The Handbook of Mites of Economic Plants

Identification, Bio-ecology and Control

FSC
www.fsc.org
MIX
Paper from
responsible sources
FSC® C013604

The Handbook of Mites of Economic Plants

Identification, Bio-ecology and Control

Vincenzo Vacante

Professor of General and Applied Entomology,
Mediterranean University of Reggio Calabria, Italy

www.cabi.org

CABI is a trading name of CAB International

CABI
Nosworthy Way
Wallingford
Oxfordshire OX10 8DE
UK

Tel: +44 (0)1491 832111
Fax: +44 (0)1491 833508
E-mail: info@cabi.org
Website: www.cabi.org

CABI
745 Atlantic Avenue
8th Floor
Boston, MA 02111
USA

Tel: +1 (617) 682-9015
E-mail: cabi-nao@cabi.org

A catalogue record for this book is available from the British Library, London, UK.

Library of Congress Cataloging-in-Publication Data

Vacante, Vincenzo, author.
Mites of economic plants : identification, bio-ecology and control / Vincenzo Vacante, Professor of General and Applied Entomology, Mediterranean University of Reggio Calabria, Italy.
pages cm
Includes bibliographical references and index.
ISBN 978-1-84593-994-6 (alk. paper)
1. Mites--Identification. 2. Mites--Ecology. 3. Mites--Control. I. Title.

QL458.V33 2015
595.4′2--dc23

2014046557

ISBN-13: 978 1 84593 994 6

Commissioning editor: Joris Roulleau
Editorial assistant: Alexandra Lainsbury
Production editor: James Bishop

Typeset by SPi, Pondicherry, India
Printed and bound in the UK by CPI Group (UK) Ltd, Croydon, CR0 4YY

In honour of Edgard Van Daele, acarologist of Gand University, Belgium, who patiently introduced me to Acarology

Contents

[1]For each genus, descriptions are given of the diagnostic characteristics, economic importance and injurious species. For each species listed, the information given includes common name(s), diagnostic (morphological) characteristics, geographical distribution, host plants, bio-ecology, symptomatology and damage caused (where known), and control methods, management and resistance.

Preface

More than 99.7% of world food is produced in terrestrial ecosystems and less than 0.3% in aquatic environments, and in the future more food will come from the land and less from the oceans. Unfortunately, in terrestrial ecosystems, numerous organisms (mites, insects, pathogens and vertebrate pests) either consume or compete with (weeds) food production, causing a worldwide loss of over 40% of the food produced, despite the application of about 3 billion kg of pesticides (Pimentel, 2002). Among pests, insects and mites cause worldwide yield losses of 13% (Pimentel, 1979) and necessitate the use of strict control methods. In this sense, knowledge of the problem represents a fundamental step towards the appropriate management of food and plant production.

Among the Acari, many morphological, physiological and behavioural adaptations for feeding on plants favour species included in the orders Trombidiformes and Sarcoptiformes. Thus, over 4000 species of Trombidiformes belonging to about 300 genera are recorded as being obligate phytophages (Amrine and Stasny, 1994; Bolland *et al.*, 1998; Migeon and Dorkeld, 2006–2013), and include various important pests of economic plants. There is a smaller number of species from the Sarcoptiformes that are commonly responsible for postharvest losses (Cao *et al.*, 2002). Each environment is characterized by its acarine fauna, which depends on various adaptive mechanisms and the level of development of the human economy, and can be enriched by the accidental introduction of exotic species, favoured nowadays by intense commercial exchange and the facility with which people can be moved.

The feeding activity of pest mites causes local damage and other general alterations. Knowledge of their feeding mechanisms (Jeppson *et al.*, 1975; Lindquist, 1985, 1986, 1996; Westphal and Manson, 1996) and of the symptoms of their feeding helps us to understand the nature and type of damage, to identify the pest responsible and to rationalize its control. The symptoms caused by mite feeding are correlated with the morphological and functional adaptations and the bio-ecology of the injurious species, and also with the organ that is attacked, the host plant and its phenology. Jeppson *et al.* (1975) believe that 'the term plant disease in its broadest meaning includes all injuries or abnormalities generated from sources outside the plant regardless of the cause'. There is, then, a link between mites and the diseases that they sometimes transmit, and the damage can take either one defined dimension, or a local or wider or more general dimension that involves the entire plant as a result of the injection of systemic or persistent toxins (Oldfield, 1996b; Westphal and Manson, 1996) or viruses (Oldfield and Proeseler, 1996; Childers *et al.*, 2001, 2003c). The reaction of a plant to mite attack therefore consists of a range of external symptoms resulting from mechanical damage (colour changes, shape alterations, reductions in growth, and losses of flowers and production) and also of internal alterations of a biochemical nature. Understanding these reactions requires analysis of the mutual interactions between the host plant and pest (Tomczyk and Kropczyńska, 1985).

The extent of damage varies with mite species and the environment, and in some situations it becomes economically unacceptable. Different examples, variously dated in time, illustrate the economic importance of this phenomenon. Thus, loss in citrus fruit production ranges from 2.5% caused by mite damage in the USA (LeClerg *et al.*, 1965) to 30% caused by insect and mite pests in India (Pruthi and Mani, 1945; Butani, 1979a,b). On apple, the red spider mite, *Panonychus ulmi* (Koch), can induce a reduction of fruit weight ranging from 3.1 to 55% (Light and Ludlam, 1972). The tomato russet mite, *Aculops lycopersici* (Tryon), produces devastating effects on tomato (Vacante, 1982a). Similarly, severe damage is caused by the coconut flower and nut mite, *Aceria guerreronis* Keifer, to coconut palm (Moore and Howard, 1996) and queen palm (Ansaloni and Perring, 2004). The extent of damage caused by the coconut mite to coconut may vary in different countries. In Mexico, the damage may affect from 80 to 100% of coconut trees (Ortega *et al.*, 1967; Olvera-Fonseca, 1986), in Cuba it may affect 42–65% (Suárez, 1991a) of trees, in Costa Rica 90% (Schliesske, 1988), in Sri Lanka 5–100% (Fernando *et al.*, 2002a) and in India 20–70% (Muthiah and Bhaskaran, 2000; Nair and Koshy, 2000; Nair, 2002). The tomato red spider mite, *Tetranychus evansi* Baker *et* Pritchard, causes crop losses of up to 90% in south-east Africa (Sibanda *et al.*, 2000; Saunyama and Knapp, 2003) and West Africa (Duverney and Ngueye-Ndiaye, 2005). The spruce spider mite, *Oligonychus ununguis* (Jacobi), is a serious pest of conifers worldwide, and in 1957 seriously damaged over 800,000 acres of Douglas fir forests in Montana and Idaho (USA) (Furniss and Carolin, 1977).

The cost of control is variable, as is the amount of damage caused, which depends on species, environmental conditions and the control strategies adopted. These aspects may also influence the agrarian economy of a region and so generalizations as to control costs cannot be made. Indeed, control of the citrus rust mite, *Phyllocoptruta oleivora* (Ashmead), in Florida (USA) requires an annual expenditure of US$75–100 million (McCoy, 1996b), while the pest poses no problem in other important citrus areas such as Italy, where its presence is not well documented. Equally important in the case of citrus is the reddish black flat mite, *Brevipalpus phoenicis* (Geijskes), which is responsible for transmission on the American continent of 'lepra explosiva' or 'leprosis' (Childers *et al.*, 2001, 2003c), a feared viral disease caused by Citrus leprosis virus (CiLV), the prevention of which costs approximately US$100 million a year in Brazil

alone (Rodrigues *et al.*, 2003), but which is non-existent and does not spread from the mite in the Mediterranean region. In the lemon groves of Italy and Spain, up to three acaricide treatments are carried out annually with a cost per hectare of approximately €450 per treatment (Vacante, 2010). In Mediterranean greenhouses, the cost of chemical control of the two-spotted spider mite on pepper and aubergine is about €400–500/1000 m^2, while that of the tomato russet mite is €60–100/1000 m^2. The use of biological control and/or integrated pest management (IPM) limits the technical problems involved, and costs depend on the crop and environmental conditions. Thus, in citriculture, in deciduous fruit trees and in vineyards, the action of natural enemies of pest mites has a relatively low or inconsistent cost because of the presence of natural predators and/or their introduction through programmes of classical biological control. In the case of vegetables or different protected crops, the cost of biological control could be higher, as for the phytoseiid mites *Phytoseiulus persimilis* Athias Henriot, at €13–25/2000 individuals, and *Euseius californicus* (McGregor), at €32–35/2000 individuals, in the Mediterranean region; this is because supplemental natural enemies are artificially reared and released in open field crops and greenhouses by the technique of augmentative biological control in quantities averaging from up to 2000 or over 15,000/1000 m^2 of crop surface. Furthermore, pest control may have toxicological and ecological consequences, as in the case of the use of chemicals such as dithiocarbamates, aldicarb and other substances, whose social cost is difficult to quantify.

As already mentioned, possession of the appropriate knowledge of mite pests helps to rationalize their control and to optimize the production of economic plants. The book *Mites Injurious to Economic Plants* by Jeppson *et al.* (1975) responded to this need and has assisted different generations of students, researchers and technicians. Jeppson *et al.*'s book is a pioneering work that successfully provides an introduction to economic acarology and management approaches to pest mites. This book was inspired by the work of Jeppson *et al.*, and its main goal is to present an update to the knowledge that has been acquired since 1975 on international issues concerning mites that are injurious to economic plants. The work is intended for students, researchers and technicians interested in plant production for food and feed, the production of fibre and/or wood, and plants of the natural environment or used as ornamentals. The huge number of host plants of the different species of mites that have been examined has required selection of the most important plants of economic interest, but at the same time has pointed the interested reader in the direction of further study via consultation of other literature sources.

The practical aims of the book have favoured a more extensive discussion of certain topics at the expense of others – which, while also interesting in terms of knowledge of the Acari, are not so 'interesting' or are 'irrelevant' to plant defence. From this viewpoint, the book may seem unbalanced in its arrangement into chapters. However, the need to determine the direction of the discussion dictated this choice. In addition, the treatment has evaluated the need for there to be sufficient consideration of the intensive development that there has been of knowledge of a taxonomic and bio-ecological nature, etc. The risk that the adoption of any new taxonomic status might lead to some confusion (to the non-expert) suggested that the bibliographic information should not be changed, except in those cases that are highly significant, e.g. the synonymy between the two-spotted spider mite, *Tetranychus urticae* Koch, and the carmine spider mite, *Tetranychus cinnabarinus* (Boisduval). The same choice was largely made for natural enemies, such as phytoseiid mites and other predators.

The book is divided into two parts: the first provides an introduction to mites; the second discusses injurious mite species. In particular, the first part of the book covers mite systematics, their external and internal morphology, the fundamental aspects of mite bio-ecology, mite collection and laboratory study, the damage caused by mites, the means and the strategies of mite control and the history of economic acarology; it ends with discussion of a classification scheme for the higher categories of the subclass Acari and a key for the identification of the major taxa. The second part of the book describes fundamental aspects of the morphology, bio-ecology and systematics of the most important mite families from a phytopathological point of view: the Histiostomatidae, Acaridae, Erythraeidae, Penthaleidae, Eriophyidae, Phytoptidae, Diptilomiopidae, Tuckerellidae, Tetranychidae, Tenuipalpidae, Tarsonemidae, and five more minor families – the Dolichocybidae, Siteroptidae, Pygmephoridae, Scutacaridae and Microdispidae. For each species, information is presented on diagnostic characteristics, geographical distribution, host plants, bio-ecology, symptomatology, damage caused and control methods, strategies and management (chemical, physical, biological, cultural, integrated, etc.).

The contents of the book can be considered exhaustive, as they represent a synthesis and up-to-date selection of the most important references on the subject – although it must be added that further input by different specialists could improve it further.

Vincenzo Vacante
Reggio Calabria, July 2014

Acknowledgements

The need to present a comprehensive account of the main species of mites harmful to the most important plants of economic interest has meant that the author has dealt with many different taxonomic groups. In this endeavour, he has been assisted by the involvement of specialists in the various systematic groups from throughout the world. These renowned acarologists have assisted in the realization of the book by providing a significant amount of literature and/or critically reviewing the chapters within their competence. Accordingly, I sincerely thank the following for helping me in the bibliographic research and/or giving me permission to use different figures: Prof. Tetsuo Gotoh, Laboratory of Applied Entomology and Zoology, Faculty of Agriculture, Ibaraki University, Japan; Dr Z.-Q. Zhang, Landcare Research, Auckland, New Zealand; Prof. Eddy Ueckerman, ARC Plant Protection Research Institute (ARC-PPRI), Pretoria, South Africa; Prof. Enrico de Lillo, Department of Biology and Chemistry of Agro-Forestry and Environment (Di.B.C.A.), Entomological and Zoological Section, Faculty of Agriculture, University of Bari, Italy; Prof. G.W. Krantz, Emeritus Professor of Zoology at Oregon State University (USA); Prof. Carlos Luis Vásquez Freitez, Departamento de Ciencias Biológicas, Decanato de Agronomía, Universidad Centro Occidental Lisandro Alvarado (UCLA), Tarabana, Venezuela; and Dr Maria Navajas, Institut National de la Recherche Agronomique, Campus International de Baillarguet, Montferrier-sur-Lez, France. Moreover, I sincerely thank Prof. Uri Gerson, Emeritus Professor of Entomology at The Hebrew University of Jerusalem, Rehovot, Israel, for critically reviewing the chapter on the Tenuipalpidae, and Dr Bruce R. Halliday, Research Fellow (Acarology), CSIRO Ecosystem Sciences, Canberra, Australia, for critically reviewing the chapters on the Erythraeidae and Penthaleidae, and for the significant bibliographic support offered to me during the writing of the book. I am also very grateful to Dr Alain Migeon, Institut National de la Recherche Agronomique, Campus International de Baillarguet, Montferrier-sur-Lez, France, for having sent to me a great deal of bibliographic support. Further, I thank Prof. Carlos H.W. Flechtmann, CNPq-Brazil Researcher, Universidade de São Paulo-ESALQ, Departamento de Entomologia e Acarologia, Piracicaba, Brazil, for critically reviewing most of the work, for the large amount of bibliographic material sent to me and for the various suggestions made. Some of the specialists who have assisted me have been extraordinarily available, including Carlos Flechtmann, Bruce Halliday, Uri Gerson, and different colleagues from the French School of Montpellier (France) under the direction of Prof. Serge Kreiter; this has given me a singular experience that has transcended the usual boundaries of scientific cooperation and publishing. Indeed, it provided a reason for me to meet 'special' colleagues who were able to give something without expecting to get anything. A special thanks is addressed once again to Serge Kreiter for advice given me in setting up the project proposal.

I sincerely thank Rachel Cutts and Alexandra Lainsbury of CABI for help given to me during the various stages of the writing of the book. In particular, without the competent support afforded by Alexandra Lainsbury, the work would never have seen the light of day. Lastly, I thank Everild Haynes, a freelance copy-editor working for CABI, who has reviewed the text with great linguistic, technical and scientific competence, improving it significantly in many places.

PART I INTRODUCTION

The first part of the book covers mite systematics, their external and internal morphology, the fundamental aspects of mite bio-ecology, mite collection and laboratory study, the damage caused by mites, the means and the strategies of mite control and the history of economic acarology in ten chapters. It ends with a chapter that discusses a classification scheme for the higher categories of the subclass Acari and a key for the identification of the major taxa.

1 Introduction to the Acari

Systematic Position and Ecological Role

The Acari, or mites, are small animals; the adults range from 300 to 500 μm in body length, except for some eriophyoids which are approximately 100 μm long and certain ticks whose females measure about 30,000 μm long. They belong to the phylum Arthropoda, a monophyletic taxon of terrestrial and marine invertebrates (Weygoldt, 1998) which are characterized by jointed legs and a chitinous exoskeleton. Two groups of Arthropoda, the Pantopoda and the Chelicerata, lack antennae and mandibles, and molecular analyses have shown that the Chelicerata is a sister taxon of a clade including the Crustacea and the Insecta (Telford and Thomas, 1998). From a systematic point of view, the Chelicerata is a subphylum that includes the class Arachnida, a primarily terrestrial group that includes the subclass Acari along with several orders (Amblypygi, Araneae, Opiliones, Palpigradi, Pseudoscorpionida, Ricinulei, Schizomida, Scorpiones, Solifugae and Uropygi) (Weygoldt and Paulus, 1979a,b). Members of the subclass Acari lack opisthosomatic spinnerets and conspicuous primary segmentations (or tergites arranged segmentally). The Acari includes the superorders Acariformes and Parasitiformes, and is morphologically differentiated from other relatively simple arachnids (Krantz, 2009a). Adults of the Acari usually have four pairs of legs, with the exception of eriophyoids, which have two pairs. Their biological cycle progresses through the stages of egg, larva and nymph. The larva is six legged and the nymph eight legged. Another characteristic that sets the Acari apart from the other Arachnida is that the phase which hatches from the egg has fewer legs than subsequent phases, while the eggs of spiders, scorpions, pseudoscorpions and opilionids all hatch into eight-legged organisms.

Up to 1999, approximately 55,000 species of Acari have been described, but the group is likely to grow to number over a million species (Walter and Proctor, 1999). By 2011, a total of 54,617 mite and tick species had been reported (Zhang, 2011). This high number of species can be explained by assuming that if in each of the 400,000 plants currently known there lives almost one species-specific mite, then the number of plant-feeding mites could be as many as 400,000 species. Similarly, if each of the 60,000 known species of vertebrate animals harbours almost two species-specific mites, the number of species-specific mites might be at least 120,000 species (Hoy, 2011). That there could be such a high number of species is not surprising when one considers that at present our knowledge of the Acari is small and fragmentary. A bibliographic analysis of the acarological papers published in *Zootaxa* from 2001 to 2005 gives a good representation of the strong development that there has been in this subject. In this period of time, the list of new taxa described and published in this journal only includes 16 genera and 185 species (Fan, 2006). Krantz (2009b) reported that the large number of species that have been described has led to the designation of about 5500 genera and 1200 subgenera in 540 families.

The small size and the strong evolutionary plasticity of the Acari have permitted these singular organisms to colonize most of the available aquatic and terrestrial habitats, where they exhibit a great ability to exploit the available trophic resources owing to their different and varied dietary regimes – phytophagous, parasitic, predaceous, mycophagous, saprophagous, coprophagous and necrophagous. In short, the Acari may be found worldwide in any environment capable of supporting life. As such, they can be collected from regimes as diverse as windswept arctic tundra, hot desert sands, icy depths or the hair follicles of eyebrows (Krantz, 2009b). For humans, the Acari form a significant component of the fauna of the cultivated plants and forests of any region of the world, either as pests (Jeppson *et al.*, 1975; Keifer *et al.*, 1982; Helle and Sabelis, 1985; Lindquist *et al.*, 1996; Zhang and Liang, 1997; Bolland *et al.*, 1998; Amrine *et al.*, 2003) or as beneficial organisms (Huffaker *et al.*, 1970; McMurtry *et al.*, 1970; Gerson *et al.*, 2003), etc. Furthermore, these little animals play a fundamental role in the litter and humus layers of forests, grasslands and agricultural soils, where they recycle minerals (Balogh, 1972). Bader (1989), in an interesting booklet written in German, discussed the fundamental features of the Acari. Walter and Proctor (1999) very effectively summarized the ecological importance and the economic role of these small and interesting animals. They wrote that: 'The average mite is minuscule, barely perceptible to even the sharpest eyes. When enough are present, however, mites can exert efforts disproportionate to their size. When a crop disappears under a blanket of silk or stored food turns into a seething mass of hairy motes, then the numbers of mites will be legion. If the mite is an itch mite burrowing into one's skin or a tick injecting virulent rickettsiae into one's blood, then "enough" can be a single mite'.

Origins and Phylogenesis

Incomplete conservation, inadequate sampling and taxonomic problems make it very difficult to assess the exact evolution of the Acari over a long period of time. In addition, species such as the Acari are very small in size and have very little tough exoskeleton, so are not suitable for preservation. So to better uncover the essential features of the origins and phylogenesis of the Acari, one must make some assumptions. In particular, the first fossil records of marine invertebrates date back about 600 mya

and it is estimated that these evolved in a few million years. In the previous 2,500 million years, life on earth appeared to be exclusively represented by algae and bacteria. The abundant availability of a monoculture of algae and eukaryotic cells formed in that time could have led to the largest explosion of evolutionary activity so far known (Gould, 1981). During the Permian period, there was a major decrease in the number of families of animals living at the bottom of the sea (Valentine, 1970), and Schopf (1974) reported that the later Permian was the time when the earth's continents came together to form the (only) supercontinent of Pangaea. This union produced a decrease in the surface of the earth covered by shallow seas, and consequently a decrease in the habitat available to shallow water invertebrates. Likewise, the increase in the abundance of fauna recorded after the Permian can be explained by the subsequent increase in shallow seas produced by the separation of Pangea into smaller continents. An interesting analysis of the history of marine animal diversity has recently been published by Stanley (2007). In addition to the above-mentioned fossil records of marine invertebrates 600 mya, since the early Silurian early fossil records appeared of vascular plants, followed by records of pteridophytes in the Carboniferous, and records of gymnosperms to angiosperms in the Cretaceous and Tertiary (Niklas *et al.*, 1983). The earliest phytophagous Insecta date from the Carboniferous (Strong *et al.*, 1984), and it is well known that there is reciprocal evolution between plants and insects that represents an important mechanism for increasing the wealth of wildlife (Begon *et al.*, 1986).

It is commonly believed that the first fossil records of the Arachnida (and the Acari) originate in the late Silurian–early Devonian periods (*c.*425 mya) in the form of a pulmonate trigonotarbid (Arachnida, Trigonotarbida), which is now extinct. Dunlop and Selden (2009) report that the oldest mite fossil is from the Devonian (410 mya), while the age of scorpion and spider fossils is 428 and 312 mya, respectively. The advent of the Acari probably also relates to the latter part of the evolution of the arthropods, with diversification having occurred about 200 million years earlier in the Cambrian period (Gould, 1989; Walter and Proctor, 1999). Fossil records show that the terrestrial Arachnida acquired respiratory organs of different types at different times during the transition from the marine environment on to land, as in the scorpions, trigonotarbids, etc. Formally, this pattern of developmental behaviour contrasts with the hypothesis of the monophyletic origin of the Arachnida (Bergstrom, 1979). The fossil records now available seem to indicate that the Acari had achieved a certain amount of diversity by the early to mid-Devonian (Hirst, 1923a; Woolley, 1961; Norton *et al.*, 1988; Evans, 1992; Subías and Arillo, 2002), which suggests that they colonized terrestrial environments as early as the late Silurian. Moreover, in the early Devonian, all fossil Acari now known belonged to the superorder Acariformes (Walter and Proctor, 1999), and the fossil records of the Parasitiformes dating to the late Cretaceous (Witalinski, 2000; Poinar and Brown, 2003); the parasitiform Trigynaspida may date to as early as the upper Triassic (*c.*220 mya) (Kim, 2004). Studies of mitochondrial phylogeny have shown that the orders and classes of spiders, scorpions, mites and ticks diversified in the late Palaeozoic (Jeyaprakash and Hoy, 2008, 2009). Recently, Sidorchuk and Bertrand (2013) described *Sellnickiella balticae* Sidorchuk *et* Bertrand (Labidostomatidae), thus highlighting the presence in Eocene Baltic amber (more than 40 Ma) of the genus *Sellnickiella* Feider *et* Vasiliu and a second species of the genus *Labidostoma* Kramer (*Labidostoma* (*Pseudocornutella*) *electri* Sidorchuk *et* Bertrand), after the congeneric species previously described by Dunlop and Bertrand (2011). The report pointed out that notwithstanding the few fossil mite descriptions, the discovery of these mite carnivores in Baltic amber suggests that Palaeogene mite biodiversity was comparable to current mite diversity (Sidorchuk and Bertrand, 2013).

The fossil records now available seem to show that after the extinction events registered in the Permo–Triassic and after the statement of modern lineages, a significant adaptive process may have occurred in the Acari during the late Mesozoic and early Cenozoic periods (60–100 mya). At this time, there was a strong morphological and ecological diversity among the Acari that permitted significant exploitation of the many co-evolving plants and animals, and there was a consistent diversification of angiosperm plants and Insecta (Lindquist, 1975; Krantz, 2009c). Specialized plant-parasitic mites, e.g. the eriophyoid *Triasacarus fedelei* Lindquist *et* Grimaldi and *Ampezzoa triassica* Lindquist *et* Grimaldi, existed as early as the Triassic (Schmidt *et al.*, 2012). Acari are often associated with Insecta as phoretic or parasitic mites in fossil records dating from the Cretaceous about 20–40 million years earlier (Poinar, 1992; Poinar *et al.*, 1993, 1997). The same phenomenon has been observed between parasitic Acari and vertebrate hosts in the mid-Tertiary (Poinar, 1995). Mites of the suborder Oribatidae have been found in fossils of the late Carboniferous (Labandeira *et al.*, 1997). The oldest oribatid fossils are from the Devonian (Norton *et al.*, 1988), and some of them belong to modern families (Subías and Arillo, 2002). All mites described from the Carboniferous, Jurassic and Cretaceous periods are included in modern families, with one exception (Subías and Arillo, 2002; Arillo *et al.*, 2012).

Regardless of the above information, in the last half century a need has developed to understand whether mites are characterized by a condition of dichotomy between acariforms and non-acariforms, or not. In short, we need to know whether in the course of the evolution of the Arachnida an event has resulted in a monophylogenous origin of the Acari or whether they result from the convergence of two or more groups that have evolved independently of one other and are now similar. Woolley (1961) notes that 'no known connections with other arachnids can be ascertained from the fossil record, but the chelicerae and other morphological features indicate that the Acari probably had an arachnid origin in opilionid-like ancestors, but developed with inherent peculiarities, principally the gnathosoma. They are at least diphyletic, probably polyphyletic, in origin, with two main groups, the older Anactinochitinosi (Notostigmata, Mesostigmata, Ixodides) and the younger Actinochitinosi (Trombidiformes, Sarcoptiformes)'. Following Jeyaprakash and Hoy (2008, 2009), an estimate of the evolutionary age of mites places their chelicerate ancestors in the Precambrian and their divergence as mites to at least 424 mya.

Studies of the morphological, developmental and behavioural aspects of mites from various research support different positions

on their evolutionary origin (Zakhvatkin, 1952; Lindquist, 1984; Van der Hammen, 1989; Walter and Proctor, 1999). Examination of 40 characters of 12 extant orders of arachnids (including the Acari), and in particular the examination of the common characters – possession of a movable gnathosoma and the existence of exapod larva and of the three octopod nymphal instars – suggest that the mites are derived from a common ancestral stock (Lindquist, 1984) and that monophyletic Acari and Ricinulei are sister groups (Kaestner, 1980). However, spermatological evidence from the Chelicerata does not encourage support for a common origin of acariforms and non-acariforms (Alberti, 2000).

2 External Morphology

Integument

In the Acarina, the developing integument appears first as undifferentiated tissue, covered by a thin layer of 'cuticulin' and separated from the underlying 'epidermis layer' by a 'Schmidt layer'. As development proceeds, the anatomical organization of the integument undergoes further transformation and shows from the outside inwards an 'epicuticular layer' (the epicuticle), a 'procuticular layer' (the procuticle), which consists of an outer 'exocuticular layer' and an inner 'endocuticular layer', an underlying Schmidt layer, an epidermal layer and, finally, a basal lamina set below the epidermis. This anatomical organization underlies the external 'cerotegument'; this consists of an outer 'cement layer' and an underlying 'wax layer' that are set on the outer epicuticle and the underlying inner epicuticle (Fig. 2.1).

Together, the epicuticular complex – the cerotegument and the outer and inner epicuticle – provides multilayered protection. The cerotegument is a secreted layer that is formed by the epidermis, from which 'pore canals' travel towards the inner epicuticle, which includes an underlying cuticulin layer and a covering cuticulin layer. Overall, the cerotegument may have various consistencies, but its cement layer is dense and frequently sculptured. Together, the wax and cement layers of the cerotegument are involved in the regulation of water loss and/or water intake (Alberti *et al.*, 1981; Norton *et al.*, 1997).

In the different regions of the body, tanning by orthoquinones can induce differential sclerotization of the two layers of the procuticle. The secretion of granules which coalesce with 'microfibres' occurs before the secretion of new cuticle below the endocuticular layer, and the microfibres consolidate and form laminations (Brody, 1970). The width of these laminations progressively decreases towards the cuticle surface. The pore canals in the endocuticular layers are first organized in a helical pattern, but as they move towards the surface they gradually become linear, sometimes with branching, and finish at the micropores of the cuticulin layer below the epicuticle (Norton *et al.*, 1997).

In addition to having micropores, the body surface of mites shows a variety of 'macropores' that are involved in different functions (secretory and sensory – including proprioception), and belong to three fundamental categories of cuticular apertures: poroidotactic (lyrifissures or cupules), setal (openings for the insertion of tactile, olfactory and chemosensory hairs and pegs) and adenotactic (openings of secretory structures, such as solenostomes or structures with a calyx) (Athias Henriot, 1969a,b). These topics will be discussed more extensively in subsequent paragraphs and other information can be found in Krantz (2009a).

Body Division

Primary segmentation is largely absent in the Acari and instead a segmental fusion or 'tagmosis' has become established. Some aspects of mite ontogenesis assist in a better understanding of this phenomenon. Several factors influence the establishment of somatic derivatives for most Acari, among these the posteroventral idiosomatic curvature or 'caudal bent' (Evans, 1992). Several interpretations of somatic origins and structural homologies in the Acari have been based on setae, pores or gland patterns. According to some specialists (Zakhvatkin, 1952; Van der Hammen, 1970; Sitnikova, 1978; Jesionowska, 1991), primitive segmentation in the Acari ranged from 12 to 22 segments, but observations carried out on caeculid mites by Coineau (1974a,b), supported by the analysis of Grandjean (1954), have led to a mapping in which the archetypal adult acariform is characterized by 16 somites plus a cheliceral segment (Fig. 2.2). The larval idiosoma (the part of the body that follows on from the first region, or gnathosoma) includes 11 somites, and one segment is added posteriorly at each post-larval moult, according to the phenomenon of 'anamorphosis'. The larva itself has six idiosomal segments (C, D, E, F, H and Ps), and one terminal segment (AD) is added to its idiosoma when it moults into a protonymph, another segment (AN) is added to the protonymph idiosoma when it moults into a tritonymph, and a third segment (PA) is added to the tritonymph idiosoma when it moults into an adult. These changes can affect the numbers of genital papillae, the setae of the genital and anal fields, and the cupules; the latter number four pairs in the larva, five in the protonymph and six in the other biological stages.

Fusion of the primitive body segments just described has led to the formation of the two major body regions or 'tagmata' of mites mentioned above: the anterior 'gnathosoma' and the posterior 'idiosoma' (Figs 2.3 and 2.4). The gnathosoma is derived from the first two anterior segments and has the primitive somatic primary organs for feeding. The idiosoma is derived from the remaining segments. The two body regions are separated by a 'circumcapitular furrow', which is clear in some groups, but is absent in others owing to secondary somatic development. Other furrows can occur in several groups of Acari, such as a 'dorsosejugal furrow' which is set between the second and third pairs of legs of acariform mites (Fig. 2.4), and a 'postpedal furrow' that may be set behind the pair of fourth legs. In some groups, e.g. the Prostigmata, a certain somatic plasticity can complicate this scheme with the appearance of other furrows (the 'abjugal furrow', 'disjugal furrow', etc.).

Gnathosoma

The gnathosoma, or 'capitulum', is the first region of the body; it is very complex and specialized, and is characterized by important adaptations (sensory and feeding organs, etc.) (Alberti and Coons, 1999). It includes the chelicerae and the palpi (Figs 2.3, 2.5D–E, 2.6A–I, 2.7A–H), which are the fundamental mite feeding organs and are variously adapted to chewing, piercing, tearing and sucking. Ontogenetically, the gnathosoma derives from the cheliceral segment and the paired biramous appendages of the second segment (Fig. 2.2); the cheliceral segment is also homologous with the first antennal segment of insects (Manton, 1977; Telford and Thomas, 1998). The palpal coxae are derived from primitive paired protopods and contribute to the formation of the 'subcapitulum' and 'hypostome', and the gnathosomatic floor and the walls, respectively; moreover, they form a channel for the paired chelicerae that are set above it and an outlet for the buccal cavity that opens below the chelicerae. The malar processes ('corniculi' or 'rutella') are sometimes set alongside the lateral lips of the hypostome. An 'epistome' or 'gnathotectum' can be extended over the gnathosomatic dorsum to partially protect the chelicerae. A 'subcheliceral plate' constitutes a base for the chelicerae and provides attachment for the muscles of the preoral 'labrum', an underlying expansion of the dorsal pharyngeal wall (Fig. 2.5A–B) (Evans, 1992; Alberti and Coons, 1999). Depending on the taxonomic group, the labrum can develop the function of a prepharyngeal valve, which is an adaptation to prevent food loss, as in the Ixodida (Sonenshine and Gregson, 1970) and other acarine groups; it can also help to prevent the ingestion of solid food particles (Evans and Loots, 1975). The buccal cavity opens into the 'pharynx', a suction pump for food whose action is related to the constrictor and dilatator muscles. The shape of the pharynx varies in the different orders, e.g. it is triradiate in cross section in the Mesostigmata and reniform in the acariform mites (Alberti and Coons, 1999); the form of the pharynx can be taken into account in the taxonomy of families and genera. The buccal cavity may show paired ducts of idiosomatic 'salivary glands' below the subcheliceral plate (Sonenshine and Gregson, 1970; Evans, 1992), or other adaptations. In many acarine species, the salivary glands produce enzymes for the preoral digestion of food (Moss, 1962), while other glandular systems are involved in silk production or osmoregulation (Alberti, 1973; Brody *et al.*, 1976).

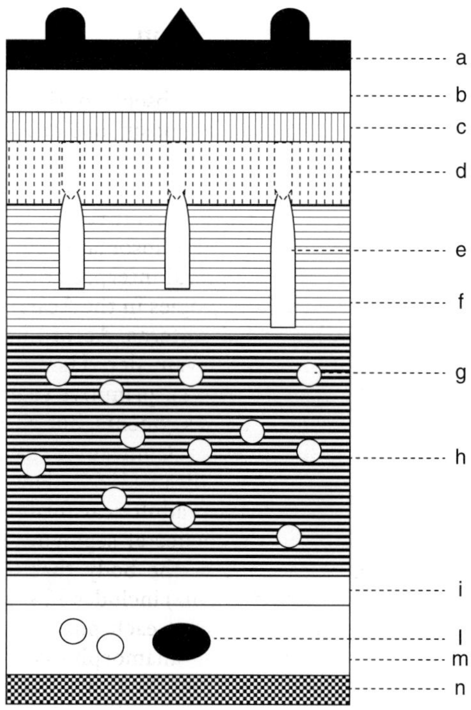

Fig. 2.1. Schematic drawing of acarine cuticle: a, cement layer; b, wax layer; c, outer epicuticle; d, inner epicuticle; e, pore canal; f, exocuticle; g, evolving pore canal; h, endocuticle; i, Schimdt layer; l, nucleus; m, epidermis; n, basal lamina (according to Norton *et al.*, 1997).

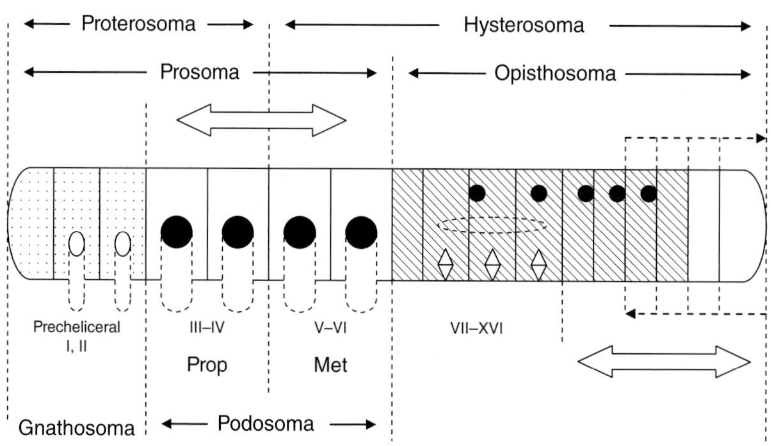

Fig. 2.2. Diagram of somatic segments of the archetypal adult acariform (according to Coineau, 1974b). The segments XII–XVI correspond to the idiosomal segments H, Ps, AD, AN and PA, respectively (see text). Prop, propodosoma; Met, metapodosoma.

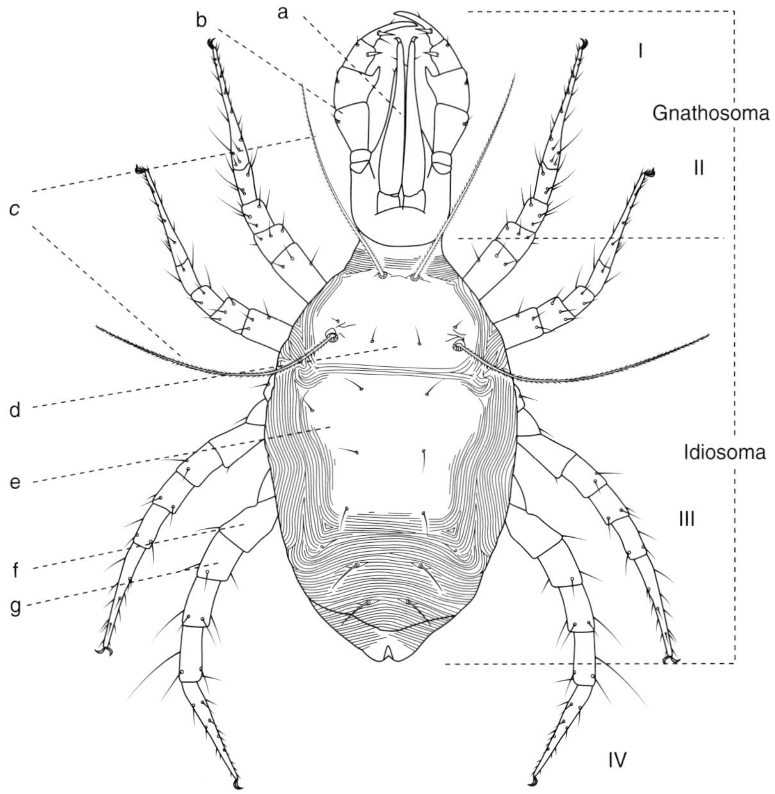

Fig. 2.3. *Cunaxa capreolus* (Berlese). Female: a, chelicerae; b, palpus; c, bothridial setae; d, prodorsal shield; e, opisthosomal shield; f, basifemur; g, telofemur; I–IV, legs (from Vacante and Nucifora, 1985).

Chelicerae

The chelicerae are placed dorsally in relation to the opening mouth and commonly consist of three segments called the 'cheliceral base', 'digitus fixus' and 'digitus mobilis'. The first segment is basal and bears the digitus fixus articulated with the distal digitus mobilis dorsally. From an ontogenetic point of view, the chelicerae derive from the endopodal appendages of the first segment. These organs evolve from a leg-like to a mandible-like appendage resulting from a gradual series of fusions and setal modifications. The basal cheliceral segment derives from the endopodal trochanter of the first somite, the attached digitus fixus from the fusion of the femur, genu, tibia and tarsus, and the terminal digit from the pretarsus (Grandjean, 1947). Cheliceral morphology shows a strong variation within the subclass Acari, as a result of the large number of evolutionary adaptations that have occurred and the variable feeding habits (Akimov *et al.*, 2002). The mandibulate form occurs in many parasitiform and acariform mites (Fig. 2.6A–D), while in phytophagous and parasitic species there are often cheliceral modifications that result in styliform, hood-like or finely toothed organs adapted to pierce plant or animal tissues or attack bacterial films (Fig. 2.6E–I) (Krantz, 2009a). The bases of the chelicerae may be partially or completely fused to each other to form the 'stylophore', as in the Tetranychoidea, while the digitus mobilis is modified into a stylet that can pierce the (plant) cell wall. The chelicerae are characterized by extrusion and retraction movements. Extrusion movements depend fundamentally on hydrostatic pressure produced by the contraction of the dorsoventral idiosomatic muscles, while retraction involves the action of retractor muscles arising from the dorsal gnathosoma or from the 'idionotum'. In some groups, the chelicerae are also involved in sperm transfer (Krantz, 2009a).

Palpi

The palpi are postoral appendices that consist of a variable number of segments – the trochanter, femur, genu, tibia, tarsus and apotele – which may be from one or two to six (Figs 2.6E, 2.7A–H). The two coxae merge with each other and contribute to form the subcapitulum, before which the mouth opens, bounded dorsally by the labrum and at the sides by various structures (the lateral labra, rutella, malae) (Fig. 2.5A–E). Ontogenetically, the palpi derive from the leg-like endopodal appendages of the second somatic segments (Fig. 2.2). Their function is primarily related to searching for and handling food, with the distal segments usually leading to chemosensory and thigmotactic sensory receptors which act as a support for the feeding activity, with various modifications related to various evolutionary adaptations. Modifications can occur in the number of free segments, which are one or two in some Astigmatina and Prostigmata, three or four in the Ixodida, and five in many Mesostigmata and Oribatida (Krantz, 2009a). In the Tetranychidae, the palpi contain part of

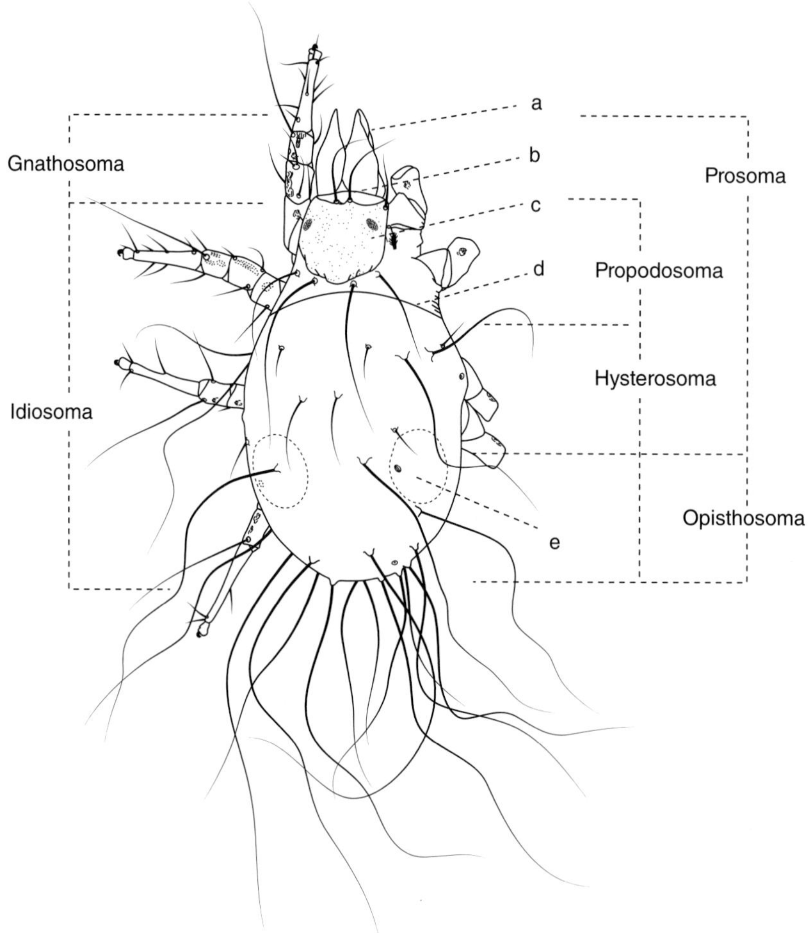

Fig. 2.4. Different body areas of *Tyrophagus neiswanderi* Johnston *et* Bruce. Female: a, chelicerae; b, circumcapitular furrow; c, prodorsal shield; d, dorsosejugal furrow; e, opisthonotal gland (from Vacante, 1988).

the silk gland, whose secretions reach a distal hollow eupathidium (also eupathidion) or 'spinneret' (Alberti and Crooker, 1985). Palpus chaetotaxy (the use of setal position and form in taxonomy) is briefly discussed under the family Tetranychidae.

Idiosoma

The idiosoma is the part of the body that follows on from the gnathosoma. It shows a high diversity in form, sclerotization and chaetotaxy, which is difficult to summarize and variable in the different systematic groups. Commonly are referred to as part of the idiosoma are the various external organs (respiratory, reproductive, secretory, osmoregulatory) and sensory receptors.

Idiosomal regions

A simple pattern divides the idiosoma into the anterior 'propodosoma' and the posterior 'hysterosoma', distinct or not by a sejugal furrow, as in the acariform mites. The region bearing the legs derives from four embryonal somites and is also called the 'podosoma'; the region of the body deriving from the first two somites includes the first two pairs of legs and represents the 'propodosoma', while the last part, which is derived from two other somites, bears the other two pairs of legs and is referred as the 'metapodosoma' (Coineau, 1974a,b; Krantz, 2009a; Alberti and Coons, 1999) (Figs 2.2–2.4). The part of the hysterosoma excluding legs III and IV constitutes the opisthosoma and derives, according to systematic groupings, from six to 13 embryonal somites; it bears the genital and anal openings (Coineau, 1974a,b). Its shape is ovoid to fusiform and presents a cuticle that is variously striate, with furrows or covered by 'shields' or plates of a shape and size that is variable in the different systematic groups. Male and female genital openings open both anteriorly or posteriorly. In addition to the terminology used above, other old terms can be found in the literature, such as 'prosoma', which refers to the gnathosoma plus podosoma, and 'proterosoma' which refers to the gnathosoma plus the propodosoma. Moreover, the idiosoma can bear one or more sclerotized shields or plates that are weakly tanned or strongly developed and heavily tanned, according to group (Fig. 2.8A–D). Dorsal and ventral shields may be lacking in many Prostigmata (Fig. 2.9A–D). The pattern of ornamentation of the dorsal shield is often used in taxonomic studies.

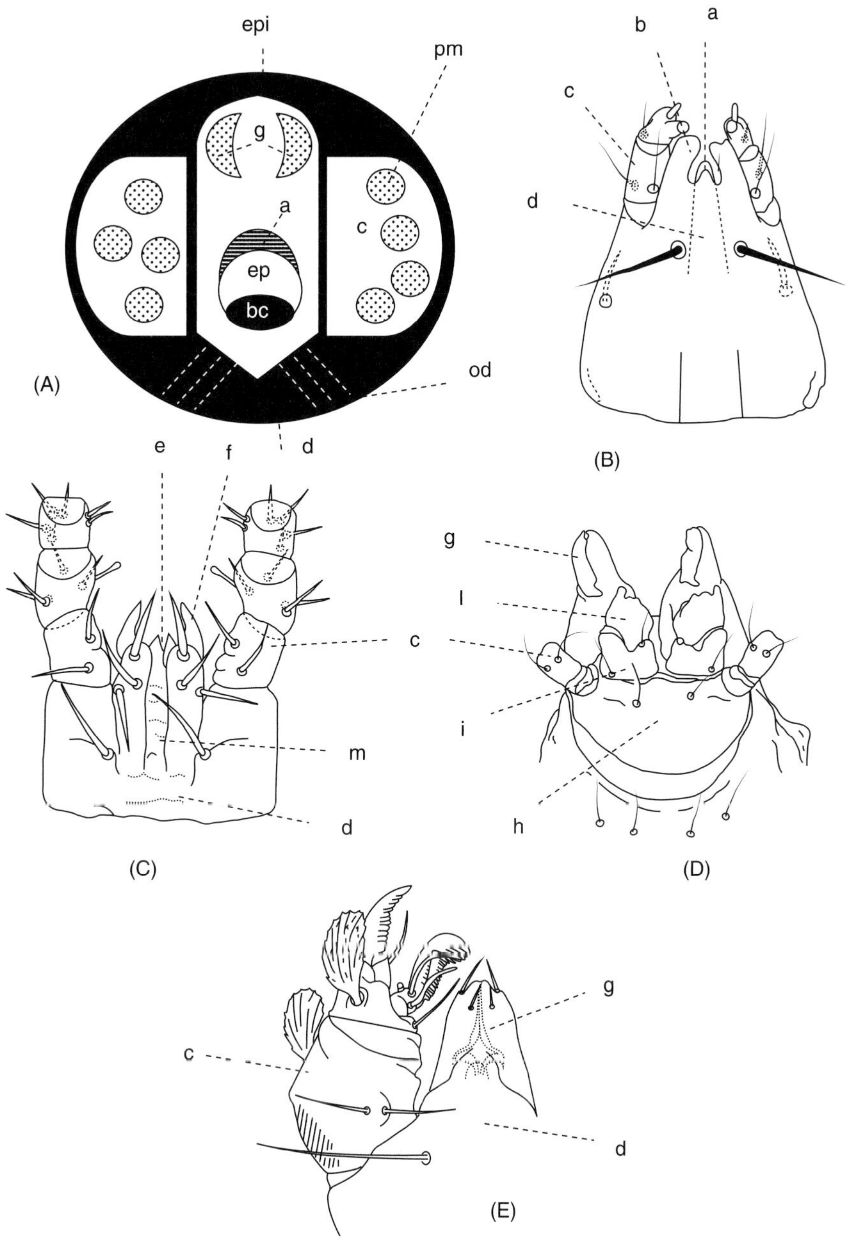

Fig. 2.5. (A) Diagram of the acarine gnathosoma in cross section (new drawing from Vacante). (B–E) Ventral view of the gnathosoma of four acarine species: (B) *Tyrophagus* sp., female (from Vacante, 1989a); (C) *Proctolaelaps pigmaeus* (Müller), female; (D) *Micreremus gracilior* Willmann, female; and (E) *Cheletogenes ornatus* (Canestrini *et* Fanzago), female (drawings C–E from Vacante and Nucifora, 1985). Key: a, labrum; b, mala; bc, buccal cavity; c, palpus; d, hypostome; e, internal mala; ep, epipharynx; epi, epistoma; f, corniculus; g, chelicerae; h, mentum; i, gena; l, rutellum; m, deutosternum; od, oral dilators; pm, palpal muscle.

The ventral idiosoma may be sclerotized or not. Variation in the ventral shield characterizes the morphology of different taxa and species. In the Acariformes, the shield remnants, commonly called 'epimera', mark the attachment for the legs, and in several groups they are modified to form the 'epimeral plates'. Epimera can be distinguished by 'epimeric furrows' that extend internally to form typical 'apodemes' (Evans, 1992) (Fig. 2.10A–B). Leg articulations and genital and anal openings have a surrounding sclerite (hardened part) that can be variously sclerotized and of different shapes and sizes (Fig. 2.10C–D).

Legs

Adult and nymphs of Acari, except in some Prostigmata and Astigmatina, have four pairs of jointed legs, and the larval instar

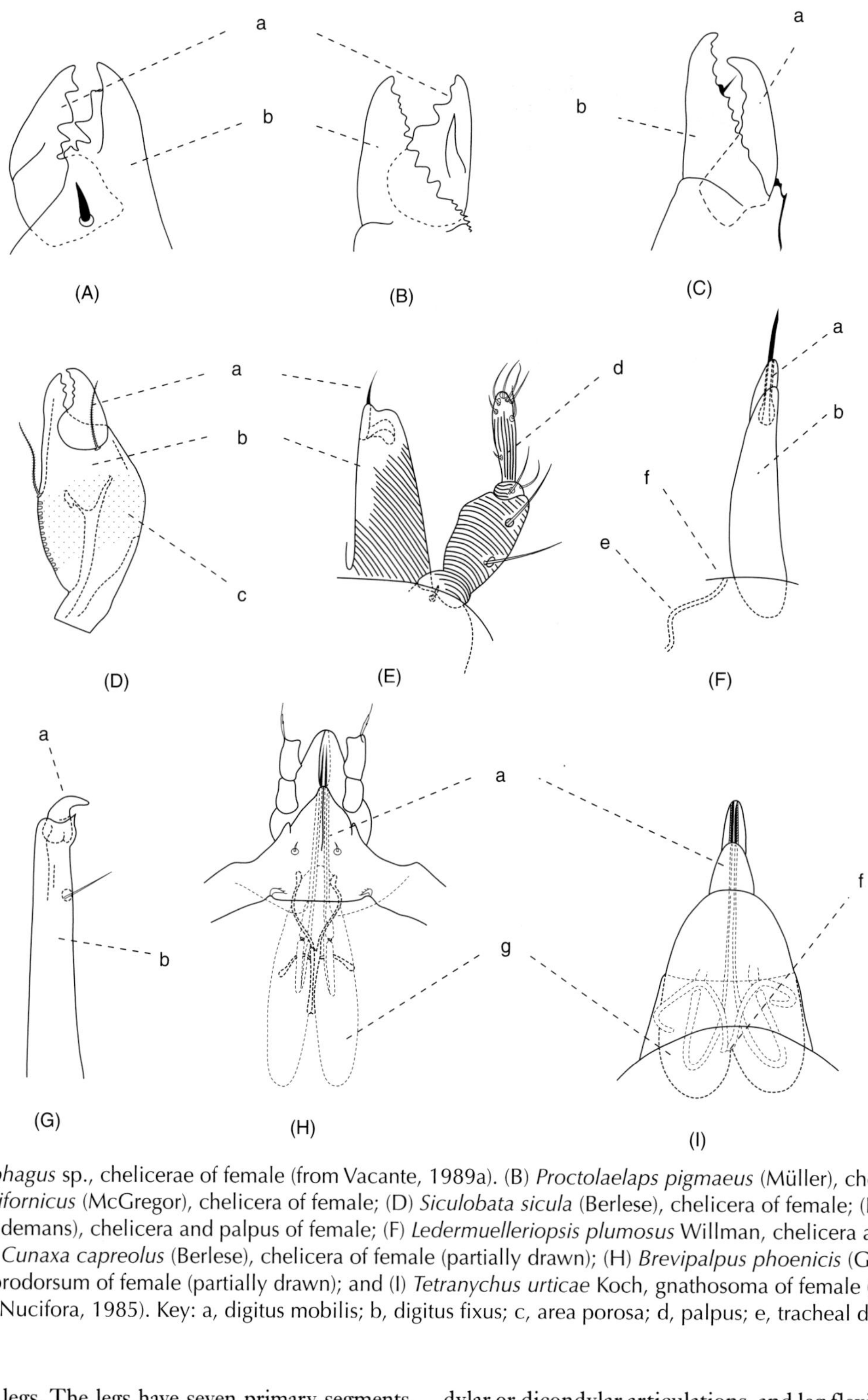

Fig. 2.6. (A) *Tyrophagus* sp., chelicerae of female (from Vacante, 1989a). (B) *Proctolaelaps pigmaeus* (Müller), chelicera of female; (C) *Neoseiulus californicus* (McGregor), chelicera of female; (D) *Siculobata sicula* (Berlese), chelicera of female; (E) *Triophtydeus triophthalmus* (Oudemans), chelicera and palpus of female; (F) *Ledermuelleriopsis plumosus* Willman, chelicera and tracheal duct with stigmata; (G) *Cunaxa capreolus* (Berlese), chelicera of female (partially drawn); (H) *Brevipalpus phoenicis* (Geijskes), gnathosoma and prodorsum of female (partially drawn); and (I) *Tetranychus urticae* Koch, gnathosoma of female (drawings B–I from Vacante and Nucifora, 1985). Key: a, digitus mobilis; b, digitus fixus; c, area porosa; d, palpus; e, tracheal duct; f, stigma; g, stylophore.

has three pairs of legs. The legs have seven primary segments, called in proximal to distal order the coxa, trochanter, femur, genu, tibia, tarsus and pretarsus (Figs 2.11A–E, 2.12A–E). As the order is always given as proximal–distal, the last segment (the most distal) could formally be called a 'posttarsus' instead of a pretarsus; however, in almost every text, the term used is pretarsus.

The leg segments are variously organized in different taxa. Segment movement is led by intersegmental monocondylar or dicondylar articulations, and leg flexion and extension are supported by two different physiological mechanisms. Extension exploits the hydrostatic pressure resulting from contraction of the dorsoventral idiosomatic muscles, while flexion results from the action of intrinsic and extrinsic intersegmental muscles (Krantz, 2009a). According to systematic group, the coxae may be free (Fig. 2.10D) or fused with the ventral podosoma (Figs 2.8D, 2.9C, 2.10A–B) and the femur

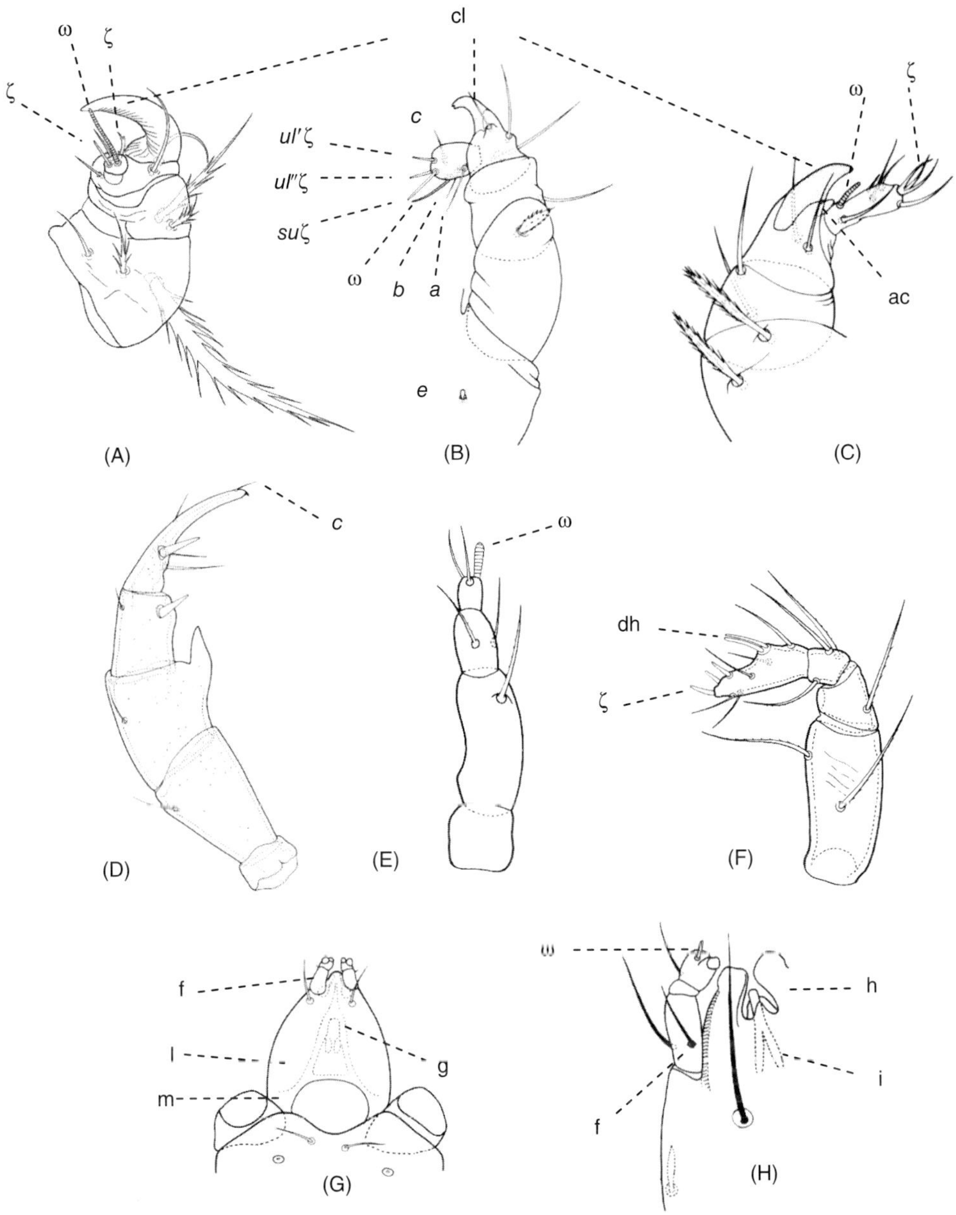

Fig. 2.7. Palpi of acarine species. (A) *Cheyletiella blakei* Smiley, female (from Patti and Vacante, 1985, produced by Vacante). (B) *Bryobia praetiosa* Koch, female; (C) *Ledermuelleriopsis plumosus* Willmann, female; (D) *Cunaxa capreolus* (Berlese), female; (E) *Brevipalpus obovatus* Donnadieu, female; (F) *Humerobates rostrolamellatus* Grandjean, female; (G) *Tarsonemus waitei* Banks, female (drawings B–G from Vacante and Nucifora, 1985). (H) *Tyrophagus neiswanderi* Johnston *et* Bruce, male (from Vacante, 1988). Key: *a*, *b*, *c*, simple setae; ac, accessory claw; cl, claw; dh, fused solenidion and eupathidium (double horn); e, supracoxal seta; ω, solenidion; ζ, *su*ζ, *ul′*ζ and *ul″*ζ, eupathidia; f, palpus; g, chelicera; h, mala; i, labrum; l, gnathosomal capsule; m, ventral apodemes.

may be divided into the basifemur and the telofemur (Fig. 2.3). The various leg segments bear a number of setae, tactile or specialized (solenidia, eupathidia, famuli, microsetae, trichobothria), commonly corresponding to the pattern of ontogenetic appearance in the taxon, and the tactile setae can be organized in whorls or 'verticils', and set in dorsal, ventral or lateral positions. In the Mesostigmata (and other groups), the distribution of setae on each leg segment can be examined by a formula that is useful in systematic and phylogenetic analysis (Evans, 1963). The system was applied by Grandjean (1935c, 1940, 1941, 1942, 1946a, 1958) to the leg chaetotaxy of the Oribatida and successively reviewed by Norton (1977) and variously applied to other acarine groups. In this system, the leg faces with respect to the central body axis are identified as a 'paraxial face' and an 'antiaxial face', and the schematic verticils of setae of a leg segment are represented by one dorsal (*d*),

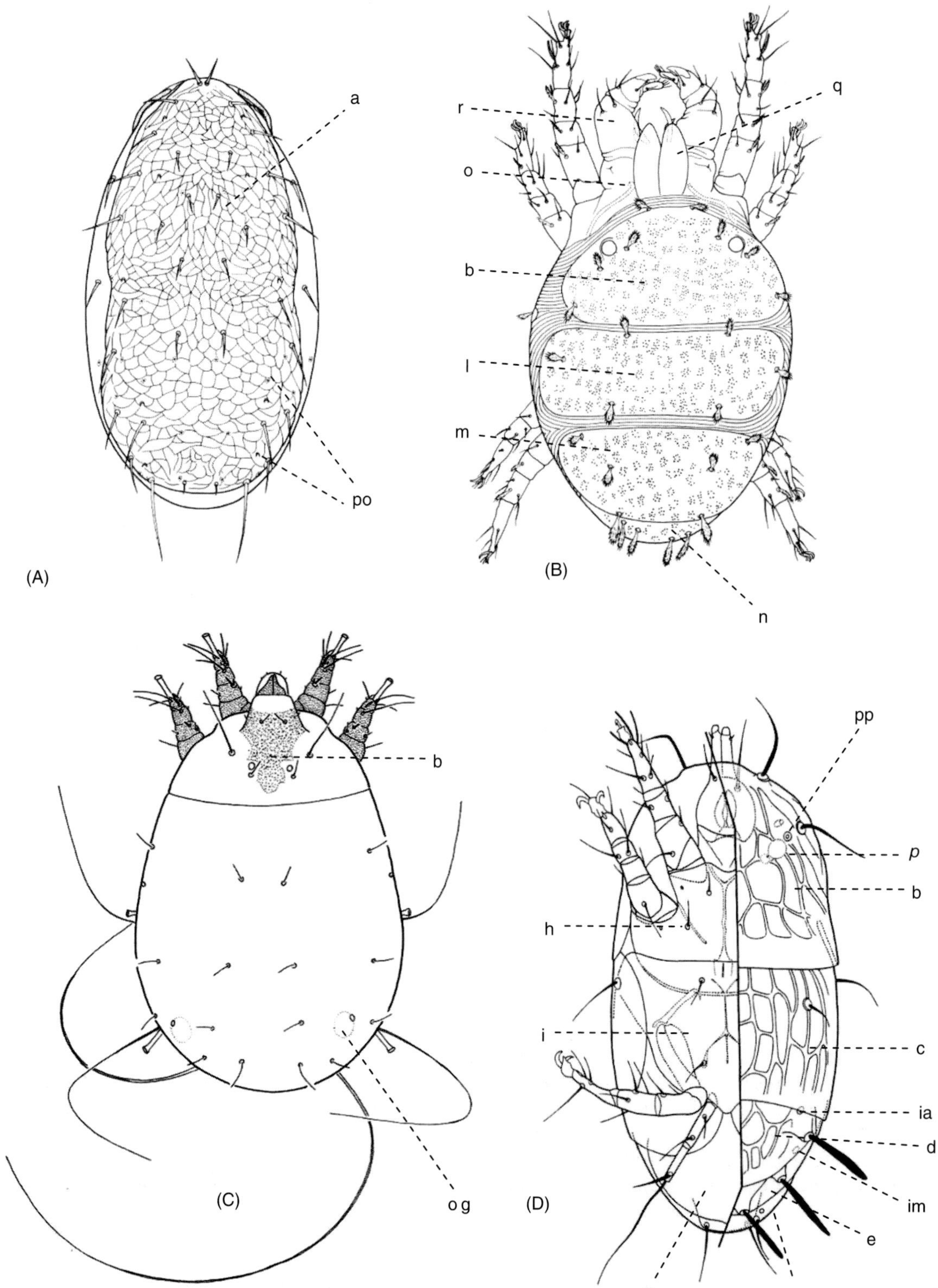

Fig. 2.8. (A) *Neoseiulus californicus* (McGregor), female partially drawn; (B) *Ledermuelleriopsis plumosus* Müller, female; (C) *Hemisarcoptes malus* (Shimer), female (drawings from Vacante and Nucifora, 1985). (D) *Daidalotarsonemus vandevriei* Suski, dorsal (at right) and ventral (at left) views of female (from Nucifora and Vacante, 2004, produced by Vacante). Key: a, dorsal shield; b, prodorsal shield; c–f, tergites C, D, EF and H; g, aggenital plate; h, coxisternal plates I and II; i, coxisternal plates III and IV; ia, im, cupules; l, metapodosomal shield; m, opisthosomal shield; n, supranal shield; o, stigma; og, opisthonotal gland; *p*, bothridial seta; po, pores; pp, prodorsal pit; q, chelicerae; r, palpus.

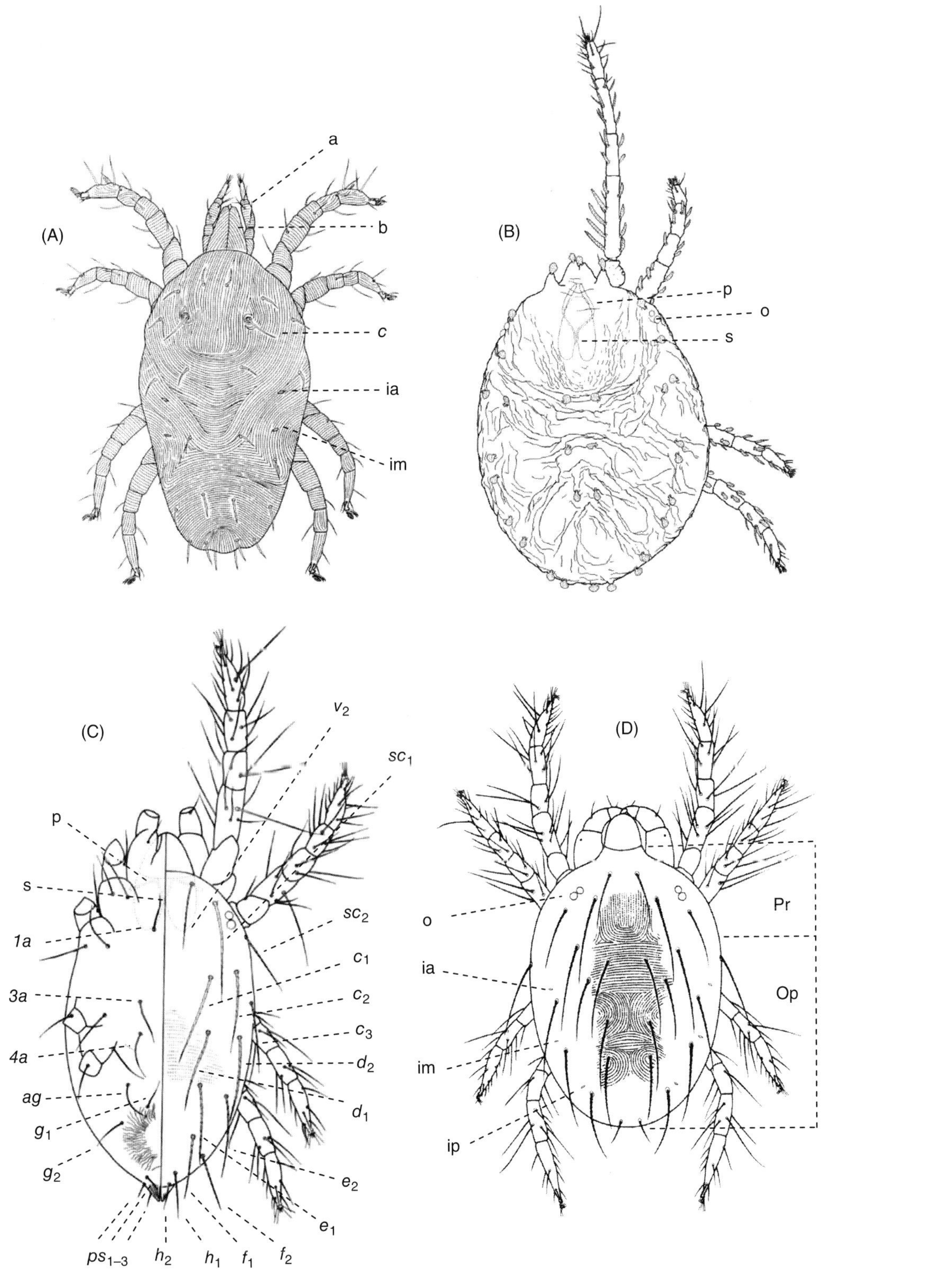

Fig. 2.9. (A) *Tydeus australensis* Baker, female (from Vacante and Nucifora, 1985). (B) *Bryobia cavalloroi* Vacante, female (from Vacante and van Eyndhoven, 1989). (C) *Eotetranychus rubiphilus* (Reck), dorsal (at right) and ventral (at left) views of female (from Vacante, 1985a). (D) *Tetranychus urticae* Koch, female (from Vacante, 2010). Key: a, palpus; *ag*, pregenital seta; b, chelicerae; *c*, bothridial seta; c_1–c_3, setae on segment C; d_1–d_2, setae on segment D; e_1–e_2, setae on segment E; f_1–f_2, setae on segment F; g_1–g_2, genital setae; h_1–h_2, setae on segment H; ia, im, ip, lyrifissures; o, ocelli; Op, opisthosoma; p, peritreme; Pr, prodorsum; ps_{1-3}, setae on the pseudanal segment; s, stigma; sc_1, internal scapular seta; sc_2, external scapular seta; v_2, external vertical seta; *1a*–*4a*, coxisternal setae.

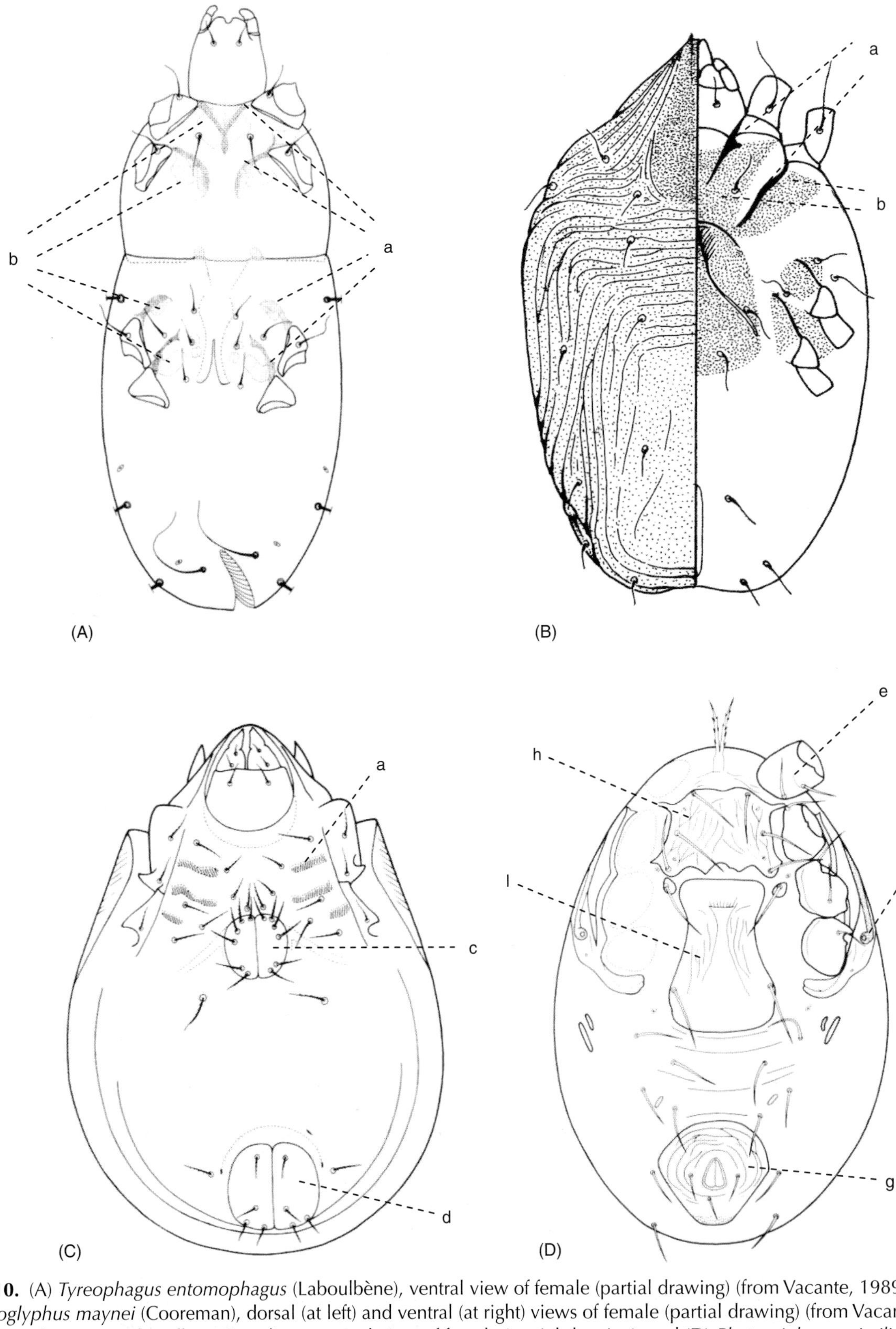

Fig. 2.10. (A) *Tyreophagus entomophagus* (Laboulbène), ventral view of female (partial drawing) (from Vacante, 1989a). (B) *Euroglyphus maynei* (Cooreman), dorsal (at left) and ventral (at right) views of female (partial drawing) (from Vacante, 1997a). (C) *Humerobates rostrolamellatus* Grandjean, ventral view of female (partial drawing); and (D) *Phytoseiulus persimilis* Athias Henriot, ventral view of female (partial drawing) (drawings C–D from Vacante and Nucifora, 1985). Key: a, epimera; b, apodemes; c, genital plate; d, anal plate; e, coxa; f, stigma; g, ventrianal shield; h, sternal shield; l, epigynial shield.

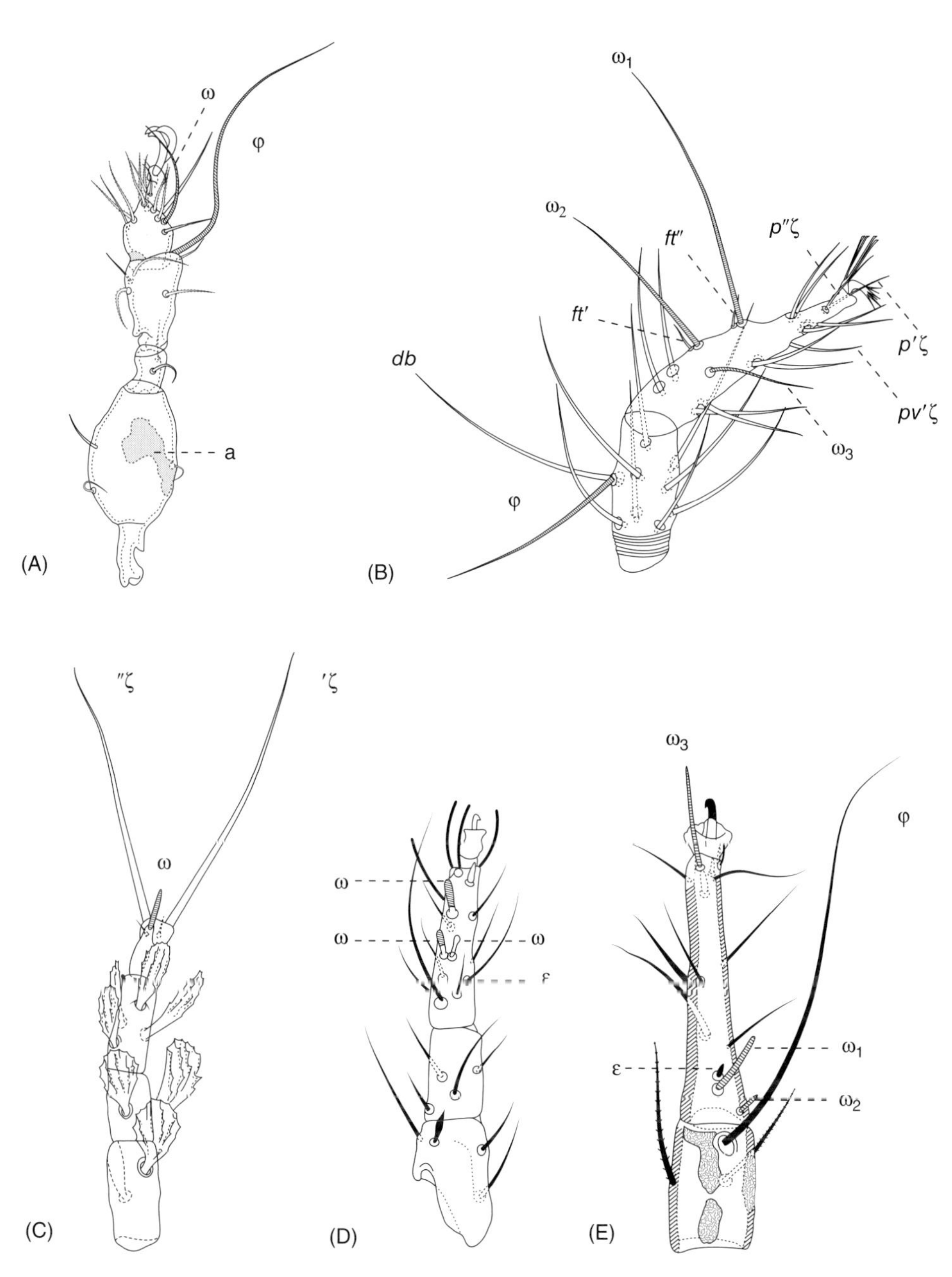

Fig. 2.11. Legs I of acarine species. (A) *Micreremus gracilior* Willmann, female (femur, genu, tibia and tarsus); (B) *Tetranychus urticae* Koch, female (tibia and tarsus); (C) *Cheletogenes ornatus* (Canestrini *et* Fanzago), female (femur, genu, tibia, tarsus) (drawings A–C from Vacante and Nucifora, 1985). (D) *Tarsonemus parawaitei* Kim. Qin *et* Lindquist, female (femur, genu, tibiotarsus) (from Nucifora and Vacante, 2004, produced by Vacante). (E) *Tyrophagus neiswanderi* Johnston *et* Bruce, female (tibia and tarsus) (from Vacante, 1988). Key: a, area porosa; *db*, trichobothridial seta; ω ($ω_1$–$ω_3$), tarsal solenidia (chemosensory setae); φ, tibial solenidion; ε, famulus; *ft*, tactile setae of duplex setae (*ft*′ and $ω_2$, *ft*″ and $ω_1$); ζ, eupathidia.

two lateral (*l*′, *l*″) and two ventral setae (*v*′, *v*″), respectively; taking into account that the paired setae are both paraxial (indicated by ′) and antiaxial (indicated by ″), the setae *l*′ on leg I are paraxial, whereas the setae on leg IV are antiaxial (Figs 2.13 and 2.14). The setal notation of the tarsi includes a greater number of elements, and from the proximal to the distal part of the segment there may be four pairs of dorsal setae, called fastigial (*ft*′, *ft*″), tectal (*tc*), iteral (*it*) and proral (*p*); the ventrolateral setae are the primiventral pair (*pv*′, *pv*″), the unpaired subunguinal seta (*s*), the anterolateral pair (*a*′, *a*″) and the unguinal pair (*u*) (Fig. 2.15) (Norton, 1977).

Usually, the tarsi and pretarsi of legs I and II have various sensory structures (see section on sensory receptors). The pretarsus includes a 'basilar sclerite' and other structures, such as a median 'empodium' (an unpaired structure arising between the tarsal claws), a pad-like or sucker-like 'pulvillus' (a membranous, pad-like structure associated with the claws) or paired 'claws'. When the pretarsus bears median elements, it is called

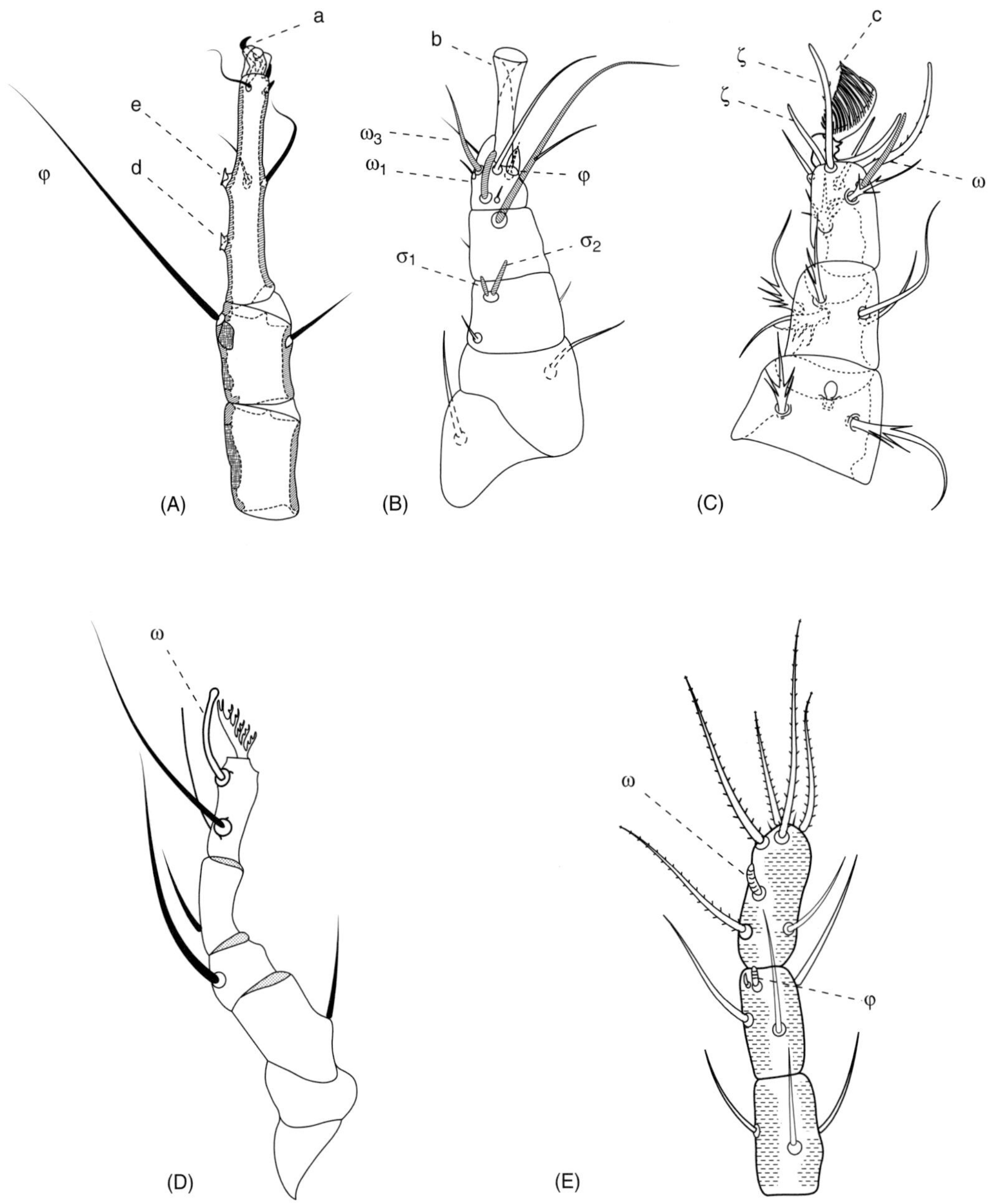

Fig. 2.12. Legs of acarine species. (A) *Tyrophagus neiswanderi* Johnston *et* Bruce, leg IV (genu, tibia and tarsus) of male (from Vacante, 1988). (B) *Hemisarcoptes malus* (Shimer), leg I (trochanter, femur, genu, tibia, tarsus) of female (from Vacante and Nucifora, 1985). (C) *Cheyletiella blakei* Smiley, leg I (genu, tibia, tarsus) of female (from Patti and Vacante, 1985, produced by Vacante). (D) *Aceria neocynarae* (Keifer), leg I of female (new drawing from Vacante). (E) *Pronematus ubiquitous* (McGregor), leg I (genu, tibia, tarsus) of female (from Vacante and Nucifora, 1985). Key: a, claw; b, sucker-like ambulacrum; c, empodium; d, e, tarsal suckers; ω, φ, σ, solenidia; ζ, eupathidia.

an 'ambulacrum'. The basilar sclerite acts as a fulcrum articulated with the tarsus by condylophores and is related to the extensor and depressor muscles arising in the tibia or in the tarsus (Krantz, 2009a). There can be strong diversity in the structure of the ambulacrum; it may be absent, it may have a claw-like or pad-like empodium or the empodium may be absent, paired claws may be absent, there may be a sucker-like pulvillus, etc. (Figs 2.11A–E, 2.12A–E, 2.16A–I).

Respiratory Organs

Respiration in the Acari can be 'cuticular', but is more commonly based on an internal branched transport system that includes spiracular tegumental openings called 'stigmata' or 'stigma'. The number, presence and position of the stigmata vary in different systematic groups and these variations are

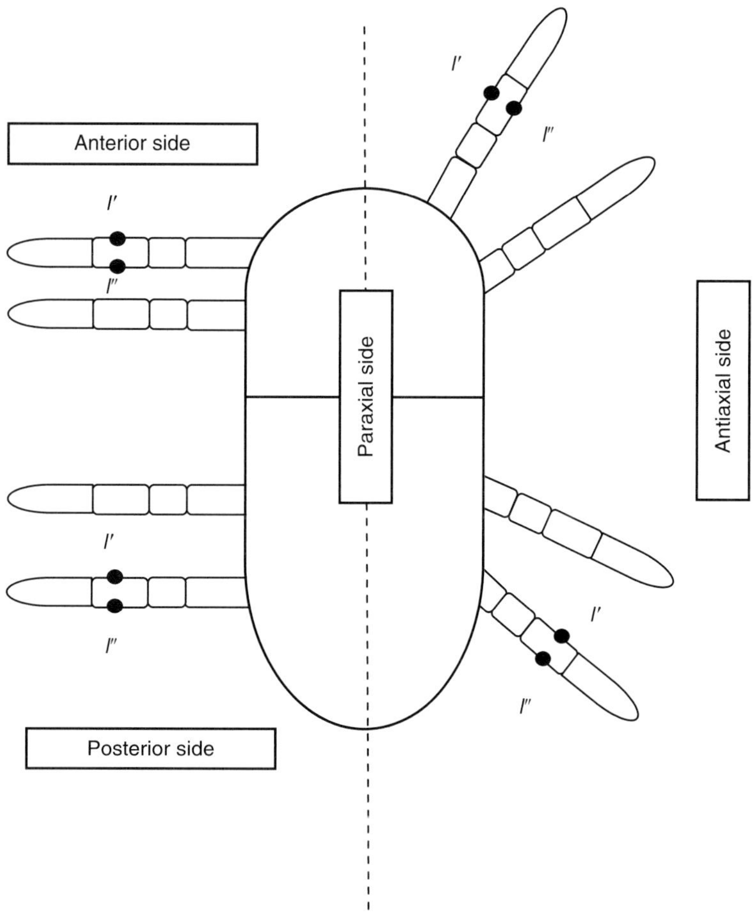

Fig. 2.13. Schematic drawing of an acariform mite showing the paraxial and antiaxial positions of the leg setae, *l′* and *l″* (according to Norton, 1977).

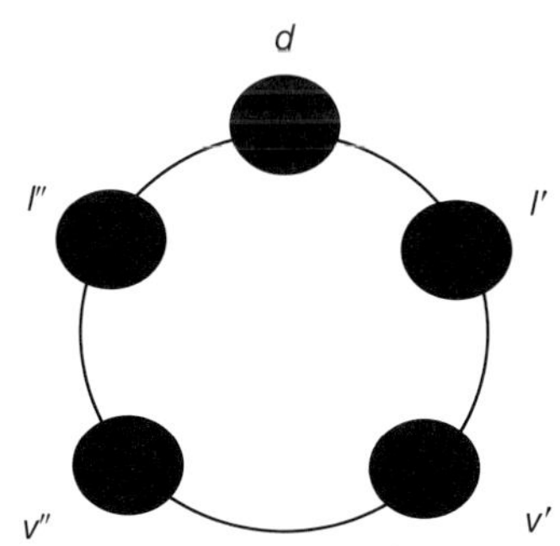

Fig. 2.14. Schematic drawing of the verticil (whorl) of five setae (following Norton, 1977). Key: *d*, dorsal; *l*, lateral; *v*, ventral; ′ indicates paraxial and ″ indicates antiaxial.

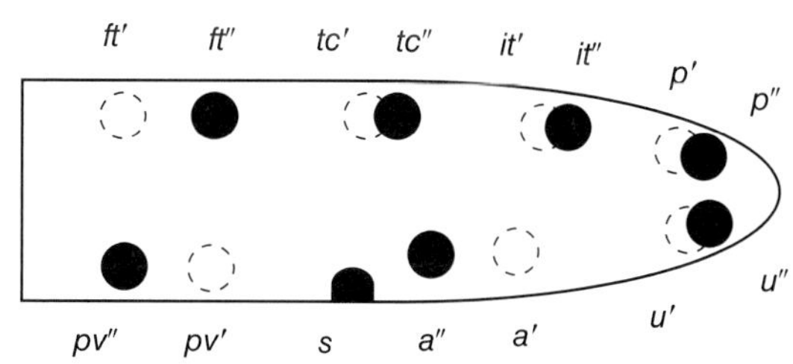

Fig. 2.15. Schematic drawing showing the nomenclature and notation of the tarsal setae (according to Norton, 1977). Key: *a*, anterolateral setae; *ft*, fastigial setae; *it*, iteral setae; *p*, proral setae; *pv*, primiventral setae; *s*, subunguinal seta; *tc*, tectal setae; *u*, unguinal setae; ′ indicates paraxial and ″ indicates antiaxial.

used in taxonomy for the identification of acarine orders and suborders. In all orders of the superorder Parasitiformes, the stigmata are set in the median idiosoma (Fig. 2.10D), while in the suborder Prostigmata of the superorder Acariformes, paired stigmata are located anteriorly (Figs 2.8B, 2.9B–C), behind or between the cheliceral bases (Fig. 2.6F, H and I), or in relation to the lateral humeral angles of the propodosoma (Fig. 2.17A). The stigmatic opening can be connected to tubular 'peritremes' (Figs 2.9B–C, 2.18A–F), which are sometimes extended throughout most of the propodosoma (Fig. 2.17B), or are arranged simply with ducts connecting to a tracheal system (Figs 2.8B, 2.17A). Moreover, in the different acarine groups, there

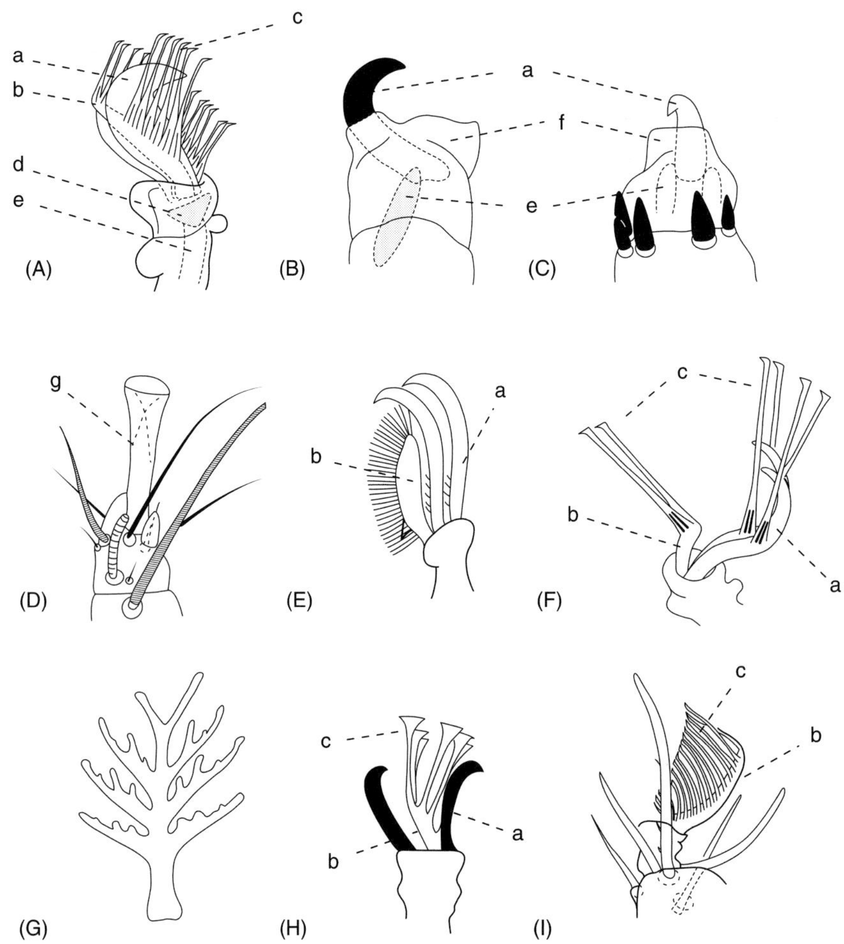

Fig. 2.16. Pretarsi of acarine species. (A) *Bryobia cavalloroi* Vacante *et* Van Eyndhoven, pretarsus IV of female (from Vacante and Van Eyndhoven, 1989). (B) *Tyrophagus tropicus* Robertson, pretarsus I of female; (C) *Michaelopus corticalis* (Michael), distal end of tarsus and pretarsus III of female; (D) *Hemisarcoptes malus* (Shimer), distal end of tibia and tarsus I of female (drawings B–D from Vacante, 1989a). (E) *Tydeus australensis* Baker, pretarsus I of female (from Vacante and Nucifora, 1985). (F) *Bryobia kissophila* Van Eyndhoven, pretarsus I of female (from Vacante, 1985a). (G) Empodium of *Aculops pelekassi* (Keifer) (new drawing from Vacante). (H) *Eryngiopus siculus* Vacante *et* Gerson, pretarsus I of female (H) (from Vacante and Gerson, 1987). (I) *Cheyletiella blakei* Smiley, distal end of tarsus and pretarsus I of female (from Patti and Vacante, 1985, produced by Vacante). Key: a, claw; b, empodium; c, tenent hairs (setae or seta-like processes); d, basilar sclerite; e, condylophore; f, ambulacrum; g, sucker-like ambulacrum.

may be morphological and physiological variation of the respiratory system that involves different structures (encircling plates, microvilli, areae porosae, brachytracheae, ostia, maculae, aeropyles) (Figs 2.11A, 2.18G–H) (Krantz, 2009a).

Reproductive Organs

The reproductive organs of mites can be discussed in terms of their external and internal morphology, and their physiology and ecology, mainly in relation to strategies of sperm transfer (Alberti and Coons, 1999). In this section, the idiosomal external features of the reproductive organs are briefly discussed. Sperm transfer may take place either directly (direct sperm transfer) or indirectly (indirect sperm transfer). Direct sperm transfer involves the entry of the male copulatory organ or 'aedeagus' (Fig. 2.19A–L) into the female copulatory orifice, which can be located either between the ovipositor and the anal opening (Fig. 2.20A), as in several Prostigmata (Alberti and Crooker, 1985), or in a special extragenital structure, the 'bursa copulatrix' (Fig. 2.20D), at the rear of the idiosoma, as in the Astigmatina and some Prostigmata. The bursa is connected to the internal seminal receptacle, or 'receptaculum seminis', from which two ducts lead into the ovaries (Evans, 1992). The function of the genital opening is fundamentally that of oviposition, and for this reason, it is called an 'oviporus'. The morphology of the receptaculum seminis varies according to systematic group (Fig. 2.21A–N). In several Prostigmata, as the Tetranychidae (Pritchard and Baker, 1955; Gutierrez and Helle, 1985; Vacante, 1985a; Van Eyndhoven and Vacante, 1985) or the Eriophyidae

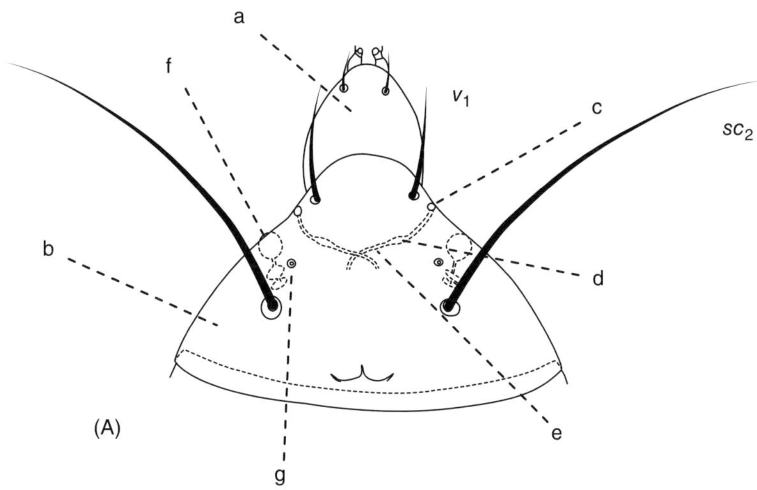

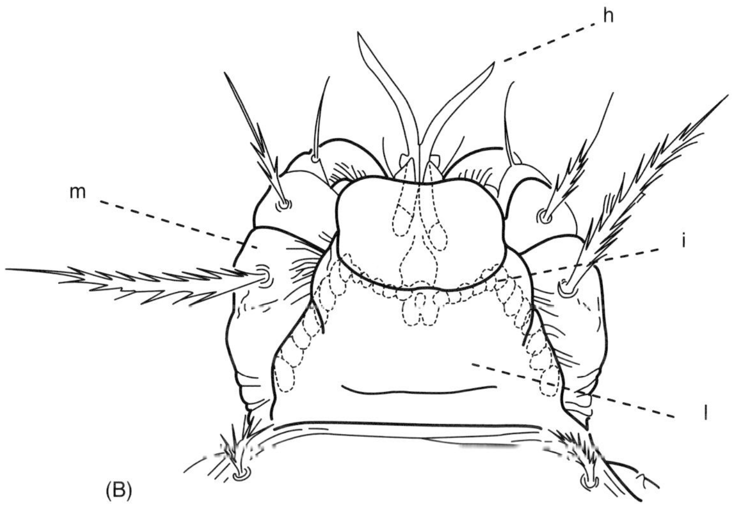

Fig. 2.17. (A) Gnathosoma and prodorsum of female of *Tarsonemus waitei* (Banks) (from Vacante and Nucifora, 1985, partially modified). (B) Gnathosoma of female of *Cheyletiella blakei* Smiley (from Patti and Vacante, 1985, produced by Vacante). Key: a, gnathosomal capsule, b, prodorsum, c, stigma, d, atrium, e, tracheal trunk, f, bothridial sensillum or pseudostigmatic organ, g, pit, h, chelicera; i, peritreme; l, gnathosoma; m, palpus; sc_2, external scapular seta; v_1, external vertical seta.

(Amrine *et al.*, 2003), the morphology of the aedeagus and the receptaculum seminis (spermatheca) is used in specific or generic taxonomy.

Non-copulatory sperm transfer uses various strategies, ranging from a transfer of sperm during male/female pairing (Walter and Proctor, 1999) to the production by the male of 'spermatophores' that are deposited on a support surface. The morphology of spermatophores varies from elongate to sculpted, and from small to very small, and is distinctive by family and genus (Fernandez *et al.*, 1991). The deposition of spermatophores can be 'proximal', when the females are close to or paired with the males, or 'dissociative' if occurs in the absence of females, as in the Eriophyidae (Oldfield *et al.*, 1970, 1972; Oldfield and Michalska, 1996). Spermatophores can also be deposited in variable numbers, singly or at high numbers in masses or organized in rows. In general, presence of the male assists the possibility of the contact between females and spermatophores by various means (physical, chemical or behavioural). Indirect sperm transfer can occur by 'tocospermy', in which the male uses one or both chelicerae to transfer its sperm into the female genital opening, or by 'podospermy', which uses a cheliceral sperm transfer organ or 'spermatodactyl' (Fig. 2.21O) to transfer sperm into solenostomes or sperm induction pores (Athias Henriot, 1969c), variously located in the idiosoma and analogous to the bursa copulatrix of the Astigmatina (Krantz, 2009a).

The reproductive behaviour of Acari, especially sperm transfer and mating, can be influenced by pheromones that are emitted by receptive biological stages (adult and immature females). Three types of these sex pheromones are known: 'arrestans', 'attractans' and 'contact pheromones' (Sonenshine, 1985). The arrestans pheromones are emitted by immature females and act by stimulating precopulatory guarding by adult males. The attractans pheromones are produced by adult females and act by attracting the males. The contact pheromones

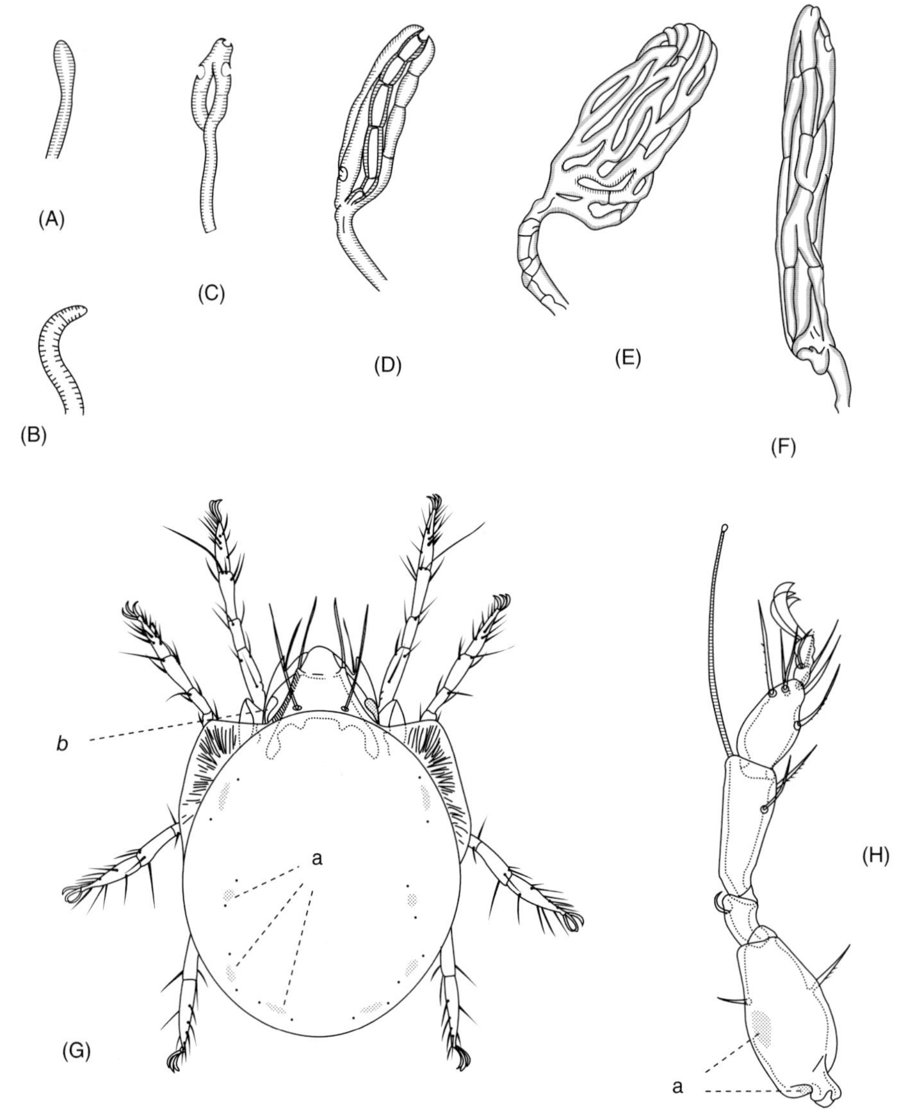

Fig. 2.18. (A–F) Peritremes of acarine species. (A) *Panonychus citri* (McGregor), female; (B) *Petrobia tunisiae* Manson, female; (C) *Bryobia rubrioculus* (Scheuten), female; (D) *Tetranycopsis horridus* (Canestrini *et* Fanzago), female; (E) *Aplonobia histricina* (Berlese), female (drawings A–E from Vacante, 1985a). (F) *Bryobia cavalloroi* Vacante *et* Van Eyndhoven, female (from Vacante and Van Eyndhoven, 1989). (G–H) Areae porosae. (G) *Humerobates rostrolamellatus* Grandjean, female; and (H) *Micreremus gracilior* Willmann, female (drawings G and H from Vacante and Nucifora, 1985). Key: a, area porosa; *b*, bothridial seta.

are emitted by adult females and stimulate copulatory behaviour in the males. The phenomenon has been investigated in several groups of acari (Krantz, 2009a).

Sensory Receptors

The idiosoma has different types of sensory receptors (chemoreceptors, mechanoreceptors and photoreceptors) commonly with a setal structure (Evans, 1992; Alberti and Coons, 1999). According to Grandjean (1935a), tactile and chemosensory setae may or may not contain an optically active iodophilic material, 'actinopilin', which is birefringent in polarized light and constitutes a layer surrounding part of the basal setal nerve cell. The Acariformes have 'anisotropic' setae that are optically active, and for this reason they also have the name Actinotrichida, while the Parasitiformes have 'isotropic' setae that are optically inactive, so they also have the name Anactinotrichida (Van der Hammen, 1961; Evans, 1992). The cuticular surface also has different pore-like openings, some of which have a sensory function (Krantz, 2009a).

Both cuticular and intracuticular pore-like organs may receive inputs from changes in physical tension within the body (Alberti and Crooker, 1985). These proprioceptors have various shapes, but they are commonly small membrane-covered clefts

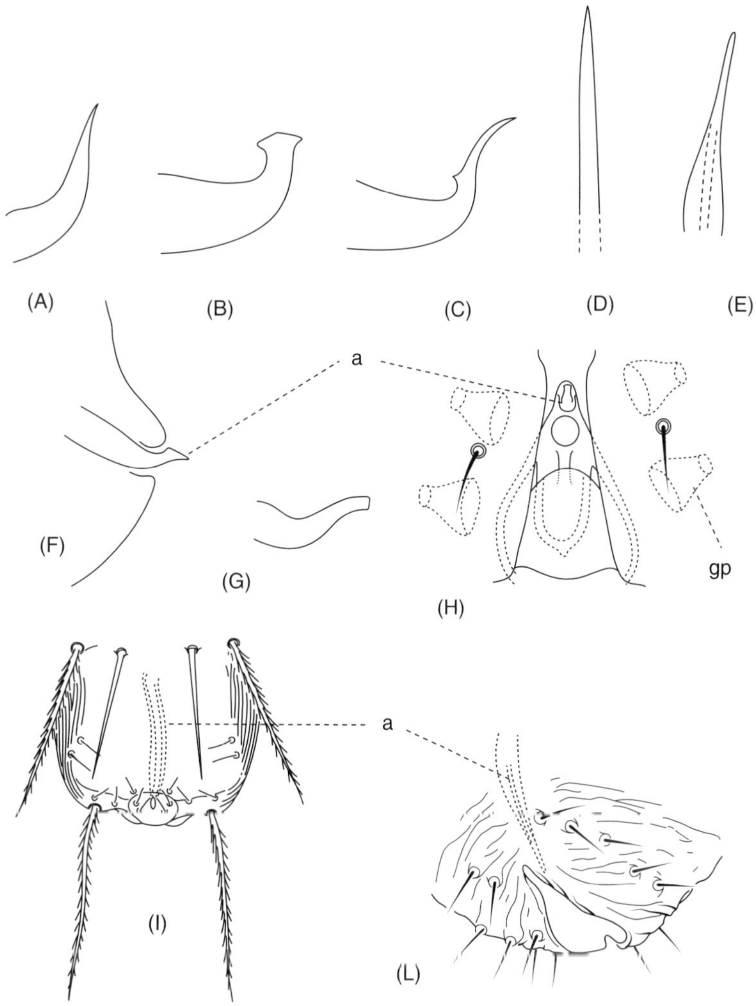

Fig. 2.19. Aedeagi of acarine species. (A) *Panonychus citri* (McGregor); (B) *Tetranychus urticae* Koch; (C) *Amphitetranychus viennensis* (Zacher); (D) *Petrobia tunisiae* Manson; (E) *Bryobia vandaelei* Vacante; (F) *Eotetranychus rubiphilus* (Reck) (drawings A–F from Vacante, 1985a). (G) *Tyrophagus neiswanderi* Johnston *et* Bruce (from Vacante, 1988). (H) *Tyrophagus palmarum* Oudemans (from Vacante, 1989a). (I) *Cheyletiella blakei* Smiley, male opisthosoma (from Patti and Vacante, 1985, produced by Vacante and partially modified). (L) *Bryobia pyrenaica* Van Eyndhoven *et* Vacante (from Van Eyndhoven and Vacante, 1985). Key: a, aedeagus; gp, genital papilla.

that are referred to as 'slit organs', 'poroids', 'cupules' or 'lyrifissures' (Alberti, 1998) (Figs 2.8A, D, 2.9A, D, 2.20C).

Commonly, chemosensory receptors have a setal appearance; they are usually anisotropic and located on distal segments of legs I and II and on the palpi. Inside, they have a sheath of actinopilin, and externally, their structure (and morphology) vary significantly. Among the variations found are the 'eupathidia', which are spine like, apically rounded or divided, and set on the distal segment of tarsi I and II and at the end of the palptarsus (Figs 2.7A–C and F, 2.11B–C, 2.12C). They probably have a single terminal pore and serve as contact or gustatory receptors or chemoreceptors (Alberti, 1998; Alberti and Coons, 1999). Their presence on the legs and palpi has been associated with a gustatory–mechanosensory function (Evans, 1992). Another receptor, whose function is not clear, is the famulus, which is a simple or peg-like seta, sometimes falcate, branched, etc. (Fig. 2.11D–E). The eupathidia have a terminal pore, which is usually small and set on tarsus I in the Acariformes, and sometimes on tarsus II. In many cases, it is invaginated into pits (depressions) (Alberti and Coons, 1999). Another type of setal receptor is the solenidia; these are common on the tarsi and on tibiae I and II (sometimes on femur III and IV) and on the palptarsus of Acariformes (Figs 2.7A–C, E, H, 2.11A–E, 2.12A–E). They are rounded, pointed or peg like, and lack the actinopilin that is present in the eupathidia and famuli; they are also birefringent and sometimes striated, as in the Tetranychidae. Their function is probably olfactory (Evans, 1992). Other probable olfactory receptors are the 'sensilla basiconica' of various Mesostigmata, and the 'Haller's organ' of the Ixodida.

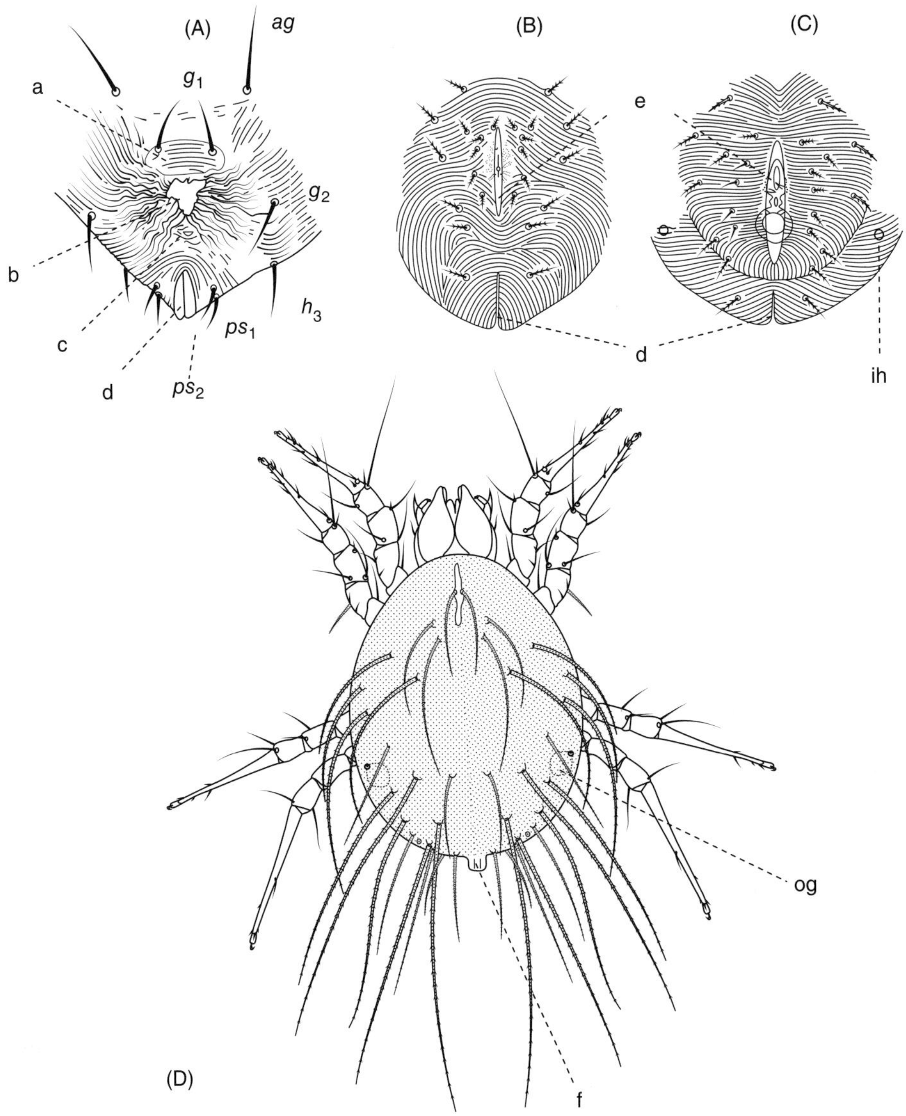

Fig. 2.20. (A) Posteroventral body surface of female of *Tetranychus urticae* Koch; (B–C) posteroventral body surface of female (B) and male (C) of *Triophtydeus triophthalmus* (Oudemans); and (D) *Glycyphagus domesticus* (De Geer), female (from Vacante and Nucifora, 1985). Key: a, genital coverflap; ag, pregenital seta; b, ovipositor; c, copulatory orifice; d, anal opening; e, progenital aperture; f, bursa copulatrix; g_1–g_2, genital setae; h_3, seta on segment H; ih, lyrifissure; og, opisthonotal gland; ps_{1-2}, setae on the pseudanal segment.

Most mechanoreceptors are simple tactile setae, with an isotropic or anisotropic shaft, a membranous socket and basal receptor cells that send signals from the setae to the central nervous system. Morphologically, the tactile setae are usually simple and spinose, but they may also have different shapes, such as expanded or variously ornate, etc. (Fig. 2.22A–D). On the base of leg I of the Astigmatina there is a supracoxal seta (Fig. 2.22E–G) that is variable in form – from a short spine-like seta in those species inhabiting wetter habitats to a many-branched seta in those species inhabiting drier environments; the latter type of seta seems to cover gland openings and reduce the loss of water. The shape, number and position of the tactile setae can form a pattern that is used in the taxonomy of various groups. A particular type of tactile setae are the 'trichobotria', which are specialized sensilla variously present on the body, legs and appendages of some acarine groups. Prodorsal tricho-bothria, called also the 'bothridial setae' or 'bothridial organs' or 'pseudostigmatic organs' are located in cup-like cavities or 'bothridia', and consist of a long and specialized seta of various shapes (simple, clavate, pectinate, etc.) (Evans, 1992) (Figs 2.3, 2.9A, 2.18G, 2.22 L–N). Despite the variable form of the both-ridial setae, they are similar to the acariform tactile setae and have a solid core of actinopilin. Their main function is anemo-tactic (Evans, 1992), and they may also be an acoustic organ

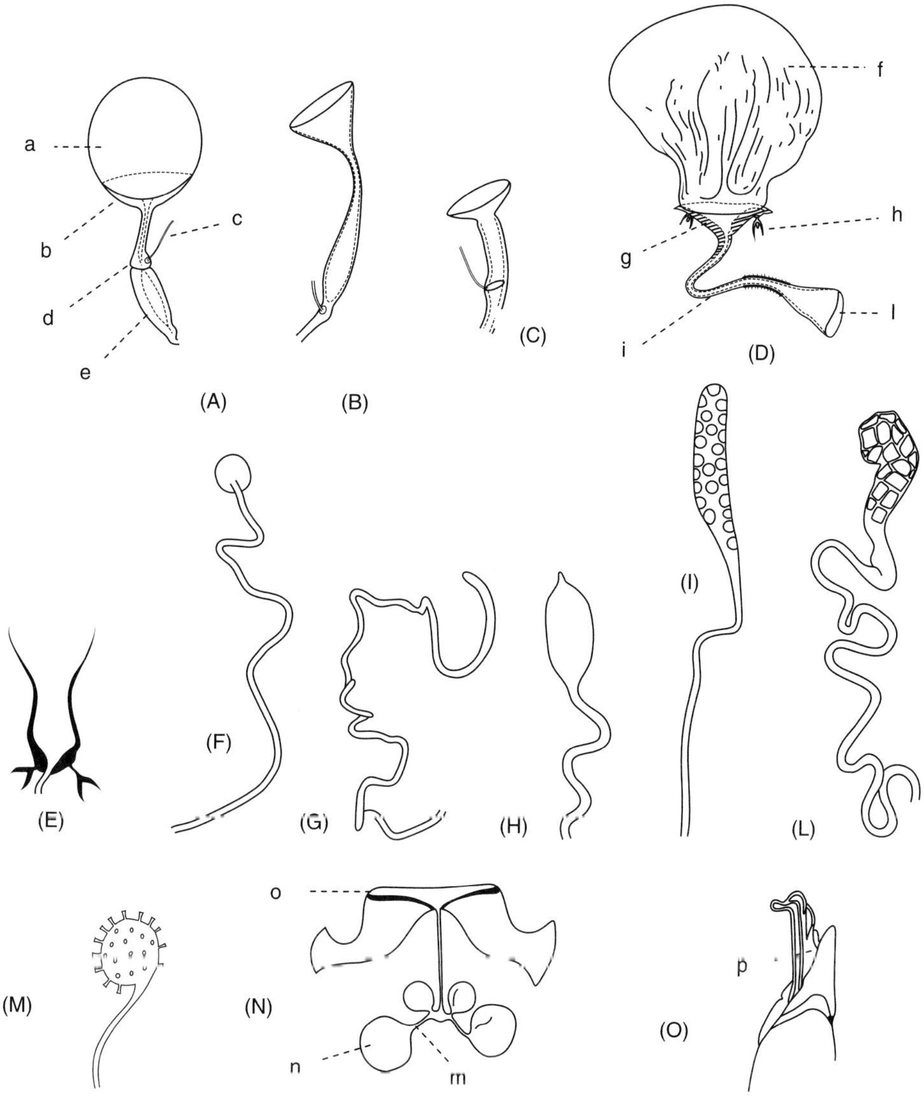

Fig. 2.21. (A–N) Receptacula seminis of acarine species. (A) *Phytoseius finitimus* Ribaga; (B) *Phytoseiulus persimilis* Athias Henriot; (C) *Euseius stipulatus* (Athias Henriot) (drawings A–C from Vacante and Nucifora, 1987). (D) *Tyrophagus neiswanderi* Johnston *et* Bruce (from Vacante, 1988). (E) *Thyreophagus entomophagus* (Laboulbène) (from Vacante, 1989a). (F) *Bryobia watersi* Manson; (G) *Bryobia nikitensis* Livshitz *et* Mitrofanov; (H) *Bryobia parietariae* Reck; (I) *Bryobia urticae* Sayed; (L) *Bryobia vasiljevi* Reck (drawings F–L from Vacante, 1985a). (M) *Brevipalpus phoenicis* (Geijskes) (from Vacante and Nucifora, 1985). (N) *Aceria neocynarae* (Keifer) (from Vacante and Tropea Garzia, 1990). (O) Chelicera of male of *Amblyseius californicus* (McGregor) with spermatodactyl (sperm finger) (from Vacante and Nucifora, 1985). Key: a, vesicle; b, cervix; c, minor duct; d, atrium; e, major duct; f, spermathecal sac; g, base of spermathecal sac; h, sclerite of oviduct; i, spermathecal duct; l, copulatory opening; m, spermathecal tube; n, spermatheca; o, anterior apodeme; p, spermatodactyl.

(Coineau *et al.*, 1997), act as vibroreceptors and anemoreceptors (Alberti *et al.*, 1994) and have a geotaxic function (Aoki, 1973). Moreover, trichobothria could be involved as capacitors in deflecting electromagnetic signals produced by their host or food substrate (Wergin *et al.*, 2000a), or in maintaining balance (Evans, 1992).

Several acarines have eyes, eyespots and photoreceptive organs. One or two pairs of simple 'photoreceptors' or 'ocelli' are set on the anteromedian or anterolateral prodorsum (the dorsal surface of the propodosoma) in various groups (Evans, 1992; Alberti and Coons, 1999; Coons and Alberti, 1999). Each ocellus consists of a cuticular lens, biconvex or convex–concave, and an underlying photoreceptor, with axons forming the optic nerve (Evans, 1992). Tetranychid mites have lateral prodorsal ocelli divided into two parts (Fig. 2.9B–D), each functioning differently from the other; the anterior part represents a biconvex eye and may act as a scanning-point detector for green and UV light, while the posterior part receives signals in the

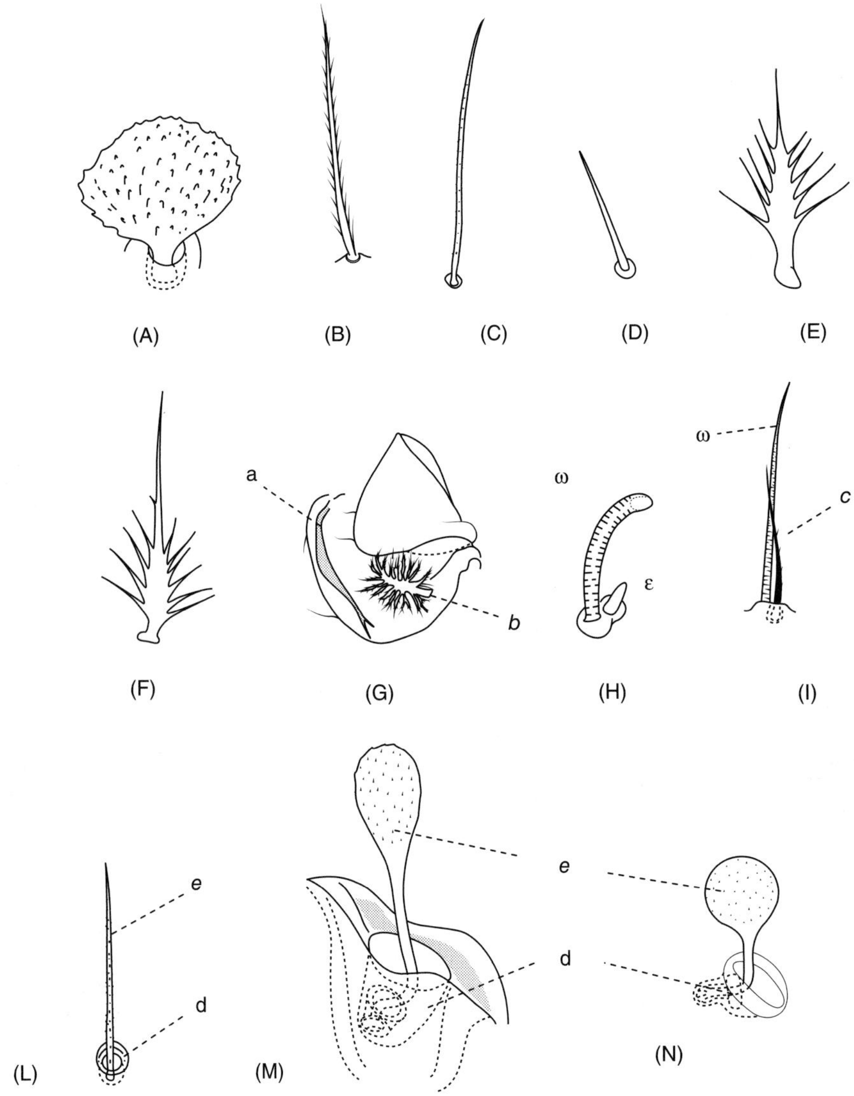

Fig. 2.22. (A) First dorsal idiosomal seta of *Bryobia cavalloroi* Vacante *et* Van Eyndhoven (from Vacante and Van Eyndhoven, 1989). (B) Opisthosomal seta of male of *Cheyletiella blakei* Smiley (new drawing from Vacante). (C) Dorsal idiosomal seta of female of *Eotetranychus rubiphilus* (Reck) (from Vacante, 1985a). (D) Seta *f* of tergite EF of female of *Tarsonemus waitei* Banks (from Vacante and Nucifora, 1985). (E) Supracoxal seta of *Tyrophagus palmarum* Oudemans; (F) supracoxal seta of *Tyrophagus tropicus* Robertson; (G) Grandjean's organ and supracoxal seta of *Glycyphagus domesticus* (De Geer); (H) solenidion and famulus of tarsus I of female of *Thyreophagus cooremani* Fain (drawings from Vacante, 1989a). (I) Duplex setae of tarsus IV of *Bryobia praetiosa* Koch (from Vacante, 1985a, partially modified). (L–N) Bothridial setae of *Triophtydeus triophthalmus* (Oudemans) (L), *Humerobates rostrolamellatus* Grandjean (M) and *Micreremus gracilior* Willmann (N) (drawings from Vacante and Nucifora, 1985). Key: a, Grandjean's organ; *b*, supracoxal seta; *c*, tactile seta; d, bothridium; *e*, bothridial seta; ω, solenidium; ε, famulus.

near UV that are probably non-directional (Alberti and Crooker, 1985). Other variations may occur in different groups.

Notation of mechanosensory and chemosensory setae

There has been a certain confusion around the designation of mechanosensory and chemosensory setae and their derivative structures on the appendages of actinotrichid mites. This problem was investigated by Grandjean (1935b, 1939a, 1958), who indicated a solution for identifying the homologies between the structures in the various taxa. In the Grandjean system, the setae are indicated by Greek symbols, with the solenidia of the femur, genu, tibia and tarsus designated by θ (theta), σ (sigma), φ (phi) and ω (omega), respectively. Famuli and eupathidia are designated by ε (epsilon) and ζ (zeta), respectively (Figs 2.11A–E, 2.22 H–I).

Secretory Organs

The secretory organs discussed in this section concern external secretions. It is very difficult to summarize the amount of information that is available on this topic and at the same time to present an account that is consistent in character.

Pore-like openings located in the acarine cuticle are common among the secretory organs and may be connected by ducts to subcuticular gland cells (Alberti and Seeman, 2005). The nature of the products that are secreted is heterogeneous and varies from cement like to waxy (Evans, 1992; Coons and Alberti, 1999) to sex pheromones (Hoy and Smilanick, 1979) or to defensive secretions as a means of protection from predators (Kerfoot, 1982).

Some oribatid mites have lateral 'opisthonotal glands', which are easily seen as variously coloured spots below the transparent integument (Figs 2.4, 2.8C, 2.20D) (Evans, 1992; Alberti and Coons, 1999). These oil glands have a duct, with a pore and a cuticular flap. The secretions of these glands are of varying chemical nature (they may be monoterpenes, hydrocarbons, esters, aromatics, etc.) and may have various functions, such as acting as pheromones (alarm, aggregation, sexual, defence, etc.). Neryl formate, a compound secreted by the opisthonotal glands of some Oribatida, has been identified as a key alarm pheromone, and is also found in the injurious acarid mite *Rhizoglyphus robini* Claparède (Mizoguchi *et al.*, 2003). Raspotnig *et al.* (2011) found that the secretions from the opisthonotal glands of four species of oribatid mites contain the iridoid monoterpenes chrysomelidial and epi-chrysomelidial and the diterpene β-springene, and discuss the chemosystematic discrimination of morphologically similar, syntopic or cryptic species based on their oil gland secretion profiles. Other secretory organs, such as the 'sensilla sagittiformia' are known for ixodid mites. A detailed treatment of this topic can be found in Alberti and Coons (1999).

Osmoregulatory Organs

Acarine osmoregulation relies on a pair of 'urstigmata' or 'organs of Claparède'. Serially homologous genital papillae are present in nymphs and adults whose larvae have Claparède's organ (Oudeman's Rule), and tydeid mites may retain the urstigmata beyond the larval stage. In the larval stage, the organs of Claparède are located ventrally on coxae I or II or between them and have been retained as homologous to the genital papillae (Fig. 2.19H). According to Vercammen-Grandjean (1975), in representative Prostigmata,the urostigmata and the genital papillae play a respiratory function, but an alternative interpretation concludes that these organs play a role in osmoregulation (Alberti, 1979; Alberti and Bader, 1990). There is also a diversity of other structures that might be related to the function of osmoregulation.

3 Internal Morphology

Like other Arthropoda, the internal morphology of mites primarily involves the idiosoma, which includes the 'haemocoel', a cavity that extends also to the legs and surrounds the internal organs that constitute the different systems of the body (circulatory, nervous, reproductive, glandular, digestive and excretory).

Circulatory System

The haemocoel is filled with 'haemolymph', a free-flowing liquid whose circulation is facilitated by body movement and the dorsoventral contraction of the idiosomal muscles. The haemolymph contains various substances, such as amino acids (Boctor, 1972), lipids (Woodring and Galbraith, 1976), sugars (Aboul-Nasr and Bassal, 1971), etc. and several cellular elements, the 'haemocytes', which are adapted to different functions, such as clotting, macrophagy and tissue dissolution, the development of which may involve germinal cells (Evans, 1992; Alberti and Coons, 1999; Coons and Alberti, 1999). The Ixodida and some of the larger sized Gamasida may also have a 'heart', which is set in the anterior opisthosoma and is hexagonal or rhomboid in shape and flattened dorsoventrally. This has a single chamber with two dorsal ostia and is related anteriorly to the long aorta.

Nervous System

The nervous system consists primarily of a 'synganglion', which is a nervous mass extended around the oesophagus and basically corresponds to the central nervous system. It is characterized by a protocerebral 'supraoesophageal ganglion' and a tritocerebral 'suboesophageal ganglion' (Alberti and Crooker, 1985; Nuzzaci and Alberti, 1996) (Fig. 3.1). The synganglion has an outer 'cortex' and an inner 'neuropil' but lacks a deutocerebral portion owing to the absence of antennae, the lack of which is the cause of this secondary loss (Evans, 1992). The nervous system of mites does not have separate dorsal or ventral portions, but there are dorsal and ventral ganglia related by commissures (circumoesophageal, etc.) that are located in the prosoma. In some groups, such as the Ixodoidea, the cortex has a 'perineurium', 'glial cells' and 'neuronal cell bodies' which are involved in the nerve impulse and in the secretion of neuromediators (Coons *et al.*, 1974). Studies carried out on ixodids have shown that the central nervous system includes neurosecretory cells that produce different substances that mediate various physiological processes (nymphal cuticular apolysis, moulting, oviposition, salivation, etc.) (Hughes, 1954; Ioffe, 1965; Binnington and Obenchain, 1982; Hänel, 1986). In parasitiform mites, a degree of sexual dimorphism may depend on neurosecretory development (Krantz, 2009a). The cholinergic system (acetylcholine, choline acetylase, cholinesterase) and synaptic action are basic to nerve function.

Supraoesophageal ganglion

Much of the knowledge on the supraoesophageal ganglion is from studies that have been carried out on the Ixodoidea, Astigmatina and some Prostigmata. This ganglion may be variously connected with other nervous organs. In *Dermacentor variabilis* (Say), for example, an 'optic lobe' and 'cheliceral lobes' have been found, with a posterior 'stomodaeal bridge' and 'pedipalpal ganglia' (Ivanov and Leonovich, 1983). In the Astigmatina, these pedipalpal ganglia serve the palpi and the pharyngeal muscles (Hughes, 1959). In the two-spotted spider mite, *Tetranychus urticae* Koch, the nervous structure is different from that found in ticks. In *T. urticae*, an anterior unpaired 'rostral nerve' reaches the capitulum and the pedipalpal ganglia originate in the ventral suboesophageal part of the synganglion (Alberti and Crooker, 1985).

Suboesophageal ganglion

Arising from the suboesophageal synganglion are the pedal ganglia and an unpaired posterior 'opisthosmatic ganglion'; from the latter an unpaired abdominal nerve may originate that extends posteroventrally beyond the podosoma. Numerous peripheral nerves, which are variable in the different suborders and in their nomenclature (e.g. genital, scutal, stomodaeal, splanchnic, tritosternal, rectal), are connected to the synganglion (Hughes, 1959; Woodring and Galbraith, 1976; Woolley, 1988).

Reproductive System

The female and male reproductive systems of mites are composed of different organs that are structured in different ways (paired, fused or multiple) and are variable in the higher taxa.

Female system

The female reproductive system includes one or two 'ovaries' and 'oviducts', the 'uterus', the 'progenital chamber', the 'vagina' and the 'genital aperture' (Figs 3.1 and 3.2A). In various

Acariformes, sperm transfer occurs via an extrudable aedeagus that transfers sperm directly through the female genital aperture (Alberti and Crooker, 1985), while in the Astigmatina, the sperm may be transferred to the female through a 'bursa copulatrix', which is related to a 'seminal receptacle', and the ovary and the genital aperture functions as an 'oviporus' – an opening for the discharge of eggs (Hughes, 1959). In some Mesostigmata, sperm transfer by extragenital insemination occurs through two 'induction pores' or 'solenostomes' (Athias Henriot, 1968) that are connected to a 'sacculus vestibulus' and to a 'tubulus annulatus', which leads the sperm into the organ complex. In podospermous Mesostigmata, the organ complex may have two different forms, which are designated the type A and type B complex. The phytoseiid mites have a type B complex, structured in a solenostome, a major duct, an atrium, a calyx and a vesicle; a minor duct originates from an enlargement of the major duct and ends in a lyrate organ that is related to an ovary (Alberti and Di Palma, 2002) (Fig. 3.2B).

The number of ovaries varies between acarine groups. The Parasitiformes, Trombidiformes and several acariform Oribatida have an unpaired ovary, while the Astigmatina may have two ovaries. The oviducts are commonly paired, but a single oviduct occurs in some Mesostigmata and Trombidiformes (Evans, 1992). The Eriophyidae and Tarsonemidae have an unpaired oviduct, probably as a result of an adaptation that is related to its small size. The oviducts may secrete substances involved in eggshell production (Woodring and Galbraith, 1976). The vagina is next to the oviducts and may develop in an 'ovipositor' that is variously structured in different acarine groups; it may be a tube-like eversion of the vaginal wall, as in the Tetranychidae, or it may be more complex.

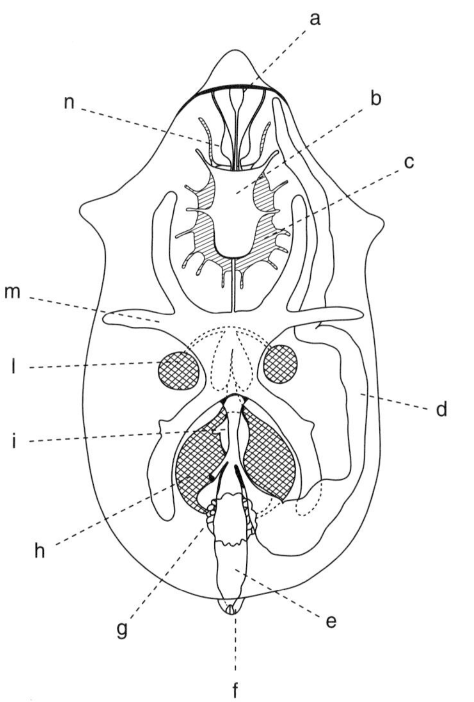

Fig. 3.1. *Caminella peraphora* Krantz *et* Ainscough, internal aspect of female (from Evans, 1992). Key: a, oesophagus; b, supraoesophageal ganglion; c, suboesophageal ganglion; d, Malpighian tubule; e, anal atrium; f, anus; g, ovary; h, oviduct; i, colon; l, accessory gland; m, gastric caeca; n, salivary gland.

Male system

The male reproductive system consists of paired or fused 'testes' and 'vasa deferentia', a variable number of 'accessory glands' and an 'ejaculatory duct' (Fig. 3.2C). A 'seminal vesicle' may occur

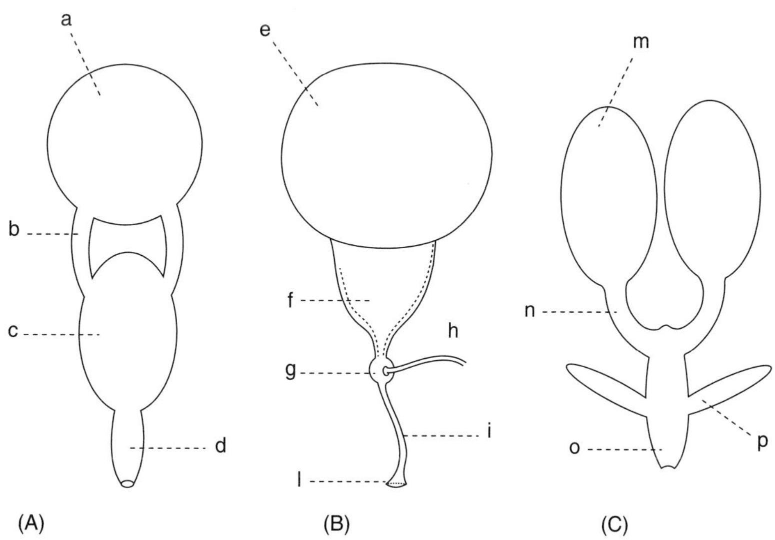

Fig. 3.2. Schematic drawings of the female and male reproductive systems (produced by Vacante). (A), female reproductive system; (B) type B complex of a phytoseiid mite; and (C) male reproductive system. Key: a, ovary; b, oviduct; c, uterus; d, vagina; e, vesicle; f, calyx; g, atrium; h, minor duct; i, major duct; l, solenostome; m, testis; n, vas deferens; o, ejaculatory duct; p, accessory gland.

in some groups. Some of the Mesostigmata, Astigmatina and Trombidiformes have paired testes, while other groups, e.g. small Prostigmata, may have unpaired testes or still others elongate, acinous or multi-lobed testes. In direct insemination, a muscular sperm pump interposed between the vasa deferentia and the ejaculatory duct may push the sperm into the aedeagus and from this into the female seminal receptacle (Alberti and Crooker, 1985; Evans, 1992). The accessory glands may play different roles, such as the production of the seminal fluid matrix (Woodring and Galbraith, 1976), or biochemicals involved in spermiogenesis or in the prevention of desiccation of the spermatophore stalk (Evans, 1992). All acarine spermatozoa are flagellated, but they are variable in form – from ribbon like or vacuolate to small or simple; their ultrastructure tends to be taxon specific at subcohort levels (Alberti, 2000), and it may be a potential tool for the examination of systematic relations (Alberti, 1984).

Glandular System

The 'glandular system' discussed here is a non-integumental system, i.e. the complex of glands that is variously located in the acarine body but not in the integument. The main glands are the 'salivary glands', the 'coxal glands' and the 'podocephalic glands'. Some acarine groups also have 'tracheal glands' and 'silk glands'.

Salivary glands

The structure and role of the salivary glands vary in the different acarine taxa. The Parasitiformes have a simpler gland type, consisting of paired acinous glands set in the prosoma, while the Acariformes have a slightly more complex organization (Figs 3.1 and 3.3). The eriophyoid mites have 'prosomal glands', which are glandular structures associated with the mouthparts; these consist of three glands that pass their secretions through ducts between the stylets of the mouthparts at the base of the labrum. One pair of these glands is retained as the podocephalic glands, while the unpaired gland represents the 'tracheal gland' (see below). The function of the secretions of these glands is unclear, but they are probably involved in the production of saliva or of a lubricant substance facilitating the action of the stylet (Nuzzaci and Alberti, 1996). The Tetranychidae have a tracheal gland set in the dorsal anterior part of the propodosoma, and its duct crosses the stylophore medioventrally and carries the secretion to between the cheliceral stylets or into the infracapitulum (Alberti and Crooker, 1985).

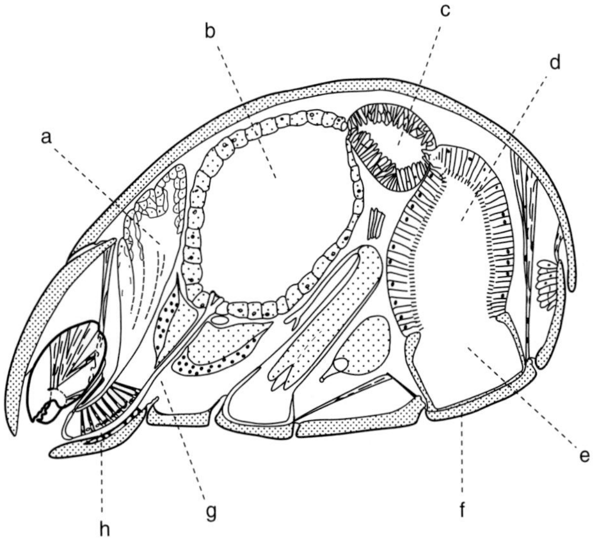

Fig. 3.3. Schematic midsagittal section of *Euzetes* sp. (from Evans, 1992). Key: a, salivary gland; b, ventriculus; c, colon; d, postcolon; e, anal atrium; f, anal plate; g, oesophagus; h, pharynx.

Coxal glands

The coxal glands are probably involved in osmoregulation or the ion/water balance and excretion, and no solid excretory products have been found in them (Alberti and Crooker, 1985). Mills (1973) notes that in the female these glands may secrete a pheromone additive that attracts the males to female webbing (see section on water balance below). Regardless of their function, the coxal glands have a nephridial origin and consist of a 'coelomic sacculus' and a 'coiled duct' or 'labyrinth'; they may be present in different body regions, but are commonly found in the coxal regions. Several species bear coxal glands in the prosoma, near the legs; in the Mesostigmata, they may be set in the opisthosoma near a long duct opening in coxa I (Alberti *et al.*, 1996). In the Oribatida and Astigmatina, the typical structure consists of the 'supracoxal complex' located at the base of leg I; this covers a pair of coxal glands which open by a small slit at the base of the 'supracoxal seta' (Fig. 2.22G). In the Tetranychidae, the sacculus seems to be absent and only the tubulus has been found. The tubulus is divided into three parts (proximal, middle and distal parts); the proximal and middle parts double back on one another, and the distal part leads through a duct to the podocephalic canal (Alberti and Crooker, 1985).

Podocephalic glands

All Acariformes have a pair of lateral 'podocephalic canals' (Van der Hammen, 1980), which are connected to the propodosomatic glands, collect product from the glands and carry it to the gnathosomatic region (Evans, 1992; Alberti and Coons, 1999). The structure of these canals may vary; sometimes they are like an open external gutter, while in other cases they are clearly internal, as in the Prostigmata. In the Astigmatina, they are short, and in the Oribatida they are long. The two podocephalic ducts collect the products secreted by the coxal glands and by a variable number (from one to three) of additional gland pairs. The functions of these secretions are variable, and include endocrine substances for moulting, products of the salivary glands (Woodring, 1973) and products for silk production (Alberti, 1973). In the Tetranychidae, each podocephalic canal receives the secretion of three glands. The two canals meet at the basis of the gnathosoma to form a single canal, the 'median salivary

duct', which travels along the infracapitulum to the gnathosoma. The dorsal and anterior podocephalic glands contain numerous granules, probably of a protein nature. The dorsal glands consist of two cells and the anterior glands of four cells, as in *T. urticae* (Alberti and Crooker, 1985). These glands are retained salivary glands (Mothes and Seitz, 1981a), but their function is unclear and they may be fat bodies (Summers *et al.*, 1973).

Tracheal glands

Unpaired tracheal glands are found in the Tetranychidae and they consist of many cells related by interdigitations and containing droplets of a probable lipid nature (Alberti and Crooker, 1985). Mills (1973) considers them to be the dorsal salivary glands, while Mothes and Seitz (1981a) suggest that they produce a secretion which helps the action of the stylets.

Silk glands

Different mite families produce silk from salivary gland secretions (Gerson, 1985a), e.g. the Tydeidae (Schuster, 1972), Cunaxidae (Walter and Kaplan, 1991) and Cheyletidae (Furman and Sousa, 1969). Silk production assumes considerable importance in several tetranychid species. In these mites, the silk glands consist of a unicellular gland located in each palpus which extends back to the central nervous mass (Alberti and Stork, 1974; Mothes and Seitz, 1981a). The gland cell is very large and contains many vacuoles filled with proteinaceous secretions. In the distal palpal segments, the droplets are contained in the reservoir-like area of the cell apex, which narrows and is connected to a terminal outlet at the tip of the spinneret. Mills (1973) reported that the podocephalic complex of *T. urticae* was infolded interiorly in silk production, while Alberti and Crooker (1985) reported that the secretions of other glands may also be involved in silk production.

Digestive System

The digestive system functions as both a 'preoral digestive system' and a 'postoral digestive system'. The preoral system is based on the secretion of salivary enzymes on to or into the food in order to predigest food and facilitate its ingestion as a liquid (Evans, 1992; Alberti and Coons, 1999; Coons and Alberti, 1999), while postoral digestion is the typical digestive process. The fundamental features of the postoral digestive system are discussed in this section.

The postoral digestive system of mites consists of various organs (Figs 3.1 and 3.3). Anteriorly, the 'buccal cavity' or 'mouth' opens into a muscular 'pharynx' which is connected posteriorly by the 'oesophagus' to the 'stomach' or 'ventriculus'. The oesophagus is narrow and, in some groups, such as the Tetranychidae, passes through a part of anterior salivary or silk gland (Alberti and Crooker, 1985) and the synganglion, from whence it continues into the ventriculus. The ventriculus may be differently shaped and sized, ranging from simple to sac like or equipped with two or more gastric 'diverticula' or 'caeca' (Evans, 1992) of complex function. In general, the Trombidiformes have a larger ventriculus and two large caeca (Woodring and Galbraith, 1976). Any feeding adaptation is generally correlated with the type and size of caeca, e.g. phytophagous mites exhibit expanded large caeca during feeding. Intracellular digestion may occur in the anterior portion of the ventriculus, and this involves vacuolated epithelial digestive cells that absorb nutrients and secrete digestive enzymes. The ventriculus opens directly into the 'colon' or 'postventricular midgut', which is sometimes divided by a 'valvular intercolon' into an anterior part and a 'postcolon' or 'anterior hindgut'. In the Tetranychidae, the midgut has a type of cell that functions in different ways at different sites: the cells absorb nutrients in the midgut and excrete them between the ventriculus and the colon (Alberti, 1973; Mothes and Seitz, 1981b). In the Astigmatina, a 'peritrophic membrane' may provide a protective wall that envelops the food bolus as it moves in the postcolon (Wharton and Brody, 1972). The colon leads to an 'anal atrium' or 'posterior hindgut', which leads to the 'anal opening'.

Excretory System

Mites excrete liquid and solid waste products through the 'Malpighian tubules' and other structures. They can also both lose water and regulate its balance. The latter capacity makes it difficult to separate the two functions.

Malpighian tubules

In the Parasitiformes and in some Astigmatina, the posterior midgut, postcolon or intercolon may bear one or two pairs of Malpighian tubules (Woodring and Galbraith, 1976) (Fig. 3.1), which collect and store guanine or uric acid (insoluble nitrogenous waste); these substances can reabsorb water and become paste like and are slowly excreted (Hughes, 1959).

Other structures

The discontinuity of the alimentary duct in some acarine species may complicate the digestive process. The complication primarily involves indigestible solid products, such as chromolipids, which are stored in the caeca until the accumulated faecal material becomes overabundant, leading to the critical situation of 'schizeckenosy', in which there is integument tearing and extrusion, and caecal lobes filled with faeces are pinched off, but without leaving any strong indications of the event (Mitchell and Nadchatram, 1969).

Water balance

The small size of mites means that there is a high surface-to-volume ratio, which affects their water balance. Thus, fluctuations in temperature and relative humidity (RH) oblige the mites to regulate

their water loss by means of various adaptations; these include simple diffusion from the body surface or mechanisms such as defaecation, excretion, the secretion of digestive fluids and the production of various other fluids or reproductive products.

Tetranychid mites use two adaptive mechanisms: a relatively impermeable cuticle, and movements of the stylophore (see Fig. 2.61), which regulate the diffusion of water in the tracheal system. The coxal glands probably act in osmoregulation and ion/water balance and excretion, as no solid excretory products have been found in them (Alberti and Crooker, 1985). Other adaptive mechanisms can also be involved, such as the webbing produced by leaf-feeding mites that conserves moisture in the adjacent leaf surface. Indeed, the RH of plant organ surfaces directly affects feeding rate and egg deposition, as does the rate of water excretion. Furthermore, a high food water content may cause problems in water excretion (Boudreaux, 1958a).

Movement of the stylophore regulates the openings of the peritremes and controls the surface available for diffusion from the tracheal system (Blauvelt, 1945). The mechanism induces an active ventilation of the tracheal system, and at lower RH represents the most important factor in controlling water balance. Mites are therefore sensitive to vapour pressure and capable of intervening in control of the rate of diffusion from the tracheae. The metabolic rate also affects water vapour loss from the tracheae, and under water stress, it is possible to reduce oxygen consumption for long periods (McEnroe, 1961).

The cuticle is also concerned in the regulation of water balance via changes in its permeability and external structure. It functions as a relatively impermeable layer and evaporation barrier and so can partially control water loss, properties that are affected by changes in the lipid components of the cuticle caused by treating mites with chloroform and silica gel (McEnroe, 1961). The strial lobes of tetranychids and the microtubercles on the annuli of eriophyoid mites are evaporative structures that provide for water loss through the cuticle. The absence of strial lobes in non-feeding diapausing tetranychid females assists in the conservation of body water (Boudreaux, 1963), as does the lack of microtubercles in the overwintering deutogynes of several eriophyoid species. Various other adaptations occur in the tetranychid mites, such as *T. desertorum* Banks, the tumid spider mite, *T. tumidus* Banks (Iglinsky and Rainwater, 1954; Nickel, 1960) and *T. urticae* (Boudreaux, 1958a). The clover mite, *Bryobia praetiosa* Koch, actively regulates water loss at RH 53–85%; below 53% RH, water loss is restricted but not regulated (Winston, 1964).

Enlarged lateral and dorsal bands of wax filaments develop in adults of the summer generations of *Abacarus hystrix* (Nalepa). These filaments increase the total surface area of summer-generation mites, but not the cuticular surface-to-volume ratio. High temperatures, but not changes in photoperiod, stimulate wax production. The significantly greater survival times of waxed compared with unwaxed mites at low RH suggest that the wax filaments have a function in the regulation of water loss (Frost, 1997). Note that it seems that the waxy nature of these structures has not yet been properly established, so they should properly be referred to as wax like, although the term wax is commonly used by different authors.

The amount of fluid intake by spider mites is about 20–25% of the total weight an hour (McEnroe, 1961). Studies of the water balance of spider mites have shown that females of *T. urticae* are capable of passing the equivalent of 25% of their weight of water through the system at intervals of 30 min (McEnroe, 1963), and that the liquid could bypass the midgut and directly reach the hindgut (McEnroe, 1961). Various advantages could derive from this adaptation, such as the direct adsorption of small molecules and ions in the hindgut, and the digestion of solid particles for a considerable time in the midgut. Moreover, the rapid passage of water through the digestive system in its excretion by that system does not require selective adsorption and excretion via the tracheal system, and the excretion of water as fluid allows feeding at high RH (McEnroe, 1963). Even the egg has adaptations for satisfying its oxygen requirements without losing too much water. In tetranychid eggs, different organs, such as an air-filled duct system that is kept open by small pillars ('micropillars'), perforation organs and an intermediate lamella are involved in embryonic gas exchange (Dittrich, 1971).

An example of mite-water relations is given by research on the net water exchange of the predatory phytoseiid mite *Phytoseiulus persimilis* Athias Henriot. Water loss by transpiration increases with environmental temperature. The transpiration rate is directly related to the saturation deficit of the air and at constant temperature is linearly dependent on the water vapour activity. Above the critical equilibrium activity (CEA), the mite can absorb water vapour from the unsaturated atmosphere. The amount of water vapour within the laminar layer of the leaf surface has periods during which it is sufficient for the phytoseiid to employ its water vapour sorption capability. *P. persimilis* also exhibits positive hygrotactic behaviour, and can absorb water vapour that can at least double the survival time of starving individuals. The supply of a prey mite (water content about 5.5 mg) every 3 days, and a water vapour activity (a_v) of 0.76 (at 20°C), can allow a survival time of about 120 days (Gaede, 1991).

4 Biology

'Biological instars' and 'reproduction' are two basic factors of mite life. They affect the ecology of mite populations, and knowledge of their characteristics is of strategic importance in the control of injurious species.

Biological Instars

The life cycle of Acari develops through the egg and six (sometimes more) biological instars that follow: the prelarva, larva, protonymph, deutonymph, tritonymph and adult. There are usually two or three nymphal instars between the larval instar and the adult. In no more than a few cases is there only a single nymphal instar, as in the ticks. Protonymphal, deutonymphal and tritonymphal instars are found in the Prostigmata, Astigmatina and some other acarine groups. The adult represents the final stage after a series of three or four moultings, although in some groups there may be post-imaginal moulting during the adult stages.

The life cycle of eriophyoid pest mites develops through the stages of egg, larva, nymph and adult, with two 'quiescent' or 'resting stages' that occur between the larva and the nymph and between the nymph and the adult (Manson and Oldfield, 1996), which are called the 'nymphochrysalis' and the 'imagochrysalis', respectively (Sternlicht and Goldenberg, 1971). The life cycle of the Kanzawa spider mite, *Tetranychus kanzawai* Kishida, has three active biological stages (larva, protonymph and deutonymph) and four quiescent stages (egg, protochrysalis, deutochrysalis and teleiochrysalis) (Oku *et al.*, 2003). André and Van Impe (2012) reported that *T. urticae* Koch lacks an imago, the name that is proposed for the last 'stase', and that its post-embryonic development is characterized by the 'stases' of prelarva (calyptostase), larva (six-legged stase), protonymph (eight-legged stase), deutonymph (eight-legged stase) and tritonymph. This last stase shows a precocious development of sexual maturity (paedogenesis). According to the approach emphasized by Grandjean (1938, 1970) and by André (1988, 1989), André and Van Impe (2012) proposed the use of the word 'stase' instead of the words 'instar', 'stade' or 'stage'. In their view, the use of the word stase (and its derivatives) is based on 'idionymy' (the existence of structures or stages that differ in discontinuous surface characters) and it is thus possible to homologize the different stases. The same words are used to define ontogenetic levels, larvae and adults.

The egg

The egg is deposited on the substrate through the genital aperture or the ovipositor. According to the group, this instar exhibits a typical morphology, size and colour. It is oval, long, flattened, or some other shape, with the external surface smooth, striate, reticulate, etc., and variously coloured, such as white, red or orange. During the process of deposition, it usually has strong elasticity.

The prelarva

The prelarva is also called the 'prolarva' or 'deutovum', and represents a non-feeding form, which is quiescent and closed within the egg chorion, and sometimes without legs and mouthparts, as in some Prostigmata (Walter and Proctor, 1999). In other groups, the prelarva bears three pairs of legs, mouthparts and setae. The first condition is referred to as 'calyptostasis' (strong inhibition) and the second as 'elattostasis' (incomplete inhibition) (Coineau, 1974a,b). A typical deutovum occurs in the Penthaleidae. A calyptostase is reported for *T. urticae* (André and Van Impe, 2012).

The larva

The larva is a typical six-legged instar, commonly lacking external genitalia. Many larval instars do not feed, but others may be active predators or parasites. The primordia of legs IV may occur in the larva of the Opiloacaridida (Walter and Proctor, 1999), while in the Eriophyoidea, legs III and IV are suppressed in all stages (Lindquist, 1996). In this last group of mites, the larva develops into the nymph through a quiescent or resting stage (Manson and Oldfield, 1996), also called the nymphochrysalis (Sternlicht and Goldenberg, 1971). The larva of the tetranychid *T. kanzawai* develops into the protonynph through the quiescent stage of the protochrysalis (Oku *et al.*, 2003).

The protonymph

The protonymph represents the first eight-legged nymphal instar that is free living and active. It may feed or not and develops in the same habitat as the later instars. The protonymph develops into the deutonymph through a quiescent stage. The protonymph of the tetranychid *T. kanzawai* develops into the deutonymph through the quiescent stage of the deutochrysalis (Oku *et al.*, 2003).

The deutonymph

This second nymphal instar is eight legged and greatly resembles the adult but does not have its external sexual characters or size.

Some groups have non-feeding deutonymphs, but in general this stage feeds actively. In the Astigmatina, there is a heteromorphic deutonymph called the 'hypopus', whose appearance is influenced by various factors (environmental, genetic, trophic, etc.). The hypopus is characterized by a typical morphology, lacks mouthparts, has a closed gut and is strongly resistant to environmental stresses; it is adapted to phoretic dispersal and attaches to animal carriers by ventral suckers (Fain, 1971). In *Tyrophagus* spp., the hypopus is absent, while in the Hemisarcoptidae family it absorbs nutrients from the haemolymph of its carrier coccinellid, the ladybird, through the ventral plate suckers (Houck, 1994). The behaviour of hypopodes of some Glycyphagidae is very unusual in that they confine themselves to the subdermal tissues of their carriers to absorb nutrients (Fain *et al.*, 1973). The active biological stage of the deutonymph of the tetranychid *T. kanzawai* develops into the adult through the quiescent stage of the teleiochrysalis (Oku *et al.*, 2003). When the tritonymph instar is absent, the deutonymph develops into the adult through a teleiochrysalis.

The tritonymph

This third nymphal instar is eight legged and uncommon. When present, it is active. It is absent in the Mesostigmata and in many Prostigmata. *T. urticae* lacks both the adult and the tritonymph and is characterized by pedogenesis (André and Van Impe, 2012).

The adult

Formally, the adult instar concludes the biological cycle. It is eight legged, with the exception of the Eriophyoidea, which are secondary tetrapods, and according to species may belong to one or both sexes, which can either be morphologically very similar except in the genital region, or show strong diversity. Body growth usually ends with the last moult and sexual maturity. However, in few cases, post-imaginal moulting has been observed, as in some Prostigmata and Astigmatina. Some groups may exhibit different 'morphofunctional' adaptations to selected food regimes. In *T. urticae*, as already noted, the adult instar is missing and is replaced by the tritonymph instar, which is able to reproduce by pedogenesis (André and Van Impe, 2012).

Reproduction

Mites reproduce sexually and exhibit a common gonochoristic condition, which may or may not be amphigonic. Parthenogenesis is also a common mode of reproduction, and pedogenesis has been observed in various families of the Acariformes, e.g. the Podapolipidae (Volkonsky, 1940), Pyemotidae (Rack, 1972), Tenuipalpidae (Baker, 1979a) and Cheyletidae (Atyeo *et al.*, 1984). Pedogenetic tritonymphs and females of *Tuckerella knorri* Baker *et* Tuttle (Tuckerellidae) were found by Ochoa (1989a), and in the same family, a male of a *Tuckerella* sp. was found to emerge directly from the deutonymph (Beard and Ochoa, 2010). In *T. urticae*, the adult instar is missing and replaced by the tritonymph instar able to reproduce by pedogenesis (see above; André and Van Impe, 2012). Obligate ovovivipary has been reported in eriophyoid mites, e.g. *Metaculus mangiferae* (Attiah) (Abou-Awad, 1981b), and other species. However, senescent females of eriophyoids can show apparent ovoviparity and/or apparent viviparity, as in *Aceria guerreronis* Keifer. This suggests that there is no true ovoviparity or vivipary in eriophyoids and that reports of their occurrence should be viewed as erroneous (Návia *et al.*, 2005a).

Some important aspects of mite biology that are related to reproduction are discussed below, including sex ratio and the chromosome complement, which is related to the genetic system that is used in reproduction, and adaptations such as the role of endosymbiont bacteria. The roles of various extrinsic and intrinsic factors in reproduction are also examined.

Sex ratio

The sex ratio is the proportion of adult males to females in a natural field population at a certain moment. Knowledge of the sex ratio helps to determine the mode of reproduction of a species and the management of pest mite populations. A method that is commonly used to evaluate the proportion of the two sexes is to measure the 'apparent sex ratio', i.e. the observed proportion of both sexes at each biological stage of the species in the field. However, various factors (differential mortality of the sexes and instars, sex dispersion, etc.) may distort the evaluation of the apparent sex ratio, and it has been suggested that instead, the 'primary sex ratio' – the proportion of each sex in the offspring of a brood at fertilization – should be considered. Theoretically, an estimate of the proportion of the sexes at the adult stage (the time when oviposition first occurs), which is designated the 'secondary sex ratio', could be a better measurement, but this is very difficult to realize. Therefore, in order to obtain an estimate of the primary sex ratio, it is considered preferable to calculate the 'tertiary sex ratio', which is the sex ratio at the time of the adult moult (Walter, 2009).

In eriophyoid mites, the percentage of females ranges from 51 to 95% (Sabelis and Bruin, 1996). In the Tetranychidae, the sex ratio is generally female biased and there are usually three times as many females as males (Wrensch, 1985). Among extrinsic factors affecting the sex ratio, the temperature (Hazan *et al.*, 1973), the population density (Wrensch and Young, 1978) and the quality of leaf on which the mite is feeding (Wrensch and Young, 1983) are very important. Intrinsic factors affecting the sex ratio include the amount of sperm transferred to the female, which varies according to the length of mating and sperm availability (Overmeer, 1972). Age also influences the sex ratio and old females generate more males or all males (Shih, 1979; Wrensch, 1979). Females fertilize a certain number of eggs daily during the oviposition period (Kongchuensin, 1983; Young *et al.*, 1986), and the first egg laid by a mated female is unfertilized and male determined (Helle, 1967), as sperm cannot reach it in time for internal fertilization to occur. Movement of the sperm towards the oocytes does not appear to be controlled by the female. The ventral part of the ovary receives less sperm than the dorsal region, and in the medioventral region the oocytes

probably escape sperm penetration (Feiertag-Koppen and Pijnacker, 1982). The sex ratio has a hereditary basis and a maternal effect has also been reported (Overmeer and Harrison, 1969; Mitchell, 1972).

Genetic systems

Reproduction in mites is characterized by a haploid–diploid genetic system that is based on haploid and diploid chromosomal complements. The main genetic systems reported for mites are 'diplodiploidy', 'haplodiploidy' and 'thelytoky'. All these systems are sexual.

The chromosomes

Mites have few and very small chromosomes, which are less than 0.003 mm long, without centromeres and usually holokinetic (Wrensch *et al.*, 1994). Their haploid complement varies by group. The Acariformes have from two to 12 chromosomes and the Parasitiformes from three to nine (Norton *et al.*, 1993; Weeks *et al.*, 2001). In the Eriophyidae, there may be eggs with two or four chromosomes as a result of a haplodiploidy condition that is characterized by $n = 2$ (Helle and Wysoki, 1983). The haploid condition of $n = 2$ is reported for the Tenuipalpidae (Helle *et al.*, 1980; Bolland and Helle, 1981), Tetranychidae (Helle and Bolland, 1967; Helle *et al.*, 1981), Tydeidae (Helle and Wysoki, 1983), Cheyletidae and Tarsonemidae (Helle *et al.*, 1984), and also in other mite groups. In the Tetranychidae, the haploid complement of n chromosomes ranges from two to seven (Helle and Pijnacker, 1985). A higher number of chromosomes ($n = 11$) has been found in a camerobiid mite (Bolland, 1983).

Diplodiploidy

This is an ancestral genetic system in which both sexes have two sets of chromosomes. The adaptation is known in various acarine groups, e.g. the Mesostigmata. The sex ratio of the species involved could be 1:1.

Haplodiploidy

In this system, the males have half the number of chromosomes of the females, and they are produced parthenogenetically from unfertilized and haploid eggs, while the females are produced from fertilized and diploid eggs. This genetic system is known in several acarine groups, including the Prostigmata and the Astigmatina. The formation of haploid males from unfertilized eggs is also called 'arrhenotoky'. Haplodiploidy has been demonstrated by karyotype analysis of over 100 bisexual tetranychid species. Only seven species out of 126 species investigated exhibited thelytoky (obligate parthenogenesis producing only females), two species exhibited both thelytoky and arrhenotoky, and 117 only arrhenotoky (Helle and Pijnacker, 1985). The haploid chromosomal complement may be obtained also by elimination or deactivation of the paternal genome, as occurs in the Phytoseiidae, where the male chromosomes are either totally or in part eliminated and the zygote acquires a haploid condition; this is called as 'parahaploidy' (Helle *et al.*, 1978).

The reproduction of the Kanzawa spider mite is arrhenotokous parthenogenesis and the haploid number of chromosomes is three (Helle and Bolland, 1967). The proportion of females ranges from 0.76 to 0.83 and the sex ratio is determined by the genotype of mothers; it appears to be controlled by several genes, without the involvement of cytoplasm (Takafuji and Ishii, 1989). The males probably use odours to determine the mating status of the females. These abilities of males may play an important role in gaining access to virgin females. The behaviour of adult females also varies with their mating status. Virgin females are more gregarious and remain on infested kidney leaves for a longer time than mated females, and this behaviour is likely to increase their mating opportunities (Oku *et al.*, 2005).

Arrhenotokous parthenogenesis is the rule in the Eriophyoidea, and the females that are not inseminated generate all-male progenies. Helle and Wysoki (1983) provided evidence of arrhenotokous parthenogenesis in eriophyoid male haploid mites. Thelytokous parthenogenesis does not occur and there is an opportunity for females to become inseminated before they disperse (Jeppson *et al.*, 1975; Lindquist and Oldfield, 1996).

Thelytoky

This genetic system involves obligate parthenogenesis and the females reproduce only parthenogenetically. It is common in many acarine groups (Walter and Proctor, 1999), but only rarely does it reach the condition of 'amphitoky' or 'deuterotoky', in which both sexes are produced parthenogenetically. Thelytoky occurs in many Prostigmata and it is well known in the Tetranychidae family (Bryobiinae subfamily) (Helle, 1985). It has not been found in the eriophyoid mites (Jeppson *et al.*, 1975; Helle and Wysoki, 1996).

Endosymbiont bacteria and mite genetic systems

Wolbachia and *Cardinium* are maternally inherited bacterial endosymbionts that are widely associated with arthropods. *Wolbachia* is an obligate intracellular Gram-negative organism that influences the reproductive characteristics of several species of Acari. In spider mites, it induces cytoplasmic incompatibility (CI), parthenogenesis and feminization (Breeuwer, 1997), which is seen in *T. urticae* as a male-biased sex ratio and low hatchability of eggs (Van Opijnen and Breeuwer, 1999). Crossing between a population infected by *Wolbachia* with a population that is either not infected by *Wolbachia* or hosts a different type of *Wolbachia* can induce partial to total reproductive incompatibility (Johanowicz and Hoy, 1998; Weeks and Stouthamer, 2003; Enigi and Schausberger, 2007; Kitajima *et al.*, 2007; Hoy and Jeyaprakash, 2008, 2009). In various mite species, haploidy is due to the action of this bacterium. Females of *Brevipalpus phoenicis* (Geijskes) are haploid, have two chromosomes and are thelytokous (Helle *et al.*, 1980; Pijnacker *et al.*, 1981; Weeks *et al.* 2001); their haploid condition is caused by a *Wolbachia* sp. (Otto and Jarne, 2001).

Similarly, *Wolbachia* bacteria are responsible for cytoplasmic incompatibility in various tetranychid mites (Breeuwer, 1997). The Kanzawa spider mite is infected by *Wolbachia*, but the bacterium has not been found in Japanese populations of this mite (Gotoh *et al.*, 2003a). *Wolbachia* sp. was detected

in the strawberry spider mite, *Tetranychus turkestani* (Ugarov *et* Nikolskii) (Breeuwer, 1997). Recently, multi-locus sequence typing was used to confirm *Wolbachia* infection status and to examine the relationship between this status and host phylogeny, mitochondrial diversity and geographical range in five spider mite species (*T. truncatus* Ehara, *T. urticae*, *T. pueraricola* Ehara *et* Gotoh, *T. phaselus* Ehara and *T. kanzawai*) from 21 populations in China. The prevalence of *Wolbachia* within the five species ranged from 31.4 to 100%, and the strains were remarkably diverse. These findings showed that *Wolbachia* was introduced to these populations at multiple distinct times, and suggested that its transmission occurs both vertically and horizontally, with horizontal transmission probably mediated by the host plants. Moreover, the distribution patterns of *Wolbachia* are quite different among different populations of the same mite species, indicating that the natural dynamics of *Wolbachia* may be affected by ecological and other factors (Zhang *et al.*, 2013).

The above discussion suggests that cytoplasmic incompatibility represents a major mechanism by which certain strains of *Wolbachia* invade and persist in host populations. However, the mechanisms that underlie the invasion and persistence of non-CI strains are little known. Zhao *et al.* (2013) established a spider mite (*T. phaselus*) population multiply infected by *Cardinium* as well as by two distinct lineages of *Wolbachia* (wCon and wOri) to study the mechanisms leading the spread of the non-CI strain of *Wolbachia* wOri. The lineage wOri induced a longevity advantage to its female hosts, stayed longer in this population and was transmitted to more offspring. When uninfected females mated with multiply infected males, they exhibited a reduced lifespan. The uninfected population is weakened by multiply infected males, and host age effects of multiple infection may represent a mechanism that leads to the spread of wOri through the host population (Zhao *et al.*, 2013).

Cardinium is associated with various tenuipalpid species, such as *B. phoenicis*, a common thelytokous species, in which it feminizes unfertilized haploid eggs and induces thelytoky (Weeks *et al.*, 2001; Groot and Breeuwer, 2006). It is also found in *B. californicus* (Banks) (Chigira and Miura, 2005). In addition, *Cardinium* sp. induces thelytoky by feminizing the unfertilized haploid eggs of *B. obovatus* Donnadieu, although some clones of *B. obovatus* contain neither *Cardinium* nor any other bacterium that could induce thelytoky. Also in some *Brevipalpus* lines, females treated with antibiotics produce no males, suggesting that in these lines thelytoky is probably a genetic property of the species itself (Groot and Breeuwer, 2006). *Cardinium* is reported from the tetranychids *Eotetranychus suginamensis* (Yokoyama) (Gotoh *et al.*, 2007a), and *E. uncatus* Garman and some phytoseiid mites as well, such as *Neoseiulus californicus* (McGregor) and other species (Enigi and Schausberger, 2007). *Cardinium* and *Wolbachia* were both found in *T. pueraricola* (Van der Geest, 2010). The bio-ecological implications that are associated with these bacteria have been extensively investigated in Japan (Nakamura *et al.*, 2011).

Biological parameters

A number of extrinsic and intrinsic factors affect the reproduction of mites (Huffaker *et al.*, 1969; van de Vrie *et al.*, 1972; Wrensch, 1979, 1985). Individual reproductive parameters (fecundity, hatchability, length of the period of oviposition and longevity, rate of development, survivorship, sex ratio) influence the intrinsic rate of increase (r_m), an important parameter that provides information on variation in the population that is related to the individual and to unit time. The designation of these parameters as strictly biological in some way can be misleading. In fact, although they are expression of the biology of species, they also affect the ecology of their populations, so it might be more appropriate to refer to them as bio-ecological parameters.

Extrinsic factors influencing biological parameters include temperature, humidity, light, level of predation, intraspecific and interspecific competition, the quantity, type and timing of pesticide applications, and different features of host plants (variety, plant and soil nutrition, plant age, etc.). Intrinsic factors affecting reproductive potential include mite strain, level of inbreeding, population density, age of females and age of the population, fertilization status, quality of mating, duration of insemination and behavioural aspects (Wrensch, 1985).

Consideration of the biological parameters that have been outlined above allows a certain understanding of the life cycles and the reproductive potential of species. Other parameters that are more directly involved in population dynamics (see next chapter) can also be considered. For example, the 'egg to egg' or 'egg to adult' developmental times, the 'mean oviposition period', the 'fecundity' and the 'mean ovipositional rate' provide basic information on the biology of various species. In particular, the egg to egg or egg to adult developmental times indicate the duration of the life cycle of a species in days. The mean oviposition period provides information on the average time in days of the oviposition of a species. The fecundity is the total number of eggs laid by a single species during its life cycle. The mean ovipositional rate refers to the number of eggs laid daily.

Fecundity is fundamental to the innate capacity to increase the population, and is affected by various environmental factors. Variations in adult offspring density are caused mainly by variations in fecundity. The ability to produce a large number of offspring under marginal conditions explains the high success of spider mites (Young and Wrensch, 1981). However, fecundity alone is inadequate either as a measure of or as a factor that is correlated with the capacity of spider mite populations with the same frequencies of the sexes to increase (Wrensch, 1985). This suggests that the sex ratio should be included in estimates of population increase (Hazan *et al.*, 1973; Wrensch and Young, 1975; Wrensch, 1979; Feldmann, 1981; Carey, 1983).

The aforementioned 'intrinsic rate of increase', r_m, provides information on variation in the population related to the individual and unit time. If the values of survival and fecundity of the population remain constant over a wide range of time, the value of the intrinsic rate of natural increase is constant and the population has a stable age structure. If the values of fecundity and survival vary over time, then the population growth rate varies over time (Begon *et al.*, 1986). Sabelis and Bruin (1996) report that r_m values observed in eriophyoid mites range from 0.05 to 0.2–0.25, and that populations double in about 3–14 days. The values found in eriophyoid mites are lower than those observed in tetranychid mites, which range from 0.13 to 0.33 at 25 ± 2°C (Sabelis 1985a, 1981). However, populations of some strains/

species of eriophyoid mite increase in a manner that is comparable to that of some spider mites (Sabelis and Bruin, 1996). Development time and fecundity are fundamental in determining the intrinsic rate of increase (Snell, 1978), and age of first reproduction affects r_m more than fecundity does (Cole, 1954). In terms of increasing r_m, small decreases in the rate of development correspond to high increases in fecundity (Lewontin, 1965). This relationship was confirmed by Wrensch and Young (1975), but Meats (1971) found that for values of r_m near zero, fecundity was more important. Snell (1978) reports that for small values of r_m, or if developmental time and fecundity are correlated, then decreases in developmental time strongly affect r_m if fecundity is large.

Other parameters that relate biological factors to population ecology and allow the evaluation of the development of populations in the environment include the 'net reproductive rate' (R_0), the 'finite rate of increase' (λ) and the 'generation time' (T). The net reproductive rate (R_0), consists of the average number of offspring of an initial individual by the end of the cohort, and is usually calculated using life tables. In annual species, this parameter indicates the total amount by which the population has increased or decreased during that period of time. If the generations overlap, or if the species breed continuously, as occurs in many mites, the situation becomes more complex. In this case, R_0 refers only to the average number of offspring produced by a single individual (Begon *et al.*, 1986). Sabelis (1985a), in a summary of the literature on the net reproductive rate of spider mites, reports values of R_0 ranging from 20.1 to 111.1 at 25 ± 1°C. The 'generation time' (T) is the average time for a generation and the 'generation time for the cohort' (Tc) is the average time between the birth of an individual and the birth of one of its descendants (Begon *et al.*, 1986). The finite rate of increase (λ) is the growth rate of a population per time period examined (usually a year). When the population lacks an age distribution or shows a stable age distribution, the value of the finite rate of increase is constant.

5 Ecology

Study of the ecology of mites is a crucial step in the process of acquiring knowledge of mites (and other organisms), and includes traditional natural history, taxonomy, modern evolutionary and behavioural ecology, and theoretical models. Such studies can also provide insight into the evolution of plant mites. Here, the focus is on those aspects that help a better understanding of the dynamics of populations and on outbreaks of pest mites in the field. Readers interested in further insights can consult the contributions of Sabelis (1985a,c, 1996) or the more recent work of Saitō (2010), and other publications.

Feeding Habits and Habitats

The feeding habits of mites involve various aspects of their morphology and lifestyle. Their morphological features (mouthparts, postoral digestive systems, etc.) usually reflect the ecological approach that these animals exhibit with respect to various food resources and habitats. From this viewpoint, morphology, habitats and behavioural patterns are interdependent and may markedly influence systematic features.

Mites can be free living, symbiotic or parasitic, and occupy various terrestrial and aquatic habitats. Terrestrial species may be arboreal or nidicolous or live in niches related to soil and/or litter. Aquatic species may be littoral, marine or freshwater species. Mites are common in all ecosystems, and are rarely lacking in any habitat. This latter characteristic is very important and has been described as an 'evolutionary revolution' in this subclass of organisms (Krantz, 2009d). Mites can also feed on a large variety of organic food or other material, including both living and dead parts of plants and animals. Several species are adapted to ectoparasitism or endoparasitism and feed on organic substrates from vertebrates or invertebrates, such as blood, tissue fluids or other parts. Many species are fungivorous, others feed on decaying leaves or other vegetable tissues, while others feed on living plant tissues. The various species may be either general or specialized feeders. Most of the mites that live and feed on plants belong to the Prostigmata, and among these there are phytophagous mites that have characteristic structures of their mouthparts, such as the Tetranychoidea and Eriophyoidea. Other mite groups, such as some of the phytophagous Penthaleidae and Acaridae, have mouthparts that are not strictly adapted to phytophagy.

Free-living species

Free-living species occupy soil habitats ('soil mites') and aboveground (aerial) habitats ('aerial mites').

Soil mites commonly inhabit the underlying rhizosphere, which hosts a rich microfauna and microflora. They are variously associated with plant roots and rhizomes, which represent a considerable food resource for organisms with different feeding habits, such as fungivores, microbivores, detritivores, herbivores and predators. Soil mites have a fundamental role in the transformation of litter into humus. According to the environment, their ecological characteristics and other variables (extraction method, depth of layer, water content, etc.), mites can be collected in numbers ranging from 50,000 to 250,000 or more from each 1 m^2 area of soil surface (Petersen, 1982); they can number up to 400,000 in winter in forest soil (Wallwork, 1959). Microarthropods in general can reach a density greater than 1,000,000 individuals m^{-2} (André *et al.*, 1994). Among the soil fauna, the Oribatida play a key role in the dispersal of ectomycorrhizal fungal spores and in the breakdown of detritus, etc. Other mites feed on algae, lichens and mosses. Soil mites also include predators belonging to several families. A restricted group of phytophages inhabits the soil, where the mites feed on root tissue, corms and bulbs in the rhizosphere, sometimes causing severe damage to ornamentals and vegetables; examples include *Rhizoglyphus* spp., acarids, etc.

Aerial mites typically include predators and phytophages. Often associated with these important groups are other mites that feed on fungi and algae. Predators are very common in the aerial parts of plants, where they feed on different instars of phytophagous or mycophagous mites and small insects. Some dozens of families of predatory mites are known (Gerson *et al.*, 2003), and among these are the Phytoseiidae, Stigmaeidae, Cheyletidae, Hemisarcoptidae and Cunaxidae. The most important predatory mites belong to the Phytoseiidae family, which includes thousands of species (2250 in 2004) (Moraes *et al.*, 2004). Phytophages include species adapted to phytophagy, pollenophagy and nectarivory. They mostly belong to the Trombidiformes, a group with over 4000 species and 300 genera (Bolland *et al.*, 1998; Amrine *et al.*, 2003; Migeon and Dorkeld, 2006–2013). The most important families are the Tetranychidae, Eriophyidae, Phytoptidae, Diptilomiopidae, Tenuipalpidae and Tarsonemidae. Other families may exhibit a partial adaptation to phytophagy, such as the Penthaleidae, Acaridae, etc.

Littoral, marine and freshwater species

These species are important from natural and ecological viewpoints, but they do not influence economic plants. Further information can be found in Krantz (2009b).

Parasitic species

These include species that are ectoparasites and endoparasites of vertebrates and invertebrates. Although they may cause serious damage to the livestock industry and to agricultural production in general, their role, except for some groups, as the Hemisarcoptidae, which are parasites and predators of armoured scales (Gerson *et al.*, 2003), is of secondary importance in plant production. Further information may be found in Krantz (2009b).

Webbing and Life Types

The webbing produced by the Arachinida is produced by different spinning organs – the abdominal spinnerets of spiders and the modified salivary glands of pseudoscorpions and mites. This adaptation is significant in the ecology of several species of the subfamily Tetranychinae (family Tetranychidae) and may negatively affect the control of these mites. The use of webbing by single species was extensively investigated by Gerson (1979, 1985b), and the interference of webbing with spider mite life type was studied mainly by Saitō (1983, 1985). Information presented here is mainly drawn from these studies.

According to Gerson (1985b), the production of webbing has various effects on the bio-ecology of mites. One important function of webbing is in the protection of eggs, e.g. in some Bdellidae (Wallace and Mahon, 1972), Camerobiidae (Bolland, 1983) and Cheyletidae (Avidov *et al.*, 1968), and in primitive Tetranychinae. In *Tetranychus* sp., the amount of webbing produced and the number of eggs laid are correlated (Hazan *et al.*, 1974). Webbing can protect eggs from mite predators, regulate the humidity around the eggs (Hazan *et al.*, 1975) and also protect mites from wind and rain (Davis, 1952; Linke, 1953). Web-producing species, such as *Tetranychus* spp., may displace other plant-feeding mites, such as *Panonychus ulmi* (Koch) (which has also been known as *Metatetranychus ulmi* Oudemans), a weak spinner (Foott, 1962), or *Bryobia* sp., a typical non-spinning mite (Georgala, 1955).

The behaviour of natural enemies in relation to webbing varies according to whether they have co-evolved with webbing and to their specialization. Thus, some species tend to avoid webs (McMurtry *et al.*, 1970), for example the phytoseiid mite *Iphiseius degenerans* Berlese (Takafuji and Chant, 1976), while others, such as *Phytoseiulus persimilis* Athias Henriot, are attracted by webs more than by eggs or exuviae – the outer layer of skin that is not recycled during a moult (Schmidt, 1976). The complex webs of *Tetranychus* mites can hinder the performance of generalist predatory mites, although this may not be true for specialist predatory mites. A comparison between the thread-cutting behaviour of three phytoseiid mites, *P. persimilis*, *Neoseiulus womersleyi* (Schicha) and *N. californicus* (McGregor), inside the complex web of *Tetranychus urticae* Koch, the two-spotted spider mite, showed no significant difference in their basic pattern of the behaviour, and showed that each phytoseiid mite could move inside the complex web without suffering from strong obstruction (Shimoda *et al.*, 2009). Webbing stops and holds sprays droplets and dust grains of pesticides, thereby protecting mite colonies, suggesting that the destruction of webs is needed during chemical control (Davis, 1952; Linke, 1953). Some pesticides induce the two-spotted spider mite to leave treated leaves by a thread, the so-called 'spindown' (Gemrich *et al.*, 1976). Mite populations treated with different acaricides form spindown threads of the same length, but the percentages of mites spinning down may be different (Gerson and Aronowitz, 1981a). Spindown significantly affects fenvalerate-induced dispersal. Mites spinning down from sprayed bean plants are favoured in their dispersal by wind (Penman and Chapman, 1983).

Webbing also affects mite finding, as in *T. urticae*, where it serves as a means of physical contact between the two sexes, and as a pheromone substrate or carrier (Penman and Cone, 1974). Guarding and aggressive behaviour that end in successful mating use webbing as a means of intervening in territoriality (Potter *et al.*, 1976a). Webbing is strongly related to mite locomotion and dispersal. Some tetranychids spin silk threads or 'lifelines' (Saitō, 1979a) and use them for dispersing by wind and animals (Fleschner *et al.*, 1956). The phenomenon has been investigated in greenhouses, where *T. urticae* forms silk ropes when leaving infested plants containing many mites (Hussey and Parr, 1963a). Mites usually live in their web (Foott, 1962), which represents the size of the colony. Silk production is affected by humidity and temperature (Hazan *et al.*, 1974; Saitō, 1977), structure of the substrate and the type of host plant (Gerson and Aronowitz, 1981b), but does not depend strictly on feeding (Hazan *et al.*, 1974). Uninterrupted light induces continuous spinning, and consequently an imbalance between webs and eggs.

The Tetranychinae have evolved different behavioural patterns in response to webs, which has suggested the use of web characters to classify their mode of life into 'life types' (Saitō, 1983, 1985). Two types of spinning behaviour (weaving and dragging), four forms of web (complicated, nest, egg cover and guy rope), eight types of use (in oviposition, while quiescent, in the deposition of faeces, as a footing, as a means of mooring during quiescence, as a lifeline, as parachute threads and as ballooning threads) and a series of other supposed functions have been suggested for this group (Saitō, 1983). Thus, nest webs and complicated web threads can be used for the deposition of faeces, and complicated web threads are used in ovideposition (Hazan *et al.*, 1974) or during quiescence. Threads are also employed for mooring of quiescent stages (Ewing, 1914), or as lifelines on the lower leaf (Saitō, 1977). In some species, when populations leave the substrate the females can use lifeline threads as ballooning threads (Fleschner *et al.*, 1956; Wanibuchi and Saitō, 1983). Saitō (1983, 1985) described the web pattern of each species in relation to its life pattern with the aim of highlighting the relationship between the functions mentioned above and the evolution of individual species. The spider mite has eight patterns of webbing (structure of the web; density of the web; use of the web as a site of oviposition, as an egg cover produced by females, as a site for quiescent stages, as a site for defecation and as a site of feeding and walking; and spinning behaviour during walking). Based on the eight web types, it is possible to classify many tetranychines into the following three 'life groups': LW (little web type), CW (complicated web type) and WN (web nest type).

The LW life type is the basic type and includes four subtypes that are separated in relation to spinning no threads or webs

(LW-f), having only webbed egg covers (LW-j), spinning silk threads while walking on the substrate and weaving egg guy ropes (LW-c) and spinning lifelines during walking and while attaching eggs and moulting stages to the tips of leaf hairs (LW-s).

The life type CW includes five subtypes. The subtype CW-p occurs in *Eotetranychus* sp., which produces silk threads during walking, and threads are accumulated on the substrate. The species uses the web as footing. The subtype CW-r is similar to CW-p, but the mites use the web as a site on which to deposit faeces. The subtype CW-u is characterized by a complex and irregular web on the substrate. A high population density is associated with the presence on the web threads of faecal pellets, eggs and cast skins. Serious pests such as *T. urticae* belong to this subtype. The subtype CW-b produces a dense and stratiform rather than a complex web, and the eggs are usually laid on the substrate under the web. The web may serve as a footing for moving and for defence. In some cases, eggs are laid on the web. The subtype CW-a is similar to CW-b. Mites move easily between webs, which facilitates their movement and serves as a footing. Eggs and quiescent stages occur on the web. This behaviour always occurs and probably serves as protection against predators.

The WN life type includes five subtypes. The subtype WN-u is similar to subtype CW-p. Mites feed and walk under the web. Mites of the subtype WN-t deposit faeces near the margin of their web nest, whereas mites of the WN-s subtype deposit their faeces on their nest web threads. The subtype WN-c includes mites that construct an extremely densely woven roof over the depressions on the lower leaf surface. Weaving is limited to the construction of the nest and egg covers. The subtype WN-r is characterized by mites that construct cells for each individual. Thus, a juvenile stage or an adult constructs a nest web around three or four leaf hairs, within which it lives.

In addition, webbing may affect the relationship between predators and pests (Gerson, 1985b; Gerson *et al.*, 2003).

Adaptations of Some Eriophyoids

Some eriophyoid species exhibit particular adaptations, such as 'web spinning', 'wax secretion' and 'liquid secretion'. There are no findings that are able to explain the ecological role of these adaptations. Information presented on this topic is taken from the review of Manson and Gerson (1996). As previously mentioned (see Chapter 3), the waxy nature of the wax secretion has not been established, so they should really be referred to as wax like, although to avoid confusion the term wax is retained.

Web spinning

Web spinning is common in the Tetranychinae and its general features have been summarized above. In addition to the Tetranychinae, Manson and Gerson (1996) have reported eight web-spinning eriophyoid species. Knorr *et al.* (1976) first reported web spinning in *Aculops knorri* Keifer, a species living on the upper leaflet surfaces of the fruit tree *Lepisanthes rubiginosa* (Roxb.) Leenth in Thailand. Populations of *A. knorri* live under the webbing, usually along the leaflet midribs and veins. Another species, *Cisaberoptus kenyae* Keifer, is found on mango (*Mangifera indica* L.) in tropical and subtropical areas, and its populations live under white leaf coatings on the upper leaf surfaces, usually along the midribs. The white coating consists of irregular strands of web of various forms and of bunches, which are not separate as in *A. knorri* (Hassan and Keifer, 1978). *Aceria gersoni* Manson lives under patches of loose white-to-greyish webbing on the undersides of the leaflets (pinnules) of the tree fern (*Dicksonia squarrosa* (Forst.f.) Swartz) in New Zealand (Manson, 1984b). *Aberoptus platessoides* Meyer and *Cisaberoptus pretoriensis* Meyer live under a waxy layer on the leaf bases, petioles and tender twigs of *Ochna pretoriensis* Phill in South Africa (Meyer Smith, 1989a). Curiously, web spinning was reported in Japan for *Aceria tulipae* (Keifer), *Trisetacus juniperinus* (Nalepa) and *A. eucricotes* (Nalepa) (Nemoto, 1991), even though webbing was not previously known in these mites. Protogynes, deutogynes and two forms of the males of *Aceria inusitata* Britto *et* Návia occur under patches of webbing on pau-brasil (*Caesalpinia echinata* Lam.) leaves in Brazil (Britto *et al.*, 2008).

Wax secretion

Wax secretion is known in several eriophyoid mites, and according to its typology the species involved can be classified as either 'species producing longitudinal wax ridges' or 'species producing flocculent wax'. All three families of the Eriophyoidea include wax-producing species, usually living as vagrants on host leaf surfaces. Jeppson *et al.* (1975) speculated that dorsal white wax stripes, or flocculent wax, are probably produced by some opisthosomal microtubercles, especially the granules of dorsal ridges. The presence of flocculent wax could assist in water conservation or in protection against predators. In some cases, such as in the cereal rust mite, *Abacarus hystrix* (Nalepa), the bands or stripes were associated with the enhancement of mite survival at lower relative humidity (RH) and with an increase in the total surface drag during dispersal (Frost and Ridland, 1996; Frost, 1997).

Various genera of eriophyoid mites include species producing longitudinal wax ridges. The genera *Calacarus* Keifer, *Abacarus* Keifer, *Acamina* Keifer and *Apodiptacus* Keifer all have three opisthosomal longitudinal wax-bearing ridges. The genus *Neocalacarus* ChannaBasavanna has five opisthosomal longitudinal wax-bearing ridges, and the prodorsal shield has a pattern of wax-bearing lines. The genus *Callyntrotus* Nalepa has about six low dorsal ridges with microtubercles bearing hair-like, wax-producing spinules. The genus *Retracrus* Keifer has the body protected by wax, which is arranged longitudinally on the opisthosoma and arises from the prodorsal shield tubercles as wax pencils (Amrine *et al.*, 2003). Species of the genera *Calepitrimerus* Keifer and *Epitrimerus* Nalepa bear wax ridges (or other wax formations).

Species producing flocculent wax belong to various genera. *Trimeroptes aleyrodiformis* (Keifer) is an undersurface leaf vagrant on sweetgum (*Liquidambar styraciflua* L.) and bears dorsal and lateral masses of wax that appear as a unique mass (Keifer, 1940c). *Dialox stellatus* Keifer is an undersurface leaf vagrant

on coconut in the Philippines, and produces stellate masses of white-like miniature mealybugs (Keifer, 1962c). *Floracarus eugenifoliae* Mohanasundaram is an undersurface leaf vagrant on *Eugenia* sp. in India, and has wax covering the entire body (Mohanasundaram, 1981a). *Porcupinotus humpae* Mohanasundaram lives on *Cassia* sp. in India, and bears long waxy filaments arranged in rows on the dorsal opisthosomal plates. There are a median and two lateral ridges on each side of the opisthosoma, and waxy filaments come from the ridges, with the dorsal filaments longer then the shorter lateral filaments (Mohanasundaram, 1984). *Aceria cerostructor* (Flechtmann) lives in Brazil under small patches of a wax-like layer produced by females bearing spatulae on leaves of jatobá (*Hymeanaea courbaril* L.). The mite scrapes off the leaf epidermis underneath the wax-like shelters and lives in a wet environment (Flechtmann, 2001a).

Liquid secretion

Liquid secretion has been reported by Manson and Gerson (1996) only for two eriophyoid mites, *Hoderus globulus* (Mohanasundaram) and *Hyborhinus kallarensis* Mohanasundaram, both from the Indian Continent. The first species lives on an unidentified shrub, and during feeding it secretes a clear liquid that forms a shining globule covering the body (Mohanasundaram, 1981b). *H. kallarensis* is an undersurface leaf vagrant on *Flacourtia ramontchi* L'Herit and produces a clear secretion on its body during feeding (Mohanasundaram, 1986).

Chemical Communication

As in insects, chemical compounds are used in communications between mites and/or with their environment, including their host plants. This section discusses chemical communication between individuals of the same species (by pheromones) and between different species (by allelochemicals). Different types of pheromones have been reported in the literature, and allelochemicals can be classified into three groups, 'allomones', 'kairomones' and 'synomones', which have distinct actions.

Pheromones

In the 1970s, the strong attraction of males of different tetranychid mites, e.g. *T. urticae*, towards females was shown to be due to the action of chemical compounds with sex-attracting capabilities, which were named pheromones (Cone *et al.*, 1971a). Methods for the preparation of pheromone extracts and bioassay procedures were developed (Cone, 1979) in association with investigations on the bio-ethology of some pest species. Thus, male behaviour towards quiescent deutonymphs was described by Cone *et al.* (1971b), while the quantification of the male response to the female deutonymph was investigated by Penman and Cone (1972). Later, it was found that the deutonymphal web aided males to find females, with a female sex pheromone acting as an attractant until the first male contact, and thereafter acting as an arrestant with a range of about 1.8–2.3 mm (Penman and Cone, 1974).

Different release sites have been suggested for pheromones (dorsal lyrifissures, basal portion of body), but it seems that the moulting fluid, or the fluid contained between the old and new cuticle, plays an active role in attracting the male (Cone, 1985). Research on the chemical nature of pheromones showed that farnesol was one component of the male sex attractant system in the two-spotted spider mite (Regev and Cone, 1975a), and that its content was correlated with egg deposition by the female (Regev and Cone, 1976).

Further studies on pheromones in different mite groups demonstrated that there were different types of pheromones, including sex pheromones, e.g. rosefuran, a sex pheromone from an acarid mite, *Caloglyphus* sp. (Mori *et al.*, 1998) and alarm pheromones, such as the isopiperitenone of the acarid mites *Tyrophagus similis* Volgin (Kuwahara *et al.*, 1987) and *T. neiswanderi* Johnston *et* Bruce (Kuwahara *et al.*, 1989). Mites of the cohort Astigmatina (= Astigmata, Acaridida), or astigmatic mites, have been found to use a single chemical compound (or different compounds) to achieve two of the three communication functions of pheromones (alarm, sex and aggregation) (Kuwahara, 2010). Each of these communication functions emerges from the conditions the mites are facing, and is based on the functional characteristics of the chemical concerned, e.g. the dose–response relationship, the active dose range and the amount of the chemical compound(s) in the opisthonotal gland. In the astigmatic mite *Rhizoglyphus setosus* Manson, all three communication functions can be achieved using products of the opisthonotal glands (Kuwahara, 2010).

Allomones

Allomones benefit the organism that produces them by a modification of the behaviour of the organism that receives them, although sometimes they do not have much effect on the receiver. Chemicals emitted by the opisthonotal glands of the Astigmatina are anti-predator or antifungal allomones. The relationships between pest mites and their predators may be led by 'infochemical' cues to alert the prey to the presence of their predators. These chemicals are usually produced by the exuviae, excreta, etc. of predators and recognized by the prey. Such infochemicals can act both as chemical signals emitted by alarmed, damaged or dead conspecifics in the prey population, and as allelochemicals emitted by other potential prey (Kats and Dill, 1998; Grostal and Dicke, 2000; Dicke and Grostal, 2001). The role of infochemicals in interactions between the cassava green mite, *Mononychellus tanajoa* (Bondar), and the fungal pathogen *Neozygites tanajoae* Delalibera, Humber *et* Hajek has been investigated by Hountondji *et al.* (2010).

Kairomones

Kairomones produce a benefit to the organism that receives them and a disadvantage to the one that produces them. Schmidt (1976) observed that females of the phytoseiid *P. persimilis* spent

a lot of time in and around the webbing of their prey, *T. urticae*. Later, it was found that this type of behaviour by predatory mites depended on kairomones (Hislop and Prokopy, 1981; Hoy and Smilanick, 1981; Sabelis and Van de Baan, 1983). These are volatile substances, extractable in solvents, that are emitted by the prey and, when present in an airstream, affect the behaviour of predators (Sabelis *et al.*, 1984).

Synomones

Synomones are the subject of intensive research because of the implications that they have for pest control, so they are treated quite extensively here. Plant tissues that are injured by phytophages produce specific volatiles or 'synomones', also called 'herbivore-induced plant volatiles' (HIPVs) (Hunter, 2002a; Frost *et al.*, 2008), that deter pests or attract predators (Cortesero *et al.*, 2000; Farmer, 2001). The chemical signals are emitted only as a result of pest feeding and provide 'a chemical burglar alarm' (Dicke, 1995). Spider mite damage triggers the emission of a blend of around 60 attractant volatiles that entice predators (Dicke *et al.*, 1999). The chemical spectrum of synomones emitted by a plant species differs according to the pest, the host plant variety and the genetics of the predators. These volatiles increase the response of uninfested leaves to spider mites, and stimulate the action of defence genes in neighbouring plants. An example is that different cucumber accessions differ in the jasmonic acid (JA)- and herbivore-induced plant volatiles that are perceived by *P. persimilis* (Kappers *et al.*, 2010).

HIPVs may prime the damaged plants (and neighbouring plants) to activate direct and indirect plant defences (Frost *et al.*, 2008). Plants become more defensive against phytophages after exposure to HIPVs, and this is a type of priming of, or preparation by, the plant that perceives an HIPV signal to respond to the phytophage. Research carried out by Choh and Takabayashi (2010) on the priming of two induced indirect defences, the HIPV-mediated induction of predator attraction and the secretion of extrafloral nectar (EFN), an alternative food source for natural enemies of phytophages, showed that on lima bean plants (*Phaseolus lunatus* L.) the priming of HIPV-exposed plants attracts the phytoseiid predator *P. persimilis* and induces the secretion of EFN, which functions to protect the plants before and after they have been fed on by its prey, *T. urticae*.

The Kanzawa spider mite, *Tetranychus kanzawai* Kishida, causes either red or white feeding scars on lima bean plants. Leaves showing red scars produce more salicylates than leaves bearing white scars. Volatiles emitted by the leaves differ as well, as does the response of predators to the plant-emitted volatiles (Matsushima *et al.*, 2006). The phytoseiid *Neoseiulus womersleyi* (Schicha) responds to mixtures of three synthetic compounds produced by the *T. kanzawai*-induced tea plant volatiles and to a mixture of four synthetic compounds similar to the volatiles of bean plants infested by *T. urticae* (Ishiwari *et al.*, 2007). However, although predatory mites can perceive these volatiles, it is unknown whether they enable recognition of the infested plant from long distances in the field.

Some mites suppress or delay the onset of the emission of the plant hormones JA and salicylic acid (SA). In particular, on tomato, some genotypes of *T. urticae* and *T. evansi* Baker *et* Pritchard induce and others suppress the JA and SA responses simultaneously. *Aculops lycopersici* (Tryon) uses the SA response to suppress the JA response. Study of the genes expressed in these mites and in the host plant, in combination with behavioural assays, genetic manipulation, expression analyses and metabolomics, allow an understanding of these plant defence mechanisms (Alba *et al.*, 2012a). *T. evansi* exhibits direct exploitative competition with *T. urticae*, and indirect competition via the direct and indirect defences induced in the host plant. It also locally suppresses direct defences (proteinase inhibitors) to below their usual greenhouse levels in undamaged tomato plants and promotes its own reproduction as well as that of *T. urticae*. However, *T. evansi* protects the leaf area that is subject to plant defence suppression by a dense web that is impenetrable to *T. urticae*, as well as to the natural enemy employed to control *T. urticae*, i.e. *P. persimilis*. Furthermore, *T. evansi* induces the plant to produce volatiles that act as attractants to this predatory mite (Janssen *et al.*, 2012). The defence mechanisms of plants create a selective environment that pressurizes herbivores to adapt. Spider mites sometimes develop resistance to constitutive and induced plant defences. It has been observed that *T. evansi* strongly suppresses induced defences in tomato plants, although how it does this, and which part of the plant's metabolism is affected, is still unknown. Suppressor mites exhibit a strong negative impact on the accumulation of jasmonates and salicylate as well as on the downstream defence genes, while causing a high degree of feeding damage. Moreover, the fitness of defence-inducing mite genotypes increases when it shares the leaflet with suppressor mites. However, suppressor mites do not suppress phytohormone accumulation induced by these co-inhabiting inducer mites, they only suppress the expression of downstream defence genes. This suggests that the suppression operates downstream of phytohormone accumulation but may feed back on to hormone biosynthesis (Alba *et al.*, 2012b). In the field, *T. evansi* is often found together with defence-inducing genotypes of *T. urticae*. Within leaflets, the induction of defences mostly concerns the feeding site, while the suppression effects are manifested beyond the feeding site, albeit not equally in all directions. The suppression of plant defences can backfire on *T. evansi* during natural infections, but it also produces large amounts of webbing that protect the feeding site not only from predators but also from invading competitors (Ataide *et al.*, 2012).

Predators may be attracted to and the plants influence the composition of the blend of synomones more than pests. It has been observed that the predator orientation to prey needs the contemporaneous reception of two chemical cues coming from different sites (Hagen, 1987), one group corresponding to plant volatiles and the other emitted by pests or their faeces. Using HIPV (synthetic methyl salicylate, MeSA) as a bait in hop yards in Washington State attracted a variety of predatory insects (*Stethorus* sp.) (James, 2003). Later, James *et al.* (2010) evaluated MeSA in the biological control of mites, and found that in a hop plot baited with 'controlled release sachet' (CRS) dispensers of synthetic MeSA, larger populations of spider mite predators (*Stethorus* spp., *Orius tristicolor* (White), *Geocoris pallens* Stål, *Deraeocoris brevis* (Uhler), Chrysopidae,

Hemerobiidae, Nabidae and Thripidae) were attracted than in unbaited plots. The predators controlled spider mite populations in MeSA-baited hops and chemical treatments were not applied. In addition, the direct distribution of natural MeSA (oil of wintergreen) to hop plants in canola oil or rosemary/peppermint oil pesticide formulations induced a greater attraction of two mite predators (*O. tristicolor* and *Stethorus* spp.). This suggested that the MeSA-mediated stimulation of the plants to produce predator-attracting volatiles was the likely mechanism involved. The employment of synthetic or natural HIPV/plant-signalling compounds like MeSA as 'herbivore-induced plant protection odours' (HIPPOs) could be a new technique for improving the biological control of mites in various crops (James *et al.*, 2010).

The staphylinid beetle *Oligota pygmaea* Sol. and the coccinellid *Stethorus gilvifrons* Mulsant showed a significant response to tea leaves infested by the red spider mite, *Oligonychus coffeae* (Nietner). Chemical analysis of the volatiles emitted from infested and healthy leaves showed the presence of seven compounds, of which β-ocimene and α-farnesene were present in higher quantity in the infested leaves than in the healthy leaves. MeSA and *cis*-3 hexenyl acetate were found only in the infested leaves. Infestation by mites triggered the emission of a greater amount of volatiles responsible for the attraction of predators. Benzaldehyde, MeSA, α-ocimene and α-farnesene induced a higher response in *O. pygmaea*, whereas in *S. gilvifrons* there was a higher response to methyl salicylate and benzaldehyde (Sachin *et al.*, 2008).

Studies were carried out on three types of leaves of *Castanea mollissima* Blume: uninfested; infested by the spruce spider mite, *O. ununguis* (Jacobi); and infested by the spider mite but without the predatory phytoseiid mite *Amblyseius finlandicus* (Oudemans). The different kinds and varied contents of the emitted volatiles (caryophyllene, α-farnesene, 1-octene, benzoic acid ethyl ester, 1,6,10-doclecatrien-3-ol, 3,7,11-trimethyl, and 4-bromo-2-bentenl [?4-bromo-3-pentenol) changed the chemical fingerprint of the host plant's odour so that *O. ununguis* had difficulty in recognizing its host plants, but these volatiles could also attract *A. finlandicus* (Sun and Yin, 2002). *O. ununguis* is also attracted to odours from the leaves (all uninfested by *O. ununguis*) of *Quercus acutissima* Carruth., *Q. variabilis* Bl., *C. mollissima* and *Q. dentata* Thunb. However, the odours emitted from the leaves of *Q. dentata* are less attractive to this spider mite than those from *Q. acutissima*, *Q. variabilis* and *Q. mollissima*, and the leaf odours of *Pinus tabulaeformis* Carr. and *Picea koraiensis* Nakai are repulsive to the mite. *O. ununguis* prefers the odours emitted by infested leaves, while *A. finlandicus* could be attracted to the odours from both leaves injured by *O. ununguis* and uninjured leaves, but exhibited a greater preference for the former. There was a significant difference between the response of *A. finlandicus* to leaves infested with spider mites and their products, from that of infested leaves without spider mites or their products (Sun *et al.*, 2002).

Finally, subterranean tulip bulbs attacked by the eriophyoid mite *A. tulipae* emit odours that attract the predatory mite *Neoseiulus cucumeris* Oudemans (Aratchige *et al.*, 2004). The phenomenon suggests a new approach for exploiting the natural enemies occurring in the soil (Gerson, 2014).

Mating and/or Sperm Transfer

Mating behaviour varies according to mite species and groups. This section discusses the mating or sperm transfer of the main pest mites groups (Tetranychidae, Eriophyoidea, Tarsonemidae, Tenuipalpidae), focusing attention on male and female behaviours and the ecological implications.

Tetranychidae

Both sexes of tetranychid mites exhibit precopulatory behaviour, followed by the act of mating. The precopulatory behaviour of the male seems to be more complex and takes place in several phases, 'aggression', 'attraction', 'guarding', 'webbing' and 'territoriality' (Cone, 1985).

Studies on the aggression between males of the two-spotted spider mite show that it does not cause a specific dominance but a series of random confrontations. Guarding females is a form of ephemeral territoriality, and the procurement and retention of territory depends on fighting capacity (Potter *et al.*, 1976b). Males spin their webs over the quiescent deutonymph. The webbing consists of strands of silk over the females and radiating outwards by 2–3 mm. This 'guarding position' allows the males to detect changes in their immediate areas, e.g. the arrival of another male or the onset of ecdysis (Cone *et al.*, 1971a; Saba, 1974). Just before the moult of the deutonymph, the male remains on or at least touches the female. When the deutonymphal exoskeleton splits, it helps the female to free herself from the exuvium. Mating occurs when the posterior part of the exuvium is shed. The males usually emerge before the females, thus ensuring that all of the females are mated (Laing, 1969; Saba, 1974). The brief availability of females, the excessive number of males and the concentration of male activity on quiescent deutonymphs intensifies male competition for mating (Potter and Wrensch, 1978).

Female precopulatory behaviour is less complex than that of the male. Males find females more quickly if females deposit webbing on the substrate before they become quiescent; the opposite occurs when female deutonymphs are placed on a clean substrate. This suggests that males find deutonymphs because they have left webbing as a cue. The same mechanism could be used by predators (Penman and Cone, 1972).

During the mating process, the male slips under the female from behind and clasps her legs with his front pair. This pushes the opisthosoma upwards to bring the extruded aedeagus in contact with the female genital opening, and the sperm are transferred in a period of 0.5–2.5 min (Cagle, 1949; Laing, 1969; Banerjee, 1979). A complete first mating causes a totally ineffective second mating (Boudreaux, 1963; Helle, 1967), probably because the female accepts only a certain amount of sperm. The findings of a study on the correlation of the duration of the first mating prior to an interruption with the results of a second mating show that second matings are about 75% effective after 30 s, 60% after 60 s, 47% after 120 s and only 31% after 150 s (Potter and Wrensch, 1978).

Eriophyoidea

Mating *sensu stricto* does not occur in the Eriophyoidea, which adopt only a sperm transfer. Detailed information on this can be found in the comprehensive review of the 'mating' behaviour and the population mating structure of eriophyoids by Oldfield and Michalska (1996). The topic has also been extensively treated by Michalska *et al.* (2010) in a updated review titled the 'Behavioural studies on eriophyoid mites'.

Female eriophyoids reared without males generate males by arrhenotoky (Putman, 1939; Bailey and Keifer, 1943; Oldfield *et al.*, 1970). In these mites, no physical mating between sexes occurs and the females are inseminated through spermatophores (8–25 μm long, with a head diameter of 9–19 μm) that are deposited by males on the plant surface (Oldfield *et al.*, 1970; Sternlicht and Goldenbergh, 1971). Michalska and Boczek (1991) observed male attraction to quiescent deutonymphs and spermatophores deposition around them. As in *Tetranychus* spp., the males of some free-living eriophyoid mites, e.g. *Aculus robiniae* (Nalepa), guard pre-emergent females for some time before their emergence, adopting a clear territoriality. According to Oldfield and Michalska (1996), there are eight subfamilies of the three families of eriophyoid spermatophore-producing species. The spermatophores of more than 20 species are known (Oldfield *et al.*, 1970; Sternlicht and Goldenbergh, 1971; Carèsche and Wapshere, 1974; Chandrapatya, 1984; Michalska and Boczek, 1991). Females pick up sperm from either single spermatophores (at the most two) or several spermatophores according to the mating system, which may be 'polygynous' or 'polygamous'. In the first case, males mate several times during their life cycle and females only once; in the second case, both males and females mate several times. Sperm storage may be 'asymmetrical' or 'symmetrical'. Females of *A. fockeui* (Nalepa *et* Trouessart) pick up sperm from only a single spermatophore (Oldfield and Newell, 1973a) and store it asymmetrically in just one of the paired spermathecae (Oldfield, 1973). Most species storing sperm asymmetrically inhabit dicotyledonous plants, and they usually show the predominance of male attendance of pre-emergent females. Symmetrical sperm storage involves both spermathecae and multiple inseminations by females (Oldfield, 1999).

Males of some species, e.g. *A. fockeui*, deposit spermatophores in the absence of females, which do not require any help from males in picking up the sperm from a spermatophore. In some species, e.g. *Aceria sheldoni* (Ewing) (Sternlicht and Goldenberg, 1971; Sternlicht and Griffiths, 1974), *Epitrimerus pyri* (Nalepa) and *Aculus schlechtendali* (Nalepa) (Oldfield, 1989), the two sexes exhibit a 'sexual independence' that leads to a complete 'sex dissociation', and the males deposit isolated spermatophores (Oldfield *et al.*, 1970; Michalska and Boczek, 1991; Michalska and Shi, 2004).

Some eriophyoid females pick up sperm after encountering a spermatophore, whereas others change their behaviour when they approach a spermatophore, thus showing a form of recognition (Oldfield *et al.*, 1972). There is a progressive decrease in the attractiveness of ageing spermatophores to females (Oldfield *et al.*, 1970; Oldfield and Newell, 1973a), and this has been related to a decrease of their pheromone concentration. Free-living species, e.g. *A. schlechtendali* and *A. fockeui*, and gall makers, e.g. *Acalitus essigi* (Hassan) (Michalska, 1997) and *A. sheldoni* (Sternlicht and Griffiths, 1974), can deposit spermatophores in groups. Vagrant species deposit a high number of spermatophores on leaves, along the mid-vein and near its junction with the lateral veins (Oldfield *et al.*, 1970; Michalska, 1997). Interspecific attraction of spermatophores is characterized by different behaviours (Oldfield, 1989). For example, females of *A. fockeui* and *A. schlechtendali* pick up the spermatophores of congeneric males. Females of *Eriophyes pyri* (Pagenstecher), *Phyllocoptruta oleivora* (Ashmead), *A. fockeui* and *A. schlechtendali* did not visit the spermatophores of males of different genera. *E. pyri* shows some interest in the spermatophores of *Aculus* spp. (*A. fockeui* and *A. schlechtendali*).

The rate of spermatophore deposition varies with eriophyoid species and environmental conditions. An *A. fockeui* male produces on average 30 spermatophores a day, while the male of *P. oleivora* produces 16 spermatophores a day (Oldfield *et al.*, 1970) and *A. sheldoni* a few spermatophores a day (Sternlicht and Goldenberg, 1971). In *A. fockeui,* the rate of spermatophore deposition depends on the age of the male (Oldfield and Newell, 1973a) and the time of day (Michalska, 2005). The host plant and its quality also affect the output of spermatophores, as in males of *Cecidophyopsis hendersoni* (Keifer), which deposit many fewer spermatophores on an older yucca leaf than on a young one. In this mite, the presence of females can stimulate spermatophore deposition, and no response to or impediment of spermatophore placement occurs in the presence of competitors (Michalska and Shi, 2004). The same behaviour has been observed in mature males of *A. fockeui* (Michalska, 2005). Spermatophore deposition on a particular patch may also rely on the probability that any receptive female will seek spermatophores within that patch. A lack of females and other cues, and the presence of many other males, might force a male to conserve sperm (Michalska, 2000).

Eriophyoid males may be 'solitary' or 'joint guarding' and 'non-guarding' of quiescent female nymphs. *Aculops allotrichus* (Nalepa) males deposit spermatophores around quiescent female nymphs and attend for many hours until female emergence (Michalska, 1999). Males of *A. fockeui* and *A. essigi* are non-guarding and make short visits to pre-emergent females and occasionally deposit spermatophores around them (Oldfield and Michalska, 1996; Michalska and Mańkowski, 2006). This behaviour seems to be related to a single or relatively few inseminations in a female's life cycle (Michalska and Mańkowski, 2006).

Tarsonemidae

Information on tarsonemid mites has been summarized by Lindquist (1986). Adult males of this family exhibit a precopulatory behaviour with immobile pharate females contained within the larval cuticle. The male engages his genital capsule to the pharate female and carries it on her back, where it is usually borne in various positions and attached on the opisthosoma. This behaviour may continue also for 24 h, until the emergence of the adult female. Males are usually very selective in the choice of pharate females (Nucifora, 1963; Suski, 1972). Copulation occurs immediately after the adult females emerge,

before they disperse. Copulation is sometimes observed between males and old females (Suski, 1972). In *Polyphagotarsonemus latus* (Banks), copulation lasts from 15 s to 2 min (Nucifora, 1963), while in *Tarsonemus pallidus* Banks it lasts from 6 to 15 min (Karl, 1965a). During copulation, the males and females are arranged at the same level with the gnathosoma facing in opposite directions in a straight line. Regenfuss (1973) called this position 'retroconjugate'. The males hold on using legs III and IV, and the females use legs III and IV. The genital capsule is affixed to the posterior part of the female's idiosoma, and the penis and the accessory stylets are introduced into the female's copulatory opening.

Nucifora and Vacante (1986) carried out observations on the mating behaviour of *T. waitei* Banks, a common species on citrus and other plants and feeding on *Alternaria* sp. They observed that males waited impatiently for several hours for female emergence, but that they did not move from the side occupied by the pharate female. The males appeared to be excited and often enacted a series of movements which simulated a fight for conquering the awaited female. Two or more males may fight from 3 to 7 min, attaching themselves with their fourth pair of legs, which are moved as a pincer. This fight does not hurt the contenders. Only the strongest male mates (first) with the female. This helps the pharate female to get out the larval skins. During the mating, the couple is often disturbed by other males, and in some cases the mating is interrupted. Usually, only one mating occurs, but when other copulations occur these last from a few seconds to 1–2 min. Sometimes, males attempt to mate with newly deposited eggs, and in some cases they may both attempt to copulate.

Tenuipalpidae

Tenuipalpid mites accomplish mating only irregularly. Several species lack males and others have only sporadic males. For example, males of *Brevipalpus obovatus* Donnadieu are rare and this species reproduces by thelytokous parthenogenesis (Helle and Bolland, 1972; Helle *et al.*, 1980). When males do mate, they do not fertilize the female and some 'male' specimens are clearly intersex (Pijnacker *et al.*, 1981; Weeks *et al.*, 2001). The same phenomenon occurs in other species.

Dispersal

Mites move within and/or between their habitats. These movements may involve single individuals or entire populations and are accomplished through a 'simple locomotion or crawling', a 'passive dispersal' or a 'phoresy' (Jeppson *et al.*, 1975; Kennedy and Smitley, 1985; Schliesske, 1990; Michalska *et al.*, 2010). In addition to the mode(s) of dispersal adopted, the survival of species also depends on the ability to easily find a suitable habitat for its development and, in the case of phytophagous species, a suitable host plant. This need relates dispersal to the phenomena of 'host acceptance' and 'host preference and host specificity'. Only the first topic is discussed in this section as it is more directly related to dispersal; the second topic is discussed later on in the chapter in the section on 'Relationships between mite pests and their host plants'.

The dispersal of pest species is closely related to crop infestations, and consists of the movement of individuals from the sites in which they developed. It can be characterized as intra-plant movements toward uninfested plant parts and inter-plant movements toward uninfested plants (Kennedy and Smitley, 1985). Intra-plant dispersal has been investigated in some tetranychid mites, e.g. *T. urticae*, in which part of the population of pre-reproductive females moves from the sites on which they developed regardless of their density (Hussey and Parr, 1963a; Coates, 1974). Physiological differences between disperser and non-disperser tetranychid mites are not known (Kennedy and Smitley, 1985). Dispersal in the Tetranychidae is not usually passive, and may represent a response to particular conditions. The dispersal of the natural enemies of mite pests has been extensively treated by Gerson *et al.* (2003).

Simple locomotion or crawling

Within the same habitat, a single movement may be caused by overcrowding, or by searches for better food resources, sites of hibernation or aestivation or appropriate microclimates for survival. Diurnal and seasonal changes in temperatures and humidity may induce vertical movement in populations of mite pests, as *T. urticae*, soil mites, Oribatida, etc.

Michalska *et al.* (2010) note that at least seven eriophyoid mite species show ambulatory dispersal only, including *Phytoptus avellanae* Nalepa, *Cecidophyopsis vermiformis* (Nalepa) (Burgess and Thompson, 1985) and *Phyllocoptes gracilis* (Nalepa) (van Dinther, 1951). Populations of the tomato rust mite, *Aculops lycopersici*, are mostly spread by humans and insects (aphids), but they can also disperse quickly to higher stems and leaves, and slowly to adjacent plants (Grahl and Leuprecht, 1998; Tanaka and Shibao, 2003). The citrus rust mite, *Phyllocoptruta oleivora* (Ashmead), is usually spread by wind (Bergh and McCoy, 1997), but at night and on cloudy days, populations are found on both leaf surfaces and tend to disperse across the foliage; they are easily observed during the most sunny days, with the highest levels of attack occurring on the outside of the canopy (Jeppson *et al.*, 1975).

Simple locomotion or crawling of tetranychid mites occurs between different parts of the same host plants and/or between plants. In modern agriculture, fields of a single crop may be identified as host plant aggregations, and mites leave infested plants at high densities and move towards uninfested plants through intertwined foliage and over the ground. For example, *T. urticae* has been observed moving from weeds on the margins of fields of maize and infests plants by crawling between them (Brandenburg and Kennedy, 1982a; Margolies and Kennedy, 1985). Some mites may drop actively or passively from plants and crawl between them over the soil surface, dispersing towards illumination with a rate of crawl that varies in different soil types, ranging from 5 cm h^{-1} to 6 m h^{-1} (McGregor, 1913; Parker, 1913). At high population densities and where there is serious plant damage, *T. urticae* exhibits a positive phototactic response in the dispersal phase (not exhibited by sedentary

individuals), which probably depends on food shortage and on decreasing RH (McEnroe and Dronka, 1971), and induces the mites to move up the plants and aggregate at high density at the periphery (Suski and Naegele, 1963). Inter-plant dispersal of *T. urticae* in glasshouses occurs by the formation of masses of mites at the apices of infested plants, which then disperse by dropping from the plant on webs, but only in still air, thus showing that the mites leave the plant by crawling (Hussey and Parr, 1963a). In the field, the formation of these aggregations is not a prerequisite to aerial dispersal, and aggregation behaviour plays a minor role in aerial dispersal; its function is unknown (Kennedy and Smitley, 1985). Observations made at low mite density have shown that populations of *T. urticae* and the citrus red mite, *Panonychus citri* (McGregor), occupied uninfested plants by crawling (McEnroe and Dronka, 1971; Wanibuchi and Saitō, 1983).

The tenuipalpid *Brevipalpus phoenicis* (Geijskes) moves less than 1 cm a day (Alves *et al.*, 2005a). Females of this mite leave infested plants and walk through the stems in search of new plants, not stopping for food for more than 3 days (Moraes and Flechtmann, 2008).

It is probable that in tarsonemid mites, e.g. *Polyphagotarsonemus latus*, inter-plant dispersal may be effected directly by females, which move between adjacent plants by walking in search of new food. However, intra-plant dispersal to new growth is frequently accomplished by males, which carry pharate females, as in *P. latus* (Lavoipierre, 1940; Moutia, 1958; Nucifora, 1963); in other tarsonemid mites, such as *Tarsonemus waitei* Banks, the males only rarely carry pharate females (Nucifora and Vacante, 1986).

Migrations of the species complex *Penthaleus major* (Dugès), variously named the blue oat mite or winter grain mite, to grain fields may take place from fence rows or other uncultivated areas. Fescue appears to be the favoured host in such situations in Texas (Jeppson *et al.*, 1975). In Australia, Weeks *et al.* (2000a) estimated the rates of movement of adult *P. major* using mark–release–recapture techniques, and mean square dispersal distances were used to estimate diffusion coefficients. In pasture, the diffusion coefficients were in the range 0.3–1.3 and 90% of the population moved <5–11 m in a 10-day period, or 7–16 m within their adult lifetime. Releases of mites into pea/wheat crops adjacent to the pasture indicated that there was directional movement towards the more favoured pea host.

Passive dispersal

Several ecological factors induce mites to adopt passive dispersal, including the depletion of food resources, overcrowding and desiccation. Air currents, rain, irrigation water, agricultural operators, harvesting tools and different animals may be involved in passive dispersal. The role of air currents has been known for some time and mites have been captured at altitudes as high as 10,000 ft (Coad, 1931, in Fleschner *et al.*, 1956). This mechanism is largely adopted by tetranychid and eriophyoid mites (Kennedy and Smitley, 1985; Lindquist and Oldfield, 1996; Michalska *et al.*, 2010) and also by their phytoseiid predators (Sabelis and Dicke, 1985). However, while aerial dispersal represents a real possibility for carrying mites great distances (Johnson, 1969), most of these attempts fail (Kennedy and Smitley, 1985). Observations carried out on the distribution of *T. urticae* in groundnut fields showed that even though aerial dispersal occurred from surrounding maize fields, the majority of individuals were not carried far. Moreover, although mites were widespread on the whole crop, the highest populations and damage were found close to the areas where dispersal had occurred (Brandenburg and Kennedy, 1982a).

Eriophyoid and tetranychid mites exhibit different predispersal behavioural patterns. In the tetranychids, web production is actively involved in population dispersal (Gerson, 1985b). The efficiency of passive dispersal requires a high density of mite populations and a high number of new sites to colonize, and there is a high risk that the new sites may not be reached. Tetranychids have various adaptations that compensate for the low probability of reaching a suitable host plant, including a wide host range, mating before dispersal and an increased number of female eggs; on deteriorating hosts, the mortality rate of males may be increased and the males may be smaller and less abundant than females (Mitchell, 1970; Wrensch and Young, 1978, 1983).

The aerial dispersal of *T. urticae* involves mostly adult females, few nymphal stages and usually no males. Females assume a dispersal posture in which legs I are raised. Wind carries the mites at a rate of 1.5 m s^{-1}, and there is a positive relationship between wind velocity and the number of mites assuming a dispersal posture (Brandenburg and Kennedy, 1982a). Moreover, dispersal posturing requires both wind and light, and according to the positive phototactic response investigated by Suski and Naegele (1963) and McEnroe and Dronka (1971), it occurs usually in populations in the upper portions of the plants. The directional orientation is characterized by a negative response to light, with mites facing away from the light source. It seems that there is not a directional response to wind, although wind can cause a negative response in which a vertical orientation is induced with the mites facing downwards. Dispersal on updraughts would facilitate horizontal displacement.

On maize, it has been observed that mites exhibit the greatest aerial dispersal after their populations have spread throughout the whole canopy (Silberman, 1983). Similarly, Margolies and Kennedy (1985) found that on infested maize crops, the aerial dispersal of *T. urticae* starts when mite populations infest the upper leaves of plants. This behaviour is explained by the fact that until mites spread on to the upper leaves they are protected from wind by the surrounding plants. An assessment of the aerial dispersal of snow-borne mites demonstrated the presence of 88 eriophyoid mites and 75 mites from other groups in 249 kg of snow from 11 major snowfalls (Zhao and Amrine, 1997a). Several mites were alive when trapped. These results suggest that many mites, especially eriophyoids, possess a high potential for winter and long-distance aerial dispersal. Michalska *et al.* (2010) report that at least 30 species of eriophyoid mites use wind, rain and different animal carriers (spiders, insects, bats, humans). Among these modes, aerial dispersal is considered to be the most important (Lindquist and Oldfield, 1996), but at the same time, it is the most risky mode of dispersal. In fact, the possibility of specialized animals finding a suitable host carrier

is very low, although this mode is advantageous for species with wide host ranges, e.g. *Abacarus hystrix* (Nalepa) (Sabelis and Bruin, 1996). Recently, it has been suggested that species adopting aerial dispersal may represent a complex of highly specialized types (Skoracka and Kuczyński, 2006a; Skoracka, 2008). Furthermore, aerial dispersal can also be regarded as a mode to increase the probability of attaching to a carrier, as in *Aceria tulipae* (Keifer) (Gibson and Painter, 1957) or *Calepitrimerus vitis* (Nalepa) (Duffner *et al.*, 2001).

A wind speed of 23 km h^{-1} is not enough to trigger the dispersal of *B. phoenicis*, and even wind speeds of 30 and 40 km h^{-1} can trigger the dispersal of less than 1% of mites on fruits. However, mite dispersal studies using sticky traps under field conditions also showed that the dispersal *B. phoenicis* is limited when compared with that of other mite species from citrus groves (Alves *et al.*, 2005a). In tarsonemid mites, e.g. *P. latus*, several factors, such as air currents, rain, irrigation water, agricultural operators, harvesting tools and various animals allow passive dispersal (Vacante, 2000; Benuzzi, 2007). Movement of the erythraeid mite *Balaustium medicagoense* Meyer *et* Ryke, a potential pest of cereal crops (Halliday, 2001a), winter crops and pastures in Australia (Halliday, 2001a; Arthur *et al.*, 2010), may also occur when over-summering eggs are moved by summer winds (Arthur and Umina, 2010). The dispersal of *Penthaleus* spp. relies on different mechanisms, including the transportation of aestivating eggs or mites on grain stubble or leaves, on soil adhering to implements that are moved about, or on forage or straw carried from infested fields during livestock feeding operations. Debris transported by wind disperses the aestivating eggs of *P. major*, and the adults can also achieve a local distribution (Jeppson *et al.*, 1975).

Phoresy

Phoresy is the dispersal of an animal by other animals. In mites, it occurs in the deutonymphal or adult stages, and it has been reported for various mite groups (Binns, 1982). The phenomenon has been defined as a temporary symbiosis between a 'phoretic animal', which is usually smaller, not feeding and inactive, and a 'carrier animal', which is usually larger, that transports it (Walter and Proctor, 1999) from a disadvantaged to a more appropriate habitat. Camerik (2010) revised the concept of phoresy by adapting Athias-Binche's (1994) classification in which phoresy may be 'facultative' or 'obligatory'. Facultative phoresy is usually related to the density of populations and/or environmental perturbations. Obligatory phoresy occurs when a species living in a relatively stable environment adopts a seasonal migration cycle using a coexisting carrier. Many examples of phoresy involve the hypopodes (heteromorphic deutonymphs) of acarine species from various taxa (Mesostigmata, Astigmatina) that have phoretic relationships with arthropods and sometimes also with vertebrates. The hypopus may attach to its carrier by several mechanisms, which include the 'anal pedicel' (Uropodina), 'posteroventral suckers' (in different Astigmatina) and 'posteroventral claspers' (Glycyphagidae), etc. In different families (Tarsonemidae, Pyemotidae, Scutacaridae, etc.), adult females are phoretic on a variety of arthropod carriers and sometimes on vertebrates (Evans, 1992; Palevsky *et al.*, 2001). Further information can be found in Krantz (2009b).

Among the pest mites, although phoresy has been variously reported for different eriophyoid species (Massee, 1928; Gibson and Painter, 1957; Shvanderov, 1975; Waite and McAlpine, 1992), these mites do not possess phoretic morphological adaptations (anal pedicels, suckers, claspers) or adaptations that enable them to select a proper carrier. This suggests that the use of carriers by eriophyoid mites is casual (Lindquist and Oldfield, 1996). The phoretic mode of dispersal seems to be more efficient than aerial dispersal, and it would guarantee a greater possibility of finding a specific host (Sabelis and Bruin, 1996). However, it could require the ability to distinguish between host-specific carriers and non-host-specific carriers. Waite (1999) reports honeybees as vectors of the lychee erinose mite, *Aceria litchii* (Keifer).

In tetranychid mites, phoresy is fundamentally unknown, although it is widely known that farm machinery, farmers and various tools can spread populations of *T. urticae*, *P. citri* and other spider mites from infested crops to uninfested crops.

Inter-plant dispersal of the broad mite, *P. latus*, is accomplished by phoresy using the tobacco whitefly, *Bemisia tabaci* (Gennadius) (Flechtmann *et al.*, 1990), and the greenhouse whitefly, *Trialeurodes vaporariorum* (Westwood) (Parker and Gerson, 1994). Palevsky *et al.* (2001) have determined that there is a specificity in the association between the mite – which also attaches to a citrus pest, the whitefly *Dialeurodes citri* (Asmehad), and other aleyrodids – and its phoretic hosts. The hypopodes of acarid mites are carried by insects. Thus, the hypopodes of *Mycetoglyphus fungivorus* (Oudemans), which have been reported from mushrooms and various other economic plants (Mahunka, 1961; Manson, 1983; Nakao and Kurosa, 1988; Nakao, 1989), are probably carried by ants (Türk and Türk, 1957), while those of *Sancassania berlesei* (Michael), which have been reported from edible fungi (Hussey *et al.*, 1969; Clancy, 1981), are carried by stored products insects (*Tenebrio molitor* L., etc.) (Hughes, 1976). The hypopodes of different Histiostomatidae are carried by insects. Adults, juvenile stages and hypopodes of *Histiostoma feroniarum* (Dufour), which have been recorded on edible fungi (Hussey *et al.*, 1969) and are responsible for serious crop losses in various mushroom species (Kim and Hwang, 1996; Greenslade and Clift, 2004), have been collected from associations with the weevil *Otiorhynchus cribricollis* Gyll. on strawberry in a protected crop in Sicily (Vacante, 1989b). Hypopodes of *H. sapromyzarum* (Dufour), which are commonly found on decaying fungi (*Lactarius*, *Russula*, *Tricholoma*, *Amanita*, *Scleroderma*) (Hughes, 1976) and are also recorded on cultivated mushrooms (Boness, 1968), have been collected on decaying hyacinth bulbs and on damp timber. The hypopus of this species is carried by several beetle species, flies and myriapods (Hughes, 1976).

Host acceptance

Host acceptance is important in the passive dispersal of mites via host plants. It occurs in eriophyoid or tetranychoid mites, and undoubtedly also in other mite groups, in which it has not

been investigated as much. Eriophyoid mites may be transported both on 'hosts' or 'non-hosts', and their ability to discriminate between these host types may represent a selective advantage. However, only a few species are able to distinguish between 'accepting' and 'non-accepting' hosts (Michalska *et al.*, 2010). Behavioural observations and differences in reproduction on various hosts allow the evaluation of host acceptance, and such investigations have been carried out in the eriophyids *Epitrimerus pyri* (Nalepa) (Bergh and Weiss, 1993) and *Abacarus hystrix* (Skoracka and Kuczyński, 2006a), and also in various other eriophyoid species. *Aculus schlechtendali*, for example, exhibits different feeding behaviour (time of probing and feeding) on different apple varieties (Kozłowski and Boczek, 1987a; Kozłowski, 1995). Gibson (1974) observed that *A. hystrix* preferred to feed only on three of 11 grass species tested. Further studies on the same mite showed that it exhibits a different activity and occupies various sites on the various grass species (Skoracka *et al.*, 2007). The different behaviours observed on hosts and non-hosts and the tendency towards dispersal when mites occupy non-hosts suggested the possibility of an adaptive discrimination (Skoracka and Kuczynski, 2006a). Another example is given by *Aceria tosichella* Keifer, which behaves differently according to the quality of the grass host plant, and disperses when the quality of host plant decreases as a result of feeding damage, thus exhibiting non-preference for the host (Nault and Styer, 1969).

The same sort of behaviour has been reported for tetranychid mites, which disperse passively by air currents and show randomized landings. Byrne *et al.* (1982a) reported that in spider mites host plant selection could be examined as host plant acceptance and not as host finding. According to Sabelis (1985a), 'host plant selection', 'host plant change' and 'reproductive success' are all involved. Host plant selection has been shown using Y-tube olfactometers (Sabelis and Van de Baan, 1983). Dabrowski and Marczak (1972) found that spider mites that have landed or arrived on host plants start piercing the substrate and according to its nature either continue feeding or move away. Byrne *et al.* (1982a) reported host plant preference in the tetranychid mite *Mononychellus tanajoba* (Bondar). Behavioural observations correlated with survival and oviposition on different host plants showed that spider mites might evaluate the host plant quality by initial piercing before oviposition starts in order to determine the best host plant for reproductive success. The ability of mites to locate host plants far away increases reproductive success somewhat, and natural selection should improve the ability to find proper host plants. Moreover, natural selection should favour genotype coding for a greater ability to find suitable host plants, which is also characterized by a wide range of proper host plants (Sabelis, 1985a). There are various scientific and technical aspects to this, such as the response of spider mite populations forced to feed on particular host plants, or how spider mites adapt to new host plants through selection. Further information on this topic and on the compromises that underlie reproductive success are discussed by Sabelis (1985a).

Skorupska and Boczek (1984) compared ten apple varieties and found that 'Landsberg' was the least attractive to the hawthorn spider mite, *Tetranychus viennensis* Zacher. Among seven plum varieties, 'Emma Leppermann' and 'Althann's Gage' were some of the least attractive. The preferences established for *T. viennensis* do not agree closely with those of the European red mite, *Panonychus ulmi* (Koch) (Skorupska and Boczek, 1984). When *T. viennensis* was transferred from hawthorn (*Crataegus*) to apple, pear or plum in the laboratory, several reproductive parameters did not differ between mites transferred from hawthorn to another plant and those reared continuously on one plant, and the mite adapted to the new host plant by the F_2 or F_3 generation. The adaptability of *T. viennensis* is much faster than that of *T. urticae,* which takes up to ten generations to adapt to a new host plant (Skorupska and Boczek, 1985).

Host acceptance was examined in the tarsonemid mite, *Steneotarsonemus spinki* Smiley. Research carried out on 70 plant species growing in or near rice fields showed that none of these plants, including 44 monocotyledons, hosted the mite, which normally lives on rice (Ho and Lo, 1979). Návia *et al.* (2010b) reported other congeneric host plants.

Dispersal and crop infestation

Dispersal affects both the life cycles and the population dynamics of mites injurious to economic plants, and so the monitoring of dispersal facilitates the management of mite infestations (Kennedy and Smitley, 1985). Some aspects of dispersal depend closely on the agricultural techniques that are used, such as the use of grafts with scions infested by eriophyoid mites or young plants in the nursery spreading populations of tetranychids (*T. urticae* and other species) throughout crops. The intercropping of susceptible plants facilitates mite dispersal from infested to uninfested plants. Good agricultural practice suggests the planning of suitable crop rotations and/or arranging different crops over a wide area, taking into account the different components of the agroecosystem. The topic was extensively examined from the viewpoint of spider mite dispersal in a maize/groundnut system (Brandenburg and Kennedy, 1982a; Margolies and Kennedy, 1985). Dispersal may also affect the management of mite strains resistant to acaricides. Various studies on dispersal have shown that the immigration of susceptible individuals into a crop subject to selective pressure for resistance limits the development of resistance (Comins, 1977; Georghiou and Taylor, 1977a,b; Taylor and Georghiou, 1979; Tabashnik and Croft, 1982). Some insecticides and acaricides stimulate the dispersal of tetranychid mites from treated plants (Gerson and Aronowitz, 1981a).

In females of *T. urticae*, permethrin causes an immediate arresting effect; a small number of females die, although the majority recover in 1–3 h (when sprayed) or 18–36 h (when immersed) and egg deposition is not affected. The females abandon treated leaves in less than 72 h, and migrate and disperse into the areas of the plant that have not been treated, but the repellent effect ceases after 72 h, and the females feed and oviposit eggs regularly. Eggs dipped in permethrin develop normally, as do the larvae and other phases (Flechtmann and Zem, 1980). Hence, the planning of control measures in a restricted area is more difficult than in a wide area.

The acarid mite *Tyrophagus similis* Volgin is introduced into cucumber greenhouses in England via straw or manure (Hussey *et al.*, 1969). Research in Japan on the horizontal and vertical distribution

of acarid populations in agricultural soils in greenhouses indicates that the mite inhabits not only the cultivated soil, but also the uncultivated soil in the periphery of the cultivated area inside the greenhouse and the soil immediately outside. The main control sites for *T. similis* are therefore in both the cultivated and peripheral soils of the greenhouse (Kasuga and Amano, 2005).

Seasonal Cycles

Seasonal adaptations in the life cycles of mites indicate that more primitive or less derived (plesiotypic or plesiomorphic) traits of seasonality involve complex 'systems of seasonal control' (SSC), characterized by different dormant stages leading the control of perennial, semivoltine and monovoltine development. Conversely, less primitive or more derived (apotypic or apomorphic) traits exhibit simple SSC with a reduced number of dormant stages for the control of monovoltine and polyvoltine development. The greater part of the Prostigmata show evolutionary transformations that range from plesiotypic to apotypic traits of seasonality, and only some taxa have fixed apotypic traits (Belozerov, 2010).

Regardless of the comparative and evolutionary evaluations, specific adaptations, environmental conditions and host plants, nature leads the seasonal cycles of mite species. There is not a single behavioural pattern. Some species are active on the host plant throughout the year, while others interrupt their activity in the winter or summer months. The stopping of activity allows mite survival in limiting conditions, and it assumes a state of 'diapause'. According to the species, diapause may occur in the winter or the summer months and involve different biological instars, such as fertilized females or eggs, in addition to morphofunctional adaptations.

Diapause

Diapause allows the survival of an organism under unfavourable environmental conditions and has been described as a physiological state of arrested development over a certain period and independently of environmental conditions (Dickson, 1949). Conversely, Beck (1980) defines diapause as a state of suppressed development that is genetically determined, which can be controlled by environmental factors. However it is defined, the phenomenon occurs in both winter or summer periods in different acarine groups. It is closely related to the climatic conditions and represents an adaptive response to weather conditions supported by a genetic basis. Diapause may occur in both eggs and adult females. It differs from quiescence owing to the involvement of physical conditions, and usually ends as environmental conditions become favourable again (Veerman, 1985). Diapause has been investigated in different mite groups, including the Tetranychidae (Veerman, 1985) and Penthalaeidae, but there is less information available on the Eriophyoidea (Jeppson *et al.*, 1975; Manson and Oldfield, 1996).

From a practical point of view, knowledge of factors inducing or governing the end of diapause may be very important in the management of different pest species in the field. For example, prediction of the period of hatching of the winter eggs of *P. ulmi* from the sums of effective temperatures accumulated by the mites assists in their control (Cranham, 1972, 1973; Herbert and McRae, 1982). Various examples are discussed by Veerman (1985).

Tetranychidae

Diapause in tetranychid mites involves different aspects of biology and physiology (Veerman, 1985). Diapause and non-diapause forms may differ in coloration (Veerman, 1974), morphology and structure of the eggs (Lees, 1961) and adult females (Pritchard and Baker, 1952a; Boudreaux, 1958a). Variations on oxygen consumption (McEnroe, 1961; Geispitz and Orlovskaya, 1971) and the accumulation of sorbitol (Sømme, 1965) confirm the involvement of physiological features. Cold hardiness and RH are also reported (see Veerman, 1985) to be involved. Diapausing and non-diapausing females of *T. urticae* exhibit different tolerance to a number of pesticides (Bondarenko, 1958; Parr and Hussey, 1966; den Houter, 1976; Hürková and Weyda, 1982). The genetic basis of diapause, and the diapause adaptations of both eggs and adults of some species play an important ecological role.

Genetic basis of diapause

The genetic basis of diapause has mainly been investigated in spider mites populations. In tetranychid mites, there is a strong diversity in photoperiodic responses that provides the means to adapt to climatic changes, e.g. in greenhouses. Usually, non-diapause characteristics are dominant (Geispitz, 1968; Helle, 1968). The process of spider mite adaptiveness arises from inter-population selection rather than from intra-population variability. However, intra-population variability cannot be underestimated, and is based on sex-linked variants and elevated spontaneous mutation rates (Helle, 1965a). The diapause response is sex linked in arrhenotokous species, acting in selection against unfavourable mutations only in diploid females and not in haploid males. This sex-linked mechanism influences several aspects of mite ecology and reduces the elimination of deleterious genes (Helle, 1968). Cross-fertilizing populations have both diploid sexes and can lead recessive characters, which are also deleterious, being facilitated by the heterozygous condition, which protects them from the pressure of natural selection. This phenomenon helps populations to store important adaptive characters. In contrast, in arrhenotokous species, this possibility is more restricted, and deleterious characters, dominant or recessive, that also occur at low frequency, are subjected to natural selection. In this case, hemizygous males immediately highlight the harmful characters. In general, positive mutations in arrhenotokous species are less involved in the dispersive process (genetic drift) than species with diploid sexes (Helle, 1965a).

Research on the chilling requirement of early- and late-hatching winter eggs of *P. ulmi* in England showed differences in the intensity of diapause that had a genetic basis (Cranham, 1973). The exposure of winter eggs susceptible to cold for four successive winters increased resistance to the cold in Nova

Scotia. Populations surviving at −34.4°C showed an increase of winter eggs of up 91% in the fourth winter population. Crosses between susceptible and resistant forms indicated that the resistance factor was recessive (MacPhee, 1961).

Winter diapause of eggs

Selected references on the dormant stages of tetranychid mites that refer to a true diapause stage indicated that almost 18 species had diapause eggs. Five of these species had eggs that exhibited aestival (summer) diapause and the rest showed hibernal (winter) diapause (Veerman, 1985).

Some tetranychids (*Panonychus*, *Oligonychus* and *Bryobia* spp.) and other phytophagous mites produce two kinds of eggs. The first type hatches shortly after laying in the time during which the mite is active; the second type hatches after a long period, and these eggs are usually referred to as 'diapause eggs'. Weather and other environmental changes may induce females to lay diapausing eggs, which allow the survival of the species through the unfavourable conditions, and hatch when the environmental conditions return to favourable. Generally, the outer egg covering or 'shell layer' of winter and summer eggs appears identical, but the two kinds of eggs exhibit different responses to environmental stimuli according to the presence (winter eggs) or absence (summer eggs) of cement in the shell layer of the egg in contact with the substrate, and also because the winter eggs are held in the female body until the stage of embryonic development (Beament, 1951). The availability of food resources influences the laying of diapause eggs, but its influence is secondary to that of temperature and photoperiod conditions. At medium temperatures and daily photoperiods of 6–13 h, only 'winter females' of *P. ulmi* develop, while with a photoperiod of 15–16 h or continuous light, only 'summer females' develop. The response is independent of light intensity, but some parts of the spectrum (ultraviolet, blue and blue green) are active, with the maximum activity in the blue region, while radiation of above 550 μm wavelength (orange, red, the infrared region) are inactive. Winter eggs in diapause do not hatch at 18–25°C and diapause can be stopped by chilling the eggs at 1–9°C for 150–200 days (Lees, 1953a,b). The upper limit of temperature occurs between 10 and 18°C and the lower limit between 1 and −5°C (Lees, 1953a). According to Hueck (1951), light provides a stimulus to the fully developed embryo to break the egg shell, with 86% of eggs hatching during daylight.

Low temperatures are an important direct mortality factor for winter eggs of tetranychid mites, e.g. *P. ulmi* (Lienk and Chapman, 1958a; MacPhee, 1961; Sømme, 1966) and *Bryobia rubrioculus* (Scheuten) (MacPhee, 1963, 1964). Winter eggs of both species are killed by freezing, but show a considerable adaptation to supercooling (the temperature at which freezing occurs at a rapid cooling rate) (Sømme, 1966). The mean supercooling point of two strains of *P. ulmi* in Canada was at −31°C and −37°C (MacPhee, 1961), and the lowest supercooling points found in Norway were −31.4°C and −33.2°C (Sømme, 1965). The mean supercooling point for *B. rubrioculus* is −32°C (MacPhee, 1963).

Further studies on the influence of temperature on winter eggs of *P. ulmi* in England show that the threshold temperature for complete development to eclosion (hatching) is 7°C. The processes of termination of diapause and early morphogenesis are concurrent. The diapause develops steadily without interruption and late phases can also be completed above the threshold of morphogenesis or in a range of temperatures above the threshold. The gradually declining heat sum needed to complete diapause provides adjustment of the mechanism controlling the time of hatch. In this way, the rate of development is either slowed or speeded up according to environmental temperatures. Premature hatching may be prevented if the weather is warm, but can occur faster if the cold is prolonged (Cranham, 1972).

The RH during incubation does not affect the percentage hatch of *P. ulmi* (Andersen, 1947).

Winter diapause in adults

Almost 25 species of tetranychid species exhibit winter diapause of female adults, but only two species have summer diapause of female adults (Veerman, 1985). Diapause females of *T. urticae* are positively geotactic and negatively phototropic. They leave the plants and search crevices in wood, bark, ground clover, earth clods, fence posts, etc. Diapause is not a constant characteristic of a given species, but may vary widely between populations in different countries or within a population after a number of generations (Veerman, 1985). Overwintering females can differ in colour and morphology from the summer forms. Tetranychid overwintering females are often reddish or orange in colour, and more deeply coloured than the summer forms. In the past, the changes of colour occurring in the tetranychids were attributed to the alteration of leaf metabolism (Reiff, 1949), but other research has indicated that the rapidity of these changes in colour depends either on temperature (Linke, 1953; Parr and Hussey, 1966) or on changes in the metabolism of the leaves of the host plants (Jeppson *et al.*, 1975). However, the orange and red pigments of tetranychid mites are hydroxy-keto carotenoids, which are usually absent in the leaves (Veerman, 1974). Whatever the cause, the colour change involves the stopping of feeding and migration from plants to various winter shelters. In the summer forms of the genera *Tetranychus* and *Eotetranychus*, the cuticular ridges bear semicircular, triangular or elongate lobes according to the species, while in the diapause (winter) forms these ridges usually lack lobes (Pritchard and Baker, 1952a; Dosse, 1964a). Exposure to winter conditions (13°C and 8 h daylight) induces changes in integumentary striation (Parr and Hussey, 1966) that are related to the need of the species to conserve water content during overwintering (Boudreaux, 1958a).

Three factors lead the diapause in the tetranychids: the photoperiod, the temperature and food resources. Their effects seem to be cumulative and are sometimes reversed, indicating that the phenomenon may be influenced by the endocrine activity. This helps to explain why old instars are more likely to show a tendency to diapause, probably because they produce greater amounts of hormones (Parr and Hussey, 1966). The photoperiod is the predominant factor in promoting diapause, but the importance of all three promoting factors varies (Veerman, 1985).

Light influences diapause in terms of the duration of the photoperiod or the total light energy, as light intensity over a certain threshold. The first studies on the role of photoperiod in the induction of diapause were carried out by Lees (1950)

and Miller (1950) on *P. ulmi* and by Bondarenko (1950) on *T. urticae*. Subthreshold values of light probably equate to darkness. The dark phase of the illumination cycle is also important in diapause in terms of the absolute length of the phases rather than the light/dark ratio. Mites are insensitive to high-intensity radiation and diapause is influenced by radiation in the green, blue and near ultraviolet (Lees, 1953a,b, 1955). The intensity threshold observed in *T. urticae* is about equal to 3.5 lux (0.3 ft candles) (Bondarenko, 1958), and β-carotene is necessary for the photoperiodic induction of diapause (Veerman and Helle, 1978). Albino mutants show a lower incidence of diapause than the wild type. Phenotypically wild type mothers obtained by crossing the two types give a high incidence of diapause. Small amounts of carotenoids, transferred by the hybrid female to the eggs, are sufficient for photoperiodic induction of diapause in albino offspring (Veerman, 1980a,b).

Diapause females do not transfer the capacity to go into diapause to their progeny. Photoperiod acts directly on the mites and the induction depends on the day length experienced by the mites and not on a gradual change in day length (Lees, 1953a). The critical day length for the photoperiodic conclusion of diapause is equal to that for the induction of diapause. Sensitivity to photoperiod gradually disappears some months after the initiation of diapause, allowing maintenance of the sensitivity to photoperiod until the arrival of winter temperatures. Diapause usually ends before midwinter (Glinyanaya, 1972; Veerman, 1977a). The variability of the photoperiod reaction is well known in spider mites, and a possible role of the circadian system in photoperiodic induction in *T. urticae* has been demonstrated (Vaz Nunes and Veerman, 1982). However, circadian-based models for the photoperiodic clock were unable to explain the results obtained with spider mites (Veerman and Vaz Nunes, 1984), and the 'hourglass timer oscillator counter model' was developed. This model is based on the assumption that the circadian system may influence the expression of a photoperiod response, but that it is not involved in the measurement of time (Vaz Nunes and Veerman, 1982). It has been observed that in *T. urticae* both diapause induction and diapause termination are under photoperiodic control. Resonance experiments (the testing of different light–dark cycles) with different European strains highlighted the involvement of the circadian system in the photoperiodic termination of diapause. In all strains the τ-value of the rhythm involved in diapause termination was shorter than that of the rhythm involved in diapause induction; this could indicate either that there is a change in the period length of a circadian oscillator involved in both the induction and termination of diapause, or that different circadian oscillators affect diapause induction and diapause termination in *T. urticae*. Other studies have indicated that the photoperiodic clock controlling diapause termination is probably the same as the one controlling diapause induction. Temperature appeared to affect the resonance rhythm of diapause termination differently in different strains of the spider mite, resulting in the rhythm either being compensated for by the temperature or not (Koveos and Veerman, 1996).

The role of temperature in relation to the light and dark phase has been variously investigated, e.g. for *P. ulmi* (Lees, 1953a) and for *T. urticae* (Veerman, 1977b). In *P. ulmi*, the dark phase is temperature sensitive, but high temperatures during the light phase have no effect. The same behaviour has been observed in *T. urticae*, in which no thermoperiodic effect on diapause induction was observed (Veerman, 1977b). In addition to the influence of temperature during the hours of darkness, temperature may influence the incidence of diapause and so affect the rate of development of mites (Veermer, 1985). Studies on developmental stages sensitive to photoperiod induction have been carried out on different spider mites. In *T. urticae*, sensitivity to photoperiod is already present in the larva (Parr and Hussey, 1966; Veerman, 1977b), but the greatest sensitivity was observed in the protonymph (Veerman, 1977b). Diapause is concluded only by a phase of chilling before warming. A English strain of *T. urticae* ended diapause after exposure to temperatures below 10°C for 100 days (Lees, 1953a). In Russia, *T. urticae* concludes diapause after 55 days at 3–6°C (Bondarenko, 1958). Females kept at 25°C and 16 h of daylight resume their feeding activity and start to lay eggs within 7 days without pre-chilling. Some 75, 73 and 90% of similar females kept at 2.7 and 13°C, respectively, become active in 4–10 days at 21°C and with 16 h of daylight. This behaviour shows the possibility of reversion to feeding forms without chilling. The differences that have been demonstrated in diapause emergence are probably related to environmental factors rather than strains (Parr and Hussey, 1966).

Diapausing females resist low temperatures very well. At −27°C only 10–15% of females die, but at −32°C there is total mortality (Bondarenko, 1958). Mites require low temperatures and high humidity. Individuals of *T. urticae* kept at 6°C and 75 or 90% RH survive for more than 8 months, which is about twice the survival at 40% RH. Mites survive better outside than inside greenhouses, and winter RH is lower inside greenhouses than outside (Parr and Hussey, 1966). Stagnant water usually hinders the survival of diapausing mites (Helle, 1962). Diapausing females of *T. urticae* survive after 100 h of submersion in water, but non-diapausing females die after 10 h (Tibilova, 1932, in Jeppson *et al.*, 1975). After winter chilling, the mites return to activity when it becomes warm. In *T. urticae*, the conclusion of diapause is related to increase of the temperature, while other factors, such as light intensity, photoperiod and food resource availability, are secondary (Jeppson *et al.*, 1975). Takafuji and Hinomoto (2008) investigated the distribution and geographic variation in diapause capacity among populations of *T. urticae* and the Kanzawa spider mite using populations from various countries in East and South-east Asia that were reared at three temperatures and under short-day conditions. They found that populations of *T. urticae* from northern Japan had a high capacity for diapause and that the mites showed a clinal decrease in diapause capacity. In general, all the populations investigated showed a very low capacity for diapause. Observations on populations of *T. kanzawai* from various countries did not show a clinal decrease in diapause capacity.

Ito and Saitō (2006) studied the effects of host plant species on the induction of diapause in the Kanzawa spider mite and found that the relationship between diapause incidence and performance was consistent with the prediction of theoretical studies that a short development time or high fecundity delayed the timing of diapause induction.

The third factor inducing diapause is the quality of food. The different behaviour of *T. urticae* populations on older

leaves from that on green leaves confirms the importance of this factor (Bondarenko, 1958; Parr and Hussey, 1966; Gould and Kingham, 1965). However, food usually has only a secondary importance in diapause induction (Parr and Hussey, 1966) and becomes fundamental only when photoperiod and temperature are critical.

Summer diapause

Summer diapause allows different mite species to survive hot and dry summers. The most common adaptation is the deposition of heat-resistant or 'aestivating' eggs on various substrates (soil, plant parts, etc.). However, some species, such as *Eotetranychus yumensis* (McGregor) (Jeppson *et al.*, 1975) and *T. urticae* (Geispitz and Orlovskaya, 1971) do have summer diapause females.

Females do not lay both non-diapausing and diapausing eggs simultaneously, and they usually deposit diapausing eggs when they are stimulated by winter changes. However, there are several examples of summer diapause females in tetranychid mites. These include the brown mite, *B. rubrioculus*, which lays diapausing eggs on the twigs of deciduous plants in early summer, which go through the summer and winter and hatch in the spring. The clover mite, *Bryobia praetiosa* Koch, lays aestivating eggs in late spring. The brown wheat mite, *Petrobia latens* Müller, lays heat-resistant eggs on ground litter during rising summer temperatures. The production of diapausing winter eggs of this species is not influenced by photoperiod in England (Lees, 1961).

In general, few data on the factors influencing the initiation of the aestival female diapause are available. Thus, the identity of aestivating females reported by Srivastava and Mathur (1962) in India is unclear, as is that of orange-red diapausing females obtained by Geispitz and Orlovskaya (1971) in the field and in the laboratory at 28–30°C under constant illumination.

Eriophyoidea

A number of Eriophyoidea have a simple pattern of development, while other species have a winter diapause in overwintering females that usually differ in colour and morphology from the summer forms. The phenomenon is common in those species inhabiting trees and is referred to as 'deuterogyny'; the summer females are called 'protogynes' and the winter females 'deutogynes' (Figs. 5.1 and 5.2) (see next section for further information). The deutogynes of eriophyoids living on deciduous plants are darker brown than the protogynes. The eriophyoid diapause mainly involves species inhabiting deciduous plants in the northern hemisphere. There is very little knowledge on species living on non-deciduous plants and in the tropics, but these species do not lay resistant eggs and go through the summer and winter as a female stage. The influence of photoperiod or thermoperiod on the appearance of diapausing females is variable. In the pear rust mite, *Epitrimerus piri* (Nalepa), temperature and exposure to light determine the rate of emergence of deutogynes from the overwintering sites, but not the stage of bud development (Bergh, 1992). In general, eriophyoid mites exhibit a considerable capacity to survive adverse conditions. Many species are able to survive the harsh low temperatures of winter either in the egg stage or through diapause. The survival of 50–76% of the deutogynes of *Aculus schlechtendali* (Nalepa) during the winter has been observed in Poland (Kozłowski and Boczek, 1987b). The eggs of *Aceria tulipae* (Keifer) can endure temperatures of −31°C for 2.5 min (Jeppson *et al.*, 1975). Mites can usually survive lower temperatures than their host plants, and this characteristic enables them to survive when they are associated with plant products that are transported for extended periods of time at low temperatures, for example, in fresh fruits, which are usually transported at temperatures from 2 to 6°C (Návia *et al.*, 2010a). The effect of winter on the host plant influences its seasonal cycles and consequently induces deutogyne development. In some cases, the development of deutogynes is related to leaf drop of the host plant, or to leaf maturity or to leaves being old or dry.

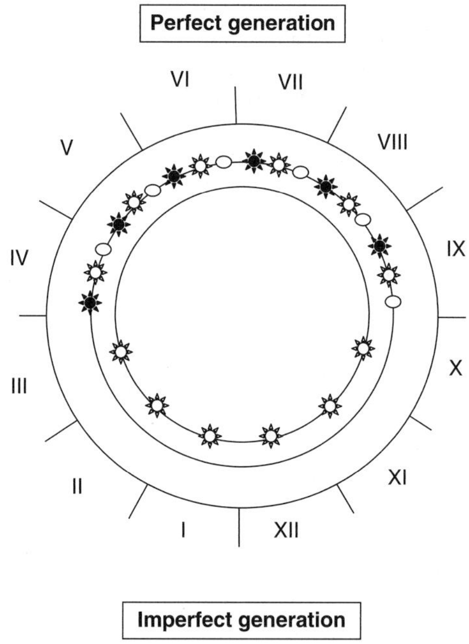

Fig. 5.1. Schematic representation of perfect and imperfect generations of eriophyoid mites. The perfect generation involves males (white), protogynes (black) and eggs (oval and white). The imperfect generation involves only deutogynes (dotted). Months are indicated as Roman numerals. According to species and environmental conditions, variations of this cycle are possible.

Deuterogyny

Deuterogyny is known in all three families of the Eriophyoidea, and the greatest number of deuterogynous species is included in the Eriophyidae. The Rhyncaphytoptidae have various deuterogynous species, and the Nalepellidae only one (Jeppson *et al.*, 1975). Although there is a non-structural deuterogyny, the most common type of deuterogyny is characterized by males and two types of females that are anatomically distinct: the protogynes or 'primary females' and the deutogynes or

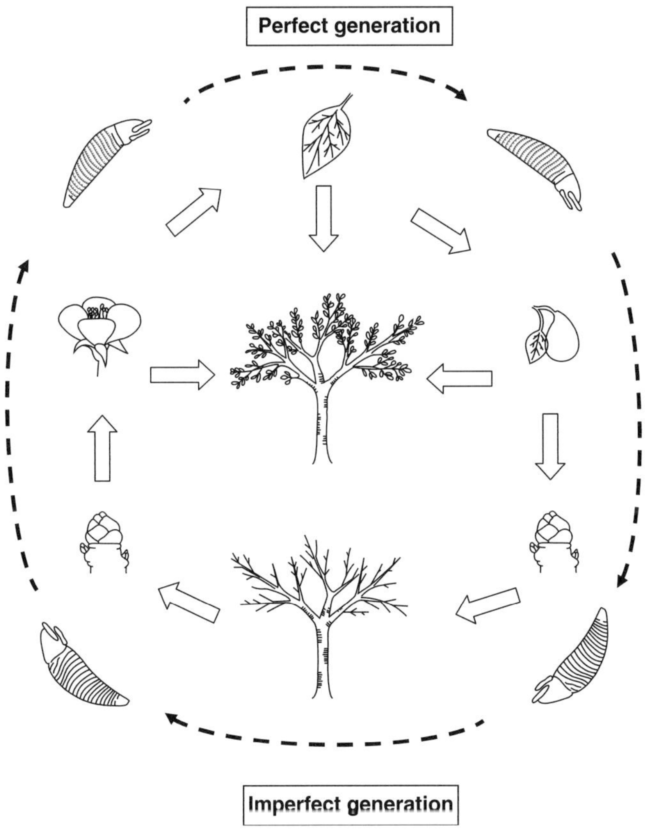

Fig. 5.2. Schematic representation of host plant phenology in relation to perfect and imperfect generations of eriophyoid mites. The top of the diagram shows the development of a perfect generation characterized by protogynes (outer circle, males and eggs are not drawn) in the spring and summer. The bottom of the diagram shows the development of an imperfect generation (outer circle) characterized by deutogynes from autumn to spring. According to species and environmental conditions, variations of this cycle are possible.

'secondary females' (Figs 5.1, 5.2). Males are commonly present during the least cold part of the year and their spermatophores are used by both types of females (Jeppson *et al.*, 1975). Protogynous females and males differ only in their genitalia. The protogynes are more or less contemporary with the males, lay eggs and have a lifespan of some weeks. They reproduce as long as environmental conditions are favourable. Protogynes and males are the retained primary forms of deuterogynous species, and together represent the 'perfect generation'. In addition, the protogynes are comparable to females of other species, and this is not always the case for the deutogynes. Because of this, and because they lack males, the deutogynes are said to represent the 'imperfect generation'.

Deutogynes differ from the protogynes in various features, mainly in the characteristics of their microtubercles, which are often reduced. A series of rules (almost seven) allows the presence of deutogynes to be detected (Jeppson *et al.*, 1975). These authors propose that the microtuberculation patterns on secondary females inhabiting deciduous plants in the northern hemisphere are correlated with the locations of aestivation and/or hibernation of eriophyoids. They classify these hibernating types into four groups. The first group includes deutogynes that are strongly microtuberculate over the entire opisthosoma, with the microtubercles more widely spaced and prominent than in protogynes. Different Eriophyinae and some Phyllocoptinae species belong to this group. They leave leaf galls in late summer/early autumn and retire to terminal buds of bark crevices, depending on the family. The second group has microtubercles entirely around the opisthosoma, but weakened and thinned out and more frequent dorsally. Some Phyllocoptinae belong to this group, and they inhabit lateral buds or bud clusters. The third group includes various species of Nalepellidae, Phyllocoptinae and Eriophyinae, and is characterized by deutogynes with microtubercles only on the lower side of the opisthosomal sternites. They leave the feeding and breeding parts of the host plant in late summer or in early autumn and retire to bark crevices. The fourth group includes some Eriophyinae and Phyllocoptinae. In this group, the deutogynes totally or almost totally lack microtubercles on the opisthosomal annuli, with the exception of some elongate microtubercles on the distal ventral part. They first aestivate and later overwinter, leaving galls or leaves during the hot summer period, to inhabit either dry masses of bud hairs or dry bark crevices.

Aceria inusitata Britto *et* Návia presents a rare case of deuterogyny that is characterized by two forms of females and two forms of males, all developing in colonies under the same 'nest' (Britto *et al.*, 2008). A similar behaviour has been observed in *Trisetacus kirghisorum* Shevchenko. In this species, a female 'summer form' (protogyne) occurs during the first year with a similar male, while a female 'winter form' (deutogyne) occurs during the second year along with a similar male. The second year forms are dispersive, entering the pollen tubes of new developing fruits of *Juniperus semiglobosa* Regel in Kyrgyzstan (Shevchenko and De-Millo, 1968). Deutogyne reproduction needs winter chilling and increasing spring temperatures. Good weather induces protogynes to produce eggs, from which the primary forms originate and gradually disappear. In this phase, the populations consist of nymphs, males and protogynes. The growth habit of the host plant and the behaviour of mite species lead to the appearance of new deutogynes at various times in the spring and summer (Jeppson *et al.*, 1975). A different type of behaviour characterizes rust mites and leaf vagrants, and erineum-making and gall-making eriophyoids.The former seem to be conditioned by leaf hardening, while the latter do not show any urgency to leave these microenvironments and may be found within them over 2 or 3 months. Leaf fall at the end of the season induces the protogynes to leave these sites. According to their various biological adaptations, variations in the life history of deuterogynous forms are known, and this topic has been extensively discussed by Jeppson *et al.* (1975).

Penthaleidae

Different Penthaleidae exhibit an adaptation of diapause that is usually found in the eggs. In some species, such as the red-legged earth mite, *Halotydeus destructor* (Tucker), other biological instars are also involved.

Diapause of eggs

Penthaleus major (Dugès) lays both winter eggs and aestivating eggs. In Australia, mite populations are active between April and October and develop two or three generations before entering diapause throughout the summer months. The life cycle is completed with the hatching of the diapause eggs in the autumn, when conditions are optimal, following cool temperatures and rainfall (Narayan, 1962; Wallace, 1970a; Ridsdill-Smith and Annells, 1997). The blue oat mite, *Penthaleus falcatus* Qin *et* Halliday (part of the *P. major* species complex), lays diapause eggs in early winter, although the first appearance of these eggs is slightly later than in *P. major* (Umina *et al.*, 2004). There is a constant rate of production of diapause eggs, although the trend is less clear than in *P. major*. The production of diapause eggs throughout a season will reduce population growth, and may contribute to the relative abundance of *Penthaleus* increasing less dramatically in the early spring than does *H. destructor* (Weeks and Hoffmann, 2000). The species develops three generations a year, passing the summer as a diapausing egg in the cadaver of the adult female (mother) on the soil surface. Long-distance passive dispersal probably occurs when these female cadavers are blown by the wind (Ridsdill-Smith, 1997).

Ecological Features of Biological Cycles

Life cycles can be examined both from biological and ecological points of view. Their diversity is related to natural selection, the environment in which each species lives and intra- and inter-specific relationships. Species can be studied as individuals, but it is usually necessary both to count individuals and to assess the development of their populations. Fertility, mortality and population increase are a reflection of the life cycle of each species, and their assessment through 'life tables', which summarize mortality and survival, are an important tool in the understanding of population dynamics. Life tables often allow the definition of survival curves, i.e. decreases in the numbers of individual offspring and the probability of their survival, and the production of fertility schedules. The basic features of the biological cycle are 'mortality and birth' and '*r*/*k* selection', a concept that underlies species interactions with the habitat that may make significant changes to the biological cycle. Both of these concepts assist in introducing the basic features of the population dynamics.

Mortality and birth

Data on the mortality and birth of the individuals of a population provide important information about the effective status of the population itself, especially its dynamics. Life tables, as well as the biological parameters of net reproductive rate (R_0), intrinsic rate of increase (r_m), generation time (T) and finite rate of increase (λ) (see Chapter 4, last section), are fundamental to this discussion, which demonstrates that the biological and ecological aspects of populations are closely related and that a rigid distinction between them can only have a didactic justification.

r/*K* Selection

The concept of *r*/*K* selection was originally proposed by MacArthur and Wilson (1967) and was further developed by Pianka (1970). The letters *r* and *K* are derived from logistic equations that represent the development of populations and indicate that individuals subject to *r*-selection are favoured by their high reproductive capacity, i.e. they have a high value of *r*, the rate of population increase, and that those subject to *K*-selection are favoured by their capacity to maintain their populations to their carrying capacity (*K*). The individuals on which *K*-selection acts live in a more stable or predictably seasonal habitat. They are generally larger, reproduce later and are iteroparous (have repeated reproductive cycles and produce offspring in successive groups); moreover, they have lower reproductive allocation, the progeny are larger, less numerous and need more parental care. In short, they use more resources on survival than on reproduction and, due to high competition, their lifespan is usually short (Begon *et al.*, 1986). In contrast, individuals subject to *r*-selection live in unpredictable or ephemeral habitats. These populations have periods of rapid growth with low competition, interspersed with negative periods with high mortality, depletion of food resources, etc. Individuals subject to *r*-selection are small sized, mature early, are probably semelparous (have only one brood), allocate more resources to reproduction, are numerous and have smaller progeny. In short, they invest very little in survival, which varies significantly according to the environment (Begon *et al.*, 1986).

The concepts outlined above occur commonly in the bio-ecology of mites and are particularly useful for understanding their relationship with the factors that most affect their populations. Mite species living in an unstable and resource-rich environment usually develop an *r* demographic strategy and the *r* value at or near the end of the selection spectrum permits a large number of progeny. Typically, systematic groups adopting an *r* strategy have many generations a year and produce a high number of progeny each generation. Several Tetranychidae living on vegetables develop *r* strategies. In relatively stable habitats characterized by limited food resources, populations of mite species at or near the *K* end of the spectrum approach the carrying capacity of the habitat, optimize the resources available and produce only a few progeny, but with a high survival potential. This strategy occurs in various acarine groups, for example in several eriophyoids, which are characterized by a frequently non-deleterious association with their host plants that allows a relatively stable and perennial resource, e.g. *Aceria sheldoni* and other species. The Mesostigmata also include species with *k* strategies.

Population Dynamics

The population dynamics of injurious mites (and other mites) depend on the size of their populations and their ability to change over time. The dynamics are assessed by analysis of the processes of birth, death, immigration and emigration. The population density assumes significant importance, particularly in intraspecific competition, and can act on the birth and death

rates to adjust the population to a density that is characterized by a stable birth rate that is equal to the rate mortality. This density is known as the 'carrying capacity' and is indicated by the symbol *K*, i.e. the amount of the population that the environmental resources can support without an increase or a decrease. The phenomenon is represented by the logistic equation:.

$$dN/dt = rN\,[(K - N)/K]$$

where *N* is the population size, *t* the time, *r* the intrinsic rate of natural increase in the absence of competition and *K* the carrying capacity. Under well-defined laboratory conditions (e.g. a microbial culture) and from a theoretical point of view, the increase in population density assumes a sigmoid pattern, increasing up to the value of the carrying capacity. However, in reality, the behaviour of a pest population is also influenced by other factors, such as interspecific competition, the action of predators and parasites, and other environmental factors (physical, chemical, etc.), each of which may be more important than the effects of intraspecific competition. Normally, the density values of pest populations oscillate above and below the value of *K* in the wake of factors that are internal to the population, such as intraspecific competition, and external factors (abiotic and biotic), such as the resistance of plant hosts, the presence of natural enemies, interspecific competition, etc. (Begon *et al.* 1986). However, it is possible that other factors (such as agricultural practices) can increase pest density values far above the values of *K* and precipitate outbreaks of pest populations.

Below, a brief treatment is given of the various mechanisms that underlie the behaviour of populations of phytophagous mites (and other pests): the ecological concept of feedback; cyclic dynamics; and eruptive dynamics.

The influences on population dynamics of abiotic factors (temperature, RH, rain, light) and biotic factors (competition, natural enemies, the relationships between mite pests and their host plants, and agricultural practices) are treated in the two separate (main) sections that follow on from this discussion of population dynamics per se.

Feedback

The relationship between the rate of population growth of a pest and the increase of the density of the population defines the 'feedback' of the population. If the rate of population growth increases with increasing population density, the feedback is positive; vice versa, if the population growth declines with increasing density, the feedback is negative. The latter condition is generally detected as a result of a deterioration in the quality and quantity of the trophic resource or of the action of natural enemies, and can lead the population towards an equilibrium density. The association between positive feedback and potential exponential growth of population means that explosion of the population can occur (Berryman, 1999).

Cyclic dynamics

The trend of density over time in pest populations determines whether the dynamics are 'cyclic' or 'eruptive'. Most citrus pests have cyclic population dynamics which are characterized by predictable density fluctuations that regularly take the population above the economic threshold values, depending on the time delay that is induced by a negative feedback (Turchin, 1990). Among the causes of this behaviour is a delay in the response of natural enemies, as occurs with some citrus mites (Huffaker, 1958). In addition to natural enemies, and of particular importance, are density-dependent processes, such as the availability and/or integrity of trophic resources (Hunter, 2002b). Several abiotic factors (temperature, rain, wind, etc.) can also influence cycles of population density and need to be borne in mind when analysing the possible density dependence of pest dynamics (Hunter and Price, 1998). In citrus, the climate affects the population dynamics of phytophagous species such as the citrus red mite (Jeppson *et al.*, 1975), and the two-spotted spider mite, which generally develops cyclic dynamics showing three peaks of population, depending on the weather and plant phenology (Vacante, 2010). Deutogynous eriophyoid mites generate cyclic dynamics with population peaks that are usually in the spring and/or summer.

Eruptive dynamics

If the frequency of explosion of the density of a pest is not constant over time, the population dynamics is called eruptive. The prediction of this phenomenon and the identifying environmental causes are not simple. Physical factors that are aperiodic and unpredictable, such as temperature, rain, wind, etc., and escape from low-density regulation are involved. If natural enemies normally regulate the population of an injurious species by keeping it at a low density, then an escape from this regulatory process, precipitated by some other factor, can cause an eruptive population increase until other forces (competition, etc.) stabilize or reduce the pest density once again (Hunter, 2002b). In citrus pests characterized by cyclical dynamics, such as the two-spotted spider mite or the citrus red mite, this happens after the use of non-selective pesticides such as methomyl, imidacloprid or deltamethrin. Unpredictable environmental factors (fires, drought, no weed control, etc.) may also allow escape from a regulation of low density and determine eruptive dynamics in pest species.

Abiotic Factors Influencing Population Dynamics

The most important abiotic factors influencing the dynamics of mite populations are temperature, relative humidity, rain and light. Adaptations to weather variations play a key role in this. The small size and the delicate structure of the mites' body expose them to severe ecological pressure – to extreme annual cycles in temperature, to extreme weather conditions in continental regions and to large fluctuations of humidity in tropical regions. In response, mites have established specific adaptive mechanisms to mitigate these pressures. Several mechanisms have been adopted, including diapause, migration to protected niches and the stimulation of plant growth to produce galls or

succulent parts (Jeppson *et al*., 1975). Adaptation to cyclic weather modifications needs structural and behavioural changes. For instance, high summer temperatures and low humidity are easily overcome by sheltering in moist ground crevices or under bud scales, or by developing more resistant forms. In some species, females stop their development or produce resistant eggs. As a result, the populations of each species develop optimally within a range of more suitable climatic conditions.

Temperature

Temperature is probably the abiotic factor that has been most studied in relation to mite population development and is the most important abiotic factor affecting this. High or low temperatures may have the same influence according to the mite species, its life cycle and season. Temperature acts on different biological stages, eggs, adults, etc., whose tolerances vary with species and life cycle. Mori (1961) reported on the temperature preferences of *Panonychus citri* (McGregor), *Amphytetranychus viennensis* (Zacker), *T. urticae* Koch and *Bryobia rubrioculus* Scheuten, which he found to be within the ranges of 25–28, 25–30, 13–35 and 21–24°C, respectively. Increases of temperature influences the potential progeny output exponentially. For example, at 15.5°C, a tetranychid female can produce 20 individuals a month, but at 21°C the figure is 12,000 individuals and at 26.5°C it is 13,000,000 individuals (Jeppson *et al*., 1975). Temperature is the major factor affecting development of the citrus red mite (Keetch, 1968). Temperatures of 40.5°C, or several days of hot dry weather (5% RH and 32°C), together with strong wind, commonly produce a high mortality. Similar conditions can be observed during the summer in California, but can occur also in the winter.

Extended periods of high RH are also unfavourable for the development of mite populations, and susceptibility to extreme conditions of both temperature and RH limits mite distribution and affects the seasonal trends of populations (Jeppson *et al*., 1975). This phenomenon may cause the temporary disappearance of mite pests over large areas for several years. A number of examples on the influence of temperature and RH on (r_m) have been reported for the Tetranychidae by Wrensch (1985). Studies carried out on the biology of a severe tetranychid pest of date in Iraq, *Oligonychus afrasiaticus* (McGregor), at 20, 25, 30 and 35°C, 50–60% RH, and under a photoperiod of light 16 h:dark 8 h, showed that no development occurs at 15 and 40°C. At 20°C, the incubation period was 7.6 days and it reached a minimum at 35°C (2.7 days); the development of the larva, protonymph and deutonymph took 5.3, 4.3 and 4.5 days at 20°C, and 1.9, 1.3, and 1.4 days at 35°C, respectively. Fecundity at 20°C was 12.5 eggs/female and at 35°C was 27.2 eggs/female. Female and male longevity were 33.2 and 29.1 days at 20°C, and 11.5 and 10.1 days at 35°C, respectively (Al-Jboory and Al-Suaide, 2010). There are many other such studies on various mite species in different countries.

Temperature also influences eriophyoid mites. Spring temperatures affect the emergence of the filbert big bud mite, *Phytoptus avellanae* (Nalepa) (AliNiazee, 1980, 1998). The nymphs leave the big buds at daily temperatures ranging from 15 to 20°C, but not at lower temperatures (Özman and Toros, 1997a). High temperatures cause greater mortality of *P. avellanae* as well as very low rainfall, or heavy rain (Jeppson *et al*., 1975; Özman and Toros, 1997a; Gantner, 2001). High temperatures and low humidity affect the survival of *Cecidophyopsis vermiformis* (Nalepa) nymphs and adults (Özman and Toros, 1997a). High temperature and RH increase the rate of development and stimulate the reproduction of the olive bud mite, *Aceria oleae* (Nalepa). Warm and humid climatic conditions are the most important factors favouring a population increase (Al-Azzazy, 2000; Abou-Awad *et al*., 2005). The fecundity of the pink tea mite, *Acaphylla theae* (Watt *et* Mann), is significantly correlated with effective temperature. The rate of increase in mite population reduces when the temperature becomes lower than 18°C or higher than 27°C, and the most favourable RH is between 75 and 90% (Yin *et al*., 2003).

The egg to adult developmental time of the broad mite, *Polyphagotarsonemus latus* (Banks), on papaya at 25, 28 and 31°C and 55, 67 and 97% RH varies from 2.81 and 4.93 days. The intrinsic rate of increase (r_m) ranges from 0.03 to 0.4, according to RH and temperature (de Coss *et al*., 2010).

Furthermore, global climatic change (and change in temperatures) is assumed to have an impact on species distribution. Agricultural pests that often threaten food security at a large scale may have different effects under changing environmental conditions if their distribution ranges can expand. For example, the Tetranychidae includes over 1200 species, several of which are considered among the worst agricultural pests worldwide. As ectotherms, they rely on external heat sources to regulate their body temperature and biochemical reactions. Hence, the ultimate limits of their distribution are strongly determined by climate, and increasing temperature is likely to have significant effects on their global distribution.

Relative humidity

Atmospheric humidity levels influence the development of mite populations according to mite species and their ecological adaptations. From a physiological point of view, this phenomenon is related to water balance. Boudreaux (1958a) first found a negative influence of high humidity on the fecundity and survival of pest mites. In some tetranychids, low RH over several days can cause high mortality (Jeppson *et al*., 1962), while in others, high levels of humidity hinder moulting and feeding, and lifespans become shorter (Harrison and Smith, 1961). The optimum development of populations of the two-spotted spider mite, *T. urticae*, occurs at low humidity (Mori and Chant, 1966a,b), and this adaptation has suggested that the pest could be controlled by the use of lower temperatures and relatively high humidity (Fritzsche, 1960). At a high population density and with severe plant damage, *T. urticae* in the dispersal phase exhibits a positive phototactic response (which is not exhibited by sedentary individuals) that is probably dependent on food shortage and decreasing RH, which induces the mites to move up the plant and to aggregate at high density on the periphery (Suski and Naegele, 1963). Such behaviour could represent a response to variations in RH (McEnroe and Dronka, 1971).

Optimum atmospheric humidity for the desert spider mite, *Tetranychus desertorum* Banks, probably occurs above 85% and independent of temperature (Nickel, 1960).

In some eriophyoids, low RH over several days may cause high mortality (Hobza, 1970). Conversely, hatching of the eggs of the citrus bud mite, *Aceria sheldoni* (Ewing), is optimal at 25°C and 98% RH, but reduces with lower RH (35–40%) (Sternlicht, 1970). In southern California, the development of this mite and that of the citrus rust mite, *Phyllocoptruta oleivora* (Ashmead), needs high atmospheric humidity and occurs only on citrus located on the immediate coast, although citrus is also cultivated in inland areas (Jeppson *et al.*, 1975). The ivy bud mite, *Phytoptus hedericola* Keifer, lives in the buds and petiole bases of English ivy (*Hedera helix* L.) in California (Keifer, 1943) and in other countries. It is adapted to conditions of high humidity of the coastal environments, and in California lives near the ocean, never appearing in the interior valleys (Jeppson *et al.*, 1975). Low humidities (and high temperatures) limit the survival of *Cecidophyopsis vermiformis* nymphs and adults (Özman and Toros, 1997a). High humidities (and temperatures) increase the rate of development and stimulate the reproduction of *Aceria oleae* (Nalepa) (Al-Azzazy, 2000; Abou-Awad *et al.*, 2005). The population increase of *Acaphylla theae* (Watt *et* Mann) is inhibited when the RH is lower than 50% or higher than 95% (Yin *et al.*, 2003). In southern California, the peach mosaic vector mite, *Eriophyes insidiosus* (Keifer *et* Wilson), prefers lower interior humidities and lives further back from the ocean coast. The same behaviour is exhibited by the apple rust mite, *Aculus schlechtendali* (Nalepa), which prefers higher humidities, and Bailey's rust mite, *Calepitrimerus baileyi* Keifer, which replaces the apple rust mite in interior apple orchards characterized by lower atmospheric humidities (Jeppson *et al.*, 1975).

Rain

Prolonged rains do not usually facilitate the development of mite populations, but it is difficult state a general behavioural pattern. It is well known that eggs of the European red mite, *Panonychus ulmi* (Koch), are not killed by immersion in water for 48 h, but other biological instars are killed, with immersion causing their metabolism and other biological activities (hatching of eggs, feeding, moulting, ovipositing, etc.) to stop (Herne, 1968). During rain, most mites move to the undersides of leaves or to other sheltered sites. However, powerful rains (Asiatic monsoons) may cause a high reduction of mite populations. Rain and other weather factors act on the development of mite populations, as in the case of the tea red spider mite, *Oligonychus coffeae* (Nietner), which suffers high levels of mortality under heavy rain (Das, 1959) or the Kanzawa spider mite, *Tetranychus kanzawai*, which stops its development with a combination of wind and rain (Osakabe, 1965).

Higher mortality of *P. avellanae* is due to high temperatures and very low rainfall, or to heavy rain (Jeppson *et al.*, 1975; Özman and Toros, 1997a; Gantner, 2001). In Munnar (Kerala, India), heavy rainfall, high temperature and high RH significantly reduce the density of the purple tea mite, *Calacarus carinatus* (Green) (Muraleedharan *et al.*, 1994); in Taiwan, the mite populations increase gradually from June until October, but are adversely affected by rainfall (Shiao, 1976). The rate of increase of the *Acaphylla theae* population is negatively correlated with precipitation, rainfall and leaf age (Yin *et al.*, 2003).

Females of *Brevipalpus phoenicis* washed by rain from the leaves of infested plants can withstand immersion in water for more than 15 days, and can then return to the plant of origin or move on to other plants (Moraes and Flechtmann, 2008).

Light

The influence of light on the development of mite populations is well known, and has been largely investigated in tetranychid mites. These mites show a positive phototaxis and the adult females are characterized by a more sensitive and positive response to white light in summer than in winter (Suski and Naegele, 1968). The wavelength determines the response, with the maximum response in the near-ultraviolet region (375 μm), with a smaller peak observed in the yellow green region (525–550 μm) and a negative or indifferent response at over 600 μm (Naegele *et al.*, 1965). The anterior pairs of eyes of tetranychids have point scanning detectors containing ultraviolet and green photoreceptors. *T. urticae* responds to the plane of light polarization (Naegele *et al.*, 1965; McEnroe and Dronka, 1971). At high population densities and on severely damaged plants, the mite exhibits a positive phototactic response in the dispersal phase that is not exhibited by sedentary individuals, and probably depends on food shortage and decreasing RH, which induce the mite populations to move up the plant and to aggregate at high density at the periphery (Suski and Naegele, 1963). This behaviour could represent a response to variations of RH (McEnroe and Dronka, 1971). From 15 to 30°C, the response to light occurs before that to temperature (Mori, 1961, 1962a), and summer adults of *T. urticae* are more sensitive to light than winter adults. No differences between the sexes have been observed (Mori, 1962a,b).

There is a relationship between starvation and the response of mites to light, and the pattern of response depends on RH. Between the first and the fifth hours of starvation, there is either a photo-indifferent or a photonegative response, according to the humidity. Low levels of humidity promote a negative or indifferent response, while high values promote an indifferent or positive response. Between the fifth and the eleventh hours of starvation, the mites exhibit a photosensitive, photonegative or phototelotactic response according to the humidity. Between the eleventh and the twentieth hours of starvation, mites show a positive phototelotactic response according to the humidity (Suski and Naegele, 1968). The mechanism of these reactions is the result of an interaction between young leaves characterized by high humidity and usually set at the periphery of the canopy, and the old leaves that have been attacked by mites and are dry and have low transpiration and humidity. Damaged leaves increase mite starvation. Light and humidity stimulate the mites to leave the injured leaves and to search for new ones, usually in the periphery of plants and with strong transpiration (and high humidity). In this phase, the mites develop a positive light response. But as soon as they encounter a side of the plant

or a leaf with rising humidity they become photonegative. If the gradient of humidity is produced by factors independent of the transpiration of new leaves, and the mites do not exhibit a behavioural pattern aimed towards finding a new food resource, they establish under the humid conditions and move at random towards a new site. When the mites do not find a site with rising humidity, there is a positive response that induces aggregation in the outermost part of canopy, which enables the mites to leave the plant by dispersal. If they reach a new and undamaged plant, they become photonegative or photo-indifferent to increases of humidity, or other stimuli, and search for food resources (Suski and Naegele, 1968).

The responses of eriophyoid mites to light are not well known, but they differ from tetranychids in some adaptations and also in their inability to move in the same way. The extent of damage caused by the Lucerne bud mite, *Aceria medicaginis* (Keifer), at 20°C varies with differences in the growth of the plants under different photoperiods. With a short photoperiod of 10.3 h, and where there are no axillary shoots on either infested or uninfested plants, the extent of damage is higher than with a longer photoperiod of 13.2 h, and there are axillary shoots on all of the plants. The leaves of infested plants show chlorotic patches (mainly at 20 and 25°C), agate-like banding (at 29°C) and a waxy mottle (at 15°C) (Ridland and Halloran, 1981).

Biotic Factors Influencing Population Dynamics

Two types of biotic factors, internal and external, influence the population dynamics of mites. The internal factors are changes resulting from competition between individuals of the same species. The external factors include competition between individuals of different species, natural enemies and the influence of host plants. In addition, various aspects of agricultural practices may act as external biotic factors.

Competition

Competition between individuals of the same species ('intraspecific competition') or between individuals belonging to different species ('interspecific competition') plays a key role in the regulation of population dynamics.

Intraspecific competition

Intraspecific competition may be referred to as 'an interaction between individuals, caused by a common need for a resource available in limited quantities and causes a decrease in survival, growth and/or reproduction of individuals interested in competition' (Begon *et al.*, 1986). The phenomenon has certain characteristics. The first is that intraspecific competition results in a reduction of resources, which is responsible for a decrease in survival and fecundity. In addition, the resource may develop limited availability as a result of exploitation by other individuals, or because of the physical interference that individuals can exercise when together. If all the individuals of a population belong to the same species, there can be certain reciprocal and/or unilateral behaviour without prejudice to the inputs arising from the different hereditary characteristics of individuals. In this case, intraspecific competition can positively influence the fitness of some individuals (Wall and Begon, 1985). The effects of intraspecific competition are more marked at high density and are density dependent. Theoretically, the competition can lead to a stable equilibrium by acting on birth and mortality rates and leading the population to a stable density (the carrying capacity, *K*) and characterized by a birth rate that is equal to the rate of mortality. *K* represents the amount of the population that the environment can support, without decreasing or increasing.

However, natural populations have different behaviours influenced by various factors, and intraspecific competition is just one of these factors. In general, intraspecific competition tends to determine the amount of the population, which has a net rate of growth placed at highest value at intermediate densities below the carrying capacity, and also a sigmoid curve of population growth. In reality, we must also take into account the influence on populations of natural enemies, other competitor species, and several physical factors that may be more important than intraspecific competition (Begon *et al.*, 1986).

Population density affects reproductive parameters. It has been observed that high population density in *Tetranychus* sp. induces lower egg production, a higher amount of non-viable eggs and mortality of the immature stages (Davis, 1952). In spider mites, high population density also reduces longevity (Attiah and Boudraux, 1964). Wrensch and Young (1978) reported that in *Tetranychus* sp., high density induced marked reductions in developmental rate, survival and sex ratio as well.

Under crowded conditions, the two-spotted spider mite, *T. urticae*, shows a strong tendency to leave damaged leaves and/or fruits and moves, according to its phototropic and geotropic adaptations, to other parts of the plant where there is a lower population density, or finds means to disperse to new host plants and areas. Populations of the broad mite, *Polyphagotarsonemus latus*, usually develop at high density on young tissues, but when the resource is depleted, which coincides with the development and hardening of the infested organs, move onto new young and non-infested leaves, flowers and fruits.

The availability of the trophic resource, the deterioration of feeding sites and changes in physical factors play a fundamental role in intraspecific competition. *Tetranychus* sp. reared on banana squash exhibits a reduction of fecundity (also with an increase in the population), an increase of non-viable eggs (also with an increase in the population) and mortality of the immature stages (Jeppson *et al.*, 1975). The six-spotted mite, *Eotetranychus sexmaculatus* (Riley), and the citrus red mite, *Panonychus citri*, leave severely damaged leaves (Ebeling, 1934). The age of the leaf and the amount of injury caused by the feeding activity (and by high population density) influence the population of *P. citri*. The number of eggs laid on young and medium-aged leaves is greater than that on old leaves, but feeding injury influences oviposition (Henderson and Holloway, 1942) and does not represent a regulatory factor in the density of populations. Furthermore, it is well known that there is a direct relationship between population density and the nitrogen of the host plant (Schwartz, 1984). Lead damage reduces

oviposition by *Oligonychus punicae* (Hirst) and induces an increase in the mortality of different stages (McMurtry and Johnson, 1966). In general, food supply plays a key role and induces mite populations to self-limit in the absence of other limiting factors (Davis, 1952; Fleschner, 1952; Watson, 1964). However, there is no general rule about the response of populations to overcrowding (Boudreaux, 1963). Thus, with inadequate food resources, the European red mite, *Panonychus ulmi*, lays premature diapause eggs, but the two-spotted spider mite leaves the host plant and may start premature diapause in the early autumn (McEnroe, 1969).

Interspecific competition

Interspecific competition occurs if 'the fecundity, survival or growth of individuals of a species decrease as a result of the use of resources or the interference by individuals of other species' or 'if two or most organisms use a source that is inadequate for all or if the resource is available in limiting quantities' (Begon *et al.*, 1986). It acts differently on the population dynamics of the various species concerned, and may influence both species distribution and their evolution. Interspecific competition may be realized by interference and by exploitation, and is commonly asymmetric, with different consequences for the two (or more) species involved (Lawton and Hassell, 1981). Competition for a resource may also affect other resources. The use of mathematical models, such as the Lotka–Volterra interspecific competition model (Lotka, 1925; Volterra, 1926), allows definition of the conditions of coexistence of competitors that lead to 'competitive exclusion'. In general, stronger interspecific competitors win over the weaker ones. If interspecific competition is more important than intraspecific competition, the population density determines the outcome between species, whereas if interspecific competition is less important than intraspecific competition, the species coexist. In the absence of competitors, the niche of a species is defined as 'fundamental', while the presence of competitors makes the niche 'realized' and its nature is defined by the competitor species involved.

Sabelis (1991) produced a detailed review of data on the life histories of Tetranychidae at 25 ± 2°C that highlighted a consistent interspecific variation. He noted that the values of the intrinsic rate of population increase (r_m) varied from *c.*0.1 to 0.3 daily. Phylogenetic factors, host-plant related factors and/or co-evolutionary interactions with other organisms (competitors, predators, pathogens) could cause this variation, acting as constraints to the evolutionary change of one or more life-history components. To highlight their importance, Sabelis (1991) assumed the absence of these constraints in order to evaluate hypotheses on how life-history components evolved under *r*-selection. The available data show that life-history components (with the exclusion of sex ratio among offspring) covary strongly, and that an increase in r_m derives from a combination of factors (faster development, higher rate of oviposition and higher fecundity). The progress in development rate becomes slower when the fecundity and the oviposition rate enter high-value ranges, pointing to the existence of a constraint to speeding up development. To explain the absence of a relationship between the sex ratio among offspring and the other life-history components, Sabelis (1991) suggest that spacing and aggregating within a web are two alternative ways of decreasing predation risk, and that these strategies have different consequences for the mating structure of the population (and hence also for the evolution of sex ratio). Competitive interactions between the tetranychid mites *T. urticae* and *Eotetranychus carpini* (Oudemans) have been investigated under laboratory conditions on red raspberry leaves (cv. Meeker). When placed as the first colonizer on red raspberry leaves, *T. urticae* caused extinction of the *E. carpini* secondary populations, but when the two species were placed on the leaves in the reverse order, *E. carpini* caused only detrimental effects on *T. urticae*, but not population extinction. The leaf surface used by *E. carpini* for egg deposition decreases with an increase in the number of *T. urticae* eggs. Red raspberry shows an induced response to *T. urticae* and *E. carpini* feeding after 2 weeks of infestation. The population growth of each species is reduced on plants previously infested by conspecifics, but the inducible response by one species has no effects on the other species (Bounfour and Tanigoshi, 2001a). In the field, the operational problem of interspecific competition occurs frequently in different plant species of economic importance that are contemporaneously infested by different phytophagous mites. Some examples confirm this. On citrus, the two-spotted spider mite may compete with the citrus red mite, but the latter commonly prevails because it exhibits a greater population density. In the same crop, the greater importance of intraspecific competition leads to coexistence between the citrus bud mite, *Aceria sheldoni*, and the two-spotted spider mite. An example of use of the same niche and of exclusion competition occurs in some tetranychid mites of deciduous fruit trees. In spring, the brown fruit tree mite, *Bryobia rubrioculus*, commonly hatches before the European red mite, *Panonychus ulmi*, and depresses its development. However, to avoid summer conditions, the brown fruit tree mite starts to aestivate in late spring, whereas the European red mite continues to be active. Later, a third species, the two-spotted spider mite, may infest the leaves occupied by the European red mite and the web made by the two-spotted spider mite may limit the development of the European red mite (Lienk and Chapman, 1958b).

Mites as competitors of mites in other groups or as alternative prey

Mites from different groups, such as eriophyoid and tetranychid mites (or other pest mites), can inhabit the same host plant together and consume the same food resource. Competition between pest mites in different groups is not well known, and when it occurs may either be 'direct', involving physical interaction(s) and/or displacement, or 'indirect', and expressed through another organism or trophic level. There is no evidence for direct competition between eriophyoids and tetranychids, but indirect competition has been studied (Dunley and Croft, 1996). The use of mites of other groups as alternative prey to increase the population of predatory mites for the control of outbreaks of mite pests has also been suggested.

INDIRECT COMPETITION. Indirect competition has been extensively investigated on apple infested by the European red mite, *Panonychus ulmi*, a tetranychid, and the apple rust mite, *Aculus*

schlechtendali, an eriophyoid. It has been found that the reproduction of *P. ulmi* in the field was suppressed on apple leaves where rust mites were present or had previously fed. A positive response was observed when *P. ulmi* was released from competition by *A. schlechtendali* (Croft and Hoying, 1977). Studies by Santos (1984) on the action of the stigmaeid *Zetzellia mali* (Ewing) on *P. ulmi* and *A. schlechtendali* indicate that high rust mite populations reduced the tetranychid populations. In commercial apple orchards, the use of non-selective pesticides allowed the apple rust mite to develop to levels that reduced *P. ulmi* reproduction in the absence of natural enemies (Croft and McGroarty, 1977). A moderate to high density of populations of *A. schlechtendali* early in the season negatively affected the development of *P. ulmi* populations (Hoyt *et al.*, 1979). Indirect competition can act through both the effects on the host plant and the avoidance of predation (Dunley and Croft, 1996).

As regards the effects on the host plant, the preliminary feeding activity of *A. schlechtendali* on leaves could play an important role in inducing a physiological change in the plant, e.g. the induction of a corky tissue layer inhibiting *P. ulmi* from feeding. This preconditioning is probably explained by the different feeding behaviours of eriophyoids and tetranychids, which have differently sized oral stylets. Tetranychid stylets are about 130 μm long (Jeppson *et al.*, 1975), while those of eriophyoids range from 7 to 20 μm (McCoy and Albrigo, 1975; Hislop and Jeppson, 1976; Easterbrook and Fuller, 1986). The short stylets of rust mites do not destroy mesophyll and feeding does not kill the plant tissues (McCoy and Albrigo, 1975; Easterbrook and Fuller, 1986; Royalty and Perring, 1988). In contrast, tetranychid stylets penetrate the epidermal layer and feed on the mesophyll, killing the plant cells during feeding (Jeppson *et al.*, 1975; Tanigoshi and Davis, 1978; De Angelis *et al.*, 1982). Feeding by eriophyoids can induce the formation of callus in the epidermal layer, thus limiting probing by tetranychids.

The occurrence of avoidance of predation in indirect competition may be due to the influence of the population density of eriophyoids and tetranychids on natural enemies (Croft and McGroarty, 1977). For example, it has been found that in apple orchards, rust mites are the primary alternate prey for different predators (Hoyt, 1969a; Sabelis and van Rijn, 1996). Early-season eriophyoid (rust mite) populations increase more rapidly than spider mite populations, which allows the predatory mite populations to develop to high density in mid-season before the tetranychid (spider mite) populations increase. This behaviour allows the suppression of spider mite outbreaks (Hoyt *et al.*, 1979). In the absence of apple rust mites, it is necessary for there to be a higher density of predators to control spider mites (Croft and McGroarty, 1977).

MITES AS ALTERNATIVE PREY. The possibility of using eriophyoid mites as alternative prey to increase the population of predatory mites (mainly phytoseiids) for the control of outbreaks of spider mites in the field has been variously proposed (Collyer, 1964b,c; Herbert and Sanford, 1969; Hoyt, 1969a,b). The topic has been extensively examined from the viewpoint of an ecological analysis by Sabelis and van Rijn (1996), who deal with the five main assumptions supporting the phenomenon, but demand a more rigorous approach.

Natural enemies

Natural enemies play a key role in the population dynamics of mites. They include pathogens, other mites, insects and spiders, and according to various factors (density of populations, weather, use of chemicals, etc.) can exert different levels of control. No parasitoids of mites are known. The action of natural enemies is essential both for the biological equilibrium of natural systems and for the management of agricultural ecosystems. It represents a strong scientific challenge involving phytoiatric and ecological implications. The topic has been extensively investigated in the last half century and is supported by a rich literature that is difficult to summarize. The works of McMurtry *et al.* (1970), Jeppson *et al.* (1975), Gerson *et al.* (2003), Hoy (2011) and Gerson (2014) are fundamental to the topic.

The pathogens of arthropods (including mites) comprise viruses, bacteria, fungi and microspora (Van der Geest, 1985, 2010; McCoy, 1996a; Boucias and Pendland, 1998; Poinar and Poinar, 1998; Van der Geest *et al.*, 2000; Bruin and Van der Geest, 2009). In addition, almost three dozen mite families are involved in the biological control of pests, some of which play an important ecological role in the control of mite pest populations (Gerson *et al.*, 2003). Predatory insects of mites belong to several different orders (Jeppson *et al.*, 1975; Chazeau, 1985; Perring and McMurtry, 1996). The bio-ecological characteristics of these predators have clear practical implications in the interpretation and evaluation of the role of the various predatory species in the field and their interrelationships, and can lead to a classification of the most important predators. A small number of beneficials appear to be highly effective in controlling mite pests, while others seem to be less efficient. For mite pests of deciduous fruit species, Croft and Brown (1975) proposed a first tentative classification of predators, and reported as the most efficient in descending order Phytoseiidae and Coccinellidae (*Stethorus* spp.), followed by Araneidae, Thripidae, Stigmaeidae, Hemiptera, Neuroptera and Diptera. According to McMurtry *et al.* (1970), Chazeau (1985) reported that few species of insects played a key role in the biological control of pest mites in Europe and North America. The most important of these insects include *Stethorus punctillum* Weise, *Stethorus picipes* Cas., *Blepharidopterus angulatus* (Fall.), *Anthocoris musculus* Say, *Haplothrips faurei* Hood and *Scolothrips sexmaculatus* (Per.). However, there is in the literature conflicting information from Chazeau (1985) that limits the interpretation of Croft and Brown (1975), and a detailed treatment of the topic shows that there are interactions between species that cannot be underestimated in both ecological and practical terms. Known cases include the compatibility of *Leptothrips mali* (Fitch) with *Stethorus punctum* (LeConte) and *Orius insidiosus* (Say) when feeding on *Panonychus ulmi* (Parrella *et al.*, 1980). Interactions between various predators in the field are very common and ecological interpretation of these allows control strategies to be better defined. A typical interaction, 'intraguild predation', occurs when two natural enemies share a host (or prey), and at least one also feeds on the other (Rosenheim *et al.*, 1995). This phenomenon is well known among the natural enemies, whose interactions can be competitive without predation, and even increase the extent of control that can be achieved, by either facilitating or hindering it. The level and severity of intraguild

predation also has a bearing on the establishment, success and survival of exotic natural enemies (Gerson, 2014).

Viruses

Although viruses have been reported in the literature since the 1960s when a 'non-inclusion virus disease' in tetranychid mites was found on citrus in Florida (Muma, 1955) and California (Munger *et al.*, 1959; Smith *et al.*, 1959), up to now, no viruses have been commercially produced for use in mite control. There are also some disadvantages associated with their use as mite control agents, including the need for them to be ingested, which results in a slow speed of action, and their sensitivity to ultraviolet (UV) light, which requires that they be specially formulated in order to preserve their effectiveness.

Munger *et al.* (1959) reported a disease affecting the citrus red mite (*Panonychus citri*), which was later shown by Gilmore (1965) and Shaw *et al.* (1968) to be important in the natural control of this mite pest. That the disease was viral was first suspected by Smith *et al.* (1959), who found it to be a very small non-inclusion form; diseased mites could be identified by the presence of birefringent irregular bodies from a few to 50 μm in size (Smith and Cressman, 1962). Estes and Faust (1965) examined the nucleic acid composition of the virus. In further investigations, Reed and Hall (1972) found rod-shaped virus particles in infected mites; they also found spherical particles, but these were more common in healthy than infected mites, and subsequent studies showed that these were not related to the virus disease (Reed and Desjardins, 1978). In California and Arizona, infected mites were common in citrus orchards, and the appearance of the disease appeared to be related to suitable environmental conditions (Shaw *et al.*, 1968; Reed, 1981). Reed *et al.* (1972a), like Smith and Cressman (1962) earlier, demonstrated the presence of birefringent bodies in infected mites, and found them both inside the midgut and associated with faecal pellets (Reed *et al.*, 1972a). The amount of these particles that were present allowed an assessment of the incidence of the disease in mite populations, and Reed *et al.* (1972b) proposed the use of a portable polarizing unit to facilitate this operation, but the success of this technique in the field was uncertain. The virus did not affect phytoseiid mites feeding on citrus red mites (Shaw *et al.*, 1967), and predaceous phytoseiid mites were shown to play only a secondary role to the virus disease in controlling citrus red mite populations in southern California (McMurtry *et al.*, 1979). The virus is inactivated by sunlight after aqueous application in the field (Reed *et al.*, 1973); its survival is also influenced by temperature (Reed, 1974). The role of the virus in mite control has been examined by Rath (1991).

A virus disease has been reported of the European red mite (*Panonychus ulmi*) on walnut in California (Steinhaus, 1959). A virus was also suggested by Putman and Herne (1966) as the cause of extensive epizootics of *P. ulmi* in peach orchards in Ontario, Canada (Putman and Herne, 1966), in which Bird (1967) identified a rod-shaped non-inclusion virus. Most of the infected mites showed in birefringent spheroidal inclusions with a radiating crystalline structure in the midgut (Putman, 1970a). Inoculum is deposited on the leaves by infected mites (probably by excreta or oral secretions) and is ingested by uninfected mites, although it is unstable and is inactivated if exposed to water (and in aqueous suspensions). In the field, the disease develops only at a high density of mite infestations. Introduction of the virus can give good mite control.

An unidentified virus was found in a greenhouse population of *T. urticae* in California. It has been evaluated as highly lethal under laboratory and greenhouse conditions, but is not produced commercially (Hoy, 2011). Three types of icosahedral virus-like particles were found in diseased females of the phytoseiid *Metaseiulus occidentalis* (Nesbitt) (Poinar and Poinar, 1998). In the laboratory, they showed reduced oviposition and exhibited a paralysed appearance (Hoy and Jeyaprakash, 2008, 2009). Virus-like particles have also been found in the eggs inside females of *Neoseiulus cucumeris* (Steiner, 1993) and *Phytoseiulus persimilis* (Bjørnson *et al.*, 1997).

Bacteria

Bacteria do not normally drastically affect the population dynamics of mites. Table 5.1 lists the main bacterial species that have been found to be associated with or investigated in Tetranychidae and Tenuipalpidae. The earliest species reported in literature is *Bacillus thuringiensis* Berliner, whose toxic β-exotoxin has been evaluated in tetranychids (Krieg, 1968; Hall *et al.*, 1971; Royalty *et al.*, 1990) and eriophyoid mites (Sulaimanov *et al.*, 2000). This represents one of the few examples in which bacteria have been shown as a potential acaricide but, unfortunately, *B. thuringiensis* is toxic for several organisms, including beneficial mites such as *Phytoseiulus. persimilis* (Guo *et al.*, 1993a). Another bacterium, *Acaricomes phytoseiuli* Pukall, Schumann, Schütte, Gols *et* Dicke, induces a change in response to prey-induced plant volatiles in laboratory populations of *P. persimilis* (Pukall *et al.*, 2006; Schütte, 2006).

Endosymbiont bacteria of the genera *Wolbachia* and *Cardinium* were discussed in Chapter 4 (see Endosymbiont bacteria and mite genetic systems). *Spiroplasma* sp. was found in *T. urticae*, but its action is unknown (Enigi and Schausberger, 2007). The efficacy of the *Proteus* strain RRLJ, isolated from tea (*Camellia sinensis* (L.) O. Kuntze) plantations in India, has been evaluated on the pink tea mite *Acaphylla theae* (Bezbaruah *et al.*, 1996). Putative gut bacteria were found also in *T. urticae* and in some phytoseiid mites, but their biological role is not really known (Hoy and Jeyaprakash, 2008, 2009). Preparations from bacteria are currently being investigated as a means of mite control, but there is no news of their possible use in the field. Eight genera of bacteria (*Bacillus*, *Nocardia*, *Corynebacterium*, *Erysipelothrix*, *Lactobacillus*, *Arachnia*, *Micrococcus*, *Staphylococcus*) (and fungi) were isolated from dead specimens of *Tyrophagus similis* on spinach in greenhouses in Japan (Kikuta and Takagi, 2001). Acaricidal activity was confirmed in 77.8% of the bacterial isolates and in 100% of the fungal isolates. The microorganisms isolated were deemed to be effective in the biological control of the spinach mites.

Fungi

There are some quite early reports of fungi affecting mites. The first report of a fungus infecting eriophyoids was from Spears and Yothers (1924), which found a *Hirsutella* sp. in a population of the citrus rust mite, *Phyllocoptruta oleivora*, on grapefruit in

Table 5.1. Families and species of bacteria reported in pest mites.

Family and species	Host pest mite	Reference
Bacillaceae		
Bacillus thuringiensis Berliner	Spider mites	Payne *et al.*, 1993
	Tetranychus urticae Koch	Jung *et al.*, 2007
Pseudomonadaceae		
Pseudomonas putida Trevisan	*Tetranychus urticae*	Aksoy *et al.*, 2008
Rickettsiaceae		
Wolbachia sp.	*Amphitetranychus quercivorus* (Ehara et Gotoh)	Gotoh *et al.*, 1995
	Bryobia sp.	Weeks (in van der Geest, 2010)
	Eutetranychus orientalis (Klein)	Breeuwer and Jacobs, 1996
	Oligonychus biharensis (Hirst)	
	Tetranychus kanzawai Kishida	
	Tetranychus neocaledonicus André	
	Tetranychus turkestani (Ugarov *et* Nikolskii)	
	Tetranychus urticae	
	Tetranychus yusti McGregor	
Flexibacteraceae		
Cardinium sp.	*Brevipalpus californicus* (Banks)	Chigira and Miura, 2005
	Brevipalpus phoenicis (Geijskes)	Groot and Breeuwer, 2006
Spiroplasmataceae		
Spiroplasma sp.	*Tetranychus urticae*	Enigi and Schausberger, 2007

Florida. The same fungus was later described as *Hirsutella thompsonii* Fisher (Fisher, 1950). About 30 years later, also in the USA, a population of the citrus red mite, *Panonychus citri*, was severely affected by a probable *Neozygites* sp. (Fisher, 1951). In the last 50 years, there has been extensive investigation of fungi in mites, and a wide range of pathogenic species have been reported in literature. About 58 species of fungi infect at least 73 species of mites included in the Acaridida, Oribatida, Actinedida, Gamasida and Ixodida (Chandler *et al.*, 2000).

Most of the known species of acaropathogenic fungi belong to the order Entomophthorales, phylum Zygomycota. This order includes six families, but the pathogens of arthropods are mainly ascribed to the families Neozygitaceae, Entomophthoraceae and Ancylistaceae. The Neozygitaceae are obligate pathogens of insects and mites, the Entomophthoraceae are obligate pathogens of arthropods and the Ancylistaceae may infect both arthropods and vertebrates. The characteristics of the nuclei, the formation of resting spores and the discharge of primary conidia are used in the identification of these fungi (Van der Geest, 2010). Other acaropathogenic species are included in the phylum Ascomycota and in the Deuteromycetes (fungi imperfecti). In the Ascomycota, the Hypocreales, Capnodiales and Eurotiales, and the class Sordariomycetes, include most important species. The activity of fungal entomopathogens may also involve plant pathogens (Ownley *et al.*, 2010).

Tables 5.2–5.4 list the main species of acaropathogenic fungi reported in literature, their hosts and the references for each. The most common species developing on tetranychids and eriophyoids are in the genera *Hirsutella* and *Neozygites* (Chandler *et al.*, 2000). Despite the numerous reports and the evidence of the positive use of acaropathogenic fungi in field, is not always possible to report consistent results on the activity of different species as they are strongly affected by manipulation and by the ecological conditions in the field on various crops. The case of *H. thompsonii* provides an example. The mass production of this fungus using industrial methods was developed using semisolid and submerged fermentations to produce mycelial and conidial formulations (McCoy, 1981, 1996a; McCoy *et al.*, 1975). However, under field conditions, even though conidiation of the inoculum on the leaves began 48 h after spraying, the use of the fungus was unsuccessful because of hyphal lysis in storage (McCoy, 1981; McCoy, 1996a). Cold storage of the fungus arrests this lysis, but refrigeration is expensive and so does not encourage pathogen production (McCoy, 1996a). The fungus is registered as a mycoacaricide on citrus, and one commercial product that uses it is Mycar®. Under weather conditions that were appropriate for the growing of the propagules, the use of the fungus in the field was effective, but there was no difference in percentage fruit injury between the fungal and standard chemical treatments. Several factors combined stopped the commercial production of *H. thompsonii* in 1985, including the sensitivity of conidia to (free) water, and problems over fungal survival in the field and stability in storage and transport (McCoy, 1996a).

Despite some failures, the contribution of entomophthoraceous fungi is sometimes very important in the regulation of natural mite populations, with the response depending on some key factors, such as the density of the mite population, temperature and humidity (Saba, 1973a, 1974; Kenneth *et al.*, 1972; Humber *et al.*, 1981; Brandenburg and Kennedy, 1982b). The use of these fungi in the field requires that the most suitable strains are identified and/or selected, and that their bio-ecology is known and their industrial production is possible. The situation is similar to that with pesticides, except that the use of pathogens is more acceptable to the farmer than a pack containing predatory mites or various entomophagous fungi. The main steps involved

in commercializing a microbial pesticide and the potential barriers to commercialization are discussed by Falcon (1985), Lacey *et al.* (2001) and Montesinos (2003).

The use of acaropathogenic fungi represents a scientific challenge that involves various researchers and economic interests. It can be assumed that in the near future, studies on entomopathogenic fungi will be further intensified, as well as biological control by arthropods, as a result of improved techniques for their management and production, and of a gradual restriction of the use of traditional chemical means. A greater knowledge of these organisms can improve their use and their action in the field. With this aim, several studies have investigated details of the bio-ecology of various acaropathogenic fungi, for example the role of infochemicals in the interactions between the cassava

Table 5.2. Families and species of fungi belonging to the Entomophthorales (Zygomycota) reported on pest mites.

Family and species	Host pest mite	Reference
Ancylistaceae		
Conidiobolus sp.	*Bryobia* sp.	Balazy *et al.*, 2008
Conidiobolus obscurus (Hall *et* Dunn) Remaudiere *et* Keller	*Mononychellus tanajoa* (Bondar)	Chandler *et al.*, 2000
	Tetranychus urticae Koch	
Conidiobolus thromboides Drechsler	*Eotetranychus sexmaculatus* (Riley)	Selhime and Muma, 1966
	Mononychellus sp.	Bartkowski *et al.*, 1988
	Mononychellus tanajoa	Delalibera *et al.*, 1992
	Panonychus citri (McGregor)	
	Tetranychus evansi Baker *et* Pritchard	Fisher, 1951
	Tetranychus pacificus McGregor	Humber *et al.*, 1981
	Tetranychus tumidus Banks	Steinhaus and Marsh, 1962
	Tetranychus turkestani (Ugarov *et* Nikolskii)	Carner and Canerday, 1968
	Tetranychus urticae	Chandler *et al.*, 2000
Basidiobolaceae		
Basidiobolus sp.	*Tetranychus urticae*	Chandler *et al.*, 2000
Entomophthoraceae		
Entomophthora sp.	*Panonychus citri*	Fisher, 1951
	Tetranychus pacificus	Steinhaus and Marsh, 1962
Zoophthora radicans (Brefeld) Batko	*Tetranychus urticae*	Chandler *et al.*, 2000
Neozygitaceae		
Neozygites sp.	*Tetranychus evansi*	Humber *et al.*, 1981
	Tetranychus ludeni Zacher	Ramaseshiah, 1971
	Tetranychus neocaledonicus André	
	Tetranychus tumidus Banks	Saba, 1973a
	Tetranychus urticae	Ramaseshiah, 1971; Keller and Wuest, 1983
Neozygites abacaridis Miętkiewski *et* Balazy	*Abacarus hystrix* (Nalepa)	Miętkiewski and Balazy, 2003
	Aculodes sp.	Balazy *et al.*, 2008
	Aculodes mckenziei (Keifer)	Miętkiewski and Balazy, 2003
	Aculus fockeui (Nalepa *et* Trouessart)	
Neozygites floridana (Weiser *et* Muma) Reumadière *et* Keller	*Bryobia* sp.	Miętkiewski *et al.*, 1993
	Eutetranychus banksi (McGregor)	Weiser and Muma, 1966
	Oligonychus gossypii (Zacher)	Yaninek *et al.*, 1996
	Oligonychus hondoensis Ehara	Nemoto and Aoki, 1975
	Oligonychus pratensis (Banks)	Saba, 1973a; Smith and Furr, 1975
	Panonychus citri (McGregor)	Smith and Furr, 1975
	Tetranychus ludeni	
	Tetranychus tumidus	
	Tetranychus urticae	Kenneth *et al.*, 1972; Buergo *et al.*, 1986; Smith and Furr, 1975
Neozygites tanajoae Delalibera, Humber *et* Hajek	*Mononychellus tanajoa* (Bondar)	Delalibera *et al.*, 2004; Hountondji *et al.*, 2010
	Oligonychus gossypii (Zacher)	Delalibera *et al.*, 2004
Neozygites tetranychi (Weiser) Reumadière *et* Keller	*Tetranychus urticae*	Weiser, 1968

Table 5.3. Species of fungi belonging to the Clavicipitaceae (Hypocreales, Ascomycota) reported on pest mites.

Species	Host pest mite	Reference
Beauveria bassiana (Balsamo) Vuillers	*Aceria guerreronis* (Keifer)	Mikunthan and Manjunatha, 2010
	Panonychus citri (McGregor)	Shi and Feng, 2006; Shi *et al*., 2008
	Panonychus ulmi (Koch)	Afifi *et al*., 2010a
	Phyllocoptruta oleivora (Ashmead)	Alves *et al*., 2005b; Shi *et al*., 2008; Afifi *et al*., 2010a
	Polyphagotarsonemus latus (Banks)	Peña *et al*., 1996
	Tetranychus evansi Baker *et* Pritchard	Bugeme *et al*., 2008
	Tetranychus neocaledonicus André	Sreenivas *et al*., 2005
	Tetranychus truncatus Ehara	Shi *et al*., 2008
	Tetranychus turkestani (Ugarov *et* Nikolskii)	
	Tetranychus urticae Koch	Dresner, 1949; Chandler *et al*., 2005; Shi *et al*., 2008; Afifi *et al*., 2010a
Hirsutella sp.	*Amrineus cocofolius* Flechtmann	Van der Geest *et al*., 2002
	Epitrimerus goniathrix Micos *et* Flechtmann	
	Mononychellus sp.	Bartkowski *et al*., 1988
	Notostrix attenuata Keifer	Van der Geest *et al*., 2002
	Propilus syagris Correa-Gondim, Flechtmann *et* Moraes	
	Retracrus johnstoni Keifer	
	Tarsonemus sp.	
Hirsutella danubiensis Balazy, Mietk., Tkaczuk, Wegenst. *et* Wrzosek	*Tetranychus urticae*	Balazy *et al*., 2008
Hirsutella gregis Minter	*Abacarus hystrix* (Nalepa)	Minter *et al*., 1983
Hirsutella kirchneri (Rostrup) Minter, Brady *et* Hall	*Abacarus hystrix*	Minter *et al*., 1983
	Eutetranychus orientalis (Klein)	Sztejnberg *et al*., 1997
	Panonychus citri (McGregor)	Sztejnberg *et al*., 1997
	Phyllocoptruta oleivora	Cabrera and Dominguez, 1987b
Hirsutella necatrix Minter	*Abacarus hystrix*	Minter *et al*., 1983
Hirsutella nodulosa Petch	*Aceria guerreronis*	Cabrera and Dominguez, 1987a
	Phyllocoptruta oleivora	Cabrera and Dominguez, 1987b
	Phytonemus pallidus (Banks)	Miętkiewski *et al*., 2000
	Polyphagotarsonemus latus	Peña *et al*., 1996
Hirsutella thompsonii Fisher	*Abacarus hystrix*	Lewis *et al*., 1981
	Acalitus vaccinii (Keifer)	Baker and Neunzig, 1968
	Acaphylla theae (Watt)	Selvasundaram *et al*., 2001
	Aceria sp.	McCoy and Selhime, 1974
	Aceria cynodoniensis (Sayed)	McCoy, 1981
	Aceria guerreronis	Hall *et al*., 1980; Sreerama Kumar, 2010
	Aceria sheldoni (Ewing)	Gomez and Nasca, 1983
	Aculops lycopersici (Tryon)	Chandler *et al*., 2000
	Brevipalpus phoenicis (Geijskes)	Cabrera, 1978; Rosas Acevedo and Sampedro Rosas, 2000; Rossi-Zalaf and Alves, 2006
	Calacarus heveae Feres	Tanzini *et al*., 2000
	Colomerus novahebridensis Keifer	Hall *et al*., 1980
	Dolichotetranychus floridanus (Banks)	Humber, 1992
	Epitrimerus goniathrix	Van der Geest *et al*., 2002
	Eutetranychus banksi (McGregor)	McCoy and Selhime, 1974
	Eutetranychus orientalis	McCoy and Selhime, 1974; Gerson *et al*., 1979; Dhooria and Butani, 1984
	Eutetranychus sexmaculatus (Riley)	McCoy and Selhime, 1974
	Mononychellus tanajoa (Bondar)	Yaninek *et al*., 1996
	Notostrix formosae (Xue *et* Hong)	Van der Geest *et al*., 2002

Continued

Table 5.3. Continued.

Species	Host pest mite	Reference
	Oligonychus gossypii (Zacher)	Yaninek *et al.*, 1996
	Oligonychus ilicis (McGregor)	Gardner *et al.*, 1982
	Panonychus citri (McGregor)	McCoy and Selhime, 1974; Rath, 1991
	Phyllocoptruta oleivora	Spears and Yothers, 1924; Fisher, 1950; McCoy and Kanavel, 1969; McCoy and Selhime, 1974; Yen, 1974; Van Brussel, 1975; Cabrera, 1977; Gomez and Nasca, 1983; McCoy, 1996a; Aghajanzadeh *et al.*, 2006
	Polyphagotarsonemus latus	Peña *et al.*, 1996
	Rhynacus sp.	Cabrera *et al.*, 1987
	Retracrus elaeis Keifer	Urueta, 1980
	Tetranychus turkestani (Ugarov *et* Nikolskii)	Chandler *et al.*, 2000
	Tetranychus urticae Koch	Gardner *et al.*, 1982
	Vasates destructor (Keifer)	Cabrera, 1984
	Vasates mckenziei Keifer	Miętkiewski *et al.*, 2000
Hirsutella vandergeesti Bałazy, Miętk. *et* Tkaczuk	*Tarsonemus lacustris* Schaarschmidt	Balazy *et al.*, 2008
Metarhizium anisopliae (Metschnikoff) Sorokin	*Aceria guerreronis*	Mikunthan and Manjunatha, 2010
	Brevipalpus phoenicis	Albuquerque *et al.*, 2000; Magalhães *et al.*, 2005
	Eutetranychus orientalis	El-Hady, 2004
	Tetranychus evansi	Bugeme *et al.*, 2008
	Tetranychus neocaledonicus	Hanchinal and Manjunatha, 2000; Sreenivas *et al.*, 2005

green mite, *Mononychellus tanajoa* (Bondar), and its pathogen, the fungus *Neozygites tanajoae* Delalibera, Humber *et* Hajek (Hountondji *et al.*, 2010). McCoy *et al.* (2009) have also recently published a substantial account of the pathogens infecting insect and mite pests of citrus.

Microspora

From a systematic point of view, the Microsporidia (phylum Microspora) are traditionally included in the Protozoa. However, recent studies consider them to be reduced fungi that are able to replicate within the host's cells and infect via unicellular spores (Van deer Geest, 2010). The first microsporidian species described in mites was *Nosema steinhausi* Weiser, which was found associated with *Tyrophagus putrescentiae* (Schrank) (Weiser, 1956). Microsporidian (*Microsporidium*) infections have been reported from several mite species, mainly oribatids. They are often found in mass rearings of phytoseiids as well, and affect the efficacy of these predators (Beerling and Van der Geest, 1991a,b; Bjørnson and Keddie, 1999). However, it does not seem that they have any potential for the biological control of mites (Van der Geest *et al.*, 2000, 2010).

Mites

The first review on predatory mites involved in the biological control of mites injurious to plants of economic importance was from McMurtry *et al.* (1970), who discussed the predatory species involved in the biological control of tetranychids. A few years later, a more general approach to the topic was published by Jeppson *et al.* (1975). Other specialists have also examined the role of mites in the biological control of pests (Gerson and van de Vrie, 1979; Hoy *et al.*, 1983; Xin, 1985), but the first extensive contribution on acarine biocontrol agents was by Gerson and Smiley (1990), followed a few years later by those of Petanović (1993), Karg (1994), Lo (1996) and Grout (1998), and the book of Xin *et al.* (1998) on mites as biological control agents of insect pests in China; the publications by Karg (1994) and Grout (1998) specifically addressed predatory mites. The use of mites in biological pest control has been extensively reviewed by Gerson *et al.* (2003), who reported at least 34 families of mites that were variously involved. Recently, Hoy (2011) treated the most important predatory mites.

Predatory mites of pest mites are included in several families (Anystidae, Ascidae, Bdellidae, Camerobiidae, Cheyletidae, Cunaxidae, Erythraeidae, Laelapidae, Parasitidae, Phytoseiidae, Stigmaeidae, Tarsonemidae, Trombidiidae, Tydeidae) (Tables 5.5–5.7), among which the Phytoseiidae (Table 5.6) and the Stigmaeidae (Table 5.7) play a key role.

PHYTOSEIIDAE. The Phytoseiidae (see Table 5.6) are the most important family of predatory mites, both because there are a high number of species involved and they have the appropriate ecological traits. In the last half century, the family has been

Table 5.4. Families and species of fungi belonging to the Ascomycota (Capnodiales, Sordariomycetes and Eurotiales) and the Basidiomycota reported on pest mites.

Family and species	Host pest mite	Reference
Davidiellaceae (Capnodiales)		
Cladosporium sp.	*Tetranychus ludeni* Zacher	Godse and Patil, 1977
Mycosphaerellaceae (Capnodiales)		
Mycosphaerella citri Whiteside	*Phyllocoptruta oleivora* (Ashmead)	Hernández and Escobio, 1988
Plectosphaerellaceae (Sordariomycetes)		
Verticillium lecanii (Zimmerman) Viégas	*Abacarus hystrix* (Nalepa)	Lewis *et al.*, 1981
	Acaphylla theae (Watt)	Selvasundaram *et al.*, 2001
	Aceria guerreronis (Keifer)	Mikunthan and Manjunatha, 2010
	Brevipalpus phoenicis (Geijskes)	Albuquerque *et al.*, 2000
	Eutetranychus orientalis (Klein)	Sewify and Mabrouk, 1990; El-Hady, 2004
	Tetranychus neocaledonicus André	Sreenivas *et al.*, 2005
	Tetranychus pacificus McGregor	Steinhaus and Marsh, 1962
	Tetranychus urticae Koch	Gams, 1971; Kenneth *et al.*, 1971
Trichocomaceae (Eurotiales)		
Paecilomyces sp.	*Brevipalpus phoenicis*	Albuquerque *et al.*, 2000
Paecilomyces eriophyes (Massee)	*Cecidophyopsis ribis* (Westwood)	Taylor, 1909
	Eriophyes padi Nalepa	Leatherdale, 1965
	Phytoptus avellanae Nalepa	Del Guercio, 1911
Paecilomyces fumosoroseus (Wize) Brown *et* Smith	*Acaphylla theae* (Watt)	Selvasundaram *et al.*, 2001
	Polyphagotarsonemus latus (Banks)	Peña *et al.*, 1996
Paecilomyces terricola (Miller *et al.*) Onions *et* Barron	*Tetranychus urticae*	Kenneth *et al.*, 1971
Exobasidiomycetidae (Basidiomycota)		
Acaromyces ingoldii Boekhout *et al.*	*Eutetranychus orientalis*	Sztejnberg *et al.*, 2004; Paz *et al.*, 2007a,b
	Panonychus citri (McGregor)	
	Phyllocoptruta oleivora	
	Tetranychus urticae	
Aschersonia sp.	*Brevipalpus phoenicis*	Albuquerque *et al.*, 2000
Meira argovae Boekhout *et al.*	*Eutetranychus orientalis*	Sztejnberg *et al.*, 2004; Paz *et al.*, 2007a,b
	Panonychus citri	
Meira geulakonigii Boekhout *et al.*	*Phyllocoptruta oleivora*	
	Tetranychus urticae	

extensively investigated and almost half of the bibliographic references on the agents of biological control concern phytoseiids. There is a priority interest in this group of mites, as witnessed by the many studies that there have been on the systematic and bio-ecological aspects of the group, and on biological control programmes and/or integrated pest management (IPM). There has been a considerable increase in the numbers of described species, from about 1500 in 1986 (Moraes *et al.*, 1986) to about 2250 in 2004 (Moraes *et al.*, 2004). Chant and McMurtry (2007) and Kreiter and Tixier (2010) reported that about 2300 species belonging to 84 genera have been described in the family. Demite *et al.* (2014b) reported over 2450 valid species, several added from collections made in countries that have been little investigated (Guanilo *et al.*, 2008; Tixier *et al.*, 2012; Kreiter *et al.*, 2013), and a database has been compiled for all species (Demite *et al.*, 2014a,b).

Molecular markers have been used by various authors to identify phytoseiid species (e.g. Li *et al.*, 2012). Gerson (2014) noted that 'two confounding trends are meanwhile impacting on the understanding and use of these species'. The first trend is the recognition of more taxa within the same species, and the difficulty of identifying the various species with only morphological criteria. Various authors (Walter and Campbell, 2003; Okassa *et al.*, 2009) have proposed the use of molecular methods to resolve this problem, and others have suggested that such methods also be used to resolve issues such as misidentifications and synonymies (Tixier *et al.*, 2011, 2012; Bowman and Hoy, 2012). The second trend identified by Gerson (2014) is the existence in common phytoseiid species of strains or populations differing in traits such as response to aridity and temperature, and in their biocontrol efficacy for prey, a phenomenon that was hypothesized at the time for *Phytoseiulus persimilis* in Mediterranean greenhouses. Some other examples are also discussed by Gerson (2014). However, it should be noted that most taxonomic studies are still based on traditional morphological examination, although with hopes both for injurious and beneficial species of the increased use of molecular tools to identify races, biotypes and cryptic species.

Results from morphological and molecular (12S rDNA) analyses of specimens of *Phytoseiulus longipes* Evans collected

Table 5.5. Predatory mites by family (Anystidae, Ascidae, Bdellidae, Camerobiidae, Cheyletidae, Cunaxidae, Erythraeidae, Laelapidae, Parasitidae, Tarsonemidae, Trombidae and Tydeidae) with their prey pest mites.

Family and species	Prey	Reference
Anystidae		
Anystis agilis (Banks)	*Bryobia praetiosa* Koch	Lord, 1949
	Panonychus citri (McGregor)	Mostafa *et al*., 1975
	Tetranychus urticae Koch	Sorensen *et al*., 1976
Anystis baccarum (L.)	Tetranychidae	Lange *et al*., 1974a,b
Anystis wallacei Otto	*Halotydeus destructor* (Tucker)	Otto and Halliday, 1991
	Penthaleus major (Dugés)	
Ascidae		
Lasioseius bispinosus Evans	*Rhizoglyphus robini* (Claparède)	Leśna *et al*., 1995
Lasioseius chaudhri (Wu *et* Wang)	*Tarsonemus fuzhouensis* Lin *et* Zhang	Zhang and Lin, 1991
Lasioseius parberlesei Tseng	*Steneotarsonemus spinki* Smiley	Tseng, 1984
Bdellidae		
Bdella depressa Ewing	*Bryobia praetiosa*	Snetsinger, 1956
	Petrobia latens (Müller)	
	Tetranychus urticae	
Bdella longicornis (L.)	Tetranychidae	Sorensen *et al*., 1983
Spinibdella sp.	Pest mites	Goh and Lange, 1989
Camerobiidae		
Neophyllobius spp.	Eriophyidae, Tenuipalpidae	De Leon, 1967; Bolland, 1983
Cheyletidae		
Cheletomimus berlesei (Oudemans)	*Aceria sheldoni* (Ewing)	Sternlicht (1969a)
Cheyletia sp.	*Aceria ficus* (Cotte)	Baker, 1939
Cheyletus eruditus (Schrank)	*Acarus siro* L.	Solomon, 1969
	Lepidoglyphus destructor Schrank	Barker, 1991
Cheyletus malaccensis Oudemans	*Tyrophagus putrescentiae* (Schrank)	Yousef *et al*., 1982
Cheyletus cacahuamilpensis Baker	*Dolichotetranychus floridanus* (Banks)	Mohamed *et al*., 1982
Hemicheyletia bakeri (Ehara)	Tetranychidae	Kanavel and Selhime, 1967
	Tetranychus urticae	Laing, 1973
Nodele calamondin Muma	*Lepidoglyphus destructor* Schrank	Barker, 1991
Cunaxidae		
Cunaxa capreolus (Berlese)	*Eutetranychus orientalis* (Klein)	Zaher *et al*., 1975
Cunaxa lukoschusi Smiley	Tenuipalpidae	Corpuz-Raros and Garcia, 1995
Cunaxoides olivieri (Schuft)	*Calepitrimerus vitis* (Nalepa)	Schruft, 1969, 1971
Cunaxa setirostris (Hermann)	*Oligonychus iseilemae* (Hirst)	Sathiamma, 1995
Erythraeidae		
Balaustium putmani Smiley	*Aculus schlechtendali* (Nalepa)	Cadogan and Laing, 1977
	Bryobia arborea Morgan *et* Anderson	Putman, 1970b
	Panonychus ulmi (Koch)	
	Tetranychus urticae	
Laelapidae		
Geollaelaps aculeifer (Canestrini)	*Rhizoglyphus robini*	Leśna *et al*., 1995, 1996
	Tetranychus urticae Koch	Lobbes and Schotten, 1980
	Tyrophagus putrescentiae (Schrank)	Lobbes and Schotten, 1980
Stratiolaelaps miles (Berlese)	*Tyrophagus putrescentiae*	Enkegaard *et al*., 1997
Parasitidae		
Parasitus fimetorum (Berlese)	*Halotydeus destructor*	James *et al*., 1995
	Rhizoglyphus robini	Leśna *et al*., 1995
Tarsonemidae		
Acaronemus destructor (Smiley *et* Landwehr)	Tetranychidae, Tenuipalpidae	Smiley and Landwehr, 1976
Dendroptus sp.	*Aculus schlechtendali*	Villanueva and Harmsen, 1996
Dendroptus sp.	Tydeidae (eggs)	Mitrofanov *et al*., 1986
Trombidiidae		
Allothrombidium pulvinum Ewing	Tetranychidae	Zhang, 1992

Continued

Table 5.5. Continued.

Family and species	Prey	Reference
Tydeidae		
Homeopronematus anconai (Baker)	*Aculops lycopersici* (Massee [Tryon])	Hessein and Perring, 1986
	Brevipalpus lewisi McGregor	Hessein and Perring, 1988
Pronematus ubiquitous (McGregor)[a]	*Aceria ficus* (Cotte)	Baker, 1939, 1968; Abou-Awad *et al.*, 1999a
	Aculops lycopersici	Rice, 1961; Abow-Awad, 1979
	Brevipalpus phoenicis (Geijskes)	Borthakur, 1981
	Eutetranychus orientalis	Rasmy, 1970a
	Phyllocoptruta oleivora (Ashmead)	Muma *et al.*, 1961
Tydeus californicus (Banks)	*Aceria sheldoni*	Baker and Wharton, 1952
Tydeus goetzi Schruft	*Colomerus vitis* (Pagenstecher)	Schruft, 1972a
	Calepitrimerus vitis	
Tydeus sp.	*Phyllocoptruta oleivora*	Chen *et al.*, 1988

[a]This species is now assigned to another family (André and Fain, 2000), but to avoid confusion, its old classification has been retained here.

in Argentina, Brazil, Chile and South Africa, are very promising. Two of these populations feed and develop on *Tetranychus evansi* Baker *et* Pritchard, whereas the two others do not. Populations from Argentina and Brazil feed and develop on *T. evansi* and are morphologically closer to each other than to the two other populations. Genetic distances among the four populations of *P. longipes* are very low, suggesting that despite their different feeding habits, they belong to the same species. However, the populations associated with *T. evansi* showed some genetic differentiation from those that do not use this pest for their development. Knowledge of this phenomenon may ensure the success of biological control programmes and suggests that strains adapted to both crops and prey species should be developed (Tixier *et al.*, 2010a).

Except for some species, phytoseiids have a medium-sized body and, with the exception of some species, they are off-white or amber in colour. They are usually very mobile and feed mostly on mites, but sometimes on small insects, nematodes, fungi and plants – on which they consume pollen and extrafloral exudates. The exploitation of these different food resources identifies different specializations supported by significant ecological adaptations, including aggregating ability and prey density dependence (Zhang and Croft, 1994). In some phytoseiids, pollen may be the main diet or it may provide subsistence nutrients, thus maintaining and allowing predator development at low prey density. The availability of pollen can affect the outcome of intraguild predation contests between phytoseiids, although not all phytoseiids feed on pollen of the same plant (Gerson *et al.*, 2003), and the pollen of some plants may be refused by one species but consumed by another (Kolokytha *et al.*, 2011). The availability of pollen can have unexpected results, as shown by the results of studies carried out by Aguilar-Fenollosa *et al.* (2011) on the effect of pollen of plants associated with Spanish citrus groves on the two-spotted spider mite. Specialist phytoseiids (Types I and II, see below and Table 5.6) were consistently collected in plots planted with *Festuca arundinacea* Schreber, and produced a better control of *T. urticae*. Conversely, in citrus plots with wild flower cover, more pollen-feeding generalists occurred, and these competed with the specialists and reduced their numbers, resulting in inadequate pest control. In a study by Weintraub *et al.* (2009), populations of *Amblyseius swirskii* Athias Henriot in greenhouses were increased by spraying plants with supplementary pollen using a hand-held electrostatic pollen applicator. The early phytoseiids living on plants were probably generalized types with omnivorous food habits, but the exploitation of the various resources available promoted various life styles and 'body plans'. For instance, the predation of spider mites producing abundant webbing has probably independently established itself in at least four of the 16 tribes included in two of the three subfamilies reported for the family. However, some 'body plans' are more closely related to the morphology of the host plant than to the food source (McMurtry, 2010).

The relationships between phytoseiid mites and the biocontrol of pests (mites and insects) have been extensively discussed by Gerson *et al.* (2003) and Gerson (2014). Other information is given on the relationships of the Phytoseiidae with the Tetranychidae by Helle and Sabelis (1985) and with the Eriophyoidea by Sabelis (1996). Systematic features are discussed by Chant and McMurtry (2007). Pérez-Sayas *et al.* (2010) successfully used the PCR-Multiplex technique for prey detection in citrus mites in Spain, and the role of main species on the defence of cultivated plants in Spain has been examined by Ferragut *et al.* (2009). Hoy (2011) discussed the use of phytoseiids in integrated mite management.

When food is available and the temperature and RH are optimum, many phytoseiid species complete their biological cycle within a week (Tanigoshi, 1982). Fecundity normally ranges from 30 to 40 eggs per female, but may be higher and is related to RH. Most species are adapted to a genetic system of 'parahaploidy' or 'pseudo-arrhenotoky', in which the females require mating to deposit eggs (Schulten, 1985). In the field, the sex ratio is sometimes greater than 0.75 females per male (Sabelis, 1985c). Overwintering occurs either by reproductive diapause (Veerman, 1992) or through resistance to low temperatures. The main factors influencing diapause seem to be daylight and low temperatures, and resistance to low temperatures

has been extensively investigated in order to select phytoseiid strains adapted to field conditions or to commercial manipulation. This has significant practical implications, as in the case of strains of *Phytoseiulus persimilis* reared in central and northern Europe and used in Mediterranean greenhouses. Examples of phytoseiid species that have different strains have been reported by Gerson (2014). The dispersal of phytoseiids may be active and achieved by moving or may be via air streams. The first method has been investigated in *P. persimilis* (Takafuji, 1977) and *Neoseiulus californicus* (McGregor) (Raworth *et al.*, 1994), and the second in *Amblyseius fallacis* (Garman) (Johnson and Croft, 1981) and *Metaseiulus occidentalis* (Hoy, 1982a). Dispersal behaviour may be influenced by various contrasting factors, such as the depletion of food resources and the presence of tactile stimuli and/or volatiles that indicate the presence of food resources even at low density, as in *P. persimilis*

Table 5.6. Types and species of predatory Phytoseiidae and their prey mites.

Types and species	Prey	Reference
Type I		
Phytoseiulus longipes Evans	*Oligonychus punicae* (Hirst)	Badii and McMurtry, 1983
	Tetranychus spp.	
Phytoseiulus macropilis (Banks)	*Tetranychus* spp.	McGregor and McDonough, 1917
Phytoseiulus persimilis Athias Henriot	*Tetranychus* spp.	Chant, 1961; Bravemboer and Dosse, 1962; Smith *et al.*, 1963
Type II		
Galendromus occidentalis (Nesbitt)	*Eotetranychus pruni* (Oudemans)	Petrushov, 1987
	Homeopronematus anconai (Baker)	Flaherty and Hoy, 1971
	Panonychus citri (McGregor)	Grafton-Cardwell *et al.*, 1997
	Tetranychus spp.	McMurtry, 1982; Wilson *et al.*, 1984
	Tetranychus mcdanieli McGregor	Hoyt, 1969a
	Tetranychus urticae Koch	Grafton-Cardwell *et al.*, 1997
	Typhlodromus pyri Scheuten	MacRae and Croft, 1993
Neoseiulus californicus (McGregor)	*Tetranychus urticae*	Greco *et al.*, 1999
Neoseiulus fallacis (Garman)	*Aculus schlechtendali* (Nalepa)	Welty, 1995
	Panonychus ulmi (Koch)	
	Tetranychus kanzawai Kishida	Lo *et al.*, 1990
	Tetranychus urticae	Smith *et al.*, 1963; Welty, 1995; Hamstead, 1970; Lee and Lo, 1990
Neoseiulus idaeus Denmark *et* Muma	*Mononychellus tanajoa* (Bondar)	Moraes *et al.*, 1994
	Oligonychus gossypii Zacher	Yaninek *et al.*, 1992
Neoseiulus longispinosus (Evans),	*Polyphagotarsonemus latus* (Banks)	Hariyapa and Kulkarni, 1988
Neoseiulus womersley (Schicha)	*Tetranychus kanzawai*	Lo *et al.*, 1990
	Tetranychus urticae	Mori and Saitō, 1979; Waite, 1988
Type III		
Amblyseius channabasavanni Gupta	*Raoiella indica* Hirst	Daniel, 1981
Typhlodromalus aripo De Leon	*Mononychellus tanajoa*	Bellotti *et al.*, 1999
Typhlodromus doreenae Schicha	*Brevipalpus lewisi* McGregor	James and Whitney, 1991
Typhlodromalus manihoti (Moraes)	*Mononychellus tanajoa*	Toko *et al.*, 1996
Typhlodromus pyri Scheuten	*Aculus schlechtendali*	Dicke *et al.*, 1990a
	Bryobia arborea Morgan *et* Anderson	Herbert, 1962
	Eotetranychus carpini (Oudemans)	Schruft, 1985
	Calepitrimerus vitis (Nalepa)	Hluchy *et al.*, 1991
	Colomerus vitis (Pagenstecher)	Mathys, 1959; Engel, 1990
	Panonychus ulmi	Collyer, 1964b; Schruft, 1985
Type IV		
Euseius addoensis (Van de Merwe *et* Ryke)	*Panonychus citri*	Keetch, 1972
	Tetranychus urticae	Schwartz, 1990
Euseius scutalis (Athias Henriot)	*Brevipalpus californicus* (Banks)	El-Halawany *et al.*, 1993
	Eutetranychus orientalis (Klein)	Abdel-Samad *et al.*, 1996
Euseius victoriensis (Womersley)	*Calepitrimerus vitis*	James and Whitney, 1993
	Colomerus vitis	James, 1989
	Tegolophus australis Keifer	Smith and Papacek, 1991
	Tetranychus urticae	James, 1990

Table 5.7. Predatory species of the Stigmaeidae and their prey mites.

Species	Prey	Reference
Agistemus spp.		
Agistemus africanus Gonzalez	*Panonychus ulmi* (Koch)	Pringle, 1975
Agistemus arcypaurus Gonzalez	*Brevipalpus phoenicis* (Geijskes)	Oomen, 1982
Agistemus collyerae (Gonzalez)	*Aculus fockeui* (Nalepa *et* Trouessart)	Castagnoli and Nannelli, 1987
Agistemus cyprius Gonzalez	*Aculus* spp.	Villaronga *et al.*, 1993
Agistemus exsertus Gonzalez-Rodriguez	*Aceria sheldoni* (Ewing)	Sternlicht, 1969a
	Aceria dioscoridis (Soliman *et* Abou-Awad)	Rasmy *et al.*, 1987
	Aceria ficus (Cotte)	*El-Halawany et al.*, 1990a
	Aculops lycopersici (Massee [Tryon])	Osman and Zaki, 1986
	Aculus schlechtendali (Nalepa)	*El-Halawany et al.*, 1990b
	Colomerus vitis (Pagenstecher)	Osman *et al.*, 1991
	Eutetranychus orientalis (Klein)	Elbadry *et al.*, 1969
	Panonychus citri (McGregor)	Ehara, 1964a; Yue and Childers, 1994; Yue and Tsai, 1995
	Tetranychus cinnabarinus (Boisduval)	Elbadry *et al.*, 1969
Agistemus fanari Dosse	Tetranychidae	Dosse, 1967
Agistemus fleschneri Summers	*Aculus schlechtendali*	Nelson *et al.*, 1973
	Brevipalpus phoenicis	Oomen, 1982
	Eotetranychus sexmaculatus (Riley)	Muma, 1975
	Eutetranychus banksi (McGregor)	Zaher and El-Badry, 1961
	Panonychus ulmi	Nelson *et al.*, 1973
	Tetranychus cinnabarinus	Zaher and El-Badry, 1962
	Tetranychus kanzawai Kishida	Osakabe, 1963
	Tetranychus urticae Koch	Nelson *et al.*, 1973
Agistemus floridanus Gonzalez-Rodriguez	*Eotetranychus sexmaculatus*	Muma, 1975
	Fungitarsonemus peregrinus Attiah	Muma, 1975
	Phyllocoptruta oleivora (Ashmead)	Muma and Selhime, 1971
Agistemus hystrix Gupta	Tenuipalpidae	Borthakur *et al.*, 1997
Agistemus longisetosus Gonzalez-Rodriguez	*Brevipalpus* sp.	Collyer, 1964a
	Bryobia rubrioculus (Scheuten)	
	Panonychus citri	
	Panonychus ulmi	
Agistemus orbicularis Ehara *et* Oomen-Kalsberek	*Brevipalpus phoenicis* (Geijskes)	Oomen, 1982
Agistemus terminalis (Quayle)	*Panonychus citri*	Ehara, 1964a; Inoue and Tanaka, 1983
	Brevipalpus phoenicis	Oomen, 1982
Zetzellia spp.		
Zetzellia graeciana Gonzalez	*Aculus schlechtendali*	White and Laing, 1977
	Panonychus ulmi (Koch)	Inserra, 1970; Nucifora and Vacante, 1978; Vacante and Tropea Garzia, 1987a
Zetzellia javanica Ehara *et* Oomen-Kalsbeek	*Brevipalpus phoenicis*	Oomen, 1982
Zetzellia mali (Ewing)	*Aculus cornutus* (Banks)	Rice *et al.*, 1976
	Aculus schlechtendali	Hoyt, 1969a; Santos, 1976
	Bryobia praetiosa Morgan *et* Anderson	Lord, 1949
	Calepitrimers vitis (Nalepa)	Schruft, 1969
	Colomerus vitis	Liguori, 1987
	Diptacus gigantorhynchus (Nalepa)	Amano and Chant, 1990
	Panonychus ulmi	Santos, 1976; Vacante and Tropea Garzia, 1987a
	Tetranychus urticae Koch	Santos, 1976
Zetzellia maori Gonzalez	*Brevipalpus phoenicis*	Oomen, 1982
Zetzellia talhouki Dosse	*Acalitus phloeocoptes* (Nalepa)	Bayan, 1988
	Aculus fockeui	
	Diptacus gigantorhynchus	
	Phyllocoptes abaenus Keifer	
	Spider mites	Dosse, 1967

(Charles and White, 1988). Mites generally occupy surfaces exposed to air currents so that they can be dispersed, and different wind velocities are probably needed for the dispersal of different species, as suggested by the results of investigations in vineyards (Tixier *et al.*, 1998). The localization of prey is led by various factors, among which volatiles seem to play a fundamental role. Research has indicated the action of cues from the pest colonies (Hoy and Smilanick, 1981) and of various volatiles emitted by plants damaged by spider mites such as *T. urticae* and *Panonychus ulmi*, which emit specific kairomones (Sabelis and Van de Baan, 1983) (see the earlier section on kairomones). The production of synomones (i.e. attractants for predatory mites) may differ according to factors such as the age of the damaged leaves (Takabayashi *et al.*, 1994a) and the cultivar, as shown in apples (Takabayashi *et al.*, 1994b) (see the earlier section on synomones).

The analysis of their different feeding habits and bio-ecological and morphological traits, including the length of the dorsal setae (McMurtry and Croft, 1997), has permitted the categorization of many phytoseiid species into four groups or types. Further research by various specialists has shown that this categorization could be used to select phytoseiid mites for various purposes, including biological control programmes (Croft *et al.*, 1998a,b; Luh and Croft, 1999). According to their biological features, these groups are referred to as Types I–IV (see Table 5.6):

- **Type I**. This type includes phytoseiids with long dorsal setae, which are usually specialized predators of webbing spider mites (mainly *Tetranychus* spp. and sometimes *Oligonychus* spp). It is represented by *Phytoseiulus* spp.
- **Type II**. This type also includes phytoseiids with long setae that are selective predators of spider mites, but not only *Tetranychus* spp. They include species that produce dense webbing, e.g. *Galendromus* spp., some *Neoseiulus* spp. and a few *Typhlodromus* spp. Their prey mites include species that produce little webbing (e.g. *Panonychus ulmi*, etc.) and other pest mites. They may also feed on pollen and plant exudates.
- **Type III**. This type includes phytoseiids with short mid-dorsal and lateral setae. They are usually generalists that prefer prey other than spider mites, such as tarsonemid mites or thrips, and they consume pollen, honeydew and fungi as well.
- **Type IV**. Species of this type have short mid-dorsal and lateral setae. They usually include only *Euseius* species, and they live better on pollen than on animal prey.

Various examples of these types of phytoseiids are discussed in the relevant chapters on pest mites in Part II of the book. The possibility that further types of predatory mites may exist, possibly more specialized for certain prey, was advanced by Adar *et al.* (2012), who proposed that the ability to feed on plants be added as a cross-type trait of phytoseiid lifestyle types. Recently, McMurtry *et al.* (2013) proposed a new classification of phytoseiid lifestyles with many subtypes. The discovery of a phytoseiid whose main food source is the coffee leaf rust fungus (Oliveira *et al.*, 2014) suggests that more lifestyles may be discovered.

STIGMAEIDAE. The Stigmaeidae are little mites, red to yellow in colour, and are flattened and slow. They live in the soil and on plants, where they feed on different trophic resources, including eggs and motile forms of tetranychid, eriophyoid and tenuipalpid mites. Like the phytoseiid mites, some genera (*Agistemus*, *Zetzellia*) of the Stigmaeidae (see Table 5.7) are important predators of different pest mites. However, their bio-ecology is less known than that of the Phytoseiidae. A generation develops in more or less a month and the reproduction is arrhenotokous. Their populations usually develop significantly with either rationalized or no spraying of chemicals. Reports on the predaceous species of this family can be found in Gerson *et al.* (2003), Santos and Laing (1985) and Thistlewood *et al.* (1996). Various examples of these predatory mites are also discussed in the relevant chapters on pest mites in Part II of the book.

Spiders

Spiders are generalist predators of insects and pest mites (Lord, 1949; Parent, 1973), and carry out 'wasteful killing', i.e. they capture and kill more prey than they can feed on (Maloney *et al.*, 2003). Despite this, it is thought that they play a positive ecological role, although evidence of their actual importance is lacking (Jeppson *et al.*, 1975). They are present irregularly in the field and require a high density of prey for development, but in natural habitats they may represent a stable assemblage that includes species able to remove large numbers of insects (Riechert and Lockley, 1984).

Studies have been carried out on the spider community in different crops, including citrus (Carroll, 1980; Mansour and Whitcomb, 1986; Benfatto *et al.*, 1992; Yan *et al.*, 1992; Rodriguez-Almaraz and Contreras-Fernandez, 1993; Dippenaar Schoeman, 1998), peach (Putman, 1967) and apples (Parent, 1973). A lycosiid spider feeding on *Tetranychus cinnabarinus* (Boisduval) has been reported from South Africa (Coates, 1974). In Israel, the potential of four species of spiders in the control of *T. cinnabarinus* has been investigated by Mansour *et al.* (1995). There have been no actual reports on the use of spiders as biological control agents of pests, with the exception of one on their use as biological control agents of pests in apple orchards and rice paddies (Maloney *et al.*, 2003), but Riechert and Lockley (1984) review the potential of spiders as biological control agents.

Insects

Predatory insects of pest mites are included in the orders Coleoptera, Dermaptera, Diptera, Hemiptera (Heteroptera), Neuroptera and Thysanoptera (McMurtry *et al.*, 1970; Jeppson *et al.*, 1975; Chazeau, 1985; Perring and McMurtry, 1996). Predatory insects are commonly generalists (except for *Stethorus* spp., some thrips, etc.) and the interaction of their populations with pests and other arthropods in the field is often unclear (Blommers, 1994; Letourneau *et al.*, 2009). Despite their ecological characteristics and adaptation to intraguild predation, predatory insects are more beneficial than detrimental. Their large size forces them to consume a large amount of prey and to develop at high density of population, often facilitated by the ability to feed on alternative foods (pollen, honeydew). Hoy (2011) summarized the positive and negative aspects of generalist

predatory insects with respect to phytoseiid mites. The positive aspects include a large consumption of prey per unit time (a high functional response), a long lifespan (inducing delayed numerical response), a high rate of reproduction (potentially with a greater numerical response), the ability to consume non-prey foods, and the possibility that some species can be mass reared for augmentative release. The negative aspects include a longer generation time than pest mites, the need for large amounts of prey (with a high risk of damage to crops), cannibalism (intraguild predation), sensitivity to chemicals, lack of suitability for use in classical biological control, and the expense of mass rearing. The advantages related with the presence of the predatory insects in the field suggest the use of a strategy of 'conservation biological control', a strategy that aims to integrate beneficial insects back into crop systems for natural pest control.

The most important predatory families of the Coleoptera are the Anthicidae, Coccinellidae and Staphylinidae, in addition to some species (*Saula japonica* Gorham) of the Endomychidae. Common predatory families of the Neuroptera are the Chrysopidae, Coniopterygidae and Hemerobiidae. Hemipterans feeding on spider mites belong to the families Anthocoridae and Miridae; of secondary importance are the Nabidae (*Himacerus apterus* (Fabr.) in Europe) and the Lygaeidae (*Geocoris* sp. in North America) (McMurtry *et al.*, 1970). The most important Thysanoptera belong to the families Aeolothripidae, Phlaeothripidae and Thripidae. Among the Diptera, the most important family is the Cecidomyiidae family; other species of secondary importance are included in the families Syrphidae (*Toxomerus geminatus* Say), Dolichopodidae and Empididae (*Drapetis micropyga* Melander) (Fleschner and Ricker, 1953a; McMurtry *et al.*, 1970). Some Dermaptera of the Labiduridae family (*Labidura riparia* (Pallas)) prey upon the two-spotted spider mite in South Africa (Coates, 1974). The ant *Tapinoma melanocephalum* (Fabricius) (Hymenoptera, Formicidae) preys on *T. urticae* in greenhouses in Florida, but its phytophagous habit renders its use unlikely (Osborne *et al.*, 1995). Although some predatory insects are mass reared and released in augmentative biological control programmes (*Orius* spp., *Macrolophus caliginosus* (Wagner), *Feltiella acarisuga* (Vallot), *Chrysoperla* spp., etc.), fundamentally they are used in conservation programmes.

COCCINELLIDAE. The Coccinellidae have a wide geographical distribution, and occur in various ecosystems, including tropical rain forests, dry savannahs and various crops. The adults and larvae are predators of aphids, scale insects, mealybugs and mites, and few are phytophagous. The most important species are included in the genus *Stethorus* Weise, which numbers over 60 predaceous species of pest mites. Almost 35 species are predators of tetranychid mites, and to a lesser extent of tenuipalpid mites (Chazeau, 1985; Obrycki and Kring, 1998; Biddinger *et al.*, 2009). The Coccinellidae are commonly referred to as ladybird beetles and species included in the Stethorini tribe are important in the biological control of pest mites (and other pests) (see Table 5.8). The use of ladybird beetles in conservation biological control programmes requires the availability of overwintering habitats and refuges that are free from factors that will reduce their populations (e.g. pesticides), in either the crop concerned and/or the surrounding environment. Ladybird beetle populations that are conserved by these means may contribute to the suppression of tetranychid populations. Reports of the successful use of ladybird beetles in biological control include the control of pest mites of apple, citrus and other crops in various countries. The existence of many cryptic species makes the systematics and taxonomy of these predators difficult, and molecular diagnostic tools are needed. The best utilization of the potential of ladybird beetles in the biological control of mites requires better knowledge of their requirements, such as alternative foods, refuges for dormancy and host-finding mechanisms, and freedom from non-selective pesticides (Biddinger *et al.*, 2009). With this aim, knowledge of their ecology and behaviour are fundamental (Hodek *et al.*, 2012).

About 40% of coccinellids prey on spider mites of economic importance. Their sex ratio is usually one male for one female, and in temperate countries the various species overwinter as adults. Two or three generations a year have been observed, and the generation time is always longer than that of the prey (Chazeau, 1985). The adults fly actively and eat the entire prey, showing a preference for large mobile individuals. Females producing eggs eat twice as many prey as males. An increase of prey consumption has been observed with an increase of temperature, and there is a positive correlation between the numbers of prey eaten and eggs laid (Chazeau, 1974). Adults of *Stethorus punctum* (LeConte) can find *Panonychus ulmi* on apple trees at a low density (Hull *et al.*, 1977). The same behaviour has been observed on citrus (Haney *et al.*, 1987) and other crops, and has been attributed to an olfactory mechanism (Colburn and Asquith, 1970), while a visual mechanism has been reported in other studies (Biddinger *et al.*, 2009). *Stethorus punctillum* has larger prey range than *S. punctum* and feeds on various pest mites, including the spruce spider mite, *Oligonychus ununguis* (Jacobi), on ornamentals (Wheeler *et al.*, 1973) and the European red mite, *P. ulmi*, on apple (Pasqualini and Malavolta, 1985) or on strawberry (García Marí and González-Zamora, 1999).

Coccinellids have also been found associated with eriophyoid mites, for example *Stethorus nanus* LeConte feeds on rust mites (Yothers and Mason, 1930) and *Delphastus pusillus* (LeConte) probably feeds on *Phyllocoptruta oleivora* (Muma *et al.*, 1961).

Some species of ladybird beetles appear to be more selective in their choice of the prey. In general, these important predators show a clear capacity to aggregate on mite colonies and to actively disperse when prey density decreases, but their establishment on the mite colonies needs a minimum density of pest mites (Chazeau, 1985). A review of the bio-ecology and behaviour of ladybird beetles was recently published by Biddinger *et al.* (2009), and further updates may be found in Hodek *et al.* (2012).

ANTHICIDAE. The anthicid beetle, *Ischyropalpus nitidulus* (LeConte), preys on the eriophyoid *Platyphytoptus sabinianae* Keifer on pines in California (Landwehr, 1977).

STAPHYLINIDAE. The genus *Oligota* Mannerheim is the most representative of this family. It is cosmopolitan and includes over 170 species living on various habitats and feeding on different resources. Chazeau (1985) reports almost ten important predatory species of tetranychids that are widespread in various areas of the world. Larvae and adult prey on tetranychid mites, usually more readily on the motile forms than on the eggs. *Oligota*

Table 5.8. Coleoptera, Coccinellidae, genus *Stethorus* Weise: main predatory species and prey mites.

Species	Prey	Reference
Stethorus aethiops Weise	*Mononychellus tanajoa* (Bondar)	Yaseen *et al.*, 1982
	Tetranychus lombardini Baker *et* Pritchard	
	Tetranychus neocaledonicus André	
	Tetranychus urticae Koch	
Stethorus bifidus Kapur	*Bryobia* sp.	McMurtry *et al.*, 1970
	Panonychus ulmi (Koch)	Collyer, 1964a
	Tetranychus lambi Pritchard *et* Baker	McMurtry *et al.*, 1970
	Tetranychus urticae	Collyer, 1964a
Stethorus chengi Sasaji	*Panonychus citri* (McGregor)	Chen and Zhao, 1994
	Tetranychus urticae	Wen and Lee, 1981
Stethorus comoriensis Chazeau	*Oligonychus coffeae* (Nietner)	Chazeau, 1971a
	Tetranychus neocaledonicus	
Stethorus darwini (Brétes)	*Mononychellus* spp.	Yaseen *et al.*, 1982
	Panonychus ulmi	Lorenzato, 1987
	Tetranychus evansi Baker *et* Pritchard	Paschoal, 1970
Stethorus exspectatus Chazeau	*Oligonychus*, *Panonychus*, *Schizotetranychus* spp.	Chazeau, 1983
	Tetranychus fijiensis Hirst	
	Tetranychus lambi	
Stethorus exsultabilis Chazeau	*Oligonychus*, *Panonychus*, *Schizotetranychus* spp.	Chazeau, 1983
	Tetranychus fijiensis	
	Tetranychus lambi	
Stethorus fenestralis Houston	*Tetranychus kanzawai* Kishida	Houston, 1980
	Tetranychus lambi	
	Tetranychus urticae	
Stethorus fijiensis Kapur	*Oligonychus* sp.	Chazeau, 1979
	Tetranychus lambi	
	Tetranychus marianae McGregor	
	Tetranychids	Swaine, 1971
Stethorus fuerschi Chazeau	*Oligonychus chazeaui* Gutierrez	Chazeau, 1971a
	Tetranychus roseus Gutierrez	
Stethorus gilvifrons (Mulsant)	*Eutetranychus hirsti* Pritchard *et* Baker	Afshari *et al.*, 2007
	Eutetranychus orientalis (Klein)	
	Oligonychus afrasiaticus (McGregor)	
	Oligonychus coffeae	Sarmah and Bhattacharyya, 2002
	Oligonychus sacchari (McGregor)	Afshari, 1999
	Panonychus citri, *Eotetranychus* sp.	Handoko, 2004
	Panonychus ulmi	Hajizadeh *et al.*, 1993
	Tetranychus atlanticus McGregor	Georgis *et al.*, 1976
	Tetranychus turkestani (Ugarov *et* Nikolskii)	Ahmed and Ahmed, 1989
	Tetranychus urticae	Mathur, 1969
	Tetranychids	McMurtry *et al.*, 1970
Stethorus griseus Whitehead	Tetranychids	Chazeau, 1979
Stethorus japonicus Kamiya	*Panonychus citri*	McMurtry *et al.*, 1970
	Panonychus mori Yokoyama	Kishimoto and Adachi, 2008
	Tetranychus kanzawai	Mori *et al.*, 2005
	Tetranychus urticae	
	Tetranychus viennenis Zacher	Kishimoto and Adachi, 2008
Stethorus jejunus Casey	*Mononychellus tanajoa* (Bondar)	Yaseen *et al.*, 1982
	Tetranychus lombardinii	
	Tetranychus neocaledonicus	
	Tetranychus urticae	
Stethorus loi Sasaji	*Panonychus citri*	Wen *et al.*, 1993
	Tetranychus urticae	Wen and Lee, 1981
	Tetranychids	Chang and Leu, 1986

Continued

Table 5.8. Continued.

Species	Prey	Reference
Stethorus madecassus Chazeau	*Oligonychus* spp., *Tetranychus neocaledonicus*, *Tetranychus* sp.	Chazeau, 1971b
Stethorus pauperculus Weise	*Oligonychus indicus* (Hirst)	Kapur, 1948
	Oligonychus neocaledonicus	Puttaswamy and ChannaBasavanna, 1977
	Tetranychus ludeni Zacher	Puttaswamy and ChannaBasavanna, 1980
Stethorus punctillum Weise	*Eotetranychus buxi* (Garman)	Creary, 2009
	Eotetranychus carpini Oudemans	Laffi, 1982
	Eotetranychus tiliarium Hermann	Jäckel *et al.*, 2000
	Oligonychus afrasiaticus (McGregor)	Idder and Pintureau, 2008
	Oligonychus bicolor (Banks)	Cinti *et al.*, 1995
	Oligonychus ununguis (Jacobi)	Putman, 1955
	Panonychus citri	Tian, 1995
	Panonychus ulmi	Putman, 1955
	Tetranychus bioculatus (Wood-Mason)	Taleb and Sardar, 2007
	Tetranychus mcdanieli McGregor	Roy *et al.*, 2002
	Tetranychus shoenei McGregor	Creary, 2009
	Tetranychus urticae	Putman, 1955
	Tetranychus viennensis Zacher	Yiğit and Uygun, 1986
	Tetranychids	Plaut, 1965
Stethorus punctum picipes Casey	*Oligonychus punicae* (Hirst)	McMurtry *et al.*, 1969
	Panonychus citri	McMurtry *et al.*, 1970
	Tetranychus tumidus Banks	Perez *et al.*, 2004
	Tetranychids	McMurtry *et al.*, 1970
Stethorus punctum (LeConte)	*Panonychus ulmi*, Tetranychids	McMurtry *et al.*, 1970
	Tetranychus mcdanieli	Robinson, 1953
	Tetranychus pacificus McGregor	
Stethorus siphonulus Kapur	*Eutetranychus banksi* (McGregor)	Raros and Haramoto, 1974
	Oligonychus exsiccator (Zehntner)	
	Panonychus citri	Huang *et al.*, 1988
	Tetranychus neocaledonicus	Chazeau, 1979
	Tetranychus piercei McGregor	Lui and Lui, 1986
	Tetranychus tumidus Banks	Raros and Haramoto, 1974
	Tetranychus urticae	
Stethorus tridens Gordon	*Panonychus citri*	Guanilo and Martinez, 2007
	Tetranychus evansi	Britto *et al.*, 2009
	Tetranychus urticae, Tetranychids	Gordon, 1982
Stethorus utilis Horn	*Eotetranychus hicoriae* (McGregor)	Tedders, 1983
	Eotetranychus sexmaculatus (Riley)	McMurtry *et al.*, 1970
	Eutetranychus banksi	
	Mononychellus caribbeanae McGregor, *Mononychellus tanajoa*, Tetranychids	Yaseen *et al.*, 1982
	Oligonychus pratensis (Banks)	Ehler, 1974
	Panonychus citri, *Tetranychus urticae*	Mora Morin, 1991
Stethorus vagans (Blackburn)	*Bryobia praetiosa* Koch	Evans, 1943
	Oligonychus exsiccator	Swezey, 1923
	Oligonychus thelytokus Gutierrez	Chazeau, 1979
	Oligonychus spp.	Chazeau, 1979; Houston, 1980
	Panonychus ulmi	Walters, 1976
	Tetranychus lambi	Chazeau, 1979
	Tetranychus marianae McGregor	
	Tetranychus neocaledonicus	
	Tetranychus urticae	

flavicornis (Boisduval *et* Lacordaire) and *O. oviformis* (Casey) respectively develop in 28 and 21–30 days, have a fecundity of 40–50 eggs and over 300 eggs, and an adult longevity of 35 and 30 days (Collyer, 1953; McMurtry *et al.*, 1970). The larvae of *O. oviformis* consume from 200 to 300 mites, and the adult ten mites a day (Badgley and Fleschner, 1956). *O. pygmaea* (Solier) preys on *Bryobia arborea* Morgan *et* Anderson (= *Bryobia rubrioculus*), *T. urticae*, *Oligonychus yothersi* (McGregor), and *Panonychus ulmi* (Frank *et al.*, 1992). Moreover, this staphylinid is an important predator of the red spider mite, *Oligonychus coffeae*, on tea (Perumalsamy *et al.*, 2009).

CHRYSOPIDAE. This family has a worldwide distribution and numbers over 1200 species in 80 genera (Winterton and de Freitas, 2006). They are mainly known as predators of aphids, and while some species can prey on spider mites, they are not specialized mite predators. *Chrysopa carnea* Stephens develops on *Panonychus ulmi*, *C. lateralis* Guérin feeds on *Eotetranychus sexmaculatus* and *Chrysopa* spp. on *Panonychus citri* and *P. ulmi* (McMurtry *et al.*, 1970). In the field, the Chrysopidae usually prey on tetranychids after the populations of their former prey (aphids) decline (Fleschner, 1958b). They operate at low levels of attack and show good searching capacity (Chazeau, 1985). *Chrysopa* sp. has been observed feeding on rust mites (Yothers and Mason, 1930) and on free-living gall mites on pomes and stone fruits (Schliesske, 1992), but their impacts on eriophyoid populations are irrelevant.

Chrysperla spp. are mass reared in various countries and released in different crops. Of these, *C. carnea* and *C. rufilabris* (Burmeister) are used in North America, and *C. nipponensis* (Okamoto) and *C. externa* (Hagen) in Latin America and Asia, but the status of the species in North America is unclear (Tauber *et al.*, 2000). Releases of *Mallada basalis* (Walker) in Taiwan on various crops against spider mites have given promising results (Ho, 2000), and this chrysopid has been also been evaluated as a potential predator of *P. citri*, *T. urticae* and *T. kanzawai* on papaya (Cheng *et al.*, 2010).

CONIOPTERYGIDAE. Coniopterygids are generalists, and sometimes play a role in spider mite control. *Conwentia pineticola* Enderlein preys on the European red mite in England but its presence is irregular. *C. psociformis* (Curtis) feeds on the same pest mite but is polyphagous (Collyer, 1951); in Mediterranean citriculture, this species develops usually on high density of populations of the citrus red spider mite. *Coniopteryx vicina* Hagen feeds on *Eotetranychus sexmaculatus* on citrus in Florida, but prefers whiteflies as prey (Muma, 1967; McMurtry *et al.*, 1970). It has been found associated with *Phyllocoptruta oleivora* on citrus at high density, but does not have much impact on the populations of the pest (Muma, 1955), even though adults feed actively on *P. oleivora* (Muma *et al.*, 1961). The species develops slowly when feeding only on rust mites and has a high mortality when feeding on insect prey (Muma, 1967).

ANTHOCORIDAE. Anthocorids live in a variety of habitats and usually overwinter as adults. Their sex ratio ranges from 1:1 to 1:1.5 or 2 and according to temperature their development takes up to a month. The genera *Anthocoris* Fallén, *Cardiastethus* Fieber, *Orius* Wolff and *Wollastoniella* Reuter include generalist predators, and almost 17 species feed on tetranychid mites (Chazeau, 1985; Lattin, 1999). Their prey include Heteroptera, Homoptera, Thysanoptera, Lepidoptera, larvae of Diptera and Coleoptera, and spider mites, which are not usually a suitable prey. The youngest larval instars feed on small prey, while the adults and more developed larvae attack larger prey. Diets of tetranychids usually allow the development of *Orius* spp. (Tawfik and Ata, 1973a,b). Some *Orius* spp. also prey on eriophyoid mites. For example, *O. vicinus* (Ribaut) feeds on the apple rust mite, *Aculus schlechtendali*, both in apple orchards (Heitmans *et al.*, 1986) and in the laboratory, where it preys on *Panonychus ulmi* and *T. urticae* as well (Wearing and Colhoun, 1999); the predator is also common on galls of *Eriophyes fraxinivorus* Nalepa (Fauvel *et al.*, 1975). *O. tristicolor* (White) preys on *Tetranychus pacificus* McGregor, a motile prey, both in the light and without light (Askari and Stern, 1972). *O. insidiosus* (Say) preys on *T. urticae*, *P. ulmi* and *P. citri* in various crops (Iglinsky and Rainwater, 1950; Isenhour and Yeargan, 1981). *Anthocoris nemorum* (L.) locates its prey by smell (Mpkagiannis, 1982).

MIRIDAE. The Miridae are mainly adapted to phytophagy, but many species are facultative predators. Most species overwinter as eggs embedded in the bark of trees or in soft plant tissues. In tropical areas, the generations overlap and no diapause occurs. Chazeau (1985) lists eight species (*Atractotomus mali* (Meyer), *Blepharidopterus angulatus* (Fall.), *Campylomma verbasci* (Meyer), *Deraeocoris nebulosus* (Uhler), *Diaphnocoris pellucida* (Uhler), *Hyaliodes harti* Knight, *Malacocoris chlorizans* Panz., *Pilophorus perplexus* (D. *et* S.)), living in North America and Europe and feeding on *Panonychus ulmi*, *T. urticae*, *Bryobia praetiosa* and *Bryobia* sp. Mirids are important in the control of the European red mite on apple orchards in England (Collyer, 1952, 1953). Some species may reach maturity when reared on *P. ulmi* only (Lord, 1971). *Macrolophus caliginosus* (Wagner) is mass reared and commercialized in various countries and preys on aphids, whiteflies and mites, e.g. *T. urticae* on tomatoes (Hansen *et al.*, 1999). *Hyaliodes vitripennis* (Say) feeds on European red mites in apple orchards in Canada (Quebec) (Chouinard *et al.*, 2006).

PHLAEOTHRIPIDAE. This family includes several predators of spider mites. Chazeau (1985) lists four important species belonging to the genera *Haplothrips* Amyot *et* Serville, *Leptothrips* Hood and *Cryptothrips* Uzel, but there are more. The development of *H. faurei* Hood from first nymph to adult takes 8–10 days at 24°C; during its development, the predator consumes 143 eggs of *Panonychus ulmi*, and female longevity is over 16 days. *H. faurei* feeds on tetranychid eggs but not on motile forms (Putman, 1965); it also feeds on the peach silver mite, *Aculus cornutus* (Banks), an eriophyoid, but its contribution to the control of the pest population is unclear. *H. subtilissimus* (Haliday) preys on eriophyoid mites on pome and stone fruit trees (Schliesske, 1992), but its contribution to mite control is uncertain; in the USA, it preys on *P. ulmi* on apple (Hoy, 2011). The development of *Leptothrips mali* (Fitch) on *P. ulmi* takes 25.6 days at 23.9°C, and female longevity ranges from 36 to 54 days (Parrella *et al.*, 1982); this predator feeds on the tomato russet mite, *Aculops lycopersici*, but does not give good control of the pest population (Bailey and Keifer, 1943; Anderson, 1954). *Xylaplothrips fuliginosus*

(Schille) preys on eriophyoid mites on pome and stone fruit trees (Schliesske, 1992), but its contribution to their control is unclear.

AEOLOTHRIPIDAE. The predatory species of this family are included in the genus *Aeolothrips* Haliday, and are usually associated in the field with *Panonychus ulmi* and *T. urticae* (Putman and Herne, 1966). *A. intermedius* Bagnall is probably the species that has been most investigated. Its reproduction is arrhenotokous and the sex ratio is of 1:1. It feeds mainly on pest thrips and also on spider mites, but needs to consume plant tissue to reach sexual maturity. At 26°C it preys on up to 68 adults of *P. ulmi* during its development (Bournier *et al.*, 1978, 1979).

THRIPIDAE. This family numbers several predatory species of spider mites. Chazeau (1985) lists about nine species included in the genera *Parascolothrips* Mound, *Perissothrips* Hood and *Scolothrips* Hinds. *Scolothrips* spp. are obligate predators of spider mites, and Mound (2011) recently discussed the species of this genus as predators of leaf-feeding mites. *S. sexmaculatus* (Pergande) has been most investigated. The species is arrhenotokous and after mating the sex ratio is 1:8 males to females. It specializes more on preying on spider mites than do other Thysanoptera, and is able to use the silk of tetranychids as traces. At 23.9°C and with *T. pacificus* as prey, its development takes 23.1 days, the fecundity is 220 eggs, female longevity is 49 days, and a female consumes 68 eggs during its development (Gilstrap and Oatman, 1976). *S. sexmaculatus* has been found associated also with the tomato russet mite, but there were no reports of its predation on this eriophyoid (Abou-Awad, 1979). *S. takahashii* Priesner feeds on the hawthorn spider mite, *Amphitetranychus viennensis* (Zacher), and exhibits type II functional responses (i.e. a decelerating rate of feed intake) at various temperatures, with females more voracious than males (Li *et al.*, 2007a).

The western flower thrips, *Frankliniella occidentalis* (Pergande), is typically a severe pest on various plants, but it can prey on spider mites on cotton (Trichilo and Leigh, 1986). In Australia, the cotton-leaf feeding thrip, *F. schultzei* (Trybom), feeds on spider mites (Milne and Walter, 1997), and *F. schultzei*, *Thrips imaginis* Bagnall and *T. tabaci* Lindeman feed on the two-spotted spider mite (Wilson *et al.*, 1996).

CECIDOMYIIDAE. Cecidomyiidae of the genera *Arthrocnodax* Rübsaamen, *Feltiella* Rübsaamen, *Lestodiplosis* Kieffer and *Therodiplosis* Kieffer may be associated with pest mites in the field (Oomen, 1982; Chazeau, 1985), and about ten species feed as vermiform larvae on tetranychid mites (Chazeau, 1985). The better known species belong to the genera *Feltiella* and *Therodiplosis*, which complete their life cycles when they feed on spider mites. Gagne (1995) included eight species in the genus *Feltiella*, and not 24 as previously reported, and synonymized the genera *Feltiella* and *Therodiplosis*. At 23°C, the life cycle of *Feltiella* sp. takes 18 days, with a pre-oviposition period of 2–3 days, a sex ratio of 1:5.6 males to females and a lifespan of 5–6 days for adults reared on raisins. The second and third larval instars prey on 9–12 eggs a day of *Tetranychus marianae* McGregor and on the same numbers of adults (Moutia, 1958). In the laboratory, the first, second and third instar larvae of *F. acarisuga* (Vallot) prey on 38, 60 and 87 eggs of *T. urticae*, and the fecundity is 33 eggs at 26.7°C and 85% RH (Mo and Liu, 2007). In the Mediterranean region, *F. acarisuga* usually occurs on ornamentals and vegetables, cultivated both in the field and in the greenhouse, and can exert good control of the two-spotted spider mite (Vacante, 1981). The species is mass reared for commercial releases in greenhouses. In Canada, the predator developed in from 10 days at 27°C to 34 days at 15°C on greenhouse vegetables infested by *T. urticae*. At 25°C the fecundity is 46 eggs. Releases of 1000 individuals ha^{-1} controlled spider mite population on tomato, pepper and cucumber (Gillespie *et al.*, 2000).

Arthrocnodax occidentalis Felt feeds on eriophyoids in fig trees and on other species of the same group (Baker, 1939). In citrus crops, a larva preyed on 380 mites in 17 days (Quayle, 1912, in McMurtry *et al.*, 1970). *A. fraxinella* (Meade) feeds on free-living eriophyoids on pome and stone fruits, and controls their populations (Schliesske, 1992). Other unidentified cecidomyiids are reported as predators of eriophyoids, e.g. *Phyllocoptruta oleivora* (Muma *et al.*, 1961), *Phyllocoptes didelphis* Keifer (Beer, 1963) and the tomato russet mite (Bailey and Keifer, 1943).

Relationships between mite pests and their host plants

Plants are the habitat of herbivores, including phytophagous mites. The mites live and feed, and also interact with their natural enemies, on plants, the last being a major, interactive component of biological control strategies (Cortesero *et al.*, 2000). Jeppson *et al.* (1975) discussed the effects of host plants on phytophagous mites in terms of the relationship between the mite pest and the host plant, and also in relation to agricultural practices. Gerson *et al.* (2003) noted that both their hereditary characteristics and environmental factors affect plants, and consequently their pests and the pests' natural enemies, directly as well as indirectly. Several direct effects (plant facies (aspect or appearance) and growth pattern, plants as shelter, plant species and cultivars, leaf surface topography and glandular hairs, the presence of domatia, age of leaf plant, fertilizers, plant health, pesticides) are involved in the relationship of plants to mite pests, whereas relatively few indirect effects (agricultural chemicals and plant health, plant resistance or direct plant defence, indirect plant defence, etc.) are known to be involved. In order to simplify the discussion that follows, the two possibilities are integrated.

The main effects of host plants on their relationships with mite pests are 'plant resistance', 'host preference' and 'plant architecture' (plant facies and growth habit, leaf surface topography, glandular hairs, domatia, age of leaves). The relationships between mite pests and the host plant also involve the natural enemies of the pest, and these three trophic levels – the pest, its natural enemies and the host plant – are fundamental to the defence of plants. Plants also influence the interactions between the phytophages and their natural enemies (Price *et al.*, 1980; Sabelis *et al.*, 1999; Gerson, 2014), and an understanding and manipulation of plant attributes may enhance biological control (Cortesero *et al.*, 2000). Furthermore, knowledge of the 'synomones' (specific volatiles emitted by plant tissues damaged by pests and attracting predators) involved can help to optimize the process of plant defence.

Host plant resistance

Host plant resistance can have fundamental importance in the relationship between the mite pest and its host plant. This topic is also discussed further in Chapter 9 (Control). Genetic diversity is common in plants, and resistance to pests has been reported in many crops. For example, resistance to spider mites has been reported in many crops (almonds, avocado, barley, bean, cassava, castor bean, chrysanthemum, citrus, cotton, cucumber, aubergine, geranium, grapes, hops, maize, groundnuts, potatoes, roses, soybean, strawberry, sugarbeet, sunflower, tea, tobacco, tomatoes) (De Ponti, 1985; Harman *et al.*, 1996; Bynum *et al.*, 2004; Lopez *et al.*, 2005). From a practical point of view, this shows that it is possible to find sources of resistance in any crop. The strategy requires the identification and use of differences in resistance between existing cultivars, breeding for resistance and monitoring the decline of resistance in the field (De Ponti, 1982). Most of the available information is on spider mites, with much less on eriophyoid mites. Up until 1980, breeding for resistance to spider mites had taken place on only a few plant species, such as cassava, cotton, cucumber and strawberry (De Ponti, 1985). However, an absolute resistance to spider mites does not occur, and the evaluation of partial resistance needs long-term breeding programmes. Hence, the approach to control is via the integration of different control measures into an integrated pest management system (De Ponti, 1985). The discussion here covers the general features of the topic, focusing attention on 'tolerance' and 'resistance'. Further details can be found in De Ponti (1977, 1985) for spider mites, and in Westphal *et al.* (1996) for eriophyoid mites.

Like pest resistance, tolerance can be common (Gaunt, 1981) and consists of 'the ability of a host to endure exploitation by such an organism' (De Ponti, 1977). Environmental factors and yield potential of cultivars may affect their pest tolerance (Schuster *et al.*, 1980; Gaunt, 1981). Tolerance levels are evaluated from 'injury levels' or 'damage thresholds', with higher tolerance of a cultivar relating to a higher damage threshold. Tolerance and resistance can be differentiated only if pest (e.g. spider mite) populations and yields are simultaneously evaluated (Shanks and Barritt, 1975; De Ponti, 1978; Schuster *et al.*, 1980). Tolerance alone does not reduce mite infestations, but in combination with resistance may become important (De Ponti, 1985).

There are various different definitions of resistance. Painter (1968) stated that host plant resistance can be 'expressed by the relative damage to the host plant in comparison to a susceptible host plant'. De Ponti (1977) stated that 'resistance is the complex of characters of a host reducing the reproduction of a phytophagous organism'. Daly (1983) stated that resistance was 'an innate ability of a plant to prevent the establishment or to limit the subsequent activities of an invading organism'.

Painter (1951) described plant resistance as developing through one or more of three mechanisms: 'tolerance', 'non-preference or antixenosis' and 'antibiosis'. Tolerance is the capacity of the plant to resist pest attack, but does not affect plant density or yield quality and quantity. Non-preference is the capacity of the plant to be uninteresting to pests, thus discouraging their presence on a host plant, as occurs in several varieties of sorghum resistant to the Banks grass mite, *Oligonychus pratensis* (Banks) (Foster *et al.*, 1977a). Antibiosis implies an adverse effect of the plant on the pests, usually through repellency, a decrease in oviposition, aborted progeny or early death. This commonly involves the production of secondary substances by plants (alkaloids, cyanogenic compounds, glycosides, lectins, resins, tannins, terpenoids, repellents, deterrents and/or poisons). For example, in hop varieties, susceptibility to *T. urticae* has been related to chemical differences based on farnesol contents (Regev and Cone, 1975b), and these were also associated with the differential susceptibility of strawberry varieties to tetranychid mites (Regev, 1978), although Gunson and Hutchins (1982) found the opposite. Cucurbitacin C was found to be associated with the resistance of cucumber to *T. urticae* (Dacosta and Jones, 1971; Gould, 1978a). Males of the two-spotted spider mite are attracted to 10 ppm synthetic nerolidol but not to 100 ppm nerolidol or the tested rate of geraniol. Citronellol at 10 ppm is a strong attractive to males of *T. urticae* (Regev and Cone, 1980). Resistance to *T. urticae* in *Nicotiana* spp. seems to be due both to non-preference and antibiosis, probably resulting from alkaloids (Patterson *et al.*, 1974). Viscid secretions from trichomes, which are toxic and entrap mites, were also associated with resistance. Resistance to *T. urticae* in tomatoes is related to avoidance of and the toxicity of sesquiterpenoids (Patterson *et al.*, 1975). Tomato resistance to mites seems to be due to various factors, such as pubescence (a hair or downy covering), exudates and essential oils that repel and reduce the mite infestation. However, some tomato varieties exhibit low pubescence in association with a high tolerance (Rodriguez *et al.*, 1972; Van Haren *et al.*, 1987; Guo *et al.*, 1993b; Fernandez-Munoz *et al.*, 2000). Chemical analysis of strawberry leaves showed that the total phenols and amino acids were higher in 'Sweet Charlie' than in the 'Camarosa' variety, with the opposite for total sugar content; the increase in total phenol content suppresses infestation by the two-spotted spider mite, whereas the increase of total sugars stimulates its development. Moreover, the leaf trichomes of 'Sweet Charlie' have a higher density and are longer and sharper pointed than those of the 'Camarosa' variety, and the dense and long-haired leaves of 'Sweet Charlie' hinder the mite (Afifi *et al.*, 2010b).

Resistance commonly depends on multiple factors with a polygenic inheritance, which reduces the chance of increased fitness in a resistant genotype (De Ponti, 1985). Plant breeding combined with pest breeding can help to solve this problem (Gould, 1978b).

As well as the factors discussed above, direct plant defence may include mechanisms such as deficient diets and the presence of physical and chemical barriers (a thick epidermis, spines or glandular hairs, etc.) (Gerson *et al.*, 2003). In its effects on natural enemies, the relationship between plant resistance and biological control is controversial and may be either disruptive (Hare, 1992) or hindering (Moraes and McMurtry, 1987; Popov and Khudyakova, 1989). Furthermore, genetically modified (GM) crops do not seem to directly influence the attractiveness of host plants to natural enemies (Hoy *et al.*, 1998).

Among the tetranychid mites, several varietal responses are reported for different host plants, e.g. species of the genera *Cucumis* and *Citrullus* in the Cucurbitaceae (Tulisalo, 1972; Gould, 1978b, 1979; De Ponti and Garretsen, 1980) or tomatoes in the Solanaceae (Rodriguez *et al.*, 1972; Patterson *et al.*, 1975;

Van Haren *et al.*, 1987; Guo *et al.*, 1993b; Fernandez-Munoz *et al.*, 2000). On Algerian ivy and beans cultivated under similar environmental conditions, spider mites exhibited very different intrinsic rates of increase (r_m) (Gerson and Aronowitz, 1980). Different cultivars of strawberry affected mite success (Dabrowski *et al.*, 1971; Dabrowski and Bielak, 1978; Afifi *et al.*, 2010b). In New Guinea, species and hybrids of *Impatiens* showed resistance to the two-spotted spider mite (Al-Abbasi and Weigle, 1982). Cotton biotypes and crosses to them were resistant to spider mites (Leigh and Hyer, 1963), but the mites adapted easily to new hosts (Jesiotr and Suski, 1976; Jesiotr, 1980).

Plant resistance to eriophyoids includes a 'constitutive resistance' involving morphological and chemical features, and an 'induced resistance' (Westphal *et al.*, 1996). The resistance of *Ribes* to *Cecidophyopsis ribis* (Westwood) and that of some solanaceous plants to *Aceria cladophthirus* (Nalepa) are typical examples of the latter form of resistance.

Resistance can be induced by inoculating bacteria, fungi and viruses into the plant or by the feeding of phytophages on the host plants (Karban and Carey, 1984; Karban, 1986; Karban and Meyers, 1989; Karban and English-Loeb, 1990). Chemical and physical changes induced in plants by the feeding of herbivores may be associated with a lower attachment of pests. This phenomenon may involve phytoalexins, which hinder feeding by herbivores. Examples of mite feeding inducing resistance to mites in crops have been reported for various crops, e.g. cotton (Karban and Carey, 1984), and in vineyards (Karban and English-Loeb, 1990). This possibility could be exploited in the near future in IPM programmes. With this aim, English-Loeb *et al.* (1998) investigated genotypic variation in constitutive and induced resistance in grapes against the Willamette spider mite, *Eotetranychus willamettei* (McGregor), and the Pacific spider mite, *Tetranychus pacificus*. Early-season damage by Pacific mites induces resistance against subsequent Willamette mite populations, but early-season damage by Willamette mites did not induce resistance against subsequent Willamette mite populations. The induction occurs in several cultivars of the Old World species *Vitis vinifera* L., as well as the North American species *Vitis californica* Bentham. Phylogenetic relationships among grape genotypes give little help in understanding the variation that is found in induced resistance. In contrast, phylogenetic relatedness among grapes does help in understanding constitutive resistance to *T. pacificus*; cultivars of *V. vinifera* are inclined to be susceptible, whereas North American species are resistant. *E. willamettei* performed well on some Old World cultivars and two North American species of *Vitis* native to California. The negative correlation between constitutive resistance and the strength of induced resistance in these grape genotypes is unclear. Grape genotype, history of mite damage (induction), the mite species causing previous damage and phylogenetic relatedness among grapes all affect the variation in constitutive and induced resistance in grapes against these two spider mites. The mite genotype could play an important role (English-Loeb *et al.*, 1998). *Citrus* spp. contain toxic furanocoumarins and terpenoids, etc. Unlike other spider mites, which cannot develop on citrus plants, the citrus red mite, *Panonychus citri*, is able to develop on these host plants, and on other non-toxic hosts such as the Japanese pear (*Pyrus pyrifolia* (Burm.) Nak.). It was hypothesized that *P. citri* adapted to citrus plants by evolving a detoxification system against the allelochemicals contained in citrus leaves. The efficacy of this detoxification system was examined in relationship to *P. citri* survival on citrus plants using piperonyl butoxide, a well-known inhibitor of cytochrome P450s. The addition of critical concentrations of piperonyl butoxide has a lethal effect on *P. citri* larvae and adults reared on citrus plants, but does not affect larvae or adults reared on Japanese pear. This adaptation shows that *P. citri* adapted to citrus plants by evolving detoxification systems composed of cytochrome P450s and/or other detoxifying enzymes (Takeyama *et al.*, 2006).

Host plant resistance may be important for the control of the earth mites (Penthaleidae). Several screening programmes have been developed in Australia to identify resistant plant varieties, mainly against *Halotydeus destructor* (Ridsdill-Smith, 1997). Subclover varieties, legume species, annual medics (*Medicago* spp.) and canola accessions have been screened for damage resistance to feeding by this mite (Gillespie, 1993, 1994; Lake and Howie, 1995; Moritz and McDonald, 1995; Ridsdill-Smith *et al.*, 1995). Biochemical and physical mechanisms of resistance to *H. destructor* have been identified for subclover, which has antifeedant properties, with the basis of resistance being antixenosis (Jiang and Ridsdill-Smith, 1996). Increased strength of the cotyledon results in reduced damage because the mites cannot penetrate the epidermis of the resistant varieties. Levels of the volatile compound 1-octen-3-one have also been correlated with reduced mite feeding (Jiang *et al.*, 1996). It is unknown whether these mechanisms also reduce feeding damage by *Penthaleus* spp. With gene transfer technology now available, work has also started on the identification of plant proteins that may confer resistance to mite attack (Higgins and Ridsdill-Smith, 1994; Weinman *et al.*, 1995).

Despite the aspects reported above, breeding and selection methods are usually fundamental to the development of pest-resistant cultivars. Details of these can be found in various plant breeding textbooks (e.g. Russell, 1978; Simmonds, 1979) or in publications on the development of pest-resistant cultivars (De Ponti, 1982). Research carried out by Gould (1979) showed that resistance to host plant resistance is possible in spider mites, as a result of the extensive genetic variation that occurs in their populations, for instance in *T. urticae*. The phenomenon is more common when crops are cultivated over large geographical regions for many years; the continuity in time and space gives mites the opportunity to develop resistance (Hoy, 2011).

Host preference and host specificity

Ecological studies define host availability as a factor that influences the evolution of the use of the host resource (Strong *et al.*, 1984; Begon *et al.*, 1986; Bernays and Chapman, 1994). Plant hosts that are short lived, uncommon, unpredictable and difficult to locate induce herbivores to generalize; conversely, hosts that are long lived, common in occurrence, predictable and easy to locate induce herbivores to specialize (Jaenike, 1990). Host strain selection, and then host specialization, is more likely for pest mites living on perennials than for mites living on annual or herbaceous plants (Magalhães *et al.*, 2007). Thus, eriophyoids inhabiting short-lived host plants (e.g. annuals) must disperse more frequently than those on long-lived host plants.

On short-lived plants, a generalist strategy is more likely to evolve, with wind-dispersed species frequently showing a low host specificity (Nault and Styer, 1969). There is then a close relationship between pest mites (or other pests) and their host plants that is characterized by the adaptive conditions of 'host preference' and 'host specificity'. These two adaptations involve different factors, internal and external to the host plant and mites, and primarily aimed to optimize both exploitation of the trophic resource and reproduction, i.e. survival.

Host preference is based on the ability of a pest mite to utilize more host plants and plants belonging to different botanical families. Thus, the two-spotted spider mite has been recorded on about 1059 plant species (Migeon and Dorkeld, 2006–2013), both natural and cultivated, but in Mediterranean greenhouses it shows an obvious host preference for some crops, e.g. for aubergine rather courgette or tomato, and on citrus crops it prefers lemon or clementine rather than orange trees. The citrus red mite, *Panonychus citri*, has been recorded on 105 host plants (Migeon and Dorkeld, 2006–2013). It usually develops on various citrus crops, and on lemon it develops more or less constantly where there is new growth and new fruits (Jeppson *et al.*, 1975), but it can also infest strawberry, ornamentals, etc., without causing severe damage. The broad mite, *Polyphagotarsonemus latus*, is injurious to many plant species in about 57 botanical families (Gerson, 1992), both herbaceous plants and trees, including citrus, but in Mediterranean greenhouses it primarily infests pepper and the gerberas (Vacante, 2000; Vacante and Benuzzi, 2007). Knowledge of their host preference may help in the monitoring of pest mites in the field (although other pest mites may also be present).

Host specificity relates the pest mite to a single plant species or to a closely related host plant species, and occurs in many eriophyoid species. This adaptation closely relates mite pests and host plants, over a long time, and involves a co-evolutionary process. Because of this, the relationship between some mites, e.g. Eriophyoidea, and their host plants is not always characterized by any easily evaluated harm to the plant. Skoracka *et al.* (2010) analysed the different factors influencing host specialization and host specificity in the Eriophyoidea, and the role of variations in the patterns of association between the herbivores and their hosts. They examined the importance of mediation by plant chemical cues, plant chemical similarities and correlations between host phylogeny, chemistry and geography. Host availability and the dispersal capacities of pests also play an important role. The eriophyoids, for example, are almost totally dependent on wind currents for their dispersal (Nault and Styer, 1969), and the probability of finding a particular host during air dispersal is very low (Sabelis and Bruin, 1996); nevertheless, wind remains the dominant means of inter-plant dispersal for these mites (Zhao and Amrine, 1997b; Zhao, 2000). Deuterogyny seems to be fundamental for the species inhabiting long-lived, stable and predictable hosts (Lindquist and Oldfield, 1996), and most eriophyoids live on perennial plants (Oldfield, 1996a). Eriophyoids living on annual plants use adaptive strategies that allow them to survive over the year through the availability of host plants, various overwintering possibilities, etc. The degree of intimacy with the host environment can influence the pattern of host use by the parasite, and can be characterized as either a weak association of the parasite with its host or as a strong and intimate association (gall-inducing eriophyoid mites) (Magalhães *et al.*, 2007). Gall-inducing and refuge-seeking eriophyoids are physiologically more intimately related with their hosts than vagrants (Lindquist, 1996; Lindquist and Oldfield, 1996), and species living in shelters are less subject to predation (Sabelis, 1996). Populations of these mites are mainly limited by intraspecific and interspecific competition for food resources, which leads to specialization on host plants. Only a very few of the several thousand Eriophyoidea species described are known as generalists, and most of these few are vagrant species limited to either monocotyledonous or dicotyledonous hosts (Oldfield, 1996a). The preference to remain at the natal site (passive philopatry), and host-associated mating (which often arises as a by-product of philopatry), have been suggested as promoting the evolution of host specialization in the Acari (Magalhães *et al.*, 2007).

The food preferences in tarsonemid mites has been little investigated (Lindquist, 1986). *Tarsonemus* spp. showing a certain food preference are *T. granarius* Lindquist and *T. waitei*, both reported from stored grain (Sinha, 1964; Sinha and Wallace, 1966), and *T. myceliophagus* Hussey, reported from mushroom cultures (Hussey and Gurney, 1967), all probably feeding on *Alternaria* and *Penicillium*. The genus *Steneotarsonemus* includes various species showing food preferences or a food preference restricted to one host plant species or a species group. These include *S. ananas* (Tryon), living on pineapple, *S. bancrofti* (Michael) on sugarcane (*Saccharum*), *S. phyllophorus* (Ewing) on bamboo (*Phyllostachys*), and *S. hyaleos* Beer on saltmarsh grass. *S. spinki* is host specific to rice (Ho and Lo, 1979) and has only recently been found on other congeneric plants (Návia *et al.*, 2010b).

Regardless of the ecological implications, from a practical point of view, knowledge of host specificity helps weed control.

Plant architecture

The relationships between pest mites and host plants are also mediated by plant architecture, including plant facies and growth habit, e.g. the relationships between the bud size in lemon and attack of the citrus bud mite, *Aceria sheldoni*, and the presence or absence of basal leaf glands on peach and attacks by the plum rust mite, *Aculus fockeui* (Nalepa) (Jeppson *et al.*, 1975). Usually, the distribution of natural enemies seems to be habitat specific rather than plant specific. However, in several cases, facies and plant growth patterns may affect the natural enemies, as for *Phytoseiulus persimilis*. In this species, variations in plant structures, such as surface area, vestiture and the numbers, kinds and proportions of plant parts, affected its foraging on cucumber (Gontijo *et al.*, 2012). Schmidt (2014) reviewed the interaction between the leaf surfaces of crops and the natural enemies, with special regard to the influence on foraging. Further details on this subject may be found in Gerson (2014).

Other factors also affect the distribution on plants of phytophagous mites, such as the plant's branching pattern, its leaf morphology, leaf age, etc. The main leaf features affecting pest populations are veins, non-glandular hairs, glandular trichomes and domatia. Many pest mites live on the undersides of leaves, where they are protected from sunlight and air currents, which

prevents the drying of their bodies, and their distribution on the leaf undersides depends on the leaf morphology. Thus, leaf surface structure affects mite distribution, with hairy leaves hosting higher densities of mites than do the smooth leaves of co-occurring host plants (Walter, 1992). Hairs along the veins offer shelter for mites. Several plants bear tiny leaf axillary invaginations, often associated with small tufts of hairs. These axillary pits are usually referred to as 'acarodomatia'. Plant domatia host a large number of mites that are characterized by predatory, fungivorous and microbivorous feeding habits; furthermore, plants with leaf domatia host more predators, usually belonging to the families Phytoseiidae, Cheyletidae, Cunaxidae and Stigmaeidae (O'Dowd and Willson, 1989, 1997; Walter, 1996; Walter and Proctor, 1999). The phytoseiids *Amblyseius herbicolus* Chant and *Iphiseiodes zuluagai* Denmark *et* Muma live on coffee and feed on each other's larvae. Domatia on detached coffee leaves were either glued closed or left open, and larvae and adults of both predators were released. The survival of larvae of both species was significantly higher on leaves with open domatia than on leaves with closed domatia (Ferreira *et al.*, 2011). In addition, *I. zuluagai* preferred to place its eggs inside domatia. The ecological role of these structures in the defence of plants against the attack of phytophages has been examined by various researchers (Walter and O'Dowd, 1992; Grostal and O'Dowd, 1994; Agrawal and Karban, 1997). The age of the leaf (or plant) affects the mite pest–host plant relationship. Young leaves have high nutrient quality (Slansky and Scriber, 1985) and possess few of the protective secondary chemicals that are usually metabolized later. The age of a citrus leaf and the amount of injury caused by mite-feeding activity influence the populations of citrus red mite, *P. citri*. The number of eggs laid on young and medium-aged leaves is greater than on old leaves, but feeding injury also influences oviposition (Henderson and Holloway, 1942) and cannot represent a regulatory factor in the density of populations.

Synomones

These are specific volatiles emitted by plant tissues damaged by pests and attracting predators. The topic has been discussed above in the section on 'Chemical communication'.

Agricultural practices

Even though mites injurious to economic plants have been known for a long time and are widespread, they came to represent a severe problem in agricultural production only after World War II. A classic example is some of the *Panonychus* and *Tetranychus* spp., which have become very common, while other species of other genera are either less or not harmful. This phenomenon depends mainly on the changes that have occurred in agricultural practices (e.g. the use of new varieties and bud selection, increased plant vigour, monoculture, and the use of pesticides, natural enemies, inert dust and soil organisms) (Jeppson *et al.*, 1975). Overall, such practices can lead to populations of the pest mites developing to above the environmental carrying capacity (K) of the agroecosystems. Other more extensive approaches include the manipulation of ground cover, weeds and nearby plants (Gerson *et al.*, 2003; Gerson, 2014).

New varieties and bud selection

Reduction in the genetic diversity of cultivated plants plays an important role in the population dynamics of pests, in consequence of a decrease in the genotypic variation of the resistance and the availability of a greater amount of resources for the strains of harmful organisms adapted to a particular genotype. The characters that have been selected for may be attractive to pests (Speight *et al.*, 1999). As noted above, plant species and cultivar may influence mite development, e.g. the citrus red mite develops better on lemon than on orange (Fleschner, 1958a). Jeppson *et al.* (1975) reported that in citrus, new varieties and selections were replacing old bud plant stocks, but there was very little information on the relationship of these changes to mite infestations. In nurseries, the citrus red mite attacks Troyer citrus seedlings before other citrus varieties. In Israel, the European red mite develops on the apple cultivar 'Oregon Spur' better than on 'Delicious' (Palevsky *et al.*, 1996). In addition, plant species and cultivar influence the spectrum of synomones produced by spider mite-injured leaves that attract predators to their colonies (Dicke *et al.*, 1990a; Takabayashi *et al.*, 1991). The large number of references on the influence of new varieties of plants and bud selection on pest mites does not allow an overview of the topic to be presented here, although many examples could be given. The influence of citrus rootstocks on pest mites is little known. Recently, Silva *et al.* (2015) evaluated the influence of rootstocks on the population densities of the citrus rust mite *Phyllocoptruta oleivora* (Ashmead) and the spider mite *Tetranychus mexicanus* (McGregor) on the sweet oranges 'Pera CNPMF D-6' and 'Valencia Tuxpan'. Counting the adults of both mites monthly from June 2011 to February 2013 showed that the rootstocks influence the density of population of both mites. Lower density of population of citrus rust mite were observed on 'Pera CNPMF D-6' sweet orange grafted on the hybrid TSKC × CTTR - 002 and on 'Swingle' citrumelo in comparison with the hybrid LVK × LCR - 010, 'Red' rough lime and 'Santa Cruz' rangpur lime as rootstocks. Likewise, lower density of populations of *T. mexicanus* were observed on 'Valencia Tuxpan' sweet orange grafted on the hybrid HTR-051 in comparison to 'Indio' citrandarin, 'Sunki Tropical' mandarin and LVK × LCR - 010 as rootstocks. Density of populations of *P. oleivora* and *T. mexicanus* on the sweet oranges 'Pera CNPMF D-6' and 'Valencia Tuxpan' are influenced in some periods of the year by some rootstocks, highlighting the influence of some genotypes on both pest mites.

Plant vigour

Plant vigour is directly related to the optimization of agricultural practices (fertilization, pruning, weed control, etc.). In the second half of the last century, this became very important and was extensively investigated in different cultures (Post, 1962). The inputs derived from fertilizing and other agronomic management techniques influence the 'net primary productivity' of crops and affect the flow of matter and energy through the biotic communities that constitute them. The rates of net primary

productivity passing through every possible energy pathway are governed by the use and transfer efficiencies from one level to another (Begon *et al.*, 1986). The 'production efficiency', which corresponds to the percentage of assimilated energy that is incorporated into new biomass, demonstrates very well the influence of primary productivity on the structure of ecosystems. Invertebrates have a higher production efficiency than vertebrates, with 40.7% efficiency observed in social insects, of which 38.8% is in pests, and 47.0% and 55.6% in predators and parasitoids (detritivores and carnivores), respectively (Humphreys, 1979). The primary productivity of crops increases with decreasing latitude, as a result of increased light intensity, increased average temperatures and increased duration of the season of growth (Cooper, 1975).

The optimal conditions for each crop depend on its area of distribution, which can affect crop primary productivity and consequently the populations of pests. In general, the number of pest species increases with increase in productivity (Brown and Davidson, 1977). However, primary productivity may also be directed towards an increase in the growth rates of animal populations (Huston, 1979), leading to the potential limitation of competitive exclusion. This phenomenon has been reported in various agroecosystems (citrus, apple, grapes, etc.), where infestations of mites and other pests (aphids, aleyrodids, etc.) coincide with the appearance of new vegetation and a greater supply of sap and/or as a result of the application of potassium and phosphorus/potassium fertilizers (McNeill and Southwood, 1978; Onillon *et al.*, 1986). In conclusion, the exploitation of the resource may on the one hand stimulate the growth rates of pest populations and on the other hand regulate the numbers of their species. This occurs in many plants of economic importance, where at certain times of the year various pests (including mites) reach densities that require control.

Soil conditions and structure, and cultivation practices, all affect the amounts of nutrients carried from the roots to the leaves. In turn, nutrient availability influences both plant physiology and the nutrition of pest mites. The osmotic pressure of plants can be increased by two or three times as a result of high fertilizer rates, and excess soluble elements (silicon, magnesium, calcium) in plants facilitate the development of spider mite populations. The relationships between plant mineral nutrition and the status of the two-spotted spider mite and the European red mite have been investigated by various researchers. The first research on the effects of plant mineral nutrition on mites was carried out by Rodriguez (1951, 1958, 1960), who also studied the relationship between minerals and B-complex vitamins and mite populations in tomato leaves (Rodriguez and Rodriguez, 1952). Leroux (1954) studied the influence of various levels of nitrogen, phosphorus and potassium in the nutrient solution on the fecundity of the two-spotted spider mite on cucumber. The effects of increasing the nitrogen levels of apple leaves on mite and predator populations were examined by Lord and Stewart (1961). Morris (1961) analysed the development of the clover mite, *Bryobia praetiosa*, in relation to the nitrogen (N), phosphorus (P) and potassium (K) nutrition of its host plant. The effects of cultural measures on European red mite development were examined by Post (1962), and Storms (1969) investigated the relationships between mineral nutrition and the reproduction of *T. urticae* on apple rootstocks in gravel culture. Huffaker *et al.* (1969) show that plant mineral nutrition affects the ecology of tetranychid mites, and that besides N, K, P, Ca (calcium), Mg (magnesium), Zn (zinc) and Co (cobalt) are also important in mite nutrition. In particular, the first research carried out on orchards pointed out that high nitrogen levels induced shorter developmental time, fecundity and longevity of the European red mite (Breukel and Post, 1959). A positive relationship between nitrogen, phosphorus and potassium levels in the nutritional solution and leaves of grain sorghum in the greenhouse and the development of *T. urticae* was found by Flechtmann and Malavolta (1976). In the absence of nitrogen or phosphorus, only small populations of mites developed, and in the absence of potassium, no mites were observed. Moreover, minimum levels of 2.10% N, 0.18% P and 1.20% K in the leaves were required to allow a large mite population to develop.

Schwartz (1984) found a direct relationship between the density of populations of the citrus red mite and the nitrogen nutrition of the citrus plants. Glutamine and glutamic acid in the leaves could facilitate mite reproduction, but in particular environments, the soil nitrogen and the leaf nitrogen may not be strictly correlated and the different mite species can show different reactions to nutrients. Thus, *Panonychus ulmi* and *T. urticae* can react differently to the amounts of nutrients. Reproduction of the European red mite increases with higher nitrogen levels, while that of the two-spotted spider mite increases at low nitrogen levels; however, the level of nitrogen also been observed to have no influence (Hukusima, 1958). In the last few decades, the positive correlation between high soil nitrogen concentrations and the density of *Tetranychus* spp. has been well documented (for example by van de Vrie *et al.*, 1972; Suski and Badowska, 1975; Wermelinger *et al.*, 1985; Wilson *et al.*, 1988; Wood and Reilly, 2000). The main results from some of these studies are reported below.

Wilson (1994) tested the fitness of cotton for the development and reproduction of the two-spotted spider mite, and found that the mite development time and fecundity increased in relation to nitrogen content. Mites prefer to feed on cotyledons more than older leaves, and as plants develop and the leaves grow old, the nitrogen content decreases. Chow *et al.* (2009) studied the influence on spider mite population density of nitrogen levels lowered from the recommended cut flower standards. They found that the mean numbers of *T. urticae* instars and eggs per flower shoot were twice as high on roses fertilized with 100% vs 33% or 50% N; fertilization using 33% N and 100% N affected total yield, which was not affected on roses fertilized with 50% N.

Wermelinger *et al.* (1985) evaluated the development rate, fecundity, pre-oviposition period, oviposition rate and survivorship of *T. urticae* in relation to different levels of nitrogen concentration on apple trees and bush beans. Female weight, fecundity and oviposition rates decreased in nitrogen-deficient treatments and were positively correlated with the N content of the leaves. Weight, oviposition rate and fecundity of the mites were higher on beans than on apples. Wermelinger and Delucchi (1990) studied the two-spotted spider mite on apple trees fertilized with three different nitrogen levels (plain water, 224 ppm and 448 ppm). All received 1 mM phosphorus (14 ppm) and 6 mM potassium (84 ppm) after 2 months. A slight increase was

found in the sex ratio (males:females) as well as female fecundity in 'high N' treatments. When the nitrogen levels were in the range 1.8–3.0% N, the sex ratio increased from 0.64 to 0.76. In the first 7 days, mites from the 'high N' treatment had a 23% higher sex ratio than those from 'low N' treatments. A relationship between increasing fecundity and higher sex ratio was also observed in the highest N treatment. The content of phenolic compounds in the leaves of micropropagated apple trees increases with N and P deficiency. Juvenile development and oviposition of *T. urticae* are both positively correlated with leaf N; female fecundity is positively correlated with N and carbohydrate contents of leaves, and negatively correlated with the phenolic content. The longevity of the two-spotted spider mite is not affected by different levels of N, P and K, although minor differences have been observed with K. The intrinsic rate of increase (r_m) declines gradually with N and P deficiency, and r_m is clearly correlated with leaf N and especially with phenolic content. Nitrogen shortage plays a key role in the population growth, significantly affecting net reproduction (R_0) and the generation time (T). So, between 1.5 and 3% leaf N, r_m increases by a factor of four and R_0 by 11. Moreover, a strong deficiency of N in leaves prolongs the doubling time of the population fourfold (Wermelinger *et al.*, 1991).

Chen *et al.* (2007b) studied ivy geranium as host to the two-spotted spider mite in response to six different combinations of nitrogen (28, 112, 224, 336 and 448 ppm) and phosphorus (1.12, 4.48, 8.96, 17.92 and 35.84 ppm). Initially, two rates of N (112 or 336 ppm) and three of P (4.48, 8.96 and 17.92 ppm) were applied. No difference in the number of mites per host plant or the amount of plant injury was found between the two different N fertilization rates. The rates of P application did not affect the mite population until week 8. From week 8, plants fertilized with 4.48 ppm P were of lower quality (not saleable) than plants fertilized with 8.96 and 17.92 ppm. Higher rates of P application therefore allow more compensation for mite-feeding injury on ivy geranium and can have a positive effect on infestation by tetranychid mites within a certain concentration range; it is possible that N or K could also have such an effect on further exploration.

Wood and Reilly (2000) assessed the role of cultural practices (leaf nitrogen concentration, leaf water status, crop load) in pecan orchard trees (*Carya illinoinensis* (Wangenh.) K. Koch), in order to evaluate damage by the pecan leaf scorch mite, *Eotetranychus hicoriae* (McGregor). They found that pecan leaf scorch mite damage increase as leaf N concentration increase as well as when water availability increase, causing reduction in pecan leaflets. The pecan leaf scorch mite causes the greatest foliar damage when high N, high water status and a heavy crop load are combined. Mature trees also exhibit greater leaf damage with high N concentrations. Cultural practices can affect crop outcome both positively and negatively and should be considered in an integrated pest management system.

Wilson *et al.* (1988) studied the Pacific spider mite using four different nitrogen rates applied to potted 'Thompson Seedless' grape plants in the form of ammonium sulfate at rates of 0.0, 2.0, 4.5 and 9.0 g per pot. The results showed that between the lower, middle and upper leaf positions, the upper leaf position contained the highest amount of nitrogen and the lower position the least. The immature development time of the Pacific spider mite decreased as the leaf nitrogen increased, whereas adult fecundity increased. Adult longevity, immature survivorship and the duration of oviposition were not affected by leaf nitrogen content.

English-Loeb (1990) documented the fertility of the two-spotted spider mite when feeding on bush beans treated with nitrogen concentrations ranging from 0 to 157.5 ppm, in three experiments that were assigned different N ranges; the control treatment was 105.0 ppm N. Plants assigned to the different ranges of nitrogen received 100 ml of water every 6 days, and the majority of all rates were below standard and the plants under nitrogen stress. Changes in plant biomass were used as a continuous measure of N stress. Fecundity was greatest at intermediate nitrogen stress levels in two of the experiments, but declined with increasing stress in the third. Root nodulation increased with decreasing N concentration, with the initial development of nodules coinciding with a peak in fecundity. Overall then, there was a non-linear response, with the intermediate N treatments performing better than the high and low N treatments.

Potassium can help plants to resist disease, regulates over 60 enzyme systems in the plant and aids in overall plant vigour. Applied in adequate amounts, it promotes the growth of plant fibres, creating long and strong shoots. Damage caused by Willamette spider mites causes a reduction in stomatal conductance over time, and potassium could then favour the mites and their survival by providing a disease-free, drought-resistant plant. Feeding by the mites presumably does not affect plant health as quickly as it would without K because of the plant's ability to resist stress. Mite populations are therefore favoured by this macronutrient (Geddes, 2010).

Other studies have shown that nitrogen and phosphorus at low and moderate levels induce spider mite populations to increase, whereas high concentrations cause decreases (Rodriguez and Rodriguez, 1987; Walde, 1995). Recently, Sharma and Bhardwaj (2010) examined the effect on *P. ulmi* populations infesting the apple cultivar 'Delicious' of three levels of N, P and K, each at the recommended dose (N1, P1, K1), 25% above this (N2, P2, K2) and 25% below it (N3, P3, K3). The lowest mite population was found at N3P2K2 (i.e. 25% less than recommended N and 25% more than recommended P and K). A slight manipulation in the recommended doses of NPK fertilizers therefore helps to control the development of the *P. ulmi* population on apple trees. In Brazil, Moreira *et al.* (1999) evaluated the effects of different levels of NPK on the infestation of *Aculops lycopersici* in tomato plants; four levels of N (0, 60, 120 and 180 kg ha^{-1}), three levels of P (0, 80 and 160 kg ha^{-1}) and two levels of K (0 and 120 kg ha^{-1}) were used. Fertilization with 120 kg ha^{-1} of K induced higher infestation. Mite infestation and levels of P applied to the soil showed a direct relationship. In the high K treatment (120 kg ha^{-1}), the infestation increased with P levels. At lower levels of N (60 kg ha^{-1}) and higher levels of N (180 kg ha^{-1}), the infestation increased, whereas at 120 kg ha^{-1} N, it was reduced. However, the NP interaction showed a low effect on the infestation at higher levels of N. The levels of N and P applied to the soil were positively and significantly correlated with the content found in the tomato leaves. No significant correlation was observed for K.

Plant water stress has long been related to population outbreaks of agricultural pests. Research on this topic has failed to

highlight a positive relationship between plant water stress and pest outbreaks. Many studies have examined the impact of plant water stress on the nutritive value of plants, but only a few of them have investigated how plant water stress can alter the microenvironment experienced by spider mites, which usually live within the thin layer of air immediately surrounding the leaf surface (Stavrinides *et al.*, 2012). It has been shown that water stress enhances spider mite success in strawberry (Dabrowski *et al.*, 1971) and peppermint (Hollingsworth and Berry, 1982) plants. Mite feeding damages the leaf epidermis and the cuticle, and the consequent water stress causes an accumulation of carbohydrates that positively affects the development of spider mites (De Angelis *et al.*, 1982, 1983a,b). Stressed and well-irrigated bean plants develop higher populations of two-spotted spider mites than do moderately stressed plants (English-Loeb, 1989, 1990). Recently, Stavrinides *et al.* (2012) investigated the relationships between plant water stress and leaf temperature, and between leaf temperature and spider mites and predatory mites in California vineyards. It was observed that leaf temperature increased by 0.8°C for every 1.0°C increase in ambient air temperature, and by 5.3°C for every 1 MPa drop in leaf water potential. Peak population densities of the Pacific spider mite increased with increasing frequency of leaf temperatures above 31°C. Conversely, peak population densities of the Willamette spider mite showed no relationship with the frequency of leaf temperatures above 31°C. The relationship between population densities of the two spider mites and leaf temperature is consistent with their upper thresholds for development, which are 40°C for the Pacific spider mite and 31°C for the Willamette spider mite. The population density of phytoseiid mites decreased with increasing frequency of leaf temperatures above 31°C, and showed no association with the population density of either spider mite species.

Finally, plants may affect natural enemies through their symbiotic associations with arbuscular mycorrhizal fungi and/or rhizobacteria. The growth rate of *Phytoseiulus persimilis* was enhanced when feeding on *T. urticae* feeding on arbuscular mycorrhizal bean plants (Hoffmann *et al.*, 2011). Moreover, mycorrhizal bean plants attacked by the two-spotted spider mite attracted more predators than non-mycorrhizal plants (Schausberger *et al.*, 2012). However, this behaviour was observed on plants grown in pots, and its validity in the field must be confirmed. Greater numbers of *P. persimilis* occurred on cucumbers treated with plant growth promoting rhizobacteria than on untreated plants, and more prey (*T. urticae*) were located thereon by the predator (Tomczyk and Burda, 2005).

Monoculture

The variation in 'quality' of plants in space and time results in a mosaic of the spatial–temporal population dynamics of pests. Commonly, the type of crop influences pest population dynamics. In monocultures, pest populations usually have higher birth and immigration rates and lower death and emigration rates; they also exhibit a limited tendency to spread and migrate, but are favoured in finding the host. The search for a food resource normally results in energy savings, which are profitably employed in the reproductive process, leading to a low probability of failure of couplings and higher birth rates. In short, monocultures offer real opportunities for colonization by pests and the stimulation of increased rates of immigration (Root, 1973; Power, 2001; Hunter, 2002b). Unfortunately, monocultures also provide natural enemies with a smaller amount of resources (alternative prey or hosts, nectar, pollen, shelters, etc.) (Root, 1973), and in this regard the biodiversity planning of crops is not always easy. Examples of the involvement of monocultures in the development of mite populations usually occur in agricultural practice, and protected crops probably represent the most extreme case.

Ground cover, weeds and nearby plants

Secondary plants, different plants or a secondary crop, grown along with the primary crop, may affect pests and their natural enemies (Parolin *et al.*, 2012). Many phytoseiids (and other natural enemies) occur on wild and cultivated plants near or within crops, and may serve as sources of predators (Mailloux *et al.*, 2010). This idea was developed by using banker plants, which are intended to provide and sustain a reproducing and releasing system of natural enemies within a crop, thus providing long-term pest suppression (Huang *et al.* 2011; Parolin *et al.*, 2013). Moreover, ground cover plants (adventitious or planted) between rows of cultivated plants allow a lowering of ground temperature and increased humidity. In citrus crops in China, the billygoat-weed (*Ageratum conyzoides* L.) modified the climate within orchards, reducing summer temperatures by about 5°C and RH by 5% (Huang *et al.*, 1983). In South-east Asia and south China, the weed often invades the crops, resulting in reduced yields. However, intercropping this weed in citrus orchards may increase the population of the phytoseiid mite *Amblyseius newsami* (Evans), a natural enemy of the citrus red mite, *Panonychus citri*, and be able to keep the population of *P. citri* at low levels. In fact, *A. conyzoides* releases volatile allelochemicals into the air in the intercropped citrus orchard, which influence the olfactory responses of *A. newsami* and *P. citri*. It has been observed that at 25°C, fresh leaves of *A. conyzoides*, its essential oil, and the major constituents of this (demethoxy-ageratochromene, β-caryophyllene, α-bisabolene, and *E*-β-farnesene) attract *A. newsami* and slightly repel *P. citri*. Spray application of *A. conyzoides* essential oil emulsion in a citrus orchard not intercropped with *A. conyzoides* increased the population density of *A. newsami* per leaf so that it reached the same level as in a citrus orchard intercropped with *A. conyzoides*. Unfortunately, this effect could not be maintained beyond 48 h because of the volatility of the essential oil. Conversely, in the citrus orchard intercropped with *A. conyzoides*, the *A. conyzoides* plants continuously emit volatile allelochemicals and maintain the population of *A. newsami* for a long time (Kong *et al.*, 2005).

In temperate regions, this technique may offer overwintering shelters for pests and natural enemies, and facilitate pest development in the spring. It has been observed that high infestations of *T. urticae* develop more easily on pears (Gotoh, 1997) and peaches (Meagher and Meyer, 1990) in orchards with ground cover than in crops where the weeds have been chemically treated. In the field or in protected crops, the two-spotted

spider mite usually also lives on weeds, and this adaptation could be used for the monitoring and release of phytophage predators (Dubitzki, 1981, in Gerson *et al.*, 2003). Research in apple orchards (Croft and McGroarty, 1977; Alston, 1994), on citrus (Huang *et al.*, 1983) and in a vinyl house vinery (Takahashi *et al.*, 1998) indicate that ground cover plants provide prey, shelters and alternative foods for natural enemies, thereby increasing pest control. Other research in apple orchards has been less encouraging (Nyrop *et al.*, 1994) and pointed to a greater complexity of the situation.

The use of sown weed strips allows the making of 'ecological compensation areas' between agroecosystems that occupy up to 5–10% of the agricultural surface. They offer overwinter shelters and are dispersal sites for natural enemies, ensuring an increase in biodiversity. According to the environmental conditions, it is possible to employ various mixtures of annual and perennial weeds (Nentwig *et al.*, 1998).

Commercial, uncultivated or weedy vegetation surrounding cultivated crops may be a source of pests, which more or less regularly invade cultivated crops, e.g. windbreaks may provide different species of pests and predators (Solomon, 1981). Usually, species of natural enemies are related to the trees of their origin. The importance of the vegetation surrounding agrosystems for the predatory mites (Phytoseiidae) that are associated with vineyards is well documented (Kreiter *et al.*, 2006).

The effects of windbreaks of arboreal legumes on the mite population has been evaluated in Brazil in the control of *Oligonychus ilicis* (McGregor) on coffee. The windbreak consisted of parallel lines of shrubby/arboreal pigeon pea (*Cajanus cajan* Millsp.), bracatinga (*Mimosa scrabrella* Benth.), leucaena (*Leucaena leucocephala* (Lam.) De Wit) and acacia (*Acacia mangium* Willd.). However, these did not affect the numbers of pest mites or of predatory mites, especially of the phytoseiid mite *Euseius citrifolius* Denmark *et* Muma, on coffee plants (Pedro Neto *et al.*, 2010).

Pesticides

Chemicals employed in agroecosystems may cause ecological disturbance of varying intensity, depending on the crop, the chemical nature of the pesticide(s) and the conditions of its use. Some chemicals create easily reversible effects and their impact on crops may be unnoticed thanks to the resistance and resilience of the agroecosystems involved. Among chemicals, pesticides may cause severe ecological disturbance and encourage the development of pest mite infestations through different mechanisms. Wrensch (1985) notes that every chemical manipulation made in favour of agricultural production (water, fertilizers, pesticides, etc.) could favour mite reproduction.

Historically, the impact of pesticides on crops was first feared by Gilliat (1935). Successively, the suppression of natural enemies and the outbreaks of pest mites following the spread of the use of chemicals on crops were reported by various researchers (Cutright, 1944; Lord, 1949; Clancy and Pollard, 1952). Ripper (1956) examined the effects of pesticides on the important families of predatory mites, and Eichmeier and Guyer (1960) investigated the role of gibberellins on mite development. The first extensive review on the effects of pesticides on natural enemies (mainly phytoseiid mites) was from McMurtry *et al.* (1970). At the same time, Huffaker *et al.* (1970) pointed out that organic pesticides destroy natural enemies, thus freeing spider mites from their control. Outbreaks of pest mites may be related to the use of chemicals (pesticides and fertilizers), the destruction of natural enemies and the escape of pest mites from predators in time and space (Huffaker *et al.*, 1969, 1970).

Pesticides may affect the population of phytophagous mites by directly stimulating their reproductive rate, as in the case of the old DDT (an organochlorine) (Klostermeyer and Rasmussen, 1953; Balevski, 1960; Post, 1962; Huffaker and Flaherty, 1966). This phenomenon is called 'hormoligosis' and occurs when living organisms are treated with low concentrations of toxic materials. It identifies the response of the organism to low quantities of stressors, e.g. chemicals or radiation (Luckey, 1968). From a physiological point of view, hormoligosis is related to the activity of detoxifying enzymes (against chemicals) or DNA repair mechanisms (against radiation). This induced physiological activity allows the organism to deal with higher quantities of chemicals later. Hormoligosis has been observed in the two-spotted spider mite and the phytoseiid *Amblyseius victoriensis* (Womersley) treated with imidacloprid; they both respond by increasing their egg production (James, 1997; James and Price, 2002). Sprayings of pyrethroids (Gerson and Cohen, 1989) suppress the populations of various predators, or induce physiological effects in the plants or pests that favour mite development. Easterbrook (1996) reports that the inability to use effective acaricidal fungicides and the large use of pesticides such as pyrethroids have a detrimental effect on predatory mites. It is possible that the effects of chemical sprays on mite populations result from a combination of factors (McMurtry *et al.*, 1970; Jeppson *et al.*, 1975). Pesticides could indirectly affect phytophagous mites and their natural enemies by influencing plant growth. Their action can induce an increase in or the reduction of photosynthesis and yield (Jones *et al.*, 1986), hinder or promote vegetative growth (Lloyd and Krieg, 1987) and influence the metabolism of secondary chemicals (Lydon and Duke, 1989). Regardless of the factors involved, the phenomenon has long been known to be the major contributor to the general increase of spider mites on fruit trees, cotton and other plants that are usually treated with pesticides (Clancy and Pollard, 1952; Klostermeyer and Rasmussen, 1953; Tew and Groves, 1957; Lochar, 1958). Another factor that has been associated with outbreaks of pest mites, e.g. the Pacificus spider mite, is sulfur (Smith, 1950; English-Loeb *et al.*, 1986; Hanna *et al.*, 1997; James *et al.*, 2002).

Natural enemies

Natural enemies (mainly predators) are fundamental to the biological control of pest mites on crops. In addition to the facts reported above, natural enemies may be absent in a crop grown in new areas or may be unable to survive the new habitat. This calls for research on the natural enemies of phytophagous mites in areas of the native host plants or in areas where the natural enemies of the pests are active. Specific programmes on the introduction and acclimatization of new natural enemies are being conducted worldwide in different crops. The topic is

divided into phases of introducing, culturing and establishing these biological agents, and has been extensively treated by Gerson *et al.* (2003).

Inert dusts

Inert dusts can affect the development of tetranychid populations mainly in crops grown near dusty roads. Bartlett (1951) observed the influence of dusts on hymenopteran parasitoids, and Fleschner (1958c) observed the effects on the biological control of avocado pests. Dusts help in the rearing of mites on citrus in the laboratory, and it seems that they have different functions, such as providing bases for the development of webs on a smooth leaf surface and/or being lethal towards predators. They may increase the permeability of the cuticle of predatory mites, inducing water evaporation and desiccation, which causes death; they may also hinder predators in searching for or in finding prey (Jeppson *et al.*, 1975). In addition to dust, the influence is known of poor soil conditions such as compaction, extremes of pH and shallow soil on mite populations and their predators (Flaherty *et al.*, 1992).

Soil organisms

Soil fumigation (or other treatments) can induce changes in the soil fauna and flora, thereby affecting plant growth and vigour, and consequently the development of pest mite populations. It seems that various soil organisms, such as *Streptomyces griseus* (Kyanshy) Waksman *et* Henrici, may negatively influence the development of tetranychid mites. There is a need to better understand the relationship between toxins and/or antibiotics produced by these organisms and the development of mite pests in orchards (Harries, 1963) and citrus crops (Jeppson *et al.*, 1966).

6 Collecting and Detecting, Preserving, Preparing and Rearing

The study and examination of mites, both live and dead, is carried out for identification purposes and/or to gain further knowledge of their bio-ecology. To accomplish these tasks, protocols for collecting, preserving, preparing and rearing the different mite categories and groups of mites have been developed over time. According to their bio-ecology, the study and collection of mites living on plants of economic importance involves the host plant, the soil, supports of a different nature around the plant, and sometimes the three habitats together. Different analytical tools, such as light microscopy (using the dissecting microscope, phase contrast microscope or Nomarski interference microscope), electron microscopy (using the scanning electron microscope and the transmission electron microscope) and physico-chemical and biochemical methods (electrophoretic analysis, molecular methods, etc.) can be employed. The aspects of mite collection treated in this chapter do not include sampling techniques and methods for estimating populations, which are covered in Chapter 9, on mite control.

Collecting and Detecting

Collection methods vary depending on the behaviour of species and the needs of the collector (laboratory rearing, systematic study, bio-ecological observations, etc.). For identification purposes, it is desirable to collect the largest possible number of specimens, preferably from all biological stages, noting their colour, the damage symptoms caused and the host plant, as well as the date, location of collection and name of the collector. This chapter covers the techniques that are used for the collection of mites injurious to economic plants, in particular arboreal and terrestrial species. For beneficial mites, information can be found in Gerson *et al.* (2003), which is also applicable to pest species.

Collecting and detecting from plants

Various techniques allow the collection and detection of immature and adult stages of mites on plants (Jeppson *et al.*, 1975; Sabelis, 1985b; Upton, 1991; Amrine and Manson, 1996; Perring *et al.*, 1996). The collection of arboreal mites can usually be achieved using both the hand collection and mass collection techniques described below.

Hand collection

Hand collection is carried out using several methods. Many phytophagous or predaceous mites living in the canopy can be collected by shaking the foliage with a stick or by hand on to a white or black support (tray, paper, etc.), and then using a needle or brush to collect them. The material that is manually collected from the plant is generally kept in sealed plastic or paper bags, which are stored in the refrigerator at 10–15°C and examined within a short time in the laboratory through a dissecting microscope. This direct examination avoids the deterioration or death of the specimens that have been collected as a result of curing or predation. The mites can be kept alive and used in the laboratory for biological studies and various other kinds of analyses (morphological, biochemical, etc.).

Field collection of mites from different organs of plants (leaves, branches, fruits, portions of bark) can also be carried out using pins mounted on sticks of various sizes, pipettes or fine hair brushes and a hand lens. Mites collected in this way can either be kept alive for a short time in empty vials or transferred to vials containing a preservative liquid such as 70% ethanol or Oudeman's fluid. This method is not suitable for all systematic groups and is normally used for species that are not very mobile and of middle or large size. Mushrooms infested by mites, e.g. Histiostomatidae, may be arranged on moist paper in containers at room temperature and examined periodically as the fungi decay (Walter *et al.*, 1998), using a fine brush to collect the specimens from the paper. Leaves and other vegetative parts can be immersed in 70% alcohol for 2–3 days; the alcohol can then be passed through a sieve equipped with white or black filter paper to enable the mites to be seen.

The collection of eriophyoid mites is especially difficult because their small size and their particular host–plant interactions. Jeppson *et al.* (1975) suggested collecting vagrant species from leaves and fruits using pins and brushes. Alternatively, the leaves and fruits can be placed in a small container in syrup of sorbitol mixed with 25% solution of isopropyl alcohol with a few iodine crystals and a very small amount of potassium iodide (otherwise the iodine crystals will never dissolve). The material that has been collected can then be observed through a dissecting microscope. It is also possible to wash the leaves and other plant parts in hot water and then separate the mites out by pouring the water through a sieve.

Amrine and Manson (1996) report various basic techniques for collecting Eriophyoidea, such as 'beating or washing vegetation', 'scanning vegetation', the 'gall technique', the 'special erinea technique' and the 'dried material technique'. The beating or washing of vegetation technique is used for vagrant or free-living species and involves beating the vegetation over a No 16 screen mesh sieve, washing the sieve with AGA (alcohol/glycerine/acetic acid) fluid (ethanol 240 ml, glacial acetic acid 30 ml, glycerine 30 ml, sorbitol 3.8 g) over a funnel and collecting the washings in a vial. The fluid is then examined under a dissecting

microscope, vacuum filtered and the mites transferred to a work slide. The scanning vegetation technique involves the direct examination of leaves, buds, stems, corms, etc. under a dissecting microscope and the transfer of live mites to a work slide. When the mites are abundant, the vegetation can be pressed, dried and stored in a plant collection. Mummified mites can be collected and mounted after some years of storage (Jeppson *et al.*, 1975). The gall technique involves either dissecting galls and removing the mites individually or placing the galls into glass tubes or vials. After several days, it is possible to wash the migrating mites into a vial containing AGA fluid. The special erinea technique is applied to erinea, such as those of *Acalitus fagerinea* (Keifer), or to dried material. This method allows the collection of eriophyoids from dried galls, erinea, rolled leaf edges, buds, etc. The collected material is immersed in slightly heated water for a variable time, and the water is then filtered for the recovery of the eriophyoids.

Further details on the collection and detection of eriophyoid mites can be found in an updated review from Monfreda *et al.* (2010), who reported seven distinct methods of extraction, indicating procedures, chemicals or solutions used, suitable devices and suggested uses.

Mass collection

The method uses different devices, some of which, although designed to solve various problems, do not seem to be very practical. For highly mobile and medium or large-sized mites, an aspirator or 'Pooter' can be used. This consists of a vial made of glass or some other transparent material, with a rubber bung and two tubes, one of which is used as a sucking tube equipped with a filter so that the mites being collected are not sucked up by the operator, and the other with a thin tip that is used for collecting specimens from the substrate concerned. This equipment forms part of the 'vacuum samplers' that are used for removing mites from plants in the field, sometimes with low efficacy; for instance in the case of mites inhabiting crevices or spider mites producing webs. Despite these limitations, the vacuum sampling method is largely employed for collecting arthropods in various environments, such as grasslands or agricultural fields (Stewart and Wright, 1995). It is also possible to use a portable, battery-operated vacuum device, which is basically a portable vacuum cleaner with a nozzle that is adapted for collecting mites and insects. This is commercially available as the 2820GA Heavy Duty Hand-Held Vac/Aspirator from BioQuip Products of Rancho Dominguez, California.

Sieves equipped with a 2.5–3.0 mesh cm^{-1} are used to collect mites from litter and other substrates. The sifted material is shaken on to a white support and the fallen mites collected using various tools (an aspirator, brush, forceps). This method can also be applied also to leaves, which are beaten on to a sieve connected through a funnel to a container or vial containing alcohol or wetted plaster of Paris (Gerson *et al.*, 2003). Another device for collecting mites from leaves is the 'leaf-brushing machine', which was used by Henderson and McBurnie (1943) to determine populations of the citrus red mite, *Panonychus citri* (McGregor), and its predator mites. The device was also used by Morgan *et al.* (1955) to estimate the pest and predatory mites present on leaves in an apple orchard, but it does not allow a distinction to be made between live and dead mites, or the leaf side on which they occur, and may give rise to unclear variations between leaves within a sample (Gerson *et al.*, 2003). It is also possible to employ the 'Berlese funnel' (see below), able to examine large amounts of vegetable material, although with low efficiency. The Berlese funnel also allows the examination of mushrooms infested by mites.

Different types of traps can be used to collect arboreal mites, such as plankton nets, plates coated with Tanglefoot (an insect glue) or greased slides (Hoy, 1982a; Charles and White, 1988), arboreal emergence traps or photocollectors (Franklin *et al.*, 1997), trap plants (Walter *et al.*, 1998), and a technique in which the mites in the material that has been collected are excited in a microwave oven so that they fall to the bottom of a plastic bag (James *et al.*, 1992).

Two type of traps are used for sampling and monitoring eriophyoid mites. The first type consists of sticky tape or other kinds of adhesive traps that collect mites during their movements over the plant surface. This has been used for species such as *Aceria guerreronis* Keifer on coconut (David and Varadarajan, 2001), the wheat curl mite, *Aceria tosichella* (Keifer), on wheat spikes (Harvey and Martin, 1988) and *Calepitrimerus vitis* (Nalepa), the grape rust mite (Duffner and Schruft, 1998; Duffner, 1999; Bernard *et al.*, 2005; Walton *et al.*, 2007). The second type of trap consists of sticky glass slides, plates coated with grease (Perring *et al.*, 1996; Duffner *et al.*, 2001) or detergent-water pan traps (Zhao and Amrine, 1997b), which are used to trap mites during passive aerial dispersal in the environment. Double-sided adhesive tape is used as a method for monitoring the emergence of *Phytoptus avellanae* (Nalepa), the hazel big bud mite (Webber, 2007).

Another method, 'canopy fogging', allows the collection of canopy arthropods (and mites) by fumigating individual trees or parts of the canopy with an insecticide, e.g. pyrethrin, used as a mist or fog. Mites that fall from the canopy are collected on sheets and then examined with a dissecting microscope (Majer and Recher, 1988).

Collecting and detecting from substrates

The collection and detection of the terrestrial mites from substrates can be effected through 'hand collection', 'funnel extraction', 'flotation separation' and 'trapping' (Walter and Krantz, 2009).

Mites living on the soil surface can be collected using a hand-operated vacuum apparatus to suck up specimens living on plants (weeds), soil, rocks, wood, tree trunks, etc. It is usually preferable to employ mechanical suction, because it avoids any toxicity caused by biological and chemical agents and/or allergic reactions due to mites, fungi, etc. The specimens collected on the filter can either be observed directly with a hand lens or dissecting microscope or they can first be washed with water (or not) and poured on to a coloured mat and then observed.

The use of the Berlese–Tullgren funnel allows the collection of mites (also nematodes and other microarthropods) from the ground vegetation and the extraction of living mites from soil and litter (Krantz, 1978; Upton, 1991; McSorely and

Walter, 1991; Walter and Krantz, 2009). The funnel extraction technique was developed by Antonio Berlese, and originally used copper coils heated by hot water or steam to produce a temperature and moisture gradient in the substrate. The design was subsequently improved with the replacement of the hot water with an overhead incandescent light bulb. Many modifications are reported in the literature (see Walter and Krantz, 2009) which aimed to improve the apparatus as a quantitative tool. The modern version consists of a funnel which has at the top (larger side) a screened support on which the substrate (soil, litter, vegetable matter, etc.) lies. The bottom (smaller) end is usually connected to a vial. The upper surface of the substrate is exposed to a heat source that produces a gradient of humidity that encourages the arthropods that are present to move below towards the vial, into which they fall and are collected either in a preservative mixture (70–80% ethanol with 1–3% glycerine, alcohol, etc.) or on a mixture of moistened plaster of Paris mixed with charcoal on which they can be kept alive (Walter and Krantz, 2009). The use of alcohol may hamper specimen mounting and some of the media used, e.g. polyethylene glycol or glycerine/glacial acetic acid (Krantz, 1978), are not safe. The number of specimens collected depends on their activity and the dryness of substrate, and the preservative mixture contains particles of soil, various detritus and/or arthropods as well as the mites. The use of a layer of wide mesh cloth or fine screening located under the original sample can prevent such items from falling into the funnel.

It is also possible to use other techniques, such as autosegregators, but flotation techniques seem to be better (Walter and Krantz, 2009). The technique of specific gravity flotation exploits both the specific gravity of mites in respect to soil particles and the marked affinity of their cuticles for petroleum derivatives (McSorley and Walter, 1991). These properties permit the separation of mites from their substrate by mixing them with a concentrated sugar or salt solution (Hart and Fain, 1987; Pugh, 1996). The mites are collected in suspension and the various soil particles are deposited at the bottom of the container; the mites may be then be collected using a Buchner funnel. Another flotation technique, 'hydrocarbon flotation', is based on the tendency of liquid hydrocarbons, e.g. kerosene, to stick to the arthropod cuticle (Walter *et al.*, 1987; Geurs *et al.*, 1991; Kethley, 1991). Different types of traps may also be used (Walter and Krantz, 2009).

Preserving

After collection, mites are usually kept in small vials or glass containers containing 70–80% alcohol with the addition of 5% glycerol to prevent evaporation of the alcohol (Evans *et al.*, 1961). This mixture is sufficient to kill, fix and preserve them. Oudeman's fluid (a mixture of 87 parts of 70% alcohol, five parts of glycerol and eight parts of glacial acetic acid) can be used as an alternative, and has the advantage in some cases of long preservation and clarifying (to varying degrees according to the species) the preserved specimens. For the Tetranychidae, AGA fluid has been suggested, which is a modification of Oudeman's fluid, consisting of 8:1:1:1 isopropyl or ethyl alcohol (70%), glacial acetic acid, glycerol and sorbitol (or sugar) (Jeppson *et al.*, 1975). Before preserving, it is necessary to immerse the mites in hot water, forcing them to extend their legs and other appendages in order to facilitate their subsequent morphological examination; it is possible to fix the mites in a solution of methanol and acetic acid in the quantity of two parts each to one part distilled water, and then transfer them to the preservative medium within a week (Evans, 1992; Saitō and Osakabe, 1992).

Preparing

The preparation of mites for light or electron microscopic study or for other analyses aiming to identify them to species or subspecies needs the use of distinct protocols.

This section discusses the procedures that are required for the use of light microscopic analysis. These include the clearing/maceration of specimens, and according to the species and the aim of the investigation, the preparation of either temporarily or permanently mounted specimens. Sometimes, the study of heavily sclerotized or large mites requires dissection of the specimens in order to separate the dorsum and the venter and remove chelicerae and the gnathotectum. More detailed information can be found in Singer (1967), Krantz (1978), Gutierrez (1985b), Evans (1992), Amrine and Manson (1996) and Walter and Krantz (2009).

Clearing/maceration

The microscopic study of the mites by optical phase contrast and interference microscopy requires that the specimens are cleared and mounted on slides. It is possible to use different media (Krantz, 1978), including lactophenol, which is produced by adding in sequence 50 parts lactic acid, 25 parts phenol crystal and 25 parts distilled water. To mitigate the aggressiveness of lactophenol on the most sensitive mites, an aqueous solution of 50–95% lactic acid is used. Tetranychidae may be treated with pure lactic acid coloured with lignin pink (Evans and Browning, 1955). Hardened specimens may be treated with Nesbitt's fluid (chloral hydrate 40 g, hydrochloric acid 2.5 ml, distilled water 25 ml), André's fluid (chloral hydrate 40 g, glacial acetic acid 30 cm^3, water 30 cm^3) (Gutierrez, 1985b) or with a 5–10% solution of KOH. Eriophyoids may be treated with Keifer's booster (sorbitol 3 g, chloral hydrate 7.5 g, iodine crystals 1 g, water 15 ml, concentrated HCl, 1 ml) or lactic acid (Amrine and Manson, 1996). In one way or another, depending on the size and nature of the species, it is possible to clear a specimen in a week at room temperature. Heating the mites in the clearing medium on a hot plate produces a rapid maceration within a period ranging from a few minutes to approximately 1 h. Cleared specimens must be washed with distilled water before mounting on slides.

Temporary mounting

Lactic acid can be used for temporary mounts, applying a drop at the centre of a glass slide and placing the mite at the centre

with the help of an appropriately sized needle or spatula. It is possible to intervene with the same tools to better guide the specimen, paying attention to the position of the legs so that they do not settle under the body. It is then possible to place a coverslip of the appropriate shape and size over the mite, which must be carefully set down in a suitable position; the outcome depends on the experience of the operator, the size of specimens and the slip covering the specimen. Some researchers use a concave glass slide covered partially with the coverslip in order to guide the specimen under study. This method was suggested for the study of particular groups, such as the Oribatida (Grandjean, 1949). It also may be possible to use a porous carbon block moistened by 75% ethanol slowly delivered by a capillary tube; the specimen is disposed on the surface of this block and studied under incident light (Grandjean, 1949; Evans *et al.*, 1961).

Permanent mounting

Permanent mounts require the use of a slide, some drops of medium and a coverslip of different sizes and shapes depending on the species or groups of the mite under study. Common media for permanent mounts are Hoyer's medium, lactophenol media and resin-based media (Krantz, 1978; Saitō *et al.*, 1993; Upton, 1993; Amrine and Manson, 1996), which are included, respectively, in the three classes of water-miscible mixtures, polyvinyl alcohol (PVA) solutions and resin-based permanent mounts (Upton 1991, 1993).

Water-miscible mixtures include Hoyer's medium (Singer, 1967; Krantz, 1978), which is widely used and basically derives from Berlese fluid, which contains gum arabic and chloral hydrate, and is usually used for clear, weakly sclerotized specimens; it has good optical properties (refractive index =1.2) but it is not always considered to be a permanent medium. Because of its toxicity, its use requires precautions to be taken so that contact with the skin and eyes is avoided. Hoyer's medium quickly degrades specimens that have been clarified with lactic acid and not sufficiently washed with distilled water. It is produced by mixing in sequence the following:

Distilled water	25 ml
Gum arabic	15 g
Chloral hydrate	100 g
Glycerine	10 ml

Arabic gum from a crystalline source and not from powdered form must be used, and the mixture filtered several times. Gilberto Moraes has found a 'way to beat the system' in the preparation of suitable gum arabic: he spread a thin layer of gum arabic powder on a glass surface (maybe a large Petri dish or just a flat piece of glass), and then sprayed it with 70% ethanol and allowed it to dry. Most of the liquid (alcohol) does not take part in the reaction and evaporates quickly, and so the preparation dries fast. When completely dry, the gum arabic is now in the form of a thin continuous layer, which can be scraped off the glass and results in gum arabic flakes. These perform very well in the process of preparing Hoyer's medium (Flechtmann, personal communication). Pritchard and Baker (1955) stated that among the available mounting media, Hoyer's solution was better for the Tetranychidae.

For the PVA or lactophenol media, which can be used as an alternative to Hoyer's medium, Heinze's PVA medium is commonly employed. According to Evans (1992), this is produced by mixing the following:.

Polyvinyl alcohol	10 g
Distilled water	40–60 g
Lactic acid (85–92%)	35 ml
Phenol 1% aqueous solution	25 ml
Glycerol	10 ml
Chloral hydrate	100 g

Resin-based media include Euparal (eucalyptus oil and paraldehyde) and Canada balsam. The opinions of acarologists on their use is divided, because of the need to macerate specimens and dehydrate those that have already been treated (Saitō *et al.*, 1993). Euparal has good optical properties (refractive index =1.483), but when it sets, the preparation becomes powdery or polymerizes. Other synthetic resins (such as dimethyl hydantoin formaldehyde – DMHF or DMHFR) have good optical properties and do not require dehydration of the mounted specimens (Bameul, 1990).

Keifer's F-medium may be employed in the mounting of eriophyoids. This is made from sorbitol 3 g, gum arabic powder 1 g, iodine crystals 0.02 g and 4% formalin solution 5 ml, to which is added 24 h or more later, chloral hydrate 14 g, glycerine 1 ml, potassium iodide 0.1 g and, if necessary, more 4% formalin (Jeppson *et al.*, 1975).

Slide preparation

Slide preparation requires knowledge of the procedure as well as considerable manual skill honed over time. The protocol varies according to the media used for clearing/macerating the specimens. Thus, specimens cleared or macerated in alcohol or Nesbitt's fluid, or maintained in water, may be directly mounted into aqueous mounting media, whereas specimens treated with lactic acid or potassium hydroxide must first be washed in water. The procedures are divided into several steps, some of which require the help of a dissecting microscope, and they can be summarized in the basic steps discussed below. Further details can be found in Walter and Krantz (2009).

In the first step, a droplet of mounting medium is placed at the centre of a clean slide. The specimen is placed in the medium using the proper tools (e.g. watchmaker's forceps) and pressed gently on to the slide with a small pin in order to reduce its displacement when the coverslip is applied. Any air bubble contained within the medium droplet is removed and then forceps are used to place a clear coverslip – usually circular and 12–15 mm in diameter (larger coverslips are available for larger specimens) – over the droplet medium. The coverslip should be applied to the margin of the droplet medium and gently helped into place. Further measures to place it correctly may be necessary to better accommodate the arrangement of the legs and the dorsoventral position of the body. If the medium does not dry, it is possible to gently warm the slide over an alcohol flame or on a hot plate.

The methodology for mounting the specimen in a special position has been investigated in a number of cases, such as in tetranychid males, whose identification requires that they be

mounted sideways to see the male aedeagus in a lateral position. It is advisable to spread a small drop of Hoyer's medium on the slide, placing the mite in the centre and on its lateral side and then to dry in a drying oven at 40°C for 3 h, or for a longer period at room temperature. Next, a new drop of Hoyer's medium is placed on the specimen and on the first layer of semi-dry Hoyer's medium with the aim that the two layers combine (Henderson, 2001).

Finally, the slide is placed in an oven at 45–50°C to dry the medium, for a time that is variable depending on the season and the thickness of the mount. When water-soluble media are used, drying in a hot oven at 40–50°C is required for 1 or 2 weeks to fully dry the slide. If the dried slides are not to be kept in a low-humidity environment, the slides must be sealed around the coverslip. For this purpose, many acarologists use insulating paint (Glyceel, Euparal or Glyptal), with which the coverslip is ringed using a small paint brush, and the coats are applied several times (Travis, 1968, 1972; Fain, 1980). Permanent mounts need both to be dried thoroughly and to be marked to identify the location of the mite on the slide. Each slide must also bear two labels, one showing the collection date and one the identification information. The collection data include locality, data of collection, habitat, host and name of collector. The identification information includes the name of the species and of the identifier, sex and position of the specimen, the year and the mounting medium used.

Electron Microscopy

Electron microscopy has enabled significant advances in the knowledge of mites (and other organisms) since the 1950s. Two types are usually used: transmission electron microscopy (TEM) and scanning electron microscopy (SEM). TEM is used to investigate mite ultrastructure. The possible applications of this method have been discussed by Alberti and Coons (1999), and several important studies on the ultrastructure of tetranychid and eriophyoid mites (and other mites) have been carried out during the last few decades. However, the use of TEM is highly specialist and expensive. Easier and less expensive is SEM, which allows the observation of surface structures only. Crooker *et al.* (1985) treated the procedure for tetranychids (and other mites). Alberti and Nuzzaci (1996) treated both TEM and SEM techniques extensively, focusing on the conventional procedures with particular regard to eriophyoid mites. Both TEM and SEM techniques use an electron beam that passes through a microscope that is structured as a column and is able to maintain a vacuum (in order to avoid the presence of gas molecules which may interfere with electrons). The microscope includes various elements, such as the electron source, the magnetic objective lens and the specimen chamber. The use of a vacuum means that dried specimens (free of water or other evaporating substances) must be used. SEM requires that the surface of the specimens is clear so that it can emit secondary electrons to form an image on a screen. For this purpose, the specimens are coated with a gold layer and mounted on a metal support using silver paint or conducting carbon as an electron-conducting glue (Alberti and Nuzzaci, 1996). TEM is based on the observation of ultrathin sections. Materials are embedded with epoxy resins and cut with glass or diamond tools using an ultramicrotome. The sections must allow the penetration of electrons and so need to be less than 100 nm thick. They are placed on metal grids and stained with metal solutions. Metal atoms bind differentially with the internal structures of tissues, and the scattered electrons form the image. Manipulations of living organisms and an intact structure give better results. From this point of view, fixation is a very important step. In TEM, the results are evaluated by photography as a two-dimensional view (Alberti and Nuzzaci, 1996).

In this section, only the essential steps that characterize electron microscopy are discussed, and it is emphasized that the use of this interesting technology requires special expertise and the availability of expensive tools. Its application is primarily aimed at research that is fundamentally not compatible with the solution of practical problems (at least in the short term), and the risk of artefacts and misidentification is great. Some of the artefacts that have occurred in the preparation of Tetranychidae for SEM and TEM are reported by Weyda (in Crooker *et al.*, 1985). Low temperature SEM (LTSEM) has been positively employed to obtain low magnification images of mites (Amrine *et al.*, 1994; Duffner *et al.*, 1998; Otto, 1999) and their feeding sites (Spieser *et al.*, 1998). LTSEM allows the avoidance of deformities to soft-bodied mites and produces images that are similar to those of live specimens (Wergin *et al.*, 2000a,b).

Other new possibilities of analysis are given by the technique of confocal laser scanning microscopy (CLSM). This method provides an extended depth of focus and 3D visualization and has been positively tested in the study of the internal genitalia and external morphology of eriophyoid mites (Chetverikov, 2012). Below, we deal briefly with main steps of TEM, SEM and LTSEM. The reader interested in more specialized information can consult the contributions of Crooker *et al.* (1985) for tetranychids (and other mites) and Alberti and Nuzzaci (1996) for eriophyoids.

Transmission electron microscopy (TEM)

This technique allows observation of biological material at the molecular level (and sometimes at the atomic level). Protocols for mite studies are well described in the literature. Protocols for tetranychids are discussed by Crooker (in Crooker *et al.*, 1985), and consider the implementation of the different preparatory phases: fixation, dehydration and embedding, specimen orientation, sectioning and staining. A list of compounds used as fixatives, buffers, embedding media and stains is given. For eriophyoids, the protocol requires the application of the precise steps of fixation and embedding, trimming and sectioning, staining of ultrathin sections and 'photography'. The fixation and embedding step includes six phases (prefixation, rinsing, postfixation, rinsing, dehydration and embedding), in each of which specific chemicals and tools are used, apart from the other steps. For more information see Alberti and Nuzzaci (1996).

Scanning electron microscopy (SEM)

This technique allows observation of the surface structures of organisms only. It includes the steps of fixation, cleaning and drying,

mounting and coating. Specific applications of SEM, such as the study of fresh eriophyoid mites (Nuzzaci and Vovlas, 1976), or of mites reconstituted from dry plant material, are known (Alberti and Nuzzaci, 1996). A common protocol for producing images with SEM requires killing the specimens in liquid nitrogen, 5% KOH or, alternatively, boiling water (Evans and Loots, 1972; Krantz, 1978). Crooker (in Crooker *et al.*, 1985) suggested a similar protocol for Tetranychidae. Mites are usually fixed in glutaraldehyde or osmium tetroxide and dehydrated before critical point drying. This method can lead to some inconvenience, e.g. damage to specimens, and depending on the species of mite, other procedures, e.g. dehydration in 100% ethanol, transferring to acetone and drying under a fume hood, can be used. The specimens are then treated with gold, gold–palladium, platinum or carbon. In another treatment, the specimens are transferred from 100% ethanol into a solution of hexamethyldisilazane (HMDS), taking in to account the use of a fume hood to avoid its high toxicity.

Low-temperature scanning electron microscopy (LTSEM)

The manipulation of mites (chemical fixation, dehydration and/or through drying) for SEM may induce artefacts. To avoid this, alternative preparation procedures have been developed, such as the use of sputter-coat vacuum drying (Ebrahim *et al.*, 1996) or imaging the specimens quickly *in vacuo* before they could become deformed or collapse. Although mites may be lost from the plant surface when specimens are plunged into liquid nitrogen (LN_2), low-temperature SEM has been successfully used to produce low magnification images of mites (Amrine *et al.*, 1994; Duffner *et al.*, 1998; Otto, 1999) and their feeding sites (Spieser *et al.*, 1998). The potential of a field emission SEM equipped with a low-temperature stage to image the structural details of frozen, hydrated mites and their plant substrates has been evaluated by Wergin *et al.* (2000b); the main steps of this technique are summarized below (from Wergin).

Samples of leaves, fruits or other plant organs harbouring the mites are excised and placed on a flat holder of copper plates containing a thin layer of Tissue Teck, which acts as a cryo-adhesive upon freezing. Samples are frozen conductively using a styrofoam box, placing the holders on a surface of a pre-cooled (−196°C) brass bar half submerged in LN_2. After 20–30 s, the holders with the frozen samples are transferred to LN_2 and then stored in an LN_2 dewar. This allows their future use or cryo-transfer under vacuum to the cold stage in the pre-chamber of the cryosystem. The frozen samples are etched for about 10 min in the pre-chamber by raising the temperature to −90°C, and surface contaminants, such as condensed water vapour, are removed. The temperature is then lowered to below −110°C, and the samples are coated using a magnetron sputter head with a platinum target, and then transferred to the pre-cooled (−170°C) cryostage in the SEM. The specimens that were previously coated, analysed and stored in LN_2 are reintroduced on to the cold stage of the microscope and slowly etched at −98°C, and surface contaminants removed (Wergin *et al.*, 2000b).

Further studies have shown that, in general, LTSEM has more advantages than 'ambient temperature scanning electron microscopy' (ATSEM) (Achor *et al.*, 2001). LTSEM has been applied to various different mite species, e.g. the honeybee tracheal mite, *Acarapis woodi* (Rennie) (Ochoa *et al.*, 2005).

Molecular Methods

Knowledge of DNA structure has been an important innovation in various fields of biology, including general and applied acarology (Klompen, 2010). The development of analytical techniques and methods of 'manipulation' of DNA, e.g. polymerase chain reactions (PCR) (Mullis and Faloona, 1987), has allowed the amplification of copies of a gene or a genomic region, and thereby provided a series of previously unimaginable possibilities, such as the identification of species, genes and genotypes, or investigations of ecological and evolutionary aspects.

According to Navajas and Návia (2010), the following advances have significantly characterized the development of molecular methods: (i) the development of PCR (Mullis and Faloona, 1987); (ii) the use of evolutionarily conserved sets of PCR primers (Simon *et al.*, 1994; Ben-Ali *et al.*, 2000; Navajas and Fenton, 2000); (iii) the advent of hypervariable microsatellite loci (Goldstein and Schlötterer, 1999); and (iv) the possibility of routine DNA sequencing. Genetic markers are related to the diversity in DNA sequences, and allow the assessment of genetic diversity at all taxonomic levels (Ben-Ali *et al.*, 2000; Navajas and Fenton, 2000; Behura, 2006). Molecular methods are used to differentiate properly between species (Caterino *et al.*, 2000; Tautz *et al.*, 2003), and give information in population studies by assisting in the estimation of gene flow (Nishimura *et al.*, 2010) and identifying haplotypes and lineages or predicting migration and colonization mechanisms (Sunnucks, 2000; Garant and Kruuk, 2005). They can also be used to support phylogenetic hypotheses (Navajas and Návia, 2010; Tixier *et al.*, 2010b).

In acarology, molecular methods have been variously applied. The electrophoretic analysis of isoenzyme patterns has allowed the separation of species (Houck, 1989) and strains or populations (Guichou *et al.*, 2010). For example, it was possible to differentiate strains of predatory phytoseiid mites resistant to pyrethroids that were released in permethrin-treated apple orchards from native mites that were already present by their pyrethroid resistance and allozyme markers (Navajas *et al.*, 2001a). Random amplification of polymorphic DNA (RAPD) analysis does not require prior knowledge of DNA sequences and allows the separation of cryptic species, as in the case of three *Typhlodromalus* spp. (phytoseiids) that were discriminated by using random amplified polymorphic DNA markers (Edwards *et al.*, 1998). The use of PCR in restriction fragment length polymorphism (RFLP) analysis allows the preparation of profiles of the DNA banding patterns of eriophyoid mites (Fenton *et al.*, 1995).

Molecular markers have allowed the taxonomic status and dispersion paths to be determined in spider mites; for instance, the intraspecific diversity of the cassava green mite, *Mononychellus progresivus* Doreste, has been examined in African (Benin and Congo) and South American (Columbia and Brazil) populations.

Comparisons based on mitochondrial and ribosomal DNA sequences and the results of several cross-breeding experiments highlighted a single introduction of the species in the two African populations. The crossing experiments also showed partial hybrid sterility, suggesting a genetic incompatibility consistent with differences between the populations that had been detected by sequence data (Navajas *et al.*, 1994). In another study, analyses of nucleotide sequences of a fragment of cytochrome oxidase subunit I (COI) showed phylogenetic relationships among 20 species of tetranychids (Navajas *et al.*, 1997). Matsuda *et al.* (2012) examined the cytochrome c oxidase subunit I gene of mitochondrial DNA and the internal transcribed spacer and 28S regions of the nuclear ribosomal RNA gene for 17 *Oligonychus* spp. The genetic distances (*p*-distances) of the three DNA regions highlighted a range of intraspecific divergence below the range of interspecific divergence, allowing the discrimination of the 17 species correctly according to their morphological classification.

Boubou *et al.* (2012) used approximate Bayesian computation (ABC) and variation in multi-locus genetic markers to reconstruct the complex historical demography of the tomato red spider mite, *Tetranychus evansi* Baker *et* Pritchard, a cryptic invasive pest. By distinguishing among multiple pathways and timings of introductions, evidence was found for the 'bridgehead effect', in which one invasion serve as a source for subsequent invasions. Tomato red spider mite populations in Europe and Africa resulted from at least three independent introductions from South America and involved mites from two distinct sources in Brazil that corresponded to highly divergent mitochondrial DNA lineages. In *Tetranychus urticae* Koch, genome sequencing has allowed better understanding of the mechanisms of adaptation of this species to phytophagy, as well as related characteristics, and has implications for the development of plant protection strategies (Grbić *et al.*, 2011).

Although in recent years the application of molecular methods has been widely standardized to the point that now it is commonly employed in laboratories, its use requires a particular approach, characterized by accuracy and adequate technical tools, as the possibility of errors and approximations is very real. Taxonomic misidentification of the specimens used to obtain DNA sequences is a growing problem that has been reported for different groups of organisms, including mammals, birds, arthropods, fungi and bacteria, which threatens the utility of the sequences deposited in public DNA databases. Misidentifications and insufficient annotations in public DNA accessions not only have consequences for the reliability of the phylogenetic and ecological conclusions inferred from such sequences and cause obstacles to sequence-based species identification, but also affect plant protection and pest management. Evidence of misidentifications in molecular DNA public databases in tetranychid mites of the genus *Tetranychus* was produced by Mendonça *et al.* (2010), who reviewed and analysed 105 GenBank accessions matching ITS2 rDNA sequences and 137 *COI* mtDNA sequences of 11 morphologically close species of *Tetranychus sensu stricto*. They included in the data set the sequences available in GenBank of four red-coloured species that could be easily misidentified with the aforementioned 11 species, as well as ITS2 and *COI* sequences of 18 new *T. urticae* samples from various countries (Brazil, France, Spain and Greece) which had been unambiguously identified by morphological criteria. Numerous examples were found in the database of mistaken identities in the *Tetranychus sensu stricto* group, as well as several of the taxonomically important DNA sequences that were publicly available being incorrectly named (Mendonça *et al.*, 2010).

Collecting and storing mites

Mite specimens are usually preserved in 70% ethanol in order to recover DNA, but better results have been obtained by killing and preserving specimens in absolute alcohol or using either live mites or mites preserved in 95–100% ethanol kept at 4°C. It is also possible to store freshly collected mites under dry conditions at −20°C (Evans *et al.*, 2008).

Monfreda *et al.* (2010) described four methods for collecting and/or storing eriophyoid specimens for DNA extraction, and found that: (i) fresh mites that had been collected by hand picking under the dissecting microscope or by the wash and sieve method could be directly used for DNA extraction; (ii) fresh mites could be killed and stored under dry conditions by freezing and keeping them at −20°C; (iii) fresh mites could also be killed in 95% ethanol and stored at 4°C or −20°C; and (iv) after soaking in warm water, mites stored in dry leaves could be removed from the leaves under the dissecting microscope.

Corriveau *et al.* (2010) tested eight storage possibilities for *Balaustium* sp. (Erythraeidae) fixed in 95% ethanol. After 8 months of storage, all specimens stored in ethanol at concentrations of 75 or 95% and at temperatures ranging from ambient to −20°C gave good quality DNA. Fixation in ethanol and continued dry and cold storage performed adequately, but storage in Koenike's fluid, clearing in lactic acid prior to ethanol storage, and the use of a lysis buffer did not give good quality DNA.

DNA extraction

The availability of high-quality DNA is fundamental for molecular analyses, but extracting DNA from mites is difficult because of their small size, and the low quantity of extracted DNA is often not suitable for amplification (Návia *et al.*, 2005b). Each extraction generally uses from five to 25 specimens (Fenton *et al.*, 1996, 2000; Návia *et al.*, 2005b), but sometimes up to 50 (Fenton *et al.*, 1993, 1995). Because of the risk of analysing multiple species, recent population genetic research has used only single mite extracts (Carew *et al.*, 2004; Evans *et al.*, 2008). The extraction of DNA from single mites and the preservation of each exoskeleton for anatomical study so that the two can be correlated has been reported by Dabert *et al.* (2008) and Skoracka and Dabert (2010). The first three methods for collecting and storing eriophyoid specimens described above allow the use of a single mite for DNA extraction, but it is unknown whether the fourth method is suitable. Dry mite extractions involve the use of about 20–30 mites using the DNeasy Tissue Kit (QIAGEN Inc.) with some modification in the method (Monfreda *et al.*, 2010).

In addition to DNA sequencing methods, future scenarios could use serological methods based on the recognition of pests' antigens by antibodies. This method could be used for rapid mite identification for quarantine purposes. In addition, enzyme immunoassays and molecular methods for the detection of mites (in particular eriophyoids) will help in distinguishing the symptoms of mite infections from phytoplasmal diseases (Monfreda *et al.*, 2010).

Rearing

Summarizing the information on rearing methods for mites is not simple, as the techniques vary according to the systematic groups and the aims that are involved. A comprehensive review has been published by Krantz (1978), and detailed information is given for predatory mites by Gerson *et al.* (2003). Descriptions of rearing methods for tarsonemids may be found in Liang (1980) and Xu *et al.* (1994), and in Oomen (1982) for tenuipalpids.

Below, brief accounts are given of the rearing techniques used for the most important groups of injurious mites: eriophyoids and tetranychids.

Eriophyoidea

The rearing of eriophyoid mites probably presents the greater difficulties, owing to their high degree of host specificity and the adaptation of several species to reproduce only on young and undifferentiated cells of the host plant. In general, species inhabiting buds or causing galls are less easily reared, while leaf or fruit vagrant species are reared more easily (Oldfield and Perring, 1996).

Several methods for rearing vagrant species are reported in the literature. Species reproducing on leaves or fruits may be reared in the laboratory or greenhouse indefinitely if their growth requirements are met, e.g. *Phyllocoptruta oleivora* (Ashmead) and *Aculops pelekassi* (Keifer) of leaves and fruits of citrus (Reed *et al.*, 1964) or *Calacarus citrifolii* Keifer of potted rough lemon seedlings (van der Merwe and Coates, 1965). The tomato russet mite, *Aculops lycopersici* (Tryon), a leaf vagrant species on annual plants, may be reared on seedlings in the greenhouse by taking care to provide new plants periodically (Rice and Strong, 1962). Leaf vagrant species on perennial plants, such as *Aculus schlechtendali* (Nalepa) or *Epitrimerus pyri* (Nalepa), are reared on seedlings or grafted plants (Oldfield *et al.*, 1970). The grass mite, *Aceria tulipae* (Keifer), is reared on greenhouse-grown seedlings of wheat and maintained by periodic transfer to new plants. In some cases, e.g. the wheat curl mite, *Eriophyes tulipae* Keifer, the mites move from culture plants to seedlings by the stimulation of a fan (Harvey and Martin, 1980). It is possible to rear vagrant eriophyoids by other methods, including enclosing them in a ring of lanolin applied to the leaf surface (Slykhuis, 1967), folding the leaf along the mid vein to enclose the mites and holding the leaf folded with a clip (Staples and Allington, 1956), and using self-watering transparent acrylic cages for confining the mites (or insects) on leaves (Tashiro, 1967). Another method uses watering platforms that provide watering to leaves maintained in rearing cages (Beavers and Oldfield, 1970). A method suggested by del Rosario and Sill (1958) based on rearing chambers gives even more confinement; it uses five rearing cages in sequence and allows the rearing of large colonies of *A. tulipae* in pure culture from single eggs or adults.

The rearing of bud mites and gall mites is more difficult and only a few examples are available. In research on the citrus bud mite, *Aceria sheldoni* (Ewing), citrus seedlings that had been grown hydroponically were used, and single mites placed in the aerial parts of plants and settled on terminal or axillary buds (Sternlicht, 1967). Knowledge of the biology of the mite, such as its attraction to yellow cellophane discs rather than to green, red or blue (Sternlicht, 1969b) and its thigmotaxis and negative phototropism, allowed the study of mites in glass rings, with yellow discs inside them, that were attached to the top of citrus fruit and touched the peel on only one side. Red cellophane allowed the aeration of the disc and the mites reproduced under the yellow cellophane discs (Sternlicht, 1970). *Aceria chondrillae* (Canestrini) is transferred to new plants of *Chondrilla juncea* L. by drying out the galls on infested plants in order to force the mites to move out of them (Cullen *et al.*, 1982). The rearing of some eriophyoid mites can be effected by transferring single mites to buds that have been artificially opened to allow their introduction.

Other studies on rearing methods for eriophyoids have been carried out by Gispert *et al.* (1997) for *Eriophyes insidiosus* (Keifer *et* Wilson) and by Haq (2001) for *Aceria guerreronis*. Interesting results were obtained from studies on the rearing of *A. guerreronis* in *in vitro* culture on immature leaf tissues of coconut (*Cocos nucifera* L.) (Wickramananda *et al.*, 2005; De Silva and Fernando, 2008).

Tetranychidae

The rearing of tetranychids involves relatively fewer difficulties than that of eriophyoids because of their greater size and the facility with which they may be handled. Three systems are mainly employed in the laboratory: detached leaf and leaf disc cultures, rearing in closed cells and rearing on plants (Helle and Overmeer, 1985a). Each of these has advantages and disadvantages, and their use depends on the specific needs of the study and/or operation to be carried out.

Detached leaf and leaf disc cultures offer many advantages, such as the easy observation of the cultures under the dissecting microscope, ease of handling of the specimens and the possibility of producing uniform material. The detached leaf system consists of a primary leaf (for instance *Phaseolus vulgaris* L., *Phaseolus limensis* MacFadyen or *Ricinus communis* L.) collected with a short portion of petiole and placed with the upper surface facing up and pressed on to wet cotton wool placed in a Petri dish or similar dish. However, some species must be reared with the undersurface of the leaf uppermost, e.g. *Tetranychus fijiensis* Hirst. The leaf must be carefully placed above the cotton so as to prevent the mites from escaping from under it. Equally important is the reduced length of the petiole. The cotton wool can be moistened either with tap water or, for the first time, with a solution containing nutrients (Helle, 1962).

More water is added according to the rate of evaporation. When the tetranychids increase, it is possible to transfer the new specimens to other fresh detached leaf cultures. Several tetranychid species are reared using this system (Helle and Overmeer, 1985a).

The leaf disc method is similar to the detached leaf method, differing in using only part of a leaf. *Panonychus citri* (McGregor) and *Panonychus ulmi* (Koch) are reared by this method in leaf disc cultures of citrus and apple, respectively. Grass mites are also reared by this technique (Congdon and Logan, 1983).

Rearing in closed cells gives a perfect isolation of mites, but it is more laborious. Its use is suggested for mites that move easily and/or use leathery leaves. A plate of plexiglas is used as a base, on which is laid some layers of moistened filter paper, with the leaf above them. Above the leaf is placed another plate of plexiglas of the same size, but with one or more holes (cells), within which the mites are reared. Cover slips are used to close these cells (Helle and Overmeer, 1985a). The method does not allow the control of humidity, but the mites are easily observed and manipulated.

Rearing on plants is usually on potted host plants. For *T. urticae*, young bean plants (French beans, castor beans, lima beans, etc.) with two primary leaves are usually used, and these are infested with a few mites. When the mite populations reach a high density, the old plants are replaced with new fresh plants, which are inoculated with leaves collected from the old infested culture. Other mites may require different techniques. *P. ulmi* may be reared on apple or plum (cultivar Myrobalan) seedlings or potted rootstocks. Mites living on particular host plants (*Asparagus*, *Buxus*, *Pinus*, etc.) may be reared on entire plants or on branches of these plants kept in a glass beaker with water (Helle and Overmeer, 1985a). Other techniques are used for rearing various other mite species, e.g. *Olygonychus ununguis* (Jacobi) is reared on conifers (Reeves, 1963; Wanibuchi and Saitō, 1983).

7 Mites and Plant Damage

Pest mites *sensu stricto* belong to the Tetranychoidea and Eriophyoidea, both of which have mouthparts that are able to pierce plant tissues and suck out their contents. Other mite groups (Acaridae, Erythraeidae, Penthaleidae, etc.) have different mouthparts that are not strictly adapted to phytophagy, but are nevertheless also able to cause severe damage. The feeding mechanism, together with the plant response, determine the damage typology. Plant damage by mite pests has been extensively investigated in the Tetranychoidea and Eriophyoidea, but less so in the Tarsonemidae and other minor phytophagous groups. Sometimes, the symptoms can be easily taken for boron deficiency, or they may be confused with virus symptoms or even with herbicide action, such as on the new leaves of papaya infested by the broad mite, *Polyphagotarsonemus latus* (Banks).

The main part of this chapter is devoted to an account of the damage to plants that is caused by the main phytophagous groups of mites (Tetranychoidea, Eriophyoidea) and others (Tarsonemidae, minor groups), with details of the feeding mechanisms that they use, the symptoms caused by feeding and the resulting plant damage. The last part of the chapter discusses the transmission of plant diseases – mainly plant virus diseases – by mites.

Tetranychoidea

The Tetranychoidea include two important families of pest mites, the Tetranychidae and the Tenuipalpidae. This family includes 1275 described species in about 77 genera and represents the most important family of pest mites, together with the Eriophyoidea, which are described next (Bolland *et al.*, 1998; Migeon and Dorkeld, 2006–2013). They have a worldwide distribution and association with higher plants, including economic species, and often cause severe injury or the death of the host plant. The role of the Tetranychoidea in the transmission of plant viruses has been largely investigated in American citriculture, and most of the reported work on the damage caused concerns the Tetranychidae, except for a few tenuipalpids.

Feeding mechanism

The morphological and physiological adaptations of the Tetranychidae to plant feeding are very similar to those of the Tenuipalpidae (discussed by Beard *et al.*, 2012) and Tuckerellidae. The feeding mechanism involves actively the stylophore, the chelicerae (Fig. 2.6H, I)[a] and the palpi (Fig. 2.7B, E)[b]. The stylophore is mobile (retractable and extrudable) and lies on the dorsomedial surface of the infracapitulum; it consists of the bases of the chelicerae fused together, curves ventrally and is divided by a median and vertical septum and into two basal cheliceral compartments, each of which hosts a stylet that is attached at the base to a cheliceral lever that activates it in a vertical plane. The chelicerae have a uniformly long and thick digitus mobilis, except at the distal end, where it is sharpened and slightly notched, and if this is protracted and juxtaposed with the one from the other side, it forms an empty tube. The stylophore pushes the chelicerae into plant tissues and when it is retracted usually leaves an opening through which the plant's cellular fluids may be pushed out by the turgor pressure of the cell. Immediately behind the stylets, the palpi press the plant surface and the pharyngeal pump located on the distal side sucks in the cellular contents. The stylets penetrate to a depth of between 70 and 120 μm, and once they are inserted into the plant tissue they push and withdraw continuously, thereby causing mechanical damage. As soon as a mite has consumed the part of the tissue fist attacked, it moves and looks for a site that has not yet been attacked. The depth of damage depends on the length of stylets, the feeding time and the mite population density (Jeppson *et al.*, 1975; Sances *et al.*, 1979a; Mothes and Seitz, 1982; André and Remacle, 1984; Lindquist, 1985). At a microscopic level, a significant volume of the plant's cellular contents passes through the intestinal tract of mites, which was estimated by McEnroe (1963) to be 1.2×10^{-2} ml h^{-2} per mite, which corresponds to approximately 50% of the mass of an adult female of *Tetranychus urticae* Koch. The mite can empty 100 photosynthetically active foliar cells within 1 min (Leisering, 1960). After feeding, thylakoid granules, derived from thylakoid grana in the plant cell and made up of 45–50% protein, 50–55% lipid and small amounts of RNA and DNA, are found in the mite intestinal tract (Mothes and Seitz, 1981b; Noggle, 1983). Water and other small quantities of substances are excreted directly (McEnroe, 1963). On the plant, it is sometimes difficult to separate the influence of mechanical damage caused by the mite from that caused by local toxins (Jeppson *et al.*, 1975).

Symptoms of feeding

The main feeding activity of tetranychids usually occurs on the leaves of the plant, but it can also involve other plant organs, such as the cotyledons, fruits, flowers, shoot tips, etc. The symptoms caused result from the removal of the contents of the cells of the palisade tissue and of the disappearance of the chloroplasts. Coalescence of the cell residues leads to the appearance of small amber-coloured masses. Only the punctured cells of the palisade layer are damaged, with the surrounding cells left

undamaged along with the leaf ribs. The small punctures are light in colour and after exposure for some time assume the appearance of irregular white or greyish spots. The density of the punctures is a function of the size and duration of the attack and of the injurious species. The density of punctures on the leaf surface can be high, e.g. *Eutetranychus banksi* (McGregor) produced 873 feeding punctures cm^{-2} on the leaves of Valencia orange (Hall and Simms, 2003).

Tetranychoid mites give variable symptoms as a result of feeding, according to the species, anatomical structure of the infested organ, the weather (during or after the infestation) and the reaction of the host plant to the attack. The symptoms of feeding are most common on the leaves, particularly on the lower surface, although some mite species may feed on both leaf surfaces. On citrus, colonies of the two-spotted spider mite, *T. urticae*, usually develop on a well-defined surface of the lower leaves and fruits. The oriental red mite, *Eutetranychus orientalis* (Klein), the red citrus mite, *Panonychus citri* (McGregor), and the yellow citrus mite, *E. cendanai* Rimando, may develop on both leaf surfaces, but they usually prefer the upper leaf surface (Jeppson *et al.*, 1975). The resulting colour often characterizes the mite attack, and there may be leaf curling and necrosis of young leaves, stems and growing tips. High levels of attack may induce defoliation. On citrus, the damaged areas can be yellowish (*T. urticae*), grey (*E. orientalis*, *Meyernychus emeticae* (Meyer), *Schizotetranychus baltazari* Rimando) or yellow bronze (*Oligonychus coffeae* (Nietner)). Necrosis of young leaves (*M. emeticae*, *E. sexmaculatus* (Riley)) and of stalks and shoots (*E. orientalis*, *E. yumensis* (McGregor)), or even leaf burning and defoliation (*E. africanus* (Tucker), *E. orientalis*, *E. sexmaculatus*, *E. yumensis*, *O. coffeae*) and/or fruit drop (*E. yumensis*) (Vacante, 2010) can occur. Different tetranychids (*P. citri*, *E. lewisi* McGregor, *E. yumensis*) and tenuipalpids (*Brevipalpus californicus* (Banks), *B. obovatus* Donnadieu, *B. phoenicis* (Geijskes)) cause citrus russets. The same pest mite can cause symptoms that differ in severity on different host plants, e.g. *Tetranychus pacificus* McGregor on pear, apple, grape, etc.

Although silk production in tetranychid mites has different functions (Gerson, 1985b), webbing may be an aspect of the symptomatology of mite attack (see Chapter 5, 'Webbing and Life Types').

Damage

Jeppson *et al.* (1975) used the phrase 'plant disease' rather than 'plant damage' and noted that 'the term plant disease in its broadest meaning includes all injuries or abnormalities generated from sources outside the plant regardless of the cause'. In the case of mites, there is a link between mites and the diseases that they sometimes transmit, and the damage that they cause can either be defined, or local, or wider, or more general in character and involve the entire plant as a result of the injection of systemic or persistent toxins and/or viruses. The reaction of the plant to mite attack therefore consists of a range of external symptoms resulting from mechanical damage (colour changes, alterations in shape, reductions in growth and loss of flowers and production) and internal alterations or changes involving biochemical mechanisms. Understanding these reactions requires the analysis of the mutual interaction between the host plant and the pest (Tomczyk and Kropczyńska, 1985).

With the exception of some tenuipalpids, e.g. *B. phoenicis*, which transmits the virus disease known as 'lepra explosiva' or 'leprosis' (Childers *et al.*, 2001; Childers and Rodrigues, 2011), plant damage caused by the Tetranychoidea is basically due to feeding activity – stylet penetration into plant tissues and the sucking out of the cells contents, often associated with the injection of saliva. The length of the stylets, feeding time and the density of the attack, as well as the host plant, determine the depth at which the damage occurs (Sances *et al.*, 1979a; Mothes and Seitz, 1982). The damage involves morphological, histological and cytological aspects, as well as chemical and physiological changes and the associated impact on growth, flowering and cropping (Tomczyk and Kropczyńska, 1985).

Cytological and histological aspects

The mechanism by which the stylets penetrate the epidermis and the mesophyll layers is partially known. On young fruit trees, *Panonychus ulmi* (Koch) may pass the stylets between cells (Avery and Briggs, 1968a), while in leaves of the apple, *Tetranychus mcdanieli* McGregor inserts the stylets through the walls of the epidermal cells (Tanigoshi and Davis, 1978). Damage may occur on the epidermal cells, which may show punctures (Summers and Stocking, 1972), or appear flattened or deformed, as a result of the punctures or osmotic effects. Damage to the spongy parenchyma may lead to destruction of the stomatal apparatus (Sances *et al.*, 1979a). Mite feeding reduces the number of cells and infested leaves may show a reduced thickness, which on bean plants may be as much as 50% (Mothes and Seitz, 1982) and on apple up to 35–55% (Tanigoshi and Davis, 1978).

On strawberry leaves, a low density of two-spotted spider mite causes fundamental damage to the spongy mesophyll tissue, but little damage to the lowest parenchyma cell layer, while a high density of the mites causes severe damage to the palisade parenchyma (Sances *et al.*, 1979a; Kielkiewicz, 1981). When *Bryobia rubrioculus* (Scheuten) feeds on the upper side of almond leaves, symptoms of damage may occur in the first two layers of the palisade parenchyma, and the cells of the spongy parenchyma may be undamaged (Summers and Stocking, 1972).

Damage may also involve both directly punctured cells and adjacent cells. The damaged cells are deformed, lack protoplasts and their walls are smaller. Starch grains are sometimes observed within damaged cells, as well as the walls being broken and the spaces containing air. Amber-coloured protoplasts or dense material, and other structural changes, are observed within the cells or in the intercellular spaces that separate them. These modifications may depend on the breakdown of tonoplasts and the resulting contact between the vacuolar sap and the cytoplasm (Tanigoshi and Davis, 1978). According to the time of feeding, the surrounding cells may present various types of damage (coagulation of the cytoplasm, destruction of the nucleus, modification of the wall structure, degeneration of the chloroplasts) (Mothes and Seitz, 1982).

The effects of feeding by *T. urticae* and *Eotetranychus carpini* (Oudemans) on red raspberry (cv. Meeker) physiology were investigated under greenhouse conditions by comparing the chlorophyll fluorescence and chlorophyll content of infested leaves with those of the most recently expanded and uninfested leaves of infested plants. The results show that cellular injury occurs before the appearance of damage and that the primary site of injury in spider mite feeding may be the plastoquinone pool (and within that, QA, the primary quinone receptor of photosystem II), which plays a major role in electron transport during photosynthesis (Bounfour *et al.*, 2002).

Cytological alterations resulting from feeding by *Oligonychus biharensis* (Hirst) on cassava leaves were examined using transmission electron microscopy (TEM). The severity of damage to the cassava leaves is directly related to the feeding intensity of the mites and is a function of time. Infested cassava leaves show a reduction in thickness as a result of the disruption and reduction of the mesophyll layers. Cells exhibit coagulated protoplasts, a reduction in the number of cells and chloroplasts, alterations in cell structure, increased space in the spongy layer, extensive disruption of the mesophyll cells and even a reduction of chloroplasts in adjacent unpunctured cells (Kaimal *et al.*, 2011).

The importance of the mechanical action of the stylets versus that of the saliva injected before the sucking out of the cellular contents is not really known. However, the process involves both the host plant and pest mite. The mite saliva could cause changes in the adjacent cells that are not damaged directly by feeding. The cells that are damaged by stylets may also induce changes in the water balance of the mesophyll, and uninjured cells may be separated by damaged cells, which make empty spaces between them. In addition, the destruction of liposome membranes may cause autolysis. According to Mothes and Seitz (1982), tetranychid damage could be due to other factors as well, such as the formation of a vacuum which induces the cells to burst, the influence of mite saliva on adjacent undamaged cells and damaged plant tissue influencing the undamaged cells.

Effects of saliva

The role of saliva injected by mites into plant tissues is complex and sometimes unclear. Unfortunately, the small size of mites does not allow an adequate amount of saliva to be obtained for chemical analysis, so indirect investigations have to be made, focusing mainly on the damaged organs and the plant reactions. Studies on *T. urticae* by Sances *et al.* (1979a) show that damage to strawberry plants is by mechanical means and that no toxic substances are involved in the feeding action of the mite. However, Mothes and Seitz (1981a), in a study of the prosomal glands of *T. urticae*, found that the salivary glands take part in the feeding action and in the digestive process. The salivary secretions contain mucous substances, glycoproteins and a liquid component. The use of various markers, such as $^{14}CO_2$ or ^{32}P, has highlighted that during mite feeding, injected substances are translocated to various parts of the plant, such as the roots and young shoots of plum and apple fed on by *Panonychus ulmi* (Avery and Briggs, 1968a) and the growing parts of bean plants infested by spider mites (*T. urticae*) (Storms, 1971). The saliva injected by mites into plant tissues may contain proteolytic enzymes and plays a fundamental role in the degradation of plant cell contents. Moreover, the saliva also could contain plant hormone-type substances that could cause a specific reaction in the host plant. A high hormone content could stimulate physiological processes in the plant that would compensate for the losses caused by mite damage (Avery and Briggs, 1968a; Storms, 1971). Increases of gibberellin-like substances have been found in leaves infested by *P. ulmi*, associated with decreases of auxin-like substances, with the exception of indole-3-acetic acid (IAA), which is higher in damaged leaves. This process may involve monophenolic substances (Avery and Briggs, 1968a). Overall, phenolic compounds increase in strawberry leaves infested by the two-spotted spider mite, decreasing in palisade parenchyma and increasing in the upper and lower epidermis and in the leaf veins. This may interfere in the host plant–mite pest relationship (Kielkiewicz, 1981).

Chemical and physiological changes

Mite feeding causes changes in the chemical composition of the host plant, including both organic and inorganic compounds. The contents of nitrogen and phosphorus decrease after a long period of *P. ulmi* feeding on apple, but only small changes or none have been observed in potassium, calcium and magnesium contents (Herbert and Butler, 1973a; Golik, 1975). Decreases in potassium, nitrogen and phosphorus have been found on chrysanthemum infested by *T. urticae*. The exact contribution of uptake by the roots from the soil and the role of damaged cells in the decrease in the mineral contents is largely unknown. The process has been little investigated and could also involve other factors, such as the ageing process of the cells and changes in their potassium content (Tomczyk and Kropczyńska, 1985). The main organic compounds involved in mite feeding activity are proteins, carbohydrates and chlorophyll. The organic compounds of saliva have been briefly mentioned above. Changes in the amounts of these compounds could affect plant physiology. A reduction of protein contents and decrease of activity of amylase and proteolytic enzymes have been reported in cotton plants infested with the red spider mite, *Tetranychus telarius* (L.) (Zukova, 1963), as well as an increase in amino acids and soluble sugars in chrysanthemum (Kolodziej *et al.*, 1975). Changes in chlorophyll concentration have also been documented (van de Vrie *et al.*, 1972). Further studies show that any reduction in chlorophyll content depends on the length of feeding activity and the number of mites feeding (Tomczyk and van de Vrie, 1982; De Angelis *et al.*, 1983b). However, other studies have reported minor changes (Sances *et al.*, 1979b, 1982a; Kolodziej *et al.*, 1979). Any reduction in chlorophyll content probably results from mechanical damage of the chloroplasts during feeding, and will vary according to the mite and host plant species. Feeding on the upper side of the leaf causes palisade damage and the greatest reduction in chlorophyll content (Summers and Stocking, 1972; Sances *et al.*, 1982b), whereas feeding on the underside causes damage to the spongy mesophyll, which usually has fewer chloroplasts (Sances *et al.*, 1979b).

Infested plants may exhibit significant changes in transpiration and photosynthesis. Damaged plants lose water by transpiration, which increases up to a certain level and then decreases (Sances *et al.*, 1979b, 1982a), depending on the damage to the protective surface of the leaf and disturbance to the stomatal openings (Sances *et al.*, 1979a). During the day, the transpiration is stomatal whereas during the night it is cuticular (De Angelis *et al.*, 1982). Severe infestations of *Panonychus citri* lead to substantial changes in the transpiration and photosynthesis of plants. In particular, transpiration increases during high-density attacks and decreases with lower mite density, and the quantity of chlorophyll may be reduced to 60% (Wedding *et al.*, 1958). Heavy infestations of the oriental red mite, *Eutetranychus orientalis*, associated with poor irrigation, drought and strong ventilation, may induce defoliation (Jeppson *et al.*, 1975). Water stress may increase the metabolic changes, such as hydrolytic processes and the degradation of starch to simple sugars and of proteins to amino acids (De Angelis *et al.*, 1983a). Mite density and the length of feeding affect the assimilation of CO_2, which is correlated with the chlorophyll content of the leaves. However, this correlation is sometimes unclear (Hall and Ferree, 1975; Sances *et al.*, 1979b;), and other factors may also be involved in the decrease in CO_2 assimilation (Tomczyk and Kropczyńska, 1985). The use of ^{14}C markers has assisted in studying carbon metabolism in photosynthesis and highlighted significant changes in the synthesis of soluble sugars, amino acids, organic acids and starch in chrysanthemum and apple infested by mites (Kolodziej *et al.*, 1975; Borichenko and Manolov, 1982).

Impact on growth, flowering and cropping

Damaged plant tissues and anomalies in physiological processes in mite-infested plants may cause changes in growth, flowering and yield. However, although retardation of the growth of the damaged organs is common (Tomczyk and Kropczyńska, 1985), infested plants do not always show a growth reduction (Zwick *et al.*, 1976; Hoyt and Tanigoshi, 1983). A low mite population density and a short feeding time may conversely stimulate plant growth (Avery and Briggs, 1968b; Storms, 1971) according to their effects on the level of growth regulators and the compensatory capacity of the plants (Tomczyk and Kropczyńska, 1985). Infested plants usually have a reduced total leaf area as a result of the decrease in leaf area growth rate (Avery, 1962; Avery and Briggs, 1968b; Summers and Stocking, 1972) and the number of leaves per plant (Avery, 1962). Shoots have reduced dry matter and so may be shorter (van de Vrie *et al.*, 1972). Mite feeding may also affect the roots, which appear reduced, and there may be fewer branches and less dry matter (Avery and Briggs, 1968b). Infestations on fruit trees may induce a reduction of flowering or fruiting of 75% in the following year (Golik, 1975). Damage may occur at the early stage of fruit development, and the early dropping of immature or ripening fruits has been reported (van de Vrie *et al.*, 1972; Golik, 1975; Hoyt and Tanigoshi, 1983). The effects on fruiting may appear after 2 or 3 years and at various intensities, depending on the plant species. A low density of *P. ulmi* causes yield losses on pear (Hoyt and Tanigoshi, 1983), whereas on strawberry the numbers and size of the fruits increase (Oatman *et al.*, 1982). The storage quality of apples from infested plants is lower (Zwick *et al.*, 1976) but grapes from plants that had or had not been attacked by *T. pacificus* had the same sugar contents (Laing *et al.*, 1972). The fruits of damaged plants are sometimes more coloured (Golik, 1975). In Iraq, *Oligonychus afrasiaticus* (McGregor) caused 50–80% yield loss of dates in years of dry, dusty and stormy weather (Al-Jboory and Al-Suaide, 2010).

Eriophyoidea

The Eriophyidae is the most economically important family of the Eriophyoidea and the second family of the most severe plant pests after the Tetranychidae (Lindquist and Amrine, 1996). Leaf vagrants and gall formers cause different morphological alterations to infested plants (Jeppson *et al.*, 1975; Keifer *et al.*, 1982). According to Amrine (see Huang, 2008), the leaf vagrants include 43.6% of all eriophyoid species (1618 vagrant species out of a total of 3708), while the gall formers make up 15.8%, the erineum formers 11.4% and the bud mites 8.9%. However, altogether, the gall, erineum and bud mites – which may be referred to as refuge-creating mites – make up 36.1% of the eriophyoids, which is almost as great a proportion as the vagrants. Other assessments have noted that the Eriophyoidea include 2263 vagrants, 462 refuge-seeking and 941 refuge-inducing species; moreover, at least half of the species adopt the vagrant life strategy and the other half the refuge-creating or refuge-seeking strategy (Skoracka *et al.*, 2010). The damage caused by these mites is due to the mechanical action of their mouthparts and to the injection of saliva containing various biochemical agents. In addition, various eriophyoid species are vectors of viruses causing severe diseases (see the last part of this chapter on the transmission of plant diseases).

Some of the species of this important group of phytophagous mites are very injurious, e.g. *Phyllocoptruta oleivora* (Ashmead), *Aculops lycopersici* (Tryon) and others. In several species, the damage is less and understanding it requires appropriate ecological analysis, e.g. the citrus bud mite, *Aceria sheldoni* (Ewing). This citrus pest has long been feared worldwide, but the results of the research on lemon carried out in Italy by the author have shown that the citrus bud mite does not negatively influence the lemon harvest and that the interaction between the mite populations and the host plant takes the form of a mutual symbiosis in which the mite uses the buds as a resource and repays the host plant positively by regulating its production system. The case is not unique in this mite group, and its ecological study represents a serious challenge that is largely underestimated by eriophyoid specialists in the Mediterranean region.

Feeding mechanism

The feeding mechanism of the Eriophyoidea involves the subcapitulum or rostrum, the palps and the cheliceral stylets. The cheliceral complex is placed in a subcapitular groove, covered by the free edges of the cheliceral sheath. The cheliceral stylets

are connected to the underlying motivator, which provides basal support (see Fig. 16.3). They are rigid, separated from each other for their entire length and do not form a food channel; they provide only an alternative longitudinal movement. On the sides of the cheliceral stylets there are two auxiliary stylets of uncertain function. In addition, a single long stylet, called the labrum and conformed below as a gutter, is connected to the pharynx and propelled by strong muscles (Lindquist, 1996; Nuzzaci and Alberti, 1996; Westphal and Manson, 1996). After selecting a suitable site for feeding, the mite stops, puts its rostrum on to the host surface and contracts the telescopic segments of the palps, which allows the protrusion of cheliceral stylets over a short distance. In a few seconds, the stylets mechanically penetrate the epidermal cells (Keifer, 1959a; Shevtchenko and Silvere, 1968; Krantz, 1973; McCoy and Albrigo, 1975; Hislop and Jeppson, 1976; Nuzzaci, 1976; Westphal *et al.*, 1990; Westphal and Manson, 1996). At the same time, three glands (two paired and one unpaired) produce secretions (Nuzzaci and Alberti, 1996). Of these, the paired glands are probably salivary glands (Thomsen, 1988; Nuzzaci and Alberti, 1996), while the function of the unpaired gland is unclear. The saliva is injected into the plant cell through the cheliceral stylets and the labrum, and the cell contents are sucked up through a canal formed by the infracapitular stylets (see Fig. 16.3) and the labrum (Nuzzaci and Alberti, 1996). It is difficult for the mite to maintain its cheliceral stylets within the wound, given that they do not participate in the establishment of a food tube. It is possible that they go backwards and that the action of the terminal segments of the palp allow the consumption of food from the plant. The mechanism of movement of the food in the gnathosoma has not been definitively clarified, but it is possible that the labrum is adapted in the hole in the plant cell wall and acts as a funnel carrying food along its ventral gutter to the pharynx. The regular contractions of the pharyngeal pump allow active ingestion (Westphal *et al.*, 1990; Westphal and Manson, 1996).

There is no agreement on the mechanism of insertion of the cheliceral stylets into the cell wall. One hypothesis is that the cheliceral stylets are inserted into the cell during the action of feeding (Jeppson *et al.*, 1975; Nuzzaci, 1979); another is that the stylets are inserted into the cell wall after it has been perforated and then the labrum is fixed to the hole in the cell wall to actively suck out the cell contents (Westphal and Manson, 1996). It seems that the limited time that is available to the mite does not allow the saliva that is excreted to dissolve the cell wall enzymatically, only to notch it. The eriophyoid first ejects the saliva on to the leaf surface to digest the plant cuticle and the cellulose of the cell wall enzymatically, and after that pierces the site with its chelicerae and sucks out the cell contents within 10–20 min (Thomsen, 1988). However, the process of probing and cell penetration needs only a few seconds, and so the saliva ejected on to the plant surface may not have time to be effective in its action on the cell wall before the chelicerae are inserted (Westphal and Manson, 1996).

The feeding mechanisms and mouthparts of the different families of the Eriophyoidea vary. In the Eriophyidae and Phytoptidae, the mouthparts are 7–30 μm long (Keifer, 1959a; Jeppson *et al.*, 1975; McCoy and Albrigo, 1975; Krantz, 1978, 2009a; Westphal, 1983; Easterbrook and Fuller, 1986; Royalty, 1987) and not penetrate the epidermis of plant leaves and fruits as far as the parenchyma. The mouthparts of the Diptilomiopidae can be 50–70 μm long and are capable of reaching the mesophyll tissue but do not cause clear damage to the host plant (Keifer, 1959b; Jeppson *et al.*, 1975; Krantz, 1978, 2009a).

The cheliceral stylets of eriophyid gall mites penetrate only the cell wall, to a depth of about 2 μm (Westphal, 1972, 1977a). On fruits of *Citrus sinensis* (L.) Osbeck, feeding adults of the pink citrus rust mite, the eriophyid *Aculops pelekassi* (Keifer), produce about 20 punctures, each approximately 1 μm in diameter in the epidermal cells (which are 10 μm long and 7 μm wide). The depth of penetration is about 20 μm and reaches as far as the second and third layer of the fruit epidermis (Tagaki, 1981).

Symptoms of feeding

The eriophyoid mites are highly specific and live only on susceptible host plants with which they are closely associated (Jeppson *et al.*, 1975). Their feeding activity is strictly related to the host plants and causes a great variety of symptoms. The leaf vagrant and gall former species induce different morphological alterations (Jeppson *et al.*, 1975; Keifer *et al.*, 1982). Common feeding symptoms of eriophyoid mites include various types of 'galls' and 'distortions' (Westphal, 1992; Westphal and Manson, 1996). Plant galls are abnormal structures characterized by various shapes, size, colours, etc. Among the mites, only eriophyoids and some tropical tenuipalpids (Oldfield, 2005) are gall formers. Eriophyoids induce galls with a typical form that are placed on a specific green organ of the plant (stems, leaves, petioles, buds, flowers, fruits and seeds), but not on the roots (Westphal, 1992). The different effects caused by species living on the same host plant (Boczek and Griffiths, 1994) suggests that gall morphology is determined by the pest mites rather than the host plant (Kane *et al.*, 1997). It has been observed that *Cecidophyopsis* spp. cause the same type of galls on different *Ribes* spp. (Amrine *et al.*, 1994); likewise, three *Aceria* spp. induce similar galls on different maple leaves (Rančić and Petanović, 2008). However, the same eriophyoid may cause different symptoms on different plant species or on different developmental stages of the same host plant (Westphal, 1980; Oldfield, 1984; Soika and Kielkiewicz, 2004).

A single gravid female (sometimes also the immatures) may induce the development of a gall (Westphal *et al.*, 1990). The galls are a localized growth reaction exhibited by the host plant, although the feeding activity of the eriophyoid can also severely affect the growth patterns of the host plant via morphogenetic mechanisms. The gall-forming process generally develops in two steps, first there is local inhibition of the growth and differentiation of the infested plant organ, and next the adjacent tissues grow to shelter the mites in their gall, associated with the formation of nutritive tissues (Němec, 1924; Westphal, 1977a).

According to their morphology, Küster (1911) classified the galls caused by various organisms as 'organoid galls' and 'histoid galls', respectively. Organoid galls are abnormal modifications of plant organs and histoid galls have more complex arrangements of the internal tissues. The histoid galls are divided into the 'kataplasmas', which are usually less differentiated, and

the 'prosoplasmas', which are more complex. The recent description of gall typology by Larew (1982) includes among eriophyoid galls both 'simple' and 'complex' kataplasmas, which are subcategorized by the thickness of the nutritive layer at gall maturity or the alignment and general type of nutritive cells. Galls may also be divided according to the adaptation of the mites to live externally to the epidermis or inside the plant tissues (Mani, 1964). A further classification based on the plant organ that is infested and the degree of damage complexity classifies galls as 'leaf galls', 'stem galls', 'bud galls', 'fruit galls' and 'other distortions' (which include other distortive and non-distortive alterations and toxaemias) (Westphal, 1992; Westphal and Manson, 1996). The structure of the various types of galls are described by Jeppson *et al.* (1975) and by Keifer *et al.* (1982), and the latter authors published a number of very interesting pictures of eriophyoid feeding symptoms.

Leaf galls

Leaf galls include 'erinea', 'blister galls', 'roll galls', 'vein galls' and 'pouch galls'.

ERINEA. The erinea are felt-like masses consisting of an unusual development of plant hairs or 'trichomes'. According to the eriophyoid species, these hairs may be unicellular or multicellular, elongate, globular, lobate or ramified (Fig. 7.1). The erinea may be set either on the lower or upper leaf surface. They have various sizes and sometimes occupy the whole leaf, or cause distortions, and may exhibit a characteristic coloration (Westphal, 1977a). Abnormal hairs may be associated with leaf rolling, pouch galls, witches' brooms or inflorescence galls. Erinea may or may not be associated with other leaf modifications, such as leaf bulging, convex swellings on both sides of the leaves, different colours, etc. (Westphal and Manson, 1996). The grape mite, *Colomerus vitis* (Pagenstecher), causes erinea that are associated with leaf bulging.

BLISTER GALLS OR POCKET GALLS. Some eriophyoid mites reach the internal leaf tissues through a stomatal aperture either in the lower or the upper leaf surface. Their feeding activity induces abnormal growth of the mesophyll tissues associated with an increase of the lacunae and swellings of the leaf lamina (Westphal and Manson, 1996). The pear leaf blister mite, *Eriophyes pyri* (Pagenstecher), causes blister galls.

ROLL GALLS. The feeding activity of some eriophyoid mites on the leaf margin may cause rolling of various size and shapes (Fig. 7.2A), depending on the species, and consist of a well-defined serration or affect the entire length of the leaf margin. Roll galls can develop upwards or downwards, be thickened or not and have one or more windings (Westphal and Manson, 1996). *Aceria stefanii* (Nalepa) causes roll galls on pistachio leaves.

VEIN GALLS. The feeding activity of *Aceria macrotrichus* (Nalepa) on the lateral veins on the upper leaf surface of the European hornbeam (*Carpinus betulus* L.) causes a vein elongation forming sinuous prominent galls (Westphal and Manson, 1996).

POUCH GALLS. These are very common galls of various sizes (from a few to over 10 mm wide), shapes (nail, bladder, spindle or finger, woolly bead, rounded, etc.) and colours (red, yellow, green, etc.) (Fig. 7.2B). Depending on the species, they develop on both leaf surfaces or on only one leaf surface (Westphal and Manson, 1996). *Aceria brachytarsa* (Keifer) causes pouch galls on walnut.

Stem galls

Few eriophyoids cause stem galls. Their formation involves the mite penetrating into the bark of young twigs through lenticels or splits near the insertion of bud scales, as in Norway maple (*Acer platanoides* L.) infested by *Aceria heteronyx* (Nalepa). The feeding activity of the mite causes outgrowth of the internal tissues, with a central cavity covered by nutritive tissues. These galls are usually small, close and coalescing. They persist in time, becoming woody and turning brown (Westphal and Manson, 1996). *Acalitus phloeocoptes* (Nalepa) causes stem galls on plum twigs.

Bud galls

Bud galls include 'big buds', 'bud proliferation and witches' brooms' and 'inflorescence galls'.

BIG BUDS. The feeding activity of some eriophyoid mites induces the infested buds to develop to several times their normal size. These buds consist of thickened scales containing fleshy nutritive excrescences and may host hundreds of mites. The buds remain closed and fail to develop leaves (Westphal and Manson, 1996). The filbert bud mite, *Phytoptus avellanae* Nalepa, on hazelnut (*Corylus maxima* Mill.) (Fig. 7.2C) and *Cecidophyopsis ribis* (Westwood) on blackcurrants are typical examples.

BUD PROLIFERATION AND WITCHES' BROOMS. Some eriophyoid mites infest buds causing anomalies of growth pattern in different parts of the plant (Figs 7.2E–F, 7.3A–C). The feeding activity of these mites induces the proliferation of adventive buds or the development of dormant buds, accompanied by other symptoms, such as leaf reduction and short internodes, inducing the formation of cauliflower-like galls, such as the stunted twigs with crowded buds caused by *Aceria mangiferae* Sayed on young mango trees. Further formation of branchlets, usually short and thin, produces a typical malformation, called witches' brooms. These witches' brooms may be typical, such as those caused by *Aceria celtis* (Kendall) on *Celtis occidentalis* L., or cause a different kind of damage, such as that on the common chrysanthemum infested by *Paraphytoptus chrysanthemi* Keifer in North America and the British Isles (Westphal and Manson, 1996).

INFLORESCENCE GALLS. Some plants infested by eriophyoid mites show typical inflorescence galls. The phenomenon is due to the feeding activity of the mites, which inhibits the differentiation of the floral organs and induces the modification of various parts of the host plant into curving and rolling leaf-like formations. *Aceria peucedani* (Canestrini) causes this type of gall on

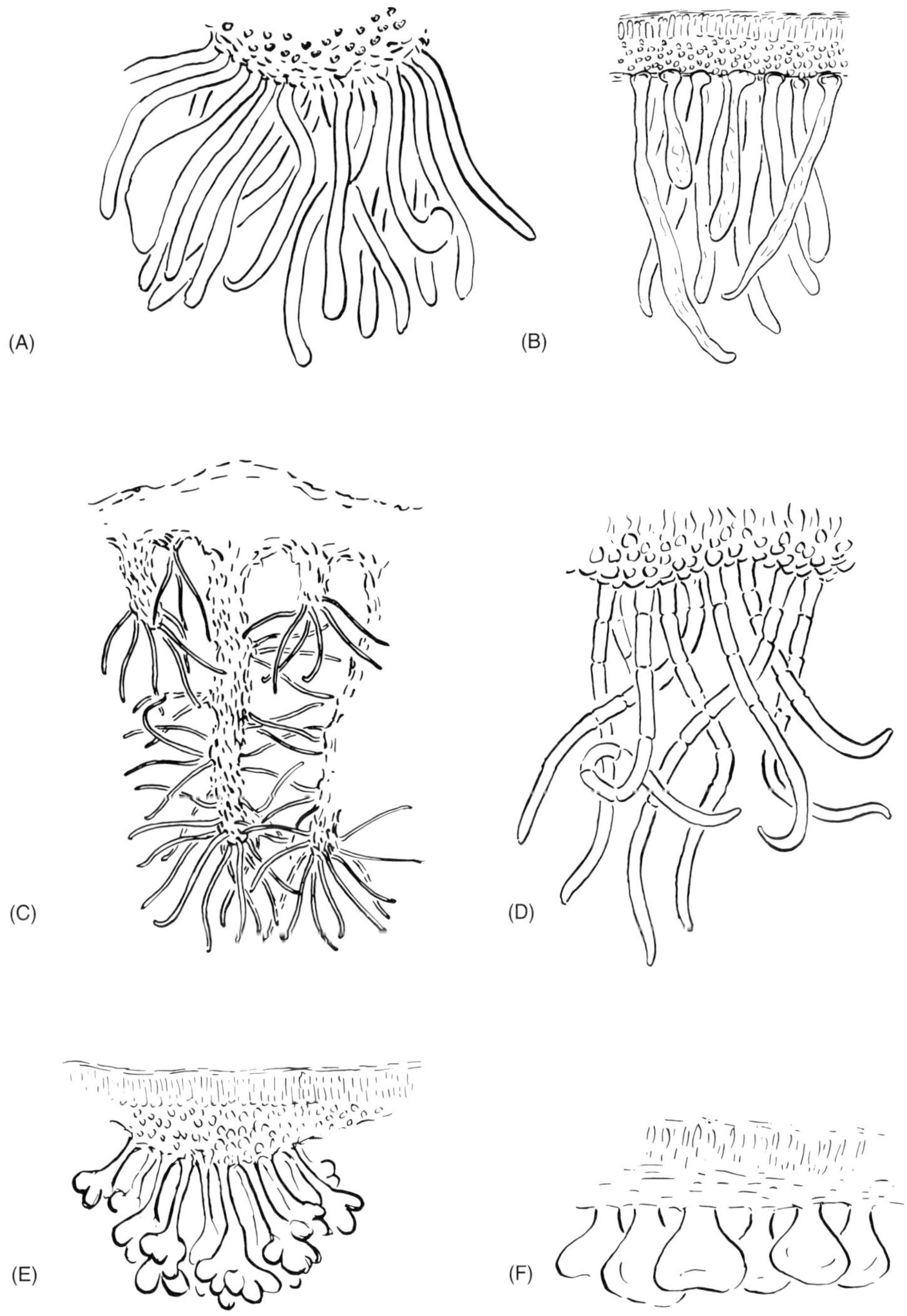

Fig. 7.1. (A) Unicellular hair-like erineum on petiole gall caused by *Aceria caulis* (Cook) on *Juglans nigra* L.; (B) unicellular erineum caused by *Phyllocoptes calisorbi* Keifer on *Sorbus* leaves; (C) erineum caused by *Aceria erinea* (Nalepa) on English walnut; (D) multicellular hair-like erineum on litchi leaf caused by *Aceria litchii* (Keifer); (E) compound capitate erineum on alder leaf caused by *Acalitus brevitarsus* (Fockeu); and (F) simple capitate erineum on sugar maple caused by *Aceria modesta* (Hodgkiss) (from Jeppson *et al*., 1975).

carrot and pimpinella (*Pimpinella saxifraga* Linnaeus), and *Aceria fraxinivora* (Nalepa) causes inflorescence galls consisting of masses of small leaves on *Fraxinus* spp., which are easily visible and pendulous on the tree during the winter. *Aceria aloinis* (Keifer) damages the leaves and flowers of *Aloe* sp. (Denmark and Keifer, 1979).

Fruit galls

These anomalies are uncommon and are produced by *Trisetacus quadrisetus* (Thomas), which causes 'berry' galls on junipers (Fig. 7.2D). The feeding activity of the mite induces slight swelling and hypertrophy of internal tissues of the female cones,

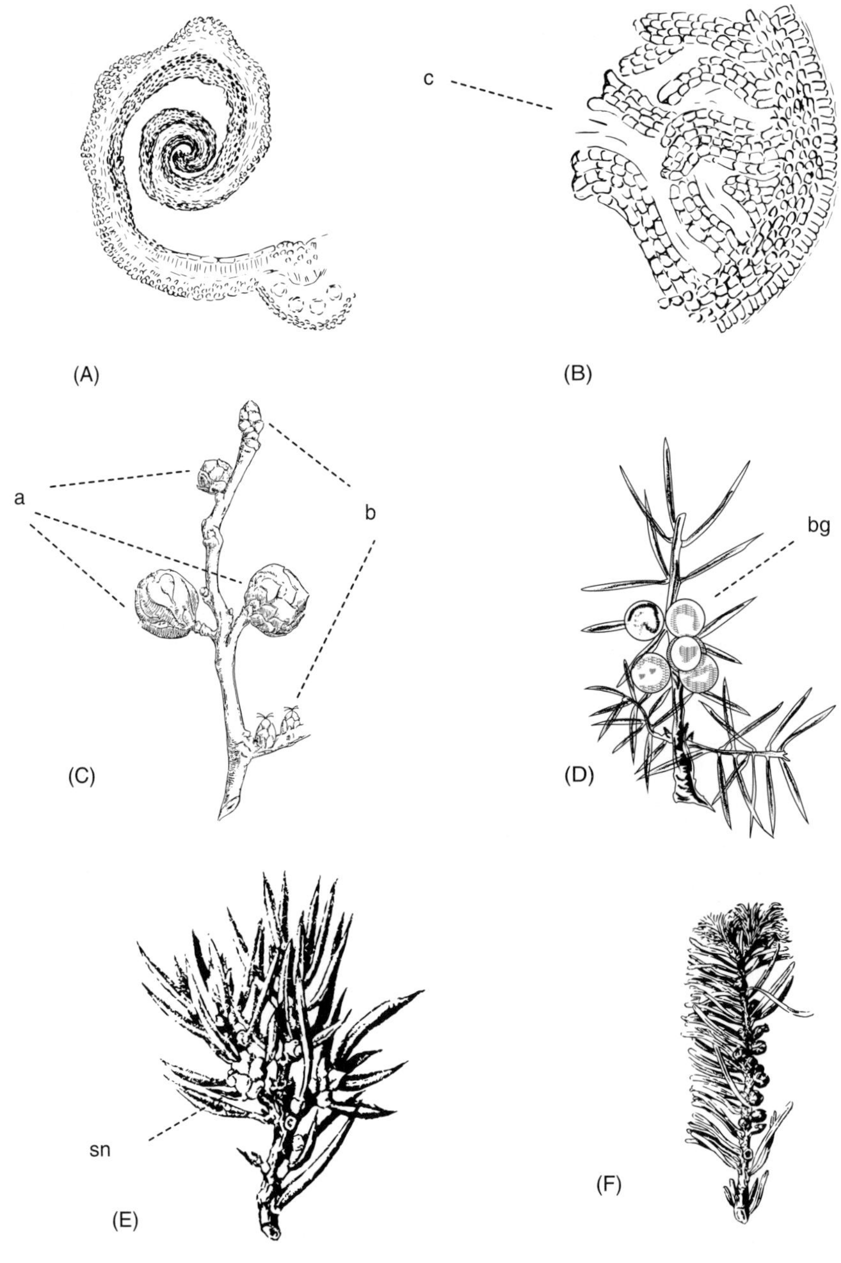

Fig. 7.2. (A) Cross section of leaf roll gall on pistachio caused by *Aceria stefanii* (Nalepa); and (B) interior of pouch gall caused by *Aceria brachytarsa* (Keifer) on walnut (from Jeppson *et al.*, 1975). (C) Big buds and normal buds caused by *Phytoptus avellanae* (Nalepa) on a hazelnut branch (from Targioni Tozzetti, in Berlese, 1925). (D) Berry galls on juniper caused by *Trisetacus quadrisetus* (Thomas); and (E) swellings of needles on juniper caused by *Trisetacus juniperinus* (Nalepa) (from Houard, 1903, and Schlechtendal, 1911). (F) Alterations caused by *Trisetacus floricolus* (Trotter *et* Cecconi) (from Houard, 1903). Key: a, big buds; b, normal buds; bg, berry galls; c, inside; sn, swelling of needles.

hindering the fusion of the different parts of the cones with the formation of an opening at the top of the berry (Westphal and Manson, 1996).

Other distortions

In addition to the symptoms or malformations discussed above, other different distortions are associated with the feeding activity of eriophyoids. Some malformations have a short-lasting effect on plant development, causing the death of the mites and not impeding plant growth (Westphal *et al.*, 1996). Other malformations may have a more long-lasting effect. Usually, this phenomenon characterizes free-living eriophyoids that are not confined within galls and are dispersed on the host plant. These species cause distortions of various organs, but do not induce the formation of new morphological structures. The symptoms usually

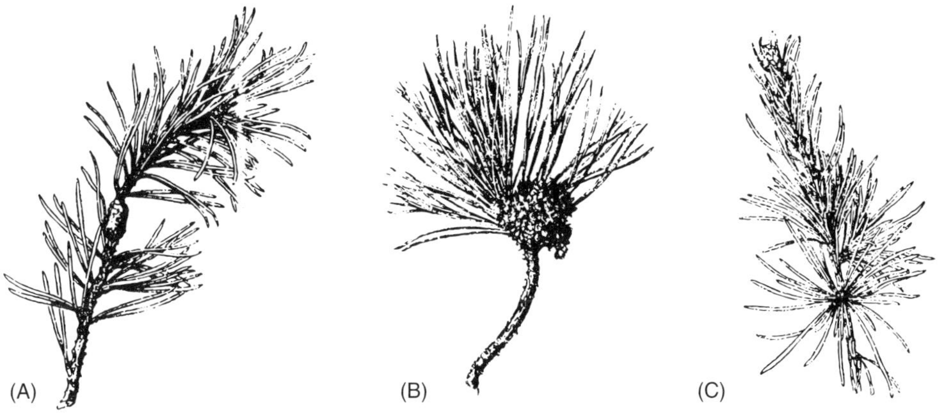

Fig. 7.3. (A) Knots on a pine twig caused by *Trisetacus pini* (Nalepa) (from Schlechtendal, 1911). (B) Bud proliferation on pine caused by *Trisetacus cembrae* (Tubeuf); and (C) alterations on pine caused by *Trisetacus laricis* (Tubeuf) (C) (from Tubeuf, 1910).

involve changes in the appearance of the epidermis of green leaves, young fruits, young stems or bud bracts and are referred to as rust, russeting, browning, bronzing, silvering, chlorotic spots, reddening, etc. These symptoms could be primarily described as a discoloration of the plant organs, but they may also be associated with other symptoms. Petanović and Kielkiewicz (2010b) reported a comprehensive list of these distortions, which included 'brown scarification and necrosis of infested buds, mortality of buds, witches' broom or multiple branching owing to the death of primary growing points, distortion of veins, chlorotic spots, concentric ring blotch, mottle effects, hairless spots on the lower surfaces of attacked leaves, hardening and late ripening, irregular streaks, defoliation (if the collapse of lower leaf mesophyll follows extensive epidermal damage), lignifications, yellowish white areas, stunting, twisting, curling, desiccation of terminal buds or whole terminal shoots, distorted fruits, and premature drop'.

The most common non-distortive effects are 'toxaemias' and 'other non-distortive' feeding effects, which are usually considered to be specific and distinct types of alterations (Oldfield, 1996b). In addition, Petanović and Kielkiewicz (2010b) reported a series of 'complex and mixed symptoms'.

TOXAEMIAS. Toxaemias are macroscopic visual changes on plant epidermal tissues caused by agents in the saliva injected by some eriophyoids during their feeding activity. The short cheliceral stylets of these mites do not reach the cells of the tissues located below the epidermal layer, and the alterations of the tissues underlying the epidermal cells that result suggests that toxaemias are caused by a transfer of a toxicogenic agent(s). The alterations appear on leaves as chlorotic areas associated or not with veins and sometimes with distorted effects, such as wrinkling of the leaf surface, etc. On fruits, toxaemias may cause striking colour changes. Eriophyoid species causing distortions (leaf, bud, stem galls) do not produce toxaemias or other non-distortive effects on plants (Oldfield, 1996b).

Several eriophyoids cause toxaemias. *Aculus fockeui* (Nalepa *et* Trouessart) induces toxaemias on developing green tissues and russeting on the epidermal tissue of mature leaves (Oldfield, 1996b). *Calepitrimerus vitis* (Nalepa) causes brown scarification and necrosis on the winter buds of grapes, death of the buds and witches' broom, and the leaves turn yellow and drop (Carmona, 1973). *Phyllocoptes gracilis* (Nalepa) produces leaf blotches of chlorotic spots or rings on tayberry (Jones *et al.*, 1984). The feeding activity of *Aceria medicaginis* (Keifer) on young leaves of *Trifolium* spp. induces toxaemias (Ridland and Halloran, 1980a). *Aceria tulipae* (Keifer) causes 'kernel red streak' on maize (Nault *et al.*, 1967), leaf stunting, twisting and curling of tulip bulbs, onions and garlic (Conijn *et al.*, 1996; Perring, 1996), and premature bud drop or 'bull head' blooms on *Camellia japonica* L. (Subirats and Self, 1972). Toxaemias may include 'red berry disease', caused by the feeding activity of *Acalitus essigi* (Hassan), in which the ripening of the drupelets of blackberry (*Rubus* spp.) is prevented (Hamilton, 1948). Further details on toxaemias are given in the discussions of single eriophyoid species (see Chapters 16–18).

OTHER NON-DISTORTING FEEDING EFFECTS. The feeding activity of leaf vagrant species on leaves and other green plant organs may cause non-distortive feeding symptoms, such as russeting, bronzing, silvering and discoloration of leaves, stems and fruits.

The citrus grey mite, *Calacarus citrifolii* Keifer, causes 'concentric rings blotch' on citrus foliage. The symptoms consist of minute chlorotic or necrotic spots which gradually coalesce at the mid-veins. The production of resin gives these blotched areas a dark brown colour associated with darkish rings of resinous tissues. In addition, the mite secretes white wax on the upper side of the leaves (Doidge, 1925; Dippenaar, 1958a,b,c).

The tomato russet mite, *Aculops lycopersici*, lives on various solanaceous plants, and its feeding symptoms vary according to the host plant. Thus, it causes severe russeting on the leaves and stems of tomato, while on potato stems it produces little bronzing, with leaf symptoms similar to those of the tomato. Leaves of the aubergine show slight russeting and crinkling. Infested tomatillo plants do not have russeting, while *Convolvulus* leaves become silvery (Rice and Strong, 1962).

The citrus rust mite, *Phyllocoptruta oleivora*, feeds on the leaves and fruits of different species of citrus. Its cheliceral stylets are about 7 μm long and infestation affects the epidermal

cells at a depth of 6–12 μm (McCoy and Albrigo, 1975). The mite destroys the stomatal guard cells on the lower leaf surface, inducing uncontrolled water loss. The epidermal cells turn brown and sometimes the lower leaf surface shows collapse of the mesophyll tissue, associated with yellow chlorotic patches which gradually become necrotic spots. The leaf drying may induce leaf abscission. On leaves that are severely infested, the upper leaf surface is slightly shiny and appears rough and dull bronze (McCoy, 1976). The fruits are usually more rusty than the leaves. Fruit damage first occurs on the epidermal layer, and afterwards also on cells of the underlying layer. It involves both the oil glands and parenchymatous structures, and appears as dark groups of cells surrounded by clear and healthy cells. The browned cells contain lignin but not lipid, callose or tannin. Ethylene emission has been observed. Some cells are punctured up to 26 times (McCoy and Albrigo, 1975). Bronzing and peel shrinkage occur on 'Valencia' orange fruits infested late in the season, and they have less juice, more soluble solids and acids, and higher concentrations of acetaldehydes and ethanol (McCoy *et al.*, 1976a).

The apple rust mite, *Aculus schlechtendali* (Nalepa), damages the leaves and fruits of apple. The mite feeds on the upper and lower leaf surfaces, damaging mesophyll tissue and causing callus tissue to form over the parenchyma in regions where the epidermis is destroyed. The hypodermal layer undergoes accelerated cell division, resulting in the formation and subsequent rupturing of periderm tissue (Easterbrook and Fuller, 1986; Royalty and Perring, 1996). The feeding activity of *A. schlechtendali* on the lower leaf surface causes epidermal cell collapse and the rupture of cell walls and the plasmalemma; this induces the degeneration of the cells of the spongy parenchyma associated with the thickening of cell walls and intercellular deposition of lignin-like compounds (Kozlowski and Zielinska, 1997; Kozlowski, 1998). Leaf damage appears as browning or rusting of the undersides during the summer, and is most obvious on leaves of the last shoots. High levels of attack may cause premature leaf fall. Damage is most severe on young trees in nurseries and in newly planted orchards (Easterbrook, 1996). In spring, the feeding activity of the apple russet mite on developing fruitlets causes russeting (Ciampolini *et al.*, 1976; Easterbrook and Fuller, 1986). The damage occurs from full bud until early July. Cells of the epidermal layer are damaged and cork gradually formed with russeting of the fruit. The lignification of the thickened cell walls is in the regions adjacent to the epidermal cells destroyed by the feeding activity of the eriophyoid on the leaves and fruits. Besides lignin formation, phellogen or 'cork cambium' forms within the attacked tissues (Easterbrook and Fuller, 1986; Kozlowski, 1998).

The coconut flower and nut mite, *Aceria guerreronis* Keifer, lives on coconut flowers and feeds on young nutlets. The green surface of the young nutlets shows triangular, pale yellowish white or whitish marks located at tightly adpressed bracts when the nuts are about 5 cm long. An increase of infestation enlarges the blemishes, which become brown and corky, sometimes forming deep fissures with exudate of gum. Irregular growth of the fruit causes cracking; damage is usually only to a part of the fruit and the kernel is partially reduced in size. Nuts about 20 cm long host a low density of mites under the bracts (Hall and Espinosa, 1981). Infested fruits turn brown and corky in appearance with sometimes extensive cracking reaching deep into the mesocarp. Fruits may be small and deformed as a result (Mariau, 1986; Schliesske, 1988; Howard and Moore, 2008).

Oxycenus maxwelli (Keifer) causes silvering of leaves and premature leaf drop on olive trees (Hatzinikolis, 1982). The feeding activity of *Calacarus carinatus* (Green) produces bronzing on tea leaves, followed at a high density of attack by desiccation and premature leaf drop (Shiao, 1976).

COMPLEX AND MIXED EFFECTS. Symptoms that are difficult to separate from those of pathogens transmitted by eriophyoids or by mixed infestation and/or infection can be attributed to eriophyoids. Several other symptoms (chlorosis, phyllody, flower virescence, internode shortening, enations, fasciation, axillary bud growth, necrotic spots, wart-like structures, blastomania, witches' brooms, stunting, multiple inflorescences and vegetative and floral malformations) may be confused with those caused by other agents (thrips, fungi, viruses, phytoplasmas, microelement deficiency). Toxaemias and non-distortive feeding effects are only known from the macroscopic visual changes of the host plant infested by the mites. Eriophyoid mites can be either vectors of diseases or prefer infested/infected plants, but many such symptoms have been proved to be caused exclusively by eriophyoid mites (Oldfield, 1996b). Petanović and Kielkiewicz (2010b) treat the topic extensively.

Damage

Eriophyoids feed on young and tender tissues, frequently on meristems, because of the high nutritional value of their cells. This adaptation forces them to inhabit particular plant organs, such as buds, sheaths, leaf veins, fruit calyxes, etc. Moreover, these microhabitats provide protection from natural enemies (Royalty and Perring, 1996). The feeding activity causes cellular and histological effects, which may or may not be associated with chemical and physiological disorders. Depending on the species and the mite population density, the attack can lead to slowdown in growth, leaf, flower and fruit drop, and death of branches or whole plants. The various responses do not have a general pattern and the long-term effects of eriophyoid infestation on host plants range from tolerance (without visible damage) to total damage (with plant death). The feeding activities of leaf vagrants and gall formers result in different morphological alterations (Jeppson *et al.*, 1975; Keifer *et al.*, 1982).

Cytological and histological effects induced by vagrant species

The small size of the mouthparts of eriophyoids forces them to feed on epidermal tissues. The vagrant species attack a single cell and leave shortly afterwards to search for another cell. At a high mite density, this behaviour causes the destruction of epidermal tissue. In the areas where the epidermal cells are destroyed, a layer of callous tissue rich in lignin is formed above the parenchyma. Russeting, bronzing, silvering or discoloration due to the feeding activity of mites arises from an oxidation process within the layer of injured cells. The destruction of

individual epidermal cells requires multiple punctures by many mites or continuous feeding by a single mite. Commonly, the surrounding cells do not undergo any damage, but in some cases the changes that occur involve cells that are not directly damaged by mites (Jeppson *et al.*, 1975; McCoy and Albrigo, 1975; Westphal, 1977a,b; Royalty and Perring, 1988, 1996).

Depending on the vagrant and host plant species, the plant tissues that are damaged vary. *Phyllocoptruta oleivora* and *Aculus schlechtendali* are both leaf vagrants and damage mesophyll tissue. However, on citrus, *P. oleivora* does not feed directly on mesophyll cells, rather, the damage that it causes to epidermal cells on the abaxial leaf surfaces causes the mesophyll cells to collapse (McCoy and Albrigo, 1975), resulting in leaf drop (McCoy, 1976) and fruit abscission (Ismail, 1971). On apple, the feeding activity of *A. schlechtendali* causes the formation of callus tissue over the parenchyma where the epidermis is destroyed. The hypodermal cells increase their division, and formation and rupturing of the periderm tissue occurs (Easterbrook and Fuller, 1986). Massive feeding of *A. schlechtendali* on the lower leaf surface leads to the degeneration of the cells of parenchyma accompanied by the thickening of the cell walls and intercellular formation of lignin-like compounds (Kozlowski and Zielinska, 1997; Kozlowski, 1998). Other leaf vagrant eriophyoids do not damage mesophyll cells, e.g. *Aculops lycopersici* (Royalty and Perring, 1988) and the grass rust mite, *Abacarus hystrix* (Nalepa) (Gibson, 1974). Both of these species generally feed in depressions around the veins on the upper leaf surface. In these areas, they find a suitable microhabitat that is characterized by greater nutritional content. After the destruction of the cells around the veins, both mite species move and feed on exposed leaf surfaces. Parenchymal cells are not attacked and the water loss through the destroyed epidermis causes desiccation and plant death (Royalty and Perring, 1996). Recently, Pećinar *et al.* (2007) found that *Leipothrix dipsacivagus* Petanović et Rector caused significant cellular damage and changes in the mesophyll organization of the curled flower-stem leaves of *Dipsacus laciniatus* L. Infested leaves have hypertrophied epidermal cells and a mesophyll consisting of two to three layers of palisade parenchyma and several layers of non-differentiated parenchyma.

Cytological and histological effects induced by gall-forming species

Cecidogenesis – the process of gall formation – has been extensively investigated in the gall mite *Aceria lycopersici* (Wolffenstein), and in the rust mite *Thamnacus solani* Boczek *et* Michalska, both reared on woody nightshade (*Solanum dulcamara* L.). The feeding activity of these two eriophyoids stimulates a different reaction on this host plant, and comparison of their behaviour has allowed an understanding of the process. On susceptible plants, *A. lycopersici* induces a 'compatible interaction' which is characterized by the formation of witches' brooms, while on resistant plants, it causes an 'incompatible interaction' or hypersensitive response (HR), which is associated with small necrotic lesions and leads to the death of the mite. As a result, the mite develops only on plant varieties upon which it can induce galls. In resistant plants, callose (1,3-β-glucan) does not appear around the mite puncture wounds and the damaged cells collapse within 45 min. The neighbouring epidermal and mesophyll cells do not show hyperplasia, their nuclei shrink and they become plasmolysed. The cells may collapse and form a necrotic spot, but no erinea form and the mite population does not develop (Westphal *et al.*, 1981, 1989, 1990). The hypersensitive response and the absence of galls are associated with the production of several novel proteins referred to as pathogenesis-related proteins (Bronner *et al.*, 1991a). *T. solani* exhibits a compatible interaction with *S. dulcamara*, causing vein russeting and the withering of leaves. The mite kills only the punctured cell, which turns brown and dies, and does not damage the surrounding cells. It then moves on from this cell and seeks a new cell to attack, which exhibits the same limited damage. Severe damage to the host plant depends on there being a high density of populations. This compatible interaction gradually causes leaf russeting as a cumulative effect of severe attack (Westphal and Manson, 1996).

Unlike the rust mites, the gall-forming eriophyoids feed on one epidermal cell for a long time (hours or days). The damaged cell often dies and the neighbouring epidermal and parenchymal cells show morphological alterations that result in various abnormalities (galls, erinea, witches' brooms, etc.). After the penetration of the stylets, a cone-shaped cell wall thickening, called a 'feeding puncture', forms around the puncture, which has been reported for several gall mites (Němec, 1924; Westphal, 1977a; Thomsen, 1988). This has a callosed disc-like structure and appears 10–20 min after stylet penetration (Bronner and Westphal 1987; Bronner *et al.*, 1989). Callose deposition is a widespread phenomenon and is not specific to stylet penetration; for example, it also occurs following superficial mechanical wounding caused by microinjections (Nims *et al.*, 1967) or by the action of fungi (Aist, 1976). The vacuoles of the epidermal and palisade cells surrounding the attacked cells become smaller and the pH inside the vacuoles increases, as does that of the vacuoles of the damaged cells which exhibit a clear alkalinization (pH > 8) within 30 min that is associated with alterations in membrane permeability (Westphal, 1982) and nuclear structure (Bronner *et al.*, 1989). The nucleus swells, occupies a central position and shows progressive DNA-denaturation, with a lower contrast and appearing to be optically empty. The nuclei and nucleoli of the neighbouring cells swell and mitosis increases, thereby providing more cells for mite feeding (Westphal *et al.*, 1981). In infestation by *A. lycopersici*, this mechanism results in erinea or witches' brooms (Westphal *et al.*, 1981, 1990). Hyperplasia of the glandular trichomes and epidermal hairs has also been observed (Westphal, 1985), and forms a favourable microhabitat for mites.

Callose formation plays an important role in cecidogenesis. The suction cone and the nucleus are characterized by chitosan accumulation. The chitosan may be introduced into the cell by the chitinous cheliceral stylets of the mite, because higher plants do not synthesize this derivative of chitin (Bronner *et al.*, 1989), and it seems that it may trigger a modification of cell permeability and alterations in the nuclear DNA. The high affinity of chitosan for DNA molecules and polygalacturonate in plant–pathogen interactions (Hadwiger, 1999) suggests that its accumulation within the nucleus and in feeding punctures

on the walls stimulates the activity of neighbouring cells. Callose is involved in the formation of the callosic disc-like feeding punctures of the compatible interaction (Thomsen, 1988), and is probably also used in the regeneration of the pierced plasmalemma of attacked cells and in other cellular functions. The cellular biochemistry and metabolic responses induced in mite-injured plants have been extensively discussed by Petanović and Kielkiewicz (2010a).

Like callose, the layers of nutritive tissues used by gall mites for feeding are important (Thomsen, 1975; Westphal *et al.*, 1981; Bronner and Westphal, 1987; Bronner *et al.*, 1989). The punctured cells undergo a similar process to that of injured epidermal cells and die, but the maintenance of nutritive tissues (for the mite) and the formation of galls require continuous puncturing (Mani, 1964; Westphal and Manson, 1996). If gall formers are removed, the nutritive tissue turns into regular parenchyma cells (Rohfritsch, 1975). For example, when *Eriophyes tiliae* (Pagenstecher) is removed from the leaves of *Tilia platyphylla* Scopoli and gall formation stops, the gall cells transform into epidermal cells with stomata and mesophyll cells containing chloroplasts (Thomsen, 1975); the epidermal cells also contain anthocyanins.

The macro-morphology of the gall is specific to the mite species. Conversely, the micro-morphology of galls is very similar. It begins with repeated anticlinal and periclinal divisions of the lower epidermal cells, which lead to the formation of both nutritive and mesophyll tissues, but no differentiation into palisade and spongy tissues. The hypertrophied parenchyma cells and nutritive tissue contain starch and proteins, but not lipids. The vascular bundles of the galls relate to those of the healthy leaf surface, but there is no direct connection between the bundles and the nutritive tissue. Fibres and tannins have been observed in gall phloem and in the lignified vascular bundle sheaths. Many unicellular and cylindrical hairs may grow on the gall surface (Petanović and Kielkiewicz, 2010a). Other cytological and histological features of the gall include hyperplasia of the epidermal and parenchymal cells adjacent to the damaged cells (Anthony *et al.*, 1983). Hypertrophy of the nuclei and nucleoli associated with hyperplasia of the epidermal cells has also been reported (Moha, 1969), as well as hypertrophy and hyperplasia of the epidermal and mesophyll cells and polyploidy in the nuclei of the nutritive cells of the galls (Wcislo, 1977). Polyploidy and multiple nuclei have been observed in the epidermal cells of galls too (Kendall, 1930; Hesse, 1971; Westphal, 1974). Different gall mites induce fragmentation of the vacuoles, increased amounts of endoplasmic reticulum, increased numbers of mitochondria and ribosomes, and plasmid alteration (Westphal, 1970, 1977a).

Some gall mites induce the formation of different gall types, such as *Eriophyes goniothorax* (Nalepa) (Boczek and Griffiths, 1994), and *Phytoptus tetratrichus* (Nalepa), which causes edge rollings on the small-leaved linden (*Tilia cordata* Mill.) and warty galls on the upper sides and erinea on the undersides of the leaves of the silver linden (*Tilia tomentosa* Moench.) (Soika and Kielkiewicz, 2004).

The feeding activity of eriophyoid mites modifies the quality of plant tissues and affects the behaviour of other herbivorous arthropods that feed successively on the same host plants. Thus, severe infestation of *Aculus schlechtendali* on apple trees inhibits the development of the European red spider mite, *Panonychus ulmi* (Croft and Hoying, 1977). In other compatible interactions, e.g. between *T. solani*, *P. oleivora*, *Aculops lycopersici* and susceptible host plants, the quality of the food resource within infested organs favours further development of the mites (Royalty and Perring, 1996; Westphal and Manson, 1996; Westphal *et al.*, 1996). Infestations of *Aceria cladophthirus* (Nalepa) trigger the incompatible interaction (or HR) in resistant *S. dulcamara* plants, which induces the death of *A. cladophthirus* and, successively, increases the mortality of *T. solani* (Westphal *et al.*, 1991). Curiously, the HR does not protect *S. dulcamara* plants from later infestation by the two-spotted spider mite, *T. urticae* (Westphal *et al.*, 1992).

Chemical and physiological changes

The feeding activity of eriophyoid mites may induce changes in the physiological processes of the host plants, but the effects of both vagrant and non-vagrant eriophyoids on the metabolism of the host plants are variable. Alterations in photosynthesis, concentrations of plant hormones and hormone regulators, and nutrient contents are well known. However, most of the data concern leaf vagrant mites, because the effects of gall-forming mites have not been extensively investigated. The topic is very complex, involving various factors, such as eriophyoid genotypes (or strains), mite population density and period of feeding, and host plant species, cultivar, organ age and environmental conditions (Petanović and Kielkiewicz, 2010a).

The feeding activity of leaf vagrant eriophyoids may reduce leaf gas exchange and thereby affect physiological processes. On peach (*Prunus persica* (L.) Stokes), 3000 *Aculus fockeui* feeding days per leaf reduce the CO_2 assimilation rate and stomatal conductance by about 20%, and there is a negative relationship between gas exchange and damage to the epidermis of the adaxial leaf surface (Andersen and Mizell, 1987). A high density of *A. fockeui* on *Prunus cerasifera* L. reduces photosynthesis by more than 70% and respiration by more than 50% (Zawadzki, 1975). On tomato, 450 *Aculops lycopersici* days cm^{-2} reduce leaflet net photosynthesis by 50% (Royalty and Perring, 1989). On apple, a density of *A. schlechtendali* of 4000–5000 mites per leaf causes direct damage to the leaf epidermis and indirect damage that is mainly correlated with changes in the mesophyll cells of leaves and reduction of the net CO_2 exchange by as much as 65%. Furthermore, the stomata are partially open and affect the transpiration process, causing a decrease of guard cell turgor and the desiccation of epidermal and parenchyma cells (Spieser *et al.*, 1998).

Galled leaves within a heavily galled portion of *Prunus americana* Marsh. infested by *Phytoptus emarginatae* (Keifer) reduce the photosynthetic surface of the nearby leaves as a result of the reallocation of resources within the infested shoot (Wilson and O'Dowd, 1990). Galls induced by *Phytoptus cerasicrumena* Walsh reduce the photosynthetic rate of galled leaves of *Prunus serotina* Poit. *et* Turpin by 47% and the photosynthesis of nearby ungalled leaves within the same shoot by 24% (Larson, 1998).

The presence of hormonal substances in the saliva of eriophyoid mites has been discussed by various authors (Jeppson *et al.*, 1975; Westphal, 1992; Boczek and Griffiths, 1994).

Lipophobic substances have been reported to be secreted by the mouthparts of *Aceria caulobia* (Nalepa), along with IAA and cytokinin-like compounds, suggesting the presence of plant growth regulatory substances in the mite saliva (de Lillo and Monfreda 2002, 2004). The feeding activity of eriophyoids may cause alterations in the concentrations of plant hormones and hormone regulators, usually associated with galling, growth abnormalities and damage. The introduction into the host plants of auxin 'protectors' (*o*-dihydroxyphenols, coumarins, polyphenols, gibberellins) that bind with peroxidase and IAA oxidase (indole-acetic-acid-oxidase), which usually degrades IAA, suggested that there was an abnormal accumulation of IAA in the damaged cells that was stimulating unusual growth and gall formation. Research on *Ziziphus jujuba* Lamk. and *Z. Mauritania* Lamk. infested by *Aceria cernuus* (Massee) highlighted IAA regulation in the galls (Purohit *et al.*, 1979; Tandon and Arya, 1980, 1982; Tandon, 1982). The citrus bud mite, *Aceria sheldoni*, induces auxin regulation and plant growth abnormalities in the buds of lemon, *Citrus limon* (L.) Burman. The feeding activity of the mite stimulates an increase in the level of phenols in the bud tissues with a simultaneous decline in auxin activity, with alterations in ribonuclease (RNase) activity. Infested tissues do not show any inhibition of IAA-oxidizing enzymes, but the auxin produced is oxidized, inducing inhibition of growth. The increase of polyphenol oxidase that is found is probably due to the conversion of polyphenols to quinones (Ishaaya and Sternlicht, 1969, 1971), which react with proteins to form melanin and other tannins, which are typically brown, as are the lemon buds damaged by the mite. On the apple cultivar 'Pioneer', a density of *A. schlechtendali* of 22–27 cm^{-2} of leaf surface induces an increase in the ratio of total phenolic compounds to carbohydrates from 3:1 in control to 5:1, and a decrease in the palatability of the injured host plant to mites (Kozlowski, 1998). On blackberry, a density of *Epitrimerus gibbosus* (Nalepa) of 600–1000 per leaf changes the nutritional value and palatability of the infested leaves after 3 weeks (Shi and Tomczyk, 2001). In the leaves of a 'Thornfree' blackberry cultivar, an increase in the concentration of phenolics is associated with decreases in carbohydrate concentration, whereas in leaves of a 'Thornless' evergreen cultivar, an increase was found in the soluble protein concentration. As an appropriate balance between feeding inhibitors and stimulators is a very important factor in the eriophyoid–plant interaction, it is believed that a rise in the ratio of total phenolics to total carbohydrates/proteins in infested blackberry leaves results in a decrease in successful mite feeding.

The localized necrotic lesions caused by *Aceria lycopersici* are associated with an increase of peroxidase activity which is involved in the oxidative cross-linking of proline-rich proteins in cell walls (Bronner *et al.*, 1991b). This mechanism renders the cell walls indigestible to enzymes and characterizes the hypersensitive response of host plants to pathogens (Dixon *et al.*, 1994). Other enzyme proteins, such as chitinase or 1,3-β-glucanase, accumulate in response to the feeding activity of *A. lycopersici* (Bronner *et al.*, 1991a). Several enzymatic proteins are not only induced locally but also systemically in the leaves of tomato plants (*Solanum lycopersicum* L.), in response to short-term feeding by *A. lycopersici*. The local induction of peroxidase (POX, hydrogen peroxide (H_2O_2) oxidoreductase) activity was much more evident than the induction of polyphenol oxidase (PPO) or lipoxygenase (LOX) (Stout *et al.*, 1996).

Alterations in the nutrient contents were observed in leaves of *P. cerasifera* infested by *A. fockeui*, e.g. a reduction in carbohydrate content and increased nitrogen, Ca^{2+}, Na^{+}, K^{+} and chlorophyll contents (Zawadzki, 1975). Leaves of *Jasminum auriculatum* Vahl infested by *Aceria jasmini* ChannaBasavanna showed a reduction of carbohydrate to 21% less than in uninfested leaves, associated with a reduction of 31% of the soluble sugars, probably as a result of the inhibition of photosynthesis (Rajagopal *et al.*, 1970). The contents of nitrogen, PO_4^{2-}, and K^{+} of the sheath leaves of sugarcane infested by *Abacarus sacchari* ChannaBasavanna were respectively 23, 16 and 37% greater than those of uninfested leaves (Sithanantham *et al.*, 1975). Studies on the effects of feeding by *Aceria acnisti* Keifer on *Acnistus cauliflorus* Schott. and by *Phyllocoptes bougainvilleae* Keifer on *Bougainvillea spectabilis* Willd. on the host plants show that *A. acnisti* induces the formation of galls on both leaf surfaces, with a tenfold increase in the number of galls associated with an increase of 50.6% in fresh weight and 55.63% in dry weight of the leaves. Moreover, the content of six macronutrients increased markedly, and potassium and phosphorus increased 299.37% and 105%, respectively. The micronutrients iron and copper increased 173% and 81.8%, respectively. These results indicate that plants invest large amounts of their resources in gall production, although it should be noted that the nutrient analyses do include the mites living on the plant tissues. In *Bougainvillea* leaves infested by *P. bougainvilleae*, there was a negligible decrease in nitrogen content, a 19% decrease in calcium and a 9.8% decrease in manganese, but there were increases in the other elements analysed. In both of these latter cases, there was an increase in leaf copper contents: 81.8% in *A. cauliflorus* infested by *A. acnisti* and 100% in *B. spectabilis* infested by *P. bougainvilleae* (Flechtmann and Berti-Filho, 1994).

Gall mites may induce different changes in the metabolism of galled leaves of walnut, apple and pear trees. In walnut leaves, the walnut erineum mite, *Aceria tristriatus* (Nalepa), induces no changes in soluble proteins and a marked decrease in carbohydrate, with a slight decrease of carbohydrates in the morphologically unaltered leaf tissues surrounding the mite erinea. The concentration of phenolics remains almost unchanged. A slight increase in the concentration of these compounds occurs within the leaf tissue between the galls. The erinea occupy at most 15–20% of the total leaf surface, but a reduction of five times occurs in the rate of photosynthesis (Tomczyk and Boczek, 2006). On apple leaves, infestations of the apple leaf blister mite, *Eriophyes mali* (Nalepa), induce an increase in soluble protein associated with a lack of changes in total carbohydrates and an increase in phenolics (Tomczyk and Boczek, 2006). On pear leaves, the pear leaf blister mite, *E. pyri*, causes a large increase in soluble proteins in slightly galled leaves but a rapid decrease in strongly galled leaves (Tomczyk and Boczek, 2006). Leaves of the linden (*T. cordata*) galled by *E. tiliae* show a decrease in the content of carbohydrates, with a non-uniform distribution of ^{14}C among the leaf blade, midribs and galls (Boczek, 1974). Studies on the composition of carotenoids in the leaf blade of *T. cordata* and in galls caused by *E. tiliae* var. *rudis* Nalepa showed that seven carotenoids were present in the leaf blade but missing in the galls, whereas three carotenoids

were located specifically within the galls. This indicates that the feeding activity of *E. tiliae* promotes the biosynthesis of some carotenoids and hinders that of others (Czeczuga, 1975).

Eriophyoid galls contain a variable amount of tannins. Within each gall type, apart from the nutritive tissue layers, there are numerous cells accumulating *o*-dihydroxyphenols, which are commonly considered to be plant defence compounds. The deterrent, antifeedant or toxic effects of condensed tannins on many arthropod herbivores are well known, as exemplified by the finding that differences in tannin content between alder genotypes at the early stages of gall formation can significantly delay gall development and differentiation of the defensive response. Another known function of tannins and phenolic compounds, especially those located in the outer layer of the galls, is protection of the gall formers from parasitoids, predators and pathogenic infections (Mound, 1994; Hartley, 1998). The gall and leaflet tissues of ash produced by *Aceria fraxini* (Garman) have few tannins, whereas the galls of willow caused by *Eriophyes aenigma* Walsh (= *Aculops aenigma* (Walsh)) and the galls of filbert caused by *Phytoptus avellanae* have many tanniferous cells in their walls, but their nutritive tissues have few tannins. The nutritive cells of very young alder galls induced by *Eriophyes laevis* (Nalepa) contain tannins, and the nutritive tissue accumulates more tannins as the alder gall matures. Week-old pouch galls induced by *E. laevis* on the leaves of European alder, *Alnus glutinosa* Gaertn., have the same structure as the galls on the leaves of *Alnus incana* (L.) Moench., but differ in the presence of tannins in the vascular bundle sheaths in undifferentiated cell layers. Tannin deposits did not appear within bundle sheaths and cells derived from the mesophyll until the galls of *A. incana* were 3 weeks old. Comparison of the internal structure of 3-week-old galls show that the galls of *A. incana* are more differentiated and have more tannin deposits than the galls of *A. glutinosa* (Kane *et al.*, 1997).

Impact on growth, flowering and cropping

The feeding activity of several eriophyoids has impacts on the growth, flowering and cropping of economic plants.

On lemon, the extent of blossom and fruit abscission significantly increases with the distortion caused by the feeding of the citrus bud mite, *Aceria sheldoni*. Distortion of the fruit is significantly linked to that of the blossoms from which they develop. It is probable that the increase in blossom and fruit abscission may depend on a decrease in auxin activity and other biochemical changes in the axillary buds infested by the mite (Phillips and Walker, 1997). On navel orange, the mite causes a reduction in fruit size and the number of fruits (Schwartz, 1972).

The tomato russet mite, *Aculops lycopersici*, develops on the stems, leaves, flowers and fruits of tomato plants (and other plants). The stems become bronzed or russeted, and may crack longitudinally or check. The leaves become brown and paper like and usually wilt. Severe infestations cause leaf drop, even before the leaves change in colour. The flowers lose turgor, cease to develop, show russeting of the sepals and fall (Jeppson *et al.*, 1975). Most yield loss is due to mite feeding on the flower stalks and pedicels, which dry and cause flower bud death (Kamau *et al.*, 1992). Severely infested fruits do not develop, and gradually become russeted and roughened (Jeppson *et al.*, 1975).

The pink citrus rust, *Aculops pelekassi*, develops on the leaves, green stems and fruits of citrus. Its feeding activity affects the diameter, volume and weight of citrus fruits, compromising their marketing as fresh fruit. The sugar content of the juice is higher in the damaged fruit, and the increased concentration of soluble solids most likely occurs through water loss caused by mite feeding (Tono *et al.*, 1978). A density of 200 or more mites per leaf is destructive (Childers *et al.*, 2007).

The citrus rust mite, *Phyllocoptruta oleivora*, feeds on the leaves and fruits of all citrus species. The result of infestations on the lower (abaxial) surfaces of the leaves differs from that on the upper (adaxial) side, and in citrus fruits; attacks on the upper side are confined to the epidermal cells and show symptoms similar to the russeting of immature fruit (Albrigo and McCoy, 1974). In severe attacks, the cuticle of the upper surface may lose its shine and become bronze with yellowish spots in areas that have been 'degreened' by the release of ethylene during the wounding process (McCoy and Albrigo, 1975), while on the lower surface the attack can lead to appreciable mesophyll collapse visible as degreened yellowish spots that later become necrotic (Albrigo and McCoy, 1974). Severe attack on the lower leaf surfaces can increase transpiration and stimulate defoliation (McCoy, 1976). Damage to fruits varies depending on the variety and stage of fruit development, and results in russeting or bronzing. A high density of the eriophyoid population causes reduced size of the fruits (Yothers and Mason, 1930; Schwartz, 1975a), fruit water loss (McCoy *et al.*, 1976a; Allen, 1979), fruit drop (Allen, 1978) and alterations in juice quality (McCoy, 1976). Although the aesthetic damage that is caused has a commercial impact, the total content of soluble solids does not appear to be affected and the fruits that result can be used for industrial applications without any problem (Chandramani *et al.*, 2004).

Severe attacks of the grape mite, *Colomerus vitis*, on grapevines significantly reduces the growth of shoots and canes and the number of dead buds increases (Smith and Schuster, 1963).

On peach trees, *Aculus fockeui* causes a reduction in the weight, sugar content and acidity of next year's fruit. Severe injury to the peach leaves causes a reduction in the sugar content of fruits in the same year, and damaged trees show lower vigour as a result of postharvest defoliation, resulting in lower fruit quality in the next year (Kondo and Hiramatsu, 1999a).

On *Medicago sativa* L., infestations by the alfalfa bud mite, *Aceria medicaginis*, reduce the length of the petioles at each stem node, as well as the leaf surface, shoot weight and number of axillary shoots (Ridland and Halloran, 1980b, 1981).

The dry bulb mite, *Aceria tulipae*, infests shallots, onions, garlic and tulips. Severe infestations induce a mean weight reduction of 19%, and the chronic impact of mite infestation is more severe than the acute damage that occurs during a short infestation (Santos and Lima, 1976). Severely infested bulbs have a 20% reduction in emergence and a 23% yield reduction (Larraín, 1986).

The coconut flower and nut mite, *Aceria guerreronis*, lives on coconut flowers and feeds on the young nutlets. Infested fruits turn brown and corky in appearance, sometimes with

extensive cracking reaching deep into the mesocarp. The fruits that result may be small and deformed (Mariau, 1986; Schliesske, 1988; Howard and Moore, 2008), and infestation may also affect copra yield, causing reductions of about 30% in American and African countries (Hernández Roque, 1977; Julia and Mariau, 1979a; Moore, 1986) and about 40% in India (Nair and Koshy, 2000; Muthiah and Bhaskaran, 2000). A later study in India (in Kerala) by Haq and Sobhal (2010) demonstrated that infestation by *A. guerreronis* reduced the weight of copra from highly infested coconuts by 32%, but showed that when mite infestation was low, the weight loss was only 7%. In Venezuela, early dropping of the coconuts induced a reduction in copra yield of about 70% (Doreste, 1968a), and in Tanzania and the African coastal islands of Mafia, Zanzibar and Pemba, the reduction in copra yield ranged from 10 to 100% (average 21%) (Seguni, 2002).

A density of 4000–5000 apple rust mites (*Aculus schlechtendali*) per leaf affects flower formation and the quality of apple yield. Fresh fruit weight, soluble carbohydrate content and skin colour intensity show a stronger decrease in the 'Jonagold' than in the 'Golden Delicious' variety, indicating the possibility that the latter cultivar is relatively resistant to the mite (Spieser *et al.*, 1998). Easterbrook and Palmer (1996) found that according to the cultivar, the level of infestation affected flower formation and the quality of apple yield.

Tarsonemidae

Only a restricted number of the Tarsonemidae are feeders on higher plants, and they belong to five genera. Other tarsonemid species probably feed on fungi and/or algae; their mouth appendages seem to be unable to penetrate the tissues of higher plants (Jeppson *et al.*, 1975).

Feeding mechanism

The feeding mechanism of the Tarsonemidae involves various mouthparts, including the stylophore, the gnathosomal capsule, the palpi, the chelicerae and the pharyngeal complex. The stylophore is formed by the fused cheliceral bases, and is integrated with the infracapitulum to form the gnathosomal capsule. This latter is usually ovoid, sometimes prolonged, and composed of modified cheliceral bases and rostral elements. The palpi are simple and are adapted to tactile function or in some cases to eating solid food particles – certain species are capable of swallowing whole fungal spores. The chelicerae are simple, stylet like, retractable and, according to the feeding habit, of various length (Fig. 2.7G). Each stylet is attached proximally to a 'cheliceral lever', which relates it to the retractor and protractor muscles. The pharyngeal complex lies in the ventral part of the gnathosomal capsule, and consists of sclerotized parts, muscular sheaths and a pair of gland-like structures, probably salivary glands. A great diversity of pharyngeal structures occurs among the various species of the family, which seem to be more suited to penetrating the cell walls of fungi and of soft tissues rather than ligneous tissues such as leaves and mature shoots (Jeppson *et al.*, 1975; Lindquist, 1986; Nuzzaci *et al.*, 2002).

Symptoms of feeding

The symptoms resulting from feeding by tarsonemids vary according to the species and are sometimes unclear or doubtful. Only a restricted number of species included in the genera *Steneotarsonemus* Beer, *Phytonemus* Lindquist, *Polyphagotarsonemus* Beer *et* Nucifora, *Hemitarsonemus* Ewing and *Tarsonemus* Canestrini *et* Fanzago are plant feeders. Some tarsonemid mites initiate fungal infection (*Steneotarsonemus* spp.).

The pineapple leathery pocket mite, *Steneotarsonemus ananas* (Tryon), causes brown spots on the white tissues at the base of the leaves of pineapple, especially in young plants (Meyer Smith, 1981a). In South Africa, the presence of the mite is associated with 'leathery pocket disease', in which there is hardening of the walls of the seed cavities and nectary ducts of the pineapple fruitlet (Petty, 1975). The hardening is probably due primarily to the fungus *Penicillium funiculosum* Thom (Rohrbach, 1979, in Meyer Smith, 1981a).

The feeding activity of the West Indian sugarcane mite, *Steneotarsonemus bancrofti* (Michael), causes transparent brown crater-like depressions on the young sugarcane stalk, and the plant tissues appear to be scabby or scarred. As the cane grows, these areas successively become opaque, brown and rough (Holloway, 1936; Butani, 1959).

Steneotarsonemus furcatus De Leon causes severe distortion of the leaves and shortened internodes on the ornamental grass *Paspalum* spp., and on *Maranta leuconeura* E. Morren, resulting in stunted plants (Jeppson *et al.*, 1975).

The bulb scale mite, *Steneotarsonemus laticeps* (Halbert), feeds mainly on epidermal surfaces of the scales of bulbs of *Narcissus* and other Amaryllidaceae. On developing shoots, it causes distortion, stunting and death of leaves and flowers. Scar-like tissues on the leaves and flower stems cause bronze streaks and transverse cracks. Infested bulbs have yellowish brown areas (Doucette, 1936; Carmona, 1966; Jeppson *et al.*, 1975).

The rice panicle mite, *Steneotarsonemus spinki* Smiley, has a probable Asian origin and was first described on the plant hopper, *Sogata orizicola* Muir, in Louisiana in 1960 (Smiley, 1967). It is a rice pest in China, the Philippines, Taiwan (Smiley *et al.*, 1993), India, Kenya (Rao and Das, 1977), Cuba (Ramos and Rodríguez, 2001), the Dominican Republic (Ramos *et al.*, 2001), Japan (Shikata *et al.*, 1984), Korea (Schall, 2004), Haiti and Columbia (Ochoa, 2004, in NPAG, 2008). In addition, the rice panicle mite has been reported in Belize, Costa Rica, Guatemala, Haiti, Honduras, Nicaragua, Panama, the Philippines, Sri Lanka, Thailand (Návia *et al.*, 2006a; EPICA, 2007; GPDD, 2007), the USA (Texas, Louisiana and Arkansas), Mexico (Arriaga, 2007) and Puerto Rico (USDA, 2007). Reports of the mite in the Americas regions have been recently discussed by Návia *et al.* (2010b). This tarsonemid feeds on rice (*Oryza sativa* L.), on the invasive *O. latifolia* Desv. and on other *Oryza* spp. (USDA NRCS, 2007). Symptoms, such as necrotic bands and spots, develop on the rice grains and leaf sheaths when

populations are large, and populations greater than 450 mites cm^{-2} have been recorded (Ramos and Rodríguez, 2001).

Steneotarsonemus spirifex (Marchal) causes spiral malformations of the rachis, blind spikelets and incompletely emerged inflorescences of cereals (Evans *et al.*, 1961).

Phytonemus pallidus (Banks) infests various economic plants, such as strawberries and watercress, and many ornamental flowers and shrubs (Van Eyndhoven and Groenewold, 1959). On pikake, *Jasminum sambac* (L.) Aiton, the feeding activity of this mite results in fewer flowers per cluster and often severe abortion of the clusters and buds, poorly formed flowers, with twisted, distorted and smaller leaves. The shoots have elongated internodes (Boyle and Haramoto, 1956). Infested strawberry leaves have a roughened and wrinkled upper leaf surface, with irregular folding and fluting of the leaf margins. The veins bulge upward like blisters, and the petioles fail to elongate, and bear leaflets that fail to unfold completely. These small leaflets become pale yellowish green with a hard brittle texture and turn brown or silvered on exposure to the sun. The flowers and fruits become brown near the inner bases of the sepals, and may turn black and dry (Dustan and Matthewman, 1932).

The broad mite, *Polyphagotarsonemus latus* (Banks), feeds on tender tissues and is common on the young leaves, flowers and immature fruits of a large variety of crops in subtropical and tropical countries, and in the greenhouses of temperate regions. On *Gerbera*, the lower leaf surface becomes rigid and rolled under at the edges and turns bronze, and the flowers have distorted and discoloured rays. Overall, the feeding activity of the mite causes curling and crinkling of the leaves and patches of blisters. High mite population densities stop the plants growing and cause death (Smith, 1939a). On potato plants, feeding symptoms consist of blackish spots on the undersurfaces of young leaves, which turn gradually reddish. The plant becomes rosetted, showing prominent leaf hairs and leaves wrinkled at the edges. The plant withers and the buds are killed (Mann *et al.*, 1920). On pepper seedlings, the infestation prevents flower and fruit development, later causing the dropping of flowers and silvering of fruits (Kulkarni, 1923).

P. latus also damages young leaves and fruits of citrus, particularly of lemon. Its feeding activity causes distortions and curlings of the leaves, sometimes so pronounced as to prevent the development of the shoot. The lower surface of the young leaves can develop a shining and a silver colour. The fruits are infested as soon as they start to develop and until they reach the size of a walnut, and the mite populations develop more frequently on young fruits in the shade and inside the apical part of the canopy. The attack causes a silvering of the pericarp with irregular contours and sometimes over the entire fruit surface, and with the progress of the infestation and the development of the fruit, the pericarp develops russeting and an irregular shape, with a variable colour from white to silver to brown (Nucifora, 1961; Jeppson *et al.* 1975; Vacante, 2010).

On tomato, the stems of terminal shoots and the undersurfaces of the young leaves infested by the broad mite become shiny bronze or brownish. The first type of injury consists of browning of the epidermal cells, but later the cells of developing tissues collapse and die. The young expanding leaves become narrow, stiff, twisted or crumpled, fail to develop and dry up. The stem of the young plant becomes swollen, roughened or russeted and greyish green. On cotton, the first symptoms occur on the lower leaf surfaces, and the leaves become rigid and roll under at the edges. Mature leaves may split or crack open, with various shapes (Hambleton, 1938).

Cucumber plants (cultivars 'M450' and 'Kfir') infested by *P. latus* produce more ethylene than uninfested plants. The broad mite feeds on epidermal cells, inducing their disappearance or collapse. The leaves of infested plants are thicker, smaller and harder, and have additional layers of more compact cells without mesophyll differentiation (Grinberg *et al.*, 2004). The plants also show growth inhibition, have fewer leaves and a decreased leaf area. These aberrations involve the entire leaf tissue, and severe infestation induces a complete loss of the epidermis and an increase in mesophyll cell size and number. The epidermal tissue appears to have collapsed, and the mesophyll cell walls are thick and distorted. The morphological, structural and ultrastructural changes that are induced by *P. latus* are associated with an induction of the genes involved in plant defence pathways (Grinberg *et al.*, 2005). Mite feeding induces the expression of the defence pathway genes *BGL2*, *LOX1*, *LOX2* and *PRX*, and the transcript levels increase between 24 and 72 h post infestation. In contrast, mechanical wounding does not induce these transcripts, with exception of a slight induction of *LOX1*. The gene *ACO1* shows a uniform expression pattern. Feeding by the broad mite causes an induction of both the salicylic acid (SA) and the jasmonic acid (JA) pathways (Grinberg *et al.*, 2004), and infested leaves exhibit the induction of genes related to the JA/ethylene and SA pathways, such as lipoxygenase (*LOX*) and β-1,3 glucanase (*BGL2*), and also an induction of oxidative stress-responsive transcripts, such as those of peroxidase (*PRX*). The transcript levels of ACC oxidase (*ACO*), which is involved in ethylene biosynthesis, remained relatively constant (Grinberg *et al.*, 2005).

The fern mite, *Hemitarsonemus tepidariorum* (Warburton), feeds on various species of ferns in greenhouses in the USA and the UK (Beer, 1954). Mite feeding causes little depressions on the frond surface, which successively become brown speckles. Attacked fronds also show distortions and swellings (Cameron, 1925; Pritchard, 1951a).

Tarsonemus myceliophagus Hussey is a serious pest in mushroom culture (Hussey and Gurney, 1967).

Damage

A close relationship between symptomatology and economic damage usually occurs on flowers and ornamental plants infested by tarsonemid mites (and other pests), because the first symptoms are sufficient to cause damage. On fruit plants, although symptoms and damage also occur on the foliage, the damage is mainly (but not exclusively) related to yield losses, e.g. *Polyphagotarsonemus latus* infestations on lemon fruits are similar to those caused by the eriophyoid *Phyllocoptruta oleivora*, and may result in a reduction of commercial value of up to 100%. In South Africa, lemon losses of between 30 and

70% have been reported from *P. lata* (Meyer Smith and Schwartz, 1998e). In the same country, *Steneotarsonemus ananas* causes the development of leathery pocket disease on pineapples causing up to 16% of the fruit damage (Petty, 1977).

The damage caused by *Steneotarsonemus spinki* on rice can be direct, due to toxin injection during feeding, or indirect, through the spreading of phytopathogens (Návia *et al.*, 2010b). When population levels of the mite are low, *S. spinki* is typically found feeding on the internal surfaces of leaf sheaths, but at high density the mite populations spread on to all plant parts, including the kernel (Chow *et al.*, 1980; Ramos and Rodríguez, 2001). In China, the Philippines and Taiwan, the mite has caused substantial crop losses, ranging from 30 to 90% (USDA, 2007). In addition to reducing the value of the rice by scarring the kernel, the mite affects yield by transmitting the fungus, *Acrocylindrium oryzae* Sawada, and the mycoplasma-like organism, *Spiroplasma citri* Saglio *et al.*, both of which cause rice sterility (Chow *et al.*, 1980). Ramos and Rodríguez (2001) estimated that the combined activities of feeding and disease transmission by *S. spinki* lowered rice yield in Cuba by 30–60%.

Infestations of *Hemitarsonemus tepidarum* on ferns can stop plant growth and cause the plant to die (Cameron, 1925; Pritchard, 1951a).

Other Pest Mites

In addition to the mites discussed above that are equipped with typical mouthparts which relate them closely to phytophagy, other mites in different families (Acaridae, Erythraeidae, Histiostomatidae, Penthaleidae and others), although characterized by different mouth structures, may also severely damage economic plants.

Feeding mechanism

The mouthparts of mites lacking stylet-like chelicerae (present in some Penthaleidae) can sometimes lacerate the tender tissues of young leaves, bulbs, etc. and suck up the internal contents. Various morphological structures have been reported that enable this to occur. The Acaridae have the dorsal surface of the gnathosoma not extended and forming a small lobe between the cheliceral bases. The chelicerae are chelate, or pincer like, dorsally free and laterally compressed and with a large basal region stretching forward and forming the digitus fixus articulated with the digitus mobilis. Both chelicerae move in a vertical plane and are often toothed distally. The chelicerae are located under the epistome and over the labrum, and are partially involved in the mouth structure (Griffiths, 1970; Hughes, 1976) (Figs 2.4, 2.5A and B, 2.6A). The chelicerae of the Erythraeidae are retractable into the idiosoma, and the palp is equipped with a thumb-claw process bearing a long straight chela (pincer) (Walter *et al.*, 2009); *Balaustium medicagoense* Meyer *et* Ryke feeds on plants using its adapted mouthparts to probe the leaf tissue and in this way obtain food (Arthur and Umina, 2010). The chelicerae of the Histiostomatidae are laterally compressed, not chelate, but with fixed comb-like digits and bearing many fine teeth. The movable digit is short, fused to the base of the fixed digit, and with few short distal teeth (Hughes, 1976; O'Connor, 2009) (Fig. 12.1C). The Penthaleidae have chelate chelicerae, separate at the base and each bearing a seta; the fixed digit is regressed, commonly slender and sometimes finger like at the tip, and the movable digit is sickle like to prong like, and smooth and sometimes strongly stylet like (Baker, 1990, 1995) (Fig. 15.2A). The penthaleid *Penthaleus major* (Dugès) penetrates the epidermal cells of the host plant with chelicerae and removes the cellular contents, making a hole larger than 3 μm (Qin and Halliday 1996a,b). The movable digits of *Penthaleus* spp. can be moved separately or together, but do not form a channel for transporting fluid. The subcapitulum is applied to the plant surface and the cell contents are sucked out using a pharyngeal pump (Nuzzaci and de Lillo, 1991a).

Symptoms of feeding

The feeding symptoms of the erythraeid *Balaustium medicagoense* on canola are distorted and cupped cotyledons, which may have a leathery appearance. Common damage on cereals, grasses and pulses is 'silvering' or 'whitening' of the infested leaves, similar to that caused by red-legged earth mites or blue oat mites (Arthur and Umina, 2010).

The feeding symptomatology of the blue oat mite, *Penthaleus major*, is similar to that of tetranychid mites. The removal of the plant cell contents typically results in silvery to grey patches on the leaf surface. In the paddock, this is often mistaken for frost damage. The leaves appear greyish or silvery and at high levels of attack for several days the tips of the leaves show a scorched aspect, turn brown and the plant may die. Many of the infested plants do not die, but become stunted and produce little forage or grain (Jeppson *et al.*, 1975).

On *Gerbera*, the symptoms of the feeding activity of the acarid *Tyrophagus neiswanderi* Johnston *et* Bruce involve the young leaves and consist of more or less extensive curling of the internal parts of the leaf surface, especially in the vicinity of the central veins, accompanied by distortion of the margins of the leaf, and morphological alterations of the flowers. On expanded leaves, the attack can also involve irregular and jagged tears of the surfaces, with the leaf margins more or less curved and regular. On cucumber, melon and pumpkin, the young leaves show similar alterations, while on expanded leaves alterations of the tissues are more frequent in the areas adjacent to the peduncle (Vacante, 1988, 1997b; Vacante and Benuzzi, 2007). Other examples of symptom occur in other families.

Damage

A few examples of the damage that may be caused by miscellaneous groups of pest mites are given here to illustrate the diversity of the topic, which will be extensively treated in the chapters on each single family. The damage caused by *Penthaleus major* on young plants is more severe than that on large, healthy ones. Small grains suffer two types of damage, which give

either a reduced amount of forage throughout the winter or reduced yields of grain in the spring and summer. The reduction in forage probably has a greater importance in those countries where winter grazing is used by farmers who rely on the small grains for fall and winter pasturing of livestock (Jeppson *et al.*, 1975). Mite damage reduces the available feed for livestock and severely damages young seedlings; most damage occurs at the establishment phase and this may lead to entire crops needing resowing (Ridsdill-Smith and Pavri, 1998).

Infestations of *T. neiswanderi* on young plants in nurseries and/or post transplanting can cause serious economic damage. The economic impacts of mite infestation of ornamental plants may be similarly severe. For example, infested *Gerbera* flowers are not marketable (Vacante, 1988, 1997b; Vacante and Benuzzi, 2007).

The histiostomatid *Histiostoma feroniarum* (Dufour) has been collected on decaying food material such as wet grain or flour (Hughes, 1976) and on edible fungi (Hussey *et al.*, 1969), and is responsible for serious crop losses to *Agaricus bisporus* (Lange) Pilát, *A. bitorquis* (Quél.) Sacc. and *Lentinula edodes* (Greenslade and Clift, 2004).

Transmission of Plant Diseases

According to Jeppson *et al.* (1975), 'The term plant disease in its broadest meaning includes all injuries or abnormalities generated from sources outside the plant regardless of the cause'. This definition of plant diseases therefore includes mite species that inject toxins or growth regulators into the host plant and also species that transmit virus diseases or spread fungal diseases to the host plant. The relationship between mites and plant disease can be divided into four distinct mechanisms: the causation of local injuries, the injection of systemic or persistent toxins, virus transmission and the transmission of other diseases. The first two of these have already been covered in this chapter. This section examines the other two: virus diseases and diseases caused by fungi or other organisms that are carried by mites or facilitated by their feeding activities.

Species belonging to the Eriophyoidea, Tetranychidae and Tenuipalpidae are known to vector one or more plant viruses (Robertson and Carroll, 1988; Skaf and Carroll, 1995; Oldfield and Proeseler, 1996; Smidansky, 1996; Rodrigues *et al.*, 2000, 2003; Kitajima *et al.*, 2010; Childers and Rodrigues, 2011). However, only the Tenuipalpidae and the Eriophyoidea include important vectors of plant pathogens. There is a high degree of host specificity between these mites and the viruses that they transmit (Oldfield and Proeseler, 1996). No virus is transmitted by more than one eriophyoid or occurs in more than one host plant, and eriophyoid-borne viruses are not transmitted by any other group of mites or insects. This suggests that both mite and virus find a mutual benefit in the transmission and that their association derives from a process of selection, which through a mutualistic virus–mite association has led to a form of co-evolution (Sabelis and Bruin, 1996). Here, a brief treatment is given of the general features of the common diseases that are transmitted. The mite vectors themselves are discussed in the later chapters on separate mite families.

Viruses and virus diseases transmitted by Tenuipalpidae

Brevipalpus spp. transmit the disease of leprosis, which is also known as lepra explosiva, and also passion fruit green spot virus (PFGSV), coffee ringspot virus (CoRSV), and orchid fleck virus (OFV), all of which belong to the Rhabdoviridae family (Childers and Rodrigues, 2011). In addition, tenuipalpids may transmit diseases of ornamental plants (*Annona*, *Brunfelsia*, *Clerodendron*, *Hedera*, *Hibiscus*, *Ligustrum*, *Malvaviscus*, *Pelargonium*, *Pittosporum*, *Salvia*, *Schefflera* and *Thunbergia*) (Kitajima *et al.*, 2003b).

Leprosis/lepra explosiva

Leprosis, or lepra explosiva, is a severe virus disease of citrus reported from the American Continent (Argentina, Bolivia, Brazil, Central America, Colombia, Costa Rica, Honduras, Mexico, Panama, Paraguay, Venezuela) (Kitajima *et al.*, 1972; Rodrigues *et al.*, 1997; Childers *et al.*, 2003a; Bastianel *et al.*, 2010; Childers and Rodrigues, 2011). Currently, the known vector of citrus leprosis in Brazil is the red and black flat mite, *Brevipalpus phoenicis* (Rodrigues *et al.*, 2003). However, in the past, both *B. californicus* in Florida and *B. obovatus* Donnadieu in Argentina have been reported as vectors of citrus leprosis (Frezzi, 1940; Vergani, 1945; Knorr, 1968), but morphological studies of the specimens collected from these earlier reports highlighted the presence of both *B. californicus* and *B. phoenicis* spp. in the Florida samples and of *B. obovatus* and *B. phoenicis* in the Argentinian samples (report from Ochoa, R., US Department of Agriculture, Agricultural Research Service, Systematic Entomology Laboratory, Baltimore, Maryland, 2002, in Childers and Rodrigues, 2011). These earlier reports did not define the vector status of the tenuipalpids involved either. Recently, *B. californicus* and *B. phoenicis* were reported as vectors of citrus leprosis in Guatemala (Palmieri *et al.*, 2007), but no direct transmission tests have been carried out to confirm that *B. californicus* is a vector, and neither are there any available specimens for either *B. phoenicis* or *B. californicus* from citrus samples collected in that country. At present, the only confirmed vector of citrus leprosis in the western hemisphere is *B. phoenicis* (Childers and Rodrigues, 2011).

In Florida, citrus leprosis has been known since the late 1800s, but there have been no reports of it since 1960 (Morishita, 1954; Childers *et al.*, 2001), and later research has not found the virus to be present in Florida and Texas (Childers *et al.*, 2003c). There is then some confusion on the presence and distribution of the virus, and there are few clear differences between damage resulting from feeding and infection with the leprosis virus (Childers *et al.*, 2003a). Until a few years ago, the agent responsible for the disease was identified as citrus leprosis rhabdovirus (CiLV), which occurred in two forms, the more common CiLV-C, which infects the cytoplasm of infected plant cells, and CiLV-N, which occurs in the cell nucleus; the two forms produce the same symptoms (Rodrigues *et al.*, 2003). Recently, the virus associated with the cytoplasm was identified as the type member of the new virus genus, *Cilevirus*, and the virus associated with the nucleus was included in the new genus '*Dichorhabdovirus*', both belonging to the family Rhabdoviridae

(Bastianel *et al.*, 2010). The cytoplasmic type is referred to as *Citrus leprosis virus C* (CiLV-C) and is vectored by *B. phoenicis* (Rodrigues *et al.*, 2003). If the mite vector is not controlled, CiLV-C can kill a tree within 3 years. This *Brevipalpus*-borne virus is one of the most serious emerging exotic diseases threatening sweet orange production within the USA, the Caribbean and potentially other citrus-producing countries in Africa, Asia, Australasia and Europe. Citrus species, especially oranges (*Citrus sinensis* (L.) Osbeck), can be infected by citrus leprosis viruses. Mandarins and hybrids such as 'Murcott' are considerably less susceptible. Annual costs of prevention of the disease in Brazil alone are about US$100 million (Rodrigues *et al.*, 2003). The damage from the disease negatively affects yield and more than US$60 million per season are spent on chemical sprays to control the mite vector (Rodrigues, 2006). Bastianel *et al.* (2010) report that in Brazil the annual costs of the disease to growers are around US$80 million dollars. The virus mainly attacks sweet orange, but also citrange, citron, 'Cleopatra' mandarin, grapefruit, lemon, mandarin, sour orange and tangor (Lovisolo, 2001). The main symptoms are lesions to the fruits, leaves, twigs and small branches, with early fruit drop, leaf drop and death of the twigs and branches until the vitality of the trees declines (Rodrigues *et al.*, 2003).

Passion fruit green spot virus

The passion fruit green spot virus causes a cytoplasmic viral disease vectored by *Brevipalpus phoenicis* that is reported only from Brazil. The symptoms include green spots and patches of green tissue on mature leaves, and the flowers show necrotic lesions that encircle the stems. High population densities of *B. phoenicis* cause leaf and fruit drop. Mature yellow fruits show green spotting along with patches of green spotting on the leaves. The most serious damage results from the necrotic lesions that girdle the stems and kill the plants (Kitajima *et al.*, 2003a,b; Moraes *et al.*, 2006; Childers and Rodrigues, 2011).

Coffee ringspot virus

The coffee ringspot virus causes a nuclear type of viral disease of coffee in Brazil and Costa Rica, and is vectored by *Brevipalpus phoenicis* (Chagas *et al.*, 2003). Symptoms include localized ring spot lesions on both leaves and berries. The disease may cause severe leaf and fruit drop with accompanying reduced coffee berry yields. According to Chagas *et al.* (2003), a similar disease occurs in the Philippines.

Orchid fleck virus

The orchid fleck virus is a nuclear type of virus reported from Australia, Brazil, Denmark, Germany, Japan, Korea and the USA (Kondo *et al.*, 2006), which is transmitted by the false spider mite, *Brevipalpus californicus*, to several orchid genera (Kondo *et al.*, 2003). The virus has been classified in the new genus '*Dichorhabdovirus*' in the family Rhabdoviridae. The virus particles accumulate in the nucleus of infected orchid plants (Kondo *et al.*, 2006; Bastianel *et al.*, 2010), and give rise to the symptoms of chlorotic or necrotic spots and rings on the leaf fronds of several orchid genera (Kondo *et al.*, 2006). The vector is infective for a long time, and the actual distribution of the virus is likely to be much greater than indicated above (Childers and Rodrigues, 2011).

Virus diseases transmitted by Tetranychidae

Few tetranychids are vectors of diseases, and among the viruses that they transmit only the transmission of *Potato virus Y* (PVY) is well known (see below) – that of other viruses is either uncommon or has not been sufficiently investigated. These latter include the mosaic symptoms associated with the feeding of *Tetranychus sinhai* Baker on wheat, maize and rye, which consist of darkening, followed by yellowing and withering in patches along the midrib. However, no data have confirmed the role of a virus, and the symptoms could be due to the injection of toxin by the mites (Slykhuis, 1963, 1965). The brown wheat mite, *Petrobia latens* (Müller), has been reported to transmit a virus-like particle on barley in Montana (Robertson and Carroll, 1988).

Potato virus Y

This widespread virus is transmitted by *Tetranychus* sp. to different Solanaceae (potato, tomato, tobacco and other plants) and displays different severity and symptoms in the various growing areas in which it is found. The first symptoms are observed 19 days after infection as a blotchy mottling on the top leaves, which spreads outward from the veins. On the lower leaf surface, a fine necrosis appears along the veins, followed by necrosis of the upper leaf surface. The necrosis extends from the petioles to the stems, which show brown lesions. The leaves may become necrotic and withered. In the years that follow, the plants show little necrosis, but exhibit leaf drop, have brittle leaves and stems, short internodes and mottled, twisted and bunched leaves. However, despite the plant appearing dwarfed and rosetted, it produces normal tubers (Thomas, 1969; Oldfield, 1970).

Virus diseases transmitted by Eriophyoidea

Virus transmission by the Eriophyoidea has been known since the 1930s (Amos *et al.*, 1927), but has been difficult to investigate because of confusion in several cases between the effects of mite feeding and the symptomatology of virus infection. The situation is further complicated by the small size of these mites and because it is not possible to verify whether the symptoms that develop depend on the injection of mite toxins or a virus (Jeppson *et al.*, 1975). The assumption that an eriophyoid mite is a vector of plant viruses must be completed by verifying the following: that there is a correlation between the presence of the mites and the disease in nature; that the disease symptoms do not depend on the presence of the mites; and that the transmission of the disease symptoms to healthy plants requires that the mites have fed on diseased plants (Jeppson *et al.*, 1975). About 20 years later, Oldfield and Proeseler (1996) reported that about a dozen plant diseases were transmitted by eriophyoid

mites, but that the nature of the pathogens transmitted was largely unknown. Further studies showed that the pathogens were viruses in the genera *Rymovirus*, *Tritimovirus* (Potyviridae) or *Allexivirus* (Comoviridae) (Jones, 2000; Kumar *et al.*, 2003a, 2005; Kang *et al.*, 2007). It is now known that 12 pathogens are transmitted by nine eriophyoid vectors, but research on these has only just begun and many aspects await elucidation, such as an understanding of the mechanisms of virus transmission (Young *et al.*, 2007) and virus detection within the mite body (Latha and Doraiswamy, 2008). Among the disease agents transmitted by eriophyoids, those infecting dicotyledonous plants are less well known, while those that usually develop on monocotyledons are better known as viruses. The relationship between eriophyoid vectors and disease agents is usually highly specific (Oldfield and Proeseler, 1996).

A summary is given below of the main features of the major diseases caused by viruses transmitted by eriophyoids, and following Oldfield and Proeseler (1996), these are categorized as 'cereal pathogens' and 'pathogens of woody dicotyledonous plants'.

Cereal pathogens

The most important viruses transmitted by eriophyoids on cereals are *Wheat streak mosaic virus*, *Wheat spot mosaic virus*, *Ryegrass mosaic virus* and *Agropyron mosaic virus* (Oldfield and Proeseler, 1996).

WHEAT STREAK MOSAIC VIRUS. The wheat curl mite, *Aceria tosichella* Keifer, is the only known vector of the *Wheat streak mosaic virus* (WSMV) (Slykhuis, 1955). Immature and adult instars transmit the virus to wheat and other wild and cultivated graminaceous (poaceous) species, and the adults can acquire the pathogen as immatures or adults. The virus persists in the wheat curl mite for several days (Slykhuis, 1955; Orlob, 1966). The disease caused is widespread and probably occurs in all areas of wheat cultivation in the Palaearctic and Nearctic regions. The symptoms are light green to yellow streaks and dashes, followed by general chlorosis, stunting and, depending upon the degree of severity, poorly filled seed heads and shrivelled kernels. At temperatures ranging from 20 to 25°C, the symptoms on wheat appear 6–8 days after inoculation (Styer and Nault, 1996).

WHEAT SPOT MOSAIC VIRUS. The *Wheat spot mosaic virus* (WSpMV) is transmitted by *Aceria tosichella*. Symptoms consist of an initial chlorotic spotting or ring spot, followed by general chlorosis, stunting and, sometimes, plant death. This symptomatology has been observed in several Poaceae, including maize, barley and wheat (Slykhuis, 1956; Nault *at al.*, 1970; Styer and Nault, 1996). Studies carried out on the persistence of the virus in the vector showed a lack of transovarial passage and retention after moulting of both WSpMV and WSMV (Nault and Styer, 1970). WSpMV is not sap transmissible (Styer and Nault, 1996).

RYEGRASS MOSAIC VIRUS. *Ryegrass mosaic virus* (RgMV) is transmitted by the cereal rust mite, *Abacarus hystrix*. It is a rod-shaped virus that is sap transmissible but not seed borne and causes mild chlorosis, chlorotic mottle, chlorotic streaks or leaf necrosis (Oldfield and Proeseler, 1996). Elevated amounts of feeding are necessary to cause damage on Italian ryegrass (Gibson, 1974) as a result of the probable absence of phytotoxins from the mite saliva. High densities of attack cause growth retardation and cuticular damage produces wilting (Frost and Ridland, 1996).

AGROPYRON MOSAIC VIRUS. *Abacarus hystrix* is an inefficient vector of *Agropyron mosaic virus* (AgMV) to *Agropyron repens* L. (= *Elymus repens* L. (Gould)), or couchgrass, a weed or forage grass (Slykhuis, 1969), but it causes a secondary disease of wheat and *A. repens* in the USA, Canada, the UK, Finland and Germany (Slykhuis, 1980).

OTHER VIRUSES. *Aceria tosichella* is a vector of three other plant disease viruses apart from WSMV and WSpMV: *Brome streak mosaic virus* (BrSMV) (Stephan *et al.*, 2008), *Triticum mosaic virus* (TriMV) (Seifers *et al.*, 2009), and High Plains disease (HPD), which affects wheat and maize (Mahmood *et al.*, 1998; Seifers *et al.*, 2002; Skare *et al.*, 2003, 2006). It also transmits kernel red streak of maize (Nault *et al.*, 1967; Styer and Nault, 1996; CABI, 2002). The dry bulb mite, *Aceria tulipae*, transmits the important onion mite-borne latent virus (OMbLV) and the shallot mite-borne latent virus (ShMbLV) (van Dijk *et al.*, 1991; Perring, 1996; Oldfield and Proeseler, 1996). Infection by both of these viruses can be asymptomatic (Ostojá-Starzewski and Matthews, 2009). In addition, it has recently been found that *A. tulipae* transmits an *Allexivirus* (GarV-B1) in garlic plants (Kang *et al.*, 2007).

Pathogens and diseases of woody dicotyledonous plants

The most important viruses transmitted by eriophyoids on woody dicotyledonous plants include *Blackcurrant reversion virus*, *Fig mosaic virus*, *Peach mosaic virus* (PMV), *Cherry mottle leaf virus*, *Rose virus 1* and *Pigeon pea sterility mosaic virus* (PPSMV).

BLACKCURRANT REVERSION VIRUS. The *Blackcurrant reversion virus* (BRV) is transmitted by the blackcurrant big bud mite, *Cecidophyopsis ribis* (Jones, 2000; Susi, 2004). Not much was known about the disease up to 1980 (Slykhuis, 1980), although the evidence for its transmission by the mite was clear (Thresh, 1963). Different *Cecidophyopsis* spp. are probably involved in transmission of the disease (Lemmetty *et al.*, 2004). All active instars of *C. ribis* transmit the virus after 3 h of feeding on infected bushes, but the optimum acquisition needs about 50 h. The virus is transmitted after 48 h feeding on healthy blackcurrant bushes (Jacob, 1976). Reversion has been reported for different host *Ribes* species, including *R. nigrum* L., *R. alpinum* L. and *R. spicatum* Robson (Bremer and Heikinheimo, 1979). Medium-to-high levels of infestation cause young buds to become galled, fail to flower and frequently to turn into 'big buds' that fail to differentiate into flowers or true leaves. Buds that are less attacked may develop secondary buds that give rise to grossly deformed or asymmetrical foliage, abnormal flowers

and fruits. Higher levels of attack induce reduction of plant vigour and the leader growth is replaced by many short twiggy side shoots (Thresh, 1964a).

FIG MOSAIC VIRUS. *Fig mosaic virus* (FMV) is transmitted by the fig bud mite, *Aceria ficus* (Cotte), only to *Ficus carica* L. (Oldfield and Proeseler, 1996). One single immature instar or adult could acquire the virus after as little as 5 min of feeding activity on infected plants, and transmit it within 5 min of feeding activity on healthy plants. At 20–30°C, virus acquisition was almost twice as great as at 5–10°C. Acquisition was best from the terminal buds, followed by the lower surface of leaves showing symptoms. Moulting did not impede the ability of the mite to transmit the virus and the adult could inoculate the virus for 6–10 days. Virus transmission does not occur transovarially (Proeseler, 1969, 1972a). The first symptoms of infection on fig seedlings are observed 5–7 days after feeding by inoculated mites and chlorotic spots develop first on the leaves and show cleared veins, chlorotic and mosaic areas. The chlorotic spots are often bordered by a rust band. The fruits bear mosaic symptoms and the trees become stunted and with leaves of reduced size. The virus is graft transmissible but not sap transmissible. The disease is widespread in commercial fig orchards, because of the horticultural practice of propagating from rooted cuttings (Bradfute *et al.*, 1970; Plavsic and Milic, 1980; Appiano, 1982).

PEACH MOSAIC VIRUS. The peach mosaic vector mite, *Eriophyes insidiosus* (Keifer *et* Wilson), is the vector of the *Peach mosaic virus* (PMV) (Gispert *et al.*, 1998), the agent of peach mosaic disease (Wilson *et al.*, 1955). On peach, the symptoms of this disease consist of vein clearing and irregular chlorosis, stunted growth of the leaves, colour breaking on the blossoms and rosetting of branches. In south-western USA and Mexico, these symptoms also occur on certain other *Prunus* fruit trees (Oldfield and Proeseler, 1996). The pathogen is not transmitted transovarially (Wilson *et al.*, 1955). The adults can inoculate the virus for at least 2 days after removal from infected plants, and transmit the virus when they feed for 0.5–1 h on peach seedlings, but not when they feed for only 15 min (Oldfield and Proeseler, 1996). In southern California, the disease spreads during the growing season (Jones and Wilson, 1952) as a result of the dispersal of some of the vector mites during those months. In the USA, peach mosaic disease is also known on plum, apricot and almond, but the vector mite has not been collected on these plants (Oldfield and Proeseler, 1996).

CHERRY MOTTLE LEAF VIRUS. *Eriophyes inaequalis* (Wilson *et* Oldfield) transmits a flexuous virus of the Closterovirus group, the *Cherry mottle leaf virus* (CMLV), which causes cherry mottle leaf (CML) disease (James and Mukerji, 1993), a graft-transmissible disease occurring in sweet cherry-growing districts of western North America from northern California to south-western Canada. The symptoms of the disease are irregular chlorotic mottling, edge tattering and small leaves. The fruits seem to be normal but lack flavour and often ripen late (Oldfield and Proeseler, 1996). The disease is commonly found in orchards with natural stands of *Prunus emarginata* (Douglas) (McLarty *et al.*, 1951), but can sometimes spread through sweet cherry orchards in countries where *P. emarginata* is absent. *E. inaequalis* collected on infected *P. emarginata* can transmit the disease to peach (a symptomless host) or to sweet cherry (Oldfield, 1970). Mites collected on *P. emarginata* in California transmitted the disease to sweet cherry in the laboratory. In British Columbia (Canada), the disease occurs in apricots (Hansen, 1978), but the mite does not reproduce on either sweet cherry or apricot (Oldfield and Proeseler, 1996). This suggests the existence of other mite vectors of the disease.

ROSE ROSETTE VIRUS. The wild rose bud and fruit mite, *Phyllocoptes fructiphilus* Keifer, is vector of a virus that was formerly designated *Rose virus 1* (RSV-1), an *Ilarvirus* (Jesse *et al.*, 2006; Tzanetakis *et al.*, 2006), which causes rose rosette disease (RRD) on the wild multiflora rose, *Rosa multiflora* Thunb. ex J. Murray, a weedy scrambling shrub, and on ornamental roses (see also Chapter 8). Recently, however, the virus has been formally identified as *Rose rosette virus* (RRV), an *Emaravirus* (Laney *et al.*, 2011; Hong and Hansen, 2012). The disease occurs mainly in the western and middle states of the USA and in some Canadian provinces, where many *Rosa* varieties are resistant (Amrine, 1996). The symptoms are a red or purplish vein mosaic on the underside of leaflets, reduction in shoot growth, early breaking and stunting of the auxiliary buds, shortened internodes and the development of bright red lateral shoots and foliage, characterized by small, wrinkled and deformed leaflets (Doudrick, 1983; Amrine and Hindal, 1988). The symptoms commonly occur first on auxiliary buds of new shoots, and gradually move towards the shoot apex. Young plants become dwarfed. Symptoms on ornamental roses consist of yellow mosaic on the leaves, thorniness of the stems, the development of lateral shoots larger than the parent stem, misshapen leaves and the early development of lateral buds, with witches' brooms (Amrine and Hindal, 1988). The disease can be transmitted both by grafts and by the feeding activity of *P. fructiphilus* to *R. eglanteris* L., *R. suffulta* Greene, *R. woodsii* Lindley, *R. multiflora* Thunb. and *R. rubrifolia* Villars (Allington *et al.*, 1968). Single mites can transmit the rosette pathogen, which can be transferred to healthy roses from symptomatic roses. The symptoms persist after mite control (Amrine *et al.*, 1988).

PIGEON PEA STERILITY MOSAIC VIRUS. In several countries of the Indian subcontinent, *Aceria cajani* ChannaBasavanna transmits *pigeon pea sterility mosaic virus* (PPSMV), the agent of sterility mosaic disease (SMD) to pigeon peas (*Cajanus cajan* (L.) Huth); the viral nature of SMD has only recently has been documented (Kumar *et al.*, 2001, 2003a; Latha *et al.*, 2007). The disease exhibits three types of symptoms: complete sterility, partial sterility and ring spot. In complete sterility, there is severe mosaic on the leaflets and the plants lack flowers and pods; this manifestation of the disease occurs if the infection begins early (usually before 45 days old). In partial sterility, there is mild mosaic on the leaflets of a few branches and only these branches do not produce flowers and pods; this occurs if the infection takes place at the later stages of plant development (usually beyond 45 days). Ring spot is characterized by green spots surrounded by a chlorotic halo, which appear on leaves,

and occurs only in certain genotypes and disappear with plant maturity. Some SMD-tolerant cultivars show this symptomatology without any effect on flowering (Reddy and Nene, 1979; Ghanekar *et al.*, 1992; Reddy *et al.*, 1990a, 1998).

OTHER (NON-VIRUS) PATHOGENS. As well as spreading viruses, mites can spread fungal spores and other non-viral organisms causing plant diseases. Usually, mite behaviour limits the possibility of disease spread, but herbivory does assist in fungal infection by carrying pathogen spores or by creating wound sites for fungal penetration (Agrios, 1980; Hatcher and Paul, 2001), and mite species from different families (Siteroptidae, Acaridae, Tenuipalpidae, Tarsonemidae and Eriophyidae) can increase the rate of some diseases or be closely associated with some pathogens. Some examples are reported below.

Stewart's bud rot – or central bud rot of carnations – is caused by the fungus *Fusarium poae* (Peck) Wollenweber, which is carried as spores attached externally to the body surface of the siteroptid mite *Siteroptes cerealium* Kirchner (Cooper, 1940). *S. cerealium* may also be involved in 'silver top' disease of grasses in North America, Europe and Asia (Hardison, 1959; Suski, 1973, 1984). The symptoms of central bud rot of carnations are a moist, brownish, decayed mass of the inner organs, with rotted stamens, stylets and petals. Externally, the fungus appears as a heavy or sparse white cottony growth (Cooper, 1940). In South Africa, *F. poae* has been found in association with the siteroptid mite *Siteroptes avenae* (Muller). The presence of sporothecae containing microsporidia of the fungus suggests that the mite is involved in *Fusarium* glume spot on wheat (Kemp *et al.*, 1996).

The robine bulb mite, *Rhizoglyphus robini* (Claparède), is closely associated with the spread of fungal diseases caused by *Fusarium*, *Stromatinia* and *Pseudomonas*. The mite feeds on the fungus responsible for *Stromatinia* rot of gladioli and can also be a vector of *Pseudomonas* (Forsberg, 1965; Meyer Smith, 1981a). *R. robini*, together with the mould mite *Tyrophagus putrescentiae* (Schrank), can transfer fungal pathogens that have become attached to the outsides of their bodies to bulbs; the pathogens include *Aspergillus niger* van Tieghem, *Haematonectria haematococca* (Berk. *et* Broome) Samuels *et* Rossman, *Rhyzopus stolonifer* (Ehrenb.) Vuill. and *Penicillium chrysogenum* Thom. In contrast, the pathogens *Aspergillus flavus* J.H. Friedrich Link and *Aspergillus ochraceus* Wilh. are transferred from the mites via their digestive tracts along with their food (Abdel-Sater and Eraky, 2001). The bulb mite, *Rhizoglyphus echinopus* (Fumoze *et* Robin), is often associated with *Fusarium* spp. in gladioli corms and *Freesia*, and with bacterial or fungal rots of dahlia tubers. Attacked freesia and gladiolus corms are usually soft and mushy and may be unsuitable for planting. If only a few bulbs are attacked, any that are soft should be rejected (Hussey *et al.*, 1969).

In Venezuela, *Brevipalpus phoenicis* moves to the scabs caused by the fungus *Elsinoë fawcettii* Bitanic *et* Jenkins (Elsinoaceae) on fruit of sour orange. This pathogen produces a disorder known as 'halo scab', with symptoms ranging from spotting of the leaves to death of the plants; the leaves attacked by the scab fungus alone do not drop, but leaves attacked by both the fungus and the mite drop, resulting in defoliation and death of the plant (Knorr and Malaguti, 1960; Chiavegato and Kharfan, 1993). In Honduras, *B. phoenicis* is associated with *E. fawcettii* on citrus (Evans *et al.*, 1993). A similar interaction between two mite species and the fungus *Sphaceloma fawcetti* Jenkins (= *Elsinoë fawcettii* Bitanic *et* Jenkins) has been reported in Costa Rica (Ochoa *et al.*, 1994). In Brazil, the mite prefers citrus fruits already infected by sweet orange scab, a fungal disease caused by the fungus *Elsinoë australis* Bitancourt *et* Jenkins (Chiavegato and Kharfan, 1993).

The tarsonemid *Steneotarsonemus spinki* causes grain discoloration of rice and facilitates the entry of the fungi *Helminthosporium* and *Sarocladium oryzae* (Sawada) (= *Acrocylindrium oryzae* Sawada), and has also been found to carry the mycoplasma-like organism (MLO), *Spiroplasma citri* Saglio *et al.* It seems likely that the mite, *A. oryzae* and the MLO are all involved in rice sterility (Chow *et al.*, 1980; Cardenas *et al.*, 2003). Ramos and Rodríguez (2001) reported that in Cuba the combined activities of mite feeding and disease transmission lowered rice yield by 30–60%.

The feeding of the pineapple leathery pocket mite, *Steneotarsonemus ananas*, on trichome cells of the pineapple epidermis helps the development of the fungus *Penicillium funiculosum* Thom, the causal agent of inter-fruitlet corking, leathery pocket, and fruitlet core rot/black spot (Petty *et al.*, 2002). The dry bulb mite, *Aceria tulipae*, is implicated in the spread of fungus-causing rot of garlic bulbs in the field (Jeppson *et al.*, 1975).

Research carried out on the association between *Aceria mangiferae* and *Fusarium mangiferae* Britz, Wingfield *et* Marasas in mango, and the cereal rust mite, *Abacarus hystrix*, and rusts of *Puccinia* spp., shows that both mites bear the spores of the two pathogens on their body integument. *A. mangiferae* carries the pathogen's spores into the mango bud (the sole port of entry for the fungus) and the frequency and severity of fungal infection increase in the presence of the mite. However, further research is needed to enhance our understanding of the direct and indirect (plant-mediated) interactions between plant pathogens and eriophyoid mites in different plant–pathogen systems (Gamliel-Atinsky *et al.*, 2010).

8 Mites as Biocontrol Agents of Weeds

Weeds compete in agriculture in various ways. They may directly affect the management of cultivated plants and reduce the yield, or act in other ways, such as by limiting the utilization of land for grazing. Weeds can be indigenous, exotic or cryptogenic (neither demonstrably native nor introduced) (Carlton, 1996). Herbicides comprise about 50% of all the agrochemicals sold worldwide (Woodburn, 1995), which compares with 30% for insecticides. The biological control of weeds could represent a desirable solution to this problem, and the use of arthropods to accomplish this was proposed in the 1830s (Goeden, 1988). The strategy is to employ natural enemies to reduce the density of weeds to an acceptable level (Van Driesche *et al.*, 2008). This type of control is usually applied against invasive weeds and requires the importation of natural enemies from their areas of origin. These natural enemies are herbivores (mites, insects, etc.) and must not damage other (accompanying) plant species. For this reason, host specificity tests are carried out before they are used (McEvoy, 1996; Sheppard *et al.*, 2003; Del Fosse, 2004) – tests that need to be improved as a result of increasing regulatory requirements and the decreasing tolerance of risk to indigenous species (Briese, 2005; Carruthers and D'Antonio, 2005; Sheppard *et al.*, 2005).

Weed biocontrol methods were categorized by McFadyen (1998) into four types: classical or inoculative; inundative or augmentative; conservation; and broad spectrum (for more information see Gerson *et al.*, 2003). The prediction of the effectiveness of a weed natural enemy is a fundamental step of the strategy. Harris (1973) proposed a scoring system to individualize effective species for weed control, which was successively revised by Goeden (1983) and Cromroy (1983). The system proposed by Cromroy includes 12 categories and each candidate is scored by a numerical value. Scores of more than 27 indicate that the candidate might be suitable for weed control. The system has since been slightly modified by Gerson *et al.* (2003).

Currently, only five mite species are officially used for weed control internationally (Briese and Cullen, 2001). One of these belongs to the Tetranychidae, three to the Eriophyidae, and one to the Galumnidae. Results from research carried out in various countries indicate that the outcome is not always clear.

Tetranychidae

Only a few tetranychid mites are suitable as weed control agents because of the low host specificity of many species, which is characterized by polyphagy, including feeding on weeds, and unclear reproductive isolation (Hill and Stone, 1985). Only three cases are well known and investigated: the control of prickly pear (*Opuntia inermis* DC), gorse (*Ulex europaeus* L.) and water hyacinth (*Eichhornia crassipes* (Martius) Solms-Laubach).

The first tetranychid mite used for weed control was *Tetranychus desertorum* Banks, which fed on cacti in Texas and was accidentally introduced into Australia in the 1920s (Dodd, 1929, 1940). In Queensland, the mite established on prickly pear, an introduced weed cactus that decreases the value of land for grazing. The mite-infested epidermis becomes corky and whitish grey, and the affected segments dry and drop off. Mite feeding killed seedlings and young plants, and the upper growth of old plants was destroyed. The tetranychid reduced 75% of a weeded surface of about 250 ha in 2 years. Unfortunately, the later introduction of the pyralid *Cactoblastis cactorum* (Bergroth) displaced the mite as a control agent (Mann, 1970). Later on, Seeman and Beard (2011) reported that *T. desertorum* did not occur in Australia and had been probably mistaken for *T. ludeni* Zacher.

The history of the biological control of gorse is more complex. This weed is a spiny leguminous, woody shrub of European origin that had been introduced into South and North America, Australia and New Zealand, where it became a weed of pastures. The introduction of the tetranychid *T. lintearius* Dufour (Longworth, 1987), which is specific to *Ulex* in Europe, was suggested for its biological control (Van Eyndhoven, 1967a). The mite was introduced into New Zealand from the UK in 1989 and was released in Australia in 1998 after tests on 32 plant species (Ireson *et al.*, 2003). However, some specialists placed *T. lintearius* (as *T. lintearicus* Dufour, 1832) in synonymy with *T. telarius* (L.) (Pritchard and Baker, 1955), and this last species has since been identified as the two-spotted spider mite, *T. urticae* Koch (Baker and Pritchard, 1960). This poses the need to evaluate the reproductive isolation of *T. lintearius* both from *T. urticae* and from the closely related species of *T. turkestani* (Ugarov *et* Nikolskii) (Hill and O'Donnell, 1991a), and also to evaluate its host specificity on gorse (Hill and O'Donnell, 1991b). The introduction of *T. lintearius* into New Zealand in 1989 has not solved the problem of gorse as a weed, for the mite often fails to provide sufficient control, either leaving the plants too soon or being unable to develop sufficiently large populations; it also suffers from predation by the coccinellid *Stethorus bifidus* Kapur (Richardson and Hill, 1998). According to Cromroy's (1983) scoring system, the mite was rated at 40, and so should be suitable for use in biological control. It is now present throughout New Zealand, is in Australia and has been released in Oregon and Hawaii (Markin *et al.*, 1996; Ireson *et al.*, 2003).

The role of some tetranychid mites has been evaluated in the control of water hyacinth, a serious weed that has spread in

warm regions of the world. It is a free-floating, perennial plant that forms rich mats on waterways and lakes (Harley, 1990). Trials carried out with the tetranychids *Oligonychus silvestris* Gutierrez, *T. macfarlani* Gutierrez, *T. neocaledonicus* André and *T. tumidus* Banks show that these mites reproduce on the weed and that the infested leaves turn yellow. This suggests that although spider mites do not consistently control stands of water hyacinth, in some areas they could be employed to delay its growth (Pieterse, 1972). In Kerala (India), the role of *O. biharensis* (Hirst) has been evaluated as a potential control agent of water hyacinth (Haq and Sumangala, 2003).

Eriophyidae

Eriophyid mites have a high potential as classical biological control agents of weeds. In the last two decades, various possible candidates have been listed. Boczek and Petanović (1996) reported 93 eriophyoids from 102 weed species in Poland and Yugoslavia. Rosenthal (1996) discussed species of *Aceria*, *Epitrimerus* and *Aculus* as candidates for the biological control of weeds. Thirteen species were evaluated for future release, but only four, *Aceria genistae* (Nalepa), *Aceria* sp., *Cecidophyes rouhollahi* Craemer and *Floracarus perrepae* Knihinicki *et* Boczek, have been authorized for introduction (Smith *et al.*, 2010). Previously, three species, *Aceria chondrillae* (Canestrini), *Aceria malherbae* Nuzzaci and *Aculus hyperici* (Liro), had been introduced into different countries (Briese and Cullen, 2001).

Aceria chondrillae lives on rush skeletonweed, *Chondrilla juncea* L., a herbaceous perennial native to Eurasia that has been introduced into Argentina, Australia and the USA, where it successfully competes with winter crops and colonizes rangelands. The weed is apomictic and has a number of forms that grow in various parts of its range. These forms have adaptive capacities that affect the activity of the mite in different ways. *A. chondrillae* has spread into almost all native stands of its host plant, and is very common in Western Europe, where up to 50% of the plants are infested. The mite feeds on both vegetative plant parts and flower buds, inducing the formation of clusters of hyperplastic leafy galls, in which the mites feed and reproduce. Mite feeding causes destruction of the buds, reduces seed production and weakens the plants. Plants that have been infested for a long time exhibit early senescence and stunting, and die (Carèsche and Wapshere, 1974). *A. chondrillae* was introduced into Australia in 1971 (Cullen *et al.*, 1982) and into the USA a few years later (Sobhian and Andres, 1978). Research in the field has shown that the effectiveness of the mite depends on its population size (Cullen *et al.*, 1982). Small weed stands allow mite survival in the winter, whereas the destruction of most of the stand leads to mite elimination from the field. In dense weed stands, the mite populations increase quickly, and produce more drastic effects (Carèsche and Wapshere, 1974). The first introduction of the mites into North America failed because of the unsuitable climate of California and the mite strain that was used (Sobhian and Andres, 1978). The introduction of another strain was successful, but no correlation was observed between mite populations and weed control (Supkoff *et al.*, 1988), probably because of the dry inner climate of California (USA) (Rosenthal, 1983). In contrast, the introduction of *A. chondrillae* into Idaho, a more humid US state, was clearly successful (Prather *et al.*, 1989). A study of the effects on three Australian forms of rush skeletonweed, referred to as forms A (narrow leaf), B (intermediate leaf) and C (broad leaf) (Hull and Groves, 1973), of three mite strains (French, Italian and Greek) (Carèsche and Wapshere, 1974) allowed a better analysis of the mite–weed relationship. The findings showed that the French plants were attacked by all mite strains, while the Greek mite strain established only on forms A and B, with form A being severely infested. The Greek mite strain was introduced successfully into Australia (Cullen and Moore, 1983); it has also been introduced into California (via Australia), where it does not produce galls on American strains of rush skeletonweed, whereas the Italian strain does (Sobhian and Andres, 1978). The variability observed in the experimental use of the Greek mite strain maintained on the rush skeletonweed form B suggests that the mite–plant relationship is affected by many environmental factors, such as temperature, relative humidity and light intensity (Cullen and Moore, 1983).

Aculus hyperici is native of Europe and lives on St John's wort (*Hypericum perforatum* L.), a European perennial and competitor of various plant species, which also affects the skin of livestock through production of the toxin hypericin. In Australia, the mite also lives, at low densities, on the native small St John's wort (*H. gramineum* G. Forst.). *A. hyperici* feeds on the rosette growing points and flowering stems, and causes dwarfing, impairment of plant vigour and plant death (Mahr *et al.*, 1997). The mite was introduced into Australia in 1991, and its effects on root systems, stems and the production of seed capsules were positively evaluated (Jupp and Cullen, 1996). However, it has been observed that the phenotypic variability in weed susceptibility to the eriophyoid may have a negative effect on mite colonization. Among the phenotypes investigated, an intermediate form was resistant to the mite (Jupp *et al.*, 1997). An evaluation of the distribution and status of the weed in Australia gave unclear results (Briese and Cullen, 2001).

Aceria malherbae is a Mediterranean species living on field bindweed, *Convolvulus arvensis* L., a perennial weed of the Convolvulaceae family and competitor of cultivated crops in temperate regions. Mite feeding on the bindweed causes leaf furrowing along the veins, characterized by thickenings, wrinklings and twistings that host the mites. The leaves, petioles and stems of infested plants bear small tubercle-like galls. The leaves turn yellow or red, infested stems do not produce flowers, and bud swelling appears to be irregular (Nuzzaci *et al.*, 1985). The mite lives on the host plant throughout its growing season (from March to November) and overwinters on rhizomes in the soil (Boldt and Sobhian, 1993). It also reproduces and forms galls on several North American species of *Calystegia*, also from the Convolvulaceae, and various North American bindweed ecotypes are differentially susceptible to the mite. Thus, in New Jersey, 96% of bindweed plants were infested, in California and Texas, there were intermediate levels of infestation of 87 and 75%, respectively, and in Nebraska, there was 53% infestation (Rosenthal and Platts, 1990). A Greek strain of *A. malherbae* was released into the USA in 1987, and introductions into South Africa took place in 1989 and 1994 (Rosenthal

and Platts, 1990; Craemer, 1995). The mite has established successfully in the USA (Boldt and Sobhian, 1993) and Canada (McClay *et al.*, 1999), and is sold by Praxis (Allegan, Michigan) for the biological control of the weed (IPM Practitioner, 2000).

Also interesting is the action of *Phyllocoptes fructiphilus* Keifer in the USA. This is an endemic eriophyoid that is a disease-associated/transmission factor in the mortality of the multiflora rose (*Rosa multiflora* Thunberg ex J. Murray), an Asian thorny shrub rose introduced into the USA as a rootstock and an ornamental, which became a weed by the 1950s (Amrine, 1996). The mite vectors a virus, which was formerly designated *Rose Virus 1* (RSV-1), an *Ilarvirus* (Jesse *et al.*, 2006; Tzanetakis *et al.*, 2006), but has since been formally identified as *Rose rosette virus* (RRV), an *Emaravirus* (Laney *et al.*, 2011; Hong and Hansen, 2012) (see also Chapter 7). The virus causes rose rosette disease (RRD) on *R. multiflora* and ornamental roses, mainly in the western and middle states of the USA and in some Canadian provinces, where many *Rosa* varieties are resistant (Amrine, 1996). After grafting RRD-infected buds on to healthy plants, RRD appears some 30–90 days later, whereas after infestation by infected mites (collected from infested plants), only 17 days are needed before the appearance of RRD. Both grafting and mite infestation give 100% transmission, and rose mortality may reach 98% in some years. From a practical viewpoint, the grafting method is used for disease transmission, but the method is laborious, and the use of *P. fructiphilus* has been evaluated as an alternative (Epstein and Hill, 1999).

Other eriophyoid species that have been investigated for weed control purposes are *Cecidophyes rouhollahi* Craemer and *Floracarus perrepae* Knihinicki *et* Boczek (Smith *et al.*, 2009).

Galumnidae

This mite family includes about four or five species living on water hyacinth and other species (*Pontederia cordata* L., etc.) of the Pontederiaceae in various countries. About four galumnid species have been reported feeding on leaves of water hyacinth in India (Sumangala and Haq, 1995), but a better known species is *Orthogalumna terebrantis* Wallwork, a native of South America. The immature stages of this mite excavate longitudinal tunnels in the false leaves of the water hyacinth (Del Fosse, 1978a,b), producing up to 25–50 feeding tunnels over a 2 cm^2 surface (250–500 tunnels per leaf) and reaching a density of 20,000 mites m^{-2} of weed. The leaf surface corresponding to the tunnels shows yellowish brown streaks, which turn brown and dry out, resulting in the death of the mites living therein (Perkins, 1973). The adults leave the galleries through holes on the upper leaf surface. Overlapping galleries form broad dry patches on the leaf surface, and the leaves collapse within a month of the attack (Sumangala and Haq, 1995), although at least 200,000 tunnels m^{-2}, or 75,000 mites m^{-2}, or 10,000 tunnels per plant, are needed to collapse many leaves (Cordo and DeLoach, 1976). In addition to direct damage due to feeding and tunnelling, *O. terebrantis* causes indirect damage to the plant by attracting the mottled water hyacinth weevil, *Neochetina eichhorniae* Warner (Curculionidae), which lays eggs on the plant. *N. eichhorniae* is also used for the biological control of the water hyacinth. The fungus *Acremonium zonatum* (Saw.) Gams (Hyphomycetes), a pathogen of *Ficus* spp., which produces zonate leaf spot disease in the water hyacinth, also damages the plant after penetrating the leaves through the holes produced by the adult mites when they emerge, but is not observed at weevil feeding points. The combined activity of the mite and weevil produce a better control of the water hyacinth than either agent alone, giving a reduction in plant density of about 50% within a year, which is superior to the sum of their single effects (Del Fosse, 1978a,b). The mite has been introduced into India, Central America and Africa (Waterhouse, 1994), but its association with the fungus *A. zonatum*, because it is injurious to *Ficus* spp., has not permitted its introduction into Fiji (Kamath, 1979). According to Cromroy's (1983) scoring system, *O. terebrantis* scores 31, but despite this evaluation and its demonstrated direct and indirect control of water hyacinth, the mite is not much used for biological weed control (Charudattan, 1986).

9 Control

Outbreaks of mite populations can have a serious impact on crops of economic plants, including yield losses, the high costs of mite control and/or undesirable ecotoxicological effects. In order to control such outbreaks, there is a need to understand the bio-ecology of mite pests and the factors regulating the development of their populations. Population monitoring, the assessment of damage and knowledge of the appropriate means and strategies for control are fundamental to making rational decisions. It is not possible to base decisions on knowledge acquired elsewhere, or to standardize control measures and strategies worldwide. Each mite species must be examined according to its environment, as this can variously influence the behaviour of its populations.

As with other injurious pests, the control of mites harmful to economic plants is achieved by physical, chemical, biological and other means – cultural, integrated pest management (IPM), etc. – according to the control strategy that is adopted. In this chapter, the general features of the strategies and means that are usually employed are discussed: physical control, chemical control, biological control (including the use of natural enemies) and IPM. Host plant resistance is discussed as a biological strategy mainly related to IPM. The use of genetically modified organisms (GMOs) is not considered as, in the opinion of the author, they lack any ecological foundation and are therefore considered to be environmentally harmful to serious agriculture.

Physical Control

Physical control includes controlled atmosphere treatment (CAT), controlled atmosphere temperature treatment (CATT), irradiation and thermal treatment.

Controlled atmosphere treatment (CAT)

CAT consists of modifications of carbon dioxide concentration made in well-defined environments that are lethal to mite pests. The method has been applied to the apple rust mite, *Aculus schlechtendali* (Nalepa), and to the European red mite, *Panonychus ulmi* (Koch). Apple storage at 2.8°C in an atmosphere of 1% O_2 and 1% CO_2 kills mite eggs in 5.3 months without any negative effects to the fruit. However, the length of storage exposure required for sufficient mortality is a disadvantage (Lidster *et al.*, 1981).

Controlled atmosphere temperature treatment (CATT)

CATT combines a constant temperature with a large amount of CO_2. For example, the application of a temperature of 35°C and 50% CO_2 for 48 h to strawberry mother plants gives over 99.8% mortality of the strawberry tarsomenid mite, *Phytonemus pallidus* (Banks), and no harmful or irreversible results of CATT have been observed on either the vitality of the mother plants or runner production from those plants (van Kruistum *et al.*, 2012).

Irradiation

The use of this method is very limited because it does not kill pests speedily and requires protocols to be approved by government regulatory agencies. The dose of radiation that controls most mites is about 300 Gy, which is higher than that required to control other pests. Irradiation has been evaluated for the control of some tetranychid mites, including the two-spotted spider mite, *Tetranychus urticae* Koch (Dohino and Tanabe, 1994; Lester *et al.*, 1997), and *Oligonychus biharensis* (Hirst) (Majumder *et al.*, 1996); for eriophyoid mites, few or no data are available (Molins, 2001). Irradiation and cold treatment were applied alone and in combination against the grape flat mite, *Brevipalpus chilensis* Baker. Treatments with caesium 137 at 0.25, 0.5 and 1 kGy and kept for 14 days at 4°C and then at room temperature did not cause total mortality during the 20 days following treatment (Sazo and Gerstle, 1989). The effects of 0.5, 1.0 and 1.5 kGy of gamma irradiation, combined with cold storage (24 h pre-cooling and 15 days at 0–2°C and 90% RH), on egg viability, mortality and reproduction of *B. chilensis* were evaluated by Jadue *et al.* (1997). All treatments resulted in significantly greater adult mortality than the control, and higher dosages had stronger and more rapid effects. A few adults surviving the 0.5 kGy treatment produced eggs, but none hatched. There was total mortality of eggs, juveniles and adults after exposure to 0.5–1.5, 1.0–1.5 and 1.5 kGy, respectively (Jadue *et al.*, 1997). On 'Thompson Seedless' grapes, the irradiation doses required to cause 90% mortality of adults, nymphs and eggs were 1307, 970 and 328 Gy, respectively. An irradiation dose of 200 Gy combined with 15 days of cold treatment stopped mite reproduction. The organoleptic properties of the grapes irradiated at 600 Gy were not affected, nor were ripeness parameters altered relative to soluble solids and acidity, and no phytotoxicity was detected in either berries or the stem (Castro *et al.*, 2004).

Thermal treatment

Various thermal treatments have been suggested for the containment of eriophyoids and other mites. Hot water treatment (HWT) of bulbs (55°C for 10–20 min, or 60°C for 10–15 min) can reduce populations of *Aceria tulipae* (Keifer) (the onion, garlic and tulip mite), but treatment may affect bulb germination. Drying the bulbs in the postharvest phase can also reduce populations (MacLeod, 2007). An exposure time of 15–45 min to hot water at temperatures between 40 and 43°C is effective in killing the broad mite, *Polyphagotarsonemus latus* (Banks), a tarsonemid that feeds on a wide variety of horticultural crops. Plants have to be immersed in the hot water for long enough to allow penetration into areas such as the meristematic tissues where the mites live, but at the same time not damage the plants. However this practice is a short-term solution without any residual effects, and after transplanting, a reinfestation may occur (Cloyd, 2010). Hot water treatment of dormant bulbs of narcissus is applied from time to time in the control of the bulb scale mite, *Steneotarsonemus laticeps* (Halbert), although the treatment could have consequences for the following season's flower growth, and there is a fine line between eradicating the pest and damaging the bulbs (Winfield, 1967; Hussey *et al.*, 1969; Lane, 1984). The method is therefore unsuitable for bulbs destined for forcing or sale. There is clear disagreement on the duration of treatment, but the solution to the problem seems to be given by Wallis (1967) and Gratwick and Southey (1986), who report that warm storage of narcissus bulbs infested by *S. laticeps* before hot water treatment reduces bulb damage. In addition, this technique improves growth and flowering. Bulbs stored in this way have a lower rate of respiration, which makes the developing flowers, leaves and roots more tolerant to the hot water treatment. During the warm season, this condition may occur naturally, usually during July–August, and from 4 to 6 weeks after lifting, when the flowers of most varieties are fully formed (Tompsett *et al.*, 1977). According to the variety, locality and season, the safest time to apply the treatment is mid-July in south-west England, and in the eastern counties it is late July to mid-August (Lynch, 1993). Despite the above reports, the control of bulb scale mites in lifted amaryllis bulbs (*Hippeastrum*) by hot water treatment is not always satisfactory. Growers apply a great variety of time and temperature combinations. The influence of bulb size, water temperature and initial bulb temperature on the time until a temperature of 43°C is reached in the bulb was investigated. Larger bulbs, a low initial bulb temperature and a low water temperature all increased the time taken for the bulb to reach 43°C. The effects are described in a model, and it is suggested that growers underestimate the influence on the duration of the initial bulb temperature and bulb size and overestimate the effect of water temperature (Doorduin *et al.*, 1997).

The immersion of strawberry plants in water at 43.5°C for 30 min gives full mortality of the cyclamen mite, *Phytonemus pallidus*, but treatment of the plants with water-saturated air at 43.5°C for 1 h is more efficient. The plants must be loose on a screen or slatted boxes, or stacked to allow the penetration of the water vapour, but after treatment the plants should be dried before packing, and planted as soon as possible (Munger, 1933; Smith, 1939a). However, the application of treatment to strawberry plants might retard their growth, and further studies were needed so that the risk of damage could be reduced (Hussey *et al.*, 1969). Later, it was observed that during the summer the shortest exposure times to kill all *P. pallidus* adults in hot water at temperatures of 44, 46 and 48°C were 6, 2.5 and 1.25 min, respectively. Hot water treatment for 6.5 min at 46°C kills all mites on infested runners, and this exposure time includes the period needed to raise the temperature of the plant material to that of the water bath. From the middle of September and later, the mites became more heat tolerant as they prepared for hibernation, and longer exposure periods could be necessary to kill mites on frigo (cold-stored) plants (Hellqvist, 2002).

Cold storage (15 days storage at 0–2°C and 90% relative humidity, RH) delayed and reduced the hatching of 1- to 4-day-old eggs of the tenuipalpid *Brevipalpus chilensis*, with lesser effects on eggs older than 4 days, probably because of their more developed embryos. Juveniles are highly susceptible to cold storage, while adults show relatively less susceptibility. Cold storage also reduces oviposition by *B. chilensis* females (Jadue *et al.*, 1996).

Chemical Control

The chemical control of injurious mites is carried out using chemical compounds – acaricides – belonging to different classes of chemical. In addition, various insecticides and fungicides may have a combined acaricidal and insecticidal, or acaricidal and fungicidal, action. According to the mite species and the environmental conditions, these compounds, must be spread on crops either one or more times during a crop season. Various factors may affect (and sometimes limit) the effectiveness of chemical treatments, such as the weather conditions, the development of resistant strains, the introduction of new pest species, and the ecotoxicological risks related to treatment. The strategy does not provide a complete resolution to the problem, but may protect crops in terms of giving an economically acceptable yield and quality of produce.

Acaricides

Historically, the oldest chemical treatment used against mites is sulfur, which was used either as a suspension in water or as a powder at the end of the 19th century (Hubbard, 1885), or as lime sulfur at the beginning of the last century (Yothers, 1915). Another older treatment was the use of petroleum oils from 1920, followed by that of synthetic organic compounds, including dinitrophenol, which were widely used between 1930 and 1950, either alone or in combination with oils. Only from 1940 did a series of acaricides began to appear in the world market. In addition, many insecticides are effective acaricides, but the development of resistance to these, and other limitations (toxicological effects, etc.), have largely resulted in the banning of their use. Similarly, different fungicides also have acaricidal properties.

Acaricides can be inorganic, organic or synthetic chemicals (see Table 9.1). According to their phytoiatric action, these

Table 9.1. Chemical types and classes of the main acaricides. The table also lists compounds that are banned from use, which are included only because of their historical importance.

Type	Class	Acaricide
Inorganic	Element	Sulfur
Organic	Triterpenoids	Azadirachtin
	Macrocyclic lactone glycosides	Avermectin, milbemectin
	Essential oils from plants	Various compounds
Synthetic organic	Benzoylacetonitriles	Cyflumetofen
	Bridged diphenyls	Benzoximate
	Carbamates	Aldicarb, benomyl, carbaryl, carbofuran, carbosulfan, methomyl, propoxur and other compounds
	Carbazates	Bifenazate
	Dinitrophenols	Dinocap, dinobuton, bynapacril and other compounds
	Diphenyl carbinols	Chlorobenzilate, bromopropylate, dicofol and other compounds
	Dithiocarbamates	Zineb, maneb, mancozeb and other compounds
	Formamides	Amitraz, chlorodimeform, formetanate
	Methoxyacrylates	Fluacrypyrim
	Naphthoquinones	Acequinocyl
	Natural hydrocarbon-based substances	Petroleum oils, paraffin oils
	Organobromines	Methyl bromide
	Organochlorines	DDT, dienochlor, endosulfan and other compounds
	Organophosphates	Diazinon, dichlorvos, ethion, methamidophos, pirimiphos-methyl, phosalone and other compounds
	Organotins	Azocyclotin, cyhexatin, fenbutatin oxide
	Oxazoles	Etoxazole
	Pyrazoles	Fenpyroximate, cyenopyrafen, tebufenpyrad
	Pyrethroids	Bioresmethrin, fenpropathrin, bifenthrin and other compounds
	Pyrroles	Chlorfenapyr and pyridaben
	Quinazolines	Fenazaquin
	Quinoxalines	Chinomethionate
	Sulfur bridged	Chlorfenson, propargite, tetradifon and other compounds
	Tetrazines	Clofentezine
	Tetronic acids	Spirodiclofen, spiromesifen
	Thiazolidines	Hexythiazox, flubenzimine

compounds are classified into insecticides/acaricides, fungicides/acaricides and acaricides. Over 200 compounds are known in almost 26 chemical classes, but technical, economic and/or ecotoxicological limitations shrink their availability to a few tens of compounds (Table 9.1).

The use of acaricides is subject to the specific legislative constraints of each country, and permissions to use various compounds in different countries may be different. In addition, the number of acaricides available varies according to newly proposed compounds and the suppression of others as they are gradually not authorized. Any information on the legality of use or approval of an acaricide – including that provided in this chapter – needs to be completely up to date to be correct. Thus, the EU Pesticides Database (European Commission, 2014) lists 103 acaricides but only 33 are approved. Data on acaricides and their toxicity for the USA are available from the US Environmental Protection Agency (US EPA, 2014) and EXTOXNET (The Extension Toxicology Network (see EXTOXNET, 2015). It should be noted that the European and North American legislative constraints may differ on several points.

The toxicity of each acaricide is assessed by the use of a dose–response curve or a concentration-response curve, which is obtained spraying test mites with increasing concentrations of an acaricide and recording the mortality in a particular time interval (Yu, 2008). Estimates of toxicity are given by the LD_{50}, the dose needed to kill 50% of the mite population tested, and the LC_{50}, the concentration needed to kill 50% of the mite population tested. An oral LD_{50} refers to an acaricide introduced through the mouth, while a dermal LD_{50} refers to an acaricide introduced through the integument. Several variables may affect the response of the mite population tested.

Acaricides may enter in the mite body as stomach poisons, contact poisons, fumigants or systemic poisons (Yu, 2008). Some acaricides exhibit a specific action, such as tetradifon which displays ovicidal activity, while others have both ovicidal and adulticidal activities. According to its chemical nature and interactions with the mite species, host plant and environment, each acaricide exhibits a different mode of action, persistence and safety period, and the use of each requires its own dosage. Inappropriate use of these compounds may also induce the development of resistance. Research carried out by Gotoh *et al.* (2008) in Japanese

pear orchards demonstrated another outcome. They observed that the mite population in pear orchards remained low – even after inoculation with adult females, suggesting that natural resistance may have developed. They also determined that a compound extracted from the pear leaves had ovicidal activity against eggs of *Panonychus citri* (McGregor), deterred moulting to the protonymph and was similar to hexythiazox. These results strongly suggested that the moulting deterrent extracted from pear leaves was in fact hexythiazox, an acaricide in use on pear trees, rather than a natural pear product, a conclusion that was supported by the occurrence of moulting deterrence in a hexythiazox-spray year, but not in a non-spray year.

Another classification of the mode of action of acaricides has been given by the Insecticide Resistance Action Committee (IRAC, 2012), which reports that acaricides act on three main targets: nerve and muscle targets, respiration targets and growth and development targets. The mode of action of yet other acaricides (e.g. benzoximate, bifenazate, chinomethionate, dicofol) is unknown.

Acaricides that act on nerve and muscle targets include five groups. The first group is acetylcholinesterase (AChE) inhibitors, which inhibit AChE, the enzyme that terminates the action of the excitatory neurotransmitter acetylcholine at nerve synapses, and thereby causes hyperexcitation. They include the carbamate, methomyl, and the organophosphate, pirimiphos-methyl. The second group is the GABA (γ-aminobutyric acid)-gated chloride channel antagonists. GABA is the major inhibitory neurotransmitter in insects, and GABA antagonists block the GABA-activated chloride channel, causing hyperexcitation and convulsions. These acaricides include cyclodiene organochlorines, such as endosulfan. The third group is the sodium channel modulators. Sodium channels are involved in the propagation of action potentials along nerve axons, and channel modulators keep the sodium channels open, causing hyperexcitation and sometimes nerve blocking. Pyrethrins and pyrethroids (e.g. bifenthrin, halfenprox) belong to this group. The fourth group (formally Group 6) consists of chloride channel activators. Glutamate is an important inhibitory neurotransmitter in insects and chloride channel activators allosterically activate glutamate-gated chloride channels (GluCls), causing paralysis. They include the avermectins, milbemycins (e.g. abamectin, milbemectin). The fifth group (formally Group 19) consists of octopamine receptor agonists, and these activate octopamine receptors, leading to hyperexcitation. Octopamine is the insect equivalent of adrenaline, the fight-or-flight neurohormone. These acaricides include the formamidines (e.g. amitraz).

Acaricides that act on respiration targets number several active ingredients that are able to interfere with mitochondrial respiration by the inhibition of electron transport and/or oxidative phosphorylation. These are generally fast to medium-fast acting, and they include five groups. The first group (formally Group 12) are inhibitors of mitochondrial ATP synthase, and inhibit the enzyme that synthesizes ATP. They include diafenthiuron, some organotin miticides (azocyclotin, fenbutatin oxide) and propargite. The second group (formally Group 13) are uncouplers of oxidative phosphorylation via disruption of the proton gradient. These are protonophores that short-circuit the mitochondrial proton gradient so that ATP cannot be synthesized, and they include chlorfenapyr. The third group (formally Group 20) are the mitochondrial complex III electron transport inhibitors that are able to inhibit the electron transport complex III, thus preventing the utilization of energy by cells. These include acequinocyl and fluacrypyrim. The fourth group (formally Group 21) are the mitochondrial complex I electron transport inhibitors that are able to inhibit electron transport complex I, thereby preventing the utilization of energy by cells. They include METI (mitochondrial electron transport inhibitors) acaricides, such as fenazaquin, pyridaben and tebufenpyrad. The fifth group (formally Group 25) are the mitochondrial complex II electron transport inhibitors that are able to inhibit electron transport complex II, and these too prevent the utilization of energy by cells. They include cyenopyrafen and cyflumetofen.

Acaricides with growth and development targets are usually slow acting, with target proteins that are not always known. They are divided into three groups. The first group (formally Group 10) are the mite growth inhibitors, which have an incompletely defined mode of action that leads to growth inhibition. They include clofentezine, hexythiazox and etoxazole. The second group (formally Group 15) are inhibitors of chitin biosynthesis, type 0, and these also are characterized by an incompletely defined mode of action that leads to the inhibition of chitin biosynthesis. They include the benzoylureas, such as flucycloxuron and flufenoxuron. The third group (formally Group 23) are inhibitors of lipid synthesis that are able to inhibit the acetyl coenzyme A carboxylase, part of the first step in lipid biosynthesis. These include the tetronic and tetramic acid derivatives, such as spirodiclofen.

Further information on this topic may be found in Yu (2008), Tomlin (2009), Kramer *et al.* (2012) and Gowariker *et al.* (2014). The discussion below covers the main acaricides, including those compounds that have been banned from use in different countries, so as to give a complete framework of their historical importance and to avoid the risk of them being proposed for use in other countries.

Inorganic acaricides

Inorganic acaricides really only include sulfur, a fungicide/acaricide that was used as a water suspension or powder at the end of 19th century (Hubbard, 1885) and as lime sulfur at the beginning of the last century (Yothers, 1915). It is effective against several eriophyoid mites, and can reach populations of mites protected in various shelters. Sulfur also controls some tenuipalpid (*Brevipalpus phoenicis* (Geijskes)) and tarsonemid (*Polyphagotarsonemus latus*) (Cranham *et al.*, 1962) mites, but is less effective on *Tetranychus* and *Panonychus* spp. (Childers and Enns, 1975a; Jeppson *et al.*, 1975; Croft, 1976), with the exception of the Atlantic spider mite, *T. turkestani* (Ugarov *et* Nikolskii), and the desert spider mite, *T. desertorum* Banks (Jeppson *et al.*, 1975). Sulfur compounds can be phytotoxic during hot weather, and in California vineyards some spider mites, as well as phytoseiids, have developed resistance to sulfur (Hoy and Standow, 1981, 1982). However, this fungicide/acaricide has a very low toxicological impact on mammals and, where possible, its use must be preferred to that of other acaricides.

Organic acaricides

Botanicals – or plant-derived compounds (pyrethrin, azadirachtin, nicotine, rotenone, etc.), various essential oils from plants, and products from microorganisms (avermectins, milbemectin, etc.) are organic compounds. They are used mainly to bypass the technical and ecotoxicological limitations of the traditional acaricides (Isman, 2006; Isman *et al.*, 2007; Yanar *et al.*, 2011). Among these organic compounds, the triterpenoid azadirachtin, the macrocyclic lactone glycoside avermectin and some essential oils are of particular interest. The use of some of these compounds, e.g. rotenone, is banned in various countries.

TRITERPENOIDS. The triterpenoids are terpenoid derivatives of triterpene molecules, which have a C_{30} skeleton and six isoprene units. The most important triterpenoid is azadirachtin, an insecticide/acaricide that is extracted from the seeds of the neem tree, *Azadirachta indica* A. Juss. Azadirachtin blocks the moulting hormone ecdysone, but it degrades within 100 h when exposed to light and water and has a very low toxicity to mammals. The effects of pure and formulated azadirachtin preparations on *Tetranychus urticae* have been evaluated in various countries (Sundaram and Sloane, 1995; Mansour *et al.*, 1997; Martínez-Villar *et al.*, 2005).

MACROCYCLIC LACTONE GLYCOSIDES. This class of organic compounds includes the insecticides/acaricides avermectin and milbemectin. Avermectin is the most effective acaricide and is a fermentation product of the Gram-positive soil bacterium *Streptomyces avermitilis* (ex Burg *et al.*) Kim and Goodfellow. It is usually used as a mixture of two homologues, both biologically active, and exhibits a translaminar activity and a short residual activity to phytoseiid mites. It blocks the action of the neurotransmitter GABA at the neuromuscular junction. Avermectin controls a range of eriophyoid mites, including *Epitrimerus pyri* (Nalepa) on pear (Campbell and Easterbrook, 1985; Undurraga and Dybas, 1988), *Aculus schlechtendali* on apple (Sterk, 1994) and *Phyllocoptruta oleivora* (Ashmead) on citrus (French, 1982). It is also effective against the tetranychids *Eutetranychus banksi* (McGregor) (Childers *et al.*, 1989), *Panonychus ulmi* (Sterk, 1994) and *Tetranychus urticae* (Undurraga and Dybas, 1988). Resistant strains of spider mites are known (Beers *et al.*, 1998; Stumpf and Nauen, 2002).

ESSENTIAL OILS. Essential oils are concentrated hydrophobic liquids containing volatile aromatic compounds extracted from plants. Many of these have been investigated as repellents, oviposition deterrents or acaricides, mainly against the two-spotted spider mite, *Tetranychus urticae*. They include garlic extracts (Boyd and Alverson, 2000), essential oils from *Artemisia absinthium* L. and *Tanacetum vulgare* L. (Chiasson *et al.*, 2001), a *Chenopodium*-based botanical (Chiasson *et al.*, 2004), wild tomato leaf extracts (Antonious and Snyder, 2006), hot pepper extracts (Antonious *et al.*, 2006), essential oils from members of the Lamiaceae (Çalmasur *et al.*, 2006) and ethanolic extracts of leaves and seeds of *Datura stramorium* L. (Kumral *et al.*, 2009). Essential oils were also extracted from lemon eucalyptus, peppermint, citronella Java, thyme red, caraway seed, clove leaf, and pennyroyal. Of these, lemon eucalyptus was found to be effective against *T. urticae* and toxic to females of the phytoseiid *Neoseiulus californicus* (McGregor) (Han *et al.*, 2010). Extracts from the flowers and leaves of *Lolium perenne* L., from the flowers of *Anthemis vulgaris* L. and from the flowers and leaves of *Chenopodium album* L. gave significantly higher mortality rates than azadirachtin and synthetic pesticides against *T. urticae* (Yanar *et al.*, 2011). Some compounds, for example soybean oils, are registered as insecticides/acaricides.

Synthetic organic acaricides

Most acaricides are synthetic organic compounds (Table 9.1). Several of these are no longer used in various countries for both ecotoxicological reasons and/or because of widespread resistance to them. The treatment below aims to provide information on both the compounds that are still used and those that are not – owing to their historical importance.

BENZOYLACETONITRILES. This class includes the new acaricide cyflumetofen, which is characterized by a contact action and is effective against all stages of tetranychids only and probably selective for beneficial organisms (which are usually known as beneficials). Its registration is pending in various countries.

BRIDGED DIPHENYLS. The most important acaricide in this class is benzoximate, which acts both as a contact poison and a stomach poison. It is active against spider mites (tetranychids) and is moderately toxic to phytoseiids. Its use is banned in various countries.

CARBAMATES. The carbamates are compounds that are derived from carbamic acid and they usually act as inhibitors of cholinesterase. Some of them are fungicides/acaricides (benomyl) and others are insecticides/acaricides (aldicarb, carbaryl, carbofuran, carbosulfan, methomil, propoxur, etc.). They are widely characterized by a hormoligotic effect (in which sublethal quantities of a pesticide are stimulatory to the pest) and by a high ecotoxicological impact. In general, the carbamates are much more active against eriophyoids than against tetranychids (Childers *et al.*, 1996). For example, benomyl was more effective on *Aculus schlechtendali* than on *Panonychus ulmi* and other tetranychids (Karg *et al.*, 1973), and is toxic to phytoseiid and stigmaeid mites (Childers and Enns, 1975a; Croft, 1975). The use of these compounds is widely banned.

CARBAZATES. The carbazates include bifenazate, which is effective against tetranychid mites. Bifenazate and its active metabolite, diazene, are highly toxic and specific, both orally and topically, to all life stages of *Tetranychus urticae* and *Panonychus citri*. All stages of development of the phytoseiid mites *Phytoseiulus persimilis* Athias Henriot and *Neoseiulus californicus* survived treatment with both bifenazate and diazene (Ochiai *et al.*, 2007).

DINITROPHENOLS. Dinitrophenols are nitrated phenols, and they include various fungicides/acaricides, such as dinocap, dinobuton, bynapacril and other compounds. Programmes of several sprays of these compounds against pathogenic fungi have also given good control of pest mites, especially eriophyoid mites.

Dinocap controlled the eriophyoids *Aculus schlechtendali* on apple (Hoyt, 1962), *Epitrimerus pyri* on pear (Laffi, 1983), *Phyllocoptruta oleivora* on citrus (Spencer and Selhime, 1954), *Aculops lycopersici* (Tryon) on tomatoes (Osman, 1979), and other pest mites. A mixture of dinocap and binapacryl was effective against *Panonychus ulmi* on apple, but resistant strains (Cranham and Helle, 1985) and a low selectivity for phytoseiids were found (Cranham and Easterbrook, 1984). Binapacryl was effective against *Oligonychus coffeae* (Nietner) and *Brevipalpus phoenicis* on tea (Mukerjea, 1968). In general, dinobuton was less effective against spider mites. The use of these compounds is widely banned.

DIPHENYL CARBINOLS. Diphenyl carbinols are aromatic organic compounds that include chlorobenzilate, bromopropylate, dicofol and other acaricides. These compounds have been widely employed against eriophyoids and other pest mites. Chlorobenzilate and bromopropylate have given good control of the citrus bud mite, *Aceria sheldoni* (Ewing) (Brown and Jesser, 1981), and both compounds controlled free-living rust mites (Childers *et al.*, 1996). A mixture of chlorobenzilate and petroleum oils controlled the citrus red mite, *P. citri* (Brown, 1980). Dicofol has been widely used since 1952 for the control of eriophyoids and tetranychids (Kirby and Tew, 1952; March, 1958; Childers *et al.*, 1996), and the control of some tenuipalpids and tarsonemids (Jeppson *et al.*, 1975). The compound is toxic to different natural enemies, and resistant strains have developed in spider mites (Cranham and Helle, 1985). The use of diphenyl carbinols is widely banned.

DITHIOCARBAMATES. The dithiocarbamates are analogues of carbamates in which both oxygen atoms have been replaced by sulfur atoms. The most important of these compounds are zineb, maneb and mancozeb, all of which are fungicides/acaricides and are more or less effective against different eriophyoid mites, and effective against the tarsonemid broad mite, *Polyphagotarsonemus latus* (Liu *et al.*, 1984), but not against tetranychid and tenuipalpid mites (Boykin and Campbell, 1982; Childers, 1990). Zineb and maneb have been used to control the citrus rust mite, *Phyllocoptruta oleivora* since the end of the 1950s (Johnston *et al.*, 1957), but control by zineb has gradually failed (Childers, 1990), and mancozeb and maneb are more effective (Swirski *et al.*, 1969), although their use alone has been considered insufficient (Childers, 1990). A further problem is that the dithiocarbamates are toxic for predatory mites (Kashio and Tanaka, 1979; Blommers *et al.*, 1986) and not all are allowed.

FORMAMIDES. Structurally, the formamides are amides derived from formic acid. The most common of these compounds is amitraz, an insecticide/acaricide effective both against tetranychid mites, mainly of the genera *Panonychus* and *Tetranychus*, and rust mites such as *Aculus schlechtendali* on apple (Bostanian *et al.*, 1981a), and *Phyllocoptruta oleivora* (Childers and Sorrell, 1984) and *Aceria sheldoni* (Brown and Jesser, 1981) on citrus, as well as other eriophyoids. Amitraz is also effective against tetranychids on citrus (Childers and Konsler, 1980; Brown and Jesser, 1982). It is very toxic to phytoseiids (Cranham and Easterbrook, 1984) and its use is not authorized in several countries. Other formamide compounds are chlorodimeform and formetanate.

METHOXYACRYLATES. This class includes the fluacrypyrim, an acaricide proposed for use against spider mites on fruit and registered in Japan. It inhibits the mitochondrial electron transport at complex III of the respiratory chain, and its toxicity seems low.

NAPHTHOQUINONES. Naphthoquinones include the acequinocyl, the first acaricide identified as affecting the enzyme complex III in mitochondria, a mechanism of action differing from that of the conventional acaricides. Acequinocyl has low mammalian toxicity, little cross-resistance and acts mainly as a contact poison and partly via ingestion. It seems selective to arthropods and less so to aquatic fauna. The authorization for its use is pending in various countries.

NATURAL HYDROCARBON-BASED SUBSTANCES. A broad range of natural hydrocarbon-based substances and refined petroleum products, each having a different chemical composition, are well known. This class includes petroleum oils and paraffinic or narrow-range oils, both of which are insecticidal/acaricidal compounds. Petroleum oils were first used in crop protection around 1870, initially based on kerosene emulsions and later on lighter lubricating oils. In 1920, the development of oil-and-water emulsions resulted in significant practical benefits, such as better coverage of the canopy, etc. (De Ong *et al.*, 1927). The development of synthetic pesticides that occurred after World War II resulted in petroleum oils being relegated to lower importance as pesticides. From 1970, the control failures associated with the irrational use of synthetic molecules rekindled interest in these oils (Davidson *et al.*, 1991). Oils from a narrow distillation range (paraffinic, summer or narrow-range oils) began to appear on the market at the same time. These compounds have innovative features and were initially applied against insect and mite pests and sooty mould on citrus in Florida and California (Trammel, 1965; Riehl, 1969). The narrow-range oils are selective compounds (DeBach *et al.*, 1971; DeBach and Rosen, 1991) usually employed in citrus IPM programmes (Riehl, 1969, 1983). Advantages associated with their use depend on their physico-chemical properties and on the technique of distillation. Further details may be found in Vacante and Bonsignore (2012). The insecticidal activity of oils also stems from their persistence on vegetation (Davidson *et al.*, 1991). Regardless of their category, all oils act as contact poisons and by asphyxia. They penetrate into the breathing apparatus of the insect/mite and occlude the trachea, preventing breathing and causing the suffocation in a few hours. They also act as repellents, reducing oviposition and trophic activity. In general, the use of oils is sometimes treated with mistrust owing to the risk of phytotoxicity and because of their limited insecticidal efficacy. However, it should be noted that the extent of their phytotoxicity depends on the physico-chemical properties of the oils and their correct use. For instance, there may be problems if oils are applied during winter frosts, or in summer when the temperature is above 32–34°C, or in flooded citrus orchards and/or when the trees are exposed to strong south-east winds. Mixing the oils with sulfur, or treating the plants with sulfur a few days before spraying, increases the danger of phytotoxicity. There is a similar risk from adding organotins to the oil mixture (Vacante, 2009a).

Petroleum oils are effective against the eriophyoid mites, including the citrus bud mite, *Aceria sheldoni* (Boyce and Korsmeier, 1941; Sale, 1988) and the citrus rust mite, *Phyllocoptruta oleivora* (Selhime, 1984), and against tetranychids, for example, the citrus red mite, *Panonychus citri* (Jeppson, 1977; Vacante *et al.*, 1980) and the Texas citrus mite, *Eutetranychus banksi* (Dean, 1980). The use of these compounds has been recently banned in several countries, and is possible to use paraffinic oils as an alternative, employed from time on deciduous and fruit trees and on ornamentals against different tetranychid and eriophyoid mites (Jeppson *et al.*, 1975).

ORGANOBROMINES. The most important organobromine acaricide is methyl bromide, which originates from both natural and synthetic sources. Marine organisms living in the ocean are estimated to produce 1–2 billion kg annually (Gribble, 1999). Small quantities of methyl bromide are also produced by some terrestrial plants belonging to the Brassicaceae family. Methyl bromide is a broad-spectrum fumigant that is used as an acaricide, fungicide, herbicide, insecticide, nematicide and rodenticide (US EPA, 1986). Fumigation using methyl bromide has been applied for mite disinfestation of fresh plant products. Methyl bromide is extremely toxic and its use requires special safety facilities, which are not available for large amounts of commercial material in some countries. It has been employed for the mite disinfestation of several vegetable products, especially fresh fruits (Katayama *et al.*, 2001). Narcissus bulbs infested by eriophyoid and other mites were treated after 10–14 days in order to kill the eggs (Mackie *et al.*, 1942). Methyl bromide fumigations were applied successfully in the control of *Steneotarsonemus laticeps* on bulbs of narcissus, amaryllis and daffodil (Wiese and Meyer Smith, 1967, in Meyer Smith, 1981a; Hussey *et al.*, 1969; Gurney and Gandy, 1974), and in the control of the cyclamen mite, *Phytonemus pallidus*, on strawberry planting stock for runner production. However, because this compound depletes the stratospheric ozone layer, the amount produced in and imported into various countries has been reduced incrementally until it was phased out to comply with the Montreal Protocol on Substances that Deplete the Ozone Layer and the Clean Air Act (CAA). So, despite its effectiveness, the use of methyl bromide is prohibited in various countries (Collier *et al.*, 2009).

ORGANOCHLORINES. Organochlorines are organic compounds containing at least one covalently bonded chlorine atom. They include insecticides/acaricides such as endosulfan and DDT, and acaricides such as dienochlor. These compounds act by disrupting the sodium/potassium balance around the nerve fibres. Dienochlor was widely employed in the past against spider mites on ornamentals and other crops. Endosulfan has been mainly used against eriophyoids, such as *Epitrimerus pyri* on pear (Westgard, 1969), *Aculus schlechtendali* on apple (Hoyt, 1962) and other mites, including *Polyphagotarsonemus latus*. It has little effect on tetranychid mites (Childers *et al.*, 1996). The use of these compounds is banned in various countries, except that DDT is used as an insecticide in different regions of the world to control vectors of malaria.

ORGANOPHOSPHATES. Organophosphates are esters of phosphoric acid. The evaluation of their acaricidal activity began in 1946 (Dean and Newcomer, 1948), and they include several compounds (diazinon, dichlorvos, ethion, methamidophos, parathion, pirimiphos-methyl, phosalone, TEPP (tetraethyl pyrophosphate) and others) that are widely banned in various countries/regions, for example in the European Union (EU) where only pirimiphos-methyl is authorized, but with restrictions. With some exceptions, these compounds are very toxic to vertebrates, but relatively few are persistent in the environment. Their action is by inhibiting cholinesterases in the nervous system. Widespread resistance to organophosphates occurs in tetranychids (Cranham and Helle, 1985) and probably in eriophyoids (Morgan and Anderson, 1958; Baker, 1979b; Abou-Awad and El-Banhawy, 1985).

ORGANOTINS. Organotins are compounds based on tin with hydrocarbon substituents, and they include mainly cyhexatin, azocyclotin, and fenbutatin oxide. They have a long persistence against eggs and motile forms of spider mites and rust mites, but a low toxicity to phytoseiids. Cyhexatin has been widely used against free-living rust mites, including *Aculus schlechtendali* (Lienk *et al.*, 1978; Bostanian *et al.*, 1983) and spider mites on various crops, but since 1987 it has been withdrawn in several countries. Fenbutatin oxide has been shown to control *Phyllocoptruta oleivora* and *Eutetranychus banksi* (Childers and Konsler, 1980), and fenbutatin oxide and azocyclotin also gave good control of *A. schlechtendali* (Bostanian *et al.*, 1983; Sterk, 1994) and *Calepitrimerus vitis* (Nalepa) (Horváth, 1989). Strains that are resistant to organotins are known and only fenbutatin oxide is authorized in various countries/regions, e.g. in the EU.

OXAZOLES. Oxazoles are heterocyclic aromatic organic compounds with an oxygen and a nitrogen separated by one carbon. They include etoxazole, an insecticide/acaricide effective against eggs and motile forms of spider mites and acting as a chitin biosynthesis inhibitor. Resistant strains are largely controlled by a single major locus. The use of etoxazole is authorized in various countries.

PYRAZOLES. Pyrazoles are heterocyclic organic compounds with a ring structure composed of three carbon atoms and two nitrogen atoms in adjacent positions. These compounds include acaricides (fenpyroximate, tebufenpyrad) and insecticides/acaricides (cyenopyrafen). Fenpyroximate is effective both against eriophyoid and tetranychid mites (Konno *et al.*, 1990). On apple, this compound gives good control of *Aculus schlechtendali* and *Panonychus ulmi* (Sterk, 1994). Its selectivity depends on the rate of application that is used. Tebufenpyrad controls both eriophyoids and tetranychids (Merriam *et al.*, 1990), and its toxicity to phytoseiids is moderate to high (Sterk, 1994). Cyenopyrafen is not authorized in various countries.

PYRETHROIDS. The pyrethroids are organic compounds that are similar to the natural pyrethrins produced by the flowers of pyrethrums (now classified as *Chrysanthemum* or *Tanacetum*): *Chrysanthemum cinerariifolium* (Trevir.) Vis. = *Tanacetum cinerariifolium* (Trevir.) Sch. Bip. and *Chrysanthemum coccineum* Willd. (= *Tanacetum coccineum* (Willd.) Grierson). Pyrethroids have increased stability in sunlight, and the class includes several

compounds that are usually not selective but have a long persistence. They are axonic poisons and cause paralysis of an organism by keeping the sodium channels open in the neuronal membranes. The first pyrethroids were developed as insecticides. Of these, fenvalerate is moderately active against *Aculus schlechtendali*, but very toxic to phytoseiids, but induces a resurgence in the population of the tetranychid *Panonychus ulmi* (Zwick and Fields, 1978), as does permethrin (Kapetanakis *et al.*, 1986). Sprays of permethrin, cypermethrin and deltamethrin on tea caused higher development of the eriophyoid *Acaphylla theae* (Watt *et* Mann) (Muraleedharan and Varatharajan, 1985). Other pyrethroids, as bioresmethrin, fenpropathrin and bifenthrin exhibit good acaricidal activity against eriophyoid and tetranychid mites in the short term, but they all cause population resurgence (Childers and Sorrell, 1982; Easterbrook, 1985; French and Villareal, 1990a). All pyrethroids are very toxic to natural enemies, and many of them are banned in various countries.

PYRROLES. Pyrroles are heterocyclic aromatic compounds characterized by a five-membered ring. The class includes the insecticides/acaricides chlorfenapyr and pyridaben. Pyridaben is a mitochondrial electron transport inhibitor that blocks cellular respiration and is effective both against eriophyoid mites (*Aculops pelekassi* (Keifer), *Aculus schlechtendali*, *Phyllocoptruta oleivora*) and tetranychid mites (Hirata *et al.*, 1988; Sterk, 1994). Its toxicity to phytoseiids is high (Sterk, 1994). The use of chlorfenapyr is not authorized in various countries.

QUINAZOLINES. The chemical structure of quinazolines is characterized by two fused six-membered simple aromatic rings, a benzene ring and a pyrimidine ring. The class includes fenazaquin, an acaricide that is especially effective against the tetranychid *Panonychus ulmi* but much less so against the eriophyoid *Aculus schlechtendali*. Its toxicity to phytoseiids is high (Sterk, 1994).

QUINOXALINES. The quinoxalines are heterocyclic compounds containing a ring complex characterized by a benzene ring and a pyrazine ring. The main compound included in this class is the fungicide/acaricide chinomethionate. The use of this compound was proposed in 1960 but it is now widely banned. Chinomethionate is effective against a range of eriophyoid mites, such as *Epitrimerus pyri* (Downing and Moillet, 1969), *Aculus schlechtendali* (Karg *et al.*, 1973), *Aculops pelekassi* (Mijuskovic and Kosac, 1972), and *Phyllocoptruta oleivora* (Brown and Jesser, 1981), and tetranychid mites, such as *Panonychus ulmi* (Downing, 1966), *P. citri* (Morse and Jones, 1983), and *T. urticae* (Vacante, 1983). Resistant strains have developed in various countries.

SULFUR-BRIDGED COMPOUNDS. The most important sulfur-bridged compounds are tetradifon, chlorfenson and propargite. Tetradifon and chlorfenson are mainly effective against eggs. Tetradifon was proposed as an acaricide in 1954. It is ineffective against eriophyoids and acts as an ovicide and against the larval stages of tetranychids. Propargite controls both the eggs and motile forms of spider mites on apple and citrus (Tuttle and Mullis, 1981; Childers *et al.*, 1989) and of eriophyoids, including *Aculops pelekassi* (Mijuskovic and Kosac, 1972) and *Phyllocoptruta oleivora* (Brown and Jesser, 1981) on citrus, *Aculus schlechtendali* on apple (Lienk *et al.*, 1978) and *Calepitrimerus vitis* on vine (Horváth, 1989). All of these compounds have low persistence in the environment, but resistant strains have been reported. Their use is banned in several countries.

TETRAZINES. Tetrazines are unstable compounds consisting of a six-membered aromatic ring bearing four nitrogen atoms. The main compound of this class is clofentezine, a very effective growth regulator of mites, active against eggs and immature stages of tetranychids but usually less effective against eriophyoids (Childers *et al.*, 1996). Clofentezine is selective against phytoseiids (Easterbrook, 1984a; Sterk, 1994). Resistant strains of *Panonychus ulmi* (Thwaite, 1991; Grosscurt *et al.*, 1994; Pree *et al.*, 2002) and *T. urticae* (Herron *et al.*, 1993) have been reported.

TETRONIC ACIDS. Spirodiclofen is the most important compound of this class. It is an insecticide/acaricide acting on lipid metabolism and affecting the production of energy by all biological instars of mites. It is effective against eriophyoid mites, such as *Calepitrimerus vitis* (de Lillo *et al.*, 2005, 2007), *Colomerus vitis* (Pagenstecher), and *Aculus schlechtendali*, and some tetranychids and tarsonemids (Guerra *et al.*, 2002; Soika and Łabanowski, 2004; Łabanowska, 2006a; Raudonis, 2006). It is moderately toxic to phytoseiids. Another tetronic acid is spiromesifen.

THIAZOLIDINES. Thiazolidines are heterocyclic compounds characterized by a five-membered saturated ring bearing a thioether group and an amine group in the 1 and 3 positions, respectively. The most important thiazolidine is the growth regulator hexythiazox, which is effective against tetranychid mites, including *Panonychus citri* and *Eutetranychus banksi*, but not against eriophyoids such as *Phyllocoptruta oleivora* (Childers, 1988), *Aculus schlechtendali* (Sterk, 1994) or *Epitrimerus pyri* (Riedl and Shearer, 1988). It is selective to phytoseiids (Easterbrook, 1984a; Sterk, 1994). Resistant strains of *P. ulmi* (Thwaite, 1991; Grosscurt *et al.*, 1994), *P. citri* (Yamamoto *et al.*, 1995) and *T. urticae* (Herron *et al.*, 1993) have been reported. Another compound belonging to this class is flubenzimine, but this is not authorized in various countries.

Commercial products

Like other pesticides or agricultural chemicals, acaricides are available as commercial products with labels containing information such as the commercial name, manufacturer, active ingredient and type of formulation, product description, characteristics and authorized applications, dose and method of use, toxicology and safety period.

The product or trade name identifies the product being marketed, and is needed because chemical names are usually long and complex, and for the purposes of trade, registration and legislation, acaricides need short names that are distinctive and easy to use. Official names are assigned by the International Organization for Standardization (ISO), while the manufacturer (and distributor) is (are) responsible for the qualitative

and quantitative standards of the product and technical information on it, such as the dose(s) and use(s) suggested on the label. The 'active ingredient' (a.i.) is the actual chemical constituent that is effective against mites. A commercial product never contains 100% a.i., and the percentage of the a.i. present must be shown on the label; the remainder of the product is a bulking agent. The formulation describes the physical and chemical state of the product, i.e. the a.i. plus the excipient. There are different types of formulations and these are usually shown in abbreviated form and in capital letters.

The product description indicates the type of product (acaricide, insecticide, etc.) and the nature of its action (systemic, by contact, ingestion, inhalation). There may also be an indication as to whether the product is specific or not. The product characteristics and authorized applications indicate the crops and mite species against which the agricultural authorities (Department of Agriculture, the Ministry of Agriculture, etc.) have authorized the use of the acaricide and its allowed percentage for use. The label of a commercial product shows the dose and the method of distribution suggested by the manufacturer and authorized by the agricultural authorities. It also reports any potential risks of phytotoxicity, any sensitive plant varieties and the phenological stages that can be treated. The use of the appropriate dose and tools of application affect the results of treatments, and choices on these must be made after consideration of the intrinsic characteristics of each acaricide, the mite species against which it is to be used and its bio-ecology and population density, and plant phenology. The doses and protocols used may influence the selection of resistant strains.

Toxicological information is usually shown on the manufacturer's label according to a toxicological classification. This may differ in different countries, but it gives information on the effects of the a.i. on humans, the soil and aquatic flora and fauna, with information also included on bees and on beneficials. It is also possible to obtain information from sources such as the EU Pesticides Database (European Commission, 2014), the environmental databases of the US EPA (2014) or other databases. The safety period is the time between the spreading of the acaricide in the field (or elsewhere) and product harvesting; it varies according to the a.i. and must be shown on the product label. In addition to knowledge of the effectiveness of acaricides in the laboratory from a practical point of view, it is very important to know about their effectiveness in the field, which depends on the chemical nature of the acaricide, its formulation, dose, distribution and the environmental conditions. Each acaricide has its own effects on the different biological stages of mites and on various taxonomic groups, so their effectiveness is generally restricted to a single group of mites or to different mite groups (Childers *et al.*, 1996). Thus, some acaricides are mainly used against eriophyoids and others against tetranychids, etc. The effectiveness of acaricides is also related to the presence of any sources of reinfestation and to the general pest management of the crop, such as the use of pesticides that stimulate mite development. The label reports if the product is inflammable, hygroscopic, volatile, irritant, etc., and also gives information on whether it can be mixed with other products and on its phytotoxicity.

Side effects

The toxicological and ecotoxicological properties of a chemical acaricide, such as its selectivity towards beneficial species, can affect the control strategies employed (Overmeer, 1985b; Childers *et al.*, 1996). An extensive review of knowledge that has been gained on the influence of chemicals on the most important groups of beneficial mites has been published by Gerson *et al.* (2003), and further updates on phytoseiid mites have been reported recently by Gerson (2014). The formulation, solvents used, application methods, persistence of active ingredients, mode of action, effects of chemicals on plants and biology of beneficials all affect the impact of chemicals on beneficials. Increased fecundity (Grout *et al.*, 1997), the removal of competitors and alternative prey, and repellence and dispersal (Hislop *et al.*, 1981) also play a key role.

The side effects of various substances that are commonly used in crop defence (Duncombe, 1972; Coates, 1974; McCoy, 1977a,b; Beattie *et al.*, 1991; Smith and Papacek, 1991) and their stimulation of the development of mite populations are well known. The negative roles of organophosphates (Attiah and Wahba, 1973), synthetic pyrethroids (Gerson and Cohen, 1989) and carbamates have been widely verified. For example, carbamates control major mite pests, such as the citrus rust mite (Johnston *et al.*, 1957), but at the same time assist the development of the two-spotted spider mite (Boykin and Campbell, 1982; Childers, 1990) and inhibit beneficials (Buskovskaya, 1976; Kashio and Tanaka, 1979). The carbamate aldicarb is effective against eriophyoid and tetranychid mites injurious to citrus (French and Timmer, 1981; Childers *et al.*, 2007) and other host plants, but has high mammalian toxicity and persistence in plants. Moreover, its use involves a risk of groundwater contamination and for this reason has been restricted (Childers *et al.*, 1996). The indirect (plant-mediated) and direct impacts of imidacloprid on the fecundity and longevity of the boxwood mite, *Eurytetranychus buxi* (Garman), on boxwood (*Buxus sempervirens* L.) have been evaluated by Szczepaniec and Raupp (2013). Females of the boxwood mite that consumed imidacloprid-treated plants laid more eggs than those feeding on untreated plants, but their longevity did not change. Fecundity was not affected when the mites were directly sprayed with imidacloprid but consumed the foliage of untreated plants. Populations of *E. buxi* were greater on plants treated with imidacloprid, and on landscape plants a high density of populations of *E. buxi* persisted into a second year. Furthermore, the absence of predators of spider mites on treated and untreated boxwoods in the field suggested that the eruptive effect of imidacloprid on *E. buxi* stems more from indirect changes in plant quality than from a loss of top-down regulation by the natural enemies of *E. buxi* (Szczepaniec and Raupp, 2013).

Understanding the side effects of the chemicals used in agriculture against injurious mites (and other pests) is then crucial both for pest control and for the conservation of the environment and human health. The detailed information that exists today, as well as the availability of new principles, assists in the implementation of suitable IPM programmes (Childers *et al.*, 1996, 2007; McCoy, 1996b; Hoy, 2011). Knowledge of the side effects of chemicals is considered a prerequisite for the

implementation of an IPM programme (Croft, 1990; Kostiainen and Hoy, 1996). The working group 'Pesticides and Beneficial Organisms' of the International Organisation for Biological Control (IOBC) periodically publishes reports on the influence of new chemicals (insecticides, acaricides, fungicides, herbicides) on beneficials, and Sterk *et al.* (1999) published a report on the influence of 20 substances on three important species of phytoseiid predatory mites. It should be noted that the information that is available on the side effects of pesticides is affected by different variables, some of which are of a technical nature, including the methods of assessment and the availability of new substances, and others that are closely related to the bio-ecology and environment of the species under investigation, its prey and its natural enemies. Several examples confirm the great importance of these aspects. For example, Kreiter *et al.* (2010) used the methodology of the French AFPP/CEB Guideline No. 167 (Kreiter and Sentenac, 2004), a sequential procedure for testing the side effects of commercial products on *Typhlodromus pyri* Scheuten, a phytoseiid mite reported in French vineyards and orchards, using both laboratory and field trials. Laboratory trials of more than 120 pesticides showed that most of the insecticides/acaricides tested were toxic, while most of the fungicides were non-toxic or slightly toxic. The use of different databases can also lead the choice of acaricide, e.g. the environmental databases of the US EPA (2014).

Resistance to chemicals

The first report of the resistance of mites to chemicals was in 1937 on a systemic selenium compound (in Cranham and Helle, 1985), but the phenomenon has acquired serious phytoiatric importance only since the early 1950s (Huffaker and Kennett, 1953), by which time it had gradually spread worldwide (Swirski *et al.*, 1967; Jeppson *et al.*, 1975; Herne *et al.*, 1979; Omoto *et al.*, 1994; Gerson *et al.*, 2003). Mite resistance to chemicals has been reviewed by a number of authors: Helle (1965b); Helle and van de Vrie (1974); Jeppson *et al.* (1975); and Georghiou and Saito (1983). Cranham and Helle (1985) and Helle (1985) provided extensive information on spider mites, and Childers *et al.* (1996) and Messing and Croft (1996) on eriophyoids. Mite resistance to acaricides is now of economic importance and almost 745 cases have been reported involving species of the Tetranychidae (691 cases), Acaridae (37 cases), Eriophyidae (13 cases) and Tenuipalpidae (4 cases) (Whalon *et al.*, 2010). The subject has most recently been discussed by Hoy (2011).

According to Messing and Croft (1996), the 'resistance' of mites to chemicals can be defined as 'the increase of non-susceptible genotypes in a population that was initially susceptible as a response to a selection pressure of chemicals'. Resistance can be confused with 'tolerance', i.e. the physiological mode of action of a chemical that is initially not toxic to most of the population. Genetic and bio-ecological factors and chemical control strategies influence the development of resistance to chemicals in mites. A decreased response (active ingredient(s)–target site interactions) and/or reduced exposure (e.g. penetration, distribution, metabolism and excretion) are mechanisms (adaptations) that allow mites to survive the application of lethal doses of insecticides/acaricides. The use of acaricides involves cross-resistance, and the inheritance of resistance is not necessarily common to each locality. Each selection process possibly leads to a different mechanism of resistance, depending on the location and past selection history, and may yield different cross-resistance patterns. The development of pesticide resistance is an example of microevolution in arthropod pests. Population genetics, which is affected by many factors, including the biological and ecological traits of arthropod pests, pesticide properties and pesticide application patterns, underlies the evolution (Osakabe *et al.*, 2010). It is therefore difficult to summarize exhaustively the existing knowledge on pesticide resistance, as it is characterized by constant change, as exemplified in *Tetranychus urticae* by the identification of several genes that are involved in the digestion, detoxification and transport of xenobiotics. These include several genes for cysteine peptidase that lead to proteolytic digestion, dozens of genes for cytochrome P450 and for esterases (carboxylesterases), as well as the gene for acetylcholinesterase and a new group of glutathione-*S*-transferases (GSTs). In addition, several proteins (ABC transporters) have been identified as involved in resistance to toxic substances. Here, the sequencing of the genome of *T. urticae* has allowed a better understanding of the mechanisms of adaptation to phytophagy and the ability to develop resistance to acaricides by detoxification mechanisms (Grbić *et al.*, 2011; Dermauw *et al.*, 2012).

From a genetic viewpoint, resistance to chemicals represents a response to selection, and indicates that resistant alleles are widespread in mite populations and that pesticide treatments select for those individuals carrying resistant alleles. The most extensive studies have been carried out in the tetranychid mites, mainly in *Tetranychus* (Helle, 1962; Schulten, 1968; Overmeer, 1967; Cranham, 1974; Osakabe, 1974) and *Panonychus* spp. (Inoue, 1980; Cranham, 1982), with the early studies carried out on *T. urticae*. Genetic analysis adapted to allow for the arrhenotokous reproduction of *T. urticae* have shown that single major genes control specific types of resistance (Stone, 1968). Arrhenotoky plays an important role in Prostigmata that are adapted to phytophagy, permitting their survival and success when they are homozygous for recessive alleles carrying the characteristic of resistance to certain chemicals. Further studies have led to the identification of a mutation for resistance to organophosphates (OPs) at the locus *OP* hosting the gene *R*, which controls decreased sensitivity of acetylcholinesterase (AChE) (Schulten, 1968). At the same time, it was found that a mutation induced resistance to tetradifon (Overmeer, 1967). Later, the genes for resistance were mapped and visible mutations studied in *Tetranychus pacificus* McGregor and *T. urticae* (Helle, 1985). Two types of resistance to OPs were found by a combined genetic and biochemical analysis. The first involves an alteration in the AChE of resistant *T. urticae* (Smissaert, 1964), while the second is based on a detoxification process (Matsumura and Voss, 1964). Overall, the main resistance mechanisms in mites include the decreased penetration of acaricides through the cuticle, the greater degradation of acaricides via esterases, and the activity of cytochrome P450-dependent monooxygenases and GSTs. In addition, a change in the target site has been reported (Knowles, 1997). The phenomenon of resistance is complicated by cross-resistance

and multi- (multiple) resistance, the action of minor genes and negative cross-resistance (Cranham and Helle, 1985). Cross-resistance is the ability of a chemical to induce resistance to itself and to other compounds, while multi-resistance is resistance of the same individuals to different chemicals because of the simultaneous presence of different mechanisms of resistance.

In eriophyoid mites, it is possible to distinguish between resistance and tolerance (Messing and Croft, 1996). In these mites, acaricide resistance is not as strong as it is in tetranychid mites (Cranham and Helle, 1985; Croft and van de Baan, 1988), and the suspected cases of pesticide resistance are few. Different factors have slowed the development of resistance, such as the carrying out of only minor control compared with that for other mites, reduced exposure to chemicals, etc. (Messing and Croft, 1996). Rotation of the use of different acaricides in the field also reduces the selection pressure for resistance (Dennehy and Omato, 1994). Several eriophyoid mites are tolerant to some of the acaricides that are active against tetranychids and other mites, and some species are tolerant to acaricides that are effective against other eriophyoids (Herne *et al.*, 1979). This last phenomenon also occurs in tetranychids. Indeed, different plants affect the susceptibility of *T. urticae* to different pesticides, inducing different levels of pesticide metabolizing enzymes. The response of the phytophages to their host plants may also lead to an increase in their metabolism of pesticides because the mechanisms that they use to detoxify plant allelochemicals could act to detoxify pesticides as well (Yang *et al.*, 2001).

The results of various studies have confirmed the involvement in pesticide resistance of genetic and bio-ecological factors. Resistance of *Panonychus citri* to abamectin was mainly associated with increasing activities of AChE and mixed-function oxidases (oxidoreductases), and resistance to fenpropathrin was attributed to changes in the activities of carboxylesterase, AChE and GST (Meng *et al.*, 2002). Quantitative and qualitative changes were recorded in the general esterases of the red spider mite, *Oligonychus coffeae*, on tea in India. Mites in conventional plantations developed a greater tolerance to synthetic pesticides, possibly due to a higher quantity and activity of general esterases connected with detoxification (Sarker and Mukhopadhyay, 2006). A significant correlation between carboxylesterase activity and acaricide susceptibility was found through statistical analysis (Chen *et al.*, 2006a). Moreover, carboxylesterase and AChE might be involved in resistance to phoxim in the citrus red mite, *P. citri* (Chen *et al.*, 2009).

Screening of field strains of *T. urticae* exposed to bifenazate for several seasons showed several cytochrome b mutations, and investigations on the levels of resistance in bifenazate susceptible and resistant strains showed distinct mutations in the cd1- and ef-helices aligning the mitochondrial cytochrome b Q_o (quinol-binding) pocket. Some of these mutations caused cross-resistance to the acaricidal Q_o inhibitor acequinocyl, and investigations on cross-resistance to acequinocyl allowed the establishment of a correlation between the cytochrome b genotype and acequinocyl resistance (Van Nieuwenhuyse *et al.*, 2010). Analyses of genetic linkage between the loci for hexythiazox and etoxazole resistance using crossing experiments showed that etoxazole resistance was largely controlled by a single major autosomal locus in Japanese *T. urticae* populations, while hexythiazox resistance was controlled by more than one locus, and one hexythiazox resistance locus was tightly or completely linked to the etoxazole resistance locus. The construction of genetic linkage maps of microsatellite loci in *T. urticae* and mapping of the etoxazole resistance gene will contribute to the efficiency of linkage analysis among resistance genes. Population structure should largely affect the development rate and also the spatial distribution of acaricide resistance genes. The analyses of a fine- and large-scale gene flow of *T. urticae* in greenhouse roses and apple orchards using microsatellite markers highlighted different gene flow in *T. urticae* populations between the two environments, with limited gene flow in the glasshouse and successful migration in the orchard (Osakabe *et al.*, 2010).

Two strains of the tomato red spider mite, *Tetranychus evansi* Baker *et* Pritchard, from Malawi showed tolerance to chlorpyrifos and resistance to bifenthrin, but were susceptible to two contemporary acaricides (abamectin and fenpyroximate). Cloning of a 3.1 kb fragment (domains IIS5 to IVS5) of the *T. evansi para* gene from pyrethroid-resistant and pyrethroid-susceptible strains revealed a single non-synonymous mutation in the resistant strain that resulted in an amino acid substitution (M918T) within the domain II region of the *para*-type sodium channel. This mutation is novel to mites, and confers high levels of resistance to pyrethroids in several insect species in which it has always been associated with another mutation (L1014F). Diagnostic tools were developed that allowed sensitive detection of this mutation in individual mites (Nyoni *et al.*, 2011).

Under laboratory conditions, *P. citri* developed resistance to fenpropathrin in the 9th to 12th generation, with an LD_{50} 10.97–17.11 times that of the initial generation. The strain selected from the laboratory fenpropathrin treatment was further treated with ten other (acaricidal) active ingredients, and this highlighted cross-resistance with five of them, but a negative resistance towards flufenoxuron and the four other compounds. The findings show that the use of fenpropathrin can be combined with that of flufenoxuron and isocarbophos (Meng *et al.*, 2000).

Methods for the measurement of resistance in the field and by bioassay have been variously discussed by Busvine (1980), Helle and Overmeer (1985b), Robertson *et al.* (2007) and Hoy (2011). Common methods include slide-dip bioassays, leaf-dip or leaf-spray bioassays, whole-plant bioassays and field trials (Robertson *et al.*, 2007).

From a practical point of view, the stability of pesticide resistance in field populations, the delaying and reversing of resistance development, and the monitoring of resistance in the field may be fundamental (Cranham and Helle, 1985). In field populations, if resistance to an acaricide family is widespread and stable, its use must be stopped. New resistant biotypes evolve under the combined influence of natural selection for fitness and chemical selection for resistance. The 'co-adaptive' process related to natural selection acts more slowly than the response to chemical selection. This suggests that the process refers not just to a gene *R* but to its genetic background (McEnroe and Naegele, 1968; Cranham and Helle, 1985). Resistant strains can have lower fertility and/or fecundity (McEnroe and Naegele, 1968; Schulten, 1968; Inoue, 1980;

Cranham, 1982), lower development rates (Schulten, 1968), and the general disadvantages of homozygous *R* individuals with respect to heterozygotes (Cranham, 1982). When chemical selection is stopped early, natural selection will favour non-resistant populations, and reversion to a susceptible population becomes possible. However, when resistance develops and becomes widespread, such a reversion is unlikely when/if the use of acaricide is discontinued (Cranham and Helle, 1985). The delay and reversal of the development of resistance can be achieved by limiting the selection pressure, for instance by using fewer acaricide treatments, by using acaricides with a short persistence, or by creating 'refugia' in which acaricides are not used (Cranham and Helle, 1985). The above discussion indicates that the management of resistance in the field is difficult and sometimes unproductive. To rationalize the approach to control, a number of genetic models have been developed. These include the 'mixture model', the 'rotations model', the 'mosaic model' and the 'high dose approach model' (Tabashnik, 1990; Hoy, 1998). The mixture and rotation models are adopted more than the other two. The mixture model assumes that cross-resistance towards a second chemical is not present in the mite population concerned and, for this reason, two chemicals belonging to different pesticide classes can be used. However, this approach does not totally exclude the development of cross-resistance. The rotation model provides for the rotation of pesticide classes, to take advantage of the fact that the resistance allele frequency for a chemical will decline during the period of time when another chemical is being sprayed instead. Databases on arthropod pesticide resistance (e.g. the Arthropod Pesticide Resistance Database (APRD) from Michigan State University, which includes reports of resistance cases from 1914 to the present; see http://www.pesticideresistance.com/index.php) can lead the choice(s) of acaricide(s).

Biological Control

As for other organisms, the ecology of mite populations is influenced both by abiotic factors, such as climatic conditions, and biotic factors, such as the biotic potential of the species, its interaction with the host plant, its natural enemies (predators, pathogens) and different agricultural practices (Jeppson *et al.*, 1975). Among these factors, natural enemies are very important. Their action is at the basis of biological control, a strategy that generally assures economic, toxicological and ecological advantages. Even so, a biological control strategy has not been established conclusively for all pests and it may have some limitations, such as the use of generalist natural enemies, a lack of natural enemies for all pests, the presence of different pests, etc. (Gerson *et al.*, 2003). In these cases, the significant benefits associated with the implementation of a biological control strategy call for intensified efforts. Despite the comments on generalists above, their importance has been documented by Stiling and Cornelissen (2005), who observed that the addition of two or more natural enemies, especially if they were generalists, could increase pest mortality by 13% and decrease pest abundance by 27.2%, in comparison with single releases or releases of specialist natural enemies. This phenomenon may depend on the ability of generalists to survive on the crop when the population density of the target pest is very low, so that they are still present if the pest population rises again (Palevsky *et al.*, 2013).

In general, biological control includes both natural biological control (*sensu* Costanza *et al.*, 1997) and biological control due to natural enemies that are artificially released into the environment. The separation of these two components in natural and artificial ecosystems is not easy. From a strictly technical point of view, biological control refers as a priority (at least in this discussion) to natural enemies that are artificially introduced, although in many cases the natural enemies that are already present in nature exert more pronounced control of pests than those that are artificially introduced, e.g. on citrus. The estimated world market value of natural biological control is about US$ 400,000 × 10^6 billion (Costanza *et al.*, 1997), while that of induced (artificial) biological control is greater (van Lenteren, 2005).

Induced biological control manipulates the interspecific relationships within agroecosystems and the antagonism between living organisms, and in this sense, the strategy assigns a central role to natural enemies (DeBach, 1964; Van Den Bosch *et al.*, 1982), although it does not exclude intraspecific actions and the use of biotechnological means (pheromones, autocidals, lethal genes, etc.). A more extensive definition includes behaviour-modifying chemicals (BMCs) (food attractants) (Jones, 1987) and the sterile insect technique (SIT), in which male insects (or, in this case, mites) are sterilized by chemicals or radiation and released so that females lay only infertile eggs (Hendrichs *et al.*, 1991). A further definition suggests the genetic improvement of natural enemies (Hoy, 1990). Rosen (1986a) stated that 'we cannot deny that when successful, biological control is an ideal form of pest management permanent, inexpensive and non-hazardous. It is by far the most promising, and most desirable alternative to chemical control available wholesale in our arsenal'. Gerson *et al.* (2003) give a realistic and modern definition of the biological control of pests by mites based on mite actions in reducing pest density and/or the extent of pest damage to below the economic injury level. The choice includes both natural and applied reductions, and utilizes three basic strategies: the importation, conservation and augmentation of natural enemies, which strategies lead on naturally to the use of integrated pest management (IPM). The importation, conservation and augmentation of natural enemies are briefly discussed below, followed by an account of the main species of predatory mites and insects that prey upon mites, and their production and release. IPM is treated separately and forms the last section of this chapter.

Importation

This strategy is also known as 'inoculative release' or 'seeding' and is the introduction and release of exotic natural enemies in order to control introduced pests. It aims to establish a stable relationship between the pest populations and the natural enemies. Moreover, it is necessary that the natural enemies occupy the same habitats as their prey, and as a result are able to live and reproduce on the prey and/or alternative food. Many examples involve phytoseiid mites (McMurtry, 1982), and other mites, throughout the world. The most significant project

is the importation into Africa of different phytoseiid mites to control the cassava green mite, *Mononychellus tanajoa* (Bondar) (Gerson *et al.*, 2003).

Conservation

Conservation aims to remove from the environment any factors adverse to natural enemies, such as insecticides (Huffaker *et al.*, 1970). The negative effects of these factors may be eliminated by different approaches, such as assaying all agricultural chemicals for their effects on natural enemies (Sterk *et al.*, 1999), using only pesticides with little or no effects on natural enemies (Pickett, 1965; Bostanian *et al.*, 2010), introducing strains of phytoseiid mites that are genetically improved and resistant to chemicals (Hoy, 1992; Duso *et al.*, 2010), using ground cover crops among plants as shelter for the arthropod natural enemies (Liang and Huang, 1994; Duso *et al.*, 2010) or planting windbreaks for the production of pollen as an alternative food for natural enemies (Grout and Richards, 1992a; Barbar *et al.*, 2010; Duso *et al.*, 2010).

Augmentation

In this strategy, the populations of natural enemies are manipulated by different releases. The approach requires that the natural enemies are able to prey on the pests and a lack of limiting factors, such as unsynchronized phenology, a high pest population density, or chemical treatments. These requirements suggest that the pest populations are treated by other methods before the introduction of natural enemies, and that the natural enemies are released either before the pest arrives or develops a high population density. According to the predatory species used, there may be different behaviours, such as the ability to consume pollen before the arrival of the prey or the need to consume prey. Markkula and Tiittanen (1976) described the 'pest-in-first' method, in which the plants are first evenly infested with the pest mites and then the predator mite is introduced, but in the case of cucumbers infested with the spider mite *T. urticae* and the predator mite *Phytoseiulus persimilis*, found that the pest-in-first method required more labour and a greater supply of predators.

Natural enemies

The predatory mites are the most important natural enemies of pest mites, followed in descending order by insects and pathogens. The role of these little arthropods as biocontrol agents of other mite pests has been known for a long time. Gerson *et al.* (2003) discussed about 30 families of beneficial mites involved in the biological control of injurious mites and/or other pests. Here, we list the main species, focusing attention on the fundamental features of their practical use: mass production, storage, shipment and release in the field. The so-called 'biopesticides' are included among the natural enemies, treating them as acarine biocontrol agents (ABAs) that are mass reared and commercially offered for the control of pests and/or the level of damage caused by pests (Gerson *et al.*, 2003). Further insights may be found in Gerson (2014).

Main species

Tables 9.2 and 9.3 list mites and insects that are predators of pest mites and are commercially available in Europe and North America. At the beginning of this decade – and excluding the Laelapidae *Hypoaspis aculeifer* (Canestrini) (= *Gaeolaelaps aculeifer* (Canestrini)) and *Hypoaspis miles* (Berlese) (= *Stratiolaelaps miles* (Berlese)), there were overall 17 phytoseiid and five insect species (Thripidae, Coccinellidae, Cecidomyiidae, Chrysopidae) that were marketed commercially (van Lenteren, 2005). However, of these, only four or five phytoseiid mites are widely used. Gerson *et al.* (2003) reported a more complete list of predatory mites based on different references (Zdárková, 1991; IPM Practitioner, 2000) and including four families and over 20 species. Around seven species are offered from more than ten suppliers, and of these, the phytoseiid *P. persimilis* is available from about 49 suppliers. The high number of suppliers demonstrates the significance and importance of this predacious species.

Mass production

Finney and Fisher (1964) described the mass production of natural enemies as a 'skillful and highly defined processing of an entomophagous species through insectary procedures which results in economical production of millions of beneficial insects'. The method is used in the augmentation strategy described above and involves various technical and commercial aspects. It usually involves rearing natural enemies on their natural hosts, under the appropriate environmental conditions and using procedures that are economic. Sometimes, rearing on the natural host is expensive or involves the risk of the transmission of pests or other organisms. To avoid this problem, it has been proposed that natural enemies are reared on artificial diets. However, this technique may have quality problems (Singh, 1984). The strategy that is adopted aims to standardize the production of the maximum number of beneficials at the lowest cost and in the shortest time – which will allow beneficials to be easily available in the field at affordable cost. Currently, over 20 species of predatory mites are mass reared commercially by different companies worldwide.

Mass rearing is developed through three gradual steps from the laboratory to the insectary. The first step occurs in the post-collection period and aims to avoid the introduction of diseases or new pests. In the second step, better conditions are identified for mass rearing. The third step is the mass rearing process per se (Gerson *et al.*, 2003). Generally, it is necessary to produce uninfested host plants and pure (i.e. predator free) pest populations. According to the aims and the species, the mass rearing of predatory mites may be carried out in artificial 'arenas' of various types (McMurtry and Scriven, 1965, 1975; and other authors), on detached leaves (Overmeer, 1985a) or on leaf discs (Brodeur and Cloutier, 1992). It is possible to use alternative prey, different types of cages (Tanigoshi *et al.*, 1975a; Fournier *et al.*, 1985; Vacante *et al.*, 1989a) or artificial diets

Table 9.2. Mites and insects predatory of pest mites available commercially in Europe in 2000, listed by species, indicating their origin (endemic, en; or exotic, ex), pest type/species and year of first use (from van Lenteren, 2005).

Species	Origin	Pest (prey)	Origin	Year
Mites				
Amblyseius fallacis (Garman) (= *Neoseiulus fallacis* (Garman))	ex	Mites	ex	1997
Amblyseius largoensis (Muma)	ex	Mites	ex	1995
Hypoaspis aculeifer (Canestrini) (= *Gaeolaelaps aculeifer* (Canestrini))	en	*Rhizoglyphus* spp.	en	1996
Hypoaspis miles (Berlese) (= *Stratiolaelaps miles* (Berlese))	en	*Rhizoglyphus echinopus* (Fumouze *et* Robin)	en	1994
Kampimodromus aberrans (Oudemans)	en	*Panonychus ulmi* (Koch)	en	1960
Metaseiulus occidentalis (Nesbitt)	ex	Mites	en	1993
Neoseiulus barkeri Hughes	en	Mites	en	1990
Neoseiulus californicus (McGregor)	ex	Mites	en/ex	1995
Neoseiulus cucumeris (Oudemans)	en/ex	Mites	en/ex	1990
Phytoseiulus longipes Evans	ex	*Tetranychus urticae* Koch	en	1990
Phytoseiulus persimilis Athias Henriot	ex	*Tetranychus urticae*	en	1968
Typhlodromus pyri Scheuten	en	Mites	en	1985
Insects				
Scolothrips sexmaculatus (Pergande)	en	Mites	en	1990
Stethorus punctillum Weise	en	Mites	en	1995
Feltiella acarisuga (Vallot)	en	Mites	en	1990
Chrysoperla carnea (Stephens)	en/ex	Mites	en	1987
Chrysoperla rufilabris (Burmeister)	ex	Mites	en	1987

Table 9.3. Predatory mites that have been commercialized worldwide, listed by family, species, number of suppliers and the prey pest (from Gerson *et al*., 2003).

Family	Species	No. of suppliers	Pest (prey)
Cheyletidae	*Cheyletus eruditus* (Schrank)	1	Grain mite
Laelapidae	*Hypoaspis aculeifer* (Canestrini) (= *Gaeolaelaps aculeifer* (Canestrini))	1	Bulb mite
	Hypoaspis miles (Berlese) (= *Stratiolaelaps miles* (Berlese))	5	House fly
	Stratiolaelaps scimitus Womersley	1	Larvae of Sciaridae, soil pupae of thrips
Macrochelidae	*Glyptholaspis confusa* (Foà)	1	House fly
	Macrocheles spp.	1	House fly
Phytoseiidae	*Amblyseius californicus* McGregor	23	Spider mites
	Amblyseius cucumeris Oudemans	20	Spider mites
	Amblyseius degenerans Berlese	3	Spider mites
	Amblyseius fallacis (Garman) (= *Neoseiulus fallacis* (Garman))	6	Spider mites
	Galendromus annectens (De Leon)	3	Spider mites
	Galendromus helveolus (Chant)	5	Spider mites
	Galendromus occidentalis (Nesbitt)	22	Spider mites
	Mesoseiulus longipes (Evans)	11	Spider mites
	Neoseiulus scyphus (Schuster *et* Pritchard)	1	Spider mites
	Neoseiulus setulus (Fox)	1	Cyclamen mite
	Phytoseiulus macropilis (Banks)	3	Spider mites
	Phytoseiulus persimilis Athias Henriot	49	Spider mites
	Typhlodromus athiasae Porath *et* Swirski	1	Spider mites
	Typhlodromus mcgregori Chant	1	Spider mites
	Typhlodromus pyri Scheuten	1	Spider mites
	Typhlodromus rickeri Chant	1	Spider mite

(McMurtry and Scriven, 1965; Ochieng *et al.*, 1987; and other authors). In addition, the mass rearing of predatory mites can be effected in greenhouses (Vacante and Benuzzi, 2004), in apple trees (Field *et al.*, 1979), in the field (Ireson and Webb, 1996) and in other environments. Despite the importance of mass rearing, there is often no detailed knowledge of many techniques as this is strictly guarded by the manufacturers, and in some cases, development projects have involved acarologists from different countries, as in the recent case of *Neoseiulus californicus* (Palevsky *et al.*, 2010b). Recently, Nguyen *et al.* (2014) have developed an artificial diet for *Amblyseius swirskii* Athias Henriot.

In addition to its technical aspects, the mass rearing of beneficials has political and economic implications, which can influence the technical decisions made and the planning of interventions, especially at long distances. Production for use at short distances may involve methods of mass rearing and manipulations that are less stressful and often have a lower overall cost. In contrast, the production of beneficials for use a long distance away, or more generally for globalized marketing, necessarily requires more sophisticated methods and means, and is consequently more expensive. In short, the two possibilities are supported by different ecological, ethical and socio-economic visions. In fact, there seems to be no need for large production complexes, operating even thousands of miles away, with strains of beneficials that do not always fit the requirements, but rather a need for the widespread distribution of small insectaria operating with suitable strains and not aiming for a great economic return.

Quality control

Quality control of beneficials is a crucial step for mass rearing, and consequently for the success of biological control. Indeed, many field experiments have concluded negatively because of the poor state of the beneficials introduced. There are technical, economic and political aspects to this, and sometimes problems arising from conflicting interests. Here, only the technical aspects are discussed, based on information on mite beneficials that has been drawn mainly from Gerson *et al.* (2003), and partly from van Lenteren (2005), as well as personal experience.

The mass rearing of predatory mites may be affected by different technical and biological (genetic, etc.) problems. These include inadequate feeding (Dicke *et al.*, 1989; James, 1993) and plant health and the accidental introduction of pests (Nakao *et al.*, 1990; Mégevand *et al.*, 1993), the development of invading predators (Hoy *et al.*, 1982a; Ho and Chen, 1992a), the 'founder effect' (van Lenteren and Woets, 1988) – a type of genetic drift in which there is little genetic variability in the new mass reared population, genetic drift (Roush, 1990; Mégevand *et al.*, 1993), inbreeding depression (Poe and Enns, 1970; Roush, 1990), directional selection (Roush, 1990; Roush and Hopper, 1995; Mégevand *et al.*, 1993) and various diseases (for references see Gerson *et al.*, 2003, and Gerson, 2014).

The problems noted above confirm the need for constant monitoring of the quality of mass reared beneficials. However, the operation is not easy, it can be expensive and require specialization that is not easily available. Gerson *et al.* (2003) reported a monitoring method based on 'fluctuating asymmetry' (FA), in which measurements are made of the differences between the two halves of bilaterally symmetrical organisms – differences that express hereditary and environmental stress during development (Parsons, 1990). A loss of genetic variation results in increasing values of FA (Leary and Allendorf, 1989) and any stress registered by the developing organism will increase its FA. The method is not expensive and specific analyses have been developed (Palmer, 1994; Swaddle *et al.*, 1994), although as of now there are only four quality control guidelines known for predatory mites.

Harvesting

According to the bio-ecology of different predatory species, various methods and tools of harvesting have been reported in the literature, such as the use of aspirators to collect phytoseiids (Theaker and Tonks, 1977). Other methods include the use of low-vacuum suction pumps or cooling at 4°C (McMurtry and Scriven, 1975), dry cotton wads (Jedlicková, 1992), humid granular media (Pickett *et al.*, 1990) or container traps with glue on the openings (Vacante and Benuzzi, 2004). When beneficials are produced for use at a short distance, leaves of host plant with the predator and its prey can be collected and kept in plastic bags – this is usually at the end of the rearing process, when the predatory populations are predominant. The systems that are used in commercial rearings are unknown.

Storage and transport

The storage and transport of beneficials are related, with the aim of using mites that are as fresh as possible. The optimization of the two steps contributes to the success of the trade. Efficient storage allows predatory mites to be released when they are needed (usually by growers a short distance away) and also to preserve their integrity during shipment for release at long distances (usually in the trade). The long-distance transport of predatory mites usually requires storage at low temperatures and within a humid carrier, in order to reduce their movement and slow down their metabolism. Short transport involves the application of different techniques and tools, but is usually easier.

Storage has been investigated for main predatory species, such as *Phytoseiulus persimilis*, for which it has been found that storage at 8–10°C for almost 10 days does no damage (Nakao *et al.*, 1990). This predator can also be maintained at 7.5°C for 4–6 weeks if the available food and moisture are sufficient (Morewood, 1992). Young females of *P. persimilis* tolerated 18 days exposure to either 5 or 10°C and starvation without affecting their quality (Luczynski *et al.*, 2008). Mated females of the Japanese strain of *N. californicus* survived for several weeks at 100% RH and 5°C without affecting the quality of their progeny (Ghazy and Amano, 2014).

In related studies, Kawashima and Jung (2010) monitored the winter survival of phytoseiids in an apple orchard in Japan using artificial shelters placed on the ground, and Szabó and Pénzes (2013) reported the transfer of ground litter from Hungarian orchards containing large numbers of the overwintering phytoseiid *Amblyseius andersoni* (Chant) to young orchards.

Release techniques

The release of predatory mites can be carried out either by hand or mechanically, and on small or large surfaces, according to the importation or augmentation strategy that is being used. Manual release is generally used over small cultivated areas, but it may be negatively affected by the behaviour and stress of the worker. The augmentation strategy is commonly based on the field introduction of large numbers of predators. The quantities of predatory mites released on to plants depend on the crop and its level of infestation. For mechanical release, e.g. of *Phytoseiulus persimilis*, the predatory mites are mixed with vermiculite or some other substrate and immobilized by chilling. A mechanical dispenser was used to distribute *P. persimilis* in a rose greenhouse and reduced *T. urticae* infestations by about 50% (Casey and Parrella, 2005). Two types of mechanical blowers used to disperse phytoseiids in a greenhouse both produced better coverage than manual release (Opit *et al.*, 2005), and the use of a motorized leaf blower to release *P. persimilis* against *T. urticae* on hop was four times faster than hand distribution (Shaw and Wallis, 2007). Another technique is the use of a rotating metering plate, which allows mites to fall down through openings (Giles *et al.*, 1995; Abenavoli *et al.*, 2009); the results of this research also pointed out the greater reliability of mechanical release (Giles *et al.*, 1995).

Other release methods include the use of various packaging systems. For example, a bran-based *Neoseiulus cucumeris* (Oudemans) culture is contained in waxed paper packs with a particular porosity. The packs are placed on plants and the phytoseiid regularly distributes within the crop, as in its use for the control of western flower thrips, *Frankliniella occidentalis* (Pergande), in the greenhouse (Bennison and Jacobson, 1991; Vacante and Tropea Garzia, 1993). In the biological control of the spider mite *Oligonychus perseae* Tuttle, Baker *et* Abbatiello on avocado trees, *Neoseiulus californicus* was blown with a mist blower (Takano-Lee and Hoddle, 2001). Over maize, aerial releases of phytoseiid mites were carried out with light aircraft using maize grits as carrier. The mites and carrier were introduced into a mechanized and refrigerated system, metered out on to a rotor and pneumatically released from the aircraft. The resulting distribution is about uniform, but wind may negatively affect the results (Pickett *et al.*, 1987). A variant was proposed on cassava in Africa, called the Airborne Insect Release System (AIRS), which is characterized by a particular packaging and delivery system (Drukker *et al.*, 1993). In another method, called 'pest-in-first' (see under 'Augmentation' above), the prey is released before the introduction of the predator, and the predator is introduced into the crops only after the development of a small non-damaging population of pests (Huffaker and Kennett, 1956).

In addition to their release, it is necessary to control the establishment of predatory mites in the field, as well as the effects of any post-colonization evolution (PE) (Murray, 1985; Secord and Kareiva, 1996). Various ecological factors may be involved, among which global warming may play a significant role, particularly as the response to an increase of temperature can be different for pests and natural enemies, and the magnitude and direction of this difference will determine the result of the biological control strategy. From a biological control viewpoint, it is necessary to understand the relationship between the effect of warming on higher trophic levels and herbivore population outbreaks, an approach that has been considered only recently (Montserrat and Guzmán, 2010).

Legislation and economics

Like chemical control, biological control may be regulated by specific legislation, but there is no common legislation for all countries of the world. Here, we must distinguish between legislative measures aimed at facilitating the application of the strategy in a country, e.g. the European laws that cover the extension of biological control in all countries of the EU, from those that are primarily aimed at facilitating the introduction of natural enemies to and their exchange between different countries, as in the trading of beneficials. The latter does not exclude the risk of introducing non-native biological control agents (Sheppard *et al.*, 2003). Biological control and IPM can be complementary in the EU, as witnessed by the legislation on 'Methods of Organic Production' (EEC Regulation 2092/91 (Annex II B), which was superseded in June 2007 by Regulation No 834/2007) and 'Agricultural Production Methods Compatible with the Requirements of Environmental Protection and the Maintenance of the Countryside' (Regulation 2078/92 and beyond). The EU has set out the basic principles of IPM (Directive 2009/128/CE), assigning a central role to biological means and horticultural management. Further details may be found in van Lenteren (2005).

It has been estimated that biological control by arthropods and nematodes has a world market value of US$400,130 × 10^6 billion (van Lenteren, 2005). In 1997, the worldwide turnover of natural enemies of all producers was calculated to be US$25 million and in 2000 it had reached about US$50 million – an annual growth of 15–20% (Bolckmans, 1999). About 85 commercial producers of natural enemies are reported worldwide. Of these, 25 are active in Europe, 20 in North America, six in Australia and New Zealand, five in South Africa, about 15 in Asia (Japan, Korea, India, etc.) and about 15 in Latin America. Some 80% of natural enemies are used in greenhouses. The phytoseiids *Phytoseiulus persimilis* and *Neoseiulus cucumeris* account for 12% of the total market (Bolckmans, 1999). Other significant experiences with augmentative biological control programmes in Europe involve 20,000 ha of strawberries treated with *P. persimilis* to control *T. urticae* and 40,000 ha of vineyards treated with *Amblyseius andersoni* (Chant) against the same pest (Sigsgaard, 2006).

The average price per mite varies according to the species and country. Thus, the average price per mite of *Amblyseius cucumeris* [= *Neoseiulus cucumeris*] is US$0.00190 in the USA (Cranshaw *et al.*, 1996) and US$0.0094 in Europe (van Lenteren *et al.*, 1997); for *P. persimilis*, the average price per mite is US$0.01924 in the USA (Cranshaw *et al.*, 1996) and US$0.1013 in Europe (van Lenteren *et al.*, 1997). It has been calculated that releases of *P. persimilis* on commercial strawberries in Florida have cost US$250–375 ha^{-1}, while acaricides have cost to US$925 ha^{-1} (Decou, 1994). On almond in California, the use of pesticide-resistant *Galendromus occidentalis* (Nesbitt) produced a saving to growers of US$60–110 ha^{-1} annually, with benefit: cost ratios from 14:1 to 34:1 for different strategies

(Headley and Hoy, 1987). The annual economic benefits were more than US$20 million (Hoy, 2000). Yaninek (2007) reported that from 1983 to 1997 about US$9.5 million were spent on the biological control programme for the cassava green mite. The funds were used to operate laboratories in Bénin, to support a network of collaborators in 20 countries, and to pay for foreign exploration services, international quarantine work and training. National programmes in Africa were provided with basic laboratory and field equipment, funds for personnel and supplies. Funds also were provided for training, including training workshops for local scientists and staff, as well as training for scientists from Africa to obtain doctorates and Master of Science and Bachelor of Science degrees in universities in Africa, South America, North America and Europe. Yaninek (2007) attributed the successful programme to creating 'a legacy of biological control training, infrastructure and experiences that is still evident today' in African subsistence agriculture (Hoy, 2011).

It should be noted however, that regardless of the above data, it is difficult to present a comprehensive overview of the topic of mass rearing when it is largely managed by commercial producers who (legitimately) do not disclose all of their business data.

Integrated Pest Management (IPM)

The strategy of IPM was proposed around the start of the 20th century, but the paper published by Stern *et al.* (1959) over 50 years ago was a significant step in defining the concept in a scientifically appropriate manner. Later, Flint and Van den Bosch (1981) adopted what is probably the most comprehensive definition when they reported that integrated pest control

> is based on the factors of plant resistance and natural mortality, such as that caused by natural enemies, and take control techniques that disturb as little as possible to these factors. The method does not exclude the use of chemicals but uses them as a last resort after a systematic monitoring of pest populations and if the natural control factors will pose the need. An integrated program considers all possible control measures, including the option not to act, and assesses the potential interaction between different techniques, cultivation practices, the development climate, the various pests and the crop.

The strategy aims to control mite populations (and other pests) using all possible means (chemical, biological, biotechnical, horticultural, etc.) that do not hinder the action of indigenous and artificially introduced natural enemies and are in agreement with the ecological, toxicological and economic requirements of agricultural production. Prevention of the introduction of exotic pests is also fundamental. The method relates economic and ecological aspects, defines a strong phytoiatric vision, and results in multiple technologies that can be applied to and are adapted to specific areas of cultivation. IPM is a fundamental prerequisite for the realization of 'integrated production', i.e. a holistic approach to the production system. The farm represents a base unit and the agroecosystem a central point of intervention, in which the conservation and improvement of soil fertility and the diversity of the environment are essential. Biological, technical and chemical methods must be balanced and respectful of environmental protection, income and social needs (El-Titi *et al.*, 1995).

According to Vacante and Bonsignore (2012), the following components (see Fig. 9.1) are involved in the implementation of an IPM programme:

- prevention and quarantine;
- agricultural practices;
- monitoring of climate conditions;
- identification of pests and their natural enemies;
- sampling and monitoring of pests, natural enemies and diseases;
- decision tools (economic injurious levels, economic thresholds, action thresholds; mite days);
- crop consultants and training;
- chemical control;
- biological control.

These are each discussed below, with the exception of chemical and biological control, which have been addressed in the first part of the chapter.

Prevention and quarantine

The accidental introductions of pests (and diseases) can represent a serious problem, usually due to the increasing ease of transport, displacement of people and goods, and other factors, including the illegal introduction of infected and/or infested plant material. This means that there is a constant demand in agricultural production for the prevention of pest outbreaks, along with long-term and economic pest control. This need occurs across all control strategies, including IPM. It is possible to limit the risk of pest introduction by formulating plant quarantine measures (Burger and Ulenberg, 1990). In addition, quarantine measures – inspections at departure and entry points and during post-entry quarantine and quarantine treatments – play a fundamental role the interception of pest mites and preventive measures to reduce the risk of pest mite introduction.

Prevention also includes the removal of sources of new infestation, the correct choice of rootstocks and scion cultivars that are less susceptible to mite pests and diseases, appropriate irrigation systems and enhancement of the activities of natural enemies (Vacante and Bonsignore, 2012).

Quarantine

To prevent the accidental introduction of exotic species, agricultural production is subject to international regulatory frameworks and guidelines (Dieckmann *et al.*, 1994; FAO, 2001a,b, 2006). Prevention of the accidental introduction of exotic pests requires that all possible mechanisms are put in place that assist in intercepting adventive (i.e. not native) species in time and in eradicating them at first outbreak. This aim is achieved through quarantine, the protocol for which is mainly based on inspection at departure and entry points and on interventions during post-entry quarantine. The usual course of action is to suggest quarantine treatments, but interception and

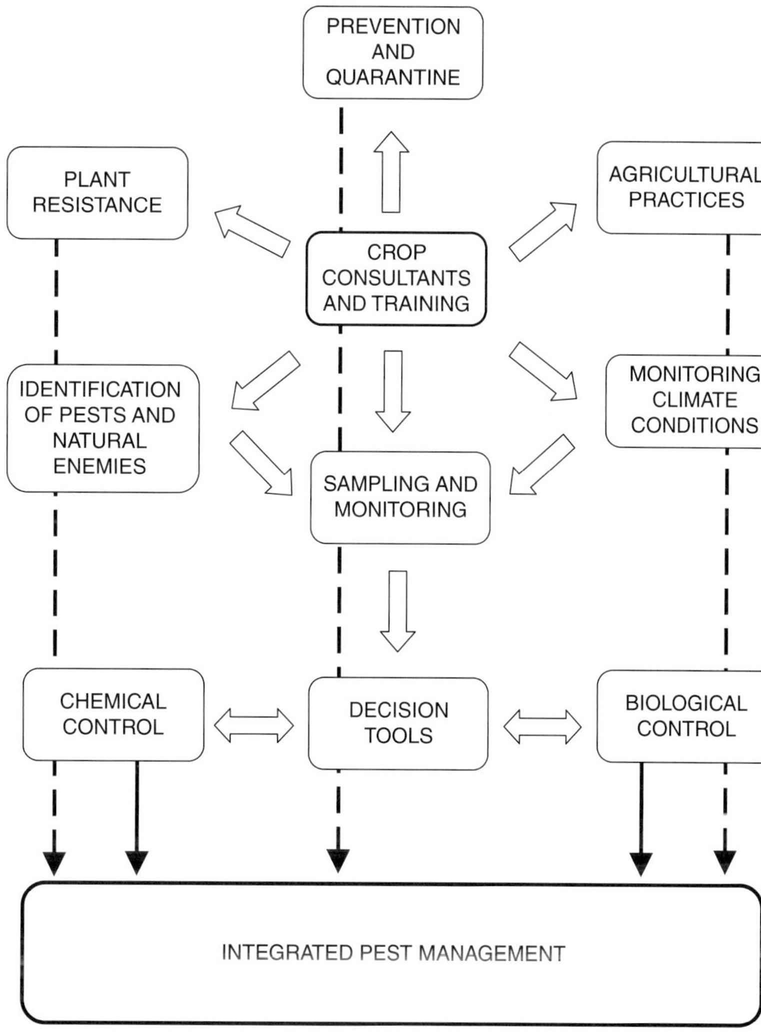

Fig. 9.1. Factors and steps involved in the implementation of an IPM programme.

preventive measures to reduce risk may also be important. Several cases involving mites have been reported in the literature, among which the most recent seems to be that of the problem of the rice mite, *Steneotarsonemus spinki* Smiley, in South and North America, against which it has been suggested that the use of quarantine measures be reinforced in order to avoid or to delay its introduction (Návia *et al.*, 2010b). This topic has been extensively discussed for the eriophyoid mites by Návia *et al.* (2010a), but the general features are applicable to other pest mites, and information given here is drawn mainly from this work.

The small size of mites, and their bio-ecological characteristics, make it very difficult to inspect goods for their presence at entry or departure points, so that the inspections that are carried out are sometimes insufficient; there is then a need to integrate inspections with other quarantine measures (Návia *et al.*, 2007a). For example, eriophyoids are very small and may shelter inside cracks and crevices, below bracts and sheaths or even inside plant organs such as buds, bulbs, etc. This amplifies the problem of their detection, so it has been suggested that there is a need also to look for symptoms resulting from mite feeding (plant abnormalities, discolorations, galls and blisters, necrotic spots, etc.). Infested plants or sampled goods must be examined in the laboratory to identify the mite responsible. In the USA, in states lacking trained quarantine inspectors, a distance diagnostic system based on digital imaging (the Remote Pest Identification Program) has been implemented by the Plant Protection and Quarantine Program of the US Department of Agriculture (USDA). Such a system could also be applied in other countries of the world.

For post-entry quarantine inspections, goods are officially confined after entry for observation and further inspection (FAO, 2006). This measure means that inadvertent pest introduction through the importation of plant material for propagation is avoided. The time needed depends on the quarantine pest. Thus, for eriophyoid mites that develop within plant organs, a long period may be required. Fruit species from arid regions are usually imported as propagation materials (budwood, cuttings, rooted plants), and glasshouse confinement for one to three seasons may be recommended in order to detect latent infestations of eriophyoid mites (Verma and Kapur, 1990).

At least 10 months of confinement have been required for *Ribes* fruits (Dieckmann *et al.*, 1994). The complete dissection of plant material assures that it is not infested by mites. The use of tissue culture allows the disadvantage of quarantine to be by-passed, and may be applied when it is necessary to introduce a plant of economic importance that is usually infested by eriophyoid mites into a new habitat/country/continent. It is quite easy to make sure tissue-cultured material is eriophyoid free and carry it to the required destination – it takes a little longer to have a grown plant, but prevents the carrying of an eriophyoid mite.

Modelling the future distribution of a species constitutes one of the most important tools for biodiversity management and pest risk assessment. To address the need of invasive risk assessment, Genevest *et al.* (2012) modelled the potential distribution of five species of Tetranychidae (*Tetranychus evansi*, *T. kanzawai* Kishida, *T. urticae*, *Panonychus ulmi* and *P. citri*) under current and future climate conditions using mechanistic models based on the physiological characteristics of the species. Contrary to the commonly used correlative models, mechanistic models (NicheMapper and Climex) are assumed to have a better predictive capacity because they explicitly take into account the functional and physiological traits of the species that might constrain their range. They are useful for predicting the physiological consequences of different environmental conditions as a function of climate, vegetation and topography in order to model the species range limits independently of a species' present range. As they rely on mechanisms rather than correlations, these models should be more reliable in terms of their predictions under climate change and their transferability under new environmental conditions, such as during an invasion process. The potential global distributions of the above five mite pests modelled in this way show that *T. urticae* is cosmopolitan, while the other species either have more restricted distribution ranges or have very recently expanded their distribution and are regarded as invasive, as in the case of *T. evansi* (Genevest *et al.*, 2012).

T. evansi is native from South America, has recently expanded its geographic distribution in different continents and is likely to keep spreading under the current warming trends. Its potential climatic niche at the global level under current conditions, as well as under two contrasting climatic scenarios for 2080, has been modelled using different correlational species distribution models (SDMs) to draw maps of potential distributions and areas of uncertainties that arise due to differences in projection methods. In the Mediterranean basin, two haplotypes were distinguished that were consistently found in different geographic regions. For each haplotype, the species climatic niches were modelled, along with their overlap and differentiation along major temperature and precipitation gradients. The potential effects of climate change on each haplotype and their distributions were then explored to put into perspective the ecological and agricultural consequences of niche differentiation under a context of climate change (Migeon *et al.*, 2012). Similarly, molecular tools have helped to place the problem of *T. evansi* in a more rational context. Analyses of two DNA sequence-based markers – the cytochrome c oxidase subunit I of mitochondrial DNA (868 bp of *COI*) and the internal transcribed spacer region of nuclear ribosomal DNA (1137 bp of ITS) were carried out on samples spanning the current geographical distribution to study the earliest stages of the invasive process. The results highlighted two different patterns of invasion: the first consisted of populations in the Mediterranean basin that are a result of multiple cryptic introductions, and the second of emerging invasions of Africa and Asia, each likely to be the result of the introduction of propagules from one or limited sources (Boubou *et al.*, 2011).

QUARANTINE TREATMENTS. The problem of quarantine treatments awaits proper resolution, as few treatments have achieved the high levels of control that are required (MacDonald and Mills, 1994). The main means employed and/or proposed are methyl bromide fumigation (see 'Chemical Control'), controlled atmosphere treatment (CAT), controlled atmosphere temperature treatment (CATT), irradiation and the thermal treatment (see 'Physical Control').

Interception of pest mites

Exotic species arrive at and colonize new countries mainly by international trade. It is therefore very important to intercept them at the border, before they can establish in new geographical areas (National Research Council, 2002). Knowledge of the pests that are associated with specific commodities and should be intercepted is very important when conducting a pest risk assessment (PRA). Interception is basic, as is information on the frequency with which it occurs so that an entry risk evaluation can be made (FAO, 2006). Examples of the eriophyoids that have been intercepted at borders are reported by Návia *et al.* (2010a).

Preventive measures to reduce the risk of introduction of pest mites

Preventive measures to reduce the risk of introduction of exotic species is related to biosecurity, and specifically with all policy and regulatory frameworks that aim to manage risk for food, agriculture and environment (FAO, 2001a). The primary purpose of biosecurity is the protection and mitigation of factors associated with invasions of exotic species. This involves the pest risk management, e.g. the analytical process of identifying risk mitigation options and evaluating the implications of these, such as their efficiency, feasibility and impacts (FAO, 2001b). Different options allow the mitigation of risks in plant biosecurity. They include market restrictions, inspection at departure and entry points, post-entry quarantine, quarantine treatments, transport conditions and pre-harvest and postharvest procedures for material originating from pest-free or low-incidence areas. Phytosanitary procedures should be chosen according to the risk level of the importation of different commodities from various countries (Návia *et al.*, 2010a).

There is also the problem of whether or not exotic natural enemies can be viewed as 'benevolent invaders'. This has recently been addressed by Palevsky *et al.* (2013), who reviewed the efficacy of biological control efforts with phytoseiids in case studies in which exotic and indigenous species were used against pests on indigenous host plants and on crops that had been either historically or recently introduced. They concluded that on some crops (citrus, apple, grape and cassava) exotic phytoseiids have enhanced biological control without negatively affecting indigenous species of natural enemies, except in the

case of *Euseius stipulatus* (Athias-Henriot) on citrus, which displaced *Euseius hibisci* (Chant), an inferior biocontrol agent of *Panonychus citri* in a restricted part of coastal California. *Phytoseiulus persimilis* used on invasive gorse is the only example (so far reported) of a negative effect of an established exotic phytoseiid on biological control (Palevsky *et al.*, 2013).

Host plant resistance

Host plant resistance has been discussed in Chapter 5 (Ecology), under 'Relationships between mite pests and their host plants', where its positive implications were also evaluated in relation to plant defence against phytophages. Varieties resistant to mite pests have been developed for a number of cultivated plants, but are not widely available for each, and each case must be individually related to the existing technical possibilities. In general, the use of resistant varieties represents a valid choice for the implementation of IPM.

Agricultural practices

Chapter 5 (Ecology) highlighted the influence of agricultural practices on outbreaks of mite populations in crops. The phenomenon has been investigated in depth for spider mites (less so in eriophyoids), where a number of factors have been indicated as involved, including the introduction of new varieties, bud selection, plant vigour, monoculture, ground cover, the presence of weeds and nearby plants, use of pesticides, natural enemies, the presence of inert dusts and soil organisms. In some mite groups, e.g. the Penthaleidae, it has been found that crop rotations, trap and border crops and weed management play an important role (Umina and Hoffmann, 2004), as do the modification of tillage practices and use of mixed cropping (Umina, 2008). The correct management of these factors helps in the implementation of control strategies, including IPM.

Monitoring climate conditions

Monitoring climate conditions helps the management of cultural practices (irrigation, tillage, etc.) and mite pest control, including chemical treatments, e.g. in the case of sulfur sprays and other compounds, and the release of natural enemies. Information is usually acquired from an agro-meteorological station, but in the absence of this, data from a few miles away can be used after correction for attitude. The data are input into meteorological models, with threshold and maximum temperatures for degree-day (°D) models. These allow prediction of the timing of the developmental stages of mite pests with an acceptable accuracy (between 1 day and 1 week). The data needed for preparing pest predictive models can be reconstructed using a meteorological model (Vacante and Bonsignore, 2012).

Identification of pests and their natural enemies

The correct identification of pests and their natural enemies is an absolute prerequisite for the implementation of a biological control programme (Rosen, 1986b). Consequently, it is not possible to implement an IPM programme without complying with this requirement. For pest mites, the case of three cryptic species of penthaleids (*Penthaleus major* (Dugès), *P. falcatus* Qin *et* Halliday and *P. tectus* Halliday) provides an example of the problem. Electrophoretic analysis indicates that these three *Penthaleus* spp. differ in clonal diversity (Weeks and Hoffmann, 1999). The main morphological differences between them are in the lengths and numbers of setae on the dorsal surface of the body. All three species live in Australia, but differing in their distribution, plant hosts, timing of diapause egg production and susceptibility to pesticides. The control strategy for each needs their exact identification and must consider all of these aspects (Umina *et al.*, 2004). Comparable cases have been reported for other mite groups, including predatory species. In the latter case, Tixier *et al.* (2010a) investigated the differentiation of specimens of *Phytoseiulus longipes* Evans collected from four countries, two of which fed on *Tetranychus evansi* and two of which did not; they demonstrated small morphological and genetic differences between them by which they could be identified.

Sampling and monitoring

In the last 50 years, sampling methods have attracted great research interest. Sabelis (1985b) gave an extensive review for the tetranychid mites and Perring *et al.* (1996) for the eriophyoid mites. Sampling of the Penthaleidae has been discussed by Umina (2008). The main findings are discussed below.

Tetranychidae

Sampling methods allow an evaluation of the abundance of populations. In agricultural practice, the population size indicates whether or not a critical level, above or below a defined threshold, has been reached, in order to determine whether control measures should be implemented. Formally, absolute counts should give a better measure, but they are impractical, especially in the case of eriophyoids. So, chosen elements of the population are sampled in order to reduce the counting effort. Often, only mobile stages are counted, while in other cases only females or eggs are counted. The presence or absence of mites in sampling units has been used to assess the population as well (Wilson *et al.*, 1983). Another concept, the 'sampling unit', allows rationalization of the sampling method. In this, the habitat is subdivided into parts of the same size. These units must cover the whole population and must not overlap. Leaves are largely used as sampling units, as well as other plant organs, such as buds, fruits, twigs, etc. The choice of sampling unit depends on mite bio-ecology and plant phenology. Variation in the number of elements per sampling unit is low when the population elements are distributed homogenously in the habitat. In this case, the sample size may be small. Conversely, when the spatial distribution is irregular or clumped, the variation is large and a very large sample size is needed. The habitat is usually divided into strata so as to reduce the variation within a stratum. Stratified random sampling may strongly reduce the sampling effort required but does not reduce the precision (Cochran, 1963).

The distribution of mite populations between and within plants in the field may vary significantly. Commonly, variation in the number of tetranychid mites on leaves is largest between plants in the field (Croft *et al.*, 1976; Mowery *et al.*, 1980). This phenomenon could be explained as the consequence of the reduced ability of both tetranychids and phytoseiids to disperse into trees (Croft *et al.*, 1976). Different factors can be used to explain the heterogeneity of a habitat. Indeed, within a crop, all plants (and their sites) do not have the same biological characteristics or nutritional and physical parameters. Even within a greenhouse crop that is apparently homogeneous, there is a diversity of habitat that involves a variation in the number of mites over the area. On diseased plants, and on plants transplanted into terrain that is less suitable, more exposed to the wind, or not irrigated and/or fertilized, develop smaller populations of mites than on plants kept in the same field under optimal conditions. In addition, there may be differential mite development over a habitat. Some authors hold that the within-plant distribution/abundance of tetranychid mites does not vary with geographical direction (Herbert and Butler, 1973b; Wilson *et al.*, 1984), but this conclusion does not take into account the non-homogenous exposure of the plant to sun and the resulting influence on the temperature-dependent growth rates of the populations of some mites (Flaherty and Huffaker, 1970). Populations of some tetranychid mites begin to develop in the lower parts and centres of trees (van de Vrie, 1964), while other tetranychids are more fecund on young leaves (Watson, 1964; Tanigoshi *et al.*, 1975b).

The start date of sampling is usually related to the bioecology of the mites and the phenology of the host plants. The interval between successive sampling dates is proportional to the time constant of population growth, corresponding to the inverse of the intrinsic rate of increase (r_m day^{-1}). Spider mite populations generally increase twofold over a week (Sabelis, 1985b). The statistics of sampling methodologies have been extensively examined by Cochran (1963) and Seber (1973), and further advances in sampling theory are due to Wilson *et al.* (1983) and other authors. Different sampling methods for field populations of tetranychid mites have been reported in the literature (van de Vrie, 1966; Huffaker *et al.*, 1970; Jeppson *et al.*, 1975; Poe, 1980). These are fundamentally destructive and the most common ones are outlined below.

DIRECT COUNTING. Direct counting consists of collection in the field of leaf samples, which are observed under a binocular microscope to count the mites present. The method is expensive, but has high precision. Storage of the leaves at about 5°C reduces the escape of mites from the leaves. Alternatively, it is possible to count the mites directly on the leaves using a hand lens (van de Vrie, 1966). In order to reduce effort, it has been suggested that only adult females are counted (Jeppson, 1951), or that mites are counted only on portions of the leaves (Hoskins *et al.*, 1938; Baten and Hutson, 1943; Michelbacher, 1959). Other possibilities are the counting of mite-free leaves only (Pielou, 1960), knocking mites from the foliage (Summers and Baker 1952; Boudreaux, 1953), or recording the proportion of infested leaves from specific plant strata (Wilson *et al.*, 1983).

IMPRINT COUNTING. In imprint counting, the leaves are pressed between the layers of a folded sheet of absorbent paper. The mites that are crushed appear as imprints. However, the method does not allow two similar species to be distinguished and it does not allow for the coalescence of imprints at a high population density (Venables and Dennys, 1941; Poe, 1980).

USE OF A BRUSHING MACHINE. In this method, the infested leaves are introduced between rotating brushes to brush off the mites, which then fall on to a revolving disc. The disc bears a sticky coating and is divided into sections (Henderson and McBurnie, 1943; Klostermeyer and Rasmussen, 1956). However, the type of leaf surface may affect the results that are obtained (van de Vrie, 1966; Poe, 1980).

WASHING. In this method, the mites are washed from the leaves (Jones and Prendergast, 1937; Newell, 1947; Henderson, 1960; Leigh *et al.*, 1984). It is possible to use a machine that separates mites from silk webbing and soil particles. The leaves are treated with a hypochlorite solution, which dissolves the silk and separates the mites from the leaves and any silt. The mites are counted on a filter paper under a binocular microscope (Wilson *et al.*, 1983). Further to the method, modifications have been suggested by Zacharda *et al.* (1988) and by Jedliková (1997).

BEATING FOLIAGE. This method is applied to the Bryobiinae, which do not typically produce webbing on leaves. Leaves and branches are beaten over a funnel and the mites are collected in a bag below the funnel (Summers and Baker, 1952). The method is also efficient for collecting phytoseiid mites.

Eriophyoidea

Variations in intra-plant distribution are well known in the Eriophyoidea. For example, species living on tea – *Acaphylla theae* (Watt), *Acaphyllisa parindiae* Keifer and *Calacarus carinatus* (Green) (Muraleedharan *et al.*, 1988a) – and on citrus – *Phyllocoptruta oleivora* (Peña and Baranowski, 1990) – are irregularly distributed on the plants. Populations of *P. oleivora* show an aggregated distribution pattern within trees, between trees and between sampling areas. When mite density increases, the sample sizes decrease. With densities of 10 mites cm^{-2}, a single sample unit per fruit has an average of 25% variation (Hall *et al.*, 1991, 1994). The citrus bud mite, *Aceria sheldoni*, has a crumpled distribution on lemon twigs (Walker *et al.*, 1992). Eriophyoids may feed only on specific plant tissues (buds, leaf axils), selected according to age and/or nutritional status, as occurs on lemon infested by *A. sheldoni*, whose density increases on young twigs (Walker *et al.*, 1992). On small fruit of grapefruit trees, the citrus rust mite occupies the north-east quadrant of the tree in the northern hemisphere, and as a result of this behaviour, the concept of subsampling was proposed (Dean, 1959b). This was used later to define the 'station trees' in a grapefruit grove; at these stations, the trees are periodically sampled, with sampling consisting of either six fruits (four on the east and north sides in the outer and inner canopy and two at the top centre) or three leaves. Three lens counts are made per fruit or leaf (Smith, 1980). In contrast, on Monterey pine, *Pinus radiata* D. Don, sampling of *Platyphytoptus sabinianae* Keifer showed that there were no differences in mite densities on leaves at different levels (Landwehr and Koehler, 1980).

Two types of sampling are reported for eriophyoid mites: non-destructive sampling, which is usually carried out *in situ*, and destructive sampling, which is based on removing mites and infested parts (leaves, twigs, buds, fruits) from plants (Perring *et al.*, 1996).

NON-DESTRUCTIVE SAMPLING. This method does not change the micro-environment, but is generally too time-consuming to use routinely for eriophyoid mites, although it is appropriate for the study of animals with reduced dispersal capabilities (Perring *et al.*, 1996). The first application of non-destructive methods to mites was done by Yothers and Miller (1934), who used a counting tool consisting of a 0.5 inch square cut in a piece of paper to calculate densities of *Phyllocoptruta oleivora* on citrus. They placed these 'tools' on both leaf surfaces and on fruits, using them as a 'series', and counting almost 25 series in a 40 acre grove (five series in each of five sections). They found a positive relationship between the numbers of mites counted in the squares and the numbers of squares hosting mites; 50% occupation of squares by mites was considered a threshold. Allen (1976b) used a piece of clear plastic upon which a grid of 1 cm^2 had been etched; the whole of this can be examined using a 10× hand lens. The grid contained 25 equal squares of area 4 mm^2. When there was low infestation, mites were counted on the whole area of 1 cm^2, but when mites were present on five diagonal squares or on the centre square, they were considered to be at high density. Count data allowed calculation of the mite time units (mite days) by computing the cumulative surface over which the mite population was developing. Rogers *et al.* (1994) applies the modified Horsfall and Barratt system for estimating citrus rust mite populations in an integrated crop management programme. Hall *et al.* (2005) investigated the consequence of reducing sample size on the accuracy and precision of estimates of the citrus rust mite. Childers *et al.* (2007) suggested a sampling method based on rust mite density (rust mites cm^{-2}) in late spring or summer on fruit trees selected at random and within uniformly distributed areas throughout a 10–40 acre block representing a single variety with uniform horticultural practices. They also suggested that sampling adjacent trees should be avoided. Fruits should be sampled at random, representing the four quadrants of the tree and midway in the canopy. One fruit surface area should be examined midway between the sun and shade areas. The number of rust mites cm^{-2} should be recorded and averaged for 10 acres, represented by 20 trees with four fruits per tree, or 80 readings for 10 acres. Six rust mites cm^{-2} would be a threshold measurement at which pesticide treatment may be required within 10–14 days. Ten rust mites cm^{-2} would be an action threshold at which treatment would be required as soon as possible. On fresh fruit, it is necessary to monitor mite populations every 10–14 days, and an average of two mites cm^{-2} would be used as an action threshold.

Gispert *et al.* (1989) sampled the tomato russet mite, *Aculops lycopersici* (Tryon), by placing tomato leaflets between two glass microscope slides. The slide on the lower leaflet surface was etched with three 1 cm^2 squares placed in the medial, central and distal parts of the leaflet, and mites were counted on these areas.

The non-destructive sampling of gall-forming mites is much more difficult and cannot be exhaustive.

DESTRUCTIVE SAMPLING. Destructive sampling involves removing parts of plants (leaves, twigs, buds, fruits) and transporting them to the laboratory to count the mites that are present under a binocular microscope. To prevent mite movement or escape, the sample may be treated with Breck® Super Hold Hair Spray (Peña and Baranowski, 1990). Destructive sampling requires knowledge of the plant tissues that are infested by mites and of plant phenology. Various methods have been reported in the literature.

Nascimento *et al.* (1982) reports that populations of the citrus rust mite, *Phyllocoptruta oleivora*, can be accurately sampled by taking five leaves per tree from 1.5% of the trees in the citrus grove and counting the number of mites in an area of 2 cm^2 on the upper surface of each leaf, 2 cm from the leaf base. Oliveira *et al.* (1982) established that the population densities of *P. oleivora* were approximately the same all over a tree, and therefore the samples could be taken from any cardinal point. At any level of infestation, the results obtained by counting the mites present on a predetermined limited area of a leaf (about 4.9 cm^2 on the lower surface of the leaf at about 2 cm from the petiole) were as accurate as counting the numbers on both surfaces of the entire leaf. A leaf-brushing machine provided fairly accurate results when the infestation level was low, but not when it was high. If a precise determination of the number of mites is required, the limited area method (see above) is to be preferred, but any of the methods can be used for comparative studies.

A subsampling method was applied for *Abacarus hystrix* (Nalepa) and *Aculodes mckenziei* (Keifer) on bluegrass, *Poa pratensis* L., in which all tillers were clipped within a 15 cm^2 surface area and investigated in the laboratory, where a subsample of 25 tillers chosen at random was examined under the stereomicroscope (Smilanick and Zalom, 1983). On mango infested by *Eriophyes mangiferae* (Sayed) and *Cisaberoptus kenyae* Keifer, Sternlicht and Goldenberg (1976) proposed the sampling of ten 10–15 cm long twigs from ten trees, each of different varieties; the buds, leaves and twigs were then examined in the laboratory under the stereomicroscope.

Other techniques involve the removal of vagrant mites from tissues before counting them under a microscope, e.g. *Aceria tulipae* from wheat spikes. Green spikes are collected in the field and placed separately on the sticky surface of a 2 × 12 cm piece of transparent tape. Mites are trapped on the tape and leave dried spikes. After 2 weeks, the spikes are removed and the mites counted (Harvey and Martin, 1988). This method can also be used for galls or deformed plant material hosting mites (Perring *et al.*, 1996). In addition, it is possible to brush eriophyoids from leaves using a mite brushing machine (Henderson and McBurnie, 1943), as for *Aculus schlechtendali* on apple (Hossain, 1992). Another method is the alcohol-based technique, in which the infested plant tissues are washed in 80% ethanol or propanol, and the mites counted under a dissecting microscope (Elliot *et al.*, 1987). Various modifications of this method have been applied to different eriophyoids (Janarthanan *et al.*, 1971; Zacharda *et al.*, 1988). Estimation of the eriophyid mite populations on ryegrass was effected by

Gibson (1975) using alcohol and ultrasonic radiation. In population studies of the blackcurrant gall mite, *Phytoptus ribis* Nalepa, specimens were sampled by macerating the gall tissues (Smith, 1961); after release of the mites, the plant matter was removed and the filtrate observed under microscope. An electrostatic method has also been applied to remove mites from open galls (Stone, 1981).

Estimation of population size and mite dispersal in the Tetranychidae and Eriophyoidea

ESTIMATION OF POPULATION SIZE. For the tetranychids, the sample size depends on the distribution of the mites in the habitat sampled. Prior sampling of the population supplies information on the distribution, and savings on the censuses that follow justify the time spent on doing this; quantitative knowledge of mite distribution is fundamental to the determination of sample size (Sabelis, 1985b). Three sampling methods are well known: numerical sampling, which has implications for populations of known or unknown size and distribution and for the variance–mean relationship; presence–absence sampling; and binomial sampling. Reductions in sampling time can be obtained by using sequential sampling (Wald, 1945), particularly when this is applied in conjunction with binomial sampling (Ingram and Green, 1972; Sterling, 1975; Zalom *et al.*, 1984a). Sequential sampling is usually appropriate when sampling for management decisions (Sabelis, 1985b), a proposal that was first discussed by Wald (1945) and Onsager (1976), and later also by Wilson *et al.* (1983).

In different eriophyoid species, indirect estimates of population size allow time to be saved (Perring *et al.*, 1996). One such technique is the plant damage-based method that was developed by McCoy *et al.* (1976a) to rate *Phyllocoptruta oleivora* infestations on citrus fruit. This uses a scale of 1–4 based on fruit damage, and allows the correlation of damage with the phytochemicals that are involved in fruit or juice quality. Evensen *et al.* (1980) used the spectral reflectance of *P. oleivora* damaged surfaces on citrus fruit as a measure of damage. Rogers *et al.* (1994) adapted the Horsfall–Barratt system (Horsfall and Barratt, 1945), a grading system for measuring plant disease, to evaluate numbers of *P. oleivora* on citrus. The system adopts a standardized visual comparison key for density assessment and reduces monitoring in the field by 50–75%. Hall *et al.* (2007) investigated binomial sampling to estimate rust mite densities on orange fruit and concluded that it seemed to be a viable alternative to absolute counts of mites per sample for a grower using a low management threshold such as two or three mites per sample. On hazelnut, the intra-plant distribution of *Phytoptus avellanae* (Nalepa) and *Cecidophyopsis vermiformis* (Nalepa) may be evaluated by rating the number of galls per tree on a scale ranging from 1 (0–2 galls per tree) to 5 (4 or more galls per shoot) (Burgess and Thompson, 1985).

ESTIMATION OF MITE DISPERSAL. Mites disperse from plant to plant mainly by walking, wind or phoresy. Sometimes, they are dispersed by infested cuttings, by workers or on various tools. Sampling methods to evaluate mite movement by walking and through aerial trapping are important for obtaining knowledge of their bio-ecology and for applying rational control strategies. The phenomenon is discussed below for different eriophyoid species.

Walking. The movement of vagrant eriophyoid mites has been variously investigated. Oldfield (1969) wrapped 2 inch bands of Parafilm® around the plant stems and applied a ring of Stickem® around the middle of the bands to evaluate the movement in spring of *Eriophyes emarginatae* Keifer on *Prunus virginiana* L. Mites trapped on the bands were removed and placed on a slide with toluene to dissolve the Stickem®. The mites were mounted for identification after the toluene had evaporated. Sticky-band traps were used to monitor the emergence and behaviour of the pear rust mite, *Epitrimerus pyri* (Bergh, 1992), and a degree-day model was used to predict the emergence of deutogynes from overwintering sites (Bergh and Judd, 1993). A mechanical method was applied to sampling the emergence of *Phytoptus ribis* from blackcurrant buds (Smith, 1961).

Aerial trapping. The dispersal of eriophyoid (and other) mites from plant to plant relies heavily on the wind. Davis (1964b) investigated the wind dispersal of *Rhynacus breitlowi* Davis, an eriophyoid mite living on the undersurface of the evergreen magnolia (*Magnolia grandiflora* L.), its only known host. The mite appeared to be adapted to wind dispersal, and no carrier arthropods were observed. At high RH, the down on the underside of the magnolia leaf matted down, but at low RH it sprang erect and launched the eriophyoids into the air, where they were carried in the air flow at velocities above 11.26–19.31 km h^{-1}. Nault and Styer (1969) suggested developing methods for capturing eriophyoids in flight, such as using sticky-coated glass slides. *Aceria tulipae* was captured with silicone grease-coated slides placed in a holder and facing the wind (Pady, 1955). Different substances have been used by researchers to optimize the method, such as petroleum jelly or vaseline (Slykhuis, 1955; Somchoudhury *et al.*, 1985). Cleaning of mites in 5% KOH has also been proposed (Nault and Styer, 1969). Wind velocity and temperature may induce variations in the response obtained to the use of sticky-coated traps (Staples and Allington, 1956).

Penthaleidae

MONITORING. Monitoring of the red-legged earth mite, *Halotydeus destructor* (Tucker), requires inspections of the pasture and crop from autumn to spring. Crops should be visited regularly in the first 3–5 weeks after sowing, taking into account that the mite commonly feeds on the leaves in the morning or on overcast days. During the warmer hours of the day, *H. destructor* gathers in the lower parts of the plants, sheltering in leaf sheaths and under debris, and takes refuge in cracks on the ground to avoid extreme temperatures. Disturbances during feeding induce the mite to drop on to the ground and seek shelter (Umina, 2008). The use of the provisional model temerite helps to individualize the proper timing to treat females before they lay diapause eggs in the spring (Ridsdill-Smith and Pavri, 2010).

MITE DISPERSAL. The dispersal of *Penthaleus* spp. relies on different mechanisms, such as the transportation of aestivating eggs or mites on grain crop stubble or leaves, on soil adhering

to implements that are moved about, or on forage or straw that is carried from infested fields in livestock feeding operations. Debris transported by the wind disperses the aestivating eggs, and the adults can achieve a local distribution. Migrations to grain fields may take place from fence rows or other uncultivated areas. Fescue grass appears to be the favoured host in such a situation in Texas (Jeppson *et al.*, 1975). Estimates of adult movement rates in the red-legged earth mite and the blue oat mite, *Penthaleus major*, using mark–release–recapture techniques and mean square dispersal distances to estimate diffusion coefficients, indicate that in pasture, coefficients were in the range 0.3–1.3 for both species. Some 90% of the population moves <5–11 m in a 10-day period, or 7–16 m within the adult lifetime. Releases of mites in adjacent pea/wheat crops indicated directional movement towards the more favoured pea host. However, there was no directional movement when adjacent plots of peas and lupins were compared, even though lupins are poor hosts. Broad border sprays or border cultures will be needed to prevent mite movement from adjacent paddocks (Weeks *et al.*, 2000a).

Decision tools for pest management

Decision tools of pest management include the economic injury level (EIL), the economic threshold (ET), the action threshold (AT), the control decision threshold (CDT) and the mite day. The EIL and ET were both proposed in the 1960s (Stern *et al.*, 1959) and enabled assessment of the real injury caused by pest populations and the tolerance to some levels of attack or even the theoretical role of certain pests. Regrettably, the relevant data are not available for all crops and mites. The relationship between the EIL and the overall equilibrium position of population dynamics of a pest permits definition of the status of the pest. The AT has been extensively discussed by Moffitt and Baritelle (1990) and the CDT by Wilson *et al.* (1983). The mite day has been examined by Andrews and La Pre (1979). These thresholds can be examined from both the entomological (or pest) and the economic points of view (Tomerlin, 2005), and herein the latter is discussed first. A further tool, the growing degree-days (GDDs), also called degree-days (DDs or °D), is also applied to predict plant and pest (including mites) development rates.

Economic injury level (EIL)

The EIL can be defined as 'the lowest population density of a pest capable of producing a loss, i.e. the amount of harm that justifies the economic cost of control' (Higley and Pedigo, 1996; Peterson and Higley, 2002). EIL values have been reported for various mite pests, e.g. for the broad mite, *Polyphagotarsonemus latus*, it corresponds to 9–15 mite days (Peña, 1990) (see below). The EIL for the persea mite, *Oligonychus perseae*, has been based on the percentage of leaf area damaged (PLAD). The action threshold (AT) (see below) has been assessed as 50–100 mites per leaf, and the mean annual cumulative mite days (CMDs) of levels of infestation of 100 and 250 mites per leaf as *c.*13,500 ± 700 per leaf. Using the linear regression equation PLAD = 0.0009 CMDs + 2.42, leaf damage has been described as a function of CMDs, and gave an estimated EIL of *c.*15 PLAD (Maoz *et al.*, 2011a).

The EIL is an indicator of the need for control and its use is not practical if the mite pest always causes economic damage, or if the estimation of damage, or control, are difficult or impossible, e.g. in the case of the pink citrus rust mite, *Aculops pelekassi*, or the broad mite, against which control measures are undertaken as soon as infestation is first seen on the fruits.

Changes in the methods for calculating EIL are related to operations intended to improve environmental sustainability, and if the environmental costs are included in the cost of managing the pest, the EIL can be described as the 'economic injury level environmental (EILE)' (Peterson and Higley, 2002).

Economic threshold (ET)

The ET can be defined as 'the density of the pest against which control measures should be undertaken in order to avoid reaching the EIL'. This means that pest populations have become large enough to cause an economic level of damage and they must have a minimum density in order to justify control. A certain pest density would thus lead to an increase of damage that is equal to the cost of prevention, and population density should decline after the use of a pesticide (or other means) in order to maximize profit. Knowledge of pest density, supported by appropriate sampling and monitoring methods, assists the practitioner in taking control measures before the population reaches the EIL. Pesticide use should include marginal costs equal to marginal benefits, and marginal costs that rise with the dose, and profits should be not be negative (Higley and Pedigo, 1996; Regev, 2002).

These conditions have to be assessed in any IPM programme, as control may require the use of more chemicals. This defines the ET as 'the lowest level of the population that justifies the use of pesticides when their amount is calculated to maximize profits'. Each specific ET involves the choice of a control strategy and relates to its optimal use and the timeliness of the action. If during the growing season more interventions are required, the use of pesticides is identified with a dynamic decision-making process and the calculation of ET must take into consideration the host plant phenology and its differential susceptibility to attack. In calculating the ET, using only the direct benefits of pesticides will bias the estimate towards lower values. We generally tend to identify the mean density of the pest with ET, assigning importance only to the private marginal cost; when considering the social cost (the sum of the costs of the external effects of pesticides and the market cost), profits becomes positive only at a high pest density. This indicates that the social costs are moving towards a higher ET and a lower consumption of pesticides (Regev, 2002). Manzur (1989) reports that *Phyllocoptruta oleivora* showed a negative binomial distribution on grapefruit tree leaves and that the economic thresholds of pest density were set at 0.5 mites per leaf for no insecticide treatment and greater than 2.0 mites per leaf requiring treatment.

Action threshold (AT)

The application of the AT uses only the population threshold and the dosage rates of the chemicals used. The AT corresponds to 'pest density prior to taking an action' (Moffitt and Baritelle, 1990) and is based on a discrete choice decision problem, where the choice consists of either not treating or applying a fixed dosage. Pest management is thus to apply a certain dosage if the

pest density exceeds a certain level and not to treat at a lower level. The method is simple and its application requires only that a farmer calculates whether the pest population exceeds a specific level, and if it does, treats it with a specific dosage. However, one cannot underestimate the risk arising from the arbitrariness of the AT and the dosages used, and any decisions with a serious economic calculation must be reported (Moffitt and Baritelle, 1990). The fixed dosage related to the threshold equates to the expected value of a marginal product when a treatment equals the cost of the chemicals, and might differ from that associated with a change in the average population size. Childers *et al.* (2007) found that for *P. oleivora*, 10 rust mites cm^{-2} would be an action threshold where treatment would be required as soon as possible. In fresh fruit, it is necessary to monitor mite populations every 10–14 days and an average of 2 mites cm^{-2} as an action threshold.

In the Korea Republic, the density of *Panonychus citri* in commercial satsuma mandarin (*Citrus unshiu* Marcow.) groves was determined by counting the number of citrus red mites per leaf for 2 years. The increase in sample size, regardless of tally threshold, had minimal effects on the precision of the binomial sampling plan. The increase in sampling size also had minimal effect on the precision of the estimated mean regardless of tally thresholds. $T = 1$ was chosen as the best tally threshold for estimating densities of *P. citri* based on the precision of the model. The binomial model with $T = 1$ provided reliable predictions of mean densities of *P. citri* observed in these commercial groves. A binomial sequential sampling procedure was developed for classifying the density of *P. citri*, and a binomial sampling programme for decision-making citrus red mite population levels based on an AT of two mites per leaf was obtained (Song *et al.*, 2001).

Control decision threshold (CDT)

The CDT may be defined as the number of mites per leaf or the percentage of infested leaves. The use of these parameters allows an easy approach to the implementation of IPM of pest mites in different crops, e.g. of the European red mite, *Panonychus ulmi*, in apple orchards (Walgenbach, 1993). On aubergine, it has been found that the optimum threshold for application of a chemical spray to control *Tetranychus marianae* McGregor is four mites per leaf during the dry season and eight mites per leaf in the wet season (Reddy *et al.*, 2013). However, a (control) decision threshold level is not the same as an injury threshold level. For instance, it has been found that healthy, properly thinned apple trees can host more than ten mites per leaf, and in several cases apple trees can tolerate up to 30 mites per leaf without any reduction in fruit yield or quality (Walgenbach, 1993). Therefore, the CDT alone does not explain the interaction between the mite populations and the host plant, and so the role of the physiological time has been examined. On cotton infested by *Tetranychus* spp., this seems to be the degree-days above 12°C starting from the date of planting (Wilson *et al.*, 1983). Other explanations have also been advanced.

Mite days

Andrews and La Pre (1979) found on almond foliage that there was a negative correlation between the chlorophyll content and the number of *Tetranychus* spp. 'mite days' per leaf, where one mite day is one mite (in this case a spider mite) feeding for 1 day, and 10 mite days might be one mite feeding for 10 days or ten mites feeding for 1 day. The effects of mite days depend on the season, the time of year and the host plant. Similar peaks of density of spider mites per leaf do not necessarily determine the same amount of crop loss. Indeed, the same number of mite days can be reached in a different range of times according to the number of mites per leaf. The concept of mite days is applied to different crops to evaluate the potential damage that can be caused by mite pests. On almond, reduced development, growth and yield in the following season were recorded after first-season infestation levels of a certain number of mite days (Welter *et al.*, 1984).

Growing degree-days (GDSS)

The temperature influences the developmental rate of living organisms and, depending on the species and the environment, a precise amount of heat is required to develop from one point of an organism's biological cycles to another. It follows that the amount of heat required to complete the development of an organism does not vary. This accumulated heat is usually called the physiological time and is employed as a reference for the development of organisms. Physiological time is usually expressed and approximated in units called degree-days (°D). The use of °D allows pest control to be rationalized and has been investigated by various researchers (Baskerville and Emin, 1969; Andrewartha and Birch, 1973; Allen, 1976a; Wilson and Barnett, 1983; Zalom *et al.*, 1983). Growing degree-days (GDDs) are an extension of degree-days (°D), and are calculated using the following equation:.

$$GDD = (T_{max} + T_{min})/2 - T_{base}$$

where T_{max} and T_{min} are the daily maximum and minimum temperatures, and T_{base} is a base temperature below which plant growth or pest development is zero. GDDs are typically measured from the winter low and are calculated each day and are accumulated by adding each day's contribution as the season progresses. This tool is extensively applied in pest control in various countries. In the USA, the University of California (UC IPM Online, 2012a; and many other specific UC IPM Pest Management Guidelines, variously dated and referenced under 'UC IPM Online') and Cornell University's Crop and Pest Management Guidelines (Cornell University, 2012) have provided detailed information on various pest and plant models, including UC-recommended degree-day models.

Crop consultants and training

Crop consultants are specialists in pest management employed by agricultural companies or Departments of Agriculture. Some consultants operate privately, and they are usually members of professional organizations. Their activities and organization vary in different countries and also according to the socio-economic situation. Thus, in the USA, there are various professional organizations, including the National Alliance of Independent Crop Consultants (NAICC), which has over 450 members in 40 states and foreign countries. Certified Professional Crop Consultants (CPCCs) and the Independent Certified Professional Crop Consultants (CPCC-I) programmes

in the USA are submitted to and administered by NAICC (Hoy, 2011). There are other kinds of organizations in other countries. Crop consultants generally provide a comprehensive service to farms, from pest control to crop production aspects such as economy, agronomy, etc. For IPM implementation, consultants identify the pests and their natural enemies, lead the monitoring of their populations and evaluate the potential risks of pest attacks. They then identify the best management options, according to the results from the most recent research, from chemical spraying to the release of natural enemies and/or implementation of improved agricultural practices, etc. In addition to this, crop consultants must be able to inform the farmer clearly and in real time up-to-date IPM strategies, using all possible means, including personalized reports, written reports, and printed and web-based newsletters. In cases in which an IPM strategy is applied for the first time, crop consultants are also involved in the training of the farmer. There are no strict guidelines for the setting up of a farmer and crop consultant relationship, which may involve a university, Department of Agriculture, private company, etc. In all cases, the relationship between the consultant and the farmer is not limited to a short term.

Practical experience of IPM

Overall, IPM aims to control mite populations (and other pests) by all the means discussed above, and others – chemical, biological, biotechnical, cultural, etc. – on condition that they do not hinder the action of indigenous and artificially released natural enemies, and are in agreement with the ecological, toxicological and economic requirements of the strategy and agricultural production in general.

Various IPM projects have been developed in different countries, and the results that have been obtained from these have independently confirmed a common need to extend the application of the strategy of other countries, with the primary aim of limiting the damage that can be caused by traditional crop protection. Hoy (2011) has recently published an extensive review on integrated mite management that covers the most important North American crops, including apple, almond, citrus and ornamentals. Similar information is available from other regions of the world, as for citriculture in the Mediterranean region (Vacante and Gerson, 2012) and protected crops worldwide (van Lenteren, 2005). The use of IPM will be developed more broadly in the discussion on each pest mite that follows in Part II of the book.

10 Brief History of the Main Published Works on the Mites of Economic Plants

The mites of economic plants are mainly included in the superfamily Eriophyoidea and the families Tetranychidae and Tenuipalpidae. Other families, e.g. the Tarsonemidae and the Penthaleidae, have relatively few injurious species. Summarizing the history of these mite groups according to their economic importance is very difficult because of the very large number of references. This brief history covers only the main works on economic acarology, and the references that are included on systematic and taxonomic aspects highlight the importance of basic knowledge in the intervention that is applied. The discussion is arranged by geographic area. The Mediterranean region is taken to include the North African countries, the Middle East, Turkey and Cyprus; the northern Mediterranean countries are included in the section on Europe.

Europe

The European history of acarology follows for long stretches of time the world history of the discipline, in conjunction with North American work. In the Greek and Roman periods, the word 'Acari' occurs many times in the writings of Hippocrates, Plutarch, Aristophanes and Pliny (Oudemans, 1926). However, until the 17th century, reports on Acari were still fragmented. The publication in 1738 of the first edition of the *Systema Natura* by Carl Linnaeus (1707–1778) inaugurated the birth of acarology. For the first time, the generic name of *Acarus* was used, and in the tenth edition of *Systema Natura*, no fewer than 30 mite species were reported, all included in the genus *Acarus*, whose type species was named *Acarus siro*. In 1834, the first tenuipalpid mite, *Trombidium caudatus*, was described in France (Dugès, 1834).

During the following century, various authors (DeGeer, 1778; Latreille, 1806–1809; Leach, 1815; Dugès, 1839; Michael, 1884) treated the Acari and defined schemes for their classification. Despite these reports, acarology acquired the status of a true discipline only between the 19th and early 20th centuries, thanks to the contributions of several authors (Mégnin, 1876; Kramer, 1877; Berlese, 1882–1902, 1925; Michael, 1884; Canestrini, 1891a; Oudemans, 1906; Reuter, 1909; Thor, 1929; Vitzthum, 1940–1943; Trägårdh, 1946). In 1875, Donnadieu first used the family name Tétranychidés. In the same paper, he established the genera *Tenuipalpus* and *Brevipalpus* (Tenuipalpidae), placing both in the Tetranychidae (Donnadieu, 1875). After this, the name Tetranychidae was adopted by the most important acarologists of the time (Kramer, 1877; Murray, 1877; Trouessart, 1891; Berlese, 1897; Oudemans, 1897, 1910). Then, in 1877, Giovanni Canestrini, Filippo Fanzago and Paul Matthias Kramer independently proposed the family Tarsonemidae (Canestrini and Fanzago, 1877).

Later on, the work of the Italian Antonio Berlese (1863–1927) played a significant role in the development of modern acarology. Berlese's studies concerned both systematics and taxonomy (*Acari, Myriopoda et Scorpiones Hucusque in Italia Reperta*, volumes 1–101, 1882–1902; *Acarotheca Italica*, 1913) and economic acarology (*Acari Dannosi alle Piante Coltivate*, 1886; *Insetti delle Case e dell'Uomo e Malattie che Diffondono*, 1917; and 'Gli affini degli insetti', Chapter 1 in *Gli Insetti, loro Organizzazione, Sviluppo, Abitudini e Rapporti coll'Uomo, Volume Secundo*, 1925). Berlese was responsible for the origins of economic acarology, and his work led the subject worldwide for many years. In 1913, he was also the first to use the name Tenuipalpini (Berlese, 1913), from which the family name Tenuipalpidae would originate (Sayed, 1950a).

In Austria, in the second half of the 19th century, the Hungarian Alfred Nalepa (1856–1930) published nearly 90 papers on eriophyoid mites, describing dozens of species and 12 new genera (Amrine and Stasny, 1994). His catalogue of the eriophyoids that were described includes 322 species and 652 host plants (Nalepa, 1929), and was used worldwide during the following half century. In the same period, in Italy, Giovanni Canestrini published about 20 papers on eriophyoid mites (see Reference list), and described about 53 species (Amrine and Stasny, 1994). A number of papers on spider mites were published in the first half of the 20th century, among them those of Stanley Hirst (1920) in England, of Hermann Graf Vitzthum (1929a) in Germany and of Dirk Cornelis Geijskes (1939) in Holland. References on mites published from 1926 to 1966 can be found in Prasad (1988), and on various websites (e.g. http://www.miteresearch.org/wishlist.php/A).

Together with Antonio Berlese, the French François Grandjean (1882–1975) made a significant contribution to 20th century acarology both in Europe and the world. Grandjean was a wild and brilliant figure in the long list of French acarologists (Bertrand *et al.*, 2008). His 241 works on the systematics, morphology, ontogeny and phylogeny of the Oribatida and other groups (Van der Hammen, 1972–1976) were strongly innovative in the study of mites and opened the way towards modern acarology.

In the 1950s, acarology developed still further. Thus, from the years 1950 to 1960 over 230 eriophyoid species were described, 83 in Finland by Liro and Roivainen, 64 in Hungary by Farkas and 91 in Poland by Boczek (Amrine and Stasny, 1994). Boczek is co-author of the *Generic Key to World Fauna of Eriophyid Mites* (Boczek *et al.*, 1989) and author of several papers on eriophyoid mites living on plants (Boczek, 1964a,b 1966). From 1960 to 1992, Manuela M. Carmona published several papers on mites injurious to cultivated plants in Portugal, including 'O complexo Acarina nas culturas portugueas'

(Carmona and Dias, 1980). At the same time, the French Jean Gutierrez began to study tetranychid mites of cultivated plants in various tropical and subtropical areas, especially in the French colonies; in this work, 53 new species were described and the use of DNA tools was first introduced, in collaboration with Maria Navajas, in the phylogenetic study of Tetranychidae (see References under both authors). This work has strongly influenced modern acarology (for references see Bolland *et al.*, 1998).

Evangelos N. Hatzinikolis published several papers on tetranychoid mites in Greece, among which were 'Les acariens phytophages et leurs hôtes en Grèce' (Hatzinikolis and Papanou, 1975) and 'Tenuipalpid mites of olive trees in the Mediterranean area (Acarida-Tenuipalpidae)' (Hatzinikolis and Kolovos, 1985). In France, Amédée H.M. Rambier (1917–2007) worked mainly on mites infesting vineyards and orchards and published about 40 papers, including 'Les acariens nuisibles en cultures fruitiers' (Rambier, 1958a). In Italy, Pierantonio Rota (1923–1996) published the 'Osservazioni sugli acari tetranychidi dannosi alle piante coltivate ed ornamentali in Italia' (Rota, 1962), and Alfio Nucifora was co-author of 'Revisione dei generi della famiglia Tarsonemidae (Acarina)' (Beer and Nucifora, 1965). In Bulgaria, Balevski published *Tetranychid Mites Occurring on Fruit Crops* in 1967. In the 1970s, Gerd Alberti began his work on mites in Germany. His researches with the electron microscope was fundamental to better knowledge of the anatomy, morphology and physiology of mites. He published many papers, alone or as co-author (see Helle and Sabelis (1985) *Spider Mites. Their Biology, Natural Enemies and Control*; and Lindquist *et al.* (1996) *Eriophyoid Mites: Their Biology, Natural Enemies and Control*). Some years later, the Italian Franco Laffi (1981) published *Acari di Interesse Agrario*, a comprehensive work on economic acarology. At the same time, Balevski *et al.* (1982) published 'Acari of cultivated plants' in Bulgaria and the Dutch W. Helle and Maurice W. Sabelis edited their volume on *Spider Mites* (see above), the first worldwide update on injurious tetranychid mites since that of Jeppson *et al.* (1975).

In the 1980s, the writer of this book began a series of studies in Italy, either alone or in collaboration, on the pest mites of the main Mediterranean crops, and published papers on pest mites on olive trees (Vacante, 1984), protected crops (Vacante, 1985b), citrus (Vacante and Nucifora, 1986), apple (Vacante and Tropea Garzia, 1987a) and grape (Vacante and Tropea Garzia, 1987b). Vacante *et al.* (1989b) also published a paper on citrus mites in the Mediterranean region. In 1998, Enrico de Lillo and James W. Amrine Jr compiled a computer database on the Eriophyoidea (de Lillo and Amrine, 1998). At the same time, Bolland *et al.* (1998) produced a detailed list of tetranychid species and genera that had been described worldwide in the second half of the 1900s, and reported 1200 species and 71 genera. Some years later, de Lillo (2004) compiled a database on the systematics and taxonomy of the European Eriophyoidea within the Fauna Europaea. Papers on advances on eriophyoid studies carried out both in Europe and other countries may be found in the recent work edited by Edward A. Ueckermann, *Eriophyoid Mites: Progress and Prognoses* in a special (combined) issue of *Experimental and Applied Acarology* (Ueckermann, 2010), which indicates the high value of the Polish and Serbian acarological schools. Vacante published the monograph *Citrus Mites: Identification, Bionomy and Control* in 2010 and, in France, Alain Migeon and Franck Dorkeld (2006–2013) from INRA (Institut National de la Recherche Agronomique) produced a database on the tetranychid species of the world.

North America

Since the beginning of the 1900s, the efforts of North American acarologists have led acarology to become an independent discipline (Banks, 1904a; Jacot, 1925). Pioneering figures, such as Nathan Banks (1868–1953), Ernest A. McGregor and Henry E. Ewing played a key role in this. Banks studied different mite families (Tetranychidae, Stigmaeidae, Cunaxidae, Tarsonemidae) and in 1915 published the first English handbook on mites (*The Acarina or Mites*, 1915). McGregor is the author of several papers on spider mites and his research was fundamental to both the systematics and the taxonomy of the Tetranychidae and for economic acarology. In his paper on 'Mites of the family Tetranychidae', he listed 102 species and 15 genera (McGregor, 1950). Ewing studied predaceous and parasitic mites from different families, including the Trombiculidae, bee mites in the family Tarsonemidae, and Tetranychidae (Ewing, 1909, 1913, 1929a,b). It was Ewing who first showed the possibility of using the male aedeagus in the taxonomic analysis of species included in the genus *Tetranychus* Dufour.

Edward W. Baker (1914–2005), one of the most important American acarologists, started working at the US Department of Agriculture in 1944. He was interested in the different mite families that were associated with agriculture, livestock and stored grain, and also carried out studies on predatory, saprophytic and parasitic mites. His work was fundamental to the development of acarology in North America and worldwide. He described 700 species, 60 genera and nine new families. In 1945, Baker published his first work on the Tenuipalpidae, and this was followed by 17 other important papers that he published, as either author or co-author, up to 1987 (for references see Mesa *et al.*, 2009). Some years later, McGregor (1949) reported 13 Nearctic tenuipalpid species and Baker published with George W. Wharton (1914–1990) an *Introduction to Acarology* (Baker and Wharton, 1952), a basic text used worldwide for several decades. In the same year, Arthur E. Pritchard (1915–1965) wrote with Baker 'A guide to the spider mites of deciduous fruit trees' (Pritchard and Baker, 1952a), and 'The false spider mites of California (Acarina: Phytoptipalpidae)' (Pritchard and Baker, 1952b). In 1953, they published 'The family categories of tetranychoid mites, with a review of the new families Linotetranidae and Tuckerellidae' (Baker and Pritchard, 1953a), followed in 1955 by *A Revision of the Spider Mite Family Tetranychidae* (Pritchard and Baker, 1955), in which they reported 204 species and 18 genera. The *Guide to the Families of Mites* (Baker *et al.*, 1958) was published in 1958, as was *The False Spider Mites (Acarina: Tenuipalpidae)*, the first major revision of the Tenuipalpidae, in which 143 species and 14 genera were reported (Pritchard and Baker, 1958). Donald M. Tuttle studied spider mites and false spider mites of the American continent and in 1968 published with Baker *Spider Mites of Southwestern United States and a Revision of the Family*

Tetranychidae (Tuttle and Baker, 1968); nine of the papers therein concern the tetranychid mites.

From 1938 to 1982, i.e. over a 45-year period, Hartford H. Keifer (1902–1986) worked at the California Department of Agriculture. He published about 80 papers and described 711 species and 113 eriophyoid genera (Amrine and Stasny, 1994). His work is the most comprehensive review and compilation of systematic and other information ever published on these mites worldwide (Lindquist and Amrine, 1996). In 1975, Keifer published with E. Baker and Lee R. Jeppson *Mites Injurious to Economic Plants* (Jeppson *et al.*, 1975), an innovative updated review of the systematics, bio-ecology and control of the main pest mites of the world. In this book, Newkirk and Keifer (1975) published a synoptic key to the groups and genera of eriophyoids. Some years later, Davis *et al.* (1982) compiled a world *Catalogue of Eriophyid Mites (Acari: Eriophyoidea)*, the second after that of Nalepa (1929). In the same year, Keifer published with other authors *An Illustrated Guide to Plant Abnormalities Caused by Eriophyid Mites in North America* (Keifer *et al.*, 1982). In 1986, Evert E. Lindquist revised the genera of the tarsonemid mites (Lindquist, 1986). Ten years later, Lindquist would be editor, together with the Dutch Maurice W. Sabelis and Jan Bruin, of the book *Eriophyoid Mites*, the most comprehensive work on the morphology, bio-ecology and control of these mites. In 1994, James W. Amrine Jr and Terry A.H. Stasny (1994) published the *Catalog of Eriophyoidea (Acarina: Prostigmata) of the World* and in 2003, together with the Brazilian Carlos H.W. Flechtmann, the *Revised Keys to World Genera of Eriophyoidea (Acari: Prostigmata)* (Amrine *et al.*, 2003).

The research that has been carried out in North America on applied acarology during the last three decades has been intense and involved many authors, so it is very difficult to summarize their efforts. Among the major figures are Martin H. Muma, Carl C. Childers, Clayton W. McCoy, Marjorie A. Hoy and Ronald Ochoa. Marjorie A. Hoy (2011) recently published *Agricultural Acarology: Introduction to Integrated Mite Management*, and is one of the most qualified contributors worldwide. Further reports on other important acarologists may be found in Helle and Sabelis (1985), Lindquist *et al.* (1996) or Mesa *et al.* (2009), but this list is not exhaustive.

Central and South America

In the last 60 years, a qualified community of acarologists has become active in Central and South America.

In 1940, the paper 'Tres ácaros dañinos para los cultivos argentinos' was published in Argentina by Blanchard (1940). This was followed in 1953 by the 'Acaros de interes economico para nuestros cultivos' (Vergani, 1953) and some years later by the 'Lista de la especies de Tetranychidae (Acari) de la Republica Argentina' (Rossi Simons, 1961).

In the 1950s, the 'Second list of insects and other arthropods of economic importance in Uruguay' was published in Uruguay (Ruffinelli and Carbonell, 1954). In 1958, Cromroy published the paper 'A preliminary survey of the plant mites of Puerto Rico' (Cromroy, 1958), and Beer and Lang (1958) 'The Tetranychidae of Mexico'. Contemporaneously, Donald De Leon studied the Tenuipalpidae of Mexico, British Guiana, the Dominican Republic and Puerto Rico (for references see Mesa *et al.*, 2009). At the same time, the 'Contribución al conocimiento de los ácaros del manzano en Chile central' was published in Chile (González, 1961), followed some years later by 'Acaros planticolas del género *Tenuipalpus* en Chile (Acarina: Tenuipalpidae)' (González, 1968a); the same author later published 'Insectos y acaros de importancia agricola y cuarantenaria en Chile' (González, 1989).

For Venezuela, Guagliumi (1967) compiled information for the publication *Insects and Arachnids of the Common Plants in Venezuela Reported in the Period 1938–1963*, and Ernesto Doreste (1968b) compiled the *Primera lista de acaros de importancia agricola en Venezuela*. At the same time, Estebanes-Gonzalez and Baker (1968) published the paper 'Arañas rojas de Mexico (Acarina: Tetranychidae)'. In Paraguay, Aranda (1969) published a thesis entitled 'Contribução para o conhecimento de ácaros de plantas do Paraguay'. Various works appeared in Colombia, including the 'Lista preliminas de acaros de importancia en Colombia' (Zuluaga Cardona, 1971), *Acaros de Importancia Economica en Plantas Cultivadas en Colombia* (Madrigal, 1974) and, more recently, *A Catalog of the Tenuipalpidae (Acari) of the World with a Key to the Genera* (Mesa *et al.*, 2009). This last work is particularly interesting and lists 891 species, of which 217 are from Neotropical regions and 180 from Oriental regions.

In 1977, Roberto H. González and Carlos H.W. Flechtmann published 'Revision de los acaros fitofagós en el Perú y descripción de un nuevo genero de Tetranychidae (Acari)'. Andrews and Poe (1980) published a paper on the 'Spider mites of El Salvador, Central America (Acari: Tetranychidae)'. In 1982, Bernal and Piñeiro described *Nuevos Acaros Encontrados sobre Citrus en la Zona de Salto* in Uruguay. In Costa Rica, Doreste (1988) published *Acarologia*, an extensive book on acarology. In the same country, in 1990, Ronald Ochoa *et al.* reported on the mites of mango in 1990, and later published *Acaros Fitofagos de America Central: Guia Ilustrada* (Ochoa *et al.*, 1991a). Alone, or as co-author, Ochoa also investigated the Tenuipalpidae of Costa Rica, Honduras and Bermuda (for references see Mesa *et al.*, 2009) and has, in addition, published other works on pest mites, e.g. on the Tuckerellidae, etc. In 1991, Estebanes-Gonzalez and Rodriguez-Navarro (1991) examined the pest mites of vegetables in Mexico. In 2002, Lérida Almaguel Rojas published *Curso Introductorio a la Acarología Aplicada* in Cuba.

In Central and South America (and in other regions), an important contribution was made by the work of the Brazilian Carlos H.W. Flechtmann. His prolific work includes several papers on false spider mites of the South American Continent and Sri Lanka (for references see Mesa *et al.*, 2009), and over 40 papers on Tetranychidae. He is co-author with H.R. Bolland and Jean Gutierrez of the *World Catalogue of the Spider Mite Family (Acari: Tetranychidae)* (Bolland *et al.*, 1998), with James W. Amrine Jr and Terry Stasny of the *Revised Keys to World Genera of Eriophyoidea (Acari: Prostigmata)* (Amrine *et al.*, 2003), and of the *Catalogue of the Eriophyid Mites (Acari: Eriophyoidea)* (Davis *et al.*, 1982), which have played a fundamental role in the development of modern acarology. In addition, Flechtmann

has made many other contributions to the field of applied acarology, including the following publications: *Ácaros de Importância Médico Veterinária* (Flechtmann, 1973a); *Elementos de Acarologia* (Flechtmann, 1975); *Ácaros de Importância Agricola* (Flechtmann, 1977a); *Acaros em Produtos Amazenados e na Poeira Domiciliar* (Flechtmann, 1986); *A Residência para o Alérgico* (Flechtmann *et al.*, 1998); and *Manual de Acarologia* (Moraes and Flechtmann, 2008). Other well-known Brazilian acarologists are Denise Návia and Gilberto J. de Moraes.

Mediterranean Region

In this region, the first significant contributions to the history of economic acarology appeared in the early 20th century.

In 1932, the French Marc André published the paper 'Contribution à l'étude du "Bou-Faroua" Tétranyque nuisible au dattier en Algérie' (André, 1932a), and this was followed in 1959 by Claire Athias Henriot's 'Acarologie appliquée et agronomie algérienne'.

Mohamed A. Zaher (2007) reports in his brief history of acarology in Egypt that despite previous references 'a real contribution of Egyptian acarologists began in 1930 by the two pioneers Hassan and Sayed'. In particular, M. Taher Sayed published several papers on the Tetranychoidea and discussed the main injurious species in Egypt (Sayed, 1950b). Some years later, Fahmy (1956) examined 'The economic importance of mites affecting fruit trees, with special reference to citrus, in Egypt', and thus began a systematic study of most important pest mites, e.g. *Tetranychus* spp. (Attiah, 1969). Later, Zaher *et al.* (1984) published a report on *Phytophagous Mites in Egypt (Nile Valley and Delta)*. Many other papers on the bio-ecology of pest mites have been published by Egyptian authors.

Pest mites have also been investigated in Turkey, where various papers on economic acarology appeared in the 1950s and 1960s, including 'Citrus mites in Turkey' (Düzgünes, 1952), 'Türkiyede yeni bulunan akarlar (Mites recently found in Turkey)' (Düzgünes, 1963) and 'The phytophagous mites of the family Tenuipalpidae in Turkey' (Düzgünes, 1965a). In 1958, Millet gave an account of the pest mites in Lebanon, and Cuenod an account of the mites injurious to citrus in Tunisia. In 1959, Georghiou discussed the 'Plant-feeding mites of Cyprus' and Millet discussed those mites in Tunisia. Additional records of insect pests from Lebanon and Syria have been reported by Talhouk (1961). Saba (1964) reported on the injurious mites in the Near East, and some years later El-Haidari (1981) published 'New records of mites and insects infesting date palms in the Near East and North Africa'. There were various papers on mites from Libya, where investigations were made of the diseases and pests of date palms (Martin, 1958, 1959), mites of other crops (Damiano, 1964, 1968a,b) and citrus mites (Vacante and Di Martino, 1987). The most important works from Morocco were *Maladies, Troubles et Ravageurs des Agrumes au Maroc* (Chapot and Delucchi, 1964), and 'Les acariens nuisibles aux plantes cultivées au Maroc' (Saba, 1973b). Phytophagous mites in Morocco were examined again later by El-Jaouani (1988a,b).

In the first half of the 20th century, there were various contributions from Israeli authors on the systematics, bio-ecology and control of mite pests (Klein, 1936; Bodenheimer, 1951; Swirski and Amitai, 1958). Later, in a PhD thesis, Dubitzki (1981) reported on the 'Spider mites (Acari: Tetranychidae) infesting wild plants in Israel' and Uri Gerson (2008) on 'The Tenuipalpidae: an under-explored family of plant-feeding mites'. Gerson works on systematic, bio-ecological and applied acarology, and has investigated the mites of date palms (Gerson *et al.*, 1983), mango (Gerson, 1986), protected crops (Gerson, 1992) and citrus (Gerson, 1971, 2003). His book *Mites (Acari) for Pest Control* (Gerson *et al.*, 2003) is one of the basic contributions to the biological control of pest mites.

Former USSR

Various research has been recorded in the former USSR since the end of World War II. In Armenia, A.T. Bagdasarian (1957) published *Tetranychoid Mites (Superfamily Tetranychoidea). Fauna of the Armenian S.S.R.* In Georgia, from 1941 to 1959, G.F. Reck published, either alone or as co-author, over 24 papers on the Tetranychidae (for references see Bolland *et al.*, 1998), including *A Key to the Tetranychoid Mites* (Reck, 1959). The work of Reck concerns mainly the systematics and the taxonomy of Tetranychoidea, although it includes some work on applied acarology, e.g. 'Mite pests on cultivated plants' (Reck, 1941a).

In Kazakhstan, the systematics and taxonomy of the Tetranychidae were investigated by Boris Aronovich Wainstein, who published about 18 papers – or more substantial publications – from 1954 to 1983 (for references see Bolland *et al.*, 1998), including the *Tetranychoid Mites of Kazakhstan (With Revision of the Family)* (Wainstein, 1960). Wainstein also published papers on economic acarology, such as 'Fauna of spider mites injurious to orchards in South Kazakhstan' (Wainstein, 1954), and 'Notes on the biology of fruit mites' (Wainstein, 1958).

However, the most significant works appear from the 1960s in Crimea by V.I. Mitrofanov, I.Z. Livshits and Z.I. Strunkova, all specialists on the Prostigmata and other mite groups. Mitrofanov is author or co-author of over 23 important works on the Tetranychidae (especially the Bryobiinae), among which there are papers on economic acarology that include: 'Tetranychoid mites of the fauna of the USSR, damaging conifers' (Mitrofanov, 1967), 'A key for determination of tetranychoid mites on coniferous trees' (Mitrofanov *et al.*, 1975), and 'A contribution to the fauna of plant-feeding mites (Acariformes, Tetranychoidea) of Crimea' (Mitrofanov and Sharonov, 1983). The taxonomic publications of Mitrofanov include the *Keys to the Tetranychid Mites (Tetranychidae, Bryobiidae) Fauna of the USSR and Adjacent Countries* (Mitrofanov *et al.*, 1987). Livshits is author of almost 14 papers on the systematics and taxonomy of the Tetranychidae and Tenuipalpidae. He is also co-author of the book *Plant-inhabiting Mites: An Illustrated Key to Families* (Livshits and Mitrofanov, 1975), and has published various papers on economic acarology, including: 'Brown fruit mite *Bryobia redikorzevi* Reck, 1947 (Taxonomic position, morphology and biology)' (Livshits, 1960a), and 'Contribution to morphology and biology of fruit mites injuring cultures (New species of tetranychid mites on plum)' (Livshits, 1960b). Strunkova published several papers on the Tetranychidae of Tadzhikistan and

central Asia, and some works on applied acarology, including 'Phytophagous mites of "Tigrovaya Balka" nursery (Tadzhikistan)' (Strunkova, 1971) and *Tetranychid Mites – Pests of Fruit Crops in the Hissar Valley of Tadzhikistan* (Strunkova, 1972). Mitrofanov and Strunkova (1979) wrote *A Key to False Spider Mites*, in which they recognized 365 species and 33 genera of Tenuipalpidae.

Studies on eriophyoid mites were carried out in the former USSR by various different acarologists, among whom were V.G. Shevchenko and S.I. Suchareva, and the works of many other applied acarologists are reported in the literature.

Africa

References on pest mites can be found in the numerous papers of Meyer Smith and other South African specialists. South African acarology started in the early 1900s (Tucker, 1926), but the most intense period began in the 1960s with Magdalena K.P. Meyer Smith (1931–2004), one of the most important acarologists of the century. Meyer Smith had a deep knowledge of the systematics and taxonomy of a number of African mite families, including the Eriophyoidea, Tetranychidae and Tenuipalpidae, and published, alone or as co-author, a number of important papers on the systematics and taxonomy of African fauna, including *A Revision of the Tetranychidae of Africa (Acari) with a Key to the Genera of the World* (Meyer Smith, 1974) and *African Tetranychidae (Acari: Prostigmata) – with Reference to the World Genera* (Meyer Smith, 1987). In her revision of the Tenuipalpidae of Africa (Meyer Smith, 1979), she reported 504 species and 21 genera. Meyer Smith and Edward A. Ueckermann described 181 eriophyoid species (see Amrine and Stasny, 1994). In 1981, Meyer Smith published the *Mite Pests of Crops in Southern Africa* (Meyer Smith, 1981a), a fundamental work on economic acarology for this region. The contributions from the South African school were actually made by Edward A. Ueckermann, a prestigious acarologist known worldwide. In Mozambique, Rodrigues (1968) published a paper on the 'Acarina de Mocambique catalogo das especies relacionadas com a agricultura'. There are also many further reports from other African acarologists working in economic acarology.

Asia

The state of Asian economic acarology is mixed. In 1961, Talhouk reported additional records of pests from Syria, including mites. Various works on mite pests have been published in Iraq, the most important of which were *A Preliminary List of Mites of Iraq* (El-Haidari, 1965), two overviews of the mite fauna (Mohamed and El-Haidari, 1965, 1968), and a thesis by Al-Jboory (1987) on 'Taxonomic studies of false spider mites (Acari: Tenuipalpidae) in central Iraq'. There are various publications on the mites of Iran, including *A Checklist of Economically Important Insects and Other Enemies of Plants and Agricultural Products in Iran* (Farahbakhsh, 1961), 'Phytophagous mite fauna of Iran' (Khalil-Manesh, 1973), *Injurious Mites of Agricultural Crops in Iran* (Khanjani and Haddad Irani-Nejad, 2006) and 'The 20 years research of acarology in Iran' (Sepasgosarian, 1977). In 1983, Sepasgosarian published the first world list of the Tenuipalpidae, reporting 541 species and 21 genera, and in four following addenda, increasing the number of species to 697 and the number of genera to 25 (see Mesa *et al.*, 2009). In 1971, Faizyar published the *List of Plant Pests (including Insects and Mites) and Diseases of Economic Importance in Afghanistan*. In 1975, Soliman and Mahfood published an account of phytophagous and predaceous mites from southern Yemen, and Hasan published various papers on false spider mites from Pakistan (for references see Mesa *et al.*, 2009).

Economic acarology in India has a long history. In 1940, Rahman and Sapra published 'Mites of the family Tetranychidae from Lyallpur with descriptions of four new species', followed in 1945 by Pruthi and Mani with *Our Knowledge of the Insect and Mite Pests of Citrus in India and their Control*. About 15 years later, Puttarudriah and ChannaBasavanna (1959) published 'A preliminary account of phytophagous mites of Mysore'. In 1974, Prasad produced *A Catalogue of Mites of India*. The same year, Chaudhri *et al.* published *Taxonomic Studies of the Mites Belonging to the Families Tenuipalpidae, Tetranychidae, Tuckerellidae, Caligonellidae, Stigmaeidae and Phytoseiidae*. Some years later, ChannaBasavanna *et al.* (1979), and then Mallik *et al.* (2003), produced bibliographies on the plant-feeding mites of India. Nassar and Ghai (1981) published a paper on 'Taxonomic studies on tetranychoid mites infesting vegetable and fruit crops in Delhi and surrounding areas'. Gupta produced a *Handbook: Plant Mites of India* (Gupta, 1985), a paper on 'The mites of agricultural importance in India with remarks on their economic status' (Gupta, 1991) and a comprehensive study for the Zoological Survey of India: *Arachnida: Plant Mites (Acari)* (Gupta, 1992). Sadana published various works on economic acarology, including *Plant Feeding Mites of India (Tetranychoidea)* (Sadana, 1985) and *False Spider Mites Infesting Crops in India* (Sadana, 1997). In 1994, Gupta and Gupta published *A Taxonomic Review of Indian Tetranychidae (Acari: Prostigmata) with Description of New Species, Redescriptions of Known Species and Keys to Genera and Species*. In 1997, Gupta and Chatterjee published another paper for the Zoological Survey of India 'Acari: plant mites'. There was also some serious research done on eriophyoid mites: ChannaBasavanna (1966) reported 61 species, Mohanasundaram described 230 eriophyoid species and Chakrabarti *et al.* 92 species, primarily from India (see Amrine and Stasny, 1994). Later, Chhillar *et al.* (2007) published *Agricultural Acarology* and Thakur *et al.* (2010) discussed the host range, distribution and morphometrics of predatory mites associated with the phytophagous mites of fruit crops in Himachal Pradesh.

In China, there has been much activity in acarology in recent decades. Several reports noted here are drawn from a commemorative volume dedicated to Jie-Liu Xin (1909–1994), a pioneer of acarology in China (Zhang and Hong, 2010). During the last century, the main groups of pest mites were extensively investigated in China. Hong and Zhang (1996b) treated nearly 200 species included in 74 genera, and later Hong *et al.* (2010a) reported three families, 187 genera and 932 species. En-Pei Ma published various papers on tetranychid mites (for references see Bolland *et al.*, 1998) and also on economic acarology, e.g. 'Tetranychid mites' (Ma and Yuan, 1992).

Tetranychid mites have also been studied by H.-F. Wang, who published alone, or as co-author, several papers, among which are 'Tetranychid and tenuipalpid mites infesting fruit trees in the north of China' (Wang *et al.*, 1981), and 'Acari; Tetranychoidea' (Wang and Cui, 1992). In 2010, the Tetranychidae family in China included 28 genera and 212 species (Hong *et al.*, 2010b), and the Tenuipalpidae family nine genera and 56 species (Zhang, 2010). In 1999, Lin and Zhang published the book *Tarsonemidae of China (Acari: Prostigmata): Annotated and Illustrated Catalogue and Bibliography*, followed in the 2002 by another book, *Tarsonemidae of the World: Key to Genera, Geographical Distribution, Systematic Catalogue and Annotated Bibliography*. The list of Chinese tarsonemids now includes 16 genera and 105 species (Lin and Zhang, 2010). In the last two decade various works on economic acarology have been published, including: the *Economic Insect Fauna of China* (Kuang, 1995); *An Illustrated Guide to Mites of Agricultural Importance* (Zhang and Liang, 1997); and *Mites of Greenhouses: Identification, Biology and Control* (Zhang, 2003). The work of Zhi-Qiang Zhang has played a key role in the development of Chinese acarology.

In Taiwan, there have been various works published on economic acarology since the 1960s. The most important include 'Tetranychoid mites infesting tea in Taiwan' (Lo, 1968a), 'Tetranychoid mites infesting fruit plants in Taiwan' (Lo, 1968b) and 'Tenuipalpid and tetranychid mites infesting citrus in Taiwan and a life history study of the citrus green mite, *Schizotetranychus baltazarae* Rimando' (Lo and Hsia, 1968). From 1970 to 1990, Y.-H. Tseng published several important works, including 'Systematics and distribution of phytophagous and predatory mites on grape in Taiwan. Part I. Phytophagous mites' (Tseng, 1974), 'Mites associated with the conifers of Taiwan, (1) Tetranychoidea, with a checklist of the world' (Tseng, 1980), 'Mites associated with weeds, paddy rice and upland rice fields in Taiwan' (Tseng, 1984), and *A Monograph of the Mite Family Tetranychidae (Acarina: Trombidiformes) from Taiwan* (Tseng, 1990).

Pest mites have been investigated also in Korea, where Han published 'Studies on mites (II). On some mites of the apple trees' (Han, 1969), and 'Studies on mites (III). Mites on persimmon and citrus' (Han, 1970).

In the Philippines, Rimando published 'The tetranychoid mites of the Philippines' in 1962 (Rimando, 1962a).

In Japan, the most important acarologist is Shōzō Ehara (1928–2008), who from 1954 to 1996, either alone or as co-author, published over 36 works on tetranychid mites, several of which concern aspects of economic acarology. Among his papers are 'A guide to the tetranychid mites of agricultural importance in Japan. Approaches to biological control' (Ehara, 1975a), and the *Illustrations of the Mites and Ticks of Japan* (Ehara, 1980). His scientific work on the false spider mites is similarly important and includes almost ten papers on Japanese and Brazilian fauna (see Mesa *et al.*, 2009). Alone or with co-authors, Ehara created many new taxa (205 species, 4 genera, 1 subgenus, 1 tribe) (Ehara *et al.*, 2009). Another important Japanese publication was *Mites. An Introduction to Classification, Bionomics and Control of Acarina* (Sasa, 1965). Other highly qualified Japanese acarologists are Tetsuo Gotoh and Yutaka Saitō, who are both world-renowned specialists.

Australia and New Zealand

Bruce Halliday of the Australian National Insect Collection of CSIRO (the Commonwealth Scientific and Industrial Research Organisation) has summarized the history of Australian acarology over the last 250 years, discussing the main stages from Linnaeus up to 2001, focusing attention on the work of the most important Australian acarologists, who included Henry Tryon, Herbert Womersley, Ronald Vernon Southcott and Frederick Hugh Sherston Roberts (Halliday, 2001a). Among these researchers, Womersley (1889–1962) played a significant role in the development of Australian acarology. He published over 150 papers on the taxonomy of various mite groups, including the Tetranychidae (Womersley, 1940). In the 1950s and 1960s, J.J. Davis published several papers on the same topic (for references see Bolland *et al.*, 1998). Halliday himself is a most eminent Australian acarologist, and has produced several fundamental works on Australian Acarina, including *Mites of Australia: A Checklist and Bibliography* (Halliday, 1998) and the recent list compiled as part of the Australian Biological Resources Study database: Higher Taxon Acari: Mites (Halliday, 2013). His contributions on economic acarology are also important and include extensive discussion of the red-legged earth mite Halliday (2001b). Another, well-known and respected Australian acarologist is David Walter of Alberta University and co-editor of *A Manual of Acarology* (Krantz and Walter, 2009).

David Charles M. Manson has published several papers on tetranychid and tenuipalpid mites from New Zealand. His works include 'Mites of the families Tetranychidae and Tenuipalpidae associated with citrus in South East Asia' (Manson, 1963), and Manson has described 54 eriophyoid species (Amrine and Stasny, 1994). Collyer has published various papers on mites in New Zealand, including 'Phytophagous mites and their predators in New Zealand orchards' (Collyer, 1964a). Zhi-Qiang Zhang, from China, who is mentioned above in the section on Asia, works at Ladare Research in Auckland, New Zealand.

Worldwide

Several very important texts were published during the second half of 1900 by a number of authors, and these are singled out here as of particular note (they do include one or two that have already been mentioned above): Radford, 1950; Hughes (T.E.), 1959; Evans *et al.*, 1961; Sasa, 1965; Flechtmann, 1975; Krantz, 1978; Woolley, 1988; Van der Hammen, 1989; Evans, 1992; Alberti and Coons, 1999; Coons and Alberti, 1999; Walter and Proctor, 1999; and Krantz and Walter, 2009.

The account that has been presented in this chapter shows that over a period of about 250 years acarology has become an autonomous discipline, which has been divided into different specialisms. It has been applied to solve problems related to human economy (agricultural practices, stored food, human health, etc.) and shows close links with the evolution of the socio-economic characteristics of human populations and the impact of conditions resulting from intensive agricultural practices. Like other zoological and/or biological science disciplines, the development of acarology has been influenced by the availability of means

and economic resources and technological innovation, such as electron microscopy, biochemical analysis, statistical computing, etc. The large number of researchers who are today engaged in the study of different branches of acarology worldwide, including pest control, is further confirmation of the development of the discipline.

The results of all of this research are reported in various specialized journals. These include: *Acarologia*, the *Acarology Bulletin*, *Experimental and Applied Acarology*, the *International Journal of Acarology*, *Systematic and Applied Acarology* and the *Journal of the Acarological Society of Japan*, as well as numerous entomological journals.

Further information on acarology may be found at the following websites, which are a small selection of the many that are available:.

- The 'Acarology' home page hosted by the UK's Natural History Museum: http://www.nhm.ac.uk/hosted_sites/acarology/
- Acarological Society of America: https://sites.google.com/site/acarologicalsociety/home
- *Experimental and Applied Acarology* journal articles: http://www.springerlink.com/content/100158/

11 Classification of and Key for the Identification of Major Taxa

The major taxa included here are the higher categories of the subclass Acari – the superorders Parasitiformes and Acariformes, and the key to families that is presented identifies the lower categories of the superorder Acariformes and its two orders Trombidiformes and Sarcoptiformes.

Higher Categories of the Subclass Acari

At present, there is no a definitive version of the higher classification of the Acari, because of a lack of knowledge of the fossil record and the existence of many unclear relationships between groups. The latest studies consider the Acari to be a subclass of the class Arachnida (Krantz, 1978; Johnston, 1982; Lindquist, 1984; Evans, 1992; Walter and Proctor, 1999; Zhang, 2003). Zhang (2003) divides the subclass into three superorders (Opilioacariformes, Parasitiformes and Acariformes) and seven orders; within these three superorders, the Opilioacariformes include the order Opilioacaridida, the Parasitiformes include the orders Holothyrida, Mesostigmata and Ixodida, and the Acariformes include the orders Prostigmata, Astigmata and Oribatida. Lindquist *et al.* (2009) recognize only two superorders, the Parasitiformes and the Acariformes. Hoy (2011) follows the classification of Evans *et al.* (1961) and considers the Acari to be a subclass including two orders, the Parasitiformes and the Acariformes. Here, the classification proposed by Lindquist *et al.* (2009) is adopted. Further information on pest mite families is given in the key to families that follows, and in Chapters 12–23.

Parasitiformes

The Parasitiformes have 1–4 dorsolateral or ventrolateral stigmata set posteriorly to coxae II. Usually, all of the coxae are free and movable (Fig. 2.10D). Tarsus I bears a dorsal cluster of solenidiform setae subdistally, and tarsi II–IV have a peripodomeric fissure and slit organ. This superorder includes four orders (the Opilioacarida, Olothyrida, Ixodida and Mesostigmata). The order Mesostigmata is divided into three suborders (the Sejida, Trigynaspida and Monogynaspida). The suborder Monogynaspida is divided into five cohorts; of these, the cohort Gamasina includes ten superfamilies, among which the Ascoidea and the Phytoseioidea (with the family Phytoseiidae) number very important species in biological control.

Acariformes

The main morphological feature of the Acariformes is the absence of stigmata posteriorly to coxae II. Moreover, the coxae are integrated with the podosoma and sometimes form coxisterna (Fig. 2.9C). Tarsus I bears paired dorsal setae distally and subdistally, and tarsi II–IV lack a peripodomeric fissure and slit organ. This superorder is divided into two orders, the Trombidiformes and the Sarcoptiformes.

The order Trombidiformes is divided into two suborders, the Sphaerolichida and the Prostigmata, which include about 38 superfamilies. The Sphaerolichida have two monogeneric families and about 16 species; they are closely related to the Prostigmata but do not have any economic importance. The suborder Prostigmata shows various evolutionary adaptations, and the species that are known are terrestrial, aquatic and marine predators, phytophages, saprophages, paraphages and parasites (Walter *et al.*, 2009). The Prostigmata have an internal respiratory system and paired stigmata, set at or near the cheliceral bases (Fig. 2.6F, H and I) or at the humeral angles of the propodosoma (Fig. 2.17A). The chelicera may be external or set within the gnathosomatic capsule; they may be chelate or lacking a digitus fixus and with the digitus mobilis transformed into a stylet. The cheliceral bases may form a stylophore or be fused to the subcapitulum (Fig. 2.6 H and I). The palp segments may be blended or reduced (Fig. 2.7G). In different groups, the palptarsus may be displaced to the base of the tibia and form the 'thumb-claw complex' (Fig. 2.7B). The prosoma and opisthosoma may bear plates or dorsal shields. The latest estimates indicate that the Prostigmata include about 17,170 described species, 1348 genera, 36 superfamilies and 131 families, which are included in four supercohorts (the Labidostomatides, Eupodides, Anystides and Eleutherengonides) (Walter and Proctor, 1999). The supercohort Eupodides is divided into five superfamiles, among which is the Eupodoidea, which includes the Penthaleidae family, which has some important pest mites, and the Eriophyoidea, which numbers three important families of pest mites, the Phytoptidae, Eriophyidae and Diptilomiopidae. The supercohort Eleutherengonides is divided into two cohorts, the Raphignathina and the Heterostigmatina. The Raphignathina includes five superfamiles, among which is the Tetranychoidea, with the families Tetranychidae, Tenuipalpidae and Tuckerellidae, all of which include pest mites. The cohort Heterostigmatina is divided into eight superfamilies, among which the Tarsonemoidea (with the family Tarsonemidae), the Pygmephoroidea (with the families Siteroptidae and Pygmephoridae) and the Scutacaroidea (with the families Scutacaridae and Microdispidae) all include several pest mites.

The order Sarcoptiformes is divided into two suborders, the Endeostigmata and the Oribatida. The suborder Oribatida is divided into five supercohorts, among which is the Desmonomatides, which has three cohorts, the Nothrina, Brachypilina and Astigmatina. The cohort Astigmatina includes

ten superfamilies, among which the Hemisarcoptoidea, with the family Hemisarcoptidae, includes important natural enemies of armoured scale insects, the Glycyphagoidea (with the family Glycyphagidae), the Acaroidea (with the family Acaridae) and the Histiostomatoidea (with the family Histiostomatidae). The Acaridae and the Histiostomatidae include several mite pests.

We present below a key to identify the adults of the pest mite families of the superorder Acariformes and the orders Trombidiformes and Sarcoptiformes. The key is fundamentally drawn from Lindquist *et al.* (2009), integrated with the suggestions of Walter *et al.* (2009) for the Trombidiformes and O'Connor (2009) for the cohort Astigmatina. The Parasitiformes have been included only to help in the building of the key. The suborder Sphaerolichida of the order Trombidiformes is not included in the key because (as already noted) its species do not have any economic importance. The family Ereynetidae is included following the account by Vázquez-Rojas and Estrada-Venegas (2010), who collected *Ereynetes amplectorus* Hunter and *Ereynetes faini* (Hunter) in soil samples inside garlic crops in several localities of Mexico. However, as there is no information on harmfulness of these species, they are not treated further in this text.

Key for the Identification of Families of Superorder Acariformes and Orders Trombidiformes and Sarcoptiformes (Adults)

1. • With 1–4 pairs of dorsolateral or ventrolateral stigmata posterior to coxae II; coxae free and commonly movable; tarsi II–IV with peripodomeric fissure associated with slit organ; tarsus I with dense cluster of solenidiform setae subdistally ....superorder **Parasitiformes**

   • Without visible stigmata posterior to coxae II; coxae integrated with venter of podosoma and often forming coxisterna; tarsi II–IV lacking peripodomeric fissure and slit organ; tarsus I with sparse pairings of dorsal setae distally and subdistally....................................superorder **Acariformes**...2

2. • Chelicerae rarely chelate, digitus fixus often regressed and digitus mobilis commonly a hook, knife, needle or stylet-like structure; cheliceral bases sometimes fused medially; palpi simple or modified into thumb-claw process, sometimes reduced; ambulacra of at least legs II and III with 2 lateral claws and with median empodium (rarely lacking) pad like or rayed and often with tenent hairs; opisthosoma lacking lateral glands; opisthosomatic setal row *c* commonly with 2 pairs of setae (c_1–c_2), sometimes with 3 pairs or hypertrichous; 1 pair of stigmata opening between bases of chelicerae or on anterior prodorsum commonly present and sometimes associated with peritremes dorsally on the cheliceral bases or on the anterior margin of prodorsum (order **Trombidiformes**); tracheal system with 1 pair of stigmata opening between bases of chelicerae or on anterior prodorsum commonly present; prodorsum commonly with 4 or fewer pairs of setae or hypertrichous, sometimes including 1–2 pairs of bothridial sensilla; chelicerae rarely chelate, commonly with digitus fixus sheath like or regressed; coxal fields contiguous or II–III distinct.........................suborder **Prostigmata...4**

   • Chelicerae chelate, commonly dentate, rarely attenuate or stylet like; cheliceral bases separate; palpi simple, lacking thumb-claw process; ambulacra of legs I–IV with 1 or 3 claws, empodium claw like or sucker like, not pad like, rarely rayed; opisthosoma commonly with pair of lateral glands; opisthosomatic setal row *c* commonly with 3–4 pairs of setae or hypertrichous; tracheal system lacking or arising from bases of legs or as brachytracheae on various parts of the legs or idiosoma; stigmata and peritremes not present between cheliceral bases or on prodorsum (order **Sarcoptiformes**); prodorsum without specialized sensory organs other than setiform setae; genital aperture exposed or partially covered by paragenital flaps, inversely 'V'-, 'U'-, or 'Y'-shaped, with usually 2 pairs of genital papillae, sometimes reduced or modified; anal aperture lacking of conspicuous paired plates; idiosoma weakly sclerotized, with epimeral plates undeveloped or weakly formed between bases of legs; pretarsus with empodial claw and fleshy pulvillus or sucker like and true paired claws lacking; palpi 2- rarely 3-segmented, males with a sclerotized aedeagus, and often with a pair of copulatory adanal suckers (suborder **Oribatida**, cohort **Astigmatina**)..3

3. • Chelicerae non-chelate, laterally compressed; often attenuate, commonly with many fine teeth; genital valves of female fused to body posteriorly, open anteriorly forming a transverse oviporus; genital papillae on both sexes often as large rings on ventral surface, not arising from a furrow and not associated with genital opening; body larger and not dorsoventrally flattened ....family **Histiostomatidae** Berlese

   • Chelicerae chelate, with few teeth, rarely chelicerae vestigial or digitus fixus lacking; female oviporus longitudinal with genital valves fused anteriorly to body, free posteriorly; genital papillae associated with genital opening and never as large rings; prodorsal sclerotization as rectangular shield, sometimes incised posteriorly, or prodorsal sclerotization lacking; dorsal setae variable, but not normally both elongate and heavily barbed; cuticle smooth or rarely with small rounded protuberances; leg III of male similar to leg IV, or leg III enlarged and bearing a large empodial claw and reduced tarsal setation....................family **Acaridae** Latreille

4. • Adults and deutonymphs often densely hypertrichous; larvae commonly strongly heteromorphic, protonymphs and tritonymphs calyptostatic; terrestrial forms with 1 or 2 pairs of bothridial setae on a prodorsal sclerite (cohort **Parasitengonina**); idiosoma and appendages hypertrichous; idiosoma lacks glandularia; chelicerae with digitus mobilis stylet form and retractable, either independently or with entire gnathosoma (subcohort **Erythraeina**); prodorsal area bearing 2 pairs of bothridial organs (superfamily **Erythraeoidea**); chelicerae retractable into idiosoma; idionotal setae usually setiform..............family **Erythraeidae** Robineau-Desvoidy

   • Adults and nymphs with or without dense hypertrichy, bothridial setae and prodorsal sclerite present or lacking, without glandularia; larvae more or less similar to other stages; alternating calyptostasis rare.................................**5**

5. • Females with one pair of stigmata and tracheae, set anterolaterally on prodorsum, and 1 pair of capitate bothridial setae (commonly lacking in males); peritremes lacking; opisthosoma with 4 dorsal tergites, sometimes overlapping (cohort**Heterostigmatina**)..13
• Females and males with 1 pair of stigmata set between cheliceral bases or on posterodorsal gnathosoma and often opening into peritreme of various shapes (sometimes lacking); prodorsal trichobothria, when present, of various shapes and on both sexes..6
6. • One or 2 pairs of prodorsal bothridial setae, set in distinct bothridia; peritremes lacking; palptibia without claw-like setae or thumb-like palptarsus; post-pedal furrows lacking (supercohort **Eupodides**, in pars).......................................7
• Prodorsal bothridial setae absent................................8
7. • Chelicerae freely articulating, digiti variously structured; tarsus I with a famulus and at least 1 recumbent solenidion; idiosomatic chetome sometimes with postlarval hypertrichy; naso commonly present (superfamily **Eupodoidea**); anal opening dorsal or dorsoterminal; intercoxal region bearing 15 or more pairs of setae; femora IV similar to others.......family **Penthaleidae** Oudemans
• Chelicerae closely contiguous or fused medially, digitus fixus regressed and digitus mobilis stylet like to whip like; tarsus I lacking famulus, solenidia erect; idiosoma never hypertrichous; naso lacking (superfamily **Tydeoidea**); tibia I with deeply recessed solenidion in sac-like structure connecting to surface by narrow duct; legs and prodorsum often with net-like ornamentation; adults with 2 pairs of genital papillae.................family **Ereynetidae** Oudemans
8. • Legs III present and legs IV commonly present in postlarval stages; genital papillae lacking in all stages; ovipositor lacking; dorsal shield various (cohort **Raphignathina**); cheliceral bases fused into movable stylophore deeply retractable into idiosoma; cheliceral digitus mobilis elongated, whip like, recurved basally within stylophore (superfamily **Tetranychoidea**); both sexes with legs III and IV; legs ending in an empodium with a pair of lateral claws; opisthosoma variously shaped..11
• Legs III and IV lacking in all stages; legs I and II ending in a branched empodial featherclaw; opisthosoma elongated, annulated, vermiform or fusiform (supercohort **Eupodides**, in pars; superfamily **Eriophyoidea**); legs I and II with distal empodial featherclaw; opisthosoma elongated, cuticle annulated, vermiform or fusiform........................9
9. • Prodorsal shield without setae or with 1 pair of setae; spermathecal tubes short; tibia I lacks solenidion 10
• Prodorsal shield with 1, 3, 4 or 5 setae; spermathecal tubes commonly long; solenidion on tibia I present or absent ....family **Phytoptidae** Murray
10. • Gnathosoma commonly small in respect to body; when large, with chelicerae straight or slightly curved; terminal segments of palps short and truncated and enclosing the short oral stylet; female genital coverflap commonly with ridges; featherclaw entire rarely divided ....family **Eriophyidae** Nalepa
• Gnathosoma large in respect to body; chelicerae abruptly curved and bent down near base; palps attenuated, enclosing the long oral stylet; female coverflap commonly smooth, less often with ridges; featherclaw deeply divided medially or entire.......................family **Diptilomiopidae** Keifer
11. • Posterior setae of idiosoma (*h*) forming a fan-like transverse row of 5 or more pairs of flagellate or bipectinate setae; dorsal and lateral opisthosomatic setae expanded, fan like, the anterior row (*c*) including 5 or more pairs ....family **Tuckerellidae** Baker *et* Pritchard
• Posterior setae of idiosoma slender, simple, not forming a fan-like row; dorsal and lateral opisthosomatic setae of various shapes, the anterior row (c) including at most 4 pairs..12
12. • Tarsi I and II lacking peg-shaped or bulbous solenidia, with 1–2 long solenidia, slender, commonly associated with a short or very little seta to form duplex setae; stylophore attachment to idiosoma without ribbed collar; palpi 5-segmented, with thumb-claw process; palptarsus with or without one of distal eupathidia enlarged as spinneret ....family **Tetranychidae** Donnadieu
• Tarsi I and II with distal peg-shaped or bulbous solenidia, and lacking solenidia associated with a seta to form duplex setae; stylophore attachment to idiosoma with ribbed collar; palpi linear, with 5 or fewer segments, without thumb-claw process; opisthosoma with at most 13 pairs of setae dorsally and laterally (c_1 commonly present, c_4 absent, 2 pairs of *h* setae); anal valves with 1–3 pairs of simple pseudanal setae............................family **Tenuipalpidae** Berlese
13. • Legs IV of female with distinct femur and genu 14
• Legs IV of female with fused femur and genu, tibia and tarsus with 2–3 setae, and lacking claws and empodium (superfamily **Tarsonemoidea**); female with 4 pairs of legs, leg IV 3 segmented, commonly more slender than legs II–III; males with 4 pairs of legs, legs IV commonly 3- or 4-segmented, and inserted ventrally; male genital capsulecaudal.. ....family **Tarsonemidae** Canestrini *et* Fanzago
14. • Tarsus I with paired claws; trochanters I, II, IV and usually III without setae; stigmata and associated tracheae anterolaterally on prodorsum absent; leg IV of male with pretarsus, empodium, and paired or reduced claws like those on legs II–III (superfamily **Dolichocyboidea**); prodorsum and tergite C of females normal in form, lacking wing-like expansions; female genital aperture small, flanked by 1–3 pairs of minute or vestigial genital setae, or these are absent; tarsi I–IV with smooth flexible empodium between claws; trochanter III lacking seta; setation of genua maximally 2–1–1–1 ....family **Dolichocybidae** Mahunka
• Tarsus I with single claw or without claws; trochanters of legs I–IV usually each with a seta; leg IV of male with claw variously shaped; trochanter IV quadrangular, longer than wide, different from subtriangular trochanter III...15
15. • Prodorsum of female not covered posteriorly by tergite C and usually with 3 pairs of setiform setae; coxisternal plates together with 4–6 pairs of setae; femur I with 4 setae (superfamily**Pygmephoroidea**)...................................16

• Prodorsum of female commonly covered to greater or lesser extent by tergite C and with 1–2 pairs of setiform setae; coxisternal plates I–II together with maximum of 4 pairs of setae; femur I with 3 setae (superfamily **Scutacaroidea**)..17

16. • Femur I of female with all setae slender, and attenuate (seta *d* setiform); coxisternal plates I–II commonly with 4 pairs of setae; prodorsum of female with stigmata often extended into longitudinal peritremes ....family **Siteroptidae** Mahunka

• Femur I of females with seta *d* thickened, rigid, and blade like, rod like, or hooked. Coxisternal plates I–II together with 4–6 pairs of setae; prodorsum of females with stigmata round or oval, not extended into elongate grooves ....family **Pygmephoridae** Cross

17. • Tergite C of female expansive, covering prodorsum entirely, forming a roof over gnathosoma; prodorsum usually with 2 pairs of short setiform setae; idiosoma broadly oval or rounded, compact, distance between insertions of legs II–III similar to that between legs III–IV; seta *d* on femur I short, stout, usually branched; aedeagus apparently annulated or striated ....family **Scutacaridae** Oudemans

• Tergite C of female covering posterior region of prodorsum to variable extent, not covering entire prodorsum and gnathososma; prodorsum with 1 pair of setiform setae, and a second pair (v_2) rudimentary or absent; idiosoma oval; distance between insertions of legs II–III at least twice that between legs III–IV; seta *d* on femur I attenuated, smooth or barbed; aedeagus apparently smooth ....family **Microdispidae** Cross

PART II MITES INJURIOUS TO ECONOMIC PLANTS

The second part of the book describes the major families of mites that are injurious to economic plants. Chapters 12–22 are each devoted to a particular family and the genera within, while the last chapter covers five families of more minor importance. For each genus, descriptions are given of the diagnostic characteristics, economic importance and injurious species. For each species listed under that genus, the information given includes common name(s), diagnostic (morphological) characteristics,[1] geographical distribution, host plants, bio-ecology, symptomatology and damage caused (where known), and control methods, management and resistance.

[1]Under the 'Diagnostic (morphological) characteristics' of each species, there is normally a description given of the 'Female', but there is only an entry for the 'Male' if the male is known to exist or there is some reference to it being unknown.

12 The Family Histiostomatidae Berlese

Morphological Characteristics, Systematics and Bio-ecology

The body of mites of the family Histiostomatidae is large and not dorsoventrally flattened. The gnathosoma is strongly modified and adapted to feed on minute particles in suspension. The chelicerae are laterally compressed, not chelate, with the digitus fixus comb like and equipped with many fine teeth. The digitus mobilis is short, fused to the base of the digitus fixus, and with few short distal teeth. The palpi have a freely movable and flattened distal segment, with long tarsal solenidion and eupathididal setae (Figs 12.1C and 12.4B). The prodorsum has four pairs of setae (*ve*, *vi*, *sce*, *sci*) and the hysterosoma ten pairs of dorsal setae (c_1, c_2, c_3, d_1, d_2, e_1, e_2, f_2, h_1, h_2); one pair of setae (h_3) is commonly set ventrally. The opisthonotal glands are commonly located between setae e_1 and e_2. The bursa copulatrix is set between the setae h_1 (Fig. 12.1A). The genital valves of the female are fused to the body medially, and form a transverse oviporus; the genital papillae of both sexes are often seen as large rings on the ventral surface and are not closely associated; the posterior pair is set on the ventral opisthosoma (Figs 12.1B and 12.2A) (Hughes, 1976; O'Connor, 2009).

The deutonymph has gnathosomatic remnant that is either well developed or reduced. The idiosoma is well sclerotized, and a sejugal furrow separates the propodosomatic and hysterosomatic sclerites. The ventral coxal surface has a posterior median apodeme, and normal chaetotaxy, with the exception of setae *4b*, which are absent. The attachment organ is commonly developed and bears all structures (Figs 12.3B and 12.7B). The anterior legs are also commonly developed; legs III and IV sometimes have some segments fused and secondary articulation in the midtarsus. The articulations of the trochanter–femur of legs III–IV are anteriorly directed in repose. The pretarsi have a small membranous ambulacrum and empodial claw; empodium IV is sometimes missing (Figs 12.3B–D, 12.7B–C) (Hughes, 1976; O'Connor, 2009). The terminology follows mostly that of Houck and O'Connor (1990) and Fashing and O'Connor (1984).

The Histiostomatidae family (formerly the Anoetidae) belongs, with the Guanolichidae, to the superfamily Histiostomatoidea and numbers 59 genera and over 567 species (Schatz *et al.*, 2011). Histiostomatid mites live in wet substrates, and use modified mouthparts to filter organic material and microorganisms from water film (Hughes, 1953; O'Connor, 2009); their genital papillae seem to be an adaptation to regulate their osmotic balance in this aquatic environment (Witalinski *et al.*, 2002) and their size varies with the osmotic pressure of that environment. Many species are also found on wet vegetable matter where decomposition is taking place, and some species are found on decaying mushrooms, on cow dung, or on the resins and saps exuding from the damaged bark of trees. Many species live in fully aquatic environments, and they can inhabit water-filled tree holes and leaf axils, often using insects as dispersal agents. Other species have developed an intimate association with animal hosts (O'Connor, 2009). *Histiostoma laboratorium* Hughes is a common pest of *Drosophila* rearing (Hughes, 1976), and Ashburner (1989) has treated extensively the Histiostomatidae associated with *Drosophila* cultures. Fashing (2010) described two new adaptations for dispersal in this family.

Some species belonging to the genus *Histiostoma* Kramer have some economic importance, feeding on various bulbs and vegetable substrates (Hughes and Jackson, 1958; Qin and Rohitha, 1996) or developing on mushrooms, where they can cause either direct or indirect damage (Hussey *et al.*, 1969; Hill and Deahl, 1978; Greenslade and Clift, 2004).

The Genus *Histiostoma* Kramer

Diagnostic characteristics

The genital opening of the female consists of a transverse slit lying between the anterior pair of chitinous rings. The posterior chitinous rings are set at the same level as coxae IV, or between coxae III and IV. The male lacks genital sense organs, and has the aedeagus barely protrusible. All setae of tarsus I are thickened as spines (except for the setae *d*), and the solenidion φ of tibiae I and II is short and inconspicuous. The attachment organ of the hypopus is well developed and with all structures. The suckers are often present on coxae I and II. The legs III and IV are commonly directed anteriorly (Hughes, 1976).

Economic importance

This genus includes about 80 species, some of which have a certain economic importance, feeding on various bulbs and vegetable substrates (Hughes and Jackson, 1958; Qin and Rohitha, 1996) or developing on fungal fruiting bodies, where they feed on necrotic tissues and may spread pathogenic organisms in commercial mushroom houses (Hussey *et al.*, 1969; Hill and Deahl, 1978; Greenslade and Clift, 2004). The species of economic importance are *Histiostoma feroniarum* (Dufour), *H. sapromyzarum* (Dufour) and *H. heinemanni* Hill *et* Deahl. The mites have been collected on decaying food material such as grain or flour that was too wet (Hughes, 1976) and on edible fungi (Hussey *et al.*, 1969), and they are responsible for serious losses of crops of

Agaricus bisporus (Lange) Pilát, *A. bitorquis* (Quél.) Sacc. and *Lentinula edodes* (Berk.) Pegler (Greenslade and Clift, 2004). *H. sapromyzarum* has been recorded feeding on edible fungi in Europe (Boness, 1968), but there is no information available on damage in other countries, such as Australia, where the mite has been recorded but any damage that it causes is unknown (Greenslade and Clift, 2004). *H. heinemanni* develops high populations in commercial mushrooms and although not directly injurious, its presence is undesirable and it is difficult to remove before marketing (O'Connor, 2009).

Histiostoma feroniarum (Dufour) (Figs 12.1–12.3)

Common name

Unknown.

Diagnostic characteristics

FEMALE. The body is pear shaped, 525 μm long and 338 μm wide, whitish in colour and contains chalky crystals. The cheliceral digitus fixus is attenuate, with about 7–9 large spines and about six small teeth near the apex. The cheliceral setae are each shorter than the digitus fixus. A long solenidion and a short eupathidial seta are set on the terminal segment of the palpus. The subcapitulum has a pair of subcapitular setae. A sejugal furrow divides the dorsal idiosoma into the prodorsum and hysterosoma. The prodorsum has a slightly ornamented sclerite with a pattern of rectangular or irregular areas. All the prodorsal setae are filiform. The internal vertical setae (*vi*) are set at the anterior side of the prodorsum, and the external vertical setae (*ve*) in the lateral area of the sclerite. The scapular setae are set around the median area of the prodorsum, and the internal scapular setae (*sci*) are shorter than the external

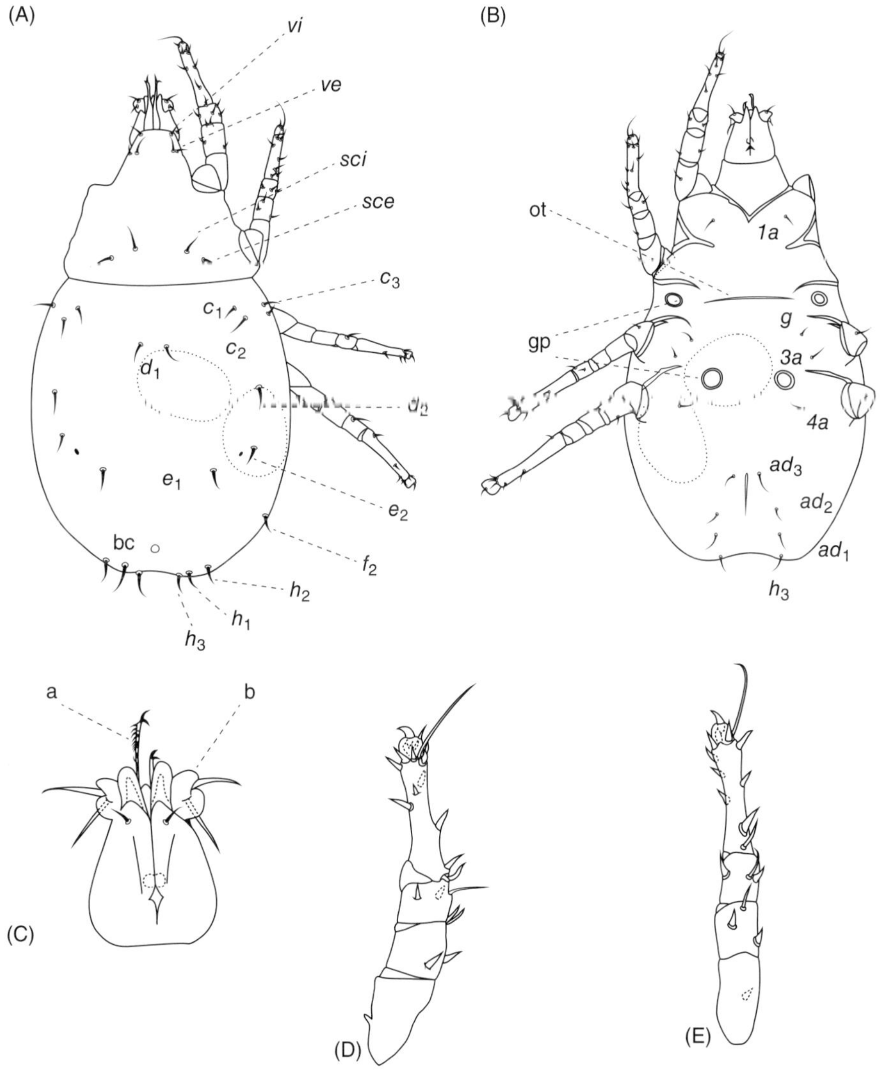

Fig. 12.1. *Histiostoma feroniarum* (Dufour). Female: (A) dorsal and (B) ventral view; (C) ventral view of gnathosoma; (D) legs I and (E) II (from Hughes, 1976). Key: a, chelicerae; b, modified palpus; bc, bursa copulatrix; gp, genital papilla; ot, transverse oviporus (genital opening). The rest of the labels designate setae, and an explanation of the setal notation is given in the text.

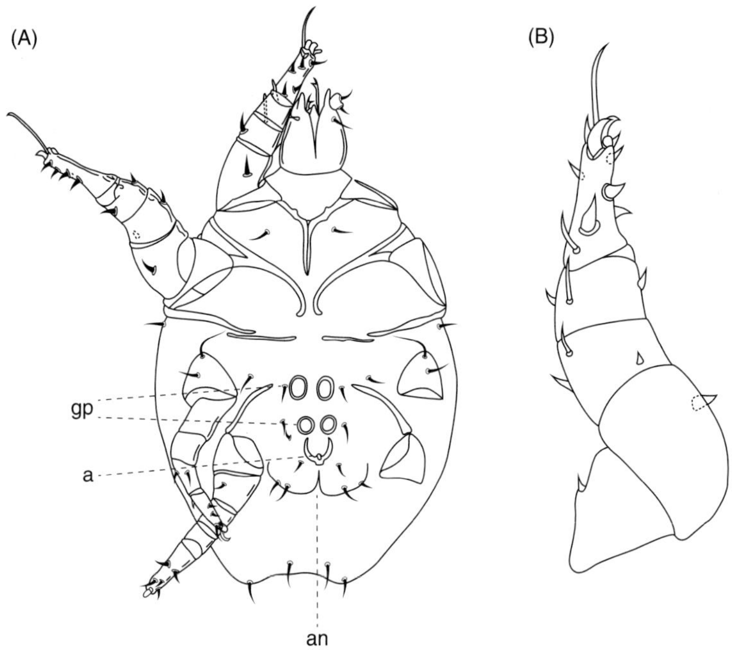

Fig. 12.2. *Histiostoma feroniarum* (Dufour). Male: (A) ventral view; (B) leg II (from Hughes, 1976). Key: a, aedeagus; an, anus; gp, genital papillae.

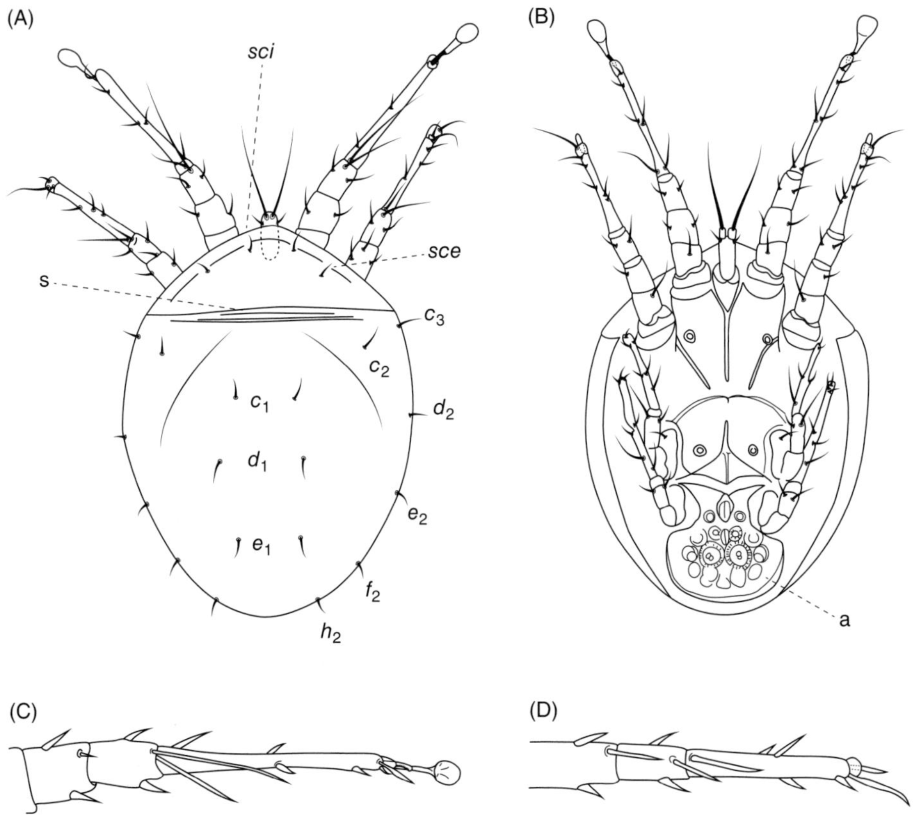

Fig. 12.3. *Histiostoma feroniarum* (Dufour). Hypopus: (A) dorsal and (B) ventral view; (C) leg I and (D) leg II (from Hughes, 1976). Key: a, attachment organ; s, sejugal furrow. The rest of the labels designate setae, and an explanation of the setal notation is given in the text.

scapular setae (*sce*). The hysterosoma has ten pairs of filiform dorsal setae (c_1, c_2, c_3, d_1, d_2, e_1, e_2, f_2, h_1, h_2), and the pair h_3 are inserted ventrally. The opisthonotal glands open between the setae e_1 and e_2. The bursa copulatrix is set between the setae h_1. The coxal apodemes are well developed, and the coxal apodemes 1 are fused medially in the form of a V. The genital papillae are circular or broadly oval, with the anterior pairs located between apodemes of coxae II and III, and the posterior pair between coxae IV. The genital opening is a transverse slit, set between the posterior apodemes 2. The anal opening consists of a longitudinal slit set in the posterior part of the venter. The coxal setae (*1a*, *3a*, *4a*) are hair like; the genital setae are set laterally to the genital opening, and three pairs of setae (ad_1, ad_2, ad_3) surround the anal opening. The leg setae are spine like except for one seta set on the apices of legs I and II which are slender and long. The claws are short and stout (Qin and Rohitha, 1996).

MALE. The body is elongate oval, 300 μm long and 135 μm wide. Its shape and the chaetotaxy of the gnathosoma and dorsal idiosoma are similar to those of female. The coxal apodemes are better developed than in the female. The coxal apodemes 1 are fused medially in form of a Y, the coxal apodemes 2 extending almost to the midline and directed posteriorly in the middle, and the coxal apodemes 3 extending almost to the midline but not directed posteriorly. The genital papillae are circular or broadly oval, with an anterior pair set between the coxal apodemes 4, and a posterior pair at the midcoxal field between coxae IV. The genital opening is set posteriorly between coxae IV. The anal opening is a longitudinal slit set posteriorly to genital opening. The coxal setae (*1a*, *3a*, *4a*) are hair like, and the genital setae (*g*) are posterolateral of the genital opening. Three pairs of setae (ad_1, ad_2, ad_3) are set laterally to the anal opening, of which ad_2 and ad_3 are set on a lobe. The legs are similar to those of the female but are shorter and stouter (Qin and Rohitha, 1996).

DEUTONYMPH. The body is 190 μm long and 136 μm wide. The gnathosomal base lacks setae. The palpi are short, each with a short basal seta and a long apical solenidion. A sejugal furrow divides the dorsal idiosoma into the prodorsum and dorsal hysterosoma. The prodorsum is triangular in outline, without sclerites. All prodorsal setae are short and hair like. The internal vertical setae (*vi*) and external vertical setae (*ve*) are absent, and the internal scapular setae (*sci*) and external setae (*sce*) are set on the anterolateral prodorsum. The hysterosomal setae are short and hair like. The opisthonotal gland and the bursa copulatrix are apparently absent. Ventrally, the coxal apodemes are well developed, and the coxal apodemes 1 are fused medially in form of a Y. The coxal setae (*1a*, *3a*, *4a*) are in the form of small conoids. The legs are slender, with fewer setae than in adults, with legs III and IV smaller than legs I and II (Qin and Rohitha, 1996).

Geographical distribution

H. feroniarum is found worldwide (Hughes, 1976).

Host plants and other biological hosts and substrates

Substrates include all kinds of decaying vegetable matter, wet grain and decaying mushrooms (Hughes, 1976), including on sewage bacterial beds, feeding on the zoogloecal film covering the clinker surface (Baker, 1964b). In the soil, the mites are found among the microfauna associated with gladiolus corms and feeding on diseased and decayed plant tissues (Poe, 1971). They are also found associated with crops of edible fungi (Hussey *et al.*, 1969), including mushrooms species such as *Agaricus bisporus* (J.E. Lange) Imbach, *A. bitorquis* (Quél.) Sacc. (Greenslade and Clift, 2004), *Lentinula edodes* (Berk.) Pegler (Kim and Hwang, 1996; Greenslade and Clift, 2004) and *Pleurotus ostreatus* (Jacq.) P. Kumm. (Kim and Hwang, 1996). All biological instars, including adults and hypopi (deutonymphs) have been collected on the weevil *Otiorrhynchus cribricollis* Gyll. on strawberry (Vacante, 1989b). *H. feroniarum* has also been found on tubers of *Sandersonia aurantiaca* (Hook.) (Qin and Rohitha, 1996) and in house dust (Learner and Chawner, 1998).

Bio-ecology

H. feroniarum reproduces by parthenogenesis and develops two types of females, one generating males and the other females (Hughes and Jackson, 1958). Scheucher (1957) reported that unfertilized eggs produced males. Between 20 and 25°C the mite reproduces in 2–4 days (Cooreman, 1944), but the optimum temperature is between 25 and 30°C and, under these conditions, the life cycle takes 3–3.5 days (Scheucher, 1957). The mite has been collected on all kinds of decaying vegetable matter, including wet grain and decaying mushrooms (Hughes, 1976). It is the most common mite to be found in sewage bacterial beds where it feeds on the zoogloecal film covering the clinker surface, helping to prevent the beds from becoming clogged by too much film (Baker, 1964b). *H. feroniarum* has also been collected in soil among microfauna living associated with gladiolus corms and feeding on diseased and decayed plant tissues. Populations develop in response to either an increase in disease or to precipitation, and infestations appear to have been sourced from outside the soil because the attack is initiated from above ground in the neck region of the host before proceeding to the corm (Poe, 1971). *H. feroniarum* is also recorded on edible fungi (Hussey *et al.*, 1969) and is responsible for serious crop losses to various mushrooms species, including *A. bisporus*, *A. bitorquis* (Greenslade and Clift, 2004), *L. edodes* (Kim and Hwang, 1996; Greenslade and Clift, 2004), and *P. ostreatus* (Kim and Hwang, 1996). Adults, juvenile stages and hypopi have been collected in association with the weevil *Otiorrhynchus cribricollis* Gyll. on strawberry cultivated in the greenhouse in Sicily (Vacante, 1989b). The mite has been collected on tubers of *Sandersonia aurantiaca* in New Zealand, but the plants were grown under strictly clean conditions, and the exact mode of entry of the mites into the *Sandersonia* cultures was not clear; it may have been by phoresy on insects, from the sphagnum moss commonly employed for packing the tubers, or from the potting medium (Qin and Rohitha, 1996). The mite has also been found in house dust (Learner and Chawner, 1998). The hypopi attach to many species of arthropods and are particularly plentiful on

filter beds where the zoogloecal film has been allowed to dry. Moulting occurs in 2–3 days when the hypopi come in contact with a film of water (Baker, 1964a; Learner and Chawner, 1998).

Symptomatology and damage

As already mentioned, *H. feroniarum* has been reported in mushroom production areas (Hussey *et al.*, 1969). In Korea, it caused severe damage to the lamellae of the shiitake mushroom, *L. edodes*, and the oyster mushroom, *P. ostreatus* (Kim and Hwang, 1996), and in Australia produced serious crop losses to *A. bisporus*, *A. bitorquis* and *L. edodes*, especially if associated with the red pepper mite (*Pygmephorus*) (Greenslade and Clift, 2004). The histiostomatid is commonly collected in baled straw and can survive in compost during pasteurization, and its presence on the compost is not selective for mushroom mycelium (Clift and Terras, 1994). The mite is a vector of disease organisms and also a pest of garlic (Learner and Chawner, 1998). Its populations probably develop in abundance on mushrooms if they are stimulated by favourable ecological conditions (temperature, relative humidity, trophic resources, etc.). It is probable that the species feeds on decaying parts of mushrooms as a result of its own attack and that of other organisms (pathogens, insects, etc.). The presence of *H. feroniarum* is feared owing to the risk of disease transmission to the trouble that its populations can cause at the commercial level.

Tuber rot of *S. aurantiaca* represents an economic problem in New Zealand because of the importance of this species in the cut flower and tuber export industry. The cause of this tuber rot, with which *H. feroniarum* has a strong association, is under research (Qin and Rohitha, 1996).

Control

The control of *H. feroniarum* is based on the application of preventive and curative methods (physical and chemical). These are discussed below for mite problems in cultivated mushrooms.

PREVENTION. Good hygiene should be maintained in mushroom houses, using appropriate air filters. Flies should be controlled, because they can be vectors of mites. Control of weed moulds is also important, because they are a preferred food for mites (EPPO, 2000).

PHYSICAL CONTROL. Pasteurization at 58–60°C for 6–8 h should suppress mites in compost. The casings should be disinfected at 65–70°C for 3 h. If mites occur despite these measures, chemical control can be applied (EPPO, 2000).

CHEMICAL CONTROL. Dichlorvos fogs may be applied between flushes. From spawning until 10 days after casing, dicofol may be used (EPPO, 2000). In addition to dicofol, other acaricides, such as abamectin and hexythiazox, have produced significant reduction of tarsonemid populations on mushrooms (Babar *et al.*, 2005) and can be applied also in the control of *H. feroniarum*.

Histiostoma sapromyzarum **(Dufour) (Fig. 12.4)**

Common name

Unknown.

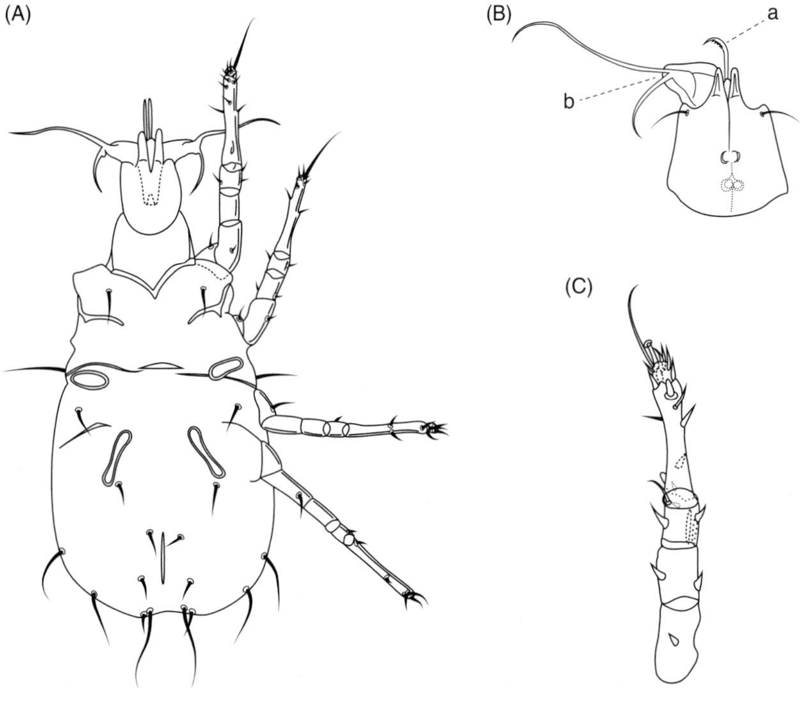

Fig. 12.4. *Histiostoma sapromyzarum* (Dufour). Female: (A) ventral view; (B) ventral view of gnathosoma; (C) leg I (from Hughes, 1976). Key: a, chelicerae; b, modified palpus.

Diagnostic characteristics

FEMALE. The general morphological features of *H. sapromyzarum* are similar to those of *H. feroniarum*. The idiosoma is 300–650 μm long. The palpus has the distal tip flattened and entire, not bilobed, and one of the setae arising from this lobe is more than twice as long as the other. The ventral body surface bears two pairs of chitinous rings, oval and constricted in the middle, resembling the sole of a shoe. The legs are more slender that in *H. feroniarum*, and the chaetotaxy of leg I is similar in both species (Hughes, 1976).

MALE. The male is similar to the female. The chitinous rings are kidney shaped, the convex side directed externally (Hughes, 1976).

DEUTONYMPH. The deutonymph closely resembles that of *H. feroniarum* (Hughes, 1976).

Geographical distribution

H. sapromyzarum is found worldwide (Hughes, 1976).

Host plants and other biological hosts and substrates

Hosts/substrates include decaying fungi (*Lactarius*, *Russula*, *Tricholoma*, *Amanita*, *Scleroderma*) (Hughes, 1976), cultivated mushrooms (Boness, 1968) and decaying hyacinth bulbs and damp timber.

Bio-ecology

The bio-ecology of this species has not been much investigated, but is probably similar to that of *H. feroniarum*. The hypopus is carried by several beetle species, flies and myriapods (Hughes, 1976).

Symptomatology and damage

The damage caused by *H. sapromyzarum* to mushrooms or to plants of economic importance is not well defined, but is probably similar to that caused by *H. feroniarum*.

Control

Currently, control of this species is unnecessary, but if it becomes necessary, it will be possible to apply the same protocol as for *H. feroniarum*.

Histiostoma heinemanni Hill *et* Deahl (Figs 12.5–12.7)

Common name

Unknown.

Diagnostic characteristics

FEMALE. The idiosoma is pyriform, 279–467 μm long and 147–267 μm wide. A sejugal furrow dorsally divides the prodorsum from the hysterosoma. The anterior part of the prodorsum is sculptured and the prodorsal setae are subequal (i.e. not quite equal) in length. The internal vertical setae (*vi*) are not enlarged

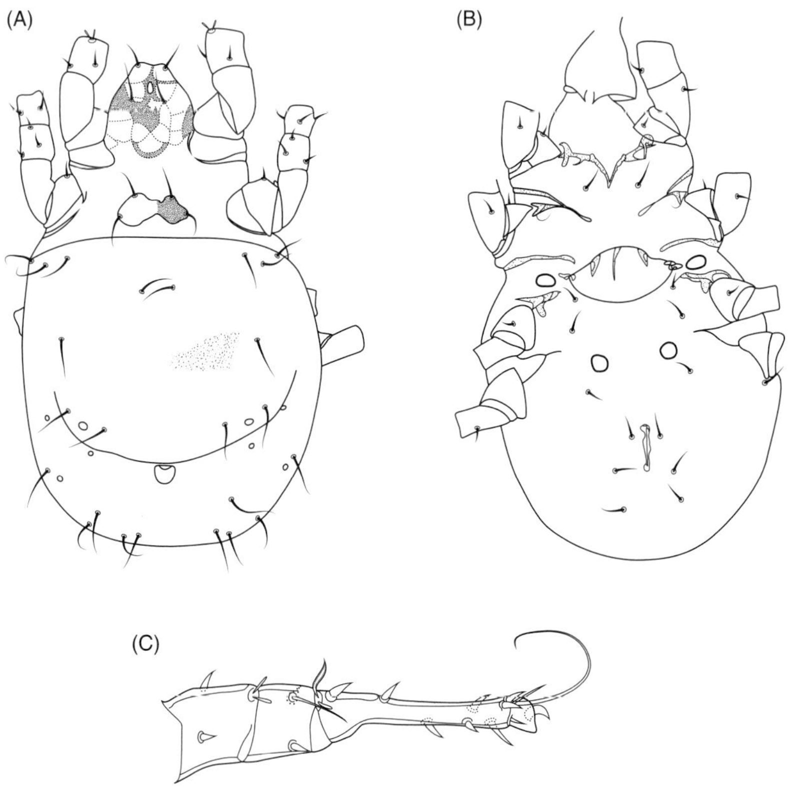

Fig. 12.5. *Histiostoma heinemanni* Hill *et* Dehal. Female: (A) dorsal and (B) ventral view; (C) leg I (from Hill and Dehal, 1978).

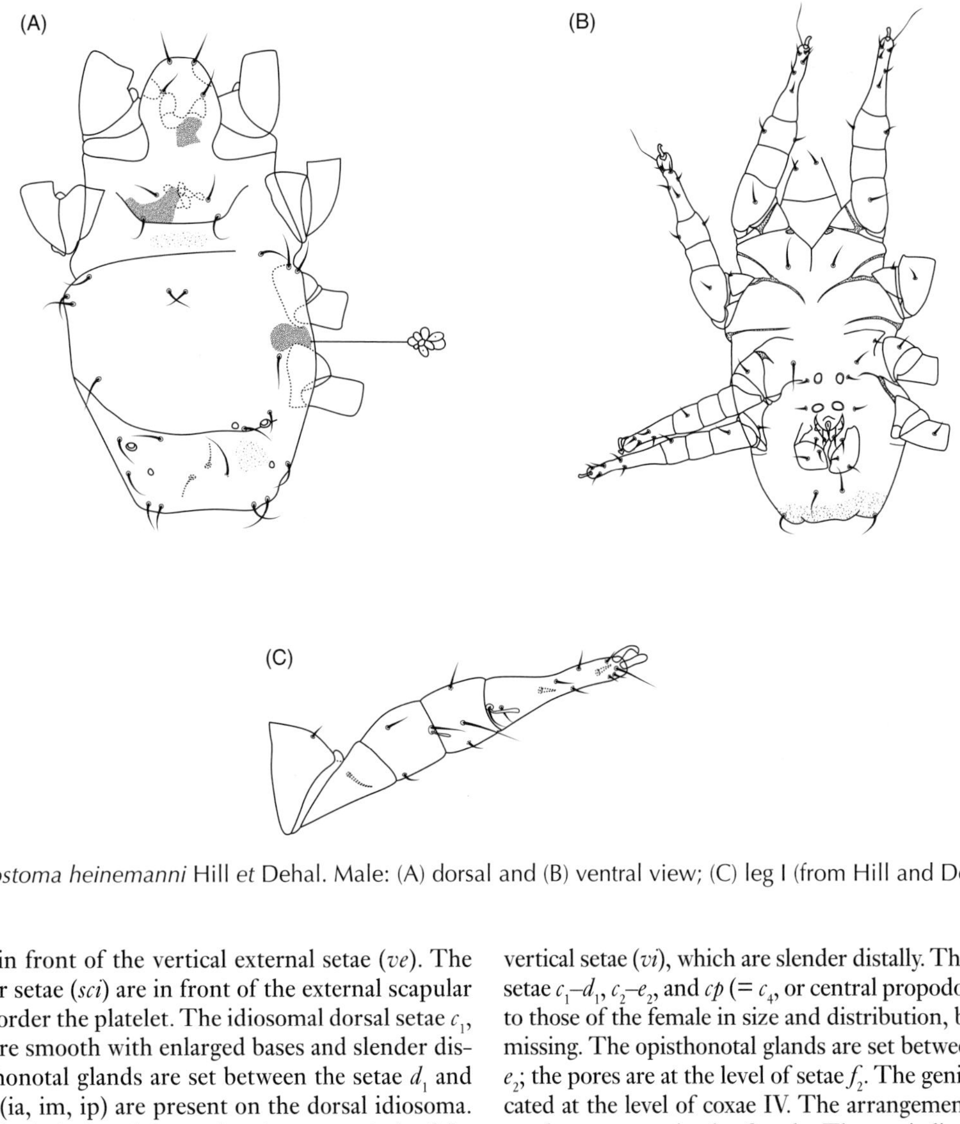

Fig. 12.6. *Histiostoma heinemanni* Hill *et* Dehal. Male: (A) dorsal and (B) ventral view; (C) leg I (from Hill and Dehal, 1978).

basally and are in front of the vertical external setae (*ve*). The internal scapular setae (*sci*) are in front of the external scapular setae (*sce*) and border the platelet. The idiosomal dorsal setae c_1, h_2, c_2, h_3 and c_3 are smooth with enlarged bases and slender distally. The opisthonotal glands are set between the setae d_1 and e_2. Three pores (ia, im, ip) are present on the dorsal idiosoma. The bursa copulatrix is dorsal, located at about one-sixth of the length of the hysterosoma from the caudal end. The coxal apodemes 1 are 'Y'-shaped. The transverse genital opening is located between coxae II and III. Two pairs of almost circular genital papillae, the anterior pair set between coxae II and III, and the posterior pair at the level of coxae IV. Three pairs of coxal setae (*1a*, *3a*, *4a*) on the coxal fields, and three pairs of setae (ad_1, ad_2, ad_3) are set laterally to the anal opening. The legs have five free segments, with the tarsi relatively long, and a stout claw originating from the short pretarsus. All the setae are spine like (Hill and Deahl, 1978).

MALE. Two types of males are known. Homeomorphic males have normal-sized spines on tarsus II, whereas heteromorphic males bear two enlarged spines on tarsus II. The idiosoma of the heteromorphic males is 255–319 μm long and 128–160 μm wide. The idiosoma of the homeomorphic males is 230–287 μm long and 121–134 μm wide. The cuticle of the idiosoma is covered with small projections. The anterior part of the idiosoma is sculptured. The internal vertical setae (*vi*) are in front of external vertical setae (*ve*). The sculptured platelet is bordered by internal scapular setae (*sci*) and external scapular setae (*sce*). The prodorsal setae are subequal in length, and enlarged basally except for the internal vertical setae (*vi*), which are slender distally. The dorsal idiosomal setae c_1–d_1, c_2–e_2, and *cp* (= c_4, or central propodosomal) are similar to those of the female in size and distribution, but the setae h_1 are missing. The opisthonotal glands are set between the setae e_1 and e_2; the pores are at the level of setae f_2. The genital papillae are located at the level of coxae IV. The arrangement of the setae and apodemes are as in the female. The genitalia are located about one-sixth of the length of the idiosoma from the caudal end; there are two pairs of internal setae, the shorter ones located more dorsally, and four pairs of anal setae. The tarsi are broader and shorter than in the female. Chaetotaxy and solenidiotaxy are as in the female, except for tarsus I, which bears ten setae and two solenidia (Hill and Deahl, 1978).

DEUTONYMPH. The idiosoma is 103–217 μm long and 74–160 μm wide. The gnathosoma is elongate. One pair of flagella is longer than the gnathosoma; an anterolateral pair of short setae is present. The prodorsum is broad, and the anterior part has an interrupted furrow. The internal vertical setae (*vi*) are in front of the external vertical setae (*ve*). The transverse furrow is complete, posteriorly with a broken-line area. The idiosoma is oval shaped, with the posterior broadly rounded and with a smooth surface. The dorsal setae are short, setiform and subequal in length. The apodemes 1 are fused medially, and the sternum is free posteriorly. The apodemes 2 are connected with apodemes 3 by a fine sclerotized extension from the latter, although the apodemes 3 are free; the posterior and anterior of the posterior sternum are free. The coxal setae (*1a*, *3a*, *4a*) are in the form of small conoids (Hill and Deahl, 1978).

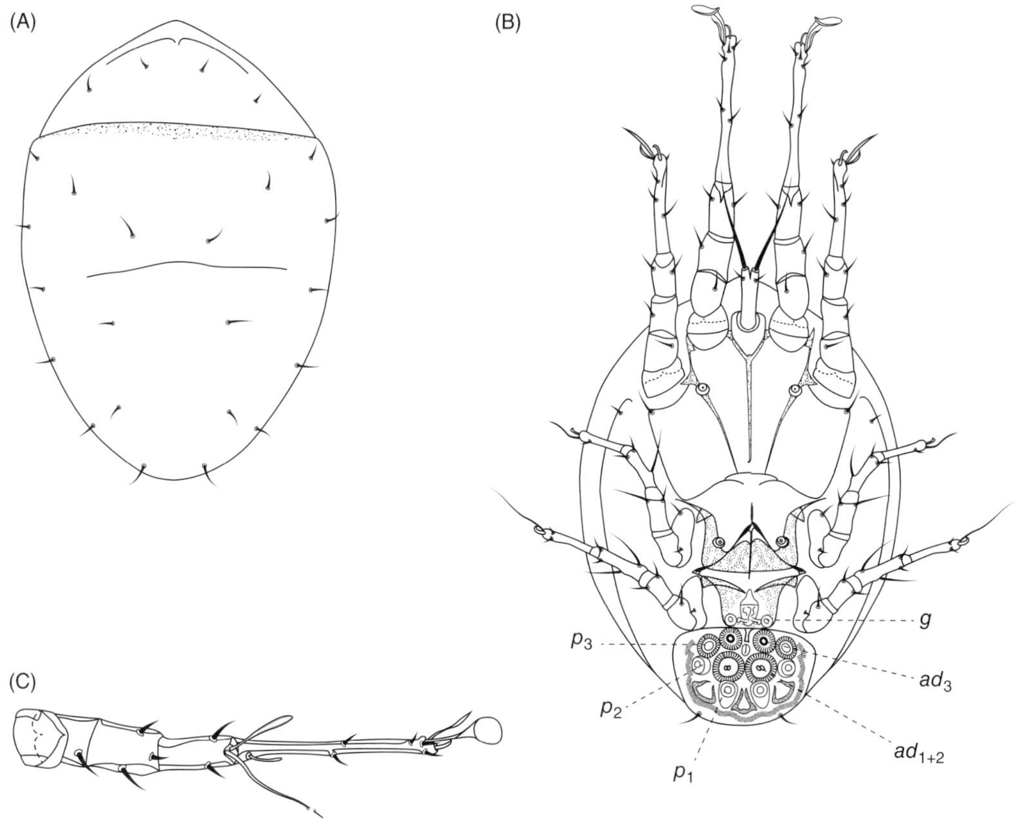

Fig. 12.7. *Histiostoma heinemanni* Hill *et* Dehal. Hypopus: (A) dorsal and (B) ventral view; (C) leg I (from Hill and Dehal, 1978). The labels designate setae, and an explanation of the setal notation is given in the text.

Geographical distribution

H. heinemanni is found in the states of Pennsylvania and Maryland in the USA (Hill and Deahl, 1978).

Host plants and other biological hosts and substrates

The species is found in commercial mushroom beds and compost piles (Hill and Deahl, 1978).

Bio-ecology

H. heinemanni has been recorded in commercial mushroom beds and compost piles in Pennsylvania and Maryland. The mite could reproduce by arrhenotokous parthenogenesis, developing a life cycle very similar to that of *H. feroniarum*, except for the presence of both homeomorphic and heteromorphic males (Hill and Deahl, 1978) (only heteromorphic males are known in *H. feroniarum*) (Scheucher, 1957). Hill and Deahl (1978) collected many specimens of *H. heinemanni* in mushroom houses that were decaying as a result of infection by *Verticillium malthousei* Ware, *Pseudomonas tolaasii* Paine and various secondary saprophytic invaders. In rearing experiments carried out in the laboratory on potato dextrose and some decaying potato put in Petri dishes and maintained at 27°C, some hypopi entered the quiescent stage within a few hours of being transferred into rearing dishes. Tritonymphs occurred on the second day and the first eggs were observed on the fourth day. The females select dry and compacted material on which to lay the eggs, either singly or in small groups. The first larvae occur on the fifth day and the first protonymphs on the sixth day. When there is no overpopulation or food depletion, the hypopus stage is omitted and the first tritonymphs occur on the seventh day. Increasing numbers of hypopi have been found in dishes kept for 4–6 weeks (Hill and Deahl, 1978).

Symptomatology and damage

H. heinemanni apparently does not feed on mushroom tissue but is probably attracted by the presence of the microorganisms infecting the mushrooms. However, by feeding on the microorganisms, the mite may be important in the spread of diseases (Hill and Deahl, 1978). High populations develop in commercial mushroom beds and although they do not cause direct damage, their presence is undesirable, and is difficult to remove before marketing (O'Connor, 2009).

Control

See *H. feroniarum*.

13 The Family Acaridae Latreille

Morphological Characteristics, Systematics and Bio-ecology

The body of the members of the Acaridae is soft, of medium length, and pale whitish to brownish coloured. Some species are weakly pigmented. The integument is smooth, rough or finely furrowed; sometimes completely sclerotized or ornamented with cuticular tubercles. A sejugal furrow separates the idiosoma into the propodosoma and hysterosoma (Fig. 13.1A).

The dorsal surface of the gnathosoma is not extended and forms a small lobe between the cheliceral bases. The chelicerae are chelate and dorsally free, laterally compressed and with a large basal region stretching forward and forming the digitus fixus articulated with the digitus mobilis; at the base of this articulation there is a small cheliceral seta (*cha*) (Fig. 13.1C). Both chelicerae move in a vertical plane and are often toothed distally. The chelicerae are located above the labrum or true epistome, and partially involved in the mouth structure. The labrum is grooved and extends posteriorly and its side walls are related to the ventral area of the gnathosoma; on it are inserted the pharynx dilator muscles. The ventral gnathosoma, called the hypostome or infracapitulum, originates mainly from the fusion of the palpcoxae; it has two whip-like subcapitular ventral setae (*m*) and extends anteriorly into two inner lobes or malae. The palpi have two freely articulating segments (articles); each coxa bears dorsally a palpal supracoxal seta (*elcp*) (Fig. 13.1D). The first free palpal segment has two simple setae and the distal segment bears one simple seta, a solenidion and a button-like ventral eupathidium. Several species have a Grandjean's organ – a thin cuticular plate embracing the base of gnathosoma – on the anterolateral propodosoma. From the base of this organ, a lateral sclerite extends around the base of leg I, and at the posterior end of this sclerite is set the supracoxal seta (*scx*) (Fig. 13.1A and E). In several species, there is a crack or pore between the lateral sclerite and the supracoxal seta, the opening of which is similar to those of the opisthonotal glands (Grandjean, 1939b; Griffiths, 1970; Hughes, 1976).

The prodorsum usually bears a prodorsal sclerite or shield that is variously sized, shaped and structured. The idiosomal setae may vary in number, size and shape, but are distributed according to the same pattern for the group. The prodorsum bears four pairs of prodorsal setae (*vi*, *ve*, *sci*, *sce*); the inner vertical setae (*vi*) are medioanterior and the external vertical setae (*ve*) are lateral; the scapular setae (*sci*, *sce*) are posterior. The hysterosoma has 12 pairs of setae (c_1, c_2, c_p (third pair of setae in first series or row), c_3 (fourth pair of setae in first series or row), d_1, d_2, e_1, e_2, f_2, h_1, h_2, h_3), four pairs of lyrifissures (ia, im, ip, ih) and one pair of opisthonotal glands. The gland opening (gla) is commonly posteriorad (towards the rear) of d_2 (Figs 13.1A and B) (Hughes, 1976; Griffiths *et al.*, 1990).

Ventrally, the coxae are fused with ventral idiosoma and each is bounded by the sclerotized apodeme. The medial apodemes of coxae I are fused together along the midline and forming a sternum. Coxa I has one seta (*1a*), coxa III two setae (*3a*, *3b*) and coxa IV one seta (*4a*). The genital opening is located between coxae III and IV, with two pairs of genital papillae covered by the genital valves and 1–3 pairs of genital setae (g_1–g_3). Genital folds are commonly present. The anal opening is posteriorad of the genital opening, and has 1–6 pairs of setae (ad_3, ps_3, ps_2, ad_2, ps_1, ad_1) (Fig. 13.1B). The copulatory opening is posteriorad of the anal opening, connecting a small shield, and leads by a thin or progressively broadened spermathecal duct into the sac-like inner part of the spermatheca or receptaculum seminis. The base of the inner part of the spermatheca is often sclerotized at its tips with a pair of 'Y'-, 'V'- or 'U'-shaped sclerites of the oviducts. The male has its aedeagus and anal suckers lateral to the anal opening or on tarsi IV (Hughes, 1976; Griffiths *et al.*, 1990; Klimov and O'Connor, 2003; Fan and Zhang, 2004).

The legs are six-segmented, with the last five articles free and the following chaetotaxy: coxae: 1, 0, 2, 1; trochanters: 1, 1, 1, 0; femora: 1, 1, 0, 1; genua: 2 + 2 solenidia (σ_1 and σ_2), 2 + 1 solenidion (σ), 1 + 1 solenidion (σ), 0; tibiae: 2 + 1 solenidion (φ), 2 + 1 solenidion (φ), 1 + 1 solenidion (φ), 1 + 1 solenidion (φ); tarsi: 12 + 3 solenidia (ω_{1-3}) + 1 famulus (ε), 12 + 1 solenidion (ω), various numbers, various numbers. Solenidia ω_{1-3} of tarsus I are set in distal, median and proximal positions, respectively. On the tibiae, the solenidion φ is dorsal, as well as the solenidia σ on genua I–III (Fig. 13.2C–E) (Grandjean, 1939b; Griffiths, 1970).

The sexes commonly differ in the genital regions, but have minimal sexual dimorphism, although they are variable between species and commonly characterized by the males, which in some species are homeomorphic and heteromorphic with respect to the females (Michael, 1903; Türk and Türk, 1957; Woodring, 1969a). This adaptation is called 'andropolymorphism', in which there are different body shapes, setal lengths and shapes of leg III, and there are four types of males: 'omotype homeomorphic', 'heteromorphic', 'bimotype bimorphic' and 'pleomorphic' (Woodring, 1969a). In the genus *Rhizoglyphus*, the adult homeomorphic males are similar to the adult females, from which they differ in a few characters. They have the genital opening set between coxae IV, have an aedeagus, one pair of anal suckers and three pairs of pseudanal setae (ps_3, ps_2, ps_1); moreover, they lack adanal setae ad_1, ad_2 and ad_3, the solenidion φ on tibia IV is spine like and tarsus IV bears two suckers. The leg chaetotaxy is as in the adult female excepting tarsi IV. The

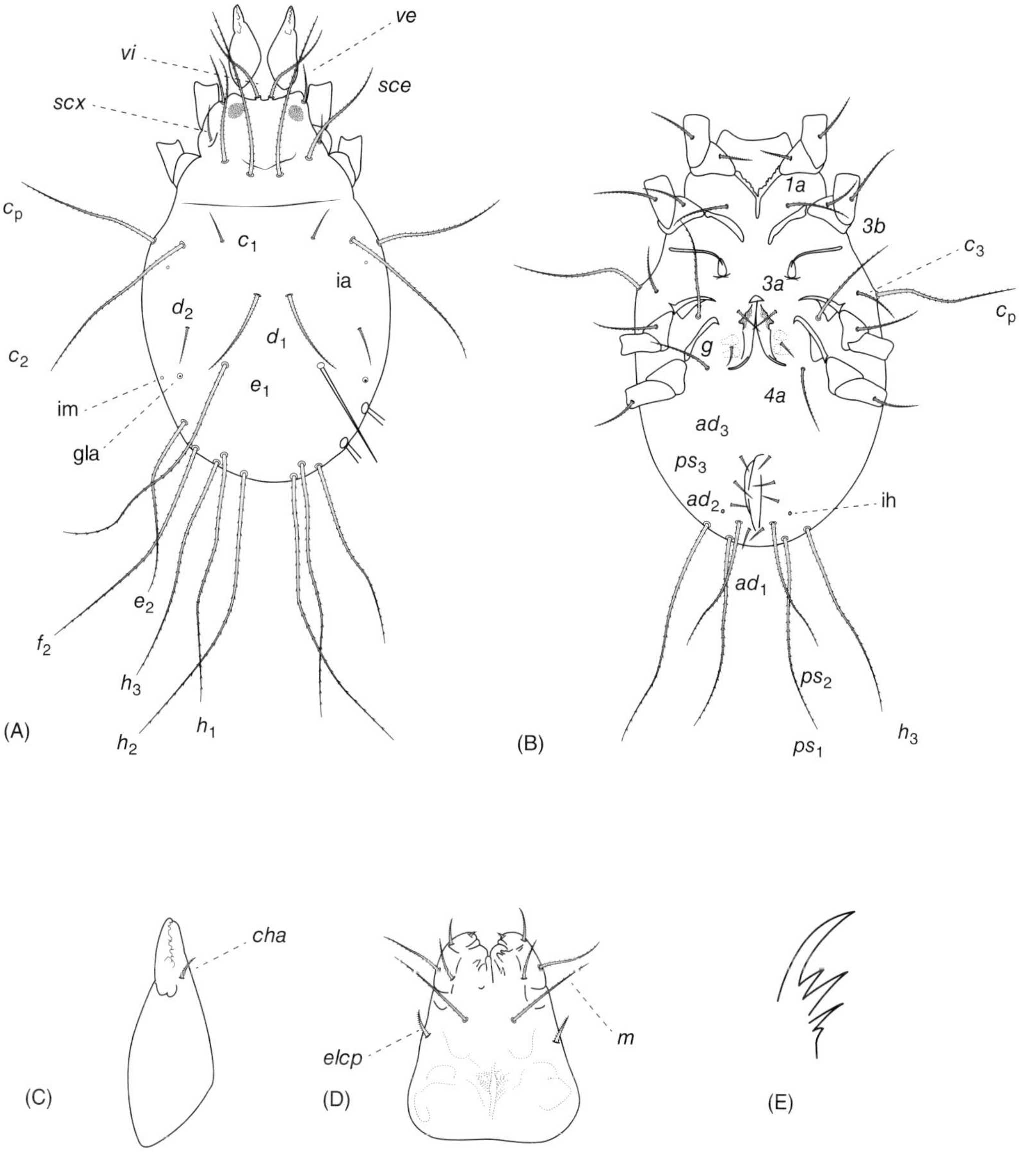

Fig. 13.1. *Tyrophagus communis* Fan *et* Zhang. Female: (A) dorsal and (B) ventral view; (C) chelicerae; (D) subcapitulum; (E) Grandjean's organ (from Fan and Zhang, 2007). Key: gla, gland opening; ia, ih and im, lyrifissures. The rest of the labels are setal notations, which are explained in the text.

adult heteromorphic male is similar to the adult homeomorphic male but both sides or one side of leg III are about doubled in size. The body setae are slightly longer. The leg chaetotaxy is as in adult homeomorphic male excepting tarsi III (Klimov and O'Connor, 2003; Fan and Zhang, 2004).

The life cycle develops through the egg, a six-legged larva, protonymph and/or tritonymph stages. In some species, the protonymph may develop into the deutonymph or hypopus, a heteromorphic stage which does not eat and allows the survival and dispersal of the species. Several *Tyrophagus* spp. have abandoned deutonymphal dispersal and colonize many habitats. Other genera, e.g. *Rhizoglyphus*, have retained the hypopus stage.

The Acaridae number over 541 species and 88 genera (Schatz *et al.*, 2011). They are widely spread in various habitats and play various ecological roles – as scavengers, coprophages, micophages or phytophages (Zakhvatkin, 1941; Johnston and Bruce, 1965; Hughes, 1976; O'Connor, 2009). Some species are injurious to ornamentals and vegetables or are associated with mushrooms. Seven of these species are included in the genus *Tyrophagus* Oudemans: *T. communis* Fan *et* Zhang, *T. curvipenis* Fain *et* Fauvel, *T. longior* (Gervais), *T. neiswanderi* Johnston *et* Bruce, *T. perniciosus* Zakhvatkin, *T. putrescentiae* (Schrank) and *T. similis* Volgin. Five other species are included in the genera *Mycetoglyphus* Oudemans (*M. fungivorus* (Oudemans)), *Rhizoglyphus* Claparéde (*R. robini* (Claparède),

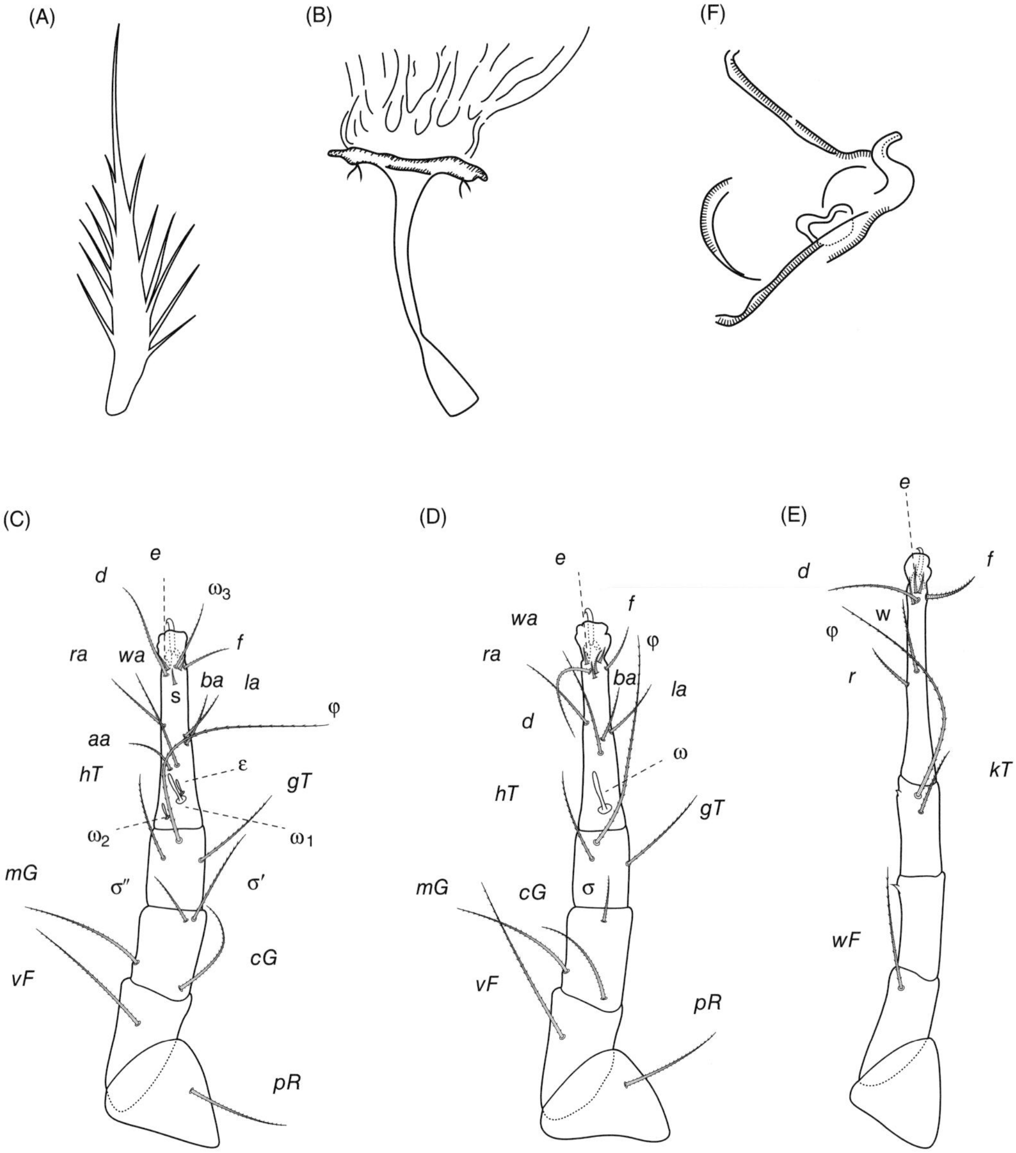

Fig. 13.2. *Tyrophagus communis* Fan *et* Zhang. Female: (A) supracoxal seta; (B) copulatory opening and spermatheca; (C) leg I, (D) leg II and (E) leg IV. Male: (F) aedeagus (from Fan and Zhang, 2007). The labels are all setal notations, which are explained in the text.

R. echinopus (Fumoze *et* Robin), *R. setosus* Manson) and *Sancassania* Oudemans (*S. berlesei* (Michael)). *M. fungivorus* in particular is injurious to various plants. *R. robini*, *R. echinopus* and *R. setosus* damage the underground organs of various plants in the field and store (Ho and Chen, 1987; Zhang, 2003; Fan and Zang, 2004). *S. berlesei* is reported from stored food and other substrates (Hughes, 1976), as well as on edible fungi in various countries (Zakhvatkin, 1941; Hussey *et al.*, 1969; Clancy, 1981; Greenslade and Clift, 2004); it has also been collected on melon in greenhouses (Laffi and Ponti, 1997) and on fruiting bodies and hyphae of *Lentinus edodes* (Berk.) Singer [=*Lentinula edodes* (Berk.) Pegler)] in Japan along with the acarid *Histiogaster rotundus* Woodring (Okabe, 2006), but there is no information available on the damage done. Some (and other Astigmatina) cause human allergies (Hage-Hamsten *et al.*, 1985; Terho *et al.*, 1985). In addition to direct damage to crops, unidentified *Tyrophagus* spp. (or other acarid mites) can pose access problems to markets for horticultural crops and stored foodstuffs when intercepted by an importing country. Bass and Hays (1976) observed that *T. putrescentiae* may feed in the laboratory on eggs of the fire ant and on organic debris.

The key to the genera described in this chapter that is presented below partly follows that of O'Connor (1998).

Selected websites

Images for Acaridae. Available at: https://www.google.it/#q=acaridae+pictures

Spinnen Forum Wiki. Available at: http://wiki.eu-arachnida.de/index.php?title=Tyrophagus

Key for the Identification of Genera

1. • External vertical setae (*ve*) set on the same transverse line as internal vertical setae (*vi*) or slightly posterior; hysterosoma with 12 pairs of setae, some of which may be elongate and finely barbed..2
   • External vertical setae (*ve*) set approximately midway between internal vertical setae (*vi*) and scapular setae; hysterosoma often with fewer than 12 pairs of setae; setae rarely elongate and barbed...3
2. • External vertical setae (*ve*) less than 1/3 the length of internal vertical setae (*vi*); male aedeagus long and straight; living in decaying materials, sometimes in stored products... ....genus *Mycetoglyphus* Oudemans
   • External vertical setae (*ve*) more than 1/3 the length of internal vertical setae (*vi*); male aedeagus usually short and often curved; tarsi I–II more than twice as long as the basal width; proral setae thinner than unguinal setae and similar in length; Grandjean's organ finger like; in a variety of field and storage habitats.............................genus *Tyrophagus* Oudemans
3. • External vertical setae (*ve*) lacking, or if present (some species), then less than 10 μm long; hysterosoma with maximum complement of dorsal setae (12 pairs); internal scapular setae (*sci*) present although usually short (absent in one species); tarsi I–II with seta *fi'* spine like and closely associated with ω_1; in soil, decaying vegetable material, fungi or in intact bulbs and corms..................................... ....genus *Rhizoglyphus* Claparède
   • External vertical setae (*ve*) present on lateral sides of prodorsal shield approximately halfway between internal vertical setae (*vi*) and scapular setae, length of *ve* setae greater than 10 μm; tarsi I–II each with seta *fi'* filiform, not associated with solenidion ω_1; female with copulatory opening separated from posterior end of anus by a distance greater than half the length of the anus; in decaying organic matter or often associated with scarabaeoid beetles......................genus *Sancassania* Oudemans

The Genus *Tyrophagus* Oudemans

Diagnostic characteristics

The idiosoma is oblong, 300–600 μm long, and whitish to semi-transparent in colour. The prodorsum has four pairs of setae and the hysterosoma 12 pairs of setae; all dorsal setae are barbed. The external vertical setae (*ve*) are long and set on anterior lateral sides of the prodorsal shield. The internal scapular setae (*sci*) are longer than the external scapular setae (*sce*). The hysterosomal setae c_1 and d_1 or d_2 are shorter than the distance to the setae in the next row. The supracoxal setae (*scx*) are commonly pectinate, rarely smooth. The distal end of the Grandjean's organ is finger like. The legs are light brown, with tarsi I–IV slender, more than twice as long as wide. The dorsal distal spine *e* on tarsi I–II is slender. The proral setae on tarsi I–II are thin, longer than unguinal setae. Tarsus I with two long distal setae and one distal solenidion (ω_3); tarsi II–III each with two long distal setae; tarsus IV with two long distal setae in the female (*d*, *f*) and one long distal seta in the male (Fan and Zhang, 2007).

Economic importance

Tyrophagus spp. are the most abundant and economically important mites of the family Acaridae inhabiting stored food and products (Hughes, 1976). Some species are injurious to economic plants, ornamentals and vegetables grown both in the open field and in greenhouses. The most injurious species are *T. neiswanderi*, *T. perniciosus*, *T. similis* and *T. longior*. Another important species is the mould mite, *T. putrescentiae* (Schrank), but systematic implications make it difficult to report its biological features and aspects involving its control. Damage to cucumber leaves by *T. cucumeris* Karg has been reported in glasshouses in Germany (Karg, 1970).

Tyrophagus communis Fan et Zhang (Figs 13.1–13.2)

Common name

Unknown.

Diagnostic characteristics

FEMALE. The idiosoma is 432 μm long and 285 μm wide. The prodorsal shield has eyespots, nearly pentagonal with apex towards the rear margin. Grandjean's organ is finger like, with the basal lobe ornamented with small spiniform teeth. The setae *scx* expand from the base to pectinations, and the shaft bears 8–14 pectinations of various length. The setae d_1 are long – about 2.7 times longer than the setae c_1 and 2.4 times longer than the setae d_2; the setae d_2 are about 1.1 times longer than the setae c_1. Coxal plates I are beyond the apex of the prosternal apodeme; coxal plates II are broadly triangular, with a very slightly concave posterior margin, and the distal point of the triangle stands slightly proud of the apodeme apex. The spermathecal duct narrows rapidly from the copulatory opening for a distance about 1.2 times the distance between the sclerites of the oviducts; it forms a neck and then gradually widens to the base of the spermathecal sac over a distance about 1.5 times the distance between the sclerites of the oviducts. The sclerotized base of the spermathecal sac is nearly flat or bends slightly forwards. The solenidion ω_1 of tarsus I is slightly widened at the apex, and the solenidion ω of tarsus II is also slightly widened at the apex; the setae *w* and *r* of tarsus IV are setiform (Fan and Zhang, 2007).

MALE. The idiosoma is 429 μm long and 280 μm wide. The prodorsal shield and Grandjean's organ are as in the female. The eyespots, coxal plates I and II, solenidia of legs I (ω_1) and

II (ω) are as in the female; the setae d_1 are long, about 2.8 times as long as the setae c_1 and 2.3 times as long, as the setae d_2; the setae d_2 are about 1.2 times longer than the setae c_1. The aedeagus is 'S'-shaped, with the distal one-third reversely curved and tapering from base to apex, and the lateral arms turning outwards. The setae *w* and *r* of tarsus IV are setiform (Fan and Zhang, 2007).

Geographical distribution

T. communis is found worldwide. In particular, the mite is known in Africa, Argentina, Australia, Brazil, Chile, China, (mainland, Hong Kong), Taiwan, Cook Islands, Crete, Ecuador, Fiji, Greece, Germany, India, Indonesia, Italy, Jamaica, Japan, Madagascar, Malta, the Netherlands, New Zealand, Papua New Guinea, Philippines, Samoa, Singapore, Solomon Islands, Spain, Thailand, Tokelau Islands, Tonga, Turkey, the UK, the USA, Vanuatu and West Africa (Fan and Zhang, 2007).

Host plants and other biological hosts and substrates

The list of plants or vegetable products hosting this mite includes agar tissue cultures of orchids, almonds, asparagus, aubergines, bananas, bark scrapings, beans in the kitchen cupboard, buffet, *Caladium*, the calyx end of oranges, caper seeds, capsicum, chaff, water vine (*Cissus antarctica* Vent.) seeds, citrus, lemon (*Citrus limon* (L.) Burm.), cocoa beans, coconuts, Arabian coffee (*Coffea arabica* L.), elephant ear (*Colocasia esculenta* (L.) Schott) bulbs, copra, *Cryptes baccarum* [?], cucumber leaf, cultures of *Tribolium*, curry leaves, cycad palm seed, *Dahlia* bulbs, damp grain, debris in containers, decaying coconut, dormant bulbs, dried figs, dried food, dried fruit, dried lychees, dried vegetable leaves, dust in grain silos, feijoas, figs, foodstuffs in household pantries, *Freesia* bulbs, garlic, gladioli bulbs, grain, groundnuts hyacinth, *Iris* bulbs, kangaroo diet, kentia palm, lemon, bay laurel (*Laurus nobilis* L.) leaves, lychees, matua on nutmeg [?], mixed grain horse fodder and young stock foods, maize, molactrate molasses blocks, mould in hazel nuts, muscari bulbs, *Narcissus* bulbs, *Nipa* palm leaves, onions, orange, orchid plants, Orchidaceae, palm seeds, persimmons, pimentos, pine tree seeds, pineapple, plant roots, *Plumeria* sp., potato, pollen from hives, *Ranunculus* bulbs, raspberry jam, red dates, rice plant, roses, *Saintpaulia*, seeds of sycamore (*Acer pseudoplatanus* L.), seeds and spices, soil, soybeans, stored sorghum, strawberries, taro, *Theobroma cacao* L. (cacao), tomatoes, tulip bulbs, vanilla beans, vanilla pods, *Virgilia divaricata* Adamson, water chestnuts (*Scirpus tuberosus* Desf.), wheat grains ex silo, wheat stores, ya (Chinese white) pear and yams (Fan and Zhang, 2007).

Animals and their products hosting the mite include beehives, *Calolampra* (Blaberidae), cheese, Cheshire cheese, cockroach colonies, dead larvae in wood infested by the citrus borer, the dead cerambycid *Prionoplus reticularis* White, dried milk powder, dried shellfish, dust beneath banana leaf matting, fish houses, Gouda cheese, human urine, 'Kai-iwi' foil-wrapped cheese, lab culture of *Psocoptera*, *Locusta migratoria* (L.), mealworm cultures, millipedes (*Polyconoceris alaskis*?), *Nauphota* (cockroach), passalid beetle, rabbit, rooster feathers 'Tintinhull', sheep faeces, silverfish cultures, *Tineola biselliella* Hum.

The mite has also been collected from fungal cultures (*Sporotrichum*), mushrooms, cracks and the floor, dust beds, glass surfaces in a house, houses, a kitchen in a farmhouse, pallet and bag scrapings and a spoon from the kitchen of a house (Fan and Zhang, 2007).

Bio-ecology

Fan and Zhang (2007) pointed out that the taxonomic concept of *T. putrescentiae* involved two closely related species, one common and the other rare. The neotype designated by Robertson (1959) for *T. putrescentiae* is the rare species, while the most common mite is a new species that Fan and Zhang (2007) described as *T. communis* Fan *et* Zhang. However, Klimov and O'Connor (2010) believed that 'the prevailing usage of the name *T. putrescentiae* comprises almost exclusively the common species, the name *T. communis* is a junior synonym of eight previously named taxa with extant types, and proposed a new name for the rare species: *Tyrophagus fanetzhangorum* Klimov *et* O'Connor (2009)'. According to this interpretation, they 'proposed to conserve the prevailing usage of the name *T. putrescentiae* by designation of a new neotype from a culture currently maintained in a research institution, which was cited in many published works and started from specimens collected close to the type locality of *T. putrescentiae*'. Later, Klimov and Tolstikov (2011) reported that *T. putrescentiae* Schrank corresponds to *T. communis* Fan *et* Zhang (2007) and that *T. fanetzhangorum* is a probable synonym. Unfortunately, the vision of Klimov and O'Connor (2010) does not simplify the topic and suggests that the more linear choice of Fan and Zhang (2007) should be followed.

According to findings of Fan and Zhang (2007), *T. communis* has a worldwide distribution and has been recorded from various plants, animals and their substrates. This highlights a wider geographic distribution than *T. putrescentiae* (*sensu* Fan and Zhang, 2007) and identifies the species with *T. putrescentiae auctorum*. The aspects discussed above do not take account of any bibliographic information previous to the study of Fan and Zhang (2007).

Symptomatology and damage

Zhang (2003) described the symptomatology and the damage for ornamentals and vegetables in greenhouses. They could be similar for other crops, but there is no adequate information available.

Control

For the control of *T. communis*, see the general guidelines suggested below for the other acarid species – *T. longior*, *T. neiswanderi* and *T. similis*.

Tyrophagus curvipenis Fain *et* Fauvel (Fig. 13.3)

Common name

Unknown.

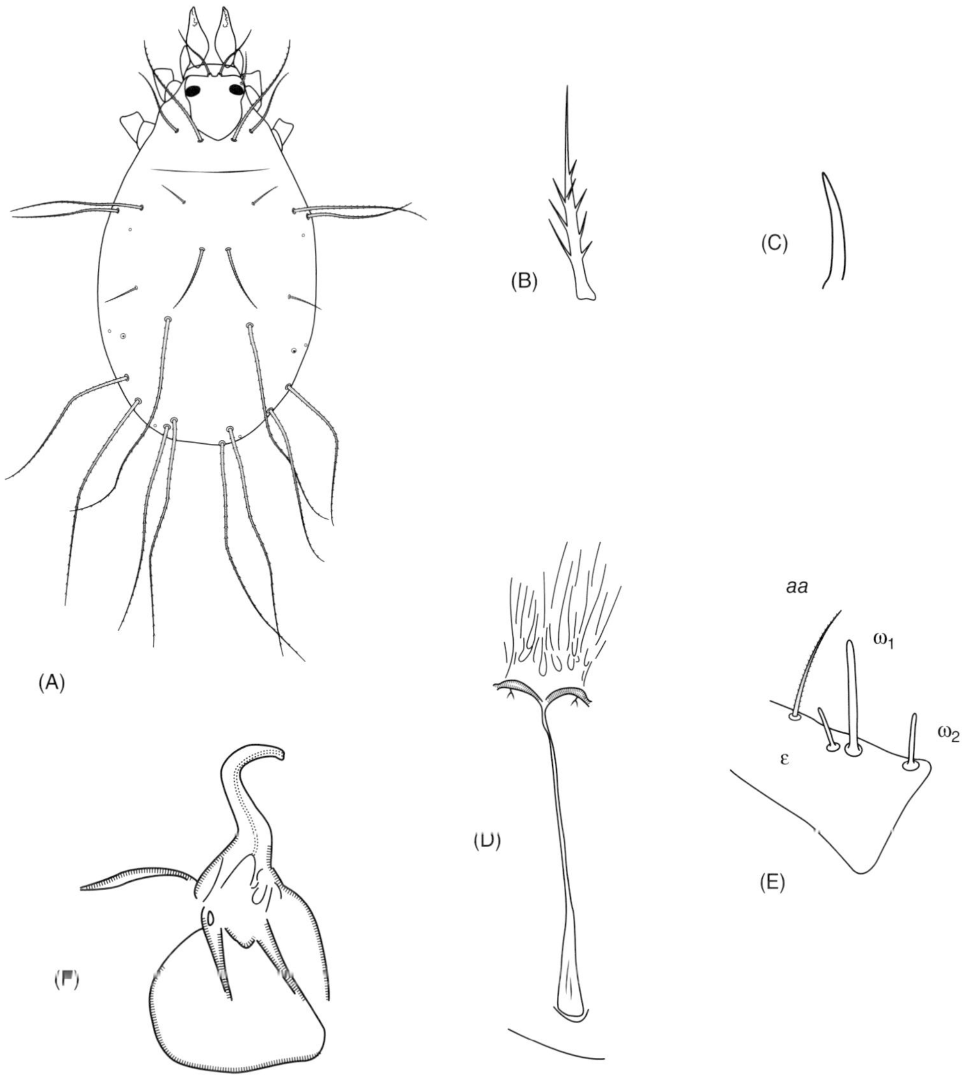

Fig. 13.3. *Tyrophagus curvipenis* Fain *et* Fauvel. Female: (A) dorsal view; (B) supracoxal seta; (C) Grandjean's organ; (D) copulatory opening and spermatheca; (E) solenidia, famulus and seta of tarsus I. Male: (F) aedeagus and genital papillae (from Fan and Zhang, 2007). The labels are all setal notations, which are explained in the text.

Diagnostic characteristics

FEMALE. The idiosoma is 367 μm long and 212 μm wide. The prodorsal shield has prominent eyespots, nearly pentagonal, with the lateral side behind the eyespots concave. Grandjean's organ is finger like, smooth, with the basal lobe bearing 2–3 spiniform teeth. The setae *scx* are slender, with 6–8 pectinations. The setae d_1 are about 3.0 times longer than the setae c_1 and 2.6 times longer than the setae d_2; the setae d_2 are about 1.2 times longer than the setae c_1. The coxal plates I extend just beyond the apex of the apodeme; the coxal plates II are broadly triangular, extending distally beyond the apex of apodeme 2, with one-third of the posterior margin slightly concave. The spermathecal duct narrows gradually from the copulatory opening for a distance about 1.5 times the distance between the sclerites of the oviducts and then extends to the base of the spermathecal sac over a distance about 2.8 times the distance between the sclerites of the oviducts; the base of the spermathecal sac is 'r'-shaped. The solenidion ω_1 of tarsus I is slender, cylindrical and slightly widened at the apex; the solenidion ω of tarsus II is slender and almost cylindrical. The setae *w* and *r* of tarsus IV are setiform (Fan and Zhang, 2007).

MALE. The idiosoma is 321 μm long and 205 μm wide. The prodorsal shield is similar to that of the female. Grandjean's organ is as in the female. The coxal plates I and II, and the solenidia ω_1 and ω of tarsi I and II are as in the female. The setae d_1 are about 3.0 times longer than the setae c_1 and 2.0 times longer than the setae d_2; the setae d_2 are about 1.5 times longer than the setae c_1. The aedeagus is 'S'-shaped, the distal two-fifths is reverse curved, tapering from the base to the tip

and with the lateral arms turning outwards. The setae *w* and *r* of tarsus IV are setiform (Fan and Zhang, 2007).

Geographical distribution

T. curvipenis is found in Australia, France, New Zealand (Fan and Zhang, 2007) and Portugal (Fain and Fauvel, 1993; Fan and Zhang, 2007).

Host plants and other biological hosts and substrates

Plants and their products on which this mite has been found include apples, the bark of apricot, cherry and nectarine, barley, the bird-of-paradise flower, blackcurrant stems, blackcurrants, capsicum, cherry foliage, clover roots, *Coleophora* tubes, cucumber leaves, *Cymbidium* sp. pollinia, garlic, Granny Smith apple calyxes in sooty mould in store, grapefruit (unsprayed fruit/leaves), hydrangea leaf, kiwi fruit, lemons, mandarin, mummified apricot fruit, mummified loquat fruit, Nashi (Japanese pear) leaves, onions, passion fruit, persimmon, poplar (*Populus nigra* L. var. *italica*) galls, shallots, strawberries, tree onion bulbs (Fan and Zhang, 2007). The mite has also been found on algae covering the wooden structures of orchids (Fain and Fauvel, 1993).

Animals and animal products on which the mite has been found include coccid scale on *Buxus* sp., dead eggs and embryos of *Miomantis caffra* Saussure, dead larva of the citrus borer in wood, dead larva of *Stathmopoda*, dead *Rattus norvegicus* Berkenhout, *Disonycha* and *Agasicles* beetles, dried milk powder, egg cluster of Avondale spider, eggs of *Ceroplastes destructor* (Newstead), honeybees, honeybee hive, nests of the bee *Exoneura concinnula* Cockerell, *Rattus exulans* (Peale), sparrow, starling (*Sturnus vulgaris* L.) and swallow, scrapings from pallets and logs, wasp (*Vespula germanica* (Fabricius)) (Fan and Zhang, 2007).

Bio-ecology

T. curvipenis is known from a wide range of habitats (Fan and Zhang, 2007), but there is no information available on its bio-ecology, any damage caused and its economic importance.

Tyrophagus longior (Gervais) (Fig. 13.4)

Common name

This is a seed mite.

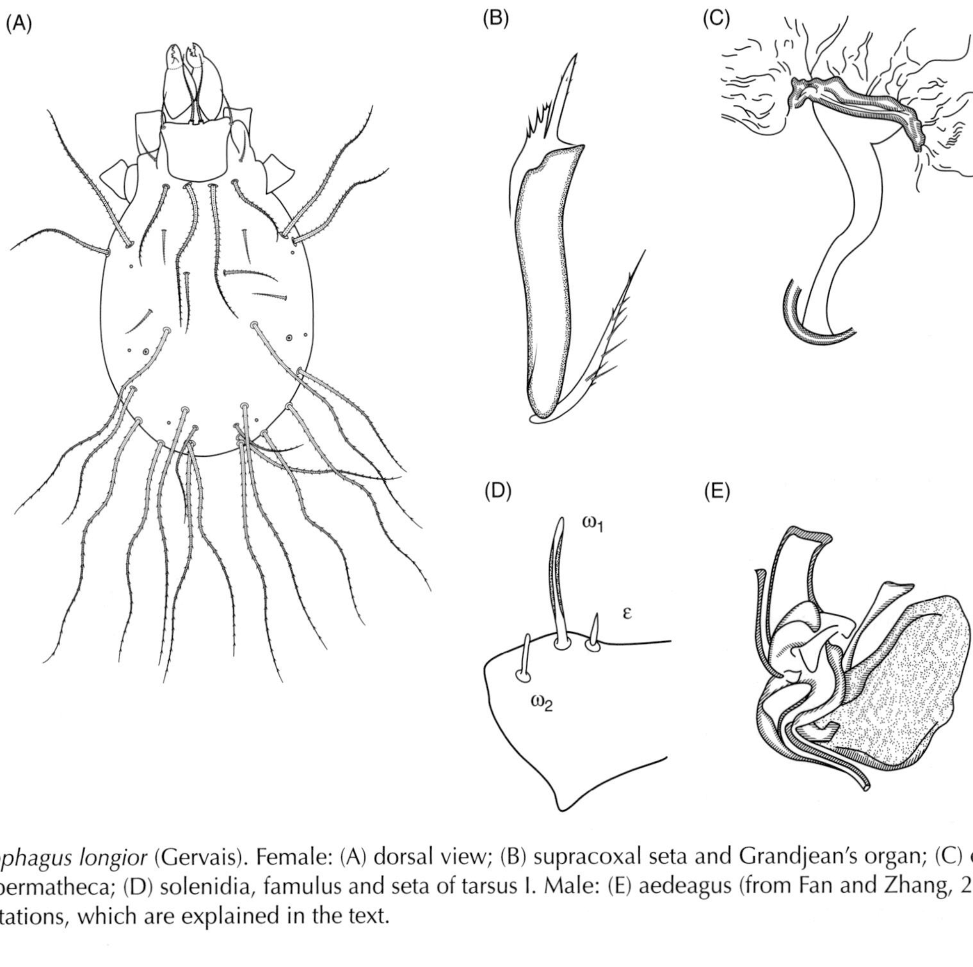

Fig. 13.4. *Tyrophagus longior* (Gervais). Female: (A) dorsal view; (B) supracoxal seta and Grandjean's organ; (C) copulatory opening and spermatheca; (D) solenidia, famulus and seta of tarsus I. Male: (E) aedeagus (from Fan and Zhang, 2007). The labels are all setal notations, which are explained in the text.

Diagnostic characteristics

FEMALE. The idiosoma is 525 μm long and 344 μm wide. The prodorsal shield lacks eyespots, is nearly rectangular, with the lateral sides slightly concave and the posterior broadly round. Grandjean's organ is finger like and barbed, with the basal lobe bearing one large tridentate and 2–3 small spiniform teeth. The setae *scx* are slender, tapering from the base to the apex, with 8–12 short pectinations. The setae d_1 are about 1.3 times longer than the setae c_1 and 1.5 times longer than the setae d_2; the setae d_2 are about 1.0 times as long as the setae c_1. The coxal plates I are triangular and do not reach to the posterior apex of the prosternal apodeme; the coxal plates II are triangular, with the posterior margin slightly sinuous and not reaching the apex of the apodeme 2 distally. The spermathecal duct is broad, widening gradually from midway to the base of the spermathecal sac, about 1.5 times longer than the distance between the sclerites of the oviducts, and the base of the spermathecal sac is nearly flat or bends slightly backwards. The solenida ω_1 and ω of tarsi I and II are slender, almost cylindrical but taper slightly from the base to the apex. The setae *w* and *r* of tarsus IV are spiniform (Fan and Zhang, 2007).

MALE. The idiosoma is 563 μm long and 366 μm wide. The prodorsal shield is similar to that of the female. The coxal plates I and II, and the solenidia ω_1 and ω of tarsi I and II, are as in the female. The setae d_1 are about 1.4 times longer than the setae c_1 and 1.6 times longer than the setae d_2; the setae d_2 are about 0.9 times the length of the setae c_1. The aedeagus has one major curve, narrowing gradually in the basal half, the distal end is straight or slightly reverse curved, and the lateral arms turn inwards. The setae *w* and *r* of tarsus IV are spiniform (Fan and Zhang, 2007).

Geographical distribution

T. longior is found worldwide (Hughes, 1976).

Host plants and other biological hosts and substrates

Host plants and their products on which *T. longior* has been found include acorns of *Quercus* sp., avocadoes, banana, barley, barn dust, bulked wheat, coconut, a combine and other harvesters, cucumber, culms and panicles of wheat, cucumber leaves, *Cymbidium*, *Dahlia*, damaged peas, damaged cucumber leaves, fig leaves, garlic (*Allium sativum* L.), gladioli bulbs, grassland, grass seed, mallow leaves, mandarin, *Musa sapientum* L., nectarine bark, onion, pomelo, ornamentals, rotting coconut, rotting mango, ryegrass seed, shallots, soybean flour, stored apple, stored food products, stored grain, stored hay, stored pollen, strawberry, tamarillo, tomato chutney, weed, wheat (Fan and Zhang, 2007). Lange and Bacon (1958) reported the mite on spinach, and Hughes (1976) reported it on stored home-produced grain, hay and straw, haystacks and grain stacks standing in the open field, barley, grass cleaning plant [tools], cucumber plants, beetroots, tomatoes and *Cyclamen* seed. Other economic plants on which the mite is found are cucumber (Chmielewski, 1979, 1984; Hussey, 1985; Voigt, 1990; Laffi and Ponti, 1997), ornamentals (Czaikowska *et al.*, 1988; Buxton, 1989), melon, strawberry, tomato (Vacante, 1997b) and orchids (*Cymbidium* spp.) (Ciampolini *et al.*, 1985).

Animals and their products on which *T. longior* has been found include cheese (Robertson, 1961), stock fish (Gigja, 1964), bee frames, beehives and surveyed bees and *Bombus* nests, hives of *Apis cerana indica* Fabricius, *Apis mellifera* L., *A. dorsata* Fabricius and *A. florae* Fabricius, honey, bird's nests (including one in a poplar tree) and nests of seabirds (*Eudyptula minor albosignata* Finsch and *Pachyptila turtur* (Kuhl)) and the dipper (*Cinclus aquaticus* (Bechst.)) and the starling (*Sturnus vulgaris*), cheese shelves, dead *Prionoplus reticularis* Lav., ham, house dust, humans, litter in a densely populated urban area, scrapings from pallets and bags (dried milk powder), ham in a meat works, scrapie-infected sheep, shepherds and wedding cake (Fan and Zhang, 2007).

Bio-ecology

In several European countries, *T. longior* has been found on cucumber (Chmielewski, 1979, 1984; Hussey, 1985; Voigt, 1990; Laffi and Ponti, 1997) and ornamentals (Czaikowska *et al.*, 1988; Buxton, 1989) in greenhouses. Vacante (1997b) collected the mite on melon, strawberry and tomato in protected crops of the Mediterranean area but did not observe any damage. On orchids (*Cymbidium* spp.) in greenhouses, the mite prefers *C. clauboda* Goliath and vectors fungi that prevent the flowers from opening, while on other *Cymbidium* spp. the problem is secondary (Ciampolini *et al.*, 1985). In California, the mite damages spinach in spring and autumn crops (Lange and Bacon, 1958), and in the USA it has been found associated with *Histiostoma* and *Rhizoglyphus* mites.

At 23°C and 87% relative humidity (RH), the life cycle of the seed mite takes from 2 to 3 weeks (Hughes, 1976). At 20°C on wheat grain, the mite populations increase from 5 to 2200 mites before a population crash when the RH is 90%, while at 75% RH the population peaks at 1000 mites and then decreases (Parkinson, 1990). Dry, warm weather and rapid growing conditions often circumvent more severe damage by the mite (Lange and Bacon, 1958). The mite produces an alarm pheromone (Noguchi *et al.*, 1998).

Symptomatology and damage

T. longior has caused periodic damage to autumn- and spring-planted spinach in most areas of California, and its attacks are severe in coastal areas. The mites occur in the crowns of the plants, feeding on the young, interfolded leaves. Damage at this early stage appears as a black stippling, later progressing to small, circular holes, then to blackened areas. The petioles develop normally, but the leaf blades become stunted, disfigured and in general have a ragged, unthrifty appearance. As many as several hundred mites may occur together in the interfolded leaves. The mites require considerable moisture and as the leaves grow out they migrate down to the new growth. In some years, damage has been quite widespread in the spinach areas, but usually damage occurs only in localized portions of fields and not over large acreages. Damage often follows a row, in effect running with the rows, or it may occur over a circular area,

usually in the low part of a field. The damage often appears similar to that caused by certain chemicals applied for weed control (Lange and Bacon, 1958).

The mite can also swarm up on to young plants of cucumber, eating minute holes in the youngest leaves, which then become quite large as the leaves expand. They may cause a proportion of the infested plants to go 'blind' (Hussey, 1985). The development of populations commonly needs a high RH, but the attack can occur on young plants in nurseries, in greenhouses and in the open field. On cotyledons, the damage consists mainly of alterations to the epidermal tissue of the upper leaf surface, where little lesions can give part of leaf a russeting aspect; similar alterations can also involve the stems at the level of cotyledon insertion. On orchids (*Cymbidium clauboda* and other species), the mite vectors fungal spores that prevent the flowers from opening, while on other *Cymbidium* spp. the damage is secondary (Ciampolini *et al.*, 1985). On *Verbena*, *Lavatera* and *Delphinium ajacis* L., distortion and death of the growing points and holing of young leaves has been reported. No organic matter, e.g. straw or manure, is associated with either infestation (Buxton, 1989).

Control

The monitoring of seed mite populations is fundamental to control, which requires both preventive and curative actions.

MONITORING. Early monitoring of mite presence on crops is crucial to the outcome of the control. The effectiveness of a mite trap for detecting the mite storage pests has been positively tested (Wakefield and Dunn, 2005). Generally, an average of five mites or more per central leaf indicates that control is required, and this is usually best when the plants are small as coverage is very difficult with the usual commercial equipment at economic rates. After the damage has been done, applications of chemicals are of little value (Lange and Bacon, 1958).

PREVENTION. Prevention requires meticulous care in the nursery, reducing the RH and the use of peat-free compost. Substrates for the production of seedlings should be properly sterilized before use, e.g. by heat. Lange and Bacon (1958) reported that applications of excessive amounts of organic fertilizer or discing under large amounts of plant debris may lead to mite damage under certain conditions.

CHEMICAL CONTROL. Parathion dust at 40–50 lb acre^{-1} applied by ground equipment has proved in the past to be satisfactory under most conditions. Sprays were effective only on small plants and 75–100 (US) gal acre^{-1} of spray were usually necessary (Lange and Bacon, 1958). Trials carried out in the laboratory showed the effectiveness of etrimfos, heptenophos, methacrifos, profenofos and propargite, while cycloprate, fenbutatin oxide and isoprothiolane failed to give complete mortality; several compounds reputedly effective against pest mites or ticks were ineffective against *T. longior* (Stables, 1980). On cucumbers, pirimiphos-methyl sprays were an effective and less persistent alternative to drenches and sprays of parathion (Hussey, 1985). On *Verbena*, *Lavatera* and *D. ajacis*, a high-volume spray of dichlorvos gave good control of the mite at one site (Buxton, 1989).

Tyrophagus neiswanderi Johnston *et* Bruce (Fig. 13.5)

Common name

Pollen-cap mite.

Diagnostic characteristics

FEMALE. The idiosoma is 493 μm long and 327 μm wide. The prodorsal shield has a pentagonal shape, with prominent eyespots, and lateral sides slightly concave and posterolateral sides strongly concave. Grandjean's organ is finger like, with a basal lobe with two basally confluent and one separate spiniform teeth. The setae *scx* have a shaft widened at the point where the pectinations begin, with 8–11 moderate or short pectinations. The setae d_1 are about 1.6 times the setae c_1 and 1.7 times longer than the setae d_2; the setae d_2 are about 0.9 times the length of the setae c_1. Coxal plates I reach to the posterior apex of the prosternal apodeme; coxal plates II narrow sharply along the distal three-fifths, then become contiguous with the apex of apodeme 2, with two-thirds of the posterior margin strongly concave. The spermathecal duct narrows gradually from the copulatory opening to the base of the spermathecal sac, and is about 2.6 times longer than the distance between the sclerites of the oviducts; the base of the spermathecal sac is triangular and funnel shaped. The solenida ω_1 and ω of tarsi I and II are slender, almost cylindrical. The setae *w* and *r* of tarsus IV are setiform (Fan and Zhang, 2007).

MALE. The idiosoma is 397 μm long and 242 μm wide. The prodorsal shield and Grandjean's organ are as in the female. Coxal plates I and II, and solenidia ω_1 and ω of tarsi I and II, are as in the female. The setae d_1 are about 1.5 times longer than the setae c_1 and 1.3 times longer than the setae d_2; the setae d_2 are about 1.1 times longer than the setae c_1. The aedeagus has one major curve, with the distal end almost straight, tapering from base to tip and lateral arms turning outwards. The setae *w* and *r* of tarsus IV are setiform (Fan and Zhang, 2007).

Geographical distribution

T. neiswanderi is found in Australia (Fan and Zhang, 2007), China (Teng *et al.*, 1988), Germany (Fan and Zhang, 2007), Italy (Vacante, 1988), Japan (Nakao and Kurosa, 1988), Mexico (Estebanes-Gonzalez, 1997), the Netherlands (Fan and Zhang, 2007), New Zealand (Martin and Workman, 1985; Fan and Zhang, 2007), Poland (Czaikowska *et al.*, 1988), Serbia and Montenegro (Petanović, 2001), Spain (Sánchez-Ramos *et al.*, 2007), South Africa (Meyer Smith and Rodrigues, 1966; Fan and Zhang, 2007), Switzerland (Fischer, 1993), the UK (Griffiths, 1979; Fan and Zhang, 2007) and the USA (Johnston and Bruce, 1965; Fan and Zhang, 2007).

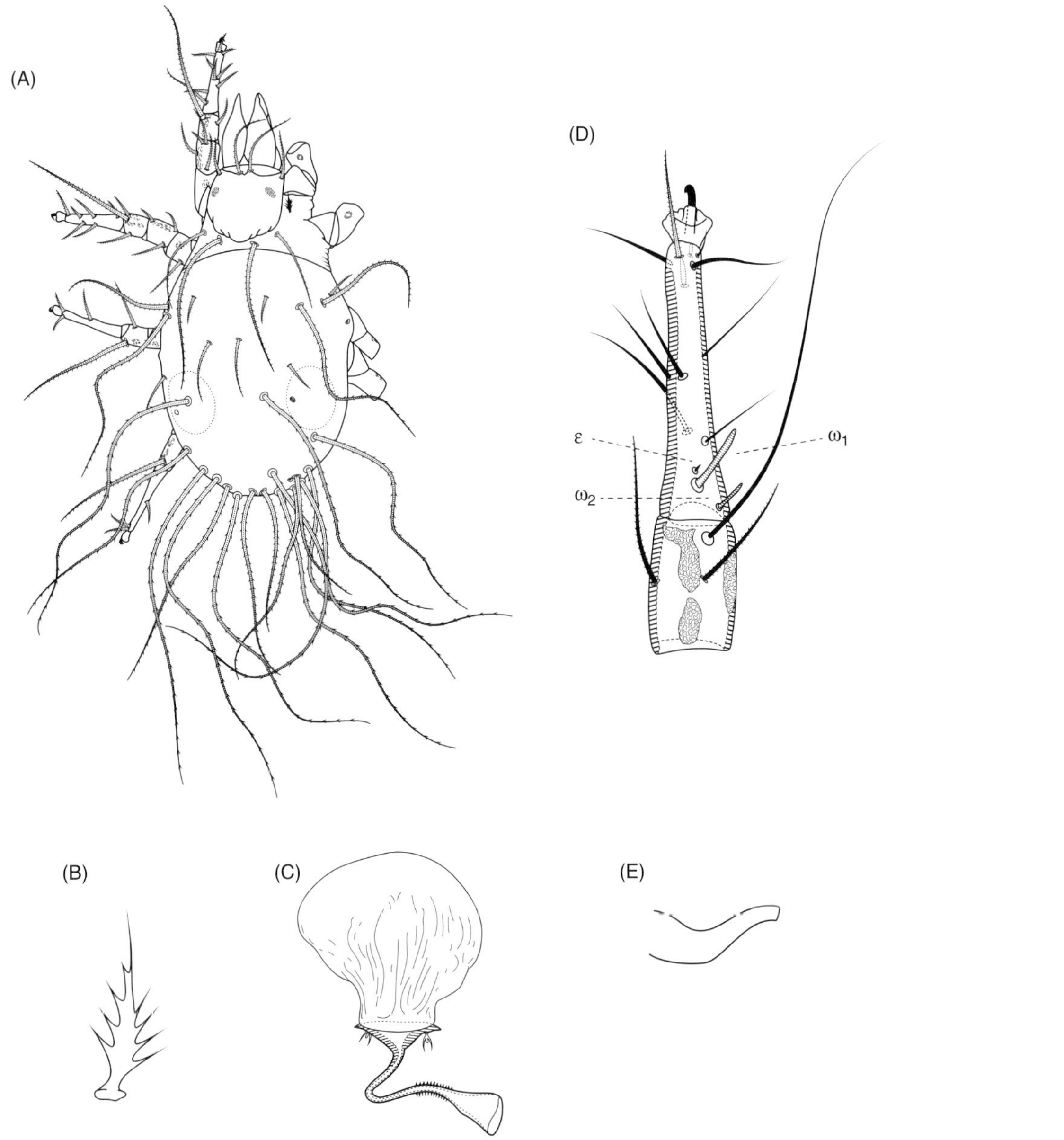

Fig. 13.5. *Tyrophagus neiswanderi* Johnston *et* Bruce. Female: (A) dorsal view; (B) supracoxal seta; (C) copulatory opening and spermatheca; (D) tibia and tarsus I. Male: (E) aedeagus (from Vacante, 1988). The labels are all setal notations, which are explained in the text.

Host plants and other biological hosts and substrates

The list of host plants and vegetable products on which *T. neiswanderi* has been found includes buds on gooseberry cultures, bulbs of Amaryllidaceae (*Amaryllis*, *Hippeastrum*), capsicum (*Capsicum frutescens* (L.)), cucumber (including cucumber fruits and cucumber in the greenhouse), *Cymbidium* orchid flower buds, leaves, pollen caps, miniature plants, *Epiphyllum* sp., flower bulbs, garlic, *Geranium*, grape leaves with galls, grapevines, *Hibiscus* sp., lemon, maize, *Narcissus* bulbs, onions and diseased onions, oranges, orchid flowers, orchid house floors and orchid pots, pepino (*Solanum muricatum* Ait.), *Phalaenopsis* orchid flower buds, *Prunus* leaves, soil and plant material (various), tomato in the greenhouse and *Zantedeschia* bulbs (Fan and Zhang, 2007).

The mite is a pest of various economic plants in greenhouses, such as cucumber (Johnston and Bruce, 1965; Griffiths and Southgate, 1967; Chmielewski, 1979, 1984; Nakao and Kurosa, 1988; Kurosa and Nakao, 1993; Vacante and Benuzzi, 2007), melon and pumpkins (Vacante, 1997b), the growing tips of chrysanthemum cuttings (Hughes, 1976), *Cymbidium* (Martin and Workman, 1985), *Phalaenopsis* (Kadono and Endo, 1996), flowers of *Gerbera* and cyclamen, bulbs of *Narcissus*, *Tulipa*,

Hippeastrum and hyacinth, corms of *Crocus* and *Freesia* (Czajkowska *et al.*, 1988; Vacante, 1988; Czajkowska and Kropczynska, 1991; Vacante and Benuzzi, 2007). The mite has also been found in stored products and nests (Sánchez-Ramos *et al.*, 2007), in the respiratory tract of humans, in the short-tailed bat and in scale culture rooms (Fan and Zhang, 2007).

Bio-ecology

The pollen-cap mite inhabits various organic substrates, stored food and cultivated plants, such as ornamentals and vegetables grown in greenhouses. At 25°C, the development from egg to adult takes from 2 to 3 weeks, and is faster on corms of *Freesia* and *Crocus* than on tulip or other plants (Czajkowska and Kropczynska, 1991). At temperatures ranging from 10 to 31°C, and with an RH of 90 ± 5%, the pre-oviposition period, the fecundity and the daily fecundity are adversely affected by extreme temperatures, while the oviposition period increases as the temperature decreases. Male and female longevity increase as the temperature decreases, and longevity in males is greater than in females. A non-linear Lactin model describes the effect of temperature on the intrinsic rate of natural increase of mite populations, and the optimum temperature for development is predicted at 26.6°C. At this temperature, the population doubling time is 2.8 days; the lower and upper thresholds are 7.4 and 31.7°C, respectively (Sánchez-Ramos *et al.*, 2007). Also, this mite produces a pheromone alarm (Kuwahara *et al.*, 1989).

Symptomatology and damage

The symptomatology and damage of *T. neiswanderi* are similar to those of other *Tyrophagus* spp. On *Gerbera*, the symptoms involve the young leaves, which curl more or less extensively over the internal parts of the leaf surface, especially near the central veins; the curling is associated with distortion of the margins of the leaf, and morphological alterations of the flowers. On expanded leaves, the attack induces irregular and jagged tears of the surfaces, with margins that are more or less curved and regular. On cucumber, melon and pumpkins, the young leaves show similar alterations, while on expanded leaves, alterations of the tissues are more frequent in areas adjacent to the peduncle (Vacante, 1997b; Vacante and Benuzzi, 2007). Attack on young plants in nurseries and/or post-transplanting can lead to serious economic damage. Equally severe may be the economic impacts on ornamentals. Infested gerbera flowers are not marketable.

Control

Control is via preventive and curative actions. Prevention is mainly based on physical methods, such as high temperatures for soil sterilization. The curative actions are based on the use of chemical and/or biological means.

SOIL STERILIZATION. Research carried out on the tolerance of *T. neiswanderi* and other acarid mites to high temperatures (40°C for 24 h, 45°C for 3 h and 50°C for 1 h) showed that the tolerance of *T. similis* to high temperature was lower than that of *T. neiswanderi*. *T. similis* may not have developed high-temperature tolerance because, in contrast to *T. neiswanderi*, the species dwells in the soil, where temperature fluctuation is not generally strong (Kasuga and Honda, 2006a). The benefit of this control method for *T. neiswanderi* is only partial.

CHEMICAL CONTROL. The need to maintain high standards of quality make the use of chemical means on ornamental and flower plants a priority. The choice of acaricides is not easy for reasons of efficiency, cost and phytotoxicity, and many chemicals have been tested. Thus, on orchids in New Zealand, it has been observed that carbaryl, deltametrin, dimethoate and methiocarb initially reduce pollen-cap mite numbers, but after 14 days only methiocarb sustains this reduction to a similar level to that achieved by the predator mite *Hypoaspis* sp. Dimethoate gave the least effective control of the pollen-cap mite and the greatest reduction in *Hypoaspis* sp. populations. The ability of these four pesticides, plus taufluvalinate, acephate and methamidophos, to reduce movement of the pollen-cap mite was low, and after 48 h only methiocarb reduced the number of mites moving up the flower stem (Workman and Martin, 2002).

BIOLOGICAL CONTROL. Releases of *Hypoaspis* spp. (Laelapidae) into the orchid media hindered the development of *T. neiswanderi* populations, thereby preventing injury to the flowers (Workman and Martin, 2002).

Tyrophagus perniciosus Zakhvatkin (Fig. 13.6)

Common name

Unknown.

Diagnostic characteristics

FEMALE. The idiosoma is 572 µm long and 387 µm wide. The prodorsal shield is nearly rectangular, lacks eyespots, with the lateral sides slightly concave and the posterior broadly round. Grandjean's organ is finger like, and the basal lobe bears 3–5 spiniform teeth. The setae *scx* are strong, tapering from the base to tip or slightly widened in the basal two-thirds, with 16–22 moderate or short pectinations. The setae d_1 are about 2.9 times longer than the setae c_1 and 2.7 times longer than the setae d_2; the setae d_2, are long about 0.9 times the length of the setae c_1. Coxal plates I do not reach the apex of the prosternal apodeme; coxal plates II are triangular, not extending posteriorly beyond the apex of apodeme 2, with two-thirds of the posterior margin slightly concave. The spermathecal duct is broad, widening gradually from midway to the base of the spermathecal sac, which is broadly round and bends backwards. The solenidion ω_1 of tarsus I is stout and clavate; the solenidion ω of tarsus II is stout and slightly clavate. The setae *w* and *r* of tarsus IV are spiniform (Fan and Zhang, 2007).

MALE. The idiosoma is 419 µm long and 296 µm wide. The prodorsal shield and Grandjean's organ are as in the female. The coxal plates, solenidia ω_1 and ω of tarsi I and II are as in the female. The setae d_1 are about 4.0 times longer than the setae c_1

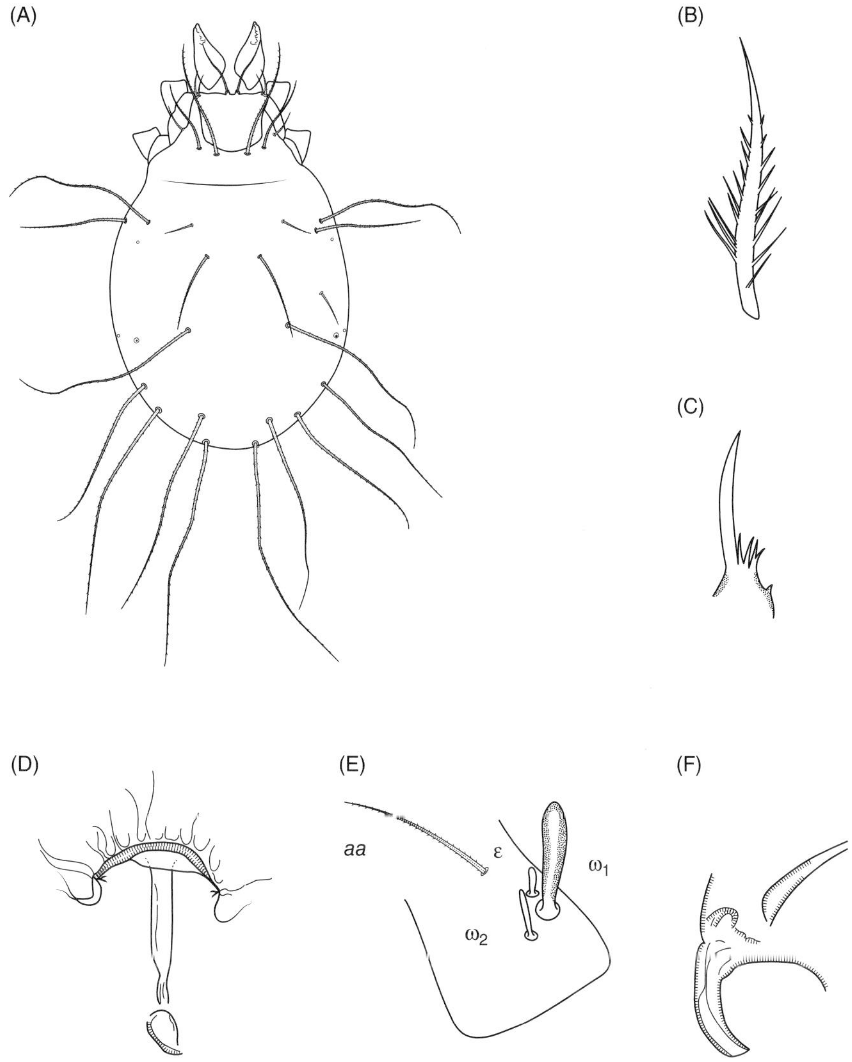

Fig. 13.6. *Tyrophagus perniciosus* Zakhvatkin. Female: (A) dorsal view; (B) supracoxal seta; (C) Grandjean's organ; (D) copulatory opening and spermatheca; (E) solenidia, famulus and seta of tarsus I. Male: (F) aedeagus (from Fan and Zhang, 2007). The labels are all setal notations, which are explained in the text.

and 3.4 times longer than the setae d_2; the setae d_2 are about 1.2 times longer than the setae c_1. The aedeagus has one major curve, narrowing gradually in the basal quarter and almost straight in the distal three-quarters; its distal end is nearly truncated, and the internal diameter of the distal three-quarters is broad, with the lateral arms turning inwards. The setae *w* and *r* of tarsus IV are spiniform (Fan and Zhang, 2007).

Geographical distribution

T. perniciosus is found in Australia (Hughes 1976; Fan and Zhang, 2007), Bulgaria (Hughes, 1976), Germany (Kazhdaya, 1996), Japan (Nakao and Kurosa, 1988), Kazakhstan (Sadieva, 1984), the Netherlands (Jonge, 1988), Russia (Zakhvatkin, 1941), Turkey (Cobanoğlu and Bayram, 1998), the UK (Hughes, 1976) and the USA (USDA, 1981).

Host plants and other biological hosts and substrates

Zakhvatkin (1941) reported *T. perniciosus* in dust granaries and warehouses, and associated with oats and barley. Hughes (1976) reported the mite on cheese and dirty seed in a budgerigar cage. Sadieva (1984) observed the mite feeding on the mycelium of *Sporidesmium mucosum* var. *pluriseptatum* P. Karst. *et* Har. on cucumber. Fan and Zhang (2007) produced an extensive list of various hosts of *T. perniciosus*, including cauliflower seedlings, melon, oil-producing seeds and probably many other plants stored in granaries and warehouses, pumpkin, soil and dead plant residues, spinach, tomato (seed and newly germinating plants), various grain products (cereals, etc.). The mite has also been collected on mushrooms (*Morchella conica* Persoon and *M. esculenta* (L.) Pers.), and it occurs on protected crops (Zhang, 2003) and is a pest of cucumbers, melons, pumpkins and spinach (Sadieva, 1984; Nakao and Kurosa, 1988; Nakao, 1991).

Animals and their products on which *T. perniciosus* has been found include bird cages (e.g. budgerigar cages), humans (causing acariasis), nests of house martins (*Delichon urbica* (L.)) and starlings (*Sturnus vulgaris*).

Bio-ecology

The bio-ecology of *T. perniciosus* is little known. According to the information discussed above, the mite may occur in protected crops (Zhang, 2003), where it causes damage to cucumbers, melons, pumpkins and spinach (Sadieva, 1984; Nakao and Kurosa, 1988; Nakao, 1991). Leal *et al.* (1988) have identified its pheromone alarm.

Symptomatology and damage

The symptomatology and damage caused by *T. perniciosus* is as for other *Tyrophagus* spp.

Control

Control methods are as for the most important *Tyrophagus* spp.

Tyrophagus putrescentiae (Schrank) (Fig. 13.7)

Common name

This is a mould mite.

Diagnostic characteristics

FEMALE. The idiosoma is 492 μm long and 314 μm wide. The prodorsal shield is nearly pentagonal, with eyespots and lateral sides slightly concave. Grandjean's organ is finger like, and the

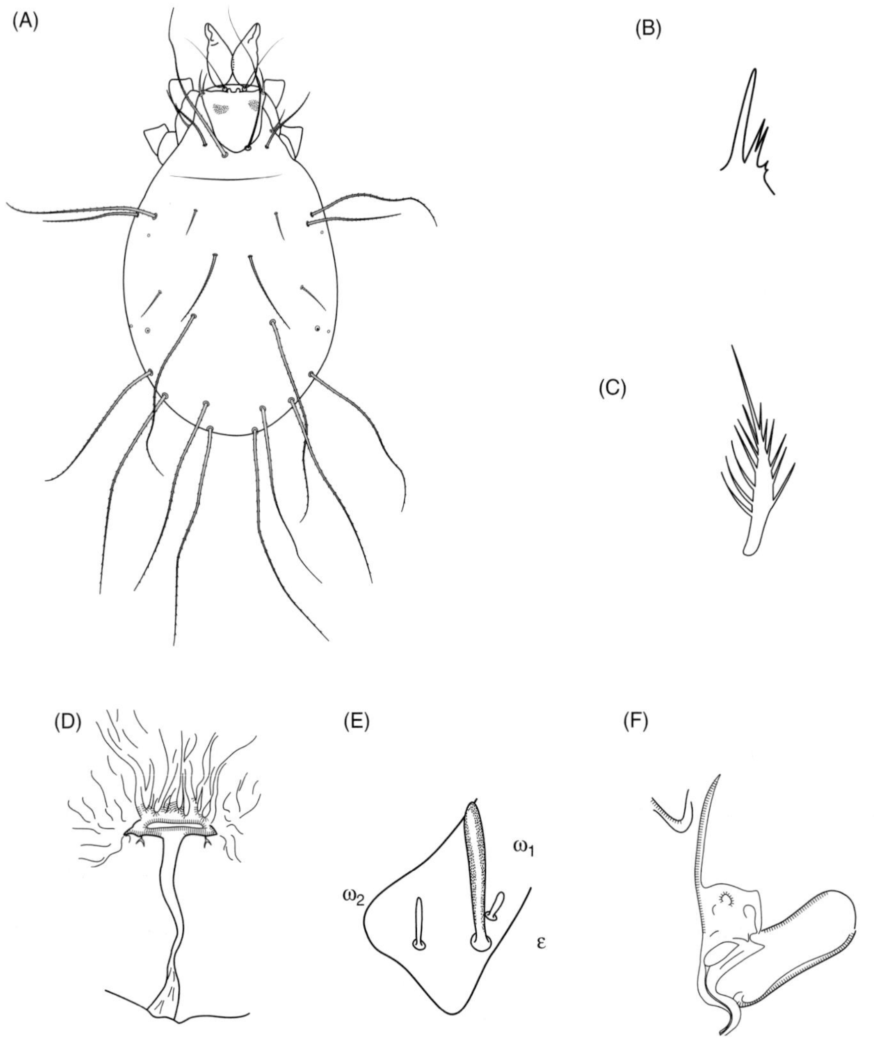

Fig. 13.7. *Tyrophagus putrescentiae* (Schrank). Female: (A) dorsal view; (B) Grandjean's organ; (C) supracoxal seta; (D) copulatory opening and spermatheca; (E) solenidia, famulus and seta of tarsus I. Male: (F) aedeagus (from Fan and Zhang, 2007). The labels are all setal notations, which are explained in the text.

basal lobe bears 1–2 large and two small spiniform teeth. The setae *scx* are widened at the bases of the pectinations, long and number 10–14. The setae d_1 are about 2.1 times longer than the setae c_1 and 2.4 times longer than the setae d_2; the setae d_2 are long, about 0.9 times the length of the setae c_1. Coxal plates I are just beyond the apex of the prosternal apodeme; coxal plates II have a sinuous posterior border, narrowing sharply medially and becoming contiguous with the apex of their apodeme. The spermathecal duct narrows rapidly from the copulatory opening for a distance about 0.6 times the diameter of the distance between the sclerites of the oviducts, then gradually widens to the base of the spermathecal sac over a distance about 2.2 times that between the sclerites of the oviducts; the base of the spermathecal sac is flat. The distal two-thirds of solenidion ω_1 of tarsus I obviously widens; the solenidion ω of tarsus II slightly widens at the apex. The setae *w* and *r* of tarsus IV are setiform (Fan and Zhang, 2007).

MALE. The idiosoma is 378 μm long and 233 μm wide. The prodorsal shield is as in the female. Grandjean's organ is finger shaped and basally has 1–4 spiniform teeth. Eyespots, coxal plates I and II, solenidia ω_1 and ω are as in the female. The setae d_1 are about 2.0 times longer than the setae c_1 and 1.9 times longer than the setae d_2; the setae d_2 are about 1.0 times the length of the setae c_1. The aedeagus is 'S'-shaped, with the distal one-third reverse curved, tapering from the base to the tip, with the lateral arms turning outwards. The setae *w* and *r* of tarsus IV are setiform (Fan and Zhang, 2007).

Geographical distribution

According to what was previously discussed under *T. communis*, the records of *T. putrescentiae* before 2007 are doubtful because of the validity of the two species *T. communis* and *T. putrescentiae*. *T. putrescentiae* (as recently validated) is distributed in Australia, China, Ecuador, Germany, Japan (Fan and Zhang, 2007), the Netherlands (Oudemans, 1924; Fan and Zhang, 2007), New Zealand, Taiwan and the USA (Fan and Zhang, 2007).

Host plants and other biological hosts and substrates

T. putrescentiae has been collected on banana, sour orange (*Citrus aurantium* L.), garlic, *Gladiolus*, humus, hyacinth, muscari bulbs, mushroom, *Narcissus* bulbs, orange, orchids, palm seed, *Ranunculus* sp. bulbs and tulip bulbs. The mite has also been collected from the burrows and nests of mutton birds (*Puffinus tenuirostris* (Temminck)) and dead tench (*Tinca* L.) (Fan and Zhang, 2007).

Bio-ecology

In their analysis of Robertson's (1959) concept of *T. putrescentiae* (Schrank), Fan and Zhang (2007) pointed out a mixture of two species now described as *T. putrescentiae* (Schrank) and *T. communis* Fan *et* Zhang. *T. putrescentiae* differs from *T. communis* in that two-thirds of the posterior margin of coxal plates II is strongly concave and the solenidion ω_1 is widened in the distal two-thirds. Unfortunately, this interpretation, although formally correct, has no available bibliographical data on the two species. Hence, the bio-ecology of this mite is unknown.

Symptomatology and damage

See *T. communis*.

Control

See *T. communis*.

Tyrophagus similis Volgin (Fig. 13.8)

Common name

The French fly.

Diagnostic characteristics

FEMALE. The idiosoma is 522 μm long and 324 μm wide. The prodorsal shield lacks eyespots, is obviously punctate, nearly rectangular in shape, with lateral sides slightly concave and the posterior broadly round. Grandjean's organ is finger like, and the basal lobe bears three obvious spiniform teeth. The setae *scx* are slender, tapering from the base to tip, with 12–16 short pectinations. The setae c_1, d_1 and d_2 are micro-setae, much shorter than others; the setae d_1 are about 1.0 times the length of the setae c_1 and 1.1 times longer than the setae d_2, the setae d_2 are about 0.8 times the length of the setae c_1. Coxal plates I do not reach to the posterior apex of the prosternal apodeme; coxal plates II do not extend posteriorly beyond the apex of apodeme 2, with the posterior margin sinuous and distal, one-third becoming concave and narrow. The spermathecal duct is broad, widening gradually from the copulatory opening to the base of the spermathecal sac, about 1.4 times the distance between the sclerites of the oviducts; the base of the spermathecal sac is broadly round, bending backwards. The solenidion ω_1 of tarsus I is stout and obviously widened at the apex; the solenidion ω of tarsus II is strong and widened at the apex. The setae *w* and *r* of tarsus IV are spiniform (Fan and Zhang, 2007).

MALE. The idiosoma is 526 μm long and 372 μm wide. The prodorsal shield is laterally concave and posteriorly convex. Grandjean's organ is as in the female. Coxal plates I and II, and the solenidia ω_1 and ω, are as in the female. The setae d_1 are about 1.0 times the length of the setae c_1 and 1.1 times longer than the setae d_2; the setae d_2 are about 0.9 times the length of the setae c_1. The aedeagus has one major curve, narrowing gradually in the basal half and almost cylindrical in the distal half, with the distal end truncated, the internal diameter tapering rapidly from the distal end to mid-length, and the lateral arms turning inwards. The setae *w* and *r* of tarsus IV are spiniform, set anteriorly or close to the distal sucker (Fan and Zhang, 2007).

Geographical distribution

T. similis is found in Australia (Robertson, 1959; Halliday, 1998; Fan and Zhang, 2007), Belgium (Hughes, 1976), China (Li,

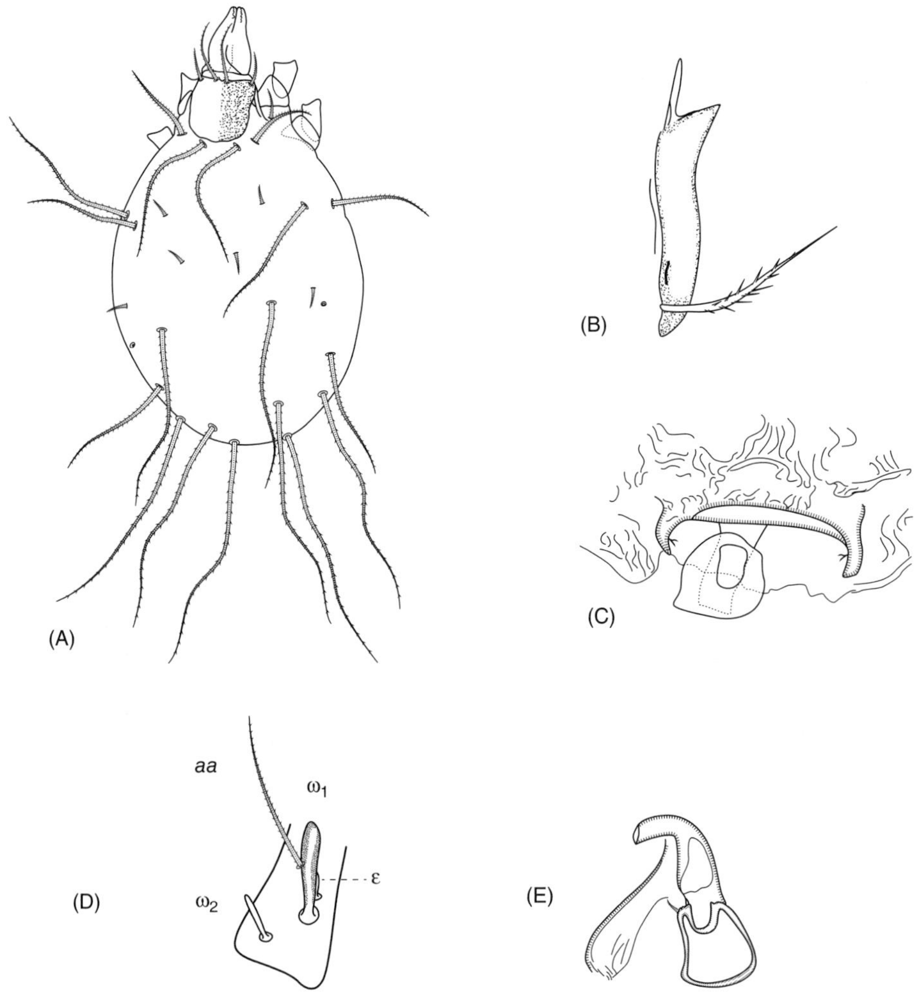

Fig. 13.8. *Tyrophagus similis* Volgin. Female: (A) dorsal view; (B) Grandjean's organ and supracoxal seta; (C) copulatory opening and spermatheca; (D) solenidia, famulus and seta of tarsus I. Male: (E) aedeagus (from Fan and Zhang, 2007). The labels are all setal notations, which are explained in the text.

1999), Germany (Robertson, 1959), the Faroe Islands (Hallas and Solberg, 1989), France (Giustina, 1981), Iceland (Hughes, 1976; Hallas and Gudleifsson, 2008), Iran (Hajiqanbar *et al.*, 2002), Italy (Laffi, 1980), Japan (Nakao and Kurosa, 1988), Mexico (Estebanes-Gonzalez and Rodriguez-Navarro, 1991), the Netherlands (Hughes, 1976; Fan and Zhang, 2007), New Zealand (Robertson, 1959; Ramsay and Paterson, 1977; Fan and Zhang, 2007), Romania (Cindea, 1978), South Africa (Fan and Zhang, 2007), Sweden (Bostrom *et al.*, 1997), the UK (Robertson, 1959; Fan and Zhang, 2007), the USA (Walter *et al.*, 1986) and Yemen (Al-Safadi, 1987).

Host plants and other biological hosts and substrates

Hughes (1976) reported *T. similis* on grassland, soil, *Phlox*, spinach, mushrooms, old hay, chaff, in an eider duck nest and in a *Bombus ionellus* (Kirby) nest. Fan and Zhang (2007) listed the following plants and plant-associated products/substrates: acorns (*Quercus* sp.), agricultural soil, barn dust, bean plumules, cucumber, dahlia, garlic, kiwi fruit, *Lilium* bulbs, maize, melon, melon seedbeds, moss, mushrooms, *Narcissus* bulbs, orchid, pasture, pumpkin, rotten straws of rice, sandy soil in pots of *Rama* ryegrass, seeds and debris from a tent, spinach, spinach buds, stored hay, sugarbeet fields and tea tree scrub. The list of animals and their products on which *T. similis* has been found includes bat debris, black-backed gull colony nests and debris, a dead *Rhopoea* larva (grub), a larva of *Zenarge turneri* Rohwer, nematode culture, the nematode *Heterodera avenae* (Wollenweber), nests of seabirds, starlings (*Sturnus vulgaris* L.), the Norway rat (*Rattus norvegicus* (Berkenhout)), Polynesian rat (*Rattus exulans* (Peale)), rabbit (*Oryctolagus cuniculus* L.) and white heron (*Egretta alba modesta* (Gray)).

Bio-ecology

In addition to the hosts and substrates listed above, the French fly also occurs on vegetables and ornamentals in the greenhouse (Zhang, 2003) and has been reported in several European countries on leaves of cucumber, French beans, *Phlox*, spinach

(Hussey *et al.*, 1969; Hughes, 1976; Ippolito and Triggiani, 1988), stems of *Cineraria* and bulbs of *Narcissus* (Czajkowska *et al.*, 1988). In Japan, *T. similis* is known as a pest of melon, watermelon, cucumber, pumpkin, tomato, sweet pepper, Welsh onion and maize, and in early spring may severely damage spinach in the greenhouse (Nakao and Kurosa, 1988; Nakao, 1989, 1991). The mite is introduced into cucumber greenhouses in straw or manure, and in the warm environment becomes active and wanders upon the newly set-out plants. Attacks decrease naturally within a few weeks, after one or two generations have developed (Hussey *et al.*, 1969).

The biology of *T. similis* is known from laboratory cultures (Al-Safadi, 1987, 1990, 1991; Kasuga and Amano, 2000a). In Iceland, the French fly has the deutonymphal stage that is responsible for dispersal. The eggs laid by females from the midsummer generation may overwinter and hatch as winter generation eggs the next spring. The formation of the deutonymph takes place in the cold winter; the summer generation does not develop this stage. In this environment, the species has a male:female ratio of 1:3 (Hallas and Gudleifsson, 2008).

On wheatgerm at 25°C, development from egg to adult takes less than 10 days (Al-Safadi, 1991). At 10, 15, 20 and 25°C, and with a 14 h light (14L):10 h dark (10D) photoperiod, the developmental time of immature stages is shorter at 25°C, and the highest fecundity occurs at 10°C (663.4 eggs/female). At 10°C, the intrinsic rate of natural increase (r_m), the net reproductive rate (R_0) and the mean generation time (T) are 0.051, 190.05 and 102.03, respectively; at 15°C, they were 0.085, 190.04 and 61.43, respectively; at 20°C, they were 0.127, 125.37 and 37.98, respectively, and at 25°C, they were 0.160, 36.57 and 22.46, respectively. The species exhibits the highest growth potential at 25°C, but at this temperature the fecundity is the lowest. A limited survival of females occurs at 35°C and at higher temperatures. The survival of females was investigated at 53, 66, 76, 87 and 100% RH, 20°C and with a 14L:10D photoperiod; limited survival was observed at 53 and 66% RH (Kasuga and Amano, 2000a).

On spinach in greenhouses in central Japan, the mite density in the soil decreases from May to August, and increases rapidly in late autumn, remaining at a high level from December to February; it then increases in April and rapidly decreases in May. The high temperatures registered in greenhouses from spring to early autumn is the main cause of the population decrease. The numbers of mites on spinach buds increase after the increase of those on and in the soil. The attack on spinach buds occurs mostly in late autumn and early spring (Kasuga and Amano, 2003).

In the greenhouse, the mite inhabits either cultivated or uncultivated soil at the periphery of the cultivated area inside the greenhouse and in the soil immediately outside. Its density in the cultivated and peripheral soils increases in the cooler months and decreases in the warmer months. Only small numbers of *T. similis* are collected from the adjacent outside soil. In the cultivated area, most of the mites inhabit the soil at depths of 0–5 cm throughout the year, and their numbers decrease with depth (Kasuga and Amano, 2005). In England, the mite is introduced into cucumber greenhouses in straw or manure, and in the warm environment they become active and wander on to the newly set-out plants. Attacks decrease naturally within a few weeks, after one or two generations have been developed (Hussey *et al.*, 1969). The species produces an alarm pheromone (Kuwahara *et al.*, 1987).

In Yemen, there is a positive correlation between the density of *Gamasellodes/Protogamasellus* spp. (predator mites) and that of the most abundant mite species present (*T. similis*) (Al-Safadi, 1987). Eight genera of bacteria (*Bacillus*, *Nocardia*, *Corynebacterium*, *Erysipelothrix*, *Lactobacillus*, *Arachnia*, *Micrococcus*, *Staphylococcus*) and fungi have been isolated from dead mites. Acaricidal activity has been confirmed in 77.8% of the bacterial isolates and 100% of the fungi isolates. The microorganisms isolated are deemed effective in biological pest control of the mite on spinach in the greenhouse (Kikuta and Takagi, 2001).

Symptomatology and damage

T. similis infests cucumber, French beans, spinach, tomatoes, sweet peppers, lettuce and other vegetables (Hussey *et al.* 1969; Hughes, 1976; Ippolito and Triggiani, 1988; Nakao, 1989, 1991). It also attacks stems of *Cineraria*, bulbs of *Narcissus* (Czajkowska *et al.*, 1988) and *Phlox* (Hughes, 1976). In addition to infestations on cucumbers, tomatoes, sweet peppers and other plants, serious damage has been recorded in Japan to spinach in greenhouses in the early spring (Nakao, 1989, 1991). In nurseries of cucurbits, the feeding activity of the mite produces many little russeting patches with reddish yellow edges, scattered on the upper leaf surface of cotyledons, and similar alterations can be observed on the stems at the cotyledon insertion level. Adult plants exhibit deformations and perforations of the leaf tissue between the main veins, frequently close to the leaf petiole (Laffi, 1980; Laffi and Ponti, 1997). Nakao (1991) summarized the symptoms of feeding activity of acarids (and hence also those of *T. similis*) on melon, cucumber and pumpkin as numerous small holes and yellowish spots on young leaves, which later become deformed. On watermelon, tomato and sweet pepper, the leaves become lustrous, discoloured and deformed, while on Welsh onion the seedlings become dwarfed and stunted.

In infestations of spinach (van den Bruel, 1940; Ippolito and Triggiani, 1988; Nakao, 1989; Laffi and Ponti, 1997), *T. similis* causes serious damage to greenhouse crops in Japan in the cooler months (Kasuga and Amano, 2000b; Nakao, 1989). The mite populations initially increase in number on or in cultivated soils rich in organic matter, then invade the spinach buds and feed on the young leaves (Kasuga and Amano, 2003). As the plants grow, the damaged leaves begin to show small holes and deformation. The mite may also damage germinating spinach seeds. Mite populations in the soil are thought to have invaded the germinating seeds and fed on their embryos and albumen. As a result, young roots and buds were injured, and some plants did not develop. The mites affected the number of growing plants, and mite numbers on the germinating seeds increased as those in the soil increased. Temperature probably affects the movement of mites to seed, with low temperatures possibly hindering mite activity (Kasuga and Amano, 2006; Kasuga and Honda, 2006a,b).

Control

Control requires preventive and curative actions. The main preventive measures are disinfestation of the soil and agronomic

management. The curative actions are mainly the use of chemical and biological methods. Early monitoring of the mite populations in the field is fundamental to detect the early symptoms of attack, but this is commonly delayed and so the problem is not solved satisfactorily.

MONITORING. On spinach, to prevent damage by *T. similis* during seed germination, the monitoring and control of mites in the soil before sowing are important. As direct counting of *T. similis* numbers in the soil is difficult, the Tullgren funnel or acarid bait trap (Kasuga *et al.*, 2005) might be useful for monitoring numbers. Kasuga and Amano (2003) report that the *T. similis* population in the soil of a spinach greenhouse remains low during the hot season, increases rapidly in the autumn, then remains high during the cool season. Thus, the mite population should first be monitored in early autumn, between the harvest of the preceding crop and the sowing of the next. If a large number of mites is found in the soil before sowing, the application of acaricides to the soil, or heat treatment (e.g. hot water or steam) of the soil, will be necessary to prevent damage by *T. similis* during the seed germination period (Kasuga and Amano, 2006).

SOIL STERILIZATION. Soil sterilization treatments by different techniques (solar heat, hot water and steam) are positively applied in Japan. Nevertheless, a few acarid mites can survive in greenhouses, e.g. as a result of uneven heat treatment, or on the outskirts of greenhouses, and in untreated areas, which do not permit a lasting effect over a long period. Soil sterilization should probably not be done in early summer, while a short-term steam disinfection in spring or autumn will probably suppress mite infestation. Use of this method will most likely spread because it is economic and practical if applied for prevention before mites can reproduce or as an emergency control measure after an outbreak (Matsumura *et al.*, 2005). Most *T. similis* inhabit the soil at depths of 0–5 cm throughout the year, and their numbers decrease with depth, so the main control point of *T. similis* is the 0–5 cm soil layer in the cultivated and peripheral soils of the greenhouse (Kasuga and Amano, 2005). Kasuga and Honda (2006a) investigated the high-temperature tolerance of *T. similis* to evaluate the possibility of controlling its populations in agricultural soil by heat treatment. No eggs and adults of *T. similis* survived treatments at 40°C for 24 h, 45°C for 3 h, or 50°C for 1 h, although the heat tolerance of eggs was relatively higher than that of adults. This indicates that hot water soil sterilization or steam soil sterilization have great potential for the control of mite populations, and that these methods should be effective for *T. similis* even though less hot water or steam are used for treatment than in the case for plant pathogens.

CULTURAL PRACTICES. Different organic substrates can influence the increase of population of *T. similis*. In Japan, an evaluation has been made of the effect of various foods/substrates on the fecundity of females of *T. similis*: three organic wastes (cattle faeces, sawdust and rice straw), an organic fertilizer (rapeseed meal), three composts (made from cattle faeces, rice husks and rice straw), three fungi (*Fusarium oxysporum* von Schlechtendal, *Pythium aphanidermatum* (Edson) Fitzp. and *Rhizoctonia solani* (Cooke) Wint.) and eight vegetables (lettuce, cucumber, komatsuna, spinach, pak choi, garland chrysanthemum, Welsh onion and tomato). Rapeseed meal, fungi and most vegetables were suitable, while organic wastes and composts were unsuitable (Kasuga and Honda, 2006c). This suggests that the use of organic fertilizers such as rapeseed meal should be reduced during the mite season, that plant residues should be removed from the soil before spinach cultivation, and that the use of immature organic materials should be reduced because they increase fungi that are suitable for *T. similis*.

BIOLOGICAL CONTROL. Release of the laelapid mite *Hypoaspis aculeifer* Canestrini has been suggested for the biological control of the French fly. The bio-ecology of this predator has been investigated by Kasuga *et al.* (2006).

CHEMICAL CONTROL. Smoking with parathion or spraying both plants and substrates with the same chemical if high infestations occur has been reported on cucumber (Hussey *et al.* 1969; Nakao 1989, 1991), but usually, mite infestation in protected crops can be controlled by soil sterilization (Misayo *et al.*, 2005). In Italy, the best chemicals for lowering populations of *T. similis* on spinach were benzoximate, endosulfan and fenpropathrin, but none of them gave good results in controlling the eggs and they all had little selectivity on predaceous mites (Ippolito and Triggiani, 1988). In Japan, dichlorvos did not reduce the numbers of mites in the greenhouse, even though it was highly toxic under laboratory conditions, which indicates that the mites present in the soil and on spinach buds had little direct contact with the chemicals. This highlights the fact that once crop damage by mites is detected, it is usually too late to use chemicals, and that mites living in buds are in some way protected from the action of agrochemicals (Kasuga and Amano, 2003). On spinach in Japan, DDVP EC (dichlorvos) gave high acaricidal activity in all populations of *T. similis*, but had only a low control effect under field conditions (Kasuga and Amano, 2000b), due to the chemical particles not sufficiently reaching mites in the soil and on spinach plants (Kasuga and Amano, 2002).

The Genus *Mycetoglyphus* Oudemans

Diagnostic characteristics

The prodorsal external vertical setae (*ve*) are short and smooth, less than a quarter of the length of the prodorsal vertical internal setae (*vi*), and set posteriorly to setae *vi*. The internal scapular setae (*sci*) are longer than the external scapular setae (*sce*). The dorsal tarsal setae *e* and the ventral tarsal setae *p*, *q*, *s*, *u* and *v* are spine shaped. On tibia I, the solenidion σ_1 is never more than twice as long as the solenidion σ_2. The leg I of the male is not enlarged and the femur does not bear a ventral apophysis. The male has a long aedeagus (Hughes, 1976).

Economic importance

The genus includes only two species: *Mycetoglyphus fungivorus* (Oudemans) and *Mycetoglyphus sevastianovi* Kadzhaja

(O'Connor, 2008). The first species has been collected on mushrooms growing and on other substrates (Hughes, 1976) and the second species has been recorded in natural soil near Odessa (Ukraine) (Kadzhaya, 1988). Damage caused by *M. fungivorus* to economic plants is similar to that of *T. similis*; severe attacks on spinach in greenhouses have been reported (Nakao, 1989; Zhang, 2003).

Mycetoglyphus fungivorus (Oudemans) (Fig. 13.9)

Common name

Unknown.

Diagnostic characteristics

MALE. The general aspect is similar to that of *T. similis*. The body is 400–600 μm long and with well-tanned appendages. The setae *ve* are very short, less than a quarter as long the setae *vi*, set posteriorly to the bases of *vi*, never as far back as the mid-point of the lateral edge of the prodorsal shield. The idiosomal dorsal setae d_2 are short, about 6% of the length of the idiosoma; the setae c_1 are 1–1.5 times longer than the setae d_2 and the setae d_1 are 1.5 to twice as long as the setae d_2. The supracoxal seta is curved and slightly pectinate. The solenidia ω_1 and ω of tarsi I and II are club shaped; on tarsi I–III, there is a distal seta *e* and five ventral spines, slightly unequal in size; on tarsus IV, the suckers are set in the basal half of the article, and the setae *r* and *w* are distal to them. The aedeagus is very long, curved and arising from a basal plate (Hughes, 1976).

FEMALE. The idiosoma is 500–600 μm long and the general aspect is as for the male (Hughes, 1976).

HYPOPUS. The body is 250 μm long and yellow–brown in colour, with a widely rounded posterior margin. The prodorsal shield has an almost straight anterior edge lacking any sign of a rostrum. The setae *vi* are lacking. The gnathosomal base is completely concealed by the prodorsal shield. Ventrally, the sternum is extended posteriorly, so that coxal plates I and II are completely separate from one another. The sucker plate is well removed from the posterior edge of the body and is almost circular in outline. Tarsus I bears a long expanded and three lanceolate setae, and tarsus IV has two lanceolate setae (Zakhvatkin, 1941).

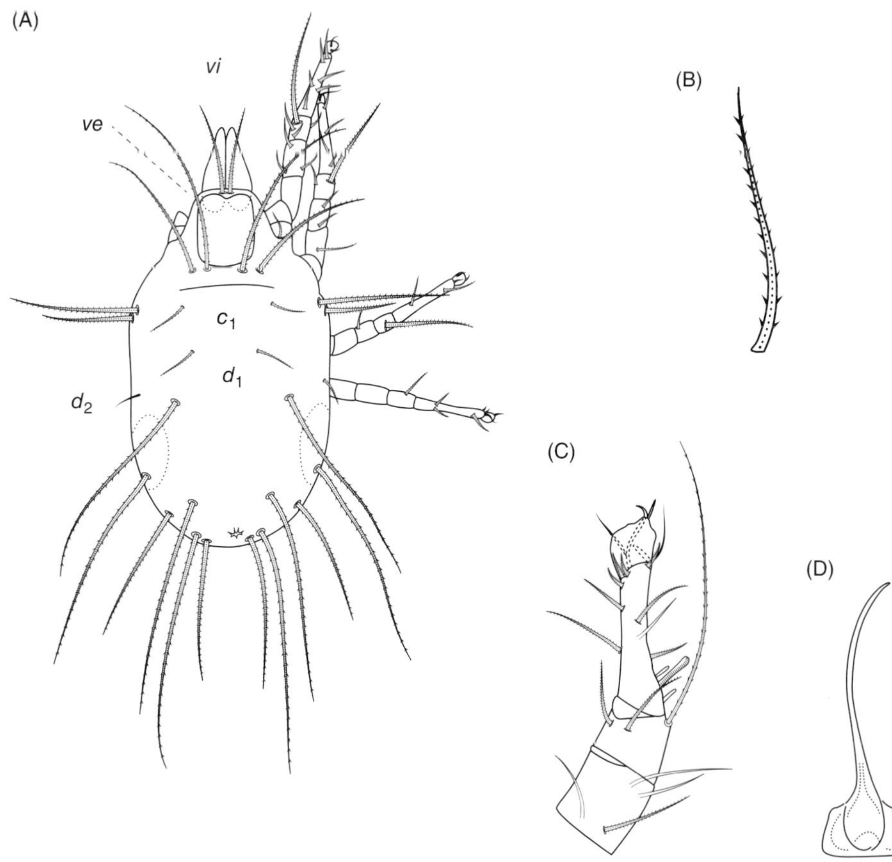

Fig. 13.9. *Mycetoglyphus fungivorus* (Oudemans). Female: (A) dorsal view; (B) supracoxal seta; (C) leg I. Male: (D) aedeagus (from Hughes, 1976). The labels are all setal notations, which are explained in the text.

Geographical distribution

M. fungivorus is found in Africa, England, Germany, Hungary (Hughes, 1976), Japan (Nakao and Kurosa, 1988), Russia and the USA (Hughes, 1976).

Host plants and other biological hosts and substrates

M. fungivorus has been reported on grassland and beneath stacks of hay and straw (Griffiths, 1960), on moles' nests and nests of the ant *Camponotus ligniperda* (Latreille), with hypopi probably carried by the ants (Türk and Türk, 1957). The acarid has also been collected on growing mushrooms, celery waste, lettuce, decaying radishes, wet decaying wood debris (Mahunka, 1961), *Phaseolus* sp. (pole beans) (Manson, 1983) and spinach (Nakao and Kurosa, 1988; Nakao, 1989).

Bio-ecology

The bio-ecology of the mite is not known.

Symptomatology and damage

As for *Tyrophagus* spp.

Control

See *Tyrophagus* spp.

The Genus *Rhizoglyphus* Claparède

Diagnostic characteristics

Until a short time ago, there had been no revision of this economically important genus, and about 52 names were in use under the genus *Rhizoglyphus* (Díaz *et al.*, 2000). Recently, Fan and Zhang (2004) examined different *Rhizoglyphus* specimens, mainly from Australasia and Oceania, and described 11 species. In addition, they provided keys to adult females and males of *Rhizoglyphus* from Australasia and Oceania, with lists of hosts and distribution details for each species.

The idiosoma is oblong and whitish to semi-transparent. There are four pairs of prodorsal setae and 12 pairs of hysterosomal setae. The setae *ve* are small or vestigial, set about halfway along the lateral margins of the prodorsal shield. The setae *sci* are shorter than the setae *sce*, and very small, vestigial or lacking. The setae *scx* are smooth, large or small. Grandjean's organ is bifurcate or simply bladed. The legs are reddish brown. Tarsi I and II have a proximal conical spine (*ba*) set near the bases of the solenidion ω_1 and a famulus (ε). Tarsus I bears four long terminal setae and one apical solenidion ω_3; tarsus II has four long terminal setae; tarsus III has three long terminal setae, and tarsus IV bears two long terminal setae in the female and one long terminal seta in the male. Some tarsal setae are slightly flattened apically (Fan and Zhang, 2004).

The homeomorphic male differs from the female in that the genital opening is set between coxae IV, an aedeagus is present, there are one pair of anal suckers and three pairs of setae (ps_3, ps_2, ps_1); the setae ad_1, ad_2 and ad_3 are lacking. On tibia IV, the solenidion φ is spine like, and tarsus IV bears two suckers. The setae on the legs are as in the adult female, except for tarsi IV. The heteromorphic male is similar to the homeomorphic male but both sides or one side of leg III are about doubled in size. The body setae are slightly longer. The setae on the legs are as in the homeomorphic male, except for tarsi III (Fan and Zhang, 2004).

The hypopus has a highly sclerotized body that is dorsoventrally flat. The gnathosoma is reduced and the chelicerae are lacking. Excepting setae *vi* and h_2, the idiosomal setae are small. The setae *scx* are set ventrally. There are two pairs of genital papillae and one pair of genital setae. The ventral hysterosomal area forms an anal attachment organ, which bears six pairs of attachment structures (modified setae). Genu I lacks solenidion σ. On tarsus I, solenidion ω_3 is proximal. There is no seta *ba*. Some apical tarsal setae are foliate (Fan and Zhang, 2004).

Economic importance

Mites of the genus *Rhizoglyphus* are usually associated with the bulbs and corms of many plants and are found in different substrates. Some species cause damage to crops and ornamental plants, both in greenhouses and in the field, and their presence may also involve exported/imported horticultural products (carrots, corms of taro, flower bulbs) (Fan and Zhang, 2004). Thirteen *Rhizoglyphus* spp. are reported as pests of crops or have been described from agricultural settings, usually in close association with cultivated plants (Díaz *et al.*, 2000). However, only *R. robini* (Claparède) and *R. echinopus* (Fumoze *et* Robin) are traditionally known as severe pests of economic plants. Recently, the list of injurious *Rhizoglyphus* spp. included *R. setosus* Manson, found damaging various bulbs and plants of economic importance in different countries. In particular, *R. robini* has been collected on bulbs, leaves and seeds of various cultivated plants, many ornamentals and several vegetables. The mite damages mainly bulbs of various ornamentals and corms of *Freesia* in storage, greenhouses and in the field, inducing distorted growing tips and leaves, and reducing the bulb to a rotten pulp (Hussey *et al.*, 1969; Manson, 1972; Hughes, 1976; Meyer Smith, 1981a; Ascerno *et al.*, 1983; Okabe and Amano, 1991; Zhang, 2003). *R. echinopus* also infests the bulbs and roots of ornamental plants and causes similar damage (Manson, 1972; Fan and Zhang, 2003; Zhang, 2003). Both *R. robini* and *R. echinopus* have quarantine importance, as in exported horticultural products (e.g. carrots) and ornamental bulbs (Fan and Zhang, 2003). *R. setosus* is associated with the underground organs of various plants (lilies, leeks, garlic, onions, etc.) and different substrates (Ho and Chen, 1987; Chen *et al.*, 2002a; Fan and Zhang, 2004), and like *R. robini* and *R. echinopus*, has a quarantine importance. In addition to direct damage, the presence of bulb mites has been associated with pathogenic fungi and bacterial development.

Rhizoglyphus robini **(Claparède) (Figs 13.10–13.11)**

Common name

The robine bulb mite or bulb mite.

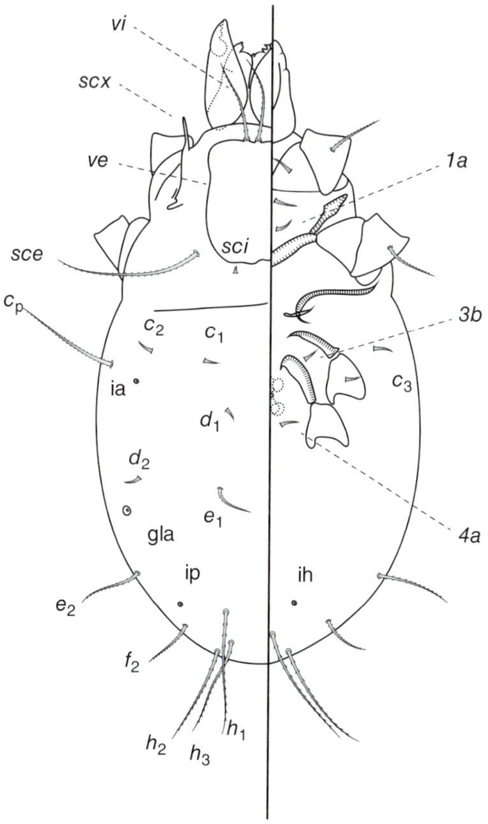

Fig. 13.10. *Rhizoglyphus robini* (Claparède). Female: dorsal (left) and ventral (right) view (modified from Fan and Zhang, 2004). Key: gla, gland opening; ia, ih and ip, lyrifissures. The rest of the labels are setal notations, which are explained in the text.

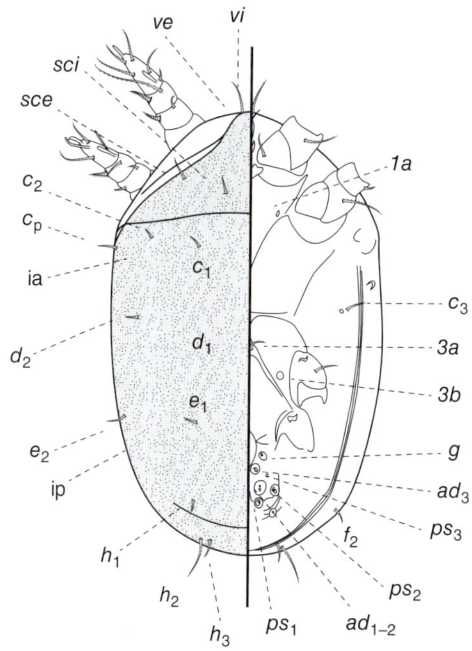

Fig. 13.11. *Rhizoglyphus robini* (Claparède). Hypopus. Dorsal (left) and ventral (right) view (modified from Fan and Zhang, 2004). Key: ia, ip, lyrifissures. The rest of the labels are setal notations, which are explained in the text.

Diagnostic characteristics

FEMALE. The idiosoma is 676–934 μm long, 482–650 μm wide, whitish and with a smooth integument. The legs are short and brownish. The prodorsal shield is evenly punctulate. The setae *vi* are blunt, the setae *sci* are about one-third the length of genu I, and the ratio *sci-sci:sci-sce* ranges from 0.7 to 0.8 (i.e. the setae *sci* are not close to *sce* as the ratio is <2). Grandjean's organ is not bifurcate or bifurcate to small branches. The setae *scx* are slender and pointed. The setae c_1 are from 0.8 to 1.0 times the length of the setae d_2; the setae c_3 are about as long as the setae c_1; the setae d_1 are from 0.8 to 1.0 times the length of the setae d_2; the setae d_2 are far from the opisthonotal glands (gla); the setae e_1 are from 2.3 to 3.5 times the length of d_2; the setae e_2 are from 0.8 to 1.5 times the length of the setae f_2. The anal opening has six pairs of small setae, subequal in length, less than one-half of the length of genu IV. The copulatory opening has an incomplete circular shield and the spermathecal duct is long and narrow. The base of the inner part of the spermatheca is minute; the sclerites of the oviducts are 'Y'-shaped, with a distance between them from 6 to 8 μm, and the basal part is about one-half the length of the branches (Fan and Zhang, 2004).

HOMEOMORPHIC MALE. The body is 603–671 μm long. The setae *sci* are 7–25 μm long. The setae *vi* are from 2.4 to 3.0 times as long as genu I. The setae c_1 are from 0.7 to 1.0 times the length of the setae d_2; the setae c_3 are about as long as or slightly longer than the setae c_1; the setae d_1 are from 0.7 to 1.0 times the length of the setae d_2, the setae d_2 are far from gla; the setae e_1 are from 2.1 to 3.3 times longer than the setae d_2, and the setae e_2 are from 0.9 to 1.3 times the length of the setae f_2. The ratio *sci-sci:sci-sce* ranges from 0.6 to 0.8. The aedeagus is conical, and the canal of the aedeagus lacks lumps. The anal discs lack radiating lines, and their diameter is about one-third of the length of the setae ps_2. The setae ps_1 are from 3.4 to 4.0 times longer than genu IV, and more than three times as long as the setae ps_2. The setae ps_2 are from 0.8 to 1.1 times the length of genu IV; the setae ps_3 are thick and pointed. The ratio σ′:σ″ ranges from 0.8 to 0.9. The solenidion of tibia IV is slightly shorter than the seta *kT* (Fan and Zhang, 2004).

HETEROMORPHIC MALE. The idiosoma is 512–721 μm long and 375–456 μm wide. The prodorsal shield is evenly punctulate, with the posterior margin concave, and the distance between the setae *ve* ranges from 77 to 102 μm. The setae *vi* are from 2.2 to 2.6 times as long as genu I. The setae c_1 are from 0.7 to 0.8 times the length of the setae d_2; the setae c_3 are longer than the setae c_1; the setae d_1 are from 0.7 to 0.8 times the length of the setae d_2; the setae d_2 are far from gla; the setae e_1 are from 2.6 to 3.3 times longer than the setae d_2, and the setae e_2 are from 1.0 to 1.4 times longer than the setae f_2. The ratio *sci-sci:sci-sce* ranges from 0.8 to 1.1. The aedeagus is conical, and the canal of the aedeagus lacks lumps. The anal discs lack radiating lines, and their diameter is about one-third the length of the setae ps_2.

The setae ps_1 are long, 3.0–3.9 times longer than genu IV, and about three times as long as the setae ps_2. The setae ps_2 are from 0.5 to 1.1 times the length of genu IV, and ps_3 is thick and pointed. The ratio σ′:σ″ is from 0.8 to 0.9. The solenidion of tibia IV is shorter than the seta *kT* (Fan and Zhang, 2004).

HYPOPUS. The gnathosoma is 21–22 µm long and 14–15 µm wide. The apical solenidia are 31–32 µm long. The palpal supracoxal setae (*elcp*) are 9–10 µm long; the infracapitular setae (*m*) are 7 µm long. The idiosoma is 302–306 µm long and 192–195 µm wide. The prodorsal shield is 82–92 µm long. The setae *vi* are 1.9 times the length of genu I. The ratio c_1–c_1:d_1–d_1:e_1–e_1 is 1.3:1.0:1.2–1.3, and h_2:h_3 is 2.2–2.5. The anal disc ad_1 + ad_2 attachment structures are small. The solenidion ω_2 of tarsus I is less than one-half as long as ω_1. The solenidion of genu II is about one-half the length of genu II. The solenidion of tibia II is about two times the length of tibia II. Tarsus II is from 2.3 to 2.8 times the length of tibia II. The solenidion of genu III is less than one-third the length of genu III (Fan and Zhang, 2004).

Geographical distribution

R. robini is found worldwide (Hughes, 1976; Fan and Zhang, 2003, 2004).

Host plants and other biological hosts and substrates

Host plants and their families, and other biological hosts and substrates are listed below.

Amaryllidaceae: bulbs of shallot (*Allium ascalonicum* L.) (Fan and Zhang, 2003), onion (*Allium cepa* L.) (Manson, 1972; Gerson *et al.*, 1985; Bu and Li, 1998; Díaz *et al.*, 2000; Fan and Zhang, 2003), *A. chinense* G. Don (Díaz *et al.*, 2000), Welsh onion (*A. fistulosum* L.), *A. porrum* (L.) (Chen and Lo, 1989), garlic (*A. sativum* L.) (Manson, 1972; Tseng, 1979; Díaz *et al.*, 2000; Fan and Zhang, 2003), chives (*A. schoenoprasum* L.) (Bu and Li, 1998), *A. tuberosum* (Rottl. ex Sprengel) (Díaz *et al.*, 2000), scallion (*Allium* sp.) (Bu and Li, 1998), amaryllis (*Amaryllis* sp.) (Manson, 1972, Meyer Smith, 1981a; Fan and Zhang, 2003), bulbs (Michael, 1903; Womersley, 1941; Zakhvatkin, 1941), *Crinum* sp. (Manson, 1972; Fan and Zhang, 2003), *Hypiastrum* sp. (Fan and Zhang, 2003), magic lily (*Lycoris squamigera* Maxim.), *Lycoris* sp. (Manson, 1972; Fan and Zhang, 2003), *Narcissus* sp. (Manson 1972; Díaz *et al.*, 2000; Fan and Zhang, 2003), bulbs of *Nerine* sp. in the shade a house, fairy lily (*Zephyranthes*) (Fan and Zhang, 2003).

Apiaceae: rotting basal material of *Aciphylla* sp. (Manson, 1972; Fan and Zhang, 2003), carrot (*Daucus carota* L.) (Rossetto and Camargo, 1966; Manson, 1972; Fan and Zhang, 2003).

Araceae: pinellia (*Pinellia ternata* (Thunb.) Makino) (Bu and Li, 1998).

Asparagaceae: *Arthropodium cirrhatum* (G. Forst.) R. Br. (Manson, 1972; Fan and Zhang, 2003), rotting roots of *Asparagus* sp., *Galtonia* sp., hyacinth (*Hyacinthus* sp.) (Fan and Zhang, 2003).

Asteraceae: *Dahlia* sp. (Claparède, 1869; Womersley, 1941; Manson, 1972; Fan and Zhang, 2003).

Brassicaceae: roots of swedes (*Brassica napus* L.) (Fan and Zhang, 2003).

Cycadaceae: rotting seeds of *Cycas revoluta* Thunb. (Fan and Zhang, 2003).

Iridaceae: *Freesia* sp. (Díaz *et al.*, 2000; Fan and Zhang, 2003; Vacante, unpublished data), *Gladiolus* (Manson, 1972; Díaz *et al.*, 2000; Fan and Zhang, 2003; Vacante, unpublished data), *Iris* sp. (Manson, 1972; Fan and Zhang, 2003).

Liliaceae: oriental lily (*Lilium speciosum* Thunb.) (Fan and Zhang, 2003), *Lilium* sp. (potted) (Manson, 1972; Fan and Zhang, 2003), bulb of *Tulipa* sp. (Fan and Zhang, 2003; Vacante, unpublished data).

Mushroom (in compost) (Fan and Zhang, 2003).

Nothofagaceae: *Nothofagus* sp. (Fan and Zhang, 2003).

Poaceae: barley (*Hordeum* sp.) (Fan and Zhang, 2003), *Lolium longiflorum* Thunb. (Díaz *et al.*, 2000), rice with husk (*Oryza sativa* L.), bamboo shoot (*Sasa* sp.) (Tseng, 1979), *Secale cereale* L. (Díaz *et al.*, 2000), seeds of maize (*Zea mays* L.) (Fan and Zhang, 2003).

Primulaceae: *Auricula* sp. (on bulbs)? (Fan and Zhang, 2003).

Solanaceae: stem and damaged root of potato (*Solanum tuberosum* L.) (Claparède, 1869; Manson, 1972; Fan and Zhang, 2003).

Stored products (Michael, 1903).

Animal hosts include a restricted number of species: humans (Fan and Zhang, 2003); Fringillidae (*Fringilla coelebs* L.); Passeridae (*Passer montanus* L.); Talpidae (*Scalops aquaticus* L.); Turdidae (*Turdus philomelos* C.L. Brehm) (Fain, 1988).

Bio-ecology

The robine bulb mite is primarily associated with bulbs, corms and tubers or roots of plants, and has been found in seeds and the lower parts of plants. It is also common in compost and soil that is rich in organic matter. Its life cycle develops through the larva, protonymph (proceeding towards a hypopus), tritonymph and adult stages. The hypopus stage does not feed, resists unfavourable environmental conditions and employs phoresy with insects (narcissus flies, sciarids, etc.) to spread. The mite severely damages gladiolus both in the field and in storage (Wang, 1983). In Taiwan, it infests 10% of bulbs within 7 days of planting, and the optimum temperature for development is 24–28°C (Liu and Tzeng, 1994).

Different kinds of food influence the reproduction of the robine bulb mite, and temperature affects its development. At temperatures from 13 to 26°C, the life cycle takes from 7 to 27 days and at 25°C it takes 9 days (Garman, 1937; Zakhvatkin, 1941; Woodring, 1969b); at 27 ± 1°C, egg to adult development takes 12.1 days (Ostovan and Kamali, 1996). At 15, 20, 25 and 30°C, and at room temperature (14–34°C), the life cycle of *R. robini* takes 10.6 ± 2.9, 10.1 ± 2, 9.15 ± 3.1, 8.0 ± 1.48 and 14.6 ± 7.7 days, respectively. The incubation period of the eggs and the developmental period of the nymphs are influenced by temperature. Though the males have a shorter lifespan than the females, the duration of life, irrespective of sex, gradually decreases with increase in temperature from 15 to 30°C (Raut and Ruma Sarkar, 1991). At 27 ± 1°C on potato tubers, the pre-oviposition, oviposition and post-oviposition periods are long – 1.0, 22.5 and 8.2 days, respectively, while the reproductive rate

(eggs day^{-1}) and the fecundity are 58 and 730 eggs (Ostovan and Kamali, 1996). The threshold for development is 11.8°C and the thermal constant is 184.8 degree days. On garlic at 27°C, the fecundity is about 400 eggs, male longevity is longer than female longevity and the intrinsic rate of increase is 0.128 (Liu and Tzeng, 1994). Li *et al.* (2008) reported a threshold temperature and an effective accumulated temperature of 10.01°C and 171.72 degree-days, respectively. To reach the larval stage, the eggs require 43.5 degree-days above 11.7°C and from 9.6 to 23.9 days to reach the adult stage, depending on the temperature. The fecundity ranges from 99.0 to 203.5 eggs and the rate of eggs day^{-1} from 4.8 to 8.9. The longevity of the female and male ranges from 20.8 to 34.4 and 12.9 to 42.1 days, respectively (Liu and Tzeng, 1994). In the Kunming region of China, the mite can develop 12 generations a year (Li *et al.*, 2008).

The moulting rate of hypopi of *R. robini* is highest, and requires the least number of days, at 24°C and 100% RH, and no moulting occurs below 93% RH. Placing hypopi in distilled water inhibits the process somewhat, and moulting is generally better at lower temperatures than at the higher temperature of 35°C (Capua and Gerson, 1983). *R. robini* can complete its life cycle utilizing filter paper as a sole dietary source. However, the complete cellulase complex that is necessary for crystalline cellulose degradation rarely occurs in arthropods, and in this case the mite probably derives nutrients from the fungal mycelia growing on the filter paper rather than from the filter paper itself (Wooddy and Fashing, 1993).

Different predatory mites living in the soil prey on the robine bulb mite. In Egypt, the ascid *Protogamasellus minutus* Hafez, El-Badry *et* Nasr (Afifi *et al.*, 1987) and the digamasellid mites *Dendrolaelaps sayedi* Hassan, Afifi *et* Nawar (Hassan *et al.*, 1987) were investigated. In the Netherlands, the action of different soil mites collected in association with *R. robini* has been evaluated, such as the laelapid *Hypoaspis aculeifer*, the ascid *Lasioseius bispinosus* Evans and the parasitic mite *Parasitus fimetorum* (Berlese) (Leśna *et al.*, 1995). The first species gave promising results (Leśna *et al.*, 1996, 2000; Conijn *et al.*, 1997).

Symptomatology and damage

R. robini damages stored bulbs as well those planted in the field or in greenhouses. The mite penetrates inside bulbs and corms through the basal plate and the external layers. Although it feeds on any decaying organic material, it prefers healthy bulbs to rotten ones (Meyer Smith, 1981a). It has been observed that bruised bulbs and fungus-infected bulbs help its establishment more than healthy bulbs. For example, on *Fusarium*-infected bulbs, the mite populations show exponential growth after 14 days, and on healthy bulbs exponential growth of the population is reached after 60–90 days. The mite is attracted to and penetrates *Fusarium*-infested bulbs more rapidly than healthy ones (Okabe and Amano, 1990, 1991). The attack produces dark brown streaks on the roots externally around the neck, and distorted growing tips, leaves and flowers. On damaged bulbs, the population density of *R. robini* increases rapidly, transforming the whole bulb to a rotten pulp. In this phase, it possible to see thousands of mites and eggs in the damaged tissue.

On lilies, the stems can be attacked and break. Severe attacks induce small and stunted plants with few or no roots. The use of a combination of fungicides and acaricides to study the dynamics of the interactions between *R. robini* and root rot (*Pythium ultimum* Trow) on lily bulbs highlighted that a low mite population allowed suppression of the pathogen (Ascerno *et al.*, 1983). On onions, the mite population gathers in the roots, producing plant collapse and arrest of the growth of the bulbs.

The robine bulb mite is closely associated with the spread of the fungus diseases due to *Fusarium*, *Stromatinia* and *Pseudomonas*; the mite feeds on the fungus responsible for *Stromatinia* rot of gladioli and can be a vector of *Pseudomonas* (Forsberg, 1965; Meyer Smith, 1981a). Baker (1983) reported that viable *Fusarium* and *Cylindrocarpon* propagules could be found in the faecal pellets of *R. robini* feeding on a substrate of these pathogens, and noted the possible role of the mite as a vector of these diseases. Similarly, the acquisition and retention of *Pseudomonas marginata* (McCulloch) Stapp by *R. robini* has been widely established (Poe *et al.*, 1979). In the laboratory, dissemination of the pathogen occurs for a maximum of 3 days, which would reduce the mite's role as a vector to restricted conditions. Disinfestation of surface mites does not influence transmission of pathogen (Noble *et al.*, 1978), so the propagules are likely to be carried within the digestive system. However, other mechanisms, such as the soil fauna, may be involved (Beute and Benson, 1979) as well, and their role in disease transmission by mites must be evaluated. It is, then, possible that mites play only a minor role in the spread of diseases.

Nevertheless, on bulbs, the mite transfers (together with *Tyrophagus putrescentiae*) the fungal pathogens *Aspergillus niger* van Tieghem, *Haematonectria haematococca* (Berk. *et* Broome) Samuels and Rossman, *Rhizopus stolonifer* (Ehrenb.) Vuill. and *Penicillium chrysogenum* Thom attached to the outside of their bodies, whereas *Aspergillus flavus* J.H. Friedrich Link and *A. ochraceus* Wilh. are transferred through the mite's digestive tract along with their food (Abdel-Sater and Eraky, 2001). The mite has been collected on 85.9% of commercial bulbs in central and southern Chile (Rojas and Klimov, 2007).

Control

Control involves the use of chemical, physical, cultural and biological methods, applied alone or together. Cleaning of the environment, the handling and storage of bulbs and vegetable material, and the use and marketing of healthy material, free from fungal infections and the presence of mites are all very important. The integration of rigid prophylaxis with biological, physical and chemical means defines an IPM strategy in which the monitoring of mite populations could play a key role.

MONITORING. The results of research on the monitoring of mite populations do not seem to be conclusive. However, the use of some types of traps might improve this situation. An acarid bait trap constructed from moistened, folded filter paper containing a small quantity of dry yeast allows the number of *R. robini* in agricultural soils to be monitored. Under laboratory evaluations, the trap captured 65% of the acarid adults by 3 days after application. Field trials carried out in Welsh onion field soil using both the bait trap and Tullgren funnel traps showed that numbers of *R. robini* obtained from the soil did not differ significantly between the two trap types, suggesting that

efficient monitoring of *Rhizoglyphus* is possible using bait traps (Kasuga *et al.*, 2005).

PHYSICAL CONTROL. The high-temperature tolerance of *R. robini* (and other acarids) has recently been reviewed to explore the possibility of controlling the mite in agricultural soils by heat treatment at 40°C for 24 h, 45°C for 3 h and 50°C for 1 h. The tolerance of *R. robini* is similar to that of *Tyrophagus similis*, and both have lower tolerances than other acarids. *T. similis* and *R. robini* may not have developed high-temperature tolerance because the two species dwell in soil where temperature fluctuations are not severe (Kasuga and Honda, 2006a). This confirms the effectiveness of control strategies already used successfully by farmers. In lilies, mite populations are controlled using acaricides, e.g. pirimiphos-methyl in combination with cultural measures such as hot-water treatment (2 h at 39°C or 41°C) and the storage of bulbs at −2°C. In peat growing media, the combination of hot-water treatment and biological control by the predatory mite *Hypoaspis aculeifer* can replace acaricide application and control *R. robini* on lily bulbs during the propagation phase (Conijn *et al.*, 1997). Flooding the soil with water for 5 days kills 96.1% of the populations in gladiolus and all mites are controlled after 14 days (Wang and Lin, 1986).

CULTURAL PRACTICES. In garlic cultures in Brazil, a rather large amount of organic matter (compost) added to the soil before 'seeding' the garlic induced the development of *Rhizoglyphus*. In particular, sugarcane bagasse is readily available and cheap, but as it takes quite some time to compost totally; it has been used 'semi-raw' on garlic fields, which results in the field giving off a slightly sweet odour. This attracts insects, mainly Coleoptera, from a long distance, and many of these carry hypopi of *Rhizoglyphus*, which end up developing into adults and damaging the garlic plants. A reduction in the amount of organic matter used and the use of totally composted bagasse solved this problem. Repetitive ploughing of the soil in between subsequent cultures and exposing the mites to sunlight also helped to eliminate them. So, a preventive cultural procedure to avoid *Rhizoglyphus* problems in the soil could be the use of only a moderate amount of totally composted organic matter, and no excess of moisture (Flechtmann, personal communication).

BIOLOGICAL CONTROL. Biological control is provided mainly by releases of the laelapid mite *H. aculeifer*. The study of this predator began in the 1980s in propagation lily bulbs and other crops. Experiments carried out in the greenhouse and the field showed that the initial predator-to-prey ratios required to achieve suppression (*c.*1:2) or elimination (*c.*3:1) in the soil were much higher than those required for bulb mite elimination when lily bulb scales were embedded in a medium of vermiculite (*c.*1:20) (Leśna *et al.*, 2000). *H. aculeifer* is used in various countries. The biology and potential use in biological control (and bioindication) of free-living mesostigmatic mites in Australia has been reviewed by Beaulieu and Weeks (2007).

CHEMICAL CONTROL. *R. robini* is commonly sensitive to various active ingredients including azocyclotin, cyhexatin, methidathion, phosmet, pirimiphos methyl, profenfos, prothiofos and sulfur sprayed on to the soil or directly on to bulbs, or used as a soaking solution for bulbs. However, strains that are resistant to one or more chemicals have been recorded in various countries.

Using the filter paper method, Kuwahara (1986) evaluated the action of 41 organophosphorus insecticides, and found that strains collected in the field in Japan showed a high level of resistance to a wide range of active ingredients, involving a cross-resistance and major resistance factors. Shinkaji *et al.* (1986) evaluated the susceptibility of *Rhizoglyphus* spp. to dimethoate and disulfoton on rakkyo (*Allium chinense*) and Chinese chive (*A. tuberosum*). Wang and Lin (1986) reported that metasystox and neoron failed to control *R. robini* on gladiolus. Chen and Lo (1990) observed high multiple resistance to organophosphorus insecticides, including diazinon, dimethoate, fenitrothion, malathion, methyl parathion, omethoate, parathion and prothiophos. Chen (1990a) improved a method for determining the susceptibility of the robine bulb mite to chemicals; the same author studied the resistance, biology and control of the mite (Chen, 1990b). Kato *et al.* (1993) tested, using a filter paper method, 32 pesticides on females of the mite collected in the field on Welsh onion (*A. fistulosum*) and found that all collected populations showed a high susceptibility to prothiofos, methidathion, fenthion and pirimiphos methyl. Among the older chemicals, methyl bromide (which is no longer allowed) is very effective at a rate of 32 g m^{-3} (Meyer Smith, 1981a) or 50 g m^{-3}, as is dazomet at 35g m^{-2} (Wang and Lin, 1986). Dip treatment using benzoximate, bromopropylate or demeton-methyl, or fumigation for 48 h with phosphine from aluminium phosphite tablets (1 tablet m^{-1} at 25–38°C or 2 tablets m^{-1} at 15–20°C) were also effective (Wang, 1983). Similarly, dipping the bulbs of gladiolus before planting and applying a prothiofos solution twice on to the soils 7 and 28 days after planting is effective (Liu, 1995).

Rhizoglyphus echinopus (Fumoze *et* Robin) (Fig. 13.12)

Common name

This is a bulb mite.

Diagnostic characteristics

FEMALE. The idiosoma is 791–860 μm long. The prodorsal shield is evenly punctulate. The seta *scx* is thick and 45–50 μm long. Grandjean's organ has a distinct forked tip. The dorsal idiosomal setae are relatively long. The setae *vi* are thick and pointed, and the setae *sci* 1–2 times the length of genu I. The ratio *sci-sci*:*sci-sce* ranges from 0.9 to 1.1. The setae c_1 are from 0.8 to 1.2 times the length of the setae d_2; the setae c_3 are about as long as the setae c_1; the setae d_1 are from 0.6 to 1.1 times the length of the setae d_2; the setae d_2 are far from the gla; the setae e_1 are from 1.2 to 1.5 times the length of the setae d_2, and the setae e_2 from 0.9 to 1.3 times the length of the setae f_2. The anal

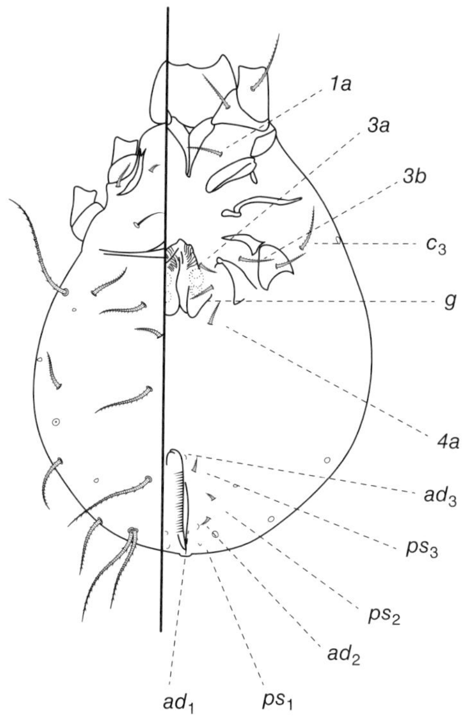

Fig. 13.12. *Rhizoglyphus echinopus* (Fumoze *et* Robin). Female: dorsal (left) and ventral (right) (modified from Fan and Zhang, 2004). The labels are all setal notations, which are explained in the text.

opening has six pairs of small setae, and three pairs are longer than the others; the setae ps_1 are less than one-quarter of the length of genu IV, the setae ps_2 are less than one-half of the length of genu IV, and the setae ps_3 about one-half of the length of genu IV; the setae ad_1 and ad_3 are small. The copulatory opening has a small circular shield; the spermathecal duct is broadly expanded from the copulatory opening to the inner part of the spermatheca. The inner part of the spermatheca is large and 'Y'-shaped; the sclerites of the oviducts are widely expanded, and the distance between them is from 97 to 122 μm, with the basal part subequal to the branches (Fan and Zhang, 2004).

HOMEOMORPHIC MALE. The body is 590–756 μm long. The seta *scx* is thick and 45–50 μm long. Grandjean's organ has a distinctly forked tip. The dorsal idiosomal setae are relatively long, and the setae *sci* are long, from 45 to 95 μm. The setae *vi* are from 2.2 to 3.2 times as long as genu I. The setae c_1 are from 0.8 to 1.0 times the length of the setae d_2; the setae c_3 are about as long as or slightly shorter than the setae c_1; the setae d_1 are from 0.5 to 1.0 times the length of the setae d_2; the setae d_2 are far from the gla; the setae e_1 are from 1.0 to 1.7 times the length of the setae d_2, and the setae e_2 are from 0.7 to 1.3 times the length of the setae f_2. The ratio *sci–sci:sci–sce* ratio ranges from 1.2 to 1.5. The aedeagus is terminally truncated, and its canal lacks lumps. The anal discs lack radiating lines, and their diameter is less than one-third the length of the setae ps_2. The setae ps_1 are from 4.5 to 4.9 times longer than genu IV, and more than three times longer than the setae ps_2. The setae ps_2 are from 0.8 to 1.4 times the length of genu IV, and the setae ps_3 are slightly blunt. The ratio σ′: σ″ is from 1.1 to 1.2. The solenidion φ on tibia IV is pointed and about as long as the ventral seta *kT* (Fan and Zhang, 2004).

Geographical distribution

R. echinopus is found worldwide (Fan and Zhang, 2003, 2004).

Host plants and other biological hosts and substrates

Host plants and their families and other biological hosts and substrates are listed below.

Amaryllidaceae: *Allium bakeri* Regel (Díaz *et al.*, 2000), *A. cepa* L. (Sandhu, 1976; Bu and Li, 1998; Díaz *et al.*, 2000; Fan and Zhang, 2003), corms of garlic (*A. sativum* L.) (Manson, 1972; Díaz *et al.*, 2000; Fan and Zhang, 2003), bulbs of amaryllis (*Amaryllis* sp.) (Fan and Zhang, 2003), daffodils (*Narcissus* sp.) (Manson, 1972; Díaz *et al.*, 2000; Fan and Zhang, 2003).

Apiaceae: carrot (*Daucus carota* L.) (Rossetto and Camargo, 1966).

Araceae: *Pinellia ternata* (Thunb.) Makino (Bu and Li, 1998).

Asparagaceae: hyacinths (*Hyacinthus* sp.) (Manson, 1972; Fain, 1988; Díaz *et al.*, 2000; Fan and Zhang, 2003), roots of *Lachenalia pendula* Aiton (Fan and Zhang, 2003).

Bulbs (Manson, 1972).

Convolvulaceae: sweet potato seed in budgerigar cage (*Ipomoea batatas* (L.) Lam.) (Fan and Zhang, 2003).

Gesneriaceae: gloxinia (*Sinningia speciosa* (Lodd.) Hiern) (Fan and Zhang, 2003), gloxinia (*Sinningia*) (Manson, 1972).

Iridaceae: *Freesia* sp. (Díaz *et al.*, 2000), *Gladiolus* (Manson, 1972; Díaz *et al.*, 2000; Fan and Zhang, 2003; Vacante, unpublished data), iris (*Iris*) (Manson, 1972; Fan and Zhang, 2003).

Liliaceae: *Lilium longiflorum* Thunb. (Díaz *et al.*, 2000), *Tulipa* sp. (Manson, 1972; Díaz *et al.*, 2000; Fan and Zhang, 2003). Lily bulb, rice straw (Tseng, 1979).

Paeoniaceae: paeony plants (*Paeonia* sp.) (Manson, 1972; Fan and Zhang, 2003).

Plant material (Manson, 1972).

Poaceae: straw of rice (*Oryza sativa* L.) (Fan and Zhang, 2003), stored wheat (Bu and Li, 1998).

Solanaceae: *Capsicum* (Díaz *et al.*, 2000), *Solanum* sp. (Díaz *et al.*, 2000).

Stored food (Hughes, 1961).

Zingiberaceae: *Curcuma domestica* Valeton (Díaz *et al.*, 2000).

The animal hosts include the Ctenophthalmidae (Insecta) *Palaeopsylla minor* (Dale) ex *Talpa europaea* L. (Fain, 1988).

Bio-ecology

The biological cycle of *R. echinopus* is similar to that of *R. robini*. This mite also lives on bulbs and corms and has been collected on several ornamentals, vegetables and organic substrates, including animals. At 27°C on potato tubers, the development from egg to adult takes 13.5 days, the pre-oviposition period is 0.9 days, the oviposition period 31.5 days and the post-oviposition period 13.4 days. The reproductive rate (eggs day^{-1}) is 43 and the

fecundity 620 eggs (Ostovan and Kamali, 1996). In the laboratory, at 20, 25, 27 and 30°C and 95% RH, the highest survival rate is at 30°C, and this decreases markedly at the lower rearing temperatures. The developmental period becomes shorter with rise in temperature and each stage becomes longer as development progresses. The threshold temperature is 9.44 and the thermal constant is 105.52 degree-days. As the temperature increases, adult longevity and the sex ratio increase, and the incubation period decreases. Temperature does not affect fecundity (Sakurai *et al.*, 1992). At temperatures from 25 to 34°C, the larva, protonymph and tritonymph live for 4.0, 2.5 and 1.2 days, respectively; adult longevity is 45.8 days (Jiang and Ge, 1994). At 25°C and 95% RH on potato tubers, potato starch and fish powder, the mortality is 15.8, 33.3 and 29.4%, respectively. On the same foods, fecundity is 202, 161 and 132 eggs and female longevity is 38.6, 20.8 and 32.8 days, respectively. Only on filter paper does the mite fail to reach the tritonymphal stage (Chen *et al.*, 2006b). Both adults and nymphal stages can overwinter in soil and on damaged host plants. According to the region, in China the mite has from nine generations a year (on *A. fistulosum*) (Jia *et al.*, 2002) to 10–12 generations a year (Hu *et al.*, 2005, in Fan *et al.*, 2010). In Jiaxiang, Shandong Province, its populations increase by the end of March to the beginning of April and peak in May and August (Jia *et al.*, 2002), while in Wuhan, Hubei Province, two population peaks have been observed, from late September to late October and from late March to mid-May (Hu *et al.*, 2005, in Fan *et al.*, 2010).

Symptomatology and damage

The symptomatology of attack and damage are similar to those of *R. robini*. For *R. echinopus*, attack of the roots of *Freesia* and *Gladiolus* also induces the formation of brown streaks, and interior mining is often observed. Corms show distorted growing tips and leaves. While this mite is not a key pest of bulbs, it often causes severe losses, as the slightest injury or bruising of the bulb will enable the mites to breed. Once *R. echinopus* is established, it extends the damage and reduces the bulb to a rotten pulp. In storage, the mite readily wanders from bulb to bulb and may act as a vector of different pathogenic fungi. Attack by this mite is similar to that of other common pests of bulbs in that it is associated with mechanical damage or unsatisfactory conditions of storage (Hussey *et al.*, 1969). The bulb mite is often associated with *Fusarium* spp. in gladiolus corms and *Freesia*, and with bacterial or fungal rots of dahlia tubers. When bulbs are intended for forcing, a representative sample should be examined. Attacked freesia and gladiolus corms are usually soft and mushy, and may even be unsuitable for planting. If infestation is widespread, the bulbs should be rejected, and when only a few bulbs are attacked, any that are soft should be rejected (Hussey *et al.*, 1969). The passage of viable propagules of *Verticillium albo-atrum* Reinke *et* Berthold through the digestive system of *R. echinopus* when the mites fed on colonies, conidia and microsclerotia of the fungus highlights the possible role of mite in fungal epidemiology (Price, 1976). Gray *et al.* (1975) found that vectoring of smoulder (*Sclerotinia narcissicola* P.H. Gregory) in *Narcissus* by *R. echinopus* had little importance because the incidence of the disease in commercial stocks was already high. The effects of vectoring are also negligible in soils with a high inoculum of pathogens, as occurs in some muck soils (Abawi and Lorbeer, 1971). In Chile, the mite has been found in 14.1% of commercial bulbs, and is predominant in central areas (Rojas and Klimov, 2007).

Control

The control of bulb mites follows the same protocol as for *R. robini*, based on the application of physical, chemical and biological methods, alone or together. Also, in this case, control requires a rigorous commercial and agronomic management of the bulbs. The health of the storage environment and use of healthy material are of primary importance.

PHYSICAL CONTROL. Hot-water treatment gives good results. Immersion in water for 1 h at 38°C is sufficient for all types of bulbs, except probably large lily bulbs. The treatment is unsuitable for forcing tulips, and should be given to narcissi at the correct developmental stage (Hussey *et al.*, 1969).

BIOLOGICAL CONTROL. See *R. robini*.

CHEMICAL CONTROL. The mite easily develops strains that are resistant to various chemicals, so the active ingredients used may exhibit a variable action, resulting in the highest known toxicity of old acaricides as cyclodiene GABA (γ-aminobutyric acid) antagonists (dieldrin, endrin and aldrin), some organophosphates (chlorpyrifos, diazinon and azinphos ethyl) and carbamates (carbofuran), anticholinesterases and the thiazolidine flubenzimine. Amitraz, fenazaflor, fenazaquin and oxythioquinox are less toxic than the cyclodienes, carbamates and organophosphates, but sodium channel agonists (DDT and pyrethroids) and several acaricides are inactive (Gencsoylu *et al.*, 1998). Fumigation with methyl bromide (3 1b 42.5 m^{-3}) gives good control of *R. echinopus* with slight phytotoxicity. The application of granular acaricides, e.g. terbufos, in planting trenches (60 kg ha^{-1}) gave good control (Liu, 1995), and soil treatments with dazomet are effective (Lee and Leu, 1997).

Rhizoglyphus setosus Manson (Fig. 13.13)

Common name

This is a bulb mite.

Diagnostic characteristics

FEMALE. The idiosoma is 465–686 μm long and 301–459 μm wide. The prodorsal shield is evenly punctulate, with the posterior margin concave. The cheliceral setae (*cha*) are branched distally. The setae *vi* are blunt, the setae *sci* are about one-third of the length of genu I, and the ratio *sci-sci*:*sci-sce* ranges from 0.6 to 0.8. Grandjean's organ is bifurcate to two minute branches. The setae *scx* are slender and pointed. The setae c_1 are 0.9–1.3 times the length of the setae d_2; the setae c_3 are about as long as c_1; the setae d_1 are 1.6–2.0 times the length of d_2; the

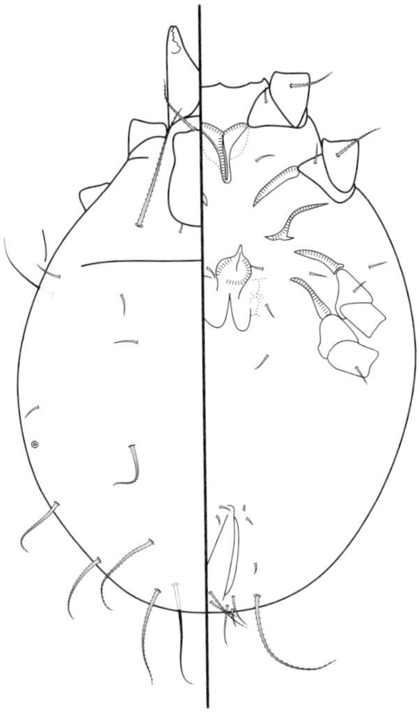

Fig. 13.13. *Rhizoglyphus setosus* Manson. Female: dorsal (left) and ventral (right) view (modified from Fan and Zhang, 2004).

setae d_2 are not close to the gla; the setae e_1 are 2.8–3.3 times the length of d_2, and the setae e_2 are 0.9–1.4 times the length of the setae f_2. The anal opening has six pairs of setae, two pairs of them very long and whip like, with ps_1, ps_2 and ps_3 about one-half the length of genu IV; the setae ad_1 are more than three times as long as ps_3; the setae ad_2 shorter, about one-half as long as the setae ad_1, and the setae ad_3 are shorter than ps_1 (Fan and Zhang, 2004).

HOMEOMORPHIC MALE. The idiosoma is 526–624 μm long and 372–441 μm wide. The prodorsal shield is evenly punctulate, posteriorly often concave. The setae *vi* are 2.4–3.0 times the length of genu I. The setae c_1 are 0.8–1.2 times the length of the setae d_2; the setae c_3 are about as long as or slightly shorter than c_1; the setae d_1 are 1.1–1.7 times the length of d_2; the setae d_2 are not close to the gla; the setae e_1 are 2.1–4.1 times as long as d_2; the setae e_2 are 1.0–1.5 times as long as f_2. The ratio *sci-sci*:*sci-sce* is 0.7–0.8. The aedeagus is straight and the canal lacks lumps. The anal discs lack radiating lines, and their diameter is about one-half of the length of setae ps_2. The ratio σ':σ" ranges from 0.8 to 0.9. The solenidion φ of tibia IV is slightly shorter than the ventral seta *kT* (Fan and Zhang, 2004).

HETEROMORPHIC MALE. The idiosoma is 562–566 μm long and 345–390 μm wide. The prodorsal shield is evenly punctulate, posteriorly slightly concave. The setae *vi* are long 2.5–2.60 times the length of genu I. The setae c_1 are long, 0.9–1.4 times the length of the setae d_2; the setae d_1 are long, 1.1–1.3 times the length of the setae d_2; the setae d_2 are not close to the *gla*; the setae e_1 are 3.5 times as long as the setae d_2; the setae e_2 are 1.2 times as long as the setae f_2. The ratio *sci-sci*:*sci-sce* is 0.8–0.9. The aedeagus is straight, and the canal lacks lumps. The anal discs lack radiating lines. The setae ps_1 are 0.3 times the length of genu IV. The setae ps_2 are 0.4 times of length of genu IV, and the setae ps_3 are shorter than ps_2. The ratio σ':σ" is from 0.8 to 0.9. The solenidion φ on tibia IV is slightly shorter than the ventral seta *kT* (Fan and Zhang, 2004).

HYPOPUS. The idiosoma is 291–301 μm long and 222 μm wide. The gnathosoma is 32–37 μm long and 22 μm wide; the apical solenidia are 32–35 μm long. The palpal setae are 14–15 μm long, and the infracapitular setae are 7 μm long. The prodorsal shield is 85–87 μm long. The setae *vi* are 30–35 μm long, from 2.0 to 2.2 times the length of genu I; the setae *ve* are 8 μm long and the distance between the setae *ve-ve* is 15 μm. The setae *sci* are 8 μm long and the distance between the setae *sci-sci* is 37 μm; the setae *sce* are 10 μm long and the distance between the setae *sci-sce* is 30 μm; the ratio *sci-sci*:*sci-sce* is 1.2. The ratio c_1-c_1: d_1-d_1: e_1-e_1 is 1.6:1.0:1.3 and h_2:h_3 is 1.8–2.1. The anal disc attachment structures ad_1 + ad_2 are medium sized and the disc diameter is 10 μm. The solenidion ω_2 of tarsus I is about one-half the length of the soleniodion ω_1. The solenidion σ of genu II is about two-thirds the length of genu II. The solenidion φ of tibia II is about two times the length of tibia II. Tarsus II is three times the length of tibia II. The solenidion σ of genu III is about as long as genu III (Fan and Zhang, 2004).

Geographical distribution

R. setosus is found worldwide (Fan and Zhang, 2004).

Host plants and other biological hosts and substrates

Host plants and their families and other biological hosts and substrates are listed below.

Amaryllidaceae: *Allium ascalonicum* L. (Chen *et al.*, 2002a), *A. cepa* L. (shallots) (Fan and Zhang, 2004), *A. cepa* (onion) (Almaguel Rojas, 2002; Chen *et al.*, 2002a; Martínez *et al.*, 2004; Suárez, 2004), *A. fistulosum* L. (Chen *et al.*, 2002a), *A. porrum* (L.) (Martínez *et al.*, 2004), *A. sativum* L. (Chen *et al.*, 2002a; Martínez *et al.*, 2004; Suárez, 2004), *A. tuberosum* L., *Hippeastrum equestre* (Aiton) Herb., *Lycoris aurea* (L'Hér.) Herb., *Nerine bowdenii* W. Watson (Chen *et al.*, 2002a).

Apiaceae: carrot (*Daucus carota* L.) (Rossetto and Camargo, 1966; Chen *et al.*, 2002a; Martínez *et al.*, 2004).

Araceae: *Colocasia esculenta* (L.) Schott (Almaguel Rojas, 2002; Chen *et al.*, 2002a; Martínez *et al.*, 2004), taro (*Colocasia* sp.) (Fan and Zhang, 2004), *Xanthosoma* sp. (Suárez, 2004).

Arecaceae: tubers of *Caladium*, bulbs of *Caladium* (Manson, 1972), seeds of *Caryota mitis* Lour. (Fan and Zhang, 2004), *Cocos nucifera* L. (Suárez, 2004), *Spathiphyllum kochii* Engl. *et* K. Krause (Chen *et al.*, 2002a).

Asparagaceae: *Hosta sieboldiana* (Hook.) Engl., *Liriope graminifolia* (L.) Baker (Chen *et al.*, 2002a), *Polianthes tuberosa* L. (Chen *et al.*, 2002a; Martínez *et al.*, 2004; Suárez, 2004).

Asteraceae: *Aster chinensis* L. (Suárez, 2004), *Callistephus hortensis* (Cass.) Cass. (Martínez *et al.*, 2004), *Dahlia coccinea* Cav. (Martínez *et al.*, 2004; Suárez, 2004), *Gerbera* sp., *Helianthus annuus* L., *Zinnia elegans* Jacq. (Martínez *et al.*, 2004).

Bulbs of green onions, gladiolus, lilies, leeks and garlic (Ho and Chen, 1987).

Convolvulaceae: *Ipomoea batatas* (L.) Lam. (Manson, 1972; Almaguel Rojas, 2002; Chen *et al.*, 2002a; Fan and Zhang, 2004; Martínez *et al.*, 2004; Suárez, 2004).

Cucurbitaceae: *Momordica charantia* L. (in yacht stores) (Fan and Zhang, 2004).

Cycadaceae: *Cycas revoluta* Thunb. (Fan and Zhang, 2004).

Dioscoreaceae: *Dioscorea alata* L. (Martínez *et al.*, 2004; Suárez, 2004), yam (*Dioscorea* sp.) (Fan and Zhang, 2004).

Euphorbiaceae: *Euphorbia heterophylla* L. (Gutierrez and Gonzalez, 1993), *Manihot esculenta* Cranz (Martínez *et al.*, 2004; Suárez, 2004).

Fabaceae: *Inocarpus edulis* J.R. Forst. *et* G. Forst. (Fan and Zhang, 2004).

Fungus growth on frangipani (Manson, 1972).

Gentianaceae: *Eustoma russellianum* Salisb. (Chen *et al.*, 2002a).

Iridaceae: bulbs of *Gladiolus* sp. (Liu, 1996), *Gladiolus communis* L. (Almaguel Rojas, 2002; Martínez *et al.*, 2004; Suárez, 2004), *Gladiolus hybrida* L. (Chen *et al.*, 2002a).

Lamiaceae: *Mentha nemorosa* Willd. (Martínez *et al.*, 2004).

Liliaceae: *Lilium longiflorum* Thunb. *Lilium* oriental hybrids (Chen *et al.*, 2002a), *Lilium* sp. (bulbs) (Liu, 1996).

Malvaceae: *Althaea rosea* (L.) Cav. (Martínez *et al.*, 2004), *Pachira macrocarpa* (Cham. *et* Schltdl.) Walp. (Chen *et al.*, 2002a), cocoa (*Theobroma cacao* L.) (Suárez, 1991a, 2004).

Musaceae: *Musa* spp. (Suárez, 2004).

Orchidaceae: rotting *Calanthe rubens* Rdl. (Fan and Zhang, 2004).

Orobanchaceae: *Orobanche ramosa* L. (Martínez *et al.*, 2004).

Palm seed (Fan and Zhang, 2004).

Pandanaceae: *Pandanus odorus* Ridley (Chen *et al.*, 2002a).

Poaceae: *Echinochloa colonum* (L.) Link (Gutierrez and Gonzalez, 1993), *Saccharum officinarum* L. (Martínez *et al.*, 2004; Suárez, 2004), *Zea mays* L. (Chen *et al.*, 2002a).

Polygonaceae: *Polygonum multiflorum* (Thunb.) Haraldson (Chen *et al.*, 2002a).

Rubiaceae: *Coffea arabica* L. (Suárez, 2004).

Solanaceae: *Capsicum annum* L., *Nicotiana tabacum* L. (Martínez *et al.*, 2004), *Solanum lycopersicum* L. (Suárez, 2004), *S. tuberosum* L. (rot on potato) (Almaguel Rojas, 2002; Fan and Zhang, 2004; Martínez *et al.*, 2004; Suárez, 2004).

Talinaceae: *Talinum triangulare* (Jacq.) Willd. (Suárez, 2004).

Taro with Chinese food (Manson, 1972).

Unknown bulbs (Manson, 1972).

Zingiberaceae: *Curcuma roscoeana* Wall. (decaying rhizomes) (Akiyama *et al.*, 1997).

R. setosus has also been reported on mealybugs on pineapples (Manson, 1972), and in the upper supralittoral zone (Fain and Schuster, 1983).

Bio-ecology

R. setosus was first described from bulbs in Taiwan (Ho and Chen, 1987), and was later reported as injurious to onions (Chen *et al.*, 2002b). The mite lives associated with underground organs of various plants and different substrates in tropical and subtropical countries. The general features of its bio-ecology are similar to those of *R. robini*. At 23–25°C the developmental time of biological cycle is from 9 to 11.5 days, the threshold of development is 11.9°C and the development of a generation needs 183.4 degree-days. Populations living in the soil infest the underground organs and the plants, and the first motile forms are observed within 10 days of massive germination. Higher densities are found within 35–40 days (Almaguel Rojas, 2002). At 28°C, the developmental time from egg to adult is 8 days. The fecundity is 202.14 eggs, with a rate of 7.95 eggs day^{-1}. The longevity of the female and male is 23.79 and 23.30 days, respectively. The sex ratio is 3.31 females:1 male. The peak of oviposition occurs on day 6 after mating (Chen *et al.*, 2002a).

Molecular techniques applied to identify various *Rhizoglyphus* species and the nucleotide sequences of the partial 16S rDNA gene of the six bulb mites show that the average nucleotide sequence divergence between *Caloglyphus* and *Rhizoglyphus* is 15%, and that the divergence between *R. setosus* and the other four *Rhizoglyphus* spp. is 13% (Yeh, 1999). Through the amplification of single bulb mites by nested PCR, species-specific primers have been identified to detect *R. robini* and *R. setosus* (Li *et al.*, 2010). The alarm pheromone (neryl formate) is not species specific among *Rhizoglyphus* spp. (Akiyama *et al.*, 1997), and the female sex pheromone (*S*-isorobinal) (Mizoguchi *et al.*, 2005) is well known.

Symptomatology and damage

Populations of *R. setosus* occur within onion bulbs kept as stock and in the field. Mites inhabiting the soil infest the underground organs of onions and other cultivated host plants, and the first motile forms are observed within 10 days from massive germination, and reach higher densities within 35–40 days (Almaguel Rojas, 2002). *R. setosus* penetrates the bulb via the neck, and from there it spreads to the outermost scales. Damaged bulbs show brown rotting tissue externally around the neck. The leaves may be delayed in development and may appear abnormal. Thousands of mites and eggs may be observed in damaged tissues. Severely infested plants may be small and stunted and have few or no roots. Sometimes, the mites aggregate on the roots, causing the plant to collapse and inhibiting bulb growth (Almaguel Rojas, 2002; Chen *et al.*, 2002b).

Control

Control requires both preventive and curative measures. Prevention is based on the use of healthy seeds and on scrupulous hygiene during storage. Other options are as for other *Rhizoglyphus* spp. Chemical control does not seem to give definitive solutions.

CHEMICAL CONTROL. Sprayings of chemicals are suggested when 2–3 mites plant^{-1} are detected from 5 days of transplantation and until 45 days. Chemical sprayings on the leaves are usually ineffective (Almaguel Rojas, 2002). Trials carried out in Taiwan show that *R. setosus* is more tolerant to several organophosphorus, carbamate, pyrethroid and other compounds than *R. robini*. Among the compounds tested, azocyclotin,

cyhexatin, demeton-*S*-methyl, methidathion, phosmet, profenofos and prothiophos are toxic to *R. setosus*. Both *R. setosus* and *R. robini* were tolerant to avermectin and to some synthetic pyrethroids (Chen and Lo, 1989). As for *R. robini*, dip treatment using benzoximate, bromopropylate, demeton-methyl or fumigation for 48 h with phosphine from aluminium phosphite tablets (1 tablet m^{-1} at 25–38°C or 2 tablets m^{-1} at 15–20°C) were effective (Wang, 1983). Similarly, dipping bulbs of gladiolus before planting and using a prothiofos solution twice on the soil 7 and 28 days after planting is effective (Liu, 1995).

The Genus *Sancassania* Oudemans

Diagnostic characteristics

The most important steps in the history of this genus up to the 1980s were discussed by Timms *et al.* (1981). Further updates may be found in Klimov and O'Connor (2003), who included in the genus *Sancassania* Oudemans the genera *Caloglyphus* Berlese, *Isoglyphus* Zakhvatkin, *Mycetosancassania* Klimov and *Rhizoglyphoides* Volgin.

Knowledge on the systematics of this genus seems not to be conclusive and in order to avoid creating confusion, the genus is defined here using the suggestions of Hughes (1976). The setae *ve* are short and fine, set near the middle of the lateral edge of the prodorsal shield, or are absent. The setae *sce* are longer than the setae *sci*, which are well developed. The hysterosoma has a complete set of setae, and the longer setae may be expanded at the base. The posterior side of the male body does not form a projecting opisthosomal plate. On legs I and II, the seta *ba* is not thickened to form a short conical spine and is well separated from the solenidion ω_1. Tarsus I bears the seta *aa*. The setae *e* at the distal end of tarsi I–III are shaped like a spine; the setae *ra* and *f* are usually curved and may have the distal ends expanded into leaf-like plates. Five ventral terminal spines are present, of which *p*, *q*, *u* and *v* are about equal in size, and *s* is a little larger. Heteromorphic males and hypopi are common (Hughes, 1976).

Economic importance

Some species of *Sancassania* (mainly those previously belonging to the genus *Rhizoglyphoides*) live on fruiting bodies of mushrooms and only reproduce several weeks each year when the mushrooms develop; during the remaining time they are inactive in the soil, waiting for the next mushroom season (Klimov and O'Connor 2003). The best known and most common species is *Sancassania berlesei* (Michael), which has been reported from time on edible fungi in different countries (Hussey *et al.*, 1969; Clancy, 1981). The mite feeds on hyphae and sporocarps of all sizes of the button mushroom, *Agaricus bisporus* Lange, but is not common in mushroom houses (Greenslade and Clift, 2004). In Italy, *S. berlesei* has been found on melon in greenhouses (Laffi and Ponti, 1997), but the damage produced is unknown. In Iran, *S. rodionovi* (Zakhvatkin) is a serious pest of the mushrooms *A. bisporus* and *Pleurotus ostreatus* Kummer (Kheradmand *et al.*, 2007).

Sancassania berlesei (Michael) (Fig. 13.14)

Common name

This acarid mite has no particular common name

Diagnostic characteristics

FEMALE. The idiosoma is rounded and from 800 to 1000 µm long. The idiosomal dorsal setae are shorter than in the homomorphic male; the setae h_1 are not as long as the setae e_1, slightly pectinate and with a truncate distal tip. There are six pairs of anal micro-setae, two pairs of which are grouped on either side of the anterior margin of the anus, with four pairs around the posterior margin. The genital sense organs are large and conspicuous. Apart from secondary sexual characters, the leg chaetotaxy is the same as the male. The bursa copulatrix is terminal and surrounded by a small sclerotized plate; it is connected with the sac-like receptaculum seminis by a narrow tube (Hughes, 1976).

HOMOMORPHIC MALE. The idiosoma is 600–900 µm long. The cucticle is smooth, colourless and shining, with pale brown appendages. In damp surroundings, the body is almost spindle shaped and widest between the third and fourth pairs of legs. The gnathosoma is narrow and tapering, and the chelicerae with toothed shears and a conspicuous mandibular spine. Grandjean's organ is a truncate spine with small pectinations. The prodorsal shield is oblong, with a slightly concave or irregular posterior margin and straight sides. All the dorsal idiosomal setae, excepting the setae *vi*, are almost completely smooth, of various lengths and thicker at the base. The setae *ve* are fine and set midway along the lateral sides of the prodorsal shield. The setae *sce* are three to four times longer than setae *sci*. The setae *scx* are slightly pectinate and more than half the length of the setae c_1. The setae c_1 are short, d_1 are two or three times longer than c_1, and d_2 and c_2 are 1.5 to twice as long as c_1; the setae e_1, h_1 and e_2 are long, and the setae h_1 extend well beyond the end of the body. The setae ps_2 are two to five times longer than ps_3, and the setae ps_1 are longer than ps_2 and, like them, stretch beyond the posterior margin. The anal suckers are round and conspicuous. The solenidion ω_1 of tarsus I has a slightly expanded distal end and arises from the same chitinous depression of the famulus ε. The solenidion ω_3 is of even diameter, while the seta *d* extends well beyond the end of the claw; the seta *e* is a conspicuous stout spine and *f* and *ra* are falcate, with their distal ends expanded into transparent leaf-shaped blades. The copulatory suckers of tarsus IV arise from the distal half of the segment; the seta *f* is long and slender, slightly expanded at the tip, and *r* and *w* are stout spines (Hughes, 1976).

HETEROMORPHIC MALE. The idiosoma is from 700 to 1000 µm long. The idiosomal setae are longer than those of the homomorphic male and tend to be more obviously thicker towards their bases. The third pair of legs is conspicuously thickened and each ends in a stout apophysis (Hughes, 1976).

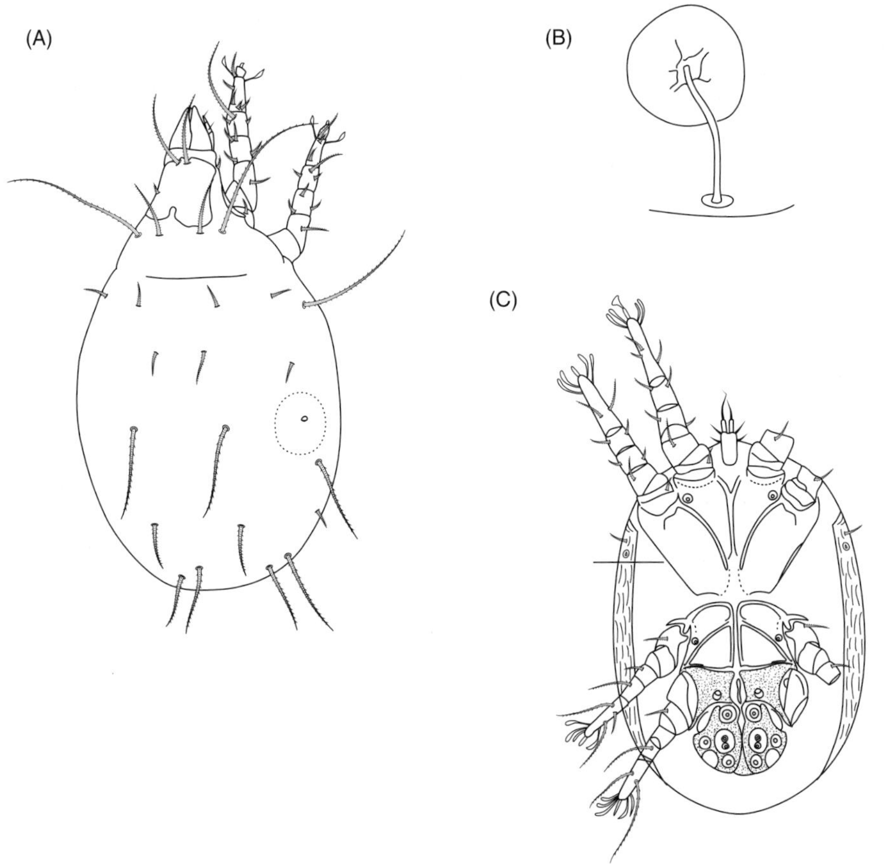

Fig. 13.14. *Sancassania berlesei* (Michael). Female: (A) dorsal view; (B) copulatory opening and spermatheca. Hypopus: (C) ventral view (from Hughes, 1976).

HYPOPUS. The idiosoma is from 250 to 350 μm long and dark brown in colour, with an arched surface, except that the anterior part of the propodosoma is covered with smooth cuticle. The propodosoma is almost triangular in shape and ends anteriorly in a rostral projection that is broadly angled, with the setae *vi* at its apex, which covers the basal region of the gnathosoma. The setae *vi* arise from the tip of this apex and two pairs of scapular setae form a curved transverse row across the prodorsum. The hysterosoma is four to five times longer than the propodosoma and bears fine inconspicuous setae. Ventrally, the sternal shield is well defined laterally. The posterior edge of the sternal shield does not touch the anterior edge of the ventrogenital shield. The sternal apodeme does not reach the posterior ends of apodemes 2. The coxal fields II are open, with a short unsclerotized distance between the anterior and posterior apodemes; the apodemes do not reach the posterior edge of the sternal shield. The coxal setae *3b* and *4a* are conoidal. The claw of tarsi I and II is encircled by five curved leaf-shaped setae, of which two are smaller than the remaining three (Hughes, 1976).

Geographical distribution

S. berlesei is found worldwide (Hughes, 1976).

Host plants and other biological hosts and substrates

Host plants and substrates include the following: rotting potato tubers (Zakhvatkin, 1941), yams and copra (Womersley, 1941), *Neurospora crassa* Shear *et* Dodge (Ascomycetes) (Pimental *et al.*, 1960), wheatgerm, prematurely baled lucerne hay (Hughes, 1961), damp and mouldy wheat, copra, linseed and groundnuts, insect cultures, deep litter of broiler houses associated with the laelapid mite *Androlaelaps casalis* (Berlese) (Hughes, 1976), edible fungi (Zakhvatkin, 1941; Hussey *et al.*, 1969; Clancy, 1981), hyphae and sporocarps of the button mushroom, *Agaricus bisporus* (Greenslade and Clift, 2004), melon (Laffi and Ponti, 1997).

Bio-ecology

S. berlesei cannot reproduce parthenogenetically and mates after the final moult (Timms *et al.*, 1981). With wheatgerm as food, under saturated conditions and at 22°C, the life cycle takes from 8 to 9 days. With high humidity, the female lays 1034 eggs in 39 days – as many as 213 eggs in 24 h. A small number of larvae pass through the hypopal stage even though they are supplied with abundant food and water (Hughes, 1961). In contrast to the above figures, Rodriguez and Stepien (1973) reported a

fecundity of 145 eggs, and Timms *et al.* (1981) of 260 eggs, respectively. With *Neurospora crassa* as food, at 27°C and 73% RH, each mite consumes between 0.01 mg and 0.05 mg a day, the eggs hatch in 24–48 h and the life cycle completes in 9 days (Pimental *et al.*, 1960).

This mite has been cultured monoxenically on 15 fungal isolates of the genera *Aspergillus*, *Cladosporium*, *Fusarium* and *Pencillium* growing on malt-extract agar. Fungus feeding could serve as more than a simple means of survival for the mite (Daneshvar *et al.*, 1977). At 25°C and 90% RH, and with a yeast diet as food, the development from egg to adult takes 158.8 h (Timms *et al.*, 1981), while with a xenic diet, it takes 193.3 h (Rodriguez and Stepien, 1973). The similar species *S. rodionovi*, reared at 25°C on button mushrooms, *A. bisporus*, and on oyster mushrooms, *Pleurotus ostreatus*, completes its immature development in 17.35 ± 0.58 and 20.17 ± 0.88 days, respectively. On button mushrooms, the rate of development increases gradually from 10 to 35°C. The developmental zero is estimated to be 3.5°C with a thermal constant of 357.14 degree-days (Kheradmand *et al.*, 2007).

The hypopi are carried by stored product insects (*Tenebrio molitor* L., etc.) (Hughes, 1976), and by various Scarabaeidae. They are also found on the abdomen and wings of the beetles *Oryctes nasicornis* L. and *Polyphylla fullo* L. (Zakhvatkin, 1941), and on the termite *Eutermes exitiosis* Hill (Womersley, 1941).

Symptomatology and damage

The mite has been recorded in various countries on edible fungi (Zakhvatkin, 1941; Hussey *et al.*, 1969; Clancy, 1981). It attacks hyphae and sporocarps of all sizes of *Agaricus bisporus*, but in Australia is not a common species in mushroom houses (Greenslade and Clift, 2004). In Italy, the mite has been collected on mushrooms (Zakhvatkin, 1941) and on melon in greenhouses (Laffi and Ponti, 1997), but on this last crop the damage caused is unknown.

Control

Control is by spray application of acaricides. On edible fungi, the mite has high levels of attacks, while on vegetables it is of secondary importance.

14 The Family Erythraeidae Robineau-Desvoidy

Morphological Characteristics, Systematics and Bio-ecology

The Erythraeidae have a large and reddish body. The chelicerae are retractable into the idiosoma, and the palp bears a 'thumb-claw' process. The prodorsum has a median longitudinal 'crista metopica' and two pairs of sensilli, with two sensilli set at each distal part (Fig. 14.1). The 'idionotal setae' are commonly setiform (Walter *et al.*, 2009). The biological cycle of these mites develops throughout three active stages (larva, deutonymph and adult), with the protonymph and tritonymph quiescent. The larva is heteromorphic to the deutonymph and adult, and in only a few species are the post-larval biological stages correlated with their larval stage. Their identification therefore needs the use of separate keys (Welbourn and Young, 1987). The family numbers eight subfamilies (Erythraeinae, Callidosomatinae, Abrolophinae, Balaustiinae, Phanolophinae, Myrmicotrombidiinae, Leptinae, Augustsonellinae) (Walter *et al.*, 2009) and over 771 species in 55 genera (Zhang *et al.*, 2011).

The Erythraeidae inhabit a broad range of edaphic and arboreal habitats worldwide (Walter *et al.*, 2009). Their larvae are parasites of various arthropods, and may attach to a site on the host by a cementing substance, forming a superficial cone. This cone enhances the mite's attachment and surrounds a funnel through which the victim's fluids are sucked (Åbro, 1988). The post-larval stages are usually predators and rarely all biological stages prey the same organism, which is then both host and prey. The male deposits spermatophores and each female lays several hundred eggs in batches. Most species develop a single generation a year (Gerson *et al.*, 2003).

The subfamily Balaustiinae has prodorsal pit-shaped 'urnulae', which are paired dorsal protrusions on the idiosoma (Lindquist, 2001; Yoder *et al.*, 2006), and post-larval instars that live in arboreal and litter habitats worldwide (Southcott, 1961; Smiley, 1966, 1968; Welbourn and Jennings, 1991; Gabryś, 2000; Halliday, 2001b; Saboori, 2001). The deutonymphs and adults of the genus *Balaustium* von Heyden are reported as general predators of small arthropods, and also feed on pollen grains; the larvae are free-living predators (Childers and Enns, 1975b; Cadogan and Laing, 1977; Childers and Rock, 1981; Welbourn and Jennings, 1991; Welbourn, 1995). *Balaustium putmani* Smiley is an important predator of pest insects and mites in fruit orchards in North America (Putman and Herne, 1966; Putman, 1970b; Cadogan and Laing, 1982). *Balaustium* spp. are known to bite humans, sometimes causing mild to severe dermatitis (Newell, 1963; Rack, 1973; Southcott, 1976; Ido *et al.*, 2004). Grandjean (1946) observed that all active instars of *Balaustium* feed on pollen grains and Newell (1963) found that *Balaustium* may feed on plants. *B. murorum* (Hermann) can feed on pollen from different sources; no attraction–aggregation pheromone or alarm pheromone is emitted by fed mites, and pollen feeding is more important for larvae that emerge in the early spring when other food options are scarce (Yoder *et al.*, 2006). *B. medicagoense* Meyer *et* Ryke is reported from South Africa (Meyer and Ryke, 1959a) and Australia (Southcott, 1976), and is a significant pest of cereal crops (Halliday, 2001b), winter crops and pastures in Australia (Arthur *et al.*, 2010).

Selected websites

CSIRO (*Commonwealth Scientific and Industrial Research Organisation*) science image, colour photographs of various Australian mites. Available at: www.scienceimage.csiro.au

Government of Western Australia Department of Agriculture and Food, information on Australian mites and their control. Available at: https://www.agric.wa.gov.au/pests-weeds-diseases/pests/mites-spiders

The Genus *Balaustium* von Heyden

Diagnostic characteristics

The prodorsum has one pair of anterolateral eyes and a single pair of sensory structures near the eyes (urnulae) (Gabryś, 2000; Halliday, 2001b).

Economic importance

The economic importance of the genus *Balaustium* is related to the feeding habits of predatory species and to the implication that some species are involved in human dermatitis. Also, in the last decade, the feeding habits of *B. medicagoense* have been demonstrated to cause extensive damage to cereal crops (Halliday, 2001b) and to be an emerging pest of winter crops and pastures in Australia (Halliday, 2001b; Arthur *et al.*, 2010).

Balaustium medicagoense Meyer *et* Ryke (Fig. 14.1)

Common name

The *Balaustium* mite.

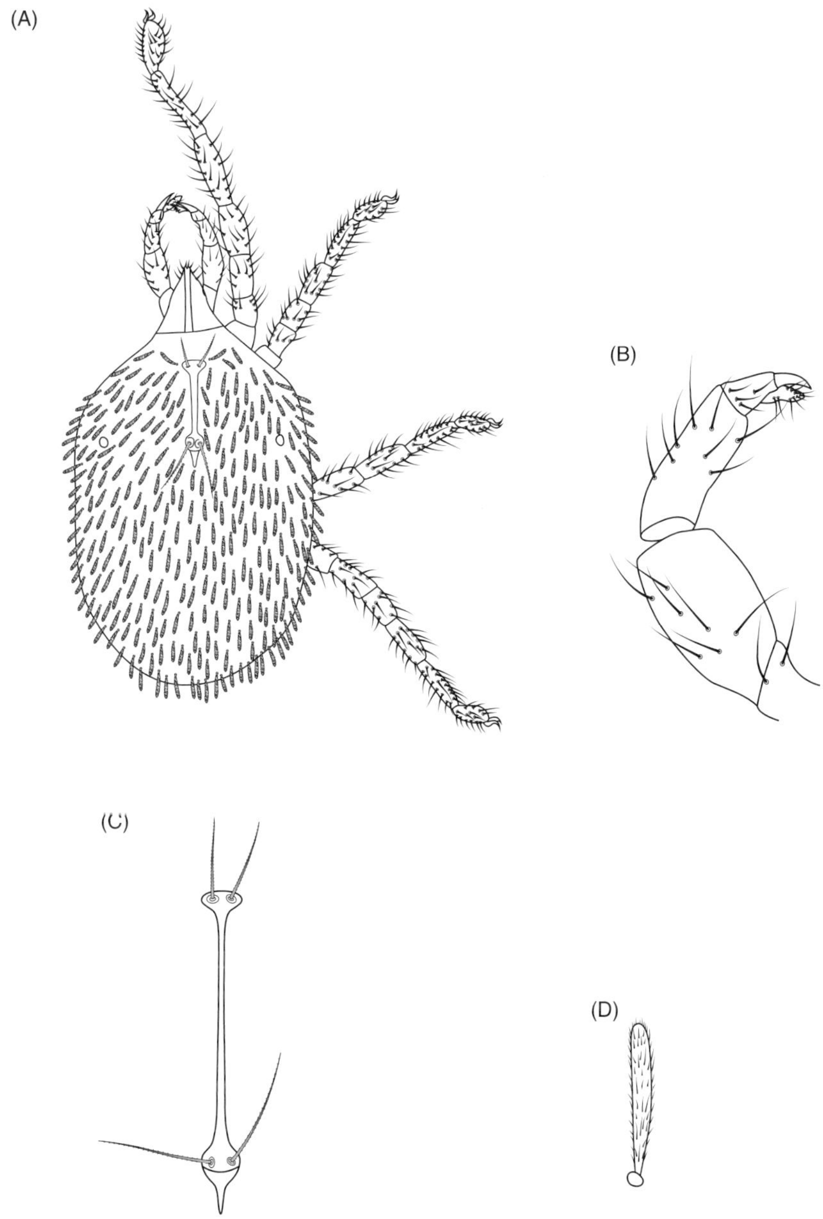

Fig. 14.1. *Balaustium medicagoense* Meyer *et* Ryke. Female: (A) dorsal view; (B) palpus; (C) crista metopica; (D) dorsal seta (from Meyer and Ryke, 1959a).

Diagnostic characteristics

FEMALE. The body is large and oval shaped, dark red or brown in colour, and the legs are bright red. The body (excluding the gnathosoma) is about 1300 μm long and 850 μm wide. The palp is five-segmented and the tibia has a strong claw extending beyond the palptarsus; the latter bears about ten short setae and the inner margin of the tibial claw has a small proximal tooth. The prodorsum has a well-developed crista metopica, lacking a plate; each of the anterior and posterior sensory areas of the crista has two sensory setae, of which the anterior pair measures 71 μm and the posterior pair 100 μm. The anterior area is rounded in front but the posterior area extends backwards and ends in a sharp point. One pair of anterolateral eyes and a single pair of sensory structures are set near the eyes (urnulae). The suture between the propodosoma and the hysterosoma is missing and the dorsum has plumose setae, stout and blunt distally. The genital area bears many setae. The first pair of legs is slightly longer than the body, and each tarsus has two claws. The ventral surface of the coxae and trochanters is ornamented with a large number of rounded papillae (Meyer and Ryke, 1959a; Halliday, 2001b).

MALE. The male has been found but is rare and probably non-functional (Arthur and Umina, 2010).

Geographical distribution

B. medicagoense is found in Australia (Southcott, 1976; Halliday, 2001b; Arthur *et al.*, 2010) and South Africa (Meyer and Ryke, 1959a; Halliday, 2001b).

Host plants

The list of host plants of the *Balaustium* mite includes a restricted number of botanical families, among which the Fabaceae and the Poaceae have the highest number of species.

Asteraceae: capeweed (*Arctotheca calendula* (L.) Levyns) (Halliday, 2001b; Arthur *et al.*, 2010), smooth cat's ear (*Hypochoeris glabra* L.), bristly ox-tongue (*Picris echioides* L.) (Arthur *et al.*, 2010).

Brassicaceae: canola (*Brassica napus* L.) (Halliday, 2001b; Arthur *et al.*, 2010), mustard crops (*Brassica* sp.), wild radish (*Raphanus raphanistrum* L.) (Arthur *et al.*, 2010).

Fabaceae: lupins (*Lupinus* spp.), burr medic (*Medicago polymorpha* L.) (Arthur *et al.*, 2010), clover (Halliday, 2001b; Arthur *et al.*, 2010), dryland lucerne (Halliday, 2001b), lucerne (*Medicago sativa* L.) (Meyer and Ryke, 1959a; Arthur *et al.*, 2010), peas (Halliday, 2001b), field peas (*Pisum sativum* L.) (Arthur *et al.*, 2010), *Trifolium* sp. (Meyer and Ryke, 1959a).

Oxalidaceae: *Oxalis* sp. (Halliday, 2001b).

Poaceae: oats (*Avena sativa* L.) (Halliday, 2001b; Arthur *et al.*, 2010), wild oats (*Avena* spp.), great brome grass (*Bromus* spp.), couch (*Cynodon dactylon* (L.) Pers.), African lovegrass (*Eragrostis curvula* (Schrad.) Nees), barley grass (*Hordeum leporinum* Link) (Arthur *et al.*, 2010), barley (*Hordeum vulgare* (L.)) (Halliday, 2001b; Arthur *et al.*, 2010), kikuyu (*Pennisetum clandestinum* Hochst. ex Chiov.), phalaris (*Phalaris* spp.), tall wheatgrass (*Thinopyrum ponticum* (Podp.) Z.-W. Liu *et* R.-C. Wang), wheat (*Triticum* spp.) (Arthur *et al.*, 2010).

Rosaceae: apple (*Pyrus malus* L. = *Malus domestica* Borkh.), peach (*Prunus persica* (L.) Batsch) (Meyer and Ryke, 1959a).

The mite has also been collected on boards in reservoirs, from grass, herbage on the shore, mixed weeds (Halliday, 2001b), pasture (Halliday, 2001b; Arthur *et al.*, 2010) and roadside weeds (Halliday, 2001b).

Bio-ecology

The *Balaustium* mite is sometimes misidentified with the red-legged earth mite and blue oat mites (*Penthaleus* spp., Penthaleidae). Its feeding habits are both carnivorous (James, 1995; James *et al.*, 1995) and phytophagous (Hoffmann *et al.*, 2008; Micic *et al.*, 2008). The mite has been proposed as a potential candidate for the biological control of *Halotydeus destructor* (Tucker). Further details of this last aspect may be found in Halliday and Paull (2004) and Halliday (2005a). Unfortunately, in Australia, *B. medicagoense* extensively damages cereal, lupin and canola crops (Agriculture Western Australia, 2000). Recently, the mite has been identified as an emerging pest of winter crops and pastures in Australia, and has been shown to cause severe damage to wheat and lupins, but only moderate damage to canola, oats and pasture (Arthur *et al.*, 2010). It infests a variety of economic plants, but prefers grasses, cereals and weeds, particularly barley, wheat, oats, barley grass and capeweed. In pastures, *B. medicagoense* prefers grasses and weeds over clovers and other medics. The species feeds on plants and also preys on small invertebrates, such as various collembolan species and other mites, and for this reason it was originally thought to be primarily a predator; it has only recently been confirmed to feed on plants (Arthur and Umina, 2010).

In Australia, *B. medicagoense* occurs throughout the Mediterranean climate areas of the south, and it has been collected almost exclusively during the southern winter and spring from March to November (Halliday, 2001b). It has been found in Western Australia, Victoria, New South Wales and South Australia, generally restricted to the coastal areas and not occurring too far inland or in the drier mallee areas of Victoria and South Australia; it has also been recorded in Tasmania (Arthur and Umina, 2010).

Halliday (2005a) found that the sex ratio of the *Balaustium* mite is heavily biased towards females (159 females:4 males); however, the significance of this aspect seems unclear, although further research on its biology may show that the mite is usually thelytokous. The *Balaustium* mite reproduces asexually, and its populations have a limited number of genetic or 'clonal' variants. Males are rare and probably non-functional. The populations are made up of clones that can respond differently to environmental and chemical conditions. Eggs are laid in batches of up to 30. The development from larva to adult takes about 5–6 weeks, and adults have been found to live for at least 8 weeks in the field (Arthur and Umina, 2010). In Australia, the mite populations are active in the cool, wet part of the year. They usually emerge in March–April and can be found until December, with a peak in the autumn and spring when they are most active. The species develops two generations a season: in autumn/winter and spring/early summer. The larvae of the second generation hatch in August–September. The mite has a longer active season than other earth mites, and is still active in late spring/early summer. Its biology is characterized by summer diapause at the egg stage during December and by inactivity until late summer/early autumn. *B. medicagoense* is most active in the warmer hours of the day (unlike the red-legged earth mite and the blue oat mite), where it can be observed on the leaves, frequently near the apex of plants (Arthur and Umina, 2010).

Like other pest mites, long-range dispersal is thought to occur via the movement of eggs from the soil adhering to livestock and farm machinery, or through the transportation of plant material. Movement may also occur when over-summering eggs are moved by summer winds (Arthur and Umina, 2010).

Symptomatology and damage

Usually, the *Balaustium* mite is not a severe pest, and it was not until 1997 that it was reported as such in Western Australia, where it was found attacking cereal, lupin and canola crops (Agriculture Western Australia, 2000). Plants were reported as severely bleached and wilted to the point of death, with as many as 20–50 mites per seedling in a barley crop. It was suspected of attacking lucerne in 1979, and was found feeding on lucerne in large numbers in 1985. Preliminary observations showed that the mite feeds on some plant species in South Africa, although it has not been reported as a pest there (Halliday, 2005a). *B. medicagoense* feeds on plants using its

adapted mouthparts to probe the leaf tissue of plants and suck up sap. It can produce little damage, but at high attack density and with plants stressed by other environmental conditions, serious damage can occur. Feeding damage in canola consists of distorted and cupped cotyledons, which may have a leathery appearance. Typical damage to cereals, grasses and pulses is 'silvering' or 'whitening' of the infested leaves, similar to that of red-legged earth mites or blue oat mites. However, the *Balaustium* mite damage differs in that it tends to attack the edges and apices of plants. Adult mites cause the majority of the feeding damage to plants (Arthur and Umina, 2010).

Control

Control requires the monitoring of mite populations and cultural practices. Chemical control, as well as biological control by predators or parasites, are not very effective (Arthur and Umina, 2010).

MONITORING. From autumn to spring, susceptible pastures and crops should be inspected for early detection of the mite and its damage. After sowing, the crops need to be monitored regularly in the first 3–5 weeks, paying attention to crops sown into paddocks that were pasture the previous year; weeds present in the paddocks before the crop is sown should also be examined and monitored for *Balaustium* mites. Usually, cold and/or wet field conditions make it difficult to find mites in the crops, but the mites are easily observed on the leaves during the warmest hours of the day, particularly near or on the apex of plants (Arthur and Umina, 2010).

A productive method for sampling mites uses a D-vac, based on the vacuum principle. A standard petrol-powered garden blower/vacuum machine is employed, with a sieve placed over the end of the suction pipe to trap mites vacuumed from plants and the soil surface (Arthur and Umina, 2010).

CULTURAL PRACTICES. Horticultural management can be effective at controlling the *Balaustium* mite. Mite damage increases when plants are stressed from adverse conditions, such as prolonged dry weather or waterlogged soils. Consequently, ideal conditions for seedling growth enable plants to tolerate a higher density of mites. Early control of summer weeds within and around paddocks, especially capeweed and grasses, can help to prevent mite outbreaks. Rotating crops or pastures with non-host crops can also limit pest colonization, reproduction and survival. In particular, before planting a susceptible crop (cereals, canola, etc.), a paddock could be sown to a broad-leaved plant (vetch) that is not attacked by mite (Arthur and Umina, 2010).

CHEMICAL CONTROL. The mite exhibits a high natural tolerance to common active ingredients and at present no chemical has been registered for its control in Australia (Arthur *et al.*, 2010). The mite may therefore prove difficult to control in the field, as reflected by documented cases of chemical control failures across southern Australia (Arthur *et al.*, 2010). This suggests that alternative control strategies not based on chemicals should be used against the *Balaustium* mite (Arthur and Umina, 2010).

15 The Family Penthaleidae Oudemans

Morphological Characteristics, Systematics and Bio-ecology

The Penthaleidae have a soft body, black to dark green in colour, with the legs long and red coloured. The chelicerae are chelate, separate at the base and each possessing a cheliceral seta (*cha*); the fixed digit is regressed, commonly slender and sometimes finger like at the tip; the movable digit is sickle like to prong like and smooth, and sometimes strongly stylet like (Fig. 15.2A). The palpi are four segmented and lack a tibial claw (Fig. 15.2B). The stigmata are set at the base of the chelicerae and do not have external peritremes. The ventral gnathosoma has four pairs of setae, the distal pair are the adoral setae (or_{1-2}) and the proximal pair, the longer, subcapitular setae (sbc_{1-2}). The prodorsum has an anteromedian projection, the 'naso', a pair of internal vertical setae (v_1) and a pair of trichobothrial internal scapular setae (sc_1) (Figs 15.1, 15.3). The idiosoma (particularly on the ventral surface) and the legs are hypertrichous. Four pairs of lyrifissures (ia, im, ip, ih) may be present, with the fourth pair located on the ventral surface. The anal opening is terminal, dorsoterminal or dorsal, surrounded by three pairs of pseudanal setae (ps_{1-3}) (Figs 15.1, 15.3). Two pairs of genital papillae are present in the adult, the eugenital setae (*eu*) are arranged in six pairs and the genital shields possess a variable number of genital setae (*gs*) and are surrounded by a variable number of aggenital setae (*ags*) (Fig. 15.2C). The 'rhagidial organs', which are fields of modified recumbent dorsal solenidia, are present on tarsi I–II only in *Halotydeus* and *Penthaleus* (Fig. 15.2D) and on tarsi I–III on *Chromotydeus*. The leg apotele has two hooked claws and a median pad-like empodium with radiating hairs (Jeppson *et al.*, 1975). The setae of the body and legs have been described and identified by Baker (1990, 1995).

The family Penthaleidae belongs to the superfamily Eupodoidea, and numbers five genera and 16 species (Zhang *et al.*, 2011). The strong stylet-like movable cheliceral digits highlight the clear adaptation of these mites to phytophagy (Nuzzaci and de Lillo, 1991a; Di Palma, 1995; de Lillo *et al.*, 2001a). Only two species of this family have been traditionally considered to have any economic importance, the red-legged earth mite, *Halotydeus destructor* (Tucker), and the blue oat mite, *Penthaleus major* (Dugès), respectively (Jeppson *et al.*, 1975). The red-legged earth mite feeds on vegetables, legumes, tobacco and various grasses (Meyer and Ryke, 1960; Wallace and Mahon, 1971; Jeppson *et al.*, 1975; Halliday, 1991; Qin and Halliday, 1996a; Umina *et al.*, 2004). The blue oat mite is reported from temperate and tropical regions (Wallace and Mahon, 1971; Jeppson *et al.*, 1975; Halliday, 2005b) and feeds on wheat (*Triticum aestivum* L.), lettuce, oats (*Avena sativa* L.), lucerne (*Medicago sativa* L.), clover, capeweed (*Arctotheca calendula* (L.) Levyns), pasture, grass, weeds (Qin and Halliday, 1996b), thick-bladed grasses, ryegrasses (*Lolium* sp.), phalaris (*Phalaris aquatica* L.) and peas (Umina *et al.*, 2004); it has also been collected on dark loam (Qin and Halliday, 1996b). However, all these records must be interpreted with caution, because some records may be based on misidentifications.

In Australia, the genus *Penthaleus* Koch includes three cryptic species. In addition to the old species *P. major*, Weeks *et al.* (1995) and Qin and Halliday (1996b) described a new blue oat mite species, *P. falcatus* Qin *et* Halliday. This new description highlighted problems in previous studies because the morphological descriptions for *P. major* resembled those of *P. falcatus* (Qin and Halliday, 1996a). Moreover, Weeks and Hoffmann (1999) reported another *Penthaleus* species of economic importance, found only in Australia and South Africa and later described as *P. tectus* Halliday (Halliday, 2005b). Electrophoretic analysis showed that the three *Penthaleus* spp. differ in clonal diversity (Weeks and Hoffmann, 1999), and that the main morphological differences were in the length and number of setae on the dorsal surface of the body. *P. major* has long setae arranged in four to five longitudinal rows, whereas *P. falcatus* has a higher number of short setae scattered in an irregular array, and *P. tectus* has dorsal setae of intermediate length and number. Another species, *P. minor* (Canestrini), can be easily distinguished from the other three *Penthaleus* spp. by its small, short cheliceral movable digits and a shorter palp with a truncate apex to the tarsus (Qin and Halliday, 1996a). This species has not been found in Australia since the original specimens were collected in 1962, but has, in addition, been found in Europe and North America (Thor and Willmann, 1941; Qin and Halliday, 1996a; Weeks and Hoffmann, 1999). *P. minor* is uncommon, primarily living in moss and lichen, and is thought to be of little agricultural importance (Qin and Halliday 1995, 1996a,b). The contemporaneous presence of the original three *Penthaleus* spp. (*P. falcatus*, *P. major*, *P. tectus*) complicates their management in the field.

Penthaleus spp. and *H. destructor* have different ranges of distribution, plant host preferences and responses to pesticides (Umina and Hoffmann, 1999; Weeks and Hoffmann, 1999; Robinson and Hoffmann, 2001), but despite these differences, all of the species are currently treated identically in terms of control. This is of concern, particularly as suppression or eradication of one species of earth mite may result in another species increasing in relative abundance (Weeks and Hoffmann, 2000). To this restriction is added the inability to use the previous bibliographic information because of identification problems between the *Penthaleus* spp.

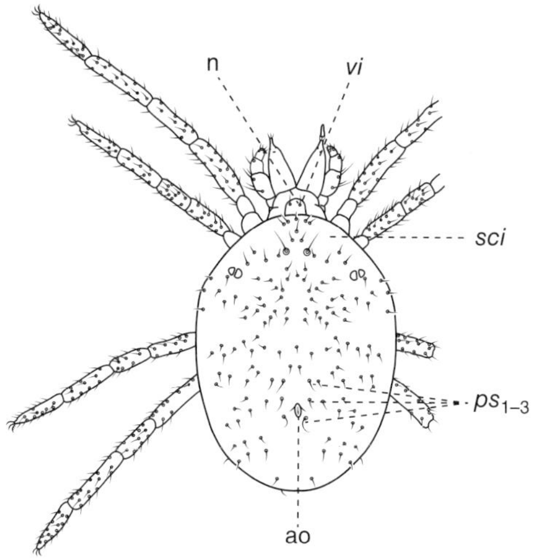

Fig. 15.1. *Penthaleus* sp. Dorsal view of adult (from Livshits and Mitrofanov, 1975). Key: ao, anal opening; n, naso; *sci*, inner scapular setae; ps_{1-3}, pseudanal setae; *vi*, internal vertical setae.

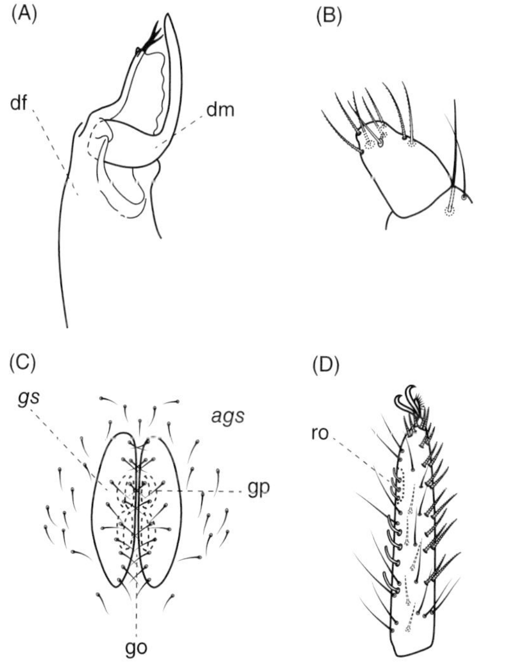

Fig. 15.2. *Penthaleus* sp. (A) chelicera; (B) distal segment of palpus; (C) genital region; (D) tarsus I (from Livshits and Mitrofanov, 1975). Key: ags, aggenital setae; df, digitus fixus; dm, digitus mobilis; go, genital opening; gp, genital papillae; gs, genital setae; ro, rhagidial organ.

Selected websites

Information on the blue oat mite from the State Government of Victoria Department of Environment and Primary Industries. Available at http://www.depi.vic.gov.au/agriculture-and-food/pests-diseases-and-weeds/pest-insects-and-mites/blue-oat-mite

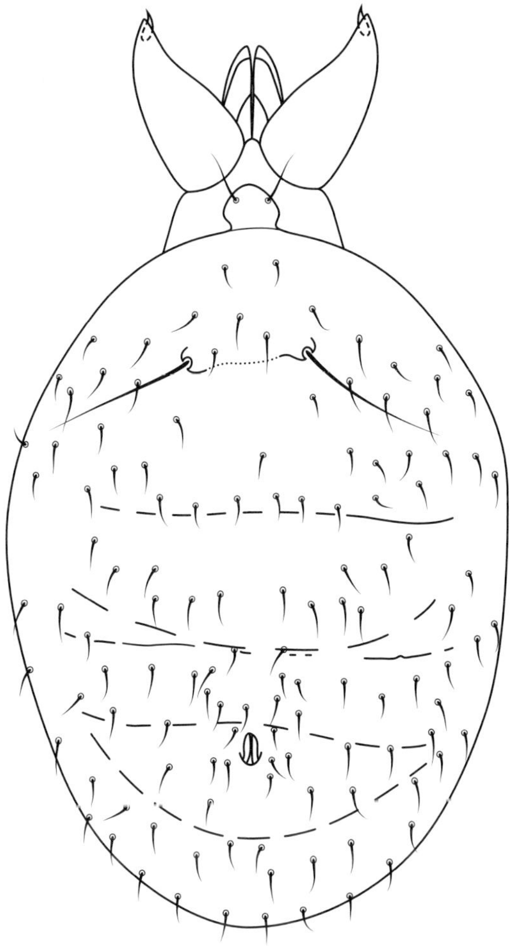

Fig. 15.3. *Penthaleus major* (Dugès). Dorsal view of female (from Walter *et al.*, 2009).

Photos and information on the red-legged earth mite and the blue oat mite. Available at: https://www.google.it/search?tbm=isch&source=univ&sa=X&ei=tQjtUu7YBeLhywP264KICg&ved=0CCwQsAQ&biw=1280&bih=551&q=Halotydeus%20destructor%20pictures

Key for the Identification of Genera (Adults)

1. • Anal opening dorsal....................genus *Penthaleus* Koch
   • Anal opening terminal...............genus *Halotydeus* Berlese

The Genus *Penthaleus* Koch

Diagnostic characteristics

The anal opening is dorsal (Qin and Halliday, 1996b).

Economic importance

Infestations of *P. major* on young plants can be severe, and on small grains the feeding activity of the mite causes a reduction of the amount of forage available through the winter and reduces yields of grain in the spring and summer. The reduction in forage probably has a greater importance in those countries where farmers use winter grazing and rely on the small grains for the autumn and winter pasturing of livestock (Jeppson *et al.*, 1975). Also, *P. falcatus* is a pest of economic importance, commonly injurious to oats, wheat, canola, lucerne and pastures (Weeks and Hoffmann, 1999; Robinson and Hoffmann, 2001), and high densities of attack by *P. tectus* can cause extensive crop damage (Umina and Hoffmann, 1999).

Penthaleus major (Dugès) (Fig. 15.3)

Common name

Blue oat mite, winter grain mite or pea mite.

Diagnostic characteristics

FEMALE. The body is round globular, dark blue to black, and the legs are red–orange coloured. The idiosoma is 730–960 μm long and 560–920 μm wide. The gnathosomal adoral setae (or_{1-2}) are curved and smooth, and only one pair is clearly visible. The subcapitular setae (sbc_{1-2}) are smooth. The palpus setal formula is 0-4-3-9, all tibial setae are smooth, and the solenidion is erect and pointed. The movable digit is 75–85 μm long and the fixed digit three tined; the cheliceral seta (*cha*) is about 23 μm long. The naso is oval, 40–58 μm long and 58–78 μm wide. The internal vertical setae (v_1) are smooth, 50–60 μm long, set in a separate socket in the anterior half of the naso. The trichobothria (sc_1) are smooth and 85–90 μm long. The anterior dorsal idiosoma bears 4–5 setae posterior to the naso. The 30–33 hysterosomal setae are scattered but arranged roughly into six longitudinal rows. There are three pairs of lyrifissures (ia, im, ip), elongate and oval. The anus is dorsal, tube like, bordered by two valves, surrounded by three pairs of pseudanal setae (ps_{1-3}), commonly longer and more curved than the other dorsal setae. The ventral idiosoma is neotrichous, with all the setae smooth. Coxae I have 7–8 pairs of setae (some probably of intercoxal origin), and coxae II–IV bear 2–5 pairs of setae. The region posterior to coxae II bears 61–91 setae. The genital shields each have 11–13 genital setae (*gs*), 33–45 μm long, and are surrounded by 25–45 pairs of aggenital setae (*ags*). The eugenital setae (*eu*) are arranged in six pairs, 40–50 μm long. The lyrifissures (ih) are located on the ventral surface. The femora are undivided, and genua I and II bear the papilla. The rhagidial organ formula (tibia–tarsus) is: leg I, 1-3; leg II, 1-3. The famulus formula is: leg I, 1-1; leg II, 0-1, and apparently lacking on tibia II. The solenidion formula (genu–tibia–tarsus) is: leg I, 3(4)-5(6)-4; leg II, 2-2(3)-1; leg III, 2-3-1; leg IV, 3-3(4)-0. All leg setae are smooth, except the ventral setae on the tarsi, which are ciliate. The setal number is variable (Qin and Halliday, 1996b).

MALE. Unknown (Qin and Halliday, 1996b).

Geographical distribution

P. major is found in Argentina, Australia, Canada, France, Germany, Morocco (Qin and Halliday, 1996b), Iceland (Gudleifsson *et al.*, 2002), British Isles, France, Germany, Norway (including Svalbard and Jan Mayen), Poland, Switzerland (Baker and Beron, 2004) and South Africa (Tucker, 1925; Meyer and Ryke, 1960).

Host plants

P. major feeds primarily on Poaceae, such as wild oats (*Avena fatua* L.) and thick-bladed grasses within pastures (Weeks and Hoffmann, 1999; Robinson and Hoffmann, 2001), but its list of host plants also includes other plants in different botanical families. These are the: Asteraceae – capeweed and lettuce (Qin and Halliday, 1996b); Fabaceae – clover, lucerne (Qin and Halliday, 1996b) and peas (Umina *et al.*, 2004); and Poaceae – oats (Qin and Halliday, 1996b), phalaris, ryegrasses (Umina *et al.*, 2004) and wheat (Qin and Halliday, 1996b). The blue oat mite has also been collected on dark loam, grass, pasture, thick-bladed grasses (Umina *et al.*, 2004) and weeds (Qin and Halliday, 1996b).

Bio-ecology

Penthaleus major is a diploid species reproducing by thelytokous parthenogenesis (Weeks *et al.*, 1995; Weeks and Hoffmann, 1999). Like other *Penthaleus* spp., its life cycle develops through the stages of egg, prelarva (deutovum), larva, protonymph, deutonymph and adult (Jeppson *et al.*, 1975). In Australia, populations of *P. major* are active between April and October (autumn, winter, spring) and develop two or three generations a year, aestivating as eggs. Jeppson *et al.* (1975) reported two generations per season. The diapause eggs hatch in the autumn when conditions are optimal following cool temperatures and rainfall. The generation time of *Penthaleus* spp. in the field is around 8–10 weeks, and in the laboratory *P. major* develops a generation in less than 7 weeks (Weeks and Hoffmann, 1999). In the cold temperate countries of the North Atlantic, such as Iceland (Gudleifsson *et al.*, 2002) or northern Norway (Johansen, 1986), where the summer temperatures are comparable to the winter temperatures of southern Australia, the blue oat mite is a summer pest on pastures and hayfields; it is usually active in May, June and early July and disappears in midsummer (Gudleifsson *et al.*, 2002). Sometimes, high population densities also occur in August and September. In Iceland, the mite develops only a summer generation and a winter generation. As in the winter development observed in Australia, the summer generation takes about 64 days, excluding egg development; the mite commonly hibernates as the protonymph and no diapause forms occur in the colder months (Hallas, in Umina *et al.*, 2004).

The female lays the eggs on the sheath leaves and stems and on and in soil near the base of the plants, fastening them on to the leaf surface by a mucilaginous substance that it produces and depositing the eggs singly or, commonly, in large numbers close together. *P. major* produces two types of eggs, 'winter eggs', which have a short period of development, and 'diapause eggs', which have an extended development period. From the

time of emergence in the autumn, a fraction of all of the eggs produced by *P. major* are diapause eggs, and their production continues throughout the period of activity of adults in the field. This begins early in the season, and is likely to have occurred in the first generation of the mite (James and O'Malley, 1993) (Fig. 15.4). Different clones of the species may differ in the timing of diapause egg production, which provides another potential selection factor contributing to the maintenance of clonal diversity within this group (Umina *et al.*, 2004). These differences in the timing of diapause egg production are likely to contribute to the distribution and abundance of the species. Robinson and Hoffmann (2001) mapped the distribution of *Penthaleus* spp. within south-eastern Australia. For instance, *P. major* has been collected from Mt Buller (in the alpine region of Victoria) before the onset of snow (Weeks and Hoffmann, 1999). In such alpine regions, the period that mites are active is curtailed because of cold temperatures and snow cover, and here, the early production of diapause eggs would be an adaptive strategy ensuring survival into the next season.

Like other *Penthaelus* spp., the larvae of *P. major* are active soon after hatching and begin to feed on the sheath leaves or tender shoots near the ground (Jeppson *et al.*, 1975). As adults, they feed higher up on the plants at night or on cloudy days, leaving the plants as the sun rises and seeking protection during the hot part of the day on the moist soil surface, perhaps under foliage (Gudleifsson *et al.*, 2002). They may even dig into the moist, cool soil under extreme conditions, or cover themselves with dry soil and a little foliage, and dig into the soil in search of more moisture and a better temperature. At sunset and thereafter, the plants are invaded by the feeding mites, and it is possible to observed them at all hours of the night with a searchlight. When disturbed, they drop to the ground (Jeppson *et al.*, 1975).

Penthaleus populations can reach densities over 150,000 mites m^{-2} in the field (Kanda and Hirai, 1990), but densities are commonly below this level. The blue oat mite lives singly, or sometimes in small groups of 5–10 individuals, probably because it is not attracted by plant volatile compounds (André, 1932b). As already noted, young mites commonly feed on the sheath leaves or tender shoots near the soil surface, while the adults, probably because they have stronger and larger mouthparts, feed on the more mature plant tissues. The early life stages of *Penthaleus* spp. are also likely to feed on lower plants and microflora found on the soil surface (McDonald *et al.*, 1995). The presence of microflora would allow populations to survive, develop and reproduce when higher plants are unavailable or are unsuitable as hosts (Maclennan *et al.*, 1998).

The dispersal of *Penthaleus* spp. occurs via different mechanisms, such as the transportation of diapause eggs or mites on grain stubble or leaves, on soil adhering to implements that are moved about, or on forage or straw that is carried from

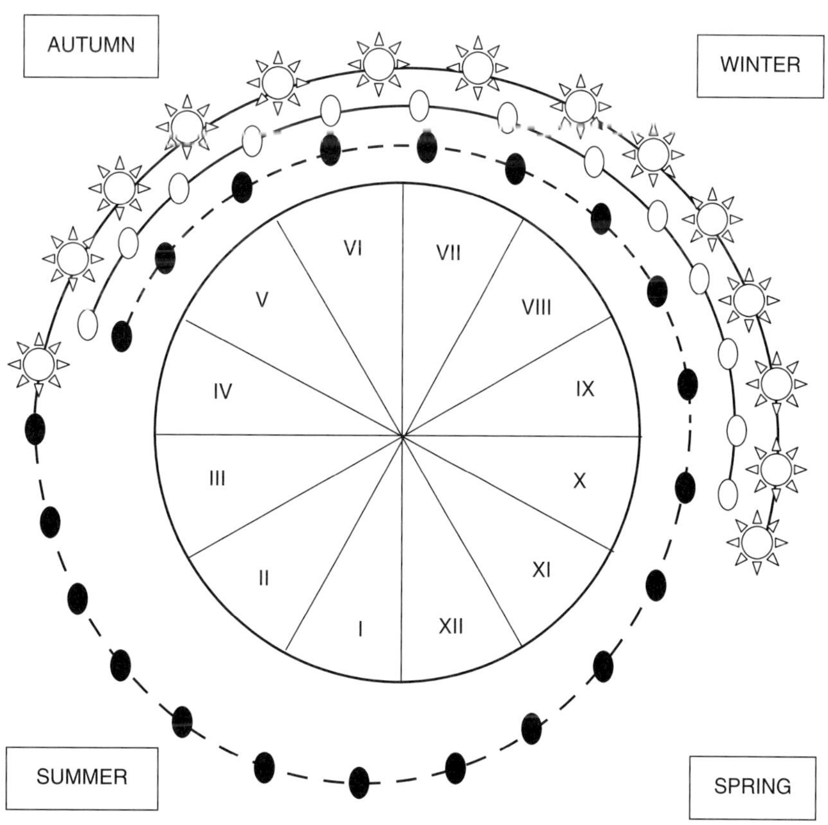

Fig. 15.4. Schematic drawing of the life cycle of *Penthaleus major* (Dugès) in the Australian climate. Diapause eggs are shown in black, winter eggs in white and adults as white stars. I–XII represent the months of the year.

infested fields in livestock feeding operations. Diapause eggs can be carried in the debris transported by wind, and adults can also achieve a local distribution by wind. Migrations to grain fields may take place from fence rows or other uncultivated areas. Fescue grass appears to be the favoured host in such situations (Jeppson *et al.*, 1975). Rates of the movement of *P. major* adults have been estimated in Australia using mark–release–recapture techniques. Mean square dispersal distances are used to estimate diffusion coefficients; in pasture, these are in the range 0.3–1.3 and 90% of the population moves <5–11 m in a 10-day period, or 7–16 m within their adult lifetime. Releases of mites into adjacent pea/wheat crops indicate directional movement towards the more favoured pea host (Weeks *et al.*, 2000a).

Natural enemies of *Penthaleus* include fungal pathogens and predators. Several fungi have been identified as pathogenic to blue oat mites and considered for biological control (Hastings, 1994). *Penthaleus* spp. make good hosts for fungal pathogens because they are soft bodied and inhabit environments with humid microclimates (Ferro and Southwick, 1984; Evans, 1992). The fungus *Neozygites acaracida* (Petch) (Neozygitaceae) is prevalent among *Penthaleus* spp. during wet winters and could be responsible for 'population crashes' (James, 1995), although no data are available to support this conjecture. Predators include the larvae of *Chrysopa* (Chrysopidae) and the predatory mite *Balaustium* (Erythraeidae), but their importance in limiting the development of *Penthaleus* populations is inconsistent (Swan, 1934; Chung *et al.*, 1963; Greenup, 1967). At least 19 predators and one pathogen are known to attack mites in southern New South Wales (Australia); the most important predators are included in the mite families Anystidae, Bdellidae, Erythraeidae, Parasitidae and Cunaxidae (James, 1995). It has also been postulated that the European earwig, *Forficula auricularia* L. (Forficulidae), may contribute to regulating populations of earth mites and other micro-invertebrate pests in pastures (Weiss and McDonald, 1998).

Symptomatology and damage

The blue oat mite feeds on a variety of pasture plants and on virtually all winter crops, mainly on the dorsal surface of cotyledons and leaves. The mites penetrate the epidermal cells of the host plant with their chelicerae and remove the cellular contents, making a hole that is probably larger (>3 μm) than that made by *H. destructor* because it has larger mouthparts (Qin and Halliday, 1996a,b). The movable digits of *Penthaleus* spp. can be moved separately or together, but do not form a channel for transporting fluid. The subcapitulum is applied to the plant surface and the cell contents are sucked out using a pharyngeal pump (Nuzzaci and de Lillo, 1991a).

The symptomatology of attack by the blue oat mite differs from that of tetranychid mites. The removal of the cell contents typically results in silvery to grey patches on the leaf surface. In the paddock, this is often mistaken for frost damage. The leaves appear greyish or silvery, and at high levels of attack over several days, the tips of the leaves show a scorched aspect and turn brown, and the plant may die. Many of the infested plants do not die, but become stunted and produce little forage or grain. The damage on young plants is more severe than on large, healthy ones. Small grains suffer two types of damage, which result in a reduced amount of forage throughout the winter and reduced yields of grain in the spring and summer. The reduction in forage probably has a greater importance in those countries where winter grazing is adopted by farmers who rely on the small grains for autumn and winter pasturing of livestock (Jeppson e*t al.*, 1975).

Control

The control of *P. major* is complicated by the simultaneous presence in the field of the three cryptic *Penthaleus* spp., different from each other in distribution, host plants, timing of diapause egg production and susceptibility to pesticides. Each control strategy must consider these aspects. Chemicals are currently commonly used in control programmes, but in the short term, important benefits could also be given by cultural management, the conservation of natural enemies and improved timing of pesticide application. The integration of these practices, combined with the development of a simple field-based kit to distinguish the different mite species, may be crucial (Umina *et al.*, 2004), and would allow the implementation of an integrated pest management (IPM) protocol.

CULTURAL PRACTICES. Cropping practices markedly affect the occurrence of and damage caused by *Penthaleus* spp. Injury can be prevented by crop rotation – not planting more than 2 years in succession. Crops such as cotton, maize, clover or sorghum may be employed in such rotations (Jeppson *et al.*, 1975). Other strategies, such as trap and border crops and weed management, require a thorough understanding of pest–host associations. The mite survives and reproduces from one season to the next when confined to pasture. Canola and a common weed, bristly ox-tongue (*Picris echioides* L.), are suitable hosts for *H. destructor* and *P. falcatus*, whereas *P. tectus* and *P. major* fail to persist on these plants. A mixture of wheat and oats sustains *P. tectus* and *H. destructor*, but not *P. falcatus* or *P. major*. Lentil plants are generally a poor host for all mite species (Umina and Hoffmann, 2004).

Control methods for blue oat mites may often depend on low dispersal rates, yet there are no experimental estimates of these rates. The mite does not have directional movement when adjacent plots of peas and lupins are compared, even though lupins are poor hosts. This indicates that broad border sprays or border cultures are needed to prevent mite movement from adjacent paddocks (Weeks *et al.*, 2000a). However, a 5 m wide wheat border around a canola crop is unsuccessful in preventing the movement of mites into the crop (Merton *et al.*, 1995), although recent dispersal data suggest that the border should have been wider (Weeks *et al.*, 2000a). Appropriate grazing management can be used to reduce *Penthaleus* populations significantly to below damaging thresholds (Grimm *et al.*, 1995), possibly because shorter pasture has a reduced relative humidity (RH), which increases mite mortality and limits food resources (Ridsdill-Smith, 1997).

Weed management, in particular the management of broad-leaved weeds, reduces mite density (Swan, 1934), as weeds and/or other non-vascular flora can provide an alternative food source, particularly for juvenile stages, and a more suitable

environment for mites within a crop or pasture (McDonald *et al.*, 1995; Maclennan *et al.*, 1998). Although cultivation (Tucker 1925; Merton *et al.* 1995), summer pasture burning (Swan, 1934; Wallace, 1961), maintaining fallow land (Newman, 1923; Swan, 1934) and flooding of summer pastures (James and O'Malley, 1991) can reduce the mite density, they are not commonly practical for current farming methods.

BIOLOGICAL CONTROL. Wallace introduced the anystid mite *Anystis wallacei* Otto, a predator of *Penthaleus* spp. reported from France, into Western Australia in 1965 (Michael *et al.*, 1991). The predator established successfully and reduced pest numbers. Because of a very low migration rate (Michael *et al.*, 1993), further attempts have been made to release *A. wallacei* into new areas, but the benefits have yet to be demonstrated. No other information on the practical use of natural enemies is available.

CHEMICAL CONTROL. Chemical control requires the correct identification of the mite species concerned and knowledge of the main features of its bio-ecology. There is a linear relationship between the number of diapause eggs laid by mites and the timing of treatments during the season. This may reflect a fairly constant production of diapause eggs throughout the season, and the low number of these eggs produced by *P. major* towards the end of spring supports this hypothesis (Weeks and Hoffmann, 2000). The pesticides that are used do not kill diapause or winter eggs (James, 1996), and their use aims mainly to control juvenile stages and adults. A cost-effective strategy for controlling *H. destructor* is to spray adults just before they lay diapause eggs in the spring (Ridsdill-Smith and Pavri, 1998). This 'spring-spraying' approach appears to be successful against *H. destructor*, but is not effective in the control of the *Penthaleus* spp. (James and O'Malley, 1993; Umina and Hoffmann, 2003), because the mites differ considerably in their timing of diapause egg production. In eastern Australia, *Penthaleus* populations do not have the large spring peak in numbers seen in *H. destructor* (Weeks and Hoffmann, 2000), so it is not necessary to control them with chemicals in the spring. Spraying *Penthaleus* within 2–3 weeks of emergence in the autumn is likely to be the most effective strategy for reducing the density of these mites in crops and pastures throughout the season. Pesticide border sprays may prevent the movement of *Penthaleus* spp. into crops from adjacent paddocks harbouring large populations. For wheat and canola crops, border spraying with omethoate repeated over two field seasons were effective at widths greater than 15 m (unpublished data, Umina *et al.*, 2004).

Various active ingredients have been proposed in the past, including several organophosphorus (malathion, parathion, dimethoate) and organic sulfur acaricides (Chada, 1956). Now, later on, the majority of pesticides suggested for *Penthaleus* spp. are organophosphates, such as omethoate and methidathion. Other chemicals that are available are pyrethroids, such as bifenthrin and alpha-cypermethrin, which are recommended in many instances because of their increased specificity to pest earth mites and their reduced effect on several acarine predators (James *et al.*, 1995). Laboratory tests have shown that *P. falcatus* has a much higher tolerance to omethoate, methidathion, bifenthrin and endosulfan (now banned in several countries and also in Australia) than *P. major*, *P. tectus* and *H. destructor*. Control strategies that are based on one species may then not be adequate for other species, so control failures may not be the result of poor application (Umina and Hoffmann, 1999). The species also respond differently to alpha-cypermethrin, phosmet, chlorpyrifos and lambda-cyhalothrin, with *P. falcatus* generally the most tolerant and *H. destructor* the least tolerant species (Robinson and Hoffmann, 2000). Bifenthrin, methidathion, endosulfan and phosmet are the most effective against *P. major* and *P. tectus*, while it may be difficult to control *P. falcatus* with any of the pesticides, particularly omethoate, unless relatively high concentrations are used. As chemical control is the only short-term management option available, pesticide resistance in earth mites could pose an emerging problem (Umina *et al.*, 2004).

Penthaleus falcatus Qin *et* Halliday

Common name

Blue oat mite, winter grain mite.

Diagnostic characteristics

FEMALE. General features and colour are as for *P. major*. The idiosoma is 710–1100 μm long and 570–840 μm wide. The adoral setae (*or*) are curved and smooth (only one pair is clearly visible); the subcapitular setae are smooth. The palpus setal formula is 0-5(4)-3-8; all tibial setae are smooth, and the solenidion is erect and pointed. The movable digit is 80–105 μm long, the fixed digit is three tined, and the cheliceral seta (*cha*) is 25–30 μm long. The naso is oval, 60–75 μm long, and 75–80 μm wide. The internal vertical setae (v_1) are smooth, 35–55 μm long and inserted in a separate socket in the anterior half of the naso. The trichobothria (sc_1) are smooth and 65–80 μm long. The anterior dorsal idiosoma posterior to the naso bears 7–9 dorsal setae. The hysterosoma is neotrichous, and bears 70–120 setae, smooth and arranged in three groups; of these setae, about 45 form a transverse band between the trichobothria and the lyrifissure ia, about 34 setae are set between the lyrifissures ia and im, and about 29 setae are set in the area posterior to lyrifissure im. The lyrifissures (ia, im, ip) are elongate and oval. The anus is dorsal, tube like, bordered by two valves, and surrounded by three pairs of pseudanal setae (ps_{1-3}), longer and curved than the other dorsal setae. The ventral idiosoma is neotrichous, and all of the setae are smooth. Coxae I have 8–10 pairs of setae, coxae II–IV bear 2–5 pairs of setae. The ventral idiosoma posterior to coxae II has about 65 setae. The genital shields each have 11–12 genital setae (*gs*). The eugenital setae (*eu*) are arranged in six pairs. The genital shields are surrounded by about 30 pairs of aggenital setae (*ags*). The lyrifissures ih are located on the ventral surface. The femora are undivided, and the genua I and II have papillae. The rhagidial organs are erect or semi-erect, with the following formula (tibia–tarsus): leg I, 1-3; leg II, 1-2. The famulus formula is: leg I, 1-1; leg II, 0-1. The solenidia formula (genu–tibia–tarsus) is: leg I, 5-5-3; leg II, 2(3)-3-0; leg III, 2-3-0; leg IV, 2(3,4)-3-0.

All of the leg setae are smooth, except for the ventral setae on the tarsi, which are ciliate. The setal number is variable (Qin and Halliday, 1996b).

MALE. Unknown (Qin and Halliday, 1996b).

Geographical distribution

P. major is found in Australia (Qin and Halliday, 1996b).

Host plants

By family, the host plants include: Asteraceae – capeweed (Qin and Halliday, 1996b), bristly ox-tongue, broadleaved weeds (*Hypochoeris* spp.) (Weeks and Hoffmann, 1999), cat's ear (*Hypochoeris* spp.) (Umina *et al.*, 2004); Brassicaceae – canola (*Brassica napus* L.) (Umina *et al.*, 2004); Fabaceae – clover, lucerne (Qin and Halliday, 1996b), peas (*Pisum* spp.) (Umina *et al.*, 2004); Papaveraceae – poppies (Umina *et al.*, 2004); Poaceae – oats (Qin and Halliday, 1996b), thick-bladed grasses (*Lolium* sp.) (Umina *et al.*, 2004), wheat (Qin and Halliday, 1996b).

The mite has also been collected on ex-pasture, grass and under stone (Qin and Halliday, 1996b).

Bio-ecology

Qin and Halliday (1996b) reported that 'some records of *P. major* in the literature may actually refer to *P. falcatus*, and many specimens previously identified as *P. major* are actually *P. falcatus*. Further study is needed to clarify the confusion between these two species'. In Australia, *P. falcatus* is widespread in New South Wales and has been recorded from South Australia, Victoria and Tasmania (Qin and Halliday, 1996b).

The general features of the bio-ecology of *P. falcatus* are as those of *P. major*. The mite is diploid, reproduces by thelytokous parthenogenesis and no males have been reported (Weeks *et al.*, 1995; Weeks and Hoffman, 1999; Robinson *et al.*, 2002). A generation time (from adult to adult) in the laboratory takes more than 9 weeks (Weeks and Hoffmann, 1999). In the field, the diapause eggs are laid in early winter, although the first appearance of these eggs is slightly later than in *P. major* (Umina *et al.*, 2004). There is a constant rate of production of diapause eggs, although the trend is less clear than in *P. major*. Producing diapause eggs throughout a season will reduce population growth, and may contribute to the relative abundance of *Penthaleus* increasing less dramatically in early spring compared with *H. destructor* (Weeks and Hoffmann, 2000).

Symptomatology and damage

The symptomatology and damage of this blue oat mite are as for *P. major*. Qin and Halliday (1996b) reported that 'the pest status of *P. falcatus* needs further attention because it occurs in pasture and also appears to feed on crops such as wheat and oats'. The mite mainly infests canola and broadleaved weeds and in south-eastern Australia it is – with *P. major*, *P. tectus* and *H. destructor* – a pest of economic importance, with damage most commonly found on oats, wheat, canola, lucerne and pastures (Weeks and Hoffmann, 1999; Robinson and Hoffmann, 2001).

Control

Control depends on the spray application of pesticides. The use of other methods follows the same considerations as indicated for *P. major*.

CHEMICAL CONTROL. Currently, chemical control is the most practicable strategy. Unfortunately, the spring spraying that is effective in the control of *H. destructor* (James and O'Malley, 1993; Umina and Hoffmann, 2003) is ineffective in the control of *P. falcatus*, because its populations (and those of other *Penthaleus* spp.) increase following spring spraying, and the treatment must be effected within 2–3 weeks of emergence in the autumn (Umina *et al.*, 2004). *P. falcatus* exhibits a higher tolerance to omethoate, methidathion, bifenthrin and endosulfan than *P. major*, *P. tectus* and *H. destructor*. A control strategy based mainly on one species may then be inadequate for the control of other species, and the failure of chemical treatments may not be due to their poor implementation (Umina and Hoffmann, 1999). The different penthaleid species are also differentially sensitive to alpha-cypermethrin, phosmet, chlorpyrifos and lambda-cyhalothrin (Robinson and Hoffmann, 2000), with *P. falcatus* commonly the most tolerant and *H. destructor* the least tolerant. Bifenthrin, methidathion, endosulfan and phosmet are the most effective against *P. major* and *P. tectus*, while it may be difficult to control *P. falcatus* with any of the pesticides, particularly omethoate, unless relatively high dosages are used. As chemical control is the only short-term management option available, pesticide resistance in earth mites could pose an emerging problem.

Penthaleus tectus Halliday

Common name

Blue oat mite, winter grain mite.

Diagnostic characteristics

FEMALE. The idiosoma is 821–976 μm long and 706–882 μm wide. The subcapitulum is roughly triangular, with the surface of the integument finely spiculate; two pairs of smooth and pointed adoral setae are inserted near the tips of attenuate membranous malae, with or_2 about half the length of or_1. The subcapitular setae are smooth, with sbc_1 25 μm long and sbc_2 15 μm long. The palpus has a finely spiculate surface throughout and with the following chaetotaxy: trochanter, 0; femur/genu, 4; tibia, 3; tarsus, 8 + 1 solenidion; all tarsal setae are plumose, and the solenidion is short, erect and pointed. The setae of other palpal segments are smooth and pointed. The cheliceral fixed digit is unsclerotized, ending in a trifid fork, with a smooth dorsal seta, and the movable digit is sclerotized, elongate, narrow and pointed. The naso is oval, 37–50 μm long and 66–71 μm wide. The setae *vi* are 45–60 μm long, with their edges smooth to very slightly roughened, and they are set in separate sockets in the anterior half of the naso. One pair of trichobothria (sc_1) is 84 μm long, and there are about 30 pairs of smooth pointed setae, 33–59 μm long, some of which are set asymmetrically or apparently

unpaired, especially posterior to the anus. Three pairs of lyrifissures (ia, im, ip) are present. The anus is protuberant and tube like in living specimens, flattened to a pair of valves, surrounded by three pairs of long, strongly curved pseudanal setae (ps_{1-3}). The ventral idiosoma bears about 90 irregularly scattered smooth, pointed setae, plus about 15 pairs of shorter aggenital setae (*ag*) and six pairs of eugenital setae (*eu*). The femora are undivided. The rhagidial organs have the formula: tarsus I, 2; tibia I, 1; tarsus II, 1; tibia II, 1; all are recumbent and lying in a pit. The famulus has the formula: tarsus I, 2; tibia I, 1; tarsus II, 1; tibia II, 0. The ventral and lateral tarsal setae are ciliate, but all of the other setae are smooth and pointed (Halliday, 2005b).

MALE. Unknown (Halliday, 2005b).

Geographical distribution

P. tectus is found in South Africa (Weeks and Hoffmann, 1999) and Australia (Halliday, 2005b).

Host plants

By family, the host plants include: Brassicaceae – canola (Halliday, 2005b); Fabaceae – lucerne (Umina *et al.*, 2004), oats (Umina *et al.*, 2004), perennial ryegrass (*Lolium perenne* L.), wheat (*Triticum* sp.) (Weeks and Hoffmann, 1999), wheat (*T. aestivum* L.) (Umina *et al.*, 2004), wild oats (Weeks and Hoffmann, 1999).

The blue oat mite has also been collected on thick bladed grasses (Umina *et al.*, 2004), and unidentified pasture plants (Halliday, 2005b).

Bio-ecology

This species was recorded as *Penthaleus* 'species x' by various specialists (Weeks and Hoffman, 1999; Umina and Hoffmann, 1999; Robinson and Hoffmann, 2000; Umina and Hoffman, 2003), and first described by Halliday (2005b). It belongs to the blue oat mites group and occurs in limited areas of western Victoria and in northern New South Wales in Australia (Qin and Halliday, 1996a; Weeks and Hoffmann, 1999). Its distribution is disjunct, and separate populations have been recorded in western Victoria and north-eastern New South Wales (Robinson and Hoffmann, 2001). The mite has also been reported from Cape Town, South Africa (Weeks and Hoffman, 1999).

P. tectus seems to prefer wheat and thick-bladed grasses (Weeks and Hoffmann 1999; Robinson and Hoffmann 2001). Its bio-ecology is similar to that of *P. major*. It reproduces by thelytokous parthenogenesis and no males have been reported from Australia (Weeks *et al.*, 1995; Weeks and Hoffman, 1999; Robinson *et al.*, 2002). The diapause response of *P. tectus* is somewhat later than in *P. major* and *P. falcatus*, but earlier than in *H. destructor* (Umina *et al.*, 2004).

Symptomatology and damage

P. tectus was previously misidentified as *P. major* and the symptomatology of its attack and the damage caused are similar to those of *P. major*. In south-eastern Australia, high densities of this mite can cause extensive crop damage (Umina and Hoffmann, 1999).

Control

Control is similar to that of other *Penthaleus* spp. Pesticide applications may produce inconsistent results because the different *Penthaleus* species vary in pesticide tolerance, and the species has not always been accurately identified (Umina and Hoffmann, 1999; Robinson and Hoffmann, 2000).

The Genus *Halotydeus* Berlese

Diagnostic characteristics

The anal opening is terminal. The idiosoma has 11 pairs of dorsal setae. The palp is slender and its setal formula is 0-2-3-8. Leg III lacks a rhagidial organ (Qin and Halliday, 1996a).

Economic importance

The genus *Halotydeus* has only a few species, two of which have a European range of distribution (Baker and Beron, 2004), and it includes only one pest species, *H. destructor*, commonly called the red-legged earth mite. This mite infests vegetables, legumes, tobacco and various grasses (Meyer and Ryke, 1960; Wallace and Mahon, 1971; Jeppson *et al.*, 1975; Halliday, 1991; Qin and Halliday, 1996a; Umina *et al.*, 2004), and is a major pest of pastures, crops and vegetables in regions of Australia, South Africa and New Zealand characterized by cool, wet winters and hot, dry summers (Ridsdill-Smith, 1997).

Halotydeus destructor (Tucker) (Fig. 15.5)

Common name

As well as having the name of the red-legged earth mite, *H. destructor* is also called the black sand mite.

Diagnostic characteristics

FEMALE. The idiosoma is 480–1100 μm long, 450–890 μm wide and black coloured; the gnathosoma and the legs are fresh red or orange coloured. The eyes are pale white, the setae white, and the anus and the surrounding area orange in colour. The genital shield is white or cream. The setae or_1 are 12–16 μm long, and or_2 5–10 μm long; the subcapitular setae are branched, sbc_1 are 27–35 μm long, and sbc_2 12–20 μm long. The palpus is 168–204 μm long, with the solenidion semi-erect. The chelicerae are 115–190 μm long, with the movable digit 40–60 μm long and 28–60 μm wide; the seta *cha* is missing. The naso is oval to broadly oval or hemispherical, 20–30 μm long and 25–35 μm wide. The setae v_1 are 12–20 μm long, set at the side of the naso and branched. The setae v_2 and sc_2 are both plumose, with v_2 22–30 μm long and sc_2 22–35 μm long. The trichobothria sc_1 are 35–55 μm long; the opisthosomal setae c_1, c_2, d_1, e_1, f_1 and f_2 are plumose and 17–32 μm long. The lyrifissures ia, im, ip are 7–8 μm long and 5–6 μm wide. The ventral idiosomal setae are

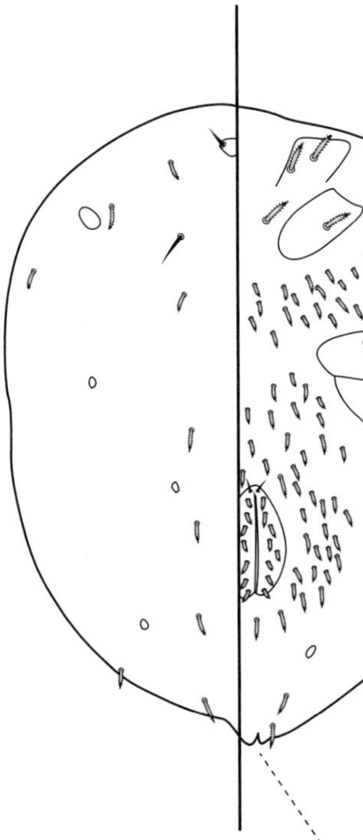

Fig. 15.5. *Halotydeus destructor* (Tucker). Female: dorsal (left) and ventral (right) view (modified from Qin and Halliday, 1996a). Key: ao, anal opening.

17–40 μm long. The coxa setal formula is 3-1-3-3, or occasionally 3-1-4-4 (*3d* and *4d* are present). The ventral chaetome is neotrichous and bears 46–64 setae between the coxae II and III and 20–26 setae between coxae IV (anterior to the aggenital setae). Each genital shield bears 7–11 setae (*gs*), 10–25 μm long. There are six pairs of eugenital setae (*eu*), 17–30 μm long, with the anterior two pairs close to each other and posterior four pairs close to each other. There are 18–25 pairs of aggenital setae (*ags*), 20–25 μm long. The seta h_2 is dorsal. Genua I and II have papillae. The leg setal formulae is variable (Qin and Halliday, 1996a).

MALE. The male is like the female, except that the genital shield is orange in colour (Qin and Halliday, 1996a).

Geographical distribution

H. destructor is found in Australia (Qin and Halliday, 1996a; Ridsdill-Smith, 1997), Malawi (Meyer Smith, 1981a), New Zealand (Ridsdill-Smith, 1997) and southern Africa (Tucker, 1925; Meyer and Ryke, 1960; Goldsmid, 1962; Jeppson *et al.*, 1975; Meyer Smith, 1981a; Ridsdill-Smith, 1997; Qin and Halliday, 1996a).

Host plants

By family, the host plants include: Asparagaceae – *Asparagus* sp. (Qin and Halliday, 1996a); Asteraceae – bristly ox-tongue (Umina *et al.*, 2004), capeweed (Qin and Halliday, 1996a; Umina *et al.*, 2004), cat's ear (Umina *et al.*, 2004), flat weed, *Gazania krebsiana* Less., *Osteospermum* sp., *Senecio* sp., variegated thistle (Qin and Halliday, 1996a); Boraginaceae – Paterson's curse (Qin and Halliday, 1996a); Brassicaceae – canola (Umina *et al.*, 2004); Fabaceae – clovers (*Trifolium* spp.), lucerne (Qin and Halliday, 1996a; Umina *et al.*, 2004), peas (*Pisum* spp.) (Umina *et al.*, 2004), subclover (*Trifolium subterraneum* L.) (Qin and Halliday, 1996a); Iridaceae – *Watsonia* sp. (Qin and Halliday, 1996a); Malvaceae – *Malva* sp. (Qin and Halliday, 1996a); Oxalidaceae – *Oxalis* sp. (Qin and Halliday, 1996a); Papaveraceae – poppies (*Papaver* spp.) (Umina *et al.*, 2004); Poaceae-barley (Qin and Halliday, 1996a), oats (Qin and Halliday, 1996a; Umina *et al.*, 2004), phalaris, ryegrasses (*Lolium* spp.), wheat (Umina *et al.*, 2004).

Other hosts include moss (Bryophyta), medic, unidentified plants (Qin and Halliday, 1996a), and thick-bladed grasses (Umina *et al.*, 2004).

Bio-ecology

H. destructor was accidentally introduced into Australia from South Africa before 1917 (Otto and Halliday, 1991), and the native range of the Australian population is the Western Cape Province of South Africa, mainly the area around Cape Town (Qin, 1997). During the last century, the mite has spread throughout southern Australia, and is now a serious pest of many crop and pasture plants in the country (Ridsdill-Smith, 1991). The red-legged earth mite has a large geographical distribution and lives on a wide range of herbaceous plants. Jeppson *et al.* (1975) reported that it lives on annual, broad-leaved plants and grasses such as in permanent pastures containing clovers, and on young wheat, potatoes, tobacco, peas, beets and tomatoes; it is frequently found on seedlings in hot-beds and greenhouses and on ornamental annuals. However, the above reports could not be confirmed.

Allozyme electrophoresis has shown that the mite is diploid, and reproduces sexually (Weeks *et al.*, 1995). The life cycle develops through the egg, larva, protonymph, deutonymph, tritonymph and adult stages; four to five nymphal stages are known. The eggs hatch in 2.5–8.5 days and each nymphal stage requires 5 days for development; thus, 20 or more days may elapse before the immature stages are completed and the adults start laying eggs. The adult lives from 25 to 50 days (Jeppson *et al.*, 1975). *H. destructor* is fundamentally a soil mite and spends 90% of its time on or near the soil surface, moving up on to plants only to feed. When pasture is grazed short, the density of populations is lower, but the same number of eggs is laid. Populations of *H. destructor* increase with high RH and mortality is probably greater at low RH in short pasture. The species develops three generations a year, passing the summer as a diapausing egg in the cadaver of an adult female on the soil surface. Long-distance passive dispersal probably occurs when these female cadavers are blown by the wind (Ridsdill-Smith, 1997).

The red-legged earth mite feeds at any time during the day or night depending on the moisture level. It penetrates the leaf epidermis and sucks up the cell contents; other mites are attracted to the volatiles released from the damaged leaves by feeding mites, and they commonly feed in aggregations (3–36 individuals)

(Ridsdill-Smith, 1997). The mites are, therefore, gregarious when feeding; when not feeding. they congregate in hollows on the ground or beneath leaves. Although *H. destructor* prefers moist soil, it avoids free water. It is sensitive to touch stimuli and selects its habitat according to its texture, food and other tactile surroundings. The females do not produce webbing and populations can be observed in the field from the onset of the winter rains until early summer, when they decrease as a result of high temperature and desiccation (Swan, 1934; Norris, 1948; Goldsmid, 1962). In contrast, the males do produce webbing, which holds the spermatophores; they spin silk webbing in which they deposit small globules of water-like fluid and the spermatophores, which are mushroom-like structures placed singly at intervals along the threads.

The reproductive success of *H. destructor* varies considerably on different host plants. Its populations reach a higher density and lay more eggs on an annual pasture at a site with high rainfall than on one with lower rainfall (Ridsdill-Smith, 1997). Research carried out on the foraging behaviour of the mite in annual pasture in winter and spring in south-western Australia showed that the main feature of *H. destructor* foraging behaviour is the presence of aggregations ranging in size from 3–36 individuals on the adaxial surfaces of the upper canopy of the pasture. Most aggregating individuals are feeding, while most solitary and paired individuals are searching. There is no evidence that individuals have aggregated for reproductive behaviour. The adult sex ratio is similar to that of the population, even though the juvenile stages are under-represented in aggregations. During winter, the distribution of numbers of *H. destructor* on subclover, capeweed and grasses – mainly *Lolium rigidum* (Gaudin)–in the upper canopy of pasture is similar to the estimated ground cover of these plant species. However, individuals show host-plant selection for subclover when feeding. During spring, when the pasture plants are flowering, there are more *H. destructor* in capeweed flowers than in aggregations on subclover foliage. Capeweed flowers have been shown to enhance the rate of *H. destructor* reproduction in a laboratory experiment. Most *H. destructor* are found on or near the soil surface, with 10% at any one time on the upper canopy of the pasture. The mite is polyphagous and behaves as a selective opportunist feeder in pasture (Gaull and Ridsdill-Smith, 1996).

The red-legged earth mite produces two types of eggs, 'winter eggs' and 'diapause eggs'. The winter eggs are bright yellow or orange coloured and are laid in masses in a single layer. They are largely deposited on the lower surface of leaves located in damp sites or in contact with the soil. These eggs have a smooth and glossy surface but may become whitish when dry, owing to the secretion used to stick the eggs to the leaf surface. Over 18°C, the mite slows its activity and begins to produce the diapause eggs. The female does not lay these eggs, but retains them in its body which, when it dies, protects them from desiccation. Diapause eggs that lose moisture become flattened or concave on one side and have a thicker shell than winter eggs, which is an additional protection against hot summer conditions. When they are exposed to a water surface after a dry period, diapause eggs assimilate moisture and assume the appearance of winter eggs. Adequate, but not excessive, moisture is essential for hatching. A high level of moisture causes the eggs to continue to swell and eventually to burst. The habitat of this species is therefore limited to light sandy soils (Jeppson *et al.*, 1975). The production of diapause eggs is influenced by the increasing maturity of the host plants in the spring, and occurs within a temperature range of 15–70°C and RH 10–100%, even when the eggs are immersed in water. The most effective conditions are 32 days at 50°C and 50% RH. In atmospheres with a vapour pressures of about 40 mm Hg, diapause development is successful at all combinations of temperatures and RH. Thus, at 20% of RH, diapause development is slower than at higher or lower RHs, whereas tolerance to high temperatures is greatest at 20%. Moreover, diapause eggs are not harmed either by complete immersion in water or by periodic desiccation and remoistening (Wallace, 1970a,b). The highest number of diapause eggs are laid by the third generation of *H. destructor* in mid–late spring, but the first two generations produce largely winter eggs (Wallace, 1970a; James and O'Malley, 1993; Ridsdill-Smith and Annels, 1997). Moreover, climatic factors, e.g. photoperiod, induce diapause (Ridsdill-Smith and Annels, 1997), and Wallace (1970b) pointed out the importance of the increasing maturity of the host plants.

Moistened diapause eggs develop within a temperature range of 5–20.5°C. The rate of development of the deutovum stage increases with temperatures up to 20°C and then rapidly decreases. Eggs may tolerate temperatures between 25 and 31.5°C for up to 53 days without undergoing morphogenesis, but this treatment retards subsequent development at 16°C. Exposure to alternating temperatures accelerates the rate of development, but if the higher temperature exceeds 20.5°C, development is retarded. Air-dry eggs may tolerate temperatures up to 75°C, but moist eggs are killed by exposures of 1 day at 45°C. High temperature tolerance, together with the limitations on development imposed by temperature, ensure that diapause eggs do not hatch before early autumn in the event of unseasonal summer rain (Wallace, 1970a).

The range of temperature and RH also influences the distribution and fluctuations of *H. destructor* populations. The mite cuticle is permeable to water, and individuals that are exposed to dry air lose their moisture rapidly. Feeding mites are less affected by air humidity, because moisture lost by evaporation is replaced by the water present in the juices imbibed from the leaf tissues, but non-feeding mites must remain in damp, sheltered situations to survive. Survival at a certain temperature is longer as humidity is increased, the optimum humidity being near saturation. In atmospheres with the same saturation deficit and different temperatures, the survival period decreases with increasing temperature, although within the range of 18–26°C, temperature changes have no effect. Above 30°C, the temperature becomes the major factor influencing survival rate. At 34°C, the mite dies within a few hours, and at 50°C in a fraction of a minute (Solomon, 1937). The survival of medium-sized nymphal stages of the red-legged earth mite (mainly tritonymphs and deutonymphs) stored at low temperature (1.5°C) in sealed plastic boxes was more than 50% after 12 days of storage, with some mites surviving for up to 8 weeks. Adding fresh subclover leaves to the storage box increased the survival rate of mites from 12 to 28%, 19 days after the storage started. Mites stored for 2 weeks at low temperatures showed feeding activity in a screening experiment that was

similar to that of mites collected directly from the field (Liu and Ridsdill-Smith, 2000).

Natural enemies include the bdellid mite *Odontoscirus virgulatus* (Canestrini *et* Fanzago), which was collected in South Africa at high density in association with the red-legged earth mite, but their relationship is not clear (Womersley, 1933; Meyer and Ryke, 1959b). In 1965, Wallace introduced into Western Australia the anystid mite *Anystis wallacei* Otto, which had been reported as a predator of blue oat mites in France (Michael *et al.*, 1991); however, technical difficulties have hindered the practical release of this predator. Recently, several predatory mites from the families Bdellidae, Anystidae, Erythraeidae and Parasitidae were reported from southern New South Wales (Australia). The trends in their seasonal abundance were not clear, but bdellids were the most common predators (58.5%), followed by erythraeids (34.6%), anystids (4.9%) and parasitids (2%) (James, 2000). Research carried out in crops and pastures in the Western Cape Province of South Africa on predatory mites that could have potential for introduction into Australia as biological control agents of *H. destructor* has shown the presence of 56 species of predatory mites belonging to 14 families. Twelve species occur in significant numbers in South Africa and are not known from Australia. Most of these could be generalist predators, and are therefore unlikely to be approved for introduction into Australia, but the anystid *Chaussieria capensis* Meyer *et* Ryke has potential and has been studied further. Unfortunately, the research did not identify any other species considered to be suitable as biological control agents (Halliday, 2005a).

Symptomatology and damage

Usually, *H. destructor* lives on annual, broadleaved plants and grasses such as permanent pastures with various botanical species, and it frequently infests seedlings in hotbeds and greenhouses, and ornamental annuals. In regions of Australia, South Africa and New Zealand characterized by cool, wet winters and hot, dry summers, it is a severe pest of pastures, crops and vegetables (Ridsdill-Smith, 1997) and commonly occurs in regions with mixed pasture species grown in rotation with grain crops (Ridsdill-Smith and Pavri, 2000).

The red-legged earth mite feeds on lower plants on the soil surface and its motile forms penetrate the epidermis of host plants with movable digits, and sucks up the cell contents. Other mites are attracted to the volatiles released from the leaves damaged by feeding mites, and they usually feed in aggregations. Although the mites feeds on a wide range of plant species, its reproductive success varies considerably on different plant species. Mites are more abundant and lay more eggs on an annual pasture at a site with high rainfall than one with lower rainfall. The rate of oviposition is probably resource limited. When pasture is grazed short, the mites are less abundant, but the same number of eggs is laid (Ridsdill-Smith, 1997). The ability of *H. destructor* to utilize the foliage of a range of plant species of differing suitability for food, while living mostly on the soil surface in niches favourable for their survival, has enabled it to become very abundant (Ridsdill-Smith and Pavri, 2000).

The feeding activity of the mite causes silvery or whitish blemishes along the main veins of leaves. The whole surface becomes progressively bleached, and the leaves wilt rapidly and shrivel so that the infested plants appear scorched. Seedlings may be killed outright, as though from frost damage. Sections of silvered leaves of *Sonchus oleraceus* L. show that the cells of the palisade parenchyma are empty, including chloroplasts. In these areas, the damaged epidermis collapses, leaving fine breaks in the cuticle. The silvery appearance results from the replacement of the cell contents by air (Jeppson *et al.*, 1975).

Control

The monitoring of populations is an important step in control, which is mainly provided by spray applications of acaricides according to the susceptibility of the different biological stages of the mite. The chemical impact on natural enemies may be reduced by using chemicals with the least impact, e.g. cyclodienes, and by reducing the number of treatments. Natural enemies living in windbreaks and roadside vegetation suppress *H. destructor* in adjacent pasture paddocks. When pesticides with residual activity are applied as border sprays to prevent mites moving into a crop or pasture, beneficial insect numbers may be reduced, thereby protecting *H. destructor* populations (Umina, 2008). Host plant resistance and cultural practices could also limit the development of mite populations. The implementation of an IPM protocol is by integrating the means available, according to the information available, the correct identification of the mite species and knowledge of its bio-ecology (see below).

MONITORING. Monitoring of mite populations needs inspections of the pastures and crops from autumn to spring. The crops should be monitored regularly in the first 3–5 weeks after sowing, taking into account that the mite commonly feeds on the leaves in the morning or on overcast days. In the warmer hours of the day, *H. destructor* gathers in the lower part of the plants, sheltering in the leaf sheaths and under debris, and takes refuge in cracks on the ground to avoid higher temperatures. Disturbances during feeding induce the mite to drop to the ground and seek shelter (Umina, 2008). The females can be killed before they begin to lay diapause eggs in the spring. The predictive model TIMERITE® facilitates the proper timing of sprays (Ridsdill-Smith and Pavri, 2010).

PLANT RESISTANCE. The development of host plant resistance is an important aspect of the current approach to controlling earth mites. Several screening programmes have been carried out in Australia to identify plant varieties that are resistant to mites, mainly *H. destructor* (Ridsdill-Smith, 1997). Subclover varieties, legume species, annual medics and canola accessions have been screened for resistance to damage from *H. destructor* feeding (Gillespie, 1993; Lake and Howie, 1995; Moritz and McDonald, 1995; Ridsdill-Smith *et al.*, 1995). Biochemical and physical mechanisms of resistance to *H. destructor* have been identified for subclover, which has antifeedant properties, with the basis of resistance being antixenosis – the inability of a plant to serve as a host for a particular arthropod or its non-preference for that plant (Jiang and Ridsdill-Smith, 1996). In subclover, increased strength of the cotyledon results in reduced

damage because mites cannot penetrate the epidermis of the resistant varieties. Levels of the volatile compound 1-octen-3-one have also been correlated with reduced mite feeding (Jiang *et al.*, 1996). It is unknown whether these mechanisms also reduce feeding damage by *Penthaleus* spp. With gene transfer technology now available, work has also begun to identify plant proteins that may confer resistance to mite attack (Weinman *et al.*, 1995). These studies are yet to reach the field-testing stage. Unfortunately, resistance to *H. destructor* may not necessarily confer resistance to all three (or any) *Penthaleus* spp. Moreover, the future management of this pest will require the integration of plant resistance with other control methods (Ridsdill-Smith, 1997).

CULTURAL PRACTICES. Cultural practices, such as crop rotations, the use of trap and border crops, and weed management, require a thorough understanding of pest–host associations. Canola and the common weed, bristly ox-tongue, are suitable hosts for *H. destructor* and *P. falcatus*, whereas *P. tectus* and *P. major* fail to persist on these plants. A mixture of wheat and oats sustained *P. tectus* and *H. destructor*, but not *P. falcatus* or *P. major*. Lentils were generally a poor host plant for all mite species. These findings show that earth mites differ in their ability to persist on different plant types, highlighting the importance of distinguishing mite species before implementing control strategies (Umina and Hoffmann, 2004). Thus, for *H. destructor*, the cultivation of lentils could reduce the risk of mite population development on a next susceptible crop (canola, etc.). Hot stubble burns also help to control the *H. destructor* population by decreasing the density of over-summering eggs. Clean fallowing and controlling weeds (especially thistles and capeweed) around crops and pasture reduce mite populations as well, by destroying important breeding sites for *H. destructor*. Previous attacks at high population densities suggest avoiding sowing pastures with a high clover content. Grazing management maintains mite populations to below damaging levels, as it produces a shorter pasture and a lower RH, which are responsible for mite mortality and reduced food resources. The modification of tillage practices, the use of trap or border crops and mixed cropping are all further measures that can reduce infestation to below the economic threshold, particularly when applied together with other measures (Umina, 2008).

BIOLOGICAL CONTROL. Natural enemies of *H. destructor* include generalist predators and some fungal pathogens. The first group includes *Chrysopa* spp. (Chrysopidae), *Hemerobius* spp. (Hemerobiidae), *Armonia* sp. and *Diomus* sp. (Coccinellidae), phytoseiid mites (*Amblyseius* spp., *Typhlodromus* sp.), *Walzia australica* Womersley and *Erythracarus* sp. (Anystidae), *Balaustium murorum* (Hermann) (Erythaeidae), and parasitid, bdellid, cunaxid and stigmaeid mites. In the second group, infection of *H. destructor* by the pathogen *Neozygites acaracida* (Petch) (Neozygitaceae) ranges from 0 to 50%, especially during wet winters (James, 1995; Ridsdill-Smith, 1997; Umina *et al.*, 2004). The introduction into Australia of the anystid *A. wallacei* from France (Michael *et al.*, 1991) appeared satisfactory, although the predator posed several problems that inhibited its practical application. The predator was found to spread very slowly at the sites into which it was released (Michael *et al.*, 1993), although feeding tests carried out later showed that the release of 100 *A. wallacei* m^{-2} could kill 16,000 red-legged earth mites in one pest generation (Michael, 1995). However, various problems have hindered its practical release. Research carried out in southern New South Wales on natural populations of *H. destructor* and *Penthaleus* spp. and predatory mites of the families Bdellidae, Anystidae, Erythraeidae and Parasitidae at six pesticide-free grassland sites and three pesticide-treated improved pasture sites show that natural enemies may play a significant role and that the development and use of more specific insecticides in pastures represents a way of minimizing the impacts on earth mite predators and improving biological control (James, 2000; Umina, 2008). At the moment, there is no information available on artificial biological control.

CHEMICAL CONTROL. A common problem is mite tolerance to different acaricides. Thus, the red-legged earth mite is tolerant to omethoate, but no tolerance to endosulfan has been detected, although this chemical is not used heavily (and is now banned in Australia and other countries). Some tolerance to alphamethrin has also been observed. This situation suggests that alternative methods for controlling this pest will eventually need to be considered (Hoffmann *et al.*, 1997).

Traditionally, chemicals should be used in the autumn as soon as possible after egg hatching is complete, because the eggs are the most difficult stage to control by acaricides. Small and delicate plants are particularly susceptible to injury by a few mites. To achieve early control with non-persistent acaricides such as the organophosphorus compounds, the time of distribution must be carefully determined. The more persistent acaricides may remain sufficiently toxic to the young mites as they emerge from the eggs or at the moulting stage and thus provide adequate control over an extended period. Injurious infestations of agricultural crops may be prevented by clean cultivation, especially during autumn and early winter, as weed hosts serve as breeding areas (Wright, 1961).

Control methods in pest earth mites and other mites often depend on them having low dispersal rates. Estimates of adult movement rates in *H. destructor* and *P. major*, using mark–release–recapture techniques and mean square dispersal distances to estimate diffusion coefficients, indicate that in pasture, coefficients were in the range of 0.3–1.3 for both species; 90% of the population moves <5–11 m in a 10-day period, or 7–16 m within their adult lifetime. Releases of mites in adjacent pea/wheat crops indicated directional movement towards the more favoured pea host. However, there was no directional movement when adjacent plots of peas and lupins were compared, even though lupins were poor hosts. Broad border sprays or border culturing will be needed to prevent mite movement from adjacent paddocks (Weeks *et al.*, 2000a).

Chemical control aimed at interrupting the life cycle at the most effective stage, by spraying in the spring before the mites enter diapause, showed that numbers of *H. destructor* at the beginning of the winter growing season were suppressed by up to 85%, but that this control level was not sustained, as the numbers of *H. destructor* and other pests were similar to or higher than those in the unsprayed areas by spring. Spring spraying is, therefore, a useful means of suppressing *H. destructor* in early

autumn, but its effects disappear relatively quickly over time. This strategy also has little impact on the control of *P. major* and other invertebrate pests (Gower *et al.*, 2008).

Controlling the first generation of mites in the autumn before the deposition of eggs helps to avoid the need for a second treatment. Chemicals employed at or after sowing should be used within 3 weeks of the first sampling of mites, before egg deposition. Chemicals with persistent residual activity can be distributed, either pre-sowing or at sowing, to kill emerging mites and protect the seedlings, which are usually the most susceptible to damage. Leaf treatment of the emerged crop is carried out and is commonly effective. Systemic pesticides are usually employed as seed dressings, but if mite density is high, plants may suffer severe damage before the chemical has much effect (Umina, 2008).

One accurately timed spring treatment with an effective chemical can significantly reduce populations of *H. destructor* in the following autumn. This option allows mites to be killed before diapause egg deposition in mid–late spring. Knowledge of climatic factors, and the use of the predictive model (TIMERITE®), help to identify the optimum date for treatments (Ridsdill-Smith and Pavri, 2010). The same treatments are not usually effective in the control of other mites. Also, massive use of the spring spraying technique can stimulate resistance to the product used, and to prevent this the selective rotation of products with different modes of action is suggested (Umina, 2008).

16 The Family Eriophyidae Nalepa

Morphological Characteristics, Systematics and Bio-ecology

The Eriophyidae are very small, worm-like or fusiform mites, approximately 200 μm long. The eriophyid body consists of the gnathosoma, the prodorsum and an extended and annular opisthosoma (Fig. 16.1). The prodorsal shield lacks anterior setae (*vi* or *ve*) and the scapular setae (*sc*) may be present or absent (Fig. 16.2). The gnathosoma is usually prognathous and directed anteriorly; it has two stumpy palpi, the infracapitulum and the chelicerae. The palpi are parallel to the infracapitulum, and each consists of a base and three segments. The mouth is surrounded by a series of stylets derived from the digits of the chelicerae, the oral stylet or the labrum, and the infracapitulum or rostrum, which has a 'U'-shaped cross section. The walls of the infracapitulum form a membranous sheath in front and externally have edges that overlap dorsally, concealing from seven to nine stylets (Fig. 16.3). The stylets include one pair of cheliceral shafts divided towards the apex into a dorsal digitus fixus and a ventral digitus mobilis that are separated from the labrum, and one pair of auxiliary stylets, also known as the inner infracapitular stylets. The basal segments of the chelicerae are close to one other and to the motivator, which provides a basal support for the chelicerae (Nuzzaci and Alberti, 1996). The opisthosomal surface has a series of transverse rings, or annuli, that never have a subdorsal seta c_1; the other setae are generally present, although, occasionally, one or two pairs of setae (c_2, *d*, *e*) and the setae h_1 are lacking. The coxal region bears setae but plate I may lack the setae *1b* (Fig. 16.1). The Eriophyidae have two pairs of legs; the leg chaetotaxy is complete, but sometimes the femoral setae I and II, genual seta II, tibial seta I and both tarsal setae *ft′* and *u′* of legs I and II are absent; tibia I is without a solenidion (Fig. 16.4). The tarsal empodium (Fig. 16.4) is sometimes thickened or of varied shape, and is rarely deeply divided. The female genital coverflap or epigynium is usually adorned, often with one or two rows of linear striae, and sometimes with granules (Fig. 16.5); the female spermathecal tubes are always short, either projecting laterally or, caudally, diagonally (Fig. 16.6). For a comprehensive treatment of the morphology of the Eriophyidae, the works of Lindquist (1996), Lindquist and Amrine (1996) and Amrine *et al.* (2003) should be consulted, and for updated treatments of their bio-ecology, damage and control, consultation is suggested of the contributions edited by Lindquist *et al.* (1996) and Ueckermann (2010).

The Eriophyidae has six subfamilies (Aberoptinae Keifer, Nothopodinae Keifer, Ashieldophyinae Mohanasundaram, Cecidophyinae Keifer, Eriophyinae Nalepa, Phyllocoptinae Nalepa) and 12 tribes (Lindquist and Amrine, 1996; Amrine *et al.*, 2003). According to a conservative estimate, the eriophyoid fauna of the world could include from 35,000 to 50,000 species, and some estimates are as high as 250,000 species (Amrine *et al.*, 2003). About 2500 species of all known Eriophyoidea belong to the Eriophyidae (85% of total) (Oldfield, 1996a). A database developed by de Lillo and Amrine with all available papers regarding eriophyoids published between 1830 and 2007 includes 3957 eriophyoid species and 7699 host plant species (de Lillo and Amrine, 2006; Skoracka *et al.*, 2010). Later, Zhang *et al.* (2011) reported 274 genera and 3790 species. DNA analyses – as yet, few concerning the eriophyoids –will expand knowledge on the population genetics, taxonomy, systematics and evolutionary relationships of many species of this family (Lindquist, 2001). An example of the use of this type of analysis is given by *Aceria guerreronis* Keifer, for which host associations and colonization history have been explained by the use of mitochondrial and nuclear markers (Návia *et al.*, 2005b).

Females are commonly more numerous than males in this family, and in some species males are unknown (Manson and Oldfield, 1996). Males are similar to females, but smaller and without a genital coverflap, and with different genitalia (Lindquist, 1996). In some cases, as in *Aceria inusitata* Britto *et* Návia (Britto *et al.*, 2008) and *Trisetacus kirghisorum* Shevchenko (Shevchenko and De-Millo, 1968), there are two types of males, one similar to a protogyne and the other to a deutogyne. The life cycle develops through the stages of egg, larva, protonymph and adult (Lindquist, 1996; Manson and Oldfield, 1996). A quiescent stage occurs between larva and nymph, and between nymph and adult, which is commonly called the 'nymphochrysalis' and 'imagochrysalis', respectively (Sternlicht and Goldenberg, 1971). In some species, the female produces hatched larvae (Abou-Awad, 1981b; de Lillo, 1991). The Eriophyidae are fundamentally bisexual, with a common number of chromosomes ($n = 2$) and reproduction is by arrhenotokous parthenogenesis (Helle and Wysoki, 1996). The male lacks an aedeagus and sperm transfer occurs via a spermatophore deposited on the host plant (Sternlicht and Goldenberg, 1971). In general, males seem to show interest in quiescent female nymphs for a single or a few inseminations (Michalska and Mańkowski, 2006). In some species, the males are attracted to 'deutonymphs' and deposit spermatophores that encircle them (Michalska and Boczek, 1991). Males of the free-living species *Aculus robiniae* (Nalepa) guard the pre-emergent females for about an hour before their emergence, but sex pheromones are probably not involved, and the male attraction is to old cast skin, the female emergence site and freshly moulted females (Oldfield and Michalska, 1996). Several factors influence spermatophore deposition, such as male age (Oldfield and Newell, 1973a), time of day (Michalska, 2005), the host plant and its quality, the presence of females (Michalska and Shi, 2004) or rivals (Michalska, 2000) and sperm competition (Michalska, 2005). In some

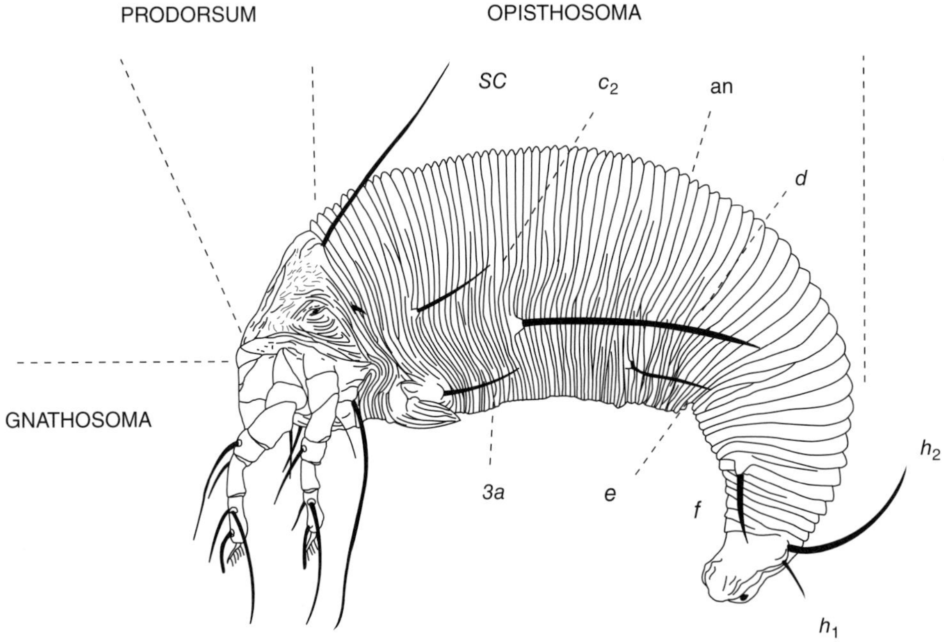

Fig. 16.1. Morphological features of eriophyoid mites. Lateral view of adult of *Eriophyes neocynarae* Keifer (from Vacante and Tropea Garzia, 1990). Key: an, annuli. The rest of the labels are setal notations, which are explained in the text.

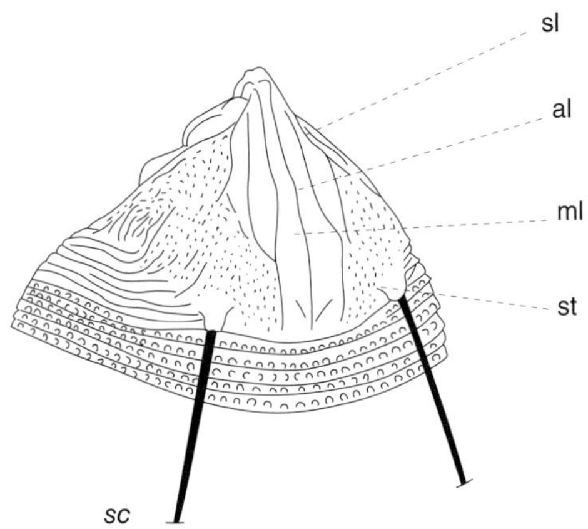

Fig. 16.2. Prodorsal shield of *Eriophyes neocynarae* Keifer (from Vacante and Tropea Garzia, 1990). Key: al, admedian line; ml, median line; *sc*, scapular seta; sl, submedian lines; st, scapular tubercle.

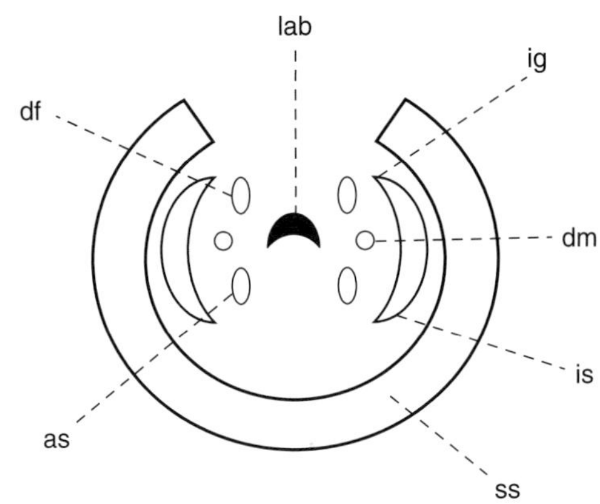

Fig. 16.3. Diagram of transverse section of an eriophyoid mite near the apices of the stylets showing cheliceral and associated structures (from Vacante, 2010). Key: as, auxiliary stylet; df, digitus fixus; dm, digitus mobilis; ig, infracapitular guide; is, infracapitular stylet; lab, labrum; ss, stylet sheath.

species, the deposition of a spermatophore may be induced by an emitted attractant (Oldfield *et al*., 1970, 1972). In free-living species, spermatophores frequently occur in groups, and in vagrant species, spermatophores are numerous along the mid-vein and the junctions of the lateral veins of the leaves (Oldfield *et al*., 1970). Research on the interspecific attraction of spermatophores has shown that females of different species self-inseminate from the spermatophores of congeneric males (Oldfield, 1989). In several species of Eriophyidae, sexual 'independence' of males and females has been observed, with males able to deposit spermatophores in the absence of females, and females not requiring the presence of males to pick up sperm from a spermatophore (Sternlicht and Goldenberg, 1971; Michalska and Boczek, 1991; Michalska and Shi, 2004). This phenomenon is also called 'sex dissociation' and is widespread in several arthropods, including many prostigmatic mites. Sex dissociation can be complete or incomplete, and a high population density may favour the phenomenon (Sabelis and Bruin, 1996).

Eriophyidae can develop two types of life cycle. The first type has the female corresponding structurally to the male, when it is called a 'protogyne'; females and males together represent the 'perfect' or 'primary form'. The female of the second type represents the 'deutogyne' or 'secondary form' (deuterogyny is discussed further in Chapter 5). The deutogynes have reduced or absent microtuberculation, or the microtubercles can have a different shape. In some cases, they have narrower dorsal annuli, and

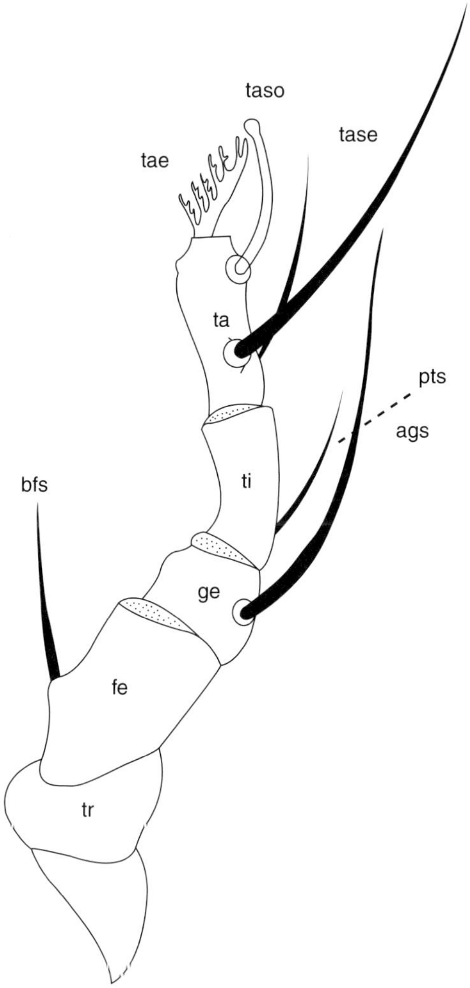

Fig. 16.4. *Eriophyes neocynarae* Keifer. Leg I of female. Key: ags, antaxial genual setae; bfs, basiventral femoral setae; fe, femur; ge, genu; pts, paraxial tibial setae; ta, tarsus; tae, tarsal empodium; tase, tarsal setae; taso, tarsal solenidion; ti, tibia; tr, trochanter (drawing by Vacante).

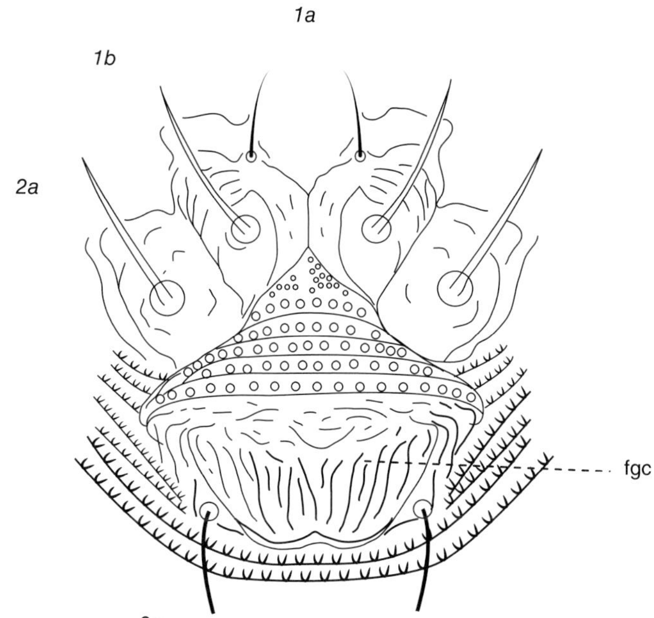

Fig. 16.5. *Cecidophyopsis vermiformis* (Nalepa). Coxal and genital regions (from Keifer, 1944). Key: *1a*, proximal seta on coxisternum I; *1b*, anterolateral seta on coxisternum I; *2a*, proximal seta on coxisternum II; *3a*, proximal seta on coxisternum III; fgc, female genital coverflap.

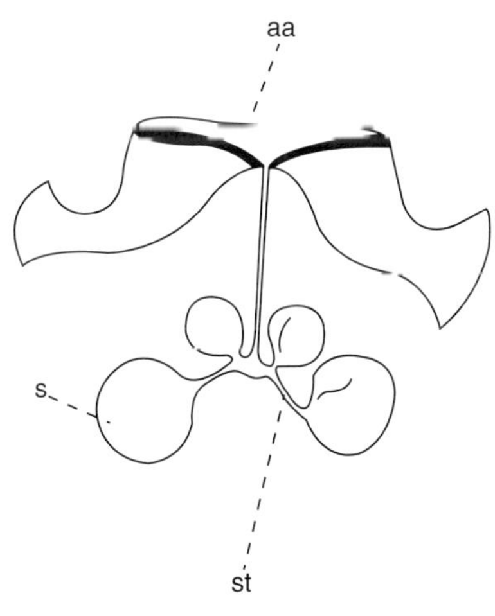

Fig. 16.6. *Eriophyes neocynarae* Keifer. Internal genitalia of female (from Vacante and Tropea Garzia, 1990). Key: aa, anterior apodeme; s, spermatheca; st, spermathecal tube.

any annuli, furrows or protuberances occurring in the protogynes are lacking in the deutogynes. Other differences can be observed in the prodorsal shield (Jeppson *et al.*, 1975; Manson and Oldfield, 1996). From the ecological point of view, deutogynes survive through adverse conditions. The modification of their opisthosoma represents a means for overwintering deutogynes to conserve body fluids by transforming the cuticle to one that is more resistant to water loss (Krantz and Ehrensing, 1990). According to their adaptations, various species can develop different behaviours. In leaf vagrants and rust mites, leaf hardening and the rising summer temperatures stimulate the production of the deutogynes in late spring or summer. In other cases, the onset of cool temperatures in the autumn stimulates the formation of deutogynes. The deutogynes move from the leaves into crevices on twigs, under bud scales or around lateral buds for aestivation and hibernation. They are inseminated before wintering, but produce eggs after a period of winter cold, followed by rising spring temperatures (Jeppson *et al.*, 1975). The wintering deutogynes emerge in spring and lay eggs that give rise to the primary forms (Manson and Oldfield, 1996).

Knowledge of phenology and developmental biology is fundamental to the control of eriophyid mites. Indeed, a better knowledge of some of these characteristics helps to establish the timing of their management, as in the control of refuge-seeking and gall-inducing Eriophyidae, which live in these structures for a long time. Knowledge of physical environmental factors (temperature, relative humidity (RH), etc.) and of the developmental stages of the host plant allows prediction/modelling of the emergence time of different species of Eriophyidae (Balaji and Hemavathy, 2007; Webber *et al.*, 2008).

Over 50 Eriophyidae are injurious to economic plants and about 30 cause severe damage; they include several pests of broadleaved plants, vectors of plant pathogens and nearly all gall-forming species. Eriophyid mites induce cytological, biochemical and physiological changes in infested plants. Host plant responses are variable, and largely dependent on mite–plant interactions. The changes that occur are related to mite genotype, density of the mite populations, feeding time and plant characteristics (species, cultivar, age) and locality (Petanović and Kielkiewicz, 2010a). Moreover, the alterations caused by eriophyid mites to plants can be both specific and non-specific (Petanović and Kielkiewicz, 2010b). About half of the injurious species are leaf vagrants, while other species (most species of the genera *Aceria* and *Eriophyes*) produce galls on leaves, green twigs, flower buds, vegetative buds or fruits. A few species attack monocotyledons, conifers, other gymnosperms and ferns, but most live on dicotyledonous plants (Oldfield, 1996a). The effects of feeding on host plants can be examined both as direct damage caused by the mites, which concerns the structure (morphology) and functioning of the eriophyid mouthparts, and as plant damage at the cellular level, which is related to early events and the differential responses (Westphal and Manson, 1996). On the host plants, the symptoms include various kinds of galls (leaf galls, stem galls, bud galls, fruit galls) and other distortions. Attacks by many Eriophyidae that do not induce gall formation or other distortions can give rise to toxaemias and non-distortive feeding effects on epidermal tissue (Oldfield, 1996b). Web spinning and the production of wax-like and liquid secretions are reported in some species (Manson and Gerson, 1996; Gondim and Moraes, 2003; Britto *et al.*, 2008).

Eriophyoid mites have a strong relationship with their host plants and reproduce on a precise range of plants. Knowledge of the mechanisms of host specificity enables understanding of mite–plant interactions, their coevolution and the diversity of the group, and also the use of rational control strategies such as the biological control of weeds (Skoracka *et al.*, 2010). This last possibility is supported by a number of considerations. In particular, the available data show that 80% of eriophyoids are reported on only one host plant, 95% on one genus and 99% on one family. The Diptilomiopidae include the highest number of monophagous species and the Phytoptidae the fewest. Non-monophagous species tend to develop on closely related plant hosts. Vagrant eriophyoids have a higher number of monophagous species than refuge-seeking and refuge-inducing species. Monophagous species develop in the same proportion on both annual and perennial host plants, but many species infesting annuals have wider host ranges than those infesting perennials; they also develop in the same proportion on evergreen and deciduous plants. Non-monophagous species have a wider geographical distribution (Skoracka *et al.*, 2010). These characteristics suggest that several species of eriophyoids have high potential for the biological control of weeds. In the last decade, 13 species have been evaluated for this purpose, but only four have been authorized for introduction. Another three species were introduced earlier, but it is not clear how they have contributed to reduction of the target weed population (Smith *et al.*, 2010). Despite the positive premise, it seems that several ecological factors (natural enemies, resistant plant genotypes, etc.) seriously limit the use of this biological strategy.

Various eriophyid species are vectors of plant pathogens. The nature of most of these pathogens was uncertain until a few years ago (Oldfield and Proeseler, 1996), and only recently has it been found that most of them are viruses belonging to the families Potyviridae (*Rymovirus*, *Tritimovirus*) and Comovirinae (*Allexivirus* (now classified as belonging to the Alphaflexiviridae) and *Nepovirus* (now classified as belonging to the Potyviridae)) (Jones, 2000; Kumar *et al.*, 2003a, 2005; Kang *et al.*, 2007). Classic pathogens of cereals include the *Wheat streak mosaic virus* (WSMV), the 'Wheat spot mosaic pathogen' (WSpM), *Ryegrass mosaic virus* (RgMV) and *Agropyron mosaic virus* (AgMV). *Allium* spp. are infected by various viruses. On woody dicotyledonous plants, viruses such as the *Blackcurrant reversion virus* (BCRV), *Fig mosaic virus* (FMV), *Peach mosaic virus* (PMV), *Cherry mottle leaf virus* (CMLV), the formerly designated '*Rose virus 1* (RSV-1)' and '*Pigeonpea sterility mosaic virus*' (PPSMV) have been reported.

In addition to the transmission of viruses, the eriophyoids may have a role in the epidemiology of some fungal plant pathogens. Thus, *Aceria mangiferae* (Hassan) vectors spores of the fungal pathogen *Fusarium mangiferae* Britz in the bud of mango, a typical port of entry for the fungal pathogen. The frequency and severity of infection were increased in the presence of the mite (Gamliel-Atinsky *et al.*, 2010).

The dispersal of Eriophyidae and the colonization of a new habitat occurs via active ambulation, the wind, the rain and several carriers (insects, phytoseiids, mole crickets, humans, spiders). The phenomenon involves various factors, such as host acceptance, feeding and the longevity of the dispersing mites. Knowledge of the genetic and environmental factors influencing host acceptance and detection of the mechanisms effecting mite transfer is important, as is knowledge of the genes responsible for resistance in original hosts (Michalska *et al.*, 2010).

Natural enemies of Eriophyidae include insects (dipterans, coleopterans, neuropterans, hemipterans, thysanopterans), mites (phytoseiids, stigmaeids, cheyletids, cunaxids, tarsonemids, tydeids) (Perring and McMurty, 1996) and pathogens (McCoy, 1996a). Among the predatory mites, phytoseiids and stigmaeids are the most important (Sabelis, 1996) and stigmaeids prey preferably upon eriophyids (Thistlewoood *et al.*, 1996). In general, refuges of eriophyoid mites are not perfect for predatory mites. Vagrant species are generally more vulnerable to predation than are refuge-forming or refuge-seeking eriophyoids (Michalska *et al.*, 2010).

The account of eriophyid mites presented in this chapter covers genera including mite species of economic importance and is mainly taken from Amrine *et al.* (2003). Abbreviations used for the accompanying figures are been drawn from Lindquist (1996) and Amrine *et al.* (2003) (see Table 16.1). When possible, details are given of a website for each species that will show coloured photos of the damage that it causes.

Selected website

Acari: Plant-Feeding Mites. Systematic Entomology Laboratory, US Department of Agriculture Agricultural Research Service, Beltsville, Maryland. Available at: http://www.sel.barc.usda.gov/acari/frames/plantfeed.html

Table 16.1. Abbreviations used in the illustrations (from Amrine *et al.*, 2003).

AD	Anterodorsal mite	GM	Genital region, male
AL	Anterolateral mite	GN	Gnathosoma
AV	Anteroventral mite	IG	Internal genitalia, female
C	Coxal region	L1, L2	Legs I, II
CG	Coxal–genital region	LM	Lateral mite
CO	Caudal opisthosoma	LO	Lateral opisthosoma
cx	Coxa	PD	Posterodorsal mite (telosoma)
D	Dorsal mite	PM	Posterolateral mite (telosoma)
DO	Dorsal opisthosoma	PV	Posteroventral mite (telosoma)
em	Empodium	T	Tarsus
fm	Femur	ti	Tibia
ge	Genu	tr	Trochanter
GF	Genital region, female	V	Ventral mite

Key for the Identification of Genera

1\. • Less than ten genera without scapular setae (*sc*) on prodorsal shield; female genitalia appressed to coxae, separating coxae more than normal, and in lateral view commonly projecting from venter; female genital apodeme bent up and short, like a heavy transverse line; ridges on female coverflap arranged in two irregular ranks; coxae I usually narrowly connate at centre line, sternal line short; curved lines often outlining tubercles of coxal setae, particularly setae *1a* ....subfamily **Cecidophyinae** Keifer...3

• Prodorsal shield rarely lacking scapular setae (*sc*); female genitalia not appressed to or separating coxae, and in lateral view lying more on a level with venter; female genital apodeme usually extending moderately forward, does not appear as a heavy transverse bar; female coverflap variably sculptured, ridges rarely arranged in 2 ranks; sternal line commonly present; coxae usually lacking curved lines outlining setiferous tubercles...2

2\. • Body vermiform, dorsal and ventral annuli subequal, at least on anterior ½ to ⅔ of opisthosoma; prodorsal shield lacking frontal lobe, or with a slight projection over gnathosoma base; when frontal lobe is present, it is narrow, basally flexible, and combined with narrow annuli...subfamily **Eriophyinae** Nalepa...5

• Body. more fusiform; prodorsal shield with a broad-based and rigid frontal lobe over gnathosoma; opisthosoma divided into broad dorsal annuli, and narrow, microtuberculate ventral annuli; if frontal lobe lacking or only a slight one present, then dorsal and ventral annuli are different, at least in larger dorsal microtubercles; if annuli subequal, and a broad frontal lobe is absent, then annuli are as broad as genital coverflap is long.... ....subfamily **Phyllocoptinae** Nalepa...8

3\. • Prodorsal shield without scapular tubercles and setae (tribe **Cecidophyini** Keifer); vermiform bud mites; frontal lobe absent or strongly indistinct; coxal setae *1b*, tibial setae and setae *e* present; genu II with setae...genus ***Cecidophyopsis*** Keifer

• Prodorsal shield with scapular tubercles and setae ....tribe **Colomerini** Newkirk *et* Keifer...4

4\. • Scapular setae (*sc*) projecting to rear; tibial seta lacking ....genus ***Cosetacus*** Keifer

• Scapular setae (*sc*) projecting diagonally forward or straight ahead, occasionally backwards or laterally; prodorsal shield with frontal lobe; all leg setae present; opisthosomal setae h_1 absent..........................genus ***Colomerus*** Newkirk *et* Keifer

5\. • Scapular tubercles and setae more or less ahead of rear prodorsal shield margin, setae directed forward or up; if scapular tubercles and setae near rear shield margin, then setae directed forward, or have longitudinal basal axes (tribe **Eriophyini** Nalepa); prodorsal shield without frontal lobe; dorsal opisthosoma evenly arched in cross section; dorsal and ventral opisthosomal annuli subequal for entire length; coxal setae *1b* present.............genus ***Eriophyes*** von Siebold

• Scapular tubercles on, or very near, rear prodorsal shield margin, with transverse basal axes, setae directed to rear, commonly divergently...tribe **Aceriini** Amrine *et* Stasny..6

6\. • With tibial setae..7

• Without tibial setae; femur I without setae; coxae I often fused across midline, with sternal line weak or lacking ....genus ***Acalitus*** Keifer

7\. • Posterior opisthosomal annuli continuous, subequal dorsoventrally...genus ***Aceria*** Keifer

• Posterior opisthosomal dorsal annuli wider; opisthosoma with all setae..........................genus ***Paraphytoptus*** Nalepa

8\. • Empodium divided (tribe **Acaricalini** Amrine *et* Stasny); body fusiform; one middorsal ridge on dorsal opisthosoma; scapular setae (*sc*) short; coxal setae *1b* lacking; genu II without setae; tibial setae present............................... ....genus ***Acaphylla*** Keifer

• Empodium entire...9

9\. • Scapular tubercles present or absent; scapular setae (*sc*) absent (tribe **Calacarini** Amrine *et* Stasny); dorsum with three ridges; middorsal ridge narrow, extends as far as lateral ridges; tibial setae present ....genus ***Calacarus*** Keifer

• Scapular tubercles and setae (*sc*) present.................10

10\. • Opisthosoma in dorsal view with lateral lobes or pointed projections from some or all annuli, or with a plate behind prodorsal shield with lateral extensions.................................... ....tribe **Tegonotini** Bagdasarian...12

• Opisthosoma in dorsal view with annuli evenly downcurved over lateral margins; dorsum varying from evenly arched in cross section to flattened, ridged or furrowed 11

11\. • Scapular tubercles well formed, often plicate, set ahead of rear prodorsal shield margin, directing scapular setae (*sc*) forward, up or centrally; if tubercles and setae are near rear shield margin, then tubercles are subcylindrical and bent forward or the alignment of their bases is longitudinal or diagonal to the body.....tribe **Phyllocoptini** Nalepa...13

• Scapular tubercles on or very near the rear shield margin, directing scapular setae (*sc*) to rear, commonly divergently; scapular tubercles either subcylindrical, or the alignment of their bases is transverse to the body.......................tribe **Anthocoptini** Amrine *et* Stasny...17

12. • Dorsal area above opisthosomal setae *f* depressed; scapular setae (*sc*) set near rear prodorsal shield margin and directed posteriorly.......................genus ***Oxycenus*** Keifer
• Dorsal area above opisthosomal setae *f* not depressed; scapular setae (*sc*) variously directed but not posteriorly; tibiae with setae...........................genus ***Tegonotus*** Nalepa

13. • Dorsal opisthosoma with ridges or evenly rounded...14
• Opisthosoma with a wide middorsal longitudinal furrow; opisthosomal setae *e* present; sternal line on coxae I present............................genus ***Phyllocoptruta*** Keifer

14. • Dorsal opisthosoma with 3 longitudinal ridges..............15
• Dorsal opisthosoma evenly rounded..........................16

15. • Dorsal opisthosoma with middorsal ridge finishing in a large furrow before termination of subdorsal ridges...genus ***Calepitrimerus*** Keifer
• Dorsal opisthosoma with middorsal ridge fading with subdorsal or lateral ridges, the shallow subdorsal furrows often causing ridges to be weak...genus ***Epitrimerus*** Nalepa

16. • Scapular tubercles set ahead of rear shield margin, directing scapular setae (*sc*) forward or up and centrally; when they are set near rear margin, *sc* directed forward; sometimes *sc* on rear margin, and tubercles with longitudinal or oblique basal axes directing *sc* up, central or convergently to the rear; genu II with seta......genus ***Phyllocoptes*** Nalepa
• Scapular tubercles set on rear shield margin, directing scapular setae (*sc*) anteriorly or medioposteriorly; scapular tubercles with basal axes convergently anteriorly, directing *sc* convergently to the rear; opisthosomal annuli differentiated ....genus ***Vasates*** Shimer

17. • Opisthosoma with middorsal longitudinal ridge.........22
• Dorsal opisthosoma evenly rounded (a slight middorsal ridge sometimes present, anteriorly).............................18

18. • Coxae without setae *1b*; tibiae with setae; coxae I narrowly joined near setae *1a*.......................genus ***Metaculus*** Keifer
• Coxae with setae *1b*...19

19. • Dorsal opisthosoma with varying patterns of annuli, never uniform; first 3–10 annuli posterior to prodorsal shield margin narrow and microtuberculate, forming complete annuli around opisthosoma; the continuing dorsal annuli form broad bands until caudal annuli...genus ***Heterotergum*** Keifer
• Dorsal opisthosoma with a uniform pattern of annuli 20

20. • Prodorsal shield with distinct frontal lobe; opisthosomal dorsal annuli large, smooth, abruptly contrasting with narrow ventral annuli beyond opisthosomal setae *f*...genus ***Anthocoptes*** Nalepa
• Prodorsal with frontal lobe variable; opisthosomal dorsal annuli, when large, gradually narrowing to and not contrasting with posterior annuli; tibiae with setae.....................21

21. • Frontal lobe large and rounded, not acuminate; sometimes with 2–4 small spines or spinules projecting forward from under front edge.......................genus ***Aculus*** Keifer
• Frontal lobe more acuminate, may end in a sharp point; never bearing spinules under front edge; body fusiform; opisthosomal dorsal annuli usually wider than ventral annuli; empodia usually bearing fewer than 7 rays...genus ***Aculops*** Keifer

22. • Opisthosoma with middorsal ridge shorter than subdorsal ridges and ending in a dorsal furrow..genus ***Abacarus*** Keifer
• Opisthosoma with middorsal ridge stronger than lateral ridges but fades caudally, and not ending in a furrow ....genus ***Tegolophus*** Keifer

The Genus *Cecidophyopsis* Keifer

Diagnostic characteristics

The body is worm like. The gnathosoma is small, downcurved, with the first visible segment longer than the second. The prodorsal shield is broadly rounded anteriorly, not projecting over the gnathosoma and concealing it in dorsal view; the frontal lobe is absent or very indistinct, and the scapular tubercles and setae are missing. The legs have all setae, including genu II setae. The opisthosoma is circular in cross section, with opisthosomal annuli similar dorsoventrally and not divided into dorsal and ventral, also completely microtuberculate. The female genitalia are close to the coxae, with coxae separated somewhat; the coxal tubercles, especially the second pair, are partially surrounded by subcircular elevations; the female genital coverflap has numerous uneven longitudinal furrows, partly in two ranks; the apodeme is shortened in ventral view (Keifer, 1959a; Amrine *et al.*, 2003).

Economic importance

The genus includes about 20 species, living on dicotyledons, monocotyledons and conifers. The type species, *Cecidophyopsis vermiformis* (Nalepa) (Fig. 16.7), is widespread on European hazelnut (*Corylus avellana* L.) associated with *Phytoptus avellanae* (Nalepa), the main causal agent of gall formation on hazel and filbert (*Corylus* spp.), but only as a harmless inquiline (Massee, 1930a; Jeppson *et al.*, 1975). This interpretation is commonly accepted, with *P. avellanae* itself considered to be a severe pest of hazelnut worldwide (Özman, 2000) (see Chapter 17). *Cecidophyopsis ribis* (Westwood) damages buds and transmits the reversion disease of blackcurrant (*Ribes nigrum* L.) (Oldfield and Proeseler, 1996). *Cecidophyopsis selachodon* Van Eyndhoven produces similar big bud deformation of redcurrant (*Ribes rubrum* L.) (Van Eyndhoven, 1967b). *Cecidophyopsis psilaspis* (Nalepa) causes big bud on *Taxus baccata* L. in Europe and North America (Amrine and Stasny, 1994), and *Cecidophyopsis malpighianus* (Canestrini *et* Massalongo) produces hypertrophy of flower buds and distortion of flowers on *Laurus nobilis* L. in the Mediterranean region and North America (Castagnoli, 1996; Jeppson *et al.*, 1975). Among these, *C. ribis* is the economically more important species.

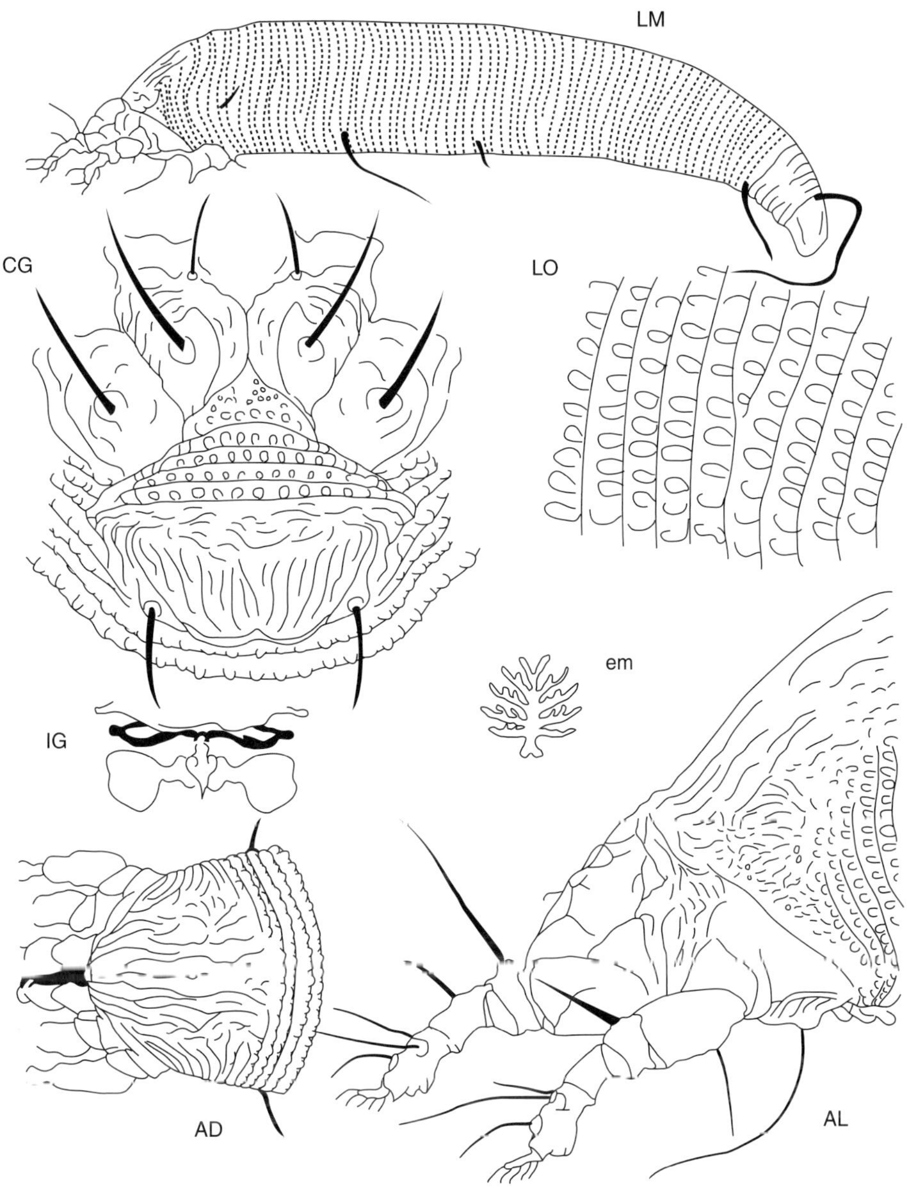

Fig. 16.7. *Cecidophyopsis vermiformis* (Nalepa) (from Keifer, 1944). See Table 16.1 for key.

Cecidophyopsis ribis (Westwood) (Fig. 16.8)

Common name

Blackcurrant big bud mite or blackcurrant gall mite.

Diagnostic characteristics

FEMALE. The body is worm like, 230–240 μm long and light yellowish in colour. The prodorsal shield has a slight central frontal lobe over the gnathosoma base, lacks scapular tubercles or setae, and shows a pattern of strong longitudinal lines, with submedians forked just laterally to admedians. The sides of the shield are granular. All opisthosomal annuli are microtuberculate and the microtubercles are pointed. There are no setae h_1. The genitalia are appressed to the coxae and the genital coverflap has about 14 longitudinal ribs in two uneven ranks. The empodium is 5- or 6-rayed (Jeppson *et al.*, 1975).

MALE. Similar to female, but smaller.

Geographical distribution

Worldwide (CABI, 1996).

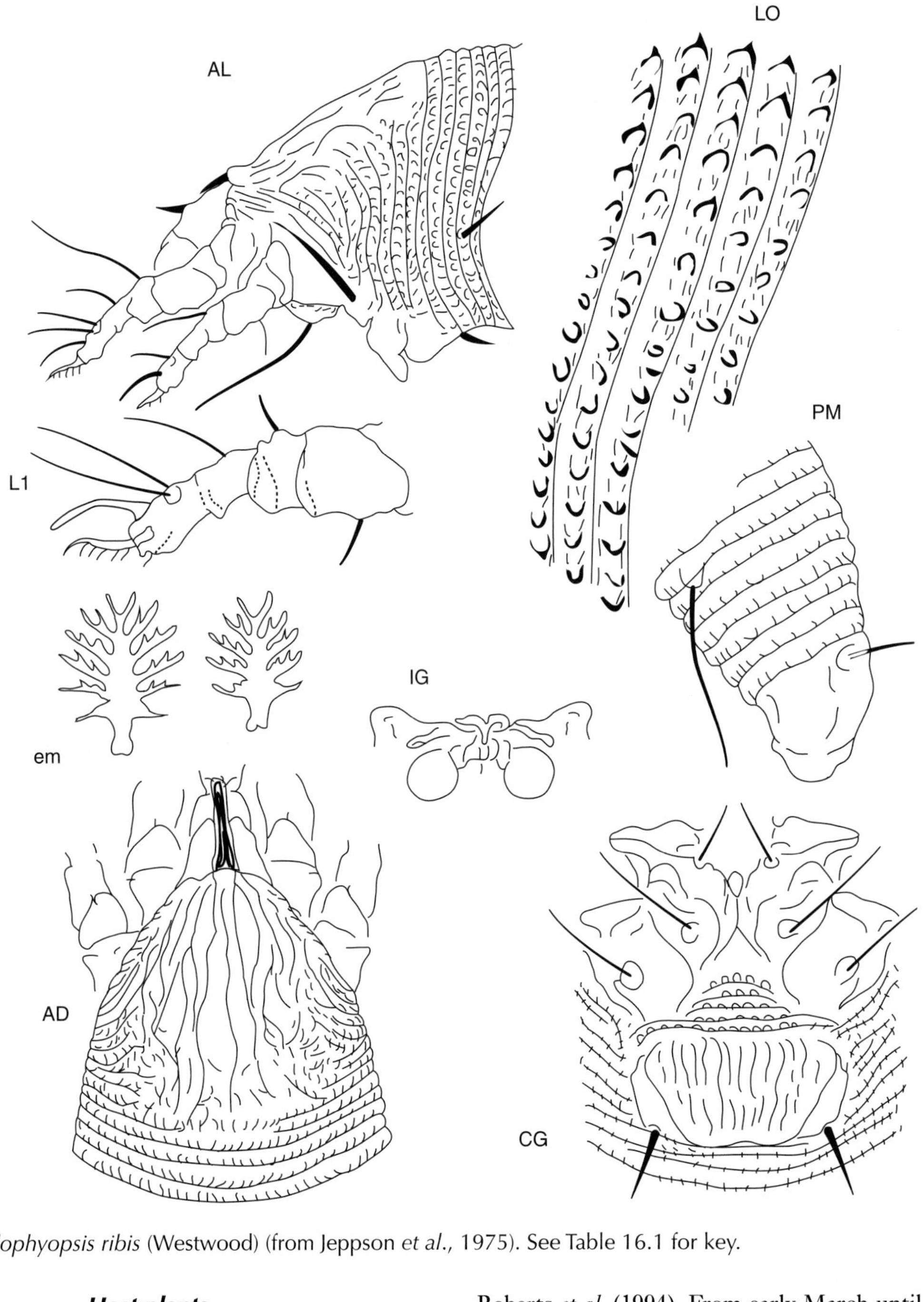

Fig. 16.8. *Cecidophyopsis ribis* (Westwood) (from Jeppson *et al.*, 1975). See Table 16.1 for key.

Host plants

Grossulariaceae: blackcurrant (*Ribes nigrum* L.) (Jeppson *et al.*, 1975), redcurrant (*R. rubrum* L.), gooseberry (*R. uva-crispa* L.) (Easterbrook, 1980), *R. alpinum* L. and *R. spicatum* Robson (Bremer and Heikinheimo, 1979).

Bio-ecology

C. ribis lives on blackcurrant in both free-living and bud-confined modes. Its ultrastructure has been investigated by Roberts *et al.* (1994). From early March until mid-June, when the temperature rises above 5°C, the mites begin to migrate from old dried-out buds and move on to outer bud scales, shoot and leaves, and also on to the upper surface of the petiole and main stem, where they feed and reproduce. The population density reaches a peak in the latter part of April or in May. The hardening of the lower leaves of the shoots stimulates *C. ribis* to migrate towards the apex of the leaves, on which it develops at high density. From late May until late July, the eriophyoid colonizes new buds, where it lays eggs and reproduces intensely, reaching a density that may exceed 10,000 mites per

bud. Some 3–4 days after the colonization of new buds, egg laying begins, with hatching about 3 days later. The infested buds become 'big buds', usually as early as July, and are found at high density in September to November; sometimes, their development continues until the next spring. The eggs are found throughout the year, but mainly in the spring and autumn. The mite is resistant to low temperatures (−7 to −2°C). The galls contain several thousand mites and roughly double the number of eggs during the early part of the migration, when the internal tissues of the galls are green and succulent. The internal tissues of the galls became progressively dry and chlorotic as migration progresses, and the numbers of mites and mite emergence dwindle and eventually cease. The mite populations in the buds develop at high density from summer to spring, with peaks in early autumn. The blackcurrant big bud mite develops 2–7 generations a year and overwinters as fertilized females (Massee, 1928; Savzdarg, 1957; Collingwood and Brock, 1959; Smith, 1961; Thresh, 1963, 1964b,d; Kanervo and Makinen, 1967; van de Vrie, 1967; Dobrivojevic and Petanović, 1982; Csapo, 1992; Gajek and Boczek, 1998; Cross and Ridout, 2001; Xia *et al.*, 2007).

In autumn, *C. ribis* can leave the big buds, and some individuals move to the extremities of deformed leaves or on to the stems. They may fasten on to the plant using their opisthosomal appendages, and wave their bodies in the air; they can also pose erect on buds and leap as high as 5 cm into the air (about 200 times its length). In this way, the mite can either be dispersed by wind or moving objects or leap into the air and drift to a new site on the leaves and move into new buds. On the plant, *C. ribis* can leap as far as 400 μm (Smith, 1960). Under experimental conditions, the mites leave the buds in increasing numbers, which are correlated with increasing wind velocities up to 39 km h^{-1}, but above this speed, the mites show a decreasing tendency to leave the buds (see Lindquist and Oldfield, 1996). According to Massee (1928), this behaviour can be explained as an attempt by the mite to attach itself to a passing vector, possibly confirmed by reports of the mite on currant aphids, honeybees and lady beetles, spiders, dipterans and coleopterans (Massee, 1928; Smith, 1960; Behrens, 1964; van de Vrie, 1967; Shvanderov, 1975; Duffner *et al.*, 2001).

Natural enemies include pathogens, insects and mites. Among the pathogens reported is the fungus *Paecilomyces eriophyes* (Massee) (Taylor, 1909). Insects include the eulophid *Tetrastichus eriophyes* Taylor, whose larvae prey on a high number of mites but do not have a relevant practical importance (Collingwood, 1959a). Among the phytoseiid mites reported as natural enemies are *Amblyseius aberrans* Oudemans, *A. finlandicus* Oudemans (Schausberger, 1992) and *A. andersoni* (Chant) (Gajek and Niemczyk, 2001).

Symptomatology and damage

Feeding by the blackcurrant big bud mite damages buds and transmits *Blackcurrant reversion virus* (BCRV) (Jones, 2000; Susi, 2004), the causal agent of blackcurrant reversion disease, the cause of which was not clear until 1980 (Slykhuis, 1980), although the evidence for its transmission by the mite was obvious (Thresh, 1963). All active instars of *C. ribis* can transmit the virus after 3 h of feeding on infected bushes, but optimum acquisition requires about 50 h. The virus is transmitted after 48 h of mite feeding on healthy bushes (Jacob, 1976). Non-gall-forming mites indistinguishable from *C. ribis* develop on redcurrant and gooseberry cultivars as well as blackcurrant in Great Britain, and mites from each host can reproduce on the others (Easterbrook, 1980). The reversion disease has also been reported for different plant hosts of *C. ribis*, including blackcurrant (*R. nigrum*), *R. alpinum* and *R. spicatum* (Bremer and Heikinheimo, 1979).

Medium to high levels of infestation cause young buds to become galled and to fail to flower, and frequently to turn into 'big buds' that fail to differentiate into flowers or true leaves. Buds that are less attacked may develop secondary buds that give rise to grossly deformed foliage and abnormal flowers. High levels of attack produce a general reduction of plant vigour and leader growth is replaced by many short twiggy side shoots (Thresh, 1964b). As a result of the mite feeding on meristems, tissues appear in the leaf epidermis and mesophyll that have hypertrophied cells, which are binucleate and may have a hypertrophic nucleus and nucleolus (Westphal, 1977a).

Damage caused by mite feeding and the symptoms of reversion disease are often confused, although they each have distinct characteristics. When mite control has occurred, the disease does not appear in the following year; instead, disease symptoms are observed in the second and third year after virus infection. The virus disease makes the plant more susceptible to mite attack, with the result that fewer mites reach healthy buds than reach buds on bushes with reversion virus. Healthy and 'reverted blackcurrants' differ in the hairiness of the leaves and young stems. These hairs, mainly those of the stipules and those set at the base of the petioles, impede mite dispersal in the spring. The lack of hairs on reverted bushes helps mite migration and the infestation of new buds; the reversion breaks down the plant's natural resistance, thus increasing colonization and reinfestation by mites. Current varieties differ in the hairiness of leaves and stems, and those that are more resistant to mites (and so to the virus) have the most hair (Thresh, 1964a, 1967).

The reversion infection can be detected earliest from the characteristics of the leaves arising from the buds on the previous year's growth. The reverted leaves tend to be narrower and smaller than normal, flattened at the base (owing to a reduction in the depth of the basal sinus). Infected leaves from extension growth are flatter than healthy ones. In many varieties, the healthy leaves have lots of pointed serrations along the leaf margin, with 13 or more serrations on one side and five or more sub-main veins from the midrib. Reverted leaves bear fewer than 12 broad, rounded serrations with less than five sub-main veins. However, the diagnosis can be complicated by varietal differences and feeding damage by insects or the effects of pruning. Infections by the most severe virus strains stop fruit production, while infections due to the less severe virus involve only a decrease in fruit production. Healthy blackcurrants have magenta flower sepals but appear grey because of their numerous hairs, while the sepals of reverted bushes are lacking hairs and appear to be magenta. This difference is best appreciated before the first flowers open. The same phenomenon occurs in all currant varieties, but it is most obvious in the hairy types (Thresh, 1964a,b,d, 1967).

Control

Control of *C. ribis* is provided by both preventive and curative treatments, based on cultural and sanitary methods and on physical and chemical treatments, integrated with appropriate monitoring in the field.

MONITORING. Monitoring is based on regular observation of damaged flowers, leaves and buds and reversion symptoms. In spring, it is necessary to observe mite migration (Łabanowska and Gajek, 1999) and symptoms of attack. In young groves in the first year, it is useful to inspect the bushes for reversion at the end of July; in the second year, the monitoring must be done in May, June and July. Inspection should also be done for galled buds during the first winter on lifting (Jeppson *et al.*, 1975).

Mite emergence from galls is preceded by swelling of the galls. Research in England on first emergence shows that the phenomenon is often associated with the first day after 1 March with a maximum air temperature ≥16°C. Accumulated temperature sums above a threshold of 4°C from Julian day 46 allowed prediction of the seasonal timing of emergence, and the average accumulated temperatures for the first and for 5 and 50% emergences were 122, 199 and 316 degree-days with mean errors in the predictions of 3.1, 1.3 and 7.2 days, respectively. The accumulated temperature model predicted the first and the 5 and 50% emergences to within 4, 4 and 5 days, respectively. There was great day-to-day variability in the mean number of mites that emerged, with a positive correlation between the number of mites emerging and the mean or maximum daily temperature; numbers were suppressed on days with rainfall. There is also great variability in the timing of emergence of mites from individual galls and the variance of the number of mites emerging is related to the mean according to Taylor's power law. The emergence of mites has a strong diurnal rhythm, controlled by both light and temperature, with virtually no mites emerging between 23.00 h and 09.30 h (Cross and Ridout, 2001).

PLANT RESISTANCE. Methodologically, studies on plant resistance commonly involve the transfer of *C. ribis* from blackcurrant to other *Ribes* plants to highlight mite-resistant *Ribes* to use for crossing with blackcurrant (Anderson, 1971; Proeseler, 1973; Knight *et al.*, 1974; Easterbrook, 1980; Potapenko, 1985). Some cultivars of *R. nigrum* (Kippen's seedling, Taylor's seedling) are not susceptible to colonization by *C. ribis*, and research carried out on resistance to the mite shows that a dominant gene (*Ce*) from *R. grossularia* L. controls resistance in *R. nigrum*. However, a deficit of mildew resistance has been found in blackcurrant progenies of parents heterozygous for resistance both to the American gooseberry mildew, *Sphaerotheca morsuvae* (Schweinitz) Berkeley *et* Curtis, and to *C. ribis*. The genes involved are Sph_2 from the Swedish cultivar 'ojebyn' and *ce* from *R. grossularia*, respectively (Keep, 1985). In gooseberry varieties, resistance is based on antixenosis, and in redcurrant on antibiosis (Herr 1988, 1991). In general, a low quantity of phenolic compounds is found in the buds of resistant varieties (Herr, 1986), as well as different terpenoids (Brennan *et al.*, 1992). The high density of covering hairs in some resistant varieties (Thresh, 1967) could hinder bud infestation by the mite.

Different research programmes are involved in pursuing the resistance strategy, e.g. in Poland, where there is a small fruit breeding programme (Zurawicz *et al.*, 2009), and in England, where the development of molecular markers to more effectively link genotype with preferred phenotypes is continuing in blackcurrant, and the first genetic linkage map has recently been completed. Several phenotypic traits have been added to the map, and the development of markers linked to these traits, both agronomically and quality related, is continuing (Brennan *et al.*, 2008). A PCR-based marker linked to resistance to the blackcurrant gall mite has been developed (Brennan *et al.*, 2009). Evaluation of the yield and quality of a number of blackcurrant cultivars has shown that some of them exhibit some resistance to the blackcurrant gall mite (Kahu *et al.*, 2009).

CULTURAL PRACTICES. Nursery and field practices help to depress the prevalence of reversion disease. Propagation must occur only from plants that are free of big bud or reversion symptoms. The planting of cuttings must be done away from infested and infected currants. Fruiting plantations with certified bushes should be planted away of sources of mite infestation and sprayed to eliminate mite populations before the bushes start to crop. Some have suggested positive steps to remove infested swollen buds during winter (Proeseler, 1973; Jeppson *et al.*, 1975; Oldfield and Proeseler, 1996), whereas others doubt the effectiveness of this strategy. In particular, research carried out in Poland showed that the removal of infested buds as the only method of gall mite control was not sufficient to protect an integrated block effectively against this pest (Łabanowska and Gajek, 1999; Gajek *et al.*, 2004).

BIOLOGICAL CONTROL. Biological control alone does not solve the problem of *C. ribis*. However, different phytoseiid mites can contribute to the implementation of integrated pest management (IPM) of *C. ribis* in the field. With this aim, the bio-ecology of different phytoseiid species has been investigated. At 25°C and with *C. ribis* as prey, the egg to egg developmental time of *Amblyseius aberrans* takes 7 days and the mean ovipositional rate is 1.55 eggs day^{-1}; under the same conditions, *A. finlandicus* has an egg to egg developmental time of 6.1 days and the mean ovipositional rate is 1.72 eggs day^{-1} (Schausberger, 1992). At 26°C, the egg to egg developmental time of this last mite lasts 9.1 days, the mean ovipositional rate is 2.3 eggs day^{-1} and the intrinsic rate of population increase (r_m) is 0.175 (Dicke *et al.*, 1990b). At 18°C and with *C. ribis* as prey, the mean oviposition period of overwintered females of *Typhlodromus pyri* Scheuten lasts 26.7 days, the fecundity is 13.8 eggs and the daily oviposition is 0.6 eggs (Zemek, 1993a); under the same conditions, the egg to egg developmental time of the first generation of the progeny takes 37.1 days, a mean oviposition period lasts 50.2 days, the fecundity is 17.8 eggs and the daily oviposition is 0.4 eggs (Zemek, 1993b).

PHYSICAL CONTROL. The effectiveness of hot-water treatment of currant cuttings in warm water for 40 min at 40°C is unclear, having had both positive (Jeppson *et al.*, 1975; Tuovinen *et al.*, 2008) and negative (Thresh, 1964c; Smolarz and Pala, 1982) results. In Finland, the use is suggested of healthy mother plant stocks

produced by hot-water treatment and the implementation of strict hygiene measures; this is essential to guarantee freedom from gall mites when plants leave the nursery (Tuovinen *et al.*, 2008).

CHEMICAL CONTROL. Chemical control can be applied during mite dispersal in the spring when bushes start to flower (Oldfield and Proeseler, 1996; Gajek and Boczek, 1998). Its application is very important in the nursery when growth starts and can be followed by four more treatments 10–14 days apart. Groves producing berries can be treated at first flowering, 3 weeks after first flowering, or when the last flowers open and again 2 weeks later (Thresh, 1964d; Jeppson *et al.*, 1975). Postharvest treatments were not effective (Nielsen, 1986). Applications of sulfur and lime sulfur (Massee, 1954; Thresh, 1964d), repeated at short intervals, are most effective, taking into account their possible phytotoxicity on some cultivars. Other chemicals that can be used are endosulfan, benomyl and fenazox, sprayed 3–4 times during spring, while the use of the carbamates aldicarb and endrin has been suggested in nurseries (Dicker *et al.*, 1972; Easterbrook and Cross, 1987; Nielsen, 1987; Oldfield and Proeseler, 1996). Aldicarb gives good control but it is highly toxic to mammals (Easterbrook and Cross, 1987). In Poland, two treatments with endosulfan gave good control (Gajek and Niemczyk, 2001). In China, the best strategy for mite control consists of spraying solutions of dicofol after the young leaves reach the length of 0.5 cm (after the spring shoots have sprouted) (Xia *et al.*, 2007). Cyhexatin failed to give control of the mite (Easterbrook and Cross, 1987). The pyrethroid fenpropathrin reduced populations of *C. ribis* (Nicholls *et al.*, 1986; Cross, 2006), but abamectin has little effect on mite control (Nielsen, 1987).

Combining the use of one or two early sulfur sprays with the use of varieties resistant to the gall mite (e.g. Ben Hope) or to reversion virus (e.g. Ben Gairn) gives a very high degree of control of both the gall mite and reversion virus, which is sustainable in the long term, and gives much better results than either using resistant varieties or sulfur sprays alone. Sprays of sulfur can be phytotoxic to blackcurrant, causing leaf discoloration and significant loss in yield. Tebufenpyrad is being investigated as an alternative to sulfur that can be used later in the season at times when sulfur is phytotoxic (Cross, 2006).

INTEGRATED PEST MANAGEMENT (IPM). Control of the mite population and the prevention of reversion disease provide two fundamental steps. The heat therapy of cuttings and chemical treatment in nurseries are required. The choice of environment for new plantings, the pruning of mite-infested branches and chemical control in commercial groves are important (Proeseler, 1973; Oldfield and Proeseler, 1996). The use of resistant cultivars is a prerequisite (Gajek and Nowacki, 1996).

IPM has been very poorly developed for blackcurrant, mainly because of the need for the frequent application of sprays of broad-spectrum acaricides to control the blackcurrant gall mite. The breeding of new blackcurrant cultivars resistant to the gall mite or to reversion disease offers realistic possibilities for the development of IPM. In 1998, a model that would predict the migratory activity of the mite to allow the timing of sprays was still under development, but it was predicted that the inspection of plantations and rogueing of infested or infected bushes may still be required (Cross and Easterbrook, 1998). Later, the same researchers reported that IPM approaches for the most important pests of blackcurrant now include a forecasting model for blackcurrant gall mite migration that employs daily maximum and minimum air temperatures to improve the timing of acaricidal sprays for gall mite control. Early results indicate that the combined use of host plant resistance and sulfur sprayings provide an effective strategy for both gall mite and reversion virus management (Cross and Harris, 2003). In Poland, spray application of endosulfan once, twice or three times during bloom did not affect the phytoseiid mites *T. pyri* and *A. andersoni*. The increase in populations of this last phytoseiid after endosulfan sprays during 3 successive years suggests that this species could be useful for efficient regulation of the two-spotted spider mite on blackcurrant (Gajek and Niemczyk, 2001).

The Genus *Cosetacus* Keifer

Diagnostic characteristics

These are bud mites with a worm-like body and narrow, completely microtuberculate annuli, approximately equal dorsoventrally. The gnathosoma is of moderate size and the oral stylets are short. The prodorsal shield is subsemicircular in dorsal anterior outline. There is no frontal lobe over the gnathosoma base, and the scapular tubercles are set near the rear shield margin, directing the setae divergently to the rear. The legs have the usual setae, including femoral setae; the seta on tibia I is missing. The coxae have three pairs of setiferous tubercles, not outlined or particularly pronounced; coxae I are approximately central but lacking a sternal line and hardly contiguous. The opisthosoma has setae c_2, d, e and f; the setae h_1 are absent. The female genitalia are somewhat protruded from the body, overlapping the rear part of coxae II, transversely subelliptical in shape, and the coverflap has rather faint longitudinal ribs in two transverse ranks. The internal genital apodeme is folded back, describing a transverse convex curve in ventral view (Keifer, 1966b; Amrine *et al.*, 2003).

Economic importance

This genus has a palaeotropical geographical range and has only two species. The type species, *Aceria camelliae* Keifer (= *Cosetacus camelliae* (Keifer)), was collected in buds, especially flower buds of camellia (*Camellia japonica* L.) in California, and causes damage to buds and floral parts (Keifer, 1945; Keifer *et al.*, 1982).

Cosetacus camelliae (Keifer) (Fig. 16.9)

Common name

The camellia bud mite.

Diagnostic characteristics

FEMALE. The body is worm like, 180–200 μm long, 35–40 μm wide and white in colour. The gnathosoma is 23 μm long and

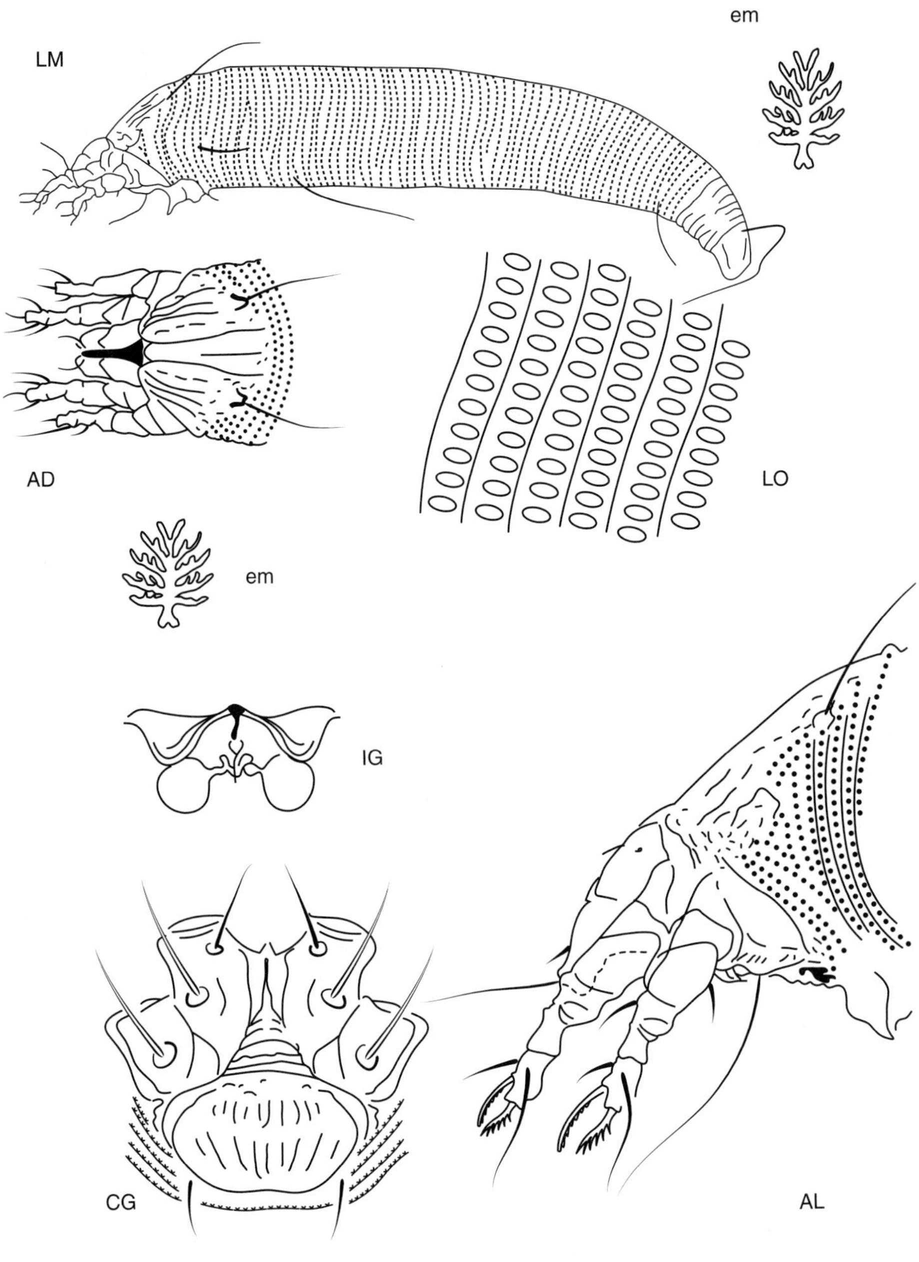

Fig. 16.9. *Cosetacus camelliae* (Keifer) (from Keifer, 1945). See Table 16.1 for key.

curved down, with a short apical seta. The prodorsal shield is 29 μm long and 30 μm wide, with central longitudinal lines; the rest is obscured by granulations, and granular laterally. The scapular tubercles are 20 μm apart, near the hind margin; the scapular setae (*sc*) are 35 μm long and project backwards. Leg I is 26 μm long, with the tibia 5 μm long and lacking the usual seta; the tarsus is 6 μm long, with the tarsal solenidion 6.5 μm long, curved and with a small knob; the empodium is 6-rayed. Leg II is 26 μm long, the tibia 4 μm long and the tarsus 6.5 μm long with a tarsal solenidion 8.5 μm long. The coxae are not particularly spread apart, though the genitalia are situated closer than usual. The opisthosoma has 55–60 annuli and is completely microtuberculate, with some ventral reduction in the number of annuli; the opisthosomal microtubercles are not pointed. The opisthosomal setae c_2 are 17 μm long, set on about annulus 5; the setae *d* are 50 μm long, set on about annulus 11; the setae *e* are 8.5 μm long, set on about annulus 24; the setae *f* are 18 μm long, set on about annulus 7 from the rear; the setae h_1 are absent. The genitalia are somewhat appressed to the coxae, with the apodeme shortened accordingly in ventral view; they are 19 μm wide and 9 μm long, and the coverflap has a double rank of about 10 furrows each; the seta is 3 μm long (Keifer, 1945).

MALE. The male is similar to the female, but smaller (Flechtmann, 1998).

Geographical distribution

California (Keifer, 1945), the former Yugoslavia (Petanović, 1997), Spain (Mansilla Vázquez, 1991a), Portugal (Mansilla Vázquez *et al.*, 2003), Brazil (Flechtmann, 1998) and Queensland (Australia) (Nash, 1991).

Host plants

Theaceae: camellia (*Camellia japonica* L.) (Keifer, 1945; Jeppson *et al.*, 1975).

Bio-ecology

The camellia bud mite lives on *C. japonica*, a species native to Asia, but the pest has not yet been reported from that continent (Jeppson *et al.*, 1975). It overwinters under the outer scales of the flower buds, and in Spain the first ovipositions occurred in April on the internal face of flower bud scales, when the mite moves to the auxiliary and terminal buds (Mansilla Vázquez, 1991a).

Symptomatology and damage

C. camelliae lives under the scales of the vegetative and flower buds of *C. japonica*, causing browning of the flower parts and buds. Its feeding induces premature bud drop or failure of the buds to open fully, producing enlarged 'bull head' blooms (Keifer, 1945; Gibson, 1967; Subirats and Self, 1972; Jeppson *et al.*, 1975; Keifer *et al.*, 1982; Meyer Smith, 1996a). In Spain, the first symptoms under the outer scales of the flower buds occur in February as browning of the outer scales, and the feeding activity of the mite causes premature fall of buds and late opening of flowers (Mansilla Vázquez, 1991a).

Control

Spray applications of amitraz, propargite, dinocap plus colloidal saffron, dicofol plus tetradifon and summer oils have been reported to be effective against *C. camelliae* infestations (Pérez Otero *et al.*, 2004).

The Genus *Colomerus* Newkirk *et* Keifer

Diagnostic characteristics

The diagnostic characteristics of this genus were first described by Newkirk and Keifer (1971) and later adopted by Amrine *et al.* (2003); recently, the genus has been redescribed by Chandrapatya *et al.* (2014), and this last revision is adopted here. The idiosoma is worm like, with the opisthosomal annuli subequal dorsoventrally and microtuberculate, but in some species smooth on the few posterior-most opisthosomal annuli; the opisthosomal setae h_1 are absent. The anterior lobe is present, varying from indistinguishable to distinctly triangular or round and broad based; the scapular tubercles are set variably from very near the posterior shield margin to well anterior to the posterior shield margin, directing the scapular setae diagonally forward or straight ahead (occasionally backwards or laterally); the gnathosoma is short. Coxae I are widely separate, with a moderate or short internal coxisternal apodeme (in some species, anterior coxisternal regions are totally separated and internal coxisternal apodeme not seen); legs have all the usual setae, with the empodia entire and 4- to 6-rayed. The genital opening is somewhat appressed to coxisternum II (four coxigenital semiannuli anterior to genital coverflap); the coverflap has longitudinal ridges distinctly arranged in one or two transverse rows, or with some (shorter) ridges in two rows and some (longer) ridges running along most of the length of the genital coverflap, constituting a single row; the genital apodemes are usually visible as a narrow dark band in ventral view, but sometimes appear to constitute a pair of subtriangular structures, depending on the position of focus (Chandrapatya *et al.*, 2014).

Economic importance

This genus has almost 23 species, variously reported on different host plants, but this number could vary according to the revision of some species that are now included in other genera. A key to the identification of these species was produced by Chandrapatya *et al.* (2014), which mentioned that the injury caused by *Colomerus* spp. might vary, thus causing some species to produce more than one type of damage.

The main types of damage are disturbance to the development of new leaves by bud damage (*C. oculivitis* (Attiah) and *C. vitis* (Pagenstecher)), fruit deformation (*C. bucidae* (Nalepa)), the formation of leaf erinea (*C. bucidae*, *C. holodisci* (Keifer), *C. spathodeae* (Carmona), *C. vitis* and other species), leaf outgrowth (*C. tricaseri* Meyer *et* Ueckermann), 'witch´s broom' by damage to inflorescences (*C. antidesmae* Meyer *et* Ueckermann), the formation of leaf galls (*C. bucidae*, *C. neopiperis* (Wilson) and other species), leaf distortion (*C. spathodeae*, *C. vitis* and other species) and fruit necrosis (*C. novahebridensis* Keifer) (Chandrapatya *et al.*, 2014).

Several species are known to infest ornamental plants, but only three species are known from major crops: *C. oculivitis* and *C. vitis* from grapevine and *C. novahebridensis* from coconut. *C. oculivitis* and *C. vitis* have been reported as injurious to their host plant, but only the grape bud mite or grape erineum mite, *C. vitis*, plays a major economic role, causing leaf erinea and bud damage, and this species has a wide geographical distribution (Jeppson *et al.*, 1975; Duso and de Lillo, 1996; Craemer and Saccaggi, 2013).

Other species are secondary pests on different host plants. *C. gardeniella* (Keifer) was reported from petiole bases of gardenia (*Gardenia jasminoides* Ellis) in Mexico (Keifer, 1964), but its presence, as well as that of *C. codiaeum* Keifer, was not associated with any type of damage on plants from which type specimens were collected (Chandrapatya *et al.*, 2014).

C. novahebridensis was collected under bracts of coconut (*Cocos nucifera* L.) in the New Hebrides islands (now Vanuatu)

(Keifer, 1977); Chandrapatya *et al.* (2014) studied specimens collected from under the bracts of coconut fruit, which usually caused the appearance of scanty triangular brown patches of damaged tissue on the fruit surface next to the bracts under which the colonies of mites had developed. The mite is usually not considered a pest, although West African cultivars growing in the Philippines and Malaysia are sometimes significantly damaged (Howard *et al.*, 2001). Chandrapatya *et al.* (2014) generally found this species at very low levels, causing little damage, but on a few occasions, damage was slightly more extensive, and the mite apparently caused premature fruit drop (fruit necrosis).

C. neopiperis causes galls of pepper leaves (*Piper* sp.) in Mexico (Wilson, 1970). *C. holodisci* makes erinea on the cream bush (*Holodiscus microphyllus* Rydb.) in California (Keifer, 1970). The African flame erineum mite or tulip tree erineum mite, *C. spathodeae*, is reported from African flame or tulip trees (*Spathodea campanulata* Beauv.) in Angola (Carmona, 1965) and South Africa (Meyer Smith and Craemer, 1999). The mite causes yellow erineum patches on the lower leaf surfaces and often curling of the whole leaf (Meyer Smith, 1996a).

Colomerus vitis (Pagenstecher) (Fig. 16.10)

Common name

Grape bud mite, grape erineum mite.

Diagnostic characteristics

FEMALE. The body is worm like, 160–200 μm long, 40 μm thick and pale yellowish in colour. The gnathosoma is 21 μm long and curved down. The prodorsal shield is 27 μm long, 32 μm wide, blunt, triangular and with numerous longitudinal lines and an ocellus-like lateral spot; the scapular tubercles are 15 μm apart, ahead of the rear margin; the scapular setae (*sc*) are 18 μm long, projecting ahead and centrally. The opisthosoma has 65–70 annuli, completely microtuberculate, with the microtubercles slightly pointed. The setae c_2 are 19 μm long, set on about annulus 7; the setae *d* are 38 μm long, set on about annulus 2; the setae *e* are 42 μm long, set on about annulus 40; the setae *f* are 15.5 μm long, set on about annulus 4 from the rear; the seta h_1 is missing. The genitalia are somewhat appressed to coxae II, the anterior apodeme is shortened in ventral view, 20 μm wide, 10 μm long; the genital coverflap has 16 or more furrows, interrupted across the centre; the seta is 4 μm long (Keifer, 1944).

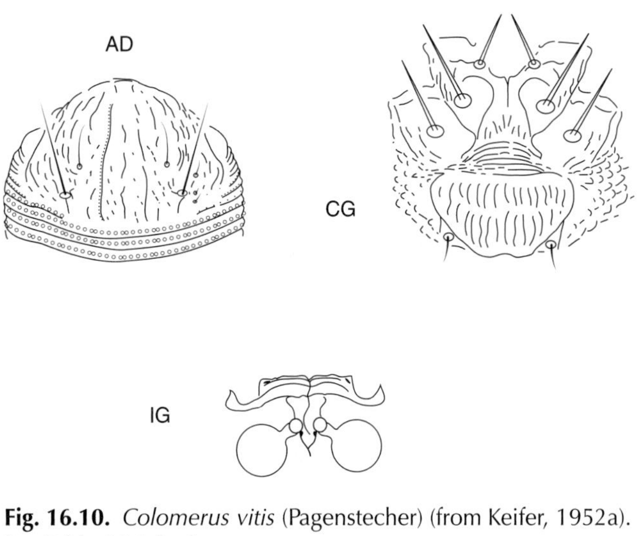

Fig. 16.10. *Colomerus vitis* (Pagenstecher) (from Keifer, 1952a). See Table 16.1 for key.

MALE. The male is smaller than the female, 140–160 μm long and 36 μm wide (Keifer, 1944).

Geographical distribution

Worldwide (Jeppson *et al.*, 1975).

Host plants

Vitaceae: grapes (*Vitis vinifera* L.) (Keifer, 1944; Smith and Stafford, 1948; Jeppson *et al.*, 1975).

Bio-ecology

The grape bud mite lives on grapevines (*V. vinifera*) and is characterized by three distinct strains: the 'bud strain', 'erineum strain' and 'leaf-curling strain' (Smith and Stafford, 1948). Recently, the use of DNA-based markers has indicated that the bud and the erineum strains are separate species, although no morphological differences between them have been described (Carew *et al.*, 2004). Here, lacking more information on this aspect and in order to avoid confusion, the old systematic concept of the species is used.

The bud strain has been recorded from Australia, California, Chile, Crimea, Egypt, Israel, South Africa and Spain (Duso and de Lillo, 1996). The females overwinter inside the outer part of the buds and as the buds swell, the mites attack the primordia, laying single eggs or small groups of eggs. Shoot development helps the mite to move on to the axils of young leaves; it also actively crawls from bud to bud. The mite is found first on the stipular scales, but as its population increases in density, it moves deeper into the new bud scales and colonizes the primordia. Attack to cluster primordia usually occurs from midsummer, and the population density increases from summer to November, with the largest density of eggs in July. The ovipositional rate is about one egg a day and the developmental time is about 20 days (Kido and Stafford, 1955; Smith and Schuster, 1963). In Spain, 10–12 generations a year have been observed (Castillo, 1990). The highest percentage of infested buds is in summer to autumn, and basal and median buds (from the 7th to the 10th) are more attacked than apical buds (Kido and Stafford, 1955; Smith and Schuster, 1963). In Crimea, the mite has 4–6 generations a year (Svetov, 1985).

The erineum strain – which causes erinose – has worldwide geographical distribution. Females overwinter between the outer scales and in bark crevices. On swelling of the buds, they move into the unfolding leaves, where they reproduce and cause the first erinea; this behaviour induces the most severe damage on the oldest leaves. The great availability of food and the higher

RH occurring in leaf patches stimulate great mite development. Between late spring and early summer, *C. vitis* infests the apical leaves, causing new erinea, with the number of erinea increased in those countries where summer pruning is applied. At the end of the summer, the females move to buds and/or other overwintering sites (Smith and Stafford, 1948; Mathys and Hugi, 1961; Mathez, 1965; Baggiolini *et al.*, 1969). A grapevine genetic map that was constructed by crossing two cultivars ('Welschriesling' × 'Sirius') of grapevine was used as the basis for a quantitative trait locus (QTL) analysis for susceptibility to erinose showed that three biological races of erinose commonly infected vineyards every year (Santiago *et al.*, 2009).

The leaf-curling strain causes leaf curl and sparse abnormal hairs on the areas colonized, but does not produce erinose. Its geographical distribution is well defined (California, Chile, Egypt, Hungary, Romania, South Africa) (Duso and de Lillo, 1996), and its biology is similar to that of the erineum strain (Kido, 1981; Barnes, 1992).

In spring, mite activity begins with temperatures of about 15.5°C. The developmental period from egg to adult takes from 10 to 14 days. The oviposition rate is about one egg a day, and the eggs can be laid in groups of 2–10 or more. Nymphs appear in 5–9 days after egg deposition in early April and the adults appear about 14 days after hatching (Kido and Stafford, 1955; Collingwood and Brock, 1959). In Switzerland, the first generation of the erineum strain develops in 25 days, but the development of the later generations is faster (Mathez, 1965).

The natural enemies of *C. vitis* include insects (anthocorids, chrysopids, cecidomyiids, coccinellids) and phytoseiid, stigmaeid and tydeid mites (see Table 16.2). The table includes only the predatory mites observed preying strictly on *C. vitis* and not those species only collected in the vineyard.

Symptomatology and damage

The economic importance of *C. vitis* is unclear and its presence in vineyards requires careful consideration.

The bud strain does not cause a typical erinea. Its feeding activity on epidermis induces a cellular hypertrophy, accompanied by 'polyp' formation and scar tissue (Smith and Schuster, 1963). The main symptoms are the scarification of green bark, short basal internodes, flattened shoots, death of distal and dormant buds, and stunting or death of the principal growing points of the bud along with the development of the lateral shoots, also known as witches' broom growth. The basal leaves are crinkled and show coalescent venation and cut leaf margins, and flower clusters drop early (Smith and Stafford, 1948; Kido and Stafford, 1955; Smith and Schuster, 1963; Gärtel, 1970; Castillo, 1990). However, there are differing opinions on these symptoms. Kido and Stafford (1955) reported that the level of bud infestation was not correlated with yield, whereas in Crimea 20–30% of damaged shoots, and sometimes up to 80–100%, have been observed, with retarded growth on susceptible varieties. The feeding activity of the mite on the axillary buds, and on the young forming leaves and the side shoots, may cause small leaves and young shoots to dry out; although there is some growth regeneration, the yield is reduced, and in the absence of control measures, the plants dry out in 3–4 years (Svetov, 1985).

Table 16.2. Predatory insects and mites of *Colomerus vitis* (Pagenstecher).

Families and species	Reference(s)
INSECTS	
Cecidomyiidae	
Arthrocnodax vitis Rubsaamen	Ocete and Skuhrava, 1996; Sasso and Viggiani, 2001
MITES	
Phytoseiidae	
Amblydromella dalii (Rather)	Rather, 2008
Amblydromella viniferae (Rather)	Rather, 2008
Amblyseius aberrans (Oudemans)	Duso and de Lillo, 1996; Crespan and Cestari, 2009
Amblyseius addoensis (van der Merwe *et* Ryke)	Dennill, 1986
Amblyseius andersoni (Chant)	Duso and de Lillo, 1996
Amblyseius californicus McGregor	Gonzalez, 1983a
Amblyseius loxtoni Schicha	James and Whitney, 1993
Amblyseius victoriensis (Womersley)	James, 1989
Phytoseius finitimus Ribaga	Duso and de Lillo, 1996
Phytoseius plumifer (Canestrini *et* Fanzago)	Duso and de Lillo, 1996
Typhlodromus doreenae Schicha	James and Whitney, 1993
Typhlodromus exhilaratus Ragusa	Castagnoli and Liguori, 1986
Typhlodromus occidentalis Nesbitt	Smith and Schuster, 1963
Typhlodromus phialatus Athias Henriot	Castillo, 1990
Typhlodromus pomi Parrot	Kido and Stafford, 1955
Typhlodromus pyri Scheuten	Mathys, 1959; Maixner, 1990
Typhlodromus reticulatus (Oudemans)	Buchanan *et al.*, 1980
Typhlodromus rhenanoides Athias Henriot	Lara Benítez and Ocete Rubio, 1993
Typhlodromus saevus van der Merwe	Dennill, 1986
Typhlodromus talbii Athias Henriot	Camporese and Duso, 1995
Stigmaeidae	
Agistemus exsertus (Gonzales)	Osman *et al.*, 1991
Zetzellia mali (Ewing)	Liguori, 1987
Tydeidae	
Tydeus caudatus Dugès	Duso and de Lillo, 1996
Tydeus goetzi Schruft	Schruft, 1972b
Tydeus grabouwi Meyer *et* Ryke	Dennill, 1986

The feeding activity of the erineum strain on the undersides of young leaves causes dense, long, narrow, bent, thin-walled and mainly unicellular trichomes that are plaited together to develop. The expanded leaves have trichomes with 2–6 cells that are horn like. Mature leaves have erineous trichomes that are thin-walled, elongate, unicellular, rounded at the apex and with the appearance of the trichomes of young leaves. After the opening of the bud has just finished, the undersurface of the leaves shows felty patches or 'trichomogenic erinea'. These patches develop in 10–30 days, first appearing whitish and then yellow. On the upper leaf surface in the corresponding areas, blister-like whitish or reddish swellings appear that are slightly dry and become brown. The distal part of the leaf, above the upper lateral incision, is mainly intersected by patches (Batchelor, 1952; Slepyan *et al.*, 1969). Clear differences distinguish the erineous trichomes from those of the uninjured epidermis, taking into account that at the boundaries of erinea they maintain the aspect of undamaged leaves. In particular, erineal trichomes are thin walled, elongate, non-partitioned and have convexly round apexes, while normal leaf trichomes show acuminate apexes. When leaf unfolding is complete, the trichomes of the erinea appear to be a fatty mass, and at the end of the growth phase, tannic substances appear which give a rusty brown colour to the erinea. The oxidizing of these tannic substances results in a gradual browning, and at the conclusion of the growth phase the browning trichomes dry out.

Simultaneously with the transformation of the trichomatous integument just described, the epidermal cells are involved in the process of erineum formation. The cuticle either does not develop fully, or is completely lacking, and the walls are thickened. The stomata opposite the erinea are normal or slightly deformed. At the sites of erineum development, the mesophyll chlorenchyma shows big changes. The mesophyll cells of the juvenile grape leaves are tightly closed, without intercellular spaces. In the first phase of leaf development, the palisade chlorenchyma has cells with the same columnar appearance and meristematic character as the cells of the upper and lower epidermis. The cells of the columnar chlorenchyma have large nuclei and a large column of cytoplasm. In the process of normal leaf growth, the structure of the mesophyll changes and the palisade and spongy chlorenchyma become distinct; many intercellular spaces develop and the chlorenchyma cells enlarge, losing their rectangular outlines. During erineum formation, the mesophyll assumes a different structure. The cells of the columnar chlorenchyma undergo one or two periclinal divisions, which result in a local invagination of the leaf blade towards the upper side of the leaf. The cells of the spongy chlorenchyma are involved in different types of divisions. The structural and growth mechanisms of erinea show that these are examples of cataplasmic galls (Slepyan *et al.*, 1969). Erinea rarely develop on petioles, peduncles or buds, even if the environmental condition are favourable (Mathez, 1965).

There is no agreement on the real danger of the erineum strain of *C. vitis*. In Switzerland, stunted shoot growth and failure of the fruits to ripen have been found (Baggiolini *et al.*, 1969). In Azerbaijan, yield losses of up to 30–40% are reported (Sadigov, 1990). In Turkey, an experimental infestation of grapevine (cv. 'Kabarcek') with 10, 11–30, 30–50 and 50–100 mites per plant, before and after blossoming, showed the highest number of mites at the beginning of August on plants infested with 11–30 mites. When plants were infested after blossoming, there was a significant yield loss of about 500–700 g. Control measures against *C. vitis* should therefore be implemented when 11–30 mites are seen on a plant before blossoming (Bahadroglu and Avgin, 2003). In England, *C. vitis* has not been regarded as an important problem because it rarely decreases vine productivity and not all cultivars are susceptible; nevertheless, the mosaic virus 'Vine Panaschure' has been reported as being transmitted by the mite (Savage, 1978). Recent research in Switzerland on the grape variety 'Muscat' showed that the most severe infestation of the erineum mite (about 3% of leaves had more than 60% of the leaf area damaged) had almost no effect on transpiration rates. Photosynthesis and stomatal conductance rates slightly decreased on heavily infested leaves, while the chlorophyll index was unaffected. The infestation did not affect shoot growth and the amount of foliation. Neither was individual plant damage correlated with the overwintering mite population or with plant damage in the previous year. In conclusion, the impact of the mite on photosynthesis was negligible (Linder *et al.*, 2009). In Italy, *C. vitis* does not reach levels of infestation that are sufficient to require specific control measures (Laffi, 1996). However, the various structural disorders of the leaf at the sites of erineum formation facilitate infection by *Plasmopara viticola* Berlese *et* de Toni, and the mycelium of the fungus is more abundant among erineous trichomes than in the adjacent part of the leaf (Zeck, 1955; Smith and Schuster, 1963; Slepyan *et al.*, 1969).

As already noted, the leaf curl of strain of *C. vitis* does not cause erinea. The mites mainly infest hairy and young leaves, which, in summer, results in the leaves becoming curled downward or rolled up and bearing a few irregular trichomes. These symptoms are associated with stunted growth and scarring of the shoots, necrosis and hypoplasia on the leaf undersurface (Smith and Stafford, 1948; Gärtel, 1972; Kido, 1981; Schwartz, 1986; Barnes, 1992). Surveys of unburst buds from vineyards in Australia show that when the grape bud mite numbers in unburst buds reach 100–500 per bud, the apical meristems of primary buds, and commonly also of secondary buds, die, thus preventing bud burst. The remaining living scale tissue appears distinctly scarred. A related condition, that of retarded shoot growth and leaf distortion, which was previously attributed to 'restricted spring growth' (RSG), is now attributed to misdiagnosed spring feeding damage by another mite, the grape rust mite (*Calepitrimerus vitis* (Nalepa)). Nevertheless, clustered high infestations of *Colomerus vitis* can cause bud axis necrosis, bud burst failure, shoots with short basal internodes and short, thin, zigzagged shoots with absent fruit clusters; which were also all previously considered to be caused by RSG (Bernard *et al.*, 2005).

Control

The need for control of *C. vitis* may vary in different areas of cultivation and relates to factors such as environmental conditions, varieties, etc. Control is provided by the use of resistant varieties, cultural management and/or spray application of chemicals.

PLANT RESISTANCE. Control is not necessary for all grape varieties, owing to differential varietal susceptibility. American varieties are commonly resistant to injury (Jeppson *et al.*, 1975). Some Russian varieties ('Dekabr'skii' and 'Pamyati Verderevskogo') are also resistant to the mite (Malyuganov, 1990). In Hungary, *C. vitis* seems to prefer some varieties ('Olasz Rizling', 'Oporto' 'Rizlingszilvani', 'Tramini', 'Zefir' and 'Zengo') (Dellei and Szendrey, 1991). Screening of different grape varieties in India revealed that 'Anab-e-Shahi', 'Hussaini', 'Khalili' and 'Sahibi' varieties are resistant to the mite, whereas 'Himrod' proved to be very susceptible (Rather, 2008).

CULTURAL PRACTICES. Research in South Africa on the effects of pruning of the 'Chenin Blanc' grape variety shows that the greatest proportions of populations of *C. vitis* occur in the basal ten buds on canes. The distribution of mites along the canes is explained in terms of their distribution within mature overwintering buds in which the basal 7–10 nodes are already present and infested. Although, when infested, the basal ten buds all contain a high density of mite populations, the damage caused to the buds is greatest in the buds at the basal two nodes and decreases significantly in the buds set at nodes 4 and 6, and 8 and 10, on mature overwintering canes. Comparison of various pruning techniques showed that vines severely infested by *C. vitis* should be pruned back to six buds for one season, thereby encouraging the healthier buds at nodes 5 and 6 to produce the next season's growth, and reducing the yield lost to *C. vitis* from 56 to 35%. During the subsequent year, the basal four canes are removed to reduce the *C. vitis* populations in the vineyard, using the usual pruning technique of leaving only two buds per bearer; an effective chemical control programme against *C. vitis* is regularly applied during the spring (Dennill, 1991).

The proper management of surrounding vegetation can improve the biological control of grapevine mite pests by providing refuges, hibernating sites and alternative food sources for predatory mites (Rather, 2008).

BIOLOGICAL CONTROL. Biological control depends on natural enemies present in the field. In Spanish vineyards, various predatory phytoseiids were collected, including *Euseius stipulatus* (Athias Henriot), *Kampimodromus* sp., *Neoseiulella litoralis* (Swirski *et* Amitai), *Phytoseiulus persimilis* Athias Henriot, *Typhloseiella isotricha* (Athias Henriot), *Typhlodromus phialatus* Athias Henriot, *Typhlodromus rhenanoides* Athias Henriot; a tydeid mite, *Orthotydeus caudatus* (Dugès), and some predatory insects, such as the cecidomyiid *Arthrocnodax vitis* Rubsaamen were also found (Ferragut *et al.*, 2008). Twenty phytoseiid mites inhabit vineyards in Greece, and among these *Phytoseius finitimus* (Ribaga) seems to be the most dominant species, reaching population densities of 75–100% in most regions and grape varieties; other phytoseiids present are *Neoseiulus californicus* (McGregor), *Typhlodromus exhilaratus* Ragusa, *Typhlodromus perbibus* Wainstein *et* Arutunjan, *Typhlodromus kerkirae* Swirski *et* Ragusa and *Seiulus finlandicus* Oudemans (Papaioannou-Souliotis *et al.*, 1999a). However, the presence of predatory mites in vineyards is strictly correlated with the management of the crop, especially with the use of chemical control. In Germany, populations of *Typhlodromus pyri* Scheuten were significantly correlated with those of *C. vitis* in vineyards, and a high density of this predator was observed on the undersides of leaves from May to August; sprays of fenbutatin oxide and parathion-ethyl were the least toxic to this phytoseiid mite, while trichlorfon and fenvalerate were the most toxic. Cross resistance has been observed between parathion, azinphos-methyl plus demeton-*S*-methyl and acephate (Maixner, 1990). Resistance to deltamethrin, lambda-cyhalothrin and chlorpyriphos-ethyl was observed in some populations of *T. pyri* and *Amblyseius andersoni* (Chant) in vineyards in south-west France (Bonafos *et al.*, 2007).

Investigations of the basic parameters of the bionomics of various phytoseiid mites under laboratory conditions confirmed the importance of some species. For example, at 25°C and with *C. vitis* as food, the developmental time (egg to egg) of *Amblyseius aberrans* (Oudemans) is 10 days (Daftari, 1979), while the daily oviposition of *A. victoriensis* (Womersley) at 20°C and with *C. vitis* as food is 1.1 eggs (James, 1989). Under the same conditions, the developmental time (egg to egg) of *Typhlodromus talbii* Athias Henriot is 21.3 days, the mean oviposition period is 21.7 days, the fecundity is 5.1 eggs, the daily ovideposition is 0.24 eggs and the intrinsic rate of population increase (r_m) is 0.03 (Camporese and Duso, 1995).

CHEMICAL CONTROL. The decision making that supports chemical control is complicated by the different assessments of the economic risk caused by the grape bud mite in different countries, which is sometimes very low, as in California (Barnes, 1992), or high, as in South Africa (de Klerk, 1985). Although crop losses have been related to severe attacks of *C. vitis* in California (Smith and Stafford, 1948), Israel (Harpaz and Bernstein, 1960) and South Africa (Dennill, 1986), the damage may have been due to microelement deficiency (Barnes, 1958) or to both microelement deficiency and mite attack (Gärtel, 1970). The influence of both factors is unclear, and in South Africa the damage has been reduced by spraying with acaricides and not by the distribution of microelements (Whitehead *et al.*, 1978). The capacity of the principal and lateral growing points to produce grape clusters influences yield loss, and the greatest injury occurs in those varieties in which only the primary growing points produce fruits (May and Webster, 1958; Harpaz and Bernstein, 1960). In Crimea, spraying with DNOC (dinitro-*o*-cresol; now banned) at the beginning of April before bud burst has proved the most effective measure and eliminates damage to the shoots (Svetov, 1985).

In infestations by the erineum strain of *C. vitis*, it has been observed that leaves with 20–30 erinea develop normally (Kido, 1981), and although spring infestations or young vine infestations are regarded as economically important (Baggiolini *et al.*, 1969; Barnes, 1992), a relationship between infestation level and yield loss has not been found (Hluchy and Pospisil, 1992). In Switzerland, it has been observed that acaricide sprays can be abandoned at moderate pest infestations (Linder *et al.*, 2009).

Usually, the application of chemical control must coincide with the presence of the mite in the erinea. When the

mite occurs in the buds, it is more protected than on the leaves, and the effectiveness of chemicals is reduced. Acaricide treatments must therefore be applied just as the mite begins to disperse, leave its overwintering sites, infest the leaves and begin the erineum formation. This occurs when the infested buds begin to swell (Jeppson *et al.*, 1975). In South Africa, the new buds are infested 1 week after burst, and 3 weeks later about 50% of the resulting mite population is protected inside the buds. This suggests spraying vineyards immediately after the buds have burst (Dennill, 1986). The chemicals that have been employed include methiocarb (Whitehead *et al.*, 1978; de Klerk, 1985), endosulfan (Whitehead *et al.*, 1978), chlorobenzilate (Whitehead *et al.*, 1978; Baum, 1984; Bernstein, 1988), clofentezine (Bernstein, 1988), bromopropylate (Whitehead *et al.*, 1978; Baum, 1984; Bernstein, 1988), carbosulfan (Baum, 1984; Bernstein, 1988), dicofol (Khalilov, 1972), furathiocarb (Bernstein, 1988), propoxur (Whitehead *et al.*, 1978; de Klerk, 1985), methyl-demeton plus chlorphenamidine and formetanate (Gärtel, 1972).

However, the best results are obtained using sulfur. In Egypt, two sprays of a mixture of dichlorvos and micronized sulfur applied at 21-day intervals (starting at budding), followed by three sprays of micronized sulfur at 21-day intervals gave the best control of the mite populations (Osman, 1975a). Colloidal sulfur applied when the new shoots reach 5 cm, followed by a repeated dose at berry growth, are suggested in Crimea (Mitrofanov and Antipov, 1973). Wettable sulfur is effective at doses of 10,000 g a.i. ha^{-1}. It is very important to spray early at the 'wool stage' and justify the use of higher doses at this period in cases of high infestation (Goebel *et al.*, 2001). A new formulation based on sulfur and terpenes was developed to reduce the quantity of sulfur applied per hectare. This formulation was tested on powdery mildew (*Uncinula necator* (Schw.) Burr.) and on *C. vitis* and shown to have a similar order of efficacy. The results of these tests should improve the use of sulfur application in IPM programmes in viticulture (du Fretay *et al.*, 2001). In Hungary, mite infestation was better suppressed by the application of sulfur than by the action of predatory mites (Mikulas, 2001).

Extracts of *Daphne gnidium* L. (Thymelaeaceae) and *Anagyris foetida* L. (Papilionaceae) cause a decrease in erineum density and seem effective against *U. necator* (Ocete and Pérez, 1996). Preventive treatments, particularly for organic vineyards, are recommended during the very early wool stage of bud development so as to avoid problems with product phytotoxicity (Crespan and Cestari, 2009).

The Genus *Eriophyes* von Siebold

Diagnostic characteristics

The dorsal opisthosoma is evenly arched in cross section; the opisthosomal annuli are subequal dorsoventrally for the entire length. The female genitalia are not shortened longitudinally and the internal apodeme has a normal anterior length (Amrine *et al.*, 2003).

Economic importance

Eriophyes is an old genus (von Siebold, 1851) including a large number of species (many more than 260), recorded from over 50 dicotyledonous families; only a few species live on monocotyledonous, conifer and fern families. About a quarter of the species infest one genus of the Rosaceae, and the genus *Prunus* includes about half of this number of species (Oldfield, 1996a). All species occur either in North America or Eurasia, and cause elongated leaf galls or leaf erinea, or live in buds. Various *Eriophyes* spp. are specialists and the bio-ecology of the genus is characterized by one or more host plants of the same genus.

Several *Eriophyes* spp. inhabit fruit trees. The pear leaf blister mite, *E. pyri* (Pagenstecher), causes blisters on the leaves and damages the fruit buds of pear (Jeppson *et al.*, 1975). *E. pseudoinsidiosus* (Wilson) has been found in pear leaf blisters (Wilson, 1965), and is recorded from California, Utah and Wisconsin in the USA, Chile (González, 1985), Iran (Fatemi, 1986) and Italy (de Lillo, 1988), but it is not economically important. *E. mali* Burts causes blisters on apple leaves but does not live on pear (Burts, 1970). The pear and apple leaf edgeroller, *E. marginemtorquens* (Nalepa), lives on pear and apple and causes rolling of the leaf edges and misshapen leaves. These rollings occur on to the upper leaf surface (Jeppson *et al.*, 1975). The peach mosaic vector, *E. insidiosus* (Keifer *et* Wilson), lives on drupaceous Rosaceae and transmits the *Peach mosaic virus* (PMV) to peaches (Wilson *et al.*, 1955), while *E. inaequalis* (Wilson *et* Oldfield) lives on lateral buds of the only known wild host, *Prunus emarginata* (Douglas) (Wilson and Oldfield, 1966) and transmits *Cherry mottle leaf virus* (CMLV), the agent of Cherry mottle leaf disease (Oldfield, 1970). *E. prunandersoni* (Keifer) has been found on desert peach (*Prunus andersoni* Gray) in California and causes an erineum in a baggy deformation on what appears to be the upper surface (Keifer, 1939d). *E. geratopus* (Keifer) has been found in buds of sierra plum (*Prunus subcordata* Benth.) in California (Keifer, 1965b) and *E. hybridicola* (Keifer) inhabits the buds and petiole bases on new growth of a *Prunus* hybrid known as elephant heart, and also on other *Prunus* hybrids and on chickasaw plum (*P. angustifolia* Marsh.) in California (Keifer, 1957). Of these *Eriophyes* spp., only *E. pyri*, *E. insidiosus* and *E. inaequalis* play an economic role.

Other eriophyoid mites, including *E. amygdali* Bagdasarian, *E. armeniacus* Bagdasarian, *E. insidiosus*, *E. padi* Domes, *E. similis* (Nalepa), are reported from commercial stone fruit and nut trees or from ornamental plants and shade trees. Of these, *E. padi* causes long galls on the leaves of Palaearctic *Prunus* species, such as *P. domestica* L., bird cherry (*P. padus* L.) and blackthorn (*P. spinosa* L.) in Poland (Szulc, 1966), in Armenia and Austria (de Lillo and Beron, 2004) and on almond (*P. dulcis* (Mill.) D.A. Webb) in Crete (Greece) (Hatzinikolis, 1969a), but has no economic importance. The mite is also reported on mahaleb cherry (*P. mahaleb* L.), prostrate cherry (*P. prostrata* Labill.) and black cherry (*P. serotina* Ehrh.) from Iran (Kamali *et al.*, 2001). *E. similis* is largely spread in Europe on European plum; in particular, it is reported from Armenia (Bagdasarian, 1970), Austria, Belgium, Bosnia, the British Isles (de Lillo and Beron, 2004),

Bulgaria (Nachev, 1983), the Czech Republic, Denmark, Estonia (de Lillo and Beron, 2004), Finland (Roivainen, 1947), France (Harranger, 1983), Germany (Engel, 1973; Schliesske, 1983), Hungary (de Lillo and Beron, 2004), Iran (Kamali *et al.*, 2001), Italy (de Lillo and Beron, 2004), Norway, Poland, Slovenia, Switzerland (de Lillo and Beron, 2004), Sweden (Roivainen, 1950) and Yugoslavia (Petanović and Dobrivojevic, 1987). It has also been reported from the Near East (de Lillo and Beron, 2004). *E. similis* causes long purse gall on leaves of Palaearctic *Prunus* spp., but does not usually have any economic importance, although a high number of galls on leaves and fruits (ten or more galls per fruit) has sometimes been found in Germany, resulting in economic loss (Engel, 1973).

Ornamental plants and shade trees host *E. leiosoma* (Nalepa), which causes erinea on *Talia* in Europe; the mite produces discoloured areas on the upper leaf surface, which appear thick, white and hairy on the underside. The hairs are similar to those of young leaves and have large nuclei and nucleoli (Westphal, 1975). *E. diversipunctatus* (Nalepa) inhabits the aspen (*Populus tremula* L.) in Europe, and *E. laevis* (Nalepa) causes bead-like galls on leaf surfaces of *Alnus* spp. in North America, Europe and Iran. *E. tiliae* (Pagenstecher) is widespread on *Tilia* spp. It causes nail galls 5–12 mm long, which appear erect on the upper side of the leaves, irregularly arranged and have long simple hairs near the gall hole on the lower leaf surface. The mite overwinters in buds, where if conditions allow, it can reproduce and cause damaged areas. In spring, a single female induces gall formation on young leaves. The galls that are revisited or characterized by a high reproductive rate appear older (Thomsen, 1976). Some species inhabit genera of the Cupressaceae (*Ephedra*, *Callitris* and *Juniperus*) as vagrants, or in cecidia on needles and fruits in France and Algeria (Boczek and Shevchenko, 1996).

Eriophyes pyri (Pagenstecher) (Fig. 16.11)

Common name

Pear leaf blister mite, pear bud mite.

Diagnostic characteristics

FEMALE. The body is slender, worm like and from 200 to 240 μm long. The prodorsal shield has a variable median line, usually present only on the rear half. The submedian lines are also variable, but the longer ones run back and curve out in front of the scapular tubercles. Lateral lines of granules run back subparallel to the lower prodorsal shield margin; some then join the portion of body annuli below the scapular tubercles. The lower prodorsal shield angle ocellar lobes have granules. The scapular tubercles are set not far ahead of the rear prodorsal shield margin, and they direct the scapular setae (*sc*) up and divergently ahead. Granules of the opisthosomal annuli are unpointed, elongate–elliptical and usually reach the annulus margin. The coxae have lines of granules (Jeppson *et al.*, 1975).

MALE. The male is like the female but smaller.

Geographical distribution

The blister strain is found worldwide (Jeppson *et al.*, 1975). The bud strain is recorded from South Africa and California (Burts, 1970).

Host plants

Rosaceae: pear (*Pyrus communis* L.) (Burts, 1970; Jeppson *et al.*, 1975; Easterbrook, 1996) and (ornamental) mountain ash (*Sorbus aucuparia* L.) (Soika and Łabanowski, 1999a).

Bio-ecology

The identification of *E. pyri* (Pagenstecher, 1857) has, in the past, posed some difficulties resulting from the systematic vision of Nalepa (1929), which 'considered the *pyri* complex on the various hosts to represent a series of subspecies and varieties' (Jeppson *et al.*, 1975). The subsequent discrimination of different species and redescription of *E. pyri* was based on early drawings of the species that showed subparallel and rather close admedian prodorsal shield lines running back to the rear shield and there turning inward to meet centrally. This morphological pattern was then taken as the basic indicator of the original *E. pyri* (Jeppson *et al.*, 1975).

The pear leaf blister mite lives on pear and is characterized by two strains, referred to as the 'blister strain' and the 'bud strain'. The blister strain is widespread in all pear-growing areas of the world and causes 'blisters' on the pear leaves, while the bud strain is less common, although is potentially a serious pest that has been recorded from South Africa and California (Burts, 1970; Jeppson *et al.*, 1975; Easterbrook, 1996).

The blister strain overwinters under bud scales (preferably under the second and third layers), sometimes in depressions or grooves more or less protected by the pubescence of the bud scales. The populations observed in the Pacific Northwestern region of the USA have similar behavioural characteristics to the mite populations of the northern countries of the world. Overwintering populations are inactive and as the pear buds swell, they begin to feed and lay the eggs at the bases of the scales. The mite penetrates into the buds and then moves on to young leaves and blossoms. Their feeding activity on young leaves within the buds at the cluster-bud stage causes the first blister, soon after which the overwintering mites infest the young leaves and leaves from 1/20 to 1/5 inch long (0.12–0.5 cm). From initial feeding to the appearance of the blister takes 3–7 days, and the first blister neither opens nor shows the presence of host mites. A feeding period of 4–6 days is necessary because from one to four epidermal cells (commonly located near the centre of the blister) become hypertrophic, dying after 6–15 days and producing small holes in the epidermis. As the attack increases, several epidermal cells die and permit the entrance into the blister of one adult mite. Initially, only one female inhabits the blister, where it feeds and begins to reproduce. The feeding activity of mites at the margins of the blister enlarge the cavity. From September to November the mites migrate to buds. The fecundity ranges from seven to 21 eggs, and at 10–17°C and 18–24°C, egg development requires 18 and 5–8 days, respectively. A generation develops in 23–36 days and there are two or

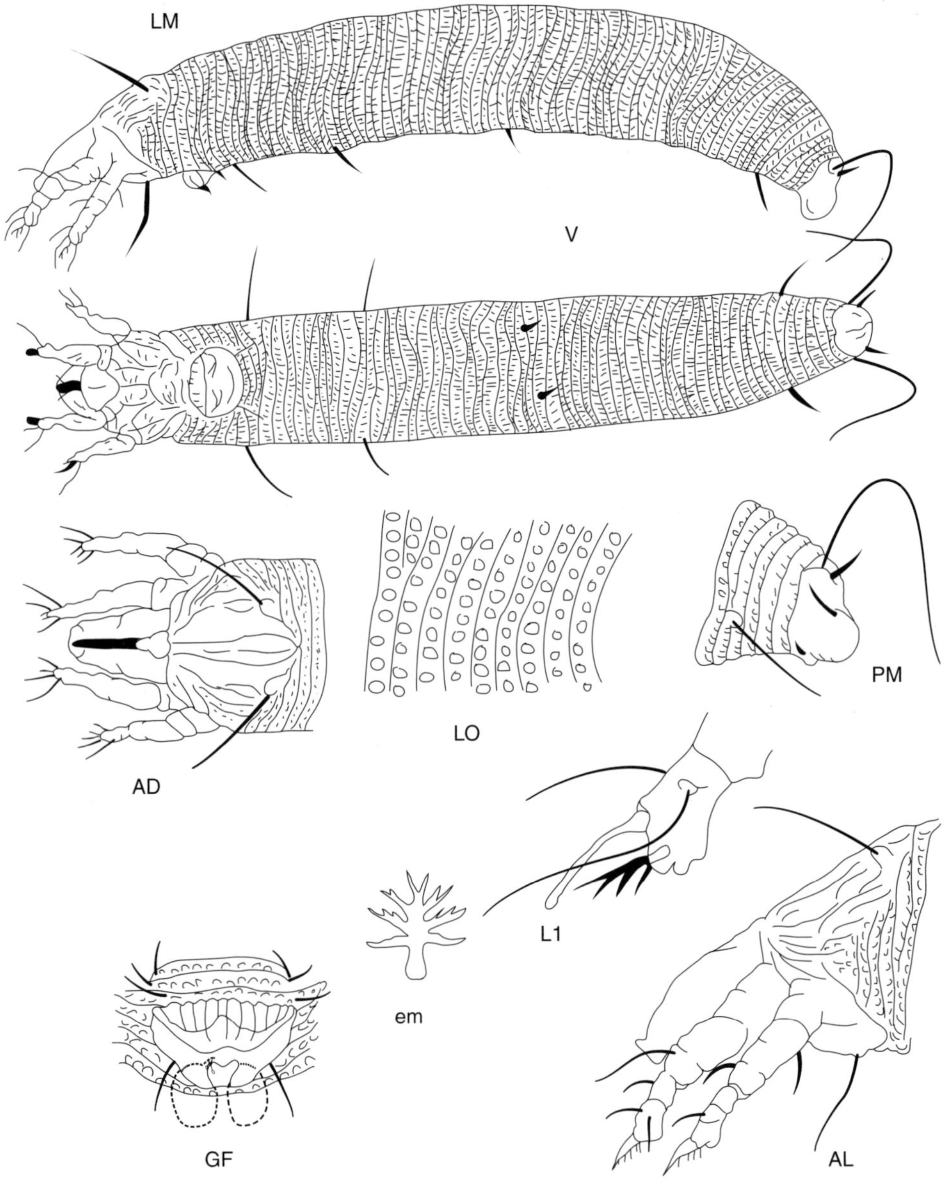

Fig. 16.11. *Eriophyes pyri* (Pagenstecher) (from Keifer, 1952a). See Table 16.1 for key.

three generations a year. The developmental time from egg to adult takes 18 days in May and 10–12 days in June (Parrott *et al.*, 1906; Borden, 1932; Prescott, 1934; Heriot, 1935; Venables and Heriot, 1937). The mite needs a high RH, and in drier weather it leaves the blister through the opening and colonizes another environment; it can survive outside the blister for up to 4 days.

The strain of *E. pyri* causing damage to fruit buds, and abnormal falling of the blossoms and young fruits, only sporadically produces leaf blisters. In South Africa, Stubbings (1947) found that the bud form of the mite differed from the blister form in various characteristics (ratio of body width to body length, number of opisthosomal annuli, and placement of the opisthosomal setae), but observations by Wilson on populations in the north-western states of North America (Jeppson *et al.*, 1975) and by Meyer Smith (1981a) in South Africa showed no such morphological differences. Bud strain populations enter the outer bud scales from August to early October. During this period, the mites are pinkish and cluster about the leaf buds behind the bases of the leaf stems on new growth. In the fruit bud scales, they become white and tend to cluster. Ovideposition occurs soon in the buds and the mites reproduce throughout the winter (a period when the blister form is inactive) from September to June, moving deep into the bud. Initially, the mite feeds on the tissues at the base of the bud scales, but later it feeds on the floral parts of the bud. In the spring, coinciding with the live buds opening, the mite leaves the cluster buds and moves to young tender new growth, developing on the leaf axils. It feeds on new growth until June, and at this time it ceases oviposition and most other activity. During June, July

and August, the mite is either not active or only slightly active; only in August does it begin to be active and to migrate to fruit buds (Parrott *et al.*, 1906; Borden, 1932; Prescott, 1934; Heriot, 1935; Venables and Heriot, 1937).

Symptomatology and damage

E. pyri causes physiological and biochemical changes in infested leaves (Tomczyk and Boczek, 2006). Initially, the attack involves few plants, or even a few branches, but it can sometimes spread rapidly to other plants. Damage caused by the blister strain on pear is common and usually is not as economically important as that caused by the bud strain. The blisters or galls that are caused on the undersides of pear leaves are at first small greenish pimples with a reddish tinge and are not really distinguishable from the leaf epidermis. Gradually, they increase in size and become more reddish, sometimes appearing as coloured spots. With the drying up of the damaged tissues, the spots become dark brown or black and the damaged areas on the undersurface of the leaf appear corky and raised, with the epidermis wrinkled. According to its age, each blister has an irregular shape and is variously sized, with the largest about ½ inch (0.31 cm) in diameter. Each blister has a tiny opening that allows the mite to enter or leave the blister. High levels of attack on the leaves of some varieties cause reddening, swelling and blistering. When unfolded, the leaves are wrinkled and unshapely. On the leaf surface, the blisters are usually set in a row on each side of the midrib. This distribution seems to be due to pubescence and to the degree of folding of the leaf at the time of migration of the mites from under the bud scales. As the attack increases, the blisters become contiguous and form dark brown patches that may rupture, usually along the margins. The feeding activity of the mite may cause light-coloured pimples on stems and around the calyx of the fruit; in midsummer, there is yellowing of the foliage, with light brown or dark green spots on the upper leaf surfaces and brown ones on the undersurface. Along each side of the midrib, dark bands contrast with the yellow of the remaining leaf surface (Burts, 1970; Jeppson *et al.*, 1975; Meyer Smith, 1981a). Severe damage may occur on dwarf leaves and on trees with few leaves, with a 25% loss of crop production (Morgan *et al.*, 1962). In South Africa, the damage mainly involves young trees, particularly 'Kieffer' and 'Bon Chrétien' pear varieties. Sometimes, the mite causes fruit drop (Meyer Smith, 1981a).

High levels of infestation of the bud strain induce the fruit buds to become brown early in the winter months and often to flare open. In the cold season, the buds drop. The damage becomes more severe during blooming, when many of the fruit buds fail to open. Other fruit buds open but only have one or two blossoms or a weak bloom that may drop or fail. The feeding activity of the mite on the floral part of the bud may cause misshapen and badly russeted fruit. The fruits derived from buds on old spurs may show russeting and be round (Jeppson *et al.*, 1975; Meyer Smith, 1981a).

Control

Control is provided by the spray application of chemicals, but resistance to some active ingredients has been reported.

PLANT RESISTANCE. Variable susceptibility of various pear cultivars to bud attack has been reported from South Africa. The most susceptible cultivars are 'Doyenne du Comice', 'Clapp's Favourite', 'Bon Chrétien', 'Buerre Hardy' and 'Louis Bonne', while 'Buerre Bosch', 'Josephine de Malines', 'Winter Nelis' and 'Glou Morceau' cultivars are less susceptible and 'Kieffer' and 'Le Conte' are almost completely resistant (Meyer Smith, 1981a).

CHEMICAL CONTROL. Blister strain populations live within the blisters during summer and spring and are free living only in the spring, before they move into the blister. Rimes (1968) reported that spray application of dicofol at petal fall controlled both the blister and the bud strains of *E. pyri*. Carbaryl and diazinone used during the summer controlled both strains (Jeppson *et al.*, 1975). Spray applications of carbaryl in the summer (Morgan *et al.*, 1962) and of endosulfan (Rammer *et al.*, 1963), fenpyroximate (Holighaus, in Easterbrook, 1996) and sulfur (Holighaus and Dahlbender, 1992; Holighaus, in Easterbrook, 1996) were effective in controlling the blister strain. Downing and Moilliet (1974) found that endosulfan plus oil, or ethion plus oil, gave good control of the blister strain when used at the delayed dormant stage, as did lime sulfur distributed at the dormant stage. Treatment during dormancy with either lime sulfur or lime sulfur plus dormant oil were effective (Jeppson *et al.*, 1975). The bud strain must be treated before the mite penetrates the buds, and carbaryl (Rammer *et al.*, 1963; Rimes, 1968; Meyer Smith, 1981a), dicofol (Rimes, 1968) or diazinone were effective (Jeppson *et al.*, 1975) against it. For trees and shrubs, the crop and pest management guidelines of Cornell University (2012) suggest sprayings of abamectin, carbaryl, fenazaquin, fenpyroximate, horticultural oil or milbemectin. On callery pear (*Pyrus calleryana* Decne.), injury is cosmetic and it may not be necessary to treat this.

Eriophyes insidiosus (Keifer *et* Wilson) (Fig. 16.12)

Diagnostic characteristics

FEMALE. The body is worm like, 160–190 μm long and dull yellowish white in colour. The gnathosoma is curved down. The prodorsal shield has a long central line and a complete median line. Beside this median line, the admedians gradually diverge to the rear and slightly recurve centrally at the margin. The submedian line runs back to the dart-shaped mark on the outer side of the tubercle, the inner line of which forms the diagonal line across in front of the scapular tubercles. Laterally, the prodorsal shield lines curve up and forward above the rear coxa. The scapular tubercles are 13 μm apart and the scapular setae (*sc*) are 17 μm long, projecting up and forward. The opisthosoma has 50–55 annuli, which are microtuberculate, and the granules of the elliptical opisthosomal annuli touch the margins of the annuli. The setae c_2 are 25 μm long, set on about annulus 7 behind the shield; the setae *d* are 28 μm long, set on about annulus 19; the setae *e* are 6 μm long, set on about annulus 27; the setae *f* are 18 μm long, set on annulus 5 from the rear; the setae h_1 are present. The genital coverflap is 9 μm long and 17 μm wide, with basal granules and about 12 longitudinal ribs, and the seta 6 μm long. The coxae are coarsely granulate.

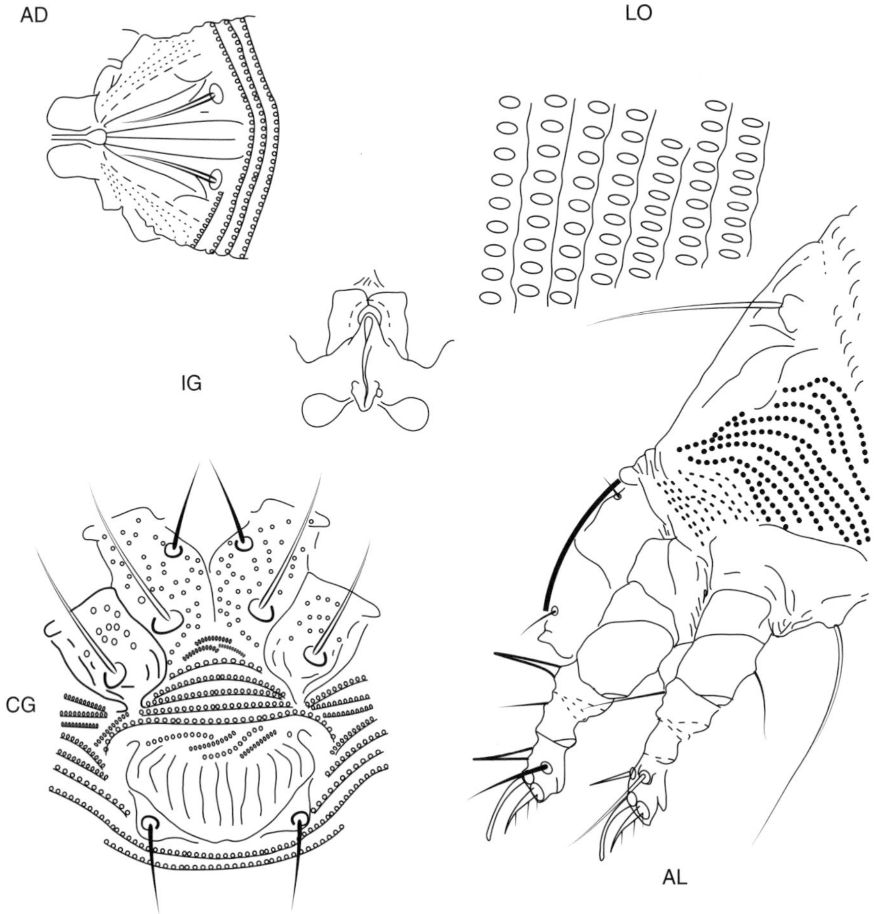

Fig. 16.12. *Eriophyes insidiosus* (Keifer *et* Wilson, 1955). See Table 16.1 for key.

Tibia I is shorter than tarsus I and not much longer than it is thick. The empodium is 4-rayed (Keifer and Wilson, 1955).

MALE. The male is like the female but smaller.

Common name

Peach mosaic vector mite.

Geographical distribution

South-western USA (Keifer and Wilson, 1955; Wilson *et al.*, 1955; Oldfield, 1970) and Mexico (Oldfield *et al.*, 1995).

Host plants

Rosaceae: cultivated *Prunus* spp. (peach, plum?, apricot?), wild *Prunus* (*P. ortulana* L.H. Bayley, *P. mexicana* S. Watson and *P. munsoniana* Wight *et* Hedrich), *Amygdalus* (almond?) (Keifer and Wilson, 1955; Wilson *et al.*, 1955; Oldfield, 1970; Jeppson *et al.*, 1975; Oldfield *et al.*, 1995; Oldfield and Proeseler, 1996).

Bio-ecology

The peach mosaic vector mite lives on plants of the genus *Prunus* (and also of the genus *Amygdalus*) and does not make leaf blisters. It inhabits buds, commonly developing colonies in lateral twig or limb buds, which gives rise to a perennial stunted bud (Jeppson *et al.*, 1975). The mite is a strict bud inhabitant, and rarely occurs in leaf axils or on the external surface of infested buds. However, on ornamental peach varieties, the population density can reach 1000 mites per bud. *E. insidiosus* is common on commercial peach varieties, where it is not easily observable; it also occurs in orchards in which mosaic is present. In northern Mexico, it has been collected on wild *P. munsoniana*, and in several central Mexican states, it develops at high density on 'Criollo' varieties of peach (Oldfield *et al.*, 1995). In freestone varieties in the south-western USA, it is commonly found in adventitious buds near the base of the trunk or on large branches; in California, all biological stages and both sexes are found throughout the year, and populations are present at low density in the dormant season, reproduction continuing at a favourable temperature (Oldfield and Proeseler, 1996). From midsummer onward, the infested buds often show strong desiccation, which causes a heavy reduction in the mite population. Dead buds apparently host low mite densities from summer until the end of the dormant season. In California, the wild host is absent and the species spreads via ornamental and commercial peaches. However, it is known that peach mosaic can be spread naturally by its vector from the symptomless wild host (*P. ortulana*, *P. mexicana* and

P. munsoniana). The mite has been found also on wild *Prunus* spp. in countries where peach mosaic disease is absent (Oldfield and Proeseler, 1996).

Symptomatology and damage

E. insidiosus feeds on buds of *Prunus* spp. and commercial peach and transmits the *Peach mosaic virus* (PMV) (Gispert *et al.*, 1998), which is responsible for peach mosaic disease (Wilson *et al.*, 1955). On peach, the symptoms of this disease are vein clearing and irregular chlorosis, stunted growth of the leaves, colour breaking of blossoms and russeting of branches. These symptoms also occur on certain other *Prunus* fruit trees in south-western USA and Mexico (Oldfield and Proeseler, 1996). The pathogen is not transmitted transovarially (Wilson *et al.*, 1955). Adult mites can inoculate the virus for at least 2 days after removal from infected plants, and transmit after feeding for 0.5–1 h on peach seedlings but not after feeding for 15 min (Oldfield and Proeseler, 1996). In southern California, the disease spreads during the growing season (Jones and Wilson, 1952) as a result of the dispersal of some vector mites during these months. In the USA, peach mosaic disease is known on plum, apricot and almond as well as peach, but the vector mite has not been collected from these plants (Oldfield and Proeseler, 1996).

Mite damage to peach following feeding activity consists of swelling and retardation of the buds, although these symptoms do not occur on buds of wild Nearctic *Prunus* that the mite inhabits (Oldfield and Proeseler, 1996). The direct damage caused by the mite to its host plants is not very significant.

Control

Control is based on the spray application of chemicals and prevention. A single application of diazinon at petal fall suppressed the development of mite populations and the spread of peach mosaic (Jones *et al.*, 1970). Similarly, the removal of infected trees is effective (Oldfield *et al.*, 1995).

Eriophyes inaequalis (Wilson *et* Oldfield) (Fig. 16.13)

Common name

Unknown.

Diagnostic characteristics

FEMALE. The body is worm like, 191 μm long, 48 μm wide, and yellowish white in colour. The gnathosoma is curved downward evenly. The prodorsal shield has a median line that is often broken; the admedian lines are complete, gradually diverging to the rear but recurving slightly centrad (to or towards the centre) at the posterior shield margin; a distinct diagonal line arises laterad (towards the side or laterally) to the dorsal tubercle and runs anteriorly and medially nearly to the admedian line; this diagonal line and two distinct submedian lines arise from a single point on the posterior shield margin and diverge anteriorly, forming a dart-shaped mark; granules are oriented in lines curving downward to the rear. The dorsal tubercles are 13 μm apart, the seta is 15 μm long, and projected forward. The opisthosoma has 65 annuli; the microtubercles are short, subconical, tipped slightly backwards, irregularly spaced, 3–4 per 10 μm arc distance on any annulus between the lateral and first ventral setae. The setae c_2 are set on annulus 6 behind the shield, and are 19 μm long; the setae *d* are set on annulus 19, and are 25 μm long; the setae *e* are set on annulus 34, and are 9 μm long; the setae *f* are set on annulus 6 from the rear, and are 16 μm long; the setae h_1 are present. The genital coverflap is 21 μm wide and 16 μm long, with 10 longitudinal furrows; the seta is 9 μm long. The empodium is 4-rayed (Wilson and Oldfield, 1966).

Geographical distribution

Western North America (Wilson and Oldfield, 1966).

Host plants

Rosaceae: *Prunus emarginata* (Douglas) (Wilson and Oldfield, 1966), sweet cherry (*Prunus avium* (L.) L.) (Oldfield, 1970).

Bio-ecology

E. inaequalis lives in the lateral buds of its only known wild host, *Prunus emarginata* (Douglas), in western North America (Wilson and Oldfield, 1966); it can also occur on sweet cherry plants, causing cherry mottle leaf disease (Oldfield, 1970). The adult retains the uneven nymphal distribution of opisthosomal microtubercles (Jeppson *et al.*, 1975). A single swollen bud can host a thousand mites and reproduction continues as temperatures allow during the year (Oldfield and Proeseler, 1996).

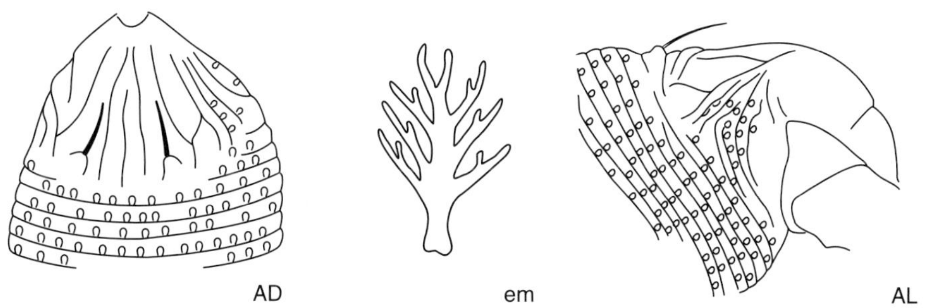

Fig. 16.13. *Eriophyes inaequalis* (Wilson *et* Oldfield) (redrawn and adapted from Wilson and Oldfield, 1966). See Table 16.1 for key.

Symptomatology and damage

Direct damage caused by *E. inaequalis* is of secondary economic importance; the symptoms are swelling and retardation of the buds of *P. emarginata* in north-western USA and south-western Canada. The most important damage caused by this mite is as a transmitter of the graft-transmissible disease 'Cherry mottle leaf disease', which is caused by a flexuous virus [formerly thought to be] of the closterovirus group, (James and Mukerji, 1993), but now formally designated as *Cherry mottle leaf virus* (CMLV) and belonging to the *Betaflexiviridae*. CMLV occurs in the sweet cherry-growing districts of western North America from northern California to south-western Canada (Oldfield and Proeseler, 1996). The main symptoms consist of irregular chlorotic mottling, edge tattering and little leaves. The fruits seem normal but lack flavour and often ripen late. The disease is common in orchards with natural stands of *P. emarginata* (McLarty *et al.*, 1951), but can sometimes spread through sweet cherry orchards in countries where *P. emarginata* is absent. *E. inaequalis* collected from infected *P. emarginata* can transmit the disease to peach (a symptomless host) or to sweet cherry (Oldfield, 1970). Mites collected on *P. emarginata* in California transmitted the disease to sweet cherry in the laboratory. The disease occurs on apricots in British Columbia (Canada) (Hansen, 1978), but the mite does not reproduce on either sweet cherry or apricot (Oldfield and Proeseler, 1996). This suggests the existence of other mite vectors.

Control

Control is provided by spray applications of acaricides and the use of mite- and virus-free grafts.

Eriophyes laevis (Nalepa) (Fig. 16.14)

Common name

Alder bead gall mite.

Diagnostic characteristics

FEMALE. The body of the protogyne is worm like, 225–250 μm long, and yellowish in colour. The prodorsal shield pattern is missing, or the admedians are only slightly indicated. The scapular tubercles are a little ahead of the rear prodorsal shield margin, and direct the short scapular setae (*sc*) up and diagonally centrad. These setae appear pointed more towards the front on the protogyne form. The opisthosomal annuli on the protogyne are sparsely granulated; the distribution varies from general to being limited to the sides and venter. The genitalia are shallow and bowl shaped, with the coverflap unmarked. Between the non-ornamented coxae, the sternal line is moderately long. The legs are slender, with the tibia and tarsus about the same length, and the setae are ⅓ of the way along the tibia. The empodium is 4-rayed, with the rays at least partially thickened basally. The deutogyne is reddish and differs from the protogyne in the setting of the dorsal setae and the fewer annular granules, which are limited to the venter and sometimes only to the rear venter (Jeppson *et al.*, 1975).

MALE. The male is similar to the protogyne, with the granuli on the annuli limited to the sides and venter (Jeppson *et al.*, 1975).

Geographical distribution

East Palearctic, Nearctic, North Africa and Oriental regions (Keifer, 1952a; Jeppson *et al.*, 1975; de Lillo and Beron, 2004).

Host plants

Betulaceae: Alder (*Alnus* spp.) (Jeppson *et al.*, 1975), including European alder (*A. glutinosa* Gaertn.), *A. incana* (L.) Moench. (Kane *et al.*, 1997), Oregon alder (*A. oregona* Nutt.) (Keifer *et al.*, 1982), white alder (*A. rhombifolia* Nutt.) (Hesse, 1971, Keifer *et al.*, 1982), hazel alder (*A. rugosa* (Du Roi) K. Spreng.), mountain alder (*A. tenuifolia* Nutt.) (Keifer *et al.*, 1982).

Bio-ecology

The alder bead gall mite infests alder worldwide. It reproduces by deuterogyny and attacks the seedlings and causes bead galls on the leaves. After overwintering, the deutogynes stimulate the formation of galls and lay eggs, from which are derived the protogynes and the males of the first generation. The larva is folded back on itself in the eggs. The eggs may change their shape and colour a few days after their deposition and become more evident; they hatch within a few days. Embryonic and post-embryonic development take 20 and 30 days, respectively. Two or three generations occur during the vegetative period of the host plant before the leaf changes occur and deutogynes appear when the day length decreases; these are the overwintering forms. Galls may host more than 400 mites of all stages (Shevtchenko, 1957, 1961). The age of the plant tissue is important for the clumped distribution of galls. The highest density of the mites was observed on the middle and lower foliage (Vuorisalo *et al.*, 1989). The marked positive phototaxis of the deutogynes leads to the emergence of the mite from galls. Dispersal occurs when the mite migrates to its winter habitat on old twigs in summer and autumn, and lasts 6–12 weeks (Shevtchenko, 1957, 1961). The mite has a poor dispersal capacity and tends to stay near to the foliage that it infested in the previous year, moving to the upper new foliage only later the growing season (Vuorisalo *et al.*, 1989).

Symptomatology and damage

E. laevis infests alder seedlings and causes bead galls on both leaf surfaces. The galls are variable in colour – from green to yellow to red or brown – and their openings commonly occur on the lower leaf surface; they are usually grouped together along the midrib, mainly on the old shoots within branches (Vuorisalo *et al.*, 1989). The underside opening of the galls may be due to the fact that the embryonic leaf in the bud is folded (Jeppson *et al.*, 1975). On *A. rhombifolia*, the cells attacked by the mite show polyploidy (Hesse, 1971). The nutritive cells of very young alder galls contain compounds that stain positively

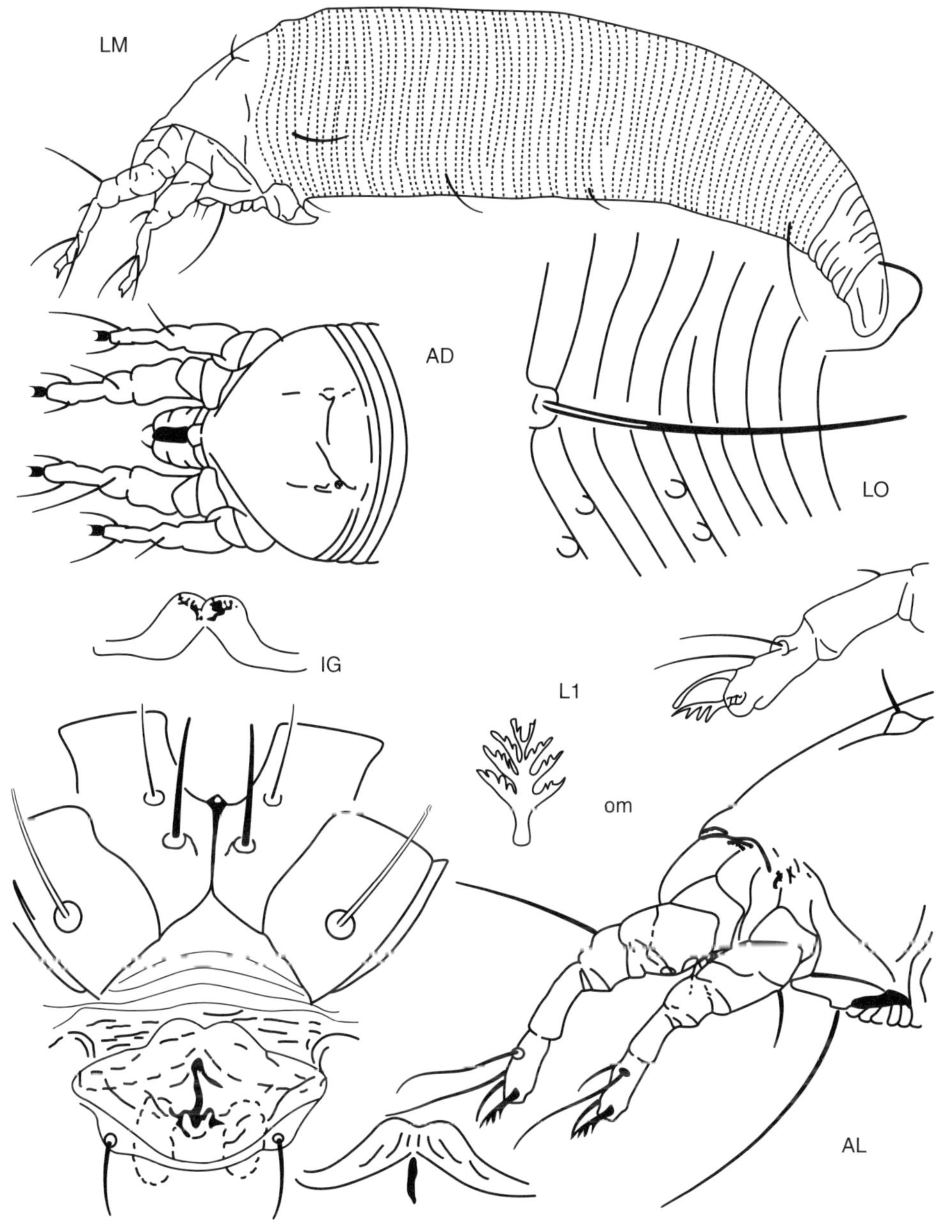

Fig. 16.14. *Eriophyes laevis* (Nalepa) (from Keifer, 1952a). See Table 16.1 for key.

for tannins, and the nutritive tissue accumulates more tannins as the alder gall matures (Kane *et al.*, 1997). Using both safranin and phloroglucinol staining procedures, young (1 week old) pouch galls caused by *E. laevis* on the leaves of *A. glutinosa* are very similar in structural architecture to the galls on leaves of *A. incana*, except for the presence of tannins in the vascular bundle sheaths located in undifferentiated cell layers. Tannin deposits did not appear within bundle sheaths and cells derived from mesophyll until galls of *A. incana* were 3 weeks old. Comparison of the internal structure of 3-week-old galls shows that galls of *A. incana* are more differentiated and have more tannin deposits than galls of *A. glutinosa*. Taking into account that condensed tannins are commonly considered to be plant defence compounds, the tannin deposits may be an adaptive deterrent or antifeedant or have toxic effects (Kane *et al.*, 1997). The mite may cause damage to the seedlings.

The removal of a portion of the leaves of short shoots with both high and low gall density to investigate the influence of galls on the growth of short shoot leaf area in *A. glutinosa* showed that the relative growth of leaf area in short shoots was negatively affected by a high gall density of *E. laevis*. However, the artificial removal of leaves positively affected leaf area growth,

and the growth of leaf area did not differ between high gall density short shoots with leaf removal and non-infested short shoots with no leaf removal. This phenomenon may involve the opposite effects of leaf removal and gall infestation (Vuorisalo *et al.*, 1989).

Control

Control is usually unnecessary.

Eriophyes emarginatae (Keifer) (Fig. 16.15)

Common name

Chokecherry finger gall mite.

Diagnostic characteristics

FEMALE. The body is worm like, 240 µm long, 60 µm wide and light yellow to amber in colour. The gnathosoma is quite short and curved down. The prodorsal shield is smooth; the scapular tubercles are 22.5 µm apart, and the scapular setae (*sc*) are 11 µm long, projecting dorsocentrad. The sternal ridge is forked. Coxal setae III are 42 µm long. The opisthosoma has about 60 smooth annuli, generally showing a slight ventrad increase. The seta c_2 is a little behind the seta *3a*, 24 µm long and set on annulus 8. The setae *d* are 28 µm long, set on annulus 20; the setae *e* are 17 µm long, set on annulus 36; the setae *f* are 32 µm long, set on annulus 5 from the rear; the setae h_2 are 55 µm long; the setae h_1 are present. The genitalia are 26 µm wide and 17 µm long; the coverflap is smooth, and the seta is 18 µm long. The empodium is 4-rayed (Keifer, 1939a).

MALE. As female.

Geographical distribution

California (USA) (Keifer, 1939b), Iran (Babaei *et al.*, 2010), Serbia and Montenegro (Glavendekić *et al.*, 2005).

Host plants

Rosaceae: *Prunus* spp. (Keifer *et al.*, 1982), almond (*P. dulcis* (Mill.) D.A. Webb) (Babaei *et al.*, 2010), *P. emarginata* (Douglas) Eaton (Keifer, 1939a).

Bio-ecology

The chokecherry finger gall mite was first collected in small finger-like, upper surface leaf galls on *P. emarginata* in California (Keifer, 1939a), and subsequently from four different *Prunus* spp. (Keifer *et al.*, 1982). Recently, *E. emarginatae* was reported from almond in Iran (Babaei *et al.*, 2010). The female of this species has the characteristics of a deutogyne; no protogyne form has been found, but the male has the characteristics of a typical protogyne (Oldfield, 1969). The tertiary sex ratio (of adults collected in the field) is 86% female (Oldfield, 1969). New females pick up spermatophores before entering diapause; males do not overwinter with females (Oldfield, 1969, 1973). In the spring, emerging females produce both sexes (Keifer, 1942; Oldfield, 1969). The species is univoltine. New-generation females, resembling deutogynes, seek wintering sites and enter diapause from the moment they emerge from the nymphal skin, shortly after the summer solstice; they remain in diapause until the following spring (Oldfield, 1969). Leaf galls are formed at the start of the season and are often produced by a single female, which may produce a maximum of about 50 first-generation progeny. In 3 successive weeks of examination, a mean of 47.52 and 47 progeny, respectively, was developed in two sets of 24 galls occupied by single females. Dispersal of the mite was presumed to be ambulatorial (Oldfield, 1969), but is probably aerial, as in the vast majority of eriophyoid mites.

Symptomatology and damage

E. emarginatae causes finger galls on Nearctic *Prunus* spp. but no anomalies on *Prunus* fruit species introduced from the Old World to North America (Oldfield, 1969). Overwintered females collected from very young galls in the spring and transferred on potted seedlings of *Prunus virginiana* L. var. *demissa* (Nuttal) induce gall formation. Females produced in the current year do not cause galls, probably because, being deutogynes, they not induce galls until they have overwintered (Oldfield, 1969). Galled leaves within a heavy galled portion of shoots of the wild plum (*P. americana* Marsh.) strongly reduced the size of the photosynthetic area of the neighbouring leaves in comparison with leaves that did not develop near galls. This suggests the reallocation of resources within the mite-attacked shoot (Wilson and O'Dowd, 1990).

Control

Control is usually unnecessary.

The Genus *Aceria* Keifer

Diagnostic characteristics

The body is worm like and circular in cross section, or nearly so. The gnathosoma has various sizes, but is usually small; if it is large, the chelicerae are evenly curved. The prodorsal shield is usually subtriangular, with scapular tubercles set in subdorsal positions on the rear shield margin and directing the scapular setae (*sc*) caudad (towards the posterior or caudally). The legs have all the usual setae and a simple empodium. The opisthosomal annuli are the same above and below, usually more numerous dorsally and nearly always microtuberculate. The female genitalia are at a moderate distance behind the coxae; the coverflap is furrowed or smooth, usually with a single row of longitudinal scoring; the anterior female apodeme is of normal length in ventral view (Keifer, 1944; Amrine *et al.*, 2003).

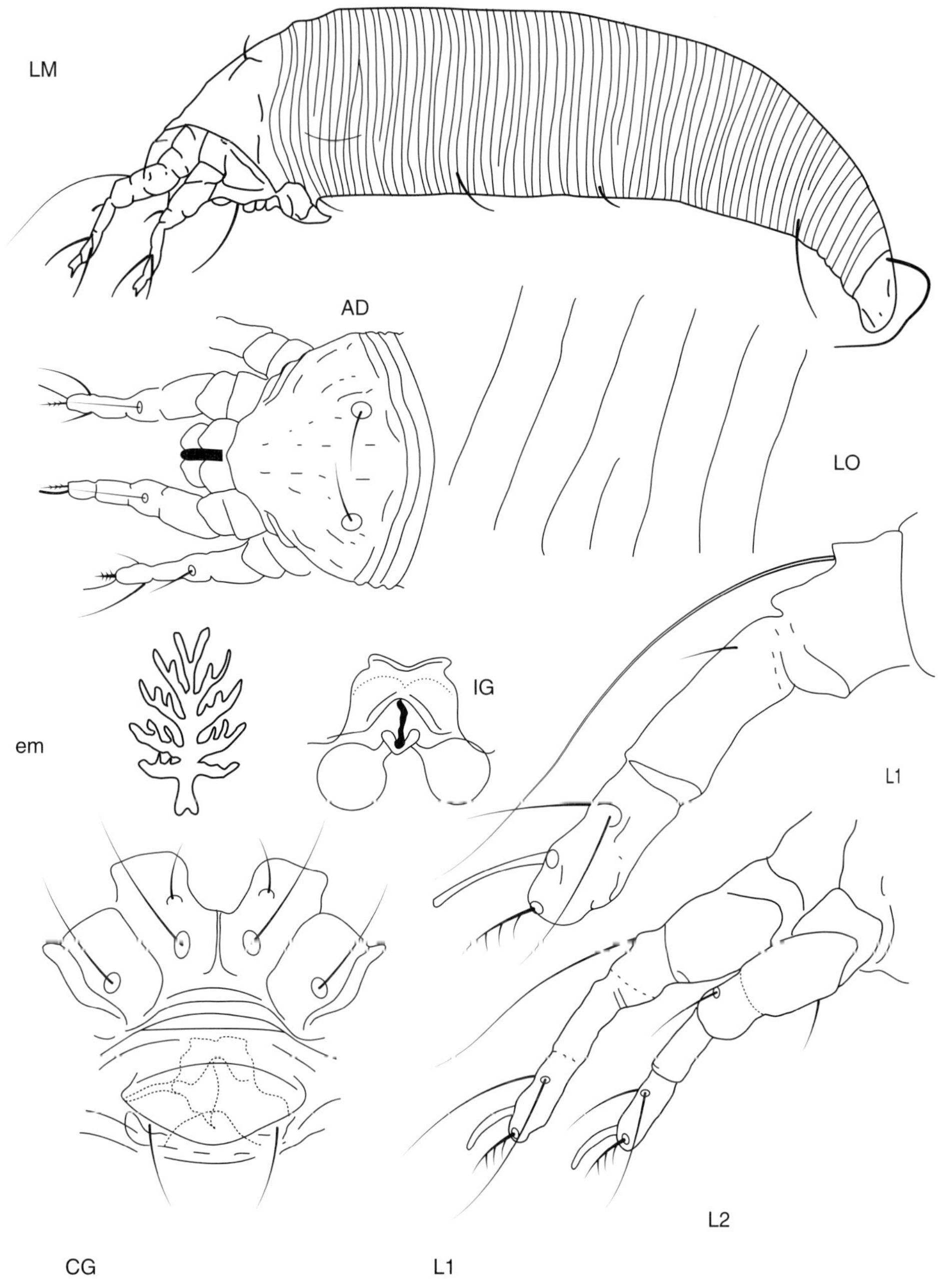

Fig. 16.15. *Eriophyes emarginatae* (Keifer) (from Keifer, 1939a). See Table 16.1 for key.

Economic importance

Amrine *et al.* (2003) include in the genus *Aceria* Keifer several species belonging to the genera *Artacris* Keifer, *Cisaberoptus* Keifer, some *Eriophyes* von Siebold 1851 prior to Keifer 1944 and *sensu* Newkirk and Keifer 1971, some *Phytoptus* Dujardin prior to Nalepa 1897, and *Trichostigma* Gerber. The genus *Aceria* includes over 900 species, but less than 20 of these are key pests of economic importance that cause either direct damage or transmit virus disease. Among these, the wheat curl mite, *A. tosichella* Keifer, causes curled leaves of wheat and transmits different viruses, such as the *Wheat streak mosaic virus* (WSMV) and the 'High Plains virus' (HPV) of wheat (*Triticum aestivum* L.) (Styer and Nault, 1996). The dry bulb mite, *A. tulipae* (Keifer), damages tulip bulbs (*Tulipa* spp.), shallots (*Allium cepa* L. var. *aggregatum*), onions (*A. cepa*) and garlic (*A. sativum* L.), and transmits some important viruses (van Dijk *et al.*, 1991; Oldfield

and Proeseler, 1996; Perring, 1996). The fig bud mite, *Aceria ficus* (Cotte), damages buds, leaves and fruits of fig (*Ficus carica* L.) (Baker, 1939; Ebeling and Pence, 1950) and transmits the *Fig mosaic virus* (FMV) (Serrano *et al.*, 2004). The olive bud mite, *A. oleae* (Nalepa), induces deformations on leaves and fruits of olive (*Olea europaea* L.) in Mediterranean countries (Jeppson *et al.*, 1975), Iran (Sepasgosarian, 1973) and South Africa (Meyer Smith, 1981b). The litchi erineum mite, *A. litchii* (Keifer), forms a fine red erineum on the lower leaf surface of litchi (*Litchi chinensis* Sonn.), often involving and curling the whole leaf and damaging flower buds, shoots and fruits (Jeppson *et al.*, 1975). The citrus bud mite, *A. sheldoni* (Ewing), deforms buds, leaves, flowers and fruits of citrus (Vacante, 2010). The coconut flower and nut mite, *A. guerreronis* Keifer, damages the fruits of coconut palm (*Cocos nucifera* L.) (Moore and Howard, 1996). The carrot bud mite, *A. peucedani* (Canestrini), damages the carrot (*Daucus carota* L.) (Scott, 1965). The hibiscus erineum mite, *A. hibisci* (Nalepa), attacks the Chinese red hibiscus (*Hibiscus rosa sinensis* L.) and other hibiscus species (Welbourn *et al.*, 2008). *A. cajani* ChannaBasavanna lives on pigeon pea (*Cajanus cajan* (L.) Huth) in several countries of the Indian subcontinent and transmits the *Pigeon pea sterility mosaic virus* (PPSMV) (Seth, 1962; Kumar *et al.*, 2001, 2003a; Latha *et al.*, 2007). The mango bud mite, *A. mangiferae* (Hassan) (= *Aceria mangiferae* Sayed), stunts and brooms twigs of mango (*Mangifera indica* L.), with bud proliferation and marked damage to young plants (Jeppson *et al.*, 1975). Many other species inhabit several plants of economic importance but are of secondary economic importance, and their presence in the field does not justify control; however, some of these species are treated to help the crop consultants to optimize the management of their populations in different environments (forest, nursery, etc.).

To facilitate the treatment of the most important *Aceria* spp., the same approach is used as adopted by Jeppson *et al.* (1975) for the genus *Eriophyes* – many species of which are now included in *Aceria*. The morphological features of *Aceria* spp. and their host plants allow the common species to be allocated to seven groups, which are referred to as the: 'Walnut group', 'Oak group', 'Fruit and nut trees other than the walnut group', 'Grasses and monocotyledonous crops group', 'Vegetables, field crops and berries group', 'Ornamental plants and forest and shade trees group' and 'Maple trees group'. These groups are described separately below.

Walnut group

This group includes species living on walnut, oak and hickory. According to Jeppson *et al.* (1975), it is a homogenous group characterized by 3-rayed empodia. Damage caused by mites of this group includes twig brooming, leaf galls, leaf erinea and leaf edge rolling, but usually is of only modest economic importance. The most common species are *A. brachytarsa* (Keifer), *A. caryae* (Keifer), *A. caulis* (Koch), *A. cinereae* (Keifer), *A. erinea* Nalepa, *A. neobeevori* Keifer, *A. tristriata* (Nalepa) and *A. vaga* Keifer. These eight species are described below.

Aceria brachytarsa (Keifer) (Figs. 16.16-16.17)

Common name

California black walnut pouch gall mite.

Diagnostic characteristics

FEMALE. The protogyne is 195–210 μm long, worm like and yellowish in colour. The prodorsal shield shows an unclear pattern, and a lateral ocellar spot at its rear angle. The opisthosomal microtubercles extend into short spinules, giving the mite a spiny appearance. The genitalia have a smooth coverflap. The empodium is 3-rayed. The deutogyne is about the same size as the protogyne and is reddish in colour (Keifer, 1939c; Keifer *et al.*, 1982).

MALE. The male is smaller than the female and yellowish in colour. The opisthosomal microtubercles are pointed (Keifer *et al.*, 1982).

Geographical distribution

California (Keifer, 1939c; Jeppson *et al.*, 1975; Keifer *et al.*, 1982) and Iran (Khanjani and Haddad Irani-Nejad, 2006; Xue *et al.*, 2009a).

Host plants

Juglandaceae: California walnut (*Juglans californica* S. Watson), Hind's walnut (*J. hindsii* (Jeps.) R.E. Sm.) (Keifer *et al.*, 1982; Khanjani and Haddad Irani-Nejad, 2006; Xue *et al.*, 2009a), black walnut (*J. nigra* L.) (Keifer, 1939c; Jeppson *et al.*, 1975).

Bio-ecology

The California black walnut pouch gall mite lives on walnut trees. Overwintering females have been observed in terminal buds in late winter and in new galls in spring. Males and protogynes were found in May, and lived in galls through August (Keifer *et al.*, 1982).

Symptomatology and damage

The mites produce galls on the upper leaf surface that are semiglobular, often solitary, set at random between the lateral veins and on the midrib, and about 3–6 mm across. The colour of these galls ranges from green to yellowish green with reddish or brown shades. The lower gall orifice bears some hairs and the internal surface of the galls has lobes deriving from consolidations of groups of papillae, with many divisions in each papilla. During summer and with gall maturity, these lobes and the surrounding tissues become red, and this colour causes the colour of the deutogyne (Jeppson *et al.*, 1975, Keifer *et al.*, 1982).

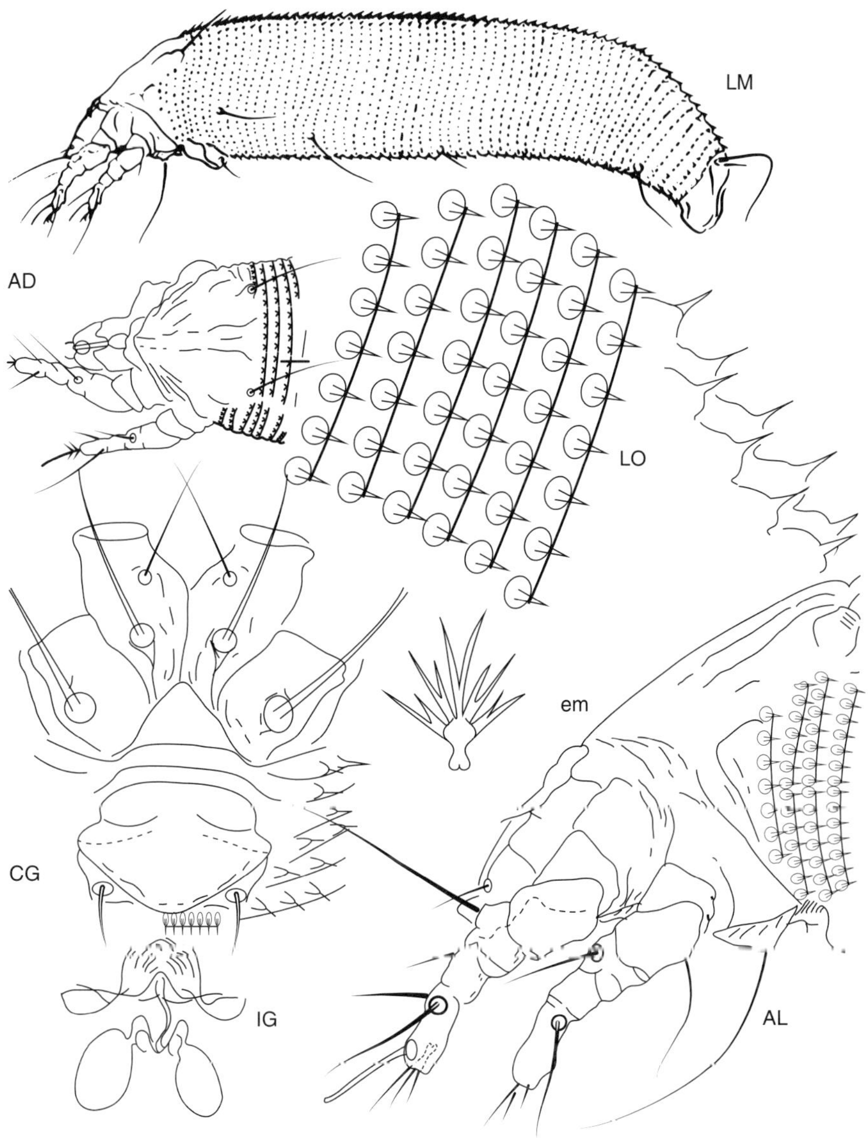

Fig. 16.16. *Aceria brachytarsa* (Keifer), protogyne (from Keifer, 1939c). See Table 16.1 for key.

Control

This is usually unnecessary.

Aceria caryae (Keifer) (Fig. 16.18)

Common name

Pecan leaf edgeroll (or edge roll) mite.

Diagnostic characteristics

FEMALE The body is worm like, 190–245 μm long, 45–50 μm wide and reddish in colour. The gnathosoma is somewhat curved down. The prodorsal shield is almost entirely smooth, except towards the rear, where the median and admedian lines are indicated; the scapular tubercles are 17.5 μm apart, set on the rear margin; the scapular setae (*sc*) are 20.5 μm long, projecting backwards. The opisthosoma has about 65 annuli, there is some ventral annulus reduction and the microtuberculation is very obscure, except immediately behind the front and on the ventral rear. The setae c_2 are 8 μm long, set on about annulus 5; the setae *d* are 10 μm long, set on about annulus 20; the setae *e* are 8 μm long, set on about annulus 37; the setae *f* are 14 μm long, set on about annulus 5 from the rear; the setae h_1 are present. The genitalia are 16 μm wide and 9.5 μm long; the coverflap is smooth, the seta is 4 μm long, and situated on pointed tubercles. The empodium is 3- rayed (Keifer, 1939e).

Geographical distribution

Texas (Keifer, 1939e; Jeppson *et al.*, 1975; Keifer *et al.*, 1982) and Poland (Soika and Łabanowski, 2006).

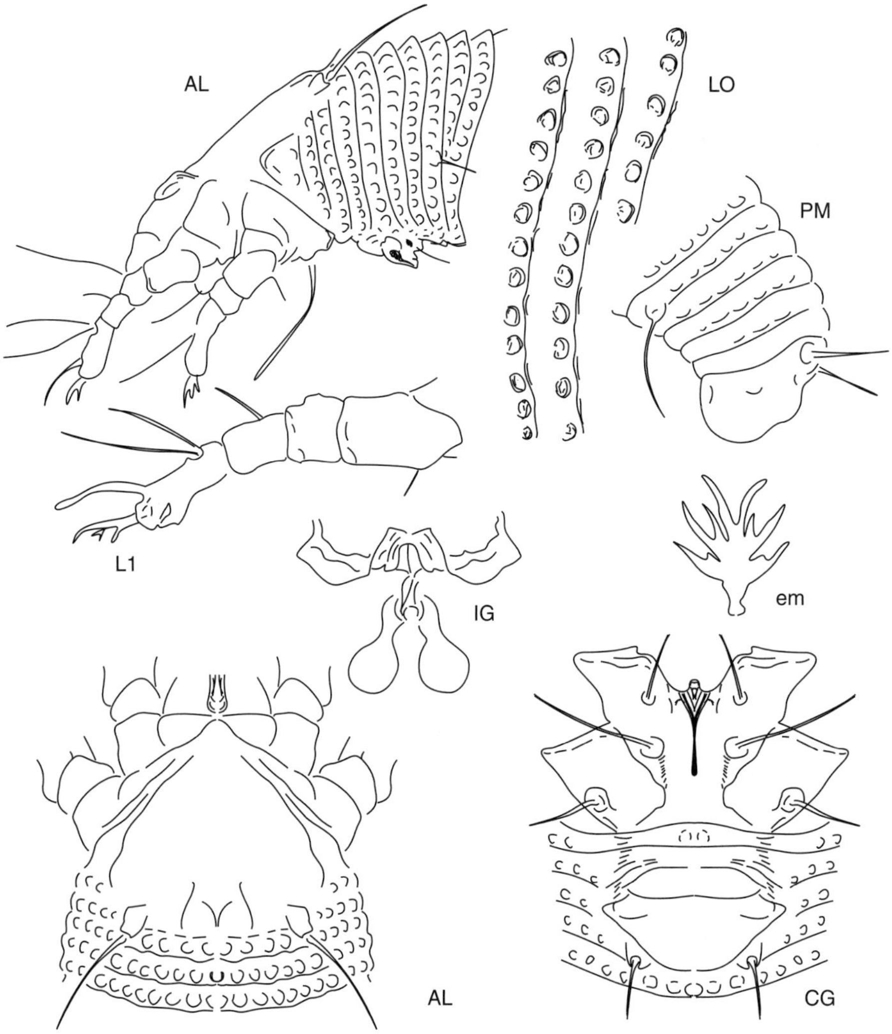

Fig. 16.17. *Aceria brachytarsa* (Keifer), deutogyne (from Keifer, 1939c). See Table 16.1 for key.

Host plants

Juglandaceae: pecan (*Carya illinoinensis* (Wangenh.) C. Koch) (= *Carya pecan* (Marsh.) Engl. *et* Graebn.) (Keifer, 1939e; Jeppson *et al.*, 1975) and shagbark hickory (*C. ovatus* (Mill.) K. Koch) (Soika and Łabanowski, 2006).

Bio-ecology

This species could represent the deutogyne of the same species of which *Aceria vaga* Keifer could be the protogyne (Jeppson *et al.*, 1975). It has been reported from pecan in Texas (Keifer, 1939e; Jeppson *et al.*, 1975) and from shagbark hickory in a nursery in Poland (Soika and Łabanowski, 2006).

Symptomatology and damage

The mites produce leaf spots on pecan and roll the leaf edges. This roll is full of spongy tissue in which the mites live (Keifer, 1939e; Jeppson *et al.*, 1975).

Control

This is usually unnecessary.

Aceria caulis (Cook) (Fig. 16.19)

Common name

Black walnut petiole gall mite.

Diagnostic characteristics

FEMALE. The body is slender, worm like and 150–210 μm long. The protogyne is worm like and reddish in colour. The prodorsal shield has two parallel ridges at the rear edge. The opisthosomal annuli have pointed microtubercles and the forms recoverable later from the galls (erinea) have rounded and humped microtubercles. The genital coverflap is smooth. The empodium is

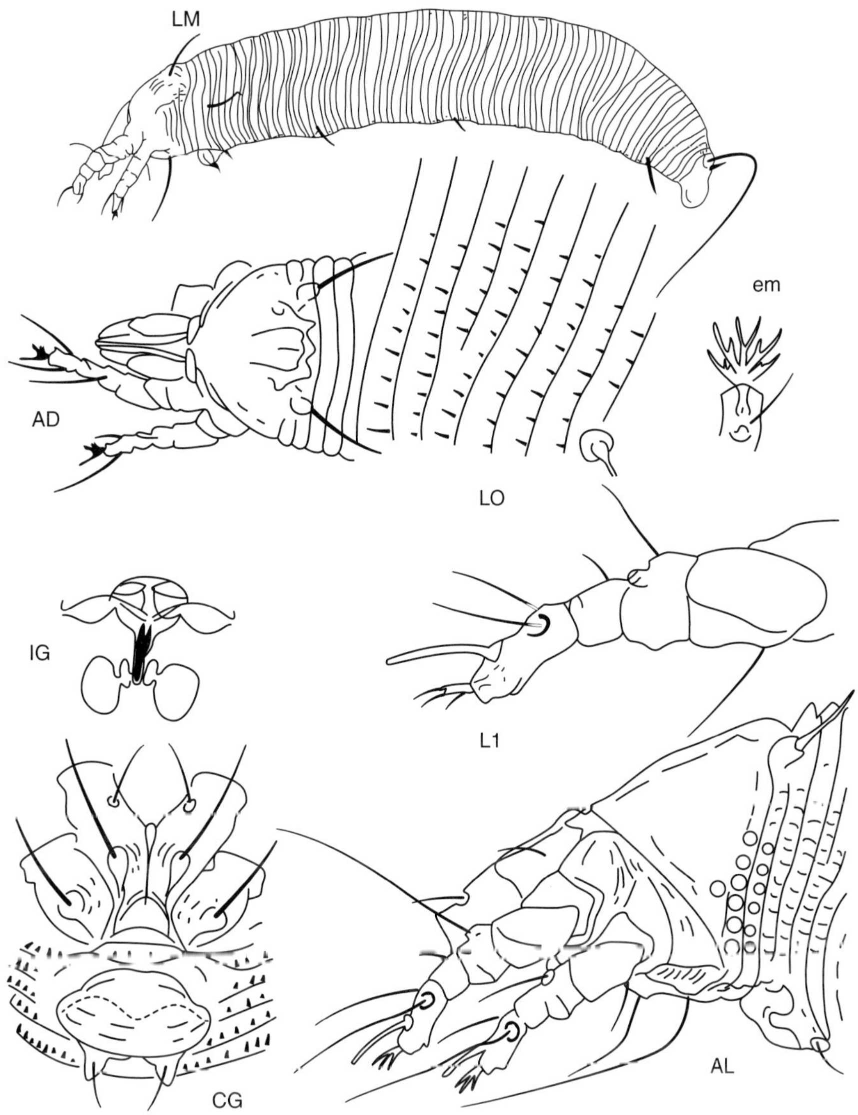

Fig. 16.18. *Aceria caryae* (Keifer) (from Keifer, 1939e). See Table 16.1 for key.

3-rayed. The deutogyne has microtubercles that are rounded and thickened posteriorly (Keifer, 1940a; Jeppson *et al.*, 1975; Keifer *et al.*, 1982).

MALE. The male is smaller than the female, with opisthosomal pointed microtubercles (Keifer *et al.*, 1982).

Geographical distribution

New Jersey, Massachusetts, and probably Ontario (Canada) and Pennsylvania, south Florida and Texas (Jeppson *et al.*, 1975; Keifer *et al.*, 1982).

Host plants

Juglandaceae: black walnut (*Juglans nigra* L.) (Jeppson *et al.*, 1975; Keifer *et al.*, 1982).

Bio-ecology

A. caulis develops alternate generations, with deutogynes overwintering in the buds. During the summer, both deutogynes and protogynes may be collected in erineum hairs (Keifer *et al.*, 1982).

Symptomatology and damage

The mite produces prominent erineum-covered swellings on petioles consisting of a solid hard mass, which are sometimes solitary and are of various sizes, varying in colour from greenish, through pink and crimson to reddish brown. The exterior surface of these erinea is covered by a rich mass of silky, unicellular hairs. The erineal papillae have no division or cell limits. The development of erineal galls causes the envelopment of the stem and leaf, with deformation of the petiole, which curls back, and a failure of leaflet development (Jeppson *et al.*, 1975; Keifer *et al.*, 1982).

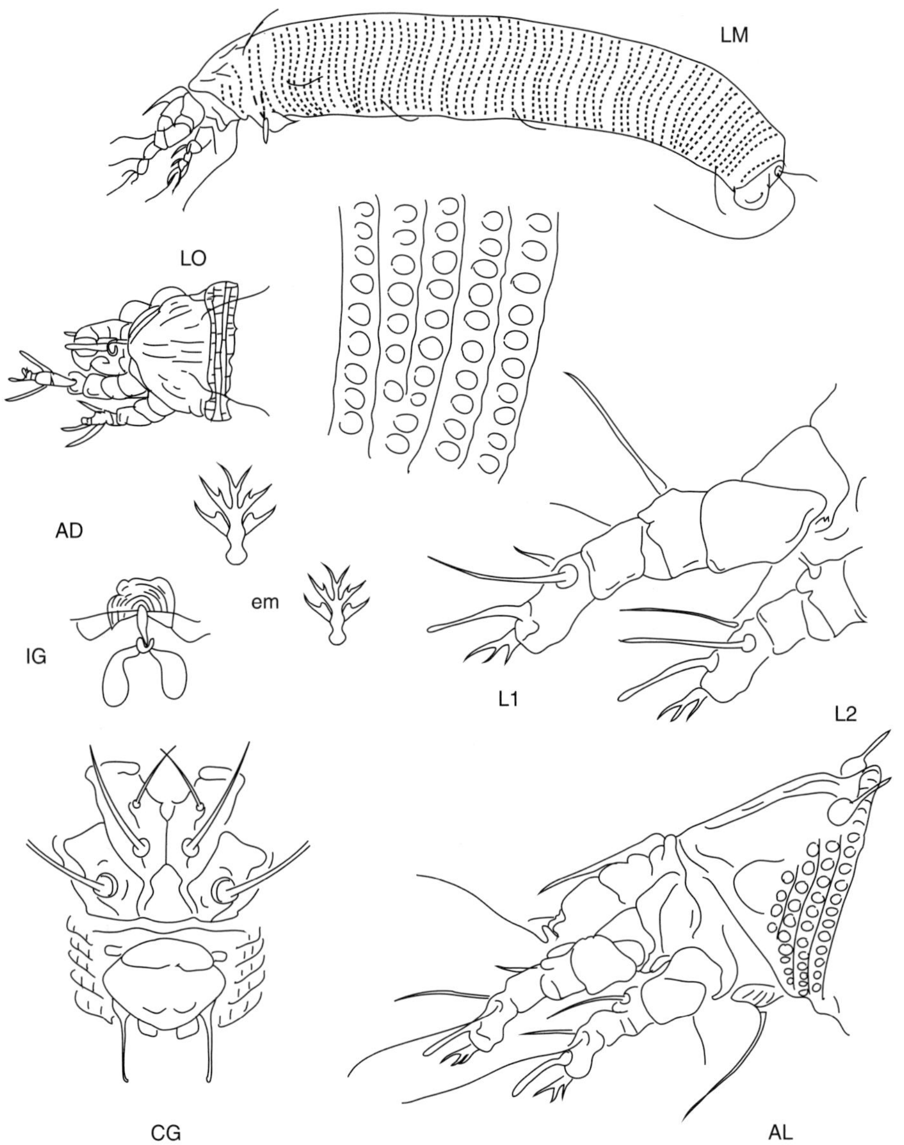

Fig. 16.19. *Aceria caulis* (Cook) (from Keifer, 1940a). See Table 16.1 for key.

Control

This is usually unnecessary.

Aceria cinereae (Keifer) (Fig. 16.20)

Common name

Butternut erineum mite.

Diagnostic characteristics

FEMALE. The body is worm like, 140–240 μm long, 35-40 μm wide and light yellowish in colour. The gnathosoma projects forward and curves down. The prodorsal shield is smooth; the scapular tubercles are 17.5 μm apart, set on the rear margin; the scapular setae (*sc*) are 14 μm long, projecting backwards. The sternal line is long, obscurely forked. The coxae are smooth. The opisthosoma has 53–61 annuli, entirely microtuberculate except for the last 6 or 7 dorsal annuli; the annuli are 3.65 μm broad; the microtubercles are rather broad, rounded or subtriangular, with 3–4 tubercles to 6 μm. The seta c_2 is 11 μm long, set on about annulus 4; the setae *d* are 9 μm long, set on about annulus 14; the setae *e* are 9 μm long, set on about annulus 29; the setae *f* are 17 μm long, set on about annulus 4–5 from the rear; the setae h_1 are present. The genitalia are 15.5 μm wide, 8.5 μm long, the coverflap is smooth, and the seta is 5.5 μm long on 4.5 μm spurs. The empodium is 3-rayed (Keifer, 1940a; Keifer *et al.*, 1982).

Geographical distribution

New York and north-eastern states of the USA (Keifer, 1940a; Jeppson *et al.*, 1975; Keifer *et al.*, 1982).

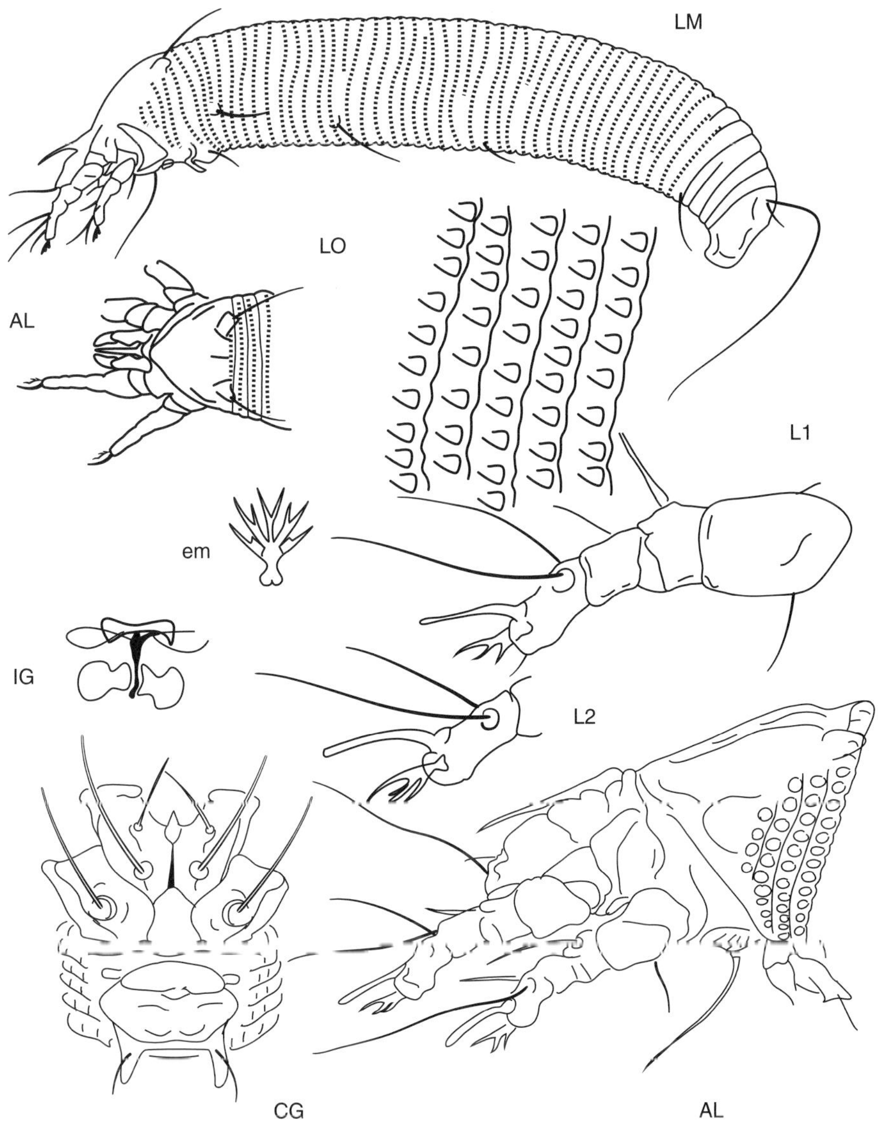

Fig. 16.20. *Aceria cinereae* (Keifer) (from Keifer, 1940a). See Table 16.1 for key.

Host plants

Juglandaceae: butternut (*Juglans cinerea* L.) (Keifer, 1940a; Jeppson *et al.*, 1975; Keifer *et al.*, 1982).

Bio-ecology

A. cinereae is reported from butternut in the north-eastern states of the USA (Keifer, 1940a; Jeppson *et al.*, 1975; Keifer *et al.*, 1982).

Symptomatology and damage

The mites cause undersurface leaf erineum patches of felted hairs set alongside the midrib, usually extended on the surface between the lateral veins and tending to turn together. On the upper leaf surface, the corresponding area appears swollen and puckered. The colour of these erineum patches ranges from light brown to reddish brown (Keifer, 1940a; Jeppson *et al.*, 1975; Keifer *et al.*, 1982).

Control

This is usually not necessary.

Aceria erinea Nalepa (Fig. 16.21)

Common name

Persian walnut erineum mite.

Diagnostic characteristics

FEMALE. The body is worm like, slender, about 225 μm long, 37 μm wide and yellowish white in colour. The gnathosoma projects

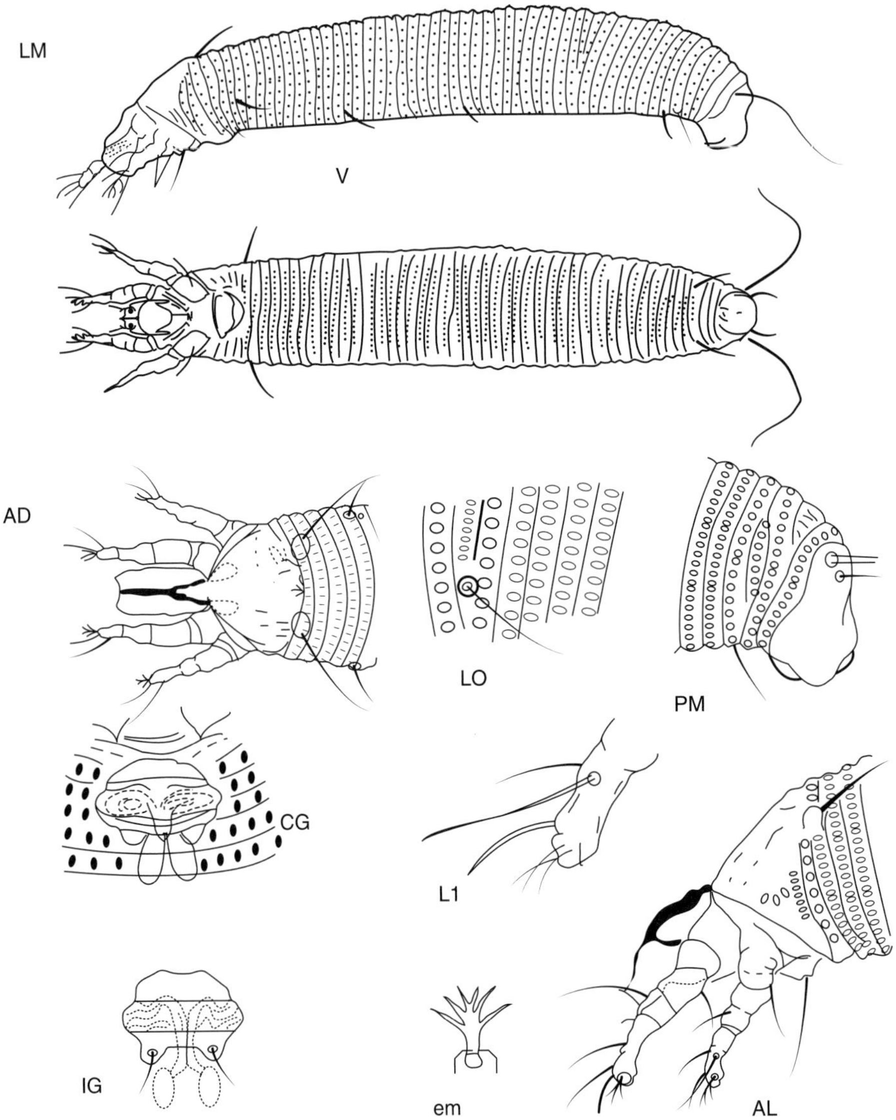

Fig. 16.21. *Aceria erinea* Nalepa (from Keifer, 1938a). See Table 16.1 for key.

forward and down. The prodorsal shield is subtriangular, gently curved above, smooth except for a slight indication of three lines between the scapular tubercles and a few large side granulations; the scapular tubercles are set on the rear margin, 17 μm apart, and the scapular seta (*sc*) are 17 μm long and retrorse. Coxae I are contiguous and the sternal line is simple. The opisthosoma has 50–60 annuli, there is a small ventrad reduction in annulus numbers, the annuli are 3 μm wide and heavily microtuberculate; the microtubercles fade on the last 4–5 annuli. The setae c_2 are set on about annulus 5, a little behind the genital seta, which is 7 μm long. The setae *d* are 7 μm long, set on about annulus 16. The setae *e* are 5 μm long, set on annulus 29. The setae *f* are 9 μm long, set on about annulus 5 from the rear. The setae h_2 are about 60 μm long; the setae h_1 are present. The genitalia are somewhat lenticular, with the coverflap smooth, 14.5 μm wide, and 9.5 μm long; the glands are small and short stalked; the seta is 3–4 μm long. The empodium is 3-rayed (Keifer, 1938a; Jeppson *et al.*, 1975; Keifer *et al.*, 1982).

Geographical distribution

Worldwide.

Host plants

Juglandaceae: Persian walnut (*Juglans regia* L.) (Nalepa, 1891; Keifer *et al.*, 1982).

Bio-ecology

A. erinea is an old European species (Nalepa, 1891) living on Persian walnut. During summer, the females move from the erinea to the terminal buds to overwinter. During autumn, it is usually possible to collect mites in both buds and erinea (Keifer *et al.*, 1982). In Germany, the eriophyoid *Anthocoptes juglandis* Domes has been found on the lower surface of leaves of Persian walnut, as a vagrant or in erinea of *A. erina*, or, from the end of

July, on new buds. The species seemed to cause little damage to the tree (Domes, 1998). Natural enemies of *A. erinea* include the phytoseiid mite *Amblydromella denmarki* Rather, reported from India (Rather, 1984).

Symptomatology and damage

The mites cause felty erineum patches on the lower leaf surface, with convex swellings on the upper leaf surface. These patches consist of a large and solitary concavity bearing felty yellowish hairs hosting the mite populations, set laterally to the midrib between the lateral veins and showing well-defined edges. Each patch bears short, unicellular hairs. Secondary veins appear as thickened partitions. Although each leaf bears only a few patches, they are easily identifiable (Jeppson *et al.*, 1975; Keifer *et al.*, 1982; Westphal and Manson, 1996).

Control

Commonly, the damage is not severe enough to warrant control measures, but the application of dicofol or another specific acaricide has been recommended in case of an unusually heavy attack (Laffi and Montermini, 1985). The crop and pest management guidelines of Cornell University (2012) suggest treating from late May through to mid-June at a threshold of 367–707 GDDs (growing degree-days) with abamectin, carbaryl, fenazaquin, fenpyroximate, milbemectin or neem oil. Individual infested leaves can also be removed by hand when possible. However, the personal experience of the author in the Mediterranean area suggests that these choices are excessive for single trees, although they could probably be justified in nurseries and/or extensive crops.

Aceria neobeevori Keifer (Fig. 16.22)

Common name

California black walnut catkin gall mite.

Diagnostic characteristics

FEMALE. The body is worm like, 175–250 μm long, 40-55 μm wide and yellowish in colour. The gnathosoma projects forward and a little down. The prodorsal shield has an obscure pattern

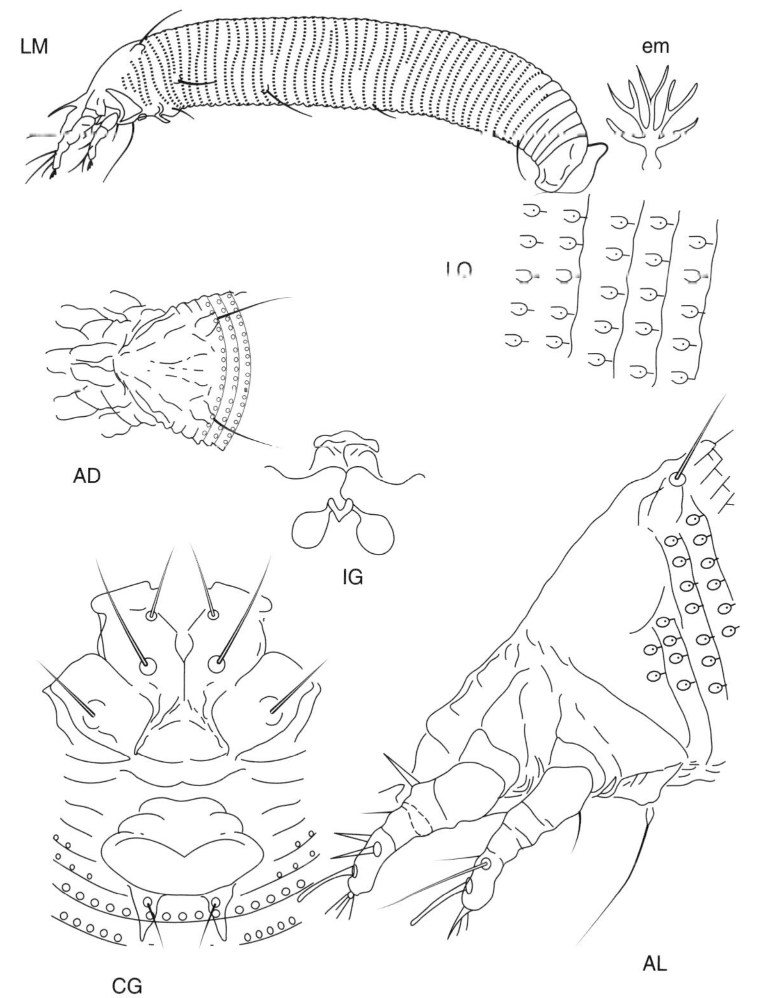

Fig. 16.22. *Aceria neobeevori* Keifer (from Keifer, 1952d). See Table 16.1 for key.

but the median and admedian lines are apparent on the rear of the shield in most cases; there is no lateral granular area. The scapular tubercles are rather large and set on the rear margin, 25 µm apart; the scapular setae (*sc*) are 22 µm long, projecting backwards. The coxae are smooth; coxae I are broadly connate. The opisthosoma has about 65–70 annuli, with microtubercles; each microtubercle is prolonged into a short fine spine. The setae c_2 are 12 µm long, set on about annulus 8; the setae *d* are 20 µm long, set on about annulus 17; the setae *e* are 16 µm long, set on about annulus 34; the setae *f* are 23 µm long, set on about annulus 5 from the rear; the setae h_1 are present. The genitalia are 20 µm wide, 12 µm long, and the coverflap is essentially smooth; the setae are 8 µm long, on elongate and pointed tubercles. The empodium is 3-rayed (Keifer, 1952d).

Geographical distribution

Southern California (Keifer, 1952d; Jeppson *et al.*, 1975).

Host plants

Juglandaceae: black walnut (*Juglans nigra* L.) (Keifer, 1952d; Jeppson *et al.*, 1975).

Bio-ecology

A. neobeevori is reported from black walnut in southern California (Keifer, 1952d; Jeppson *et al.*, 1975).

Symptomatology and damage

The mites form hanging galls of the male catkins, which then enlarge and remain alive on the tree for most of the season (Keifer, 1952d; Jeppson *et al.*, 1975).

Control

This is usually not necessary.

Aceria tristriata (Nalepa)

Common name

Persian walnut leaf gall or blister mite.

Diagnostic characteristics

FEMALE. The protogyne of this mite is long, 200–240 µm. The opisthosomal microtubercles are pointed, more so ventrally. Coxae I are elongate. The genital coverflap bears no markings. The deutogyne has no dorsal microtubercles and these are flattened ventrally (Jeppson *et al.*, 1975).

Geographical distribution

Worldwide (Jeppson *et al.*, 1975; de Lillo and Beron, 2004).

Host plants

Juglandaceae: Persian walnut (*Juglans regia* L.) (Jeppson *et al.*, 1975).

Bio-ecology

A. tristriata is widespread on Persian walnut. Its biological cycle includes both deutogyne and protogyne forms (Jeppson *et al.*, 1975).

Symptomatology and damage

The Persian walnut leaf gall mite causes small, brown, hard pustules, about 1 mm in diameter. These galls occur along midribs and the larger lateral veins; during high infestations they also develop elsewhere on the leaf. Severe attacks cause twisted and misshapen leaves (Jeppson *et al.*, 1975). The feeding activity of the mite sometimes malforms young fruit (Boesewinkel, 1977).

Control

This is usually not necessary.

Aceria vaga Keifer (Fig. 16.23)

Common name

Unknown.

Diagnostic characteristics

FEMALE. The body is elongated, worm like, 200–252 µm long, and dull light yellowish in colour. The gnathosoma is curved down. The prodorsal shield has no pattern; the scapular tubercles are 23 µm apart; the scapular setae (*sc*) are 23 µm long, projecting caudad and a little outward. Coxae I are moderately connate, with a heavy junction line. The opisthosoma has 55–50 annuli, entirely microtuberculate; the microtubercles are ahead of the annulus margins and pointed. The setae c_2 are 10 µm long, set on annulus 4 behind the shield; the setae *d* are 20 µm long, set on about annulus 16; the setae *e* are 8.5 µm long, set on about annulus 31; the setae *f* are 11 µm long, set on annulus 4 from the rear; the setae h_1 are 8.5 µm long. The genitalia are 15 µm wide and 12 µm long; the coverflap lacks ribs; the seta is 10 µm long, set on elongate, pointed tubercles. The empodium is 3-rayed (Keifer, 1962a).

Geographical distribution

Brazil (Keifer, 1962a) and California (APHIS, 1979).

Host plants

Juglandaceae: pecan (*Carya illinoinensis* (Wangenh.) K. Koch) (Keifer, 1962a), Northern California walnut (*Juglans hindsii* (Jeps.) R.E. Sm.) (APHIS, 1979).

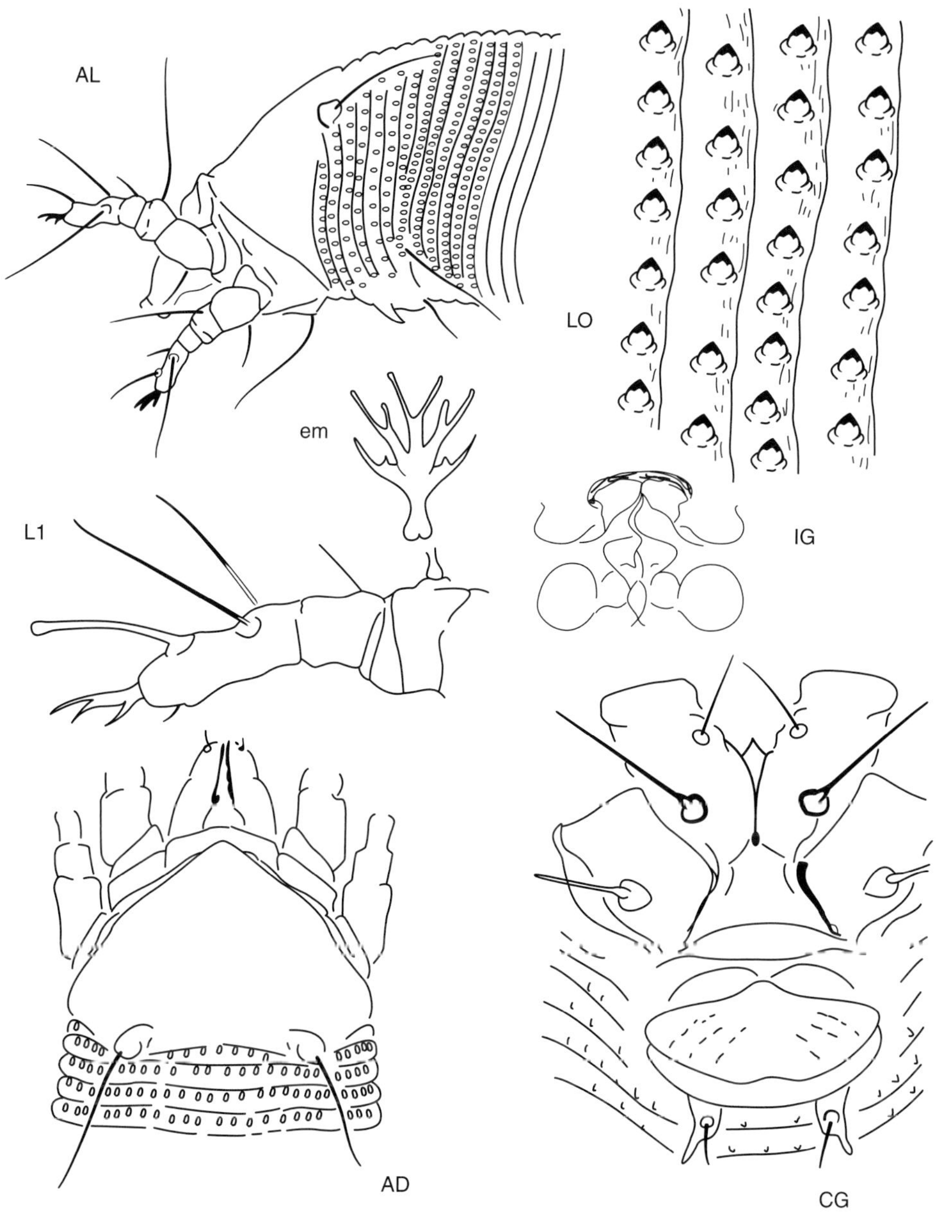

Fig. 16.23. *Aceria vaga* Keifer (Keifer, 1962a). See Table 16.1 for key.

Bio-ecology

A. vaga has been reported from pecan in Brazil (Keifer, 1962a) and from Northern California walnut in California (APHIS, 1979). The mite could be associated with 'Pecan bunch disease' (Keifer, 1962a; Jeppson *et al.*, 1975).

Symptomatology and damage

The mites could cause leaf spots or discoloration, rolled leaf edges and 'bunch disease'.

Control

This is usually not necessary.

Oak group

The species included into this group have 3-rayed empodia. The most important are the three erineum mites, *Aceria ecantyx* Keifer, *Aceria ilicis* (Canestrini) and *Aceria triplacis* Keifer. These three species are described below.

Aceria ecantyx Keifer (Fig. 16.24)

Common name

Unknown.

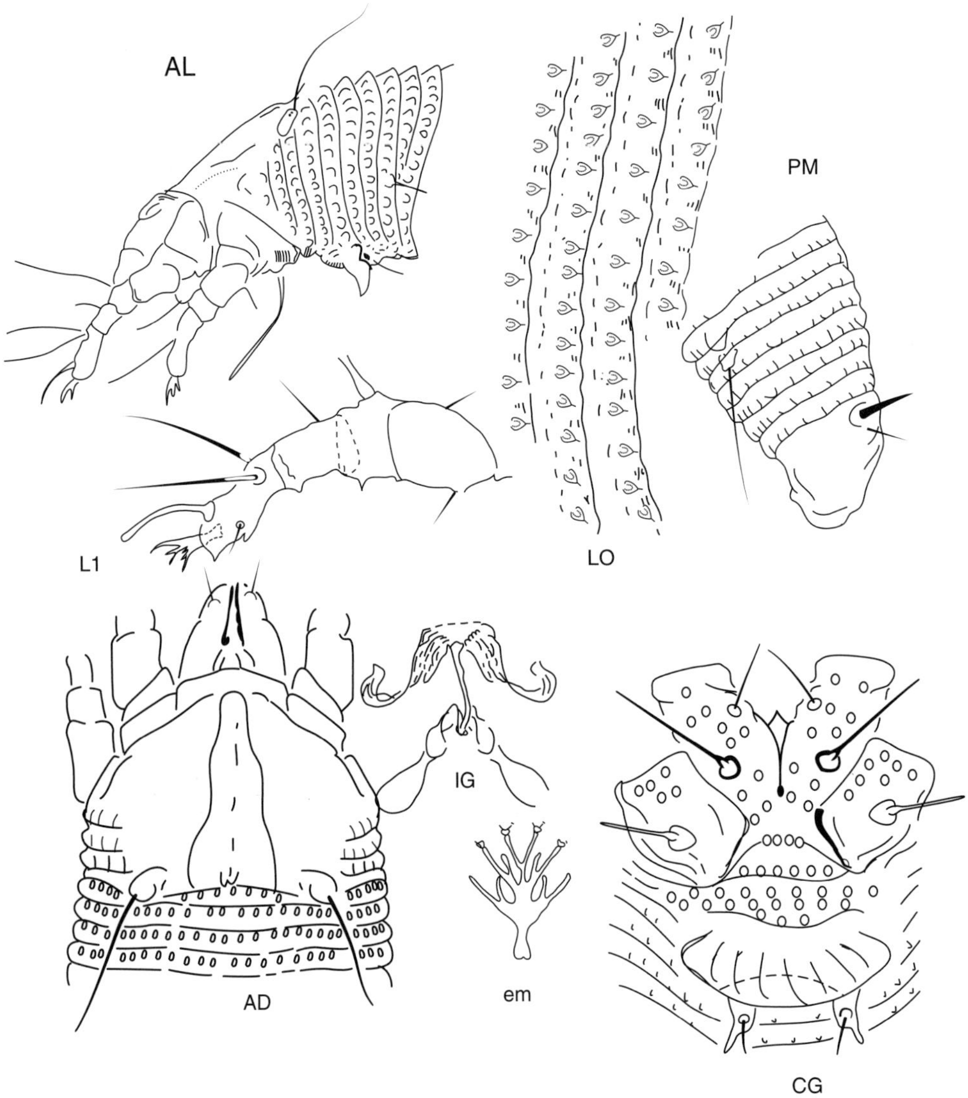

Fig. 16.24. *Aceria ecantyx* Keifer (from Keifer, 1969b). See Table 16.1 for key.

Diagnostic characteristics

FEMALE. The body is worm like, 160–200 μm long and light yellowish white in colour. The gnathosoma projects forward and is curved down. The prodorsal shield is subtriangular in dorsal view. The median shield line is faint, present only on the rear part of the shield; a short central line at the rear margin is enclosed by a pair of curved lines. The admedian lines are complete and slightly sinuate, gradually diverging to the rear. The submedian lines are absent. The shield has a pair of irregular longitudinal lines laterally, an ocellar spot above the rear coxa and three partial annuli below the scapular tubercles. The opisthosoma has about 52 annuli, with microtubercles that bear a point or small spinule each, and set just ahead of the margins of the annuli. The setae c_2 are 21 μm long, set on annulus 5; the setae *d* are 31 μm long, set on annulus 17; the setae *e* are 7 μm long, set on annulus 33; the setae h_1 are 8 μm long. The genital coverflap has about 10 longitudinal ribs that tend to be broadened, with more or less irregular microtubercles just ahead of the coverflap. The empodium is 3-rayed (Keifer, 1969b).

Geographical distribution

Portugal (Keifer, 1969b).

Host plants

Fagaceae: holly oak (*Quercus ilex* L.) (Keifer, 1969b).

Bio-ecology

The mites live in the buds of holly oak in Portugal (Keifer, 1969b).

Symptomatology and damage

Unknown.

Control

This is usually not necessary.

Aceria ilicis (Canestrini)

Common name

Unknown.

Diagnostic characteristics

FEMALE. The rear margin of the genitalia has a posterior curve or convexity. The empodium is 3-rayed (Jeppson *et al.*, 1975).

Geographical distribution

Europe, Near East, North Africa (de Lillo and Beron, 2004) and Iran (Khanjani and Haddad Irani-Nejad, 2006; Xue *et al.*, 2009a).

Host plants

Fagaceae: kermes oak (*Quercus aegilops* auct.) (= *Quercus coccifera* L.), holly oak (*Q. ilex* L.) (de Lillo and Beron, 2004), Vallonea oak (*Q. ithaburensis* Decne.), European oak (*Q. pedunculata* Ehrh.) (= *Quercus robur* L.) (Khanjani and Haddad Irani-Nejad, 2006; Xue *et al.*, 2009a).

Bio-ecology

The mite lives on oaks in Europe, the Near East, North Africa and Iran (Khanjani and Haddad Irani-Nejad, 2006; Xue *et al.*, 2009a; de Lillo and Beron, 2004).

Symptomatology and damage

A. ilicis causes erinea on the lower blade of the leaves of holly oak (Jeppson *et al.*, 1975). The papillae of these erinea are partially joined at the base and branched at the tip.

Control

This is usually not necessary.

Aceria triplacis Keifer (Fig. 16.25)

Common name

Unknown.

Diagnostic characteristics

FEMALE. The body is elongate and worm like, 190–205 μm long, 35 μm wide and dull dark yellowish in colour. The gnathosoma is curved down. The prodorsal shield is subtriangular in outline. The median line beginning just behind the front, is complete to the rear; the admedians are subparallel to the median, gradually diverging to the rear, sometimes slightly recurving. The submedian and lateral lines are obscure; there are some side granulations. The scapular tubercles are 16 μm apart, with the scapular setae (*sc*) 18 μm long, and strongly diverging to the rear. The coxae I have some large granulations and a moderately long sternal line at the junction. The opisthosoma has 55–60 annuli, is completely microtuberculate and the microtubercles are moderately large, low and rounded. The opisthosomal setae c_2 are 23 μm long, set on annulus 5; the setae *d* are 35 μm long, set on annulus 16; the setae *e* are 8 μm long, set on annulus 29; the setae *f* are 22 μm long, set on annulus 5 from the rear. The setae h_1 are 6.5 μm long. The genitalia are 10 μm long and 18 μm wide; the coverflap has 9–10 curving longitudinal ribs. The empodium is 3-rayed (Keifer, 1960).

Geographical distribution

New York State, eastern USA (Keifer, 1960; Keifer *et al.*, 1982) and China (Xue *et al.*, 2009b).

Host plants

Fagaceae: white oak (*Quercus alba* L.) (Keifer, 1960; Keifer *et al.*, 1982).

Bio-ecology

A. triplacis lives on white oak in New York State and in the eastern USA (Keifer, 1960; Keifer *et al.*, 1982).

Symptomatology and damage

The mites cause shallow erineum pockets on the leaves (Keifer, 1960; Keifer *et al.*, 1982).

Control

This is usually not necessary.

Fruit and nut trees other than the walnut group

This group includes *Aceria biopsidia* Keifer, *A. diospyri* Keifer, *A. ficus* (Cotte), *A. litchii* (Keifer), *A. jasmini* ChannaBasavanna, *A. mangiferae* (Hassan), *A. oleae* (Nalepa), *A. pistaciae* (Nalepa), *A. rossettonis* Keifer, *A. sheldoni* (Ewing), *A. stefanii* (Nalepa) and several other species. The eleven named species are described below.

Aceria biopsidia Keifer (Fig. 16.26)

Common name

Unknown.

Diagnostic characteristics

FEMALE. The body is worm like, 105–120 μm long, 30–35 μm wide and with a colour that is probably light yellowish white in life. The gnathosoma projects ahead and down. The prodorsal

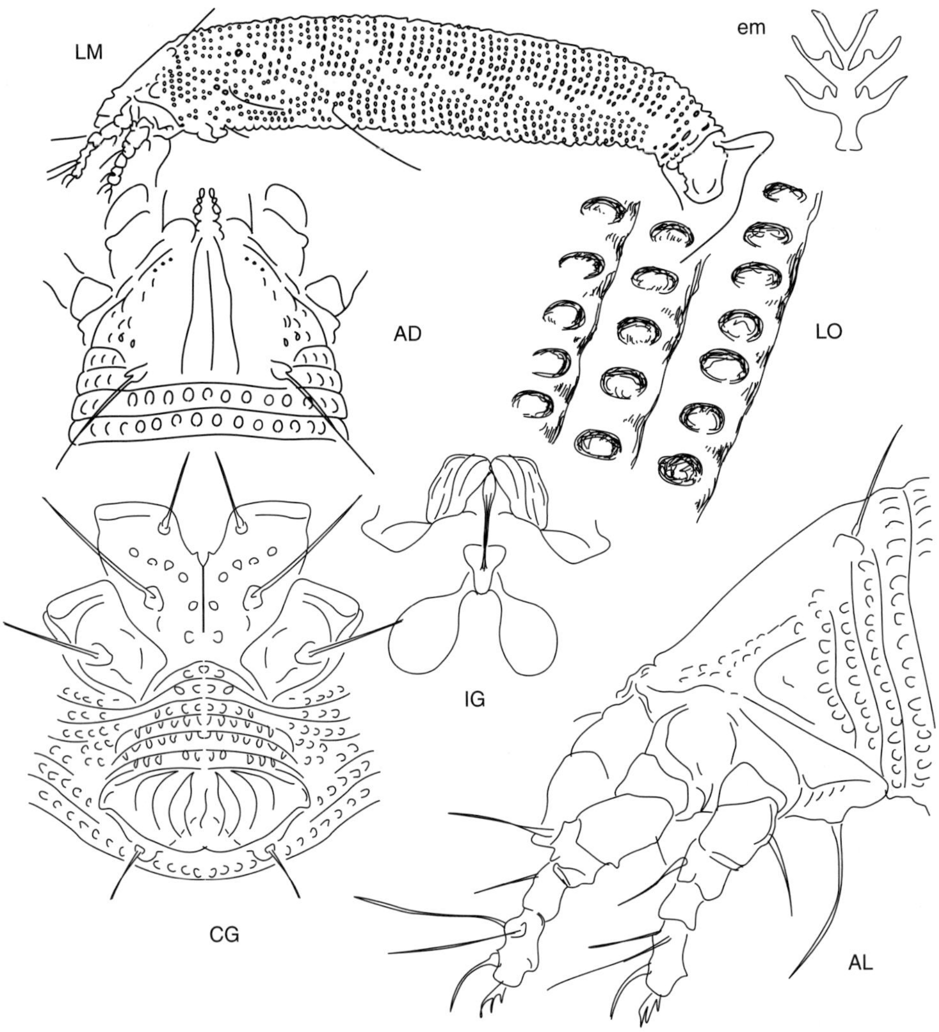

Fig. 16.25. *Aceria triplacis* Keifer (from Keifer, 1960). See Table 16.1 for key.

shield pattern consists of lines of dashes and granules, with the lines broken. The median line is present on the rear half as a broken line, ending at the rear shield margin in a dart-shaped mark. The admedian lines are complete from the base of the chelicera, broken, sinuate, with the lines curving inward at the rear 2/5, almost meeting medianly; the remainder of the admedian lines continue and diverge at the rear margin. The submedian lines of granules are more or less obscure due to a confusion of granules; three submedian lines are more or less apparent ahead of the scapular tubercles. The shield shows bands of granules and obscure lines laterally. The scapular tubercles are 16 μm apart, the scapular setae (*sc*) are 24 μm long and diverging. The coxae are ornamented with granulations. The opisthosoma has about 56 annuli, with the microtubercles rounded off outwardly, touching the annulus margins above and slightly ahead of the margins laterally and below. The microtubercles fade dorsally on the last 7–9 annuli. The opisthosomal setae c_2 are 12 μm long, set on annulus 5 behind the shield; the setae *d* are 37 μm long, set on annulus 18; the setae *e* are 5 μm long, set on annulus 32. The microtubercles of the last 5 annuli are faint, small and on the annulus margins with an anterior line from each. The setae h_1 are absent. The genitalia are 15 μm wide and 12 μm long; the coverflap has coarse granules basally, and about 12 longitudinal ribs; the seta is 6 μm long. The empodium is 4-rayed (Keifer, 1969b).

Geographical distribution

Venezuela (Keifer, 1969b).

Host plants

Myrtaceae: guava (*Psidium guajava* L.) (Keifer, 1969b).

Bio-ecology

A. biopsidia lives in buds of guava in Venezuela (Keifer, 1969b).

Symptomatology and damage

The mites cause bud damage (Jeppson *et al.*, 1975).

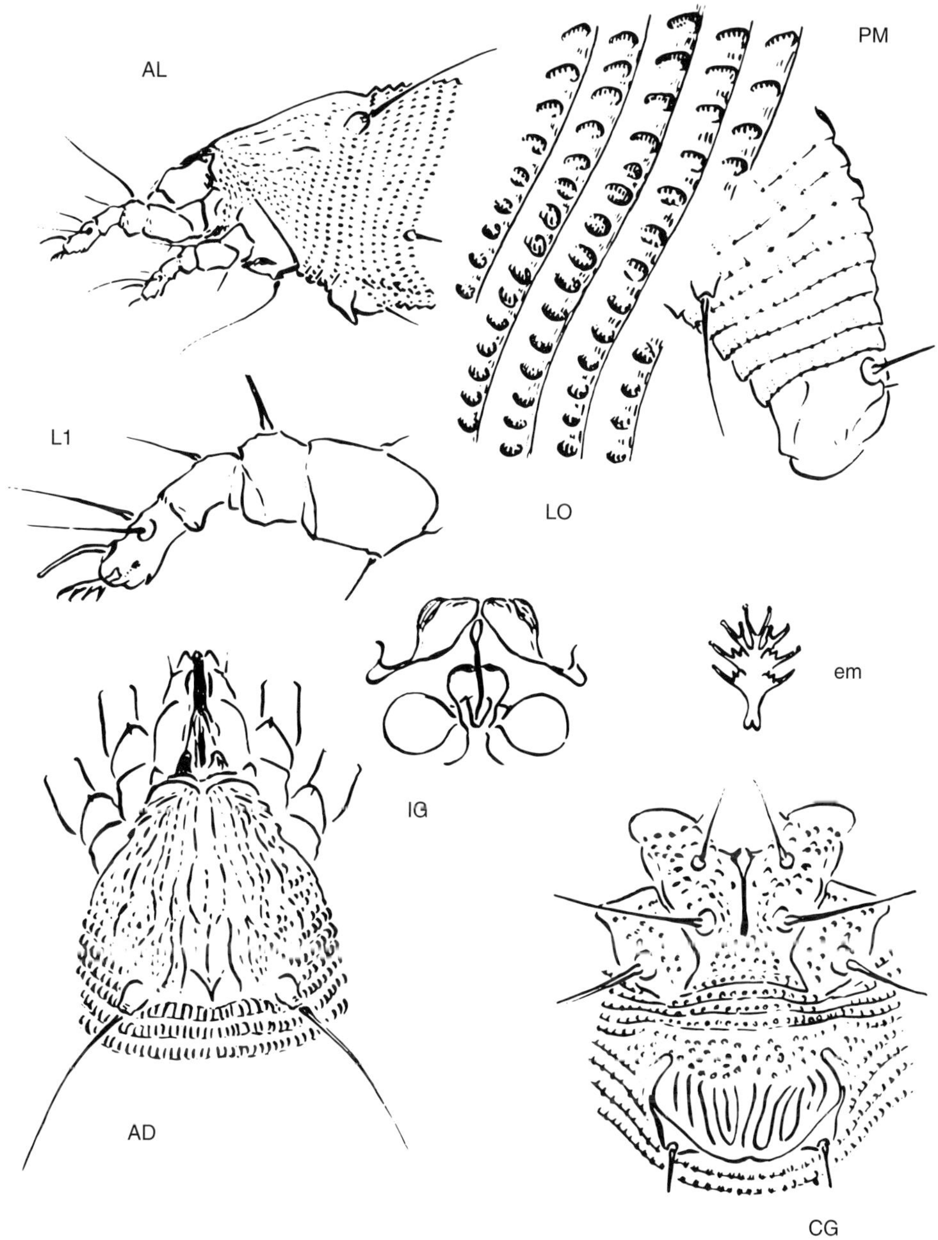

Fig. 16.26. *Aceria biopsidia* Keifer (from Keifer, 1969b). See Table 16.1 for key.

Control

Currently, this seems not to be necessary.

Aceria diospyri Keifer (Fig. 16.27)

Common name

Persimmon bud mite.

Diagnostic characteristics

FEMALE. The body is worm like, 140–170 μm long, 30 μm wide and light yellowish white in colour. The gnathosoma is somewhat downcurved. The prodorsal shield has a series of longitudinal dotted lines, and the sides are granular; the scapular tubercles are 17.5 μm apart, set on the rear margin, with the scapular setae (*sc*) 16 μm long, projecting backwards. The coxae are microtuberculate, with no definite sternal line. The opisthosoma has about 70 annuli, showing a ventrad reduction in number, and with microtubercles not pointed. The setae c_2 are 12 μm long, set on about annulus 7; the setae *d* are 40 μm long, set on about annulus 22; the setae *e* are minute, set on about annulus 40; the setae *f* are 13 μm long, set on about annulus 6 from the rear; the setae h_1 are present. The genitalia are 17.5 μm wide, 11.5 μm long, the coverflap has 12–14 furrows, and the seta is 8 μm long. The empodium is 5-rayed (Keifer, 1944).

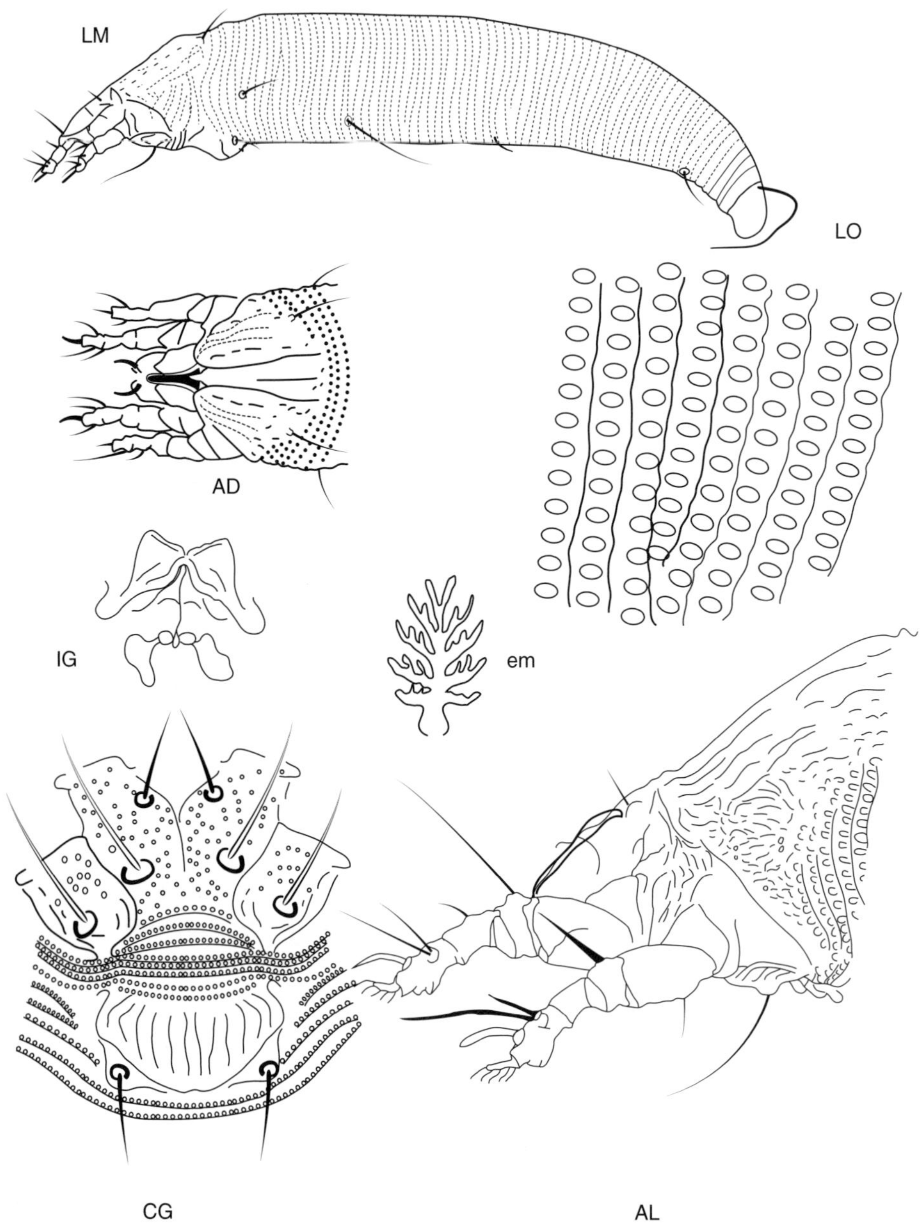

Fig. 16.27. *Aceria diospyri* Keifer (from Keifer, 1944). See Table 16.1 for key.

Geographical distribution

Brazil (Rossetto *et al.*, 1971) and southern California (Keifer, 1944; Jeppson *et al.*, 1975; Keifer *et al.*, 1982).

Host plants

Ebenaceae: persimmon (*Diospyros kaki* L.) (Keifer, 1944; Rossetto *et al.*, 1971; Jeppson *et al.*, 1975; Keifer *et al.*, 1982).

Bio-ecology

A. diospyri lives under fruit buttons and under the bud scales of persimmon.

Symptomatology and damage

The persimmon bud mite lives under the fruit buttons and bud scales of persimmon. It blackens the fruit surface and stem under the fruit buttons, resulting in premature fruit drop (Keifer, 1944; Rossetto *et al.*, 1971; Jeppson *et al.*, 1975; Keifer *et al.*, 1982).

Control

Rossetto *et al.* (1971) reported that in Brazil three spray applications of acaricides at monthly intervals, beginning after petal fall, controlled the mite infestation and increased fruit setting in the

'Fuyu' variety by 16, 10 and 25% in three consecutive crop seasons. Other reports on control of the mite are not known.

Aceria ficus (Cotte) (Fig. 16.28)

Common name

Fig bud mite.

Diagnostic characteristics

FEMALE. The body is worm like, 160–202 μm long and yellowish in colour. The prodorsal shield has a median line on the rear ²/₃, ending in a dart-shaped mark. The admedian lines are complete and recurve slightly at the rear. There is a fork in the first submedian line just in front of the scapular tubercles. The opisthosomal microtubercles are elliptical and rounded off. The coxae bear fine granules. The genital coverflap has about 8 longitudinal ribs. The empodium is 5-rayed (Jeppson *et al.*, 1975; Keifer *et al.*, 1982).

MALE. The male is similar to the female, and about 140 μm long.

Geographical distribution

Worldwide (Jeppson *et al.*, 1975).

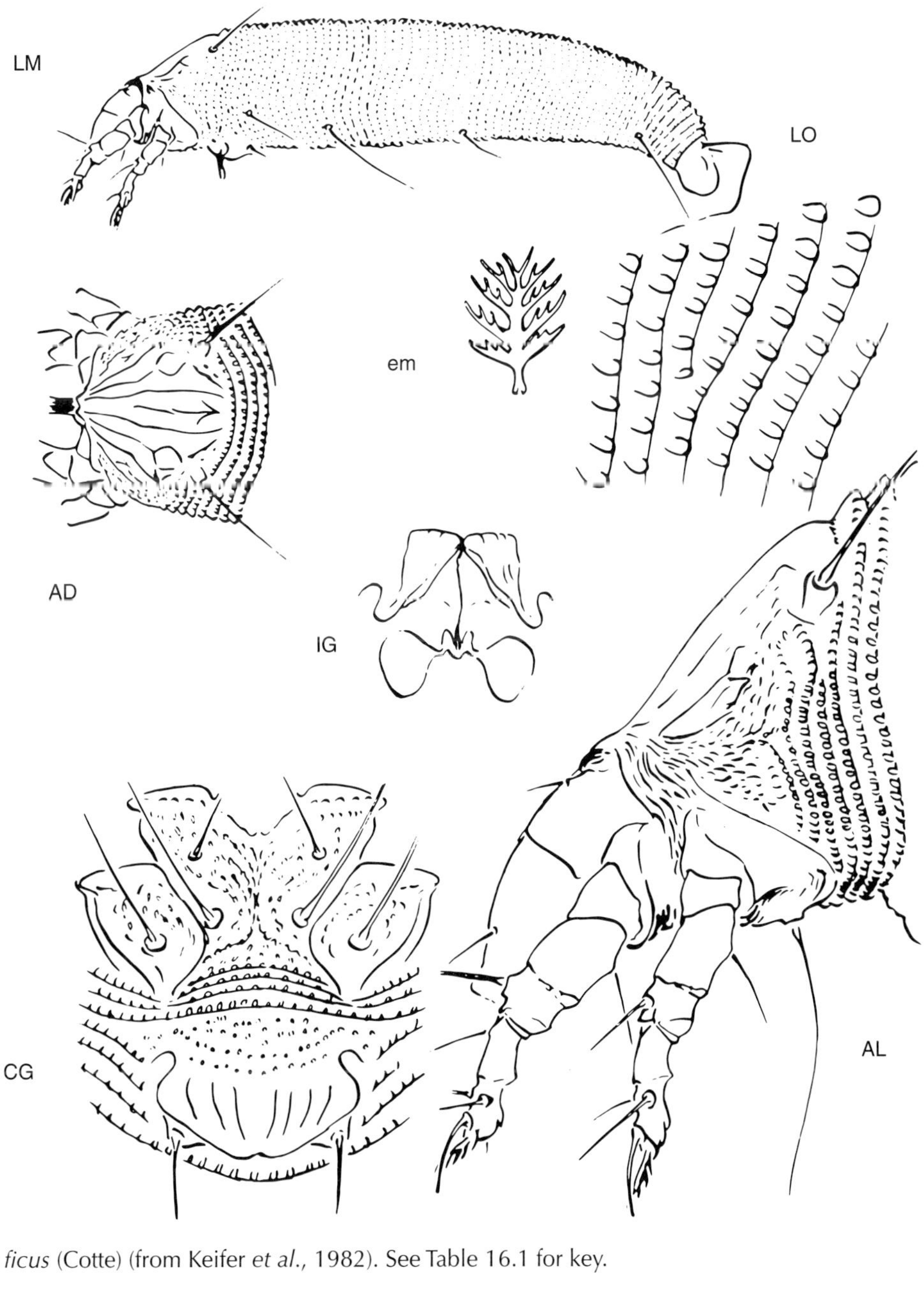

Fig. 16.28. *Aceria ficus* (Cotte) (from Keifer *et al.*, 1982). See Table 16.1 for key.

Host plants

Moraceae: fig (*Ficus carica* L.) (Cotte, 1920; Jeppson *et al.*, 1975).

Bio-ecology

The fig bud mite is a European species (Cotte, 1920) living on common fig and characterized by worldwide distribution (Jeppson *et al.*, 1975). It overwinters in and around buds of the fig, where it is possible to find all biological stages. In spring, bud breaking stimulates the mites to move to the stems and leaves, where they lay eggs and begin to reproduce. Oviposition mainly occurs on the lower leaf surface hairs, and the biological cycle from egg to adult takes 5–7 days. The females usually prefer buds and tender leaves, but are also present at high density on the undersurface leaf hairs of older leaves and under apical fruit scales. Optimum nitrogen and phosphorus nutrition induce a greater increase of mite populations, which also develop more rapidly on non-viruliferous plants (Baker, 1939; Flock and Wallace, 1955; Jeppson *et al.*, 1975; Keifer *et al.*, 1982). Abou-Awad *et al.* (1999b) found that the fig bud mites preferred immature leaves to mature leaves, while mature leaves were preferred for oviposition throughout the season.

Natural enemies include predatory mites and insects. Predatory mites include cheyletids (Baker, 1939; Flock and Wallace, 1955), phytoseiids – *Typhlodromus* sp. (Baker, 1939; Flock and Wallace, 1955), and *Phytoseiulus plumifer* (Canestrini *et* Fanzago) (Rasmy and El-Banhawy, 1974); stigmaeids – *Agistemus exsertus* (Gonzalez) (El-Halawany *et al.*, 1990a); and tydeids – *Pronematus ubiquitus* (McGregor) (Baker, 1939). Predatory insects include the black hunter, *Leptothrips mali* (Fitch) (Thripidae) and some Chrysopidae (Baker, 1939; Flock and Wallace, 1955).

Symptomatology and damage

A. ficus causes direct damage to fig and transmits *Fig mosaic virus* (FMV) (Serrano *et al.*, 2004). Feeding by mites carrying no virus damages epidermal cells, causing russeting or surface browning, bud blasting and impedes new growth; it also causes bud and leaf distortion, leaf chlorosis and, sometimes, defoliation of the entire tree. The mite does not produce galls, but under experimental conditions may stop new growth. In the field, it is impossible to observe these symptoms alone because the virus is present everywhere in all fig varieties. Hence, no evaluation can be made of the loss of fruit solely as a result of mite damage in the field. The female of *A. ficus* can produce virus symptoms in fig in less than 10 days of feeding (Baker, 1939; Ebeling and Pence, 1950; Jeppson *et al.*, 1975; Meyer Smith, 1981b; Keifer *et al.*, 1982).

The capacity of *A. ficus* to transmit FMV to the fig was first observed by Flock and Wallace (1955), and its role as vector has also been investigated by Vashisth and Nagaich (1968) and Proeseler (1969, 1972a). The virus may be transmitted by a single mite (immature instars and adults) after a feeding activity of 5 min on infected plants and within 5 min of feeding activity on healthy plants. Usually, the transmission of virus occurs within a few hours of acquisition. Virus acquisition is from buds and the lower surface of leaves showing symptoms at 20–30°C or 5–10°C. Moulting does not impede the ability to transmit the virus and the adult becomes inoculative from 6 to 10 days old. Virus transmission does not occur transovarially (Proeseler, 1969, 1972a). The first symptoms of infection on fig seedlings are observed 5–7 days after feeding by inoculative mites.

The disease is widespread in commercial fig orchards because of the horticultural practice of propagating from rooted cuttings. The first disease symptoms are chlorotic spots on tender leaves, followed by vein clearing and the development of chlorotic areas and foliar mosaic. The chlorotic spots are often bordered by a rust band. The fruits also bear mosaic symptoms and trees become stunted and bear leaves of reduced size.

The virus is graft transmissible and not sap transmissible. The leaf parenchyma cells of mosaic-infected fig plants show polymorphic double-membraned bodies 120–160 nm in size (Bradfute *et al.*, 1970; Plavsic and Milic, 1980; Appiano, 1982). The pathogen is graft transmissible to various moraceous plants, including many *Ficus* spp., *Cudrania tricuspidata* Bureau (Condit and Horne, 1933; Burnett, 1961, 1962) and *Morus indica* L. (Vashisth and Nagaich, 1965). The mite transmits the virus only to *Ficus carica* (Oldfield and Proeseler, 1996).

Control

Spray applications of acaricides reduce mite populations and increase the propagation of pathogen-free nursery stock. The production of plants by shoot tip culture helps to limit the problem (Muriithi *et al.*, 1982). Control measures are usually not applied.

Aceria litchii (Keifer) (Fig. 16.29)

Common name

Litchi erineum mite or lychee erinose mite.

Diagnostic characteristics

FEMALE. The body is worm like, 110–135 μm long, 30 μm wide and yellowish to reddish in colour. The gnathosoma is somewhat downcurved. The prodorsal shield has definite longitudinal central lines and is granular laterally; the scapular tubercles are 16 μm apart, set on the rear margin; the scapular setae (*sc*) are 17 μm long, projecting caudad. The coxae are strongly tuberculate; the setae I are ahead of the junction of coxae I. The opisthosoma has 55–60 annuli, all microtuberculate. The opisthosomal setae c_2 are 18 μm long, set on about annulus 8; the setae *d* are 40 μm long, set on about annulus 21; the setae *e* are 5 μm long, set on about annulus 32; the setae *f* are 12 μm long, set on about annulus 6 from the rear; the setae h_1 are present. The genitalia are 15 μm wide, 10 μm long; the coverflap has about 16 curved furrows, and the seta is 9 μm long. The empodium is 5-rayed (Keifer, 1943).

Geographical distribution

Australia (Cull and Paxton, 1983; Waite and McAlpine, 1992), Brazil (Raga *et al.*, 2008), China (Xu and Li, 1996), Florida (Dekle, 1957, 1960; Mead, 1993), Hawaii (Keifer, 1943; Nishida

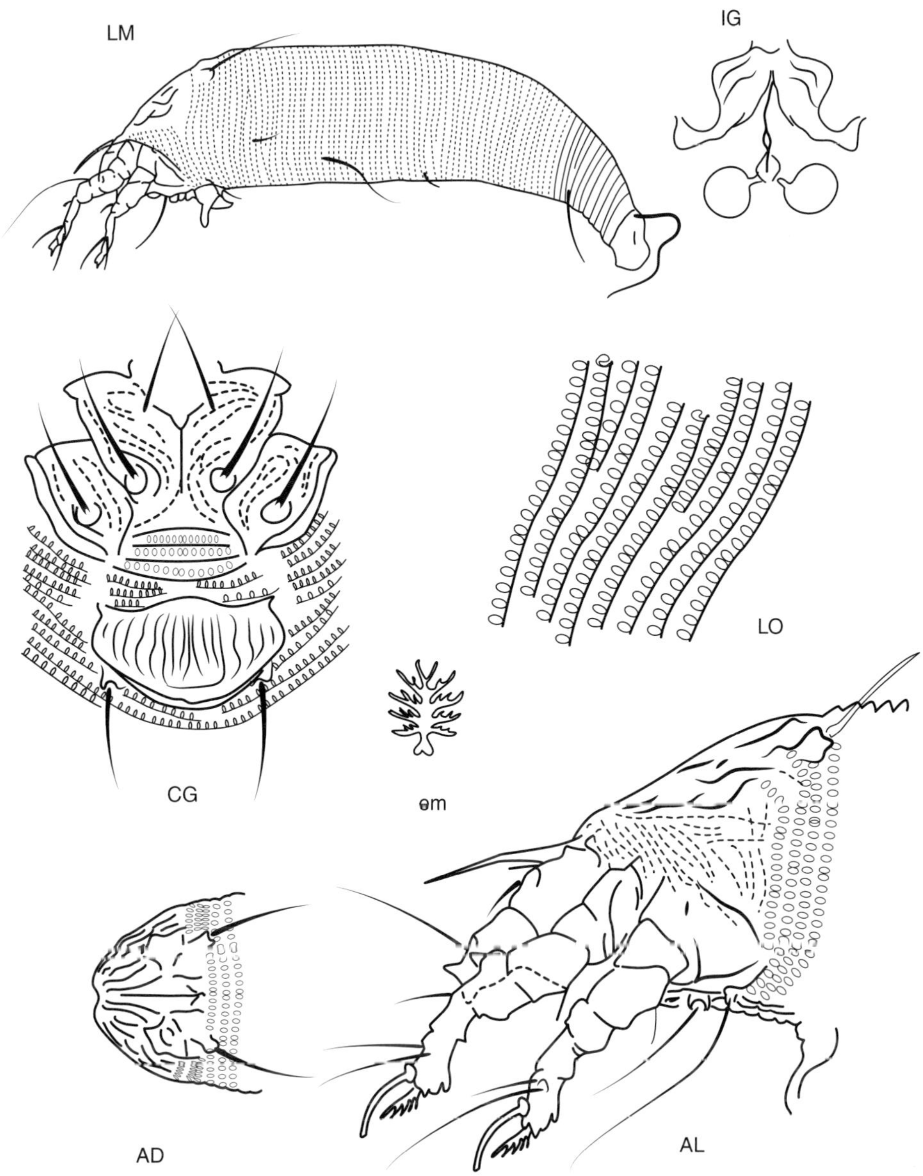

Fig. 16.29. *Aceria litchii* (Keifer) (From Keifer, 1943). See Table 16.1 for key.

and Holdoway, 1955; Jeppson *et al*., 1975), India (Chakrabarti and Mondal, 1982; Sharma *et al*., 1986), Japan (Mito and Uesugi, 2004), Pakistan (Alan and Wadud, 1963) and Taiwan (Huang, 1967; Huang *et al*., 1990).

Host plants

Sapindaceae: longan (*Euphoria longana* Lam.) (= *Dimocarpus longan* Lour.) (Huang, 1967), lychee (US litchi) (*Litchi chinensis* Sonn.) (Keifer, 1943; Jeppson *et al*., 1975).

Bio-ecology

The litchi erineum mite infests the lychee in various countries (Jeppson *et al*., 1975), and also infests the longan in Taiwan (Huang, 1967). The mite was introduced into Hawaii from Asia on *L. chinensis* many years ago (Návia *et al*., 2010a). Since then, two introductions of the mite have been registered into Florida from Hawaii on dormant trees. One occurred in 1955, and was eradicated in 1960 (Dekle, 1957, 1960), but there was a second infestation in 1993 (Mead, 1993). No litchi erineum mites have been found in Florida since 1993. The litchi erineum mite had previously been listed as a quarantine pest for Brazil but has recently been reported in the country (Raga *et al*., 2008).

Females lay eggs singly at the bases of the erineal papillae. The pre-ovipositional period lasts 1.5 days. Egg development takes on average 2.5 days; the first nymph develops for 2 or 3 days and the second nymph for 6 days. The life cycle takes from 13 to 19 days. The peaks of population density occur at each new flush of plant growth. High temperatures, high humidities and

heavy rainfall prevent mite development (Nishida and Holdoway, 1955; Alan and Wadud, 1963). In China, the mite develops 15–16 generations a year and overwinters as the adult. The population density increases with rising temperature and reaches a peak in June and July (Xu and Li, 1996). In southern Taiwan, the highest population density of mites has been observed in yellowish brown erinea, followed by reddish brown and black erinea (Wen *et al.*, 1991). The relationship between the population densities of *A. litchii* and the climatic factors showed that in India the eriophyoid populations were largest in April–May and smallest in November–December. Populations are significantly influenced by temperature and RH. A quadratic model based on maximum temperature, minimum temperature and RH explained about 94% of the population variation (Sharma *et al.*, 1986). In Taiwan, the fluctuations in population density appeared more closely correlated with rainfall than with temperature (Huang, 1967). In China, litchii erineum mite populations showed good spatial dependence over short distances in a range of 13–18 m. At the beginning of winter, the population of *A. litchii* showed a clumped distribution at high densities, and about 61% of the total variation was due to the spatial dependence found at this scale. The range and magnitude of the spatial structure of mite populations varies considerably from time to time (Zhou and Li, 2001).

In Queensland (Australia), up to 23% of honeybees (*Apis mellifera* L.) collected from flowering lychee trees severely infested with the litchi erineum mite carried live mites which were picked up as the bees foraged (Waite and McAlpine, 1992). New evidence further indicating honeybees as vectors of the lychee erinose mite has been found (Waite, 1999). In southern Taiwan, surveys using sticky plates showed that most dispersal occurred in March and August (Wen *et al.*, 1991). Xu and Li (1996) reported five natural enemies of the mite from China. In Queensland (Australia), 17 predators were collected associated with the pest mite, including ten species of phytoseiid mites, six other predaceous mites and a cecidomyiid, *Arthrocnodax* sp. In China, nine phytoseiid species were collected in lychee orchards (Waite and Gerson, 1994).

Symptomatology and damage

The litchi erineum mite develops on young leaves, flower buds, shoots and fruits. It induces the formation of a fine red erineum on the undersurface of the leaf, often involving and curling the whole leaf. The leaves dry and fall, resulting in stunted plants. The erineum usually appears first when the leaflets unfold. Initially, this erineum is greener than the normal plant, but within 2 days it becomes a thick silvery white. In 3 or 4 days, it changes from light brown to deep reddish brown. The leaves become twisted. An erineum of this colour is drying and in this phase the mite leaves it and moves to a newer part of the tree. Flower bud development is depressed, and severe damage occurs during growth (Jeppson *et al.*, 1975).

In Taiwan, *A. litchii* infests lychee and longan and is a serious pest of the former. Surveys of lychee trees showed that although the mite was present throughout the year, damage reached a peak (83.04%) in May and was lowest (20.84%) in September. The annual average ranged from 36.37 in the most northerly of the districts surveyed to 55.17 in the most southerly (Huang, 1967). In southern Taiwan, 11.11–73.59% (24.59% on average) of lychee plants are damaged by mite (Wen *et al.*, 1991).

Control

Control is provided by spray applications of acaricides and cultural practices.

CULTURAL PRACTICES. Control of the litchi erineum mite is expensive and difficult, including pesticide application and the pruning of trees that can reach 15-20 m in height. Consequently, the introduction of the litchi erineum mite into Brazil could discourage lychee production that was in the process of expansion (Návia *et al.*, 2010a). In southern Taiwan, a combination of pruning and spraying with an acaricide such as sulfur is used as an effective method of control (Wen *et al.*, 1991).

CHEMICAL CONTROL. A number of active ingredients have been applied in various countries. Among these the following have been recorded: azocyclotin, bromopropylate, carbaryl (Mishra, 1980; Prasad and Bagle, 1981), chlorobenzilate (Mathur and Tandon, 1974; Mishra, 1980; Prasad and Bagle, 1981; Prasad and Singh, 1981), cyhexatin (Mishra, 1980; Prasad and Bagle, 1981), dicofol (Mathur and Tandon, 1974; Mishra, 1980; Prasad and Bagle, 1981; Prasad and Singh, 1981; Sharma, 1985), dimethoate, methyl demeton (Sharma and Rahman, 1982), monocrotophos (Prasad and Bagle, 1981; Sharma and Rahman, 1982), phosalone (Prasad and Bagle, 1981), phosphamidon (Sharma and Rahman, 1982), propargite (Mishra, 1980; Prasad and Bagle, 1981), sulfur-bridged compounds (Prasad and Bagle, 1981) and sulfur WP (Wen *et al.*, 1991).

Aceria jasmini ChannaBasavanna

Common name

Jasmine erineum mite.

Diagnostic characteristics

FEMALE. The prodorsal shield lacks lines or a pattern. The empodium is 4-rayed (ChannaBasavanna, 1966).

Geographical distribution

India (ChannaBasavanna, 1966).

Host plants

Oleaceae: jasmines (*Jasminum auriculatum* Vahl, *J. pubescens* (Retz.) Willd.) (ChannaBasavanna, 1966).

Bio-ecology

A. jasmini lives on jasmines in India (ChannaBasavanna, 1966). Its bio-ecology has not been documented.

Symptomatology and damage

The jasmine erineum mite causes felt-like hairy outgrowths on leaf surfaces, on stems and on flower buds. Damage consists of stunted growth and flower suppression (ChannaBasavanna, 1966; Jeppson *et al.*, 1975). *A. jasmini* alters the nutrient content of its host plant. The total carbohydrate in leaves of *J. auriculatum* infested by the mite was 21% less than in non-infested leaves, and soluble sugars were reduced by 31% (Rajagopal *et al.*, 1970).

Control

Currently, this seems not to be necessary.

Aceria mangiferae (Hassan) (Fig. 16.30)

Common name

Mango bud mite.

FEMALE. The body is worm like, up to 225 µm long and yellowish in colour. The prodorsal shield bears noticeably elongate scapular tubercles behind the shield. The prodorsal shield is indistinct anteriorly but between the scapular tubercles there is a pattern of strong longitudinal ridges. The dorsal microtubercles behind the shield are elongate. The opisthosomal microtubercles are elongate dorsally and rounded ventrally. The genital coverflap has longitudinal ribs arranged in two uneven ranks. The empodium is 6-rayed (Jeppson *et al.*, 1975; Keifer *et al.*, 1982).

MALE. Similar to female, but smaller. The empodium is 6-rayed (Návia and Flechtmann, 2000a).

Geographical distribution

Southern Asia (Jeppson *et al.*, 1975), Brazil (Rossetto *et al.*, 1967; Návia and Flechtmann, 2000a), Costa Rica (Abreu *et al.*, 1987), Egypt (Hassan, 1944; Sayed, 1946a), El Salvador (Ochoa *et al.*, 1994), Florida (Denmark, 1983), Guatemala, Honduras (Ochoa *et al.*, 1994), India (ChannaBasavanna, 1966), Iran (Arbabi *et al.*, 1999; Xue *et al.*, 2009a), Israel (Sternlicht and Goldenberg, 1976), Mexico (Barajas, 1998), Nicaragua, Panama (Ochoa *et al.*, 1994), Spain (Peña and Ferragut, 1994), South Africa (Meyer Smith and Craemer, 1999) and Venezuela (Doreste, 1984a).

Host plants

Anacardiaceae: mango (*Mangifera indica* L.) (Hassan, 1944; Sayed, 1946a).

Bio-ecology

The mango bud mite was reported the first time on mango in Egypt (Hassan, 1944, Sayed, 1946a) and may be misidentified with *Tegonotus mangiferae* (Keifer) and *Metaculus mangiferae* (Attiah), all living on the same host plant. *A. mangiferae* lives under the terminal buds of the mango and is common in many areas of its cultivation (Jeppson *et al.*, 1975). It lays eggs where it feeds in groups of two or three, or even more, and insemination occurs soon after female emergence; unfertilized females generate only male offspring, while both males and females are generated by fertilized females (Abou-Awad *et al.*, 2011a). Abou-Awad (1981a) reported that in the laboratory at 25–27°C and 70% RH, the life cycle of *A. mangiferae* takes at least 15 days, and that in the field, the mite develops about eight generations a year in Egypt. A later report by Abou-Awad *et al.* (2011a) set the number of generations a year higher, at 16 and 17 over 2 successive years (2003 and 2004). Abou-Awad *et al.* (2011a) also reported that in the laboratory at 25, 30 and 35°C, the female life cycle took 9.68, 7.95 and 6.94 days, respectively, and male development took 9.20, 7.10 and 5.81 days, respectively. The fecundity is 16.92, 19.92 and 26.70 eggs during a reproduction period of 13.60, 13.86 and 15.53 days, respectively. At 25°C and 60% RH, the net reproduction rate (R_0), mean generation time (T), intrinsic rate of increase (r_m) and finite rate increase (e^{rm}) are 10.97, 21.50, 0.111 and 1.11, respectively. At 30°C and 55% RH, the same parameters are 13.47, 17.73, 0.146 and 1.15, respectively, while at 35°C and 50% RH, they are 18.46, 17.35, 0.168 and 1.18, respectively. Under these conditions, the sex ratio ranges from 2.00 to 2.75 females to one male (Abou-Awad *et al.*, 2011a). However, Mahgoob (2006) reported that temperature and RH had varying but not significant effects on the mite population in various Egyptian governorates.

There are differing reports on the periods of peak population density of *A. mangiferae* in Egypt. Abou-Awad (1981a) reported population peaks in late May and October, while Wahba *et al.* (1986) and Mahgoob (2006) reported that the highest mite population occurred during September–January or February, and the lowest during March–June. The population peaks also varied by cultivar, which may explain some of these discrepancies. On cultivar 'Alphonso' in Egypt, the first peak occurs in May, the second peak in August and a third peak in November (Abou-Awad *et al.*, 2011a). In the same country, on 'Timour' cultivar the mite developed two annual population peaks, one in February and the other in September, according to the climatic conditions, time of budding and the new growth cycle in the autumn (Zaher and Osman, 1970). Observations on numerical changes in the vertical distribution of *A. mangiferae* on the cultivar 'Alphonso' showed that over 2003 and 2004, mite density exhibited a gradual increase and decrease from February, reaching its peak during May, October and May and August, and then started declining from November onwards. At the top of the trees, the terminal and lateral buds – especially the terminal buds – had significantly more mango bud mites than the middle and bottom parts of the trees (Abou-Awad *et al.*, 2011a).

In Egypt, the predator most frequently associated with the mango bud mite is the phytoseiid *Amblyseius swirskii* Athias Henriot (Abou-Awad, 1981a). Also reported is the cheyletid *Cheletogenes ornatus* (Canestrini *et* Fanzago) (Bharadwaj and Banerjee, 1973).

Symptomatology and damage

A. mangiferae damages severely young trees, causing stunted twigs and a large number of crowded buds (ChannaBasavanna,

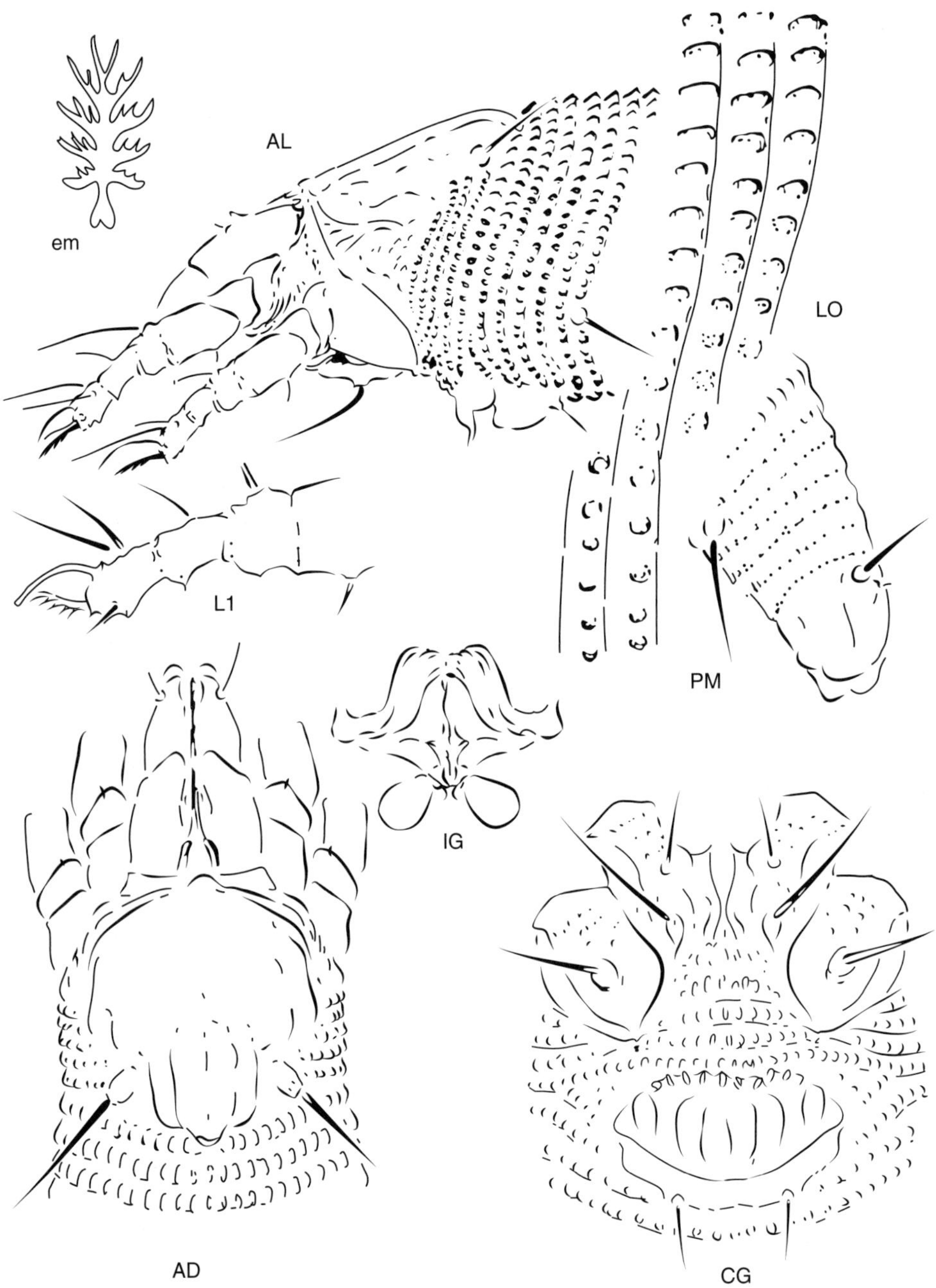

Fig. 16.30. *Aceria mangiferae* (Hassan) (from Keifer *et al.*, 1982). See Table 16.1 for key.

1966). Mite feeding brooms the twigs and induces bud proliferation, causing marked damage to young plants (Jeppson *et al.*, 1975). In Venezuela, the mite damages both old and young mango trees, and as a result is a pest of economic importance in nurseries (Doreste, 1984a), as well as in several other countries that cultivate mango, such as Israel (Wysoki, 1997), Egypt (Mahgoob, 2006) and Brazil (Sao Jose *et al.*, 2000). Attacks of *A. mangiferae* on mango in Florida result in malformed inflorescences and appear to be associated with *Fusarium* spp.; together, these were thought to be responsible for the formation of floral and foliar galls called 'mango malformation disease' (Denmark, 1983; Ochoa *et al.*, 1994). However, further investigations have shown that *A. mangiferae* does not produces mango malformation itself, but may carry *Fusarium mangiferae* Britz, *F. sterilihyphosum* Britz and *F. proliferatum* (Matsush.) Nirenberg ex Gerlach and Nirenberg, which are the causal agents of mango malformation (Kumar and Beniwal, 1992a,b; Peña *et al.*, 2005; Marasas *et al.*, 2006).

Studies in India on the effects of the mite on buds showed that variations in the percentage of malformed panicles on trees (0–81%) related to the variety of mango concerned, and that the population density in the buds was not significant (Bindra and Bakhetia, 1969; Reis *et al.*, 1970). In Israel, flower buds of 15 varieties examined monthly for malformation of the

inflorescence caused by *A. mangiferae* highlighted that some varieties ('Faizanson', 'Irwin', 'Maya', 'Palmer' and 'Dasheri') were infected only occasionally, and that not all infected trees had malformation. Varietal differences for resistance following inoculation with *Gibberella fujikuroi* (Sawada) Wollenw. have been observed (Israel Ministry of Agriculture,1975). *A. mangiferae* has no direct effect on inflorescence malformation, but its interaction with another biotic factor such as a fungus is not excluded (see above for the involvement of *Fusarium* spp.). Varieties only slightly infested throughout the year ('Carrie', 'Faizanson', 'Irvin', 'Maya', 'Mistakawi', 'Palmer', 'Rogni', 'Early Gold', 'Gedong', 'Mulgoba') were free of mites (Sternlicht and Goldenberg, 1976).

Control

Control is provided by the cultivation resistant varieties, cultural practices and spray applications of acaricides. Results from biological control studies are unclear.

PLANT RESISTANCE. The susceptibility to mite infestations of several mango varieties was investigated in various countries. Studies in Egypt on the relationships between the malformation of inflorescences and presence of *A. mangiferae* on five mango varieties show that 'Zebda' and 'Mabrouka' are more and less susceptible to the mite, respectively. 'Taimour', 'Company' and 'Hindi' varieties are moderately susceptible. 'Zebda' is the only variety that does not have malformed inflorescences. The other four varieties had the following order of susceptibility: 'Company', 'Taimour', 'Hindi' and 'Mabrouka'. There was an obvious relationship between mango malformation and mite populations on the 'Mabrouka', 'Hindi' and 'Taimour' varieties. No such relationship has been observed on the other two varieties (Wahba *et al.*, 1986). Further investigations in the same country on 11 mango varieties show that the cultivars 'Zebda', 'Taimour', 'Ewais' and 'Arnaba' are the most susceptible, while 'Mabrouka', 'Company', 'Excellent Succari' and 'White Succari' are the least susceptible. 'Mesk', 'Geolck' and 'Alphonse' varieties have moderate susceptibility to mite infestation. Buds of some highly mite-susceptible cultivars have lower values of total carbohydrates and total soluble sugars than some moderately susceptible or susceptible cultivars. Phenol contents are higher in most of the susceptible cultivars, and lower in most of the moderately susceptible and low susceptibility cultivars. The total soluble proteins and amino acids were higher in some susceptible cultivars. Susceptibility was positively correlated with the total soluble proteins and amino acids, and negatively correlated with the total carbohydrates and total soluble sugars. Correlations with the phenol content showed no clear trend. The mango cultivars had varying degrees of floral malformation, with 'Company', 'Alphonse' and 'Geolck' the most susceptible, and 'Zebda' the most resistant. There is a negative correlation between the mango bud mite and the incidence of malformation disease caused by *G. fujikuroi*. Zebda harboured the highest number of mites but exhibited the lowest percentage of malformation, whereas the opposite was observed with 'Company' (Mahgoob, 2006).

The incidence and severity of mango malformation on cultivar 'Tommy Atkins' was evaluated in Brazil. About 10% of the total area cultivated with mango trees was selected in the main counties of Bahia State, and results indicated that flower malformation amounted to 54% in Bahia State mango orchards, and that there was 17% severe malformation (Sao Jose *et al.*, 2000).

CULTURAL PRACTICES. Cultural practices involve the pruning of infested branches, usually associated with spray application of chemicals, e.g. oil and dicofol. Application of this method reduces mite populations and increases fruit yield in Egypt, where as well as in India, malformed inflorescences are a greater economic problem than in Florida (Denmark, 1983). The same technique has been proposed for use in Spain, with intervals between treatments of 7 days (Peña and Ferragut, 1994).

BIOLOGICAL CONTROL. The acaropathogen *Hirsutella thompsonii* Fisher has been investigated in Mexico (Barajas, 1998) but no information on the further use of the fungus is available.

CHEMICAL CONTROL. A number of pesticides have been employed in various areas of mango cultivation. They include abamectin (Abou-Awad *et al.*, 2011a), binapacryl (Srivastava, 1974), chlorfenapyr (Abou-Awad *et al.*, 2011a), chlorobenzilate (Rai *et al.*, 1966; Srivastava, 1974), diazinon, dichlorvos (Srivastava, 1974; Doreste, 1984a), dicofol (Rai *et al.*, 1966; Srivastava, 1974; Mishra *et al.*, 1996), dimethoate (Srivastava, 1974; Doreste, 1984a), dinocap (Mishra *et al.*, 1996), endosulfan (Rai *et al.*, 1966; Srivastava, 1974), methoxyfenozide (Abou-Awad *et al.*, 2011a), monocrotophos, phosphamidon (Srivastava, 1974; Doreste, 1984a) and sulfur (Doreste, 1984a; Mishra *et al.*, 1996; Abou-Awad *et al.*, 2011a). Pesticides and plant hormones (miraculan, planofix) have been positively employed to increase the growth parameters of the plant (Mishra *et al.*, 1996). Among the fungicides that have been tested are aluminium phosphide (Bharadwaj and Banerjee, 1973), and the biopesticides include azadirachtin (Abou-Awad *et al.*, 2011a).

Aceria oleae (Nalepa) (Fig. 16.31)

Common name

Olive bud mite.

Diagnostic characteristics

FEMALE. The body is small and worm like. The prodorsal shield has a rather obscure central pattern with lines only on the rear half. A median line and two admedian lines can occur at the rear prodorsal shield margin, with a curved line extending back just centrad of the scapular tubercles. The coxae have granules, and the sternal line is short and bifurcate at both extremities. The opisthosomal microtubercles are subcircular and rounded off. The genital coverflap has two transverse lines of granules across the base, with longitudinal ribs that are irregular and tending to be broad; there are about 12 ribs. The empodium is 4-rayed (Jeppson *et al.*, 1975).

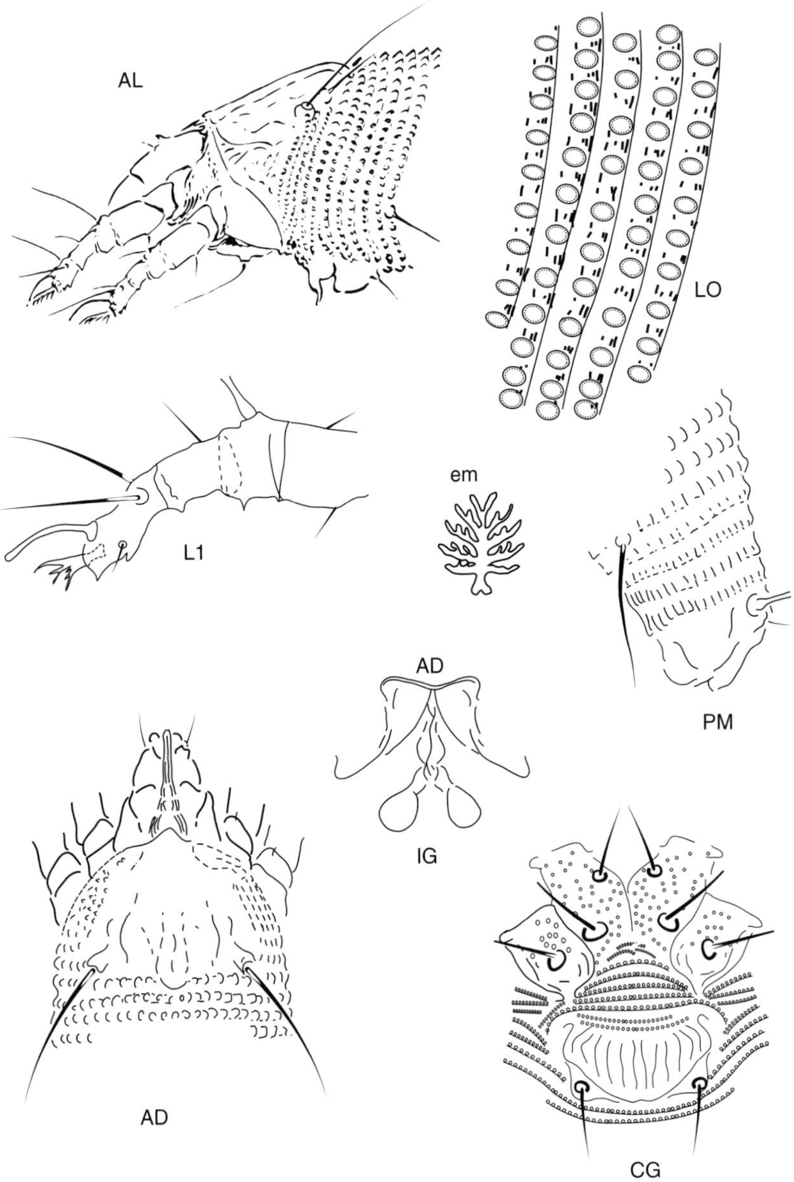

Fig. 16.31. *Aceria oleae* (Nalepa) (from Jeppson *et al.*, 1975). See Table 16.1 for key.

Geographical distribution

Cyprus (Georghiou, 1959), Egypt (Avidov and Harpaz, 1969), Israel (Harpaz, 1955), Italy (Graniti, 1954), Jordan (Castagnoli, 1981), Greece (Kavadas, 1927), Iran (Sepasgosarian, 1973; Xue *et al.*, 2009a), Libya (Avidov and Harpaz, 1969), Morocco (Saba, 1973b), Spain (Castagnoli, 1981) and South Africa (Meyer Smith, 1981b).

Host plants

Oleaceae: olive (*Olea europaea* L.) (Jeppson *et al.*, 1975; Meyer Smith, 1981b).

Bio-ecology

The olive bud mite lives on all varieties of olive in the Mediterranean region (Jeppson *et al.*, 1975) and in South Africa (Meyer Smith, 1981b). The mite is common in warm areas and may be associated in the same host plant with other eight eriophyoid mites: *Aceria olivi* Zaher *et* Abou-Awad, *Aculops benakii* (Hatzinikolis), *Aculus olearius* Castagnoli, *Shectchenkella oleae* (Nachev), *Oxycenus maxwelli* (Keifer), *Oxycenus niloticus* Zaher *et* Abou-Awad, *Tegolophus hassani* (Keifer) and *Ditrymacus athiasella* Keifer. *A. oleae* overwinters as the female on the underside of leaves. In spring, the females move to buds and young leaves and start to reproduce. In the flowering phase, the mite populations migrate to the buds and the calyxes and ovaries of the flowers and then to the young fruits (maximum diameter of 0.5 cm), occupying the lower side and the part protected from the sepal residue. As the fruit grows, the mite populations move to the leaves, where they live during the summer. The population density is very low during winter and higher during the inflorescence period.

At 21–25°C, the developmental period from egg to adult takes 13.9 days, the ovipositing period lasts a maximum of 15–16

days and the number of generations a year ranges from 12 to 15 (Hatzinikolis, 1971). In Egypt, the mite develops a maximum of 12 generations a year (Abou-Awad *et al.*, 2005). At 20, 25 and 31°C, the female life cycle lasts 13.86, 9.36 and 6.58 days, respectively, while the male develops faster. Insemination occurs soon after female emergence. Fertilization is essential for the maximum reproduction of females, as unfertilized females deposit fewer eggs than fertilized females. Unfertilized females also produce only male offspring, while both males and females are produced by fertilized females. The oviposition period is shortest at 20°C and longest at 31°C. Females lay an average of 7.4, 15.0 and 21.4 eggs during an average oviposition period of 7.60, 10.00 and 10.40 days and survive for 3.40, 2.20 and 1.40 days before death at 20, 25 and 31°C, respectively. The life table parameters show that at 31°C and 80% RH, the *A. oleae* population increases 16.70 times in a generation time of 13.50 days (Abou-Awad *et al.*, 2005).

Populations of *A. oleae* are affected by climatic conditions, predation, shady and sunny zones, leaf age and vertical distribution (Abou-Awad *et al.*, 2005). Both high temperature and RH increase the rate of development and stimulate mite reproduction (Al-Azzazy, 2000; Abou-Awad *et al.*, 2005). Natural enemies include phytoseiid, stigmaeid and tydeiid mites. *Neoseiulus cydnodactylon* (Shehata *et* Zaher) and *Agistemus olivi* Romeih have been reported in Egypt (Abou-Awad *et al.*, 2005).

Symptomatology and damage

The simultaneous presence of at least eight to nine eriophyoid species on the same host plant in the same region does not facilitate correct description of the symptomatology of and damage caused by *A. oleae*. However, analysis of the bio-ecology of the different species compared with the phenology of the host plant helps to explain the situation, and indicates that there is a succession of plant organs infested and a probable rational use of the trophic resource.

A. oleae damages leaves, inflorescences and fruits of olive trees. Attacks to young leaves cause irregularly lobed margins showing humps and curvings on the rachis. Expanded leaves bear hollows on the underside of the leaf and a corresponding chlorotic protuberance on the upper leaf surface. Mature leaves may have irregular greenish patches due to the destruction of silvery stellate hairs. These patches become brown as necrosis increases. At high density, the mites also feed on the leaf margins, inducing deformities (Harpaz, 1955; Hatzinikolis, 1969b). External morphological changes correspond to internal alterations in the epidermis and mesophyll (Graniti, 1954). Dead cells scattered on the upper leaf surface induce a rusty coloured leaf tissue. The upper layer cells are small, sometime triangular, with irregular walls and granular contents (Harpaz, 1955).

The feeding activity of the mite causes russeting and splitting of flower buds, which wither and drop; the flowers and the entire inflorescence drop early, as well as the young fruits. The mite infests fruits with a maximum diameter of 0.5 cm, and fruit damage first appears as silvering and afterwards as browning, and at the end as deformation. The silvered areas show brown droplets of exuded gum. A high density of mite populations may be observed at the stem end of the fruit or under the sepal remains and most damage occurs on the stem. Usually, the fruits that develop early are very little affected (Harpaz, 1955; Avidov and Harpaz, 1969; Hatzinikolis, 1969b). A higher density of populations usually occur in later spring, when it is possible to find thousands of individuals on inflorescences that are able to affect fruit setting.

Abou-Awad *et al.* (2005) report that field and laboratory studies indicate that the olive bud mite can be disastrous on shrubby and young olive trees, but the situation is not clear cut and needs further investigation. The degree of sensitivity of olive trees to the pest varies according to cultivar (Elhadi and Birger, 1999).

Control

Severe damage to production in the field or to young plants in the nursery does not usually occur. However, it is sometimes necessary to treat plants with a high density of mites in early spring and also to treat plants being grown for the production of table olives. In this case, it is possible to use sulfur (Avidov and Harpaz, 1969). Field studies in Israel show that spray applications of amitraz or abamectin reduce fruit damage by more than 85%. To prevent fruit damage, one early spray before flowering of either of the two chemicals is sufficient, but to prevent leaf damage, more than one spray is needed (Elhadi and Birger, 1999). One spring treatment with abamectin is the most successful method against the mites (Abou-Awad *et al.*, 2005).

Aceria pistaciae (Nalepa) (Fig. 16.32)

Common name

Pistachio bud mite.

Diagnostic characteristics

FEMALE. The body is worm like and up to 170 μm long. The prodorsal shield has central marked longitudinal lines. The median lines extends on to the rear half. The admedian lines are nearly complete and gradually diverge. The first submedian line runs

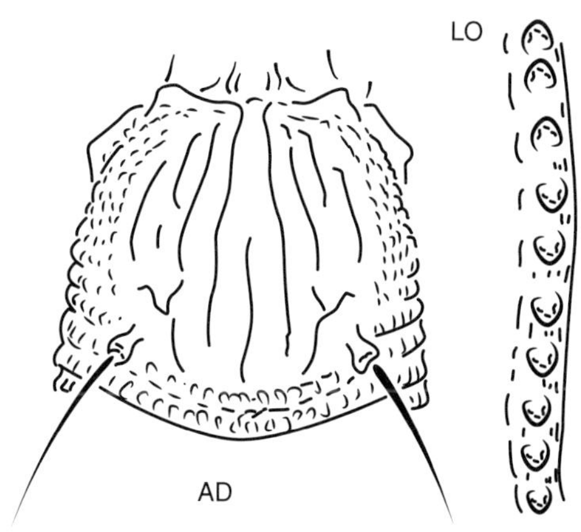

Fig. 16.32 *Aceria pistaciae* (Nalepa) (from Jeppson *et al.*, 1975). See Table 16.1 for key.

back towards the scapular tubercle and ends in a diagonal line extending inwardly across in front of the scapular tubercle. There are two more short submedian lines, and lateral granules. The opisthosoma has subcircular microtubercles, slightly ahead of the annuli margins, and pointed. The coxae have coarse granules. The genital coverflap has about 6 rather broad longitudinal ribs. The empodium is 4-rayed (Jeppson *et al.*, 1975).

Geographical distribution

West Palaearctic region (de Lillo and Beron, 2004), Iran (Mehrnejad and Daneshvar, 1991; Xue *et al.*, 2009a) and the Nearctic region (Jeppson *et al.*, 1975).

Host plants

Anacardiaceae: *Pistacia* spp., *P. mutica* Fischer *et* C.A. Mey, *P. vera* L. and *P. terebinthus* L. (Mehrnejad and Daneshvar,1991; Xue *et al.*, 2009a).

Bio-ecology

A. pistaciae inhabits the leaves, buds and flowering and fruit clusters of pistachio species. In Iran, it has been collected on *P. mutica*, *P. vera* and *P. terebinthus* (Mehrnejad and Daneshvar, 1991; Xue *et al.*, 2009a). Its bio-ecology is not well documented.

Symptomatology and damage

The pistachio bud mite causes flower stalk brooming and some leaf deformation on certain pistacio species (Jeppson *et al.*, 1975). An average density of <21 mites on a cluster of either flowers or fruits causes fruit dropping in the early spring and some leaf deformation in June. Heavy infestation causes deformation of the twigs, buds, leaves and flower clusters, with the male flowers modified and female flowers not fertilized (Mehrnejad and Daneshvar, 1991). Further investigations in Iran suggest that this mite is a minor pest in the pistachio orchards of Kerman province (Mehrnejad and Ueckermann, 2001). Witches' brooms on *P. terebinthus* are caused by *A. pistaciae*. Swellings at the lower ends of affected twigs have been anatomically compared with the structure of normal twigs, and *Uromyces terebinthi* (DC) G. Winter has occasionally been found on the abnormal twigs (Gualaccini, 1950).

Control

This seems not to be necessary.

Aceria rossettonis Keifer (Fig. 16.33)

Common name

Cashew bud mite.

Diagnostic characteristics

FEMALE. The body is worm like, 150–185 μm long, 40–50 μm wide and light yellowish white in colour. The gnathosoma tapers to a slender apex, projecting forward and down. The prodorsal shield is subsemicircular in anterior outline, slightly acuminate. The prodorsal shield has lines and granules; the median line is present on the rear 1/3; the submedian lines are complete, subparallel, close for the first 2/3, further apart alongside the median; the first submedians are subparallel to the admedians, undulating, somewhat broken centrally, ending between the base of the scapular tubercles with the admedian line a little ahead of the rear margin. The second and third submedian lines are close to the first but ending or becoming faint after the anterior ¼. The sides of the shield above the coxae bear lines of granules and about five partial annuli. The scapular tubercles are 17 μm apart, directing the scapular setae (*sc*) divergently to the rear; the scapular setae are 24–29 μm long. The coxae are ornamented with granules, and the sternal line is short. The opisthosoma has about 63 annuli, and is completely microtuberculate; the microtubercles are slightly acuminate, oval or elliptical, more or less set ahead of the rear annulus margins. The setae c_2 are on about annulus 5, 13 μm long; the setae *d* are 4 μm long, on annulus 17; the setae *e* are 11 μm long, on annulus 34; the setae *f* are 17 μm long; the setae h_1 are 3 μm long. The genitalia are 12 μm long, 18 μm wide; the coverflap has 8–10 longitudinal ribs, and the seta is 8 μm long. The empodium is 5- and 6-rayed (Keifer, 1969a).

Geographical distribution

Brazil (Keifer, 1969a; Arruda and Aquino, 1970).

Host plants

Anacardiaceae: cashew (*Anacardium occidentale* L.) (Keifer, 1969a; Flechtmann, 2001b).

Bio-ecology

Four eriophyoid mites – *Aceria rossettonis*, *Calacarus decoratus* Flechtmann, *Mesalox abathus* Keifer and *Vimola globosa* (Keifer) – are reported on cashew (Flechtmann, 2001b). Among these, the cashew bud mite, *A. rossettonis*, attacks buds and flowers and lowers nut production in Brazil (Keifer, 1969a; Arruda and Aquino, 1970; Jeppson *et al.*, 1975). The bio-ecology of the mite has not been documented.

Symptomatology and damage

Unknown.

Control

This seems not to be necessary.

Aceria sheldoni (Ewing) (Fig. 16.34)

Common name

Citrus bud mite.

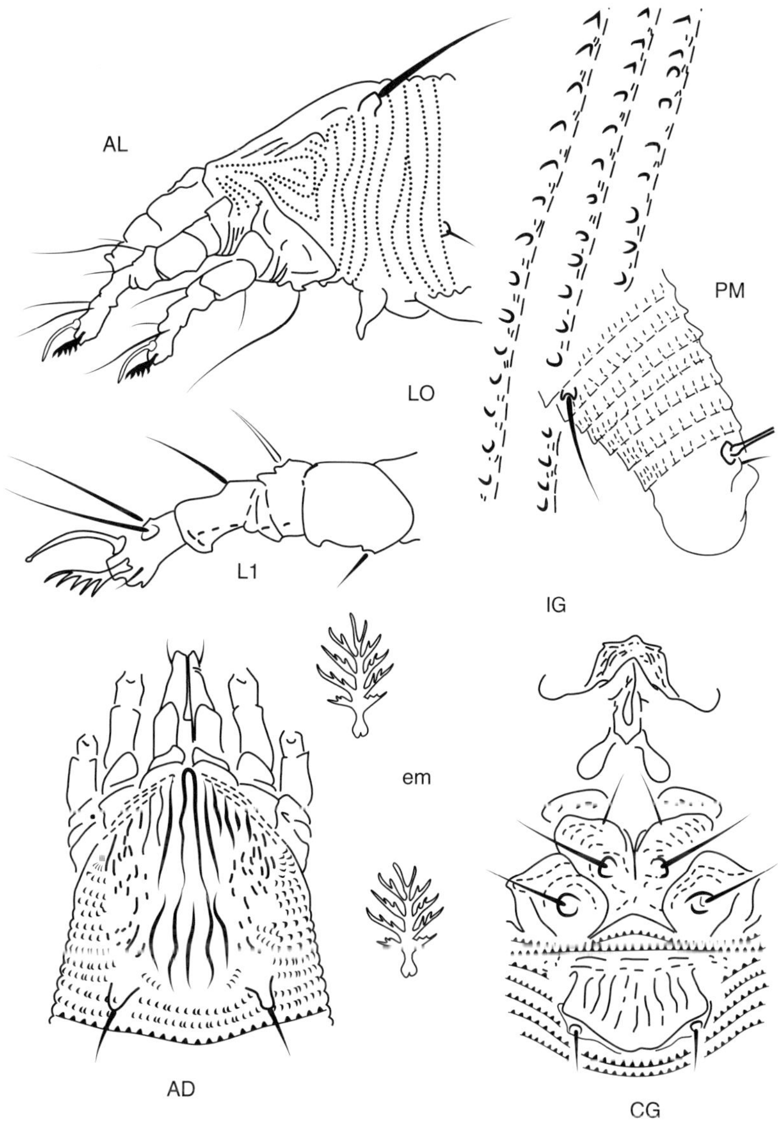

Fig. 16.33. *Aceria rossettonis* Keifer (from Keifer, 1969a). See Table 16.1 for key.

Diagnostic characteristics

FEMALE. The body is vermiform, 170–180 μm long and from hyaline white to yellowish or pink–orange in colour. The prodorsal shield has a variety of markings, with indistinct or quite marked lines; in its most defined condition, the median line is broken and ends as a dart-shaped mark close to the rear shield margin. The admedian lines are complete; the submedian lines extend back towards the scapular tubercles in front of which they encounter the transversal curved lines. Coxae I have a short sternal line and are marked with granules. The opisthosomal microtubercles are elliptical and touch the edge of the annuli. The genital coverflap has 10–12 longitudinal ribs. The empodia are 5-rayed (Jeppson *et al.*, 1975).

MALE. The body is worm like, with external features similar to those of the female, except for the genital structures.

Geographical distribution

Worldwide (Jeppson *et al.*, 1975).

Host plants

Rutaceae: *Citrus* spp. (Jeppson *et al.*, 1975; Vacante, 2010).

Bio-ecology

The citrus bud mite lives on different species of citrus (lemon, orange, lime, grapefruit, tangerine, bergamot, citron, etc.). Of these, it prefers lemon for its large buds which offer greater protection, as well as some varieties of orange ('Valencia' and 'Navel'). It is common in coastal citrus-growing areas and is probably present in every lemon-growing area of the world that has an adequate RH for its development. Populations develop

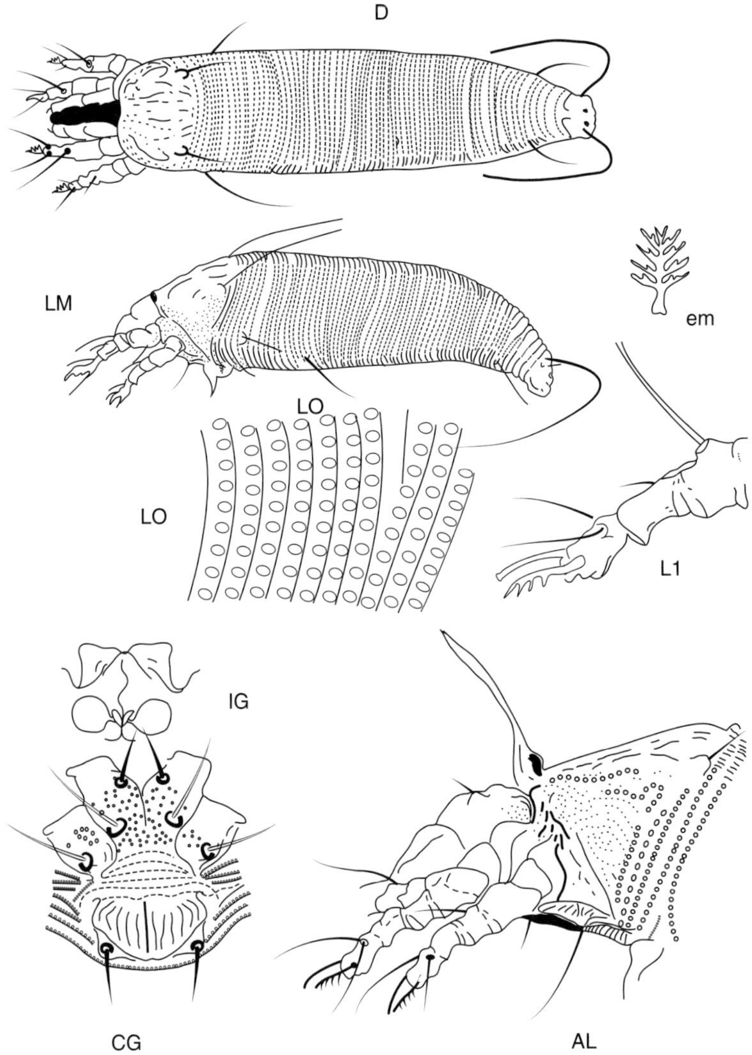

Fig. 16.34. *Aceria sheldoni* (Ewing) (from Keifer, 1952a). See Table 16.1 for key.

inside both dormant buds on old wood and flower buds or under the fruit rosette; they shelter in flowering buds as soon as fruit growth begins; individuals sometimes shelter in the points of contact between two young lemon fruits, avoiding those sites most exposed to light (Jeppson *et al*., 1975). During the year, varying population peaks are recorded from one season to another depending on various ecological factors (growth of new vegetation, climatic trends, etc.) and the highest densities are observed in the buds in the coldest months, with the lowest in spring–summer (Schwartz, 1975b; Vacante, 1986; Vacante *et al*., 2007). The population dynamics of *A. sheldoni* appear to be stable compared with those of other mites (Searle, 1978). As soon as populations of the citrus bud mite become exposed during new bud growth, they move on in search of new buds, spreading over the plant. Dispersal occurs throughout the year, reaching a peak during spring and at the beginning of summer, coinciding with the growth of new vegetation and increased air currents; nursery material, insects and birds facilitate mite dispersal (Jeppson *et al*., 1975). The mites search for new buds during the day even when there is a low population density, but this is more marked at high densities; reduced migration is observed in the morning or evening.

The reproduction is arrhenotokous parthenogenesis. During its life, the male deposits from 25 to 100 spermatophores (2–15 per day), randomly distributing them on the plant (Sternlicht, 1970). At 20–27°C, the males deposit the first spermatophores 1–5 days after eclosion and deposition continues irregularly for up to 31 days. Females of *A. sheldoni* produce only males when they use spermatophores that have been exposed for 1 h to a temperature of −7 or 41°C, and an RH above 30% is necessary for the viability of spermatophores. A small percentage of males are found on lemon in the winter period and a high percentage in spring and autumn (Sternlicht and Goldenberg, 1971). The eggs hatch in 3–14 days and the length of a generation (egg to egg) is 12–33 days; hatching is optimal at 25°C and 98% RH but reduces at lower RH (35–40%). The threshold for embryonic development is 9°C, and for completion of the life cycle it is 12.5°C. The mean fertility is six eggs (4–8) and this increases up to 12 (5–19) if the female feeds from the bud during larval development. Survival of the population is

influenced by extreme temperature values; for example, at 30 and −15°C, 50% of a population dies within 30 min (Sternlicht, 1970). In Australia, development from egg to adult takes 10 days in the field in summer and approximately 15 days in autumn. Mortality begins at 43.3°C and is complete at 47.7°C (Boyce and Korsmeier, 1941). In Israel, it has been calculated that the citrus bud mite can develop 15 generations a year on fruit (Sternlicht, 1970).

Natural enemies include the fungus *Hirsutella thompsoni* Fisher var. *synematosa*, which develops on the citrus bud mite (Searle, 1974; McCoy, 1981, 1996a,b; Gomez and Nasca, 1983). Among the predatory mites, a number of stigmaeids have been recorded in different regions of the world, such as *Agistemus africanus* (Meyer *et* Ryke) (Searle, 1974; Searle and Meyer Smith, 1998), *Agistemus collyerae* Gonzalez (Vacante, 1986), *Agistemus exsertus* Gonzalez-Rodriguez (Sternlicht, 1969a, 1970), *Agistemus tranatalensis* Meyer (Searle, 1974; Searle and Meyer Smith, 1998), *Eupalopsellus brevipilis* (Meyer *et* Rike) (Searle, 1974; Searle and Meyer Smith, 1998) and *Zetzellia graeciana* Gonzalez and *Z. mali* (Ewing) (Vacante, 1986). The cheyletid mites that have been recorded include *Chelotegenes ornatus* (Canestrini *et* Fanzago) (Searle, 1974; Vacante, 1986; Searle and Meyer Smith, 1998) and *Cheletomimus berlesei* (Oudemans) (Sternlicht, 1969a, 1970). Among the phytoseiid mites, *Typhlodromips swirskii* (Athias Henriot) (Sternlicht, 1969a, 1970), *Typhlodromus pyri* Scheuten (Mijuskovic, 1973a) and other species are known.

Symptomatology and damage

A. sheldoni infests buds of all ages, including those that are dormant in old wood. Adults, immature stages and eggs live under the axillary buds, at the base of the petioles near the buds, under the bracts, inside flowering or fruiting buds and under the sepals. The trophic activity of the citrus bud mite within the axillary leaf buds of lemon is the principal cause of damage and structural modification. Attacked buds, recognizable by the presence of brownish or black bracts, often die and encourage multiple gemmation on the infested bud; the percentage of infested axillary buds varies with the age of the shoot. Mite populations progressively attack the occasional buds that emerge in reaction to the initial attack, resulting in a widespread emergence of buds, which impedes shoot growth and flower development and negatively affects vegetative growth. Even small amounts of damage reduce the vegetative integrity of the plant and branches emerging from damaged buds may turn out to be shorter, larger and more flattened than normal; in lemon trees, rosetted growths occasionally derive from bud proliferation (Ebeling, 1959; Jeppson *et al.*, 1975). In the northern hemisphere, south-facing trees are infested to a greater extent than north-facing ones, with the levels of infestation between buds on different shoots usually greater than those recorded among buds of the same shoot (Walker *et al.*, 1992). The opposite occurs in the southern hemisphere (Searle and Meyer Smith, 1998).

Leaves originating from damaged buds have unusual shapes, being curled and twisted, and divided and divergent at the tip; flowers are misshapen and stunted with deformed or aborted reproductive organs. The altered blossoms fall and may produce fruit that falls early or has a peculiar shape (Jeppson *et al.*, 1975). Distortion of the fruit results from the attack on the embryonic tissue of the fruit in the bud; once it emerges, the fruit is not affected by further attack from the mite (Walker *et al.*, 1992). Depending on the amount of the attack and blossom damage, lemons may acquire a rounded rather than ellipsoid shape. The symptoms on oranges (cultivars 'Valencia' and 'Navel') are the same as those observed on lemons, except for a less evident deformity of the branches, leaves and fruit. Oranges may flatten vertically, and have crests and scars on the skin or small openings in the stilar cicatrix (Jeppson *et al.*, 1975).

Trophic activity of the citrus bud mite causes an increase in the phenol levels of the bud tissue with a simultaneous fall in auxin activity (attributed to inhibition by phenols), and increases in the activity of polyphenoloxidase, peroxidase and RNase (Ishaaya and Sternlicht, 1969, 1971). The extent of blossom and fruit abscission increases significantly with the distortion caused by mite attack. Distortion of the fruit is significantly linked to that of the blossoms from which they develop. It is probable that the increase in blossom and fruit abscission may depend on a decrease in auxin activity and other biochemical changes in the axillary buds infested by the mite (Phillips and Walker, 1997).

A density of between one and three mites per bud results in delayed development of the foliage (Ishaaya and Sternlicht, 1969). The level and economic importance of the attack depends on the age of the crop, the host plant and variety, and the extent of infestation. A high density attack on young groves may seriously disrupt the development of lemon and orange plants (Di Martino, 1985; Searle and Meyer Smith, 1998). Swarms of mites may significantly delay the vegetative development of 'Navel' orange. The degree and quality of vegetative shoots determines the flowering time and fructification of the following season. The fall in lemon production resulting from infestation may prove to be greater than that of 'Navel' orange (Schwartz, 1972). Mite swarms also affect 'Navel' orange in New Zealand (Harty *et al.*, 2004). However, Vacante and Nucifora (1984) have established that in lemon groves in eastern Sicily, up to 70% infestation of buds by mites guarantees a satisfactory gross marketable production. Findings in California did not demonstrate significant differences in terms of volume, quality or value in production between lemon groves treated once or twice a year with oils and untreated lemon groves; indeed, in four out of six cases, greater production levels were recorded in untreated lots, possibly owing to oil phytotoxicity (Hare *et al.*, 1999). Vacante *et al.* (2007) reported similar results in two lemon groves in southern Italy.

Control

Control may involve spray applications of acaricides, insecticides and/or fungicides. Biological control alone is inadequate, and no classical biological or microbial control programmes have been initiated (McCoy, 1996b). Some selective chemicals, such as the petroleum oils, which are selective for the beneficials (phytoseiids, stigmaeids, etc.), allow the implementation of integrated pest management (IPM) (Rosen, 1986a; Vacante, 1986, 2010; McCoy, 1996b). As regards control, when necessary, the mite can be controlled by most acaricides as well

as by mineral oils, but the cost of such control may exceed its economic benefit (Gerson and Vacante, 2012).

SAMPLING AND MONITORING. Monitoring may be done by randomly observing four shoots of the last vegetative growth on 10% of the plants (Cavalloro and Prota, 1983). In the Mediterranean region, sampling and monitoring can be conducted from late August to September in lemon subjected to the technique of water stress in July for the production of out-of-season green lemons; in other lemon groves, it is commonly done in winter from November to January; observations are made on the buds of 100 shoots (García Marí, 2009).

THRESHOLDS. In the Mediterranean region, Vacante (2010) suggested that on adult plants a reasonable threshold for treatment might be more than 70% of infested buds in August–September and/or in winter; while on young groves it was less than 70% of infested buds. In Spain, García Marí (2009) suggested a threshold of 20% of infested shoots from March to June.

CHEMICAL CONTROL. During the last 50 years, a number of active ingredients have been applied and/or proposed, and a detailed account is given by Childers *et al.* (1996). Some of the most effective chemicals have now been banned and are no longer available. The active ingredients that have been tested include organochlorines (Jeppson *et al.*, 1955, 1958; Attiah *et al.*, 1973; Mathew, 1973; Schwartz, 1975c; Brown and Jesser, 1981; Di Martino, 1985; Costilla *et al.*, 1987; Lacasa *et al.*, 1990a), oil emulsion (Schwartz and Riekert, 1967; Mijuskovic, 1973a,b; Di Martino, 1985), sulfur as polysulfide sulfur or lime sulfur (Sternlicht, 1966; Mathew, 1973; Jeppson *et al.*, 1975; Fourie, 1988) and petroleum oils (Boyce and Korsmeier, 1941; Boyce *et al.*, 1942; Jeppson *et al.*, 1975; Vacante, 1986, 1995, 2009a; Atkins *et al.*, 1987; Sale, 1988; Salas and Macian, 2003). These are sprayed in summer or in winter, according to their efficiency, the environmental conditions and the legislative requirements of the country concerned.

In nurseries and young crops, one treatment per season may be necessary. In adult crops that are heavily infested, two applications of acaricide are effected, one in late winter or early spring and the other in summer. In adult crops with a medium infestation, only one application is necessary (Searle and Meyer Smith, 1998). The treatments must be carried out when there is the highest density of the citrus bud mite population in the buds, before vegetative growth. Winter treatments with petroleum oils are also effective against the armoured scale insects, and in summer it is necessary to use summer petroleum oils in those regions, like Sicily, where it is possible to use a method for the summer production of lemons based cn a limited irrigation step, and the oil can be applied after the first irrigation and before vegetative growth (Di Martino, 1985; Vacante, 1986, 1995, 2009a). McCoy (1996b) reports that two applications a year are carried out in May and June and/or from September to November. In Greece, Papaioannou-Souliotis *et al.* (1999b) regard the periods from mid-May to mid-July and from mid-August to mid-September as the most effective times for treatment.

Aceria stefanii (Nalepa)

Common name

Pistachio leaf edge roll mite.

Diagnostic characteristics

FEMALE. The body is worm like, up to 160 μm long. The prodorsal shield has central longitudinal lines and laterally is broadly granular. The median prodorsal shield line occurs on the rear $^2/_3$ and ends in a dark-shaped mark at the rear margin; the admedian lines are slightly sinuate, complete and gradually diverging, and short submedian lines are located anteriorly. The coxae have lines of granules. The opisthosomal annuli have pointed microtubercles ahead of the annuli margins. The genital coverflap has about 10 longitudinal ribs. The empodium is 4-rayed (Jeppson *et al.*, 1975).

Geographical distribution

Europe, Near East (Nalepa, 1898; de Lillo and Beron, 2004) and Iran (Mehrnejad and Daneshvar, 1991).

Host plants

Anacardiaceae: *Pistacia* spp. (Nalepa, 1898; Jeppson *et al.*, 1975; de Lillo and Beron, 2004), lentiscus (*P. lentiscus* L.) (Mehrnejad and Daneshvar, 1991; Mehrnejad and Ueckermann, 2001; Xue *et al.*, 2009a), *P. mutica* Fischer *et* C.A. Mey, pistachio (*P. vera* L.).

Bio-ecology

A. stefanii lives on *Pistacia* spp. in different European countries, the Near East (Nalepa, 1898; Jeppson *et al.*, 1975; de Lillo and Beron, 2004) and Iran (Mehrnejad and Daneshvar, 1991; Mehrnejad and Ueckermann, 2001; Xue *et al.*, 2009a).

Symptomatology and damage

The pistachio leaf edge roll mite rolls the pistachio leaflets over the upper surfaces into tight rolls. These rolled leaflets become thin and stick like. Sometimes, entire branches may appear to be without leaflets (Jeppson *et al.*, 1975).

Control

This is usually not necessary.

Grasses and monocotyledonous crops group

The most common species of this group are *Aceria aloinis* (Keifer), *A. bakkeri* Keifer, *A. cynodoniensis* Sayed, *A. guerreronis* Keifer, *A. merwei* (Tucker), *A. sacchari* Wang, *A. slykhuisi* Hall, *A. tosichella* Keifer, *A. tulipae* (Keifer) and *A. zeasinis* Keifer. Several other species are known. The ten named species are described below.

Aceria aloinis (Keifer) (Fig. 16.35)

Common name

Aloe wart or aloe gall mite.

Diagnostic characteristics

FEMALE. The body is worm like, 190–220 µm long, 45–50 µm wide and whitish to purplish in colour. The gnathosoma is somewhat downcurved. The prodorsal shield has the median line extended to the rear and the admedians complete; the sublaterals are distinct, curved, inter-branched. The scapular tubercles are 27.5 µm a part, set on the rear margin; the scapular setae (*sc*) are 35 µm long, projecting backwards. The opisthosoma has 68–75 annuli, completely microtuberculate, the tubercles slightly pointed. The setae c_2 are 34 µm long, set on about annulus 6; the setae *d* are 52 µm long, set on about annulus 21; the setae *e* are 50 µm long, set on about annulus 40; the setae *f* are 33 µm long, set on about annulus 7 from the rear; the setae h_1 are present. The genitalia are 25 µm wide, 18.5 µm long; the coverflap is basally tuberculate with 7–9 furrows, and the seta 10 µm long. The empodium is 6-rayed (Keifer, 1941).

Geographical distribution

California (Keifer, 1941), Florida (USA) (Denmark and Keifer, 1979), Hawaii (Garrett and Haramoto, 1969) and South Africa (Meyer Smith, 1981b, 1996a).

Host plants

Xanthorrhoeaceae: *Aloe arborescens* Mill., *A. dichotoma* Masson, *A. nobilis* Haw. and other aloe species (Meyer Smith, 1996a), aloe

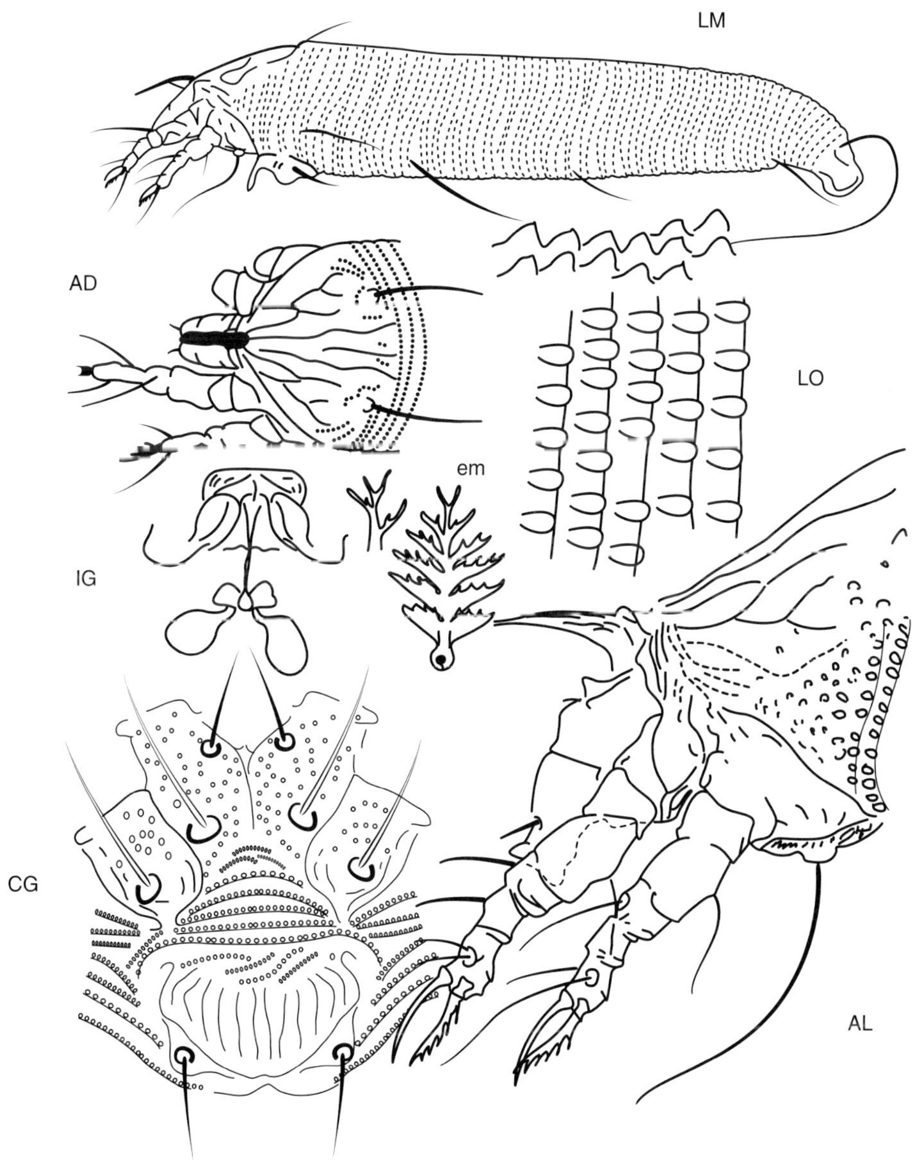

Fig. 16.35. *Aceria aloinis* (Keifer) (from Keifer, 1941). See Table 16.1 for key.

(*A. spinossissima* Hort ex A. Berger), *Haworthia* (Keifer, 1941; Garrett and Haramoto, 1969; Denmark and Keifer, 1979).

Bio-ecology

A. aloinis lives on aloe and *Haworthia* in California (Keifer, 1941), Hawaii (Garrett and Haramoto, 1969) and Florida (Denmark and Keifer, 1979), and on a number of aloe species in South Africa (Meyer Smith, 1996a).

Symptomatology and damage

The mites cause large warty deformations on the leaf bases (Keifer, 1941; Jeppson *et al*., 1975). These gall-like outgrowths are small to large, rounded to irregular thickenings and are yellowish, brownish or green tinged with yellow. In South Africa, *A. arborescens* and its varieties are frequently infested (Meyer Smith, 1996a). The mite also damages the flowers (Denmark and Keifer, 1979).

Control

The adaptation of aloe gall mite to living in galls makes the use of chemicals difficult and suggests that prevention is a better approach. Regular observations of aloes allow detection of the first symptoms of attack and the infested portions of the leaves should then be cut out and burned. Acaricides could be used on wounds. Careful examination of plants before planting also helps in prevention (Meyer Smith, 1996a). The acaricides that have been applied include dicofol (Denmark and Keifer, 1979).

Aceria bakkeri Keifer (Fig. 16.36)

Common name

Kenya rice mite.

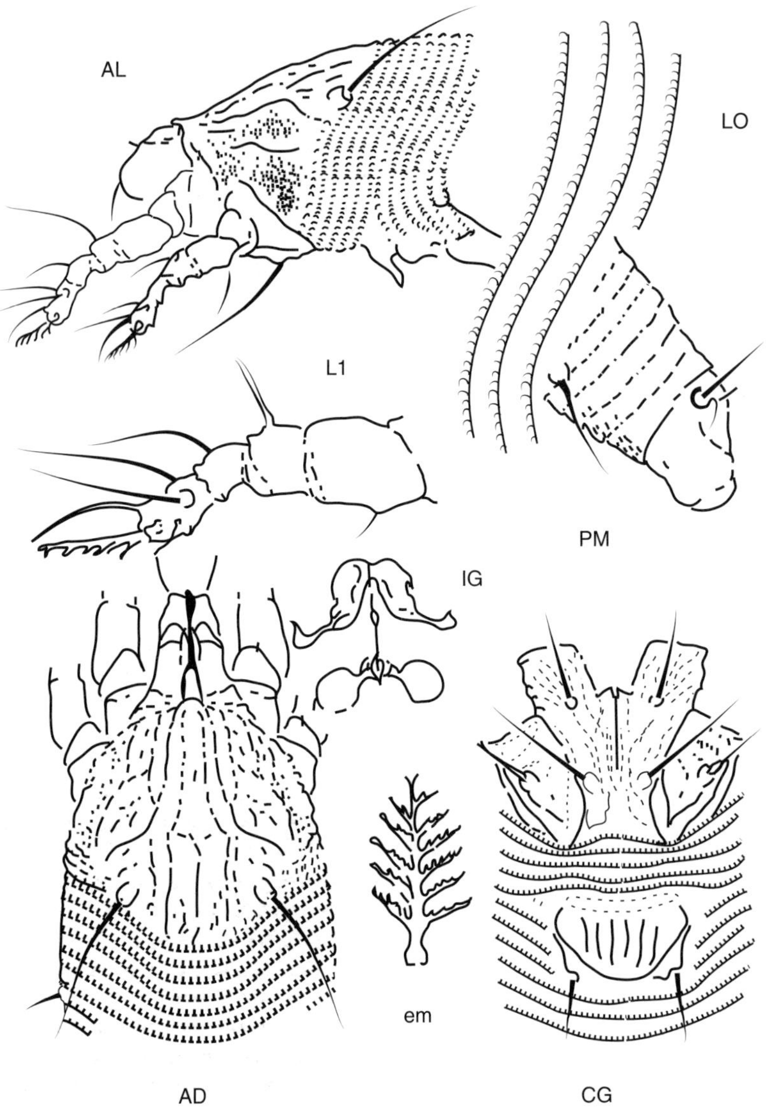

Fig. 16.36. *Aceria bakkeri* Keifer (from Keifer, 1969a). See Table 16.1 for key.

Diagnostic characteristics

FEMALE. The body is elongate, worm like, 170–185 µm long, 35 µm wide, and in life probably coloured light yellowish white. The gnathosoma is curved down. The prodorsal shield is subsemicircular in anterior outline, with the rear margin rather strongly convex between the scapular tubercles. The shield pattern is lines of granules, interspersed with granulations and short dashes; the median line is present on the rear 2/5; the admedians are complete from just above the base of the chelicera subparallel from about halfway, gently curving outward to break at 3/5, then slightly sinuate to the rear margin curving inward at the margin. The first and second submedians are close to admedian, subparallel, extending back and running together at just before halfway, the line forking again in front of the scapular tubercle, the inner branch weak and broken, the outer stronger end bending abruptly outward ending lateral to the tubercle. There is a lateral line above the coxae enclosing a band of granulations, and about three partial microtuberculated annuli below the scapular tubercles. The scapular tubercles are 17 µm apart; the scapular setae (*sc*) are 23 µm long, diverging to the rear. The coxae are rather elongate, with lines of dashes and granulations. The opisthosoma has about 64 annuli. The microtubercles are rounded off, elongate dorsally, especially anteriorly, becoming shorter laterally and ventrally, usually reaching the annulus margins. The setae c_2 are on annulus 8 behind the shield, 27 µm long; the setae *d* are 51 µm long, set on annulus 22; the setae *e* are 11 µm long, set on annulus 36; the setae h_1 are 3 µm long. The genitalia are 17 µm wide, 12 µm long; the coverflap has about 6 longitudinal ribs, and the seta is 12 µm long. The empodia are 6- to 7-rayed (Keifer, 1969a; Jeppson *et al.*, 1975).

MALE. The empodium is 5-rayed (Jeppson *et al.*, 1975).

Geographical distribution

Kenya (Keifer, 1969a), Cuba (de la Torre and Martínez, 2004) and the Dominican Republic (Behysi *et al.*, 2002).

Host plants

Poaceae: rice (*Oryza sativa* L.) and unidentified grass (Keifer, 1969a; Jeppson *et al.*, 1975).

Bio-ecology

A. bakkeri is reported inside the leaf sheaths of rice and from an unidentified grass in Kenya (Keifer, 1969a; Jeppson *et al.*, 1975).

Symptomatology and damage

The Kenya rice mite seems not to cause any visible damage in Cuba (de la Torre and Martínez, 2004).

Control

This seems not to be necessary.

Aceria cynodoniensis Sayed (Fig. 16.37)

Common name

Bermuda or couch grass mite.

Diagnostic characteristics

FEMALE. The body is worm like, 170–210 µm long, 40 µm wide and whitish cream in colour. The gnathosoma is downcurved. The prodorsal shield is semicircular anteriorly; it is marked with a distinctive pattern of two scalloped, parallel ridges running the full length of the shield, with a straight and short ridge in between. The scapular tubercles are 23 µm apart; the scapular setae (*sc*) are 45 µm long. The coxae are granular, the junction line between coxae I is paralleled by lines of granulations. The opisthosoma has about 60 annuli, with microtubercles elliptical, pointed and located ahead of margins. The setae c_2 are 35 µm long, set on annulus 7 behind the shield; the setae *d* are 32 µm long, set on about annulus 21; the setae *e* are 7 µm long, set on annulus 38; the setae *f* are set on about annulus 4 from the rear and 27 µm long; the setae h_1 are 2.5 µm long. The genitalia are 18 µm wide and 10.5 µm long; the coverflap has about 11–12 narrow longitudinal ribs, with the median ribs twice as long as the lateral ribs; the setae are 8.5 µm long. The empodium is 6-rayed (Keifer, 1960; Jeppson *et al.*, 1975; Keifer *et al.*, 1982).

MALE. The empodia are 5-rayed (Jeppson *et al.*, 1975; Keifer *et al.*, 1982).

Geographical distribution

Australia (Naumann, 1993), Egypt (Sayed, 1946c), Greece (Kapaxidi *et al.*, 2008), New Zealand (Jeppson *et al.*, 1975), South Africa (Meyer Smith, 1968; Meyer Smith, 1981b) and the USA (Arizona, California, Florida, Georgia) (Keifer, 1960; Tuttle and Butler, 1961; Davis, 1964a; Barké and Davis, 1971), but probably wider.

Host plants

Poaceae: *Cynodon bradleyi* Stent (Meyer Smith, 1968), Bermuda grass (*C. dactylon* (L.)) (Keifer, 1960; Jeppson *et al.*, 1975).

Bio-ecology

A. cynodoniensis lives on terminal leaf sheaths, under broadened leaf bases of Bermuda grass (Keifer, 1960; Jeppson *et al.*, 1975), and on *C. bradleyi* (Meyer Smith, 1968). In Florida, the mite develops one generation in 7–10 days and its populations are active most of the year. All biological instars live under the leaf sheath and the populations increase during late spring and summer, with a density ranging from a few to a hundred or more motile forms per sheath (Keifer *et al.*, 1982). Dispersal occurs by wind, insects (mole crickets) and via grass clippings. Infestations usually develop in the taller grass (rough areas, along canals, fence rows, etc.) (Short and Buss, 2005). The populations are less rich where flood irrigation is applied (Keifer *et al.*, 1982). The variety of grass is an important factor in

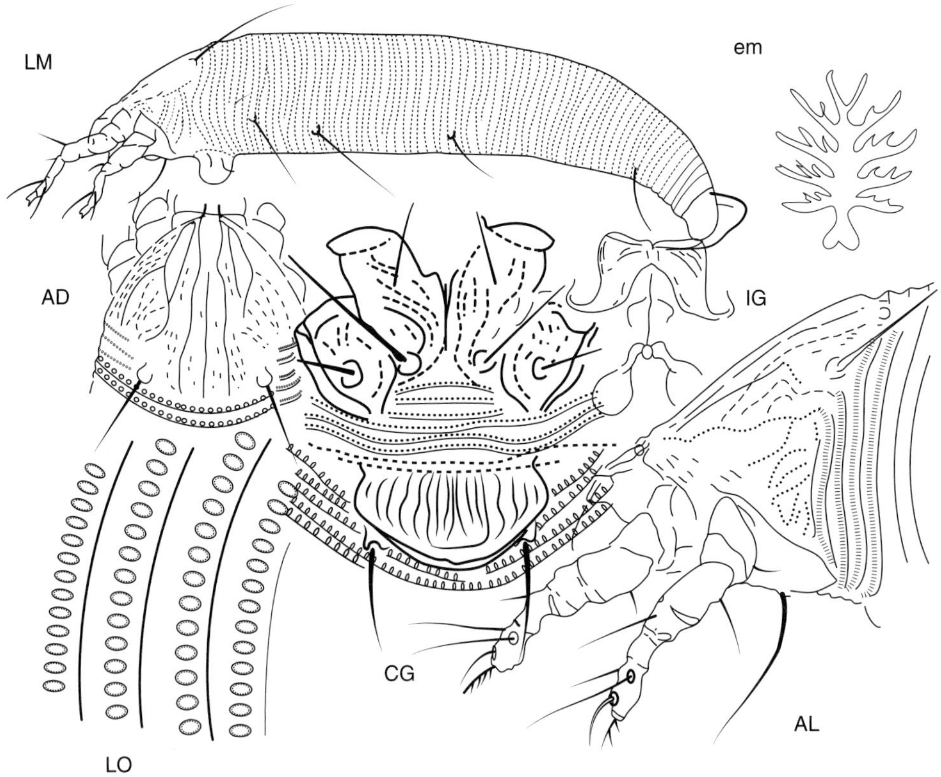

Fig. 16.37. *Aceria cynodoniensis* Sayed (from Keifer, 1960). See Table 16.1 for key.

susceptibility to attack, and the mite shows high specificity for the grass species (Skoracka, 2006). Soil fertilization increases infestation (Tuttle and Butler, 1961).

Symptomatology and damage

Feeding by the Bermuda grass mite causes swollen leaf sheaths, which become closely packed, thickened and bunched at the stem node, and the leaf blades become stunted. Infested stems have greatly deformed and enlarged nodes and shortened internodes. On Bermuda grass grown in lawns, the damage is evident in the spring, and then browning and thinning out of the grass follow (Keifer, 1960; Jeppson *et al*., 1975; Keifer *et al*., 1982; Kapaxidi *et al*., 2008). In general, all common varieties of Bermuda grass are susceptible to attack and plant water or nutrient stressed plants show severe damage. This aspect has suggested the potential use of the mite in the biological control of this weed. On strains of Bermuda grass from *C. dactylon* parentage, which are usually cultivated as important golf course grasses, the feeding activity of the mite can cause damage (Cromroy, 1983), suggesting that *A. cynodoniensis* be considered as a pest.

Control

Control can be needed on strains of Bermuda grass of *C. dactylon* parentage, which is usually cultivated as an important golf course grass (Cromroy, 1983).

PLANT RESISTANCE. *A. cynodoniensis* develops on some strains of Bermuda grass of *C. dactylon* parentage, failing on hybrids from crossings with the African species *C. transvalensis* Davy (Cromroy, 1979, 1983). The ability of *A. cynodoniensis* to damage some varieties of *C. dactylon* has been sufficient to stimulate plant breeders to develop commercial varieties that are resistant to the mite (Reinert *et al*., 1978, 2008). However, because the plant is invasive in some regions, it would be interesting to find mites that damage the invasive varieties, but not the commercial varieties. This has involved the efforts of several researchers (Rodgers and Baltensperger, 2004; Henry *et al*., 2005), but the problem is as yet not solved.

CULTURAL PRACTICES. Close mowing of attacked areas removes a great part of the mite populations. Free irrigation and fertilization reduce damage (Tuttle and Butler, 1961).

BIOLOGICAL CONTROL. The action has been evaluated of the acaropathogenic fungus *Hirsutella thompsonii* Fisher (McCoy, 1981), but there are no other known reports on the biological control of this mite.

CHEMICAL CONTROL. Among the chemicals that have been used for control are diazinon and sulfur (Tuttle and Butler, 1961).

Aceria guerreronis Keifer (Fig. 16.38)

Common name

Coconut flower and nut mite.

Diagnostic characteristics

FEMALE. The body is worm like, 205–255 μm long, wider just behind the prodorsal shield, and light yellowish white in colour. The gnathosoma curves down. The prodorsal shield is subsemicircular in anterior outline, with a pattern of strong lines and lines of granules, of which those located centrally are solid, but showing granular thickenings; the median line develops from $^{1}/_{3}$ of the way to the rear margin; the admedian lines are complete from the base of the chelicera, subparallel, closer ahead of the median, farther apart alongside the median and slightly recurved at the rear. The first submedian line develops from the side of the base of the chelicera, subparallel to the admedian and following its curves, ending in a transverse line at $^{2}/_{3}$ of the way in front of the scapular tubercle, this line faintly extending across the centre of the shield. The second submedian line runs to the outside first, extending back by $^{1}/_{3}$, then abruptly outcurving for a distance and recurving back, ending laterad of the scapular tubercle. The area in front of the scapular tubercles has some short dashes. Laterally, the shield has a band of granules above the coxae, some lateral lines and dashes, and partial annuli below the scapular tubercles. The opisthosoma has about 56 annuli to the annulus bearing the seta *f*; all annuli are completely microtuberculate, these structures noticeably elongate, reaching, or nearly reaching, the annulus margins and thickened at the rear end. The opisthosoma is thicker across the genital area. The seta

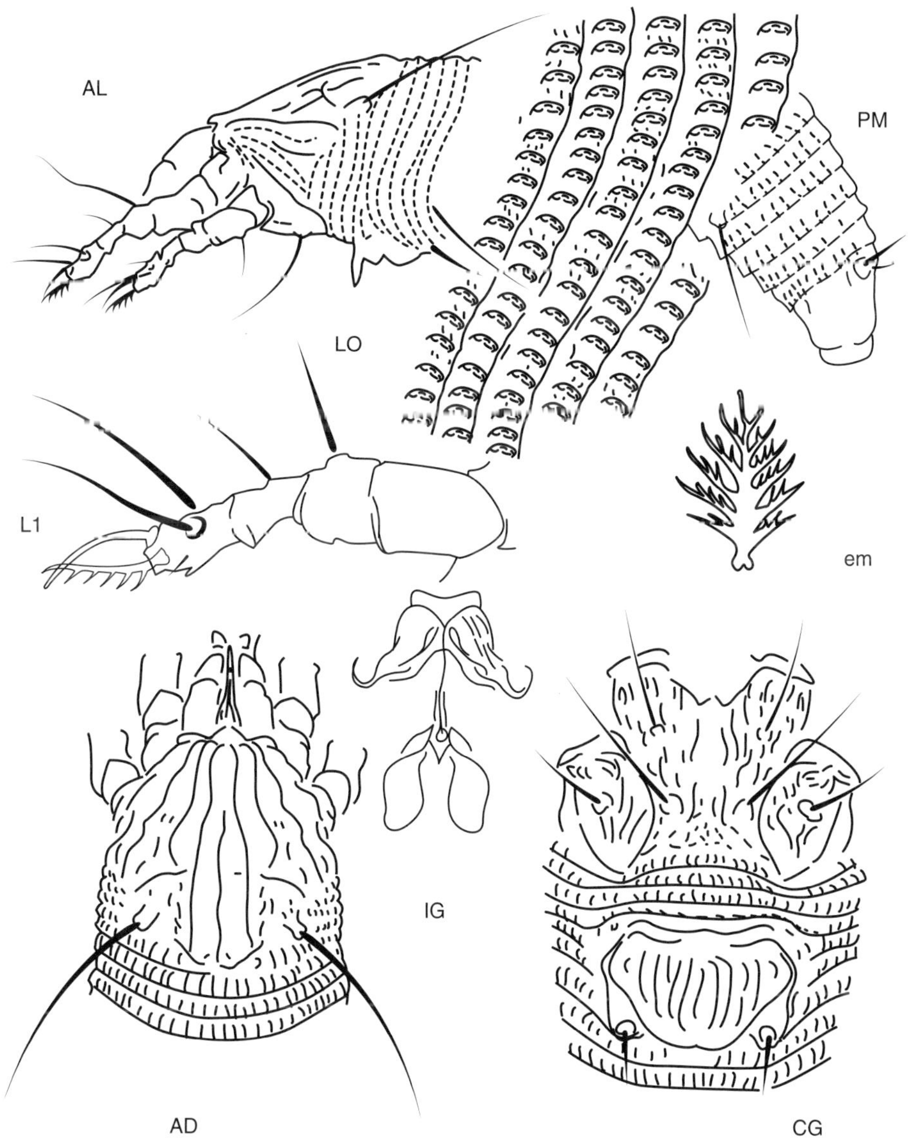

Fig. 16.38. *Aceria guerreronis* Keifer (from Keifer, 1965b). See Table 16.1 for key.

c_2 is 37 μm long, set on annulus 7 behind the shield; the setae *d* are 50 μm long, set on annulus 20; the setae *e* are 12 μm long, set on annulus 35; the setae *f* are 24 μm long, set on annulus 5 from the rear; the setae h_1 are 3 μm long. The genitalia are rather deep bowl shaped, 21 μm wide and 17 μm long; the coverflap has about 9–12 longitudinal ribs, and the seta is 16 μm long. The empodium is 6-rayed (Keifer, 1965b).

MALE. The male is similar to the female, but smaller.

Geographical distribution

Worldwide. The mite has been reported from the main areas of coconut palm cultivation (Moore and Howard, 1996; Fernando *et al.*, 2002b; Nair *et al.*, 2005).

Host plants

Arecaceae: palmyra palm (*Borassus flabellifer* L.) (Ramaraju and Rabindra, 2002), coconut palm (*Cocos nucifera* L.) (Keifer, 1965b; Návia *et al.*, 2005b), Weddell palm (*Lytocaryum weddellianum* (H. Wendland) Toledo) (= *Cocos weddelliana* H.A. Wendland) (Flechtmann, 1989a), queen palm (*Syagrus romanzoffiana* (Cham.) Glassman) (Ansaloni and Perring, 2004).

Bio-ecology

A. guerreronis is a severe pest of coconut palms that probably originated in tropical America (Návia *et al.*, 2005b) and spread to most of the coconut production areas of the world. It also lives on the fruits of palmyra palm, a plant that is native to southern and South-east Asia, and in India (Ramaraju and Rabindra, 2002) and Sri Lanka (de Moraes, in Návia *et al.*, 2010a). In Florida, the mite was discovered on coconut in 1984 (Howard *et al.*, 1990). In California and Florida, it was reported on Brazilian ornamental queen palm seedlings in nurseries (Ansaloni and Perring, 2004). However, surveys on *S. romanzoffiana* leaves and fruits carried out in Brazil showed that the queen palm was not a host for *A. guerreronis* in its areas of natural occurrence (Návia, 2004). In Brazil, the mite has also been found on *L. weddellianum* (Flechtmann, 1989a).

The coconut mite is a tropical and subtropical species characterized by a worldwide geographical distribution (Moore and Howard, 1996; Fernando *et al.*, 2002b; Nair *et al.*, 2005). The species was described from Mexico (Keifer, 1965b), but it was present in different areas in central and South America before 1965 (Robbs and Peracchi, 1965; Ortega *et al.*, 1967; Zuluaga Cardona and Sánchez Potes, 1971; Arruda, in Návia *et al.*, 2010a). It was reported on coconut in Florida in 1984 (Howard *et al.*, 1990), and in California in 1997 (Ansaloni and Perring, 2004). In West Africa, the mite has been known since 1967 (Cabral and Carmona, 1969; Mariau, 1969; Seguni, 2002). In India and Sri Lanka, *A. guerreronis* was found in the 1990s (Fernando *et al.*, 2002a; Sathiamma *et al.*, 1998), and it was found later in the Middle East (Oman) (de Moraes, in Návia *et al.*, 2010a). The coconut mite has a strong economic and social impact and represents a serious problem in South-east Asia (Fernando *et al.*, 2002a; Haq, 2007). It has already spread into Malaysia, India, the Philippines and Sri Lanka (Muthiah *et al.*, 2001; Návia *et al.*, 2006d). Studies on the molecular and morphological diversity, geographical range and expansion history of the species may allow limitation of its spread (Návia *et al.*, 2005b). Molecular analysis highlighted the American origin of the mite and suggested that its original host plant was a non-coconut palm. In contrast to the diversity found in the Americas, all samples from Africa and Asia were identical or very similar, confirming that the mite had spread into these regions recently (Návia *et al.*, 2005b).

A. guerreronis develops on the meristematic zone of the fruits covered by the perianth, usually referred to as the tepals or bracts (Moore and Howard, 1996). Within a few weeks to a month, the mites move between the upper and lower tepals, until they reach the fruit surface protected by the perianth (Ortega *et al.*, 1967; Mariau and Julia, 1970; Hall and Espinosa, 1981; Moore and Alexander, 1987a; Howard and Abreu, 1991). The development of life cycle from egg to adult takes about 10 days (Mariau, 1977), and spermatophores have benn found beneath the perianth. The fruits are susceptible to mite infestation during the entire time of their development, but 10- to 13-month-old fruits rarely harbour a high density of mites (Hall and Espinosa, 1981; Moore and Alexander, 1987a). The coconut mite reproduces by arrhenotoky. Uninseminated females lay eggs that give rise only to male offspring (Ansaloni and Perring, 2004). Ovoviviparity or viviparity may occur, but the conditions influencing this strategy are not known (Návia *et al.*, 2005a,b). Patterns of diel activity have been reported (Haq *et al.*, 2002; Sumangala and Haq, 2005). Observations on the seasonal abundance of *A. guerreronis* in India pointed out its persistent nature and the occurrence of population peaks in the summer months (April–May) (Nair, 2002).

Although the mite is adapted to tropical and subtropical climates, its populations can survive for short periods of frost and cold temperatures (Howard *et al.*, 1990). Infestation can be more severe in dry climates or during the dry season of wetter climates (Zuluaga Cardona and Sánchez Potes, 1971; Griffith, 1984).

Trees situated near the sea are attacked less than those inland, both because of the strong sea winds blowing the mites off the nuts and because of salt spray deposits on the nuts; monthly treatment of the inflorescences of inland trees with sprays of seawater almost halved the rate of infestation (Mariau, 1977). In St Lucia (West Indies), mites were not found in unfertilized flowers but were present within a few weeks of fertilization. Nuts remained vulnerable to colonization for a number of months, and mites were present on nuts for up to 13 months. Protection of nuts for the first few months after fertilization is of utmost importance. On the tree, much migration could be due to mites walking in large numbers across nut surfaces, and they could move from one inflorescence to another if the two were in contact. Increased nut damage appeared to encourage mite migration (Moore and Alexander, 1987a).

In general, wind appears to be the main factor in dispersing mites from one region to another; they cannot be transported with the nuts as these cease to be attractive to the pests 2–3 months before harvest, owing to their lack of young growing tissues (Mariau, 1977). In Costa Rica, the wind is important in large-area dispersal, while transmission between bunches is by

phoresis and rain (Schliesske, 1988). Dispersal occurs also through flower-visiting insects, ants and bats (Moore and Alexander, 1987a). Touching inflorescences allow mite dispersal which results from negative geotaxis and from mites moving from older to younger inflorescences (Moore and Alexander, 1987a). The mites walk at a rate of 20–100 μm s^{-1} (Moore and Howard, 1996).

Natural enemies of the coconut mite include pathogens and predatory mites. The pathogens include the fungi *Hirsutella thompsonii* Fisher (Julia and Mariau, 1979b; Hall *et al.*, 1980) and *Verticillium lecanii* (Zimmerman) Viégas (Julia and Mariau, 1979b). The predators include the bdellid mite *Bdella distincta* Baker, the phytoseiid mites *Amblyseius largoensis* Muma, *Neoseiulus mumai* Denmark and *Neoseiulus paspalivorus* DeLeon (Howard *et al.*, 1990), two unidentified phytoseiids and a tarsonemid mite (Julia and Mariau, 1979b). Two species of *Lupotarsonemus* (*Tarsonemus*?) were found in the Ivory Coast and New Guinea, but appeared to be of only minor significance in reducing populations (Hall *et al.*, 1980).

Symptomatology and damage

A. guerreronis lives on coconut flowers and feeds on young nutlets. Keifer (1965b) reported that the mites damaged the floral bracts and scarred the young developing nuts. A more extensive description of damage by Hall and Espinosa (1981) points out that the green surface of nutlets shows triangular, pale yellowish white or whitish marks located at tightly adpressed bracts when the nuts are about 5 cm long. Increases in infestation enlarge the blemishes, which become brown and corky, sometimes forming deep fissures with gum exudate. Irregular growth of the fruit causes cracking; damage is usually only to a part of the fruit and kernel, which are partially reduced in size. Nuts about 20 cm long host a low density mite population under the bracts (Hall and Espinosa, 1981). Infested fruits turn brown and corky in appearance, sometimes with extensive cracking reaching deep into the mesocarp. The resulting fruits may be small and deformed (Mariau, 1986; Schliesske, 1988; Howard and Moore, 2008). In the Ivory Coast, the mite affects the size and copra content of the nuts but does not cause premature nut fall, as had previously been thought (Mariau, 1986).

The extent of damage caused by the mite varies in different countries. In Mexico, damage may range from 80 to 100% of coconut trees (Ortega *et al.*, 1967; Olvera-Fonseca, 1986), in Cuba from 42 to 65% (Suárez, 1991b), in Sri Lanka from 5 to 100% (Fernando *et al.*, 2002a), in India from 20 to 70% (Muthiah and Bhaskaran, 2000; Nair and Koshy, 2000; Nair, 2002) and in Costa Rica it is 90% (Schliesske, 1988). Environmental conditions, varieties and horticultural choices may affect copra yield. Thus, in American and African countries reductions in copra yield of about 30% have been reported (Hernández Roque, 1977; Julia and Mariau, 1979a; Moore, 1986) and they were about 40% in India (Muthiah and Bhaskaran, 2000; Nair and Koshy, 2000). In Venezuela, early dropping induced about 70% reduction of copra yield (Doreste, 1968a) and the reduction was 10–100% (average 21%) in Tanzania and the African coastal islands (Mafia, Zanzibar and Pemba) (Seguni, 2002).

The coconut mite feeds on the meristematic tissue of queen palm seedlings in California and Florida nurseries (Ansaloni and Perring, 2004), causing necrosis, leaf distortion and plant death. The phenomenon severely limits the ornamental palm business in Florida and California (Návia *et al.*, 2010a).

Control

Control of *A. guerreronis* is provided by preventive and curative measures. Prevention through quarantine may play an important role, as well as chemicals, plant resistance and cultural practices. Physical and biological means seem to be less effective. Integrated pest management (IPM), based on the integration of selected means can provide practical control of the mite (Reddy and Naik, 2000).

MONITORING. Monitoring systems help in mite control. The use of the sellotape embedding technique has been suggested, in which transparent sellotape is pressed gently over the mite colony to embed the whole population, including eggs, nymphs and adults, on to the sellotape. The populations are counted using the microscope (Girija *et al.*, 2001).

PREVENTION. The species has a strong history of expansion and quarantine measures may help to limit its spread. Thus, in India, a lack of quarantine measures is thought to have been responsible for the introduction of the mite (Prasad and Ranganath, 2000).

PLANT RESISTANCE. The current emphasis of research on control of the coconut mite is to identify and develop mite-resistant coconut varieties (Shalini *et al.*, 2007). The arrangement of the bracts of growing coconuts influences the pattern of attack by *A. guerreronis*. Mite infestation is worst when the bract is less tightly adpressed to the nut and where bracts overlap one another, so it is concluded that physical space is necessary for mite infestations to become established. The tightness of adpression, as in some Cambodian cultivars, may be an advantageous characteristic to select for when replanting (Mariau, 1977; Moore, 1986; Moore and Alexander, 1990). A number of studies have indicated that there are various degrees of susceptibility in different coconut varieties. Thus, the incidence of attack on 'Yellow Dwarf' coconuts was greater than on 'West African Tall' varieties. Moreover, a Cambodian variety was immune, probably because the floral parts adhered firmly to the nut (Mariau, 1977). Further research showed that the 'West African Tall' and 'Malayan Yellow Dwarf' varieties were very susceptible to the mite, whereas the hybrid between these two, as well as other varieties such as 'Malayan Tall', 'Tahiti Tall' and 'Cameroon Red Dwarf', performed better (Julia and Mariau, 1979b). Resistance to the mite, assessed as damage to the nuts, was greater in 'Malay Tall', 'Tahiti Tall' and 'Cameroon Red Dwarf' than in 'Malay Yellow Dwarf', 'West African Tall' and 'Malay Green Dwarf' varieties. The tolerant hybrid PB121 was derived from two susceptible parents, 'West African Tall' and 'Malay Yellow Dwarf' (Julia and Mariau, 1979a). The 'West African Tall' or 'Grand Ouest African' varieties were always more susceptible than other African or foreign varieties. Even within a variety or hybrid, considerable individual differences in susceptibility were observed between trees standing together; this led to the conclusion

that coconut resistance might be improved by genetic means (Mariau, 1986). Up to 90% of trees and fruits were damaged in the variety 'Atlantic Tall', with lower infestation of 'Malayan Yellow Dwarf' (Schliesske, 1988).

On the cultivars 'West Coast Tall' and 'Dwarf Green', infestation significantly affected kernel weight, pulp thickness and nut water volume. Infestation reduced pulp thickness particularly during the tender stage. The 'Dwarf Green' cultivar showed a greater reduction in pulp thickness due to infestation than 'West Coast Tall'. The acidity level in coconut water is reduced due to infestation. However, the maximum acidity is observed in infested 'Dwarf Green' plants (Meenakshi and Vadivel, 2004). In the Ivory Coast, comparisons made between dwarf × tall hybrids (90% PB121, 10% PB111 and PB141) and 25% local 'West African Talls' (WAT) showed that the hybrids were superior to WAT with regard to precocity, number of nuts ha^{-1} and copra per nut (on average, 80% higher). Hybrids are more tolerant to *A. guerreronis* but more susceptible to nut fall caused by the pathogenic fungus *Phytophthora heveae* Thompson than WAT (Taffin *et al.*, 1991). Different responses to attack by *A. guerreronis* were shown by nine 'Jamaica Tall' genotypes grown in St Lucia; some trees were highly susceptible whereas others were reliably resistant. Reduced attack was associated with a rounded nut shape and green nut colour (Moore and Alexander, 1990).

CULTURAL PRACTICES. Management efforts include a programme to develop and popularize eco-friendly nutrient management, including organic manuring and botanical pesticides (Moore, 2000; Pushpa and Nandihalli, 2008). Studies of a hybrid crop suffering from water shortage and subsequently irrigated showed that drought aggravated the situation by causing the nuts to develop more slowly and thus remain susceptible to the mite for a longer period (Mariau, 1977, 1986). Crops that are rationally fertilized may be less infested (Mariau, 1977), and an increase in nutrients can worsen the levels of infestation. There are some associations between nutrient status (levels of the major nutrients, N, P and K) and levels of damage by *A. guerreronis*. Damage generally increases with increasing levels of N, and the level of K may be associated with decreasing mite damage (Moore *et al.*, 1991). In St Lucia, the major scope for yield increases did not lie with control, but rather in improved agronomic practices, which, if allied to replanting with improved and mite-resistant cultures, could at least double the current yields (Moore *et al.*, 1989).

BIOLOGICAL CONTROL. The current emphasis of research is to survey for promising predators (Lawson-Balagbo *et al.*, 2008) and for acaropathogenic fungi as prospective mycoacaricides (Rosas Acevedo and Sampedro Rosas, 2006; Edgington *et al.*, 2008). The action of the acaropathogen fungus *H. thompsonii* has been assessed with unclear results in different countries, e.g. in Mexico (Espinosa Becerril and Carrillo Sanchez, 1986; Lampedro and Luis Rosas, 1989; Beevi *et al.*, 1999), Cuba (Cabrera and Dominguez, 1987a) and St Lucia (Moore *et al.*, 1989); the fungus is also reported from Jamaica and the Ivory Coast (Hall *et al.*, 1980). The pathogenic fungus *Hirsutella nodulosa* Petch is reported from Cuba (Cabrera and Dominguez, 1987a). Further research on the acaropathogenic fungus *Fusarium semitectum* Berkeley *et* Ravenel, *Fusarium* sp. isolate GM15, *Lecanicillium lecanii* R. Zare *et* W. Gams, *Beauveria bassiana* (Balsamo) Vuill., *Metarhizium anisopliae* (Metchnikoff) Sorokin and *Trichoderma viride* Pers. showed that 23 days after spraying, *Fusarium* sp. isolate GM15 and *F. semitectum* were the most effective (Mikunthan and Manjunatha, 2010). On predatory mites, Aratchige *et al.* (2007) reported that 68% of coconuts infested by *A. guerreronis* had a perianth gap 100 µm wider than in uninfested fruits, and enabled the phytoseiid mite *Neoseiulus baraki* Athias Henriot to enter beneath the perianth and prey on the coconut mite.

Despite its limits, classical biological control is considered to be a promising strategy against populations of *A. guerreronis* (Moraes and Zacarias, 2002). Molecular analysis has shown the American origin of *A. guerreronis* and suggested that biological control agents for the species should be found in the Americas (Lawson-Balagbo *et al.*, 2007a,b, 2008). Critical to success in finding effective agents for biological control is the determination of the historical range of the mite (Návia *et al.*, 2010a). At present, international efforts supported by WOTRO Science for Global Development, a division of the Netherlands Organization for Scientific Research (NWO), are directed to the development of a classical biological control programme in Africa and Asia. The search for natural enemies will be focused in Brazil, Colombia, Mexico and Venezuela; the target countries for their introduction will be Benin, Tanzania and Sri Lanka. Researchers working on this project include those working at the ESALQ/University of São Paulo-Brazil, the University of Amsterdam in the Netherlands, the International Institute of Tropical Agriculture (IITA) in Benin, the Coconut Research Institute in Sri Lanka, and collaborators from Colombia, Mexico and Venezuela (Návia *et al.*, 2010a).

PHYSICAL CONTROL. In St Lucia, a sticky material has been used for physical control of the mite, with unclear results (Moore *et al.*, 1989).

CHEMICAL CONTROL. In general, chemical control is not practicable for the coconut mite (Moore and Howard, 1996), because it is scarcely effective (Moraes and Flechtmann, 2007) and limits the development of predators. Chemicals are also expensive and difficult to apply because mite colonies develop at the tops of the palms (Hernández Roque, 1977; Ramaraju *et al.*, 2002). Application of an acaricide every 2 months considerably reduced losses (by about 90%). However, the use of a non-selective chemical six times a year is not advisable, and the use of biological control is recommended from time to time (Mariau, 1977). Acaricides must be applied several times, and in most production areas of the world, farmers cannot afford the continuous sprayings of chemicals that are required (Moore and Howard, 1996; Muthiah and Bhaskaran, 2000; Ramaraju *et al.*, 2002).

Active ingredients used in the different countries of the world include fungicides, insecticides and acaricides. The most common chemicals are abamectin (Mikunthan and Manjunatha, 2010), carbophenothion (Ortega and Banda Torre, 1967; Ortega *et al.*, 1967), carbosulfan (Sujatha *et al.*, 2003), chlorpyrifos (Ortega and Banda Torre, 1967; Ortega *et al.*, 1967), cyhexatin

(Hernández Roque, 1977; Mariau, 1977; Julia and Mariau, 1979b), DDT (Olvera-Fonseca, 1986), dicofol (Nair, 2002), dicrotophos (Hernández Roque, 1977), fenpyroximate (Sujatha *et al.*, 2003), mancozeb (Julia and Mariau, 1979b), methyl demeton (Nair, 2002), phosphamidon (Ortega and Banda Torre, 1967; Ortega *et al.*, 1967), monocrotophos (Ortega and Banda Torre, 1967; Ortega *et al.*, 1967; Hernández Roque, 1977; Mariau, 1977; Julia and Mariau, 1979b; Nair, 2002; Sujatha *et al.*, 2003), quinomethionate (Hernández Roque, 1977; Mariau, 1977; Julia and Mariau, 1979b), BHC (correct name HCH), polimar and palmarol (Olvera-Fonseca, 1986), triazophos (Sujatha *et al.*, 2003), and vamidothion (Moore *et al.*, 1989). Stem injection of vamidothion was ineffective (Moore and Alexander, 1987b).

Active ingredients with low impact include monthly treatment of the inflorescences of inland trees with sprayings of seawater, which almost halved the rate of infestation (Mariau, 1977), botanical insecticides (Pushpa and Nandihalli, 2008), among which are a neem oil/garlic mixture (Nair, 2002) and salt (sodium chloride) (Olvera-Fonseca, 1986). Azadirachtin gives an intermediate action (Mikunthan and Manjunatha, 2010).

Aceria merwei (Tucker) (Fig. 16.39)

Common name

Unknown.

Diagnostic characteristics

FEMALE. The body is narrow and cylindrical in shape, and 200 μm long. The opisthosoma has about 80 annuli and the empodium is 6-rayed. The tarsal solenidion is very long (Tucker, 1926; Meyer Smith, 1981b).

Geographical distribution

South Africa (Tucker, 1926; Ryke and Meyer, 1960; Meyer Smith, 1981b; Meyer Smith and Craemer, 1999).

Host plants

Poaceae: sugarcane (*Saccharum officinarum* L.) (Tucker, 1926; Meyer Smith, 1981b; Meyer Smith and Craemer, 1999).

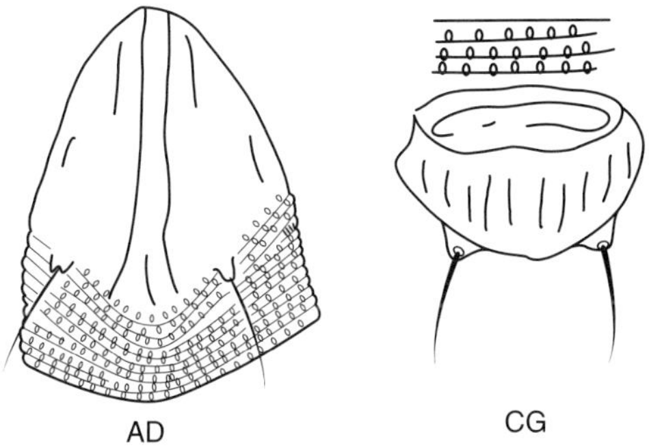

Fig. 16.39. *Aceria merwei* (Tucker) (from Meyer Smith, 1981b). See Table 16.1 for key.

Bio-ecology

The mite lives in the buds of sugarcane (Tucker, 1926; Meyer Smith, 1981b; Meyer Smith and Craemer, 1999). Its bio-ecology is not known.

Symptomatology

The feeding activity of the mite causes stunted growth (Tucker, 1926; Meyer Smith, 1981b), but it seems to be of secondary economic importance.

Control

This seems not to be necessary.

Aceria sacchari Wang

Common name

Sugarcane blister mite.

Diagnostic characteristics

FEMALE. The body is worm like, 180–260 μm long, and light yellowish in colour. The prodorsal shield has central long lines, of which the median and admedian are complete. The coxae have many lines. The opisthosomal microtubercles are elongate elliptical, touching the margins of the annuli. The genital coverflap has 8–12 longitudinal ribs. There are two forms, which are referred to as the tick and slender form; the slender form has 6-rayed empodia and the thick form has 7-rayed empodia (ChannaBasavanna, 1966; Jeppson *et al.*, 1975).

Geographical distribution

Bangalore (India) (ChannaBasavanna, 1966), Iran (Kamali *et al.*, 2001; Xue *et al.*, 2009a) and Taiwan (Wang, 1964; Huang, 2008). The mite may be spreading into other countries (see Bio-ecology below).

Host plants

Poaceae: sugarcane (*Saccharum officinarum* L.) (Wang, 1964; ChannaBasavanna, 1966, 1996; Jeppson *et al.*, 1975; Kamali *et al.*, 2001).

Bio-ecology

Until recently, the number of eriophyoid mites reported from sugarcane in various countries of the world was eight: *Abacarus sacchari* ChannaBasavanna, *Abacarus delhiensis* ChannaBasavanna, *Abacarus queenslandiensis* Ozman-Sullivan, Amrine *et* Walter, *Aceria merwei* (Tucker), *Aceria sacchari* Wang, *Catarhinus sacchari* Kuang, *Cathetacarus spontaneae* (Mohanasundaram) and *Diptacus sacchari* (Xin *et* Dong) (Özman-Sullivan *et al.*, 2006). Recently, another has been added: *Abacarus doctus* Návia, Flechtmann, Lindquist *et* Aguilar (Návia *et al.*, 2011).

So currently, there are four *Abacarus* spp. (*A. doctus*, *A. delhiensis*, *A. queenslandiensis* and *A. sacchari* reported from sugarcane (*Saccharum* spp.).

The new species, *Abacarus doctus*, is a vagrant species that is found on the inner surface of the sugarcane leaves in Costa Rica, and can cause reddish or bronzed spots to spread randomly on the inner leaf surfaces which can extend to cover almost the entire leaf sheath. At a distance, such damage symptoms could be confused with those caused by rust fungi, which is commonly associated with sugarcane (Návia *et al.*, 2011). Three other species, *Abacarus delhiensis*, *A. sacchari* and *Cathetacarus spontaneae*, are leaf vagrants on *Saccharum spontaneum* L. in India, and exhibit symptoms of damage that are either not apparent or not stated (ChannaBasavanna, 1966; Mohanasundaram, 1984). Six species (*Abacarus sacchari*, *Abacus merwei*, *Aceria sacchari*, *Cathetacarus spontaneae*, *Catarhinus sacchari*, *D. sacchari*) are leaf vagrants on sugarcane proper (*S. officinarum* L.). Of these, *Abacarus sacchari* causes occasional chlorotic blotches on leaves and may be a vector of plant viruses; it is distributed in India, Brazil, Réunion, Cape Verde, Venezuela and the French Antilles (also known as the French West Indies) (ChannaBasavanna, 1966; Flechtmann and Aranda, 1970; Gutierrez, 1982; Gupta, 1985; Meyer Smith, 1989b; Cermeli and Aponte, 1990; Flechtmann *et al.*, 1999b). *Aceria merwei* is a bud mite that causes stunting and dwarfing, and has been reported from South Africa (Tucker, 1926; Nalepa, 1929; Ryke and Meyer, 1960; Meyer Smith, 1981b). *Aceria sacchari* causes leaf sheath 'blisters', and is probably the vector of streak virus; the mite has formally been reported from India, Indonesia, Iran, Java (Indonesia), Taiwan and Australia (Van Hall, 1924; Box, 1953; Muthukrishnan, 1956; Wang, 1964; ChannaBasavanna, 1966, 1996; Sithanantham *et al.*, 1973; Jeppson *et al.*, 1975; Mohanasundaram, 1981c; Singh, 1982; Baco *et al.*, 1991, 1992; Kamali *et al.*, 2001). Symptoms of damage in the remaining species are either not apparent or not stated. The last mite, *Abacarus queenslandiensis*, is a leaf vagrant on *Saccharum* spp. hybrids in Australia and causes discoloration of the leaves and may be a virus vector (Özman-Sullivan *et al.*, 2006). Among the species discussed above, the sugarcane blister mite is economically most important, although it is usually a minor pest.

Aceria sacchari lives on the inner surface of the leaf sheaths of sugarcane (*Saccharum* sp.) and is well known in India (ChannaBasavanna, 1966) and Taiwan (Wang, 1964). The records of the mite in Queensland (Australia) (Box, 1953), Java (Van Hall, 1924) and other South Asian countries are doubtful. Halliday (2010) reviewed the historical records of some Eriophyidae infesting rice (*Oryza sativa* L.) and sugarcane in South-east Asia and Australia, and pointed out that the names *Aceria saccharini* Wang, *Aceria sacchari* Wang, *Eriophyes saccharini* (Wang), *Aceria saccharium* ChannaBasavanna and *Abacarus sacchari* ChannaBasavanna have all been applied to these mites, but the identity of the species that occur in Australia has not been definitely established. Inevitably, the situation involves taxonomic and nomenclatural confusion that affect both pest management and biosecurity. Appropriate studies on the taxonomy and bio-ecology of *A. sacchari* are, therefore, needed, in order to determine the exact identity and distribution of the species in different countries. To avoid increasing the existing confusion, until the problem is resolved, the information reported here is as summarized by ChannaBasavanna (1996).

The mite lives where the tissues become disrupted to a spongy form. Infestation is severe in the ratoon crops, and in India the highest population density occurs in the rainy season from June to October (Singh, 1966), while in Taiwan, mite populations increase in the dry season (Wang, 1964). Eggs are laid in the spongy tissues of the blister, and at room temperature the development from egg to adult takes about 15 days (Puttarudriah and Usman, 1957). Both forms, the thick and the slender, can give rise to both forms, but the slender form seems to be more adapted for dispersal (Mohanasundaram, 1981c). Entry of the mite is most frequent into the fourth to the tenth leaf sheath from the apex. Leaf sheaths 7–10, respectively, from the apex host a higher population density at 4, 6, 8 and 10 months of age of the host plant (Sithanantham, 1981).

Symptomatology and damage

The feeding activity of *A. sacchari* causes large spots 1–2 cm in diameter on both sides of the sheaths. These spots first appear watery and afterwards become dark red. Hypertrophied spherical leaf blisters contain mite populations (Wang, 1964; ChannaBasavanna, 1966; Jeppson *et al.*, 1975). Damaged tissues show both anatomical and chemical changes. The anatomical changes are in the shape and size of the cells of the blister tissues (Agarwal and Kandasami, 1959). Direct effects of the mite on the chemical composition of the sugarcane leaf sheath include a higher content of potassium, nitrogen and phosphorus in the blister tissues, but lower contents of calcium and magnesium (Agarwal and Kandasami, 1959; Sithanantham *et al.*, 1975). The damage to the leaf sheath facilitates infection by saprophytic fungi, such as *Gleosporium*, *Fusarium* and *Alternaria* (Muthukrishnan, 1956).

The mechanisms of susceptibility of sugarcane varieties have been investigated (Sithanantham and Velayutham, 1981). Various morphological characters are associated with susceptible varieties, such as opening leaves with the blades curved in the middle, which are probably more receptive to mites being moved by air currents. Sheath moisture and thickness favour the development of the mites that are present at high levels. The positive influence of sheath thickness on infestation has been known for some time (Agarwal, 1969). The size and density of the blisters vary among varieties. Thus, the 'CO 419' variety shows larger and more numerous blisters than other varieties (Sithanantham *et al.*, 1975). The mite has been shown to have a role in the transmission of the sugarcane streak virus (Sithanantham *et al.*, 1972), expression of which may be related to varietal differences (Sithanantham *et al.*, 1973).

Control

There are only a few reports on the control of this mite, which is carried out by spray applications of chemicals that are registered for the infested crop (Singh, 1982; Baco *et al.*, 1992).

Aceria slykhuisi Hall

Common name

Buffalo grass mite.

Diagnostic characteristics

FEMALE. The body is worm like and 220 μm long. The opisthosoma has 65–70 annuli, with dorsal elongate microtubercles, more bead like laterally, slightly pointed ventrally and ahead of the margins of the annuli. The coxae are granular. The genital coverflap has about 7–8 ribs. The empodium is 6-rayed, sometimes 7-rayed (Jeppson *et al.*, 1975).

Geographical distribution

Kansas and the midwestern USA (Hall, 1958; Jeppson *et al.*, 1975).

Host plants

Poaceae: buffalo grass (*Buchloe dactyloides* (Nutl.) Engelm.) (Hall, 1958, 1967; Jeppson *et al.*, 1975).

Bio-ecology

A. slykhuisi is reported from buffalo grass in Kansas and midwestern areas of North America (Hall, 1958, 1967; Jeppson *et al.*, 1975).

Symptomatology and damage

The buffalo grass mite has been associated with witch brooming of pistillate plants (Hall, 1967). It seems to be injurious to turf grass sites, but few reports have been found in the literature.

Control

Control is provided by the use of resistant varieties, e.g. the 'Bowie' cultivar, which has an excellent resistance to the buffalo grass mite, and good resistance to leaf spot (caused by *Helminthosporium* spp.) and dollar spot (caused by *Sclerotinia homoeocarpa* F.T. Bennett); it also has moderate resistance to ergot (caused by *Claviceps purpurea* (Fr.) Tul). The use of the 'Bowie' variety is suggested for low- to medium-maintenance turf grass sites in the northern and central Great Plains arid regions of the USA (Severmutlu *et al.*, 2005a). 'Cody' cultivar buffalo grass has shown excellent winter hardiness and drought tolerance, and has excellent resistance to mealybug and to the buffalo grass mite. 'Cody' is recommended for low- to medium-maintenance turf grass sites in the northern, southern and central Great Plains and southern arid regions of the USA (Severmutlu *et al.*, 2005b).

Aceria tosichella Keifer (Fig. 16.40)

Common name

Wheat curl mite.

Diagnostic characteristics

FEMALE. The body is worm like, 200–230 μm long, and the colour in life is light yellowish white. The gnathosoma projects forward and down. The prodorsal shield is generally acuminate anteriorly, with a small frontal lobe over the base of the gnathosoma. The median shield line is present on the rear ½ or ⅔; the admedian lines are complete from the frontal lobe, gradually diverging to the rear margin and somewhat sinuate; the submedian line from the front margin is lateral to the admedian, running back towards the scapular tubercle, curving laterally in front of the tubercle, and with a broken inner fork on the inside of the tubercle. Laterally, the shield has two longitudinal lines, the upper less distinct and obscurely branched, ending in granules below the tubercle; the lower line is from the base of the gnathosoma to granulations below the tubercle; there are bands of granulations above the coxae and dipping between the coxae. The scapular tubercles are 22 μm apart; the scapular setae (*sc*) are about 60 μm long, diverging to the rear. The coxae are ornamented with short lines and lines of granules; there is a moderately long sternal line set between coxae I. The opisthosoma has about 66 annuli, which are completely microtuberculate; the microtubercles tend to be slight, especially dorsally and ventrally, and just behind the genitalia; the microtubercles are not particularly ahead of the margins of the annulus ventrally; the ventral microtubercles are ahead of the caudal opisthosoma, each with a thin anterior line. The setae c_2 are 44 μm long, set on annulus 9; the setae *d* are 50–60 μm long, set on annulus 21; the setae *e* are 50–60 μm long, set on annulus 42. The caudal opisthosoma has 5 annuli, with fine microtubercles on the margins, dorsally projecting over the margins as short points, laterally and ventrally, each with a fine line ahead of the margin, especially ventrally. The setae h_1 are 4–5 μm long. The genital coverflap has about 10 longitudinal ribs. The empodium is 7- to 8-rayed (Keifer, 1969c).

Geographical distribution

Worldwide (Oldfield and Proeseler, 1996; Harvey *et al.*, 2002; de Lillo and Beron, 2004). The mite is spreading in South America, where it has been reported from Argentina (Návia *et al.*, 2006b), Brazil (Pereira *et al.*, 2009) and Uruguay (Castiglioni and Návia, 2010), and also in China (Hong *et al.*, 2006). It has also been reported from Australia (Halliday, 2004) and Iran (Malik *et al.*, 2003a).

Host plants

Poaceae: wheat (*Triticum sativum* Lam.), maize, sorghum, barley, oat, rye, pearl millet and a large number of grasses of minor economic importance, as well as various weeds (Keifer, 1969c; Jeppson *et al.*, 1975; Oldfield and Proeseler, 1996; Harvey *et al.*, 2002). The mite has been collected on about 120 species of monocotyledons (Nault and Styer, 1969), and also on ryegrass (*Lolium multiflorum* Lam.) and brome (*Bromus unioloides* (Willd.) Kunth) (Castiglioni and Návia, 2010).

Bio-ecology

Several reports of *Aceria tosichella* Keifer have been made under *Aceria tulipae* (Keifer). This misidentification does not facilitate the use of references and results in confusion.

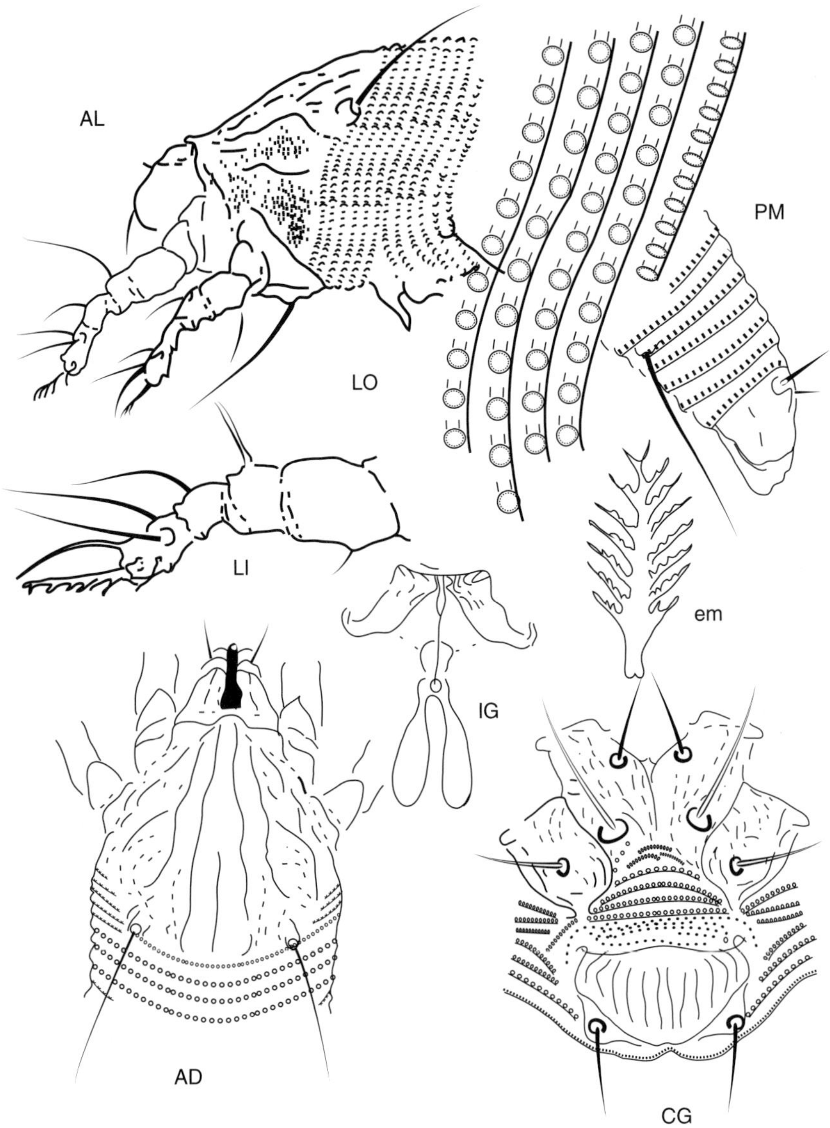

Fig. 16.40. *Aceria tosichella* Keifer (from, Keifer, 1969c). See Table 16.1 for key.

However, the species living on wheat, *A. tosichella*, does not develop on onion as does *Aceria tulipae*, and the two mites are separate species (Razvyazkina, 1966). Moreover, *Aceria tritici* Shevtchenko (Shevtchenko *et al.*, 1970) is synonymous with *A. tosichella* (Amrine and Stasny, 1994). The use of molecular markers has recently indicated that *A. tosichella* may be a species complex in Australia (Carew *et al.*, 2009).

The wheat curl mite was first described from wheat in Yugoslavia (Keifer, 1969c). It is a widespread pest of wheat and maize (Oldfield and Proeseler, 1996; Harvey *et al.*, 2002) and has been reported from the main wheat production areas in North America, Europe, Asia, the Middle East, Africa and Oceania (de Lillo and Beron, 2004). *A. tosichella* has been collected on about 120 species of monocotyledons (Nault and Styer, 1969), but it lives mainly on wheat and may occur on maize, sorghum, barley, oat, rye, pearl millet and a large number of grasses of minor economic importance, as well as on various weeds (Jeppson *et al.*, 1975). In Uruguay, *A. tosichella* has been found also on ryegrass (*L. multiflorum*) and brome (*B. unioloides*) (Castiglioni and Návia, 2010).

In north-central Washington State, during the vegetative stage of development, the highest density of mites was always found in the youngest leaf of a wheat or barley tiller. The mite populations were concentrated at the base of the leaves near the collar and the ligule. Mites continued to move to the succeeding terminal leaves on each tiller as it developed, and then into the wheat head as it emerged from the flag leaf. In club winter wheat, the mite populations stabilized, while in common (awned) winter and spring wheat and barley, the density increased quickly as the heads emerged from the flag leaf (Gillespie *et al.*, 1997). In Tibet, the mite develops 6–7 generations a year; damage to wheat and barley leaves was observed in January and June–July, and to grains in February and July–August. Temperature and RH affect mite development significantly (Lin *et al.*, 1987).

A. tosichella behaves differently according to the quality of its grass host plant. It stays on hosts during periods of active

plant growth and disperses when host plant quality is deteriorating because of heavy feeding damage. All biological stages overwinter on the wheat leaf (Nault and Styer, 1969). Adults and eggs can survive the lowest temperatures of the North American winter (Slykhuis, 1955; Briones and Sill, 1972).

At the beginning of a new dispersal, the mite exhibits non-preference for the host plant (Michalska *et al.*, 2010). Although wheat is an annual plant, it has early and late varieties, so *A. tosichella* commonly overwinters on new annual growth occurring late in the season or on winter wheat, and then disperses to summer wheat when it appears (Skoracka, in Smith *et al.*, 2010). The landing efficiency of the wheat curl mite increases on cultivars that have high densities of leaf trichomes (Harvey *et al.*, 1990). Mites may occur for a limited period on different grasses growing between harvest and the emergence of volunteer wheat. A few mites developing on volunteer wheat may be sufficient to begin an infestation on wheat. In this way, volunteer wheat emerging just before or soon after harvest can play an important role in the epidemiology of *Wheat streak mosaic virus* (WSMV, a virus that is transmitted by *A. tosichella*) in the Central Plains States of the USA (Connin, 1956). In the Northwest and Northern Plains States, there is a large incidence of WSMV because of the overlap of the winter and spring wheat crops (Edwards and McMullen, 1988).

Although dispersal of the wheat curl mite may be via aphids, thrips and greenbugs (for references see Michalska *et al.*, 2010), air dispersal appears to be the principal mechanism (Slykhuis, 1955; Staples and Allington, 1956; Gibson and Painter, 1957; Nault and Styer, 1969), and this involves high mortality of the mite populations (Jeppson *et al.*, 1975). However, the high number of grasses reported as hosts for the mite suggest that it is also transported on contaminant grasses accompanying potted ornamental plants infested by mites (Návia *et al.*, 2006b). Another hypothesis is that *A. tosichella* is disseminated through seeds. At high mite populations in the field, or when mites start aerial dispersal at harvest time, life stages could be associated with seeds during their transport, and so colonize new areas that are being sown. The interception of *A. tosichella* and *Aceria zeala* Keifer on wheat and maize germplasm seeds at the Embrapa Laboratory of Plant Quarantine in Brazil (Návia and Flechtmann, 2008) supports this possibility. In north-central Washington State, the mites disperse to spring wheat and barley while the plants are tillering, well after the seedling stage, when the plants are most susceptible to damage from mite feeding and infection by WSMV. Cool moist summers, hailstorms and warm winters create a favourable set of environmental conditions for the first outbreak of wheat curl mites and WSMV (Gillespie *et al.*, 1997).

Mite biology depends on its dispersal and available hosts, and differences are recorded from one region to another (Styer and Nault, 1996). At the end of the wheat crop in the Central Plains of the USA, the mites move on to volunteer wheat and wild grass (Connin, 1956). In the northern regions of North America, the winter and spring wheat crops allow mite survival (Slykhuis, 1955; Edwards and McMullen, 1988), while in the South Central States of the USA, the mite populations develop on wheat and irrigated maize that overlap in the spring and autumn (Wiese, 1987). In the Corn Belt States of the USA, the mites move from wheat to maize in early summer, where high populations develop until maturation, after which mite dispersal and the infestation of new wheat occur (Nault and Styer, 1969).

Symptomatology and damage

A. tosichella has been collected on about 120 species of monocotyledon plants (Nault and Styer, 1969). It lives mainly on wheat, but it can also occur on maize, sorghum, barley, oat, rye and pearl millet, and on a large number of grasses of minor economic importance, as well as various weeds (Jeppson *et al.*, 1975). Recently, it has also been found on ryegrass and brome (Castiglioni and Návia, 2010). The feeding activity of the mite causes direct damage to wheat and other host plants and transmits various viruses (*Wheat streak mosaic virus*, the Wheat spot mosaic pathogen and *High Plains virus*) (Oldfield and Proeseler, 1996; Styer and Nault, 1996; Malik *et al.*, 2003b). In addition to these viruses/pathogens, two other new viruses transmitted by *A. tosichella* have been reported, the *Brome streak mosaic virus* (BSMV) from Europe (Stephan *et al.*, 2008) and *Triticum mosaic virus* (TMV) from the USA (de Wolf and Seifers, 2008; Seifers *et al.*, 2009); however, no information is available on the losses caused by these viruses. *A. tosichella* also transmits Kernel red streak (Nault *et al.*, 1967; Styer and Nault, 1996; CABI, 2002). The direct plant damage and virus diseases caused by this mite are summarized below.

DIRECT DAMAGE. On wheat plants, mite infestation begins inside the whorl of the young leaf, where the mites settle on the adaxial surface and feed on tissue between veins in the grooved part bearing the bulliform cells (Orlob, 1966). In normal development, the growing leaves unroll after the enlargement of these cells (Esau, 1960). However, the mites feed only on the bulliform cells and consequently prevent the growing leaf from uncurling. The growth of the curled leaf traps inside it the subsequent leaf and causes curled, looped and trapped leaves. This leads to moisture conditions that facilitate the development of mites (Styer and Nault, 1996). In Tibet, *A. tosichella* damages wheat, highland barley, maize and some weeds. Damaged leaves appear curved in shape and the heart leaf appears 'R'-shaped. Infested wheat grains are easily broken and the grain weight and yield decrease significantly (Lin *et al.*, 1987). On maize, mite feeding activity causes kernel red streak (Styer and Nault, 1996). Yield losses to wheat crops due to wheat curl mite infestations can reach 30% (Harvey *et al.*, 2002).

WHEAT STREAK MOSAIC VIRUS. The wheat curl mite is the only known vector of the *Wheat streak mosaic virus* (WSMV) (Slykhuis, 1955), a widespread virus that is probably present in all areas of wheat cultivation in the Palaearctic and Nearctic regions. The symptoms of WSMV transmitted by *A. tosichella* are light green to yellow streaks and dashes on the leaves, followed by general chlorosis, stunting and, depending on the degree of severity, poorly filled seed heads and shrivelled kernels. At 20–25°C, the symptoms on wheat appear 6–8 days after inoculation (Styer and Nault, 1996). Immature and adult stages transmit the virus, and the adults can acquire the pathogen as immatures or adults. The virus persists in the wheat curl mite for several days (Slykhuis, 1955; Orlob, 1966). Large numbers of viral particles

occur in the midgut of the mites and persist undegraded at this site for about 5 days. The particles are frequently observed in the haemocoel and in the salivary glands, but not in the hindgut (Paliwal, 1980).

WHEAT SPOT MOSAIC DISEASE. Symptoms of wheat spot mosaic disease, which is thought to be caused by the *Wheat spot mosaic virus* (WSpMV), are transmitted by *A. tosichella*. The symptoms are initial chlorotic spotting or ring spot, followed by general chlorosis, stunting and sometimes plant death. This symptomatology has been observed in several Poaceae, including maize, barley and wheat (Slykhuis, 1956; Nault *at al.*, 1970; Styer and Nault, 1996). Large amounts of WSMV accumulate in the mid-gut of *A. tosichella* feeding on infected plants. Masses of flexuous rod-shaped virus particles persist in the midgut, without degradation, for at least 5 of the 6–9 days of the adult life of the mites. WSMV particles are also found in the body cavity and in the salivary glands of mites, showing that the virus probably circulates (Paliwal, 1980). The transmission of WSMV from wheat to sorghum by the mite is sporadic, inefficient and probably of limited epidemiological significance (Harvey and Seifers, 1991). Studies on the persistence of virus in the vector showed a lack of transovarial passage but that the virus was retained after moulting in both WSMV and the WSpMV (Nault and Styer, 1970). WSpMV is not sap transmissible (Styer and Nault, 1996).

HIGH PLAINS VIRUS. The *High Plains virus* (HPV) is reported in the Great Plains of the central USA (Jensen *et al.*, 1996), Nebraska (Mahmood *et al.*, 1998) and Kansas (Seifers *et al.*, 2002; Skare *et al.*, 2003, 2006). It causes economic damage on wheat and maize (Jensen *et al.*, 1996).

KERNEL RED STREAK. Kernel red streak (KRS) is a not virus disease of maize. It is characterized by streaks of colour ranging from a deep red in yellow kernels to a pink or purple in white kernels. The presence of red pigment within the pericarp forms these streaks. Sometimes, the entire pericarp may bear this pigment and rarely is it present in the endosperm. The symptomatology is directly related to mite density and may depend on a phytotoxic salivary substance (Nault *et al.*, 1967). No economic damage is attributed to KRS (Styer and Nault, 1996).

VIRUS DISEASE YIELD LOSSES. The main damage resulting from wheat curl mite infestation is due to the transmission of WSMV and HPV (Oldfield and Proeseler, 1996; Malik *et al.*, 2003b). WSMV causes higher yield losses in North and South America, Europe, the Middle East, Oceania and Asia (Oldfield and Proeseler, 1996; Sanchez Sanchez *et al.*, 2001; French and Stenger, 2003; Truol *et al.*, 2004). HPV is reported in the Great Plains of central USA (Jensen *et al.*, 1996), Nebraska (Mahmood *et al.*, 1998) and Kansas (Seifers *et al.*, 2002; Skare *et al.*, 2003, 2006) and causes economic damage on wheat and maize (Jensen *et al.*, 1996). Mixed infections of both viruses (WSMV and HPV) make it difficult to estimate the losses caused by each virus in the USA. Losses from HPV infection in maize are estimated by the Australian Quarantine and Inspection Service (AQIS, 2000) to be about 75%. In Kansas, more severe damage occurs. In the USA, losses due to WSMV amounted to US$30 million in 1949 (Fellows and Sill, 1955); yield losses in 1974 were 30 million bushels and in 1981 they were 21 million bushels (Martin *et al.*, 1984); in 1988, a 13% loss was estimated (Christian and Willis, 1993). A pest risk analysis (PRA) for *A. tosichella* and WSMV/HPV was carried out in South America that considered as risk areas the non-affected regions of Argentina, and all territories of Brazil, Uruguay and Paraguay, showed that the risk of introduction, establishment and economic importance of these pests in the risk areas was high (Návia *et al.*, 2007a,b). In Argentina, WSMV has spread to the main wheat production areas (Truol *et al.*, 2008). Severe WSMV epiphytes have been observed in wheat (Truol and Sagadin, 2008a; Truol *et al.*, 2008). The presence of HPV was also confirmed in mixed infections with WSMV in Argentina (Truol and Sagadin, 2008b). At present, the pathosystem *A. tosichella* and WSMV/HPV is a threat to wheat production in Argentina because all of the wheat cultivars commonly used for short and intermediate cycles are susceptible (Bainotti and Vanzetti, 2008).

Control

Control aims to limit feeding damage and – mostly – virus transmission. Resistant varieties, cultural practices (tillage), environmental factors (Jiang *et al.*, 2005; Coutts *et al.*, 2008), the control of volunteer wheats (Thomas and Hein, 2003) and spray applications of acaricides all contribute in various ways. Biomolecular methods can limit virus transmission (Stenger *et al.*, 2006; Young *et al.*, 2007).

MONITORING. Monitoring allows measurement of the numbers of *A. tosichella* on wheat cultivars in the field in order to help in its management. By excising milk-stage wheat spikes and placing them on strips of transparent sticky tape, it is possible to count the numbers of specimens of *A. tosichella* that crawl out of the drying heads and become stuck on the tape (Harvey and Martin, 1988).

PLANT RESISTANCE. Resistance to the wheat curl mite has been investigated in order to limit mite development and transmission of virus disease. Harvey *et al.* (1994) reported that the use of varieties resistant to *A. tosichella* in Kansas reduced the incidence of the virus by an average of 76%. In the 1980s, breeders begin to use the resistance of rye (*Secale cereale* L.) to the mite. In this period, some new lines of hard, red winter wheat were developed that carried resistance to the mite contributed by the wheat–rye translocator cultivar 'Salmon'. A representative line from these trials reduced the incidence of WSMV by 74% (Martin *et al.*, 1984). A close relative of wheat, *Aegilops squarrosa* L., also exhibits resistance to the mite and the characters seems to be readily transmitted to wheat (Thomas and Conner, 1986; Thomas and Whelan, 1991).

The transferring of the genes for WSMV resistance from various grasses involved *Agropyron elongatum* (Host) P. Beauv., a wild plant resistant to both the mite and the virus disease. Three lines were developed by crossing wheat with *A. elongatum* (Jiang *et al.*, 1993). The cross of the 'Rescue' cultivar × *A. elongatum* gave a resistant variety of hard, red spring

wheat (Whelan and Conner, 1989). Later, *Agropyron intermedium* (Host) P. Beauv. was investigated for the transfer of resistance to WSMV (Friebe *et al.*, 1996). A collection of wild 'Timopheevi' wheat was also evaluated for resistance to disease and arthropod pests (Brown Guedira *et al.*, 1996). Research on the relationships between plant morphology and resistance to the mite shows that new cultivars developed for areas where WSMV is a production constraint would probably benefit from having leaves with a low trichome density (Harvey *et al.*, 1990).

The prospects for global improvement of wheat also include studies on the genetics and gliadin protein characteristics of a wheat–alien translocation (Whelan and Lukow, 1990). Further, genetics research allows the identification of a *Triticum–Lophopyrum* non-compensating translocation line and the detection of *Lophopyrum* DNA using a wheatgrass-specific molecular probe (Kim *et al.*, 1992), as well as the cytological and molecular characterization of a chromosome interchange and addition lines in the 'Cadet' cultivar involving chromosome 5B of wheat and 6Ag of *L. ponticum* (Podp.) (Kim *et al.*, 1993a). Genetic segregation and detection of spontaneous wheat–alien translocations are being investigated as well (Thomas *et al.*, 1997).

From a practical point of view, research on resistant varieties allows the commercialization of resistant wheat varieties. The 'TAM 107' and 'KS85H22' varieties, both resistant to the wheat curl mite, host significantly lower numbers of mites in the spikes and have a lower incidence of WSMV than susceptible varieties (Harvey and Martin, 1988). 'TAM 107' reduces the incidence of WSMV in the wheat (Harvey *et al.*, 1995). The 'Caucho' variety showed a high degree of resistance to the mite (Wood *et al.*, 1995), as did the 'Amigo' variety (Sebesta *et al.*, 1995). The 'TAM 110' variety is resistant to some biotypes of *A. tosichella* and *Puccinia graminis* Pers. f. sp. *tritici* Eriks. *et* Henn. and *Erysiphe graminis* DC. f. sp. *tritici* Marchal (Lazar *et al.*, 1997). However, despite the commercial availability of resistant varieties, the problem is still wide open and being investigated (Chen *et al.*, 2003; Malik *et al.*, 2003a; Thomas *et al.*, 2004; Harvey *et al.*, 2005; Li *et al.*, 2005, 2007b).

CULTURAL PRACTICES. The destruction of volunteer wheat and sowing in the next year help in mite control. This practice allows the harvesting of the new crop when the mite populations have declined (Wiese, 1987), although there are obstacles resulting from the use of volunteer wheat as winter grazing and also to the planting date, which is managed according to soil moisture availability (Martin *et al.*, 1984).

CHEMICAL CONTROL. Several active chemical ingredients have been tested over time (Kantack and Knutson, 1958), including applications of the systemic insecticides carbofuran and disulfoton with wheat seed at sowing. Carbofuran treatment controlled the mite during the autumn, reduced the incidence of the virus in spring and resulted in increased grain yields; disulfoton was relatively ineffective (Harvey *et al.*, 1979). Imidacloprid seed treatment did not consistently control WSMV or its vector (Harvey *et al.*, 1998). In general, there is no known effective chemical control of the mite (Whelan and Conner, 1989), primarily because the part of the host plant where the mite populations live shelters them from chemicals (Kantack and Knutson, 1958).

Aceria tulipae (Keifer) (Fig. 16.41)

Common name

Dry bulb mite.

Diagnostic characteristics

FEMALE. The body is spindle formed, 210–250 μm long and whitish in colour. The gnathosoma is gently bent down. The prodorsal shield is a little curved above, subtriangular, slightly overlying the gnathosoma base; the shield has five diverging lines and the shield sides have curved lines and a band of granulations above the edge; the scapular tubercles are moderate in size and on the rear margin, placed 23 μm apart; the scapular setae (*sc*) project backwards and are over 60 μm long. The opisthosoma has 85–90 annuli, with some ventrad reduction in number; the annuli are about 2 μm wide, the microtubercles partly rest on the rear edge of each annulus; the last 5–6 annuli have microstriations. The setae c_2 are 35.5 μm long, set on about annulus 10; the setae *d* are 54 μm long, set about on annulus 27; the setae *e* are about 35 μm long, set on annulus 50; the setae *f* are 31 μm long, 5 or 6 annuli from the rear; the setae h_2 are about 80 μm long; the setae h_1 are 4.5 μm long. The genitalia are 24 μm wide, 17.5 μm long and subcordate; the coverflap is longitudinally furrowed, and the seta is 12 μm long (Keifer, 1938a).

Geographical distribution

Worldwide (de Lillo and Beron, 2004).

Host plants

Amaryllidaceae – shallot (*Allium cepa* L. var. *aggregatum*), onion (*A. cepa*), garlic (*A.* sativum L.); Liliaceae – tulips (*Tulipa* spp.) (Keifer, 1938a; Jeppson *et al.*, 1975).

Bio-ecology

The dry bulb mite was described in California on bulbs that had originated from Holland (the Netherlands) (Keifer, 1938a) and was for a long time misidentified with *Aceria tosichella* Keifer (Razvyazkina, 1966). Usually, it feeds between the bulb layers of shallot, onion, garlic and tulips, scarifying and drying the surfaces (Keifer, 1938a).

At 19°C and RH over 90%, pre-oviposition lasts 3.7 ± 0.48 days, the incubation period is 7.5 ± 1.43 days and the oviposition period is 25.1 ± 9.13 days; the fecundity is 44.7 ± 17.44 eggs, the daily oviposition is 1.76 ± 0.05 eggs and adult female longevity is 28.8 ± 10.70 days. At 18–20°C and at over 90% RH, the life cycle takes about 17 days (Wahba *et al.*, 1985). Eggs can endure temperatures of −31°C for 2.5 min (Jeppson *et al.*, 1975). In the laboratory, the shortest development time occurs at about 25°C. The upper lethal temperature and the development threshold for the eggs are estimated as about 45°C and 6°C, respectively. An RH close to 100% is required for a high percentage of egg hatching, but water condensation on leaves is harmful. Host transpiration has an important regulatory effect on the humidity conditions favouring mite survival

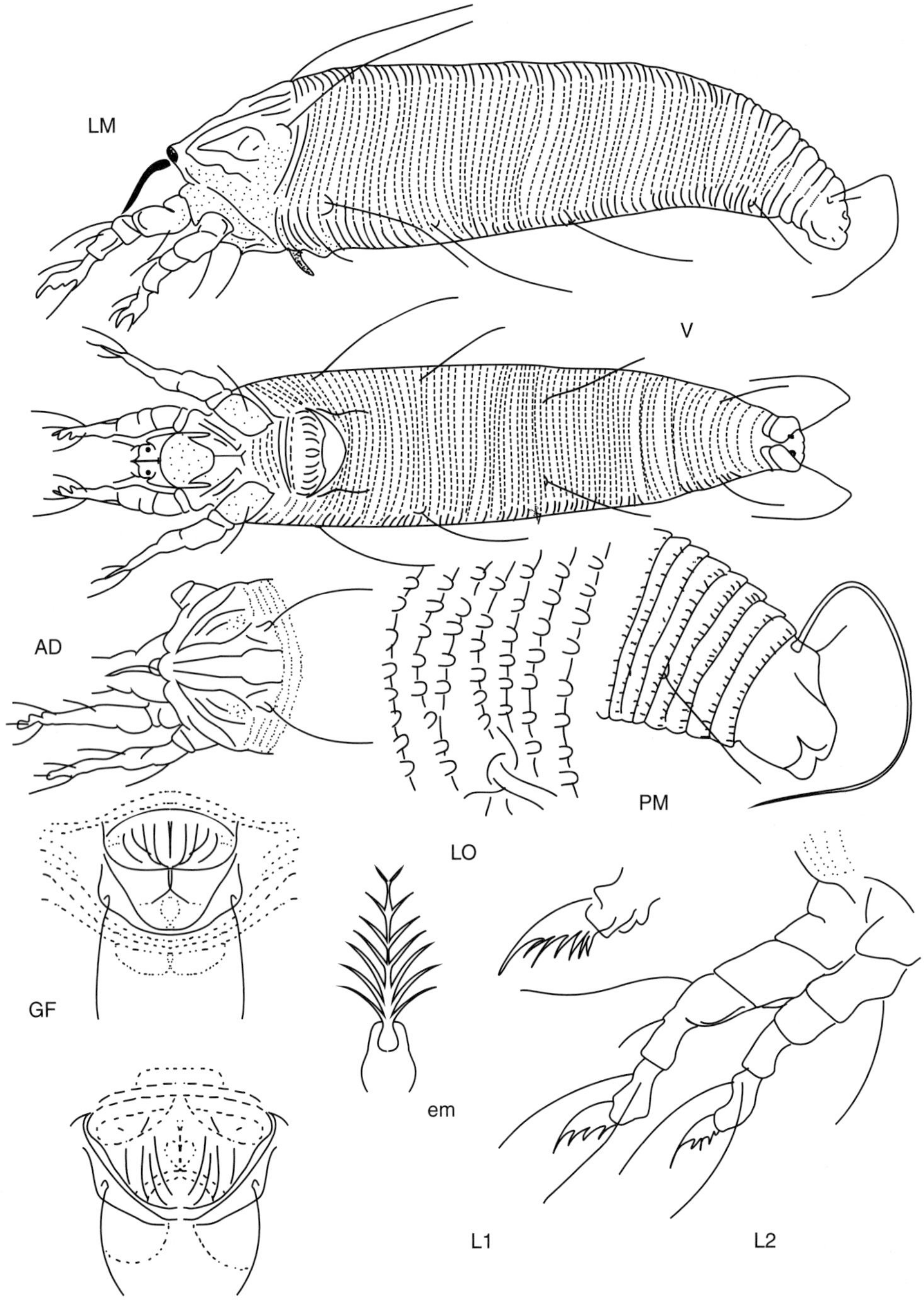

Fig. 16.41. *Aceria tulipae* (Keifer) (from Keifer, 1938a). See Table 16.1 for key.

(Courtin *et al.*, 2000). Generally, the mites survive to lower temperatures than their host plants. This characteristic enables them to survive when they are associated with plant products transported for extended periods of time at low temperatures. Investigations of stored garlic bulbs in Egypt showed that all stages of the mite were present inside the bulb in a quiescent state, but became active and reproduced rapidly if the bulb was planted and germinated. Stored bulbs thus appear to be the main source of infestation for garlic plants in the following year (Wahba *et al.*, 1985).

The emission of ethylene by tulip bulbs (*Tulipa gesneriana* L., cultivar 'Yokohama') in response to the dry bulb mite indicates the occurrence of a wound response (Leśna *et al.*, 2004), and ethylene is a putative factor mediating changes in bulb metabolism, which most probably affects the success of attempts to control *A. tulipae* with predatory mites. A detailed study on the involvement of signalling molecules in the plant defence response to eriophyoids is needed to understand the mechanism. Bulbs that are infested by eriophyoids produce volatiles that attract phytoseiids to them (Aratchige *et al.*, 2007), and bulb damage

triggers the production of ethylene, which causes the widening of the space between scales at the apex of the bulb to such an extent that it becomes penetrable to predatory mites (Leśna *et al.*, 2004).

Natural enemies of *A. tulipae* include the phytoseiid mite *Neoseiulus cucumeris* (Oudemans) (Leśna *et al.*, 2004).

Symptomatology and damage

The dry bulb mite causes direct damage to infested plants and transmits several important viruses such as the *Onion mite-borne latent virus* (OMbLV) and the *Shallot mite-borne latent virus* (ShMbLV) (van Dijk *et al.*, 1991; Perring, 1996; Oldfield and Proeseler, 1996). Infection by both viruses can be asymptomatic (Ostojá-Starzewski and Matthews, 2009). Recently, it has been found that *A. tulipae* transmits an allexivirus (GarV-B1) to garlic plants (Kang *et al.*, 2007). The mite attacks both bulbs in storage and plants in the field. Plants show stunted, twisted, curled and yellow-mottled leaves, which is similar to viral symptomatology (Lange, 1955). Mite feeding activity damages upper epidermal cells of the bulb scales, which collapse, resulting in scarification (Smalley, 1956). Infested garlic cloves have some tissue breakdown which is visible as one or different brownish sunken spots (Manson, 1970; Scalopi *et al.*, 1971). Severe infestation may induce a mean weight reduction of 19%, and the chronic impact of mite infestation is more severe than acute damage occurring over a short infestation (Santos and Lima, 1976). Severely infested bulbs have a 20% reduction in emergence and a 23% yield reduction (Larraín, 1986). *A. tulipae* is also involved in the spread of a fungus-causing rot of garlic bulbs in the field (Jeppson *et al.*, 1975).

The economic importance of the pest led to the performance of a pest risk analysis (PRA) in the UK to assess the invasive potential of the dry bulb mite and the economic consequences of its establishment. Even though the PRA considered that the mite was capable of establishing both outdoors and in storage facilities, the PRA concluded that significant damage was not expected in the UK and recommended no phytosanitary measures (MacLeod, 2007).

Control

Control is provided by the use of chemical and physical means before storage and just before planting. The action of other methods of control is not certain.

PLANT RESISTANCE. In trials in Brazil, the most resistant garlic cultivar 'Chinês' showed an average infestation of 16.4%, followed by 'Cateto Roxo' and 'Seleção Jetibá' with 36.5 and 37.8% infestation, respectively. The most susceptible cultivar was 'Gigante Inconfidente' with 72.7% infestation (Fornazier *et al.*, 1987). At present, there does not seem to be a mite-resistant garlic cultivar available on the market.

CULTURAL PRACTICES. Chemical fertilizer and organic compost influence the productivity and reaction of garlic plants to *A. tulipae*. The highest percentage of plants with symptoms of attack by the mite (21.9%) occurred in plots treated with the fertilizer formulation NPK 4-14-8 + B + Zn + Mg at 1500 kg/ha^{-1}. In contrast, the percentages of plants with symptoms of attack in the plots that received 10, 30, 40 and 50 t of organic compost ha^{-1} were 4.2, 4.6, 4.2 and 3.1, respectively, which did not differ from the control (2.8%). The lowest intensity of blight pustules occurred in plots treated with 30 t of organic compost ha^{-1} and the highest number in the treatment with inorganic fertilizer formulation 4-14-8 (Pereira *et al.*, 1987).

BIOLOGICAL CONTROL. Leśna *et al.* (2004) showed that ethylene played an important role in the biological control of *A. tulipae* on *T. gesneriana* (cultivar 'Yokohama') using the phytoseiid mite *N. cucumeris*. In the absence of ethylene, no predation by *N. cucumeris* was observed. However, when tulips were exposed repeatedly to ethylene, they seemed to be well protected from bulb mite damage by *N. cucumeris*. Ethylene increased the distance between the inner bulb scales, thereby facilitating entry of the predator and hence stimulating the control.

PHYSICAL CONTROL. Soaking garlic cloves in water at 55 and 60°C for 10 min is effective in controlling the mite, but affects the sprouting capacity. The best results were obtained by soaking the cloves in water for 2 h and then dipping them for 10 min in a solution of dicofol or chlorobenzilate mixed with dimethoate and mineral oils (Almaguel *et al.*, 1986). MacLeod (2007) reported that the hot water treatment of bulbs at 55°C for 10–20 min, or at 60°C for 10–15 min could reduce mite populations, although it might damage bulb germination. A postharvest treatment in which the bulbs are dried can also reduce populations. It has been observed that dipping the bulbs first in water and then in acaricide solution also proved effective. Controlled atmosphere methods also showed promising results (Conijn, 2006).

CHEMICAL CONTROL. Generally, dipping of garlic bulbs into chemicals before storage and just before planting is suggested (Doreste, 1963, 1965, 1966). In the past, bulbs in storage were fumigated with methyl bromide and phosphine, but the methyl bromide had an adverse effect on clove sprouting (Cosenza *et al.*, 1981). Also suggested was the use of methyl bromide fumigation of bulbs in storage along with sulfur dusting of plants (Lange, 1955; Mann and Lange, 1960). Malathion was also used for bulbs in storage (Cosenza *et al.*, 1981).

A number of chemicals have been evaluated in different countries, either directly sprayed on to plants or applied by drip irrigation (Depaoli *et al.*, 1983), either alone or mixed. These include alcohol (Lorenzato, 1984), aldicarb (Depaoli *et al.*, 1983), binapacril (Lorenzato, 1984), carbaryl (Depaoli *et al.*, 1983), bromopropylate (Hafez and Maksoud, 1984), carbofuran (Larraín, 1986), chlorobenzilate (Hafez and Maksoud, 1984; Almaguel *et al.*, 1986), common soap, cyhexatin (Lorenzato, 1984), cypermethrin, dichlorvos plus phosalone (Depaoli *et al.*, 1983), dicofol (Hafez and Maksoud, 1984; Almaguel *et al.*, 1986), fenbutatin oxide (Hafez and Maksoud, 1984), formol, mancozeb (Lorenzato, 1984), maneb (Hafez and Maksoud, 1984), mercuric chloride sublimate, methyl parathion plus mineral oil (Lorenzato, 1984), mineral oil (Lorenzato, 1984; Almaguel *et al.*, 1986), mineral oil plus soap (Lorenzato, 1984), monocrotophos (Hafez and Maksoud, 1984), omethoate, pirimiphos-ethyl

(Almaguel *et al.*, 1986), propargite, sodium hypochlorite (Lorenzato, 1984), thiometon (Chiavegato *et al.*, 1968), triazophos (Almaguel *et al.*, 1986), wettable sulfur, wettable sulfur plus soap or plus mineral oil (Lorenzato, 1984), zineb (Giannotti, 1971).

Good control was also afforded by immersion for 24 h in a mixture containing neutral soap of coconut and wettable sulfur or neutral soap of coconut and mineral oil (Lorenzato, 1984).

Aceria zeasinis Keifer (Fig. 16.42)

Common name

Corn sheath mite.

Diagnostic characteristics

FEMALE. The body is worm like, 255–270 μm long and light yellowish white in colour. The gnathosoma is curved down. The prodorsal shield is subsemicircular anteriorly; the median line has elongate dashes on the rear ¾, with a pair of transverse cross lines at about ⅔. The admedian lines are complete, gently sinuate from the base of the chelicera, most widely expanded around a short cross line at about ⅔, gently recurving at the rear; there are irregular dashes within the rear area and between the admedians and scapular tubercles. The first submedian is from near the anterior end of the admedian, subparallel to the admedian to an area in front of the scapular tubercle, then abruptly bending outward and ending laterally to the scapular tubercle; the second submedian is from a short distance back to the first, curving back to parallel with the first for a short distance, then

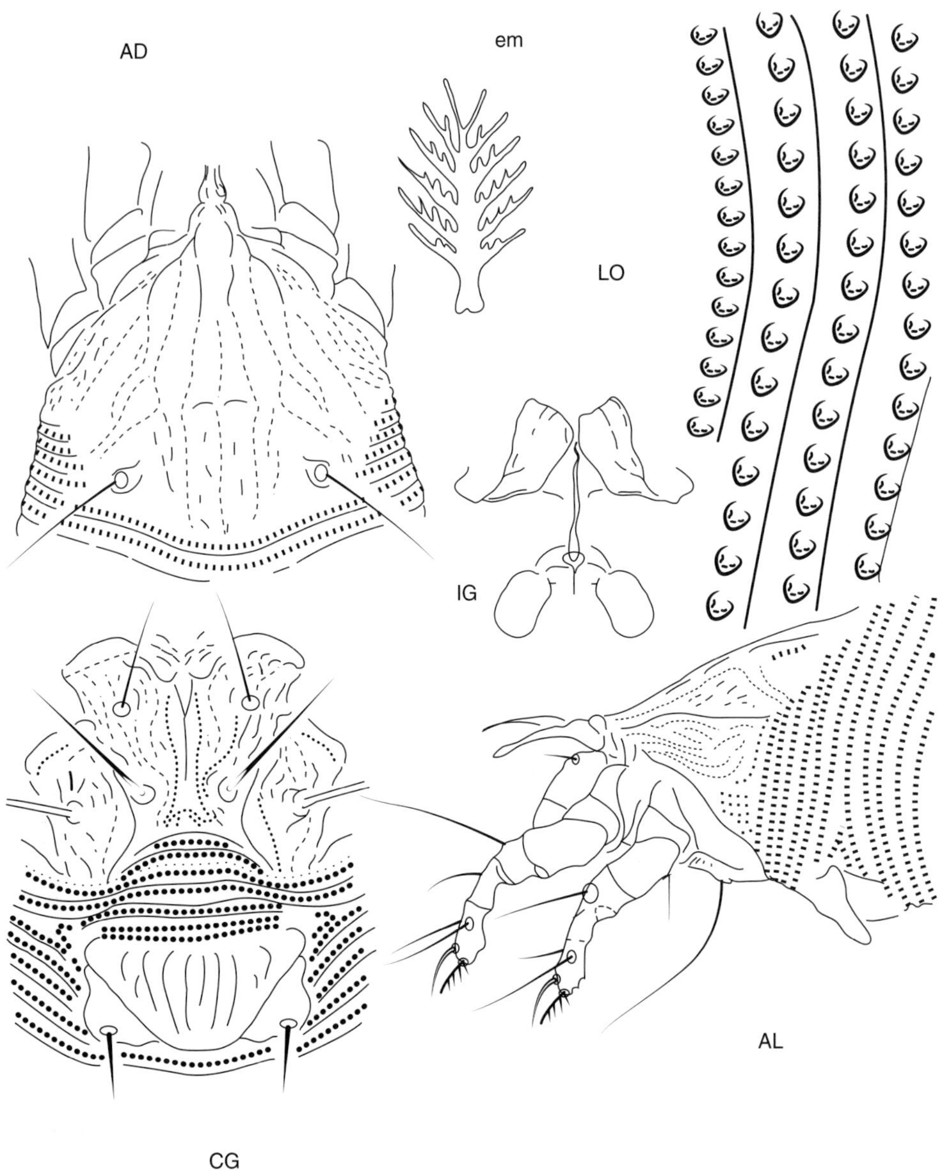

Fig. 16.42. *Aceria zeasinis* Keifer (Keifer, 1962c). See Table 16.1 for key.

usually forked and fading to the rear lateral to the scapular tubercle. The admedian and submedian lines are more or less composed of short dashes. Laterally, the prodorsal shield is generally granular, with a series of partial annuli below the scapular tubercles. The scapular tubercles are 22 µm apart; the scapular setae (*sc*) are 53 µm long, diverging to the rear. The coxae are elongate, well ornamented with lines of dots and dashes, and with a gently sinuate line of dots lying on each side of coxal junction. The opisthosoma has about 70–75 annuli, completely microtuberculate; the microtubercles are rounded apically and more elongate dorsally and ventrally, ahead of the rear annulus margins. The setae c_2 are 44 µm long, set on about annulus 9; the setae *d* are 60 µm long, set on about annulus 25; the setae *e* are 14 µm long, set on about annulus 42; the setae *f* are 24 µm long, set on annulus 5 from the rear; the setae h_1 are 3 µm long. The genitalia are moderately bowl shaped, 23 µm wide and 15 µm long; the coverflap has about 10 longitudinal ribs. The empodium is 6-rayed (Keifer, 1962c).

Geographical distribution

Alabama (Keifer, 1962c), Cuba (de la Torre and Martínez, 2004) and India (Rather, 1983).

Host plants

Poaceae: maize (*Zea mays* L.) (Keifer, 1962c; Rather, 1983; de la Torre and Martínez, 2004).

Bio-ecology

A. zeasinis is a tropical species reported from maize (Keifer, 1962c; Rather, 1983; de la Torre and Martínez, 2004). The mite lays its eggs near the leaf veins and at 26–31°C hatching takes 5–8 days. Fecundity ranges from 7 to 18 eggs, and egg deposition ceases over 33°C. In India, *A. zeasinis* develops on maize during the growing season and has 3–5 generations. Natural enemies include *Stethorus* sp. and a phytoseiid mite (*Euseius vignus* Rishi *et* Rather) and can suppress the populations of the mite (Rather, 1983).

Symptomatology and damage

Damage gives rise to blistered and wrinkled foliage, and at high infestations, the leaves become distorted and the plants appear stunted (Rather, 1983). Attacks of the mite on greenhouse maize may induce severe stunting and distort growth (Keifer, 1962c; Jeppson *et al.*, 1975).

Control

Few reports are available on the control of *A. zeasinis*. Active ingredients tested during the last few decades include binapacryl, carbaryl, chinomethionate, dimethoate, endosulfan and parathion (Rather, 1983).

Vegetables, field crops and berries group

The most important species of this group of eriophyid mites are *Aceria gastrotrichus* (Nalepa), *A. granati* (Canestrini *et* Massalongo), *A. hibisci* (Nalepa), *A. lycopersici* (Wolffenstein), *A. medicaginis* (Keifer), and *A. peucedani* (Canestrini). *A. neocynarae* (Keifer) is a secondary species living on artichoke (*Cynara scolymus* L.) and either causing no damage (Keifer, 1939c) or inducing curling of some of the leaves at high levels of attack (Zaher and Abou-Awad, 1979; Vacante and Tropea Garzia, 1990). The first six species named above are described in more detail below.

Aceria gastrotrichus (Nalepa)

Common name

Sweet potato leaf gall mite.

Diagnostic characteristics

FEMALE. The body is worm like and 145–160 µm long. The prodorsal shield is sometimes concave on the side towards the front, and the front edge is broad and emarginate. The opisthosoma has elliptical microtubercles. The genital coverflap has many fine longitudinal ribs. The coxae have transverse lines. The empodium is 4- or 5-rayed (ChannaBasavanna, 1966; Jeppson *et al.*, 1975).

Geographical distribution

Java (Indonesia) (Nalepa, 1918), India (ChannaBasavanna, 1966), the Philippines (Amalin *et al.*, 1991), southern Asia and the East Indies (Jeppson *et al*, 1975).

Host plants

Convolvulaceae: sweet potato (*Ipomoea batatas* (L.) Lam.) (Nalepa, 1918; Jeppson *et al.*, 1975; Amalin *et al.*, 1991), *I. staphylina* Roemer *et* Schultes (ChannaBasavanna, 1966).

Bio-ecology

A. gastrotrichus is reported from sweet potato (Nalepa, 1918; Jeppson *et al.*, 1975; Amalin *et al.*, 1991) and *I. staphylina* (ChannaBasavanna, 1966). Its bio-ecology has not been documented.

Symptomatology and damage

The feeding activity of this mite causes the development of extensive irregularly shaped outgrowths (galls) on leaves and stems of sweet potato, with severe deformation of leaves and stems (ChannaBasavanna, 1966; Jeppson *et al.*, 1975).

Control

This seems not to be necessary.

Aceria granati (Canestrini *et* Massalongo) (Fig. 16.43)

Common name

Pomegranate leaf curl mite.

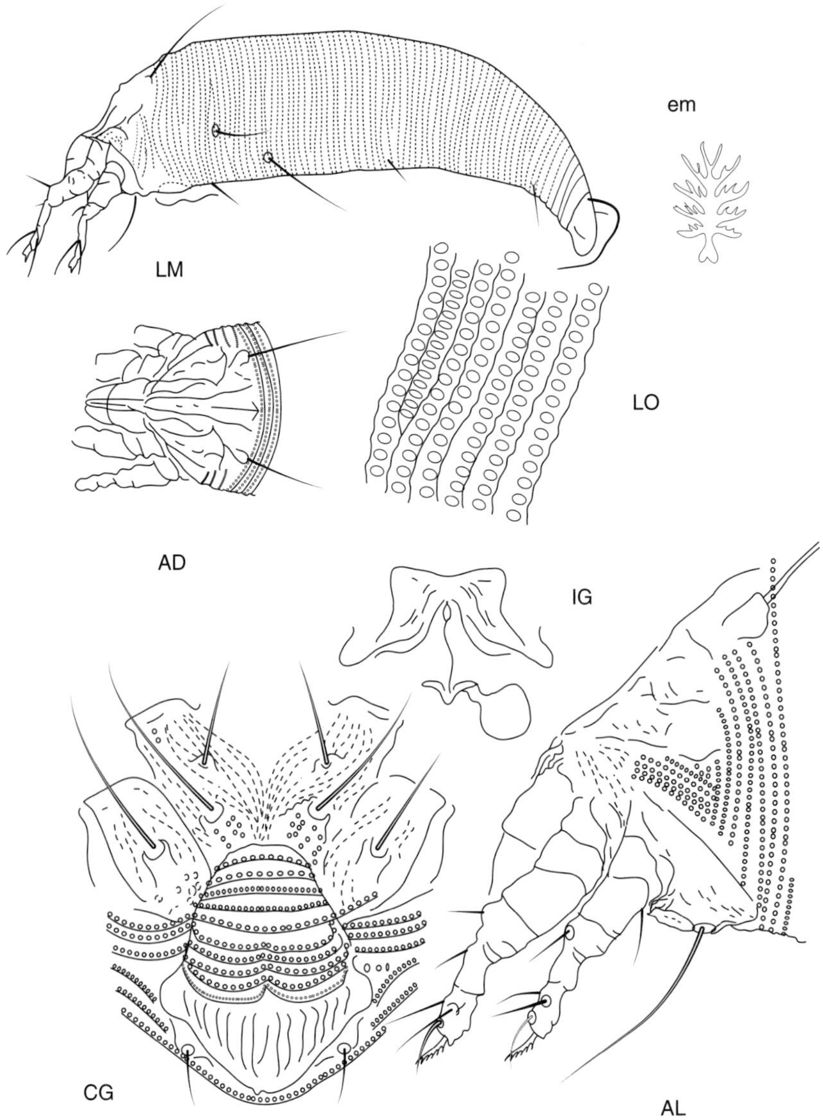

Fig. 16.43. *Aceria granati* (Canestrini *et* Massalongo) (from Keifer, 1952a). See Table 16.1 for key.

Diagnostic characteristics

FEMALE. The body is worm like and 165–180 μm long. The prodorsal shield has a median line that is faint anteriorly and ends in a dart-shaped mark posteriorly. The admedian lines are complete and gradually diverge; the first submedian lines are short; the second submedian lines curve back towards the admedians before the scapular tubercles and then abruptly angle out past the front of the tubercles. There are heavy granules on the coxae obscuring the sternal line. The opisthosomal microtubercles are sometimes elliptical and set ahead of the margins of the annuli. The genital coverflap has about 5 granular transverse posteriorly convex lines and 14–16 longitudinal ribs. The empodium is 4-rayed (Jeppson *et al.*, 1975; Keifer *et al.*, 1982).

Geographical distribution

Worldwide (Jeppson *et al.*, 1975).

Host plants

Lythraceae: pomegranate (*Punica granatum* L.) (Canestrini and Massalongo, 1894; Jeppson *et al.*, 1975).

Bio-ecology

This species was described from pomegranate in Italy by Canestrini and Massalongo (1894). It is probably native to the south European and Mediterranean area, and followed pomegranate planting in the tropical and subtropical regions, so that it now has a worldwide geographical distribution (Jeppson *et al.*, 1975).

In India, occurrence of the mites starts from January and reaches a peak during the first fortnight of March; populations decrease from the second fortnight of May, probably affected by rainfall. An increase in morning RH seems to favour mite development. Evening RH, rainfall and number of rainy days have a negative relationship with the mite population and hinder mite development. Of these abiotic factors, evening and

morning RH have an influence of 53 and 36%, respectively. Karuppuchamy (1994) also found maximum edge-rolling symptoms by mites during the second fortnight of June, although the correlation between incidence and weather parameters was non-significant. Peaks of mite activity on leaves were observed in March to April (Ananda *et al.*, 2009).

Symptomatology and damage

The pomegranate leaf curl mite rolls the leaves from the edge towards the lower surface to form a tube in which it lives. The leaves are rolled strongly and appear to be leaflets. The twig develops normally, showing that the terminal twig is not injured. Damage may involve each young leaf (Jeppson *et al.*, 1975; Meyer Smith, 1996a).

Karuppuchamy (1994) reported that in Tamil Nadu (India), the maximum edge-rolling symptoms occurred during the second fortnight of April (18.18%) to the first fortnight of August (26.14%) in hill areas and from the first fortnight of February (41.53%) to the first fortnight of June (31.02%) on the plains.

Control

Plants grown for fruit production do not usually require any control. In nurseries and/or in the production of ornamental plants, it may be necessary to use some acaricides.

Aceria hibisci (Nalepa) (Fig. 16.44)

Common name

Hibiscus erineum mite.

Diagnostic characteristics

FEMALE. The body is worm like, robust and 160–185 µm long. The prodorsal shield is subtriangular, with strong lines only to the rear portion. The median line is hardly indicated; the admedian lines are located on the rear half of the shield as outwardly convex arches, ending posteriorly. Some lateral shield lines are present, with an ocellar spot at the rear prodorsal shield angles. The opisthosoma has laterally strong annuli dorsally and weaker annuli ventrally. The microtubercles are dorsally elongate and reach the margins, and are bead like ventrally. The dorsal opisthosoma lacks microtubercles on the rear ¼. The coxae lack ornamentation. The genital coverflap has 8–10 ribs. The empodium is 4-rayed (Jeppson *et al.*, 1975).

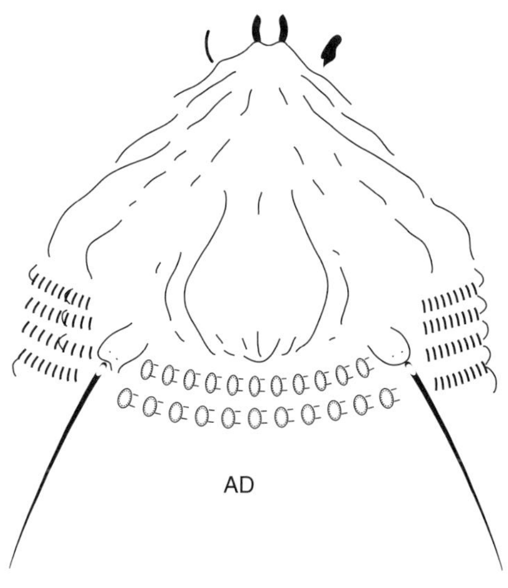

Fig. 16.44. *Aceria hibisci* (Nalepa) (from Jeppson *et al.*, 1975). See Table 16.1 for key.

Geographical distribution

Brazil (see under Bio-ecology), Cook Islands and Tonga (Jeppson *et al.*, 1975), China (Zhao and Kuang, 1996), Cuba (de la Torre and Martínez, 2004), Dominica (Welbourn *et al.*, 2008), Fiji Islands (Nalepa, 1906), Guadeloupe (Flechtmann and Etienne, 2001), Hawaii (Hara *et al.*, 1996), Jamaica (Welbourn *et al.*, 2008), Martinique (Flechtmann *et al.*, 1999b), New South Wales, Queensland (Australia) (Carson and Gough, 2000, 2007), Puerto Rico (Welbourn *et al.*, 2008), Réunion (Quilici *et al.*, 1997) and the Samoan Islands (Nalepa, 1909a).

Host plants

Malvaceae: okra (*Abelmoschus esculentus* (L.) Moench) (Jeppson *et al.*, 1975), *Hibiscus* spp. (Nalepa, 1906; Jeppson *et al.*, 1975; Hara *et al.*, 2001), Chinese red hibiscus (*H. rosa-sinensis* L.), wax mallow (*Malvaviscus arboreus* Cav.) (Welbourn *et al.*, 2008), blue mahoe (*Talipariti elatum* (Sw.) Fryxell) (de la Torre and Martínez, 2004).

Bio-ecology

The hibiscus erineum mite was described on Chinese red hibiscus from the Fiji islands (Nalepa, 1906), and although it prefers this host plant, it can also develop on other hibiscus species. In Cuba, it has been collected on blue mahoe (de la Torre and Martínez, 2004). In Réunion and some Caribbean countries, the mite occurs only on *Hibiscus* sp. In Hawaii, the *H. rosa-sinensis* cultivars 'Apricot', 'Empire', 'Pink Hibiscus', 'Itsy Bitsy Peach Monch', 'Zham Chinese' and 'Apple Blossom' are less susceptible to mite infestation than 'Chinese Red', 'Herman Shierman', 'Orange hibiscus', 'Nii Yellow' and 'Kardinal' (Hara *et al.*, 2001). *A. hibisci* also causes irregular erineum pockets on okra (Jeppson *et al.*, 1975). In Puerto Rico, wax mallow growing side by side with heavily infested hibiscus plants did not show galls or signs of infestation by the mite (Welbourn *et al.*, 2008). The report of this mite in Brazil is unclear. Indeed, Jeppson *et al.* (1975) consider *Aceria esculenti* Keifer, reported on okra from Brazil (Braga, 1957), to be synonymous with *A. hibisci*, although they do not give any explanation. In addition, Perring (1996) reported *A. hibisci* from 'Eastern Europe', but in the Fauna Europaea (de Lillo and Beron, 2004) this report was not confirmed. The vegetative propagation of hibiscus plants facilitates mite dispersal via infested plants or plant material. The wind, as with most other plant-feeding mites, could play a role as well (Welbourn *et al.*, 2008).

In Dominica, Jamaica and Puerto Rico, predatory mites have been collected on galls of *A. hibisci* (Welbourn *et al.*, 2008). On Réunion, *Phytoseius intermedius* Evans *et* McFarlane was found associated with hibiscus erineum mite on *Hibiscus* sp. (Quilici *et al.*, 1997).

Symptomatology and damage

The feeding activity of *A. hibiscus* damages the leaves, stems, twigs and flowers of hibiscus. Puckered bumps appear on the leaf edge and both leaf surfaces. The underside of these bumps (erinea) has different compartments and is usually covered with a velvety growth that is pale yellowish in colour. Deformations occur also on the leaf axils. Unopened flowers may be damaged. Infested flowers bear deformations and the galls are usually located in the internal part of the flowers. The lower part of the leaf stem has a corky tissue. Gall development and leaf deformation is more severe when the attack is on young leaves and vegetative buds. To begin with, the new galls are small and a few millimetres wide, greenish in colour and slightly raised; the developed galls are about 0.2 cm long and increase in size and length as mite feeding proceeds (Welbourn *et al.*, 2008). On okra, the mite causes irregular erineum pockets, which induce misshapen leaves (Jeppson *et al.*, 1975).

Control

Control is provided by cultural practices and chemical sprays.

CULTURAL PRACTICES. Pruning of infested branches is a temporary control strategy, resulting in new growth on the same (frequently infested) plant (Welbourn *et al.*, 2008).

CHEMICAL CONTROL. Chemical control is by the use of acaricides registered for hibiscus. Sprays affect the new growth but the damaged leaves persist on the plants. This suggests that damaged plants should be pruned after chemical treatment (Welbourn *et al.*, 2008).

Aceria lycopersici (Wolffenstein)

Common name

Tomato erineum mite.

Diagnostic characteristics

FEMALE. The body is worm like and 150–200 μm long. The prodorsal shield is subtriangular and bears complete median and submedian lines, with a broad lateral granular band on the sides. The coxae lack markings. The opisthosomal microtubercles are elliptical, thick posteriorly, and reach the margins of the annuli. The posterior distal opisthosomal annuli lack microtubercles and are smooth dorsally. The genital coverflap has 10–12 longitudinal ribs. The empodium is 4-rayed (Jeppson *et al.*, 1975).

Geographical distribution

Pantropical and temperate areas (in greenhouses) (Jeppson *et al.*, 1975).

Host plants

Solanaceae: pepper (*Capsicum annuum* L.), Chinese boxthorn (*Lycium chinense* Mill.), broadleaf nightshade (*Nicandra physaloides* (L.) Gaertn.), *Nicotiana glutinosa* L., tobacco (*N. tabacum* L.), *Petunia hybrida* hort. ex E. Vilm., Chinese lantern (*Physalis alkekengi* L.), *Solanum atropurpureum* Schrank, bitter nightshade (*S. dulcamara* L.), *S. luteum* Miller, tomato (*S. lycopersicum* L.), black nightshade (*S. nigrum* L.) and potato (*S. tuberosum* L.) (Wolffenstein, 1879; Jeppson *et al.*, 1975; Westphal *et al.*, 1981).

Bio-ecology

A. lycopersici was first described from tomato in southern Spain (Wolffenstein, 1879), and has been misidentified in the past with *Phytoptus lycopersici* Wolffenstein, *Aceria calacladophora* (Rolfs) and *Aceria cladophthira* (Nalepa), and probably also with *Aceria lycopersici* (Massee) (Amrine and Stasny, 1994). It lives in holotropical countries and in more temperate areas in greenhouses, and its host range includes wild and cultivated solanaceous plants, such as pepper, tobacco, tomato and potato (Jeppson *et al.*, 1975; Westphal *et al.*, 1981). Tomato erineum mite infestations usually occur in the dry season, but it possible to observe outbreaks in rainy seasons as well (Costa and Gonçalves, 1950).

Natural enemies include the phytoseiid mite *Amblyseius gossypii* El-Badry (Abou-Awad, 1983).

Symptomatology and damage

Some reports on the tomato erineum mite stated that infestations frequently occurred in the dry season, mainly on aubergine and tomato, causing a consistent hair-like growth on leaves and stems that was sometimes similar to white mould (ChannaBasavanna, 1966; Jeppson *et al.*, 1975). Later, it was found that the symptomatology of and damage caused by tomato erineum mite were more complex. The feeding activity of the mite causes misshapen leaves, rolled leaf edges, erinea, wart-like alterations on the leaves, leaf enations, leaf necrotic spots, short stem internodes, axillary bud growth, stem fasciation and flower virescence (Westphal *et al.*, 1981; Anthony *et al.*, 1988). The mite feeds on epidermal cells, which produce callose around the punctures (Westphal *et al.*, 1981). After a few minutes, alterations appear to the organelles. Within the cells, the nucleoli are enlarged, optically clear and move towards the side of the puncture. Chitosan or other hexosamines induce alterations in the nucleic acids (Bronner *et al.*, 1989). Vacuoles of the epidermal and palisade cells surrounding the injured cells decrease in size and the pH of their contents increases (Westphal, 1982). The nuclei and nucleoli of these adjacent cells enlarge and intensify mitosis. These cells form a trophic resource for the mite and cause the erinea (Westphal *et al.*, 1981, 1990; Anthony *et al.*, 1988). These alterations are associated with hyperplasia of the glandular trichomes and epidermal hairs (Westphal, 1985), and produce a favourable microhabitat

for the mite (Royalty and Perring, 1996). Cellular biochemical and metabolic responses induced by the mite on injured plants are further described by Petanović and Kielkiewicz (2010a).

Control

There are few reports on the control of this mite. In the past, the use of sulfur has been suggested (Rolfs, 1907; Massee, 1939).

Aceria medicaginis (Keifer) (Fig. 16.45)

Common name

Alfalfa broom mite or lucerne mite.

Diagnostic characteristics

FEMALE. The body is worm like, 150–180 μm long, 40–50 μm wide and light yellow in colour. The gnathosoma is somewhat downcurved. The prodorsal shield has an obscure pattern, and some acuminate tubercles at the side; the scapular tubercles are 27 μm apart on the rear margin; the scapular setae (*sc*) are 24 μm long, projecting backwards. The opisthosoma has about 60 annuli, the dorsal half-annuli are faintly microtuberculate anteriorly, practically non-tuberculate on the posterior third; the ventral half-annuli are strongly set with acuminate microtubercles, contrasting sharply with the dorsum alongside the body. The setae c_2 are 46 μm long, set on about annulus 7; the setae *d* are 63 μm long, set on about annulus 19; the setae *e* are 23.5 μm long, set on about annulus 36; the setae *f* are 27 μm long,

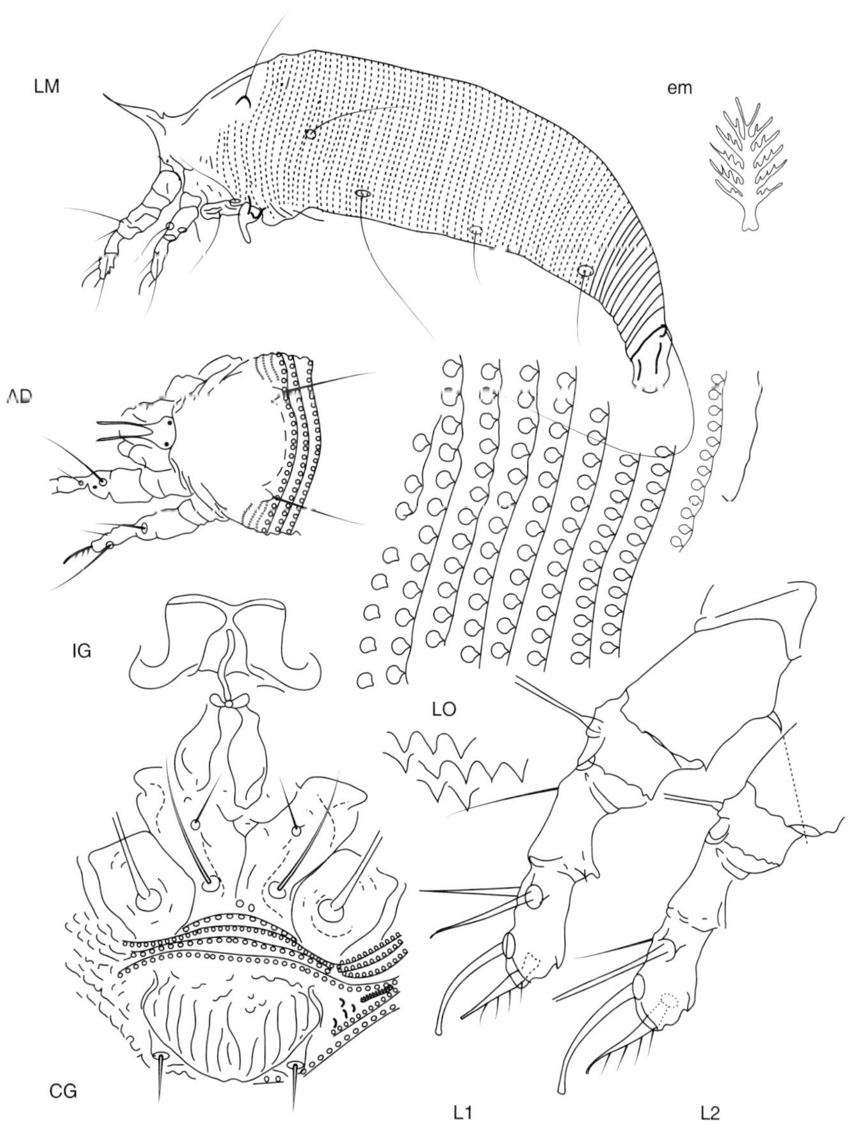

Fig. 16.45. *Aceria medicaginis* (Keifer) (from Keifer, 1941). See Table 16.1 for key.

set on about annulus 6 from the rear; the setae h_1 are present. The genitalia are 21 µm wide, 15 µm long; the coverflap has 10–12 furrows, and the seta is 16 µm long (Keifer, 1941).

MALE. The body is 130–170 µm long and 35-45 µm wide (Keifer, 1941).

Geographical distribution

Arizona (Keifer, 1965c), Australia (Stubbs and Meagher, 1965; Ridland and Halloran, 1980a), Belgium, Bulgaria (de Lillo and Beron, 2004), California (Keifer, 1941), Greece (Emmanouel and Papadoulis, 1987), India (ChannaBasavanna, 1973), Italy (Majorana, 1970), the Near East (de Lillo and Beron, 2004), Oklahoma (Hall, 1967), Spain and Yugoslavia (de Lillo and Beron, 2004).

Host plants

Fabaceae: sweet vetch (*Hedysarum coronarium* L.), *Medicago* spp., alfalfa (lucerne) (*M. sativa* L.), crimson or Italian clover (*Trifolium incarnatum* L.), woods clover (*T. pinetorum* Greene), subterraneum clover or subclover (*T. subterraneum* L.) (Keifer, 1941, 1965c; Majorana, 1970; Jeppson *et al.*, 1975; Ridland and Halloran, 1979, 1980a).

Bio-ecology

The lucerne broom mite lives on Lucerne and other Fabaceae, such as the sweet vetch, woods clover and other *Trifolium* spp. In Greece, high densities of *A. medicaginis* were observed on lucerne from mid-September to mid-March, with the highest mean numbers 38 and 82 for every 100 g dry weight. The mite infests mainly the growing tips of the plant, and its density decreases after the crop has been cut; consequently, it may become a serious pest in lucerne if it is not often cut (Emmanouel and Papadoulis, 1987); the highest density occurs during the cooler and more humid period of the year (Lykouressis and Emmanouel, 1991). At 38°C, the mite could survive for 2 weeks (Stubbs and Meagher, 1965; Jeppson *et al.*, 1975). In California, the peak population of *A. medicaginis* was observed between May and September (Lehman, 1971).

Symptomatology and damage

The infested plants show a symptomatology similar to that of virus diseases, characterized by many witches' brooms. The plants become dwarfed, with a considerable increase in the crown shoots. The stems are fine and show erect bunchy foliage, with small leaves, which are rounded, puckered and slightly chlorotic. The severity of the symptoms depends on the density of the mite population (Stubbs and Meagher, 1965; Jeppson *et al.*, 1975). The development of leaf malformations is probably related to the stage of the leaf when mite attack occurs. The mite infests the leaves just before petiole extension, when cell division has just about ceased (Ridland and Halloran, 1980a). If mites do feed on younger leaves set at apex that have a high rate of cell division, an inhibition of cell division can occur at the feeding points, which induces abnormal development of leaflets (Ridland, 1996).

The feeding activity of the mite causes significant reduction in the growth of seedlings of the annual species *T. incarnatum* and *T. subterraneum*, but does not influence the growth of the perennial species *T. fragiferum* L., *T. pratense* L. and *T. repens* L. *A. medicaginis* does not reproduce on either annual or perennial species of *Trifolium*. Leaf symptoms on the annual species are similar to those caused by bean yellow mosaic virus on annual clovers. The symptoms on *T. incarnatum* are not evident until 10–12 days after infestation, which is the time taken for undifferentiated leaves to develop fully. This shows that the presumed salivary toxin of the mite is of short-lived activity in the plant or is not readily translocated beyond the leaf upon which feeding takes place, or both (Ridland and Halloran, 1980a).

At 25 and 29°C, a density of ten mites/seedling reduces the dry weight of the tops, stem length and the development of axillary shoots within 22 days of infestation. The numbers of mites increase 16 times at 29°C, nine times at 25°C and six times at 20°C, but no increase occurs at 15°C. The extent of damage by *A. medicaginis* at 20°C varies with the differences in growth of the plants under different photoperiods. With a short photoperiod of 10.3 h, and with no axillary shoots on either infested or uninfested plants, the extent of damage is higher than with a longer photoperiod of 13.2 h with the axillary shoots developed on all seedlings. The leaves of infested plants show chlorotic patches (mainly at 20 and 25°C), agate-like banding (at 29°C) and a waxy mottle (at 15°C) (Ridland and Halloran, 1981).

Control

Few reports are available about the control of this mite. Some studies have been made of physiological tolerance to the mite. The occurrence of symptomless plants hosting high mite numbers among the perennial species *Medicago* spp., *M. coerulea* Less., *M. falcata* L., *M. glutinosa* M.B., *M. hemicycla* Grossh. and *M. quasifalcata* Sinsk. suggested the presence of physiological tolerance to the mite (Ridland and Halloran, 1979). Research on the Australian cultivar *M. sativa* 'Hunter River' infested with *A. medicaginis* and harvested 23 days later showed that the seedlings least affected by infestation had the lowest number of axillary shoots and the highest rate of growth of the primary shoot. Thus, selection for greater seedling vigour within this cultivar could increase tolerance to *A. medicaginis* (Ridland and Halloran, 1980b).

***Aceria peucedani* (Canestrini) (Fig. 16.46)**

Common name

Carrot bud mite.

Diagnostic characteristics

FEMALE. The body is worm like and about 210 µm long. The prodorsal shield lines are granular. The median line extends back posteriorly from ⅓. The admedian lines are close to each other before the median line, but then diverge as two convex arcs extending back posteriorly. The prodorsal shield is granular laterally. The coxae have coarse granules. The genital coverflap

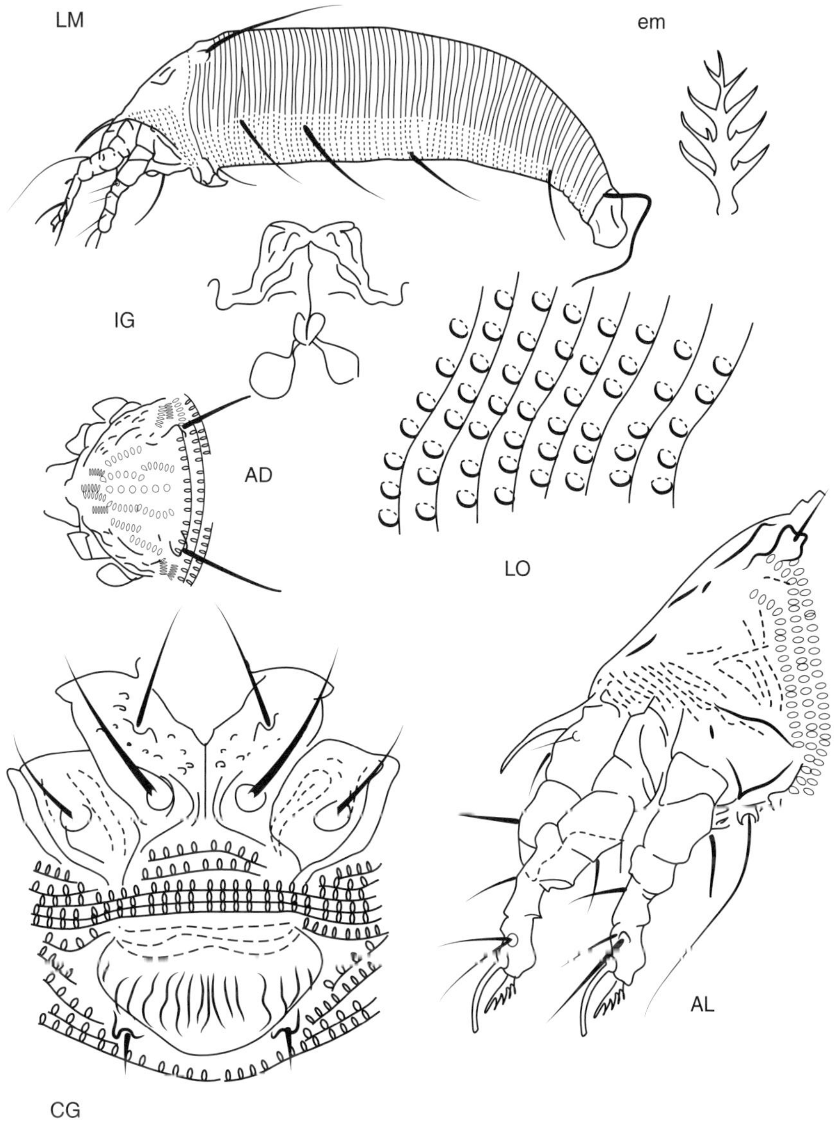

Fig. 16.46. *Aceria peucedani* (Canestrini) (from Keifer, 1952a). See Table 16.1 for key.

has about 12 longitudinal ribs. The empodium is 4- to 5-rayed (Jeppson *et al.*, 1975).

Geographical distribution

Austria, Bosnia and Herzegovina, Finland, France, Germany, Hungary, Italy, Luxemburg, Russia, Slovenia and Yugoslavia (de Lillo and Beron, 2004), and California (Keifer, 1952a).

Host plants

Apiaceae: carrot (*Daucus carota* L.), *Orlaya grandiflora* (L.) Hoffm. (Nalepa, 1929), *Peucedanum venetum* (Spreng.) Koch (Canestrini, 1891b), *Pimpinella saxifraga* L., *Seseli glaucum* L., *S. hippomarathrum* L. and *Torilis infesta* Koch (Nalepa, 1929).

Bio-ecology

A. peucedani is a pest of carrot (Keifer, 1952a; Scott, 1965; Jeppson *et al.*, 1975) and was described for the first time in Italy from the herb plant *P. venetum* by Canestrini (1891b), and later reported on different wild and cultivated Apiaceae, including the carrot (Nalepa, 1929). On carrot, the development of mite populations begins in early summer, when the blooms for seed production appear. In this phase, it is possible to observe at the apex of the plant congregations of mites ready for air dispersal. The mite is dispersed via infested carrot roots that are harvested and stored for later planting, on which the mites feed and reproduce (Scott, 1965; Jeppson *et al.*, 1975).

Symptomatology and damage

On carrot, the mite causes drying of the seed heads. Observations have been made of discoloured flower heads with damage to the developing seeds (Keifer, 1952a). A high mite population density induces deformation of the leaves and umbels, which may turn brown and show discoloured petals. The infested plants produce little or no seed and attacked seedlings develop small roots. The infestation has a negative effect on root growth,

and kills at the growing point at high mite density. Moderate attacks reduce the seed yield by 18%, and higher levels of attack by 53% (Scott, 1965).

Control

Control uses both preventive and curative measures, the latter by spray applications of acaricides.

PREVENTION. Prevention involves field isolation of at least ¼ mile in order to prevent mite dispersal between fields. Isolation of root fields from seed production fields is also a suitable strategy (Scott, 1965).

CHEMICAL CONTROL. Chemicals tested in the past include various insecticides/acaricides, such as demeton, diazinone, phorate and disulfoton (Scott, 1965).

Ornamental plants and forest and shade trees group

This group includes almost 20 *Aceria* species. Among these, *A. cajani* ChannaBasavanna is the most injurious pest, and is discussed extensively. The other species are sometimes injurious, but their role must be examined in relation to environmental variables. Nine species – *A. loewi* (Nalepa), *A. fraxinivora* (Nalepa), *A. genistae* (Nalepa), *A. lantanae* (Cook), *A. ligustri* (Keifer), *A. proteae* Meyer, *A. tenella* (Nalepa), *A. arceosae* (Briones *et* McDaniel), *A. tumisetus* Meyer *et* Ueckermann – are reported from ornamental plants. The first six of these species are treated extensively below. The last three have only a naturalistic importance. Of these, *A. tenella* is a European species (de Lillo and Beron, 2004) living on European hornbeam (*Carpinus betulus* L.), *A. arceosae* is a Nearctic species first reported from South Dakota (Briones and McDaniel, 1976) and later found on caragana (*Caragana arborescens* Lam.) in nurseries in Poland (Łabanowski and Soika, 1998; Soika and Łabanowski, 2002), and *A. tumisetus* was reported from *Gazania* spp., an indigenous species of Asteraceae in South Africa (Meyer Smith and Ueckermann, 1992) and today grown worldwide as many hybrids. *A. tumisetus* is commonly called the gazania gall mite and causes open cup domes on leaves, with motile forms occurring in the abaxial indumentum (Meyer Smith, 1996a).

On flowering plants, six *Aceria* species are reported – *A. caliberberis* Keifer, *A. georghioui* Keifer, *A. paradianthi* Keifer, *A. barbertoni* Meyer *et* Ueckermann, *A. diastolus* Meyer *et* Ueckermann and *A. dianthi* (Lindroth). The first three of these are of some economic importance and are treated more extensively below. The other species play no such role. Of these, *A. barbertoni*, commonly called the gerbera erineum mite, lives on gerbera (*Gerbera jamesonii* H. Bol. ex Adlam.) in South Africa (Meyer Smith and Ueckermann, 1992). Its feeding activity on indigenous gerbera causes a whitish woolly erineum on the lower leaf surfaces and leaf distortion (Meyer Smith, 1996a). *A. diastolus* is reported from *Plumbago auriculata* Lam. in South Africa (Meyer Smith and Ueckermann, 1992). This ornamental plant is native to South Africa, but its cultivation has spread to Australasia and Europe. The feeding activity of *A. diastolus* causes rolling of leaf edges, deformation of the inflorescences and distal growths (Meyer Smith, 1996a). *A. dianthi* is a leaf vagrant on *Dianthus deltoides* L. in Finland and causes stunting of plants (Lindroth, 1904). Three of the flowering plant species listed above live on carnation and can be distinguished morphologically by their empodia, which is 5-rayed in *A. dianthi*, 6-rayed in *A. paradianthi* and 7-rayed in *A. georghioui*.

Also included in this group is *A. peyerimhoffi* (Nalepa), which infests flax (*Linum usitatissimum* L.), a plant cultivated for its fibre and oil; this is described below.

Two other species in the group live on poplars, *A. parapopuli* (Keifer) and *A. populi* (Nalepa), and one lives on *Acer* spp., *A. macrorhyncha* (Nalepa). The first of these, *A. parapopuli*, is better known and is described below. *A. macrorhyncha* is widespread on *Acer* in Europe and has also been reported in North Africa (de Lillo and Beron, 2004) and Iran (Khanjani and Haddad Irani-Nejad, 2006; Xue *et al.*, 2009a). It has been found on European sycamore maple (*Acer pseudoplatanus* L.) in nurseries and city greeneries in Poland (Soika and Łabanowski, 2000) and Iran (Xue *et al.*, 2009a). *A. populi* is a European species living on poplars, which has also been reported in the East Palaearctic and Nearctic regions, and in North Africa (de Lillo and Beron, 2004).

Aceria cajani ChannaBasavanna (Fig. 16.47)

Common name

Unknown.

Diagnostic characteristics

FEMALE. The body is worm like, robust, 120–130 μm long and pinkish or violet in colour. The gnathosoma is obliquely inclined forward. The prodorsal shield has lines that are light centrally, and complete median and admedian lines; short longitudinal lines are set laterally. The scapular tubercles are located at the rear margin and 13 μm apart; the scapular setae (*sc*) are 35–40 μm long, diverging to the rear. The coxal ornamentation is unclear, and the sternal line is moderately long. The opisthosoma has about 68 annuli, uniformly microtuberculate, with the exception of those of the dorsum of the rear 6–10 annuli, which are almost bare; the microtubercles are oval, occupying the entire width of the annuli, sparsely distributed dorsally and more closely located on the lateral and ventral sides. The setae c_2 are 25–28 μm long, set on annulus 10; the setae *d* are 45–50 μm long, set on annulus 23; the setae *e* are 13–14 μm long, set on annulus 42; the setae *f* are 16–18 μm long, set on annulus 6; the setae h_1 are 4 μm long, and the setae h_2 are 65–71 μm long. The genital coverflap has about 12 longitudinal lines, and the seta is 18–21 μm long. The empodium is 6-rayed (ChannaBasavanna, 1966).

MALE. The male is similar to the female and about 100 μm long (ChannaBasavanna, 1966).

Geographical distribution

The Indian subcontinent and, in particular, India, Bangladesh, Sri Lanka (ChannaBasavannaa, 1966; Kannaiyan *et al.*, 1984;

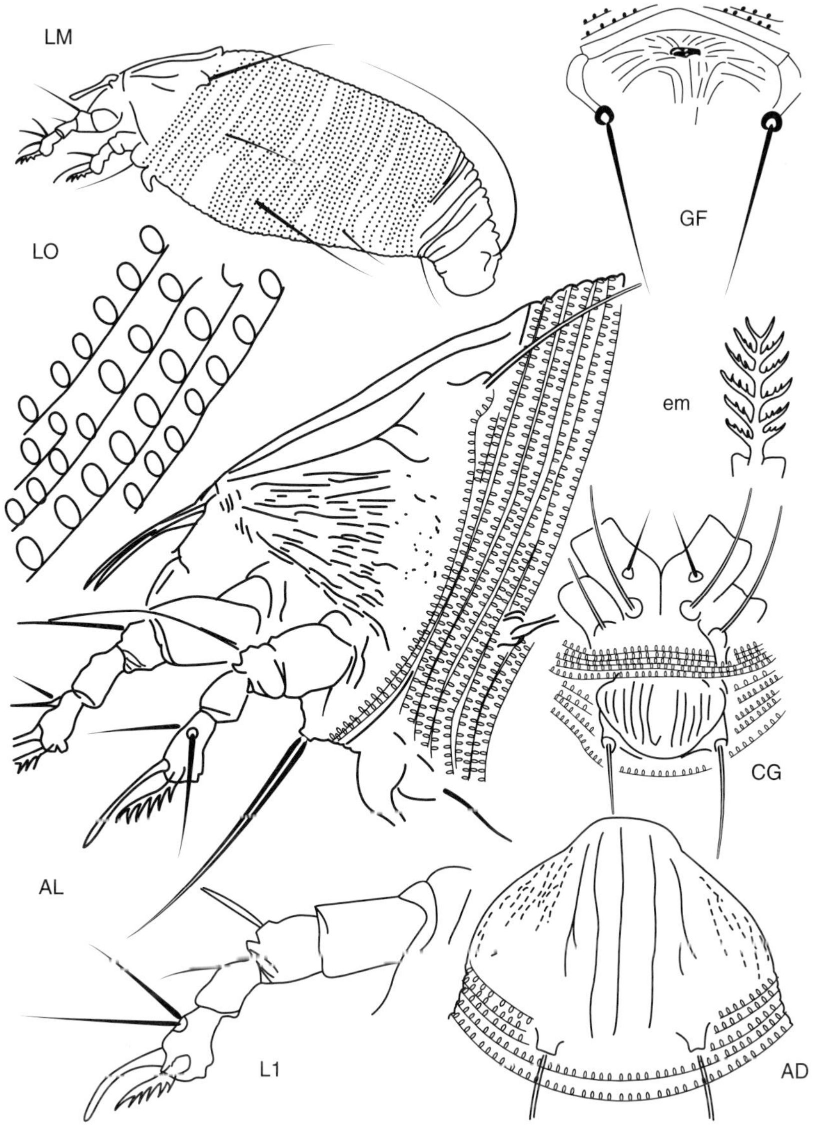

Fig. 16.47. *Aceria cajani* ChannaBasavanna (from ChannaBasavanna, 1966). See Table 16.1 for key.

Nene *et al.*, 1984; Kumar *et al.*, 2001), Myanmar (Su, 1931; Kumar *et al.*, 2001) and Nepal (Dahal and Neupane, 1991).

Host plants

Fabaceae: pigeon pea (*Cajanus cajan* (L.) Huth) (ChannaBasavanna, 1966), *C. cajanifolius* (Haines) Maesen and showy pigeon pea (*C. scarabaeoides* (L.) Thouars var. *scarabaeoides*) (Dahal and Neupane, 1991).

Bio-ecology

A. cajani lives on pigeon pea in several countries of the Indian subcontinent. In these countries, the pigeon pea is cultivated as a green manure crop, nurse crop and cover crop, as a windbreak hedge, as a host for lac insects and as food for silkworms. The stalks are used for fuel, thatch and basketry (Duke, 1983). ChannaBasavanna (1966) reports that *A. cajani* lives on tender leaves amid the surface hairs. Dhar and Rathore (1994) report that the mite usually lives on the terminal parts of the plants, and on vegetative buds in the lower, middle and upper parts. It survives on pigeon pea and its wild relatives, such as showy pigeon pea and *C. cajanifolius*. High population densities do not cause visible damage to the leaves, and are usually not noticed. The male deposits stalked spermatophores that are attached to leaf by an enlarged base. The pre-oviposition period lasts 24 h or more and the female lays 1–3 eggs daily. The eggs are milky white in colour, and usually attached to trichomes at the growing tips of the plants. The eggs hatch in 4–5 days, nymphal instar II takes from 48 h to more than 5 days, and the total life cycle is completed in less than 2 weeks (Oldfield *et al.*, 1981; Sheila *et al.*, 1988; Reddy *et al.*, 1990a). The mite is mainly spread by wind (Nene *et al.*, 1983; Mehan and Reddy, 1990), which carries it in the downward direction for up to 2 km; mite density is variable within a distance of 150 m from the source (Nene *et al.*, 1983; Reddy *et al.*, 1990b). In Bangalore,

India, *A. cajani* is present throughout the cropping period, but its density is low at the later stages of the crop. The populations increase from 45 to 60 days after sowing, reaching a peak on the 120- to 150-day-old crop between mid-October and mid-November (Prabhuswamy *et al.*, 1995). In Coimbatore, India, the populations increase from mid-August to the end of September, and at Navsari, mites occur throughout the crop period, reaching a maximum density during February (AICRP, 1996).

Symptomatology and damage

The feeding activity of *A. cajani* causes direct damage and transmits a viral disease to infested plants. The direct damage consists of slight leaf yellowing and the infested branches do not produce flowers and fruits (ChannaBasavanna, 1966). The mite transmits *Pigeon pea sterility mosaic virus* (PPSMV) (Seth, 1962; Janarthanan *et al.*, 1972; Reddy *et al.*, 1989), the causal agent of Sterility mosaic disease (SMD), the viral nature of which has only recently been documented (Kumar *et al.*, 2001, 2003a; Latha *et al.*, 2007). It is estimated that in Nepal and India, the mite causes annual pigeon pea grain losses worth US$282 million (Reddy *et al.*, 1993a).

STERILITY MOSAIC DISEASE. SMD has been extensively discussed by Oldfield and Proeseler, 1996, Perring, 1996, Dharmaraj *et al.*, 2004, Jones *et al.*, 2004 and Srinivasa *et al.*, 2004. *A. cajani* on pigeon pea across the Indian subcontinent is one species and no other *Aceria* species, and probably no *A. cajani* biotypes that differ in vectoring ability, are involved in the transmission of the agent of the SMD (Kumar *et al.*, 2001). The disease is reported from India, Bangladesh, Nepal, Sri Lanka (Kannaiyan *et al.*, 1984; Nene *et al.*, 1984; Dahal and Neupane, 1991) and Myanmar (Su, 1931). In India, SMD was first reported by Mitra (1931) and described by Alam (1933), and is widespread in all pigeon pea-growing states (Janarthanan *et al.*, 1973; Mali *et al.*, 1977; Kannaiyan *et al.*, 1981, 1984; Reddy *et al.*, 1988; Zote *et al.*, 1991; AICRP, 1994; Singh *et al.*, 1999).

Three types of symptoms are reported, which are referred to as 'complete sterility', 'partial sterility' and 'ring spot' (Reddy and Nene, 1979; Ghanekar *et al.*, 1992; Reddy *et al.*, 1998). Complete sterility consists of severe mosaic on the leaflets, and the plants lack flowers and pods; it occurs if the infection begins at the early stage (usually before 45 days old). Partial sterility has mild mosaic on the leaflets on a few branches, and only these branches do not produce flowers and pods; it occurs when infection takes place at later stages (usually beyond 45 days). In ring spot, green spots surrounded by a chlorotic halo appear on leaves; this occurs only in certain genotypes and disappears with plant maturity (Reddy *et al.*, 1990a). Some SMD-tolerant cultivars show this symptomatology without any effect on flowering (Reddy *et al.*, 1998). Further information on the transmission biology and relationships between the virus and the mite can be found in Kulkarni and Kumar (2002) and Latha and Doraiswamy (2004, 2008).

Five mites cm^{-2} of leaf surface are correlated with disease spread in the field, while one mite per leaf results in very low disease incidence (Dhar *et al.*, 1998). Increase in the mite population is associated with a simultaneous increase in the extent of disease (Prabhuswamy *et al.*, 1995), which spreads in the downwind direction for up to 2 km from the inoculum source. The mite density on plants differs considerably within a distance of 150 m from the source of inoculum. Spread in the upwind direction is very limited (usually less than 200 m). Commonly, the incidence and spread of the disease depends on the vector population in the field (Reddy *et al.*, 1989).

A single individual of *A. cajani* may transmit the disease, and both nymphs and adults transmit efficiently (Janarthanan *et al.*, 1972). The efficiency of a single mite in transmitting PPSMV of 20–60% (Reddy *et al.*, 1989; Kulkarni, 2003) has been reported as up to 53% by one individual but 100% when there are more than five mites per plant. The PPSMV is acquired by the eriophyoid after a minimum acquisition access period (AAP) of 15 min and inoculated after a minimum inoculation access period (IAP) of 90 min. No latent period is known. Starvation of *A. cajani* before, or following, PPSMV acquisition reduces the minimum AAP and IAP periods to 10 and 60 min, respectively; mites retain the virus for up to 13 h. None of the mites that develops from eggs taken from PPSMV-infected leaves transmits the virus, indicating that it is not transmitted transovarially. This suggests a semi-persistent mode of transmission of PPSMV by *A. cajani* (Kulkarni and Kumar, 2002). Nene *et al.* (1990) report an AAP of 5–10 min and an IAP of 30 min. A minimum AAP of 15 min and IAP of 25 min with no latent period can occur (Kulkarni, 2003). More than five mites per seedling with at least 15 h of AAP and more than 5 h of IAP results in 100% PPSMV transmission. Once the mite has acquired the virus, it retains it for up to 13 h of starving, but starvation beyond 3–4 h affects virus transmission significantly. The mite prefers SMD-infected plants for its multiplication and survival. Younger leaves of infected plants supported a higher mite population than the mature leaves. Mite density is positively correlated with PPSM virus concentration in the leaves. The minimum number of vector mites required for successful transmission of disease is one to five. Also, mites do not prefer to survive on infected plants for longer than 4–6 weeks after disease symptoms have been expressed (Narayanaswamy, 2004).

Infection at the early stages, within 45 days of sowing, causes complete sterility, sometimes resulting in complete yield loss. In older plants (beyond 45 days), sterility symptoms occur in a few branches, causing partial sterility, and the reduction in yield is not correlated with the percentage of infected plants (Reddy and Nene, 1981). Large reductions in pod length, pod width, number of grains per pod and 1000 grain weight (Singh and Rathi, 1994, 1997), shoot and root weight, nodule weight and nodule number (Prameela *et al.*, 1990) have been reported. Losses could be huge in ratoon and perennial pigeon pea and estimated yield losses due to SMD would be more than 2 lakh (hundred thousand) t in India alone (Kannaiyan *et al.*, 1984).

Control

Control relies upon various technical possibilities, such as the growing of resistant varieties, cultural choices and spray applications of chemicals (Reddy *et al.*, 1993b).

PLANT RESISTANCE. The topic has been summarized by Srinivasa *et al.* (2004), and much information is available on resistant or tolerant varieties. Thus, it is well known that some varieties

(Bihar, ICPL 151, Hyd 3C) are resistant to mite (Reddy *et al.*, 1993b). Other varieties (ICPL 87108, ICPL 90028, ICPL 88025) are free from both the disease and the mite (Patel *et al.*, 1995). Pigeon pea lines ICP 8852, ICP 11278 (variant 1), ICP 87119, ICP 7035 and Hyd 3C are tolerant of the disease (Srinivas and Reddy, 1995; Rangaswamy *et al.*, 1997; Reddy *et al.*, 1998). There are also later reports on pigeon pea varieties resistant to the mite and the virus (Lakshmikantha and Prabhuswamy, 2003; Kumar *et al.*, 2005).

CULTURAL PRACTICES. These include breaking the pigeon pea cycle by crop rotation, changing the sowing date, etc. Other cultural practices are destruction of the infected plants during the crop season and of voluntary diseased plants during the off season (Reddy *et al.*,1993b).

CHEMICAL CONTROL. Spray applications of acaricides at the early stages of the crop significantly reduce the mite populations and disease transmission (Anonymous, 1994–1996, 1996–1998; Reddy *et al.*, 1998). More details on disease management may be found in Kandulna *et al.* (2005).

Chemicals tried in various countries include aldicarb (Rathi, 1979, 1983), carbofuran (Rathi, 1979; Mehan and Reddy, 1990; Reddy *et al.*, 1990a), carbosulfan (De *et al.*,1996), chinomethionate (Reddy *et al.*, 1990a; Ghanekar *et al.*, 1992), dicofol (Reddy *et al.*, 1990a; Ghanekar *et al.*, 1992; AICRP, 1996, 1998; Chinnaiah and Mohanasundaram, 1995; De *et al.*,1996; Reddy *et al.*, 1998; Narayanswamy, 2004), ethion (AICRP, 1996, 1998; Reddy *et al.*, 1998), metasystox (Rathi, 1983), monocrotophos (Arjunan *et al.*, 1989; AICRP, 1996, 1998; Reddy *et al.*, 1998), phorate (Arjunan *et al.*, 1989; De *et al.*, 1996), profenophos (Narayanswamy, 2004), tetradifon (Reddy *et al.*, 1990a; Ghanekar *et al.*, 1992) and wettable sulfur (Lakshmikantha, 1992; Natarajan and Beevi, 1992; Singh, 1992; AICRP, 1996, 1998; Reddy *et al.*, 1998).

Biopesticides that have been tried include neem seed kernel extract plus neem oil emulsion (Chinnaiah and Mohanasundaram, 1995).

Aceria loewi (Nalepa)

Common name

Lilac bud mite.

Diagnostic characteristics

FEMALE. The body is 160–170 µm long. The prodorsal shield is somewhat acuminate anteriorly and has a faint pattern that seems to be of longitudinal lines centrally, with submedian lines arching back towards the scapular tubercles. The scapular tubercles are placed near the rear margin, but direct the scapular setae (*sc*) towards the front. The opisthosoma is completely microtuberculate. The genital coverflap has longitudinal ribs (Jeppson *et al.*, 1975).

Geographical distribution

Europe and the Nearctic region (Nalepa, 1890; Łabanowski and Soika, 1995; de Lillo and Beron, 2004; Šefrová and Laštůvka, 2005).

Host plants

Oleaceae: *Syringa* spp. (Šefrová and Laštůvka, 2005), Hungarian lilac (*S. josikaea* J. Jacq. ex Rchb.) (Łabanowski and Soika, 1995) and lilac (*S. vulgaris* (Shr.)) (Šefrová and Laštůvka, 2005).

Bio-ecology

A. loewi is an old European species described on lilac by Nalepa (1890), and it feeds on buds. Recently, the lilac bud mite was reported on *Syringa* spp. in the Czech Republic (Šefrová and Laštůvka, 2005) and on lilac and Hungarian lilac in nurseries and in city greeneries in Poland (Łabanowski and Soika, 1995). Dispersal occurs by spiders, dipterans, aphids and coleopterans, and is not selective with respect to carriers (Shvanderov, 1975).

Symptomatology and damage

The feeding activity of *A. loewi* on buds causes drying of the lilac buds. The shoots become dwarf and many lateral growths develop, appearing as witches' broom (Hughes, 1959; Jeppson *et al.*, 1975). Meyer Smith (1996a) also reports enlarged buds.

Control

This is not usually necessary.

Aceria fraxinivora (Nalepa) (Fig. 16.48)

Common name

Ash spangle gall mite.

Diagnostic characteristics

FEMALE. The body is worm like and about 180 µm long. The prodorsal shield is smooth and bears a short median ridge posteriorly. The empodium is 4-rayed (Jeppson *et al.*, 1975; Keifer *et al.*, 1982).

Geographical distribution

Europe, the Nearctic region, North Africa (Nalepa, 1909b; Keifer, 1952a; Jeppson *et al.*, 1975; Keifer *et al.*, 1982; de Lillo and Beron, 2004) and Iran (Xue *et al.*, 2009a).

Host plants

Oleaceae: ash (*Fraxinus excelsior* L.), Oregon ash (*F. oregona* Nutt.) and flowering ash (*F. ornus* L.) (*F. latifolia* Benth. = *F. oregona* Nutt.) (Nalepa, 1909b; Keifer, 1952a; Keifer *et al.*, 1982; Anthony *et al.*, 1983; de Lillo and Beron, 2004).

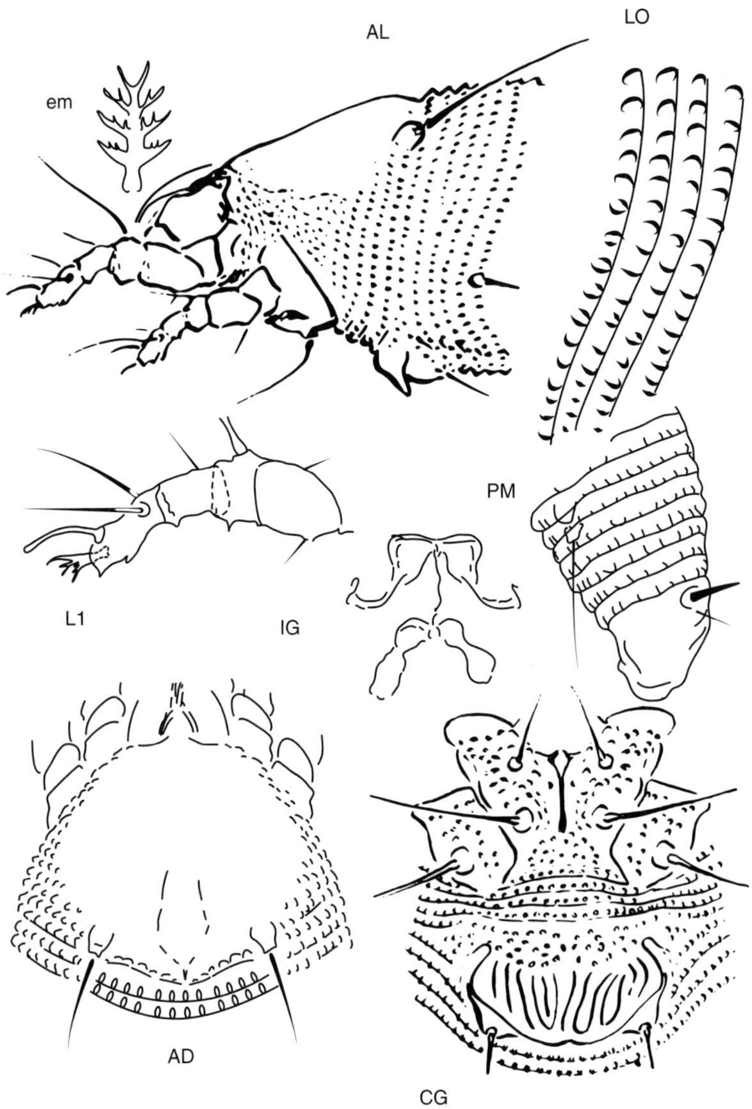

Fig. 16.48. *Aceria fraxinivora* (Nalepa) (from Keifer *et al.*, 1982). See Table 16.1 for key.

Bio-ecology

The ash spangle gall mite lives on the inflorescences of various ash species. Natural enemies include the cecidomyiid *Arthrocnodax fraxiniella* (Meade), the anthocorid *Orius vicinus* Ribaut and some phytoseiid mites, *Kampimodromus aberrans* (Oudemans) and *Typhlodromus* sp. (Fauvel *et al.*, 1975).

Symptomatology and damage

A. fraxinivora causes inflorescence galls, which form pendulous masses of abnormal, small leaves on attacked twigs during the cold season (Westphal and Manson, 1996). On *F. ornus*, there is hyperplasia of the epidermal and parenchymal cells around the damaged cells (Anthony *et al.*, 1983). The galls that develop are markedly brownish and commonly remain attached to the twigs for a long time. Galls are observed through the year and can be very numerous in some plant parts (Keifer *et al.*, 1982).

Control

This is usually not necessary.

Aceria genistae (Nalepa)

Common name

Broom erineum mite.

Diagnostic characteristics

FEMALE. The body is fusiform, 129–240 μm long, 45–54 μm wide and white or pale cream in colour. The gnathosoma is curved down. The prodorsal shield is subsemicircular, with ornamentation consisting of a median line, admedians and a submedian line, usually of a broken nature; the median and admedian

lines run the full length of the shield, whereas the submedians run just over half the shield length and usually end in a small circular area containing dash-like markings; the lateral shield areas and the area posterior of the circular areas have dash-like markings; the scapular tubercles are located on the rear shield margin, 19–21 µm apart; the scapular setae are 16–19 µm long, directed posteriorly. The opisthosoma has about 66–89 microtuberculate annuli, with rounded microtubercles. The setae c_2 are set on about annulus 13, and are 16–26 µm long. The setae *d* are set on about annulus 32, and are 37-46 µm long; the setae *e* are set on about annulus 51, and are 4-6 µm long; the setae *f* are set on about the 7th annulus from the rear, and are 16-21 µm long. The setae h_1 are 3–6 µm long. The genitalia are 17–19 µm wide and 7–10 µm long. The coverflap has longitudinal markings; the seta is 6–9 µm long. The empodium is 5-rayed (Manson, 1989).

MALE. The male is similar to the female, but smaller.

Geographical distribution

Australia (Sheppard *et al.*, 2006), Central Europe (Nalepa, 1891), Great Britain (NBN, 2015), Italy (Canestrini, 1892; Castagnoli, 1978), New Zealand (Manson, 1989; Smith *et al.*, 2010), South Africa (Craemer *et al.*, 1996), Spain (Roivainen, 1953) and the USA (Smith *et al.*, 2010).

Host plants

Fabaceae: Scotch broom (*Cytisus scoparius* (L.) Link), *Genista cinerea* (Vill.) DC, *G. corsica* (Lois.) DC, hairy greenweed (*G. pilosa* L.), Dyer's broom (*G. tinctoria* L.), *Sarothamnus purgans* Gren. *et* Godr., Spanish broom (*Spartium junceum* L.), European gorse (*Ulex europaeus* L.) and *U. parviflorus* Pourr. (Nalepa, 1891, Manson, 1989; Meyer Smith, 1996a; Sheppard *et al.*, 2006; Smith *et al.*, 2010).

Bio-ecology

A. genistae is an old European species (Nalepa, 1891) living on various Fabaceae (Meyer Smith, 1996a), whose geographical distribution is expanding. The broom erineum mite now also infests the Spanish broom, a shrub native to Europe and Asia and now imported into other countries of the world as a garden ornamental; it has become an invasive weed in California, Hawaii and South Africa (Meyer Smith, 1996a).

The high potential of the mite as a classical biological control agent of weeds has suggested that a as strain of *A. genistae* living on Scotch broom in France (Smith *et al.*, 2010) should be authorized for introduction into New Zealand. The same strain has been introduced into Australia (Sheppard *et al.*, 2006; Sagliocco, 2008). The mite has been accidentally introduced in the USA and is fairly widespread on *C. scoparius* between Mount Vernon, Washington State, and Portland, Oregon (Smith *et al.*, 2010).

Symptomatology and damage

A. genistae causes growth deformities of the shoots. On *G. tinctoria*, a whitish erineum covers the tips of the shoots (Roivainen, 1953). In the USA, the mite seems to damage *C. scoparius* in the field, reducing the flowering and other plant tissues. Host specificity is currently being evaluated using the mite from *Cytisus* in Washington State, and so far this strain has not formed galls on gorse (*U. europaeus*) or French broom (*Genista monspessulana* (L.) L.A.S. Johnson) in transfer experiments (Smith *et al.*, 2010). A different strain of the mite has possibly been reported on *U. europaeus* and *G. monspessulana* in California (Chan and Turner, 1998). Infested plants had little or no symptoms in the field; however, when gorse seedlings were brought into a greenhouse, they developed stunted shoot tips and reduced, thickened scale leaves, which were attributed to mite infestation. This suggests that natural enemies, cooler outdoor temperatures or some other factor may be preventing mite populations from damaging gorse in the field. In New Zealand, the same mite produces galling on *U. europaeus* but does not attack *C. scoparius* (Manson, 1989). It seems that there are different host-specific *A. genistae* biotypes or cryptic species associated with *C. scoparius* and *U. europaeus* (Smith *et al.*, 2010).

The broom erineum mite causes a general stunting of *S. junceum*, with multiple stem growth and dense witches' brooms (Meyer Smith, 1996a). It infests young apical shoots, inducing the formation of excessive hairiness, thickening of the axis, shortening of the internodes and witches' brooms (Petanović and Kielkiewicz, 2010b). *Spartium* witches' broom disease was found to be associated with the phytoplasma 16SrX group strain in Italy (Marcone *et al.*, 1996) and Spain (Torres *et al.*, 2002). X-disease-related phytoplasmas have been identified in ornamental trees and shrubs (forsythia, spiraea, cypress and willow), showing witches' broom, stunting and fasciation symptoms (Paltrinieri *et al.*, 1998).

Control

Control is sometimes necessary in nurseries or on young ornamental plants. On adult plants it is usually not necessary (but see the discussion on biological control above).

Aceria lantanae (Cook)

Common name

Lantana flower gall mite.

Diagnostic characteristics

FEMALE. The prodorsal shield lacks a definite pattern. The opisthosomal annuli have rounded microtubercles and the empodium is 4-rayed (Jeppson *et al.*, 1975).

Geographical distribution

Brazil (Flechtmann, 1973b), Cuba (Cook, 1909; de la Torre *et al.*, 2004), the Caribbean, Florida, Central and South America and Texas (Keifer and Denmark, 1976; Jeppson *et al.*, 1975).

Host plants

Verbenaceae: lantana (*Lantana camara* L.) (Cook, 1909; Flechtmann, 1973b; Jeppson *et al.*, 1975; Keifer and Denmark, 1976), other

host plants of the *Lantana* section (= section *Camara*) (Palmer and Pullen, 1995; Day *et al.*, 2003; Sanders, 2006).

Bio-ecology

A. lantanae lives on lantana, a verbenaceous plant native to Central and South America, Florida, Mexico and islands of the West Indies. The taxonomy of this weed is unclear, and members of this group have been distributed globally, and are continuing to hybridize in the field (Spies, 1984). Over 650 cultivars have been selected, and both ornamental and wild forms are spread worldwide; several cultivars have become weeds in various countries, including Australia, eastern and southern Africa, southern Asia and the Pacific Islands. In tropical and subtropical countries, lantana is a widespread weed, while in more temperate areas it is mainly an ornamental. The mite has been collected on three other host plants of the *Lantana* section (= section *Camara*) (Palmer and Pullen, 1995; Day *et al.*, 2003; Sanders, 2006).

Eight eriophyoid species have been described from *L. camara* and, with the exception of *A. lantanae*, six of these species are leaf vagrants (Amrine and Stasny, 1994). Moreover, only *A. lantanae*, *Rhynacus kraussii* Keifer (Keifer, 1962a; Keifer and Denmark, 1976), *Shevtchenkella stefneseri* Craemer and *Paraphytoptus magdalenae* Craemer were reported from the original geographical range of *L. camara* (Craemer, 1996). Among the above reported eriophyoids, *A. lantanae* has been suggested as a possible biological control agent in those countries where lantana is a weed (Keifer and Denmark, 1976). A detailed review of this topic is given by Smith *et al.* (2010). The lantana flower gall mite is characterized by two biotypes (probably species) that are morphologically indistinguishable (Craemer and Neser, 1990; Craemer, 1996), one common in South America (Flechtmann, 1973b; Moura *et al.*, 2008) and the other in Central America (Cook, 1909; Keifer and Denmark, 1976; Craemer and Neser, 1990; Baars and Neser, 1999; Urban *et al.*, 2001a).

After the North American experience (Keifer and Denmark, 1976), the second biotype of the lantana flower gall mite has been introduced in South Africa, where a mixed culture of possible biotypes from Florida and Cuba is now established on certain lantana varieties (Smith *et al.*, 2010). The results of tests being conducted in quarantine in South Africa will be important for the application for permission to introduce the mite into quarantine in Australia (Smith *et al.*, 2010).

Symptomatology and damage

The infested flowers become a mass of small leaf-like outgrowths. A very similar species on the same host plant causes simple leaf galls and occurs in Central and South America (Jeppson *et al.*, 1975). Distortion of the flower buds and flowers, producing a mass of very small green leaves, was observed in Brazil; these are often large. On the leaves, the mite causes small crinkle galls that can be seen on both sides of the leaves (Flechtmann and Harley, 1974). Keifer and Denmark (1976) also report the distortion of flower buds and flower galls by the mite. Moreover, *A. lantanae* may cause smaller galls on the leaf surfaces (Meyer Smith, 1996a).

The first biotype of *A. lantanae* causes 'crinkle' galls on leaves (Flechtmann, 1973b; Moura *et al.*, 2008), while the second causes galls on flower and shoot buds (Cook, 1909; Keifer and Denmark, 1976; Craemer and Neser, 1990; Baars and Neser, 1999; Urban *et al.*, 2001a). The latter biotype produces a dense witches' broom that may inhibit further flowering and growth (Craemer and Neser, 1990; Baars and Neser, 1999). Sometimes, both types of galls have been found on the same host plant.

Research in South Africa on host specificity of the inflorescence-galling biotype of *A. lantanae* from Florida showed that galls formed only on varieties of *L. camara* (Urban *et al.*, 2001a,b, 2004; Mpedi and Urban, 2003, 2005). South African and Australian lantana varieties that were both infested by a Florida strain of *A. lantanae* displayed a flower suppression of 10–95% and 0–30%, respectively (Urban *et al.*, 2004). This suggests that more lantana varieties should be used (Craemer and Neser, 1990; Urban *et al.*, 2001b, 2004). Abiotic environmental conditions may also reduce the potential efficacy of the mite. In fact, although high population densities of *A. lantanae* have sometimes been found in Florida (Baars and Neser, 1999), gall induction requires the availability of differentiating buds on rapidly growing plants and high humidity, which must occur simultaneously in order for severe damage to be caused. Details on the ontogenesis of the leaf and leaf galls induced by *A. lantanae* in *L. camara* may be found in Moura *et al.* (2005, 2009).

Control

Control is usually unnecessary, but in those regions where lantana is cultivated as an ornamental it is possible to use acaricides, among which dicofol has been suggested in the past (Keifer and Denmark, 1976).

***Aceria ligustri* (Keifer) (Fig. 16.49)**

Common name

Privet mite.

Diagnostic characteristics

FEMALE. The body is worm like, 150–190 μm long, 35-40 μm wide and yellowish in colour. The gnathosoma is curved down. The prodorsal shield is rounded anteriorly, with an obscure central pattern, and the sides granular; the scapular tubercles are 21 μm apart on the rear margin; the scapular setae (*sc*) are 24 μm long, and projecting caudad. The coxae are somewhat tuberculate, and coxae I touching. The opisthosoma has 60–65 annuli, densely microtuberculate; the last 10–12 annuli are broader, the tubercles finer and more elongate. The opisthosomal setae c_2 are 20 μm long, set on about annulus 10; the setae *d* are 50 μm long, set on about annulus 23; the setae *e* are 12 μm long, set on about annulus 39; the setae *f* are 23 μm long, set on about annulus 4 from the rear; the setae h_1 are present and of moderate size. The genitalia are 20 μm wide, 11 μm long; the coverflap has about 10 furrows, and the seta is 11 μm long. The empodium is 4-rayed (Keifer, 1943).

MALE. The body is 120–140 μm long (Keifer, 1943).

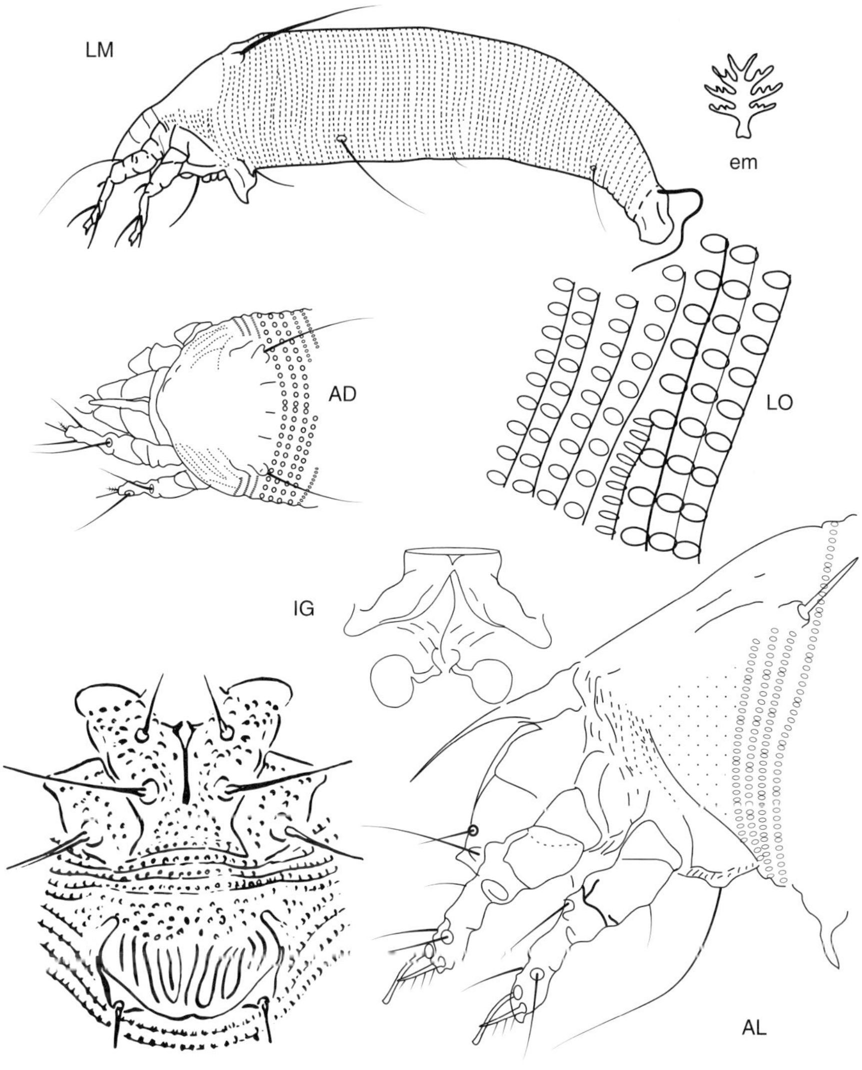

Fig. 16.49. *Aceria ligustri* (Keifer) (from Keifer, 1943). See Table 16.1 for key.

Geographical distribution

California (Keifer, 1943), Hungary (de Lillo and Beron, 2004), Iran (Hajizadeh and Hosseini, 2003a), Serbia and Montenegro (Glavendekić *et al.*, 2005) and Poland (Łabanowski and Soika, 1998).

Host plants

Oleaceae: unspecified *Ligustrum* sp. (Glavendekić *et al.*, 2005), privet (*L. ovalifolium* Hassk.) (Keifer, 1943), hedge privet (*L. ovalifolium*) (Hajizadeh and Hosseini, 2003a), wild privet (*L. vulgare* L.) (Łabanowski and Soika, 1998).

Bio-ecology

A. ligustri was originally described as *Eriophyes ligustri* from privet in California (Keifer, 1943). The mite is also reported on an unspecified *Ligustrum* sp. from Serbia and Montenegro (Glavendekić *et al.*, 2005), on wild privet in city greeneries from Poland (Łabanowski and Soika, 1998) and on hedge privet shrubs from Iran (Hajizadeh and Hosseini, 2003a). The mite lives on the leaves and buds of privet as a vagrant and feeds on the sap of plant leaf cells. In Iran, the annual activity of privet mite starts in early spring and continues to early summer; during this time, curling of the leaves is considerable and they may be so tightly rolled that the infested twigs appear leafless. In the early summer, owing to increasing temperature, the mites migrate to aestivation sites under the privet bud scales. The species has a long summer and winter diapause. In late winter, the mites lay their eggs in the privet buds (Hajizadeh and Hosseini, 2003b).

Symptomatology and damage

A. ligustri forms a moderately plentiful population in the host buds (Keifer, 1943). It lives as a vagrant on the leaves and

buds of *Ligustrum* hedges, where it can cause curling, bronzing and rusting of infested leaves. It is potentially a pest that can cause considerable damage in urban areas (Návia *et al.*, 2010a).

Control

Control is not usually needed, but if the host plant is cultivated as an ornamental, it may be possible to use an acaricide. The crop and pest management guidelines of Cornell University (2012) suggest treating the infested trees in the last 10 days of May through to mid-June, with 298–802 GDDs (growing degree-days) and to mid-July, with 1266–1515 GDDs, using abamectin, carbaryl, fenazaquin, fenpyroximate, horticultural oil, milbemectin or neem oil.

Aceria proteae Meyer (Fig. 16.50)

Common name

Protea witches' broom mite.

Diagnostic characteristics

FEMALE. The body is worm like, 211–268 μm long and 43 μm wide. The gnathosoma projects ahead and down. The prodorsal shield is subsemicircular, with a small frontal lobe. The median line is present, complete or sometimes broken, and ends in an inconspicuous dart-shaped mark; the admedian lines are complete and gradually diverge; the first 2 submedian lines join at ½ and from there curve slightly inwards and end near the rear

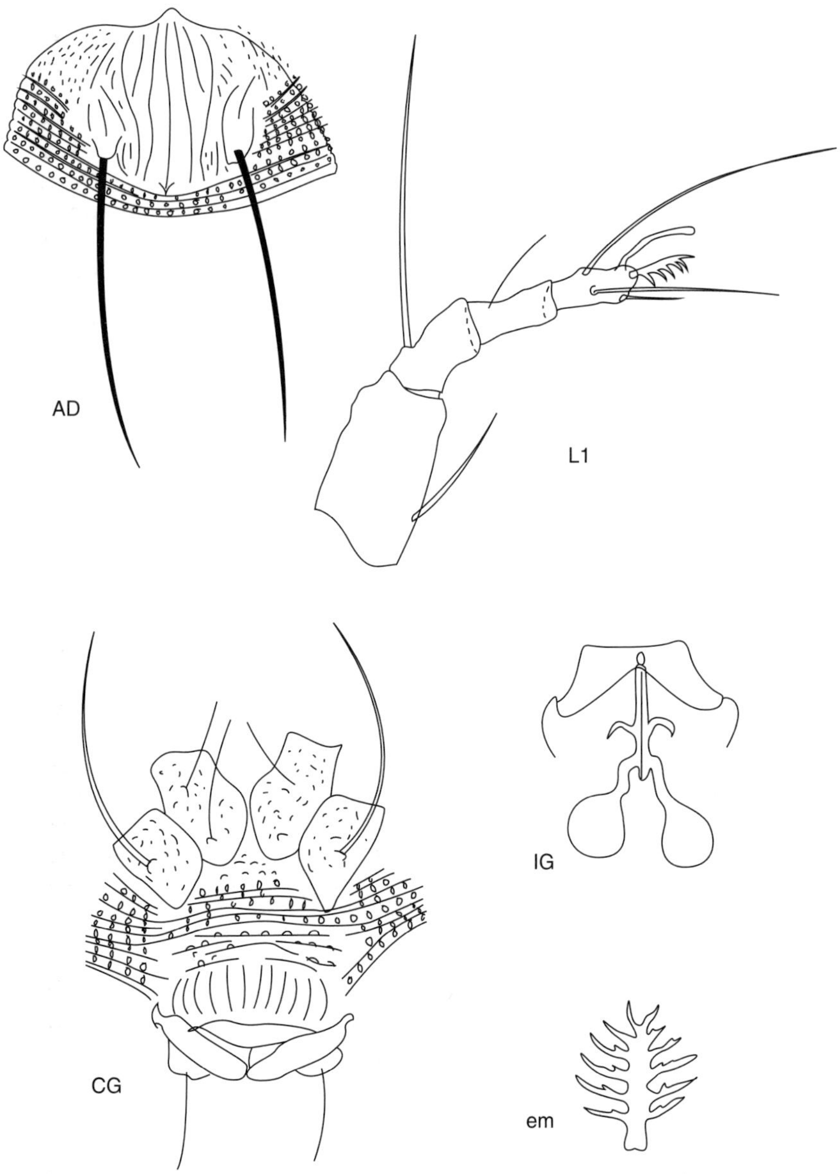

Fig. 16.50. *Aceria proteae* Meyer (from Meyer Smith, 1981b). See Table 16.1 for key.

margin; laterally, there are short dashes and granules. The scapular tubercles are 25 µm apart, just craniad to the rear margin; the scapular setae (*sc*) are 55 µm long, directed backwards to about annulus 20; some partial annuli are located below the tubercles. The ventral opisthosoma has about 75 microtuberculate annuli, with the microtubercles elliptical and located ahead of the annuli margins. The coxae are narrowly contiguous anteriorly and granulate. The genital coverflap has 10–12 longitudinal ribs. The empodium is 6-rayed (Meyer Smith, 1981b).

MALE. The male is like the female, but 169–175 m long and 50 m wide. The empodium is 5-rayed (Meyer Smith, 1981b).

Geographical distribution

South Africa (Meyer Smith, 1981b).

Host plants

Proteaceae: *Leucadendron* sp., *Leucospermum cordifolium* (Salisb. ex Knight) Fourc., *L. cuneiforme* (Burm. F.) Rourke, *L. linifolium* (Jacq.) R. Br., *Mimetes cucullatus* (L.) R. Br., *Protea caffra* Meisn., *P. cynaroides* (L.) L., *P. laurifolia* Thunb., *P. lepidocarpodendron* (L.) L., *P. nerifolia* R. Br., *P. nitida* Mill. and *P. repens* (L.) L. (Meyer Smith, 1981b, 1996a).

Bio-ecology

A. proteae lives on several plants belonging to the Proteaceae family in South Africa (Meyer Smith, 1981b, 1996a). Its bio-ecology has not been documented.

Symptomatology and damage

The feeding activity of this mite causes malformed growth, probably as a result of the injection of toxin or mycoplasma into the tissues. The buds divide repeatedly and do not develop a flower-bearing shoot but hundreds or more tiny shootlets that form a witche's broom the size of a man's fist or wider. Clusters may sometimes die off. The mites cause severe damage to proteas. High levels of attack may reduce flower production and necessitate the replanting of entire plantations. The popular giant protea, *P. cynaroides*, is very susceptible to mite infestations, whereas *P. repens* is less susceptible (Meyer Smith, 1996a).

Control

Control is provided by spray applications of acaricides and pruning of the infested part(s) of the plants (Meyer Smith, 1996a).

Aceria caliberberis Keifer (Fig. 16.51)

Common name

Unknown.

Diagnostic characteristics

FEMALE. The body is worm like, 170–195 µm long, 40–45 µm wide and light yellow in colour. The gnathosoma is curved down. The prodorsal shield has an obscure central pattern that is mostly missing, with the admedian lines distinct on the rear half; laterally, two diverging longitudinal lines suddenly converge to the rear of the shield, with two diagonal connecting lines; the lateral granules are nearly absent; the scapular tubercles are 21 µm apart; the scapular setae (*sc*) project backwards 30 µm. The opisthosoma has 50–55 annuli, evenly set with rounded microtubercles. The opisthosomal setae c_2 are 40 µm long, set on about annulus 6 behind the shield; the setae *d* are 60 µm long, set on about annulus 15; the setae *e* are 14 µm long, set on about annulus 25; the setae *f* are 35 µm long, set on about annulus 6 from the rear; the setae h_1 are present. The genitalia are 21 µm wide and 16 µm long; the genital coverflap has about 12 longitudinal ribs, and the seta is 45 µm long. The empodium is 5-rayed (Keifer, 1952c).

Geographical distribution

California (Keifer, 1952c), Poland (Łabanowski and Soika, 1998) and Serbia and Montenegro (Glavendekić *et al.*, 2005).

Host plants

Berberidaceae: barberry (*Berberis californica* Jepson) (Keifer, 1952c), Japanese barberry (*B. thunbergii* DC.) (Łabanowski and Soika, 1998), shining netvein barberry (*Mahonia dictyota* (Jeps.) Fedde) (Keifer *et al.*, 1982) and an imprecisely specified host plant (Glavendekić *et al.*, 2005).

Bio-ecology

The mite is reported from barberry in California (Keifer, 1952c) and recently in Serbia and Montenegro on an imprecisely specified host plant (Glavendekić *et al.*, 2005) and on Japanese barberry in Poland (Łabanowski and Soika, 1998). It lives on leaf folds and buds, and in California has been collected in November from shining netvein barberry at 500–1800 m altitude (Keifer *et al.*, 1982).

Symptomatology and damage

A. caliberberis makes an irregular, sinuate fold in the leaves from the upper surface. This fold is usually longitudinal, but may occur on any part of the leaf from the midrib to the edge (Keifer, 1952c). Another leaf distortion is uncommon, and consists of the leaf margins curled upward, folded, and twisted across the upper surface towards the midrib. The fold can involve the entire leaf, with one or two windings forming a tube. The leaves can become crumpled and brown in colour (Keifer *et al.*, 1982).

Control

Control is not usually needed, but if the host plant is cultivated as an ornamental, an acaricidal spray may be necessary.

Aceria georghioui Keifer (Fig. 16.52)

Common name

Unknown.

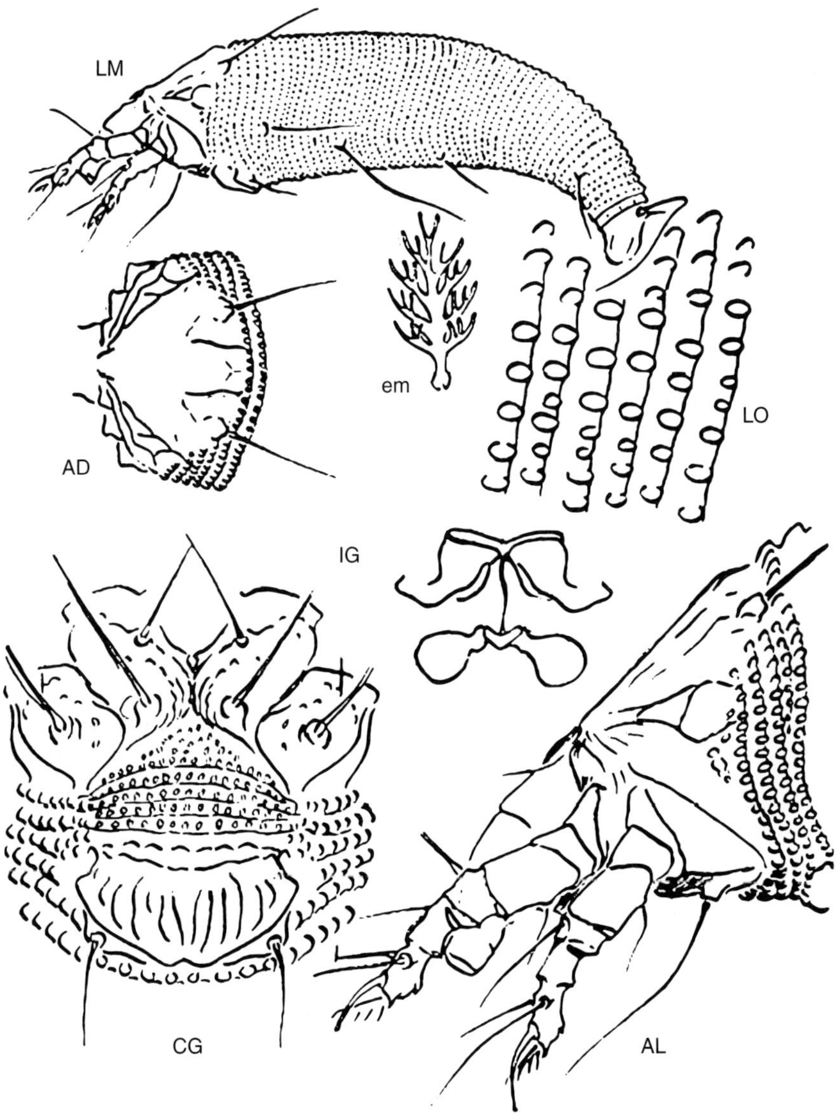

Fig. 16.51. *Aceria caliberberis* Keifer (from Keifer, 1952c). See Table 16.1 for key.

Diagnostic characteristics

FEMALE. The body is worm like, 210–220 µm long, and dull yellow in colour. The gnathosoma projects diagonally down. The prodorsal shield is subtriangular; the median line is complete, faint anteriorly, and ends close to the rear shield margin in a dark shaped mark; the admedian lines are complete, and gradually diverging. The lateral angles of the prodorsal shield are somewhat bulging, with lines and granules. The scapular tubercles are 21 µm apart, and the scapular setae (*sc*) are 20 µm long. The coxae are generally lined with a pattern of granules. The opisthosoma has about 85 annuli, completely microtuberculate, and the microtubercles are rounded and located ahead of the rear annulus margins. The setae c_2 are 53 µm long; the setae *d* are 60 µm long, set on annulus 27; the setae *e* are 12 µm long, set on annulus 46; the setae *f* are 28 µm long, set on annulus 9 from the rear; the setae h_1 are 6 µm long. The genital coverflap has 3 or 4 basal cross lines of granules and about 14–16 longitudinal ribs. The empodium is 7-rayed (Keifer, 1959b).

Geographical distribution

California (Hoy, 2011), Cyprus (Keifer, 1959b) and Italy (de Lillo and Beron, 2004).

Host plants

Caryophyllaceae: carnation (*Dianthus* sp.) (Keifer, 1959b; de Lillo and Beron, 2004; Hoy, 2011).

Bio-ecology

The mite is reported from carnation in Cyprus (Keifer, 1959b), Italy (de Lillo and Beron, 2004) and California (Hoy, 2011).

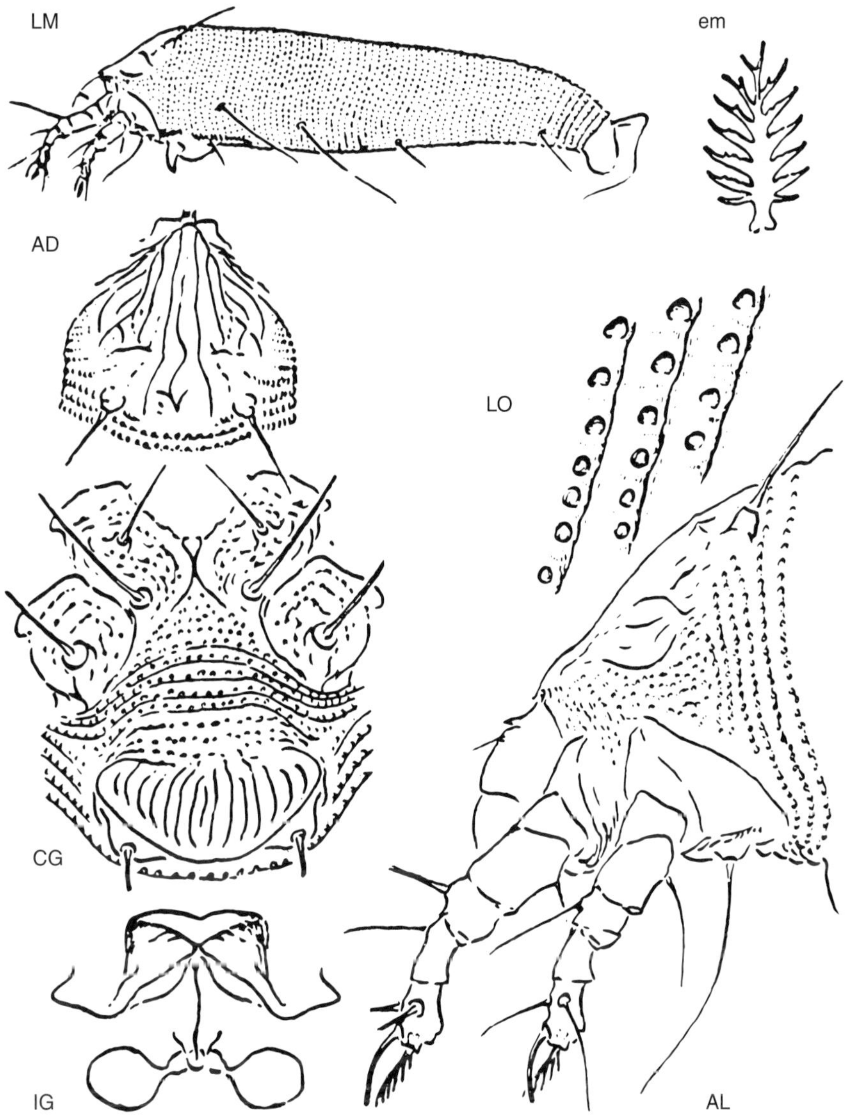

Fig. 16.52. *Aceria georghioui* Keifer (from Keifer, 1959b). See Table 16.1 for key.

Symptomatology and damage

The mites are said to live in the leaf axils and cause discoloration and unthriftiness of the plant (Keifer, 1959b; Jeppson *et al.*, 1975).

Control

Spray applications of various chemicals, such as chlorobenzilate, diazinon, dicofol and parathion, repeated several times, have been suggested in the past (Pritchard, 1957).

Aceria paradianthi Keifer (Fig. 16.53)

Common name

Carnation bud mite.

Diagnostic characteristics

FEMALE. The body is worm like, 200 μm long, 50–55 μm wide and yellow in colour. The gnathosoma is downcurved. The pattern of the prodorsal shield is a network; the median line starts from a fork from the admedian lines at ⅓; at ⅔ a transverse line runs across the shield from in front of the scapular tubercles on each side, and two curved lines connect the median line to the admedian lines just ahead of the rear margin; the admedian lines curve close together in front and ⅓ ahead of the median line; the first submedian line extends to a point laterad of each scapular tubercle, where it connects by a cross line to the transverse line; the second submedian line is broken; the shield has a well-defined longitudinal granular area above the coxae; the scapular tubercles are 25 μm apart; the scapular setae (*sc*) are 25 μm long, projecting backwards. The coxae are somewhat granular, with coxae I hardly connate centrally. The opisthosoma has about

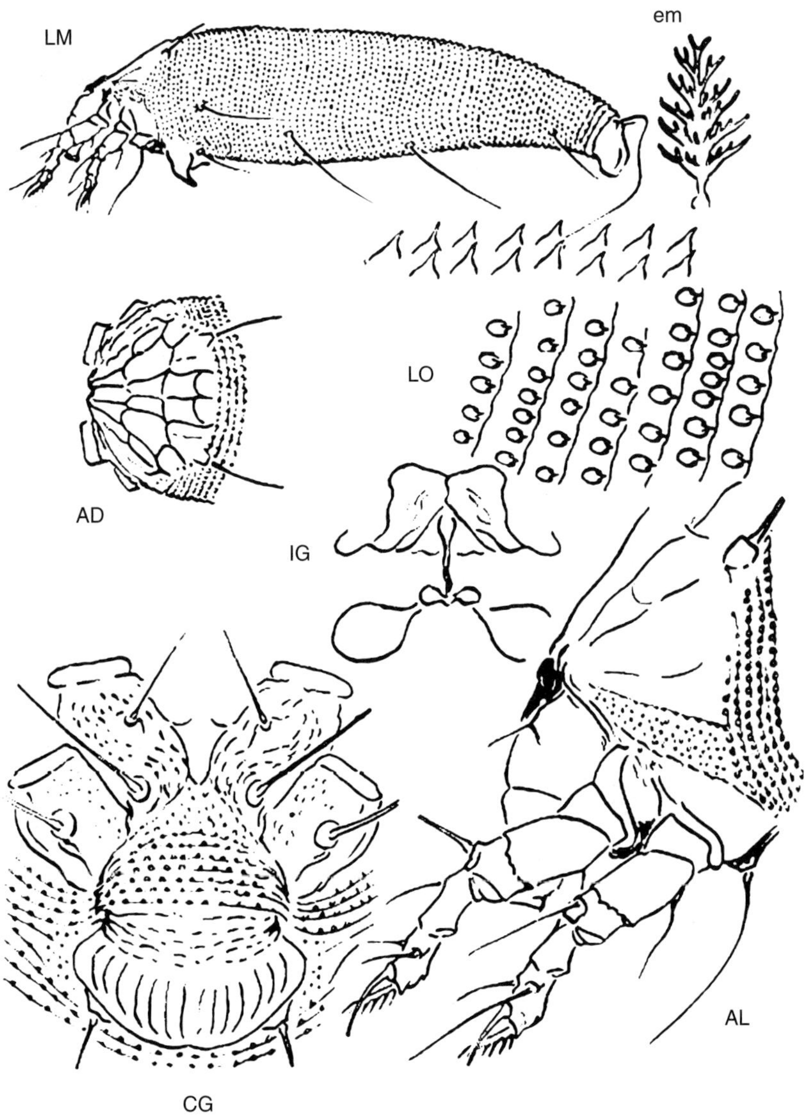

Fig. 16.53. *Aceria paradianthi* Keifer (from Keifer, 1952c). See Table 16.1 for key.

85 annuli; the microtubercles are equally developed dorsally and ventrally and each formed into a point. The setae c_2 are 45 μm long, set on about annulus 10; the setae *d* are 55 μm long, set on about annulus 29; the setae *e* are 45 μm long, set on about annulus 50; the setae *f* are 43 μm long, set on annulus 8 from the rear; the setae h_1 are present. The genitalia are 28 μm wide, 17 μm long; the genital coverflap has about 14 longitudinal furrows; the seta is 25 μm long. The empodium is 6-rayed (Keifer, 1952c).

Geographical distribution

California (Keifer, 1952c), Italy (de Lillo and Beron, 2004), Japan (Mito and Uesugi, 2004), Poland (de Lillo and Beron, 2004) and South America (Jeppson *et al.*, 1975).

Host plants

Caryophyllaceae: carnation (*Dianthus* sp.) (Keifer, 1952c).

Bio-ecology

The carnation bud mite lives between the leaf bases and stems of carnation, being especially numerous towards the lower parts of the plant, which shows a tendency towards stunting and basal sprouting (Keifer, 1952c).

Symptomatology and damage

Infested plants are said to have a water-soaked appearance. Clippings indicate that a certain amount of yellowing results from the mites' activities (Keifer, 1952c).

Control

The economic importance of this mite is unclear, and consequently so is the need for its control.

Aceria peyerimhoffi (Nalepa)

Common name

Flax leaf edgeroller mite.

Diagnostic characteristics

FEMALE. The body is worm like and about 160 μm long. The prodorsal shield has central lines. The dorsal opisthosomal annuli are wider than the ventral annuli and more strongly granular. The genital coverflap has fine lines (Jeppson *et al.*, 1975).

Geographical distribution

North Africa (Jeppson *et al.*, 1975).

Host plants

Linaceae: flax (*Linum usitatissimum* L.) (Jeppson *et al.*, 1975).

Bio-ecology

The mite lives on flax in North Africa (Jeppson *et al.*, 1975). Its bio-ecology is not well documented.

Symptomatology and damage

The mites cause flax leaves to roll upward (Jeppson *et al.*, 1975).

Control

The economic importance of this mite is unclear, and consequently so is its control.

Aceria parapopuli (Keifer) (Fig. 16.54)

Common name

Cottonwood or poplar bud gall mite.

Diagnostic characteristics

FEMALE. The body is spindle formed, about 180 μm long, 55 μm wide and reddish in colour. The gnathosoma projects ahead and down. The prodorsal shield is nearly smooth, with a faint central indication of a network pattern, and a few lateral tubercles; the scapular tubercles are 26 μm apart, on the rear margin; the scapular setae (*sc*) are 38 μm long, projecting backwards. The opisthosoma has 48–54 annuli, microtuberculate except for the last 10–12 dorsal annuli, which are smooth. The setae c_2 are 23 μm long, set on about annulus 5; the setae *d* are 61 μm long, set on about annulus 15; the setae *e* are 15 μm long, set on about annulus 27; the setae *f* are 23 μm long, set on about annulus 4 from the rear; the setae h_1 are present. The genitalia are 21.5 μm wide, 10.5 μm long; the genital coverflap is smooth, and the seta is 21 μm long. The empodium is 4-rayed (Keifer, 1940a).

MALE. The body is 170 μm long and 50 μm wide (Keifer, 1940a).

Geographical distribution

North America (Keifer, 1940a; Jeppson *et al.*, 1975; Keifer *et al.*, 1982) and Iran (Sepasgosarian, 1973; Xue *et al.*, 2009a).

Host plants

Salicaceae: Canadian poplar (*Populus* sp.) (Keifer, 1940a), mountain cottonwood (*P. angustifolia* E. James), balsam poplar (*P. balsamifera* L.) and intermediate hybrids (Floate and Whitham, 1993), and white poplar (*P. alba* L.), Fremont's cottonwood (*P. fremontii* S. Watson), black poplar (*P. nigra* L.) (Sepasgosarian, 1973; Xue *et al.*, 2009a).

Bio-ecology

A. parapopuli was first reported on Canadian poplar from Montana (Keifer, 1940a) and later from the western part of North America, mainly from the Great Plains and in the Intermountain Region (Jeppson *et al.*, 1975). Primary hosts include the mountain cottonwood, the balsam poplar and intermediate hybrids (Floate and Whitham, 1993). In Iran, the mite has been collected on white poplar, Fremont's cottonwood and black poplar (Sepasgosarian, 1973; Xue *et al.*, 2009a). The mite overwinters in galls and leaves in spring to attack new vegetative and reproductive buds. The infested buds form woody, cauliflower-like galls also over 3 cm in diameter. These galls are usually active for a year but may persist for 15 years. During the summer, new generations are produced every 2 or 3 weeks. Dispersal occurs mainly by wind (McIntyre and Whitham, 2003).

Symptomatology and damage

The cottonwood or poplar bud gall mite causes woody galls to be made from buds, and stops twig growth at that stage. The development of these galls occurs for several years and the mites move to new tissues formed in them; the galls may reach a diameter of an inch or more and then die, but they stay on the branch or twig. Strong attacks result in galls that may erupt on several buds along twigs, inducing a growth retardation of the tree and twig deformation. Branches may die early as well. Ornamental poplar may fail to develop satisfactorily (Keifer, 1940a; Jeppson *et al.*, 1975). The galls are brown, tinged with a greenish and whitish powdery deposit. Old galls can be observed at any season (Keifer *et al.*, 1982). Ecological research over several years showed that the plant genotype and hybridization influence the population dynamics of *A. parapopuli* (McIntyre and Whitham, 2003).

Control

The economic importance of this mite is moderate. However, in nursery and on ornamental poplars, control may sometimes be necessary and require spray application of acaricides.

Maple trees group

This group includes the most common eriophyoid mites inhabiting maple trees and making leaf erinea. Of these, *Aceria*

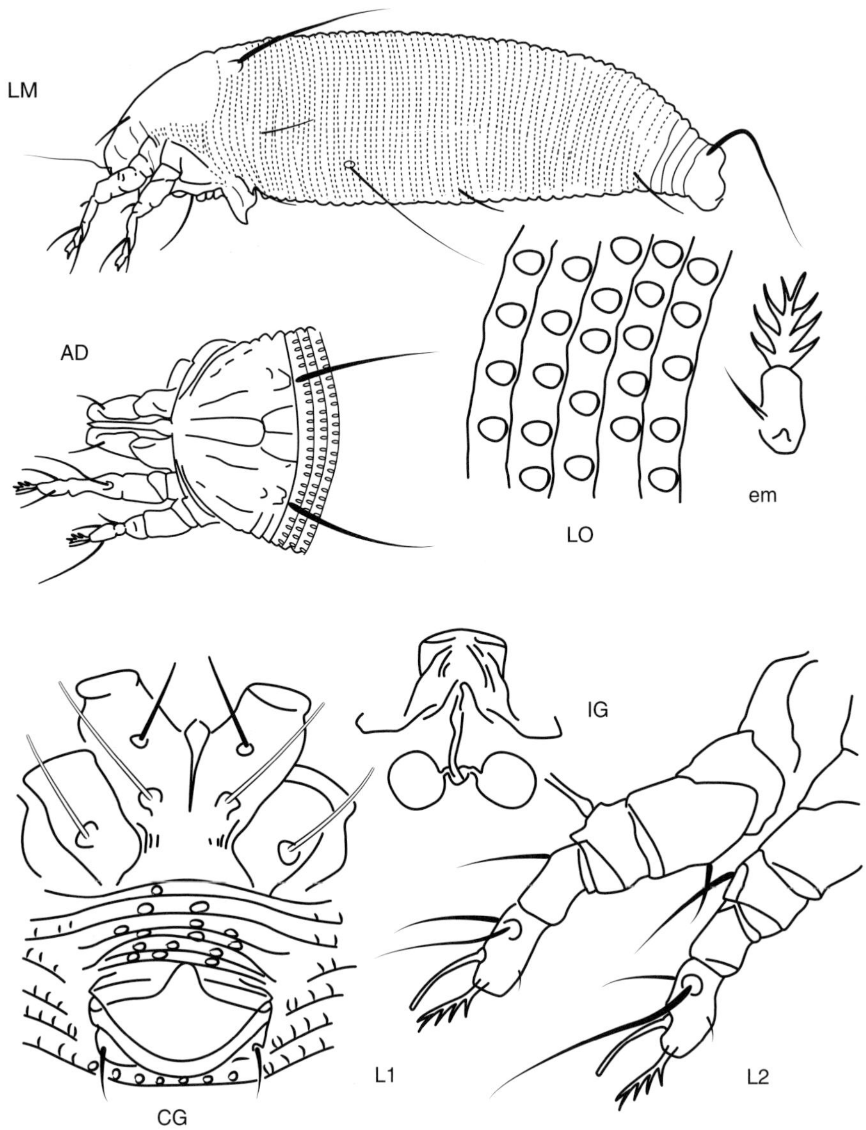

Fig. 16.54. *Aceria parapopuli* (Keifer) (from Keifer, 1940a). See Table 16.1 for key.

elongata (Hodgkiss) lives on sugar maple (*Acer saccharum* Marshall) in North America and *Aceria pseudoplatani* (Corti) on European sycamore maple (*Acer pseudoplatanus* L.) in Europe. These two species are described below. *Aceria aceris* (Hodgkiss) lives on silver maple (*Acer saccharinum* L.) in North America and causes greenish or reddish leaf erinea on the undersurface. *Aceria major* (Hodgkiss) lives on red maple (*Acer rubrum* L.) in eastern North America. The protogyne has strong prodorsal shield lines and the deutogyne has a few transverse dorsal annuli creases. The mite causes white to red leaf erinea on the undersurface. *Aceria modesta* (Hodgkiss) lives on sugar maple in eastern North America and makes green erinea on the leaf undersurface (Hodgkiss, 1930; Jeppson *et al.*, 1975). *Aceria negundi* (Hodgkiss) is reported from box elder (*Acer negundo* L.) in North America and is commonly referred to as the box elder erineum mite. The deutogyne has the prodorsal shield not overhanging the gnathosoma and the opisthosoma is characterized by a strong dorsal–ventral differentiation of annuli. The mite makes erineum pockets that are dull tan or grey in colour on the undersurface of the leaves, protruding out above the upper surface (Hodgkiss, 1930; Jeppson *et al.*, 1975; Keifer *et al.*, 1982). Several other mites are known.

Aceria elongata (Hodgkiss) (Fig. 16.55)

Common name

Crimson erineum mite.

Diagnostic characteristics

FEMALE. The protogyne is worm like, 216 μm long and yellowish white in colour. The prodorsal shield lacks many markings,

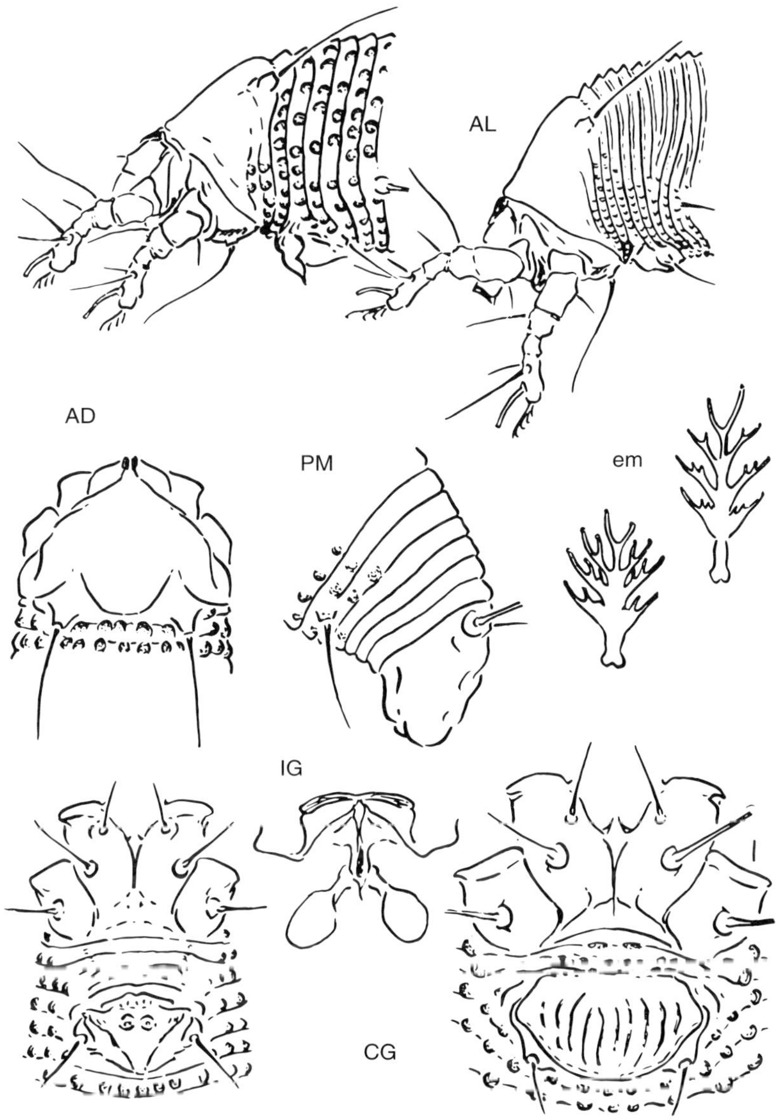

Fig. 16.55. *Aceria elongata* (Hodgkiss) (from Jeppson *et al*., 1975). See Table 16.1 for key.

but has an inverted 'V'-shaped cross line in front of each scapular tubercle. The opisthosomal microtubercles are irregularly placed, coarse, rounded and evenly present on the dorsal and ventral annuli. The dorsal posterior opisthosomal area lacks microtubercles. The genital coverflap has about 10 longitudinal ribs. The empodium is 4-rayed. The deutogyne lacks dorsal opisthosomal microtubercles, but has more annuli ventrally, with a central transverse crease in each dorsal half annulus (Hodgkiss, 1930; Jeppson *et al*., 1975; Keifer *et al*., 1982).

MALE. The males have microtubercles as the protogyne (Keifer *et al*., 1982).

Geographical distribution

East Coast of North America (Hodgkiss, 1930; Jeppson *et al*., 1975; Keifer *et al*., 1982).

Host plants

Sapindaceae: sugar maple (*Acer saccharum* Marshall) (Hodgkiss, 1930; Jeppson *et al*., 1975; Keifer *et al*., 1982).

Bio-ecology

The crimson erineum mite lives on sugar maple on the East Coast of North America. It develops both protogyne and deutogyne instars (Hodgkiss, 1930; Jeppson *et al*., 1975), and its life history shows alternation of generations, with the deutogynes overwintering. The highest population peaks develop in the summer, with erineum growth most abundant then (Keifer *et al*., 1982).

Symptomatology and damage

The mite causes upper leaf surface erinea which are initially greenish and then crimson or purplish in colour. The erinea

are composed of tiny rounded capitate papillae containing red fluid. At high population density of the mite, the erineum occupies a consistent leaf surface, causing distortion and leaf drop (Hodgkiss, 1930; Jeppson *et al.*, 1975; Keifer *et al.*, 1982).

Control

Control is not usually necessary.

Aceria pseudoplatani (Corti)

Common name

Unknown.

Diagnostic characteristics

FEMALE. The deutogyne lacks microtubercles on the dorsal annuli. The protogyne is worm like and 190 μm long. The prodorsal shield has 3 central lines. The opisthosoma has about 65 annuli, strongly granular, and the annuli are broader and smoother caudally (Jeppson *et al.*, 1975).

Geographical distribution

Europe (de Lillo and Beron, 2004).

Host plants

Sapindaceae: European sycamore maple (*Acer pseudoplatanus* L.) (Jeppson *et al.*, 1975).

Bio-ecology

A. pseudoplatani lives on European sycamore maple. Its bio-ecology has not been documented.

Symptomatology and damage

The mites make small yellowish undersurface erineum patches near leaf veins, commonly located basally (Jeppson *et al.*, 1975).

Control

Control is not usually necessary.

The Genus *Paraphytoptus* Nalepa

Diagnostic characteristics

The gnathosoma is short and curved downward; the anterior prodorsal shield lacks spine-like processes. The opisthosoma has all the standard setae; posteriorly, the dorsal annuli are wider (Amrine *et al.*, 2003).

Economic importance

This genus includes over 30 species mainly living on dicotyledonous in north temperate areas and Palaeotropics (Oldfield, 1996b). *Paraphytoptus paradoxus* Nalepa induces browning of leaves of absinthe (*Artemisia absinthium* L.) in Europe. *Paraphytoptus pannolus* Keifer causes rust on leaves and invades flowers of giant ragweed (*Ambrosia trifida* L.) in Virginia (USA) (Keifer, 1962a). Only the chrysanthemum rust mite, *Paraphytoptus chrysanthemi* Keifer, has some economic importance, causing some brooming on the leaves of common *Chrysanthemum* in North America and in the British Isles; it occurs also in India, where brooming has not been observed (Westphal and Manson, 1996).

Paraphytoptus chrysanthemi Keifer (Fig. 16.56)

Common name

Chrysanthemum rust mite.

Diagnostic characteristics

FEMALE. The body is elongate and spindle formed, 130–150 μm long, 40 μm wide and light yellowish. The gnathosoma is curved down. The prodorsal shield has a distinct median line, with the admedian and two submedian lines present, and the sides granular; the scapular tubercles are 22.5 μm apart, on the rear margin; the scapular setae (*sc*) are 47 μm long, projecting caudad. The coxae are somewhat lined and granular, and coxae I are touching. The opisthosoma has about 65–70 annuli ventrally; dorsally, the broader dorsal annuli begin on about annulus 50 and the total is slightly less than 60. The opisthosomal setae c_2 are 20.5 μm long, set on about annulus 10; the setae *d* are 55 μm long, set on about annulus 25; the setae *e* are 17.5 μm long, set on about annulus 42; the setae *f* are 17.5 μm long, set on about annulus 6 from the rear; the seta h_1 are present. The genitalia are 21.5 μm wide and 12.5 μm long; the genital coverflap has 14–16 somewhat irregular ridges, and the seta is 16 μm long. The empodium is 5-rayed (Keifer, 1940a).

MALE. The male has not been studied (Keifer, 1940a).

Geographical distribution

British Isles and North America (Jeppson *et al.*, 1975), China (Hong and Zhang, 1996b) and Serbia and Montenegro (Petanović, 1997; Petanović and Stanković, 1999).

Host plants

Asteraceae: *Chrysanthemum* sp. (Keifer, 1940a; Breakey and Batchelor, 1950; Miles, 1964; Jeppson *et al.*, 1975; Petanović, 2004).

Bio-ecology

The chrysanthemum rust mite was first found around the leaf axil buds and the terminal flowers of chrysanthemum in California (Keifer, 1940a). It lives and feeds among hairs on the

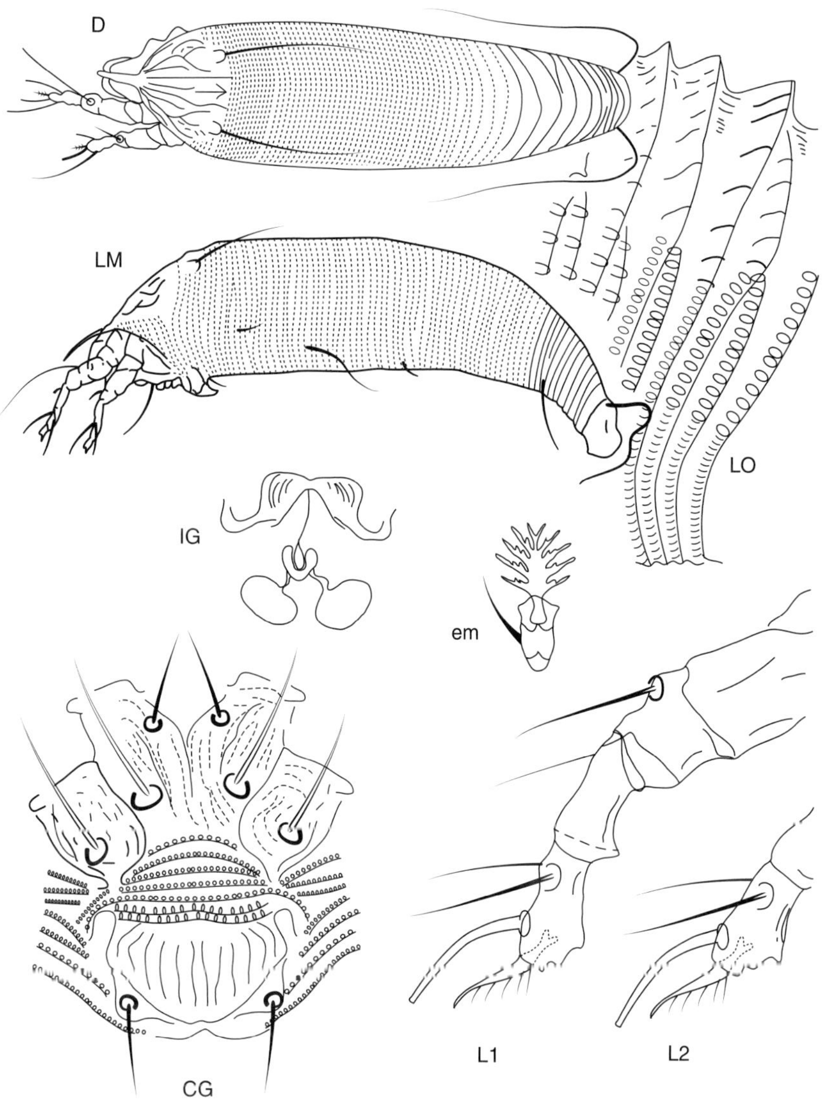

Fig. 16.56. *Paraphytoptus chrysanthemi* Keifer (fromKeifer, 1940a). See Table 16.1 for key.

lower leaf surfaces and on the green stems of chrysanthemums. The mite has a wide range (Jeppson *et al.*, 1975).

Symptomatology and damage

The feeding activity of *P. chrysanthemi* causes stunting and backward curling of the apical leaves and reddishness. The leaves become brittle and the stems have shortened internodes. Bud clustering appears as brooming. The infestation may cause flower phyllody with partial petal colouring and the flower parts tend to revert to leaf-like structures (Breakey and Batchelor, 1950; Miles, 1964). Petanović (2004) reported that the chrysanthemum rust mite caused deformed buds, hairy leaves and rust on *Chrysanthemum*.

Control

Control is provided by monitoring mite attack and by spray applications of acaricides that are admissable for use on chrysanthemum. The control of wild Asteraceae around crops helps to suppress mite oubreaks.

The Genus *Acalitus* Keifer

Diagnostic characteristics

The dorsal pedipalp genual setae (*d*) and the seta h_1 are very tiny or absent. The standard femoral I seta are lacking. The ventral spine on femora I is in place of the seta; the seta is missing from the shortened tibia I. The coxae have prominent granules, and the coxae I are not well separated centrally; the first setiferous coxal tubercles moved ahead towards the front end of the coxae I. The genital coverflap has coarse granules but lacking longitudinal ribs (Keifer, 1965b; Amrine *et al.*, 2003).

Economic importance

This genus includes at least 80 species. The type species, *Acalitus ledi* Keifer, produces yellow erineum patches on the leaf undersurface of Labrador tea (*Ledum glandulosum* Nutt.) in California (Keifer, 1965b). Four species are economically important. These are the cotton blister mite, *Acalitus gossypii* (Banks), which causes severe blistering on cotton (*Gossypium* spp.) (Jeppson *et al.*, 1975), the almond and plum bud gall mite, *A. phloeocoptes* (Nalepa), which produces irregular galls on different *Prunus* fruit and nut plants (Jeppson *et al.*, 1975), the redberry mite, *A. essigi* (Hassan), which lives on wild and cultivated berries and causes delayed drupelet ripening – usually called 'redberry disease' (Hanson, 1930; Borgman, 1950; Keifer *et al.*, 1982) – and the blueberry bud mite, *A. vaccinii* (Keifer), which damages blueberries (*Vaccinium* spp.). These four species are further described below.

Other *Acalitus* species include the alder erineum mite, *A. brevitarsus* (Fockeu), which causes white or orange erinea on the leaf undersurfaces on different alders in Europe and North America. European erinea have various capitate papillae, from simple in western Europe to compound–capitate in Turkey (Moha, 1969), whereas in North America the alder erinea consist of simple, tangled filaments (Jeppson *et al.*, 1975). The boysenberry bud mite, *A. orthomerus* (Keifer), distorts buds of blackberry (*Rubus* sp.), dewberry (*R. macropetalus* Dougl. ex Hook) and boysenberry (*Rubus ursinus* var. *loganobaccus* (L.H. Bailey)) in California (Keifer, 1951) and Maryland. The mite causes bud distortion and shoot stunting that often kill the buds. Buds of blackberry from Maryland hosting specimens of the mite show symptoms of 'redberry disease' (Keifer *et al.*, 1982). These two species have a little economic importance, as do other *Acalitus* species.

Acalitus gossypii (Banks) (Fig. 16.57)

Common name

Cotton blister mite.

Diagnostic characteristics

FEMALE. The body is worm like, 225–250 μm long and whitish in colour. The prodorsal shield has a strong median line only on the rear ¾. The admedian lines are complete, gradually diverging, and slightly recurved to the rear. The first submedian line is short, the second gently curves back, almost reaching the scapular tubercle. The opisthosomal microtubercles are rounded off and bead like, and located ahead of the annulus margins. The coxae usually have granules, with coxae I centrally joined. Granules from the coxae extend back on to the female genital coverflap, on which the granules are in transverse rows. A slight line runs across the flap just inside the rear margin. The empodium is 4-rayed (Jeppson *et al.*, 1975).

Geographical distribution

Antigua, Barbados, Brazil, Colombia, Florida, Guadeloupe, Haiti, India (Bihar, Gujarat, Haryana, Maharashtra, Tamil Nadu), Jamaica, Martinique, Montserrat, Mustique, Nevis, Pakistan, Peru, Puerto Rico, St Lucia, St Vincent, Venezuela and the Virgin Islands (CABI, 1990a). The presence of the mite in Brazil has been well documented by Hambleton and Sauer (1938), Braga (1957) and Rosetto (1972), and has been recently confirmed (Policápio da Costa Gonçalves *et al.*, 2008).

Host plants

Malvaceae: cultivated and wild cotton, *Gossypium* spp., cotton (*Gossypium barbadense* L.), globe mallow (*Sphaeralcea orcutti* Rose) (Jeppson *et al.*, 1975; Policápio da Costa Gonçalves *et al.*, 2008).

Bio-ecology

A. gossypii is structurally indistinguishable from *A. sphaeralceae* Keifer (Keifer, 1965b; Jeppson *et al.*, 1975), but *A. sphaeralceae* has never been found infesting cotton, although cotton is cultivated in the vicinity of its native environment. Further investigations have shown that *A. sphaeralceae* is a synonym of *A. gossypii* (Amrine and Stasny, 1994). *A. gossypii*, the cotton blister mite, lives on cultivated and wild cotton, and there is a wide range of susceptibility to the mite within the genus *Gossypium*. In the American tropics, it could infest other native malvaceous species. In the south-western USA, the mite infests the globe mallow, *S. orcutti*, and it has been reported in Brazil on *G. barbadense* (Hambleton and Sauer, 1938; Braga, 1957; Rosetto, 1972; Policápio da Costa Gonçalves *et al.*, 2008) and other *Gossypium* spp. Indeed, *A. gossypii* was a pest of cotton in Brazil in the past, when *G. barbadense* was cultivated, but later on, the cultivated varieties all came from *Gossypium hirsutum* L. and/or *Gossypium herbaceum* L., and, so far, are all non-susceptible to *A. gossypii*. However, *G. barbadense* can still be found in many backyards of residences where people, mainly in smaller cities and where they still have gardens and backyards, like to have a few of these plants for ornamental purposes, mainly because of their flowers; quite often, severe attacks of *A. gossypii* can be seen on these plants (Carlos Flechtmann, personal communication). Flechtmann says 'The cotton plant breeder should bear this in mind – if by any chance he comes up with a susceptible variety, the source of *A. gossypii* is everywhere around!'.

The cotton blister mite has a short life cycle, it does not overwinter and reproduces throughout the year. The best development of its populations occurs in humid tropical and subtropical regions, but it may adapt to semi-arid conditions (Jeppson *et al.*, 1975).

Symptomatology and damage

The cotton blister mite infests all parts of the cotton plant, with the exception of the larger stems and roots. Its feeding causes hairy deformations (erinea) on the leaves and flowers, in which the mite population develops. The first injuries appear on young fruiting buds in leaf axils, followed by distortion and crumpling of the infested parts. The leaves show silvery spots and fine short white hair, which are frequently on the lower surface. The palisade cells of the injured leaves develop to less than their normal length (only one-third). Damaged plants show crumpled leaves, distorted growth and a reduced number of fruiting branches. When the fruiting branches develop normally, small leaves and short petioles occur on the main stem. Damaged squares (floral

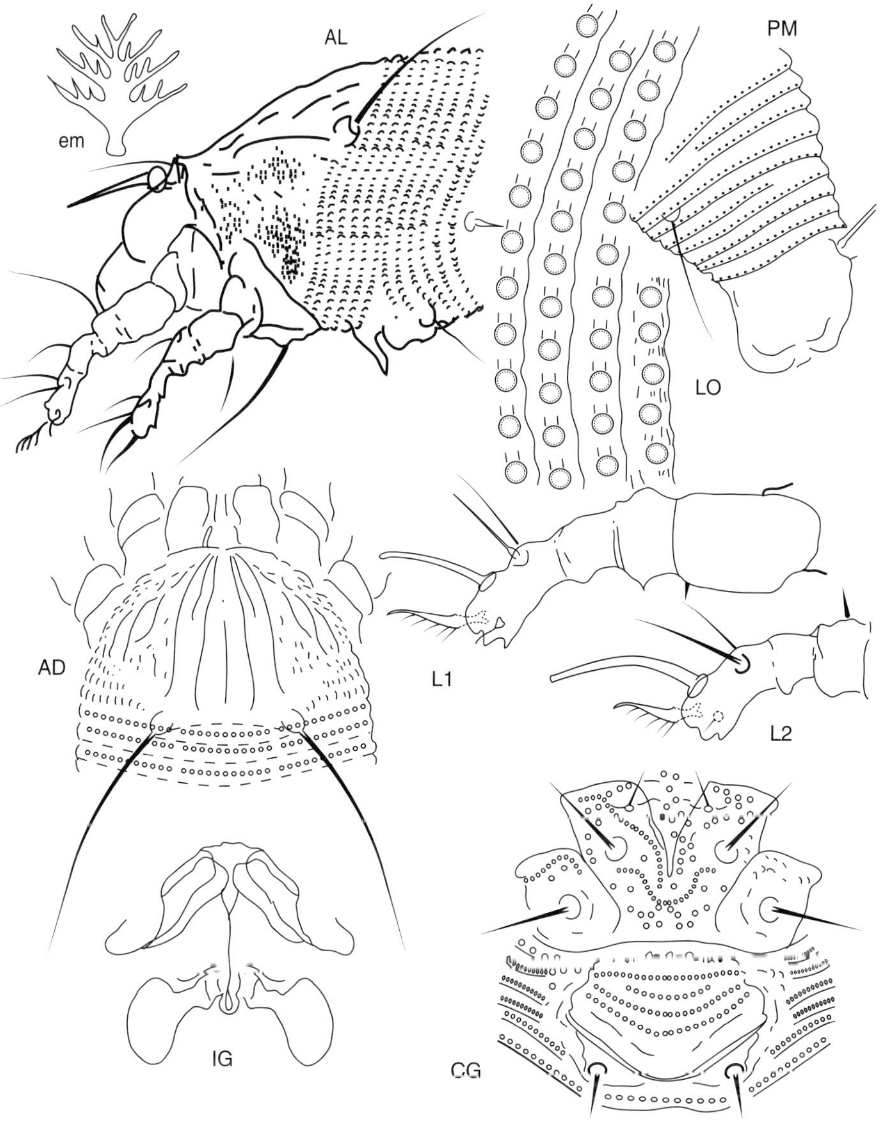

Fig. 16.57. *Acalitus gossypii* (Banks) (from Keifer, 1965b). See Table 16.1 for key.

buds with bracts) do not open into normal flowers. In addition, abnormal shedding of squares occurs. Plants that are highly infested can die (Thaker and Desa, 1929; Fife, 1937). On globe mallow, *A. gossypii* causes crumpled leaves, and lives down in the succulent tissues of the creases (Jeppson *et al.*, 1975). Damage by *A. gossypii* may be confused with that by *A. puttarudriahi* ChannaBasavanna, which causes hairy and felt-like outgrowths on Levant cotton in India (Jeppson *et al.*, 1975).

Control

Control is by using resistant varieties, cultural practices and spray applications of acaricides.

PLANT RESISTANCE. The occurrence of resistant varieties is well documented, e.g. 'Sea Island' cotton is very susceptible to the mite, while 'Upland' cotton is less susceptible (Jeppson *et al.*, 1975).

CULTURAL PRACTICES. Cultural practices are a combination of field clean-up with a cotton-free period. During the host-free period (from 2 to 4 weeks), the growers uproot and burn the old plants and destroy wild cotton (Jeppson *et al.*, 1975).

CHEMICAL CONTROL. Sulfur applications give better control in dry climate, and are less successful in humid countries such as India (Jeppson *et al.*, 1975).

Acalitus phloeocoptes **(Nalepa) (Fig. 16.58)**

Common name

Almond and plum bud gall mite.

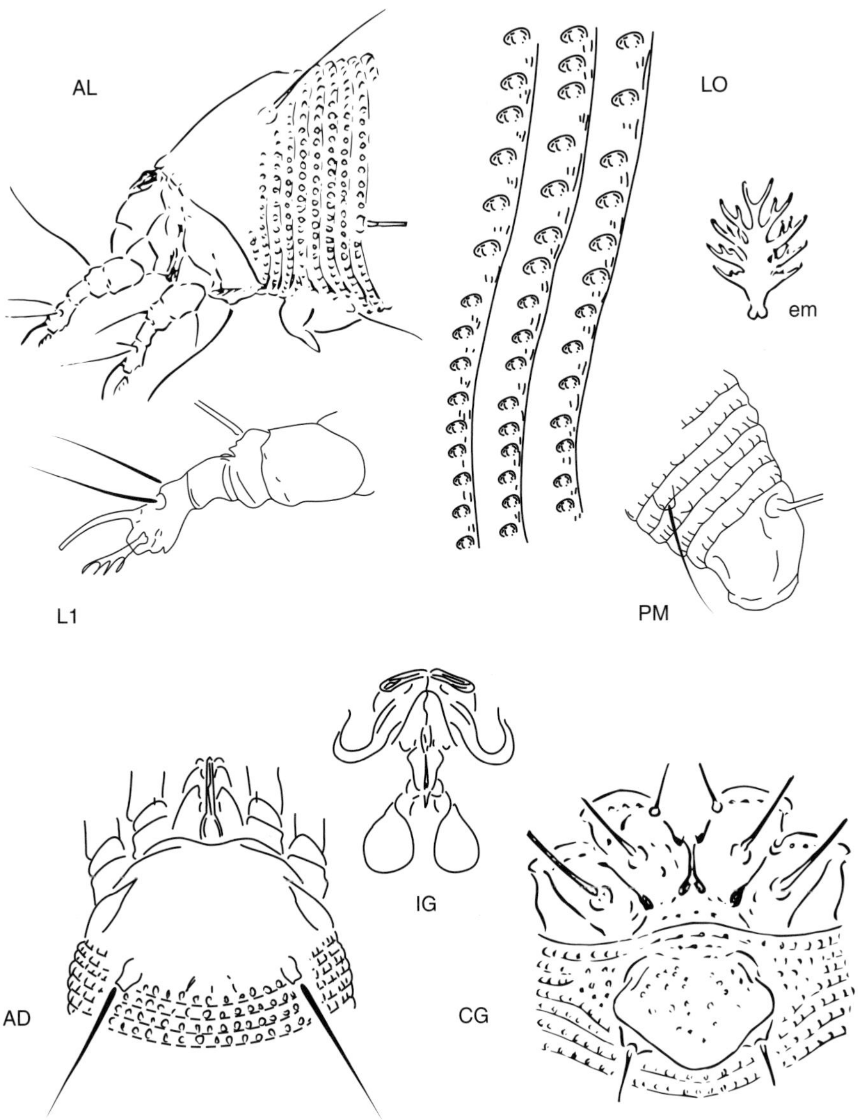

Fig. 16.58. *Acalitus phloeocoptes* (Nalepa) (from Jeppson *et al.*, 1975). See Table 16.1 for key.

Diagnostic characteristics

FEMALE. The body is worm like, 135–150 μm long and whitish in colour. The prodorsal shield lacks markings. The opisthosomal microtubercles are oval or rounded, smaller and acuminate ventrally, and all located ahead of the annulus margins. There is a strong sternal line between coxae I, which ends in a fork that has thick lines. The genital coverflap has indistinct granules on its surface. The empodium is 5-rayed (Jeppson *et al.*, 1975).

Geographical distribution

Worldwide. Common in areas of *Prunus* cultivation.

Host plants

Rosaceae: *Cotoneaster* spp., almond (*Prunus amygdalus* Batch.), apricot (*P. armeniaca* L.), myrobalan plum (*P. cerasifera* Ehrh.), European plum (*P. domestica* L.), peach (*P. persica* L.), blackthorn (*P. spinosa* L.) (Nalepa, 1890; Di Stefano, 1971; Talhouk, 1971; Sternlicht *et al.*, 1973; Boczek, 1974; Jeppson *et al.*, 1975; Dzhadaibaev and Dyagilev, 1977).

Bio-ecology

A. phloeocoptes was described from European plum in Austria by Nalepa (1890). It causes galls around buds on different *Prunus* fruit and nut plants, such as almond, apricot, European plum, myrobalan plum, peach and *Cotoneaster* spp. (Nalepa, 1890; Di Stefano, 1971; Talhouk, 1971; Sternlicht *et al.*, 1973; Boczek, 1974; Jeppson *et al.*, 1975; Dzhadaibaev and Dyagilev, 1977).

Overwintering forms are slender and cuneate, and summer forms robust. Winter females are darker and slender, while summer females are pearly white. The field sex ratio shows a preponderance of females, and the male population in winter is

0.3%, in spring 5% and in summer and autumn 1.5% (Sternlicht *et al.*, 1973). Females overwinter in galls, 1 or 2 years old, which crack in the early spring so that the mites can move towards new buds. Mites leave old galls, penetrate new buds and start to feed, inducing the basal bud cells to proliferate. The migration occurs during the night and can last for 2 months. From April to May, polylocular soft galls of various shapes and sizes develop around infested buds. Each gall contains an average of six 'locules', separated from one another by parenchymatous tissue (Jeppson *et al.*, 1975). In Spain, the adults emerge from winter galls in mid-March, and the emergence ends in the second fortnight of May (Duran *et al.*, 2006). In Kazakhstan (former USSR), it has been observed that some adults leave the galls in spring and feed on the buds and young leaves, while others remain in the galls and reproduce there until the summer. Eggs are thus laid on the shoots and bud scales and also in the galls. A single gall may contain up to 600 mites. Occasionally, the galls are formed on leaves (Dzhadaibaev and Dyagilev, 1977). In May, each locule hosts one female, which deposits about one egg per day for 20–25 days. Hatching of the eggs occurs during the first part of June, and in this period the parenchymatous cells lining the locules turn violet red. The first adults are observed in mid-June, and a life cycle during July and August take about 3 weeks. The locules contain 4000-5000 mites and many eggs. Egg deposition decreases in late August and stops in late October (Talhouk, 1977). On European plum and blackthorm in Poland, the galls are well developed at the end of June. One-year-old galls host about 30 times as many mites as first-year galls. On blackthorn, the galls are broader than those on plum and host 2–4 times as many mites (Boczek, 1974). Dispersal occurs by wind, insects and birds (Jeppson *et al.*, 1975).

Natural enemies include insects and mites. The insects include Chalcidoidea and Thysanoptera. *Tetrastichus sajoi* (Szelenyi), a chalcidoid, develops in older galls, where their larvae prey on many mites (Mezei, 1995). Mite natural enemies include Anystidae, Phytoseiidae and Stigmaeidae (Philipp, 1957; Talhouk, 1963; Sternlicht *et al.*, 1973). In Israel, there is strong mite predation by natural enemies during the spring migration (Sternlicht *et al.*, 1973).

Symptomatology and damage

The feeding activity of *A. phloeocoptes* causes galls around buds on different *Prunus* fruit and nut plants. On almond in Syria and Lebanon, the galls on the bark of young twigs are spherical, with a diameter ranging from 1.3 to 1.8 mm. The galls formed around the buds are large and irregular in shape. The damaged buds do not develop regularly, producing weak leaves and no flowers. The mite also causes deformation of fruit spurs. Mite infestation affects the physiology of the plant, hindering the formation of fruit buds and inducing loss of plant vigour. The damage appears progressively and irreversibly in 3–6 years (Talhouk, 1971; Jeppson *et al.*, 1975).

On plum, mite feeding induces the formation around the buds of small subspherical galls, from 1.3 to 1.8 mm in diameter. These can be single or grouped around the buds, and become woody, deformed fruit spurs, although the plum plants may recover from mite infestation (Jeppson *et al.*, 1975). Infested plants have short terminal internodes and appear thinned and weakened (Sternlicht *et al.*, 1973). In Krasnodar (former USSR), infestation of the buds leads to loss of growth, drying up and death of the tree (Gushina *et al.*, 1976). On some cultivars of peach, the mite causes tubercle-shaped galls at the base of buds on old (1–2 years) twigs, death of shoots and twigs, and loss of yield (Di Stefano, 1971). The gall locules are parenchymatous, consisting of four rows of isodiametric nucleated cells delimited by very small intercellular spaces, and vacuolate with cellulose walls. Externally to the paranchymatous cells there are layers of typical parenchyma, with large intercellular spaces. The cells are isodiametric or with an elongate palisade form. Beyond these parenchyma cells there are various layers of cellulose-walled cells, which may become dead cork cells constituting the gall outside walls. Most external gall layers consist of dead flattened cells, with cutin on the outside and suberized inside. Young vascular tissues are located in the direction of the locules, but do not reach them. Between these tissues and the cavity of locules there are damaged cells. The vascular tissue contains bacteria and crystals (Jeppson *et al.*, 1975).

Control

Control is provided mainly by preventive and curative measures – the use of resistant varieties, cultural practices and, later, the spray applications of chemicals.

PLANT RESISTANCE. Plant resistance to the almond and plum bud gall mite is well known. A strain of almond growing in Lebanon is immune to attack (Talhouk, 1971; Jeppson *et al.*, 1975). Among 13 species of *Cotoneaster* examined in Kazakhstan, six were very susceptible to the mite (*C. lucida* Schltdl. is one of these) and two were resistant (*C. integerrimus* Medik. and *C. oliganthus* Pojark.) (Dzhadaibaev and Dyagilev, 1977). In Spain, the plum varieties 'Red Beaut' and 'Santa Rosa' showed greater infestations than 'Golden Japan', while the cultivars 'Mayeros' and 'Beauty' were not colonized by the mite (Lacasa *et al.*, 1990a).

CULTURAL PRACTICES. The use of uninfested material for planting is suggested and the importance of cutting off infested shoots during the winter is emphasized (Dzhadaibaev and Dyagilev, 1977).

CHEMICAL CONTROL. Chemical control measures are effective only when they are synchronized with the migration of the adults and the start of gall formation. In Israel, the best results were obtained with spray applications of endosulfan in combination with naled or dichlorvos (Sternlicht *et al.*, 1973). The same strategy was suggested in Italy, where the action of various chemicals (demeton, dieldrin, endosulfan, ethoate-methyl, phenthoate, parathion, schradan) has been evaluated (Di Stefano, 1971). In Lebanon, one treatment in early May with methyl demeton gave good control of mite populations (Talhouk, 1971). In Krasnodar (former USSR), sprays with dicofol after flowering, at an interval of 1 month, were more effective than wettable sulfur applied twice at an interval of 17 days or three times at intervals of 10 days (Gushina *et al.*, 1976). In Kazakhstan,

sprays of the bushes with dimethoate at the time of migration from the buds, with colloidal sulfur during oviposition and with malathion or formothion during bud hatching, gave positive results. A fourth application of one of the sprays was made when the mites were entering hibernation (Dzhadaibaev and Dyagilev, 1977). In Hungary, spray applications of clofentezine controlled the mite (Mezei and Czepo, 1988). In Spain, winter applications of pesticides were ineffective, as well as a mixture of dicofol plus tetradifon applied at the end of April, but endosulfan applied at the beginning of April reduced the infestation significantly (Lacasa *et al.*, 1990a); in the same country, 3–8 applications of sulfur effectively controlled the pest, and the application of carbaryl at 70% hatching effectively controlled mite infestation (Duran *et al.*, 2006). In China, sprayings of propargite, dimethoate and bifenthrin produced positive results (Wei *et al.*, 1991).

Acalitus essigi (Hassan) (Fig. 16.59)

Common name

Redberry mite.

Diagnostic characteristics

FEMALE. The body is worm like, 120–140 μm long, 28–30 μm wide and whitish or very light amber in colour. The gnathosoma is rather small. The prodorsal shield has median and admedian lines and a long submedian line; there are many short lines and a lateral clear or marked lobe, with side granulations. The scapular tubercles are 17.5 μm apart, set on the rear margin; the scapular setae (*sc*) are 19.5 μm long, projecting backwards. The coxae I are fused with each other, granular, as are the coxae II.

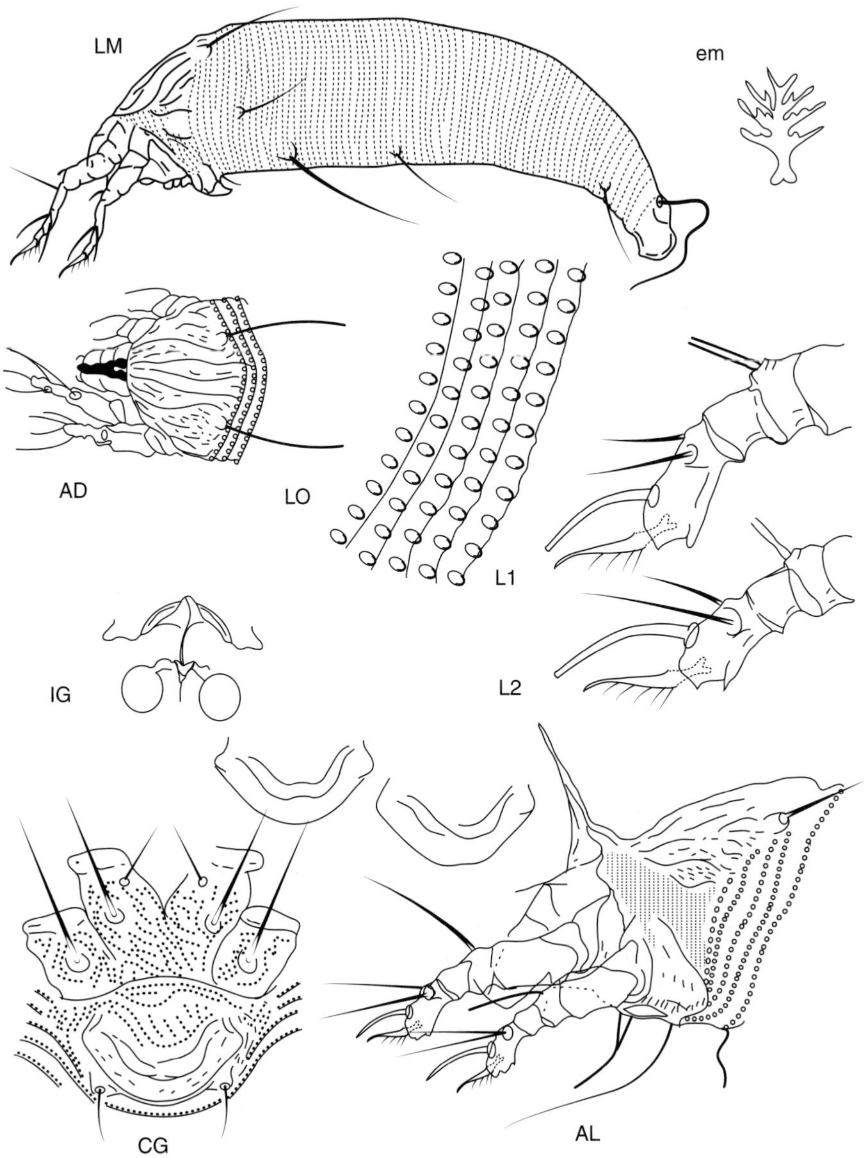

Fig. 16.59. *Acalitus essigi* (Hassan) (from Keifer, 1941). See Table 16.1 for key.

The opisthosoma has about 65–70 annuli, some ventrad reduction in annulus number, and the tubercles are pointed. The opisthosomal setae c_2 are 13.5 μm long, set on about annulus 7; the setae *d* are 44 μm long, set on about annulus 19; the setae *e* are 30 μm long, set on about annulus 38; the setae *f* are 12.5 μm long, set on about annulus 6 from the rear; the setae h_1 are present and minute. The genitalia are 16.5 μm wide and 9.5 μm long; the genital coverflap has two concentric, irregular half rings and basal granules; the seta is 8 μm long. The legs are rather short and stocky; the empodium is 4-rayed (Keifer, 1941).

MALE. The body of the male is 105–135 μm long and 25–30 μm wide (Keifer, 1941).

Geographical distribution

Australia (Davies *et al.*, 2001a; Scott *et al.*, 2008), Belgium (Paternotte, 2000), California (Hassan, 1928; Keifer, 1941), Chile (Gerding, 1992), England (Alford, 1979), Germany (Schliesske, 1985a; Roeser, 1996), Hungary (Szendrey *et al.*, 2003), New Zealand, North America (Jeppson *et al.*, 1975), Poland (Łabanowska and Suski, 1990), Switzerland (Baillod *et al.*, 1989b) and Yugoslavia (Zivanovic, 1973; Petanović, 1998).

Host plants

Rosaceae: blackberry (*Rubus* spp.), *Rubus anglocandicans* A. Newton, European blackberry (*R. fruticosus* L.), *R. laudatus* A. Berger, whitebark raspberry (*R. leucodermis* Douglas), loganberry (*R.* ×*loganobaccus* L.H. Bailey), Himalaya blackberry (*R. thyrsanthus* Focke), *R. ulmifolius* Schott, dewberry or wild dewberry (*R. ursinus* Cham. *et* Schldl.), boysenberry (*R. ursinus* × *idaeus* L.) (Hanson, 1930; Keifer, 1941; Borgman, 1950; Keifer *et al.*, 1982, Scott *et al.*, 2008; Davies *et al.*, 2001a,b).

Bio-ecology

A. essigi can be misidentified with *A. orthomerus* (Keifer), a pest of boysenberry, native blackberry and dewberry in California (Keifer, 1951; Jeppson *et al.*, 1975). However, *A. essigi* differs by lacking longitudinal ribs on the coverflap. Yet, *A. essigi* can also be misidentified with *Phyllocoptes gracilis* (Nalepa).

The redberry mite lives on wild and cultivated berries, such as European blackberry, Himalaya blackberry, dewberry or wild dewberry, whitebark raspberry, boysenberry, raspberry, loganberry and blackberry (*Rubus* spp.) (Hanson, 1930; Keifer, 1941; Borgman, 1950; Keifer *et al.*, 1982; Davies *et al.*, 2001a). In south-west Australia, the mite has been collected on the fruits of three species of weedy blackberry (*R. anglocandicans*, *R. laudatus* and *R. ulmifolius*) and it has been hypothesized that it could be used in the biological control of *Rubus* spp., especially those that are not susceptible to *Phragmidium violaceum* (Schultz) G. Winter (Uredinales), the rust fungus that is being released against species of European blackberry (Scott *et al.*, 2008).

The redberry mite survives at temperatures below −6°C in the northern hemisphere, and egg deposition begins around the buds in early March, increases in September and then decreases until it stops in early December. During the year, the mite has many life cycles. In the state of Washington, *A. essigi* lives during the year in and around buds, and also occurs on berries at drupelet bases and around the core and stem ends. It does not have a general autumnal migration from the berries, and the greater part of the population remains until berry decomposition. In the autumn, the part of population that has left berries, together with the populations remaining along the canes, move towards the new axillary cane buds. The mites live in and around these buds until the development of the berries the next spring, when they move to the developing flowers, green berries and the bases of the leaves. Many mites do not find suitable sites for overwintering, and during the autumn and winter they die in high numbers; during these periods, predatory mites cause high mortality of the redberry mite (Jeppson *et al.*, 1975). Sexual 'independence' of males and females has been confirmed (Michalska and Boczek, 1991).

Studies on the occupation of microhabitats and aggregation patterns of *A. essigi* on European blackberry in Australia show that the mite is a refuge-inhabiting species. It resides in bud and leaf axil microhabitats on both primocanes and fructocanes, and also occupies berry and bract microhabitats on fructocanes. Population fluctuations within the different microhabitats are evident across seasons. From summer to winter, populations significantly decline in bract and leaf axil microhabitats, but significantly increase within bud microhabitats where overwintering takes place as slowly reproducing colonies. Live fruits and young shoots were also identified as overwintering sites. Mite populations display an aggregated distribution both within and between individual blackberry canes. Within primocanes, *A. essigi* is aggregated in the lower 20% of the cane length. On fructocanes, mite aggregation is in the lower 20% and, especially, in the upper 20% of the cane length. In spring, *A. essigi* emerges from bud overwintering sites and colonizes shoots mainly in the lower third of the previous season's primocanes, suggesting that there is limited dispersal away from overwintering sites. It is proposed that biotic factors such as tissue age, microhabitat morphology and limited ambulatory dispersal capabilities are responsible for the aggregation patterns of the redberry mite (Davies *et al.*, 2001a). In Australia, no mites have been found on unopened flower buds or open flowers of blackberry, and infestation of fruit starts during the green stage of fruit development and significantly increases during the red fruit stage. The use of a sticky barrier to prevent the mite moving up the pedicel of open flowers significantly reduces its populations within the fruit. Very low levels of aerial dispersal on to fruit cannot be discounted, but non-aerial or crawling dispersal via the fruit pedicel is the dominant method of blackberry fruit infestation (Davies *et al.*, 2001b). According to Hanson (1933), the dispersal occurs by wind. Natural enemies include some phytoseiid mites.

Symptomatology and damage

Although *A. essigi* causes some cane bud damage, the main drawback is delayed drupelet ripening, which is usually called 'redberry disease'. Infested berries host the mite populations at drupelet bases and around the core and stem ends. During its feeding activity, the mite injects a toxin that prevents the drupelet from colouring. The berries have uneven greenish red or red drupelets, and only a part of infested berries show

the normal dark brown or black drupelets. Highly infested berries turn completely red. The extent of the damage is proportional to the density of the attack. Infested berries gradually fade and dry out during the autumn and winter (Jeppson *et al.*, 1975).

Control

Control is based on preventive cultural choices and curative chemical treatments. The effectiveness of the sticky tape method has been evaluated for monitoring.

MONITORING. The effectiveness of the sticky tape method of eriophyoid mite extraction was evaluated in Australia with and without the addition of a new method – the use of a supplementary synthetic pyrethroid spray – for estimating the critical abundance of eriophyoid mites from within architecturally complex plant structures. *A. essigi* was sampled from within the complex fruit of *R. fruticosus*. Sticky tape extraction supplemented with a synthetic pyrethroid spray was significantly the most successful method of mite extraction, with, on average, over twice the number of mites extracted from blackberry fruit using this method compared with sticky tape alone. Visual counts of *A. essigi* on the receptacle region of blackberry fruit were correlated with the number of mites extracted using the sticky tape method both with and without a spray treatment. However, due to the low numbers that were assessed visually, the use of visual methods alone appears to be limited (Davies *et al.*, 2002).

CULTURAL PRACTICES. The removal of canes before the autumnal applications of acaricides lowers the density of mite populations and helps in control (Jeppson *et al.*, 1975). Moreover, early varieties of berries that are harvested before the increase of mite populations usually have reduced or no damage (Texas berry variety), whereas late-maturing berry varieties show greater damage, because mite migration from the winter cane buds to the new fruit occurs late. In New Zealand, the red berries appear about 2 weeks before the main crop ripens (Jeppson *et al.*, 1975). Research carried out in Belgium shows that 'Silvan' (a blackberry–raspberry hybrid) is an early, high-yielding cultivar with large fruits and good flavour, but is very thorny. 'Loch Ness' is a thornless cultivar that ripens 12 days later than 'Silvan', and is susceptible to attack by *A. essigi*. 'Black Satin', which ripens 11 days later than 'Loch Ness', has large but acid fruits (Verlinden, 1995).

CHEMICAL CONTROL. Two or three chemical sprayings before flowering are traditionally suggested (Krczal, 1966, 1969; Alford, 1979; Bolay *et al.*, 1989). Low population densities can be controlled by sulfur alone, avoiding treating at hot temperatures – sulfur and oil must be used only in winter. High infestations can be treated with a lime sulfur application in the autumn or with lime sulfur and oil combinations in late winter or a delayed dormant application; these are all effective (Essig, 1925; Massee, 1932; Hanson, 1933; Edwards *et al.*, 1935; Hamilton, 1950). In West Germany, the control of *A. essigi* was entrusted to sprayings of wettable sulfur or endosulfan. Either chemical as a wettable powder or in an emulsion spray was effective if applied three times, first when the new shoots reached a length of 40 cm, again when they reached 50–60 cm and last when the plants were in full bloom (Krczal, 1969). In Yugoslavia, a winter oil wash was recommended, followed in spring by three applications of endosulfan, to which a sulfur compound may be added for disease prevention (Zivanovic, 1973). In England, endosulfan was an effective alternative to endrin for the control of *A. essigi*. Three pre-flowering sprayings of endosulfan applied at about 14-day intervals reduced the incidence of redberry disease to an acceptable level; two sprays about 4 weeks apart were generally less effective. Endosulfan residues in fruits at harvest were below detectable levels; dicofol, vamidothion and chlorpyriphos were ineffective (Alford, 1979). In Germany, Belgium and Poland, the fungicide tolylfluanid was effective against *A. essigi* on blackberries and raspberries. The chemical is safe to all of the most important beneficial organisms in integrated pest management (IPM) (e.g. bees, beneficial mites and anthocorids) and has a favourable toxicological profile; it is very rapidly degraded in soil and water, and because it is strongly adsorbed by soil particles has no potential to migrate into the deeper soil layers (Merabet *et al.*, 2002).

Acalitus vaccinii (Keifer) (Fig. 16.60)

Common name

Blueberry bud mite.

Diagnostic characteristics

FEMALE. The body is worm like, 185–210 μm long, 57 μm wide and whitish in colour. The gnathosoma is bent down. The prodorsal shield has an indistinct pattern, with the submedian lines indicated to the rear, and the side with a band of tubercles; the scapular tubercles are 20 μm apart, located on the rear margin; the scapular setae (*sc*) are 25 μm long, projecting caudad. The coxae are contiguous, with setae I near the apex of coxae I. The opisthosoma has about 65 annuli. There is a considerable ventrad annulus reduction. The opisthosomal setae c_2 are 19 μm long, set on about annulus 7; the setae *d* are 33 μm long, set on about annulus 20; the setae *e* are 35 μm long, set on about annulus 35; the setae *f* are 13 μm long, rather stout, 6 annuli from the rear; the setae h_2 are about 40 μm long; the setae h_1 are absent. The genitalia are 18 μm wide, 11 μm long; the genital coverflap has about 10 ridges; the seta is 13 μm long. The empodium is 6-rayed (Keifer, 1939c).

MALE. The body of the male is 150–170 μm long and 40 μm wide (Keifer, 1939c).

Geographical distribution

Eastern North America (mainly coastal areas), from Canada to Florida and Texas (Keifer, 1939c; Jeppson *et al.*, 1975; Keifer *et al.*, 1982).

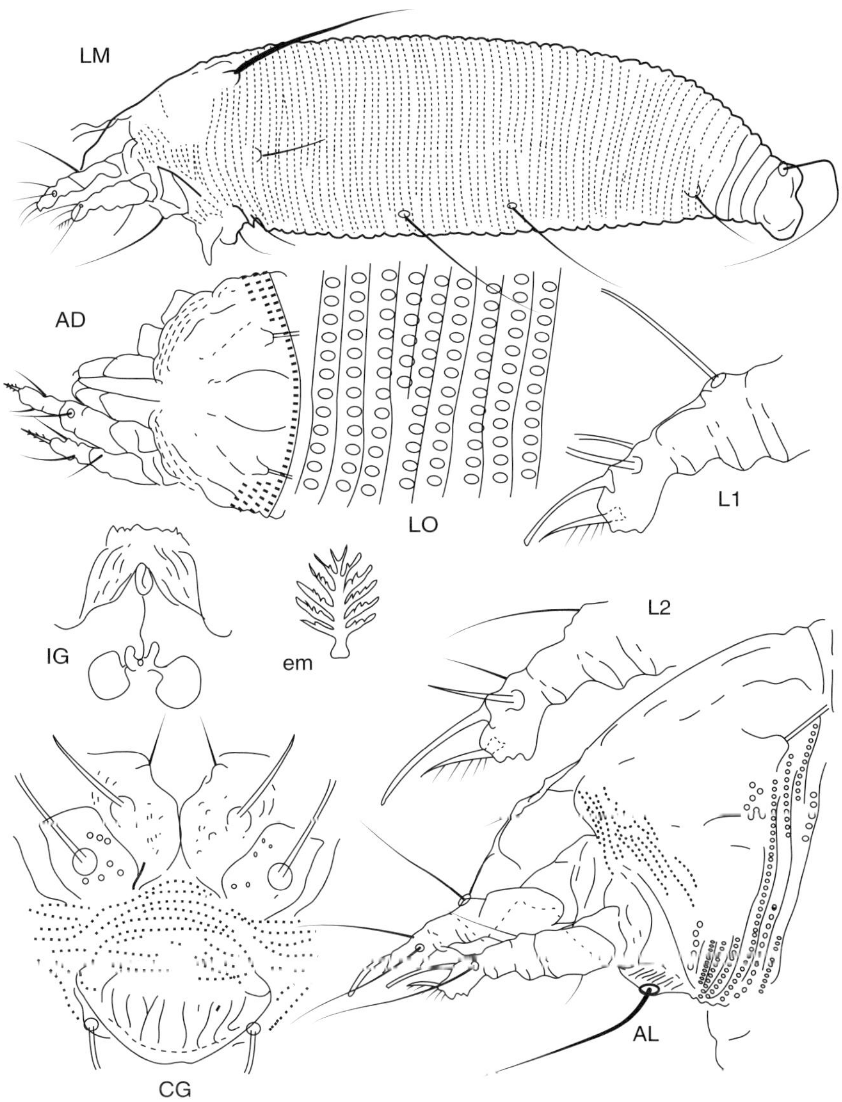

Fig. 16.60. *Acalitus vaccinii* (Keifer) (from Keifer, 1939c). See Table 16.1 for key.

Host plants

Ericaceae: black huckleberry (*Gaylussacia baccata* (Wangenh.) K. Koch), *Vaccinium* spp., blueberries (*V. corymbosum* L.) (Keifer, 1939c; Fulton, 1940; Jeppson *et al.*, 1975; Keifer *et al.*, 1982; Amrine and Stasny, 1994).

Bio-ecology

A. vaccinii is native to eastern North America, common in the coastal areas, and spread from Canada to Florida and Texas (Keifer, 1939c; Jeppson *et al.*, 1975; Keifer *et al.*, 1982). It lives on ericaceous plants, mainly on the genus *Vaccinium*, and is commonly referred to as blueberries and huckleberries. Eight species of eriophyoids have been reported from *Vaccinium* spp. (Amrine and Stasny, 1994). The mite also lives on black huckleberry. The females overwinter in the outer fruit bud scales, but less under the basal nodes and inside the buds (Fulton, 1940). In early spring, when the blossom buds swell, the mites move to the bases of all bud scales and stay on the clusters, which consequently develop into rosettes. Later, the mites reproduce between the scales of the rosettes and in other shelters. In early summer, they leave the dried rosettes and move to sheltered sites; in late summer, they move to new distal fruit buds. Mites also live between the corolla and calyx of flowers and sometimes on developing fruits. In North Carolina, the numbers of *A. vaccinii* start to decrease in early summer with rosette drying, and the fall in numbers continues through the autumn and winter, reaching its lowest in February. The spring flush of bush growth induces breeding and the mites reach a peak population during the spring. Migration is accomplished as in other eriophyoids, by wind and insects. Mites within flowers can ride away on bees that are collecting pollen. The blueberry bud mite lays eggs through most the year, and the number of eggs is

proportional to the population density. The two nymphal stages grow for 4 or 5 days. The maximum development time from egg to adult is about 15 days at 19°C (Edwards *et al.*, 1935; Darrow *et al.*, 1944; Baker and Neunzig, 1970).

Predatory mites and insects are reported in association with *A. vaccinii* (Baker and Neunzig, 1970). In Michigan (USA), tydeid predatory mites were found in spring, and both tydeid and phytoseiid mites during summer (Isaacs and Gajek, 2003).

Symptomatology and damage

The mite damages fruiting bodies. On wild plants, the damage consists of unnatural succulence and epidermal roughening or blistering of the fruit bud scale bases, inducing buds to hang in tight rosettes around the fruit stems. These rosettes are conspicuous and make visual surveying for the mite relatively easy during spring. The bases of the attacked fruit stems form watery blisters and retain the juvenile red colouring of growing tissues. Surface succulence and roughening of bud scales caused by the mite are similar to some extent to the inside of some leaf galls that other eriophyoids make on other plant hosts. This rosetting may be retained as a pseudogall (Keifer, 1941).

Damage to the fruiting bud scales may not prevent the berry clusters from developing relatively normal berries, often on shortened stems. In severe attacks, blistering extends to the calyx and petal parts, and the berry is deformed. Blistered petals tend to fuse with the calyx. On wild plants, highly infested clusters remain through the summer as drying rosettes. Badly blistered and curled fruit stems make no fruit. Fruiting clusters on cultivated plants are more severely injured than on wild plants. High infestation may totally deform every berry cluster on a plant. The vegetative bush parts show relatively little damage, except for some delayed leaf growth (Jeppson *et al.*, 1975).

Control

Plant resistance to the mite, cultural practices and biological control do not result in sufficient control, so this is mainly based on spray applications of acaricides.

PLANT RESISTANCE. Several species of *Vaccinium* grown in North Carolina (*V. arkansanum* Ashe, *V. darrowi* Camp, *V. myrsinites* Lam., *V. stamineum* L. and *V. tenellum* Ait.) are totally free of mites (Neunzig and Galletta, 1977).

CULTURAL PRACTICES. Plants in deep shade are less injured than plants cultivated on dry sand flats or in open woods and at cleaning margins, which are most likely to have rosetting. As cultivated plants fruit better when grown in little or no shade, it may be that apparent differences in susceptibilities of different blueberry varieties are at least partially the result of location and climate (Jeppson *et al.*, 1975).

BIOLOGICAL CONTROL. The acaropathogenic fungus *Hirsutella thompsonii* Fisher is effective in control of the blueberry bud mite but its use needs high temperatures and RH (Baker and Neuzing, 1968). Fungus varieties *thompsonii* and *vinacea*, characterized by vinaceous colonies, affect *A. vaccinii* (Samson *et al.*, 1980).

CHEMICAL CONTROL. Chemical control is effective during mite migration and in the postharvest period (Bailey and Bourne, 1946; Tomlinson, 1950). Applications of postharvest summer oil produced good control (Tomlinson, 1950), as well as diazinone and chlorobenzilate (Bailey and Bourne, 1946). The mite is tolerant of tetradifon, dicofol, propargite and chlorobenzilate (Herne *et al.*, 1979). A mixture of abamectin and oil provided good control, but pyridaben and fenpropathrin were less effective. Abamectin and oil alone also provided a high level of control of the blueberry bud mite. A comparison between postharvest applications of endosulfan, delayed-dormant application of oil, and a combination of both treatments, showed that endosulfan was most effective, followed by oil alone, and on a par with the combination of endosulfan and oil (Isaacs *et al.*, 2004).

The Genus *Acaphylla* Keifer

Diagnostic characteristics

The prodorsal shield is triangular and the frontal lobe overhanging the gnathosoma; the scapular tubercles are set ahead of the rear margin, and the scapular setae point up and forward. The gnathosoma is small, projecting down. The coxae have seta I missing. Legs I all have a seta; legs II lack a patellar seta. The opisthosomal dorsal annuli are smooth, of moderate width, forming a longitudinal middorsal ridge; the ventral annuli are microtuberculate. The opisthosomal setae are as usual except for no seta h_1. The genital coverflap has longitudinal furrows. The empodia are divided (Keifer, 1943; Amrine *et al.*, 2003).

Economic importance

The genus *Acaphylla* includes about seven species. Among these, the economically more important species are *A. theae* (Watt *et* Mann), which is injurious to the leaves of tea (*Camellia sinensis* (L.) O. Kuntze) (Watt and Mann, 1903), *A. steinwedeni* Keifer, which causes rust on Japanese camellia (*Camellia japonica* L.) leaves (Keifer, 1943) and tea, and *A. theavagrans* Kadono, which is injurious to *Thea sinensis* L. (Kadono, 1992). These three species are described below. Other species are *A. acromia* (Nalepa), an alder rust mite occurring in Finland, north Russia, Sweden and Germany (de Lillo and Beron, 2004), *A. eugeniae* Chandrapatya, a vagrant on the upper leaf surface of the myrtaceous species *Eugenia crumini* DC. in Thailand (Boczek and Chandrapatya 1996), *A. subacromius* (Roivainen) and *A. syzigii* Channabasavanna.

Acaphylla theae (Watt *et* Mann)

Common name

Pink tea mite.

Diagnostic characteristics

FEMALE. The body is fusiform, blunt anteriorly, 170–194 μm long, 70–75 μm wide and orange in colour. The prodorsal shield has an indented frontal lobe that projects over the gnathosoma, and a centrolongitudinal ridge derived from the segmented median line and the admedian lines; the pattern has a few longitudinal and transverse lines and lateral granules. The scapular tubercles are well ahead on the rear margin, and the scapular setae (*sc*) are 5 μm long, directed centrad and upward. The opisthosoma bears a middorsal ridge, and there are 29–30 dorsal annuli without microtubercles, and 68–70 ventral annuli with small, bead-like microtubercles touching the rear annular margins. The setae h_1 are lacking, and the tubercles are tiny and vestigial. The coverflap has numerous broken longitudinal lines and granules basally and 4–5 irregular longitudinal lines apically. On coxae I, the seta *1b* is absent, on leg II, the antaxial genual seta *l′* is lacking, and the empodia are divided and 3-rayed (Lee *et al.*, 2014).

MALE. The male is smaller than the female, about 150 μm long and 55 μm wide; the bilobed frontal lobe is plainly visible (Lee *et al.*, 2014).

Geographical distribution

Australia (CSIRO, 2004), China (Zhao and Hou, 1993; Hong *et al.*, 2006), Georgia (Russia) (Demokidov, 1916), India (Watt and Mann, 1903; Das and Sengupta, 1958; Rao, 1970), Indonesia (Bernard, 1909), Iran (Khanjani and Haddad Irani-Nejad, 2006; Xue *et al.*, 2009a), Italy (Ferrari *et al.*, 1999), Korea (Lee *et al.*, 2014), Portugal, Spain (Mansilla Vázquez *et al.*, 2003), South-east Asia (du Pasquier, 1932). The CABI (1988) map includes Asia, the Georgian SSR (now Georgia), India (Assam, Tamil Nadu, Tripura, West Bengal), Indonesia, Java, Sumatra, Malaysia, Taiwan, the USSR and Vietnam.

Host plants

Theaceae: *Camellia sasanqua* Thumb. (Lee *et al.*, 2014), tea (*C. sinensis* L.) (Watt and Mann, 1903; Hong *et al.*, 1998; Lee *et al.*, 2014).

Bio-ecology

Five species of eriophyoid mites are reported on tea (ChannaBasavanna, 1996). Of these, *A. theae* is the most economically important, but its exact identity was doubtful until a few years ago. Indeed, the genus *Acaphylla* included three structurally similar species, *A. theae*, injurious to tea (Watt and Mann, 1903), *A. steinwedeni* Keifer, causing rust on Japanese camellia (*C. japonica* L.) (Keifer, 1943) and *A. indiae* Keifer, causing some leaf rusting on tea (Keifer, 1954). The first species was described by Watt and Mann as *Phytoptus theae*, and some years afterwards transferred by Nalepa (1929) into the genus *Eriophyes* Von Siebold. After the description of the genus *Acaphylla* by Keifer (1943), whose type species (*A. steinwedeni*) had been reported from common camellia, Das and Sengupta (1958) transferred *P. theae* to the genus *Acaphylla* and synonymized *A. steinwedeni* with *A. theae*. Some years afterwards, Huang (1974) reported *A. steinwedeni* from Taiwan. Jeppson *et al.* (1975) stated that while there were no obvious morphological differences between these two species, no experimental evidence supported or contradicted the synonymy either. Amrine and Stasny (1994) report that if *A. steinwedeni* occurs on both tea and camellia it may be synonymous with *A. theae*. ChannaBasavanna (1996) agreed with the conclusion of Das and Sengupta (1958), and Meyer Smith (1996a) with that of Jeppson *et al.* (1975). However, rearings of *A. theae* on tea, Japanese camellia and oil tea camellia (*Camellia oleifera* Abel.) showed that the mite survived only on tea leaves and failed to complete a generation on Japanese camellia and oil tea camellia (Hong *et al.*, 1998). Currently, some authors (Amrine and Stasny, 1994; Lee *et al.*, 2014) still regard *A. steinwedeni* and *A. indiae* as synonyms of *A. theae*. Until the problem is finally solved, and to avoid further confusion, this author prefers to distinguish *A. theae* from *A. steinwedeni*.

A. theae is spread through all tea-growing tracts of several Asian countries. It feeds on both leaf surfaces, petioles and green stems of the tea plant. Development from egg to adult takes 8.0–8.7 days in March, 5.6–7.0 days in July–August, and 12.0–13.5 days in December–January. The average adult longevity is 12 days, the female lays from 2–4 eggs a day and the fecundity is 30–40 eggs. The eggs are laid mainly on the undersurface of the leaf, along the midrib and by the lower leaf veins (Das and Sengupta, 1958). The fecundity is significantly correlated with effective temperature. The rate of increase of populations reduces when the temperature is less than 18°C or higher than 27°C, and the most favourable RH is between 75 and 90%. Population increase is inhibited when the RH is lower than 50% or higher than 95%. The rate of increase of populations is negatively correlated with precipitation, rainfall and leaf age. Population increase is not affected by mite density when the initial density is lower than 56.4 per leaf; however, the rate of increase is adversely affected by density when the initial density is higher than 102.2 mites per leaf. Sunshine has no significant effect on population dynamics (Yin *et al.*, 2003).

Rearings of *A. theae* at 18, 22, 26, 30 and 34°C on tea (*C. sinensis*), Japanese camellia (*C. japonica*) and oil tea camellia (*C. oleifera*) showed that the mite only survived on tea leaves and failed to complete a generation on Japanese camellia and oil tea camellia. Most mites died during the 1st instar, but some reached the 2nd instar on *C. japonica* (Hong *et al.*, 1998). At 28 ± 2°C, the intrinsic rate of increase is 0.1680, the net reproductive rate is 12.39 and the mean generation time is 15.49 days (Loganathan and Muraleedharan, 1997). Lee *et al.* (2014) also reported *A. theae* on *C. sasanqua*, but it is not known whether the mite develops on this host plant. In China, *A. theae* develops over 20 generations a year in tea plantations (Lu, 1993). The mean number of mites per leaf is higher at the top of tea bushes, and lower in the middle and bottom portions. Mite populations and rainfall are correlated negatively, whereas a positive correlation occurs between mite populations and maximum temperatures (Muraleedharan *et al.*, 1988a). In north-eastern India, the density of the populations of *A. theae* on leaves increases from March to June, reaching outbreak proportions, but it decreases in dry weather and may not recover its former numbers even though there are wetter periods later on (Das and Sengupta, 1958).

In China, fluctuations in the population density of *A. theae* have one or two peaks. When there are two peaks, the population density of mite during the first peak is greater than that during the second peak. The first maximum peak date is positively correlated with total rainfall and RH in January (Lu and Lou, 1995). In Tamil Nadu (India), maximum populations of *A. theae* occur in January and February and a second smaller peak occurs in August–September. The mite population density is negatively correlated with rainfall and RH (Radhakrishnan *et al.*, 1987). Mite populations also prefer unshaded tea plants (Pande and Nandi, 1983).

Natural enemies of *A. theae* include various acaropathogens (fungi and bacteria) and predatory mites and insects. Pathogens include the fungi *Verticillium lecanii* (Zimmerman) Viégas, *Paecilomyces fumosoroseus* (Wize) Brown *et* Smith and *Hirsutella thompsonii* Fisher (Selvasundaram *et al.*, 2001), and a bacterial strain of *Proteus* (Bezbaruah *et al.*, 1996). Different phytoseiids have been found on tea plants in India, such as *Amblyseius deleoni* Denmark *et* Muma (Rao *et al.*, 1970), *A. herbicolus* (Chant) (Muraleedharan and Chandrasekharan, 1981), *Amblyseius* spp. (Rao *et al.*, 1970; Muraleedharan and Chandrasekharan, 1981), *A. obtusus* (Koch) (Muraleedharan and Chandrasekharan, 1981) and *A. rhabdus* Denmark (Rao *et al.*, 1970). Other predatory mites reported from India are acarids (*Acarus* sp.) (Muraleedharan and Chandrasekharan, 1981), cunaxids (*Cunaxa* sp.), stigmaeids (*Agistemus* sp.) (Borthakur and Das, 1988), tydeids (*Tydeus* sp.) (Muraleedharan and Chandrasekharan, 1981) and eupalopsellids (*Exothorhis caudata* Summers) (Borthakur and Das, 1988). The insects include some coccinellids (*Stethorus* sp.) and chrysopids (*Chrysoperla nipponensis* (Okamoto)) collected in China (Zhao and Hou, 1993), and the cecidomyiid *Lestodiplosis oomeni* Harris reported from Indonesia (Harris, 1982). In China, the natural enemies of *A. theae* had no significant effect on their populations (Yin *et al.*, 2003).

Symptomatology and damage

The pink tea mite feeds on the tender parts of tea – both leaf surfaces, petioles and green stems. The infested leaves become pale, often leathery, and the veins and margins of the undersurface become pinkish. Young tea plants are more susceptible to attack than adult plants (Das and Sengupta, 1958; Das, 1965). Lee *et al.* (2014) reported the mite as a vagrant on the lower leaf surface and living under the floral bud scales; sometimes causing leaf rust. Leaves that are severely infested become crinkled, and the underside browns, giving bushes a sickly appearance. A high attack density retards bush growth. The mite prefers 'Assam' tea varieties over 'Chinese' varieties. Watt and Mann (1903) reported a sometimes huge loss. Field experiments in India indicated crop losses of 5.55% (Muraleedharan and Selvasundaram, 1996).

Control

Cultural practices, plant resistance and biological control all hinder mite outbreaks. However, control usually requires the sampling and monitoring of populations, and spray applications of chemicals led by threshold determination.

SAMPLING AND MONITORING. The sampling and monitoring of *A. theae* populations are fundamental for control. Statistical analysis show that mite infestation indexes are positively correlated with observed population densities. Based on the statistical analysis of 38 groups of data, a 3-grade method and two 4-grade methods of classifying infestation by *A. theae* were developed. Mite infestation indexes calculated using these methods were significantly positively correlated with observed population densities. Population densities calculated by transforming the mite infestation indexes obtained by the two 4-grade methods were the same as the observed values of population density, and as accurate as that provided by a 6-grade method. Compared with the 6-grade method, the 3-grade and 4-grade methods reduced both time and effort (Yin *et al.*, 1995). The positive correlation of population peaks with total rainfall and RH helps enable a contingency table to be made to predict mite occurrence for previous years based on this correlation. Predicted values agreed well with observed values. The method has the advantages of simplicity, convenience and a longer range (Lu and Lou, 1995).

THRESHOLDS. The application of chemical control may be led by an economic threshold of 3–4 mites cm^{-2} leaf surface. Damage is not apparent in the field until that level of mite density is reached, but this may not happen at the same time in different parts of the same bush, especially if superficial spraying has resulted in the treatment of only the tops of the bushes. When the threshold density on a leaf has been exceeded, the mites in excess of this disperse to fresh leaves because the original leaf is too badly damaged for them all to feed on it properly (Banerjee, 1971).

PLANT RESISTANCE. Research on the morphology and biochemistry of varieties of the tea plant has been carried out over the last two decades. Chen (1994) examined the effects of these physical and chemical features of the tea leaf surface on the damage caused by *A. theae* in China. Xu *et al.* (1996) investigated the morphological parameters of tea varieties resistant to pink mite by scanning electron microscopy, and found that the leaves of resistant tea varieties had denser pubescence and stronger cuticularization on the lower surface and lower stomatal density than susceptible varieties. This study also found that there were higher contents of total amino acids, theanine, glutamic acid, aspartic acid and caffeine, and lower contents of reducing sugars and water-soluble sugars in the more resistant varieties. Furthermore, the shoots of resistant varieties had stronger glutamate dehydrogenase and phenylalanine ammonia lyase activities. It is suggested that the morphological characters inhibit mite feeding activity and the biochemical characters contribute to antibiosis.

Further studies by Chen *et al.* (1996a) demonstrated that the hair densities on the undersurface of leaves and the caffeine and amino acid contents in young shoots of resistant cultivars were significantly higher than in susceptible cultivars, while the stomatal densities and reducing sugar contents were lower than those in susceptible cultivars. There were no significant differences between the cultivars in the polyphenol and soluble protein contents in young shoots, but there were differences in the

contents of theanine and other amino acids. High contents of caffeine and many amino acids inhibited the feeding activity of the mite, while a high glucose content promoted it. Pubescent leaves with a low stomatal density had high contents of amino acids and caffeine and low contents of reducing sugars that promoted resistance to the mite.

Yet more studies showed that in resistant varieties, the total free amino acid contents – mainly theanine, glutamic acid, alanine, glycine, aspartic acid, serine and valine (etc.) – were significantly higher in resistant than in susceptible varieties. The average activities of glutamine dehydrogenase and phenylalanine ammonia lyase in the former were 2.3 and 3.2 times as much as those in the latter. Each (all) of the amino acids mentioned above inhibited the feeding activity and the reproduction of the mite in bioassays. The contents of amino acids obtained from the leaf surface of resistant varieties by water rinsing were more than those from susceptible varieties. It was suggested that the resistance of tea to the mite might be increased by enhancing the content of free amino acids in tea shoots by the application of nitrogen fertilizer (Chen *et al.*, 2000). Research has also been done on host plant suitability of *A. theae* at different temperatures (Hong *et al.*, 1998).

CULTURAL PRACTICES. Cultural and chemical control methods are recommended for mite control (Lu, 1993). In a fertilizer and shading experiment, shaded tea plants hosted fewer mites (although they did not necessarily have less mite damage) than did unshaded tea plants, while fertilizer treatments had little or no effect, implying that mite numbers were affected more by light distribution and leaf temperature than by chemical changes to the plants resulting from the use of fertilizers (Banerjee, 1969).

BIOLOGICAL CONTROL. The action of three entomogenous fungi (*V. lecanii*, *P. fumosoroseus* and *H. thompsonii*) has been tested in the control of the pink mite both in the laboratory and field. The mite was susceptible to all three pathogens and the fungal formulation used gave satisfactory control of populations from the 5th day after application; its efficacy was 100% on the 10th day after application. Mycelial growth was observed on dead pink mites within 5–7 days of the date of application. Application of the three pathogens at 3500 g formulation ha^{-1} significantly reduced the population density of pink mites in the field (Selvasundaram *et al.*, 2001). A strain of *Proteus* RRLJ 16 cultured in a normal bacteriological medium without iron stress produced an insect-repellent compound that was effective when sprayed on tea plants infested with *A. theae*. Reinfestation was not observed on treated plants for 1 year (Bezbaruah *et al.*, 1996).

CHEMICAL CONTROL. A number of active ingredients have been employed on tea, but regardless of their effectiveness, there are legal constraints on their use. Thus, investigations have been made of the levels of residues of dicofol, endosulfan and malathion in India (Singh and Agnihotri, 1984) in relation to the suggested clearances by the US Environmental Protection Agency on residue tolerance levels on tea for dicofol, endosulfan, ethion and phosalone (Muraleedharan and Varatharajan, 1988).

Chemicals that have been tested include aldoxycarb (Murthy *et al.*, 1979), aramite (Das and Sengupta, 1958), binapacryl (Haq and Ali, 1973), chlorobenzilate (Das and Sengupta, 1958; Banerjee, 1978), bifenthrin (Lu, 1993), cyhexatin (Muraleedharan and Kandasamy, 1982), cypermethrin, deltamethrin (Muraleedharan and Varatharajan, 1985), dicofol (Haq and Ali, 1973; Banerjee, 1978; Ram, 1978; Murthy *et al.*, 1979; Muraleedharan and Kandasamy, 1982; Das, 1983; Das and Gope, 1983; Gope and Borthakur, 1988), dicrotophos (Haq and Ali, 1973), diflubenzuron (Chen *et al.*, 1992), dimethoate (Banerjee, 1978), dinobuton (Banerjee, 1978; Murthy *et al.*, 1979), dithane (Lu, 1993), endosulfan (Wu *et al.*, 1995), ethion (Banerjee, 1978; Ram, 1978; Murthy *et al.*, 1979; Muraleedharan and Kandasamy, 1982; Das, 1983; Das and Gope, 1983; Gope and Borthakur, 1988), fenpropathrin (Banerjee, 1978; Murthy *et al.*, 1979; Muraleedharan and Varatharajan, 1985), fenvalerate (Banerjee, 1978; Murthy *et al.*, 1979; Muraleedharan and Varatharajan, 1985), lime sulfur (Das and Sengupta, 1958; Haq and Ali, 1973), malathion (Das and Sengupta, 1958), monocrotophos (Murthy *et al.*, 1979; Muraleedharan and Kandasamy, 1982), permethrin (Muraleedharan and Varatharajan, 1985), phosalone (Ram, 1978; Murthy *et al.*, 1979), pirimiphos-methyl (Murthy *et al.*, 1979), propargite (Wu *et al.*, 1995), pyridaben (Chen *et al.*, 1993; Lu, 1993), quinalphos (Lu, 1993), sulfur (Das and Sengupta, 1958; Rao, 1970), tetradifon (Banerjee, 1978; Das and Gope, 1983; Gope and Borthakur, 1988), thiometon (Banerjee, 1978) and tridemorph (Chandrasekaran, 1980). Biopesticides include neem-based formulations (Muraleedharan and Selvasundaram, 1996).

Acaphylla steinwedeni Keifer (Fig. 16.61)

Common name

Orange camellia rust mite.

Diagnostic characteristics

FEMALE. The body is spindle formed and blunter anteriorly, 175–190 μm long, 60 μm wide and orange in colour. The prodorsal shield has a central longitudinal elevation with a bell outline, and the frontal lobe bilobed, overhanging the gnathosoma; coarse tubercles are located laterally above the coxae; the scapular tubercles are 25 μm apart, set well ahead of the rear margin; the scapular setae (*sc*) are 4.5 μm long, directed cephalodorsad. The coxae have seta I missing. The opisthosoma has moderately broad, smooth dorsal annuli, forming a central longitudinal ridge; the ventral annuli are microtuberculate except near the confluence with the dorsal annuli; there are 30 dorsal annuli and about 60–65 ventral annuli. The setae c_2 are about 12 μm long, set on about ventral annulus 5; the setae *d* are 30 μm long, set on about ventral annulus 21; the setae *e* are 23 μm long, set on about ventral annulus 38; the setae *f* are 17 μm long, set on about ventral annulus 6 from the rear; the setae h_1 are absent. The genitalia bear fine tubercles basally, have longitudinal scoring and are 24 μm wide and 20 μm long; the genital coverflap has 6–8 longitudinal ridges; the seta is 9 μm long. The empodium is 3-rayed and bifurcate (Keifer, 1943).

MALE. The body is 140–160 μm long (Keifer, 1943).

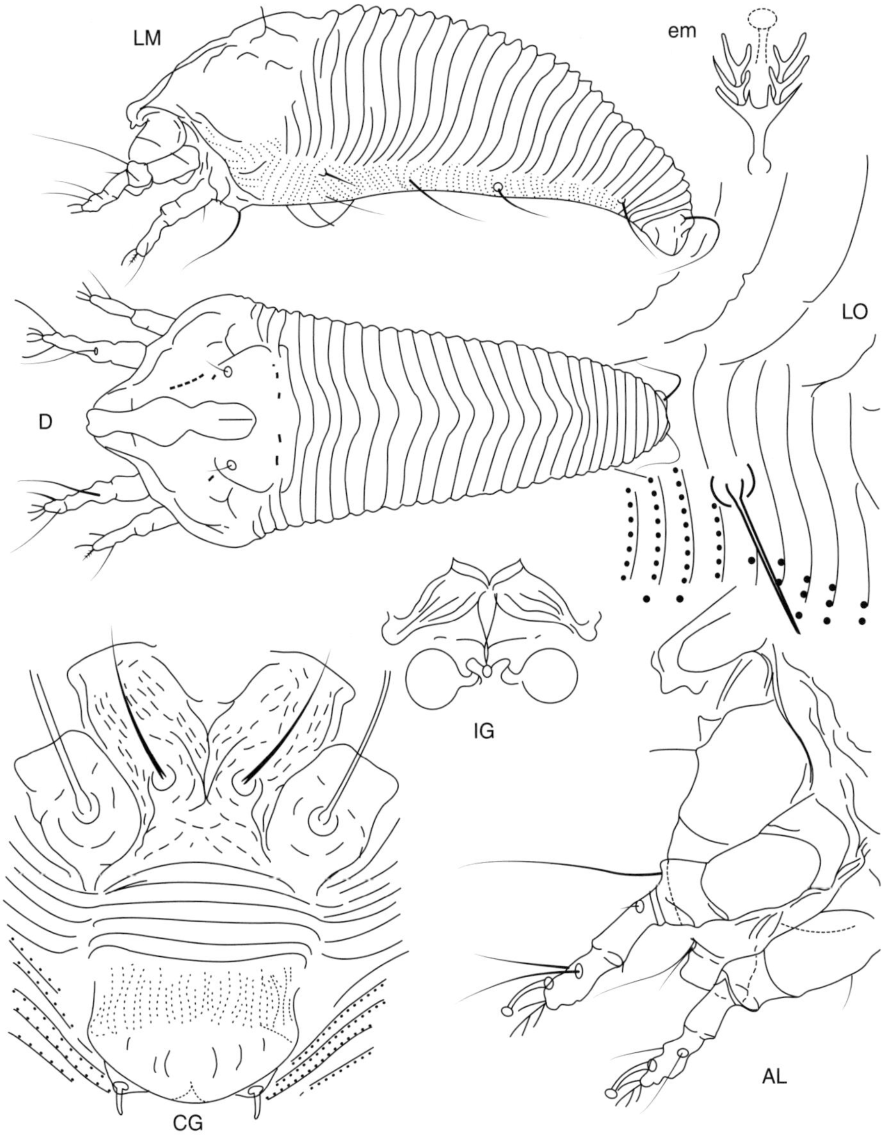

Fig. 16.61. *Acaphylla steinwedeni* Keifer (from Keifer, 1943). See Table 16.1 for key.

Geographical distribution

China (Kuang, 1995; Hong and Zhang, 1996b), Malaya (Hall, 1954), Spain (Mansilla Vásquez *et al.*, 2002, 2003), Taiwan (Chao, 1971; Huang, 1974, 1978) and the USA (Alabama, California, Florida, Louisiana) (Keifer, 1943; Jeppson *et al.*, 1975; Olivier and Cancienne, 1980; Keifer *et al.*, 1982). The mite has also been reported from Japan, India and the (former) USSR (Biota Taiwanica, 2011).

Host plants

Theaceae: Japanese camellia (*Camellia japonica* L.) (Keifer, 1943; Jeppson *et al.*, 1975; Keifer *et al.*, 1982), dian shan cha (*C. reticulata* Lindl.) (Biota Taiwanica, 2011), tea (*C. sinensis* (L.) Kuntze) (Hall, 1954; Chao, 1971; Yeh and Huang, 1975).

Bio-ecology

A. steinwedeni is reported from Japanese camellia leaves in various countries. The mite probably originates from Asia but there is as yet no information on this. It is morphologically similar to *A. theae* (Watt *et* Mann) and in addition to its original host plant also infests tea plants (Hall, 1954; Yeh and Huang, 1975). It has a haploid number of chromosomes of $n = 2$ (Helle and Wysoki, 1996), the threshold for development is 5.37°C and the sum of effective temperatures for development is 170.07 degree-days. At 21°C, mite development takes 11 days, and in the Nanjing region in China it can develop 22–23 generations a year. The intrinsic rate of population increase is 0.1691, the net reproductive rate is 6.2496, the finite rate of increase is 1.1843 and the mean generation time is 10.8357 days. No responses have been observed to different photoperiods, and diapause

could not be induced. At 33°C, the eggs hatch and post-embryonic development occurs, but at 35°C hatching does not occur. The lower cooling point is −25.44°C (Kuang and Zhao, 1988). In Taiwan, the mite developed 28 generations a year in the laboratory, and the duration of the developmental stages varied from about 6.2 days at 28.4°C to about 27.3 days at 14.3°C (Chao, 1971). One generation takes 9.5 days at 23.1°C and 18.2 days at 15.9°C. Over one generation, each female laid an average of 3.8 eggs, with 89.5% hatching per female, and the mite may develop up to 30 generations a year (Yeh and Huang, 1975).

In Louisiana, *A. steinwedeni* reaches the adult stage in 7–8 days. Females begin ovipositing on the third day after the final moult, and lay an average of 9.4 eggs in 7 days. Populations are largest in the spring when new growth of the food plant begins. The lowest populations occur in the summer and early autumn (Olivier and Cancienne, 1980). In Taiwan, *A. steinwedeni* occurs in tea plantations throughout the year, and outbreaks generally occur from late October to early December, with peak numbers in early November. The population density is highest on pubescent varieties with round-shaped leaves, on the east-facing side of plants, in the middle of plants, on middle-aged leaves and around the primary veins in the central parts of individual leaves (Yeh and Huang, 1975).

Symptomatology and damage

The feeding activity of the orange camellia rust mite causes considerable rusting, mainly in more protected places with higher humidity (Jeppson *et al.*, 1975). In Louisiana, *A. steinwedeni* infests the Japanese camellia throughout the state. Large populations cause foliar discoloration and scabbing, which result in loss of aesthetic value and consequently in economic losses; the same symptoms are attributed to another eriophyoid, *Calacarus carinatus* (Green) (= *Calacarus adornatus* (Keifer)) (Olivier and Cancienne, 1980). However, the damaging symptoms produced by the two mites on leaves are easily distinguished from one another, as the upper-surface discoloration of infested leaves to dark brown to purplish brown occurs only with *A. steinwedeni* (Biota Taiwanica, 2011). Host plants of *A. steinwedeni* in Taiwan include tea and *C. reticulata*. The mite is reported as a vagrant, inhabiting mostly the undersurface of tea and camellia leaves and causing yellow to yellowish brown spots. It also infests buds and young shoots (Biota Taiwanica, 2011). The mite has been reported as an important pest of tea in Taiwan, particularly in plantations in the north of the island. It generally infests new growth, causing the young leaves to curl and become brown, and thus reducing the quality and quantity of the yield (Chao, 1971).

Control

As for *A. theae*.

Acaphylla theavagrans Kadono

Common name

Tea rust mite.

Diagnostic characteristics

FEMALE. The body is fusiform, 201–215 μm long, 68–85 μm wide and orange to orange yellow in colour. The gnathosoma is curved ventrally. The prodorsal shield is subtriangular in anterior outline, with the frontal lobe over the gnathosma base; the median line is lacking, and the admedians run in zigzags from the frontal lobe to a point slightly posterior to the transverse level through the scapular setae, and curve inwardly towards each other at the rear; transverse lines run from the admedian lines at anterior ⅓ to the lateral shield margin; there are irregularly curved lines in the area surrounded by the admedian lines and the transverse lines from the admedians. The scapular tubercles are located ahead of the rear shield margin; the scapular setae (*sc*) are 4 μm long and directed upwardly. The opisthosoma has 30–34 smooth dorsal annuli and 69–79 ventral microtuberculate annuli. The setae c_2 are 14 μm long; the setae *d* are 53 μm long; the setae *e* are 43 μm long; the setae *f* are 20 μm long. The genitalia are 18 μm long and 25 μm wide, the coverflap has 6–10 anterior ribs and is microtuberculate basally. The empodium is 3-rayed (Kadono, 1992).

MALE. The male is similar to the female, except for the genitalia (Kadono, 1992).

Geographical distribution

China (Hong and Zhang, 1996a), Japan (Kadono, 1992; Mizuta, 2000), Korea (Lee *et al.*, 2014) and Taiwan (Huang, 1980).

Host plants

Theaceae: tea (*Camellia sinensis* (L.) Kuntze) (Kadono, 1992; Mizuta, 2000; Lee *et al.*, 2014).

Bio-ecology

The tea rust mite is a leaf vagrant on tea (Kadono, 1992). At 15, 18, 21, 24 and 27°C and on mature tea leaves, development from egg to egg takes 22.4, 16.5, 11.4, 9.3 and 7.7 days, respectively. The developmental zero and the total effective temperature for one generation have been estimated to be 9.1°C and 137.2 degree-days, respectively. Oviposition occurs at all the above temperatures from 15 to 27°C, and the number of eggs per female increases with increasing temperature. Adult longevity and the pre-ovipositional period are shortened with increasing temperature. The fecundity zero, estimated by regression analysis on the average number of eggs per ovipositing female daily at five different temperatures, is 8.7°C. The intrinsic rate of natural increase is higher at higher temperatures, and is 0.037 and 0.181 at 15°C and 27°C, respectively (Mizuta, 2000).

In Japan, the seasonal occurrence of the pest shows two peaks, one in spring and the other in autumn. The spring peak is higher than the autumn peak in tea fields of arc-shaped bush formation for 'sencha' (a type of green tea) and 'gyokuro' (shaded green tea) production, and the autumnal peak is higher than that in spring in tea fields of natural-shaped bush formation for gyokuro production (Matsuda *et al.*, 2001).

Research carried out in the laboratory on the adult density at different levels using different tea varieties and lines showed that the adult density of the mite was highest on the 4th to 5th leaves from the terminal leaf. Females laid more eggs on the middle leaves. The difference in the adult density in relation to the leaf order is mainly caused by a difference in the oviposition rate as the number of eggs laid per 20 females for 3 days was the highest on the 5th leaf (Mizuta, 2002). Among the eight Japanese tea cultivars, the number of eggs laid per 20 females for 3 days varied from 34.8 on 'Kuritawase' to 97.8 on 'Ooiwase' and was significantly different among the varieties. Large differences in the numbers of eggs were found among nine introduced tea lines, ranging from 3.5 on 'Miya taiwan-yamacha' to 63.5 on 'Chinshindapan'. The intrinsic rate of natural increase varied from 0.1559 on 'Miyataiwan-yamacha' to 0.1973 on 'Chinshindapan' (Mizuta, 2002).

Natural enemies include the cecidomyiid *Feltiella acarisuga* (Vallot) (Abe *et al.*, 2011).

Symptomatology and damage

The tea rust mite is a vagrant on the lower leaf surface of tea, and causes leaf rust (Lee *et al.*, 2014). In Fukuoka Prefecture, Japan, it is one of the most serious pests in tea production (Matsuda *et al.*, 2001).

Control

Control of the tea rust mite requires spray applications of acaricides. Damage by the pest is prevented successfully by the application of acaricides before the first sprouting. Chlorfenapyr, dicofol, fenpyroximate, milbemectin and pyridaben are reported as most effective in suppressing mite damage (Matsuda *et al.*, 2001).

The Genus *Calacarus* Keifer

Diagnostic characteristics

The gnathosoma is quite large, downcurved and recurved, and the dorsal pedipalp genual seta (*d*) is long. The legs have all the setae except for the patellar seta of legs II, which is missing. The prodorsal shield lacks scapular tubercles and setae and has a pattern of curved wax-bearing lines or narrow ridges. The opisthosoma has a central and one or two lateral wax-bearing ridges; the dorsal annuli are not much less numerous than the ventral annuli. The genital coverflap has fine but obscure scoring (Keifer, 1940c; Amrine *et al.*, 2003).

Economic importance

This genus includes over 40 species, but only two of these play an economic role. The most important species are the citrus grey mite or citrus blotch mite, *Calacarus citrifolii* Keifer, living on citrus and other plants (Jeppson *et al.*, 1975; Vacante, 2010), and the purple or ribbed tea mite, *C. carinatus* (Green), which is injurious to tea in southern Asia (Jeppson *et al.*, 1975). Two minor pests live on papaya (*Carica papaya* L.): the papaya leaf edgeroller, *Calacarus brionesae* Keifer, reported from the Philippines and causing edge rolling of the leaves and white spots along the rolls (Jeppson *et al.*, 1975); and *C. flagelliseta* Flechtmann, de Moraes *et* Barbosa, which causes upward curling of leaves, drying and necrosis in Brazil (Flechtmann *et al.*, 2001). The coffee leaf rust mite, *C. coffeae* Keifer, lives on the underside of coffee bush leaves in the Kivu area of Zaire (Jeppson *et al.*, 1975). The pepper purple mite, *C. capsica* Chakrabarti *et* Mondal, is reported from West Bengal (India) (Chakrabarti and Mondal, 1980) and South Africa (Meyer Smith and Craemer, 1999).

Calacarus citrifolii Keifer (Fig. 16.62)

Common name

Citrus grey mite or citrus blotch mite.

Diagnostic characteristics

FEMALE. The body is fusiform, solid, 185–200 µm long, 70 µm wide and grey to purplish in colour. The gnathosoma is curved downwards. The prodorsal shield has a broad and rounded frontal lobe, and the scapular setae (*sc*) are absent. The shield pattern is a network, with admedian lines without small vertical dashes along them. The median line is present on the frontal lobe, and slightly indicated on the rear ½ of the shield. The admedian lines are strongly sinuate and undulating, arising from the sides of the frontal lobe, arching back to a lateral cross line at about ⅓, converging centrally to a slight cross line at about ½, then curving out ¾, where they meet another lateral line and then fork. The inner fork lines of the admedians meet centrally a little ahead of the rear shield margin; the outer part of the bifurcation arches back and recurves outwardly at the rear margin. The submedian line, with branches, forms three large cells on the outside of the admedian. The opisthosoma has about 60 dorsal annuli and 65–70 ventral annuli; with 5 white longitudinal wax bands, gradually diminishing in distinctness caudally. The setae c_2 are 40 µm long, set on about ventral annulus 11; the setae *d* are 45 µm long, set on about the ventral annulus 27; the setae *e* are 40 µm long, set on about ventral annulus 46; the setae *f* are 25 µm long, set on ventral annulus 8 from the rear; the setae h_1 are absent. Coxae I are narrow and meet centrally, and are ornamented with many curved lines. The genital coverflap has many fine longitudinal lines placed in 2 ranks. The empodia are 5-rayed (Keifer, 1955a; Jeppson *et al.*, 1975).

MALE. The male is unknown (Keifer, 1955a).

Geographical distribution

Angola (De Carvalho and Cardoso, 1972), Iran (Khanjani and Haddad Irani-Nejad, 2006; Xue *et al.*, 2009a), Kenya, Mozambique, Nigeria, Zambia, Zimbabwe (Meyer Smith, 1996a; Bedford, 1998) and South Africa (Keifer, 1955a; Meyer Smith, 1996a; Bedford, 1998).

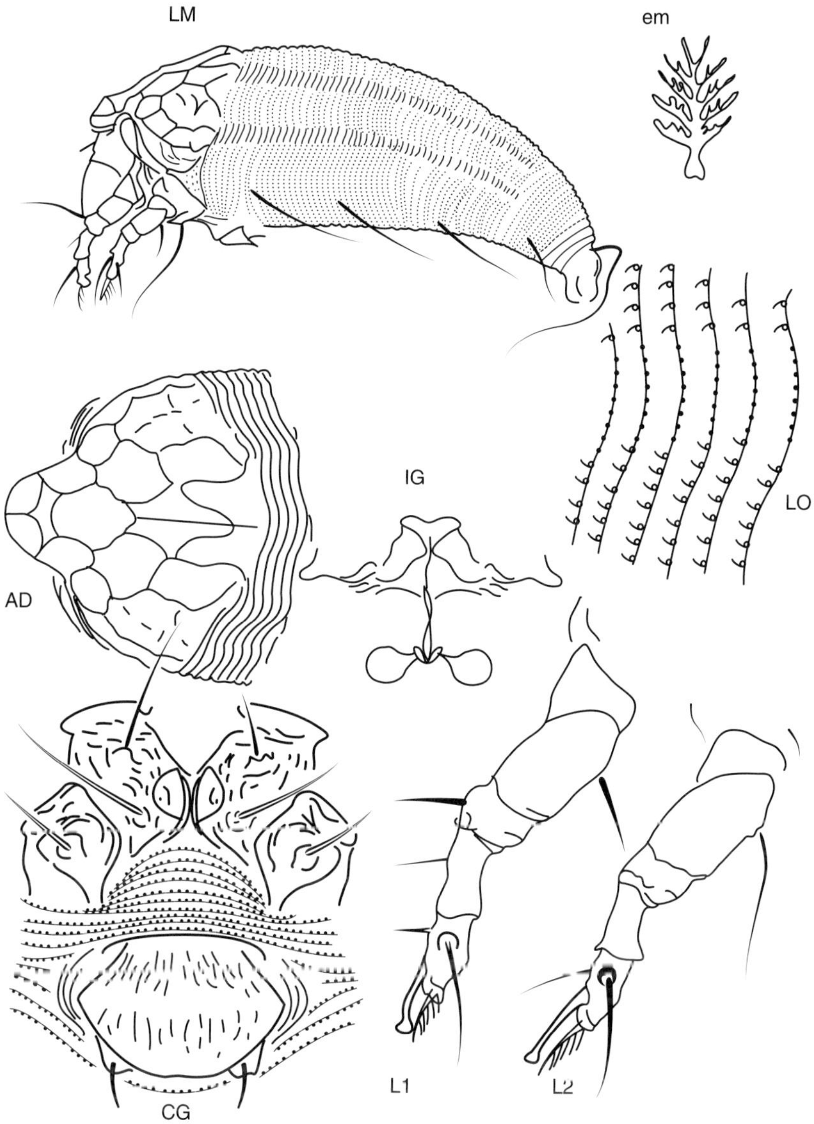

Fig. 16.62. *Calacarus citrifolii* Keifer (from Keifer, 1955a). See Table 16.1 for key.

Host plants

The citrus grey mite has been reported on 28 plant species belonging to 21 families (Amrine Jr, in Flechtmann *et al.*, 2001); however, records may involve more than one *Calacarus* species (Jeppson *et al.*, 1975). The main economic plants of importance are listed here by family: Araceae – common calla (*Zantedeschia aethiopica* (L.) Spreng.); Caricaceae – papaya (*Carica papaya* L.); Caryophyllaceae – *Dianthus* sp.; Euphorbiaceae – poinsettia (*Euphorbia pulcherrima* Willd. ex Klotzsch); Musaceae – banana (*Musa paradisiaca* L.); Passifloraceae – common passion fruit (*Passiflora edulis* Sims), granadilla (*P. quadrangularis* L.); Rosaceae – peach (*Prunus persica* (L.) Batsch); Rutaceae – *Citrus* spp.; Solanaceae – *Brunfelsia* sp. (van der Merwe and Coates, 1965; Jeppson *et al.*, 1975; Meyer Smith, 1996a; Schwartz, 1971, in Bedford, 1998; Khanjani and Haddad Irani-Nejad, 2006; Xue *et al.*, 2009a).

Bio-ecology

The citrus grey mite is a vagrant on young leaves, twigs and fruits of various economic plants including citrus. On citrus leaves, it normally prefers the upper leaf surface, where the immature and adult stages secrete a white wax that partly protects them and appears as a greyish powder when looking at the leaves obliquely to incident light. The mite is active throughout the year (van der Merwe and Coates, 1965).

C. citrifolii reproduces by parthenogenesis, and at 20 and 27°C the development of a generation (from egg to egg) takes 13 and 7 days, respectively. The female begins to lay after 24 h, and at 27°C lays on average 3 eggs daily for a total of 33 eggs in 10 days; at 30°C the fecundity is 43 eggs (van der Merwe and Coates, 1965). Oviposition occurs in the hours of darkness, with little light or on cloudy days, mainly in depressions over the oil glands. At 27°C, the eggs hatch in 5–6 days

and at 20°C in 13 days. The mite produces several generations a year and wintering stages are not known. Nymphal stages I and II live 1–2 and 3–5 days, respectively. The climatic conditions and the vegetative state of the crop can influence the population dynamics of the mite, whose seasonal fluctuations are generally unpredictable (Dippenaar, 1958a). The population density can reach 200 mites per leaf and the highest peaks are observed in the warmest months (van der Merwe and Coates, 1965).

Symptomatology and damage

On most host plants, infestation by the citrus grey mite induces the appearance of leaf browning or necrotic yellow blotches. In particular, on *Brunfelsia* sp, the mite causes concentric ring spots (Meyer Smith, 1996a). On papaya, as well as on poinsettia, the feeding activity of *C. citrifolii* induces bronzing of the leaves (Meyer Smith, 1996a; Meyer Smith and Craemer, 1999). The symptoms of attack on banana are fine, reddish brown russet on the peel of the fruit (Meyer Smith, 1996a). On *Citrus*, it transmits a serious disease, commonly known as 'concentric ring blotch' that was reported for the first time in South Africa (Doidge, 1925) and has been extensively studied by Dippenaar (1958a,b,c); this is described below.

CONCENTRIC RING BLOTCH. The disease is caused by the transmission of toxic saliva from the mite, although it has been observed that high levels of attack are not always associated with disease (Rossouw and Smith, 1963), which affects mainly the young tissues and those in strong growth – the young leaves, shoots, branches and fruits, while the mature leaves and shoots can be exempt. It becomes particularly serious in nurseries and young plants, and the symptoms are seen more frequently on the parts of the canopy exposed to the sun. The disease is discontinuously present over years. The leaves and fruits show blotched areas marked by concentric rings, hence the term 'concentric ring blotch', that vary significantly in number, form, size and colour; other characteristics include a general chlorosis of the leaf surface, partial or general defoliation, dieback of shoots and shedding of fruit reaching maturity. On the leaves, there are two types of spots, one consisting of an atypical chlorotic blotch that is concentrically marked and the other a necrotic spot with or without concentric markings, surrounded by a halo of chlorotic tissue. The blotches on the fruits are map-like outlines or circular, green to whitish in colour, with the damaged area depressed, of variable size and expanding with the development of the fruit until it reaches a diameter of 50 mm and above. On shoots, concentric point-like blotches, either individual or several, are observed. The young lesions are chlorotic or green, and clearer than the surrounding surface. From these areas on shoots and branches, necrosis, longitudinal cracks and abundant emissions of rubber may arise. A high density of attack can lead to death of the entire shoot or the branch (Dippenaar, 1958a,b; Jeppson *et al.*, 1975; Bedford, 1998). In concentric alterations of the infested leaves, a high concentration of microorganisms such as *Spiroplasma* has been found (Kotze *et al.*, 1987). On orange, attack by the mite may initially go unnoticed (Grout and Schutte, 2004).

Control

Biological control is inadequate and mite infestations are usually treated with chemicals.

CHEMICAL CONTROL. In the nursery, it is possible to treat every 3 or 4 months, while in the field, a first treatment can be carried out at petal fall and a second at a later stage (Dippenaar, 1958c). On 'Navel' orange and lemon, young plants of 'Valencia' orange and cultivars with an average age of over 10 years, only one treatment in the autumn generally allows the implementation of integrated pest management (IPM) programmes for both the citrus bud mite and the citrus blotch mite (Bedford, 1968, 1998). Among the chemicals used, in addition to older acaricides such as different organochlorines and some other organohalogens, various formulations of sulfur (dust, lime and wettable) have been reported as effective against both mobile forms and eggs of the mite (Dippenaar, 1958a,b,c; Krause *et al.*, 1996; Meyer Smith, 1996a; Bedford, 1998). A good response to some organophosphates has been reported on banana (Jones, 1979).

Calacarus carinatus (Green) (Fig. 16.63)

Common name

Purple or ribbed tea mite.

Diagnostic characteristics

FEMALE. The body is robust and fusiform, 190–240 μm long, 79–93 μm wide and a dull deep purple; it has shield lines and dorsal ridges with white wax. The prodorsal shield has a cellular pattern; the frontal lobe is round, and the scapular tubercles are apparently present, but lack scapular setae. The opisthosoma has 52–68 dorsal annuli, 5 dorsal longitudinal ridges, is smooth and has 69–79 ventral annuli, with small bead-like microtubercles set on the rear annular margins. The setae h_1 are lacking. The coverflap is smooth, with 11 short faint lines. The coxae are smooth, the empodia are 5-rayed and the tarsal solenidia are strongly knobbed (Lee *et al.*, 2014).

Geographical distribution

Australia, Cambodia (CABI, 1974), China (Hong *et al.*, 2006), Georgia (USSR) (Tulashvili, 1930), Hungary (de Lillo and Beron, 2004), India (Assam) (Das and Sengupta, 1962), India (south India) (Anstead, 1911), Indonesia (Bernard, 1909), Italy (de Lillo and Beron, 2004), Japan (CABI, 1974), Kenya (IPPC Secretariat, 2005), Korea (Lee *et al.*, 2014), Krasnodar (USSR) (Maslennikov, 1979), Laos (CABI, 1974), Mauritius (Moutia, 1958), New Zealand, Pacific Islands, Peninsular Malaysia (CABI, 1974), Poland (de Lillo and Beron, 2004), Portugal (Mansilla Vasquez *et al.*, 2003), Sri Lanka (Green, 1890), South Africa (Immelman, 1983), South-east Asia (du Pasquier, 1933), Spain (Manzilla Vásquez, 1991b; Mansilla Vásquez *et al.*, 2003), Taiwan (Huang, 1974), USA (California, Florida, Georgia, Louisiana) (Keifer, 1940c; Jeppson *et al.*, 1975; Olivier and Cancienne, 1980; Keifer *et al.*, 1982) and Vietnam (du Pasquier, 1933).

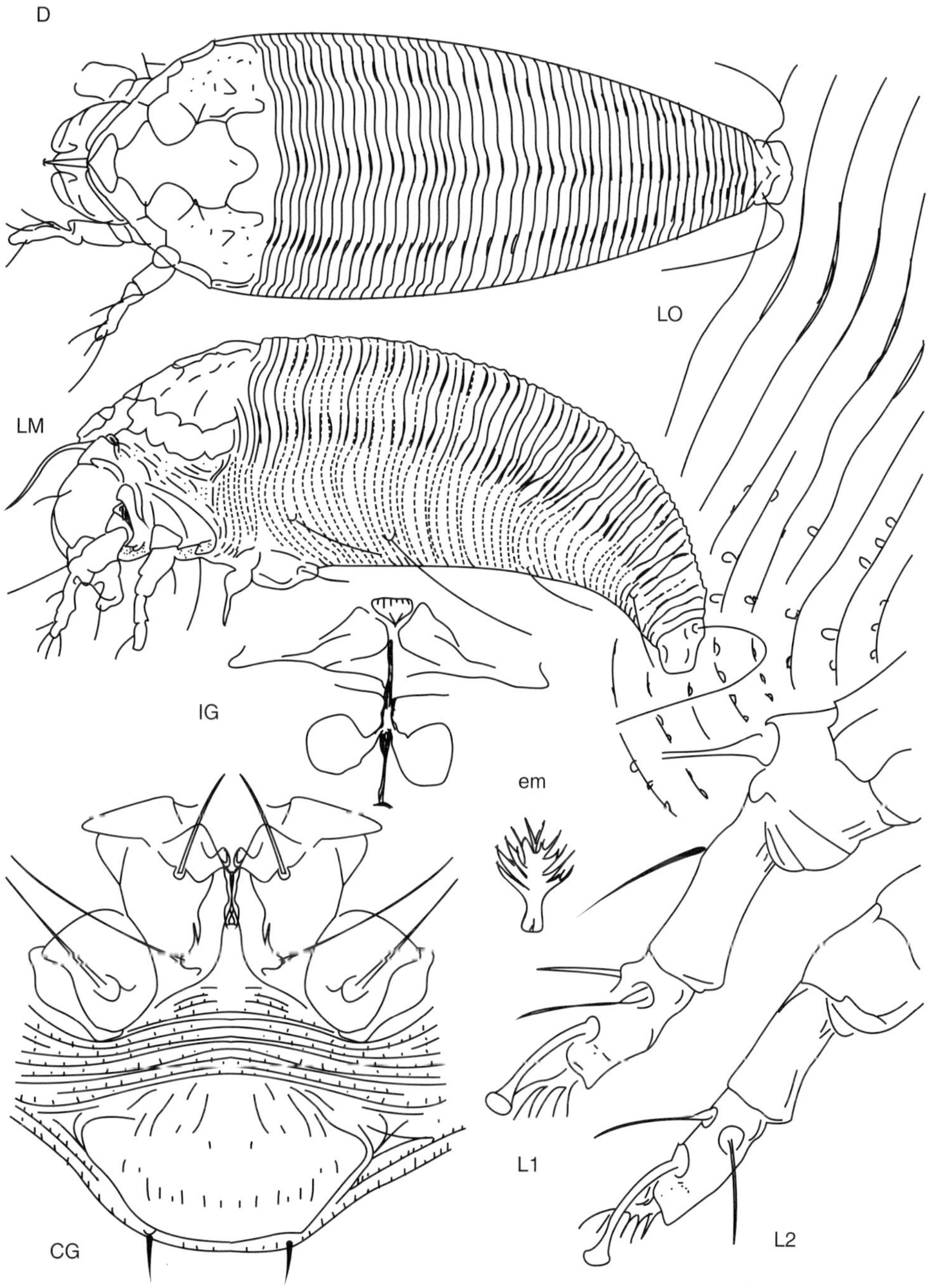

Fig. 16.63. *Calacarus carinatus* (Green) (from Keifer *et al.*, 1982). See Table 16.1 for key.

Host plants

The main economic plants of importance are listed here by family: Adoxaceae – cranberry bush (*Viburnum opulus* L.) (Keifer, 1940c); Araceae – *Spathiphyllum* sp.; Solanaceae – chilli (*Capsicum annum* L.) (Moutia, 1958); Theaceae – *Camellia* (Keifer, 1955b), *C. caudata* (Wallich) (Das and Sengupta, 1962), Japanese camellia (*C. japonica* L.) (Keifer, 1940c), *C. kissi* (Wallich), tea (*C. sinensis* (L.) Kuntze) (Keifer, 1940c; Lee *et al.*, 2014).

Bio-ecology

C. carinatus is a leaf vagrant on tea and other plants in southern Asia and other countries. It has also been reported on cranberry bush, Japanese camellia (Keifer, 1940c), *Camellia* (Keifer, 1955b), chilli (Moutia, 1958), teas (Das and Sengupta, 1962) and *Spathiphyllum* sp.

The purple tea mite usually lives on tender parts of the plant, including young leaves of tea. Feeding occurs on both

leaf surfaces but is much commoner on the upper surface of older leaves (Shiao, 1976). On tea bushes, the mite prefers the leaves at the lower level of the bush; leaves at the top and middle levels harboured significantly less dense populations of mites. The population density is greater on bushes in the third and fourth years from pruning than on bushes in the first half of the pruning cycle (Muraleedharan *et al.*, 1994). On tea leaves, the populations are present throughout the year, but the rains reduce them to presence only on the lower levels. Commonly, population development starts in October and becomes injurious before the cold weather; from late February to early March, the population increases and develop until the new rains (King, 1937). In south India, peaks occur in January, April, May and November, and the percentages of the mites on the upper, middle and lower thirds of the tea plants were 45.85, 42.4 and 11.75, respectively (Muraleedharan and Chandrasekharan, 1981). In Munnar, south-west India, at 1500 m above sea level, the population density of the mite on tea leaves is high during January to April, and heavy rainfall, high temperature and high RH induce a significant decrease (Muraleedharan *et al.*, 1994). In Louisiana, mite populations on *C. japonica* are largest in the spring when new growth of the food plant begins; the lowest populations occur in summer and early autumn (Olivier and Cancienne, 1980). In Taiwan, the population increases gradually from June until October but is adversely affected by rainfall (Shiao, 1976).

In India, the egg to egg developmental period on tea in July–August is 6.0–7.5 days (Das and Sengupta, 1962). In Louisiana on *C. japonica*, the mite reached the adult stage in 7–8 days. Females began ovipositing on the third day after the final moult and laid an average of 7.5 eggs in 8 days (Olivier and Cancienne, 1980). At 28.5 and 16.5°C, a generation lasted 20 and 50 days, respectively (Shiao, 1976). Most eggs are laid along veins. The female lays 1–4 eggs a day, with a fecundity of 21 eggs and an oviposition period of 10–13 days (King, 1937). At 23°C, the egg hatches in 6.5–8 days. The first nymphal instar covers a period of 1–3 days, the second nymphal instar develops about 2 days before moulting. The developmental period from egg to adult takes 10–12 days. The life cycle on tea develops in about 13.5 days in January, 9 days in March and April, and 6–8 days in July and August (Das and Sengupta, 1962).

Natural enemies include the cecidomyiid *Lestodiplosis oomeni* Harris (Harris, 1982) and some predatory mites, including the phytoseiids *Amblyseius rhabdus* Denmark, *A. deleoni* Muma *et* Denmark (Muraleedharan and Chandrasekharan, 1981), *A. ovalis* (Evans) (Moutia, 1958; Rao *et al.*, 1970) and *Amblyseius* sp. (Muraleedharan and Chandrasekharan, 1981). In addition, some tydeids (*Tydeus* sp.) and an acarid mite (*Acarus* sp.) have been reported (Muraleedharan and Chandrasekharan, 1981).

Symptomatology and damage

The purple tea mite is an important tea pest in different countries of the world and is very common in southern Asia. In Java (Indonesia), it causes leaf tea damage more severe than that in India. Damage occurs mainly during droughts, usually involves the leaves and may result in bush defoliation. 'Assam' tea varieties show greater damage than 'Chinese' varieties, and higher mite densities are observed on mature leaves. Severely infested leaves turn bronze and both surfaces show a dusty aspect due to the presence of mite skins and wax. Transparent and pale spots or blisters appear on the leaves (Jeppson *et al.*, 1975). Infested leaves initially show a brownish discoloration, which is more apparent at the margins, and subsequently turn completely brown, dry up and are shed (Das and Sengupta, 1962; Shiao, 1976). In Louisiana, *C. carinatus* infests the Japanese camellia, and large populations cause foliar discoloration and scabbing, which result in losses in aesthetic value and economic value (Olivier and Cancienne, 1980).

Control

Control is provided by both spray applications of acaricides and cultural practices. The choice of selected acaricides and bioacaricides, associated with cultural practices, allows the implementation of integrated pest management (IPM).

CULTURAL PRACTICES. In Kenya, a marked response has been found to N, the highest annual return amounting to 23 kg of made tea kg^{-1} N at 90 kg N ha^{-1}. Bushes of tea receiving no fertilizer are heavily infested by *C. carinatus*, whereas none are found on fertilized bushes. Recovery from pruning is also much faster on fertilized bushes (Othieno *et al.*, 1981). Populations of *C. carinatus* on tea are significantly reduced by applying permethrin and propargite combined with nitrogenous fertilizers (2% urea or NPKS) as a top dressing. Nitrogenous fertilizers alone also reduce the mite populations, NPKS being superior to urea (Sudoi, 1990).

CHEMICAL CONTROL. Spray applications of acaricides prevent serious damage to tea leaves (Jeppson *et al.*, 1975). Before the introduction of organic acaricides, kerosene emulsion, soap solutions and sulfur preparations were employed for control of this pest mite (ChannaBasavanna, 1996). Most of the active ingredients listed in the literature are no longer allowed, which highlights the current importance of the problem. Chemicals that have been tested include aldoxycarb (Murthy *et al.*, 1979), carbophenothion (Ananthakrishnan, 1962, 1963; Mukerjea, 1967; Banerjee, 1978; Muraleedharan, 1981), chlorobenzilate (Ananthakrishnan, 1963; Cranham, 1966; Mukerjea, 1967), cyhexatin (Shiao, 1976), dicofol (Das and Sengupta, 1962; Ananthakrishnan, 1963; Cranham, 1966; Mukerjea, 1967; Banerjee, 1978; Ram, 1978; Murthy *et al.*, 1979; Das, 1983; Das and Gope, 1983; Muraleedharan *et al.*, 1988b), dimethoate (Ananthakrishnan, 1962, 1963; Mukerjea, 1967; Banerjee, 1978; Muraleedharan, 1981), dinobuton (Ananthakrishnan and Krishnamurthy, 1964), ethion (Ananthakrishnan, 1962, 1963; Das and Sengupta, 1962; Mukerjea, 1967; Banerjee, 1978; Ram, 1978; Murthy *et al.*, 1979; Muraleedharan, 1981; Das, 1983; Das and Gope, 1983), fenpropathrin, flucythrinate, fluvalinate (Gopal *et al.*, 1987), monocrotophos (Murthy *et al.*, 1979), omethoate, oxydemeton-methyl (Ananthakrishnan, 1962, 1963; Mukerjea, 1967; Banerjee, 1978; Muraleedharan, 1981), spiromesifen (Sharma *et al.*, 2005), permethrin (Sudoi, 1990), phosalone (Ram, 1978; Murthy *et al.*, 1979), pirimiphos-methyl (Murthy *et al.*, 1979), propargite (Sudoi, 1990), sulfur (Jeppson *et al.*, 1975; Muraleedharan *et al.*, 1988b), tetradifon (Ananthakrishnan, 1963; Cranham, 1966;

Mukerjea, 1967; Banerjee, 1978), tridemorph (Chandrasekaran, 1980). Recently, the use of fenazaquin has been proposed (Turner, 2011). Bioacaricides include neem formulations (Subaharan and Regupathy, 2006).

The Genus *Oxycenus* Keifer

Diagnostic characteristics

The gnathosoma is moderately large, with a short recurved apical portion of oral stylet. The prodorsal shield is subtriangular, and the scapular tubercles are set on the rear shield margin, directing the scapular setae (*sc*) towards the rear and slightly diverging. The opisthosomal dorsal annuli are moderately broad, with 2 mid-dorsal ridges extending caudad about 13 dorsal annuli, then ending abruptly at a caudal depression which is situated just before the termen. Laterally, the dorsal annuli project irregularly beyond the main body line, those projecting being approximately the 1st, 4th and 5th, 7th and 9th. Each dorsal annulus covers 3 or 4 ventral annuli laterally (Keifer, 1961a; Amrine *et al.*, 2003).

Economic importance

This genus has three species. The type species, *Oxycenus maxwelli* Keifer, has a worldwide distribution, having followed the diffusion of olive over the world. *O. diospyrosis* Ghosh, Mondal *et* Chakrabarti lives on *Diospyros malanoxylon* Roxb. in India and *O. niloticus* Zaher *et* Abou-Awad has been collected in Egypt. Combined infestation by both *Eriophyes olivi* Zaher *et* Abou-Awad and *O. niloticus* cause malformation of the tender leaves of the developing terminal shoots of olive (Zaher and Abou-Awad, 1980). The species that is best known is *O. maxwelli*.

Oxycenus maxwelli (Keifer) (Fig. 16.64)

Common name

Olive leaf and flower mite.

Diagnostic characteristics

FEMALE. The body is flattened, wedge shaped, 140–160 µm long, 60 µm wide and yellow to amber or orange in colour. The gnathosoma projects down. The prodorsal shield is acuminate over the gnathosoma, with an open network pattern; the scapular tubercles are located on the rear margin; the scapular setae (*sc*) are 11 µm long, projecting to the rear. The opisthosoma is roof like, the dorsal ridge is serrate in side view, with a part of the dorsal annuli prominently lobed laterally, and a posterior notch at the end of the dorsal ridge; there are 19–21 dorsal annuli and about 55 ventral annuli. The opisthosomal setae c_2 are ahead of the genital seta, 15 µm long, set on annulus 7; the setae *d* are 36 µm long, set on annuli 18–19; the setae *e* are 15 µm long, set on annuli 33–35; the setae *f* are 25 µm long, set on annuli 4–5 from the rear; the setae h_2 are 53 µm long; the setae h_1 are absent. The genitalia are 26 µm wide, 16 µm long; the genital coverflap has 18–20 ridges; the seta is 13 µm long (Keifer, 1939a).

MALE. The male body is 125–145 µm long and 50 µm wide. The genitalia are 18 µm wide, 13 µm long; the seta is 12 µm long (Keifer, 1939a).

Geographical distribution

Argentina (Becerra *et al.*, 2007), Australia (Knihinicki and Boczek, 2002), Brazil (Reis *et al.*, 2011), California (USA) (Keifer, 1939a; Sibbett *et al.*, 1976), Chile (Peralta, 1998), Egypt (Attiah, 1970), Iran (Soroosh, 2003; Xue *et al.*, 2009a), Italy (Pegazzano, 1971; Vacante, 1984), Malta (Mifsud *et al.*, 2006), Montenegro (Hrncic *et al.*, 2009), Spain (González *et al.*, 2000) and South Africa (Meyer Smith, 1990a).

Host plants

Oleaceae: olive (*Olea europaea* L.) (Keifer, 1939a).

Bio-ecology

O. maxwelli inhabits leaves, flower stalks and young fruits of olive (*O. europea*). The eriophyoid is principally an upper surface leaf feeder but on occasion numerous mites may be found on the undersurface (Keifer, 1939a). In Tuscany (Italy), *Aculus olearius* Castagnoli, *O. maxwelli* and *Ditrymacus athiasellus* Keifer are common on olive, and the first species predominates. The three eriophyoids have similar biology, and the highest peak of population occurs during flowering, when the mites migrate from the leaves to the flowers and affect the rate of fruit set (Castagnoli and Souliotis, 1982). *O. maxwelli* is very susceptible to low temperatures, and heavy mortality occurs on leaves during periods of cold weather (Russo, 1973).

Natural enemies reported in Australia are some phytoseiid mites (*Euseius elinae* (Schicha)) and ladybirds (*Stethorus* spp.) (Spooner-Hart *et al.*, 2007).

Symptomatology and damage

The mite feeds on both leaf surfaces, the flowers and developing fruits. The leaves become spotted and misshapen, the flowers turn brown, wither and fall prematurely, and the fruits assume a leaden colour and shrivel round the punctures. Premature fall of flowers and spotting of the leaves of olive occur usually in spring (Russo, 1973). In Tuscany (Italy), the eriophyoids *A. olearius*, *O. maxwelli* and *D. athiasellus* cause the same damage. Because the three species live on the same host plant, the amount of damage depends on their common density. At the end of April, a density of 2–3 mites per leaf influences the shape and the consistency of the leaves. During flowering, 20–25 mites per inflorescence cause no economic losses. A significant yield loss requires a density of several hundred mites per inflorescence, which happens much less frequently in Italy than in other Mediterranean countries (Castagnoli and Souliotis, 1982). Jeppson *et al.* (1975) reported that the mite generally caused no damage, except at high population density. In California, as a rule, the mite is not a major problem. It feeds on stem and bud

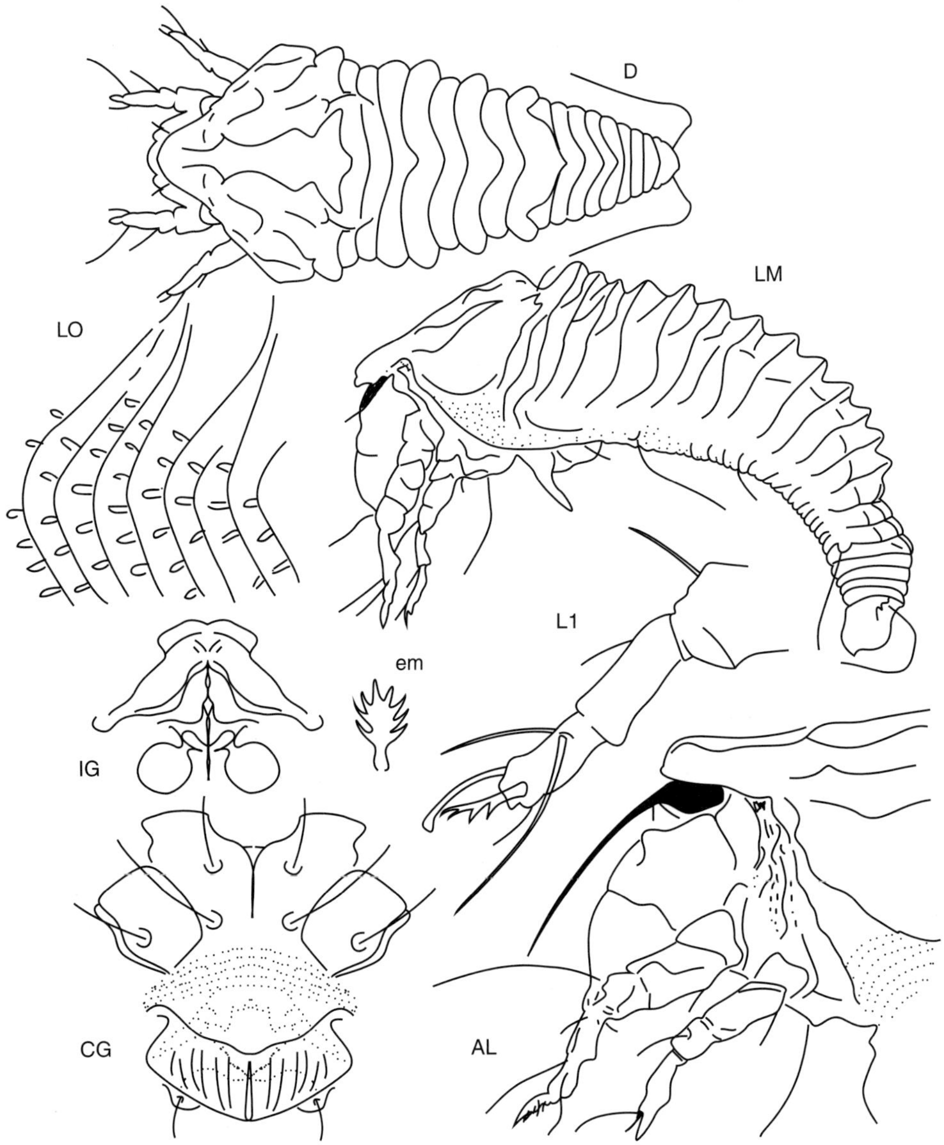

Fig. 16.64. *Oxycenus maxwelli* (Keifer) (from Keifer, 1939a). See Table 16.1 for key.

tissues and on the upper surface of leaves. Symptoms of damage consist of sickle-shaped leaves, dead vegetative buds in the spring, discoloration of flower buds, bud drop, blossom blasting, inflorescence abscission and reduced shoot growth (UC IPM Online, 2009a). Silvering and distortion of young leaves were observed in Brazil, but without serious damage to the plant and without affecting yield (Reis *et al.*, 2011).

Control

The real economic damage caused by the olive mite is not clear, and there is reason to suspect that the damage attributed to the mite is partially connected with feeding by other species. Regardless, it is a monophagous – or closely oligophagous – species, and it is possible that its relationship with the host plant is marked by a mutualistic symbiosis whose contours need to be better clarified. From this point of view, the necessity for control could be secondary, certainly for oil production but probably less so for olive production. If strictly necessary, the common acaricides can be used, but insecticides or minerals oils used against other pests give a sufficient control. In Egypt, sulfur has given good control (Attiah, 1970). In California, treatment effected before August against scale insects assured some control of the olive bud mite. Spray applications of trichlorophon, phosmet, diazinon and methidathion with oils were effective and have been considered as alternatives to the use of parathion and carbaryl (Sibbett *et al.*, 1976). The University of California IPM Pest Management Guidelines (UC IPM Online, 2009a) reported that in California the olive mite is generally not managed in olives unless fruit set and the crop have been below normal for several years. If crop yield has been increasingly poor for several years in a row, shoot tips and developing flower buds should be examined in the spring

for the presence of olive mites, and treatment administered before bloom if large populations are present. Sulfur sprays are acceptable for use on organically certified crops. Wettable sulfur in applied pre-bloom is effective, but application at temperatures above 32.2°C should be avoided. Dusting sulfur is less damaging than wettable sulfur at hot temperatures.

The Genus *Tegonotus* Nalepa

Diagnostic characteristics

The prodorsal shield has the frontal lobe not emarginate and the scapular setae (*sc*) directed up, medially or laterally. The tibiae bear setae (Amrine *et al.*, 2003).

Economic importance

This genus has over 46 vagrant species living in the Holarctic and Palaeotropical regions, mainly recorded from dicotyledonous families. Only a few species have been found on monocotyledons and about three on coniferous hosts (Oldfield, 1996b). *Tegonotus acutilobus* (Nalepa) deforms leaves of dogwoods (*Cornus sanguinea* L. and *C. mas* L.), *T. mangiferae* (Keifer) lives on mango leaves and *T. convolvuli* (ChannaBasavanna) on sweet potato leaves (*Ipomoea batatas* L.). The latter two mites play the major economic role.

Tegonotus mangiferae (Keifer) (Fig. 16.65)

Common name

Mango rust mite.

Diagnostic characteristics

FEMALE. The body is spindle formed, flattened, 160–180 μm long, 60 μm wide and light amber in colour. The gnathosoma projects down. The prodorsal shield is flattened, with an obscure pattern and a diamond-shaped area towards the front centre, with granular ridges running over the frontal lobe, which is also wrinkled in lateral view; the sides of the shield have longitudinal lines; the scapular tubercles are 20 μm apart, located well ahead of the rear margin; the scapular setae (*sc*) are 3 μm long, projecting up. The coxae are granular, and coxae I are broadly contiguous. The opisthosoma has 6 or 8 anterior dorsal annuli with characteristic lateral points; there is a rather acute central ridge; the ventral annuli are microtuberculate; there are about 21 dorsal annuli and 55–65 ventral annuli. The setae c_2 are 18 μm long, set on about ventral annulus 6; the setae *d* are 26 μm long, set on about ventral annulus 20; the setae *e* are 7 μm long, set on ventral annulus 33; the setae *f* are 13 μm long, set on about ventral annulus 5 from the rear; the seta h_1 are absent. The genitalia are 20.5 μm wide, 15 μm long; each side of the genital coverflap has diagonal centrally converging furrows, the base of the coverflap has lines of granules; the seta *3a* is 12 μm long. The empodium is 6-rayed (Keifer, 1946).

MALE. The male body is 130–150 μm long and 50 μm wide (Keifer, 1946).

Geographical distribution

Worldwide on mango cultivation.

Host plants

Anacardiaceae: mango (*Mangifera indica* L.) (Keifer, 1946; Jeppson *et al.*, 1975; Návia and Flechtmann, 2000a).

Bio-ecology

T. mangiferae is a vagrant on leaves of mango and may be misidentified with *Aceria mangiferae* (Hassan) and *Metaculus mangiferae* (Attiah), all living on the same host plant (Keifer, 1946; Jeppson *et al.*, 1975).

Symptomatology and damage

The mite is a rust producer on the undersides of the leaves (Keifer, 1946), and causes no apparent damage (Návia and Flechtmann, 2000a).

Control

T. mangiferae is a rather easily controlled by sulfur applications (Jeppson *et al.*, 1975).

Tegonotus convolvuli (ChannaBasavanna)

Common name

Sweet potato rust mite.

Diagnostic characteristics

FEMALE. The body is spindle formed, flattened, 185–200 μm long and yellowish in colour. The prodorsal shield is semicircular anteriorly, with the frontal lobe apically roughened, rather small and projecting. The prodorsal shield has a few lines; the admedian lines are subparallel and close. The scapular tubercles are close to each other and well ahead of the rear shield margin. The scapular setae (*sc*) are 3 μm long and directed up. There are some curved lines on the coxae; the sternal line is moderately long, forking at the rear. The opisthosomal dorsal annuli have a rather low, longitudinal mid-dorsal ridge, and the lateral lobes end in small lateral teeth. The microtubercles are on the ventral annuli only, and set slightly ahead of the annulus margins. The genitalia are shown as having numerous longitudinal ribs on the coverflap (Jeppson *et al.*, 1975).

Geographical distribution

India (ChannaBasavanna, 1966; Mohanasundaram, 1983, 1991).

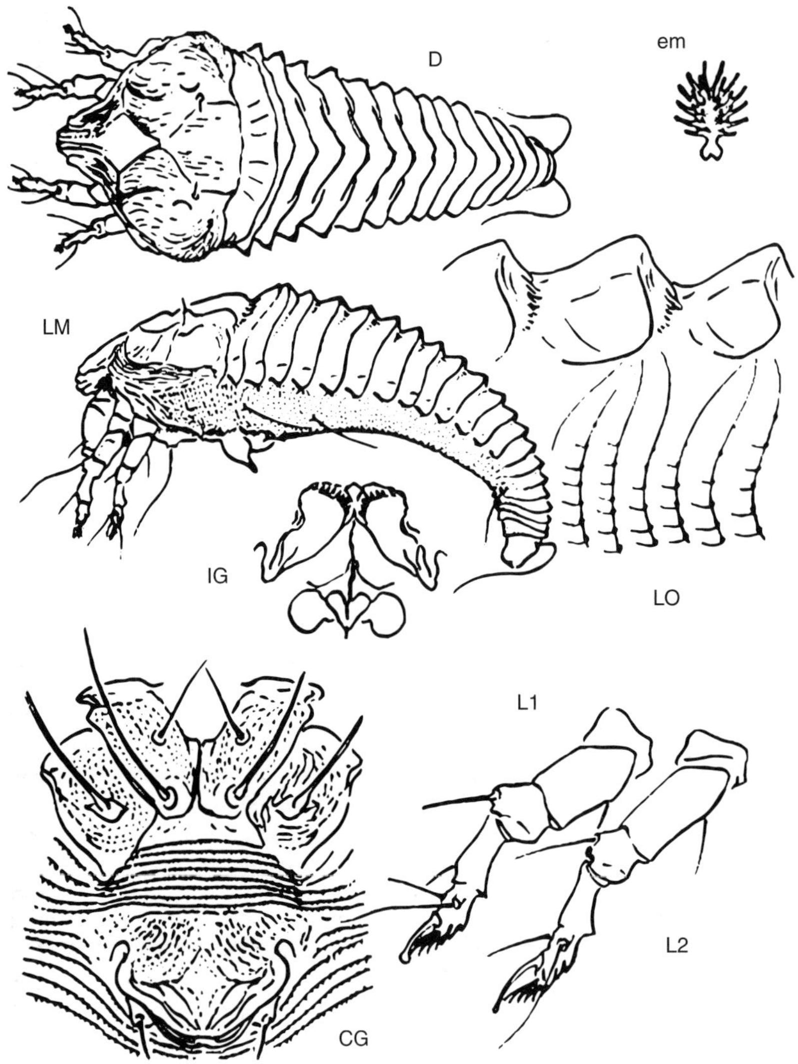

Fig. 16.65. *Tegonotus mangiferae* (Keifer) (Keifer, 1946). See Table 16.1 for key.

Host plants

Convolvulaceae: sweet potato (*Ipomoea batatas* (L.) Lam.) (ChannaBasavanna, 1966; Mohanasundaram, 1983), *I. carnea* Jacq. (Mohanasundaram, 1991), *I. purpurea* (L.) Roth, *I. sepiaria* J. Koenig ex Roxb. (David, 1959).

Bio-ecology

T. convolvuli lives on the upper leaf surfaces of sweet potato in India (ChannaBasavanna, 1966; Mohanasundaram, 1983). During the cropping cycle, it develops on both young and old plants, and may live on some wild Convolvulaceae, such as *I. sepiaria*, *I. purpurea* (David, 1959) and *I. carnea* (Mohanasundaram, 1991).

Symptomatology and damage

The feeding activity of the mite causes the leaves to become bright brown and at high density causes stunting of plant growth. During the cropping cycle, the eriophyoid infests both young and old plants (David, 1959).

Control

Among the host plants, *I. sepiaria* is an alternate host plant for the mite, offering refuge to its populations in the absence of the sweet potato. It has been suggested that this weed plant should be removed from the cultivated environment before planting the crop (David, 1959).

The Genus *Calepitrimerus* Keifer

Diagnostic characteristics

The gnathosoma is short and directed ventrad. The prodorsal shield projects anteriorly over the gnathosoma base; the scapular

setae (*sc*) are present. The legs are 6-jointed, and the empodium is simple; all setae are present. The opisthosoma has all setae, but there are no h_1 setae; it is somewhat flattened dorsoventrally, and elongate wedge shaped in dorsal view; the opisthosomal ventral annuli are not much more numerous than the dorsal annuli; the principal part of the opisthosomal dorsum is longitudinally concave, flanked laterally by a ridge on each side which begins somewhat behind the shield, ending shortly before the anal lobes; the middorsal anterior ½ or ⅔ of the opisthosoma consists of a large sharp carina, ending abruptly to the rear (Keifer, 1938b; Amrine *et al*., 2003).

Economic importance

The genus *Calepitrimerus* includes over 76 species, reported mainly from the Holarctic region, India and China. The most important species is *C. vitis* (Nalepa), which is widespread in the main viticultural areas of the world and causes rust and distortion on vine/grape (*Vitis vinifera* L.) foliage (Jeppson *et al*., 1975). Another rust mite is *C. baileyi* Keifer, reported from California and South Dakota (Keifer, 1938b; Jeppson *et al*., 1975), Iran (Hajizadeh and Hosseini, 2004; Xue *et al*., 2009a), New Zealand (Manson, 1984a), and Poland (Kozłowski, 1979). In the inner regions of California, *C. baileyi* replaces the apple rust mite, *Aculus schlechtendali* (Nalepa), and is probably better adapted to the hot and dry conditions there, where it causes rust on apple (*Malus* × *domestica*) leaves (Jeppson *et al*., 1975). *C. muesbecki* Keifer causes some rust on avocado leaves (*Persea americana* Mill) in Florida (Keifer, 1940b). *C. azadirachtae* ChannaBasavanna lives on leaves of the neem tree (*Azadirachta indica* A. Juss.) (ChannaBasavanna, 1966). Three species are also reported from Cupressaceae (Boczek and Shevchenko, 1996).

Calepitrimerus vitis (Nalepa) (Figs 16.66 and 16.67)

Common name

Grape rust mite.

Diagnostic characteristics

FEMALE. The body is spindle formed, about 150 µm long and yellowish. The prodorsal shield is subtriangular, with a moderately acute and rounded off frontal lobe. The admedian shield lines are complete from the sides of the frontal lobe, running back to and forking between the scapular tubercles. The scapular setae (*sc*) are moderately long and project diagonally up and centrad. The empodium of the protogyne is 4-rayed. The central opisthosomal ridge runs back to about 4/5 of the opisthosomal length, and has a moderate rear slope. The genital coverflap has about 10 longitudinal or radial ribs. The deutogyne lacks microtubercles, except on the ventral distal part of the opisthosoma. The empodium is 5-rayed (Jeppson *et al*., 1975).

MALE. The male is like the female but smaller.

Geographical distribution

Worldwide (Jeppson *et al*., 1975).

Host plants

Vitaceae: grapes (*Vitis vinifera* L.) (Keifer, 1942).

Bio-ecology

The grape rust mite is a pest of vine and is widespread mainly in temperate vineyards. Its biological cycle develops through the summer form or protogyne and overwintering deutogynes (Keifer, 1942). It is a vagrant species and its optimum development occurs in spring and summer at temperatures ranging from 22 to 25°C and at 40–60% RH (Maltshenkova, 1969). In the laboratory at 23°C and RH 70%, the developmental cycle lasts 13 days and under natural conditions it takes 20–30 days (Kozłowski, 1993). The development from larva to adult takes 6 days and adult longevity is about 1 month (Carmona, 1978). Deutogyne forms overwinter in cracks along twigs, at insertions between shoots and branches, or also in hair masses on the underside of outer bud scales, in hairy sites around the growing points (Maltshenkova, 1969; Carmona, 1973; Jeppson *et al*., 1975). Many deutogynes overwinter on the basal buds than on other buds (Carmona, 1978). At bud swelling in spring, the deutogynes appear and feed on buds, depositing eggs that generate males and protogynes. The progeny develop mainly on the basal area of new shoots. In the Mediterranean region, the mite populations develop on the lower surface of young leaves of shoots and clusters from mid-May to October. In early summer, the mites also move into the new buds, within which they increase during the summer months. In Portugal, the deutogynes are observed in late summer and move to overwintering sites in September and October (Carmona, 1973, 1978). The same behaviour occurs in Spain, where the deutogynes emerge from overwintering sites in spring. The protogynes appear in the second half of May when the deutogynes disappear. The mite populations increase during June and at the beginning of summer increase more quickly. The highest density is reached in August or at the beginning of September when the first deutogynes appear. Mite density rapidly declines in September and becomes very low in October. The central buds on canes tend to shelter a large number of females. Large numbers are also found at the cane base and in the first buds (Pérez Moreno and Moraza-Zorrilla, 1998).

In Hungary, the deutogynes emerge from their overwintering sites in spring and the protogynes start to appear at the beginning of May, and gradually replace the deutogynes. The process lasts until the end of May when the deutogynes disappear and reappear at the beginning of August until September. They overwinter in the buds at the cane base with the 2-year-old wood and take refuge continuously in their hibernation shelters until the end of October. Most deutogynes move to the buds during September. There is no connection between the length of the moving period and annual infestation (Gabi and Mészarós, 2000, 2003).

Two main peaks of populations were observed in Portugal, in late spring and midsummer (Carmona, 1978). In Hungary,

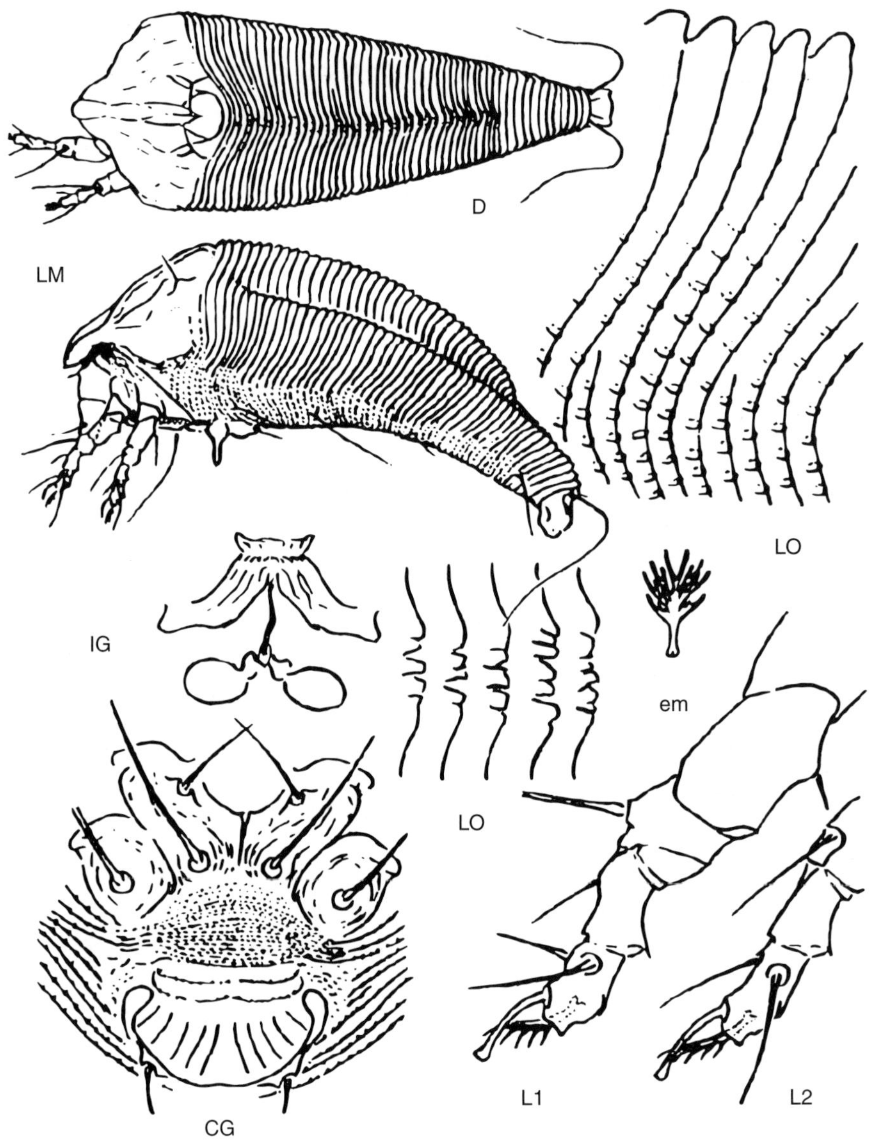

Fig. 16.66. *Calepitrimerus vitis* (Nalepa), protogyne (from Keifer, 1952a). See Table 16.1 for key.

the mite population increases slightly during the summer and a rapid increase occurs in August, with the highest density of mites in the middle of August or at the beginning of September. The mite density rapidly declines in September and is very low in October (Gabi and Mészarós, 2000). According to the environmental conditions, a variable number of generations per year has been reported; thus, 3–4 generations are reported from Switzerland and France (Mathys, 1959; Bonnemaison, 1969, in Carmona, 1978), 5–12 in Moldavia (Maltshenkova, 1969), and 4 in Poland, with a peak in late August (Kozłowski, 1993). In central Hungary, *C. vitis* develops two generations a year. In neglected vineyards, the pest can multiply considerably within 3–4 years (Gabi, 1994).

Correlation and regression analyses of temperature and rainfall data show that outbreaks of grape rust mite depend on warmer conditions in the August of the preceding year (Kast *et al.*, 2004). The passive dispersal of the grape rust mite occurs mainly via air currents, rain and human activity, which also transport other arthropods, including the phytoseiid mite *Typhlodromus pyri* Scheuten (Duffner *et al.*, 2001; Gabi and Mészáros, 2003).

Natural enemies of *C. vitis* include predatory insects and mites. The insects include Thysanoptera, Anthocoridae and Cecidomyiidae (James and Whitney, 1991). The mites belong to the Phytoseiidae, Stigmaeidae, Cunaxidae and Tydeidae. Phytoseiids include *Amblydromella dalii* Rather (Rather, 1984), *Amblyseius aberrans* (Oudemans), *A. andersoni* (Chant) (Duso and de Lillo, 1996), *A. finlandicus* Oudemans (Hluchy *et al.*, 1991), *A. victoriensis* (Womersley) (James and Whitney, 1993), *T. doreenae* (Schicha) (Bernard *et al.* 2005), *T. pyri* (Mathys, 1957a; Hluchy *et al.*, 1991; Pérez Moreno and Moraza, 1997) and *T. talbii* Athias Henriot (Camporese and Duso, 1995). The stigmaeids include *Zetzellia mali* (Ewing), the cunaxids *Haleupalus olivieri* Schruft and the tydeids *Pronematus staerki* Schruft and *Tydeus goetzi* Schruft (Schruft, 1969, 1972b).

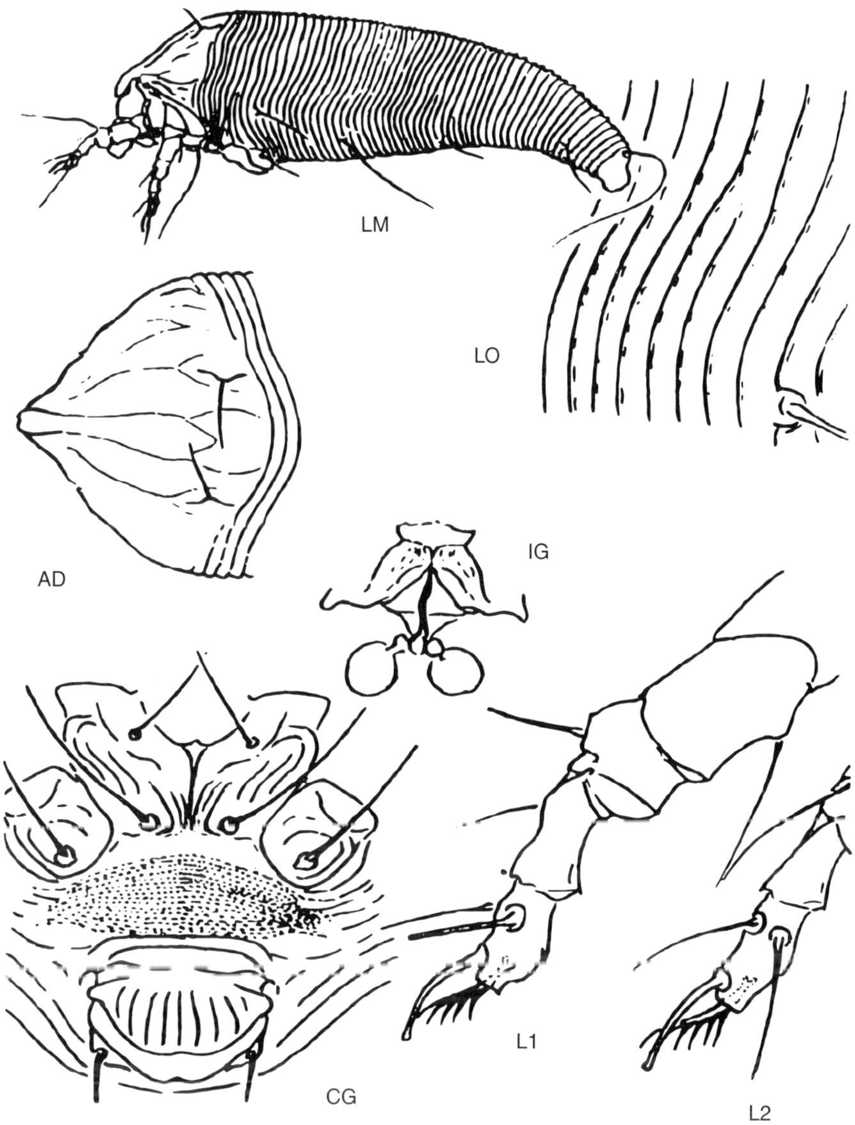

Fig. 16.67. *Calepitrimerus vitis* (Nalepa), deutogyne (from Keifer, 1952a). See Table 16.1 for key.

The phytoseiid *T. pyri* has the best potential for reducing the density of the grape rust mite population (Hluchy, 1993).

Symptomatology and damage

Symptoms and damage affect buds, shoots, leaves and clusters of grapes. Inside the outer scales, infested buds show brown scarification and necrosis. The primary growing point and the buds may be dead before sprouting. The shoots dry and show shortened internodes. The lateral shoots have a stunted development and take the form of witches' brooms. Developing leaves show leaf rolling and ripples, as a result of irregular unfolding of the leaf, with cracks on the margins, chlorotic spots and necrosis, which may conclude with premature drop. According to the grape variety, the expanded leaves are green to reddish brown or yellow. Infestation on clusters before blossoming may induce deformation, accompanied by reduction of size and premature flower drop. Attacks are more frequent in young vineyards than on adult crops and may involve serious damage. Reductions in the number and size of bunches, flower dropping and russeting of the rachis and bunches cause yield loss (Mathys, 1959; Schruft, 1962, 1966; Garcia, 1966; Barnes, 1970; Carmona, 1973, 1978). The problem is common in the northern Mediterranean (Mathys, 1959; Garcia, 1966; Baillod and Guignard, 1986; Kreiter and Planas, 1987; Rota, 1991; Laffi, 1994, 1996; Duso, 1997; de Lillo *et al.*, 2001b).

The symptomatology may be partially confused with the symptoms of the attack of other mites or insects, fungi, viruses or microelement deficiency (Kreiter and Planas, 1987). Leaf distortion and retarded shoot growth have been identified in Australia as damage due to feeding by extremely high populations of overwintered deutogynes of *C. vitis*. This damage was hitherto known in Australia as 'restricted spring growth' (RSG), a syndrome comprising several growth abnormality symptoms,

none with a clearly identified cause or a successful treatment. Chemical treatments against *C. vitis* used to eliminate RSG selectively show that severe leaf distortion is associated with >400 *C. vitis* deutogynes per spur, while >1000 per spur has the added effect of severely retarding shoot growth. A 43.0–47.2% shoot length reduction occurs for 'Cabernet Sauvignon' and 27.1–32.8% for 'Sauvignon Blanc' when 4–6 leaves are separated. Symptoms are most prominent up to 8–9 separated leaves; however, a 24.7–30.4% shoot length reduction is still evident at flowering, and 12.8% around fruit set (Bernard *et al.*, 2005).

Control

The economic importance of this rust mite is not clear, owing to the fact that in some countries, including a wide area of south Mediterranean viticulture, and on some cultivars, such as the Italian 'Nero d'Avola' (not genetically manipulated), mite development does not represent a severe problem and does not often require control. Conversely, in the northern Mediterranean region, central Europe, North America, Australia, South Africa, etc., mite infestation may be enhanced by various conditions (environmental, weather, varieties, phytoiatric protocols, etc.) and control may be needed, as provided by the application of preventive and curative measures.

SAMPLING AND MONITORING. Various means and strategies can be employed to sample and monitor *C. vitis* populations. Some of them have a strictly scientific value and are well adapted to laboratory studies; others are more practical but require more extensive validation in the field.

For the short-term prediction of *C. vitis*, it is possible to use microscopic observations, the washing out and washing off method, trapping with blue tapes, migrating sheets and vaseline streaks, or observations of puncture marks. The washing out and off method, developed by a comparison of the usual soaking methods, enables the overwintering mite and thrip populations to be determined, together with those actually causing damage to the leaves. Reliable farm predictions can be made using a mite screen (Szabo, 1992).

The sticky trap method for monitoring mite migration was suggested by Duffner and Schruft (1998) and Duffner (1999), and has been recently discussed by Bernard *et al.* (2005) and Varner *et al.* (2006). It involves trapping the emerging forms from buds during the spring. Double-sided sticky tape (changed every 3 days) is located above the base of each spur and below the node 1 buds. Transparent plastic sheets allow counting of the mites at 50–70× magnification. Walton *et al.* (2007) monitored eriophyoids in grapevines using tapes winding around each cane, and observing the samples of sticky tapes (which were replaced every 15 days) in the laboratory.

A standardized method for sampling *C. vitis* based on a washing and sieving technique was developed by Pérez Moreno and Moraza-Zorrilla (1998). Leaf samples are immersed in a 70% ethanol solution for 5 min to kill mites. The leaves are then washed under tap water and the mites are collected in a 25 μm sieve, then transferred to a Petri dish using a solution of 70% ethanol and 5% glycerine. After evaporation of the ethanol and glycerine embedding, the mites are counted at 70× magnification. A similar protocol is applied to buds and basal canes to examine deutogyne distribution in woody shoots.

Duffner (1999) collects *C. vitis* from leaves and buds and places the vegetable material in a plastic container covered by a detergent solution in water (0.2% wetting agent). The suspension is shaken strongly for a few minutes and after 2 h it is agitated again. The leaves are then washed under tap water and mites are collected using two sieves. Eriophyoids retained by the finest sieve are stained using 1% methylene blue. Mite counts are made on the sieves, after eliminating water with a vacuum pump and absorbent paper.

Gabi and Mészáros (2001) developed a washing method to count the population of *C. vitis* on grapevine leaves and buds, cut up into small pieces and shaken in tap water. The shaking solution is treated with azur II eosin dye and the leaf suspension is sieved through a two-phase vacuum sieve. Mites are counted at 50× magnification on the sieving dish.

de Lillo *et al.* (2001b) suggested washing the plant material in tap water with added 2% sodium hypochlorite and 0.2% of a dish detergent. The suspension is filtered using sieves and the filtered material is observed in the laboratory. The method seems to be of interest for extracting free-living mites from plant surface, buds or in several plant deformations, but it does not work for strictly gall-making populations, whose specimens live in tight spaces.

Overwintering deutogynes and mite density during the growing season are partially correlated, and a density of 20–25 deutogynes per bud could be associated with damage (Carmona, 1978). Later, it was also found that there was a correlation between the numbers of females overwintering under the bark at the base of shoots and under bud scales. These results suggest sampling of the buds set from the 4th to 8th nodes, and spraying at a density of 1–2 females per bud (Baillod and Guignard, 1986). An economic injury level of 280 mites per leaf has been suggested for late summer (Hluchy and Pospisil, 1992).

Sequential–enumerative and binomial sampling programmes have been developed to evaluate probable mite density. To characterize the spatial distribution of *C. vitis*, the aggregation indexes of Taylor and Iwao were calculated. The best correlation of the population density was with the Taylor index, with $b = 1.79$. To develop an enumerative sampling plan, the sample size based on Taylor's index was calculated with a 25% precision. A binomial sampling scheme was based on the relationship between mite numbers per leaf and the percentage of mite-infested leaves (Pérez Moreno and Moraza, 1996).

CULTURAL PRACTICES. Different cultural practices can hinder mite development. Good results are usually obtained by using planting stock free of vine leaf mites, the correct choice of planting site and vine variety, the distribution of a uniform level of nutrients, the adoption of a wise cultivation method and the regular execution of shoot selection (Szabo, 1992). Taking into account that the greater part of the population of *C. vitis* lives on the young leaves of the growing shoots, the population density may be reduced by pruning the growing shoots in the middle of summer. The predatory mite *Z. mali* lives mainly on the older leaves in the middle and lower level of the shoots, and its density is not affected by this pruning. One of the reasons for the mass multiplication of *C. vitis* is the cultivation of grapevine without any summer pruning (Koleva and Jenser, 1996).

In Slovakia, the control method referred to as 'Culen's trap' has been validated for 6 years, and its use has given both economic and ecological benefits. In the post-bloom period, three reserve leaves, including sheath leaves, are left beyond the anticipated length of the yearling shoots to allow the mites to migrate towards the younger leaves to reproduce, thus setting up a 'trap' for them. Only at the time when the mites set off for a downward migration to hibernate in August will the trap including the leaves simply be cut off (Culen, 2001).

BIOLOGICAL CONTROL. Biological control involves the use of various predatory species, mainly phytoseiid and stigmaeid mites. Some phytoseiid mites, such as *T. pyri*, have been introduced into different countries (Koleva *et al.*, 1996; Pérez Moreno and Moraza, 1997; Voigt *et al.*, 1998). However, the conclusions drawn from this research differ. In central Europe, a population of *T. pyri* (resistant to organophosphorus compounds and sulfur) reduced the population densities of *C. vitis* and *Tetranychus urticae* Koch in vineyards below the economic threshold within two seasons (Hluchy, 1993) and a density of at least two predatory mites per leaf was effective (Wegner Kib, 2003). Conversely, in Spain, the high temperatures and low humidity hinder the development of the phytoseiid (Pérez Moreno and Moraza, 1997); similarly, the high temperatures in Italy caused a decline in the activity of the natural phytoseiids (Varner *et al.*, 2006). *T. pyri* is probably better adapted to vineyards of temperate areas and increases in temperature depress its reproduction and consumption of prey (Sengonca *et al.*, 2003).

In Australian vineyards, *C. vitis* is controlled by natural populations of *A. victoriensis*, *Typhlodromus* sp. (James and Whitney, 1991; Bernard *et al.* 2005) and *T. doreenae* (Bernard *et al.*, 2005). Natural enemies alone cannot carry out adequate control, but their presence is essential in integrated pest management (IPM) protocols. This suggests the need for a strict protection of existing biodiversity and/or the enrichment of the existing complex with the periodic introduction of one or more species. Knowledge of the environment and of the bio-ecology of the species involved helps to determine the colonization pattern, the choice of operating protocols (winter or spring predatory releases) and the role of natural vegetation, which may sometimes have controversial results (Tixier *et al.*, 2006). The introduction into vineyards of *T. pyri* or *Kampimodromus aberrans* (Oudemans) reduced the populations of *C. vitis* to low levels, but *K. aberrans* releases also reduce the populations of phytophagous mites and predators (Duso and Vettorazzo, 1999).

PHYSICAL CONTROL. In this method, dormant infested cuttings are incubated in water baths for up to 60 min at 30–52°C. Treatment at 52°C for 30 min totally eliminates the mite population, but does not reduce the viability of several varieties (excepting the cultivar 'Bluefrankish' and the rootstocks 5BB and 5C). The secondary buds of these varieties stay viable and develop into healthy shoots after incubation for 2 months (Dulinafka and Szendrei, 1995; Dulinafka *et al.*, 1995; Szendrey *et al.*, 1995; Laffi, 1996).

CHEMICAL CONTROL. Chemical control must be effected when the buds swell, and the green tip and shoots reach the length of 5–10 cm (Garcia, 1966). One treatment in spring and another in summer, or twice in summer, are effective. If only a single treatment is possible, it must take place in the summer (Gabi, 1994). Usually, spraying in spring and early summer is sufficient, but a spraying too early in the spring is less effective as every mite female has not yet left its overwintering site (Györffyné *et al.*, 1994). Spraying can be done twice in August. The shoots that are longer in the check that is made in autumn after treatment(s)are also longer in the spring of the following year. Vineyards can be protected from *C. vitis* by using acaricides during the summer before the departure of the deutogyne female to buds for overwintering (Injac *et al.*, 1995). Spray application in the spring at the time of early shoot growth (first two leaves open) is also effective (Antonacci *et al.*, 2000).

The list of chemicals that have been tested in various countries is very long. Active ingredients include acrinatrin, amitraz (Antonacci *et al.*, 2000), bifenthrin (Fleury, 1992; Sitti *et al.*, 2008), bromopropilate (Kreiter and Planas, 1987; Horváth, 1989; Redl and Hiebler, 1991; Gabi, 1994; Györffyné, 1994; Laffi, 1994; Pérez Marín and Ortega Sáenz, 1994; Vanek and Ackermann, 1996; Antonacci *et al.*, 2000; de Lillo *et al.*, 2004, 2007; Sitti *et al.*, 2008), deltamethrin (Györffyné, 1994), demephion (Reinecke *et al.*, 1986), dichlofluanid (de Lillo *et al.*, 2004), dicofol (Reinecke *et al.*, 1986; Fleury, 1992; Laffi, 1994), dimethoate (Reinecke *et al.*, 1986), endosulfan (Redl and Hiebler, 1991; Szabo, 1992; Laffi, 1994; Injac *et al.*, 1995), ethion (Pérez Marín and Ortega Sáenz, 1994), fenazaquin (Györffyné, 1994; Györffyné *et al.*, 1994; Sekulic *et al.*, 1997; Gueguen, 1998, Antonacci *et al.*, 2000; de Lillo *et al.*, 2004, 2007; Sitti *et al.*, 2008), fenbutatin oxide (Kreiter and Planas, 1987; Horváth, 1989; Györffyné, 1994; Scannavini *et al.*, 2008), fenoxycarb (Voigt, 1998), fenpyroximate (Antonacci *et al.*, 2000; de Lillo *et al.*, 2004, 2007), fenpropathrin (Szabo, 1992), methamidophos, methidathion, parathion-methyl (Reinecke *et al.*, 1986), propargite (Kreiter and Planas, 1987; Horváth, 1989; Laffi, 1994; Injac *et al.*, 1995; Vanek and Ackermann, 1996), pyridaphenthion (Pérez Marín and Ortega Sáenz, 1994), quinoxifen (Hernández Riesco *et al.*, 2002), spirodiclofen (de Lillo *et al.*, 2004, 2007), tolylfluanid (de Lillo *et al.*, 2004) and trichlorfon (Reinecke *et al.*, 1986). Mixtures of endosulfan plus mineral oils and diazinon plus mineral oils (Redl and Hiebler, 1991), and hexythiazox plus fenazaquin have also been used (Scannavini *et al.*, 2008; Sitti *et al.*, 2008).

Among the low-impact acaricides, sulfur alone plays an important role (Horváth, 1989; Redl and Hiebler, 1991; Goebel *et al.*, 2001; Wegner Kib, 2003) in the control of *C. vitis*. Others have also emphasized the important role of sulfur (Bernard *et al.*, 2005; Walton *et al.*, 2007) applied both as wettable sulfur (Goebel *et al.*, 2001) and mixed with an oil-based product (Varner *et al.*, 2006). Early sulfur sprayings at the 'wool stage' can also justify the use of higher doses at this period in cases of high infestation (Goebel *et al.*, 2001).

Recently, various kinds of oils have been tested, among which are the mineral or paraffin oils (Hernández Riesco *et al.*, 2002; Martín *et al.*, 2005), fish oils (Hernández Riesco *et al.*, 2002; Martín *et al.*, 2005) and vegetable oils, such as refined rapeseed oil (Hernández Riesco *et al.*, 2002; Wegner Kib, 2003; Martín *et al.*, 2005). There is also a type of mineral oil called polithiol, whose phytotoxicity on growing grapevines restricts its use only to preventive applications, which must be carried out

no later than the wool stage (Scannavini *et al.*, 2008). A water-soluble, modified starch dextrin produced by high-temperature acid hydrolysis of natural starches, referred to as nontox, forms a thick film on its surface that entraps, immobilizes and destroys small insects and mites. The dried film flakes off, and the residues are washed off by rain or watering (Boh *et al.*, 1995).

INTEGRATED PEST MANAGEMENT. A holistic view of the vineyard based on sampling and monitoring of the pest populations and diseases, and associated with a rational cultural management and preventive and curative methods, such as the use of selective chemicals and natural enemies, allow the implementation of IPM strategy (Baggiolini *et al.*, 1976). Here, the introduction into vineyards of predatory mites is a positive choice (Duso and Vettorazzo, 1999).

The Genus *Epitrimerus* Nalepa

Diagnostic characteristics

The body is spindle shaped rather than globose. The opisthosoma has all setae and the middorsal ridge fades simultaneously with the subdorsal or lateral ridges; the shallow subdorsal furrows often cause ridges to be weak. The femora have setae (Amrine *et al.*, 2003; Amrine and Stasny, 2005).

Economic importance

This widespread genus has about 168 species, mostly in the Holarctic region, with a few in the Tropics and China. The greater part of the species inhabit dicotyledons, but there are a few on monocotyledons and less than 20 species on conifers (Oldfield, 1996a). These live on more or less 50% of the Pinaceae and 50% of the Cupressaceae of the genera *Abies*, *Picea*, *Taxus*, *Pseudotsuga*, *Cupressus* and *Juniperus* (Boczek and Shevchenko, 1996). The most important species are the pear rust mite, *Epitrimerus pyri* (Nalepa), which causes rust on leaves and fruits of pear (*Pyrus communis* L.) (Jeppson *et al.*, 1975), the coffee rust mite, *E. congoensis* Keifer, which causes rust on coffee leaves (*Coffea arabica* L.) in Zaire (Keifer, 1960) and the bald cypress rust mite, *E. taxodii* (Keifer), which is a vagrant on needles and leaves and causes characteristic white skins cast as short streaks on the foliage of bald cypress (*Taxodium distichum* Rich.) in California (Keifer, 1939e). Another species, the Douglas fir needle vagrant, *E. pseudotsugae* Keifer, is a vagrant on the needles of Douglas fir (*Pseudotsuga taxifolia* (Lamb.) (=*P. menziesii* (Mirb.) Franco)) along the Pacific coast of North America, and its populations are most numerous when new shoots are growing but the mite has done no observable damage (Keifer, 1946; Jeppson *et al.*, 1975). *E. pungiscus* Keifer has been found as a needle vagrant on blue spruce (*Picea pungens* Engelm.) in Arizona (Keifer, 1962b) and was found later in Finland on different spruce species; in Finland, the mite overwinters as the egg, the nymphs appear in spring and there are 4–5 generations a year. The mite lives at low population density on natural stands of *Picea abies* (L.) H. Karst., and develops at high densities on nursery trees of this conifer and also on exotic members of *Picea* (Löyttyniemi, 1975). Other *Epitrimerus* spp. inhabiting coniferous trees and other host plants in forests of temperate regions of the world are *E. sabinae* Xue *et* Hong, living on needles of *Juniperus* in China (Xue and Hong, 2005), *E. pini* Kuang *et* Li, living on needles of *Pinus* in China (Kuang *et al.*, 1994), *E. gemmicolus* (Nalepa), associated with buds of *Taxus* in Europe, *E. piceae* (Soika) living on *P. abies* in Europe (Soika and Łabanowski, 1999b), *E. subacromius* Rovainen living on silver birch (*Betula pendula* Roth) and downy birch (*B. pubescens* Ehrh.) in Europe (Łabanowski and Soika, 1998) and *E. trilobus* (Nalepa) on elder (*Sambucus nigra* L.) in Palaearctic and Nearctic regions (Łabanowski and Soika, 1995; Soika and Łabanowski, 1999a; de Lillo and Beron, 2004). The species that is most economically important is *E. pyri*.

Epitrimerus pyri (Nalepa) (Figs 16.68–16.69)

Common name

Pear rust mite.

Diagnostic characteristics

FEMALE. The body is flattened and fusiform, 145–160 μm long and yellowish in colour. The prodorsal shield has the frontal lobe rather small and acuminate, rounded distally; its pattern is plain, composed of fine lines or granules, without the median line. The admedian lines are nearly complete, close and subparallel anteriorly, abruptly diverging between the proscapular tubercles. The lateral shield lobes are quite prominent, with finely serrate edges, and these lobes are a definitive feature of the mite. The scapular setae (*sc*) are short and directed upward and inward. The opisthosoma has microtuberculate annuli, and the dorsum bears a very shallow subdorsal longitudinal furrow on each side. The genital coverflap has 2 or 3 basal cross lines, and about 12 ribs. The deutogyne is slightly smaller than the protogyne, with a larger anterior frontal lobe than the protogyne and the prodorsal annuli devoid of tubercles. The empodium is 4-rayed (Jeppson *et al.*, 1975; Amrine and Stasny, 2005).

MALE. The male is similar to the female but smaller (Jeppson *et al.*, 1975).

Geographical distribution

Worldwide.

Host plants

Rosaceae: quince (*Cydonia oblonga* Mill.) (Easterbrook, 1978; Bergh and Weiss, 1993), *Pyronia* sp. (Easterbrook, 1978), *Pyrus calleryana* Decne (Bergh and Weiss, 1993), pear (*P. communis* L.) (Jeppson *et al.*, 1975; Easterbrook, 1978).

Bio-ecology

E. pyri is a worldwide pest of pear. It is a leaf vagrant, overwintering as deutogyne forms under the bark, in crevices and

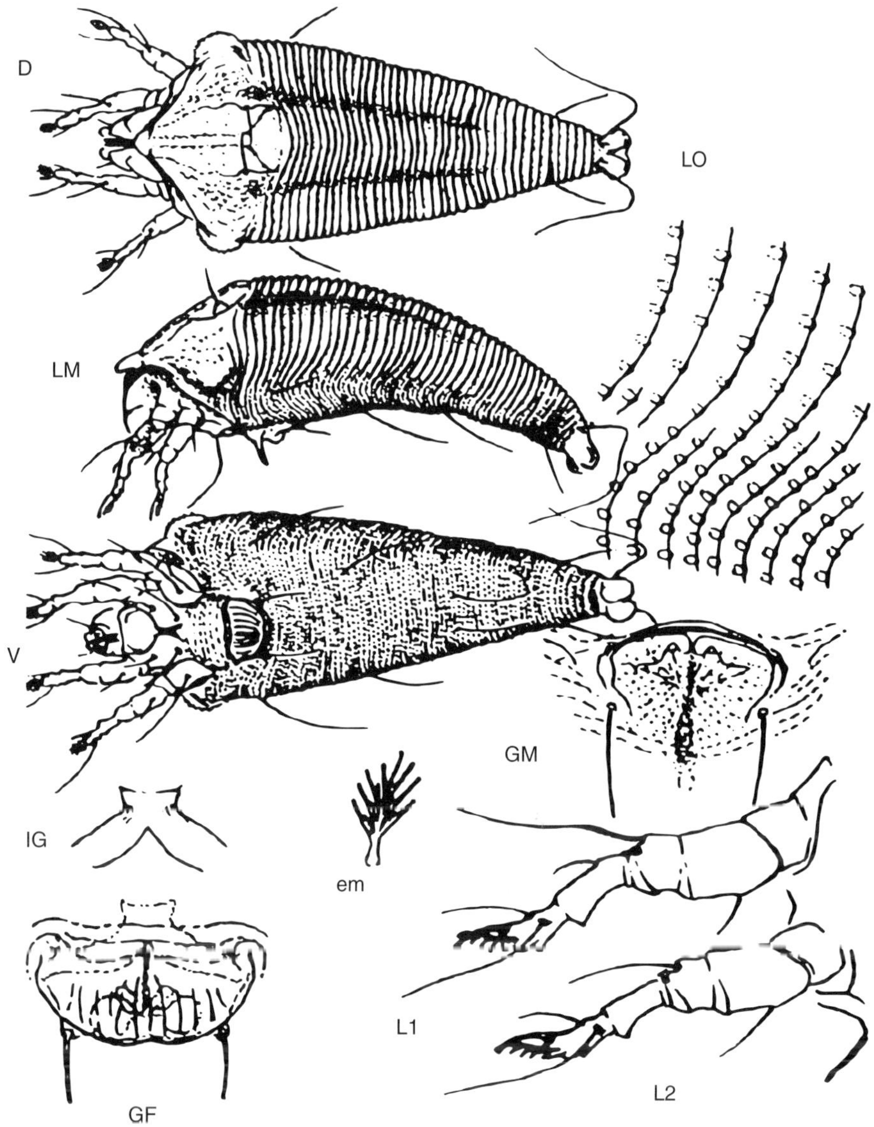

Fig. 16.68. *Epitrimerus pyri* (Nalepa), protogyne (from Keifer, 1952a). See Table 16.1 for key.

under the scales of dormant buds (Westigard, 1975; Easterbrook, 1978; Herbert, 1979). Most of the population is found on spur systems or on 2- to 3-year-old branch wood and few on 1-year-old shoots. Emergence from these sites and the invasion of fruit buds occur mainly between the green cluster stage (stage D) and full bloom (stage F2) (Easterbrook, 1996). Initially, the mite lives on the scars derived from the abscission of the outer scales of the fruit bud, where it feeds and lays eggs; afterwards, it moves on to the leaves, flower receptacles and young fruitlets (Easterbrook, 1978). In England, the density of the population living on flowers and fruits in this period is similar to that living on leaves in the period until late May and early June. Later, mite density on the fruits decreases compared with the leaves (Easterbrook, 1978; Baillod *et al.*, 1991). In Yugoslavia, deutogynes emerge at the beginning of flowering and lay eggs at the base of the rosettes; the protogynes settle on the fruits and feed below the calyx. Higher mite densities occur on the fruits in May–June. Later, these forms follow the growth of the annual shoots, causing defoliation in young orchards. In varieties ripening in August and later, russeting of the fruits occurs (Injac *et al.*, 1988).

Temperature and exposure to light determine the rate of emergence of deutogynes from overwintering sites, and not the stage of bud development. The pest moves up the branches in response to light to find feeding sites in opening buds (Bergh, 1992). As already noted, in England, most of the mites are found to migrate from their hibernation sites under loose bark or in small, permanently dormant buds on spurs to fruit buds between the green cluster and petal fall stages. Peak numbers of mites are observed in mid-June on the fruits and in early August on the leaves, with the return to hibernation sites beginning in July (Easterbrook, 1978). In Canada, overwintering is as deutogynes

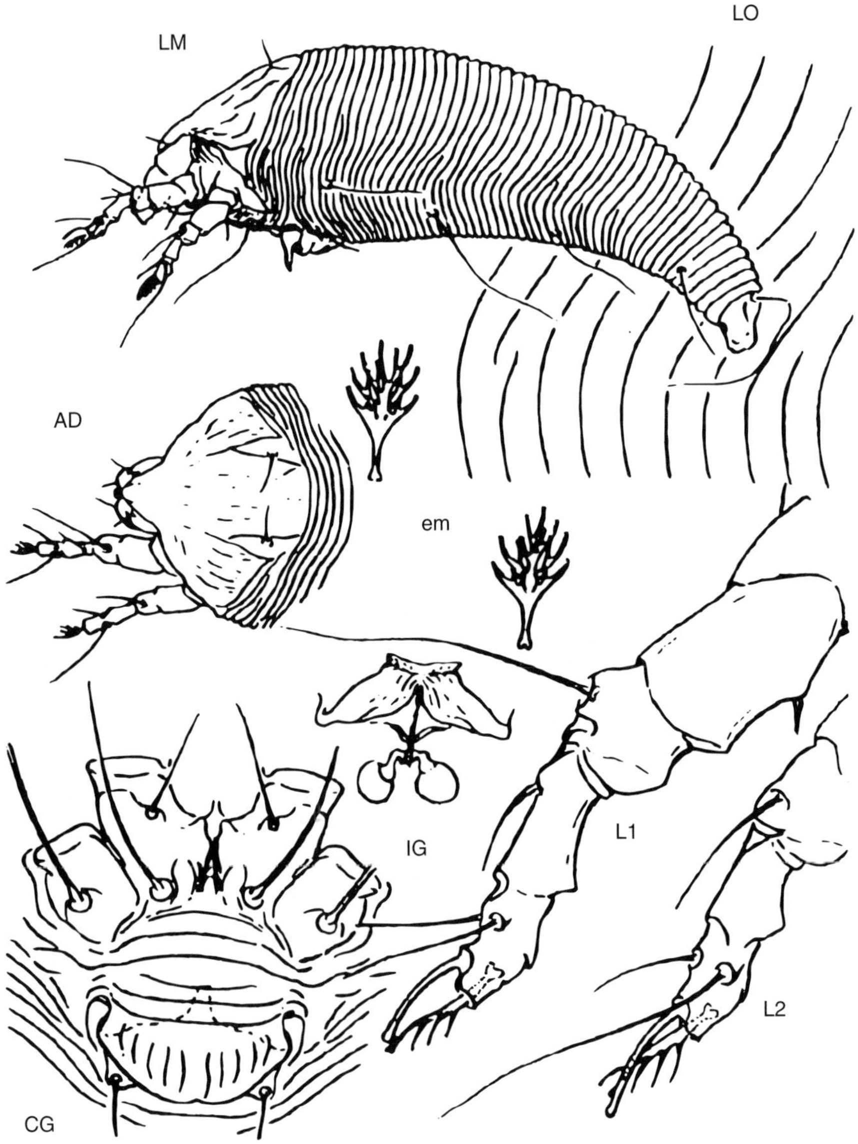

Fig. 16.69. *Epitrimerus pyri* (Nalepa), deutogyne (from Keifer, 1952a). See Table 16.1 for key.

in a semi-active state in old leaf scar crevices and under bud chips of the leaf clusters. The deutogynes become active within the bud in mid-April, as soon as the weather is warm, and there are three generations in the growing season (as indicated by peaks in the numbers of eggs in May, June and July). The first-generation eggs are laid within the swelling bud. The mites feed on the ventral surface of the leaves, causing bronzing, and on the calyx end of the fruit, resulting in russeting of the epidermal tissue. Deutogynes of the third generation begin to enter hibernation by mid-August (Herbert, 1979). In Bulgaria, *E. pyri* overwinters in the buds and in cracks in the bark, is very resistant to adverse winter weather conditions and resumes activity in late March or early April during migration from their winter quarters to the leaves; populations therefore increase very slowly and reach a peak density of 120–150 mites per leaf only in August (Nachev, 1976). Higher populations of overwintering *E. pyri* occur on 2-year-old shoots, mainly in folds, growth scars and under the first scale of buds on the spurs (Badowska-Czubik *et al.*, 2002).

At 10, 15 and 20°C, the mean longevity and pre-oviposition period of deutogynes decrease with increasing temperature. The average rate of oviposition increases with temperature, but lifetime fecundity does not differ among temperatures. At 20°C, the protogynes begin ovipositing sooner than deutogynes and lay more eggs at a faster rate, but the average longevity of both is the same. There is a significant linear relationship between temperature and the developmental rates of eggs, nymphs and generations. Threshold temperatures for development of eggs, nymphs and generations are 2.2, 3.3 and 3.8°C, respectively. Degree-day (± sem) requirements for median development of eggs, nymphs and generations are 93.5 ± 2.5, 78.7 ± 3.4 and 144.9 ± 3.6 above their respective threshold temperatures (Bergh, 1994). At 20°C, the egg to egg development time is 11.0 days, the mean oviposition period is 20.8 days, the fecundity

is 60.4 eggs and the mean ovipositional rate is 2.9 eggs a day (Bergh, 1994). Easterbrook (1978) gave an egg to egg development time of about 11.2 days. There is a sexual independence of male and females (Oldfield, 1989).

Increases of temperature after blooming can induce a rapid development of populations. The density of mites on fruits shows a peak in mid-June, both in England (Easterbrook, 1978) and Oregon (Westigard, 1975). The higher development of populations on leaves starts in late June or July (Easterbrook, 1978; Herbert, 1979), with a population peak in late July or early August also of over 1000 mites per leaf.

In England, the mites are dispersed by wind but are found only on quince, *Pyronia* and species of *Pyrus* (Easterbrook, 1978). In Bulgaria, a large number of mites is carried away by wind or rain (Nachev, 1976).

Symptomatology and damage

The pear rust mite infests both pear leaves and fruit in spring. The damage is due to feeding by the offspring of deutogynes and their generations, which causes browning of undersides of the leaves and russeting on fruits. High levels of attack produce severe damage, while a lower population density usually causes damages only on the calyx end of the fruit (Jeppson *et al.*, 1975).

In southern Oregon, the mite reaches the fruit during the bloom period, and remains there during the entire growing season. The proportion of the population at this site decreases during the summer, whereas that on fruit cluster leaf tissue increases. The population density is higher throughout the entire season on fruits of the upper parts of the tree than on those of the lower parts. Although a density of five or more mites per fruit in June or July may cause economic damage, the data indicate that higher densities could be tolerated earlier in the season. Fruit varieties with a clear skin ('Doyenne du Comice', 'Bartlett') suffer the most injury (Westigard, 1975). Damage of over 5–10% of the fruit surface may involve the downgrading of fruits or prevent their marketing as fresh fruit, with variable levels of damage according to variety and the environment. In Canada, downgrades of up to 50% of the fruit are reported from fresh marketing to processing (Herbert, 1979). Also in Canada, serious difficulties result from the hard flesh that is found beneath the skin of russeted pears (Morgan and Arrand, 1969). In Switzerland, the cultivar 'Williams' Bon Chrétien' is severely damaged by the pear rust mite, with 90% of fruits downgraded (Baillod *et al.*, 1991).

First browning of the leaf tissue occurs along the midrib near the base of the leaf and extends gradually with increase of the mite population on the ventral surface. The symptoms involve all leaves, including the spur cluster leaves. These leaves may fall early and in some cases (2- to 4-year-old pear orchard) by up to 43% (Injac *et al.*, 1988).

On 1-year-old pear trees in pots, a population over 2000 mites per leaf reduces the length of new shoots by 8% and their dry weight by 14% (Easterbrook, 1978), or causes the cessation of terminal growth (Westigard, 1975). In Poland, the highest density of the pest has been found on 'd'Anjou' and 'Conference' varieties, and the lowest on 'Hnidzik', 'Albigowianka' and 'Carola' (Badowska-Czubik *et al.*, 2002).

Control

Control is achieved by the sampling and monitoring of populations and the use of different methods, either alone or integrated.

SAMPLING AND MONITORING. A method derived from that of Oldfield (1969) to study the movement of the vagrant eriophyoid mites was developed by Bergh (1992), which observed that mite emergence was stimulated by light. Active mites are positively phototactic but are not affected by bud development, and the rate of mite emergence at 15°C is significantly less than that at 20°C.

Bergh and Judd (1993) found a base threshold temperature for emergence of 6.2°C. The relationship predicts that median (50%) emergence of mites requires 62 ± 1 degree-days (± SEM) above the base temperature. This degree-day model has been validated using field phenological data collected at one and four sites, respectively. Using a 6.0°C threshold temperature for emergence, and daily minimum–maximum air temperatures beginning on 1 March, the model accurately (± 2 days) predicts the 50% emergence point.

PLANT RESISTANCE. The ovipositional response of deutogynes of *E. pyri* and the rate of development of nymphs differ according to the host plant. Among 14 *Pyrus* hosts, leaves from 'Clapp's Favorite' cultivar of *P. communis* exhibited the strongest egg-laying response and the fastest rate of nymphal development, whereas egg laying was least and nymphal development slowest on leaves from *P. calleryana* Decne. Among three non-*Pyrus* hosts, *E. pyri* oviposits and completes nymphal development on leaves of apple and quince, but will not deposit eggs on leaves of apricot. Antixenosis appears to be the primary mechanism of resistance, although antibiosis may be operating to a lesser extent (Bergh and Weiss, 1993).

CULTURAL PRACTICES. The effects of over-tree irrigation on the population density of and damage by *E. pyri* on pear trees were evaluated in southern Oregon. Damage to fruit by *E. pyri* in under-tree irrigated plots was twice that measured in over-tree irrigated plots (Westigard *et al.*, 1979).

CHEMICAL CONTROL. Spray applications of acaricides against *E. pyri* should be carried out early in the growing season (Nachev, 1976). The numbers of mites living on fruits and the amount of russeted fruits are related. A mite density averaging or exceeding 5–10 mites per fruit may result in over 5% of fruits russeted (Easterbrook, 1978; Herbert, 1979). Summer treatments against deutogynes encounter more difficulties because of their resistance and their protection in dry bark crevices (Jeppson *et al.*, 1975). The timing of spring spray applications is critical and could be completed by the cluster-bud stage. The need to prevent leaf and fruit damage may suggest other interventions.

Various chemicals have been tested for cluster-bud treatment. These include abamectin (Campbell and Easterbrook, 1985; Undurraga and Dybas, 1988), amitraz (Campbell and Easterbrook, 1985), benzoximate (Downing and Moilliet, 1975), binapacryl (Westigard and Berry, 1964), carbaryl (Westigard and Berry, 1964; Jeppson *et al.*, 1975), chinomethionate (Downing and

Moilliet, 1969), chlorobenzilate (Joos *et al.*, 1974), cyhexatin (Bulla, 1975; Downing and Moilliet, 1975), dichlofluanid (Laffi, 1999), dicofol (Westigard and Berry, 1964; Joos *et al.*, 1974; Jeppson *et al.*, 1975; Laffi, 1983; Injac *et al.*, 1988), DNOC (dinitro-*o*-cresol; now banned) (Injac *et al.*, 1988), dofenapyn (Easterbrook and Campbell, 1986), endosulfan (Downing and Moilliet, 1974; Joos *et al.*, 1974; Jeppson *et al.*, 1975), ethion (Downing and Moilliet, 1974), fenvalerate (Riedl and Hoying, 1980), maneb (Westigard, 1969), mineral oil (Joos *et al.*, 1974), pirimiphos-methyl (Easterbrook, 1984b), polysulfide (Bulla, 1975), pyrethroids (Campbell and Easterbrook, 1985), tolylfluanid (Laffi, 1999), vamidothion (Easterbrook, 1984a) and zineb (Westigard, 1969). The use of various mixtures, such as mineral oils plus abamectin (Campbell and Easterbrook, 1985), diazinon (Joos *et al.*, 1974; Bulla, 1975), endosulfan (Joos *et al.*, 1974; Bulla, 1975) or ethion (Downing and Moilliet, 1974; Bulla, 1975) has also been reported.

Chemicals with low impact include colloidal sulfur (Shvanderov, 1984) and flowable or lime sulfur (Downing and Moilliet, 1974; Joos *et al.*, 1974; Jeppson *et al.*, 1975). Plant growth regulators, such as paclobutrazol and chlormequat chloride, have been suggested (Campbell *et al.*, 1989).

Particle films (HPF technology) for the protection of plants consisting of microscopic mineral particles have been proposed. They are sprayed on to plants as a liquid, which evaporates leaving a film on the plant surface. Tests on apple and pear trees have shown that HPF is effective against a broad range of insect pests and some diseases (Stanley *et al.*, 1998; Puterka *et al.*, 2000).

INTEGRATED REST MANAGEMENT. IPM of the pear rust mite must take place within a holistic programme of control of various pests and diseases of the orchard, primarily based on the use of selective chemicals. The economic advantages of the method have already been extensively evaluated in North America.

Selective chemical control programmes have been tested in Oregon pear orchards with the goal of maximizing the effect of biological control agents for the suppression of pear pests. The programmes used pre-bloom oil sprays and other selective chemicals for various insect pests and, where needed, half the usual rate of cyhexatin for the control of *E. pyri* and other mites. Beneficial increases in predators occurred in orchards, gave commercially acceptable control of the psyllid and lowered acaricide use by 75%. The costs of control were about US$300 ha^{-1} in the selective programmes and about US$700 ha^{-1} in the standard programmes. The density and period of colonization by pests and predators varied from site to site and appeared to depend on previous treatment history within the study area and on the nature and management of the surrounding vegetation (Westigard *et al.*, 1986).

Recently, pear growers in California made a rapid transition from organophosphate insecticides to a mating disruption-based codling moth (*Cydia pomonella* L.) control programme. The reduction in use of broad-spectrum insecticides resulted in a decrease of pest populations with a decrease in pesticide use for their control. However, there have been sporadic increases of secondary pests such as the pear rust mite. Savings were minimal during the first 2 years of transitioning into the mating disruption programme, but became significant beginning in the third year. From the third year on, savings were US$247–511 ha^{-1} annually. Fluctuations in savings were due to variability in insecticide applications for psyllids, mites and secondary pests (Varela and Elkins, 2008).

The Genus *Phyllocoptruta* Keifer

Diagnostic characteristics

The prodorsal shield protrudes over the gnathosoma base, with a distinct frontal lobe; the scapular tubercles are well ahead of the rear margin. The opisthosoma is divided into dorsal and ventral annuli by the lateral doubling of the annuli of the opisthosoma, which is broadly flat or concave, this area tapering and disappearing caudad, and flanked on each side by a subdorsal longitudinal ridge; the dorsal half-annuli are not microtuberculate, and the ventral half-annuli are microtuberculate. The setae are as usual for Eriophyidae. The female coverflap entirely covers the genitalia. The legs are 6-jointed, the empodium simple, and coxae I contiguous (Keifer, 1938a; Amrine *et al.*, 2003).

Economic importance

This genus has about 34 species, among which are the citrus rust mite, *Phyllocoptruta oleivora* (Ashmead), which is distributed worldwide and causes severe damage to fruits and leaves of *Citrus* spp. (Jeppson *et al.*, 1975; Vacante, 2010), the banana rust mite, *P. musae* Keifer, which produces spotting on bananas (*Musa* spp.) in Queensland (Australia) (Keifer 1955a) and India (Mohanasundaram *et al.*, 1989), and *P. sakimurae* Keifer, which lives in shell grooves on pineapple fruit (*Ananas sativus* Schult.) in Hawaii (Keifer, 1966a); this mite is also reported on *Aralia* in the USA (Návia *et al.*, 2010a). Only the citrus rust mite causes severe damage.

Phyllocoptruta oleivora (Ashmead) (Fig. 16.70)

Common name

Citrus rust mite.

Diagnostic characteristics

FEMALE. The body is fusiform, flattened, 150–165 µm long, 53 µm wide and yellow to straw coloured. The prodorsal shield has a moderate frontal lobe, with a transverse furrow across the front edge. The admedian lines rise from each side of the frontal lobe, and curve out to the area just past the lobe base; from there, they curve inward and meet a slight cross line at ⅓, then curve slightly outward past the inner side of the scapular tubercles at ⅔, and end in transverse arcs ahead of the rear shield margin. The scapular tubercles are moderately large, well ahead of the rear shield margin, 23 µm apart; the scapular setae (*sc*) are 9 µm long and project dorsoanteriorly. The opisthosoma has a broad longitudinal trough delimited by 2 longitudinal ridges, with about 31 dorsal annuli and 58 ventral annuli. Microtubercles are

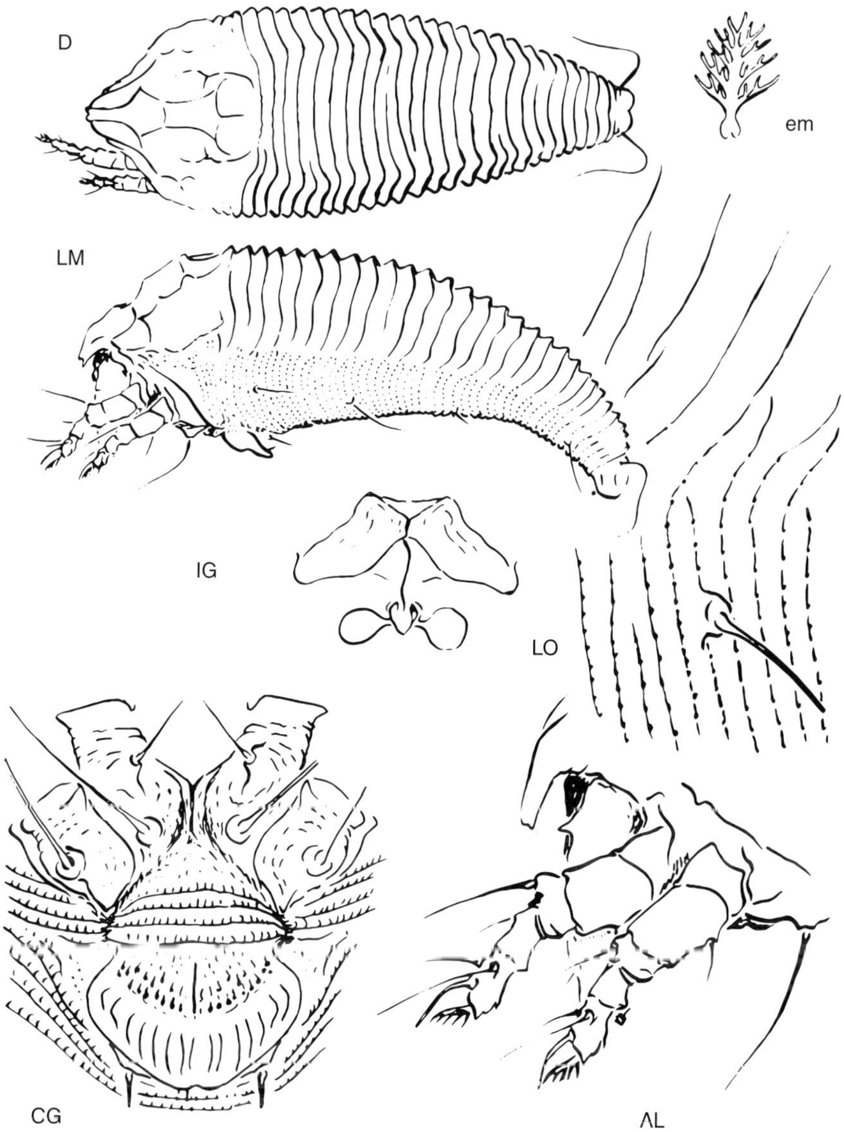

Fig. 16.70. *Phyllocoptruta oleivora* (Ashmead) (from Keifer, 1938a). See Table 16.1 for key.

present only on the ventral annuli, and each dorsal annulus covers 2 or 3 ventral annuli. The setae c_2 are 25 µm long, set on about ventral annulus 5; the setae *d* are 35 µm long, set on ventral annulus 17; the setae *e* are 8 µm long, set on about ventral annulus 33; the setae *f* are about 15 µm long, set on ventral annulus 5 from the rear; the setae h_1 are very short, and the setae h_2 are nearly 40 µm long. The genital coverflap is granular at the base, with a median longitudinal line, and has 14–16 ribs apically. The empodia are 5-rayed (Keifer, 1938a; Jeppson *et al.*, 1975).

MALE. The body is fusiform, 135 µm long, 54 µm wide, with the same external features as the female except for the genital structures (Keifer, 1938a; Jeppson *et al.*, 1975).

Geographical distribution

Worldwide (Jeppson *et al.*, 1975).

Host plants

Rutaceae: *Citrus* spp. (Jeppson *et al.*, 1975; Vacante, 2010).

Bio-ecology

The citrus rust mite is a severe pest of citrus in many areas of the world, and damages fruit quality by its feeding activities (Jeppson *et al.*, 1975). It is a vagrant species, living on the leaves and fruits of all citrus species and varieties and reproducing by arrhenotokous parthenogenesis (Swirski and Amitai, 1958). The mite is ovoviviparous (Hall, 1967). The male deposits an average of 16 spermatophores a day (Oldfield *et al.*, 1970). The female lays eggs in the depressions of the pericarp of the fruits or leaves. At 32°C, the embryonic development is completed in 3 days and the nymphal stage I in 1.8 days; at 22°C, the egg develops in 5.5 days, nymphal stage I in 4.3 days and nymphal stage II in 6.4 days. In the summer, a generation develops in

7–10 days and in winter in about 14 days. The female lives less than 20 days and lays a total of 20–30 eggs (Yothers and Mason, 1930; Ebeling, 1959). At constant temperatures between 14 and 31°C, the development threshold is 29°C (Allen *et al.*, 1995). At 32°C and with 88% RH, the lifespan of the female is 3.2 ± 0.673 days, with an intrinsic rate of (population) increase (r_m) of 0.350 individuals per female daily. The thermal threshold of development is 11°C and oviposition ends at 10°C. The mortality from egg to adult is 53.06% at 17°C and decreases to 25.81% and 26% at 30 and 32°C. The optimum temperature for development is between 30 and 32°C (Ebrahim, 2000). Development from egg to adult requires 330.07 degree-days with a temperature of between 23 and 32°C and a photoperiod of 12/12 h light/dark (Li *et al.*, 1989).

The citrus rust mite is active in all seasons and can develop a large number of generations during the year, for instance approximately 40 generations have been observed in Suriname (Van Brussel, 1975) and in Israel up to 28 or more (Swirski and Amitai, 1958). Although the mite develops well with high values of RH, and particularly after rain (Pratt, 1957; Dean, 1959a), in South Africa high values of RH induce an increase in the mortality of nymphs and a decrease of oviposition (Van Brussel, 1975). In Florida, the mite is active throughout the year and populations increase in the spring and during the summer (Griffiths and Thompson, 1957); the population density increases in May–July and then declines in late August, but can increase again in late October or early November (Childers *et al.*, 2007).

The behaviour of *P. oleivora* on exposure to direct sunlight and in different parts of the canopy is unclear. Indeed, Van Brussel (1975) reports that the mite does not tolerate direct sunlight and usually prefers the lower surface of leaves and areas in the shade of the fruits; on young plants, it can be found on the lower surface of the leaves and on the upper surface of trees older than 1 year. McCoy (1979) found that populations migrated to the new shoots and settled on the lower surface of the leaves before moving towards the upper surface to reproduce and then on to the fruits. Allen and McCoy (1979) analysed the distribution of populations on individual fruits and over the canopy and reported that the citrus rust mite refused direct exposure to sunlight. Populations are generally more abundant on fruits and leaves on the outer margins of the tree canopy. The northern bottom of the tree canopy is preferred, and the least favourable conditions for population increase occur in the southern top of the tree canopy (Childers *et al.*, 2007). In Israel, it has been observed that the mite prefers the lower surface of the leaves and the outside of the canopy rather than the internal part, and prefers the fruits over the leaves; it also develops better in citrus groves with a high density of transplantation and attacks the fruits on high branches and not on low branches and lemons, in spite of the presence of other species and varieties of citrus (Swirski, 1962). At night and on cloudy days, pest populations are found on both leaf surfaces; they tend to disperse across the foliage and are easily observed during the day on sunny green surfaces, showing the highest levels of attack on the outside of the canopy (Jeppson *et al.*, 1975).

Mite dispersal is facilitated by the wind, with shifts of up to 135 m away, and with higher peaks between mid-September and mid-October in Florida (Bergh and McCoy, 1997).

Several natural enemies develop on the citrus rust mite, including fungi, molluscs, insects and mites. The insects include a number of coniopterigids, cecidomyiids, thrips and coccinellids (Table 16.3), and the mites include a number of stigmaeids, cheyletids and tydeids, and several phytoseiids (Table 16.4).

Symptomatology and damage

The citrus rust mite is a serious problem in all citrus areas of the world, with the exception of a number of regions such as New South Wales (Australia), where it is not always harmful to

Table 16.3. *Phyllocoptruta oleivora* (Ashmead). Natural enemies (fungi, molluscs and insects).

Families and species	Reference(s)
FUNGI	
Clavicipitaceae	
Beauveria bassiana (Balsamo) Vuillemin	Alves *et al.*, 2005a
Hirsutella kirchneri (Rostrup) Minter, Brady *et* Hall	Cabrera and Dominguez, 1987b
Hirsutella nodulosa Petch	Cabrera and Dominguez, 1987b
Hirsutella thompsonii Fisher	Spears and Yothers, 1924; McCoy and Kanavel, 1969; McCoy, 1996a
Exobasidiomycetidae	
Acaromyces ingoldii Boekhout *et al.*	Paz *et al.*, 2007a
Meira argovae Boekhout *et al.*	Paz *et al.*, 2007a
Meira geulakonigii Boekhout *et al.*	Sztejnberg *et al.*, 2004; Paz *et al.*, 2007a
Mycosphaerellaceae	
Mycosphaerella citri Whiteside	Hernández and Escobio, 1988
MOLLUSCS	
Orthalicidae	
Drymaeus dormani (Binney)	Bledsoe and Minnick, 1982
INSECTS	
Cecidomyiidae	
Cecidomyiids	Muma *et al.*, 1975
Feltiella sp.	Villanueva *et al.*, 2006
Lestodiplosis sp.	Villanueva *et al.*, 2006
Coccinellidae	
Pentilia castanea Mulsant	Van Brussel, 1975
Coniopterygidae	
Coniopteryx westwoodi Fitch	Muma, 1967
Semidalis vicina Hagen	Muma, 1967
Psychidae	
Cryptothelea gloverii (Packard)	Villanueva *et al.*, 2005
Thripidae	
Leptothrips mali (Fitch)	Griffiths and Thompson, 1957

Table 16.4. *Phyllocoptruta oleivora* (Ashmead). Natural enemies (mites).

Families and species	References
Cheyletidae	
Cheletogenes ornatus (Canestrini *et* Fanzago)	Rezk and Gadelhak, 1996
Hemicheyletia wellsi (Baker)	Feres *et al.*, 2005
Phytoseiidae	
Amblyseius herbicolus (Chant)	Argov, 1993; Matioli *et al.*, 1998; Argov *et al.*, 2002
Euseius sp.	Matioli *et al.*, 1998
Euseius citri (van der Merwe *et* Ryke)	Schwartz and Meyer Smith, 1998
Euseius citrifolius Denmark *et* Muma	Gravena *et al.*, 1993
Euseius elinae (Schicha)	Argov, 1993; Argov *et al.*, 2002
Euseius mesembrinus (Dean)	Abou-Setta and Childers, 1989; Flores *et al.*, 1996
Euseius stipulatus (Athias Henriot)	McMurtry *et al.*, 1992; Argov *et al.*, 2002
Euseius victoriensis (Womersley)	Smith and Papacek, 1991; Argov, 1993; Argov *et al.*, 2002
Galendromus helveolus (Chant)	Caceres and Childers, 1991
Iphiseiodes zuluagai Denmark *et* Muma	Gravena *et al.*, 1993; Matioli *et al.*, 1998
Iphiseius degenerans (Berlese)	Harpaz and Rosen, 1971; Palevsky *et al.*, 2003
Iphiseoides quadripilis (Banks)	Villanueva and Childers, 2007
Neoseiulus cucumeris (Oudemans)	Zhang *et al.*, 2003a
Typhlodromalus peregrinus (Muma)	Peña, 1992
Typhlodromips swirskii (Athias Henriot)	Harpaz and Rosen, 1971; Palevsky *et al.*, 2003
Typhlodromus athiasae Porath *et* Swirski	Harpaz and Rosen, 1971
Typhlodromus rickeri Chant	Argov *et al.*, 2002
Stigmaeidae	
Agistemus cyprius Gonzalez	Palevsky *et al.*, 2003
Agistemus exsertus Gonzalez-Rodriguez	Rezk and Gadelhak, 1996
Agistemus floridanus Gonzalez	Muma and Selhime, 1971
Tydeidae	
Tydeus sp.	Chen *et al.*, 1988

orange (Beattie *et al.*, 1991), or Italy, where it has been reported (de Lillo and Beron, 2004) but up to now is not severely injurious.

The mite feeds on leaves and fruits of all citrus species, but prefers lemon. On the upper leaf surface, the damage is to the epidermal cells and appears as slightly brownish patches to the back. In severe cases, the cuticle loses its lustre, takes on a bronze-like colour and may show patches of yellowish cells in russeted areas that have been degreened by ethylene release during the attack (Yothers and Mason, 1930; McCoy and Albrigo, 1975). On 'Valencia' orange and 'Marsh' grapefruit, as a result of the damage, the leaf surfaces present a wound periderm, consisting on the leaves of 'Sunburst' tangerine of three layers (phellem, phelloderm and phellogen); on the leaves of other cultivars that do not form phellem, the phelloderm consists of up to seven layers of cells. The lower leaf surface may present a mesophyll collapse that appears first as yellow degreened patches and later as necrotic spots (Albrigo and McCoy, 1974). The probability of defoliation increases with the amount of damage, but the relationship between the size of the area involved and that of the leaves does not seem to have an effect on leaf abscission (McCoy, 1976).

The mite attacks early fruits of the size of a pea, and symptoms are seen when the size has not quite reached that of a walnut. At harvest, fruits infested in the early, mid or late stage of growth and development show different levels of injury, and symptoms that are designated 'sharkskin', 'russet' or 'bronzing'. The stylar feeding of the mite on fruits causes the collapse of the epidermal cells, probably as a result of the inoculation of saliva. As this is a vagrant species characterized by continuous probing of the leaf surface and fruit, it destroys a large quantity of cells that develop the appearance of brownish black patches (Albrigo and McCoy, 1974). Affected tissues emit ethylene, which can stimulate an early degreening of the leaves and fruits (McCoy and Albrigo, 1975). The discoloration of the fruit surface appears to be associated with the formation of lignin and likely oxidation of cytoplasmic substances. According to variety and the maturity of the fruits, the injuries caused may be different. If it occurs in the growth phase before fruit maturity, it leads to a breaking up of the dead epidermis and the formation of a peridermal wound beneath the epidermis. The development of cracks may occur in the epidermis but the epidermal layer does not break up into separate patches. The cracks form a rough texture that cannot be polished, while the oxidized cell contents give the fruit a brownish black appearance. This relatively early injury phase is call a 'russet condition'. On grapefruit, lemon, lime and orange, the 'russet condition' occasionally assumes the sharkskin appearance (Albrigo and McCoy, 1974). This develops when the fruit is damaged at a very early stage and is the result of epidermal damage followed by cracking when the fruit grows further, which may result in a sloughing off to expose a smooth wound periderm. In bronzed fruits, the damage occurs later when little additional fruit growth will occur; the epidermal cells die and become brownish, but there is not usually cracking, and the fruits will polish because the cutin and the wax layer are still present (Albrigo and McCoy, 1974). Severe attacks result in smaller fruits (Yothers and Mason, 1930; Schwartz, 1975a), fruit water loss (McCoy *et al.*, 1976a; Allen, 1979), fruit drop (Allen, 1978) and alterations in juice quality (McCoy, 1976). Although the aesthetic damage to fruits does cause commercial damage, the content of total soluble solids does not appear to be altered and so there are no drawbacks to their industrial use (Chandramani *et al.*, 2004).

Examination of the frequency distribution of the damaged fruits has shown that those sites in the northern quadrant present a higher average area of damage, followed by the eastern, southern and western quadrants. The distribution frequency of the damaged surface of the fruit changes with the level of average damage. With low average injury, most of the fruits show no damage, while increasing the average amount of injury to the fruits decreases the proportion of fruit without damage (Yang *et al.*, 1995). On orange, a positive linear correlation between the cumulative mite population and the percentage damaged surface of the fruit was found, allowing the estimation that a single mite damages 0.048% of the surface of a fruit every day (Sanchez Salas and Padron Chavez, 1985). Increased damage to fruits can lead to a small increase in fruit drop, with a slight negative relationship between the size of fruits and the amount of damage (Yang *et al.*, 1994).

Control

Control of the citrus rust mite is provided by spray applications of chemicals in a context of integrated pest management (IPM) programmes (McCoy, 1996b). The choice of strategy depends on the aim of production. Production destined for fresh consumption normally requires more control treatments than that destined for juice production, with there being a priority for cosmetic appearance in fresh fruit consumption, while for processing for juice the development and abscission of the fruits are not affected until 50–75% of the fruit surface is damaged, so there is less justification for the chemical control of rust mites on fruit grown for processing (Allen and Stamper, 1979; Rogers *et al.*, 2013). In Florida, production destined for the fresh market is treated with pesticides 3–4 times a year in April, June, August and October, while production designated for processing is treated from up to twice, once with petroleum oil and once with acaricides (McCoy, 1985; Browning, 1992). Allen (1981) reports that there is no single economic threshold value, as there is a range of damage thresholds equivalent to the cost of the treatment which depends on the value of production, time of the pest attack, harvest and temperature.

SAMPLING AND MONITORING. Nascimento *et al.* (1982) reported that populations of the citrus rust mite could be accurately sampled by taking five leaves per tree from 1.5% of the trees in the grove and counting the number of mites in an area of 2 cm^2 on the upper surface of each leaf, 2 cm from the leaf base. Oliveira *et al.* (1982) established that population levels were approximately the same all over a tree, and therefore the samples could be taken from any cardinal point. At any level of infestation, the results obtained by counting the mites present on a predetermined limited area of a leaf (about 4.9 cm^2 on the lower surface of the leaf at about 2 cm from the petiole) were as accurate as counting the numbers on the whole of both surfaces of the entire leaf. A leaf-brushing machine provided fairly accurate results when the infestation level was low, but not when it was high. If a precise determination of the number of mites is required, the 'limited area' method is to be preferred, but any of the methods can be used for comparative studies. Manzur (1989) reported that *P. oleivora* showed a negative binomial distribution on grapefruit tree leaves and that the economic thresholds of pest density were set as 0.5 mites per leaf for no insecticide treatment and >2 mites per leaf requiring treatment.

Rogers *et al.* (1994) applied the modified Horsfall–Barratt system for estimating citrus rust mite populations in an integrated crop management programme. In South Africa, the first discovery of the mite on leaves allows treatments (Schwartz and Meyer Smith, 1998). Hall *et al.* (2005) examined the consequence of reducing sample size on the accuracy and precision of estimates of citrus rust mite. Hall *et al.* (2007) investigated binomial sampling for estimating rust mite densities on orange fruit and concluded that a binomial sampling for a general estimate of mites densities seemed to be a viable alternative to absolute counts of mites per sample for a grower using a low management threshold such as two or three mites per sample. In Florida, Childers *et al.* (2007) suggested one sampling method based on rust mite density (rust mites cm^{-2}), in late spring or summer, in processed fruit trees selected at random and within uniformly distributed areas throughout a 10–40 acre block representing a single variety with uniform horticultural practices. The sampling of adjacent trees should be avoided and fruits should be sampled at random, representing the four quadrants of the tree and taken midway in the canopy. One fruit surface area should be examined midway between the sun and shade areas. The number of rust mites cm^{-2} should be recorded and averaged for 10 acres, represented by 20 trees with four fruits per tree or 80 readings for 10 acres. In Pakistan, an adhesive tape method was evaluated (Aghajanzadeh and Malik, 2007).

THRESHOLDS. Six rust mites cm^{-2} would be a planning threshold where pesticide treatment might be required within 10–14 days. Ten rust mites cm^{-2} would be an action threshold where treatment would be required as soon as possible. For fresh fruit, it is necessary to monitor mite populations every 10–14 days and use an average of 2 mites cm^{-2} as an action threshold (Childers *et al.*, 2007).

BIOLOGICAL CONTROL. The most important mycopathogen is probably *Hirsutella thompsonii* Fisher, a fungus of the family Clavicipitaceae reported for the first time in association with *P. oleivora* (Spears and Yothers, 1924) and described as a new species by Fisher (1950). The pathogen exhibits a clear pathogenicity to the citrus rust mite (McCoy and Kanavel, 1969), but its effectiveness requires values of RH that do not hinder the development of propagules on leaves and fruit (McCoy, 1996b; Cabrera, 2004), and high levels of population of the citrus rust mite that unfortunately usually coincide with damage to the plant (McCoy, 1981). The action of a protein (Irsutellina A) extracted from the substrate of a culture of *H. thompsonii* is able to inhibit the fecundity of *P. oleivora* (Omoto and McCoy, 1998). The effectiveness of the fungus has been verified in several citrus areas throughout the world, including Florida (McCoy *et al.*, 1971; McCoy and Selhime, 1974; McCoy and Couch, 1982), China (Yen, 1974; Chen and Chen, 1986; Chen *et al.*, 1987), Suriname (Van Brussel, 1975), Cuba (Cabrera *et al.*, 1981), Argentina (Gomez and Nasca, 1983), Brazil (Almeida *et al.*, 1981; Santos da Silva *et al.*, 1981), Thailand (Saowanit *et al.*, 1999) and Iran (Aghajanzadeh *et al.*, 2006), and the results are comparable to those of chemical control but

the impossibility of the continued and rapid development of the mycopathogen in the field has suggested that its mass production be stopped (McCoy, 1996a).

Other species are *H. kirchneri* (Rostrup) Minter, Brady *et* Hall, *H. nodulosa* Petch (Cabrera and Dominguez, 1987b), *Beauveria bassiana* (Balsamo) Vuillemin (Clavicipitaceae) (Alves *et al.*, 2005a) and *Mycosphaerella citri* Whiteside (Mycosphaerellaceae) (Hernández and Escobio, 1988), but various factors suggest that their use is not practical.

Three species of the family Exobasidiomycetidae (*Meira argovae* Boekhout *et al.*, *M. geulakonigii* Boekhout *et al.* and *Acaromyces ingoldii* Boekhout *et al.*) that have been recently described (Boekhout *et al.*, 2003) develop on *P. oleivora*, on tetranychid mites (e.g. *Panonychus citri* (McGregor)) and on *Sphaerotheca fusca* (Fr.) Blumer, a pathogenic fungus of cucumber (Sztejnberg *et al.*, 2004). Paz *et al.* (2007a) have evaluated the role of these mycopathogens in the laboratory for the control of *P. oleivora* and found that after 14 days they produced mortality rates of 87.1, 91.1. and 89.1%, respectively. These results encouraged a verification in the field of the action of *M. geulakonigii* on grapefruit infested by *P. oleivora*. One application of the pathogen and monthly applications gave 87.4 and 87.2% of non-damaged fruits respectively, while in the control the percentage of damaged fruits was 77.2%. The fungus is a beneficial endophyte of grapefruits that colonizes the fruit's peel, thus protecting it from attack by *P. oleivora* (Paz *et al.*, 2007b). Unfortunately, at present, there are no indications of the practical use of these pathogens.

For predators, Schicha (1987) and Smith and Papacek (1991) reported that in Australia the phytoseiid mite *Euseius victoriensis* (Womersley) controlled an attack of *P. oleivora* on fruit by less than 5%. In Israel, the phytoseiids *Amblyseius herbicolus* (Chant), *E. victoriensis*, *E. elinae* (Schicha), *E. stipulatus* (Athias Henriot) and *Typhlodromus rickeri* Chant have been introduced from different geographical areas (Argov, 1993; Argov *et al.*, 2002). In China, *Neoseiulus cucumeris* (Oudemans) has been released at the rate of 900,000 larvae ha^{-1} for the control of *P. citri* and *P. oleivora*, and its efficacy against mites on citrus trees was 85–95% 30 days after release and the pests were controlled for 6 months (Zhang *et al.*, 2003a). In Florida, the Psychidae *Cryptothelea gloverii* (Packard) preyed on the citrus rust mite and consumed the eggs and adults of both *P. oleivora* and *P. citri* (Villanueva *et al.*, 2005) and two species of Cecidomyiidae of the genera *Feltiella* and *Lestodiplosis* were both efficient predators of *P. oleivora* eggs, larvae and nymphs (Villanueva *et al.*, 2006).

Unfortunately, different obstacles limit the use of these biological means and the control of citrus rust infestations is still entrusted to the use of chemicals, sometimes in an IPM context.

CHEMICAL CONTROL. Chemical control is needed, especially in wetlands and crop varieties for the fresh market (McCoy *et al.*, 1989). Important considerations here are the easy development of the mite populations, the rapid increase in damage, the small size of the mite which makes monitoring difficult and the lack of natural enemies (McCoy, 1996b). The influence of biotic and abiotic factors and cultural practices (Jeppson *et al.*, 1975; McCoy, 1977a; Zamorra and Nasca, 1985; Smith and Papacek, 1991) can influence the population density of the citrus rust mite; also, the side effects of many chemicals can stimulate the development of populations (McCoy, 1977a,b; Smith and Papacek, 1991; Beattie *et al.*, 1991) or the selection of resistant strains (Swirski *et al.*, 1967; Jeppson *et al.*, 1975; Herne *et al.*, 1979; Omoto *et al.*, 1994).

A detailed account of active ingredients reported in the literature is given by Childers *et al.* (1996). The most effective substances are included into the avermectins (French, 1982, 1984; McCoy *et al.*, 1982; Childers, 1985; Zamora *et al.*, 1989; French and Villarreal, 1990b; Moreira, 1992; Iskander, 1993; Chiaradia and Cruz, 1996; Smith *et al.*, 1998; Bergh *et al.*, 1999; Lu, 2000; Grout and Stephen, 2002; Andrade *et al.*, 2009), dithiocarbamates (Schwartz, 1977a; Childers, 1990; Ibrahim, 1992; Iskander, 1993; Drishpoun *et al.*, 1995; Rezk and Gadelhak, 1996; Kalaisekar *et al.*, 2000, 2003; Ying *et al.*, 2000; Ran *et al.*, 2001; Grout and Stephen, 2002), organochlorine (Dean, 1979; French and Gage, 1983; Oliveira *et al.*, 1985; Childers, 1990), organotin (Childers and Konsler, 1980; French, 1982; Brown and Jones, 1983; McCoy *et al.*, 1990; Oliveira Guassu *et al.*, 1993; Childers, 1994b; Drishpoun *et al.*, 1995; Moraes *et al.*, 1995b; Wu, 1995), narrow range petroleum oils (Selhime, 1984; Knapp, 1991; Iskander, 1993; Ibrahim, 1994; Rae *et al.*, 2000; Knapp *et al.*, 2001; Cai *et al.*, 2001; Lei *et al.*, 2003) and sulfur (Griffiths and Thompson, 1957; Brown and Jones, 1983; Ibrahim, 1992; Moraes *et al.*, 1995b; Chiaradia and Cruz, 1996; Scarpellini and Santos, 1997; Scarpellini and Clari, 1999; Kalaisekar *et al.*, 2000, 2003; Huang, 2005).

Avermectins are widely used in various regions of the world either alone or in combination with petroleum oils, and are considered partially selective towards the beneficials (McCoy *et al.*, 1982). Sulfur applications were proposed between the end of the 19th century (Hubbard, 1885) and the beginning of the last century (Yothers, 1915), but sulfur is not very selective for natural enemies (McCoy, 1977a; Smith and Papacek, 1991). Medium-range petroleum oils have been used worldwide since 1940 (Jeppson *et al.*, 1975) and are effective, but have a modest persistence and are largely selective (McCoy *et al.*, 1976b; Vacante, 1986). An alternative approach is to use 435-66 FC, FC 455-88, or 470 petroleum oils as a fungicide for sooty mould control and to suppress pest mites (Childers *et al.*, 2007). Acaricides can be used in combination with other pesticides (insecticides, fungicides) and it is a good rule to assess their mutual interactions and their selectivity (McCoy, 1996b). Recently, Bell *et al.* (2005) reported that a new acaricide tetronic acid derivative, spirodiclofen, in combination with horticultural oil spray provided excellent initial and residual control of the mite, equal or superior to that of avermectins and pyridazinone derivatives. Nevertheless, the selectivity of this last acaricide is as yet unclear.

INTEGRATED PEST MANAGEMENT. The above-mentioned chemical control methods require that control is placed in an IPM context (McCoy, 1996a), with particular attention to pertinent technical aspects (choice of chemicals, threshold values, sampling techniques, etc.). Equally important is the canopy density, which affects the ability of populations to increase in a short period of time (Childers *et al.*, 2007). In China, in the context of prevention, irradiation is suggested as a quarantine treatment (Hu *et al.*, 2004b).

The Genus *Phyllocoptes* Nalepa

Diagnostic characteristics

The prodorsal shield has the frontal lobe over the gnathosoma. The scapular tubercles are usually ahead of the rear shield margin, directing the scapular setae (*sc*) forward or up and centrad; if near the rear margin, the setae *sc* are directed forward; in some species, the setae *sc* are located on the rear margin, but the scapular tubercles show longitudinal or oblique basal axes directing *sc* up, central or convergently to the rear. The opisthosomal annuli are variable (Amrine *et al.*, 2003).

Economic importance

The genus *Phyllocoptes* has a worldwide geographical distribution and includes over 190 species, mostly recorded from dicotyledonous plants (Oldfield, 1996a). About ten species live on conifers of the genera *Taxodium* and *Juniperus* both in Europe and North America, inhabiting needles and stems and causing discoloration (Boczek and Shevchenko, 1996). Many species show considerable morphological variation, with, for example, some having broad dorsal annuli and no microtubercles, while others are worm like and bear well-developed microtubercles. Species of the genera *Phytocoptes* Donnadieu and *Phytocoptes* Nalepa, some species of the genera *Phytoptus* Dujardin (prior to 1898) and some species of *Eriophyes* Von Siebold are now included in the genus *Phyllocoptes* (Amrine *et al.*, 2003).

The species that are economically most important are the dryberry or blackberry leaf mite, *P. gracilis* (Nalepa), a European and north American species, which causes deformed leaves, bud injury and dry berry in raspberry, blackberry, loganberry and other berries (*Rubus* spp.) (Jeppson *et al.*, 1975), and *P. fructiphilus* Keifer, a north American species (Keifer, 1940a) and vector of rose rosette disease (RRD) in *Rosa* spp. (Allington *et al.*, 1968; Jesse *et al.*, 2006; Tzanetakis *et al.*, 2006), which has been proposed for the biological control of wild roses (Epstein and Hill, 1999; Amrine, 1996). Another species proposed for weed control is *P. cruttwellae* Keifer, which has been suggested for use against *Chromolaena odorata* (L.) King *et* Robinson (Cromroy, 1983).

Several other *Phyllocoptes* spp. live on plants of economic importance, but their phytopathological roles are secondary. These species include *P. abaenus* Keifer, which is widespread in Europe, Australia, the Near East and Nearctic and Neotropical regions on almond (*Prunus dulcis* (Mill.) D.A. Webb), apricot (*P. armeniaca* L.), mahaleb cherry (*P. mahaleb* L.), sweet cherry (*P. avium* L.), European plum (*P. domestica* L.) and myrobalan plum (*P. cerasifera* Ehrh.) (Jeppson *et al.*, 1975; de Lillo and Beron, 2004). In late summer, populations of *P. abaenus* are sometimes observed at high density mainly on the undersurface of leaves, associated with patches of hairs along the midrib, corresponding to its erinea (Delley, 1973). The species overwinters as deutogyne forms between bud scale scars and at the base of lateral buds. Its economic importance is secondary (Schliesske, 1983). *P. pyrivagrans* Kadono causes browning of the under leaf surface of pear in Japan (Kadono, 1985). *P. pruni* Soliman *et* Abou-Awad lives on European plum in Egypt and Iran (Soliman and Abou-Awad, 1978; Xue *et al.*, 2009a). *P. goniothorax* (Nalepa) is an erineum former on pomaceous Rosaceae in the Nearctic region and Europe, with hawthorn as the typical host (Jeppson *et al.*, 1975; de Lillo and Beron, 2004). *P. calisorbi* Keifer lives on mountain ash (*Sorbus californica* Greene) in California, forming erineum patches on the undersides of the leaflets. These patches are at first white, later turning brown. Shrubs in the shade are the ones that are most affected (Keifer, 1965a). *P. nageli* Keifer is a needle vagrant on Douglas fir (*Pseudotsuga menziesii* ((Mirb.) Franco) in coastal areas of central California and north to Oregon, and probably also other north American states, but is not injurious (Keifer, 1965d; Jeppson *et al.*, 1975). *P. graniti* Keifer makes underside erineum patches on the leaves of almond in Italy. The erineum consists of capitate papillae (Keifer, 1959a). The bougainvillea rust mite, *P. bougainvilleae* Keifer, was originally described from *Bougainvillea spectabilis* Willd. in Brazil (Keifer, 1959c), and was later collected on paper flower (*B. glabra* Choisy var. *sandeina*) in Kenya and described as *Vittacus bougainvilleae* by Abou-Awad and El-Banhawy (1991), under which name it was also reported from Florida (Halbert and Mead, 1995) and California. The species causes extensive edge rolling and rusting of the flowers and leaves, with the damage most severe in the white and pink bougainvillea varieties.

Phyllocoptes gracilis (Nalepa) (Fig. 16.71)

Common name

Dryberry mite, blackberry leaf mite.

Diagnostic characteristics

FEMALE. The body is long from 115 to 130 μm, more worm like than fusiform, and light yellow in colour. The prodorsal shield is subtriangular, with an acuminate, apparently notched apex. The scapular setae (*sc*) are moderately long and directed diagonally backwards and centrad. The opisthosomal annuli are approximately subequal dorsoventrally and entirely microtuberculate. The genital coverflap has about 10 longitudinal ribs. The empodium is 5-rayed (Jeppson *et al.*, 1975).

MALE. The male is like the female, but smaller.

Geographical distribution

Europe and North America (British Columbia, Washington, Oregon, California) (Jeppson *et al.*, 1975).

Host plants

Rosaceae: *Rubus* spp., including wild and cultivated raspberry, blackberry, loganberry, Himalaya berry and thimbleberry (Jeppson *et al.*, 1975).

Bio-ecology

P. gracilis infests several *Rubus* spp., including raspberry, blackberry, loganberry, Himalaya berry and thimbleberry (Jeppson

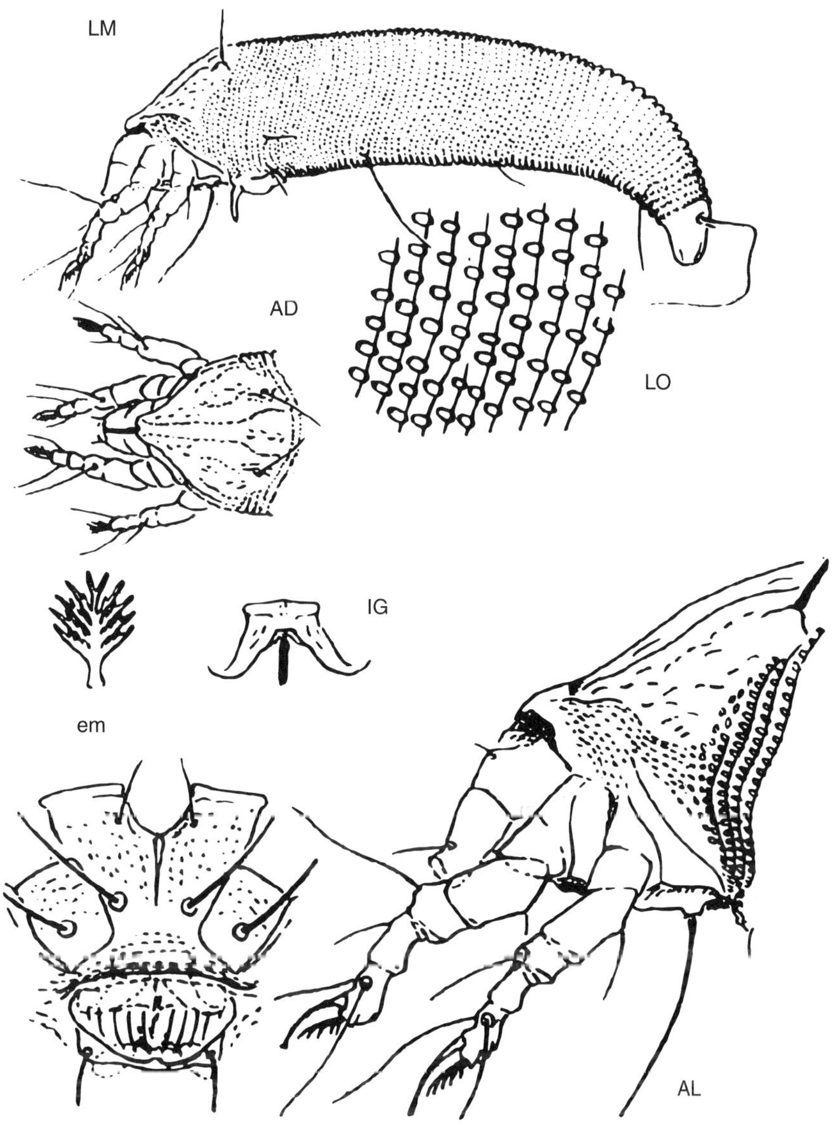

Fig. 16.71. *Phyllocoptes gracilis* (Nalepa) (from Keifer, 1952a). See Table 16.1 for key.

et al., 1975). In Scotland, the dryberry mite is widespread on cultivated and wild raspberries. Its feeding activity causes damage to leaves and fruits, but usually only on raspberries grown in sheltered sites, where mites tend to aggregate. Females overwinter in the buds, under bud scales or in petiole scars (Gordon and Taylor, 1976). On raspberry in Yugoslavia, mated adult females overwinter in the bud axilla in colonies ranging from 40 to a few hundred individuals. In spring, when the temperature increases to above 11°C, the females became active, and lay their first eggs when the temperature is about 25°C (Dobrivojevic and Petanović, 1985). In Scotland, high rates of mortality were observed in buds on raspberry canes. The mites are most abundant in buds 60–140 cm above soil level, but mortality decreases with the height and size of the bud; the number of living mites is greatest at 120 cm. Mortality seems to be independent of population density and is probably the result of being crushed between bud scales (Macintyre and Foster, 1977). Also in Scotland, the mites disperse in the spring to the emerging foliage of the fructocanes, where they feed and oviposit. Populations reach a maximum on the fructocanes in mid-July, when the first fruits are maturing, and on the primocanes in late September. Migration to the overwintering sites begins in late September. In spring, the overwintering females lay 1–2 eggs in 12 h (Massee, 1932; Breakey and Batchelor, 1957). Embryonic development lasts about 7 days (Dobrivojevic and Petanović, 1985). The life cycle takes 14 days at 25°C and several generations develop a year (Domes, 1957; Gordon and Taylor, 1976; Dobrivojevic and Petanović, 1985). On wild and cultivated raspberry in the southern Urals region of the ex-USSR, the mite is a specialized pest of raspberry, and is distributed mainly with planting material; it has the same bio-ecology as observed in European countries (Klement'eva, 1979).

There is little information on natural enemies, but among these some phytoseiid mites have been reported (Breakey and Batchelor, 1957; Gordon and Taylor, 1976). In Yugoslavia, *Phytoseius macropilis* (Banks) significantly reduced numbers of the mite in the field (Dobrivojevic and Petanović, 1985).

Symptomatology and damage

Feeding activity of the dryberry mite causes damage to berries and leaves, with a symptomatology that is variable according to the *Rubus* species involved, timing and duration of attack, and environmental factors. Infested berries show rapid development of drupelets, early ripening and drying. The distal growing point may die off, and lateral buds may develop (Domes, 1957; Gordon and Taylor, 1976). Fruits of red raspberry of the 'Lloyd George' variety become hardened, misshapen and do not ripen normally. In red raspberry 'Washington' variety, there is drupelet dying and drying at the apex, and drying spreads towards the margin; this symptomatology is referred to as 'sun scald'. The damage may affect most of the yield, with a loss ranging from 50 to 70%. On loganberry, the infestation causes killing of the early fruits after petal fall. Infested berries are reddish, but afterwards turn brown and dry. Fruits that develop later are unaffected, while mature berries may appear withered, mainly distally. A similar symptomatology occurs on boysenberries. The feeding activity of the mite on red and yellow garden raspberries in Europe destroys the 'hair felt' and the fruits show pale spots (Jeppson *et al.*, 1975). Leaves of raspberry that are attacked early during their development show pale green areas, sometimes coalescing and appearing first as large and irregular chlorotic spots and afterwards as necrotic areas that are reddish in colour. A high mite population density deforms and twists the growing leaves, while the expanded leaves show less marked damage (Domes, 1957; Gordon and Taylor, 1976). In tayberry (a blackberry–raspberry hybrid), the known symptoms are close to toxaemias, and include leaf blotches of chlorotic spots or rings related to mite feeding. Leaves are downcurved, narrowed and can die back. Mite feeding activity may cause multiple branching following the death of the terminal growing points (Jones *et al.*, 1984). On raspberries with a high mite population density occurring over several years, a reduction of plant vigour and fruit quality is induced. The phenomenon is more frequent in sheltered areas (Gordon, 1981). On cultivated blackberry in Yugoslavia, damage to raspberry took the form of light green to yellow spots on the leaves, progressive depigmentation of the foliage, shoot proliferation, reduced photosynthesis and the production of small and deformed fruits (Dobrivojevic and Petanović, 1985).

Control

Despite relatively resistant varieties being recommended as sources of planting material (Klement'eva, 1979), control is mostly provided by spray applications of acaricides. A first treatment can be applied in spring, as the cane buds enlarge, when the mites leave their winter sites, and before the beginning of their feeding activity on expanding leaves. A second treatment may be done in summer as soon as the flower buds form and before they open. This treatment controls the mite populations before they penetrate into the blossoms and damage the young fruits (Jeppson *et al.*, 1975; Gordon and Taylor, 1977; Jones *et al.*, 1984).

PLANT RESISTANCE. Cultivars differ in the extent to which they support overwintering populations and show symptoms of infestation. For example, no mites are found on the 'Malling Promise' variety, whereas the 'Malling Jewel' variety appears to be a good host to the mite (Gordon and Taylor, 1976). In the southern Urals region, heavy infestation leads to death of the leaves. The most susceptible variety may have up to 400 mites per leaf. Certain varieties are not infested, whereas infestation on others reached 10, 25, 50 and even over 50%, this highest infestation being accompanied by the deformation of leaves and shoots (Klement'eva, 1979).

CHEMICAL CONTROL. A number of active ingredients are reported in the literature as suitable for use. They include acaricides, insecticides and fungicides, such as azinphos-methyl (Stenseth, 1973), benomyl (Gordon and Taylor, 1977), chlorobenzilate (Jeppson *et al.*, 1975), demeton-*S*-methyl, dicofol, dimethoate, methomyl (Stenseth, 1973), mineral oils (Jeppson *et al.*, 1975), spirodiclofen (Linder *et al.*, 2008), lime sulfur (Jeppson *et al.*, 1975) or wettable sulfur (Linder *et al.*, 2008) and vamidothion (Gordon and Taylor, 1977). Sulfur provides good control (Jeppson *et al.*, 1975; Linder *et al.*, 2008).

Phyllocoptes fructiphilus Keifer (Fig. 16.72)

Common name

Wild rose bud and fruit mite.

Diagnostic characteristics

FEMALE. The body is spindle formed, 140–170 μm long, 43 μm wide and amber to light yellow in colour. The gnathosoma is rather large. The prodorsal shield has a network pattern, which is granular laterally; the scapular tubercles are 20.5 μm apart, slightly ahead of the rear margin; the scapular setae (*sc*) are 16 μm long, directed dorsocentrad. Coxae I are contiguous. The opisthosoma is totally microtuberculate, the posterior ventral annuli are striated, the dorsal and ventral annuli are subequal in number, with about 55 dorsal annuli and about 60 ventral annuli. The seta c_2 is 21 μm long, set on about ventral annulus 8; the setae *d* are 46 μm long, set on about ventral annulus 20; the setae *e* are 32 μm long, set on about ventral annulus 36; the setae *f* are 25 μm long, set on about ventral annulus 5 from the rear; the setae h_1 are present. The genitalia are 23 μm wide, 14.5 μm long; the genital coverflap has 6 or 7 ridges, and the seta is 32 μm long. The empodium is 5-rayed (Keifer, 1940a).

Geographical distribution

Western half of North America (Jeppson *et al.*, 1975).

Host plants

Rosaceae: California rose (*Rosa californica* Cham. *et* Schltdl.) (Keifer, 1940a), *R. eglanteris* L., multiflora rose (*R. multiflora*

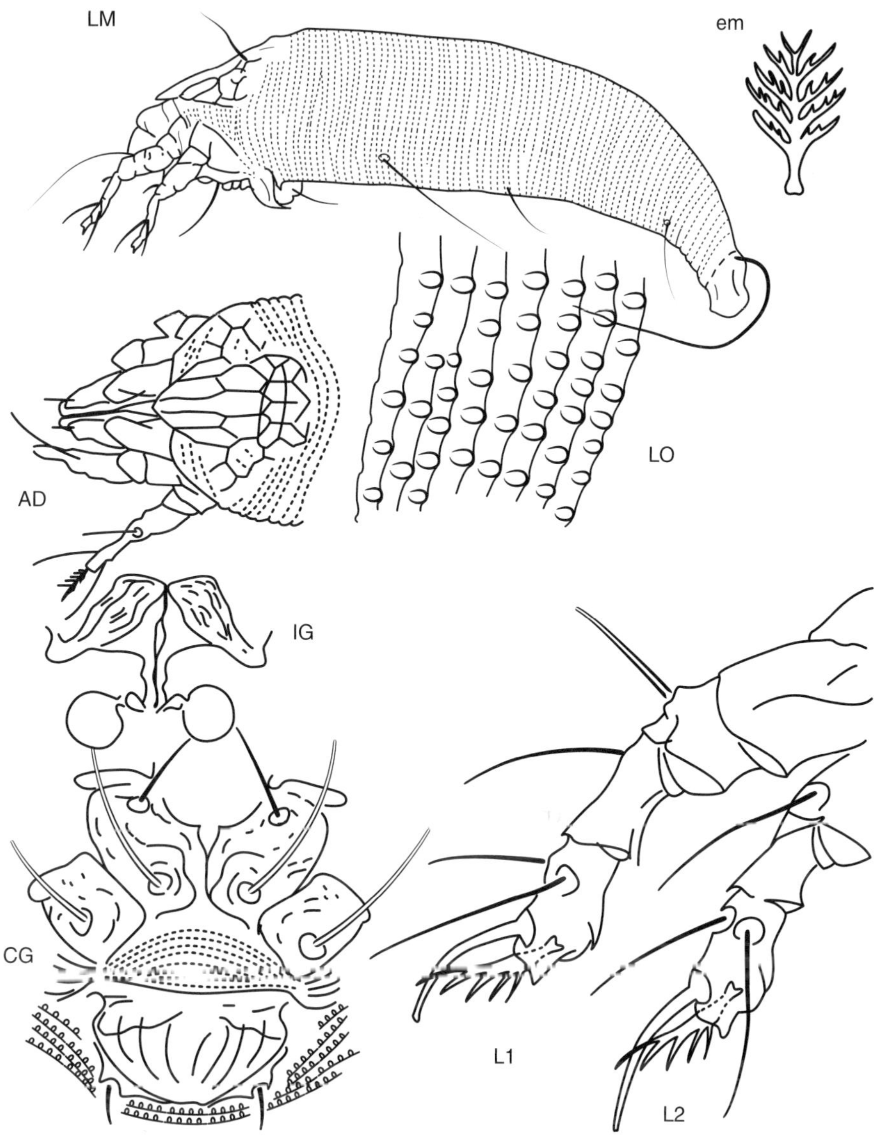

Fig. 16.72. *Phyllocoptes fructiphilus* Keifer (from Keifer, 1940a). See Table 16.1 for key.

Thumberg ex J. Murray), *R. rubrifolia* Villars, *R. suffulta* Greene, Woods's rose (*R. woodsii* Lindley), (Allington *et al.*, 1968) and various ornamental roses (Amrine, 1996).

Bio-ecology

The wild rose bud and fruit mite was found the first time in the fruits around the seeds and at the petiole bases of California rose in California (USA) (Keifer, 1940a); since then, the mite has been reported from several states of the western half of North America (Jeppson *et al.*, 1975).

The species reproduces easily in the laboratory on multiflora rose, a plant introduced to North America from east Asia that has become an injurious weed in the central states of the USA. At 23°C, the developmental time from egg to egg takes 14.9 days, the mean ovipositional period is 12.1 days, the fecundity is 7.2 eggs and the daily oviposition is 0.6 eggs (Kassar and Amrine, 1990). During the growing season of the host plants, all biological stages of *P. fructiphilus* occur in protected places on the plants between the leaf petioles and axillary buds, mainly towards the apex of the shoot. The mite overwinters in any protected sites of the plant characterized by appressed green surfaces. Later, it penetrates into the open end of the fruit and settles among the seeds. *P. fructiphilus* also lives between bud scales. In spring, the population density is low and it increases slowly until mid-June to early July; from the summer, it increases further until September or October, when the populations move towards various sites to overwinter, although the mite can be active until December (Allington *et al.*, 1968; Jeppson *et al.*, 1975).

Symptomatology and damage

P. fructiphilus vectors the virus that was formerly designated '*Rose virus 1*' (RSV-1), an Ilarvirus (Jesse *et al.*, 2006; Tzanetakis

et al., 2006), which has recently been formally identified as *Rose rosette virus* (RRV), an *Emaravirus* (Laney *et al.*, 2011; Hong and Hansen, 2012). This virus causes Rose rosette disease (RRD) on the wild multiflora rose, *Rosa multiflora*, a weedy scrambling shrub, and on ornamental roses (see also Chapters 7 and 9). The disease has spread mainly in the western and middle states of the USA and in some provinces of Canada, although many varieties of rose are resistant (Amrine, 1996). This is probably the only example of an eriophyoid mite controlling a weed indirectly by means of a plant pathogen. As noted above, the identity of the causal agent of RRD was unproven until recently, although spherical virus-like particles had been found in the cytoplasm of cells of diseased plants that measured 120–150 nm and had a trilaminar wall; they were similar to other particles of eriophyid-borne diseases, such as 'fig mosaic' and the 'wheat spot' (Gergerich and Kim, 1983; Gergerich *et al.*, 1983).

The symptoms of RRD are a red or purplish vein mosaic on the underside of leaflets, reduction of growth, early breaking and stunting of axillary buds, shortened internodes and the development of bright red lateral shoots with small, wrinkled and deformed leaflets (Doudrick, 1983; Amrine and Hindal, 1988). In a new shoot, the symptoms commonly occur first on axillary buds, and develop gradually towards the shoot apex. Young plants become dwarfed. The symptoms on ornamental rose are yellow mosaic on the leaves, thorniness of the stems, the development of lateral shoots larger than the parent cane, deformed foliage and the early development of lateral buds, with witches' brooms (Amrine and Hindal, 1988).

The disease may be transmitted both by grafts and by the feeding activity of *P. fructiphilus* to *R. eglanteris*, *R. suffulta*, *R. woodsii*, *R. multiflora* and *R. rubrifolia* (Allington *et al.*, 1968). Single individuals can transmit the rosette pathogen, which may be transferred to a healthy rose from a symptomatic rose; the symptoms persist after mite control has been applied (Amrine *et al.*, 1988). Graft transmission of the pathogen to rooted cuttings of *R. multiflora* takes from 45 to 80 days and is only 46% successful, whereas graft transmission to large vigorously growing transplants takes from 30 to 60 days and is 100% successful. The graft transmission characteristics show that the agent resides in the roots of multiflora rose. Laboratory transmission of RRD by *P. fructiphilus* to transplants is 92.3% successful and symptoms appear after 17–24 days. Transmission by mites to rooted cuttings is unsuccessful (Amrine *et al.*, 1988). Multiflora roses inoculated by the mite show symptoms after 17–30 days and witches' brooms develop after 30–90 days. The disease spreads from infected canes to the roots, and plants usually die after 1–2 years (Amrine and Hindal, 1988). The spread of RRD is best described by a logistic population growth model. Peak susceptibility to graft transmission of the disease in the field occurs in Iowa from late May to early July. Populations of *P. fructiphilus* are low until mid-July and reach peak levels at the end of August, with their host range limited to *R. multiflora* and hybrid ornamental roses. There is no evidence for seed or soil transmission of the causal agent and no remission of symptoms is obtained after either heat therapy or tetracycline treatment. A significant decrease in stored starch corresponds to a major decrease in sucrose and increases in fructose and glucose in infected tissue. It has been suggested that the pathogen may have potential for the biocontrol of multiflora rose (Epstein and Hill, 1995). Monthly observations of *R. multiflora* bushes carried out in Indiana showed that the percentage of plants affected by RRD increased in about 2 years, from 30.3 at the time of planting to 86.9%. In the same period, the population mortality rate also increased from 1.1 to 44.3%. During this period, infestation by *P. fructiphilus* increased from 47 to 100 and 30 to 87% from spring to autumn of each of the 2 years, respectively, and the density of mites also increased from 12–39 to 113–133 per tip. Mite density was 15 times higher on diseased than on healthy plants. Mite density is adversely affected by a cool and wet spring (Amrine *et al.*, 1990). The viral agent and the mite are both natives of North America and restricted to roses. Because of its major effect on *R. multiflora*, in the USA, the use of the mite has been suggested for controlling this wild (weed) plant (Amrine, 1996; Epstein and Hill, 1999).

Control

The mite can be a pest of ornamental roses, and for this reason its presence may require chemical control (Amrine, 1996). However, many varieties are not susceptible (Amrine, 2002). Public concern over the risk of damage to ornamental plants has prevented efforts to further distribute the disease or the mite (Amrine 1996, 2002; Epstein and Hill, 1999).

Genus *Vasates* Shimer

Diagnostic characteristics

The scapular tubercles have the basal axes converging anteriorly, directing the scapular setae (*sc*) convergently to the rear. The opisthosomal annuli are differentiated (Amrine *et al.*, 2003).

Economic importance

This genus has over 43 species. The most important species is the maple bladder-gall mite, *Vasates quadripedes* Shimer, causing rounded galls on leaves of red maple (*Acer rubrum* L.) and silver maple (*A. saccharinum* L.) in North America (Jeppson *et al.*, 1975; Keifer *et al.*, 1982); it has also been reported in Europe (Petanović, 1997; Soika and Łabanowski, 1999a,b; Ripka *et al.*, 2002; Wurzell, 2002; Ellis and Ulenberg, 2005; Glavendekić *et al.*, 2005) and Iran (Kamali *et al.*, 2001; Xue *et al.*, 2009a). Another species is the finger gall mite, *V. aceriscrumena* (Riley), which causes finger galls on leaves of sugar maple (*A. saccharum* Marsh.) in the eastern USA and Canada; these galls are solitary, elongate, pointed or truncate at the apex, or spindle shaped (Keifer *et al.*, 1982). *V. allotrichus* (Nalepa) causes leaf rolling on black locust (*Robinia pseudoacacia* L.), a tree native to the USA that has spread into temperate North America, Asia and Europe; the mite is also reported from Serbia and Montenegro (Šefrová and Laštůvka, 2005). *V. aegyptiacus* Abou-Awad has been found in the terminal buds of mango (*Mangifera indica* L.) in Egypt, where it is associated with *Metaculus mangiferae*

(Attiah) and causes malformations such as russeting in the buds and stunting in the inflorescences (Abou-Awad, 1980). Except for the *V. quadripedes*, which has an inconsistent economic role, there is no information on the economic importance of the other *Vasates* spp.

Vasates quadripedes Shimer (Figs 16.73 and 16.74)

Common name

Maple bladder-gall mite.

Diagnostic characteristics

FEMALE. The body is spindle shaped, 170–220 μm long, and whitish in colour. The prodorsal shield is produced anteriorly, and the scapular setae (*sc*) are directed backward. The deutogyne has a smooth dorsal shield and the genital coverflap; the dorsal opisthosoma lacks microtubercles. The empodium is 5-rayed and palmate. The protogyne has a pattern on the prodorsal shield, and the whole opisthosoma has microtubercles. The genital coverflap is ribbed. The empodium is 5-rayed and not palmate (Keifer *et al.*, 1982).

MALE. The male is smaller than the female, with a pattern on the prodorsal shield, and the empodium 4-rayed (Keifer *et al.*, 1982).

Geographical distribution

England (Wurzell, 2002), Hungary (Ripka *et al.*, 2002), Iran (Kamali *et al.*, 2001; Xue *et al.*, 2009a), Poland (Soika and Łabanowski, 1999a,b), the Netherlands (Ellis and Ulenberg, 2005), eastern North America (Keifer, 1959b; Jeppson *et al.*,

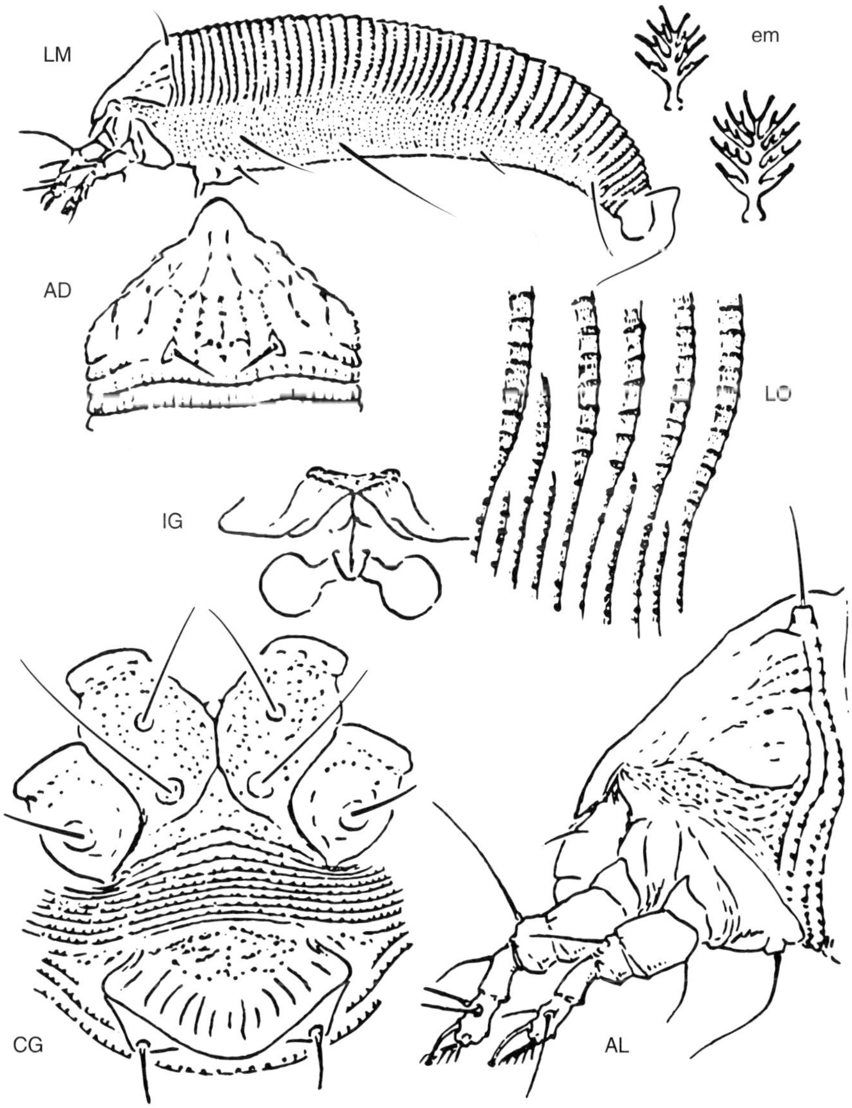

Fig. 16.73. *Vasates quadripedes* Shimer, protogyne (from Keifer, 1959b). See Table 16.1 for key.

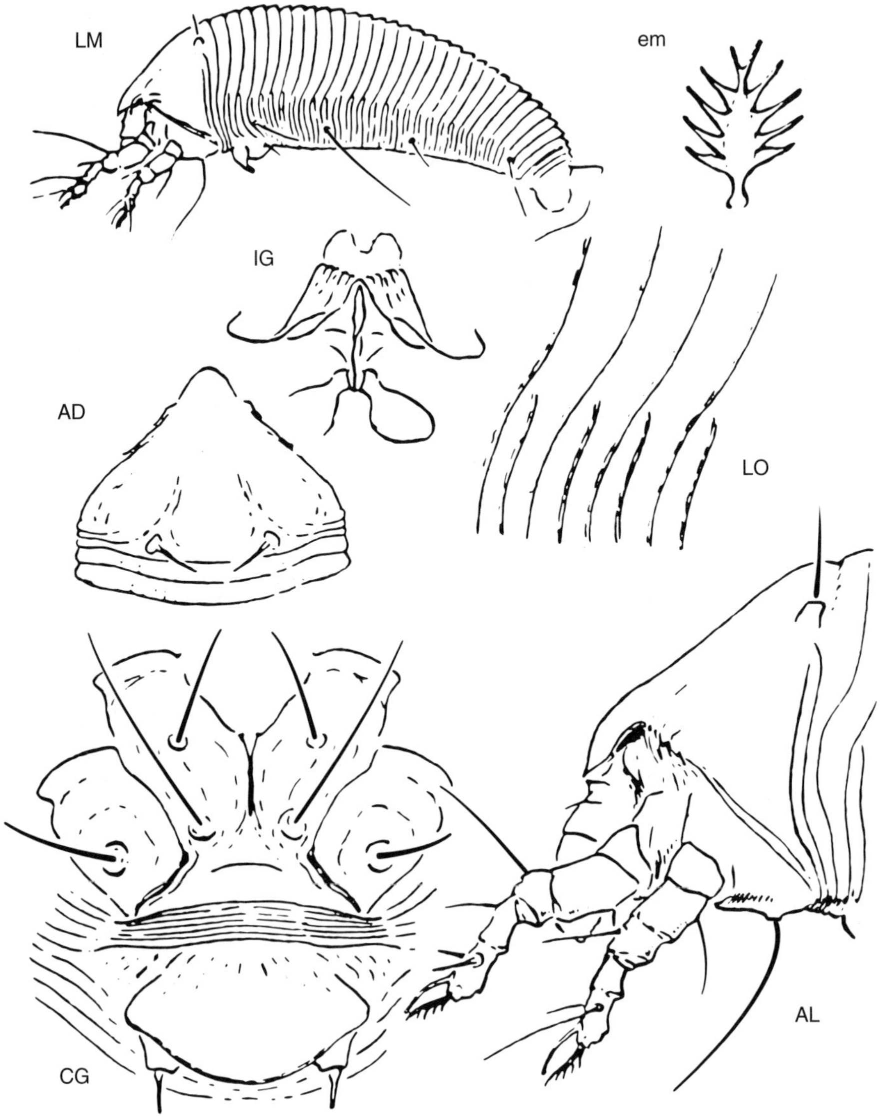

Fig. 16.74. *Vasates quadripedes* Shimer, deutogyne (from Keifer, 1959b). See Table 16.1 for key.

1975; Keifer *et al.*, 1982), Serbia and Montenegro (Glavendekić *et al.*, 2005) and the former Yugoslavia (Petanović, 1997).

Host plants

Sapindaceae: red maple (*Acer rubrum* L.), silver maple (*A. saccharinum* L.) (Keifer, 1959b; Jeppson *et al.*, 1975; Keifer *et al.*, 1982; Wurzell, 2002; Ellis and Ulenberg, 2005), black maple (*A. saccharum* Marshall) (Soika and Łabanowski, 1999b).

Bio-ecology

The maple bladder-gall mite lives on red maple, silver maple (Jeppson *et al.*, 1975; Keifer *et al.*, 1982; Wurzell, 2002; Ellis and Ulenberg, 2005) and black maple (Soika and Łabanowski, 1999b). On these host plants, the mite develops a typical alternation of generations. Deutogyne forms overwinter in bark cracks or wounds and do not feed until the following spring. During the summer, the protogyne forms inhabit leaf galls that form on the upper leaf surface of red maple and silver maple on the expanding spring leaves. They start to migrate to the trunk and branches as the galls turn from green pepper to dark brown or black from late July. This migration from the galls may occur until early October, and the mite populations spread on to many parts of the tree (Jeppson *et al.*, 1975).

Symptomatology and damage

The galls that are produced on the expanding spring are globular and protrude from the upper leaf surface; they are green pepper like, with a hole on the undersurface. The galls start to grow in spring and sometimes cause leaf deformation (Jeppson *et al.*, 1975). On silver maple, the galls may be rounded or elongate, from 1.5 to 5 mm long and with a short stem. They

are numerous at the basal part of the leaf between the veins. Externally, the galls are wrinkled and glossy and change from yellowish green or dark green to pink, brown or black. Internally, they are hollow and the exit hole is on the underside leaf surface and has unicellular whitish hairs (Keifer *et al.*, 1982). Wurzell (2002) reported that the number of galls on some leaves of silver maple, especially on the lower branches, may appear to pile on top of one another, occupying the whole leaf surface. The galls are roughly spherical, obconical, irregularly or kidney shaped, smooth, wrinkled or knobbly, with a texture ranging from slightly glossy to matt; they are lime green to dark green when young, with shades of pink and dingy red in the summer before turning dark brown or coal black in the autumn. On red maple, the galls are smaller, from 1 to 2 mm long, bead like and lacking a basal stem. The leaves appear curled, form cylindrical rolls and drop early (Keifer *et al.*, 1982). High levels of attack can disfigure silver maple trees but infestations rarely require control (Jeppson *et al.*, 1975).

Control

The strong degree of deuterogyny has played a role in the rapid expansion of this mite in Europe, because it enables the species to hibernate on the trees. This results in plant material that leaves a nursery in early spring already being infected. From this point of view, prevention may play a significant role, especially in the nursery (Ellis and Ulenberg, 2005). Several acaricides have been suggested for mite control, such as dicofol, chlorobenzilate, ethion or chlorobenzide. However, their use is rare and any sprayings are ineffective against mites sheltered inside the galls (Weidhaas and Reeves, 1963). The presence of galls, even in considerable numbers, does not usually harm trees, and the problem is largely one of unsightliness. There are registered products available but should the problem be severe enough to warrant control. They should be applied in spring before the leaf buds open (Antonelli, 1992). Depending on the variety of maple, the crop and pest management guidelines of Cornell University (2012) suggest treating in late April through to early May with 58-148 GDDs, and to mid–late May with 98-155 GDDs, using carbaryl, fenazaquin, fenpyroximate, fluvalinate, horticultural oil or milbemectin.

The Genus *Metaculus* Keifer

Diagnostic characteristics

The tibial setae are present. The coxae I are narrowly joined near the setae *1a*, and the coxal setae *1b* are absent (Keifer, 1962b; Amrine *et al.*, 2003).

Economic importance

This genus numbers over ten species reported from Africa, Europe and Asia. The species type, *Metaculus syzygii* Keifer, is a rust mite on the undersides of the leaves of *Syzygium cordatum* (Hochst.) in Congo (Zaire) (Keifer, 1962b). *M. mangiferae* (Attiah) damages the terminal buds and inflorescences of mango (*Mangifera indica* L.) in various countries, including India (Jeppson *et al.*, 1975), Egypt (Attiah, 1955; Abou-Awad *et al.*, 2011a) and Thailand (Carlos Flechtmann, personal communication). *M. rapistri* Carmona wrinkles and yellows the leaves of *Rapistrum rugosum* (L.) All. in Portugal (Carmona, 1969). *M. londaensis* Mohanasundaram, *M. sapindiphagus* Mohanasundaram, and *M. foveolatae* Mohanasundaram are reported from India, and *M. foveolatae* was found on a medicinal plant called 'maa krathuep rong' (*Ficus foveolata* Wall. ex Miq.) (Mohanasundaram, 1983). *M. autumnali* Meyer is reported from *Dombeya autumnalis* I. Verd., *M. capassae* Meyer from *Lonchocarpus capassa* Rolfe and *M. magalismontani* Meyer from *Bequaertiodendron* in Transvaal and *M. natalensis* Meyer from *Rhus* in Natal (South Africa) (Meyer Smith, 1990b). Of these species, only *M. mangiferae* causes economic damage.

Metaculus mangiferae (Attiah) (Fig. 16.75)

Common name

Mango rust mite.

Diagnostic characteristics

FEMALE. The body is fusiform, 170–240 μm long and light yellowish in colour. The prodorsal shield is subtriangular, with a pattern consisting of a network of lines. The median line develops on the rear ⅔. The admedian lines are irregular owing to cross lines and angles with the lateral lines. There is a cross line on the shield at ¼, a second just behind the ½ point, and a third just ahead of the rear margin. The submedian lines make a network of cells outside the admedian lines, and there is a row of lateral cells around the shield margins. The scapular tubercles are located at the rear shield margin, directing

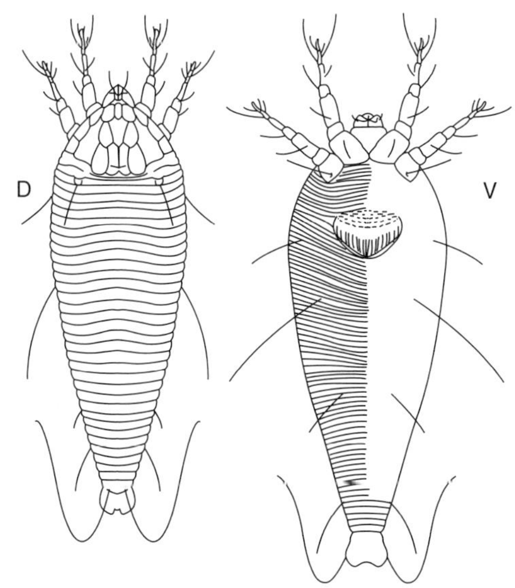

Fig. 16.75. *Metaculus mangiferae* (Attiah) (from Attiah, 1955). See Table 16.1 for key.

the scapular setae (*sc*), which are about 19 μm long, caudad. The opisthosoma is microtuberculate and there are 28–35 dorsal annuli not much wider than the ventral annuli (about 80). The opisthosomal dorsal annuli have a shallow subdorsal furrow for a short distance behind the shield. The setae c_2 are 24 μm long, set on annulus 18 anteriorly; the setae *d* are 56 μm long, set on 14 annulus behind; the setae *e* are 19 μm long, set on annulus 20 behind; the setae *f* are 30 μm long, set on 23 annulus behind and on the 5th annulus from the rear; the setae h_1 and h_2 are present, and h_2 are 64 μm long. The genital coverflap has granular cross lines basally, and about 24 longitudinal ribs; the seta is 14 μm long. The empodium is 4-rayed (Attiah, 1955; Jeppson *et al.*, 1975).

MALE. The males are similar to the females but smaller, about 170 μm long (Attiah, 1955).

Geographical distribution

Angola (Meyer Smith, 1990b), Egypt (Attiah, 1955; Abou-Awad *et al.*, 2011a), India (Jeppson *et al.*, 1975) and Thailand (Carlos Flechtmann, personal communication).

Host plants

Anacardiaceae: mango (*Mangifera indica* L.) (Attiah, 1955; Jeppson *et al.*, 1975; Meyer Smith, 1990b; Abou-Awad *et al.*, 2011a).

Bio-ecology

M. mangiferae is a vagrant species, which may be misidentified with *Aceria mangiferae* (Hassan) and *Tegonotus mangiferae* (Keifer); all three species live on mango. At 25–29 °C and 70% RH on excised young terminal leaves of mango it develops a generation in about 7.5 days. Females produce 1–3 larvae daily. The mite overwinters under the scale leaves of the buds and its density increases in spring and early summer, with peaks in late April and early June (in Egypt) (Abou-Awad, 1981b). Abou-Awad (1981b) reports that the mite reproduces by obligate ovovivipary. However, senescent females of eriophyoids show only apparent ovoviparity and/or apparent viviparity, as occurs in *Aceria guerreronis* Keifer. In general, there is no true ovoviparity or vivipary in eriophyoids and reports concerning their occurrence should be considered erroneous (Návia *et al.*, 2005a). At 25, 30 and 35°C, the female life cycle takes 6.29, 4.90 and 4.44 days, respectively, while the development of the male takes 5.20, 4.27 and 3.74 days, respectively. Insemination occurs after female emergence, and the mating process maximizes reproduction, producing lower numbers of larvae from unmated females than mated females. At 25, 30 and 35°C, the fecundity is 20.53, 24.50 and 32.20 larvae, during a reproduction period that averages 15.93, 17.40 and 17.06 days, respectively. The reproduction period is shorter at 25°C and 60% RH and longer at 30°C and 50% RH. At 25°C and 60% RH, the net reproduction rate (R_0), the mean generation (T), the intrinsic rate of increase (r_m) and the finite rate of increase (e^{rm}) are 13.02, 18.42, 0.139 and 1.14, respectively; at 30°C and 55% RH, the same parameters are 15.04, 16.22, 0.167 and 1.18, respectively, while at 35°C and 50% RH, they are 22.35, 16.05, 0.193 and 1.21, respectively. At 25°C and 60% RH and 30°C and 55% RH, the sex ratio is 1.72 females:1 male, and at 35°C and 50% RH it is 2.30:1 (Abou-Awad *et al.*, 2011a).

The distribution of *M. mangiferae* in buds shows three annual peaks of seasonal abundance in Egypt. Flowering buds are a source of infestation for the new terminal buds of the new year, and the increases occur during January–February, June and October. In March and December, the population suddenly decreases. Populations of *M. mangiferae* also reach the leaves in three annual peaks; the first peak occurs in January–February, the second in May and the third (the highest peak) occurs in October (Abou-Awad *et al.*, 2011a).

The mango rust mite develops about 11–16 generations a year on buds and leaves. The longest generation is throughout February–March and March–April or April–May, and lasts for about 35–49 days for both buds and leaves. Population development starts in April or May and gradually increases until it reaches a peak in June for both buds and leaves; it then fluctuates and declines from July to October, and tails off in November and December. A comparative study of population densities at different levels in the tree showed that rust mite density was significantly higher in the lateral buds than in the terminal buds. In the bottom branches of the trees up to a height of 150–200 cm above ground level, the buds host relatively fewer mites compared with the top level (upper 80–100 cm of the branches) and the middle level (in between the top and the bottom). Individuals prefer hibernation under the scale leaves of the vegetative buds in the period from the end of October to December. At the beginning of the growing season in January and February, a remarkable number of the first instar nymphs appear in the swollen vegetative buds. These buds are the main source of infestation for the new leaves in spring and autumn. The density of the mango rust mite on the upper and lower leaf surfaces at the three vertical tree levels shows that the distribution is significantly higher on the upper leaf surfaces, especially the top level (Abou-Awad *et al.*, 2011a).

Natural enemies include some phytoseiid mites, such as *Amblyseius swirskii* Athias Henriot (Abou-Awad, 1981b) and *Typhlodromus mangiferus* Zaher *et* El-Brolossy (Abou-Awad *et al.*, 2010b).

Symptomatology and damage

The mango rust mite causes russeting of the terminal leaves, buds and inflorescences. Young seedlings show brooming consisting of a crowding of short shootlets called a 'bunchy top'. The growth of buds and inflorescences stops and they dry off. High levels of attack induce arrested terminal buds. The damage to inflorescences results in a compact mass of malformed flowers. The infested panicle produces flower buds when fruit setting has concluded. Malformed inflorescences produce a low percentage of hermaphrodite flowers, while the fruits either drop or are pea sized. Malformations usually occur on seedlings when the plants are 5 months old. Trees from 4 to 8 years old may have about 50% of damaged shoots. Late-blooming varieties have low damage (Jeppson *et al.*, 1975; Abou-Awad, 1981b). At high levels of infestation, the leaves become curled and roll around themselves (Abou-Awad *et al.*, 2011a).

Control

Biological control does not really work, so control is entrusted to spray applications of acaricides.

CHEMICAL CONTROL. Chemical control of the mite is easy because it lives in more-or-less exposed conditions. Sulfur gives good results (Attiah, 1955; Singh *et al.*, 1961). Recently, abamectin, chlorfenapyr, sulfur, methoxyfenozide and azadirachtin have been tested (Abou-Awad *et al.*, 2011a).

The Genus *Heterotergum* Keifer

Diagnostic characteristics

The body is fusiform, larger anteriorly and tapering caudally. The gnathosoma is small, and the chelicerae are nearly straight. The prodorsal shield is triangular in dorsal view, and the frontal lobe over the gnathosoma is acuminate and short; the scapular tubercles are large, located on the rear margin; the scapular setae (*sc*) point backwards and are long. The legs have all the standard segments and setae. The opisthosoma is evenly curved dorsally, the first 4–6 narrow microtuberculate annuli are behind the shield completely encircling the opisthosoma, followed by a series of broad dorsal annuli, which become gradually reduced caudally. The first 4–6 body annuli and the subsequent ventral annuli are completely microtuberculate; the broad dorsal annuli have obscure elongate microtubercles. The female genitalia are a moderate distance behind the coxae; the internal anterior apodeme has a normal length (Keifer, 1955a; Amrine *et al.*, 2003).

Economic importance

This genus is widespread and includes 12 species at present. The type species, *Heterotergum gossypii* Keifer is reported on cotton (*Gossypium hirsutum* L.) from Brazil (Keifer, 1955a). A number of other American species are known. These are *H. olneyae* Keifer collected on *Olneya tesota* Gray in Arizona (Keifer, 1963a), *H. schlingeri* Wilson *et* Oldfield living on *Franseria chenopodiifolia* Benth. in northern Mexico (Wilson and Oldfield, 1966), *H. tuttlei* Keifer reported on *Trixus californicus* Kelley from Arizona (Keifer, 1961a), *H. wilsoni* Keifer found on *Simmondsia californica* Gray in California (Keifer, 1956), *H. yumensis* Keifer living on *Acacia greggii* Gray in Arizona (Keifer, 1966b) and *H. zexmeniae* Keifer detected on *Zexmenia* sp. in Mexico (Keifer, 1973). *H. artemisiae* Hong *et* Kuang is reported from *Artemisia lavandulaefolia* DC. in China (Hong and Kuang, 1989) and *H. shevtchenkoi* Bagdasarian from *Pistacia mutica* Fischer *et* C.A. Mey in the former USSR (Bagdasarian, 1972). Some species, such as *H. crispae* Meyer living on *Euclea crispa* (Thunb.) Guerke, and *H. munduleae* Meyer living on *Mundulea sericea* (Willd.) A. Chev., are reported from South Africa (Meyer Smith, 1992), while *H. flacourtiae* Abou-Awad *et* Nasr is known from *Flacourtia indica* Merr. in Egypt (Abou-Awad and Nasr, 1983). Among the above species, only *H. gossypii* has some economic importance.

Heterotergum gossypii Keifer (Fig. 16.76)

Common name

Cotton rust mite.

Diagnostic characteristics

FEMALE. The body is 150–215 μm long and light yellow in colour. The gnathosoma projects down. The prodorsal shield is triangular, and the frontal lobe is short and acute; the pattern is obscure, laterally granular above the coxae; the scapular tubercles are 22 μm apart on the rear margin; the scapular setae (*sc*) are 61 μm long, projecting backwards. Coxae I are broadly contiguous. The opisthosoma has about 14 broad dorsal annuli bearing obscure elongate microtubercles; the first dorsal annulus is separated from the shield by 4–6 narrow microtuberculate annuli extending around the opisthosoma from the respective dorsal annuli. The caudal 5–6 body annuli are similar dorsoventrally. There are 63-68 dorsal annuli, completely microtuberculate. The setae c_2 are 7.5 μm long, set on about ventral annulus 5; the setae *d* are 50 μm long, set on about ventral annulus 22; the setae *e* are 65 μm long, set on ventral annulus 38; the setae *f* are 16 μm long, set on ventral annulus 5 from the rear; the setae h_1 are present. The genitalia are 20 μm wide and 16 μm long; the genital coverflap has obscure microtubercles basally and about 12 longitudinal furrows; the seta is 13 μm long. The empodium is 5-rayed (Keifer, 1955a).

MALE. The male is similar to the female but smaller (Keifer, 1955a).

Geographical distribution

Brazil (Keifer, 1955a).

Host plants

Malvaceae: cotton (Keifer, 1955a).

Bio-ecology

H. gossypii has been collected on cotton plants in Brazil (Keifer, 1955a).

Symptomatology and damage

The feeding activity of the mite causes bronzing of mature leaves and tip blighting of young leaves (Jeppson *et al.*, 1975).

Control

Control is not usually necessary.

The Genus *Anthocoptes* Nalepa

Diagnostic characteristics

The opisthosomal dorsal annuli are broad, smooth and abruptly different from the narrow annuli beyond the setae *f* (Amrine *et al.*, 2003).

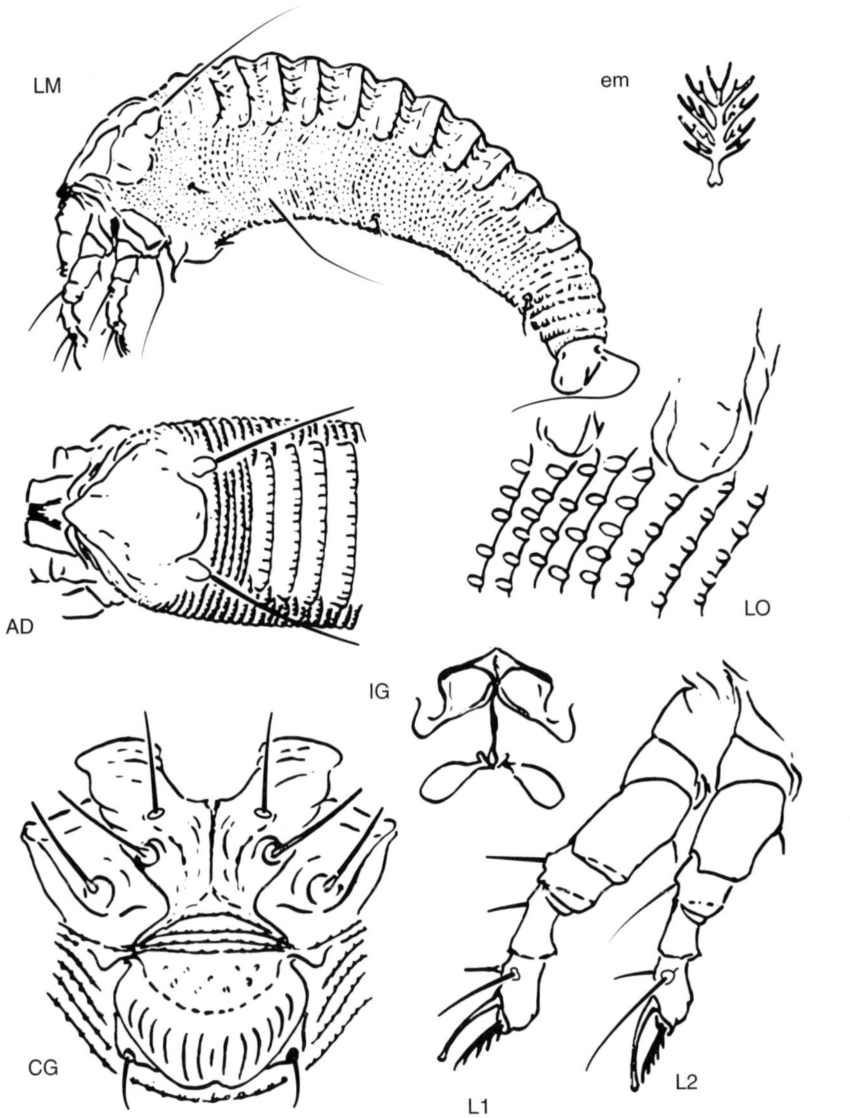

Fig. 16.76. *Heterotergum gossypii* Keifer (from Keifer, 1955a). See Table 16.1 for key.

Economic importance

This genus includes over 40 species living on plants belonging to about 20 dicotyledonous families and one monocotyledonous family. Among these, *Anthocoptes bakeri* Keifer is reported from honey locust (*Gleditsia triacanthos* L.) in eastern North America (Jeppson *et al.*, 1975) and Italy (de Lillo and Beron, 2004). Different European species inhabit ornamentals, such as *A. galeatus* (Nalepa), which was recently reported from European white elm (*Ulmus laevis* Pall.) in nurseries in Poland (Łabanowski and Soika, 1998), *A. loricatus* (Nalepa) reported from filbert (*Corylus avellana* L.) in Europe and the Near East (de Lillo and Beron, 2004) and causing non-serious leaf rust, and *A. transitionalis* Hodgkiss reported as an alien invasive species in Montenegro and Serbia (Glavendekić *et al.*, 2005). The most important species seems to be *A. bakeri*.

Anthocoptes bakeri Keifer (Fig. 16.77)

Common name

Honey locust rust mite.

Diagnostic characteristics

FEMALE. The body is robust, spindle formed, 120–140 μm long and dull yellowish in colour. The gnathosoma projects down. The prodorsal shield is curved laterally and with the centre somewhat raised and outlined, extending from the frontal lobe to the rear margin and expanding to the rear; the frontal lobe is short and broadly rounded. The scapular tubercles are widely spaced, 32 μm apart on the rear margin; the scapular setae (*sc*) are 25 μm long, diverging to the rear. Coxae I are narrowly

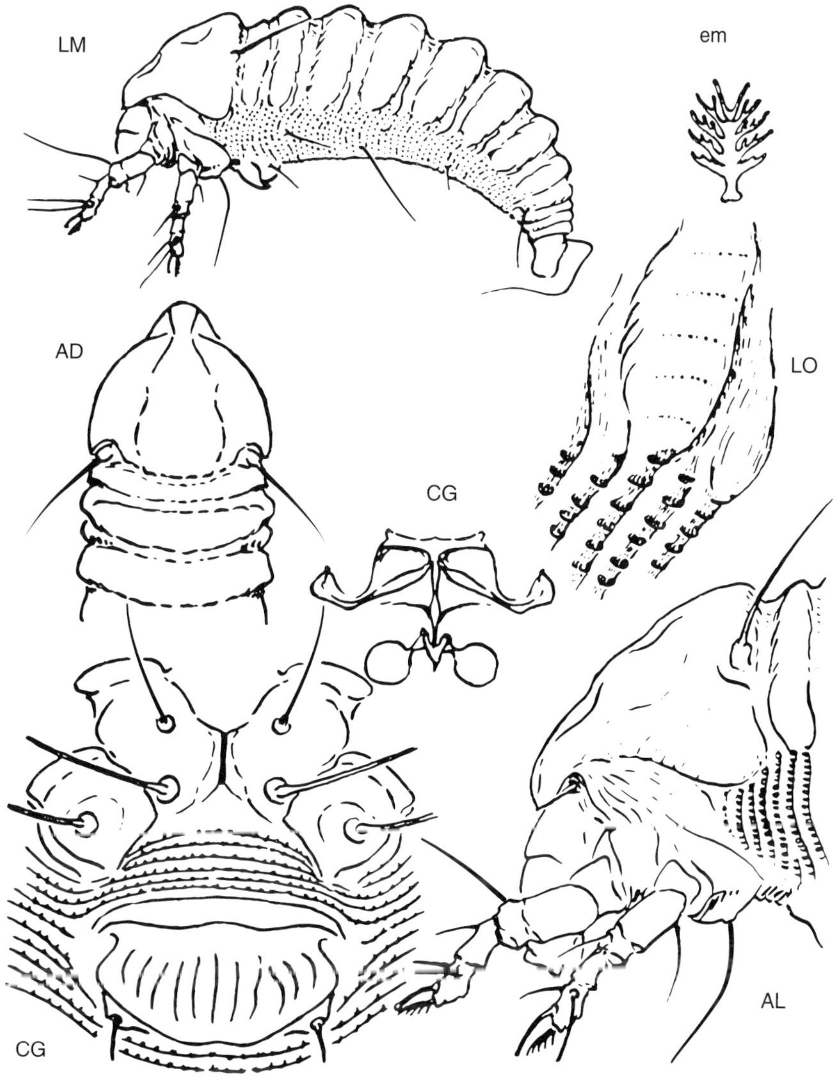

Fig. 16.77. *Anthocoptes bakeri* Keifer (from Keifer, 1959c). See Table 16.1 for key.

contiguous centrally. The opisthosoma has about 13 dorsal annuli, the first narrow and depressed, followed by 7–9 broad rounded dorsal annuli, there are no distinct microtubercles on the dorsal annuli; there are about 50–55 narrow ventral annuli heavily set with elliptical microtubercles touching the rear margin of each ventral annulus. The setae c_2 are 16 µm long, set on about ventral annulus 7; the setae *d* are 40 µm long, set on about ventral annulus 20; the setae *e* are 17 µm long, set on about ventral annulus 35; the setae *f* are 23 µm long, set on ventral annulus 5 from the rear; the setae h_1 are 2.5 µm long. The genitalia are 20 µm wide and 15 µm long; the genital coverflap has 10–12 longitudinal furrows. The empodium is 5-rayed (Keifer, 1959c).

Geographical distribution

Italy (de Lillo and Beron, 2004), Maryland (Keifer, 1959c) and eastern North America (Jeppson *et al.*, 1975).

Host plants

Fabaceae: honey locust (*Gleditsia triacanthos* L.) (Keifer, 1959c; Jeppson *et al.*, 1975).

Bio-ecology

A. bakeri lives on underleaf surfaces of honey locust in eastern North America (Keifer, 1959c; Jeppson *et al.*, 1975) and Italy (de Lillo and Beron, 2004).

Symptomatology and damage

The feeding activity of the honey locust rust mite sometimes causes serious leaf damage (Jeppson *et al.*, 1975).

Control

Control is usually not necessary.

The Genus *Aculus* Keifer

Diagnostic characteristics

The body is usually spindle formed, and is generally somewhat dorsoventrally flattened. The gnathosoma is relatively small, projecting diagonally down; the chelicerae are short, nearly straight; the oral stylet is short, recurving well below the base of the chelicera. The prodorsal shield is subtriangular, with the frontal lobe overhanging the gnathosoma; the protogynes have a pair of forward-directed small spines from the anterior point of the lobe, or from the underside. The scapular tubercles are located on the rear margin, well spaced, projecting backwards and directing the scapular setae (*sc*) divergently to the rear. The opisthosoma of protogynes usually has broader dorsal annuli that are strongly differentiated from the ventral annuli; often with a slight subdorsal furrow extending a short distance caudad from the rear shield margin. The setae c_2, *d*, *e* and *f* are present. The female genital coverflap usually has furrows; the internal apodeme is of a normal length and is broad. Deutogynes lack frontal lobe spines, and have the dorsal and ventral annuli less differentiated and the microtubercles more or less suppressed (Keifer, 1959b).

Economic importance

This genus includes over 270 species, some of which cause economic damage. The plum rust mite, *Aculus fockeui* (Nalepa *et* Trouessart), has a worldwide geographical distribution and damages the leaves of plum, peach, nectarine, almond and cherry orchards (*Prunus* spp.) (Oldfield, 1996a; Kondo and Hiramatsu, 1999a). The apple rust mite, *A. schlechtendali* (Nalepa), is a widespread pest causing damage to the terminal growth and rusts of the leaves and fruits of apple (*Malus domestica* Borkh.) (Easterbrook, 1996). The hedge privet rust mite, *A. ligustri* (Keifer), causes damage to the leaves of hedge privet (*Ligustrum ovalifolium* Hassk) in California (Keifer, 1938a); in Iran, the same mite is also reported on *L. vulgare* L. and *Ligustrum* sp. (Hajizadeh and Hosseini, 2004; Xue *et al.*, 2009a). The azalea bud and rust mite, *A. atlantazaleae* (Keifer), infests native creeping azalea in eastern North America and causes russeting (Keifer, 1940c; Meyer Smith, 1996a). The shagbark hickory rust mite, *A. caryfoliae* Keifer, is reported from shagbark hickory (*Carya ovata* Mill.) leaves in eastern North America, and may cause severe leaf rusting (Keifer, 1961a). Some species, such as *A. hyperici* (Liro), have been proposed for biological control of St John's wort (*Hypericum perforatum* L.), a perennial European weed, also found in eastern Australia, where it is toxic to livestock and competes with favourable pasture plants (Rosenthal, 1996). Many other species occur on plants of economic importance, but their presence does not usually require control. *A. fockeui*, *A. schlechtendali* and *A. ligustri* are the most economically important species and are described below.

Aculus fockeui (Nalepa *et* Trouessart) (Figs 16.78 and 16.79)

Common name

Plum rust mite.

Diagnostic characteristics

FEMALE. The body is fusiform. The protogyne form is 160–175 μm long. The prodorsal shield is subtriangular, with the frontal lobe slightly acute. The prodorsal shield pattern consists of lines, of which only the central line on the frontal lobe is clear. A marked transverse line occurs at ¼. The admedian lines are barely visible, and starting from the transverse line run back divergently. The opisthosoma has about 30 dorsal annuli and 50 ventral annuli. The dorsal microtubercles are elongate and faint. On the ventral annuli, these microtubercles are elliptical and touch the annulus margin. The genital coverflap has basally curved transverse lines, and about 16 longitudinal subapical ribs. The empodium is 4-rayed.

The deutogyne is smaller than the protogyne. The frontal lobe is thinner than on the protogyne and without a small pair of spines. The opisthosoma has narrower annuli, especially dorsally. The dorsal microtubercles are unclear but shorter than on the protogyne (Jeppson *et al.*, 1975).

MALE. The male is similar to the female, but smaller.

Geographical distribution

Worldwide.

Host plants

Rosaceae: almond (*Prunus amygdalus* Batch.) (Attiah, 1970; Rice and Jones, 1978), apricot (*P. armeniaca* L.) (Boczek *et al.*, 1984), sweet cherry (*P. avium* L.) (Kozłowski, 1983; Oldfield, 1984), myrobalan plum (*P. cerasifera* L.) (Boczek, 1974; Schliesske, 1977a,b; Oldfield, 1984), sour cherry (*P. cerasus* L.) (Keifer, 1952a; Boczek, 1974; Schliesske, 1977a; Boczek *et al.*, 1984), *P.* × *cistena* (Schliesske, 1977a), European plum (*P. domestica* L.) (Putman, 1939; Keifer, 1952a; Proeseler, 1972b; Boczek, 1974; Schliesske, 1977a,b, 1981; Boczek *et al.*, 1984), *P. insistitia* (= *P. domestica* subsp. *interstitia* (L.) Bonnier *et* Layens) (Schliesske, 1977b), *P. italica* Borkh. (= *P. domestica* subsp. *interstitia* (L.) Bonnier *et* Layens) (Schliesske, 1977c), mahaleb cherry (*P. mahaleb* L.) (Keifer, 1952a; Proeseler, 1972b; Schliesske, 1977a,b; Oldfield, 1984), peach (*P. persica* (L.) Batsch) (Schliesske, 1977a; Boczek *et al.*, 1984; Oldfield, 1984), *P. pseudocerasus* Lindl. (Schliesske, 1977a), *P. spinosa* L. (Schliesske, 1977b), *P. triloba* Lindl. (Schliesske, 1977a).

Bio-ecology

A. fockeui is structurally indistinguishable from *A. cornutus* (Banks), and successful interbreeding experiments have showed synonymy between the two mites (Oldfield, 1984). The mite is a vagrant on about 17 wild and cultivated *Prunus* spp. (Schliesske, 1977a), including various commercial stone fruit and nut trees in orchards and in nurseries, such as almond, apricot, mahaleb cherry, sour cherry, sweet cherry, European plum, myrobalan plum and peach (Putman, 1939; Attiah, 1970; Proeseler, 1972b; Boczek, 1974; Jeppson *et al.*, 1975; Schliesske, 1977a, 1981; Rice and Jones, 1978; Boczek *et al.*, 1984; Oldfield, 1984).

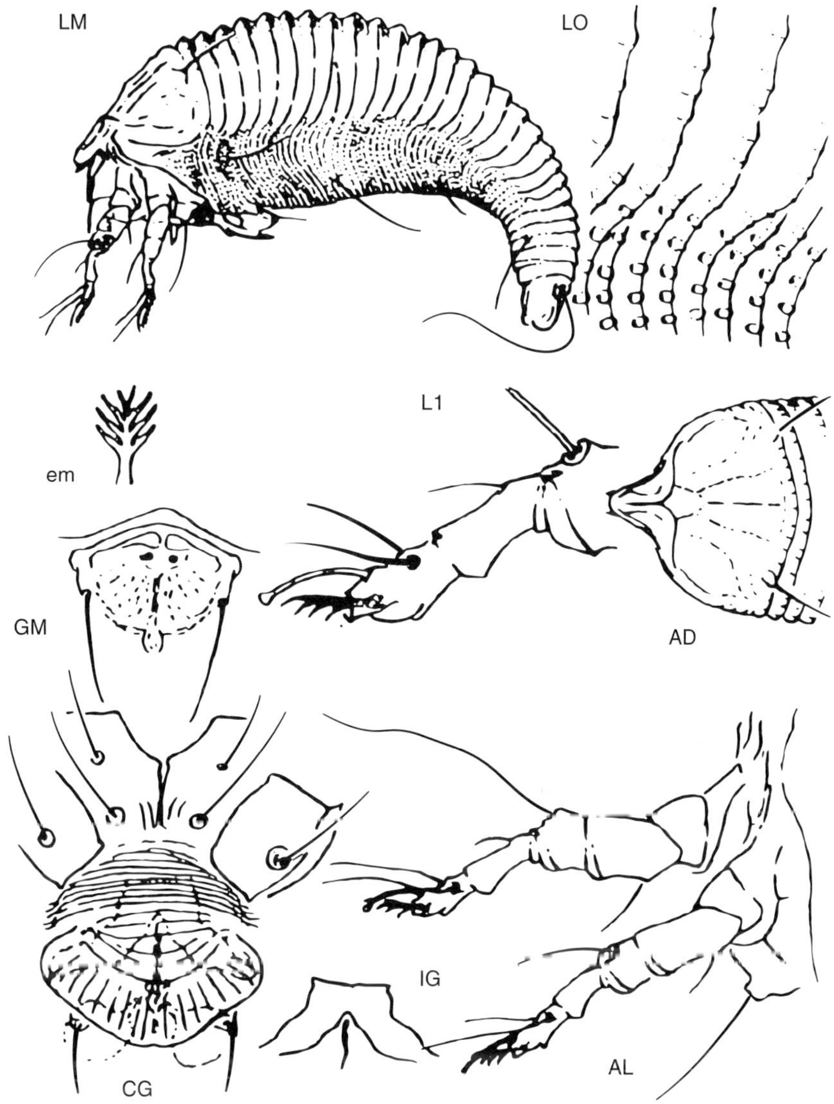

Fig. 16.78. *Aculus fockeui* (Nalepa *et* Trouessart), protogyne (from Keifer, 1952a). See Table 16.1 for key.

The plum rust mite overwinters as deutogynes in various niches on or near the last growth, protected in bark crevices, behind active buds and under bud scales. The expanding of buds in the spring induces the deutogynes to move towards the underside of new leaves, where they begin to feed and lay eggs, commonly along the midrib (Putman, 1939; Boczek *et al.*, 1984). The first generation consists of males and protogynes, and spreads towards the tips of shoots (Schliesske, 1981). The mite usually lives on the lower leaf surface near the midrib and disperses as its populations increase on both leaf surfaces (Barké *et al.*, 1972). These develop until midsummer and active forms are found on the new sucker growth as long as they are fresh. With their hardening, mature leaves induce the appearance of new deutogynes, which after insemination move towards the overwintering sites (Massee, 1932; Gilmer and McEwen, 1958). In Poland, on European plum, sour cherry and myrobalan plum, the deutogynes are first observed in the latter half of August (Boczek, 1974). In Germany, on European plum, the deutogynes are present in early June as 70% of females, but by mid-July less than 30% of females are deutogynes. The percentage of deutogynes increases again from early August, and by early November there are generally no protogynes present (Proeseler, 1972b). In Yugoslavia, the migration of deutogyne females occurs in late April and the first half of May, but only in a very small number of individuals, depending on the temperatures at this time of year. The population density starts to increase in June, reaching a maximum in the second half of July and in August. The peak population ranges from 285.0 to 370.4 mobile forms cm^{-2} of leaf area (Pesic *et al.*, 1998). In China, the adults start to overwinter in mid-November (Huang *et al.*, 1994).

The plum rust mite reproduces by arrhenotoky and unfertilized eggs develop only into males, which last represent 20–30% of the adult population (Putman, 1939). Pre-hibernation

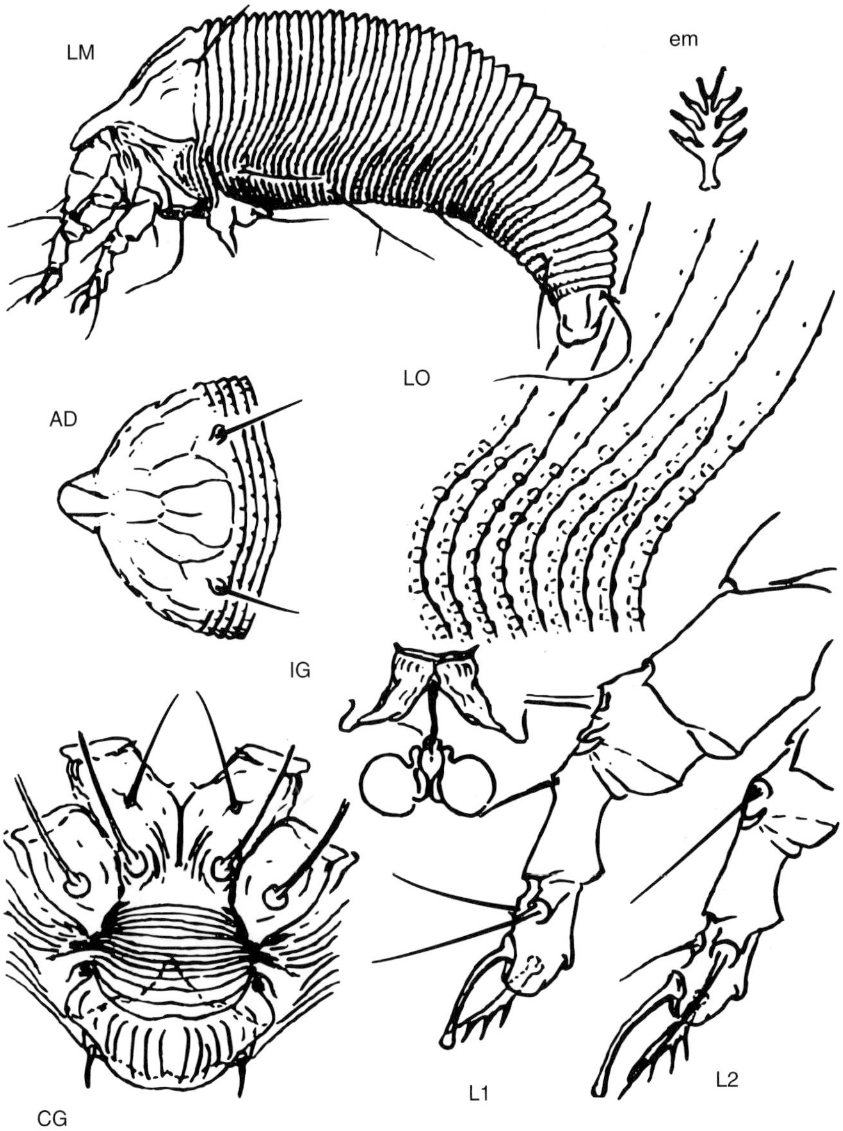

Fig. 16.79. *Aculus fockeui* (Nalepa *et* Trouessart), deutogyne (from Keifer, 1952a). See Table 16.1 for key.

insemination occurs by newly emerged deutogynes visiting spermatophores. Spermatozoa are detected in the spermathecae of deutogynes in autumn on mature leaves and in overwintering sites. Spermatophores contain about 50 spermatozoa (Oldfield and Newell, 1973a), and are frequent along the mid-vein of the adaxial leaf surface. The deutogynes visit single spermatophores deposited on the leaves by males of the same generation, and after diapause they generate, like protogynes, both sexes, mostly females for several days and then only males (Oldfield and Newell, 1973b). The male visits the pre-emergent females and deposits a single spermatophore in her vicinity, so that pre-emergent females are often encircled with spermatophores from different males (Michalska and Boczek, 1991). In the absence of females, males start to deposit spermatophores less than 1 day after eclosion, and at 26°C they deposit about 30 spermatophores a day for about 2 weeks (Oldfield *et al.*, 1970). Uninseminated protogynes can walk directly to a spermatophore or walk past one or more spermatophores until they bump one with their legs. They kick the spermatophore and walk in circles until it is placed ahead. The female commonly mounts a spermatophore within a minute of access to it. The sac of the spermatozoa is transferred through the genital orifice, up the spermathecal duct and into one of the pair of the spermathecae (Oldfield *et al.*, 1972) or both (Oldfield, 1999; Michalska and Mańkowski, 2006). The protogynes visit the spermatophores after eclosion and start to deposit eggs, laying 3–4 eggs a day over about 2 weeks (Oldfield and Newell, 1973a). Protogynes inseminated from newly deposited spermatophores produce higher numbers of female progeny (Oldfield *et al.*, 1970), while insemination with 3-day-old spermatophores produces mostly males, showing that the viability of the spermatozoa may decrease in older spermatophores (Oldfield and Michalska, 1996).

Studies on the interspecific attraction of spermatophores showed that females of *A. fockeui* searched for and mounted

spermatophores both from males of the same species and congeneric males (of *A. schlechtendali*), and readily accomplished self-insemination from spermatophores from both, although there was no evidence of fertilization. Females do not visit spermatophores deposited by males of different genera (Oldfield, 1989).

Populations of *A. fockeui* are active from 22 to 31°C but stop activity at 21°C (Barké *et al.*, 1972). On plum and sour cherry at 23°C, egg development lasts 4–4.5 days, and the larval nymphal stage lasts 3–3.3 days, but at 33°C it takes 2.5 days; egg to egg development takes 7.2 days. At 28°C (without the pre-oviposition period), egg to egg development lasts 5.8–6.1 days, the fecundity is 30.8–41.3 eggs and daily oviposition is 1.1–1.5 eggs (Boczek *et al.*, 1984). On nectarine and peach leaflets with 70% RH and below 20°C, the mite stops all activity, and by 33°C, the adult begins to slow down and ceases all activity. At 29°C, at least 34% of the generation time is spent in the egg stage. Adult longevity decreases with increasing temperature. The fecundity is highest (43.16 eggs per female) at 29°C. At this temperature, the population on nectarine leaves multiplied 30.04 times in a generation time of 18.49 days, and on peach leaves the population increased 13.87 times in a generation time of 19.18 days (Abou-Awad *et al.*, 2010a). In China, the development of *A. fockeui* occurs from 17 to 29°C; the egg stage lasts 3–5 days and the nymphal stage 5–7 days. A generation develops within 11 days. Effective accumulated temperature and temperature threshold are 131.4°D and 12.6°D, respectively; at 26–27°C, the adult lifespan is 3–11 days. The mite aggregates on the lower surface of leaves (Huang *et al.*, 1994). In Georgia (USA), field populations of *A. cornutus* on peach below 50 mites per leaf occur more commonly on the lower leaf surface near the mid-vein; at above 100 mites per leaf, the pest spreads to all areas and both leaf surfaces. The mites are active on a leaf at temperatures between 22 and 31°C, with nymphs active up to 32.5°C. Below 24°C, the mites are sluggish, and their activity stops at 21°C (Barké *et al.*, 1972). In Florida, the mite first appears in significant numbers in the last week of May on unsprayed peach trees, peaks in early June, and disappears in early July. Between 50 and 3400 mite days per leaf accumulate per tree, the percentage of leaf damage varies between 0 and 88% of the adaxial leaf surface (Andersen and Mizell, 1987). In Japan, on peach leaves (cv. Shigahakutou), *A. fockeui* appears in June, and the numbers of the mite increase up to 704 per leaf in July (Kunugi *et al.*, 1993).

In Poland, in midsummer, the mite develops about 11 generations a year on European plum (Boczek *et al.*, 1984); in Germany, there are 12 generations a year (Schliesske, 1981). Usually, the presence of expanded leaves and an increase in natural enemies slows population growth down (Putman, 1939), while summer rains and an increase of humidity induce the persistence of populations (Schliesske, 1981). Dispersal occurs by wind, tree transportation, rain and human activities (Barké *et al.*, 1972; Schliesske, 1977c).

Natural enemies include the fungus *Sporothrix schenckii* Hektoen *et* Perkins (Schliesske, 1981), the anthocorid *Orius insidiosus* Say (Putman, 1939) and several mites, among which are the phytoseiids *Typhlodromus pyri* Scheuten and *Amblyseius finlandicus* Oudemans (Collyer, 1964b,c), *A. andersoni* (Chant) and *A. stipulatus* Athias Henriot (Castagnoli *et al.*, 1984), and the stigmaeids *Zetzellia mali* (Ewing) (Rice *et al.*, 1976), *Z. talhouki* Dosse (Bayan, 1988) and *Agistemus collyerae* (Gonzalez) (Castagnoli *et al.*, 1984).

Symptomatology and damage

On young leaves of *Prunus* stone fruit trees, the feeding activity of *A. fockeui* causes a toxaemia reported as 'yellow spot' of peach in California (Wilson and Cochran, 1952), 'chlorotic fleck' of myrobalan plum in New York (Gilmer and McEwen, 1958) or 'asteroid spot' of *Prunus* fruit trees in Europe (Vukovits, 1961). The feeding activity of the mite on mature leaves produces silvering on peaches and rusting on other stone fruit species. Toxaemia on *Prunus* reported from North America and Europe consists of reactions of the host plants to a salivary phytotoxin secreted by *A. fockeui* (Oldfield, 1996b). The elimination of mites from symptomatic plants induces the disappearance of these symptoms (Costa *et al.*, 1962).

On young leaves of 'Levell' peach seedlings, mite feeding develops yellow spot, usually near to the veins. Each spot measures about 1 mm in diameter and has a circular to irregular shape, at first showing diffuse margins and later sharp margins. The coalescence of spots causes a mottled effect. The toxins injected into the vascular system that move a short distance seem to cause chlorosis near the veins. Symptomatic leaves have altered palisade cells and spongy parenchyma cells; the disintegrated contents include refractive bodies that stain green with methyl green and acid fuchsin and not the red colour of normal cells (Wilson and Cochran, 1952). The saliva of *A. fockeui* affects different peach varieties differently. On 'Salberta' peach seedlings, red borders occur and current-season stems have yellow areas that become raised and darker in time. Vein-associated chlorosis occurs rarely on 'Hale' and 'Elberta' varieties of peach, which show little yellow spots and wider red spots with central necroses and drop out. The 'Hale' variety may show puckered leaves with longitudinal rolling. Peach varieties with leaf glands do not develop yellow spots (Oldfield, 1996b).

Chlorotic fleck symptoms develop on young leaves of myrobalan plum, but not on the mature leaves upon which the mites have been feeding. They first appear when mites feed on embryonic spring leaves, when symptoms develop from 13 to 15 days after the start of feeding. The flecks consist of well-defined circular yellow areas, 1–2 mm in diameter. Small spots often occur but wider spots show well-defined margins and may have concentric rings. Leaves bearing about 50 spots are wavy or slightly twisted on their longitudinal axes. High levels of attack may cause rosette shoots and leaves that are prevented from expanding to their regular size. Twigs bear green spots, more or less ovoid and usually larger than leaf spots (Hildebrand, 1945; Gilmer and McEwen, 1958).

In Romania, *A. fockeui* transmits *Plum pox virus* (PPV), a *Potyvirus*, in plum orchards. In the cultivars 'Centenar', 'Pescarus', 'd'Agen', 'Stanley' and 'Tuleu Gras', the transmission rate ranges from 20 to 60%, while in the susceptible cultivars 'Vanat Romanesc' and 'Vanat de Italia', the transmission rate ranges from 40 to 80% (Isac *et al.*, 1998).

In plum nursery trees in Yugoslavia, infested trees show a change in coloration, mild inward rolling and brittleness of leaves, brooming and stunted growth. The 'Cacanska Rodna'

cultivar is one of the most susceptible to the plum rust mite. Nursery trees measured at the end of the season showed an average height from 198.0 to 227.4 cm, with an average diameter from 1.21 to 1.31 cm. The average growth increment between July (when the mite population density rapidly increases) and at the end of the season is statistically significant, but the coefficient of correlation between tree height and population density is less than 1%, and the growth of trees is not significantly affected by mite attacks (Pesic *et al.*, 1998).

In peach, there is an inverse relationship between the CO_2 assimilation rate, stomatal conductance and transpiration rate of the leaves and mite days of feeding by *A. fockeui*. In particular, 3000 mite days per leaf reduce the CO_2 assimilation rate and stomatal conductance by about 20%. There is a negative relationship between these parameters and the damage to the adaxial leaf epidermis. The mite is primarily an epidermal feeder, and its feeding may inhibit leaf gas exchange by affecting stomatal function. *A. fockeui* is at least tenfold less debilitating to leaf gas exchange of peach than the two-spotted spider mite, *Tetranychus urticae* (Koch), and may not ordinarily need to be controlled by chemical sprays (Andersen and Mizell, 1987). Low levels of populations do not affect photosynthetic processes, but high levels cause silvering and reduce photosynthesis by more than 70% and respiration by more than 50%. Leaves of *P. cerasifera* injured by the plum rust mite show reduction in carbohydrate content and increased nitrogen, Ca^{2+}, Na^+, K^+ and chlorophyll contents. The injury varies according to the species of *Prunus* (Zawadzki, 1975).

Usually, the European plum is more severely damaged than peach. In a plum nursery in Canada, injury was reported on terminal shoots, associated with curling and dwarfism of leaves, but sour cherry was free of injury (Putman, 1939). In Germany, the European plum and the mahaleb cherry are affected in spring by chlorotic fleck, and later by browning of the leaves (Proeseler, 1972b). In Poland, a density of 100 mites cm^{-2} of leaf surface causes rusting of prune leaves (Kozłowski, 1979). On European plum, a density of 370 mites cm^{-2} stops growth and induces leaf drop and twig wilting (Zawadzki, 1975). There is a positive correlation between increase of mite density and the regular growth of the host plant (Iacob *et al.*, 1977). In Japan, on peach, the weight and sugar content of fruit from trees not sprayed with miticide decreased to 15 and 8%, respectively, in comparison with sprayed trees (Kunugi *et al.*, 1993). Also in Japan, damaged leaves had lower photosynthetic rates than healthy leaves. At fruit harvest in August, no significant relationships were found between leaf injury levels. The sugar content of fruits was negatively correlated with injury levels, and the percentage leaf drop in September and leaf injury levels at fruit harvest were positively correlated. Trees characterized by 70–80% of leaf drop in September had enlargement of the floral bud volume suppressed (defoliation in undamaged trees was 5–10%). Damaged trees had smaller ovaries, and fruit enlargement was suppressed. In damaged trees, the weight, sugar content and acidity of next year's fruit are lower than in undamaged ones. Severe peach leaf injuries reduce the sugar content of fruit in the same year, and damaged trees show lower vigour due to postharvest defoliation, resulting in lower fruit quality in the next year (Kondo and Hiramatsu, 1999a).

Control

Cultural measures significantly contribute to reducing the risk of attack by *A. fockeui* (Stamenkovic *et al.*, 1995). However, mite control is currently entrusted to the use of acaricides.

CHEMICAL CONTROL. Chemical control is related to mite bioecology. It is possible spray shortly before flowering and again at petal fall (Delley, 1973). During the decline in mite numbers, treatments reduce the mite population so sharply that it becomes unable to recover much before the autumn (Schliesske, 1977a). Treatments with petroleum oils are suggested in winter or at bud burst (Delley, 1972). Two applications during the growth period gave good protection to plum seedlings (Iacob *et al.*, 1977).

Other active ingredients tested include azinphos-methyl (Schliesske, 1977a), bromopropylate (Iacob *et al.*, 1977; Doreste, 1979), chlordimeform (Iacob *et al.*, 1977), diazinon (Stamenkovic *et al.*, 1995), dicofol (Iacob *et al.*,1977; Doreste, 1979), endosulfan (Delley, 1972, 1973; Stamenkovic *et al.*, 1994, 1995), methyl-*S*-demeton (Schliesske, 1977a; Baker, 1979b; Doreste, 1979), mevinphos (Schliesske, 1977a), propargite, pyridaben (Stamenkovic *et al.*, 1995), wettable sulfur (Doreste, 1979; Stamenkovic *et al.*, 1995), white oil emulsion and white oil emulsion with parathion or endosulfan (Delley, 1972). Sprayings with fluvalinate (Kondo and Hiramatsu, 1999b) or other pyrethroids induce resurgence of *A. fockeui* (Ashihara *et al.*, 2004; Gyorffy, 2006). Like other pest mites, it is possible to use only acaricides labelled for the mite and crop concerned in each country.

Aculus schlechtendali **(Nalepa) (Fig. 16.80)**

Common name

Apple rust mite.

Diagnostic characteristics

FEMALE. The protogyne has a fusiform body, 160–175 μm long and yellowish. The prodorsal shield is subtriangular, with bulging sides and a pattern of granular lines; the frontal lobe is moderately acute. The median line occupies the rear ⅓. The admedian lines are complete, sinuate, gradually diverging to the rear. The opisthosomal dorsal annuli have rather faint and elongate microtubercles. Each dorsal annulus covers about 2 ventral annuli, which have bead-like microtubercles on the annulus margins. The genital coverflap has about 3 cross lines of granules and 8–10 longitudinal subapical ribs. The empodium is 4-rayed.

The deutogyne is smaller than the protogyne and should have a thinner frontal lobe lacking spinules and an indistinct shield pattern. The opisthosomal dorsal annuli should be narrower than those on the protogyne (Keifer, 1946; Jeppson *et al.*, 1975).

MALE. The male is similar to the female, but smaller.

Geographical distribution

Worldwide.

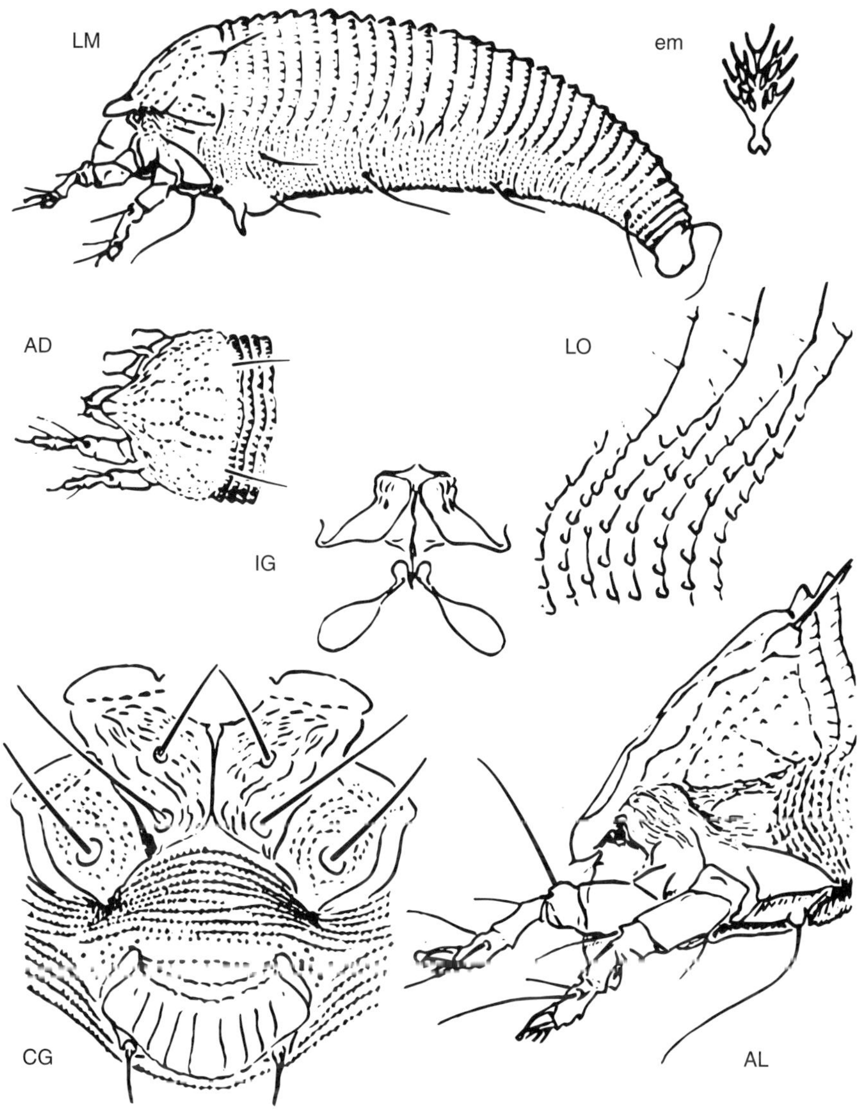

Fig. 16.80. *Aculus schlechtendali* (Nalepa) (from Keifer, 1946). See Table 16.1 for key.

Host plants

Rosaceae: apple (*Malus domestica* Borkh.) (Hoyt, 1969b; Herbert, 1974; Jeppson *et al.*, 1975; Easterbrook, 1979; Schliesske, 1985b; Kozłowski and Boczek, 1987b).

Bio-ecology

The apple rust mite is a leaf vagrant on apple in most cultivated areas of the world. It overwinters as deutogynes, and on young trees the highest number of deutogynes is found on 1-year-old shoots, between the vegetative buds and shoots, while on old trees, the greater part of the population overwinters under loose bark, in tiny crevices and in dormant buds on older wood and spurs (Herbert, 1974; Easterbrook, 1979; Sapozhnikova, 1982; Schliesske, 1985b; Kozłowski and Boczek, 1987b; Funayama and Takahashi, 1992). On old apple trees, deutogynes have been collected from ten sites and the highest numbers were from under the edge of bud scales. Higher survival has been found among deutogynes congregated in higher numbers. In diapausing deutogynes, the mortality ranges from 25 to 50% and a further 30–36% mortality occurs after diapause and migration to new growth (Kozłowski and Boczek, 1987b); in contrast, Schliesske (1985b) reported 77% mortality. Deutogynes leave their overwintering sites in spring, and the pattern of their emergence varies with temperature. The first deutogynes begin to emerge at the bud break stage (Fleckinger stage B; EPPO, 1989), with a peak during the mouse ear to pink bud stage (Fleckinger stage C3 to E), and emergence commonly concludes before flowering starts (Easterbrook, 1979, 1984b; Kozłowski and Boczek, 1987b). From the overwintering sites, the deutogynes move to the green bud scales and later to the leaves, sepals and flower receptacles, where they feed and lay eggs, initially at the mouse ear (C3) stage. The protogynes and

males are derived from these eggs. Studies on the interspecific attraction of spermatophores show that females of *A. schlechtendali* search for and mount spermatophores both from males of the same species and congeneric males (of *A. fockeui*), and readily accomplish self-insemination from spermatophores from both, although there was no evidence of fertilization. Males and females are sexually independent. Females do not visit spermatophores deposited by males of different genera (Oldfield, 1989).

There is a seasonal variability in the production of the diapausing deutogynes that is related to different photoperiods (Sapozhnikova, 1982). Protogynes lay a mean of 87 eggs at 22°C and deutogynes lay 33 eggs at 10°C and 26 eggs at 16°C (in total from 67 to 100 eggs). At 22°C, a generation develops in 10 days, and at 16°C in 16 days (Easterbrook, 1979). The optimum temperature for development is in a range of 23–28°C (Kozłowski and Boczek, 1989). In a cool maritime climate, the species develops three generations a year (Herbert, 1974), but in warmer regions there are from six to eight (Croft and McGroarty, 1977). In Poland, the mite develops five generations a year (Kozłowski and Boczek, 1989).

Peaks of mite density occur in June or July or until August (Hoyt, 1969b; Herbert, 1974; Easterbrook, 1979; Gólya and Kozma, 2001). In China, peaks are reported in mid-May to mid-June (Li and Cai, 1996). The population density can reach 2000 mites per leaf and decreases with leaf deterioration and the appearance of deutogynes. The new deutogynes appear from July (Easterbrook, 1979; Kozłowski and Boczek, 1989) to late June (Funayama and Takahashi, 1992). In the growing season, the increase of density mostly depends on the RH and the temperature. Decrease in populations is mainly caused by low humidity (<60%), a permanently high temperature (<30°C) and stronger winds and intensive cloudbursts (due to their mechanical effects) (Gólya and Kozma, 2001). The density may vary during the year according to different factors, both environmental and technical. Acaricide treatments effected in early summer may make the populations that develop at high levels during the autumn low. This phenomenon has been observed in the Netherlands, where 40 mites cm^{-2} have been detected in autumn (van Epenhuijsen, 1981), while in England, 680 mites per leaf have been detected in early October (Easterbrook, 1996). In China, infestations are more severe with moderate temperature, low humidity and less rain, and on varieties with more pubescence on the leaves (Li and Cai, 1996). In hot areas, mite density may decline during summer and there may be two peaks in autumn (Hoyt, 1969a). A summer diapause is reported from hot areas (Sapozhnikova, 1982).

A. schlechtendali is a competitor of other pest mites on apple, such as the European red mite, *Panonychus ulmi* (Koch). There is an indirect competition that acts at two levels, through effects on the host plants and via avoidance of predation (Dunley and Croft, 1996). Competition via plant defence consists of the different effects of competition between eriophyoids and tetranychoids for the same food resource. The feeding activity of eriophyoids promotes the formation of a callous in the epidermal layer, thus limiting the likelihood of spider mite probing. Conversely, the feeding activity of tetranychids destroys the plant tissue, reducing photosynthesis in the mesophyll (Croft and Hoying, 1977). This has been variously reported (Hoyt, 1969a; Solomon, 1975; Croft and McGroarty, 1977; Hoyt *et al.*, 1979) and is referred to as 'competition for predator-avoidance', a method of indirect competition between eriophyoids and tetranychids (Dunley and Croft, 1996). More details on this topic are given by Sabelis and van Rijn (1996). A mortality of 29–36% of deutogynes occurs during the spring migration (Kozłowski and Boczek, 1987b). *A. schlechtendali* is dispersed by wind (Duffner *et al.*, 2001). Its natural enemies include predatory insects and mites (see Table 16.5).

Table 16.5. Natural enemies (insects and mites) of *Aculus schlechtendali* (Nalepa).

Families and species	Reference(s)
INSECTS	
Anthocoridae	
Orius vicinus Ribaut	Heitmans *et al.*, 1986
Chrysopidae	
Chrysoperla carnea (Stephens)	Khan and Fent, 2005
Dolichopodidae	
Medetera spp.	Rathman *et al.*, 1988
Miridae	
Hyaliodes vitripennis (Say)	Chouinard *et al.*, 2006
Malacocoris chlorizans Panzer	Easterbrook, 1996
Thrips	Parrella *et al.*, 1981; Schliesske, 1985b
MITES	
Anystidae	
Anystis baccarum (L.)	Cuthbertson *et al.*, 2003
Stigmaeidae	
Zetzellia mali (Ewing)	Hoyt, 1969b; Santos and Laing, 1985
Phytoseiidae	
Amblyseius andersoni (Chant)	Dicke *et al.*, 1990b
Amblyseius fallacis (Garman)	Croft and McGroarty, 1977
Amblyseius finlandicus (Oudemans)	
Galendromus flumenis (Chant)	Croft and Luh, 2004
Galendromus occidentalis (Nesbitt)	
Metaseiulus citri (Garman et McGregor)	
Typhlodromus arboreus (Chant)	AliNiazee, 1979
Typhlodromus caudiglans Schuster	Croft and Luh, 2004
Typhlodromus occidentalis Nesbitt	Hoyt, 1969a
Typhlodromus pyri Scheuten	Karg, 1972; Easterbrook *et al.*, 1985; Genini and Baillod, 1987

Symptomatology and damage

The apple rust mite is reported as an apple pest in most fruit-growing areas, depending on the cultivar (Spieser *et al.*, 1998; Cuthbertson and Murchie, 2006). These outbreaks seem largely to depend on pesticide availability and on the use of fungicides lacking acaricidal activity and non-selective pesticides (Easterbrook, 1996), including pyrethroids (Croft and Slone, 1998). The feeding activity of *A. schlechtendali* damages the leaves and fruits. The mite feeds on the upper and lower leaf surfaces, damaging the mesophyll tissue and causing callus tissue to form over the parenchyma in regions where the epidermis is destroyed. The hypodermal layer undergoes accelerated cell division, resulting in formation and subsequent rupturing of the periderm tissue (Easterbrook and Fuller, 1986; Royalty and Perring, 1996). The feeding activity of *A. schlechtendali* on the lower leaf surface causes epidermal cell collapse and rupture of cell walls and plasmalemma; this induces the degeneration of the cells of spongy parenchyma associated with the thickening of cell walls and intercellular deposition of lignin-like compounds (Kozłowski and Zielinska, 1997; Kozłowski, 1998).

Leaf damage appears as browning or rusting of the undersides of the leaves during the summer, and is most obvious on leaves of the last shoots. Mite density, host plant variety and weather conditions influence the appearance of the symptoms, which are usually appreciable during July. Sometimes, it is possible to observe longitudinal upward rolling of the leaf edge, and higher infestations may cause silvery white blotches on the upper leaf surface (Lienk *et al.*, 1981). High levels of attack may cause premature leaf fall. The damage is most severe on young trees in nurseries and in newly planted orchards (Easterbrook, 1996). High population densities induce reductions in crop yield the following year (Parent, 1979). High numbers of deutogynes (and their offspring) can damage the spur leaves surrounding the blossom clusters, which show necrotic areas on the undersides of the leaves. Although the importance of these spur leaves in the development of the fruit is clear, there is no real evidence of reduction in fruit set, even at high levels of infestation (Easterbrook, 1996). In the Netherlands, chlorosis and curling of the leaves have been observed, and severe infestation of flower buds in the spring retards development (van Epenhuijsen, 1981).

In spring, the feeding activity of the apple russet mite on developing fruitlets causes russeting (Ciampolini *et al.*, 1976; Easterbrook and Fuller, 1986). The damage occurs from full blossom until early July. Cells of the epidermal layer are damaged and cork formation and russeting of the fruit gradually happen. Lignification occurs of the thickened cell walls in the regions adjacent to the epidermal cells destroyed by feeding activity of the mites on leaves and fruits. Besides lignin formation, phellogen or 'cork cambium' forms within the attacked tissues (Easterbrook and Fuller, 1986; Kozłowski, 1998).

The number of mites on fruitlets is correlated with damage to fruits, seen as russeting on the calyx end and in other parts. According to the level of infestation, the damage ranges from little spots or streaks of russet, usually around the calyx, to large rough russeting that may even be associated with cracking. The calyx end is most damaged, and from this site the alterations spread to other parts of fruit, sometimes severely. As a consequence of high levels of attack in the summer, the different cultivars, such as 'Jonagold' or 'Boskoop', often show different reductions in the colour of their fruits (von Palm, 1985; Sterk and Highwood, 1992). However, the effects of rust mite on the 'Golden Delicious' variety are unclear (Höhn and Höpli, 1990). There are few reports of fruit damage in North America (Easterbrook, 1996).

The time of probing and feeding by the apple rust mite varies on different apple varieties (Kozłowski and Boczek, 1987a; Kozłowski, 1995). In the susceptible apple cultivar 'Pionier', a density of 22–27 mites 1 cm^{-2} of leaf surface increases the ratio of the total phenolic compounds to carbohydrates from 3:1 in control to 5:1 in infested leaves, which finally decreases the palatability of the injured host plant to mites (Kozłowski, 1998). Levels of attack from 4000 to 5000 mites per leaf cause direct damage to the leaf epidermis and indirect damage that is mainly correlated with changes in the mesophyll cells of the leaves and a reduction of net CO_2 exchange by as much as 65%. The stomata are also partially open, which affects transpiration, causing a decrease of guard cell turgor and desiccation of epidermal and parenchyma cells. Infestation affects flower formation and the quality of apple yield. Fresh fruit weight, soluble carbohydrate content and skin colour intensity show a stronger decrease in 'Jonagold' than in the 'Golden Delicious' variety, raising the possibility that the latter is relatively resistant to mite attack (Spieser *et al.*, 1998). However, it has been found that at densities of 14,523 and 31,750 mite days per leaf, the size, weight and colour intensity of fruits decreases (Easterbrook and Palmer, 1996). Moreover, the 'Golden Delicious' variety is more susceptible to moderate and high mite densities in the summer (Angeli *et al.*, 2007), and it is possible that various factors (environmental, agricultural, etc.) lead the response of this variety.

Control

Control measures involve spray applications of acaricides, alone or according to integrated pest management (IPM) programmes. Monitoring and threshold determination of mite populations are fundamental tools. Mechanical reduction of overwintering stages results in a 50% reduction of the initial population in May/June. Neither winter nor summer prunings reduce the levels of mite infestation (Koschier, 1997).

The suitable management of natural enemies is fundamental to reducing mite populations (Croft and McRae, 1993; Walde *et al.*, 1997; Cuthbertson and Murchie, 2004). The use of selective pesticides is a prerequisite in IPM programmes (Hardman *et al.*, 2003), as is damage assessment to establish threshold levels (Spieser *et al.*, 1998). Early in the season, the populations of spider mites are low and populations of *A. schlechtendali* represent an alternative food resource for predaceous mites. This occurs in the early part of the growing season, during the emergence of the deutogynes from overwintering sites, before egg deposition by the European red mite, *P. ulmi*. Survival of the phytoseiids needs the application of selective or non-disruptive chemicals. Occasionally, some acaricide treatment may be required to balance apple rust mite and predator populations (Easterbrook, 1996).

THRESHOLDS. Studies carried out on thresholds in Michigan showed that a density of 200 mites per leaf indicate a requirement for chemical control (Mailloux, 1984), although a tolerance level of 200 mites per leaf for 10–14 days has been found (Croft, 1975), populations up to 300 mites/leaf produce little crop damage (Hoyt, 1969a). Population increases above 50 mites cm^{-2} induce increases in transpiration and decreases in photosynthesis (Kozłowski, 1980).

CHEMICAL CONTROL. Chemical treatments can be effected prebloom and in the summer (Jeppson *et al.*, 1975; van Epenhuijsen, 1981; Easterbrook, 1984a). The list of active ingredients that have been tested in various countries is very wide. It includes aldicarb (Lienk *et al.*, 1978), amitraz (Lienk *et al.*, 1978; Bostanian *et al.*, 1981a; Easterbrook, 1984a,b), abamectin (Sterk, 1994; White *et al.*, 1996; Raudonis *et al.*, 2007), azocyclotin (von Palm, 1985; Sterk, 1994), benomyl (Karg *et al.*, 1973), benzoximate (Bostanian *et al.*, 1981b), bifenthrin (Easterbrook, 1985; Schliesske, 1985b; AliNiazee, 1989), brofenprox (Sterk, 1994), dithiocarbamates (Westigard and Hoyt, 1965), bromopropylate (van Epenhuijsen, 1981; Bostanian *et al.*, 1983), captan (Karg *et al.*, 1973), carbaryl (Hoyt, 1962; Westigard and Hoyt, 1965; Easterbrook, 1984a; Young *et al.*, 1990), chinomethionate (Westigard and Hoyt, 1965; Karg *et al.*, 1973), chlorpyrifos (Easterbrook, 1984a; Young *et al.*, 1990), clofentezine (Easterbrook and Campbell, 1986; Pfeiffer *et al.*, 1989), cycloprate (Bostanian *et al.*, 1983; Lienk *et al.*, 1978), cyhexatin (Lienk *et al.*, 1978; Bostanian *et al.*, 1983; Easterbrook, 1984a; Bostanian and Vincent, 1985; Schliesske, 1985b), dichlofluanid (Karg *et al.*, 1973), dicofol (Hoyt, 1962; Westigard and Hoyt, 1965; Lienk *et al.*, 1978; von Palm, 1985; Schliesske, 1985b), dinocap (Hoyt, 1962; Young *et al.*, 1990), endosulfan (Hoyt, 1962; van Epenhuijsen, 1981; Schliesske, 1985b), fenazaquin (Sterk, 1994), fenbutatin oxide (Bostanian *et al.*, 1983; Bostanian and Vincent, 1985; Rheinland-Pfalz Ministerium für Landwirtschaft, Weinbau und Forsten, 1991), fenpropathrin (Easterbrook, 1985), fenpyroximate (Rheinland-Pfalz Ministerium für Landwirtschaft, Weinbau und Forsten, 1991; Sterk, 1994), fenvalerate (Zwick and Fields, 1978), flubenzimine (Bostanian and Vincent, 1985), flucycloxuron (Sterk, 1994), flucythrinate (Bostanian and Vincent, 1985; Easterbrook, 1985), flufenoxuron (Sterk, 1994), fluvalinate (Bostanian and Vincent, 1985; Easterbrook, 1985), formetanate (Lienk *et al.*, 1978; Bostanian *et al.*, 1983), lambda-cyhalothrin (Angeli *et al.*, 2008), lime sulfur (Westigard and Hoyt, 1965), malanoben (Bostanian *et al.*, 1983), mancozeb (Easterbrook, 1984a), methomyl (Lienk *et al.*, 1978), mineral oils (Fernandez *et al.*, 2006), oxamyl (Lienk *et al.*, 1978), permethrin (Kapetanakis *et al.*, 1986), phenoxypyrazole (Sterk, 1994), pirimiphos-methyl (Easterbrook, 1984a; Easterbrook and Buss, 1988), propargite (Lienk *et al.*, 1978), pyridaben (Sterk, 1994), spirodichlofen (Angeli *et al.*, 2008), sulfur (Westigard and Hoyt, 1965; Easterbrook, 1984a; von Palm, 1985), tebufenpyrad (Sterk, 1994), tetradifon (Hoyt, 1962; Easterbrook, 1984a), tetrazine (Pap *et al.*, 1994), thiodan (Westigard and Hoyt, 1965) and triazophos (Bostanian *et al.*, 1983).

Of the chemicals reported above, several acaricides and/or insecticides (fenpyroximate, flucycloxuron, flufenoxuron, pirimiphos-methyl, pyridaben) and fungicides (sulfur) are highly effective (Childers *et al.*, 1996). Pre-bloom sprays with sulfur or lime sulfur are effective (Westigard and Hoyt, 1965). Resistant strains to synthetic pyrethroids and organophosphates are known (Sterk and Highwood, 1992). A strong suppressive effect of dichlofluanid plus bitertanol is reported (Ciglar and Baric, 2000). The use of many of the active ingredients cited above is no longer allowed, and also for this mite only acaricides labelled for apple should be used.

Modifications of carbon dioxide concentration in closed environments, usually referred to as controlled atmosphere treatment (CAT), kill the mite eggs in 5.3 months. The method of application demands that the apples are stored at 2.8°C in an atmosphere of 1% O_2 and 1% CO_2, and does not cause detrimental effects to fruits. However, the length of storage exposure is a disadvantage (Lidster *et al.*, 1981).

Aculus ligustri (Keifer) (Fig. 16.81)

Common name

Hedge privet rust mite.

Diagnostic characteristics

FEMALE. The body is fusiform, brownish yellow in colour and the protogyne is 185–200 μm long. The prodorsal shield is subtriangular, with a moderately acute frontal lobe over the gnathosoma. The median line occurs on the rear shield ½. The admedian lines are nearly complete, arising on the side of the frontal lobe, describing gentle in-and-out curves, and reaching nearly to the rear margin. The admedians receive a cross line at about ¼, and a second at ½. The opisthosoma has about 36 dorsal annuli bearing faint and elongate microtubercles, and about 66 ventral annuli, with bead-like microtubercles. The genitalia have 1 or 2 basal transverse curved lines, and about 9 longitudinal ribs. The empodium is 4-rayed.

The deutogyne is about the same size, but browner and with reduced microtubercles, none of which are present on the dorsal annuli (Jeppson *et al.*, 1975).

MALE. The body of the male is 130–150 μm long (Jeppson *et al.*, 1975).

Geographical distribution

Belgium (Witters *et al.*, 2003), California (USA) (Keifer, 1938a, 1952a), China (Xue *et al.*, 2012), Italy (de Lillo and Beron, 2004), Hungary (Gólya *et al.*, 2001), Iran (Hajizadeh and Hosseini, 2003b), Ohio (USA) (Boggs *et al.*, 2004), Poland (Soika and Łabanowski, 1998a), Serbia and Montenegro (Petanović and Stanković, 1999) and Virginia (USA) (Weidhaas, 1979).

Host plants

Oleaceae: privet (*Ligustrum* sp.) (Keifer, 1938a; Jeppson *et al.*, 1975), Amur privet (*L. obtusifolium* Siebold *et* Zucc. (= *L. amurense* Carrière)), California privet (*L. ovalifolium* Hassk.) (Keifer, 1952a; crop and pest management guidelines of Cornell University, 2012).

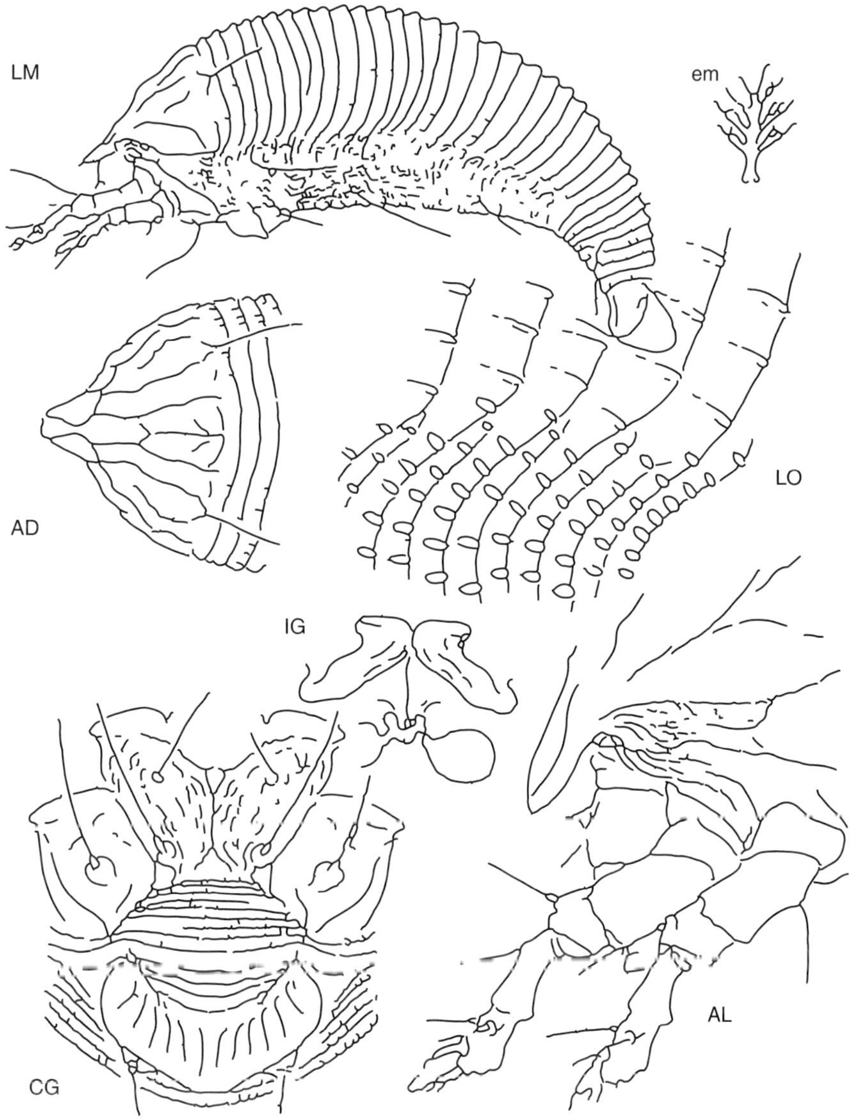

Fig. 16.81. *Aculus ligustri* (Keifer) (from Keifer, 1952a). See Table 16.1 for key.

Bio-ecology

A. ligustri was first collected on privet in California and described as *Phyllocoptes ligustri* (Keifer, 1938a). In the USA, the mite develops on hedge privet, but does not infest privet trees (Jeppson *et al*., 1975). The mite populations live on both leaf surfaces and the spring increase in temperature induces them to leave overwintering sides and to move towards the leaves. At this time, it is possible to observe a variety of females differing in microtuberculation. In late spring, all forms move towards the lateral bud scales or buds and do not return to the leaves. Only deutogynes survive until late winter, when they move around and among the lateral buds, starting to feed on the green bud scale bases. They are probably inseminated in spring (Jeppson *et al*., 1975). The crop and pest management guidelines of Cornell University (2012) report that the mites may be present from May through into November, and 'California', 'Amur' and 'Regal privet' are commonly infested. In Iran, the privet rust mite lives on forest and ornamental privet shrubs, and it starts its annual activity in early spring. The mite populations increase during the spring, and vagrant mites are observed on both leaf surface. Leaves might be tightly rolled in such a way that infested twigs appears to be leafless. In late spring, because of increasing temperature, the mites migrate to aestival positions under privet bud scales (Hajizadeh and Hosseini, 2003b).

Natural enemies of the hedge privet rust mite include some predacious mites.

Symptomatology and damage

Keifer (1938a) observed browning or scarification of the leaf surface as a result of mite infestation in California. In Iran, the hedge privet rust mite caused browning, rusting and curling of

leaves of forest and ornamental privet shrubs (Hajizadeh and Hosseini, 2003b). The main symptoms are premature leaf drop, a silvery stipple to the foliage and cupping downward of the leaves (crop and pest management guidelines of Cornell University, 2012). In Ohio, the hedge privet rust mite has produced damaging localized infestations in many areas of the state (Boggs *et al.*, 2004). In Belgium, on *Ligustrum* hedges, the mite has caused considerable damage (Witters *et al.*, 2003).

Control

Although control is usually unnecessary, in nursery and in young stands it may be necessary to apply chemical control. According to privet variety, the crop and pest management guidelines of Cornell University (2012) suggest treating from the last 10 days of May to mid-June at 298-802 GDDs and until mid-July at 1266–1515 GDDs. It is possible to use abamectin, carbaryl, fenazaquin, fenpyroximate, horticultural oil, milbemectin and neem oil. The chemicals must be sprayed on to both leaf surfaces.

The Genus *Aculops* Keifer

Diagnostic characteristics

Aculops spp. are fusiform mites with a short oral stylet. All regular setae are present. The frontal lobe is produced over the gnathosoma base, of small or moderate size, acuminate to rounded or terminating in a sharp or spine-like point. There is no pair of small spines projecting forward from the lower front of the frontal lobe margin. The scapular tubercles are usually subcylindrical, projecting back over the rear shield margin, with transverse axes on the rear margin, directing is the scapular setae (*sc*) to the rear, usually divergently. The opisthosoma on non-gall makers is clearly divided laterally into broader dorsal annuli and narrower ventral annuli; this distinction is less clear on most gall formers, the deutogynes tend to have clearer dorsoventral distinction. The opisthosomal microtubercles are round, elliptical or produced as spinules, either located ahead of the annulus margins dorsally, extending anteriorly from the margins, or as spinules bent either ahead or caudad from the ridges. The genitalia are not closely appressed to the coxae; the interior female apodeme extends forward from the base (Keifer, 1966a).

Economic importance

This genus has a worldwide distribution and includes over 170 species, of which almost 111 live on more than 30 dicotyledonous plants (Oldfield, 1996a). They are fundamentally rust mites and leaf vagrants, and some species cause severe damage to economic plants. The type species, *Vasates populivagrans* Keifer, is a deuterogynous species causing rust on leaves of fremont cottonwood (*Populus fremonti* S. Watson) in California (Keifer, 1953). The pink citrus rust mite, *Aculops pelekassi* (Keifer), causes serious russeting of fruit and leaves and growth distortion of orange, mandarin and clementine (Vacante, 2010). The tomato russet mite, *A. lycopersici* (Tryon), damages leaves, stems and fruits of tomato (*Lycopersicon esculentum* (L.) Karsten ex Farw.) and to lesser extent tobacco (*Nicotiana tabacum* L.), potato (*Solanum tuberosum* L.), pepper (*Capsicum* spp.), *Petunia* sp. and other plants (Perring, 1996). The olive yellow spot mite, *A. benakii* (Hatzinikolis), damages leaves, flowers and young fruits of olive (*Olea europaea* L.) (Hatzinikolis, 1969b). The honey locust rust mite, *A. gleditsiae* (Keifer), lives on leaves of honey locust (*Gleditsia triacanthos* L.) (Keifer, 1959a) in eastern North America and Europe, and causes russeting of the lower leaf surface. The European *A. allotrichus* (Nalepa) (de Lillo and Beron, 2004) causes rolled leaves and rust of black locust (*Robinia pseudoacacia* L.); early severe attack in nurseries produces strong defoliation of tips, short internodes and stops development. The lilac rust mite, *A. massalongoi* (Nalepa), is reported from North America, Eurasia and temperate regions, where it lives on the lower leaf surface of lilac (*Syringa vulgaris* (Shr.)), causing russeting (Jeppson *et al.*, 1975; Meyer Smith, 1996a). The cottonwood rust mite, *A. lobuliferus* (Keifer), first collected on eastern cottonwood (*Populus deltoides* Bartr.) in Mississippi (Keifer, 1961a), is a leaf vagrant and rust mite damaging nursery-grown young trees of poplar and cottonwood (Jeppson *et al.*, 1975). *A. magnolivora* (Keifer) lives on the lower leaf surface of *Magnolia grandiflora* L. It was first reported from North America (Keifer, 1939e) and then also from Europe (de Lillo and Beron, 2004); it causes yellowing, browning and premature dropping of leaves (Davis, 1964a). Only a few of the above reported species (*A. pelekassi* and *A. lycopersici*) play a key role in the phytopathogical scenario. Many other species live on economic plants but they have no economic importance.

Aculops pelekassi (Keifer) (Fig. 16.82)

Common name

Pink citrus rust mite, Japanese citrus mite.

Diagnostic characteristics

FEMALE. The body is spindle formed, 140–150 μm long and pinkish in colour. The gnathosoma projects diagonally down. The prodorsal shield is subtriangular in anterior outline, with bulging sides and with moderately acute but terminally rounded, frontal lobe over the gnathosoma. The shield pattern is a more or less obscure network of lines. The median lines is present on the rear ½ or ⅔ and receives transverse lines at the ⅓ and ⅔ positions. The line at a ⅓ extends laterally around to the rear shield margin. A subparallel line, below the transverse line at a ⅓, runs from the side of the frontal lobe to the shield margin above the coxae. The admedian lines are complete, extending from the sides of the frontal lobe to the rear margin, somewhat sinuate and furthest apart at the rear; admedians also receive the cross lines at ½ and ⅔, which connect them with the medians. The scapular tubercles are 26 μm apart, on the rear margin; the scapular setae (*sc*) are 9 μm long and diverging to the rear. The opisthosoma has about 36 dorsal annuli and 50 ventral annuli. The microtubercles on the dorsal annuli are faint; they are bead like on the ventral annuli, and the rest are on the

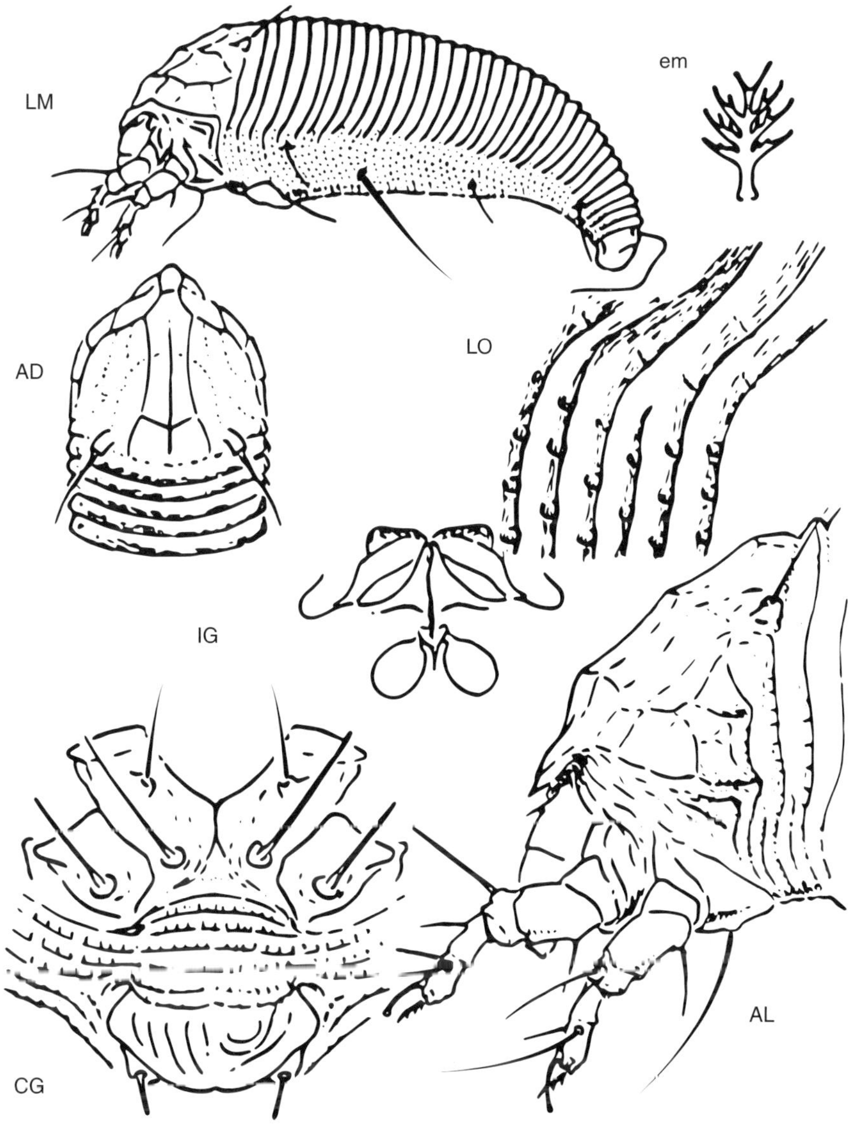

Fig. 16.82. *Aculops pelekassi* (Keifer) (from Keifer, 1959b). See Table 16.1 for key.

ring margins. The setae c_2 are 26 μm long, set on ventral annulus 6; the setae *d* are 40 μm, set on ventral annulus 16; the setae *e* are 7 μm long, set on ventral annulus 30; the setae *f* are 21 μm long, set on ventral annulus 6 from the rear; the setae h_1 are 2 μm long. The genital coverflap has 2 basal transverse lines with granules; the longitudinal ribs may have 2 forms, either about 10 longitudinal ribs, or transversely curved ribs. The empodia are 4-rayed (Keifer, 1959b).

MALE. The male is like the female, except for the genital structures (Keifer, 1959b).

Geographical distribution

China (Kuang *et al.*, 2005), Croatia (Mijuskovic and Kosac, 1972), Florida (Denmark, 1962; Burditt *et al.*, 1963), Greece (Keifer, 1959b), Iran (Hajizadeh and Hosseini, 2004; Xue *et al.*, 2009a), Italy (Costantino, 1962), Japan (Ehara, 1964a,b), Paraguay (Flechtmann and Aranda, 1970), Taiwan (Huang and Wang, 1997) and Thailand (Keifer and Knorr, 1978).

Host plants

Rutaceae: *Citrus* spp. (Keifer, 1959b; Jeppson *et al.*, 1975; Vacante, 2010.)

Bio-ecology

A. pelekassi may be easily confused with *Phyllocoptruta oleivora* (Ashmead). However, the first pest is pink and the second yellow, as well as the eggs of the first species being translucent white and being laid scattered, while those of the second species are

opaque and laid in the hollows along the ribs. Populations of *P. oleivora* also live on mature leaves and are found over the leaf surface, while those of *A. pelekassi* prefer young leaves and are more numerous on the margins of leaves, where they are easily dispersed by the wind. Also, *P. oleivora* is not stimulated by light, while *A. pelekassi* reacts to the stimulation of fluorescent light (Burditt *et al.*, 1963).

The pink citrus rust mite is a vagrant on leaves, twigs and fruits of *Citrus*. The sexes are separate and the sex ratio changes seasonally; in Japan, there are 84% of females in October and 100% in December (Ashihara *et al.*, 2004). Also in Japan, the maximum oviposition (21.8 eggs per female) occurs at 25°C and stops when the temperature drops to 15°C. The developmental time from egg to adult varies with temperature and ranges from 14.9 days at 20°C to 6.3 days at 30°C in Japan (Seki, 1979); in the summer (with variable temperatures), it requires 6–7 days and in the spring 8–13 days, while with a constant temperature of 35°C, it takes 5 days and at 18°C 18–19 days. In Croatia, the mite develops 18–22 generations a year in the field between May and October. The female lives 5–32 days and its fecundity is 55 (average 31.5) (Mijuskovic and Tomašević, 1975). At 27°C, the female lifespan is 6.4 ± 11.22 days and the intrinsic rate of increase (r_m) is 0.192 individuals per female daily. The mortality from egg to adult is 45.59% at 27°C and 32.25% at 17°C. The optimum conditions for mite development are between 22 and 27°C and 75–77% RH. The thermal threshold for development is 11°C, and at 10°C oviposition stops (Ebrahim, 2000). The developmental zero and effective heat units from egg to adult are 10.6°C and 119 degree-days (Seki, 1979).

Although the species can reproduce in all seasons (Nucifora, 1969; Jeppson *et al.*, 1975), it does not normally reproduce during the winter and shelters between the bracts of buds (Ciampolini and Rota, 1963; Seki, 1983), under the rosettes of the fruits, in dips on the plants and in branch injuries (Martelli, 1964), under the empty follicles of scale insects (Nucifora, 1967) and in the stylar scars of the fruits, either as the adult female or other biological stages (Nucifora , 1967; Di Martino, 1985; Vacante, 1995). In the late spring, the mite begins to reproduce and completes the first generation in the buds, while the subsequent generations develop on young sprouts, new leaves and fruits from the month of June (Ashihara *et al.*, 2004). The density of the mite populations on the leaves increases rapidly in early summer, and reaches a peak in late July. Summer generations of the pink citrus rust mite move from the leaves to the immature fruits, on which they feed, reproduce and cause russeting to the fruits. Adult males begin to crawl from the leaves and fruits to the buds in early October (Seki, 1983).

In Japan, the relationship between the overwintering population density and the succeeding seasonal prevalence of *A. pelekassi* on satsuma mandarin trees showed that during the winter the population density of the mite was higher on succulent shoots. The density of the overwintering population gradually decreases, varying according to the weather conditions. The timing of monitoring to estimate the population density of overwintering mites is just before the sprouting of the spring shoots, and allows prediction of the later development of the populations. The seasonal behaviour of the mite populations shows two parts, which are referred to as the first half (until July) and the second half (after August). The seasonal occurrence of the first half has some relationship with the overwintering population density, while the relationship of the overwintering population density with the second half is unclear. The dependable forecasting of occurrence of the mite based on its wintering population density is difficult. The population dynamics of *A. pelekassi* displays four patterns: low density through the season, peaking in the first half, peaking in the second half, and peaking in both the first and second halves (Ohnishi *et al.*, 2008).

In Florida, *A. pelekassi* starts to feed on small fruits during April or May; and at the end of June or early July, the pink citrus rust mite populations rapidly decrease and the citrus rust mite, *Phyllocoptruta oleivora* (Ashmead), starts to increase. In Japan, high densities of populations were related to higher spring temperatures and low precipitation in early summer; similar conditions occur in Florida (Childers and Achor, 1999). The ban on the use of some dithiocarbamates effective against eriophyoids (Johnston *et al.*, 1957; Martelli and Di Martino, 1963; Swirski *et al.*, 1967) has recently posed a problem with this pest in some areas of eastern Sicily (Italy), where the lemon orchards were largely treated with these chemicals, which were also applied to control the disease ‘mal secco’, caused by the fungus *Phoma tracheiphila* (Petri) Kantschaveli *et* Gikachvili.

The dispersal of the pink citrus rust mite relies on weather conditions, on the use of infested propagating materials, on the transport of infested fruit and on farmers who involuntarily transport it from one crop to another. High population densities of the pest were observed in the sunspot areas of fruit at midday, or in the afternoon as they dispersed. This behaviour differs from that observed with *P. oleivora*, in which there is avoidance of direct sunlight. The pink citrus rust mite forms a golden yellow to pinkish dust patch on the fruit during the dispersal phase as a result of their large aggregations (Childers and Achor, 1999).

Natural enemies include pathogens and predatory mites. The most important fungus is *Hirsutella thompsonii* Fisher (McCoy, 1996b). The mites include several phytoseiids, such as *Amblyseius andersoni* (Chant) (Kim and Paik, 1999), *A. eharai* Amitai *et* Swirskii (Ashihara *et al.*, 2004), *A. fallacis* (Garman) (Kim and Paik, 1999), *A. longispinosus* (Evans) (= *Neoseiulus longispinosus* (Evans)) (Ashihara *et al.*, 2004), *A. womersleyi* Schicha (Kim and Paik, 1999), *Iphiseiodes quadripilis* (Banks) (Villanueva and Childers, 2007), *Typhlodromus pyri* Scheuten and *T. occidentalis* Nesbitt (Kim and Paik, 1999). Some ascids are also reported, including *Proctolaelaps pygmaeus* (Müller) (Ehara, 1964b) and stigmaeid mites such as *Agistemus terminalis* (Quayle) (Ashihara *et al.*, 2004).

Symptomatology and damage

Potentially, this mite is more harmful than *P. oleivora* and attacks all citrus, among which it seems to prefer orange, mandarin and clementine. The symptomatology can be confused with viral diseases. The damage to citrus plant is similar to that caused by *P. oleivora* and affects leaves, green stems and fruits. In the early spring, the mites move from overwintering sites and develop on new leaves, where they feed on the lower surface. The feeding activity of pink citrus rust mite produces along the

margins and the median ribs of the developing leaves irregular brown blemishes, different degrees of distortion, curling under the margins, crinkling of tissues, burn and leaf dieback. Densities of 200 or more mites per leaf are destructive. Later, the mites move on to the fruits and back to the leaves (Childers *et al.*, 2007; Rogers *et al.*, 2013).

The green stems show corky peridermal alterations which, according to the level of attack, may become rusty and run with cracks. The damage is most common in young nurseries. In the field, these symptoms are more rare, because the attack can affect only fruits (Burditt and Reed, 1963; Nucifora, 1967; Jeppson *et al.*, 1975; Pennisi *et al.*, 1975; Di Martino, 1985; Vacante, 1995, 2009b; Vacante and Bonsignore, 2009; Gerson and Vacante, 2012).

On fruit of sweet orange (*Citrus sinensis* (L.) Osbeck), each feeding adult of *A. pelekassi* produces about 20 punctures (each about 1 μm in diameter) per epidermal cell (10 μm long and 7 μm wide). The depth of penetration is about 20 μm and reaches the second and third layer of the fruit epidermis (Tagaki, 1981). Callus formation on the peel surface of *Citrus unshiu* Marcowich fruit occurs at early infestation in August, when the cells are dividing. Late attacks from September to October produce bronzing injury. Generally, the injury during the middle stage of fruit growth results in both types of symptoms near the oil glands (Kato, 1977). Diameter, volume and weight of damaged fruits are less than those of undamaged fruits. In particular, the sugar content of juice is higher in damaged fruit, showing that the concentration of soluble solids has been increased through water loss caused by mite attack (Tono *et al.*, 1978).

At the beginning of the attack, the fruits lose their natural shine and are scattered with a thin and delicate brown powder consisting of thousands of mite individuals. The lemon fruits present a green bronze colour that turns after to silver and a more or less intense reddish brown. The attack may be confused with that of the broad mite, *Polyphagotarsonemus latus* (Banks). Nevertheless, the eriophyoid (i.e. *A. pelekassi*) infests the fruit at all stages of development, while the tarsonemid (*P. latus*) infests it until it reaches the size of a walnut; also, *P. latus* searches for fruit sites in the shade within the foliage, while *A. pelekassi* prefers those exposed to direct sunlight. Moreover, while the skin surface of ripe fruit infested by *P. latus* is cracked and desquamated, the skin of fruit infested by *A. pelekassi* is not desquamate (Nucifora, 1967). Orange and tangerine fruits initially show an easing of brightness and the skin gradually becomes a green bronze with tones that are more or less intense depending on the severity of the attack in the glandular concavities. Mandarins that are infested early take on at ripening a leather colour that becomes dark and then may become olivaceous or fuliginous. The symptoms can be confused with those of the red and black flat mite, *Brevipalpus phoenicis* (Geijskes) *sensu lato* (see Chapter 21). However, the feeding activity of the tenuipalpid causes pericarp dimples with crater-shaped cracks, while that of *A. pelekassi* usually induces longitudinal cracks that sometimes intersect with one other. When the attack affects only a part of the fruit, spots may occur that are more or less numerous and dark orange, delicate and with undefined contours. These spots gradually converge on each other and lead towards the part of the fruit that has been most attacked (Nucifora, 1967).

Control

Control requires the strict monitoring of mite populations. Two factors should trigger the need to monitor the mite on the leaves and fruit. Monitoring should be initiated following mild winter temperatures and/or low rainfall during the previous year or into early spring, or following a recent infestation of the mite in the field (Childers and Achor, 1999). The guidelines for pink citrus rust mite control in the citrus groves of the Mediterranean region were established during the EC Experts' Meeting held in Siniscola and Muravera (Sardinia, Italy) on 20–22 October 1982, and suggest general inspection of the fruit on 10% of the plants sampled. The threshold should be kept within the limits of 2–3% of infested fruit (Cavalloro and Prota, 1983). In North American citriculture, a sampling method and thresholds were proposed for processed fruit and fresh fruit (see *P. oleivora*) (Childers *et al.*, 2007). An analysis of the use of binomial sampling for estimating rust mite densities on orange fruit concluded that binomial sampling for a general estimate of mite densities seemed to be a viable alternative to absolute counts of mites per sample for a grower using a low management threshold such as two or three mites per sample (Hall *et al.*, 2007). In Japan, presence–absence density estimation has been investigated for the pink citrus rust mite on Satsuma mandarin fruits (Ikemiya *et al.*, 2007). Also in Japan, the combined application of chemical control in early July and in early September was more effective than a single application in early July. Using these findings and references, a pest control guidance (The Pink Citrus Rust Mite Control Manual) has been proposed to support decision making for the timing of control and selection of more effective acaricides or insecticides (Sugiura *et al.*, 2008). In addition to the above tools, the implementation of an integrated pest management (IPM) programme requires the use of selective chemicals. No practical experience of the use of biological control alone has been reported.

CHEMICAL CONTROL. Several acaricides, such as acequinocyl, chlorobenzilate, etoxazole, hexythiazox and kelthane, are ineffective (Martelli and Di Martino, 1963; Costantino, 1968; Nucifora, 1969; Di Martino and Benfatto, 1973; Ashihara *et al.*, 2004). Chlorpyrifos, fluazinam, pyrimidifen, quinalphos and tebufenpyrad controlled adults, with pyridaben being the more effective against adults and eggs (Qu *et al.*, 1997). Amitraz, chlorfenapyr, fenbutatin oxide and milbemectin had high acaricidal activity (Sugiura *et al.*, 2008). The activity of amitraz and pyridaben has also been confirmed (Qu *et al.*, 1997; Tsuchiya, 2002; Sugiura *et al.*, 2008). Spirodiclofen mixed with petroleum oils provided excellent initial and residual control of the mite (Childers *et al.*, 2007), equal or superior to spraying with horticultural oil, abamectin and pyridaben (Bell *et al.*, 2005).

The use of sulfur, as lime sulfur, sulfur dust or wettable sulfur, is well known (Mijuskovic, 1973a; Seki, 1983; Childers *et al.*, 2007), and it is also used in a mixture with zineb or ziram (Ciampolini and Rota, 1963; Martelli and Di Martino, 1963; Nucifora, 1967; Di Martino and Benfatto, 1973; Tsuchiya, 2002); however, the sulfur negatively influences the populations of natural enemies (McCoy, 1977a; Smith and Papacek, 1991). Petroleum oils can also be used alone or with acaricides (Childers *et al.*, 2007).

Aculops lycopersici (Tryon) (Fig. 16.83)

Common name

Tomato russet mite.

Diagnostic characteristics

FEMALE. The body is robust, fusiform, from 150 to 180 μm long and whitish to yellowish in colour. The prodorsal shield has a broad and short frontal lobe, precipitous anteriorly, and topped by a transverse line that extends back along the sides to the rear margins. Laterally, this line has several lower branches that form lateral cells above the coxae. The median line is faintly present on the rear ⅔. The admedian lines arch back from the anterior transverse line, curve inward at ½, extend towards the rear subparallel to each other on each side of the median line, and end by flaring out towards the scapular tubercles. The dorsal setae are short. The opisthosoma has about 27 dorsal annuli and 60 ventral annuli. The microtubercles are pointed and located on the ventral annuli margins. The genitalia have faint basal granules and 8–10 longitudinal ribs. The empodium is 4-rayed (Jeppson *et al.*, 1975; Keifer *et al.*, 1982).

MALE. The male is like the female and 140–150 μm long (Keifer *et al.*, 1982).

Geographical distribution

Worldwide. Common in the main areas of tomato cultivation.

Host plants

These are listed by family: Convolvulaceae–bindweed (*Convolvulus arvensis* L.), morning glory (*Ipomoea* sp.); Grossulariaceae – wild

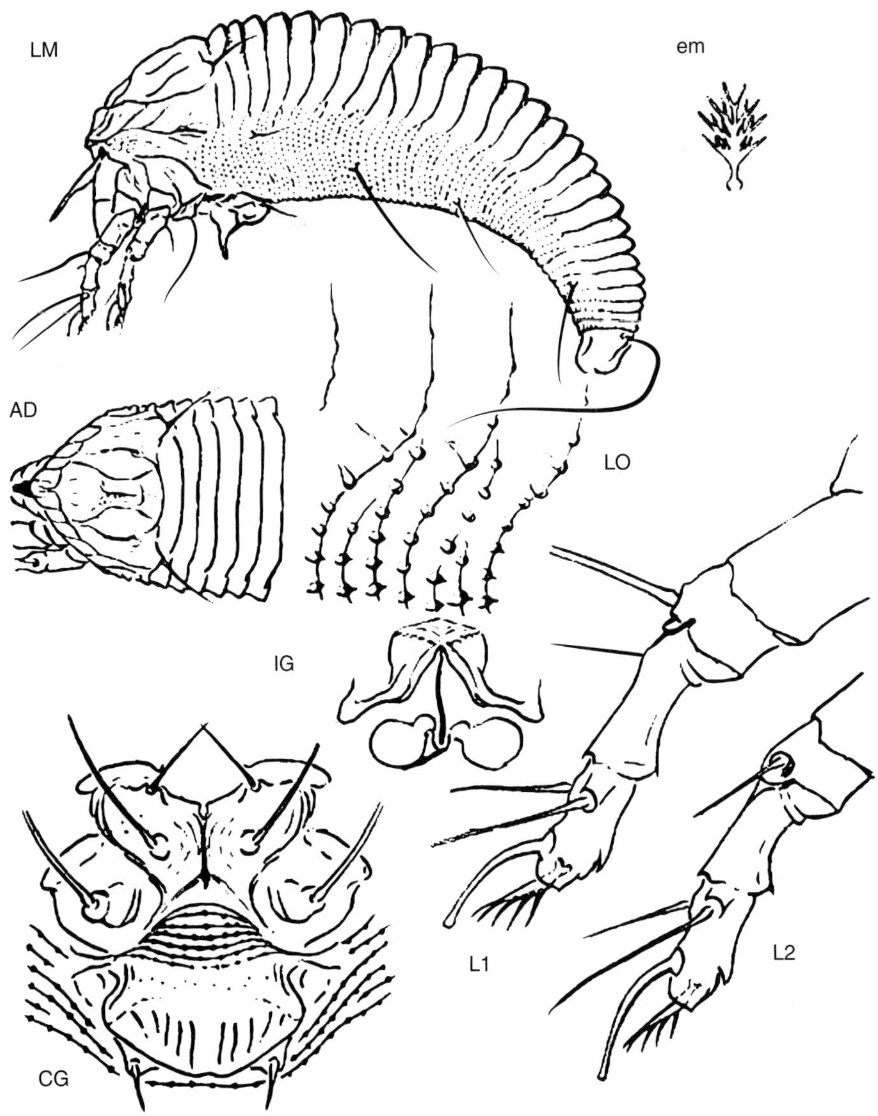

Fig. 16.83. *Aculops lycopersici* (Tryon) (from Keifer,1940c). See Table 16.1 for key.

blackcurrant (*Ribes americanum* Mill.), gooseberry (*R. uva-crispa* L.).); Rosaceae – blackberry (*Rubus fruticosus* L.); Solanaceae – amethyst-flower (*Browallia speciosa* Hook.), bell pepper (*Capsicum annuum* L.), bird pepper (*C. frutescens* L.), *Datura* sp., large thorn-apple (*D. ferox* L.), downy thorn-apple (*D. inoxia* Mill.), common thorn-apples (*D. stramonium* L.), hairy thorn-apple (*D. wrightii* Regel), tomato (*Lycopersicon esculentum* (L.) Karsten ex Farw.), tobacco (*Nicotiana tabacum* L.), petunia (*Petunia* sp.), *Physalis* sp., cape-gooseberry (*P. peruviana* L.), tomatillo (*P. philadelphica* Lam.), small flower nightshade (*Solanum americanum* Mill.), horse nettle (*S. carolinense* L.), greenspot nightshade (*S. douglasii* Dunal), aubergine (*S. melongena* L.), black nightshade (*S. nigrum* L.), Jerusalem cherry (*S. pseudocapsicum* L.), popolo (*S. sodomeum* L.), potato (*S. tuberosum* L.), red nightshade (*S. villosum* Mill) (Keifer, 1940c; Bailey and Keifer, 1943; Rice and Strong, 1962; Sepasgosarian, 1973; Perring and Farrar, 1986; Perring, 1996; Kim *et al.*, 2002a).

Bio-ecology

Aculops lycopersici may be confused with the tomato erineum mite, *Aceria lycopersici* (Wolffenstein), which causes patches of white erineum on the stems and leaves of tomato plants. It is a vagrant species living mainly on cultivated and wild Solanaceae, and on some Convolvulaceae, Rosaceae and Grossulariaceae. The original area of the tomato russet mite and its original solanaceous host plants are unknown, but its habitat might have been subtropical and the origin Australian (Bailey and Keifer, 1943). The tomato russet mite is a severe pest of tomato, infesting sometimes avbergine, potato, tobacco, pepper, and petunia, occurring also on blackberry and on a number of wild plants (Bailey and Keifer, 1943; Rice and Strong, 1962; Sepasgosarian, 1973; Perring and Farrar, 1986; Perring, 1996; Kim *et al.*, 2002a). High densities of population may develop on round red fruiting varieties of tomato or on petunia, both plants sometimes hairy or glandular and producing a favourable habitat for mite development. Varieties of petunia that are very sticky are not affected by the mite, and potato plants may host many mites but not show as much damage as tomato. The glabrous surface of most nightshades and daturas hinders mite development. There is individual variation among black nightshade plants in respect of mite reproduction (Bailey and Keifer, 1943). Severe damage usually occurs only on tomato and is rare on other plants (Kay, 1986). This behaviour can be explained by the phenomenon of 'solanum stimulation' in which the tomato russet mite can kill tomato plants by feeding and reproducing rapidly on the plants (Bailey and Keifer, 1943; Anderson, 1954), and on tomatillos, potato, aubergine, black nightshade and horsenettle (Rice and Strong, 1962); other hosts ensure survival of the mite populations (Perring, 1996).

The tomato russet mite is haplodiploid, and the haploid number of chromosomes is $n = 2$. Males develop from haploid eggs and females from diploid eggs (Helle and Wysoki, 1983). The female lives longer than the male, and adult longevity decreases with increasing temperature (Haque and Kawai, 2003). Bailey and Keifer (1943) observed a sex ratio of 64% males and 36% females. The mite feeds on all green plant tissues and its populations develop both in greenhouses and in the field. It is a warm-weather species and has no dormant stages. Winter cold may kill all mite stages in the field but not in protected crops. In southern areas of North America and in greenhouses, the mite commonly survives during winter on black nightshade growing near tomato crops. Morning glory and bindweed can also harbour the mites, and the latter develops a silvery sheen and may fall without russeting (Rice and Strong, 1962). In California, continuous reproduction has been observed in the field from early May to November or until frost. Reproduction is continuous as long as conditions permit, and only slows down at minimum survival temperatures for the host plants. Under these conditions, the mites are slow, their populations are sparse and they live mainly in leaf axils, stem crevices or curved leaves. The greenhouse represents a better reproduction environment for *A. lycopersici* at any time of the year. The females lay eggs irregularly among the hairs or in the slight crevices on both leaf surfaces, or on the smooth fruits in heavy attacks. Higher populations occur on the outer margin of infested plants. At a high level of attack, the mites crawl between plants, assuming the appearance of a fine yellowish dust. They are able to resist direct sunlight without apparent damage (Bailey and Keifer, 1943). Holdaway (1941) found more abundant mite populations during dry months, and Gispert *et al.* (1989) reported a higher mite density on water-stressed than well-irrigated tomato plants. Bailey and Keifer (1943) found all biological stages of the tomato russet mite on the leaves of protected outdoor nightshade plants in California. In areas where the winter hosts, including tomato, are not hampered by frost, the pest populations will survive and cause earlier injury to tomato crops in the field.

The first studies on mite biology detected optimum conditions for development near 26.5°C and 30% RH. Increases in temperature and decreases of RH were appropriate for development, while photoperiod seemed less important. The females lay eggs on leaflets near to naturally protected areas (leaf hairs, crevices, veins) (Rice and Strong, 1962). Recently, Kim *et al.* (2002a) found that the optimum development was between 15 and 30°C. Studies in Japan on tomato leaflets at temperatures between 15 and 27.5°C showed that survival rates from egg to adult were over 69%, and 53% at 30°C (Haque and Kawai, 2003). The highest survival rate (89.9%) of *A. lycopersici* was at 23°C (and 53% RH), and at 26°C it was 87.1% (and 75.5% RH). The optimal temperature for mite development should be within 26–29°C and the optimal RH within 53–75% (Xu *et al.*, 2006). The developmental rates increase linearly as temperature increases from 15 to 27.5°C (Haque and Kawai, 2003), and the duration of mite development decreases as the rearing temperature increases from 14 to 35°C, and increases as the RH increases from 30 to 100% (Xu *et al.*, 2006). At 21°C and 30% RH, egg to egg development takes 6.5 days, the mean oviposition period is 32 days, the fecundity is 50 eggs and the mean oviposition rate is 1.6 eggs a day. At 27 and 32°C and 30% RH, the mean oviposition period is 16 and 12 days, the fecundity is 47 and 24 eggs, and the mean oviposition rate is 2.9 and 2.0 eggs a day, respectively (Rice and Strong, 1962). Bailey and Keifer (1943) observed an egg to egg developmental time of 6.5 days at 21°C. At 25°C, the egg to egg development time is 6.9 days, the mean oviposition period is 19.12 days, the fecundity is 16 eggs and the mean oviposition rate is 0.8 eggs a day (Abou-Awad, 1979). The egg to egg development time at 27°C and 30% RH

is 6.0 (in Flechtmann, 1977a) and at 26–28°C is 7.0 days (Kamau, 1977), respectively. Under optimum conditions of temperature and RH, the fecundity can reach up to 53 eggs per female, the incubation period takes 2 days, the first nymph instar lasts 1 day, the second nymph instar 2 days and the life cycle develops in 6 or 7 days (Rice and Strong, 1962). Kim *et al.* 2002a) reported an egg to adult developmental time of 6–11 days. Haque and Kawai (2003) reported that the highest fecundity (51.7 eggs per female) occurred at 25°C, whereas Xu *et al.* (2006) found the highest fecundity at 26°C (53% RH) and 53% RH (26°C), with 44.3 eggs and 42.2 eggs laid per female, respectively. Moreover, the duration of oviposition decreases with increasing temperature and increases with increasing humidity (Xu *et al.*, 2006). At temperatures between 15 and 27.5°C, the development from egg to adult requires a total of 81.2 degree-days above a developmental zero of 10.5°C (Haque and Kawai, 2003). Other research has shown that the developmental zero of the egg, the nymph, and from egg to adult, are 10.51, 9.02 and 9.15°C, respectively. A total of 105.56 degree-days above a developmental zero of 9.15°C are required to complete development from the egg to adult emergence (Xu *et al.*, 2006). Haque and Kawai (2003) reported that the highest intrinsic rate of natural increase (0.253 a day) occurred at 25°C, and Xu *et al.* (2006) found that the highest intrinsic rates of natural increase (0.2645 and 0.2669, respectively) occurred at 26°C (53% RH) and 53% RH (26°C).

In Switzerland, the mite develops several generations during a growing season and at 25°C its population doubles in less than 3 days; it survives strong variations in temperature and RH (Fischer and Mourrut-Salesse, 2005). In Egypt, peak infestations on tomato usually occur during June and July (Osman, 1975b). In Turkey, *A lycopersici* appears on tomato at 26.1 and 28.3°C, when the first green tomato is observed in the field. The population density reaches the highest level when the green tomatoes are abundant, and also increases during harvest, when it is dense on the lower and middle leaves (Hıncal *et al.*, 2002). On tomato in Tunisia, *A. lycopersici* generally develops in the field at the beginning of July. One month later, the peak can reach 78 mobile forms per leaf (Grissa and Sahraoui, 2007). In Japan, on tomatoes in the greenhouse, the mite is first detected in mid-July and its peak density is observed in early October (Tanaka and Shibao, 2003).

The tomato russet mite abandons the injured and drying parts of infested plants by crawling and can move about ½ inch in 2 min. When there are no fresh tissues, it crawls to the tips of the stems and leaves, and up hairs, etc. At these sites, it rears up on the posterior part of its opisthosoma and claws at the air as it awaits some means of spread. The mite may also drop from dying plants (Bailey and Keifer, 1943). It is usually spread by human activity on tools, boxes, clothes, irrigation water, etc. and by insects (aphids). It disperses to higher stems and leaves quickly, but moves slowly to adjacent plants (Grahl and Leuprecht, 1998; Tanaka and Shibao, 2003).

Natural enemies include mycopathogens, such as the fungus *Hirsutella thompsonii* Fisher (Cabrera, 1984), predatory insects, such as the phlaeothripid *Leptothrips mali* (Fitch) (Bailey and Keifer, 1943; Anderson, 1954) and the thripid *Scolothrips sexmaculatus* (Pergande) (Abou-Awad, 1979), and various mite species. These last include some phytoseiids, such as *Amblyseius gossipi* Elbadry (Abou-Awad, 1983), *A. victoriensis* (Womersley) (James, 1989) and *Euseius concordis* (Chant) (Moraes and Lima, 1983), one stigmaeid, *Agistemus exsertus* (Gonzalez) (Osman and Zaki, 1986), and some tydeids, such as *Pronematus ubiquitus* (McGregor) (Carmona, 1970; Abou-Awad, 1979; Hıncal *et al.*, 2002) and *Homeopronematus anconai* (Baker) (Hessein and Perring, 1986).

Symptomatology and damage

The tomato russet mite has undergone a host expansion, having lived originally on wild solanaceous plants native to the Americas and later having associated with tomato (Oldfield, 1996a). On tomato, the mites feed on the leaves, stems and fruits, causing severe browning and curling of the leaves, and russeting of the fruits (Keifer, 1940c). The damage starts at the base of plant and spreads gradually towards the top, slowly affecting the stems, leaves, flowers and fruits. The stems become bronzed or russeted, crack longitudinally and check. The leaves become brown and paper like and usually wilt. Severe infestations cause leaf drop, even before the leaves turn colour. Loss of leaves may induce sunburning of the lower fruits, and under such conditions, weeds around tomato plants protect the fruits (Bailey and Keifer, 1943). Flowers lose their turgor, cease to develop, show russeting of the sepals and fall (Jeppson *et al.*, 1975). According to the variety and environmental factors, the early symptoms may occur on leaves. Much of the loss of yield is due to mite feeding on the flower stalks and pedicels, which dry and cause flower bud death (Kamau *et al.*, 1992). This explains why some varieties show most damage and have the lowest density of mites on the leaves. The fruits are attacked only in the severe cases, and in the field the mites are mainly observed on the green fruits around the stem; usually, a few individuals reproduce. Severely infested fruits do not develop, and gradually become russeted and roughened (Bailey and Keifer, 1943; Jeppson *et al.*, 1975). The damage first occurs on individual plants or in small areas of the crop, where brown patches due to mite attack may be observed. From these 'focal points', the infestation spreads in the crop in the direction of the prevailing wind; it is also spread by cultural practices (first picking, etc.) (Bailey and Keifer, 1943).

The pest lives preferentially on the leaf upper surfaces near the depressions surrounding the veins, and its populations reach a high density near the leaf petioles. Once it has destroyed the cells around the veins, it moves to more exposed areas and begins to feed there. Its feeding activity does not damage the parenchymal cells, and the drying and death of plants are due to the high density of mite attack, which results in water loss through the destroyed epidermis. The sites surrounding the veins enable the tomato russet mite to access the greater nutritional content of the vascular and meristematic tissues beneath the epidermis near veins and petioles, and can also provide a more favourable habitat for the mites (Royalty and Perring, 1988, 1996). It is possible to quantify feeding by the mite over time by analysing the regression relationship between injury to the tomato leaf epidermal cells and mite days. The proportion of epidermal cells damaged on a tomato leaflet and the feeding rate of mites are positively correlated (Royalty and Perring, 1988). This relationships can be explained by the

reduction of gas exchange between the leaflet and the surrounding atmosphere that is caused by accumulated damage to the epidermis, accompanied by an increase of leaflet temperature and vapour pressure deficit (Royalty, 1987; Royalty and Perring, 1989). Increased feeding activity can also be explained by increased mite density, which stimulates the pest to probe more cells. Mites can punctuate the cells, feed for a short time and move to new cells, resulting in an increase in the number of damaged cells (Royalty and Perring, 1996).

In California, a greater frequency of damage symptoms occurs in early and middle plantings than in late plantings, while no consistent relationship has been found between heat unit accumulation from seedling emergence to the first detected damage symptoms (either by year or by category of seedling emergence date). However, a relationship has been observed for most years and all categories of seedling emergence dates between the expression of damage symptoms and the stage of plant growth. The most likely stages of plant growth at which damage symptoms are expressed are when most plants have pink fruits or mature green tomatoes >2.5 cm (Zalom *et al.*, 1986a). The same behaviour has been reported from South Africa, where the mite causes field losses of up to 55% (Daiber, 1985).

Mite damage seems not to depend upon salivary toxins. It has been found that 800 mite days feeding by the tomato russet mite on a 50 mm^2 surface of leaflet is insufficient to cause russeting symptoms that clearly depend on saliva injected by the mite (Royalty and Perring, 1988). The linear relationship between mite day feeding cm^2 and a reduction in photosynthesis seem to indicate that the increase in the number of damaged cells during a given period of time results from a decrease in gas exchange (Royalty and Perring, 1989). Severe damage caused by *A. lycopersici* is probably a physiological response to mechanical damage of the mite and not a response to salivary toxins (Royalty and Perring, 1996).

Research in tomato on the effects of mechanical and chemical damage and three types of biotic damage on the activities of four foliar proteins of the tomato plant showed that putative plant defences were a proteinase inhibitor, polyphenol oxidase, peroxidase and lipoxygenase. Various damage treatments cause differential induction of these proteins, and different subsets of the four proteins are induced by different types of damage, so plants can to respond differentially to different types of damage (Stout *et al.*, 1994).

Short-term feeding of *A. lycopersici* on the leaves of tomato plants is associated locally and systemically with the induction of enzymatic proteins, such as peroxidases (POX, hydrogen peroxide (H_2O_2) oxidoreductase), which activity occurs locally more than that of polyphenol oxidase (PPO) or lipoxygenase (LOX). The induction of POX results in over eight times more damaged leaflets than undamaged leaflets on the 3rd leaf, while the increase is less than twofold for PPO and LOX. POX and PPO induction is host-plant systemic, while LOX is leaf systemic. Moreover, the systemic induction of POX is much greater than the induction of PPO or LOX. It seems that the high activity of POX induces lignin formation shortly after mite feeding (Stout *et al.*, 1996).

Severe damage to plants other than tomato is rare (Kay, 1986). This phenomenon can be explained by 'solanum stimulation' (Anderson, 1954; see above). Potato stems show little bronzing, but the leaves may dry and the plants die. On aubergine, high levels of infestation do not cause the plant to die, but the leaf shows leaf distortion and crinkling, and varieties with many small glandular epidermal hairs do not russet (Jeppson *et al.*, 1975).

Control

Control is provided by preventive and curative measures. In addition to prevention, the monitoring of mite populations to detect early attack, cultural practices and appropriate choice and use of chemicals are fundamental.

PREVENTION. Prevention consists mainly of weeding the crop and the surrounding areas, with particular attention to wild and cultivated solanaceous plants. The presence of infested Solanaceae from the previous crop is very dangerous. In addition, actions must be avoided that might inadvertently result in transfer of the mites by workers, tools, etc. from infested crops to mite-free crops. Plants should also be inspected before transplanting.

MONITORING. Usually, the tomato russet mite has a patchy distribution in the field and the level of group aggregation is correlated with different factors, such as its habits, environmental variables, etc. This facilitates the detection and monitoring of its populations and the assessment of density. Plant protection from the tomato russet mite requires knowledge such as the effects of temperature and humidity on the growth rate, in order to understand mite population dynamics (Haque and Kawai, 2003). Research on tomato has shown that adults and nymphs of *A. lycopersici* have a similar distribution on leaflets. About half of the population infests the leaflets and the other half the petioles. Most of the mites lives on the upper leaf surface. On petioles, the mite population is high at the posterior part and large on the leaflets near the stem. The percentage of mite population on each part of the leaflet to the total number on the whole leaf does not vary, irrespective of the population density. On leaflets, the mite density is higher in the middle than at the periphery, and it is also higher at the bottom than at the top. On leaflets, the mite density is uniform when the attack is high. This behaviour suggests a method for estimating the densities of the mite population on the whole leaf, in which the mite numbers on the second and third leaflets and several parts of the petiole are counted (Kawai and Haque, 2004). According to the theoretical sampling formula of Iwao–Kuno, Hirata *et al.* (2007) proposed a theoretical sampling number for different densities of *A. lycopersici*. Xu *et al.* (2008) estimated the density of tomato russet mites in different parts of a tomato plant at different stages and analysed the spatial distribution pattern by different methods. They found that the tomato russet mite exhibited a gathering distribution in tomato in the field; its basic distribution element is the individual group, and its group aggregation depends mainly on its habits and environmental factors. This suggests a theoretical sampling number for different mite densities according to the formula of Iwao–Kuno.

PLANT RESISTANCE. Lines resistant to tomato russet mite have been investigated in various countries, but no line appears to be currently available commercially. On a resistant genotype of bittersweet nightshade (*S. dulcamara* L.), the feeding activity of the tomato russet mite induces the development of a local incompatible interaction (HR) (*sensu* Westphal and Manson, 1996) that prevents further development of its populations. Damaged cells lack callose and lignin and contain polyphenol oxidation compounds (Westphal *et al.*, 1981). Moreover, cells around the HR surfaces have a high amount of phytoalexin and strong hydrolytic activity (Bronner *et al.*, 1991a) and oxidative enzymes (Bronner *et al.*, 1991b); they also lack lignin (Westphal *et al.*, 1981; Bronner *et al.*, 1991a) and callose is present on the cell walls. However, it is unclear whether the hydrolytic enzymes affect the mites directly or indirectly (by toxin production). Several other abiotic and biotic stresses (e.g. pathogen infection, mechanical wounding) trigger HR that may or may not be associated with hydroxyproline-rich glycoproteins (HPRG) formation. Therefore, the response of a resistant plant genotype to feeding by *A. lycopersici* cannot be considered to be a specific defence reaction (Westphal *et al.*, 1996).

Leaves of tomato cultivars resistant to *A. lycopersici* contain high levels of 2-tridecanone (alternative name tridecan-2-one) and other methyl ketones (Gonçalves *et al.*, 1998). On *L. esculentum* and *L. hirsutum* Dunal, the different levels of tridecan-2-one and undecan-2-one, the density and type of trichomes and the size of leaves affect the distribution of the mite (Leite *et al.*, 1999). Tomato trichomes affect and hinder the natural enemies and the host plant is therefore more resistant to the natural enemies than to the mite pest (Simmons and Gurr, 2005). Also, the presence of toxic secondary metabolites, such as the steroidal glycoalkaloid α-tomatine, can intoxicate predators (Roddick, 1974).

Difference in susceptibility to damage by the tomato russet mite among varieties within and between species in the genus *Lycopersicon* have been investigated in Japan (Kitamura and Kawai, 2006). In China, Wang *et al.* (2008a,b) examined mite development in the field and the laboratory on several tomato lines characterized by more hair (YZ618, YZ619) or less hair (YZ419, YZ507, YZ504), as well as wild lines (YZ7, YZ5), yellow-leafed lines (YZ401) and conventional lines (YZ406, YZ412, YZ413, YZ515). High population densities developed on some lines (YZ504, YZ507, YZ419), while on other lines (YZ7, YZ618, YZ619) low population densities developed. Serious differences were found in survival rates, development duration and egg production on different lines. The highest survival rate occurred on the lines with less hair (YZ504, YZ507 and then YZ419), and the lowest on YZ7, a wild line. The highest mite fecundity occurred on YZ504, followed by YZ507 and YZ419, and the lowest on YZ7. The data showed that the YZ7 and YZ618 lines were resistant to the mite, while other lines (YZ504, YZ419, YZ507) were susceptible to the mite (Wang *et al.*, 2008a). Multi-haired plants (YZ618) hosted an amount of mites that was significantly lower than those with fewer hairs (YZ504, YZ413), and mite distribution varied on these three tomato lines. Fewer mites occurred on the upper part of YZ618, and the amount of mites on the upper part of YZ504 was higher than those on the middle and lower parts, while the amount of mites on the upper and middle parts of YZ413 was higher than that on the lower part. Moreover, the enzyme activities of PPO, POX and ascorbic peroxidase (ASP) in leaves increased after infestation by mites, and the mite-resistant YZ618 was most sensitive in this respect. Correlation analysis showed that PPO was the mite-resistant biochemical factor induced by mite infestation, and that ASP was the original mite-resistant biochemical factor. No significant relationship between POX and mite resistance was established (Wang *et al.*, 2008b).

CULTURAL PRACTICES. Fertilization and irrigation may affect tomato russet mite development. Irrigation management early in the first phase of tomato growth could help to prevent mite development later in the growing phase (Gispert *et al.*, 1989). The levels of NPK can affect the development of *A. lycopersici* under field conditions. Plants fertilized with 120 kg ha^{-1} K are more infested by the mite. There is a direct relationship between mite infestation and the levels of P applied to the soil. At a K level of 120 kg ha^{-1}, infestation increases with P levels. At N levels of 60 and 180 kg ha^{-1}, the infestation increases, whereas at 120 kg ha^{-1}, it is reduced. However, the NP interaction has a low effect on infestation at higher levels of N. The levels of N and P applied to the soil are positively and significantly correlated with the content found in the tomato leaves. No significant correlation was observed for K (Moreira *et al.*, 1999).

BIOLOGICAL CONTROL. Relatively few studies have been carried out with pathogens, and first report concerns probably *Bacillus thuringiensis* Berliner (Royalty and Perring, 1987), for which the exotoxin of *B. t.* subsp. *thuringiensis* has been tested with no encouraging results on tomato and potato in Uzbekistan (Russia) (Sulaimanov *et al.*, 2000). In the last three decades, most studies have concerned the phytoseiid mites and some predatory insects, but biological control of the tomato russet mite awaits adequate solutions. The first investigations were probably carried out in Brazil, where the action of the phytoseiid mite *Euseius concordis* (Chant) was evaluated as a predator of the tomato russet mite (Moraes and Lima, 1983). The difficulty of introducing natural enemies into Canada to control *A. lycopersici* effectively was reported by Shipp *et al.* (2001). In Switzerland, investigations of the role of the phytoseiid *Amblyseius andersoni* (Chant) gave unclear results (Fischer *et al.*, 2005). Feeding trials in Australia with indigenous phytoseiid mites (*Typhlodromips montdorensis* (Schicha), *Typhlodromalus lailae* (Schicha), *Amblyseius lentiginosus* Denmark *et* Schicha, *Neoseiulus wearnei* (Schicha), *Euseius victoriensis* (Womersley) and *Neoseiulus cucumeris* (Oudemans) showed that some species developed on tomato russet mite; however, they may not be suitable candidates as biocontrol agents because they do not cope well with trichomes on the tomato plant (see Plant resistance above) (Steiner and Goodwin, 2005). The actions of *Neoseiulus californicus* (McGregor) and *N. cucumeris* were not satisfactory on greenhouse tomatoes in France (Trottin-Caudal *et al.*, 2005). Gerson and Weintraub (2007) report that trials carried out with several phytoseiids (*N. californicus*, *N. fallacis* (Garman), *N. cucumeris*, *Euseius concordis*, *A. victoriensis*, *T. lailae*) indicate that currently none of them produce adequate control of *A. lycopersici*. Further research by Momen and Abdel-Khalek (2008) pointed out that *Typhlodromips swirskii* (Athias Henriot) and *Typhlodromus athiasae*

Porath *et* Swirski fed on mobile stages of the tomato russet mite, but *Paraseiulus talbii* (Athias Henriot) did not as it never reached the adult stage when fed on the tomato russet mite. Some tydeid mites are effective natural enemies of tomato russet mite on tomato in greenhouse in Japan, and spot-spraying of acaricides is recommended to preserve them for suppressing *A. lycopersici* damage (Tanaka and Shibao, 2003). The response of an unidentified predatory midge, *Feltiella* sp., was evaluated on tomato in the greenhouse in Australia (Steiner and Goodwin, 2005), and gave a promising response, but its off-plant pupation and difficulties in mass rearing were obstacles to its commercial use.

CHEMICAL CONTROL. Severe attacks in the pre-harvest phase cause higher crop losses (Zalom *et al.*, 1986a), and prophylactic pesticide applications may be necessary to control recurrent outbreaks (Perring and Farrar, 1986). Many active ingredients are reported from various countries, some of which show mite-resistant strains, such as methamidophos (Abou-Awad and El-Banhawy, 1985). Insecticides/acaricides tested include abamectin (Royalty and Perring, 1987; Haji *et al.*, 1988; Silva *et al.*, 1988), azocyclotin (Blanck *et al.*, 1954; Kamau, 1977; Kay, 1986; Kay and Shepherd, 1988), bifenthrin (Perring and Trumble, 1984), binapacryl (Ramalho and Leao Veiga, 1980), bromopropylate (Blanck *et al.*, 1954; Kamau, 1977; Ramalho and Leao Veiga, 1980; Oliveira and Sponchiado, 1983; Costilla, 1991; Baradaran Anaraki and Daneshvar, 1992), clofentezine (Haji *et al.*, 1988), chlorfenethol, chlorfensulfide (Ramalho and Leao Veiga, 1980), chlorobenzilate (Blanck *et al.*, 1954; Kamau, 1977; Ramalho and Leao Veiga, 1980; Cermeli *et al.*, 1982; Oliveira and Sponchiado, 1983), cyhexatin (Blanck *et al.*, 1954; Kamau, 1977; Cermeli *et al.*, 1982; Kay, 1986; Royalty and Perring, 1987; Kay and Shepherd, 1988; Silva *et al.*, 1988), cypermethrin (Bourdouxhe and Collingwood, 1982; Perring and Trumble, 1984; Abou-Awad and El-Banhawy, 1985; Kim *et al.*, 2002a), deltamethrin (Bourdouxhe and Collingwood, 1982), dichlorvos (Ramalho and Leao Veiga, 1980), dicofol (Blanck *et al.*, 1954; Kamau, 1977; Ramalho and Leao Veiga, 1980; Cermeli *et al.*, 1982; Oliveira and Sponchiado, 1983; Abou-Awad and El-Banhawy, 1985; Royalty and Perring, 1987; Haji *et al.*, 1988; Kay and Shepherd, 1988; Costilla, 1991; Monkman, 1992; Hıncal *et al.*, 2002), endosulfan (Blanck *et al.*, 1954; Kamau, 1977; Cermeli *et al.*, 1982; Kay and Shepherd, 1988), fenarimol (Abou-Awad and El-Banhawy, 1985), fenazaquin (Kim *et al.*, 2002a), fenbutatin oxide (Kay and Shepherd, 1988), fentin acetate, fentin hydroxide (Nunez and Maeso, 1983), fenvalerate (Bourdouxhe and Collingwood, 1982; Oliveira and Sponchiado, 1983), flubenzimine (Haji *et al.*, 1988; Kay and Shepherd, 1988), fluvalinate (Cheremushkina *et al.*, 1991), formothion (Ramalho and Leao Veiga, 1980), lambda-cyhalothrin (Cheremushkina *et al.*, 1991), methamidophos (Ramalho and Leao Veiga, 1980; Haji *et al.*, 1988), mineral oils (Monkman, 1992), monocrotophos (Kay and Shepherd, 1988), oxythioquinox (Blanck *et al.*, 1954; Kamau, 1977), parathion-methyl (Oliveira and Sponchiado, 1983), petroleum oils (Singh *et al.*, 2000b), profenofos (Oliveira and Sponchiado, 1983), propargite (Ramalho and Leao Veiga, 1980; Cermeli *et al.*, 1982), pyridaphenthion (Abou-Awad and El-Banhawy, 1985), quinomethionate (Messiaen, 1971; Hıncal *et al.*, 2002), sulfur (Bailey and Keifer, 1943; Tuft and Anderson, 1953; Anderson, 1954; Blanck *et al.*, 1954; Harpaz, 1955; Rice and Strong, 1962; Smith, 1955; Kamau, 1977; Cermeli *et al.*, 1982; Vacante, 1985b,c, 2000; Royalty and Perring, 1987; Haji *et al.*, 1988; Kay and Shepherd, 1988; Berlinger *et al.*, 1988; Silva *et al.*, 1988; Baradaran Anaraki and Daneshvar, 1992; Monkman, 1992; Hıncal *et al.*, 2002; Kitamura and Honda, 2006; Vacante and Benuzzi, 2007), sulprofos (Kay and Shepherd, 1988) and tetradifon (Ramalho and Leao Veiga, 1980; Haji *et al.*, 1988; Hıncal *et al.*, 2002).

Fungicides include anilazine (Nunez and Maeso, 1983), binapacryl, captafol (Blanck *et al.*, 1954; Kamau, 1977), chlorothalonil (Nunez and Maeso, 1983), dinocap (Blanck *et al.*, 1954; Osman, 1975b; Kamau, 1977; Baradaran Anaraki and Daneshvar, 1992), folpet (Blanck *et al.*, 1954; Kamau, 1977; Nunez and Maeso, 1983), metalaxyl (Perring and Trumble, 1984), and triforine (Nunez and Maeso, 1983).

Despite its negative impact on beneficials, the use of sulfur must be privileged because of its reduced toxicological impact on mammals.

INTEGRATED PEST MANAGEMENT. In north-western Europe, the climate permits control of *A. lycopersici* mainly by means of biologically based IPM (Khromova, 2001), whereas in southern Europe, the implementation of an IPM strategy is reduced because the milder climatic conditions favour a wider availability of alternative host plants for the tomato russet mite (Yoldas *et al.*, 1999). Experiences in Italy on tomato in greenhouses showed that a reasonable use of sulfur allowed the implementation of IPM (Vacante and Benuzzi, 2007). Many bibliographic reports confirm the worldwide and broad consensus on the use of sulfur. Attacks occurring early in the season usually require more than one spray with sulfur, which gives good control for 3 or 4 weeks. Unfortunately, lack of knowledge of the problem and commercial interests assign an insufficient action to control by sulfur, e.g. in Mediterranean greenhouses. References to its use in cold Mediterranean greenhouses (Vacante, 1985b,c, 2000; Vacante and Benuzzi, 2007, and others) indicate the opposite. Horticultural practices and weed control are a part of the strategy.

Aculops benakii (Hatzinikolis)

Common name

Olive yellow spot mite.

Diagnostic characteristics

FEMALE. The body is fusiform, 165–175 μm long and pinkish in colour. The prodorsal shield has a long acuminate frontal lobe over the gnathosoma, and a network pattern. A median line occurs on the rear ¼, with a cross line at about ⅓ and a transverse line curving back from in front of the scapular tubercles and crossing the prodorsal shield just ahead of the rear margin. The scapular setae (*sc*) are 21 μm long. There are many lateral shield cells. The opisthosoma has about 20 dorsal annuli and 68 ventral annuli. The microtubercles on the ventral annuli are bead

like on the annulus margins. The last annuli are completely microtuberculate and contrasting with the anterior part of the opisthosoma. The genitalia have transverse basal lines, and about 14 longitudinal ribs on the coverflap (Jeppson *et al.*, 1975).

Geographical distribution

Greece (Hatzinikolis, 1968a).

Host plants

Oleaceae: olive (*Olea europaea* L.) (Hatzinikolis, 1968a, 1969b, 1972).

Bio-ecology

The olive yellow spot mite lives on the underside of olive leaves under the stellate hairs (Hatzinikolis, 1968a). At 20–25°C, the egg to egg development time is 14.0 days, the mean oviposition period is 10.5 days and the fecundity is 12–33 eggs (Hatzinikolis, 1979). In Greece, the mite can develop from 12 to 15 generations a year.

Symptomatology and damage

The possible simultaneous presence on olive trees of different eriophyoid mites hinders the proper description of the symptomatology and damage caused by this species (and others living on olive). Regardless, olive eriophyoids attack in succession the buds, shoots, leaflets, inflorescences and fruits. According to Hatzinikolis (1969b), expanded leaves show lighter green hollows on the upper surface associated with a corresponding chlorotic protuberance on the undersurface, which gradually become brownish. However, other eriophyoids, such as *Oxycenus maxwelli* (Keifer) and *Tegolophus hassani* (Keifer), may cause the same symptomatology. On the inflorescences and young fruits, the feeding activity of *Aculops benakii* causes russeting and splitting of flower buds, which gradually become whiter and drop, and also early drop of flowers (and of inflorescences) and the deformation and/or dropping of young fruits (Hatzinikolis, 1969b, 1972).

Control

The correlation between mite density and damage is unclear, and like the citrus bud mite, *Aceria sheldoni* (Ewing), the role of this mite could regulate fruit setting. The lack of other host plants shows a convergent evolution that could justify a mutualistic symbiosis between the mite and its host plant. In the cultivation of olives for oil, the mite is usually a secondary problem.

Aculops gleditsiae (Keifer) (Fig. 16.84)

Common name

Honey locust rust mite.

Diagnostic characteristics

FEMALE. The body is elongate, spindle formed, 175–190 μm long, 60 μm wide and yellowish in colour. The gnathosoma is curved down. The prodorsal shield has a frontal lobe with a narrow rounded point; the pattern is a network with numerous curved and cross lines that form cells; the median line is a short mark at the central rear; the admedian lines are distinct on the anterior ⅔; there are lateral granulations below the margin. The scapular tubercles are 37 μm apart, and the scapular setae (*sc*) are 16 μm long. The opisthosoma has about 28 dorsal annuli and 60–65 ventral annuli; the dorsal annuli have elongate microtubercles that project unevenly slightly beyond the rear annulus edge; the microtubercles on the ventral annuli are small and rounded. The setae c_2 are 17 μm long, set on ventral annulus 9; the setae *d* are 54 μm long, set on ventral annulus 24; the setae *e* are 14 μm long, set on ventral annulus 41; the setae *f* are 18 μm long, set on ventral annulus 4 from the rear; the setae h_1 are present. The genitalia are 23 μm wide and 20 μm long, and the coverflap has about 12 furrows; the seta is 28 μm long. The empodium is 5-rayed (Keifer, 1959a; Jeppson *et al.*, 1975).

Geographical distribution

Hungary, Italy (de Lillo and Beron, 2004), Virginia (Keifer, 1959a) and various areas of eastern USA (Jeppson *et al.*, 1975), Yugoslavia (de Lillo and Beron, 2004) and Serbia and Montenegro (Glavendekić *et al.*, 2005).

Host plants

Fabaceae: honey locust (*Gleditsia triacanthos* L.) (Keifer, 1959a; Jeppson *et al.*, 1975).

Bio-ecology

The honey locust rust mite lives on honey locust in Virginia (Keifer, 1959a) and other areas of the eastern USA (Jeppson *et al.*, 1975). Its bio-ecology is not well documented.

Symptomatology and damage

A. gleditsiae causes severe rusting on young honey locust leaves and high attacks may be associated with dying trees (Keifer, 1959a; Jeppson *et al.*, 1975).

Control

Control it is not usually necessary.

The Genus *Abacarus* Keifer

Diagnostic characteristics

The body is more or less elongate. The gnathosoma has a moderate size. The prodorsal shield is subtriangular, with an acuminate frontal lobe over the gnathosoma; the scapular tubercles are located on the rear margin, directing the scapular setae (*sc*)

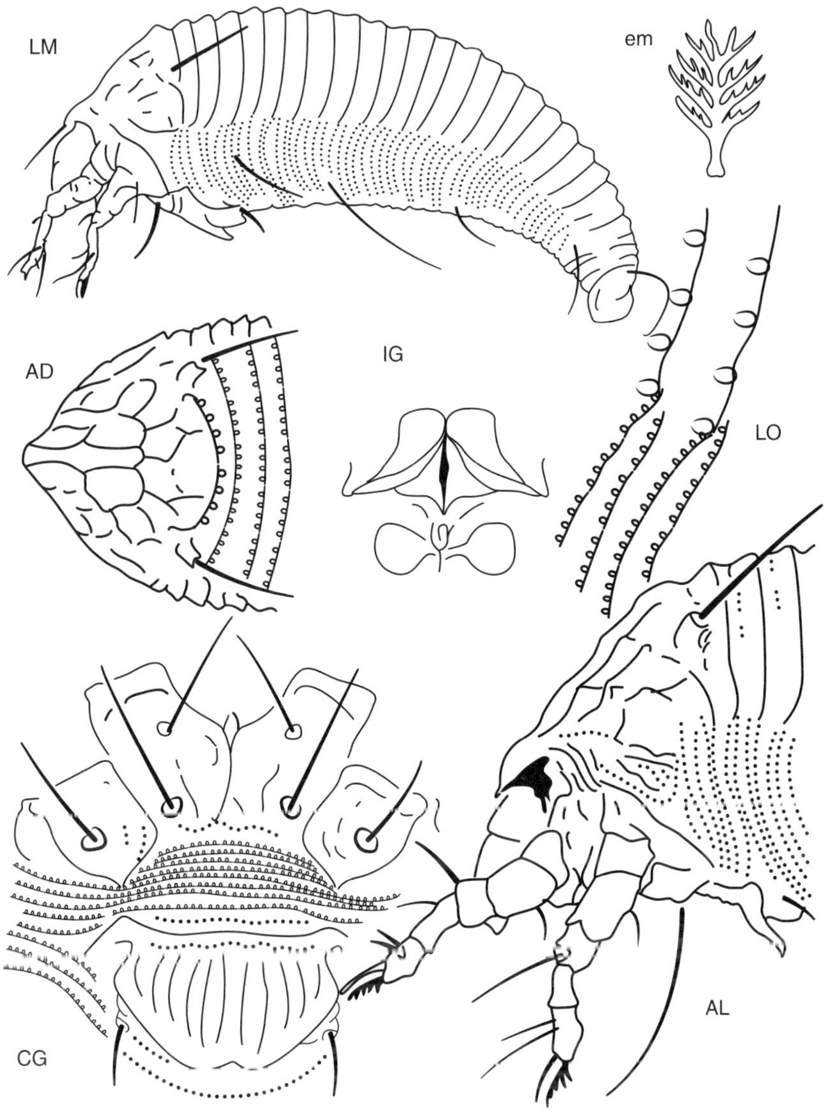

Fig. 16.84. *Aculops gleditsiae* (Keifer) (from Keifer, 1959a). See Table 16.1 for key.

caudad. The opisthosomal dorsal annuli are nearly as numerous as the microtuberculate ventral annuli, with the microtubercles suppressed or absent dorsally; the dorsal annuli are formed into three dorsal longitudinal wax-bearing ridges, the middorsal opisthosomal ridge being shorter than the subdorsal ridges and ending in a dorsal furrow. The empodium is simple (Keifer, 1944; Amrine *et al.*, 2003). In *Abacarus doctus* Návia, Flechtmann, Lindquist *et* Aguilar, Návia *et al.* (2011) observed an unusual typical seta on tibia II, set in a dorsal or lateral position.

Economic importance

This genus *Abacarus* includes over 60 species living both on dicotyledonous and monocotyledonous plants, and it has a worldwide geographical distribution. The type species, *Calepitrimerus acalyptus* Keifer, is a vagrant on the upper side of the leaves and on stems of the snow bush (*Ceanothus cordulatus* Kellogg) in California (Keifer, 1939e). Different species are reported as injurious to economic plants, such as the cereal rust mite, *A. hystrix* (Nalepa), causing damage to leaves and growth of oat (*Avena pubescens* (Hudson) Dumort.), barley (*Hordeum* spp.), wheat (*Triticum aestivum* L.), rye (*Secale cereale* L.) and other Poaceae; the mite also transmits *Ryegrass mosaic virus* (Jeppson *et al.*, 1975; Oldfield and Proeseler, 1996). The coffee mite, *A. afer* Keifer, causes rust on the lower leaf surface of leaves of coffee (*Coffea arabica* L.) in Zaire (Keifer, 1962b). The rice rust mite, *A. oryzae* Keifer, is often associated with plants dwarfed by 'tungo' disease in the Philippines (Keifer, 1963a). *A. sacchari* ChannaBasavanna lives along the laminar furrows on the upper surface of tender leaf blades of sugarcane (*S. officinarum* L.) in India (ChannaBasavanna, 1966) and other tropical countries,

such as Brazil (Flechtmann, 1970), Réunion (Gutierrez, 1982) and the Cape Verde Islands in the Atlantic Ocean off the coast of Africa (Meyer Smith, 1989b); it is probably not injurious. Currently, there are four *Abacarus* spp. (*A. doctus* Návia, Flechtmann, Lindquist *et* Aguilar, *A. delhiensis* ChannaBasavanna, *A. queenslandiensis* Ozman-Sullivan, Amrine *et* Walter and *A. sacchari*) reported from sugarcane (*Saccharum* spp.). *A. doctus* is reported from Costa Rica and is a vagrant species on the inner surface of leaves; high levels of populations can cause reddish or bronzed spots to spread randomly on the inner leaf surfaces, which can extend to cover almost the entire leaf sheath. At a distance, these damage symptoms could be confused with those caused by rust fungi, which are commonly associated with sugarcane (Návia *et al.*, 2011). Among the species mentioned above, only the cereal rust mite causes severe damage.

Abacarus hystrix (Nalepa) (Fig. 16.85)

Common name

Cereal rust mite.

Diagnostic characteristics

FEMALE. The body is 180–200 μm long, 35–40 μm wide, whitish in life and has a central longitudinal stripe of wax, enclosed by a lateral band on each side. The prodorsal shield has the frontal lobe narrowed above, a spine below. The shield has a central ridge, laterally granulate; the scapular tubercles are 23.5 μm apart, on the rear margin; the scapular setae (*sc*) are 20 μm long, projecting caudad. The opisthosoma has the ventral annuli

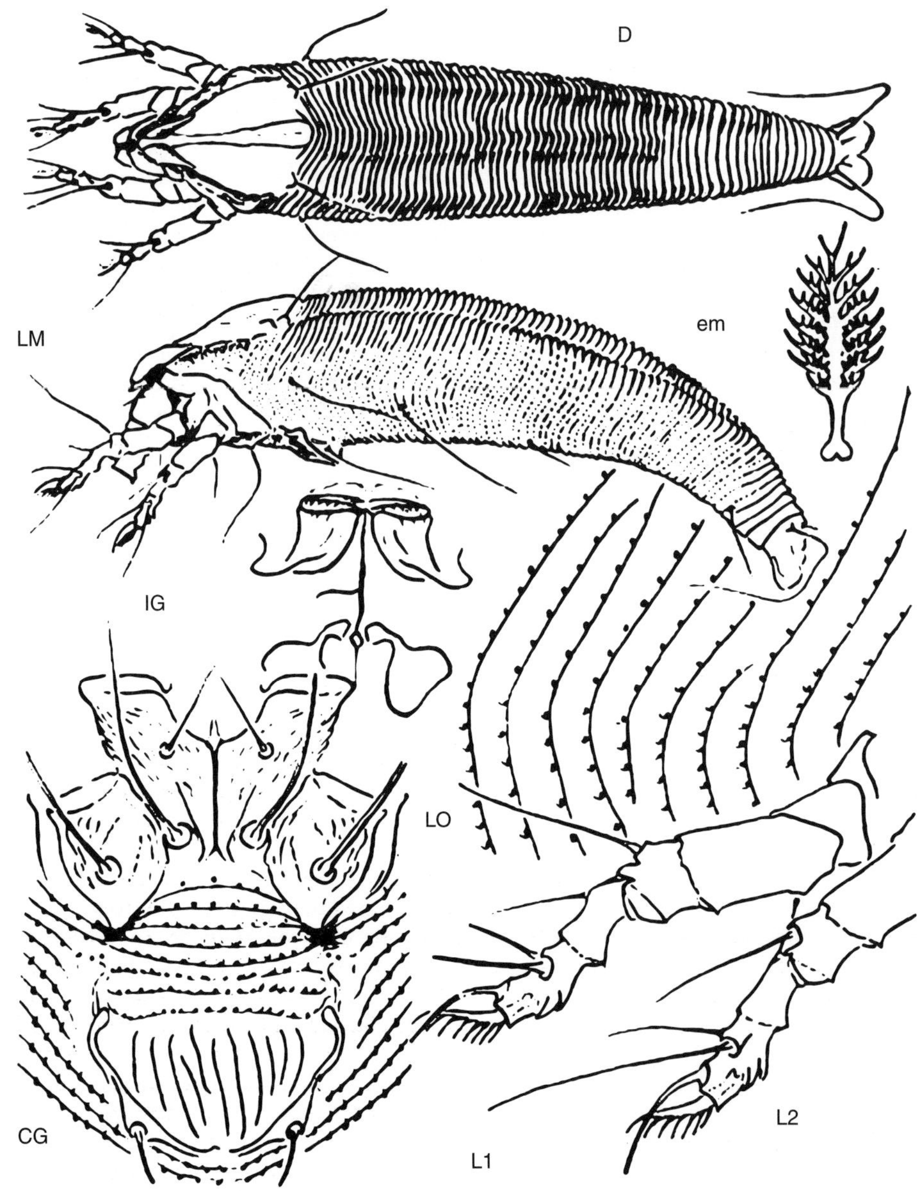

Fig. 16.85. *Abacarus hystrix* (Nalepa) (from Keifer, 1944). See Table 16.1 for key.

microtuberculate, the microtubercles are suppressed or absent on the dorsal annuli; the central longitudinal wax-bearing ridge ends in the dorsal trough on about the 46th dorsal annulus; supralateral wax-bearing ridges extend further caudad; the dorsal annuli are about as numerous as the ventral annuli, of which there are 60–65. The setae c_2 are 35 µm long, set on about ventral annulus 6; the setae *d* are 36 µm long, set on about ventral annulus 21; the setae *e* are 23 µm long, set on about ventral annulus 38; the setae *f* are 27 µm long, set on about annulus 4 from the rear; the setae h_1 are present. The genitalia are deep basin shaped externally, 21 µm wide, 16 µm long, and the coverflap has about 12 furrows. The empodium is 8-rayed (Keifer, 1944).

MALE. The body of the male is 170–190 µm long and 35-40 µm wide (Keifer, 1944).

Geographical distribution

Worldwide.

Host plants

The range of host plants of the cereal rust mite includes 60–70 species of Poaceae. The most common host plants are listed: *Agropyron* spp. (Jeppson *et al.*, 1975), couchgrass (*A. repens* (L.) P. Beauv.) (Nault and Styer, 1969; Kozłowski, 2001), colonial bentgrass (*Agrostis tenuis* Sibthorp) (Proeseler, 1972c), false oat-grass (*Arrhenatherum elatius* (L.) P. Beauv. ex J. Presl and C. Presl) (Kozłowski, 2001), oat (*Avena sativa* L.), cocksfoot (*Dactylis glomerata* L.) (Jeppson *et al.*, 1975), *Elymus* spp. (Jeppson *et al.*, 1975), quack grass (*E. repens* (L.) Gould) (Skoracka and Kuczynski, 2006b), *Festuca* spp. (Jeppson *et al.*, 1975), tall fescue (*F. arundinacea* Schreb.) (Frost and Ridland, 1996), English bluegrass (*F. pratensis* Huds.) (Gibson, 1974; Jeppson *et al.*, 1975), barley (*Hordeum vulgare* (L.)) (Jeppson *et al.*, 1975), *Lolium* spp. (Jeppson *et al.*, 1975; Frost and Ridland, 1996), Italian ryegrass (*L. multiflorum* Lamarck) (Gibson, 1974; Jeppson *et al.*, 1975; Salm *et al.*, 1994), English ryegrass (*L. perenne* L.) (Gibson, 1974; Jeppson *et al.*, 1975; Kozłowski, 2001), rice (*Oryza* spp.), *Phleum* spp., timothy hay (*P. pratense* L.), *Poa* spp. (Jeppson *et al.*, 1975), rye (*Secale cereale* L.) (Meyer Smith, 1989b), *Triticum* spp. (Jeppson *et al.*, 1975) and wheat (*T. aestivum* L.) (Skoracka and Boczek, 2001; Skoracka *et al.*, 2007).

Bio-ecology

The cereal rust mite was first reported as widespread on perennial grasses in the northern hemisphere (Keifer, 1944). It is a vagrant species that has been collected from 60–70 species of grasses, including economically important species such as wheat (Jeppson *et al.*, 1975; Skoracka and Boczek, 2001; Skoracka *et al.*, 2007), barley, oat and rice (Jeppson *et al.*, 1975), and grasses involved in pasture production. Grasses hosting the cereal rust mite include couchgrass (Nault and Styer, 1969; Kozłowski, 2001), *Agropyron* spp. (Jeppson *et al.*, 1975), colonial bentgrass (Proeseler, 1972c), false oat-grass (Kozłowski, 2001), cocksfoot (Jeppson *et al.*, 1975), quack grass (Skoracka and Kuczynski, 2006b), *Elymus* spp. (Jeppson *et al.*, 1975), tall fescue (Frost and Ridland, 1996), English bluegrass (Gibson, 1974; Jeppson *et al.*, 1975), *Festuca* spp., Italian ryegrass (Gibson, 1974; Jeppson *et al.*, 1975; Salm *et al.*, 1994), English ryegrass (Gibson, 1974; Jeppson *et al.*, 1975; Kozłowski, 2001), *Lolium* spp. (Jeppson *et al.*, 1975; Frost and Ridland, 1996), timothy hay, *Phleum* spp. and *Poa* spp. (Jeppson *et al.*, 1975), and rye (Meyer Smith, 1989b). In Ohio, *A. repens* is the major host plant of this mite (Nault and Styer, 1969); Gibson (1974) reported that the mite established only on English ryegrass, Italian ryegrass and English bluegrass. In Australia, the cereal rust mite has been collected mostly on *Lolium* spp. and sometimes on tall fescue (Frost and Ridland, 1996). In South Africa, the cereal rust mite lives on rye (Meyer Smith, 1989b) and Italian ryegrass (Salm *et al.*, 1994). In Poland, the most infested grasses are quack grass and English ryegrass and least infested is false oat-grass, while *Abacarus hystrix* and *Aculodes mckenziei* (Keifer) are the dominating eriophyoid species (Kozłowski, 2001).

Despite this range of host plants, *A. hystrix* is unable to infest other graminaceous plants (timothy, maize, barley, oat and wheat) from ryegrass (Gibson, 1974), probably owing to the existence of biotypes adapted to specific host plants of the same botanical family. Because of its wide host range, *A. hystrix* has long been considered to be a host generalist. However, host specialization between some host populations of *A. hystrix* has been demonstrated. Studies on the development time, survivorship, longevity and fecundity of *A. hystrix* living on *L. perenne* and *E. repens* support the hypothesis that populations of the cereal rust mite from *E. repens* and *L. perenne* are highly specialized in their host use and may be regarded as host races or separate species (Skoracka and Kuczynski, 2006b). When exchanged on to each other's hosts, *A. hystrix* females collected from different host plants showed no acceptance of and failed to colonize the novel host plant (Skoracka and Kuczyński, 2006a; Skoracka *et al.*, 2007). Crossing experiments have shown that in addition to host fidelity, which acts as a pre-zygotic barrier, post-zygotic barriers between *A. hystrix* host populations also exist (Skoracka, 2008). A molecular genetic study pointed out that host populations of this mite exhibited more than 20% sequence divergence in the cytochrome oxidase subunit I (COI) gene (Skoracka and Dabert, 2010). Thus, we can expect that over its huge host range, *A. hystrix*, an ostensible generalist, may comprise many specialized 'species'.

On Italian ryegrass, the cereal rust mite feeds on bulliform cells at the base of grooves on the adaxial leaf surface, using the subcapitulum to remove the cuticular leaf wax before feeding. The feeding activity of the mite does not damage the mesophyll cells. Leaf development raises the eggs and immature stages higher into the canopy and the adults move gradually downward towards the plant crown, where they usually feed on tender tissues (Gibson, 1974). This behaviour causes a vertical age stratification. Females overwinter in the crowns of the plant, and there are no diapause stages, although during winter mites reduce their activity in the crowns of the host plants. In the spring, mites congregate on the tender tips, and occupy the grooved upper surface. Few mites inhabit the smooth undersurfaces or the old leaves.

The reproduction of *A. hystrix* is arrhenotokous (Skoracka, 2008). Females lay eggs in the leaf grooves and over a 6-day period their fecundity varies from 51 to 81 eggs. At room temperature,

the eggs hatch in 5–6 days. Mite populations decrease after cutting of the first hay crop. When hot weather occurs after cutting, the mite density remains low until late August or early September, and increases after this. Burning stubble in September does not reduce mite density (Pepper, 1942; Mulligan, 1960). At 20°C on quack grass, the egg to egg development time is 12 days, while at 25°C it is 8.5 days (Boczek and Chyczewski, 1975).

At constant temperature of 19.5–20.5°C and 94 ± 1% RH, egg to adult development takes 7.98 days. The life expectancy of females is longer than that of males (9.72 and 5.41 days, respectively). The sex ratio is 0.80 females:1 male. The reproductive output for females is age dependent, and daily egg production reaches a peak of 3.83 eggs on the 5th day. The net reproductive rate (R_0) is 10.12 female progeny per female per generation, the generation time (T) is 11.31 days, the intrinsic rate of increase (r_m) is 0.20 female progeny per female per day, and the finite rate of increase (λ) is 1.23 female progeny per day. The mite has a great potential capacity for rapid population increase when colonizing new hosts, and its density is low. The population of the cereal rust mite on quack grass may rapidly build up to very high densities and can be a reservoir population, which may easily disperse and infest other, including cultivated, grasses (Skoracka and Kuczynski, 2004).

On quack grass, the population density of *A. hystrix* is dependent both on location of the leaf on the shoot and date of collection. The highest number of individuals per leaf is recorded on the first day of June and the lowest on the first day of August. The highest population density occurs on the upper leaves. The densities on the median and lower leaves did not differ significantly. However, they were significantly different from the group on the upper leaves. A significant correlation was found between population density and unevenness of distribution. At a low population density, the mites are distributed evenly across the host plant, but at a high density the selectivity increases and population distribution changes towards a clumped pattern (Skoracka *et al.*, 2003).

The mite walks actively between leaves and spreads over about a 1 m of sward during the summer and autumn (Gibson, 1981); dispersal basically depends on the action of the wind (Nault and Styer, 1969; Gibson, 1981; Frost, 1997). In Great Britain, the mites colonize healthy ryegrass plants from June to October (A'Brook, 1975), while in Germany, on turf grass (*L. perenne*, *A. tenuis* and *D. glomerata*) cut at 2- to 3-week intervals from late May until the end of August, the cereal rust mite populations start to develop in July and peaks occur during the autumn and winter, with lower populations during the spring and summer (Proeseler, 1972c). Other studies on the population dynamics of the cereal rust mite show that a higher density occurs in autumn, with a strong host growth. Lower levels of populations were observed during winter, and increases during spring and early summer. Migration occurs in late spring and summer, and low herbage quality causes a decrease in the population in summer (Gibson, 1976a; Lewis and Heard, 1980; Frost, 1993a). In Northern Ireland, the mite densities on *L. multiflorum* or *L. perenne* increase in early summer, decline in July and early August, increase in late summer and autumn, and then decline from October onwards. When plots are cut 2–3 times during the year, the mite density varies in autumn depending on the stage of grass growth when it is first cut. Differences appear to be related to the regrowth of grass and not to the removal of mites by cutting or to immigration. Apparent preferences for particular cultivars are eliminated when allowance is made for growth pattern and vigour. Mite populations are similar in plots cut at 3- or 5-week intervals, but cutting to 3 cm instead of 6 cm greatly reduced the population density (Mowat, 1978).

Among the natural enemies of *A. hystrix*, the presence of different acaropathogenic fungi is best documented. These include *Hirsutella thompsonii* Fisher (Lewis *et al.*, 1981), *H. kirchneri* (Rostrup), *H. necatrix* Minter, *H. gregis* Minter (Minter *et al.*, 1983) and *Hirsutella* spp. (Lewis *et al.*, 1981) (Clavicipiteae), *Neozygites abacaridis* Mietkiewski *et* Baazy (Neozygitaceae) (Miętkiewski and Baazy, 2003), *Paecilomyces eriophyes* (Massee) (Lewis *et al.*, 1981) (Trichocomaceae) and *Verticillium lecanii* (Zimmerman) Viégas (Plectosphaerellaceae) (Lewis *et al.*, 1981).

Symptomatology and damage

The feeding activity of the cereal rust mite causes direct damage to host plant and transmits *Ryegrass mosaic virus* (RgMV), a rod-shaped virus which is sap transmissible but not seed borne and causes mild chlorosis, chlorotic mottle, chlorotic streaks or leaf necrosis (Oldfield and Proeseler, 1996). Elevated amounts of feeding are necessary to cause damage on Italian ryegrass (Gibson, 1974) as a result of the probable absence of phytotoxins from the mite saliva. High densities of attack cause growth retardation and cuticular damage produces wilting (Frost and Ridland, 1996).

RYEGRASS MOSAIC VIRUS. RgMV is reported from the British Isles and several European countries (Slykhuis, 1958) and north-western North America (Paliwal and Tremaine, 1976). Nymphs and adults of the cereal rust mite feeding on diseased plants acquire the virus in 2 h and can transmit it to healthy plants. A feeding activity up to 12 h allows a greater transmission. The mite is inoculative up to 24 h, and after 2 h of feeding on immune grass it loses its infectivity (Mulligan, 1960). RgMV can spread easily and in the first year of the sward may occur on more than 70% of ryegrass plants; up to 5 years old, all plants may be infected (Heard and Chapman, 1986). Infected plants of *L. multiflorum* host fewer mites than healthy plants. Multiplication is slower on infected plants and mites disperse easily from infected host plants (Gibson, 1976b). The first foci are due to airborne mites in the summer, and intra-sward transmission occurs with increase in the mite population during autumn and spring (Gibson, 1980, 1981). Cutting may transmit the virus within plots (Gibson and Plumb, 1976). The primary cause of yield losses in *Lolium* seems to be related to a decrease in net canopy photosynthesis and increase in dark respiration, associated with a decrease in tillering (Jones *et al.*, 1977). High N levels induce the greatest effects of RgMV and damage is most severe in spring and autumn during higher herbage production (A'Brook and Heard, 1975).

Studies on wheat colonization by quack grass and ryegrass adapted cereal rust mites highlight a low fecundity and female survival of the ryegrass population and indicate no success in wheat colonization. In contrast, fecundity and survival of the

quack grass strain are similar on wheat and the familiar host, or even higher on wheat. The phylogenetic similarity of quack grass and wheat could be affecting the patterns of host colonization. *A. hystrix* is the only vector of RgMV and this behaviour could explain the inability of RgMV to infest wheat successfully. In conclusion, quack grass and ryegrass adapted strains of the cereal rust mite have different physiological host ranges, and the phylogenetic relationships between host plants appear to be drivers for host specialization in the mite (Skoracka, 2009).

Damage on timothy hay consists of discoloured areas at the tip of the plant where the mites congregate and feed. On ryegrass, the feeding activity of the mites and the virus disease cause reduction in size, vigour and yield. High levels of attack hinder unfolding of the blade tips, deforming the plants and stopping their development; sometimes, the new growth dies. Diseased grass usually bears green to yellow mottling and streaking, and in some ryegrass strains, a brownish necrosis is observed (Mulligan, 1960).

RgMV induces losses in pasture production. Annual ryegrass is more susceptible to the virus than perennial ryegrass, and in susceptible varieties the yield losses may reach 27% (Wilkins and Hides, 1976), while in mixed swards there is retardation of competitiveness (Catherall, 1987). In Scotland, the field losses are between 2 and 10% (Holmes, 1980). Plants at advanced stages of infection show a decrease in total soluble carbohydrate and of digestibility values (Holmes, 1979). The reduction of ryegrass in a field is usually compensated by increases in weed growth, and the total herbage yield is not influenced. A selection pressure for resistance in the field could explain the existence of old plants in the field being infected with more difficulty (Gibson and Heard, 1974). This behaviour also helps to explain why RgMV epidemiology reaches a maximum in a few years and then decreases, e.g. in Australian *Lolium* pastures, where 39% infection has been observed 2 years after sowing and 10% infection after 10 years (Frost and Ridland, 1996).

Mixed infections with *Ryegrass spherical virus* or *Barley yellow dwarf virus* may occur (Catherall, 1987). Furthermore, *A. hystrix* is an inefficient vector of *Agropyron mosaic virus* (AgMV) to *A. repens* (Slykhuis, 1969; Catherall and Chamberlain, 1975), a secondary disease of wheat and *A. repens* in the USA, Canada, England, Finland and Germany (Slykhuis, 1980).

Control

All choices of control that hinder an increase in mite populations help to prevent the development of RgMV (Gibson, 1981). Before replanting, it is necessary to destroy the old plants because they host both virus and vector (Gibson, 1981; Oldfield and Proeseler, 1996). Control is provided by cultural practices and spray applications of pesticides. Natural enemies and resistant varieties are not envisaged to be a definitive solution.

PLANT RESISTANCE. Different types of resistance of ryegrass to RgMV are known, such as polygenic inherited resistance to infection (effective against all virus strains), post-infection resistance inherited by two recessive genes (effective against specific virus strains) and tolerance (Salehuzzanian and Wilkins, 1984; Catherall, 1986). Polygenic resistance to RgMV found in the perennial ryegrass cultivar 'S.23' has been transferred to Italian ryegrass (RgMV-resistant cultivar Bb 2113) (Wilkins, 1987). Some degree of resistance to the virus has been found by Frost (1993b).

CULTURAL PRACTICES. Cutting and/or grazing affect the population dynamics of *A. hystrix*. Heavy cutting and strip grazing remove a greater part of the mite population, including females living at the base of the sward (Frost, 1993a), and retards its development. Moreover, the removal of herbage during summer exposes the mite population to lethal temperatures. Cutting to 4 cm hinders the development of the mite population and the spread of RgMV, but the removal of cut grass is ineffective (Gibson, 1976a). Postharvest burning induces a decrease of mite population over 6 months (Smilanick and Zalom, 1983) but damages plant crowns and roots (Pepper, 1942). The sowing of mixed perennial and annual cultivars of ryegrass causes a decrease of RgMv infection but does not affect the mite population, resulting in less productive perennial ryegrass (Lewis *et al.*, 1985).

CHEMICAL CONTROL. Spray applications of pesticides prior to or during periods of peak population growth are most effective (Frost and Ridland, 1996). Chemical treatments are suggested in crops with a previous infestation and/or when 25% of the plants show curled tips of the new leaf blades. Aldicarb distributed to seedbeds (Gibson, 1974, 1981; Lewis, 1982) or sprayings of amitraz, chlorfenvinphos, chlorpyrifos, dimethoate, endosulfan, fenitrothion (Mowat, 1985), fenpropathrin (Lewis, 1982), and pirimiphos-methyl (Mowat, 1985) are reported to be effective. The use of some of the active ingredients cited above is not currently allowed in various countries, and the choice of pesticides must fall back on principles labelled for this mite and the crop concerned.

The Genus *Tegolophus* Keifer

Diagnostic characteristics

Leaf vagrant mites with the scapular tubercles located on the rear shield margin; the tubercles have transverse axes and direct the scapular setae (*sc*) straight back or diverging. The opisthosoma has a middorsal longitudinal ridge, stronger than the lateral ridges but fading caudally, not ending in a furrow. The coxal setae *1b* are present (Keifer, 1961a; Amrine *et al.*, 2003).

Economic importance

This genus has over 70 species, which mostly live on dicotyledons. The type species, *Epitrimerus califraxini* Keifer 1938 (= *Tegolophus califraxini* Keifer 1961), is a leaf vagrant on flowering ash (*Fraxinus dipetala* Hoak *et* Arn.) in California, probably causing some leaf deformation and stunting (Keifer, 1938b). *Tegolophus australis* Keifer causes rust of leaves and blemishes of the fruits of orange and grapefruit in Australia (Keifer, 1964; Beattie *et al.*, 1991). *T. brunneus* Flechtmann is a

vagrant species living on leaves and fruit of orange and mandarin in Brazil, and producing rust on leaves but mainly on fruits (Flechtmann, 1999). However, studies carried out on *Citrus sinensis* L. Osbeck orange trees in the State of Amazonas, in northern Brazil, showed that despite a mite population representing 13% of the population of phytophagous and mycophagous mites, it did not represent any danger to citrus (Bobot *et al.*, 2011). *T. perseaflorae* Keifer damages flowers and reduces fruit production of avocado (*Persea americana* Mill.) in Brazil (Keifer, 1969a; Arruda and Aquino, 1970). *T. hassani* (Keifer) was reported early from olive tree leaves (*Olea europaea* L.) in Egypt (Keifer, 1959b), Armenia, Greece, Iran and Italy. On olive, the mite is usually associated with other eriophyoids, but its damage is not well documented. At 31°C and 80% RH, the population of *T. hassani* multiplied 9.92 times in a generation time of 14.42 days, and up to 15 generations a year may be developed (Abou-Awad *et al.*, 2005). Young plants show disorders in growth or stop developing (Attiah, 1970). Young leaves have irregular lobed margins, with humping and curving of the rachis. Expanded leaves bear lighter green hollows on the upper side and chlorotic protuberances on the lower surfaces, which gradually turn brownish (Hatzinikolis, 1969a). Infestation of the inflorescences and small fruits causes russeting and splitting of flower buds, with early drop of flowers and inflorescences, associated with the dropping and deformation of small fruits (Hatzinikolis, 1969b, 1972). *T. zizyphagus* (Keifer) causes rust on leaves of jujube (*Ziziphus jujuba* Mill.) in California (Keifer, 1939d). *T. braziliensis* Keifer is a rust mite living on a tropical forage legume (*Desmodium* sp.) in Brazil (Keifer, 1969a, 1978). *T. ipomoeifoliae* Keifer curls the edges of the leaves up, and causes discoloration on sweet potato (*Ipomoea batatas* Lam.) in Brazil (Alves *et al.*, 1972; Keifer, 1972). Other injurious species are formally reported in the literature, but the damage they cause is sometimes unclear and their presence does not usually require any control. Only *T. australis* is discussed here.

Tegolophus australis Keifer (Fig. 16.86)

Common name

Brown citrus rust mite.

Diagnostic characteristics

FEMALE. The body is fusiform, 180–190 μm long, 75 μm wide, rough and brownish in colour. The prodorsal shield is coarsely pitted, with the frontal lobe moderately acuminate over the gnathosoma. The shield surface is rough and largely obliterates the basic pattern, although the admedian lines are faintly discernible, with an amphora-shaped median boss with handles. The rear edge of the shield between the scapular tubercles is partially furrowed. The scapular tubercles are small, 42 μm apart; the scapular setae (*sc*) are 5 μm long and do not reach across the first dorsal annulus. The opisthosoma has about 22 dorsal annuli, broken laterally, and 50–55 ventral annuli, with more small and elongate ventral microtubercles. The first dorsal annulus is rather broad and rough with longitudinal lines; the second dorsal annulus is narrow dorsally; the central ridge begins at the third dorsal annulus and fades on the 16th dorsal annulus. The setae c_2 are 8 μm long, set on about ventral annulus 5; the setae *d* are 55 μm long, set on about ventral annulus 16; the setae *e* are 4 μm long, set on about ventral annulus 29; the setae *f* are 14 μm long, set on about ventral annulus 5 from the rear; the setae h_1 are absent. The genital coverflap bears in the basal part a pair of subcircular areas containing short dashes, and about 16 longitudinal ribs (Flechtmann, 1999, reports 18 ribs). The empodia are 4-rayed (Keifer, 1964).

MALE. The male is unknown (Keifer, 1964).

Geographical distribution

Australia (Keifer, 1964).

Host plants

Rutaceae: grapefruit (*Citrus paradisi* Mcfad.) (Beattie *et al.*, 1991), orange (*C. sinensis* (L.) Osbeck) (Keifer, 1964).

Bio-ecology

The brown citrus rust mite was first reported on orange (*Citrus sinensis*) in Australia (Keifer, 1964). In New South Wales, its infestations do not represent a severe problem for orange and grapefruit, and the dynamics of the population is marked by peaks clashing with an increase and decrease in seasonal temperature and RH recorded in summer/autumn and spring/early summer and with stasis of the reproduction of natural enemies. The decreases observed in winter are generally associated with lower values of RH and temperature, particularly in July when the average temperature is 10°C. The density of mobile forms does not appear to be affected by climatic adversities (Beattie *et al.*, 1991).

Natural enemies include mites and insects. Predatory mites include various phytoseiids, such as *Amblyseius deleoni* Muma *et* Denmark, *A. lentiginosus* Denmark *et* Schicha (Beattie and Gellatley, 1983), *Euseius elinae* (Schicha) and *E. victoriensis* (Wormsley) (Smith and Papacek, 1991). Predatory insects include the coccinellids *Halmus chalybeus* (Boisduval) and *Serangium bicolor* Blackburn (Beattie and Gellatley, 1983).

Symptomatology and damage

T. australis is a minor pest of orange and grapefruit. The symptoms of its attack may be confused with those of the citrus rust mite, *Phyllocoptruta oleivora*; however, in New South Wales, *P. oleivora* prefers to shelter inside the fruit and inner fruit surface, producing a slightly rough and greyish brown injury, with deeper brown margins (Hely, 1968). In contrast, feeding injury by the brown citrus rust mite consists of smooth and brown blemishes on the fruit located on the outer part of the canopy. In smooth blemishes, the rind looks as though it has been polished with dark tan boot polish (Jeppson *et al.*, 1975).

Control

The need for control is rare, although it is possible to use both biological and chemicals methods.

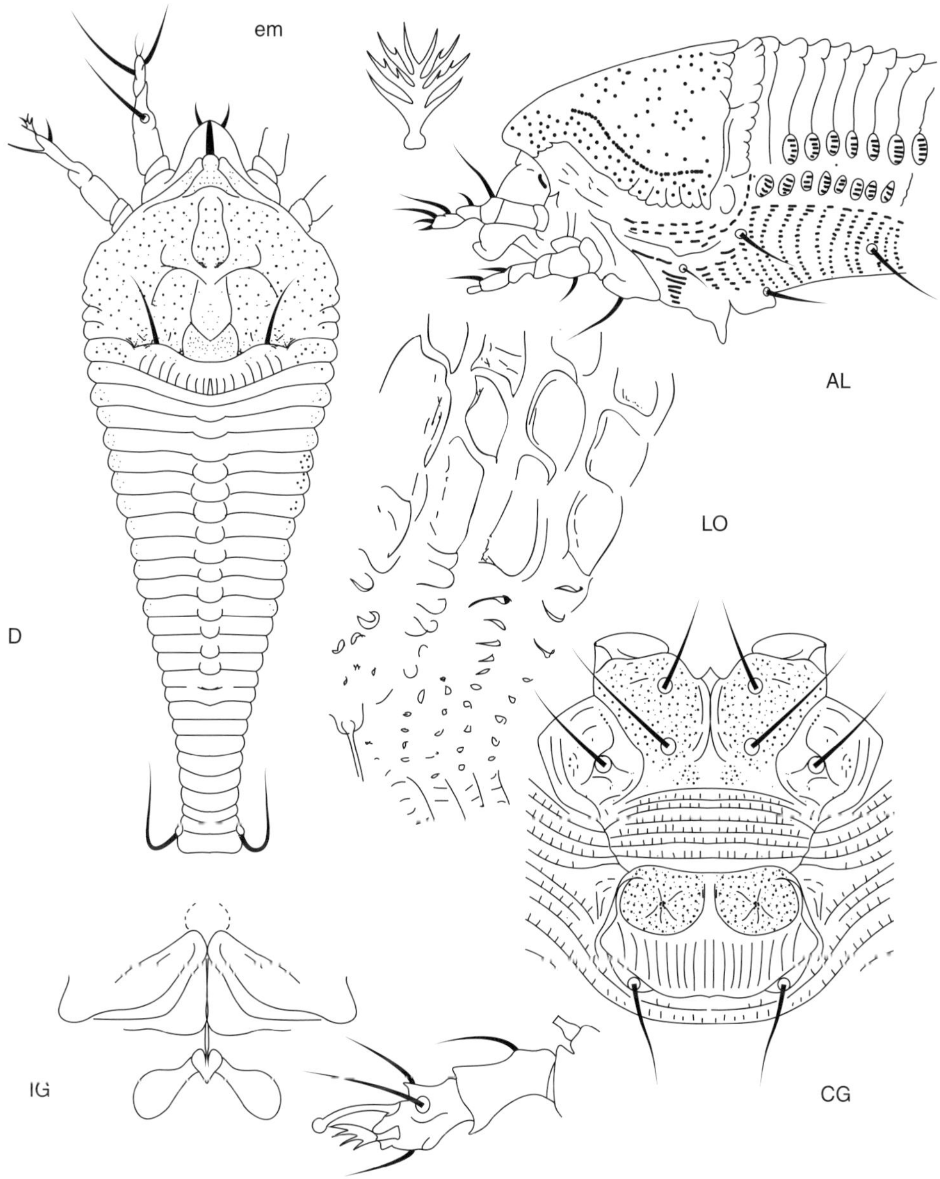

Fig. 16.86. *Tegolophus australis* Keifer (from Keifer, 1964). See Table 16.1 for key.

BIOLOGICAL CONTROL. The phytoseiid mite *E. victoriensis* produces a good control of the pest, and in spring a density of 10–20 individuals per 100 leaves increases to just 100 individuals per 100 leaves in midsummer, bringing the amount of infested fruits to below the economic threshold of 10%. The encouragement of host plants for the production of pollen as food increases the phytoseiid population, and when Rhodes grass (*Chloris gayana*) was allowed to flower between rows, the windblown pollen served as a supplementary food source. (Smith and Papacek, 1991).

CHEMICAL CONTROL. Trials conducted on Valencia late orange in Queensland with avermectin mixed with petroleum oils gave good control of both *P. oleivora* and *T. australis* for 20 weeks; some organotin was also effective (Smith *et al*., 1998).

17 The Family Phytoptidae Murray

Morphological Characteristics, Systematics and Bio-ecology

The general morphological characteristics of the Phytoptidae are the same as those of other eriophyoid mites (see Chapter 16, Eriophyidae), and details of their external morphology and setal notation given here are fundamentally based on Lindquist (1996) and Amrine *et al.* (2003). The family is distinctive in having on the prodorsal shield from one to five setae, with anterior setae present (paired or unpaired *vi* and/or paired *ve*). The gnathosoma can be variously sized and is often large, and the chelicerae are straight or slightly and evenly curved; the palp is commonly short and truncate and enclosing the oral stylet. The legs have the usual setae and frequently the solenidion φ on tibia I. All empodia are undivided. The opisthosoma bears the usual setae, and the setae h_1 are sometimes long. The female genital coverflap lacks ridges; the anterior female apodeme always extends forward for a moderate distance, and the spermathecal tubes are commonly long, sometimes extending diagonally forward and then recurving caudally (see Fig. 17.1).

The family Phytoptidae includes five subfamilies: Prothricinae Amrine, Novophytoptinae Roivainen, Nalepellinae Roivainen, Phytoptinae Murray and Sierraphytoptinae Keifer; the Nalepellinae include the tribes Pentasetacini Boczeck *et al.*, Trisetacini Farks and Nalepellini Roivainen. The Sierraphytoptinae includes the tribes Sierraphytoptini Keifer and Mackiellini Keifer (Lindquist and Amrine, 1996; Amrine *et al.*, 2003). The phytoptids are 'considered as the earliest primitive eriophyoid members and retain a greater array of plesiomorphic attributes than other families and are difficult to define apomorphically' (Lindquist *et al.*, 2009). Cladistic analysis seems to indicate that the family may be paraphyletic (Hong and Zhang, 1996a). The Phytoptidae includes 20 genera and 164 species (Zhang *et al.*, 2011), primarily infesting ancient host plants such as the Araucariaceae, Coniferaceae, Palmaceae, Cyperaceae, Fagaceae, Corylaceae and Betulaceae (Lindquist and Amrine, 1996). About ten species may cause potentially or usually severe damage to forest or cultivated plants. Many species inhabit forest environments and some may cause economic damage in nurseries or young plants that have just been transplanted. Few species are harmful to plants of agricultural interest. The various species colonize different microhabitats of the host plant, such as cones, buds, sheath and leaf surfaces. Some species form galls – 'big buds' (Jeppson *et al.*, 1975) or twig and bark galls (Schliesske, 1985c; Shevchenko *et al.*, 1993). It is not known whether these mites transmit viruses. Several pictures of damage and abnormalities caused by eriophyid mites in North America (and in other countries) can be found in Keifer *et al.* (1982) and on specific websites. The key presented below for the identification of genera (and subfamilies and tribes) is taken from Amrine *et al.* (2003), as are the abbreviations used in the figures (see Table 16.1).

Key for the Identification of Genera

1. • Prodorsal shield with 3 (1 *vi* and 2 *ve*) or 1 (*vi*) anterior setae; scapular setae (*sc*) present or absent; spermathecal tubes from 3 to 5 times longer than the spermatheca; opisthosoma either vermiform with setae c_1 present, or more robust and fusiform, with setae c_1 absent.............................subfamily **Nalepellinae** Roivainen...3
   • Prodorsal shield with 2 anterior setae (*ve*), interior vertical setae (*vi*) lacking and setae *sc* rarely minute or absent; spermathecal tubes short...**2**
2. • Opisthosomal setae c_1 present; body vermiform; opisthosomal annuli narrow and subequal dorsoventrally; setae *sc* pointing up if short, forward if long (subfamily **Phytoptinae** Murray); femora and genua not fused; prodorsal shield without gland; setae *sc* short or long, but not little; dorsal annuli with microtubercles distributed normally ....genus ***Phytoptus*** Dujardin
   • Opisthosomal setae c_1 present or absent; body more fusiform and often flattened; opisthosomal annuli broad or with dorsoventral differentiation (subfamily **Sierraphytoptinae** Keifer); opisthosomal setae c_1 absent...tribe **Mackiellini** Keifer...5
3. • Opisthosomal setae c_1 present (tribe **Trisetacini** Farkas); body elongate, vermiform, with narrow and subequal opisthosomal annuli............................genus ***Trisetacus*** Keifer
   • Opisthosomal setae c_1 absent...tribe **Nalepellini** Roivainen...**4**
4. • Body elongate and vermiform; opisthosomal annuli subequal dorsoventrally.........................genus ***Setoptus*** Keifer
   • Body fusiform and robust; opisthosomal annuli with dorsoventral differentiation; dorsal opisthosoma evenly arched ....genus ***Nalepella*** Keifer
5. • Prodorsal shield with 2 setae *ve* on humeral angles; scapular tubercles and setae lacking.................................**6**
   • Prodorsal shield with 2 *ve* and 2 *sc* setae; prodorsal shield setae *sc* and *ve* with bulbous, enlarged tubercles, directing *ve* forward and *sc* backward............... genus ***Retracrus*** Keifer
6. • Shield setae *ve* on anterior margin at humeral angles; dorsal annuli much broader than ventral annuli...genus ***Propilus*** Keifer
   • Shield setae *ve* not as far forward as above, prominent eye-like structures posterolaterally; dorsal and ventral annuli subequal...........genus ***Palmiphytoptus*** Návia *et* Flechtmann

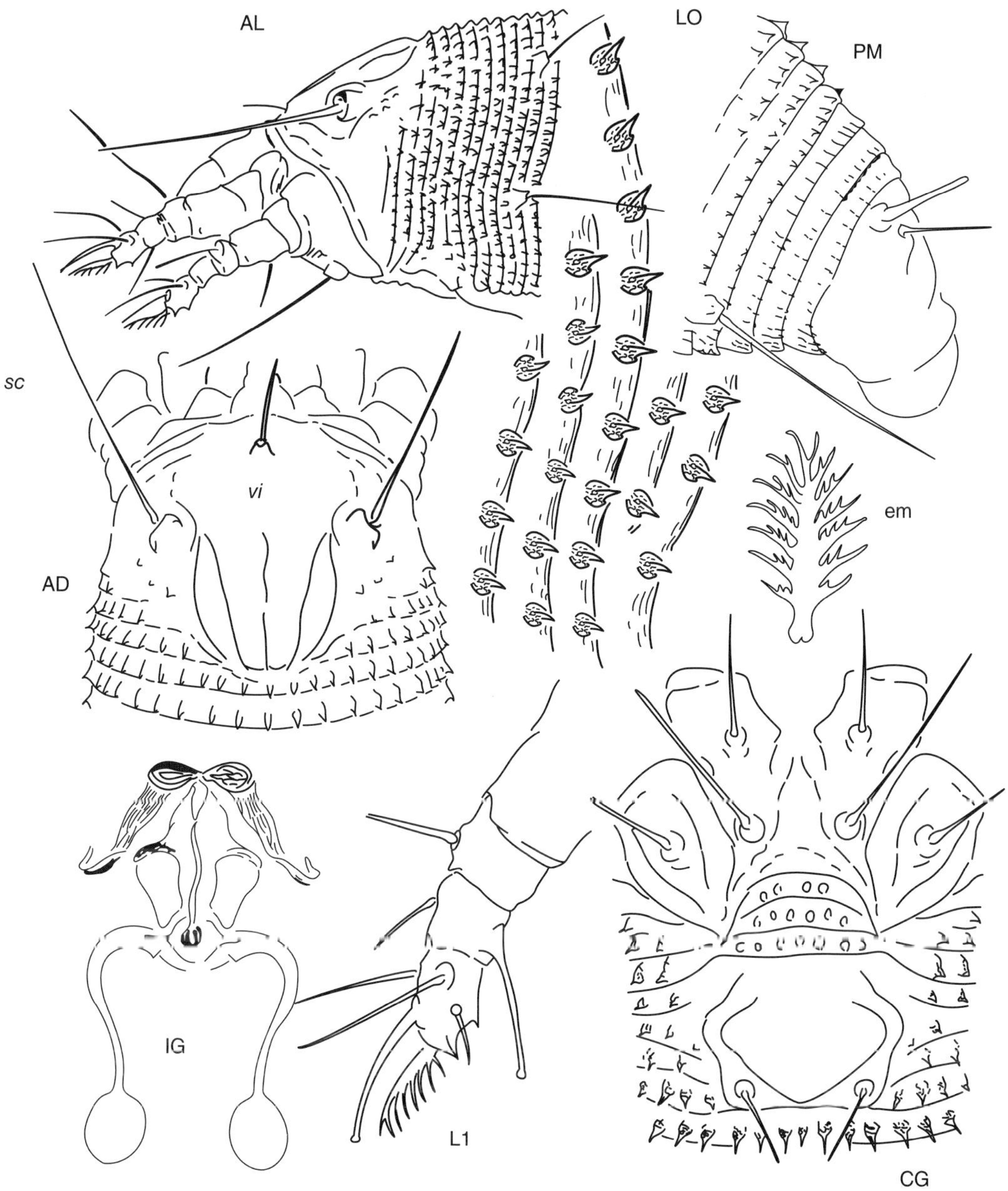

Fig. 17.1. *Trisetacus pini* (Nalepa) (from Keifer, 1963b). Key: sc, scapular seta; *vi*, internal vertical seta; for key to rest of labels, see Table 16.1.

The Genus *Trisetacus* Keifer

Diagnostic characteristics

The body is elongate and worm like. The gnathosoma has a moderate size and is downcurved. The prodorsal shield is broad, approximately semicircular, not projecting over the gnathosomal base, with three setae (1 *vi* and 2 *sc*), and seta *vi* may be small. The opisthosomal annuli are completely microtuberculate, the setae c_1 are present and the annuli are narrow and subequal. The female genital coverflap is smooth, and the seminal vesicles are on long stalks (Keifer, 1952b; Amrine *et al.*, 2003).

Economic importance

The genus *Trisetacus* numbers over 50 Holarctic species, 24 of which are reported from Europe (de Lillo and Beron, 2004); they are commonly known as needle sheath mites on pines or

bud mites on other conifers (Jeppson *et al.*, 1975). They are needle vagrants, bud mites or gall mites, and induce various disorders on Pinaceae, Cupressaceae and Taxodiaceae of the genera *Abies*, *Pinus*, *Larix*, *Pseudotsuga*, *Cupressus*, *Chamaecyparis*, *Sequoia*, *Cedrus*, *Thuja* and *Juniperus*. The disorders include chlorosis, browning, distorted and stunted needles, rosette galls, cortical galls and undeveloped buds, as well as destruction of the seeds (Boczek and Shevchenko, 1996) (see Figs 7.2D–F, 7.3A–C). Different species live on foliage, at needle bases, or under the sheath; and sometimes on fruits, vegetative buds and male cones (Castagnoli, 1996). Females usually overwinter in the feeding organs and become active when new growth develops; in this phase, they migrate to colonize new organs and reproduce. Generally, no damage is recorded for vagrant eriophyoids in forests, and outbreaks of their populations sometimes involve only nurseries and Christmas tree plantations (Boczek, 1962; Löyttyniemi, 1969, 1975; Siewniak, 1971; Marshall and Lindquist, 1972; Boczek *et al.*, 2002). The species that are adapted to confined microhabitats (galls, buds, fruits, scales, needle sheaths) can influence their host plant markedly (Castagnoli, 1996), e.g. *Trisetacus* spp. on coniferous trees. In temperate forests, over 30 species are reported, six of which live on *Abies*, four on *Juniperus*, four on *Larix*, six on *Picea* and ten on *Pinus* (Castagnoli *et al.*, 2010). The most common species inhabiting forest and ornamental plants, their host plants and (in parentheses) their microhabitat are listed here. The mites living on *Abies* are *T. abietis* Postner (needles), *T. abietivagrans* Kadono (buds), *T. grosmanni* Keifer (buds), *T. laricinae* Smith (male flower buds), *T. neoabietis* Smith (needle vagrant) and *T. relocatus* Bagnyuk *et* Shevtchenko (under shoot scales). The species reported from *Juniperus* are *T. juniperinus* (Nalepa) (needles), *T. neoquadrisetus* Smith (berries), *T. quadrisetus* (Thomas) (berries), *T. kirghisorum* Shevchenko (berries) and *T. thujivagrans* Smith (needles). The species reported from *Larix* are *T. ehamanni* Keifer (needle sheaths), *T. laricinae* Smith (buds), *T. laricis* (Tubeuf) (buds) and *T. pini* (Nalepa) (galls on twigs). The species living on *Picea* are *T. grosmanni* (buds), *T. neopiceae* Smith (needles), *T. piceae* (Roivainen) (buds), *T. pini* (galls on twigs), *T. relocatus* (under shoot scales) and *T. ucluelетensis* Smith (needles). The species living on *Pinus* are *T. alborum* Keifer (needle sheaths), *T. campnodus* Keifer *et* Saunders (needle sheaths), *T. cembrae* (Tubeuf) (buds), *T. ehamanni* (needle sheaths), *T. gemmavitians* Styer, Nielsen *et* Baldestron (buds), *T. grosmanni* (buds), *T. neoabietis* (needles), *T. pini* (galls on twigs), *T. strobi* Smith (needles) and *T. silvestris* Castagnoli (needle sheaths) (Castagnoli *et al.*, 2010).

In temperate areas, the most common species is *T. pini*, which is injurious to young plants in nurseries and transplanted stock, and in extreme cases can cause the death of all branches of a tree growing at the forest margin. *T. juniperinus* causes enlarged and deformed buds, russeting and/or branch apex folding. The buds can become dried, as can brachyblasts and/or parts of the branches. Irregular proliferations of axillary buds, witches' brooms, blastomania (similar to witches' brooms) and deformed cones result in the production of fewer seeds.

Other species, such as *T. laricinae*, *T. abietivagrans*, *Trisetacus bagdasariani* Bagnyuk, *T. piceae* and *T. grosmanni*, may induce the death of buds and hinder the development of shoots (Roivainen, 1951; Keifer, 1959d; Smith, 1979; Kadono, 1981; Bagnyuk, 1984), but there are no reports of population outbreaks limiting the growth of trees. The species developing on juniper berries could have similar behaviour (Smith, 1978), except for *T. kirghisorum* which infested fruits of *Juniperus semiglobosa* Regel and destroyed its seeds in Kyrgyzstan and Uzbekistan (Shevchenko, 1962). *T. batonrougei* Smith and *T. neoquadrisetus* Smith are common in North America and injurious to juniper berries, transforming them into gall-like structures; when all of the berries are infested, the seeds can be destroyed. *T. chamaecypari* Smith, known on Cupressaceae in Ontario (Canada) (Smith, 1977), and recently reported as injurious to several Cupressaceae in England, feeds on the cones, leaves and terminal buds of its host plants (Ostojá Starzewski and Halstead, 2006). *T. cembrae* (Tubeuf) produces bud proliferation on Swiss stone pine (*Pinus cembra* L.) in Europe, while in Italy, *T. dorsospinosus* Castagnoli, *T. etruscus* Castagnoli, *T. halepensis* Castagnoli and *T. silvestris* (Castagnoli, 1973) do not usually cause severe damage. *T. piceae* probably exhibits similar behaviour. Relatively few eriophyoid mites are reported on from South America (Schliesske, 1985c; Flechtmann and Návia, 1998), but among these there is a needle vagrant species, *T. calvus* Návia *et* Flechtmann, which has been collected on evergreen cypress (*Cupressus sempervirens* L.) in Brazil (Návia and Flechtmann, 2000b).

It can be seen from the above discussion that different species of *Trisetacus* do not give rise (with some exceptions) to a constant phytopathological problem associated with any precise economic importance. Their presence does, however, imply a general condition of risk that must be taken into account in planning activities related to their host plants. It seems that only *T. pini* and *T. juniperinus*, and some other species (e.g. *T. chamaecypari*, see above), may be injurious in nurseries and to young plants, and it is difficult to highlight attacks that undermine the vitality of adult plants and/or the structure of a forested area. None the less, the production and/or handling of certain host plant species (*Juniperus*, *Abies*, etc.) can stimulate the development of populations of these mite species, and adequate knowledge of them would then be required.

Trisetacus pini (Nalepa) (Fig. 17.1)

Common name

Twig gall, twig knot mite.

Diagnostic characteristics

FEMALE. The prodorsal shield has lines set to the rear ½, and the median line extends forward to a point even with the scapular tubercles; the anterior internal vertical seta (*vi*) is rather prominent. The opisthosomal annuli have strongly spinulate microtubercles. The empodium is 7-rayed (Keifer, 1963b).

Geographical distribution

Nearctic, Palaearctic (Austria, British Isles, Czech Republic, Danish, Finland, France, Germany, Greece, Hungary, Italy, Poland, Russia (north), Slovenia, Ukraine) and Oriental regions (de Lillo and Beron, 2004), Anatolia (Turkey) (Yüksel, 1999) and Japan (Huang, 1971).

Host plants

Pinaceae: Scots pine (*Pinus sylvestris* L.) (Nalepa, 1887; Keifer, 1963b).

Bio-ecology

T. pini was collected by Nalepa (1887) on Scots pine (*Pinus sylvestris* L.) in central Europe and Italy (South Tyrol and the north). Keifer (1938a) identified as *T. pini* a mite collected on *Pinus torreyana* Parry, *P. radiata* Don. and *P. sabiniana* Dougl. in California from inside the sheaths of the youngest needles, between the sheath and the needle and also living between the needles. In California, the mite overwintered as females and the males appeared in the spring. In 1963, a study of some specimens of a mite collected inside galls on *P. sylvestris* in Germany showed that the specimens of *T. pini* from California differed from those described in Europe (Keifer, 1963b). In the Baltic regions, *T. pini* develops about two generations a year. Migrating females are smaller than other females and have slightly fewer opisthosomal annuli, but the phenomenon is unclear (Shevchenko *et al.*, 1993).

Symptomatology and damage

The twig knot mite feeds within the basal sheath of a needle cluster and causes premature shedding of the needles and a weakening of young stands. Infested young twigs are crowded and swollen, and the terminal buds are killed. Damage caused by the mite may be misidentified with abiotic disorders such as air pollution injury or drought and these need to be taken into consideration when diagnosing the problem (Yüksel, 1999). The feeding activity of *T. pini* leads to the formation of galls and knots on pine twigs (Nalepa, 1887; Shevchenko *et al.*, 1993; Petanović *et al.*, 1996; Skoracka *et al.*, 2005) (see Fig. 7.3A). The twig swelling, or knots, are supposed to persist several years and turn out successive broods of mites. In Germany and Baltic regions, the mite may cause serious damage to young pines (Kruel, 1963; Shevchenko *et al.*, 1993). A high number of galls can also involve the death of all branches of a tree, but the phenomenon has been observed only in the margins of forests (Lewandowki, in Castagnoli *et al.*, 2010). Large-scale outbreaks of mite in young pine plantations in Anatolia (Turkey) caused in some areas up to 80% mortality. The mite killed about 550,000 young Crimean pine trees with outbreaks in pine plantations in the last 2-3 months of 1998 (Yüksel, 1999). In Japan, *T. pini* is a very destructive pest to larch trees, and in some areas the damage of buds caused by the mites is as high as 61% (Huang, 1971).

Control

Control is not usually required under natural conditions, but it may be necessary in nurseries or in young stands. In this case, the spread of the mite from infested plants located nearby must be stopped and some acaricide labelled for pines should be sprayed in the nursery.

Trisetacus juniperinus (Nalepa) (Fig. 17.2)

Common name

Juniper gall mite.

Diagnostic characteristics

FEMALE. A distinguishing feature of *T. juniperinus* is that the microtubercles form a pattern on the front part of the dorsal opisthosoma but are lacking from two subdorsal areas that extend longitudinally back from the rear dorsal shield margin. There is also a gland pit present at the rear central part of the prodorsal shield margin (Jeppson *et al.*, 1975).

Geographical distribution

East Palaearctic region (Austria?, Czech Republic, Germany, Hungary, Italy, Poland, Ukraine), Nearctic region (Denmark, 1970; de Lillo and Beron, 2004) and Iran (Baradaran and Arbabi, 2005; Xue *et al.*, 2009a).

Host plants

Cupressaceae – *Chamaecyparis*, *Cupressus* (Castagnoli *et al.*, 2010), evergreen cypress (*Cupressus sempervirens* L.) (Nuzzaci and Monaco, 1977), *Juniperus* spp. (Denmark, 1970; Castagnoli *et al.*, 2010), *J. procumbens* (Endl.) Miq. (Soika and Łabanowski, 1999a; Łabanowski and Soika, 2000), *J. silicicola* (Small) L.H. Bailey, *J. virginiana* L. (Denmark, 1970) and ornamental junipers (Jeppson *et al.*, 1975); Pinaceae – *Cedrus* sp. (Denmark, 1970).

Bio-ecology

The juniper gall mite lives in the apical and subapical buds of ornamental Cupressaceae (*Juniperus*, *Cupressus*, *Chamaecyparis*). In Mediterranean countries, the populations of *T. juniperinus* develop throughout the year (Castagnoli and Simoni, 2000), and on adult host plants the mite also lives in the young reproductive organs (Guido *et al.*, 1995). The mite leaves the buds when they dry out and migrates to new buds. The continuous presence of buds on the host plant allows the species to migrate gradually during the year. The presence of fast-growing plant tissue could be related to the high level of damage observed in nurseries and young stands, whereas in natural environments the populations of *T. juniperinus* generally occur at low densities, probably because of the different activity of meristems and the action of natural enemies. The development of populations of the juniper gall mite may also be related to the selection of new canker-resistant clones (Castagnoli *et al.*, 2010). The long-lasting effect of mite attack can be interpreted as a trade-off between the plant resources devoted to growth and to the defence reactions of plants according to the 'growth/

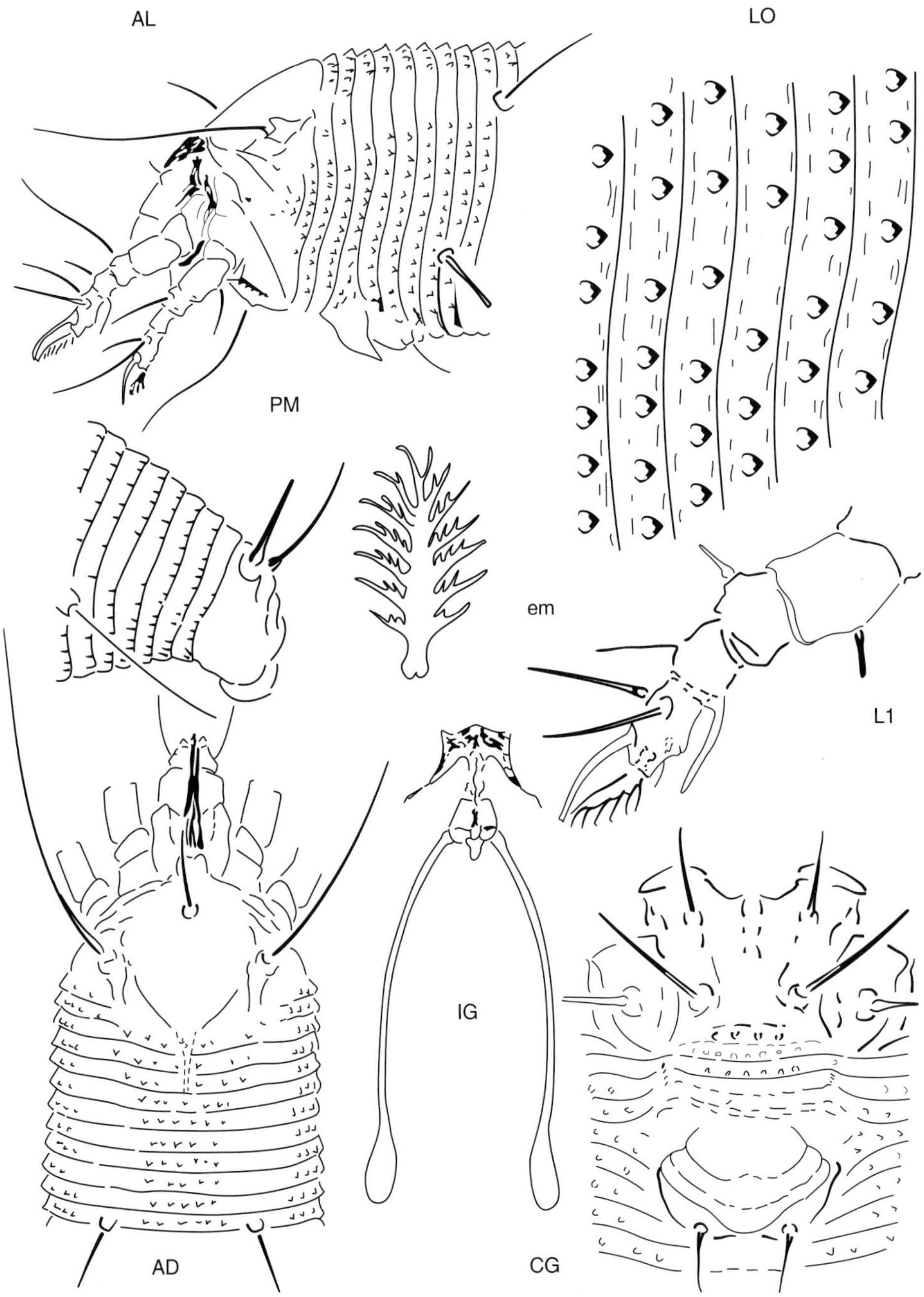

Fig. 17.2. *Trisetacus juniperinus* (Nalepa) (from Jeppson *et al.*, 1975). See Table 16.1 for key.

differentiation balance' hypothesis formulated by Herms and Mattson in 1992 (Castagnoli *et al.*, 2010).

Symptomatology and damage

The symptoms of attack may range in severity, and different types of injury can coexist on the same host plant. The mites cause the needles of ornamental junipers to become basally swollen and crowded, and the terminal buds of low ornamental junipers with pointed projecting needles are sometimes killed (Jeppson *et al.*, 1975). In Italian nurseries and young groves of evergreen cypress, *T. juniperinus* is considered a severe pest (Nuzzaci and Monaco, 1977), especially for the commercial clone 'Bolgheri', in which it disturbs the apical growth of the tree. The mite produces enlarged and deformed buds, russeting and/or branch apex folding. The buds can become dried, as can the brachyblasts and/or parts of the branches. Irregular proliferations of axillary buds, witches' brooms, blastomania and deformed cones result in the production of fewer seeds (Fig. 7.2E). The damage is negatively correlated with the average increment in the height of the plant (Castagnoli *et al.*, 2002, 2010).

Control

Natural infestations rarely occur and the mite outbreaks arise exclusively in nurseries and in young transplanted trees. Control on mature trees is not necessary. The main cause of *T. juniperinus* outbreaks in nurseries is associated with the susceptibility of different varieties and the proximity of infested adult plants. Castagnoli *et al.* (2002) developed a rating system based on the type of mite injury and this is useful for analysing plant damage and mite bio-ecology. The severity of each damage symptom is rated from zero to four (highest intensity), and it is possible to measure damage intensity by the calculation of a global damage index (GDI) per plant obtained by adding the mean intensities of the five categories of damage. Damage is correlated negatively with the average increment in plant height (Castagnoli *et al.*, 2002, 2010). Healthy rootstock and graft material should be used to limit damage and maintain plant growth. In case of attack, a pesticide should be applied as soon as possible (Simoni *et al.*, 2004). In the nursery, acaricides such as abamectin, and/or petroleum oils are sprayed.

The Genus *Setoptus* Keifer

Diagnostic characteristics

The body is elongate, worm like, somewhat clavate anteriorly. The gnathosoma is large, and the chelicerae evenly curved. The prodorsal shield is longer than wide, with a broad, transverse front edge and a narrower transverse rear edge, with 3 setae – 1 internal vertical seta (*vi*) and 2 scapular setae (*sc*). The narrow abdominal annuli are completely microtuberculate, and a shallow sublateral furrow is present; the opisthosoma lacks the setae c_1 and the annuli are subequal dorsoventrally; the remaining setae are in the usual positions. The female genitalia have a smooth coverflap and long-stalked genital glands. The empodium is simple (Keifer, 1944; Amrine *et al.*, 2003).

Economic importance

The genus has about 14 species (Amrine *et al.*, 2003) living on conifers (Oldfield, 1996a), and spread over North America, Europe and Russia. *Setoptus* spp. produce needle sheath stunting of *Pinus* (Boczek and Shevchenko, 1996). The following species have been recorded in temperate forests: *S. jonesi* (Keifer), living on needles, *S. koraiensis* Kuang *et* Hong, living in an unlisted microhabitat, *S. pini* Boczek living on needles, *S. strobacus* Keifer living on needle sheaths, *S. thunbergi* Hong living in an unlisted microhabitat and *S. multigranulatus* Castagnoli living on needle sheaths (Castagnoli *et al.*, 2010). *S. pineae* (Castagnoli) lives in the Mediterranean region. *S. viator* (Flechtmann *et* Návia) and *S. pineae* were both collected after being beaten off needles of *Pinus oocarpa* Schiede ex Schlectendahl in Brazil (Flechtmann and Návia, 1998; Návia and Flechtmann, 2000b). *S. strobacus* is economically the more important species.

Setoptus strobacus Keifer (Fig. 17.3)

Common name

White pine sheath mite.

Diagnostic characteristics

FEMALE. The body is long and worm like, being 200–225 μm long and 58 μm wide. It is light yellowish white in colour, sometimes slightly red anteriorly. The gnathosoma is curved down. The prodorsal shield has scattered small granules. The internal vertical seta (*vi*) is 21–25 μm long. The scapular tubercles are 27 μm apart and the scapular setae (*sc*) are 72 μm long. The opisthosoma has about 87 annuli, which are completely microtuberculate and the microtubercles are spine like. The setae c_2 are 15 μm long, set on annulus 4; the setae *d* are 40 μm long, set on annulus 24; the setae *e* are 18–20 μm long, set on annulus 42; the setae h_2 are 27 μm long and the setae h_1 are 10 μm long. The genitalia are 28 μm wide, 24 μm long, the coverflap is unmarked and the setae *3a* are 15–20 μm long. The coxae have some spinules distributed over the surface. The empodium is 10-rayed (Keifer, 1966a).

Geographical distribution

Eastern North America (Keifer, 1966a; Jeppson *et al.*, 1975).

Host plants

Pinaceae: eastern white pine (*Pinus strobus* L.) (Keifer, 1966a; Jeppson *et al.*, 1975).

Bio-ecology

S. strobacus lives on eastern white pine in eastern North America (Keifer, 1966a; Jeppson *et al.*, 1975). During the phase of elongating candles, its populations inhabit mainly the needle sheath and feed on the elongating needles; they feed openly on mature needles (Thomas and Rajotte, 2011).

Symptomatology and damage

The white pine sheath mite was initially associated with needle sheaths and considered responsible for stunting or deformation of the young needles (Keifer, 1966a), but later on severe rust and needle stunting were recorded (Lindquist and Amrine, 1996). During the phase of elongating candles, the mites inhabit the needle sheath and feed on the elongating needles, inducing the needles to become stunted. When the mites feed on mature needles, they induce them to turn silver or brown (Thomas and Rajotte, 2011).

Control

In natural environments, the damage potential of *S. strobacus* is only moderate and control is usually not necessary, but in nurseries and in young stands it may sometimes be justified. Preventive measures, such as the transplanting of pest-free

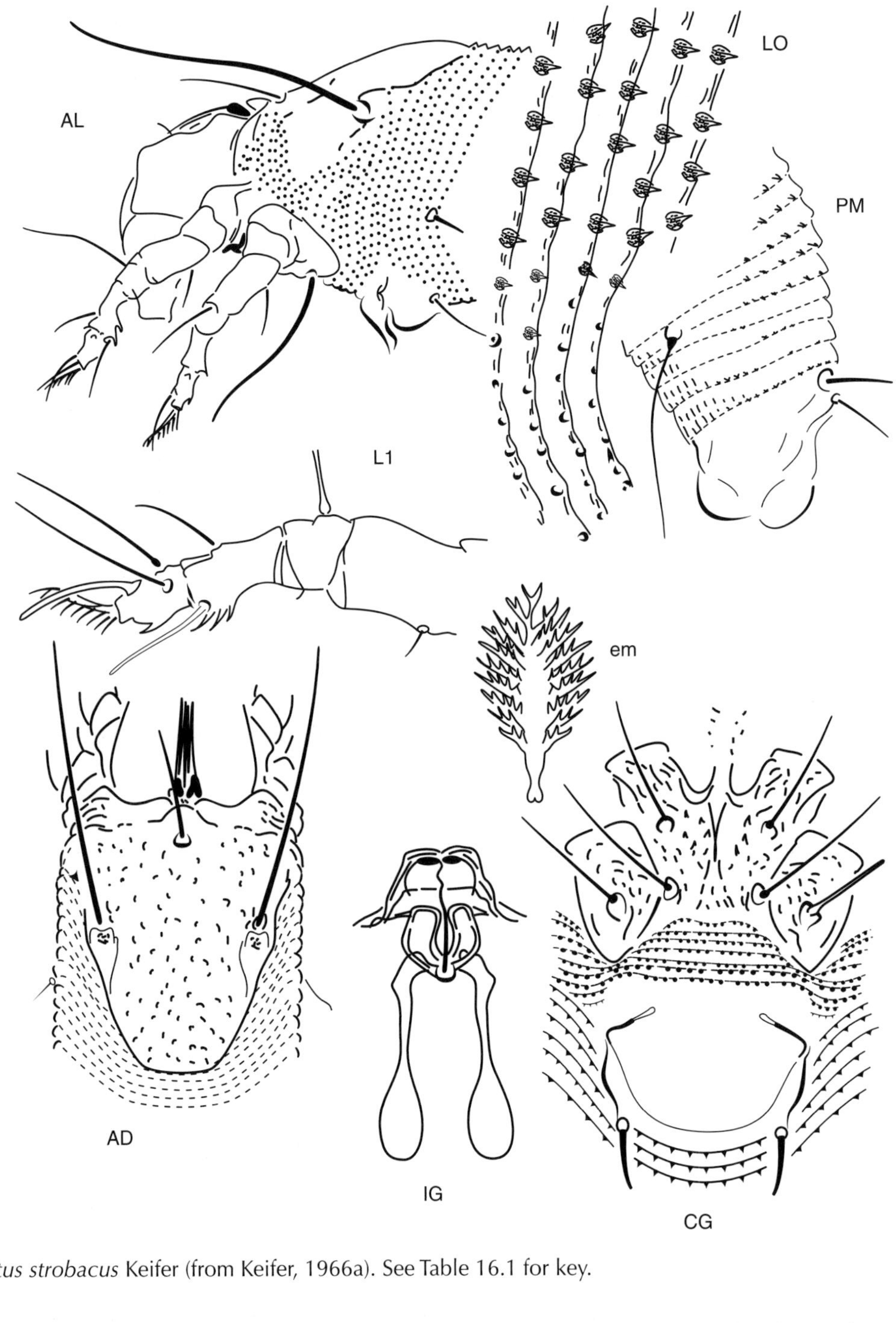

Fig. 17.3. *Setoptus strobacus* Keifer (from Keifer, 1966a). See Table 16.1 for key.

plants, and curative measures, e.g. spray applications of biorational products (e.g. insecticidal soap, pyrethrum and rotenone) and chemicals, are employed. Monitoring leads control.

MONITORING. From early March to May and from late August to October, 10–20 trees acre^{-1} are examined, either choosing trees that have been previously damaged or at random, and looking at the exposed needle base of last year's growth. Monitoring should be continued until temperatures rise, looking gently and pulling elongating needles from the sheath and observing the needle bases. During autumn, monitoring helps to intercept population resurgence. During the growing season, it is possible to use the growing degree-days (GDDs), taking into account that overwintering eggs hatch at 0–15 GDDs (Thomas and Rajotte, 2011).

CHEMICAL CONTROL. The threshold level for control (and also for *Nalepella* spp., see next section) is at least 80% sampled shoots infested by mites, and at least eight mites on a single needle in the plantation (on all sides of the needle). It is possible to apply dormant oil in early spring before bud break. Eggs may have already hatched, but the oil will control active mites if coverage

is thorough, although it will remove the blue colour on some spruces. A first acaricide spraying may be applied in mid-April to mid-May, followed by a second application if warranted 1–2 weeks later (Thomas and Rajotte, 2011).

The Genus *Nalepella* Keifer

Diagnostic characteristics

The body is fusiform and robust. The gnathosoma is moderately sized. The prodorsal shield is subtriangular, with the frontal lobe over the gnathosoma, and characterized by 3 setae, 1 internal vertical seta (*vi*) and 2 scapular setae (*sc*); the scapular tubercles are well ahead of the rear margin and directing the long setae ahead. The opisthosomal setae c_1 are lacking and the dorsal opisthosoma is evenly arched. The empodium is simple (Keifer, 1944; Amrine *et al.*, 2003).

Economic importance

The genus *Nalepella* numbers about 16 Holarctic species, spread in Europe, Russia, North America and China. They are needle vagrants on *Abies*, *Picea*, *Larix*, *Tsuga* and *Torreya*, and cause browning, russeting and needle dropping (Boczek and Shevchenko, 1996). In coniferous temperate forests, there are nine species, seven of which were collected on *Abies*, one on *Larix*, three on *Picea* and one on *Taxus*. The species recorded for *Abies* are *N. danica* Boczek, Harding *et* Shi, *N. ednae* Keifer, *N. longoctonema* Hu *et* Krantz, *N. octonema* Keifer, *N. shevtchenkoi* Boczek, *N. triceras* (Börner) and *N. tsugifoliae* Keifer, all collected from needles. On *Larix*, only *N. triceras* has been recorded, also found on needles. *N. haarlovi* Boczek, *N. haarlovi picea-abietis* Löyttyniemi and *N. shevtchenkoi* live on *Picea*, and *N. tsugifoliae* on *Taxus*, all on the needles (Castagnoli *et al.*, 2010). *N. pini* Kuang *et* Luo is reported from China (Kuang *et al.*, 2005). The most economically important species are *N. tsugifoliae*, *N. haarlovi*, *N. halourga* Keifer and *N. longoctonema* Hu *et* Krantz.

Nalepella tsugifoliae Keifer (Fig. 17.4)

Common name

Hemlock rust mite.

Diagnostic characteristics

FEMALE. The body is robust, fusiform, 180–250 μm long, 90 μm wide and orange yellow in colour. The frontal lobe projects a short distance over the gnathosoma, and the prodorsal shield surface has fine short longitudinal streaks. The scapular tubercles are papilla like, 53 μm apart, located ahead of the rear margin; the scapular setae (*sc*) are strong, 115–140 μm long, projecting well ahead. The internal vertical setae (*vi*) are 20–25 μm long and point forward. Coxa I barely touch each other and are heavily spined on the proximal half; there are also spines on the suboral plate ahead of the coxae. The opisthosoma has about 60–70 dorsal annuli and 90–105 ventral annuli; the annuli are microtuberculate, with microtubercles that project over the margin of the annulus but do not form definite spinules. The setae c_2 are 20 μm long, set on about annulus 9; the setae *d* are 40 μm long, set on about annulus 32; the setae *e* are 60 μm long, set on about annulus 53; the setae *f* are 40 μm long, set on about annulus 8 from the rear. The setae h_1 are present. The genitalia are 38 μm wide and 25 μm long; the coverflap is smooth, and the setae *3a* are about 40 μm long (Keifer, 1953).

MALE. The body is 200 μm long and 90 μm wide (Keifer, 1953).

Geographical distribution

Eastern North America (Keifer, 1953; Jeppson *et al.*, 1975).

Host plants

Pinaceae – *Abies*, *Larix*, *Picea*, *Pseudolarix* (Taylor, 1970), Canada hemlock (*Tsuga canadensis* (L.) Carrière) (Keifer, 1953), fir, hemlock and Colorado blue spruce (Thomas and Rajotte, 2011); Taxaceae – *Taxus* (Taylor, 1970).

Bio-ecology

The hemlock rust mite lives in needles of Canada hemlock in Virginia and Maryland, where it is active early in the spring (Keifer, 1953), and has also been reported on needles of other Pinaceae, such as *Abies*, *Larix*, *Pseudolarix*, *Picea* and Taxaceae, such as *Taxus* (Taylor, 1970). In Pennsylvania, the mite is frequent on fir and hemlock, and like other *Nalepella* spp. overwinters as eggs on the needles of host plants. The eggs hatch in early spring, and during late spring and summer the population density is very low; it increases in late summer when temperatures begin to moderate. The hemlock rust mite lays overwintering eggs in autumn and may be active into December (Thomas and Rajotte, 2011).

Symptomatology and damage

N. tsugifoliae lives in the field and in nurseries on freshly matured needles of conifers and its populations develop early in spring. In March, it is possible to observe chlorosis and browning of the leaves, and at high mite density there is severe damage to the hemlock needles (Keifer, 1953; Jeppson *et al.*, 1975). In nurseries, premature dropping of the previous season's needles can occur. When chlorosis appears, the mite leaves the attacked needles and moves towards younger leaves. A density of 20–25 mites per needle involves evident damage; in summer, the population density decreases as a result of dispersal (Eidit, 1966). Chlorotic or rusty-brown-appearing needles, premature needle drop and a silvered, dusty, or bleached-out appearance on Colorado blue spruce are reported in Pennsylvania (Thomas and Rajotte, 2011).

Control

Monitoring should be done by looking for yellowish needles in spring. Cultural management includes not overfertilizing conifers. Control measures are not usually necessary in the field, but they can be carefully evaluated in nurseries. Among biorational pesticides, it is possible to use abamectin, horticultural oil and insecticidal soap. Traditional acaricides include bifenthrin, dicofol, fenbutatin oxide and lambda-cyhalothrin (Krischik

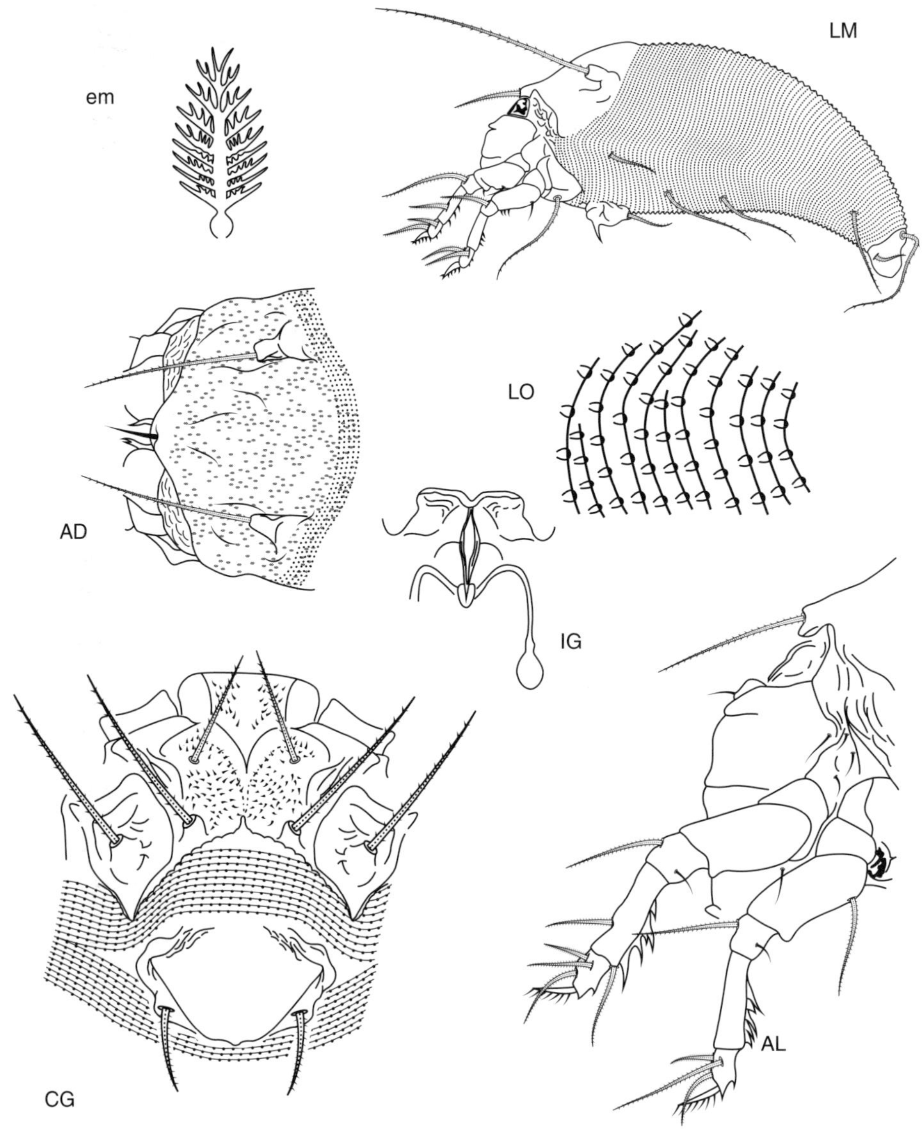

Fig. 17.4. *Nalepella tsugifoliae* Keifer (from Keifer, 1953). See Table 16.1 for key.

and Davidson, 2013a). The crop and pest management guidelines of the Cornell University (2012) suggested treating infested trees from late March through to the end of May, with 7–450 GDDs, using abamectin, carbaryl, dimethoate, fenazaquin, fenpyroximate, horticultural oil, milbemectin or neem oil.

Nalepella haarlovi Boczek

Common name

Unknown.

Diagnostic characteristics

FEMALE. The body is 272–340 μm long, 121 μm wide and dull yellow in colour. The gnathosoma is downcurved. The prodorsal shield surface has fine, short longitudinal streaks; the frontal lobe projects a short distance over the gnathosoma. The scapular tubercles are 62 μm apart and somewhat ahead of the rear shield margin; the scapular setae (*sc*) are 155 μm long, and directed forwards. Coxae I are spinulate proximally, with the posterior smooth. The opisthosoma has 42 dorsal annuli and about 105 ventral annuli. The ventral annuli are microtuberculate, and the microtubercles end in a fine spine. The dorsal annuli have very indistinct and pointed microtubercles. The setae c_1 are absent. The setae c_2 are 88 μm long, located laterally, on ventral annulus 28; the setae *d* are 44 μm long, set on ventral annulus 45; the setae *e* are 82 μm long, set on ventral annulus 63; the setae *f* are 48 μm long, located laterally on ventral annulus 7 from the rear; the setae h_1 are 10 μm long. The genitalia are 20 μm long and 33 μm wide, the coverflap is smooth, and the seta is 64 μm long (Boczek, 1962).

Geographical distribution

Danish mainland and Finland (Boczek, 1962; de Lillo and Beron, 2004).

Host plants

Pinaceae: white spruce (*Picea abies* (L.) H. Karst.), Sitka spruce (*P. sitchensis* (Bong.) Carrière), (Boczek, 1962; Löyttyniemi, 1969, 1971).

Bio-ecology

N. haarlovi lives on needles of Sitka spruce and white spruce in northern Europe. It overwinters as diapause eggs protected on the needles and laid by females of the two last generations from late August and through to the end of September. Most of these eggs survive in the field until spring and break diapause after a specific exposure to cold. Hatching occurs after 17 days at 10°C, 10 days at 15°C and 7 days at 20, 25 or 30°C. Other biological stages disappear before the end of November. The mite develops 4–8 generations a year (Löyttyniemi, 1971).

Symptomatology and damage

N. haarlovi is a very common needle vagrant but until now has only been found on forest trees, on which it turns the needles of the trees yellow brown, sometimes with a fatal result for the weaker individuals (Boczek, 1962). It damages *Picea*, especially in nurseries. High levels of mite populations cause yellowish brown needles and then leave the tree; sometimes death of young plants may occur. In Finnish nurseries, the mite caused the death of about 600,000 *P. abies* seedlings in a year (Löyttyniemi, 1969).

Control

Control is unnecessary in the field, but may sometimes be required in nurseries or in young stands.

Nalepella halourga Keifer (Fig. 17.5)

Common name

Spruce rust mite.

Diagnostic characteristics

FEMALE. The body is robust, 185–205 µm long, about 80 µm wide and deep purplish red. The gnathosoma is large and curved down. The median line of the prodorsal shield is faint, especially anteriorly; the admedian lines are visible anteriorly; the central rear of the prodorsal shield shows fine dashes; the submedian lines are indicated ahead of the scapular tubercles; the sides of the prodorsal shield have a faint network of lines. The internal vertical seta (*vi*) is 40 µm long; the scapular tubercles are 64 µm apart; the scapular setae are 108 µm long. Coxae I have spinules on the inner angle and on the anterior diagonal area; these coxae are not connate. The opisthosoma has about 43 narrow dorsal annuli to ring-bearing seta *f* and about 70 ventral annuli to the same annulus; all parts of the ventral annuli bear fine microtubercles on the rear margins; they are bead like or laterally pointed on the dorsal annuli. The setae c_2 are 40 µm long, set on ventral annulus 9 behind the prodorsal shield; the setae *d* are 50 µm long, set on ventral annulus 26; the setae *e* are 40 µm long, set on ventral annulus 42; the setae *f* are 40 µm long, set on annulus 6 ahead of the rear. The setae h_1 are 12 µm long. The genitalia are 28 µm wide and 32 µm long; the coverflap has faint lines, and the setae *3a* are 50 µm long. The empodium is 10-rayed (Keifer, 1965d).

Geographical distribution

Eastern North America (Keifer, 1965d).

Host plants

Pinaceae: blue spruce (*Picea pungens* Engelman) (Keifer, 1965d; Marshall and Lindquist, 1972; Thomas and Rajotte, 2011).

Bio-ecology

The spruce rust mite is a needle vagrant on blue spruce in Vermont (Keifer, 1965d), Pennsylvania (Thomas and Rajotte, 2011) and other eastern North American states. Its bio-ecology is similar to that of other *Nalepella* spp. In seedlings artificially infested with a few mites, it has been calculated that there is a population density of 20–40 mites per needle 2 months later, and groups of eggs are observed at bases of the needles (Marshall and Lindquist, 1972).

Symptomatology and damage

No symptoms of damage were described by Keifer (1965d), but later damage similar to that caused by *N. tsugifoliae* was reported for this mite. High attacks of *N. halourga* to black spruce were observed in greenhouses in Canada. Infested plants become yellowish brown and show a coated aspect owing to a high density of mites (Marshall and Lindquist, 1972).

Control

Control may be necessary in nurseries and in young stands.

Nalepella longoctonema Hu et Krantz

Common name

Unknown.

Diagnostic characteristics

FEMALE. The body is fusiform, 258 µm long, 111 µm wide and white or yellowish white in colour. The gnathosoma is hypognathous and with blunt spinules suborally. The prodorsal shield has a short and broad frontal lobe overhanging the base of the gnathosoma, and the internal vertical seta (*vi*) is 39 µm long. Anteriorly, the median and admedian lines are distinct, divergent posteriorly and crossing with short, submedian lines curving posteriorly towards the shield axis; paired transverse lines arise anteriorly from the admedian lines and more or less parallel to the leading edge of the shield; there is a pair of strong anteriorly directed scapular setae (*sc*) about 100 µm long in tuberculate

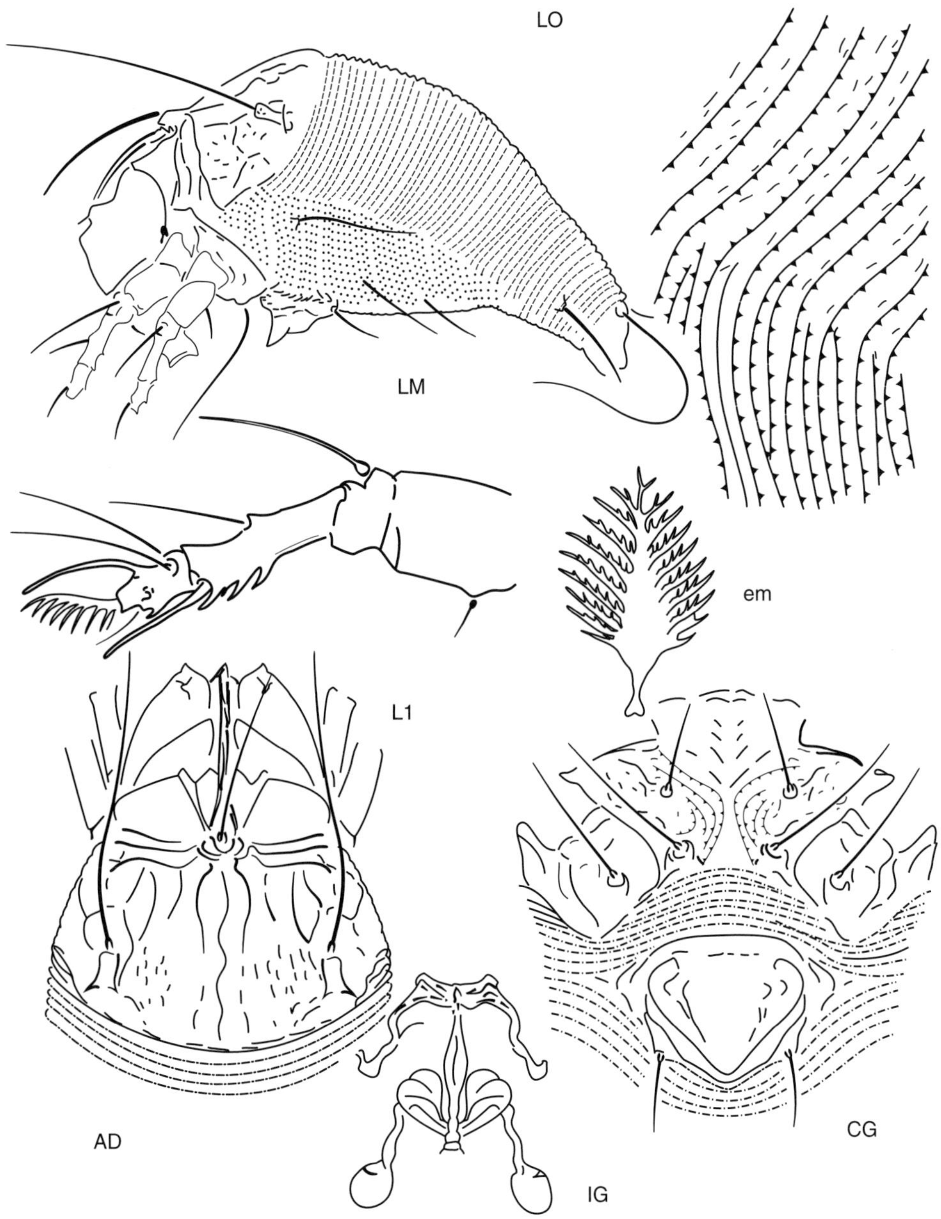

Fig. 17.5. *Nalepella halourga* Keifer (from Keifer, 1965d). See Table 16.1 for key.

insertions, each of which is flanked by a field of minute spines. The opisthosoma has 27–30 broad dorsal annuli and 94–101 narrow ventral annuli anterior to setae *f*; the annuli are ornamented with acuminate microtubercles, weak laterally. The setae c_2 are 57 µm long, 14–15 annuli from coxae II; the setae *d* are 74 µm long and inserted 22–25 annuli behind setae c_2; the setae e_2 are about 74 µm and inserted 43-47 annuli behind the setae c_2; the setae *f* are on annulus 6; the setae h_2 are about 142 µm long and h_1 about 7 µm long. The genital coverflap is 15 µm long, 29 µm wide and non-ornamented; the setae *3a* are 67 µm long. The empodium is 8-rayed (Hu and Krantz, 1991).

MALE. The body is 206 µm long and 90 µm wide (Hu and Krantz, 1991).

Geographical distribution

Oregon and other North American areas (Hu and Krantz, 1991).

Host plants

Pinaceae: grand fir (*Abies grandis* (Dougl. ex D. Don) Lindl.), noble fir (*A. procera* Rehder), (Hu and Krantz, 1991).

Bio-ecology

N. longoctonema has been collected in large numbers from all stages on plantation plantings of noble fir and grand fir in Oregon and other areas of North America. Feeding mites

were observed on the underside of the needles. All biological stages are present in the winter at temperatures ranging from −6 to 16°C, and in the laboratory, the mite develops without difficulty at 4°C (Hu and Krantz, 1991).

Symptomatology and damage

The feeding activity of *N. longoctonema* causes chlorotic spots on needles of *Abies*, and at high levels of attack, the needles turn brown and drop in late summer (Hu and Krantz, 1991).

Control

Under natural conditions control is unnecessary.

The Genus *Phytoptus* Dujardin

Diagnostic characteristics

The prodorsal shield lacks a gland and has 2 internal vertical setae (*vi*) and 2 scapular setae (*sc*); the scapular setae (*sc*) are short or long, but not minute. Microtubercles are normally distributed on the dorsal annuli. The femora and genua are separated and not fused (Amrine *et al.*, 2003).

Economic importance

This genus numbers about 40 Holarctic species, mostly living on dicotyledonous (about 20) and monocotyledonous (about 14) plants and a very small number on conifers (Oldfield, 1996a). In Europe, about 16 species have been reported, including the filbert bud mite, *Phytoptus avellanae* (Nalepa), a pest of filbert, and the ivy bud mite, *P. hedericola* Keifer, infesting English ivy (*Hedera helix* L.); both mites have worldwide distribution. *Phytoptus potentillae* Chen, Wei *et* Nan has recently been described in China (Chen *et al.*, 2005b).

Phytoptus avellanae (Nalepa) (Fig. 17.6)

Common name

Filbert big bud mite, nut gall mite.

Diagnostic characteristics

FEMALE. The body is wormlike, 220-260 µm long, 45 µm thick, and light yellow. The gnathosoma is bent down. The prodorsal shield has admedian lines extending from the chelicera base to the rear margin between the setae. The prodorsal shield has 4 setae (2 *ve* and 2 *sc*); the external vertical setae (*ve*) are 14 µm apart and 5.5 µm long; the scapular tubercles are 14.5 µm apart, ahead of the rear margin. The scapular setae (*sc*) are 10.5 µm long, and projecting forward. The opisthosoma has about 70 annuli, both ventrad and dorsad reduction in number, the last 8 or 10 annuli are partially or entirely smooth above, and the microtubercles are pointed. The setae c_1 are 36 µm long, set on about annulus 10; the setae c_2 are 17.5 µm long, set on about annulus 9; the setae *d* are 17.5 µm long, set on about annulus 22; the setae *e* are 12 µm long, set on about annulus 38; the setae *f* are 25 µm long, set on about annuli 7 from the rear; the seta h_1 is present. The genitalia are 26.5 µm wide, 11.5 µm long; the coverflap is smooth, and the seta *3a* is about 25 µm long, and the glands are clavate. The empodium is 4- to 5-rayed (Keifer, 1940b).

MALE. Not studied (Keifer, 1940b).

Geographical distribution

Worldwide.

Host plants

Betulaceae: hazel, filbert, cobnut (*Corylus avellana* L.) (Massee, 1930a; Jeppson *et al.*, 1975; AliNiazee, 1980).

Bio-ecology

P. avellanae feeds and reproduces in the subterminal and terminal buds and leaves of hazel, filbert and cobnut. Other eriophyoids (*Aculus comatus* (Nalepa), *Tegonotus depressus* Nalepa, *Coptophylla lamimani* (Keifer) and *Anthocoptes loricatus* Nalepa) live on the lower leaf surface of filbert (Jeppson *et al.*, 1975), and of these, *A. comatus* is the most harmful (Krantz, 1973). Moreover, *P. avellanae* is often associated in the buds with another eriophyoid, *Cecidophyopsis vermiformis* (Nalepa), and together they cause swelling of the buds. *C. vermiformis* is easily misidentified with *P. avellanae* (Jeppson *et al.*, 1975), and the two have been incorrectly considered to be one species. Knowledge of their life cycles was not conclusive until some time ago. In fact, initially, it was thought that *P. avellanae* had a simple life cycle based on a single nymph emerging from 'winter big buds' – the swollen buds where the mite overwinters, and which it leaves in spring to move to the axillary buds, where it feeds and reproduces until the next spring (Massee, 1930a; Krantz, 1979; AliNiazee, 1980). Following Jeppson *et al.* (1975), a nymph of *P. avellanae*, which is very different from the adult, would live on the leaves during the summer.

It has also recently been observed that the life cycle of *P. avellanae* involves both gall and vagrant forms (Özman and Toros, 1997a,b), and that each has a different life cycle and produces typical damage. Following Özman (2000), the form of *P. avellanae* inducing gall formation has a simple life cycle involving a single nymph (which is similar to the adult) that develops with two nymphal stages. In spring, the second nymphal stages leave the winter buds and move to the new axillary buds, colonizing the outer part of the bud, where they presumably diapause until late spring, when they start moulting to adults. This behaviour contrasts with assertions that the mite moves to the core tissue (Krantz, 1979; Manson and Oldfield, 1996; Özman and Toros, 1997a). The buds occupied only by *P. avellanae*, or where its populations are predominant, begin to swell during the summer when new adults begin to reproduce, and swelling continues until autumn, resulting in 'spring' or 'medium big buds' (MBB in Fig. 17.7) (Krantz, 1979; Manson and Oldfield,

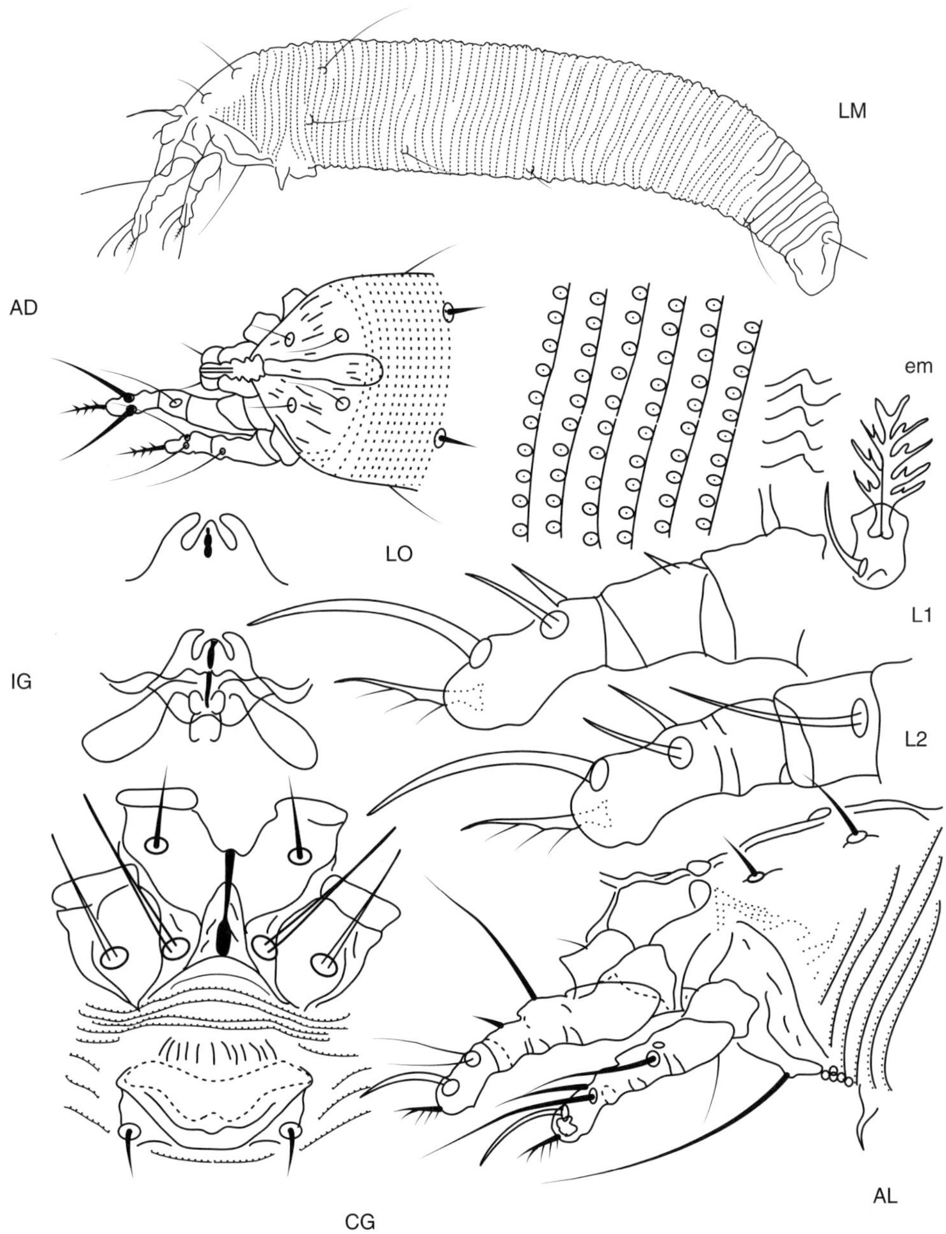

Fig. 17.6. *Phytoptus avellanae* (Nalepa) (from Keifer, 1940b). See Table 16.1 for key.

1996), which swell further to form the winter big buds (see Fig. 7.2C and WB in Fig. 17.7). The development of populations occurs within the same bud during the year, with an overlap of generations that is difficult to calculate (Özman and Toros, 1997a). Arzone (1983) reported that in north Italy, *P. avellanae* completed six generations a year.

The vagrant form of *P. avellanae* evolves through two nymphal forms. The first form, also called the 'normal nymph', is similar to the adult and evolves into the adult through two nymphal stages. It lives on unexposed sites such as big buds, generative/reproductive buds (i.e. flower buds) and on the lower leaf surfaces at the start of spring (Özman and Toros, 1997a; Özman, 2000). The other nymphal form, which is also called a '*Tegonotus*-like nymph', differs from the adult in that it has a flat body, broad dorsal annuli, and lateral projecting fleshy points (Jeppson *et al.*, 1975; Manson and Oldfield, 1996). These vagrant forms live on exposed surfaces, along the veins on leaf undersurfaces, and migrate to catkins, female flowers, vegetative buds and shoot tips in summer and autumn. Their life cycle can develop without the presence of big buds (Özman and Toros, 1997a). They move into terminal buds at the same time as they moult into the adult stage (Jeppson *et al.*, 1975; Alford, 1984); some adults penetrate into vegetative buds and remain on the outer scales until the next spring. Other adults colonize the spring buds or winter big buds and remain on the outer scales, and when the gall forms invade the outer scales,

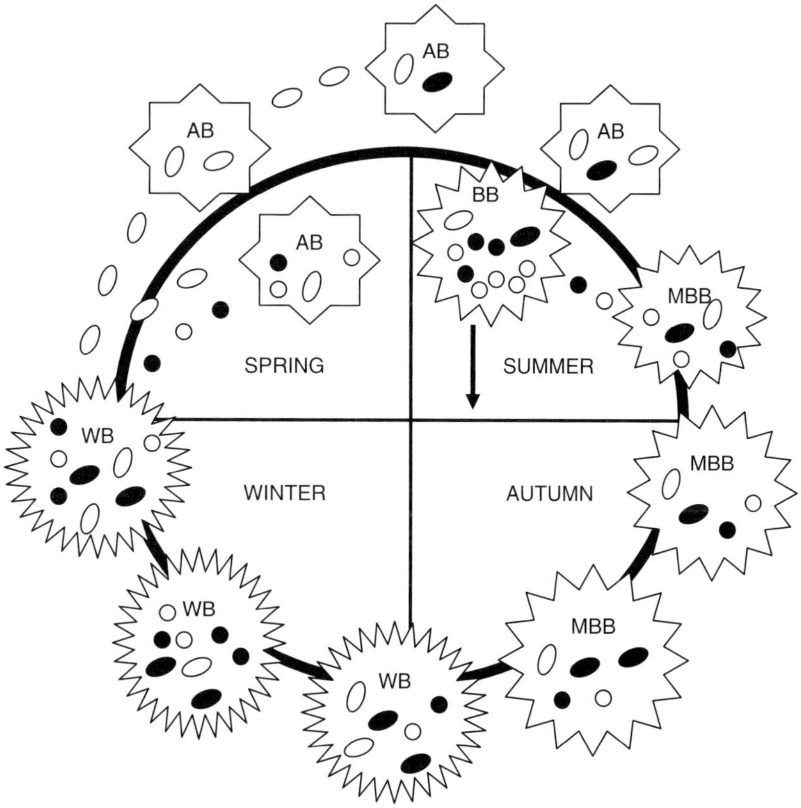

Fig. 17.7. Schematic drawing of gall formation by *Phytoptus avellanae* (Nalepa) and *Cecidophyopsis vermiformis* (Nalepa). Adults of *P. avellanae* are designated by black ovals and nymphs by white ovals; adults of *C. vermiformis* are designated by black circles and nymphs by white circles. Key: AB, axillary buds; BB, (summer) big buds; MBB, medium big buds; WB, winter (big) buds. The arrow indicates the fall of BBs during the summer.

they diffuse to the inner part of the big buds. In spring, the vagrant form migrates earlier than the gall form (Özman and Toros, 1997a).

The life cycle of *C. vermiformis* develops along with *P. avellanae*. In early spring, both mites leave winter buds and move towards the axillary buds (AB in Fig. 17.7). Adult females (Krantz, 1979; AliNiazee, 1998) and nymphs (Özman and Toros, 1997a) of *C. vermiformis* penetrate the core of the axillary buds, while nymphs of *P. avellanae* colonize the outer part of the buds, where they presumably diapause until late spring, when they start moulting to adults. *C. vermiformis* increases its density and induces the formation of 'big buds' (BB in Fig. 17.7). In late summer, these big buds open too soon, desiccate and fall, and *C. vermiformis* leaves them and searches for vegetative buds, catkins and female flowers; it then moves into spring (medium big) buds hosting *P. avellanae*, where it overwinters at low density (Özman and Toros, 1997a). In contrast, *P. avellanae* living in big buds probably disappears with the big bud and does not migrate into medium (big) buds (Krantz, 1979) – unlike *C. vermiformis*, which leaves the big buds (Özman and Toros, 1997a). *P. avellanae* populations already living in the medium (big) buds continue to develop, and the buds swell further to form winter buds (Krantz, 1979; Özman and Toros, 1997b). The density of *P. avellanae* is high in both the winter (big) buds and the spring medium (big) buds, while that of *C. vermiformis* is higher in the summer big buds (Krantz, 1979). Moreover, although the numbers of *C. vermiformis* in winter big buds are lower in the autumn, its populations develop and the densities of both species reach similar values; commonly, *C. vermiformis* prevails in the spring (Özman and Toros, 1997c). The behaviour of *P. avellanae* and *C. vermiformis* can be described as temporal niche partitioning (Sabelis and Bruin, 1996). The two species can be found together in the same bud during the year, but either one or the other always dominates in terms of density. The dominance by either species in newly infested buds probably occurs only if bud infestation relates to the colonizing of core tissues (Krantz, 1979). The reasons for this coexistence are unknown. In summer, the faster population development of *C. vermiformis* induces the early opening and fall of the *C. vermiformis*-dominated buds and the *P. avellanae*-dominated buds represent a solution for overwintering. This may result in a frequency-dependent mechanism of coexistence. *C. vermiformis* is the best competitor, but its populations cannot develop at high density so as not to hinder their chances of overwintering (Sabelis and Bruin, 1996).

The phase of emergence is influenced by weather and plant cultivar (Özman and Toros, 1997a). *P. avellanae* moves only short distances from the overwintering buds, and its dispersal in the environment is facilitated by wind, insects or birds (Vidal-Barraquer *et al.*, 1966); the time of emergence is related to the country and cultivar and occurs in early spring or spring.

C. vermiformis has a second emergence time in late summer. Thus, the emergence of *P. avellanae* starts in mid-April in Turkey (Özman and Toros, 1997a) and in early spring in Australia (Snare and Knihinicki, 2000). In North America, it commonly occurs from April to May (AliNiazee, 1998). In Europe, emergence has been observed from March to July in England (Massee, 1930a; Alford, 1984), from April to June in central and north Italy (Viggiani and Bianco, 1974; Mozzone *et al.*, 1994) and in north-eastern Slovenia (Beber, 1994), from April to later May in Sicily (south Italy) (Nucifora *et al.*, 1979a), from April to May in Turkey (Özman and Toros, 1997a) and from April to July in Yugoslavia (Stamenkovic *et al.*, 1997). The time of peak numbers of emerging big bud mites also varies in the different countries. In Slovenia, the peak occurs in mid-June (Beber, 1994), while in Turkey it occurs from the last week of April and the last week of May (Özman and Toros, 1997a). However, in Yugoslavia, it was found that the maximum mite number varied in different years, occurring at the beginning, in the middle and at the end of the emergence period (Stamenkovic *et al.*, 1997).

The influence of the phenology of the host plant on the time of big bud mite emergence is difficult to discern from the available data, except it seems to vary with season, environment and cultivar. In Romania, it occurs at the same time as the emergence of the third leaf (Ioachim and Bobarnac, 1997), in central Italy it happens when there are 2–3 leaves per shoot (Viggiani and Bianco, 1974), while in Turkey the highest emergence coincides with 3–4 leaflets on shoots about 2–3 cm long (Özman and Cobanoğlu, 2001). Jeppson *et al.* (1975) mention that emergence occurs with the formation on terminal shoots of the 5th and 6th buds, while Krantz (1979) observes that emergence starts when the infested buds reach the pre-blasting stage in late winter and early spring. The most convincing interpretation of the ecological factors underlying emergence is probably from Burgess and Thompson (1985), who highlight a close synchronization between mite emergence and shoot development. Research on six cultivars of all levels of susceptibility indicated that mites colonized only those buds that developed during the emergence phase.

The weather conditions also play an important role in the emergence of big bud mites (Jeppson *et al.*, 1975; AliNiazee, 1980, 1998; Özman and Toros, 1997a) and among these are reported the spring temperature (AliNiazee, 1980, 1998). *P. avellanae* nymphs leave the big buds at a daily temperature range of 15–20°C, but not at lower temperatures (Özman and Toros, 1997a). The buds offer the mites adequate protection from limiting weather conditions, and emergence exposes them to the possibility of a high level of mortality (up to 90%) (Vidal-Barraquer *et al.*, 1966; Jeppson *et al.*, 1975), with greater losses due to high temperatures and very low rainfall, or as a result of heavy rain (Jeppson *et al.*, 1975; Özman and Toros, 1997a; Gantner, 2001). High temperatures and low humidity also limit the survival of *C. vermiformis* nymphs and adults (Özman and Toros, 1997a).

The natural enemies of *P. avellanae* include fungi, insects and mites, and are summarized in Table 17.1.

Symptomatology and damage

P. avellanae commonly lives together with *C. vermiformis*, but the main causal agent of gall formation is *P. avellanae*, and *C. vermiformis* is only a harmless inquiline (Massee, 1930a; Jeppson *et al.*, 1975). This interpretation of the relationship between the two mite species is now commonly accepted, and *P. avellanae* is considered to be a severe pest of hazelnut worldwide (Özman, 2000).

Until a short time ago, it was believed that *P. avellanae* damaged both vegetative and generative buds (Massee, 1930a; Vidal-Barraquer *et al.* 1966; Jeppson *et al.*, 1975; Krantz, 1979; AliNiazee, 1980), but recently it has been found that the mite feeds only on generative buds and induces the formation of big buds (Özman and Toros, 1997b). In spring, the second-stage nymphs leave the winter buds and move to new axillary buds, remaining in the more external part of the bud. The infested new buds begin to swell in the summer when new adults reproduce. The swelling continues during the summer and produces the 'spring or medium big buds', which consist of an aggregation of thickened scales bearing fleshy nutritive excrescences (Westphal and Manson, 1996); the buds are spherical, with a diameter of about 10 mm, and harbour thousands of mites in winter and spring.

The vagrant form of *P. avellanae* infests catkins, leaves and the apexes of shoots; the greatest damage occurs in female flowers and fruits, which can fall (Özman and Toros, 1997b). Formerly, it was believed that adults developing from the second type of nymphs (*Tegonotus*-like nymphs) and living on leaves during the summer moved to terminal buds to cause big buds (Jeppson *et al.*, 1975), but recently it was observed that the vagrant form did not produce big buds (Özman and Toros, 1997b).

The injury caused by *C. vermiformis* is not clear. The mite is involved in the formation of summer and winter big buds, and in the emergence phase during spring and autumn, it lives on female flowers and the tips of catkins, where it can produce catkin deformation (Özman and Toros, 1997a,b; Özman-Sullivan, 2006).

When gall development is completed, the galls are no longer joined to the twig and they normally fall. When the number of galls is high, the twig remains partially damaged, which can inhibit development and cause withering. The presence of galls also prevents the emergence of leaves, and consequently the process of photosynthesis. The most serious damage to twigs occurs on young plants.

The galls are distributed only on buds formed during mite migration, and at this time they are young and tend to occur within nodes five to eight from the base of the shoots. The newest elongating twigs are attacked soon after they begin to grow (Burgess and Thompson, 1985). Krantz (1979) reported that the loss due to big gall formation by *C. vermiformis* ranged from 3 to 5%, while that due to gall formation by *P. avellanae* ranged from 18 to 20%. In general, the damage varies with cultivar, environment and cultural practices. The amount of infestation varies from one country to another – from 0.1–0.5% in Poland (Maziarz, 1984) and Spain (Vidal-Barraquer *et al.*, 1966) to 80–90% in England (Massee, 1930a) and Georgia (ex USSR) (Tavamaishvili, 1990). Commonly, the level of infestation ranges from 20 to 30% (Arzone, 1975; Krantz, 1979; AliNiazee, 1998; Özman and Cobanoğlu, 2001).

Bud injury can produce yield loss and has economic implications (Ecevit *et al.*, 1992; Stamenkovic *et al.*, 1997); however,

Table 17.1. *Phytoptus avellanae* (Nalepa). Natural enemies (fungi, insects and mites).

Families (or cohorts) and species	References
FUNGI	
Clavicipitaceae	
Verticillium lecanii (Zimmerman) Viegas	Özman and Cobanoğlu, 2001
Trichocomaceae	
Paecilomyces eriophyes (Massee) Leath.	Del Guercio, 1911
Penicillium funiculosum Thom	Santamarina *et al.*, 1988
(Not assigned to a family)	
Cephalosporium spp.	Pesante, 1962
INSECTS	
Anthocoridae	
Orius minutus L.	AliNiazee, 1994; Özman and Cobanoğlu, 2001
Cecidomyiidae	
Arthrocnodax coryligallarum (Targioni Tozzetti)	Hovasse, 1930; Pesante, 1962
Chalcididae	
Tetrastichus eriophyes Taylor	Pesante, 1962
MITES	
Alycina (cohort)	Özman-Sullivan *et al.*, 2005
Eupodina (cohort)	
Phytoseiidae	
Kampimodromus aberrans (Oudemans)	Pesante, 1962; Ragusa, 1974; Arzone, 1983; Nicotina and Viggiani, 1985; Tsolakis *et al.*, 2000; Özman-Sullivan, 2006
Phytoseiulus persimilis Athias Henriot	Ragusa, 1974; Nicotina and Viggiani, 1985; Tsolakis *et al.*, 2000
Phytoseius plumifer (Canestrini *et* Fanzago)	Daneshavar and Khosrowshahi, 1994
Typhlodromus cryptus Athias Henriot	Ragusa, 1974; Nicotina and Viggiani, 1985; Tsolakis *et al.*, 2000
Typhlodromus doreenae Schicha	Webber, 2007
Typhlodromus intercalaris Livshitz *et* Kuznetsov	Ragusa, 1974; Nicotina and Viggiani, 1985; Tsolakis *et al.*, 2000
Typhlodromus rhenanus (Oudemans)	
Typhlodromus sternlichti Swirski *et* Amitai	
Typhlodromus tilarum Oudemans	Michelatti *et al.*, 1994
Stigmaeidae	
Agistemus spp.	AliNiazee, 1994; Copping, 2001; Özman and Cobanoğlu, 2001
Zetzellia spp.	AliNiazee, 1994; Özman and Cobanoğlu, 2001
Tarsonemidae	
Tarsonemus spp.	
Tydeidae	
Tydeus spp.	

reports of the effects of mite infestation on yield vary, and the available data indicate little or no effect on yield (Alford, 1984), serious yield losses (AliNiazee, 1998) or significant yield loss (Özman-Sullivan and Akça, 2005). Some authors report that the degree of bud damage or bud loss gives an equivalent reduction in yield (Krantz, 1979; AliNiazee, 1998). Big bud mite harmfulness has also been assessed as a percentage of infested buds (Stamenkovic *et al.*, 1997). Some authors (Bergoughoux *et al.*, 1978; Westphal and Manson, 1996) report that infested big buds do not develop into vegetative or fruiting branches, with a resulting yield loss, while others report that crops are reduced by severe infestations (Thompson, 1977). Research by Planes *et al.* (1965) shows that there may be up to a 20% reduction in yield due to mite-induced pistillate flower distortion in compound buds. Viggiani and Bianco (1974) indicated a 15% bud loss (infestation rate) as an economic threshold and 20% infestation as an economic injury level (EIL). Alford (1984) reports that big buds have little or no effect on yield. These conflicting data indicate that the role of the big bud mites on the development of twigs and leaves has not been fully investigated.

Burgess and Thompson (1985) developed a rating scheme based on plant response to mite attack. They calculated the intra-plant distribution of *P. avellanae* and *C. vermiformis* by rating the density of galls per tree and produced a scale ranging from 1 (0–2 galls per tree) to 5 (4 or more galls per shoot). In North America, a rating scale of 1–5 is used to rate the severity of infestation and the frequency with which insecticides are applied year after year; *P. avellanae* was rated at 4.0, while *C. vermiformis* was rated at 2.0 (AliNiazee, 1994).

Control

The control of *P. avellanae* is provided by the use of various means, such as mechanical, biological and chemical, and the use of resistant cultivars, all led by the monitoring of populations and the appropriate decision tools.

MONITORING. Double-sided adhesive tape is a simple method for monitoring the emergence of *P. avellanae* while avoiding interference with the physiology of the mite; it also circumvents the limitations of the method of Cross and Ridout (2001), which is based on cutting off the shoot above the gall. The different tapes need to be placed in the same position relative to the big bud above or below the big bud, and the sampling done from the top, middle and lower canopy; sampling from each quarter of the tree does not appear to be important (Webber, 2007).

DECISION TOOLS. Peak emergence and the phenology of the host plant should both be used to calculate the timing of control. It is important to assess the development of the new bud as well as the number of emerging mites. Control should be applied before the new bud has developed to the stage that is susceptible to the attention of big bud mites and their moving towards bud entry. In New Zealand, an appropriate time to apply control measures would be before the new buds reach a size of 0.5 × 0.5 mm (width × height) or at 50% mite emergence. Here, the use of an 'accumulated heat sum model' provides acceptable prediction of the start of emergence. Such a forecast would be very useful for growers or a technician to establish a start date for monitoring and visual inspection of the growth of the hazelnut (Webber, 2007).

Accumulated degree-days alone do not give an accurate prediction of the accumulated percentage emergence. However, a regression model involving leaf number, bud height, bud width, degree-days with a lower threshold temperature of 6°C from start of emergence, and Julian date, permits determination of the approximate accumulated percentage emergence of *P. avellanae*. This would be used by growers to determine the time to start the control. These two models can be used in conjunction with one other (Webber, 2007).

In Italy, Viggiani and Bianco (1974) found that 15% of bud losses represented the economic threshold, and 20% the EIL. Commonly, levels of attack higher than this involve serious economic loss (Bergoughoux *et al.*, 1978).

Regardless of the reliability of the above values, any decision tool used to determine when to control the mite populations must take into account that the hazelnut is not a traditional grove, and in several countries represents a transition between cultivated environments and forest. From this standpoint, notwithstanding the requirement of sound economic management of control, preferred methods of control are marked by having a lower ecological impact.

PLANT RESISTANCE. Resistant varieties allow the limitation of mite outbreaks (AliNiazee, 1994, 1998). Susceptibility to *P. avellanae* is heritable and controlled by many genes with a predominantly additive gene action (Thompson, 1977). The level of resistance or tolerance to the mite probably depends on the structure of the meristems and bud primordia (Burgess and Thompson, 1985), although the actual resistance mechanism is unknown (Mehlenbacher, 1994) and has involved various research (Hassan, 1928; Averna Saccà, in Pesante, 1961; Vidal-Barraquer *et al.*, 1966). For hazel in the Mount Etna region of Sicily (Italy), several resistant cultivars were found and selected ('Agghirara', 'Carrello', 'Tonda Gentile Romana', etc.), as well as a resistant clone ('Selvaggiola tardiva'), whose buds cause the death of skin tissues rather than the hyperplastic formations that are fundamental for eriophyoid development (Nucifora *et al.*, 1979b). Resistant cultivars commercially available are known worldwide and include 'Barcellona' (Lagerstedt, 1979; Mehlenbacher, 1994; Crawford, 1995), 'Ennis' (Crawford, 1995), 'Tonda di Giffoni' (Mehlenbacher, 1994; Crawford, 1995; Baldwin *et al.*, 2007) and 'Tonda romana' (Mehlenbacher, 1994; Crawford, 1995).

MECHANICAL CONTROL. Cutting off all infested twigs and branches in the autumn and burning them is a method that has been used for a long time (Del Guercio, 1911; Matons, 1920; Schimitschek, 1939; Bardia y Bardia, 1958; Pesante, 1962). The collection and burning of infested buds controlled mite infestation in Turkey (Sureya, 1933), and significantly reduced (by up to 65%) the number of galls in the next year in Spain (Vidal-Barraquer *et al.*, 1966). At present, it is considered that big bud mites are easy to control without the use of pesticides by carefully removing and burning infested buds from the young trees (Eastmond and Eastmond, 2003). However, this is laborious, uneconomic and inefficient (Vidal-Barraquer *et al.*, 1966), and is also associated with a reduction of fruiting connected with the removal of galls (Nucifora *et al.*, 1979a).

BIOLOGICAL CONTROL. Biological control alone seems to be unable to exercise adequate control of populations of *P. avellanae*. Various natural enemies of *P. avellanae* are known, such as pathogens, and predatory insects and mites, but no studies have been initiated to exploit their action in the field. Some examples of studies that have been done are mentioned briefly. The mycopathogen *Verticillium lecanii* (Zimmerman) Viegas is able to produce up to 99.5% of control of *P. avellanae* (Özman and Cobanoğlu, 2001), but extensive applications of the fungus in the field are not known. In north Italy, although 90% of galls were colonized by 1–2 *Tetrastichus eriophyes* Taylor (Eulophidae) and 60% of galls by 2–6 *Kampimodromus aberrans* (Oudemans) (Phytoseidae) (Arzone, 1983), chemical treatments were usually applied. The positive role of phytoseiid mites cannot be denied, but their usefulness in the control of *P. avellanae* is uncertain according to various authors. Thus, AliNiazee (1980) reported that in North America phytoseiid mite species did not play an economic role, while in north Italy, Michelatti *et al.* (1994) found that *Typhlodromus tilarum* Oudemans was able to limit mite populations under the injury threshold. Özman-Sullivan (2006) has indicated that *K. aberrans* is a potential biological control agent of *P. avellanae* in Turkey, and Webber (2007) has reported that *T. doreenae* Schicha appears to have good potential as a biological control agent for hazelnut big bud mites in New Zealand.

CHEMICAL CONTROL. Several chemicals can be sprayed against *P. avellanae* in the spring during mite emergence from the overwintering bud galls (AliNiazee, 1980), when the mites are most susceptible. The success of chemical control also depends on the correct timing of the treatments (Özman-Sullivan and Akça, 2005), and spraying must coincide with the peak

emergence period. Timing of treatments also relates to the time of year, leaf formation and the monitoring of mite emergence. The different persistence of chemicals sometimes makes the control unsatisfactory (Cross and Ridout, 2001) and helps to explain the good results obtained from the use of chemicals such as wettable sulfur, which is characterized by a persistence on the bushes for 7–10 days (Van Dinther, 1952), or endosulfan, which persists for 21 days after treatment (Stamenkovic *et al.*, 1997).

Sulfur has been used for a long time in various forms: as lime sulfur (Massee, 1930a), sulfur wettable powder (Van Dinther, 1952) and sulfur dust (Alkan, 1959). Three treatments with sulfur as wettable powder at intervals of 15–20 days beginning when the 3rd leaf of the bud emerged allowed Vidal-Barraquer *et al.* (1966) to reduce the number of galls in the following year by 72%. Recently, it was observed that one treatment with sulfur 80% wettable powder is as effective as endosulfan (Özman-Sullivan and Akça, 2005). In New Zealand, the distribution of sulfur at 2, 50 and 80% of mite emergence reduced the number of emerging and moving big bud mites compared with the controls (Webber, 2007). Endosulfan has also been used from time (Pesante, 1961; Krczal, 1963; Planes *et al.*, 1965; Martí Fabregat and del Rivero, 1966; Vidal-Barraquer *et al.*, 1966; Minetti *et al.*, 1986). One treatment with endosulfan at peak emergence (Arzone, 1977), or two (Viggiani and Bianco, 1974) or three (Vidal-Barraquer *et al.*, 1966) treatments at the release of at least the third leaflet on the twigs (or when the terminal shoots begin to form the fifth and sixth buds) ensured good control (Massee, 1930a; Planes *et al.*, 1965; Vidal-Barraquer *et al.*, 1966; Jeppson *et al.*, 1975). However, sulfur has low mammalian toxicity and low harmful effects on beneficials (Nucifora *et al.*, 1979b; Michelatti *et al.*, 1994; Özman-Sullivan and Akça, 2005), while endosulfan is very toxic to entomophagous insects (Arzone, 1975).

INTEGRATED PEST MANAGEMENT. Sampling and monitoring techniques, natural biological control and the use of safe pesticides are at the basis of IPM programmes developed in North America and Europe for hazelnut (AliNiazee, 1998). In Oregon, an expert system (HAZLPEST) has been developed for insect and disease management based on different sub-programmes that include 'Insect identification' (incorporating tetranychid and eriophyoid mites), 'Pests monitoring and control' and 'Insecticide selection and disease management' (Drapek *et al.*, 1990). Oregon State University has developed an IPM programme for monitoring the movement of *P. avellanae* during March and April by placing 'Tanglefoot' on the twigs surrounding buds and checking for mites with a 20× hand lens. The use of pesticides is suggested when the action threshold is reached, normally with a consistent mite movement and usually occurring with bud break (Pscheidt *et al.*, 2014).

Resistant varieties help the implementation of IPM in hazelnuts, and increasing numbers of hazelnut growers are adopting IPM programmes (AliNiazee, 1997; Özman-Sullivan and Akça, 2005), with better returns for the growers because of the reduction in pesticide contamination and improvement in orchard environment (AliNiazee, 1994; Mozzone *et al.*, 1994).

Phytoptus hedericola Keifer (Fig. 17.8)

Common name

Ivy bud mite.

Diagnostic characteristics

FEMALE. The body is worm like, 150–180 µm long, 35-40 µm thick, and light yellow. The gnathosoma projects forward and down. The prodorsal shield has a pattern of longitudinal lines and 4 setae, 2 external vertical setae (*ve*) and 2 scapular setae (*sc*). The scapular tubercles are 17.5 µm well ahead of the rear margin; the scapular setae (*sc*) are 15 µm long, and projecting forward; the external vertical setae (*ve*) are 7.5 µm long, and 20.5 µm apart. The opisthosoma has about 60 annuli which are strongly microtuberculate, with the microtubercles elongate, and a slight dorsoventrad increase in annuli number and the dorsal rear without tubercles. The setae c_1 are about 30 µm long, set on annulus 11; the setae c_2 are 15 µm long, set on about annulus 10; the setae *d* are 25 µm long, set on about annulus 23; the setae *e* are 15 µm long, set on about annulus 35; the setae *f* are 15 µm long, set on about annulus 5 from the rear. The seta h_1 is present. The genitalia are 17.5 µm wide and 10 µm long; the coverflap is smooth except for slight basal tuberculation, and the seta *3a* is about 5 µm long. Coxae I are connate. The empodium is 5-rayed (Keifer, 1943; Jeppson *et al.*, 1975).

MALE. The male has not been seen (Keifer, 1943).

Geographical distribution

Australia (Knihinicki and Boczek, 2002), California (Keifer, 1943), Serbia and Montenegro (Glavendekić *et al.*, 2005) and South Africa (Meyer Smith and Craemer, 1999).

Host plants

Araliaceae: English ivy (*Hedera helix* L.) (Keifer, 1943; Jeppson *et al.*, 1975).

Bio-ecology

P. hedericola was first collected from the buds and petiole bases of English ivy in California (Keifer, 1943), and today has a worldwide distribution. The mite is adapted to conditions of high humidity of coastal environments, and in California lives near the ocean, never appearing in the interior valleys (Jeppson *et al.*, 1975).

Symptomatology and damage

The ivy bud mite causes leaf deformation and growth stunting. The damage to dwarf potted ivy is severe (Keifer, 1943).

Control

Chemical control may sometimes be necessary on young plants.

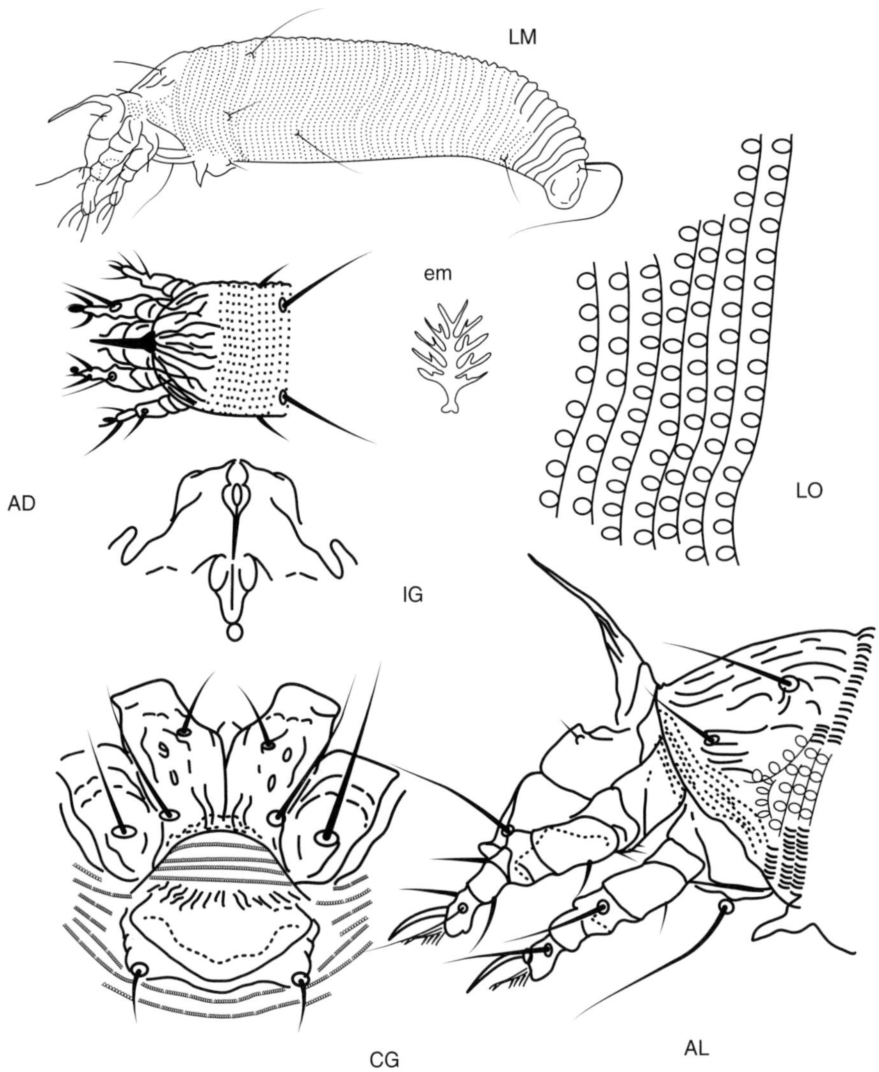

Fig. 17.8. *Phytoptus hedericola* Keifer (from Keifer, 1943). See Table 16.1 for key.

The Genus *Propilus* Keifer

Diagnostic characteristics

The prodorsal shield is broad, with the frontal lobe overhanging the gnathosoma base, with 2 external vertical setae (*ve*) on the humeral angles, and tubercles and scapular setae (*sc*) lacking. The opisthosomal dorsal annuli are much broader than the ventral annuli. All the usual opisthosomal setae are present. The female genital coverflap lacks definite ornamentation; the spermathecal tubes recurve from an anterior extension of tubes from the spermathecal pore, with the pore located at the central genital rear (Keifer, 1975b; Amrine *et al.*, 2003).

Economic importance

This genus includes about six species (*Propilus alternatus* Návia *et* Flechtmann, *P. gentyi* Keifer, *P. pellitus* Návia *et* Flechtmann, *P. spinosus* Keifer, *P. syagris* Gondim Jr, Flechtmann *et* Moraes, and *P. tavaresi* Návia *et* Flechtmann), all variously associated with Palmae in South America (Návia *et al.*, 2007c).

P. alternatus has been collected on *Mauritia flexuosa* L. in Manaus, Amazonas (Brazil) on the upper leaf surface (Návia and Flechtmann, 2005). *P. gentyi* has been found in San Alberto, Bucaramanga, Department of Santander, Colombia, on prickle palm (*Aiphanes* sp.); the mites rust the underside of the fronds (Keifer, 1975b). *P. pellitus* has been collected in Cotia, São Paulo (Brazil) on the upper leaf surface of toledo (*Lytocaryum hoehnei* (Burret)) (Návia and Flechtmann, 2002). *P. spinosus* has been found on *Aiphanes* sp. in Bucaramanga, Santander (Colombia) (Keifer, 1975b) and on *Bactris setosa* Martius and *Syagrus romanzoffiana* (Cham.) Glassm. in Brazil on the upper leaf surface, causing russeting of the affected tissue (Návia *et al.*, 2007c). *P. syagris* lives on *S. romanzoffiana* (Cham.) in Pariquera-Açu, São Paulo (Brazil) (Gondim *et al.*, 2000), and on *Astrocaryum aculeatissimum* (Schott) Burret and *B. vulgaris* Barb. Rodr. (Návia *et al.*, 2007c). *P. tavaresi* was found on the upper leaf surface of *Elaeis oleifera* (Kunth) in Cortés in

Manaus, Amazonas, and the specimens were collected under a dense trichoma layer along the midrib on the upper leaf surface of *E. oleifera* (Návia and Flechtmann, 2005). All species may be harmful, but at present cannot be included *sensu stricto* among the pest mites. Monitoring and chemical control of other pests in the production and marketing of ornamental Palmae can be sufficient to control *Propilus* spp.

The Genus *Palmiphytoptus* Návia *et* Flechtmann

Diagnostic characteristics

The prodorsal shield has prominent eye-like structures posterolaterally, and the external vertical setae (*ve*) are not located on the anterior margin. The dorsal and ventral opisthosomal annuli are subequal (Amrine *et al.*, 2003).

Economic importance

The genus includes the species *Palmiphytoptus oculatus* Návia *et* Flechtmann and *P. barbosae* Návia *et* Flechtmann. *P. oculatus* has been collected from the terminal shoots and leaves of *Bactris setosa* Martius in Cananéia, São Paulo (Brazil) and from *Astrocaryum aculeatissimum* (Schott) Burret (Návia and Flechtmann, 2002). *P. barbosae* lives on the upper leaf surface of *A. acaule* Mart. in Manaus, Amazonas (Brazil) and the colonies were found under a dense trichoma layer along the midrib on the upper leaf surface (Návia and Flechtmann, 2005). Currently, none of these species play a serious economic role, but their presence should be monitored within the general management of the defence of their host plants.

The Genus *Retracrus* Keifer

Diagnostic characteristics

The body is fusiform, and widest across the centre of the shield. The gnathosoma has a moderate size, the chelicerae are evenly bent down, and there is a short form oral stylet. The prodorsal shield is broad and long, with a short broad frontal lobe over the gnathosoma; the lateral projecting lobe is just outside the anterior setiferous tubercles; the scapular setae (*sc*) and the external vertical setae (*ve*) bear bulbous, enlarged tubercles; the setae *ve* direct forward and *sc* backwards. The opisthosoma is divided laterally into dorsal and ventral annuli. The opisthosomal setae c_1 are lacking. The empodium is simple (Keifer, 1965d; Amrine *et al.*, 2003).

Economic importance

The genus includes the pacaya mite, *Retracrus johnstoni* Keifer, and the oil palm mite, *R. elaeis* Keifer, both living on *Palmae* in Central and South America (Keifer, 1965d, 1975a; Ochoa *et al.*, 1994; Gondim and Moraes, 2003; Návia *et al.*, 2007c). The first mite is also recorded from Egypt (El-Halawany *et al.*, 2001). An unidentified specimen of the genus *Retracrus* was first collected on peach palm (*Bactris gasipaes* Kunth) in Peruvian Amazonia (Peru) (Couturier *et al.*, 1996).

Retracrus johnstoni Keifer (Fig. 17.9)

Common name

Pacaya mite.

Diagnostic characteristics

FEMALE. The body is covered by abundant wax, placed longitudinally on the opisthosoma, and the white wax pencils from the prodorsal shield tubercles, the anterior resembling antennae; it is clear water white, 170–180 μm long, 70–90 μm wide and about 60 μm thick, broadest across the centre of the prodorsal shield. The gnathosoma is 36 μm long and curved down. The dorsal pedipalp genual seta (*d*) is 12 μm long. The prodorsal shield is 75 μm wide and 90 μm long, subquadrate, with no design of lines. The lobes below the scapular tubercles ahead of the dorsal annuli are pointed. The anterior lateral lobes are rounded. The anterior setiferous tubercles are 45 μm apart, the external vertical setae (*ve*) are 20 μm long; the scapular tubercles are 40 μm apart, the scapular setae (*sc*) are 14 μm long. The opisthosoma has about 13 dorsal annuli ahead of the rear fused area, their edges are somewhat irregular and projecting laterally; there are about 42 ventral annuli from seta c_2 to seta *f*. Anteriorly, the ventral annuli are unclear laterally. The microtubercles of the ventral annuli become stronger caudad, especially on the venter after seta *e*. The setae c_2 are above seta *3a* and 26 μm long, the setae *d* are missing; the seta *e* are 15 μm long, and 13 ventral annuli ahead of seta *f*; the setae *f* are 25 μm long, on annulus 4 from the rear. The seta h_1 is lacking. The genitalia are 20 μm long, and 25 μm wide; the coverflap lacks ribs, but has a central base longitudinal line; the seta *3a* is 18 μm long. The coxae are unornamented and the empodium is 6-rayed (Keifer, 1965d).

MALE. The male is like the female, but smaller.

Geographical distribution

Brazil (Santana *et al.*, 1994; Santana and Flechtmann, 1998; Návia *et al.*, 2007c), Chiapas (Mexico) (Keifer, 1965d), Costa Rica (Schliesske 1988; Ochoa *et al.*, 1994; Návia *et al.*, 2007c), Egypt (El Halawany *et al.*, 2001) and Peruvian Amazonia (Couturier *et al.*, 1996).

Host plants

Arecaceae: *Astrocaryum aculeatissimum* (Schott) Burret (Návia *et al.*, 2007c), peach palm (*Bactris gasipaes* Kunth) (Couturier

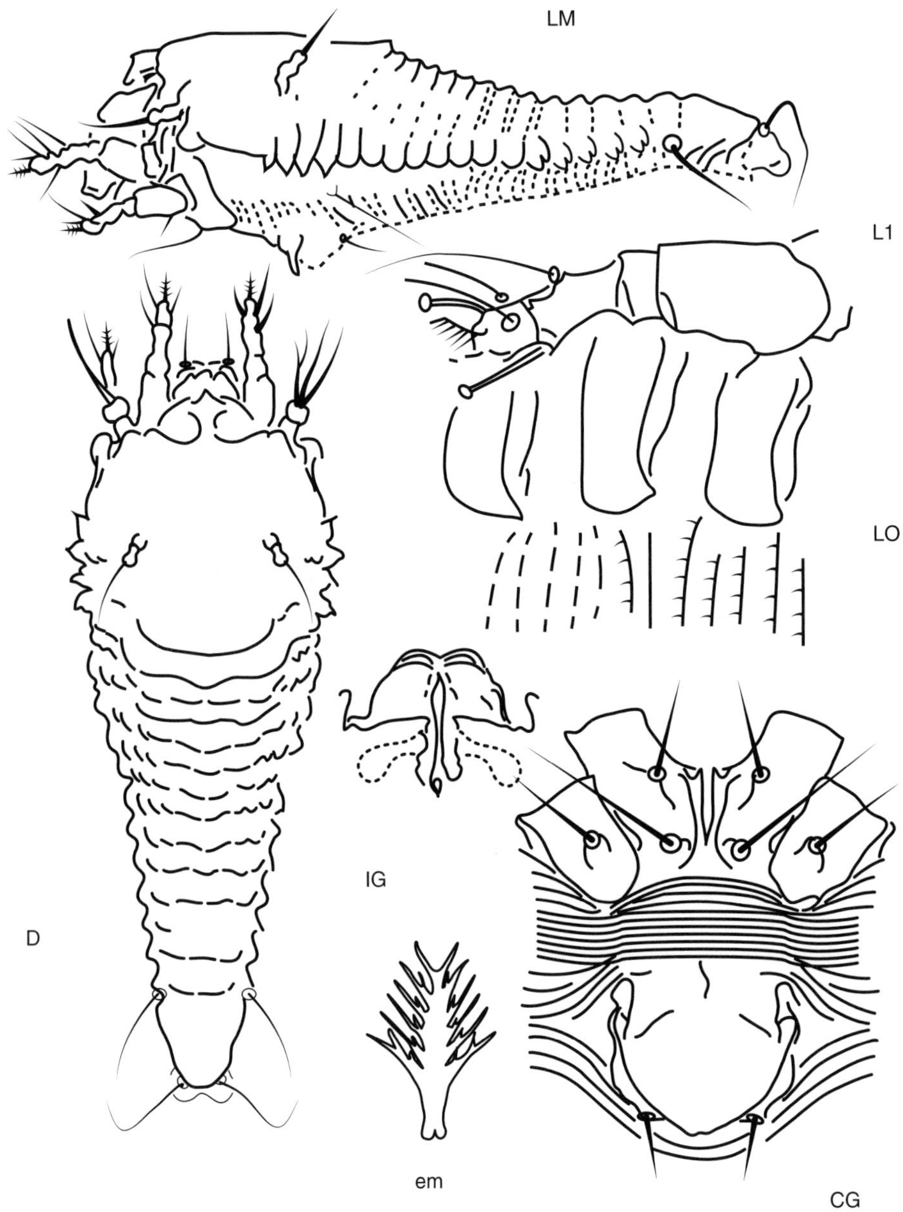

Fig. 17.9. *Retracrus johnstoni* Keifer (from Keifer, 1965d). See Table 16.1 for key.

et al., 1996; Návia *et al*., 2007c), *B. setosa* Martius (Návia *et al*., 2007c), *Chamaedorea* sp. (Ochoa *et al*., 1994), *Chamaedorea* sp. (var. Chiapas) (Keifer, 1965d), pacaya (*C. costaricana* Oerst.) (Ochoa *et al*., 1994), good luck palm (*C. elegans* Martius) (Schliesske 1988), coconut palm (*Cocos nucifera* L.) (Santana *et al*., 1994), African oil palm (*Elaeis guineensis* Jacquin), *Euterpe* sp., assai palm (*E. edulis* Martius), cabbage palm (*E. oleracea* Martius), açaí (*E. precatoria* Martius), *Geonoma gamiovora* Barb. Rodr., ouricana palm (*G. pohliana* Martius), *G. schottiana* Martius, tree of life (*Mauritia flexuosa* L.) (Návia *et al*., 2007c), date palm (*Phoenix dactylifera* L.) (El-Halawany *et al*., 2001), *Scheelea* sp., queen palm (*Syagrus romanzoffiana* (Cham.) Glassman) (Santana and Flechtmann, 1998).

Bio-ecology

R. johnstoni was intercepted in quarantine in Texas by D. Johnston of the US Department of Agriculture on *Chamaedorea* sp. (var. Chiapas); it had originated from Tuxtla or Chiapas (Mexico), about 150 miles north of the Guatemalan border (Keifer, 1965d). The pacaya mite is currently reported on about 20 plant species of the Arecaceae family (Keifer, 1965d; Schliesske 1988; Ochoa *et al*., 1994; Couturier *et al*., 1996; El-Halawany *et al*., 2001; Návia *et al*., 2007c). In Brazil, the mite has been collected on peach palm, coconut palm, African oil palm, queen palm and other Arecaceae (Santana *et al*., 1994; Santana and Flechtmann, 1998; Návia *et al*., 2007c); in Costa

Rica, the pacaya mite was found on good luck palm (Schliesske 1988), pacaya (Ochoa *et al.*, 1994) and peach palm (Návia *et al.*, 2007c), and in Egypt on date palm (El-Halawany *et al.*, 2001). *R. johnstoni* is sedentary and gregarious. Like other eriophyoids, high levels of populations appear as powdery puffs along the midrib of the leaves and on the lamina (Ochoa *et al.*, 1994). The incidence of the mite on date palm in Egypt is high in all governorates (El-Halawany *et al.*, 2001).

On sections of leaves of the palm tree under laboratory conditions each immature stage of *R. johnstoni* lasts about 7 days, corresponding to about 20.5 days for immature development. The sex ratio is about 2 females per male. The fecundity is slightly over 5 eggs per female. Immature and adult mites secrete a layer of white material, which covers their bodies. The spermatophores consist of a head of about 6 × 6 µm on a base about 6 µm long (Gondim and Moraes, 2003).

Natural enemies include the pathogenic fungi *Cladosporium* sp. and *Hirsutella* sp., both collected in Brazil (Van Der Geest *et al.*, 2002).

Symptomatology and damage

R. johnstoni lives on the underside of leaves and produces characteristic black spots that are appreciable on both surfaces (Keifer, 1965d). On pacaya and *Chamaedorea* sp., the damage occurs first at the edges of leaves and consists of dark and longitudinal patches. Around the bark patches a chlorotic halo forms, and green, chlorotic and necrotic areas are observed on the leaf surface. Leaves that are severely infested show diffuse necrosis and folding of the edges; white spots corresponding to mite colonies occur on the lower surface. Sometimes, the leaves turn completely yellow, but the mite populations remain in their definite patches. The distribution of damage on the plant is irregular and can be on both the upper and lower leaves (Ochoa *et al.*, 1994). Under laboratory conditions, dark spots appear on the lower surface of leaflets of the palm tree and yellowish spots on the corresponding regions of the upper surface, and the leaf spots become progressively larger and coalescent; the leaf dies and falls prematurely (Gondim and Moraes, 2003).

Control

Control can be necessary, and the species of this family are susceptible to sulfur-based products (Ochoa *et al.*, 1994).

Retracrus elaeis Keifer (Fig. 17.10)

Common name

Oil palm mite.

Diagnostic characteristics

FEMALE. The body is 154–166 µm long, about 40 µm wide and is white with wax. The gnathosoma projects down. The prodorsal shield has 4 pointed lobes laterally. The frontal lobe is bluntly acuminate from above, downcurved in side view, with a notch just under the top. The shield design is of faint longitudinal lines, and the shield is somewhat raised at the rear. The external vertical setae (*ve*) are located on pronounced tubercles; the setae have a bulbous base, and a beyond base are 15 µm long, stiff, with a terminal bulb. The scapular tubercles are 30 µm apart; the scapular setae (*sc*) are 15 µm long, with a bulbous base, stiff and a slight terminal bulb. The opisthosoma has about 14 dorsal annuli and approximately 46 ventral annuli. The seta c_1 is lacking. Dorsal annuli lack indications of microtubercles but are somewhat lobed laterally; the first dorsal annulus has a point on the lateral lobe. The ventral annuli are broader anteriorly but narrow towards the rear; anteriorly, the ventral annuli do not reach up to the dorsal annuli but are complete towards the rear. The microtubercles on the dorsal annuli are elongate and touching the rear ring margins, stronger to the rear; the microtubercles are confined to the central ventral area anteriorly but reach the dorsal annuli towards the rear. The seta c_2 is 25 µm long, set on about ventral annulus 3 behind the shield. The seta *d* is missing; the seta *e* is 13 µm long, set on ventral annulus 28. The ventral microtubercles are elongate, especially on the last ventral annulus; the seta *f* is 19 µm long; stiff and laterally placed. The genitalia are 22 µm wide, and 15 µm long, and deep bowl shaped; the coverflap has a transverse row of short basal longitudinal lines, otherwise with no ornamentation; the seta *3a* is 19 µm long. The empodium is divided into about 7 rays (Keifer, 1975a).

MALE. The body of the male is 144 µm long (Keifer, 1975a).

Geographical distribution

Colombia (Keifer, 1975a; Genty and Reyes, 1977), Costa Rica (Schliesske, 1988; Ochoa *et al.*, 1994; Návia *et al.*, 2007c) and Peruvian Amazon (Couturier *et al.*, 1996).

Host plants

Arecaceae: peach palm (*Bactris gasipaes* Kunth) (Ochoa *et al.*, 1994), pacaya (*Chamaedorea costaricana* Oerst.) (Návia *et al.*, 2007c), coconut palm (*Cocos nucifera* L.) (Genty and Reyes, 1977; Schliesske, 1988), African oil palm (*Elaeis guineensis* Jacquin) (Keifer, 1975a; Návia *et al.*, 2007c),.

Bio-ecology

The oil palm mite was first described from African oil palm in Colombia by Keifer (1975a). The mite has been reported on the same host plants in Costa Rica (Návia *et al.*, 2007c), and on other host plants, such as peach palm (Ochoa *et al.*, 1994) and pacaya in Costa Rica (Návia *et al.*, 2007c), and coconut palm in Colombia (Genty and Reyes, 1977) and Costa Rica (Schliesske, 1988). *R. elaeis* is a solitary mite and develops higher density populations on the lower and middle leaves. The symptoms of attack appear first at these sites. On peach palm, the higher densities of populations occur on higher, second and third leaves. Mite dispersal takes place via the wind and various mechanical actions, such as during pruning when a loss of 10–20% of the populations has been observed. No significant mobility was observed in the field or during daylight, and mite displacement probably occurs during the

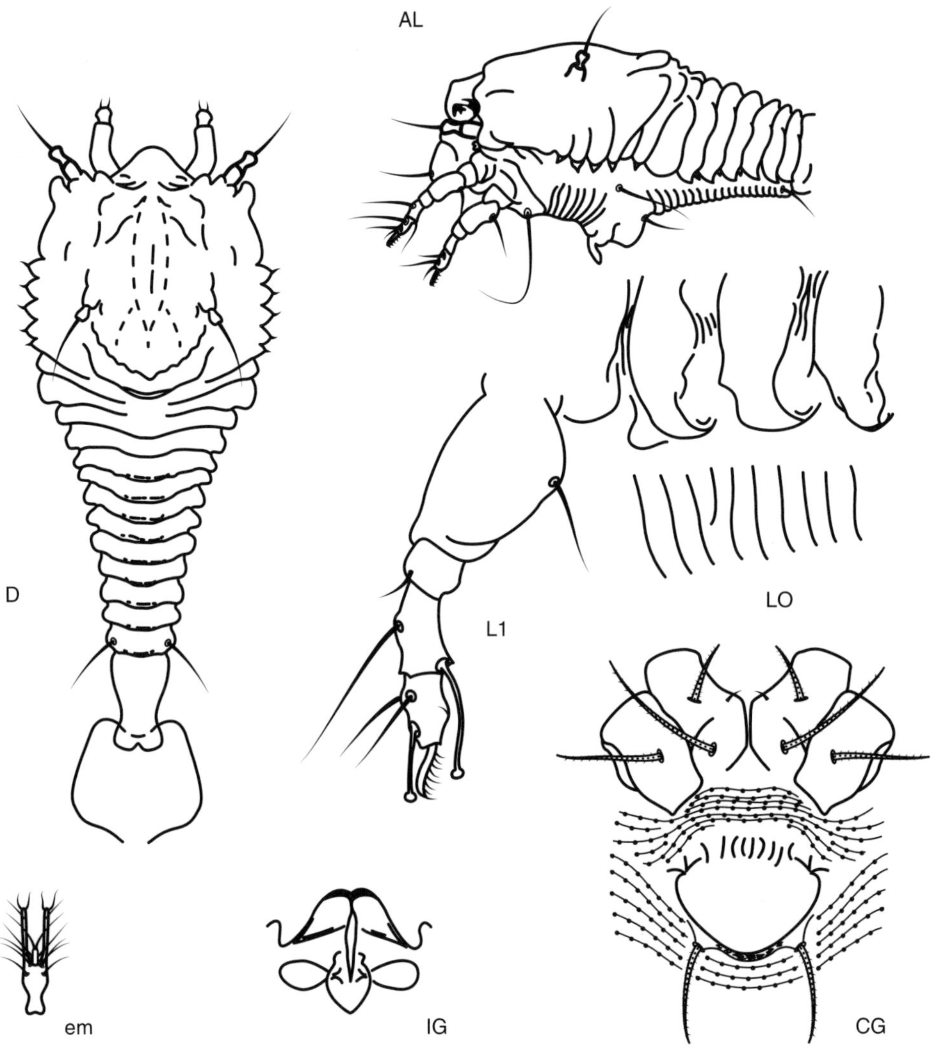

Fig. 17.10. *Retracrus elaeis* Keifer (from Keifer, 1975a). See Table 16.1 for key.

night (Ochoa *et al.*, 1994). Natural enemies include the entomopathogenic fungus *Hirsutella thompsonii* Fisher (Urueta, 1980).

Symptomatology and damage

R. elaeis speckles the fronds extensively with small black spots which tend to turn yellow or orange (Keifer, 1975a). Damage to the host plant may be mistaken for potassium or microelement deficiency, waterlogging or the presence of algae and fungi such as *Cercospora* sp., *Helminthosporium* sp., *Phyllosticta*, etc. (Ochoa *et al.*, 1994).

The feeding activity of the mite on oil palm initially produces small black patches on the lower and middle leaves. With maturing of the plant, these lesions turn yellow and join up, and the whole leaf becomes yellow or orange, followed successively by chlorosis and death of the plant (Ochoa *et al.*, 1994). On oil palm, mite attack can result in a serious loss of production, as recorded in Colombia, where a 50% of loss has been registered on badly attacked palms (Genty and Reyes, 1977).

On peach palm, the patches are yellow, the older leaf presents the most marked symptoms, and necrosis is observed at the tips and edges of the leaves. There is a differential susceptibility to the mite, with spineless varieties more susceptible than spined varieties (Ochoa *et al.*, 1994). In Brazil, it has also been observed that the mite causes numerous spots on leaves that are initially dark, gradually becoming yellowish or orange. Some infested leaves can dry completely. Attacked plants can have their yield impaired (Návia *et al.*, 2007c).

Control

Pathogens (Urueta, 1980) and chemical means have been used in control, including endosulfan, which gives good results (Genty and Reyes, 1977). The mite is susceptible to sulphur-based products (Mariau and Genty, 1987; Ochoa *et al.*, 1994).

18 The Family Diptilomiopidae Keifer

Morphological Characteristics, Systematics and Bio-ecology

The morphological characteristics of the family Diptilomiopidae are similar to those of the Eriophyidae. The prodorsal shield has two or no setae, the scapular setae (*sc*) are present or absent, and the unpaired setae *vi* and *ve* are largely absent. The gnathosoma is sharply bent towards the base, with cheliceral stylets folded in the same way and long oral stylets. The opisthosoma frequently lacks the setae c_1; the remaining setae are present, or sometimes any one of the setae c_2 or *d* or setae h_1 are absent. The chaetotaxy of the coxal plate is complete, plate I is sometimes without the setae *1b* and rarely with the setae *1a*. The leg chetotaxy is complete but may be missing the basiventral femoral setae I and II, the antaxial genual seta of genu II, the paraxial tibial seta of tibia I and both the paraxial and fastigial tarsal setae (*ft'*) or the paraxial and unguinal tarsal setae (*u'*) of legs I and II; the tibia I lacks a solenidion; the tarsal empodium may be thick and is commonly divided. The genital coverflap sometimes has one or two rows of ridges and spots or semilunar granules. The spermathecal tubes of the female genitalia are shorter than the diameter of the spermathecal sacs (Figs 18.1–18.3) (Lindquist 1996b; Lindquist and Amrine, 1996, Amrine *et al.*, 2003).

The family includes the subfamilies Diptilomiopinae Keifer, which is distinguished by having a usually deeply divided empodium, and Rhyncaphytoptinae Roivainen, which has an entire empodium (Lindquist and Amrine, 1996; Amrine *et al.*, 2003). The mites are leaf vagrants and rarely cause severe damage to their host plants (Jeppson *et al.*, 1975). Most species inhabit dicotyledonous plants, and only a few genera live on monocotyledons, mainly in the tropics (Oldfield, 1996a).

The Diptilomiopinae include the blackberry leaf mite, *Rhynacus abronius* (Keifer), collected on the undersurface of blackberry (*Rubus vitifolius* Cham. *et* Schltdl.) leaves along the Pacific Coast of North America (Keifer, 1939e). *Trimeroptes aleyrodiformis* (Keifer) is a vagrant on the undersurface of the leaves of American sweet gum (*Liquidambar styraciflua* L.) in North America, and has dorsal and lateral masses of wax (Keifer, 1940c). The big-beaked hickory mite, *Apodiptacus cordiformis* Keifer, which is also known as the white wax vagrant on the underside of hickory leaves in the eastern USA, lives on walnuts and other plant species as well (Keifer, 1960). The genus *Apodiptacus* includes several other species too: *A. acutissimae* Lin *et* Kuang, living on sawtooth oak (*Quercus acutissima* Carruth.) (Lin and Kuang, 1997), *A. castaneae* Liu *et* Kuang on *Castanea* sp. in China (Liu and Kuang, 1998), *A. liquidambarus* Boczek *et* Davis collected on American sweetgum (*Liquidambar styraciflua* L.) in Georgia (USA) (Boczek and Davis, 1990), and *A. rubi* Kuang on raspberry (*Rubus idaeus* L.) in China (Kuang, 1998). *Diptilomiopus assamica* Keifer causes rust of lemon (*Citrus limon* (L.) Burman) leaves in Australia (Knihinicki and Boczek, 2002) and India (Keifer, 1959c; Chakrabarti and Mondal, 1983; Dhooria *et al.*, 2005). *Diptacus gigantorhynchus* (Nalepa) inhabits *Prunus* and other rosaceous plants in the northern hemisphere and other regions. Other diptilomiopid mites are *Davisella globosa* (Keifer), a leaf vagrant on cashew (*Anacardium occidentale* L.) leaves in Brazil (Keifer, 1969a), the coconut frond vagrant *D. stellatus* (Keifer), collected on the underside of coconut (*Cocos nucifera* L.) leaves in the Philippines and forming stellate masses of white wax (Keifer, 1962c), and the rhododendron leaf vagrant *Vimola tampae* (Keifer), reported from rhododendron leaves in Florida (Keifer, 1963b). The Rhyncaphytoptinae include the corn rust mite, *Catarhinus tricholaenae* Keifer, reported from maize (*Zea mays* L.) and *Tricholaena rosea* Nees in Brazil, which probably causes rust on maize leaves (Keifer, 1959b; Flechtmann and Santana, 1997). The fig leaf mite, *Rhyncaphytoptus ficifoliae* Keifer, is the commonest species of the genus *Rhyncaphytoptus* Keifer. None of the above species cause severe damage to economic plants, and the most important mites of this family belong to the genera *Diptilomiopus* Nalepa, *Diptacus* Keifer and *Rhyncaphytoptus*.

Key for the Identification of Genera

1. • Empodium divided, usually deeply.................................. subfamily **Diptilomiopinae** Keifer....................................2
   • Empodium entire (subfamily **Rhyncaphytoptinae** Roivainen); coxal setae *1b* present; both legs with femoral setae; opisthosomal dorsal annuli larger than ventral; dorsum evenly rounded; empodia not lobed, with lateral rays; frontal lobe not emarginate.............genus ***Rhyncaphytoptus*** Keifer
2. • Prodorsal shield without scapular setae (*sc*); scapular tubercles present; tibial setae lacking; both legs without genu; coxal setae *1b* lacking......genus ***Diptilomiopus*** Nalepa
   • Prodorsal shield with scapular setae (*sc*) (may be small); tibial setae present; coxae with three setae; frontal lobe without filament; dorsal opisthosoma with a weak central ridge; some species produce wax............genus ***Diptacus*** Keifer

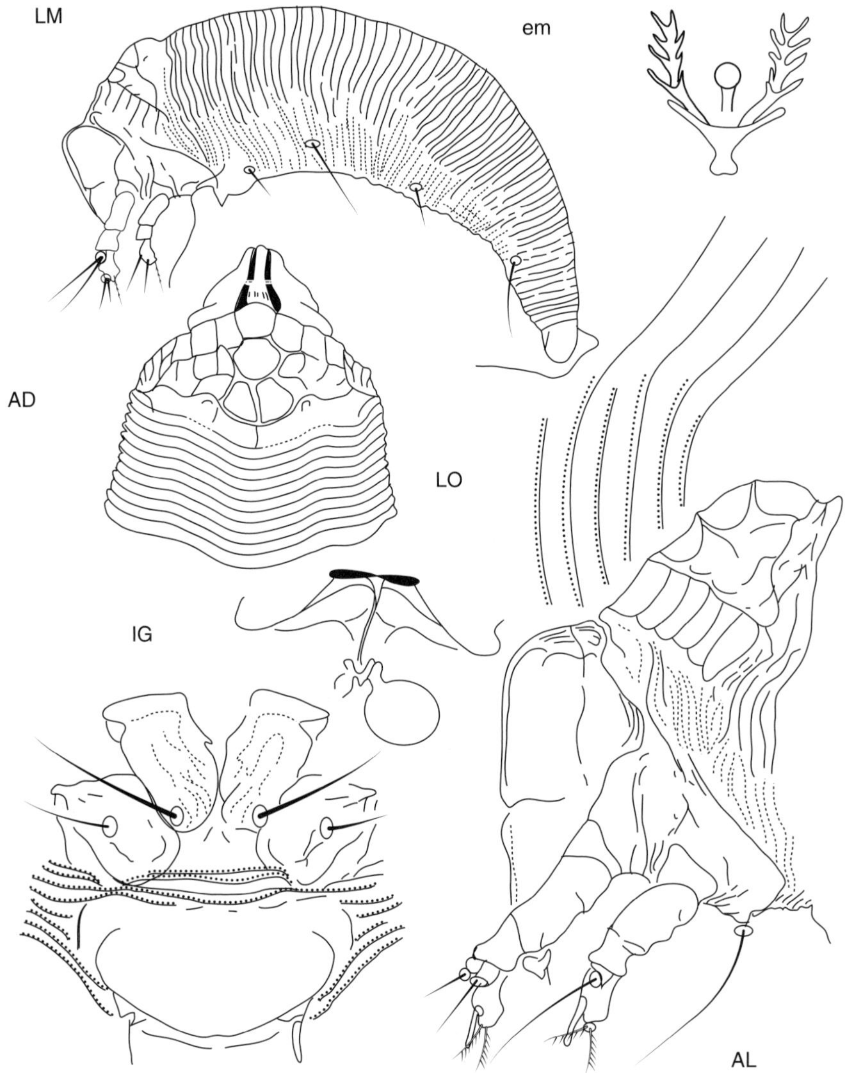

Fig. 18.1. *Diptilomiopus assamica* Keifer (from Keifer, 1959c). See Table 16.1 for key.

The Genus *Diptilomiopus* Nalepa

Diagnostic characteristics

Prodorsal shield without scapular setae (*sc*); coxae lack setae *1b*; both legs without genu (Amrine *et al.*, 2003).

Economic importance

The genus *Diptilomiopus* has over 90 species, mostly tropical and reported from Africa, Asia, the Indo–Malaysian region and Australia. The type species, *D. javanicus* Nalepa, is a leaf vagrant on *Hemigraphis confinis* T. Anders. in Africa, South Asia and Australia (Nalepa, 1916). The most important species is *D. assamica* Keifer, which causes rust of lemon leaves in Australia (Knihinicki and Boczek, 2002) and India (Keifer, 1959c; Chakrabarti and Mondal, 1983; Dhooria *et al.*, 2005). Other species are *D. bengalensis* Chakrabarti *et* Mondal, which causes rust and distortions to leaves of *Gardenia* in India (Chakrabarti and Mondal, 1979), *D. davisi* Keifer, a vagrant on the leaf undersides of macadamia nut (*Macadamia* spp.) in Queenland (Australia) (Keifer, 1969a), *D. jevremovici* Keifer, an undersurface leaf vagrant causing rust on coffee leaves (*Coffea arabica* L.) in Zaire (Keifer, 1960), and *D. ficus* Attiah, a leaf vagrant on figs in Egypt (Attiah, 1967a). Only *D. assamica* is of some economic importance.

Diptilomiopus assamica Keifer (Fig. 18.1)

Common name

Unknown.

Diagnostic characteristics

FEMALE. The body is fusiform, 215–230 μm long, 60 μm wide and light yellowish white in colour. The gnathosoma is moderately

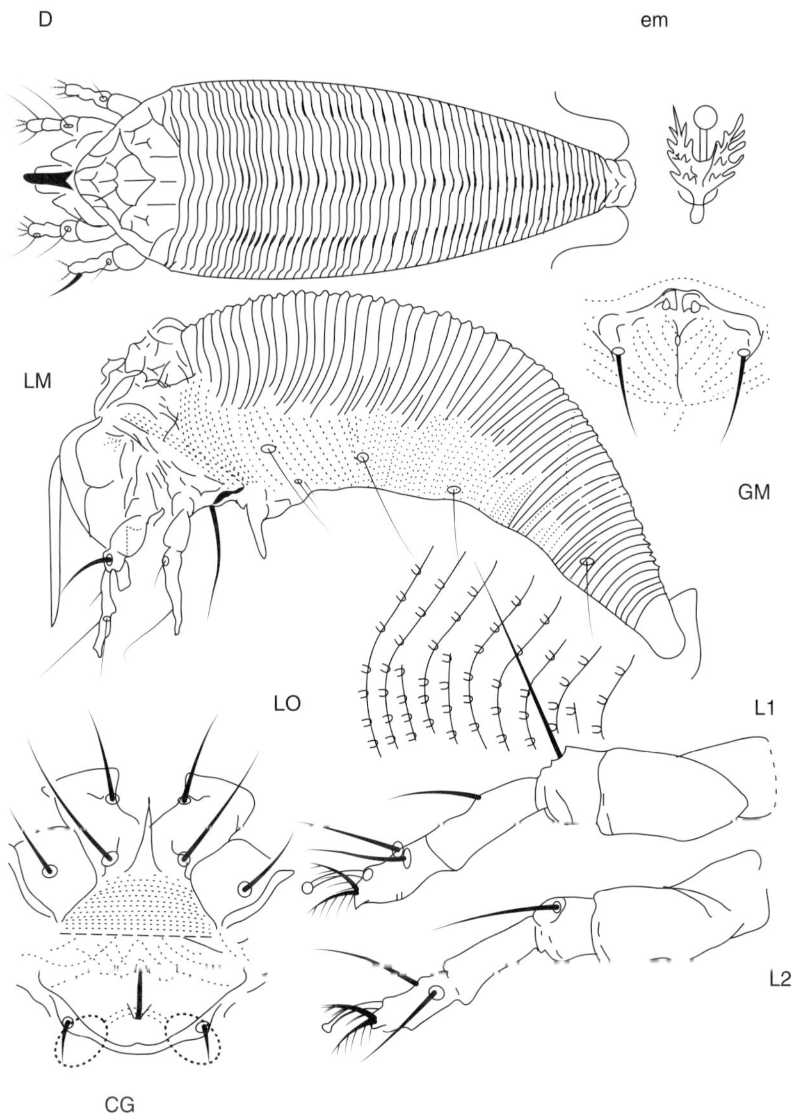

Fig. 18.2. *Diptacus gigantorhynchus* (Nalepa) (from Keifer, 1939a). See Table 16.1 for key.

attenuate. The prodorsal shield has a network of raised lines. There is almost no anterior shield frontal lobe, a median shield line is present at the rear only, admedian lines are complete, strongly angled at cross-line connections and joined by three transverse lines, the latter two joining the median line; the lower front and sides of the shield have a pattern of vertically elongate cells. The scapular tubercles, minus scapular setae (*sc*), are present within the rear shield margin, 20 μm apart. The opisthosoma has about 55–60 dorsal annuli and 70–75 ventral annuli; a shallow longitudinal furrow extends on each side of the middorsum, fading caudally; there are small ventral microtubercles on the ventral annuli touching the ring margins, which are absent dorsally. The setae c_2 are missing; setae *d* are 18 μm long, set on about ventral annulus 26; setae *e* are 15 μm long, set on about ventral annulus 45; setae *f* are 23 μm long, set on about ventral annulus 9 from the rear; setae h_1 are absent. The genital coverflap is smooth. The empodia are 5-rayed (Keifer, 1959c).

MALE. Unknown (Keifer, 1959c).

Geographical distribution

Australia (Knihinicki and Boczek, 2002) and India (Keifer, 1959c; Chakrabarti and Mondal, 1983; Dhooria *et al.*, 2005).

Host plants

Rutaceae: lemon (*Citrus limon* (L.) Burman), orange (Keifer, 1959c; Chakrabarti and Mondal, 1983; Knihinicki and Boczek, 2002; Dhooria *et al.*, 2005).

Bio-ecology

D. assamica is a leaf vagrant mite, commonly associated with other eriophyoids (*Aculops pelekassi* (Keifer), *Tegolophus australis*

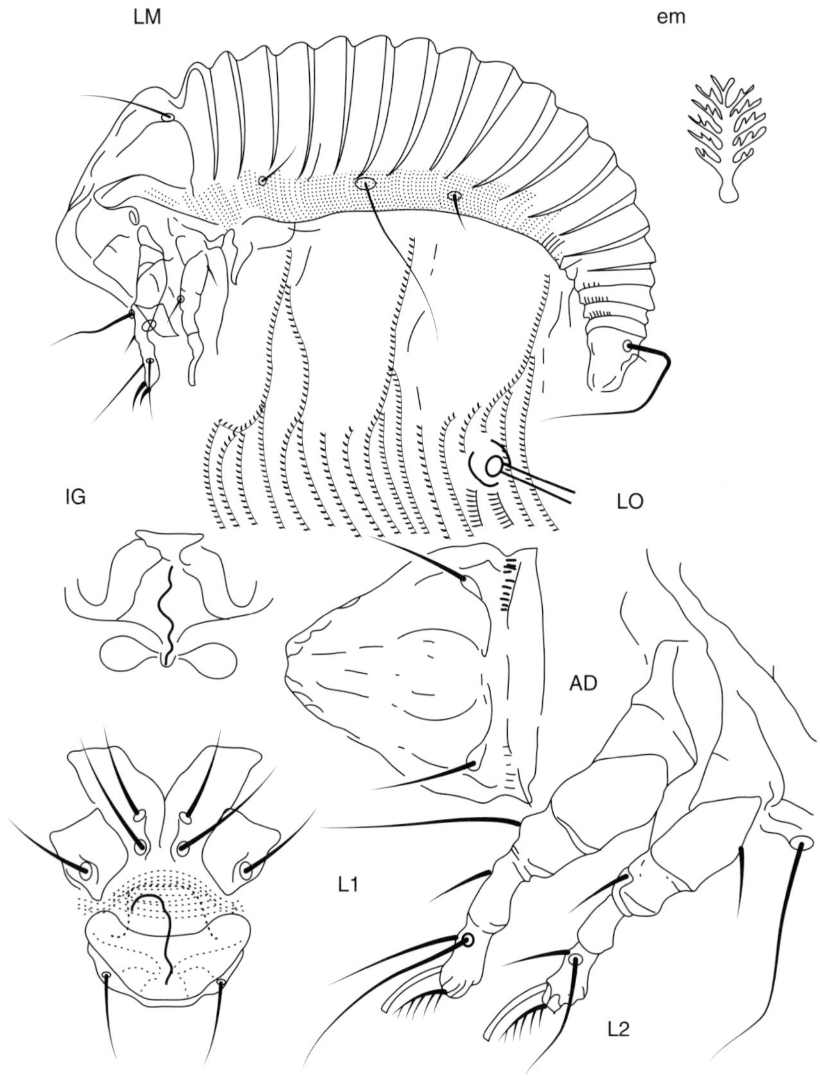

Fig. 18.3. *Rhyncaphytoptus ficifoliae* Keifer (from Keifer, 1939a). See Table 16.1 for key.

Keifer, *Cosella fleshneri* (Keifer)) and with *Phyllocoptruta oleivora* (Ashmead) on citrus (Seki, 1983; Smith and Papacek, 1991). It is reported from lemon and several varieties of orange in India (Keifer, 1959b) and Australia (Knihinicki and Boczek, 2002).

Symptomatology and damage

Keifer (1959c) reported *D. assamica* as a rust mite not causing serious damage. The leaves can show diffused russeting but control is unnecessary (McCoy, 1996b), and the pest status of the mite on citrus is fundamentally unknown (Gerson, 2003).

Control

Control is not usually necessary.

The Genus *Diptacus* Keifer

Diagnostic characteristics

The prodorsal shield has scapular setae (*sc*), which may be small. The frontal lobe is rounded distally. The dorsal opisthosoma has a weak central ridge. The coxae have 3 setae, and the tibial setae are present. Some species produce wax-like secretions (Amrine *et al.*, 2003).

Economic importance

The genus *Diptacus* is widespread and includes over 40 species, many of which have a Holarctic geographical range and

live mainly on dicotyledonous plants (Oldfield, 1996a). The most common species seems to be *D. gigantorhynchus* (Nalepa), living on leaves of *Prunus* and other rosaceous plants, but the damage that it causes needs to be well documented. *D. camarai* Carmona induces rust on leaves of laurel (*Laurus nobilis* L.) and *D. swensoni* Keifer causes browning on leaves of holly (*Ilex aquifolium* L.) (Keifer, 1959b), but even when the density of its populations are high, the damage is slight (Jeppson *et al.*, 1975). Many other species are reported in the literature but the damage that they may cause is unknown.

Diptacus gigantorhynchus (Nalepa) (Fig. 18.2)

Common name

Big-beaked plum mite.

Diagnostic characteristics

FEMALE. The body is robust and spindle formed, 200–210 µm long and dull light purplish or greyish purple in colour. The gnathosoma is large. The prodorsal shield is declivitous but a little overlying the gnathosoma base, with the frontal lobe very short. The shield design is a network of lines, of which the median line is complete, and broken by cross lines; the admedian lines more or less curve to join with the cross lines. The front and side cells are set in the lines. The rear prodorsal shield margin is elevated above the adjacent opisthosoma. The scapular tubercles are 50 µm apart, ahead of the rear margin, and the scapular setae (*sc*) are 6.75 µm long, projecting dorsoanteriorly. The opisthosoma has a shallow longitudinal subdorsal furrow, about 60 dorsal annuli, finely microtuberculate, and 80 ventral annuli, more heavily microtuberculate. The setae c_2 are 35 µm long, set on annulus 16; the setae *d* are 55 µm long, set on annulus 36; the setae *e* are 55 µm long, set on annulus 57 ; the setae *f* are 35 µm long, set on annulus 12 from the rear; the setae h_2 are 63 µm long and the setae h_1 are present. The genitalia are 37 µm wide and 20 µm long; the coverflap has no ridges; the seta is 9 µm long. The empodium is partly divided and 5-rayed (Keifer, 1939a; Jeppson *et al.*, 1975).

MALE. The body is 200 µm long, 65 µm wide. The genitalia are 25 µm wide, 18 µm long, and the seta is 8.5 µm long (Keifer, 1939a).

Geographical distribution

Armenia (Bagdasarian, 1970), Austria (Nalepa, 1892), Belgium (Paternotte and Soenen, 1982), Bulgaria (Nachev, 1978), California and Georgia (USA) (Keifer, 1939a, 1952a), Canada (Amano and Chant, 1986), Chile (Peralta, 1991), Finland (Roivainen, 1947), Germany (Schliesske, 1977b), Hungary (Györffyné, 1992), Italy (de Lillo and Beron, 2004), Lebanon (Bayan, 1989), Poland (Kozłowski, 1983), Spain (Roivainen, 1953), Sweden (Roivainen, 1950), Switzerland (Delley, 1972) and the USSR (Shvanderov, 1984).

Host plants

Rosaceae – almond and stone fruit (Rice and Jones, 1978), *Malus* spp. (Nachev, 1978; Amano and Chant, 1990; Schliesske, 1992), pear (*Pyrus*) (Shvanderov, 1984), plums (*Prunus* spp.) (Keifer, 1939a, 1952a; Schliesske, 1977b, 1992; Paternotte and Soenen, 1982; Kozłowski, 1983; Bayan, 1988, 1989; Peralta, 1991), myrobalan (*Prunus cerasifera* Ehrh.) (Schliesske, 1977b), sour cherry (*P. cerasus* L.) (Keifer, 1939a, 1952a), plum (*P. domestica* L.) (Nalepa, 1892; Keifer, 1939a, 1952a; Schliesske, 1977b), *P. domestica* subsp. *insititia* (L.) C.K. Schneid, aka mahaleb cherry (*P. mahaleb* L.) (Schliesske, 1977b), peach (*P. persica* (L.) Batsch) (Keifer, 1939a, 1952a; Paternotte and Soenen, 1982), blackthorn (*P. spinosa* L.) (Schliesske, 1977b), flowering almond (*P. trilobus* Lindl.) (Keifer, 1939a, 1952a), ornamental plum (Keifer, 1939a, 1952a; Schliesske, 1977b), blackberry (*Rubus vitifolius* Cham. *et* Schltdl.) (Keifer, 1952a); Vitaceae – native grape (*Vitis californica* Benth.) (Keifer, 1952a).

Bio-ecology

D. gigantorhynchus is a leaf vagrant species first described from plum in Austria (Nalepa, 1892). The mite seems to prefer rosaceous hosts, especially plums (Keifer, 1939a, 1952a; Schliesske, 1977b, 1992; Paternotte and Soenen, 1982; Kozłowski, 1983; Bayan, 1988, 1989; Peralta, 1991), but it has also been reported from different additional host plants, such as peach (Keifer, 1939a, 1952a; Paternotte and Soenen, 1982), flowering almond, Royal Ann cherry, ornamental plum (Keifer, 1939a), blackberry, native grape (Keifer, 1939a, 1952a), *Malus* spp. (Nachev, 1978; Amano and Chant, 1990; Schliesske, 1992), pear (Shvanderov, 1984), almond and stone fruit (Rice and Jones, 1978). In Germany, *D. gigantorhynchus* is the most common eriophyoid mite living on *Prunus* spp. grown as windbreaks and plants growing wild (Schliesske, 1977b).

The male deposits spermatophores mainly on the abaxial surface, commonly in the area between the mid-vein and the lateral veins (Oldfield *et al.*, 1970). The mite overwinters as deutogynes sheltered in bark crevices (Keifer, 1952a). In south Italy and Sicily, deutogynes have been found under the scales or in the axils of the buds of peach, from which they move in spring to feed on the new leaves. The same behaviour is reported from countries of central Europe, such as Switzerland (Delley, 1972) and Hungary, where overwintering colonies are commonly embedded between the eggs of the European red mite, *Panonychus ulmi* (Koch), and form mixed populations with dust mites and *Aculus fockeui* (Nalepa *et* Trouessart) (Györffyné, 1992). The big-beaked plum mite is found frequently associated on the same host plant with the latter mite. In plum orchards and nurseries in Switzerland, peaks of populations of *D. gigantorhynchus* occur from July to September (Delley, 1972, 1973). In Nova Scotia (Canada), peak numbers are observed in June or July (Herbert, 1974; Amano and Chant, 1990)

The natural enemies include phytoseiid and stigmaeid mites. The phytoseiids include *Typhloseiopsis citri* (Garman *et* McGregor) (Rice *et al.*, 1976), *Typhlodromus pomi* (Parrot), *Phytoseius macropilis* (Banks) and *Amblyseius finlandicus* (Oudemans) (Amano and Chant, 1986, 1990). The stigmaeids are *Zetzellia mali* (Ewing) (Amano and Chant, 1990) and *Z. talhouki* Dosse (Bayan, 1988).

Symptomatology and damage

The damage caused by *D. gigantorhyncus* is not well documented. Keifer (1939a) reported that the mite causes no apparent damage. Further studies on mite damage to the leaves of plum in orchards and nurseries in Switzerland highlighted withering of the leaves and, indirectly, an influence on both the size and quality of the fruits (Delley, 1972); on *P. domestica* and *P. insititia* greying and browning of the foliage are reported (Delley, 1973). Similarly, in Sweden and Spain, browning of leaves of *P. spinosa* has been observed (Roivainen, 1950, 1953).

Control

Control of this mite is not usually necessary. Moreover, the insecticidal treatments suggested for application to plum during the winter or at bud burst have no action on *D. gigantorhynchus*. Phosphorus compounds appeared to help an increase of eriophyoid species, while positive results were obtained with endosulfan, applied shortly before flowering and again at petal fall (Delley, 1972, 1973). In the USSR, infestations of *D. gigantorhynchus* on pear were controlled by spraying a suspension of colloidal sulfur, which was not phytotoxic and does not harm natural enemies (Shvanderov, 1984).

The Genus *Rhyncaphytoptus* Keifer

Diagnostic characteristics

The prodorsal shield has the frontal lobe not emarginate, and it is not separated from the 1st annulus. The dorsum is evenly curved, and the ventral annuli are more numerous than the dorsal annuli. The empodium has lateral rays (Amrine *et al.*, 2003).

Economic importance

Many of the described species of the family Rhyncaphytoptinae belong to the genus *Rhyncaphytoptus*, and are mainly reported from deciduous trees widely distributed in the Holarctic region. Various species live on Fagaceae of the genera *Castanea*, *Fagus* and *Quercus*, others on Rosaceae, and others on plants of the genera *Acer*, *Ulmus* or *Salix* (Oldfield, 1996a). The fig leaf mite, *R. ficifoliae* Keifer, is a vagrant on leaves of fig (*Ficus carica* L.) (Keifer, 1939a) in subtropical countries. *R. amplus* Keifer is a vagrant on the undersurface of leaves of Norway maple (*Acer platanoides* L.) in Maryland and causes rust on the leaves (Keifer, 1961b). The sycamore big-beaked leaf mite, *R. platani* Keifer, is a vagrant on the undersurface of leaves of *Platanus* sp. in California (Keifer, 1939b) and in central European countries, such as Hungary (Ripka *et al.*, 2005), Poland, where it lives on London plane (*Platanus acerifolia* (Aiton) Willd.) (Soika and Łabanowski, 1998b) and Germany, where it causes complete defoliation of this host plant (Schliesske and Billen, 1989). The elm leaf mite, *R. ulmivagrans* (Keifer), is a vagrant on the lower leaf surface of *Ulmus* sp. in California and causes no apparent damage (Keifer, 1939d). Usually, none of the above mites cause severe damage, and only the fig leaf mite is interesting from the phytopathological point of view. The mite does not cause any damage to the fig plants, but its wide presence on this host plant can lead to a false evaluation of its role and the conclusion that it requires chemical control.

Rhyncaphytoptus ficifoliae Keifer (Fig. 18.3)

Common name

Fig leaf mite.

Diagnostic characteristics

FEMALE. The body is elongate, 180–190 μm long, 45 μm wide and amber in colour. The prodorsal shield is nearly smooth, flanked by a curved line, anteriorly extending over the gnathosoma base; the scapular tubercles are 33 μm apart, slightly ahead of the rear margin, and the scapular setae (*sc*) are 18 μm long. The opisthosoma has 19–20 broad dorsal annuli, somewhat scale like and with spine-like microtubercles along the rear edge, but no subdorsal furrows; there are 75 ventral annuli as usual, the rear few distinct from the anterior annuli. The setae c_2 are 13 μm long, set on annulus 12; the setae *d* are 40 μm long, set on annulus 33; the setae *e* are 16 μm long, set on annulus 46; the setae *f* are 27 μm long, set on annulus 5 from the rear; the setae h_2 are 67 μm long; the setae h_1 are present. The genitalia are 27 μm wide and 14 μm long; the coverflap is smooth, and the seta is 14 μm long. The empodium is 6-rayed (Keifer, 1939a).

Geographical distribution

Worldwide.

Host plants

Moraceae: common fig (*Ficus carica* L.) (Keifer, 1939a).

Bio-ecology

R. ficifoliae is a vagrant mite on the undersurface leaf hairs of the fig tree. The mite is viviparous (Abou-Awad *et al.*, 1999b) and its immature stages are more or less covered with white flocculent wax (Keifer, 1939a). Its population overwinters as deutogynes between the bark crevices and under the hardened milky layer formed at the leaf scars. The deutogynes move towards the new leaves in late March and early April. They lay eggs on the leaves, and the hatching nymphs produce males and protogyne females which mate and reproduce. In the Mosul region of Iraq, the fig leaf mite began to appear on the leaflets of fig trees in April and reached peak numbers in the 3rd and 4th weeks of June. Population densities then decreased gradually and the mites disappeared completely in the 2nd and 3rd weeks of October. The new deutogynes appeared before the leaves fell. There was a positive correlation between temperature, predators and mite density, and a negative correlation with RH (Al-Mallah and Mohammad, 1989). On fig trees in

Egypt, the population of *R. ficifoliae* increased by 16.50 times in a generation time of 14.61 days (Abou-Awad *et al.*, 1999b).

Natural enemies include the tydeid *Pronematus ubiquitus* (McGregor), the phytoseiid *Amblyseius swirskii* Athias-Henriot and the stigmaeid mite *Agistemus exsertus* Gonzalez (Abou-Awad *et al.*, 1999b).

Symptomatology and damage

No damage was noted by Keifer (1939a) in California, as well as in other countries. Conversely, Jafari *et al.* (2015) reported that *Eotetranychus hirsti* Pritchard *et* Baker (Tetranychidae) and *R. ficifoliae* were serious pests of fig trees in the south and western regions of Iran, including Lorestan Province, which had negative effects on the yield of this tree, especially in summer.

Control

Control is usually unnecessary. However, Abou-Awad *et al.* (1999b) reported that one summer application of abamectin was the most successful control method.

19 The Family Tuckerellidae Baker *et* Pritchard

Morphological Characteristics, Systematics and Bio-ecology

The Tuckerellidae are usually referred to as 'peacock mites'. They are bright red in colour and have evident white setae, with the caudal setae long and plumose. The body consists of the propodosoma and hysterosoma. The dorsal surface shows the prodorsum, corresponding to the anterodorsal part of the propodosoma, and the opisthosoma, formed from the hysterosoma excluding legs III and IV. The prodorsum has four pairs of setae: the vertical internal (*vi*), vertical external (*ve*), scapular internal (*sci*) and scapular external (*sce*) setae (Fig. 19.1). The opisthosoma has 36 pairs of dorsal palmate setae and five or six pairs of flagelliform caudal setae set on four distinct dorsal plates (C, D, EF, H). The first plate (C) has two pores and a setal row (c_1–c_7), the second plate (D) has two pores and a setal row (d_1–d_5), the third plate (EF) has four pores and two setal rows (e_1–e_4) and (f_1–f_2) and the last plate (H) has two pores and a caudal row of setae (h_1–h_5 + h_7 or h_1–h_8). Setal rows *c*, *d* and *e* form an 'L' pattern, whereas rows *f* and *h* form a transverse pattern (Quirós-Gonzalez and Baker, 1984). *Tuckerella revelata* Beard *et* Walter has the *d* row consisting of six setae (instead of five) and an *e* row of three setae (instead of four) (Beard and Walter, 2005). The palp is elongate and bears a 'thumb-claw complex'. The distal end of the peritreme is similar to that of tetranychids (Fig. 19.2B). The female genital area has two pairs of aggenital setae (ag_1, ag_2), four pairs of genital setae (g_1–g_4) and three pairs of pseudanal setae (ps_1–ps_3) (Fig. 19.2C). The male has only one pair of genital setae (g_1). Both sexes have three pairs of intercoxal setae (ic_1, ic_3, ic_4) (Fig. 19.1). The coxa bears 2-2-1-1 setae, the trochanter 1-1-2-1 setae, the femur 7-7-2-1 setae, the genu 5-4-1-0 setae and the tibia 8-5-5-4 setae; tibia I has one dorsal solenidion. Tarsi I–IV bear 14, 11, 8 and 8 tactile and sensorial setae. The pretarsi have two true claws and a pad-like empodium with tenent hairs (Fig. 19.2D–E) (Quirós-Gonzalez and Baker, 1984).

The Tuckerellidae retain the ancestral prostigmatan condition of three nymphal stages during their development, but only the female developmental sequence retains a tritonymphal stage. Thus, adult females develop from a tritonymphal stage and males directly from the deutonymphal stage (Beard and Ochoa, 2010). *T. knorri* Baker *et* Tuttle exhibits paedogenesis, with eggs developing and emerging from tritonymphs (Ochoa, 1989a).

The family Tuckerellidae belongs to the superfamily Tetranychoidea, and includes only the genus *Tuckerella* Womersley, with nearly 30 species worldwide (Walter *et al.*, 2009). One species (*T. litoralis* Collier) has been collected in moss on boulders bordering a saltwater harbour soil (Collyer, 1969); others have been collected from the soil, probably associated with underground plant organs (*T. knorri* Baker *et* Tuttle, *T. hypoterra* McDaniel, Morihara *et* Lewis, *T. coleogynis* Jorgensen) (Jorgensen, 1967; Baker and Tuttle, 1975; McDaniel *et al.*, 1975) or from aerial habitat on some 25 plant families (Jorgensen, 1967; Zaher and Rasmy, 1969; McDaniel *et al.*, 1975; Ehara, 1975b; Jeppson *et al.*, 1975; Meyer Smith, 1981a; Ochoa, 1989b; Meyer Smith and Ueckermann, 1997). Three of these mites are reported on economic plants in China, e.g. *T. hainanensis* Lin *et* Fu, collected on *Coffea arabica* L., *T. jianfengensis* Lin *et* Fu, reported from *Anona muricata* L. (Lin and Fu, 1997a) and *T. xinglongensis* Lin *et* Fu, living on *Polyscias fruticosa* (L.) Harms var. *plumata* and *Camellia sinensis* (L.) Kuntze (Lin and Fu, 1997b). Four species (*T. knorri*, *T. pavoniformis* (Ewing), *T. nilotica* Rasmy *et* Abou-Awad, *T. ornata* (Tucker)) are reported on citrus worldwide (Vacante, 2010) and also on other plants, but they do not cause severe economic damage and do not usually require control measures, with the exception of *T. knorri* and *T. pavoniformis*, which damage citrus and other plants in Central America (Ochoa *et al.*, 1991a, 1994). Prasad (1973) reports that *Tuckerella* mites feed mainly on the stems of their host plants, and some species on fruits or sprouts, but this behaviour is ignored by most authors; the problem appears to be complex and needs further investigation.

Selected website

'Peacock mites', Tuckerellidae. Information from the Systematic Entomology Laboratory, US Department of Agriculture Agricultural Research Service, Beltsville, Maryland. Available at: http://www.sel.barc.usda.gov/acari/content/tuck1.html

The Genus *Tuckerella* Womersley

The diagnostic characteristics of the family can be applied to the genus *Tuckerella* (Meyer Smith and Ueckermann, 1997).

Tuckerella ornata (Tucker) (Figs. 19.1–19.2)

Common name

Unknown.

Diagnostic characteristics

FEMALE. The body is oval and dark red in colour, about 400 μm long and 200 μm wide. The dorsum is reticulate, the prodorsum has 4 pairs and the opisthosoma 18 pairs of leaf-shaped and silvery white setae; the caudal opisthosoma bears 5 pairs of

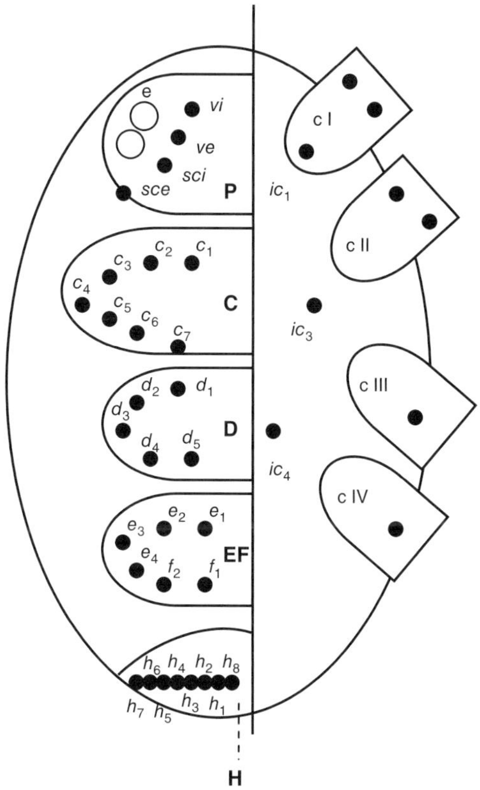

Fig. 19.1. Tuckerella ornata (Tucker). Diagram of dorsal (left side) and ventral (right side) showing the chaetotaxy of the female idiosoma according to Quirós-Gonzalez and Baker (1984). Key: e, eyes; c I–IV, coxae I–IV; C, D, EF, H, dorsal plates; P, prodorsum. The rest of the labels are setae, and the setal notation is explained in the text (from Vacante, 2010).

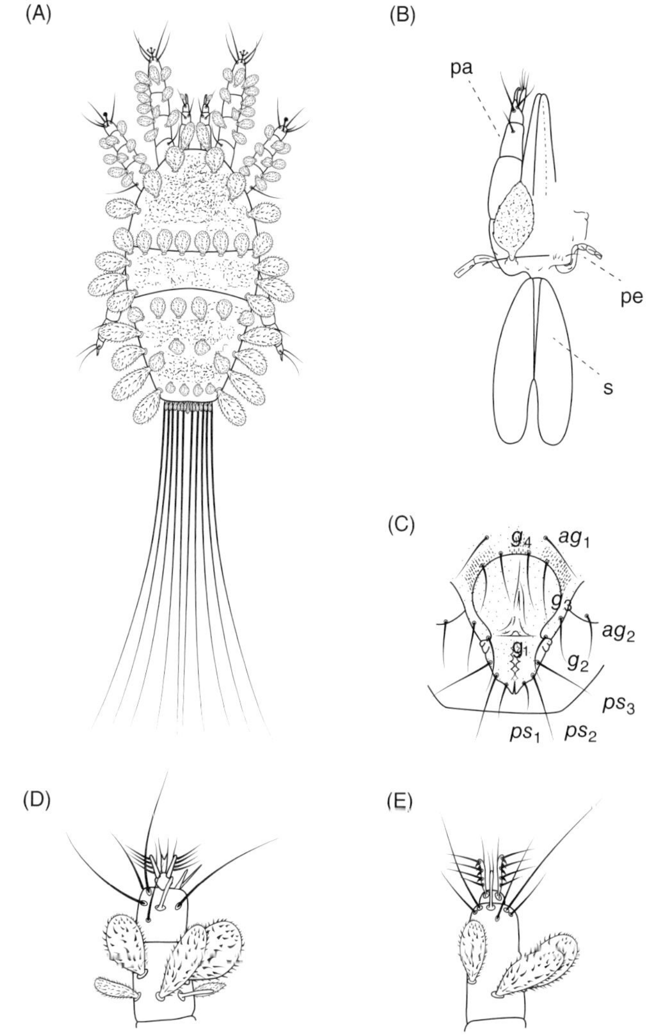

Fig. 19.2. *Tuckerella ornata* (Tucker). Female: (A) adult; (B) gnathosoma and stylophore; (C) genito-anal region; (D) tarsus and tibia I; (E) tarsus and tibia II (from Baker and Pritchard, 1953a). Key: pa, palp; pe, peritreme; s, stylophore. The rest of the labels are setae, and the setal notation is explained in the text.

flagellate caudal setae; the opisthosomal setae f_2 are situated slightly anterior to the setae f_1, and the setae f_{1-2} are about equal in length (Meyer Smith, 1981a; Meyer Smith and Ueckermann, 1997). The last 4 leaf-shaped prodorsal setae have the outer pairs smaller than the inner pair and placed caudally of the inner pair. The anterodistal solenidion on tarsus I is as long as but slenderer than the posterodistal solenidion. The tarsus II bears a slender solenidion anterodistally. The leg IV has slender setae dorsally (Baker and Pritchard, 1953a).

MALE. Unknown (Baker and Pritchard, 1953a).

Geographical distribution

Brazil (Flechtmann, 1979), California (Baker and Pritchard, 1953a), China (Lin, 1982), Costa Rica (Ochoa, 1989b), Florida (Baker and Pritchard, 1953a; De Leon, 1955), Guatemala (Baker and Pritchard, 1953a), Hawaii (Garrett and Haramoto, 1967), Kenya, Namibia (Meyer Smith and Ueckermann, 1997), Philippines (Baker and Pritchard, 1953a; Corpuz-Raros, 1989a), Réunion (Meyer Smith and Ueckermann, 1997), South Africa (Tucker, 1926; Baker and Pritchard, 1960; Meyer Smith and Ueckermann, 1997), Thailand, Venezuela (Quirós-Gonzalez and Baker, 1984) and Zimbabwe (Meyer Smith and Ueckermann, 1997).

Host plants

These belong to several families:

Anacardiaceae: mango (*Mangifera indica* L.) (Garrett and Haramoto, 1967; Lin, 1982; Meyer Smith and Ueckermann, 1997), Brazilian pepper (*Schinus terebinthifolius* Raddi) (De Leon, 1955).

Apocynaceae: Natal plum (*Carissa grandiflora* (E. Mey.) A. DC.) (= *Carissa macrocarpa* (Eckl.) A. DC.) (De Leon, 1955).

Aquifoliaceae: dahoon holly (*Ilex cassine* L.) (De Leon, 1955).

Asparagaceae: *Protasparagus* sp. (Meyer Smith and Ueckermann, 1997).

Asteraceae: *Pluchea dioscoridis* (L.) DC. (Meyer Smith and Ueckermann, 1997).

Burseraceae: gumbo-limbo (*Bursera simaruba* (L.) Sarg.) (De Leon, 1955).

Calophyllaceae: mammee fruit (*Mammea americana* L.) (Baker and Pritchard, 1953a).

Caricaceae: papaya (*Carica papaya* L.) (Garrett and Haramoto, 1967).

Celastraceae: *Maytenus heterophylla* (Eckl. *et* Zeyh.) N.K.B. Robson (Meyer Smith and Ueckermann, 1997).

Combretaceae: *Combretum hereroense* Schinz. (Meyer Smith and Ueckermann, 1997).

Cupressaceae: cypress, Chinese arborvitae (*Thuja orientalis* L.) (= *Platycladus orientalis* (L.) Franco) (Corpuz-Raros, 1989a).

Ebenaceae: *Euclea schimperi* (A. DC.) Dandy (Meyer Smith and Ueckermann, 1997).

Euphorbiaceae: *Hevea* cuttings (Baker and Pritchard, 1953a), rubber tree (*H. brasiliensis* Müll. Arg.) (Lin, 1982).

Fabaceae: *Bauhinia* sp. (Baker and Pritchard, 1953a), wiliwili (*Erythrina sandwicensis* O. Deg.) (Garrett and Haramoto, 1967), *Faidherbia albida* (Delile) A. Chev. (Meyer Smith and Ueckermann, 1997), tamarind (*Tamarindus indica* (L.)) (De Leon, 1955), *Xanthocercis zambesiaca* (Baker) Dumaz-le-Grand (Meyer Smith and Ueckermann, 1997).

Lauraceae: avocado (*Persea americana* Mill.) (De Leon, 1955).

Malpighiaceae: acerola (*Malpighia emarginata* A. DC.) (Barbosa *al.*, 2003).

Malvaceae: Chinese hibiscus (*Hibiscus rosa-sinensis* L.) (Lin, 1982), cacao tree (*Theobroma cacao* L.) (Flechtmann, 1979).

Moraceae: *Ficus thonningi* Blume (Meyer Smith and Ueckermann, 1997).

Myrtaceae: *Psidium* sp. (Corpuz-Raros, 1989a), guava (*Psidium guajava* L.) (De Leon, 1955).

Oleaceae: *Forsythia* sp. (Meyer Smith and Ueckermann, 1997).

Orchidaceae: *Phalaenopsis* sp. (Baker and Pritchard, 1953a).

Penaeaceae: *Penaea cneorum* Meerb. subsp. *cneorum* (Meyer Smith and Ueckermann, 1997).

Phyllanthaceae: *Bridelia micrantha* (Hochst.) Baill. (De Leon, 1955).

Pittosporaceae: *Pittosporum* sp. (Meyer Smith and Ueckermann, 1997).

Rubiaceae: *Breonadia salicina* (Vahl) Hepper and Wood (Meyer Smith and Ueckermann, 1997).

Rutaceae: *Citrus* (Baker and Pritchard, 1960), lemon (*C. limon* (L.) Burm.) (Meyer Smith, 1981a; Ochoa, 1989b; Meyer Smith and Ueckermann, 1997), pomelo (*C. maxima* (Burman) Merril), musk lime (*C. mitis* Blanco) (= *Citrofortunella mitis* (Blanco) J. Ingram *et* H.E. Moore) (De Leon, 1955; Garrett and Haramoto, 1967), sweet orange (*C. sinensis* (L.) Osbeck) (Meyer Smith, 1981a; Meyer Smith and Ueckermann, 1997).

Sapotaceae: zapodilla (*Achras zapota* (L.)) (De Leon, 1955).

Theaceae: tea *Camellia sinensis* (L.) Knutze (Meyer Smith and Ueckermann, 1997).

Vitaceae: *Rhoicissus* sp. (Meyer Smith and Ueckermann, 1997).

Bio-ecology

T. ornata has a wide range of host plants that includes almost 29 botanical families. The mite was first collected on orange fruit in the Western Cape Province and Transvaal in South Africa (Tucker, 1926), and later on several economic plants, such as zapodilla (De Leon, 1955), tea (Meyer Smith and Ueckermann, 1997), papaya (Garrett and Haramoto, 1967), sweet orange (Meyer Smith, 1981a; Meyer Smith and Ueckermann, 1997), musk lime, pomelo (De Leon, 1955; Garrett and Haramoto, 1967), lemon (Meyer Smith, 1981a; Ochoa, 1989b; Meyer Smith and Ueckermann, 1997), rubber tree, Chinese hibiscus (Lin, 1982), acerola (Barbosa *et al.*, 2003), mango (Garrett and Haramoto, 1967; Lin, 1982; Meyer Smith and Ueckermann, 1997), avocado (De Leon, 1955), *Phalaenopsis* sp. (Baker and Pritchard, 1953), guava, tamarind (De Leon, 1955) and the cacao tree (Flechtmann, 1979).

The bio-ecology of the species is not well documented, but could be similar to that of other congeneric species. Natural enemies include the phytoseiid mite, *Amblyseius largoensis* (Muma) (Corpuz-Raros, 1989a).

Symptomatology and damage

Tucker (1926) reported the mite was probably associated with scarring of citrus fruit, but Meyer Smith (1981a) pointed out that its population density was generally low and that the injury caused was not clear.

Control

This is not usually necessary.

Tuckerella knorri Baker *et* Tuttle (Fig. 19.3)

Common name

Ornamented mite.

Diagnostic characteristics

FEMALE. The body is oval, about 400 µm long and well reticulated; the colour in life is red with white setae. The prodorsal setae *ve* are about two-thirds the length of setae *vi*. The opisthosoma has the setae f_{1-2} subequal in length, the setae c_4 is about half as long as the distances between their bases and those of the setae *sce*; there are 6 pairs of flagellate caudal setae; the posterior opisthosoma has irregular reticulations. The distal solenidion of tarsus I is longer than the proximal solenidion (Baker and Tuttle, 1975).

MALE. Unknown (Baker and Tuttle, 1975).

Geographical distribution

China (Lin, 1982), Costa Rica (Ochoa, 1989b; Ochoa *et al.*, 1991a, 1994), Iran (Kamali, 1990), Philippines (Corpuz-Raros, 1989a) and Thailand (Baker and Tuttle, 1975).

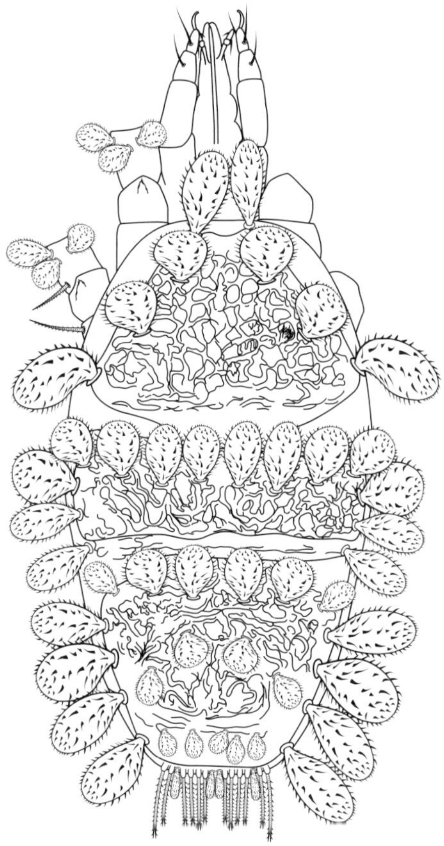

Fig. 19.3. *Tuckerella knorri* Baker *et* Tuttle. Female (from Baker and Tuttle, 1975).

Host plants

The host plants include several families: Anacardiaceae - mango (*Mangifera indica* L.) (Ochoa, 1989a,b; Ochoa *et al.*, 1991a, 1994); Annonaceae - Brazilian pawpaw (*Anona muricata* L.) (Lin, 1982); Asteraceae - *Elephantopus scaber* L. (Corpuz-Raros, 1989a); Caricaceae - papaya (*Carica papaya* L.) (Lin, 1982); Cupressaceae - cypress, Chinese arborvitae (*Thuja orientalis* L.) (= *Platycladus orientalis* (L.) Franco) (Corpuz-Raros, 1989a); Lauraceae - avocado (*Persea americana* Mill.) (Lin, 1982); Pandanaceae, *Pandanus odoratissimus* L. f. (Baker and Tuttle, 1975); Rutaceae - lemon (*Citrus limon* (L.) Burm. (var. 'Mesina')), sweet orange (*C. sinensis* (L.) Osbeck) (Ochoa, 1989a,b; Ochoa *et al.*, 1991a, 1994); Sapindaceae - lychee (*Litchi chinensis* Sonnerat) (Lin, 1982); Sapotaceae - sapodilla (*Achras zapota* (L.)) (= *Manilkara zapota* (L.) P. Royen) (Baker and Tuttle, 1975; Lin, 1982); Theaceae - tea (*Camellia sinensis* (L.) O. Kuntze) (var. 'assamica') (Lin, 1982).

Bio-ecology

T. knorri was first described on *Pandanus odoratissimus* L. roots and on sapodilla in Thailand (Baker and Tuttle, 1975). It is reported on plant species from 11 botanical families, including a number of economic plants, such as papaya, Brazilian pawpaw, tea, lychee, avocado (Lin, 1982), lemon, sweet orange, mango (Ochoa, 1989a,b; Ochoa *et al.*, 1991a, 1994), Chinese arborvitae and cypress (Corpuz-Raros, 1989a).

The life cycle of *T. knorri* develops through three nymphal stages, and paedogenetic tritonymphs lay the eggs (Ochoa, 1989a). On lemon, the mite lives on branches, where it feeds and moves in small depressions or cracks. It uses the crevices derived from its feeding activity for protection and oviposition (Ochoa *et al.*, 1994). When predatory mites approach *T. knorri*, it moves the flagellate setae forward (Ochoa, 1989b).

Symptomatology and damage

In Costa Rica, *T. knorri* is a pest of citrus and is often associated with the fungus *Sphaceloma fawcettii* Jenkins (Ochoa, 1989b). The mite is considered to be a causative agent of the cracking of citrus fruits (Aguilar and Gonzalez, 1990), producing irregular cracking of the epidermis of fruit infected by *S. fawcettii*, which changes the light brown colour of the fungus to brownish black. When the fungus is widespread, the ornamented mite is common around the lesions, occasionally changing the colour of the fungus to reddish brown. The combination of *T. knorri* and of the fungus causes more severe damage than that caused by the fungus and the tenuipalpid *Brevipalpus phoenicis* (Geijskes) (Ochoa, 1989b). In Costa Rica, heavy infestation of *T. knorri* on lemon caused a significant reduction in yield. On sweet orange (var. 'Valencia'), the symptoms are occasionally mistaken for fungal disease (Ochoa *et al.*, 1994).

Control

Control is applied in some countries, for example Costa Rica.

Tuckerella pavoniformis (Ewing) (Fig. 19.4)

Common name

Peacock spider mite.

Diagnostic characteristics

FEMALE. The body is oval and about 400 µm long, 200 µm wide and red with whitish orange setae. The dorsum is reticulate, the prodorsum has 4 pairs of setae and the opisthosoma 18 pairs of foliaceous or palmate setae. The prodorsum has the last pair of lateral palmate setae (*sce*) larger than the inner pair (*sci*) and the anterior pair (*vi*) is as wide as long. The caudal region of the opisthosoma has 6 pairs of long flagellate and pectinate setae. The anterodistal solenidion of tarsus I is very short compared with the posterodistal solenidion. Tarsus II bears one short solenidion (Baker and Pritchard, 1953a).

MALE. Unknown (Baker and Pritchard, 1953a)

Geographical distribution

Australia (Womersley, 1940), Brazil (Mineiro *et al.*, 2005), California (USA) (McGregor, 1950; Baker and Pritchard, 1953a), China, Taiwan (Wang, 1981a; Ma *et al.*, 1984), Costa Rica

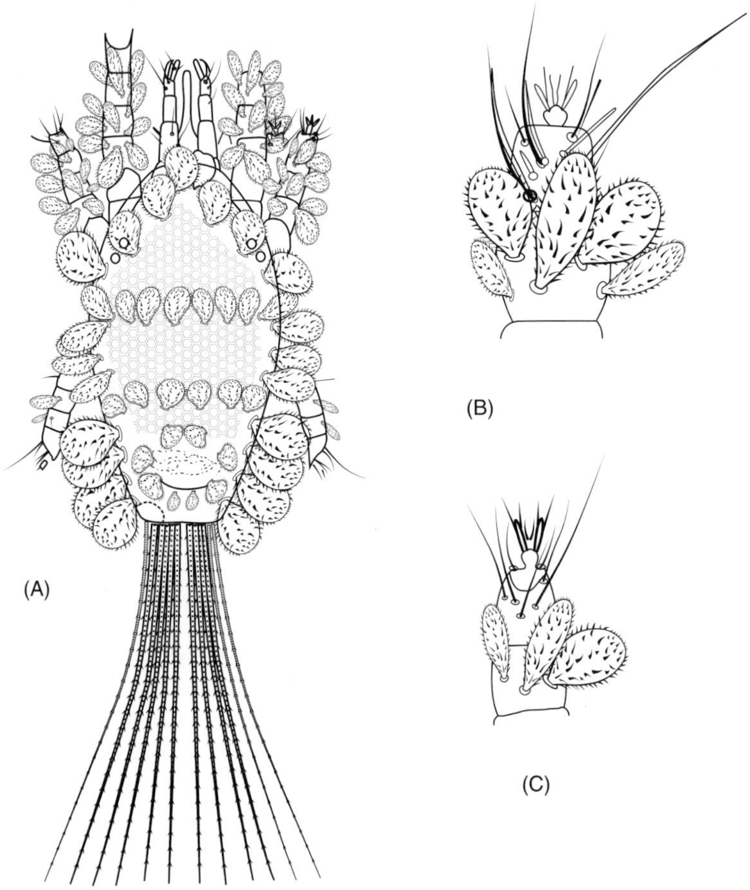

Fig. 19.4. *Tuckerella pavoniformis* (Ewing). (A) female; (B) tibia and tarsus I; (C) tibia and tarsus II (from Baker and Pritchard, 1953a).

(Ochoa, 1989b; Ochoa *et al.*, 1991a, 1994), Florida (Jeppson *et al.* 1975), Georgia (USSR) (Reck, 1959), Hawaii (Ewing, 1922), Japan and Formosa (Vietnam) (Ehara, 1966a) and Mauritius (Jeppson *et al.*, 1975).

Host plants

The host plants include several families: Caricaceae – papaya (*Carica papaya* L.) (Baker and Pritchard, 1953a; Jeppson *et al.*, 1975); Casuarinaceae – beach she-oak (*Casuarina equisetifolia* J. *et* G. Forst.) (Ehara, 1966a); Cupressaceae – cypress pine (Womersley, 1940); Lauraceae – avocado (*Persea americana* Mill.) (Ochoa, 1989b; Aguilar and Murillo, 2008); Malpighiaceae – acerola (*Malpighia glabra* L.) (Aguilar and Murillo, 2008); Malvaceae – hibiscus (Ewing, 1922; McGregor, 1950; Baker and Pritchard, 1953a); Myrtaceae – eucalyptus (Womersley, 1940, on an *Apiomorpha* gall; McGregor, 1950), Cattley guava (*Psidium cattleianum* Sabine) (Ochoa *et al.*, 1991a); Oleaceae – privet (Womersley, 1940; McGregor, 1950; Jeppson *et al.*, 1975); Pinaceae – pine (McGregor, 1950; Jeppson *et al.*, 1975), Japanese black pine (*Pinus thunbergii* Parlatore) (Ehara, 1966a); Rutaceae – citrus (McGregor, 1950); Sapindaceae – lychee (*Litchi chinensis* Sonnerat) (Mineiro *et al.*, 2005); Theaceae – tea (Jeppson *et al.*, 1975).

Bio-ecology

T. pavoniformis has a wide range of host plants, including various economic plants, such as papaya (Baker and Pritchard, 1953a; Jeppson *et al.*, 1975), citrus (McGregor, 1950), lychee (Mineiro *et al.*, 2005), avocado (Ochoa, 1989b; Aguilar and Murillo, 2008), tea (Jeppson *et al.*, 1975) and Cattley guava (Ochoa *et al.*, 1991a). The peacock spider mite also occurs on a variety of ornamental plants. Like *T. knorri*, it lays its eggs in the cracks of host plants, where adults and other mobile forms hide (Ochoa *et al.*, 1994). The bio-ecology of the species is not well documented.

Symptomatology and damage

T. pavoniformis is a pest of fruits and citrus plants in Central America (Ochoa *et al.*, 1994), but up to now the mite has never been associated with severe damage to citrus. On Cattley guava, the symptoms produced by the mite are characterized by cracks in fruits that can cover the entire surface (Ochoa *et al.*, 1991a)

Control

At present, control is used only in Central America (Ochoa *et al.*, 1994).

20 The Family Tetranychidae Donnadieu

Morphological Characteristics, Systematics and Bio-ecology

The bodies of the Tetranychidae are soft and more or less ovoid or round. The body is from 350 to 1000 μm long, variously coloured (red, orange, green or yellow), and consists of a gnathosoma and idiosoma. The dorsal disjugal and the ventral sejugal sutures divide the idiosoma into an anterior propodosoma and a posterior hysterosoma. The dorsal surface of the body consists of the prodorsum, corresponding to the anterodorsal part of the propodosoma, and the opisthosoma, i.e. the hysterosoma without legs III and IV (Fig. 2.9D).

The gnathosoma has a stylophore and two chelicerae. The chelicerae have a long and evenly thick digitus mobilis, except at the distal end, where they are pointed; if they are protracted and one side is juxtaposed with that of the other side, an empty tube is formed. The cheliceral bases are consolidated, curve ventrally and form the stylophore, which is able to protract and retract and has at its base two stigmata that are associated with the two peritremes (Fig. 2.6I), and lead anteriorly to the membranous prodorsum. The lower surface of the stylophore matches the back medial surface of the infracapitulum, and is conformed as a longitudinal hollow. The ventral surface of the infracapitulum shrinks to the distal end and constitutes the buccal cone, which hosts the mouth. The palp consists of 4–5 segments, including the 'thumb-claw', and bears simple setae, eupathidia and solenidia. The palp chaetotaxy is usually constant, with the exception of the palptarsus. One palpal supracoxal seta (*e*) is set at the base of the palpi. The trochanter lacks setae, and the femur and genu each have one seta, while the tibia has a dorsal claw and three setae; the palptarsus (thumb) bears three simple setae (*a*, *b*, *c*), three eupathidia (*ul'*, *ul''*, *su*) and one solenidion (ω) (Lindquist, 1985) (Fig. 2.7B). The eupathidium *su* (spinneret) is thicker than the other eupathidia and is involved in web production.

The prodorsum has three or four pairs of setae: the internal vertical setae (v_1), the external vertical setae (v_2), the internal scapular setae (sc_1) and the external scapular setae (sc_2). Between the scapular setae, there are two lateral prodorsal ocelli divided into two parts (Fig. 2.9B–D), each functioning differently from the other; the anterior part represents a biconvex eye and may act as a scanning-point detector for green and UV light, and the posterior part receives near-UV signals and is probably non-directional (Alberti and Crooker, 1985). Other variations may occur in the different groups. The integument can be striate, dotted or reticulate, or furnished with 1–4 projections that are ornate with setae (subfamily Bryobiinae). The setae may be arranged on tubercles (Fig. 2.9B).

The opisthosoma bears up to 14 pairs of dorsal setae, which are named according to the standard notation proposed by Grandjean (1939c). Commonly, only the setae of segments C (c_{1-3}), D (d_{1-3}), E (e_{1-3}), F (f_{1-2}) and H (h_{1-3}) are evident as dorsal setae. Except for the setae h_{2-3}, the opisthosomal setae are generally similar to one another and also to the prodorsals; they can be finely barbulate, pointed, setiform, clavate, spatulate or lanceolate, and all or partially set on tubercles. The integument is commonly similar to that of the prodorsum. The opisthosomal ventral region of the female has one pair of aggenital setae (*ag*), the genital opening, two pairs of lateral genital setae (g_{1-2}), and the anal opening, with 1–3 pairs of pseudanal setae (ps_{1-3}) (Figs 2.9C, 2.20A). The orifice (copulatory orifice) of the bursa copulatrix is located between the anal and genital openings, joined inside through a thin duct to the receptaculum seminis (Fig. 2.21F–L). In the male, the aedeagus is placed between the genito-anal valves (Fig. 2.19A–F, L); the shape and size of this organ are characteristics that are used in systematics for the discrimination of species (Lindquist, 1985).

The ventral podosomal region commonly bears three pairs of simple or intercoxal ventral setae (*1a*, *3a*, *4a*) (Fig. 2.9C). The legs consist of five articulating segments (trochanter, femur, genu, tibia and tarsus), which are not subdivided, fused or reduced. The general features of the setal notation of the legs are discussed in Chapter 2 (External Morphology). Further insights on the setal notation of tetranychids may be found in Lindquist (1985) and Gutierrez (1985a). Commonly, the maximum number of coxisternal setae for the four legs is 2-2-1-1. The supracoxal setae (*e*) are set dorsally on coxae I. The trochanter bears 1-1-1-1 setae (*v'*). On the femur and genu, there are additional setae present, and it is not possible to determine the correct notation of the two segments. The standard chetotaxy for the tibia of the Tetranychinae subfamily is 9 (+ 1 solenidion) -7-6-7, while in the Bryobiinae subfamily it ranges from 8 (+ 1 solenidion) -5-5-5 to 15 (+ 15 solenidia) -11-11-11, with some variation in the males. The tarsal chetotaxy of the Tetranychinae is variable and ranges from (i) 15 (+ 3 solenida) -14 (+2 solenidia) - 9 or 10 (+1 solenidion) - 10 (+1 solenidion) to (ii) 13 (+3 solenidia) -11 (+2 solenidia) - 9 (+1 solenidion) - 9 (+1 solenidion). In the Bryobiinae, there is greater variation; tarsi I and II bear duplex setae (Figs 2.11B, 2.22I) formed by a solenidion and a tactile seta. Further insights on the setal notation of legs may be found in Lindquist (1985) and Gutierrez (1985a). A short and flexible pretarsus (Fig. 2.16A, F) bears a pair of lateral structures derived from the true claws and a median empodium. These structures may change in different systematic groups (Lindquist, 1985).

The Tetranychidae family includes two subfamilies, Bryobiinae Berlese and Tetranychinae Berlese. The Bryobiinae

include three tribes (Bryobiini Reck, Hystrichonychini Pritchard and Baker and Petrobiini Reck), as do the Tetranychinae (Eurytetranychini Reck, Tenuipalpoidini Pritchard and Baker and Tetranychini Reck) (Gutierrez, 1985a; Meyer Smith, 1987; Bolland *et al.*, 1998). The family has a worldwide distribution and so far 1275 species belonging to 77 genera have been reported on about 3607 plants (Pritchard and Baker, 1955; Gutierrez, 1985a; Meyer Smith, 1987; Bolland *et al.*, 1998; Migeon and Flechtmann, 2004; Migeon and Dorkeld, 2006–2013). However, the exact number of species is unclear because the taxonomy of this group is not finally resolved, and additional species may be reported in new habitats and on new host plants. In addition, some species may include cryptic species. Members of this family have never been examined by cladistic analysis to analyse the history and relationships among the taxa, as in the case of the first phylogenetic analysis and taxonomic review of species of the genera *Aponychus*, *Paraponychus* and *Stylophoronychus* by Hernandes and Feres (2013). The results of this study may indicate a new systematic structure for the group.

The life cycle of the Tetranychidae develops through the biological stages of egg, larva, protonymph, deutonymph and adult, with a quiescent stage between each instar before moulting. André and Van Impe (2012) report that *Tetranychus urticae* Koch lacks an imago, a new name proposed for the last stase, and its post-embryonic development has the stases of prelarva (calyptostase), larva (six-legged stase), protonymph (eight-legged stase), deutonymph (eight-legged stase) and tritonymph. This last stase shows a precocious development of sexual maturity (paedogenesis). The process of losing a stase during ontogeny was named 'meristasy' by André (1988). According to the approach emphasized by Grandjean (1938, 1970) and by André (1988, 1989), André and Van Impe (2012) proposed the use of the word stase instead of the words instar, stade or stage. In their vision, the word stases is based on 'idionymy', and thus it is possible to homologize them. Thus, the same words are used to define ontogenetic levels, larva and adults (see also Chapter 4, Biology).

Regardless of this clarification and nomenclature, the life cycles of tetranychid mites vary according to both the species and ecological factors (population, climate, host plant, etc.). So, in tropical and subtropical areas, different species are active on their host plants throughout the year, and may stop their reproduction when temperatures are lower than their development threshold. Conversely, in temperate areas, some species overwinter in diapause. Diapause involves various aspects of biology and physiology (Veerman, 1985), and diapause and non-diapause forms may differ in coloration (Veerman, 1974), morphology and structure of the eggs (Lees, 1961) and of the adult females (Pritchard and Baker, 1952a; Boudreaux, 1958a). Variations in oxygen consumption (McEnroe, 1961; Geispitz and Orlovskaya, 1971) or in the accumulation of sorbitol (Sømme, 1965) confirm the involvement of physiology in diapause. Cold hardiness and RH are also reported to be involved (see Veerman, 1985). Diapausing and non-diapausing females of *T. urticae* exhibit a different tolerance to a number of pesticides (Bondarenko, 1958; Parr and Hussey, 1966; den Houter, 1976; Hürková and Weyda, 1982). Diapause has a genetic basis, and in tetranychid mites there is a strong diversity in photoperiodic responses that provides the means to adapt to changes in the weather, e.g. in greenhouses. Non-diapause characteristics are usually dominant (Geispitz, 1968; Helle, 1968). Almost 18 species of tetranychid mites produce diapause eggs. Five of these species have eggs that undergo aestival (summer) diapause and the rest undergo hibernal (winter) diapause. Moreover, in almost 25 species of tetranychids, winter diapause occurs in female forms, while aestival diapause occurs in only two species of female forms (Veerman, 1985).

The sexes are generally dimorphic, with the males small and pear like, and the females broader and oval. Males are absent in several species (mainly the Bryobiinae). Most species are arrhenotokous, and unmated females lay haploid eggs producing males. Within tetranychid populations, inbreeding is common between females and sons and between brothers and sisters. However, mostly there is no expression of deleterious mutations and/or death of the populations, as there is adequate genetic variability to accommodate to changes of conditions (Helle, 1968; Helle and Sabelis, 1985). A mating male places its body under the female and bends the idiosoma upward to insert the aedeagus into the copulatory orifice of the bursa copulatrix of the female to transfer the sperm. Mated females lay haploid and diploid eggs, which generate haploid males and diploid females, respectively. Mated females produce progeny with a sex ratio of 3:1 females to males. Development of the male is faster than that of the female. Both sexes exhibit precopulatory behaviour, followed by the act of mating. Male precopulatory behaviour seems more complex and is separated into several phases: aggression, attraction, guarding, webbing and territoriality (Cone, 1985). Males are stimulated to guard female deutonymphs by the emission of a contact sex pheromone. Different males may guard the same deutonymph and may fight between themselves, but they rarely cause severe damage; the stronger and aggressive male usually mates with the young female (Potter *et al.*, 1976a). Adult females also produce a sex pheromone and may mate more times until they have sufficient sperm to fertilize their eggs (Cone *et al.*, 1971a,b; Sonenshine, 1985; Royalty *et al.*, 1992, 1993; Rasmy and Hussein, 1994). In the laboratory, a male could inseminate 70 females and contribute to the generation of about 1145 daughters (Krainacker and Carey, 1989).

Tetranychid mites are commonly known as 'spider mites' owing to the tendency of many species of the subfamily Tetranychinae to make webs. In these mites, the silk glands consist of a unicellular gland located in each palpus and extending back to the central nervous mass (Alberti and Storch, 1974; Mothes and Seitz, 1981b). These cells are very large and contain many vacuoles filled with proteinaceous secretions. In the distal palpal segments, the droplets are contained in the reservoir-like area of the cell apex, which narrows and is connected with a terminal outlet at the tip of the spinneret. Mills (1973) reported that only the podocephalic complex of *T. urticae* was involved in silk production, but Alberti and Crooker (1985) suggested that the secretions of other glands might be involved as well. Webbing can protect eggs from mite predators, regulate the humidity around the eggs (Hazan *et al.*, 1975) and protect mites from wind and rain (Davis, 1952; Linke, 1953). Webbing species, such as *Tetranychus* spp., may displace other plant-feeding

mites, such as *Panonychus ulmi* (Koch), a weak spinner (Foott, 1962), or *Bryobia* sp., a typical non-spinning mite (Georgala, 1955). The Tetranychinae have evolved different behavioural patterns in response to the webs, and this has suggested that web characters are used to classify their mode of life into 'life types' (Saitō, 1983, 1985).

The dispersal of tetranychid mites occurs by locomotion between different parts of the same host plants and/or between plants, and by passive dispersal. Air currents, rain, irrigation water, agricultural operators, harvesting tools and various animals may be involved in passive dispersal.

Population dynamics are influenced by biotic and abiotic factors. Biotic factors include intrinsic and extrinsic factors, for example, density of populations, species of host plant, natural enemies (mites, insects, pathogens), etc. Abiotic factors include climate, fertilizers, pesticide treatments, etc.

The tetranychid mites are typical plant feeders and include the mites that are more harmful to economic plants. About 100 species in almost ten genera may cause economic damage to herbaceous and arborescent plants in crop and forest environments. A few species (*Eutetranychus orientalis* (Klein), *Oligonychus coffeae* (Nietner), *O. punicae* (Hirst), *Panonychus citri* (McGregor), *P. ulmi* (Koch), *Tetranychus kanzawai* Kishida, *T. pacificus* McGregor, *T. urticae*) are major pests in various countries and crops. *T. urticae* has worldwide distribution and is probably the tetranychid that is best known and feared. Other species, such as *Tetranycopsis horridus* (Canestrini *et* Fanzago), *Paraplonobia myops* (Pritchard *et* Baker), etc. cause damage that is either not well documented or sporadic. Some Bryobiinae, such as *Aplonobia* spp. and *Petrobia harti* (Ewing), are often associated with economic plants in that they lay their eggs in the wood, but they do not cause damage.

Plant damage caused by tetranychid mites is fundamentally direct and relates strictly to the feeding mechanism adopted by the mites, which actively involves the stylophore, the chelicerae and the palps. The main feeding activity of tetranychids occurs on leaves, but it may also involve other plant organs, such as cotyledons, fruits, flowers, tips of shoots, etc. The feeding results in the removal of the contents of cells of palisade tissue and the disappearance of the chloroplasts. The small punctures have a light colour, and with exposure to the air, assume the appearance of irregular white or greyish spots. The density of the punctures is a function of the size and duration of the attack and of the injurious species. Symptoms of feeding may vary according to species, anatomical structure of the infested organ, weather (during or after the infestation) and the reaction of the host plant to the attack. They are most common on leaves, particularly on the lower surface, but feeding by different mite species may involve both leaf surfaces. Cytological and histological aspects, the effects of mite saliva, and chemical and physiological changes determine the quality and amount of the mite impact on growth, flowering and cropping, i.e. the damage.

Natural enemies include pathogens and predator mites and insects. Control is provided by the use of beneficials, chemicals and other technical means. According to species, host plants and the environmental conditions, it is possible to integrate these means to implement an integrated pest management (IPM) strategy. Tetranychid mites may exhibit resistance to various acaricides. The phenomenon has been extensively investigated in *Tetranychus* (Helle, 1962; Overmeer, 1967; Schulten, 1968; Cranham, 1974; Osakabe, 1974) and *Panonychus* species (Cranham, 1982; Inoue, 1980). Genetic analysis of the arrhenotokous reproduction of *T. urticae* shows that single major genes control specific types of resistance (Stone, 1968). The arrhenotokous mechanism of reproduction permits the survival (and success) of the character of resistance to various chemicals via a state of homozygous recessive allele carriers. The role of mutations in resistance to organophosphates (Schulten, 1968) or in the induction of resistance to tetradifon (Overmeer, 1967) is well documented. The resistance genes have been mapped and the visible mutations in *T. pacificus* and *T. urticae* have been analysed (Helle, 1985). Major resistance mechanisms in mites include decreased penetration of acaricides through the cuticle, the greater degradation of acaricides via esterases, and the activity of cytochrome-P450-dependent monooxygenases and glutathione-*S*-transferases. In addition, a change in the target site is known (Knowles, 1997). The phenomenon is complicated by 'cross-resistance' and 'multi (multiple) resistance', the action of minor genes and negative cross-resistance (Cranham and Helle, 1985). Further information is presented in Chapter 9 (Control).

Further insights on the external anatomy, chaetotaxy and systematics of this family of mites can be found in Lindquist (1985) and Gutierrez (1985a). The key to subfamilies, tribes and genera given below follows Bolland *et al.* (1998). The taxa reported comply with the spider mites web of Migeon and Dorkeld (2006–2013). In the sections that follow on 16 tetranychid genera, under 'Host plants', only those of economic importance are reported for each mite species. Complete lists are available in Migeon and Dorkeld (2006–2013).

Selected websites

Alain Migeon and Franck Dorkeld (2006–2015) Spider Mites Web: a comprehensive database for the Tetranychidae. Available at: http://www1.montpellier.inra.fr/CBGP/spmweb/

Acari: Plant-Feeding Mites. Systematic Entomology Laboratory, US Department of Agriculture Agricultural Research Service, Beltsville, Maryland. Available at: http://www.sel.barc.usda.gov/acari/frames/plantfeed.html

Key for the Identification of Genera (Based on Females)

1. • Empodium with tenent hairs; female with 2 or 3 pairs of pseudanal setae (ps_{1-3}) and male with 5 pairs of genital (g_{1-2}) and pseudanal setae (ps_{1-3}) .... subfamily **Bryobiinae** Berlese...**2**

• Empodium without tenent hairs or absent; female with 1 or 2 pairs of pseudanal setae (ps_{1-2}) and male with 4 pairs of genital and pseudanal setae (ps_{1-2}, g_{1-2}) .... subfamily **Tetranychinae** Berlese...5

2. • True claw uncinate; empodium pad like (tribe **Bryobiini** Reck); opisthosoma with 12 pairs of dorsal setae; prodorsum with 4 anterior projections and 4 pairs of setae; empodia II–IV with more than 1 pair of tenent hairs ....genus ***Bryobia*** Koch
   • True claw pad like; empodium pad like or uncinate....3
3. • True claw and empodium pad like (tribe **Hystrichonychini** Pritchard and Baker).......................4
   • True claw pad like and empodium uncinate (tribe **Petrobiini** Reck); prodorsum without projections over gnathosoma; body with a maximum of 15 pairs of dorsal setae; 3 pairs of medioventrals; setae h_{2-3} in ventral position; empodium curved distally and with 2 rows of tenent hairs; true claw without hair-like processes ....genus *Petrobia* Murray
4. • Prodorsum with 4 pairs of setae; opisthosoma with 12 pairs of dorsal setae......genus *Tetranycopsis* Canestrini
   • Prodorsum with 3 pairs of setae; opisthosoma with 10 pairs of setae..................genus *Aplonobia* Womersley
5. • Empodium when present is claw like; tarsus I with loosely associated setae or with 1 pair of duplex setae; when tarsus I possesses 2 pairs of duplex setae, tarsus II is devoid ....tribe **Eurytetranychini** Reck...6
   • Empodium claw like or split distally; tarsus I with 2 pairs of duplex setae and tarsus II with 1 pair.....................7
6. • Empodial claw apparently lacking; two pairs pseudanal setae (ps_{1-2}); 10 pairs of dorsal opisthosomal setae............. ....genus ***Eutetranychus*** Banks
   • Empodial claw small; 2 pairs of para-anal setae (h_{2-3}) ....genus ***Eurytetranychus*** Oudemans
7. • Opisthosomal setae f_1 in marginal position or absent........tribe **Tenuipalpoidini** Pritchard and Baker
   • Opisthosomal setae f_1 in normal dorsal position......... ....tribe **Tetranychini** Reck...8
8. • Two pairs of setae h (h_{2-3})..9
   • One pair of setae h..15
9. • Empodium claw like...10
   • Empodium split near the middle into 3 pairs of hairs......13
10. • Empodium split into 2 claw-like structures...............11
   • Empodium a single claw-like structure.................12
11. • Distal segment of the palpus with 3 simple setae (a, b, c), 3 eupathidia (ul', ul'', su) and 1 solenidion (ω) ....genus *Schizotetranychus* Trägårdh
   • Distal segment of the palpus with 2 simple setae (a, c), 3 eupathidia (ul', ul'', su) and 1 solenidion (ω ....genus *Stigmaeopsis* Banks
12. • Empodial claw as long or longer than proximoventral hairs, which are at right angles to the claw......................... ....genus *Panonychus* Yokoyama
   • Empodial claw shorter than proximoventral hairs, which are at less than right angles to the claw............................ ....genus *Allonychus* Pritchard *et* Baker
13. • Opisthosoma with longitudinal striae between the e_1 setae; dorsal body setae serrate...................................... .... genus *Mononychellus* Wainstein
   • Opisthosoma with transverse striae..........................14
14. • Dorsal body setae much shorter than intervals between their bases.....................genus ***Platytetranychus*** Oudemans
   • Dorsal body setae as long or longer than the intervals between their bases............genus ***Eotetranychus*** Oudemans
15. • Empodium claw like with proximoventral hairs; duplex setae of tarsus I distal and adjacent; 2 pairs of anal setae (ps_{1-2}); opisthosoma with 10 pairs of dorsal setae; all legs or most of them with empodial claws as long or longer than the proximoventral hairs ....genus ***Oligonychus*** Berlese
   • Empodium split distally, usually into 3 pairs of hairs; duplex setae of tarsus I well separated..........................**16**
16. • Empodial spur generally visible; peritreme recurved distally..................................genus *Tetranychus* Dufour
   • Empodial spur absent; peritreme anastomosed distally ....genus *Amphitetranychus* Oudemans

The Genus *Bryobia* Koch

Diagnostic characteristics

The prodorsum has 4 pairs of setae, and 4 prominent lobes over the gnathosoma. The opisthosoma has 12 pairs of setae; the setae h_{2-3} are present. The empodia II–IV have more than 1 pair of tenent hairs (Bolland *et al.*, 1998).

Economic importance

This genus has worldwide distribution and includes 130 species (Migeon and Korkeld, 2006–2013) living on herbaceous and arborescent plants. Nomenclatural problems or misidentification make the identification of some species difficult. Worldwide, *Bryobia graminum* (Schrank) is reported on 27 herbaceous and fruit plants, including *Prunus dulcis* Miller, *Pyrus communis* L. and *Citrus aurantium* L. (Bolland *et al.*, 1998; Migeon and Dorkeld, 2006–2013), and it is occasionally injurious. *B. kissophila* Van Eyndhoven also occurs worldwide; the species is monophagous on *Hedera helix* L. (Van Eyndhoven, 1955) and has also been reported on *Buxus sempervirens* L. (Vierbergen, 1990) and *Ipomoea batatas* L. (Baker and Pritchard, 1962). The Palaearctic *Bryobia ribis* Thomas lives on *Ribes alpinum* L., *Ribes rubrum* L. (Thomas, 1896; Rambier, 1958a), *Ribes* sp. (Vierbergen, 1990), and *R. uva-crispa* L. (Migeon and Dorkeld, 2006–2013). The polyphagous *B. vasiljevi* Reck is reported on 22 host plants, including *Fragaria* sp., *Malus domestica* Borkh. (Migeon and Dorkeld, 2006–2013), *Lolium perenne* L. (Gutierrez and Schicha, 1983), *Trifolium* sp., *Vicia sativa* L. (Hatzinikolis and Emmanouel, 1991), *Medicago sativa* L. (Gutierrez and Schicha, 1983; Hatzinikolis and Emmanouel, 1991), *Triticum aestivum* L. and other plants (Migeon and

Dorkeld, 2006–2013); the mite may cause bleaching of the blade resulting, together with dry climate, in severe damage (Manson, 1967a). The clover mite, *B. praetiosa* Koch, and the brown mite, *B. rubrioculus* (Scheuten), cause the most important damage and are described below.

Bryobia praetiosa Koch (Fig. 20.1)

Common name

Clover mite.

Diagnostic characteristics

FEMALE. The body is 770–785 µm long, dark red in colour, sometimes greenish, with lighter edges and legs. The prodorsal lobes are subtriangular and well separated from each other; the inner lobes are higher than the outer lobes and the base is wider than high; the outer lobes are higher than wide. All the dorsal setae are arranged on tubercles, spatulate serrate and longer than wide. The peritremes end in an anastomosis about 50 µm long. Pretarsus I bears 2 true claws, with a pair of tenent hairs each and a median empodium with 2 tenent hairs. Tarsi III and IV have 2 duplex setae, with the solenidion longer than the tactile seta (Vacante, 1985a).

MALE. Unknown (Vacante, 1985a).

Geographical distribution

Worldwide (Bolland *et al.*, 1998; Migeon and Dorkeld, 2006–2013).

Host plants

The clover mite is reported on 268 wild and cultivated plant species belonging to 70 families, with the highest number of host plants (in decreasing order) in the families Asteraceae (31), Poaceae (27), Rosaceae (23) and Fabaceae (20) (Migeon and Dorkeld, 2006–2013). Families and species of economic importance are listed below:

Amaranthaceae: spinach (*Spinacea oleracea* L.) (Bolland *et al.*, 1998; Migeon and Dorkeld, 2006–2013).

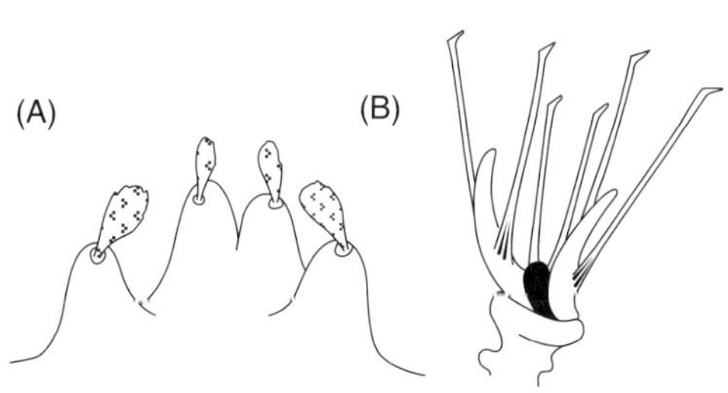

Fig. 20.1. *Bryobia praetiosa* Koch. Female: (A) prodorsal projections; (B) pretarsus I (modified from Vacante, 1985a).

Apiaceae: carrot (*Daucus carota* L.) (Bolland *et al.*, 1998; Migeon and Dorkeld, 2006–2013).

Anacardiaceae: pistachio nut (*Pistacia vera* L.) (Bolland *et al.*, 1998; Migeon and Dorkeld, 2006–2013).

Asteraceae: sunflower (*Helianthus annuus* L.) (Zaher *et al.*, 1982).

Brassicaceae: black mustard (*Brassica nigra* (L.) W.D.J. Koch) (McGregor, 1950), cabbage (*B. oleracea* L.), field mustard (*B. rapa* L.) (Meyer Smith, 1974).

Cannabaceae: common hop (*Humulus lupulus* L.) (Bolland *et al.*, 1998; Migeon and Dorkeld, 2006–2013).

Caryophyllaceae: carnation (*Dianthus caryophyllus* L.), baby's-breath (*Gypsophila paniculata* L.) (Tuttle and Baker, 1964).

Cucurbitaceae: watermelon (*Citrullus lanatus* (Thunb.) Matsum. and Nakai), cantaloupe (*Cucumis melo* L.) (McGregor, 1950), garden cucumber (*C. sativus* L.) (Bibby and Tuttle, 1959; Tuttle and Baker, 1964), crookneck squash (*Cucurbita moschata* Duchesne), field pumpkin (*C. pepo* L.) (Tuttle and Baker, 1968).

Fabaceae: Bermuda grass (*Cynodon dactylon* (L.) Pers.) (Zaher *et al.*, 1982), lucerne (*Medicago sativa* L.) (McGregor, 1950; Bibby and Tuttle, 1959; Rossi Simons, 1961; Tuttle and Baker, 1964; Hatzinikolis, 1968b, 1969c), bean (*Phaseolus vulgaris* L.) (McGregor, 1950; Rossi Simons, 1961), garden pea (*Pisum sativum* L.) (McGregor, 1950; Rossi Simons, 1961), Egyptian clover (*Trifolium alexandrinum* L.) (Zaher *et al.*, 1982), small clover (*T. dubium* Sibth.), red clover (*T. pratense* L.), Dutch clover (*T. repens* L.) (Tuttle and Baker, 1964), bird vetch (*Vicia cracca* L.) and fava bean (*V. faba* L.) (Rossi Simons, 1961).

Grossulariaceae: cultivated currant (*Ribes rubrum* L.) (Thomas, 1896).

Juglandaceae: English walnut (*Juglans regia* L.) (McGregor, 1950).

Moraceae: fig (*Ficus carica* L.) (Bolland *et al.*, 1998; Migeon and Dorkeld, 2006–2013).

Passifloraceae: purple granadilla (*Passiflora edulis* Sims) (Meyer Smith, 1974), giant granadilla (*P. quadrangularis* L.) (Bolland *et al.*, 1998; Migeon and Dorkeld, 2006–2013).

Poaceae: barley (*Hordeum vulgare* L.) (McGregor, 1950), sugarcane (*Saccharum officinarum* L.), cereal rye (*Secale cereale* L.) (Bibby and Tuttle, 1959; Tuttle and Baker, 1968), wheat (*Triticum aestivum* L.) (McGregor, 1950).

Rosaceae: quince (*Cydonia oblonga* Miller) (Rossi Simons, 1961), loquat (*Eriobotrya japonica* (Thunb.) Lindl.), strawberry (*Fragaria* (L.) Mill.) (= × *ananassa* Duchesne ex Rozier) (Ehara, 1956a), woodland strawberry (*F. vesca* L.) (McGregor, 1950; Bibby and Tuttle, 1959; Rossi Simons, 1961), apple (*Malus domestica* Borkh.) (McGregor, 1950; Ehara, 1956a; Bibby and Tuttle, 1959; Rossi Simons, 1961), apricot (*Prunus armeniaca* L.) (Rossi Simons, 1961), sweet cherry (*P. avium* (L.) L.) (McGregor, 1950; Rossi Simons, 1961), European plum (*P. domestica* L.) (Rossi Simons, 1961; Gupta and Gupta, 1994; Gupta and Chatterjee, 1997), sweet almond (*P. dulcis* (Mill.) D.A. Webb) (Bibby and Tuttle, 1959; Georghiou, 1959; Rossi Simons, 1961), peach (*P. persica* (L.)

Batsch) (McGregor, 1950; Rossi Simons, 1961), black cherry (*P. serotina* Ehrh.) (McGregor, 1950; Bibby and Tuttle, 1959; Rossi Simons, 1961; Gupta and Gupta, 1994; Gupta, 1995; Gupta and Chatterjee, 1997), pear (*Pyrus communis* L.) (McGregor, 1950; Bibby and Tuttle, 1959; Rossi Simons, 1961; Gupta and Gupta, 1994; Gupta, 1995; Gupta and Chatterjee, 1997).

Rutaceae: tangerine (*Citrus reticulata* Blanco) (Rossi Simons, 1961), sweet orange (*C. sinensis* (L.) Osbeck) (McGregor, 1950).

Sapindaceae: lychee (*Litchi chinensis* Sonn.) (Bolland *et al.*, 1998; Migeon and Dorkeld, 2006–2013).

Solanaceae: tomato (*Solanum lycopersicum* L.), aubergine (*S. melongena* L.), Irish potato (*S. tuberosum* L.) (Meyer Smith, 1974).

Vitaceae: grapevine (*Vitis vinifera* L.) (Ehara, 1959).

Bio-ecology

B. praetiosa lives on wild and cultivated plants. It is fundamentally a pest of ornamental flowers, vegetables, legumes, lawn grasses and cereals, but sometimes also occurs on arborescent plants, such as ornamentals, deciduous shrubs, deciduous fruit trees, and forest and other plants. The mite reproduces by thelytokous parthenogenesis induced by *Wolbachia* (Weeks and Breeuwer, 2001) and has a haploid number of four chromosomes (Helle *et al.*, 1970). *B. praetiosa* is active at high temperatures under very dry conditions. At an RH of 53–85%, it is able to regulate water loss, but at an RH below 53%, water loss is not regulated, although it is restricted. Well-fed mites prefer low RH, but somewhat desiccated mites prefer a high RH (Winston, 1964). The most favourable conditions for development occur between 21 and 24°C (Mori, 1961), and a better temperature for oviposition and egg hatching is from 8 to 18°C; the period of development decreases below 7°C, and increases above 18°C, with a little hatching above 30°C (Meyer Smith, 1981a). In warmer regions, the clover mite overwinters as the egg, immature and adult stages. In spring, the winter eggs hatch and begin the spring generation, simultaneously with the generation derived from eggs laid by emerging overwintering females. Overwintered larvae and nymphs mature in the spring and their progeny represent a delayed portion of the spring generation. The mite oversummers in the egg stage; the eggs are laid in terrain cavities and cracks, in the walls of dwellings and on the bark of trees. Hatching occurs when the temperature decreases (late September in Illinois) and continues until early winter, when all of the stages overwinter (English and Snetsinger, 1957a). In north Italy, on wall rocket (*Diplotaxis tenuifolia* (L.) DC.), overwintering eggs of *B. praetiosa* hatch in spring at the same time as overwintering adults emerge and begin to lay eggs, some of which rapidly hatch. Peak numbers are reached in May. Most of the spring eggs hatch in autumn (Laffi, 2001). At temperatures between −2 and 8°C, the eggs may hatch after a week; overwintering eggs may hatch in 12–18 h at temperatures ranging from 18 to 24°C. A small proportion of the eggs laid in spring do not oversummer and hatch in the early summer, giving continuous generations; in British Columbia (Canada), the oversummer eggs laid in May and June may hatch in late summer or at the first decrease in temperature; the peak of hatching may follow the first decrease in temperature (Anderson and Morgan, 1958; Gäbele, 1959).

During the early spring and occasionally in the autumn, houses that are near clovers, grass, weeds or other host plants may be invaded by *B. praetiosa*. Hibernating mites and mites originated from overwintering eggs penetrate into the buildings through cracks in the flour, walls, etc. to find shelter to moult or lay eggs. Although the mite is not directly harmful to human health, its presence alarms people, especially when it is present at high density (Van Eyndhoven, 1954; Jeppson *et al.*, 1975).

Occasionally, the clover mite lays eggs at the base of the trunk of arborescent plants or occurs as immature and adult stages on the leaves of the lower foliage, but does not cause any damage. This behaviour explains the long list of arborescent host plants and the risk of confusing the mite with other tetranychids and incorrectly treating its populations with acaricides. In Sicily (Italy), the clover mite has been collected on herbaceous plants (Vacante, 1985a) and grasses, under the canopy of citrus trees, or on leaves of the lower foliage of lemon (Vacante and Nucifora, 1986) and orange trees.

Natural enemies of the clover mite include the bdellid *Bdella depressa* Ewing, which alone represents over 75% of the agents of biological control in Illinois (Snetsinger, 1956). In Tasmania, the coccinellid *Stethorus vagans* Blackburn is reported (Evans, 1943; Chazeau, 1979), and in Switzerland the mirid *Malacocoris chlorizans* Panz. is found on fruit trees (Geier and Baggiolini, 1952).

Symptomatology and damage

Feeding by the clover mite causes stipplings on the leaves that are similar to the first appearance of the damage caused by leaf miners; as populations increase, the leaves turn yellow or brown and wilt. Heavy infestations may cause severe damage to various host plants (Jeppson *et al.*, 1975). In addition to cultivated plants, *B. praetiosa* is a widespread pest of cool-season turf grasses (Kramer and Cranshaw, 2009). Moreover, hibernating mites and mites derived from overwintering eggs sometimes cause serious problems in dwellings, especially when these are surrounded by or near clovers, grass, weeds or other hosts, as the mites migrate from these host plants to protected sites in dwellings to moult or lay eggs, entering through cracks in the floor, walls, etc. Although the clover mite does not adversely affect human health, its presence alarms people, especially when the mites are present at high density on walls, windows and floors (Jeppson *et al.*, 1975). In Canada, the incidence of *B. praetiosa* infestation has been reported as accompanying dermatitis lesions in a cat (Lindo and Grenn, 1968). Such clover mite problems generally end on their own with warm weather in May or early June (Cranshaw, 2012).

Control

Control is provided by prevention, horticultural practices, and use of physical and chemical means. An updated review on the topic has been produced by Gomez and Mizell (2008).

PREVENTION. The introduction of clover mites into houses can be prevented by strategies such as caulking exterior cracks around doors and windows and in the foundation. A grass-free band 18–24 in wide around dwellings also hinders mite movement. The surfaces in front of the south-, south-west- and east-facing sides of the house must be left barren or covered with mulch or pea gravel. Ornamental and flowering plants not attractive to *B. praetiosa* (zinnia, marigold, salvia, rose, chrysanthemum and petunia, or shrubs such as barberry, juniper, spruce, arborvitae) may be planted as barriers in flower beds. In addition, tight-fitting screens can be used on windows and doors (English and Snetsinger, 1957a; Houseman, 2000; Towsend, 2007; Lyon, 2008; Cranshaw, 2012). The entry of clover mites may also be prevented by spreading a fine layer of various dusts around windows and other ways into dwellings. Talc-containing baby powder, diatomaceous earth, and even baking soda or corn (maize) starch, can form an effective barrier against clover mites (Cranshaw, 2012).

CULTURAL PRACTICES. Supplemental irrigation can negatively affect clover mite populations. On lawns, the mites can be greatly reduced by providing some supplemental watering to areas where they develop, such as warm, dry areas at the base of sun-exposed walls and around evergreens. The practice can be supplemented with spot spray application of acaricides in those areas where clover mites develop (Cranshaw, 2012). Studies to quantify the effect of twice weekly 2.5 cm spring irrigations on clover mites show that the numbers of mites are about twice as low on irrigated turf grass compared with that receiving natural precipitation alone. The effects on larvae seem to be greater than on nymphs and adults (Kramer and Cranshaw, 2009).

PHYSICAL CONTROL. Within the house, a vacuum cleaner can be used to collect mites around windowsills, walls and elsewhere. It is necessary to close the vacuum bag before disposal (Houseman, 2000; Lyon, 2008; Cranshaw, 2012). A soapy rag or wet sponge can also be used to clean mites off surfaces, while wiping carefully to avoid crushing the mites and causing stains (Townsend, 2007; Cranshaw, 2012).

CHEMICAL CONTROL. Attacks of the clover mite on lawns were in the past controlled with sprays of malathion, dicofol or chlorobenzilate (English and Snetsinger, 1957a). Good control was obtained on wall rocket with deltamethrin applied immediately after the first cut (Laffi, 2001). However, the use of effective materials against this mite has been dramatically curtailed as they are no longer registered for use on turf grass (Kramer and Cranshaw, 2009). Cranshaw (2012) mentions that presently there are no highly effective chemical control agents for clover mites.

Sprays of acaricides provide temporary control of the heavy presence of *B. praetiosa* in and around houses, and physical control measures are necessary for effective control (Houseman, 2000). During the mite invasion period, a perimeter spray barrier should be made around the outside of the house, and the foundation, exterior walls up to the bottom of the first floor windows, and a strip 10–20 feet wide from the foundation out on to the grass sprayed (Houseman, 2000; Lyon, 2008). Within houses, labelled acaricides must be used, spraying lightly and applying them directly on to the mites (Houseman, 2000). Lyon (2008) reported that within houses mites could be controlled with aerosol or contact sprays of pyrethrins. Around the house, spraying can be done with bifenthrin, lambda-cyhalothrin (Cranshaw, 2012), cyfluthrin and fluvalinate (Bambara, 2003). Other chemicals labelled for outdoor use are chlorpyrifos, cyfluthrin, diazinon, fluvalinate, propoxur and pyrethrins. Spot applications are made to cracks and crevices at baseboards, around windows and doors, and between windows and screens with propoxur, diazinon or chlorpyrifos (Lyon, 2008). Mites seen on the outside of houses can be killed with a direct spray of an insecticidal soap or regular liquid dish-washing soap at the rate of 2 tablespoons gal^{-1} water. This treatment will not provide any residual control. A spray of chlorpyrifos or dicofol along the outside walls and extending about 10 ft out from the foundation may provide some relief (Townsend, 2007).

Bryobia rubrioculus (Scheuten) (Fig. 20.2)

Common name

Brown mite, brown almond mite.

Diagnostic characteristics

FEMALE. The body is 600–650 μm long, dark in colour, sometimes greenish, with light lobes and legs. The outer prodorsal lobes are subtriangular, small and separated from the inner lobes, which are thin and higher. Both lobes may have small pointed ridges. The dorsal setae are set on small tubercles and are spatulate. The peritremes end in a small anastomosis 20 μm long. Pretarsus I has 2 true claws with a pair of tenent hairs each and a median empodium with two tenent hairs.

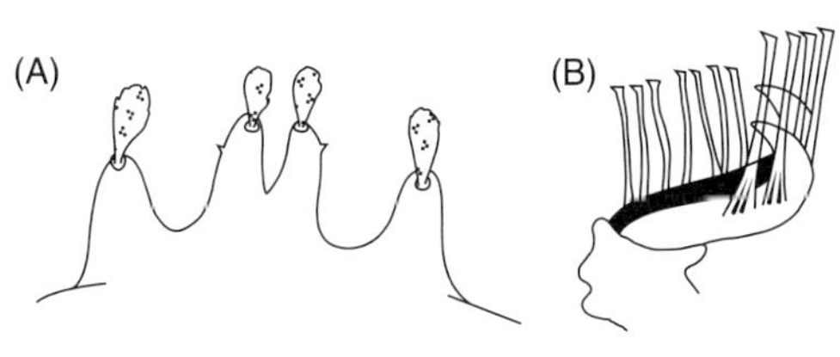

Fig. 20.2. *Bryobia rubrioculus* (Scheuten). Female: (A) prodorsal projections; (B) pretarsus IV (modified from Vacante, 1985a).

The solenidion and the tactile seta of tarsus III are of about the same length; the solenidion and the tactile seta of tarsus IV are well separated, the solenidion is proximal and shorter than the tactile seta (Vacante, 1985a).

MALE. Unknown (Vacante, 1985a).

Geographical distribution

Worldwide (Bolland *et al.*, 1998; Migeon and Dorkeld, 2006–2013).

Host plants

The whole list of host plants of brown mite includes 57 species belonging to 20 families, and about half the species are included in the Rosaceae family (Migeon and Dorkeld, 2006–2013). Families and species of economic importance are listed below:

Fabaceae: lucerne (*Medicago sativa* L.), pea (*Pisum sativum* L.), fava bean (*Vicia faba* L.) (Bolland *et al.*, 1998; Migeon and Dorkeld, 2006–2013).

Juglandaceae: English walnut (*Juglans regia* L.) (Abdullaev and Mugutdinov, 1976).

Moraceae: fig (*Ficus carica* L.) (Meyer Smith, 1974).

Rosaceae: quince (*Cydonia oblonga* Mill.) (Meyer Smith, 1974; Abdullaev and Mugutdinov, 1976), loquat (*Eriobotrya japonica* (Thunb.) Lindl.) (Hatzinikolis and Emmanouel, 1991), apple (*Malus domestica* Borkh.) (Van Eyndhoven, 1955; Rambier, 1958a; Baker and Pritchard, 1960; Tuttle and Baker, 1964), American plum (*Prunus americana* Marshall), apricot (*P. armeniaca* L.) (Rambier, 1958a; Meyer Smith, 1974), sweet cherry (*P. avium* (L.) L.) (Rambier, 1958a; Reeves, 1963; Abdullaev and Mugutdinov, 1976), sour cherry (*P. cerasus* L.) (Ehara, 1959; Abdullaev and Mugutdinov, 1976), European plum (*P. domestica* L.), sweet almond (*P. dulcis* (Miller) D.A. Webb.) (Rambier, 1958a; Hatzinikolis, 1968b, 1969c; Meyer Smith, 1974; Abdullaev and Mugutdinov, 1976), European plum (*P. insititia* (L.) C.K. Schneid.) (Hatzinikolis, 1968b), peach (*P. persica* (L.) Batsch) (Rambier, 1958a; Meyer Smith, 1974; Abdullaev and Mugutdinov, 1976), pear (*Pyrus communis* L.) (Van Eyndhoven, 1955; Rambier, 1958a; Hatzinikolis, 1968b, 1969c; Abdullaev and Mugutdinov, 1976; Osakabe *et al.*, 2000), Chinese pear (*Pyrus pyrifolia* (Burman) Nakai) (Hatzinikolis and Emmanouel, 1991; Ehara, 1999).

Rutaceae: tangerine (*Citrus reticulata* Blanco) (Bolland *et al.*, 1998; Migeon and Dorkeld, 2006–2013).

Bio-ecology

B. rubrioculus generally lives on rosaceous plants and is a pest of apples, pears, peaches and other fruit trees (Jeppson *et al.*, 1975) in the most important areas of their cultivation in the world. It reproduces by thelytokous parthenogenesis and has a haploid chromosome number of four (Helle *et al.*, 1970). The mite overwinters as (winter) eggs laid on the bark of spurs, along the undersides of twigs and forks of branches, and on the sides of the trunk and branches, commonly close together in folds and crevices. In spring, the eggs hatch and the larvae migrate to buds, where they feed on the leaves. Hatching is completed before the fruit blooms, and is most at temperatures between 19 and 26°C; it decreases at 80% RH.

On apple, the mite lays eggs during the spring on the lower leaf surface near the prominent veins, the base of the leaf, the petioles, the twig spurs and the wood of the younger shoots. Later, the eggs may also be laid along the midrib of the upper leaf surface, and in the calyx end of the apples. On almond, the hatching increases at the beginning of the bloom, with maximum rate between full bloom and petal drop. The larvae hatching from winter eggs are positively phototactic and move to the first buds opening at the tips of the shoots. The first eggs hatch in 6–18 days, and the larvae feed on the lower leaf surface, while adults feed over the whole leaf surface (Anderson and Morgan, 1958; Georgala, 1958; Dosse, 1963; Jeppson *et al.*, 1975). Mathys (1957b) found that at 25°C the egg incubation period lasted 9.96 days and at 20°C the fecundity was 20 eggs. Rearing of the brown mite on cherry at nine constant temperatures (15, 17.5, 20, 22.5, 25, 27.5, 30, 32.5 and 35°C), 60 ± 5% RH and a photoperiod of 16:8 h (light:dark) showed that the development time from egg to adult ranged from 43.36 ± 0.63 days at 15°C to 14.54 ± 0.37 days at 32.5°C. Female longevity is 14.02 ± 0.88 days at 15°C and 6.46 ± 0.67 days at 32.5°C. The highest and lowest mean number of eggs laid per female during adult life is 13.82 ± 1.07 at 20°C and 4.38 ± 0.881 at 32.5°C (Honarparvar *et al.*, 2010a). An assessment of the demographic traits of *B. rubrioculus* at eight constant temperatures (15, 17.5, 20, 22.5, 25, 27.5, 30 and 32.5 °C), 60 ± 5% RH and a photoperiod of 16:8 h (light:dark) on sweet cherry showed a significant difference between demographic parameters at the various temperatures. Net reproduction rate (R_0), generation time (T) and intrinsic rate of increase (r_m) ranged from 5.83 ± 0.72, 42.79 ± 0.83 and 0.041 ± 0.003 at 15°C to 0.67 ± 0.13, 24.15 ± 0.41 and 0.025 ± 0.002 at 32.5°C, respectively. Maximum and minimum doubling times were 18.26 ± 0.82 at 15°C and 10.67 ± 0.39 at 20°C. The fecundity ranged from 4.39 ± 0.77 eggs at 32.5°C to 29.45 ± 2.87 eggs at 15°C. The maximum survival rate (lx) is at 15°C and the minimum at 32.5°C. The optimal temperature for *B. rubrioculus* population growth is at 20°C (Honarparvar *et al.*, 2010b).

In Nova Scotia (Canada), *B. rubrioculus* develops two generations a year, and sometimes one partial third generation. Females developing two and three generations lay summer eggs on the leaves and twigs, but winter eggs (or overwinter eggs) are laid only on twigs. Females of bivoltine populations lay both summer and winter eggs, and the eggs produced by females of the third generations hatch first in the following spring (Herbert, 1965). In Lebanon, the mite develops four generations per year (Dosse, 1963), and in South Africa six generations a year, with a mean generation time from 22 to 33 days (Georgala, 1955). In California in almond orchards, the brown mite was observed primarily in the late winter and early spring. Hatching of the winter eggs begins in late February and continues throughout the blooming season. Active stages rest on the woody portions of the tree and move out to feed on the upper leaf surfaces during limited periods of the day or night. The second generation develops in May and a partial third sometimes in June (Summers, 1950; Summers and Stocking, 1972; Rice and Jones, 1978; Hoy, 1985). Damage is usually limited

because the mite typically has only two generations a year, and primarily occurs in the lower portion of the tree, which the mites first colonize in the spring. In the USA, the populations decline in June because the eggs deposited on twigs are in diapause and do not hatch until the following spring (Hoy, 1985, 2011).

In Nova Scotia, temperatures up to 36°C do not affect the adults negatively, which move to shaded sides of foliage (Herbert, 1965), while in Lebanon high temperatures induce death of the mites in July and August and so reduce its populations. Long periods of rain also reduce mite density by killing the forms that remain in water drops, washing mites from leaves and retarding the development of immature stages. Conversely, high temperatures early in the spring induce a population increase. The most important factors affecting later population development are the availability and the quality of the food resource (Jeppson *et al.*, 1975). The mean freezing point of the mite was −26°F (−32.2°C), although doubling the period of exposure in the range of 0.25 to 16 h increased mortality from 8 to 10% (MacPhee, 1963).

Natural enemies of *B. rubrioculus* include the anystid *Anystis baccarum* (L.) (Hoy, 2011), the eupalopsellid *Eupalopsellus hamedaniensis* Khanjani *et al.* (Rostami *et al.*, 2010), and various phytoseiid mites, including *Typhlodromus pyri* Scheuten, *T. rhenanus* (Oudemans), *T. corticis* Herbert, *Amblyseius finlandicus* Oudemans, *Phytoseius macropilis* (Oudemans), (Herbert, 1959), *A. cucumeris* (Oudemans) (= *Neoseiulus cucumeris* (Oudemans)) (Burrell and McCormick, 1964) and some stigmaeid mites, including *Agistemus longisetus* González-Rodríguez (Collyer, 1964a).

Symptomatology and damage

The larvae emerging during spring from the overwintering eggs move to buds, and their feeding activity causes whitish grey spots on the upper surface of young or spur leaves, especially near the base of the veins. As infestation continues, all of the foliage become lighter in colour, and the leaves that were infested earlier have a reduced size. Moulting residues are not commonly found over apple leaf surfaces, but are usually near the midrib or on the branches. Rarely, the pest causes leaf drop, but moulting residues, excreta and eggshells in the hollow of the stem end may negatively affect the macroscopic aspect of dessert peaches (Jeppson *et al.*, 1975). On almond, *B. rubrioculus* is typically a less serious pest and can cause stippling to leaves and, rarely, defoliation. Only the cells penetrated by cheliceral stylets are damaged (Summers and Stocking, 1972). High levels of attack can result in a 29.8% reduction in the leaf surface after feeding during the interval of time from petal fall until early June. The damage is limited because the mite usually develops only two generations a year, and primarily occurs in the lower portions of the tree first colonized by mites in the spring (Hoy, 2011). On citrus, the species is occasional but does not cause any damage (Vacante, 2010).

Control

Control of this mite is provided by acaricide sprays that aim to prevent damage to trees and fruit yield. There are various timing options. Control may be direct against the diapause eggs during the winter, the juvenile and adult stages of the first generation during the spring and the eggs and adults during the summer. The first strategy is partially ineffective because of the impossibility of reaching the eggs in their concealed sites, although the treatment is valuable in preventing damage by the first generation of the mite in spring. The second strategy involves pre-bloom spraying and is particularly effective when spraying(s) is/are carried out against the motile forms of the mite. This strategy prevents damage to the tree caused by the early generation of the mite, and also prevents a reduction in fruit numbers and quality. Summer spraying usually fails because of the difficulty of reaching eggs in their niches, particularly when contact acaricides are used. This difficulty may result in a need for more treatments using acaricides with a shorter residue life in order to control the mites that emerge from these eggs, or in the use of residual-type acaricides (Jeppson *et al.*, 1975). Because the mite moves off the foliage to rest, sampling must take this into account (Hoy, 2011).

Against winter eggs, long-time dormant sprays have been suggested, such as oil applied in the spray, or the use of an acaricide in combination with ovicides, or summer-type oils or acaricides applied during spring (Kremer, 1956). On almond in California, overwintering eggs were destroyed by applications of dormant sprays containing oil and an organophosphate insecticide. If dormant sprays were applied, *B. rubrioculus* populations were limited to orchards in the cooler growing areas of the Central Valley and to orchards with few pesticide applications (Hoy, 1985, 2011).

The Genus *Petrobia* Murray

Diagnostic characteristics

Not more than 15 pairs of dorsal setae; ventrally 3 pairs of medioventral setae, and setae h_{2-3} set in a ventral position. Tarsus I has 2 or more sets of duplex setae. The empodium is curved distally with 2 rows of tenent hairs and the true claw lacks hair-like processes (Bolland *et al.*, 1998).

Economic importance

This genus includes 36 species abscribed to the subgenera *Tetranychina* Wainstein (16 species), *Petrobia* Wainstein (15 species) and *Mesotetranychus* Wainstein (5 species) respectively (Migeon and Dorkeld, 2006–2013). *Petrobia* spp. are mainly associated with herbaceous plants but are sometimes collected on arborescent economic plants. For instance, *P. harti* (Ewing) is reported on 46 wild and cultivated host plants, including various *Oxalis* spp. and is sometimes collected from fruit trees on the trunk and branches, on which it lays overwintering eggs (Vacante, 2010). The brown wheat mite, *Petrobia* (*Petrobia*) *latens* (Müller), and the legume mite, *Petrobia* (*Tetranychina*) *apicalis* (Banks), may be injurious.

Petrobia latens (Müller) (Fig. 20.3)

Common name

Brown wheat mite.

Diagnostic characteristics

FEMALE. The body is 570–600 µm long, dark brown or greenish in colour and with pale yellow legs. The dorsal body setae are finely serrate, not set on tubercles and shorter than the distance between their bases. The dorsal striae are lobate. The prodorsal setae v_2 are twice as long as the other dorsal setae. The peritremes end in a large anastomosis, about 50 µm long. The receptaculum seminis is smooth and tubular, about 15 µm long. Tarsi III and IV bear duplex setae (Vacante, 1985a).

MALE. Meyer Smith (1981a) and Vacante (1985a) reported the male as unknown, but Lu (1979) and del Rivero and García Marí (1983) reported that it existed. The body is pear like and 460 µm long (Lu, 1979).

Geographical distribution

Worldwide (Bolland *et al.*, 1998; Migeon and Dorkeld, 2006–2013).

Host plants

The whole list of host plants includes 115 species belonging to 33 families. The Poaceae have the highest number of species (35), followed in decreasing order by the Asteraceae (16 species), Fabaceae (13 species) and other families (Migeon and Dorkeld, 2006–2013). Families and species of economic importance are listed below:

Amaryllidaceae: onion (*Allium cepa* L.) (Sayed, 1946b; McGregor, 1950; Bibby and Tuttle, 1959; Rossi Simons, 1961; Tuttle and Baker, 1964; Zaher *et al.*, 1982), garlic (*A. sativum* L.) (Ehara, 1956a).

Apiaceae: coriander (*Coriandrum sativum* L.) (Singh and Raghuraman, 2011), carrot (*Daucus carota* L.) (Bibby and Tuttle, 1959; Rossi Simons, 1961; Tuttle and Baker, 1964).

Asteraceae: endive (*Cichorium endivia* L.) (Bibby and Tuttle, 1959; Tuttle and Baker, 1964), chicory (*C. intybus* L.) (Meyer

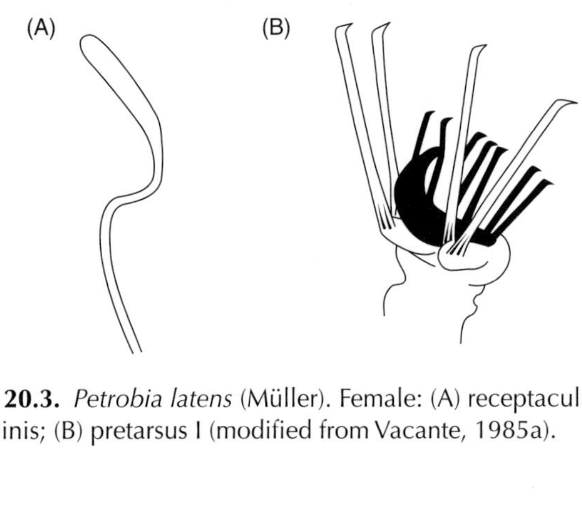

Fig. 20.3. *Petrobia latens* (Müller). Female: (A) receptaculum seminis; (B) pretarsus I (modified from Vacante, 1985a).

Smith, 1974), lettuce (*Lactuca sativa* L.) (Bibby and Tuttle, 1959; Rossi Simons, 1961; Tuttle and Baker, 1964).

Cucurbitaceae: cantaloupe (*Cucumis melo* L.) (Bibby and Tuttle, 1959; Tuttle and Baker, 1964).

Fabaceae: pigeon pea (*Cajanus cajan* (L.) Millsp.) (Prasad, 1974; Gupta and Gupta, 1994), soybean (*Glycine max* (L.) Merr.) (Bibby and Tuttle, 1959; Tuttle and Baker, 1964), lucerne (*Medicago sativa* L.) (McGregor, 1950; Bibby and Tuttle, 1959; Rossi Simons, 1961; Tuttle and Baker, 1964), kidney bean (*Phaseolus vulgaris* L.) (McGregor, 1950; Gupta and Gupta, 1994; Gupta and Chatterjee, 1997), tiny vetch (*Vicia hirsuta* (L.) Gray) (Prasad, 1974; Gupta and Gupta, 1994; Gupta and Chatterjee, 1997), garden vetch (*V. sativa* L.) (Prasad, 1974), azuki bean (*Vigna angularis* (Willd.) Ohwi and H. Ohashi) (Migeon and Dorkeld, 2006–2013).

Malvaceae: upland cotton (*Gossypium hirsutum* L.) (Bibby and Tuttle, 1959; Rossi Simons, 1961; Tuttle and Baker, 1964).

Moraceae: fig (*Ficus carica* L.) (Tuttle and Baker, 1964).

Poaceae: common oat (*Avena sativa* L.) (Meyer Smith, 1974), Bermuda grass (*Cynodon dactylon* (L.) Pers.) (Bibby and Tuttle, 1959; Tuttle and Baker, 1964; Gupta and Gupta, 1994; Gupta and Chatterjee, 1997), hare barley (*Hordeum leporinum* Link) (Rossi Simons, 1961), barley (*H. vulgare* L.) (Banks, 1917; Bibby and Tuttle, 1959; Gupta and Gupta, 1994; Gupta and Chatterjee, 1997; Ashok and Mahendra, 2001), sugarcane (*Saccharum officinarum* L.) (Zaher *et al.*, 1982), wheat (*Triticum aestivum* L.) (Womersley, 1940; McGregor, 1950; Baker and Pritchard, 1960; Tuttle and Baker, 1964; Gupta and Gupta, 1994; Ashok and Mahendra, 2001), maize (*Zea mays* L.) (Sandhu *et al.*, 1973; Gupta and Gupta, 1994; Gupta and Chatterjee, 1997).

Rosaceae: beach strawberry (*Fragaria* (L.) Mill. (= *x ananassa* Duchesne ex Rozier)) (Ehara, 1956a), apple (*Malus domestica* Borkh.) (Womersley, 1940), pear (*Pyrus communis* L.) (Gupta and Chatterjee, 1997).

Vitaceae: grapevine (*Vitis vinifera* L.) (Bolland *et al.*, 1998; Migeon and Dorkeld, 2006–2013).

Bio-ecology

The brown wheat mite lives on wild and cultivated monocotyledonous plants. Its populations commonly inhabit grasses, from which they move on to cultivated plants, such as garlic, gladiolus, grain sorghum, iris, maize, onion and wheat; at high levels of attack, the mite also spreads on to cotton, carrot, lettuce, watermelon, strawberry and lucerne, and it is sometimes found on fruit trees (Jeppson *et al.*, 1975; Meyer Smith, 1981a).

P. latens reproduces by thelytokous parthenogenesis and has a haploid number of chromosomes of four (Helle *et al.*, 1970). Its populations oversummer as diapause eggs that hatch at the beginning of winter; the larvae move on herbaceous plants, on which they begin the first part of their life cycle. Populations of *P. latens* are active until halfway through May, when diapause eggs appear and motile forms reduce and/or disappear. The female lays both non-diapause and diapause eggs, which are deposited on the soil or on any object in the area of plants. The non-diapause eggs are subspherical, lightly ribbed and with an inconspicuous central stalk. At hatching, the shell splits near the equator. The diapause eggs are glistening,

with a rounded basal portion and an expanded cap. The cap has 20–30 ridges that are expanded radially to form a small elevated 'boss' in the centre (Jeppson *et al.*, 1975). Baker and Pritchard (1953b) reported that in Oklahoma, diapause eggs serve for overwintering, but that in California, they serve for aestivating. In England, it has been observed that during rising summer temperatures, the brown wheat mite lays heat-resistant eggs on litter on the ground, and also produces diapausing winter eggs, the deposition of which is not influenced by photoperiod (Lees, 1961). The incubation of non-diapause eggs to hatching takes 7 days at 22°C. Diapause eggs remain unhatched for an indefinite period during hot, dry weather, but soil moisture and increased humidity stimulate them to hatch. Females lay 70–90 summer eggs in 3 weeks and 8–30 winter eggs over the same period.

More than 12 mm of rain may destroy most of the pests, but 6 mm or 12 mm of rain does not have a deleterious effect, and neither does rocky, gravelly or sloping ground (Fenton, 1951; Cox and Lieberman, 1960; Jeppson *et al.*, 1975; Meyer Smith, 1981a). The time from sampling the first larvae in the field and the appearance of adults ranges from 8 to 11 days (Fenton, 1951; Cox and Lieberman, 1960; Meyer Smith, 1981a), while a range of 7–20 days has been reported by Khan *et al.* (1969). The mite feeds during the day, and the best time to look for them is mid-afternoon, or at temperatures of 20–25°C (Meyer Smith, 1981a). The life cycle from egg to adult takes 8–11 days. Larval development lasts 2–3 days, and the deutonymph develops into an adult female in 2–3 days. The preoviposition period takes 1–2 days (Jeppson *et al.*, 1975). The life cycle from egg to egg takes an average of 27 days (17–36 days) in Rajasthan (India) (Khan *et al.*, 1969), while in the greenhouse it takes 20–24 days (13–116 days) (Cox and Lieberman, 1960). Adult longevity ranges in the different conditions from about 2 to 3 weeks (Fenton, 1951; Cox and Liberman, 1960; Khan *et al.*, 1969; Jeppson *et al.*, 1975; Meyer Smith, 1981a). The brown wheat mite does not produce webbing and will quickly drop to the soil when disturbed by high day temperatures or strong winds. Under these conditions, the mite takes refuge under clods of soil. Dispersal occurs by active movement through the soil or from leaf to leaf, and via irrigation water (Khan *et al.*, 1969), wind (Khan *et al.*, 1969; Meyer Smith, 1981a) or human activities (Meyer Smith, 1981a). The mite develops a variable number of generations a year, from 3–6 in Rajasthan (Khan *et al.*, 1969) to 4–9 in the greenhouse (Cox and Lieberman, 1960).

In India, mite activity begins from early December and goes on to mid-April, with a peak in March, and mite populations are most common under rainwater-fed cultivation (Khan *et al.*, 1969). A comprehensive report about *P. latens* in India has been published by Sharma and Srinivasa (2004).

Natural enemies include coccinellids, such as *Chausseria* sp. (Meyer Smith, 1981a), and predatory mites, such as the bdellid *Bdella depressa* (Snetsinger, 1956) and the aceosejid *Lasioseius terrestris* Menon *et* Ghai (Menon and Ghai, 1968).

Symptomatology and damage

The brown wheat mite is mainly a pest in dry weather, and the damage it causes is similar to that produced by drought. Infested plants appear to be dried out even with adequate soil moisture. The mite inserts its stylets into the leaf to suck the plant sap, which pierces the plant leaf cells, producing fine stipples that at a distance may give a bronzing or yellowish effect. When attack continues, the plants become pale yellow to bronze, then dry out and die (Jeppson *et al.*, 1975; Meyer Smith, 1981a). A density of 30 or more mites per plant may cause damage and require control (Meyer Smith, 1981a).

P. latens is a pest of seedling cotton and spreads on to cotton fields from adjacent wheat, other cereals and grasses. Its feeding activity on the leaves and stems of young cotton plants may induce desiccation and wilting (Smith, 1942).

On carrot, the infested leaves turn yellow early and soon dry out and become brittle; in fields, the symptoms occur first in small areas and then gradually spread through the crop fields (Jeppson *et al.*, 1975). In India, *P. latens* infests barley, coriander and wheat. It is a serious pest in dryland agriculture, and usually reaches higher densities on wheat than on barley. Coriander has also been found to be infested by the mite (Singh and Raghuraman, 2011).

The mite was found to transmit filamentous virus-like particles, a causal agent of barley in Montana (USA) and Canada (Robertson and Carroll, 1988). Like *Bryobia pratiosa*, the brown wheat mite penetrates into buildings and causes similar problems to people (Jeppson *et al.*, 1975).

Control

Favourable climatic conditions such as winter rain followed by a dry climate may stimulate an increase in brown wheat mite populations. The monitoring of populations helps to manage their control, which is provided by horticultural practices and chemical sprays. The presence of the mite in buildings may be controlled as described for *B. praetiosa*.

SAMPLING AND MONITORING. Different methods have been developed to sample and monitor the brown wheat mite. In India, sampling of populations on wheat is done by tapping plants over microscope slides that are smeared with glycerine and held in a thermocouple sampler at ground level. The slides are taken to the laboratory for counting the mites under a binocular microscope (Khan *et al.*, 1977). It is possible to place a double folded sheet of paper open next to the plant on the ground. The mites are then shaken off on to one half and squashed flat with the other half. The adults leave greenish brown spots and the juvenile stages reddish spots. It is also possible to use an enamel tray held against the plant and shake the mites on to it, and then count them. First signs of yellowing or browning in patches visible at a distance may also help in determining when to apply control, while taking into account that these signs are not due to other factors such as shortage of nitrogen or drought; it is necessary to monitor reinfestations by mites living on grasses surrounding the crops as well (Meyer Smith, 1981a).

CULTURAL PRACTICES. Late sowing of wheat avoids the cool conditions that stimulate increases in mite populations. Deep ploughing destroys the diapause eggs laid in the soil. Crop rotation of wheat with a non-host plant will prevent the development of the mite populations too (Reynolds and Swift, 1951; Chung *et al.*, 1963).

CHEMICAL CONTROL. Chemical treatments must be applied in early spring when the first damage occurs (Neal, 1979). Various chemicals have been tested on different crops in various countries. In laboratory experiments carried out in India, a number of chemicals were tested, including formothion, phosphamidon, monocrotophos, parathion, chlorpyrifos, carbophenothion, dicofol, malathion and binapacryl (Bhatia *et al.*, 1974). Also in India, Saxena and Rawat (1969) tested thiometon, dimethoate and parathion on wheat in the field, and Deol and Sandhu (1974) tested dimethoate, monocrotophos and oxydemeton-methyl. In the USA, oxydemeton-methyl was applied to cereals (except for rye), dimethoate to wheat and parathion (exceptionally) to rye (Neal, 1979). Spray applications of parathion or sulfur were suggested on garlic and onion; sulfur was effective when it was hot (Allen *et al.*, 1967, in del Rivero and García Marí, 1983). On onions in Egypt, the best results were obtained using dichlorvos alone or in combination with mancozeb, or mancozeb with dicofol. Jeppson *et al.* (1975) reported that the organophosphate compounds demeton and parathion, and Metacide® (metam-sodium, a dithiocarbamate), have given good results in the past. In South Africa, a first spray of demeton-*S*-methyl, monocrotophos or formothion was suggested, with a second treatment after 2–3 weeks when necessary (Meyer Smith, 1981a).

Despite the effectiveness of the above chemicals, applications of sulfur dusts or sprays usually suppress mite populations under warm conditions (Jeppson *et al.*, 1975).

Petrobia apicalis (Banks)

Common name

Legume mite.

Diagnostic characteristics

FEMALE. The body is subovate, the prodorsum is punctate and the rest of the body has a transverse striate pattern. The peritreme has a distal anastomosis. The dorsal body setae are very short, except for the first pair of prodorsal setae (v_2). The fourth pair of dorsal setae (f_1) and the fourth pair of marginal setae (f_2) are 4–5 times as long as the other dorsal setae (Jeppson *et al.*, 1975).

MALE. Like the female, except for the legs, which are much longer (Jeppson *et al.*, 1975).

Geographical distribution

Georgia and Louisiana (USA) (Banks, 1917; McGregor, 1950; Pritchard and Baker, 1955; Jeppson *et al.*, 1975), Hungary (Bozai, 1970) and Portugal (Carmona, 1970).

Host plants

The list of host plant includes 14 species belonging to four families. Families and species of economic importance are listed: Amaryllidaceae – garlic (*Allium sativum* L.) (Bolland *et al.*, 1998; Migeon and Dorkeld, 2006–2013); Asteraceae – *Solidago* sp. (Bolland *et al.*, 1998; Migeon and Dorkeld, 2006–2013); Fabaceae – *Lathyrus hirsutus* L., pea (*Pisum sativum* L.) (McGregor, 1950), *Trifolium campestre* Schreb., *T. dubium* Sibth., *T. incarnatum* L. (Flechtmann and Hunter, 1971), *T. repens* L. (Banks, 1917), *T. resupinatum* L., *Trifolium* sp., vetch (*Vicia sativa* L.), *Vicia* sp. (McGregor, 1950), *V. villosa* Roth (Bolland *et al.*, 1998; Migeon and Dorkeld, 2006–2013); Poaceae sp. (Carmona, 1970).

Bio-ecology

The legume mite lives mostly on herbaceous plants, both wild and cultivated. In the USA, it is active from November to May and has an aestival diapause as summer eggs (Zein-Eldin, 1956), mainly laid on rough surfaces (cracks, loose bark, etc.) or on trees, bushes and fences (Jeppson *et al.*, 1975). Diapause eggs have been found on trees at heights exceeding 12 ft (3.60 m) from the soil surface (Zein-Elden, 1956). The eggs have a white, rigid, cup-shaped receptacle covered with a lid, and the internal contents are cherry red in colour. The receptacle has 20–30 longitudinal sections, each of which corresponds to a radial section of the lid (Jeppson *et al.*, 1975).

The importance of day length as a controlling factor in the induction of diapause has not been demonstrated (Glancey, 1958; Boudraux, 1963). Zein-Eldin (1956) reported that photoperiod, temperature and food availability did not affect the initiation of diapause egg laying, and that the number of generations without diapause could be controlled endogenously. The females originated from diapause eggs may lay diapause eggs by exposing them during their development to a daily photoperiod with a photophase of 14 h (Glancey, 1958; Boudraux, 1963). The lengthening of the days during the spring induces the female to lay diapause eggs (Brooking, 1957). The role of temperature seems to be unimportant, and weakened food plants could induce diapause independently of the photoperiod (Veerman, 1985). The non-diapause eggs are almost hemispherical with a straight wrinkled and sticky base. They have a pearly, transparent and finely striated shell containing cherry red embryonic material. These eggs are laid singly or in masses on debris or on the host plant. Fecundity is 70 or more eggs and nine or more generations may be developed in a year (Jeppson *et al.*, 1975).

Dispersal occurs by crawling, floating in water and air currents, but dispersal through the spinning ability of larvae is probably more effective. By this mechanism, mites use thread to 'balloon' various distances according to the altitude and speed of the air current (Zein-Elden, 1956; Gutierrez and Helle, 1985).

Symptomatology and damage

The legume mite feeds mainly on the upper leaf surface. Leaves that are strongly infested turn greyish and are no longer suitable as forage. The plants either reduce their seed production or do not produce seeds and may die, which results in the elimination of stands of these plant species after some years of infestation (Jeppson *et al.*, 1975). The pattern of feeding punctures by the larvae results typically in straight, bent or criss-crossed lines (Brooking, 1957; Glancey, 1958; Boudreaux, 1963).

Like *Bryobia praetiosa* and *Petrobia latens*, this mite also disturbs people, entering in great numbers into buildings, residences, offices and factories during the period of summer egg deposition (Smith and Webber, 1954).

Control

See *Petrobia latens*.

The Genus *Aplonobia* Womersley

Diagnostic characteristics

The prodorsum has 3 pairs of setae and lacks lobes over the gnathosoma. Some or all of the dorsal setae are set on strong tubercles. The dorsum bears 10 pairs of opisthosomal setae. The setae f_1 are in the normal position. The coxal formula does not exceed 4-3-2-2. The opisthosoma lacks plates. Tarsus I has 2 sets of duplex setae (Bolland *et al.*, 1998).

Economic importance

This genus has a worldwide distribution and includes 31 species living on herbaceous plants but sometimes collected on arborescent fruit plants (Migeon and Dorkeld, 2006–2013). Some species lay diapausing eggs on the wood of fruit plants. *Aplonobia citri* Meyer has been collected covered by its eggs on the trunk of *Citrus sinensis* (L.) Osbeck in South Africa (Meyer Smith, 1974), and on *C. limon* (L.) Burm. f. and *Prunus persica* (L.) Batsch in New South Wales (Australia) (Gutierrez and Schicha, 1983). *A. honiballi* Meyer (Meyer Smith, 1974) has been collected on the trunk of *C. sinensis* in South Africa (Meyer Smith, 1974). *A. histricina* (Berlese) is reported on *Sphaeralcea ambigua* Gray (Migeon and Dorkeld, 2006–2013), *Oxalis pes-caprae* L. (Baker and Pritchard, 1960; Ben-David *et al.*, 2007; Zriki *et al.*, 2015), *Oxalis* sp. (Womersley, 1940), *Pyrus communis* L. (Baker and Pritchard, 1960) and an unknown host plant (Berlese, 1910). In Sicily (Italy), the mite is very common on *O. cernua* Thumb. (Vacante, 1985a; Vacante and Nucifora, 1986) under the canopy of citrus trees or their surroundings. The mite oversummers as eggs laid on the trunk and branches of citrus and fruit trees, on fences near the plants or on plastic irrigation tubes. None of the above-mentioned species play a phytopathological role. Unfortunately, they are sometimes identified as phytophagous species and treated with acaricides. This has suggested that the genus should be reported in order to focus the attention of the technician on the need to avoid any treatment against these mites.

The Genus *Tetranycopsis* Canestrini

Diagnostic characteristics

The prodorsum has 4 pairs of setae, and no prominent lobes occur over the gnathosoma. The opisthosoma has 12 pairs of dorsal setae (Bolland *et al.*, 1998).

Economic importance

This genus include eight species with a Palaearctic, Nearctic and Oriental distribution (Migeon and Dorkeld, 2006–2013). Of these, the Nearctic and Palaearctic *Tetranycopsis horridus* (Canestrini *et* Fanzago) is reported on *Achillea* sp. (Migeon and Dorkeld, 2006–2013), *Alnus* sp., *Carpinus caucasica* Grossh., *Corylus avellana* L. (Pritchard and Baker, 1955; Reck, 1956; Vierbergen, 1990; Skorupska, 1994), *C. maxima* Mill., *Thuja* sp., *Phaseolus vulgaris* L., *Juglans regia* L., *Picea glauca* (Moench) Voss and *Pinus* sp. (Migeon and Dorkeld, 2006–2013). The mite infests hazelnut trees in various countries, but it does not produce severe damage and its presence does not require control measures. *T. borealis* Ehara *et* Mori is reported on *Rubus allegheniensis* Porter, *R. chamaemorus* L. and *R. parvifolius* L. in Japan (Ehara and Mori, 1969). *T. cerasi* Strunkova has been found on *Prunus cerasus* L. in Tajikistan (Strunkova, 1969; Mitrofanov *et al.*, 1987). *T. hystriciformis* Reck is a Nearctic and Palaearctic mite that has been reported on *Eryngium campestre* L. (Reck, 1956), *Potentilla argentea* L. (Wainstein, 1956a; Baker and Tuttle, 1972), *P. arvernensis* Gandoge and *P. chinensis* Ser. (Migeon and Dorkeld, 2006–2013). *T. kuzminae* Strunkova has been reported on *P. pamiroalaica* Juz. in Tajikistan (Strunkova, 1969; Mitrofanov *et al.*, 1987). *T. matikashviliae* Reck was described from *P. dulcis* (Mill.) D.A. Webb in Georgia (USSR) (Reck, 1953) and was also reported on the same host plant in Armenia, Azerbaijan, Kazakhstan, Tajikistan and Ukraine (Reck, 1953; Mitrofanov *et al.*, 1987). *T. spiraeae* Reck is reported on *Spiraea hypericifolia* L. in Armenia, Azerbaijan (Mitrofanov *et al.*, 1987), Georgia (USSR) (Reck, 1948a; Mitrofanov *et al.*, 1987), Tajikistan (Reck, 1948a; Mitrofanov *et al.*, 1987) and Hungary (Bozai, 1970). None of the above-mentioned mites cause significant damage.

The Genus *Eutetranychus* Banks

Diagnostic characteristics

There are 10 pairs of opisthosomal dorsal setae, 2 pairs of pseudanal setae (ps_{1-2}), but the empodial claws are apparently absent (Bolland *et al.*, 1998).

Economic importance

This genus has a worldwide distribution and currently includes 33 species, some of which are pests of economic plants: *Eutetranychus africanus* (Tucker), *E. banksi* (McGregor) and *E. orientalis* (Klein). A number of minor pests has been reported on various wild and cultivated plants in different countries. *E. citri* Attiah has been collected on lime trees in Egypt (Attiah, 1967b) and on *Citrus* sp. and other host plants in India (Nassar and Ghai, 1981; Gupta and Gupta, 1994). *E. cratis* Baker *et* Pritchard has been found on *Bixa* sp., *Macaranga* sp. and orange leaves in Zaire (now the Democratic Republic of the Congo) (Baker and Pritchard, 1960), and on *Trema orientalis* Blume and *Manihot esclulenta* Crantz in Nigeria (Matthysse, 1980) and

Congo (Gutierrez and Bonato, 1994). *E. eliei* Gutierrez *et* Helle has been collected on *Artocarpus incisa* L., *Plumeria alba* L. and *Citrus* sp. in Madagascar (Gutierrez and Helle, 1971). *E. palmatus* Attiah infests different Arecaceae (Palmae) in the Near East (Attiah, 1967b; Zaher *et al.*, 1982; Kamali, 1990; Ben-David *et al.*, 2007). *E. pantopus* (Berlese) has been collected on *Albizia lebbeck* L., *Azadirachta indica* A. Juss., on leaves and fruits of *Citrus* sp., *Ficus* sp. and *Ricinus communis* L. in Egypt and Sudan (Sayed, 1946b) and on *Ficus* sp. (Berlese, 1910) and citrus (Walter *et al.*, 1995) in Australia. *E. pyri* Attiah has been collected in Egypt on *Pyrus communis* L., *Ricinus communis* L., *Cynodon dactylon* L., *C. aurantium* L., *Solanum melongena* L. and *Dalbergia sissoo* Roxburgh ex De Candolle (Attiah, 1967b; Zaher *et al.*, 1982). None of these species generally requires control, with the exception of *E. palmatus*, whose role is, however, unclear.

Eutetranychus africanus (Tucker) (Fig. 20.4)

Common name

African red mite.

Diagnostic characteristics

FEMALE. The body is brownish green in colour. The dorsal opisthosomal setae are long, slender and set on tubercles. The prodorsum is striate and the striae have weaker lobes. The dorsal opisthosomal setae c_2, d_2, e_2 and f_2 are long and slender and the setae c_1, d_1, e_1, f_1 and h_1 are lanceolate to subspatulate and vary in length from half to three-quarters of the length of the other dorsal setae. The dorsal opisthosomal setae c_2, d_2, e_2 and f_2 can vary from relatively long, slender and tapering to long and subspatulate. The dorsal opisthosomal setae c_1, d_1, e_1, f_1 and h_1 can be tapered or subspatulate and vary in length from half as long to as long as the setae. The dorsal opisthosoma is striate and bears weakly developed lobes, and there are longitudinal striae between the setae d_1 and e_1. The peritremes end simply. The receptaculum seminis is oval. Tibia II bears 6 setae (Meyer Smith, 1987).

MALE. The male is smaller than the female, yellowish green in colour and with more slender dorsal body setae, which are shorter and taper more than those of the female. The distal bent portion of the aedeagus is shorter than the dorsal margin of the shaft, which is relatively straight (Meyer Smith, 1987).

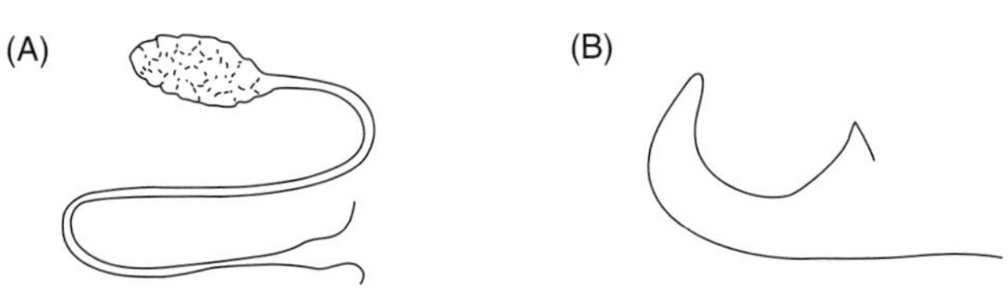

Fig. 20.4. *Eutetranychus africanus* (Tucker). (A) female, receptaculum seminis; (B) male, aedeagus (redrawn from Meyer Smith, 1987).

Geographical distribution

Australia (Walter *et al.*, 1995), Comoros (Bolland *et al.*, 1998), Egypt (Attiah, 1967b), India (Manson, 1963; Gupta and Gupta, 1994), Kenya (Toroitich *et al.*, 2009), Iran (Sheikholeslam Zadeh and Sadeghi, 2010a), Japan (Ehara and Yogi, 1998), Madagascar (Gutierrez and Helle, 1971), Mauritius (Moutia, 1958; Baker and Pritchard, 1960), Mozambique (Meyer Smith and Rodrigues, 1966), Myanmar (Burma) (Manson, 1963; Baker, 1975a), Papua New Guinea (Schicha and Gutierrez, 1985), Philippines (Baker, 1975a), Réunion (Gutierrez and Etienne, 1986), South Africa (Tucker, 1926; Baker and Pritchard, 1960; Meyer Smith, 1974, 1987) and Thailand (Uraisakul *et al.*, 1999; Kongchuensin *et al.*, 2005).

Host plants

The whole list of host plants numbers 62 species belonging to 18 families. The highest number of species belong to the Fabaceae and Rutaceae families, each including 10 species (Migeon and Dorkeld, 2006–2013). Families and species of economic importance are listed below:

Caricaceae: papaya (*Carica papaya* L.) (Gutierrez and Helle, 1971; Ehara and Yogi, 1998).

Cucurbitaceae: chayote (*Sechium edule* (Jacq.) Sw.) (Gutierrez and Etienne, 1986).

Euphorbiaceae: rubber tree (*Hevea brasiliensis* (Willd. ex A. Juss.) Müll. Arg.), cassava (*Manihot esculenta* Crantz), castor bean (*Ricinus communis* L.) (Gutierrez and Helle, 1971).

Malvaceae: upland cotton (*Gossypium hirsutum* L.) (Gutierrez and Etienne, 1986).

Moraceae: fig (*Ficus carica* L.) (Attiah, 1967b), white mulberry (*Morus alba* L.) (Zaher *et al.*, 1982; Ohno *et al.*, 2011a).

Myrtaceae: guava (*Psidium guajava* L.) (Zaher *et al.*, 1982).

Rosaceae: loquat (*Eriobotrya japonica* (Thunb.) Lindl.) (Baker and Pritchard, 1960), apple (*Malus domestica* Borkh.), European plum (*Prunus domestica* L.) (Attiah, 1967b), peach (*P. persica* (L.) Batsch) (Baker and Pritchard, 1960), *Rosa hybrida* (Ehara and Yogi, 1998).

Rubiaceae: gardenia (*Gardenia jasminoides* J. Ellis) (Ehara and Yogi, 1998).

Rutaceae: sour orange (*Citrus aurantium* L.) (Gupta and Gupta, 1994), Taiwan tangerine (*C. depressa* Hayata) (Ehara and Yogi, 1998), Kaffir lime (*C. hystrix* DC.) (Blommers and Gutierrez, 1975), lemon (*C. limon* (L.) Burm.) (Tucker, 1926), grapefruit (*C. paradisi* Macfad.) (Blommers and Gutierrez, 1975), tangerine (*C. reticulata* Blanco), sweet orange (*C. sinensis* (L.) Osbeck) (Tucker, 1926; Attiah, 1967b), orange jessamine (*Murraya paniculata* (L.) Jack) (Ehara and Yogi, 1998), hardy orange (*Poncirus trifoliata* (L.) Raf.) (Blommers and Gutierrez, 1975).

Solanaceae: aubergine (*Solanum melongena* L.) (Attiah, 1967b; Gupta and Gupta, 1994).

Vitaceae: fox grape (*Vitis labrusca* L.) (Gutierrez and Helle, 1971).

Bio-ecology

The African red mite reproduces by arrhenotokous parthenogenesis, and has a haploid number of chromosomes of two

(Bolland and Helle, 1981). Its populations feed on and produce webbing on the upper leaf surface; on the lower surface, the females stop laying and die after some days. The period of pre-oviposition lasts 1–2 days and the life cycle from egg to egg takes about 15 days. On *C. hystrix* at 24 ± 1°C and 85 ± 10 % RH, the female lays on average 4.2 eggs a day (maximum 8) and on mandarin it lays 2 eggs a day (maximum 5). The developmental time of the egg, larva, protonymph and deutonymph is 7.2, 1.3, 1.0 and 0.7 days, respectively. The distribution of populations on the leaves seems independent of the age of the leaves (Blommers and Gutierrez, 1975). Mite dispersal is hindered by rain (Moutia, 1958; Baker and Pritchard, 1960) but is resistant to moderate rain (Blommers and Gutierrez, 1975).

Various natural enemies develop on *E. africanus*. The pathogens include the fungus *Hirsutella thompsoni* Fisher. The line weaving spider *Hylyphantes graminicola* Sundevall (Linyphiidae) preys on immature and adult stages of the mite (Wipada, 1997). Predatory insects include the coccinellid *Stethorus madecassus* Chazeau and some chrysopids (Blommers and Gutierrez, 1975). Phytoseiid mites include *Amblyseius parasundi* Blommers, *A. tamatavensis* Blommers, *A. vazimba* Blommers *et* Chazeau (Blommers and Gutierrez, 1975), *Neoseiulus californicus* (McGregor), *N. longispinosus* (Evans) (Manita and Vatana, 2000; Kongchuensin *et al.*, 2005), *Phytoseiulus hongkongensis* Swirski *et* Shechter (Charanasri and Kongchuensin, 2001) and *P. persimilis* Athias Henriot (Manita and Vatana, 2000).

Symptomatology and damage

The feeding activity of the African red mite causes fine stipplings on the leaves, which drop early without turning brown (Moutia, 1958; Baker and Pritchard, 1960). The pest attacks all citrus species (Gutierrez and Helle, 1971; Bolland *et al.*, 1998) and does not prefer lemon. The immature and adult stages live exclusively on the upper leaf surface of citrus and their feeding activity causes fine whitish spots over the entire surface. The infested leaves become lead grey, and a high mite population density may present lateral necrosis of the edges similar to the effects of mineral deficiency or attack by tetranychids. The infested leaves may yellow early (Blommers and Gutierrez, 1975). *E. africanus* is an important pest of durian in Thailand, especially during the cool season (late October to early March). It feeds on the upper surface of leaves, and its populations develop rapidly during hot and dry weather conditions (IPM Thailand, 2005).

Control

Control is provided by horticultural practices and the use of biological and chemical means.

CULTURAL PRACTICES. Ground cover with weeds offers refuge to natural enemies. Water jet spraying increases humidity during the dry season and hinders mite development, and it is possible to install a sprinkler on the top of the tree(s) for water spraying for mite control (IPM Thailand, 2005).

BIOLOGICAL CONTROL. In Thailand, biological control is provided by the release of mass-reared *N. longispinosus* (IPM Thailand, 2005).

CHEMICAL CONTROL. The ETL (economic threshold level) of this mite is 25% infested (with more than five mites) mature leaves. When natural enemies are present in the field, it is better to use water jet spraying than chemicals for control. However, hexythiazox, propargite or amitraz are effective against the mite (IPM Thailand, 2005). In Madagascar, the need for control on citrus is necessary only on *C. hystrix* (Blommers and Gutierrez, 1975). In Thailand, on *C. reticulata*, pyridimifen and abamectin, followed by amitraz and propargite, were tested positively for mite control (Sukumalanand *et al.*, 1999). The action of black pepper essential oil has also been positively tested in Thailand (Pumnuan *et al.*, 2009).

Eutetranychus banksi (McGregor) (Fig. 20.5)

Common name

Texas citrus mite.

Diagnostic characteristics

FEMALE. The body is broad and robust, tan to brownish green in colour with dark brown to greenish spots and bars near the lateral margins. The legs are strong, with the distal segments pale and the proximal elements brown. The dorsal setae are variable in length, commonly broader distally and not as long as the distance between their bases. The prodorsum is striate longitudinally. The dorsal opisthosomal setae e_1 are closer together than the other dorsal setae. The opisthosomal striae are transverse, with the exception of a 'V' pattern between the dorsal setae d_1 and e_1. Tibia II has 6 setae (Pritchard and Baker, 1955; Jeppson *et al.*, 1975).

MALE. The male is smaller than the female, and similar in colour. The aedeagus is simple and abruptly turns dorsally near the distal end (Pritchard and Baker, 1955; Jeppson *et al.*, 1975).

Geographical distribution

North, Central and South America (Jeppson *et al.*, 1975; Bolland *et al.*, 1998; Migeon and Dorkeld, 2006–2013), Egypt (Zaher and El-Badry, 1961), Hawaii (Garrett and Haramoto, 1967), India (Sandhu *et al.*, 1982), Portugal (Gonçalves *et al.*, 2002) and Spain (Garcia *et al.*, 2003, 2004).

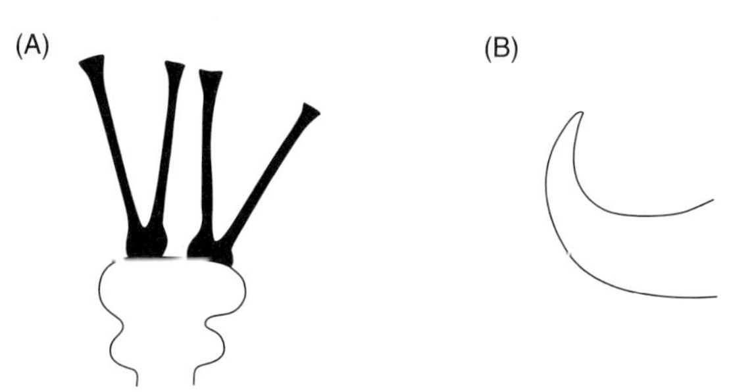

Fig. 20.5. *Eutetranychus banksi* (McGregor). (A) female, pretarsus; (B) male, aedeagus (drawing by Vacante).

Host plants

The whole list of host plants of *E. banksi* currently includes 107 species in 32 families, with the highest number of species in the families Fabaceae (22 species) and Rutaceae (11 species) (Migeon and Dorkeld, 2006–2013). Families and species of economic importance are listed below:

Aquifoliaceae: mate (*Ilex paraguariensis* A. St.-Hil.) (Aranda and Flechtmann, 1971).

Arecaceae: date palm (*Phoenix dactylifera* L.) (Bolland *et al.*, 1998, Migeon and Dorkeld, 2006–2013).

Caricaceae: papaya (*Carica papaya* L.) (McGregor, 1917; Aranda and Flechtmann, 1971; Álvarez and Alarcón, 2004; Barbosa *et al.*, 2004; Guanilo *et al.*, 2012).

Cucurbitaceae: chayote [*Sechium edule* (Jacq.) Sw.] (Andrews and Poe, 1980; Ochoa *et al.*, 1989).

Euphorbiaceae: rubber tree (*Hevea brasiliensis* (Willd. ex A. Juss.) Müll. Arg.) (Paschoal and Reis, 1968; Flechtmann and Baker, 1975; Bellini *et al.*, 2005; Demite and Feres, 2005; Hernandes and Feres, 2006; Vis *et al.*, 2006), cassava (*Manihot esculenta* Crantz) (Andrews and Poe, 1980), castor bean (*Ricinus communis* L.) (McGregor, 1914; Andrews and Poe, 1980; Garcia *et al.*, 2003; Guanilo *et al.*, 2012).

Fabaceae: groundnut (*Arachis hypogaea* L.) (Urueta, 1975), pigeon pea (*Cajanus cajan* (L.) Millsp.) (Andrews and Poe, 1980).

Malvaceae: cacao tree (*Theobroma cacao* L.) (Baker and Pritchard, 1962).

Meliaceae: neem (*Azadirachta indica* A. Juss.) (Livshits and Salinas-Croche, 1968).

Moraceae: fig (*Ficus carica* L.) (Garrett and Haramoto, 1967; Aguilar and Murillo, 2008), white mulberry (*Morus alba* L.) (Aguilar and Murillo, 2008), black mulberry (*Morus nigra* L.) (Flechtmann, 1967a; Flechtmann and Baker, 1975).

Rhamnaceae: jujube (*Ziziphus jujuba* Mill.) (Livshits and Salinas-Croche, 1968; Aguilar and Murillo, 2008).

Rosaceae: loquat (*Eriobotrya japonica* (Thunb.) Lindl.) (Livshits and Salinas-Croche, 1968), European plum (*Prunus domestica* L.) (Baker and Pritchard, 1962), sweet almond (*P. dulcis* (Mill.) D.A. Webb) (Livshits and Salinas-Croche, 1968), peach (*P. persica* (L.) Batsch) (Baker and Pritchard, 1962).

Rubiaceae: Arabian coffee (*Coffea arabica* L.) (Flechtmann and Baker, 1975).

Rutaceae: Key lime (*Citrus aurantiifolia* (Christm.) Swingle) (Rossi Simons, 1961), sour orange (*C. aurantium* L.) (Rossi Simons, 1961; Doreste, 1968b), lemon (*C. limon* (L.) Burm. f.) (Rossi Simons, 1961; Baker and Pritchard, 1962), grapefruit (*C. paradisi* Macfad.) (Rossi Simons, 1961; Baker and Pritchard, 1962), tangerine (*C. reticulata* Blanco) (Rossi Simons, 1961; Paschoal, 1970; Guanilo *et al.*, 2012), sweet orange (*C. sinensis* (L.) Osbeck) (Rossi Simons, 1961; Baker and Pritchard, 1962; Garcia *et al.*, 2003; Martínez *et al.*, 2004; Horn *et al.*, 2011), curry leaf tree (*Murraya koenigii* (L.) Spreng.) (Estebanes-Gonzalez and Baker, 1968).

Sapotaceae: sapodilla (*Achras zapota* L.) (McGregor, 1950).

Vitaceae: grapevine (*Vitis vinifera* L.) (Baker and Pritchard, 1962).

Bio-ecology

E. banksi is a pest of fruit trees, ornamental and flowering plants, and vegetables. It reproduces by arrhenotokous parthenogenesis and has a haploid number of chromosomes of three (Helle *et al.*, 1970). Its populations develop at low RH, under strong dry conditions and at temperatures close to 27°C. At low densities, the pest commonly occurs in the southern quadrants of the canopy (Dean, 1959a). On grapefruit (*C. paradisi*) leaves, the life cycle from the egg to the adult stage at 15, 20, 25, 28, 30 and 32°C and 51–84% RH takes 29.6, 17.2, 13.1, 11.6 and 9.6 days for the female and 27.7, 16.4, 12.0, 10.1, 10.8 and 8.5 days for the male. The sex ratio is female biased (<80%) at 20–30°C, but the proportion of females declines to 68% at 15°C and 61% at 32°C. The maximum net reproductive rate (R_0) at 28, 30 and 25°C is 31.26, 26.06 and 25.29, respectively (Childers *et al.*, 1991). On leaves of *C. sinensis*, the optimum range of development occurs from 28 to 31°C. At 20, 25, 30, 32.5 and 35°C, the intrinsic rate of increase is −0.0649, 0.1723, 0.1759, 0.1973 and −0.0711, the net reproductive rate is 0.30, 8.25, 7.24, 9.21 and 0.44 and the mean generation time is 19.21, 12.79, 11.74, 11.52 and 11.64 days, respectively (Badii *et al.*, 2003). In Texas, *E. banksi* occurs on citrus leaves throughout the year as eggs that are laid along the midrib and near the lateral leaf margins; there are lower egg densities laid from February to April but these are high during May and June; in Mexico, there are 22–23 generations per year (Badii *et al.*, 2003).

Pathogenic fungi and predatory insects and mites develop on populations of *E. banksi*. The fungi include *Neozygites floridana* (Weiser *et* Muma) Reumadière *et* Keller (Weiser and Muma, 1966; Muma, 1969) and *Hirsutella thompsonii* (McCoy and Selhime, 1974). The insects include the coccinellids *Stethorus atomus* Casey (McMurtry *et al.*, 1970) and *S. siphonulus* Kapur (Raros and Haramoto, 1974). Predatory mites include the phytoseiids *Euseius mesembrinus* (Dean) (Abou-Setta and Childers, 1989) and *Galendromus helveolus* (Chant) (Caceres and Childers, 1991), the ascid *Lasioseius scapulatus* Kennet (Ramirez and Esponda, 1986) and the stigmaeid *Agistemus fleschneri* Summers (Zaher and El-Badry, 1961).

Symptomatology and damage

The feeding symptomatology is similar to that of the citrus red mite, *Panonychus citri* (McGregor) (Jeppson *et al.*, 1975). The results of monthly inspections for arthropods in more than 130 citrus groves in Florida in 1956–1971 showed that *E. banksi* increased in the 16-year period (Simanton, 1976). Also in Florida, in a citrus orchard (cultivar Rhode Red Valencia) under irrigation management, the final mean density of feeding stipples on infested leaves was 873 cm^{-2} and leaf life was 380 days. There was no significant negative relationship between leaf life and mite damage. The study indicated that damage by Texas citrus mites to Valencia citrus leaves promoted little or no premature leaf abscission in irrigated trees (Hall and Simms, 2003).

Control

The control of *E. banksi* is provided by the use of biological and chemical means, which are also applied in integrated pest

management (IPM) programmes. Some clones of *H. brasiliensis* hinder mite development.

PLANT RESISTANCE. Research on different clones of *H. brasiliensis* shows that the mite develops only on clone FX 2784, on which the highest mite infestation occurs in the first 4 months of the year, especially in January (0.4 mite cm^{-2}) (Castro *et al.*, 2010a).

BIOLOGICAL CONTROL. Biological control on citrus has been extensively discussed by McMurtry (1985b).

CHEMICAL CONTROL. A number of active ingredients are reported in the literature for the chemical control of the Texas citrus mite, including avermectins (McCoy *et al.*, 1982; French and Villarreal, 1988), carbamates (French *et al.*, 1985), organochlorines (French, 1975; French *et al.*, 1985; Buergo *et al.*, 1986; Flores *et al.*, 1994; Flores Suarez *et al.*, 1996), organotins, pyrethroids (Ochoa *et al.*, 1989), phenyl pyrazoles (French and Villarreal, 1988), organophosphorus compounds (Buergo *et al.*, 1986), petroleum oils (French, 1975; Buergo *et al.*, 1986; French and Villarreal, 1988; Childers, 2010), pyridazinones (French and Bruno, 1996) and sulfur (Jeppson, 1989).

INTEGRATED PEST MANAGEMENT. Implementation of IPM strategies on citrus has been well known for some time (Knapp, 1981). Childers (2010) has reported further updates.

Eutetranychus orientalis (Klein) (Fig. 20.6)

Common name

Oriental red mite, citrus brown mite, lowveld citrus mite.

Diagnostic characteristics

FEMALE. The body is ovoid, about 400 μm long, from green to brown to dark green in colour and with dark inner blotches. The dorsal body setae are set on tubercles and vary in length and shape. The prodorsum is striate longitudinally and the striae are tuberculate. The dorsal opisthosomal striae between the setae d_1 and e_1 vary from longitudinal to 'V' shaped. The dorsal opisthosomal setae c_2, d_2, e_2 and f_2 are long and lanceolate, subspatulate or broadly spatulate; the setae c_1, d_1, e_1, f_1 and h_1 are short and spatulate or lanceolate or subspatulate. The setae c_1 are more or less in line with the setae c_2 and c_3. The setae c_1 and f_1 form a square. The receptaculum seminis is pointed distally. Tibia II has 6 setae (Baker and Pritchard, 1960; Meyer Smith, 1987).

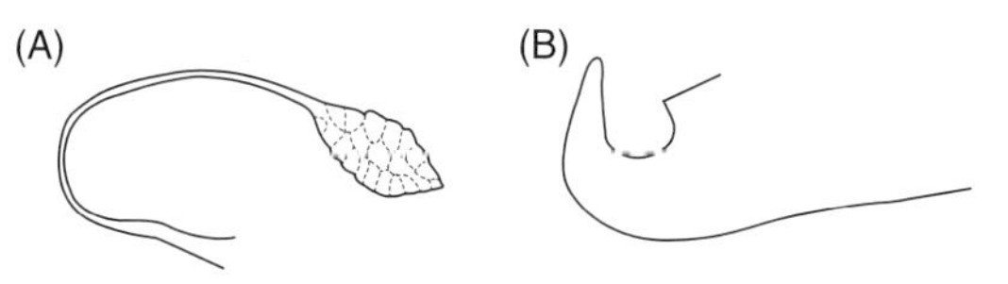

Fig. 20.6. *Eutetranychus orientalis* (Klein). (A) female, receptaculum seminis; (b) male, aedeagus (from Meyer Smith, 1987).

MALE. The male is smaller than the female, 360 μm long and with dorsal body setae shorter than those of the female and more lanceolate or subspatulate. The aedeagus has the bent portion longer than the dorsal margin of the shaft (Baker and Pritchard, 1960; Meyer Smith, 1987).

Geographical distribution

Worldwide (Bolland *et al.*, 1998; Migeon and Dorkeld, 2006–2013).

Host plants

The oriental red mite is reported on 207 host plants of 58 families. The Fabaceae harbour the highest number of species (32), followed in decreasing order by the Malvaceae (16 species), Rutaceae (13 species), Rosaceae (11 species) and other families (Migeon and Dorkeld, 2006–2013). Families and species of economic importance are listed below:

Anacardiaceae: cashew (*Anacardium occidentale* L.) (Gupta, 1992), mango (*Mangifera indica* L.) (Garcia *et al.*, 2003).

Annonaceae: custard apple (*Annona reticulata* L.) (Gupta and Gupta, 1994; Walter *et al.*, 1995), sugar apple (*Annona squamosa* L.) (Zaher *et al.*, 1982).

Arecaceae: coconut palm (*Cocos nucifera* L.) (Baker, 1975b; Gupta, 1992; Gupta and Gupta, 1994; Gupta and Chatterjee, 1997), date palm (*Phoenix dactylifera* L.) (Beccari, 1971; Chaudhri *et al.*, 1974).

Asteraceae: chicory (*Cichorium intybus* L.), sunflower (*Helianthus annuus* L.) (Chaudhri *et al.*, 1974).

Bromeliaceae: pineapple (*Ananas comosus* (L.) Merr.) (Meyer Smith, 1987).

Cannabaceae: marijuana (*Cannabis sativa* L.) (Chaudhri *et al.*, 1974).

Caricaceae: papaya (*Carica papaya* L.) (Baker, 1975b; Ehara and Wongsiri, 1975; Nassar and Ghai, 1981; Gupta and Gupta, 1994; Gupta and Chatterjee, 1997; Ehara, 2004).

Convolvulaceae: sweet potato (*Ipomoea batatas* (L.) Lam.) (Baker, 1975b).

Cucurbitaceae: cantaloupe (*Cucumis melo* L.), winter squash (*Cucurbita maxima* Duchesne), crookneck squash (*C. moschata* Duchesne), field pumpkin (*C. pepo* L.) (Georghiou, 1959).

Euphorbiaceae: rubber tree (*Hevea brasiliensis* (Willd. ex A. Juss.) Müll. Arg.) (Ehara and Tho, 1988), cassava (*Manihot esculenta* Crantz) (Gutierrez and Etienne, 1981b), castor bean (*Ricinus communis* L.) (Attiah, 1967b; Chaudhri *et al.*, 1974; Zaher *et al.*, 1982; Gupta, 1992; Gupta and Gupta, 1994; Gupta and Chatterjee, 1997; Kreiter *et al.*, 2002; Zriki *et al.*, 2015).

Fabaceae: groundnut (*Arachis hypogaea* L.) (Gutierrez and Etienne, 1981b; Gupta and Gupta, 1994), pigeon pea (*Cajanus cajan* (L.) Millsp.) (Gupta and Gupta, 1994; Gupta and Chatterjee, 1997), soybean (*Glycine max* (L.) Merr.) (Baker, 1975b), kidney bean (*Phaseolus vulgaris* L.) (Prasad, 1974), winged bean (*Psophocarpus tetragonolobus* (L.) D.C.) (Baker, 1975b).

Lythraceae: pomegranate (*Punica granatum* L.) (Zaher *et al.*, 1982).

Juglandaceae: English walnut (*Juglans regia* L.) (Georghiou, 1959; Zriki *et al.*, 2015).

Lauraceae: avocado (*Persea americana* Mill.) (Garcia *et al.*, 2003).

Malvaceae: okra (*Abelmoschus esculentus* (L.) Moench) (Chaudhri *et al.*, 1974), durian (*Durio zibethinus* Murray), short-staple cotton (*Gossypium herbaceum* L.) (Gupta and Gupta, 1994), upland cotton (*G. hirsutum* L.) (Chaudhri *et al.*, 1974), cacao tree (*Theobroma cacao* L.) (Bolland *et al.*, 1998; Migeon and Dorkeld, 2006–2013).

Meliaceae: neem (*Azadirachta indica* A. Juss.) (Rahman and Sapra, 1940a; Chaudhri *et al.*, 1974; Gupta, 1992; Gupta and Gupta, 1994; Gupta and Chatterjee, 1997; Gupta and Krishna Karmakar, 2011).

Moraceae: fig (*Ficus carica* L.) (Georghiou, 1959; Nassar and Ghai, 1981), white mulberry (*Morus alba* L.) (Chaudhri *et al.*, 1974; Nassar and Ghai, 1981; Zaher *et al.*, 1982; Gupta and Gupta, 1994; Gupta and Chatterjee, 1997), black mulberry (*M. nigra* L.) (Meyer Smith and Rodrigues, 1966).

Musaceae: banana (*Musa paradisiaca* L.) (Nassar and Ghai, 1981).

Myrtaceae: guava (*Psidium guajava* L.) (Gupta and Gupta, 1978, 1994; Gupta and Chatterjee, 1997).

Oxalidaceae: carambola (*Averrhoa carambola* L.) (Bolland *et al.*, 1998; Migeon and Dorkeld, 2006–2013).

Pedaliaceae: sesame (*Sesamum indicum* L.) (Gupta and Gupta, 1994; Gupta and Chatterjee, 1997).

Poaceae: sugarcane (*Saccharum officinarum* L.) (Baker, 1975b), maize (*Zea mays* L.) (Prasad, 1974; Abo-Korah, 1978; Gupta and Gupta, 1994; Gupta and Chatterjee, 1997).

Rhamnaceae: common jujube or ber (*Ziziphus mauritiana* Lam.) (Rahman and Sapra, 1940a; Gupta and Gupta, 1994).

Rosaceae: quince (*Cydonia oblonga* Mill.), beach strawberry (*Fragaria chiloensis* (L.) Mill.) (= × *ananassa* Duchesne ex Rozier), apple (*Malus domestica* Borkh.) (Meyer Smith, 1987), European plum (*Prunus domestica* L.) (Beccari, 1971; Gupta and Gupta, 1994; Gupta and Chatterjee, 1997), *P. dulcis* (Mill.) D.A. Webb (Rahman and Sapra, 1940a; Chaudhri *et al.*, 1974; Gupta and Gupta, 1994; Gupta and Chatterjee, 1997; Ben-David *et al.*, 2007), peach (*P. persica* (L.) Batsch) (Gupta *et al.*, 1971; Nassar and Ghai, 1981; Gupta and Gupta, 1994; Gupta and Chatterjee, 1997), pear (*Pyrus communis* L.) (Gupta *et al.*, 1971; Nassar and Ghai, 1981; Gupta and Gupta, 1994; Gupta and Chatterjee, 1997; Arora *et al.*, 2004), Chinese pear (*P. pyrifolia* (Burm. f.) Nakai) (Gupta and Gupta, 1994; Gupta and Chatterjee, 1997).

Rutaceae: calamondin (*Citrofortunella* × *microcarpa* (Bunge) Wijnands) (Ehara, 2004), sour orange (*Citrus aurantium* L.) (Nassar and Ghai, 1981; Zaher *et al.*, 1982), pomelo (also pummelo) (*C. grandis* (L.) Osbeck) (Meyer Smith, 1974), rough lemon (*C. jambhiri* (Lush.)), karna (*C. karna* Raf.), lemon (*C. limon* (L.) Burm. f.) (Baker and Pritchard, 1960; Chaudhri *et al.*, 1974; Meyer Smith, 1974; Nassar and Ghai, 1981; Gupta and Gupta, 1994; Garcia *et al.*, 2003; Ledesma *et al.*, 2011; Zriki *et al.*, 2015), citron (*C. medica* L.) (Gupta and Gupta, 1994), grapefruit (*C. paradisi* Macfad.) (Baker and Pritchard, 1960; Gupta and Gupta, 1994; Ben-David *et al.*, 2007), tangerine (*C. reticulata* Blanco) (Prasad, 1974; Gupta and Gupta, 1994; Ben-David *et al.*, 2007), sweet orange (*C. sinensis* (L.) Osbeck) (Rodrigues, 1968; Chaudhri *et al.*, 1974; Gupta and Gupta, 1994; Walter *et al.*, 1995; Ledesma *et al.*, 2011), orange jessamine (*Murraya paniculata* (L.) Jacq.) (Bolland *et al.*, 1998; Migeon and Dorkeld, 2006–2013), hardy orange (*Poncirus trifoliata* (L.) Raf.) (Ehara, 2004).

Solanaceae: eggplant (*Solanum melongena* L.) (Attiah, 1967b; Zaher *et al.*, 1982).

Vitaceae: grapevine (*Vitis vinifera* L.) (Ehara and Wongsiri, 1975).

Bio-ecology

The oriental red mite infests a wide range of economic plants, such as ornamental and flowering plants, forest plants, fruit plants, vegetables, legumes, cereals and others; between cultivated plants, it prefers the citrus. *E. orientalis* is infected by *Wolbachia* (Breeuwer and Jacobs, 1996) and reproduces by arrhenotokous parthenogenesis. It has a haploid condition of three chromosomes and a sex ratio of 1 male: 3.25 females (Helle *et al.*, 1970). On common jujube (*Z. mauritiana*, cv. Gola), *E. orientalis* deposits eggs singly in a linear fashion on the upper leaf surface, mostly along the midrib near the lateral veins. Incubation and the larval, protonymphal and deutonymphal stages last 5.38 ≥ 0.05, 2.21 ≥ 0.17, 2.09 ≥ 0.17 and 2.15 ≥ 0.04 days, respectively (Yadav *et al.*, 2003a). During summer, the mite ovideposits mostly along the midrid of the upper leaf surface, but in winter or at high density, it deposits eggs on both leaf surfaces. The developmental threshold is 11°C. Above 26°C, reproduction becomes slower; at 23°C the pre-oviposition period is 12 days, at 20–22°C it is 2.5–3 days and at 14–15°C it is 4–8 days. The egg development rate decreases at low RH. The optimum conditions for development are 21°C and 59–70% RH. The mite develops at temperatures ranging from 18 to 30°C and at 35 to 75% RH. Beyond these limits, development decreases or stops (Bodenheimer, 1951).

The longevity of fertilized females is lower (6.9 days) than that of non-fertilized females (9.8 days) (Angsumarn and Kosol, 1988). In summer, the adult lifespan is about 12 days, in spring and autumn it is 14–18 days and in winter it is more than 21 days (Bodenheimer, 1951). In summer, the periods of pre-oviposition, oviposition and post-oviposition are 1.0, 10.4 and 2.4 days, and in winter they are 2.4, 12.2 and 3 days, respectively. The egg incubation period and larval and protonymph developmental times are 4.3, 2.9 and 1.7 days in summer and 5.7, 3.7 and 2.4 days in winter, respectively (Siddig and El-Badry, 1971a,b). At 20°C, egg incubation period and immature stage development period and adult longevity are 8.06, 20.75 and 21.88 days, while at 30°C they are 3.57, 10.0 and 7.36 days, respectively (Atwa *et al.*, 1987). The egg-to-adult developmental time of the female is 22.32 days at 20°C and 12.43 days at 30°C. The development of the mite above the lower threshold temperature (6.37°C) requires 305 degree-days. The pre-imaginal mortality is 19.61, 22.76, 43.11 and 100% at 20, 25, 30 and 35°C, respectively. The longevity of the female ranges from 16.57 days at 20°C to 7.50 days at 30°C. The fecundity ranges from 14.56 to 16.33 eggs per female, and the sex ratio is 75–80% female.

The intrinsic rate of increase (r_m) ranges from 0.144 at 30°C to 0.094 individuals per female daily at 20°C. The population doubles in 4.79 days at 30°C and in 7.33 days at 20°C (Imani and Shishehbor, 2009). The net reproductive rate (R_0) is 18.85, the cohort generation time (T_c) is 17.03 days, the innate capacity for cohort increase (R_c) is 0.17 and the finite rate of increase (λ) is 1.19 mites a day (Angsumarn and Kosol, 1988). Between 18 and 30°C, the developmental time of *E. orientalis* decreases and the development rate increases with increasing temperature. Between 22.5 and 27.5°C, the lifespan of the female is 8–10 days and that of the male is 4–6 days. The females exhibit 'protection of their eggs' behaviour (Zhou *et al.*, 2006). At 30°C, female longevity is 45 days and fecundity is 318 eggs. The mean pre-imaginal mortality is 24%. The mean generation time (T) is 22.83 days, the net reproductive rate (R_0) is 154.08 and the intrinsic rate of increase (r_m) is 0.221 (Imani *et al.*, 2009).

A positive correlation between temperature and day length and the population development of *E. orientalis* on *C. aurantiifolia*, *C. reticulata*, *C. limon* and *C. paradisi* has been observed in Ludhiana, India, whereas RH, rainfall and the phytoseiid mite *Amblyseius delhiensis* (Narayanan *et* Kaur) (= *Euseius scutalis* (Athias Henriot)) are negatively correlated with a population increase on all citrus varieties except *C. reticulata* (Neena Singla and Sadana, 1998). In coastal regions, where humidity is medium, populations develop during the summer, while in regions with low humidity, populations do not increase in the spring and summer, but may develop later and cause serious damage in the autumn (Jeppson *et al.*, 1975; Jeppson, 1989). In Egypt, a generation develops in 14.5 days in the cold months, and in 11.2 days in the summer; through the year, the mite has as many as 27 generations (Siddig and El-Badry, 1971a,b). In India, there are two population peaks a year in late April and September, but no mites are found from the end of November until the third week of January (Bhumannavar and Singh, 1986). At temperatures between 21 and 27°C and an RH of 30–40%, a population peak has been observed in May–June, with a decrease in July–August following intense rainfall (Dhooria and Butani, 1984). In Punjab (northern India), population peaks were observed in September–December on Baramasi lemon and in June on kaghzi lime, while on grapefruit there were two peaks – in June and October (Kapur-Ghai and Mandeep Kaur, 2003a). In South Africa, damage is more evident in autumn, especially in dry conditions. Infestations appear between February and March and the peak is reached between March and April (Meyer Smith, 1998). Fluctuations in the populations of *E. orientalis* on guava (var. allahabadi), in relation to climatic factors and the population of a predatory mite, *E. scutalis*, were investigated in Punjab by Neena Singla (2001). The populations of *E. orientalis* were positively correlated with those of *E. scutalis*. *E. orientalis* was observed from July to September, and had a population peak in August at a median temperature of 28.18°C, RH of 81.39%, rainfall of 8.72 mm, wind velocity of 5.23 km h^{-1}, sunshine of 06.79 h and a day length of 13.00 h, when the population of *E. scutalis* was also at its peak. Also in northern India (Uttar Pradesh), the population of *E. orientalis* increases on apple from March to June (Mukherjee and Singh, 1993; Singh *et al.*, 2000a). In China, the mite populations peak twice in a year; the first peak is from the last 10 days of May to the middle 10 days of June, and the second peak is from the middle 10 days to the last 10 days of September (Zhou *et al.*, 2006).

Natural enemies include acaropathogenic fungi (Ascomycota, Clavicipitaceae and Exobasidiomycetidae), predatory mites (cheyletids, stigmaeids, tydeids, cunaxids, smaridids and phytoseiids) and insects (coccinellids, staphylinids, anthocorids, chrysopids and eolothripids) (Tables 20.1 and 20.2).

Symptomatology and damage

Among *Citrus* spp., the oriental red mite prefers sour lemon to sweet lemon, mandarin and orange (Mohamed, 1965), and lays a higher number of eggs on lemon leaves than on orange and mandarin (Rasmy, 1977). Feeding on the upper leaf surface and the fruits causes stippling, similar to that of the citrus red mite, *Panonychus citri* (McGregor). At high population densities, the trees become silver–grey and the leaves stop their growth and may fall and the shoots die back. Bare trees represent a serious problem in the nursery or in young orchards. The combined effect of insufficient water and low infestation produces as much defoliation and twig dieback as a heavy attack (Jeppson *et al.*, 1975; Jeppson, 1989). In India, *E. orientalis* is a major pest of citrus and sometimes infests pear, peach, ber (*Z. mauritiana*), cucurbits and cotton, damaging the upper surface of fully grown leaves. It has also been found on papaya in the Varanasi region in the eastern part of Uttar Pradesh. Infestations start along the midrib, with mites moving further away as the size of the population increases. The webbing produced by the mites traps dust particles, giving the upper leaf surface a dirty appearance (Singh and Raghuraman, 2011).

Control

Hot dry environmental conditions and/or irrational programmes of control can stimulate outbreaks of mite populations. In Egypt, if citrus orchards are not treated with chemicals, *E. orientalis* is no longer a problem (Rasmy, 1969). When sprays are applied against other citrus pests, it is necessary to add acaricides to control phytophagous mites (Rasmy, 1970a). In South Africa, in orchards where pyrethroids have been sprayed against other pests, attacks are more severe and an average of three or more females per leaf justifies chemical control (Meyer Smith, 1998). In Egypt, one spray application of acaricide as populations of *E. orientalis* begin to increase is sufficient to maintain control throughout the year (Wafa *et al.*, 1969). Plant resistance may hinder mite development.

PLANT RESISTANCE. The possibility of plant resistance has been widely investigated on citrus in India (Sadana and Kanta, 1972; Dhooria and Sandhu, 1973; Dhooria, 1982; Singh *et al.*, 1983; Bhumannavar *et al.*, 1988), and in other countries. In Egypt, it was observed that seedlings of sour orange and citron were the most favoured hosts as they were severely infested, followed by sweet orange, Morrakishi orange, navel orange and mandarin (Rasmy and Zaher, 1983). Certain structural and physiological characteristics of host varieties, for example the density of the oil glands and the quantity of tannic acid in the leaves, display a positive correlation with the rate of population development on

leaves; in contrast, there is a negative correlation with the thickness of the cuticle (Mohamed, 1965). Protein concentration in the leaves is positively correlated with mite infestation, and lemon (cv. 'Baramasi') contains phenols and free amino acids, which both seem to be negatively correlated with mite infestation (Jyotika and Mandeep, 2003). Studies on the biochemical components of *C. limon* (cv. 'Baramasi'), *C. aurantiifolia* (kaghzi lime) and *C. paradisi* (grapefruit) and resistance/susceptibility to *E. orientalis* confirm that protein concentration is positively correlated with mite attack. 'Baramasi' lemon contains the least amount of phenols and free amino acids, and both these constituents seem to be negatively correlated with mite infestation.

Table 20.1. *Eutetranychus orientalis* (Klein). Natural enemies (fungi, insects and mites).

Families and species	References
FUNGI	
Ascomycota	
Verticillium lecanii (Zimmerman) Viégas	Sewify and Mabrouk, 1990; El-Hady, 2004
Clavicipitaceae	
Hirsutella kirchneri (O. Rostr.) Minter *et al.*	Sztejnberg *et al.*, 1997
Hirsutella thompsonii Fisher	McCoy and Selhime, 1974; Gerson *et al.*, 1979; Dhooria and Butani, 1984
Metarhizium anisopliae (Metschnikoff) Sorokin	El-Hady, 2004
Exobasidiomycetidae	
Acaromyces ingoldii Boekhout *et al.*	Paz *et al.*, 2007a
Meira argovae Boekhout *et al.*	
Meira geulakonigii Boekhout *et al.*	
Aeolothripidae	
Scolothrips sp.	Meyer Smith, 1998
Scolothrips indicus Priesner	Lal, 1977; Dhooria, 1983
Scolothrips longicornis Priesner	Heikal *et al.*, 1990
INSECTS	
Anthocoridae	
Orius spp.	Meyer Smith, 1998
Chrysopidae	
Chrysopa spp.	Meyer Smith, 1998
Coccinellidae	
Parastethorus histrio (Chazeau)	Dhooria, 1981
Stethorus gilvifrons Mulsant	Afshari *et al.*, 2007; Imani *et al.*, 2009
Stethorus pauperculus (Weise)	Banu and ChannaBasavanna, 1972
Staphylinidae	
Oligota spp.	Meyer Smith, 1998
Oligota (*Domatium*) *oviformis* (Casey)	Banu and ChannaBasavanna, 1972
MITES	
Cheyletidae	
Acaropsellina sollers (Rohdendrof)	Nassar and Kandeel, 1986
Acaropsis aegyptiaca Wafa *et* Soliman	Wafa *et al.*, 1970
Cheletogenes ornatus (Canestrini *et* Fanzago)	Rizk *et al.*, 1983; Rezk and Gadelhak, 1996
Chiapacheylus macrocorneus Zaher *et* Soliman	Zaher and Soliman, 1971
Eutogenes punctata Zaher *et* Soliman	Hassan, 1989
Cunaxidae	
Cunaxa capreolus (Berlese)	Zaher *et al.*, 1975
Smarididae	
Smaris biscutatus Meyer *et* Rike	Meyer Smith, 1998
Stigmaeidae	
Agistemus buntex Chaudhri	Khan and Muhammad, 2005
Agistemus exsertus Gonzalez Rodriguez	Rizk *et al.*, 1983; Rezk and Gadelhak, 1996
Tydeidae	
Pronemtus ubiquitus (McGregor)	Rasmy, 1969; Rasmy, 1970b
Tydeus californicus (Banks)	Rizk *et al.*, 1983
Tydeus kochi Oudmans	Rasmy, 1970b
Tydeus munsteri Meyer *et* Ryke	Meyer Smith, 1998

Table 20.2. *Eutetranychus orientalis* (Klein). Natural enemies (Phytoseiidae).

Species	References
Amblyseius zaheri Yousef *et* El-Borolossy	Rasmy *et al.*, 2003
Cydnoseius negevi (Swirski *et* Amitai)	Abou-Awad *et al.*, 1989
Euseius alstoniae (Gupta)	Dhooria, 1983
Euseius bwende (Pritchard *et* Baker)	Meyer Smith, 1998
Euseius citri (van der Merwe *et* Ryke)	Meyer Smith, 1998
Euseius finlandicus (Oudemans)	Sharma and Sadana, 1987
Euseius gossipi (El-Badry)	El-Enany *et al.*, 1993
Euseius hibisci (Chant)	Swirski *et al.*, 1967, 1970
Euseius olivi (Nasr *et* Abou-Awad)	Abou-Awad *et al.*, 1998b; Momen and El-Borolossy, 1999
Euseius ovalis (Evans)	Shih *et al.*, 1993
Euseius rubini (Swirski *et* Amitai)	Swirski *et al.*, 1967, 1970
Euseius scutalis (Athias Henriot)	Tanigoshi *et al.*, 1990; Abdel-Samad *et al.*, 1996; Neena Singla, 2001
Euseius stipulatus (Athias Henriot)	Garzón-Hidalgo and Ferragut, 2010
Metaseiulus occidentalis (Nesbitt)	Swirski and Dorzia, 1969
Neoseiulus barkeri Hughes	Abou Awad *et al.*, 1998b; Momen and El-Borolossy, 1999
Neoseiulus californicus (McGregor)	Ibrahim *et al.*, 2005
Neoseiulus longispinosus (Evans)	Angsumarn, 1994
Paraseiulus talbii (Athias Henriot)	Abou-Awad *et al.*, 1998b; Momen and El-Borolossy, 1999
Phytoseius plumifer (Canestrini *et* Fanzago)	Nawar *et al.*, 2001a
Proprioseiopsis asetus (Chant)	Fouly, 1997
Proprioseiopsis badryi (Yousef *et* El-Borolossy)	Abou Awad *et al.*, 1998a; Momen and El-Borolossy, 1999
Proprioseiopsis cabonus (Schicha *et* Elshafie)	Abou Awad *et al.*, 1998b; Momen and El-Borolossy, 1999
Proprioseiopis lindquisti (Schuster *et* Pritchard)	Abou Awad *et al.*, 1998b; Momen and El-Borolossy, 1999
Typhlodromips deleoni (Muma)	Rasmy *et al.*, 2000
Typhlodromips swirskii (Athias Henriot)	Hoda *et al.*, 1987
Typhlodromus athiasae Porath *et* Swirski	Abou-Awad *et al.*, 1998b; Momen and El-Borolossy, 1999; Nawar *et al.*, 2001b
Typhlodromus balanites El-Badry	Abou-Awad *et al.*, 1998b; Momen and El-Borolossy, 1999
Typhlodromus citri Hassan *et* Romeih	Romeih *et al.*, 2005
Typhlodromus mangiferus Zaher *et* El-Borolossy	El-Halim *et al.*, 2004
Typhlodromus praeacutus van der Merwe	Meyer Smith, 1998
Typhlodromus pyri Scheuten	Rasmy, 1970b
Typhlodromus transvaalensis (Nesbitt)	Abou-Awad *et al.*, 1998b; Momen and El-Borolossy, 1999

This information can be used in the development of strategies (e.g. nutritional manipulation via fertilizers and breeding resistant cultivars) for the control and management of the pest mite (Kapur-Ghai and Mandeep Kaur, 2003b). The results of studies on the differential susceptibility of *C. limon* (cv. Baramasi), *C. aurantiifolia* (kaghzi lime) and *C. paradisi* (grapefruit) to *E. orientalis*, and its correlation with abiotic factors, showed that 'Baramasi' lemon is the most heavily infested, followed by kaghzi lime and grapefruit, which is the least infested. No significant variation occurs in the mite population with respect to the direction (aspect), but the mite population is positively correlated with temperature and sunshine, and negatively correlated with rainfall, irrespective of the cultivar. RH is negatively correlated with the mite population in kaghzi lime, but positively correlated in the other two cultivars (Kapur-Ghai and Mandeep Kaur, 2003a). Assessment for susceptibility to oriental red mite carried out on *Z. mauritiana* cultivars 'Gola', 'Jogia', 'Chomulocal', 'Mundia', 'Sev' and 'Umran' showed that 'Chomulocal' was the least susceptible, and 'Gola' the most susceptible, while the other cultivars were moderately susceptible (Yadav *et al.*, 2000).

BIOLOGICAL CONTROL. In spite of the numerous natural enemies of the oriental red mite that are well known throughout the world, biological control alone is insufficient and chemical control is usually necessary. Natural enemies tested include acaropathogenic fungi and predatory mites and insects.

The pathogens that have been tested against *E. orientalis* include *Verticillium lecanii* (Zimmerman) Viégas and *Metarhizium anisopliae* (Metschnikoff) Sorokin, which were studied in trials carried out at 22 and 28°C on different stages of *E. orientalis*. *M. anisopliae* was more effective than *V. lecanii* against all stages and temperature played an effective role in the susceptibility of the different stages to the pathogens. Treatment of the eggs resulted in a reduced rate of hatchability. Newly hatched larvae from the treated eggs failed to develop and died soon after emergence (El-Hady, 2004). In the laboratory, *Meira argovae* Boekhout *et al.* and *M. geulakonigii* Boekhout *et al.* produced 83.6 and 87.8% mortality after 14 days, while *Acaromyces ingoldii* Boekhout *et al.* caused 83.9 and 79.2% mortality, respectively, after 7 and 14 days (Paz *et al.*, 2007a).

Phytoseiid mites feeding on *E. orientalis* exhibit different responses, such as *Typhlodromus athiasae* Porath *et* Swirski, *Amblyseius barkeri* (Hughes) (= *Neoseiulus barkeri* Hughes) and *A. olivi* Nasr *et* Abou-Awad (= *Euseius olivi* (Nasr *et* Abou-Awad), which have a shorter developmental period than *T. transvaalensis* (Nesbitt). *T. athiasae* and *N. barkeri* have the highest oviposition rates. *E. olivi* shows a low rate of oviposition, although *T. transvaalensis* is not able to lay any eggs. Survival of immatures of *T. talbii* Athias-Henriot (= *Paraseiulus talbii* (Athias Henriot)), *T. balanites* El-Badry, *A. badryi* Yousef *et* El-Borolossy (= *Proprioseiopsis badryi* (Yousef *et* El-Borolossy)), *A. cabonus* (Schicha) (= *Proprioseiopsis cabonus* (Schicha *et* Elshafie) and *A. lindquisti* Schuster *et* Pritchard (= *Proprioseiopsis lindquisti* (Schuster *et* Pritchard)) is very low on *E. orientalis* and all phytoseiids fail to develop to adulthood (Momen and El-Borolossy, 1999). Studies on releases of *N. californicus* at three levels (40, 50 and 70 individuals/tree) at an average level of infestation of 4.64, 5.06 and 4.70 *E. orientalis*/leaf, respectively, showed that after releasing of phytoseiid mites, the population of *E. orientalis* decline gradually and the percentage reduction in its population after 4 months of releasing reached 57.84, 73.76 and 88.25%, respectively (Ibrahim *et al.*, 2005).

Research in Korea shows that both *Tetranychus turkestani* (Ugarov *et* Nikolskii) and *E. orientalis* are essential prey for the development and reproduction of the coccinellid *Stethorus gilvifrons* Mulsant and that *E. orientalis* is slightly more suitable than *T. turkestani* (Imani *et al.*, 2009).

CHEMICAL CONTROL. Active ingredients tested in the control of the oriental red mite include organochlorines (El-Kady *et al.*, 1977; Dhooria, 1983; Dhooria and Butani, 1984; El-Zemaity *et al.*, 1984; Deshpande *et al.*, 1988; Iskander *et al.*, 1993; Helal *et al.*, 1995; Rezk and Gadelhak, 1996), petroleum oils (Jeppson *et al.*, 1975; Iskander *et al.*, 1993; Ibrahim, 1994; Osman *et al.*, 1994; Osman, 1997), sulfur (Bodenheimer, 1951; Klapperich, 1957; Jeppson *et al.*, 1975; Singh *et al.*, 1986; Deshpande *et al.*, 1988; Jeppson, 1989; Meyer Smith, 1998) and other acaricides that are effective in controlling other tetranychid mites.

Yadav *et al.* (2003b) tested the efficacy of monocrotophos, ethion, dicofol, dimethoate, phosphamidon, endosulfan, cypermethrin and phosalone on common jujube infested by *E. orientalis*. Monocrotophos was the most effective, followed by phosphamidon and ethion. Cypermethrin was the least effective and the other acaricides were moderately effective. Arora *et al.* (2004) found that dimethoate, ethion, quinalphos and triazophos were equally effective even 10 days after spraying. Following the application of ethion to pear, the initial deposit of the chemical was observed to be less than its maximum residue limit of 2.0 mg kg^{-1}. However, based on an acceptable daily intake of ethion of 0.002 mg kg^{-1}, a waiting period of 10 days is recommended for the safe consumption of pear fruits following application of ethion at the recommended dose. Marquez *et al.* (2006) report that several phytosanitary treatments a year are usually required for the control of *E. orientalis* on citrus in Spain. Dicofol, propargite, hexythiazox, etoxazole and fenpyroximate are effective against the mite. Nevertheless, the compared efficacy showed differences as the assay progressed. Dicofol and fenpyroximate caused an early and persistent efficacy of nearly 100%. Hexythiazox and etoxazole showed a 95% efficacy, but this was observed 3 weeks after the treatment. Propargite showed a good early impact, but its efficacy finally decreased (Marquez *et al.*, 2006). Laboratory studies on the deterrent and toxic effects of two essential oils, those from *Majorana hortensis* Moench (*Origanum vulgare* L.) and *Rosmarinus officinalis* L., on *E. orientalis* showed that increasing concentrations of the two oils increased mortality of the mite and reduced the total numbers of eggs laid (Amer *et al.*, 2001).

INTEGRATED PEST MANAGEMENT. The rational use of insecticides, and appropriate spray application of petroleum oils and selected acaricides prevent the development of mite populations and indirectly allow realization of a preliminary condition of IPM. The role of resistant citrus cultivars may be complementary.

Eutetranychus palmatus Attiah (Fig. 20.7)

Common name

Unknown.

Diagnostic characteristics

FEMALE. The prodorsum has mediodorsal striae that are longitudinal and slightly undulating; the striae are lobate and between the setae d_1 and e_1 forming a 'V' pattern. The dorsal body setae are lanceolate, serrate and not set on tubercles; the prodorsal setae v_2 and sc_{1-2} and the opisthosomal setae f_1, h_1 and f_2 tend to be longer than other dorsal body setae; the setae e_1 form a slight rectangle with setae f_1. The spermatheca is sharply pointed distally (Meyer Smith, 1987).

Geographical distribution

Egypt (Attiah, 1967b; Zaher *et al.*, 1982), Iran (Kamali, 1990), Israel (Gerson *et al.*, 1983; Ben-David *et al.*, 2007) and the Palestinian territories (Ben-David *et al.*, 2007).

Host plants

E. palmatus is reported on six host plants, five of which belong to the Arecaceae: *Hyphaene* sp., *Phoenix canariensis* Chabaud (Ben-David *et al.*, 2007), date palm (*P. dactylifera* innnaeus) (Attiah, 1967b; Zaher *et al.*, 1982), *P. loureirii* Kunth (Ben-David *et al.*, 2007), *Washingtonia* sp. (Bolland *et al.*, 1998; Migeon and Dorkeld, 2006–2013). One species belongs to the Malvaceae (*Gossypium* sp.) (Bolland *et al.*, 1998; Migeon and Dorkeld, 2006–2013).

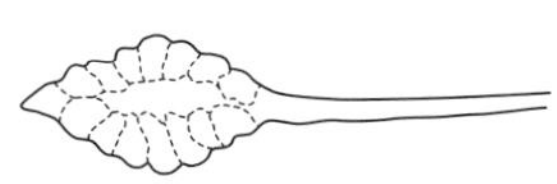

Fig. 20.7. *Eutetranychus palmatus* Attiah. Female, receptaculum seminis (from Meyer Smith, 1987).

Bio-ecology

E. palmatus lives mostly on species of the Arecaceae. Gerson *et al.* (1983) report the mite on date fronds throughout the year, and it infests the fruit during midsummer; when the dates turn yellow, it is again restricted to the fronds. The mite is solitary, produces little webbing and lives on fronds from midwinter and throughout the summer; the highest population densities occur from May to early June; later on, in July and August, the populations decrease.

Adult cumulative mite days on date fronds are always considerably higher than those of the Old World date mite, *Oligonychus afrasiaticus* (McGregor), another spider mite that lives on the same host plant. Thus, studies in Israel by Palevsky *et al.* (2010a) showed that the seasonal cumulative mite days (SCMDs) of *E. palmatus* on fronds of the cultivars 'Barhi', 'Medjool' and 'Deglet Noor' were 2100, 28,176 and 5429, respectively, while, the SCMDs for *O. afrasiaticus* were 434, 407 and 498, respectively. Infestations of *E. palmatus* on fruit strands of date palm began earlier than those of *O. afrasiaticus*, but peak numbers of the latter species were usually much higher than those of *E. palmatus.* Populations of *E. palmatus* reach a peak in late April, before those of *O. afrasiaticus*. However, the mites coexist, and reach their highest but different levels by July on the fruit strands of all cultivars; on 'Barhi', 567 *E. palmatus* were sampled per strand, and on 'Deglet Noor', 8341 *O. afrasiaticus* were sampled per strand (Palevsky *et al.*, 2010a). Despite these differences, the two mites may be confused in the field, although knowledge of their phenology allows them to be distinguished. In particular, *E. palmatus* occurs on date fronds throughout the year, but *O. afrasiaticus* infests fronds only from late April until late August. Moreover, *E. palmatus* attacks the fruits earlier in the season than *O. afrasiaticus*. Populations of *E. palmatus* usually have low densities, seldom reaching 1000 mites per fruit strand, in contrast to *O. afrasiaticus*, whose populations increase very quickly during June–August and can reach densities higher than 1000 mites per fruit strand. In addition, *E. palmatus* produces very little webbing, and is a solitary species living on the fronds throughout the year, while *O. afrasiaticus* produces large amounts of webbing that are able to host its dense colonies, which rarely occur on the fronds (Palevsky *et al.*, 2010a). The natural enemies of *E. palmatus* are as those of *O. afrasiaticus*.

Symptomatology and damage

The mite is a potential or minor pest of date palm, but the potential damage caused by it is lower than that of *O. afrasiaticus*. Feeding by *E. palmatus* causes chlorotic scarring on date fruit, and on some cultivars ('Barhi'), the blemishes caused by its feeding are clearly visible because the fruit is picked when it is still yellow. The mite may become an important pest when *O. afrasiaticus* does not occur (Palevsky *et al.*, 2010a).

Control

Despite the relatively minor pest status of *E. palmatus* compared with *O. afrasiaticus*, in Israel, growers who have experienced spider mite damage caused by high populations of *E. palmatus* very early in the season (April–May) apply acaricides to control the mite (Palevsky *et al.*, 2010a).

The Genus *Eurytetranychus* Oudemans

Diagnostic characteristics

The genus has two pairs of para-anal setae (h_{2-3}), the empodial claw is small, and tarsus I has 1 pair of associated setae (Bolland *et al.*, 1998).

Economic importance

This genus includes 17 species, living mainly on Cupressaceae, Pinaceae, Ulmaceae and Fagaceae. Only a few species live on herbaceous plants and/or shrubs.

Four species of *Eurytetranychus* have an Afrotropical distribution. These are *E. madagascariensis* Gutierrez, which lives on *Nerium oleander* L. and *Calotropis procera* (Aiton) W.T. Aiton in Madagascar (Gutierrez, 1966), *E. acaciae* Meyer, which lives on *Acacia kraussiana* (= *Acacia nilotica* subsp. *kraussiana* (Benth.) Brenan?) (Meyer Smith, 1987), *E. spathatus* Meyer on an unknown host and *E. neobuxi* Meyer, which lives on *Diplorhynchus condylocarpon* Muell. Arg. (Meyer Smith, 1974), *Tabernaemontana elegans* Stapf, *Elytropappus scaber* L.f., Levyns, *Euclea racemosa* Hiern, *Otholobium fructicans* (L.) C.H. Stirt., *Psoralea* sp. and *Myrica cordifolia* L. in South Africa (Meyer Smith, 1987).

Of the other species of *Eurytetranychus* that have been reported, *E. admes* Pritchard *et* Baker is reported on *Cupressocyparis leylandii* (A.B. Jacks. *et* Dallim.) Dallim., *Juniperus virginiana* L. (Migeon, 2003), *J. communis* L., *J. sabina* L. and *Juniperus* sp. (Pritchard and Baker, 1955; Tuttle and Baker, 1976), and also on *Calocedrus decurrens* (Torr.) Florin (Pritchard and Baker, 1955), *Calocedrus* sp. (Thewke and Enns, 1969), *Picea abies* (L.) H. Karst. and *P. glauca* (Moench) Voss (Migeon and Dorkeld, 2006–2013) from the USA (Pritchard and Baker, 1955), France (Migeon, 2003) and Hungary (Bozai, 1970). *E. cyclobalanopsis* Hu *et* Chen is reported on *Quercus glauca* (Thunb.) in China (Hu and Chen, 1994). *E. fengchengensis* Ma *et* Yuan and *E. huaqingnicus* Ma *et* Yuan are reported on *Ulmus* sp. from Palaearctic China (Ma and Yuan, 1981a). *E. furcisetus* Wainstein is reported on *Picea pungens* Engelm., *P. schrenkiana* Fisch. *et* C.A. Mey. and *Pinus silvestris* L. in Kazakhstan (Wainstein, 1956b, 1960; Mitrofanov *et al.*, 1987) and in Hungary (Bozai, 1974) and Kyrgyzstan (Mitrofanov *et al.*, 1987). *E. koreanus* Ehara inhabits the ulmaceous *Hemiptelea davidii* (Hance.) Planch. in South Korea (Ryu and Ehara, 1995). *E. mexicanus* Estebanes *et* Baker is reported on *Cupressus* sp. in Mexico (Estebanes-Gonzalez and Baker, 1968). *E. piceus* Löyttyniemi is reported on *Picea glehnii* F. Schmidt, *P. jezoensis* (Siebold *et* Zucc.) Carr. and *P. koyamai* Shiras in Finland (Löyttyniemi, 1973). *E. recki* Bagdasarian is reported on *Amblytropis* sp. and *Astracantha caucasica* (Pall.) Plodech (Bagdasarian, 1948), *Caragana*

arborescens Lam., *C. sinica* (Buc'hoz) Rehder (Wang, 1977), *Colutea arborescens* L., *Cytisus scoparius* (L.) and *Medicago sativa* L. (Bagdasarian, 1948; Wang, 1977) and on *Thymus* sp. and *Spiraea* sp. (Migeon and Dorkeld, 2006–2013) in Armenia (Bagdasarian, 1948; Mitrofanov *et al.*, 1987), Azerbaijan (Mitrofanov *et al.*, 1987), China (Wang, 1977) and Georgia (USSR) (Mitrofanov *et al.*, 1987). *E. shenyangensis* Yin, Lu *et* Lan, was collected on *Ulmus* sp. in China (Yin *et al.*, 1987). *E. ulmi* Wang was collected on *Ulmus davidiana* Planch. and *U. pumila* L. in China (Wang, 1977) and Italy (Bolland and Beron, 2004). *E. wuyishanensis* Hu *et* Chen was collected on *Castanea mollissima* Blume in China (Hu and Chen, 1994). *E. buxi* (Garman) is reported on several species of Buxaceae and is found on various European countries, Iran and the USA (see next section).

None of the mites mentioned above generally plays a severe phytopathological role, with the exception of the boxwood mite, *E. buxi* (Garman), which is sometimes injurious to boxwood, *Buxus sempervirens* L (see below).

Eurytetranychus buxi (Garman) (Fig. 20.8)

Common name

Boxwood mite.

Diagnostic characteristics

FEMALE. The body is 450–500 μm long, and yellowish green or reddish in colour. The peritreme ends in a simple bulb. The dorsal setae are as long as the distances between their insertion. The empodia are claw like, tibiae III and IV lack solenidia, and tarsus I has a single set of duplex setae that are somewhat separated (Jeppson *et al.*, 1975).

MALE. The male is smaller than the female, and pear like. The aedeagus is hook like, bent upwards and rounded at the tip.

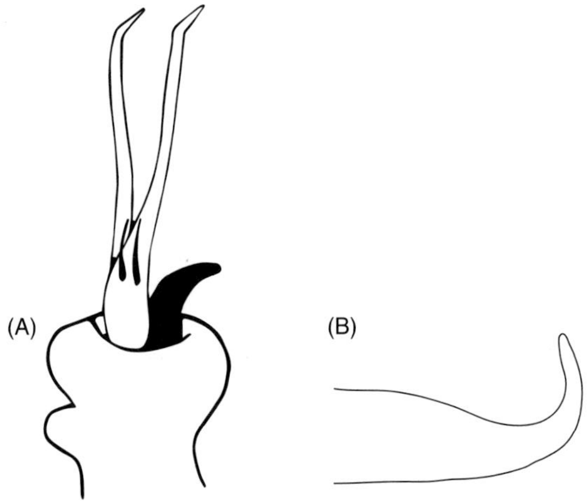

Fig. 20.8. *Eurytetranychus buxi* (Garman). (A) female, pretarsus I; (B) male, aedeagus (drawing by Vacante).

Geographical distribution

Armenia (Bagdasarian, 1948; Mitrofanov *et al.*, 1987), Azerbaijan (Mitrofanov *et al.*, 1987), Belgium (Bolland *et al.*, 1998), France (Navajas *et al.*, 1996), Georgia (USSR) (Mitrofanov *et al.*, 1987), Hungary (Bozai, 1970), Iran (Khalil-Manesh, 1973), Ireland (Curry, 1976), Italy (Canestrini and Fanzago, 1876; Berlese, 1889), the Netherlands (Helle and Bolland, 1967; Vierbergen, 1990), Poland (Łabanowski, 1993), Portugal (Carmona and Dias, 1980), Spain (Bolland, 2001), Ukraine (Mitrofanov *et al.*, 1987), the UK (Jeppson *et al.*, 1975) and the USA (Ewing, 1932; Ries, 1935; McGregor, 1950; Pritchard and Baker, 1955).

Host plants

The boxwood mite is reported on a restricted number of Buxaceae, such as the littleleaf boxwood (*Buxus microphylla* Siebold *et* Zucc.) (Pritchard and Baker, 1955), common box (*B. sempervirens* L.) (Berlese, 1889; Ries, 1935; Pritchard and Baker, 1955; Vierbergen, 1990; Łabanowski, 1993; Navajas *et al.*, 1996) and *Buxus* sp. (Pritchard and Baker, 1955; Reeves, 1963).

Bio-ecology

E. buxi lives on species of Buxaceae. It has a haplodiploid reproductive system, and the haploid number of chromosomes is five (Helle and Bolland, 1967). The species is also known to have a thelytokous strain (Ries, 1935). The boxwood mite overwinters as eggs mainly on the upper leaf surface, which hatch from early April to May. The biological cycle takes from 18 to 21 days, and larval, protonymphal and deutonymphal development lasts 2–4 days, 2–3 days and 3–7 days, respectively. The fecundity is 25–35 eggs and a female lays 3–4 eggs a day. The eggs hatch in 6–10 days. The mite usually develops at least eight generations a year, and the overwintering eggs are laid from late September to early October. Better conditions for mite development occur at high temperatures and low humidities. The lifespan of the adults is 2–5 weeks at constant temperatures (Ries, 1935; Reeves, 1963; Jeppson *et al.*, 1975). According to the guidelines of the University of Maryland Extension, the mite breeds rapidly and has 5–6 generations a summer (Malinoski *et al.*, 2009). Few natural enemies are known (Malinoski *et al.*, 2009), but they include the coccinellid *Stethorus punctillum* Weise (Creary, 2009).

Symptomatology and damage

The feeding activity of the boxwood mite causes mottling of the leaves early in the season, followed by yellowing and browning near the mid-vein on the undersurface of the leaf. The upper leaf surface shows small comma-shaped spots. At high density of attack, the leaves are entirely bronzed and may drop, leaving a scraggy looking plant. At high density of attack, the mite lives on both leaf surfaces, and usually feeds on the tender shoots and the upper surface of the young leaves (Ries, 1935; Reeves, 1963; Jeppson *et al.*, 1975). Japanese boxwood is

less susceptible to the mites, but its cultivars are very susceptible (Malinoski *et al.*, 2009).

Control

Control measures are provided by spraying against overwinter eggs before new plant growth begins in the spring and/or sprayings during summer. Both possibilities require monitoring of populations. As for other economic plants, the use of pesticides must be carefully evaluated to avoid their side effects, as occurs when using imidacloprid (Szczepaniec and Raupp, 2013).

MONITORING. Mite infestations can be monitored by beating a small branch over a white sheet of paper. The mites will fall on to the paper and move around. Damage occurs at about 25 mites per beat. The American Boxwood Company has reported that this amount of infestation allows natural enemies to become established.

CHEMICAL CONTROL. Overwintering eggs can be destroyed by a delayed dormant oil spray on the underside of the leaves. Insecticidal oil and soap are effective against light populations when summer control measures are required. The American Boxwood Company has reported that heavy summer populations can be treated with bifenthrin, dicofol, dimethoate, fluvalinate, hexythiazox, lambda-cyhalothrin, or malathion. The crop and pest management guidelines of Cornell University (2012) suggest treating in the first 2 weeks of May, at 245–600 growing degree-days (GDDs), using abamectin, bifenazate, bifenthrin, dimethoate, fenazaquin, fenpyroximate, fluvalinate, hexythiazox, horticultural oil, insecticidal soap, malathion, milbemectin or neem oil. The guidelines of the University of Maryland Extension suggest applying a dormant rate of horticultural oil to the undersides of the leaves before new plant growth begins in the spring. Light summer populations of mites can be controlled with a summer rate of horticultural oil or insecticidal soap sprays, but heavy infestations may require a residual acaricide application (Malinoski *et al.*, 2009). In Belgium, clofentezine, hexythiazox, flucycloxuron, bromopropylate, fenbutatin oxide, pyridaben and tebufenpyrad were tested on *B. sempervirens*. Effective control of up to 50% of developing overwinter eggs was possible with sprayings of clofentezione and hexythiazox at the end of April. All acaricides controlled 50–100% of developing overwinter eggs, with flucycloxuron being the best. Mixed populations of overwinter and summer eggs were best controlled with hexythiazox, followed by clofentezione and flucycloxuron (Hellemans and Goossens, 2000).

The Genus *Schizotetranychus* Trägårdh

Diagnostic characteristics

The opisthosoma has 10 pairs of dorsal setae. The empodium is claw like and split into 2 claw-like structures (Bolland *et al.*, 1998). The distal segment of the palpus has 3 simple setae (*a*, *b*, *c*), 3 eupathidia (*ul′*, *ul″*, *su*), and 1 solenidion (ω) (Saitō *et al.*, 2004).

Economic importance

The genus *Schizotetranychus* has worldwide distribution and currently includes 115 species that live on monocotyledonous and dicotyledonous plants (Migeon and Dorkeld, 2006–2013). Recently, Saitō *et al.* (2004) redescribed the genus *Stigmaeopsis* Banks and included in it seven species of *Schizotetranychus*, all with six palpal setae. Many *Schizotetranychus* spp. live on economic plants, but the damage they cause is often poorly documented or unknown and does not allow a reliable classification of their impact. Only four species are categorized as at risk of damaging economic plants: *S. andropogoni* (Hirst), *S. asparagi* (Oudemans), *S. bambusae* (Reck) and *S. hindustanicus* (Hirst). These four species are described below in separate sections devoted to each of them. A number of other species that may be found on various economic plants do not pose a serious risk, and these are briefly mentioned in the paragraphs that follow.

Several *Schizotetranychus* spp. live on citrus and other economic plants. The citrus green mite, *S. baltazarae* Rimando, is reported on *Citrus grandis* (L.) Osbeck, *C. madurensis* Lour., *C. medica* L., *C. sinensis* (L.) Osbeck (Gupta and Gupta, 1994) and *Citrus* sp. (Prasad, 1974), and also on *Murraya koenigii* (L.) Sprengel (Gupta and Gupta, 1994), *Dioscorea* sp. (Ehara and Wongsiri, 1975), *Azadirachta indica* (Gupta, 1992) and *Lansium domesticum* Corrêa (Ehara, 2004). Its geographical distribution includes China (Manson, 1963; Wang, 1981b; Wang *et al.*, 1985; Wang and Cui, 1992), India (Manson, 1963; Gupta, 1992; Gupta and Gupta, 1994), Indonesia (Ehara, 2004), Japan (Ohashi *et al.*, 2009), Myanmar (Manson, 1963), the Philippines (Rimando, 1962b), Taiwan (Lo and Hsia, 1968) and Thailand (Ehara and Wongsiri, 1975). *S. spiculus* Baker *et* Pritchard is reported on *Citrus* sp. (Baker and Pritchard, 1960) and *M. koenigii* (Migeon and Dorkeld, 2006–2013) in Kenya (Baker and Pritchard, 1960) and India (Karuppuchamy and Mohanasundaram, 1987).

A group of species including *S. andropogoni* (Hirst), *S. freitezi* Ochoa *et al.*, *S. oryzae* Rossi de Simons, *S. paezi* Alvarado *et* Fréitez, *S. pseudolycurus* Ochoa *et al.* and *S. yoshimekii* Ehara *et* Wongsiri lives on rice (*Oryza sativa* L.) in Neotropical and Oriental regions, and are included in the US Department of Agriculture's *Pest Risk Assessment for Rice Seed from the Western Hemisphere and Europe* (Cave and Randall-Schadel, 2012). Of these, *S. oryzae* has the wider geographical distribution and is reported on rice in Argentina, Brazil (Rossi Simons, 1966), Colombia (Zuluaga Cardona, 1971; Mejia *et al.*, 1998; Pantoja *et al.*, 1998), Costa Rica (Aguilar and Murillo, 2008), Suriname (Bolland *et al.*, 1998) and Venezuela (Alvarado and Fréitez, 1976); this species has also been collected on *Panicum maximum* Jacq. (Migeon and Dorkeld, 2006–2013). *S. yoshimekii* can lead to a reduction in rice yield of up to 30% (Wang, 1981a), but similar damage has not yet been confirmed.

Some *Schizotetranychus* spp. live on the pigeon pea (*Cajanus cajani* L.). These are *S. bhandhufalcki* Ehara *et* Wongsiri in Thailand (Baker, 1975b; Ehara and Wongsiri, 1975), *S. cajani* Gupta and *S. fluvialis* McGregor in India (Gupta, 1976; Gupta and Gupta, 1994) and *S. spireafolia* Garman (Migeon and Dorkeld, 2006–2013).

A group of *Schizotetranychus* species were reported on sugarcane (*Saccharum officinarum* L.), such as *S. sacharum*

Flechtmann *et* Baker in Brazil (Flechtmann and Baker, 1975) and *S. spireafolia* in India (Gupta, 1976) (also *S. andropogoni* in Thailand and India). *S. malayanus* Ehara has been collected on *Manihot* sp. in Indonesia (Ehara and Tho, 1988).

Several *Schizotetranychus* spp. live on forest plants. *S. alni* Beglyarov *et* Mitrofanov has been collected on *Alnus* sp. in eastern Russia (Beglyarov and Mitrofanov, 1973). *S. garmani* Pritchard *et* Baker is reported on *Acer* sp., *Quercus robur* L. (Migeon and Dorkeld, 2006–2013), *Populus tremula* L., *Salix caprea* L., *S. humilis* Marshall, *S. petiolaris* Sm. (Reeves, 1963) and *Salix* sp. (Pritchard and Baker, 1955); its geographical distribution includes the USA (Pritchard and Baker, 1955), Iran (Khalil-Manesh, 1973), Poland (Boczek and Kropczynska, 1964), eastern Russia (Mitrofanov *et al.*, 1987) and Switzerland (Günthart and Günthart, 1959). *S. jachontovi* Reck lives on *Quercus* sp. (Reck, 1953) in Georgia (USSR) (Reck, 1953; Mitrofanov *et al.*, 1987), Azerbaijan, eastern and western Russia and Ukraine (Mitrofanov *et al.*, 1987). *S. zhongdianensis* Wang *et* Cui lives on *Salix* sp. in China (Wang and Cui, 1992). The Nearctic and Palaearctic *S. schizopus* (Zacher) has a wide list of host plants, including *Vaccinium uliginosum* L. (McGregor, 1950) and several Salicaceae, such as *P. tremula*, *S. alba* L. (Reeves, 1963), *S. aurita* L., *S. babylonica* L. (Gotoh *et al.*, 2003a), *S. balsamifera* (Hook.) Barratt ex Anderss., *S. bicolor* Willd., *S. caprea* L., *S. daphnoides* Vill., *S. fragilis* L. (Dobosz *et al.*, 1995), *S. myrsinifolia* Salisb., *S. planifolia* Pursh, *S. purpurea* L. (Reeves, 1963), *Salix* sp. (Zacher, 1913; Reck, 1948b; Reeves, 1963; Vierbergen, 1990), *S. triandra* L. (Gotoh *et al.*, 2003a) and *S. viminalis* L. (Migeon and Dorkeld, 2006–2013).

A number of *Schizotetranychus* spp. are found on Poaceae (Bambusoideae) in Oriental, Palaearctic, Nearctic and Neotropical regions. *S chizotetranychus* mites from moso bamboo (*Phyllostachys pubescens* Mazel ex J. Houz.) have been described by Lin *et al.* (2000) and Zhang *et al.* (2000e). Species damaging bamboo forests in China include *S. elongatus* Wang *et* Cui (Cui, 1989), *S. bambusae* (Wang, 1981b; Ma *et al.*, 1984; Cui, 1989; Sun *et al.*, 1997a; Zhang *et al.*, 1997), *S. minutus* Wang (Cui, 1989) and *S. emeiensis* Wang (Cui, 1989). *S. camur* Pritchard and Baker is reported on *Arundinaria* sp. in the USA (Pritchard and Baker, 1955). *S. floresi* Rimando has been collected on *Bambusa bambos* (L.) Voss and *Arundo formosana* Hack. in the Philippines (Rimando, 1962a) and Taiwan (Tseng, 1990). *S. longirostris* Feres *et* Flechtmann and *S. paraelymus* Feres *et* Flechtmann are reported on *Bambusa* sp. in Brazil (Feres and Flechtmann, 1995), *S. saitōi* Ehara and *S. taquarae* Paschoal on *B. vulgaris* Schrad. *ex* J.C.Wendl. in Malaysia (Ehara and Tho, 1988) and Brazil (Paschoal, 1971a), respectively. *S. approximatus* Ehara lives on *B. vulgaris*, *Thyrsostachys siamensis* Gamble and other Poaceae in Malaysia (Ehara and Tho, 1988).

Schizotetranychus andropogoni (Hirst) (Fig. 20.9)

Common name

Web-spinning mite.

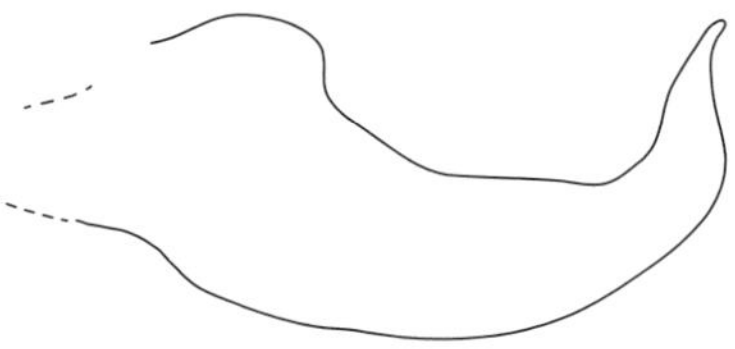

Fig. 20.9. *Schizotetranychus andropogoni* (Hirst). Male, aedeagus (from Hirst, 1926).

Diagnostic characteristics

FEMALE. The propodosoma is longitudinally striate, and the hysterosoma transversely striate. The dorsal setae are as long as the distances between their insertions, widened proximally and finely tapering distally. Tibia I has 6 tactile setae, tarsus I only 1 seta proximal to the duplex setae, and tibiae II and III have 5 tactile setae (Jeppson *et al.*, 1975). Variations are reported in leg chetotaxy of specimens collected in Oriental countries (Ehara and Wongsiri, 1975; Gupta and Gupta, 1994).

MALE. The male is smaller than the female, with relatively shorter dorsal setae. The aedeagus forms a dorsal directed sigmoid curve that is not acutely angled (Jeppson *et al.*, 1975).

Geographical distribution

Mexico (Jeppson *et al.*, 1975), India (Hirst, 1926; Gupta, 1992, 1995; Gupta and Gupta, 1994; Gupta and Chatterjee, 1997), Pakistan (Gupta, 1976) and Thailand (Baker, 1975b; Ehara and Wongsiri, 1975).

Host plants

The web-spinning mite is reported on eight species of host plants of three families. There is one species in the Asteraceae, *Zinnia* sp. (Gupta, 1984; Gupta and Gupta, 1994) and one in the Fabaceae, pigeon pea (*Cajanus indicus* Spreng. (= *Cajanus cajan* (L.)Millsp.) (Gupta and Gupta, 1994). The Poaceae includes six species that host *S. andropogoni*: *Andropogon annulatus* Forssk. (Hirst, 1920; Gupta and Gupta, 1994; Gupta and Chatterjee, 1997), rice (*Oryza sativa* L.) (Gupta, 1976, 1992, 1995; Gupta and Gupta, 1994), sugarcane (*Saccharum officinarum* L.) (Baker, 1975b; Ehara and Wongsiri, 1975; Gupta and Gupta, 1994; Gupta and Chatterjee, 1997), wild sugarcane (*Saccharum spontaneum* (L.)) (Pritchard and Baker, 1955; Gupta and Gupta, 1994; Gupta and Chatterjee, 1997), sorghum (*Sorghum bicolor* (L.) Moench) (Khan and Murty, 1956) and an unidentified poaceous species (Prasad, 1974; Gupta and Gupta, 1994).

Bio-ecology

S. andropogoni is a polyphagous mite reported from different countries of Nearctic and Oriental regions and has a haplodiploid reproductive system. In India, the web-spinning mite infests cereals and sugarcane. It is a sporadic post-monsoon pest of sugarcane throughout the country, occasionally infesting paddy in eastern and north-eastern India (Gupta and Gupta, 1994). The webbing that is produced turns brown and at a later stage

is blown off by the wind, hence dispersing the mites (Singh and Raghuraman, 2011).

Symptomatology and damage

Feeding by the mite on sugarcane, and occasionally on paddy, causes whitish patches in rows on either side of the midrib (Gupta and Gupta, 1994). It is a very detrimental pest of sugarcane in the northern part of India, causing the formation of white blotches on the leaf, which gradually dry out (Singh and Raghuraman, 2011).

Control

Sprayings of dicofol were not effective, but organophosphorus compounds suppressed mite outbreaks (Agarwal, 1957; Rattan *et al.*, 1962).

Schizotetranychus asparagi (Oudemans) (Fig. 20.10)

Common name

Pineapple mite, asparagus spider mite.

Diagnostic characteristics

FEMALE. The body is light green in colour. The propodosoma is longitudinally striate and the hysterosoma transversely striate. The striae bear semicircular lobes. The peritreme ends in a simple bulb, and the terminal sensillum of the palptarsus is longer than broad. All dorsal setae are setose and shorter than the distances between them. The setae c_1, c_2, d_1, d_2, e_1 and e_2 are about the same length and ¾ the length of the prodorsal setae v_2, sc_1, sc_2 and c_3. The setae f_1, f_2 and h_1 are slightly longer than other opisthosomal setae. Seate c_1, d_1 and f_1 are equally far apart but the setae e_1 are closer together. Tarsus I has 5 tactile setae proximal to the duplex setae (Meyer Smith, 1974).

MALE. The male is smaller than the female. The shaft of the aedeagus turns ventrad at an angle of about 90° and the distal end is approximately sigmoid; the knob has an acute anterior angulation and a relatively longer posterior projection (Meyer Smith, 1974).

Geographical distribution

Australia? (Davis, 1968a; unconfirmed), Germany (Jeppson *et al.*, 1975), Hawaiian Islands (USA) (Pritchard and Baker, 1955), Israel (Bolland *et al.*, 1998; Ben-David *et al.*, 2007), Malawi (Meyer Smith, 1974), Morocco (Saba, 1973b), the Netherlands (Oudemans, 1928), Portugal (Carmona, 1970), Puerto Rico (Pritchard and Baker, 1955), South Africa (Meyer Smith, 1974, 1987), the USA (McGregor, 1930, Pritchard and Baker, 1955) and Zimbabwe (Meyer Smith, 1974).

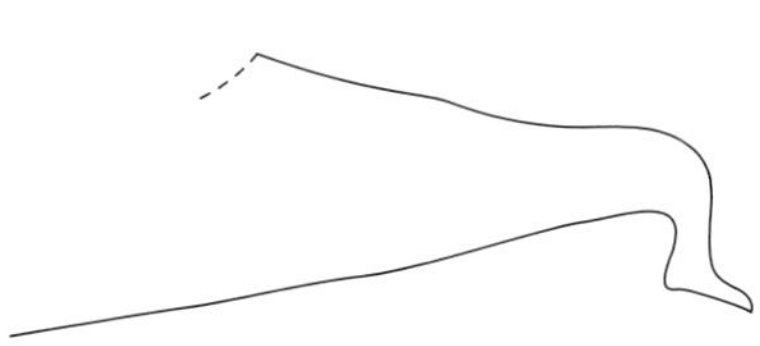

Fig. 20.10. *Schizotetranychus asparagi* (Oudemans). Male, aedeagus (from Meyer Smith, 1987).

Host plants

The list of host plants of the asparagus spider mite includes 11 species in three families: Asparagaceae – *Asparagus* sp. (Pritchard and Baker, 1955; Meyer Smith, 1974; Lee Goff, 1986), asparagus fern (*Asparagus aethiopicus* L.) (Oudemans, 1928), *A. africanus* Lam. (Meyer Smith, 1974), *A. capensis* (L.) Oberm. (Meyer Smith, 1987), *A. laricinus* (Burch.) Oberm. (Meyer Smith, 1974), asparagus (*A. officinalis* L.) (Garrett and Haramoto, 1967; Carmona, 1970; Saba, 1973b; Lee Goff, 1986), common asparagus fern (*A. setaceus* auct. non) (McGregor, 1930; Pritchard and Baker, 1955; Garrett and Haramoto, 1967; Carmona, 1970; Saba, 1973b; Meyer Smith, 1974; Lee Goff, 1986), *A. suaveolens* Burch. (Meyer Smith, 1974), *A. virgatus* Baker (Ben-David *et al.*, 2007); Bromeliaceae – *Ananas* sp. (Jeppson *et al.*, 1975); Fabaceae – *Acacia karroo* Hayne (Meyer Smith, 1974).

Bio-ecology

The asparagus spider mite has a wide distribution and has a haplodiploid reproductive system. It infests several species of Asparagaceae. The record of its presence in Australia is not confirmed (Halliday, 2000).

Symptomatology and damage

The mite usually causes severe damage to newly set pineapple plants. Infested plants remain small and produce little or no fruit. Severe infestations kill the plants (Jeppson *et al.*, 1975).

Control

Planting mite-free plants usually helps to control the pest (Jeppson *et al.*, 1975). Monitoring and control in the nursery are fundamental.

Schizotetranychus bambusae Reck (Fig. 20.11)

Common name

Bamboo spider mite.

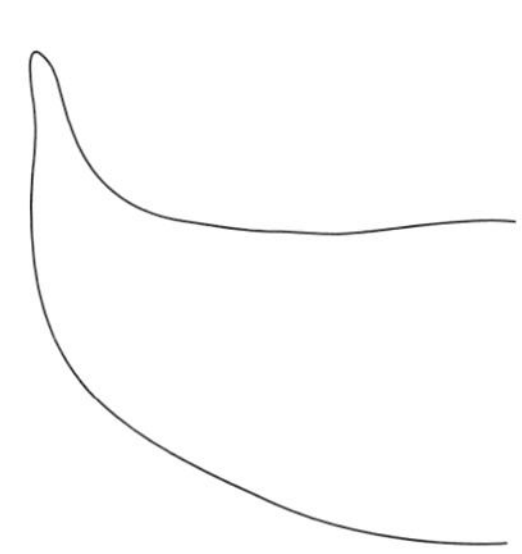

Fig. 20.11. *Schizotetranychus bambusae* Reck. Male, aedeagus (drawing by Vacante).

Diagnostic characteristics

FEMALE. The body is 365 μm long and 205 μm wide. The setae c_1, d_1, e_1 and f_1 are subequal and longer than the distances between their bases. The distal eupathidium of palptarsus ($su\zeta$) is subcylindrical with a rounded tip, twice as long as wide; two lateral eupathidia ($ul'\zeta$, $ul''\zeta$) are 7 and 9 μm long, respectively, and the solenidion (ω) is about four times as long as wide (Zhang *et al.*, 2000e).

MALE. The male is smaller than the female and pear like. The aedeagus is broad basally and gradually curves dorsally with a blunt tip (Zhang *et al.*, 2000e).

Geographical distribution

China (Wang, 1981b; Zhang *et al.*, 2000e, 2001a,c), France (Auger and Migeon, 2007), Georgia (USSR) (Reck, 1941b, 1948b; Mitrofanov *et al.*, 1987), Hainan (China) (Ma *et al.*, 1979), Japan (Ehara, 1973; Gotoh *et al.*, 2003a) and South Korea (Lee *et al.*, 1987).

Host plants

The bamboo spider mite is reported on several species of bamboo (Poaceae, Bambusoideae): *Arundinaria chino* (Franch. *et* Sav.) Makino (Auger and Migeon, 2007), *Arundinaria* sp. and *Bambusa* sp. (Ehara, 1973), *Phyllostachys* sp. (Reck, 1941b, 1948b), golden bamboo (*P. aurea* Rivière *et* C. Rivière) (Auger and Migeon, 2007), timber bamboo (*P. bambusoides* Siebold *et* Zucc.), moso bamboo (*P. edulis* (Carrière) J. Houz.) (Zhang *et al.*, 2000e, 2001a,c), black bamboo (*P. nigra* (Lodd. ex Lindl.) Munro) and ruscus bamboo (*Shibataea kumasaca* (Zoll. ex Steud.) Makino ex Nakai) (Gotoh *et al.*, 2003a).

Bio-ecology

S. bambusae is a polyphagous mite that has been extensively investigated in China, where it is recorded on moso bamboo in Hainan and Guangxi provinces (Ma *et al.*, 1984) and included among the tetranychid mites damaging bamboo forests (Cui, 1989). Lin *et al.* (2000) included *S. bambusae* in the checklist of mites from moso bamboo in Fujian. The mite is also reported in China as a serious pest of ornamental bamboos in Shandong Province (Sun *et al.*, 1997a), and as the major pest mite on moso bamboo in Sichuan Province (Liu *et al.*, 2000a).

Populations of *S. bambusae* grow well only on young bamboo leaves (Sun *et al.*, 1997a; Zhang *et al.*, 2001a). On leaves of moso bamboo from 1-, 2-, 3- and 4-year-old plants, the survival of immature stages of *S. bambusae* is 92.00, 88.46, 70.81 and 94.34%, respectively. The mean number of eggs laid per female reared on leaves from the four ages of leaves, is 41.4, 37.3, 14.7 and 42.8, respectively. The trend index of population (I), net reproductive rate (R_0), intrinsic rate of increase (r_m) and finite rate of increase (λ) are 5.2073, 5.7719, 0.1117, 1.1182 on 3-year-old bamboo leaves, and the population doubling time is the longest (6.2050 days). On the other three ages of bamboo leaves (1, 2 and 4 years old), the net reproductive rates (R_0) are 22.1223, 16.6168 and 18.2853, respectively (Liu *et al.*, 2004b).

The species has a haplodiploid reproductive system, and the females spin very loose webbing, which is quite different from the tight silky webbing produced by *Stigmaeopsis najingensis* (Ma *et* Yuan) (Zhang *et al.*, 2000e). They feed all over the leaf, which is also different from the confined feeding within web nests of *S. nanjingensis* and also have a much higher reproductive rate than *S. nanjingensis* (Zhang *et al.*, 2001a). According to Saitō (1985), the life type of *S. bambusae* is CW-b, and it has a dense web which seems to be stratiform rather than complicated. The eggs are laid on the leaf surface under the web, but when the population increases, some eggs are laid on the web, which can also be used as a footing for moving and as a defence from predators. On moso bamboo leaves, the female develops through the larval, protonymphal and deutonymphal stages, and the male through the larval and protonymphal stages. At 24–30°C, male and female development takes 6.6–6.8 and 8.0–9.1 days, respectively. Survival of the egg, larva, protonymph and deutonymph is 98.0 ± 3.3%, 95.6 ± 2.9%, 96.2 ± 4.5% and 100%, respectively. The pre-oviposition period lasts 3 days and each female lays 1–7 eggs a day during most of the oviposition period, which lasts 8–20 days. The fecundity is 50.3 ± 12.1 (27–60) eggs. Adult females live for 18.7 ± 4.4 (11–24) days. The offspring sex ratio averages 3:1 females:males (Zhang *et al.*, 2001a).

The developmental time from egg to adult ranges from 16.7 days at 21°C to 7.6 days at 33°C. The developmental thresholds for egg, larva, protonymph, deutonymph, pre-oviposition and the combined immature stages are 12.4, 15.3, 14.0, 15.2, 10.7 and 12.5°C, respectively. High temperatures (33°C) result in a decline in larval and protonymphal development. Survivorship is highest (88.3%) at 27°C; whereas at 21 and 33°C, it is 61.7 and 55.7%, respectively. After emergence, the adult has a pre-oviposition period that ranges from 1.9 days at 21°C to 1.0 days at 33°C. The highest number of eggs (23.4 eggs per female) is produced at 27°C and the lowest at 33°C (11.2 eggs per female). The longest duration of reproduction (15.5 days) occurs at 21°C, and the shortest (4.3 days) at 33°C. Mites reared at 27°C show the highest net reproductive rate (15.6), whereas at 33°C they have the highest intrinsic rate of increase (0.1686) (Liu *et al.*, 2000a). On bamboo leaves, the developmental threshold and the effective accumulated temperature of immature stages of *S. bambusae* at five constant temperatures (21, 24, 27, 30 and 33°C) are 12.53°C and 150.25 degree-days, respectively (Liu *et al.*, 2000b).

Reciprocal cross-testing between populations of *S. bambusae* reared on the bamboo species *Phyllostachys pubescens* and *Neosinocalamus affinis* (Rendle) Keng f. showed that the two populations could mate, but lay unfertilized eggs that produce males only, highlighting the existence of reproductive isolation (Liu *et al.*, 2004a).

In the Chang Ning bamboo area in Yunnan Provice, China, the mite develops 13–14 generations a year (Liu *et al.*, 2000b). Wang *et al.* (1998) report that *S. bambusae* develops about ten generations a year and feeds individually from April to November under thin webs on the back sides of leaves. On *P. pubescens*, the occurrence of a population peak in spring is obvious and coincides with the period of sensitive development of the host plant, and the development and population dynamics of the mite are seriously affected by the precipitation and higher temperature in July and August (Liu *et al.*, 2000b).

Various phytoseiid mites are associated with *S. bambusae*.

Damage

S. bambusae causes economic injury on moso bamboo only during spring and early summer in a big-harvest year, e.g. the year of the 2-year cycle when the moso bamboo changes leaves. It is rarely observed during other periods of the year, when the leaves are mature. The bamboo spider mite is not found during a small harvest year either, when there is no change of leaves (i.e. all of the leaves are old leaves) (Zhang *et al.*, 2001a). Feeding by the mite sometimes causes leaves to fall from a severely damaged bamboo, especially during dry seasons (Wang, 1981a), and death has occurred on a large scale in the moso bamboo forests in Fujian (Zhang *et al.*, 1997). The mite has also caused serious damage to bamboos in Shandong, where its biology has been investigated, especially the effects of temperature on development and reproduction (Sun *et al.*, 1997a). In Fujian, *S. bambusae* has been found in a few sites and can cause some damage, and in Zooning County, a large proportion of infested leaves, many of which were covered with brownish spots resulting from mite feeding, have been observed. Mite numbers varied from 35 to 133 per leaf and the damage index reached III or IV on a scale of 0–V (Zhang *et al.*, 2000e).

Control

Control is provided by chemical and biological means. High mortality rates of the pest were also observed resulting from the activity of predatory mites. Indiscriminate use of linden on *Phyllostachys heterocycle* (Carrière) Mitford. (= *Phyllostachys pubescens* Mazel ex J. Houz.) has in the past caused some pests to become predominant and decrease the beneficials. This suggested that integrated pest management (IPM) must be implemented in bamboo forests (Zhang *et al.*, 2007a).

BIOLOGICAL CONTROL. Research on biological control has involved studies on the bio-ecology of different phytoseiid mites, such as *Typhlodromus bambusae* Ehara (Liu *et al.*, 2002, 2006b) and *Neoseiulus cucumeris* (Oudemans) (Zhang *et al.*, 2004a,b).

The generation survival rate of *N. cucumeris* feeding on *S. bambusae* is 77.55%, and the average fecundity per female is 45.77 eggs. On the same prey, the generational survival rate of *T. bambusae* is 100% and the average fecundity per female is 46.8 eggs. The intrinsic rate of natural increase of *T. bambusae* and *N. cucumeris* feeding on *S. nanjingensis* is 0.155 and 0.154, respectively, and higher than that of their prey (0.089). On *S. bambusae*, the intrinsic rate of natural increase of *T. bambusae* and *N. cucumeris* is similar (0.152) and both are lower than that of their prey. Releases of *N. cucumeris* into moso bamboo forests in Fujian during May/July to control pest mites showed that this phytoseiid might be the better biological control agent of pest mites of moso bamboo (Zhang *et al.*, 2004a). Further releases of *N. cucumeris* to control pest mites in moso bamboo forests in Napping and Tongan cities (Fujian) showed that the densities of *S. bambusae* were significantly reduced (by 99.37%). The number of predatory mites in the biocontrol area increased by 73.45% (1 month after release) and 12.18% (2 months after release) compared with those of the control. The moso bamboo grew best in the biocontrol plots (Zhang *et al.*, 2004b).

The functional responses of *T. bambusae* to *S. bambusae* at different development stages are expressed by a Holing Type II curve. The amounts of pest mites consumed by the predator increase quickly with the initial prey density, but soon level off. At 25 ± 0.5°C, the handling time (*THz*) of *T. bambusae* to *S. bambusae* at the stages of egg, larva, protonymph, deutonymph and female is in that order, and the selection coefficients (*Q*) of *T. bambusae* to *S. bambusae* for the egg, larva, protonymph, deutonymph and female are 1.47, 1.34, 1.16, 0.90 and 0.15, respectively. The numerical response of *T. bambusae* to *S. bambusae* shows that the fecundity of the predatory mite is significantly affected by prey density. When the ratio of predator to prey is 1:1–9, the daily oviposition of the predator ranges from 0.5 to 2.0 eggs per female, and this increases with increasing prey density, but when the ratio is 1:15 or more, the daily oviposition decreases to about 1.8 eggs per female, probably due to the interference of the high prey density with predation (Liu *et al.*, 2006b).

CHEMICAL CONTROL. As for other *Schizotetranychus* spp. infesting bamboo (Yu and Shi, 1991; Shi *et al.*, 1992), injection of the culm cavity with 50 or 40% dimethoate at 1.5 ml per culm is suggested for this pest (Wang *et al.*, 1998). In general, the control achieved is more than 86.4%. Treating by root application of insecticide with fertilizer together with bamboo cavity injection gives better results than other control methods. The root application of insecticides with fertilizer is favourable to fast growth of moso bamboo, while the injection of the bamboo cavity gives a rapid effect of the insecticide compared with the root application of insecticide along with fertilizer (Chen, 2007).

INTEGRATED PEST MANAGEMENT. See *Stigmaeopsis najingensis* (Ma *et* Yuan).

Schizotetranychus hindustanicus (Hirst) (Fig. 20.12)

Common name

Citrus Hindu mite, nest-webbing mite.

Diagnostic characteristics

FEMALE. Jeppson *et al.* (1975) reported that the female was unknown. Recently, Návia and Marsaro (2010) published a detailed picture of a female collected in Brazil on citrus. A rough description of the mite based on this picture, and on the reports of Quirós de González and Geraud-Pouey (2007) and Návia *et al.* (2013), is given here. The body is similar in its general features to other *Schizotetranychus* spp. The adult and juvenile stages are yellowish or yellowish green, with dark internal spots on the sides of the idiosoma. Legs I and II and the gnathosoma can be amber in colour. The body is oval, somewhat flattened, about 430 μm long, with stubby legs. The peritreme ends in a bulb. The prodorsum is striate longitudinally and the hysterosoma transversely. All dorsal setae are broad at the base and finely setose. The prodorsal setae v_2, sc_1 and sc_2 have more or

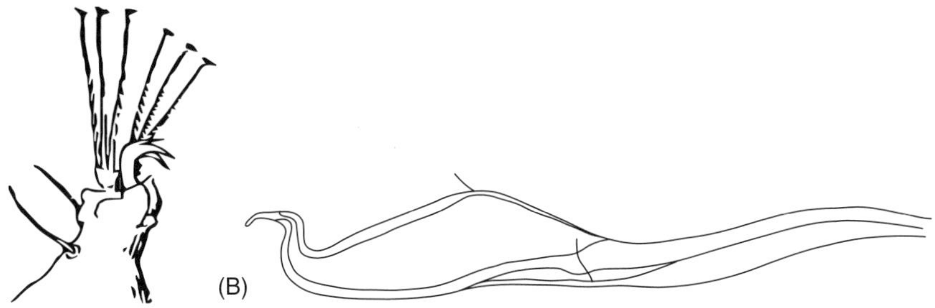

Fig. 20.12. *Schizotetranychus hindustanicus* (Hirst). Male: (A) pretarsus I; (B) aedeagus (from Hirst, 1924).

less the same length. The setae c_3 appear the longer and robust, c_1 slightly shorter and c_1, d_1 and e_1 are set more or less on the same longitudinal line, with the c_1 setae closer and the f_1 setae with at least twice as far between them as the distance between the setae e_1. The setae c_2, d_2 and e_2 are longer than c_1–e_1. The setae f_2 and h_1 are longer and robust. The cupules ia are set between the setae c_1 and c_2, but slightly below.

MALE. The body is pear shaped, paler and smaller than that of the female, with remarkably long legs (especially the first pair) and typical red eyes; it measures less than 350 μm. The dorsal body setae are about ⅔ as long as the distances between their bases; the dorsal opisthosomal setae f_1 are much further apart than the other dorsal opisthosomal setae. The aedeagus has the distal part turned dorsally to form a sigmoid curve, slightly hooked at the tip, and with the axis of the knob almost parallel with the shaft (Jeppson *et al.*, 1975; Quirós de González and Geraud-Pouey, 2007; Návia *et al.*, 2013).

Geographical distribution

Brazil (Návia and Marsaro, 2010; Návia *et al.*, 2013; Ferragut *et al.*, 2013a), India (Hirst, 1924; Gupta and Gupta, 1994) and Venezuela (Quirós de González and Geraud-Pouey, 2007; Ferragut *et al.*, 2013a; Návia *et al.*, 2013).

Host plants

The citrus Hindu mite is reported on different species of host plants from five families: Arecaceae – coconut palm (*Cocos nucifera* L.) (Bolland *et al.*, 1998; Migeon and Dorkeld, 2006–2013); Fabaceae – *Acacia* sp. (Gupta and Gupta, 1994); Meliaceae – neem (*Azadirachta indica* A. Juss.) (Cherian, 1931), *Melia azedarach* L. (Gupta and Gupta, 1994); Poaceae – grain sorghum (*Sorghum bicolor* (L.) Moench) (Gupta and Gupta, 1994); Rutaceae – *Citrus* sp. (Hirst, 1924; Gupta and Gupta, 1994; Quiros de Gonzalez and Geraud-Pouey, 2007; Ferragut *et al.*, 2013a), key lime (*C. aurantifolia* (Christm.) Swingle), sweet orange (*C. sinensis* (L.) Osbeck) and mandarin (*C. reticulata* Blanco) (Nienstaedt-Arreaza, 2007; Návia *et al.*, 2013), Tahití lime (*C. latifolia* (Yu. Tanaka) Tanaka) (Quirós de González and Dorado, 2005; Návia *et al.*, 2013), lemon (*C. limon* (L.) Burm. f.) (Nienstaedt-Arreaza, 2007; Návia and Marsaro, 2010; Návia *et al.*, 2013), Rangpur lime (*C. limonia* Osbeck) (Marsaro *et al.*, 2010; Návia *et al.*, 2013), 'Poncan' mandarin, 'Tangor Murcott' (*C. reticulata* Blanco × *C. sinensis* (L.) Osbeck) and Valencia sweet orange (Marsaro *et al.*, 2010; Návia *et al.*, 2013).

Bio-ecology

Among its host plants, *S. hindustanichus* seems to prefer citrus, on which it has a haplodiploid reproductive system. The females web over concavities or depressions of the rind of fruits, spinning fine, circular 1–2 mm diameter webs, under which they lay the eggs. The larvae and nymphs feed and defecate on the cells and tissue protected by the web, whereas the adults may or may not remain under the nest web (Nienstaedt-Arreaza, 2007; Quirós de González and Geraud-Pouey, 2007). The mite probably exhibits a social behaviour similar to that of other tetranychid mites building nests, such as *Schizotetranychus* and *Stigmaeopsis* (Návia *et al.*, 2013). On orange, lime, lemon and mandarin trees in Venezuela, the mite populations are uniformly distributed in the canopy and females begin colonization on younger leaves (Nienstaedt-Arreaza, 2007).

Studies on the bio-ecology of the citrus Hindu mite done in Venezuela on different citrus species show that the peak of population on key lime, orange and lemon leaves occurs in the spring, from March to May, with the highest densities on lime. The female life cycle has been investigated under laboratory conditions at 25 ± 2°C and 83 ± 12% RH; the developmental periods for the egg, larva, protonymph and deutonymph on orange are 7.89 ± 0.86, 2.46 ± 1.19, 1.97 ± 0.80 and 1.59 ± 0.62 days, respectively. On lime, these same developmental periods are 7.48 ± 0.67, 2.36 ± 0.85, 1.88 ± 1.19 and 2.14 ± 0.78 days, while on lemon they are 7.69 ± 0.88, 3.27 ± 1.22, 2.10 ± 0.76 and 1.76 ± 1.01 days, respectively. The development from egg to adult on orange, lime and lemon lasts 13.91 ± 3.47, 13.86 ± 3.49 and 14.82 ± 3.87, respectively. On orange, lime and lemon, the fecundity is 11.59 ± 5.35, 13.33 ± 8.79 and 10.92 ± 3.82 eggs, respectively (Nienstaedt-Arreaza, 2007).

Natural enemies include some phytoseiid mites, such as *Iphiseoides zuluagai* Denmark *et* Muma, *Galendromus annectens* (De Leon), *Euseius concordis* (Chant), some *Agistemus* spp. (Stigmaeidae) and *Bdella* (Bdellidae) (Marsaro *et al.*, 2009).

Symptomatology and damage

Damage by this mite to lemon trees (cv. 'Tahiti' and 'Galeguinho') consists of uniformly distributed circular whitish spots (1–3 mm in diameter) on the leaves and fruits, sometimes over the entire canopy (Návia and Marsaro, 2010). On citrus species in Venezuela, the spots were often observed on the fruits and upper leaf surface, but were never found on orange trees. The symptoms first appear on the upper leaf surface, along the main rib, and later extend to the entire leaf blade; attacked fruits become uniformly silvered and hard when the infestation is

extensive (Quirós de González and Geraud-Pouey, 2007). Severe damage to citrus leaves and fruits has also been observed in Brazil. Even if there are no confirmed reductions in citrus production as a result of *S. hindustanichus* infestations, there is no doubt that the mite can reduce the commercial value of fresh fruits owing to depreciation in their aesthetic quality (Návia *et al.*, 2013).

Control

There are few reports of trials on the chemical control of the citrus Hindu mite. Active ingredients tested in South America include azocyclotin, soap and mineral oil; azocyclotin seems to produce the best control, but it appears the risk of phytotoxicity is underestimated (Quirós de González and Dorado, 2005). The postharvest procedure provides immersing fruits in chlorine solution, washing fruits with a *o*-phenylphenol solution that is usually applied for cleaning citrus fruits, and brushing, drying and waxing the fruits with vegetable wax made of *Copernicia prunifera* Miller palm wax and oxidized colophony resin (Návia *et al.*, 2013).

The Genus *Stigmaeopsis* Banks

Diagnostic characteristics

The distal segment of the palp has 2 simple setae (*a*, *c*), 3 eupathidia (*ul*′, *ul*″, *su*), and 1 solenidion (ω). On the hysterosoma, 2 lines connecting the bases of the c_1, and c_4 setal pairs form a 'V' shape (these are more or less parallel in other species of Tetranychini) (Saitō *et al.*, 2004).

Economic importance

A molecular phylogeny study based on the 28S region of rDNA show that the *celarius* species group within the genus *Schizotetranychus* Trägårdh is a distinct monophyletic clade (Sakagami, 2002). These findings and a morphological analysis suggested to Saitō *et al.* (2004) that this species group should be separated from the genus *Schizotetranychus* and included in the genus *Stigmaeopsis* Banks. Saitō *et al.* (2004) re-categorized seven species of the *celarius* species group as members of *Stigmaeopsis*.

Stigmaeopsis spp. live typically on bamboo plants in different countries of the Australasian, Nearctic, Oriental and Palaearctic regions, except for *S. miscanthi* (Saitō), which lives on the poaceous species *Miscanthus sinensis* Anderss. and *Miscanthus* sp. in the Oriental (Nansei-shoto, Japan) and Palearctic regions (Japan) (Saitō, 1990a). These mites are characterized by their web-nesting habit, which is known as the WN-c life type (Saitō, 1983), in which webs are spun over depressions on the lower leaf surface and are related to a kind of sociality. Thus, the four species living on bamboo plants in Japan make different nest areas. The nest area of *S. longus* (Saitō) is the wider, followed by that of *S. celarius* Banks, *S. takahashii* Saitō *et* Mori and *S. saharai* Saitō *et* Mori, in decreasing order of width. Smaller nests effectively prevent the adults of several predator mites from intruding, and variation in nest size could be associated with different anti-intruder adaptations of this mite genus according to their sociality (Mori and Saitō, 2005). Attacks by the subsocial *Schizotetranychus nanjingensis* Ma *et* Yuan (=*Stigmaeopsis nanjingensis* (Ma *et* Yuan)) against its specific predator, the phytoseiid mite *Typhlodromus bambusae* Ehara, are very distinctive. Males of this spider mite capture predatory larvae and deutonymphs in the nest and repeatedly push themselves against them. It has been observed experimentally that three-quarters of the predator larvae introduced into the nests with one male spider mite may be either killed or forced out of the nests within the first 10 min. Spider mite males may kill the protonymphs of *T. bambusae* within 10 min, and drag or push the dead predators out of the nest. Females of *S. nanjingensis* themselves never attack and kill the predators, but sometimes help to remove the dead intruders (Zhang *et al.*, 2001d).

Three of these mites are secondary pests of Poaceae (Bambusoideae). These are *S. saharai*, reported on *Sasa kurilensis* (Rupl.), *S. senanensis* (Franc. *et* Sav.) and *Sasa* sp. in Japan (Saitō *et al.*, 2004), *Stigmaeopsis takahashii*, inhabiting the same host plants in Japan (Ehara, 1957; Saitō and Takahashi, 1982; Saitō *et al.*, 2004) and *S. tenuinidus* (Zhang *et* Zhang) on *Phyllostachys edulis* (Carrière) J. Houz. in China (Zhang *et al.*, 2000e, 2001b). The most important pest mites of this genus are *S. celarius*, *S. longus* and *S. nanjingensis*. Their feeding activity on bamboos potentially affects their aesthetic and economic value, mainly in commercial nurseries, and may also cause damage in natural stands. Sometimes, populations of *S. celarius* are identified as populations of *S. longus*, or mixed populations of both species may occur. The distribution of *S. longus* (mostly in the USA), seems more restricted than that of *S. celarius*. Regardless, the problem of the bamboo mites has an economic importance, as indicated in reports from the USA on the websites of the American Bamboo Society, Oregon State University and Washington State University.

Stigmaeopsis celarius Banks (Fig. 20.13)

Common name

Bamboo spider mite.

Diagnostic characteristics

FEMALE. The body is 417–498 μm long, strawy to greenish yellow in colour, with small blackish green spots. The propodosoma is distinct from the hysterosoma. The idiosomal dorsal setae are slender. The prodorsal setae sc_1 are ¾ as long as the distance between their bases. The bases of the hysterosomal setae c_1–f_1 are approximately in a straight line. The hysterosomal setae d_1 exceed the bases of setae f_1, but do not reach the base of the setae h_1. The genital flap and the area anterior to the flap are transversely striate. The distal segment of palpus bears 2 simple setae, 3 eupathidia (including one spinneret), and 1 solenidion. The spinneret is conical in shape and $2/3$ as long as wide (Saitō *et al.*, 2004).

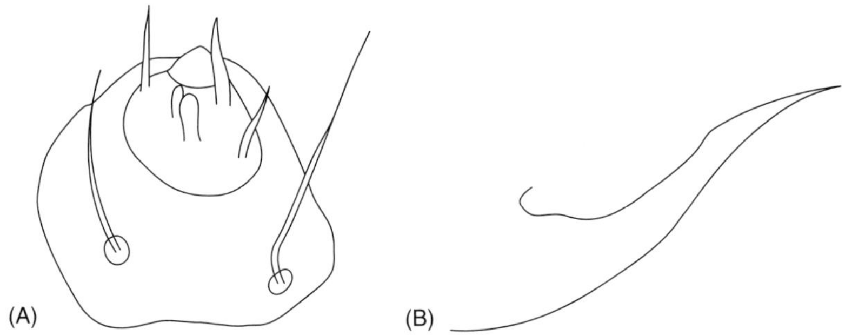

Fig. 20.13. *Stigmaeopsis celarius* Banks. Male: (A) distal segments of palpus; (B) aedeagus (from Saitō, 2004).

MALE. The body of the male is 285–317 μm long, and the aedeagus is broadly curved dorsally (Saitō *et al.*, 2004).

Geographical distribution

Australia (Gutierrez and Schicha, 1983), Belgium (Witters *et al.*, 2003), China (Ehara and Lee, 1971; Wang, 1981b; Wang *et al.*, 1985), France (Bolland *et al.*, 1998; Auger and Migeon, 2007), Hawaiian Islands (USA) (Garrett and Haramoto, 1967; Lee Goff, 1986), Japan (Yokoyama, 1932; Saitō and Ueno, 1979; Saitō and Takahashi, 1980; Saitō *et al.*, 2004), South Korea (Lee *et al.*, 1987), Japan (Nansei-shoto) (Ehara, 1966a), the Netherlands (Bolland *et al*, 1998), Romania (Gutue *et al.*, 2010), Taiwan (Tseng, 1990), the UK (Ostojá-Starzewski, 2000) and the USA (Banks, 1917; Saitō *et al.*, 2004).

Host plants

The bamboo spider mite is reported on 23 species of the Poaceae and on one of the Moraceae. The latter is *Ficus pumila* L. (Ehara, 1966a). The species of Poaceae are the following: an unnamed Poaceae sp. (Ehara and Lee, 1971), *Arundinaria humilis* Mitford (Auger and Migeon, 2007), *Bambusa* sp. (Garrett and Haramoto, 1967; Lee Goff, 1986; Ostojá-Starzewski, 2000), *Miscanthus sinensis* Anderss. (Ehara, 1966a), *Oryza* sp., golden bamboo (*Phyllostachys aurea* Rivière *et* C. Rivière) (Ostojá-Starzewski, 2000; Witters *et al.*, 2003; Auger and Migeon, 2007), yellow-groove bamboo (*P. aureosulcata* McClure) (Gutue *et al.*, 2010), Japanese timber bamboo (*P. bambusoides* Siebold *et* Zucc.) (Ehara, 1957; Auger and Migeon, 2007), moso bamboo (*P. edulis* (Carrière) J. Houz.) (Witters *et al.*, 2003; Saitō *et al.*, 2004; Auger and Migeon, 2007; Sakagami *et al.*, 2009), Makino bamboo (*P. makinoi* Hayata), black bamboo (*P. nigra* (Lodd. ex Lindl.) Munro) (Gutierrez and Schicha, 1983; Witters *et al.*, 2003), fen lü zhu (*P. viridiglaucescens* (Carrière) Rivière and C. Rivière) (Auger and Migeon, 2007), pygmy bamboo (*Pleioblastus fortunei* (Van Houtte) Nakai) (Saitō and Takahashi, 1980), ram-rod bamboo (*Pseudosasa hindsii* (Munro)), *P. japonica* (Sielbold *et* Zucc. ex Steud.) Makino ex Nakai (Banks, 1917; Saitō *et al.*, 2004), *P. viridula* Chen *et* Sheng (Auger and Migeon, 2007), sugarcane (*Saccharum officinarum* L.) (Ehara, 1966a), wild sugarcane (*S. spontaneum* L.), *Sasa kurilensis* (Rupr.), *S. nipponica* (Null.) (Saitō and Takahashi, 1980), *S. senanensis* (Franc. *et* Sav.) (Saitō and Ueno, 1979; Saitō and Takahashi, 1980), *Sasa* sp. (Ehara, 1957), ruscus bamboo (*Shibataea kumasaca* (Zoll. ex Steud.) Makino ex Nakai) (Witters *et al.*, 2003).

Bio-ecology

Stigmaeopsis celarius was first described by Banks (1917), and was later renamed as *Schizotetranychus celarius* (McGregor, 1950; Pritchard and Baker, 1955); recently, the mite was transferred to the genus *Stigmaeopsis* Banks again (Saitō *et al.*, 2004). Moreover, a complex of species was delineated within the redesignated *Schizotetranychus celarius* species, which were named *Stigmaeopsis celarius* Banks, *S. miscanthi* (Saitō) and *S. longus* (Saitō) (Saitō, 1990a). *S. celarius* has a worldwide distribution and in the USA has been reported from California, Florida and Georgia (Baker and Tuttle, 1994). Bamboo mites have also been reported from Maryland, New York, Virginia and Louisiana.

S. celarius infests a number of Poaceae (Bambusoideae). It has a haplodiploid reproductive system, and lives in colonies on the undersurface of bamboo leaves in woven nests that are separate from one other (Saitō and Ueno, 1979). According to Saitō (1985), its life type is WN-c, which includes mites that construct an extremely densely woven roof (the web nest) over depressions on the lower leaf surface. During walking, spinning is suppressed and weaving time is limited to construction of the nest and eggs covers. One or two sites outside and near the woven nest are employed for defecation. All mites living in the same nest use these fixed sites to defecate. This shelter-like web helps the mite to resist adverse weather conditions, competing herbivores and predators (Saitō and Takahashi, 1982).

The overall sex ratio in the offspring is 0.84. At a constant temperature of 25°C ± 1°C, net reproduction (R_0), the intrinsic rate of increase (r_m) and the finite rate of increase (λ) of the population are 67.6, 0.162 and 1.176, respectively. In the laboratory, the mite may have 26 generations a year (Saitō and Ueno, 1979).

Natural enemies include various predatory mites, such as the phytoseiid *Typhlodromus bambusae* Ehara, *Neoseiulus cucumeris* (Oudemans) and other mites. *T. bambusae* is able to overcome the shelter-like web constructed by *S. celarius*. However, the males and females of *S. celarius* leave the nest and return later and the males kill offspring of the phytoseiid mite (Saitō, 1984).

Symptomatology and damage

Feeding by the mite causes chlorotic spots on the upper leaf surface opposite to the mite population on the underside (Jeppson *et al.*, 1975). The damage is visible on both leaf surfaces, and the feeding site is yellow and may look similar to variegation. This damage is thought to vary slightly among the bamboo

mite species and may be an indicator of which species is present (Rosetta, 2005). A population density of 85.2 mites per leaf on bamboo in China causes a reduction in new shoot production of 200 kg ha^{-1} (Yu and Shi, in Wang *et al.*, 1998). Gerdeman and Tanigoshi (2004) report that feeding damage on bamboo is lighter in colour than the surrounding leaf tissue, and is permanent and aesthetically displeasing. The damage may first appear as light speckling or blotches on the upper leaf surface. As infestation increases, the damaged area acquires a chequered pattern sometimes resembling leaf variegation. Heavy infestations can cause green bamboos to appear yellow–green in colour. Leaf damage can impair photosynthesis and reduce plant vigour (Meredith, 2001), and damaged plants may lose their leaves more frequently (Rosetta, 2001). Infested leaves of the yellow-groove bamboo show a distinctive symptom, visible on both leaf surfaces, in which the feeding site turns white–yellow and is similar to variegation (Gutue *et al.*, 2010).

Control

The bio-ecology of the mite constrains its control, which requires rigid prevention and the application of cultural techniques and biological, physical and chemical means.

PREVENTION. Prevention requires the inspection of new bamboo plants to detect mite presence and consider placing them under quarantine for a period of time before planting in the ground. It is necessary to isolate new plant material in quarantine away from other susceptible plants, downwind if possible. New and existing plants should be monitored regularly to locate mite infestations. When and where possible, infested plants should be isolated away from non-infested plants. Work in the nursery should be done in non-infested areas before working in infested areas, taking into account that mites hitchhike on clothing, and that the infestation can be spread in this way. Repeated spot applications in infested areas have been successful in some cases (Rosetta, 2005).

CULTURAL TECHNIQUES. Dense bamboo groves should be thinned by 50% (removing older or weak culms first) to improve air circulation and make it harder for mites to establish dense colonies, cutting down all the stems in early spring before the new culms or shoots emerge. It is important to remove or destroy all of the cut canes and leaf litter on the ground. If this is done properly, new shoots should emerge mite free (MGG, 2013). The above-ground foliage of infested plants should also be mostly removed and burned or otherwise disposed of. The emerging foliage should then be treated to kill any remaining pest mites. This technique is a practical solution for heavily infested sites, particularly where chemical application presents difficulties (Rosetta, 2005).

BIOLOGICAL CONTROL. The better control option is biological control (MGG, 2013). This method usually requires low levels of the pest in order to sustain the natural enemy population. These low levels would not be desirable for plants in a nursery, which are to be sold or transported. A production nursery might use a biological control programme to suppress bamboo mites below a damaging threshold, but later clean up the plants before sale with miticide applications (Rosetta, 2005).

Biological control programmes for bamboo mites have been carried out in various countries. In China, the potential of the phytoseiid mite *N. cucumeris* as a biocontrol agent against the bamboo mites has been investigated by Zhang *et al.* (2000a, 2004a,b), and that of *T. bambusae* by Liu *et al.* (2002, 2006b) and Zhang *et al.* (2004a). In Japan, *T. bambusae* is considered a specific predatory mite of the *S. celarius* mites (Saitō, 1990b). In the USA, *N. fallacis* (Garman) develops on *S. longus* and effectively reduces infestation (Pratt and Croft, 1999). The predatory mite readily enters nests of bamboo mites through natural openings or by creating new holes. Releases of *N. fallacis* carried out on different plants in a number of nurseries have successfully suppressed· infestations by bamboo mites (and other spider mites). Good scouting and releasing the mite at a very conservative threshold (one pest mite for every five leaves) ensures success. The mite requires 80% RH, which is easily found in the canopy of bamboo grown in the Northwest USA (Rosetta, 2005).

PHYSICAL CONTROL. Where possible, plants should be sprayed with a pressure washer from time to time in order to dislodge existing bamboo mites and keep the population threshold (and symptoms) down. Hot water treatment eliminates mites from dormant rhizomes. For this treatment, rhizomes are immersed for 10 min in hot water at 50°C, then heeled into sand or sawdust and kept cool until planting time in the spring (Young and Haun, 1961).

CHEMICAL CONTROL. To reduce the movement of bamboo mites in commerce and also to reduce the potential for aesthetic and economic damage, chemical treatment may be warranted (Rosetta, 2005), but its use in natural stands of bamboo must be carefully evaluated. Also, bamboo mites are difficult to control with chemical sprayings as they are located on the leaf underside and protected by webbing. Small outbreaks can be treated by bending the canes and spraying the undersides with either horticultural oil or insecticidal soap. This technique is not usually effective in large stands that are severely infested (Mike's Garden Guide, 2013). Few acaricides have been registered specifically for bamboo; an example is bifenazate. Chemicals reported in the literature include abamectin, bifenazate, dicofol, oxythioquinox and oil. When using miticides without ovicidal activity, two applications 7–10 days apart are recommended. Silicon-based surfactants are suggested for bifenazate, and latron (modified phthalic glycerol alkyd resin and butyl alcohol) is recommended for dicofol. Oil may enhance the activity of abamectin, and oil itself has miticidal activity upon direct contact, including the ability to smother mite eggs (Rosetta, 2005).

INTEGRATED PEST MANAGEMENT. Bifenazate has been used in conjunction with releases of predatory mites in Oregon, with no apparent impact on the predatory mite *N. fallacis*. Oils and soaps, although they can kill on contact, have no residues, leaving immigrating natural enemies unharmed (Rosetta, 2005). The topic is discussed below under *S. najingensis* (Ma *et* Yuan).

Stigmaeopsis longus (Saitō) (Fig. 20.14)

Common name

Bamboo spider mite.

Diagnostic characteristics

FEMALE. The body is 378–435 μm long and 255 μm wide, strawy to greenish yellow in colour, with small blackish green spots. The propodosoma is quite distinct from the hysterosoma, with the setae sc_1 as long as the distance between their bases. The dorsal idiosomal setae are slender. The hysterosomal setae c_1–f_1 are set approximately in a straight line. The hysterosomal setae d_1 and e_1 are more than twice as long as the distance between them and the consecutive setae e_1 and f_1. The genital flap and the area anterior to the flap are transversely striate. The spinneret on the palp is conical in shape and about ½ to ⅓ as long as wide (Saitō, 1990a).

MALE. The male is smaller than the female. The aedeagus is broadly curved dorsally (Saitō, 1990a).

Geographical distribution

Japan (Saitō, 1986; 1990a; Saitō *et al.*, 2004; Gotoh and Shida, 2007) and the USA (Pratt and Croft, 1999).

Host plants

The bamboo spider mite is reported on the following five poaceous species: *Bambusa* sp. (Pratt and Croft, 1999), *Sasa* sp. (Saitō, 1990a; Saitō *et al.*, 2004), *Sasa kurilensis* (Rupl.) (Saitō *et al.*, 2004), *S. senanensis* (Franc. *et* Sav.) (Saitō, 1986, 1990a; Gotoh *et al.*, 2003a; Saitō *et al.*, 2004; Gotoh and Shida, 2007) and *S. veitchii* (Carrière) Rehder (Gotoh *et al.*, 2003a).

Bio-ecology

Stigmaeopsis longus was previously identified as the 'B strain' of *Schizotetranychus celarius* (Banks, 1917) (Saitō, 1979a; Saitō and Ueno, 1979) and as the 'long seta form' of *Schizotetranychus celarius* (Banks, 1917) (Saitō and Takabashi, 1980, 1982). Later, the mite was described as *Schizotetranychus longus* Saitō (Saitō, 1990a), but it was recently transferred to the genus *Stigmaeopsis* Banks (Saitō *et al.*, 2004).

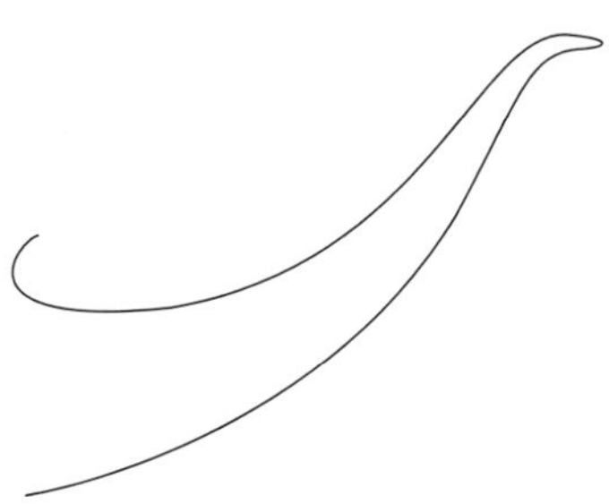

Fig. 20.14. *Stigmaeopsis longus* (Saitō). Male, aedeagus (from Saitō, 1990a).

The bamboo spider mite has a Nearctic and Palaearctic distribution and lives mainly on Poaceae (Bambusoideae) of the genus *Sasa*, which seem to be the most susceptible hosts. It is thought that leaf pubescence may deter the mite (Pratt and Croft, 1999). Females of *S. longus* use silk threads to clean the nest and the eggs. Silk-weaving behaviour does not represent a by-product of nest construction, but is rather aimed at cleaning the space and eggs. The effectiveness of the attending females is connected to the cleaning behaviour of the parental females. The ability to maintain clean habitats is fundamental to animals evolving aggregative social lives (Kanazawa *et al.*, 2010). Within the genus *Schizotetranychus*, the method of nest expansion in relation to nest size and group size, and the groups' host plant history, relate to species differences (Sato and Saitō, 2010).

The mite overwinters successfully on ornamental bamboo in western Oregon, but its long-term survival over a cold winter or multiple cold winters is uncertain, although the mite seems to survive well in the semi-enclosed nurseries that are used to grow bamboo in Oregon (Pratt and Croft, 1999).

Symptomatology and damage

The symptoms of infestation and damage caused are similar to those of other bamboo mites. Conversely to earlier reports of Young and Haun (1961), the mite could have an economic importance because it can reduce the marketability of spider mite-infested bamboo, and the potential for dissemination of the pest by nursery workers, plant collectors and homeowners is high. Moreover, the presence of the pest may hamper the development of bamboo cultivation for other purposes within the region (Pratt and Croft, 1999). At present, the mite is reported as a pest of bamboo in the USA (Rosetta, 2005).

Control

For control, see *S. celarius*. Research by Pratt and Croft (1999) on biological control shows that the phytoseiid mite *N. fallacis* may be a good candidate for the biological control of *S. longus* in temperate humid climates. According to their findings, the webbing and faeces of the pest were similarly attractive to this predatory mite, but webbing was more attractive than eggs and, when searching, faeces were more attractive than eggs. Releases of other phytoseiid mites are proposed by suppliers of biological control agents.

Stigmaeopsis nanjingensis (Ma *et* Yuan)

Common name

Bamboo spider mite.

Diagnostic characteristics

FEMALE. The body is 500–620 μm long and 240–310 μm wide. The opisthosomal setae c_1 are less than $^1/_3$ as long as the setae d_1. The distance between the setae d_1 is equal to or less than that between the setae e_1. The terminal eupathidium of the palptarsus $su(\zeta)$ is conical, 4 μm long and 3 μm wide at

the base; the 2 lateral eupathidia *ul′*(ζ) and *ul″*(ζ) are 5 μm long; the solenidion (ω) is 4 μm long, and about 3 times as long as wide (Zhang *et al.*, 2000e).

MALE. The body is 410 μm long and 210 μm wide. The aedeagus is gradually narrowing and slightly bent distally to form a sigmoid shape (Zhang *et al.*, 2000e).

Geographical distribution

China (Ma and Yuan, 1980; Yu and Shi, 1991; Zhang *et al.*, 1999a; Zhang *et al.*, 2000e) and Italy (Pellizzari and Duso, 2009).

Host plants

S. nanjingensis is reported on seven poaceous plants: *Phyllostachys* sp. (Ma and Yuan, 1980), moso bamboo (*P. edulis* (Carrière) J. Houz.) (Yu and Shi, 1991; Zhang *et al.*, 1999a, 2000e; Saitō *et al.*, 2004), black bamboo (*P. nigra* (Lodd. ex Lindl.) Munro) (Pellizzari and Duso, 2009), sulphur bamboo (*P. sulphurea* (Carrière) Rivière and C. Rivière) (Pellizzari and Duso, 2009), zao zhu (*P. violascens* (Carrière) Rivière and C. Rivière), running gant bamboo (*P. vivax* McClure) (Saitō *et al.*, 2004) and arrow bamboo (*Pseudosasa japonica* (Siebold *et* Zucc.)) (Pellizzari and Duso, 2009).

Bio-ecology

S. nanjingensis is known in the Oriental and Palaearctic regions. In China, the mite was first described from bamboo leaves in Nanjing (Ma and Yuan, 1980). Later, it was reported on moso bamboo in Zhejiang, where frequently it damaged bamboo (Yu and Shi, 1991). In Fujian, *S. nanjingensis* is a major pest on the moso bamboo, and is more common and widespread than other *Stigmaeopsis* spp. (Zhang *et al.*, 2000e).

Yu and Shi (1991) gave a preliminary account of the life history and biology of this mite in Zhejiang. Zhang *et al.* (1999a) studied its biology in Fujian. Like other *Stigmaeopsis* spp., *S. nanjingensis* spins web nests on the lower surface of new bamboo leaves and exhibits social behaviour. Its populations overwinter as females, which spin the web nests and begin to lay eggs in late February. The mite feeds actively and reproduces in May, aestivates from June to mid-July, and from late July to late October reproduces again. Populations *S. nanjingensis* increase in autumn and decrease during winter. Monoculture, crowding of bamboos and the removal of ground vegetation are major causes of mite outbreaks (Zhang *et al.*, 1999b, 2004d).

Natural enemies include various phytoseiid mites, either indigenous or introduced into China, such as *N. cucumeris* (Zhang *et al.*, 2000a, 2004a), *N. longispinosus* (Zhang *et al.*, 2000b,c) and *T. bambusae* Ehara (Zhang *et al.*, 2000c, 2004a).

Symptomatology and damage

In China, damage to bamboo leaves is caused by mixed populations of three pest mites, *S. nanjingensis*, *Aponychus corpuzae* Rimando (Tetranychidae) and *Aculus bambusae* Kuang (Eriophyidae), which have different population dynamics (Zhang *et al.*, 2001e). Of these, *S. nanjingensis* is the most severe pest of moso bamboo (*Phyllostachys pubescens* = *P. edulis*), a plant that is very important in the country as material for furniture, scaffolding and various manufactured products. Mite damage to the bamboo forests affects the economy, ecology and biodiversity of Fujian Province (Zhang *et al.*, 2000c), where mite outbreaks have become particularly severe since 1980, when the natural bamboo forests were replaced by artificial monocultures to increase bamboo production (Zhang *et al.*, 2003b). In Europe, where bamboos are cultivated as ornamentals, damaged leaves persist on the plants and affect their aesthetic value negatively.

As it goes on, the feeding activity of *S. nanjingensis* causes the leaves to become yellow and brown, and then to fall; it also affects the physiology and the metabolism of the plants, resulting in a reduction of leaf surface and loss of new culms. The presence of 1–3 mite nests per leaf causes a reduction of 11% in the chlorophyll content of bamboo leaves, 4–10 nests per leaf causes a 24% reduction and 11–20 nests per leaf causes 57% or more reduction. The feeding activity of *S. nanjingensis* also reduces the sugar content of bamboo leaves by 2% with light damage, 13% with moderate damage and 55% with severe damage. For total sugar content in bamboo leaves, light damage induces a reduction of 12%, moderate damage a reduction of 19% and severe damage a reduction of 57% (Zhang *et al.*, 1998). Unlike normal bamboo plants, which change their leaves every other year, damaged plants might change their leaves every year, or even twice a year. A reduction of leaf surface of 30% is associated with light damage and a reduction of 58% with severe damage (Zhang *et al.*, 1998). An acre of lightly damaged forest lost 20–30% of new culms (300–450), resulting in a loss of 3750 kg of shoots, while moderately damaged forest lost 30–40% new culms (450–600) and severely infested forests seem burned when observed at a distance (Zhang *et al.*, 2000c).

Control

Control is provided by the use of cultural techniques and biological and chemical means, usually by injections into joints on the culm. The integration of these technical possibilities is fundamental to the implementation of an integrated pest management (IPM) programme. Sampling methods allow the evaluation of mite presence in the forest, damage index per leaf and average damage index per plant.

SAMPLING AND MONITORING. Sampling methods require the observation of 9–30 leaves at random collected from different parts of the plant (for a forest, 15 plants within a range of 7–20 plants) and counting under the microscope of the number of mites per leaf. A damage index is visually assigned to each leaf and an average index is calculated for each plant. The damage index per leaf is estimated by the proportion of leaf area covered by webbing and is grouped into six classes ranging from class 0 (no web nests per leaf) to class V (>17 web nests per leaf or 20% of the leaf surface covered) (Zhang *et al.*, 2000d).

CULTURAL TECHNIQUES. Since 1980, the change from polyculture (750 bamboo plants ha^{-1}) to monoculture (over 2000 bamboo plants ha^{-1}) has caused outbreaks of populations of *S. nanjingensis*

in Fujian. Natural polyculture forests had 5–10 times as many natural enemies (especially phytoseiids) as artificial monocultures, where the lower under-plants are cut, soils are turned and plant growth hormones are used. Monoculture is ecologically unstable and the weakened forests are more susceptible to pest outbreaks (Zhang *et al.*, 2000c). Mite damage in the monoculture forests is on average twice (35%) as high as that in polyculture forests (17.5%, i.e. less than an injury level of 20%). The relative contribution of each mite species to total damage varies among sites, and *S. nanjingensis* is the major pest. Mite damage is also inversely correlated with the altitudes of individual forests across sites (Zhang *et al.*, 2000d). Interplanting broad-leaved trees with moso bamboo at ratios of about 2:8 or 3:7 restores biodiversity and establishes a sustainable and ecological pest management system in bamboo forests. Moreover, a healthy culture of bamboo forests has increased natural resistance to pests and diseases, and an increased compensatory ability (Zhang *et al.*, 2000c). The use of organic fertilizers (fermented sugarcane by-products) mixed with N, P, K, Mn, Zn and Si has a positive effect on the growth of bamboo plants. These fertilizers are mixed with 'imitate compound biopesticides' and applied to root systems of bamboo plants during spring, 40–60 days after releasing predatory mites (*N. longispinosus*) (Zhang *et al.*, 2000c).

BIOLOGICAL CONTROL. Biological control programmes for bamboo mites have been carried out over the last two decades. Different phytoseiid mites have been tested and some species have given a good response.

The potential of the phytoseiid mite *N. cucumeris* as a biocontrol agent against *S. nanjingensis* has been investigated in China. At 27–28°C, and with females and eggs of *S. nanjingensis* as food, the developmental time from egg to egg of the phytoseiid is 7.7–7.8 days, and is as long as its life cycle on its normal prey in the laboratory, *Tyrophagus putrescentiae* (Schrank). The pre-oviposition period of *N. cucumeris* is 3 days, with a daily rate of 1–4 eggs over a period of 7–18 days and a total fecundity of 35.8 eggs. The number of prey consumed increases with prey density and the number of eggs laid is directly correlated with the number of prey consumed. Females prey on twice as many female spider mites as male and deutonymph predators. *N. cucumeris* females are unable to invade intact web nests of *S. nanjingensis* but may invade and like to stay and lay eggs in broken nests with existing openings or holes (Zhang *et al.*, 2000a). Studies on the response of *A. longispinus* to the web nest of *S. nanjingensis* show that this predatory mite responds to stimuli associated with webbing, mite feeding damage and other residues in the web nests (Zhang *et al.*, 2000b).

The results of studies in China on the indigenous phytoseiid *T. bambusae* and the introduced phytoseiid *N. cucumeris* to evaluate their potential as a biocontrol agent against *S. nanjingensis* in moso bamboo forests are promising. On *S. nanjingensis* as food, the generation survival rate and average fecundity per female of *N. cucumeris* are 90.40% and 38.12, while for *T. bambusae*, these figures are 95.23% and 44.5, respectively. The intrinsic rate of natural increase of *T. bambusae* (0.155) is very similar to that of *N. cucumeris* (0.154) and higher than that of their prey, *S. nanjingensis* (0.089). *N. cucumeris* was released into moso bamboo forests to control pest mites during May–July, and it has been shown that this species may be a better biological control agent of pests of moso bamboo (Zhang *et al.*, 2004a).

CHEMICAL CONTROL. On moderately infested surfaces, chemical control reduced pest mite densities by 94–98% in 30 days, with selective injections of solutions of systemic pesticides such as methamidophos, omethoate, fenvalerate, methomyl or chlorfluazuron diluted by 10–30 times into joints on the culm. Some pesticides killed natural enemies (Zhang *et al.*, 2000c).

INTEGRATED PEST MANAGEMENT. The implementation of an IPM programme is based in the long term on three main actions: changing monoculture bamboo forests into polyculture; promoting healthy bamboo culture; and increasing the presence of natural enemies of pest mites (Zhang *et al.*, 2000c). The use of a systems simulation approach clearly shows that biodiversity is necessary for the sustainable regulation of large-scale forests, such as moso bamboo plantations (Tsuji *et al.*, 2011). From a practical point of view, the proposal is economically advantageous and results in 1.4–1.6 pest mites per leaf in IMP forests, while in untreated forests there are as many as 160 pest mites per leaf. After 1 year, there were yield increases of 98–467% and increases in the number of culms of 840–975 acre^{-1}. The diameter of the new culms in the treated area increased by 2.0–4.0 cm and the height by 2.0–4.2 cm compared with untreated areas (Zhang *et al.*, 2000c). In the long term, IPM programmes should regard the diversification of the bamboo forests as a priority; where this choice conflicts with the short-term productivity, other options must be encouraged, such as the conservation of indigenous phytoseiid mites (e.g. *T. bambusae* and *N. longispinosus*) or the release of mass-reared predators; the use of selective pesticides, in combination with organic fertilizers, might also be used for spot control. IPM, along with integrated forest management, should be implemented, especially in regard to diversifying plant stands, proper nutrition input and sustainable harvesting schedules (Zhang *et al.*, 2000d).

The Genus *Panonychus* Yokoyama

Diagnostic characteristics

The empodial claw is as long or longer than the proximoventral hairs, which are inserted at right angles to the claw (Bolland *et al.*, 1998).

Economic importance

This genus has worldwide distribution and currently includes 16 species living on wild and cultivated plants (Migeon and Dorkeld, 2006–2013).

Three *Panonychus* spp. are reported on Poaceae (Bambusoideae) in Japan (Palaearctic region). These are: *Panonychus akitanus* Ehara, living on *Sasa senanensis* (Franc. *et* Sav.) (Ehara and Gotoh, 1987; Gotoh *et al.*, 2003a) and *Sasa* sp. (Ehara, 1978), *P. bambusicola* Ehara *et* Gotoh living on *S. borealis* (Hack.)

Makino and Shibata (Ehara and Gotoh, 1991; Gotoh *et al.*, 1996a) and *S. senanensis* (Ehara and Gotoh, 1991; Gotoh and Shida, 2007); and *P. pusillus* (Ehara *et* Gotoh) reported on *S. chartacea* (Makino) Makino and Shibata (Ehara and Gotoh, 1987). *P. caricae* Hatzinikolis and *P. hadzhibejliae* (Reck) both infest fig plants (*Ficus carica* L.) in Greece (Hatzinikolis, 1984), and in Armenia, Azerbaijan (Mitrofanov *et al.*, 1987) and Georgia (USSR) (Reck, 1947; Mitrofanov *et al.*, 1987). *P. inca* Vis *et* Moraes and *P. lishanensis* Tseng both live on apple (*Malus domestica* Borkh.) in Peru (Vis and Moraes, 2002; Guanilo *et al.*, 2012) and Taiwan (Tseng, 1990). *P. osmanthi* Ehara *et* Gotoh lives in Japan on three species of Oleaceae, *Osmanthus fortunei* Carrière (Kitashima and Gotoh, 1997), *O. fragrans* Lour. (Ehara and Gotoh, 1996) and *O. heterophyllus* (G. Don) P.S. Green (Gotoh *et al.*, 2003a). *P. globosus* Tseng is reported on *Achyranthes aspera* L. and *Vitis* sp. in Taiwan (Tseng, 1974). *P. spinigerus* (Lucas) lives on *Misopates orontium* (L.) Raf. in Algeria (Lucas, 1848). Other species infest ornamental plants, e.g. *P. thelytokus* Ehara *et* Gotoh develops on *Ulmus davidiana* Planch. in Palaearctic Japan (Ehara and Gotoh, 1992; Gotoh *et al.*, 2003a,b), and other mites live on different host plants. A number of species are polyphagous on wild and cultivated plants, such as the raspberry red mite (*P. caglei* Mellot), the citrus red mite (*P. citri* (McGregor)), *P. elongatus* Manson, the European red mite (*P. ulmi* (Koch)) and *P. mori* Yokoyama. *P. mori* is reported in Japan on *Aphananthe aspera* Thunb. (Gotoh *et al.*, 2003a), Japanese quince (*Chaenomeles japonica* (Thunb.) Lindl. ex Spach) (Toda *et al.*, 2000), *Cocculus orbiculatus* (L.) DC. (Gotoh *et al.*, 2003a), *Morus* sp. (Toda *et al.*, 2000), white mulberry (*Morus alba* L.) (Yokoyama, 1929; Ehara and Gotoh, 1992; Gotoh *et al.*, 2003a,b), peach (*Prunus persica* (L.) Batsch) (Gotoh *et al.*, 2003a) and Asian pear (*P. pyrifolia* (Burm. f.) Nakai) (Ehara and Gotoh, 1992; Gotoh *et al.*, 2003a).

Among the above-mentioned species, only the citrus red mite, *P. citri*, and the European red mite, *P. ulmi*, are severe pests of economic plants, although the raspberry red mite, *P. caglei*, and *P. elongatus* are potential pests. All four species are further discussed below.

Panonychus caglei Mellot

Common name

Raspberry red mite.

Diagnostic characteristics

FEMALE. The general aspect of this mite is as for *P. citri* (see below), with the body deep red in colour and the dorsal tubercles light red to pink. The distal sensillum of the palp is about 1¼ to 1½ times as long as wide. The peritremes are straight distally. The dorsal setae of the idiosoma are large, pubescent and set on strong tubercles. The striae on the genital flap are longitudinal and anterior to the setae of the flap and transverse posterior to them. Tarsus I bears 1 sensory and 5 tactile setae proximal to the duplex setae and 3 sensory and 4 tactile setae distal to them (Mellot, 1968).

MALE. In the male, the distal sensillum of the palp is small, conical and slightly longer than wide. The dorsal setae are long, pubescent and set on strong tubercles. The aedeagus is dorsally directed and sigmoid (Mellot, 1968).

Geographical distribution

Brazil (Paschoal, 1970), China (Ma and Yuan, 1981b; Wang *et al.*, 1985; Wang and Cui, 1992), Japan (Nansei-shoto) (Ohno *et al.*, 2011a), Taiwan (Tseng, 1990) and the USA (Cagle, 1962; Mellot, 1968).

Host plants

This mite is reported on 14 host plants from seven families: Asteraceae – *Artemisia argyi* H. Lév. *et* Vaniot (Ma and Yuan, 1981b); Cucurbitaceae – field pumpkin (*C. pepo* L.) (Paschoal, 1970), *Trichosanthes ovigera* Blume (Ohno *et al.*, 2011a); Fabaceae – *Pueraria montana* (Lour.) Merr. (Cagle, 1962); Malvaceae – *Malva pusilla* Sm. (Cagle, 1962); Moraceae – *Broussonetia papyrifera* (L.) Vent. (Ma and Yuan, 1981b); Rosaceae – *Potentilla norvegica* L. (Mellot, 1968), *Rosa* sp. (Mellot, 1968; Ma and Yuan, 1981b), *Rubus* sp. (Cagle, 1962; Wang and Cui, 1992), *R. allegheniensis* Porter, *R. chingii* Hu (Ma and Yuan, 1981b), American dewberry (*R. flagellaris* Willd.), raspberry (*R. idaeus* L.) (Cagle, 1962; Mellot, 1968); Urticaceae – *Boehmeria nivea* (L.) Gaudich. (Ma and Yuan, 1981b; Ohno *et al.*, 2011a).

Bio-ecology

P. caglei is a polyphagous mite reported on wild and cultivated plants. On raspberry, the mite feeds on both leaf surfaces, with the immature forms mostly on the lower surface and the adults on the upper surface. In the early spring, the raspberry red mite lays most of its eggs on the lower leaf surface, while by the third to ninth generations, 75–80% of the eggs are laid on the upper leaf surface. In autumn, many eggs are laid on the lower leaf surface and also on the canes. Some of the eggs laid on the canes in autumn may hatch in the same period as the summer eggs. During mild winters, the eggs may hatch sooner, but colder winters do not allow hatching, which begins from the last part of March. Initially warm winters followed by cold periods cause high mortality of the mite populations, which induces low levels of infestation in spring. The mite develops nine or more generations a year, according to the influence of temperature on the eggs and immature development stages (Cagle, 1962; Jeppson *et al.*, 1975).

Symptomatology and damage

On raspberry, the feeding activity of *P. caglei* causes the formation of lines or short dashes on the upper leaf surface (Jeppson *et al.*, 1975).

Control

This mite is a potential pest but there is no information on its control.

Panonychus citri (McGregor) (22.15)

Common name

Citrus red mite.

Diagnostic characteristics

FEMALE. The body is 350–400 µm long and deep red to purplish in colour. The legs I are shorter than the body. All the dorsal setae are set on strong tubercles of the same colour as the body, and longer than the distance between their bases. The prodorsal vertical setae v_2 are shorter and the scapular setae sc_1 longer. The opisthosoma is striate transversally and the striae bear conical and narrow lobes; ventrally, the lobes are broad and round. The dorsal opisthosomal setae f_2 and h_1 are of equal length and both as long as about ⅓ of setae f_1. The peritremes end in a simple bulb. Tarsus I bears 3 tactile setae and 1 solenidion proximal to the duplex setae (Vacante, 1985a).

MALE. The male is smaller than the female, and light in colour, sometimes orange. The aedeagus is sigmoid, with a broad base and the distal end pointed and turned upward (Vacante, 1985a).

Geographical distribution

Worldwide, and widely spread in the Palaearctic region (Bolland *et al.*, 1998; Migeon and Dorkeld, 2006–2013).

Host plants

The citrus red mite is reported on 105 host plants of 37 families. Of these, the Rutaceae harbour the higher number of species (23) (Migeon and Dorkeld, 2006–2013). Families and species of economic importance are listed below:

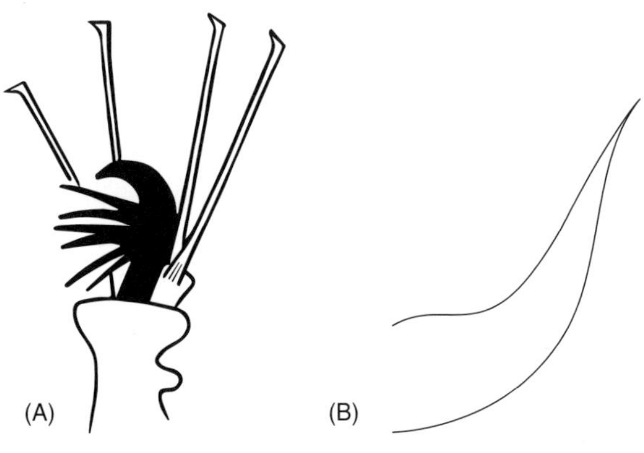

Fig. 20.15. *Panonychus citri* (McGregor). (A) female, pretarsus I; (B) male, aedeagus (modified from Vacante, 1985a).

Arecaceae: coconut palm (*Cocos nucifera* L.) (Bolland *et al.*, 1998; Migeon and Dorkeld, 2006–2013).

Caricaceae: papaya (*Carica papaya* L.) (Garrett and Haramoto, 1967; Lee Goff, 1986; Gupta and Gupta, 1994; Ho *et al.*, 1997; Álvarez and Alarcón, 2004).

Cucurbitaceae: garden cucumber (*Cucumis sativus* L.) (Gupta, 1984), field pumpkin (*C. pepo* L.) (Bolland *et al.*, 1998; Migeon and Dorkeld, 2006–2013).

Ebenaceae: Japanese persimmon (*Diospyros kaki* L.f.) (Bolland *et al.*, 1998; Migeon and Dorkeld, 2006–2013).

Euphorbiaceae: castor bean (*Ricinus communis* L.) (McGregor, 1950).

Fabaceae: soybean (*Glycine max* (L.) Merr.) (Meyer Smith, 1974; Ashihara, 1987).

Lauraceae: avocado (*Persea americana* Mill.) (Ehara, 1969a; Guanilo *et al.*, 2012).

Meliaceae: neem (*Azadirachta indica* A. Juss.) (Flechtmann and Baker, 1970).

Moraceae: white mulberry (*Morus alba* L.) (Ehara, 1956b; Gupta and Gupta, 1994; Ohno *et al.*, 2011a), black mulberry (*M. nigra* L.) (Garrett and Haramoto, 1967; Lee Goff, 1986).

Musaceae: banana (*Musa* × *paradisiaca* L.) (Gupta and Gupta, 1994).

Myrtaceae: guava (*Psidium guajava* L.) (Ho *et al.*, 1997).

Oleaceae: sweet osmanthus (*Osmanthus fragrans* Lour.) (Kitashima and Gotoh, 1995; Wang *et al.*, 2012).

Oxalidaceae: carambola (*Averrhoa carambola* L.) (Ho *et al.*, 1997).

Rhamnaceae: common jujube (*Ziziphus mauritiana* Lam.) (Ho *et al.*, 1997).

Rosaceae: loquat (*Eriobotrya japonica* (Thunb.) Lindl.) (Ho *et al.*, 1997), Virginia strawberry (*Fragaria virginiana* Duchesne) (Othman and Zhang, 2003), apple (*Malus domestica* Borkh.) (Gupta, 1981; Gupta and Gupta, 1994), sweet almond (*P. dulcis* (Mill.) D.A. Webb), cherry laurel (*P. laurocerasus* L.) (McGregor, 1950; Navajas *et al.*, 1996), *P. persica* (L.) Batch (Flechtmann and Hunter, 1971; Gupta, 1992, Gupta and Gupta, 1994; Ho *et al.*, 1997; Gotoh and Higo, 1997), black cherry (*P. serotina* Ehrh.) (Ashihara, 1987), pear (*Pyrus communis* L.) (Ehara, 1955), Chinese pear (*P. pyrifolia* (Burm. f.) Nakai) (Gotoh and Higo, 1997), China rose (*Rosa chinensis* Jacq.) (Othman and Zhang, 2003), Japanese rose (*R. multiflora* Thunb. ex Murr.) (Davis, 1968c; Gotoh *et al.*, 2003a).

Rutaceae: calamondin (*Citrofortunella microcarpa* (Bunge) Wijnands) (Bowman and Bartlett, 1978; Vierbergen, 1994), key lime (*Citrus aurantiifolia* (Christm.) Swingle) (Doreste, 1968b; Lee Goff, 1986), sour orange (*C. aurantium* L.) (Rodrigues, 1968; Gupta and Gupta, 1994; Kasap, 2011), *C. junos* Siebold ex Tanaka (Gotoh *et al.*, 2003a), lemon (*C. limon* (L.) Burm. f.) (McGregor, 1950; Ho *et al.*, 1997; Aguilar and Murillo, 2008; Ledesma *et al.*, 2011; Zriki *et al.*, 2015), *C. macracantha* Hassk. (Vierbergen, 1994), pomelo (*C. maxima* (Burm.) Merr.) (Ho *et al.*, 1997; Gotoh *et al.*, 2003a), citron (*C. medica* L.) (Gotoh *et al.*, 2003a), grapefruit (*C. paradisi* Macf.) (McGregor, 1950; Lee Goff, 1986; Guanilo *et al.*, 2012), tangerine (*C. reticulata* Blanco) (McGregor, 1950; Lee Goff, 1986; Gotoh *et al.*, 2003a,b; Guanilo *et al.*, 2012; Zriki *et al.*, 2015), sweet orange

(*C. sinensis* (L.) Osbeck) (McGregor, 1950; Lee Goff, 1986; Martínez *et al.*, 2004; Ben-David *et al.*, 2007; Ledesma *et al.*, 2011), hyuganatsu (*C. tamurana* hort. ex Tanaka) (Gotoh *et al.*, 2003a), *C. trifoliata* L. (Ehara, 1955), Indian wood apple (*Limonia acidissima* L.) (Guanilo *et al.*, 2012).

Sapotaceae: sapodilla (*Manilkara zapota* (L.) P. Royen) (Migeon and Dorkeld, 2006–2013).

Theaceae: tea (*Camellia sinensis* (L.) Kuntze) (Bolland *et al.*, 1998; Migeon and Dorkeld, 2006–2013).

Vitaceae: grapevine (*Vitis vinifera* L.) (Ho *et al.*, 1997).

Bio-ecology

P. citri is a polyphagous mite and a severe pest of *Citrus* despite the fact that these plants contain toxic furanocoumarins, terpenoids and other compounds. Unlike other spider mites that cannot develop on citrus plants, *P. citri* is able to do so; it can also develop on other non-toxic hosts such as Japanese pear. The mite has adapted to citrus plants by evolving detoxification systems composed of cytochrome P450s and/or other detoxifying enzymes (Takeyama *et al.*, 2006).

The citrus red mite reproduces by arrhenotokous parthenogenesis and virgin females lay eggs hatching only into males, while mated females produce both sexes. The karyotype is $n = 3$ or $n = 4$. The chromosomes are holokinetic, with mixoploidy in the karyotypes (Zou *et al.*, 2002). The mite has diapausing and non-diapausing forms that are characterized by morphological differences (Okasabe and Saitō, 1991) and the diapause occurs through winter eggs (Shinkaji, 1979; Takafuji and Morimoto, 1983). The ratio of females to males on leaves and fruits of lemon is 2.7:1 and 3.1:1, respectively (Chiavegato, 1988).

On citrus, *P. citri* develops on leaves, twigs and mature or green immature fruits, and it prefers light green maturing growth, feeding commonly in the greatest numbers on the upper surface of young and medium aged leaves. It has been observed that 60% of populations live in the exterior canopy and 80% on the upper side of the leaves. The mite likes the green to the yellow fruits, and in winter in California on mature fruits of sweet orange or clementine, the population densities are higher than on the leaves (Mico *et al.*, 1992). On lemon, the species develops more or less constantly where there is new growth and new fruits (Jeppson *et al.*, 1975). In Korea, it has been observed that on citrus in autumn the density of populations of citrus red mite is higher on fruits than on leaves, according to the fruit maturation level (Song *et al.*, 2003), but laboratory experiments show that lemon leaves are more suitable than fruits for the development of the pest (Chiavegato, 1988). The age of leaf and amount of injury caused by the feeding activity influence the mite populations. The number of eggs laid on young and medium-aged leaves is greater than on old leaves, but feeding injury influences oviposition (Henderson and Holloway, 1942) and this cannot represent a regulatory factor in the density of populations. Furthermore, there is a direct relationship between the density of populations and the nitrogen of the host plant (Schwartz, 1984).

At 26°C, the life cycle takes 14 days and at 10°C is five times longer; furthermore, at 10°C, it is 9 times longer than at 27°C. The range of development at high temperatures is balanced by a life cycle that is longer under colder conditions (English and Turnipseed, 1941; Shinkaji, 1954). On lemon and Valencia orange fruits at 23.9°C and 56% RH, the eggs hatch in 7–9 days, and the larval, protonymphal and deutonymphal stages last 2, 2 and 1–3 days, respectively. The males hatch before the females and oviposition reaches the maximum within 7–8 days after the females hatch. Under these conditions, the life cycle from egg to egg takes 17 days, and the longevity of male and female is 22 and 27 days, respectively (Keetch, 1968). At 25 ± 1°C, 58 ± 5% RH and LD (light:dark hours) of 24:0, and with leaves and fruits of lemon, 28.9 and 13.0 days are necessary for the female lifespan, and 9.6 and 10.9 days for the male lifespan, respectively. The number of eggs per female was 77.2 and 37.7 on leaves and fruits, respectively (Chiavegato, 1988). The development threshold for the complete cycle is 10.82°C and the thermal constant is 151.61 degrees-day (Delrio and Monagheddu, 1986). In the laboratory, on leaves of Washington navel orange at 15, 20, 25, 30 and 35 ± 1°C, the developmental period for females is 37.2, 16.6, 12.2, 9.8 and 9.0 days, respectively. The total and daily egg production are highest at 25°C, followed by 20°C and 30°C. The lowest total and daily egg production occur at 15°C. The development threshold for female eggs and the egg to adult stages are 9.22°C and 9.77°C, respectively, and the total effective temperature is 100 and 192.30 degree-days, respectively. The intrinsic rate of increase (r_m) at 30, 25, 20 and 15°C is 0.148, 0.160, 0.111 and 0.042 females female^{-1} daily, respectively (Kasap, 2009).

The instantaneous rate of reproduction is 0.1723 at 30°C and 0.1688 at 25°C, and no reproduction occurs at 15°C (Muraoka and Shiyomi, 1986). At 25°C, the intrinsic rate of increase (r_m) is 0.209 day^{-1} and the net reproduction rate (R_0) is 46.61 (Gotoh *et al.*, 2003b).

Temperature is the major factor affecting mite development (Keetch, 1968). Temperatures of 40.5°C, or several days of hot, dry weather (5% RH and 32°C) together with strong wind, commonly produce a high mortality. Similar conditions may be observed during the summer in California, but can occur also in the winter. Extended periods of high RH are also unfavourable for the development of mite populations. The susceptibility to extreme conditions of both temperature and RH limits the distribution of the mite and affects the seasonal trends in its populations (Jeppson *et al.*, 1975).

Study of the direct influence of citrus irrigation on citrus red mite development has shown than the direct effect of irrigation on yield is greater than physiological effects on the trees and on mite development (Hare *et al.*, 1989). The direct and interacting effects of four seasons of uncontrolled citrus red mite populations with 36 other combinations of grove management practices have been examined by Hare *et al.* (1992) to determine the long-term deleterious consequences of not controlling mite populations on Navel orange. Effectiveness of the acaricide, densities of mites per leaf, differential acaricide application, fertilization level and differential irrigation all caused yield reductions and/or produced reductions in

fruit size. Total yield and average fruit size did not differ consistently because of other interactions between effective *P. citri* treatments and other cultural practices. Marketing conditions continue to favour large fruit; thus, increases in average fruit size on trees where citrus red mite populations are not effectively suppressed continue to be at least a compensating, if not beneficial, consequence of withholding acaricide applications.

The dynamics of populations vary from season to season, depending on climatic conditions, chemical control of different citrus pests and other factors. Sometimes, the mite seems disappear for several years. McGregor (1956) reported that in the interior areas of Florida and California, the period of heaviest occurrence of the citrus red mite is autumn and early winter. Jeppson (1977) says that two peaks of populations occur, in spring or early summer and in autumn or early winter. In South Africa, the maximum number of mites has been observed during spring and autumn (Smith, 1953; Stofberg, 1959; Brodrick, 1965), and in some countries from November to January (Van Rooyen, 1966). In Spain, the seasonal trend in mite populations showed an increase during August–September, followed by a descent or maintenance of the population levels up to the winter (García Marí *et al.*, 1983). In Italy, the populations of *P. citri* commonly increase in the spring and autumn, with a pause or lowering of reproduction in summer and winter, but in some countries, where environmental factors permit, mite development can occur in those seasons (Vacante, 1995, 2009b).

Normally, the citrus red mite does not spread uniformly in orchards, but its presence varies from tree to tree. The influence on predators of dust covering trees alongside orchard roads is well known (Van Rooyen, 1966). Furthermore, the use of different insecticides for scale insects or for the control of other pests can either directly (Schwartz, 1978; Jones and Parrella, 1984a) or indirectly (Keetch, 1968; Grout and Richards, 1992b) stimulate the elimination of predacious mites and citrus red mite development. Because most of the hosts of *P. citri* are woody plants, mainly citrus trees usually planted at intervals of various metres, the mite is likely to face more risks (e.g. starvation) by dispersing between host plants than other spider mite species that live on both herbaceous and woody plants, such as the two-spotted spider mite, *Tetranychus urticae* Koch. A limited gene flow between patches (host plants) can lead to differentiation of populations even within a small area. It has been observed that *P. citri* populations maintain a higher level of variation between trees (or patches of trees) within groves than between groves at the local level, although the gene diversity tended to be smaller with decreasing distance between populations at the geographical level. These mechanisms are related to the dispersal behaviour of spider mites (Osakabe *et al.*, 2005). The species is dispersed by wind and by labourers and farm implements such as harvesting or other machines.

Natural enemies include pathogens and predatory insects and mites. The pathogens include a non-inclusion virus and several fungi. Several predatory insects and mites are reported in the USA (Quayle, 1938; DeBach *et al.*, 1950; Fleschner and Ricker, 1953a,b; Ebeling, 1959; McMurtry, 1969, 1977a, 1985b; McMurtry *et al.*, 1970, 1979; Yue and Childers, 1994; Goldarazena *et al.*, 2004; Villanueva *et al.*, 2004, 2005). The insects include coccinellids, staphylinids, endomychids, chrysopids, coniopterygids, reduviids and Aeolothripidae (Table 20.3). The most important predatory mites are Phytoseiidae, but also recorded are some stigmaeid and bdellid mites (Table 20.4).

Symptomatology and damage

The feeding symptomatology of the red spider mite may be confused with that of the oriental citrus mite, *Eutetranychus orientalis* (Klein). Feeding from the upper leaf surface results in the removal of all the cytoplasmic contents except for some starch grains (Albrigo *et al.*, 1983). This produces on the leaves, mostly in the upper surface, little emptied spaces looking like stipplings, more or less numerous according to the density of the attack, and consisting of light coloured spots with a grey or silvery appearance. The green immature fruits become pale, and mature fruits turn a pale straw yellow. Fruits that are damaged while they are still green will ripen and colour, but if the mite infestation occurs on mature fruits, the markings are more permanent. High levels of attack before fruit maturity may cause fruit fall (Stofberg, 1959; Keetch, 1968; Jeppson *et al.*, 1975). The influence on the quality of the fruit is commonly of minor importance (Jeppson *et al.*, 1975).

The damage is severe when high levels of attack are associated with a high transpiration rate owing to a low RH and wind speed or a deficiency in leaf moisture owing to drought, a poor root system, or other influences that limit the a supply of adequate moisture to the leaves. The combined influence of mite attack and weather may produce heavy leaf and fruit drop, twig dieback and even the death of large limbs (Jeppson *et al.*, 1975). The twigs without leaves are bare and show the leaf petioles still attached; they commonly die back as result of the defoliation. The defoliation rate during winter may reach 13.5, 20.6, 53.1 and 72.6% at densities of 1, 3, 6 or >10 mites per leaf (Choi and Kim, 1998).

Severe infestations reduce the yields in the following years. The use of one or more pesticides on Tahiti lime trees can have a negative impact on the populations of citrus red mite and their associated predators to the detriment of yield. The magnitude of the impact is dependent upon the treatment regime and application date. Spider mites increase because of reduced predator populations, and use of these pesticides, especially in combination, is to be discouraged (Childers and Abou-Setta, 1999).

Like other spider mites, *P. citri* may induce allergies in workers (Burches *et al.*, 1996), and is the most common sensitizing allergen in adolescents living in rural areas with citrus orchards; the sensitization is significantly associated with the prevalence of atopic dermatitis (Cho *et al.*, 2000). On skin prick tests, the most common sensitizing allergen was from the citrus red mite (20.6%), followed by *Dermatophagoides pteronyssinus* (Trouessart) (18.5%), *D. farinae* Hughes (14.6%) and the cockroach (9.8%) (Kim *et al.*, 2002c).

Table 20.3. *Panonychus citri* (McGregor). Natural enemies (virus, fungi and insects).

Families and species	References
VIRUSES	
Non-inclusion virus	Muma, 1955; Munger *et al.*, 1959; McMurtry *et al.*, 1979; Rath, 1991
FUNGI	
Ancylistaceae	
Conidiobolus thromboides Drechsler	Fisher, 1951
Clavicipitaceae	
Beauveria bassiana (Balsamo) Vuillemin	Shi and Feng, 2006
Hirsutella thompsonii Fisher	Rath, 1991
Hirsutella kirchneri (O. Rostr.) Minter *et al.*	Sztejnberg *et al.*, 1997
Entomophthoraceae	
Entomophthora sp.	Fisher, 1951
Exobasidiomycetidae	
Acaromyces ingoldii Boekhout *et al.*	Paz *et al.*, 2007a
Meira argovae Boekhout *et al.*	Paz *et al.*, 2007a
Meira geulakonigii Boekhout *et al.*	Sztejnberg *et al.,* 2004; Paz *et al.*, 2007a
Neozygitaceae	
Neozygites floridana (Weiser *et* Muma)	Smith and Furr, 1975
INSECTS	
Aeolothripidae	
Aeolothrips sp.	Guanilo and Martinez, 2007
Scolothrips takahashii Priesner	Kim and Choi, 2000; Kishimoto and Takagi, 2001
Anthocoridae	
Orius sauteri (Poppius)	Kim and Choi, 2000
Chrysopidae	
Ceraeochrysa cincta Schneider	Guanilo and Martinez, 2007
Chrysopa pallens (Rambur)	Kim and Choi, 2000
Chrysoperla externa Hagen	Guanilo and Martinez, 2007
Mallada basalis (Walker)	Cheng *et al.*, 2009
Coccinellidae	
Halmus chalybeus (Boisduval)	Jamieson *et al.*, 2005
Paraotcthorus histrio Chazeau	Guanilo and Martinez, 2007
Propylea japonica (Thunberg)	Kim and Choi, 2000
Stethorus spp.	McMurtry *et al.*, 1970; McMurtry, 1977a, 1985b; García Marí *et al.*, 1983; Vacante, 1995; Jamieson *et al.*, 2005, 2008; Jamieson and Stevens, 2009
Stethorus japonicus Kamiya	Kishimoto and Takagi, 2001
Stethorus punctillum Weise	Kim and Choi, 2000
Stethorus tridens Gordon	Guanilo and Martinez, 2007
Coniopterygidae	
Coniopteryx (*Holoconiopterix*) *turneri* Kimmins	Van Rooyen, 1966
Coniopteryx vicina Hagen	McMurtry *et al.*, 1970
Conwentzia sp.	Van Rooyen, 1966
Conwentzia barrette (Banks)	McMurtry *et al.*, 1970
Conwentzia capensis Tjeder	Meyer Smith and Schwartz, 1998a
Conwentzia psociformis (Curtis)	García Marí *et al.*, 1983; Vacante, 1995
Semidalis fuelleborni Enderlein	Van Rooyen, 1966
Semidalis meridionalis Kimmins	Meyer Smith and Schwartz, 1998a
Endomychidae	
Saula japonica Gorham	Nakao *et al.*, 1977
Staphylinidae	
Oligota spp.	McMurtry *et al.*, 1970; McMurtry, 1977a, 1985b; Garcia Marí *et al.*, 1983; Kishimoto and Takagi , 2001; Guanilo and Martinez, 2007
Oligota kashmirica benifica Naomi	Kim and Choi, 2000
Oligota pygmea Solier	González, 1969
Oligota yasumatsui Kistner	Kim and Choi, 2000
Reduviidae	
Ploearia culiciformis (de Gen.)	García Marí *et al.*, 1983

Table 20.4. *Panonychus citri* (McGregor). Natural enemies (Bdellidae, Cheyletidae, Phytoseiidae and Stigmaeidae).

Families and species	References
Phytoseiidae	
Amblyseius chungas Denmark *et* Muma	Guanilo and Martinez, 2007, 2009
Amblyseius eharai Amitai *et* Swirski	Tanaka and Kashio, 1977; Beattie, 1978
Amblyseius herbicolus (Chant)	
Amblyseius largoensis (Muma)	Rodriguez and Ramos, 2000b, 2006; Jamieson *et al.*, 2005
Amblyseius newsami (Evans)	Kong *et al.*, 2005
Euseius addoensis (van der Merwe *et* Ryke)	Meyer Smith, 1981a
Euseius citri (van der Merwe *et* Ryke)	
Euseius hibisci (Chant)	McMurtry, 1969
Euseius scutalis (Athias Henriot)	Kasap and Sekeroglu, 2004
Euseius stipulatus (Athias Henriot)	Vacante, 1986, 1995; Vacante and Nucifora, 1986; Vacante *et al.*, 1989b; Ferragut *et al.*, 1988, 1992; Abad *et al.*, 2006
Metaseiulus occidentalis Nesbitt	Kim *et al.*, 2002b
Neoseiulus californicus (McGregor)	Katayama *et al.*, 2006; Guanilo and Martinez, 2007
Neoseiulus cucumeris (Oudemans)	Zhang *et al.*, 2004c, 2005
Neoseiulus fallacis (Garman)	Kim *et al.*, 2002b, 2006
Neoseiulus womersleyi (Schicha)	Kim and Paik, 1999; Kim and Choi, 2000; Kim *et al.*, 2002b
Phytoseiulus longipes Evans	Badii *et al.*, 2010b
Phytoseiulus macropilis (Banks)	Rodriguez and Ramos, 2006
Phytoseiulus persimilis Athias Henriot	Jamieson *et al.*, 2005; Guanilo and Martinez, 2007
Typhlodromus athiasae Porath *et* Swirski	Kasap, 2011
Typhlodromus (*Anthoseius*) *evectus* (Schuster)	Guanilo and Martinez, 2007
Typhlodromus (*Anthoseius*) *transvaalensis* (Nesbitt)	Guanilo and Martinez, 2007
Typhlodromina subtropica Muma *et* Denmark	
Bdellidae	
Bdellodes longirrostris (Hermann)	Garcia Marí *et al.*, 1983
Cheyletidae	
Cheletogenes sp.	Guanilo and Martinez, 2007
Mexecheles sp.	
Cunaxidae	
Armascirus sp.	Guanilo and Martinez, 2007
Stigmaeidae	
Agistemus spp.	Goldarazena *et al.*, 2004
Agistemus cyprio (González)	García Marí *et al.*, 1983
Agistemus exsertus González-Rodríguez	Yue and Childers, 1994
Agistemus fanari Dosse	Dosse, 1967
Agistemus longisetus González-Rodríguez	Jamieson *et al.*, 2005, 2008; Jamieson and Stevens, 2009
Agistemus terminalis (Quayle)	Inoue and Tanaka, 1983
Zetzellia sp.	García Marí *et al.*, 1983

Control

The implementation of a reliable sampling and monitoring strategy for populations allows the rationalization of control, which is usually based on the use of chemicals and on the action of indigenous natural enemies, but sometimes those introduced from other countries. Biological control alone, through inundative releases of natural enemies, has been positively evaluated in different countries, but is not at present a practical strategy in the field. The integration of these methods with rational cultural management of orchards allows the implementation of IPM programmes. There have been some studies on the resistance of different varieties.

SAMPLING AND MONITORING. The sampling and monitoring of the citrus red mite populations on citrus crops have been variously investigated. In the Mediterranean region, the fortnightly counting has been suggested of mobile forms present on four developed leaves of the newest vegetation sampled at random from 10% of the plants, keeping the threshold within the limits of three mites per leaf of new vegetation, or 20–40% of leaves infested (Cavalloro and Prota, 1983). Jones and

Parrella (1984b) and Zalom *et al.* (1986b) discussed the presence–absence method and Rodriguez and Ramos (1998) calculated on Marsh grapefruit seedlings that the proportion of infested leaves related strongly to the average density of mites for any phase. The values of R^2 are 0.88, 0.85 and 0.85, respectively, for females, juveniles and the total population. Using the threshold value of 0.30 females per leaf and the results obtained, it is possible to apply a presence–absence sampling plan. The method provides a rapid estimation of the citrus red mite population based on 16% of infested leaves in a 30-leaf sample. Hare and Phillips (1992) found no significant relationship between the variation in density and duration of citrus red mite populations peaking at >19 adult females per leaf and variation in total yield, size or grade of lemon fruits. They observed no effects of high mite populations in the first year upon crop yield, grade, or size in the second year, and adopted a conventional treatment threshold for *P. citri* on lemon of two adult female mites per leaf. The data do not support the premise that growers who observe that threshold are achieving an economic return.

The density of citrus red mites in commercial satsuma mandarin (*C. unshiu*) groves in Korea was assessed by counting the number of mites per leaf. It was observed that this increase in sampling size had minimal effects on the precision of the binomial sampling plan and of the estimated mean, regardless of tally thresholds. When $T = 1$ is chosen as the best tally threshold for estimating densities of *P. citri* based on the precision of the model, the binomial model with $T = 1$ provides reliable predictions of mean densities of the citrus red mite. A binomial sequential sampling procedure has been developed for classifying the density of mites, and a binomial sampling programme for decision making on the *P. citri* population level based on an action threshold of two mites per leaf has been achieved (Song *et al.*, 2001). Later, Song *et al.* (2003) developed sequential sampling plans for adults of the citrus red mite, developing a decision-making model for the citrus red mite population based on different action threshold levels (2.0, 2.5 and 3.0 mites per leaf) with 0.25 precision. The maximum number of trees and required number of trees sampled on a fixed sample size plan with 2.0, 2.5 and 3.0 thresholds with 0.25 precision are 19, 16 and 15, and their critical T values are 554, 609 and 659, respectively. Later on, the spatial distribution of pre-adult and adult *P. citri* on Washington and Valencia orange crops was determined in Turkey. A precision sequential sampling plan for *P. citri* was applied and it was observed that a high infestation rate and low coefficient of variation occurred on the outside of the canopy, pointing to this area as adequate for a sampling unit. However, because of the ease of sampling and time efficiency, the bottom outside of the canopy is the best sampling unit for the pest. Sample size requirements increase with an increasing coefficient of variation (Kasap *et al.*, 2004).

The monitoring of populations is difficult because variables such as climate, mite location and chemical control can affect the economic threshold. McMurtry (1985b) reports that successful monitoring of mite populations is based more on experience than on the use of economic thresholds, and that more citrus red mites can be tolerated in California during spring and summer than during the autumn, when the dry winds may cause stress to the trees. In South Africa, an average of 2–3 adult females per leaf represents the economic threshold, but the use of chemicals must be related to climate conditions. Integrated control is the obvious solution to the citrus red mite problem. *P. citri* very seldom reaches pest proportions in unsprayed orchards or orchards under integrated control (Meyer Smith and Schwartz, 1998a). In this context, selective acaricides must be used.

PLANT RESISTANCE. Different research has been done at different times in various countries. However, development planning for citrus does not seem to always take into account the available information on varietal resistance. Studies in the field and laboratory in China on the development and population dynamics (population density, female adult longevity and innate capacity of increase) of the citrus red mite, on 14 citrus varieties show that 'Jin Dan' and 'Ponkan NL' cultivars are resistant to the mite, whereas 'Taiwan bampeiyu' pummelo, 'Diangjiang' white-flesh pummelo and 'Beibei 447' sweet orange are susceptible to the citrus red mite. Other cultivars exhibit moderate resistance. Among sweet orange and mandarin cultivars, the former show high susceptibility to the mite, while the latter (excepting 'Zaojin Wase') exhibit low susceptibility (Lei *et al.*, 2004). Other observations carried out in China on the population dynamics of citrus red mite on 20 citrus germplasm samples indicated that cv. 'Beibeiyou' is susceptible, and cultivars 'Jinju' and 'Juibingle' are resistant to the mite (Hu *et al.*, 2004a).

BIOLOGICAL CONTROL. Releases of the phytoseiid mite *Neoseiulus cucumeris* (Oudemans) carried out in China on to potted citrus seedlings showed that predator females released at predator:prey ratios as low as 1:54–58 provided control of the citrus mite in 5 weeks. Field trials showed that a single preventive release of *N. cucumeris* using a controlled release system provided adequate control of citrus red mite and also significantly reduced the rate of occurrence of black fruits. Biological control cost a third less than conventional chemical control (Zhang *et al.*, 2001f). Later, *N. cucumeris* was defined as a desirable biological control agent to suppress *P. citri* at lower population stages in citrus orchards (Zhang *et al.*, 2005).

Releases of the phytoseiid mite *Amblyseius fallacis* (Garman) (= *Neoseiulus fallacis* (Garman)) at a ratio of 16:1 (*P. citri*: *N. fallacis*) effectively suppressed the citrus red mite population in Korea. Releases of the *N. fallacis* once or twice at a ratio of 20:1 (*P. citri*:*N. fallacis*, at a density of 0.5 and 1.0 *P. citri* per leaf) did not effectively suppress the citrus red mite. Investigations in a plastic film house on the density changes of *P. citri* and *N. fallacis* on a citrus tree at a release ratio of 10:1 (at a density of *P. citri* of approximately 0.1 per leaf) showed that releases of *N. fallacis* twice at an interval of 10 days resulted in high densities of *N. fallacis* that effectively suppressed *P. citri* populations (Kim *et al.*, 2003).

In laboratory studies in Israel, the fungi *Meira geulakonigii* Boekhout *et al.* and *Acaromyces ingoldii* Boekhout *et al.* (Exobasidiomycetidae) caused 86.7 and 87.5% mortality of *P. citri* after 14 days, respectively, and *M. argovae et al.* caused

72.7% mortality (Paz *et al.*, 2007a). An indigenous population of *N. californicus* (McGregor) in conventionally controlled citrus groves in central Japan suppressed the population of *P. citri* on Satsuma mandarin (Katayama *et al.*, 2006).

Although these results highlight the possibility of a radical turning point in the biological control of the citrus red mite (and also of other herbivorous mites), the technique is not used extensively in the field at present.

CHEMICAL CONTROL. A number of acaricides have been used against *P. citri* during the last three decades. They include acid benzhydroxamic derivates (Benfatto and Lanza, 1980; Lanza *et al.*, 1980; Vacante *et al.*, 1980), amidines (Benfatto and Lanza, 1980; Lanza *et al.*, 1980; García Marí *et al.*, 1983; Meyer Smith and Schwartz, 1998a), avermectins (McCoy *et al.*, 1982; French and Villarreal, 1988; Benfatto, 1994; Peña *et al.*, 1999; Jamieson and Stevens, 2009), carbazates (Ochiai *et al.*, 2007), formamides (Ran *et al.*, 2008), organochlorines (Lanza *et al.*, 1980; Vacante *et al.*, 1980; García Marí *et al.*, 1983, 1988; French *et al.*, 1985), organosulfurs (Benfatto and Lanza, 1980; Lanza *et al.*, 1980; Vacante *et al.*, 1980; García Marí *et al.*, 1983, 1988), organotins (Benfatto and Lanza , 1980; Lanza *et al.*, 1980; García Marí *et al.*, 1983; Vacante, 1995; Meyer Smith and Schwartz, 1998a; Ran *et al.*, 2008), petroleum oils (Jeppson *et al.*, 1975; Benfatto and Lanza, 1980; Lanza *et al.*, 1980; García Marí *et al.*, 1983; Ohkubo, 1983; French and Villarreal, 1988; Benfatto, 1994; Herron *et al.*, 1995; Walker and Aitken, 1996; Meyer Smith and Schwartz, 1998a; Nohara *et al.*, 2000; Ripa *et al.*, 2006; Jamieson and Stevens, 2009; Childers, 2010), phenyl pyrazoles (Yamada *et al.*, 1987; French and Villarreal, 1988; Jamieson and Stevens, 2009), pyridazinones (Benfatto, 1994; Peña *et al.*, 1999; Shi and Feng, 2006), pyrazoles (Benfatto, 1994), quinazolines (Longhurst *et al.*, 1992; Curkovic and Araya, 2004; Ripa *et al.*, 2006), sulfur-bridged compounds (Ran *et al.*, 2008; Jamieson and Stevens, 2009), tetrazines (Santaballa *et al.*, 1994) and tetronic acid derivates (Izquierdo *et al.*, 2002), either alone or as mixtures.

In addition to acaricides, various detergents have been tested, such as sodium benzene sulfonate (Ripa *et al.*, 2006), Quix (a neutral liquid detergent) and Nobla (an alkaline powder) (Curkovic and Araya, 2004), extracts from plants, such as *Artemisia annua* L. (Zhang *et al.*, 2009a), *Boenninghausenia sessilicarpa* H. Lev. (Yang *et al.*, 2009), *Ginkgo biloba* L. (Pan *et al.*, 2006), *Humulus scandens* (Loureiro) Merrill, *Laggera pterodonta* (DC.) Benth. and *Rabdosia* sp. (Yang *et al.*, 2007), and repellents, such as extracts of wild rice (Wan *et al.*, 2006).

INTEGRATED PEST MANAGEMENT. According to what has been discussed above, the implementation of a IPM programme requires the use of selective pesticides and/or acaricides that are able to preserve natural enemies, and the correct agronomic management of orchards. The long-term solution to citrus red mite control consists of priority elimination of the use of disruptive agrichemicals, but until this happens effective miticides are needed to control the mite (Jamieson and Stevens, 2009). Usually, sprays of petroleum oils are applied once or twice a year. Alternatively, it is possible to employ organotins. IPM is strictly related to the control of other pests and to horticultural practices. In the Mediterranean region, the phytoseiid mite *Euseius stipulatus* (Athias Henriot) provides excellent control of citrus red mite populations (Garzón-Hidalgo and Ferragut, 2010), but it is necessary to preserve its presence in citrus groves by using selective acaricides or petroleum oils (Vacante, 1986). The implementation of an IPM strategy then requires assessment of the variation of susceptibility to different acaricides (Gotoh *et al.*, 2004a). The risk associated with the use of disruptive pesticides, such as neonicotinoids, synthetic pyrethroids and/or other compounds, is very high, as shown by experiences in the Satsuma mandarin orchards in Japan (Masui and Ikeda, 2003) and by other studies in different countries.

Panonychus elongatus Manson (Fig. 20.16)

Common name

Unknown.

Diagnostic characteristics

FEMALE. The body is 410 μm long, and bright red in colour. The anterior margin of the stylophore is rounded. The peritremes end in a bulb-like swelling. The dorsal body setae are long, slender, tapering, finely pubescent and mostly set on prominent tubercles. The podosomal venter bears 4 pairs of midventral setae. Coxae I and II each bear 2 setae; coxae III and IV each have 1 seta. There are 2 pairs of anal setae, 2 pairs of pseudanal setae and 2 pairs of genital setae. The species is very similar to *P. citri*, but the body setae of female *P. citri*, except for the prodorsal setae v_2, are all slightly smaller than the corresponding body setae of *P. elongatus* (Manson, 1963).

MALE. The body of the male is 273 μm long, and bright red in colour. The terminal sensillum of the palptarsus is about twice as long as broad. The peritremes end in a simple bulb. The distal part of the aedeagus is elongate, slender, tapering, directed posteriorly and similar to that of *P. citri*, but distinctly longer (Manson, 1963).

Geographical distribution

Australia (Manson, 1968), China (Ma and Yuan, 1980; Wang and Cui, 1992), South Korea (Han, 1970), Myanmar (Manson, 1963), Papua New Guinea (Schicha and Gutierrez, 1985), Taiwan (Tseng, 1990) and Thailand (Manson, 1963; Baker, 1975b).

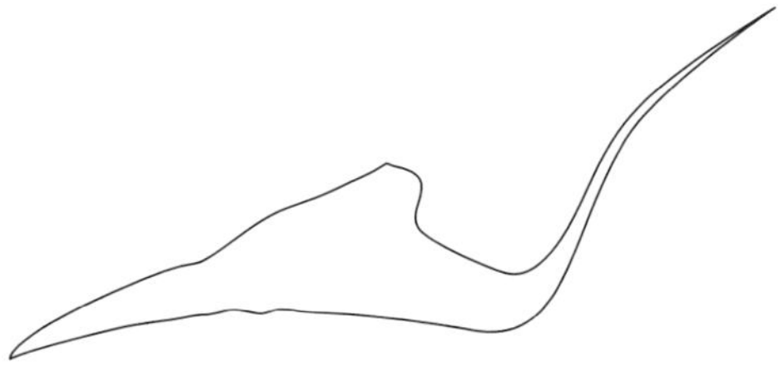

Fig. 20.16. *Panonychus elongatus* Manson. Male, aedeagus (from Manson, 1963).

Host plants

The list of host plants of *Panonychus elongatus* includes 23 species from 10 families.

Caricaceae: papaya (*Carica papaya* L.) (Bolland *et al.*, 1998; Migeon and Dorkeld, 2006–2013).

Fabaceae: *Centrosema plumieri* Benth., *Phaseolus* sp., *Pueraria phaseoloides* (Roxb.) Benth., *Teramnus uncinatus* (L.) Sw. (Davis, 1968c).

Meliaceae: neem (*Azadirachta indica* A. Juss.), *Cedrela odorata* L. (Davis, 1968c), *Melia azedarach* L. (Ma and Yuan, 1980), *Toona ciliata* M. Roem. (Davis, 1968c).

Menispermaceae: *Cocculus orbiculatus* (L.) DC (Bolland *et al.*, 1998; Migeon and Dorkeld, 2006–2013).

Moraceae: *Cudrania tricuspidata* (Carrière) Bureau, *Ficus* sp., *Morus* sp., (Bolland *et al.*, 1998; Migeon and Dorkeld, 2006–2013).

Oleaceae: *Syringa* × *persica* L. (Ma and Yuan, 1980).

Passifloraceae: purple granadilla (*Passiflora edulis* Sims) (Davis, 1968c).

Rosaceae: pear (*Pyrus communis* L.) (Baker, 1975b), *Rosa* sp. (Bolland *et al.*, 1998; Migeon and Dorkeld, 2006–2013).

Rutaceae: *Citrus* sp. (Manson, 1963, 1968; Davis, 1968c), lemon (*C. limon* (L.) Burm. f.) (Davis, 1968c), tangerine (*C. reticulata* Blanco) (Ma and Yuan, 1980), sweet orange (*C. sinensis* (L.) Osbeck) (Davis, 1968c), hardy orange (*Poncirus trifoliata* (L.) Raf.) (Davis, 1968c).

Taxaceae: *Taxus cuspidata* Siebold *et* Zucc. (Bolland *et al.*, 1998; Migeon and Dorkeld, 2006–2013).

Bio-ecology

P. elongatus has sometimes been misidentified with *P. citri* (Davis, 1968c). The mite is reported from the Australasian, Oriental and Palaearctic regions. It has a haplodiploid and reproductive system; its biological stages are similar to those of *P. citri*. Its bio-ecology has not been investigated.

Natural enemics include several phytoseiid mites. *Amblyseius largoensis* (Muma), *A. tamatavensis* Blommers, *Neoseiulus longispinosus* (Evans), *Phytoseius hawaiiensis* Prasad, *P. hongkongensis* (Swirski *et* Shechter), *P. rubiginosae* Schicha, *Proprioseiopsis peltatus* (van der Merwe) and *Typhlodromips deleoni* (Muma) (Schicha and Gutierrez, 1985) have been recorded from Papua New Guinea, and *Euseius ovalis* (Evans) (Shih *et al.*, 1993) from Taiwan.

Symptomatology and damage

Probably as for *P. citri*.

CONTROL. This is usually unnecessary.

Panonychus ulmi (Koch) (Fig. 20.17)

Common name

European red mite.

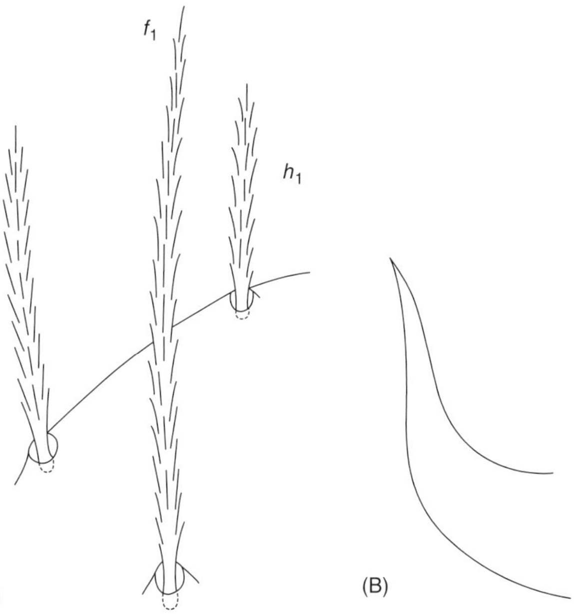

Fig. 20.17. *Panonychus ulmi* (Koch). (A) female, dorsal opisthosomal setae h_1, f_2 and f_1; (B) male, aedeagus (from Vacante, 1985a).

Diagnostic characteristics

FEMALE. The body is reddish in colour, and the dorsal body setae are strong and set on prominent tubercles that are white at the bases. The dorsal opisthosomal setae h_1 and f_2 are $\frac{1}{3}$ and $\frac{2}{3}$ as long as the dorsal opisthosomal setae f_1. The genital plate has transverse striae and its anterior region longitudinal striae. The peritremes end in a simple bulb. The species is similar to *P. citri* (Vacante, 1985a).

MALE. The male is smaller than the female, light in colour, sometimes orange. The aedeagus is sigmoid, with a broad base and the distal end pointed and turned upward (Vacante, 1985a).

Geographical distribution

Worldwide, common in the Palaearctic region (Bolland *et al.*, 1998; Migeon and Dorkeld, 2006–2013).

Host plants

The whole list of host plants of European red mite numbers 138 species of 38 families, with the highest number of species (44) in the Rosaceae family (Migeon and Dorkeld, 2006–2013). Families and species of economic importance are listed below:

Anacardiaceae: mango (*Mangifera indica* L.) (Gupta, 1992).

Apiaceae: carrot (*Daucus carota* L.) (Bolland *et al.*, 1998; Migeon and Dorkeld, 2006–2013).

Betulaceae: filbert (*Corylus avellana* L.) (Rambier, 1958a).

Cucurbitaceae: winter squash (*Cucurbita maxima* Duchesne), field pumpkin (*C. pepo* L.) (Bolland *et al.*, 1998; Migeon and Dorkeld, 2006–2013).

Ebenaceae: Japanese persimmon (*Diospyros kaki* L.) (Carmona and Dias, 1980).

Fabaceae: lucerne (*Medicago sativa* L.) (Gupta, 1992), red clover (*Trifolium pratense* L.), garden vetch (*Vicia sativa* L.) (Carmona and Dias, 1980).

Fagaceae: European chestnut (*Castanea sativa* Miller) (McGregor, 1950).

Grossulariaceae: golden currant (*Ribes aureum* Pursh), European blackcurrant (*R. nigrum* L.), cultivated currant (*R. rubrum* L.) (Rambier, 1958a), red flowering currant (*R. sanguineum* Pursh) (Bolland *et al*., 1998; Migeon and Dorkeld, 2006–2013).

Iridaceae: *Gladiolus* sp. (Carmona and Dias, 1980).

Juglandaceae: English walnut (*Juglans regia* L.) (McGregor, 1950; Hatzinikolis, 1968c).

Moraceae: fig (*Ficus carica* L.) (McGregor, 1950; Rambier, 1958a; Gupta and Gupta, 1994), white mulberry (*Morus alba* L.) (Gupta and Gupta, 1994), black mulberry (*M. nigra* L.) (Wafa *et al*., 1967).

Poaceae: oat (*Avena sativa* L.), Aleppo grass (*Sorghum halepense* (L.) Pers.), wheat (*Triticum aestivum* L.) (Gupta and Gupta, 1994), maize (*Zea mays* L.) (Bolland *et al*., 1998; Migeon and Dorkeld, 2006–2013).

Rosaceae: quince (*Cydonia oblonga* Mill.) (Hatzinikolis, 1968c), loquat (*Eriobotrya japonica* (Thunb.) Lindl.), beach strawberry (*Fragaria chiloensis* (L.) Mill.)] (= × *ananassa* Duchesne ex Rozier), woodland strawberry (*F. vesca* L.) (Bolland and Beron, 2004), apple (*Malus domestica* Borkh.) (Oudemans, 1931d; Vergani, 1953; Rambier, 1958a, 1964; van de Vrie, 1964; Hatzinikolis, 1968c; Kreiter *et al*., 2002; Gotoh *et al*., 2003b; Martínez *et al*., 2004; Zriki *et al*., 2015), apricot (*Prunus armeniaca* L.) (Hatzinikolis, 1968c; Gupta and Gupta, 1994), sweet cherry (*P. avium* (L.) L.) (Rambier, 1958a), *P. divaricata* Ledeb., European plum (*P. domestica* L.) (McGregor, 1950; Rambier, 1958a; Hatzinikolis, 1968c, 1969c; Tuttle and Baker, 1976; Gupta and Gupta, 1994), sweet almond (*P. dulcis* (Miller) D.A. Webb.) (Vergani, 1953), peach (*P. persica* (L.) Batsch) (McGregor, 1950; Vergani, 1953; Rambier, 1958a; Rossi Simons, 1961; Wafa *et al*., 1967; Hatzinikolis, 1968c, 1969c; Gupta, 1992; Gupta and Gupta, 1994), pear (*Pyrus communis* L.) (McGregor, 1950; Vergani, 1953; Rambier, 1958a; Wafa *et al*., 1967; Hatzinikolis, 1968c; Tuttle and Baker, 1976; Martínez *et al*., 2004), Chinese pear (*P. pyrifolia* (Burm. f.) Nakai) (Gupta and Gupta, 1994), dog rose (*Rosa canina* L.) (Gupta and Gupta, 1994), Japanese rose (*R. multiflora* Thunb.), American red raspberry (*Rubus idaeus* L.) (McGregor, 1950; Rambier, 1958a), black raspberry (*R. occidentalis* L.) (McGregor, 1950), whitebeam (*Sorbus aria* (L.) Crantz), *S. aucuparia* L. (Reeves, 1963).

Rubiaceae: gardenia (*Gardenia jasminoides* J. Ellis) (Bolland *et al*., 1998; Migeon and Dorkeld, 2006–2013).

Rutaceae: key lime (*Citrus aurantiifolia* (Christm.) Swingle), sour orange (*C. aurantium* L.), pomelo (*C. maxima* (Burm.) Merr.) (Zhou *et al*., 1999; Ran *et al*., 2009).

Solanaceae: tomato (*Solanum lycopersicum* L.) (Gupta and Gupta, 1994).

Theaceae: tea (*Camellia sinensis* (L.) Kuntze) (Bolland *et al*., 1998; Migeon and Dorkeld, 2006–2013).

Vitaceae: fox grape (*Vitis labrusca* (L.)) (Reeves, 1963), grapevine (*V. vinifera* L.) (Rambier, 1958a, 1964; Arthofer, 1976; Carmona and Dias, 1980; Tixier *et al*., 2003).

Bio-ecology

P. ulmi infests deciduous fruit trees, grape, ornamentals, forest trees, legumes, cereals, forages, vegetables and other economic plants and shrubs. Among the economic plants, the mite commonly infests apple, pear, plum and quince, and may also occur on peach, walnut, cherry, almond, grape, raspberry, hawthorn, mountain ash, elm, rose and chestnut (Jeppson *et al*., 1975). It is the most important pest mite on deciduous fruit trees in Europe, the eastern USA and Canada, Argentina, Tasmania, New Zealand and Japan (van de Vrie, 1985). *P. ulmi* is a severe pest of grapevines in the whole northern region of Europe (Schruft, 1985), and in Pennsylvania it has become the most important pest on grapes (*V. labrusca*) (Ramsdell and Jubb, 1979).

The mite reproduces by arrhenotokous parthenogenesis and the haploid condition is $n = 3$ (Helle and Bolland, 1967). The sex ratio is 63% females and 37% males on grapes (*V. labrusca*) (Jeppson *et al*., 1975). On deciduous fruit trees, the European red mite overwinters as eggs that are laid in groups from August to October or November (depending on the climate) on bark areas at the base of the buds and spurs, in bud wounds, and points that mark the beginning of new growth. Hatching begins in spring when a total of 195 degree-days have accumulated, when the earliest blossom buds show pink (Lees, 1953a; Jeppson *et al*., 1975). The duration of diapause and hatching are synchronized with local climatic conditions (Tsugawa *et al*., 1966; Cranham, 1973; Herbert and McRae, 1982). The larvae hatching from overwintering eggs move on to leaves, and the immature stages usually live on the undersurfaces of the leaves, while adults can live on both surfaces. On grapevines, the overwintering eggs are laid on the 1-year-old canes near the buds and on the head of the trunk, The hatched larvae move towards the young leaves. The populations feed mainly on the lower leaf surface, and the infestation is very harmful in early spring and late summer (Schruft, 1985). On deciduous fruit trees, the summer eggs are laid on the lower leaf surfaces from spring to September or October. The life cycle requires 3–4 weeks and there are 5–6 generations a year in Canada and as many as 9–10 in Virginia (Parent and Beaulieu, 1957). On grapevines, the mite develops 4–6 overlapping generations a year. Populations living on grapevines represent a specific biotype of the fruit tree populations, and transfer from one to the other crop is unsuccessful (Götz, 1956, in Schruft, 1985).

At 23.5°C, the egg develops in 5 days and at 13°C in 20 days. The development from egg to adult takes 4 days at the higher temperature and 19–22 days at the lower temperature. The average lifespan of the female takes about 19 days and it oviposits 10–90 eggs. At 25°C ± 1°C, the intrinsic rate of increase (r_m) is 0.180 day^{-1} and the finite rate of increase (λ) is 1.197 (Rabbinge, 1976). Herbert (1981) reports an intrinsic rate of increase (r_m) of 0.185 day^{-1} and a finite rate of increase (λ) of 1.203. The mite exhibits an LW-c life type, characterized by the spinning of silk threads while walking on the leaf and the weaving of egg guy ropes, but it does not produce other web structures. These spinning threads are used as lifelines that assist locomotion on the leaf surface and prevent the mite from falling off the plant (Saitō, 1979a, 1985).

Dispersal of *P. ulmi* occurs mainly by transport of the overwintering eggs to new plantings on nursery stock. However,

the mite may be dispersed locally by wind. When populations increase greatly, the mite spins a silken thread, attached on the one end to a leaf or twig, and uses the other end to support itself. The thread falls 2–5 inches, suspended in the air, until a breeze breaks it, forming a 'ballooning' exercise (Jeppson *et al.*, 1975).

Population dynamics are influenced by the weather, cultural practices, predators, and chemical treatments and the host plant (Jeppson *et al.*, 1975). More information on this topic can be found in the first part of the book. The summer weather conditions affect food availability and the amount of overwintering eggs deposited by the females and, consequently, the populations in the following spring. The development of high population densities during the summer may be reduced either by internal competition due to restricted availability of the food resource or by heavy development of natural enemies, which lead to the deposition of only a few overwintering eggs, while under good conditions orchards may be characterized by a high deposition of overwintering eggs. The first stages after hatching are very sensitive to low temperatures, and during this period if night temperatures fall below freezing, the mite populations may be significantly reduced. Overwintering eggs do not develop at 7°C or below, although overwintering eggs from different areas may have a different response to temperature. However, within the lethal range of each egg population, mortality increases by about 10% for each doubling of the exposure period. Resistance to freezing temperatures has been increased by selection (Tsugawa *et al.*, 1961; Jeppson *et al.*, 1975). Bhardwaj and Sharma (2010) observed a partial correlation between all stages of mites and meteorological parameters that indicated a negative correlation with rainfall and a significant positive correlation with humidity.

Cultural practices (fertilization, pruning, cultural choices, etc.) may facilitate the development of populations of *P. ulmi* (Post, 1962). On apple leaves with a high nitrogen content, the mite lays a higher number of eggs and has a faster development (Breukel and Post, 1959). Deficiency and/or an excess of nitrogen or potassium induce in grapevines an increase of *P. ulmi* populations and higher egg production (Besse and Götz, 1963). According to Chaboussou (1966), high infestations of spider mites usually involve a phenomenon of 'trophobiosis', i.e. improvement of the basic nutrition of plants. The influence of agricultural practice has been extensively discussed by van de Vrie *et al.* (1972).

Increase of temperature and a high content of foliar carbohydrates, amino acids and some mineral elements favour the development of *P. ulmi* (Ramsdell and Jubb, 1979; Schreiner, 1984; Wermelinger *et al.*, 1992). Conversely, rain, high RH, a high level of potassium and of phenolic compounds in the leaves hinder an increase in the populations of *P. ulmi* (Schreiner, 1984; Borgo and Giorgessi, 1987; Wermelinger *et al.*, 1992). In addition, on vines, the frequency and level of infestation by the European red mite may vary depending on the genotype determining the leaf characteristics; tomentose leaf surfaces are often associated with a higher density of pest mites (Schreiner, 1984; Rilling, 1989).

Spray applications of insecticides against other pests may synergize *P. ulmi* development, either as a means of the reduction of predators or induction of the development of mite itself (McMurtry *et al.*, 1970; Hoyt *et al.*, 1978). Some host plants seem less suitable for mite development than others. Thus, in the same area, the mortality of young stages may be higher on walnut than on pear or plum. Moreover, the life cycle on walnut is shorter but associated with fewer eggs, and the lifespan of females is shorter than on pear or plum (Doreste, 1964; Jeppson *et al.*, 1975). Biotic factors influencing the ecology of *P. ulmi* also include the apple rust mite, *Aculus schlechtendali* (Nalepa). This eriophyoid plays a beneficial role in apple orchards in Washington (and elsewhere) because it is a trophic resource that helps the development of the phytoseiid *Metaseiulus occidentalis* (Nesbitt) early in the growing season (Hoyt, 1969a). Eriophyoid populations must become relatively large before they produce severe damage to apple leaves, and so help the phytoseiid populations to increase so that they can reduce the populations of *P. ulmi* and other tetranychids on apples later in the growing season (Hoyt, 1969a; Hoyt *et al.*, 1979).

Natural enemies of the European red mite include a non-inclusion virus and various predatory mites and insects (Tables 20.5 and 20.6). Further information on this topic may be found in Jeppson *et al.* (1975), McMurtry *et al.* (1970), Chazeau (1985), Gerson (1985a), Van der Geest (1985, 2010), van de Vrie (1985), Gerson *et al.* (2003) and Hoy (2011).

Symptomatology and damage

The feeding activity of the European red mite induces the leaves to become at first pale and then bronze in colour; photosynthetic and respiratory processes are reduced and transpiration increases. Highly infested leaves turn brown and fall. The greatest damage to trees is usually associated with prolonged infestations. Thus, if infestation occurs in June, the normal number of flower buds in the following year may not develop (Jeppson *et al.*, 1975). On apple and pear, the leaf symptomatology is associated with strong changes in the spongy parenchyma (Avery and Briggs, 1968a,b). The intensity of the reaction to mite feeding depends on the plant species and variety. On apple, reduced crop yield may occur in the following year, but the effects vary with apple cultivar, tree vigour, time of damage, crop load and the weather (Hoyt *et al.*, 1979). Beers and Hull (1990) estimated the effect of *P. ulmi* on apples in Pennsylvania, allowing 1000 cumulative mite days to occur in the early season (early May to mid-June), mid-season (mid-June to August), or late season (August to mid-October). Mid-season injury induced the greatest reduction in mean fruit weight at harvest, as well as reduced bloom and fruit load the following season. Late-season injury caused reduced bloom the next year. However, Hoyt *et al.* (1979) found that in Washington State 'improved red color and higher soluble solids were positively correlated with *P. ulmi* populations in one test. No other significant effects from mite feeding damage were observed on fruit quality or yield'.

Hoyt (1969b) reported that robust trees cultivated with sufficient fertilizer might be little influenced by feeding by either *P. ulmi* or other spider mites. Moreover, they reported that 'In Washington, apples are thinned extensively, both chemically and by hand, and this thinning may overcome any potential for reduced yields in the second year'. Yield losses induced by mites on apple are 'more likely to be due to reduced fruit growth the year the population occurs rather than to reduced bloom and fruit set the following year'. 'Red Delicious' usually has higher red mite populations, but 'Golden Delicious' is more susceptible

Table 20.5. *Panonychus ulmi* (Koch). Natural enemies (virus and insects).

Families and species	References
VIRUSES	
Non-inclusion virus	Steinhaus, 1959; Putman and Herne, 1966
INSECTS	
Aeolothripidae	
Aeolothrips fasciatus (L.)	Putman and Herne, 1966
Aeolothrips malaleucus (Hal.)	Putman and Herne, 1966; Horsburgh and Asquith, 1968
Scolothrips indicus Priesner	Reddy and Jagadish, 1977
Scolothrips sexmaculatus(Pergande)	McMurtry *et al.*, 1970
Anthocoridae	
Anthocoris musculus Say	McMurtry *et al.*, 1970
Anthocoris nemoralis (F.)	Steiner, 1980
Anthocoris nemorum (L.)	McMurtry *et al.*, 1970
Anthocoris sp.	
Orius sp.	
Orius insidiosus Say	
Orius laevigatus (Fieber)	Carayon and Steffan, 1959
Orius majusculus (Reuter)	
Orius minutus (L.)	McMurtry *et al.*, 1970
Orius vicinus (Ribaut)	Fauvel *et al.*, 1975
Cecidomyiidae	
Arthrocnodax sp.	Readshaw, 1975
Chrysopidae	
Chrysopa sp.	McMurtry *et al.*, 1970
Chrysopa carnea Stephens	
Coccinellidae	
Parastethorus histrio (Chazeau)	Aguilera, 1987
Parastethorus nigripes (Kapur)	Walters, 1976
Stethorus bifidus Kapur	Collyer, 1964a; McMurtry *et al.*, 1970
Stethorus darwini (Brethes)	Lorenzato, 1987
Stethorus gilvifrons (Mulsant)	Hajizadeh *et al.*, 1993
Stethorus punctillum Weise	Putman, 1955; McMurtry *et al.*, 1970
Stethorum punctum punctum (Leconte)	McMurtry *et al.*, 1970
Stethorus vagans (Blackburn)	Walters, 1976
Coniopterigidae	
Conwentzia pineticola End.	Collyer, 1951
Conwentzia psociformis (Curt.)	
Semidalis albata End.	McMurtry *et al.*, 1970
Hemerobiidae	
Hemerobius sp.	UC IPM Online, 2012e
Miridae	
Atractotomus mali (Meyer)	Lord, 1971
Blepharidopterus angulatus (Fall.)	McMurtry *et al.*, 1970
Campylomma verbasci (Meyer)	
Deraeocoris nebulosus (Uhler)	Horsburgh and Asquith, 1968
Diaphnocoris pellucida (Uhler)	McMurtry *et al.*, 1970
Hyaliodes harti Knight	Horsburgh and Asquith, 1968
Malacocoris chlorizans Panz.	Geier and Baggiolini, 1952
Pilophorus perplexus (D. *et* S.)	Lord, 1971
Phlaeothripidae	
Haplothrips faurei Hood	Putman and Herne, 1966; McMurtry *et al.*, 1970
Haplothrips subtilissimus (Hal.)	Parrella *et al.*, 1981
Leptothrips mali (Fitch)	Horsburgh and Asquith, 1968
Staphylinidae	
Oligota sp.	McMurtry *et al.*, 1970
Oligota flavicornis (Boisduval *et* Lacordaire)	
Syrphidae	
Toxomerus geminatus Say	McMurtry *et al.*, 1970

Table 20.6. *Panonychus ulmi* (Koch). Natural enemies (spiders and mites).

Families and species	References
SPIDERS	
Theridiidae and Linyphiidae	
Webbing spiders	Gerson, 1985a
MITES	
Anystidae	
Anystis baccarum (L.)	Cuthbertson and Murchie, 2012
Erythraeidae	
Balaustium putmani Smiley	Putman, 1970b
Phytoseiidae	
Amblyseius andersoni (Chant)	Schruft, 1985; Calis *et al.*, 1988
Amblyseius potentillae (Garman)	Van de Vrie and Kropczyńska, 1965
Euseius finlandicus (Oudemans)	Schruft, 1985
Euseius hibisci (Chant)	Hoy, 2011
Kampimodromus aberrans (Oudemans)	Schruft, 1985; Kreiter *et al.*, 2000
Metaseiulus occidentalis (Nesbitt)	Hoyt *et al.*, 1979
Neoseiulus californicus (McGregor)	Jung *et al.*, 2010
Neoseiulus fallacis (Garman)	Welty, 1995
Phytoseius finitimus Ribaga	Vacante and Tropea Garzia, 1987b; Duso, 2006
Phytoseius plumifer (Canestrini *et* Fanzago)	Kreiter *et al.*, 2000; Duso and Fontana, 2002
Typhlodromus athiasae Porath *et* Swirski	Mansour *et al.*, 1993
Typhlodromus caudiglans Schuster	Downing and Moilliet, 1967
Typhlodromus exhilaratus Ragusa	Vacante and Tropea Garzia, 1987b; Duso, 2006
Typhlodromus pyri Scheuten	Collyer, 1964b; Kreiter *et al.*, 2000
Stigmaeidae	
Agistemus africanus Gonzalez	Pringle, 1975
Agistemus flechneri Summers	Nelson *et al.*, 1973
Agistemus longisetosus Gonzalez	Collyer, 1964a
Zetzellia graeciana Gonzalez	Inserra, 1970; Nucifora and Vacante, 1978; Vacante and Tropea Garzia, 1987a
Zetzellia mali (Ewing)	Parent and LeRoux, 1956; Vacante and Tropea Garzia, 1987a

to damage. The following year's crop may be reduced in terms of both fruit size and amount of fruit. Eggs may be deposited on the calyx end of fruit, causing downgrading and processing problems (Alston and Reding, 2011a).

In general, pear trees exhibit a low tolerance to the European red mite, and a low population density may induce severe leaf burning and yield losses (Hoyt and Tanigoshi, 1983). The feeding activity of *P. ulmi* on pear causes the leaves to lighten in colour and gradually become mottled or stippled. Prolonged infestation by a high density of populations may induce leaf bronzing, and under high temperatures, portions of the leaf blade or its entire surface may turn brown and dry. This damage is usually called 'mite burn'. High temperatures following feeding damage can induce leaf burn to appear even after mites have been controlled. Severe damage and leaf drop can negatively affect fruit size and the returned crop load. Among pear cultivars, the red strains are more tolerant to mite infestation than the green strains, and 'Bartlett' and 'Comice' are more tolerant to leaf burn and leaf drop than 'D'Anjou' and 'Bosc' according to the UC Pest Management Guidelines for pear (UC IPM Online, 2012e) and Alston and Reding (2011b).

Damage on stone fruits is intermediate compared with that on pear and apple. Light infestations cause leaf speckling and heavy infestations can cause bronzed and pale leaves, undersized and poor quality fruit, reduced tree vigour and decreased yield the year after significant June and July infestations because fewer fruit buds are formed. Strong infestations can induce leaf drop and tree stress, which can lead to reduced winter hardiness (Alston and Reding, 2011b).

The European red mite infests also the vineyard (table and wine grapes). The feeding activity of the mite may cause severe damage to fruits and leaves, which become yellow, brown and leaden, and may drop early. Moreover, high levels of populations of European red mite negatively affect the photosynthesis process and induce loss of water, with a decrease of the quality and quantity of the yield. In spring and summer, the populations of *P. ulmi* cause brittling and deformation of the shoots on which they live. The bunches of grapes may be seriously damaged. At the end of summer, the foliage of infested vineyards changes in colour and appears brown red. The phenomenon may be a consequence of treatments with pyrethroids, neonicotinoids, or other active ingredients sprayed against other pests.

On walnut, the feeding activity of *P. ulmi* first causes light stippling of the leaves. Prolonged high infestations gradually induce the leaves to turn bronzed. Leaf drop does not usually occur, but severe and repeated bronzing of leaves for several years can negatively affect nut yields. On this host plant, damage of *P. ulmi* is commonly rare, and it is more abundant in cooler, coastal areas than in hot, inland orchards; see the UC Guidelines of Pest Management, Walnut (UC IPM Online, 2009f).

On almond, the feeding activity of *P. ulmi* causes leaf stippling. Prolonged infestation causes leaves to pale and appear

bronzed and burned at the tips and margins. Almond trees that are not stressed by drought or by other factors can tolerate high infestation levels (in excess of 50 mites per leaf) for prolonged periods without inducing leaf drop. However, these levels of attack can cause leaf drop in stressed trees. At low levels, *P. ulmi* represents an alternative food for mite predators. Generally, the mite does not reach damaging levels on this host plant. Populations often decline when temperatures turn hot; see the UC Guidelines of Pest Management for Almond (UC IPM Online, 2012c).

Control

Control requires primarily a reliable methodology for sampling and monitoring the mite populations. Spray applications of pesticides, integrated with the action of natural enemies and agricultural practices are the best option. The choice of selective pesticides and the implementation of rational cultural practices are fundamental.

SAMPLING AND MONITORING. The Phenology Model Database of the University of California for the European red mite (UC IPM Online 2003a) uses the model of Herbert (1981) and reports a developmental threshold of 10.6°C. The eggs, larvae, protonymphs and deutonymphs of the female require accumulations of 68.6, 25.8, 22.5 and 28.0 degree-days, respectively. The eggs, larvae, protonymphs and deutonymphs of the male require accumulations of 69.9, 24.1, 19.2 and 22.0 degree-days, respectively. The generation time of the female (egg to adult), the pre-oviposition period and the generation time (egg to egg) require accumulations of 144.9, 16.9 and 161.8 degree-days, respectively. The generation time of the male (egg to egg) requires an accumulation of 135.2 degree-days.

The sampling and monitoring of populations of the mite may be based on observations of both wood and foliage. The main information for the most important crops that are infested is reported below.

Almond. The UC Pest Management Guidelines for almond (UC IPM Online, 2013a) suggest spur sampling during the dormant season to monitor the eggs of the European red mite. With this aim, 35–50 trees are selected randomly from each orchard or plot to be sampled. Major scaffolds are also selected randomly, and two or three spurs or twigs clipped from the inside of each canopy up to a total of 100. Some 20 of the spurs are examined for mite eggs with a hand lens or binocular microscope and presence or absence recorded. A figure of 20% of spurs with mite eggs indicates the need for a delayed dormant oil spray. During the growing season, monitoring must be carried out once a week (along with looking for other pests). No treatment thresholds are known, but almond trees may tolerate a high density of *P. ulmi* per leaf.

Apple. Sampling methods for winter eggs on wood differ mainly in the sampling unit chosen. The method suggested by Vogel and Bachmann (1956) is based on the branches of old wood (≥2 years). Baillod and Fiaux (1975) sampled pairs of spurs on 2-year-old wood, while Touzeau (1973) considered basal pairs of buds on 1-year-old wood. Goonewardene and Kwolek (1975) sampled nodes on both sides of bud scale scars, and Fauvel *et al.* (1978) the first ten buds at the base of a 1-year-old twig. All of these methods provide a relative estimate of egg abundance. In Greece, models have been developed to predict when the eggs will hatch so that pesticide applications can be made at an appropriate time (Broufas and Koveos, 2000). Wogel and Wilbotz (in Bassino *et al.*, 1973) reported that the presence of 400–500 overwintering eggs (linear) m^{-1} of wood during winter required adequate control in the spring. The UC Pest Management Guidelines for apple (UC IPM Online, 2009b,c) suggest that 25–100 fruit spurs should be collected during winter from trees throughout the plot. Less than 10% of infested spurs do not commonly constitute a phytopathological problem during the season. More than 10% of infested spurs requires treatment (oil spraying before bloom). Similarly, the Utah Pests Fact Sheet by Alston and Reding (2011b) for apple, pear, peach, cherry, plum and prune suggests interventions against overwintering eggs during the delayed dormant period. With this aim, it is suggested that orchards are surveyed from late autumn through to early spring to detect overwintering eggs on small limbs and twigs near buds and spurs. Control generally improves the closer the eggs are to hatching.

For the foliage, Bassino *et al.* (1973) suggested threshold levels of 400, 700 and 100 mites per 100 leaves sampled randomly on 50 trees in May, June to mid-August and from mid-August, respectively. From a practical point of view, it is possible to simplify management by using the threshold values of 65% of infested leaves in May, 75% in June–August and 45% from August. Jeppson *et al.* (1975) reported that during summer, 2–5 mites per leaf represented a threshold level. Vacante and Calabretta (1980) obtained good control of *P. ulmi* on apple by treating trees at petal fall (80% in cv. 'Delicious') with cyexatin.

The UC Pest Management Guidelines for apple (UC IPM Online 2009b,c) suggest a threshold of 10–30 mites per leaf during summer, depending on various parameters, such as age, variety, condition of the tree and the density of predatory mites. The Utah Pests Fact Sheet of Alston and Reding (2011b) suggests interventions in the post-bloom periods, so the motile forms of *P. ulmi* should be monitored beginning at petal fall to evaluate whether early summer acaricide treatments are necessary. A reliable and easy-to-use sampling method for all spider mites is the presence–absence or binomial method. Values reported for apple and pear are lower than for cherry and peach. A detailed example may be found in Beers and Hoyt (1993).

Grapevines. On grapevines, the relationship between the number of red spider mites per leaf and the percentage of leaves infested by mites leads to an estimation of the risk of damage (Baillod *et al.*, 1979). In spring, the economic threshold level is 60–70% of infested leaves, while in summer, 30–45% of infested leaves allows severe infestation to be avoided in late summer and autumn. A sequential sampling evaluation was proposed by Baillod and Schlaepfer (1982). The thresholds proposed by Baillod *et al.* (1979) correspond to a maximum value of 2–5 motile forms per leaf, which was estimated by Girolami (1981) as too low to cause economic damage. In fact, Girolami *et al.* (Girolami, 1981; Girolami *et al.*, 1989) observed that 20 motile forms of *P. ulmi* per leaf would not harm yield and for this reason a threshold of 10–20 motile forms per leaf was proposed depending on factors such as the host plant, environmental conditions and density of predators. Moreover, the binomial distribution of the European red mite suggested to Girolami and Mozzi (1984) a method of sampling based on the cumulative number of mites

per leaf. This method is, however, hindered by thresholds that are considered too high, although it has been observed that higher population densities of *P. ulmi* do not adversely affect the physiology of the vine. For instance, it has been observed that the net photosynthesis of vines infested with more than 3500 mites day per leaf (maximum densities of 60 mites per leaf) is not altered (Candolfi *et al.*, 1993a). In addition, yield, quality of the grapes and vegetative growth of three varieties were not affected by an accumulated density over 36,000 mite days per leaf (Candolfi *et al.*, 1993b). From a practical point of view, counts should be made of spider mites and their predators on about ten leaves per hectare of homogeneous vineyard counting spider mites and their predators (Duso and Girolami, 1986). The risk of early attacks of *P. ulmi* in the spring can be assessed by counting the eggs wintering in the vicinity of buds on a sample of 100 buds per hectare made by pruning branches (Baillod *et al.*, 1989a). The threshold proposed in south Italy is ten eggs per bud (Laccone *et al.*, 1993); unfortunately, this value was assigned empirically and was probably too low and so impractical.

Pear. The UC Pest Management Guidelines for pear (UC IPM Online, 2012e) suggest five distinct periods of monitoring, dormant to delayed-dormant, finger bud to petal fall, late spring, summer and harvest/postharvest periods, respectively.

During the dormant to delayed-dormant period, the guidelines suggest collecting 100 fruit spurs, one from the treetop and one at eye level from 50 trees in a 20-acre orchard, and observing the bases of buds and spurs with a hand lens. If the infestation of the spurs exceeds 10%, it is necessary to monitor *P. ulmi* during the growing season. Conversely, if spur infestation is less than 10%, there probably will not be a problem in the coming year, provided that dormant oil treatment was done. If no dormant treatment has been applied, *P. ulmi* should be monitored during the growing season if three or more spurs are infested in the sample of 100.

Sampling during the finger bud to petal fall period should be done after egg hatch by examining 100 flower clusters from 50 trees (one from the treetop and one at eye level). If more than five clusters are infested, narrow-range oil should be sprayed. If more than ten clusters are infested, an acaricide mixed with oil should be sprayed. If the amount of infested clusters ranges between one and five, oil should be added to the first codling moth spray (UC IPM Online, 2012e).

In late spring, monitoring of populations of *P. ulmi* is done by collecting two samples in each of 20 trees, one from the top of the tree and one shoot sample from eye level. For each shoot, five leaves should be observed. There is no treatment if ten or fewer motile forms of *P. ulmi* or eggs are counted per 100 leaves sampled. It is necessary to spray narrow range oil if 11–50 *P. ulmi* or eggs are counted. If more than 50 *P. ulmi* and eggs are counted, it is suggested that a mixture of acaricide and oil should be sprayed (UC IPM Online, 2012e).

During the summer, five leaves should be observed from each of the 20 top shoots and the numbers of *P. ulmi* counted. The above thresholds help to determine treatment needs only if the mite occurs (UC IPM Online, 2012e).

A lack of predators during harvest and following harvest suggests the use of the lower thresholds to determine whether a postharvest treatment is necessary. A bottom sample would be 50–150 mites and eggs per 100 leaves, while a top sample would be 50–200 mites and eggs per 100 leaves (UC IPM Online, 2012e).

Walnut. The UC Pest Management Guidelines for walnut (UC IPM Online, 2009f) suggest examining leaf and growth scars on twigs for eggs. During spring, checks should be made for the development of *P. ulmi* populations and stippling or bronzing on leaves, especially in the shady, central parts of the canopy. Damage by this mite is uncommon and no threshold levels are available to lead control. No treatment should be applied for low to moderate levels of infestation of *P. ulmi* because its populations are important in maintaining predators of other mites.

BIOLOGICAL CONTROL. The first studies on biological control of the European red mite on apple in Europe were carried out by Dosse (1960) in Germany, who pointed out the positive role of the phytoseiid mite *Typhlodromus pyri* Scheuten. The same predatory mite was investigated by Collyer (1964b) in England, while van de Vrie and Kropczyńska (1965) released the phytoseiid mite *Amblyseius potentillae* (Garman) in the Netherlands. Research on different predators of *P. ulmi* on apple have also been done in the USA and Canada. Ballard (1954) studied the biology of *T. fallacis* (Garman) (= *Neoseiulus fallacis* (Garman)). Horsburgh and Asquith (1968) reported 30 predators of *P. ulmi* in Pennsylvania, among which the ladybird beetle, *Stethorus punctillum* Weise. was very active in control. Poe and Enns (1969) studied the phytoseiid mites of orchards in Missouri. Knisley and Swift (1972) quantitatively examined the mite fauna associated with apple in New Jersey. Croft and Jorgensen (1977) evaluated the potential of *T. mcgregori* Chant in the control of the European red mite on apple orchards in California and central Utah. Apart from releases of predatory mites in orchards, results from various work have pointed out that indigenous phytoseiid mites can reduce and maintain pest mites in apple orchards below the economic threshold levels (van de Vrie, 1985). The lack of quantitative data on the predatory system suggested to Rabbinge (1976) a basic model be made in the Netherlands of the interaction between the European red mite and the phytoseiid mite *A. potentillae*. Verification of the model in the field showed a correspondence in the pattern of population fluctuations of prey and predator and allowed use of the model for mite management (van de Vrie, 1985). In Michigan, Dover *et al.* (1979) constructed a dynamic model for the interaction of the phytoseiid mite *Amblyseius fallacis* (= *Neoseiulus fallacis*) and *P. ulmi* in apple orchards, and this allowed prediction of the population development of the predator and pest mites during the season.

From a practical point of view, sprays of non-selective pesticides and fungicides against other apple pests (codling moth, aphids, etc.) and diseases may suppress the survival of predatory mites. The development of resistance of predators to the pesticides, documented first by Croft and Jeppson (1970), allows this to be avoided. Various examples are reported in the literature, e.g. the *T. pyri* strain in English apple orchards treated with carbaryl and organophosphates (OPs) (Cranham and Solomon, 1981). This possibility suggests the mass production of resistant predatory mites, such as OP-resistant *Metaseiulus occidentalis* (Nesbitt) reared in Australia for 3000 ha of apple and peach (Field *et al.*, 1979). Hoy (1982b) reared 62 million carbaryl-OP-resistant *M. occidentalis*.

It is a good strategy to avoid the application of chemical sprays (e.g. lime sulfur) to control the apple rust mite (*A. schlechtendali*), because the populations of this eriphyoid mite allow the development of phytoseiid mites early in the

season, which later in the season control the *P. ulmi* populations. The same protocol has been proposed for walnut; see the UC Pest Management Guidelines, Apple (UC IPM Online, 2009b,c).

On almond in California, *M. occidentalis* preys the immature and adult stages, but is unable to break through the egg shell so it is not as effective in controlling *P. ulmi* as it is other mite species. The brown lacewing, *Hemerobius* sp., is an effective predator, but is not always present; see the UC Pest Management Guidelines for almond (UC IPM Online, 2013a).

Much work has concerned the control of *P. ulmi* in vineyards, where several predatory insects and mites have been reported. Here, the research in the Mediterranean region on phytoseiid mites is fundamental. Among these, *Kampimodromus aberrans* (Oudemans), *T. pyri* and *Phytoseius plumifer* (Canestrini *et* Fanzago) are the three main species controlling mite pests in French vineyards (Kreiter *et al.*, 2000). In north Italian vineyards, the main phytoseiid mites are *T. pyri*, *K. aberrans* and *Amblyseius andersoni* (Chant), while in central and southern Italy, the most common phytoseiids are *T. exhilaratus* Ragusa, *P. finitimus* Ribaga and *K. aberrans* (Vacante and Tropea Garzia, 1987b; Duso, 2006); Duso and Fontana (2002) have also reported *P. plumifer* from north Italy.

In the apple orchards near the Etna volcano in Sicily (Italy), the stigmaeid *Zetzellia graeciana* Gonzalez may efficiently suppress populations of *P. ulmi* (Nucifora and Vacante, 1978; Vacante and Tropea Garzia, 1987a), as does *Z. mali* (Ewing) (Vacante and Tropea Garzia, 1987a). Even in orchards, the response of natural enemies depends on the management of the crop and the surrounding environment, especially their biodiversity, the use of other cultural and phytosanitary techniques and the release of predators.

CHEMICAL CONTROL. The choice of acaricides and timing of their application are the two fundamental steps in chemical control. Acaricides must be selective for natural enemies and assure good control of eggs and motile stages of *P. ulmi*. Timing of the application of chemicals relates to the plant phenology and bio-ecology of the pest mite. It is possible to treat overwintering eggs during the winter, spray hatching overwintering eggs and young larvae during the spring or apply treatments during the summer (Jeppson *et al.*, 1975). Some protocols of chemical control that basically concern the USA are discussed below. However, each choice must be related to the country concerned, and in particular must comply with its laws.

The hatching of eggs will vary from site to site, and this variability may affect pest management programmes directed toward controlling newly hatched larvae and nymphs. The optimal timing of sprays to control *P. ulmi* larvae varies with the local climate and the genetic constitution of the population (Hoy, 2011). Winter sprays of petroleum oils (2%) against overwintering eggs or eggs near to hatching (1%), when the eggs are more sensitive to chemical applications, are effective (Jeppson *et al.*, 1975; van de vrie, 1985; Bhardwaj and Bhardwaj, 2010). The UC Pest Management Guidelines for apple (UC IPM Online, 2009b,c) consider the delayed dormant period to be the better timing for treatment and suggest using narrow-range oils at concentrate doses of 6.0 gal and a dilute dose of 1.6 gal. For dilute application, the rate is per 100 gal water to be applied in 300–500 gal water acre^{-1}, according to the label; for concentrate applications, 80–100 gal water acre^{-1} or less can be sprayed if the label allows. During spring–summer, a number of acaricides can be applied to apple foliage, such as avermectin, clofentezine, fenbutatin oxide, pyridaben, hexythiazox, fenpyroximate, etoxazole, bifenazate and dicofol. On pear in a dormant to delayed dormant state, narrow range oils, or dormant flowable emulsion or dormant plus can be sprayed. From petal fall to harvest narrow-range oil, abamectin, pyridaben, fenpyroximate, acequinocyl, etoxazole, clofentezine, hexythiazox and bifenazate can be sprayed. Postharvest, narrow-range oils can be used. Alston and Reding (2011b) suggest a similar protocol in their Utah Pests Fact Sheet. In India, Bhardwaj and Bhardwaj (2010) used horticultural mineral oil, hexithiazox, fenpyroximate and propargite on the foliage. In Serbia, spirodiclofen compared successfully with bifenthrin, clofentezine, fenazaquin and fenpyroximate (Marcic *et al.*, 2010).

On walnut, a dilute, delayed-dormant oil treatment should be applied, but oils should not be applied to walnut during the dormant season. Foliage sprays of abamectin, bifenazate, clofentezine, fenbutatin-oxide, hexythiazox, propargite and narrow-range oils, according to their properties, can be used to see the UC Pest Management Guidelines for walnut (UC IPM Online, 2009f). On almond, the best time for treatment in an IPM programme is an oil spray during the dormant season, as determined by a dormant spur sample, see the UC Pest Management Guidelines for almond (UC IPM Online, 2013a).

The distribution of pesticide applied may reduce the ecotoxicological impact in apple orchards. In Pennsylvania, only alternate rows of trees are sprayed at any one time to control the codling moth; in this way, ladybird beetles (*S. punctillum*) are preserved and can control the European red mite (Hull and Beers 1985).

INTEGRATED PEST MANAGEMENT. The success of an IPM programme depends on different actions, such as the correct use of decision tools, the use of selective pesticides, the availability of pesticide-resistant strains of predatory mites and the application of a holistic view of the defence of orchards that is able to take into account the control of other pests and diseases of orchards and cultural control (correct fertilization, irrigation, reduction of dust, etc.). Various examples of the successful implementation of IPM programmes reported in different countries are strictly related to this ability by growers, suppliers of beneficials and consultants. Recently, it has been suggested that the release of the anystid *Anystis baccarum* (L.) in Bramley orchards in the UK can be used as a potential alternative control component for incorporation into IPM strategies. It has been demonstrated that *A. baccarum* feeds upon economically important invertebrate pest species including the European fruit tree red spider mite and has a level of compatibility with sprays of pesticides (Cuthbertson and Murchie, 2012).

The Genus *Allonychus* Pritchard *et* Baker

Diagnostic characteristics

The opisthosoma has 2 pairs of setae *h* (h_{2-3}). The empodium is of a single claw-like structure, shorter than the proximoventral hairs, which are at less than right angles to the claw (Bolland *et al.*, 1998).

Economic importance

This genus includes eight species living on wild and cultivated plants, mainly in Nearctic and Neotropical regions (Migeon and Dorkeld, 2006–2013). Six are further discussed below as having potential economic importance. The other two are *Allonychus bambusae* Lo, recorded on *Bambusa* sp. in Taiwan (Lo, 1969) and *A. wuyinicus* Gao *et* Zou on Poaceae in China (Gao and Zou, 1994).

A. brevipenis Feres lives on *Triplaris weigeltiana* (Rchb.) Kuntze in Brazil (Feres, 1992).

A. braziliensis (McGregor) is a potential pest of ornamentals and fruit tropical plants. The list of host plants includes *Cecropia peltata* L. (Flechtmann, 1971), true cinnamon (*Cinnamomum verum* J. Presl) (Flechtmann and Baker, 1975), quince (*Cydonia oblonga* Mill.) (McGregor, 1950), *Dombeya* sp. (Flechtmann and Baker, 1975), *Ficus* sp. (Flechtmann, 1971), the rubber tree (*Hevea brasiliensis* Müll. Arg.) (Chiavegato, 1968), *Hibiscus elatus* Swartz (Livshits and Salinas-Croche, 1968), mango (*Mangifera indica* L.) (Paschoal, 1970), manioc (*Manihot esculenta* Crantz) (Guerrero and Bellotti, 1980), wild banana (*Musa acuminata* Colla), banana (*M. sapientum* L.) (McGregor, 1950), *Nectandra* sp., avocado (*Persea americana* Mill.) (Livshits and Salinas-Croche, 1968), *Stenotaphrum dimidiatum* (L.) Brongn. (Migeon and Dorkeld, 2006–2013) and the European pear (*Pyrus communis* L.) (Paschoal, 1968a, 1970; Flechtmann, 2004). Its geographical distribution includes Brazil (McGregor, 1950; Chiavegato, 1968; Paschoal, 1970; Flechtmann, 2004), Colombia (Guerrero and Bellotti, 1980), Cuba (Livshits and Salinas-Croche, 1968; Suárez, 2004), Nicaragua (Jeppson *et al.*, 1975) and Venezuela (Doreste, 1968b).

A. dorestei Baker *et* Pritchard is another potential pest collected on coconut palm (*Cocos nucifera* L.) in Mexico (Tuttle *et al.*, 1976) and banana (*Musa* × *paradisiaca* L.) in Venezuela (Baker and Pritchard, 1962) and Peru (Guanilo *et al.*, 2012).

A. littoralis (McGregor) is a potential pest reported on *Bixa orellana* L. (Migeon and Dorkeld, 2006–2013), *Bursera simaruba* (L.) Sarg. (Baker and Pritchard, 1962), *Gossypium* sp. (McGregor, 1955), *M. esculenta* (Doreste, 1984b), *P. americana* (Baker and Pritchard, 1960, 1962), *Quercus rugosa* Née (Tuttle *et al.*, 1974) and *Sterculia tragacantha* Lindl. (Baker and Pritchard, 1962). Its geographical distribution includes Mexico (Tuttle *et al.*, 1976), Costa Rica (Doreste, 1984b), Ecuador (McGregor, 1955; Baker and Pritchard, 1962) and Guatemala and Honduras (Baker and Pritchard, 1962).

A. reisi Paschoal is reported on *Azalea* sp. (Paschoal, 1970), *Banisteria* sp., *C. oblonga* (Migeon and Dorkeld, 2006–2013), *Ficus elastica* Roxb. (Paschoal, 1970), *M. esculenta*, matisia (*Matisia cordata* Humb. *et* Bonpl.) (Urueta, 1975), azalea (*Rhododendron indicum* (L.) Sweet) (Paschoal, 1970) and the cacao tree (*Theobroma cacao* L.) (Urueta, 1975). Its geographical distribution includes Brazil (Paschoal, 1970; Feres *et al.*, 2005) and Colombia (Urueta, 1975).

A. querci Baker *et* Pritchard has been collected on *Quercus* sp. (Baker and Pritchard, 1962) and *Hibiscus tiliaceus* L. (Flechtmann, 1976a). Its geographical distribution includes Brazil (Paschoal, 1971b), Honduras (Baker and Pritchard, 1962) and Mexico (Tuttle *et al.*, 1976).

No species of this genus causes severe damage to economic plants.

The Genus *Mononychellus* Wainstein

Diagnostic characteristics

The dorsal body setae are serrate and the opisthosoma is longitudinally striate between the setae e_1. There are two pairs of para-anal setae (h_{2-3}). The empodium is split near the middle into 3 pairs of hairs (Bolland *et al.*, 1998).

Economic importance

This genus is widespread in Nearctic, Neotropical and Afrotropical regions, and to a lesser extent in the Palaearctic region. Currently, it includes 30 species living on wild and economic plants, some of which are potential pests, such as *Mononychellus caribbeanae* (McGregor), living on *Cornus* sp. (Migeon and Dorkeld, 2006–2013), *Dalbergia sissoo* Roxb. (Livshits and Salinas-Croche, 1968), *Erythrina berteroana* Urb. (Tuttle *et al.*, 1976), *Gossypium* sp. (McGregor, 1955), cassava (*Manihot esculenta* Crantz) (McGregor, 1950; Álvarez and Alarcón, 2004; Martínez *et al.*, 2004; Suárez, 2004), and other plants in several countries in Nearctic and Neotropical regions.

Other *Mononychellus* species are pests of economic plants, such as *M. planki* (McGregor), reported on 58 host plants in Nearctic and Neotropical regions, *M. progresivus* Doreste, infesting *Manihot* spp. in Neotropical and Afrotropical regions, and *M. tanajoa* (Bondar), injurious to *Manihot* species and other plants in Afrotropical, Nearctic and Neotropical regions (Migeon and Dorkeld, 2006–2013).

M. progresivus and *M. tanajoa* have a Neotropical origin and were largely confused with one other in the past; for this reason, they are referred to as 'cassava green mites'. Inevitably, some confusion occurs in their geographical distribution, bio-ecology and control. The work of Bolland *et al.* (1998) and Migeon and Dorkeld (2006–2013) has helped to razionalize this problem. *M. planki*, *M. progresivus* and *M. tanajoa* are further discussed below.

Mononychellus planki (McGregor) (Fig. 20.18)

Common name

Green mite.

Diagnostic characteristics

FEMALE. The body is elliptical, and about half as long again as it is wide. The median part of the prodorsum and the dorsal surface surrounding the dorsal hysterosomal setae are irregularly reticulate, but sometimes this dotting is reduced. The dorsal setae are slender, slightly enlarged distally and serrate. The peritreme ends in a bulb. The prodorsal setae v_2 are shorter and sc_1 longer. The opisthosomal setae c_1–f_1 are as long as the intervals between them and not different from the other dorsal setae. Tarsus I has 2 pairs of duplex setae, approximate and distal;

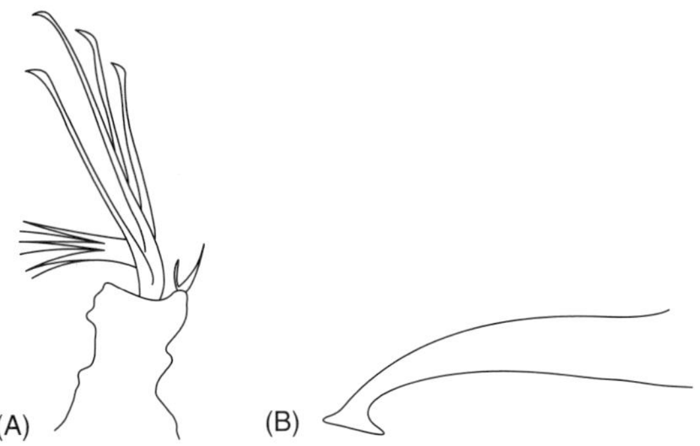

Fig. 20.18. *Mononychellus planki* (McGregor). (A) female, pretarsus I; (B) male, aedeagus (modified from McGregor, 1950)

tibia I has 9 tactile and 1 sensory setae (McGregor, 1950; Pritchard and Baker, 1955).

MALE. The body is smaller and narrower than in the female. The aedeagus ends distally in a triangular barb, with the posterior tip rounded and little extended and the anterior tip sharply pointed and directed more forward than downward (McGregor, 1950).

Geographical distribution

Argentina (Pritchard and Baker, 1955), Brazil (Costa and Correa, 1960; Flechtmann and Baker, 1970; Feres *et al.*, 2005; Furtado *et al.*, 2006; Guedes *et al.*, 2007; Mendonça *et al.*, 2011a), Colombia (Zuluaga Cardona, 1971; Urueta, 1975), Costa Rica (Fréitez, 1974), Cuba (Livshits and Salinas-Croche, 1968; Martínez *et al.*, 2004; Suárez, 2004), Ecuador (McGregor, 1955), El Salvador (Andrews and Poe, 1980), Guadeloupe (Flechtmann and Etienne, 2002), Honduras (Ochoa *et al.*, 1991a), Mexico (Tuttle *et al.*, 1974), Paraguay (Aranda and Flechtmann, 1971), Puerto Rico (USA) (McGregor, 1950; Pritchard and Baker, 1955), Trinidad and Tobago (Pritchard and Baker, 1955), the USA (Baker and Tuttle, 1994) and Venezuela (Doreste, 1968b; Fréitez and Alvarado, 1976).

Host plants

The whole list of host plants numbers 58 species of 12 families. Of these, the Fabaceae have the highest number of species (33), followed by the Malvaceae (11 species) (Migeon and Dorkeld, 2006–2013). Families and species of economic importance are listed below:

Asteraceae: sunflower (*Helianthus annuus* L.) (Fréitez and Alvarado, 1976).

Cucurbitaceae: winter squash (*Cucurbita maxima* Duchesne) (Suárez, 2004), field pumpkin (*C. pepo* L.) (Flechtmann, 1967a).

Euphorbiaceae: cassava (*Manihot esculenta* Krantz) (Comroy, 1958; Flechtmann, 1967a; Doreste, 1968b; Zuluaga Cardona, 1971), castor bean (*Ricinus communis* L.) (Paschoal, 1970).

Fabaceae: groundnut (*Arachis hypogaea* L.) (Flechtmann, 1967a; Flechtmann and Baker, 1970), pigeon pea (*Cajanus cajan* (L.) Millsp.) (Pritchard and Baker, 1955; Andrews and Poe, 1980; Suárez, 2004), soybean (*Glycine max* (L.) Merr.) (Flechtmann, 1968; Livshits and Salinas-Croche, 1968; Zuluaga Cardona, 1971; Guedes *et al.*, 2007), bean (*Phaseolus vulgaris* L.) (Paschoal and Reis, 1968; Flechtmann and Baker, 1970; Zuluaga Cardona, 1971), perennial soybean (*Neonotonia wightii* (Wight *et* Arn.) Lackey) (Paschoal and Reis, 1968; Flechtmann and Baker, 1970; Zuluaga Cardona, 1971).

Malvaceae: okra (*Abelmoschus esculentus* (L.) Moench) (Paschoal, 1968a), Creole cotton (*Gossypium barbadense* L.) (Aranda and Flechtmann, 1971), Levant cotton (*Gossypium herbaceum* L.) (Costa and Correa, 1960; Flechtmann and Baker, 1970).

Rubiaceae: gardenia (*Gardenia jasminoides* J. Ellis) (Bolland *et al.*, 1998; Migeon and Dorkeld, 2006–2013).

Moraceae: black mulberry (*Morus nigra* L.) (Flechtmann, 1967a).

Theaceae: tea (*Camellia sinensis* (L.) Kuntze) (Flechtmann, 1967a).

Bio-ecology

M. planki is a polyphagous mite reported as a pest of soybean (Flechtmann, 1968; Livshits and Salinas-Croche, 1968; Zuluaga Cardona, 1971; Guedes *et al.*, 2007), cotton (Costa and Correa, 1960; Flechtmann and Baker, 1970; Aranda and Flechtmann, 1971) and cassava (Comroy, 1958; Flechtmann, 1967a; Doreste, 1968b; Zuluaga Cardona, 1971), probably as a consequence of the cultural practices that are used (extensive monoculture, varieties, etc.), misuse of pesticides to control other pests and climatic conditions. In the state of Rio Grande do Sul in Brazil, the occurrence of spider mites on soybean has been reported as sporadic and probably associated with dry periods (Roggia, 2007). However, severe attacks by *M. planki*, *Tetranychus desertorum* Banks, *T. gigas* Pritchard *et* Baker and *Polyphagotarsonemus latus* (Banks) on soybean have been reported in Rio Grande do Sul in recent years (Guedes *et al.*, 2007), and two mites, *T. urticae* Koch and *M. planki*, are more widespread than in previous seasons, as a result of a continuous series of cropping seasons characterized by drought, which favours large populations of these mites (Roggia *et al.*, 2008).

The green mite reproduces by arrhenotokous parthenogenesis and has a haploid condition of three chromosomes (Flechtmann, 1983, in Helle and Pijnacker, 1985). The mites do not spin webs and live on both the lower and upper leaf surfaces (Roggia *et al.*, 2008). In a study by Siqueira (2011) on different soybean varieties, the developmental period of the female from egg to adult takes from 10.26 to 12.23 days on 'BRS 134' and 'Jackson', respectively. The shortest male developmental period (9.89 days) occurs on the 'BMX Titan RR' variety and the longest (12.39 days) on the 'Jackson' variety, respectively. The lower immature survival rate (43%) occurs on the 'BRS 245 RR' variety and the higher (92%) on the 'BRS 284' variety, respectively. Female longevity ranges from 8.17 days on the 'BRS 134' variety to 11.64 days on the 'BRS 284' variety, and male longevity ranges from 7.29 days on 'BRS 245

RR' to 18.93 days on 'BRS 284'. The highest fecundity (29.05 eggs per female) has been observed on the 'BRS 284' variety and the lowest (11.85 eggs per female) on the 'BRS 134' variety. The highest population density has been observed at the R3-4 phenological phases. In the 'IAC 100' variety, populations of *M. planki* increased at the end of the study, probably as a result of the early maturation of 'BRS 284' and 'BMX Titan RR'. The 'BMX Titan RR' and 'BRS 284' varieties seem the most attacked (Siqueira, 2011). Climatic factors determine mite population fluctuations on soybean, and the population peak occurs during the initial period of development of soybean pods (Roggia, 2007).

Natural enemies include the predatory mites *Neoseiulus anonymus* (Chant *et* Baker), *N. californicus* (McGregor), *Phytoseiulus fragariae* Denmark *et* Schicha, *P. macropilis* (Banks), *Proprioseiopsis cannaensis* (Muma) and *Galendromus annectens* (De Leon) (Roggia *et al.*, 2009), and the fungus *Neozygites floridana* Weiser *et* Muma (Roggia *et al.*, 2009; Duarte *et al.*, 2010).

Symptomatology and damage

The green mite feeds on the undersurface of the leaves, chiefly along the major veins, producing a very fine webbing. The upper leaf surfaces appear mottled with white (King and Saunders, 1984). On soybean, the feeding activity of *M. planki* induces chlorosis on both leaf surfaces, evolving to a greyish colour, with the leaves showing an aged aspect later (Roggia *et al.*, 2008). However, on this host plant, the symptomatology of attack may overlap with that of other mites. Indeed, in Brazil, *M. planki* may be associated on soybean with other spider mites such as *T. urticae*, *T. desertorum*, *T. ludeni* Zacher and *T. gigas* (Flechtmann, 1972; Návia and Flechtmann, 2004; Guedes *et al.*, 2007).

Control

Control of the green mite is provided by cultural practices and spray application of chemicals.

CULTURAL PRACTICES. The use of two soybean cultivars (transgenic glyphosate tolerant and non-transgenic), each treated by five weed management systems (without management, hand management and herbicide treatment with bentazone + sethoxydim, bentazone or sethoxydim), and a spray of glyphosate on the transgenic soybean, show that the weed management system is more determinant of the populations of *M. planki* and *T. gigas* on soybean than the soybean cultivar used. The mite populations have an inverse correlation with weed infestation and weed density (Roggia, 2007).

CHEMICAL CONTROL. Various active ingredients have been tested in the control of the green mite, such as imidacloprid and betacyfluthrin, either sprayed alone or mixed. Tests with imidacloprid resulted in an increase of oviposition rate in the first 7 days, but the average egg production did not differ significantly from the control. Comparisons of fecundity between mites treated with imidacloprid plus betacyfluthrin and imidacloprid alone showed that the fecundity of untreated females did not differ statistically from that of the treated mites (Siqueira, 2011).

Mononychellus progresivus Doreste (Fig. 20.19)

Common name

Cassava green mite.

Diagnostic characteristics

FEMALE. The dorsal body setae are long, setose and tapering, sometimes somewhat widened; the opisthosomal setae c_1, d_1 and e_1 are progressively longer towards the rear; the setae c_1 are about half as long as the distances to the bases of setae d_1; the latter setae are about ¾ as long as the distances to the bases of setae e_1 (Meyer Smith, 1987).

MALE. The shaft of the aedeagus turns ventrally with a small angle, gradually tapering, and distally shows 2 small projections (Meyer Smith, 1987).

Geographical distribution

Angola (Yaninek and Herren, 1988), Bahamas (Doreste, 1982), Brazil (Bolland *et al.*, 1998), Benin (Leuschner, 1980; Navajas *et al.*, 1996), Burkina Faso (Yaninek and Herren, 1988), Burundi (Girling *et al.*, 1978), Cameroon (Yaninek and Herren, 1988), Central African Republic (Gutierrez, 1987a), Colombia, Comoros (Bolland *et al.*, 1998), Congo (IITA, 1976; Bonato and Gutierrez, 1993; Gutierrez and Bonato, 1994), Congo (Democratic Republic of the Congo, formerly Zaire) (Girling *et al.*, 1978), Costa Rica (Doreste, 1982), Ethiopia (Nyiira, 1982), Gabon (Flechtmann, 1982), Ghana (Yaninek and Herren, 1988), Guyana (Doreste, 1982), Ivory Coast (Gutierrez, 1987b), Kenya (Girling *et al.*, 1978), Liberia (Yaninek and Herren, 1988), Malawi (Itimu and Sauti, 1982), Mozambique (Byrne *et al.*, 1983), Nigeria (Akinlosotu and Leuschner, 1981), Panama (Doreste, 1982), Paraguay (Aranda and Flechtmann, 1971), Puerto Rico (Nyiira, 1982), Rwanda (Nyiira, 1978), Sierra Leone (Yaninek and Herren, 1988), Sudan (Byrne *et al.*, 1983), Suriname (Bolland *et al.*, 1998), Tanzania (Girling *et al.*, 1978), Togo (Leuschner, 1980), Uganda (Lyon, 1973), Trinidad and Tobago (Nyiira, 1982; Bolland *et al.*, 1998), Venezuela (Doreste, 1981; Agudelo-Silva and Silva, 1986), Zambia and Zimbabwe (Yaninek and Herren, 1988).

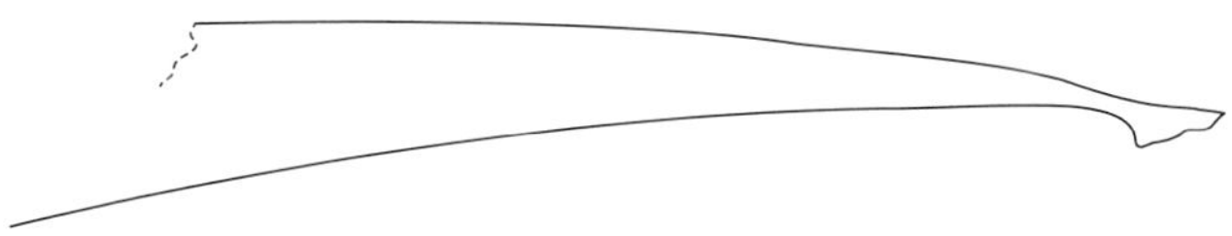

Fig. 20.19. *Mononychellus progresivus* Doreste. Male, aedeagus (from Meyer Smith, 1987).

Host plants

The cassava green mite is reported on a restricted number of species (six) of the Euphorbiaceae belonging to the genus *Manihot*: *Manihot* sp. (Meyer Smith, 1987), *M. caerulescens* Pohl, *M. carthagenensis* (Jacq.) Müll. Arg., manicoba rubber (*M. dichotoma* Ule), cassava or manioc (*M. esculenta* Crantz) (Aranda and Flechtmann, 1971; Akinlosotu and Leuschner, 1981; Doreste, 1981; Agudelo-Silva and Silva, 1986; Bonato and Gutierrez, 1993; Gutierrez and Bonato, 1994; Navajas *et al.*, 1996) and *M. heptaphylla* Ule (Meyer Smith, 1987).

Bio-ecology

M. progresivus lives in Neotropical and Afrotropical regions on plants of the genus *Manihot*. The mite has been widely reported as *Mononychellus tanajoa* (Bondar), also infesting cassava in Africa, and for this reason both species have been called 'cassava green mites'. The need to avoid confusion around the designation of *Mononychellus* suggested that the mites infesting cassava in Africa be referred to as the *Mononychellus* complex. However, Gutierrez (1987b) reported that only a single species was in fact introduced into Africa. This mite was named *M. tanajoa* for some years, but since the description of *M. progresivus* by Doreste (1981), it should have been designated by this newer name. The description of the first African *Mononychellus* was based only on the study of a female specimen, and only later were both sexes described and the male aedeagi of specimens compared; this highlighted the existence of one species, named *M. progresivus*. This conclusion is in agreement with that of Macfarlane (1984), who reported that 'the majority of records are probably *M. progresivus*'. Moreover, Gutierrez (1987b) believes that there is only an intraspecific rather than an interspecific complex, as has been suggested by various specialists. The two mites do not show genetic variability by clear morphological variations, although crosses between strains may have different levels of incompatibility. In addition, *M. tanajoa* has been reported only recently in some countries of the Afrotropical region, such as Benin (Agboton *et al.*, 2011) and Rwanda (Night *et al.*, 2011). The use of molecular markers allows the examination of intraspecific diversity in specimens of the cassava green mite collected in Africa (Benin and Congo) and America (Columbia and Brazil). Comparisons of mitochondrial and ribosomal DNA sequences, and the results of cross-breeding trials, show that the two African populations appear to be identical and similar to the Colombian population, while the Brazilian population is different. This supports the hypothesis of a single introduction of *M. progresivus* as the source of the two African populations. Moreover, crossing experiments have shown partial hybrid sterility, suggesting genetic incompatibility that is consistent with differences detected by sequence data (Navajas *et al.*, 1994).

The cassava green mite reproduces by arrhenotokous parthenogenesis. Its populations usually live around the growing points of plants, on buds, young leaves and stems, and do not produce webbing (Bellotti, 1985). At five constant temperatures (16, 22, 26, 31 and 36°C) on the cassava variety '1M20' and at 70 ± 10% RH and a 12 h light:12 h dark (12L:12D) photoperiod, the mite develops in the temperature range of 22–36°C, and the lower thermal threshold is 13°C. The shortest development time occurs at 31°C and is 7.2 days. Maximum fecundity occurs at 26°C (42.1 eggs). The highest intrinsic rate of increase (r_m) (0.289) was observed at 31°C (Bonato *et al.*, 1995a), with r_m values of 0.180 and 0.246 at 24 and 27°C, respectively, reported by Yaninek *et al.* (1989). Rearing of *M. progresivus* at three constant RH values (30, 60 and 90% RH), which were obtained using saturated salt solutions and airtight boxes placed in an air-conditioned room at 26 ± 1°C, and a photoperiod of 12L:12D h (illuminance 3500 lux), show that low (30% RH) and high (90% RH) air humidity have a negative effect on the life history traits of the mite compared with medium air humidity (60% RH). The strongest effect was obtained at 90% RH, when there was no egg hatching (Bonato *et al.*, 1995b).

Outbreaks of populations of *M. progresivus* in the field depend on various factors. The RH influences the population dynamics of the mite and explains in part the decrease in populations observed at the middle to the end of the dry season and the virtual absence of mites during the wet season (Bonato *et al.*, 1995b). The influence of pesticide treatments on population outbreaks is reported by Braun *et al.* (1987), who observed increased populations in treated plots immediately after the initiation of permethrin application; these remained significantly higher than in untreated plots during the whole treatment period.

Natural enemies include various predators, such as the phytoseiid mite *Amblyseius limonicus* Garman *et* McGregor (= *Typhlodromalus limonicus* (Garman *et* McGregor)) and the staphylinid *Oligota minuta* Cameron (Braun *et al.*, 1987).

Symptomatology and damage

The feeding activity of different *Mononychellus* spp. may overlap in the field, making it difficult to define the symptoms of and the damage attributable to each mite, but infested leaves usually lose their green colour, show yellow spots and take on a mottled, bronzed and mosaic-like appearance associated with deformations. Severe infestations induce stunting of plants, shoots lose their natural colour and stems scarify, turning rough and brown and later dying back. The necrosis of the leaves and stems develops from the top to the bottom (CIAT, 1976a; Nyiira, 1976).

During dry periods, populations of cassava green mite usually increase, inducing the reduction of root yields. Yield losses of 8 and 17% were reported in Venezuela in treated and untreated plots, respectively (Doreste and Aponte, 1978). In Colombia, yield reductions of 73% were observed in susceptible varieties and 15% in resistant varieties (Byrne *et al.*, 1982b). Moreover, yield reductions of 20–53% were attributed to a complex of four spider mites, *M. progresivus* with *M. mcgregori* (Flechtmann *et* Baker), *Tetranychus urticae* Koch and *Oligonychus peruvianus* (McGregor) (CIAT, 1977).

Control

Much of the available information on attacks in the field refers to the simultaneous presence of *M. progresivus* and *M. tanajoa*, so only

specific work on the subject can be discussed here. Thus, information on biological control is very limited and the topic is better discussed under *M. tanajoa* in the next section. Notwithstanding, the control methods used for *M. progresivus* include varietal resistance, cultural practices and biological and chemical means. Sampling and monitoring of the populations are a preliminary step to implementing control.

SAMPLING AND MONITORING. Sampling plans have been developed for *M. progresivus* on cassava in Africa. An analysis of intra-plant distribution showed a preference of the mite for the apical and middle parts of the plant. Thus, the first fully developed leaf in the middle part may be chosen as the common sampling unit. On this, the mite has an aggregated distribution. Binomial and enumerative sampling plans permit reliable estimates of the density of populations of *M. progresivus*, and 40 fully developed leaves provide satisfactory estimates of mite densities (Bonato *et al.*, 1995c).

PLANT RESISTANCE. Resistance to the green cassava mite has been extensively investigated, and thousands of varieties from germplasm banks have been screened for resistance to *Mononychellus* mites. The mechanisms of resistance are based on antibiosis and ovipositional preference. On resistant varieties, females develop slowly (75%), have a short lifespan (45%) and a longer pre-ovipositional period (21%), and produce fewer eggs (59%). Juvenile mortality is greater on resistant cultivars (Byrne *et al.*, 1982a). Susceptible cultivars exhibit an average yield reduction of 73% in respect to 15% of resistant cultivars, and during the period from the third to the eighth month of plant growth the level of infestation of the resistant cultivars was found to be 1 mite cm^{-2}, while that of susceptible cultivars was 3.9 mites cm^{-2} (Byrne *et al.*, 1982b).

CULTURAL PRACTICES. Various choices can be made. Planting different cassava varieties helps to avoid genetic uniformity. It is advantageous to plant in areas characterized by a short dry season. Indeed, planting early during the rainy season allows plants to be more vigorous when mite infestations occur. Planting non-host plants between cassava fields to space them allows attacked plants to be removed more effectively at the first sign of mite infestation (Bellotti, 1985).

BIOLOGICAL CONTROL. Biological control of cassava mites is mainly done by introducing and releasing predators that are formally employed to control *M. tanajoa* in Africa (Bennett, 1974; Bennett and Yaseen, 1975; Yaseen and Bennett, 1978). Bellotti *et al.* (1982) reported almost 35 predators and pathogens, including mites, insects and fungi. The predatory insects include various *Oligota* spp. (Staphylinidae) specializing in *Mononychellus* mites, although releases of *O. minuta* Cameron in Kenya were inconclusive (Girling *et al.*, 1978). The role of phytoseiid mites, such as *T. rapax* (De Leon) and *T. limonicus*, seems more feasible (Yaseen and Bennett, 1977; Girling *et al.*, 1978). Research has also been done on the possibilities of using other phytoseiid mites for classical biological control of *Mononychellus* spp. in Africa, such as *Typhlodromus annectens* De Leon (= *Galendromus annectens*), *Amblyseius californicus* (McGregor) (= *Neoseiulus californicus* (McGregor)), *A. idaeus* (Denmark *et* Muma), *Euseius concordis* (Chant) and *Phytoseiulus macropilis* (Banks) (Mesa *et al.*, 1990). The biological control of cassava green mites is extensively discussed by Hoy (2011), who refers to experiences with *M. tanajoa* (probably *M. progresivus*) in Africa.

CHEMICAL CONTROL. Spray applications of acaricides are not generally recommended in small crops. It is good protocol to monitor the whole crop and intercept the dynamics of infestation in its progress from isolated plants towards patches of plants and later towards crop, and then to treat patches. With this aim, various acaricides have been tested, such as chlorobenzilate or binapacryl, but mite infestations can also be reduced by sprayings of water under pressure (Bellotti, 1985).

Mononychellus tanajoa (Bondar) (Fig. 20.20)

Common name

Cassava green mite.

Diagnostic characteristics

FEMALE. The dorsal body setae are broadened distally. The prodorsal setae v_2 and sc_{1-2}, and the opisthosomal setae c_2, c_3, d_2, e_2 and f_2 are as long as the longitudinal distances between their bases; the setae c_1 and d_1 are of equal length and slightly shorter than half the distance to the bases of the setae that are next behind; the setae e_1 are longer than c_1 and d_1 and reach about halfway to the bases of setae f_1. The dorsal striae have rounded lobes with basal spots (Meyer Smith, 1987).

MALE. The aedeagus turns ventrad at a slight angle and has angulations distally (Meyer Smith, 1987).

Geographical distribution

Benin (Agboton *et al.*, 2011), Bahamas (Doreste, 1982), Brazil (Bondar, 1938; Flechtmann and Baker, 1975; Gonçalves *et al.*, 2001; Mendonça *et al.*, 2011a), Colombia (Urueta, 1970, 1975; Doreste, 1982), Costa Rica (Doreste, 1984b; Aguilar and Murillo, 2008), El Salvador (Andrews and

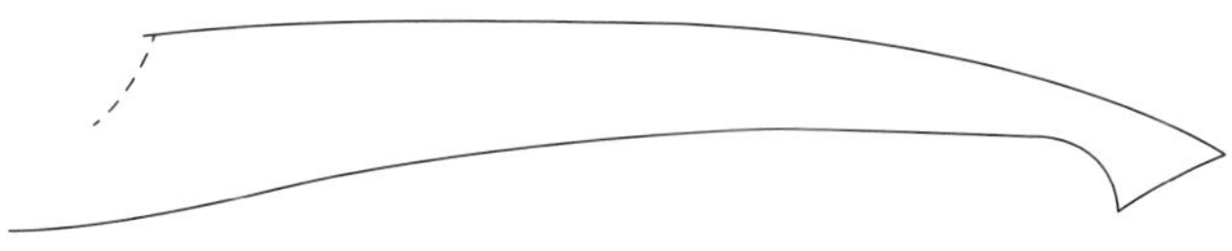

Fig. 20.20. *Mononychellus tanajoa* (Bondar). Male, aedeagus (from Meyer Smith, 1987).

Poe, 1980), Guyana (Doreste, 1982), Mexico (Tuttle *et al.*, 1976), Nicaragua (Baker and Pritchard, 1962), Panama (Doreste, 1982), Paraguay (Doreste, 1984b), Rwanda (Night *et al.*, 2011), Trinidad and Tobago (Flechtmann and Baker, 1975; Doreste, 1982), Venezuela (Doreste, 1981) and Zambia (Yaninek and Herren, 1988; Mwase and Baker, 2006).

Host plants

The list of host plants of *M. tanajoa* includes 12 species from four families: Euphorbiaceae – *Manihot* sp. (Bondar, 1938; Flechtmann and Baker, 1975), *M. caerulescens* Pohl, *M. carthagenensis* (Jacq.) Müll. Arg., *M. dichotoma* Ule (Gutierrez, 1979), cassava (*M. esculenta* Crantz) (Bondar, 1938; Urueta, 1970, 1975; Yaninek and Herren, 1988; Gonçalves *et al.*, 2001; Agboton *et al.*, 2011; Night *et al.*, 2011), *M. heptaphylla* Ule (Gutierrez, 1979); Fabaceae – *Erythrina* sp. (Andrews and Poe, 1980), *Gliricidia maculata* (Kunth) Kunth (Baker and Pritchard, 1962), *G. sepium* (Jacq.) Kunth ex Walp. (Estebanes-Gonzalez and Baker, 1968; Andrews and Poe, 1980), *Senna occidentalis* (L.) Link (Rossi Simons, 1961); Passifloraceae – *Passiflora cincinnata* Mart. (Bolland *et al.*, 1998; Migeon and Dorkeld, 2006–2013); Typhaceae – *Typha domingensis* Pers. (Aguilar and Murillo, 2008).

Bio-ecology

M. tanajoa was first collected in Brazil in the state of Bahia (Bondar, 1938) and was later reported in other countries in Afrotropical, Nearctic and Neotropical regions. The mite lives mainly on *Manihot* spp., but may also occur on other plants. It reproduces by arrhenotoky and its populations live on the lower surface of newer (cassava) leaves. The eggs are laid singly on the sides of the midrib or other veins or in concavities on the lower leaf surface, mainly in the basal half of the leaf. Most of the mites live in the upper leaves of the plant, the largest numbers being on leaves 5–10, while the top 3–4 newly flushed leaves carry few mites. After the tenth leaf, the numbers of mites fall sharply, apparently as a result of the depletion of nutrients (Yaseen and Bennett, 1977).

At 26.8 ± 2.2°C and RH 82% in the morning to 55% in the afternoon, the pre-oviposition period lasts 1–2 days. The duration of the egg, larval, protonymphal and deutonymphal stages is 3–4, 1–2, 1–2 and 2–3 days, respectively. Each active stage has a quiescent stage lasting less than a day and the total egg to adult period is 11–13 days. The males mature first and remain near the female teleochrysalis ready to mate with the emerging females. Females lay 21–65 eggs over 8–14 days and live up to 18 days (Yaseen and Bennett, 1977). In laboratory studies at 24 ± 2°C, 65 ± 10% RH and with a photoperiod of 13L:11D on *M. esculenta*, the mean generation time (T) was 24.7 days, the net reproductive rate (R_0) 25.1, the total fecundity 41.8 ± 4.4 eggs, the sex ratio (females:females + males) 0.57, the intrinsic rate of increase (r_m) 0.14 and the longevity 14.8 ± 1.0 days. The mite reproduced better on cassava (r_m = 0.14), although it also developed and reproduced on other *Manihot* spp.; it also developed to the adult stage on common bean but did not oviposit (Moraes *et al.*, 1995a). Under laboratory conditions on leaves of cassava, the incubation period of the eggs was 4.76 ± 0.01 days. The developmental stages for the larva, protochrysalid, protonymph, deutochrysalid, deutonymph and teleiochrysalid were 1.03 ± 0.02, 0.85 ± 0.02, 0.96 ± 0.02, 0.75 ± 0.01, 1.00 ± 0.02 and 0.99 ± 0.02 days, respectively. The development from egg to adult took 10.21 ± 0.11 days, and the instantaneous rate of increase (r_i) was 0.42 (Rêgo *et al.*, 2010). Samplings of cassava leaves in Brazil showed that *M. tanajoa* represented 76.3% of the mite specimens, and that the mite incidence on the cultivar 'Branca de Santa Catarina' reached 53% of those present (Montes *et al.*, 2010). Yaseen and Bennett (1977) observed that the mite dispersed in the direction of the prevailing wind over a distance of 200 m.

The population dynamics of *M. tanajoa* are affected by both biotic and abiotic factors. Dry periods influence the development of high mite populations positively. In Trinidad, the populations showed upward trends in March–April and developed peak levels by the end of May or early June; these persisted to the middle of July, when they dropped suddenly with the onset of heavy rains. A minor peak occurred during the short dry period in September–October. On cassava, 56.1–66.5% of the population is observed in June–July during the major peak, and 9.0–29.7% in September–October during the short-lived minor peak (Yaseen and Bennett, 1977).

The age and physiological condition of the host plant influences mite density. Young plants usually bear few mites, because the few newly flushed leaves provide only small surface areas and the nutrients are not fully synthesized. Similarly, old plants with depleted nutrients and retarded growth do not provide favourable conditions. Young vigorous plants about 4–8 months old produce new leaves early in the dry season and provide favourable conditions for the development of heavy mite populations (Yaseen and Bennett, 1977).

In Brazil, Teodoro *et al.* (2010) determined the relative importance of the abiotic environmental variables (rainfall, temperature and RH) in the abundance of *M. tanajoa*. The abundance of *M. tanajoa* increases and that of the predatory mite *Amblyseius largoensis* (Muma) remains constant through time. Hierarchical partitioning analysis indicates that most of the variance for the abundance of *M. tanajoa* is explained by rainfall and RH, followed by *A. largoensis* abundance and temperature. Although *A. largoensis* contributes to the regulation of the cassava green mite population, the main mechanisms explaining the abundance of *M. tanajoa* are abiotic environmental variables.

The natural enemies include predatory mites and insects and some acaropathogenic fungi (Table 20.7).

Symptomatology and damage

The feeding activity of *M. tanajoa* on the lower surface of newer cassava leaves causes small yellow spots and abnormal growth. The stems turn brownish, the epidermis ruptures and there is progressive drying from the tip to the base. Heavy infestations occur mainly during long dry periods and may result in the death of the whole plant (Flechtmann, 1977b). During the dry period, it is difficult to differentiate between the symptoms of severe infestation by this mite and those of the viral African cassava mosaic disease (Nestel, 1976).

Table 20.7. *Mononychellus tanajoa* (Bondar). Natural enemies (fungi, insects and mites).

Families and species	References
FUNGI	
Neozygitaceae	
Neozygites tanajoae Delalibera *et al.*	Moraes and Delalibera, 1992; Delalibera *et al.*, 2004; Van der Geest, 2010
INSECTS	
Ancylistaceae	
Conidiobolus obscurus (Hall *et* Dunn) Remaudiere and Keller	Chandler *et al.*, 2000
Conidiobolus thromboides Drechsler	Delalibera *et al.*, 1992
Anthocoridae	Yaseen *et al.*, 1982
Clamidiastethus sp.	
Wollastoniella gatti Ghauri	
Cecidomyiidae	
Arthrocnodax sp.	
Feltia sp.	
Feltiella sp.	
Chrysopidae	
Coccinellidae	
Stethorus aethiops Wse	
Stethorus darwini (Bréthes)	
Stethorus jejunus Casey	
Stethorus utilis Horn	
Coniopterygidae	
Coniopterix crassicornis Eb. Pet.	
Lygaeidae	
Staphylinidae	
Oligota barbadorum (Frank)	
Oligota centralis (Sharp)	
Oligota fageli Wlliams	
Oligota minuta (Cameron)	
Oligota pallidicornis Cameron	
Oligota pygmaea (Solier)	
Syrphidae	
Thysanoptera	
Scolothrips spp.	
SPIDERS	Yaseen and Bennett, 1977
MITES	
Phytoseiidae	
Amblyseius aerialis (Muma)	Yaseen and Bennett, 1977; Moraes *et al.*, 1989, 1993; Yaninek, 2007; Rêgo *et al.*, 2010
Amblyseius largoensis (Muma)	
Euseius concordis (Chant)	
Euseius hibisci (Chant)	
Galendromus annectens (De Leon)	
Neoseiulus idaeus (Denmark *et* Muma)	
Neoseiulus anonymus (Chant *et* Baker)	
Neoseiulus californicus (McGregor)	
Phytoseiulus macropilis (Banks)	
Typhlodromalus aripo De Leon	
Typhlodromalus limonicus (Garman *et* McGregor)	
Typhlodromalus manihoti Moraes	
Typhlodromalus rapax (DeLeon)	

Until the 1970s, *M. tanajoa* was not a serious pest in the neotropics, and very little information was available about its bio-ecology (Yaseen and Bennett, 1977). Later, the mite gradually became a destructive pest of cassava in South America and Africa, where it caused yield losses of up to 80%, depending (most) upon the age of the plants when the heaviest attacks occurred (CIAT, 1976b; Veiga, 1985; Yaninek and Herren, 1988).

Control

Resistant varieties, cultural practices, biological and chemical means are used in the control of *M. tanajoa*.

PLANT RESISTANCE. In the last few decades, varying degrees of varietal resistance have been reported for *M. tanajoa* (Barrios, 1972; Bennett and Yaseen, 1975). Varieties of cassava resistant to attack by this mite have been identified in East Africa (Shukla, 1977, in Flechtmann, 1977b), and in the 1970s about 42 resistant cassava varieties were reported in Bahia (Brazil) (Costa, 1973). Cassava varieties resistant to tetranychids have also been reported in Venezuela (Montaldo, 1972). Doreste *et al.* (1977) produced a list of 19 cassava cultivars with different degrees of resistance to tetranychid mites. Among these highly promising sources of resistance are the cultivars 'Ecu 133' and 'Mex 20'. Recently, the use of the oviposition rate as a tool for evaluating species of *Manihot* for resistance to the cassava green mite showed the existence of sources of resistance at the highest level in wild genotypes and also demonstrated that the oviposition rate could be used as a fast tool to select genotypes for breeding programmes for resistance to mite (Noronha *et al.*, 2010). Unfortunately, although resistant cultivars are known, Neotropical cultivars that might exhibit resistance to the cassava green mites are also susceptible to the *cassava mosaic virus* in Africa and are therefore not useful.

CULTURAL PRACTICES. Cassava plants 2–9 months old are most susceptible to mite infestation, so altering planting time is a form of cultural control; however, because subsistence farmers need to plant cassava soon after the first rains of the long wet season, altering the time of planting is difficult to do. Intercropping or mixed cropping does not produce any effect. Neither does hand removal of newly infested leaves, which is labour intensive and results in new growth on lateral shoots, which soon provide suitable new leaves for the mites (Hoy, 2011). As the densities of mite populations are associated with plant nutrient levels, studies have been made of mite response in ten cassava varieties treated with NPK. The results were inconclusive, although observations during January–March indicated that mite populations were higher in all treated plots than in the untreated check plot (Yaseen and Bennett, 1977).

BIOLOGICAL CONTROL. In agreement with the discussion above of what has been previously reported, especially on the geographical distribution of *M. progresivus* and *M. tanajoa*, and the exact identification of *M. tanajoa*, the information in reports used in this discussion may refer to *M. progresivus* but formally refer to *M. tanajoa*. Despite this drawback, it seems that classical biological programmes based on releases of phytoseiid mites were successful in the control of the cassava green mite populations on cassava in Africa. A review of this topic has been published by Yaninek and Hanna (2003). Since 1977, a specific programme has been started by the International Institute of Tropical Agriculture (IITA), the Centro Internacional de Agricultura Tropical (CIAT) and Empresa Brasileira de Pesquisa Agropecuária (EMBRAPA) in Brazil.

Research in the cassava-growing areas of the neotropics highlighted a number of predatory insects and mites and acaropathogens of cassava green mites. However, the opportunistic behaviour of insects and their adaptation to preying only at high population densities leads interest more towards the phytoseiid mites as predators. About 50 species of Phytoseiidae were collected on cassava in the neotropics (Moraes *et al.*, 1989; Yaninek, 2007). Among these, *Amblyseius aerialis* (Muma), *Euseius concordis* (Chant), *Galendromus annectens* (De Leon), *Neoseiulus idaeus* (Denmark *et* Muma), *N. anonymus* (Chant *et* Baker), *N. californicus* (McGregor), *Typhlodromus aripo* De Leon, *T. limonicus* (Garman *et* McGregor) and *T. manihoti* Moraes seemed the most interesting and were sent to Benin for further evaluation of their activity and their possible mass rearing and release in different areas of Africa. Mass-rearing methods were developed (Haug *et al.*, 1987) and between 1984 and 2001, a total of 11.6 million phytoseiids were shipped to 16 countries for release in 847 locations (Yaninek, 2007). *G. annectens*, *N. anonymus*, *N. californicus*, *N. ideaeus*, *T. aripo* and *T. manihoti* were recovered after release, and three of these predatory mites are considered to be established, *N. ideaeus*, *T. aripo* and *T. manihoti* (Yaninek *et al.*, 1998; Yaninek, 2007). *T. aripo* has been the most successful, having 'become established in more than 20 countries, where it has reduced *M. tanajoa* populations and significantly increased cassava yields by a third' (Yaninek, 2007). The phytoseiid was released directly on to cassava plants beginning in 1993 and became widely established in Africa by the end of 1995 (Hanna and Toko, 2001). It spread over an area of more than 150,000 km^2 within 2 years and appeared to be relatively specific to the cassava green mites, although it can feed and develop on other spider mites, several types of pollen (maize, castor bean and leucaena), whiteflies, cassava plant exudate and cassava leaves (Onzo, 2003; Onzo *et al.*, 2009). These alternative food resources allow the persistence of the phytoseiid even at a low density of cassava green mites. The use of leaf primordia as a refuge by *T. aripo* is an important ecological factor. Indeed, the phytoseiids take refuge in the apices of the plant during the day but move to the young leaves during the night to prey. The predators then return to the apices the next morning. This behaviour results in the protection of young foliage.

Bennett and Yaseen (1975) evaluated the effectiveness of the staphylinid *Oligota minuta* Cameron and introduced it into East Africa, but the results that were obtained did not seem conclusive. Conversely, the fungus *Neozygites tanajoae* Delalibera *et al.* has been introduced into Benin, where it seems to have established and is producing about 25% mortality of cassava green mites (Hountondji *et al.*, 2002).

CHEMICAL CONTROL. Although chemical control is not practical (mainly in depressed areas) and other measures, including

biological control, are being studied, a number of pesticides have been tested or suggested in the past, such as fosalone, arsenate, diazinone (Costa, 1975), chlordimeform or malathion (Yaseen and Bennett, 1977). Depending on the phytosanitary regulations of the country concerned, other acaricides may be used. However, predatory mites and *M. tanajoa* can be suppressed by pesticide applications, and populations of the *M. tanajoa* can recover and build up very rapidly before predatory mites reinvade the plants (Yaseen and Bennett, 1977).

The Genus *Platytetranychus* Oudemans

Diagnostic characteristics

The opisthosoma is transversely striae, and the dorsal body setae are much shorter than the distances between their bases. There are 2 pairs of setae *h* (h_{2-3}). The empodium is split near the middle into 3 pairs of hairs (Bolland *et al.*, 1998).

Economic importance

The genus *Platytetranychus* has a worldwide distribution and includes ten species, mainly Nearctic and Neotropical, some of them living on ornamentals (Bolland *et al.*, 1998; Migeon and Dorkeld, 2006–2013).

Of these, *P. libocedri* (McGregor) is a potential pest of Cupressaceae and Pinaceae in Mexico (Estebanes-Gonzalez and Baker, 1968) and the USA (McGregor, 1936; Pritchard and Baker, 1955; Tuttle and Baker, 1964, 1976), where it lives on *Abies religiosa* (Kunth) Schltdl. *et* Cham. (Estebanes-Gonzalez and Baker, 1968), *Cupressus* sp. (Pritchard and Baker, 1955), *C. arizonica* Greene (Pritchard and Baker, 1955; Tuttle and Baker, 1964), *C. sempervirens* L., *Juniperus* sp. (Pritchard and Baker, 1955), *J. communis* L. (Tuttle and Baker, 1964), *J. compacta* (Martínez) R.P. Adams (Tuttle *et al.*, 1976), *Calocedrus decurrens* (Torr.) Florin (McGregor, 1936; Pritchard and Baker, 1955), *Pinus* sp. (Estebanes-Gonzalez and Baker, 1968), *Thuja* sp. (Pritchard and Baker, 1955) and *T. occidentalis* L. (Pritchard and Baker, 1955; Tuttle and Baker, 1964, 1976). The mite also lives on *Tamarix* sp. (Tuttle and Baker, 1964).

P. pueblensis Tuttle *et al.* is reported on *Juniperus* sp., *J. compacta* and *T. occidentalis* in Mexico (Tuttle *et al.*, 1976) and the USA (Baker and Tuttle, 1994).

The arborvitae spider mite, *P. thujae* (McGregor), is a potential pest of *Juniperus* sp. (Pritchard and Baker, 1955), *J. chinensis* L. (Reeves, 1963), *J. osteosperma* (Torr.) Little (Tuttle and Baker, 1968), *J. procumbens* (Siebold ex Endl.) Miquel (Reeves, 1963), *J. virginiana* L. (Pritchard and Baker, 1955), *Platycladus orientalis* (L.) Franco (Thewke and Enns, 1969) and *T. occidentalis* in the USA (McGregor, 1950).

The honey locust spider mite, *P. multidigituli* (Ewing), is a pest of honey locust (*Gleditsia* sp., *Gleditsia triacanthos* L.) (Ewing, 1917) and aubergine (*Solanum melongena* L.) in India (Gupta and Gupta, 1976) and the USA (Ewing, 1917). This seems to be the most important pest mite of this genus and is further discussed below.

Platytetranychus multidigituli (Ewing)

Common name

Honey locust spider mite.

Diagnostic characteristics

FEMALE. The summer forms are light to dark green in colour, and overwintering females are orange red. The dorsal body setae are shorter than the distances between their insertions. The peritreme has a distal anastomosis. Tibia I has 7 tactile setae, the tibia II has 5 tactile setae, and on tarsus I there are 3 tactile setae proximal to the duplex setae (Jeppson *et al.*, 1975).

MALE. The male is smaller than the female and with the same colour. The chaetotaxy of tibiae I and II and tarsus I is the same as that of the female. The aedeagus bends sharply ventrad and forms a distal knob, the distal angulation being much stronger than the anterior angulation (Jeppson *et al.*, 1975).

Geographical distribution

India (Gupta and Gupta, 1976) and the USA (Ewing, 1917).

Host plants

Fabaceae – honey locust (*Gleditsia* sp., *Gleditsia triacanthos* L.) (Ewing, 1917); Solanaceae – aubergine (*Solanum melongena*) (Gupta and Gupta, 1976).

Bio-ecology

The honey locust spider mite is reported on honey locust (Ewing, 1917) and aubergine (Gupta and Gupta, 1976). Low temperatures induce the females to turn orange red and move to the twigs and branches, where they overwinter, mostly inactive, under bud scales, in the crevices of rough bark, at the bases of twigs or under the shells of dead scale insects. The mite becomes active at the beginning of April, when it starts to lay eggs; the first larvae occur at the start of May. The mite populations develop in May and June and high densities are observed until the end of August or September. Most of the eggs (55%) are deposited on the undersurface of leaflets, with fewer on the upper surfaces (17%) and on leaflet petioles (28%). From various reports, the pre-oviposition period and the time for each moult are about 1 day. Egg development takes 4–9 days and the development from larvae to adults takes 4–11 days in June and 8.5 days in September, with the average development period 9.5 days in June and 14.6 days in September. The fecundity is about 72 eggs and the oviposition lasts 11 days (English and Snetzinger, 1957b; Reeves, 1963; Jeppson *et al.*, 1975).

Natural enemies include minute pirate bugs, lacewings, mites, lady beetles and predatory midges (Krischik and Davidson, 2013b).

Symptomatology and damage

The feeding activity of this mite causes stippling and bleached spots on dead leaf tissue, mainly on the undersides of the leaflets, but at a low density of attack, it also causes white or yellow coloration on the upper leaf surface. Heavy infestations cause severe yellowing and browning and eventually dropping of the leaves. When damage occurs in early summer, the tree produces new leaves, resulting only in aesthetic damage and associated with only low plant risk (Jeppson *et al.*, 1975; Krischik and Davidson, 2013b).

Control

Control is provided by monitoring of the populations and the use of chemicals.

MONITORING. Monitoring requires checking for females around the bark crevices and under the bud scales during winter. Mite populations may be monitored during the growing season at the bases of leaflets on the underside. Infested leaves show stippling or browning (Krischik and Davidson, 2013b).

CHEMICAL CONTROL. Chemical control is by the spraying of various pesticides (English and Snetzinger, 1957b; Reeves, 1963; Krischik and Davidson, 2013b). A high density of overwintering females may be treated during the winter with horticultural oil. During summer, different acaricides may be used, such as abamectin, bifenazate, clofentezine, hexythiazox, horticultural oil or insecticidal soap. However, it is possible to use bifenthrin, chlorpyrifos (in the nursery), deltamethrin, dicofol, fenpropathrin, fenbutatin oxide, lambda-cyalothrin and pyridaben (Krischik and Davidson, 2013b). The side effects of chemicals may be detrimental when used against this pest. Trials carried out in Colorado showed that populations of the honey locust spider mite were as much as 3–4 times higher on *G. triacanthos* that had received an imidacloprid soil drench compared with control plants. Effects on spider mite populations or injury were not caused by either hormoligosis or phytotoxicity. The mortality of *Orius tristicolor* (White), a hemipteran predator of tetranychids, was observed to increase significantly when confined with leaves collected from plants that had received soil treatments of imidacloprid, so the adverse effects on predators may be an important reason for the increase in spider mite injury and abundance resulting from soil treatments of imidacloprid (Sclar *et al.*, 1998).

The Genus *Eotetranychus* Oudemans

Diagnostic characteristics

The dorsal body setae are as long or longer than the distances between their bases. The opisthosoma is transversely striate. There are 2 pairs of para-anal setae (h_{2-3}), and 2 pairs of anal setae (ps_{1-2}). The empodium is split near the middle into 3 pairs of hairs (Bolland *et al.*, 1998).

Economic importance

Eotetranychus is a worldwide genus that currently includes 187 species (Migeon and Dorkeld, 2006–2013), some of which are pests of economic plants. A number of species infests ornamental and shade trees, and a restricted number damages citrus, deciduous trees, berries and other economic plants. Usually, these mites are not feared as severe pests, probably because of their behaviour in living in associated colonies and being actively preyed upon by natural enemies, or on account of their wide susceptibility to chemicals (Jeppson *et al.*, 1975). Some species are potential pests of economic plants, including the *Eotetranychus* mites that are briefly described below, but they do not usually cause severe damage to their host plants.

E. ancora Baker *et* Pritchard sometimes damages apple leaves in Mauritius but its populations are usually controlled by natural enemies (Moutia, 1958).

E. caryae Reeves is a North American mite living on *Aesculus hippocastanum* L., *A. turbinata* Blume, *Carya illinoinensis* (Wangenh.) K. Koch, *C. ovata* (Mill.) K. Koch, *Castanea* sp. and *Quercus marilandica* Münchhausen. Feeding of this mite causes scorched areas dark brown in colour on the leaves, and heavy infestations induce early defoliation starting at the lower branches. Leaves remaining on the trees may be russeted or scorched (Reeves, 1963).

E. matthyssei Reeves lives on *Celtis occidentalis* L., *C. sinensis* Pers., *Robinia pseudoacacia* L., *Ulmus americana* L., *U. fulva* Michx. *et al.*, *U. rubra* Muhl. and *Quercus* sp. in the USA (Reeves, 1963). On *U. Americana*, feeding by the mite causes severe browning and cupping of the underside of leaves. The first damage occurs on the leaves close to the trunk and near the ground. The mite populations are associated with a small amount of webbing near the veins (Jeppson *et al.*, 1975).

E. pallidus (Garman) is reported on ornamental and forest plants, such as *Alnus* sp. (Pritchard and Baker, 1955), *Betula* sp. (McGregor, 1950), *Fagus* sp. (McGregor, 1950), *F. grandifolia* Ehrh. (Reeves, 1963) in the USA (Garman, 1940; Reeves, 1963). The mite lives on the lower leaf surfaces and the injured areas turn brown (Reeves, 1963).

E. querci Reeves lives on *Betula pubescens* Ehrh., *Quercus palustris* Münchh., *Q. rubra* L. (Reeves, 1963) and *Tilia japonica* (Miq.) Simonk. (Ehara, 1970) in the USA (Reeves, 1963) and Japan (Ehara, 1970). The infested leaves show distortions and discolorations and may drop, causing defoliation of trees. On *Q. palustris*, the mites first infest the leaves close to the base of mid-vein. Early infestation and high density of attack induce the leaves to turn brown and failure of development. However, low infestations do not hinder the normal development of the distal portion of the leaves, even if distortions appear around the bases of mid-veins. On the northern species *Q. borealis* (= *Q. rubra* L.), only minor damage occurs, consisting of yellowed areas on the upper leaf surfaces and of depressed surfaces viewed from below, associated with a little leaf distortion (Reeves, 1963; Jeppson *et al.*, 1975).

E. tiliarium (Hermann) lives on wild and ornamental plants (especially on Malvaceae of the genus *Tilia*) in several countries of the Palaearctic region (Bolland *et al.*, 1998; Migeon and Dorkeld, 2006–2013) and in the Nearctic region (Pritchard and

Baker, 1955). On lime and linden trees (*Tilia* spp.), the mite feeds on the lower leaf surface between the veins, causing severe discoloration of leaves (Reeves, 1963).

Nine *Eotetranychus* spp. have been reported on citrus worldwide, but of these, only two species, *E. cendanai* Rimando and *E. sexmaculatus* (Riley), may actually be injurious to citrus. Also, although some of the species described above exhibit a marked polyphagy and a wide geographical distribution, they cannot be referred to as severe pests of citrus (Vacante, 2010). *E. cendanai* is treated below as a potential pest of citrus, and *E. sexmaculatus* as a reported pest of citrus and other plants, such as avocado and ornamentals.

Eleven other *Eotetranychus* spp. are considered to be pests of economic plants, and are treated below: *E. carpini* (Oudemans), *E. hicoriae* (McGregor), *E. hirsti* Pritchard *et* Baker, *E. kankitus* Ehara, *E. lewisi* (McGregor), *E. populi* (Koch), *E. pruni* (Oudemans), *E. smithi* Pritchard *et* Baker, *E. uncatus* Garman, *E. willametti* (McGregor) and *E. yumensis* (McGregor).

Eotetranychus carpini (Oudemans) (Fig. 20.21)

Common name

Yellow vine mite.

Diagnostic characteristics

FEMALE. The overwintering forms are intense yellow in colour, and the summer forms are light yellow with green spots. The peritreme ends in a simple bulb. The genital flap and the anterior area are transversely striate. Tibia I has 9 tactile setae, and tibia II has 8 tactile setae. Tarsus I bears 5 tactile setae proximal to the duplex setae (Jeppson *et al.*, 1975).

MALE. The male is smaller than the female. The peritreme ends in a simple bulb. Proximal members of the duplex setae on tarsus I are very short. The aedeagus is long, slender, tapering distally and strongly undulate near the middle (Jeppson *et al.*, 1975).

Geographical distribution

Algeria (Athias Henriot, 1959), Armenia, Azerbaijan (Mitrofanov *et al.*, 1987), China (Ma and Yuan, 1981b), Croatia (Ciglar and Baric, 1998), Finland (Oudemans, 1915), France (Delmas and Rambier, 1954; Rambier, 1958a,b; Migeon *et al.*, 2007), Germany (Zacher, 1921a), Greece (Hatzinikolis, 1970), Georgia (USSR) (Reck, 1950; Mitrofanov *et al.*, 1987), Hungary (Bognar, 1961), Italy (Vacante and Tropea Grazia, 1987b; Migeon *et al.*, 2007), Lebanon (Dosse, 1967), Mexico (Beer and Lang, 1958), Morocco (Saba, 1973b), the Netherlands (Oudemans, 1905; Vierbergen, 1990), Poland (Boczek and Kropczynska, 1964; Dobosz *et al.*, 1995), Romania (Iacob, 1978), eastern Russia (Mitrofanov *et al.*, 1987), Slovenia (Migeon *et al.*, 2007), Switzerland (Günthart and Günthart, 1959), Tunisia (Kreiter *et al.*, 2002), Turkey (Onucar and Ulu, 1988), Ukraine (Mitrofanov *et al.*, 1987), the UK (Hirst, 1920) and the USA (McGregor, 1917).

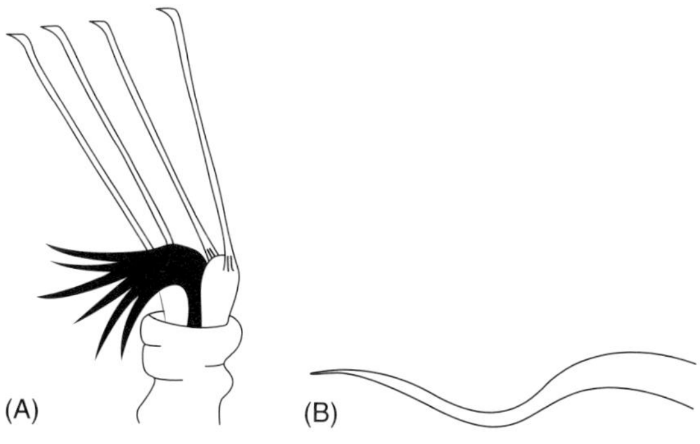

Fig. 20.21. *Eotetranychus carpini* (Oudemans). (A) female, pretarsus I; (B) male, aedeagus.

Host plants

The yellow vine mite is reported on 39 host plants of 14 families, with the highest number of species (10) included in the family Rosaceae (Migeon and Dorkeld, 2006–2013). The families and species of economic importance are: Betulaceae – European hornbeam (*Carpinus betulus* L.) (Oudemans, 1905; Reck, 1950; Dobosz *et al.*, 1995; Migeon *et al.*, 2007); Rosaceae – apple (*Malus domestica* Borkh.) (Ewing, 1913; Ma and Yuan, 1981b), sweet cherry (*Prunus avium* (L.) L., European plum (*P. domestica* L.) (McGregor, 1917), Japanese rose (*Rosa rugosa* Thunb.), American red raspberry (*Rubus idaeus* L.) (McGregor, 1950); Vitaceae – grapevine (*Vitis vinifera* L.) (McGregor, 1950; Rambier, 1958a,b; Hatzinikolis, 1970; Vacante and Tropea Grazia, 1987b; Migeon *et al.*, 2007).

Bio-ecology

E. carpini is a pest of grapevine in southern Europe. The mite lives on a number of different host plants, including the European hornbeam (*C. betulus*), which commonly occurs in stands and hedgerows bordering vineyards, where it may represent a potential mite reservoir (Malagnini *et al.*, 2012). The ability of the mite to colonize different host plants succesfully has for a long time hindered its identification. In fact, although it may live on a number of host plants (Bolland *et al.*, 1998; Migeon and Dorkeld, 2006–2013), it prefers the hornbeam and is a pest of vineyards in Europe (Delmas and Rambier, 1954; Ambrosi and Lenarduzzi, 1959; Mathys and Tencalla, 1959; Schruft, 1985). This behaviour suggested to Dosse (see Gasser and Mathys, 1960) the existence of two forms of *E. carpini* in Europe, one living on *V. vinifera* (*E. carpini* forma *vitis*) and the other on *C. betulus* (*E. carpini* forma *carpini*). In the last few decades, the opinion of Dosse has been adopted by several acarologists, among whom Jeppson *et al.* (1975) reported three distinct subspecies: *E. carpini carpini* (Oudemans), living on apple, hornbeam, hazelnut, oaks, willow, maple and alder; *E. carpini vitis* (Oudemans), living on grapes in France and Italy; and *E. carpini borealis* (Ewing), a pest of apple and pear in North America. However, experiments carried out by Bonato *et al.* (1989) showed that *E. carpini* forma *vitis* could successfully develop on hornbeams and vice versa, and a molecular analysis of specimens of *E. carpini* collected from each of these two host plants was undertaken and showed no differences (Malagnini, 2001). Recently, the use of biological and molecular

tools has allowed the resolution of this problem. Host-switch experiments to test the ability of *E. carpini* to develop on an alternative host plant, using mite populations originally collected on either *C. betulus* or *V. vinifera* plants from the same area, demonstrated the differentiation of mites collected from the two host plants. Mites collected from *C. betulus* did not survive and reproduce on *V. vinifera* and vice versa. Molecular analysis indicated a DNA-based differentiation using nucleotide sequences of the ITS1-5.8S-ITS2 region of the ribosomal DNA of individual *E. carpini* from the populations examined; sequences of individuals collected in Italy and Slovenia were analysed to estimate species variation. Significant genetic differentiation was revealed between the mites collected from the two host plants, but there was no evidence of genetic variation among specimens collected from the same host species. These results demonstrate the existence of host races of *E. carpini* (Malagnini *et al*., 2012).

The yellow vine mite reproduces by arrhenotokous parthenogenesis and has a haploid set of four chromosomes (Helle and Bolland, 1967). It overwinters as fertilized females under the bark of vines, mainly in the lower part of the plant at or above ground level or in the upper soil (Pritchard and Baker, 1952a). In March or early April, these females begin to feed on the opening buds and later move on to the lower leaf surfaces, mainly along the veins. They spin heavy webbing over the colony surface, beneath which the mite population lives. From April onward, they lay eggs either in the webbing or on the leaf surface, mainly in the areas of junction of the midrib and veins. Each female lays 30–40 eggs during its life. Egg development takes almost 7 days, depending on the climate. The first generation starts in April or May, and 4–6 generations may develop before the females move towards overwintering sites in October (Rambier, 1958b; Ambrosi and Lenarduzzi, 1959; Jeppson *et al*., 1975). The intrinsic rate of increase (for *E. carpini borealis*) ranges from 0.065 at 15°C to 0.151 at 30°C (Bounfour and Tanigoshi, 2001b).

The population dynamics is affected by heat and by alternate heavy rain and hot sun, but any of these factors induces high mortality. Biotic factors include predatory insects, such as the coccinellid *Stethorus punctillum* Weise, and a number of phytoseiid mites (Rambier, 1958b; Ambrosi and Lenarduzzi, 1959; Jeppson *et al*., 1975; Laffi, 1982). The phytoseiids *Kampimodromus aberrans* (Oudemans), *Typhlodromus pyri* Scheuten and *Phytoseius plumifer* (Canestrini *et* Fanzago) are the three main species controlling mite pests in French vineyards (Kreiter *et al*., 2000). In north Italian vineyards, the main phytoseiid mites are *T. pyri*, *K. aberrans* and *Amblyseius andersoni* (Chant), while in central and southern Italy the most common phytoseiids are *T. exhilaratus* Ragusa, *P. finitimus* Ribaga and *K. aberrans* (Vacante and Tropea Garzia, 1987b; Duso, 2006); Duso and Fontana (2002) have also reported *P. plumifer* from north Italy.

Symptomatology and damage

The feeding activity of the yellow vine mite causes stunted growth of shoots, with shrunken and deformed leaves characterized by diffuse chlorotic areas and necrosis. The lower leaves of the bud may shrivel and fall prematurely. The presence of the mite is usually associated with chlorotic areas that turn reddish (mainly in the red vines), and the leaf blade can dry out in these affected areas. Heavy infestations may induce defoliation and early attacks may negatively affect the sugar content of the grapes and the development of grapes (acini). In addition, shoot lignification is hindered (Borgo, 1988). On apple and pear leaves, the damage somewhat appears as mildew spots (Jeppson *et al*., 1975).

Control

Control is provided by sprays of chemicals and biological means. During the monitoring of populations for the European red mite (*Panonychus ulmi*) (Koch)) as suggested by Baillod *et al*. (1979), the impact of *E. carpini* was usually more marked than that of the European red mite and the threshold lower (6–10 motile forms per leaf) (Girolami *et al*., 1989).

BIOLOGICAL CONTROL. See *Panonychus ulmi*.

CHEMICAL CONTROL. In early spring, it is possible to apply lime sulfur. During the summer, sprays of acaricides give good control (Rambier, 1958b; Ambrosi and Lenarduzzi, 1959). A number of acaricides (acrinathrin, tebunfenpyrad, pyridaben, fenazaquin, flufenoxuron, dicofol) give good control of mites of the vineyard (Morando *et al*., 1998). Practical experience of the author in Mediterranean countries showed that satisfactory control was given by sulfur. Vineyards treated during spring–summer with acaricides are generally either very little or not infested.

***Eotetranychus cendanai* Rimando (Fig. 20.22)**

Common name

Citrus yellow mite.

Diagnostic characteristics

FEMALE. The body is 390 μm long, 240 μm wide, pale yellowish green in colour and with silvery setae. The dorsum of the idiosoma is longitudinally striate between the opisthosomal setae e_1. The dorsal opisthosomal setae are slender, tapering and pubescent. The dorsal opisthosomal setae c_1, d_1, e_1 and f_1 are about as long as or longer than the longitudinal distances between them. The peritremes end in a simple bulb. Tibiae I and II have 8 and 5 tactile setae, respectively; tarsus I has 4 tactile setae proximal to the duplex setae. The genital flap and the area anterior to the flap are striate transversally (Ehara, 1969a; Jeppson *et al*., 1975).

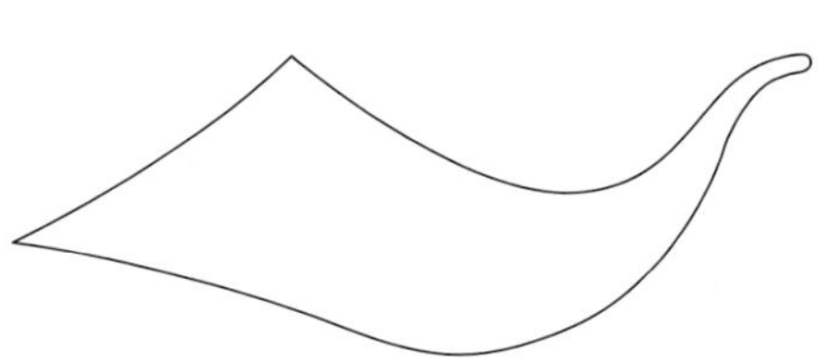

Fig. 20.22. *Eotetranychus cendanai* Rimando. Male, aedeagus (from Manson, 1963).

MALE. The body is 270 μm long and 270 μm wide. The aedeagus turns upward at almost a right angle to the shaft, and is strongly sigmoid and tapers distally (Ehara, 1969a; Jeppson *et al.*, 1975).

Geographical distribution

Cambodia (Manson, 1963; Baker, 1975a), China (Wang, 1980), Philippines (Rimando, 1962b), Taiwan (Ehara, 1969a) and Thailand (Manson, 1963; Baker, 1975a).

Host plants

The list of host plants of citrus yellow mite includes 12 species belonging to 3 families: Fabaceae – *Gliricidia sepium* (Jacq.) Kunth ex Walp., *Sesbania grandiflora* (L.) Pers. (Baker, 1975b); Poaceae – *Zoysia japonica* Steud. (Baker, 1975b); Rutaceae – *Citrus* sp. (Baker, 1975b; Ehara and Wongsiri, 1975), key lime (*C. aurantiifolia* (Christm.) Swingle) (Baker, 1975b; Thongtab *et al.*, 2002), sour orange (*C. aurantium* L.) (Rimando, 1962b), kaffir lime (*C. hystrix* DC) (Baker, 1975b), pomelo (*C. maxima* (Burm.) Merr.) (Ehara, 1969a), citron (*C. medica* L.) (Baker, 1975b; Barrion and Corpuz Raros, 1975), (*C. mitis* Blanco) (Barrion and Corpuz Raros, 1975), sweet orange (*C. sinensis* (L.) Osbeck) (Baker, 1975b) and orange jasmine (*Murraya paniculata* (L.) Jack.) (Baker, 1975b; Thongtab *et al.*, 2002).

Bio-ecology

E. cendanai is a potential pest of citrus. On *C. medica* and *C. mitis*, the ratio of males to females in the offspring of mated females averages 1:3, while the ratio from unmated females is 1:0 (Barrion and Corpuz Raros, 1975). The fertilized females produce offspring with a sex ratio of male:female of 1:1.02–1:2, and unfertilized females generate only male offspring (Thongtab *et al.*, 2002).

The citrus yellow mite feeds on the upper surface of the leaves, produces little webbing and commonly injures young trees. At 28 ± 1°C and 58 ± 5% RH on five different citrus species, the durations of the egg, larva, protonymph and deutonymph stages ranged from 4.9 to 5.8, 1.9 to 3.0, 1.9 to 2.6 and 1.8 to 2.7 days, respectively (Thongtab *et al.*, 2002). On *C. medica* and *C. mitis*, the duration of the egg, larval, protonymphal, deutonymphal and adult stages averaged 5–6, 3.5–4.5, 3–3.5, 2.5–3.5 and 19–20.5 days, respectively, in males, and 5–6, 4–4.5, 3, 4 and 15.5–18.5 days in females (Barrion and Corpuz Raros, 1975). The life cycle takes 11.2–14.0 days, and the fertilized female lays from 6.30 to 26.38 eggs (0.96–2.43 eggs a day) (Thongtab *et al.*, 2002). Mated and unmated females laid 22–23 eggs on *C. medica* and 24–25 on *C. mitis*, and 46.25% and 55.11% of these eggs hatched (Barrion and Corpuz Raros, 1975). Longevity of a fertilized female ranged from 6.55 to 10.52 days; longevity of unfertilized females was 7.52–11.62 days, and 14.35–17.73 eggs (1.40–1.97 eggs a day) are laid per female (Thongtab *et al.*, 2002). The mortality averages were 51.45% on *C. medica* and 65% on *C. mitis* (Barrion and Corpuz Raros, 1975). The net reproductive rate of increase (R_0) was 14.1, the cohort generation time (T_c) 17.37 days, the intrinsic rate of increase (r_m) 0.156 individuals day^{-1} and the finite rate of increase (λ) 1.165. The highest mortality occurs during the larval stage, followed by the nymphal and egg stages (Thongtab *et al.*, 2002).

Natural enemies include the linyphiid *Hylyphantes graminicola* (Sundevall) (Wipada, 1997) and the phytoseiid mite *Neoseiulus longispinosus* (Evans) (Thongtab *et al.*, 2002; Kongchuensin *et al.*, 2005).

Symptomatology and damage

The mite feeds on the upper surface of the leaves and produces little if any webbing. It is injurious to young trees. In the Philippines, it appeared to be a potential pest of citrus and has damaged young trees previously treated with insecticides (Rimando, 1962b).

Control

Although the mite appears to be a potential pest, at present control does not seem to be justified.

Eotetranychus hicoriae (McGregor) (Fig. 20.23)

Common name

Pecan leaf scorch mite.

Diagnostic characteristics

FEMALE. The peritremes are hooked distally. The genital flap is striate transversally and the area anterior to the flap is striate longitudinally. Tibia I and II have 9 and 8 tactile setae, respectively, and tarsus I has 5 tactile setae proximal to the duplex setae (Jeppson *et al.*, 1975).

MALE. The aedeagus is bent sharply ventrally, the bent portion is strongly sigmoid and at an acute angle to the shaft (Jeppson *et al.*, 1975).

Geographical distribution

China (Wang, 1980), India (Mallikarjunappa, 1987; Gupta and Gupta, 1994), South Korea (Lee, 1989) and the USA (McGregor, 1950).

Host plants

The list of host plants includes 13 species from 5 families: Cannabaceae – *Celtis sinensis* Pers. (Wang, 1980); Fagaceae –

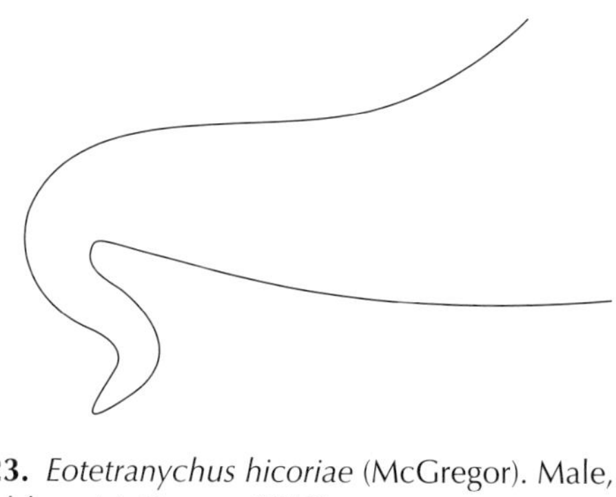

Fig. 20.23. *Eotetranychus hicoriae* (McGregor). Male, aedeagus (modified from McGregor, 1950).

Castanea dentata (Marshall) Borkh., *Fagus grandifolia* Ehrh. (Reeves, 1963), *Quercus acutissima* Carruthers, *Q. marilandica* Münchh. (Pritchard and Baker, 1955), *Q. rubra* L. (Pritchard and Baker, 1955; Reeves, 1963); Juglandaceae – *Carya* sp. (Pritchard and Baker, 1955), *C. cordiformis* (Wangenh.) K. Koch (Reeves, 1963), *C. glabra* (Mill.) Sweet (Thewke and Enns, 1969), pecan (*C. illinoinensis* (Wangenh.) K. Koch) (McGregor, 1950), *C. ovata* (Mill.) K. Koch (Reeves, 1963); Myrtaceae – guava (*Psidium guajava* L.) (Gupta and Gupta, 1994); Sapindaceae – *Aesculus hippocastanum* L. (Pritchard and Baker, 1955).

Bio-ecology

E. hicoriae is a pest of deciduous and ornamental plants. The mite lives on both leaf surfaces, but it seems to prefer the lower surface (Jeppson *et al*., 1975). It reproduces by arrhenotokous parthenogenesis and overwinters as females in the rough bark area on limbs of pecan trees; overwintering mortality may exceed 70%. In the spring, overwintering females began to oviposit within 7 days of the initiation of feeding (Jackson *et al*., 1983).

Rearing *E. hicoriae* on pecan leaf discs at 15.5, 21.1, 26.7 and 32.2°C, 30–50% RH and constant illumination (24 h light:0 h dark, or 24L:0D) showed a negative correlation between developmental time and temperature. The eggs did not hatch at 15.5°C, and at 21.1°C, and with a photoperiod of 10L:14D the females had a diapause morphology and did not lay eggs. Shortening of the photophase induced an increase in developmental time for both sexes. On pecan seedlings at the temperatures indicated above, the maximum daily oviposition was at 26.7°C. Mated and virgin females exhibited the same fecundity. There is no tritonymph stage (Micinski *et al*., 1979). Life tables constructed for the immature stages at 21.1, 26.7 and 32.2°C, and constant illumination (24L:0D), showed that immature mortality varies from 81% at 32.2°C to 96% at 21.1°C. Moreover, under two photoperiodic regimes (10L:14D and 14L:10D) at 21.1°C, the immature mortality was 48 and 46%, respectively. Intrinsic rates of increase (r_m) varied from less than 0.001 at 21.1°C and constant illumination to 0.105 at 32°C and constant illumination (Micinski *et al*., 1981). The eggs did not hatch at 18.3°C, and at 35°C, 97% of the individuals died as immatures, so these temperatures might limit population growth in the field. In the laboratory, higher mortality with increasing temperatures might be due to increasing mortality of individuals that would develop into males (Jackson *et al*., 1983).

On guava in southern India, population peaks were observed in March. During February and March, the increase in populations relates to high temperature, longer sunshine hours, low rainfall and low RH. Temperature and sunshine were positively correlated with mite populations, while rainfall and RH were negatively correlated. The correlation between predators and the tetranychid population was not significant (Mallikarjunappa and Nageshchandra, 1989). On pecan, the populations of *E. hicoriae* increased gradually during the summer and rapidly during late summer or early autumn, reaching peaks in September or early October (Ellis, 2008).

Natural enemies include predatory insects, such as *Stethorus utilis* Horn (Tedders, 1983) and several phytoseiid mites. Among these, *Amblyseius* (Athias Henriot) *delhiensis* (Narajanan *et* Kaur) (= *Euseius scutalis*), *Amblyseius tetranychivorus* (Gupta) (= *Typhodromips tetranychivorus* Gupta), *Phytoseius punjabensis* Gupta and *P. swirski* Gupta, and the cunaxid *Cunaxa setirostris* (Hermann) have been reported in south India (Mallikarjunappa and Nageshchandra, 1989). *Neoseiulus collegae* (De Leon) was reported in the Neotropical region by Mizell and Schiffhauer (1991). Other phytoseiid mites were artificially released.

Symptomatology and damage

Feeding of mites on leaflets causes scorched areas, which first appear as dark brown or liver-coloured blotches. This symptom may be observed in the first half of April. Damage occurs on the leaves of shoots of small branches, and most often begins along the midrib and spreads outward towards the leaf margins. On mature leaves, the damage may be observed from June. Old scorched areas have an irregular size and pattern on the leaves. Severe attacks induce premature defoliation, mainly starting at the lower branches. The leaves remaining on the trees may be russeted or scorched, mainly during dry and hot weather. Heavy losses of leaves in August and early September cause failure of the trees to bloom in the following spring (Pierce, 1953; Reeves, 1963; Jeppson *et al*., 1975). The damage usually begins on the lowest branches in the interior of the tree and progresses upwards through the tree; it is most severe in late summer and autumn (Ellis, 2008).

Control

Control is provided by sprays of acaricides and releases of natural enemies (Phytoseiidae). The use of selective chemicals allows the implementation of integrated pest management (IPM) programmes. Monitoring and cultural practices help in control.

MONITORING. Specific controls should be applied if mite damage appears frequently and generally on the foliage of low limbs (in 25% of leaves sampled). Spot treatments are often possible (Ellis, 2008).

CULTURAL PRACTICES. Rational fertilizer application usually hinders mite outbreaks. On pecan seedlings fertilized with N at 0, 200 or 600 ppm, the highest mite population develops on seedlings treated with N at 200 ppm (the suggested level for good orchard maintenance). This high population seems to be related to enhanced mite reproduction and not to a shorter life cycle. Moreover, generational mortalities of 65.3% at 0 ppm N, 40.6% at 200 ppm and 37.8% at 600 ppm N were found. Also, net photosynthesis (Pn) was less, compared with control seedlings, in seedlings with mites and treated with 200 ppm N, and greater in seedlings with mites that had been treated with 600 ppm N. Seedlings infested by mites and treated with 200 ppm N are shorter than seedlings that are not infested with mites; and seedlings treated with 600 ppm N that are infested with mites are taller than seedlings that are not infested with mites (Jackson and Hunter, 1983). Foliage damage to pecan by *E. hicoriae* increases with elevated leaf water status and N concentration, but is unaffected by crop load (Wood and Reilly, 2000).

BIOLOGICAL CONTROL. Experiences of the biological control of *E. hicoriae* are encouraging. Release of the phytoseiid mites *Metaseiulus occidentalis* (Nesbitt) and *Phytoseiulus persimilis* Athias Henriot suppressed populations of the pecan leaf scorch mite in pecan orchards in Georgia (USA). The first use of these predatory mites controlled a population of 4.4 mites and eggs of *E. hicoriae* after 28 days. Releases of each mite were effected at 500 and 1000 mites per tree in the centre tree of a 25-tree square plot (0.41 ha). The untreated trees were infested by 63, 240 and 38 *E. hicoriae* and eggs per leaf at 6, 10 and 24 days post release, respectively. On the treated surface, the pecan leaf scorch mites were suppressed at this high level of population 24 days after predator release. Release of the mites at 500 and 1000 *M. occidentalis* mites per tree reduced the pest mite infestation by 67 and 91%, respectively. Release of 500 and 1000 *P. persimilis* mites per tree reduced the pest mite infestation by 90 and 98%, respectively. So, predatory mite releases appear to provide an effective management tactic for pecan leaf scorch mite for pecan producers in Georgia (Dutcher, 2007).

CHEMICAL CONTROL. One or two sprays during the latter part of May were found to be effective in the USA. The first reports concerned sprays of wettable sulfur plus dinitro-*o*-cyclohexyl phenol, or of parathion or other acaricides (Pierce, 1953; Reeves, 1963; Jeppson *et al.*, 1975). Later, different chemicals, such as azinphos-methyl, carbaryl, demeton, dialifor, dicofol, dimethoate, disulfoton, endosulfan, methidathion, parathion and phosalone, were reported for pecan in the USA (Boethel and Ezell, 1978). Dicofol was the most common acaricide employed on pecan before of the gradual halting of its production, since when it has been replaced by bifenazate (Dutcher *et al.*, 2009). In India, a number of chemicals were tested on guava, including cyhexatin, dicofol, formothion, oxydemeton and oxydemeton-methyl (Mallikarjunappa *et al.*, 1989). Other tests on this host plant were carried out with cyhexatin, dicofol, formothion, malathion, oxydemeton-methyl and sulfur. Of these, the application of cyhexatin, dicofol and sulfur resulted in >90% mortality 3 days after spraying and were effective for up to 21 days. However, the effectiveness of all chemicals decreased 28 days after spraying (Mallikarjunappa and Nageshchandra, 1990). In the USA, the injection of the trunks of mature pecan trees with dicrotophos was evaluated as an effective method for controlling certain foliar pecan arthropod pests (such as the pecan leaf scorch mite) in urban plantings where conventional spray equipment was prohibited. *E. hicoriae* populations were higher in trees injected with benomyl than in trees injected twice with benomyl plus dicrotophos; a single injection of dicrotophos and benomyl did not control pecan leaf scorch mites. No chemical residues of dicrotophos were found in pecan fruits at harvest (Gentry *et al.*, 1982).

In addition, the ecological impact of pesticides on natural enemies may cause outbreaks of *E. hicoriae*. Studies on the population dynamics of *E. hicoriae* from May to October in three managed and one abandoned pecan orchards in northern Louisiana showed that the seasonal abundance of *E. hicoriae* varied in each managed orchard, primarily as a result of the insecticide programmes followed. In the abandoned orchard, the mite had low populations through the midseason and eventually reached peak levels in late August. Insecticides were detrimental to the phytoseiid populations and appeared to be partially responsible for the larger tetranychid populations found in the managed orchards (Boethel and Ezell, 1978). In the USA, chemicals used for aphid control sometimes hinder the development of mite populations. However, in the late season, these pesticides may be not adequate. In addition, some insecticides have detrimental side effects and kill beneficial non-mite organisms, thus causing mite problems to increase (Ellis, 2008).

INTEGRATED PEST MANAGEMENT. The positive evaluation of the phytoseiid mite *G. occidentals* in biological control of the pecan leaf scorch mite and the registration of the selective acaricide bifenazate (replacing dicofol) has allowed IPM of pecan leaf scorch mite in pecan orchards in the USA. Bifenazate is effective and has the additional benefit over dicofol of conserving phytoseiid mites. This newer acaricide conserves a portion of the phytoseiid mite population, and the presence of phytoseiids is similar in non-treated and bifenazate-treated trees for up to 4 weeks after treatment. However, treatment of pecan trees with bifenazate plus the release of phytoseiid mites is a more effective method of pecan leaf scorch mite control than the application of bifenazate alone. In addition, among eight chemical control treatment alternatives to dicofol, pecan trees treated with bifenazate had similar predatory mite abundance to the non-treated control (Dutcher *et al.*, 2009). Dutcher (2012) suggests a strategy for integrated control of the pecan leaf scorch mite that uses both predatory mites and miticides.

Eotetranychus hirsti Pritchard *et* Baker

Common name

Fig spider mite.

Diagnostic characteristics

FEMALE. The body is greenish yellow with black blotches all over the dorsum. The genital flap and the area anterior to the flap are both transversely striate (Jeppson *et al.*, 1975).

MALE. The empodium consists of a simple claw. The aedeagus bends ventrally, with the downturned portion slender, tapering and slightly sigmoid (Jeppson *et al.*, 1975).

Geographical distribution

Cyprus (Georghiou, 1959), India (Hirst, 1926; Nassar and Ghai, 1981; Gupta, 1992, 1995; Gupta and Gupta, 1994; Gupta and Chatterjee, 1997), Iran (Khanjani and Hadad Irani-Nejad, 2006; Daneshnia *et al.*, 2012), Israel? (Ben-David, 2008) and Pakistan (Chaudhri *et al.*, 1974).

Host plants

The list of host plants of the fig spider mite includes seven species from three families; Apocynaceae – *Nerium oleander* L. (Bolland *et al.*, 1998; Migeon and Dorkeld, 2006–2013);

Fabaceae – *Bauhinia* sp. (Gupta and Gupta, 1994; Gupta and Chatterjee, 1997); Moraceae – *Ficus* sp. (Gupta, 1992), fig (*F. carica* L.) (Hirst, 1926; Nassar and Ghai, 1981; Gupta and Gupta, 1994; Gupta and Chatterjee, 1997; Khanjani and Hadad Irani-Nejad, 2006; Daneshnia *et al.*, 2012), *F. racemosa* L. (Gupta and Gupta, 1994; Gupta and Chatterjee, 1997), sacred fig (*F. religiosa* L.) (Gupta and Gupta, 1978) and *F. semicordata* Buch.-Ham. ex Sm. (Prasad, 1974; Gupta and Gupta, 1994; Gupta and Chatterjee, 1997).

Bio-ecology

E. hirsti is a pest of figs in the fig-growing areas of Oriental and Palaearctic regions. The fig spider mite lives mainly on the lower leaf surfaces, but it may also live on the upper leaf surfaces and on fruits. In Lyallpur (now Faisalabad, Pakistan) the mite is active from May to October and when leaves die in November, the fertilized females move to the branches to overwinter in terminal buds. In February and early March, the mite resumes its activity (Rahman and Sapra, 1940a). *E. hirsti* is reported as a pest of fig in Hisar and Haryana States (India) (Putatunda *et al.*, 2002) and recently also as an alarming pest of fig in Delhi, Punjab, West Bengal and Uttar Pradesh (Singh and Raghuraman, 2011). Moreover, it is one of the major pests of fig trees in various provinces of Iran: Fars (Daneshnia *et al.*, 2012), Bushehr, Gilan, Kermanshah, Khuzestan, Kohgilouyeh and Boyer-Ahmad, Markazi, Semnan and Hormozgan, Sistan and Baluchestan, and Tehran (Beyzavi *et al.*, 2013).

In laboratory studies on fig leaves at 30 ± 2°C, 60 ± 5% RH and a photoperiod of 12L:12D, the developmental periods for the egg, larva, protochrysalis, protonymph, deutochrysalis, deutonymph and teleiochrysalis were 4.4 ± 0.113, 1.37 ± 0.089, 1, 1.37 ± 0.089, 1.03 ± 0.033, 1.33 ± 0.088 and 1 days, respectively. The longevity of the male and female were 8 and 16 days, respectively. The pre-oviposition period of females, the total pre-oviposition period of females counted from birth, and the average and the highest number of laying eggs were 1.5 ± 0.1043, 13.17 ± 0.1871, 46.21 ± 0.593 and 51, respectively. The mean generation time (T), the net reproduction (R_0), the intrinsic rate of natural increase (r_m) and the finite rate of increase (λ) were 18.48 ± 0.22, 36.97 ± 3.46, 0.1955 ± 0.0058 and 1.2159 ± 0.007, respectively (Daneshnia *et al.*, 2012).

Recently, Jafari *et al.* (2015) observed that at six constant temperatures (15, 20, 25, 30, 32 and 35°C), 50 ± 5% RH, a photoperiod of 16L:8D and on fig leaves, the total developmental time from egg to adult female of *E. hirsti* was 41.29, 24.15, 16.95, 12.35, 10.21 and 10.67 days, respectively. The developmental rate increased as the temperature increased from 15 to 32°C, but declined at 35°C. The estimated lower temperature threshold for the entire developmental time and the effective accumulative temperature using the ordinary linear model were 9.86°C and 239.48 degree-days, respectively. The estimated critical temperatures (T_{opt} and T_{max}) calculated using the Logan-6 model were 34.30 and 35.44°C for the immature development of *E. hirsti*.

Natural enemies include predatory insects and mites. In Khuzestan (Iran), the coccinellid *Stethorus gilvifrons* (Mulsant) preys on different spider mites, including *E. hirsti* (Afshari *et al.*, 2007). Also in Iran, the phytoseiid mite *Phytoseius corniger* Wainstein is considered likely to be a predator of *E. hirsti* on fig (Daneshvar, 1987). In India (Hisar, Haryana), different predatory mites, such as *Agistemus fleschneri* Summers, *Amblyseius* sp., *A. alstoniae* Gupta (= *Euseius alstoniae* (Gupta)), *A. delhiensis* (Narajanan *et* Kaur) (= *Euseius scutalis* (Athias Hennot)), *Bdella* sp., *Cheletacarus gryphus* Summers *et* Price, *Phytoseius roseus* Gupta and *Typhlodromus homali* Bindra, were found associated on different fruit crops with tenuipalpid and tetranychid mites, including *E. hirsti* (Putatunda *et al.*, 2002).

Symptomatology and damage

Feeding of the fig spider mite on leaves first causes transparent patches observed in a cross light. These turn yellowish green, and then brown and rough and dry. Premature and complete defoliation and dropping of the fruit may occur (Jeppson *et al.*, 1975; Singh and Raghuraman, 2011). In India (Haryana), the mites infest fig and jujuba trees. However, the infestation seemed moderate; mite presence was associated with pinhead white spots on the leaf surfaces (Putatunda *et al.*, 2002).

Control

Control is mainly by chemical sprays. However, the results of studies on the susceptibility of different cultivars have shown the possibility of using resistant varieties.

RESISTANT VARIETIES. Studies in Iran (Saveh) on the susceptibility of 11 fig cultivars (including 'Yellow Indigenous', 'Saveh Yellow', 'Black Indigenous', 'Varamine Black', 'Onion Yellow', 'Wild Fig', 'Garmsar Morabai', 'Siljerd Yellow' and 'American') to fig spider mites showed that the cultivar 'Wild Fig' was more susceptible than the other cultivars, followed by 'Garmsar Morabai' and 'Siljerd Yellow'. The least susceptible cultivars were 'Varamine Black', 'Yellow Indigenous' and 'Varamine Black' (Baradaran *et al.*, 2002).

CHEMICAL CONTROL. In the past, the mite was treated with sulfur sprays and other acaricides, e.g. phosphamidon and chlorobenzilate (Kanta *et al.*, 1963; Bindra and Varma, 1966). Currently, it is possible to use various acaricides, according their action and side effects, and the specific legislative constraints of each country. *E. hirsti* has long been the key pest of fig trees in Iran, where it is primarily controlled by chemical pesticides (Khanjani and Hadad Irani-Nejad, 2006).

Eotetranychus kankitus Ehara (Fig. 20.24)

Common name

Citrus yellow mite.

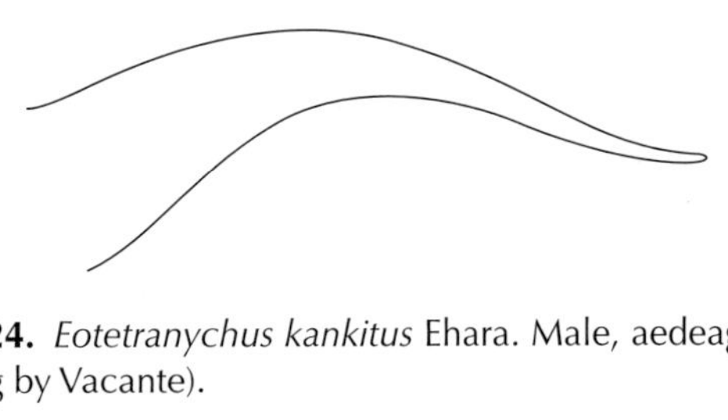

Fig. 20.24. *Eotetranychus kankitus* Ehara. Male, aedeagus (drawing by Vacante).

Diagnostic characters

FEMALE. The body is ovoid. Tibiae I and II bear 9 and 8 tactile setae, respectively; tarsi I and II have 5 and 3 tactile setae proximal to the duplex setae, respectively. The peritremes are slightly curved distally but not hooked. The genital flap is striate transversally on the posterior part and longitudinally on the anterior central part; the anterior area to the genital flap is longitudinally striate (Ehara, 1955; Jeppson *et al.*, 1975).

MALE. Empodium I is split distally in 3 pairs of ventrally direct hairs. The aedeagus is slightly down curved, tapering distally, and slightly sigmoid (Ehara, 1955; Jeppson *et al.*, 1975).

Geographical distribution

China (Wang, 1980; Wang *et al.*, 1985; Wang and Cui, 1992; Zhou *et al.*, 1999), India (Manson, 1963; Gupta and Gupta, 1994; Gupta, 1995) and Japan (Ehara, 1955).

Host plants

The list of host plants includes 15 species from ten families: Betulaceae – *Alnus* sp. (Wang and Cui, 1992); Cannabaceae – *Celtis sinensis* Fruhstorfer (Bolland *et al.*, 1998; Migeon and Dorkeld, 2006–2013); Cornaceae – *Alangium chinense* (Lour.) Harms (Bolland *et al.*, 1998; Migeon and Dorkeld, 2006–2013); Elaeagnaceae – *Elaeagnus bockii* Diels (Wang and Cui, 1992), *E. pungens* Thunb. (Wang, 1980); Lauraceae – *Litsea auriculata* Chien *et* Cheng (Bolland *et al.*, 1998; Migeon and Dorkeld, 2006–2013); Poaceae – *Eleusine indica* (L.) Gaertn. (Bolland *et al.*, 1998; Migeon and Dorkeld, 2006–2013); Rosaceae – apricot (*Prunus armeniaca* L.), European pear (*Pyrus communis* L.), *Rosa* sp. (Gupta, 1984; Gupta and Gupta, 1994), China rose (*R. chinensis* Jacq.) (Gupta and Gupta, 1994); Rutaceae – *Citrus* sp. (Ehara, 1955; Gupta and Gupta, 1994; Zhou *et al.*, 1999), mandarin (*C. reticulata* Blanco) (Prasad, 1974); Salicaceae *Salix* sp. (Wang and Cui, 1992); Vitaceae – grapevine (*Vitis vinifera* L.) (Bolland *et al.*, 1998; Migeon and Dorkeld, 2006–2013).

Bio-ecology

E. kankitus is reported on wild, ornamental and fruit plants. It is a pest of *Citrus* sp. in Japan (Ehara, 1955), India (Gupta and Gupta, 1994) and China (Wang, 1980; Chen and Li, 1985b; Li, 1990; Yang, 1998; Zhou *et al.*, 1999), and of *C. reticulata* in India (Manson, 1963; Prasad, 1974). It infests leaves, flowers, shoots and fruitlets, but young leaves are usually attacked badly (Yang, 1998). In a Dehong lemon orchard in Yunnan Province (China), *E. kankitus* is regarded as a dominant species and prevention and control methods are under investigation (Gao *et al.*, 2012).

The citrus yellow mite reproduces by arrhenotokous parthenogenesis and has a haploid number of chromosomes of four (Zou *et al.*, 2002). At 20°C, the fecundity is 87.96 eggs per female, and the oviposition rate is 2.06 eggs per female daily, whereas at 35°C the oviposition rate is 3.73 eggs per female daily and the fecundity is 55 eggs per female (Zhou *et al.*, 1999).

Life table studies of *E. kankitus* were undertaken at five constant temperatures (15, 20, 25, 30 and 35°C). At 15°C, the lowest net reproductive rate of increase (R_0), 16.64, and the longest generation time (T), 48.62 days, resulted in the lowest intrinsic rate of increase (r_m), 0.058, and the longest doubling time (11.95 days). The highest intrinsic rate of increase (0.223) and finite rate of increase (λ), 1.2493, were at 35°C, and the population was able to double in only 3.11 days (Zhou *et al.*, 1999).

In spring, the dispersal pattern of the mite is of the gregarious translocation type. There is a significant negative correlation between the population densities on the autumn and spring tips before the peak of the mite population, and also a negative correlation after the peak when numbers are declining rapidly (Chen and Li, 1985b). The generation time is negatively correlated with both temperature and humidity. The number of eggs per female is correlated with temperature and humidity. The optimal temperature interval for mite development ranges from 25 to 30°C and the optimal humidity interval is within the range 65–75% (Zhou *et al.*, 2000).

The population of *E kankitus* has a peak from late April to early May, then the population decreases rapidly, while that of its predator, the phytoseiid mite *Amblyseius nicholsi* Ehara *et* Lee, has a peak in June and the numbers remain relatively high in the autumn. The distribution patterns of *E. kankitus* and *A. nicholsi* are an aggregating distribution, and *A. nicholsi* exhibits a high reproductive rate and predation, and consequently is important in the control of *E. kankitus* (Zhi, 1997). Studies in Danjiangko Municipality, Hubei, China, showed that the unusual climate (warm winter and infrequent rainfall), the omission of orchard winter cleaning, not spraying chemicals in the early spring and the injudicious use of chemicals were reasons for the prevalence of the citrus yellow mite (Yang, 1998).

Natural enemies include the phytoseiid mite *Euseius nicholsi* (= *A. nicholsi*) (Zhi *et al.*, 1994, 1998) and the stigmaeid *Agistemus exsertus* González-Rodríguez (Wang, 1981c).

Symptomatology and damage

In Japan, the mite causes injury to citrus similar to that of the six-spotted spider mite, *Eotetranychus sexmaculatus* (Riley) (Ehara, 1964a). The injury is related to leaf age, mite abundance and soil moisture. Leaves that are infested early were able to compensate the damage more strongly than those infested later (Chen and Li, 1985a). High infestations on citrus trees cause leaves, flowers and fruits to fall prematurely, and withering of the branches (Chen, 1999). *E. kankitus* is thermophilic, producing shoot damage on citrus in China during warm, dry weather as the shoots elongate (Li, 1990).

Control

Control is provided by the monitoring of populations and sprays of chemicals, according preference to active ingredients with reduced side effects.

SAMPLING AND MONITORING. Sampling methods for estimating the population density of *E. kankitus* in China have been investigated by Wang (1985), who compared two methods: the insect density estimation technique of D.J. Gerrard and H.C. Chiang

(1970) and the sequential sampling technique of E. Kuno (1969). The practicality and effectiveness of the latter method was determined by random sampling computer simulation.

CHEMICAL CONTROL. Spraying with pyrazoles in early April is quite effective (Yang, 1998), as is spraying with organotins and pyridazinones (Chen, 1999).

Eotetranychus lewisi (McGregor) (Fig. 20.25)

Common name

Lewis spider mite.

Diagnostic characteristics

FEMALE. The body is ovoid, about 360 µm long and pale greenish amber in colour, with blackish spots laterally. The genital flap and the area anterior to the flap are both striate transversally. The peritremes are hooked distally. Tibiae I and II bear 9 and 8 tactile setae respectively, and tarsus I has 5 tactile setae proximal to the duplex setae (McGregor, 1950; Jeppson *et al.*, 1975).

MALE. The male is smaller than the female, and mustard yellow in colour. The tibia I has 9 tactile setae. The aedeagus gradually tapers distally and forms a broad sigmoid ventral bend (McGregor, 1950; Jeppson *et al.*, 1975).

Geographical distribution

Bolivia (Flechtmann, 1996), Canada (CABI, 2000), Chile (González, 1989), Colombia (Urueta, 1975), Costa Rica (Baker and Pritchard, 1962), El Salvador (Berry, 1959; Andrews and Poe, 1980), Guadeloupe (Flechtmann *et al.*, 1999a), Guatemala (Ochoa *et al.*, 1991a), Hawaii Islands (Garrett and Haramoto, 1967), Honduras (Baker and Pritchard, 1962), Libya (Damiano, 1964), Madeira (Carmona, 1992), Mexico (Tuttle *et al.*, 1974), Nicaragua (Baker and Pritchard, 1962), Panama (Ochoa *et al.*, 1991a), Peru (Bolland *et al.*, 1998; Guanilo *et al.*, 2012), South Africa (Meyer Smith, 1987), Taiwan (Ho and Shih, 2004) and the USA (McGregor, 1943a, 1950; Pritchard and Baker, 1955). The mite has also been reported in Norway, Germany and Japan (EPPO, 2006).

Host plants

The list of host plants of the Lewis spider mite includes 69 species ascribed to 26 families, with the highest number of species belonging to the Euphorbiaceae (13 species) and the Fabaceae (10 species) (Migeon and Dorkeld, 2006–2013). The families and species of economic importance are: Caricaceae – papaya (*Carica papaya* L.) (Berry, 1959; Baker and Pritchard, 1962; Andrews and Poe, 1980; Flechtmann *et al.*, 1999a); Euphorbiaceae – wild poinsettia (*Euphorbia cyathophora* Murr.) (Urueta, 1975), ghostweed (*E. marginata* Pursh), poinsettia (*E. pulcherrima* (Willd. ex Klotzsch)) (Baker and Pritchard, 1962; Andrews and Poe, 1980; Carmona, 1992; Ho and Shih, 2004; Ho *et al.*, 2004), castor bean (*Ricinus communis* L.) (McGregor, 1950; Pritchard and Baker, 1955); Moraceae – fig (*Ficus carica* L.) (Estebanes-Gonzalez and Baker, 1968); Oleaceae – European olive (*Olea europaea* L.) (Pritchard and Baker, 1955); Rosaceae – strawberry (*Fragaria* × *ananassa* Duchesne ex Rozier) (Howell and Daugovish, 2013), peach (*Prunus persica* (L.) Batsch) (Tuttle and Baker, 1964; Pérez-Santiago *et al.*, 2007a); Rutaceae – lemon (*Citrus limon* (L.) Burm. f.) (McGregor, 1943a), sweet orange (*C. sinensis* (L.) Osbeck) (McGregor, 1943a, 1950).

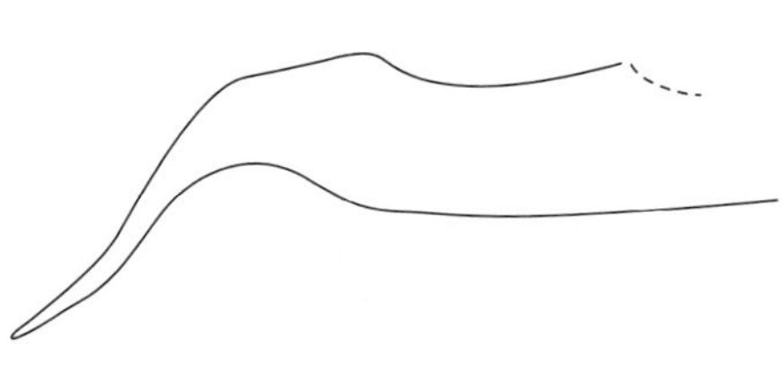

Fig. 20.25. *Eotetranychus lewisi* (McGregor). Male, aedeagus (from Meyer Smith, 1987).

Bio-ecology

The Lewis spider mite is a polyphagous species that is widely distributed, especially in the Nearctic region, where it infests a number of wild and economic plants. *E. lewisi* usually feeds on the lower leaf surfaces of the host plant, in the areas closer to the main veins, but in heavy infestations, its populations spread on to the entire lower leaf surface. On citrus, the mite lives and reproduces on the fruits, where it lays eggs in the depressions of the rind. The Lewis spider mite is a pest of lemon and sweet orange exclusively of the desert areas of California (McGregor, 1943a, 1950, 1956; Jeppson *et al.*, 1975; Jeppson, 1989) and in Israel (Jeppson, 1989). It is also known as a pest of poinsettia in greenhouses in California (Doucette, 1962; Jeppson *et al.*, 1975; Dara, 2011), Taiwan (Ho and Shih, 2004) and other countries. Recently, it was reported as the most important pest in peach orchards in north-central Mexico (Pérez-Santiago *et al.*, 2007b). Infestation on strawberries was reported from the Philippines nearly a decade ago (Raros, 2001), and it is a newly emerging pest in California commercial strawberries and raspberries (Dara, 2011; Howell and Daugovish, 2013). In the USA, the economic impact of *E. lewisi* concerns mainly poinsettia and other Euphorbiaceae. The mite develops most rapidly on *E. marginata*, poinsettias and *R. communis* in greenhouses, and when not controlled induces leaf discoloration and leaf drop, thereby lowering the sale value of the plants. It is only a minor pest of citrus (Doucette, 1962), occurring occasionally in southern California.

E. lewisi reproduces by arrhenotokous parthenogenesis and the haploid number of chromosomes is two (Helle *et al.*, 1981). On citrus, the female lays 2–3 eggs a day in the depressions of the fruit surface. The egg has a short stalk, but no threads extend from the top of the stalk to the host. The life cycle from egg to adult is 12 days for the males and 14.5 days for the females (Stiller, in Jeppson *et al.*, 1975). In a study of the development and population parameters of the Lewis spider mite on poinsettia, the developmental time of the egg was found to be 2.5–8.1 days at 16–28°C. At 30°C, the eggs did not hatch. The developmental periods for the larva, protonymph and deutonymph were 1.8–3.9, 1.4–2.9 and 2.3–4.1 days,

respectively. The developmental time from egg to adult was 8.0–19.0 days. The thermal summation was 159.16 degree-days, the low temperature threshold 8.3°C, the optimal temperature 27.8°C and the lethal temperature 28.2°C. At 20, 24 and 28°C, the longevities of the adult male were 19.4, 12.3 and 7.9 days; those of the adult female were 12.2, 16.0 and 9.6 days, respectively. The total fecundities at the same three constant temperatures were 20.7, 50.5 and 32.1 eggs, respectively. The intrinsic rate of increase was 0.0988 day^{-1} at 20°C, 0.1731 day^{-1} at 24°C and 0.1145 day^{-1} at 28°C. The mean generation time was 19.8 days at 20°C, 16.68 days at 24°C and 13.16 days at 28°C (Lai and Lin, 2005). On poinsettias in the greenhouse, the damage generally becomes obvious in October, but there have also been problems in August and September (Gilrein, 2006). In the field, the mite probably disperses by wind currents, but in greenhouses the infestation of poinsettias could depend on the introduction of infested cuttings from outside.

Natural enemies include the phytoseiid mites *Amblyseius californicus* McGregor (= *Neoseiulus californicus* (McGregor)) (Doucette, 1962), *Euseius hibisci* (Chant) and *Typhlodromalus limonicus* (Garman *et* McGregor) (McMurtry, 1985b). Recently, several mites were collected in Taiwan on poinsettia and other economic plants: *N. longispinosus* (Evans), *E. ovalis* (Evans), *Paraseiulus minutus* Athias Henriot, *Phytoseius rugatus* Tseng, and some stigmaeid and ascid mites (*Lasioseius* sp.). Predatory insects include cecidomyiids (*Oligota* sp.), thripids (*Scolothrips* spp.), and anthocorids (*Orius* sp.) (Ho *et al.*, 2004; Ho, 2006). Other phytoseiids have been artificially released.

Symptomatology and damage

Symptoms of damage vary according to the host plant. On citrus, the Lewis spider mite damages only the fruits, and its feeding results in a pale stippling of the rind. Commonly no injury is caused to the leaves; heavy infestations cause silvering on lemon and either a silvering or russeting of oranges. The abundant webbing collects dust that makes the detection of heavy infestations easy (McGregor, 1956; Jeppson *et al.*, 1975). Meyer Smith and Craemer (1999) reported similar damage to citrus.

On other host plants, feeding by the Lewis spider mite causes chlorotic stippling on leaves, which later turn either yellow or bronze; extensive leaf drop may occur. On poinsettias, the mite prefers mature leaves (Carmona, 1992). At low levels of attack, the leaves have a speckled or peppered appearance, due to a large number of clear yellow spots or yellowish patches of varying size characterized by indefinite borders and distributed over the leaf surface. Light-green to obvious chlorotic areas occur on the lower leaf surface. Intense yellow speckling can be observed on both leaf surfaces. At a high level of attack, the inter-vein surfaces turn yellow and contrast clearly with the green veins. This condition can be confused with zinc or magnesium deficiency (Ochoa *et al.*, 1991a). Severe infestation produces abundant webbing on the lower leaf surface, usually around the flower centres, and chlorotic leaves, and may induce extensive leaf drop. Damage to castor bean (*R. communis*) is similar to that to poinsettias (Doucette, 1962). On papaya, the damage to mature leaves is similar to that on poinsettias, and can be confused with alterations by hormonal herbicides. On young leaves, mite feeding causes chlorosis and distortion similar to virus diseases. In heavy infestations, young leaves have veins and lack laminas, inducing a mistaken diagnosis of a virus disease (Ochoa *et al.*, 1991a). On 'Diamante Mejorado' peach tree cultivars, *E. lewisi* may influence the concentration of total soluble sugars and starch in the leaves, bark and roots (Pérez-Santiago *et al.*, 2007b). On strawberry, the symptoms of *E. lewisi* damage are yet to be determined, because the infestation currently overlaps with that of the two-spotted spider mite, *Tetranychus urticae* Koch (Dara, 2011).

Control

According to the host plant, control involves the use of mechanical, chemical and biological means.

MECHANICAL CONTROL. This method is applied only in citrus owing to the adaptation of the mite to live more on the fruit than on the leaves. This allows the removal of most of the mite population with the harvesting of the citrus fruit (Jeppson *et al.*, 1975).

BIOLOGICAL CONTROL. The action of the phytoseiid mites *Phytoseiulus persimilis* Athias Henriot, *N. californicus*, *N. fallacis* (Garman) and *A. andersoni* (Chant) in the management of *E. lewisi* populations, and the interactions between *T. urticae* and *E. lewisi* in relation to phytoseiid efficiency, were investigated in the laboratory. When *E. lewisi* and *T. urticae* are present on the same leaf, *T. urticae* populations increase and begin displacing *E. lewisi*. *P. persimilis* did not prey upon *E. lewisi*, but the other three phytoseiid mites did prey upon the spider mites and reduce their populations. When both *E. lewisi* and *T. urticae* are present on the same leaf, *N. fallacis* and *A. andersoni* prey on both pest mites equally and reduce both populations. *N. californicus* feeds on *E. lewisi* first and reduces its population, but allows the increase of *T. urticae* populations. *P. persimilis* may be inefficient at controlling *E. lewisi* and its use may instead enhance *E. lewisi* populations (Howell and Daugovish, 2013).

CHEMICAL CONTROL. On citrus, the mite populations can be controlled by sulfur or other acaricide applications (Jeppson, 1989). On poinsettia in the greenhouse, it is important to spray the leaf lower surface on which the mite lives. One spray of demeton, dicofol or parathion gave adequate control and two applications of sulfotep aerosol (a dithio aerosol) at 5-day intervals resulted in 100% mortality. No injury to poinsettia was noted when the acaricides were applied before the bracts showed colour (Doucette, 1962). Tests carried out more recently reported acaricides such as abamectin, acequinocyl, bifenazate, fenpyroximate, hexythiazox, pyridaben and spiromesifen to be effective (Gilrein, 2006; Dara, 2011). On ornamentals, the sprays must be applied at an early stage of detection of the mite infestation, according to Gilrein (2006), who relates a low number of mites per plant with a low damage rating (based on a scale of 0–10, with 0 = no damage and 10 = completely defoliated). It is important to spray the lower leaf surface with particular care.

Eotetranychus populi (Koch) (Fig. 20.26)

Common name

Yellow poplar mite.

Diagnostic characteristics

FEMALE. The body is 360 μm long and from 180 to 210 μm wide. The peritremes are irregularly anastomosed. The dorsal body setae are longer than the distance between their bases. The genital flap and the area anterior to the flap are transversally striate. Tibia I and II have 9 and 8 tactile setae, respectively, and tarsus I has 5 tactile setae proximal to the duplex setae (Jeppson *et al.*, 1975).

MALE. The male is smaller than the female. The aedeagus is strong and only slightly undulate (Jeppson *et al.*, 1975).

Geographical distribution

Armenia, Azerbaijan (Mitrofanov *et al.*, 1987), China (Wang, 1980), France (Gutierrez and Helle, 1983), Germany (Koch, 1838), Greece (Bolland and Beron, 2004), Georgia (USSR) (Reck, 1948b; Mitrofanov *et al.*, 1987), Hungary (Bozai, 1970), India (Gupta and Gupta, 1994), Iran (Khalil-Manesh, 1973), Italy (Arru, 1961), Kazakhstan (Wainstein, 1956a; Mitrofanov *et al.*, 1987), South Korea (Lee *et al.*, 1987), the Netherlands (Oudemans, 1937; Vierbergen, 1990), Poland (Boczek and Kropczynska, 1964; Dobosz *et al.*, 1995), eastern Russia, Tajikistan (Mitrofanov *et al.*, 1987), Turkey (Düzgünes, 1965b), Ukraine (Mitrofanov *et al.*, 1987), the UK (Hirst, 1920), the USA (Garman, 1940), Uzbekistan (Mitrofanov *et al.*, 1987) and Yugoslavia (Tomašević, 1964).

Host plants

The list of host plants includes 24 species from six families: Betulaceae – *Betula papyrifera* Marshall (Bolland *et al.*, 1998; Migeon and Dorkeld, 2006–2013); Fabaceae – *Robinia pseudoacacia* L. (Migeon and Dorkeld, 2006–2013); Hamamelidaceae – *Hamamelis virginiana* L. (Bolland *et al.*, 1998; Migeon and Dorkeld, 2006–2013); Oleaceae – *Fraxinus* sp. (Bolland *et al.*, 1998; Migeon and Dorkeld, 2006–2013); Poaceae – *Elymus smithii* (Rydb.) Gould (Bolland *et al.*, 1998; Migeon and Dorkeld, 2006–2013); Salicaceae – *Populus* sp. (Koch, 1838; Reck, 1948b; Arru, 1961; Vierbergen, 1990; Gupta and Gupta, 1994), *P. alba* L. (Dobosz *et al.*, 1995), *P. balsamifera* L. (Bolland, 1999), *P. candicans* Aiton (Zacher, 1920), *P. deltoides* W. Bartram ex Marshall (Reeves, 1963), *P. fremontii* S. Watson, *P. gracilis* Grossh., *P. nigra* L. (McGregor, 1950), *P. tremula* L. (Reeves, 1963), *Salix* sp. (Zacher, 1920; Reck, 1948b), *S. alba* L. (Zacher, 1920), *S. babylonica* L. (Pritchard and Baker, 1955), *S. daphnoides* Vill. (Zacher, 1920), *S. discolor* Muhl. (Reeves, 1963), *S. fragilis* L. (Dobosz *et al.*, 1995), *S. interior* Rowlee, *S. pierothii*, *S. pseudolasiogyne* H. Lév., *S. viminalis* L. (Zacher, 1920).

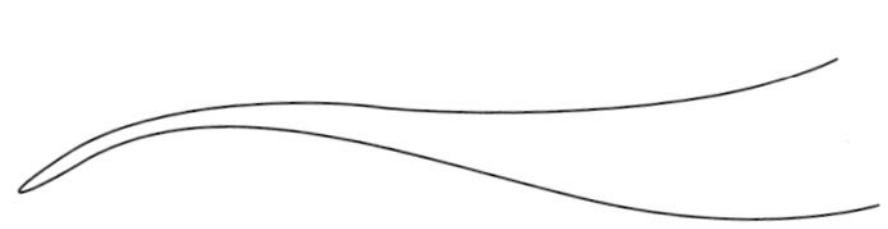

Fig. 20.26. *Eotetranychus populi* (Koch). Male, aedeagus (drawing by Vacante).

Bio-ecology

Despite its wide geographical distribution, the yellow poplar mite occurs mainly in Palaearctic countries, where it lives on ornamental plants, mainly Salicaceae. In the western Palaearctic region, the mite usually plays a secondary role as a pest of aspen, poplar and willow, and its presence does not require constant control. However, in north China, the mite is reported as an important pest of Salicaceae (Sun *et al.*, 1996), and recent studies carried out at Wuquan Mountain Park, Lanzhou, Gansu Province, indicated that the presence of *E. populi* was a very serious hazard (Liu *et al.*, 2011).

E. populi reproduces by arrhenotokous parthenogenesis and overwinters as the female on the trunk under old bark or in the forks of branches (Zacher, 1920). Sometimes, the females may be observed during mild winters among dry leaves and in the crown of the tree. From the end of March to mid-April, overwintering females move on to young leaves and start to feed and reproduce, and eggs may be found by late April. The mite reproduces on leaves until the first cold (Reeves, 1963; Tomašević, 1964). The same behaviour was observed in Tai'an City in western Shandong Province, China, with fertilized females overwintering in the fissures, galls and the holes of the willow tree. They began to appear in late March, when the average daily temperature reached 7°C and most of the willow buds unfold. The peak emergence occurred in early April. The peak occurrence in the year is from the middle of June to early July. The adult prefers crowding and webbing in the rolls of leaves and galls. According to the development at zero temperature, the effective accumulative temperature of the spider mite, and meteorological data from Tai'an City, there was an average of 11 generations, with a minimum of ten and a maximum of 12 (Sun *et al.*, 1997b).

In studies at 16, 24, 29 and 34°C, egg development lasted 7.1, 4.3, 2.5 and 2.2 days, respectively. At 24, 26 and 35°C, the immature development lasted 7.9, 4.9 and 4.4 days, respectively. At an average daily field temperature of 22°C, the eggs develop in 5–7 days, immature development lasts 3–5 days and the pre-oviposition period is 2–3 days (Reeves, 1963; Tomašević, 1964). In further experimental studies, the duration of development was shown to be negatively correlated with temperature in the range from 16 to 28°C. For the egg, the developmental zero was 8.77 ± 0.83°C and the effective accumulated temperature 100.0 ± 5.96 degree-days. The developmental zero from larva to adults was 9.64 ± 1.60°C and the effective accumulated temperature from larva to adults 113.3 ± 13.88 degree-days. The developmental zero for one generation was 8.67 ± 0.92°C and the effective accumulated temperature of one generation was 218.2 ± 14.37 degree-days (Sun *et al.*, 1996).

Symptomatology and damage

The first infestation develops on sucker growth, where the new developing leaves are heavily pubescent. The mites live on the

lower surface of these leaves and produce a layer of silk over the pubescence in order to form a covering above them. This webbing structure is reinforced by cast skins and dirt particles. Infested leaves become chlorotic or brown and are often curled (Reeves, 1963; Jeppson *et al.*, 1975). Usually, the damage does not require any control.

Control

Summer applications of dicofol or organophosphorus acaricides (Jeppson *et al.*, 1975), or other active ingredients, suppress mite outbreaks, although the need for control is infrequent. Acephate and pyridaben were succesfully tested in China (Du *et al.*, 2009).

Eotetranychus pruni (Oudemans) (Fig. 20.27)

Common name

Apple yellow mite, garden spider mite, chestnut spider mite.

Diagnostic characteristics

FEMALE. The body is yellow in colour, with lateral green spots. Overwintering females are intense yellow. The peritreme ends simply with three tubular parts. The genital flap and the anterior area are both transversally striated. Tibia I bears 8 tactile setae, 1 trobotrium and 1 solenidion, while tibia II has 8 tactile setae, and tarsus I has 5 tactile setae proximal to the duplex setae (Vacante, 1985a).

MALE. The male is smaller than the female, and lighter yellow in colour. Tibiae I and II are as in the female; the duplex setae of tarsus I are surrounded by 4 tactile setae. The aedeagus is large at the base and after a short curve downwards is straight and pointing up (Vacante, 1985a).

Geographical distribution

Algeria (Athias Henriot, 1959), Armenia (Mitrofanov *et al.*, 1987), Austria (Bohm, 1960), Azerbaijan (Mitrofanov *et al.*, 1987), Belgium (Lebdi-Grissa *et al.*, 2002), Bulgaria (Balevski, 1967), China (Ma and Yuan, 1980; Wang, 1980; Wang *et al.*, 1981), Cyprus (Georghiou, 1959), France (Rambier, 1958a), Germany (Sepasgosarian, 1956), Greece (Hatzinikolis, 1968c), Georgia (USSR) (Reck, 1948b; Mitrofanov *et al.*, 1987), Hungary (Bognar, 1961; Gyenis *et al.*, 2005), Iran (Sepasgosarian, 1975), Irish Republic (Curry, 1976), Italy (Carmona, 1960; Vacante, 1985a); Japan (Ehara, 1970), Kazakhstan (Wainstein, 1954, 1956a,b; Mitrofanov *et al.*, 1987), Moldova (Mitrofanov *et al.*, 1987), Morocco (Saba, 1973b), the Netherlands (Oudemans, 1931a; van de Vrie, 1965; Vierbergen, 1990), Poland (Boczek and Kropczynska, 1964), Romania (Iacob, 1978), eastern and western Russia (Mitrofanov *et al.*, 1987), Slovakia (Jedlicková and Kolodochka, 1994), Switzerland (Günthart and Günthart, 1959), Tajikistan (Mitrofanov *et al.*, 1987), Tunisia (Kreiter *et al.*, 2002), Turkey (Düzgünes, 1963), Ukraine (Mitrofanov *et al.*, 1987), the UK (Jeppson *et al.*, 1975) and the USA (Pritchard and Baker, 1955).

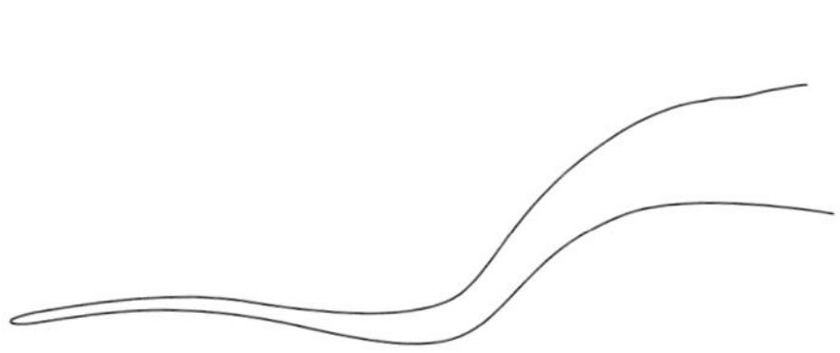

Fig. 20.27. *Eotetranychus pruni* (Oudemans). Male, aedeagus (drawing by Vacante).

Host plants

The list of host plants include 23 species belonging to 8 families; Betulaceae – filbert (*Corylus avellana* L.) (Vacante, 1985a), *C. heterophylla* Fisch. ex Trautv. (Wang, 1980); Fagaceae – *Castanea* sp., *Quercus dentata* Thunb. (Ehara, 1970), *Q. rubra* L. (Bolland *et al.*, 1998; Migeon and Dorkeld, 2006–2013); Malvaceae – *Tilia cordata* Mill. (Bolland *et al.*, 1998; Migeon and Dorkeld, 2006–2013); Rosaceae – *Cerasus vulgaris* Mill. (Georghiou, 1959), apple (*Malus domestica* Borkh.) (Sepasgosarian, 1956; Rambier, 1958a; van de Vrie, 1965; Vacante, 1985a; Kreiter *et al.*, 2002), sweet cherry (*Prunus avium* (L.) L.) (Vacante, 1985a), *P. divaricata* Ledeb., European plum (*P. domestica* L.) (Oudemans, 1931a; Rambier, 1958a; Hatzinikolis, 1968c), *P. insititia* L. (Hatzinikolis, 1968c), peach (*P. persica* (L.) Batsch) (Bolland and Beron, 2004), *P. spinosa* L., pear (*Pyrus communis* L.) (Rambier, 1958a; Vacante, 1985a), China rose (*Rosa chinensis* Jacq.) (Ma and Yuan, 1980); Salicaceae – *Salix* sp. (Wainstein, 1956b); Sapindaceae – *Acer platanoides* L. (Reck, 1950), *A. rubrum* L. (Pritchard and Baker, 1955), *A. saccharum* Marshall (Reeves, 1963), *Aesculus hippocastanum* L. (Reck, 1950; Gyenis *et al.*, 2005); Ulmaceae – *Ulmus* sp. (Bolland *et al.*, 1998; Migeon and Dorkeld, 2006–2013); Vitaceae – grapevine (*Vitis vinifera* L.) (Reck, 1948b, 1950).

Bio-ecology

E. pruni is a Palaearctic mite (which has also been reported in the USA) infesting apple, cherry, prune, grapevine, forest and ornamental trees (Jeppson *et al.*, 1975; Balevski, 1980; Telliev and Geshev, 1980; van de Vrie, 1985; Shahrokhi and Nikkhoo, 1990; AgroAtlas-Pests, 2003–2009a), and other plants.

The mite reproduces by arrhenotokous parthenogenesis, and overwinters as fertilized females, which are orange–yellow in colour, at various sites, such as in the axils of branches, under bark, inside cracks, under leaves or in grass at a distance 25 cm and less from the host plant stem. Overwintering populations may sometimes consist of a few hundred thousands of mites, and they can survive a short period of freezing to −20°C. In Russia and neighbouring countries, about 70% of the population usually dies in the frozen winter, with individuals surviving only under a blanket of snow. The mite resumes its activity in April to early May, and adults appear just before or concurrently with bud blossoming. In Transcaucasia, mites appear on the green parts of the vine in the first half of April, and mass migration to leaves occurs in the second half of the month. The mites first occupy leaves of the lower layer, then go higher. Mass oviposition occurs in the first half of May (Khalilov, 1972; Jeppson *et al.*, 1975; AgroAtlas-Pests, 2003–2009a).

E. pruni lives mainly on the lower leaf surface and lays eggs in among the webbing. In the field, the developmental cycle in the spring, summer and autumn takes about 27.4, 25.3 and 31.1 days, respectively. Jeppson *et al.* (1975) reports that the mite develops 3–4 generations a year, but according to AgroAtlas-Pests (2003–2009a), in Armenia and Azerbaijan, in Kazakhstan and in Tajikistan, it develops 8–9, 6–9 and 12 generations a year, respectively. Under greenhouse conditions, the female life cycle takes 30.5 days in the spring, 20.5 in the summer and 30 days in autumn, respectively, and the mite develops six generations a year. The egg development needs more than 40% of the total development time. The fecundity in the field and in the greenhouse is 25–61 eggs (average 37) and 27–98 eggs (average 48.1), respectively. The sex ratio (females:males) is 7:1 in the spring, 1.5:1 in the summer and 0.7:1 in the autumn (Jeppson *et al.*, 1975). In Russia and neighbouring countries, fecundity on the grapevine is about 30–100 eggs (AgroAtlas-Pests, 2003–2009a). In the Khorassan region of Iran, the mite develops 11 generations a year on apple. The females overwinter under the bark of parts of the tree in the soil and resume their activity in mid-April. The mean fecundity is 25 eggs. Populations may reach up to 500 mites per leaf after the end of June, when the temperature exceeds 35°C (Shahrokhi and Nikkhoo, 1990). In Russia and neighbouring countries, mass propagation of the mites occurs in July–August. Yellow overwintering females appear in the autumn, and the remaining population (juvenile stages, adults) dies after the first cold (AgroAtlas-Pests, 2003–2009a). Studies on the demographic traits of *E. pruni* in orchards show that the minimum generation time (*G*) is 14.2 days, calculated generation time (*T*) is 19.9 days and immature development time (*D*) is 13.7 days (Lebdi-Grissa *et al.*, 2002). The temperature threshold for development is about 12°C. The development of a generation needs 170 degree-days, and a generation lasts 15–40 days (AgroAtlas-Pests, 2003-2009a).

The optimum temperature for development is 27.5°C. Spring temperatures may affect summer and autumn populations of *E. pruni*. Thus, damp cool weather during May and June may result in low mite populations in August and September (Jeppson *et al.*, 1975). In Russia and neighbouring countries, it has been found that an air temperature of about 22–25°C and RH of 60–70% are the most favourable conditions for the mites. Low air moisture is a limiting factor (AgroAtlas-Pests, 2003–2009a). Outbreaks of *E. pruni* populations on grapevine in Bulgaria were associated with climate, cultural practices and the use of pesticides (parathion-methyl) against grape moths (Telliev and Geshev, 1980).

The distribution of the mobile forms is as an aggregate pattern, which occurs as loose colonies (mean size over 4–6 leaves). The distribution of the colonies is aggregate, but intra-colony distribution is obviously random (Luo *et al.*, 1994).

Natural enemies include predatory insects (anthocorids, coccinellids) and mites (*Typhlodromus* sp.) (Sepasgosarian, 1956). The phlaeothripid *Haplothrips subtilissimus* (Hal.) and the anthocorid *Anthocoris pilosus* Yakovlev were reported on fruit trees in Kirgizia (Kartasheva and Lesteva, 1979). In Russia and neighbouring countries (Armenia, Azerbaijan, Kazakhstan, Tajikistan), the population density of *E. pruni* in nature depends on the activity of predators, such as *T. reticulatus* Oudemans, *Phytoseius spoofi* Oudemans, *Scolothrips acariphagus* Sakh., *Scymnus* (*Stethorus*) *punctillum* Weise and *Chrysopa carnea* Stephens (AgroAtlas-Pests, 2003–2009a).

Symptomatology and damage

The feeding activity of *E. pruni* may induce premature leaf drop and reduction in fruit size (Jeppson *et al.*, 1975). Infestation of grapevine in Bulgaria caused bronzing of the leaves and other symptoms, and was deleterious to the chlorophyll system (Telliev and Geshev, 1980). Infested leaves undergo destruction of chlorophyll, and may fall off; they become yellowish on white varieties of the vine and reddish brown on black varieties. In addition, there may be a loss of grape yield by 15–60% and decreases of sugar content in grapes. The yield of apples may decrease to a third. It has been found that vine varieties with denser leaf pubescence are more resistant (AgroAtlas-Pests, 2003–2009a).

Control

Control is provided by the use of resistant varieties, and chemical and biological means. Regardless of the means, it is essential to rationalize the whole control strategy. Indeed, in addition to the influence of the climatic conditions, agronomic management and the cultivated varieties, it seems plausible that the mite outbreaks recorded in recent decades in some regions of the world are a result of the injudicious use of certain substances.

PLANT RESISTANCE. It has been found that grapevine varieties with denser leaf pubescence are more resistant, and the cultivation of resistant varieties is an important preventive control measure against the pest (AgroAtlas-Pests, 2003–2009a).

BIOLOGICAL CONTROL. The phytoseiid mite *Metaseiulus occidentalis* (Nesbitt) was released repeatedly in several apple orchards in Gansu Province, China, to control *E. pruni*. The predator became established in these orchards and played an important role in suppressing the mite populations. Chemical sprays were reduced by two or three times in these orchards compared with those in which the predator had not been released. The phytoseiid overwintered in ground crevices and under dry leaves and plant debris. The availability of food before and after overwintering, as well as the number of pesticide applications during the season, were critical in influencing the population fluctuations of the predator in the orchards (Deng *et al.*, 1990).

Introduction of an organophosphate-resistant strain of *M. occidentalis* to control *E. pruni* in the vineyards in the former USSR reduced pest numbers by about 85–96% (Petrushov, 1987). The same phytoseiid mite was introduced in New Zealand, where it became established but not did not give satisfactory mite pest control (Thomas and Walker, 1989). Further introductions of the phytoseiid in China were unsatisfactory (Z.-Q. Zhang, in Gerson *et al.*, 2003).

In apple orchards in north-west China, *E. pruni* is a less damaging spider mite than the two-spotted spider mite, *Tetranychus urticae* Koch. In China, both of these pest mites are preyed upon by the phytoseiid *Euseius finlandicus* (Oudemans), and field studies done to evaluate the impact of *E. pruni* and

E. finlandicus on *T. urticae* showed that the predator-mediated apparent competition strongly affected the population dynamics of *T. urticae*. Curiously, the addition of the apparent competitor *E. pruni* alone has little impact on *T. urticae* densities. Although the release of *E. finlandicus* alone could result in a reduction of *T. urticae* densities, the greatest reduction in *T. urticae* densities occurs when both the predator *E. finlandicus* and apparent competitor *E. pruni* are released. In apple orchards, the early introduction of both the apparent competitor *E. pruni* and *E. finlandicus* would evidently increase the population size of the predator *E. finlandicus*, and consequently significantly enhance the control of *T. urticae* (Liu *et al.*, 2006a).

CHEMICAL CONTROL. Sprays of acaricides applied during summer to control other spider mites in apple orchards are also effective against *E. pruni*. Winter sprays, are usually partially effective against overwintering females (Sepasgosarian, 1956). On grapevines in Azerbaijan, spray applications of dimethoate and dicofol were effective against the mobile forms of *E. pruni* (Khalilov, 1972). On vineyards in Bulgaria, sprays of kilakar (4-chloro- alpha -(4-chlorophenyl)- alpha -cyclopropylbenzenemethanol) and a mixture of dicofol and dinocap applied at intervals of 10–15 days gave good results, as did the application of propargite and dimethoate at high temperatures (24–25°C) (Telliev and Geshev, 1980). In a field trial in Iran on apple, dicofol, propargite, binapacryl, bromopropylate and tetradifon were equally effective against *E. pruni* (Shahrokhi and Nikkhoo, 1990). In Russia and neighbouring countries, control measures against mite infestations also include treatments by systemic acaricides at pest mass abundance (AgroAtlas-Pests, 2003–2009a).

Eotetranychus sexmaculatus (Riley) (Fig. 20.28)

Common name

Six-spotted spider mite.

Diagnostic characteristics

FEMALE. The body is ovoid, about 300 μm long and lemon yellow in colour, commonly with blackish spots organized in 3 blotches (sometimes inconspicuous or absent) along each side of the body. Tibiae I and II bear 9 and 8 tactile setae, respectively. Tarsus I has 5 tactile setae proximal to the duplex setae. The peritreme is hooked distally. The genital flap is longitudinally striate on the anterior central part and on the area anterior to the flap (Jeppson *et al.*, 1975).

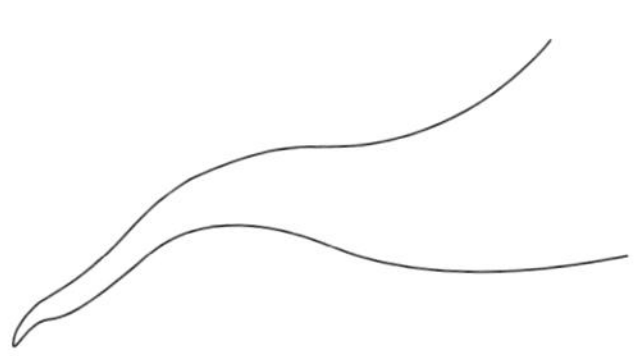

Fig. 20.28. *Eotetranychus sexmaculatus* (Riley). Male, aedeagus (modified from McGregor, 1950).

MALE Tibiae I and II of the male are similar to those of the female. The aedeagus is curved dorsally near the middle of the shaft, and distally directed caudoventrally with the tip deflexed (Jeppson *et al.*, 1975).

Geographical distribution

Australia (Froggatt, 1921), China (Wang, 1980; Wang *et al.*, 1985), Hainan Island (China) (Ma *et al.*, 1979), Hawai (Garrett and Haramoto, 1967), South Korea (Lee, 1989), India (Gupta and Gupta, 1976, 1994), Iraq (Mohamed and El-Haidari, 1965), Japan (Ehara, 1956c, 1975a), New Zealand (Manson, 1983), Peru (Guanilo *et al.*, 2012), Taiwan (Pritchard and Baker, 1955; Ehara, 1969a) and the USA (Riley, 1890; Pritchard and Baker, 1955).

Host plants

The list of host plants includes 43 species from 25 families (Migeon and Dorkeld, 2006–2013). The families and species of economic importance are: Actinidiaceae – kiwi (*Actinidia deliciosa* (A. Chev.) C.F. Liang *et* A.R. Ferguson) (Bolland *et al.*, 1998; Migeon and Dorkeld, 2006–2013); Apocynaceae – *Plumeria* spp. (Migeon and Dorkeld, 2006–2013); Ebenaceae – Japanese persimmon (*Diospyros kaki* L.f.) (Ehara, 1969b); Euphorbiaceae – poinsettia (*Euphorbia pulcherrima* Willd. ex Klotzsch) (Garrett and Haramoto, 1967), rubber tree (*Hevea brasiliensis* (Willd. ex A. Juss.) Müll. Arg.) (Bolland *et al.*, 1998; Migeon and Dorkeld, 2006–2013); Fabaceae carob tree (*Ceratonia siliqua* L.) (McGregor, 1950); Lauraceae – avocado (*Persea americana* Mill.) (McGregor, 1950); Myrtaceae – guava (*Psidium guajava* L.) (Garrett and Haramoto, 1967; Wang, 1980); Rosaceae – beach strawberry (*Fragaria chiloensis* (L.) Mill. (= × *ananassa* Duchesne ex Rozier)) (Gupta, 1984), apple (*Malus domestica* Borkh.) (Gupta and Gupta, 1994), peach (*Prunus persica* (L.) Batsch) (Gupta and Gupta, 1994); Rutaceae – pomelo (*Citrus grandis* (L.) Osbeck), lemon (*C. limon* (L.) Burm.f.), sweet orange (*C. sinensis* (L.) Osbeck) (Pritchard and Baker, 1955); Vitaceae – grapevine (*Vitis vinifera* L.) (McGregor, 1950).

Bio-ecology

The six-spotted mite is a pest of citrus, avocado and ornamentals, such as azalea, camphor, *Elaeagnus*, maple, pyracantha, etc. In Florida, it was recorded attacking oranges as early as 1885; in California, it infests citrus only in coastal areas (McGregor, 1956; Muma, 1961; Jeppson, 1989; Timmer and Duncan, 1999). In New Zealand, the mite is a key pest of avocados (Stevens, 2001; Steven, 2004; Jamieson and Stevens, 2007), while in California, it only occasionally infests this host plant, mostly near the coast in foggy areas. The mite is generally under good control in the interior areas because of predators and warm weather, but it can become a problem anywhere if trees are drought stressed or pesticides with wide side effects are employed to control other pests, according to the UC Pest Management Guidelines for the six-spotted mite on avocado (UC IPM Online, 2010). Sometimes, *E. sexmaculatus* causes severe damage to rubber trees in China (Lin *et al.*, 1985).

E. sexmaculatus lives on the lower leaf surfaces, near the midrib or the main veins and it infests the fruit only at high

densities. It produces abundant webbing that covers its populations, protecting them from predators and offering support for the eggs. The sex ratio (females:males) in the progeny of fertilized females is 2.5:1. Fertilized females reproduce by arrhenotoky and fecundity reduces as the age of the female increases. The number of females in the progeny of newly emerged and fertilized females is relatively high, but decreases with increase in age of the females (Yang, 1987). The pre-oviposition period lasts 2–3 days, and each female lays 25–40 eggs in 10–20 days (Jeppson *et al.*, 1975). The number of eggs laid by fertilized females (35.16 ± 6.90) is greater than that laid by non-mated females (20.95 ± 3.85) (Yang, 1987). In summer, egg development takes 5–8 days and the developmental stages take 2–3 days each (Jeppson *et al.*, 1975). On avocado leaves in the laboratory, no larvae hatched from eggs at 10°C. Egg development lasts 5.3 days at 30°C and 23.9 days at 13°C. The developmental time from egg to adult is 29.6 days at 18°C and 11 days at 30°C. The mean adult longevity ranges from 18.8 days at 30°C to 41.4 days at 18°C, and the fecundity ranges from 6.9 to 20.9. The threshold temperature is 12.2°C, and 202.8 degree-days are required to complete development from egg to adult (Jamieson and Stevens, 2007).

In southern California, the six-spotted spider mite is restricted to coastal areas, and although at times it can be a serious pest, it commonly does not produce severe damage (McGregor, 1956). Its population dynamics are adversely influenced by dry weather conditions and the populations reach high levels only in the more humid coastal regions. Dry winds hinder the development of the mite populations, which increase gradually during winter and more rapidly in spring and summer (Jeppson *et al.*, 1975; Jeppson, 1989). In China, the time of yearly population peak on rubber trees is largely determined by the amount of natural enemies and rainfall. The population usually begins to increase in mid-April and reaches a peak during late April and May. There is a sharp decline in the population in late May and June (Lin *et al.*, 1985). Dispersal of mites from infested plants is via silk threads.

The webbing covering the colonies probably protects the pest from predators, and the increase of prey that it represents is a source of food for predators. The action of phytoseiid mites, together with unfavourable weather conditions, reduce the mite to a pest of occasional importance on citrus in California and Florida (Jeppson *et al.*, 1975). Several natural enemies (fungi, insects and mites) develop on *E. sexmaculatus* populations (Table 20.8).

Symptomatology and damage

On citrus, the six-spotted spider mite feeds on the undersides of leaves and does not normally live on the fruit, except in high infestations. Leaves near the ground are attacked first, and only after some time may leaves more than 4–5 feet from the ground be injured (Jeppson *et al.*, 1975). Mite feeding from the lower leaf surface evacuates the spongy mesophyll cells partly or completely. Several weeks after damage, there is some recovery, and the damage appears to be the result of physical factors or carbohydrate depletion, but not the action of toxins or growth stimulants (Albrigo *et al.*, 1983). Feeding on the lower leaf surfaces of citrus produces yellow depressions on the undersides of the leaves covered by webbing. The areas of the upper surfaces corresponding to the sides of the lower surfaces where the mite colonies live become raised and yellow or yellow white, with a smooth, skinny surface. With increase in infestation, the yellowish areas converge and the leaves become entirely yellow, distorted or misshapen, and drop prematurely (McGregor, 1956; Jeppson *et al.*, 1975; Jeppson, 1989).

On avocado, the mite feeds only on the lower leaf surface, and causes irregular brown to purplish discoloration, mostly along the midrib and larger veins. It does not produce dense webbing (UC IPM Online, 2010). In New Zealand, excessive leaf drop and a subsequent reduction in tree productivity and yield are reported (Stevens, 2001; Jamieson and Stevens, 2007).

On *Plumeria*, a cultivated garden plant, feeding by the six-spotted spider mite causes brown to purple leaf discoloration. Feeding symptoms first occur on the lower leaf surface along the midrib, at the base of the leaf petiole, or along the midrib of new leaves. The mite damages the bloom clusters, induces premature leaf drop and affects plant vigour. Attacks during the course of new growth may induce abscission of developing shoot tips from the plant. Further generations of mites may spread on to the upper leaf surface, causing a blistering near the midrib, and producing webs linking the midrib to the leaf surface. At high mite density, the web spreads between the stem and petioles. When populations increase, the mites invade leaf tips and may be collected on silk threads from the edges of the uppermost leaves. Mite feeding prevents the leaf petiole glands from secreting, and damaged leaf tissues may exude latex. The mite usually prefers tender leaf growth, but it also attacks tender stem tissue, even after the shoot tips have abscised (UC IPM Online, 2007a).

Control

The sampling and monitoring of populations rationalize the use of chemical and biological means. Some cultural practices exert a positive influence.

SAMPLING AND MONITORING. For avocado, trees should be selected from the dusty and more humid parts of groves. A hand lens should be used to examine along the midrib and lateral veins on the underside of interior canopy leaves to detect the presence of brown to purplish discoloration, mite webbing and mites. The six-spotted spider mite can severely stress trees at relatively low densities by causing premature leaf drop. However, populations rarely develop more than two or three mites per leaf, and this amount does not represent a risk of damage (UC IPM Online, 2010). On plumeria, when mites are controlled early in the season, the plants produce new leaves and recover during the course of the season. During summer, plants should be regularly observed to detect webbing along the midrib on the upper leaf surface and particularly towards the leaf petiole. A hand lens helps to observe the mites and their eggs. Plants should be treated if six-spotted mites are present. The detection of webbing and mites in the autumn suggests that plants should be treated before moving them indoors for the winter. Monitoring of the plants during late autumn, winter and early spring allows the mites to be treated before they cause severe

Table 20.8. *Eotetranychus sexmaculatus* (Riley). Natural enemies (fungi, insects and mites).

Families and species	References
FUNGI	
Clavicipitaceae	
Hirsutella thompsonii Fisher	McCoy and Selhime, 1974
Entomophthoraceae	
Entomophthora sp.	Muma, 1969
INSECTS	
Aeolothripidae	McMurtry *et al.*, 1970; Steven, 2004; Jamieson and Stevens, 2007
Scolothrips sexmaculatus (Pergande)	
Chrysopidae	
Chrysopa lateralis Guérin	
Coccinellidae	
Stethorus sp.	
Stethorus utilis Horn	
Coniopterygidae	
Coniopteryx vicina Hagen	
MITES	
Anystidae	
Anystis baccarum (L.)	Jamieson and Stevens, 2007
Phytoseiidae	
Amblyseius herbicolus (Chant)	Chang and Chen, 1984
Amblyseius largoensis (Muma)	Lin *et al.*, 1985
Amblyseius maai Tseng	Chang and Chen, 1984
Amblyseius newsami (Evans)	Muma, 1970
Chelasius floridanus (Muma)	
Euseius hibisci (Chant)	Tanigoshi *et al.*, 1981; McMurtry, 1985b
Euseius mesembrinus (Dean)	McCoy and Rakha, 1985; Abou-Setta and Childers, 1987, 1989
Euseius ovalis (Evans)	Chang and Chen, 1984
Euseius tularensis Congdon	Hoy, 2011
Galendromus helveolus (Chant)	Caceres and Childers, 1991
Metaseiulus occidentalis (Nesbitt)	Solow and Sherman, 1997
Neoseiulus collegae (De Leon)	Mizell III and Schiffhauer, 1991
Neoseiulus taiwanicus (Ehara)	Chang and Chen, 1984
Neoseiulus womersleyi (Schicha)	Osakabe, 2002
Typhlodromalus limonicus (Garman *et* McGregor)	McMurtry, 1985b
Typhlodromips okinawanus (Ehara)	Chang and Chen, 1984
Stigmaeidae	
Agistemus exsertus González-Rodríguez	Wang, 1981c; Tna, 1986
Agistemus longisetus González-Rodriguez	Jamieson and Stevens, 2007
Agistemus terminalis (Quayle)	Wang, 1981c

damage. Double-sided sticky tape wrapped around the plant stem may prevent overwintering mites from crawling up the stem (UC IPM Online, 2007a). On rubber trees, from four to eight mites per leaf at the start of the population peak suggests that chemical spraying should be done (Lin *et al.*, 1985).

CULTURAL PRACTICES. On avocado, oiling or paving orchard roads to control road dust is a good practice. Dirt roads should only be driven on slowly and, when possible, a water truck or trailer used to wet dirt roads before travel during the summer. Individual backyard trees can be hosed down in the early to midsummer to remove dust and enhance biological control (UC IPM Online, 2010). Plumeria is a container plant and dust is not usually a problem, but avoiding water stress is good cultural practice for reducing mite outbreaks (UC IPM Online, 2007a).

BIOLOGICAL CONTROL. In California, the six-spotted spider mite is controlled on avocado primarily by phytoseiids, including *Typhlodromalus limonicus* (Garman *et* McGregor) and *Galendromus helveolus* (Chant). *Euseius hibisci* (Chant) plays an important role because it feeds on avocado pollen when pest mites are scarce. *Typhlodromus rickeri* Chant also preys on *E. sexmaculatus*. The coccinellid *Stethorus picipes* Casey and the thrips *Scolothrips sexmaculatus* (Pergande) are other important natural enemies (UC IPM Online, 2010). On plumeria, predatory mites may reduce populations of the pest mite naturally, but the threshold for *E. sexmaculatus* is very low and does not allow their use (UC IPM Online, 2007a).

In New Zealand, natural enemies do not appear to be able to prevent unacceptable outbreaks of *E. sexmaculatus* on avocado (Jamieson and Stevens, 2007). In addition, the fungi *Verticillium lecanii* (Zimmerman) Viégas, *Metarhizium anisopliae* (Metchnikoff)) Sorokin and *Beauveria bassiana* (Balsamo) Vuillemin were ineffective (Steven, 2003).

CHEMICAL CONTROL. It is possible to use the same acaricides against the six-spotted spider mite as those used against the citrus red mite. Petroleum oils (Ebeling, 1959; Jeppson *et al.*, 1975) or sulfur reduce populations (Jeppson, 1989). On rubber trees, sprays of tetradifon reduced the mite population by 80–90% (Lin *et al.*, 1985). In New Zealand, one or two sprays from October to December with abamectin, fenpyroximate, milbemectin or oil were effective on avocado (Steven, 2003, 2004), as were bifenazate and potassium salts combined with fatty acids in a vegetable oil base (Froud *et al.*, 2002). On plumeria, a light horticultural oil spray is effective if all of the plant surfaces are covered. Insecticidal soaps can be used as an alternative to oil but it is more likely that repeat applications will be necessary. When using soaps or oils, plants should be well watered to avoid stress, and caution must be taken not to expose the plants to excessive heat and/or direct sunlight before or after treatment (UC IPM Online, 2007a). On avocado, the botanical pesticides clove oil, garlic and cottonseed oil are not effective; neither are endosulfan, neem oil and sucrose octanoate (Steven, 2003). Trials in China show that abamectin, chlorfenapyr and lamba-cyalothrin are effective against *E. sexmaculatus*, but the mite has high resistance to azocyclotin, chlorpyrifos, fenpropathrin, propargite and pyridaben (Hao *et al.*, 2009).

INTEGRATED PEST MANAGEMENT. Implementation of IPM needs the enhancement of biological control by conserving natural enemies, the minimization of dust and avoidance of the application of non-selective pesticides toxic to natural enemies. Any applications needed should be limited to spots where the pests are most abundant (UC IPM Online, 2010). In New Zealand on avocado trees treated mainly with selective insecticides, there were more ladybirds and in some instances more predatory mites and fewer *E. sexmaculatus*. In cases where the pest mite populations were low, tydeid mites (*Orthotydeus* spp.) became more common. Some broad-spectrum insecticides appeared to have activity against *E. sexmaculatus* (Tomkins, 2002).

Eotetranychus smithi Pritchard et Baker

Common name

Unknown.

Diagnostic characteristics

FEMALE. The peritreme is hooked distally. The genital flap is transversely striate, and the area anterior to the flap is longitudinally striate. Tibiae I and II have 9 and 8 tactile setae, and tarsus I has 4 tactile setae proximal to the duplex setae (Jeppson *et al.*, 1975).

MALE. The tactile setae of tibiae I and II and of tarsus I are as in the female. The aedeagus has the proximal half with the ventral side directed dorsally; the distal part narrows and forms a fine stylet directed caudally (Jeppson *et al.*, 1975).

Geographical distribution

China (Wang, 1980; Wang *et al.*, 1985), Japan (Ehara, 1960; Gotoh *et al.*, 2003a), South Korea (Lee *et al.*, 1987), Madagascar (Gutierrez, 1967), Seychelles (Bolland *et al.*, 1998), Philippines (Corpuz-Raros, 1989b), Taiwan (Tseng, 1990) and the USA (Pritchard and Baker, 1955).

Host plants

The whole list of host plants includes 23 species from six families (Migeon and Dorkeld, 2006–2013). The families and species of economic importance are: Juglandaceae – Manchurian walnut (*Juglans mandshurica* Maxim.) (Bolland *et al.*, 1998; Migeon and Dorkeld, 2006–2013); Rosaceae – loquat (*Eriobotrya japonica* (Thunb.) Lindl.), apple (*Malus domestica* Borkh.), peach (*Prunus persica* (L.) Batsch) (Pritchard and Baker, 1955), American red raspberry (*Rubus idaeus* L.), Japanese raspberry (*R. parvifolius* L.) (Wang, 1980), wine raspberry (*R. phoenicolasius* Maxim.) (Pritchard and Baker, 1955); Vitaceae – grapevine (*Vitis vinifera* L.) (Ehara, 1960).

Bio-ecology

E. smithi lives on various host plants and seems to prefer rosaceous species, such as rose and different berries (*Rubus* spp.); it has also been reported as a pest of grapevines in Japan (Ehara, 1960, 1964a) and cotton in Tennessee (Caldwell, 1967; Jeppson *et al.*, 1975). It reproduces by arrhenotokous parthenogenesis and has a haploid number of chromosomes of five (Helle *et al.*, 1970). On cotton, it lives on the leaf surface near the midrib and produces little webbing (Jeppson *et al.*, 1975). Natural enemies include the aeolothripid *Scolothrips hartwigi* Priesner (Chazeau, 1985)

Symptomatology and damage

Damage on cotton is similar to that produced by *Tetranychus* sp. (Jeppson *et al.*, 1975).

Control

Few reports highlight the need for control. Ehara (1960) and Caldwell (1967) have suggested spray applications of acaricides.

Eotetranychus uncatus Garman

Common name

Walnut spider mite, birch spider mite.

Diagnostic characteristics

FEMALE. The body is flesh coloured or pale lemon yellow in colour, with a tinge of green and small black spots posterior to the eyes.

The peritremes are strongly hooked distally. The genital flap and the area anterior to the genital flap are both transversely striate. Tibiae I and II bear 9 and 8 tactile setae, respectively, and tarsus I has 5 tactile setae proximal to the duplex setae (Jeppson *et al.*, 1975).

MALE. The aedeagus is long, slender, tapers distally and is strongly undulate (Jeppson *et al.*, 1975).

Geographical distribution

Canada (Montréal Biodôme, 1999), China (Wang, 1981a), France (Migeon *et al.*, 2007), India (Gupta, 1985; Gupta and Gupta, 1994), Japan (Ehara, 1970; Gotoh *et al.*, 2003a), Kazakhstan (Wainstein, 1956a; Mitrofanov *et al.*, 1987), Mexico (Estebanes-Gonzalez and Baker, 1968), the Netherlands (Gutierrez and Helle, 1981), Paraguay (Aranda and Flechtmann, 1971), Poland (Boczek and Kropczynska, 1964; Dobosz *et al.*, 1995), Turkey (Yanar and Ecevit, 2005; Yanar, 2012) and the USA (Pritchard and Baker, 1952a; Ubertalli, 1955; Thewke and Enns, 1969).

Host plants

The walnut spider mite is reported on 38 host plants of 11 families, among which the Betulaceae and the Rosaceae have 11 and nine species, respectively (Migeon and Dorkeld, 2006–2013). The families and species of economic importance are: Juglandaceae – Japanese walnut (*Juglans ailantifolia* Carrière) (Ehara, 1970), English walnut (*J. regia* L.) (Bolland *et al.*, 1998; Migeon and Dorkeld, 2006–2013); Rosaceae – apple (*Malus communis* L. (= *Malus domestica* Borkh.)) (Yanar and Ecevit, 2005; Yanar, 2012), apple (*M. domestica* Borkh.) (Pritchard and Baker, 1952a; Ubertalli, 1955; Migeon *et al.*, 2007), sweet cherry (*Prunus avium* (L.) L.), cherry plum (*P. cerasifera* Ehrh.) (Migeon *et al.*, 2007), sour cherry (*P. cerasus* L.) (Reeves, 1963), European plum (*P. domestica* L.) (Skorupska, 1976), peach (*P. persica* (L.) Batsch) (Estebanes-Gonzalez and Baker, 1968); Theaceae – tea (*Camellia sinensis* (L.) Kuntze) (Gupta, 1992).

Bio-ecology

In the USA, *E. uncatus* was reported for a long time only in Massachusetts and Connecticut, where it damaged apple and stone fruits; it has also been reported in Utah and southern California (Pritchard and Baker, 1955; Jeppson *et al.*, 1975). The mite is known on a number of ornamental plants as well, such as birch (*Betula*) (Pritchard and Baker, 1955; Reeves, 1963; Dobosz *et al.*, 1995; Gotoh *et al.*, 2003a), hornbeam (*Carpinus*), linden (*Tilia*) (Reeves, 1963), alder (*Alnus*) (Reeves, 1963; Dobosz *et al.*, 1995), maple (*Acer*) (Dobosz *et al.*, 1995), elm (*Ulmus*) and other plants. Although *E. uncatus* has been reported as a serious pest of apple and stone fruits in the eastern USA, its real harmfulness has been confirmed only in a restricted number of cases that concerned especially apple orchards sprayed with fungicides (captan) and insecticides (DDT) (Oatman, 1973). In Turkey, the mite usually lives in unsprayed apple orchards (Yanar and Ecevit, 2008), and could be susceptible to the pesticides that are commonly sprayed in the apple orchards (Yanar, 2012). In Poland, *E. uncatus* occurs in 23.3% of samples of infested leaves. Most frequently, it inhibits apple and plum trees; among wild plants, it is found on cherry plum and hawthorn (Skorupska, 1989). In Canada, the mite does not cause any significant damage to birches in the wild (Montréal Biodôme, 1999).

E. uncatus reproduces by arrhenotokous parthenogenesis and has a haploid number of chromosomes of three (Bolland *et al.*, 1981). Its populations live mainly on the lower leaf surface, where they spin conspicuous webbing; they are categorized in the life type WN-u (Saitō, 1985). Females overwinter under bark or in other protected shelters on the tree. In spring, during the delayed dormant stage of bud development, the mites leave these sites and begin their activity, moving towards the spur leaves to feed before they lay eggs. At 21–22°C and 27–28°C, egg development lasts 9 and 2 days, respectively; the development of each stage takes 6 and 2 days, and the time from hatching to adult is 20 to 5 days, respectively; pre-oviposition lasts 6 to 2 days, and the adult lifespan is 14 to 8 days, respectively (Pritchard and Baker, 1952a; Ubertalli, 1955; Reeves, 1963). On apple leaf discs under controlled conditions, at 25 ± 1°C, 16L:8D photoperiod and RH 65 ± 5%, the mean developmental time of eggs for the female and male was 5.33 ± 0.15 and 4.89 ± 0.13 days, respectively. The developmental time of the female larva was 1.10 ± 0.07 days and that of the male larva 1.02 ± 0.09 days. The developmental time of the female protonymph was 1.10 ± 0.05 days and that of the male protonymph 1.07 ± 0.05 days; respective figures for the female and male deutonymphs are 1.03 ± 0.03 and 0.96 ± 0.03 days. The total developmental times from egg to adult for the female and male (also including the protochrysalis, deutochrysalis and teleiochrysalis stages) were 13.65 ± 0.30 and 11.82 ± 0.30 days, respectively. For the female, the pre-oviposition period was 2.95 ± 0.13 days and adult longevity 14.33 ± 0.19 days. The fecundity was 26.30 ± 0.69 eggs and the daily oviposition 2.75 ± 0.08 eggs. The net reproductive rate (R_0) was 17.81, the r_m 0.335 and the mean generation time (T_0) 8.58 (Yanar, 2012). On *B. platyphylla* Sukat. var. *japonica* (Miq.) Hara at 21 ± 1°C and with a 15L:9D photoperiod, the developmental time from egg to adult was 11.38 ± 0.13 days for the female and 10.30 ± 0.25 for the male. The fecundity was 39.53 ± 2.29 eggs and the daily eggs laid per female were 3.39 ± 0.15 (Gotoh, 1987a). On *B. variegata* at 26.6°C, the developmental times for the egg, larva, protonymph and deutonymph were 2.7, 2.30, 1.68 and 1.49 days, respectively. The developmental time from egg to adult is 8.01 days. At 30.6°C, the fecundity was 21.2 eggs and at 22.47°C it was 12.20 eggs. The lifespan of the adult female was 9.10 days (Lal and Mukharji, 1978).

Symptomatology and damage

The walnut spider mite feeds mainly on the lower leaf surface and induces uneven growth of both leaf surfaces, which results in cupping or crinkling of the leaves. Feeding also causes stippling, yellowing and bronzing of the leaves. Symptoms of spring infestations to new apple leaves are similar to those of frost damage. Heavy levels of attack may induce defoliation and reduction of fruit size. The feeding activity of *E. uncatus* on the lower leaf surface of alder causes yellow patches on the upper surfaces. Damage in the early growing season may induce the leaves to cup or to crinkle (Reeves, 1963; Jeppson *et al.*, 1975; Gotoh, 1987b). Feeding

of *E. uncatus* on leaves of *B. variegata* causes burning of the leaves as a toxigenic effect on the plant. It is possible that a toxic component is injected into the plant (Lal and Mukharji, 1979).

Control

The need for control is uncertain.

Eotetranychus willamettei (McGregor)

Common name

Willamette mite.

Diagnostic characteristics

FEMALE. The terminal sensillum of the palp is about 4 times as long as wide. The peritreme ends in a simple bulb. Tibiae I and II bear 9 and 8 tactile setae, respectively, and tarsus I has 5 tactile setae proximal to the duplex setae. The duplex setae (in both sexes) are more equal in length than in congeneric species (Jeppson *et al.*, 1975).

MALE. The aedeagus is long, sinuous and tapers distally (Jeppson *et al.*, 1975).

Geographical distribution

Iran (Sheikholeslam Zadeh and Sadeghi Namaghi, 2010b) and the USA (McGregor, 1917; Stavrinides *et al.*, 2010a; Stavrinides and Mills, 2011).

Host plants

The Willamette mite is reported on 11 host plants from six families: Fagaceae – *Quercus garryana* Douglas ex Hook. (McGregor, 1917), *Q. lobata* Née (Bolland *et al.*, 1998; Migeon and Dorkeld, 2006–2013); Polygonaceae – *Eriogonum jamesii* Benth. (Bolland *et al.*, 1998; Migeon and Dorkeld, 2006–2013); Rosaceae – *Amelanchier* sp., apple (*Malus domestica* Borkh.), *Purshia tridentata* (Pursh.) DC., pear (*Pyrus communis* L.), service tree (*Sorbus domestica* L.) (Bolland *et al.*, 1998; Migeon and Dorkeld, 2006–2013); Sapindaceae – *Acer negundo* L. (Bolland *et al.*, 1998; Migeon and Dorkeld, 2006–2013); Ulmaceae – *Ulmus* sp. (Bolland *et al.*, 1998; Migeon and Dorkeld, 2006–2013). Vitaceae – grapevine (*Vitis vinifera* L.) (Stavrinides *et al.*, 2010a; Stavrinides and Mills, 2011).

Bio-ecology

E. willamettei has a Nearctic distribution and is known mainly as a pest of grapevines in central and northern California, and Washington and Oregon (Jeppson *et al.*, 1975; Geddes, 2010). In Californian vineyards, the mite is found especially early in the spring (Flaherty 1969; Hanna *et al.*, 1997a), but it causes minimal damage to most grape varieties; it serves mostly as important prey for the phytoseiid *Metaseiulus occidentalis* (Nesbitt), especially early in the growing season (Hoy, 2011). Recently, it was reported (as *Eotetranychus willamettei* Ewing) on broadleaved trees in green spaces of the Mashhad region in Iran (Sheikholeslam Zadeh and Sadeghi Namaghi, 2010b).

The Willamette mite overwinters as the female (which is amber to yellow in colour), either singly or in groups, under the bark of 3-year-old or older vines and on the undersides of vines; only a few individuals reach the main trunk. When disturbed, overwintering females become active, but it is not known whether they feed after disturbance. In the spring, when the spurs swell, the mite moves toward them and waits under the bark on the 2-year-old wood until the leaves are expanded. The mite population then migrates to the swelling buds or leaves. These females are a much paler yellowish white than the winter females and start to feed in colonies on the upper leaf surface of grapevines, spinning a sparse web. The eggs are laid anywhere on the lower leaf surface. Egg development takes 7 days or less and the total juvenile development lasts about 10 days. About 15 days after migration to the leaves, some of the overwintering females leave the basal two or three leaves and move towards the young leaves. The mite populations are active on grapevine leaves until the beginning of November, when they start to overwinter (Jeppson *et al.*, 1975).

Natural enemies include the phytoseiid mite *M. occidentalis* (Flaherty 1969; Kinn and Doutt 1972; Hanna *et al.*, 1997a,b).

Symptomatology and damage

The Willamette mite lives in colonies on the upper leaf surface of grapevines, where damaged areas turn straw in colour and become scarious. On the upper leaf surfaces, this symptom appears as chlorotic green or straw bronze discoloration. Feeding initially affects the leaf surface between the larger veins, and a narrow strip of tissue close to the larger veins appears green and gives the leaf a mottled aspect. Damage on the buds or leaves is well defined owing to the rapid growth of the plant. This symptom allows the damage caused by the Willamette mite to be distinguished from that of the Pacific spider mite, *Tetranychus pacificus* McGregor, which does not cause damage in early spring but only on mature leaves in midsummer. The increase in the mite population is associated with a gradual expansion of the leaf yellow spot until the yellow centre shades into the green tissue. Damage on tender developing shoots consists of dwarfing and deformation of leaves, angularity at nodes, short internodes and spindle stems. The tips of the slow-growing axial shoot may be killed by summer or autumn. Severe infestations cause short canes, irregular fruiting, few clusters and berries, and small berries; the berries lack uniformity, and are of poor quality and delayed in maturity (Jeppson *et al.*, 1975).

In California, different grape varieties exhibit a different response to *E. willamettei*. Infestations cause relatively little harm to vigorous grape varieties ('Thompson seedless', etc.) and the presence of the mite is positively evaluated because it is important prey for the phytoseiid *M. occidentalis* (Flaherty 1969; Kinn and Doutt 1972; Hanna *et al.*, 1997a,b). However, *E. willamettei* can cause economic injury to wine grape cultivars with fewer leaves (Hanna *et al.*, 1997a,b).

Feeding damage by *E. willamettei* reduces grape leaf photosynthesis and stomatal conductance. There is a negative linear regression between feeding damage and grape leaf gas exchange, and the mite has comparable effects on field photosynthetic

rates and the stomatal conductance of grapevines. Photosynthetic rates of undamaged adjacent leaves are not affected by mite feeding damage, whereas stomatal conductance appears to increase in undamaged adjacent leaves. In greenhouse trials, significant interaction effects occur between cultivar type and stomatal conductance, but not between cultivar type and photosynthesis (Welter *et al.*, 1989b).

Control

Means of control include placing of bands of sticky material at the base of grape spurs and summer sprays of acaricides (Frazier and Smith, 1946). Unsuccessful tests have been performed of the 'induced resistance to mite feeding technique' (Karban and English-Loeb, 1990), and of genotypic variation in constitutive and induced resistance in grapes (English-Loeb *et al.*, 1998) (see below). Rational fertilization may hinder mite outbreaks.

INDUCED RESISTANCE TO MITE FEEDING. In vineyards of the San Joaquin Valley of California infested by the Willamette spider mite, outbreaks of the Pacific spider mite (*T. pacificus*), a more serious pest, are modest or inconsistent (English-Loeb and Karban, 1988). The introduction of *E. willamettei* into a commercial vineyard with a long history of infestation by *T. pacificus* significantly reduced the populations of the Pacific spider mite on rows in which the releases were made compared with the control rows. However, in the following year, no differences between the two treatments were observed (Karban and English Loeb, 1990).

PLANT RESISTANCE. Genotypic variation in constitutive and induced resistance in grapes against *E. willamettei* and *T. pacificus* have been investigated. Early-season damage by the Pacific spider mite induces resistance against subsequent Willamette mite populations, but early-season damage by the Willamette mite did not induce resistance against subsequent Willamette mite populations. Induction is known for several cultivars of the Old World species *V. vinifera* L., as well as for the North American species *V. californica* Bentham. Phylogenetic relationships among grape genotypes are of little help in understanding the variation in induced resistance. Phylogenetic relatedness among grapes did help to understand constitutive resistance to *T. pacificus*; cultivars of *V. vinifera* are inclined to be susceptible, whereas North American species are resistant. *E. willamettei* 'behaves well' on some Old World cultivars and on two North American species of *Vitis* native to California. There is no clear evidence of a negative correlation between constitutive resistance and strength of induction for these grape genotypes. Various factors affect the variation in constitutive and induced resistance in grapes against the two spider mites, such as grape genotype, previous history of mite damage (induction), the species of mite causing previous damage and phylogenetic relatedness among grapes. In this complex mechanism, the mite genotype could play an important role (English-Loeb *et al.*, 1998).

CULTURAL PRACTICES. Studies carried out on the cultivar 'Cabernet Sauvignon' on the effects of nitrogen and potassium fertilizer on the mite longevity and oviposition pointed out a non-linear response to grape N concentration. Performance is better within the two medium treatments (1000 and 500 ppm) compared to high N (1500 ppm) and low N (50 ppm) treatments. Survivorship suggests the same as days to maturation, although not statistically significant (Geddes, 2010).

CHEMICAL CONTROL. Chemical treatments are not usually necessary, because grapevines tolerate high levels of attack, with the exception of few table grapevines. *E. willamettei* (and other spider mites) in California vineyards have developed resistance to sulfur, probably due to the many applications (up to 20 times) of sulfur per season over many years to control powdery mildew (Hoy, 2011). Hanna *et al.* (1997b) have also discussed this topic, specifically with respect to the effect of sulfur (and triadimefon) on predator mites. The response by mite to sulfur and irrigation has been evaluated by Costello (2004).

***Eotetranychus yumensis* (McGregor) (Fig. 20.29)**

Common name

Yuma spider mite.

Diagnostic characteristics

FEMALE. The body is ovoid, 320 µm long and about 180 µm wide, varying in colour from straw to dark pink. The peritremes are distally hooked. The dorsal body setae are finely setose and not set on tubercles, each reaching well beyond the base of the next seta behind. The genital flap is striate transversally, and the area anterior to the flap is longitudinally striate. Tibiae I and II bear 9 and 8 tactile setae, respectively. Tarsus I has 4 tactile setae proximal to the duplex setae (McGregor, 1950; Jeppson *et al.*, 1975).

MALE. The male is smaller than the female and pale. The aedeagus forms a shallow, nearly sigmoid dorsal curve, with the distal end gradually tapering to a caudally direct point (McGregor, 1950; Jeppson *et al.*, 1975).

Geographical distribution

Mexico (Tuttle *et al.*, 1974) and the USA (McGregor, 1934, 1950; Pritchard and Baker, 1955; Tuttle and Baker, 1964).

Host plants

The whole list of host plants of the Yuma spider mite includes 27 host plants of 15 families (Migeon and Dorkeld, 2006–2013). The families and species of economic importance are:

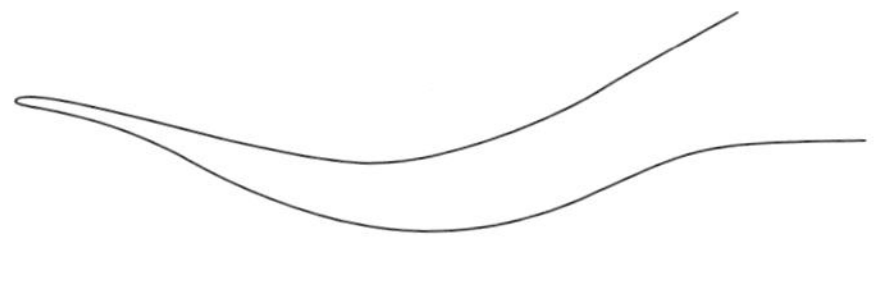

Fig. 20.29. *Eotetranychus yumensis* (McGregor). Male, aedeagus (modified from McGregor, 1950).

Euphorbiaceae – castor bean (*Ricinus communis* L.) (Bibby and Tuttle, 1959; Tuttle and Baker, 1964); Poaceae – sorghum (*Sorghum bicolor* (L.) Moench) (Bibby and Tuttle, 1959; Tuttle and Baker, 1964); Rosaceae – Japanese rose (*Rosa multiflora* Thunb.) (Tuttle and Baker, 1968); Rutaceae – lemon (*Citrus limon* (L.) Burm.f.) (McGregor, 1950; Pritchard and Baker, 1955; Tuttle and Baker, 1964), grapefruit (*C. paradisi* Macfad.) (McGregor, 1950; Tuttle and Baker, 1964), tangerine (*C. reticulata* Blanco) (McGregor, 1950; Pritchard and Baker, 1955; Tuttle and Baker, 1964), sweet orange (*C. sinensis* (L.) Osbeck) (McGregor, 1950; Tuttle and Baker, 1964); Vitaceae – grapevine (*Vitis vinifera* L.) (Tuttle and Baker, 1964).

Bio-ecology

E. yumensis is a Palaearctic mite, living on various wild and economic plants, but mainly known as a citrus pest. The mite was first observed near Yuma, Arizona, by Lauderdale in 1928, and was later described by McGregor (1934) from specimens collected on lemon leaves. Occasionally, the Yuma spider mite is a severe pest of citrus in some desert areas of the south-western USA. Its geographical distribution initially seemed to be confined to the citrus areas around Yuma, Arizona, the Coachella and Imperial Valleys in California and adjacent desert regions in northern Mexico. In the Coachella Valley, it is further limited to citrus plantings within a 20-mile radius of the northern end of the Salton Sea, although the citrus plantings extend much farther than this (Elmer, 1969). Jeppson *et al.* (1975) reported its presence in areas of Nevada. Later reports indicate that distribution of the mite does not seem to have varied from these areas (Kerns *et al.*, 2004; UC Pest Management Guidelines, Citrus, see UC IPM Online 2013b).

E. yumensis reproduces by arrhenotokous parthenogenesis and has a haploid number of chromosomes of two (Helle *et al.*, 1981). Although the mite feeds on leaves, fruits and twigs of all citrus varieties, it seems to prefer grapefruit and lemon; in the southern San Joaquin Valley, it is primarily found on mandarins during the summer. TheYuma spider mite produces quantities of webbing on the lower leaf surface of infested citrus, to which dust adheres, making it easy to detect dense mite colonies (McGregor, 1956; Elmer, 1969; Jeppson *et al.*, 1975; UC IPM Online, 2013b). Its populations seem to be restricted to a desert area, and relatively high temperatures are necessary for their development. Egg deposition does not occur at constant temperatures of 10°C, and at 15.5°C no hatching occurs if the RH is below 50%. The complete life cycle occurs between 21 and 38°C, but at 43.5°C the eggs do not hatch. This adaptation probably confines the species to hot desert areas (Elmer, 1969; Jeppson *et al.*, 1975).

Populations of the Yuma spider mite generally increase in October and November, remain high throughout the winter, and decrease in spring or early summer. During the summer, the species aestivate as adult females under the bark of the trunk and limbs, or in cracks and crevices, and it is possible to find a few eggs, but not immature stages (Elmer, 1969; Jeppson *et al.*, 1975).

Predatory insects and mites are commonly associated with populations of the Yuma spider mite; these are likely to be other *Eotetranychus* spp. (McMurtry, 1985b). In California, the six-spotted thrip, *Scolothrips sexmaculatus* (Pergande), is an effective predator of *E. yumensis*. Other natural enemies, such as the spider mite destroyer (*Stethorus picipes* Casey), minute pirate bugs (*Orius* spp.) and a phytoseiid mite (*Euseius tularensis* Congdon) play a role in the control of *E. yumensis* (UC IPM Online, 2013b). In Arizona, a number of predators prey upon *E. yumensis*, including predatory mites, lacewings, predacious thrips, etc. (Kerns *et al.*, 2004).

Symptomatology and damage

The Yuma spider mite primarily feeds on the lower leaf surface, and can cause stippling and leaf discoloration. Heavy infestation can induce some leaf abscission, mainly during periods of high winds. The mite has been associated with a condition known as 'fall dieback' (autumn dieback), which is also related to the occurrence of high winds, low humidity and lack of adequate soil moisture. Frequently, new growth results in 'whip-like' branches devoid of leaves except at the terminal; branches up to 1 inch in diameter may lose their leaves and die. However, severe leaf damage is uncommon, and economic damage to mature trees is doubtful. At high population densities, the mite infests the fruit, which is far more severe than leaf infestation. Feeding on the surface of green fruit causes a stippled and bleached appearance, though in all but severe cases the fruit colours up normally. Damage to mature fruits appears as silvering, similar to that caused by the Lewis spider mite (Elmer, 1969; Jeppson *et al.*, 1975; Kerns *et al.*, 2004; UC IPM Online, 2013b.)

Control

Control requires the monitoring of populations and, when necessary, sprays of acaricides, integrated with the action of natural enemies and cultural practices.

SAMPLING AND MONITORING. The guidelines for citrus in Arizona include monitoring for *E. yumensis* at the beginning of petal fall through to the end of July. The action threshold for the mite is unknown. Treatments of the leaves are probably unnecessary unless populations are very high. Acaricides may be sprayed on the fruit when 10% of the fruit of less than 1 inch in diameter is infested or when larger fruit averages 3–5 mites per fruit (Kerns *et al.*, 2004). The UC Pest Management Guidelines for citrus (UC IPM Online, 2013b) suggest checking for Yuma spider mites on mandarins during July and August in the San Joaquin Valley. The procedure is to look for stippling of leaves and fruit that is associated with large amounts of webbing. On bearing trees, treatments should be applied if fruit stippling is sufficient to inhibit proper fruit colouring and beneficial organisms are not already reducing mite densities. On young trees, treatment should be done if leaf drop is found (UC IPM Online, 2013b).

CULTURAL PRACTICES. In California, adequate irrigation and dust control reduce the impact of *E. yumensis* (UC IPM Online, 2013b). However, Kerns *et al.* (2004) reported that in Arizona although reducing the incidence of dust by watering dirt roads adjacent to infested groves may be useful, the technique is rarely cost-effective.

CHEMICAL CONTROL. Generally, damage from this mite is not severe enough to justify treatment. However, heavy infestations

can be controlled with sulfur, oil or other acaricides. Sulfur sprays must be applied during the autumn, winter or early spring at cool weather periods (Elmer, 1965a; Jeppson, 1989). During warm seasons, Elmer (1969) suggested that acaricides such as dicofol or chlorobenzilate should be applied. In the Coachella and Imperial valleys, it is suggested to treat if needed to prevent leaf drop. *E. yumensis* can be controlled with sulfur during the period between October and 15 March, or with acaricides during the remainder of the year (UC IPM Online, 2013b). Non-bearing trees may be treated with bifenazate or etoxazole, while bearing trees may be sprayed with a number of acaricides (acequinocyl, dicofol, fenbutatin oxide, fenproximate, hexythiazox, propargite, pyridaben, spirodiclofen) (Kerns *et al.*, 2004; UC IPM Online, 2013b).

The Genus *Oligonychus* Berlese

Diagnostic characteristics

The opisthosoma has 10 pairs of dorsal setae. The anal region has 2 pairs of pseudanal setae (ps_{1-2}) and 1 pair of setae *h*. The empodium is claw like and has proximoventral hairs. All legs or most of them bear empodial claws as long or longer than the proximoventral hairs. The duplex setae of tarsus I are distal and adjacent (Bolland *et al.*, 1998).

Economic importance

This genus has worldwide distribution and includes 205 species living on wild and economic plants (Migeon and Dorkeld, 2006–2013). Pritchard and Baker (1955) recognized five major species groups within the genus *Oligonychus*. Later, Tuttle and Baker (1968) divided the genus into six subgenera (*Oligonychus* Berlese, *Wainsteiniella* Tuttle *et* Baker, *Homonychus* Wainstein, *Metatetranychus* Wainstein, *Reckiella* Tuttle *et* Baker and *Pritchardinychus* Wainstein). Jeppson *et al.* (1975) adopted the latter systematics and in addition took into account some typical ecological adaptations, such as host plants, the leaf surface upon which the mite feeds, etc. Mites included in the subgenus *Oligonychus* usually feed on the upper leaf surface of broad-leaved plants (a few species feed on conifers), while mites of the subgenus *Metatetranychus* usually feed on the lower leaf surface. Species belonging to the subgenus *Reckiella* live on grasses and other monocotyledons. Species included in the subgenus *Pritchardinychus* are green in colour and feed on the lower leaf surface, except for some species, such as *O. biharensis* (Hirst) (Jeppson *et al.*, 1975). Unfortunately, it is not possible to apply this method to all subgenera. Later, Bolland *et al.* (1998) only partially recognized the systematic framework proposed by Tuttle and Baker (1968), as does the author.

At least 16 species of *Oligonychus* can be called major pests of economic plants: *O. afrasiaticus* (McGregor), *O. bicolor* (Banks), *O. biharensis* (Hirst), *O. coffeae* (Nietner), *O. gossypii* (Zacher), *O. ilicis* (McGregor), *O. indicus* (Hirst), *O. mangiferus* (Rahman *et* Sapra), *O. newcomeri* (McGregor), *O. oryzae* (Hirst), *O. perseae* Tuttle *et al.*, *O. platani* (McGregor), *O. pratensis* (Banks), *O. punicae* (Hirst), *O. ununguis* (Jacobi) and *O. yothersi* (McGregor). Minor pests include *O. peruvianus* (McGregor) and other species, and there are also other potential pests. These 17 named species are described below

Oligonychus afrasiaticus (McGregor) (Fig. 20.30)

Common name

Old world date mite, date palm spider mite, date palm dust mite.

Diagnostic characteristics

FEMALE. The body is 293 µm long and 182 µm wide (McGregor, 1950). The terminal sensillum of the palpus is slightly longer than broad. The opisthosoma is longitudinally striate between the fourth pair (f_1) of the dorsocentral setae (Meyer Smith, 1987).

MALE. The body is 234 µm long (McGregor, 1950). The terminal sensillum of the palpus is about 1.5 times as long as broad. The aedeagus curves dorsad and has a moderately large knob distally, about half the length of the dorsal margin of the shaft; the anterior projection of the knob is rounded, and about as long as the posterior projection; the latter projection is acute and deflexed at the tip; the axis of the knob is parallel to the shaft axis (Meyer Smith, 1987).

Geographical distribution

Algeria (André, 1932a; McGregor, 1939; Guessoum, 1986), Chad (Munier, 1952), Egypt (Zaher *et al.*, 1982), Iran (Khalil-Manesh, 1973), Iraq (Buxton, 1920), Israel (Gerson *et al.*, 1983; Ben-David *et al.*, 2007), Libya (Martin, 1958), Mali, Mauritania (Munier, 1952), Mexico (Estebanes-Gonzalez and Baker, 1968), Morocco (Saba, 1973b), Niger (Munier, 1952), Palestinian Territories (Ben-David *et al.*, 2007), Saudi Arabia (Beccari, 1971), Sudan (Baker and Pritchard, 1960) and Tunisia (Pagliano, 1951; Ben Chaaban and Chermiti, 2009; Ben Chaaban *et al.*, 2011a).

Host plants

O. afrasiaticus is reported on six host plants from three families: Arecaceae – date palm (*Phoenix dactylifera* L.) (André, 1932a;

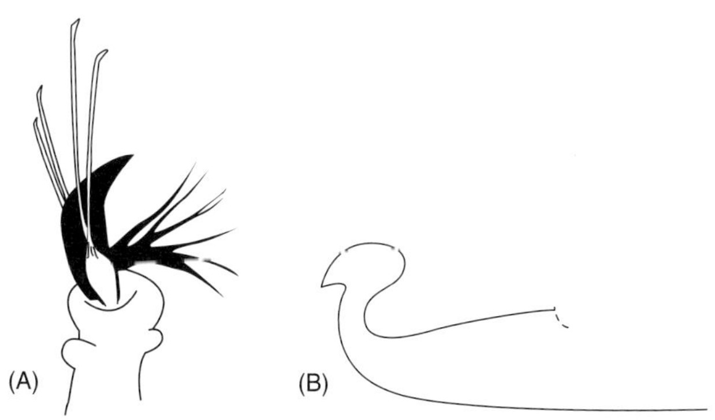

Fig. 20.30. (A) *Oligonychus* sp., pretarsus; (B) *Oligonychus afrasiaticus* (McGregor), aedeagus (from Meyer Smith, 1987).

McGregor, 1939; Baker and Pritchard, 1960; Guessoum, 1986; Palevsky *et al.*, 2004, 2005; Ben-David *et al.*, 2007; Latifian *et al.*, 2007; Ben Chaaban and Chermiti, 2009; Ben Chaaban *et al.*, 2011a); Celastraceae – Arabian tea (*Catha edulis* (Vahl) Forssk. ex Endl.) (Bolland *et al.*, 1998; Migeon and Dorkeld, 2006–2013); Poaceae – Bermuda grass (*Cynodon dactylon* (L.) Pers.) (Zaher *et al.*, 1982), *Imperata cylindrica* (L.) P. Beauv., *Lolium* sp., maize (*Zea mays* L.) (Estebanes-Gonzalez and Baker, 1968).

Bio-ecology

The old world date mite is a pest of dates in North Africa and the Near East (André, 1932a; McGregor, 1939; Baker and Pritchard, 1960; Guessoum, 1986; Palevsky *et al.*, 2004, 2005; Ben-David *et al.*, 2007; Latifian *et al.*, 2007; Idder and Pintureau, 2008; Ben Chaaban and Chermiti, 2009; Ben Chaaban *et al.*, 2011a), and may occur also on other host plants. The mite lives on the fronds, pinnae and fruits of date palms. It rarely occurs during winter and spring, and small populations are observed on the fronds in summer, simultaneously with very large numbers on the fruit strands. In addition to the damage caused by its feeding, the mite produces large amounts of webbing that covers the dates. Dust is deposited in the webbing, forming shade that prevents the dates beneath from colouring (Jeppson *et al.*, 1975). The webbing provides space and shelter for the large and dense mite populations, which rarely develop on the fronds (Palevsky *et al.*, 2010a).

O. afrasiaticus reproduces by arrhenotokous parthenogenesis. In laboratory studies at 20°C, daily egg production started on the fourth day and reached a peak (0.92 eggs) on the fifth day, while at 35°C the females started laying eggs on the second day and laying reached a peak (3.95 eggs) on the fourth day. At 20, 25, 30 and 35°C, the total reproduction rate by females was 12.58, 17.80, 19.5 and 21.74, respectively, and the net reproduction rate (R_0) was 12.5, 17.22, 18.68 and 20.05, respectively. The highest mean generation time (T) was 13.97 days at 20°C, and the lowest mean generation time was 6.04 days at 35°C. The highest rate of intrinsic increase in population ($r_m = 0.496$) occurred at 35°C, and the lowest rate (0.181) at 20°C. At 20°C, the doubling time of population was 3.83, and at 35°C it was 1.40 (Al-Sweedy *et al.*, 2006). Al-Jboory and Al-Suaide (2010) studied mite biology at 20, 25, 30 and 35°C, and 50–60% RH, with a 16L:8D photoperiod, having demonstrated that no development occurred at 15 and 40°C. At 20°C, egg development took 7.6 days and at 35°C it took 2.7 days. The development of the larva, protonymph and deutonymph took 5.3, 4.3 and 4.5 days at 20°C, and 1.9, 1.3 and 1.4 days at 35°C, respectively. At 20°C, the fecundity was 12.5 eggs, and at 35°C it was 27.2 eggs. The longevity of the female and male was 33.2 and 29.1 days at 20°C, and 11.5 and 10.1 days at 35°C, respectively. In a laboratory study at 27°C on fruits of date palms of the varieties 'Deglet Noor', 'Alig', 'Kentichi' and 'Bessr', and on 'Deglet Noor' pinnae and sorghum leaves, the average length of the life cycle of *O. afrasiaticus* ranged from 13 days on 'Alig' fruits to '10.9' days on sorghum leaves. The highest fecundity was on sorghum leaves (2 eggs per female daily) during 5.2 oviposition days, and there was low fecundity on 'Deglet Noor' pinnae and 'Alig' fruits at 0.7 eggs per female daily over 5.4 days. The longevity of *O. afrasiaticus* females ranged from 13.4 to 7.5 days on 'Deglet Noor' fruits and sorghum leaves, respectively. The intrinsic rate of increase (r_m) was highest on sorghum leaves (0.171) and 'Deglet Noor' fruits (0.166), and lowest on 'Alig' fruits (0.103) (Ben Chaaban *et al.*, 2012).

In Tunisia, the timing of the beginning of *O. afrasiaticus* infestation on fruits varies between years, ranging from the first to the third week of July. Mite populations on pinnae remain low from May through to December, not exceeding two mites per pinnae, but on fruits, the populations may reach peaks of 14 mites per fruit. At the end of August, the pest populations peak. The mite was found on leaves of *Sorghum bicolor* (L.) Moench, *Solanum melongena* L. and *Cucumis melo* L. in the orchard groundcover vegetation during early spring and autumn, and these plants may represent the source for later date palm infestation. The period spent by the mite on fruits differs from one variety to another; it is 8 weeks on 'Deglet Noor', 2–5 weeks on 'Alig', 2–4 weeks on 'Kentichi' and 2–4 weeks on 'Bessr'. The 'Deglet Noor' variety appears to be the most susceptible to *O. afrasiaticus*. Mite populations on the pinnae remain low from May through to December (Ben Chaaban *et al.*, 2011a, 2012). In Israel, *O. afrasiaticus* was rarely found during winter and spring; it occurs on fronds only from late April until late August and maintains small populations in the summer, whereas very large numbers occur on the fruit strands, with a rapid increase during June–August, in most cases averaging much more than the 1000 mites per fruit strand.

Adult cumulative mite days of *O. afrasiaticus* (on fruit strands) are always considerably higher than those of another spider mite living on palms, *Eutetranychus palmatus* Attiah, but on fronds the opposite is observed (Palevsky *et al.*, 2010a). Differences between the two spider mites are treated under *E. palmatus*. In Iraq, *O. afrasiaticus* develops six generations a year, and the peak number develops around the middle of July (Hussain, 1969). In the southern provinces of Iran, the highest population of mite has been observed from the second half of June to the beginning of August. Initiation of infestation is high in small palms 3 m tall. High populations have been observed in the 'Kabkab', 'Mazafati', 'Halileh' and 'Khasuii' varieties (Fasihi *et al.*, 2006). Further studies in the same country show that the highest population density of *O. afrasiaticus* on the fruits and pinnae of date palms occurs in the first half of August. The spatial distribution pattern of *O. afrasiaticus* is aggregated (Payandeh *et al.*, 2013). In Mesopotamia, infestations on dates are more severe in the humid environment (near the water) (Buxton, 1920). In Yemen, it has been observed that in the absence of fruits and during winter, the pest mite stayed at the bases of the fronds and in the heart of the palms. On the 'Mijraf' variety, the mite appears on date fruits in the second week of March and is present until the third week of July (Ba-Angood and Bassahaih, 2001). In Saudi Arabia, *O. afrasiaticus* is active from the first half of March until the second half of October. Infestation usually starts around mid-May to June, when the mite builds dusty, creamy silky webs on date bunches around date fruit stands, where the adults and immature stages live. Numbers of mites per 100 date fruits may reach a maximum with 9095.5 mites at the end of July; the population then gradually decreases until the middle of October to a density of 124 mites per 100 date fruits. The infestation began after the 2nd week of March and reached a mean of 3%; it reached a maximum of 80% in the 3rd week of October (Aldosari, 2009).

The dispersal of *O. afrasiaticus* is via the fig wasp, *Polistes olivaceous* (DeGeer) (Jeppson *et al.*, 1975). It also occurs via aerial dispersal of motile forms (Ben Chaaban *et al.*, 2011a).

Natural enemies include predatory insects, such as the coccinellids *Stethorus punctillum* Weise (Idder and Pintureau, 2008), and *Stethorus gilvifrons* (Mulsant) (Latifian, 2012) and phytoseiid mites. Recently, Palevsky *et al.* (2009) published a study that aimed to monitor the phytoseiids in the date palm canopy and on the orchard floor in Israel, which included a key to all species collected from date palm orchards worldwide.

Symptomatology and damage

The date dust mite infests pinnae and fruits of date palms, and in addition to damage caused by its feeding, produces large amounts of webbing that covers the dates. Dust is deposited in the webbing, forming shade that prevents the dates beneath from colouring (Jeppson *et al.*, 1975). The mite attacks the dates from their early stages of development, and dust collected in the webs plus the exuviae of different developmental stages of the date dust mite affect the date bunches, giving them a dusty appearance (Saleh and Hosny, 1979). The establishment of *O. afrasiaticus* starts from the points of contact with fruits, where it begins to spin webbing and lays its eggs. As the fruits develop through the kimri stage (which begins when the fruit colour changes from white to green and ends when the fruit turns yellow), *O. afrasiaticus* spreads out over the entire fruit cuticle (Palevsky *et al.*, 2010a). Infested dates become progressively drier at the end of the maturity and unsuitable for marketing and human consumption. Significant economic losses are then recorded.

The greatest damage occurs on the 'Deglet-Nour' date variety, which is of high commercial value (Idder and Pintureau, 2008). Research in Yemen on the effect of *O. afrasiaticus* on the physico-chemical characteristics of the dates of cultivars 'Mijraf', 'Madini' and 'Hamra' showed that the mite had a negative effect on some of these. Infested 'Mijraf', 'Madini' and 'Hamra' dates are smaller in size, malformed and unripe, and have lower contents of water and total soluble solids, particularly sugars, compared with healthy dates (Ba-Angood and Basshaih, 2000). In Israel, it was observed that the date dust mite attacked and rapidly developed on young and green fruits. Mite phenology is affected by date cultivar, but the relationship between the fruit characteristics and development of the mite population is unknown. Mite establishment on the main cultivars grown in Israel ('Medjool', 'Barhi' and 'Deglet Noor') began only when the water content of the fruit increased to ≥84%. Fruit bunch architecture and resistance to penetration, as well as sugar composition and content, are not key factors in mite establishment. Later in the season, when total sugar levels reached ≥150 mg g^{-1} fresh fruit weight, total soluble solids comprised 15% and the water content decreased to ≤75%, the mite populations declined. Annual climatic variations could be responsible for phenological asynchrony between the pest and fruit development (Palevsky *et al.*, 2005). No damage occurs on 'Deglet Noor' and 'Barhi' fruit at seasonal levels of *c.*6000 ACMDs (adult cumulative mite days), whereas substantial damage occurs when population levels exceeded *c.*20,000 ACMDs (Palevsky *et al.*, 2004). In Iran, damage by the mite under drought condition may cause significant and extensive losses, mostly on commercial date varieties. Losses of highly susceptible varieties of up to 100% were also observed (Arbabi *et al.*, 2006, 2010; Fasihi *et al.*, 2006). In Iraq, infestation by *O. afrasiaticus* causes 50–80% yield loss in years with dry, dusty and stormy weather (Al-Jboory and Al-Suaide, 2010).

Control

Various techniques to reduce pest damage have been tested, such as reducing the overwintering populations, employing physical barriers to prevent ambulatory and airborne mites from reaching fruit bunches, using indigenous phytoseiids, applying a single, properly timed seasonal acaricide and post-harvest treatments to mitigate any evident mite damage. The first three methods did not give good results. Mite damage was higher on palms (trees) that had their stems (trunks) scorched in the winter, and fruit bunch infestation was not prevented by glue barriers or dense netting. Indigenous phytoseiids were not found on fruit from mid-July to the end of August, when pest populations peaked (Palevsky *et al.*, 2004). In Tunisia, fruit bunch infestation was not prevented or even reduced by glue barriers, so infestation of the date palm by *O. afrasiaticus* might occur via the aerial dispersal of motile forms. Phytoseiid mites are scarce in ground cover (Ben Chaaban *et al.*, 2011a). In Israel, postharvest rehydration treatment appeared effective, increasing the amount of marketable fresh fruit by approximately 20% on the economically important 'Medjool' cultivar (Palevsky *et al.*, 2004).

Currently, the control of *O. afrasiaticus* is provided by the use of resistant varieties and biological and chemical means. Estimation of the difference in susceptibility of cultivars to *O. afrasiaticus* is crucial for developing efficient pest control programmes. Indeed, less susceptible cultivars can either be left unsprayed or sprayed at a low threshold (Ben Chaaban *et al.*, 2011b).

PLANT RESISTANCE. Studies in Iran on the response of 30 native date palm cultivars to infestation by *O. afrasiaticus* highlighted very highly infested cultivars ('Liloei', 'Berhee', 'Zahedi', 'Deiri', 'Ashkar', 'Berim'), highly infested cultivars ('Halavi', 'Beliani', 'Sovidani', 'Hadak', 'Shkar', 'Bent Alsabae', 'Dagal zard', 'Khazravi', 'Sayer'), moderately infested cultivars ('Bobki', 'Chabchab', 'Mashtom', 'Jahromi', 'Amobahri', 'Dagal sorkh', 'Fersi', 'Hadal') and less infested cultivars ('Khasab', 'Hamravi', 'Hasavi', 'Ashagh', 'Jozi', 'Gantar'). Bunch weight, fruit number and fruit skin thickness are significantly correlated with date palm spider mite injury (Latifian *et al.*, 2007).

The susceptibility of date palm fruits of six cultivars ('Sokary', 'Rothan', 'Moonifie', 'Sefrie', 'Cebiky' and 'Khodary') to natural infestation by *O. afrasiaticus* and the effects of infestation on the chemical composition of the fruits were investigated in Saudi Arabia. The cultivars 'Sokary' and 'Rothan' appeared to be highly susceptible, while the cultivar 'Cebiky' seemed resistant and the cultivar 'Khodary' moderately resistant. Differences between the amount of infestation and date fruit composition (contents of lipids, proteins, carbohydrates and ash) between different cultivars were insignificant, but there was a significant negative correlation between infestation and carbohydrate content (amounts of sucrose, glucose and fructose) and between infestation and protein content of the fruits (Ali and Aldosari, 2007).

The susceptibility of three cultivars ('Deglet Noor', 'Alig', 'Bessr') characterized by different ripening periods suitable for feeding by *O. afrasiaticus* was evaluated in Tunisia. The 'Deglet Noor' cultivar was the most susceptible to the mite. The largest population of *O. afrasiaticus* on these three cultivars occurred when the water content reached 83%. A high acidity (>2 mEq 100 g^{-1}) prevented mite establishment on the three cultivars. When the total sugar levels reached >420 mg g^{-1} fresh fruit, total soluble solids increased above 10% and the fruit moisture content decreased to <80%, the mite populations declined. Total phenolics and hardness of the exocarp were not key factors in mite establishment (Ben Chaaban and Chermiti, 2009). Research on four date palm varieties ('Deglet Noor', 'Alig', 'Kentichi', 'Besser') at 27 ± 1°C and 60 ± 10% RH showed that the total development time of immature females was shorter on 'Deglet Noor' fruits than on fruits of the other cultivars. The total fecundity per female is highest on 'Deglet Noor' and the lowest fecundity on 'Besser'. Comparison of intrinsic rates of natural increase (r_m), net reproductive rates (R_0) and the survival rates of immature stages of *O. afrasiaticus* on the host plants highlighted a better performance on 'Deglet Noor' fruits. The lowest intrinsic rate of natural population increase was on the 'Alig' variety (0.103 day^{-1}) (Ben Chaaban *et al.*, 2011b).

The responses of various cultivars ('Kabkab', 'Khasi', 'Haj-Mohamady', 'Zahedi', 'Gantar' and 'Khazravi', 'Piarom', 'Mezafati', 'Shahani') to mite feeding were evaluated in Iran. Some cultivars ('Khasi', 'Kabkab', 'Khazravi', 'Shahani') showed maximum infestation, while the cultivar 'Piarom' was minimally infested. A comparison of means based on infestation severity showed that 'Khasi', 'Kabkab', 'Shahani' and 'Khasravi' had maximum infestation rates, with 4.31, 3.65, 3.47 and 3.09%, respectively, while 'Piarom' and 'Zahedi' had minimum infestation rates (1.19 and 1.25%). Based on cluster analysis, the date palm cultivars were classified into two groups. The first group ('Khasi', 'Shahani', 'Kabkab', 'Khazravi', 'Haj-mohamady') are infested cultivars, and the second group ('Piarom', 'Zahedi', 'Mezafati', 'Gantar') are uninfested cultivars. The correlation between infestation severity or infestation percentage and low bunch weight was positively significant (Khajehzadeh and Latifian, 2011). In Yemen, the mite attacks almost all date palm varieties, but the 'Hamra' and 'Hajri' varieties are less susceptible, while 'Mijraf' and 'Madini' are comparatively more susceptible (Ba-Angood and Bassahaih, 2001).

BIOLOGICAL CONTROL. The action of predatory mites and insects against *O. afrasiaticus*, both artificially introduced or present in the environment, has been tested by various researchers. Releases of the phytoseiid *Neoseiulus californicus* (McGregor) in date groves in the south of Tunisia had a significant effect, which was greater at a high density of pest populations, but the quality of the fruit was better when low-scale attacks occurred as well. These promising results suggested that the effects of early release of *N. californicus* (Othman *et al.*, 2001) should be evaluated. Recently, the potential use has been evaluated of two phytoseiid mites, *Cydnodromella negevi* (Swirski *et* Amitai) and *N. barkeri* Hughes, as biological control agents against *O. afrasiaticus* (Negm, 2012).

Research on predatory insects mainly concerns the use of ladybirds. In the Wargla area of Algeria, the effectiveness of releases of the ladybird *S. punctillum*, which plays an important role in the control of the mite *O. afrasiaticus*, has been evaluated (Idder and Pintureau, 2008). In Iran, laboratory experiments on date offshoots (*P. dactylifera*) infested by *O. afrasiaticus* showed that adults of *S. gilvifrons* consumed significantly more mites than *S. gilvifrons* larvae. The predator adults preyed on 33.67 and 21.33 mites, whereas predator larvae preyed on 32.33 and 19.67 adults and larvae of 40 spider mites, respectively, in 24 h. The two life stages of the coccinellid exhibited a significant preference for adults of *O. afrasiaticus* in the presence of its larvae. This preference was more marked for adults of *S. gilvifrons*. The total number of prey consumed decreases significantly for predator adults and larvae as the number of mite larvae in the prey ratio increases. *O. afrasiaticus* larvae are an alternative prey for the predator, and adults of *S. gilvifrons* should be more efficient than their larvae in a prey assemblage with larvae and adults of the mite (Latifian, 2012). The population densities and spatial distribution pattern of *S. gilvifrons* were investigated in date palm orchards in Iran, and the highest population density of this predator on fruits and pinnae has been observed in the first half of August. The spatial distribution pattern of *S. gilvifrons* is a random distribution (Payandeh *et al.*, 2013).

CHEMICAL CONTROL. In Israel, growers have dusted fruit bunches prophylactically with sulfur since 1990, applying as many as five treatments a year, but recently the efficacy of sulfur has declined. A single treatment with the acaricides abamectin, fenbutatin oxide or hexythiazox, sprayed when the first mites are detected on the fruit, provides seasonal pest control (Palevsky *et al.*, 2004). Field tests in Saudi Arabia showed the effectiveness in descending order of abamectin, dicofol, bromopropylate, azocyclotin, tetradifon, sulphur WP, sulfur + Cu and thiometon (Al-Doghairi, 2004). Further trials in Saudi Arabia highlighted the good action of abamectin, amitraz, perpol and methidathion (Aldosari, 2009). In Iran, the action of amitraz, fenpropathrin, fenpyroximate, fenazaquin, hexythiazox, propargite and tetradifen has been evaluated. Maximum effects are recorded for tetradifen (Arbabi *et al.*, 2006, 2010). Trials in Iraq with five insecticides (alpha-cypermethrin, deltamethrin, chlofenapyr, cyfluthrin and triazophos) highlighted the good response of triazophos (Al-Dosary, 2010).

In Saudi Arabia, the botanical compound matrine gave insufficient control of the mite (Aldosari, 2009), while in Egypt, ethanolic extracts of demsisa (*Ambrosia maritima* L.), duranta (*Duranta plumeria* Jacq.) and cumin (*Cuminum cyminum* L.) exhibited ovicidal and repellent activities and significant effects on the reproduction and feeding behaviour of *O. afrasiaticus* (Fetoh and Al-Shammery, 2011).

Oligonychus bicolor (Banks) (Fig. 20.31)

Common name

Oak mite.

Diagnostic characteristics

FEMALE. The propodosoma is yellow and the opisthosoma brown in colour. The dorsal body setae are finely plumose, not set on

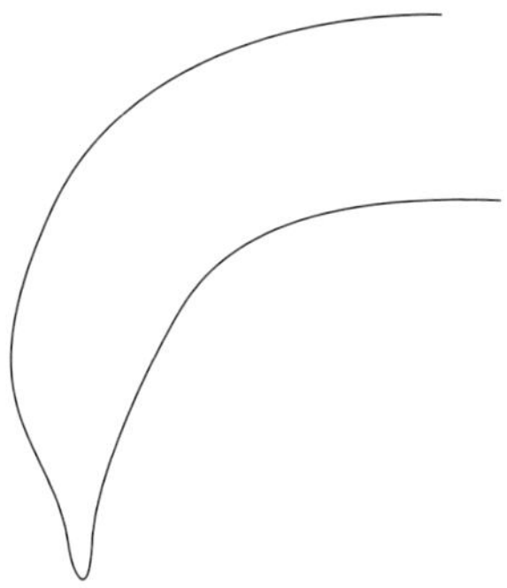

Fig. 20.31. *Oligonychus bicolor* (Banks). Male, aedeagus (drawing by Vacante).

tubercles and longer than the distances between their bases; the opisthosomal setae f_1 and f_2 are subequal in length. Tibia I has 7 tactile setae and 1 sensory seta; tarsus I bears 3 tactile setae proximal to the duplex setae, of which the proximal seta is very short (Vacante, 1985a).

MALE. Body chaetotaxy and colour of the male are as in the female. The basal part of the aedeagus is large, facing down and narrowed distally; the dorsal curve of the shaft is wider than the ventral shaft (Vacante, 1985a).

Geographical distribution

Canada (Jeppson *et al.*, 1975), Iran (Sepasgosarian, 1972), Italy (Vacante, 1985a), Portugal (Carmona and Dias, 1980), South Africa (Baker and Tuttle, 1994), Taiwan (Baker, 1975a) and the USA (Banks, 1894).

Host plants

The whole list of host plants of oak mite includes 46 species from 18 families, with the highest number included in the the Fagaceae family (13 species) (Migeon and Dorkeld, 2006—2013).

Apocynaceae: *Plumeria* sp. (Bolland *et al.*, 1998; Migeon and Dorkeld, 2006–2013).

Asteraceae: *Senecio angulatus* L.f. (Bolland *et al.*, 1998; Migeon and Dorkeld, 2006–2013).

Betulaceae: *Betula* sp., *B. pendula* Roth (Reeves, 1963), *Corylus* sp. (Bolland *et al.*, 1998; Migeon and Dorkeld, 2006–2013).

Ebenaceae: *Diospyros rumphii* Bakh. (Bolland *et al.*, 1998; Migeon and Dorkeld, 2006–2013).

Fabaceae: *Acacia confusa* Merr. (Bolland *et al.*, 1998; Migeon and Dorkeld, 2006–2013).

Fagaceae: *Castanea* sp. (Banks, 1894), *C. dentata* (Marshall) Borkh. (Reeves, 1963), *Fagus* sp. (Pritchard and Baker, 1955), *F. grandifolia* Ehrh. (Reeves, 1963), *Quercus* sp. (Banks, 1894), *Q. alba* L. (Pritchard and Baker, 1955), *Q. michauxii* Nutt. (Reeves, 1963), *Q. palustris* Münchh., *Q. phellos* L. (Pritchard and Baker, 1955), *Q. pubescens* Willd., *Q. robur* L. (Bolland and Beron, 2004), *Q. rubra* L. (Pritchard and Baker, 1955; Reeves, 1963), *Q. velutina* Lam. (Reeves, 1963).

Juglandaceae: *Carya* sp. (McGregor, 1950), *C. cordiformis* (Wangenh.) K. Koch (Reeves, 1963), *C. illinoinensis* (Wangenh.) K. Koch. (McGregor, 1950).

Lauraceae: avocado (*Persea americana* Mill.) (Bolland *et al.*, 1998; Migeon and Dorkeld, 2006–2013).

Meliaceae: *Trichilia dregeana* Sond. (Bolland *et al.*, 1998; Migeon and Dorkeld, 2006–2013).

Moraceae: *Artocarpus heterophyllus* Lam., *Morus* sp. (Bolland *et al.*, 1998; Migeon and Dorkeld, 2006–2013).

Pinaceae: *Picea* sp. (McGregor, 1950).

Proteaceae: *Hakea* sp. (Bolland *et al.*, 1998; Migeon and Dorkeld, 2006–2013).

Rosaceae: sour cherry (*Cerasus vulgaris* Mill.), quince (*Cydonia oblonga* Mill.), loquat (*Eriobotrya japonica* (Thunb.) Lindl.), sweet cherry (*Prunus avium* (L.) L.), peach (*P. persica* (L.) Batsch), Chinese pear (*P. pyrifolia* (Burm. f.) Nakai) (Bolland *et al.*, 1998; Migeon and Dorkeld, 2006–2013).

Salicaceae: *Salix* sp. (Bolland *et al.*, 1998; Migeon and Dorkeld, 2006–2013).

Sapindaceae: *Acer* sp. (McGregor, 1950), *A. coriaceifolium* H. Léveillé, *A. hippocastanum* L., *Dimocarpus longan* Lour., lychee (*Litchi chinensis* Sonn.) (Bolland *et al.*, 1998; Migeon and Dorkeld, 2006–2013).

Theaceae: tea (*Camellia sinensis* (L.) Kuntze) (Bolland *et al.*, 1998; Migeon and Dorkeld, 2006–2013).

Ulmaceae: *Ulmus americana* L. (Reeves, 1963), *Zelkova serrata* (Thunb.) Makino (Bolland *et al.*, 1998; Migeon and Dorkeld, 2006–2013).

Vitaceae: fox grape (*Vitis labrusca* L.) (Reeves, 1963), grapevine (*V. vinifera* L.) (Bolland *et al.*, 1998; Migeon and Dorkeld, 2006–2013).

Bio-ecology

O. bicolor is mostly a pest of oak, but it may also occur on other ornamentals, such as birch (*Betula*), beech (*Fagus*), chestnut (*Castanea*), elm (*Ulmus*), hickory (*Carya*), maple (*Acer*), spruce (*Picea*) and grapevine (McGregor, 1950; Reeves, 1963; Specht, 1963; Jeppson *et al.*, 1975). It lives on the upper leaf surface, where it lays eggs along the midrib and lateral veins; it sometimes occurs on the bark of young twigs as well. The oak mite overwinters as eggs in crevices and around the axils of small twigs, where the eggs remain until the following spring (May) (Jeppson *et al.*, 1975). In South Carolina, the eggs hatch as early as March (Chong *et al.*, 2012). The mite develops several generations a year, and its populations increase until autumn as long as the environmental conditions are favourable (Chong *et al.*, 2012). In Italy, on chestnut, the first generation of *O. bicolor* lasts about 5–6 weeks from the beginning of May. Later, generations overlap, and the highest density (444.4 individuals 100 cm^{-2} of leaf surface) occurs in mid-June (Cinti *et al.*, 1995). Hot and dry weather favours an increase in mite populations (Chong *et al.*, 2012).

Natural enemies include the phytoseiid mite *Euseius finlandicus* (Oudemans) and the coccinellid *Stethorus punctillum* Weise (Cinti *et al.*, 1995).

Symptomatology and damage

In the USA, the mite is an occasional pest of oak in nurseries, more during hot and dry summers (Chong *et al.*, 2012). The mite causes leaf discoloration. The damaged areas later become reddish, and at high levels of attack the damage, webbing and cast skins gives a greyish rust appearance to leaf surface (Speght, 1963). Extensive feeding can cause yellowing and bronzing of the leaves and also premature senescence (Chong *et al.*, 2012). On chestnut, although the mite did cause extended leaf discoloration, no defoliation occurred (at least at a density of 444.4 mites 100 cm^{-2} of leaf) (Cinti *et al.*, 1995).

Control

Natural stands do not usually require any mite control. However, in the nursery, it may be necessary to apply some control, such as sprays of chemicals, cultural practices and the use of mechanical means. The mite is active starting around 802 GDD (growing degree-days) and control measures can be implemented until approximately 1266 GDD (Kowalsick and Clark, 2010).

CULTURAL PRACTICES. Overhead irrigation and moving susceptible host plants away from dry and dusty roads help to reduce favourable conditions for the development of mite populations (Chong *et al.*, 2012).

PHYSICAL CONTROL. On a small number of plants, it is possible to use a forceful jet of water to dislodge the mites (Chong *et al.*, 2012).

CHEMICAL CONTROL. Nursery oak trees infested in previous years must be monitored by examining the twigs at the time of bud burst. Sprays should be applied as soon as the overwintering eggs hatch. Appropriate acaricides include abamectin, acequinocyl, bifenazate, etoxazole, hexythiazox and spiromesifen. Pyrethroids are not effective. It is possible to apply sprays of horticultural oil against overwintering eggs (Chong *et al.*, 2012).

Oligonychus biharensis (Hirst) (Fig. 20.32)

Common name

Cassava red mite (?).

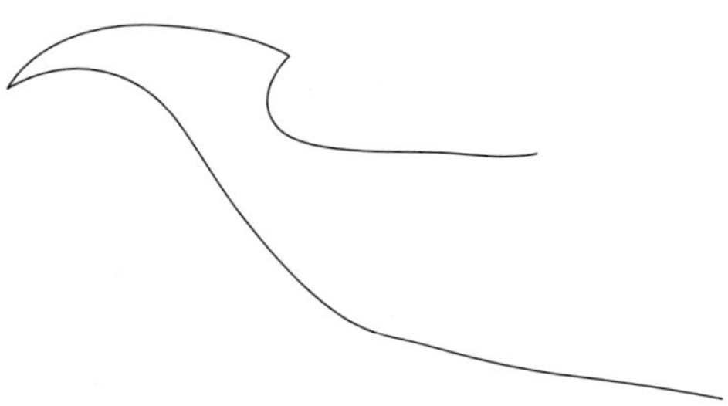

Fig. 20.32. *Oligonychus biharensis* (Hirst). Male, aedeagus (from Meyer Smith, 1987).

Diagnostic characteristics

FEMALE. The distal sensillum of the palptarsus is short and broad. The peritremes are turned backwards distally. The dorsal body setae are longer than the longitudinal distances between them. The opisthosomal striae are transverse (Meyer Smith, 1987).

MALE. The shaft of the aedeagus curves dorsally and distally forms a large knob longer than the dorsal margin of the shaft; the anterior projection of the knob is small and acute and the posterior projection long and curved downwards; the axis of the knob is parallel to the shaft axis, and the dorsal margin of the knob is convex (Meyer Smith, 1987).

Geographical distribution

Antigua and Barbuda (Ehara and Tho, 1988), Australia (Davis, 1968a), Bangladesh (Bolland *et al.*, 1998), Brazil (Flechtmann and Baker, 1970), China (Wang, 1981a; Ehara and Tho, 1988; Wang and Cui, 1992; Ji *et al.*, 2008), Fiji (Gutierrez and Schicha, 1984), French Polynesia (Gutierrez, 1981), Guadeloupe (Gutierrez and Etienne, 1988; Flechtmann *et al.*, 1999a; Flechtmann and Etienne, 2006), Hawaiian Islands (McGregor, 1950; Lee Goff, 1986), India (Hirst, 1925; Nassar and Ghai, 1981; Gupta, 1992; Gupta and Gupta, 1994), Japan (Ehara, 1966a; Gotoh *et al.*, 1996b, 2003a; Ohno *et al.*, 2011a), Malaysia (Ehara and Tho, 1988; Othman and Zhang, 2003), Mauritius (Moutia, 1958; Baker and Pritchard, 1960), Mexico (Estebanes-Gonzalez and Baker, 1968), New Caledonia (France) (Gutierrez, 1977), Papua New Guinea (Schicha and Gutierrez, 1985), Philippines (Baker, 1975a), American Samoa (Gutierrez, 1977), Western Samoa (Bolland *et al.*, 1998), South Africa (Corpuz-Raros, 1989b), Taiwan (Lo, 1968b), Thailand (Manson, 1963; Baker, 1975b), Tonga (Bolland *et al.*, 1998), Venezuela (Vásquez *et al.*, 2009) and Vietnam (Bellotti, 2009, in Bellotti *et al.*, 2012).

Host plants

O. biharensis is reported on 70 host plants ascribed to 24 families, with the highest number of species included in the Fabaceae (16 species) and Rosaceae (10 species) (Migeon and Dorkeld, 2006–2013). The families and species of economic importance are:

Anacardiaceae: mango (*Mangifera indica* L.) (Baker and Pritchard, 1960; Flechtmann and Baker, 1970; Baker, 1975b; Matsuda *et al.*, 2012).

Araceae: coco yam (*Colocasia esculenta* (L.) Schott) (Bolland *et al.*, 1998; Migeon and Dorkeld, 2006–2013).

Arecaceae: coconut palm (*Cocos nucifera* L.) (Gupta, 1980; Gupta and Gupta, 1994).

Ebenaceae: Malaysian persimmon (*Diospyros maritima* Blume) (Ehara, 1966a; Ohno *et al.*, 2011a).

Euphorbiaceae: cassava (*Manihot esculenta* Crantz) (Lal and Pillai, 1976).

Fabaceae: tamarind (*Tamarindus indica* L.) (Ehara and Tho, 1988); cowpea (*Vigna unguiculata* (L.) Walp.) (Kaimal, 2013).

Lauraceae: camphor tree (*Cinnamomum camphora* (L.) J. Presl) (Baker and Pritchard, 1960), avocado (*Persea americana* Mill.) (Ehara, 1969a).

Malvaceae: durian (*Durio zibethinus* Murray) (Baker, 1975b), cacao tree (*Theobroma cacao* L.) (Gupta and Gupta, 1994).

Myrtaceae: guava (*Psidium guajava* L.) (Estebanes-Gonzalez and Baker, 1968).

Rhamnaceae: jujube (*Ziziphus jujuba* Mill.) (Flechtmann *et al.*, 1999a).

Sapindaceae: longan (*Dimocarpus longan* Lour.) (Baker and Pritchard, 1960; Matsuda *et al.*, 2012), lychee (*Litchi chinensis* Sonn.) (McGregor, 1950; Baker, 1975b; Lee Goff, 1986).

Sapotaceae: sapodilla (*Manilkara zapota* (L.) P. Royen) (Baker, 1975b).

Rosaceae: loquat (*Eriobotrya japonica* (Thunb.) Lindl.) (McGregor, 1950; Moutia, 1958; Baker and Pritchard, 1960; Nassar and Ghai, 1981; Lee Goff, 1986; Gupta and Gupta, 1994; Ji *et al.*, 2008), apple (*Malus domestica* Borkh.), pear (*Pyrus communis* L.) (Ehara and Wongsiri, 1975), Chinese pear (*P. pyrifolia* (Burm. f.) Nakai) (Gupta and Gupta, 1994), rose de Mai (*Rosa* × *centifolia* L.) (Bolland *et al.*, 1998; Migeon and Dorkeld, 2006–2013), English rose (*Rosa hybrida*) (Baker, 1975b).

Solanaceae: aubergine (*Solanum melongena* L.) (Gupta and Gupta, 1994), Irish potato (*S. tuberosum* L.) (Gupta, 1984).

Vitaceae: grapevine (*Vitis vinifera* L.) (Bolland *et al.*, 1998; Migeon and Dorkeld, 2006–2013).

Bio-ecology

O. biharensis is a polyphagous mite living on various economic plants, including avocado, cassava, coconut, durian, grapevine, guava, loquat, lychee, mango, pear, vegetables and ornamentals. It has been reported as a pest of mango (Moutia, 1958; Jeppson *et al.*, 1975), guava (Kwee and Chong, 1990), cassava (Swaminathan, 2007) and cowpea (Kaimal, 2013), and has been evaluated as a potential control agent of water hyacinth (Haq and Sumangala, 2003).

The mite lives on the upper leaf surfaces and lays round reddish eggs bearing a short apical hair. It reproduces by arrhenotokous parthenogenesis and has a haploid number of chromosomes of two (Helle and Gutierrez, 1983). It spins a very loose web and exhibits a life type WN-u, because it always feeds and usually walks under its webs. The web serves to shelter the mites from adverse climatic conditions and from predators (Saitō, 1985).

On cassava leaves at 30 ± 2°C and 70 ± 5% RH, the pre-oviposition period, oviposition period and post-oviposition period lasted 1 day, 10.9 ± 0.75 days and 0.8 ± 0.22 day, respectively. The fecundity was 44.6 ± 4.2 eggs and 30.6 ± 1.6 eggs for mated and unmated females, respectively. The total duration of sexual development and parthenogenetic development was 7.75 ± 0.09 days and 7.0 ± 0.06 days, respectively. The sex ratio (male:female) was 1–2:10 (Kaimal and Ramani, 2011a).

The lowest finite rate of increase (λ) (1.0238) and intrinsic rate of natural increase (r_m) (0.0235) occurs at 15°C, and the population can double in 29.47 days. The highest finite rate of increase (λ) (1.3592) and intrinsic rate of natural increase (r_m) (0.3069) are at 35°C, and the population can double in 2.26 days. A high temperature is favourable for the population of *O. biharensis*. The highest net reproductive rate (R_0) (56.3000) occurred at 25°C, and the lowest mean generation time (T) (11.9721) at 35°C. The highest mean number of eggs (71.60 per female) and daily oviposition rate (4.10 eggs per female) occurred at 25°C. The lowest daily oviposition rate (0.34 eggs per female), resulting in the production of 8.80 eggs per female, was at 15°C. Longevity was lowest (19 ± 3.11 days) at 35°C and highest (98.90 ± 20.77 days) at 15°C (Ji *et al.*, 2005a). The developmental time from egg to adult was shortest (14.57 days) on longan and longest (20.71 days) on mango. The survival rate of *O. biharensis* on longan was the highest (93.8%), while it was lowest (83.8%) on lychee. The fecundity was highest (46.63 eggs) on longan and lowest (7.32 eggs) on mango. The intrinsic rate of natural increase on longan was the highest (0.1328), while it was lowest on mango. The population trend index can be ranked from the highest to the lowest in the order longan > lychee > mango (Zhang *et al.*, 2007b).

Rearing of *O. biharensis* on leaves of cowpea in the laboratory at 30 ± 2°C and 70 ± 5% RH showed that the life cycle of the mite comprised both sexual and parthenogenetic generations. The pre-oviposition, oviposition and post-oviposition periods lasted 1.5 days, 11.5 ± 0.75 days and 0.9 ± 0.25 day, respectively. The fecundity was 50.9 ± 4.7 and 40.2 ± 1.4 eggs for mated and non-mated females, respectively. The total duration of sexual development and parthenogenetic development was 6.45 ± 0.05 days and 6.0 ± 0.07 days respectively. The sex ratio (male: female) was 1–2:10 (Kaimal, 2013).

In Japan, it was observed that although the mite was polyphagous, it did not share any wild host species in the field, indicating that its source plants were completely separate in Okinawa. Such data are of great significance when considering vegetation control as part of the integrated management of these pest mites (Ohno *et al.*, 2011a). However, Kaimal (2013) suggests that cowpea may be a potential host for the successful establishment of *O. biharensis* throughout the year.

Natural enemies include the coccinellid *Stethorus vinsoni* Kapur (Moutia, 1958), the aeolothripid *Scolothrips pallidus* (Beach), and the cecidomyiid *Therodiplosis persicae* Kieffer (= *Feltiella acarisuga* (Vallot)) (Gutierrez, 1983, in Chazeau, 1985). In China, the use of the phytoseiid mite *Neoseiulus cucumeris* (Oudemans) (Ji *et al.*, 2008; Zhang *et al.*, 2009a) has been proposed, and an evaluation has been made of the predation of *O. biharensis* by the bdellid mite *Bdellodes japonicus* (Ehara) (Chen *et al.*, 2007b).

Symptomatology and damage

Feeding by *O. biharensis* produces white spots on the upper leaf surfaces of mango and dark bronzing on loquat leaves (Jeppson *et al.*, 1975). On guava, the mite feeds on the lower leaf surfaces, which become dull green and then bronzed. Populations increase during dry spells and when plants suffer from water stress, as when guava is cultivated in sandy soils (Kwee and Chong, 1990).

On cassava, the mite feeds on the upper surface of leaves and causes depletion of chlorophyll, inducing a typical rusty leathery appearance. Leaves show curling starting from the margins (Swaminathan, 2007). The feeding characteristics of and damage induced by *O. biharensis* were evaluated in the laboratory at 30 ± 2°C and 70 ± 5% RH on detached leaves of cassava. The mite developed on both surfaces of mature leaves and the petioles showed high population densities. Initial

symptoms of damage were many white spots at the points of feeding on the leaf surface. Prolonged feeding led to the formation of dark brown patches, crinkling and subsequent drying and defoliation of the infested leaves. Attack by mites affected the growth and vigour of cassava plants. Analysis of damage symptoms shows extensive bleaching and chlorosis of the leaves. The protein profile of leaves revealed five prominent bands in both infested and mite-free host plants, but mite-free samples had higher protein concentration than the infested samples. Losses of chlorophylls a and b were 86.40 ± 1.6 % and 81.03 ± 1.2 %, respectively. Chlorophyll loss is significant at 1% levels. The increase in total phenol content of infested plants was 11.67 ± 0.17 mg phenol g^{-1} plant material (Kaimal and Ramani, 2011b). The severity of damage to the cassava leaves was directly related to the feeding intensity of the mites and was a function of time. Infested cassava leaves showed a reduction in thickness as a result of disruption and reduction of the mesophyll layers. Cells exhibited coagulated protoplasts, a reduction in the numbers of cells and chloroplasts, alterations in cell structure, increases in space in the spongy layer, extensive disruption of the mesophyll cells and even reduction of chloroplasts in adjacent unpunctured cells (Kaimal *et al.*, 2011).

In India, a complex of four species of mites, *Eutetranychus orientalis* (Klein), *O. biharensis*, *Tetranychus cinnabarinus* (Boisduval) and *T. neocaledonicus* André caused significant yield reductions (17–33%) on cassava (Pillai and Palaniswami, 1983). Field studies carried out in West Bengal (India) to determine yield losses in okras infested by *E. orientalis*, *O. biharensis*, *T. cinnabarinus* and *T. neocaledonicus* highlighted a yield reduction of 17.46% (Jaydeb *et al.*, 1996).

Control

Different strategies have been evaluated to control *O. biharensis* infestations, such as the use of resistant varieties and biological, physical and chemical means.

PLANT RESISTANCE. Field experiments in Kerala (India) on the effects of infestation by *T. neocaledonicus*, *T. cinnabarinus*, *E. orientalis* and *O. biharensis* on the tuber yield of three cassava varieties (‘H 2304’, ‘H 97’, ‘M 4’) showed that the differences in tuber yield between plots with high and low mite population levels were significant for all three varieties, while the differences between moderate and high population levels were significant only for ‘M 4’ (Lal and Pillai, 1980). Moreover, the use of the cassava resistant varieties ‘CE-4’ and ‘CE-38’, combined with sprays of water and urea, and of insecticides when pest populations grew large, controlled the four pest mites (Pillai and Palaniswami, 1991).

Study of the development, longevity and reproduction of *O. biharensis* on four different cultivars of lychee (‘Feizixiao’, ‘Baitangying’, ‘Ziniangxi’ and ‘Sanyuehong’) show that the cultivar ‘Baitangying’ is the most suitable host plant and that the cultivar ‘Feizixiao’ is the most unsuitable host plant for *O. biharensis* (Chen *et al.*, 2005a). In particular, it has been observed that the total mortality rates from egg to adult on the four cultivars of lychee were 37.27, 32.45, 25.52 and 17.32%, respectively. The developmental period from egg to adult ranged from 16.97 days on ‘Sanyuehong’ to 21.05 days on ‘Feizixiao’. The average longevity of adult females ranged from 19.72 days on ‘Sanyuehong’ to 27.01 days on ‘Baitangying’. Total oviposition ranged from 68.80 eggs on ‘Baitnagying’ to 34.00 eggs on ‘Sanyuehong’. The daily oviposition rate of *O. biharensis* varied from 5.29 eggs on ‘Baitnagying’ to 1.76 eggs on ‘Feizixiao’. The net reproductive rate of increase (R_0), intrinsic rate of natural increase (r_m) and finite rate of increase (λ) on ‘Baitangying’ were the highest’, and the same three parameters (R_0, r_m and λ) were the lowest on the ‘Feizixiao’ cultivar (Chen *et al.*, 2005a).

BIOLOGICAL CONTROL. Laboratory studies have been done with various predatory mites. The bdellid mite *B. japonicus* preys on different spider mites, including *T. urticae*, *T. truncatus* Ehara, *Amphitetranychus viennenis* (Zacher) and *O. biharensis*, and at 25 ± 1°C it gives good control of *O. biharensis* (Chen *et al.*, 2007a). The functional response of female *N. cucumeris* preying on females, males or nymphs of *O. biharensis* and the numeric response of *N. cucumeris* at different temperatures has been also investigated. The results of this research indicate that *N. cucumeris* is a biocontrol agent to controlling *O. biharensis* in loquat orchard in the late autumn to early spring (Zhang *et al.*, 2009b).

PHYSICAL CONTROL. The effects of radiation on the mortality, fecundity and sterility of *O. biharensis* infesting common flower plants in Bangladesh have been studied by Majumder *et al.* (1996), and suggested as a quarantine treatment (Bhuiya *et al.*, 1999).

CHEMICAL CONTROL. Field studies on cassava in Kerala (India), to examine the economic aspects of treatments, showed that sprays of dimethoate, methyl-demeton and monocrotophos and of water had cost:benefit ratios of 1:6.42, 1:5.09, 1:2.55 and 1:2.6, respectively (Pillai and Palaniswami, 1983). Water sprays at run-off level, spraying neem oil or dimethoate are also effective (Swaminathan, 2007). Botanical acaricides include extracts of *Curcuma longa* L., *Streblus asper* Lour. (Shakhotaka), *Thunbergia laurifolia* Lindl., *Derris elliptica* (Wall.) Benth., *Derris scandens* (Roxb.) Benth., *Corchorus aestuans* L. and *Azadirachta indica* A. Juss.

Oligonychus coffeae (Nietner) (Fig. 20.33)

Common name

Tea red spider mite, red tea mite, red spider mite.

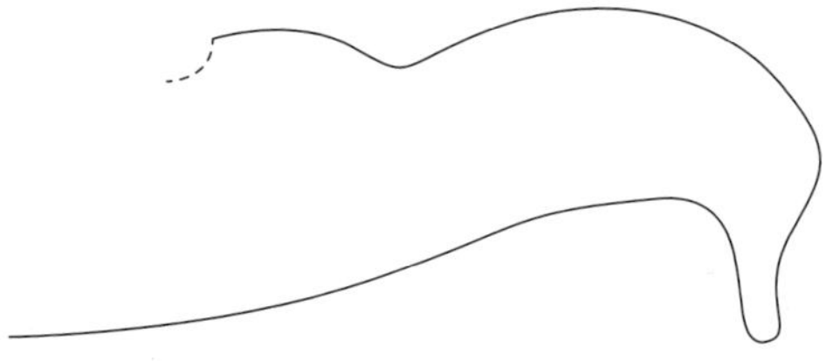

Fig. 20.33. *Oligonychus coffeae* (Nietner). Male, aedeagus (from Meyer Smith, 1987).

Diagnostic characteristics

FEMALE. The propodosoma is bright red and the hysteroma wine coloured. The peritremes end in a bulb. The dorsal body setae are serrate and longer than distances between their bases. The opisthosoma is striate mostly transversally but sometimes 'U' shaped or irregular between the dorsal setae e_1. The striae between the dorsal setae f_1 may be irregular to irregular longitudinal, and those anterior to the genital flap from longitudinal to transverse. Tibia I bears 7 tactile setae. Tarsus I has 3 tactile setae and 1 solenidion well proximal to the duplex setae, and 1 tactile setae occurs opposite the duplex setae on the ventral side of tarsus I (Pritchard and Baker, 1955; Meyer Smith, 1987).

MALE. The male is smaller than the female and lighter in colour. The palptarsus has a tiny terminal sensillum. Tarsus I bears 3 tactile setae and 2 solenidia well proximal to the duplex setae; the empodia have 5 pairs of proximoventral hairs. Distally, the aedeagus bends ventrally at a right angle to the shaft axis and gradually narrows to a slender truncate tip (Pritchard and Baker, 1955; Meyer Smith, 1987).

Geographical distribution

Australia (Pritchard and Baker, 1955), Azerbaijan (Nikolskii, 1953), Bangladesh (Andrews, 1915), Brazil (Flechtmann and Arleu, 1984), China (Ma and Yuan, 1976; Wang *et al.*, 1985; Ehara and Tho, 1988; Wang and Cui, 1992), Colombia (CABI, 1963), Congo (Gutierrez and Bonato, 1994), Congo (Democratic Republic of the Congo, formerly Zaire) (Baker and Pritchard, 1960), Costa Rica, Ecuador (CABI, 1963), Egypt (Abdel-Shaheed *et al.*, 1973), Ethiopia (CABI, 1963), Fiji (Gutierrez and Schicha, 1984), French Polynesia (Gutierrez, 1979, 1981), Hawaiian Islands (USA) (Garrett and Haramoto, 1967; Lee Goff, 1986), India (Andrews, 1928; Nassar and Ghai, 1981; Gupta, 1992; Gupta and Gupta, 1994; Gupta and Chatterjee, 1997; Babu *et al.*, 2004; Roobakkumar *et al.*, 2010), Indonesia (Kalshoven, 1951), Japan (Gotoh and Nagata, 2001; Gotoh *et al.*, 2003a, 2007b; Ohno *et al.*, 2011a), Kenya (Baker and Pritchard, 1960), Madagascar (Gutierrez, 1968), Malawi (Smee, 1943), Malaysia (Ehara and Tho, 1988; Othman and Zhang, 2003), Mauritius (Baker and Pritchard, 1960), Mozambique (Rodrigues, 1959), Myanmar (Ghosh, 1940), Nepal (Corpuz-Raros, 1989b), New Caledonia (Gutierrez, 1977), Nigeria (Matthysse, 1980), Paraguay (Aranda and Flechtmann, 1971), Philippines (Rimando, 1962a), Réunion (France) (Gutierrez, 1968), Samoa (Western) (Gutierrez, 1977), Solomon Islands (Bolland *et al.*, 1998), South Africa (Tucker, 1926; Pritchard and Baker, 1955; Baker and Pritchard, 1960), Sri Lanka (Nietner, 1861), Taiwan (Shiraki, 1919), Tanzania (CABI, 1963), Tasmania (Australia) (Gutierrez and Schicha, 1985), Thailand (Baker, 1975b; Ehara and Wongsiri, 1975), Tonga (Bolland *et al.*, 1998), Uganda (Nyiira, 1982), the USA (Pritchard and Baker, 1955), Vanuatu (Gutierrez, 1977), Vietnam (du Pasquier, 1932) and Zimbabwe (Meyer Smith, 1974).

Host plants

The whole list of host plants of *Oligonychus coffeae* includes 130 species ascribed to 43 families, with the highest number of species belonging to the Myrtaceae (16 species) and Fabaceae (13 species) families (Migeon and Dorkeld, 2006–2013). The families and species of economic importance are:

Anacardiaceae: cashew (*Anacardium occidentale* L.) (Baker and Pritchard, 1960; Meyer and Rodrigues, 1966), mango (*Mangifera indica* L.) (Baker and Pritchard, 1960; Gupta and Gupta, 1994; Matsuda *et al.*, 2012).

Annonaceae: custard apple (*Annona reticulata* L.), sugar apple (*A. squamosa* L.) (Bolland *et al.*, 1998; Migeon and Dorkeld, 2006–2013).

Aquifoliaceae: mate (*Ilex paraguariensis* A. St.-Hil.) (Aranda and Flechtmann, 1971).

Arecaceae: African oil palm (*Elaeis guineensis* Jacq.) (Baker and Pritchard, 1960).

Euphorbiaceae: rubber tree (*Hevea brasiliensis* (Willd. ex A. Juss.) Müll. Arg.), cassava (*Manihot esculenta* Crantz) (Rodrigues, 1968; Gutierrez and Bonato, 1994), castor bean (*Ricinus communis* L.) (Gupta and Gupta, 1994).

Fabaceae: derris (*Derris elliptica* (Wall.) Benth.) (Gupta and Gupta, 1994).

Lauraceae: camphor tree (*Cinnamomum camphora* (L.) J. Presl) (Das, 1959; Gupta and Gupta, 1994; Ohno *et al.*, 2011a), avocado (*Persea americana* Mill.) (Ehara, 1969a; Blommers and Gutierrez, 1975).

Lythraceae: pomegranate (*Punica granatum* L.) (Baker, 1975b).

Malvaceae: okra (*Abelmoschus esculentus* (L.) Moench) (Misra, 1913), jute (*Corchorus capsularis* L.), levant cotton (*Gossypium herbaceum* L.) (Gupta and Gupta, 1994).

Moraceae: white mulberry (*Morus alba* L.) (Gupta and Gupta, 1994).

Myrtaceae: strawberry guava (*Psidium cattleianum* Sabine), guava (*Psidium guajava* L.) (Meyer Smith, 1971).

Passifloraceae: purple granadilla (*Passiflora edulis* Sims) (Bolland *et al.*, 1998; Migeon and Dorkeld, 2006–2013).

Rosaceae: sweet cherry (*Prunus avium* (L.) L.) (Rodrigues, 1968), peach (*P. persica* (L.) Batsch) (Meyer and Ryke, 1959b; Nassar and Ghai, 1981; Gupta and Gupta, 1994; Gupta and Chatterjee, 1997).

Rubiaceae: Arabian coffee (*Coffea arabica* L.) (Nietner, 1861; Aranda and Flechtmann, 1971; Gupta and Gupta, 1994).

Sapindaceae: lychee (*Litchi chinensis* Sonn.) (Matsuda *et al.*, 2012).

Sapotaceae: sapodilla [*Manilkara zapota* (L.) P. Royen] (Migeon and Dorkeld, 2006–2013).

Theaceae: tea (*Camellia sinensis* (L.) Kuntze) (Tucker, 1926; Gupta, 1992; Gupta and Gupta, 1994; Gotoh and Nagata, 2001; Gotoh *et al.*, 2003a, 2007b; Babu *et al.*, 2004; Roobakkumar *et al.*, 2010; Vasanthakumar *et al.*, 2012).

Vitaceae: grapevine (*Vitis vinifera* L.) (Sayed, 1946b; Baker and Pritchard, 1960).

Bio-ecology

O. coffeae is a pantropical mite living on several wild and economic plants. In tropical and subtropical areas it is a severe pest of tea, as well as of coffee, mango, cotton and jute (Tucker, 1926; Meyer Smith, 1981a; Gupta, 1992; Gupta and Gupta, 1994; Gotoh and Nagata, 2001; Gotoh *et al.*, 2003a, 2007b; Babu *et al.*, 2004; Roobakkumar *et al.*, 2010). The mite is also reported

on avocado, cassava, camphor, cashew, citrus, guava, grape, mulberry, oil palm, rubber tree and ornamental plants.

The tea red spider mite reproduces by arrhenotokous parthenogenesis and has a haploid number of chromosomes of three (Helle *et al.*, 1970). It feeds on the upper leaf surface and produces little webbing. At 22°C, the female lays during its lifetime 40–50 eggs. Under these conditions, the lifespan of larva, protonymph and deutonymph is 8, 40 and 2–3 days, respectively. The developmental time from egg to adult is 14–15 days. Temperature and humidity affect the development of populations. Under constant temperatures of 34°C and at 17% RH, the eggs fail to hatch, as they do at 33°C and RH <72% as well, while at an RH just above 72% some hatch occurs. Optimal conditions of development are 20–30°C and 49–94% RH. Under these conditions, the mite may develop 22 generations a year on tea (Das, 1959, 1960; Hu and Wang, 1965; Das and Das, 1967). Gotoh and Nagata (2001) observed that the developmental time declined as the temperature increased from 15 to 32°C. The threshold temperature for development and the total effective temperature for the egg-to-egg period were 10°C and 232.6 degree-days, respectively. At 25°C, the intrinsic rate of natural increase (r_m) was 0.186. In Bangladesh, Haque *et al.* (2007) reported that eggs of *O. coffeae* developed to adults within the shortest duration of 5.30 ± 0.16 days at 30.28°C and 76.00% RH in June, but that it took 12.91 ± 0.21 days at 19.80°C and 75.41% RH in January. The highest temperature increased the developmental rate and reduced the duration of the developmental stages. Temperature exerts a significant effect on all of the developmental stages of *O. coffeae* except the deutonymph. However, RH has no significant effect except on the deutonymph. Mazid *et al.* (2013) reported that with an increase in temperature, the developmental rate of mites increased and the duration of the developmental stages decreased. They found that on tea leaves at laboratory temperatures ranging from 22.75 to 32°C, the incubation period was 7.01 ± 0.04 to 3.65 ± 0.04 days, the larval period lasted from 2.73 ± 0.02 to 1.50 ± 0.03 days, the protonymphal period lasted from 2.48 ± 0.01 to 1.39 ± 0.02 days and the deutonymphal period lasted from 2.25 ± 0.06 to 1.38 ± 0.02 days. The pre-oviposition period was longest (2.00 ± 0.04 days) at 22.75°C and shortest (1 ± 0.08 days) at 32°C. The highest fecundity (76.52 ± 0.14 eggs) was at 22.75°C and the lowest (52.30 ± 0.11 eggs) at 32°C; the maximum daily oviposition (16.06 ± 0.05) occurred at 22.75°C and the minimum (10.83 ± 0.18) at 32°C (Mazid *et al.*, 2013).

In north-east India, the tea red spider mite can be found throughout the year. During winter, its populations live on the remaining few old leaves and on small leaves at the base of the shoots of the tea bushes. These overwinter populations usually increase in early March and the highest densities are observed in late March and early April. Injuries to the leaves are severe during May and June or until the monsoon rains. The attacks are renewed after the rainy season, but never become as injurious as during the pre-monsoon period. In the cool weather, the densities of the populations are lower and damage is uncommon. On tea, pruning removes many of the old leaves and shoots, and thus a large part of the mite population (Das, 1959, 1960; Hu and Wang, 1965; Das and Das, 1967).

Coccinellids prey upon eggs and other biological stages of *O. coffeae* (Das, 1959). These include *Stethorus comoriensis* Chazeau (Chazeau, 1971a), *S. madecassus* Chazeau (Blommers and Gutierrez, 1975) and *S. gilvifrons* (Malsant) (Banerjee and Cranham, 1985). The predatory insects include chrysopids (Das, 1959; Banerjee and Cranham, 1985) and staphylinids (Banerjee and Cranham, 1985), such as *Oligota pygmaea* (Solier) (Sachin *et al.*, 2008; Perumalsamy *et al.*, 2009). Predatory mites include phytoseiids, such as *Amblyseius parasundi* Blommers (Blommers and Gutierrez, 1975), *A. coccosocius* Ghai *et* Menon and *Neoseiulus longispinosus* (Evans) (Chakraborty *et al.*, 2005; Kongchuensin *et al.*, 2005; Sahu *et al.*, 2008), and stigmaeid (Banerjee and Cranham, 1985) and cunaxid mites (Sah *et al.*, 1999).

Symptomatology and damage

Symptoms of damage on tea first appear as yellowish spotting along the midrib and leaf veins, and sometimes on the petioles. As infestation continues, these spots turn brown and may coalesce, forming large damaged areas. The whole leaf surface may be bronzed and necrotic. Defoliation may occur and growth of the tea bushes may be affected (Das, 1959, 1960; Jeppson *et al.*, 1975; Meyer Smith, 1981a). In India, *O. coffeae* is one of the major pests of tea, accounting for an estimated loss of 18% in yield and deterioration of the quality of manufactured tea (Sachin *et al.*, 2008).

Control

Different strategies to optimize the control of tea red spider mite are known, such as choice of plant cultivars, horticultural practices and the use of biological and chemical means.

PLANT RESISTANCE. Some tea clones can reduce the fecundity and decrease the rate of development of *O. coffeae* (Sudoi, 1992), and in Kenya, some ('BBLK152', '6/8' and '7/9') are classed as resistant to *O. coffeae*, while others are moderately resistant or susceptible. In Sri Lanka, there are also some tea clones ('MT18' and 'TRI 2027') that are resistant to *O. coffeae*, while others ('CY9' and 'DT 1') are susceptible (Thirugnanasuntharan and Amarasinghe, 1990). In south Assam (India), among 18 tea clonal cultivars, some were highly susceptible to the red spider mite, some were moderately susceptible and others were fairly resistant (Choudhury *et al.*, 2006). The highest fecundity of *O. coffeae* occurs at the pivotal age of 20.5 and 24.5 days in the Assam-type and the China-type leaves, respectively. The net reproductive rate is lower when the mites are reared on Assam-type than on China-type tea leaves, and the mean generation length of *O. coffeae* is shorter on the Assam-type than on the China-type. The rate of multiplication in each generation and the age-specific fecundity of *O. coffeae* reared on the China-type are higher than on the Assam-type leaves (Chakraborty *et al.*, 2007). The use of more resistant clones will reduce the development of mite populations and mite damage.

CULTURAL PRACTICES. Pruning generally removes most of the old leaves and the shoots, e.g. the janams (cataphylls), and consequently many mites. It is well known that pruned tea is less infested than unpruned tea or tea where a part of the top is removed, e.g. skiffed tea. Moreover, early pruned tea is usually

more infested than late pruned bushes. It follows that the winter removal of the leaves tends to reduce the risk of summer attacks. Preventive defoliation of bushes then helps to reduce mite infestation. Abandoned tea is heavily infested during the first summer, but the following development of the trees induces a decrease of the infestation. This phenomenon causes the mite to be absent from tea seed gardens, with the exception of some border trees (Das, 1959, 1960; Hu and Wang, 1965; Das and Das, 1967).

BIOLOGICAL CONTROL. Studies on the biological control of *O. coffeae* have looked at the response of the mite to predatory mites and insects, and some pathogens. However, none of the methods tested is currently used in the field. The most important reports are summarized below.

The interaction between the tea red spider mite and the phytoseiid *A. coccosocius* at different prey to predator ratios (1:1, 1.5:1, 6:1 and 10:1) was shown to be highly effective at all of the ratios. The phytoseiid took 7, 8, 14 and 13 days, respectively, to consume the prey entirely (Chakraborty *et al.*, 2005).

The life history and feeding potential of *S. gilvifrons*, a major predator of *O. coffeae*, were studied in the laboratory. Both the larvae and adults are voracious feeders; the larvae feed by sucking, while the adults feed by chewing the whole mite. The highest consumption is exhibited by the third instar grubs, followed by the adult beetles (Sarmah and Bhattacharyya, 2002).

Monitoring of seasonal abundance of the staphylinid *O. pygmaea* and *O. coffeae* under the prevailing field conditions in Tamil Nadu (India) shows that *O. pygmaea* exhibits a typical pattern of population dynamics, with a peak during January to March and low incidence during June to September. Peaks in the population of *O. pygmaea* coincided with the abundance of *O. coffeae* in the tea fields. Weather factors such as high temperature, low RH and low sunshine hours adversely affected the populations of *O. pygmaea*. The first to third instar larvae of *O. pygmaea* prey upon 31.0–133.2 eggs of mites a day. Third instar larva of *O. pygmaea* prey upon an average of 133.2 eggs, 46.4 hexapod larvae, 39.6 nymphs and 11.4 adults a day. Adult females prey on a greater number of red spider mites than do the males (Perumalsamy *et al.*, 2009).

The fungus *Pochonia* (*Verticillium*) *lecanii* (Zimmerman) Viegas is effective against the tea red spider mite (Devi *et al.*, 2010).

CHEMICAL CONTROL. The phenology of the tea plant and the persistence of some acaricides may make chemical control difficult. The thick canopy of the tea plant during the growing season hinders sprays of chemicals during the summer. In addition, acaricides with persistent residues inside and/or over the leaves cannot be used during the plucking season. This drawback may oblige the discarding of the leaves for some plucking rounds. However, a rational choice of acaricides allows the control of the mite populations (Das, 1960; Jeppson *et al.*, 1975). In Assam (India), prophylactic sprays of organic and inorganic chemicals administered during winter (December–February) usually help to reduce the carried-over population of *O. coffeae* and decrease the level of attack in the following years (Choudhury *et al.*, 2006).

Acaricides tested include amitraz, acrinathrin (Sah *et al.*, 1999), bifenazate (Gotoh *et al.*, 2001), bifenthrin (Gotoh *et al.*, 2001; Barindra Talukdar *et al.*, 2008b), bromopropylate (Sah *et al.*, 1999), buprofezin (Barindra Talukdar *et al.*, 2008a,b), chlorfenapyr (Gotoh *et al.*, 2001), diafenthiuron (Barindra Talukdar *et al.*, 2008b; Sarkar *et al.*, 2009), dicofol (Sah *et al.*, 1999; Jana and Somchoudhury, 2005; Chakraborty Gautam *et al.*, 2006; Rahman *et al.*, 2007a,b; Barindra Talukdar *et al.*, 2008b; Sahu *et al.*, 2008), endosulfan (Sah *et al.*, 1999), ethion (Sah *et al.*, 1999; Jana and Somchoudhury, 2005; Rahman *et al.*, 2007a,b; Barindra Talukdar *et al.*, 2008b), fenazaquin (Sah *et al.*, 1999; Sahu *et al.*, 2008), fenbutatin oxide (Gotoh *et al.*, 2001), fenpropathrin (Jana and Somchoudhury, 2005), hexythiazox (Sahu *et al.*, 2008), profenofos (Jana and Somchoudhury, 2005), propargite (Rahman *et al.*, 2007b) and spiromesifen (Chakraborty Gautam *et al.*, 2006). Some chemicals, e.g. diafenthiuron and buprofezin, were tested alone and with adjuvants (karanj oil and citronella oil) in the presence of surfactant (APSA) (Barindra Talukdar *et al.*, 2008a; Sarkar *et al.*, 2009). A mixture of ethion and sulfur controlled *O. coffeae* reasonably well (Choudhury *et al.*, 2006).

Many efforts have been focused on the study of the biorational pesticides. These include neem (*Azadirachta indica* A. Juss)-based products. A mixture of neem and ethion gave excellent control of the mite (Choudhury *et al.*, 2006). Rahman *et al.* (2007a) reported that neem formulations, alone or at reduced rates, in combination with synthetic insecticides, could play an integral role in a progressive tea plant management programme. Crude plant extracts of *Acores calamus* (L.), *Xanthium strumarium* (L.), *Polygonum hydropiper* (L.) and *Clerodendron infortunatum* (Gaertn) are suggested as safer phytopesticidal products in both organic and inorganic tea estates as one of the potent tools in integrated mite management (Sarmah *et al.*, 2009). Acaricidal activities of petroleum ether, acetone and methanol fractions of *X. strumarium*, *A. calamus* and *Pongamia pinnata* (L.) Pierre were evaluated in the laboratory. All the solvent fractions of *A. calamus* and the methanolic fraction of *X. strumarium* killed eggs, but no ovicidal action was found of *P. pinnata* extracts. All the plant extracts adversely affected post-embryonic development as well as adults of *O. coffeae* (Sarmah *et al.*, 2007). Leaf extracts of *P. hydropiper* and *Lantana camara* L., with petroleum ether, acetone and methanol, were also tested in the laboratory. The petroleum ether and acetone extracts of *P. hydropiper* showed strong acaricidal and ovicidal properties compared with the extracts of *L. camara* (Sarmah *et al.*, 2000).

A biocide (supplied by M/s. Integrated Fire Protection Pvt. Ltd., Jalpaiguri) caused high mortality of the mite after 1 h of treatment and the efficacy of the chemical continued for up to 8 days. The biocide had no effect on non-target insects, such as parasites and predators. No traces of residue or uncomfortable odours of the biocide were found in the liquor from tea samples collected from the treated bushes (Shwetaketu Dev and Chenchaiah, 2005).

INTEGRATED PEST MANAGEMENT. Biorational insecticides, especially neem formulations, alone or at reduced rates, in combination with synthetic insecticides, could play an integral role in a progressive tea plant management programme (Rahman *et al.*,

2007a). Buprofezin has been tested alone and in combination with adjuvants (karanj and citronella oils). The best result was obtained with buprofezin in both the combinations with adjuvants. Considering the residue problems in made tea, the biorational pesticide cum acaricide can be included in the IPM module of tea red spider mite in West Bengal (terai condition) (Barindra Talukdar *et al.*, 2008a). Moreover, sprays of propargite were suggested as one of the alternative acaricides in integrated red spider mite management (Rahman *et al.*, 2007b). Crude plant extracts of *A. calamus*, *X. strumarium*, *P. hydropiper* and *C. infortunatum* are suggested as safer phytopesticidal products in both organic and inorganic tea estates as one of the potent tools in integrated mite management (Sarmah *et al.*, 2009).

Oligonychus gossypii (Zacher) (Fig. 20.34)

Common name

Cotton red mite.

Diagnostic characters

FEMALE. The body is ovoid and reddish in colour. The dorsal body setae are serrate and longer than the distances between their bases. The peritremes are strongly turned backwards distally. The opisthosoma is longitudinally striate between the dorsal setae e_1 and f_1, and forms a diamond shape. Tibia I bears 9 tactile setae and tibia II bears 7 tactile setae. Tarsus I has 4 tactile setae proximal to the duplex setae, and 2 ventral setae proximal to the duplex setae (Meyer Smith, 1974; Jeppson *et al.*, 1975).

MALE. The distal sensillum of the palptarsus is about 4 times as long as broad. The aedeagus narrows distally and curves dorsally at about a right angle; it has a distal large knob with a small anterior projection and a long undulate posterior projection, with the tip directed ventrally. Tarsus I has 4 tactile setae and 3 solenidia proximal to the duplex setae. The empodia has a claw and 3 proximoventral hairs (Meyer Smith, 1974; Jeppson *et al.*, 1975).

Geographical distribution

Angola (Carmona, 1967), Benin (Carmona, 1967; Gutierrez and Etienne, 1981b), Brazil (Flechtmann and Baker, 1970; Feres *et al.*, 2005; Hernandes and Feres, 2006; Vis *et al.*, 2006), Cameroun (Meyer Smith and Bolland, 1984), Central African Republic (Gutierrez, 1987b), Colombia (Urueta, 1975), Congo (Gutierrez and Etienne, 1981b; Bonato and Gutierrez, 1993; Gutierrez and Bonato, 1994; Navajas *et al.*, 1996), Congo (Democratic Republic of the Congo, formerly Zaire) (Baker and Pritchard, 1960), Costa Rica (Byrne *et al.*, 1983), Ecuador (McGregor, 1955), Ethiopia (Matthysse, 1980), Guinea-Bissau (Hirst, 1926), Honduras (Baker and Pritchard, 1962), Kenya (Nyiira, 1982), Madagascar (Helle *et al.*, 1970), Nigeria (Baker and Pritchard, 1960), Sao Tome and Principe (Hirst, 1926), Senegal (Gutierrez and Etienne, 1981b), Sierra Leone (Hirst, 1926), Tanzania (Nyiira, 1982), Togo (Zacher, 1921a), Uganda (Nyiira, 1982) and Venezuela (Byrne *et al.*, 1983).

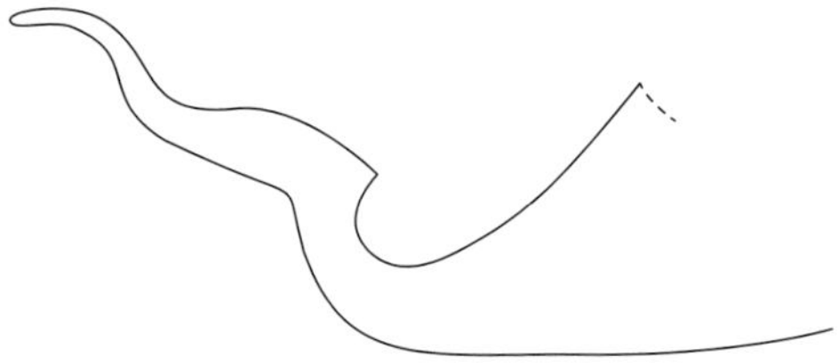

Fig. 20.34. *Oligonychus gossypii* (Zacher). Male, aedeagus (from Meyer Smith, 1987).

Host plants

The whole list of host plants includes 47 species belonging to 17 families (Migeon and Dorkeld, 2006–2013). The families and species of economic importance are: Caricaceae – papaya (*Carica papaya* L.) (Hirst, 1926); Euphorbiaceae – rubber tree (*Hevea brasiliensis* (Willd. ex A. Juss.) Müll. Arg.) (Mesa Cobo and Zuluaga Cardona, 1983; Hernandes and Feres, 2006; Vis *et al.*, 2006), cassava (*Manihot esculenta* Crantz) (Hirst, 1926; Baker and Pritchard, 1960; Gutierrez and Etienne, 1981b; Bonato and Gutierrez, 1993; Gutierrez and Bonato, 1994; Navajas *et al.*, 1996); Fabaceae – groundnut (*Arachis hypogaea* L.) (Urueta, 1975), pigeon pea (*Cajanus cajan* (L.) Millsp.) (Baker and Pritchard, 1962), sieva bean (*Phaseolus lunatus* L.) (Hirst, 1926), winged bean (*Psophocarpus tetragonolobus* (L.) DC.) (Mesa Cobo and Zuluaga Cardona, 1983); Malvaceae – okra (*Abelmoschus esculentus* (L.) Moench) (Gutierrez and Bonato, 1994), *Gossypium* sp. (Zacher, 1921a; Baker and Pritchard, 1960), cacao tree (*Theobroma cacao* L.) (McGregor, 1955; Urueta, 1975); Rosaceae – peach (*Prunus persica* (L.) Batsch) (Baker and Pritchard, 1960).

Bio-ecology

O. gossypii is common in African and South American equatorial areas, where it lives on wild and economic plants. It was originally described as a pest of cotton in Togo (Zacher, 1921a) and later in Nigeria (Baker and Pritchard, 1960). The cotton red mite also infests cassava (Hirst, 1926; Pritchard and Baker, 1960; Bonato *et al.*, 1994, 1995a,b,c), papaya (Hirst, 1926), the rubber tree (Fazolin and Pereira, 1989; Flechtmann, 1989b), and other economic plants, such as citrus, beans, peach, rose and ornamentals (Baker and Pritchard, 1960). Despite its wide polyphagy and the first record on cotton, the mite is better known as a pest of cassava.

The cotton red mite lives on both leaf surfaces and produces much webbing. It reproduces by arrhenotokous parthenogenesis and has a haploid number of chromosomes of two (Helle *et al.*, 1970). The lower thermal threshold is 11°C and development occurs in the range from 22 to 36°C, with the shortest developmental time at 31°C (8.2 days), and the maximum fecundity at 26°C (36.3 eggs). The highest intrinsic rate of increase (r_m) was observed at 31°C (0.214) (Bonato *et al.*, 1995a). The RH influences the population dynamics of the mite and may explain part of the decrease in populations observed in the middle to the end of the dry season, and the absence of mites during the wet season. Observations in the

laboratory at three constant RH values (30, 60 and 90%), at 26 ± 1°C and with a photoperiod of 12L:12D show that low (RH 30%) and high (RH 90%) air humidity have a negative effect on mite biology compared with medium air humidity (60% RH). The strongest effect occurs at 90% RH, when 96% of the immature stages die (Bonato *et al.*, 1995b).

Natural enemies include some pathogenic fungi, such as *Neozygites floridana* Weiser *et* Muma (Yaninek *et al.*, 1996), *N. tanajoae* Delalibera *et al.* (Delalibera *et al.*, 2004) and *Hirsutella thompsonii* Fisher (Yaninek *et al.*, 1996), and various phytoseiid mites, including some species mainly associated with *Mononychellus tanajoa* (Bondar) on cassava, such as *Typhlodromalus aripo* De Leon (Onzo, 2003; Onzo *et al.*, 2009) and *Neoseiulus idaeus* Denmark *et* Muma (Yaninek *et al.*, 1992).

Symptomatology and damage

The damage caused by *O. gossypii* on cassava is widely unknown. Research on cassava in Congo (Bonato *et al.*, 1994) shows that *O. gossypii* population sizes during outbreaks and their impact on yield losses are comparable to those of *Mononychellus progresivus* Doreste. In crops attacked by *M. progresivus* and *O. gossypii* in Central Africa, leaf dry matter increases during the rainy season and decreases during the dry season. Stems contain the reserves used for initiating plant growth after the dry season. Dry matter increases in the roots, which form the true storage organs. At the start of the second year of cultivation, one-third of the root biomass is used to support the recovery of plant growth and development. Mite densities fewer than 50 mobile forms per leaf (maximum number of mites) are not sufficient to cause significant loss of leaf, stem and root dry matter. Nevertheless, the mites divert part of the biomass allocated to the roots, thereby reducing the efficiency of storage root production in the infested plot (Bonato *et al.*, 1994).

In Brazil, a great abundance of the mites were registered on rubber tree crops in the states of Pará and Amazonas, leading to intense bronzing and premature fall of the leaves (Fazolin and Pereira 1989; Flechtmann, 1989b).

Control

Sampling plans have been developed on cassava in Africa. Analysis of intra-plant distribution showed a preference of *O. gossypii* for the middle and lower parts of the plant. Thus, the first fully developed leaf in the middle part was chosen as the common sampling unit. On this unit, the mite displays an aggregated distribution. Binomial and enumerative sampling plans permit reliable estimates of *O. gossypii* densities. Forty fully developed leaves provided satisfactory estimates of the densities of both mite species (*M. progresivus* and *O. gossypii*) (Bonato *et al.*, 1995c). For control options on cassava, see *M. tanajoa*. On other crops, control seems unnecessary.

Oligonychus ilicis (McGregor) (Fig. 20.35)

Common name

Southern red mite, coffee red spider mite.

Diagnostic characteristics

FEMALE. Purplish or reddish in colour, paler anteriorly with a pale spot medially. The dorsal body setae are set on small tubercles. Tibia I has 7 tactile setae; tarsus I has 3 setae proximal to the duplex setae, and the proximal member of the duplex setae is very short (Jeppson *et al.*, 1975).

MALE. The aedeagus is bent ventrally, narrow distally and rather flattened dorsoposteriorly (Jeppson *et al.*, 1975).

Geographical distribution

Australia (Knihinicki, 1999), Brazil (Calza and Sauer, 1952; Flechtmann and Baker, 1975), Italy (Rota and Biraghi, 1987; Pellizzari *et al.*, 2005), Japan (Pritchard and Baker, 1955; Ehara, 1960; Gotoh *et al.*, 2003a, 2007a), South Korea (Lee, 1989), the Netherlands (Vierbergen, 1990), Paraguay (Aranda and Flechtmann, 1971) and the USA (McGregor, 1917).

Host plants

The whole list of host plants numbers 38 species of 16 families (Migeon and Dorkeld, 2006–2013). The families and species of economic importance are:

Aquifoliaceae: *Ilex crenata* Thunb. (Reeves, 1963), American holly (*I. opaca* Aiton) (McGregor, 1917), *Ilex* sp. (Pritchard and Baker, 1955; Vierbergen, 1990).

Ericaceae: *Azalea sp.* (McGregor, 1950), *Rhododendron* sp. (Pritchard and Baker, 1955; Flechtmann and Baker, 1975; Vierbergen, 1990), koreansk azalea (*R. schlippenbachii* Maxim.).

Fagaceae: *Quercus* sp. (Bolland *et al.*, 1998; Migeon and Dorkeld, 2006–2013).

Juglandaceae: *Carya illinoinensis* (Wangenh.) K. Koch (Flechtmann and Baker, 1970), English walnut (*Juglans regia* L.) (Bolland *et al.*, 1998; Migeon and Dorkeld, 2006–2013).

Lauraceae: camphor tree (*Cinnamomum camphora* (L.) J. Presl), laurel (*Laurus nobilis* L.) (Pritchard and Baker, 1955).

Myrtaceae: *Eucalyptus* sp. (Mesa Cobo and Zuluaga Cardona, 1983), *E. camaldulensis* Dehnh., *E. grandis* W. Hill ex Maid., *E. pellita* F. Muell. (Flechtmann, 1983), *E. tereticornis* Sm. (Flechtmann, 1982), guava (*Psidium guajava* L.) (Aranda and Flechtmann, 1971).

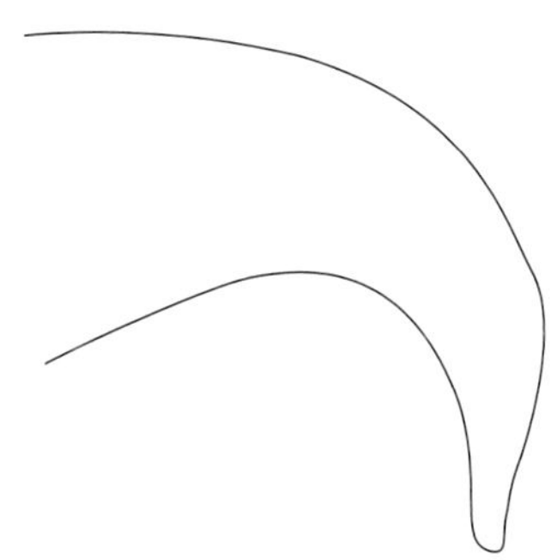

Fig. 20.35. *Oligonychus ilicis* (McGregor). Male, aedeagus (drawing by Vacante).

Pinaceae: *Picea* sp. (Bolland *et al.*, 1998; Migeon and Dorkeld, 2006–2013).

Platanaceae: American sycamore or American plane tree (*Platanus occidentalis* L.) (Bolland *et al.*, 1998; Migeon and Dorkeld, 2006–2013).

Rosaceae: quince (*Cydonia oblonga* Mill.), loquat (*Eriobotrya japonica* (Thunb.) Lindl.), pear (*Pyrus communis* L.) (Bolland *et al.*, 1998; Migeon and Dorkeld, 2006–2013).

Rubiaceae: Arabian coffee (*Coffea arabica* L.) (Calza and Sauer, 1952).

Theaceae: tea plant (*Camellia sinensis* (L.) Kuntze) (Ehara, 1960; Gotoh *et al.*, 2003a, 2007a; Matsuda *et al.*, 2012).

Bio-ecology

O. ilicis is a pest of several economic plants, such as American holly, azalea, camellia, the camphor tree, coffee, conifers, eucalyptus, guava, laurel, loquat, oak, pear, plane tree, quince, walnut, ornamentals and other plants (Ehara, 1963a; Jeppson *et al.*, 1975; Knihinicki, 1999). In Brazil, it is one of the main pest mites of coffee, although it is not reported among the primary pests (Franco *et al.*, 2009). The southern red mite lives mainly on the upper leaf surface of host plants, reproduces by thelitoky and/or arrhenotoky, and has $n = 3$ chromosomes (Flechtmann and Flechtmann, 1982; Flechtmann, in Helle and Pijnacker, 1985). Childs (1980) found that populations of *O. ilicis* were predominantly on the undersurface of *I. crenata* 'Hetzii' leaves, and Knihinicki *et al.* (1999) reported that mites also occurred on the lower leaf surface. The mite overwinters as eggs, hatching in late April and early May (in the USA). The populations peak in June at about the time of migration to new growth, and they often decline in late June and July and increase in September, when they start the deposition of overwintering eggs (Matthysse and Naegele, 1952; Mague and Streu, 1980; Knihinicki *et al.*, 1999). Between 22 and 24°C, the female lays 10–15 eggs, the life cycle takes in 2 weeks, and the female lives 15 days (Calza and Sauer, 1952). Dry climatic conditions favour an increase of populations. Conversely, rainy seasons or low damp areas hinder mite development, and control measures are usually unnecessary (Jeppson *et al.*, 1975). On coffee in Brazil, the periods of larger occurrence of *O. ilicis* coincide with low precipitation (Franco *et al.*, 2008). Dispersion occurs through webbing moved by air currents (Jeppson *et al.*, 1975).

On *I. crenata*, the southern red mite develops under constant temperatures ranging from 15.5 to 35°C. Developmental periods of all biological stages and adult longevity decrease as temperatures increase. Male development is more rapid than female development at all temperatures, except 26.5°C. The maximum rate of oviposition is affected by temperature, and the length of pre-oviposition, oviposition and post-oviposition generally decreases as the temperature increases. Mortality curves for males and females drop more abruptly as the temperature increases. The survival rate of the mite is greatest for eggs, and the highest mortality occurs in the larval stage. The net reproductive rate (R_0) is highest (7.0961) at 26.5°C. The mean generation time (T) is shortest (2.5695) at 35°C and longest (7.3119) at 21°C. The intrinsic rate of natural increase (r_m) is greatest (0.4694) at 26.5°C. The number of degree-days required for development of each immature stage at six different temperatures shows that the egg stage requires the greatest amount of heat (Childs, 1980). On *I. crenata*, the mean number of mites living in the upper portion of the tree differs significantly from the means for the lower and middle parts of the holly plants. Means for eggs did not differ with regard to height. Significant differences for zones occur for both eggs and mites, there being more eggs and mites located in the inner zone. Direction also influences egg numbers, with the most eggs and mites being found in the north and south quadrants (Childs, 1980).

Natural enemies include some acaropathogenic fungi, such as *Hirsutella thompsonii* Fisher (Gardner *et al.*, 1982) and various predators. Insects include the thrips *Scolothrips latipennis* Priesner (Oliveira, 2000), the chrysopids *Chrysopa* spp. (Oliveira, 2000) and *Chrysoperla externa* (Hagen) (Pedro Neto *et al.*, 2008), and some coccinellids, such as *Stethorus* spp. (de Oliveira, 2000). Predatory mites include some stigmaeids (*Agistemus* spp.) (Oliveira, 2000) and different phytoseiids, such as *Iphiseiodes zuluagai* Denmark *et* Muma (Oliveira, 2000; Franco *et al.*, 2008; Reis and Franco, 2010), *Amblyseius herbicolus* (Chant) (Franco *et al.*, 2008; Reis and Franco, 2010), *A. compositus* Denmark *et* Muma (Reis and Franco, 2010), *Euseius citrifolius* Denmark *et* Muma (de Oliveira, 2000; Franco *et al.*, 2008; Reis and Franco, 2010), *E. concordis* Chant (Oliveira, 2000; Franco *et al.*, 2008) and *E. alatus* DeLeon (Franco *et al.*, 2008; Reis and Franco, 2010).

Symptomatology and damage

Feeding of the southern red mite on the upper leaf surface of azalea, camellia, loquat, sycamore and other host plants causes bronzing and browning along the midrib. The upper leaf surface of walnut appears dirty and ashy grey. Symptoms are due to cells damaged by mite feeding on the epidermis and palisade layer of the upper leaf surface (Smith, 1939b; Jeppson *et al.*, 1975). Knihinicki *et al.* (1999) report that the mite prefers mature leaves rather than new shoots of azalea and feeds on both leaf surfaces. It usually damages the edges of leaves first, and high levels of infestation induce the leaves to turn grey or brown and fall prematurely. Infested coffee plants show webbing and bronzing around the leaf veins. Bronzing intensity on infested leaves of *C. arabica* cv. 'Catuai Vermelho' is proportional to the degree of internal damage. *O. ilicis* introduces its stylet into the adaxial epidermis, destroying chloroplasts in the feeding surface. The number of chlorophyll-free cells in the palisade parenchyma is proportional to the intensity of attack by the mite. The cells of the spongy parenchyma set below the feeding areas do not show alterations, having intact chloroplasts, with the exception of some leaves that suffer from heavy attack. The net photosynthetic rate declines, but only in leaves that are strongly infested, as a result of destruction of the chloroplasts in the attacked cells. However, transpiration and stomatal conductance are unaffected by red mite feeding. The concentrations of ethylene produced by the leaves are similar regardless of levels of attack, suggesting that red mite infestation does not lead to leaf abscission. Anatomical studies of *C. arabica* cv. 'Mundo Novo' leaves of plants visually recovered from red mite attack remain on plants with the scars present (Fahl *et al.*, 2007). On *C. arabica* cv. 'Catuai' leaves, a negative correlation was observed between the level of infestation and potential photosynthesis. At infestation rates of 15, 30, 60 and 120 mites

per leaf, the photosynthetic rates were reduced by 37.2, 38.7, 46.0 and 50.1%, respectively (Franco *et al.*, 2009). The *C. canephora* Pierre ex A. Froehner 'Conillon' cultivar is regarded as more sensitive to the mite than arabica coffee (Carvalho *et al.*, 2008).

Control

Control of the southern red mite is mainly based on sprays of chemicals. Biological control has been extensively investigated on coffee but has not been used commercially.

MONITORING. Research on the spatial distribution of southern red mite on *I. crenata* 'Hetzii' showed that the optimum sampling location would be in the inner zone and at the upper height in either the north or south quadrants of the tree (Childs, 1980).

CULTURAL PRACTICES. The effect of windbreaks of arboreal legumes on mite population on coffee has been evaluated in Brazil. The windbreak consisted of parallel lines of trees of pigeon pea (*Cajanus cajan* Millsp.), bracatinga (*Mimosa scabrella* Benth), leucaena (*Leucaena leucocephala* (Lam.) De Wit) and acacia (*Acacia mangium* Willd), respectively. However, arboreal legumes did not affect the number of pest mites or of predatory mites, especially of the phytoseiid mite *E. citrifolius*, on coffee plants (Pedro Neto *et al.*, 2010).

BIOLOGICAL CONTROL. Currently, it seems that no strategy of biological control alone is applied in the field. The available information concerns preparatory laboratory studies and faunistic research. The predation potential of the phytoseiid mites *I. zuluagai*, *E. citrifolius* and *A. herbicolus* on the developmental phases of *O. ilicis* has been investigated in the laboratory. The adult females are the most efficient predators, followed by the nymph, male and larval phases. Adult *O. ilicis* mites are the least predated of all of the developmental phases by the predatory mites; the larvae were the preferred stage (Franco *et al.*, 2007). Evaluation of the effects of the webbing produced by *O. ilicis* on its predation by females of the phytoseiids *I. zuluagai*, *E. citrifolius* and *A. herbicolus* were investigated in the laboratory. In the presence of webbing, the consumption of eggs, larvae and nymphs by *I. zuluagai*, and of eggs and larvae by *E. citrifolius*, was lower. For *A. herbicolus*, egg predation was lower, but larval predation did not vary significantly and the predation of nymphs and adults was higher in the presence of webbing. Predators as a whole are more efficient at consuming larvae regardless of the presence of webbing. Considering the stages of *O. ilicis* altogether, webbing reduced the predation potential of *I. zuluagai* and *E. citrifolius*, but not of *A. herbicolus* (Franco *et al.*, 2010).

In the USA and Canada, the coccinellid *Stethotus punctillum* Weise is recommended by commercial suppliers to control *O. ilicis* (Biddinger *et al.*, 2009).

CHEMICAL CONTROL. Various chemicals are effective against the southern red mite. The main acaricides/insecticides applied in past decades include abamectin (Reis *et al.*, 2004), aldicarb (Herron and Rophail, 2000), azocyclotin (Reis *et al.*, 2005), bifenthrin (Herron and Rophail, 2000), bromopropylate, chlordimeform (Wilson and Olivier, 1969), chlorpyrifos (Herron and Rophail, 2000), chlorobenzilate (Denmark, 1968; Wilson and Olivier, 1969), chloropropylate (Wilson and Olivier, 1969), cyhexatin (Poe *et al.*, 1976), dicofol (Denmark, 1968; Childs, 1980; Herron and Rophail, 2000), dicrotophos (Oliveira, 2000), ethion (Denmark, 1968), fenbutatin oxide (Teodoro *et al.*, 2005), flufenoxuron (Silveira *et al.*, 2010), hexakis (Childs, 1980), methidathion (Wilson and Olivier, 1969), methiocarb (Poe *et al.*, 1976), naled (Childs, 1980), omethoate (Herron and Rophail, 2000), propargite (Poe *et al.*, 1976; Childs, 1980; Herron and Rophail, 2000), rynaxypyr (Reis and Franco, 2010), spirodiclofen (Reis *et al.*, 2005; Silveira *et al.*, 2010), sulfur (Jeppson *et al.*, 1975; Teodoro *et al.*, 2005; Andrade *et al.*, 2010b), tau-fluvalinate, tebufenpyrad (Herron and Rophail, 2000) and tetradifon (Denmark, 1968, Wilson and Olivier, 1969). On woody ornamentals, one spray of sulfur or acaricides in mid-May provides a sufficient control (Jeppson *et al.*, 1975).

Sulfur drastically compromised populations of *I. zuluagai*, quickly leading them to extinction due to the low reproductive potential of this species compared with its prey (Teodoro *et al.*, 2005). Deltamethrin, permethrin and cypermethrin promoted a high population increase of *O. ilicis* (Oliveira, 2000). The physiological insecticide flufenoxuron is also effective in the control of *Leucoptera coffeella* (Guérin-Mèneville *et* Perrotet), the coffee leafminer, the most important pest of coffee and occurring in all coffee-producing regions in Brazil (Silveira *et al.*, 2010). It was found that copper oxychloride did not result in spider mite mortality. Instead, sprays of this compound induce an increase in mite oviposition, showing that copper oxychloride is one of the factors that is responsible for red spider mite outbreaks in coffee plantations (Oliveira, 2000; Reis and Teodoro, 2000).

The main botanical species investigated to evaluate the effectiveness of their extracts against *O. ilicis* include *Agave americana* L. (Marafeli *et al.*, 2010), *Albizia polycephala* (Benth.) Killip ex Record (Toledo *et al.*, 2010b), *Allamanda cathartica* L. (Potenza *et al.*, 2003, 2005), *Aloe vera* (L.) Burm. f. (Toledo *et al.*, 2010b), *Annona muricata* L. (Marafeli *et al.* (2010), *A. squamosa* L. (Potenza *et al.*, 2003; Carvalho *et al.*, 2008; Marafeli *et al.*, 2010), *Azadirachta indica* A. Juss (Mourao *et al.*, 2004; Andrade *et al.*, 2010b; Marafeli *et al.*, 2010), *Capsicum baccatum* L., *Chenopodium quinoa* Willd. (Marafeli *et al.*, 2010), *Dahlia pinnata* Cav. (Potenza *et al.*, 2005), *Dieffenbachia brasiliensis* H.J. Veitch (Potenza *et al.*, 2003, 2005), *Helianthus annuus* L., *Mentha sativa* L., *Persea americana* Mill. (Toledo *et al.*, 2010b), *Ruta graveolens* L. (Potenza *et al.*, 2003), *Solanum paniculatum* L. (Potenza *et al.*, 2005), *Spondias* sp. (Potenza *et al.*, 2003) and *Thymus vulgaris* L. (Toledo *et al.*, 2010b).

INTEGRATED PEST MANAGEMENT. There is no information available on the practical implementation of IPM for this mite. However, various studies have investigated the selectivity of chemicals, such as rynaxypyr or hexythiazox, towards phytoseiid species (*A. herbicolus*, *I. zuluagai* and *E. citrifolius*) (Reis and Franco, 2010), as well as the selectivity of extracts of *A. americana*, *A. indica*, *A. squamosa* and *C. baccatum* (Toledo *et al.*, 2010a).

Oligonychus indicus (Hirst) (Fig. 20.36)

Common name

Sugarcane mite, cholam mite.

Diagnostic characteristics

FEMALE. General features are as the genus as a whole (Jeppson *et al.*, 1975). The female is greyish green in colour, with black blotches on the dorsum (Gupta *et al.*, 1975).

MALE. The aedeagus is strongly upcurved, with anterior and posterior angulation of about equal size; the knob is concave dorsally (Jeppson *et al.*, 1975).

Geographical distribution

China (Ma and Yuan, 1980), India (Rahman and Sapra, 1940b; Venkatraman and Sharma, 1960; Gupta, 1992, 1995; Gupta and Chatterjee, 1997) and Pakistan (Hirst, 1923b; Gupta, 1985; Gupta and Gupta, 1994).

Host plants

The sugarcane mite is reported on 23 host plants of 5 families, with the highest number of species (17) ascribed to the Poaceae (Migeon and Dorkeld, 2006–2013). The families and species of economic importance are: Arecaceae – betel palm or arecanut (*Areca catechu* L.) (Prasad, 1974; Gupta and Gupta, 1994), coconut palm (*Cocos nucifera* L.) (Rai *et al.*, 1991; Gupta and Gupta, 1994); Musaceae – banana (*Musa sapientum* L.) (Gupta and Gupta, 1994); Poaceae – Bermuda grass (*Cynodon dactylon* (L.) Pers.) (Gupta and Gupta, 1994), rice (*Oryza sativa* L.) (Gupta and Gupta, 1994), sugarcane (*Saccharum officinarum* L.) (Venkatraman and Sharma, 1960; Gupta and Gupta, 1994), sorghum (*Sorghum bicolor* (L.) Moench) (Hirst, 1923b; Gupta and Gupta, 1994), maize (*Zea mays* L.) (Gupta and Gupta, 1994).

Bio-ecology

O. indicus is spread through India, Pakistan and China, but its distribution could be worldwide (Jeppson *et al.*, 1975). The sugarcane mite infests various economic plants, such as bananas (Prasad, 1974; Gupta, 1992, 1995; Gupta and Gupta, 1994), sugarcane (Venkatraman and Sharma, 1960; Gupta and Gupta, 1994), *Sorghum* spp. (Hirst, 1923b; Prasad, 1974; Gupta and Gupta, 1994), betel palm (Prasad, 1974; Gupta and Gupta, 1994), maize (Sandhu *et al.*, 1975), coconut (Rai *et al.*, 1991) and other plants.

Fig. 20.36. *Oligonychus indicus* (Hirst). Male, aedeagus (from Hirst, 1923b).

Fertilized females produce individuals of both sexes, and unfertilized females produce males only (Gupta *et al.*, 1975). The mite usually lives on the lower leaf surface and it spins white webs. The development rate, longevity and fecundity of *O. indicus* were studied at different temperatures (25, 27.5, 30, 32.5 and 35°C) on sugarcane, sorghum and maize. On maize, the shortest pre-oviposition period was observed at 30°C (0.21 ± 0.41 days) and the longest at 27.5°C (1.91 ± 1.04 days). The maximum number of eggs laid daily was 18 on maize at 30°C; the lowest number was no eggs laid at all. Over an entire life cycle, the maximum number of eggs laid was 193 on sorghum at 32.5°C and the lowest number laid was four on maize at 27.5 and 35°C. Fertilized females exhibited the lowest fecundity (10.44 ± 4.34 eggs) on sugarcane at 27.5°C and the highest (73.20 ± 32.87 eggs) on sorghum at 30°C. Unfertilized females had the lowest fecundity (11 ± 4.32 eggs) on maize at 32.5°C and the highest (75.92 ± 35.75 eggs) on maize at 30°C. The shortest egg development (2.12 ± 0.10 days) occurred on maize at 35°C and the longest (4.87 ± 0.83 days) on sorghum at 25°C. The minimum larval period (1 day) was observed on maize at 35°C and the maximum (2.71 ± 0.59 days) on sugarcane at 25°C. The maximum protonymphal period (3.05 ± 1.24 days) was observed on maize at 25°C. The shortest deutonymphal period (1 day) was observed on maize at 35°C and the longest (3.36 ± 0.81 days) on the same host plant at 25°C. The shortest adult longevity (5.60 ± 1.30) was observed on maize at 35°C and the longest (19 ± 6.60 days) on sugarcane at 27°C. Adults lived for a longer time on sugarcane and sorghum and for a shorter time on maize. The shortest life cycle (4.05 ± 0.73 days) was observed on maize at 35°C and the longest (11.73 ± 0.63) on sorghum at 25°C. The mite developed best on maize at 30°C, followed by sorghum (Gupta *et al.*, 1975).

All biological stages of *O. indicus* are present in the field throughout the year, and the generations usually overlap. In the Punjab (India), during the non-crop season (January and February), populations of the sugarcane mite develop on Johnson (or Aleppo) grass (*Sorghum halepense* (L.) Pers.), a wild poaceous plant; on this plant, they reach a peak in April and disperse to the developing sugarcane or sorghum grown nearby. The mite populations continue to increase and survive on Johnson grass, representing a source of infestation for the following year's crops. On sugarcane or sorghum, the mite populations develop until the onset of the monsoon rains in July, which usually kill off all motile forms of *O. indicus* except the eggs. After the monsoon rains, the mite populations continue to increase on sugarcane or sorghum and cause severe damage (Cherian, 1933; Rahaman and Sapra, 1940b). In the Punjab region of India, the sugarcane mite is a serious pest of maize during the pre-monsoon months. The counts of mites on 19 heavily infested maize varieties just before and after the rains showed that the rain affected the mite population differentially on different varieties. Populations of *O. indicus* were considerably reduced on

erect and narrow-leaved varieties compared with those on horizontal and broad-leaved varieties (Sandhu *et al.*, 1975). On arecanut in India, the incidence of *O. indicus* builds up from March up to last week of April. The peak population is associated with rise in temperature. Mite incidence was almost the same on the basal, middle and top fronds during the first week of April. The mite population gradually reduces in the top and middle fronds from May to July, with a subsequent further decline in population. The lowest population occurs during August, September, December and January. Rainfall and RH are negatively correlated with the mite population, whereas temperature is positively correlated (Yadav Babu and Manjunatha, 2007).

Dispersal occurs mainly by cane transfers and air currents. In sugarcane and sorghum fields, the spread of populations usually occurs by migration from plant to plant (Jeppson *et al.*, 1975).

Natural enemies include various spiders, such as *Uloborus* sp. (Uloboridae), *Argyrodes* sp., *Theridula* sp. (Theriididae), *Theridion* sp. and *Cheiracantheum* sp. (Chebionidae) (Manjunatha and Puttaswamy, 1989a), predatory mites and insects, such as the phytoseiid *Neoseiulus longispinosus* (Evans) (Yadav Babu and Manjunatha, 2007), the coccinellid beetles *Stethorus pauperculus* (Weise) (Venkatraman and Sharma, 1960), *Stethorus* sp. and the thrip *Scolothrips sexmaculatus* (Pergande) (Yadav Babu and Manjunatha, 2007).

Symptomatology and damage

Feeding of this mite on leaves of sorghum and sugarcane causes reddish spots, and as feeding continues, the size of these spots increases and the different spots coalesce, forming wide red patches, which may cover the whole leaf surface. Infested leaves appear red and necrotic and may dry (Jeppson *et al.*, 1975; Singh and Raghuraman, 2011). At 30 days of crop growth, attack by *O. indicus* on sorghum reduces plant height, grain yield and straw yield by up to 15.6, 57.9 and 40.4%, respectively (Manjunatha and Puttaswamy, 1990b). Despite *O. indicus* being widely distributed in India, its infestation of sugarcane is more in north India and less in south India; some 20–30% of yield loss is due to attack by *O. indicus* (Gupta, 2003).

Control

Horticultural practices, resistant varieties and the use of physical means allows mite outbreaks to be controlled without severe impact. Chemical control is provided by the use of various active ingredients (Harbans and Sudhu, 1961; Jeppson *et al.*, 1975).

CULTURAL PRACTICES. The destruction of wild plants (such as *Panicum* spp. and *Paspalum distichum* L.) surrounding the crop fields hinders mite infestations, but the use of high amounts of nitrogen fertilizers may induce heavy infestation (Harbans and Sudhu, 1961; Jeppson *et al.*, 1975).

PLANT RESISTANCE. The physiological, chemical and morphological make-up of the different sugarcane clones have a large impact on the resistance to various insect pests. So, a low number of stomata per unit has been associated with the resistance character of varieties to *O. indicus* and other pests (Agarwal, 1969).

Sugarcane varieties with soft leaves are more susceptible to infestation than varieties with harder leaves (Jeppson *et al.*, 1975). In India, tests in field plots on the resistance of 115 sorghum lines to *O. indicus* showed that several ('CSH-5', 'CSH-6', 'CSH-9', 'SPH-80', 'CSV-5', '2077B', 'IS 3687', 'SPV 106', 'SPV 135', 'SPV 192', 'SPV 220', 'SPV 222', 'SPV 224' and 'SPV 265') exhibited a high level of resistance to the mite. Mite resistance is common in tan-pigmented lines, while purple-pigmented lines are generally more susceptible. Antixenosis was apparently the dominant resistance mechanism to both *O. indicus* and the shoot fly (*Atherigona soccata* Rond.) in sorghum (Singh *et al.*, 1981a). Studies on the behavioural and physiological responses governing antixenosis in relation to *O. indicus* at the 25-day-old flag leaf and half bloom stages of 12 grain sorghum cultivars ('CSV 8R', 'CSV 14R', 'Sel 3', 'SPV 913', 'Swati', 'RS29', 'IS 2146', 'IS 2312', 'IS 5613', 'ICSV 705', 'Lakadi', 'M35-1') showed the predominant role of gustatory over visual and olfactory stimuli in adult females of *O. indicus*, but the opposite in the larvae and nymphs, for orientation and settling responses towards the host plant. On 'CSV 8R', 'SPV 913' and 'RS 29', significantly less oviposition was found consistently at all the growth stages, indicating low-to-moderate levels of antixenosis (Sridhar *et al.*, 2010).

PHYSICAL CONTROL. Canes transported from infested areas should be soaked in water at 46°C for ½ h (Harbans and Sudhu, 1961; Jeppson *et al.*, 1975).

CHEMICAL CONTROL. The application of sulfur dusts, sulfur or lime sulfur, or acaricides, suppress mite infestations in the field (Harbans and Sudhu, 1961; Jeppson *et al.*, 1975). Other active ingredients tested include cypermethrin, decamethrin, dicofol, endosulfan, ethion, fenpropathrin, malathion and tetradifon. Among these, ethion, dicofol and tetradifon proved superior over other chemicals, and considering the net income accrued due to the use of pesticides, dicofol, endosulfan, ethion and tetradifon were evaluated as the most effective acaricides (Manjunatha and Puttaswamy, 1990b).

Oligonychus mangiferus **(Rahman** ***et*** **Sapra) (Fig. 20.37)**

Common name

Mango spider mite.

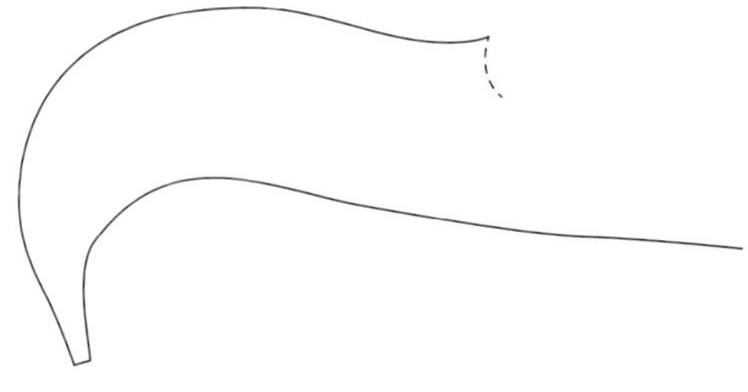

Fig. 20.37. *Oligonychus mangiferus* (Rahman *et* Sapra). Male, aedeagus (from Meyer Smith, 1987).

Diagnostic characteristics

FEMALE. Tarsus I has 4 tactile setae and 1 solenidion proximal to the duplex setae; tarsus II has 3 tactile setae well proximal to the duplex setae; the tibiae have 7(1)-5-5-5 setae (Jeppson *et al.*, 1975; Meyer Smith, 1987).

MALE. The aedeagus bends ventrally, forming an acute angle with the axis of the shaft, and the distal part abruptly narrows at the tip (Jeppson *et al.*, 1975; Meyer Smith, 1987).

Geographical distribution

Australia (Davis, 1968a), Brazil (Paschoal, 1970), Chile (Bolland *et al.*, 1998), El Salvador (Andrews and Poe, 1980), China (Ma and Yuan, 1992), Egypt (Sayed, 1946b; Mohamed, 1963; Zaher *et al.*, 1982), Hawaiian Islands (McGregor, 1950; Pritchard and Baker 1955; Lee Goff, 1986), India (Baker and Pritchard, 1960; Nassar and Ghai, 1981; Gupta, 1992, 1995; Gupta and Gupta, 1994; Gupta and Chatterjee, 1997), Iran (Sheikholeslam Zadeh and Sadeghi, 2010a), Israel (Gerson, 1986; Ben-David *et al.*, 2007), Mauritius (Moutia, 1958; Baker and Pritchard, 1960), Mozambique (CABI, 1965), Pakistan (Rahman and Sapra, 1940a), Peru (Pritchard and Baker, 1955; Guanilo *et al.*, 2012), Réunion (Gutierrez and Etienne, 1986), South Africa (Meyer Smith, 1970a, 1974, 1987), Taiwan (Baker, 1975a), Thailand (Baker, 1975b) and Yemen (Soliman and Mahfood, 1975).

Host plants

The whole list of host plants of mango spider mite numbers 55 species of 22 families (Migeon and Dorkeld, 2006–2013). The families and species of economic importance are:

Anacardiaceae: mango (*Mangifera indica* L.) (Rahman and Sapra, 1940a; McGregor, 1950; Moutia, 1958; Baker and Pritchard, 1960; Baker, 1975b; Andrews and Poe, 1980; Nassar and Ghai, 1981; Zaher *et al.*, 1982; Lee Goff, 1986; Gupta, 1992, 1995; Gupta and Gupta, 1994; Gupta and Chatterjee, 1997; Ben-David *et al.*, 2007).

Annonaceae: sugar apple (*Annona squamosa* L.) (Bolland *et al.*, 1998; Migeon and Dorkeld, 2006–2013).

Euphorbiaceae: castor bean (*Ricinus communis* L.) (Gupta and Gupta, 1994; Mendonça *et al.*, 2011a).

Lauraceae: avocado (*Persea americana* Mill.) (Gutierrez and Etienne, 1986).

Lythraceae: pomegranate (*Punica granatum* L.) (Gupta, 1976; Mendonça *et al.*, 2011a).

Malvaceae: *Gossypium* sp. (Bolland *et al.*, 1998; Migeon and Dorkeld, 2006–2013).

Meliaceae: neem (*Azadirachta indica* A. Juss.) (Gupta and Gupta, 1994).

Moraceae: fig (*Ficus carica* L.) (Nassar and Ghai, 1981; Gupta and Gupta, 1994; Gupta and Chatterjee, 1997).

Musaceae: banana (*Musa paradisiaca* L.) (Nassar and Ghai, 1981; Gupta and Gupta, 1994).

Myrtaceae: guava (*Psidium guajava* L.) (Nassar and Ghai, 1981; Gupta and Gupta, 1994).

Rosaceae: loquat (*Eriobotrya japonica* (Thunb.) Lindl.) (Moutia, 1958), peach (*Prunus persica* (L.) Batsch) (Moutia, 1958; Nassar and Ghai, 1981; Gupta and Gupta, 1994; Gupta and Chatterjee, 1997), French rose (*Rosa gallica* L.), English rose (*Rosa hybrida*) (Baker, 1975b).

Sapindaceae: lychee (*Litchi chinensis* Sonn.) (Gupta, 1991, 1992).

Vitaceae: grapevine (*Vitis vinifera* L.) (Sayed, 1946b; Nassar and Ghai, 1981; Gupta and Gupta, 1994; Gupta and Chatterjee, 1997; Domingos *et al.*, 2010).

Bio-ecology

O. mangiferus is mainly a pest of mango in tropical and subtropical countries, but may also occur on cotton, loquat, peach, pomegranate, banana, fig, guava, lychee, rose, ornamentals and other economic plants (Moutia, 1958; Mohamed, 1963; Jeppson *et al.*, 1975; Neelu Nangia *et al.*, 1990; Johann *et al.*, 2009; Domingos *et al.*, 2010; Ghoshal *et al.*, 2011).

The mango spider mite reproduces by arrhenotoky and thelytoky and has $n = 3$ chromosomes (Flechtmann, 1983, in Helle and Pijnacker, 1985). At 25 ± 2°C and 65 ± 5% RH, the incubation period averages 6.30 days for females and 6.3 days for males on mango ('Ewais' and 'Alphons' cultivars). The life cycle lasts 12.65 and 12.95 days for females and 12.05 and 12.30 days for males, respectively, on the two cultivars (Halawa and El-Sebaay, 2010). On mango, the adult lifespan is 10.11 days for females and 4.21 for males (Rai *et al.*, 1988). Development, survivorship and reproduction of *O. mangiferus* reared on mango leaves were evaluated at five constant temperatures (16, 20, 24, 28, 32°C). The development time for a generation varied from 9.20 days at 32°C to 29.07 days at 16°C. The developmental thresholds for the egg, larva, nymph, pre-oviposition and combined immature stages were estimated as 8.77, 9.37, 10.61, 11.13 and 8.91°C, and their effective cumulative temperatures are 93.46, 30.29, 44.96, 14.24 and 191.83 degree-days, respectively (Fu and Zhang, 2002). Tests using a combination of eight constant temperatures and relative humidities (RHs), viz., 7.0°C with 85% RH, 10°C with 80% RH, 15.0°C with 75% RH, 23.0°C with 70% RH, 31.0°C with 65% RH, 34.0°C with 65% RH, 36.0°C with 60% RH and 40.0°C with 55% RH, revealed that the optimal conditions for the development of *O. mangiferus* were 15.0–31.0°C and 65–75% RH. The highest temperature and the lowest RH accelerate the rate of development and induce more reproduction. The population of the mango spider mite multiplies 30.81 times in a generation time of 27.36 days at 31.0°C and 65% RH, while the same population only increases 7.46 times in a generation time of 48.07 days at 15.0°C and 75% RH. Fecundity is highest at 31.0°C and 65% RH, with 46.43 eggs per female. The highest intrinsic rate of natural increase occurs at 31.0°C (0.125 day^{-1}) (Abou-Awad *et al.*, 2011b).

In Egypt, the peak populations were observed in July on quince, and in September on pomegranate (Mohamed, 1963). Also in Egypt, the population density of *O. mangiferus* on mango has one annual peak in April (El-Halawany, 2003), while in Taiwan two population peaks were observed during the year, from February to June and October to December (Lin and Chen, 2008). The population dynamics of *O. mangiferus* on mango is influenced by weather conditions, the cultivar, natural enemies and its vertical distribution. On cv. 'Hindi', 14 and 15 generations were observed, and on 'Alphonso', 16 and 14 generations were observed over 2 successive years. Leaves at the top of

both mango cultivars were preferred by mites to those in the middle and at the bottom. 'Alphonso' was more susceptible to infestation than 'Hindi' (Abou-Awad *et al.*, 2012).

Natural enemies include predatory insects and mites. The mites include different phytoseiids, and the life-table parameters of *Neoseiulus womersleyi* (Schicha) and *Euseius ovalis* (Evans) have been investigated in the laboratory (Nguyen and Thomas Shih, 2012). In Brazil, *O. mangiferus* is often associated with *E. citrifolius* Denmark *et* Muma on the grapevine 'Italia' variety (Domingos *et al.*, 2010).

Symptomatology and damage

Feeding of *O. mangiferus* may cause severe damage to mango, consisting of drying and premature leaf drop (Jeppson *et al.*, 1975). Feeding on the leaves of mango also affects their content of organic compounds (chlorophyll, total protein, phenolics and total carbohydrate), minerals (Mg, Zn, Cu and Fe) and inorganic compounds (nitrate and nitrite). In particular, the amount of chlorophyll, total protein, total carbohydrate, Mg, Zn, Cu, Fe, nitrate and nitrite are reduced by 15.36, 19.08, 17.54, 11.93, 23.1, 32.32, 16.20, 20.19 and 23.23%, respectively, while the amount of phenolics increases by 21.21% (Ghoshal *et al.*, 2011).

On cotton and pomegranate, mites develop on the upper leaf surfaces (Moutia, 1958; Mohamed, 1963).

Control

Control of *O. mangiferus* is based on sprays of chemicals. Biological control is not applied.

CHEMICAL CONTROL. One acaricide application is sufficient to suppress the mite outbreak on mango to below the economic level when it is applied as the mite population starts to increase (Abou-Awad *et al.*, 2012). It is possible to use various acaricides. Trials carried out with bromopropylate, chlorfenapyr and fenpyroximate showed the effectiveness of bromopropylate and fenpyroximate (Lin and Chen, 2008). Suitable plant extracts include those from holy basil, lemon grass and neem (Sangsrijun and Chandrapatya, 2001).

Oligonychus newcomeri **(McGregor) (Fig. 20.38)**

Common name

Unknown.

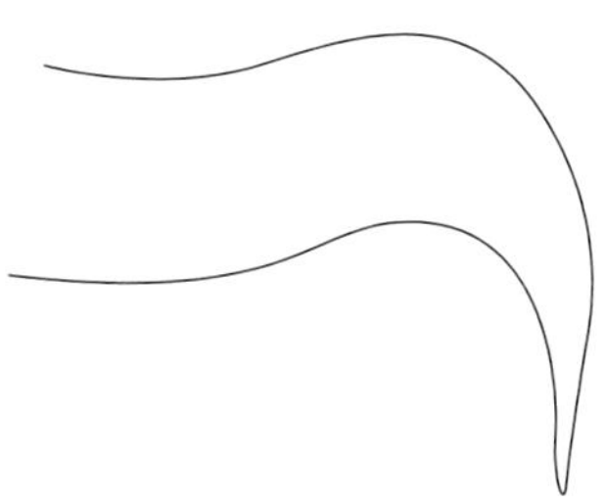

Fig. 20.38. *Oligonychus newcomeri* (McGregor). Male, aedeagus (modified from McGregor, 1950).

Diagnostic characteristics

FEMALE. The body is purple in colour, with an anterior area that is heart shaped and pinkish. The anterior legs are pale, and the posterior legs may be darker. The dorsal body setae are not set on tubercles, and the opisthosomal setae f_1 and f_2 are similar in length. Tibia I bears 7 tactile setae, tarsus I has 3 tactile setae proximal to the duplex setae, and the proximal members of the duplex setae are short (Jeppson *et al.*, 1975).

MALE. The aedeagus bends ventrally, gradually narrows to the tip and is distinctly flattened dorsoposteriorly (Jeppson *et al.*, 1975).

Geographical distribution

The USA (McGregor, 1950).

Host plants

The list of host plants includes 18 species belonging to three families: Betulaceae – *Carpinus caroliniana* Walter (Reeves, 1963); Oleaceae – *Fraxinus americana* L. (Reeves, 1963); Rosaceae – *Amelanchier* sp. (McGregor, 1950), *A. amabilis* Wiegand, downy serviceberry (*A. arborea* (F. Michx.) Fernald), *Cerasus vulgaris* Mill., service tree (*Cormus domestica* (L.) Spach [= *Sorbus domestica* L.]), cockspur hawthorn (*Crataegus crus-galli* L.), *C. pruinosa* (Wendl.f.) K. Koch, *C. stolonifera* Sarg., *Malus* sp. (Reeves, 1963), apple (*M. domestica* Borkh.), *Oxyacantha succulenta* (Schrad. ex Link) Lunell, chokecherry (*Padus virginiana* (L.)), *Pyrus canadensis* (L.) Farw. (Reeves, 1963), pear (*P. communis* L.) (McGregor, 1950), *Rosa* sp., *Sorbus* sp. (Reeves, 1963).

Bio-ecology

O. newcomeri primarily infests serviceberry and hawthorn, but it may also occur on apple, pear, *Prunus* sp. and other rosaceous plants (Pritchard and Baker, 1952a; Reeves, 1963). The mite spins little webbing and overwinters as eggs that are amber in colour. It begins its activity in early May, and by June the adults occur primarily on the upper leaf surfaces; juvenile forms and males may also be found on the lower leaf surface. Summer eggs are pale, flattened, striate above and with a dorsal stipe. According to the amount of new leaves, populations gradually increase during the summer, and have laid overwintering eggs in September. By October, motile forms disappear (Reeves, 1963; Jeppson *et al.*, 1975).

Symptomatology and damage

The upper surface of serviceberry leaves may appear entirely brown by the autumn, as well as apple leaves (Jeppson *et al.*, 1975).

Control

High levels of attack may be controlled with acaricide sprays authorized for cultures, usually during the spring or summer.

Oligonychus oryzae (Hirst) (Fig. 20.39)

Common name

Rice spider mite, paddy leaf mite, leaf mite.

Diagnostic characteristics

FEMALE. The dorsal opisthosoma is transversely striate except for a longitudinal pattern between the setae f_1. The peritreme ends in a simple bulb. Tibiae I and II have 9 and 7 tactile setae, respectively. Tarsus I has 4 tactile setae proximal to the duplex setae and 2 setae set ventrally to the duplex setae (Jeppson *et al.*, 1975).

MALE. The aedeagus has a slender, sigmoid distal tip that is broadly angled dorsally (Jeppson *et al.*, 1975).

Geographical distribution

India (Hirst, 1926; Gupta, 1992; Gupta and Gupta, 1994; Beard *et al.*, 2003), Thailand (Baker, 1975b; Ehara and Wongsiri, 1975) and Vietnam (Dam Quoc Tru, 2009).

Host plants

The list of host plants includes nine species ascribed to the Musaceae and Poaceae families: Musaceae – *Musa* sp. (Baker, 1975b), banana (*Musa* × *paradisiaca* L.) (Gupta and Gupta, 1994); Poaceae – *Bambusa* sp. (Gupta, 1992), Bermuda grass (*Cynodon dactylon* (L.) Pers.) (Gupta, 1976; Gupta and Gupta, 1994), barley (*Hordeum vulgare* L.) (Gupta and Gupta, 1994), rice (*Oryza sativa* L.) (Hirst, 1926; Gupta and Gupta, 1994; Beard *et al.*, 2003), Poaceae, *Setaria* sp. (Gupta and Gupta, 1994), maize (*Zea mays* L.) (Baker, 1975b; Ehara and Wongsiri, 1975).

Bio-ecology

The rice spider mite is a severe pest of rice in some countries of the Oriental region, and it also infests other economic plants, such as bananas (Baker, 1975b; Gupta and Gupta, 1994; Karmakar and Soma Dey, 2006) and maize (Baker, 1975b; Ehara and Wongsiri, 1975).

The mite reproduces by arrhenotokous parthenogenesis (Misra and Israel, 1968). The infestation starts in the nursery and continues to the stem elongation stage. Although the mites feed on the upper and lower surface of rice leaves, there are usually higher populations on the lower leaf surfaces than on the upper surfaces. The mites pierce the leaf tissues and suck the sap, sometimes damaging the entire leaf portion (Jalaluddin *et al.*, 2008).

Adults mate after emergence, the pre-oviposition period lasts 1–3 days and the oviposition period 6–12 days. The fertilized females lay 8–21 eggs; the unfertilized females also lay eggs, and these develop into males. The larval period lasts 2–3 days and is followed by a quiescent stage; the protonymphal period lasts 1.5–2 days and is followed by a second quiescent stage; the deutonymphal period lasts 2–3 days and is followed by another quiescent stage. The male to female ratio is 1:2.7. Males live 5–6 days and the female lifespan is 1–3 days (Misra and Israel, 1968). Under laboratory conditions, egg, larval, protonymphal and deutonymphal development takes 4.70 ± 1.09, 1.89 ± 0.30, 1.89 ± 0.18 and 2.09 ± 0.24 days for the male, and 5.90 ± 1.12, 2.23 ± 0.36, 2.27 ± 0.26 and 2.42 ± 0.28 days for the female, respectively. The total developmental time is 10.58 ± 1.50 days for the male and 12.64 ± 1.57 days for the female (Nayak *et al.*, 2008a). Jalaluddin *et al.* (2008) found that the entire life cycle could take 8–12 days, and Radhakrishnan and Ramaraju (2009) observed that on rice at 35°C, the development from egg to adult lasted 8.33 ± 0.20 days.

On banana, *O. oryzae* lives on the under leaf surface covered by a thin web, and its populations develop within the grooves of parallel veins as well as along the midrib. The mite occurs in the field throughout the year. In India, during January–February, populations of *O. oryzae* are very low; high levels occur during the second fortnight of March to the first fortnight of June and reach a peak during the first fortnight of May. High temperature and low humidity favour an increase in the mite populations, which decline with the increase of RH at the onset of the monsoon. Populations are maintained in an inconsistent manner during the rest of the year, and the lowest levels occur in the first fortnight of January, followed by the second fortnight of December (Karmakar and Soma Dey, 2006). A severe incidence of *O. oryzae* on rice has been reported in the first week of September, when the maximum temperature shot up to 36°C (Velusamy *et al.*, 1987). Higher populations were observed when temperatures reached 27.2°C and RH 87.5% (Mukherjee *et al.*, 1989). Bright, sunny weather followed by slight rainfall during September–October is optimum for the mite population increase (Misra and Israel, 1968; Jhansi Lakshmi *et al.*, 2008). On rice, the mite reproduces throughout the year and large populations develop during hot weather, when the life cycle lasts the shortest time (Jalaluddin *et al.*, 2008). Besides paddy, several weeds in and around paddy fields – *Panicum coloratum* L., *P. crus-galli* (L.) Beauv (Cherian, 1938; Nagarajan, 1957), *C. dactylon* (Misra and Israel, 1968) and *Echinochloa colona* (L.) Link – have been reported as alternate hosts for *O. oryzae*. Large numbers of mites have been collected on this last host plant, a wetland weed found with rice that may be an alternate host plant (Velusamy *et al.*, 1987).

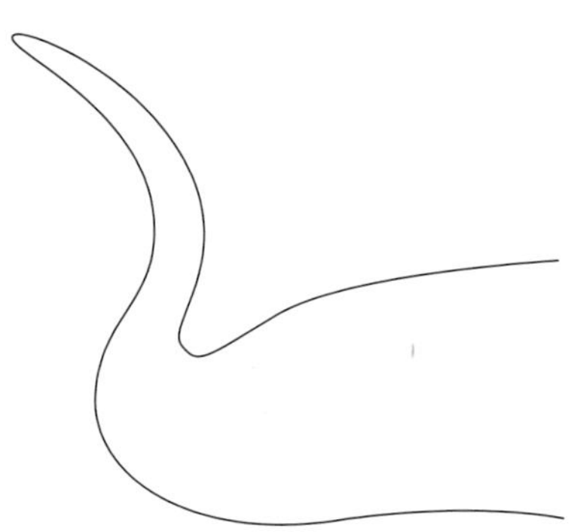

Fig. 20.39. *Oligonychus oryzae* (Hirst). Male, aedeagus (from Hirst, 1926).

Symptomatology and damage

The feeding of *O. oryzae* causes whitish patches on leaves, which later turn ash coloured and dry from the tip downward. Infested seedlings become pale, stunted in growth and die after

transplanting. Transplant crops show white patches on the upper leaf surfaces, the plants develop slightly, and in severe infestations the leaves turn greyish white and dry up (Nagarajan, 1957; Jeppson *et al.*, 1975; Velusamy *et al.*, 1987). Feeding of the mite on rice cv. 'Sambha Mahsuri' ('BPT 5204') at its late tillering stage causes scattered irregular greyish pale green patches devoid of chlorophyll on leaves that hinder photosynthesis (Singh, 2001). Infestations may reduce rice yields up to 25% (Jeppson *et al.*, 1975; Gupta, 2003). The damage is most serious during the summer months, when the temperature is high and humidity is low compared with the winter and rainy seasons (Jalaluddin *et al.*, 2008).

The mite feeds on the ventral leaf surface of banana, and feeding areas become whitish. The whitish patches, along with the eggs, cuticular exuvae of the mites and dust particles, develop characteristic symptoms. Later, the whitish patches turn brown and the leaf withers (Karmakar and Soma Dey, 2006).

Control

Horticultural practices, resistant varieties and sprays of acaricides suppress mite outbreaks.

CULTURAL PRACTICES. Various cultural practices are known, and some of these may be considered to be preventive control measures. It is good practice to chisel plough once every 4 years and to maintain the organic carbon content of the soil at optimum level. A rice-free period in which the stubble is ploughed down between crops, or planting fallow crops with pulses and green manures in endemic areas, hinders mite development. It is also good practice to maintain bunds free from weeds, and to top dress with urea and neem cake (5:1) using an LCC (leaf colour chart) based on N to avoid excess nitrogen application; top dressing can also be done with 50% potash fertilizer. Finally, it is a good practice to clean machinery before using it in an uninfested field (Jalaluddin *et al.*, 2008; TNAU, 2011).

PLANT RESISTANCE. Screening of rice varieties carried out in Tamil Nadu (India) and at IRRI (International Rice Research Institute, Philippines) for their reaction to the mite showed that most of the Tamil Nadu varieties had no mite incidence, while the IRRI varieties were affected, e.g. 'IR 62' (Velusamy *et al.*, 1987). Nine cultivars of paddy ('IR 30', 'IR32', 'IT 579', 'Dular', 'Jaya', 'Masuri', 'Jalkamini', 'CNM 31', 'TN1') were studied to ascertain any morphological basis of resistance. The maximum density of mites was found on 'IR 30', followed by 'Dular'. Physical characteristics of the leaf were not related to mite density, except for the width of the leaf (Mukherjee *et al.*, 1989). Further studies on the amount of colonization of mites in eight commercial rice varieties ('ADT 36', 'ADT 43', 'ADT 45', 'IR 50', 'CO 47', 'ASD 16', 'ADTRH 1', 'TN 1') show that the varieties 'ASD 16', 'ADTRH 1' and 'CO 47' are moderately resistant to the rice spider mite, while the 'ADT 45', 'TN 1', 'IR 50' and 'ADT 36' varieties are moderately susceptible, and the 'ADT 43' variety is the most susceptible (Radhakrishnan and Ramaraju, 2009).

BIOLOGICAL CONTROL. Predatory mites may suppress populations of the mite (TNAU, 2011), but there are few experiences of the use of this technique. The bioefficacy of the fungus *Hirsutella thompsonii* Fisher was not proven (Srimohanapriya *et al.*, 2010).

CHEMICAL CONTROL. To control severe attacks of the rice spider mite on rice, foliar spraying of acaricides is required (Jalaluddin *et al.*, 2008). Sampling 15 days after planting usually helps to catch mite populations at low levels (TNAU, 2011). Precautionary measures may be taken during the summer months for the sustainable production of banana (Karmakar and Soma Dey, 2006).

A number of acaricides and insecticides are reported in the literature. They include abamectin (Nayak *et al.*, 2008b; Radhakrishnan and Ramaraju, 2009), bassa (Mukherjee *et al.*, 1989), diafenthiuron (Jalaluddin *et al.*, 2008; Nayak *et al.*, 2008b), dicofol (Mukherjee *et al.*, 1989; Jhansi Lakshmi *et al.*, 2008; Nayak *et al.*, 2008b; Srimohanapriya *et al.*, 2010; Mahabaleshwar Hegde and Balikai, 2011), ethion, fenazaquin (Nayak *et al.*, 2008b), fenpyroximate (Jalaluddin *et al.*, 2008), hostothion, methyl demeton (Srimohanapriya *et al.*, 2010), milbemectin (Jhansi Lakshmi *et al.*, 2008; Srimohanapriya *et al.*, 2010; Mahabaleshwar Hegde and Balikai, 2011), Mit (?MIT-505 50 EC = ethion), novaluron (Srimohanapriya *et al.*, 2010), profenofos (Jhansi Lakshmi *et al.*, 2008; Srimohanapriya *et al.*, 2010), propargite (Nayak *et al.*, 2008b; Srimohanapriya *et al.*, 2010), quinalphos (Mukherjee *et al.*, 1989), spiromesifen (Jhansi Lakshmi *et al.*, 2008; TNAU, 2011) and sulfur (Nagarajan, 1957; Mukherjee *et al.*, 1989). Diafenthiuron and fenpyroximate have both been tested with wettable sulfur (Jalaluddin *et al.*, 2008). Onion extract plus neem oil plus soap, and neem seed kernel extract (NSKE) (Srimohanapriya *et al.*, 2010), or neem oil alone (Nayak *et al.*, 2008b), have also been tested. Among the above active ingredients, abamectin, dicofol, milbemectin, quinalfos and spiromesifen are reported to be effective.

Oligonychus perseae **Tuttle, Baker** ***et*** **Abbatiello (Fig. 20.40)**

Common name

Persea mite.

Diagnostic characteristics

FEMALE. The body is 319 μm long, 204 μm wide, and yellowish green in colour. The gnathosoma is short and broad. The peritreme ends in a simple bulb. The prodorsum is longitudinally striate. The opistosoma is transversely striate except for a 'V' pattern between the setae e_1. The dorsal body setae are slender, slightly expanded at the base and nude, except for sc_2, c_3, f_2 and h_1, which

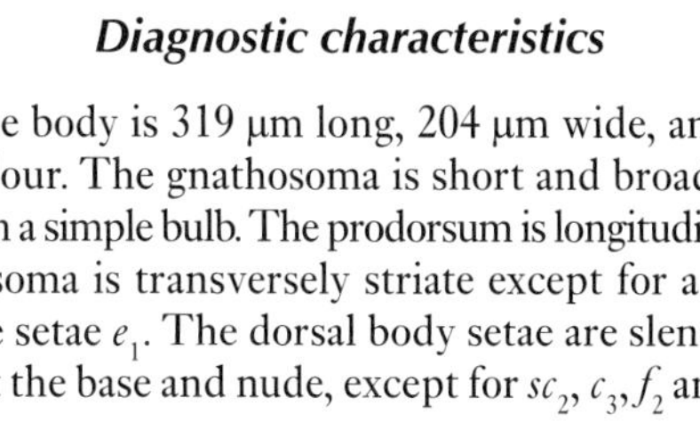

Fig. 20.40. *Oligonychus perseae* Tuttle, Baker *et* Abbatiello. Male, aedeagus (Sandoval *et al.*, 2011, courtesy of C. Vásquez, 2013).

are finely serrated and longer than the other dorsal setae. Tarsus I bears 4 tactile setae and 1 solenidion proximal to the duplex setae; the duplex setae have the proximal member much shorter than the distal member; tarsus II has 2 proximal tactile setae. Tibia I has 9 tactile setae and 1 solenidion; tibia II has 7 tactile setae (Tuttle *et al.*, 1976; Sandoval *et al.*, 2011).

MALE. The dorsal body chaetotaxy of the male is the same as that of the female. The aedeagus bends ventrally, with a longer and more slender distal portion than that of *O. peruvianus* (Tuttle *et al.*, 1976; Sandoval *et al.*, 2011).

Geographical distribution

Costa Rica (Ochoa *et al.*, 1991a), Israel (Ben-David *et al.*, 2007), Mexico (Tuttle *et al.*, 1976), Italy (Zappalà *et al.*, 2015), Portugal (Borges *et al.*, 2008), Spain (Alcázar *et al.*, 2005) and the USA (Baker and Tuttle, 1994; Aponte and McMurtry, 1997a,b; Kerguelen and Hoddle, 1999).

Host plants

The persea mite is reported on four host plants belonging to the Bixaceae, Fabaceae and Lauraceae families: Bixaceae – *Bixa orellana* L. (Tuttle *et al.*, 1976); Fabaceae – carob tree (*Ceratonia siliqua* L.) (Bolland *et al.*, 1998; Migeon and Dorkeld, 2006–2013); Lauraceae – *Persea* sp. (Tuttle *et al.*, 1976), avocado (*Persea americana* Miller) (Aponte and McMurtry, 1997a,b; Kerguelen and Hoddle, 1999; Alcázar *et al.*, 2005; Ben-David *et al.*, 2007).

Bio-ecology

O. perseae may be misidentified with *O. peruvianus* (McGregor). Both species are occasional pests in some tropical crops, but may be differentiated by the different idiosomal striation pattern between the setae e_1 and the shape of the aedeagus (Sandoval *et al.*, 2011).

The persea mite is a pest of avocado, but may also occur on a restricted number of other plants. Takano-Lee and Hoddle (2002a), following Bender (1993), Baker and Tuttle (1994) and Aponte and McMurtry (1997a,b), reported a wider list of host plants, including *Acacia senegal* L., apricot (*Prunus armeniaca* L.), camphor (*Cinnamomum camphora* T. Nees *et* C. Eberm.), *Citrus* spp., *Sonchus* spp., grape (*Vitis* spp.), *Chenopodium album* L., *Liquidambar* spp., *Asclepias* spp., nectarine and peach (*P. persica* L. Batsch), persimmon (*Diospyros* spp.), plum (*Prunus* spp.), rose (*Rosa* spp.), *Rhus* spp. and *Salix* spp. The UC Pest Management Guidelines for the avocado persea mite (UC IPM Online, 2014a) report that many ornamentals and weeds also harbour the persea mite, and in the early 1990s, when persea mites were first found in California, individual mites from heavy populations on avocado trees were observed drifting on to the leaves of adjacent stone fruit trees, although they did not feed. Later, populations were found to be reduced, and persea mites have not since been collected on stone fruit trees or fruit; neither are *Prunus* spp. known to be a host of this mite.

O. perseae lives mainly on the underside of leaves, and sometimes may also occur on the upper leaf surface. On the lower leaf surface, it feeds and reproduces in 'nests' of silken webbing, which also provide protection from some predator mites and other natural enemies. The persea mite exhibits a modification of the earlier defined life-type web nest WN-c. The greatest number of nests observed to be built by a female was 12.17 at 20°C, and the greatest number of eggs per female per nest was 5.20 at 25°C (Aponte and McMurtry, 1997a).

Persea mites reproduce by arrhenotokous parthenogenesis. In the laboratory at 25°C and 60% RH, egg development took 6.33 days. The mean lengths of the larval, protonymphal and deutonymphal stages (including quiescent periods) were 4, 4 and 3.95 days, respectively. (Mated) females oviposited one egg daily, the oviposition period ranged from 6.66 to 10.62 days, and during this time each female laid about ten eggs. The pre-oviposition time ranged from 2 to 5.33 days. The pre-oviposition period of unmated females was 5.5 days and the oviposition period 6 days, during which 6 eggs were laid. The whole biological cycle took 20.95 days (Salinas Anaya and Reséndiz García, 1995).

Aponte and McMurtry (1997a) found that on 'Hass' avocado leaves at 15, 20, 25 and 30°C, the development from egg hatch to adult took 34.89, 16.90, 13.87 and 9.81 days, respectively. The highest fecundity (45.80 eggs per female) occurred at 25°C, and the highest oviposition rate (1.84 eggs per female daily) at 30°C. The highest mortality during development was 36.92% for the combined stages at 30°C. The theoretical lower threshold temperature for development was 7.80 and 9.00°C for the combined stages of the female and male, respectively. The thermal constant for completion development was 200 and 166 degree-days for the female and male, respectively. The sex ratio ranged from 1.82:1 to 2.39:1 (female:male). The highest intrinsic rate of increase (0.1440) occurred at 30°C, the highest net reproductive rate (25.80) at 25°C and the shortest mean generation time (18.38) at 30°C (Aponte and McMurtry, 1997b).

During the winter, cool temperatures slow the development of the persea mite populations, and in California the lowest densities occur in about March. Populations gradually increase from the spring, and peaks are usually observed in July and August. Daily temperatures of 37.77°C or more on several consecutive days and low humidity suppress populations of the persea mite (UC IPM Online, 2014a). Aerial dispersal of *O. perseae* was observed by Takano-Lee and Hoddle (2002a).

Natural enemies include predatory mites and insects. The most important phytoseiid mites are *Amblyseius similoides* Buchelos *et* Pritchard, *Euseius hibisci* (Chant), *Galendromus annectens* De Leon, *G. helveolus* (Chant) (Takano-Lee and Hoddle, 2002a; UC IPM Online, 2014a), *E. scutalis* (Athias Henriot) (Maoz *et al.*, 2009, 2011b; Torres *et al.*, 2010), *E. stipulatus* (Athias Henriot) (Torres *et al.*, 2010) and *Neoseiulus californicus* (McGregor) (Takano-Lee and Hoddle, 2002a; Torres *et al.*, 2010; UC IPM Online, Avocado, 2014a). Insects include Phlaeothripidae (*Leptothrips mali* (Fitch)), Aeolothripidae (*Scolothrips sexmaculatus* Pergande)), Hemerobiidae (*Hemerobius* spp.), Chrysopidae (*Chrysopa* and *Chrysoperla* spp.), Coniopterygidae, Cecidomyiidae (*Feltiella* sp.), Staphylinidae (*Oligota oviformis* Casey) and Coccinellidae (*Stethorus picipes* Casey). Several predators are not very effective because they are hindered by the protective webbed nests of mites (UC IPM Online, 2014a).

Symptomatology and damage

Damage by *O. perseae* on 'Hass' avocado cultivar occurs mainly on the underside of the leaves along the midrib, main veins and

leaf depressions. Feeding by the mite on the lower epidermal, spongy parenchyma and palisade parenchyma destroys the cells of the leaf tissues, causing wide necrotic areas on the underside of the leaves when mite populations are high (Aponte and McMurtry, 1997a). These necrotic areas appear as distinct circular chlorotic to brown spots on the lower leaf surface, which are also appreciable on the upper leaf surface. Populations initially consist of a few individuals (i.e. the colonies), which can gradually become very numerous and produce dense webbing, appearing as a silvery spot on the underside of the leaf. Heavy attacks may be associated with numerous brown-spotted, green leaves hanging from trees and on the ground beneath infested trees. Severely infested trees appear lighter coloured and may be affected by premature leaf drop and defoliation, which leads to sunburned bark and fruits, and aborted or dropped fruits; stressed trees later have reduced yields (UC IPM Online, 2014a).

Hoddle (1998) defined a scale of damage ranging from 1 to 50% of the leaf surface. At 1% damage, there are 30 spots per leaf caused by the presence of 42 mites, while at 25% damage there are 191 spots and 267 mites per leaf. However, 50% of damage corresponds to 110 spots and 154 mites per leaf. This is because as persea mite feeding damage increases, the number of individual necrotic spots on leaves declines as the nests begin to merge. Consequently, *O. perseae* populations also decline as there is less food available. The ecological adaptations of *O. perseae* have led to its confusion with other mite species infesting avocado. For instance, persea mite damage early in the season can be confused with that caused by the six-spotted mite, *Eotetranychus sexmaculatus* (Riley). However, the webbing produced by the six-spotted mite is less dense and usually not organized in small circular patches. Moreover, feeding by the six-spotted mite causes brown to purplish irregularly shaped blotches, while the persea mite produces roundish and mostly scattered spots. Six-spotted mites usually damage the areas near the veins, while *O. perseae* often feeds throughout the lower leaf surface. In addition, the persea mite sometimes feeds on the upper leaf surface, upon which the avocado brown mite, *Oligonychus punicae* (Hirst) generally feeds, inducing the upper leaf surface to become bronzed or scorched, but with damage not usually occurring in distinct circular spots (UC IPM Online, 2014a). On avocado trees in Israel, it has been found that the mean tri-annual yield (yield over 3 years) was reduced by 20% when mite populations were as high as ≥250 mites per leaf compared with the yields obtained when plots were sprayed at a threshold of 50–100 mites per leaf (Maoz *et al.*, 2009).

Control

Control of persea mite is provided by the implementation of rational horticultural practices, the use of resistant varieties, and biological and chemical means. Monitoring of populations helps the use of decisons tools.

MONITORING. The UC Pest Management Guidelines for the persea mite on avocado (UC IPM Online, 2014a) suggested that leaves should be inspected to detect mites, their damage and natural enemies about every 7–10 days from mid-March at least through to the end of August, and perhaps October. Monitoring and treatment decision making for *O. perseae* must be related to those of other pests, such as avocado thrips, as monitoring frequency, treatment and choice of active ingredients can all be affected by management decisions for other pests (e.g. thrips). In particular, some pesticides that are usually sprayed earlier in the season against avocado thrips are also effective against persea mites, which are usually sprayed later in the season. However, some pesticides can have an adverse impact on natural enemies, and it is recommended that choices that reduce the development of pesticide resistance should be favoured. The weather also affects decision making. Thus, heavy winter rains and high winds affect the development of mite populations negatively, and daily high temperatures of 37.77°C or more, and low humidity on several consecutive days, may suppress or crash persea mite populations (UC IPM Online, 2014a).

THRESHOLDS. In California, there are no thresholds for persea mite control. Instead, it is suggested that mite densities are monitored and recorded regularly and the data from year to year compared with records of control actions and their effectiveness (UC IPM Online, 2014a).

In Israel, the adoption of an action threshold (AT) of 50–100 mites per leaf (future research is needed to refine this AT) has been provisionally suggested (Maoz *et al.*, 2011a). It has been found in Israel that mean annual cumulative mite days (CMDs) at the two higher mite densities tested (100 and 250 mites per leaf) are *c.*13,500 ± 700 per leaf. Using the linear regression equation PLAD (percentage of leaf area damaged) = 0.0009 CMDs + 2.42, to describe leaf damage as a function of CMDs, these authors estimated an EIL (economic injury level) of *c.*15 PLAD (Maoz *et al.*, 2011a).

Damage may be measured using automated image analysis software (Hoddle and Kerguelen, 1999a). Díaz *et al.* (2009) measured leaf damage using machine learning approaches for the objective segmentation and quantification of leaf areas damaged by mites in avocado leaves.

PLANT RESISTANCE. The percentage of leaf area damaged by mite feeding allowed the definition of three groups of avocado, referred to as susceptible ('Hass' and 'Gwen' cultivars), resistant ('Fuerte', 'Lamb Hass' and 'Reed' cultivars) and intermediate ('Esther' and 'Pinkerton' cultivars) susceptibility. There was no difference in mortality and rate of development of *O. perseae* of either the first or second generation reared on either 'Hass', 'Pinkerton' or 'Lamb Hass' cultivars. However, when *O. perseae* was reared on leaves collected in late spring and early summer, net reproduction and intrinsic rate of increase were higher on 'Hass' avocados in midsummer. The same behaviour occurs on 'Hass' in the field in July. Seasonal changes in the nutritional quality of leaves may then be the major factor determining the susceptibility of avocado cultivars to the persea mite (Hoddle and Kerguelen, 1999a,b; Kerguelen and Hoddle, 2000).

The UC Pest Management Guidelines for the avocado persea mite (UC IPM Online, 2014a) report that the 'Esther', 'Pinkerton' and 'Reed' varieties are of intermediate susceptibility, the 'Bacon', 'Fuerte', 'Lamb Hass' and 'Zutano' varieties are much less affected, while the 'Hass', 'Gwen' and a few other varieties are the most damaged.

CULTURAL PRACTICES. In general, the reduction of tree stress contains the effect of persea mite feeding. Appropriate horticultural practices also help to suppress the increase of populations of persea mite. These include the elimination or the reduction of other host plants growing near the avocado trees, such as susceptible ornamentals, non-commercial fruit trees and weeds. The optimization of irrigation and cultural practices maintain the flush of new growth and compensate for leaf drop caused by mites. However, trees should not be overfertilized, because an increase of foliar nitrogen may positively affect the development of persea mite populations during late spring and summer. After premature leaf drop, it is a good practice to whitewash trunks and major branches to protect the bark and wood from sunburn (UC IPM Online, 2014a).

In California, three cultural control methods were tested to reduce the vertical recolonization of avocado trees by *O. perseae*. These were: the arrangement of Tanglefoot® barriers on tree trunks, the removal of ground cover and leaf litter under tree canopies, and of the use of Tanglefoot barriers combined with ground cover removal. However, no significant differences were found among trees receiving different treatments (Takano-Lee and Hoddle, 2002a).

PHYSICAL CONTROL. On small trees, it is possible to spray the canopy (especially the undersides of leaves) with a forceful stream of water to reduce mite populations (UC IPM Online, 2014a). However, the results of trials of this method carried out in the Canary Islands were not promising (Hernández Suárez *et al.*, 2010).

BIOLOGICAL CONTROL. Biological control of the persea mite is provided by both releases of exotic predatory mites and the conservation of prevalent phytoseiid mites.

Releases of phytoseiid mites on avocados were first done in California. Six phytoseiid mites (*G. annectens*, *G. helveolus*, *G. pilosus* (Chant), *Metaseiulus occidentalis* (Nesbitt), *N. californicus* and *Typhlodromus rickeri* Chant) were released at a rate of 2000 individuals per tree, using early or late release timings in order to evaluate the effect of prey density on predator establishment rates, predator population growth and control of pest mites. The early release was made when 25% of leaves were infested by one or more adult persea mites, but 4 weeks after the release no predators were collected. The second early release was made when 75% of leaves were infested with one or more adult persea mites. Five out of six phytoseiid mites were collected after this release, but *G. pilosus* did not establish. Populations of *G. helveolus* and *N. californicus* increased following the second early release and peaked 8 and 10 weeks after release, respectively. A third phytoseiid release was made when 95% of leaves were infested by persea mites. *G. helveolus* and *N. californicus* were found from late releases, but neither exhibited increases in density as *O. perseae* populations were naturally declining (Hoddle *et al.*, 1999).

Later, three different release rates and timings of *N. californicus* were effected on avocado trees in a commercial orchard in California. Densities of predator and pest mites were monitored on trees treated with either one, two or three releases of 500, 1000 or 2000 *N. californicus*, or two applications of 5% narrow-range (NR) 415 petroleum oil. Predator releases were made when 50% (release 1), 75% (release 2) or 95% (release 3) of leaves were infested with one or more motile forms of the persea mite. Release of a minimum of 2000 *N. californicus* per tree was necessary to reduce *O. perseae* densities significantly compared with the control trees. Releasing 1000 *N. californicus* twice or 2000 *N. californicus* once provided *O. perseae* control similar to that by petroleum oil treatments and in trees treated with cumulative releases totalling more than 2000 phytoseiids per tree. This result indicated that releases totalling more than 2000 *N. californicus* per tree did not substantially improve pest mite control, and the cumulative number of predators used per tree is more important than the number of times *N. californicus* is employed for controlling the persea mite (Hoddle *et al.*, 2000a).

The contributions of *N. californicus*, *Amblyseius swirskii* Athias Henriot (= *Typhlodromips swirskii* (Athias Henriot)) and *Iphiseius degenerans* (Berlese) to the control of the persea mite on avocado in the Canary Islands have been evaluated by different researchers (see Hernández Suárez *et al.*, 2010). The results of further researches suggest that *N. californicus* is unable to establish and increase on avocados in the Canary Islands, despite the high number of predators released. To promote and improve biological control, current research is being focused on developing methods for the conservation/augmentation of *E. stipulatus* (Torres *et al.*, 2010).

The conservation of the prevalent phytoseiid mites (and release of exotic predatory mites) was tested by Maoz *et al.* (2009, 2011b) on avocado in Israel. Field trials showed that conservation of *E. scutalis* using Rhodes grass (*Chloris gayana* Kunth) as a wind-borne pollen provisioning cover crop (WPPCC) is highly effective compared with repeated artificial pollen applications. Densities of phytoseiid populations were significantly higher on trees adjacent to the Rhodes grass patches than on distant trees, whereas persea mite populations on trees adjacent to these patches were consistently lower. This study shows that the use of Rhodes grass as a WPPCC for the conservation of *E. scutalis* is both effective and sustainable. However, while the results indicate that *E. scutalis* has the potential for mite control, future studies are needed to demonstrate the efficacy of this approach in commercial orchards (Maoz *et al.*, 2011b). Other experiments showed that the conservation of this phytoseiid through the provisioning of supplemental food (e.g. maize pollen) increased its populations and enhanced persea mite control. In addition to *E. scutalis*, four generalist insect predators have been found in Israel that feed upon and rip open the nests of persea mites (Maoz *et al.*, 2009).

Some phytoseiid mites, such *N. californicus*, *G. annectens* and *G. helveolus* and larvae of *Chrysoperla* spp. are available commercially. From a practical point of view, it is suggested that a single release of 2000 phytoseiids per tree, or two releases each of 1000 phytoseiids per tree be made in spring or early summer when leaf monitoring for pest mite presence–absence detects about 50% of leaves with one or more motile forms of *O. perseae*. The most effective release strategy is to dispense predator mites and a carrier (e.g. maize grits) in small paper cups attached to branches. Four cups can be attached per tree, evenly distributed around the canopy on branches that are shaded from the sun. About 250–500 predators are added per cup depending on the release rate. The predators will disperse from

the cups (UC IPM Online, 2014a). The use of 'fresh' or 'cold-stored' predators may affect the results. The topic has been investigated by Takano-Lee and Hoddle (2002b), who compared the action of 'fresh' predators used ≤48 h after receipt from a commercial insectary and 'cold-stored' predators that were maintained at 12°C for about 14 days. Fresh and cold-stored *G. helveolus* attack *O. perseae* only after invading webbed nests. Conversely, fresh and cold stored *N. californicus* may intercept and attack migrant persea mites outside web nests, prey on persea mites through nest webbing or invade and prey upon *O. perseae* inside nests. The predatory efficacy of both *N. californicus* and *G. helveolus* was reduced following cold storage, as both species engaged in certain predatory behaviours less frequently than did fresh mites. The preying behaviours of *N. californicus* and *G. helveolus* on *O. perseae* are interpreted in relation to the chaetotaxy hypothesis, which proposes that phytoseiid invasion efficiency and the propensity for webbed nests are facilitated by dorsal setal lengths (Takano-Lee and Hoddle, 2002b).

CHEMICAL CONTROL. Chemicals may be sprayed in the early stages of a severe infestation, using narrow-range petroleum oils or acaricides. A good rule is to treat only infested trees and leave unsprayed areas or trees to conserve beneficials and provide refuges from which natural enemies and pesticide-susceptible pests can recolonize treated trees (UC IPM Online, 2014a). Acaricides commonly used are abamectin (Takano-Lee and Hoddle, 2002b; Hernández Suárez *et al.*, 2010; Maoz *et al.*, 2011a; UC IPM Online, 2014a), abamectin plus pyrethrins (Hernández Suárez *et al.*, 2010), fenpropathrin, narrow-range oil (UC IPM Online, 2014a), spirodiclofen (Hernández Suárez *et al.*, 2010; Maoz *et al.*, 2011a; UC IPM Online, 2014a) and sulfur (Hernández Suárez *et al.*, 2010).

Oligonychus peruvianus (McGregor) (Fig. 20.41)

Common name

Unknown.

Diagnostic characteristics

FEMALE. The body is ovoid and yellowish green in colour. The dorsal body setae are short, strongly lanceolate, nearly nude and broadest near the base. The peritremes end simply. The dorsal opisthosoma is longitudinally striate between the setae e_1. Tibia I has 9 tactile setae and 1 solenidion. Tarsus I bears 4 tactile setae and 1 solenidion proximal to the duplex setae and 1 single ventral tactile seta. Tibia II has 7 tactile setae. The empodium has 3 pairs of proximoventral hairs (McGregor, 1950; Pritchard and Baker, 1955).

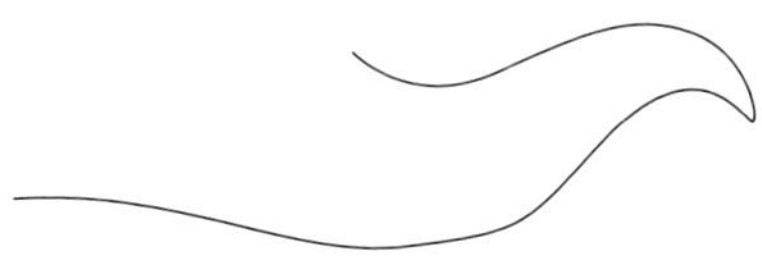

Fig. 20.41. *Oligonychus peruvianus* (McGregor). Male, aedeagus (Sandoval *et al.*, 2011, courtesy of C. Vásquez, 2013).

MALE. Tibia I has 9 tactile setae and 2 solenidia, and tibia II has 7 setae. Tarsus I has 4 tactile setae and 3 solenidia proximal to the duplex setae and 1 single ventral tactile seta. The aedeagus bends ventrally and the distal part is short (McGregor, 1950; Pritchard and Baker, 1955).

Geographical distribution

Brazil (CABI, 2009), Colombia (Thewke and Enns, 1969; Urueta, 1975), Costa Rica (Thewke and Enns, 1969), Ecuador (McGregor, 1955), Guatemala (Baker and Pritchard, 1962), Mexico (Estebanes-Gonzalez and Baker, 1968), Peru (McGregor, 1917; Guanilo *et al.*, 2012), Trinidad and Tobago (Hirst, 1922), the USA (Pritchard and Baker, 1955; Baker and Pritchard, 1962) and Venezuela (Doreste, 1982).

Host plants

The list of host plants of *O. peruvianus* includes 15 species belonging to 11 families (Migeon and Dorkeld, 2006–2013): Apiaceae – carrot (*Daucus carota* L.) (Bolland *et al.*, 1998; Migeon and Dorkeld, 2006–2013); Bixaceae – *Bixa orellana* L. (Estebanes-Gonzalez and Baker, 1968); Euphorbiaceae – cassava (*Manihot esculenta* Crantz) (Urueta, 1975; Yaseen and Bennett, 1977); Fabaceae – carob tree (*Ceratonia siliqua* L.) (McGregor, 1950), *Gliricidia sepium* (Jacq.) Kunth ex Walp. (Doreste, 1968a), *Trifolium* sp. (Baker and Pritchard, 1962); Lauraceae – avocado (*Persea americana* Miller) (Bolland *et al.*, 1998; Migeon and Dorkeld, 2006–2013); Malvaceae – *Gossypium* sp. (Pritchard and Baker, 1955); Oxalidaceae – *Oxalis* sp. (Bolland *et al.*, 1998; Migeon and Dorkeld, 2006–2013); Rubiaceae – Arabian coffee (*Coffea arabica* L.) (Doreste, 1968a); Rutaceae – *Citrus* sp. (Servicio para el Agricultor, 1973), sour orange (*Citrus aurantium* L.) (Doreste, 1968a); Salicaceae – *Salix* sp. (McGregor, 1917; Guanilo *et al.*, 2012); weeping willow (*S. babylonica* L.) (Doreste, 1968a); Vitaceae – *Vitis* sp. (Hirst, 1922), grapevine (*V. vinifera* L.) (Doreste, 1968a).

Bio-ecology

O. peruvianus may be misidentified with another occasional pest of some tropical plants, *O. perseae* Tuttle *et al.* The different idiosomal striation pattern between the setae e_1 and the shape of the aedeagus differentiate the two species (Sandoval *et al.*, 2011). The mite is a minor pest of a restricted number of economic plants that include avocado, carob, cassava, citrus, coffee, cotton, grape and some ornamentals, such as willow (McGregor, 1917, 1950; Pritchard and Baker, 1955; Doreste, 1968a; Urueta, 1975; Bellotti, 1985; Bolland *et al.*, 1998; Migeon and Dorkeld, 2006–2013).

Low levels of populations of *O. peruvianus* occur on the underside of the leaves, usually under small, compact webs resembling the egg covering of certain true spiders (McGregor, 1950; Baker and Pritchard, 1953; Estebanes-Gonzalez and Baker, 1968). These spots of white silk are commonly set alongside the midrib and lateral veins, and consist of two compact layers similar to small powder puffs. The eggs are usually deposited on the lower web, and held to it by radial threads. The lower web is in contact with the leaf surface, and each webbing

envelope contains both motile forms and eggs (up to 43) (Fréitez, 1974; Rodriguez, 1978; Salas, 1978; Ochoa *et al.*, 1991a, 1994).

Natural enemies include the phytoseiid mite *Amblyseius limonicus* Garman *et* McGregor (= *Typhlodromalus limonicus* (Garman *et* McGregor)) and a complex of minor predators (Braun *et al.*, 1989).

Symptomatology and damage

O. peruvianus lives on the underside of cassava leaves, and the corresponding sites on the upper leaf surface appear as yellow to brown dots. Severe infestations may induce drop of the basal leaves (Fréitez, 1974; Rodriguez, 1978; Salas, 1978; Ochoa *et al.*, 1991a, 1994). Despite these indications, the harmfulness of the mite has not been well defined. Few reports treat this issue and information on yield reductions of cassava of 20–53% were caused by combined injury by a complex of four pest mites, including *O. peruvianus*, and depended upon the timing and duration of infestation (CIAT, 1977).

Control

There are no special reports on the control of *O. peruvianus* because it is only a minor pest on cassava and other host plants compared with other pest mites. The main predator, *T. limonicus*, is not very effective against this mite. Tests of 2231 cassava varieties by CIAT (the International Center for Tropical Agriculture) showed that 250 had promising resistance to *O. peruvianus* (CIAT, 1976b). On cotton, repeated and heavy applications of sulfur dusts controlled *O. peruvianus* (Wille *et al.*, 1955).

Oligonychus platani (McGregor) (Fig. 20.42)

Common name

Sycamore spider mite.

Diagnostic characteristics

FEMALE. The body is greenish or brownish in colour, with black spots; the legs are light and the tarsi brown. Tibia I bears 7 tactile setae. Tarsus I has 3 tactile setae proximal to the duplex setae, and 1 ventral seta below the duplex setae. The proximal member of the duplex setae is nearly as long as the other member (Pritchard and Baker, 1955; Jeppson *et al.*, 1975).

MALE. The aedeagus bends ventrally, and the distal part tapers gradually (Pritchard and Baker, 1955; Jeppson *et al.*, 1975).

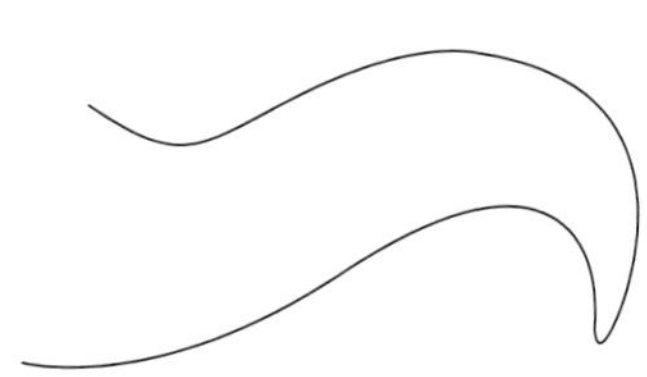

Fig. 20.42. *Oligonychus platani* (McGregor). Male, aedeagus (modified from McGregor, 1950).

Geographical distribution

France (Navajas *et al.*, 1996), Greece (Bolland and Beron, 2004), Iran (Khalil-Manesh, 1973), Mexico (Pritchard and Baker, 1955; Tuttle *et al.*, 1976) and the USA (McGregor, 1950; Pritchard and Baker, 1955; Bibby and Tuttle, 1959; Tuttle and Baker, 1964).

Host plants

The list of host plants of sycamore spider mite includes 23 species of 11 families:

Asteraceae: *Artemisia filifolia* Torr. (Baker and Tuttle, 1972), sunflower (*Helianthus annuus* L.) (Bolland *et al.*, 1998; Migeon and Dorkeld, 2006–2013), *Heteromma decurrens* (DC.) Benth. (Migeon and Dorkeld, 2006–2013).

Cupressaceae: *Cupressus* sp. (McGregor, 1950).

Fagaceae: California live oak (*Quercus agrifolia* Née) (Pritchard and Baker, 1955), emory oak (*Q. emoryi* Torr.) (Tuttle and Baker, 1968), valley oak (*Q. lobata* Née), *Quercus* sp., cork oak (*Q. suber* L.) (Pritchard and Baker, 1955).

Juglandaceae: English walnut (*Juglans regia* L.) (McGregor, 1950).

Lauraceae: camphor tree (*Cinnamomum camphora* (L.) J. Presl) (McGregor, 1950), avocado (*Persea americana* Miller) (Estebanes-Gonzalez and Baker, 1968).

Myrtaceae: *Eucalyptus* sp. (McGregor, 1950).

Nyctaginaceae: *Boerhavia* sp. (Tuttle *et al.*, 1974).

Pinaceae: Scots pine (*Pinus sylvestris* L.) (Bolland *et al.*, 1998; Migeon and Dorkeld, 2006–2013).

Platanaceae: London plane (*Platanus acerifolia* (Aiton) Willd.) (Pritchard and Baker, 1955; Navajas *et al.*, 1996), American sycamore (*P. occidentalis* L.) (McGregor, 1950), Oriental plane tree (*P. orientalis* L.) (Bolland and Beron, 2004).

Rosaceae: *Cotoneaster* sp. (McGregor, 1950), loquat (*Eriobotrya japonica* (Thunb.) Lindl.) (McGregor, 1950; Pritchard and Baker, 1955; Tuttle and Baker, 1964), toyon (*Heteromeles salicifolia* (C. Presl) Abrams) (McGregor, 1950; Pritchard and Baker, 1955), scarlet firethorn (*Pyracantha coccinea* M. Roem.) (Pritchard and Baker, 1955; Bibby and Tuttle, 1959; Tuttle and Baker, 1964).

Salicaceae: *Salix* sp. (McGregor, 1950).

Bio-ecology

O. platani has been known as a pest of sycamore trees in California since about 1924; it also infests other plants in California, including camphor, *Cotoneaster* sp., cypress, *Eucalyptus* sp., loquat, *Pyracantha* sp., *Quercus* spp., sycamore, toyon, walnut and willow (McGregor, 1950). It is a serious pest of pyracantha in southern Arizona (Butler and Abid, 1965), and has been found on avocado and oak in Texas and Mexico (Pritchard and Baker, 1955; Estebanes-Gonzalez and Baker, 1968; Jeppson *et al.*, 1975).

The sycamore spider mite lays eggs on the upper leaf surface, along the midrib, but at high levels of infestation they may be laid over the whole leaf surface. The eggs are flattened at the base, dome like, radially striate and with a dorsal stipe; when they have just been laid they are pale, but they gradually turn orange in colour. During June, July and August in the field in

southern Arizona, the developmental time from eggs to adult is 12.8, 12.1 and 9.8 days, respectively. In the laboratory, at 15, 12, and 34°C, the life cycle took 32, 10.2 and 9.6 days, respectively (Butler and Abid, 1965).

Symptomatology and damage

Feeding by *O. platani* on pyracantha causes bronzing and unsightly heavy webbing on both leaf surfaces (Butler and Abid, 1965; Jeppson *et al.*, 1975). In some areas, the susceptibility of cotoneaster to *O. platani* limits the use of this plant (Krüssmann, 1986; Wyman, 1986; Griffiths, 1994).

Control

In urban areas and/or in natural stands, control is not usually necessary. In the nursery, high levels of attack can be controlled with acaricides, generally during spring or summer.

Oligonychus pratensis **(Banks) (Fig. 20.43)**

Common name

Banks grass mite.

Diagnostic characteristics

FEMALE. The dorsal opisthosoma is transversely striate except for the longitudinal pattern between the setae f_1. The peritreme ends in a simple bulb. Tibiae I and II have 9 and 7 tactile setae, respectively. Tarsus I has 4 tactile setae proximal to the duplex setae, and 2 setae are set ventrally to the duplex setae (Jeppson *et al.*, 1975).

MALE. The distal knob of the aedeagus is about twice as wide as the stem; the axis of the knob is at a distinct angle to the axis of the shaft; the anterior projection of the knob is rounded and the posterior is acute; the tip curves slightly ventrally (Jeppson *et al.*, 1975).

Geographical distribution

Algeria (Athias Henriot, 1959), Brazil (Flechtmann and Baker, 1975; Feres *et al.*, 2009), China (Ma and Yuan, 1980), Colombia (Figueroa, 1919), Costa Rica (Ochoa *et al.*, 1991a), Egypt (Zaher *et al.*, 1982), El Salvador (Baker and Pritchard, 1962; Andrews and Poe, 1980), Hawaiian Islands (Jeppson *et al.*, 1975), Honduras (Ochoa *et al.*, 1991a), Iraq (Hirst, 1920), Madagascar (Helle *et al.*, 1970), Mexico (Baker and Pritchard, 1962), Pakistan (Chaudhri *et al.*, 1974), Senegal (Gutierrez and Etienne, 1981a,b), Thailand (Baker, 1975b) and the USA (Banks, 1912, 1914; Ewing, 1922; Pritchard and Baker, 1955; Bibby and Tuttle, 1959; Tuttle and Baker, 1964, 1976).

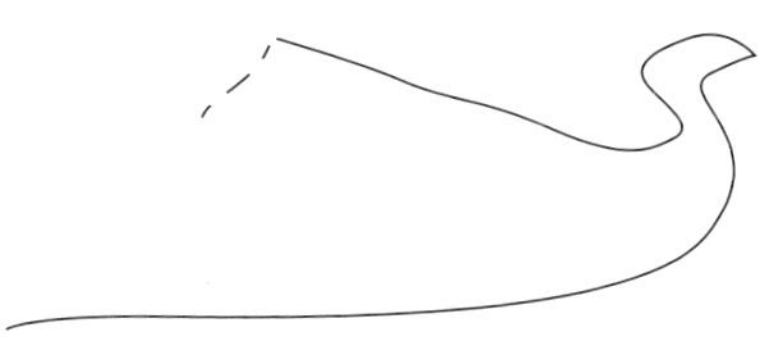

Fig. 20.43. *Oligonychus pratensis* (Banks). Male, aedeagus (from Meyer Smith, 1987).

Host plants

The whole list of host plants includes 70 species of 9 families, and the highest number of species (54) belongs to the Poaceae family (Migeon and Dorkeld, 2006–2013). The families and species of economic importance are: Arecaceae – coconut palm (*Cocos nucifera* L.), Canary Island palm (*Phoenix canariensis* hort. ex Chabaud), date palm (*P. dactylifera* L.) (Banks, 1914; Ewing, 1922; Pritchard and Baker, 1955), American cotton palm (*Washingtonia filifera* (Linden ex André) H. Wendl.) (Feres *et al.*, 2009); Cucurbitaceae – watermelon (*Citrullus lanatus* (Thunb.) Matsum. and Nakai) (Tuttle *et al.*, 1976); Malvaceae – okra (*Abelmoschus esculentus* (L.) Moench) (Chaudhri *et al.*, 1974); Poaceae – blue grama grass (*Bouteloua gracilis* (Willd. ex Kunth) Lag. ex Griffiths) (Tuttle and Baker, 1968), downy brome (*Bromus tectorum* L.) (Davis, 2008), Bermuda grass (*Cynodon dactylon* (L.) Pers.) (Pritchard and Baker, 1955; Bibby and Tuttle, 1959; Tuttle and Baker, 1964; Zaher *et al.*, 1982), barley (*Hordeum vulgare* L.) (Tuttle and Baker, 1968), African rice (*Oryza glaberrina* Steud.) (Gutierrez and Etienne, 1981b), bluegrass (*Poa pratensis* L.) (Tuttle and Baker, 1968), sugarcane (*Saccharum officinarum* L.) (McGregor, 1950; Pritchard and Baker, 1955; Baker, 1975b; Flechtmann and Baker, 1975; Lee Goff, 1986), sorghum (*Sorghum bicolor* (L.) Moench) (Bibby and Tuttle, 1959; Tuttle and Baker, 1964, 1968; Baker, 1975b; Gutierrez and Etienne, 1981b), wheat (*Triticum aestivum* L.) (Pritchard and Baker, 1955), maize (*Zea mays* L.) (Bibby and Tuttle, 1959; Baker and Pritchard, 1962; Tuttle and Baker, 1964, 1976; Baker, 1975b; Andrews and Poe, 1980).

Bio-ecology

The Banks grass mite is a severe pest of maize in the central plains states of the USA and in the states of Washington and Oregon (Hoy, 2011). In Colorado, it damages maize and is an occasional pest of turf, sorghum and wheat (Peairs, 2010; Cranshaw, 2012). In Texas, the Banks grass mite is one of the major pests of grain crops (Hobson, 1977). In Pakistan, it infests maize and other grasses (Nazir, 2009), as well as sorghum in India (Singh, 1997) and in some African countries (Gutierrez and Etienne, 1981a,b).

The Banks grass mite reproduces by arrhenotokous parthenogenesis and has $n = 4$ chromosomes (Helle *et al.*, 1970). Males develop faster than the females (Tan and Ward, 1977). The mite spins abundant webbing on the lower part of infested grasses. This webbing appears as a matting, which protects the eggs that are laid by the mite either on it or on the leaf surface. The mites overwinter as orange adult females near the bases of host plants (André, 1942; Malcolm, 1955). The populations start their activity in the early spring and the intense growth of the host plants occurring during this period usually attenuates the feeding damage. Populations pass periods of drought as immature stages on the dry leaves of the lower parts of plants and increase during late summer, when they cause severe damage to host plants. According to the temperature, the life cycle of

O. pratensis (egg to adult) takes from 8 to 25 days and the lifespan of adults is about 23 days. The mites develop 6–7 generations a year in areas of the northern USA and more in southern areas (Malcolm, 1955; Jeppson *et al.*, 1975). On maize at 25 ± 1°C, the net reproduction (R_0), the intrinsic rate of increase (r_m) and the finite rate of increase (λ) are 71.2, 0.290 and 1.340, respectively (Perring *et al.*, 1984), while on blue grama grass the same parameters are 20.1, 0.203 and 1.260, respectively (Congdon and Logan, 1983). Rearing of mites at high, medium and low vapour pressure deficits (VPD) at 25, 33, 36 and 39°C highlighted a reduced influence of humidity on their development. However, at 39°C and low VPD, egg hatching is strongly reduced. Temperature has a greater influence on longevity than humidity. The mites live longer at cooler temperatures, and at high temperatures the longevity tends to reduce at lower VPD. The intrinsic rate of natural increase (r_m) is dependent on temperature and humidity. At 25°C, r_m is similar for high, medium and low VPDs; however, at 33, 36 and 39°C, low VPD conditions cause a reduction in survival rates and fecundity, resulting in a lower r_m. The highest r_m occurs at 36°C and high VPD (Perring, 1983).

On date palm in California, all active forms of *O. pratensis* are found throughout the year; during winter the mites usually reduce their activity and populations increase in June, reaching peaks in July and August. During winter, the mites inhabit late-maturing or off-season fruit and date palm leaves, or grasses under the canopy, such as Bermuda grass (Elmer, 1965b). On fruit bunches, highest populations occur in July, when the fruit is immature and susceptible to feeding damage. Densities of *O. pratensis* on bunches vary strongly within trees. Predatory mites also live on the date fronds throughout the year, but only sporadically in January and March, and they do not occur on the bunches until June (Gispert *et al.*, 2001). In Mauritius, *O. pratensis* may be found during dry months on sugarcane growing in greenhouses and in the field on maize and wild plants (Jeppson *et al.*, 1975). In Colorado, fertilized females of Banks grass mite move into winter wheat in the autumn as their summer hosts, mainly maize, begin to dry. Winter conditions induce the mites to move to the crowns of wheat plants, where they live and feed until spring. Mites breed continuously on wheat and summer hosts until their return to winter wheat in the following autumn. According to the temperature, development of a generation takes 10–20 days (Peairs, 2010).

Severe stress of host plants results in low numbers of *O. pratensis*, probably due to reduced nutritional quality of the host. Non-stressed plants also harbour low mite numbers. On moderately stressed plants, populations tend to relate to the amount of stress. In maize, the canopy and leaf temperatures are higher in stressed plants, and canopy temperatures are significantly correlated with *O. pratensis* numbers. High canopy temperatures and low humidities may induce fast developmental rates, high survival rates and increased fecundity, resulting in a high r_m. Mite densities seem to be closely linked to microenvironmental conditions (Perring, 1983).

Light, leaf surface and gravity influence the choice of site of the adult female of *O. pratensis* on leaves of maize. When there is no light, the mites chose the abaxial side of maize leaves. When there is a light source, they stay on the leaf surface away from the light. In the absence of light and leaf-surface influences, gravity has a significant effect on the choice of the site on maize leaves by the mites; this effect also exists when there is light, and acts in concert with normal light conditions. Both heat and ultraviolet light could be responsible for the light effect (Li and Margolies, 1991). On infested leaves, the females produce more female progeny than on uninfested leaves, showing that the offspring sex ratios change in response to deteriorating food sources (Stiefel *et al.*, 1992). On undamaged sorghum leaves at growth stages 1–6 of Vanderlip (1972), *O. pratensis* has a longer lifespan (25.7 days) and ovipositional period (11.6 days) than on leaf discs at growth stages 6–8 (19.1 and 8.9 days, respectively). Females lay fewer eggs (4.3) daily at growth stages 1–6 than at stages 6–8 (Foster *et al.*, 1977b).

Heavy rain has been successful in washing the mite populations and their webbing from fruit bunches. Some growers spray water on the roads around the orchards to reduce dust (Perring *et al.*, 2000, in Hodel and Johnson, 2007).

Natural enemies include pathogens and predatory insects and mites. Pathogenic fungi include *Neozygites floridana* Weiser *et* Muma (Saba, 1973a) and *N. adjarica* (Tsintsadze *et* Vartapetov). *N. adjarica* was found infecting the Banks grass mite in field maize in Kansas (Dick and Buschman, 1995). Insects include some coccinellids, such as *Stethorus atomus* Casey (Ehler, 1974), *S. caseyi* Gordon *et* Chapin (Pollock and Michaels, 2003), *S. nigripes* Kapur (Pollock and Michels, 2002) and *S. utilis* Horn (Ehler, 1974). *S. nigripes* was introduced from Australia into California almond orchards (Hoy and Smith, 1982), but in formal releases it failed. Later, the coccinellid was found in Texas preying on *O. pratensis* (Pollock and Michels, 2002). Predatory mites include some phytoseiids, such as *Amblyseius californicus* McGregor (= *Neoseiulus californicus* (McGregor)), *Phytoseiulus persimilis* Athias Henriot and *Metaseiulus occidentalis* (Nesbitt) (Gilstrap *et al.*, 1977), and other species.

Symptomatology and damage

The mites usually attack wheat at the two- to five-leaf stage. Infested leaves first turn yellow and later become necrotic and brownish yellow. During autumn, the mites move to the crown, near the ground level of plants, where they stay through the winter. When infested, young leaves of wheat lose their colour, are curled inward, covered by webbing and wither. Sometimes, infestation of the heads may occur at their emergence from the boot. Moreover, the kernels may shrink and/or the plant not develop. Warm winters in some areas of the northern USA may facilitate mite activity. Feeding by the mite on grasses or autumn grains may cause severe damage, consisting of white stippled areas on the leaves, chlorosis and sometimes weakening of the plants. High levels of attack may induce high plant mortality. Reduction in seed production is not correlated with high numbers of mites. High infestation in grasses is common in areas of meager moisture or rainfall, and lawn grasses that are sprinkled consistently are generally not infested. The Banks grass mite most commonly damages Colorado winter wheat in the autumn in areas near maturing field maize (Jeppson *et al.*, 1975; Peairs, 2010).

Damage on sorghum plants consists of discoloration and premature wilting of leaves. Infestation first affects the underside of the lower leaves and gradually develops on the upper

leaves and head. High population densities occur along the midrib beneath the webbing, where they appear as small and white stippled spots on the leaf along the midrib. These spots increase in size, particularly on the basal portion of the leaf, and turn red or brown. Infested leaves fold downwards along the midrib nearest the base. The symptoms spread gradually from leaf to leaf. Heads show shrivelled seeds and webbing. Natural resistance of the plant to disease organisms may be affected (Ward *et al.*, 1972).

Feeding by the Banks grass mite on sugarcane causes small spots on the lower leaf surface. At high population densities, these spots appear as yellow streaks that become red as the leaves age. The edges and tips of the leaves may dry. The young internodes still in the sheath show small red blisters, which gradually turn the surface of the canes to a corroded reddish brown colour. At high levels of infestation, the internodes may appear rusty red (Jeppson *et al.*, 1975).

On maize, attacks by *O. pratensis* prematurely dry the leaves, break the stalks, reduce moisture contents of the grain, shrink kernels and reduce the overall grain yields (UC IPM Online, 2009d).

O. pratensis causes direct damage to dates by feeding on green and immature fruits, which results in hardened, shrivelled and cracked fruits. Late-season feeding causes rasping marks or bronzing on the fruit surface. Webbing produced by mites covers the fruit and collects dust, affecting fruit maturation and hindering the action of natural enemies and chemical control (Elmer, 1965b; Jeppson *et al.*, 1975; Farrar, 2000). Damaged fruits are downgraded or culled. In California, annual losses for 1997 were calculated at US$2.5 million plus control costs (California Date Commission, 1998).

On coconut palm, the mites feed along the midrib on the lower leaf surface. Feeding causes (jointly with *Raoiella indica* Hirst) yellowish patches on the infested areas (Jeppson *et al.*, 1975).

O. pratensis may be a severe pest of turf. The first symptoms consist of small and white stipplings. Damaged grass blades may show a slight purpling, which also is associated with brown wheat mite injury. Dead grass is brownish yellow in colour and the blades are stiff. Favourable conditions induce the rapid development of severe damage, which is usually related to drought stress (Cranshaw, 2012).

Control

Control of Banks grass mite is based on horticultural practices, on the use of resistant varieties, and on physical and chemical means. Biological control alone does not seem to be effective. However, reducing dust and weed hosts favour predators. According the UC Pest Management Guidelines for Corn Spider Mites (UC IPM Online, 2009d), monitoring treatments may be necessary.

MONITORING. Infestations usually begin on the lower portions of the maize plants and move upward as mite numbers increase. Older and lower leaves are randomly sampled and inspected to detect stippling on the upper surface and webbing, mites and feeding symptoms on the lower surface. Infestations reaching the ear leaf are the most damaging (UC IPM Online, 2009d).

CULTURAL PRACTICES. Fields, field margins and irrigation ditches should be kept clean of weed hosts. The reduction of dust may also hinder mite infestations (UC IPM Online, 2009d). On spring maize crops, post-emergence weed control at 40 days with or without reduction of the irrigation interval to 5 days results in minimum leaf infestation by *O. pratensis* (Nazir, 2009). On turf grass, it is suggested that all control programmes be based upon adequate water supply to the site. This includes autumn and winter watering as needed, because populations of the mites can build up during this period (Cranshaw, 2012). Growing downy brome as a winter cover crop in maize production creates favourable habitats for overwintering populations of predatory mites such as *Neoseiulus* spp. (Davis, 2008).

PLANT RESISTANCE. Although several sorghum varieties are resistant to the Banks grass mite, no antibiosis was found (Foster *et al.*, 1977b). Some varieties are more readily colonized than others, and non-preference is the mode of resistance. The mechanism of non-preference is having upright leaves; the mites prefer the undersides of bent leaves, which receive indirect sunlight and maintain a higher relative humidity. Additional research has indicated that tolerance also plays a role; resistant varieties have a higher sugar content and maintain healthy green stalks and leaves for longer, despite feeding by the mites (Hoy, 2011).

Field tests in Texas on selected sorghum lines and F_1 hybrids highlighted factors contributing to resistance to the Banks grass mite. Grain-type sorghums of similar maturity did not differ significantly in mite infestations, but some were less damaged than others, were relatively high in sugar and were 'non-senescing' types. The two sorghums that were least infested were either late maturing ('SC56-14') or both late maturing and high in tillering ('Sweet Sudan'). 'BTx618' was more resistant to damage than 'BTx 378', and this resistance was reflected in hybrids between these lines and 'SC599'. Sorghums identified in this study should be useful in the development of mite-resistant sorghum hybrids (Foster *et al.*, 1977a). The results of other research showed that the latest maturing line ('M100') and the least early-maturing lines ('60M' and 'CK60') harboured more mites per plant. Mite numbers on 'B407' (late maturing) did not differ from those on any of the other lines tested. Mite densities were positively correlated with leaf area. In contrast, mite feeding damage was significantly lower on the late-maturing lines, demonstrating that sorghum susceptibility to mite feeding increased as plants reached anthesis and caryopsis (Archer *et al.*, 1986).

Updated information concerning the screening for resistance to spider mites may be found in Singh (1997). Other studies to understand the inheritance of mite resistance and to develop mite-resistant hybrids of maize have been carried out by Bynum *et al.* (2004).

PHYSICAL CONTROL. This consists of sprayings of water on the roads around the orchards to reduce dust that deposits on dates (Perring *et al.*, 2000, in Hodel and Johnson, 2007).

BIOLOGICAL CONTROL. Currently, biological control alone does not seem to be envisaged as a solution to infestation by the Banks grass mite. Predation by *P. persimilis* and *N. californicus*

in the laboratory reduced the densities of *O. pratensis* by 60 and 28%, respectively. In general, phytoseiids preferentially fed upon the more abundant instars. Ovipositional rates of *P. persimilis* feeding on *O. pratensis* were approximately those when feeding on *Tetranychus* spp. (Pickett and Gilstrap, 1986). The persistence of *N. californicus*, *P. persimilis* and *M. occidentalis* was investigated on maize in the field, and releases of *P. persimilis* gave unsatisfactory results (Gilstrap *et al.*, 1977). In Texas and Kansas, the coccinellid *S. nigripes* is present on several native grasses from early-June to mid-July and in different maize fields. It has been found in almost every field infested with *O. pratensis*, except in fields treated for other pests. However, adult and larval densities of *S. nigripes* were highest within and beneath the webbing along the central rib of the maize leaf (Pollock and Michels, 2002). The fungus *N. adjarica* was found infecting the Banks grass mite and the two-spotted spider mite (*Tetranychus urticae* Koch) in field maize in Kansas (Dick and Buschman 1995). Epizootics were associated with high levels of attack and extended periods of high RH (>80%). Unfortunately, all *N. adjarica* infections in Kansas occur late in the field season (middle to late August), which is too late for it to serve as a dependable tool in an IPM programme (Hoy, 2011).

CHEMICAL CONTROL. The early detection of infestations on the lower leaves of young plants of maize (and other host plants), makes control effective. It is good practice to treat when maize is 60–120 cm high; sprays applied after the plants reach 120 cm in height make good coverage of the canopy difficult. It is necessary to spray some swathes of land around fields to keep mites from spreading into the field (UC IPM Online, 2009d). In Colorado, winter wheat insecticide applications to field margins bordering maize during autumn are often sufficient to prevent economic damage. Spring infestations are not common (Peairs, 2010). Strains that are resistant to chemicals are well known in Texas (Hobson, 1977).

Among the chemicals that can be used, sulfur is effective against the Banks grass mite. In the past, application of sulfur and parathion dusts has given good results. However, mite webbing near the bases of plants makes adequate coverage of plants difficult. Sulfur dusts are effective against the mite on dates, but it is necessary to distribute the chemical in the centre of the date bunches (Depew, 1960). On date palm in California, the growers make 3–5 applications of sulfur dusts (Farrar, 2000; Davies and Mauk, 1997, in Hodel and Johnson, 2007). However, bunches receiving commercial applications of sulfur harbour date mite numbers five times higher than in non-treated trees. In fact, sulfur treatment suppresses predatory mites, which may contribute to the increase in Banks mite numbers. Non-treated trees had a higher proportion of lightly infested bunches than did treated trees, while treated trees had more bunches that were heavily infested (Gispert *et al.*, 2001). In general, applications of wettable sulfur (microthiol) during July, or of hexythiazox when *O. pratensis* starts to lay eggs, give a good control of mites (Farrar, 2000; Mauk and Bellows, 1998, in Hodel and Johnson, 2007). Other chemicals that can be used include methyl parathion, permethrin (Ayyappath *et al.*, 1997), propargite and spiromesifen (UC IPM Online, 2009d). On turf grass, it is possible to use bifenthrin or lambda-cyhalothrin (Cranshaw, 2012). Recently, the use of etoxazole has been proposed on maize (Martinez *et al.*, 2013), and bifenthrin, bifenazate, dimethoate and propargite have been tested on bluegrass for seed production (Walenta, 2006).

Oligonychus punicae (Hirst) (Fig. 20.44)

Common name

Avocado brown mite.

Diagnostic characteristics

FEMALE. The female is indistinguishable from females of *Oligonychus yothersi* (McGregor) and *Oligonychus mangiferus* (Rahman *et* Sapra). The propodosoma is pinkish, and the lateral hysterosoma contains blotches of purplish brown. In older specimens, the hysterosoma may be blackish brown. The legs I and palpi are rusty pink, and the other legs pale. The dorsal body setae are strong, pale and lacking tubercles. Tibia I has 7 tactile setae and tarsus I has 4 tactile setae proximal to the duplex setae (Jeppson *et al.*, 1975).

MALE. The body is paler than that of the female. The aedeagus bends ventrally, the hook is rather broad and ends in a finger-like projection (Jeppson *et al.*, 1975).

Geographical distribution

Australia (Davis, 1968a), Brazil (Flechtmann and Baker, 1970; Feres *et al.*, 2005), Chile (Flechtmann, 1996), China (Wang, 1981a), Colombia (Urueta, 1975), Costa Rica (Baker and Pritchard, 1962; Ochoa *et al.*, 1990), Cuba (Livshits and Salinas-Croche, 1968), Egypt (Zaher *et al.*, 1982), El Salvador (Baker and Pritchard, 1962; Ortiz and Trejo, 1976; Andrews and Poe, 1980), France (Bolland *et al.*, 1998), Guatemala, Honduras (Ochoa *et al.*, 1991a), India (Hirst, 1926; Gupta, 1992), Mexico (Tuttle *et al.*, 1976), Nicaragua (Baker and Pritchard, 1962), Panama (Ochoa *et al.*, 1991a), Peru (Guanilo *et al.*, 2012), Venezuela (Fréitez and Alvarado, 1978a; Quirós de González, 1978) and the USA (McGregor, 1941b).

Host plants

The whole list of host plants of avocado brown mite includes 27 host plants of 10 families (Migeon and Dorkeld, 2006–2013).

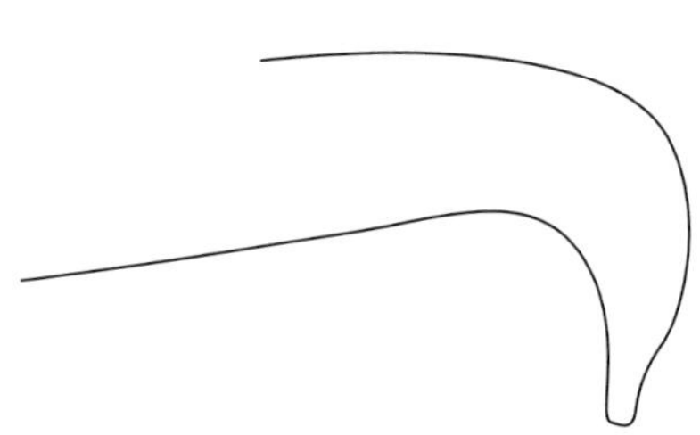

Fig. 20.44. *Oligonychus punicae* (Hirst). Male, aedeagus (from Meyer Smith, 1987).

The families and species of economic importance are: Anacardiaceae – mango (*Mangifera indica* L.) (Baker and Pritchard, 1962; Andrews and Poe, 1980; Ochoa *et al.*, 1990); Arecaceae – coconut palm (*Cocos nucifera* L.) (Gupta, 1992); Juglandaceae – English walnut (*Juglans regia* L.) (McGregor, 1950); Lauraceae – avocado (*Persea americana* Miller) (McGregor, 1941b; Andrews and Poe, 1980; Gupta and Gupta, 1994); Lythraceae – pomegranate (*Punica granatum* L.) (Hirst, 1926); Musaceae – banana (*Musa* × *paradisiaca* L.) (Urueta, 1975; Quirós de González, 1978); Rosaceae – strawberry (*Fragaria* × *ananassa* Duchesne ex Rozier) (Flechtmann and Baker, 1970), cherry laurel (*Prunus laurocerasus* L.) (McGregor, 1950), peach (*P. persica* (L.) Batsch) (Estebanes-Gonzalez and Baker, 1968); Rubiaceae – Arabian coffee (*Coffea arabica* L.) (Baker and Pritchard, 1962; Ortiz and Trejo, 1976; Andrews and Poe, 1980); Vitaceae – grapevine (*Vitis vinifera* L.) (Hirst, 1926; Baker and Pritchard, 1962; Urueta, 1975; Zaher *et al.*, 1982; Sánchez and Vásquez, 2005).

Bio-ecology

O. punicae is a pest of banana, coffee, grape, mango, pomegranate and other plants in tropical America (Baker and Pritchard, 1962; Ochoa *et al.*, 1990, 1994; Urueta, 1975; Ortiz and Trejo, 1976; Quirós de González, 1978; Andrews and Poe, 1980; Sánchez and Vásquez, 2005). In India, *O. punicae* is known as a pest of avocado (Gupta and Gupta, 1994), pomegranate (Hirst, 1926) and coconut palm (Gupta, 1992). However, Jeppson *et al.* (1975) report that on avocado the mite causes minor damage compared with that of other pest mites, such as the six-spotted spider mite. McMurtry (1985a) does not note any chemical application against its attacks on avocado, and recently Vásquez *et al.* (2008a) reported *O. punicae* as an occasional pest of grapevine in Venezuela.

The avocado brown mite reproduces by arrhenotoky and has $n = 2$ chromosomes (Helle *et al.*, 1981); it usually feeds on the upper leaf surface, but sometimes may occur on the lower leaf surface as well (McMurtry, 1985a; Tomczyk and Kropczyńska, 1985). In California, high populations may also occur during winter. However, many overwintering mites leave the tree, falling with the old leaves. Populations of *O. punicae* begin to increase in early summer, reaching a peak during late summer, and later decreasing. Sometimes, the increase in population is delayed until late summer, and the peak occurs during autumn or early winter (Jeppson *et al.*, 1975). In south China, the mite overwinters as the egg, and the eggs facultatively display diapause, both types of egg coexisting in nearly equal numbers (Kuang, 1983). At 25°C, the development of a generation takes 7 days, and at 32.5°C all biological stages die. During summer, the mite may develop two generations in a month (Ebeling, 1959).

At 25 ± 1°C on avocado leaves, the net reproduction (R_0), the intrinsic rate of increase (r_m) and the finite rate of increase (λ) were 43.1, 0.220 and 1.245, respectively (Tanigoshi and McMurtry, 1977). In further studies on avocado leaves, the time for development was 7.74, 7.78 and 9.54 days for the 'Fuerte', 'Hass' and 'Criollo' cultivars, and the average fecundity rate per female was 4.05, 2.86 and 4.38 eggs a day, respectively. The net reproduction (R_0), intrinsic rate of increase (r_m) and finite rate of increase (λ) indicate that there is a greater potential for population growth on the 'Hass cultivar' (25.81, 0.2214 and 1.24, respectively), followed by 'Fuerte' (11.96, 0.1624 and 1.17, respectively) and 'Criollo' (10.78, 0.1485 and 1.16, respectively). The same behaviour was observed for the generation time (T_g) and doubling time (t) population parameters. The 'Hass' cultivar had the fastest development rate and the population doubling time of 14.68 and 3.13 days, respectively; the corresponding results for the 'Fuerte' and 'Criollo' cultivars were 15.28 and 4.27, and 16.01 and 4.67, respectively (Cerna *et al.*, 2009).

The life cycle, fecundity and longevity of *O. punicae* have also been investigated on the leaves of six grapevine cultivars ('Tucupita', 'Villanueva', 'Red Globe', 'Syrah', 'Sauvignon' and 'Chenin Blanc') in Venezuela under laboratory conditions at 27 ± 2°C, 80 ± 10% RH and 12L:12D photoperiod. Preoviposition, oviposition and post-oviposition times in *O. punicae* females varied among the grape cultivars. The shortest pre- and post-ovipositional times were found on 'Chenin Blanc' leaves (1.2 and 0.9 days, respectively), while on other cultivars the ranges of these parameters were 1.6–2.0 to 1.2-2–5 days, respectively. Oviposition periods lasted from 6.7 days ('Villanueva' cultivar) to 16.1 days ('Sauvignon' cultivar). Average daily egg production was highest on 'Tucupita' (2.8 eggs per female daily) and lowest on 'Syrah' (0.9 eggs per female daily). Daily oviposition rate ranged from 2.0 to 6.1 eggs per female up to day 7 in females feeding on 'Tucupita' leaves, while on 'Chenin Blanc', 'Red Globe', 'Sauvignon', 'Syrah' and 'Villanueva' leaves the range was 1.1–6.1, 0.2–5.3, 0.3–5.0, 0.2–2.7 and 0.5–3.5 eggs per female, respectively. Female longevity was longest on the 'Sauvignon' (17.5 days), and shortest on the 'Villanueva' cultivar. The life cycle of *O. punicae* differed among cultivars, with average values ranging between 8.2 days on 'Tucupita' and 9.1 days on 'Syrah' (Vásquez *et al.*, 2008a, 2012). The intrinsic rate of increase (r_m) was highest on 'Sauvignon' (0.292) and 'Tucupita' (0.261), and lowest on 'Syrah' (0.146) and 'Villanueva' (0.135) (Vásquez *et al.*, 2008a). Amounts of secondary metabolites, such as alkaloids, flavonoids, tannins and polyphenols, of leaves of the six grape cultivars, as well as the thickness of the adaxial cuticle–epidermis, were assessed to determine their potential effects on mite parameters. Differences in cuticle–epidermis thickness among the six cultivars did not influence mite parameters. Secondary metabolite content did vary between the cultivars and, generally, increasing flavonoid content coincided with decreasing reproductive parameters (Vásquez *et al.*, 2008a).

On the camphor tree, a threshold temperature of 7.9°C was found for the development of *O. punicae*. The duration in degree-days above 7.9°C for females was 205.5. At 20.7 and 28°C, the intrinsic rates of increase (r_m) were 0.42 and 0.44, and the net reproduction rate (R_0) was 37.58 and 19.07, respectively (Kuang, 1983).

Populations of *O. punicae* on seedlings of *Persea indica* (L.) Spreng in the greenhouse initially infested with 20, 10 and 5 females per plant increased about 1000-, 2000- and 4000-fold, respectively, in 7–8 weeks and then declined rapidly as the leaves became strongly bronzed. The numbers of females decreased and that of males increased with time, until the males finally comprised the highest percentage of the active stages. When heavily bronzed leaves were washed to remove webbing, cast skins and other debris, and reinfested, the peak of new

mite populations was more than six times as high as those on bronzed leaves that were reinfested but not washed. On young leaves or on the lower surface of old leaves, the reproductive rate is usually lower than on the upper leaf surface of mature leaves (McMurtry, 1970). Intraspecific competition and/or predation by natural enemies may induce the decline of populations (McMurtry and Johnson, 1966; McMurtry, 1970). Females balloon from the leaves on silk threads during periods of gentle breeze, which breaks the silk threads and carries them aloft (Fleschner *et al.*, 1956). This behaviour increases migration to the undersurfaces of leaves and immature leaves, which is stimulated by reduction and deterioration of the food resource, and is important in the dynamics of the mite population (McMurtry, 1970). Suppressive effects of rainfall and natural enemies on the population of avocado brown mite were also observed in China (Kuang, 1983). Temperatures of 32–35°C or higher usually kill the mites and their eggs, as does the first cold weather in autumn or early winter (UC IPM Online, 2007b).

A number of predatory insects and mites prey upon the avocado brown mite. Insects include the coccinellid *Stethorus picipes* Casey (McMurtry and Johnson, 1966; McMurtry *et al.*, 1969), the aeolothripid *Scolothrips sexmaculatus* (Pergande) (McMurtry and Johnson, 1966; McMurtry *et al.*, 1970), the cecidomyiid *Arthrocnodax occidentalis* Felt (Fleschner, 1958b; McMurtry and Johnson, 1966), the staphylinids *Oligota oviformis* Casey (Fleschner, 1958b; McMurtry and Johnson, 1966) and *O. pygmaea* (Sol.) (Gonzalez, 1983, in McMurtry, 1985a), and the chrysopid *Chrysopa carnea* (Stephens) (McMurtry and Johnson, 1966). Predatory insects generally reduce the increase in populations of *O. punicae* late in the season (McMurtry and Johnson, 1966). Predatory mites include the phytoseiids *Amblyseius hibisci* (Chant) and *A. limonicus* Garman *et* McGregor (= *Typhlodromalus limonicus* (Garman *et* McGregor)) (McMurtry and Johnson, 1966; McMurtry, 1985a), *Typhlodromus floridanus* Muma (Tanigoshi and McMurtry, 1977; McMurtry, 1985a), *Phytoseiulus longipes* Evans (Badii *et al.*, 2010b), *Euseius mesembrinus* (Dean) (Badii *et al.*, 2010a) and *E. concordis* (Chant) (Escalona and Vásquez, 2005). The cheyletid *Cheyletus fortis* Oudemans has also been reported (Gupta and Gupta, 1992, in Gupta, 2001).

Symptomatology and damage

The avocado brown mite is a sporadic pest, mostly in the coastal growing areas of California, where damage is most noticeable from about July to September. Severe infestations tend to occur in border row trees along dirt roads, where road dust hinders mite predators. Ash deposited on leaves from wildfires reportedly also causes brown mite outbreaks (UC IPM Online, 2007b). The mites feed on the upper leaf surface of avocado, but at high population densities may feed on the lower leaf surface and fruits as well. Populations develop first along the midrib or in depressions of the leaf surface; later, they spread along the smaller veins, and at high levels of attack occupy the whole leaf surface. Leaf areas damaged by feeding become brownish or bronzed and covered by whitish hatched eggs and cast skins. The destruction of chlorophyll affects the physiology of the leaf, but when the mite population is removed, the leaf colour may return (Jeppson *et al.*, 1975; McMurtry, 1985a). Infestation does not seem to cause a large amount of damage (Tomczyk and Kropczyńska, 1985), causing only partial defoliation of avocado trees, especially of the 'Hass' variety, at about 80–100 adult females or 200–300 post-embryonic stages per leaf (McMurtry and Johnson, 1966; McMurtry *et al.*, 1969; McMurtry, 1985a). The Pest Management Guidelines of California University report that no significant damage occurs when populations are low to moderate (about 10–20 adult females per leaf), especially when *S. picipes* is active in the field. However, damage does occur if the mite averages about 50–70 adult females per leaf (about 100–200 adults and nymphs). In this case, the mites spread on the lower leaf surface and fruits, inducing partial defoliation, leaf bronzing along the midrib and then along smaller veins; finally, the entire leaf turns brown (UC IPM Online, 2007b).

Avocado leaves damaged by *O. punicae* exhibit first an increase and later a decrease in the intensity of transpiration (Sances *et al.*, 1982a). The phenomenon may be explained by the damage that has been done to the protective surface of the leaves and by the disturbance in the functioning of the stomatal openings.

The avocado brown mite damages the camphor tree and sassafras in China. As a result of injury, the chlorophyll contents of host leaves can decrease to 34–63% (Kuang, 1983).

In Venezuela, *O. punicae* is an occasional pest whose feeding on grapevines delays fruit ripening (Vásquez *et al.*, 2008a).

Control

Avocado is generally tolerant of moderate attacks of *O. punicae*, and damaging population densities occur only occasionally. Natural enemies and weather conditions usually hinder its outbreaks in California, as does the control of dust and the prohibition to use broad-spectrum pesticides for other pests (UC IPM Online, 2007b). Chemical control could, therefore, be unnecessary in the USA (McMurtry, 1985a), while in other countries, such as the Mexico, the need of treatments is more frequent (Coria-Avalos, 1993).

MONITORING. The occurrence of bronzed leaves and brown mites should be monitored during the summer through to the autumn, as should the presence of other pests (caterpillars, the persea mite), especially in coastal groves. Monitoring for the brown mite should also be done in border rows along dirt roads where trees are dusty, when trees have been sprayed earlier in the season with a broad-spectrum insecticide and after wildfires. Higher outbreaks occur after spraying a broad-spectrum insecticide to control other pests. To locate the avocado brown mite and its webbing, inspections should be done along the midrib on the upper leaf surface. The threshold is not known, and chemical applications for the avocado brown mite are rarely necessary (UC IPM Online, 2007b).

To assess the average population of *O. punicae* on avocado with 80% confidence and a relative precision of 0.25, it has been suggested in Mexico that 120 leaves in 12 trees should be examined, which is sufficient for most sampling programmes (Morales-Galván *et al.*, 2003).

CULTURAL PRACTICES. Population outbreaks of the avocado brown mite are often related to the deposition of dust on the leaves of

trees close to dirt roads or other sources of dust (Fleschner, 1958c). In this case, the control of dust improves predator activity and maintains biological control. It is advisable to drive slowly on dirt roads, and to use a water truck or trailer to wet unpaved roads and prevent airborne dust, especially during the summer months (UC IPM Online, 2007b).

PLANT RESISTANCE. Some avocado cultivars ('175PLS', '54PLS', '131PLS', '120PLS', '18PLS', '137PLS', '30PLS') exhibit some tolerance to *O. punicae* (Reyes and Salgado, 1994). The cultivars 'Colinmex' and '148PLS' are moderately susceptible (Colin and Rubi, 1992), while the cultivars 'Rincom' and 'Fuerte' are tolerant of the mite (Ortega and Ewart, 1972, in Gallegos, 1983).

BIOLOGICAL CONTROL. The biological control of *O. punicae* mainly exploits the action of natural enemies present in the environment. In California, the most important predators are the phytoseiids *A. hibisci*, *T. limonicus* and the coccinellid *S. picipes* (McMurtry and Johnson, 1966; McMurtry, 1985a). Recently, a positive laboratory evaluation has been made of the phytoseiid *E. mesembrinus* in Mexico (Badii *et al.*, 2010a). Releases of *S. picipes* on avocado timed correctly may suppress mite populations (McMurtry *et al.*, 1969), but their economic cost is unacceptable, and natural control is commonly sufficient (McMurtry, 1985a). Any control exerted by *S. picipes* is generally effected before severe damage (severe bronzing on leaves) occurs on leaves, when the predator shows a marked numerical response to an increase of the pest mite, the density of which is in the range of 10–20 adult females per leaf. When severe damage occurs, the increase of *S. picipes* is delayed (McMurtry and Johnson, 1966). Results from releases of phytoseiid mites are uncertain (McMurtry *et al.*, 1984).

CHEMICAL CONTROL. In Mexico, sprays of sulfur and propargite have been suggested three to four times during the dry season, from December and May (Coria-Avalos, 1993). The Pest Management Guidelines of California University (UC IPM Online, 2007b) suggest the use of sulfur and narrow-range oil. When temperatures exceed 32°C, sulfur sprays may cause leaf damage, and are often not effective in coastal areas where temperatures do not promote fuming action. When treating any pests, including *O. punicae*, during late summer or autumn, individual trees should be spot treated where possible (UC IPM Online, 2007b).

Oligonychus ununguis (Jacobi) (Fig. 20.45)

Common name

Spruce spider mite.

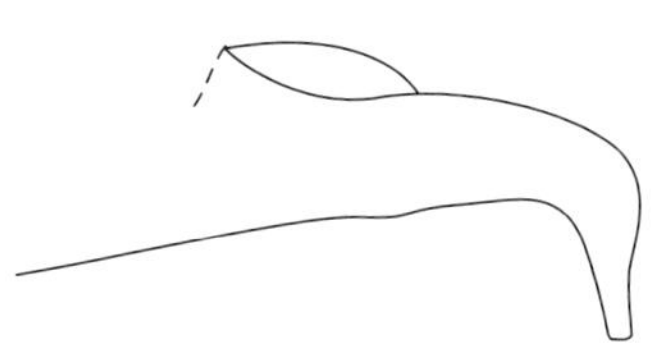

Fig. 20.45. *Oligonychus ununguis* (Jacobi). Male, aedeagus (from Meyer Smith, 1987).

Diagnostic characteristics

FEMALE. The body is orange to blackish in colour. Tibia I has 7 tactile setae. Tarsus I bears 4 tactile setae proximal to the duplex setae, and the proximal members of the duplex setae are short (Jeppson *et al.*, 1975).

MALE. The aedeagus is bent ventrally at a right angle to the shaft, and tapers gradually to the acute tip (Jeppson *et al.*, 1975).

Geographical distribution

Worldwide (Bolland *et al.*, 1998; Migeon and Dorkeld, 2006–2013).

Host plants

The whole list of host plants of spruce spider mite includes 97 host plants of 11 families, with the highest number of species ascribed to Cupressaceae (23 species) and Pinaceae (55 species) (Migeon and Dorkeld, 2006–2013).

Araucariaceae: *Araucaria angustifolia* (Bertol.) Kuntze (Bolland *et al.*, 1998; Migeon and Dorkeld, 2006–2013).

Buxaceae: *Buxus* sp. (Bolland *et al.*, 1998; Migeon and Dorkeld, 2006–2013).

Cupressaceae: incense-cedar (*Calocedrus decurrens* (Torr.) Florin) (McGregor, 1936), *Chamaecyparis* sp. (Reeves, 1963), *C. funebris* (Endl.) Franco (Meyer Smith, 1987), *C. lawsoniana* (A. Murray bis) Parl. (Dobosz *et al.*, 1995), *C. obtusa* (Siebold *et* Zucc.) Siebold and Zucc. ex Endl. (Ehara, 1962), *C. pisifera* Siebold *et* Zucc. (Czajkowska *et al.*, 2003), *Cupressus* sp (Browne, 1968; Flechtmann and Baker, 1975, Meyer Smith, 1996a), *C. arizonica* Greene (Tuttle and Baker, 1964), *C. lusitanica* Mill. (Ehara, 1966b; Urueta, 1975), *C. sempervirens* L. (Bibby and Tuttle, 1959; Tuttle and Baker, 1964; Navajas *et al.*, 1996), *Juniperus* sp. (Reeves, 1963; Tuttle and Baker, 1976), *J. bermudiana* L. (Browne, 1968), *J. chinensis* L. (Thewke and Enns, 1969; Tuttle and Baker, 1976; Dobosz *et al.*, 1995), *J. communis* L. (Reeves, 1963), *J. deppeana* Steud. (Tuttle and Baker, 1968), *J. horizontalis* Moench (Reeves, 1963), *J. rigida* Siebold *et* Zucc., *J. virginiana* L., *Platycladus orientalis* (L.) Franco (Reck, 1953; Lee Goff, 1986; Dobosz *et al.*, 1995), *Thuja* sp. (Livshits and Salinas-Croche, 1968), *T. occidentalis* L. (Ewing, 1917; Bibby and Tuttle, 1959; Tuttle and Baker, 1964, 1976; Dobosz *et al.*, 1995), *T. plicata* Donn ex D. Don (Tuttle and Baker, 1976), *Xanthocyparis nootkatensis* (D. Don) Farjon and D. K. Harder (Bolland *et al.*, 1998; Migeon and Dorkeld, 2006-2013).

Fagaceae: *Castanea* sp. (Ehara, 1962), *C. crenata* Sielbold *et* Zucc., *Quercus* sp. (McGregor, 1950), *Q. acutissima* Carruthers, *Q. alba* L. (Bolland and Beron, 2004), *Q. mongolica* Fisch. ex Ledeb., *Q. rubra* L. (Ehara, 1962), *Q. serrata* Thunb., *Q. variabilis* Blume (Bolland *et al.*, 1998; Migeon and Dorkeld, 2006–2013).

Juglandaceae: English walnut (*Juglans regia* L.) (McGregor, 1950).

Myrtaceae: *Eucalyptus* sp. (Estebanes-Gonzalez and Baker, 1968).

Pinaceae: *Abies* sp. (McGregor, 1950; Reck, 1956), *A. alba* Mill., *A. balsamea* (L.) Miller, *A. bracteata* (D. Don) Poit., *A. firma*

Siebold *et* Zucc. (Ehara, 1962), *A. fraseri* (Pursh) Poir., *A. grandis* (Douglas ex D. Don) Lindl. (Dobosz *et al.*, 1995), *A. holophylla* Maxim., *A. lasiocarpa* (Hook.) Nutt. (Tuttle and Baker, 1968), *A. nordmanniana* (Steven) Spach, *A. sachalinensis* (F. Schmidt) Mast. (Ehara, 1954), *Cedrus libani* A. Rich., *Larix* sp. (Reeves, 1963), *L. decidua* Mill., *L. kaempferi* (Lamb.) Carrière (Browne, 1968), *L. sibirica* Ledeb., *Picea* sp. (Ewing, 1921), *P. abies* (L.) Karst. (Jacobi, 1905; Ehara, 1962; Dobosz *et al.*, 1995; Czajkowska *et al.*, 2003), *P. alcoquiana* (H.J. Veitch ex Lindl.) Carrière, *P. engelmannii* Parry ex Engelm. (Tuttle and Baker, 1968), *P. glauca* (Moench) Voss (Browne, 1968; Dobosz *et al.*, 1995; Kielkiewicz and Czajkowska, 2002; Czajkowska *et al.*, 2003), *P. jezoensis* (Siebold *et* Zucc.) Carr. (Ehara, 1954), *P. mariana* (Mill.) Britton, Sterns and Poggenb., *P. omorika* (Pancic) Purk. (Dobosz *et al.*, 1995; Kielkiewicz and Czajkowska, 2002), *P. orientalis* (L.) Link (Oudemans, 1931c), *P. pungens* Engelm. (Tuttle and Baker, 1964, 1976; Kielkiewicz and Czajkowska, 2002; Czajkowska *et al.*, 2003), *P. rubens* Sarg., *P. sitchensis* (Bong.) Carrière (Browne, 1968), *Pinus* sp. (Womersley, 1940; McGregor, 1950; Hatzinikolis, 1970), *P. aristata* Engelm., *P. balfouriana* Balf., *P. banksiana* Lamb., *P. brutia* Ten., *P. cembroides* Zucc. (Estebanes-Gonzalez and Baker, 1968), *P. contorta* Douglas ex Loudon (Browne, 1968), *P. densiflora* Siebold *et* Zucc., *P. edulis* Engelm., *P. flexilis* James, *P. halepensis* Mill., *P. jeffreyi* Balf., *P. koraiensis* Siebold *et* Zucc., *P. lambertiana* Douglas, *P. leiophylla* Schiede *et* Deppe (Estebanes-Gonzalez and Baker, 1968), *P. monophylla* Torr. *et* Frém., *P. parviflora* Siebold *et* Zucc., *P. ponderosa* Lawson *et* C. Lawson (Tuttle and Baker, 1968), *P. radiata* D. Don (Browne, 1968), *P. resinosa* Aiton, *P. sabiniana* Douglas ex Douglas, *P. strobus* L., *P. sylvestris* L. (Hirst, 1924), *P. thunbergii* Parl. (Ehara, 1962), *Pseudotsuga menziesii* (Mirb.) Franco (Tuttle and Baker, 1968), *Tsuga canadensis* (L.) Carrière (Reeves, 1963), *T. dumosa* (D. Don) Eichler (Bolland *et al.*, 1998; Migeon and Dorkeld, 2006–2013).

Rosaceae: *Rosa* sp. (Suárez, 2004).

Sapindaceae: *Acer* sp. (McGregor, 1950).

Taxaceae: *Taxus cuspidata* Siebold *et* Zucc. (Bolland *et al.*, 1998; Migeon and Dorkeld, 2006–2013).

Taxodiaceae: *Cryptomeria japonica* (L.f.) D. Don (Matsuda *et al.*, 2012), *Sequoia* sp., *S. sempervirens* (Lamb. ex D. Don) Endl. (Pritchard and Baker, 1955).

Bio-ecology

The spruce spider mite is probably the most destructive conifer-feeding mite in the world (Kielczewski, 1966; Johnson and Lyon, 1976, 1988; Boyne and Hain, 1983). It infests many species of forest and ornamental plants, although it is possible that reports on some of above-mentioned host plants are those of different spider mites, e.g. in Japan, populations of *O. ununguis* reported on conifers and broadleaves are probably made up of two different species and not one common species (Sun *et al.*, 2000). In the field, the spruce spider mite may be confused with the two-spotted spider mite, *Tetranychus urticae* Koch, and the arborvitae spider mite, *Platytetranychus thujae* (McGregor). However, posteriorly, *O. unuguis* shows one consistent coloured area, while the two-spotted spider mite has two such areas and the arborvitae spider mite has several small and darkened areas (Lehman, 1997). Moreover, the egg of the spruce spider mite bears a dorsal stipe (Lehman, 2002).

O. ununguis reproduces by arrhenotokous parthenogenesis and has $n = 3$ chromosomes (Helle and Bolland, 1967). When walking, it spins silk threads that it uses as 'lifelines' on the host plant (Saitō, 1979a). These threads are usually disposed over leaf depressions, and gradually accumulate to form a 'woven roof' structure. In this way, the mites construct a nest-like web within which they feed and reproduce. Because of this behaviour, Saitō (1985) suggested that the spruce spider be included in life type WN-u. Mites belonging to this subtype always feed and walk under the web, which shelters them from adverse climatic factors and natural enemies.

Immature stages and adults of *O. ununguis* feed on needles, first near the base and then, as populations increase, over the entire needle surface to the tips. From spring until bud break, the spruce spider mites prefer to feed on the current season's growth. After this, they move to the newly developing needles, although high populations induce the mites to move back to the previous season's needles (Calkin, 1991).

The spruce spider mite overwinters as eggs laid near the base of needles and on other protected sites, but eggs are never laid on the needles (Zacher, 1921b; Geijskes, 1939). Overwintering eggs are evenly laid on all parts (top, middle, bottom) of the tree with no preference for cardinal direction. Most eggs are not laid on the ends of branches, but at some distance from the bud, and the distance from the main trunk to the oviposition site increases from the top to the bottom of the tree (Richmond and Shetlar, 1996). The egg diapause ends in the latter half of winter in south Finland (Löyttyniemi, 1970a) and in Japan (Shinkaji, 1975). In Japan, 50% hatch of the winter eggs was not found until the end of April and the beginning of May. Mites hatched from overwintering eggs feed on older growth for several generations; they will feed on new growth later when it has hardened off (Lehman, 2002). Adults complete their development in 11–23 days.

Females lay summer eggs in sheltered sites of the tree, and immature stages live mainly on the needles and shoots of the lower portion of the crown, while the adults have no preference. Fecundity is 45 eggs, and 60–80% of offspring are females. The thermal threshold is 6–7°C, and the diapause concludes after a resting period when eggs are exposed to 20°C (Serafimovski and Thalenhorst, 1962; Reeves, 1963). In Longjing (China), the spruce spider mite develops six generations a year and overwinters as eggs that are mostly deposited on branches of pine. Winter diapause begins usually in September and overwintering eggs hatch in early May. The sex ratio is 2.5 females: 1 male, and the female fecundity is 35–50 eggs. The bionomy of *O. ununguis* is closely related to temperature, and its activity and feeding depend on light and twigs. The mite is active during the day from 09:00 to 16:00 h (Jin, 2006).

Varandi *et al.* (2005) conducted laboratory and field studies on *O. ununguis* in Iran. Laboratory studies at 20 ± 1, 25 ± 1 and 30 ± 1°C, 65 ± 5% RH and under 12 h light showed that the duration of a generation, oviposition period and longevity of the mature mite were 21.49 ± 1.17, 19.2 ± 0.26 and 20.8 ± 0.6 days at 20 ± 1°C, and 15.4 ± 0.82, 15.9 ± 0.32 and 17.2 ± 0.67 days at 25 ± 1°C, respectively. Fecundity and the eggs laid daily were 23 ± 1.28 and 1.6 ± 0.26 at 20 ± 1°C, and

27 ± 1.97 and 2.3 ± 0.28 at 25 ± 1°C, respectively. The mite developed up to seven generations a year. At 30 ± 1°C, none of the immature stages of *O. ununguis* survived to maturity. The sex ratio was 3.1 females:1 male. The species overwinters as egg, and the highest population peak occurs in August and September.

On Fraser fir (*A. fraseri*) in North Carolina, the most favourable conditions for *O. ununguis* are 26°C and 50–60% RH. The peak of net reproductive rate (R_0) is at 20°C, and the rate stabilizes at 23°C and 26°C before dropping to its lowest point at 29°C. Mean generation time (T) decreases as the temperature increases. The shortest T occurs at 29°C, and at 26°C the value for T is 18.50. The maximum intrinsic rate of increase (r_m) (0.157) occurs at a constant temperature of 26°C. The development times of all biological stages were different at low (30–40%), moderate (50–60%) and high (88–98%) RH, with the most favourable response at 50–60% RH, and the least survival at high RH. Frequent and heavy rainfall are detrimental to population increase, and populations of *O. ununguis* are highest in spring and autumn (Boyne and Hain, 1983). On chestnut at 25 ± 1°C, R_0, r_m and the finite rate of increase (λ) are 24.7, 0.178 and 1.195, respectively (Saitō, 1979b).

Species and variety of host plant, especially their morphological and anatomical structures and their biochemical composition, have a significant effect on the dynamics of populations of *O. ununguis* (Przygoda, 2001; Yin and Sun, 2002; Czajkowska and Ewa, 2005; Puchalska *et al.*, 2006, 2008). Populations respond favourably to warm and dry conditions (Schread 1951; Matthysse and Naegele 1952; Stewart and Peterson, 1960; Peterson and Hildahl, 1969; McBride, 1977). Decreases in developmental time are related to increases in daily mean temperature (Garman, 1923). Pesticides applied against other pests may be detrimental to natural enemies of spruce spider mites and stimulate outbreaks (von Scheller, 1962; Fellin, 1968; Sandquist *et al.*, 1993). At high population densities and with deteriorating resources, the threads produced by mites change their function and become ballooning threads for adult females (Wanibuchi and Saitō, 1983).

Natural enemies include some pathogens (*Entomophthora* sp.) (Löyttyniemi, 1970a), and various predatory mites and insects. Phytoseiid mites are common on ornamental coniferous plants during the growing season (Czajkowska and Ewa, 2005). Among these are *Amblyseius andersoni* Chant (Kiekiewicz and Czajkowska, 2002), *A. castaneae* Wang *et* Xu (Lu *et al.*, 2008), *A. finlandicus* (Oudemans) (Sun and Yin, 2002; Sun *et al.*, 2002), *Metaseiulus occidentalis* (Nesbitt) (Shrewsbury and Hardin, 2003), *Neoseiulus fallacis* (Garman) (Kramer and Hain, 1989; Shrewsbury and Hardin, 2003), *N. collegae* (De Leon) (Mizell and Schiffhauer, 1991), *Typhlodromus americanus* Chant *et* Yoshida-Shaul (West and De Angelis, 1993), *T. bakeri* (Garman) (von Scheller, 1962), *T. laurentii* Ragusa *et* Swirski (= *T. setubali* Dosse), *T. phialatus* Athias Henriot (Kiekiewicz and Czajkowska, 2002), *T. pyri* Scheuten and *T. rhenanus* (Oudemans) (von Scheller, 1962). Anystid mites include *Anystis baccarum* (L.). Predatory insects include some Chrysopidae (Kiekiewicz and Czajkowska, 2002), the coccinellid *Stethorus punctillum* Weise (Putman, 1955; Wheeler *et al.*, 1973), minute pirate bugs, the rove beetle (*Oligota oviformis* Casey), some predaceous midges (Krischik and Davidson, 2013c) and the coniopterygid *Conwentzia pineticola* End. (Löyttyniemi, 1970a).

Symptomatology and damage

Infested needles first appear mottled, then turn yellow and finally brown. Severe infestations may induce premature drop of needles, twig dieback and sometimes even branch dieback. The first damage usually occurs in the lower part of the canopy, and extends gradually upward as the infestation increases. Severe attacks may kill seedlings, transplants and small trees; adult trees may sometimes die, especially when they are stressed. The type and extent of damage relate to the host plant and its age. The most preferred hosts seem be 3-year-old spruce trees, while trees about 2 m tall may be defoliated, but do not generally die (Jeppson *et al.*, 1975; Lehman, 2002; Krischik and Davidson, 2013c).

Feeding by *O. ununguis* causes permanent damage to conifers, and new growth is needed to repair the chlorotic needles or bare branches; on some host plants, this may require several seasons. In addition, although *O. ununguis* does not spin much silk, the accumulation of cast skins, dirt and debris make the host plants less attractive (Lehman, 2002). All of these aspects result in aesthetically unattractive trees (Cunliffe and Ryle 1923; Johnson 1958).

On young dwarf of *P. glauca* (cv. Conica), feeding by *O. ununguis* affects the needle compounds, with changes in chemical composition related to the density of the mite population. Needles of trees infested by 2–6 mites per 5 cm of twig over 8 weeks have higher amounts of soluble proteins, total phenolics and essential oil volatiles, while a population of mites that is three times larger (18 per 5 cm of twig) causes either a reduction in the amounts of these compounds or no significant difference from the control. At low mite densities (not exceeding 2–6 per 5 cm of twig for 8 weeks), mite feeding stimulates susceptible host plants to alter their metabolism, but the induced responses are suppressed as mite density increases (Kiekiewicz *et al.*, 2005).

Control

Horticultural practices, and the use of physical, chemical and biological means hinder outbreaks of the spruce spider mite. The correct implementation of monitoring and decision tools contribute to successful control. Different variables must be considered before applying sprays against *O. ununguis*, such as the host plant, age, season and cost of control. When regular monitoring indicates an increase in the mite population, control should be considered for high-value and marketable trees. It is a good practice to keep records of the spruce spider mite population levels at each farm in order to develop decision tools for each situation (Lehman, 2002).

MONITORING. Monitoring generally involves direct observations of eggs and mites in the field. In nurseries and Christmas tree farms, monitoring of *O. ununguis* should begin before symptoms appear. The amount of effort devoted to detection depends on the value of the crop. In the landscape, specimen and highly susceptible plants should have priority treatment (Lehman, 2002).

The lack of significant differences in mite location between cardinal directions within a tree means that these are irrelevant

to sampling. Sampling in the middle portion of the canopy gives a reliable estimate of mite presence. The selection of a sample from the current season's foliage for overwintering eggs and during mid- to late summer for eggs and motile stages should provide a good population assessment (Calkin, 1991). Shoots should be examined with a hand lens in early spring for overwintering eggs. Eggs are usually laid on twigs at the base of needles or around buds. On sample trees harbouring overwintering eggs, monitoring should be done every few days at the beginning of hatching, taking into account that hatching usually starts on the south side of the tree. From a phenological point of view, egg hatch is before bud break, and can be monitored from March to early May in the USA. Motile forms of *O. ununguis* can be monitored from May to early December, every 3–4 weeks (within 1 week of pesticide application) using a sheet of white paper held under a branch, which is tapped to dislodge mites; the mites, their eggs and natural enemies are observed with a lens. It should also be taken into account that only at the beginning of June do needles show fine stippling turning to bronzing (Kowalsick, 2003; Krischik and Davidson, 2013c).

Monitoring of eggs and motile forms of *O. ununguis* in the field may be integrated by the use of GDDs (growing degree-days) and by phenological indicators. In New York State, overwintering eggs hatch from 7 to 121 GDDs (Clark and Kowalsick, 1992), and the first generation of *O. ununguis* may be controlled from 100 to 200 GDDs (Orton, 1989). In Ohio, it has been observed that 50% of the overwintering eggs hatch when 171 GDDs have been accumulated, and the first generation in spring may be treated with acarides near 258 GDDs (Richmond and Shetlar, 1996). On arborvitae, hemlock, juniper and spruce, sprays of horticultural oil are suggested in April before new growth begins, and sprays of acaricides in late May through mid-June between 192 and 363 GDDs, and again in late August through mid-September, between 2375 and 2806 GDDs (Kowalsick, 2003).

Phenological indicators allow the definition of the control period for spring populations, e.g. in saucer magnolia from pink bud to early bloom, in Norway maple at blooming, and in *Amelanchier* × *grandifolia* Rehder at the beginning of blooming (Orton, 1989).

THRESHOLDS. In Pennsylvania, it has been suggested that a minimum of 15 shoots per acre are checked for eggs or mites. The detection of any eggs or mites on a shoot identifies that shoot as infested. Economic thresholds for treatment vary from an infestation rate of 40% (for trees less than waist high) down to 10% (for trees in the year of sale) (Sidebottom, 1996; Lehman, 2002). In New York State, it is suggested that a sample of 3–5 branches per plant suspected to be infested is undertaken, and that each suspected plant in the landscape is checked. According to a method suggested by the Rutgers Cooperative Extension in New Jersey, treatments should be initiated if there are more than five mites per plant on Alberta spruce or ten mites per plant on other species (Kowalsick, 2003).

Studies on the photosynthetic activity of 3-year-old white spruce (*P. glauca*) trees infested with 20, 35 or 45 female of *O. ununguis* show that the photosynthetic activity of the host plant decreases with increase in mite numbers (Puchalska, 2002). Further research shows that photosynthesis is already significantly suppressed after 3 weeks of mite feeding (14.7 mites per 5 cm of shoot length) and at a relatively low level of needle damage, e.g. mean leaf damage index (MLDI) of 2.3 and 2.6 (the MLDA was assessed using the 5-grade scale proposed by Bogatko *et al.* (1987), and the formula of Hussey and Scopes (1985)). When the population density does not exceed four mites per 5 cm of shoot length, chemical control can be delayed by up to 5 weeks, or predatory mites can be released (Puchalska, 2006).

CULTURAL PRACTICES. The effects of perennial ryegrass (*Lolium perenne* L.) planted in chestnut orchards and of the mowing of the ryegrass on the population densities of *O. ununguis* and *A. castaneae* have been investigated in China. The microclimate in the orchard was unfavourable for *O. ununguis* and beneficial for the natural enemies. The strip management of *L. perenne* had marked effects on the densities of *O. ununguis* and *A. castaneae*, and at the stage of peak abundance of *O. ununguis*, plots under alternate strip mowing showed the lowest density of the pest mite and a better ratio of *A. castaneae* to *O. ununguis* (Lu *et al.*, 2008).

Fertilization practices in forest or nursery, especially higher plant nitrogen levels, may induce increased mite populations (Thalenhorst, 1963; Löyttyniemi and Heliövaara, 1991).

PHYSICAL CONTROL. A steady stream of water applied to the plant in the landscape gives some population reduction (Stewart and Peterson, 1960).

BIOLOGICAL CONTROL. An augmentative biological control strategy, based on releases of the phytoseiid mites *N. fallacis* and *M. occidentalis*, was evaluated for reducing pest mite densities and injury, and the associated economic costs, on *J. chinensis* 'Sargentii' in an outdoor nursery. Sequential releases of predatory mites, individually and in combination, were tested and the results compared with horticultural oil and hexythiazox. Predator releases were more expensive and provided less effective suppression of *O. ununguis*, resulting in greater spider mite injury to plants compared with conventional pesticides. However, spider mite damage to plants did not differ in an economically meaningful way between treatments. Unsatisfactory levels of control seem to be related to underestimations of actual spider mite abundance based on grower perceptions and the beat sampling technique used to estimate predator release rates. These data suggest that when initial populations of the spruce spider mite are high, it is unlikely that sequential releases of predator species, individually or in combination, will suppress them. Augmentative biological control was 2.5–7.0 times more expensive than chemical control (Shrewsbury and Hardin, 2003).

CHEMICAL CONTROL. High population densities during the autumn suggest the application during the winter of dormant oil sprays (Krischik and Davidson, 2013c). This strategy was positively applied in forest nurseries in south Finland, where overwintering eggs were satisfactorily controlled by thioquinox or anthracene oil or crude petroleum. Surviving eggs on lifted

planting stock were destroyed by dipping the bundles of transplants into the control liquid, but anthracene oil caused some injury to the plants when used during bud break. Thioquinox was the best all-round acaricide for nursery use (Löyttyniemi, 1970b). During the growing season, if natural enemies are present, it is possible to treat heavy infestations with summer oil or insecticidal soap. It is good practice to use chemicals that are selective for natural enemies. For homeowners, it is possible to apply 2% oils. Acaricides/insecticides include abamectin, bifenazate, bifenthrin, clofentezine, chlorpyrifos (only in the nursery), deltamethrin, dicofol, fenpropathrin, fenbutatin oxide, hexythiazox, horticultural oil, insecticidal soap, lamba-cyhalothrin and pyridaben (Krischik and Davidson, 2013c). Kowalsick (2003) suggested the use of insecticidal soap, summer oils or malathion. Conifer monoterpene vapours of limonene, β-pinene, α-pinene and Δ3-carene are toxic to spruce spider mites (Cook, 1992).

Oligonychus yothersi (McGregor) (Fig. 20.46)

Common name

Avocado red mite.

Diagnostic characteristics

FEMALE. The body is rusty red in colour. The tibia I has 7 tactile setae; the tarsus I has 4 tactile setae proximal to the duplex setae, and the proximal members of the duplex setae are short (Jeppson *et al.*, 1975).

MALE. The aedeagus bends ventrally at a slight obtuse angle, with the bent portion as long as the distal part of the shaft; the distal bent portion is long and slender (Jeppson *et al.*, 1975).

Geographical distribution

Argentina (Pritchard and Baker, 1955; Rossi Simons, 1961, 1971), Brazil (Braga, 1957; Paschoal and Reis, 1968; Aranda, 1974; Flechtmann and Baker, 1975; Oliveira *et al.*, 2004; Feres *et al.*, 2005; Gouvea *et al.*, 2006; Vis *et al.*, 2006; Pinto *et al.*, 2012), Chile (González, 1961), China (Ma and Yuan, 1981b), Colombia (Pritchard and Baker, 1955; Urueta, 1975), Costa Rica (Baker and Pritchard, 1962; Ochoa *et al.*, 1990; Aguilar and Murillo, 2008, 2012), Cuba (Livshits and Salinas-Croche, 1968; Suárez, 2004), Ecuador (Pritchard and Baker, 1955), Hawaiian Islands (Garrett and Haramoto, 1967; Lee Goff, 1986), Mexico (Baker and Pritchard, 1962), Nicaragua (Ochoa *et al.*, 1991a), Paraguay (Aranda, 1969), Peru (González and Flechtmann, 1977; Guanilo *et al.*, 2012) and the USA (McGregor, 1914).

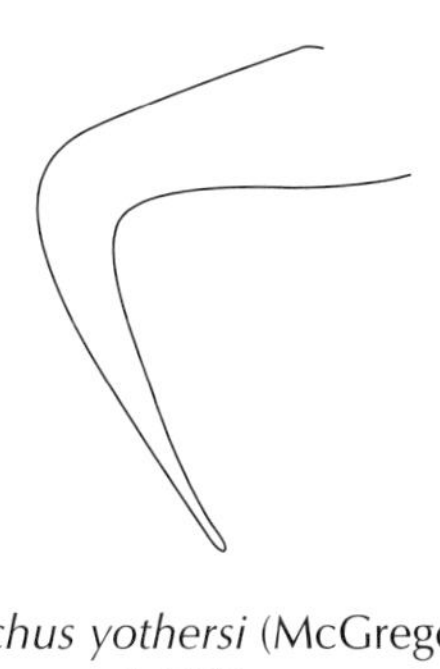

Fig. 20.46. *Oligonychus yothersi* (McGregor). Male, aedeagus (modified from McGregor, 1950).

Host plants

The whole list of host plants of avocado red mite numbers 66 species of 38 families (Migeon and Dorkeld, 2006–2013). The families and species of economic importance are:

Anacardiaceae: cashew (*Anacardium occidentale* L.), mango (*Mangifera indica* L.) (Pritchard and Baker, 1955; Paschoal and Reis, 1968; Urueta, 1975; Lee Goff, 1986; Ochoa *et al.*, 1990).

Aquifoliaceae: mate (*Ilex paraguariensis* A. St. Hil.) (Oliveira *et al.*, 2004; Gouvea *et al.*, 2006).

Euphorbiaceae: cassava (*Manihot esculenta* Crantz), castor bean (*Ricinus communis* L.) (Urueta, 1975; Guanilo *et al.*, 2012).

Caricaceae: papaya (*Carica papaya* L.) (Bolland *et al.*, 1998; Migeon and Dorkeld, 2006–2013).

Fabaceae: copaiba (*Copaifera langsdorffii* Desf.) (Flechtmann, 1967a; Flechtmann and Baker, 1975).

Fagaceae: European chestnut (*Castanea sativa* Mill.) (Paschoal and Reis, 1968).

Lauraceae: camphor tree (*Cinnamomum camphora* (L.) J. Presl) (McGregor, 1914; Rossi Simons, 1971; Ma and Yuan, 1981b), avocado (*Persea americana* Miller) (Ewing, 1921; Flechtmann and Baker, 1975; Suárez, 2004; Guanilo *et al.*, 2012).

Lythraceae: pomegranate (*Punica granatum* L.) (Bolland *et al.*, 1998; Migeon and Dorkeld, 2006–2013).

Malvaceae: cacao tree (*Theobroma cacao* L.) (Bolland *et al.*, 1998; Migeon and Dorkeld, 2006–2013).

Musaceae: banana (*Musa* × *paradisiaca* L.) (Urueta, 1975; Vargas *et al.*, 1996a).

Myrtaceae: guava (*Psidium guajava* L.) (Bolland *et al.*, 1998; Migeon and Dorkeld, 2006–2013).

Oxalidaceae: carambola (*Averrhoa carambola* L.) (Livshits and Salinas-Croche, 1968).

Rosaceae: quince (*Cydonia oblonga* Mill.) (Urueta, 1975), loquat (*Eriobotrya japonica* (Thunb.) Lindl.), apple (*Malus domestica* Borkh.) (González, 1961), peach (*Prunus persica* (L.) Batsch) (Rossi Simons, 1971).

Rubiaceae: Arabian coffee (*Coffea arabica* L.) (Pritchard and Baker, 1955; Baker and Pritchard, 1962; Urueta, 1975).

Sapindaceae: longan (*Dimocarpus longan* Lour.), lychee (*Litchi chinensis* Sonn.) (Pritchard and Baker, 1955).

Sapotaceae: star apple (*Chrysophyllum cainito* L.) (Bolland *et al.*, 1998; Migeon and Dorkeld, 2006–2013).

Theaceae: tea (*Camellia sinensis* (L.) Kuntze) (Flechtmann, 1976b).

Vitaceae: grapevine (*Vitis vinifera* L.) (Rossi Simons, 1971).

Bio-ecology

O. yothersi is a Neotropical mite that is also known in some countries of the Australasian, Nearctic and Oriental regions. It infests various tropical economic plants, such as Arabian coffee, avocado, banana, cacao, camphor, cassava, cashew, guava, loquat,

lychee, mango, mate and tea. The mite also infests apple, grapevine, peach, pomegranate and various ornamentals.

The avocado red mite reproduces by arrhenotokous parthenogenesis (Alves *et al.*, 2004). The eggs are globose, smoky amber and with a stalk at the apex. They are laid singly, initially along the midrib, and then over the entire leaf surface (Jeppson *et al.*, 1975). The mite populations are usually located on the upper leaf in the depressions formed by the ribs, under a very thin web. In severe attacks, the underside of the leaf is also colonized by mites (Cardenas, 1983). Ebeling (1959) reports that the development of the egg takes from 7 to 11 days, and that of the larva, protonymph and deutonymph lasts about 3 days each; the life cycle develops in about 14 days. At 25°C, the development of a complete generation of *O. yothersi* requires 7 days, and at 33°C all biological stages die (McGregor, 1941b). Fecundity is 35 eggs, egg development takes 7–10 days, and the life cycle develops in 14.2 days (Jeppson *et al.*, 1975).

In experimental studies on coffee plants (var. Caturra) at 15, 20, 25 or 30°C and 70 ± 5% RH, the development time from egg to adult lasted 33.8, 15.7, 12.9 and 11.7 days, respectively. The mite required 274.95 degree-days to complete its development from egg to adult, and a minimum threshold temperature of 4.8°C. At these same four temperatures, the female laid 0.87, 2.7, 2.7 and 2.3 eggs a day, respectively; the net reproduction rate (R_0) was 0.088, 15.1, 9.6 and 1.03 and the generation time (T) was 40.1, 25.4, 20.8 and 16.4 days, respectively. The intrinsic rate of increase (r_m) was 0.06, 0.10, 0.10 and 0.002, respectively (Orozco Hoyos *et al.*, 1990).

On *I. paraguariensis* leaf discs at 25 ± 1°C and 14L:10D photoperiod, egg development of male and female both took 4.9 ± 0.2 days, respectively. The larvae of the male and female lasted 2.8 ± 0.7 and 2.1 ± 0.7 days, the protonymph 1.8 ± 0.5 and 1.8 ± 0.5 days (the same), the deutonymph 1.7 ± 0.5 and 2.1 ± 0.4 days, and the adult 6.1 ± 1.9 and 6.1 ± 1.5 days, respectively (Alves *et al.*, 2004).

On leaves of avocado cv. 'Lorena' at 26 ± 3°C and 56 ± 3% RH, the development periods of the egg, larva, protochrysalis, protonymph, deutochrysalis, deutonymph and teliochrysalis averaged 4.96, 2.25, 0.76, 2.11, 0.90, 2.61 and 1.28 days, respectively. The developmental time from egg to adult was 14.34 days, and survival from egg to adult was 53%. The sex ratio was 4.8 females:1 male. The pre-oviposition time was 1.09 ± 0.90 days, the oviposition time 6.64 ± 6.75 days, daily egg deposition 1.46 ± 2.18 eggs, fecundity 14.43 eggs and female longevity 8.35 ± 7.04 days. The net reproduction (R_0), the intrinsic rate of increase (r_m) and the finite rate of increase (λ) were 70.56, 0.67 and 1.96, respectively (Reyes-Bello and Mesa-Cobo, 2011). These biological parameters differ by cultivar. On avocado leaves at 25 ± 2°C, 70% RH and under a 16L:8D photoperiod, post-embryonic development lasted 16.50 days on cv. 'Hass' and 14.9 on cv. 'Fuerte'. The reproductive period was 26 days on 'Hass' and 18 days on 'Fuerte'. Fecundity was 40 eggs on 'Hass' and 17.7 eggs on 'Fuerte'. The net reproduction (R_0) was 28.8 on 'Hass' and 8.5 on 'Fuerte'. The intrinsic rate of increase (r_m) was 0.008 on 'Hass' and 0.07 on 'Fuerte'. The low values of the intrinsic rate of increase show that the mite has a low potential to develop high levels of infestation on avocado leaves (Léon Lobos, 2003).

In Florida, *O. yothersi* is active throughout the year, while in New Jersey it overwinters as eggs and develops 5–6 generations a year. Weather and natural enemies affect the development of the population. Natural spring defoliation reduces mite populations, and the spring and summer rains often keep populations below economic levels. Populations usually increase in early summer and the peak occurs in later summer, followed by an abrupt decrease. Sometimes, an increase in populations may be delayed by rain until late summer, and the peak then occurs during autumn or early winter (Jeppson *et al.*, 1975). In Brazil, on *I. paraguariensis*, an average of 0.024 mites/leaf has been observed, with a maximum average population in August (0.181 mites per leaf). No mites were observed in October, November, January, March, April and June, and the highest population density was found on the ventral surface of the leaves and mature leaves. The peak population occurs in winter, a period of warm temperatures and low rainfall, and populations of *O. yothersi* appear to be positively affected by wind and adversely affected by moisture (Gouvea *et al.*, 2004). Cardenas (1983) reports that on coffee the mites are spread by wind and to a lesser extent, by the workers. The rains wash or drown all biological states.

The natural enemies include phytoseiid mites and some insects. Phytoseiids include *Amblyseius* sp., *A. chilenensis* (Dosse), *A. Fructicolus* (Gonzalez *et* Schuster) and *Phytoseius decoratus* (Gonzalez *et* Schuster). Moreover, some stigmaeids are reported (Vargas, 1989). Insects include some coccinellids of the genera *Scymnus* and *Stethorus* (Cardenas, 1983) and *Parastethorus histrio* (Chazeau) (Prado, 1991), and the staphylinid *Oligota pygmaea* (Solier) (Frank *et al.*, 1992).

Symptomatology and damage

Feeding of *O. yothersi* on the upper leaf surface of avocado causes first a white spotting, and as feeding increases, the area of the midrib turns a reddish brown. Severe infestations cause bronzing of the entire upper leaf surface and of a portion of lower leaf surface, and may be associated with partial defoliation. Serious damage results from only a high level of attack (McKenzie, 1935; Jeppson *et al.*, 1975).

The dorsal surface of infested leaves of mate or Paraguay tea (*I. paraguariensis*) initially shows small purple spots along the midrib, mostly at its base. These spots later coalesce, forming necrotic areas. Damaged leaves become wrinkled, with the border turning down. Great numbers of mites at all life stages, as well as exuviae and webbing, occur at scattered depressions on the upper leaf surface. The infestation affects leaf development, and most of the infested leaves remain smaller than non-infested leaves and drop prematurely (Alves *et al.*, 2004).

O. yothersi is a potential pest for eucalyptus plants and crops. Damage caused to *E. urophylla* is similar to that caused by other pest mites feeding on the leaf surface, and causes premature drop of damaged leaves (Pinto *et al.*, 2012).

Feeding of *O. yothersi* on the upper leaf surface of coffee breaks the cuticle and the epidermal cells, causing oxidation of the oils, thereby giving a brown colour to the foliage. The coloration persists even after controlling the spider mite. Severe attacks cause leaf drop. The avocado red mite is a potential pest of coffee crops, especially during prolonged periods of drought, and in coffee plantations set along roads or highways and subject to the accumulation of dust on the foliage, mainly in plantations 1–2 years old (Cardenas, 1983).

Feeding of *O. yothersi* on the leaf surface of tea first causes red spots. As infestation increases, these red spots are so numerous that the entire leaf surface becomes dark red copper. Initially, this symptom occurs on both sides of the ribs of adult leaves, but eventually it extends to the entire leaf surface and on both sides. In severe attacks, these symptoms extend even to young leaves and shoots, and in Argentina are detected between the months of June and August on isolated susceptible plants, which become a reddish colour. These patches may be formed by several plants, or over the entire planting, and may be characterized by total leaf fall. A population density of 3 motile forms cm^{-2} induces a reduction in yields of 24.76% ± 2.95 (Prat Kricun *et al.*, 2011).

Control

The harmfulness of *O. yothersi* is unclear, as it varies between different crops and countries. For instance, Benavides *et al.* (2012) consider the mite as a minor pest on coffee in Colombia, and Jeppson *et al.* (1975) report that high levels of infestation are needed for there to be serious damage by *O. yothersi* on avocado. Conversely, Rioja and Vargas (2009) treat the avocado red mite as the major leaf pest in Chile's avocado orchards, and Prat Kricun *et al.* (2011) treat it as a serious pest mite of tea in Argentina. It is probable that the population dynamics of *O. yothersi* are affected differently by environmental factors, including control measures applied to various pests. This situation does not allow a standardization of control measures.

SAMPLING. The aggregation pattern of *O. yothersi* in a mate tea orchard has been assessed through the coefficients a and b of Taylor's power law, and the proportion of infested leaves and the number of required samples estimated by a mathematical model. The proportion of infested leaves calculated by the model was shown to be a trustworthy parameter to estimate its population density. The number of samples required is small, and makes feasible the practical application of the presence–absence sampling method for *O. yothersi* on mate tea (Gouvea *et al.*, 2007).

CULTURAL PRACTICES. On tea in Argentina, the application of light pruning from May and June, at sites where there are no cold problems, represents an efficient cultural technique for reducing the mite population (Prat Kricun *et al.*, 2011).

BIOLOGICAL CONTROL. Releases of natural enemies are not known for the control of *O. yothersi*. However, an important role is attributed to natural enemies present in the field. In Colombia, the mite has not become a serious pest because the agroecosystems are quite stable with a great biodiversity, which favours the development of beneficial agents and maintains the potential pests present in equilibrium (Benavides *et al.*, 2012). However, several scientific works have evaluated the potential under laboratory conditions of different natural enemies, such as *Cydnodromus picanus* Ragusa, *Amblyseius graminis* Chant and *Metaseiulus occidentalis* (Nesbitt). For example, the intrinsic rate of increase (r_m) registered by *C. picanus* indicates its predatory potential to control *O. yothersi* (Rioja and Vargas, 2009). Similarly, the susceptibility of *O. yothersi* to *Beauveria bassiana* (Balsamo) Vuillemin strains has been demonstrated (Oliveira *et al.*, 2002).

CHEMICAL CONTROL. Sulfur dusts are active against juvenile and adults stages of *O. yothersi*, and sulfur residues deposited on webbing kill the mites that later hatch from eggs (Hamilton, 1926; McKenzie, 1935). On coffee, chemical control would be practicable under two prolonged periods of drought, the state in which there has been the most severe defoliation. In this case, spray applications are recommended of dicofol or tetradifon alone or mixed with an insecticide such as dimethoate or phosphamidon (Cardenas, 1983). Other possible compounds include abamectin, bifenazate and pyridaben (Peña *et al.*, 1999). Among the botanical acaricides, azadirachtin has been tested (Pasini *et al.*, 2003).

The Genus *Tetranychus* Dufour

Diagnostic characteristics

The peritreme is recurved distally. The anal region has 1 pair of setae *h*. The empodium is split distally, commonly into 3 pairs of hairs, and the empodial spur is commonly visible. The duplex setae of tarsus I are well separated (Bolland *et al.*, 1998).

Economic importance

The genus *Tetranychus* has a worldwide distribution, and includes 149 species living on wild and economic plants (Migeon and Dorkeld, 2006–2013). Tuttle and Baker (1968) divided this genus into seven subgenera, referred to as *Tetranychus* Dufour, *Polynychus* Wainstein, *Armenychus* Wainstein, *Pentanychus* Wainstein, *Septanychus* McGregor, *Pseudonychus* Wainstein and *Amphitetranychus* Oudemans. Bolland *et al.* (1998) do not recognize this systematic classification, and neither does the author. Recently, Seeman and Beard (2011) summarized in an exhaustive treatment the morphological and bio-ecological features of the most important species worldwide.

At least 17 species are major pests of economic plants: *T. desertorum* Banks, *T. evansi* Baker, *T. gloveri* Banks, *T. kanzawai* Kishida, *T. lambi* Pritchard *et* Baker, *T. lombardinii* Baker *et* Pritchard, *T. ludeni* Zacher, *T. macfarlanei* Baker *et* Pritchard, *T. mcdanieli* McGregor, *T. mexicanus* (McGregor), *T. neocaledonicus* André, *T. pacificus* McGregor, *T. truncatus* Ehara, *T. tumidus* Banks, *T. turkestani* (Ugarov *et* Nikolskii), *T. urticae* Koch and *T. yusti* McGregor. These species are characterized by polyphagy, which is sometimes extreme, and enables them to develop on various economic plants in different biogeographical regions. Other species are minor or potential pests, such as the four-spotted spider mite *T. canadensis* (McGregor), *T. fijiensis* Hirst, *T. schoenei* McGregor and *T. sinhai* Baker. *T. lintearius* Dufour is a severe pest of gorse (*Ulex* spp.) in Europe (Zwolfer, 1964; Girling, 1977) and is a specific pest for this host plant (Hill and O'Donnell, 1991a,b); this adaptation has suggested the use of this spider mite in the biological control of gorse. Because of

the large number of *Tetranychus* spp. mentioned above, and their phytopathological importance, further aspects of these 22 species are discussed under each individual species.

Tetranychus canadensis (McGregor) (Fig. 20.47)

Common name

Four-spotted spider mite.

Diagnostic characteristics

FEMALE. The body of summer forms is greenish yellow to orange in colour, with dark paired spots, set medially and posteriorly. Overwintering females are bright orange in colour. The dorsal striae between the opisthosomal setae e_1–e_1 and e_1–f_1 are transverse, and those between the setae f_1–f_1 are longitudinal and form an hourglass pattern; the dorsal striae generally bear small lobes or lack lobes. The ventral striae lack lobes, and the pregenital striae are weak but complete. Tarsus I has the tactile setae proximal to the duplex setae. The empodia may have very tiny dorsal spurs (Jeppson *et al.*, 1975; Seeman and Beard, 2011).

MALE. Empodia I–IV have dorsal spurs; empodium I is claw like and the other empodia end in 3 pairs of hairs directed ventrally. The knob of the aedeagus is wide, with the dorsal side convex, the anterior projection rounded or angulate, and the posterior projection acute (Jeppson *et al.*, 1975; Seeman and Beard, 2011).

Geographical distribution

Canada (McGregor, 1950; Pritchard and Baker, 1955), Hungary (Hetenyi, 1954), Poland (Boczek and Kropczynska, 1964) and the USA (Baker and Pritchard, 1953; Pritchard and Baker, 1955; Bibby and Tuttle, 1959; Tuttle and Baker, 1964, 1976; Seeman and Beard, 2011).

Host plants

The whole list of host plants of four-spotted spider mite includes 67 species of 27 families (Migeon and Dorkeld, 2006–2013). The families and species of economic importance are:

Betulaceae: common filbert (*Corylus avellana* L.) (Thewke and Enns, 1969).

Convolvulaceae: sweet potato (*Ipomoea batatas* (L.) Lam.) (Bolland *et al.*, 1998; Migeon and Dorkeld, 2006–2013).

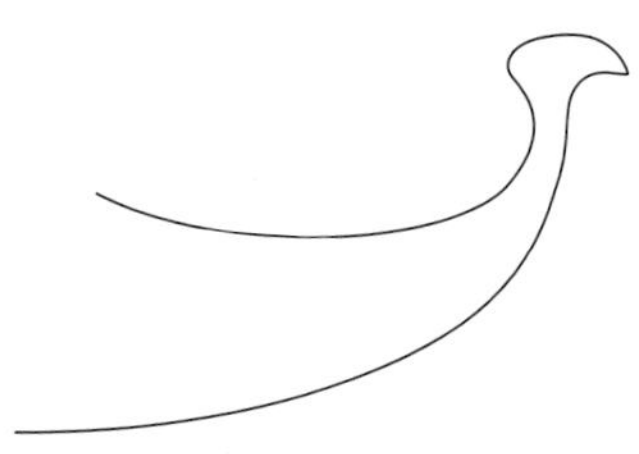

Fig. 20.47. *Tetranychus canadensis* (McGregor). Male, aedeagus (modified from McGregor, 1950).

Fabaceae: black locust (*Robinia pseudoacacia* L.) (Thewke and Enns, 1969), red clover (*Trifolium pratense* L.) (Bolland *et al.*, 1998; Migeon and Dorkeld, 2006–2013).

Fagaceae: *Fagus grandifolia* Ehrh. (Reeves, 1963).

Juglandaceae: *Carya cordiformis* (Wangenh.) K. Koch (Reeves, 1963).

Malvaceae: okra (*Abelmoschus esculentus* (L.) Moench), upland cotton (*Gossypium hirsutum* L.) (Pritchard and Baker, 1955), *Tilia americana* L. (Reeves, 1963).

Meliaceae: neem (*Azadirachta indica* A. Juss.) (Bibby and Tuttle, 1959; Tuttle and Baker, 1964).

Moraceae: white mulberry (*Morus alba* L.) (Thewke and Enns, 1969).

Oleaceae: *Fraxinus* sp. (Tuttle and Baker, 1976), *F. pennsylvanica* Marshall, *F. velutina* Torr. (Tuttle and Baker, 1964).

Poaceae: barley (*Hordeum vulgare* L.), cereal rye (*Secale cereale* L.), common wheat (*Triticum aestivum* L.), maize (*Zea mays* L.) (Bolland *et al.*, 1998; Migeon and Dorkeld, 2006–2013).

Rosaceae: Siberian crab apple (*Malus baccata* (L.) Borkh.), apple (*Malus domestica* Borkh.) (McGregor, 1950; Pritchard and Baker, 1955), European plum (*Prunus domestica* L.) (Lienk and Chapman, 1951; Pritchard and Baker, 1955), peach (*P. persica* (L.) Batsch), Japanese rose (*Rosa multiflora* Thunb.) (Thewke and Enns, 1969).

Salicaceae: *Populus* sp. (Pritchard and Baker, 1955), *P. deltoides* W. Bartram ex Marshall (Thewke and Enns, 1969).

Sapindaceae: *Acer saccharinum* L. (Thewke and Enns, 1969), *Aesculus hippocastanum* L. (Pritchard and Baker, 1955).

Solanaceae: tomato (*Solanum lycopersicum* L.) (Migeon and Dorkeld, 2006–2013).

Ulmaceae: *Ulmus* sp. (Pritchard and Baker, 1955), *U. americana* L., *U. laevis* Pall., *U. rubra* Muhl. (Reeves, 1963).

Bio-ecology

The four-spotted spider mite lives on wild and economic plants, such as okra, common filbert, barley, sweet potato, upland cotton, Siberian crab apple, apple, white mulberry, bean, European plum, peach, cereal rye, red clover and maize. It also infests ornamental plants, such as silver maple, American beech, American elm, green ash, etc. (Jeppson *et al.*, 1975; Bolland *et al.*, 1998). Despite this wide list of host plants, the restricted number of reports highlights its role as an occasional pest.

T. canadensis is similar morphologically to *T. schoenei* McGregor (see below), differing only in the shape of the aedeagus (McGregor, 1950; Pritchard and Baker, 1952a; Reeves, 1963; Jeppson *et al.*, 1975). Recently, Seeman and Beard (2011) examined specimens from Canada and the USA and concluded that the two species could be synonymous; this needs further investigation.

The mite is a cool-climate species and occurs in temperate countries. Its populations live on the lower leaf surface, and produce little webbing. The mite overwinters as females on the trunks and in the soil at the base of the host plants (Jeppson *et al.*, 1975). Its bio-ecology is not well documented.

Symptomatology and damage

Occasional damage by *T. canadensis* is reported on apple, plum and cotton (Lienk and Chapman, 1951; Pritchard and Baker, 1955).

In addition, Jeppson *et al.* (1975) recorded many other economic plants damaged by *T. canadensis*. Leaves of infested trees turn rusty brown and later drop. The damage first occurs on the top of the canopy. Feeding by *T. canadensis* on barley leaves causes darkening, followed by yellowing and withering from the bend of the leaf on to its tip and, sometimes, crinkling of the leaf edges from below (Atchenson, 1953; Wallace and Sinha, 1961; White 1964, 1965; Jeppson *et al.*, 1975).

Control

Control is not usually necessary.

Tetranychus desertorum Banks (Fig. 20.48)

Common name

Desert spider mite.

Diagnostic characteristics

FEMALE. The body is about 490 µm long, reddish in colour, and with the dorsal setae finely serrate and longer than the distances between their bases. The peritremes end in a simple hook. The dorsal striae between the setae e_1–e_1 are mixed oblique, transverse and longitudinal; the dorsal striae between the setae f_1–f_1 are longitudinal; the dorsal striae between the setae e_1 and f_1 are transverse/oblique and form a diamond shape medially. The dorsal striae are lobate, the ventral striae lack lobes, and the pregenital striae are entire. The empodia have 6 proximoventral hairs and a small dorsal spur. Tarsus I has 4 tactile setae overlapping with the proximal duplex setae, and tarsus III has 1 proximal tactile seta (Pritchard and Baker, 1955; Seeman and Beard, 2011).

MALE. The empodia of the male have a dorsal spur; empodia I–II are uncinate and III–IV have long and free proximoventral hairs. The aedeagus has a large knob that is slightly convex dorsally, with the anterior projection short and pointed and the posterior large and strongly hooked (Seeman and Beard, 2011).

Geographical distribution

Argentina (McGregor, 1943b; Baker and Pritchard, 1953; Pritchard and Baker, 1955; Rossi Simons, 1961, 1971; Flechtmann and Baker, 1975; Guanilo *et al.*, 2010), Australia (Davis, 1968b), Bolivia (Munro, 1954), Brazil (Ehara, 1966b; Aranda, 1974; Flechtmann and Baker, 1975; Holanda and Oliveira, 1992; Flechtmann, 2004; Furtado *et al.*, 2006; Guedes *et al.*, 2007; Silva *et al.*, 2008), Chile (González, 1989), China (Ma and Yuan, 1975), Colombia (Flechtmann and Baker, 1975; Urueta, 1975), Costa Rica (Vargas *et al.*, 1996b), Cuba (Suárez, 2004), Greece (Bolland and Beron, 2004), Guadeloupe (Flechtmann and Etienne, 2002), Haiti (Mann, 1969), Hawaiian Islands (Garrett and Haramoto, 1967; Lee Goff, 1986), Mexico (Pritchard and Baker, 1955; Tuttle *et al.*, 1976; Estebanes-Gonzalez and Rodriguez-Navarro, 1991), Nicaragua (Baker and Pritchard, 1962), Paraguay (Nickel, 1958; Aranda and Flechtmann, 1971), Peru (Baker and Pritchard, 1953b; Pritchard and Baker, 1955; Guanilo *et al.*, 2012), Puerto Rico (Ehara, 1966b), the USA (Banks, 1900, 1908; Ewing, 1926; McGregor, 1950; Pritchard and Baker, 1952a, 1955; Bibby and Tuttle, 1959; Tuttle and Baker, 1964; Seeman and Beard, 2011) and Venezuela (Byrne *et al.*, 1983).

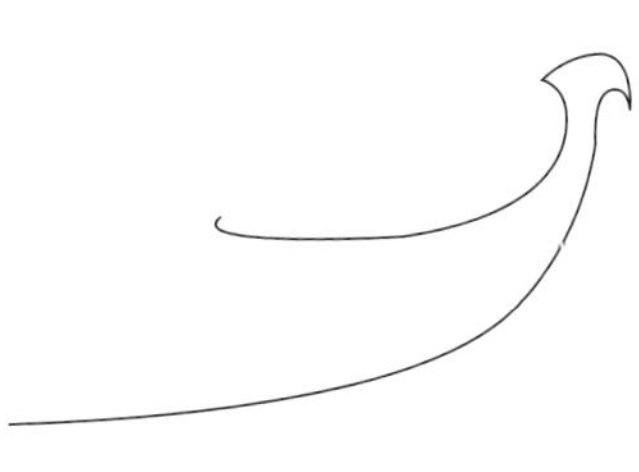

Fig. 20.48. *Tetranychus desertorum* Banks. Male, aedeagus (from Pritchard and Baker, 1955).

Host plants

The desert spider mite is recorded on 217 host plants belonging to 50 families. Among these, the Asteraceae have 32 species, followed by the Fabaceae with 27 species and the Solanaceae with 25 species (Migeon and Dorkeld, 2006–2013). The families and species of economic importance are:

Amaranthaceae: beat (*Beta vulgaris* L.) (Bibby and Tuttle, 1959; Tuttle and Baker, 1964).

Amarillidaceae: onion (*Allium cepa* L.), garlic (*A. sativum* L.) (Bolland *et al.*, 1998; Migeon and Dorkeld, 2006–2013).

Anacardiaceae: mango (*Mangifera indica* L.) (Flechtmann, 1967a).

Apiaceae: wild celery (*Apium graveolens* L.) (Pritchard and Baker, 1955; Bibby and Tuttle, 1959; Tuttle and Baker, 1964), carrot (*Daucus carota* L.) (McGregor, 1950; Pritchard and Baker, 1955), fennel (*Foeniculum vulgare* Mill.) (Bolland *et al.*, 1998; Migeon and Dorkeld, 2006–2013).

Aquifoliaceae: mate (*Ilex paraguariensis* A. St. Hil.) (Aranda and Flechtmann, 1971).

Asteraceae: chicory (*Cichorium intybus* L.) (Paschoal, 1970), sunflower (*Helianthus annuus* L.) (Pritchard and Baker, 1955), lettuce (*Lactuca sativa* L.) (Bibby and Tuttle, 1959; Tuttle and Baker, 1964).

Brassicaceae: mustard (*Brassica rapa* L.) (McGregor, 1950).

Caricaceae: papaya (*Carica papaya* L.) (Baker and Pritchard, 1962; Urueta, 1975).

Caryophyllaceae: carnation (*Dianthus caryophyllus* L.) (Bolland *et al.*, 1998; Migeon and Dorkeld, 2006–2013).

Convolvulaceae: sweet potato (*Ipomoea batatas* (L.) Lam.) (Flechtmann, 1967a).

Cucurbitaceae: watermelon (*Citrullus lanatus* (Thunb.) Matsum. and Nakai) (Bibby and Tuttle, 1959; Tuttle and Baker, 1964, 1968; Aranda and Flechtmann, 1971), cantaloupe (*Cucumis melo* L.) (Pritchard and Baker, 1955; Bibby and Tuttle, 1959; Tuttle and Baker, 1964), cucumber (*C. sativus* L.) (Pritchard and Baker, 1955; Baker and Pritchard, 1962), field pumpkin (*C. pepo* L.) (Flechtmann, 1967a), chayote (*Sechium edule* (Jacq.) Sw.) (Flechtmann and Baker, 1975).

Euphorbiaceae: cassava (*Manihot esculenta* Crantz) (Paschoal, 1968a; Guanilo *et al.*, 2012), castor bean (*Ricinus communis* L.) (Flechtmann, 1967a; Aranda and Flechtmann, 1971; Flechtmann and Baker, 1975; Urueta, 1975; Guanilo *et al.*, 2012).

Fabaceae: groundnut (*Arachis hypogaea* L.) (Guanilo *et al.*, 2012), honey locust (*Gleditsia triacanthos* L.) (Rossi Simons, 1971), soybean (*Glycine max* (L.) Merr.) (Tuttle and Baker, 1964; Urueta, 1975; Guedes *et al.*, 2007; Guanilo *et al.*, 2012), pacay (*Inga feuilleei* DC.) (Guanilo *et al.*, 2012), lucerne (*Medicago sativa* L.) (McGregor, 1950; Pritchard and Baker, 1955; Bibby and Tuttle, 1959; Tuttle and Baker, 1964; Rossi Simons, 1971; Guanilo *et al.*, 2012), bean (*Phaseolus vulgaris* L.) (Flechtmann, 1967a; Urueta, 1975; Holanda and Oliveira, 1992), white clover (*Trifolium repens* L.), fava bean (*Vicia faba* L.) (Aranda and Flechtmann, 1971; Guanilo *et al.*, 2012), vetch (*V. sativa* L.) (Rossi Simons, 1971), cowpea (*Vigna unguiculata* (l.) Walp.) (Holanda and Oliveira, 1992).

Iridaceae: *Gladiolus* sp. (Pritchard and Baker, 1955).

Juglandaceae: English walnut (*Juglans regia* L.) (Rossi Simons, 1961).

Lauraceae: laurel (*Laurus nobilis* L.) (Flechtmann, 1967a).

Malvaceae: cotton (*Gossypium herbaceum* L.), upland cotton (*G. hirsutum* L.) (Pritchard and Baker, 1955; Bibby and Tuttle, 1959; Rossi Simons, 1961; Tuttle and Baker, 1964).

Meliaceae: neem (*Azadirachta indica* A. Juss.) (Aranda and Flechtmann, 1971).

Moraceae: fig (*Ficus carica* L.) (Bolland *et al.*, 1998; Migeon and Dorkeld, 2006–2013).

Musaceae: banana (*Musa acuminata* Colla) (Flechtmann, 1967a).

Passifloraceae: granadilla (*Passiflora edulis* Sims) (Flechtmann, 1967a).

Poaceae: barley (*Hordeum vulgare* L.) (Aranda and Flechtmann, 1971), sorghum (*Sorghum bicolor* (L.) Moench) (Bibby and Tuttle, 1959; Tuttle and Baker, 1964, 1968), maize (*Zea mays* L.) (Bibby and Tuttle, 1959; Tuttle and Baker, 1964).

Rosaceae: strawberry (*Fragaria × ananassa* Duchesne ex Rozier) (Bibby and Tuttle, 1959; Tuttle and Baker, 1964; Chiavegato and Camargos, 1968; Suárez, 2004), apple (*Malus domestica* Borkh.) (Rossi Simons, 1961), European plum (*Prunus domestica* L.) (Guanilo *et al.*, 2012), peach (*P. persica* (L.) Batsch) (Flechtmann, 1967a), pear (*Pyrus communis* L.) (McGregor, 1943b; Rossi Simons, 1961).

Simmondsiaceae: jojoba (*Simmondsia chinensis* (Link) C.K. Schneid.) (Tuttle and Baker, 1964).

Solanaceae: tobacco (*Nicotiana tomentosa* Ruiz *et* Pav.) (Guanilo *et al.*, 2012), tomato (*Solanum lycopersicum* L.) (Flechtmann, 1967a; Urueta, 1975; Estebanes-Gonzalez and Rodriguez-Navarro, 1991; Guanilo *et al.*, 2012), aubergine (*S. melongena* L.) (Rossi Simons, 1971), naranjilla (*S. quitoense* Lam.) (Urueta, 1975), Irish potato (*S. tuberosum* L.) (Paschoal, 1970; Urueta, 1975).

Bio-ecology

The desert spider mite has a subtropical and tropical distribution, and is a pest in warm and humid areas (Jeppson *et al.*, 1975). The species has been mistaken for *T. ludeni* Zacher, as reported by Ehara and Masaki (1989) for Japan, Quirós de González (1981) for Central America and Seeman and Beard (2011) for Australia. Therefore, the list of host plants of the desert spider mite is probably inflated by confusion with *T. ludeni* (Seeman and Beard, 2011).

T. desertorum infests forages, cereals, vegetables, fruit plants, including various tropical species, ornamentals and some industrial plants. The main forages of economic importance that are affected are clover, lucerne and vetch. The cereals include sorghum and maize. The vegetables include a long list of plants: aubergine, barley, bean, beat, carrot, celery, chicory, cucumber, garlic, lettuce, mustard, onion, strawberry, tomato, Irish potato and sweet potato. Among the fruit plants infested are apple, cantaloupe, fig, pear, peach, European plum, and different tropical fruit plants, such as banana, cassava, chayote, jojoba, mango, mate, pacay, papaya, naranjilla, watermelon, etc. Plants of industrial importance that are infested include cotton, groundnut, soybean, sunflower and tobacco. The list of host plants also includes English walnut and various ornamentals. Wild plants are very important for the survival and development of populations of *T. desertorum* during the periods of the year when crops are absent in the field. Of these, the most well known are the evening primrose (*Oenothera speciosa* Nutt.), brome grass (*Bromus* sp.), bloodweed (*Ambrosia* sp.), bur clover (*Medicago* sp.), horehound (*Marrubium* sp.), tie vine (*Ipomoea* sp.), Johnson grass (*Sorghum* sp.), cocklebur (*Xanthium* sp.) and other species.

The desert spider mite usually lives on the lower leaf surface of the host plant. It reproduces by arrhenotokous parthenogenesis and the haploid number of chromosomes is three (Helle *et al.*, 1981). Its populations construct a highly complicated and irregular web on the leaf surface, which are included by Saitō (1985) in the subtype CW-u.

Egg development takes 2 days during the summer and 4–5 days in the winter. Each developmental stage lasts 1.0–1.5 days in the summer and 1.6–3.0 days in the winter. The intrinsic rate of increase (r_m) at 30°C is 0.46 individuals per female a day at 25–30% RH and 0.36 individuals per female a day at 85–90% RH. The mite does not survive at temperatures below 10°C and the optimum conditions for population increase are 30°C and 85–90% RH. This behaviour helps to explain the restricted economic distribution of *T. desertorum*, and suggests that it is not an important pest in central California because of the arid summer (Nickel, 1960). In Venezuela, Morros and Aponte, (1994) found that on the black bean (*P. vulgaris*) at 26.34 ± 3.92°C and 69.44 ± 19.44% RH, the life cycle (egg to adult) of *T. ludeni* lasted 9.98 and 9.25 days for females and males, respectively. The reproduction rate (R_0) was 77.42, the mean generation time (T) 19.63, the intrinsic rate of increase (r_m) 0.2526 individuals per female a day and the finite rate of natural increase (λ) 1.2874 individual per female a week. Also in Venezuela, studies by Rivero and Vásquez (2009) on *P. vulgaris* cv. 'Tacarigua' at 28 ± 2°C, 70 ± 10% RH and under a 12L:12D photoperiod showed that the female developmental period was 6.8 days, and those of the egg, larva, protonymph and deutonymph lasted 3.8, 1.4, 1.0 and 0.7 days, respectively. The pre-oviposition, oviposition and post-oviposition periods took 1.1, 8.4 and 1.3 days, respectively. The daily mean fecundity was 6.93 eggs, and female mean longevity was 10 days. The net reproduction rate (R_0) was 41.10 individuals, the intrinsic rate of increase (r_m) was 0.144 individuals per female a day, the generation time (T) was 11.15 days and the finite natural rate of increase (λ) was 1.155 individuals per female.

In Texas during the winter, *T. desertorum* lives on evening primrose and other wild plants. Plants surrounded by grass are preferred by the mite as winter hosts. From mid-March, high populations may be found in patches on bur clover, and from late April on bloodweed and seedling cotton. Heavy infestations on cotton occur in July. During the summer, the mite also lives

on tie vine, Johnson grass and other wild plants, while in October the most important host plant seems to be cocklebur, from which the mites move during November and December towards their winter host plants. The spring rains repress the development of populations of *T. desertorum*. Temperature and RH probably hinder the distribution of the species in warmer climates and arid conditions appear to be preventing the species from being a major cotton pest in California and Arizona (Jeppson *et al.*, 1975).

The dispersal of *T. desertorum* is entrusted to the air currents and crawling (Iglinsky and Rainwater, 1954; Hightower and Martin, 1956; Jeppson *et al.*, 1975).

Phytoseiid mites, thrips, anthocorids and coccinellids are the most important natural enemies (Nickel, 1960). El-Banhawy (1975) reported the phytoseiid *Amblyseius brazilli* El-Banhawy.

Symptomatology and damage

The damage caused is similar to that caused by the two-spotted spider mite, *T. urticae* Koch (Nickel, 1958). Morros and Aponte (1995a,b) evaluated the effect of two levels of mite infestation on the vegetative and reproductive stages of the black bean, *P. vulgaris*, under greenhouse and field conditions. Greater reductions of the numbers of leaves and internodes, leaf area and dry weight of vegetative organs occurred when mite infestation was in the early vegetative stage. Moraes and Flechtmann (2008) reported that this spider mite might cause severe damage to bean plants, as *T. urticae* does.

Control

Control of the winter hosts helps to prevent the development of *T. desertorum* on cotton, and where this practice fails, chemical sprays may be applied (Iglinsky and Rainwater, 1954; Hightower and Martin, 1956). The effects of different varieties of beans on the biology and fecundity of *T. desertorum* have been investigated in Colombia (Jara *et al.*, 1981). The Pest Control Guidelines for cucurbits of California University (UC IPM Online, 2012d) suggest, as for other spider mites, sprays of abamectin, dicofol and sulfur. A wider range of chemicals is available for use on cotton.

Tetranychus evansi Baker (Fig. 20.49)

Common name

Tomato red spider mite.

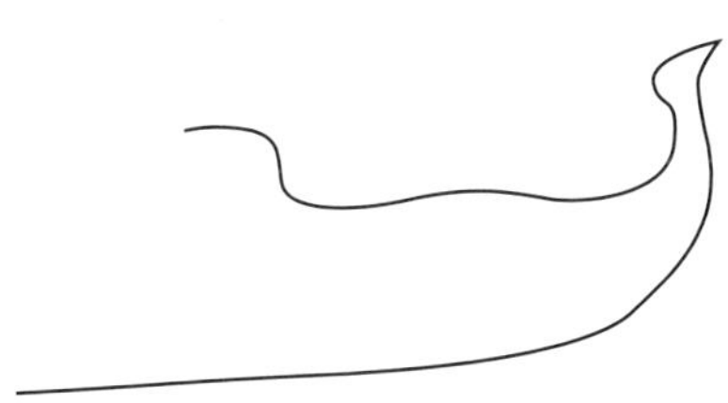

Fig. 20.49. *Tetranychus evansi* Baker. Male, aedeagus (from Meyer Smith, 1987).

Diagnostic characteristics

FEMALE. The peritreme is hooked. The dorsal striae between the setae e_1–e_1 are longitudinal and oblique, between the setae f_1–f_1 are longitudinal and between the setae e_1 and f_1 are transverse and form a diamond shape. The dorsal striae are lobate, the ventral striae lack lobes, and the pregenital striae are entire, and may be weak medially. The empodia have 1 minute dorsal spur and 6 proximoventral hairs. Tarsus I has 4 tactile setae overlapping with the proximal duplex setae; tarsus III has 1 proximal tactile seta (Seeman and Beard, 2011).

MALE. The empodia have a dorsal spur; empodia I and II are uncinate, and III–IV have proximoventral hairs that are long and free. The knob of the aedeagus has a pointed anterior projection, and a posterior projection as a blunt hook (Seeman and Beard, 2011).

Geographical distribution

Algeria (Migeon *et al.*, 2009), Argentina (Rossi Simons, 1961, 1971), Brazil (Silva, 1954; Flechtmann, 1967a), Canary Islands (Gotoh *et al.*, 2009), China (Kinmen and Matsu Islands; Ho *et al.*, 2004), Congo (Bolland *et al.*, 1998), Congo (Democratic Republic of the Congo, formerly Zaire) (Knapp *et al.*, 2003), France (Migeon, 2005), Gambia (Migeon *et al.*, 2009), Greece (Tsagkarakou *et al.*, 2007), Hawaiian Islands (Kumashiro, 1998), Israel (EPPO, 2006; Ben-David *et al.*, 2007), Italy (Bolland and Beron, 2004), Japan (Ehara and Ohashi, 2002; Ikeshima *et al.*, 2009), Kenya (Knapp, 2007), Madeira (Portugal) (Hurtado *et al.*, 2008), Malawi (Mtambo and Hoeschle-Zeledon, 2001), Mauritius (Moutia, 1958), Morocco (El-Jaouani, 1988), Mozambique (Meyer Smith, 1987), Namibia (Keizer and Zuurbier, 2001), Niger (Migeon *et al.*, 2009), Portugal (Ferreira and Carmona, 1995), Puerto Rico (USA) (Medina Gaud and Garcia Tuduri, 1977), Réunion (France) (Gutierrez, 1983), Rodrigues Island (Mauritius) (Gutierrez, 1974), Senegal (Duverney and Ngueye-Ndiaye, 2005), Seychelles (Gutierrez, 1974), Somalia (Knapp *et al.*, 2003), South Africa (Meyer Smith, 1996b), Spain (Ferragut *et al.*, 1997a), Syria (Zriki *et al.*, 2015), Taiwan (Ho *et al.*, 2004), Tanzania (Toroitich *et al.*, 2008), Tunisia (Bolland *et al.*, 1998), the USA (Wene, 1956), Virgin Islands (American) (Moraes *et al.*, 1987), Zambia (Jensen and Mingochi, 1988) and Zimbabwe (Tobacco Research Board, 1982; Blair, 1983; Meyer Smith, 1987).

Host plants

Tetranychus evansi is reported on 125 host plants belonging to 33 families, with a strong preference for solanaceous plants (43 species) (Migeon and Dorkeld, 2006–2013). The families and species of economic importance are:

Amaranthaceae: beet (*Beta vulgaris* L.) (Aucejo *et al.*, 2003).

Anacardiaceae: cashew (*Anacardium occidentale* L.) (Tuttle *et al.*, 1977).

Brassicaceae: wild radish (*Raphanus raphanistrum* L.) (Ferreira and Sousa, 2011).

Convolvulaceae: sweet potato (*Ipomoea batatas* (L.) Lam.) (Moutia, 1958; Meyer Smith, 1996b).

Euphorbiaceae: poinsettia (*Euphorbia pulcherrima* Willd. ex Klotzsch), castor bean (*Ricinus communis* L.) (Ho *et al.*, 2004).

Fabaceae: groundnut (*Arachis hypogaea* L.) (Moutia, 1958), kidney bean (*Phaseolus vulgaris* L.) (Gutierrez and Etienne, 1986).

Lamiaceae: sweet basil (*Ocimum basilicum* L.), sweet marjoram (*Origanum majorana* L.), oregano (*O. vulgare* L.), kitchen sage (*Salvia officinalis* L.) (Ferreira and Sousa, 2011).

Malvaceae: okra (*Abelmoschus esculentus* (L.) Moench) (Tuttle *et al.*, 1977), upland cotton (*Gossypium hirsutum* L.) (Wene, 1956).

Myrtaceae: guava (*Psidium guajava* L.) (Tuttle *et al.*, 1977).

Rosaceae: strawberry (*Fragaria* × *ananassa* Duchesne ex Rozier) (Ferreira and Sousa, 2011).

Solanaceae: pepper (*Capsicum annuum* L.) (Silva, 1954; Moutia, 1958; Flechtmann and Abreu, 1973; Tuttle *et al.*, 1977; Meyer Smith, 1996b; Ferreira and Sousa, 2011), tree tobacco (*Nicotiana glauca* Graham) (Oatman *et al.*, 1967), cultivated tobacco (*N. tabacum* L.) (Blair, 1983, 1989; Gutierrez and Etienne, 1986; Moraes *et al.*, 1987; Meyer Smith, 1996b; Seeman and Beard, 2011), tomato (*Solanum lycopersicum* L.) (Silva, 1954; Wene, 1956; Moutia, 1958; Baker and Pritchard, 1960; Flechtmann, 1967a; Ferragut and Escudero, 1999; Knapp *et al.*, 2003; Ho *et al.*, 2004; Duverney and Ngueye-Ndiaye, 2005; Migeon, 2007; Ikeshima *et al.*, 2009; Guanilo *et al.*, 2010; Guenaoui, 2010; Ferreira and Sousa, 2011; Seeman and Beard, 2011; Matsuda *et al.*, 2013; Zriki *et al.*, 2015), aubergine (*S. melongena* L.) (Moutia, 1958; Oatman *et al.*, 1967; Paschoal, 1968b; Denmark, 1973; Gutierrez and Etienne, 1986; Meyer Smith, 1987, 1996b; Moraes *et al.*, 1987; Ehara and Ohashi, 2002; Leite *et al.*, 2003; Ho *et al.*, 2004; Ho and Wang, 2007; Ikeshima *et al.*, 2009; Guanilo *et al.*, 2010; Ferreira and Sousa, 2011; Ohno *et al.*, 2012), Irish potato (*S. tuberosum* L.) (Moutia, 1958; Paschoal, 1968b; Denmark, 1973; Moraes and Flechtmann, 1981; Moraes *et al.*, 1987; Meyer Smith, 1996b; Ferragut and Escudero, 1999; Ikeshima *et al.*, 2009; Ferreira and Sousa, 2011; Seeman and Beard, 2011).

Bio-ecology

T. evansi is a warmth-loving mite that is injurious to a number of economic plants (including aubergine, sweet basil, kidney bean, beat, cashew, cotton, groundnut, guava, sweet marjoram, naranjilla, okra, oregano, pepper, poinsettia, wild radish, kitchen sage, strawberry, potato, tobacco and tomato), in particular solanaceous plants. It severely damages tomato (Silva, 1954; Gutierrez and Etienne, 1986; Fiaboe *et al.*, 2007a; Knapp *et al.*, 2003; Gillet *et al.*, 2006; Migeon, 2007), aubergine (Moraes *et al.*, 1987; Leite *et al.*, 2003; Gillet *et al.*, 2006), potato (Escudero and Ferragut, 2005; Gillet *et al.*, 2006), tobacco (Blair, 1989) and peppers and chillies (Silva, 1954). It has high oviposition on aubergine, tomato and nightshade, but not on cherry tomato and pepper (Wekesa *et al.*, 2011). Severe attacks of the tomato red spider mite mainly occur outdoors, and only a few reports concern its presence in glasshouses (EPPO, 2007), probably due to different varieties and control strategies applied to protected crops. Weed plants play a key role in the population dynamics of the mite, and among these, it prefers nightshade (*Solanum nigrum* L.) (Migeon, 2007), but it also has been collected on *Amaranthus blitoides* S. Watson, *Chenopodium* spp. (El-Jaouani, 1988), *Convolvulus arvensis* L., *Conyza* spp., *Diplotaxis erucoides* (L.) DC., *Hordeum murinum* L., *Lavatera trimestris* L. and *Sonchus* spp. (Ferragut and Escudero, 1999; Aucejo *et al.*, 2003).

The tomato red spider mite is native to South America and currently is an invasive species in Africa, Europe and other countries. Genetic data demonstrate two different patterns of invasion. The populations of *T. evansi* in the Mediterranean derive from multiple cryptic introductions, while the emerging invasions in Africa and Asia probably each depend on propagules from one or limited sources, respectively (Boubou *et al.*, 2011). Approximate Bayesian Computation (ABC) and variation in multi-locus genetic markers have shown that *T. evansi* populations in Europe and Africa relate to at least three independent introductions from South America and involve mites from two distinct sources in Brazil that correspond to highly divergent mitochondrial DNA lineages. Mites from south-west Brazil colonized the African continent, and from there they colonized Europe through two pathways. One pathway was a widespread invasion, not only of Europe, but also other regions in Africa, southern Europe and eastern Asia. The second pathway was a mixture with a second introduction from south-west Brazil, leading to an admixed population in southern Spain. This admixing has been also detected between invasive populations in Portugal. A third introduction from the Brazilian Atlantic region resulted in only a restricted introduction in Europe (Boubou *et al.*, 2012).

The potential hazard of mite introduction has suggested that a 'Climex model' should be developed to predict the distribution of *T. evansi*; this should be able to fit its records with exceptions for some dry areas, taking into account that dryness and excess moisture stresses may limit the spread of the mite into new areas. According to these factors, the potential distribution of the mite in North America and Eurasia seems to be mainly limited by cold stress, while the Mediterranean area and Japan could represent the climatic borders for its geographical distribution. The establishment of *T. evansi* in these latter areas could be facilitated by mild winters, and according to the model the whole Mediterranean area has the potential to be extensively colonized by the mite. The economic risk is high because the Mediterranean region is the main area of tomato cultivation in open fields in Europe (Migeon *et al.*, 2009). This issue has been further discussed by Genevest *et al.* (2012), who have addressed the need for invasive risk assessment and modelled the potential distribution of five species of Tetranychidae (including *T. evansi*) under current and future climate conditions using two mechanistic models, NicheMapper and Climex, based on physiological properties of the species.

The invasion success of *T. evansi* could depend on a particularly successful invasive population serving as a source to colonize new areas. This suggests that the biological traits that can facilitate potential invasion by the mite should be investigated (Boubou *et al.*, 2010). Janssen *et al.* (2012) hypothesized that plant- and predator-mediated competition played a decisive role in the spread of *T. evansi* into greenhouse tomato crops in the Old World. On the tomato plant, *T. evansi* competes both directly and indirectly with the two-spotted spider mite, *T. urticae* Koch, and exerts indirect competition through the direct and indirect defences induced in the plant. Feeding

by the tomato red spider mite on undamaged tomato plants suppresses proteinase inhibitors (direct defences) locally to below the normal levels, promoting both its own reproduction and that of *T. urticae*. However, *T. evansi* protects the leaf surface involved in plant defence suppression by spinning a dense web that is impenetrable to the two-spotted spider mite and its natural enemies, e.g. *Phytoseiulus persimilis* Athias Henriot. Moreover, *T. evansi* induces the tomato plant to emit volatiles that act as attractants to this phytoseiid mite (Janssen *et al.*, 2012).

How spider mites suppress the defences of the plant, and which part of the plant's metabolism is affected, are unknown. Some genotypes of *T. evansi* induce and others suppress jasmonic acid (JA) and salicylic acid (SA) responses simultaneously. Information was collected from (i) the sequencing of the expressed genes (mRNA) of these mite genotypes feeding from tomatoes with a different level of resistance, (ii) the expressed genes of the host plant on which they were feeding and (iii) other factors (behavioural assays, genetic manipulation, expression analyses and metabolomics); this allowed an analysis of the causes and the effects of these different types of plant-defence manipulations within the semi-natural context of these mite species (Alba *et al.*, 2012a). Studies on the differential induction and suppression of tomato defences by several genotypes of *T. evansi* and *T. urticae* show that suppressor mites exert a strong negative impact on the accumulation of jasmonates and salicylate as well as the downstream defence genes, while producing a high level of feeding damage. The fitness of defence-inducing mite genotypes increases when inhabiting the leaflet with suppressor mites. Suppressor mites did not suppress phytohormone accumulation caused by cohabiting inducer mites, but only suppressed the expression of downstream defence genes. Consequently, the suppression operates downstream of phytohormone accumulation and may feed back on to hormone biosynthesis (Alba *et al.*, 2012b). The assessment of the spatial characteristics of induced and suppressed defence responses during attack by *T. evansi* and *T. urticae* highlight that within tomato leaflets, the induction of defences is mostly restricted to the feeding site, while suppression occurs beyond the feeding site as well, but not in an equal manner in all directions. Defence suppression can, therefore, hinder *T. evansi*, which spins elevated amounts of webbing to protect its feeding site from predators and competitors (Ataide *et al.*, 2012). Regardless of these important ecological features, the recent invasions of *T. evansi* point to the importance of human activities in the spread of agricultural pests, and the limits of international quarantine procedures, particularly for cryptic invasions (Boubou *et al.*, 2011).

The tomato red spider mite reproduces by arrhenotokous parthenogenesis and has a haploid number of chromosomes of three (Helle *et al.*, 1981). It usually lives on the lower leaf surface, but at high humidity also spreads on to the upper leaf surface. Females lay eggs mainly on the lower leaf surfaces at the junctions of the veins, mostly on normally pubescent leaves rather than on lightly pubescent leaves (Qureshi *et al.*, 1969). The mite spins a dense and disorderly structured webbing. It does not produce denser webbing in response to predator cues, and lays most of the eggs suspended in the web, away from the leaf surface. Natural enemies, e.g. the phytoseiid *P. longipes* Evans, prey less on suspended eggs than on eggs laid on the leaf surface under the web. So, females of *T. evansi* protect their offspring by altering their oviposition behaviour in response to predator cues (Lemos *et al.*, 2010). Studies on the effect of webbing spun by *T. evansi* on the predation efficiency of the coccinellids *Cycloneda sanguinea* (L.) and *Eriopis connexa* (Germar) and of the chrysopid *Chrysoperla externa* (Hagan) show that the defensive role of *T. evansi* webbing varies according to the predator species and its developmental stage (Venzon *et al.*, 2009).

In the laboratory, and with *Solanum douglassi* Dunal as the food plant, oviposition by *T. evansi* increased with high RH. At 23°C and 40–50% RH, egg development lasted 60–72 h and the duration of the larval and nymphal stages averaged 43 and 39 h, respectively. The resting stages took about 21–25 h. The development time from egg to adult female ranged from 8.2 to 9.3 days. During the summer (averaging 22.8°C) and winter (averaging 19.4°C), the life cycle develops in 6.5 and 18.5 days, respectively. The pre-oviposition period is 1 day and the largest number of eggs is laid on the fourth day, when females may lay up to 30 eggs in a day; oviposition then decreases. The daily oviposition is 6.1 eggs over 26.7 days. The female longevity ranges from 13 to 32 days, and unfertilized females live from 27 to 39 days. In southern California, the mite reproduces constantly throughout the year, and may develop 24–30 generations in a year (Oatman *et al.*, 1967; Qureshi *et al.*, 1969). In laboratory studies, the lower thermal threshold was 10.3°C, the shortest developmental time (6.3 days) was at 36°C, and maximum fecundity at 31°C (123.3 eggs per female). The highest intrinsic rate of increase (r_m) (0.355) occurred at 31°C, and the optimal temperature for population growth seemed to be 34°C (Bonato, 1999). At 30°C, 60% RH and with 16 h light, the values of the intrinsic rate of natural increase (r_m) on potato, tomato, pepper and *S. nigrum* were 0.254, 0.236, 0.232 and 0.260 day^{-1}, respectively (Lebdi-Grissa and Mbarki, 2012).

T. evansi is associated with endosymbionts, and the association is apparently beneficial to the mite. The effect of microflora on the performance of *T. evansi* has been evaluated by comparing the oviposition of antibiotic-treated and untreated mites on tomato leaves. The rate of oviposition of untreated mites was 2.8 times higher than that of treated mites (Cavalcante *et al.*, 2010).

The survival of mite populations in the field and their invasion of cultivated solanaceous plants depends mainly on weeds such as nightshade (*S. nigrum*), and other plants, on which the mite lives and moves and from which it has the potential to invade transplanted cultivated solanaceous plants. So, Gotoh *et al.* (2010) assessed the life-history parameters on *S. nigrum* of samples of *T. evansi* collected from seven sites worldwide. Increasing temperatures from 15 and 32.5°C significantly increased the developmental rate of all of the strains, and the duration of immature development was about the same at 32.5–40°C. The rate of egg to adult development exceeded 88% at temperatures from 15 and 37.5°C. The lower thermal thresholds were 11.9–12.5°C for both egg to adult and egg to egg development. The optimum developmental temperatures ranged from 36.7 to 43.8°C and the upper developmental threshold from 45.2 to 59.4°C. The r_m values became higher with temperature, increasing from 15 to 35°C. At 25°C, r_m values ranged from 0.265 to 0.277 (relatively high for species of the genus *Tetranychus*). These results indicate that after invasion into Japan, *T. evansi* has the potential to become a serious pest

on solanaceous crops, as in Africa and Europe (Gotoh *et al.*, 2010). Another weed, American nightshade (*Solanum americanum* Miller), is considered a major source plant in Japan (Ohno *et al.*, 2012). This plant is native to the Americas and represents an alternative host plant for *T. evansi* and its natural enemies in the absence of tomato plants (Ribeiro *et al.*, 2012). Before the invasion of *T. evansi* into Spain, the two-spotted spider *mite*, *T. urticae*, and the strawberry spider mite, *T. turkestani* (Ugarov *et* Nikolskii), were the most abundant spider mites, accounting for 62.9 and 22.8% of the specimens, respectively. After the invasion, *T. evansi* has become very common, representing 60% of the total spider mites collected, and *T. urticae* has been strongly reduced (23%). This reduction took place mainly on non-crop plants, where native species have been replaced by *T. evansi*. Null model analyses provided evidence for competition structuring the spider mite community on non-crop plants after the invasion of *T. evansi*. Possible causes for this competitive displacement could be resistance to acaricides, the absence of efficient native natural enemies, manipulation of plant defences and the web type produced by *T. evansi* (Ferragut *et al.*, 2013b).

The effects of biotic and abiotic ecological factors, such as total rainfall, mean temperature, predators, height of canopy, levels of nitrogen and potassium in leaves and density of leaf trichomes on the attack intensity of *T. evansi* on aubergine (var. Natu Nobilis) were evaluated in Brazil. In some areas, higher numbers of motile forms of *T. evansi* cm^{-2} might be explained by hotter and more rainy weather. Mite population was positively correlated with total rainfall, but no significant effects were found for natural enemies, levels of N and K, and leaf trichome density on population densities of mites. Higher numbers of adults of *T. evansi* infested leaves of the apical and medium parts of aubergine plants than the lower leaves (Leite *et al.*, 2003).

Natural enemies include pathogens such as *Conidiobolus thromboides* Drechsler (Ancylistaceae) (Fisher, 1951), *Neozygites* sp. (Humber *et al.*, 1981) and *Neozygites floridana* Weiser *et* Muma (Neozygitaceae) (Guanilo *et al.*, 2010), and other species (see below), and various predatory insects and mites. Predatory insects include the coccinellids *Stethorus darwini* (Bréthes) (Paschoal, 1970), *S. tridens* Gordon (Fiaboe *et al.*, 2007b; Britto *et al.*, 2009), *S. vinsoni* Kapur (Moutia, 1958) and *Eriopis connexa* Germar, and the chrysopids *Chrysopa sanguinea* (Navás) and *C. externa* (Hagen) (Venzon *et al.*, 2009). A number of phytoseiid mites have been found associated with *T. evansi*, such as *Euseius concordis* (Chant) (Furtado *et al.*, 2007; Guanilo *et al.*, 2010), *Neoseiulus californicus* (McGregor), *N. idaeus* Denmark *et* Muma (Furtado *et al.*, 2007), *Phytoseiulus fragariae* Denmark *et* Schicha (Furtado *et al.*, 2007; Nascimento de Vasconcelos *et al.*, 2008), *P. longipes* Evans (Furtado *et al.*, 2006; Silva *et al.*, 2008; Tixier *et al.*, 2010a) and *Proprioseiopsis cannaensis* (Muma) (Furtado *et al.*, 2007). Other species were tested for the biological control of *T. evansi* under laboratory conditions and in the field (see below).

Symptomatology and damage

Feeding by the tomato red spider mite on the underside of leaves causes whitening or yellowing, followed by desiccation of the tissues, and in severe attacks, by leaf drop. Tomato plants infested by *T. evansi* turn yellow green and then brown. Plants exhibit a bleached yellow orange or russeted appearance, and may die rapidly. High temperature and low RH induce severe damage. The mite may kill its host plant within 3–5 weeks of infestation, especially tomato, potato, groundnut and aubergine plants (Qureshi *et al.*, 1969). High population densities are usually associated with the production of extensive webbing on the lower leaf surfaces.

Control

The elevated rate of reproduction of *T. evansi*, its development of resistance to pesticides and its ability to develop on weeds as alternative hosts when tomato plants are not available in the field, make its control difficult (Onyambus *et al.*, 2011). Currently, the results of studies on plant resistance do not seem to offer immediate practical solutions, and likewise the use of biological control alone. Sprays of acaricides according to their authorization of use of each country represent the better solution.

PLANT RESISTANCE. There do not seem to be commercially available tomato varieties that are resistant to *T. evansi*. However, an evaluation of tomato accessions for their resistance to the tomato red spider mite show that trichome density and plant-emitted volatiles play an important role in host-plant selection by spider mites (Murungi *et al.*, 2009). Hybrids between the cultivated tomato and the weed species *Lycopersicon hirsutum* Dunal f. *glabratum* C.H. Müll., *L. pennellii* (Correll) D'Arcy and *L. cheesmanii* Riley f. minor (Hook.) C.H. Mull. have been produced and useful levels of resistance to various pests (including pest mites) have been assessed, although these effects may be tempered by those on natural enemies (Simmons and Gurr, 2005).

The relationship was studied between zingiberene content of tomato leaves, glandular trichome density and levels of repellence to *T. evansi*. Indirect selection for high zingiberene contents increases both glandular trichome density and mite repellence. Zingiberene in glandular trichomes is therefore likely to mediate mite repellence in the tomato population. Glandular trichome densities (especially of type VI trichomes) are highly and positively correlated with leaf zingiberene contents. High zingiberene levels and their relationship with both glandular trichome densities and resistance (repellence) to *T. evansi* is significant after two backcross generations, indicating an indirect selection for high zingiberene as an effective way of selecting mite-resistant tomato genotypes (Gonçalves *et al.*, 2006). Onyambus *et al.* (2011) investigated host plant acceptance and the relative contribution of trichomes in the control of the red spider mite by comparing its survival, development and oviposition rates on eight tomato accessions. The densities of trichome types I, IV, V and VI varied among the tomato accessions, and a higher density of trichome type I corresponded to lower adult survival. The findings indicated the possible resistance of some of the tested tomato accessions against *T. evansi* that was partially associated with trichome type and density.

The *L. pennellii* accession LA716 has been used as a source of pest resistance in tomato breeding programmes because of the high contents of acylsugars in its leaflets. The relationship between high foliar acylsugar contents and repellence to *T. evansi* in plants with contrasting acylsugar levels has been investigated, and plants selected for high levels of acylsugars in

leaflets found to show mite repellence similar to that of LA716. The high correlation confirmed the association between high acylsugar contents and mite repellence (Vilela de Resende *et al.*, 2008). In addition, five African nightshade species (*Solanum sarrachoides* Sendter, *S. villosum* Miller, *S. tarderemotum* Bitter, *S. americanum* and *S. scabrum* Miller) have been assessed and their genotypes differ in their levels of resistance to *T. evansi*, which is partially associated with differences in trichome types and their densities (Murungi *et al.*, 2011).

CULTURAL PRACTICES. Weed plants play a key role in the survival of populations of *T. evansi*, and their control before and after seedling transplantation of cultivated solanaceous plants is good practice, as well as the transplantation of uninfested seedlings. Pruning and trellising of tomato may result in better mite management, less disease incidence, fewer fruit rots and reduced damage to fruits. These practices resulted in additional profit of US$18,780 ha^{-1} at Mutoko, Zimbabwe (Saunyama and Knapp, 2003). The results of studies on the population dynamics of adult females of *T. urticae* and *T. evansi* on tomato plants (cv. 'Castlemart') with their apices removed or intact showed that there was a higher final population of *T. urticae* on intact plants than on plants without apices, and a smaller population of *T. evansi* on intact plants. Both mites reached peak density earlier on intact plants than on plants lacking apices. Apex removal therefore alters the susceptibility of tomato plants to spider mite infestation, with the impact depending mainly on the mite species (Li *et al.*, 2013a).

BIOLOGICAL CONTROL. Many efforts have been made to evaluate the action of different pathogens and predators both in the field and under laboratory conditions. However, biological control alone does not currently resolve the problem of the tomato red spider mite in the field.

Attention was first focused on different isolates of *Metarhizium anisopliae* (Metschnikoff) Sorokin and *Beauveria bassiana* (Balsamo) Vuillemin to evaluate their pathogenicity against the tomato red spider mite. Despite the susceptibility of different biological stages of *T. evansi* to both entomopathogenic fungi, and a lower fecundity of fungus-treated females, no difference in egg hatchability was found between treatments (Wekesa *et al.*, 2005, 2006). Further studies on the virulence of various isolates of *B. bassiana* and *M. anisopliae* against *T. evansi* show that temperature significantly affects germination, radial growth and virulence of the various isolates. The best fungal germination occurs at 25 and 30°C, and the best fungal radial growth at 30°C. The isolates germinated and grew at all temperatures, but germination and radial growth varied with isolate and temperature. Although all isolates were virulent to *T. evansi*, virulence varied with isolates and temperature (Bugeme *et al.*, 2008).

Studies on population dynamics of *T. evansi* and its natural enemies in the field and in a protected environment on nightshade and tomato plants highlighted the positive action of *N. floridana*. This pathogen was the only natural enemy found associated with *T. evansi* under protected environment conditions and appeared in tomato plants even with a relatively low population density of *T. evansi* (less than 10 mites 3.14 cm^{-2} leaf surface); at this population density, the infection rates are higher than 50%. The application of pesticides directly affected the fungus by delaying epizootic initiation and contributing to lower infection rates than in unsprayed treatments. Rainfall did not have any apparent impact on mite populations. The results indicate that *N. floridana* can play a significant role in the population dynamics of *T. evansi*, especially under protected environmental conditions, and has the potential to control this pest in classical biological control programmes (Duarte and Delalibera, 2010). Tritrophic-level interaction experiments were conducted to evaluate the effect of selected host plants of *T. evansi* (aubergine, tomato, nightshade, cherry tomato and pepper) on *N. floridana* by evaluating the attachment of capilliconidia, presence of hyphal bodies in the infected mites, mortality from fungal infection, and the mummification of and sporulation from fungus-killed mite cadavers. The performance of *N. floridana* varied with host plants and may be an important factor in the development of pathogen epizootics (Wekesa *et al.*, 2011). In addition, field application of captan and mancozeb induced a high reduction in the sporulation and germination of *N. floridana*, while the application of propargite did not inhibit sporulation but affected germination of the primary conidia. Sprays of methomyl and abamectin have fewer effects on *N. floridana* (Wekesa *et al.*, 2008).

The bionomy and the potential of the ladybird beetle *S. tridens* as a candidate natural enemy of *T. evansi* were examined by Fiaboe *et al.* (2007b). The predator exhibits a type II functional response at the first and tenth oviposition day, with a maximum consumption per predator of 33 *T. evansi* nymphs a day at the highest prey density (Britto *et al.*, 2009). Despite the fact that *S. tridens* is often collected associated with *T. evansi* on solanaceous wild plants and the results of first studies suggesting the potential of this ladybird beetle to control *T. evansi* on tomato crops, there is no new information on its application in the field.

The first laboratory investigations with phytoseiid mites showed that *T. evansi* was an unfavourable prey for the phytoseiids *Amblyseius californicus* (= *N. californicus*) and *P. persimilis* (Moraes and McMurtry, 1985; Escudero and Ferragut, 2005). Later, it was hypothesized that two-pronged protection from natural enemies of tomato made *T. evansi* a key pest of this plant. In fact, morphological traits and/or trichome exudates may induce direct adverse effects of tomato on predators, whereas the accumulation of toxic plant compounds cause indirect prey-mediated effects. The developmental time and oviposition rate of *N. californicus* are both directly and indirectly negatively affected by the tomato plant, but the offspring sex ratio and survival of juveniles and adult females are unaffected. The direct and indirect prey-mediated adverse effects of tomato on *N. californicus* with *T. evansi* prey have similar magnitudes and are additive. These results show that *T. evansi* is a suitable prey for *N. californicus* (Koller *et al.*, 2007). Despite this conclusion, the results of releases of *N. californicus* against *T. evansi* on tomato and other host plants are uncertain. Similarly, *P. fragariae* cannot be considered a good predator of *T. evansi* (Nascimento de Vasconcelos *et al.*, 2008).

In contrast, *P. longipes* may develop on *T. evansi* at a wide range of temperatures and has the potential to control its populations (Ferrero *et al.*, 2007). Releases of four females per plant of *P. longipes* on tomato plants, each infested with eight adult

females of *T. evansi*, showed that the population densities of *T. evansi* and the damage caused were significantly lower in the plots where *P. longipes* was released (Silva *et al.*, 2010). Releases of *P. longipes* in greenhouses in the Netherlands and in Spain at an approximate rate of 1:10 predators to spider mites significantly reduced the tomato red spider mite population compared with the untreated control. The temperature and RH in greenhouses may affect the development of both spider mite and predatory mite populations (Knapp *et al.*, 2010). Further trials on tomatoes in greenhouses in Spain at high and low predator:prey ratios (1:5 and 1:25) showed that *P. longipes* was able to maintain *T. urticae* and *T. evansi* populations under a threshold of 25 mites per leaflet at a high predator:prey ratio. However, at the low predator:prey ratio, plants showed damage in the presence of both pests (Ferrero *et al.*, 2011). The results of studies in Brazil on natural populations of *P. longipes* confirmed its validity of action. Of the natural enemies found in association with *T. evansi* on the African nightshade (*S. americanum*), only *N. floridana* and *P. longipes* reduced the populations of *T. evansi*. Increases of populations of tomato red spider mites were always followed by increases in the density of one of these natural enemies. This highlights their role as regulating factors of *T. evansi* populations on *S. americanum*. Moreover, the presence of this host plant near the release sites in Africa thus might increase the establishment of these natural enemies (Ribeiro *et al.*, 2012).

CHEMICAL CONTROL. In the past, high-volume sprays of carbophenothion, chlorobenzilate and diazinon assured good control of the tomato red spider mite. Ethion, carbophenothion and dicofol dusts were also applied to reduce infestation over a 7-day period (Schuster, 1959). Lime sulfur and neem-based products, applied in appropriate concentrations and formulations, test as a viable alternative to control *T. evansi* on tomato plants (Soto *et al.*, 2010). Other acaricides positively tested include abamectin (Nyoni *et al.*, 2011), acrinatrin (EPPO, 2007), bifenazate, chlorfenapyr, cyenopyrafen, cyflumetofen (Gotoh *et al.*, 2011), dicofol, fenbutatin oxide (EPPO, 2007), fenpyroximate (EPPO, 2007; Nyoni *et al.*, 2011), hexythiazox (EPPO, 2007), milbemectin (Gotoh *et al.*, 2011), propargite (EPPO, 2007), spirodiclofen and tebufenpyrad (Gotoh *et al.*, 2011). Also effective are the mixtures dicofol plus hexythiazox and fenpyroximate and dicofol (EPPO, 2007). Tolerance has been reported to chlorpyrifos and resistance to bifenthrin (Nyoni *et al.*, 2011).

Tetranychus fijiensis Hirst (Fig. 20.50)

Common name

Unknown.

Diagnostic characteristics

FEMALE. The body is 350 µm long and orange red in colour. The peritremes are strongly 'U' shaped distally. The opisthosomal dorsal striae form a diamond-shaped pattern between the setae e_1 and f_1. The empodia have only 2 pairs of proximal hairs, the dorsal pair smaller than the ventral pair, and a strong mediodorsal spur, about half as long as the longer pair of proximal hairs. Tarsus I is attenuate, with the tactile setae on a line with the duplex setae, which are widely spaced (Pritchard and Baker, 1955).

MALE. The aedeagus is long, slender, tapering and curved dorsally, and lacking an enlarged knob (Pritchard and Baker, 1955).

Geographical distribution

Australia (Flechtmann and Knihinicki, 2002; Seeman and Beard, 2011), China (Ehara and Tho, 1988), Fiji (Hirst, 1924), India (Hirst, 1924; Manson, 1963; Daniel, 1977; Gupta, 1992; Gupta and Gupta, 1994), Kiribati (Gutierrez, 1977), Malaysia (Ehara and Tho, 1988; Othman and Zhang, 2003), Northern Marianas, Marshall Islands, Federated States of Micronesia (Gutierrez, 1980), New Caledonia (France) (Bolland *et al.*, 1981), Papua New Guinea (Schicha and Gutierrez, 1985), Philippines (Baker, 1975a), Seychelles (Gutierrez, 1974), Taiwan (Tseng, 1990) and Thailand (Manson, 1963; Baker, 1975b).

Host plants

The whole list of host plants of *T. fijiensis* includes 24 species from 10 families (Migeon and Dorkeld, 2006–2013). The families and species of economic importance are: Arecaceae – betelnut (*Areca catechu* L.) (Daniel, 1977; Gupta and Gupta, 1994; Flechtmann and Knihinicki, 2002), coconut palm (*Cocos nucifera* L.) (Hirst, 1924; Baker, 1975b; Gupta and Gupta, 1994); Caricaceae – papaya (*Carica papaya* L.) (Gupta, 1976; Gupta and Gupta, 1994); Moraceae – white mulberry (*Morus alba* L.) (Bolland *et al.*, 1998; Migeon and Dorkeld, 2006–2013); Rosaceae – peach (*Prunus persica* (L.) Batsch), pear (*Pyrus communis* L.) (Ehara and Wongsiri, 1975); Rutaceae – *Citrus* sp. (Ehara and Wongsiri, 1975; Gupta, 1992), sour orange (*C. aurantium* L.) (Prasad, 1974; Gupta and Gupta, 1994), citron (*C. medica* L.), grapefruit (*C. paradisi* Macfad.), tangerine (*C. reticulata* Blanco) (Baker, 1975b), calamondin (*Citrus* × *microcarpa* Bunge) (Othman and Zhang, 2003); Zingiberaceae – cardamom (*Elettaria cardamomum* (L.) Maton) (Gupta and Gupta, 1994).

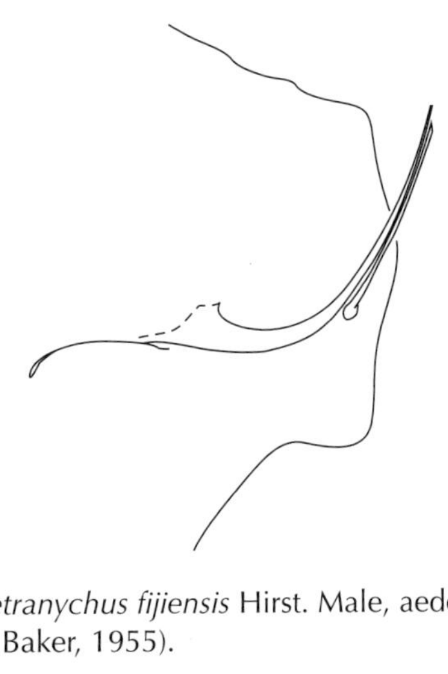

Fig. 20.50. *Tetranychus fijiensis* Hirst. Male, aedeagus (from Pritchard and Baker, 1955).

Bio-ecology

T. fijiensis is reported in Asia, Australia and Polynesia, where it infests calamondin, cardamom, citron, coconut palm, grapefruit, papaya, pear, peach, sour orange, tangerine, white mulberry and various ornamentals. Reproduction is by arrhenotokous parthenogenesis and the haploid number of chromosomes is four (Bolland *et al.*, 1981). Unmated females produce only males, and mated females produce both males and females at a ratio of 1:8 (Vatana *et al.*, 2001).

The mite feeds on the lower leaf surface, where the different stages occur in webs spun by adults (Daniel, 1979). In laboratory studies, non-inseminated females laid fewer eggs, but lived longer than inseminated females. A 50% survival rate for inseminated females was attained 14.5 days before that of non-inseminated females. The fecundity of inseminated females was roughly twice as great as that of non-inseminated females. This behaviour is an adaptation to the climate of the area of origin (the Indo-Pacific tropical area) (Bonato and Gutierrez, 1999).

In India, on coconut, high populations occur in the early summer and mid-autumn, reaching densities of 5–6 mites per leaflet (Sarkar and Somchoudhury, 1989a). On leaves of betelnut, the mite lives throughout the year; it has the maximum population in the summer and minimum in the rainy season. Egg development lasts 4–4.5 days. The larval period lasts 43–68 h, including a quiescent period of 11–20 h. The protonymph and deutonymph stages last 47–52 and 47–72 h, respectively (including quiescent periods of 4–19 and 13–23 h). The immature stages last from 9.9 to 12.1 days (Daniel, 1977). On the passion fruit leaf, the developmental period from egg to adult of the male and female is 10.7 and 12.6 days, respectively. Each unmated female could lay on average 15.1 eggs throughout her adult life, while the mated female could lay on average 20 eggs (Vatana *et al.*, 2001).

Natural enemies include predatory insects (coccinellids, cecidomyiids and chrysopids) and phytoseiid mites. The coccinellids include *Delphastus catalinae* (Horn) (Jing *et al.*, 2003), *Stethorus exspectatus* Chazeau and *S. exsultabilis* Chazeau (Chazeau, 1983). The phytoseiid mites include *Amblyseius channabasavannai* Gupta *et* Daniel (Daniel, 1979), *A. cinctus* Corpuz *et* Rimando (Vatana *et al.*, 2001), *A. largoensis* (Muma), *A. tamatavensis* Blommers (Schicha and Gutierrez, 1985), *Neoseiulus longispinosus* (Evans) (Schicha and Gutierrez, 1985; Kongchuensin *et al.*, 2005), *Euseius nicholsi* (Ehara *et* Lee) (= *Amblyseius nicholsi* Ehara *et* Lee) (Ehara *et* Lee) (Vatana *et al.*, 2001), *E. ovaloides* (Blommers), *Proprioseiopsis peltatus* Van der Merwe, *Phytoseius hawaiiensis* Prasad, *P. hongkongensis* Swirski *et* Shechter, *P. rubiginosae* Schicha, *Papuaseius dominiquae* (Schicha *et* Gutierrez) and *Typhlodromips deleoni* (Muma *et* Denmark) (Schicha and Gutierrez, 1985).

Symptomatology and damage

T. fijiensis is common on coconut in India, and does not cause severe damage (Gupta and Gupta, 1994), although Sarkar and Somchoudhury (1989a) have reported otherwise. However, the paucity of reports on control indirectly confirms the opinion of the first authors. The mite has also been recorded on *Citrus* spp., but the symptoms of attack on this crop are unknown (Vacante, 2010). It usually feeds on the lower leaf surface, and on betelnut causes yellow speckles that later coalesce and give the whole leaf a yellow colour (Daniel, 1977).

Control

Control currently does not seem to be necessary on infested crops, such as citrus, etc. However, various acaricides have been tested against outbreaks of the mite, including binapacryl, chinomethionate, cyhexatin, dicofol, fenbutatin oxide, prothoate and triazophos (Dzolkhifli and Khoo, 1989).

Tetranychus gloveri Banks (Fig. 20.51)

Common name

Cotton spider mite, Glover's spider mite.

Diagnostic characteristics

FEMALE. The body is carmine in colour, and the dorsal setae are finely serrate, long and slender. The dorsal striae are lobate; the striae between the setae e_1–e_1 and f_1–f_1 are longitudinal, and those between the setae e_1 and f_1 form a diamond shape. Between the genital area and the setae *1a* the ventral striae are lobate, and the pregenital striae are broken. The empodia are rayed, with 6 proximoventral hairs, each with a strong dorsal spur. Tarsus I has tactile setae proximal to the posterior duplex setae (Jeppson *et al.*, 1975; Seeman and Beard, 2011).

MALE. The male is smaller than the female and lighter. Empodia I–IV have a dorsal spur; the empodia I are claw like, and the other empodia end in ventrally directed hairs. The knob of the aedeagus is indented dorsally, with a globular anterior projection and a pointed posterior projection (Jeppson *et al.*, 1975; Seeman and Beard, 2011).

Geographical distribution

American Samoa (Gutierrez, 1978), Australia (?) (Seeman and Beard, 2011), Bermuda (Nyiira, 1982), Brazil (Bondar, 1930), Colombia (Urueta, 1975), Costa Rica (Ochoa *et al.*, 1991a), Cuba (Livshits and Salinas-Croche, 1968), French Polynesia (Navajas *et al.*, 1996), Greece (Bolland and Beron, 2004), Guadeloupe (Gutierrez and Etienne, 1988; Flechtmann *et al.*, 1999a; Flechtmann and Etienne, 2003, 2006), Guam Island (McGregor, 1950), Hawaiian Islands (Garrett and Haramoto, 1967), Honduras (Ochoa *et al.*, 1991a), Northern Marianas

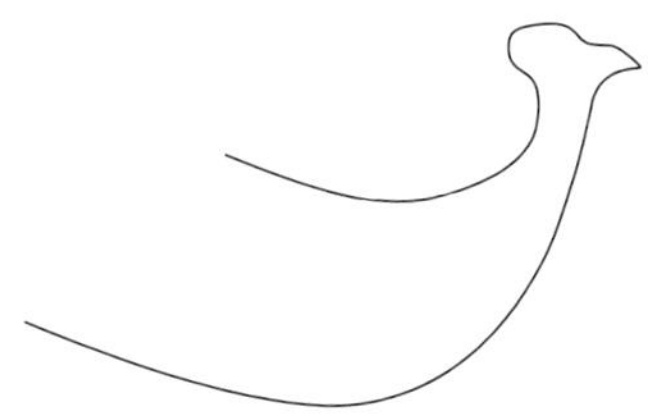

Fig. 20.51. *Tetranychus gloveri* Banks. Male, aedeagus (from Jeppson *et al.*, 1975).

(Gutierrez, 1980), Martinique (France) (Flechtmann *et al.*, 1999a), Mexico (Estebanes-Gonzalez and Baker, 1968), Panama (Baker and Pritchard, 1962), Paraguay, Puerto Rico (Banks, 1917), Suriname (Bolland *et al.*, 1998), Trinidad and Tobago (Byrne *et al.*, 1983), the USA (Banks, 1900) and Venezuela (Doreste, 1968b).

Host plants

The whole list of host plants of the cotton spider mite includes about 110 host plants from 40 families, with the highest number of host plants in the Fabaceae (15) and Malvaceae (11) (Migeon and Dorkeld, 2006–2013). The families and species of economic importance are:

Anacardiaceae: mango (*Mangifera indica* L.) (Bolland *et al.*, 1998; Migeon and Dorkeld, 2006–2013).

Apiaceae: wild celery (*Apium graveolens* L.) (Jeppson *et al.*, 1975), carrot (*Daucus carota* L.) (Bolland *et al.*, 1998; Migeon and Dorkeld, 2006–2013).

Arecaceae: coconut palm (*Cocos nucifera* L.) (Flechtmann *et al.*, 1999a).

Asteraceae: red dahlia (*Dahlia coccinea* Cav.) (Jeppson *et al.*, 1975).

Brassicaceae: cabbage (*Brassica oleracea* L.) (Flechtmann *et al.*, 1999a).

Caricaceae: papaya (*Carica papaya* L.) (Paschoal, 1970).

Convolvulaceae: sweet potato (*Ipomoea batatas* (L.) Lam.) (Livshits and Salinas-Croche, 1968).

Cucurbitaceae: chayote (*Sechium edule* (Jacq.) Sw.) (Bolland *et al.*, 1998; Migeon and Dorkeld, 2006–2013).

Euphorbiaceae: cassava (*Manihot esculenta* Crantz) (Yaseen and Bennett, 1977), *Ricinus communis* L. (Flechtmann *et al.*, 1999a).

Lauraceae: avocado (*Persea americana* Mill.) (Jeppson *et al.*, 1975).

Fabaceae: sieva bean (*Phaseolus lunatus* L.), kidney bean (*P. vulgaris* L.) (Bondar, 1930), pea (*Pisum sativum* L.) (Jeppson *et al.*, 1975), cowpea (*Vigna unguiculata* (L.) Walp.) (Paschoal, 1970).

Malvaceae: okra (*Abelmoschus esculentus* (L.) Moench) (Livshits and Salinas-Croche, 1968), Creole cotton (*Gossypium barbadense* L.), Arabian cotton (*Gossypium herbaceum* L.) (Paschoal, 1970), upland cotton (*G. hirsutum* L.) (Doreste, 1968b), *Gossypium* sp. (Livshits and Salinas-Croche, 1968).

Musaceae: banana (*Musa* x *paradisiaca* L.) (Livshits and Salinas-Croche, 1968).

Poaceae: maize (*Zea mays* L.) (Bolland *et al.*, 1998; Migeon and Dorkeld, 2006–2013).

Solanaceae: aubergine (*Solanum melongena* L.) (Jeppson *et al.*, 1975), Irish potato (*S. tuberosum* L.) (Bolland *et al.*, 1998; Migeon and Dorkeld, 2006–2013).

Bio-ecology

T. gloveri and *T. tumidus* Banks (see below) were first synonymized by Pritchard and Baker (1955). The two species were later separated by Boudreaux (1958b, 1979). The Australian records of *T. gloveri* (Bolland *et al.*, 1998; Halliday, 2000; Flechtmann and Knihinicki, 2002), *T. tumidus* (Womersley, 1942, as *Septonychus*; Baker and Tuttle, 1994) and *Allonychus braziliensis* (McGregor) (Pritchard and Baker, 1955; Jeppson *et al.*, 1975) are *Oligonychus digitatus* Davis. The first record of *T. gloveri* in Australia is dated early December 2008, when it was reported on spider lilies; in April 2009, its presence was eradicated, and other specimens have not been collected since (Seeman and Beard, 2011), so it may not exist in Australia. A major difference between *T. gloveri* and *T. tumidus* is that *T. gloveri* always lays red eggs, while mated females of *T. tumidus* lay white eggs and unmated females deposit red eggs (Boudreaux, 1958b).

The cotton spider mite lives in tropical and subtropical climates, and infests aubergine, carrot, cassava, chayote, cotton, celery, beans, beets, maize, okra, peas and Irish and sweet potato, where it causes rusty speckling and blotches on the leaves and the eventual death of the host. The mite also infests some tropical fruit plants, such as avocado, banana, coconut, mango and papaya, and several ornamentals. Despite the wide list of host plants and its reputation as a severe pest, the biology of *T. gloveri* is basically unknown.

Symptomatology and damage

The feeding of *T. gloveri* causes a reddening of the upper leaf surface. On the damaged leaf surface, there may appear either a small blotch or many blotches that often extend over the whole leaf surface. Infested plants may defoliate (Jeppson *et al.*, 1975). Despite the mite being reported as a major pest of various herbaceous plants, the damage that it causes has been properly investigated only for cotton. Feeding by the mite on cotton causes leaf senescence. Modest levels of attack reduce the number of seeds per ball, and induce a decrease in the weight, viability and oil content. A reduction in the length and maturity of the fibres and the lint index has also been observed. Infestation by *T. gloveri* can decrease crop yield by 45% (Roussel *et al.*, 1951).

Control

No reports on the control of this mite are available.

Tetranychus kanzawai Kishida (Fig. 20.52)

Common name

Kanzawa spider mite, tea red spider mite.

Diagnostic characteristics

FEMALE. The body is carmine in colour, with long and slender dorsal body setae. The opisthosomal dorsal striae are lobate;

Fig. 20.52. *Tetranychus kanzawai* Kishida. Male, aedeagus (from Jeppson *et al.*, 1975).

between the setae e_1–e_1 and f_1–f_1 the striae are longitudinal and oblique, and those between the setae e_1 and f_1 are transverse and form a diamond shape. The ventral striae lack lobes; the pregenital striae are usually entire, sometimes weak with small breaks medially. Tarsus I has 4 tactile setae proximal to the posterior duplex setae. The empodia bear 6 hairs, and the spur is minute or lacking (Jeppson *et al.*, 1975; Seeman and Beard, 2011).

MALE. The male is smaller than the female and lighter. Empodia I–II bear a dorsal spur, and empodia I are claw like; empodia II–IV have long and free proximoventral hairs. The knob of the aedeagus is dorsally convex, with the anterior projection rounded and the posterior acutely angled (Jeppson *et al.*, 1975; Seeman and Beard, 2011).

Geographical distribution

Australia (Davis, 1968b; Gutierrez and Schicha, 1983; Seeman and Beard, 2011), China (Ehara and Lee, 1971; Wang, 1981a; Wang *et al.*, 1981, 1985), Colombia (Flechtmann, 1996), Congo (Gutierrez and Bonato, 1994; Navajas *et al.*, 1996), Greece (Hatzinikolis, 1968c, 1969c, 1986), Japan (Kishida, 1927; Pritchard and Baker, 1955; Ehara, 1956b, 1960, 1966a, 1975a; Jeppson *et al.*, 1975; Gotoh *et al.*, 2003a; Takafuji and Hinomoto, 2008; Matsuda *et al.*, 2013), Korea (South) (Lee *et al.*, 1987; Lee, 1989; Takafuji and Hinomoto, 2008), India (Nassar and Ghai, 1981; Gupta and Gupta, 1994; Gupta and Chatterjee, 1997; Seeman and Beard, 2011), Indonesia (Flechtmann, 1981a; Ehara, 2004), Malaysia (Ehara and Tho, 1988), Mexico (Estebanes-Gonzalez and Baker, 1968; Tuttle *et al.*, 1974; Estebanes-Gonzalez and Rodriguez-Navarro, 1991), Papua New Guinea (Schicha and Gutierrez, 1985), Philippines (Rimando, 1962a; Takafuji and Hinomoto, 2008), South Africa (Meyer Smith, 1974, 1987), Taiwan (Ehara, 1969a; Tseng, 1990; Takafuji and Hinomoto, 2008), Thailand (Baker, 1975b; Ehara and Wongsiri, 1975; Takafuji and Hinomoto, 2008), the USA (Pritchard and Baker, 1955; Reeves, 1963; Baker and Tuttle, 1994), Vietnam (Hinomoto *et al.*, 2007; Takafuji and Hinomoto, 2008).

Host plants

This species is recorded on 183 wild and cultivated plants from 60 families, among which Fabaceae and Rosaceae have the highest number of species (24 and 19, respectively) (Migeon and Dorkeld, 2006–2013). The families and species of economic importance are:

Annonaceae: sugar apple (*Annona squamosa* L.) (Ho *et al.*, 1997).

Asteraceae: sunflower (*Helianthus annuus* L.) (Lee, 1989).

Cannabaceae: hop (*Humulus lupulus* L.) (Gotoh *et al.*, 1999).

Caricaceae: papaya (*Carica papaya* L.) (Ehara, 1966a; Ehara and Wongsiri, 1975; Schicha and Gutierrez, 1985; Gotoh *et al.*, 1996b, 1999, 2003a; Ho *et al.*, 1997; Takafuji *et al.*, 2001; Matsuda *et al.*, 2013).

Caryophyllaceae: carnation (*Dianthus caryophyllus* L.), baby's-breath (*Gypsophila paniculata* L.) (Ho *et al.*, 1997).

Convolvulaceae: sweet potato (*Ipomoea batatas* (L.) Lam.) (Ho *et al.*, 1997).

Cucurbitaceae: watermelon (*Citrullus lanatus* (Thunb.) Matsum. and Nakai) (Ehara, 1975a; Ho *et al.*, 1997), cantaloupe (*Cucumis melo* L.) (Tuttle *et al.*, 1974; Ho *et al.*, 1997; Ohno *et al.*, 2010), cucumber (*C. sativus* L.), bottle gourd (*Lagenaria siceraria* (Molina) Standl), ribbed gourd (*Luffa acutangula* (L.) Roxb.) (Ho *et al.*, 1997).

Ebenaceae: Japanese persimmon (*Diospyros kaki* L.) (Ehara, 1975a; Takafuji *et al.*, 2001).

Euphorbiaceae: poinsettia (*Euphorbia pulcherrima* Willd. ex Klotzsch) (Pritchard and Baker, 1955), manihot (*Manihot carthaginensis* (Jacq.) Müll. Arg.) (Ehara and Tho, 1988), cassava (*M. esculenta* Crantz) (Ehara, 1966a, 1969a, 2004; Flechtmann, 1981a; Schicha and Gutierrez, 1985; Ehara and Tho, 1988; Tseng, 1990; Gutierrez and Bonato, 1994; Navajas *et al.*, 1996; Takafuji and Hinomoto, 2008).

Fabaceae: groundnut (*Arachis hypogaea* L.) (Ho *et al.*, 1997), soybean (*Glycine max* (L.) Merr.) (Ehara, 1963b, 1975a; Baker, 1975b; Ho *et al.*, 1997), sieva bean (*Phaseolus lunatus* L.) (Tseng, 1990; Ho *et al.*, 1997), kidney bean (*P. vulgaris* L.) (Pritchard and Baker, 1955; Ehara, 1975a; Tseng, 1990; Ho *et al.*, 1997), garden pea (*Pisum sativum* L.) (Lee *et al.*, 1999), *Trifolium pratense* L. (Hinomoto and Takafuji, 2001; Takafuji *et al.*, 2001; Matsuda *et al.*, 2013), *T. repens* L. (Lee, 1989; Lee *et al.*, 1999; Matsuda *et al.*, 2013), fava bean (*Vicia faba* L.) (Ho *et al.*, 1997), azuki bean (*V. angularis* (Willd.) Ohwi and H. Ohashi) (Gotoh *et al.*, 1999, 2003a), cowpea (*Vigna unguiculata* (L.) Walp.) (Gotoh *et al.*, 1996b, 2003a; Ho *et al.*, 1997; Takafuji and Hinomoto, 2008; Ohno *et al.*, 2009).

Lythraceae: pomegranate (*Punica granatum* L.) (Tseng, 1990).

Moraceae: white mulberry (*Morus alba* L.) (Kishida, 1927; Ehara, 1956b, 1966a; Lee, 1989; Hinomoto and Takafuji, 2001; Gotoh *et al.*, 2003a; Ohno *et al.*, 2010, 2011a).

Oxalidaceae: carambola (*Averrhoa carambola* L.) (Ho *et al.*, 1997).

Passifloraceae: passion fruit (*Passiflora edulis* Sims) (Ho *et al.*, 1997).

Pedaliaceae: sesame (*Sesamum indicum* L.) (Ho *et al.*, 1997).

Poaceae: sugarcane (*Saccharum officinarum* L.), sorghum (*Sorghum bicolor* (L.) Moench) (Baker, 1975b), maize (*Zea mays* L.) (Ehara and Lee, 1971; Baker, 1975b; Tseng, 1990; Ho *et al.*, 1997).

Rhamnaceae: jujube (*Ziziphus jujuba* Mill.) (Ho *et al.*, 1997).

Rosaceae: loquat (*Eriobotrya japonica* (Thunb.) Lindl.) (Ho *et al.*, 1997), strawberry (*Fragaria* × *ananassa* Duch.) (Ehara, 1975a; Hinomoto and Takafuji, 2001; Takafuji and Hinomoto, 2008; Seeman and Beard, 2011), apple (*Malus domestica* Borkh.) (Pritchard and Baker, 1955; Ehara, 1956b, 1963b, 1975a; Tseng, 1990; Lee *et al.*, 1999; Hinomoto and Takafuji, 2001; Takafuji *et al.*, 2001), apricot (*Prunus armeniaca* L.) (Gupta and Gupta, 1994; Gupta and Chatterjee, 1997), sweet cherry (*P. avium* (L.) L.) (Ehara, 1963b, 1975a), peach (*P. persica* (L.) Batsch) (Ehara, 1956b, 1960, 1975a; Tseng, 1990; Gupta and Gupta, 1994; Gotoh *et al.*, 1996b, 1999; Gupta and Chatterjee, 1997; Ho *et al.*, 1997; Hinomoto and Takafuji, 2001; Takafuji and Hinomoto, 2008), pear (*Pyrus communis* L.) (Navajas *et al.*, 2001b; Seeman and Beard, 2011), Chinese pear (*P. pyrifolia* (Burm. f.) Nakai) (Ehara, 1956b, 1960, 1975a; Gotoh *et al.*, 1999, 2003a; Takafuji *et al.*, 2001; Takafuji and Hinomoto, 2008), English rose (*Rosa hybrida*) (Baker, 1975b; Lee, 1989; Tseng, 1990; Lee *et al.*, 1999; Hinomoto *et al.*, 2007).

Rutaceae: *Citrus* sp. (Ehara, 1960, 1975a), lemon (*C. limon* (L.) Burm.f.) (Estebanes-Gonzalez and Baker, 1968), pummelo (*C. maxima* (Burm. f.) Merr.), Japanese bitter orange (*C. trifoliata* L.) (Lee, 1989), orange jasmine (*Murraya paniculata* (L.) Jack) (Tseng, 1990), Japanese pepper (*Zanthoxylum piperitum* (L.) DC.) (Osakabe *et al.*, 2002).

Solanaceae: pepper (*Capsicum annuum* L.) (Tseng, 1990; Estebanes-Gonzalez and Rodriguez-Navarro, 1991; Ho *et al.*, 1997), aubergine (*Solanum melongena* L.) (Ehara, 1963b, 1975a; Ho *et al.*, 1997; Hinomoto and Takafuji, 2001; Osakabe *et al.*, 2002; Hinomoto *et al.*, 2007; Takafuji and Hinomoto, 2008; Ohno *et al.*, 2009).

Theaceae: tea (*Camellia sinensis* (L.) Kuntze) (Hotta, 1928; Ehara, 1956b, 1960, 1966a; Gotoh *et al.*, 1999, 2003a; Hinomoto and Takafuji, 2001; Navajas *et al.*, 2001b; Osakabe *et al.*, 2002; Takafuji and Hinomoto, 2008; Matsuda *et al.*, 2013).

Vitaceae: grapevine (*Vitis vinifera* L.) (Ehara, 1956b, 1975a; Ehara and Wongsiri, 1975; Ho *et al.*, 1997).

Bio-ecology

The Kanzawa spider mite is one of the most common spider mites in the East Asian region (Takafuji and Hinomoto, 2008). It lives on wild and cultivated plants, among which the economic plants include various vegetables, such as aubergine, beans, cucumber, gourds, pea, pepper and sweet potato. The mite also develops on various industrial and food crops, including cassava, cotton, groundnut, hop, Japanese pepper, jujube, maize, sesame, soybean, sugarcane, sunflower and tea. *T. kanzawai* occurs on fruiting plants, including some tropical species; examples include apple, apricot, cantaloupe, carambola, Chinese pear, pear, Japanese bitter orange, Japanese persimmon, lemon, loquat, orange jasmine, papaya, passion fruit, peach, pomegranate, pummelo, strawberry, sugar apple, sweet cherry, white mulberry, watermelon and wine grapes. The Kanzawa spider mite has also been collected on cereals and many ornamentals. However, despite its wide list of host plants, the mite is a major pest of only a restricted number of plants in Japan, southern China and Korea. Damage occurs to tea in various Asian countries (Ehara, 1960; Osakabe, 1962, 1965, 1967; Banerjee and Cranham, 1985). The mite infests cherry in China (Wang, 1981a), aubergine (Ho and Chen, 1992b) and papaya in Taiwan (Cheng *et al.*, 2009), and grapevines in Taiwan (Ho and Chen, 1994) and Japan (Ehara, 1964a; Ashihara, 1996). The mite was also recorded on strawberry in Taiwan (Chang and Huang, 1995) and cassava in the Congo (Gutierrez and Bonato, 1994) and the Philippines (Villacarlos and Vasquez, 1988). Populations of *T. kanzawai* most often live outside, but they may also occur in greenhouses, where they have been reported damaging grapevines in Japan (Ashihara *et al.*, 1992).

The Kanzawa spider mite lives mainly under leaf surfaces and on the stems of host plants, where it feeds and produces a dense web. It reproduces by arrhenotokous parthenogenesis, and the haploid number of chromosomes is three (Helle and Bolland, 1967). The tetranychid is infected by the endosymbiont *Wolbachia*, a group of maternally inherited bacteria infecting arthropods and responsible in spider mites for cytoplasmic incompatibility (Breeuwer, 1997), but it has not been found in Japanese populations of *T. kanzawai* (Gotoh *et al.*, 2003a). The proportion of females ranges from 0.76 to 0.83, and the sex ratio is determined by the genotype of the mothers, and appears to be controlled by several genes, but without the involvement of cytoplasmic genes (Takafuji and Ishii, 1989). The males probably use odours to discriminate the mating status of females. These abilities of males may play an important role in them gaining access to virgin females. However, the behaviour of adult females seems to vary with their mating status. For instance, virgin females are more gregarious and remain on infested kidney leaves for longer than mated females, and such behaviour is likely to increase the mating opportunities of virgin females (Oku *et al.*, 2005). Quiescent deutonymph females attract males, and when more males attempt to guard one female, they may become combative; the winner male then guards the female, so females could in effect choose their mates indirectly as a result of male combat during pre-copulatory mate guarding (Oku, 2010). *T. kanzawai* spins webs over leaf surfaces and usually lives under these webs. The tetranychid belongs to the subtype CW-u, in which a complex and irregular web is produced on the leaf surface (Saitō, 1985). The mite produces two types of excreta, black and yellow pellets, and uses its webs as a place to deposit these. The mite also uses the webs as refuges to escape predation by the phytoseiid *Neoseiulus womersleyi* (Schicha). It has been found that the black pellets reduce the fecundity of *T. kanzawai* adult females and the yellow faecal pellets reduce their predation by *N. womersleyi*. In this way, *T. kanzawai* avoids any negative effect on its own fitness and reduces the risk of predation (Oku, 2008).

At 15, 20, 25 and 30°C, egg development took 15.1, 8.9, 4.4 and 2.3 days, respectively, the larval period was 6.4, 3.5, 1.9 and 1.2 days, the protonymphal period 5.6, 2.8, 1.7 and 1.0 days, and the deutonymphal period 7.5, 4.0, 2.1 and 1.3 days, respectively. The corresponding adult female lifespans were 33.3, 16.8, 15.5 and 13.4 days, and the average number of eggs laid per female were 37.5, 59.4, 100.6 and 103.3. The developmental threshold temperatures for the egg, protonymphal and deutonymphal stages were 13.9, 12.6 and 12.6°C, respectively, and the corresponding temperature sums for development were 39.2, 21.4 and 18.2 degree-days (Kim *et al.*, 1993b). Following Osakabe (1967), the theoretical zero points for the development of eggs, larvae, protonymphs and deutonymphs are 8.7, 14.6, 13.4 and 13.2°C.

On tea at 15, 20, 25 and 30°C, the intrinsic rate of increase (r_m) was 0.0443, 0.1079, 0.2149 and 0.2764, respectively (Tsai *et al.*, 1989). At the same temperatures, Kim *et al.* (1993b) reported values of 0.062, 0.134, 0.252 and 0.371, respectively. The net reproductive rate (R_0) ranged from 10.85 female offspring per female at 15°C to 31.05 at 30°C, and the finite rate of increase (λ) from 1.0453 to 1.3183. The mean generation time (T) ranged from 12.4 days at 30 to 53.9 days at 15°C (Tsai *et al.*, 1989). The intrinsic rates of natural increase (r_m) varied on different host plants from 0.187 a day to 0.283 a day; on tea, hydrangea (*H. macrophylla*) and Japanese pear, the rates were 0.043–0.096 a day lower than on mulberry (Gotoh and Gomi, 2003).

The first studies on the influence of temperature and humidity showed that successful hatching occurred at 15–30°C and 46–93% RH. An RH of 20–100% has a negative effect on egg hatching and hinders the development of immature stages.

Recent studies on the influence of various temperatures and humidities showed that at 35°C and RH 60% the development of one generation took 6.23 ± 0.44 days, while at 30°C and RH 65%, it took 10.89 ± 0.5 days. At 25°C and RH 70%, the development of one generation takes 9.16 6 ± 1.14 days, at 20°C and RH 75%, it takes 15.92 ± 1.1 days, and at 15°C and RH 80%, it takes 27.49 ± 2.23 days. The mite may develop 13–16 generations during a year in Jiangxi Province, China. The starting temperature for development is 9.95 ± 0.79°C. The total effective accumulated temperature for one generation is 166 ± 3.773 degree-days (Cao *et al.*, 1998).

On tea, the mite overwinters as diapausing females, but in southern Japan, all stages may occur throughout the year, including diapausing and non-diapausing females (Osakabe, 1962, 1965; Uchida, 1980). The results of studies on the effects of host-plant species on diapause induction of the Kanzawa spider mite show that the relationship between diapause incidence and performance is consistent with the prediction of theoretical studies that a short development time or high fecundity delays the timing of diapause induction (Ito and Saitō, 2006). The population density varies with season, and high population densities are observed from March to June and from September to December in Japan, with variations according to climate (Osakabe, 1965). On the same host plant, the density of *T. kanzawai* may peak in mid-June and then decline (Mochizuki, 2002). The population dynamics of *T. kanzawai* on hydrangea in Japan exhibit two types of seasonal population trends. The first type has a population peak from May to June; the second type has a spring peak in June and an autumn peak in September–October. Populations decline after the spring peak. The predators show a delayed density-dependent numerical response, but the populations decrease even in the absence of predators, and do not affect the decrease of populations observed in June. Furthermore, the rate of development from larva to adult and the fecundity of females decrease just before the density decrease in June. Seasonal changes in plant quality seem to affect the decrease of *T. kanzawai* density that is observed in June (Gotoh and Gomi, 2000).

Natural enemies include predatory insects and mites, and are listed in Table 20.9.

Amano (2001) discussed the species structure and abundance of invertebrate natural enemies in sustainable agroecosystems, and reported that on Japanese pear the most abundant phytoseiids in descending order were *Amblyseius californicus* (McGregor) (= *Neoseiulus californicus* (McGregor)) and *A. womersleyi* (= *N. womersleyi*), while on tea they were

Table 20.9. *Tetranychus kanzawai* Kishida. Natural enemies (insects and mites).

Families and species	References
INSECTS	
Aeolothripidae	
Scolothrips indicus Priesner	Ho and Chen, 1998
Scolothrips sexmaculatus (Pergande)	McMurtry *et al.*, 1970
Scolothrips takahashii Priesner	Morishita, 2000
Anthocoridae	
Anthocoris sp.	McMurtry *et al.*, 1970
Orius sp.	
Cecidomyiidae	
Feltiella sp.	McMurtry *et al.*, 1970
Feltiella minuta (Felt)	Ho and Chen, 1998
Coccinellidae	
Parastethorus histrio (Chazeau)	Houston, 1980
Stethorus fenestralis Houston	
Stethorus japonicus Kamiya	Mori *et al.*, 2005
Staphylinidae	
Oligota flavicornis (Boisduval *et* Lacordaire)	Ho and Chen, 1998
Oligota kashmirica benefice Naomi	McMurtry *et al.*, 1970
MITES	
Phytoseiidae	
Amblyseius eharai Amitai *et* Swirski	Morishita, 2000
Euseius sojaensis (Ehara)	Osakabe *et al.*, 1987; Morishita, 2000
Neoseiulus californicus (McGregor)	Takada and Kashio, 2004
Neoseiulus fallacis (Garman)	Ho, 1990
Neoseiulus longispinosus (Evans)	Nakagawa, 1984; Nagatomo *et al.*, 1991; Ho and Chen, 1998; Kongchuensin *et al.*, 2005
Neoseiulus womersleyi (Schicha)	Shih, 1985; Kim and Paik, 1996a
Phytoseiulus persimilis Athias Henriot	Zhang *et al.*, 1996a,b
Stigmaeidae	
Agistemus fleschneri Summers	Osakabe (1963)

N. womersleyi and *N. californicus* and on hydrangea *A. eharai* Amitai *et* Swirski and *A. womersleyi*, respectively.

Symptomatology and damage

The Kanzawa spider mite infests leaves, stems and fruit (Ho and Chen, 1994; Ashihara, 1996; CABI, 2007). Feeding on leaves causes necrotic areas. The mites and webbing usually occur on the under leaf surfaces. Damaged leaves show yellowish spots, and at high levels of attack the leaves become yellow and dry. Infestation also causes discoloration of the bark and webbing on the stems. On tea, the mite infests the young shoots, inducing direct crop loss (Banerjee and Cranham, 1985).

The mite causes damage to tea in various Asian countries (Ehara, 1960; Osakabe, 1962, 1965, 1967; Banerjee and Cranham, 1985), and damage to aubergine (Ho and Chen, 1992b) and papaya in Taiwan (Cheng *et al.*, 2009), and to grapes in Taiwan (Ho and Chen, 1994) and Japan (Ehara, 1964a; Ashihara, 1996). On strawberries in China, the levels of infestation may reach 90–100%, with 2000–3000 mites per leaf, whereas on grapes in Taiwan 10% of grape clusters were infested by a low density of mites per cluster. The population density varies between grape cultivars (Ashihara, 1996), and some cultivars induce high developmental success of the mites. The Kanzawa spider mite is a minor pest of citrus, on which the symptoms and damage are unclear, and also on other cultivated plants. The r_m values seem to suggest that tea, hydrangea and Japanese pear are poor-quality plants for *T. kanzawai* compared with their performance on mulberry (Gotoh and Gomi, 2003). On tea, the economic injury level (EIL) in June is 20% of infested new leaves (Mochizuki, 2002).

Control

Control of the Kanzawa spider mite is provided by releases of natural enemies and/or spray applications of different acaricides. The results of biological control programmes in various countries appear to be good, as are those from the growing of resistant cultivars.

PLANT RESISTANCE. According to varying susceptibility of different tea varieties to mite, it has been suggested that resistant varieties could be grown (Ehara, 1960; Osakabe, 1965). The same possibility occurs on grape, in which the use of different cultivars may significantly affect the population density of *T. kanzawai*. Thus, high populations have been found on ‘Muscat Bailey A’, while only 25% of the larvae develop to adults on ‘Kychou’, 2% develop on ‘Muscat of Alexandria’ and there is no developmental success on ‘Neo Muscat’ (Ashihara, 1996).

BIOLOGICAL CONTROL. Biological control is provided by the release of both mite and insect predators. However, recently, the action of the entomopathogenic fungi *Beauveria bassiana* (Balsamo) Vuillemin and *Metarhizium anisopliae* (Metchnikoff) Sorokin was investigated under laboratory conditions, and their transmission to a population of *T. kanzawai* confirmed as a result of contact with an infected cadaver or any residue left behind on the substrate (Sanjaya *et al.*, 2013).

Different species of predatory insects were tested for biological control of *T. kanzawai* on aubergine crops that had not been sprayed with chemicals (Ho and Chen, 1998). The feeding response of the larvae of *Oligota flavicornis* (Boisduval *et* Lacordaire) to the daily supply of 10, 20, 40, 80 and 160 eggs of *T. kanzawai* was studied at 28°C and under a photoperiod of 13L:11D. One hundred and sixty spider mite eggs were considered to be the food amount to attract both the adult female and the larvae of this beetle to feed and oviposit normally (Ho and Chen, 2002). Adult females of *Scolothrips indicus* Priesner on average consumed 650 spider mite eggs during their lifespan, with a daily consumption of 57 spider mite eggs (Ho and Chen, 2001). On aubergine crops in Taiwan, populations of this thrip fluctuated with the population of *T. kanzawai*, and it was found mostly on leaves containing a few hundred spider mites. The green lacewing, *Mallada basalis* (Walker), also plays a significant role in the biological control of *T. kanzawai*. Releasing eggs or larvae of *M. basalis* at 3-week intervals suppressed 60–90% of the *T. kanzawai* population on strawberry in Taiwan. The implementation of this biological control saved *c.*US$233 ha^{-1} in control costs, and increased fruit production by 15% and that of first grade fruit by 7.7% (Chang and Huang, 1995). Moreover, the larvae of the green lacewing have high potential for augmentative biological control of mites on papaya in Taiwan (Cheng *et al.*, 2009). Recently, it has been observed that *M. basalis* has the potential for substantially suppressing populations of both *T. kanzawai* and *P. citri* on papaya at a predator:prey ratio of 1:15 or greater. However, evaluation under realistic agricultural settings is needed before specific recommendations about predator release rates can be made (Cheng *et al.*, 2012).

For predatory mites, Yang *et al.* (1991) reported that *Phytoseiulus persimilis* Athias Henriot was highly effective against spider mites in China. Releases of predators in June controlled mite infestation on grapevines in a greenhouse in Japan in plots where there were no long-term toxicity effects from pesticide spraying (Ashihara *et al.*, 1992). The release of two predators per plant in mid-March on strawberry in China suppressed *T. kanzawai* during the rest of the growing season (Zhang *et al.*, 1996a,b). In Taiwan, *P. persimilis* and *N. fallacis* (Garman) both controlled *T. kanzawai* infestation on papaya. However, *P. persimilis* was more effective than *N. fallacis* when two or three releases (= 500 predators per plant) were applied to the lower leaves (Hao *et al.*, 1996). *N. fallacis* and *P. persimilis* gave both effective control of *T. kanzawai* after release on to potted tea plants at a predator to prey ratio of 1:20 in a greenhouse in Taiwan (Ho, 1990). Mass releases of *N. womersleyi* brought about significant reductions in tetranychid populations in mulberry fields in Taiwan when their initial densities were less than or equal to 22 mites per leaf (Shih, 1985). In addition, synthetic pyrethroid (SP)-resistant *N. womersleyi* in tea fields under SP application was studied in Japan. Resistant *N. womersleyi* suppressed mite damage to new leaves, and a positive correlation was found between damage to new leaves at plucking and infestation by *T. kanzawai* when new buds sprouted. Acaricide applications were harmless to *N. womersleyi* and release or integration of predatory mites prevented damage on new tea leaves by *T. kanzawai* (Mochizuki, 2002).

CHEMICAL CONTROL. The application of chemical control depends on the crops. On citrus, no control measures are generally needed, or they are only rarely applied. On other crops, such as tea, papaya, ornamentals, etc., sprays of acaricides may be applied. These include

avermectins (Kim and Paik, 1996b), diphenyl carbinols, diphenyl ether (benzomate) (Wang and Liu, 1993), organophosphates (Ozawa, 1995), sulfur-bridged compounds (Wang and Liu, 1993; Tatara *et al.*, 1999) or synthetic pyrethroids (Ozawa, 1995). The effectiveness of chlorfenapyr/tebufenpyrad, cyenopyrafen/cyflumetofen, etoxazole, hexythiazox and milbemectin was tested recently (Ullah *et al.*, 2011). Different sprayings may be applied to tea from March until before plucking (Banerjee and Cranham, 1985). However, chemical control is hampered both by strains of *T. kanzawai* resistant to different acaricides and by economic cost. Strains of *T. kanzawai* resistant to chemicals have been reported for some time (Osakabe, 1974), and cases of resistance to chlordimeform, dicofol and phenthoate (Kuwahara, 1977), fenpyroximate, pyridaben and pyrazole (Goka, 1998), and cyflumetofen, etoxazole, fenbutatin oxide, spirodiclofen and spiromesifen (Ullah *et al.*, 2011) are well documented. The cost of chemical control of *T. kanzawai* on strawberry in Taiwan exceeds US\$233 ha^{-1} (Chang and Huang, 1995).

Tetranychus lambi Pritchard *et* Baker (Fig. 20.53)

Common name

Strawberry spider mite, banana spider mite.

Diagnostic characteristics

FEMALE. The body is carmine (Pritchard and Baker, 1955), or green or yellowish in colour (Davis 1968b; Gutierrez and Schicha, 1983; Seeman and Beard, 2011). The dorsal body setae are long and slender. The dorsal striae are lobate and transverse and/or oblique between the setae e_1–e_1 and e_1–f_1, and longitudinal between the setae f_1–f_1, forming an hourglass pattern. The ventral striae are lobate (Queensland and New Zealand specimens) or lacking lobes (some Tasmanian specimens); when the ventral striae are lobate, extending from the genital region to *1a*, the pregenital striae are entire and often lobate. Empodia have 6 hairs; spurs on empodia I–IV are minute or absent. The proximal duplex setae of tarsus I are distal to 4 tactile setae (Pritchard and Baker, 1955; Seeman and Beard, 2011).

MALE. Empodia I–IV have a minute dorsal spur; empodia I are claw like and empodia II–IV have long and free proximoventral hairs. The knob of the aedeagus has the dorsal surface more or less flat, with the anterior projection small and acute, and the posterior projection longer, slender and acute (Pritchard and Baker, 1955; Seeman and Beard, 2011).

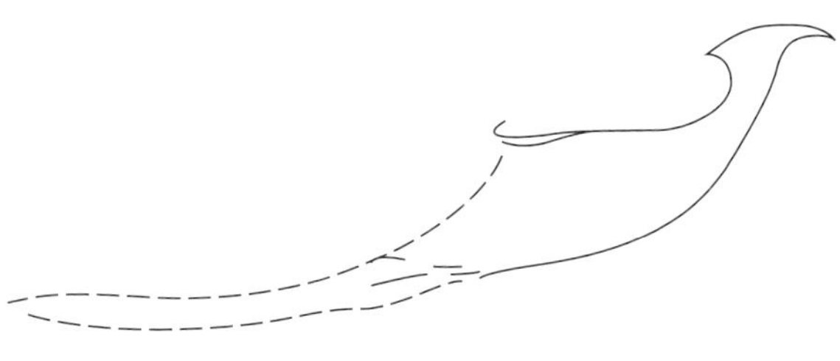

Fig. 20.53. *Tetranychus lambi* Pritchard *et* Baker. Male, aedeagus (from Pritchard and Baker, 1955).

Geographical distribution

Australia (Miller, 1966; Davis, 1968b; Seeman and Beard, 2011), Cook Islands (Gutierrez, 1980), Fiji (Swaine, 1971), French Polynesia (Gutierrez, 1980), Iran (Khalil Manesh, 1973), New Caledonia (Gutierrez, 1977), New Zealand (Pritchard and Baker, 1955; Seeman and Beard, 2011), Papua New Guinea (Schicha and Gutierrez, 1985), American Samoa (Gutierrez, 1977), Western Samoa (Bolland *et al.*, 1998), Taiwan (Baker, 1975a), Tonga (Gutierrez, 1980), Vanuatu (Gutierrez, 1977) and Wallis and Futuna (Gutierrez, 1981).

Host plants

T. lambi is reported on over 60 wild and cultivated plants belonging to 23 families, with the highest number of species (19) in the Fabaceae (Migeon and Dorkeld, 2006–2013). The families and species of economic importance are:

Arecaceae: coconut palm (*Cocos nucifera* L.) (Gutierrez and Schicha, 1984).

Caricaceae: papaya (*Carica papaya* L.) (Davis, 1968b; Seeman and Beard, 2011).

Cucurbitaceae: winter squash (*Cucurbita maxima* Duchesne) (Gutierrez and Schicha, 1983), field pumpkin (*C. pepo* L.) (Bolland *et al.*, 1998; Migeon and Dorkeld, 2006–2013).

Euphorbiaceae: cassava (*Manihot esculenta* Crantz) (Gutierrez and Schicha, 1984; Seeman and Beard, 2011).

Fabaceae: groundnut (*Arachis hypogaea* L.) (Davis, 1968b), soybean (*Glycine max* (L.) Merr.) (Davis, 1968b; Gutierrez and Schicha, 1983; Seeman and Beard, 2011), bean (*Phaseolus vulgaris* L.), red clover (*Trifolium pratense* L.), Dutch clover (*T. repens* L.) (Davis, 1968b).

Malvaceae: upland cotton (*Gossypium hirsutum* L.) (Gutierrez and Schicha, 1983).

Moraceae: white mulberry (*Morus alba* L.) (Davis, 1968b).

Musaceae: banana (*Musa* × *paradisiaca* L.) (Gutierrez and Schicha, 1983).

Rosaceae: strawberry (*Fragaria* × *ananassa* Duchesne ex Rozier) (Pritchard and Baker, 1955; Davis, 1968b; Gutierrez and Schicha, 1983; Seeman and Beard, 2011), apple (*Malus domestica* Borkh.) (Pritchard and Baker, 1955; Seeman and Beard, 2011), peach (*Prunus persica* (L.) Batsch) (Davis, 1968b), raspberry bush (*Rubus rosifolius* Sm.).

Rutaceae: Japanese bitter orange (*Citrus trifoliata* L.) (Davis, 1968b).

Solanaceae: cultivated tobacco (*Nicotiana tabacum* L.) (Davis, 1968b), aubergine (*Solanum melongena* L.) (Gutierrez and Schicha, 1984).

Bio-ecology

T. lambi infests several economic plants (apple, aubergine, banana, beans, cassava, citrus, coconut, cotton, groundnut, Japanese bitter orange, papaya, peach, pumpkin, squash, soybean, strawberry, tobacco, white mulberry). In New South Wales and Queensland (Australia), it is a pest of several leguminous plants, strawberry (Davis and Heather, 1962), banana (New South Wales Department of Agriculture, 1975; Gutierrez and Schicha, 1983; Beard, in Seeman and Beard, 2011) and cotton (Queensland Department of Agriculture and Fisheries, 2012).

The mite reproduces by arrhenotokous parthenogenesis and the haploid number of chromosomes is three (Bolland *et al.*, 1981). Its populations usually live on the lower leaf surface, where they produce a fine web. Unfertilized females of *T. lambi* lay fewer eggs but live longer than inseminated (fertilized) females. The 50% survival rate of inseminated females is attained 13.5 days before that of unfertilized females. Comparison of egg production by fertilized and unfertilized females showed that the fecundity of fertilized females is 1.25 times greater (Bonato and Gutierrez, 1999). The paucity of data on the bio-ecology of *T. lambi* suggests its role as a secondary or occasional pest.

Various predatory insects (Coccinellidae, Staphylinidae, Cecidomyiidae and Aelothripidae) were found associated with *T. lambi* (Chazeau, 1985). The coccinellids include *Parastethorus nigripes* (Kapur) (Houston, 1980), *Stethorus bifidus* Kapur (McMurtry *et al.*, 1970), *S. exspectatus* Chazeau, *S. exsultabilis* Chazeau (Chazeau, 1983), *S. fenestralis* Houston (Houston, 1980), *S. fijiensis* Kapur (Chazeau, 1979), *S. loxtoni* Britton *et* Lee (Houston, 1980) and *S. vagans* (Blackburn) (Chazeau, 1979). The staphylinids include *Oligota flavicollis* (Boisd. *et* Lac.) (Chazeau and Gutierrez, 1977). The aeolothripids include *Scolothrips pallidus* (Beach) and the cecidomyiid *Feltiella acarisuga* (Vallot) (Gutierrez, 1983, in Chazeau, 1985).

Symptomatology and damage

Feeding by *T. lambi* on the upper leaf surface of strawberry plants causes an abnormal and irregular purple colour, usually close to the principal veins. High levels of attack produce a silvering appearance on the entire leaf undersurface, leaf edge rollings, growth retardation and premature ripening of the fruit (Davis and Heather, 1962). On bananas, heavy infestations cause flecking and mottling of the leaves, and the mites also infest the fruit, causing skin blemishes and loss of market value (New South Wales Department of Agriculture, 1975; Gutierrez and Schicha, 1983). Feeding of mites on cotton causes yellowing of the upper leaf surface, and fine webbing occurs on the lower leaf surface (Queensland Department of Agriculture and Fisheries, 2012).

Control

On strawberry and bananas, chemical control may be justified, while on cotton and other economic plants the need for control is rare. Citrus damage is not documented, and control is not justified. The effectiveness of cyhexatin, dicofol, dimethoate, methidathion and profenofos was tested by Brun *et al.* (1983). Currently, various acaricides may be applied and among these, clofentezine has been proposed for bananas when mites first appear, as has bifenthrin.

Tetranychus lintearius Dufour (Fig. 20.54)

Common name

Gorse spider mite.

Diagnostic characteristics

FEMALE. The peritreme is hooked. The dorsal striae lack lobes, and are longitudinal and oblique between the opisthosomal dorsal setae e_1–e_1 and e_1–f_1, transverse between the setae e_1 and f_1, and form a diamond shape. The ventral striae lack lobes, and the pregenital striae are entire. The empodia have 6 proximoventral hairs, and dorsal spurs. Tarsus I has 4 tactile setae proximal to the proximal duplex setae (Seeman and Beard, 2011).

MALE. Empodia I–IV have dorsal spurs; empodia I and II are claw like, and empodia III–IV have free proximoventral hairs. The knob of the aedeagus is small and dorsally slightly convex, with anterior and posterior projections both rounded and small, but the posterior projection larger than the anterior projection (Seeman and Beard, 2011).

Geographical distribution

Algeria (André, 1933a), Australia (Ireson *et al.*, 1999, 2003; Davies *et al.*, 2007; Seeman and Beard, 2011), Belgium (Zacher, 1949), Chile (Norambuena and Escobar, 2004; Norambuena *et al.*, 2007), France (Dufour, 1832; André, 1933a), Hawaiian Islands (Markin *et al.*, 1996; Norambuena and Escobar, 2004; Norambuena *et al.*, 2007), Italy (Canestrini and Fanzago, 1877), New Zealand (Hill *et al.*, 1989; Ireson *et al.*, 2003; Davies *et al.*, 2007), Portugal (Hill and Stone, 1985; Ireson *et al.*, 2003; Norambuena and Escobar, 2004; Norambuena *et al.*, 2007), Spain (Ferragut and Santonja, 1989; Ireson *et al.*, 2003), the UK (Hirst, 1920; Hill and O'Donnell, 1991a; Ireson *et al.*, 2003; Davies *et al.*, 2007) and the USA (Markin *et al.*, 1996; Pratt *et al.*, 2002).

Host plants

The gorse spider mite has been reported on Fabaceae (five species) and on one species of Rutaceae: Fabaceae – spiny-broom (*Calicotome spinosa* (L.) Link) (André, 1933a), *C. villosa* (Poiret) Link, common gorse (*Ulex europaeus* L.) (Dufour, 1832; André, 1933a; Stone, 1982, 1986; Ferragut and Santonja, 1989; Hill and O'Donnell, 1991b; Ireson *et al.*, 1999, 2003; Pratt *et al.*, 2002; Norambuena and Escobar, 2004; Davies *et al.*, 2007; Norambuena *et al.*, 2007; Seeman and Beard, 2011), *U. minor* Roth (Van Eyndhoven, 1967a; Ferragut and Santonja, 1989), Mediterranean gorse (*U. parviflorus* Pourr.) (Ferragut and Santonja, 1989). Rutaceae: *Boronia* sp. (Rutaceae) (Seeman and Beard, 2011).

Bio-ecology

T. lintearius was described by Dufour (1832) on common gorse in France. Pritchard and Baker (1955) placed the species in synonymy with the polyphagous *T. telarius* (L.). However, the monophagy of Dufour's mite on gorse and the production of large webs (and other characters) suggested to Van Eyndhoven (1967a) that

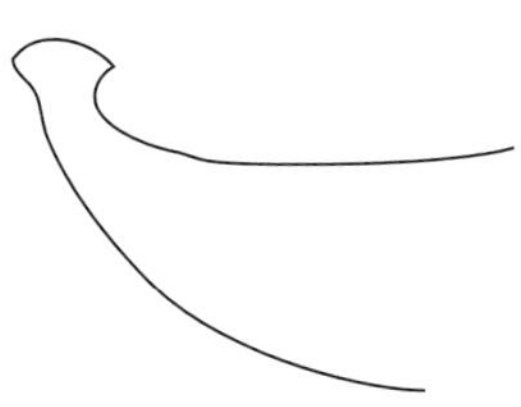

Fig. 20.54. *Tetranychus lintearius* Dufour. Male, aedeagus (drawing by Vacante).

T. lintearius should be re-established. Cross-mating experiments to highlight whether the gorse spider mite was reproductively isolated from *T. urticae* Koch and *T. turkestani* (Ugarov *et* Nikolskii) showed that normal mating might occur, but no females developed from the progeny of any crosses (Hill and O'Donnell, 1991a). In addition, investigations in the field and laboratory with various plants confirmed the mite specificity for *Ulex* (Hill and O'Donnell, 1991b).

The gorse spider mite reproduces by arrhenotokous parthenogenesis and has $n = 3$ chromosomes (Bolland *et al.*, 1981). Colonies of the mite may consist of several thousand motile forms, which spin large amounts of webbing that may extensively envelop many plants. The mite is active throughout the year and develops different generations, e.g. six generations in Britain (Hill and Stone, 1985). Experimental observations on gorse at 23°C have showed development from egg to adult takes 17.8 ± 0.17 days, female longevity is 22.4 ± 4.2 days and females make up 74% of the progeny. Daily egg deposition is 2.3 ± 0.1 eggs, and the fecundity is 52 eggs (Stone, 1982, 1986). A decrease in the quality of gorse as a food resource induces a decline in egg deposition and an increase in its development time. Mean egg deposition at 15°C ranges from 37.8 to 45.2 eggs per female over 10 days and development time from 39.2 to 43.3 days. The density of adult females also significantly affects the number of eggs laid per female over 10 days, and the mean daily egg production per female ranges from 0.78 to 0.88. The mean development time from egg to adult decreases from 40.6 to 39 days as density increases from 1 to 50 mites per spine (Millar, 1993).

Taking into account that *T. lintearius* is a severe pest of gorse in Europe (Zwolfer, 1961; Girling, 1977) and is a specific pest for this host plant (Hill and O'Donnell, 1991a, 1991b), it was released in New Zealand in 1989. Despite its wide dispersal in this country, premature abandonment of the host plant, inability to develop adequate levels of attack and predator activity, e.g. by the coccinellid *Stethorus bifidus* Kapur, may cause adequate control of gorse to fail (Richardson and Hill, 1998). The mite has also been released in Oregon and Hawaii (Markin *et al.*, 1996) and has been established in Tasmania (Australia) (Ireson *et al.*, 1999).

The predation of *T. lintearius* by a complex of predatory mites for the biological control of spider mites on crops is well known. The number of spider mite colonies and volume of gorse foliage colonized are higher when the introduced phytoseiid mite *Phytoseiulus persimilis* Athias Henriot is excluded from plots, showing that this predator is suppressing *T. lintearius* populations (Pratt *et al.*, 2003). Other predators include some coccinellids, such as *Parastethorus histrio* (Chazeau) (Ireson *et al.*, 2003) and *S. bifidus* (Peterson *et al.*, 1994).

Symptomatology and damage

Feeding by *T. lintearius* causes young branches and leaves or spines to become white with green speckles. The tips of the spines often appear entirely white (Van Eyndhoven, 1967a). The mite inserts its stylets into gorse leaves through the stomatal openings to feed on the mesophyll parenchyma. Parenchyma cells that have not been fed upon exhibit symptoms of stress, either showing deformed cell organelles aligned along the walls of empty cells or cells with no well-defined cell organelles. Externally, subsequent to feeding by the mite, infested leaves show bleaching. The leaf water potential decreases by 30% and tender shoots lose their turgor, confirming water-related stress. *T. lintearius* then has the potential to decrease vigour and retard the competitiveness of *U. europaeus* by inflicting stress (Marriott *et al.*, 2013). Studies on the impact of the mite on the growth and development of gorse showed that over a 2½ year period, the mite developed about five generations and caused a significant reduction in dry matter production of about 36% compared with bushes controlled by acaricide treatment (Davies *et al.*, 2007).

Control

No control is required.

Tetranychus lombardinii Baker *et* Pritchard (Fig. 20.55)

Common name

Crimson spider mite.

Diagnostic characteristics

FEMALE. The body is dark red with dark spots on each side. The peritreme is hooked. The dorsal striae are lobate, and between the setae e_1–e_1 they are longitudinal and/or oblique, between the setae f_1–f_1 they are longitudinal, and between the setae e_1 and f_1 they are transverse and form a diamond shape. The ventral striae are lobate between the genital region and the setae *3a*; the pregenital striae are broken and dotted. The empodia bear 6 proximoventral hairs and lack spurs; tarsus I has 4 tactile setae proximal to the proximal duplex setae (Meyer Smith, 1974; Seeman and Beard, 2011).

MALE. The body is greenish yellow. The empodia have little dorsal spurs or lack spurs; empodium I is claw like, and empodia II–IV have long and free proximoventral hairs. The knob of the aedeagus is small, supported by a thick neck, dorsally convex, with the anterior margin rounded, and the posterior projection short and rounded (Gutierrez and Schicha 1983; Seeman and Beard, 2011).

Geographical distribution

Australia (Gutierrez and Schicha, 1983), Congo (Democratic Republic of the Congo, formerly Zaire) (Meyer Smith, 1974), Guadeloupe (Flechtmann *et al.*, 1999a), Kenya (Baker and Pritchard, 1960), India (Gupta, 1985; Gupta and Gupta, 1994),

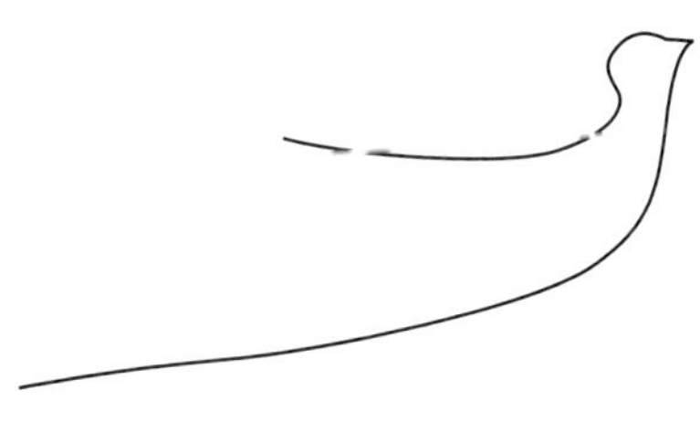

Fig. 20.55. *Tetranychus lombardinii* Baker *et* Pritchard. Male, aedeagus (from Meyer Smith, 1987).

Indonesia (Gutierrez, 1992), Madagascar (Gutierrez, 1974), Malawi, Mozambique, Namibia (Baker and Pritchard, 1960), South Africa (Meyer Smith and Rodrigues, 1966; Seeman and Beard, 2011), Zambia (Meyer Smith, 1974; Mwase and Baker, 2006) and Zimbabwe (Baker and Pritchard, 1960).

Host plants

The whole list of host plants of *T. lombardinii* includes 123 wild and cultivated plants belonging to 41 families, with the highest number of species in the families Fabaceae and Solanaceae (16 and 13, respectively) (Migeon and Dorkeld, 2006–2013). The families and species of economic importance are:

Amaranthaceae: spinach (*Spinacia oleracea* L.) (Baker and Pritchard, 1960).

Asteraceae: *Gerbera jamesonii* Bolus ex Hook. f. (Meyer Smith, 1974), garden lettuce (*Lactuca sativa* L.) (Meyer Smith and Rodrigues, 1966).

Caricaceae: papaya (*Carica papaya* L.) (Rodrigues, 1968).

Cucurbitaceae: watermelon (*Citrullus lanatus* (Thunb.) Matsum. and Nakai), African horned melon (*Cucumis metuliferus* E. Mey. ex Naud.) (Meyer Smith, 1974), winter squash (*Cucurbita maxima* Duchesne) (Meyer Smith and Rodrigues, 1966).

Euphorbiaceae: cassava (*Manihot esculenta* Crantz), castor bean (*Ricinus communis* L.) (Meyer Smith and Rodrigues, 1966).

Fabaceae: groundnut (*Arachis hypogaea* L.), pigeon pea (*Cajanus cajan* (L.) Millsp.), soybean (*Glycine max* (L.) Merr.), lucerne (*Medicago sativa* L.), kidney bean (*Phaseolus vulgaris* L.) (Meyer Smith, 1974).

Malvaceae: Arabian cotton (*Gossypium herbaceum* L.) (Meyer Smith and Rodrigues, 1966; Meyer Smith, 1974; Gupta and Gupta, 1994).

Meliaceae: neem (*Azadirachta indica* A. Juss.) (Meyer Smith, 1974).

Moraceae: fig (*Ficus carica* L.) (Meyer Smith and Rodrigues, 1966), white mulberry (*Morus alba* L.) (Meyer Smith, 1970b).

Musaceae: banana (*Musa* × *paradisiaca* L.) (Meyer Smith and Rodrigues, 1966).

Passifloraceae: passion fruit (*Passiflora edulis* Sims) (Meyer Smith, 1974).

Solanaceae: tree tobacco (*Nicotiana glauca* Graham) (Meyer Smith, 1974), tomato (*Solanum lycopersicum* L.) (Meyer Smith and Rodrigues, 1966), Irish potato (*S. tuberosum* L.) (Meyer Smith, 1974).

Poaceae: sorghum (*Sorghum bicolor* (L.) Moench) (Meyer Smith, 1965).

Bio-ecology

T. lombardinii is an Afrotropical mite that has also been reported in Australasian, Neotropical and Oriental regions. It infests a number of vegetables, tropical fruit plants and flowering and ornamental plants. Among its host plants, the mite seems to prefer cotton (Goldsmid, 1962; Meyer Smith and Rodrigues, 1966), banana and papaya (Meyer Smith, 1981a).

The crimson spider mite reproduces by arrhenotokous parthenogenesis and has $n = 3$ chromosomes (Bolland *et al.*, 1981). Its populations usually live on the lower leaf surface and at 27–33°C and 57–66% RH the development of the eggs takes 5 days. The larva is active for 1–2 days, and then becomes quiescent for about 16 h and moults in the protonymph stage. This last stage takes 1–2 days and after a quiescent period of 23 h it moults into the deutonymph which, after a quiescent period of 24 h, becomes an adult after 2 days. According to the environmental conditions, the life cycle takes 9–12 days. At 25°C and 50% RH, the female and male life cycle develops over 10 and 9 days, respectively. Female longevity ranges from 18 to 30 days, and fecundity from 70 to 110 eggs. Fertilized and unfertilized females lay the same number of eggs. The highest survival temperature is 40 and 45°C, and no survival has been observed at −3.8°C. Populations of *T. lombardinii* develop optimally under dry and warm conditions, especially in late spring and summer. The wind plays an important role in spreading the mite (Coates, 1974).

A number of predatory insects are reported as natural enemies of *T. lombardinii*. They include the coccinellids *Stethorus aethiops* We and *S. jejunus* Casey, the staphylinids *Oligota fageti* Williams and *O. pallidicornis* Cameron, the aeolothripid *Scolothrips* sp., the anthocorid *Wollastoniella gatti* Ghauri, the cecidomyiid *Therodiplosis* sp. and the coniopterygid *Coniopteryx crassicornis* Eb. Pet. (Yaseen *et al.*, 1982).

Symptomatology and damage

Jeppson *et al.* (1975) reported that any specific damage caused by this mite had not been well detailed, and neither had its biology. Meyer Smith (1981a) reported damage on some South African crops, such as banana and papaya. Feeding by *T. lombardinii* on the lower leaf surfaces of banana causes stipples along the midrib, but damage becomes important only when higher levels of infestation occur. Heavy infestation downgrades fruit, and the bunches that develop during winter and are cut during summer can be severely damaged. Populations of *T. lombardinii* may also develop on and/or in the banana bunches, where they are hidden between the fingers. At these sites, the mite develops and spins webs. Symptoms on the fruits are a stippled silver appearance, similar to that caused by the banana thrip *Hercinothrips bicinctus* (Bagnall), but the stipples caused by the thrip are much rougher. Low levels of attack do not affect the ripening of fruits, but severely attacked fruits become smoky yellow when they ripen (Meyer Smith, 1981a).

Feeding by *T. lombardinii* on the leaves of papaya causes silvering and yellowing, but the size of the leaves generally reduces the importance of the attack, and only severe infestations cause browning of leaves. Symptoms on other crops are similar to those caused by *T. cinnabarinus* (Boisduval) (Meyer Smith, 1981a).

Control

Control of *T. lombardinii* is mainly by chemical means. There are not many reports of biological control. Researches carried out under laboratory conditions showed that the predator *Neoseiulus idaeus* Denmark *et* Muma might develop and reproduce on *T. lombardinii* (Van den Berg and Markham, 1987).

CHEMICAL CONTROL. Control is provided by spray applications of various chemicals. On bananas, bromopropylate, dicofol and sulfur dust, as well as triazophos, are reported to be effective.

On tomato and cotton, chinometionate, cyhexatin, diazinon, dicofol and monocrotophos have been applied (Meyer Smith, 1981a). Currently, various acaricides registered for these crops in different countries may be used.

Tetranychus ludeni Zacher (Fig. 20.56)

Common name

Red spider mite, red-legged spider mite, bean spider mite.

Diagnostic characteristics

FEMALE. The body is dark red in colour. The peritreme is hooked. The dorsal body setae are serrate, long and slender. The opisthosomal dorsal striae are lobate, and between the setae e_1–e_1 they are longitudinal and/or oblique, between the setae f_1–f_1 they are longitudinal, and between the setae e_1 and f_1 they are transverse and form a diamond shape medially. The ventral striae set between the genital region and the setae *1a* are lobate, and the pregenital striae are obviously broken and dotted. The empodia have 6 proximoventral hairs; empodial spurs are minute or lacking. Tarsus I has 4 tactile setae overlapping with the proximal duplex setae (Gutierrez and Schicha, 1983; Seeman and Beard, 2011).

MALE. The body is orange yellow in colour. The empodia possess a dorsal spur; empodia I are claw-like, and II–IV with proximoventral hairs long and free. The distal knob of the aedeagus is very small, slightly angulate dorsally, lacking posterior projection, and with the anterior projection small and acuminate (Gutierrez and Schicha, 1983; Seeman and Beard, 2011).

Geographical distribution

Algeria (Athias Henriot, 1959), Argentina (Pritchard and Baker, 1955; Rossi Simons, 1961; Guanilo *et al.*, 2010), Australia (Pritchard and Baker, 1955; Miller, 1966; Seeman and Beard, 2011), Brazil (Flechtmann, 1967b; Flechtmann and Baker, 1975; Feres *et al.*, 2005, 2009; Furtado *et al.*, 2006; Silva *et al.*, 2008; Seeman and Beard, 2011), Canary Islands (Ferragut and Santonja, 1989), Chile (Pritchard and Baker, 1955), China (Ma and Yuan, 1975; Wang and Cui, 1992), Colombia (Urueta, 1975), Cook Islands (Dugdales, 1975), Costa Rica (Baker and Pritchard, 1962; Aguilar and Murillo, 2008, 2012), Egypt (Abdel-Shaheed *et al.*, 1973), El Salvador (Pritchard and Baker, 1955; Baker and Pritchard, 1962; Jeppson *et al.*, 1975; Andrews and Poe, 1980), Fiji (Gutierrez, 1977), France (Hardouin, 1934), French Polynesia (Gutierrez, 1980), Germany (Zacher, 1913; Oudemans, 1931b), Greece (Hatzinikolis, 1986), Hawaiian Islands (USA) (Garrett and Haramoto, 1967; Lee Goff, 1986), Honduras (Baker and Pritchard, 1962), Kenya (Jeppson *et al.*, 1975), India (Puttarudriah and Channabasavanna, 1959; Nassar and Ghai, 1981; Gupta, 1992, 1995; Gupta and Gupta, 1994; Gupta and Chatterjee, 1997), Iran (Ghasemi Moghadam *et al.*, 2012), Iraq (Mohamed and El-Haidari, 1965), Israel (Ben-David *et al.*, 2007), Japan (Ehara, 1956c, 1960, 1966a; Ehara and Masaki, 1989; Osakabe *et al.*, 2002; Gotoh *et al.*, 2003a; Ohno *et al.*, 2009; Seeman and Beard, 2011; Matsuda *et al.*, 2013), Madagascar (Helle *et al.*, 1970), Madeira Island (Portugal) (Carmona, 1992), Malawi (Meyer Smith, 1974), Mauritius (Baker and Pritchard, 1960), Mexico (Estebanes-Gonzalez and Baker, 1968), Morocco (Saba, 1973b), Mozambique (Meyer Smith and Rodrigues, 1966), Nansei-shoto (Japan) (Ehara, 1966a; Ohno *et al.*, 2009; Matsuda *et al.*, 2013), New Caledonia (France) (Gutierrez, 1980), New Zealand (Manson, 1967b), Nicaragua (Baker and Pritchard, 1962), Paraguay (Aranda, 1969; Aranda and Flechtmann, 1971), Peru (Guanilo *et al.*, 2012), Portugal (Carmona, 1960), South Africa (Pritchard and Baker, 1955; Meyer Smith, 1974), Spain (Ferragut and Santonja, 1989; Escudero and Ferragut, 2005; Hurtado *et al.*, 2008), Swaziland (Meyer Smith, 1974), Taiwan (Tseng, 1974), Thailand (Baker, 1975b), the USA (McGregor, 1919; Pritchard and Baker, 1955), Venezuela (Doreste, 1968b; Fréitez and Alvarado, 1976; Morros and Aponte, 1994), Zambia (Meyer Smith, 1974; Mwase and Baker, 2006) and Zimbabwe (Meyer Smith, 1974).

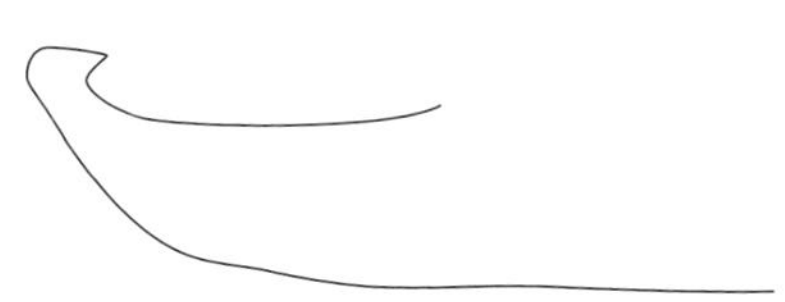

Fig. 20.56. *Tetranychus ludeni* Zacher. Male, aedeagus (from Meyer Smith, 1987).

Host plants

The red spider mite is recorded on about 325 wild and cultivated plants from 61 families. Among these, the Asteraceae, Cucurbitaceae, Fabaceae, Malvaceae and Solanaceae have the highest number of species (Migeon and Dorkeld, 2006–2013). The families and species of economic importance are:

Amaranthaceae: beet (*Beta vulgaris* L.) (Carmona and Dias, 1980), spinach (*Spinacia oleracea* L.) (Gupta and Gupta, 1994).

Apiaceae: wild celery (*Apium graveolens* L.), carrot (*Daucus carota* L.) (McGregor, 1950), fennel (*Foeniculum vulgare* Mill.), parsley (*Petroselinum crispum* (Mill.) Fuss) (Matsuda *et al.*, 2013).

Arecaceae: coconut palm (*Cocos nucifera* L.) (Bolland *et al.*, 1998; Migeon and Dorkeld, 2006–2013).

Asteraceae: artichoke (*Cynara cardunculus* L.) (Meyer Smith and Ryke, 1959; Gupta and Gupta, 1994; Seeman and Beard, 2011), African daisy (*Gerbera jamesonii* Adlam) (Davis, 1968b).

Brassicaceae: cabbage (*Brassica oleracea* L.) (Nassar and Ghai, 1981; Gupta and Chatterjee, 1997), colza (*B. rapa* L.) (Ehara and Masaki, 1989).

Caricaceae: papaya (*Carica papaya* L.) (Baker and Pritchard, 1962; Aranda and Flechtmann, 1971).

Caryophyllaceae: carnation (*Dianthus caryophyllus* L.) (Meyer Smith, 1974).

Convolvulaceae: sweet potato (*Ipomoea batatas* (L.) Lam.) (Davis, 1968b; Baker, 1975b).

Cucurbitaceae: watermelon (*Citrullus lanatus* (Thunb.) Matsum. and Nakai) (Carmona, 1960; Andrews and Poe, 1980; Gupta and Gupta, 1994), cantaloupe (*Cucumis melo* L.) (Ehara, 1956c; Carmona, 1960; Aguilar and Murillo, 2012), garden cucumber (*C. sativus* L.) (Ehara, 1956c; Doreste, 1968b; Gupta and Gupta, 1994; Ohno *et al.*, 2009, 2010; Seeman and Beard, 2011), winter squash (*Cucurbita maxima* Duchesne) (Gupta and Gupta,

1994), field pumpkin (*C. pepo* L.) (Aguilar and Murillo, 2008), bottle gourd (*Lagenaria siceraria* (Molina) Standl.) (Flechtmann, 1967a), chayote (*Sechium edule* (Jacq.) Sw.) (Paschoal, 1968a; Ochoa *et al.*, 1989; Gupta and Gupta, 1994).

Euphorbiaceae: cassava (*Manihot esculenta* Crantz) (Rodrigues, 1968), castor bean (*Ricinus communis* L.) (McGregor, 1950; Flechtmann and Baker, 1975; Urueta, 1975; Gupta and Gupta, 1994).

Fabaceae: groundnut (*Arachis hypogaea* L.) (Rodrigues, 1968), pigeon pea (*Cajanus cajan* (L.) Millsp.) (Meyer Smith, 1974; Gupta and Gupta, 1994), soybean (*Glycine max* (L.) Merr.) (Ehara, 1960; Davis, 1968b; Flechtmann and Baker, 1975; Urueta, 1975; Gutierrez and Schicha, 1983; Gupta and Gupta, 1994; Guanilo *et al.*, 2010), lucerne (*Medicago sativa* L.) (McGregor, 1950), velvet bean (*Mucuna pruriens* L.) (Paschoal, 1968a; Flechtmann and Baker, 1975), Lima bean (*Phaseolus lunatus* L.) (Gupta, 1992), kidney bean (*P. vulgaris* L.) (Meyer Smith and Ryke, 1959; Urueta, 1975; Gupta and Gupta, 1994; Morros and Aponte, 1994; Ohno *et al.*, 2009; Seeman and Beard, 2011), garden pea (*Pisum sativum* L.) (Davis, 1968b), white clover (*Trifolium repens* L.) (Gutierrez and Schicha, 1983), fava bean (*Vicia faba* L.) (Seeman and Beard, 2011), garden vetch (*V. sativa* L.) (Biasi and Santos, 1988b), cowpea (*V. unguiculata* (L.) Walp.) (Gupta and Gupta, 1994; Gupta and Chatterjee, 1997), wild cowpea (*V. vexillata* (L.) A. Rich.) (Bolland *et al.*, 1998; Migeon and Dorkeld, 2006–2013).

Geraniaceae: house geranium (*Pelargonium* × *hortorum* L. H. Bailey) (Bolland *et al.*, 1998; Migeon and Dorkeld, 2006–2013).

Juglandaceae: English walnut (*Juglans regia* L.) (Carmona and Dias, 1980).

Lamiaceae: peppermint (*Mentha* × *piperita* L.) (Paschoal, 1970; Seeman and Beard, 2011), oregano (*Origanum vulgare* L.) (Garrett and Haramoto, 1967; Lee Goff, 1986; Osakabe *et al.*, 2002), sage (*Salvia officinalis* L.) (Davis, 1968b).

Malvaceae: okra (*Abelmoschus esculentus* (L.) Moench) (Paschoal, 1970; Gupta and Gupta, 1994), creole cotton (*Gossypium barbadense* L.) (Davis, 1968b), Arabian cotton (*G. herbaceum* L.) (Flechtmann, 1967a; Gupta and Gupta, 1994), upland cotton (*G. hirsutum* L.) (Rodrigues, 1968).

Moraceae: common fig (*Ficus carica* L.) (Meyer Smith, 1974), white mulberry (*Morus alba* L.) (Gupta, 1995).

Musaceae: French plantain (*Musa* × *paradisiaca* L.) (Carmona, 1992).

Passifloraceae: passion fruit (*Passiflora edulis* Sims) (Paschoal, 1970; Seeman and Beard, 2011).

Poaceae: maize (*Zea mays* L.) (Abdel-Shaheed *et al.*, 1973).

Rosaceae: quince (*Cydonia oblonga* Mill.), strawberry (*Fragaria* × *ananassa* Duchesne) (Davis, 1968b; Meyer Smith, 1974), apple (*Malus domestica* Borkh.) (Meyer Smith and Ryke, 1959; Paschoal and Reis, 1968; Ohno *et al.*, 2010), apricot (*Prunus armeniaca* L.) (Meyer Smith, 1974), sweet cherry (*P. avium* (L.) L.) (Migeon and Dorkeld, 2006–2013), European plum (*P. domestica* L.), peach (*P. persica* (L.) Batsch) (Meyer Smith and Ryke, 1959; Aranda and Flechtmann, 1971), pear (*Pyrus communis* L.) (Carmona and Dias, 1980; Seeman and Beard, 2011).

Rutaceae: *Citrus* sp. (Gutierrez and Schicha, 1983; Gupta and Gupta, 1994; Gupta and Chatterjee, 1997), sour orange (*C. aurantium* L.) (Davis, 1968b), lemon (*C. limon* (L.) Burm.) (Meyer Smith, 1974), grapefruit (*C. paradisi* McFad.), Japanese bitter orange (*C. trifoliata* L.) (Davis, 1968b).

Solanaceae: pepper (*Capsicum annuum* L.) (Ohno *et al.*, 2010), tree tobacco (*Nicotiana glauca* Graham) (Guanilo *et al.*, 2010), gooseberry (*Physalis angulata* L.) (Gutierrez and Schicha, 1983; Guanilo *et al.*, 2010), tomato (*Solanum lycopersicum* L.) (Carmona, 1960; Gupta and Gupta, 1994), aubergine (*S. melongena* L.) (Zacher, 1913; Ehara, 1956c; Gupta and Gupta, 1994; Gupta and Chatterjee, 1997; Osakabe *et al.*, 2002; Ohno *et al.*, 2009; Guanilo *et al.*, 2010), Irish potato (*S. tuberosum* L.) (Meyer Smith, 1974; Ehara and Masaki, 1989).

Vitaceae: grapevine (*Vitis vinifera* L.) (Baker and Pritchard, 1962; Aranda and Flechtmann, 1971).

Bio-ecology

T. ludeni has often been collected in mixed populations with *T. urticae* Koch (see below) (Gutierrez and Schicha, 1983). In addition, taxonomic studies on spider mites in New Zealand showed that several specimens previously identified as *T. cinnabarinus* (Boisduval) were in fact *T. ludeni* (Zhang, 2002). Moreover, *T. ludeni* has often been confused with *T. desertorum* Banks owing to misinterpretation of the aedeagal structure. A similar misinterpretation also occurred in the description of *T. ludenensis* Attiah (1969), which was recently declared by Seeman and Beard (2011) to be a junior synonym of *T. ludeni*. In agreement with what has been stated for *T. desertorum* above, it is very likely that the list of host plants for *T. ludeni* should be reviewed.

In the tropics, the red spider mite occurs outdoors on many wild and cultivated plants, and in temperate climates, it infests various greenhouse plants (Moutia, 1958; Meyer Smith and Rodrigues, 1966; Jeppson *et al.*, 1975). *T. ludeni* usually lives on the lower leaf surfaces. On pyrethrum, the mite also lives on the young curled leaves, and when populations increase, spreads on to all leaf surfaces, producing webbing that covers the entire plant (Jeppson *et al.*, 1975). In New Zealand, it has been observed that the density of *T. ludeni* on the upper surface of the leaves of chayote may be 5–8 times as great as on the lower surfaces. The preference by *T. ludeni* for the upper leaf surface of this species may depend on the fact that the predatory mite *Phytoseiulus persimilis* Athias Henriot mostly lives on the lower leaf surface. The disparity in the within-leaf distribution between the predator and prey provides the prey with a certain refuge and reduces the searching efficiency of the predator, which is considered to be the main reason for the less effective biocontrol of *T. ludeni* on this plant (Zhang, 2002).

T. ludeni reproduces by arrhenotokous parthenogenesis and the haploid number of chromosomes is three (Helle *et al.*, 1970). The male:female ratio was 1:2.7 on French bean (*P. vulgaris*) in pots (Puttaswamy and Channabasavanna, 1979b). Adango *et al.* (2006) observed a sex ratio (% females) of 72.72 and 78.68% on amaranth (*Amaranthus cruentus* L.) and nightshade (*Solanum macrocarpon* L.), while Kaimal and Ramani (2011c) found a sex ratio (male:female) of 2:10 on *Mucuna deeringiana* (Bort) Merr.

The mite first feeds on the lower leaf surfaces or in the young curled leaves, but as the population increases, it becomes distributed over all of the leaf surfaces and produces webbing that covers the entire plant (Coates, 1974). Field studies in India on potted French bean plants at temperatures ranging from 19.3 to 28.4°C and 53–88% RH showed that the development of males and females took 11.96 and 12.48 days, respectively. Mated and unmated females had pre-oviposition periods of 1.54 and 1.43 days, and oviposition periods of 22.83 and 27.41 days

respectively; they laid a total of 165.9 and 132.0 eggs, and lived for 27.98 and 32.14 days. The adult male lifespan is 25.20 days (Puttaswamy and Channabasavanna, 1979b). On *P. vulgaris* in the laboratory at 26.34 ± 3.92 °C and 69.44 ± 19.44% RH, the mean duration of the different developmental stages for egg, larva, protonymph and deutonymph was 4.68, 1.75, 1.31 and 1.85, respectively. The total life cycle was 9.98 days for females and 9.25 days for males, respectively (Morros and Aponte, 1994).

On cotton leaves in bioclimate chambers at 20, 23, 25, 28 and 30°C, 70 ± 10% RH, and under a 12L:12D photoperiod, the development time from egg to adult ranged from 20.77 days at 20°C to 8.50 days at 30°C for males, and from 18.83 days at 20°C to 7.75 days at 30°C for females (Silva, 2002). In growth chamber experiments on amaranth (*A. cruentus*) and nightshade (*S. macrocarpon*) at 27°C, 70% ± 10% RH and under a 12L:12D photoperiod, the immature development of *T. ludeni* was shorter on amaranth (9.60 ± 0.13 days) than on nightshade (10.12 ± 0.07), and female longevity was 28.33 ± 1.27 and 24.93 ± 1.26, respectively. Fecundity was higher on *A. cruentus* (111.55 ± 10.73 eggs) than on *S. macrocarpon* (84.29 ± 10.51 eggs), but there was no difference in the daily egg laying on the two plant species (7.09 ± 1.56 and 6.74 ± 1.79, respectively) (Adango *et al.*, 2006). In laboratory studies of leaves of *M. deeringiana* at 30 ± 2°C and 70 ± 5% RH, the duration of the pre-oviposition period, oviposition period and post-oviposition period was 0.5 ± 0, 11.5 ± 0.38 and 0.5 ± 0 days, respectively. Fecundity was 83.6 ± 3.4 eggs and longevity 12.6 ± 0.37 days. The total duration of sexual and parthenogenetic reproduction was 10.04 ± 0.13 days and 9.04 ± 0.13 days, respectively, so parthenogenetic development requires a comparatively shorter time than sexual development (Kaimal and Ramani, 2011c).

On *P. vulgaris*, the reproduction rate (R_0) of *T. ludeni* was 77.42, the mean generation time (T) 19.63, the intrinsic rate of increase (r_m) 0.2526 individuals per female a day and the finite rate of natural increase (λ) was 1.2874 individuals per female a day (Morros and Aponte, 1994). On cotton, the thermal constant was 138.34 degree-days for females and 130.91 degree-days for males. At 30°C, the highest values of the following parameters were: intrinsic rate of increase (r_m), 0.418; number of eggs per females a day, 3.47; fecundity, 61.29; and reproductive rate net, 48.00. The minimum mean generation time was 9.27 days (Silva, 2002). On amaranth and nightshade at 27°C, 70% ± 10% RH and under a 12L:12D photoperiod, the intrinsic rates of natural increase (r_m) were 0.173 and 0.215, respectively; the net reproductive rates (R_0) (females per female) were 16.32 and 41.68, respectively; the mean generation times (T) were 16.20 and 17.36 days; and the finite rates of increase (λ) were 1.19 and 1.24, respectively (Adango *et al.*, 2006).

All biological stages of *T. ludeni* occur throughout the year and many overlapping generations a year are possible (Jeppson *et al.*, 1975). In India, populations of *T. ludeni* appear in the first week of April, peak in June, and their density declines sharply in July (Kumar and Sharma, 1993; Singh, 1995). Population increases are associated with periods of low rainfall and relative humidity and high temperatures (Reddy, 2002). Wind is the most important factor affecting the dispersal of *T. ludeni* (Puttaswamy and Channabasavanna, 1979a).

Natural enemies include pathogens and several predatory insects and mites, see Table 20.10.

Symptomatology and damage

Infested plants are yellowish and wilt and drop rapidly, especially during dry conditions. The leaves first turn yellow, but are soon covered by necrotic patches. In India, the mite causes severe vegetable losses (Channabasavanna, 1973). Feeding on leaves of French bean induces white stippling, which results in drying and leaf loss (Puttaswamy and Channabasavanna, 1979b). Severe damage is seen on cowpea in the Varanasi region, Uttar Pradesh, India (Singh and Mukherjee, 1989). Infested cowpea leaves first show yellow patches, which may later become necrotic. Aggregation of mites on the tips of the pods is common during periods of the day when the temperature is above 38°C (Singh and Raghuraman, 2011). Morros and Aponte (1995a,b) evaluated the effect of two levels of mite infestation on vegetative and reproductive stages of *P. vulgaris* under field (1 or 2 infested leaves per plant) and greenhouse (60 or 100 mites per plant) conditions, and found a greater reduction in the number of leaves and internodes, leaf area and dry weight of vegetative organs at the higher infestation (in the field) and when mite infestation affected the early vegetative stage (in the greenhouse). Damage to citrus by *T. ludeni* is virtually unknown.

Control

Control of *T. ludeni* involves the use of resistant varieties and spray applications of acaricides. Releases of natural enemies are sometimes possible but usually does not occur at a practical level.

CULTURAL PRACTICES. The lower growth of populations of *T. ludeni* on amaranth suggested that it could be used in integrated pest management (IPM) strategies, such as intercropping, to reduce pest density, and in the development of biopesticides for use against *T. ludeni* on vegetables (Adango *et al.*, 2006).

PLANT RESISTANCE. Of the four varieties of aubergine tested ('Madurai' ('MDU_1'), 'Coimbatore' ('CO_1'), 'Annamalai' and 'Panruti (local)'), 'Panruti' was highly resistant to *T. ludeni*, while 'Madurai', which is widely grown by farmers in India, was the most susceptible (Reddy and Baskaran 1991, 2006).

BIOLOGICAL CONTROL. Attempts at the biological control of *T. ludeni* were tried in India, in which releases of *Chrysoperla carnea* Stephens larvae (640 individuals/130 plants) were made on to resistant and susceptible aubergine varieties infested by the red spider mite. Overall, the results were evaluated positively (Reddy, 2001). Biological control of the red spider mite using this predator is common in IPM programmes for vegetable crops such as aubergine, okra and peppers, but not tomato. In fact, odours from aubergine, okra and peppers are attractive to *C. carnea*, while odours from tomato are not (Reddy, 2002).

CHEMICAL CONTROL. Restrictions on the use of chemical substances for pest control are well known. It has also been reported that *T. ludeni* from cotton has developed resistance to all organophosphate insecticides tested (Herron *et al.*, 1998).

Table 20.10. *Tetranychus ludeni* Zacher. Natural enemies (fungi, insects and mites).

Families and species	References
FUNGI	
Dematiaceae	
Cladosporium sp.	Godse and Patil, 1977
Neozygitaceae	
Neozygites sp. near *floridana*	Ramaseshiah, 1971
Neozygites floridana (Weiser *et* Muma)	Smith and Furr, 1975
INSECTS	
Aeolothripidae	
Scolothrips indicus Priesner	Reddy and Jagadish, 1977
Chrysopidae	
Chrysoperla carnea Stephens	Reddy, 2001, 2002
Coccinellidae	
Stethorus sp.	Zhang, 2002
Stethorus pauperculus Weis	Ansari and Pawar, 1992
Scymnus coccivora Ayyar	Ansari and Pawar, 1992
Staphylinidae	
Oligota flavicollis (Boisd. *et* Lac.)	Chazeau and Gutierrez, 1977
Oligota oviformis Casey	Ansari and Pawar, 1992
MITES	
Cunaxidae	
Cunaxa setirostris (Hermann)	Arbabi and Singh, 2000
Phytoseiidae	
Amblyseius multidentatus (Chant, Swirski *et* Shechter)	Kumar and Sharma, 1993; Putatunda and Anupam Tagore, 1997
Euseius alstoniae (Gupta)	Putatunda and Anupam Tagore, 1997
Euseius finlandicus (Oudemans)	
Euseius neococcineae (Gupta)	
Euseius rhododendronis (Gupta)	Krishnamoorthy, 1983
Euseius scutalis (Athias Henriot)	Putatunda and Anupam Tagore, 1997
Neoseiulus californicus (McGregor)	Escudero *et al.*, 2005
Neoseiulus fallacis (Garman)	Biasi and Santos, 1988a; Putatunda and Anupam Tagore, 1997
Neoseiulus idaeus Denmark *et* Muma	Escudero *et al.*, 2005
Neoseiulus indicus (Narayan *et* Kaur)	Putatunda and Anupam Tagore, 1997
Neoseiulus longispinosus (Evans)	Ansari and Pawar, 1992; Abhilash and Sudharma, 2003
Phytoseius jujuba Gupta	Putatunda and Anupam Tagore, 1997
Phytoseius minutus Narayanan, Kaur *et* Ghai	
Phytoseius mixtus Chaudhri	
Phytoseius roseus Gupta	
Phytoseiulus persimilis Athias Henriot	Zhang, 2002; Escudero *et al.*, 2005
Typhlodromips tetranychivorus Gupta	Puttaswamy and ChannaBasavanna, 1979c; Nangia and ChannaBasavanna, 1983; Ansari and Pawar, 1992; Kumar and Sharma, 1993
Typhlodromus homalii (Gupta)	Putatunda and Anupam Tagore, 1997
Stigmaeidae	
Agistemus industani González-Rodriguez	Arbabi and Singh, 2002

Different acaricides used on various crops included avermectins (Kumar *et al.*, 2003b), formamidines (Chiavegato *et al.*, 1975), organochlorines (Duncombe, 1972; Chiavegato *et al.*, 1975; Jagadish and ChannaBasavanna, 1983; Nangia and ChannaBasavanna, 1983; Nangia *et al.*, 1990; Kumar and Sharma, 1993; Kumar *et al.*, 2003b), organosulfur compounds (Nangia *et al.*, 1990; Kumar and Sharma, 1993), organotins (Brun *et al.*, 1983; Peter and David, 1988) and sulfur (Nangia and ChannaBasavanna, 1983; Nangia *et al.*, 1990; Kumar and Sharma, 1993; Kumar *et al.*, 2003b). Botanical acaricides include neem oil (Reddy, 2001).

INTEGRATED PEST MANAGEMENT. Host plant resistance complemented by natural enemies and biorational control agents, such as *C. carnea* and neem oil, are suitable components in an IPM programme for managing the spider mite in endemic areas. The implementation of such a strategy on aubergine may give a net profit ranging from US$77 to US$91 (Reddy, 2001).

Tetranychus macfarlanei Baker *et* Pritchard (Fig. 20.57)

Common name

Unknown.

Diagnostic characteristics

FEMALE. The body of young females is bright red and later becomes deep red (Jose and Shah, 1989a). The peritreme is hooked. The dorsal striae are lobate, and between the setae e_1–e_1 they are longitudinal, between the setae f_1–f_1 they are also longitudinal, and between the setae e_1 and f_1 they are transverse and form a diamond shape. The ventral striae set between the genital region and the setae *1a* are lobate. The pregenital striae are almost entire but weak and/or broken medially. The empodia have a dorsal spur and 6 proximoventral hairs. Tarsus I has 3 tactile setae overlapping with the proximal duplex setae (Seeman and Beard, 2011).

MALE. The empodia have a dorsal spur; empodia I are claw like, and empodia II–IV have long proximoventral hairs. The knob of the aedeagus is small, dorsally flat to slightly convex and anvil shaped, with tiny anterior and posterior projections (Seeman and Beard, 2011).

Geographical distribution

Bangladesh (Ullah *et al.*, 2012), Canary Islands (Pande *et al.*, 1989), India (Baker and Pritchard, 1960; Nassar and Ghai, 1981; Gupta and Gupta, 1994; Gupta and Chatterjee, 1997; Seeman and Beard, 2011), Madagascar (Gutierrez, 1974) and Mauritius (Baker and Pritchard, 1960).

Host plants

T. macfarlanei is reported on 29 herbaceous host plants ascribed to nine families. The Cucurbitaceae and Fabaceae both have eight species (Migeon and Dorkeld, 2006–2013). The families and species of economic importance are: Convolvulaceae – water spinach (*Ipomoea aquatica* Forssk.) (Gupta and Gupta, 1994); Cucurbitaceae – watermelon (*Citrullus lanatus* (Thunb.) Matsum. and Nakai) (Gutierrez, 1974), garden cucumber (*Cucumis sativus* L.) (Baker and Pritchard, 1960; Gupta and Gupta, 1994), winter squash (*Cucurbita maxima* Duchesne) (Bindra and Kittur, 1961; Gupta and Gupta, 1994), field pumpkin (*C. pepo* L.) (Gupta and Gupta, 1994), bottle gourd (*Lagenaria siceraria* (Molina) Standl.) (Bindra and Kittur, 1961; Gupta and Gupta, 1994), chayote (*Sechium edule* (Jacq.) Sw.)

Fig. 20.57. *Tetranychus macfarlanei* Baker *et* Pritchard. Male, aedeagus (from Meyer Smith, 1987).

(Gutierrez, 1974; Gupta and Gupta, 1994); Euphorbiaceae – castor bean (*Ricinus communis* L.) (Gutierrez, 1974); Fabaceae – groundnut (*Arachis hypogaea* L.) (Gutierrez, 1974), pigeon pea (*Cajanus cajan* (L.) Huth), soybean (*Glycine max* (L.) Merr.) (Gupta and Gupta, 1994), sieva bean (*Phaseolus lunatus* L.) (Gutierrez, 1974): Malvaceae – okra (*Abelmoschus esculentus* (L.) Moench) (Baker and Pritchard, 1960; Gupta and Gupta, 1994; Seeman and Beard, 2011), Arabian cotton (*Gossypium herbaceum* L.), American cotton (*G. hirsutum* L.) (Gupta and Gupta, 1994); Musaceae – banana (*Musa* × *paradisiaca* L.) (Migeon and Dorkeld, 2006–2013); Solanaceae – aubergine (*Solanum melongena* L.) (Nassar and Ghai, 1981; Gupta and Gupta, 1994; Gupta and Chatterjee, 1997).

Bio-ecology

T. macfarlanei is reported from tropical and subtropical countries. In India, it is known as an occasional pest of aubergine (Jose and Shah, 1989b; Gupta and Gupta, 1994), okra and cotton, and in the absence of its host plants, it may survive on weeds (Jose and Shah, 1989b).

The mite reproduces by arrhenotokous parthenogenesis and has $n = 3$ chromosomes (Bolland *et al.*, 1981). The male:female sex ratio observed on aubergine is 1:9.8 (Patil, 2005). In India, populations of *T. macfarlanei* develop on cotton from October to February and are carried over to aubergine and okra from March to June and on some weed plants from July to December (Jose and Shah, 1989b). In June, September, October and December, respectively: egg development takes 3.30, 2.75, 2.93 and 4.5 days; the larval period lasts 1.0, 1.0, 1.13 and 2.71 days; the protonymph lasts 1.27, 0.96, 1.06 and 1.67 days; the deutonymph lasts 1.13, 0.84, 0.71 and 1.0 days; and female development takes 7.73, 7.43, 7.42 and 11.81 days. In September, October and December, male development takes 6.12, 7.67 and 10.9 days, respectively. The fecundity is 137 eggs (Jose and Shah, 1989a).

In other studies in India, Thulsiram (1991) reported a daily egg depostion of 5.84 eggs over an average oviposition period of 20.52 days on cotton. Female egg development took from 3.5 to 4.5 days, and male egg development from 3 to 4.5 days. The larval, protonymphal and deutonymphal stages of the female lasted 2.65, 2.14 and 1.47 days; for males, these stages lasted 2.35, 1.46 and 1.23 days, respectively. The oviposition period lasted 20.52 days, and the longevity of male and female was 12.25 and 24.18 days, respectively. On okra, the duration of the different biological stages was more or less equal, ranging from 7.11 ± 0.33 to 8.04 ± 0.28 days, but daily fecundity, total fecundity and longevity differed according to the okra variety (Bhanderi, 1991). In March and April in south Gujarat, low temperature and low RH prolonged mite development on okra. Conversely, high temperature and high RH shortened mite development (Sejalia *et al.*, 1993).

Patil (2005) conducted laboratory and field investigations on the phytophagous mites of aubergine in Karnataka, India. *T. macfarlanei* was the major pest, and under laboratory conditions on aubergine, at temperatures ranging from 20.53 ± 1.34 to 27.38 ± 2.01°C and with 69.79 ± 5.18 and 82.25 ± 2.91% RH, male and female development from egg to adult took 10.63 ± 2.33 and 12.41 ± 2.61 days, respectively. The development period was shorter in the summer, and longer in the rainy and

winter seasons. Male longevity was 17.31 ± 2.18 days, and female longevity 20.62 ± 2.98 days. The fecundity of mated females was 47.87 ± 9.03 eggs, ranging from 55.5 eggs in the summer to 43.1 in the winter. The fecundity of unmated females was 23.67 ± 7.37 eggs, ranging from 24.9 eggs in the summer to 22.3 in the winter. The daily egg depostion of mated females was 3.76 ± 0.38 and 2.40 ± 0.58 eggs during the summer and winter, respectively. The complete life cycle of the male and female was 27.94 ± 4.34 and 33.06 ± 5.27 days, respectively (Patil, 2005). In the same region, populations of the mite on aubergine were abundant from March to May from 45 to 90 days after planting, and peaked at about 75 days (Eswara Reddy, 2000). Prasanna (2007) found that on aubergine *T. macfarlanei* developed from the first half of October (post-monsoon) until the first half of June (beginning of the monsoon). The population of the mite increased during the spring, peaked during the second half of May (pre-monsoon), and declined from the first half of June. The mite populations had a positive relationship with maximum temperature and a negative relationship with morning and evening RH. The correlations of minimum temperature (positive) and rainfall (negative) had a non-significant influence on the mite population (Prasanna, 2007). The leaves of the middle canopy harboured the maximum mite density followed by the bottom leaves (Patil, 2005; Prasanna, 2007). Weather factors and natural enemies were found not to be significantly correlated with *T. macfarlanei* populations on aubergine because of rainfall (Patil, 2005).

Predatory insects associated with *T. macfarlanei* include anthocorids, mirids (Prasanna, 2007), coccinellids (*Stethorus* sp.) (Moutia, 1958), staphylinids (*Oligota oviformis* Casey (Patil, 2005), *Oligota* sp. (Moutia, 1958 and Prasanna, 2007)), thripids (*Scolothrips rhagebianus* Preiesner) (Prasanna, 2007), cecidomyiids (*Feltiella* sp.) (Moutia, 1958). Predatory mites include different phytoseiids, such as *Amblyseius alstoniae* Gupta (= *Euseius alstoniae* (Gupta)) (Shah and Jose, 1989), *Neoseiulus longispinosus* (Evans) (Patil, 2005; Prasanna, 2007) and *A. multidentatus* (Chant, Swirski *et* Shechter) (ICAR, 1994–1996). Some stigmaeid (*Agistemus industani* González-Rodriguez) (ICAR, 1996–1998) and cunaxid mites (*Cunaxa* sp.) (Sudharma and Nair, 1999) are also reported.

Symptomatology and damage

Feeding by *T. macfarlanei* causes yellowish speckling, followed by leaf senescence. The mite is a sporadic to serious pest of aubergine in India, causing leaf fall (Jeppson *et al.*, 1975; Jose and Shah, 1989b; Gupta and Gupta, 1994). Aubergine plants that had been artificially infested with 25, 150 and 200 mites/plant were respectively 66.61, 61.83 and 60.21 cm tall at final harvest, while uninfested plants were 74.34 cm tall; the number of fruits found per plant were 9.89, 8.44, 7 and 11.44, while the yield was 489.06, 428.90, 377.54 and 556.92 g, respectively. Mite infestation under protected and unprotected conditions caused a yield loss of 129.78 and 108.97 quintals ha^{-1} (Patil, 2005).

Control

Control of *T. macfarlanei* may be required on aubergines and other crops. Currently, there are few reports on biological control, and the control of mite infestations is usually entrusted to sprays of chemicals.

BIOLOGICAL CONTROL. Reports on biological control are mainly from trials with entomopathogenic fungi, as *Verticillium lecanii* (Zimmerman) Viégas and *Metarhizium anisopliae* (Metchnikoff) Sorokin (Prasanna, 2007). There are no known extensive programmes of biological control.

CHEMICAL CONTROL. A wide list of insecticides/acaricides has been tested on different crops. They include abamectin (Prasanna, 2007), carbaryl (Kumar *et al.*, 1989), diafenthiuron (Anandkumar *et al.*, 2003; Patil, 2005; Prasanna, 2007), dicofol (Anandkumar *et al.*, 2003; Patil, 2005; Prasanna, 2007), dimethoate, endosulfan, ethion (Patil, 2005), fenazaquin (Anandkumar *et al.*, 2003; Patil, 2005; Prasanna, 2007), hexythiazox, methyl demeton, monocrotophos, phosalone (Patil, 2005), propargite (Patil, 2005; Prasanna, 2007), profenofos (Prasanna, 2007), quinalfos (Kumar *et al.*, 1989; Prasanna, 2007), spiromesifen (Patil, 2005) and sulfur (Kumar *et al.*, 1989; Prasanna, 2007). Among these active ingredients, diafenthiuron, dicofol and spiromesifen provided excellent control (Patil, 2005).

Various products of botanical/animal origin have also been tested, such as azadirachtin (Prasanna, 2007), neem oil (Srinivasa and Sugeetha, 1999; Eswara Reddy, 2000; Patil, 2005; Roopa, 2005; Prasanna, 2007), neemazal (Patil, 2005), nimbecidine, NSKE (neem seed kernel extract) (Patil, 2005; Prasanna, 2007), pongamia oil (Srinivasa and Sugeetha, 1999; Eswara Reddy, 2000; Prasanna, 2007), mahua oil (Srinivasa and Sugeetha, 1999), *Clerodendrum* aqueous leaf extract (Eswara Reddy, 2000), sweet flag rhizome extracts and others (Patil, 2005). Among these, neem oil, NSKE and sweet flag rhizome extracts are reported as effective (Patil, 2005).

Tetranychus mcdanieli McGregor (Fig. 20.58)

Common name

McDaniel mite.

Diagnostic characteristics

FEMALE. The peritreme is hooked. The opisthosomal dorsal striae lack lobes; between the setae e_1–e_1 and between the setae f_1-f_1 and e_1 and f_1 they are transverse. The ventral striae lack lobes, and the pregenital striae are sparse medially and incomplete (Seeman and Beard, 2011).

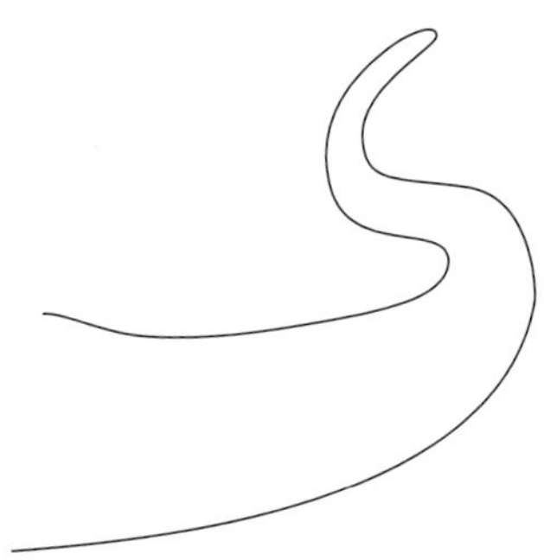

Fig. 20.58. *Tetranychus mcdanieli* McGregor. Male, aedeagus (modified from McGregor, 1950).

MALE. The dorsal striae lack lobes. Empodia I have a strong spur and are claw like; empodia II–IV have little dorsal spurs and long and free proximoventral hairs. The aedeagus lacks a knob, has a sigmoid shaft and is dorsally directed, bending anteriorly then recurving posteriorly and tapering (Seeman and Beard, 2011).

Geographical distribution

Canada (Pritchard and Baker, 1955; Jeppson *et al.*, 1975; Roy *et al.*, 1999), France (Rambier, 1982a; Navajas *et al.*, 1996) and the USA (McGregor, 1931; Pritchard and Baker, 1955; Bibby and Tuttle, 1959; Tuttle and Baker, 1964).

Host plants

The list of host plants of the McDaniel mite includes 15 species belonging to ten families (Migeon and Dorkeld, 2006–2013). The families and species of economic importance are: Rosaceae – Virginia strawberry (*Fragaria virginiana* Duchesne), apple (*Malus domestica* Borkh.), sweet cherry (*Prunus avium* (L.) L.), European plum (*P. domestica* L.) (Pritchard and Baker, 1952a; Pritchard and Baker, 1955), peach (*P. persica* (L.) Batsch), American red raspberry (*Rubus idaeus* L.) (McGregor, 1931; Pritchard and Baker, 1955; Roy *et al.*, 1999); Sapindaceae – sugar maple (*Acer saccharum* Marshall) (Bolland *et al.*, 1998; Migeon and Dorkeld, 2006–2013); Ulmaceae – American elm (*Ulmus americana* L.) (Bolland *et al.*, 1998; Migeon and Dorkeld, 2006–2013); Vitaceae – grapevine (*Vitis vinifera* L.) (Rambier, 1982a; Navajas *et al.*, 1996).

Bio-ecology

T. mcdanieli infests berries, deciduous fruit trees, grapevines and some ornamental plants in Nearctic (northern USA, southern Canada) and Palaearctic (France) regions. The mite was first described on raspberry in Michigan (McGregor, 1931) and later became a pest of tree fruits in the drier areas of the Pacific Northwest USA, Utah and California. In the central areas of Washington State, *T. mcdanieli* has mostly been investigated as a pest of apples since the 1930s. During the late 1950s and early 1960s, strong modifications to biological control induced widespread outbreaks of *T. mcdanieli*, and although it was the primary outbreak species in central Washington State during that period, it has since become relatively uncommon (Hoyt and Beers, 1993).

Populations of *T. mcdanieli* first develop on the lower leaf surfaces, and later spread on to both leaf surfaces. Young leaves infested by the McDaniel mite curl upward, and this protects mites living on the upper leaf surface. The mite produces heavy webbing that covers the leaves and traps *Panonychus ulmi* (Koch), another pest mite of apple, hindering its feeding and oviposition (Hoyt, 1969b). *T. mcdanieli* overwinters as bright orange adult females under bark scales, in various crevices and in the litter in the soil at the base of the tree. Overwintering mites are often aggregated in large masses and produce extensive webbing. In the USA, in March or April, at about the time fruit buds begin to open, these females emerge from overwintering sites and move towards weed plants, buds and new water sprout growth developing near the main trunk. They feed on new opened leaves and flower parts, and after some days start to lay eggs. While they are feeding, overwintering females change their orange colour to the usual greenish colour with dorsal spots. Eggs are first laid on the lower leaf surfaces, either within or beneath the webbing, but when the populations are high, egg deposition occurs on both leaf surfaces. Female fecundity reaches 150 eggs, laid over 5–6-weeks. Development can take 6 days at high temperatures, and at common summer field temperatures ranges from 8 to 11 days. According to the temperature, in central Washington State the mite may develop up to ten overlapping generations a year. On apple, in the early part of the season, the mites usually live close to the main trunk and limbs and their distribution covers a cone-shaped area in the centre of the tree. Although some lateral distribution may occur, the mites do not inhabit the terminal growth, because they prefer older leaves. In the absence of predators, populations peak in late July to mid-August, and then decline. Overwintering females usually appear from mid-July through to September, but if leaf quality deteriorates they may appear as early as May. The onset of the cold season induces the death of males, and overwintering females move to sheltered sites (Nielsen, 1958; Hoyt, 1969a,b; Jeppson *et al.*, 1975; Hoyt and Beers, 1993). As an adult, *T. mcdanieli* disperses by wind or walking (Hoy, 2011).

Egg to egg development took from 8.12 days at 35°C to 61.96 days at 10°C. The mite developed rapidly at 35°C. At 25 ± 1°C on leaves of bean, the net reproduction (R_0) was 75.1, the intrinsic rate of increase (r_m) was 0.201 and the finite rate of increase (λ) was 1.223. Values for the intrinsic rate of increase (r_m) ranged from 0.115 a day at 18°C to 0.431 a day at 35°C and at 40–65% RH (Tanigoshi *et al.*, 1975b). The optimum temperature for egg deposition appeared to be between 29 and 32°C. The host plants used influenced the development of the mites, the duration of the pre-ovipositional and ovipositional periods, the number of eggs laid and the proportion of mites failing to reach maturity (Nielsen, 1958).

Natural enemies include predatory insects, such as the coccinellids *Stethorus punctum* (LeConte) (Robinson, 1953) and *S. punctillum* Weise (Roy *et al.*, 1999), and mites, such as the phytoseiid mites *Metaseiulus occidentalis* (Nesbitt) and *Neoseiulus fallacis* (Garman) (Nielsen, 1958; Dondale, 1967; Jeppson *et al.*, 1975; Roy *et al.*, 1999).

Symptomatology and damage

Motile forms of *T. mcdanieli* insert their stylets through the walls of the epidermal cells. In damaged tissues of apple leaves, most of the injured cells appear deformed, lacking protoplasts and with walls shrunken, and sometimes neighbouring cells are altered (Tanigoshi and Davis, 1978). Young leaves infested by the mite curl upward. During feeding, the mite removes the chlorophyll, causing stippling, and at high levels of attack the leaves turn brown and fall, and the fruits appear sunburned. Hot, dry conditions intensify the damage. An increase in the population induces dispersal of the mites to the peripheral parts of the tree, which become brownish grey. The abundant webbing produced by the mites on leaves and fruit causes the foliage to appear matted (Jeppson *et al.*, 1975; Hoyt and Beers, 1993). Webs on the leaves may prevent spray penetration, and make chemical control difficult. Mites and webbing in the calyx end

of fruit also represent a problem as the mite is regulated by quarantine in some countries (Hoyt and Beers, 1993).

Control

Sampling and monitoring of populations, cultural practices, the management of natural enemies and application of selective pesticides are crucial for the control of the McDaniel mite. Alston and Reding (2011a) note that 'Relying primarily on biological and cultural control for spider mite management should be the goal of every orchard pest manager. In many situations, chemical control is unnecessary and may only make the mite problem worse. Use of miticides will eliminate the beneficial predatory mites, which are capable of keeping spider mite populations below economically damaging levels. Another reason to avoid chemical control is that populations of two-spotted and McDaniel spider mites have developed resistance to miticides'. These reflections further confirm that outbreaks of *T. mcdanieli* basically depend on changes induced by agricultural practices in the balance of the ecosystem.

SAMPLING. Alston and Reding (2011a) suggested that on apple in Utah, monitoring for the McDaniel spider mite and its predators should use the presence–absence or binomial sampling method proposed by Jones (1990). In this method, ten trees are randomly selected per 2–5 acre wide block. Sampling is done every 1–2 weeks from mid-June through to September (every week during the hot period). Each sampling is of 100 leaves, corresponding to ten leaves from each of ten plants. The leaves must be collected from inside the canopy, as well as from the edges. The leaves collected on each tree must be separated from those collected from other trees. The number of leaves from each tree harbouring pest and predatory mites is counted and recorded. The use of suitable (look-up) tables allows estimation of the number of mites (pest and predatory) per leaf.

THRESHOLDS. According to Jones (1990), application of the presence–absence sampling method enables one to find out whether the economic thresholds for spider mites have been exceeded. Alston and Reding (2011a) suggested the following protocol: less than five spider mites and no predator per leaf of apple and stone fruits requires sampling within 1 week; a predator:spider mite ratio of at least 1:5 requires sampling after 2 weeks; the presence of 5–10 spider mites and less than one predator per leaf requires sampling in 2–5 days. A predator:pest ratio of at least 1:10 requires a new sampling within 1 week. Lastly, an average number of spider mites per leaf more than ten and less than one predator per leaf requires the application of acaricide spray.

CULTURAL PRACTICES. Ground cover management affects orchard mite ecology. It has been found that some broadleaved weeds and apple root suckers enhance the movement of spider mites from ground cover plants to trees. Healthy grass ground cover displaces most broadleaved weeds used as reproductive hosts by spider mites and prevents an increase in their density. In spring and early summer, spider mites may develop on ground cover plants and help the development of the predator mite populations that feed on them. A low level of weeds may reduce the density of spider mites and improve the synchrony of predator and prey when they move into the trees later. However, predatory mites may be attracted in the early season by rust mites present in the trees and then there is no need for spider mites on ground cover. In brief, the movement of mites (pests and predators) may be strongly affected by ground cover management (mowing, herbicide application, cultivation, irrigation) (Alston and Reding, 2011a).

BIOLOGICAL CONTROL. Research by Hoyt (1969a,b) in commercial apple orchards in Washington State showed that when *T. mcdanieli* was the only prey available for the phytoseiid mite *M. occidentalis*, a cyclic predator–prey response developed. The potential of this predatory mite was highlighted in apple orchards in Utah and California, where *T. mcdanieli* was rarely found in unsprayed trees, but was present in sprayed orchards, along with *M. occidentalis* (Croft and Jorgensen, 1977). Knowledge on the distribution of *T. mcdanieli* females and *M. occidentalis* was regarded as relevant to the management of *T. mcdanieli* in Washington apple orchards in the 1960s (Hoyt, 1969a,b), but currently, *M. occidentalis* is regarded as a key factor for the biological control and integrated management (IPM) of the McDaniel spider mite. Its presence in apple orchards and other crops must be preserved by making choices such as the use of selective pesticides, not treating by dormant, delayed dormant and pre-bloom treatments of the lower trunk and surrounding ground cover where predators overwinter, and not controlling populations of rust mites, which are used early in the season as a food source by predators (Alston and Reding, 2011a). When predatory mites are absent from orchards, it is possible to re-establish them by cutting shoots containing both *M. occidentalis* and *T. mcdanieli* from other orchards and placing them in the trees of the orchard that is lacking predators. It is also possible to purchase predatory phytoseiid mites from commercial suppliers, and strains of phytoseiid mites tolerant to some pesticides are available (Alston and Reding, 2011a).

CHEMICAL CONTROL. Chemical control of the McDaniel mite is difficult. Overwintering forms protected in various sites prevent good results from winter treatments, and profuse webbing prevents good results from summer treatments. Moreover, *T. macdanieli* has developed resistance to various acaricides, such as dicofol, binapacryl and tetradifon (Cranham and Helle, 1985). According to their registration, several acaricides (abamectin, acequinocyl, bifenazate, clofentezine, dicofol, etoxazole, fenbutatin oxide, fenpyroximate, hexythiazox, insecticidal soap, propargite, pyridaben, spirodiclofen, sucrose octanoate esters and summer-weight oils such as horticultural mineral oil, canola oil, soybean oil and others) may be used in summer sprays (Alston and Reding, 2011a).

INTEGRATED PEST MANAGEMENT. The first outbreaks of the McDaniel mite during the late 1950s and early 1960s suggested that integrated control should be applied, and this has been successfully implemented in apple orchards in Washington State (Hoyt and Beers, 1993) and Utah (Alston and Reding, 2011b), and also in

other countries. The main component of the IPM programme is the use of selective pesticides to preserve natural and artificial populations of natural enemies of *T. macdanieli*. The strategy may be formally integrated by releases of natural enemies. Selectivity can be achieved by the choice of active ingredients, timing of applications, rates at which chemicals are applied and the placement of sprays. All of these approaches can be used in an IPM programme (Hoy, 2011). Among the natural enemies of *T. macdanieli*, the phytoseiid *M. occidentalis* plays a key role. This is an obligatory predatory mite whose development requires a consistent source of prey in apple orchards, which can be provided by the apple rust mite (*Aculus schlechtendali* (Nalepa), *T. mcdanieli* and the European red mite (*P. ulmi*)) (Hoyt, 1969a).

Knowledge on the distribution of *T. mcdanieli* females and *M. occidentalis* is fundamental to the management of *T. mcdanieli* in Washington apple orchards. *T. mcdanieli* females emerging from winter diapause in the spring lay eggs on flower parts or on the leaves (Hoyt, 1969a,b). Adult females of the phytoseiid mite overwinter on the trunk of the apple trees and also move on to these same plant organs early in the spring. During this first generation, the action of the predator exerts a strong impact on the populations of prey that can also destroy (Hoyt, 1969b). In addition, *M. occidentalis* is an important predator of *A. schlechtendali* (Hoyt, 1969a) and *P. ulmi* in Washington apple orchards. Rust mite populations are fundamental to maintaining adequate numbers of *M. occidentalis* at key times during the growing season so that they remain an effective natural enemy of *T. mcdanieli* (Hoy, 2011). The phytoseiid tolerates the profuse webbing produced by *T. mcdanieli* and responds to increases in its populations with cyclic fluctuations, which are usually indicative of effective biological control. Diapausing females of *M. occidentalis* leave the leaves in September and seek overwintering sites in crevices on the tree trunk. Cardboard bands placed on the trunks allow the phytoseiid mites to be sampled, and the counting and removal of aggregated diapausing females (Horton *et al.*, 2002).

Releases of 32, 64, 128 and 256 *M. occidentalis*/apple tree provided good control of *T. mcdanieli*. The use of the predator reduced the pest populations by over ten times, with the best results obtained by releasing 128 phytoseiids/tree (Croft and McMurtry, 1972a). The use of an index allowed prediction of the outcome of the interaction of the predator and prey in apple orchards in California. This index is based on the counting of prey and predator collected on leaves weekly. Fewer than 20 pest mites per leaf confirm effective biocontrol, and more than 20 pest mites per leaf indicate unsatisfactory biological control. The predictive value of this index demands constant monitoring, and its application in other countries and to different apple cultivars must be validated (Croft and Nelson, 1972). For further details on practical implementation of an IPM strategy see Alston and Reding (2011a).

Tetranychus mexicanus (McGregor) (Fig. 20.59)

Common name

Unknown.

Diagnostic characteristics

FEMALE. The body is carmine in colour, with the dorsal setae long, finely serrate and slender. The peritreme is hooked. The dorsal striae are lobate and transverse to oblique between the setae e_1–e_1, longitudinal between the setae f_1–f_1, and transverse between the setae e_1 and f_1 and forming an hourglass shape. The ventral striae mostly lack lobes, but low lobes are present between the setae *4a*; the pregenital striae are weak and broken medially, and dotted. The empodia have 6 proximoventral hairs, and an obvious dorsal spur; tarsus I has 4 tactile setae proximal to the proximal duplex setae (Pritchard and Baker, 1955; Seeman and Beard, 2011).

MALE. The dorsal striae bear small lobes. The empodia have a dorsal spur; empodia I are uncinate, and empodia II–IV have long and free proximoventral hairs. The knob of the aedeagus is large and dorsally convex, and with pointed posterior and anterior projections (Seeman and Beard, 2011).

Geographical distribution

Argentina (Pritchard and Baker, 1955; Rossi Simons, 1961; Guanilo *et al.*, 2010), Brazil (Flechtmann, 1967a; Paschoal and Reis, 1968; Flechtmann and Abreu, 1973; Aranda, 1974; Flechtmann and Baker, 1975; Feres *et al.*, 2005, 2009), Colombia (Urueta, 1975), Costa Rica (Ochoa *et al.*, 1991a; Aguilar and Murillo, 2008, 2012), Cuba (Livshits and Salinas Croche, 1968; Suárez, 2004), El Salvador (Andrews and Poe, 1980), Guadeloupe (Flechtmann *et al.*, 1999a; Flechtmann and Etienne, 2006), Honduras (Ochoa *et al.*, 1991a), Mexico (Beer and Lang, 1958), Nicaragua (Ochoa *et al.*, 1991a), Paraguay (Aranda, 1969; Aranda and Flechtmann, 1971), Peru (Bolland *et al.*, 1998; Migeon and Dorkeld, 2006–2013; Guanilo *et al.*, 2012), the USA (McGregor, 1950), Uruguay (Bernal and Piñeiro, 1982) and Venezuela (Quirós-Gonzalez, 2000).

Host plants

The whole list of host plants of *T. mexicanus* includes 107 wild and cultivated plants belonging to 43 families (Migeon and Dorkeld, 2006–2013). The families and species of economic importance are:

Anacardiaceae: cashew (*Anacardium occidentale* L.) (Paschoal, 1970).

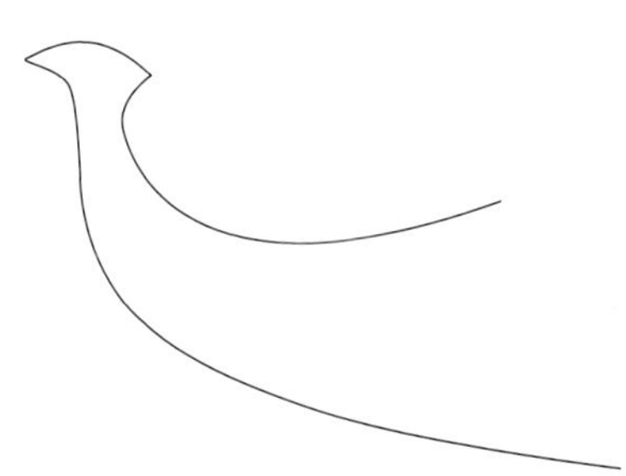

Fig. 20.59. *Tetranychus mexicanus* (McGregor). Male, aedeagus (modified from McGregor, 1950).

Annonaceae: soursop (*Annona muricata* L.) (Paschoal, 1970; Urueta, 1975), *A. purpurea* Moc. and Sessé ex Dunal (Estebanes-Gonzalez and Baker, 1968), custard apple (*A. reticulata* L.), sugar apple (*A. squamosa* L.), atemoya (*Annona* × *atemoya* Mobb.) (Livshits and Salinas-Croche, 1968).

Aquifoliaceae: mate (*Ilex paraguariensis* A. St.-Hil.) (Aranda and Flechtmann, 1971).

Arecaceae: coconut palm (*Cocos nucifera* L.) (Beer and Lang, 1958; Flechtmann, 1967a; Urueta, 1975), African oil palm (*Elaeis guineensis* Jacq.), American oil palm (*E. oleifera* (Kunth) Cortés) (Urueta, 1975).

Caricaceae: papaya (*Carica papaya* L.) (Flechtmann and Arruda, 1967; Aranda and Flechtmann, 1971; Martínez *et al.*, 2004).

Caryocaraceae: pequi (*Caryocar brasiliense* Cambess.) (Aranda, 1974; Flechtmann and Baker, 1975).

Dioscoreaceae: water yam (*Dioscorea alata* L.) (Paschoal, 1970).

Euphorbiaceae: rubber tree (*Hevea brasiliensis* (Willd. ex A. Juss.) Müll. Arg.) (Chiavegato, 1968), cassava (*Manihot esculenta* Crantz) (Paschoal and Reis, 1968), castor bean (*Ricinus communis* L.) (Guanilo *et al.*, 2010).

Fabaceae: groundnut (*Arachis hypogaea* L.) (Bernal and Piñeiro, 1982), bean (*Phaseolus vulgaris* L.) (Bolland *et al.*, 1998; Migeon and Dorkeld, 2006–2013).

Lauraceae: avocado (*Persea americana* Mill.) (Aranda and Flechtmann, 1971).

Malvaceae: Egyptian cotton (*Gossypium barbadense* L.) (Azevedo and Vieira, 2002), Arabian cotton (*G. herbaceum* L.) (Aranda and Flechtmann, 1971), cacao (*Theobroma cacao* L.) (Flechtmann, 1967a; Suárez, 2004).

Meliaceae: neem (*Azadirachta indica* A. Juss.) (Mesa Cobo and Zuluaga Cardona, 1983).

Moraceae: black mulberry (*Morus nigra* L.) (Flechtmann, 1967a).

Myrtaceae: guava (*Psidium guajava* L.) (Bolland *et al.*, 1998; Migeon and Dorkeld, 2006–2013).

Oxalidaceae: carambola (*Averrhoa carambola* L.) (Flechtmann, 1967a).

Passifloraceae: passion fruit (*Passiflora edulis* Sims) (Flechtmann and Abreu, 1973; Urueta, 1975).

Poaceae: sugarcane (*Saccharum officinarum* L.), Johnson (Aleppo) grass (*Sorghum halepense* (L.) Pers.) (Bolland *et al.*, 1998; Migeon and Dorkeld, 2006–2013).

Rosaceae: strawberry (*Fragaria* × *ananassa* Duchesne), apple (*Malus domestica* Borkh.), peach (*Prunus persica* (L.) Batsch) (Flechtmann, 1967a), Japanese plum (*P. salicina* Lindl.) (Paschoal and Reis, 1968), pear (*Pyrus communis* L.) (Paschoal, 1968a).

Rutaceae: Egyptian lime (*Citrus aurantiifolia* (Christm.) Swingle), kumquat (*C. japonica* (Thunb.) Swingle) (Fernandez, 1972), Persian lime (*C. latifolia* (Yu. Tanaka) Tanaka) (Quirós-Gonzalez, 2000), lemon (*C. limon* (L.) Brurm. f.) (Pritchard and Baker, 1955), grapefruit (*C. paradisi* Mcfad.), tangerine (*C. reticulata* Blanco) (Guanilo *et al.*, 2010), sweet orange (*C. sinensis* (L.) Osbeck) (McGregor, 1950; Aguilar and Murillo, 2008; Horn *et al.*, 2011), bitter orange (*C. trifoliata* L.) (Paschoal and Reis, 1968).

Sapindaceae: litchi (*Litchi chinensis* Sonn.) (Paschoal and Reis, 1968), Spanish lime (*Melicoccus bijugatus* Jacq.) (Flechtmann *et al.*, 1999a).

Bio-ecology

T. mexicanus is reported on several economic plants in tropical and subtropical countries, but it does not seem to play a significant role as a pest. Its outbreaks are sporadic and mostly restricted to unhealthy plants (Quirós-Gonzalez, 2000; Gerson, 2003).

The biology of *T. mexicanus* was investigated on detached leaves of three species of *Annona* (*A. coriaceae* L., *A. muricata* L. and *A. squamosa* L.) in the laboratory at 27°C, 70% RH under a 12L:12D photoperiod (Sousa *et al.*, 2010). The shortest development time of eggs was observed on *A. coriaceae* (2.4 ± 0.04 days) and the longest on *A. muricata* (3.4 ± 0.04 days) and *A. squamosa* (3.4 ± 0.08 days). The shortest larval stage was observed on *A. coriaceae* (2.5 ± 0.06 days) and the longest on *A. squamosa* (3.5 ± 0.17 days). The shortest protonymphal stage was found on *A. muricata* (1.9 ± 0.04 days) and the longest on *A. squamosa* and *A. coriaceae* (2.4 ± 0.10 days). The shortest deutonymphal stage was observed on *A. muricata* (2.3 ± 0.05 days) and the longest on *A. squamosa* (2.7 ± 0.12 days). On *A. muricata*, *A. squamosa* and *A. coriaceae*, respectively: the development from egg to adult took 10.1 ± 0.11, 12.1 ± 0.26 and 9.9 ± 0.14 days, the fecundity was 98.9 ± 7.17, 65.4 ± 5.74 and 38.9 ± 3.19 eggs, the longevity of the female was 23.0 ± 1.51, 24.9 ± 2.30 and 18.9 ± 1.33 days and the longevity of the male was 47.0 ± 1.52, 38.1 ± 3.50 and 29.5 ± 1.46 days. The percentage of males was lower that 1%. On *A. muricata*, *A. squamosa* and *A. coriaceae*, respectively, the intrinsic rate of increase (r_m) was 0.24, 0.17 and 0.21, the net reproduction (R_0) was 85.45, 31.72 and 28.30, the generation time (*T*) was 18.66, 20.74 and 15.50 and the finite rate of increase (λ) was 1.27, 1.18 and 1.24 (Sousa *et al.*, 2010).

On leaflets of peach palm (*Bactris gasipaes* Kunth) in the laboratory at 25 ± 1°C, 60.0 ± 10.0 % RH and under a photoperiod of 14 h of light, the development from egg to adult took 13 days for males and 13.6 days for females, with a survival rate of 59.3%. The female:male sex ratio was 4.7:1. After a pre-oviposition period of 2 days, the females oviposited for 12.1 days the average of 0.82 eggs per female a day. Male and female longevity was 7.1 and 16.7 days, respectively. The intrinsic rate of increase (r_m) was 0.06, the finite rate of increase (λ) 1.06 individuals per female a day, the mean generation time (*T*) 21.6 days and the net reproductive rate (R_0) 3.7 (Stein and Daólio, 2012). A study of the viability of eggs of *T. mexicanus* on the leaves of African oil palm at 15 ± 1, 20 ± 1, 25 ± 1, 30 ± 1 and 35 ± 1°C and under a 12 h photoperiod showed that at 15 and 20°C the larvae did not hatch. At 25, 30 and 35°C, the viability was >93% (Oliveira Azevedo *et al.*, 2012).

Natural enemies include some predatory mites, such as the stigmaeid *Agistemus floridanus* Gonzalez, and the phytoseiids *Euseius concordis* (Chant), *Neoseiulus anonymus* (Chant *et* Baker) (Ferla and Moraes, 2003), and *Phytoseiulus macropilis* (Banks) (Jaime de Herrero, 1984).

Symptomatology and damage

Although this mite lives on a wide list of host plants, it does not generally cause severe damage to crops. Occasionally, it reaches damaging levels on citrus in Texas, Mexico and Argentina (Jeppson, 1989; Quirós-Gonzalez, 2000; Gerson, 2003; Vacante, 2010), and on several other cropping systems in tropical and

warm subtropical areas, such as soursop (Sousa *et al.*, 2010). Water-stressed plants are probably more susceptible to attack by the mite (Quirós-Gonzalez, 2000).

Control

Various acaricides have been applied to cacao seedlings and other crops, including avermectins, dinitrophenol derivates (Nakayama *et al.*, 1987), organosulfurs (Nakayama *et al.*, 1987; Hernández *et al.*, 1988), organotins, pyrethroids (Nakayama *et al.*, 1987) and sulfur (Hernández *et al.*, 1988). Spray applications of acaricides on citrus are not justified.

Tetranychus neocaledonicus André (Fig. 20.60)

Common name

Vegetable mite.

Diagnostic characteristics

FEMALE. The body is bright red and the legs pale white in colour. The peritremes are hooked distally. The dorsal body setae are finely serrate, slender and longer than the distances between their bases. The dorsal striae are lobate; between the setae e_1–e_1 and f_1–f_1 they are longitudinal, and between the setae e_1 and f_1 they are transverse and form a diamond shape. The ventral striae lack lobes; the pregenital striae are obviously broken and dotted anteriorly. The empodia have 6 proximoventral hairs, and the spurs on the empodia are minute or lacking. Tarsi I have 4 tactile setae proximal to the duplex setae (Gutierrez and Schicha, 1983; Seeman and Beard, 2011).

MALE. The body is greenish yellow (Gutierrez and Schicha, 1983). The empodia have a minute dorsal spur; empodia I are claw like, and empodia II–IV have long and free proximoventral hairs. Morphological variations occur on the aedeagus. Seeman and Beard (2011) reported that the knob of the aedeagus was dorsally convex and posteromedially indented, with the anterior projection short and pointed and the posterior projection short and broadly rounded. Gutierrez and Schicha (1983) found that the knob of the aedeagus was berry like, and Meyer Smith (1987) reported that the anterior and posterior projections of knob were rounded and separated by a small indentation.

Geographical distribution

Angola (Meyer Smith, 1974), Australia (Davis, 1968b; Seeman and Beard, 2011), Bahamas (Pritchard and Baker, 1955), Benin (Gutierrez and Etienne, 1981b; Navajas *et al.*, 1996), Brazil (Flechtmann and Baker, 1970; Flechtmann, 1981b, 1996; Fiaboe *et al.*, 2007a; Poderoso *et al.*, 2010), Cameroon (Meyer Smith and Bolland, 1984), Canary Islands (Ferragut and Santonja, 1989), Cape Verde (Meyer Smith, 1987), China (Ehara and Lee, 1971; Ma *et al.*, 1979; Wang, 1981a), Comoros (Gutierrez, 1974), Congo (Gutierrez and Bonato, 1994), Congo (Democratic Republic of the Congo, formerly Zaire) (Baker and Pritchard, 1960), Cook Islands (Gutierrez, 1980), Costa Rica (Ochoa *et al.*, 1991a; Aguilar and Murillo, 2008), Cuba (Livshits and Salinas-Croche, 1968; Martínez *et al.*, 2004; Suárez, 2004), Egypt (Attiah, 1969; Zaher *et al.*, 1982), Fiji (McGregor, 1950), French Polynesia (Cohic, 1963; Gutierrez, 1981), Guadeloupe (Gutierrez and Etienne, 1988), Guam (Bolland *et al.*, 1998), Hawaiian Islands (McGregor, 1950; Lee Goff, 1986), Honduras (Ochoa *et al.*, 1991a), India (Khot and Patel, 1956; Manson, 1963; Nassar and Ghai, 1981; Gupta, 1992, 1995; Gupta and Gupta, 1994; Gupta and Chatterjee, 1997), Iran (Sepasgosarian, 1956), Ivory Coast (Gutierrez and Etienne, 1981b), Japan (Ehara, 1969a; Ehara and Yamaguchi, 2001; Gotoh *et al.*, 2003a, 2007c; Ohno *et al.*, 2009, 2011b; Matsuda *et al.*, 2013), Kenya (Baker and Pritchard, 1960), Kiribati (Gutierrez, 1977), Madagascar (André, 1959), Malawi (Meyer Smith, 1974), Malaysia (Doreste, 1982; Othman and Zhang, 2003), Northern Marianas (Gutierrez, 1980), Mauritius (Moutia, 1958), Mozambique (Meyer Smith and Rodrigues, 1966), New Caledonia (André, 1933b), New Zealand (Manson, 1963; Anthony, 2008 (?)), Nigeria (Matthysse, 1980), Niue (Gutierrez, 1980), Pakistan (Rahman and Sapra, 1940a), Papua New Guinea (Schicha and Gutierrez, 1985; Seeman and Beard, 2011), Peru (Flechtmann, 1996; Guanilo *et al.*, 2012), Philippines (Rimando, 1962a), Puerto Rico (McGregor, 1950), Réunion (Gutierrez, 1968), American Samoa (Gutierrez, 1976a), Western Samoa (Gutierrez, 1977), Senegal (Gutierrez and Etienne, 1981b), Seychelles (Gutierrez, 1974), South Africa (Meyer Smith, 1965), Taiwan (Ehara, 1969a), Thailand (Baker, 1975b), Tonga (Gutierrez, 1977), Trinidad and Tobago (CABI, 1990b), Tuvalu (Gutierrez, 1977), the USA (Pritchard and Baker, 1955), Vanuatu (Gutierrez, 1975), Zambia (Meyer Smith, 1974; Mwase and Baker, 2006), Venezuela (McGregor, 1950) and Zimbabwe (Baker and Pritchard, 1960).

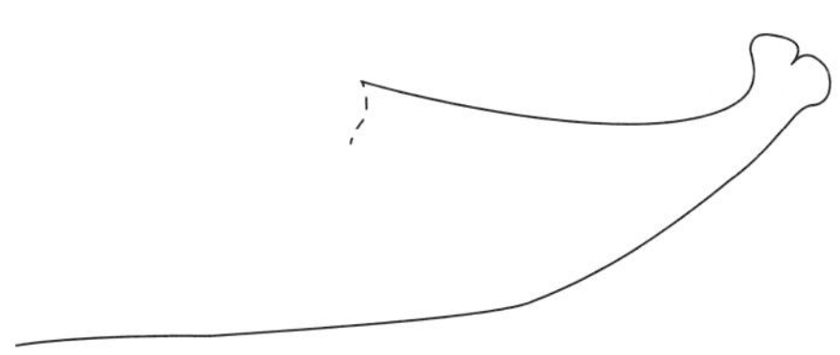

Fig. 20.60. *Tetranychus neocaledonicus* André. Male, aedeagus (from Meyer Smith, 1987).

Host plants

The whole list of host plants of the vegetable mite includes over 445 wild and cultivated plants belonging to 91 families. Most species belong to the Fabaceae (68 species), followed in decreasing order by the Asteraceae (34 species), Malvaceae (32 species), Solanaceae (26 species) and other families (Migeon and Dorkeld, 2006–2013). The families and species of economic importance are:

Amaranthaceae: beet (*Beta vulgaris* L.) (Gupta and Gupta, 1994).

Amaryllidaceae: onion (*Allium cepa* L.), Spanish garlic (*A. sativum* L.) (Bolland *et al.*, 1998; Migeon and Dorkeld, 2006–2013).

Anacardiaceae: mango (*Mangifera indica* L.) (Pritchard and Baker, 1955; Lee Goff, 1986; Gupta and Gupta, 1994).

Annonaceae: pond apple (*Annona glabra* L.), soursop (*A. muricata* L.) (Gutierrez and Schicha, 1983), sugar apple (*A. squamosa* L.) (Davis, 1968b).

Apiaceae: wild celery (*Apium graveolens* L.), coriander (*Coriandrum sativum* L.) (Prasad, 1974; Gupta and Gupta, 1994).

Asteraceae: sunflower (*Helianthus annuus* L.) (Zaher *et al.*, 1982).

Brassicaceae: cabbage (*Brassica oleracea* L.) (Rahman and Sapra, 1940a; Gupta and Gupta, 1994), Chinese cabbage (*B. rapa* L.) (Prasad, 1974; Gutierrez and Etienne, 1986), rocket salad (*Eruca vesicaria* (L.) Cav.) (Prasad, 1974; Gupta and Gupta, 1994).

Caricaceae: papaya (*Carica papaya* L.) (Pritchard and Baker, 1955; Gutierrez, 1975, 1976b; Gupta and Gupta, 1994; Ohno *et al.*, 2011a; Seeman and Beard, 2011; Matsuda *et al.*, 2013).

Caryophyllaceae: carnation (*Dianthus caryophyllus* L.) (Meyer Smith and Bolland, 1984).

Convolvulaceae: sweet potato (*Ipomoea batatas* (L.) Lam.) (Pritchard and Baker, 1955; Suárez, 2004; Ohno *et al.*, 2009, 2010).

Cucurbitaceae: ash gourd (*Benincasa hispida* (Thunb.) Cogn.) (Gupta and Gupta, 1994), watermelon (*Citrullus lanatus* (Thunb.) Matsum. and Nakai) (Rahman and Sapra, 1940a; Lee Goff, 1986; Ohno *et al.*, 2009), bur cucumber (*Cucumis anguria* L.) (Prasad, 1974), cantaloupe (*C. melo* L.) (Ohno *et al.*, 2009), garden cucumber (*C. sativus* L.) (Gutierrez and Etienne, 1981a; Gupta and Gupta, 1994), winter squash (*Cucurbita maxima* Duchesne) (Rahman and Sapra, 1940a; Gupta and Gupta, 1994; Suárez, 2004), crookneck squash (*C. moschata* Duchesne) (Rahman and Sapra, 1940a), field pumpkin (*C. pepo* L.) (Rahman and Sapra, 1940a; Gupta and Gupta, 1994), bottle gourd (*Lagenaria siceraria* (Molina) Standl.) (Gupta and Gupta, 1994), ribbed gourd (*Luffa acutangula* (L.) Roxb.) (Prasad, 1974; Gupta and Gupta, 1994), bitter melon (*Momordica charantia* L.), gac (*M. cochinchinensis* Spreng.) (Ehara, 1969a), chayote (*Sechium edule* (Jacq.) Sw.) (Paschoal, 1970), fluted gourd (*Telfairia occidentalis* Hook. f.) (Bolland *et al.*, 1998; Migeon and Dorkeld, 2006–2013).

Euphorbiaceae: poinsettia (*Euphorbia pulcherrima* Willd. ex Klotzsch) (Ohno *et al.*, 2010), cassava (*Manihot esculenta* Crantz) (Rodrigues, 1968; Gutierrez, 1975, 1976b; Gupta and Gupta, 1994; Gutierrez and Bonato, 1994; Seeman and Beard, 2011), castor bean (*Ricinus communis* L.) (Baker and Pritchard, 1960; Gupta and Gupta, 1994; Guanilo *et al.*, 2012).

Fabaceae: groundnut (*Arachis hypogaea* L.) (McGregor, 1950; Seeman and Beard, 2011), pigeon pea (*Cajanus cajan* (L.) Huth) (Davis, 1968b; Livshits and Salinas-Croche, 1968), soybean (*Glycine max* (L.) Merr.) (Gupta, 1976; Gupta and Gupta, 1994), bur medic (*Medicago polymorpha* L.), lucerne (*M. sativa* L.) (Flechtmann, 1967a; Gupta and Gupta, 1994), lima bean (*Phaseolus lunatus* L.) (Prasad, 1974; Gupta and Gupta, 1994), bean (*P. vulgaris* L.) (McGregor, 1950; Gutierrez, 1976a; Gupta and Gupta, 1994; Ohno *et al.*, 2009), pea (*Pisum sativum* L.) (Prasad, 1974; Gupta and Gupta, 1994), asparagus pea (*Psophocarpus tetragonolobus* (L.) DC.) (Gutierrez and Schicha, 1984; Ohno *et al.*, 2009), tamarind (*Tamarindus indica* L.) (McGregor, 1950), Egyptian clover (*Trifolium alexandrinum* L.), Dutch clover (*T. repens* L.), fenugreek (*Trigonella foenum-graecum* L.), mat bean (*Vigna aconitifolia* (Jacq.) Maréchal) (Prasad, 1974; Gupta and Gupta, 1994), ground bean (*V. subterranea* (L.) Verdc.) (Prasad, 1974; Gupta and Gupta, 1994; Navajas *et al.*, 1996).

Lamiaceae: basil (*Ocimum basilicum* L.) (Mendonça *et al.*, 2010), holy basil (*O. tenuiflorum* L.) (Baker, 1975b; Gupta and Gupta, 1994), perilla (*Perilla frutescens* (L.) Britton) (Ohno *et al.*, 2009).

Lauraceae: avocado (*Persea americana* Mill.) (Bolland *et al.*, 1998; Migeon and Dorkeld, 2006–2013).

Malvaceae: okra (*Abelmoschus esculentus* (L.) Moench) (Bindra and Goyal, 1966; Gupta, 1992; Ohno *et al.*, 2009; Seeman and Beard, 2011), manioc hibiscus (*A. manihot* (L.) Medik.) (Seeman and Beard, 2011), tree cotton (*Gossypium arboreum* L.), Brazilian cotton (*G. barbadense* L.) (Davis, 1968b), Arabian cotton (*G. herbaceum* L.) (Gupta and Gupta, 1994), American cotton (*G. hirsutum* L.) (Rodrigues, 1968).

Meliaceae: neem (*Azadirachta indica* A. Juss.) (Pritchard and Baker, 1955).

Moraceae: chempedak (*Artocarpus integer* (Thunb.) Merr.) (Baker, 1975b), monkey jack (*A. rigidus* Blume), fig (*Ficus carica* L.) (Prasad, 1974; Gupta and Gupta, 1994), white mulberry (*Morus alba* L.) (Gupta and Gupta, 1994; Ehara and Yamaguchi, 2001; Gotoh *et al.*, 2003a, 2007c; Ohno *et al.*, 2010; Matsuda *et al.*, 2013), black mulberry (*M. indica* L.), *M. nigra* L. (Livshits and Salinas-Croche, 1968).

Moringaceae: moringa (*Moringa oleifera* Lam.) (Kaimal and Ramani, 2007a).

Musaceae: banana (*Musa* × *paradisiaca* L.) (Migeon and Dorkeld, 2006–2013).

Myrtaceae: guava (*Psidium guajava* L.) (Seeman and Beard, 2011).

Oxalidaceae: carambola (*Averrhoa carambola* L.) (Ohno *et al.*, 2010).

Passifloraceae: passion fruit (*Passiflora edulis* Sims) (Davis, 1968b), sweet granadilla (*P. ligularis* Juss.) (Aguilar and Murillo, 2008), giant granadilla (*P. quadrangularis* L.) (Gutierrez and Etienne, 1986).

Phyllanthaceae: gooseberry tree (*Phyllanthus acidus* (L.) Skeels) (Livshits and Salinas-Croche, 1968).

Piperaceae: pepper elder (*Peperomia pellucida* (L.) Kunth), jia ju (*Piper sarmentosum* Roxb.) (Bolland *et al.*, 1998; Migeon and Dorkeld, 2006–2013).

Poaceae: African rice (*Oryza glaberrima* Steud.) (Gutierrez and Etienne, 1981a), bajra (*Pennisetum glaucum* (L.) R. Br.), sorghum (*Sorghum bicolor* (L.) Moench) (Prasad, 1974; Gupta and Gupta, 1994), maize (*Zea mays* L.) (Baker, 1975b).

Rhamnaceae: jujube (*Ziziphus jujube* Mill.) (Flechtmann *et al.*, 1999a).

Rosaceae: loquat (*Eriobotrya japonica* (Thunb.) Lindl.), apple (*Malus domestica* Borkh.), peach (*Prunus persica* (L.) Batsch) (Moutia, 1958; Lee Goff, 1986), Japanese rose (*Rosa multiflora* Thunb.) (Davis, 1968b).

Rubiaceae: Spanish tamarind (*Vangueria madagascariensis* J.F. Gmel.) (Migeon and Dorkeld, 2006–2013).

Rutaceae: Egyptian lime (*Citrus aurantiifolia* (Christm.) Swingle), sour orange (*C. aurantium* L.) (Attiah, 1969), lemon

(*C. limon* (L.) Burm. f.) (Prasad, 1974; Nassar and Ghai, 1981; Gupta and Gupta, 1994), pummelo (*C. maxima* (Burm.) Merr.), citron (*C. medica* L.) (Gupta and Gupta, 1994), sweet orange (*C. sinensis* (L.) Osbeck) (Attiah, 1969), Japanese bitter orange (*C. trifoliata* (L.) Raf.), wood apple (*Limonia acidissima* L.) (Migeon and Dorkeld, 2006–2013).

Santalaceae: sandalwood (*Santalum album* L.) (Prasad, 1974; Gupta and Gupta, 1994).

Solanaceae: pepper (*Capsicum annuum* L.) (Martínez *et al.*, 2004), tree tobacco (*Nicotiana glauca* Graham) (Meyer Smith, 1987), tobacco (*N. tabacum* L.), goldenberry (*Physalis peruviana* L.), scarlet aubergine (*Solanum aethiopicum* L.) (Flechtmann 1967a), tomato (*S. lycopersicum* L.) (Rahman and Sapra, 1940a; Gupta and Gupta, 1994; Suárez, 2004), African aubergine (*S. macrocarpon* L.), aubergine (*S. melongena* L.) (Prasad, 1974; Gupta, 1992; Gupta and Gupta, 1994; Ohno *et al.*, 2009; Matsuda *et al.*, 2013), Irish potato (*S. tuberosum* L.) (Prasad, 1974; Gupta and Gupta, 1994).

Theaceae: tea (*Camellia sinensis* (L.) Kuntze) (Bolland *et al.*, 1998; Migeon and Dorkeld, 2006–2013).

Urticaceae: China grass (*Boehmeria nivea* (L.) Gaudich.) (Ohno *et al.*, 2010).

Vitaceae: grapevine (*Vitis vinifera* L.) (Prasad, 1974; Gupta and Gupta, 1994).

Bio-ecology

The vegetable mite is a very polyphagous species reported in tropical and subtropical countries. It is injurious to a variety of economic plants, including vegetables, tuber crops, fruit crops, oil seeds, spices, ornamentals and flowering plants.

The mite reproduces by arrhenotokous parthenogenesis, and has a haploid number of chromosomes of three (Helle *et al.*, 1970). The male:female sex ratio is 1:10 (Manjunatha and Puttaswamy, 1989b). The two types of reproduction exhibit slight difference in the duration of the developmental stages. Parthenogenetic development takes 8.875 ± 0.04 days and sexual development 9.5 days for males and 10.9 ± 0.15 days for females (Kaimal and Ramani, 2007b). The 50% survival rate of inseminated females is attained 15.5 days before that of non-inseminated females, and the fecundity of inseminated females is 1.5 times as great as that of non-inseminated females (Bonato and Gutierrez, 1999).

Populations of the vegetable mite usually inhabit lower leaf surfaces and fruits. The females usually lay eggs on the lower leaf surface and rarely on the upper surface. Ovipositing females construct a silken web across the midrib and major lateral veins before they deposit their eggs (Kaimal and Ramani, 2007b). In India the mite overwinters as fertilized females on weeds, and in spring when the weather begins to warm, it moves towards cultivated plants. From this period, the populations increase and infest cultivated plants, reaching the highest densities in May to mid-July and decreasing during late July and August. Later, populations of *T. neocaledonicus* increase again in September and October, commonly on weeds. The females move to winter crops and continue their activity until December and February. Later on, the females move into the winter crops or weed hosts to overwinter (Jeppson *et al.*, 1975). On okra in Karnataka (south-west India), the highest population was observed from March to April, and the same density of motile forms occurred on the basal, middle and top leaves during the first fortnight of April. During May, the populations gradually decreased on the top and middle leaves, and the trend continued during June–July, with a subsequent decrease of the population density on the basal leaves. The lower population levels were found from June to February (Rachana *et al.*, 2009).

The length of the development period from egg to adult varies according to the host plant. In experimental studies, Gutierrez (1976b) reported that on cotton seedlings at 25°C and 50% RH, the mean egg to adult developmental time of the female was 9.2 days, and the fecundity was 80.3 eggs. Puttaswamy (1981) found that on tapioca, the egg to adult developmental time was longer (12.11 days) than on castor bean (10.48 days), amaranth (10.20 days) and mulberry (10.14 days). On cardamom, the female completed its life cycle in 10.47 days and the male in 10.01 days. The egg, larval, deutonymphal and protonymphal stages lasted on average 4.66, 1.71, 1.47 and 2.63 days for the female, and 4.65, 1.60, 1.41 and 2.35 days, for the male, respectively. Mated females each laid an average of 126.42 eggs during their oviposition period of 20.50 days, while unmated females laid an average of 87.00 eggs over 23.50 days. The lifespan of mated females was 24.00 days, that of unmated females 26.50 days and that of males 12.25 days (Puttaswamy and Reddy, 1980). On *Amaranthus viridis* (L.) at 23–26°C and 74-81% RH, the developmental period from egg to adult was 10.19 days. Females laid 147.42 eggs and their lifespan was 27.69 days (Puttaswamy and ChannaBasavanna, 1981a). Under the same conditions, the mite developed on mulberry, castor bean, cassava and *A. viridis*, but development took longer on cassava than on the other three plants. Fecundity was highest on castor bean and mulberry, and the adult lifespan longest on *A. viridis*. The results show that castor bean and mulberry are the most suitable food plants for the mite (Puttaswamy and ChannaBasavanna, 1981b).

On *Rhizophora mucronata* Lamk under laboratory conditions, Ghoshal *et al.* (2006) observed a developmental time from egg to adult of 13.5 ± 0.15 days, a fecundity of 39.8 ± 0.85 eggs, a longevity of 13.2 ± 0.23 days and a male:female sex ratio of 1:1.65. In growth chamber studies on moringa (*Moringa oleifera* Lam.), the incubation period lasts on average 3.15 ± 0.072–3.5 days for males and 3.5 ± 0.1 days for females. The male and female larval stages lasted 2–2.1 ± 0.063 days and 3.2 ± 0.08 days, respectively. The quiescence of the different biological stages lasted on average 0.5–0.9 ± 0.06 days. The protonymph stage was active for about a day and started the quiescence and moulting phases before entering into the next biological stage. The deutonymphal stage was also active for about a day. The pre-oviposition period was 1.5 ± 0.12 days for unmated females and negligible in the case of mated females, and the oviposition period lasted on average 8.4 ± 0.5089 days (Kaimal and Ramani, 2007b).

The best reproduction of the vegetable mite occurs at 30°C, and no survival has been observed at temperatures beyond 37°C; at 30 and 37°C, the mite fails to complete its life cycle. With an increase in temperature above 20°C, development was faster but the mortality of all biological stages increased. Overall, 30°C is the optimal temperature (Pande and Sharma, 1986). The optimal RH for development ranged from 45 to 50% (Pillai

and Jolly, 1986). At 25°C ± 1°C on cotton, the net reproduction (R_0) was 57.2, the intrinsic rate of increase (r_m) 0.260 individuals per female a day, and the finite rate of increase (λ) 1.297 individuals per female a day (Gutierrez, 1976b). Under appropriate conditions, the mite may develop up to 32 generations a year (Jeppson *et al.*, 1975). A negative relationship has been observed between population size and rainfall, and the responses of the mite to the different amounts of rainfall were found to vary according to the food plant (Sharma and Pande, 1983). The incidence of *T. neocaledonicus* was significantly and positively correlated with mean temperature, and there was a significant negative correlation with mean RH (Rachana *et al.*, 2009).

Natural enemies include some pathogens and several predatory insects and mites, see Table 20.11.

Symptomatology and damage

The symptomatology of *T. neocaledonicus* infestation is similar to that of *T. urticae* Koch. Feeding by the vegetable mite on leaves produces white spots that gradually coalesce. Infested leaves lose their green colour, gradually wilt, dry up and drop. The infestation adversely affects growth, flowering and fruiting. The damaged areas of some plants may turn red and a thick sheath of webbing may cover the canopy (Jeppson *et al.*, 1975). In Egypt, *T. neocaledonicus* infests leaves and fruit of mandarin and orange, causing yellowish blotches on the fruits (Attiah, 1969); however, it is a minor pest of citrus. Feeding by the vegetable mite on *M. oleifera* causes conspicuous white spots, manifested by chlorosis of the leaves. Infested leaves exhibit chlorophyll loss and subsequently dry up and are shed (Puttaswamy and Reddy, 1980). Feeding on seedlings of *Erythrina velutina* Willd., a leguminous plant used by the pharmaceutical industry, causes leaf blade reduction and crumpling symptoms (Poderoso *et al.*, 2010).

Control

Despite several natural enemies developing on populations of *T. neocaledonicus*, natural biological control of this mite is usually insufficient alone. Cultural practices and spray application of acaricides ensure better results.

CULTURAL PRACTICES. The destruction of overwintering or summer hosts decreases infestation levels by leaving the land free of plants between crop plantings (Rahman and Sapra, 1945; Khot and Patel, 1956).

CHEMICAL CONTROL. A number of acaricides have been used in the control of vegetable mites on various crops. Common active ingredients that have been tested include dicofol (Gupta *et al.*, 1972; Krishnaiah and Tandon, 1975; Singh *et al.*, 1981b; Pillai and Jolly, 1986; Peter *et al.*, 1987; Singh *et al.*, 1988), chlorobenzilate (Sidhu and Singh, 1971; Dhooria and Mann, 1980), cyhexatin (Krishnaiah and Tandon, 1975; Singh *et al.*, 1981b; Peter *et al.*, 1987; Yadav *et al.*, 1987), ethion (Peter *et al.*, 1987), propargite (Singh *et al.*, 1981b; Zhang *et al.*, 1990), tetradifon (Krishnaiah and Tandon, 1975; Dhooria and Mann, 1980), lime sulfur (Zhang *et al.*, 1990) and wettable sulfur (Anand and Chauhan, 1983; Yadav *et al.*, 1987). Chemical control is not justified on citrus.

Botanical acaricides were tested under laboratory conditions on the leaf of *Withania somnifera* (L.) Dunal in aqueous and methanolic dose. They include *Eupatorium triplinerve* Vahl, *Cassia alata* (L.) Roxb. and *Ocimum tenuiflorum* L. Highest mortality (97%) was found by 3% methanolic dose of *C. alata* (Roy *et al.*, 2011).

Tetranychus pacificus **McGregor (Fig. 20.61)**

Common name

Pacific spider mite.

Diagnostic characteristics

FEMALE. The body colour is variable (amber, salmon, orange–red, greenish yellow or other colours). The dorsum may lack spots or have as many as four blackish spots along each side, and the legs are pale (McGregor, 1950; Baker and Pritchard, 1953). The peritreme is hooked. The dorsal striae are lobate, transverse or oblique between the setae e_1–e_1, and transverse between the setae f_1–f_1 and e_1 and f_1. The ventral striae lack lobes. The pregenital striae are weak and incomplete or lacking, or with extremely weak striae laterally. The empodia have 6 proximoventral hairs, and are without dorsal spurs; tarsus I has 4 tactile setae proximal to the proximal duplex setae (Seeman and Beard, 2011).

MALE. The dorsal striae lack lobes. The empodia I are claw like and with a strong spur; empodia II–IV have long and free proximoventral hairs; empodia II bear spurs, and empodia III–IV have minute dorsal spurs. The aedeagus lacks a knob, is largely sigmoid, and shows a small and distinct projection on the anterior margin (Seeman and Beard, 2011).

Geographical distribution

Canada (Baker and Pritchard, 1953), Mexico (Beer and Lang, 1958) and the USA (McGregor, 1919; Pritchard and Baker, 1955; Tuttle and Baker, 1976; Navajas *et al.*, 1996; Stavrinides and Mills, 2009, 2011; Stavrinides *et al.*, 2010a; Seeman and Beard, 2011).

Host plants

The whole list of host plants of the Pacific spider mite includes 53 wild and cultivated plants belonging to 25 families (Migeon and Dorkeld, 2006–2013). The families and species of economic importance are: Asteraceae – sunflower (*Helianthus annuus* L.) (Bolland *et al.*, 1998; Migeon and Dorkeld, 2006–2013); Cucurbitaceae – field pumpkin (*C. pepo* L.) (McGregor, 1950); Fabaceae – lucerne (*Medicago sativa* L.) (McGregor, 1919, 1950); Malvaceae – *Gossypium* sp. (McGregor, 1950); Juglandaceae – English walnut (*Juglans regia* L.) (McGregor, 1950); Lamiaceae – horehound (*Marrubium vulgare* L.) (McGregor, 1950); Moraceae – fig (*Ficus carica* L.) (McGregor, 1950); Poaceae – maize (*Zea mays* L.) (Bolland *et al.*, 1998; Migeon and Dorkeld, 2006–2013); Rosaceae – strawberry (*Fragaria* × *ananassa* Duchesne ex Rozier), apple (*Malus domestica* Borkh.) (Seeman and Beard,

Table 20.11. *Tetranychus neocaledonicus* André. Natural enemies (fungi, insects and mites).

Families and species	References
FUNGI	
Ascomycota	
Verticillium lecanii (Zimmerman) Viégas	Sreenivas *et al*., 2005
Clavicipiteae	
Beauveria bassiana (Balsamo) Vuillemin	Sreenivas *et al*., 2005
Metarhizium anisopliae (Metschnikoff) Sorokin	Hanchinal and Manjunatha, 2000; Sreenivas *et al*., 2005
INSECTS	
Anthocoridae	
Cardiastethus sp.	Gutierrez, 1983, in Chazeau, 1985
Orius sp.	
Wollastoniella gatti Ghauri	Yaseen *et al*., 1982
Cecidomyiidae	
Feltiella acarisuga (Vallot)	Gutierrez, 1983, in Chazeau, 1985
Therodiplosis sp.	Yaseen *et al*., 1982
Chrysopidae	
Chrysoperla carnea Stephens	Sharanabasava and Manjunatha, 2001
Coniopterygidae	
Coniopterix crassicornis Eb. Pet.	Yaseen *et al*., 1982
Coccinellidae	
Brumus suturalis Fabricius	Pande and Sharma, 1984
Chilocorus sp.	
Illeis cincta (Fabricius)	
Menochilus sexmaculatus (Fabricius)	
Parastethorus histrio (Chazeau)	Chazeau, 1979
Rodolia sp.	Pande and Sharma, 1984
Stethorus aethiops Weise	Yaseen *et al*., 1982
Stethorus comoriensis Chazeau	Chazeau, 1971a
Stethorus incompletus Whitehead	Chazeau, 1979
Stethorus jejunus Casey	Yaseen *et al*., 1982
Stethorus madecassus Chazeau	Chazeau, 1971b; Gutierrez and Chazeau, 1972; Chazeau, 1974
Stethorus pauperculus (Weise)	Puttaswamy and ChannaBasavanna, 1976; Rachana *et al*., 2009
Stethorus proximus Chazeau	Chazeau, 1979
Stethorus siphonulus Kapur	
Stethorus vagans Blackburn	
Staphylinidae	
Oligota fageli Williams	Yaseen *et al*., 1982
Oligota flavicornis (Boisduval *et* Lacordaire)	Gutierrez, 1976b; Chazeau and Gutierrez, 1977
Oligota oviformis Casey	Rachana *et al*., 2009
Oligota pallidicornis Cameron	Gutierrez, 1976b; Yaseen *et al*., 1982
Thripidae	
Scolothrips sp.	Yaseen *et al*., 1982; Gutierrez, 1983, in Chazeau, 1985
Scolothrips hartwigi Priesner	Gutierrez, 1983, in Chazeau, 1985
Scolothrips pallidus (Beach)	
MITES	
Anystidae	
Anystis sp.	Singh and Singh, 1999
Phytoseiidae	
Amblyseius largoensis (Muma)	
Amblyseius multidentatus (Chant, Swirski *et* Shechter)	Putatunda and Anupam Tagore, 1997
Amblyseius tamatavensis Blommers	Schicha and Gutierrez, 1985
Amblyseius vazimba Blommers *et* Chazeau	Blommers and Chazeau, 1974
Euseius alstoniae (Gupta)	Putatunda and Anupam Tagore, 1997
Euseius concordis (Chant)	Jagadish *et al*., 1994
Euseius finlandicus (Oudemans)	
Euseius neococcineae(Gupta)	

Continued

Table 20.11. Continued.

Families and species	References
Euseius ovaloides (Blommers)	Schicha and Gutierrez, 1985
Euseius scutalis (Athias Henriot)	Putatunda and Anupam Tagore, 1997
Galendromus pilosus (Chant)	Aguilar and Salas, 1989
Neoseiulus bibens (Bloomers)	Blommers, 1976
Neoseiulus fallacis (Garman)	Putatunda and Anupam Tagore, 1997
Neoseiulus indicus (Narayan *et* Kaur)	
Neoseiulus longispinosus(Evans)	Schicha and Gutierrez, 1985; Rachana *et al.*, 2009
Neoseiulus masiaka (Blommers *et* Chazeau)	Blommers and Chazeau, 1974
Papuaseius dominiquae (Schicha *et* Gutierrez)	Schicha and Gutierrez, 1985
Phytoseiulus persimilis Athias Henriot	Rachana *et al.*, 2009
Phytoseius hawaiiensis Prasad	Schicha and Gutierrez, 1985
Phytoseius hongkongensis Swirski *et* Shechter	
Phytoseius jujuba Gupta	Putatunda and Anupam Tagore, 1997
Phytoseius minutus Narayanan, Kaur *et* Ghai	
Phytoseius mixtus Chaudhri	
Phytoseius roseus Gupta	
Phytoseius rubiginosae Schicha	Schicha and Gutierrez, 1985
Proprioseiopsis peltatus (van der Merwe)	
Typhlodromips deleoni (Muma)	
Typhlodromus homalii (Gupta)	Putatunda and Anupam Tagore, 1997

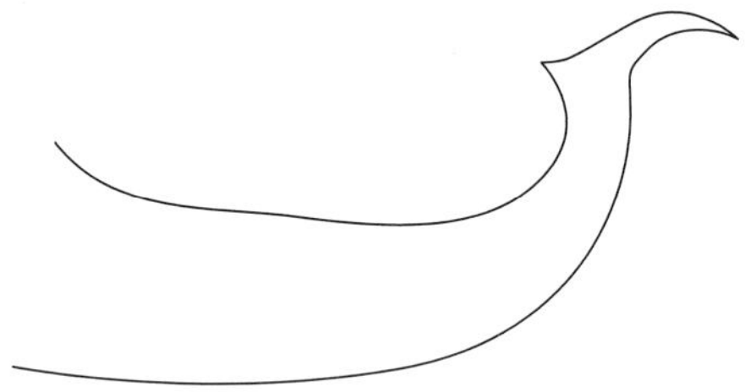

Fig. 20.61. *Tetranychus pacificus* McGregor. Male, aedeagus (modified from McGregor, 1950).

2011), apricot (*Prunus armeniaca* L.), sweet cherry (*P. avium* (L.) L.) (McGregor, 1950), European plum (*P. domestica* L.), almond (*P. dulcis* (Mill.) D.A. Webb) (McGregor, 1950; Tuttle and Baker, 1976), peach (*P. persica* (L.) Batsch) (McGregor, 1950), pear (*Pyrus communis* L.) (McGregor, 1950); Rutaceae – lemon (*Citrus limon* (L.) Burm. f.), grapefruit (*C. paradisi* Macfad.) (McGregor, 1950); Solanaceae: aubergine (*Solanum melongena* L.) (Bolland *et al.*, 1998; Migeon and Dorkeld, 2006–2013); Vitaceae – grapevine (*Vitis vinifera* L.) (McGregor, 1950; Navajas *et al.*, 1996; Stavrinides and Mills, 2009, 2011; Stavrinides *et al.*, 2010a).

Bio-ecology

The Pacific spider mite is a Nearctic mite injurious to deciduous trees including walnut, cotton, grapevines, lucerne, maize and other plants, and may be confused morphologically with *T. mcdanieli* McGregor (Jeppson *et al.*, 1975). In the USA, it has been reported on orange and lemon trees in Los Angeles County, and on grapefruit in Riverside County (McGregor, 1950, 1956), and has caused severe damage in San Joaquin Valley in California (USA), but it was rare on this crop and probably was dispersed from nearby food plants. The infestations occurred under clean crops, indicating that the mite has become established as a pest of citrus (Jeppson, 1989). Moreover, the Pacific spider mite is a major pest of almonds (Welter *et al.*, 1984) and grapevines (Hanna *et al.*, 1996, 1997a,b). In the western USA, *T. pacificus* and the Willamette spider mite, *Eotetranychus willamettei* (McGregor), are both important pests of grapevines and may coexist on the same host plant, although *T. pacificus* causes the most damage (Flaherty *et al.*, 1992).

The Pacific spider mite has a haploid number of chromosomes of three and reproduces by arrhenotokous parthenogenesis (Helle and Bolland, 1967). Its populations live on both leaf surfaces and produce webbing that may cover the entire plant. Each female lays 50–100 eggs in 2–4 weeks, and the life cycle from egg to adult takes 10–14 days (Jeppson *et al.*, 1975).

On cotton cotyledons, the average development periods in the laboratory ranged from 25.8 to 29.0 days at 15.5°C and from 6.1 to 6.7 days at 29.4°C for the females, with a slightly shorter time for the males. The total progeny at 23.8°C averaged 78.9 eggs per female and the corresponding number at 29.4°C was 68.3 eggs per female. The percentage of female offspring was 57%. The females lived for a mean of 12.71 days at 23.8°C, and 8.91 days at 29.4°C. At 24°C and 50–65% RH, the mean egg to adult development of the female took 10.5 days, pre-oviposition lasted 1.17 days, the maximum oviposition a day was 11.0 eggs and the fecundity 78.9 eggs. At 25 ± 1°C, the net reproductive rate (R_0) was 44.6, the intrinsic rate of increase (r_m) was 0.207 individuals per female a day and the finite rate of increase (λ) was 1.229 individuals per female a day. The lower developmental threshold was 11.1°C (Carey and Bradley, 1982). In laboratory studies on cotton leaves at 29°C, the lifespan of adult females was 8.68 days and the fecundity was 57 eggs per female. The immature survivorship was 89.7% and 78.7% of the offspring were females. The intrinsic rate of

increase (r_m) was 0.305, the net reproductive rate (R_0) 40.12, and the finite rate of increase (λ) 1.35; the population doubling time was 2.28 days and the mean generation time (T) 12.12 days (Bai and Carey, 1988).

In laboratory investigations on bush lima bean at 25 ± 1°C, the net reproductive rate (R_0) was 108.3, the intrinsic rate of increase (r_m) 0.290 and the finite rate of increase (λ) 1.340 (Takafuji and Chant, 1976). Under laboratory conditions on grape leaf discs at 15, 22, 28 and 34°C and under a 16L:8D photoperiod, the pre-oviposition period lasted 3.33 ± 0.37, 2.13 ± 0.22, 0.70 ± 0.12 and 0.18 ± 0.11 days, female longevity was 46.59 ± 5.08, 45.33 ± 3.85, 12.04 ± 1.20 and 5.28 ± 0.28 days, the net reproductive rate (R_0) was 13.61 ± 2.47, 26.21 ± 2.18, 35.94 ± 4.35 and 0.73 ± 0.13, the intrinsic rate of natural increase (r_m) was 0.061 ± 0.004, 0.119 ± 0.006, 0.272 ± 0.006 and −0.031, the mean generation time (T) was 43.35 ± 0.91, 27.44 ± 1.36, 13.17 ± 0.43 and 10.00 ± 0.09 and the population doubling time was 11.39 ± 0.71, 5.82 ± 0.29, 2.54 ± 0.06 and −22.11, respectively (Stavrinides and Mills, 2011). In another study under the same conditions of temperature and photoperiod, and also on grapevine foliage, immature survival was 76.49, 89.28, 92.00 and 30.18, and the time to completion of successive developmental stages was 34.61 ± 0.59 and 34.33 ± 0.67, 14.89 ± 0.23 and 15.17 ± 0.39 for the female, and 9.21 ± 0.12 and 8.95 ± 0.18, 7.06 ± 0.24 and 7.50 ± 0.69 for the male, respectively. The lower development threshold for females and males is 10.0 and 12.4°C, respectively and the upper development threshold is 40.3°C (Stavrinides *et al.*, 2010a).

In warm subtropical areas, the mites do not overwinter and develop throughout the year as well as in greenhouses, although they have high population densities only during summer and early autumn (Jeppson, 1989). In temperate areas, the females migrate in the late summer into the bark and crevices of the host plants and/or in soil to overwinter and in March move towards the host plants to reproduce. The first populations on trees are observed at the tops or the tips of the branches, and these may survive until the infestations develop severely and leaf drop occurs (Laminman, 1935; Pritchard and Baker, 1952a; Jeppson *et al.*, 1975). The overwintering females are bright orange to deep amber in colour, shiny and lack spots (Pritchard and Baker, 1952a). The presence of *T. pacificus* in Canada highlights its ability to tolerate very low temperatures. In the interior and hotter coastal vineyards of California, outbreaks of *T. pacificus* are frequent, while infestations of *E. willamettei* occur in the cooler coastal vineyards or earlier in the season in the interior vineyards (UC IPM Online, 2014b). Outbreaks of *T. pacificus* are more frequent during the hot summer months, because high temperatures allow the mite to escape control by the phytoseiid *Metaseiulus occidentalis* (Nesbitt). The results of studies conducted by Stavrinides *et al.* (2010a) confirm the field observations mentioned above, and relate *E. willamettei* damage to the cooler, coastal vineyards and early in the season in inland vineyards, and *T. pacificus* attacks to hot vineyards in inland and coastal areas. Moreover, the results of further research by Stavrinides *et al.* (2012) showed that the peak population densities of the Pacific spider mite increased significantly with increasing frequency of leaf temperatures above 31°C. Conversely, the peak population densities of the Willamette spider mite exhibit no relationship with the frequency of leaf temperatures above 31°C. The relationship between population densities of the two spider mites and leaf temperature is consistent with their upper thresholds for development: 40°C for *T. pacificus* and 31°C for *E. willamettei*. In addition, the population density of phytoseiids decreases with increasing frequency of leaf temperatures above 31°C, indicating that there is no association with the population density of either spider mite species.

Natural enemies of Pacific spider mite include pathogens and predatory insects and mites, see Table 20.12.

Symptomatology and damage

Like other spider mites, the feeding activity of the Pacific spider mite removes chlorophyll from the plant cells, causing a stippling on the leaves. The severe damage caused by low population densities on pear, citrus and other plants suggests the involvement of a toxin. Damaged leaves first appear at the top of the trees, and then turn brown and die, as though they had been scorched by fire (Jeppson *et al.*, 1975). High temperatures and low humidity increase the severity of damage.

Between economic plants, the grapevine, prune and almond are highly susceptible, with populations reaching 1400 mites per leaf on grapes (Laminman, 1935). Almond leaves damaged by *T. pacificus* exhibit an abscission layer, and leaves infested by a high density of mites appear brittle to the touch. The defoliation may be a reaction to mite feeding related to a toxic effect of saliva on the almond plant (Andrews and La Pré, 1979). Mixed populations of the Pacific spider mite and two-spotted spider mite, *T. urticae* Koch, may also occur. Feeding damage by *T. pacificus* and *E. willamettei* on grapevines is similar and leads to a loss of photosynthetic ability and leaf discoloration, followed by leaf and plant death (Jeppson *et al.*, 1975; Tanigoshi *et al.*, 2004). In particular, the first symptoms of the damage of Pacific spider mite consist of yellow spots, and increasing damage appears on the leaves as necrotic areas. High population densities cause leaf burning and bronzing and produce abundant amounts of webbing. Damage is most severe along the shoulder and tops of plants (UC IPM Online, 2014b). Flaherty *et al.* (1981) reported that loss of leaves induced a decrease in maturity and quality of fruit. The feeding activities of *T. pacificus* and *E. willamettei* cause a reduction of grape fruit sugar content and yield (Welter *et al.*, 1989a,b, 1991). Feeding on maize causes early drying of the leaves associated with grain yield reduction as a result of increasing stalk breakage, lowering of the moisture content of the grain at harvest and kernel shrinkage (Jeppson *et al.*, 1975). On citrus, the mite populations do not tend to colonize and live on the entire undersurface of the leaves and on fruits. Damage is similar to that caused by the citrus red mite, but leaf drop and dead twigs arise from lower populations (Jeppson, 1989).

Control

Cultural techniques and biological and chemical means prevent outbreaks of the Pacific spider mite. Sampling and monitoring help in control. There is no information on the economic injury level and economic threshold.

Table 20.12. *Tetranychus pacificus* McGregor. Natural enemies (fungi, insects and mites).

Families and species	References
FUNGI	
Ancylistaceae	
Conidiobolus thromboides Drechsler	Steinhaus and Marsh, 1962
Entomophthoraceae	
Entomophthora sp.	
INSECTS	
Coccinellidae	
Stethorus picipes Casey	McMurtry *et al.*, 1974
Stethotus punctum (LeConte)	Robinson, 1953
Eolothripidae	
Franklinothrips sp.	Hoddle *et al.*, 2000b
Scolothrips sexmaculatus (Pergande)	Laminman, 1935; Rice and Jones, 1972
MITES	
Anystidae	
Anystis agilis (Banks)	Sorensen *et al.*, 1976
Bdellidae	
Bdella longicornis (L.)	Sorensen *et al.*, 1983
Phytoseiidae	
Amblyseius andersoni (Chant)	Amano and Chant, 1977
Euseius citrifolius Denmark *et* Muma	Moraes and McMurtry, 1981
Euseius hibisci (Chant)	Zhimo and McMurtry, 1990
Euseius scutalis (Athias Henriot)	Bounfour and McMurtry, 1987
Euseius stipulatus (Athias Henriot)	McMurtry, 1977b; Zhimo and McMurtry, 1990; Flechtmann and McMurtry, 1992
Euseius tularensis Congdon	Zhimo and McMurtry, 1990
Galendromus annectens (De Leon)	Badii *et al.*, 1990
Iphiseius degenerans (Berlese)	Takafuji and Chant, 1976; Flechtmann and McMurtry, 1992
Metaseiulus occidentalis (Nesbitt)	Croft and McMurtry, 1972b; Rice and Jones, 1972; Hoy *et al.*, 1982a; Flechtmann and McMurtry, 1992
Metaseiulus pini (Chant)	Charlet and McMurtry, 1977
Metaseiulus validus (Chant)	
Phytoseiulus longipes Evans	Badii and McMurtry, 1983
Phytoseiulus persimilis Athias Henriot	McMurtry and Scriven, 1975; Takafuji and Chant, 1976; Amano and Chant, 1977
Typhlodromus kettanehi Dosse	McMurtry, 1977c
Typhlodromus persianus McMurtry	

SAMPLING AND MONITORING. Sampling and monitoring for spider mites and *M. occidentalis* are fundamental steps for control. In California, almond monitoring at least every week allows the predator:prey ratio to be regulated early in the growing season, before the hot weather in June (Wilson *et al.*, 1984; Zalom *et al.*, 1984a,b). Two monitoring methods for spider mites and their predators are known: the brush and count method and the presence–absence sequential sampling method (useful from mid-June). The brush and count method uses different treatment guidelines for May, June, July or August (Hoy, 1984). The guidelines for the main crops infested by Pacific spider mite in California are discussed below. Further information on other crops (walnut, cherry, etc.) may be found in the University of California (UC) Pest Management Guidelines. To predict timing of the development of Pacific spider mites, the UC Phenology Model Database for the Pacific Spider Mite (UC IPM Online, 2003b) used two phenological models. The phenological model proposed for almond stated a lower developmental threshold of 12.0°C and a degree-days accumulation required for a generation time (egg to egg) of 288.0 degree-days. The development time on water-stressed trees will be shorter (Oi *et al.*, 1989). The phenological model proposed for cotton gave a lower developmental threshold of 11.1°C and degree-day accumulations for the development of the egg, larva, protonymph, deutonymph and for egg to adult development of 52.1, 24.8, 20.4, 23.6 and 121.0 degree-days, respectively (Carey and Bradley, 1982).

Almond. The UC Pest Management Guidelines for almond (IC IPM Online, 2012a) suggest weekly monitoring from May through to August. After August, the migration of mites towards overwintering sites makes sprays unnecessary. In particular, before 1 July, it is necessary to monitor every few days those areas, often dusty or water-stressed, where mite infestations may develop first. If treatment is necessary in these areas, sampling should be extended to the whole orchard to evaluate whether a spot spray is sufficient or the entire orchard must be sprayed. After 1 July, monitoring does need to be extended

to the whole orchard, dividing it into sampling plots (areas) (that could be sprayed separately), within which a minimum of five trees should be sampled. Fifteen leaves should be selected from each tree, randomly sampling them from both the inside and outside of the canopy. Both leaf surfaces are observed under a hand lens to detect spider mites and eggs and their natural enemies. For sampling on trees not yet sprayed for mites during the season, presence–absence sampling can be done, noting the number of leaves on each tree with pest mites or their eggs, and the number of leaves with predators. Once five trees have been sampled, the total found are matched to the numbers in the 'Don't Treat' and 'Treat' columns on the UC form (UC IPM Online, 2012a). Treatments must be related to the presence of predators. No predators sampled in May, a total number of spider mite days for the season greater than ten, and an increasing mean number of spider mites per leaf each week for 3 weeks suggest that a 'lower than label rate' spray of acaricides (propargite or cyhexatin) should be done (Hoy, 1984). Detection of a predator:prey ratio greater than or equal to 1:10 leaf requires no sprays, and biological control might be expected quickly. In June, no treatment is suggested if the predator:prey ratio is greater than 1:5, and the total accumulated spider mite days is less than ten. The hotter and drier weather and the reduced ability of trees to tolerate mite damage make July and August guidelines more stringent (Hoy, 2011).

Apple. On apple, high levels of population earlier in the season – and usually from July to August – require monitoring. Sampling of five spur leaves at spaced intervals from one lateral branch located at eye level is done every 1–3 weeks from each of 20 marked trees that have been established as representative trees in a block. The leaves are brushed in a mite-brushing machine and the mites (but not eggs) counted. The average presence of ten spider mites per leaf suggests that treatment should be done, but the simultaneous presence of one predatory mite to ten pest mites suggests that sampling be continued to evaluate that pest mite populations do not increase (see UC Pest Management Guidelines for Apple at UC IPM Online, 2009b,c).

Cotton. On cotton, the timing for monitoring spider mites is between crop emergence and the first open boll. Defoliation of seedling cotton and high levels of population need treatments. In particular, from early squaring to the first open boll, if 30–50% of leaves are infested, sprays of acaricides may be necessary. Some crops, including lucerne, beans, sugarbeet and safflower grown next to cotton fields may cause more severe infestation on field margins than in the remainder of the field. This suggests that the field margins of crops should be monitored and treated separately from the remainder of the field (see UC Pest Management Guidelines for Cotton at UC IPM Online, 2013c).

Grapes. On grapevines, a prey:predator ratio exceeding 30:1 highlights the rarity of the predator. In this case, if the leaves are less than 50% infested, treatment delay is suggested; if there are 50–65% of infested leaves, treatment is suggested if the spider mite population is increasing; and if there are over 65% of infested leaves, treatment should be done rapidly. A prey:predator ratio between 30:1 and 10:1 indicates that the predator is occasional, and suggests treating after some days at 65–75% of infested leaves and rapidly at over 75% of infested leaves. A prey:predator ratio between 10:1 and 2:1 indicates that the predator is frequent; in this case, the predator increase should be monitored, and treatment done only if the percentage infestation is greater than 75% and the predator is not increasing rapidly. A ratio greater than 2:1 indicates that the predator is numerous; under these conditions, treatment may not be necessary if the pest population is dropping because of very high (>1:1) predator:prey distribution ratios, and damage should be carefully evaluated (Flaherty *et al.*, 1981; see UC Pest Management Guidelines, Grape at UC IPM Online, 2014b).

Prune. On prune, sampling must be conducted from 1 June to 15 July, until the threshold has been reached and spray application has occurred. Monitoring is not necessary if spray is applied after 15 July. In orchards, up to 40 acres of monitoring involves a 5 min search in two separate areas of the orchard, for a total sampling time of 10 min. For each area, at least 2–3 leaves should be sampled from both the inside and outside of ten plants, noting the presence–absence of prey and predators. Uncertain mite development suggests making two 5 min searches during the summer, while a consistent mite population needs only one 5 min search. Records of sampling must be noted on a monitoring form. The need for treatment is led by the following guidelines, which take into account the presence of prey and predator. Mite infestation may be low (1–20% infested leaves), low/moderate (21–39% infested leaves), moderate (40–60% infested leaves), moderate/high (61–79% infested leaves) or high (80–100%infested leaves). The predator presence may be low (less than one predator per six leaves), moderate (one predator per three leaves) or high (one or more predators per leaf). According to the above ratings, treatment must be effected if at least one 5 min search highlights low/moderate mite rating with low/moderate predator rating, or moderate/high mite rating with moderate/high predator rating (see UC Pest Management Guidelines for Prune at UC IPM Online, 2009e).

ECONOMIC INJURY LEVELS AND ECONOMIC THRESHOLD. Barnes and Andrews (1978) report that high levels of infestation by *Tetranychus* spp. do not influence growth and yield of almond trees in the same year but in the year after the infestation; yields of trees not treated with acaricides are 13–19% lower than from sprayed trees, and terminal shoot length is reduced by 11 and 48%, respectively, in orchards in their fourth and sixth years. Trunk growth is reduced by 19% in the untreated trees. Andrews and La Pré (1979) found that chlorophyll content was negatively correlated with the number of mite days per leaf. Welter *et al.* (1984) reported that *Tetranychus* mites reduced development, growth and yield in almonds in the subsequent growing season by 24 and 9%, respectively, after 517 and 424 mite days. The number of fruit was reduced by 20% after 178 mite days during the preceding growing season in trees that were 6 years old. In California, almonds are more susceptible to spider mite damage than apples and grapes. Currently, there is no precise information available on the economic injury level, but it is known

that almonds do not tolerate spider mites and may defoliate if more than 120 spider mite days accumulate per leaf. Growers do not accept tree defoliation and their control programmes are based on reducing mite days to below this value (Hoy, 2011).

According to Schruft (1985), the economic threshold in vineyards relates to the vigour and tolerance of the infested plants, the soil properties and the presence of predators and other pests. Although Laing *et al.* (1972) have not detected any correlation between the number of mites per leaf and the weight or sugar quality content of grapes, it is known that severe damage is associated with a density of 600–1400 mite per leaf (Jeppson *et al.*, 1975). Research has shown that the feeding activity of *T. pacificus* and *E. willamettei* causes reductions in both grape fruit sugar content and yield (Welter *et al.*, 1989a,b, 1991).

CULTURAL PRACTICES. Horticultural practices can enable grapevines to tolerate outbreaks of mites better. However, nitrogen fertilization of the host plants increases both the fecundity and development times of *T. pacificus* (Wilson *et al.*, 1988). It has also been observed that the irrigation of vineyards prevents outbreaks of mites, because drought stress facilitates their development (Kinn *et al.*, 1974). The same phenomenon is reported on almond, where the fecundity and development rate of *T. pacificus* increase with water stress, probably as a result of increased leaf temperature. Dehydrated leaves are warmer than well-watered leaves, and higher temperatures enhance the development of mites (Youngman *et al.*, 1988; Oi *et al.*, 1989). Recently, Stavrinides *et al.* (2012) reported that differences in heat resistance among arthropods might partly explain the differential effects of plant water stress on herbivore abundance. Moreover, spider mite outbreaks on almond are induced by early harvest, a practice applied to reduce damage caused by the navel orangeworm, *Amyelois transitella* (Walker), and induced by stopping irrigation shortly after the almond hull begins to split early in July (Micke *et al.*, 1966). Reduction of dust along roads by spraying road oil may help to prevent dusty fruits and the development of mites (Schruft, 1985). When possible, it is good practice to maintain weed or other cover in vineyard middles to further reduce dust (see UC Pest Management Guidelines for Grape at UC IPM Online, 2014b).

BIOLOGICAL CONTROL. The phytoseiid mite *M. occidentalis* is a major natural enemy of *T. pacificus* and *E. willamettei* in California vineyards (Flaherty and Huffaker, 1970b; Hoy and Smilanick, 1981; Hanna and Wilson, 1991; Flaherty *et al.*, 1992; Stavrinides and Mills, 2009; UC IPM Online, 2014b). However, this mite alone does not always adequately prevent outbreaks of spider mites, and the action of the predator must usually be integrated with sprays of selective acaricides. At 28°C, *M. occidentalis* developed 5% faster when feeding on *E. willamettei* than on *T. pacificus*. These results confirm field observations linking *E. willamettei* damage to cooler, coastal vineyards and early in the season in inland vineyards, and *T. pacificus* infestations to hot vineyards in inland and coastal areas (Stavrinides *et al.*, 2010a). Moreover, higher temperatures favour *T. pacificus* over the less damaging *E. willamettei*, and may also reduce the effectiveness of *M. occidentalis* (Stavrinides and Mills, 2011). Regular monitoring of the prey–predator distribution patterns allows estimation of the effectiveness of the action of *M. occidentalis*. This can be done by the method of the presence–absence of prey and predator on 20–30 leaves sampled in the vineyard. A prey:predator ratio exceeding 30:1 indicates the rarity of the predator, a ratio between 30:1 and 10:1 that the predator is occasional, a ratio between 10:1 and 2:1 that the predator is frequent, and a ratio greater than 2:1 that the predator is numerous (Flaherty *et al.*, 1981; UC IPM Online, Grape, 2014b).

Studies conducted in California vineyards showed that simultaneous testing of the demographic effects of pesticides on pests and natural enemies was essential for a full assessment of the impacts of pesticide application on biological control by *M. occidentalis* (Stavrinides and Mills, 2009). The impact of non-selective pesticides on phytoseiid populations may be detrimental and dramatically reduce their populations. It is necessary to avoid using disruptive pesticides such as carbaryl, dimethoate, dicofol and methomyl (UC IPM Online, 2014b). To respond to this deficiency, carbaryl- and orthophosphate-resistant *M. occidentalis* were mass reared by commercial companies using starter cultures provided by the University of California at Berkeley (Hoy *et al.*, 1982a).

Different methods of release of phytoseiid mite into orchards are known. In the common method used on almond, 100 adult females are released into every third tree in every third row (one tree in nine) early in the growing season (as soon as a few spider mites are detected). The resistant *M. occidentalis* strain multiplies and spreads throughout the orchard within a few weeks (Hoy, 2011). Vineyard releases of *M. occidentalis* are most successful when host plants (green beans) are placed directly on vines. Usually, the release of a minimum of 1000 predators per acre is suggested (UC IPM Online, 2014b). The application of reduced rates of cyhexatin and propargite (one-tenth to one-twentieth of the lowest label rate) allows low prey populations to be maintained in the orchard. The spider mite populations are reduced by about 50–80% and the population of *M. occidentalis* survives. The number of treatments depends on the density and early monitoring of the infestation (Hoy, 2011). It has also been observed that the maintenance of predator (*M. occidentalis*) to prey (*Pronematus* spp.) interactions under a regime of sulfur and cattail pollen application in a non-commercial vineyard help the biological control of *T. pacificus* (Calvert and Huffaker, 1974).

A singular method of biological control, formally very different from the traditional technique, is applied for the management of *T. pacificus* on grapes. This consists of the early introduction of *E. willamettei*, a spider mite that is not particularly injurious to vineyards, before natural infestation by *T. pacificus* has occurred. This induces a systemic reaction that reduces the later development of *T. pacificus* (English-Loeb *et al.*, 1993; Hougen-Eitzman and Karban, 1995). Unfortunately, the method cannot be used for all grape cultivars (Hanna *et al.*, 1997b).

CHEMICAL CONTROL. As for other spider mites, chemical control of the Pacific spider mite may pose the problem of the development of resistance to several acaricides (Jeppson *et al.*, 1975; Stavrinides *et al.*, 2010b; UC IPM Online, Cotton, 2013c), difficulties with their selectivity (UC IPM Online, 2012a, etc.), and the high economic cost of chemical control. For example, in California on almond during the 1970s and 1980s, in order to

prevent severe losses, growers applied propargite or cyhexatin to suppress spider mites on a preventive basis, or whenever webbing or leaf stippling was seen, and an economic analysis indicated that a single acaricide application cost US\$50.50 acre^{-1} at that time (Headley, 1983). Currently in California, grape growers treat more than 100,000 ha annually with pesticides against the two spider mites, *E. willamettei* and *T. pacificus*, at a cost of US\$20 million (CDPR, 2007).

Acaricides tested during past decades on cotton, almond, grapes and other crops include avermectins (Hoy and Conley, 1987; Leigh *et al.*, 1990), phenyl pyrazoles (Hoy and Ouyang, 1986), organochlorines (Dennehy and Granett, 1982; Leigh *et al.*, 1990), organosulfurs (Stafford and Kido, 1969; Hoy *et al.*, 1982b; Leigh *et al.*, 1990), organotins (Rice and Jones, 1972; Hoy *et al.*, 1982b), tetrazines (Hoy and Ouyang, 1986) and wettable sulfur (Stavrinides and Mills, 2009). Costello (2007) tested various formulations and timings of the distribution of sulfur, the oldest and most widely used fungicide in the vineyards of California. Sulfur was sprayed either pre-bloom (applied before grape bloom, in late May) and post-bloom. All cases showed increases in the density of the Pacific spider mites and predators, and the results did not support the hypothesis that the cause of the increase in *T. pacificus* density was the result of the negative effects of sulfur on phytoseiids or tydeids. Rather, it appears that a plant-based explanation is likely: first, because of the differences in pre-bloom versus post-bloom sulfuring; and second, because of the long lag time between the end of the sulfur applications and the corresponding increase in spider mite density (Costello, 2007). Hanna *et al.* (1997b) have also discussed this topic, specifically with respect to the effect of sulfur (and triadimefon) on predator mites.

Currently, the implementation of integrated pest management programmes in California includes the spraying of different active ingredients, e.g. for grape applications of abamectin, acequinocyl, bifenazate, clofentezine, etoxazole, fenpyroximate, hexythiazox, insecticidal soaps, narrow-range oil, neem oil, and spirodiclofen are suggested (UC IPM Online, 2014b). Soft materials such abamectin, acequinocyl, bifenazate and oils protect natural enemies.

INTEGRATED PEST MANAGEMENT. The application of cultural practices and biological and chemical control are fundamental to the implementation of IPM programmes. Releases of pesticide-resistant strains of *M. occidentalis* in combination with sprays of lower-than-label rates of selective acaricides (propargite and cyhexatin), the monitoring of predator-to-prey ratios, dust management, spot treatments of small portions of the orchard early in the season and adequate irrigation have, in the past, represented fundamental steps (Hoy, 1985, 2011). The costs and the benefits of IPM implementation on almond were discussed by Hoy (2011), who reported, according to Headley and Hoy (1986, 1987), the substantial advantages of the strategy. The method allows a saving of US\$24 acre^{-1} (with predator releases) to US\$44 acre^{-1} (without predator releases), including the expenses to pay a professional pest control consultant. Implementation of the method in the 360,000 acres of California almonds gives a calculated saving of US\$20 million year^{-1} in acaricide costs. Environmental and health benefits are not included in this estimate (Hoy, 2011).

Tetranychus schoenei McGregor (Fig. 20.62)

Common name

Schoenei spider mite.

Diagnostic characteristics

FEMALE. The body of the summer forms is light green to dark green with four dark spots, the most anterior of which are the largest (Pritchard and Baker, 1952a). The overwintering forms are orange in colour (Cagle, 1943). The peritreme is hooked. The dorsal striae between the setae e_1–e_1 and e_1–f_1 are transverse, while the dorsal striae between the setae f_1–f_1 are longitudinal and form an hourglass pattern. The dorsal lobes are small or lacking; the ventral striae lack lobes; the pregenital striae are entire, and sometimes weak medially. The empodia have 6 proximoventral hairs, and a little dorsal spur; tarsus I has 4 tactile setae proximal to the proximal duplex setae, and tarsus III has 1 proximal tactile seta (Seeman and Beard, 2011).

MALE. The body is pale yellowish green, with 4 tiny spots (Cagle, 1943). The empodia bear dorsal spurs; empodia I are claw like, and empodia II–IV have the proximal hairs long and free. The knob of the aedeagus is large and dorsally convex, with the anterior projection short and pointed, and the posterior projection sharp and hook like (Seeman and Beard, 2011). Pritchard and Baker (1955) and Jeppson *et al.* (1975) reported the anterior projection as more or less round.

Geographical distribution

USA (McGregor, 1941c; Reeves, 1963; Seeman and Beard, 2011; Szczepaniec *et al.*, 2011).

Host plants

The Schoenei spider mite is reported on 48 wild and cultivated plants belonging to 18 families, with the highest number of species (17) in the Rosaceae family (Migeon and Dorkeld, 2006–2013). The families and species of economic importance are: Fabaceae – soybean (*Glycine max* (L.) Merr.) (Thewke and Enns, 1969), *Phaseolus* sp. (McGregor, 1950); Malvaceae – upland cotton (*Gossypium hirsutum* L.) (Bolland *et al.*, 1998; Migeon and Dorkeld, 2006–2013). Moraceae: white mulberry

Fig. 20.62. *Tetranychus schoenei* McGregor. Male, aedeagus (Pritchard and Baker, 1955).

(*Morus alba* L.) (Bolland *et al.*, 1998; Migeon and Dorkeld, 2006–2013); Rosaceae – apple (*Malus domestica* Borkh.) (McGregor, 1941c; Seeman and Beard, 2011), Japanese rose (*Rosa multiflora* Thunb.) (Thewke and Enns, 1969), Allegheny blackberry (*Rubus allegheniensis* Porter), European red raspberry (*R. idaeus* L.), black raspberry (*R. occidentalis* L.) (Reeves, 1963). In addition to plants of agricultural and food importance, the mite inhabits various ornamentals, such as birch (*Betula*), beech (*Fagus*), hickory (*Carya*), willow (*Salix*) and elm (*Ulmus*).

Bio-ecology

T. schoenei is a Nearctic mite reported only from the eastern and south-eastern USA (Reeves, 1963). It is sometimes confused with *T. canadensis* (McGregor), because of their similar morphology and the overlapping distribution. It is an occasional pest of various economic plants, such as beans, apple, raspberries and cotton (Jeppson *et al.*, 1975), and is also reported on various ornamentals (Migeon and Dorkeld, 2006–2013). According to the condition of their food plants, the colour of the summer females varies from green or dark green to brown – when they feed on leaves that have turned brown.

The mite reproduces by arrhenotokous parthenogenesis. Unfertilized eggs produce males, and 80.7–86.3% of the mites reared from eggs laid by fertilized females produce females (Cagle, 1943). Populations of *T. schoenei* live on the lower leaf surface, near the veins, and produce a variable amount of webbing. The species overwinters as fertilized females, which start to lay eggs from the spring. In rearing studies in Virginia, Cagle (1943) has found that at 27.5°C, egg development takes 3 days and at 11.5°C it takes 25 days. At 25°C, the larvae and the protonymphs reach their quiescent stages in less than 1 day, while at 12–15°C this takes 4–5 and 5–6 days, respectively. At 25 and 15°C, the quiescent larval stage lasts from <1 to 3 days and the quiescent protonymphal stage lasts from <1 to 6 days. At 25–26°C and 12–15°C, the deutonymph stage feeds from <1 to 8 days and the quiescent period lasts from <1 to 12 days. At the same temperatures, the development from egg to adult takes 5 days. At 25–28°C, the pre-oviposition period takes 1 day, and the fecundity is 106 eggs, laid over a maximum of 38 days. The average daily number of eggs is 2.1–3.7. The mite develops about nine generations a year, and overwintering females may develop 5–8 generations (Cagle, 1943).

Symptomatology and damage

The damage caused to apple is similar to that by the European red mite (*Panonychus ulmi* (Koch)). Infested leaves become bronzed and the fruit fails to colour during autumn; at high levels of infestation, premature leaf drop may occur (Cagle, 1943). Severe attacks on staghorn sumac induce the leaflets to become yellow near the mid-veins of both leaf surface (Reeves, 1963).

Control

Serious infestations are sporadic, as is the need for control. Spray applications of acaricides are effective. The use of neonicotinoids to control other pests induced outbreaks of the Schoenei spider mite on elm trees (Szczepaniec *et al.*, 2011).

Tetranychus sinhai Baker

Common name

Unknown.

Diagnostic characteristics

FEMALE. The body is greenish yellow to orange in colour, and has one longitudinal black spot extending on each side. The striae between the opisthosomal setae e_1 are longitudinal (Jeppson *et al.*, 1975).

MALE. Empodium I is claw like and has a strong dorsal spur; empodium II bears a smaller spur and 3 pairs of ventral hairs directed ventrally. The aedeagus is upturned, the shaft has a short neck, and the knob is large, forming a marked angle with the neck and with the anterior portion rounded and the posterior portion acute (Jeppson *et al.*, 1975).

Geographical distribution

Canada (Baker, 1962), Poland (Boczek, 1964a) and the USA (Baker, 1962).

Host plants

The list of host plants of *T. sinhai* includes 11 species belonging to 3 families (Migeon and Dorkeld, 2006–2013). The species of most economic importance are: Asteraceae – sunflower (*Helianthus annuus* L.) (Bolland *et al.*, 1998; Migeon and Dorkeld, 2006–2013); Poaceae – barley (*Hordeum vulgare* L.) (Baker, 1962), wheat (*Triticum aestivum* L.), maize (*Zea mays* L.) (Wallace and Sinha, 1961; Sinha and Wallace, 1963); Rosaceae – black chokecherry (*Padus virginiana* (L.) M. Roem.) (Bolland *et al.*, 1998; Migeon and Dorkeld, 2006–2013).

Bio-ecology

T. sinhai is a minor or occasional pest of barley, wheat and maize (Wallace and Sinha, 1961; Sinha and Wallace, 1963), reported also on black chokecherry and wild and forage plants (Bolland *et al.*, 1998; Migeon and Dorkeld, 2006–2013). Its populations usually live on the lower leaf surface and produce little webbing (Wallace and Sinha, 1961).

Symptomatology and damage

Feeding by the mite on barley causes a darkening of leaves, followed by yellowing and withering from the bend of the leaf to its tip. The symptoms may appear along the midrib and in other patches on the leaf surface. Occasionally, the edges of the leaves crinkle from below. An increase in mite number per leaf was reflected by advanced symptoms showing on the leaves (Wallace and Sinha, 1961; Sinha and Wallace, 1963). On wheat, fewer mites usually occur but mosaic-like yellowing suggests the possibility of a transmissible virus disease (Wallace and Sinha, 1961).

Control

Control efforts have involved studies on barley and wheat resistance. No information is available on other means of control.

PLANT RESISTANCE. Sinha and Wallace (1963) examined 165 barley varieties from different parts of the world for their reaction to infestation by *T. sinhai* in Canada. Of these, 15 were found to be resistant, 47 moderately resistant, 88 moderately susceptible and 15 susceptible. In general, the barley varieties grown in arid regions appeared to be more resistant to infestation by *T. sinhai*. The most common barley varieties grown in Canada ('Parkland', 'Montcalm') were moderately susceptible. Only two North American varieties ('Canadian Lake Shore (C.I. 2750)', 'Gem (C.I. 7243)') showed high resistance to mite infestation.

In another study, 25 varieties of common wheat (*T. aestivum* L.) and 25 of durum wheat (*T. durum* Desf.) were studied in Manitoba (Canada) for their resistance to natural infestation by *T. sinhai*. Generally, the commonly grown varieties ('Selkirk', 'Thatcher', 'Ramsey', 'Stewart') were resistant, but some of the hybrid varieties ('Sask. 6070', 'Q2331-5') were susceptible (Wallace and Sinha, 1965).

Tetranychus truncatus Ehara (Fig. 20.63)

Common name

Mulberry red mite, cassava red mite (?).

Diagnostic characteristics

FEMALE. The body is carmine in colour (Ehara, 1956b). The peritreme is hooked. The dorsal striae between the setae e_1–e_1 are mixed longitudinal and oblique; between the setae f_1–f_1 they are longitudinal (sometimes combined with oblique); and between the setae e_1 and f_1 they are transverse and form a diamond shape. The ventral striae lack lobes; the pregenital striae are entire. The empodia have 6 proximoventral hairs and a minute spur. Tarsus I has 4 tactile setae proximal to the proximal pair of duplex setae (Seeman and Beard, 2011).

MALE. Empodia I–II bear a dorsal spur; empodium I is claw like, and empodia II–IV have free and long proximoventral hairs. The knob of the aedeagus is small, dorsally flat to slightly convex, with a medial indentation, with a rounded anterior projection and pointed posterior projection (Jeppson *et al.*, 1975; Seeman and Beard, 2011).

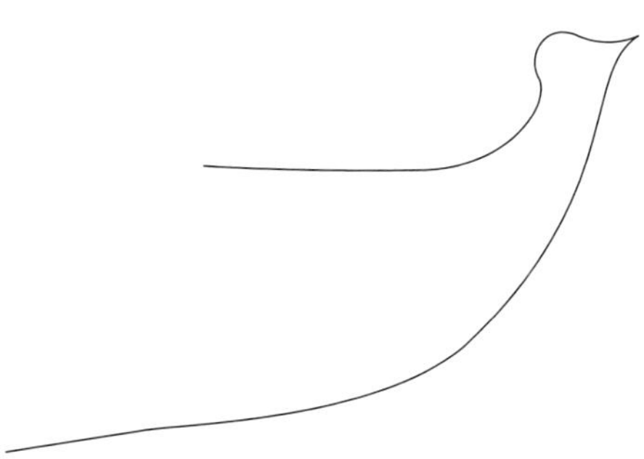

Fig. 20.63. *Tetranychus truncatus* Ehara. Male, aedeagus (from Jeppson *et al.*, 1975).

Geographical distribution

China (Ma *et al.*, 1979; Wang, 1981a; Wang *et al.*, 1985), Guam (Bolland *et al.*, 1998), India (Srinivasa *et al.*, 2012), Indonesia (Masaki, 1991), Iran (Namaghi, 2010), Japan (Ehara, 1956b; Gotoh *et al.*, 2007c; Hinomoto *et al.*, 2007; Matsuda *et al.*, 2013), South Korea (Lee *et al.*, 1987, 1999), Northern Marianas (Gutierrez, 1980), Philippines (Rimando, 1962a), Taiwan (Ho *et al.*, 1997), Thailand (Baker, 1975b; Ehara and Wongsiri, 1975; Sakunwarin *et al.*, 2003) and Vietnam (Hinomoto *et al.*, 2007).

Host plants

The whole list of host plants of *T. truncatus* includes 80 species belonging to 32 families, with the highest number of species in the Fabaceae (11 species), Cucurbitaceae (eight species), Amaranthaceae and Solanaceae (seven species) (Migeon and Dorkeld, 2006–2013). The families and species of economic importance are:

Amaranthaceae: carelessweed (*Amaranthus spinosus* L.), Chinese amaranth (*A. tricolor* L.), green amaranth (*A. viridis* L.), beet (*Beta vulgaris* L.) (Ehara, 1963b).

Amaryllidaceae: garlic (*Allium sativum* L.) (Bolland *et al.*, 1998; Migeon and Dorkeld, 2006–2013).

Araceae: cocoyam (*Colocasia esculenta* (L.) Schott) (Ho *et al.*, 1997).

Asteraceae: burdock (*Arctium lappa* L.) (Ho *et al.*, 1997), sunflower (*Helianthus annuus* L.) (Baker, 1975b).

Brassicaceae: canola (*Brassica juncea* (L.) Czern.) (Ho *et al.*, 1997), bird rape (*B. rapa* L.) (Migeon and Dorkeld, 2006–2013).

Caricaceae: papaya (*Carica papaya* L.) (Baker, 1975b).

Convolvulaceae: water spinach (*Ipomoea aquatica* Forssk.), sweet potato (*I. batatas* (L.) Lam.) (Baker, 1975b; Ho *et al.*, 1997).

Cucurbitaceae: watermelon (*Citrullus lanatus* (Thunb.) Matsum. and Nakai) (Ho *et al.*, 1997), cantaloupe (*Cucumis melo* L.) (Ehara, 1963b; Ho *et al.*, 1997), cucumber (*C. sativus* L.), pumpkin (*Cucurbita maxima* Duchesne) (Ho *et al.*, 1997), ribbed gourd (*Luffa acutangula* (L.) Roxb.) (Ehara, 1969a).

Euphorbiaceae: cassava (*Manihot esculenta* Crantz) (Baker, 1975b; Ehara and Wongsiri, 1975; Sakunwarin *et al.*, 2003), castor bean (*Ricinus communis* L.) (Bolland *et al.*, 1998; Migeon and Dorkeld, 2006–2013).

Fabaceae: groundnut (*Arachis hypogaea* L.) (Ho *et al.*, 1997), soybean (*Glycine max* (L.) Merr.) (Baker, 1975b; Ho *et al.*, 1997; Lee *et al.*, 1999), lima bean (*Phaseolus lunatus* L.), common bean (*P. vulgaris* L.) (Ho *et al.*, 1997), asparagus pea (*Psophocarpus tetragonolobus* (L.) DC) (Baker, 1975b), fava bean (*Vicia faba* L.), cowpea (*V. unguiculata* (L.) Walp.) (Ho *et al.*, 1997).

Malvaceae: okra (*Abelmoschus esculentus* (L.) Moench) (Baker, 1975b).

Meliaceae: neem (*Azadirachta indica* A. Juss.) (Baker, 1975b).

Moraceae: white mulberry (*Morus alba* L.) (Ehara, 1956b; Lee *et al.*, 1999).

Moringaceae: moringa (*Moringa oleifera* Lam.) (Bolland *et al.*, 1998; Migeon and Dorkeld, 2006–2013).

Myrtaceae: guava (*Psidium guajava* L.) (Bolland *et al.*, 1998; Migeon and Dorkeld, 2006–2013).

Pedaliaceae: sesame (*Sesamum indicum* L.) (Ho *et al.*, 1997).

Poaceae: rice (*Oryza sativa* L.), maize (*Zea mays* L.) (Baker, 1975b; Ho *et al.*, 1997).

Rhamnaceae: jujube (*Ziziphus jujuba* Mill.) (Migeon and Dorkeld, 2006–2013).

Rosaceae: Japanese pear (*P. pyrifolia* (Burm. f.) Nakai), English rose (*Rosa hybrida*) (Baker, 1975b).

Solanaceae: pepper (*Capsicum annuum* L.), Chinese boxthorn (*Lycium chinense* Mill.), gooseberry (*Physalis angulata* L.), tomato (*Solanum lycopersicum* L.) (Ho *et al.*, 1997), aubergine (*S. melongena* L.) (Ehara, 1963b; Ho *et al.*, 1997).

Vitaceae: grapevine (*Vitis vinifera* L.) (Ho *et al.*, 1997).

Bio-ecology

T. truncatus is widely distributed in South-east Asia and Indonesia, and extends eastward to Japan and Korea and westward to Iran; it is also found in some tropical and temperate countries. The mite is a pest of cotton (Gao, *et al.*, 1990b; Cao *et al.*, 1993), maize (Pang *et al.*, 2005a,b), mulberry (Sakunwarin *et al.*, 2003), vegetables (Chao and Lo, 1974; Pang *et al.*, 2004), ornamentals (Chen *et al.*, 1996b) and other plants.

Like many other tetranychid mites, this species also lives mainly on the lower leaf surface, where it feeds and produces webbing. In China, in the district of Beijing, populations of *T. truncatus* develop on *Sophora japonica* L. from May to early November, and reach a peak that is harmful from July to August. At 28°C and 67% RH, each generation develops in 6.7 days. At 27°C and 72% RH, the female and male lifespan is 7.5 and 10.7 days, respectively. The mite overwinters as adult forms shelter in the crevices of the trunk and branches, and also in modest numbers at the base of plants in the upper soil layer and leaf litter (Chen *et al.*, 1996b). In contrast, on jujube in Changzhou prefecture in Jiangsu Province, *T. truncatus* overwinters as eggs laid in the bark crevices on trunks. In March of the following year, these overwintering eggs begin to hatch, and the young mites first migrate to the ground to feed on weeds and multiply, then in April they move to jujube and cause damage during sprouting (Li *et al.*, 1998). On mulberry leaflets under laboratory conditions, *T. truncatus* completed its development within a wide range of temperatures (20, 24, 28, 31, 35°C). The development time from egg to adult ranged from 6.30 to 14.89 days. The highest immature mortality was at 20°C (39.88%) and 35°C (30.70%). The threshold temperatures for the development of the egg, larva, protonymph, deutonymph and a generation were 11.38, 9.91, 10.67, 12.51 and 11.57°C and the degree-days necessary for each stage were 62.52, 15.38, 11.36, 11.76 and 142.83, respectively. Fertilized females laid the highest number of eggs at 24°C (65.6) and 31°C (64.78), and lived for up to 28.41 days at 20°C and 9.63 days at 35°C. The survival of unfertilized eggs was longest at 24°C and egg production was slightly more at 24 and 28°C. The net reproductive rate (R_0) was highest at 24°C (37.39) and 31°C (37.00). Values of intrinsic rate of increase (r_m) and finite rate of increase (λ) were highest (0.321 and 1.378, respectively) at 31°C and lowest (0.134 and 1.143, respectively) at 20°C. The mean generation time was shortest at 35°C and the shortest time for a population to double was 2.16 days at 31°C. The highest population trend index was at 24°C (49.49) and 31°C (48.58). Overall, the temperature range 24–31°C is the most suitable for the development, survival rate and reproduction of *T. truncatus* (Sakunwarin *et al.*, 2003).

Studies in maize fields in Mongolia showed that the spatial pattern of mite distribution was aggregated, and the aggregation intensity decreased with increasing population density (Pang *et al.*, 2005a). The spatial patterns of *T. truncatus* populations on cucumber plants in a plastic greenhouse were also aggregated. The aggregation intensity is correlated with the density of the mites and declines with increase in the mite population density (Gao *et al.*, 2005).

The host plant affects the development and reproduction of *T. truncatus*. Rearings of the mite under laboratory conditions at 28°C on five different host plants (cucumber, kidney bean, soybean, aubergine, maize) showed significant differences in the duration of the various developmental stages, and in fecundity, daily egg production, net reproductive rate (R_0), finite rate of increase (λ), intrinsic rate of increase (r_m), mean generation time (T) and days needed to double the population size (DDP). The time needed by females to complete a generation on cucumber, kidney bean, soybean, aubergine and maize were 9.3, 9.3, 9.6, 11.0 and 11.6 days, respectively. The oviposition time was 9.9, 11.2, 12.9, 15.8 and 17.9 days, respectively, on aubergine, cucumber, kidney bean, maize and soybean. Both fecundity and R_0 were the highest on soybean and the lowest on aubergine. The survival rates on soybean, maize, cucumber, kidney bean and aubergine were 93.6, 91.7, 89.8, 84.3 and 61.6%, respectively. The mites showed a preference for each plant that decreased in the order soybean, maize, cucumber, kidney bean and aubergine (Pang *et al.*, 2004).

Studies conducted on the biological parameters of *T. turkestani* Ugarov *et* Nikolskii and *T. truncatus* on cotton and maize in the laboratory at 28 ± 1°C, 70 ± 10% RH and under a 16L:8D photoperiod showed that the fecundity of *T. truncatus* was greater than that of *T. turkestani* on the same host plant. Moreover, maize was the better host plant for *T. truncatus*, whereas cotton was more suitable for *T. turkestani* (Guo *et al.*, 2013). Similarly, under laboratory conditions at 28°C, the use of different cultivars of maize ('Zhongdan 2', 'Nongda 108', 'Chidan 202' and 'Badan 3') significantly affected the development and the reproduction of *T. truncatus*, influencing the duration of the egg and nymph stages, pre-oviposition, adult longevity, fecundity, daily egg production, net reproductive rate (R_0), finite rate of increase (λ), intrinsic rate of increase (r_m), mean generation time (T) and the days needed to double population size (DDP). The hatching rate on cultivars 'Zhongdan 2', 'Nongda 108', 'Chidan 202' and 'Badan 3' was 95.8, 94.0, 90.0 and 84.0%, and the survival rate on these cultivars was 90.5, 86.0, 84.0 and 72.0%, respectively. The net reproductive rate (R_0) was highest on 'Zhongdan 2' and lowest on 'Badan 3'. The mite showed a decreasing preference for each cultivar in the order 'Zhongdan 2', 'Nongda 108', 'Chidan 202' and 'Badan 3' (Pang *et al.*, 2005b).

Under laboratory conditions, *T. truncatus* developed 31 generations a year (Chao and Lo, 1974). In China, in the district of Beijing, 15–17 generations a year were observed on *S. japonica*

(Chen *et al.*, 1996b), while in Changzhou, *T. truncatus* completed 13 generations a year on jujube trees (Li *et al.*, 1998). In Thailand, the dry period (December–March) has the optimal conditions for population growth of the mite (Charanasri *et al.*, 1993), and in China also, populations seem to thrive during drier periods (Chen *et al.*, 1999).

Different coccinellids are known to be natural enemies of *T. truncatus*. These include *Stethorus chengi* Sasaji, *S. hirashimai* Sasaji and *S. loi* Sasaji (Chao and Lo, 1974), and *S. pauperculus* (Weise) (Wongkopras, 1992). Predatory insects include the staphilinid *Oligota oviformis* Casey, the cecidomyiid *Arthrocnodax occidentalis* Felt, the hemerobiid *Eumicromus sauteri* (Esben-Petersen), the thrip *Scolothrips sexmaculatus* (Pergande) and some anthocorids, such as *Orius* sp. (Chao and Lo, 1974), and *O. tantillus* (Mots.) (Navasero *et al.*, 2003). Predatory phytoseiid mites include *Amblyseius largoensis* (Muma) and *A. ovalis* (Evans) (Chao and Lo, 1974), *A. swirskii* Athias Henriot (= *Typhlodromips swirskii* (Athias Henriot)) (Chen *et al.*, 2011) and *Neoseiulus longispinosus* (Evans) (Chao and Lo, 1974; Urisakul, 1982). Other predatory mites reported included some Anystidae (*Anystis* sp.) and Bdellidae, such as *Bdella* sp. (Chao and Lo, 1974) and *Bdellodes japonicus* Ehara (Chen *et al.*, 2007a).

Symptomatology and damage

Like other spider mites, feeding damage by *T. truncatus* leads to speckling of the leaves, and then to large areas of yellowing and bronzing of the foliage (Chen *et al.*, 1999). Infested leaves of cotton show yellowish spots, often associated with webbing on the lower leaf surface. *T. truncatus* is an important pest of cotton in China (Gao *et al.*, 1990b; Cao *et al.*, 1993); in Henan, yield losses from infested cotton plants increased by 11.9, 14.2 and 28.4% at the seedling, squaring/flowering and flowering/boll stages in normal years, and by 27.2, 40.6 and 73.7% in dry years (when the proportion of red leaves increased by 1%), respectively (Gao *et al.*, 1990b). On maize plants, the vertical distribution of *T. truncatus* populations is stratiform, and the highest density of mites occurs on the middle part of the plant. Early sowing, bad growing, low maize plant density, few natural enemies and a short duration of extremely high humidity from July to August induce high population densities and severe damage (Chen *et al.*, 1999). Early symptoms on maize leaves are faded greenish spots, and with increasing attack, heavily infested leaves become yellow and dry. Severe infestation of *T. truncatus* on maize may induce a yield reduction of 15%. On *S. japonica*, peaks of mite populations from July to August in the urban district of Beijing gave about 70% of leaf damage, and over 90% damage in the suburbs (Chen *et al.*, 1996b).

Control

The application of cultural techniques and sprays of acaricides are currently the only options possible for the control of infestation by *T. truncatus*. Biological control by predatory insects and mites is a limiting factor to populations of the mite in some countries, but does not alone seem to solve the problem.

ECONOMIC THESHOLD. Gao *et al.* (1990b) found that the economic thresholds for ginned cotton yields of 50, 75 and 100 kg mu^{-1} (1 mu = 0.067 ha) were 25, 17 and 13% red leaves at the seedling stage, 21, 14 and 11% at the squaring/flowering stage, and 11, 7 and 5% at the flowering/boll stage, respectively.

CULTURAL PRACTICES. Weed control and the enhancement of soil fertility are good cultural methods for the control of *T. truncatus* on maize in Shaanxi Province, China (Chen *et al.*, 1999). Control of the mite is best achieved on jujube trees by scraping the old bark in the early spring; sticking a viscous glue ring on to the trunk of the tree in late April is also effective (Li *et al.*, 1998).

BIOLOGICAL CONTROL. Despite various records of natural enemies and the development of laboratory- and field-rearing techniques for some predators, such as the anthocorid *Orius tantilus* (Mots.) (Navasero *et al.*, 2003), the control of *T. truncatus* currently mainly depends upon chemical methods.

CHEMICAL CONTROL. Like other spider mites, the chemical control of *T. truncatus* is prevented by the presence of strains resistant to insecticides/acaricides, e.g. dicofol, monocrotophos and omethoate (Cao *et al.*, 1993). Various active ingredients have been applied in the recent past to control the mite outbreaks. They include the insecticide/acaricides methidathion and permethrin, and acaricides, such as amitraz, dicofol, propargite (Jin *et al.*, 2002) and pyridaben (Hu *et al.*, 1998). Currently, the acaricides acequinocyl, bifenazate, clofentezine, cyenopyrafen, cyflumetofen, etoxazole, fenbutatin oxide, hexythiazox and spiromesifen are considered effective for control of the mite (Ullah and Gotoh, 2013).

In addition to conventional acaricides, the study of botanical acaricides and other natural compounds is of interest. Results of field testing highlighted the effectiveness of Syrian rue (*Peganum harmla* L.) in mite control (Zhao *et al.*, 1997). Ethanolic extracts from black pepper (*Piper nigrum* L.) and turmeric (*Curcuma longa* L.) also had a strong effect on the mite (Insung, 2001). The effectiveness of sugar apple seed crude hexane and ethanol extracts was reported by Sakunwarin (2004). The toxicities of *Derris elliptica* Benth extracts and rotenone were tested by Wongthong and Pimsamarn (2007).

Application of the antibiotic liuyangmycin, a tetranactin isolated from the bacterium *Streptomyces griseolus* Waksman *et* Henrici, resulted in high mortality of *T. truncatus* (Dong, 1990). On *S. japonica*, spraying nikkomycin, liuyangmycin, propargite, washing powder and pellets of NH_4HCO_3 gave high mortality of *T. truncatus* in the spring (Chen *et al.*, 1996b).

Tetranychus tumidus Banks (Fig. 20.64)

Common name

Tumid spider mite.

Diagnostic characteristics

FEMALE. The body is light carmine in colour, marked on the back by a pair of large 'U'-shaped sooty spots. The peritremes are hooked. The dorsal body setae are slender and longer than the intervals between longitudinal bases. The dorsal opisthosomal

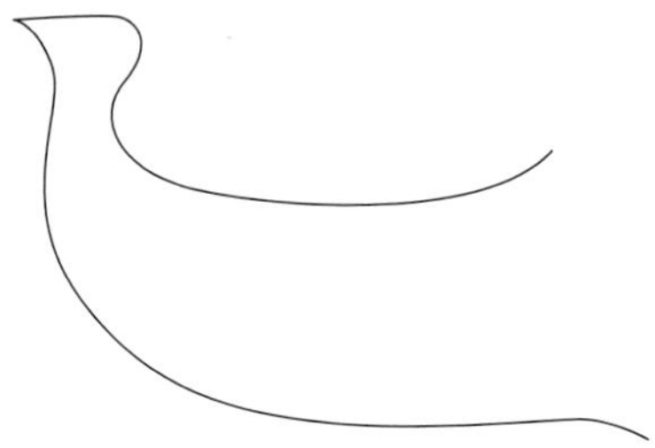

Fig. 20.64. *Tetranychus tumidus* Banks. Male, aedeagus (modified from McGregor, 1950).

striae between the setae e_1 and f_1 are longitudinal and form a diamond-shaped pattern. The tactile setae of tarsus I are proximal to the duplex setae. The empodia have large spurs (Pritchard and Baker, 1955; Jeppson *et al.*, 1975).

MALE. Empodium I is claw like and the other empodia end in 3 pairs of ventrally directed hairs. The knob of the aedeagus is at right angles to the neck, with the axis parallel to that of the shaft and a broadly rounded anterior angulation with a short acute angulation posteriorly (Pritchard and Baker, 1955; Jeppson *et al.*, 1975).

Geographical distribution

Colombia (Urueta, 1975), Costa Rica (Aguilar and Murillo, 2008), Cuba (Boudreaux, 1979; Álvarez and Alarcón, 2004; Martínez *et al.*, 2004; Suárez, 2004), Greece (Hatzinikolis, 1986), Guam (Baker and Pritchard, 1962), Hawaiian Islands (Lee Goff, 1986), Panama, Puerto Rico (Baker and Pritchard, 1962), Thailand (Kongchuensin *et al.*, 2005) and the USA (Banks, 1900; Bibby and Tuttle, 1959).

Host plants

The whole list of host plants of the tumid spider mite includes 47 species from 21 families (Migeon and Dorkeld, 2006–2013). The families and species of economic importance are:

Amaranthaceae: spleen amaranth (*Amaranthus dubius* Mart. ex Thell.) (Urueta, 1975), carelessweed (*A. spinosus* L.) (Lee Goff, 1986), green amaranth (*A. viridis* L.) (Martínez *et al.*, 2004).

Amaryllidaceae: garlic (*Allium sativum* L.) (Martínez *et al.*, 2004).

Apiaceae: celery (*Apium graveolens* L.) (Suárez, 2004).

Araceae: ape (*Alocasia macrorrhizos* (L.) G. Don) (Suárez, 2004), cocoyam (*Colocasia esculenta* (L.) Schott) (Martínez *et al.*, 2004).

Arecaceae: coconut palm (*Cocos nucifera* L.) (Suárez, 2004).

Caricaceae: papaya (*Carica papaya* L.) (Suárez, 2004).

Combretaceae: Indian almond (*Terminalia catappa* L.) (Suárez, 2004).

Asteraceae: red dahlia (*Dahlia coccinea* Cav.) (Suárez, 2004), sunflower (*Helianthus annuus* L.) (Martínez *et al.*, 2004), African marigold (*Tagetes erecta* L.) (Martínez *et al.*, 2004; Suárez, 2004), zinnia (*Zinnia elegans* Jacq.) (Martínez *et al.*, 2004).

Convolvulaceae: sweet potato (*Ipomoea batatas* (L.) Lam.) (Lee Goff, 1986; Suárez, 2004).

Cucurbitaceae: courgette (*C. pepo* L.) (Martínez *et al.*, 2004).

Euphorbiaceae: cassava (*Manihot esculenta* Crantz) (Martínez *et al.*, 2004).

Fabaceae: bean (*Phaseolus vulgaris* L.) (Álvarez and Alarcón, 2004; Martínez *et al.*, 2004; Suárez, 2004).

Malvaceae: okra (*Abelmoschus esculentus* (L.) Moench) (Martínez *et al.*, 2004; Suárez, 2004), American cotton (*Gossypium hirsutum* L.) (Bibby and Tuttle, 1959).

Musaceae: banana (*Musa* × *paradisiaca* L.) (Álvarez and Alarcón, 2004; Martínez *et al.*, 2004).

Passifloraceae: sweet granadilla (*Passiflora ligularis* Juss.) (Aguilar and Murillo, 2008).

Poaceae: maize (*Zea mays* L.) (Lee Goff, 1986; Martínez *et al.*, 2004).

Solanaceae: tomato (*Solanum lycopersicum* L.) (Suárez, 2004), quito orange (*S. quitoense* Lam.) (Aguilar and Murillo, 2008), Irish potato (*S. tuberosum* L.) (Martínez *et al.*, 2004).

Bio-ecology

T. tumidus was first synonymized with *T. gloveri* Banks (see above). Among economic plants, it damages herbaceous, tropical fruit and ornamental plants, and has been reported on citrus in south-eastern USA (Jeppson, 1989), although this report probably refers to *T. gloveri*.

The tumid spider mite reproduces by arrhenotokous parthenogenesis and has a haploid number of chromosomes of six (Helle *et al.*, 1970, 1983). Both adults and immature stages feed on leaves. The development, survivorship and reproduction of the mite were assessed on coconut palm at six different temperatures (10, 15, 20, 25, 30 and 35°C). The developmental period of immature stages ranged from 39.6 days at 15°C to 7.4 days at 30°C. The mite failed to develop beyond the larval stage at 10°C. The lower temperature developmental thresholds for egg, larva, protonymph, deutonymph and the combined immature stages were estimated as 11.1, 12.9, 12.1, 11.1 and 11.9°C, respectively. The upper temperature thresholds for development of immature stages were estimated as 25.9–35.9°C. The percentage survivorship of immature stages varied from 56.4 to 93.7% within the temperature range 15–35°C. The average longevity of adult females ranged from 48.7 days at 15°C to 7.9 days at 35°C. The average oviposition per female lifespan varied from 86.02 to 61.75 eggs within the temperature range 20–30°C. However, oviposition was greatly reduced to 19.44 and 20.54 eggs at 15°C and 35°C, respectively (Liu and Tsai, 1998).

In laboratory experiments at 20, 25 and 30°C, the minimum development threshold for the incubation period of *T. tumidus* eggs was 14°C, with a thermal constant of 35 ± 6.4 degree-days. Corresponding minimum development threshold temperatures for the larval stages were 13.7°C with a thermal constant of 26.3 ± 5.2 degree-days, for the nymphal stages, 12.9°C with a thermal constant of 50.1 ± 5.2 degree-days, and for the whole cycle from egg to adult stage, 13.9 °C with a thermal constant of 112.6 ± 8.5 degree-days (Pérez Alvarez *et al.*, 1997).

On coconut palm, the highest values of the intrinsic rate of increase (r_m) and the net reproduction (R_0) were at 30°C: 0.289 individuals per female a day and 59.67, respectively. The lowest values of net reproduction (R_0) were at 15 and 35°C: 12.74 and

12.99, respectively. The lowest value of the intrinsic rate of increase (r_m) was 0.042, at 15°C. The mean generation time (T) of the population ranged from 60.7 days at 15°C to 11.7 days at 35°C. The optimal temperature for the development of the populations was 30°C (Liu and Tsai, 1998).

Natural enemies include pathogens such as *Neozygites* spp. (Saba, 1973a) and predatory insects and mites. The predatory insects include the thripid *Scolothrips pallidus* (Beach), the cecidomyiid *Feltiella acarisuga* (Vallot) (Gutierrez, 1983, in Chazeau, 1985), the coccinellids *Stethorus punctum picipes* Casey (Pérez *et al.*, 2004) and *Stethorus siphonulus* Kapur (Raros and Haramoto, 1974), and the phytoseiid mite *Neoseiulus longispinosus* (Evans) (Kongchuensin *et al.*, 2005).

Symptomatology and damage

Feeding activity by the tumid spider mite causes reddening of the upper leaf surface. This reddened surface may appear either as a small blotch or as numerous blotches extended over the whole leaf surface. Damage may result in total defoliation of the infested plant (Jeppson *et al.*, 1975). Jeppson (1989) reported that on citrus, feeding damage to freshly flushed leaves by the tumid spider mite was similar to that of six-spotted mite, *Eotetranychus sexmaculatus* (Riley).

Control

The few reports that there are on the control of this mite indicate its secondary importance. Organochlorines, organosulfurs and organotins have been tested against the mite on *Chamaedorea elegans* Martius (Reinert and Neel, 1977). On citrus, chemical control is not justified.

Tetranychus turkestani (Ugarov *et* Nikolskii) (Fig. 20.65)

Common name

Strawberry spider mite.

Diagnostic characteristics

FEMALE. The body is variable in colour, from amber, green, brownish or almost black in the summer females to bright orange in overwintering females (Jeppson *et al.*, 1975); there is a broad spot on each side, behind the eyes and extending beyond the middle of the body, and 1 black spot for each side towards the end of the opisthosoma. The peritreme is hooked. The dorsal striae are lobate; between the setae e_1–e_1 and f_1–f_1 they are longitudinal, and between the setae e_1 and f_1 they are transverse and form a diamond shape; the ventral striae lack lobes; the pregenital striae become sparse and broken medially. The empodia have 6 proximoventral hairs, each with a minute dorsal spur or with the spur absent; tarsus I has 4 tactile setae proximal to the proximal duplex setae (Seeman and Beard, 2011).

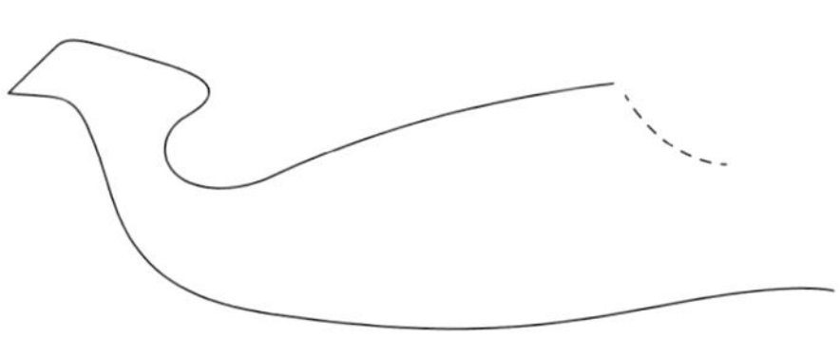

Fig. 20.65. *Tetranychus turkestani* (Ugarov *et* Nikolski). Male, aedeagus (from Meyer Smith, 1987).

MALE. Empodia I–II each have a dorsal spur; the empodia I are uncinate or claw like, and empodia II–IV have long and free proximoventral hairs. The knob of the aedeagus is large, dorsally flat with a posterior angle, the anterior projection rounded and the posterior projection bluntly pointed (Seeman and Beard, 2011).

Geographic distribution

Azerbaijan (Mitrofanov *et al.*, 1987), Bulgaria (Atanasov, 1971, 1979; Mitrofanov *et al.*, 1987), Canary Islands (Ferragut and Santonja, 1989), China (Wang, 1981a), Costa Rica (Ochoa *et al.*, 1991a), France (Rambier, 1958a, 1979; Bailly *et al.*, 2004; Hurtado *et al.*, 2008), Greece (Hatzinikolis, 1968c, 1969c, 1986), Guadeloupe (Flechtmann *et al.*, 1999a), Hungary (Bozai, 1970), Iran (Khalil-Manesh, 1973; Seeman and Beard, 2011), Iraq (El-Haidari, 1965), Israel (Bolland *et al.*, 1998; Ben-David *et al.*, 2007, 2009), Kazakhstan (Mitrofanov *et al.*, 1987), Kuwait (Bolland *et al.*, 1998), Mexico (Tuttle *et al.*, 1976), Morocco (Saba, 1973b), the Netherlands (Jeppson *et al.*, 1975; Vierbergen, 1990; Breeuwer, 1997), New Zealand (Manson, 1967b), Pakistan (Kazimi and Ghani, 1969), Poland (Boczek and Kropczynska, 1964), Portugal (Carmona, 1964), Russia (Mitrofanov *et al.*, 1987), Spain (Ferragut and Santonja, 1989; Aucejo *et al.*, 2003; Escudero and Ferragut, 2005; Hurtado *et al.*, 2008), South Africa (Meyer Smith and Ryke, 1959; Seeman and Beard, 2011), Switzerland (Günthart and Günthart, 1959), Syria (Zriki *et al.*, 2015), Tajikistan (Mitrofanov *et al.*, 1987), Tunisia (Kreiter *et al.*, 2002), Turkey (Düzgünes, 1954; Uysal *et al.*, 2001; Ozsisli and Cobanoğlu, 2011), Ukraine (Mitrofanov *et al.*, 1987), the UK (Hill and O'Donnell, 1991a,b), the USA (Pritchard and Baker, 1955; Bibby and Tuttle, 1959; Tuttle and Baker, 1964; Johanowicz and Hoy, 1996; Seeman and Beard, 2011), Uzbekistan (Ugarov and Nikolskii, 1937; Mitrofanov *et al.*, 1987) and Yugoslavia (Mijuskovic, 1981; Petanović and Filipi-Matutinovic, 1988).

Host plants

The host range of the strawberry spider mite includes about 208 wild and cultivated plants belonging to 60 families, with the highest number of species ascribed to the families Asteraceae (23 species), Fabaceae (22 species) and Rosaceae (17 species) (Migeon and Dorkeld, 2006–2013). The families and species of economic importance are:

Amaranthaceae: African spinach (*Amaranthus cruentus* L.) (Carmona, 1964).

Apiaceae: celery (*Apium graveolens* L.), Peruvian carrot (*Arracacia xanthorrhiza* Bancr.), carrot (*Daucus carota* L.) (Tuttle and Baker, 1964, 1968), parsley (*Petroselinum crispum* (Mill.) Fuss) (Carmona, 1964).

Asteraceae: chicory (*Cichorium intybus* L.) (Tuttle and Baker, 1964, 1968), sunflower (*Helianthus annuus* L.) (McGregor, 1950;

Bailly *et al.*, 2004), lettuce (*Lactuca sativa* L.) (Tuttle and Baker, 1968), dandelion (*Taraxacum officinale* F.H. Wigg.) (Tuttle *et al.*, 1976), zinnia (*Zinnia elegans* Jacq.) (Bolland *et al.*, 1998; Migeon and Dorkeld, 2006–2013).

Betulaceae: hazel (*Corylus avellana* L.) (Bolland *et al.*, 1998; Migeon and Dorkeld, 2006–2013).

Brassicaceae: canola (*Brassica juncea* (L.) Czern), pepper cress (*Lepidium draba* L.) (Bailly *et al.*, 2004).

Cannabaceae: hop (*Humulus lupulus* L.) (Ugarov and Nikolskii, 1937).

Cucurbitaceae: watermelon (*Citrullus lanatus* (Thunb.) Matsum. and Nakai) (Ugarov and Nikolskii, 1937; Tuttle *et al.*, 1976; Ben-David *et al.*, 2007, 2009), cantaloupe (*Cucumis melo* L.) (Ugarov and Nikolskii, 1937; Tuttle and Baker, 1968), cucumber (*C. sativus* L.) (Ugarov and Nikolskii, 1937; Tuttle and Baker, 1964, 1968; Hatzinikolis, 1968c, 1969c), pumpkin (*Cucurbita maxima* Duchesne), squash (*C. moschata* Duchesne) (Ugarov and Nikolskii, 1937; Carmona, 1964).

Ebenaceae: kaki (*Diospyros kaki* Thunb.) (Carmona and Dias, 1980).

Euphorbiaceae: castor bean (*Ricinus communis* L.) (Bibby and Tuttle, 1959; Tuttle and Baker, 1964, 1968).

Fabaceae: groundnut (*Arachis hypogaea* L.), soybean (*Glycine max* (L.) Merr.) (McGregor, 1950; Bibby and Tuttle, 1959; Tuttle and Baker, 1964, 1968; Carmona and Dias, 1980), liquorice (*Glycyrrhiza glabra* L.), lucerne (*Medicago sativa* L.) (Ugarov and Nikolskii, 1937; McGregor, 1950; Bibby and Tuttle, 1959; Tuttle and Baker, 1964, 1968; Bailly *et al.*, 2004; Seeman and Beard, 2011), lima bean (*Phaseolus lunatus* L.) (McGregor, 1950), common bean (*P. vulgaris* L.) (Ugarov and Nikolskii, 1937; Ozsisli and Cobanoğlu, 2011; Zriki *et al.*, 2015), Egyptian clover (*Trifolium alexandrinum* L.) (Tuttle and Baker, 1968), Italian clover (*T. incarnatum* L.) (McGregor, 1950), Chilean clover (*T. pratense* L.), Dutch clover (*T. repens* L.) (McGregor, 1950; Bailly *et al.*, 2004), fava bean (*Vicia faba* L.), black lentil (*Vigna mungo* (L.) Hepper) (Bolland *et al.*, 1998; Migeon and Dorkeld, 2006–2013).

Grossulariaceae: blackcurrant (*Ribes nigrum* L.) (Bolland *et al.*, 1998; Migeon and Dorkeld, 2006–2013).

Juglandaceae : English walnut (*Juglans regia* L.) (McGregor, 1950; Zriki *et al.*, 2015).

Lauraceae: sassafras (*Sassafras albidum* (Nutt.) Nees) (Bolland *et al.*, 1998; Migeon and Dorkeld, 2006–2013).

Malvaceae: okra (*Abelmoschus esculentus* (L.) Moench), American jute (*Abutilon theophrasti* Medik.) (Ugarov and Nikolskii, 1937), American cotton (*Gossypium hirsutum* L.) (Bibby and Tuttle, 1959; Tuttle and Baker, 1964, 1968).

Moraceae: fig (*Ficus carica* L.) (Carmona and Dias, 1980; Ben-David *et al.*, 2009), red mulberry (*Morus rubra* L.) (Tuttle and Baker, 1968).

Musaceae: banana (*Musa acuminata* Colla) (Ben-David *et al.*, 2009).

Poaceae: maize (*Zea mays* L.) (Ugarov and Nikolskii, 1937; Tuttle and Baker, 1964; Hatzinikolis, 1969c, 1986; Bailly *et al.*, 2004).

Rosaceae: quince (*Cydonia oblonga* Mill) (Ugarov and Nikolskii, 1937), strawberry (*Fragaria × ananassa* Duchesne ex Rozier) (McGregor, 1941a; Rambier, 1958a; Carmona, 1964; Johanowicz and Hoy, 1996; Escudero and Ferragut, 2005), Virginia strawberry (*F. virginiana* Mill.), apple (*Malus domestica* Borkh.) (Pritchard and Baker, 1955; Ben-David *et al.*, 2007, 2009), almond (*Prunus amygdalus* Batsch) (Ben-David *et al.*, 2009), sweet cherry (*P. avium* (L.) L.) (Ugarov and Nikolskii, 1937), European plum (*P. domestica* L.) (Carmona, 1964), peach (*P. persica* (L.) Batsch) (Ugarov and Nikolskii, 1937; Ben-David *et al.*, 2007, 2009), pear (*Pyrus communis* L.) (Carmona, 1964), red raspberry (*Rubus idaeus* L.) (Bolland *et al.*, 1998; Migeon and Dorkeld, 2006–2013).

Rutaceae: sweet orange (*Citrus sinensis* (L.) Osbeck) (Tuttle and Baker, 1968).

Solanaceae: pepper (*Capsicum annuum* L.) (Ozsisli and Cobanoğlu, 2011), tomato (*Solanum lycopersicum* L.) (Kreiter *et al.*, 2002), aubergine (*S. melongena* L.) (Ugarov and Nikolskii, 1937; McGregor, 1950; Hurtado *et al.*, 2008; Ozsisli and Cobanoğlu, 2011; Zriki *et al.*, 2015).

Vitaceae: grapevine (*Vitis vinifera* L.) (Ugarov and Nikolskii, 1937; Rambier, 1979; Carmona and Dias, 1980).

Bio-ecology

T. turkestani infests low-growing crops (vegetables and forages), fruit crops (deciduous fruits, stone fruits, citrus, tropical fruits), grapevines and various ornamentals. In southern California, the mite occurs only rarely, and this seems to be the result of straggling from other plants growing under or near the trees, but it has no economic importance on citrus (McGregor, 1956), while it is a serious pest of cotton in different countries of the world (Leigh, 1985).

The strawberry spider mite reproduces by arrhenotokous parthenogenesis, and has a haploid number of chromosomes of three (Helle *et al.*, 1970). Its populations live mainly on the lower leaf surface, where they feed and produce webbing that may cover the leaves and stems together. The mite overwinters as bright orange females, which appear in late September, while the summer forms disappear by late October. Overwintering females shelter in litter, under bark scales and in the soil (Pritchard and Baker, 1952a; Mellot and Connell, 1965; Jeppson *et al.*, 1975). The mite harbours endosymbiont *Wolbachia* bacteria, which induces cytoplasmic incompatibility (Breeuwer, 1997). Initial studies on the bio-ecology of *T. turkestani* reported a pre-oviposition time ranging from 1 to 6 days, an average daily oviposition of 7.4 eggs during midsummer and of 2–3 eggs in the summer and autumn. Female longevity is about 8 days in the summer and 33 days in autumn (Cagle, 1956). Like other spider mites, the bio-ecology of *T. turkestani* has since been investigated in more detail, including evaluation of the influence of host plants on the development and reproduction of the mite (Yasen *et al.*, 2007; Sohrabi and Shishehbor, 2008). Some results are given below from reports concerning cotton, bean, cucumber and other plants as hosts of the strawberry spider mite.

On cotton cotyledons, the mean developmental time ranged from 25.8 to 29.0 days at 15.5°C and from 6.1 to 6.7 days at 29.4°C for the female, with a slightly shorter time for the male. At 23.8°C, the fecundity was 84.6 eggs and at 29.4°C it was 73.5 eggs. The proportion of female offspring was 55%. The females lived on average 12.46 days at 23.8°C and 8.79 days at 29.4°C. At 25°C ± 1°C, the intrinsic rate of increase (r_m) was 0.203 individuals per female a day, the net reproduction (R_0)

was 46.8, and the finite rate of increase (λ) was 1.225 individuals per female a day (Carey and Bradley, 1982). At 25 and 30°C, the developmental time ranged from 7.0 to 12.1 days on cowpea (*Vigna unguiculata* (L.) Walp.), from 6.9 to 11.7 days on green gram (*V. radiata* (L.) R. Wilczek) and from 6.7 to 10.8 days on pinto bean (*P. vulgaris*). The greatest mortality occurred during the nymphal stage, followed by the larval and egg stages. Immature mortality ranged from 18.6 to 21.6% on cowpea, from 11.6 to 28.5% on green gram and from 38.3 to 42.1% on pinto bean. The adult longevity was greater for females than for males, greater on cowpea than on green gram and pinto bean and decreased with increasing temperature. The mean total fecundity ranged from 36.3 to 90.5 eggs per female and the sex ratio was 0.8–0.91 (females). The intrinsic rate of natural increase (r_m) ranged from 0.23 to 0.32 on cowpea, from 0.24 to 0.28 on green gram and from 0.19 to 0.27 on pinto bean. The generation time (*T*) decreased from 16.3 to 10.2 days with increasing temperature (Sohrabi and Shishehbor, 2008).

On cotton and maize at 28 ± 1°C, 70 ± 10% RH and under a 16L:8D photoperiod, the fecundity of *T. truncatus* Ehara (see above) was greater than that of *T. turkestani* on the same plant species. Moreover, maize was the better host plant for *T. truncatus*, whereas cotton was more suitable for *T. turkestani* (Guo *et al.*, 2013). On cucumber, the egg to adult developmental period for females ranged from 50.11 ± 0.79 days at 15°C to 7.73 ± 0.14 days at 30°C. The lowest thermal thresholds for the female and male were 11.89 and 12.21°C, respectively. The complete development of the female required an average of 140.33 ± 5.49 degree-days above the lower thermal threshold. Pre-imaginal mortality was 62.92, 16.67, 41.82 and 42.31% at 15, 20, 25 and 30°C, respectively. Mean longevity of the female ranged from 30.22 ± 2.88 days at 15°C to 5.78 ± 0.10 days at 30°C. Mean total fecundity ranged from 23.11 ± 3.14 to 49.95 ± 5.54 eggs per female. The sex ratio (% females) ranged from 69.00 ± 0.04 to 83.01 ± 0.02%. The intrinsic rate of population increase (r_m) ranged from 0.272 ± 0.010 at 30°C to 0.033 ± 0.002 (females per female a day) at 15°C. The population doubled most quickly at 30°C (2.54 ± 0.09 days) and most slowly at 15°C (20.46 ± 1.59 days). The optimal temperature for development, survival and reproduction within the above range was 30°C (Karami-Jamour and Shishehbor, 2012). The above data do not conflict significantly with the values found in experimental studies by Yu *et al.* (2000), who reported a developmental threshold temperature and an effective thermal summation of 10.74°C and 164.01 degree-days for females, and 11.56°C and 113.29 degree-days for males, respectively.

Studies conducted in the field in Iran on the seasonal abundance of the strawberry spider mite on four aubergine varieties ('Blackbeauty', 'Ghasri', 'Esfahani' and 'Inehrash') showed that infestation occurred on all varieties throughout the season, and that the highest populations are reached in June and July. Trends in population changes are correlated with various biotic and abiotic factors, average and minimum temperatures and average RH. Stepwise regression analysis showed significant influences of maximum temperature and the predator population on population changes of *T. turkestani* on the varieties studied (Nemati *et al.*, 2010b).

Canerday and Arant (1964) reported that high daily temperatures and low rainfall favoured the development of strawberry spider mite in field studies in Tennessee. Yu *et al.* (2000) observed that at 15, 20, 25 and 30°C, the mean developmental time of females took 35.15, 19.05, 11.04 and 8.68 days, respectively. Laboratory research conducted by Zhang *et al.* (2006a) highlighted the importance of temperature in the development and fecundity of *T. turkestani*. They found that the generation survival rates of strawberry spider mites reached 79.80–100% in the temperature range of 19–31°C. The net reproductive rate per generation was 44.5793 at 20 ± 1°C, 57.0307 at 25 ± 1°C, 52.6785 at 30 ± 1°C and 57.6588 at 34 ± 1°C. The period of population doubling shortened as the temperature increased; it was 4.8799 days at 20 ± 1°C and 2.3610 days at 34 ± 1°C. These data allow the conclusion to be drawn that under the climatic, field and the cultural management conditions for cotton in Xinjiang, China, the higher fecundity potential and the shorter generation of the strawberry spider mite are the main factors that cause mite outbreaks.

Experimental high temperature exposure (38, 42 and 46°C) for 2–6 h significantly affected the egg hatching rates, survival rates of eggs, larvae and adults, and developmental periods of *T. turkestani* and *T. truncatus*, but did not affect the longevity and reproduction of female adults; *T. turkestani* exhibited stronger adaptability to high temperature than *T. truncatus* (Yang *et al.*, 2013a). In Iran, studies of *T. turkestani* under laboratory conditions at different temperatures (15, 20, 25, 30, 35, 40°C), 55 ± 10 RH and a 16L:8D photoperiod showed that the upper and lower thermal thresholds were 12.4 and 41.6°C, respectively. The total effective temperature for egg to egg development was 360.74 ± 16.09 degree-days. The non-linear developmental rate Lactine-2 model estimated upper and lower developmental threshold values of 43.09 and 6.05°C and an optimal temperature of 36.5°C (Nemati *et al.*, 2010a). The mite develops 8–16 generations a year (Cagle, 1956).

Natural enemies of *T. turkestani* include the coccinellids *Stethorus gilvifrons* (Mulsant) (Ahmed and Ahmed, 1989; Afshari, 1999; Sohrabi and Shishehbor, 2007) and *Stethorus punctillum* Weise (Yang *et al.*, 2013b), the thrip *Scolothrips longicornis* Priesner (Nemati *et al.*, 2010b; Gheibi and Hesami, 2011), different phytoseiid mites, such as *Neoseiulus californicus* (McGregor) (Escudero and Ferragut, 2005; Gomez-Moya and Ferragut, 2007), *N. cucumeris* (Oudemans) (Ji *et al.*, 2005b; Zhang *et al.*, 2006b), *Phytoseiulus persimilis* Athias Henriot (Atanasov, 1979; Daneshvar and Abaii, 1994; Gomez-Moya and Ferragut, 2007; Wang *et al.*, 2013) and *Phytoseius tropicalis* Daneshvar (Nemati *et al.*, 2010b), and some stigmaeid mites, such as *Zetzellia mali* (Ewing) (Khodayari *et al.*, 2009).

Symptomatology and damage

The strawberry spider mite feeds mainly on the lower leaf surface, but the damage appears on the upper surface as necrotic areas at the point of feeding on the leaf. High infestations of the mite induce leaf drop and plant death. Large amounts of webbing result in matt leaves and stems (Jeppson *et al.*, 1975). Feeding on cotton reduces seed yields from 13 to 22%, negatively affects boll size, lowers seed viability, reduces lint production and hinders maturity. Under laboratory conditions, the mite causes reddening of the leaves and plant defoliation within 40 days (Canerday and Arant, 1964). The time between infestation and abscission of a cotyledon is directly proportional to

the damage (necrotic areas) observed on the upper leaf surface. Abscission usually occurs if necrosis extends over 70% or more of the leaf surface. The damage produced by immature stages and adult females is easily visible, but not that caused by males. The mite may inject a local toxin into the plants during the feeding process (Simons, 1964). The tolerance of cotton to mite infestations increases with crop maturity (Wilson *et al.*, 1987). On citrus, the symptoms of attack and the damage caused are unknown and control is not justified. In France, the strawberry spider mite is reported as an occasional pest of vineyards on the Mediterranean coast (Rambier, 1982b). In western Europe, populations of *T. turkestani* reach moderate levels in July. Damage to grapevines has not been documented (Duso *et al.*, 2012).

Control

Control guidelines are similar to those suggested for the Pacific spider mite (*T. pacificus* McGregor) and the two-spotted spider mite (*T. urticae* Koch). It is possible to use a specific phenological model, transgenic varieties (Li *et al.*, 2013b) and biological and chemical means.

SAMPLING AND MONITORING. To predict timing of the development of the strawberry spider mite, it has been suggested that the phenological model of Carey and Bradley (1982) is applied; on cotton, this reported a lowest developmental threshold of 12°C, and degree-day accumulations for the egg, larva, protonymph and deutonymph of 49.7, 23.2, 18.4 and 20.8 degree-days, respectively. The egg to adult development period, the pre-egg-laying adult period, and the egg to egg developmental period require 112.1, 13.8 and 125.9 degree-days, respectively. Field monitoring and degree-days may be used in conjunction (see UC Phenology Model Database for the Pacific spider mite; UC IPM Online, 2003b). The economic threshold is unknown for *T. turkestani*. However, an average of one mite per leaf of cotton has been suggested as an action threshold in Iraq when pesticides are applied for the control of other pests (Ghobrial and Dittrich, 1980). For guidelines see Pacific spider mite.

BIOLOGICAL CONTROL. The most important natural enemies that have been investigated under laboratory conditions and/or in the field for the biological control of the strawberry spider mite are the coccinellids *S. gilvifrons* (Sohrabi and Shishehbor, 2007) and *S. puntillum* (Yang *et al.*, 2013b), and the phytoseiids *N. cucumeris* (Ji *et al.*, 2005b; Zhang *et al.*, 2006b) and *P. persimilis* (Daneshvar and Abaii, 1994; Gomez-Moya and Ferragut, 2007; Wang *et al.*, 2013). Despite the encouraging results obtained in the study of predatory insects, the use of some phytoseiids seems more realistic. Field trials conducted in China on Koerle pear infested by *T. turkestani* showed that if *N. cucumeris* was released in late April to early May when the temperature was 15–30°C and there were not more than two pest mites per leaf, the pest mites could be controlled effectively, and chemical control could be reduced by 6–7 times over the whole season in the orchard (Zhang *et al.*, 2006b). In Bulgaria, Atanasov (1979) reported that *P. persimilis* controlled infestations of *T. turkestani* on strawberries, aubergine and soybean. Laboratory and greenhouse trials on plants of bean and potato, as well as simulations of the population dynamics of phytoseiid predators and phytophagous mites in Spain showed the feasibility of biological control by phytoseiid mites. If the initial population of the pest was 25 mites per plant, the minimum release dose of *N. californicus* was estimated as 20 mites per plant to control *T. turkestani*, while for *P. persimilis* the minimum release dose would be 12 mites per plant (Gomez-Moya and Ferragut, 2007).

CHEMICAL CONTROL. Guidelines for chemical control are discussed in the sections on the Pacific spider mite (see above) and the two-spotted spider mite (see below). Evaluations of the effectiveness and selectivity of acaricides are being particularly investigated, and also their importance in the development of integrated pest management strategies in different countries. The studies include investigations in Xinjiang, China, on the sublethal effects of abamectin and pyridaben on experimental populations of *T. turkestani*, which showed that a sublethal dosage of abamectin and pyridaben could decrease the developmental rate of laboratory populations of *T. turkestani* (Gu *et al.*, 2010).

INTEGRATED PEST MANAGEMENT. See the guidelines for the Pacific spider mite and the two-spotted spider mite.

Tetranychus urticae Koch (Fig. 20.66)

Common name

Two-spotted spider mite.

Diagnostic characteristics

FEMALE. The body is variable in colour; there is a green form (GF) and a red form (RF). Summer females of the red form are similar in colour to the strawberry spider mite but with one conspicuous black spot on each side, while overwintering forms are yellowish orange. The peritreme is hooked. The opisthosomal dorsal striae are lobate, and the lobes vary from triangular to semicircular; between the setae e_1–e_1 and f_1–f_1 they are longitudinal, and between the setae e_1 and f_1 they form a diamond shape. The ventral striae lack lobes, the pregenital striae are usually entire, and sometimes sparse with small breaks medially. The empodia have 6 proximoventral hairs and lack spurs; tarsus I has 4 tactile setae proximal to the proximal duplex setae (Vacante, 1985a;

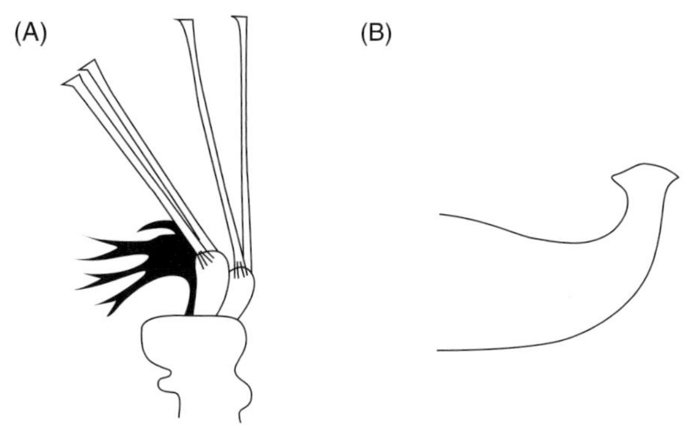

Fig. 20.66. *Tetranychus urticae* Koch. (A) female, pretarsus I; (B) male, aedeagus (drawing by Vacante). Other details of the same species are shown in Figs 2.6I, 2.9D, 2.11B, 2.19B, 2.20A.

Seeman and Beard, 2011). Following André and Van Impe (2012), the adult instar is missing in this species and is replaced by the tritonymph instar that is able to reproduce by paedogenesis.

MALE. Empodia I–II bear a dorsal spur; empodia I are claw like, and empodia II–IV have long and free proximoventral hairs. The aedeagus has a small knob, with the axis of the knob parallel to the axis of the shaft and the anterior and posterior angulations of the knob small and similar (Vacante, 1985a; Seeman and Beard, 2011).

Geographical distribution

Worldwide, reported in 112 countries of the Afrotropical, Australasian, Nearctic, Neotropical, Oriental and Palaearctic regions (Migeon and Dorkeld, 2006–2013).

Host plants

The whole list of host plants of the two-spotted spider mite consists of 1059 wild and cultivated plants belonging to 124 families. The Asteraceae and Fabaceae have 108 species each (Migeon and Dorkeld, 2006–2013). Economic plants include vegetables, cereals, citrus species, deciduous fruit plants, forages, stone fruits, tropical fruits, ornamentals and other cultivated plants. Only the most significant families and species are listed below:

Actinidiaceae: gold kiwi fruit (*Actinidia chinensis* Planch.) (Migeon and Dorkeld, 2006–2013), kiwi (*A. deliciosa* (A. Chev.) C.F. Liang and A.R. Ferguson) (Mesa Cobo and Zuluaga Cardona, 1983).

Amaranthaceae: beet (*Beta vulgaris* L.) (Bibby and Tuttle, 1959; Tuttle and Baker, 1964; Gupta and Gupta, 1994; Aucejo *et al.*, 2003; Aguilar and Murillo, 2008), spinach (*Spinacia oleracea* L.) (Prasad, 1974; Nassar and Ghai, 1981; Gupta and Gupta, 1994).

Amaryllidaceae: levant garlic (*Allium ampeloprasum* L.) (Baker and Pritchard, 1960; Livshits and Salinas-Croche, 1968; Hatzinikolis, 1969c; Gotoh *et al.*, 2007d), bulb onion (*A. cepa* L.) (Hatzinikolis, 1969c; Meyer Smith, 1974; Gupta and Gupta, 1994), Welsh onion (*A. fistulosum* L.) (Urueta, 1975), garlic (*A. sativum* L.) (Doreste, 1968b; Hatzinikolis, 1969c).

Anacardiaceae: mango (*Mangifera indica* L.) (Tuttle *et al.*, 1974; Gupta, 1992), terebinth (*Pistacia terebinthus* L.), pistachio (*P. vera* L.) (Bolland *et al.*, 1998; Migeon and Dorkeld, 2006–2013).

Annonaceae: soursop (*Annona muricata* L.), custard apple (*A. reticulata* L.), sugar apple (*A. squamosa* L.) (Martínez *et al.*, 2004).

Apiaceae: celery (*Apium graveolens* (Mill.) Pers.) (Bibby and Tuttle, 1959; Rossi Simons, 1961; Tuttle and Baker, 1964; Hatzinikolis, 1969c; Gupta and Gupta, 1994; Suárez, 2004), coriander (*Coriandrum sativum* L.) (Prasad, 1974; Gupta and Gupta, 1994), carrot (*Daucus carota* L.) (Bolland *et al.*, 1998; Migeon and Dorkeld, 2006–2013), fennel (*Foeniculum vulgare* Mill.) (Flechtmann, 1967a; Aucejo *et al.*, 2003), parsnip (*Pastinaca sativa L.*), parsley (*Petroselinum crispum* (Mill.) Fuss) (Hatzinikolis, 1969c; Urueta, 1975; Aguilar and Murillo, 2008).

Asparagaceae: *Asparagus officinalis* L. (Ehara, 1956a).

Asteraceae: Indian chrysanthemum (*Chrysanthemum indicum* L.) (Aguilar and Murillo, 2008), chrysanthemum (*C. morifolium* Ramat) (Goka and Takafuji, 1995; Lee *et al.*, 1999; Hinomoto *et al.*, 2001; Takafuji *et al.*, 2003; Takafuji and Hinomoto, 2008), endive (*Cichorium endivia* L.) (Bolland *et al.*, 1998; Migeon and Dorkeld, 2006–2013), chicory (*C. intybus* L.) (Prasad, 1974; Gupta and Gupta, 1994), cardoon (*Cynara cardunculus* L.) (Estebanes-Gonzalez and Rodriguez-Navarro, 1991), red dahlia (*Dahlia coccinea* Cav.) (Oudemans, 1937; Womersley, 1940; Rossi Simons, 1961; Hatzinikolis, 1969c; Gupta and Gupta, 1994; Ho *et al.*, 1997; Aguilar and Murillo, 2008), African daisy (*Gerbera jamesonii* Adlam) (Kielkiewicz and Witul, 1995; Witul and Kielkiewicz, 1999; Suárez, 2004), sunflower (*Helianthus annuus* L.) (Womersley, 1940; Meyer Smith and Ryke, 1959; Rossi Simons, 1961; Hatzinikolis, 1969c; Baker, 1975b; Gupta and Gupta, 1994; Guanilo *et al.*, 2012), lettuce (*Lactuca sativa* L.) (Bibby and Tuttle, 1959; Tuttle and Baker, 1964; Lee Goff, 1986; Aguilar and Murillo, 2012), zinnia (*Zinnia elegans* Jacq.) (Hatzinikolis, 1969c; Meyer Smith, 1974).

Betulaceae: hazel (*Corylus avellana* L.) (Gutierrez and Etienne, 1986).

Brassicaceae: mustard greens (*Brassica juncea* (L.) Vassilii Matveievitch Czernajew), rape (*B. napus* L.), cabbage (*B. oleracea* L.) (Bibby and Tuttle, 1959; Tuttle and Baker, 1964; Hatzinikolis, 1969c; Lee Goff, 1986; Gupta and Gupta, 1994; Guanilo *et al.*, 2012), *B. rapa* L. (Prasad, 1974), field mustard (*Sinapis arvensis* L.) (Prasad, 1974; Gupta and Gupta, 1994).

Cannabaceae: *Cannabis sativa* L. (Prasad, 1974; Gupta and Gupta, 1994), hop (*Humulus lupulus* L.) (Geijskes, 1939; Rambier, 1965; Hatzinikolis, 1969c; Sakagami *et al.*, 2009).

Caricaceae: papaya (*Carica papaya* L.) (Carmona, 1967; Urueta, 1975; Gutierrez and Etienne, 1981a; Lee Goff, 1986; Gupta and Gupta, 1994; Ho *et al.*, 1997; Martínez *et al.*, 2004; Mendonça *et al.*, 2011b; Ohno *et al.*, 2011b; Seeman and Beard, 2011; Guanilo *et al.*, 2012).

Caryophyllaceae: Sweet William (*Dianthus barbatus* L.) (Meyer Smith, 1974), carnation (*D. caryophyllus* L.) (Meyer Smith and Ryke, 1959; Hatzinikolis, 1969c; Lee Goff, 1986; Gupta and Gupta, 1994; Ho *et al.*, 1997; Gotoh *et al.*, 2007d; Guanilo *et al.*, 2012), China pink (*D. chinensis* L.) (Gasser, 1951; Rossi Simons, 1961, Rambier, 1965; Urueta, 1975; Matsuda *et al.*, 2013), *Gypsophila paniculata* (Lee Goff, 1986; Ho *et al.*, 1997; Osakabe *et al.*, 2002).

Convolvulaceae: water spinach (*Ipomoea aquatica* Forssk.) (Ohno *et al.*, 2010), sweet potato (*I. batatas* L.) (Beccari, 1971; Zaher *et al.*, 1982; Aguilar and Murillo, 2008).

Cucurbitaceae: winter melon (*Benincasa hispida* Thunb.) (Gupta and Gupta, 1994; Ho *et al.*, 1997; Ohno *et al.*, 2010; Aguilar and Murillo, 2012), colocynth (*Citrullus colocynthis* L.) Schrad.), watermelon (*C. lanatus* (Thunb.) Matsum. and Nakai) (Rahman and Sapra, 1940a; Bibby and Tuttle, 1959; Tuttle and Baker, 1964; Hatzinikolis, 1969c; Lee Goff, 1986; Gupta and Gupta, 1994; Ho *et al.*, 1997; Ben-David *et al.*, 2007; Gotoh *et al.*, 2007d; Guanilo *et al.*, 2012; Matsuda *et al.*, 2013), musk melon (*Cucumis melo* L.) (Womersley, 1940; McGregor, 1950; Bibby and Tuttle, 1959; Rossi Simons, 1961; Hatzinikolis, 1969c; Gupta and Gupta, 1994; Ho *et al.*, 1997; Lee *et al.*, 1999), cucumber (*C. sativus* L.) (Murray, 1877; Tuttle and Baker, 1964; Hatzinikolis, 1969c; Gupta and Gupta, 1994;

Ho *et al.*, 1997; Kreiter *et al.*, 2002; Álvarez and Alarcón, 2004; Seiedy *et al.*, 2010; Zriki *et al.*, 2015), seven years melon (*Cucurbita ficifolia* Bouché) (Carmona, 1967), squash (*C. maxima* Duchesne) (Carmona, 1967; Hatzinikolis, 1969c; Aguilar and Murillo, 2012), squash (*C. moschata* Duchesne ex Poir.), courgette (*C. pepo* L.) (Tuttle and Baker, 1964; Hatzinikolis, 1969c; Lee Goff, 1986; Gupta and Gupta, 1994), bottle gourd (*Lagenaria siceraria* (Molina) Standl.) (Hatzinikolis, 1969c; Gupta and Gupta, 1994), angled luffa (*Luffa acutangula* (L.) Roxb.) (Prasad, 1974; Ho *et al.*, 1997), smooth luffa (*L. cylindrica* (L.) Roem.) (Rahman and Sapra, 1940a; Hatzinikolis, 1969c; Gupta and Gupta, 1994; Ho *et al.*, 1997; Ohno *et al.*, 2011b; Aguilar and Murillo, 2012), bitter melon (*Momordica charantia* Descourt.), tinda (*Praecitrullus fistulosus* (Stocks) Pangalo) (Gupta and Gupta, 1994), chayote (*Sechium edule* (Jacq.) Sw.) (Flechtmann, 1967a; Ochoa *et al.*, 1989).

Ebenaceae: Japanese persimmon (*Diospyros kaki* Thunb.) (Carmona, 1967; Kawashima *et al.*, 2008).

Euphorbiaceae: poinsettia (*Euphorbia pulcherrima* Willd. ex Klotzsch) (Meyer Smith and Ryke, 1959; Aguilar and Murillo, 2008), rubber tree (*Hevea brasiliensis* Müll. Arg.) (Gupta and Gupta, 1994; Bellini *et al.*, 2005), cassava (*Manihot esculenta* Crantz) (Oudemans, 1931d; Livshits and Salinas-Croche, 1968; Urueta, 1975; Lee Goff, 1986; Gupta and Gupta, 1994; Ohno *et al.*, 2009; Guanilo *et al.*, 2012), castor bean (*Ricinus communis* L.) (Boisduval, 1867; Bibby and Tuttle, 1959; Rossi Simons, 1961; Tuttle and Baker, 1964; Hatzinikolis, 1969c; Gupta, 1992; Ben-David *et al.*, 2009; Guanilo *et al.*, 2012; Zriki *et al.*, 2015).

Fabaceae: groundnut (*Arachis hypogaea* L.) (Sayed, 1946b; Rambier, 1965; Hatzinikolis, 1969c; Gupta and Gupta, 1994; Ho *et al.*, 1997; Guanilo *et al.*, 2012), chickpea (*Cicer arietinum* L.) (Hatzinikolis, 1969c), soybean (*Glycine max* (l.) Merr.) (André, 1945; Bibby and Tuttle, 1959; Tuttle and Baker, 1964; Gupta and Gupta, 1994; Ho *et al.*, 1997; Lee *et al.*, 1999; Gotoh *et al.*, 2007d; Guanilo *et al.*, 2012); hyacinth bean (*Lablab purpureus* (L.) Sweet) (Baker and Pritchard, 1960; Gupta and Gupta, 1994; Ho *et al.*, 1997), lucerne (*Medicago sativa* L.) (Bibby and Tuttle, 1959; Meyer Smith and Ryke, 1959; Rossi Simons, 1961; Tuttle and Baker, 1964; Hatzinikolis, 1969c; Gupta and Gupta, 1994; Guanilo *et al.*, 2012), tepary bean (*Phaseolus acutifolius* A. Gray) (Rahman and Sapra, 1940a; Tuttle and Baker, 1964), *P. coccineus* (Smiley and Baker, 1995; Aguilar and Murillo, 2008), *P. lunatus* (Reeves, 1963; Gupta and Gupta, 1994; Ho *et al.*, 1997; Guanilo *et al.*, 2012), common bean (*P. vulgaris* L.) (Womersley, 1940; Livshits and Salinas-Croche, 1968; Hatzinikolis, 1969c; Navajas *et al.*, 1996; Ho *et al.*, 1997; Suárez, 2004; Escudero and Ferragut, 2005; Ohno *et al.*, 2011b; Guanilo *et al.*, 2012; Zriki *et al.*, 2015), pea (*Pisum sativum* L.) (Rossi Simons, 1961; Tuttle and Baker, 1968; Hatzinikolis, 1969c; Gupta and Gupta, 1994; Ho *et al.*, 1997; Gotoh *et al.*, 2003a), Egyptian clover (*Trifolium alexandrinum* L.) (Meyer Smith, 1974), golden clover (*T. aureum* Pollich), lesser hop trefoil (*T. dubium* Sibth.) (Aucejo *et al.*, 2003), ball clover (*T. glomeratum* L.), Alsike clover (*T. hybridum* L.), red clover (*T. pratense* L.) (Thewke and Enns, 1969), white clover (*T. repens* L.) (Hinomoto *et al.*, 2001; de Villiers and Pringle, 2011), fava bean (*Vicia faba* L.) (Sayed, 1946b; Hatzinikolis, 1969c; Ho *et al.*, 1997), vetch (*V. sativa* L.) (Hatzinikolis, 1969c), moth bean (*V. aconitifolia* (Jacq.) Marechal) (Gupta and Gupta, 1994), azuki bean (*V. angularis* (Willd.) Ohwi and H. Ohashi), black lentil (*V. mungo* (L.) Hepper) (Rahman and Sapra, 1940a; Gupta and Gupta, 1994), mung (*V. radiata* (L.) R. Wilczek) (Prasad, 1974; Gupta and Gupta, 1994), cowpea (*Vigna unguiculata* (L.) Walp.) (Baker and Pritchard, 1960; Tuttle and Baker, 1968; Gupta and Gupta, 1994; Ho *et al.*, 1997; Ohno *et al.*, 2011b; Moghadam *et al.*, 2012).

Grossulariaceae: blackcurrant (*Ribes nigrum* L.), redcurrant (*R. rubrum* L.) (Bolland *et al.*, 1998; Migeon and Dorkeld, 2006–2013).

Juglandaceae: English walnut (*Juglans regia* L.) (Carmona, 1967; Hatzinikolis, 1969c).

Iridaceae: gladiolus (*Gladiolus hortulanus* L.H. Bailey) (Tuttle and Baker, 1964), iris (*Iris* × *germanica* L.) (Meyer Smith, 1974).

Lamiaceae: peppermint (*Mentha* × *piperita* L.) (Gupta and Gupta, 1994), basil (*Ocimum basilicum* L.) (Bolland *et al.*, 1998; Migeon and Dorkeld, 2006–2013), rosemary (*Rosmarinus officinalis* L.) (Thewke and Enns, 1969; Aguilar and Murillo, 2008), sage (*Salvia officinalis* L.) (Bolland *et al.*, 1998; Migeon and Dorkeld, 2006–2013).

Lauraceae: bay laurel (*Laurus nobilis* L.) (Rossi Simons, 1961), avocado (*Persea americana* Mill.) (Carmona, 1967; Urueta, 1975; Suárez, 2004).

Lythraceae: pomegranate (*Punica granatum* L.) (Hatzinikolis, 1969c; Meyer Smith, 1974).

Malvaceae: okra (*Abelmoschus esculentus* (L.) Moench) (Rahman and Sapra, 1940a; Hatzinikolis, 1969c; Zaher *et al.*, 1982; Gupta and Gupta, 1994; Martínez *et al.*, 2004), Egyptian cotton (*Gossypium barbadense* L.) (Sayed, 1946b; Zaher *et al.*, 1982), Levant cotton (*G. herbaceum* L.) (Flechtmann, 1967a; Gupta and Gupta, 1994), upland cotton (*G. hirsutum* L.) (Bibby and Tuttle, 1959; Rossi Simons, 1961; Tuttle and Baker, 1964, 1968; Meyer Smith, 1974; Gupta and Gupta, 1994; Ben-David *et al.*, 2009).

Meliaceae: neem (*Azadirachta indica* A. Juss.) (Tuttle and Baker, 1968; González-Zamora *et al.*, 2011).

Moraceae: fig (*Ficus carica* L.) (Sayed, 1946b; Rossi Simons, 1961; Tuttle and Baker, 1964; Hatzinikolis, 1969c; Urueta, 1975; Abou-Awad *et al.*, 1999b), white mulberry (*Morus alba* L.) (Ehara, 1956b; Hatzinikolis, 1969c; Gupta and Gupta, 1994; Ohno *et al.*, 2010), black mulberry (*M. nigra* L.) (Carmona, 1967; Hatzinikolis, 1969c), red mulberry (*M. rubra* L.) (Tuttle and Baker, 1964).

Musaceae: Japanese banana (*Musa basjoo* Siebold *et* Zucc. ex Iinuma) (Bibby and Tuttle, 1959; Hatzinikolis, 1969c; Gutierrez and Schicha, 1983), banana (*Musa* × *paradisiaca* L.) (Meyer Smith and Ryke, 1959; Nassar and Ghai, 1981; Martínez *et al.*, 2004).

Myrtaceae: Cattley guava (*Psidium cattleianum* Sabine), guava (*P. guajava* L.) (Meyer Smith and Ryke, 1959; Nassar and Ghai, 1981; Zaher *et al.*, 1982; Gupta, 1992; Gupta and Gupta, 1994).

Oleaceae: olive tree (*Olea europaea* L.) (Carmona, 1967).

Papaveraceae: opium poppy (*Papaver somniferum* L.) (Bolland *et al.*, 1998; Migeon and Dorkeld, 2006–2013).

Passifloraceae: passion fruit (*Passiflora edulis* Sims) (Doreste, 1968b), banana passion fruit (*P. mollissima* (Kunth) L.H. Bailey) (Seeman and Beard, 2011).

Pedaliaceae: sesame (*Sesamum indicum* L.) (Abraham *et al.*, 1973; González and Flechtmann, 1977; Ho *et al.*, 1997; Guanilo *et al.*, 2012).

Poaceae: oat (*Avena sativa* L.) (Hatzinikolis, 1969c), Italian ryegrass (*Lolium multiflorum* Lam.) (Meyer Smith, 1974), Asian rice (*Oryza sativa* L.) (Suárez, 2004), common millet (*Panicum miliaceum* L.) (Thewke and Enns, 1969), sugarcane (*Saccharum officinarum* L.) (Lee Goff, 1986; Suárez, 2004), sorghum (*Sorghum bicolor* (L.) Moench) (Meyer Smith, 1974; Gupta and Gupta, 1994; Ho *et al.*, 1997), maize (*Zea mays* L.) (Bibby and Tuttle, 1959; Meyer Smith and Ryke, 1959; Tuttle and Baker, 1964; Hatzinikolis, 1969c; Aranda and Flechtmann, 1971; Gutierrez and Etienne, 1981a; Suárez, 2004; Ohno *et al.*, 2010; Guanilo *et al.*, 2012).

Primulaceae: ivy-leaf cyclamen (*Cyclamen hederifolium* Aiton), florist's cyclamen (*C. persicum* Mill.) (Boisduval, 1867; Gotoh *et al.*, 2007d), drumstick primula (*Primula denticulata* Sm.), false oxlip (*P. polyantha* Mill.) (Meyer Smith and Ryke, 1959), cowslip (*P. veris* L.) (Meyer Smith, 1974).

Rhamnaceae: jujube (*Ziziphus jujuba* Mill) (Ho *et al.*, 1997), Christ's thorn jujube (*Z. spina-christi* (L.) Desf.) (Bolland *et al.*, 1998; Migeon and Dorkeld, 2006–2013).

Rosaceae: redhaw hawthorn (*Crataegus sanguinea* Pall.), quince (*Cydonia oblonga* Mill.) (Rossi Simons, 1961; Hatzinikolis, 1969c), loquat (*Eriobotrya japonica* (Thunb.) Lindl.) (Meyer Smith and Ryke, 1959; Hatzinikolis, 1969c; Nassar and Ghai, 1981), strawberry (*Fragaria* × *ananassa* Duchesne) (Froggatt, 1921; Womersley, 1940; Ehara, 1956c; Rambier, 1958a; Bibby and Tuttle, 1959; Flechtmann, 1967a; Ochoa and Aguilar, 1989; Johanowicz and Hoy, 1996; Ho *et al.*, 1997; Gotoh *et al.*, 2007d; Ohno *et al.*, 2009; Seeman and Beard, 2011; Matsuda *et al.*, 2013), musk strawberry (*F. moschata* Duchesne), wild strawberry (*F. vesca* L.) (Oudemans, 1931b), Virginia strawberry (*F. virginiana* Duchesne), apple (*Malus domestica* Borkh.) (McGregor, 1950; Baker and Pritchard, 1960; Tuttle and Baker, 1964; van de Vrie, 1965; Hatzinikolis, 1969c; Baillod *et al.*, 1989d; Smiley and Baker, 1995; Kreiter *et al.*, 2002; Gotoh *et al.*, 2007d; Guanilo *et al.*, 2012), almond (*Prunus amygdalus* Batsch) (Meyer Smith and Ryke, 1959; Rossi Simons, 1961; Hatzinikolis, 1969c), apricot (*P. armeniaca* L.) (Rossi Simons, 1961; Tuttle and Baker, 1964; Hatzinikolis, 1969c; Zriki *et al.*, 2015), sweet cherry (*P. avium* (L.) L.) (Rossi Simons, 1961; Hatzinikolis, 1969c; Carmona and Dias, 1980; Gupta and Gupta, 1994), myrobalan plum (*P. cerasifera* Ehrh.) (Migeon and Dorkeld, 2006–2013), plum (*P. domestica* L.) (Donnadieu, 1875; Rossi Simons, 1961; Hatzinikolis, 1969c; Zaher *et al.*, 1982; Guanilo *et al.*, 2012), damson plum (*P. insititia* L.) (Oudemans, 1931a; Hatzinikolis, 1969c), peach (*P. persica* (L.) Batsch) (Donnadieu, 1875; Womersley, 1940; Rossi Simons, 1961; Livshits and Salinas-Croche, 1968; Gupta and Gupta, 1994; Ho *et al.*, 1997; Ben-David *et al.*, 2009; Zriki *et al.*, 2015), Asian plum (*P. salicina* Lindl.) (Flechtmann, 1967a), pear (*Pyrus communis* L.) (Bibby and Tuttle, 1959; Meyer Smith and Ryke, 1959; Tuttle and Baker, 1964; Hatzinikolis, 1969c; Ben-David *et al.*, 2009; Guanilo *et al.*, 2012), white rose (*Rosa* × *alba* L.) (Zaher *et al.*, 1982), dog rose (*R. canina* L.) (Uysal *et al.*, 2001), Holland rose (*Rosa* × *centifolia* L.) (Tuttle and Baker, 1964), Portland rose (*Rosa* × *damascena* Mill.) (Wang and Cui, 1992), English rose (*R. hybrida*) (Goka and Takafuji, 1995; Lee *et al.*, 1999; Hinomoto *et al.*, 2001; Hinomoto and Takafuji, 2004; Gotoh *et al.*, 2007d; Hinomoto *et al.*, 2007; Takafuji and Hinomoto, 2008), Japanese rose (*R. multiflora* Thunb.) (Davis, 1968b), tea rose (*R. odorata* (Andrews) Sweet) (Tuttle and Baker, 1964), blackberry (*Rubus fruticosus* L.), raspberry (*R. idaeus* L.) (Rossi Simons, 1961; Carmona, 1967; Badowska-Czubik *et al.*, 1977; Roy *et al.*, 1999), black raspberry (*R. occidentalis* L.) (Hatzinikolis, 1969c; Thewke and Enns, 1969).

Rubiaceae: coffee (*Coffea arabica* L.) (Garrett and Haramoto, 1967).

Rutaceae: key lime (*Citrus aurantiifolia* (Christm.) Swingle) (Doreste, 1968b), sour orange (*C. aurantium* L.) (Hatzinikolis, 1969c; Zaher *et al.*, 1982), clementine (*C. clementina* Hort. ex Tanaka) (Aguilar-Fenollosa *et al.*, 2012), lemon (*C. limon* (L.) Burm.f.) (Womersley, 1940; McGregor, 1950; Cuenod, 1958; Hatzinikolis, 1969c; Ledesma *et al.*, 2011), pomelo (*C. maxima* Merr.) (Hatzinikolis, 1969c; Meyer Smith, 1974), citron (*C. medica* L.) (Hatzinikolis, 1969c), grapefruit (*C. paradisi* Macfad.) (Cuenod, 1958; Hatzinikolis, 1969c), mandarin (*C. reticulata* Blanco) (Hatzinikolis, 1969c; Meyer Smith, 1974; Pascual-Ruiz *et al.*, 2011), sweet orange (*C. sinensis* (L.) Osbeck) (Cuenod, 1958; Hatzinikolis, 1969c; Gupta and Gupta, 1994; Martínez *et al.*, 2004; Ledesma *et al.*, 2011).

Solanaceae: pepper (*Capsicum annuum* L.) (Rossi Simons, 1961; Tuttle and Baker, 1964; Carmona, 1967; Hatzinikolis, 1969c; Suárez, 2004; Ben-David *et al.*, 2009; Guanilo *et al.*, 2012; Matsuda *et al.*, 2013), tobacco (*Nicotiana tabacum* L.) (Womersley, 1940; Bibby and Tuttle, 1959; Rossi Simons, 1961; Carmona, 1967; Hatzinikolis, 1969c), groundcherry (*Physalis angulata* L.) (Gutierrez and Schicha, 1983; Furtado *et al.*, 2006), Peruvian cherry (*P. peruviana* L.) (Meyer Smith and Ryke, 1959; Seeman and Beard, 2011), mock tomato (*Solanum aethiopicum* L.) (Paschoal, 1970), tomato (*S. lycopersicum* L.) (Bibby and Tuttle, 1959; Rossi Simons, 1961; Tuttle and Baker, 1964; Hatzinikolis, 1969c; Gupta and Gupta, 1994; Kreiter *et al.*, 2002; Ben-David *et al.*, 2007, 2009; Gotoh *et al.*, 2007d; Mendonça *et al.*, 2011b; Seeman and Beard, 2011; Guanilo *et al.*, 2012; Zriki *et al.*, 2015), aubergine (*S. melongena* L.) (Ehara, 1956a; Rossi Simons, 1961; Hatzinikolis, 1969c; Gupta and Gupta, 1994; Ho *et al.*, 1997; Gotoh *et al.*, 2007d; Ben-David *et al.*, 2009; Mendonça *et al.*, 2011b; Ohno *et al.*, 2011b; Guanilo *et al.*, 2012), naranjilla (*S. quitoense* Lam.) (Urueta, 1975), potato (*S. tuberosum* L.) (Meyer Smith and Ryke, 1959; Rossi Simons, 1961; Gupta and Gupta, 1994; Guanilo *et al.*, 2012).

Theaceae: tea (*Camellia sinensis* (L.) Kuntze) (Gupta and Gupta, 1994).

Vitaceae: grapevine (*Vitis vinifera* L.) (Boisduval, 1867; Womersley, 1940; Rambier, 1958b; Rossi Simons, 1961; Hatzinikolis, 1969c; Baillod *et al.*, 1989c; Gupta and Gupta, 1994; Ho *et al.*, 1997; Suárez, 2004).

Zingiberaceae: turmeric (*Curcuma longa* L.) (Gupta and Gupta, 1994).

Bio-ecology

The high number of synonymies (59 according to Jeppson *et al.*, 1975, and 50 according to Migeon and Dorkeld, 2006–2013) and the existence of two forms – the green form (GF) and the red form (RF) – make the definition of the pest status

of the two-spotted spider mite difficult. The two forms have been identified by various authors as two distinct species, *T. urticae* Koch and *T. cinnabarinus* (Boisduval), and only recently have morphological, biological and molecular data highlighted the synonymy of *T. cinnabarinus* and *T. urticae* (Auger *et al.*, 2013). Currently, *T. urticae* is regarded as a polymorphic species in which there is conspecific status of the green and red forms. Many reports confirm that the green form lives in cool temperate climates, and the red form inhabits warm temperate and subtropical areas. Moreover, the red form reproduces throughout the year, while the green form overwinters as the yellow-orange female on the ground or in sheltered places such as the bark (Gutierrez and Schicha, 1983). The red form has overlapping niches with *T. turkestani* (Ugarov *et* Nikolskii) in which reproductive interactions between the two species are likely to occur naturally, and there is a post-zygotic reproductive barrier between *T. turkestani* and the red form of *T. urticae* (Ben-David *et al.*, 2009).

T. urticae is a pest of a large number of economic plants, including vegetables (in the field and the greenhouse), cereals, citrus species, deciduous fruits, forages, tropical fruits, ornamentals and other plants. It reproduces by arrhenotokous parthenogenesis, and has a haploid number of chromosomes of three (Helle and Bolland, 1967). The sex ratio is 3:1 females: males (Laing, 1969). The mite is infected by endosymbiont *Wolbachia* bacteria, which in European populations are responsible for a cytoplasmic incompatibility (Breeuwer, 1997) that is not found in Japan (Gotoh *et al.*, 2003a).

Populations of the two-spotted spider mite develop usually on the lower leaf surface of young leaves, but at high densities they invade the upper leaf surface and other organs, such as the flowers, fruits and older leaves. The mite produces conspicuous amounts of webbing that may cover the entire plant at high mite population density. The complex construction of the web on the leaf surface assigns the mite to the subtype CW-u of Saitō (1985).

On leaves of strawberry in the laboratory under a diurnal temperature cycle of 15–28.3°C, the fecundity of *T. urticae* was on average 37.9 eggs per female and daily egg production on average 2.4 eggs (Laing, 1969). In a laboratory study on excised primary bean leaves, Shih *et al.* (1976), found that at 27 ± 1°C and 95 ± 5% RH the development of the egg, larva, protonymph and deutonymph took 2.3, 0.6, 0.4 and 1.9 days, respectively, and that females and males developed in 7.6 and 5.6 days, respectively. Female and male longevity was on average 19.1 and 14.6 days, respectively. The life cycle took 7.5 days. There was a quiescent period between the deutonymph and active adult female of 2.4 days. The reproductive rate was 7.97 eggs per female a day. The intrinsic rate of increase (r_m) was 0.336, the mean generation time (T) 13.6 days and the net reproductive rate (R_0) 97.4. Chahine and Michelakis (1994) reported that in laboratory studies on bean the longevity and fecundity of *T. urticae* were 10.7 days and 55.2 eggs, respectively. Also in the laboratory, Kasap (2002) found that on bean at 25 ± 2°C, 60% ± 10 RH and under a 16L:8D photoperiod, the development time of *T. urticae* was 10.9 days. The intrinsic rate of increase (r_m) was 0.265 females per female a day, the net production rate (R_0) 185.4 females per female and the mean generation time 23.2 days.

Recently, Kavousi *et al.* (2009) observed that on bean leaves the deutonymph stage, the total pre-adult stage and the pre-oviposition period, respectively, were shorter (1.22, 7.6, 0.29 days) than on leaf discs (1.73, 8.2, 0.89 days). The lifetime fecundities were 22.81 and 12.05 offspring on whole leaves and leaf discs, respectively. The intrinsic rate of increase on leaves (0.235 days^{-1}) was higher than on leaf discs (0.159 days^{-1}). Under laboratory conditions at 25 ± 1°C, 60 ± 10% RH and under a photoperiod of 18L:6D, soybean was the most favourable host for the two-spotted spider mite (mean r_m = 0.296 ± 0.012 offspring per female a day), followed by cowpea (0.242 ± 0.013), with the slowest population growth on bean (0.230 ± 0.013). These data show that the host plant may influence the time needed to reach damaging levels in a culture (Razmjou *et al.*, 2009).

On cotton cotyledons, the development period was 25.8–29.0 days at 15.5°C and 6.1–6.7 days at 29.4°C for females, and a slightly shorter time for males. At 23.8 and 29.4°C, the fecundity was 103.3 and 64.3 eggs, respectively. The proportion of female offspring was 74%, and at 23.8 and 29.4°C, female longevity averaged 14.71 and 9.71 days, respectively. At 25 °C ± 1°C the net reproduction (R_0) was 74.8, the intrinsic rate of increase (r_m) 0.219 individuals per female a day and the finite rate of increase (λ) 1.245 individuals per female a day (Carey and Bradley, 1982). On the same host plant, Gutierrez (1976b) found an intrinsic rate of increase (r_m) of 0.241 and a finite rate of increase (λ) of 1.272. Silva *et al.* (1985) reported that on cotton (*G. hirsutum*) the net reproductive rate (R_0) of *T. urticae* ranged from 34.4 to 54.5, the mean duration of one generation (T) from 17.8 and 24.9 days, the innate growth capacity (r_m) from 0.161 to 0.207 and the finite rate of increase (λ) from 1.172 to 1.230. Daily egg deposition ranged from 4.4 ± 1.0 to 7.7 ± 0.8.

Chahine and Michelakis (1994) found that the longevity of *T. urticae* on aubergine and tomato was 10.7 days, and on cucumber 8.6 days. The host plant also influences fecundity more than longevity, as 71 eggs per female were observed on aubergine 18.2 on cucumber and 10.6 on tomato. Kasap (2002) observed that the development time of *T. urticae* on cucumber at 25 ± 2°C, 60% ± 10 RH and under a 16L:8D photoperiod averaged 10.4 days. The intrinsic rate of increase (r_m) was 0.247 females per female a day, the net production rate (R_0) 110.7 females per female and the mean generation time 21.9 days.

Sabelis (1981) reported that on leaves of rose at 20°C, the pre-oviposition period took 1.7 days, and the development of the egg, larva, protonymph and deutonymph stages required 6.7, 2.8, 2.3 and 3.1 days, respectively. The development period for a generation at 15, 20 and 30°C was 36.3, 16.6 and 7.3 days, respectively. The intrinsic rate of increase was 0.0067 for each degree and the thermal threshold was 12°C. Kasap (2002) found that on rose at 25 ± 2°C, 60 ± 10% RH and under a 16L:8D photoperiod, the development of *T. urticae* lasted 11.2 days. The intrinsic rate of increase (r_m) was 0.200 females per female a day, the net production rate (R_0) 47.8 females per female and the mean generation time 22.1 days.

Silva *et al.* (2009) found that on gerbera leaf discs at 25°C, 70 ± 10% RH and 14 h photophase, the egg viability of non-mated and mated females was 96.5 and 97.1%, respectively. Larvae, protonymphs and deutonymphs originating from unmated females

lived 3.2., 1.9 and 1.6 days, respectively; when they originated from mated females, they lived 3.5, 2.0 and 1.6 days, respectively. The duration of one generation for non-mated and mated females was 18.6 and 19.7 days, respectively; other parameters differed significantly. Mated and non-mated females had a net reproductive rate of increase (R_0) of 11.5 and 24.6, respectively. The innate capacity for increase (r_m) was 0.12 and 0.17 and the finite rate of growth (λ) was 1.13 and 1.19, respectively.

Populations of the two-spotted spider mite develop both in the glasshouse and in the field; in the warmer regions of the world they are active throughout year, while in temperate climates they overwinter as diapause females. A shortened period of light, decrease in temperature and unfavourable food supply stimulate the appearance of diapause females. Diapausing females change in colour, stop feeding and laying eggs, and leave the host plants to go to various protected places to overwinter (Cagle, 1949; Gasser, 1951). Both diapause induction and diapause termination are under photoperiodic control, with the involvement of the circadian system in the photoperiodic termination of diapause (Koveos and Veerman, 1996).

In warm climates, populations of *T. urticae* continue to reproduce on their host plants, e.g. in Mediterranean greenhouses *T. urticae* infests various crops throughout the year (Vacante and Benuzzi, 2007). In relatively mild and in cooler climates, the two-spotted spider mite can overwinter on citrus as diapause forms, and although they usually leave the host plant, in the San Joaquin Valley of California they have been collected on citrus trees. Populations commonly develop on weeds or cover crops, and when these plants dry or are cut, they move to citrus, where they produce damage to leaves on the lower branches (Jeppson, 1989). Similar behaviour occurs on citrus in southern Italy, where the two spotted spider mite is mainly a pest of lemon and, in particular conditions, of other citrus species (Gerson and Vacante, 2012). In studies on leaf discs from potted mandarin trees watered with a range of concentrations of salt solutions (0–60 mM NaCl), high salt concentrations were found to have a negative effect on the two-spotted spider mite, but, remarkably, the r_m values observed at the lowest salt concentrations were significantly higher than those of the controls. This kind of effect may contribute to outbreaks of the mite (Ansaloni *et al.*, 2006). On lemon in Sicily (Italy), it has been observed that two-spotted spider mites are also active in the winter, and it has three population peaks, in July, November and February (Vacante, 1986). On clementines in Spain, populations of the mite show three peaks – during the winter, late spring and autumn. Population increases in the trees, and green cover, were apparently related to drops in humidity, while fluctuations in temperature had less effect on seasonal trends (Pascual and Ferragut, 2003).

At low population densities, *T. urticae* females usually disperse by walking to new leaves on the same host plant, and newly mated non-gravid females are the primary dispersal agents. At high levels of infestation, newly mated females that have not begun to lay eggs move to the top of a plant and may form clumps of mites, in which males and immatures are rare. Slight breezes will disperse the females, which do not appear to truly balloon because they do not have a trailing strand of silk (Li and Margolies, 1994). This behaviour commonly occurs on vegetables in the greenhouse, when plants that are severely infested become overcrowded and food resources become scarce, and individuals gather at the plant apex to form a balloon made by mites and silk threads. This balloon could be a structure facilitating group dispersion by wind or animal transport (Clotuche *et al.*, 2010).

Workers' allergies due to *T. urticae* are known (Burches *et al.*, 1996).

New infestation on vegetables and/or ornamentals in the field and in protected crops are generally the result of infested seedlings in the nursery, or are carried by workers, wind, water and various animals. Sprays of some pesticides (pyrethroids) induce hormoligosis and/or suppress the natural enemies, and stimulate mite development. In addition, horticultural practices include a complex framework of plant resistance that may variously affect the development of populations of *T. urticae*.

Natural enemies include pathogens (bacteria and fungi; Table 20.13) and predatory insects (Table 20.14) and spiders and mites (Table 20.15).

Symptomatology and damage

T. urticae feeds on tender tissues, usually on the lower leaf surface, where it sucks out cell chloroplasts and other contents; the corresponding sites of the upper leaf surface appear as whitish or yellowish stipplings. At a low density of attack, the stipplings are quite distinct from the green tissues of the leaf surface, but as mite feeding continues they may join up and become brownish or yellow–brown. In addition, mite webbing develops on the lower leaf surface, and at high population densities may cover the leaves, flowers and fruits, or the entire plant. Severe damage induces the leaves to dry and drop, and the plant may die.

In semi-tropical areas, populations of the two-spotted spider mite develop on young leaves and green or mature fruits of all citrus species. Infestation usually occurs on a limited portion of the leaves, which buckle at the site corresponding to the colonies, and the upper surface becomes raised and a yellow ochre colour. On orange, lemon and other citrus fruit, the feeding activity of the mite produces a blackish area around the navel end of the fruit that, with increasing populations, may affect the whole fruit (Lewis *et al.*, 1951; Dosse, 1964b; Gerson and Vacante, 2012). In southern Italy, infestation of lemon fruits is very common, and damage usually affects the blossom end of the lemons (Vacante, 2009b). In the San Joaquin Valley (California), damage by the two-spotted spider mite on citrus is similar to that caused by the six-spotted spider mite, *Eotetranychus sexmaculatus* (Riley); feeding activity on the undersides of the young leaves produces chlorotic areas that are visible on the upper surface, and severe injury may result in leaf drop (Jeppson, 1989). Currently, the pest status of the species on citrus is considered of minor importance.

Control

Like other spider mites, control of the two-spotted spider mite depends on prevention, sampling and monitoring of populations, horticultural practices and the use of chemical and biological means. The large number of reports that are available make it difficult to summarize the information on control and to present a uniform framework. Therefore, the general features

Table 20.13. *Tetranychus urticae* Koch. Natural enemies (bacteria and fungi).

Families and species	References
BACTERIA	
Bacillaceae	
Bacillus thuringiensis Berliner	Jung *et al.*, 2007
Pseudomonadaceae	
Pseudomonas putida Trevisan	Aksoy *et al.*, 2008
Spiroplasmataceae	
Spiroplasma sp.	Enigi and Schausberger, 2007
FUNGI	
Ancylistaceae	
Conidiobolus obscurus (Hall *et* Dunn) Remaudiere and Keller	Chandler *et al.*, 2000
Conidiobolus thromboides Drechsler	
Basidiobolaceae	
Basidiobolus sp.	Chandler *et al.*, 2000
Clavicipiteae	
Beauveria bassiana (Balsamo) Vuillemin	Chandler *et al.*, 2005
Entomophthoraceae	
Zoophthora radicans (Bref.) A. Batko	Chandler *et al.*, 2000
Exobasidiomycetidae	
Acaromyces ingoldii Boekhout *et al.*	Paz *et al.*, 2007a
Hirsutella danubiensis Balazy	Bałazy *et al.*, 2008
Hirsutella thompsonii Fisher	Gardner *et al.*, 1982
Meira argovae Boekhout *et al.*	Paz *et al.*, 2007a
Meira geulakonigii Boekhout *et al.*	
Neozygitaceae	
Neozygites floridana (Weiser *et* Muma)	Smith and Furr, 1975
Neozygites tetranychi (Weiser) Remaud. and S. Keller	Weiser, 1968

of the topic are covered and readers interested in further insights are invited to consult local pest control advisers or pest control organizations.

PREVENTION. The prevention of *T. urticae* infestation begins in the nursery with the production of seedlings free from infestation and, where possible, the use of resistant varieties. Before transplanting seedlings, it is necessary to perform careful weed control. Workers should avoid moving from one infested environment to another that is free from attack. It is good practice not to use chemicals that induce hormoligosis or are non-selective to natural enemies. In the Mediterranean area, it is useful to close greenhouses at the end of the crop cycle to kill spider mites. In plastic and wood structures, it is possible to burn sulfur after closing greenhouses.

SAMPLING AND MONITORING. Different guidelines are used for control of the two-spotted spider mite in different crops. The UC Phenology Model Database for the Two Spotted Spider Mite (UC IPM Online, 2003c) reports two models to predict the timing of the development of two-spotted spider mites which may be used in conjunction with field monitoring. The first model is based on Carey and Bradley (1982), who found a lower development threshold of the mite of 11.7°C. On cotton, development of the egg, larva, protonymph and deutonymph require the accumulation of 46.1, 21.9, 19.0 and 21.5 degree-days, respectively, while the egg to adult development, pre-egg-laying adult development and the egg to egg development require the accumulation of 108.6, 11.9 and 120 degree-days, respectively. The second model is based on Herbert (1982), who found a lower developmental threshold of 10.0°C. On apple, the egg to adult development, pre-egg-laying adult development and egg to egg development require the accumulation of 144.5, 25.3 and 169.8 degree-days, respectively.

On vegetables cultivated in the field and in greenhouses, an action threshold of 40–50% of infested leaves by one motile form of *T. urticae* is sufficiently reliable (Vacante and Benuzzi, 2007). Sampling is done by collecting 60 leaves from 3000 plants (about the number of plants transplanted to 1000 m^2 of cultivated surface), alternately collected from the top and bottom of plants, at random on one plant in every 50. On ornamental and floral plants, this threshold is lower and usually coincides with the first detection of attack. In the case of biological control, releases must be effected at the first detection of mites. The UC Pest Management Guidelines for spider mites of strawberry (2014c) suggest that for plants transplanted in autumn, ten leaflets per acre should be randomly selected in small fields and five leaflets per acre in larger fields. Mid-tier leaflets can be monitored by observing the lower leaf surface using a hand lens to count the number of mites or by using a mite-brushing machine. Observations should be recorded on a sampling form.

For citrus in the Mediterranean region, it is suggested that four shoots (newest vegetation) should be examined per plant and 20 fruits per tree examined *in situ* on 10% of plants

Table 20.14. *Tetranychus urticae* Koch. Natural enemies (insects).

Families and species	References
Aeolothripidae	
Aeolothrips melaleucus (Hal.)	Putman and Herne, 1966
Scolothrips sp.	Yaseen *et al.*, 1982
Scolothrips indicus Priesner	Kantaratanakul *et al.*, 1979
Scolothrips longicornis Priesner	McMurtry *et al.*, 1970
Scolothrips sexmaculatus (Pergande)	
Anthocoridae	
Anthocoris nemorum (L.)	McMurtry *et al.*, 1970
Orius insidiosus (Say)	Iglinsky and Rainwater, 1950; Isenhour and Yeargan, 1981
Orius laevigatus (Fieber)	Tawfik and Ata, 1973b
Wollastoniella gatti Ghauri	Yaseen *et al.*, 1982
Cecidomyiidae	
Arthrocnodax sp.	Readshaw, 1975
Arthrocnodax occidentalis Felt	Oatman *et al.*, 1981
Feltiella acarisuga (Vallot)	Vacante, 1981, 1982b; Gillespie *et al.*, 2000; Mo and Liu, 2007
Coccinellidae	
Clitostethus arcuatus (Rossi)	Liotta, 1981
Parastethorus histrio (Chazeau)	Chazeau *et al.*, 1974; Houston, 1980
Parastethorus nigripes (Kapur)	Edwards and Hodgson, 1973; Houston, 1980
Stethorus aethiops Weise	Yaseen *et al.*, 1982
Stethorus bifidus Kapur	Collyer, 1964a
Stethorus chengi Sasaji	Wen and Lee, 1981
Stethorus fenestralis Houston	Houston, 1980
Stethorus gilvifrons (Mulsant)	Mathur, 1969
Stethorus japonicus Kamiya	Mori *et al.*, 2005
Stethorus jejunus Casey	Yaseen *et al.*, 1982
Stethorus loi Sasaji	Wen and Lee, 1981
Stethorus parapauperculus Pang	Lin and Chen, 1984
Stethorus punctillum Weise	Putman, 1955
Stethorus siphonulus Kapur	Raros and Haramoto, 1974
Stethorus tridens Gordon	Gordon, 1982
Stethorus utilis Horn	Mora Morin, 1991
Stethorus vagans (Blackburn)	Chazeau, 1979; Houston, 1980
Conyopterigidae	
Conwentzia psociformis (Curtis)	Ripolles and Melia, 1980; Abad *et al.*, 2006
Formicidae	
Tapinoma melanocephalum (Fabricius)	Osborne *et al.*, 1995
Labiduridae	
Labidura riparia (Pallas)	Coates, 1974
Miridae	
Malacocoris chlorizans Panz.	Geier and Baggiolini, 1952
Staphylinidae	
Oligota fageli Williams	Yaseen *et al.*, 1982
Oligota oviformis Casey	Oatman *et al.*, 1981; Wen and Lee, 1981
Oligota pallidicornis Cameron	Yaseen *et al.*, 1982
Thripidae	
Frankliniella sexmaculatus (Pergande)	Hoy, 2011

(Cavalloro and Prota, 1983; Vacante and Bonsignore, 2012). The aggregated pattern of distribution of *T. urticae* has suggested the development in Spain of a sampling method for clementine. The method includes a random choice of both the primary units (trees in orchard) and the secondary units (leaves or fruits of each tree). The sampling plan combines the observations of symptomatic leaves in a ring 0.25 m^2 in diameter and the pest population density on the leaves (Martínez-Ferrer *et al.*, 2006).

On almond and other crops, the same UC Pest Management Guidelines suggested for the Pacific spider mite and the strawberry spider mite are applied (UC IPM, Online, 2012c). In general, closer inspections help to prevent mite outbreaks.

Table 20.15. *Tetranychus urticae* Koch. Natural enemies (spiders and mites).

Families and species	References
SPIDERS	
Lycosiidae	
Pardosa crassipalpis (Pourcell)	Coates, 1974
MITES	
Anystidae	
Anystis agilis Banks	Sorensen *et al.*, 1976
Bdellidae	
Bdella depressa Ewing	Snetsinger, 1956
Cheyletidae	
Acaropsellina sollers (Rohdendrof)	Nassar and Kandeel, 1986
Hemicheyletia bakeri (Ehara)	Laing, 1973
Erythraeidae	
Balaustium putmani Smiley	Putman, 1970b
Phytoseiidae	
Amblyseius zaheri Yousef *et* El-Borolossy	Rasmy *et al.*, 2003
Euseius mesembrinus (Dean)	Abou-Setta and Childers, 1987
Euseius stipulatus (Athias Henriot)	Vacante, 1986; Ferragut *et al.*, 1987; Grafton Cardwell *et al.*, 1997
Galendromus helveolus (Chant)	Caceres and Childers, 1991
Metaseiulus occidentalis (Nesbitt)	McMurtry, 1985b; Grafton Cardwell *et al.*, 1997
Neoseiulus californicus (McGregor)	Buergo *et al.*, 1986; Grafton Cardwell *et al.*, 1997; Ibrahim *et al.*, 2005; Abad *et al.*, 2006
Neoseiulus cydnodactylon (Shehata *et* Zaher)	El-Banhawy *et al.*, 2000
Neoseiulus longispinosus (Evans)	Kongchuensin *et al.*, 2005
Neoseilus picanus (Ragusa)	Ragusa *et al.*, 2000
Phytoseiulus longipes Evans	Grafton Cardwell *et al.*, 1997
Phytoseiulus persimilis Athias Henriot	Chant, 1961; Bravemboer and Dosse, 1962; Vacante, 1982b; Buergo *et al.*, 1986; Abad *et al.*, 2006
Phytoseius plumifer (Canestrini *et* Fanzago)	Nawar *et al.*, 2001b
Proprioseiopsis badryi (Yousef *et* El-Borolossy)	Abou-Awad *et al.*, 1998a
Typhlodromus athiasae Porath *et* Swirski	Nawar *et al.*, 2001a
Typhlodromus exhilaratus Ragusa	Ragusa, 1981
Typhlodromus phialatus Athias Henriot	Ferragut *et al.*, 1987
Stigmaeidae	
Agistemus fleschneri Summers	Nelson *et al.*, 1973
Zetzellia mali (Ewing)	Parent and LeRoux, 1956
Trombidiidae	
Allothrombium pulvinum Ewing	Hosseini *et al.*, 2005

ECONOMIC THRESHOLDS. Economic thresholds have been detected for some important crops. So, on cucumber in greenhouses, Hussey and Parr (1963b) proposed a leaf damage index (LDI) system based on a 0–5 score, ranging from 0 (no damage) to 5 (leaves blanched and starting to shrivel). Above a damage index of 1.9, crop loss occurs, and an index of 2.5 produces 40% loss after 5 weeks. On tomato, crop loss starts at an index of 2.0 (about 30% of photosynthetic surface damaged) and the maximum loss occurs at an index of 2.7 in 16 days (Scopes, 1985). On strawberry, the economic threshold in autumn transplants corresponds to an average of five mites per mid-tier leaflet in the first 4 months after transplanting. The threshold for summer transplants is the higher of an average of ten mites per mid-tier leaflet in the same period. At harvest, strawberries are more tolerant of mite feeding, and treatment thresholds increase to an average of 15–20 mites per mid-tier leaflet (UC IPM Online, 2014c). On clementine mandarins, Pascual-Ruiz *et al.* (2010) suggested relating various measurements taken weekly during the season (numbers of leaf flushes, symptomatic (chlorotic) leaves m^{-2}, adult females per symptomatic leaf (and per random leaf and per fruit) to a fruit damage index (DI) scored at the end of the season on a scale ranging from 0 (no damage) to 5 (scar covering more than 50% of fruit skin). A strong relationship was found between DI values for fruits with symptomatic (chlorotic) leaves m^{-2}. Therefore, accumulated symptomatic leaf counts taken in August and September satisfactorily predicted DI at harvest, and the economic injury level (EIL) was 15 accumulated symptomatic leaves during that period. On lemon, the action threshold is 2% of infested fruit and/or 8–10% of leaves contaminated by active colonies of the mite (Cavalloro and Prota, 1983; Vacante and Bonsignore, 2012).

CULTURAL PRACTICES. Some plant cultivars are less susceptible to spider mites and plant resistance may be used in IPM programmes. However, the possibilities are complex and do not allow easy decisions to be made. For example, cultivars and cultural practices on strawberry in California vary between production regions, and information can be sought from the University of California County Cooperative Extension office or from cooperatives before making planting decisions (UC IPM Online, 2014c).

According to Boudreaux (1958a), the performance of spider mites is optimized in hot and dry conditions, and the possibility of suppressing the development of *T. urticae* populations by high air humidity or direct contact with water was positively tested by Tulisalo (1974). In modern greenhouses, increasing the humidity level by misting plants can reduce the growth of spider mites. The absence of polyethylene plastic mulch, and surface irrigation increase humidity within the greenhouse. However, increasing humidity may increase the risk development of fungal diseases. Variation of the fertilization regime may also affect the population growth of *T. urticae* (Markkula and Tiittanen, 1969), but plant response does not encourage the application of this technique.

Short-day strawberry cultivars transplanted in the autumn are usually less tolerant of mite damage than day-neutral cultivars, especially later in the fruit-production season, but short-day cultivars transplanted in the summer are relatively tolerant of mite damage. Moreover, vernalization directly favours plant vigour. Various factors (autumn transplanting, nursery location, pre-harvest chilling, nursery harvest date and length of pre-transplant supplemental cold storage) affect plant vernalization. Plants with low amounts of chilling exhibit low vigour and often develop high levels of mite infestation. Excessive chilling increases vigour and reduces the mite population, although it has a negative effect on other production factors, such as delayed flowering, large plant size and increased vegetative runner production. Moreover, soil preparation and fumigation, the use of polyethylene plastic mulch, and proper irrigation to prevent water stress increase plant vigour as well. Like other spider mites, road dust control is also important in preventing mite outbreaks (UC IPM Online, 2014c). Hussey and Scopes (1985) reported that in greenhouses of temperate areas, the combination of interrupted lighting and vigorous growing plants should, when plants are removed at the end of the crop season, induce only the presence of the green form of *T. urticae*, and these die within a few hours. The removal of weeds and other host plants enables a subsequent crop season free of pest mites to be obtained.

BIOLOGICAL CONTROL. Biological control can effectively suppress mite outbreaks, mainly on vegetables in protected crops. Positive results have also been reported with strawberry in the field, and with other crops. Natural enemies of *T. urticae* include bacteria, entomopathogenic fungi and predatory insects and mites. Sometimes, the action of natural enemies is insufficient and must be integrated with sprays of selective acaricides. Van der Geest (2010) reported 15 pathogens variously associated with *T. urticae*. The mite is susceptible to many species of Ascomycetes, including *Beauveria bassiana* (Balsamo) Vuillemin, *Hirsutella thompsonii* Fisher, *Metarhizium anisopliae* (Metchnikoff) Sorokin, *Verticillium lecanii* (Zimmerman) Viégas, *Isaria farinosa* (Holmsk.) Fr. and *I. fumosorosea* Wize and some Entomophthorales, such as *Neozygites floridana* (Weiser *et* Muma) (Delalibera and Mascarin, 2010). Bioassays on tomato leaflets both under laboratory conditions and in experimental greenhouses on tomato plants to evaluate the action of isolates of *B. bassiana*, *H. thompsonii*, *M. anisopliae* and *V. lecanii* showed that *B. bassiana* gives the best control by reducing the numbers of adults, nymphs and eggs of the two-spotted spider mite by 97–98% (Chandler *et al.*, 2005). Bioassays of the actions of the fungi *Meira geulakonigii* Boekhout *et al.*, *M. argovae* Boekhout *et al.* and *Acaromyces ingoldii* Boekhout *et al.* showed that after 14 days, *M. geulakonigii* and *A. ingoldii* produced 81.9 and 90% mortality, respectively, while *M. argovae* caused only 46.1% mortality (Paz *et al.*, 2007a). The potential of the *M. anisopliae* isolate ICIPE78 as an alternative to acaricides for *T. urticae* management has recently been highlighted by Bugeme *et al.* (2010), while Castro *et al.* (2010b) discussed the use of *N. floridana*.

The most important predatory insects are the coccinellid *Stethorus punctillum* Weise and the cecidomyiid *Feltiella acarisuga* (Vallot). Both of these predators have been investigated and proposed as means of circumventing the limits to the use of the most common phytoseiid mite *Phytoseiulus persimilis* Athias Henriot, which is limited by high temperature, low humidity and the leaf structure of some crops, such as cucumber or tomato. Although the lady beetle *S. punctillum* is a voracious predator of spider mites, its development requires high levels of the prey population. Its release is then inappropriate in the early stages of crop infestation and releases best occur in 'hot spots' of spider mite infestation. The bio-ecology of the lady beetle and its application to the control of the two-spotted spider mite have been investigated by Rott and Ponsonby (2000a,b). The midge *F. acarisuga* feeds on various biological stages of *T. urticae*, and its role in Mediterranean greenhouses was first evaluated by Vacante (1981, 1982b) on various crops, alone or in association with the phytoseiid mite *P. persimilis*. Recently, *F. acarisuga* has been proposed for the biological control of spider mites on tomato and cucumber (Gillespie *et al.*, 1998, 2000). Unfortunately, in the Palaearctic region, the midge is parasitized by the proctotrupid *Aphanogmus fulmeki* Szelenyi (Vacante, 1981, 1982b).

The action of various phytoseiid mites appears to be more effective than that of predatory insects. Almost six phytoseiid mites (*P. persimilis*, *P. macropilis* (Banks), *Neoseiulus californicus* (McGregor), *N. fallacis* (Garman), *N. longispinosus* (Evans) and *Metaseiulus occidentalis* (Nesbitt) have been proposed for the biological control of the two-spotted spider mite in the field and the greenhouse, but only *N. californicus* and *P. persimilis* exert the most effective control of the two-spotted spider mite in appropriate conditions.

The poor response of *P. persimilis* in torrid areas characterized by low RH and high temperatures, the negative influence of poor packaging, storage and marketing of commercial preparations and the use of phytoseiid strains better adapted to the programmed temperature and humidity of their mass rearing than to field conditions have suggested that searches should be conducted for other natural antagonists, such as *N. californicus*. This phytoseiid was first described from California (McGregor, 1954), and has also been reported in Central and South America and southern Europe, where it is adapted to arid and semi-arid conditions and has retained greater tolerance to low RH and low prey density. *N. californicus* is a type II

phytoseiid generalist predator, a type that includes phytoseiid mites feeding on spider mites that produce little webbing, but not restricted to *Tetranychus* spp., and able to feed on other mites, on pollen and on plant exudates. Like the phytoseiid mites of type I, they may bear long setae (Gerson *et al.*, 2003). McMurtry and Croft (1997) include *N. californicus* in type II, but its generalist adaptations (elevated fecundity on a thrip diet, reduced inter-plant movement, shorter dorsal setae) suggested to Croft *et al.* (1998a) that it should be transferred to type III. Studies on a Japanese strain of *N. californicus* reared on eggs of *T. urticae* (red form) showed hatching of more than 97.3% of *N. californicus* eggs and more than 81.6% of newly hatched larvae developing at temperatures between 15 and 35°C. At 37.5°C, the females lay eggs, but no eggs hatch. At 40°C, there is no oviposition. The lowest threshold temperature from egg to oviposition is 10.3°C and the thermal constant is 86.2 degree-days; the maximum number of generations that could complete development in a year under field conditions are in the range 21–28. At 25°C, the fecundity is on average 41.6 eggs during a mean time of 19.4 days. At 20, 25 and 30°C, the intrinsic rates of natural increase (r_m) were 0.173, 0.274 and 0.340, respectively (Gotoh *et al.*, 2004b). However, it has been observed that reproductive females of *P. persimilis* kill more than three times the number of prey eggs killed by *N. californicus* and more than five times the number killed by *M. occidentalis*. Adults of *P. persimilis* can find a single tetranychid adult individual while traversing 47.1 cm^2, whereas *N. californicus* must encounter adult prey at 1 every 35.7 cm^2 and *M. occidentalis* at 1 every 24.5 cm^2 (Gilstrap and Friese, 1985). *N. californicus* tracked a population of the two-spotted spider mite on strawberry, but with some delay, and its populations were slightly less aggregated. These incongruities in the spatial distributions of prey and predators could depend on the use of the *T. urticae* as a temporal refuge (Greco *et al.*, 1999).

From a practical point of view, it is believed that the use of *P. persimilis* and *N. californicus* in the same crop has complementary effects and that a combination of the two predators can enhance the biological control of the two-spotted spider mite. However, the results of research on intraguild predation do not clear doubts about the interactions between the two species. Walzer *et al.* (2001) found that intraguild predation by *N. californicus* was a stronger force than food competition and finally resulted in the displacement of *P. persimilis*. They mention that previous studies showed that intraguild predation between the specialist *P. persimilis* and the generalist *N. californicus* was strongly asymmetric and favoured the generalist. Further intraguild experiments showed that *N. californicus* was a potential intraguild predator of *P. persimilis*. However, *P. persimilis* did not suffer much from intraguild predation as long as the shared prey was present. This is probably because *N. californicus* prefers to feed on the two-spotted spider mite rather than on its intraguild prey (Çakmak *et al.*, 2006).

About 30 years after the first use of *N. californicus*, the assessment of the response of this phytoseiid mite compared with that of *P. persimilis* still do not agree, in some cases giving positive results (Escudero and Farragut, 2005; Rhodes and Liburd, 2006; Weintraub and Palevsky, 2008) and in other cases less encouraging results (Oatman *et al.*, 1977; Çakmak *et al.*, 2009). Experiments conducted by the author on strawberry in a greenhouse in Sicily with natural populations of *N. californicus* were not encouraging. Regardless of the aspects discussed above, since the 1970s *N. californicus* has been used for the control of *T. urticae* in different countries on different crops. In 2000, at least 23 suppliers managed the commercial distribution of this phytoseiid (IPM Practitioner, 2000), and in 2012 the result was that it was used for the control of mites in Europe, North and South Africa, North and Latin America, Asia; consumption of the product has been estimated as large (van Lenteren, 2012).

P. persimilis belongs to the type I predator mites, having long dorsal setae that result in the mites being specialized predators of heavily webbing spider mites (Gerson *et al.*, 2003). It is a specific predator of *Tetranychus* spider mites, and in addition to *T. urticae*, it has been collected in association with other spider mites of the genera *Tetranychus*, *Eutetranychus* and *Panonychus*. Its performance is very limited on other spider mites, and sometimes it has been collected with other phytoseiid mites, such as *N. californicus* and *Iphiseius degenerans* (Berlese). The reasons for the success of *P. persimilis* lie in its specificity of action, and in the different values of the fundamental parameters of its bio-ecology compared with those of its prey, *T. urticae*. At 20°C, the oviposition times of *T. urticae* and *P. persimilis* on bean leaves are 15.7 and 22.3 days, respectively. The longevities of the prey and predator are 17.8 and 29.6 days, respectively, while the fecundities are 37.9 and 53.5 eggs, respectively. The development of a generation takes 24.0 and 17.32 days, respectively (Laing, 1968, 1969). The lowest development threshold of *P. persimilis* is 11°C, and the intrinsic rate of development is 0.011 °C^{-1}, while that of *T. urticae* is 0.0067 °C^{-1} (Sabelis, 1981). Prey density, temperature, RH and starvation affect the predacious behaviour of *P. persimilis*. At 27°C and 80% RH, the phytoseiid exerts the best action. Decreasing the moisture decreases its biological activity, and at constant temperature different RH values can dramatically affect the biological response of the predator (Mori and Chant, 1966a,b; Stenseth, 1979; Pralavorio and Almaguel-Rojas, 1980). Scopes (1985) found that at 24°C a 1:20 ratio of predator to mobile forms of *T. urticae* prey ensured control of the infestation in less than 10 days, while at a 1:200 ratio of predator:prey, control took about 16 days. At 18°C, the same ratios of predator:prey required about 20 and 28 days, respectively, and at 13°C, about 47 and 29 days, respectively.

The history of *P. persimilis* as an agent of biological control of the two-spotted spider mite started in the 1960s with its accidental introduction into Germany with orchids from Chile (Dosse, 1958). The predator was distributed to different specialists in Europe (Bravemboer and Dosse, 1962) and North America (Chant, 1961), where its effective action in the biological control of *T. urticae* was ascertained. Some years later, mass rearing of *P. persimilis* (Scopes, 1968) and large-scale control of *T. urticae* by release of *P. persimilis* were effected on cucumber (Gould *et al.*, 1969) and other crops. During the last 50 years, the use of the phytoseiid has spread to all continents. It is usually released, both in the field and the greenhouse, at amounts of 1.5–30 motile forms m^{-2} as a function of the density and the frequency of the two-spotted spider mite during early infestation (Vacante and Benuzzi, 2007), according to the inoculative pest in first or inundative methods (see Chapter 9).

For example, the UC Pest Management Guidelines for the two-spotted spider mite on strawberry (UC IPM Online, 2014c) suggest the release of 2–3 predators per plant when pest mite populations are low and an average of five predators per plant when the pest mite population is at threshold level. Widespread infestations early in the season associated with low spider mite populations can be controlled by the release of about 30,000 mites acre^{-1} (about 1.5 predatory mites per plant (= 7 m^{-2})) either as a single, large release or as three smaller releases of 10,000 acre^{-1} (= 2 predator m^{-2}), depending on the severity of the weather conditions and the spider mite population density. When mite densities increase to threshold levels, inundative release may reduce mite infestations, but at release rates exceeding 100,000 acre^{-1} (= 25 m^{-2}), once spider mite populations begin to increase, it is difficult for predators to contain the pest mite densities below the economic thresholds (UC IPM Online, 2014c).

Private companies rear the predator and transfer useful instructions on its release and management on different crops. At least 49 suppliers manage the commercial distribution of *P. persimilis* (IPM Practitioner, 2000). Summarizing the experiences on the use of *P. persimilis* worldwide and the techniques used on different crops and in different countries is very difficult. For further details, the reader may consult Hussey and Scopes (1985), Scopes (1985), Zhang (2003), Vacante and Benuzzi (2007) and Hoy (2011).

CHEMICAL CONTROL. Chemical control is provided by sprays of acaricides labelled for various crops in different countries. For example, acaricides suggested for control of the two-spotted spider mite on strawberry in California include abamectin, acequinocyl, bifenazate, etoxazole, fenbutatin oxide, hexythiazox, narrow-range oil and spiromesifen (UC IPM Online, 2014c). Like other spider mites, the selection of resistant strains may markedly affect the results of control efforts, as occurs for abamectin (Hoy and Conley, 1987), clofentezine and hexythiazox (Herron *et al.*, 1993), and other acaricides (Richter and Schulze, 1990; Rossi and Conti, 1997). Natural compounds that can be used include azadirachtin (Makundi and Kashenge, 2002), rosemary oil, stylet oil (UC IPN Online, 2014c) and other compounds. Some antibiotics, such as liuyangmycin (Dong, 1990), have also been tested. Further insights for citrus may be found in Vacante (2010). Zhang (2003) and Vacante and Benuzzi (2007) discussed the control of greenhouse crops.

INTEGRATED PEST MANAGEMENT. The implementation of an IPM programme is the best solution for the control of *T. urticae* infestations. The use of selective chemicals (acaricides, fungicides and insecticides) is a precondition (Veire and Degheele, 1992; Park *et al.*, 1995) (see Chapter 9). Moreover, the strategy must be adapted to the needs of different crops and different techniques, including the release of natural enemies and/or exploitation of the natural enemies already present in the field. The availability of resistant varieties and the application of cultural techniques are both fundamental. Insights on various crops may be found in Zhang (2003), Hoy (2011), Vacante and Bonsignore (2012) and Vacante and Gerson (2012).

Tetranychus yusti McGregor (Fig. 20.67)

Common name

Unknown.

Diagnostic characteristics

FEMALE. The body is deep carmine red in colour (Boudreaux, 1956) and has dark spots of different sizes. The peritreme is hooked. The dorsal striae are lobate; between the setae e_1–e_1 and f_1–f_1 the striae are longitudinal, and between the setae e_1 and f_1 they are transverse and form a diamond shape. The ventral striae bear weak lobes only between the genital region and the setae *4a*; the pregenital striae are almost entirely broken and strongly dotted. The empodia have 6 proximoventral hairs, and minute or absent spurs; tarsus I usually has 3 tactile setae overlapping with the proximal duplex setae (Seeman and Beard, 2011).

MALE. The body is pale green in colour (Saba, 1971). The empodia bear a small dorsal spur; empodia I are claw like, and empodia II–IV have long and free proximoventral hairs. The knob of the aedeagus is small, dorsally slightly convex and with medioposterior indentation; the anterior projection is rounded and the posterior projection pointed (Seeman and Beard, 2011).

Geographical distribution

Brazil (Paschoal, 1970), Cape Verde (Meyer Smith, 1987), Colombia (Bolland *et al.*, 1998), Cook Islands (Gutierrez, 1980), Ecuador (McGregor, 1955), El Salvador (Mancia and Cortès, 1975; Andrews and Poe, 1980), French Polynesia (Gutierrez, 1980), Greece (Hatzinikolis, 1986), Guadeloupe (Flechtmann *et al.*, 1999a), Honduras (Baker and Pritchard, 1962), Mexico (Tuttle *et al.*, 1976), Nigeria (Matthysse, 1980), Thailand (Baker, 1975b), the USA (Boudreaux, 1956) and Venezuela (Fréitez and Alvarado, 1978b).

Host plants

T. yusti is reported on 44 wild and cultivated plants belonging to 20 families, with the highest number of species (14) in the Fabaceae (Migeon and Dorkeld, 2006–2013). The families and species of economic importance are: Asteraceae – calendula (*Calendula officinalis* L.), sunflower (*Helianthus annuus* L.) (Jeppson *et al.*, 1975); Convolvulaceae – sweet potato (*Ipomoea batatas* (L.) Lam.) (Jeppson *et al.*, 1975); Euphorbiaceae – cassava (*Manihot esculenta* Crantz) (Baker, 1975b), castor bean (*Ricinus*

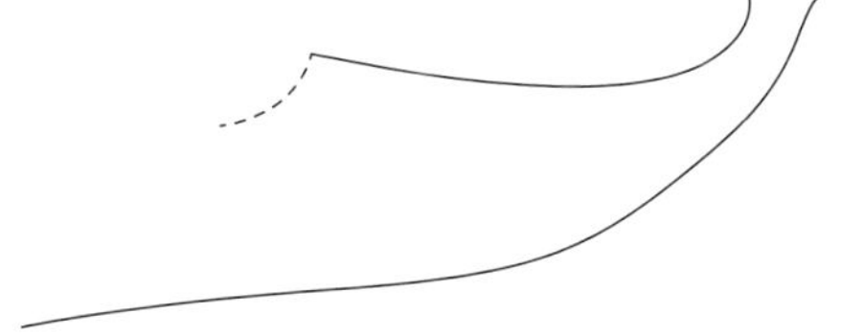

Fig. 20.67. *Tetranychus yusti* McGregor. Male, aedeagus (from Meyer Smith, 1987).

communis L.) (Flechtmann *et al.*, 1999a); Fabaceae – groundnut (*Arachis hypogaea* L.) (Jeppson *et al.*, 1975), pigeon pea (*Cajanus cajan* (L.) Huth) (Matthysse, 1980), soybean (*Glycine max* (L.) Merr.) (Jeppson *et al.*, 1975), common bean (*Phaseolus vulgaris* L.) (Mancia and Cortès, 1975; Andrews and Poe, 1980; Matthysse, 1980), pea (*Pisum sativum* L.) (Jeppson *et al.*, 1975), Dutch clover (*Trifolium repens* L.) (Jeppson *et al.*, 1975); Malvaceae – okra (*Abelmoschus esculentus* (L.) Moench) (Jeppson *et al.*, 1975), *Gossypium* sp. (Matthysse, 1980); Musaceae – banana (*Musa acuminata* Colla), banana (*Musa* × *paradisiaca* L.) (Baker and Pritchard, 1962); Poaceae – maize (*Zea mays* L.) (Bolland *et al.*, 1998; Migeon and Dorkeld, 2006–2013); Rhamnaceae – jujube (*Ziziphus jujuba* Mill.) (Flechtmann *et al.*, 1999a).

Bio-ecology

T. yusti has a tropical and subtropical distribution and is widely reported from the Nearctic region. It reproduces by arrhenotokous parthenogenesis and has a haploid number of chromosomes of three (Bolland *et al.*, 1981). Like other tetranychids, this mite also harbours *Wolbachia* bacteria (Breeuwer and Jacobs, 1996). On primary bean leaves (*P. vulgaris*) in the laboratory at 25.5°C, the development of female *T. yusti* took from 9 to 10 days, and male development took 8.5 to 9.5 days. Daily egg deposition was 4.5 eggs, and the total fecundity averaged 158 eggs. The oviposition period lasted about 35 days, and female longevity was 21 days (Saba, 1971).

Natural enemies include the phytoseiid mite *Neoseiulus fallacis* (Garman) (Smith and Newsom, 1970), the aeolothripid *Scolothrips pallidus* (Beach) (Gutierrez, 1983, in Chazeau, 1985) and the cecidomyiid *Feltiella acarisuga* (Vallot) (Gutierrez, 1983, in Chazeau, 1985).

Symptomatology and damage

Like other spider mites, damage by *T. yusti* causes speckling of the host plant foliage, followed, as infestation continues, by browning of the leaves. The mite has been reported as common on some crops, such as soybeans (Baker and Connell, 1961), but on the whole it appears to be an uncommon pest.

Control

There are no known reports in the control of *T. yusti*.

The Genus *Amphitetranychus* Oudemans

Diagnostic characteristics

The peritremes are anastomosed distally. There is 1 pair of para-anal setae. The empodium is split distally, commonly with 3 pairs of hairs, and is devoid of spurs. The duplex setae of tarsus I are well separated (Bolland *et al.*, 1998).

Economic importance

This genus includes three species. *A. viennensis* (Zacher) was first reported from rosaceous plants in Austria (Zacher, 1920), and is widely distributed in Palaearctic regions and also in China. *A. savenkoae* (Reck) was reported on *Quercus* sp. in Gruziya in Georgia (USSR) (Reck, 1956; Mitrofanov *et al.*, 1987), and Ukraine (Mitrofanov *et al.*, 1987). *A. quercivorus* (Ehara *et* Gotoh) lives on *Quercus mongolica* in Japan (Ehara and Gotoh, 1990). *A. viennensis* has some economic importance.

Amphitetranychus viennensis (Zacher) (Fig. 20.68)

Common name

Hawthorn spider mite.

Diagnostic characteristics

FEMALE. The body of summer females is bluish violet or carmine red in colour and with whitish segments. The overwintering females are bright red and the extremities are light yellow or sandy coloured. The dorsal body striae are transversal and bear little subtriangular lobes (Jeppson *et al.*, 1975; Vacante, 1985a).

MALE. The body is sandy yellow in colour and sometimes with greenish black spots. The aedeagus is bent sharply dorsally, and the knob has a small anterior projection near the base of the bent portion, and a posterior projection that is attenuated, straight and vertical (Jeppson *et al.*, 1975; Vacante, 1985a).

Geographical distribution

Armenia (Mitrofanov *et al.*, 1987), Austria (Zacher, 1920), Azerbaijan (Mitrofanov *et al.*, 1987), Belgium (Rossem *et al.*, 1961), Bulgaria (Balevski, 1967; Arnaudov, 2003), China (Ma and Yuan, 1975; Wang *et al.*, 1981, 1985; Wang and Cui, 1992; Li *et al.*, 2007a), Czech Republic (Zacharda *et al.*, 1988), France (Chaboussou and Bessard, 1954; Rambier, 1954, 1958a; Navajas *et al.*, 1996), Germany (Zacher, 1920), Greece (Bouchelos, 1962; Hatzinikolis, 1968c, 1969c, 1986), Georgia (USSR) (Mitrofanov *et al.*, 1987), Hungary (Bognar, 1961), Iran (Khalil-Manesh, 1973; Golpayegani *et al.*, 2003; Kafil *et al.*, 2007), Italy (Rota, 1962; Vacante, 1985a), Japan (Ehara, 1956a; Gotoh *et al.*, 2003a; Kishimoto and Adachi, 2006), Kazakhstan (Ugarov and Nikolskii, 1937; Wainstein, 1954, 1956a,c; Mitrofanov *et al.*, 1987), South Korea (Lee *et al.*, 1987, 1999; Lee and

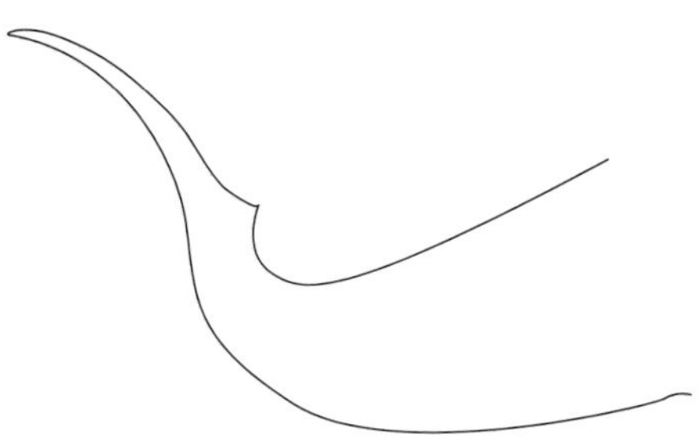

Fig. 20.68. *Amphitetranychus viennensis* (Zacher). Male, aedeagus (from Vacante, 1985a).

Chung, 2011), Lebanon (Dosse, 1967), Lithuania (Pauriene, 1970), Moldova (Mitrofanov *et al.*, 1987), the Netherlands (Rossem *et al.*, 1961; Vierbergen, 1990), Poland (Boczek and Kropczynska, 1964; Dobosz *et al.*, 1995), Portugal (Carmona, 1964), Romania (Iacob, 1978), Russia (Mitrofanov *et al.*, 1987), Slovakia (Jedlicková and Kolodochka, 1994), Spain (Alfaro, 1964; Ferragut and Santonja, 1989), Sweden (Oudemans, 1931b), Switzerland (Günthart and Günthart, 1959), Taiwan (Tseng, 1990), Tajikistan (Mitrofanov *et al.*, 1987), Turkey (Düzgünes, 1961; Uysal *et al.*, 2001), Ukraine (Mitrofanov *et al.*, 1987), the UK (Hirst, 1920) and Uzbekistan (Ugarov, 1937; Ugarov and Nikolskii, 1937; Mitrofanov *et al.*, 1987).

Host plants

The list of host plants of the hawthorn spider mite includes 45 species belonging to eight families, with the highest number of species (38) in the Rosaceae family (Migeon and Dorkeld, 2006–2013). The families and species of economic importance are: Betulaceae – hazel (*Corylus avellana* L.) (Bolland *et al.*, 1998; Migeon and Dorkeld, 2006–2013); Moraceae – fig (*Ficus carica* L.) (Bolland *et al.*, 1998; Migeon and Dorkeld, 2006–2013); Rosaceae – azarole (*Crataegus azarolus* L.) (Dobosz *et al.*, 1995), quince (*Cydonia oblonga* Mill.) (Gerini, 1971), apple (*Malus domestica* Borkh.) (Zacher, 1920; Ugarov and Nikolskii, 1937; Rambier, 1954, 1958a; Alfaro, 1964; Zacharda *et al.*, 1988; Navajas *et al.*, 1996; Arnaudov, 2003; Li *et al.*, 2007a), sweet almond (*Prunus amygdalus* Batsch.) (Vacante, 1985a), apricot (*P. armeniaca* L.) (Hatzinikolis, 1986), sweet cherry (*P. avium* (L.) L.) (Zacher, 1920; Rambier, 1958a; Vacante, 1985a), European plum (*P. domestica* L.) (Rambier, 1958a), peach (*P. persica* (L.) Batsch) (Rambier, 1958a; Carmona, 1964), blackthorn (*P. spinosa* L.) (Uysal *et al.*, 2001), pear (*Pyrus communis* L.) (Zacher, 1920; Ugarov, 1937; Rambier, 1958a; Hatzinikolis, 1968c, 1969c, 1986), Asian pear (*P. pyrifolia* (Burm. f.) Nakai), raspberry (*Rubus idaeus* L.) (Dobosz *et al.*, 1995).

Bio-ecology

The hawthorn spider mite is a pest of rosaceous plants, such as apple, almond, pear and *Prunus*, mainly in the drier climates of the Mediterranean region and central and eastern Europe (Müller, 1957; Beglyarov, 1959; Rota, 1962; Göksu, 1968; Yiğit and Uygun, 1982; Skorupska and Boczek, 1984; Skorupska and Tomczyk, 1988; Kasap, 2010). In Japan, the mite is injurious to pear (Kishimoto and Adachi, 2006). Its populations often reach high densities in crops treated with pesticides.

Navajas *et al.* (1999) found that a French strain of hawthorn spider mite was infected by *Wolbachia* bacteria, but not a Japanese strain. Before the autumn leaf drop, lack of food resources and decreasing temperatures induce fertilized females of *A. viennensis* to overwinter under bark scales, mosses and lichens on the boughs and branches of the trees (up to 50 specimens together), but not in the soil or on the weeds beneath the trees (Zacher, 1921b; Razumova, 1967). In the spring, when the mean daily temperatures are between 9 and 10°C, overwintering females emerge, and the major migration from winter shelters occurs before the beginning of blossom. Females first occupy the leaflets of young shoots, usually as a single female that begins to lay its eggs on the lower leaf surfaces. In Germany, the developmental periods for the female and male are on average 20–30 days and 8–16 days, respectively, and the mite develops 5–6 generations a year. In Turkey, the development of the female takes 12–25 days, and the mite develops 9–10 generations a year. At temperatures of 22–25°C, pre-oviposition lasts from 1 to 5 days, and fecundity ranges from 70 to 82 eggs. The mite develops a generation in 12–14.5 days, and the development of the egg, larval and nymphal stages requires 4–5, 1–2 and 0.5–3 days, respectively (Müller, 1957; Beglyarov, 1959; Göksu, 1968). The mite produces webbing on the lower leaf surface.

Laboratory studies on the development period and reproduction rate of hawthorn spider mite conducted on five apple cultivars ('Amasya', 'Golden Delicious', 'Granny Smith', 'Starking Delicious' and 'Starkrimson Delicious') at 25°C, 65 ± 10% RH and under a 16L:8D photoperiod highlighted the better performance on 'Golden Delicious', which had a short development time (10.7 days), high daily egg production (5.2 eggs) and an early reproduction peak. The highest intrinsic rate of natural increase (r_m), 0.247, was also on 'Golden Delicious', while the lowest (0.215) was on 'Starking Delicious'. The development period ranged from 7.4 to 18.8 days at 35 and 20°C for females, and from 7.9 to 17.2 days at 30 and 20°C for males. The development thresholds of the eggs and pre-adult stages were 9.72 and 9.07°C, and the total effective temperatures were 72.99 and 185.18 degree days, respectively. The mean generation time (T_0) of the population ranged from 16.13 days at 30°C to 29.15 days at 20°C. The net reproductive rate (R_0) increased from 54.33 at 20°C to 78.34 at 25°C, and decreased to 75.71 at 30°C. The highest intrinsic rate of increase (r_m) was at 30°C (0.268), the lowest at 20°C (0.136) (Kasap, 2003).

The number of males is 3–5 times fewer than that of females. The thermal threshold of development is about 10°C, and the development of one generation requires 185 degree-days. The population density in nature depends on the activity of predators (AgroAtlas-Pests, 2003–2009b).

Acceptance of the host plant and the host preference of the hawthorn spider mite may affect the mite population dynamics. The first studies done in Poland showed that among ten apple varieties, the 'Landsberg' was the least attractive to the mites; among seven plum varieties, 'Emma Leppermann' and 'Althann's Gage' were the least attractive (Skorupska and Boczek, 1984). Effects of the host plants on the reproduction of the mite were examined in a laboratory study by Skorupska and Boczek (1985). They found that when the mite was transferred from hawthorn (*Crataegus*) to another plant (apple, pear or plum), it responded with an increase in female lifespan and fecundity in the parental generation.

Studies under laboratory conditions at 25 ± 1°C, 70 ± 10% RH and with a 16L:8D photoperiod on the effects of black cherry (*P. serotina* Ehrh. cv. 'Irani'), cherry (*P. avium* cv. 'Siahe Mashhad') and apple (*M. domestica* cv. 'Shafi Abadi') on the biological parameters of *A. viennensis* showed that black cherry might be the most suitable plant for hawthorn spider mite, as it had a shorter developmental period (10.6 days), longer adult longevity (25.5 days), higher reproduction (65.6 eggs) and a higher intrinsic rate of natural increase (0.194). Cherry was the least suitable host plant (Kafil *et al.*, 2007).

In another study on the effect of peach (*P. persica*) and nectarine (*P. persica* var. *nectarina* Maxim) varieties on the biological parameters of the hawthorn spider mite at 25 ± 1°C, 60 ± 10% RH and with 16L:8D photoperiod, the best performance was on middle- and late-season ripening varieties for both peach and nectarine; the developmental periods were 10.98 and 10.77 days for middle-season ripening varieties and 11.27 and 10.82 days for late-season ripening varieties, respectively, and there was also longer adult longevity, higher reproduction (52.05, 45.13, 36.04, and 30.98 eggs per female, respectively) and higher intrinsic rates of natural increase (0.199, 0.193, 0.188 and 0.186). Early-season ripening varieties gave the poorest performance (Li *et al.*, 2009).

Low temperatures and high RH hinder population increases of the hawthorn spider mite. When these conditions occur during the spring, the development of injurious population densities during the summer is precluded. Conversely, favourable conditions during spring induce a summer increase of mite populations (Müller, 1957; Beglyarov, 1959; Göksu, 1968). The results of studies on sprayed and unsprayed apple orchards ('Golden Delicious' and 'Starking Delicious') in Turkey showed that the population densities of *Panonychus ulmi* (Koch), *A. viennensis* and *Bryobia rubrioculus* (Scheuten) generally began to increase in early May, reached higher levels from mid-June to late August, and remained up to late September. The populations of *P. ulmi* and *A. viennensis* reached higher densities in sprayed orchards. In unsprayed orchards, populations of spider mites remained very low, depending upon the presence of the phytoseiid predator *Kampimodromus aberrans* (Oudemans). In sprayed orchards, although the predatory insect *Stethorus punctillum* Weise and the predatory stigmaeid mite *Zetzella mali* (Ewing) were abundant, they did not control the spider mites on either apple cultivar (Kasap, 2010).

Natural enemies include predatory insects and mites. Predatory insects include: coccinellids – *Stethorus* spp. (Yiğit and Uygun, 1982), *S. japonicas* Kamiya (Kishimoto, 2002) and *S. punctillum* (Yiğit and Uygun, 1986; Kasap, 2010); staphylinids – *Oligota* sp. (Kishimoto, 2002); coniopterigids – *Conwentzia* sp. (Yiğit and Uygun, 1982); the mirid *Malacocoris chlorizans* Panz. (Yiğit and Uygun, 1982); the anthocorid *Orius albidipennis* Reuter; chrysopids – *Chrysopa* spp.; aeolothripids – *Scolothrips acariphagus* Sarh. (AgroAtlas-Pests, 2003–2009b), *S. takahashii* Priesner (Kishimoto, 2002; Kishimoto and Adachi, 2006); and cecidomyiids – *Feltiella* sp. (Kishimoto, 2002; Kishimoto and Adachi, 2006). Predatory mites include the phytoseiids *Amblydromella sternlichti* (Swirski *et* Amitai) (= *Typhlodromus sternlichti* Swirski *et* Amitai) (Yiğit and Uygun, 1982), *Amblyseius californicus* (McGregor) (= *Neoseiulus californicus* (McGregor)), *A. orientalis* Ehara, *A. sojaensis* Ehara (Kishimoto, 2002), *K. aberrans* (Kasap, 2010), *Phytoseius plumifer* (Canestrini *et* Fanzago) (Moeezipoor *et al.*, 2008), *T. reticulatus* Oudemans (AgroAtlas-Pests, 2003–2009b) and *T. vulgaris* Ehara (Kishimoto, 2002). Also known are the stigmaeid *Z. mali* (Kasap, 2010) and the trombidiid *Allothrombium pulvinum* Ewing (Hosseini *et al.*, 2005).

Symptomatology and damage

Early damage by hawthorn spider mite causes yellow spots on the leaves. The spots are as large as a hazelnut, and as feeding of the mite continues, the leaves appear yellowish grey. Infested trees produce very few fruit buds, and the buds that develop are small (Jeppson *et al.*, 1975). At high levels of infestation, webs cover trunks and branches, plants may defoliate and there may be a decrease in bud formation the next year. The yield of plums is reduced 2–3 times (AgroAtlas-Pests, 2003–2009b). Feeding by the hawthorn spider mite on apple induces plant physiological responses. Adult female mites feeding on growing apple seedlings caused 20% reductions in CO_2 assimilation after 7 days, which continued after that; net photosynthesis was reduced by 30% after the same period but thereafter returned to control levels. A small reduction in chlorophyll and a small increase in dry mass were recorded relative to the control plants, while transpiration increased to 50% or more than that in the controls after 4 weeks (Skorupska and Tomczyk, 1988). The hawthorn spider mite may be a vector of apple scab and apple mildew (AgroAtlas-Pests, 2003–2009b).

Control

Infestations by the hawthorn spider mite are widely associated with excessive use of pesticides, which seems to stimulate its outbreaks. These outbreaks mainly depend on the suppression of natural enemies in the field and on the capability of mites to develop resistance against pesticides (AgroAtlas-Pests, 2003–2009b). Resistant varieties and chemical and biological means may be effective in control.

PLANT RESISTANCE. Plant resistance to the hawthorn spider mite has been investigated from time, e.g. in Poland, where there are apple varieties that are able to affect the demographic parameters of *A. viennensis* and *T. urticae* koch. The scab-resistant apple varieties 'Primula', 'Pionier', 'Lodel', 'Novamac' and 'Freedom' are good hosts for both spider mites. On 'Novamac', the best host for both spider mites, the populations of *A. viennensis* and *T. urticae* Koch develop by 26-fold and 22.7-fold in 27.4 days, respectively. The least suitable varieties for *A. viennensis* are 'Primula' and 'Pionier' (11-fold population increase in 26.4 days and 11.4-fold in 27 days, respectively); these two varieties were also the least suitable for *T. urticae* (10.2-fold in 26.5 days and 8.9-fold in 26.4 days). The different apple varieties differ in the morphological structure of their leaves, and the multiplication of *A. viennensis* appears to be influenced by the number of stomata on the lower leaf surface, the thickness of the spongy and palisade parenchyma and leaf thickness (Skorupska, 1998).

BIOLOGICAL CONTROL. Studies carried out in five apple orchards in Adana province, Turkey, on the population dynamics of *T. viennensis* and its predators (*Stethorus* spp., *Conwentzia* sp., *A. sternlichti* (*T. sternlichti*) and *M. chlorizans*) showed that the predators were able to control the hawthorn spider mite when broad-spectrum insecticides were not applied repeatedly (Yiğit and Uygun, 1982).

The effects of predacious insects on the population dynamics of three spider mite species, *A. viennensis*, *T. urticae* and *P. citri* (McGregor), which are serious pests infesting the Japanese pear, were compared by experimentally introducing each spider

mite species on to Japanese pear trees in Japan. *A. viennensis* and *T. urticae* immediately decreased in density after their introduction on to the trees, because the density of the predacious insects *S. takahashii* and *Feltiella* sp. increased steeply even when spider mites were at low densities (Kishimoto and Adachi, 2006).

CHEMICAL CONTROL. The application of dormant spray oils alone or mixed with acaricides suppress the spring generation of *A. viennensis*. However, it is necessary to apply sprays of acaricides during the summer to control the mite populations adequately (Müller, 1957; Beglyarov, 1959; Göksu, 1968). Spray applications of acaricides in the spring after the mass appearance of the mites require an average air temperature of 10°C. Whitewashing of tree trunks is now an important preventive control measure against *A. viennensis* (AgroAtlas-Pests, 2003–2009b). The acaricides that are usually applied to apple for the control of *P. ulmi* are effective.

21 The Family Tenuipalpidae Berlese

Morphological Characteristics, Systematics and Bio-ecology

The tenuipalpid mites are often confused macroscopically with tetranychid mites because of the orange to red colour of their bodies but, unlike tetranychids, they are slow moving and do not produce silk webbing; for this reason, they are also called 'false spider mites' or 'flat mites'. Tenuipalpid mites are usually regarded as secondary plant pests, but the recent emergence of a number of injurious species as vectors of plant viruses (Childers *et al.*, 2003b; Childers and Rodrigues, 2011) means that an updated review is required of this relatively little known family (Gerson, 2008).

The body of a tenuipalpid mite is dorsoventrally flattened, from 200 to 400 μm long, and commonly with ridges and reticulations. It consists of the gnathosoma and idiosoma (Fig. 21.1). The dorsum of the metapodosoma (the region of the idiosoma with legs III and IV) is considered to have disappeared, and the posterior edge of the dorsum of the propodosoma (the region of the idiosoma bearing legs I and II) is apposed to the anterior edge of the dorsum of the opisthosoma (the part posterior to legs IV), or these edges are bridged by the merged 'sejugal' and 'disjugal' furrows (Mesa *et al.*, 2009).

Like the Tetranychidae, the gnathosoma of the Tenuipalpidae includes stylophore, infracapitulum, chelicerae and palpi (Figs 2.6H, and 2.7E). Anatomical studies of the morphology and functioning of the mouthparts of *Cenopalpus pulcher* (Canestrini *et* Fanzago) have highlighted the role of the various parts, such as the 'labrum', 'preoral groove', 'inferior oral commissure', 'salivary pump' and 'chelicerae' (Nuzzaci and de Lillo, 1991b). The stylophore is not well developed, is extrudable and is partially derived from fusion of the bases of the chelicerae; these have a long stylet-like digitus mobilis that is strongly recurved proximally. The base of the stylophore connects with a membranous structure, the 'sheath' or 'collar', which is characterized by typical longitudinal ribbing and is strongly developed in some genera, e.g. *Raoiella* Hirst; it is involved in the protraction and retraction of the stylophore. The dorsal infracapitulum has a 'rostral gutter' that is modified into a groove with longitudinal 'capitular lips'. Tenuipalpids do not move their stylophores as markedly as the Tetranychidae, and the gutter does not guide the sliding stylophore (Hislop and Jeppson, 1976). The movement of the stylets appears to be independent of the stylophore, and that of the collar appears to be somehow connected with stylet movement. Indeed, the collar is protracted when the stylets are protracted and retracted when the stylets are retracted.

During feeding, *Raoiella* mites repeatedly fully expand the collar out over the infracapitulum and retract it completely within the prodorsum (Beard *et al.*, 2012). The infracapitulum hosts the pharynx and the 'inferior oral commissure', and bears 1–3 pairs of apical adoral setae (ad_{1-3}), set around the oral orifice and probably involved in the location of feeding sites. The cheliceral stylets lock together in a 'tongue and groove' arrangement to form a single hollow tube. A pair of 'stylet channels' or 'cheliceral grooves' act as a sheath and guide for the stylets, and a median 'salivary duct', set deep within the rostral gutter, extends through the infracapitulum to above the 'oral orifice'. In the apical portion of the infracapitulum, the walls of the rostral gutter curve upward and envelope the stylet channels. These walls close over, touching above the stylet channels but not fusing together; they seal the stylets within a tunnel deep inside the infracapitulum, and leave only a dorsal longitudinal groove that shows where the walls of the gutter touch above the stylets; this is the stylet or cheliceral groove. In the genus *Raoiella*, and in other tenuipalpid genera, the apex of the infracapitulum has only one opening, the oral orifice, through which the stylet tube is protracted and into which the host fluid components are sucked. The apex of the infracapitulum in the Tenuipalpidae and Tuckerellidae has either vestigial or no oral flaps. The inferior oral commissure appears as a perforation in the ventral cuticle of the infracapitulum, set anteriorly to the infracapitular setae (*m*), and able to join the pharynx to the outside of the mite body. Its function may be to eliminate air bubbles ingested with food without affecting the flow in the food channel (Nuzzaci and de Lillo 1989, 1991b; de Lillo *et al.*, 2001a). In *C. pulcher* the 'food channel' derives from the labrum interlocked with the stalked lateral margins of the preoral groove, and 'the salivary channel' is formed by the cheliceral stylets which interlock when protracted (Nuzzaci and de Lillo, 1991b).

Despite the absence of the oral flaps, the infracapitulum of *C. pulcher* adheres well enough to the wounds on the leaf surface for the pharyngeal pump to operate (Nuzzaci and de Lillo, 1989). However, if oral flaps are necessary to produce a vacuum, the pharynx of tenuipalpids must act differently from that of the Tetranychidae (Beard *et al.*, 2012). The possibility that tenuipalpids feed by using the stylet tube implies that the digestive system and stylet tube are directly connected, although this connection has never been observed. Within the infracapitulum, the stylet tube is physically distant to the food channel (Nuzzaci and de Lillo, 1991b), and anatomical studies confirm a definite lack of connection. The stylet tips in *Raoiella* and other genera have strong antiaxial serrations that are used for piercing the host plant tissues, and for anchoring the mites to the leaf surface during moulting. For example, *Raoiella* anchors itself to the host leaf during moulting via the stylets being inserted into a stoma; *Brevipalpus obovatus* Donnadieu (Morishita, 1954) and *B. phoenicis* (Geijskes) (Haramoto, 1969) both anchor

themselves during moulting by inserting their fully protracted stylets into the plant tissue; likewise, *Tenuipalpus pacificus* Baker anchors itself to the leaf surface by inserting its stylets directly into the leaf epidermis. The palpi of tenuipalpids are simple, 1–5 segmented, and the penultimate segment lacks a claw (Fig. 2.7E). Respiration is by two tracheae directed anteriorly and ending in a bulb that may be related to the longitudinal folds of the invagination of the stylophore (Fig. 2.6H).

The dorsal disjugal and the ventral sejugal sutures divide the idiosoma into the anterior propodosoma and the posterior hysterosoma. The dorsal body surface is divided into the prodorsum, corresponding to the anterodorsal part of the propodosoma, and the opisthosoma, e.g. the hysterosoma without legs III and IV (Alberti and Coons, 1999; Welbourn *et al.*, 2003) (Fig. 21.1). The prodorsum bears an anterior projection or 'rostral shield', which may be of different shapes and sizes, three pairs of setae including one pair of vertical setae (v_2) and two pairs of scapular setae (sc_1, sc_2), one pair of eyes on each side, a variable ornamentation from smooth to reticulate and one pair of pores (Welbourn *et al.*, 2003).

The opisthosoma hosts the genital and anal openings, a number of plates (ventral, genital, anal) and a maximum of 15 pairs of setae, excluding the genital and aggenital setae, 13 pairs of which are dorsal (Lindquist, 1985; Welbourn *et al.*, 2003; Beard *et al.*, 2013). The maximal dorsal opisthosomal chaetotaxy consists of three pairs each in the C (c_1, c_2, c_3), D (d_1, d_2, d_3) and E (e_1, e_2, e_3) rows, and a maximum of two pairs each in the F (f_2, f_3) and H (h_1, h_2) rows; the pseudanal (*ps*) setae are ventral to ventrocaudal (Fig. 21.1). Some setae are stable (sc_1, sc_2, c_3, d_3, e_3, h_1, h_2) and present in nearly all tenuipalpid species (Beard *et al.*, 2013). The dorsal and ventral surfaces have variable reticulations.

The legs articulate ventrally with the podosoma, corresponding for pairs I–II to the posterior part of the propodosoma, and for the pairs III–IV to the anterior part of the hysterosoma. The legs are short, wrinkled and with five free segments (Fig. 21.1). Ventrally, the coxae are fused to the venter and form the coxal fields. The posterior venter includes the genital and anal regions. The coxal field of legs I has three setae (*1a*, *1b*, *1c*), and those of legs II–IV each have two setae (*2b*, *2c*; *3a*, *3b*; *4a*, *4b*). The ventral opisthosoma bears 1–3 pairs of pseudanal setae (ps_1, ps_2, ps_3), two pairs of genital setae (g_1, g_2) and one pair of aggenital setae (*ag*) (Fig. 21.2A). The aggenital setae (*ag*) are set posterior to setae *4a*, and anterior to the genital region, located between the setae *4a* and g_1–g_2, just anterior to the genital plate. The genital setae are set on the genital plate. The pseudanal setae surround the anus, sometimes set on distinct anal plates, between the genital plate and the posterior margin of the body. The females nearly always bear between the legs IV the seta *4a*, and seta *4b*, when present, is set on coxa IV (Beard *et al.*, 2013). The setal formula for legs I–IV (from coxae to tarsi) and the leg chaetotaxy are relatively simple. For example, in the genus *Brevipalpus*, the setal formula for legs I–IV (from coxae to tarsi) is: 2-2-1-1, 1-1-2-1, 4-4-2-1, 3-3-1-1, 5-5-3-3, 9(1)-10(1)-5-5. The leg chaetotaxy is: fe I–II *d*, *l'*, *v'*, *bv'*, fe III *d*, *ev'*, fe IV *ev'*; ge I–II *d*, *l'*, *l'*, ge III–IV *l'*, ti I–II *d*, *l'*, *l'*, *v'*, *v'*, ti III–IV *d*, *v'*, *v'* (Beard *et al.*, 2013). The pretarsi have two claws and the empodium has tenent hairs (Fig. 21.2B).

The males of false spider mites have an aedeagus, but they do not have a genital plate. In this region, the setal arrangement shows a different pattern from the females. The setae *4a* and *4b* are in the same location, and setae *ag*, g_{1-2} and ps_{1-3} can be in a fairly similar position to that of the female or can be quite tightly

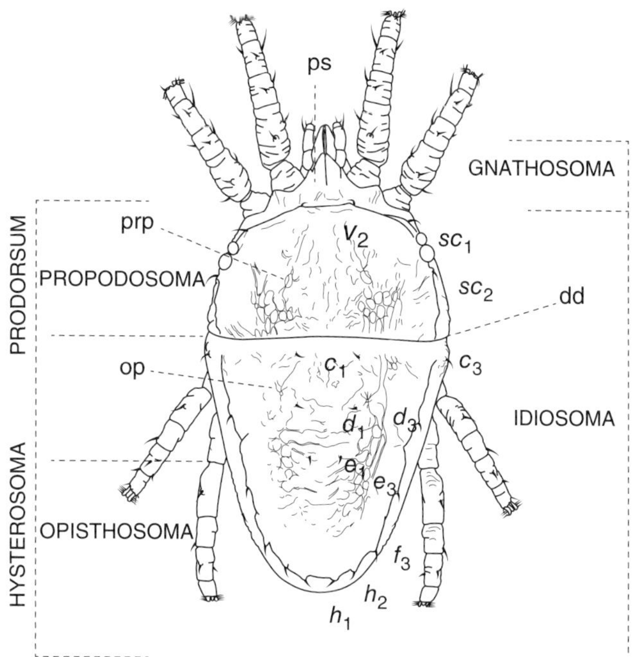

Fig. 21.1. *Brevipalpus phoenicis* (Geijskes). Dorsal view of female (from Vacante and Nucifora, 1985). Key: dd, dorsal disjugal suture; op, opisthosomal pore; prp, prodorsal pore; ps, prodorsal shield. The rest of the labels designate setae, and an explanation of the setal notation is given in the text.

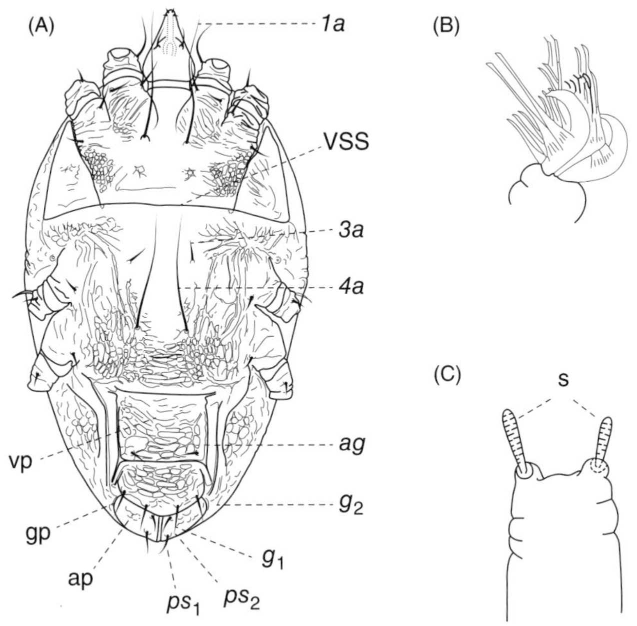

Fig. 21.2. *Brevipalpus phoenicis* (Geijskes). (A) ventral view of female (from Vacante and Nucifora, 1985); (B) pretarsus; (C) solenidia of tarsus II. Key: ap, anal plates; gp, genital plate; s, solenidion; vp, ventral plate; vss, ventral sejugal suture. The rest of the labels designate setae, and an explanation of the setal notation is given in the text.

aligned together posteriorly around the genito-anal region. The setae ps_1 of the males may be modified, tending to be thicker and sometimes longer than the other *ps* setae. The females have a spermatheca or receptaculum seminis, which may be sclerotized and exhibit a characteristic morphology, e.g. in the genus *Brevipalpus* (Beard *et al.*, 2013).

Tenuipalpids have been reported from all zoogeographic regions of the world (except Antarctica), and the highest number of species is reported from the USA (17.3%), followed by Mexico (16.3%) and Pakistan (16%) (Mesa *et al.*, 2009). Gerson (2008) reported that about 900 species in approximately 30 genera have been described, but many species may remain undiscovered on tropical and subtropical plants. Childers *et al.* (2003a) report that over 622 tenuipalpid species can be ascribed to 30 genera, and that three species (*B. californicus* (Banks), *B. obovatus* and *B. phoenicis*) develop on a total of 928 botanical species from 513 genera and 139 families. Later, Mesa *et al.* (2009) reported 891 valid species belonging to 34 genera, with the highest number of species ascribed in decreasing order to the genera *Tenuipalpus* Donnadieu (305 species), *Brevipalpus* Donnadieu (281 species), *Aegyptobia* Sayed (93 species), *Cenopalpus* Pritchard *et* Baker (59 species) and *Dolichotetranychus* Sayed (22 species). Currently, the Tenuipalpidae include over 1100 valid species belonging to 38 genera. Most of these species have been described from North America (33% of the total known Tenuipalpid fauna) and Africa (13%). Some of the available information needs further resolution. For example, the genus *Brevipalpus* is the most commonly intercepted genus of Tenuipalpidae in quarantine worldwide, but it is also the most complicated and by far the most diverse genus in the family, and the three most important species in the family worldwide, *B. californicus*, *B. obovatus* and *B. phoenicis*, are still consistently confused and misidentified. Thus, *B. californicus* species is a complex of three species groups (A, B, C), and the *B. phoenicis* species complex has been further separated into seven cryptic species, currently being described by Beard *et al.* (in preparation) (see Beard *et al.*, 2013).

Different *Brevipalpus* species, such as *B. californicus*, *B. obovatus* and *B. phoenicis*, reproduce by thelytokous parthenogenesis owing to the presence of feminizing bacterial symbionts of the genus *Cardinium* (Groot and Breeuwer, 2006). The frequency of males is low (Helle *et al.*, 1980; Weeks *et al.*, 2001; Groot and Breeuwer, 2006), and haploid females have $n = 2$ chromosomes. Other tenuipalpids, such as *B. chilensis* Baker, *B. russulus* (Boisduval) and *R. indica* Hirst have both males and females; the females are diploid ($2n = 4$ chromosomes), and the males haploid ($n = 2$ chromosomes) (Pijnacker *et al.*, 1980). Baker (1979a) reported paedogenesis in *Brevipalpus* mites. Occasional events of the sexual maturity of immatures have been observed in some *B. phoenicis* colonies that were maintained in the laboratory. There is no conclusive explanation for the occurrence of this phenomenon or of its potential association with sex-altering symbiont bacteria.

The life cycle of false spider mites includes the active stages of larva, protonymph, deutonymph and adult, which are separated from one another by a quiescent developmental stage (chrysalis). This chrysalis and the eggs are attached to the plant surface.

All tenuipalpids are phytophagous and feed on epidermal cells of the stems, leaves and fruits of various wild and cultivated plants. *Raoiella* spp. feed by inserting their stylets through stomata[l] openings in the leaf epidermis, not through the epidermis it[self] (Ochoa *et al.*, 2011; Beard *et al.*, 2012). Some species app[ear to] exhibit a wide polyphagy, such as *B. californicus*, *B.* [*obovatus*] and *B. phoenicis* (Childers *et al.*, 2003c), although thi[s statement] may be complicated by the fact that each may be a sp[ecies complex,]

and that some of these yet-to-be-named species may not be polyphagous. The feeding activity of tenuipalpids causes direct damage and sometimes transmits plant viruses that cause severe damage to economic plants, as occurs on citrus in Central and South America (Childers *et al.*, 2001, 2003a,b,c; Childers and Rodrigues, 2011); curiously, in other regions, such as the Mediterranean area (Vacante, 2009, 2010; Gerson and Vacante, 2012) or South Africa, the same species cause secondary damage (that may be due to yet-to-be-named species). This behaviour is explained by the phenomenon of clones (Weeks *et al.*, 2000b; Rodrigues *et al.*, 2003). Some species carry fungal spores of plants (Evans *et al.*, 1993; Ochoa *et al.*, 1994). Most important phytophagous species infesting economic plants belong to the genera *Brevipalpus* and *Tenuipalpus*. Currently, the widespread epidemic of the leprosis virus (and other viruses) in the Americas (Childers *et al.*, 2003a,b,c) and the introduction of invasive *R. indica* to the Americas (Flechtmann and Etienne, 2004) represent the most important economic problems.

The adaptation of mites of the genus *Raoiella* to feed via the stomata of different host plants could hypothetically involve other tenuipalpid genera also feeding via the stomata. From a practical point of view, further knowledge on tenuipalpid species feeding via the stomata and the absence of oral flaps could help in understanding the feeding mechanics and the ability of some important pests, e.g. *Brevipalpus*, to transmit plant viruses (Childers *et al.*, 2003b). This indicates that our knowledge of the bio-ecology of the family Tenuipalpidae should be reassessed, including the systematics of important genera, such as *Raoiella*, *Brevipalpus*, *Cenopalpus* and *Tenuipalpus*. The possibility of separating genera based on their feeding behaviour could have an important impact on alternatives for management and control (Beard *et al.*, 2012).

The key below to the identification of adult females of the genera including the most injurious species to economic plants follows Mesa *et al.* (2009).

Selected websites

Flat Mites of the World. Available at: http://idtools.org/id/mites/flatmites/ (see Beard *et al.*, 2013 for full details).

Acari: Plant-Feeding Mites. Systematic Entomology Laboratory, US Department of Agriculture Agricultural Research Service, Beltsville, Maryland. Available at: http://www.sel.barc.usda.gov/acari/frames/plantfeed.html

Key for the Identification of Genera (Adult Females)

1. • The setae h_2 are long and whip like, at least 4 times longer than other dorsal opisthosomal setae; 6–10 pairs of dorsal opisthosomal setae, and setae c_2 lacking; the palpus is 2- or 3-segmented, rarely 1-segmented; the setae c_1 are present in most species......................genus ***Tenuipalpus*** Donnadieu
   • The setae h_2 are not long, attenuate, similar in length to other dorsal opisthosomal setae; 5–13 pairs of dorsal opisthosomal setae; the setae c_2 are present or absent................2
2. • The setae c_2 are absent; the genital and ventral plates are separated, the former elliptical, wider than long, and the latter rectangular, usually with striate cuticle forming a frame around the plate; the palpus is 4-segmented; the lateral margin of opisthosoma has 6 or 7 pairs of setae; the setae sc_1, c_1 and e_1 are not extremely long and lanceolate...genus ***Brevipalpus*** Donnadieu
   • The setae c_2 are present...3
3. • The opisthosoma has 11 or fewer pairs of dorsal setae; the setae d_2 and e_2 are absent..4
   • The opisthosoma has 13 pairs of dorsal setae; the setae d_2 and/or e_2 are present; the dorsal opisthosomal setae are long and rod like, with clavate tips; the idiosoma is round; the palpus is 2-segmented........................genus ***Raoiella*** Hirst
4. • The genital and ventral plates are distinct and separated, the former elliptical, wider than long, and the latter rectangular, usually with a striate cuticle forming a frame around the plate; the palpus is 4-segmented (rarely 3-segmented); the idiosoma is broadly oval or slightly tapering posteriorly .... genus ***Cenopalpus*** Pritchard *et* Baker
   • The genital and ventral plates lacking, incomplete or fused, the ventral plate not rectangular and surrounded by a frame, the palpus is 1–5-segmented; the posterior end of the idiosoma has 4 pairs of dorsal setae, with setae e_2 present, and f_2 absent; the genu IV lacks setae...genus ***Dolichotetranychus*** Sayed

The Genus *Brevipalpus* Donnadieu

Diagnostic characteristics

The palpus is 4-segmented. The dorsal opisthosomal setae c_2, d_2 and e_2 are missing, and the opisthosomal setae f_2 may be present (set in a lateral position) or absent. The opisthosomal setae h_2 are not elongate. The anterior prodorsum has a broad, flat projection extending over the coxae I–II and the base of the gnathosoma. The genital and ventral plates are well developed. Two pairs of setae *ps* are usually present. Tarsi I and II usually have short and rod-like solenidia (Beard *et al.*, 2013).

Baker and Tuttle (1987) distinguished six different species groups within the genus *Brevipalpus*, stating that these groups coincided with the division proposed by Baker *et al.* (1975). The species discussed here belong to four of the groups proposed by Baker and Tuttle (1987): *Brevipalpus californicus* species group (= group IV of Baker *et al.*, 1975), *Brevipalpus obovatus* species group (= group V of Baker *et al.*, 1975), *Brevipalpus cuneatus* species group (= group I of Baker *et al.*, 1975) and *Brevipalpus phoenicis* species group (= group VI of Baker *et al.*, 1975). The main characters of each of these four species groups are listed below.

- ***Brevipalpus californicus* species group:** the opisthosomal setae f_2 are present. The dorsal central setae c_1, d_1 and e_1 have a different shape from the dorsal lateral setae c_3, d_3 and e_3. Tarsus II has 2 solenidia.
- ***Brevipalpus obovatus* species group:** the opisthosomal setae f_2 are absent. The dorsal central setae c_1, d_1 and e_1 have a different shape from the dorsal lateral setae c_3, d_3 and e_3. Tarsus II has 1 solenidion.

- ***Brevipalpus cuneatus* species group:** the opisthosomal setae f_2 are present. The dorsal central setae c_1, d_1 and e_1 have a different shape from the dorsal lateral setae c_3, d_3 and e_3. Tarsus II has 1 solenidion.
- ***Brevipalpus phoenicis* species group:** the dorsal opisthosomal setae f_2 are absent. Tarsus II has 2 solenidia.

Economic importance

This is a worldwide genus consisting of over 300 species living on wild and cultivated plants. This high number of species, however, does not correspond to the general harmfulness of the group, as only a restricted number of species (*B. californicus*, *B. chilensis*, *B. lewisi* McGregor, *B. obovatus* and *B. phoenicis*) cause direct damage by their feeding activity, and sometimes transmit virus diseases to economic plants. Among these, *B. californicus* and *B. phoenicis* have been reported as vectors of plant viruses to citrus and other crops in the Americas and some other countries (Childers and Rodrigues, 2011). In the remaining countries, the same mite species cause only direct damage to citrus and other crops.

Outbreaks of *B. phoenicis* have caused serious damage on tea in Indonesia, but this has been less of a problem in India and Sri Lanka (Oomen, 1982); the mites have also been reported infesting guava (Guerere and Gonzalez, 2000; Neena Singla, 2001; Rivero-Maldonado *et al.*, 2008) and coffee (Reis *et al.*, 2000; Teodoro and Reis, 2006) in tropical countries. Although the economic damage caused by both *B. californicus* and *B. phoenicis* cannot be denied, the recent updates on their identities suggest that both species should be treated in *sensu latu* while waiting for the complex of species that characterizes them to be adequately described. *B. lewisi* causes feeding injury on pistachios in California, and has also been reported for citrus fruits in Texas (Rice and Weinberger, 1981; Childers *et al.*, 2003c). Outbreaks of *B. phoenicis* and *B. obovatus* on 'Robinson' tangerine leaves in Florida (Childers *et al.*, 2003c) caused damage on the outer margins of the leaves and at the base adjacent to the petiole, similar to the leaf damage on tea caused by *B. phoenicis* (Oomen, 1982). *B. chilensis* infests different fruit plants (citrus, fig, almond), ornamentals, forest trees and annual weeds, and causes severe damage to grapevines in Chile (Jeppson *et al.*, 1975). *B. obovatus* exhibits a wide polyphagy (Childers *et al.*, 2003c) and is considered an important citrus pest in different areas of the world (Vacante, 2010).

Other *Brevipalpus* spp. are occasional or secondary pests. *B. essigi* Baker has a wide range of host plants, such as *Aucuba*, *Ficus*, fuchsia, *Howea*, *Pittosporum*, sage, speedwell, butterfly bush and orchids, and has been reported in California, Mexico (Baker, 1949) and New Zealand (Ashley and Manson, 1987). *B. lilium* Baker inhabits various host plants, such as azalea, croton, hibiscus, jasmine, lily, apple, sumac, brambles, berries, grapes and other plants (Jeppson *et al.*, 1975). *B. oleae* Baker and *B. olearius* Sayed live on olive trees (Baker, 1949). *B. oncidii* Baker is a pest of orchids (*Oncidium* and *Odontoglossum*) in greenhouses in California and England (Baker, 1949; Pritchard, 1951b). *B. russulus* is a pest of cactus and succulents, both in ornamentals and rangeland plants, and has been reported in different European and American countries, in Japan and in other countries. In addition to ornamental cactus, it damages a range of plants, which become uniformly reddish grey (Evans *et al.*, 1961; Jeppson *et al.*, 1975; Ashley and Manson, 1987). *B. sayedi* Baker is a pest of hickory and pecan trees in Florida, Maryland and Indiana (Baker, 1949; Jeppson *et al.*, 1975). However, none of these species is known to cause consistent economic damage.

Brevipalpus californicus (Banks) (see Beard *et al.*, 2013)

Common name

Citrus flat mite.

Diagnostic characteristics

This species belongs to the *B. californicus* species group (Baker and Tuttle, 1987). Its identity has been questioned by Beard *et al.* (2013) who, according to the different morphological structures of the prodorsum, the cuticle posterior to setae e_1–e_1 and the ventral and genital plates, distinguished three *B. californicus* group species, called A, B and C. The body of the female of *B. californicus s.l.* is ovoid, 240 µm long and 156 µm wide, red coloured and with a dark pattern in the central area (Baker, 1949). The male is similar to the female, but smaller and less common (Pritchard and Baker, 1958). The main characteristics of the group species proposed by Beard *et al.* (2013) are summarized below.

BREVIPALPUS CALIFORNICUS GROUP SPECIES A. This group is based on the type specimens of *B. californicus*. The prodorsum is smooth to wrinkled centrally with large closed cells laterally, forming reticulations. The cuticle posterior to setae e_1–e_1 has a few short, transverse to 'V'-shaped folds. The ventral plate has medium to small rounded circular cells, and the genital plate has transversely aligned elongate cells, sometimes forming transverse bands.

BREVIPALPUS CALIFORNICUS GROUP SPECIES B. The prodorsum is entirely reticulate, with large closed cells; the central reticulation is usually weak, but the cells are usually closed, and sometimes the central cells are fused to form larger cells. The cuticle posterior to setae e_1-e_1 is wrinkled, may have a few short, weak chevron ('V'-shaped) to transverse folds. The ventral plate has medium to large rounded cells, and the genital plate has transversely aligned elongate cells, sometimes forming transverse bands.

BREVIPALPUS CALIFORNICUS GROUP SPECIES C. The prodorsum is entirely reticulate, with even cells. The cuticle posterior to setae e_1–e_1 is reticulate. The ventral plate and the genital plate have medium rounded cells.

Geographical distribution

The geographical distribution based on Beard *et al.* (2013) is as follows.

BREVIPALPUS CALIFORNICUS GROUP SPECIES A. Japan, Mexico, Spain and the USA (California).

BREVIPALPUS CALIFORNICUS GROUP SPECIES B. Australia, Ecuador, Italy, Mexico, Nepal, Palestine, South Africa, Spain, Sri Lanka, Philippines and the USA (Maryland).

BREVIPALPUS CALIFORNICUS GROUP SPECIES C. Nigeria, Peru and Thailand.

Host plants

The citrus flat mite has been collected on about 316 wild and cultivated plants belonging to 67 genera of 33 botanical families (Childers *et al.*, 2003b). Only host plants based on specimens of *B. californicus* confirmed by Beard *et al.* (2013) are reported here.

BREVIPALPUS CALIFORNICUS GROUP SPECIES A. Geranicaceae – *Geranium* sp.; Rosaceae – *Malus* sp.; Rubiaceae – *Gardenia* sp.; Rutaceae – *Citrus* sp., sweet orange (*C. sinensis* (L.) Osbeck), orange.

BREVIPALPUS CALIFORNICUS GROUP SPECIES B. Annonaceae – *Chirimoya* sp.; Arecaeae – coconut (*Cocos nucifera* L.); Bignoniaceae – *Begonia* sp.; Orchidaceae – orchids; Pinaceae – *Abies* sp.; Rubiaceae – *Gardenia* sp.; Rutaceae – Mexican orange (*Choisya ternata* Kunth), *Citrus* sp., key lime (*C. aurantifolium* (Christm.) Swingle), lemon (*C. limon* (L.) Osbeck), *C. nobilis* Andrews; Theaceae – tea (*Camellia sinensis* (L.) Kuntze).

BREVIPALPUS CALIFORNICUS GROUP SPECIES C. Convolvulaceae – sweet potato (*Ipomoea batatas* (L.) Lam.); Rutaceae – *Citrus* sp.

Bio-ecology

The occurrence of three group species within *B. californicus* means that new scenarios are envisaged concerning both the identity of the species and the aspects of their bio-ecology as known today, including the direct and indirect damage that they cause to different crops worldwide. Specifically, until now, *B. californicus* had been reported as collected from 316 wild and cultivated plants belonging to 67 genera of 33 botanical families (Childers *et al.*, 2003b), but it is not possible to relate this knowledge with each group species as stated by Beard *et al.* (2013). This limitation suggests that the bio-ecology of this mite be discussed by referring to it as *B. californicus s.l.* Consequently, the information provided here is not exhaustive, although it has the fundamental aim of illustrating how this complex of forms, attributable to the old single species, lives. The same considerations need to be applied to the aspects of symptomatology and damage to be discussed, while pointing out the need to relate any knowledge closely to each form (or future species).

Feeding damage by *B. californicus s.l.* concerns mainly ornamentals such as orchids (Manglitz and Cory, 1953; Ochoa *et al.*, 1994), citrus (Lewis, 1949; Dean and Maxwell, 1967; Jeppson *et al.*, 1975; Schwartz, 1977b; Childers, 1994a; French and Rakha, 1994; Ochoa *et al.*, 1994; Childers *et al.*, 2003a,b) and tea (Cranham, 1966). Damage has also been reported to cardamoms (*Elettaria cardamomum* (L.) Maton) (Aguilar and Ochoa, 1988).

B. californicus s.l. lives on the leaves, fruits and twigs of various host plants. On citrus, it also occurs on petioles and developing buds to varying degrees (Sadana and Joshi, 1979), causing a different symptomatology according to the cultivar. On orange (cv. 'Valencia') and grapefruit, the green to ripe fruits are preferred to the twigs and leaves (Schwartz, 1970). Females of the citrus flat mite have two chromosomes and reproduce by thelytokous parthenogenesis (Helle and Bolland, 1972). The bacterial symbiont *Cardinium* sp. manipulates the reproduction of *B. californicus s.l.*, feminizing it (Chigira and Miura, 2005; Groot and Breeuwer, 2006).

In rearing experiments on leaves of seedlings of the orchid *Epidendrum cochleatum* (L.) W.E. Higgins, at temperatures between 18 and 24°C and 55% RH, egg development took about 9 days, and between 21 and 30°C the developmental time of the larval, protonymphal and deutonymphal stages required 8.6, 6.2 and 7.0 days, respectively; the developmental time of the quiescent stages was 3.6 days each. The life cycle developed in about 3 weeks, oviposition began about 3 days after the last moult and the female laid one egg per day over 25 or more days (Manglitz and Cory, 1953). On sweet orange (*Citrus sinensis* Osbeck, cv. 'Jaffa'), a decrease in temperature from 30.5 to 15.3°C increased the developmental time. Incubation, larval, protonymphal, deutonymphal and adult stages took 5.5–10.3, 4.5–13.2, 4.8–12.8, 5.1–12.8 and 14.8–54.5 days, respectively. The complete life cycle took 20.6–49.1 days. Pre-oviposition, oviposition and post-oviposition periods varied from 2.1 to 7.0, 6.9 to 27.2 and 3.4 to 20.7 days, respectively. The highest fecundity (30.5 eggs) was in generation VI and the lowest (5.2 eggs) in generation III; daily egg deposition ranged from 0.7 ± 0.1 to 1.4 ± 0.1 eggs per female. The mite develops 7 generations per year, with the 8th generation partial (Nehru and Bhagat, 2006a). On the same host plant under laboratory conditions, about 31 days were needed to complete the immature stages, including pre-oviposition. The female lifespan was, on average, 35 days, and the highest daily egg deposition ranged from 2.00 to 2.60 per female during the ovipositional period from days 34 to 40. The net reproductive rate (R_0) was 31.84, the intrinsic rate of increase (r_m) 0.086 and the finite rate of increase (λ) 1.089 females per female a day. The weekly multiplication rate of the population was 1.825 times (Nehru and Bhagat, 2006b). On *Azalea* sp. at 23 ± 1°C and 60 ± 5% RH, the life cycle from egg to adult took 26.5 days and at 27°C it took 21 days (Trindade and Chiavegato, 1994).

On citrus, the population density is commonly low and the highest densities occur in the cracks and crevices of fruit, which are also made by wind or hail; juvenile and adult stages feed around the edges of these, emphasizing the outlines. In early infestation, the mite populations are most abundant at the stem end of the fruit near or under the fruit button (Dean and Maxwell, 1967; Meyer Smith and Schwartz, 1998b). Investigations of preferential distribution of the mite on the leaves of sweet orange (cv. 'Jaffa') revealed that the tender leaf growth stage had a significantly low population of the mite (1.4) compared with the mature leaf (5.0). The preference of the mite in decreasing order is for the mature > young > tender leaf growth stage. The midrib region of tender, young and mature leaves have greater mite populations than the apex, base, margin, veins of the lamina and petiole. The abaxial leaf surface harbours the maximum mite population (Nehru *et al.*, 2005). In Egypt, populations of *B. californicus* are highest on orange and mandarin in August;

they prefer the eastern part of the canopy and the lower branches and, frequently, the fruits. The mite is present in larger populations on oranges than on mandarins (El-Halawany, 1991). The use of phosphorus compounds may cause an increase in populations (Attiah and Wahba, 1973), and light pruning and a regular succession of dry weather with light rainfall may favour their development (Baptist and Ranawerra, 1955).

Natural enemies include a number of phytoseiid mites, such as *Euseius hibisci* (Chant) (Gupta *et al.*, 1971), *E. mesembrinus* (Dean) (Badii *et al.*, 1993), *E. scutalis* (Athias Henriot) (El-Halawany *et al.*, 1993), *Galendromus helveolus* (Chant) (Chen *et al.*, 2006a), *Typhlodromus doreenae* Schicha (James and Whitney, 1993) and *T. rhenanus* (Oudemans) (Gupta *et al.*, 1971).

Symptomatology and damage

Feeding by the citrus flat mite causes direct and indirect damage to leaves, twigs, branches and fruits of different citrus species and varieties. Moreover, the feeding of *B. californicus s.l.* is associated with virus diseases on both citrus and ornamental plants. On seedlings of sour orange, feeding by the mite produces severely stunted rounded leaves, with marginal necrosis and tip burn. The infestation also results in severe stunting of new shoot growth with the formation of corky swollen buds. Leaves fail to develop from these gall-like buds. Damage has also been observed on citrus seedlings in Venezuela and in Florida and described as 'Brevipalpus gall' (Knorr *et al.*, 1960; Knorr and Denmark, 1970), and it has been reported that the affected plants do not produce new leaves (Childers *et al.*, 2003a).

In Australia, a silvering of the fruit (especially lemon) has been observed, known as 'silver mite' (Jeppson *et al.*, 1975). In Texas, injury to citrus fruit is prevalent on the inside of fruits in the lower tree canopy below 2 m height. The presence of *B. californicus s.l.* and *B. phoenicis* (Geijskes) has been associated with rind spotting of grapefruit at a high density of populations. The fruit lesions first appear as very slight yellowish and circular discoloured areas in depressions on the fruit surfaces of grapefruit or oranges. A central brown necrotic area or spot develops in these lesions and slowly becomes darker and corky. Irregular-shaped spots appear varying in size from 1 to 12 mm or larger, and most common at first on the side and stylar end of the fruit, but as feeding continues they may cover the fruit. Initially, the spots are level with the surface of the peel, but on storage they tend to become raised and darker in colour (Dean and Maxwell, 1967; French and Rakha, 1994). The irregular damaged areas on the fruit gradually become necrotic and result in fruit being rejected for the fresh market. In South Africa, the feeding activity of *B. californicus s.l.* produces brown-to-bronze coloured and corky scab-like spots on the rind of sweet orange, and on Valencia orange alterations are found in the depressions of the rind, essentially on the exposed side of the fruit; on grapefruit, these scab-like spots have an irregular shape and size and are not restricted to any particular side of the fruit rind, but they are mainly in the middle of the trees (Schwartz, 1977b; Meyer-Smith and Schwartz, 1998b). In Italy, *B. californicus* is regarded as a secondary pest that is sometimes responsible for the silvering of fruit of various citrus plants (Di Martino, 1985). Moreover, Knorr *et al.* (1960) report that feeding by *B. californicus* on the leaves and fruit of sweet orange produces small platelets of dry gum material; similar alterations are formed on twigs, thus increasing the injury, and with growth, the alterations increase and resemble scrofulous shelling of the bark.

Populations of *B. californicus s.l.* develop on the lower leaf surface of tea, causing a reddish discoloration. High levels of attack induce darkening of the lower foliage, a scorched appearance in the basal part of the leaf and a reduction in the size of new leaves. Defoliation can occur, and the tea bushes may die (Cranham, 1966).

On orchids, feeding by *B. californicus s.l.* causes silver patches on the leaves, which gradually become indented and brown. High levels of infestations induce the leaves to become chlorotic and fall off (Meyer-Smith, 1981a).

In addition to direct damage due to its feeding activity, until a few years ago *B. californicus s.l.* was thought to transmit viruses to citrus. Knorr *et al.* (1968) report that the tenuipalpid may transmit a toxin or virus to the leaves and fruit of citrus that has strongly limited the development of citrus groves in certain areas of the world. On grapefruit and orange varieties in Texas and Florida, smaller necrotic lesions form on the infested surface of leaves and fruits and appear as a 'leprosis-like spotting' or 'nail head rust' (Dean and Maxwell, 1967; Jeppson, 1989; French and Rakha, 1994). In Florida, the condition has been recorded on twigs and branches and is called 'Florida scaly bark', and in Argentina both symptoms are known as 'lepra explosiva' or 'leprosis', a severe virus disease of citrus actually reported from the American Continent (Argentina, Brazil, Central America, Colombia, Bolivia, Costa Rica, Honduras, Mexico, Panama, Paraguay, Venezuela) (Kitajima *et al.*, 1972; Rodrigues *et al.*, 1997; Childers *et al.*, 2003c; Bastianel *et al.*, 2010; Childers and Rodrigues, 2011).

At present, leprosis is the major virus disease of citrus in Brazil (Chagas *et al.*, 2003), but its only known vector is the red and black flat mite, *B. phoenicis* (Rodrigues *et al.*, 2003). In the past, *B. californicus s.l.* in Florida and *B. obovatus* Donnadieu in Argentina were both reported as vectors of citrus leprosis (Frezzi, 1940; Vergani, 1945; Knorr, 1968), but morphological studies of the specimens collected from these earlier reports confirmed the presence of both *B. californicus s.l.* and *B. phoenicis* species in the Florida samples and of *B. obovatus* and *B. phoenicis* in the Argentinian samples (Ochoa, USDA SEL, 2002, in Childers and Rodrigues, 2011). These earlier reports did not clarify the vector status of the tenuipalpids involved. Recently, *B. californicus s.l.* and *B. phoenicis* were reported as vectors of citrus leprosis in Guatemala (Palmieri *et al.*, 2007). However, no direct transmission tests have been done to prove that *B. californicus s.l.* is a vector of the disease. Moreover, there are no available specimens for either *B. phoenicis* or *B. californicus s.l.* from citrus samples collected in Guatemala. At present, the only confirmed vector of citrus leprosis in the western hemisphere is *B. phoenicis* (Childers and Rodrigues, 2011). However, feeding by *B. californicus s.l.* can transmit the *orchid fleck virus* (OFV), a nuclear type of virus reported from Australia, Brazil, Denmark, Germany, Japan, Korea and the USA (Kondo *et al.*, 2003, 2006), which has been classified in the new genus *Dichorhabdovirus* of the Rhabdoviridae. Virus particles accumulate in the nucleus of infected orchid plants (Kondo *et al.*, 2006; Bastianel *et al.*, 2010). The symptoms consist of chlorotic or necrotic spots and rings on the leaf fronds, and are also found on other plants (Kondo *et al.*, 2006). The vector is infective for a long time, so the actual virus distribution is likely much greater than known so far (Childers and Rodrigues, 2011).

Control

The need for control varies between different crops and countries. There is little information on biological control and spray applications of acaricides and cultural practices usually suppress mite outbreaks.

CULTURAL PRACTICES. On tea, *B. californicus* may also be contained by providing a host-free period or by removing infested branches of nearby infested shade trees (Cranham, 1966).

BIOLOGICAL CONTROL. Natural enemies alone are generally unable to contain the development of *B. californicus s.l.* populations. In Egypt, a decline of almost 90% in mite numbers was obtained after releasing 50 *E. scutalis* per citrus tree (El-Halawany *et al.*, 1993). Recently, a report from the Texas Invasive Species Institute (2011) notes that some phytoseiid mites (*Amblysieus linearis* Corpus, *A. longispinosus* Evans (= *Neoseiulus longispinosus* (Evans)), *Phytoseius nipponicus* Ehara) prey on all mite development stages, so they may be also used to control false spider mite populations.

CHEMICAL CONTROL. Chemical control may be necessary on some ornamentals and, according to the country, also on fruit plants. In some American countries, it is a priority, in others not. In Italy, good cultural management and/or rational use of chemicals against pests is usually sufficient to maintain low populations of false spider mites on citrus, while in Egypt chemical control appears stronger, and with this aim El-Halawany (1991) recommended that acaricides should be applied before August to the lower level and east side of trees. In South Africa, spray applications of acaricides are recommended during midsummer at the first signs of mite build-up on twigs and/or fruit (Meyer Smith and Schwartz, 1998b). Acaricides include avermectins (Iskander, 1993; French and Rakha, 1994; Iskander *et al.*, 1993), organochlorines (Jeppson *et al.*, 1975; Hanna *et al.* 1977, Iskander *et al.*, 1993), petroleum oils (Hanna *et al.* 1977; Iskander, 1993) and sulfur (Cranham 1966; Jeppson *et al.*, 1975; Ibrahim, 1992; Hoy, 2011). In Egypt, spray applications of flufenoxuron suppressed populations of *B. californicus s.l.* immatures on mandarins (Iskander and El-Atrouzy, 1991).

Brevipalpus chilensis Baker (Fig. 21.3)

Common name

Grape flat mite.

Diagnostic characteristics

FEMALE. *B. chilensis* belongs to *B. obovatus* Donnadieu species group (*sensu* Baker and Tuttle, 1987). The body is ovoid and red in colour, and about 267 µm long and 167 µm wide. The rostral shield extends beyond the base of femur I, with the central rostral projections broader than the two lateral projections and the posterior central area reticulate. The prodorsum has 3 pairs of short and lanceolate setae (v_2, sc_1 and sc_2) and 1 pair of eyes each side. The opisthosoma bears 9 pairs of dorsal, short and lanceolate setae (c_1, c_3, d_1, d_3, e_1, e_3, f_3, h_1, h_2), and 1 pair of pores. The prodorsum is reticulate centrally, with some cells fused together to form larger irregularly shaped cells; large closed cells occur laterally, forming a reticulation; the lateral cells are sometimes fused. The dorsal opisthosomal cuticle between c_1–c_1 and d_1–d_1 has large closed cells forming a reticulation; the cuticle between the setae d_1–d_1 and e_1–e_1 shows fused irregularly shaped cells; the cuticle posterior to e_1–e_1 has a series of short transverse folds. The ventral plate has small-to-medium circular cells, some of which are fused to form transversely elongate cells. The genital plate bears large, rounded cells, some of which are fused to form transversely elongate cells. The spermatheca has a long strong duct ending in a small rounded vesicle with short finger-like projections around most or all of the surface. Tarsus II has 1 solenidion (Baker, 1949; Beard *et al.*, 2013).

MALE. The male is similar to the female but smaller (Pritchard and Baker, 1958).

Geographical distribution

Argentina (Rio Negro) (Beard *et al.*, 2013), Chile (Baker, 1949), and India (Ghai and Shenhmar, 1984).

Host plants

Only host plants based on specimens of *B. chilensis* confirmed by Beard *et al.* (2013) are reported here: Actinidiaceae – kiwi fruit (*Actinidia deliciosa* (A. Chev.) C.F. Liang and A.R. Ferguson); Rutaceae – *Citrus* sp., lemon (*C. limon* (L.) Osbeck), orange; Vitaceae – grapevine (*Vitis vinifera* L.) (Beard *et al.*, 2013).

Bio-ecology

Although other species of host plants are included above, the grape flat mite is a primary pest of grapevines in Chile, although with the exception of the old grape cv. 'Ribier' it is of no economic importance in table grapes (González, 2006). Its populations live on the lower leaf surface and overwinter as fertilized females in various bark shelters. The best development occurs in the higher part of the canopy (González, 1968b). In Chile, overwintering females move up to the growing parts of the grapevine in late winter before the buds burst. They feed on

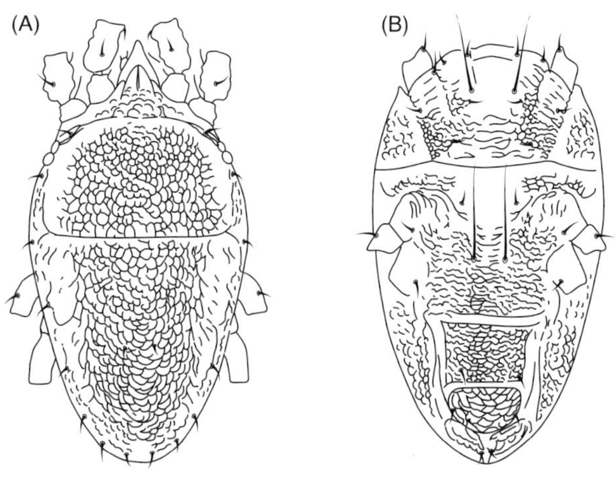

Fig. 21.3. *Brevipalpus chilensis* Baker. Female: (A) dorsal view; (B) ventral view (from Baker, 1949).

green tissue and begin to lay eggs in early October. On grapevines the mite develops 5–6 generations a year. The first generation takes place in approximately 40–50 days. The other 4–5 annual spring and summer generations appear in an overlapping fashion every 35–40 days. Females lay eggs on the lower leaf surfaces and hatching requires about 8–12 days. Males begin to appear at the end of summer when active mating occurs. In evergreen fruit trees such as on mandarins and lemons, mite activity continues throughout the winter (González, 2006).

Natural enemies include the coccinellid *Parastethorus histrio* (Chazeau) (Aguilera, 1987) and the phytoseiid mites *N. chilenesis* (Dosse) (González, 1983b), *N. californicus* (McGregor) and *T. pyri* Scheuten. Under laboratory conditions at 25 ± 2°C, 60 ± 5% RH and with a 16L:8D photoperiod, mated females of *T. pyri* consumed on average 15.7 eggs and 19.38 immature motile stages. Mated females of *N. californicus* consumed on average 28.53 eggs and 18.5 immature motile stages (Vargas *et al.*, 2005). On citrus and other host plants, *N. californicus* does not exert appropriate biological control of low flat mite populations (González, 2006).

Symptomatology and damage

The feeding activity of *B. chilensis* on grapevines induces a dark reddish discoloration, causing leaf drop and reduction in size of new growth. Spring infestation of young shoots causes the most severe damage (González, 1968b). Feeding injury of *B. chilensis* on citrus is not severe and not well known; on lemon, the damage consists of a roughened silvering of the rind (Ripa and Rodriguez, 1999). Infestations of kiwi fruit, clementines, lemons, mandarins, custard apples, figs and persimmons are well known, but do not represent a strong risk, with the exception of products for export to the USA and other countries. Pome and stone fruits are practically not infested (González, 2006).

Control

Natural enemies are unable to suppress outbreaks of *B. chilensis* (Hernera-Villamil, 1958). Physical and chemical means are suggested for the control of this mite on grapevines.

PHYSICAL CONTROL. Physical means of control include the use of irradiation and cold treatment, alone and in combination. Specimens of *B. chilensis* exposed to Cs137 at 0.25, 0.5 and 1 kGy and kept for 14 days at 4°C and then at room temperature did not exhibit total mortality during the 20 days following treatment (Sazo and Gerstle, 1989). The effects of 0.5, 1.0 and 1.5 kGy of gamma irradiation, combined with cold storage (24 h pre-cold and 15 days at 0–2°C and 90% RH), on egg viability, mortality and reproduction of *B. chilensis* were evaluated by Jadue *et al.* (1997). All treatments resulted in significantly greater adult mortality than the control, and higher dosages had stronger and more rapid effects. A few adults survived the 0.5 kGy treatment and produced eggs, but none of these hatched. There was total mortality of eggs, juveniles and adults after exposure to 0.5–1.5, 1.0–1.5 and 1.5 kGy, respectively (Jadue *et al.*, 1997). On 'Thompson Seedless' grapes, the irradiation doses required to cause 90% mortality of adults, nymphs and eggs were 1307, 970 and 328 Gy, respectively. An irradiation dose of 200 Gy combined with 15 days of cold treatment stopped mite reproduction. The organoleptic properties of 'Thompson Seedless' grapes irradiated at 600 Gy were not affected, nor were the ripeness parameters altered relative to soluble solids and acidity; no phytotoxicity was detected in either berries or the stem (Castro *et al.*, 2004).

Physical methods of control also include cold storage (15 days storage at 0–2°C and 90% RH), which delayed and reduced hatching of 1- to 4-day-old eggs, with lesser effects on eggs older than 4 days, probably because of their more developed embryos. Juveniles are highly susceptible to cold storage, but adults show a relatively lower susceptibility. Cold storage also reduces the oviposition of adult *B. chilensis* mites (Jadue *et al.*, 1996).

CHEMICAL CONTROL. Spray application of acaricides must be effected as early as possible in spring, and repeated every 15–20 days, taking into account that dormant oil sprays are not effective (Hernera-Villamil, 1958; Jeppson *et al.*, 1975). Effective acaricides include abamectin (González and Barria, 1999; González, 2001, 2006), acrinathrin (González, 2001, 2006), bifenthrin (González, 2006), chinomethionate (Curkovic *et al.*, 1994), chlorobenzilate (Jeppson *et al.*, 1975), dicofol (Jeppson *et al.*, 1975; González, 2006), petroleum oils (Curkovic *et al.*, 1994; Montano, 1995), propargite (González, 2006), pyridaben (Curkovic *et al.*, 1994), spirodiclofen (González, 2006) and sulfur (Jeppson *et al.*, 1975), applied in spring or in summer treatments. The use of some of the active ingredients cited is not permitted in various countries, and their use must adhere to the legislation in force in each country.

Brevipalpus lewisi McGregor (Fig. 21.4)

Common name

Citrus flat mite, flat mite, scab mite or bunch mite.

Diagnostic characteristics

FEMALE. *B. lewisi* belongs to the *Brevipalpus cuneatus* species group (*sensu* Baker and Tuttle, 1987). The body is ovoid and light reddish brown to bright red in colour, and about 246 μm long and 153 μm wide. The rostral shield has 2 long central rostral

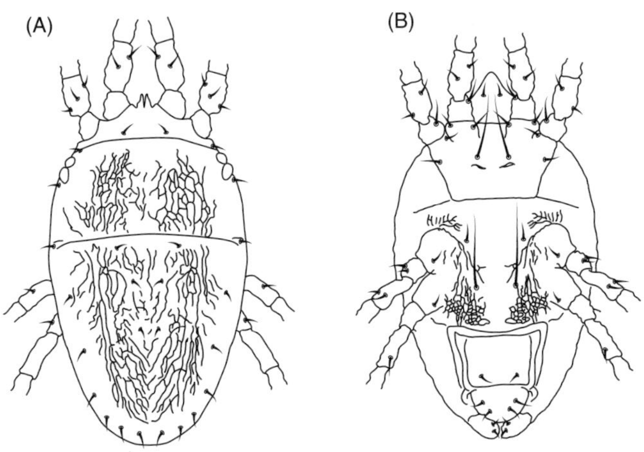

Fig. 21.4. *Brevipalpus lewisi* McGregor. Female: (A) dorsal view; (B) ventral view (from Livshits and Mitrofanov, 1967).

projections, and the medial projections are large but smaller and notched in some specimens. The prodorsum has 3 pairs of setae (v_2, sc_1, sc_2), short, slightly lanceolate and serrate, and 1 pair of eyes each side. The opisthosoma has 10 pairs of dorsal setae (c_1, c_3, d_1, d_3, e_1, e_3, f_2, f_3, h_1, h_2), short, slightly lanceolate and serrate, and 1 pair of opisthosomal pores. The prodorsum is variably wrinkled or folded centrally, appearing as areolae; laterally, longitudinal elongate cells form a reticulation, and some lateral elements may be fused or open dorsally. The opisthosoma cuticle between the setae c_1–c_1 and e_1–e_1 is smooth to wrinkled; the cuticle posterior e_1–e_1 shows a series of weak to 'V'-shaped folds, and the posterior sublateral cells form distinct bands. The ventral plate has transversely aligned elongate cells, sometimes fusing to form bands. The genital plate has transversely aligned elongate cells, usually forming bands. The spermatheca has a long and strong duct ending in a small rounded vesicle with a series of short finger-like projections around the dorsal surface. Tarsus II has 1 solenidion (Baker, 1949; Beard *et al.*, 2013).

MALE. The male is similar to the female, with the idiosomal dorsal setae shorter than the longitudinal intervals between them. The propodosoma is without mediodorsal reticulations. Hysterosomal pores are present and strongly tubular (Pritchard and Baker, 1958).

Geographical distribution

Australia (Beard *et al.*, 2013), Israel (Meyer Smith and Gerson, 1981), Mexico, the USA (Arizona, California, Washington) and the former Yugoslavia (Beard *et al.*, 2013).

Host plants

Only host plants based on specimens of *B. lewisi* confirmed by Beard *et al.* (2013) are reported here: Anacardiaceae – pistachio (*Pistacia vera* L.); Ebenaceae – *Diospyros* sp.; Juglandaceae – walnut (*Juglans* sp.); Rutaceae – *Citrus* sp., mandarin (*C. reticulata* Blanco), lemon (*C. limon* (L.) Burm. f.); Vitaceae – boston ivy (*Parthenocissus tricuspidata* (Siebold *et* Zucc.) Planch.), grapevines (*Vitis vinifera* L.).

Bio-ecology

Before the study of Beard *et al.* (2013), in addition to the host plants reported above, the list of *B. lewisi* host plants also included pomegranates (Lythraceae – *Punica* sp.) (Ebeling and Pence, 1949).

The most common infestations by the citrus flat mite reported are on citrus and grapevines. Although the mite attacks the twigs and leaves of citrus, it prefers green to ripe fruit, and indeed fruit at any stage, to the leaves. The highest population densities may occur at the stem end of citrus fruit, near and under the button. The female lays single eggs in the cracks and crevices of fruits, twigs and leaves. In central California, *B. lewisi* overwinters as an adult, but in Imperial Valley and Coachella Valley in California, it is constantly active throughout the year on citrus; on grapevines, it is inactive during the winter. Population peaks are reached in the warmest months because high temperatures and low RH do not have a negative effect on populations of *B. lewisi*. On grapevines, the mite populations live on the green organs of the vine during the summer, and move to the stems to overwinter in cracks of the bark in autumn. In Spain, the citrus flat mite moves from overwintering sites when temperatures reach above 20°C. The winter females migrate to the growing grapevine shoots over a period of 1–2 months, finishing in mid-May. Four generations develop between May and September. The most dangerous period for fruit production is at the beginning of June, when mites colonize the nodes where the bunches are situated (Rodriguez *et al.*, 1987).

In laboratory studies on the leaves of *V. champini* Planchon (cv. Dog Ridge), the development of immature stages took from 16.8 days at 34°C and 35% RH up to 27.9 days at 22°C and 70% RH. The mean fecundity ranged from 5.7 eggs per female a day at 34°C and 35% RH to 18.8 eggs per female a day at 22°C and 70% RH. The maximum value of r_m was 0.04 day^{-1} at 28°C and 35% RH. The optimal net reproductive rate (4.82) was at 22°C and 70% RH. A field study of seasonal abundance and intra-vine distribution of the mite showed that populations increased about 60 times during a season (Buchanan *et al.*, 1980).

Natural enemies include various phytoseiid mites, such as *Euseius victoriensis* (Womersley) (James and Whitney, 1993), *Phytoseius plumifer* (Canestrini *et* Fanzago) (Yousef, 1970), *Metaseiulus mcgregori* (Chant) (Rice and Weinberger, 1981), *Neoseiulus reticulatus* (Oudemans) (Buchanan *et al.*, 1980), *Typhlodromus pyri* Scheuten (Yousef, 1970), *T. doreenae* Schicha (Smith and Papacek, 1991; James and Whitney, 1993), some tydeid, such as *Homeopronematus anconai* (Baker) (Hessein and Perring, 1988) and stigmaeid mites (*Agistemus exsertus* González) (Yousef, 1970).

Symptomatology and damage

On citrus, *B. lewisi* lives on green fruits, which are preferred to ripe fruits and leaves, and the populations are abundant at the stem end of the fruit or under the button. Attacks frequently occur on areas of fruits injured by leafhoppers (Cicadellidae) or citrus thrips (Thysanoptera). On ripe fruits, these areas normally become inconspicuous, but if they are infested they develop in conspicuous scab-like isolated depressions where the mites live and the rind-blemished areas evolve as scars. High population densities may cause large alterations on the surface of the fruit. The damage may produce a total reduction of the quality of fruit. The mite does not cause damage to leaves or wood and the scab-like scars observed on most varieties of citrus rarely occur on grapefruit (Elmer and Jeppson, 1957). Infestation of tangerines by *B. lewisi* causes grade-reducing scarring on 17–28% of the fruit, with figures of 16–21% on navel oranges and 18–35% on grapefruits (Elmer, 1968). On 'Lisbon' lemon fruits, scarring may affect over 25% of the fruits (Lewis, 1949). Nowadays, *B. lewisi* is often an economically important pest of citrus (Kerns *et al.*, 2001). There are also reports of 50–90% of pomegranates being damaged by *B. lewisi*.

The feeding activity of *B. lewisi* occurs on the green parts of vines, and prevents the development of the berries. On grapevines in Bulgaria, up to 30% loss in yield has been reported (Raikov and Nachev, 1965). The mite could lower productivity for a number of crops.

Control

In south-eastern Australia, *B. lewisi* is controlled on grapevines by the phytoseiids *T. doreenae* and *E. victoriensis*, thus facilitating a reduction in acaricide applications (James and Whitney, 1993). Unfortunately, natural enemies do not always guarantee sufficient control of infestations.

CHEMICAL CONTROL. Acaricides applied in the control of *B. lewisi* include dicofol, chlorobenzilate (Jeppson *et al.*, 1975; Telliev and Geshev, 1980; Goodwin, 1982; Kerns, 2006), spirodiclofen (Kerns, 2006), petroleum oils (Goodwin, 1982), organosulfurs (Telliev and Geshev, 1980; Goodwin, 1982) and sulfur (Chakrov, 1973; Goodwin, 1982), sprayed alone or in various mixtures. On citrus, *B. lewisi* is very susceptible to sulfur, and the best results are obtained with applications in late winter and early spring; during the summer, plant damage may be prevented by dicofol and chlorobenzilate sprays (Jeppson *et al.*, 1975). In Spain, chemical treatments are recommended on grapes before June (Rodriguez *et al.*, 1987). The use of some of the above active ingredients is not allowed in certain countries.

Brevipalpus obovatus Donnadieu (Fig. 21.5)

Common name

Ornamental flat mite, privet mite.

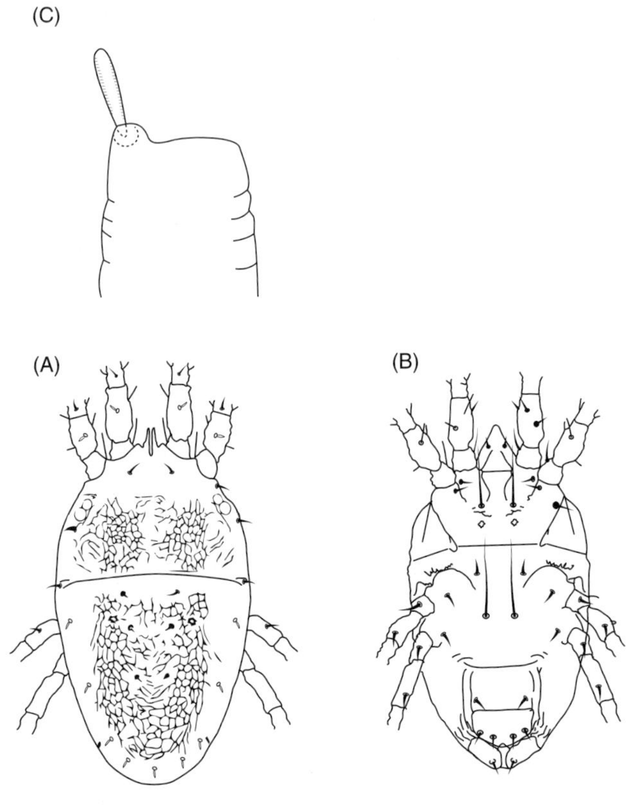

Fig. 21.5. *Brevipalpus obovatus* Donnadieu. Female: (A) dorsal view; (B) ventral view (from Livshits and Mitrofanov, 1967); (C) solenidion of tarsus II (drawing by Vacante).

Diagnostic characteristics

FEMALE. This species identifies the *B. obovatus* species group (*sensu* Baker and Tuttle, 1987). The body is 250–300 μm long, light orange to dark red in colour with various patterns of dark pigmentation. The rostral shield does not extend beyond the base of femur I, and is blunt and with a shallow division between the central and medial projections. The prodorsum has 3 pairs of marginal setae including 1 pair of vertical setae (v_2) and 2 pairs of scapular setae (sc_1, sc_2) that are lanceolate serrate, 1 pair of eyes each side, and 1 pair of prodorsal pores. The central area of the prodorsum changes from smooth to rugose, the medial area is uniformly reticulate, and the lateral area from smooth to rugose. The opisthosoma bears 9 pairs of dorsal setae (c_1, c_3, d_1, d_3, e_1, e_3, f_3, h_1, h_2) that are lanceolate serrate, and 1 pair of opisthosomal pores. The prodorsum is mostly smooth centrally, or with weak wrinkles; it is strongly reticulate laterally, with large cells posteriorly and small cells anteriorly. The dorsal opisthosoma cuticle between the setae c_1–c_1 and e_1–e_1 is weakly reticulate to weakly wrinkled; the cuticle between and posterior to setae e_1–e_1 has a series of short, transverse folds; the posterior sublateral cuticle has cells present to the body margin. The ventral plate has small rounded cells, with the central cells often fusing to form weak transverse bands. The genital plate has medium to large rounded cells, sometimes transversely aligned. The spermatheca has a long duct ending in a small rounded vesicle covered in short finger-like projections around the entire surface. Tarsus II has 1 solenidion (Jeppson *et al.*, 1975; Beard *et al.*, 2013).

MALE. The male is similar in colour to the female and smaller. The dorsomedian area of the propodosoma is irregularly striate or reticulate. The dorsocentral hysterosomal setae are very short, and the opisthosoma has 4 pairs of dorsolateral setae (Pritchard and Baker, 1958).

Geographical distribution

Argentina, Australia, Brazil, Canada, Chile, China, Colombia, Costa Rica, Democratic Republic of the Congo, Ecuador, France, Germany, Hungary, India, Iran, Israel, Italy, Jamaica, Japan, Korea, Mauritius, Mexico, Pakistan, Papua New Guinea, Thailand, Philippines, South Africa, Taiwan, the USA (California, Missouri, New York, South Carolina, Tennessee) and Venezuela (Beard *et al.*, 2013).

Host plants

The ornamental flat mite lives on a wide range of wild and cultivated host plants. Childers *et al.* (2003b) reported 451 plants belonging to 119 genera and 55 botanical families, many of which are herbs, ornamentals or shrubs. Only host plants based on specimens of *B. obovatus* confirmed by Beard *et al.* (2013) are reported here: Adoxaceae – *Viburnum* sp.; Aracaeae – *Alocasia* sp., *Colocasia* sp.; Araliaceae – *Hedera* sp.; Arecaceae – *Phoenix roebelenii* O'Brien; Asteraceae – *Chamaemelum* sp., *Chrysanthemum* sp., *Gerbera* sp., *Solidago* sp.; Berberidaceae – *Epimedium acuminata* Franch.; Bignoniaceae – *Bignonia* sp.; Buxaceae – *Buxus* sp.; Cucurbitaceae – *Cucurbita* sp.; Ebenaceae – *Diospyros* sp.;

Ericaceae – *Rhododendron* sp.; Euphorbiaceae – *Andrachne* sp.; Lamiaceae – *Mentha arvensis* L., *Origanum* sp., *Thymus* sp.; Lecythidaceae – *Barringtonia racemosa* (L.) Spreng.; Liliaceae – tulip bulbs; Malvaceae – cotton (*Gossypium herbaceum* L.), *Hibiscus* sp.; Moraceae – fig (*Ficus* sp.); Myrtaceae – *Callistemon* sp., *Kunzea baxteri* (Klotzsch) Schauer; Oleaceae – *Ligustrum* sp., *L. amurense* Carrière; Phytolaccaceae – *Phytolacca* sp.; Pittosporaceae – *Pittosporum undulatum* Vent.; Rosaceae – loquat (*Eriobotrya japonica* (Thunberg) Lindley), *Rubus* sp.; Rutaceae – *Citrus* sp., lemon (*C. limon* (L.) Burm.f.); Solanaceae – tomato (*Solanum esculentum* Dunal), potato (*S. tuberosum* L.); Theaceae – tea (*Camellia chinensis* (L.) Kuntze); Verbenaceae – *Citharexylum quadrangulare* Jacq., *Verbena bonariensis* L.; Vitaceae – grapes (*Vitis vinifera* L.) (Beard *et al.*, 2013).

Bio-ecology

The economic importance of *B. obovatus* seems to be restricted to citrus and ornamentals, and despite its wide range of host plants, it is not of economic importance on other important host plants. The mite reproduces by thelytokous parthenogenesis owing to the presence of feminizing bacterial symbionts of the genus *Cardinium* sp. Females are haploid with two chromosomes, and males are rare (Helle and Bolland, 1972; Helle *et al.*, 1980); the male mates with but does not fertilize the female, and some specimens that seem to be clearly males are intersex (Pijnacker *et al.*, 1981; Weeks *et al.*, 2001). Furthermore, some clones of *B. obovatus* contain neither *Cardinium* nor any other bacteria that could induce thelytoky. Females treated with antibiotics generate no males, suggesting that in these lines the thelytoky is probably a genetic property of the species itself (Groot and Breeuwer, 2006). Different haplotypes of *B. californicus s.l.* and *B. obovatus* are known, which although morphologically identified as these two species, differences are known in the DNA samples from the populations. These same populations showed differential capabilities to colonize different host plants, showing a closed relationship between the mite and the plant that tended towards specialization, as reported by Groot *et al.* (2005).

Populations of *B. obovatus* usually live on the lower leaf surfaces, the stems and the petioles of host plants, and overwinter as adults aggregated around the base of plants and/or in sheltered sites of the lower leaf surfaces. Eggs and immature stages may sometimes occur during the winter. In greenhouses and in mild climates, the mite reproduces throughout the year (Morishita, 1954). Jeppson *et al.* (1975) reported that the best temperatures for development were between 20 and 27°C. Goyal *et al.* (1985) investigated *B. obovatus* reared on *Solidago canadensis* L. in the laboratory at 15, 20, 25, 30 and 35°C, and found that the greatest development was at 30°C, the least at 20°C, with 15 and 35°C unsuitable. The most suitable temperature for development was 25°C, as oviposition period, fecundity and rate of hatching were all high, while mortality was low. In the laboratory on *Azalea* sp. at 23 ± 1°C and 60 ± 5% RH, the life cycle from egg to adult was 27.8 days; at 27°C, it was 21.5 days (Trindade and Chiavegato, 1994). On citrus leaves at 25 ± 2°C and 65 ± 5% RH, egg incubation lasted 3.07 days and hatching was 87.49%. The larval, protonymphal and deutonymphal stages took 2.76, 2.63 and 3.21 days, respectively (Rezk, 2001). At 20 and 27°C, the fecundity was three and 54.3 eggs, and the adult lifespan 67 and 38.1 days, respectively (Jeppson *et al.*, 1975), while at 25 ± 2°C and 65 ± 5% RH, the lifespan was 34.26 days (Rezk, 2001). In Egypt, the main population peaks on citrus have been observed in early spring and autumn, when there is an average of 0.52 ± 0.04 and 0.61 ± 0.02 mites per leaf, respectively (Rezk, 2001).

Pathogens and predatory mites develop on *B. obovatus* populations. The first include the acaropathogenic fungus *Hirsutella thompsonii* Fisher (Rosas Acevedo and Sampedro Rosas, 2000). The predatory mites number the phytoseiids *Euseius scutalis* (Athias Henriot) (Donia *et al.*, 1995) and *Neoseiulus idaeus* Denmark *et* Muma (Tamai *et al.*, 1997)], the tydeid *Tydeus californicus* (Banks), the stigmaeid *Agistemus exsertus* González and the cheyletid *Cheletogenes ornatus* (Canestrini *et* Fanzago) (Rizk *et al.*, 1983; Rezk and Gadelhak, 1996)].

Symptomatology and damage

B. obovatus infests the leaves, fruits and twigs of citrus. Its feeding activity removes the cell contents, kills the cells and injects toxic saliva into the tissues. On sweet orange leaves, this causes large chlorotic spots (Morishita, 1954). The symptoms of attack on the leaves and fruits are similar to that which has been illustrated in Texas for *B. californicus* (Banks) and *B. phoenicis* (Geijskes) (Dean and Maxwell, 1967; French and Rakha, 1994; Childers, 1994a; Childers *et al.*, 2003a). In the USA (Jeppson *et al.*, 1975), South Africa (Meyer Smith and Schwartz, 1998d) and the Mediterranean region the ornamental flat mite does not cause severe damage to citrus (Vacante, 2010; Gerson and Vacante, 2012).

On fuchsia, the feeding activity of *B. obovatus* produces faint brown flecks, consisting of sunken spots where the cell contents have been removed or the cells have been killed by the toxic saliva injected by feeding. As mite feeding continues, the flecks merge and the leaves become brownish or bronzy. High population densities of the mite induce red discoloration of the upper leaf surface, followed by leaf drop. Feeding of *B. obovatus* on anthuriums causes yellow areas on the upper leaf surfaces and brown areas on the lower leaf surfaces (Jeppson *et al.*, 1975).

In Argentina on leaves of sweet orange, feeding by *B. obovatus* has been associated with the appearance of large chlorotic spots with concentric rings of chestnut-coloured resinous substance covering up to a third of the leaf surface. Similar but smaller alterations occurred on the fruits, and ring spots are also known on twigs. At later stages of infestation, these alterations erupt to produce scaling of the bark like a scaly bark psorosis. This injury was referred to as 'lepra explosiva' or 'leprosis' (Vergani, 1945). In Venezuela, the ornamental flat mite is associated with 'halo scab' on the leaves of sour orange seedlings grown in humid structures. When *B. obovatus* and *B. phoenicis* are present on the same plant, the injury to the leaves and stems is more severe and primarily due to those two species (Knorr *et al.*, 1960). Both *B. californicus* in Florida and *B. obovatus* in Argentina were reported as vectors of citrus leprosis (Frezzi, 1940; Vergani, 1945; Knorr, 1968). However, morphological studies of the specimens collected from these earlier instances of the disease showed the presence of both *B. californicus* and *B. phoenicis* in the Florida samples, and of *B. obovatus* and *B. phoenicis* in the Argentinian samples (Ochoa, 2002, in Childers and Rodrigues, 2011). Therefore, the vector status of *B. obovatus* remains unclear based on the earlier reports. Recently, *B. californicus* and *B. phoenicis* have been reported as vectors of citrus leprosis

in Guatemala (Palmieri *et al.*, 2007). However, no direct transmission tests have been published to prove that *B. californicus* is a vector. Furthermore, there are no available voucher specimens for either *B. phoenicis* or *B. californicus* from citrus samples collected in Guatemala. Therefore, the only confirmed vector of citrus leprosis in the western hemisphere to date is *B. phoenicis* (Childers and Rodrigues, 2011).

Both *B. obovatus* and *B. phoenicis* have been identified as vectors of *Cestrum ringspot virus* (CeRSV) on lady of the night (*Cestrum nocturnum* L.), an ornamental of the Solanaceae family, in Brazil. This is a nuclear-type virus causing chlorotic ring spots on the leaves. In addition, *Solanum violaefolium* ringspot virus (SvRSV) was shown to be transmitted by both *B. obovatus* and *B. phoenicis*. This cytoplasmic type of virus caused ring spots on the leaves and was also found in Brazil (Kitajima *et al.*, 2010).

Control

Natural enemies alone are unable to control *B. obovatus*. Therefore, the control of this mite is entrusted to spray applications of acaricides, including organohalogens (Rezk, 2001), organochlorines (Jeppson *et al.*, 1975; Childers, 1994a; Rezk and Gadelhak, 1996), organotins (Rezk, 2001), petroleum oils, pyridazinones or high rates of sulfur (Jeppson *et al.*, 1975; Childers, 1994a), and pyrazoles (Rezk, 2001). Note that the use of acaricides must be permitted by the regulations of each country concerned.

Brevipalpus phoenicis (Geijskes) (see Beard *et al.*, 2013)

Common name

Red and black flat mite, reddish black flat mite, leprosis mite, scarlet mite, red crevice mite.

Diagnostic characteristics

The dorsal propodosomal reticulation pattern and the length of some of the dorsal body setae of *B. phoenicis* show some variability. Manson (1967c) noted that this variation might depend on the host plant. However, González (1975) observed this type of variation on specimens collected on the same host plant and concluded that normal variations might occur among populations. He also found other small variations on the tarsal setae on tarsus II and in the dorsal body setae of the nymph, and concluded that the phenomenon could be due to mixed populations of *B. phoenicis* and *B. californicus* (Banks), age of specimens and asymmetry in size. Variations in *B. phoenicis* nymphs have been discussed by various authors, and a detailed list of reports was presented by Welbourn *et al.* (2003), but as yet there has been no comprehensive morphological review or application of molecular analysis that has been able to construct a reliable systematic framework to help to distinguish the different species with certainty.

The body of the female of *B. phoenicis s.l.* is oval, about 300 µm long, and red coloured with a central dark pattern in old specimens (Meyer Smith, 1981a). The male is similar to the female, but smaller and rare (Pritchard and Baker, 1958). Specific characteristics are described below. Beard *et al.* (2013) recently revised the external morphology of several specimens and questioned the common concept of *B. phoenicis* that had been adopted hitherto. They distinguished within *B. phoenicis* seven groups of species, designated as groups A–G, that are waiting to be appropriately described. The main characteristics of adult females used by Beard *et al.* (2013) to distinguish the seven groups of species are reported below. When strictly necessary, the mite is referred to as *B. phoenicis s.l.*

BREVIPALPUS PHOENICIS GROUP SPECIES A. The cuticle of the central prodorsum has strong and broad areolae; the sublateral prodorsum is anteriorly mostly smooth, medially it has small rounded cells, and posteriorly it is reticulate and forms broad rounded cells. The opisthosomal cuticle from the setae c_1–c_1 to d_1–d_1 is strongly wrinkled, and can be weakly areolate; from the setae d_1–d_1 to e_1–e_1, it is wrinkled, with a few strong transverse folds; from the setae e_1–e_1 to h_1–h_1, it has transverse folds, becoming weakly reticulate towards the setae h_1–h_1. The cuticle of the sublateral opisthosoma is reticulate, forming large, uniform and rounded cells. The dorsal microplates consist of distinct individual, rounded to irregularly rounded plates, with irregular ridges on the dorsal surface, and no series of parallel ridges. Ventrally, the cuticle from the setae *4a* to the ventral plate is laterally verrucose, with separately formed rounded 'warts'; centrally, the cuticle has raised transverse bands formed by fused warts, becoming weaker towards the setae *4a*–*4a* and disappearing; the central cuticle lacks separately formed warts. The cucticle of the ventral plate is weakly verrucose; the warts are fused to form transverse bands; there are no separately formed warts centrally, but often some separately formed warts are set laterally. The genital plate is uniformly verrucose to verrucose–reticulate, with large cells formed by fused or associated warts. The spermatheca has a moderately thick convoluted duct ending in a membranous bulb.

BREVIPALPUS PHOENICIS GROUP SPECIES B. The cuticle of the central prodorsum has strong areolae; the sublateral prodorsum is anteriorly weakly reticulate to the setae v_2, and posteriorly reticulate, forming large cells. The opisthosomal cuticle from the setae c_1–c_1 to d_1–d_1 is smooth to weakly reticulate; from the setae d_1–d_1 to e_1–e_1, it is weakly reticulate and/or wrinkled, and from the setae e_1–e_1 to h_1–h_1, it usually has strong 'V'-shaped folds, becoming weaker towards the setae h_1. The cuticle of the sublateral opisthosoma is reticulate with regular cells, becoming longitudinally elongate towards the posterior. The dorsal microplates consist of distinct individual oval plates, with a series of distinct parallel ridges. Ventrally, the cuticle from the setae *4a* to the ventral plate is uniformly verrucose, with separately formed rounded warts, often weaker centrally. The cuticle of the ventral plate is uniformly verrucose, with separately formed small rounded warts. The genital plate is uniformly verrucose to verrucose–reticulate, with large cells formed by fused or associated warts. The spermathecal duct is long, narrow and convoluted, and the distal vesicle is sclerotized and oval with a thick distal stipe.

BREVIPALPUS PHOENICIS GROUP SPECIES C. The cuticle of the central prodorsum has strong areolae; anteriorly, the sublateral prodorsum has a large and smooth region and a narrow band that is weakly reticulate; posteriorly, there is some reticulation forming a few large cells. The opisthosomal cuticle from the setae c_1–c_1 to d_1–d_1, is smooth or weakly wrinkled, and with a

pair of distinct oblique folds near the setae d_1; from the setae d_1–d_1 to e_1–e_1, it is smooth or with weak irregular folds, with a few strong transverse folds, and from the setae e_1–e_1 to h_1–h_1, it has several transverse folds, abruptly becoming longitudinal folds towards the setae h_1. The cuticle of the sublateral opisthosoma is reticulate with large uniform cells (some rounded), with the lateral cells becoming longitudinally elongate towards the posterior. Ventrally, the cuticle from the setae *4a* to the ventral plate is laterally verrucose, with separately formed rounded warts; centrally it has transverse bands formed from fused warts. The cuticle of the ventral plate has bands of various orientation, with lateral and posterior bands tending to be transverse, and central bands tending to be oblique. The genital plate has irregular, narrow transverse bands. The spermathecal duct is long and moderately thick, and the vesicle is sclerotized, spherical and with a crown of minute projections.

BREVIPALPUS PHOENICIS GROUP SPECIES D. The cuticle of the central prodorsum has strong and broad areolae; anteriorly, the sublateral prodorsum has a large smooth region; medially, it has distinct small rounded cells, and posteriorly it is reticulate and forms a few rounded cells. The cuticle of the opisthosoma from the setae c_1–c_1 to d_1–d_1 and from d_1–d_1 to e_1–e_1 is almost entirely smooth, with some weak wrinkles medially; from the setae e_1–e_1 to h_1–h_1, there is a series of strong transverse folds, abruptly becoming smooth towards the setae h_1–h_1. The cuticle of the sublateral opisthosoma is anteriorly smooth, and posteriorly has large, distinctly rounded cells. Ventrally, the cuticle from the setae *4a* to the ventral plate is verrucose, with separately formed rounded warts; centrally, some weak transverse bands (formed by fused warts) may be present. The cucticle of the ventral plate shows weak transverse bands, without separately formed warts. The cuticle of the genital plate has uniform narrow transverse bands. The spermathecal duct is moderately thick.

BREVIPALPUS PHOENICIS GROUP SPECIES E. The cuticle of the central prodorsum has strong and broad areolae; the sublateral prodorsum is anteriorly mostly smooth, but posteriorly reticulate and forming broad rounded cells. The opisthosomal cuticle from the setae c_1–c_1 to d_1–d_1 is smooth to wrinkled; from the setae d_1–d_1 to e_1–e_1, it has irregular folds, and from the setae e_1–e_1 to h_1–h_1, it shows a short series of transverse folds, which become longitudinal folds towards the setae h_1–h_1. The cuticle of the sublateral opisthosoma is reticulate, with large, uniform, distinctly rounded cells, somewhat elongate towards the setae h_1. The dorsal microplates consist of distinct individual, rounded to irregularly rounded plates, with short irregular ridges on the dorsal surface arranged in apparently random directions, and no parallel ridges. Ventrally, the cuticle from the setae *4a* to the ventral plate is verrucose, with separately formed rounded warts; centrally, the cuticle shows transverse bands (formed by fused warts), which may be weakly developed or broken into elongate warts (formed by fused warts); the central cuticle lacks separately formed warts. The cuticle of the ventral plate is weakly verrucose, with few or no separately formed warts and fused to form transverse bands. The genital plate is uniformly verrucose or verrucose–reticulate, with large transversely elongate cells formed by fused or associated warts. The spermathecal duct is long, ending distally in a small rounded vesicle with a weakly formed crown of short projections.

BREVIPALPUS PHOENICIS GROUP SPECIES F. The cuticle of the central prodorsum has wrinkles or fine folds, longitudinally directed and lacking areolae. Anteriorly, the sublateral prodorsum shows weak reticulation becoming folds; posteriorly, it is reticulate and forms large cells. The opisthosomal cuticle from the setae c_1–c_1 to d_1–d_1 has strong wrinkles and folds, and may be weakly reticulate; from the setae d_1–d_1 to e_1–e_1, it has strong wrinkles and folds, becoming a series of short, narrow transverse folds, and from the setae e_1–e_1 to h_1–h_1, it has series of short, narrow transverse folds, becoming weakly reticulate towards the setae h_1. The cuticle of the sublateral opisthosoma is anteriorly reticulate with regular cells, becoming longitudinally elongate towards the posterior. The dorsal microplates consists of distinct individual rounded to oblong plates, with a series of distinct parallel ridges on the dorsal surface arranged in various directions. Ventrally, the cucticle from the setae *4a* to the ventral plate is verrucose, with separately formed rounded warts; centrally, the cuticle often shows fused warts forming elongate warts. The cuticle of the ventral plate is weakly verrucose or weakly reticulate with rounded to transversely elongate cells; the central cuticle shows some fused warts forming a few weak transverse bands. The genital plate has uniform narrow transverse bands. The spermathecal duct is long, ending in an elongate membranous bulb.

BREVIPALPUS PHOENICIS GROUP SPECIES G. The cuticle of the central prodorsum has strong and broad areolae. Anteriorly, the sublateral prodorsum shows short folds and wrinkles to setae v_2, and posteriorly it is reticulate and forms large cells. The cuticle of the opisthosoma from the setate c_1–c_1 to d_1–d_1 has some folds or wrinkles, and can be weakly reticulate; from the setae d_1–d_1 to e_1–e_1, it is wrinkled with a few transverse folds, and from the setae e_1–e_1 to h_1–h_1, it has a series of transverse folds, becoming weakly reticulate towards the setae h_1–h_1. The cuticle of the sublateral opisthosoma is reticulate with large cells, slightly longitudinally elongate. Ventrally, the cuticle from the setae *4a* to the ventral plate is laterally verrucose, with separately formed rounded warts (with some fusion of cells); centrally, the cuticle has transversely elongate warts, often fused to form short transverse bands; the elongate warts and transverse bands become weaker towards the setae *4a*–*4a* and disappear between the setae *4a*–*4a*. The cuticle of the ventral plate has irregularly shaped cells or warts, with the central cells more elongate than the lateral cells. The cuticle of the genital plate has irregular transversely elongate cells, forming irregular transverse bands. The spermathecal duct is fine, ending distally in a rounded vesicle, the outer margin of which has tiny projections.

Geographical distribution

The geographical distribution of *B. phoenicis s.l.* is worldwide, but here only the distribution based on specimens confirmed by Beard *et al.* (2013) are reported.

BREVIPALPUS PHOENICIS GROUP SPECIES A. Brazil, Costa Rica, The Netherlands, South Africa and the USA.

BREVIPALPUS PHOENICIS GROUP SPECIES B. Argentina, Australia, Bangladesh, Belgian Congo, Brazil, Burma, China, Colombia, Costa Rica, Cuba, Dominican Republic, Ecuador, El Salvador,

Ethiopia, France, Guatemala, Honduras, India, Indonesia, Israel, Malaysia, Mexico, Nigeria, Pakistan, Philippines, Puerto Rico, Spain, Sri Lanka, Thailand, Trinidad, the USA and Venezuela.

BREVIPALPUS PHOENICIS GROUP SPECIES C. The USA.

BREVIPALPUS PHOENICIS GROUP SPECIES D. Azores, Canary Islands, Democratic Republic of the Congo, Guatemala, Honduras, India, Portugal, Rwanda and Burundi, South Africa, Spain and the USA.

BREVIPALPUS PHOENICIS GROUP SPECIES E. Brazil.

BREVIPALPUS PHOENICIS GROUP SPECIES F. Spain.

BREVIPALPUS PHOENICIS GROUP SPECIES G. Argentina.

Host plants

The list of host plants of *B. phoenicis s.l.* includes almost 486 species belonging to 118 genera and 64 botanical families (Childers *et al.*, 2003b). Only host plants based on specimens of *B. phoenicis* confirmed by Beard *et al.* (2013) are reported here.

BREVIPALPUS PHOENICIS GROUP SPECIES A. Arecaceae – *Phoenix canariensis* H. Wildpret; Bixaceae – *Bixa orellana* L.; Lauraceae – *Cinnamomum camphora* (L.) J. Presl; Rubiaceae – coffee (*Coffea arabica* L.); Rutaceae – *Citrus* sp., Persian lime (*C. latifolia* (Yu. Tanaka) Tanaka); Solanaceae – *Datura stramonium* L.

BREVIPALPUS PHOENICIS GROUP SPECIES B. Many host plants of this mite are of economic importance (Beard *et al.*, in preparation). Among these, the Rutaceae is the most important family. The species include: Arecaceae – coconut (*Cocos nucifera* L.); Asteraceae – sunflower (*Helianthus annuus* L.); Caricaceae – papaya (*Carica papaya* L.); Juglandaceae – walnut (*Juglans regia* L.); Lauraceae – avocado (*Persea americana* Mill.); Malvaceae – okra, hibiscus (flowers) (*Hibiscus* spp.); Moraceae – fig (leaves), *Ficus* spp.; Musaceae – banana, *Musa* spp.; Myrtaceae – guava (*Psidium guajava* L.); Passifloraceae – passion fruit, *Passiflora* sp.; Rosaceae – loquat (*Eriobotrya japonica* (Thunb.) Lindl.); Rutaceae: *Citrus* spp., key lime (*C. aurantifolia* (Christm.) Swingle), lemon (*C. limon* (L.) Burm. f.), citron (*C. medica* L.), mandarin (*C. reticulata* Blanco), sweet orange (*C. sinensis* (L.) Osbeck), Japanese bitter orange (*Poncirus trifoliata* (L.) Raf.); Theaceae – tea (*Camelia sinensis* (L.) Kuntze); Vitaceae – grapes, *Vitis* spp. Other host families include the Amaryliidaceae, Apocynaceae, Araceae, Bignoniaceae, Euphorbiaceae, Fabaceae, Oleaceae, Rubiaceae and Sapindaceae.

BREVIPALPUS PHOENICIS GROUP SPECIES C. Caricaceae – papaya (*Carica papaya* L.).

BREVIPALPUS PHOENICIS GROUP SPECIES D. The Rutaceae is the most important host family. The species include: Araliaceae – ivy (*Hedera* sp.); Musaceae – banana (*Musa* spp.); Rubiaceae – large leaf quinine (*Chinchona* sp.); Rutaceae – *Citrus* spp., sour orange (*C. aurantium* L.), mandarin (*C. reticulata* Blanco) and bitter orange, lemon and tangerine; Solanaceae – *Datura stramonium* L.; Theaceae – tea (*Camelia sinensis* (L.) Kuntze).

BREVIPALPUS PHOENICIS GROUP SPECIES E. Euphorbiaceae – *Croton floribundus* Spreng.

BREVIPALPUS PHOENICIS GROUP SPECIES F. Scrophulariaceae – *Myoporum laetum* G. Forst.

BREVIPALPUS PHOENICIS GROUP SPECIES G. Rutaceae sweet orange (*Citrus sinensis* (L.) Osbeck).

Bio-ecology

B. phoenicis s.l. frequently lives in association with *B. californicus* (Banks) and *B. obovatus* Donnadieu on the leaves, fruits and twigs of different citrus species, and together with the latter species has been reported as an important citrus pest in different areas of the world (Lewis, 1949; Dean and Maxwell, 1967; Jeppson *et al.*, 1975; Schwartz, 1977b; Childers, 1994a; French and Rakha, 1994; Ochoa *et al.*, 1994; Childers *et al.*, 2003a). The mite is also known as a pest of coffee in Brazil (Chagas *et al.*, 2001), tea in Asia (Oomen, 1982), passion fruit (Kitajima *et al.*, 1997) and various ornamental plants worldwide (Jeppson *et al.*, 1975). On citrus, *B. phoenicis* populations usually live on fruits and the lower surface of leaves, where they aggregate along the mid-vein or major lateral veins; they normally move to another plant when resources become limited (Haramoto, 1969).

B. phoenicis s.l. reproduces by thelytokous parthenogenesis. Its haploid condition is led by an endosymbiont bacterium of the genus *Wolbachia* (Otto and Jarne, 2001), and as in other *Brevipalpus* spp., the bacterium *Cardinium* sp. induces thelytoky by feminizing unfertilized haploid eggs (Groot and Breeuwer, 2006). The haploid females have $n = 2$ chromosomes (Helle *et al.*, 1980; Pijnacker *et al.*, 1981; Weeks *et al.*, 2001), and males are rare (Gope and Das, 1992).

On leaves and fruits of 'Valencia' and 'Pera Rio' orange under laboratory conditions, at 25°C, 60% RH and with a 14L:10D photoperiod, the pre-oviposition period lasted 2.34 days, the females laid an average of 22.5 eggs each, the mean incubation period was 7.7 days, the developmental period of larvae and nymphs lasted 1.6 and 9.9 days, respectively, and the adult lifespan was 22.2 days (Chiavegato, 1986). On fruits of key lime, the thresholds for egg and post-embryonic development and the pre-oviposition period are 9.1, 15.5 and 11°C, respectively, and the sums of effective temperatures for the completion of the egg stage, post-embryonic development, pre-oviposition period and complete life cycle is 135, 145, 54 and 330.4 degree-days, respectively. Under constant temperatures, the average length of a generation ranges from 27.5 days at 24°C to 18.3 days at 30°C, while in naturally fluctuating temperatures, it varies from 30.8 days at 21–23°C to 19.4 days at 30–32°C (Prieto Trueba, 1975). In the laboratory, the life cycle on citrus took 21 days at 27.5°C and 96 days at 13.8°C. In Egypt, there are about ten generations a year (Zaher *et al.*, 1970).

Under constant experimental conditions, *B. phoenicis s.l.* had a low growth rate ($r_m = 0.127$) and high rates of survival;

the net reproduction rate was 33.2 eggs mite^{-1}, the generation time was 27.6 days and the population doubled in 5.5 days. The stable age profile of the population of eggs, immature stages and adults was 74.9, 19.24 and 6.07% (Kennedy *et al.*, 1996). On citrus fruits, the tenuipalpid has a higher reproductive rate than *B. californicus* and *B. obovatus* (Trindade and Chavegato, 1990). Under laboratory conditions at 25 ± 2°C, 70 ± 10% RH and with a 14 h photophase, *B. phoenicis s.l.* also showed better development and higher survival and fecundity on citrus fruits than on coffee leaves, according to the respective intrinsic rates of the population increase (r_m) of 0.128 and 0.090 females per female a day observed (Teodoro and Reis, 2006).

Studies on the spatial distribution of *B. phoenicis s.l.* in a 'Valencia' orange orchard in Colombia showed an inversely proportional behaviour of mites to precipitation; in particular, when the rains began, the mite population decreased. The mite also tends to move from the most humid layer of the tree towards the driest, which implies that its behaviour is dependent as much on the relative humidity as on precipitation (Solano *et al.*, 2008). On coffee plants, the mite lives mostly on the internal leaves, branches and berries, in the lower and medium parts of the plants. On the leaves, the mites were localized mostly on the lower leaf surface, mainly next to the midrib. On the berries, eggs and mites were found mainly at the crown and peduncle, and on the branches in cracks in the bark and between them (Reis et al., 2000).

In guava (var. 'Allahabadi') orchards in Punjab, India, the population of *B. phoenicis s.l.* was negatively correlated with that of the predator *Amblyseius delhiensis* Narayan *et* Kaur (= *Euseius delhiensis* (Narayan *et* Kaur)), mean temperature, RH and rainfall, but positively correlated with wind velocity, sunshine and day length; the mites lived on this crop throughout the year, and the peak populations were observed in April, when there was no population of *E. delhiensis* (Neena Singla, 2001). In experimental orchards in Venezuela, the mean mite population on guava fluctuated from 0 to 19 during the 84 DPA (days post anthesis). Mites were more abundant on sepals and the peduncles of young fruits between 14 and 35 DPA, whereas on older fruits at 49 and 84 DPA, the population density was more abundant around the exocarp. The highest mean population (19.16 mites) was found on the abaxial surface of sepals, and the lowest (0) on the peduncles of old fruit, as well as on the adaxial surface of sepals, when the fruit was at 84 DPA. This population behaviour depends on the feeding habits of *B. phoenicis s.l.*, because the sepals provide the conditions for population increase in young fruits, probably due to the special nutrition that is available and to protection from other mite predators; in addition, the anatomical characteristics of those structures probably facilitate the feeding and defence of this mite species (Rivero *et al.*, 2010).

In West Bengal, India, *B. phoenicis s.l.* is active throughout the year on the weed *Mikania micrantha* Kunth (Asteraceae), a widespread perennial creeping weed found everywhere in the fencing of cultivated fields, tea gardens, poorly managed trees of homestead plantations and on forest plants. It also acts as a source of infestation of *B. phoenicis* infesting tea, citrus, guava, papaya, pointed gourd, etc. The population of *B. phoenicis s.l.* started increasing in February/March and reached a maximum during the moderately dryer and cooler period in the month of October, as on other host plants. The mite population persisted until November, and thereafter declined sharply to a very low population during January. Significant positive correlations were observed between the mite population and minimum temperature and minimum RH (Karmakar and Saha, 2005). Also in West Bengal, the incidence of *B. phoenicis s.l.* was investigated on four cultivars of pointed gourd (*Trichosanthes dioica* Roxb.) ('IA', 'Hilli', '4A' and '13 MV') from April to November. The mite population was highest on 'Hilli' and lowest on '4A'; it was initially recorded in July, peaked from September to November, and was lower in October (Patel and Karmakar, 2005). On tea in Assam, India, the tenuipalpid was present on tea bushes throughout the year, and peak numbers occurred during May–August. The life cycle was completed in 17–23 days during April–August, and in 37–55 days during November–February (Gope and Das, 1992).

B. phoenicis s.l. moves less than 1 cm a day. A wind speed of 23 km h^{-1} is not enough to trigger mite dispersal, and wind speeds of 30 and 40 km h^{-1} are capable only of triggering the dispersal of less than 1% of mites on fruits. Dispersal studies using sticky traps under field conditions also showed that the dispersal of *B. phoenicis* was limited compared with that of other mite species from citrus groves (Alves *et al.*, 2005a).

Natural enemies include pathogenic fungi and several insect and mite predators (see Table 21.1).

Symptomatology and damage

B. phoenicis s.l. causes direct damage by its feeding activity, and may transmit severe plant diseases. Like other false spider mites, it injects toxic saliva into the fruit, leaf, stem and bud tissues of citrus, and prefers the green fruit to ripe fruit, and the fruit to leaves and twigs. On the leaves, the mite normally inhabits the lower surface along the mid-vein or principal lateral veins, causing a yellow blistering of the leaf surface opposite the feeding mite aggregations; it also feeds on the stems or green twigs (Haramoto, 1969). The damaged areas become slowly necrotic, with consequent leaf drop coinciding with high infestations (Childers, 1994a), although this phenomenon is uncommon (Childers *et al.*, 2003a). On grapefruit or orange, the injury is similar to that previously described for *B. californicus* and *B. obovatus* in Texas (Dean and Maxwell, 1967; French and Rakha, 1994; Childers *et al.*, 2003a), and in South Africa (Schwartz, 1970, 1977b; Meyer Smith and Schwartz, 1998c). Oranges infested by *B. phoenicis s.l.* are usually lighter, and their weight is inversely proportional to the degree of infestation; furthermore, affected trees may lose 50% of their yield (Rodrigues *et al.*, 2003). In Italy, Di Martino (1985) has observed greyish scabby patches and medial cracks on the apical epidermis of mandarin fruits. Many lesions are located on oleiferous glands, which, as a result, are empty and dry. The infested orange fruit shows rounded reddish patches.

On guava, the feeding activity of *B. phoenicis s.l.* affects the basal and apical tips, and the fruits. The mite induces changes in colour and texture of the fruit rind, and the intensity and development of the damage can be detected and measured by the changes in colour and texture on the fruit rind tissues. Slightly damaged fruits are more frequent than the other (greater) types of damage (Guerere and Gonzalez, 2000). The presence/absence of the sepals on the apical tip of guava fruits

Table 21.1. *Brevipalpus phoenicis* (Geijskes) *s.l.* Natural enemies (fungi, insects and mites).

Families and species	References
FUNGI	
Ascomycota	
Verticillium lecanii (Zimmermann) Viégas	Albuquerque *et al.*, 2000
Clavicipitaceae	
Aschersonia sp.	Albuquerque *et al.*, 2000
Hirsutella thompsonii Fisher	Cabrera, 1978; Rosas Acevedo and Sampedro Rosas, 2000; Rossi-Zalaf and Alves, 2006
Metarhizium anisopliae (Metschnikoff) Sorokin	Albuquerque *et al.*, 2000; Magalhães *et al.*, 2005
Paecilomyces sp.	Albuquerque *et al.*, 2000
INSECTS	
Chrysopidae	
Chrysoperla externa (Hagen)	Silva *et al.*, 2006
Coccinellidae	
Sticholotis punctata Crotch	Haramoto, 1969
MITES	
Cheyletidae	
Mexecheles hawaiiensis (Baker)	Haramoto, 1969
Phytoseiidae	
Amblyseius acalyphus Denmark *et* Muma	De Vis *et al.*, 2006
Amblyseius compositus Denmark *et* Muma	Reis *et al.*, 2007
Amblyseius largoensis (Muma)	Haramoto, 1969
Euseius alatus De Leon	Gravena *et al.*, 1994; Yamamoto and Gravena, 2001; Reis *et al.*, 2003
Euseius alstoniae (Gupta)	Kumari and Sadana, 1991
Euseius citrifolius (Denmark *et* Muma)	Gravena *et al.*, 1994; Yamamoto and Gravena, 2001; Reis *et al.*, 2003; De Vis *et al.* 2006
Euseius concordis (Chant)	Mineiro *et al.*, 2008
Euseius delhiensis (Narayan *et* Kaur)	Neena Singla and Sadana, 2000
Galendromus occidentalis (Nesbitt)	Swirski and Dorzia, 1969
Iphiseiodes zuluagai Denmark *et* Muma	Gravena *et al.*, 1994; Yamamoto and Gravena, 2001; Reis *et al.*, 2003; De Vis *et al.* 2006
Metaseiulus camelliae (Chant *et* Yoshida-Shaul)	De Vis *et al.*, 2006
Phytoseiulus macropilis (Banks)	Haramoto, 1969; Jaime de Herrero, 1984
Typhlodromips cananeiensis Gondim Jr *et* Moraes	De Vis *et al.*, 2006
Typhlodromips swirskii (Athias Henriot)	Zaher *et al.*, 1971
Typhlodromus homalii (Gupta)	Sandhu *et al.*, 1979
Stigmaeidae	
Agistemus sp.	Oomen, 1982
Agistemus brasiliensis Matioli, Ueckermann *et* Oliveira	Matioli and Oliveira, 2007
Zetzellia javanica Ehara *et* Oomen-Kalsbeek	Oomen, 1982
Zetzellia malvinae Matioli, Ueckermann *et* Oliveira	De Vis *et al.*, 2006; Mineiro *et al.*, 2008
Tydeidae	
Pronematus elongates Baker	Gupta, 2001
Pronematus ubiquitus (McGregor)	Zaher *et al.*, 1971
Pronematus sp.	Borthakur, 1981; De Vis *et al.*, 2006
Tydeus californicus (Banks)	Zaher *et al.*, 1971

during development significantly influences the mean mite density, which is 2.36 times higher on the fruits with sepals remaining (11.23 mites per fruit) than on those with the sepals removed (4.75 mites per fruit). Fruit age influences mean mite density, which increases with fruit development or age, for both kinds of fruits, with or without sepals. Fruit age influences the average degree of damage of the fruits, which increases fast and progressively in time on fruits with or without sepals but colonized early by the mite. Light and moderate damage levels are found at 24 and 38 days of age of the fruits, which then become heavily injured at 52 and 66 days old for fruits with and without sepals, respectively. The damage levels increase rapidly at 52 days of fruit growth, with 54% of the fruits with sepals showing light, moderate and heavy amounts of damage (Quiróz de González *et al.*, 2002). Studies in Venezuela on the histological alterations caused to sepals

and peduncles of small, medium and large sized fruits of guava with flat mite damage showed that the epidermal and parenchymal tissues of sepals and peduncles evolved into a necrosed superficial layer, whose thickness increased according to a combination of the factors studied. Taking into account that the epidermal and parenchymal layers have a meristematic capacity, the alterations caused by the mite could affect cellular regeneration, which is important and necessary for fruit growing, and for wound healing in the structures (Rivero-Maldonado *et al.*, 2008).

On tea, populations of *B. phoenicis s.l.* live only on the lower leaf surface on either side of the midrib, mainly around the petiole. As feeding continues, defoliation of tea bushes occurs (Gope and Das, 1992). Obviously a yield loss may occur.

VIRUS DISEASES AND OTHER DISORDERS. *Brevipalpus phoenicis s.l.* may transmit severe plant diseases such as leprosis or 'lepra explosiva' to citrus (Kitajima *et al.*, 1972; Carter, 1973; Boaretto and Chiavegato, 1994; Rodrigues *et al.*, 1997; Childers *et al.*, 2003a) in Central and South America (Argentina, Belize, Bolivia, Brazil, Colombia, Costa Rica, Guatemala, Honduras, Mexico, Panama, Paraguay, Uruguay, Venezuela) (Childers and Rodrigues, 2011). Other diseases transmitted by *B. phoenicis s.l.* are coffee ringspot and passion fruit green spot diseases.

Leprosis has been known in Florida since the late 1800s, but there has been no news of it since the 1960s (Morishita, 1954; Childers *et al.*, 2001), and subsequent research has excluded the presence of the causative virus in Florida and Texas (Childers *et al.*, 2003a; Childers and Rodrigues, 2011). There is some confusion about the presence and distribution of the virus and little clarity about the differences between feeding damage and leprosis virus infection (Childers *et al.*, 2003a). The responsible agent is the Citrus leprosis rhabdovirus (CiLV), which occurs in two forms: the more common CiLV-C, infecting the cytoplasm of infected plant cells, and CiLV-N, occurring in the cell nucleus. The prevention costs for the disease in Brazil alone are about US$100 million a year (Rodrigues *et al.*, 2003). Damage from the disease negatively affects yield and more than US$60 million a season are spent on chemical sprays to control the mite vector (Rodrigues, 2006). Bassanezi and Laranjeira (2007) recently calculated that acaricide applications for the annual control of infestations of the mite vector cost the Brazilian citrus industry US$62 million, constituting 35% of the total agrochemical costs and 14% of the total production costs in mature orchards. Unlike other *Brevipalpus* spp. found on citrus, only *B. phoenicis s.l.* transmits the virus throughout the different biological stages, but not transovarially, and each mite needs to acquire the virus separately in order to become infective. The complete nucleotide sequence and genomic organization of CiLV-C has recently been established (Pascon *et al.*, 2006). The virus is mechanically transmitted from plant to plant and from a few herbaceous plants belonging to the genera *Atriplex*, *Beta* and *Chenopodium* (Chenopodiaceae), *Gomphrena* (Amaranthaceae) and *Tetragonia* (Tetragoniaceae). Efforts to purify and characterize the virus have so far failed (Childers *et al.*, 2001). The virus mainly attacks sweet orange, but also citrange, citron, 'Cleopatra' mandarin, grapefruit, lemon, mandarin, sour orange and tangor (Lovisolo, 2001). The principal symptoms are lesions to the fruits, leaves, twigs and small branches, with early fruit drop, leaf drop and death of twigs and branches until the vitality of the trees declines (Rodrigues *et al.*, 2003). Localized lesions occur commonly, with those deriving from CiLV-N usually smaller than those due to CiLV-C (Bastianel *et al.*, 2006). The lesions result in a long-term decline in the quality of sweet orange trees, bringing about fruit drop and even death.

Studies on the incidence of leprosis on 'Valencia' and 'Natal' orange in Brazil show that the increase in disease incidence is slow and estimates of disease growth rate are quite low. The initial inoculum quantity on the branches increases significantly from one cycle to the next. After 3 years, the incidence reaches 32.0% in 'Valencia' and 6.8% in 'Natal' orange trees. Although not systemic, citrus leprosis has a polyetic performance, building up inoculum year after year, especially on the branches. Its incidence is not correlated with mite density on trees in previous data collections, but it has a significant positive correlation with disease incidence. This indicates that the presence of symptoms, rather than only the mites, must be taken into consideration when sampling for disease control (Czermainski *et al.*, 2007).

Weeks *et al.* (2000b) report that clones corresponding to isofemale lines and derived from single females of the mite collected from citrus comprise 17 genotype lines. Various other clones of *B. phoenicis s.l.* from Sao Paolo (Brazil) and from Florida, and differing along geographical lines according to a mitochondrial DNA analysis, vary in their capacity to transmit CiLV (Rodrigues *et al.* 2003), and the fitness of the mite is limited when it is displaced to new host plants. According to Groot *et al.* (2005), the phenomenon can be explained by the fact that the species consists of diverse host-specialized clones instead of one generalist form, each adapted to different environments and host plants. In a later study, Groot and Breeuwer (2006) investigated thelytokous reproduction in three *Brevipalpus* spp., including *B. phoenicis s.l.*, in 19 isofemale lines (ten of *B. phoenicis s.l.*) based on variation in their mitochondrial COI sequences. They noted that taxonomic inferences based on the COI gene sequences in *Brevipalpus* spp. – in this case *B. phoenicis s.l.* – did not assist in the classical taxonomic approach based on morphology, and that a taxonomic revision of the group was required (Groot and Breeuwer, 2006). This has now been done by Beard *et al.* (2013).

Coffee ringspot disease consists of conspicuous ring spot symptoms on leaves, berries and, less frequently, twigs of coffee, and is caused by the *Coffee ringspot virus* (CoRSV). The disease has been reported in several Brazilian states (São Paulo, Paraná, Minas Gerais and Federal District), and was recently found in Costa Rica. A similar disease is known in the Philippines, but no information exists about its relationship to CoRSV. Coffee ringspot had no economic significance until recently, when a large-scale infection was reported in Minas Gerais that resulted in yield loss (Chagas *et al.*, 2003).

Passion fruit green spot disease was first identified in 1997 after a severe outbreak in the state of São Paulo, Brazil. Mature yellow fruits of *Passiflora edulis* Simms f. *flavicarpa* Degener showed characteristic green spots 2–5 mm in diameter, and patches of green tissues, on senescent leaves. The devastating effect on passion flower is caused by necrotic lesions that encircle the stems and kill the plant. In severe cases, entire orchards of a few hectares in size have been completely destroyed. The disease was always preceded by heavy infestations of *B. phoenicis s.l.*

A viral aetiology for the disease was stated and the agent was named *Passion fruit green spot virus*. The disease was also found in other Brazilian states and spray applications of acaricides significantly reduced its incidence (Kitajima *et al.*, 2003a).

In Florida, another disorder is 'a diffuse laminar chlorosis and a marginal chlorosis of the leaves' known as '*phoenicis* blotch', but this does not represent a true leprosis or stem cankering. The disorder is similar to the early stages of leprosis, but it is not accompanied by a separation of the parts affected into resinous or corky types. The damage is linked to high populations of *B. phoenicis s.l.* and limited to plants previously defoliated by stress factors (Jeppson *et al.*, 1975). In Venezuela and Brazil, the mite exhibits a preference for sour orange fruits with scabs caused by the fungus *Elsinoe fawcetti* Bitanic *et* Jenkins (Elsinoaceae) and produces a disorder known as 'halo scab'; the symptoms range from spotting of the leaves to death of the plants; the leaves attacked by the scab fungus alone do not drop, but leaves attacked by both the fungus and the mite drop, followed by defoliation and death of the plants (Knorr and Malaguti, 1960; Chiavegato and Kharfan, 1993). In Honduras, *B. phoenicis s.l.* is associated with *E. fawcettii* on sour orange (Evans *et al.*, 1993). A similar interaction between two mite species and the fungus *Sphaceloma fawcetti* Jenkins has been reported in Costa Rica (Ochoa *et al.*, 1994). In Brazil, the mite prefers citrus fruits already infected by sweet orange scab, a disease caused by the fungus *Elsinoe australis* Bitancourt *et* A.E. Jenkins (Chiavegato and Kharfan, 1993); according to Gravena *et al.* (1994), predation by the phytoseiid *Euseius citrifolius* (Denmark *et* Muma) on *B. phoenicis s.l.* is limited on oranges with scab, thus protecting the scab for the populations of the pest mite.

Another disorder, 'Brevipalpus galls', discussed earlier for *B. californicus*, and also attributed in Venezuela to *B. phoenicis*, may cause the death of 60% of sour orange seedlings. Gall-like woody protuberances are observed on the main stem of the seedlings that seem like axils that have proliferated to the point where bud studded cushions have developed. The axils occupied by these cushions do not produce leaves, and if all the buds are replaced by cushions, the trees become devoid of leaves and soon die (Knorr *et al.*, 1960, 1968; Knorr and Denmark, 1970).

Control

The control of populations of *B. phoenicis s.l.* on citrus requires the application of quarantine, cultural practices and biological and chemical means, preferably harmonized within an integrated pest management (IPM) programme. Sampling and monitoring of the populations is fundamental to control, mainly in Central and South American citriculture, and less pressing in other regions of the world. However, the risk of accidental introduction of clones of the vector suitable for the transmission of diseases with the introduction of plant material that has not been properly examined is obvious. Similarly, the need for control arises in guava, tea and other plants.

SAMPLING AND MONITORING

Citrus. The failure of control of *B. phoenicis s.l.* on citrus in Brazil could be because of the high values of threshold used, which recommend spray applications of acaricides only when 10% of the fruits are infested with at least 30 mites cm^{-2} (Netto, 1987). Random samples should be taken from two of every 100 trees (with five fruits showing signs of attack being taken from each of the two) and treatment should begin when mites are found on 2% or more of the total number of fruits examined (Goes *et al.*, 1985). Results of studies on the spatial patterns of leprosis and of *B. phoenicis s.l.* showed that the patterns of either the diseased or the infested plants deviated from a binomial distribution. Hence, estimates of disease or mite incidence may not be precise (Bassanezi and Laranjeira, 2007). The determination of the acceptable sample size for estimating the *B. phoenicis s.l.* population was investigated in São Paulo, Brazil, in an orchard of 8-year-old orange trees (cv. 'Valencia') with 2480 plants. Samples of 1, 2, 3, 5, 10 and 100% of plants in the orchard were inspected, corresponding to 25, 50, 74, 124, 248 and 2480 plants, by walking in the direction of the planting lines. Three fruits or branches per plant were inspected. It was observed that the error percentage for estimating the mean percentage of fruits with *B. phoenicis s.l.* in an actual situation of inspection for flat mite leprosis in citrus, when only 1% of plants (25 plants) were inspected, was 50%. However, for an infestation of 10%, the variation of infested fruits percentage was 5–15%, leading farmers to underestimate or overestimate the mite population density, thus either increasing the costs with unnecessary treatments or leading to an inefficient control of the flat mite. To keep the error in the sample within an acceptable margin (20–30% of error), 105 plants must be inspected. In terms of the fruit percentage with more than ten flat mites, it was verified that for an acceptable margin of error (20–30%), 540 plants must be inspected (Lopes *et al.*, 2007). In South Africa, spray application of acaricides is recommended in the summer at detection of the first adults on fruits (Krause *et al.*, 1996). Sequential sampling for leprosis on orange trees is usually based on the binomial distribution.

Guava. On guava fruit in development, the 'linear and curvilinear regression models' relating to the variables mean mite density with average degree of damage highlighted an increase of average degree of damage with an increase in mean mite density for fruits either with or without sepals. The use of a logarithmic model to relate mean mite density to fruit age showed that mean mite density increased with age, and was higher for fruits with than without them. Sepals offer nutrients and refuges that favour the colonization and the beginning of injury by *B. phoenicis s.l.*; in addition, they are fruit structures that indicate mite presence and damage, so they can be used to make assessments of decision-making strategies for mite problems, beginning with fruit set and becoming serious as the fruit grows, especially during the dry season (Quiróz de González *et al.*, 2002).

Tea. A sampling programme for *B. phoenicis s.l.* damage and population density was developed and tested at an Indonesian tea plantation. Because of the stable coefficients of variation for the widely different mean mite densities measured during the season, a sample size of no more than 200 leaves per plot was recommended (Young *et al.*, 1995).

QUARANTINE. Quarantine methods enforce a strict quarantine regime on citrus plants that may harbour vectors of plant viruses and are probably the best way to control diseases. However, such regimes are seldom fully enforced and the diseases would,

therefore, be expected to spread from their present foci. Neither would quarantine be useful if a tenuipalpid was suspected of being wind borne (Gerson, 2008).

PLANT RESISTANCE. Plant resistance has been investigated in citrus (Sadana and Joshi, 1979), and recent studies have suggested that inheritance of resistance to leprosis may be controlled by only a few genes (Bastianel *et al.*, 2006). The role of host plant resistance to *B. phoenicis s.l.* has been investigated in various studies, including those by Grewal (1993), Bastianel *et al.* (2005, 2006) and Rodrigues (2006). Sudoi *et al.* (1994) investigated the duration of the life stages of *B. phoenicis s.l.* on eight clones of tea. The results showed that the duration of the different stages varied on different tea clones. The clones '56/89' and '54/40', but not 'SFS150' and '6/8', were preferred by the mite for oviposition and development.

CULTURAL PRACTICES. These are intended to limit the spread of leprosis and the tenuipalpid using resistant varieties, disease- and/or mite-free plants, removing or pruning infected or infested trees or parts, avoiding hedge and windbreak plants that are heavily infested or are alternate host plants for the mites, and removing susceptible weeds and other plants from the area (Maia and Oliveira, 2004).

On tea in Kenya, all mite developmental stages were positively influenced by an increase in NPKS (25:5:5:5) fertilizer. The frequency of fertilizer application did not influence the numbers of mite, tea yield and leaf nutrient level, but the yield of tea increased linearly up to 200 kg N ha^{-1} annually, after which no improvement occurred. Increases in N application rates resulted in an increase in leaf nutrient content, while the amounts of leaf phosphorus and potassium fell significantly. NPKS fertilizer application positively correlated with eggs, pre-adult and adult developmental stages. The resultant increase in nitrogen content probably led to an increase in protein content, consequently supporting the development of the mites feeding on leaves. The depressed phosphorus and potassium uptake into leaves could also influence tolerance by the tea plant to certain environmental stresses, including susceptibility to pest attack. It was concluded that the rational application of nitrogen fertilizer to tea could result in increased yields and tolerance to mites (Sudoi *et al.*, 2001a). Further studies in Kenya by Sudoi *et al.* (2001b) aimed to determine how yield losses on tea due to infestation by *B. phoenicis s.l.* related to nitrogen nutrition in plots receiving either 0, 100 or 200 kg N ha^{-1} annually. The reduction in yield losses in fertilized plots showed the beneficial effects of nitrogen in inducing tolerance to mites and possibly increasing plant vigour. Spray applications of acaricides protected the tea from mite attack and resulted in a yield advantage and a net benefit of US\$0.3, 0.2 and 0.2 kg^{-1} made tea at 0, 100 and 200 kg N ha^{-1}, respectively. Application of 100 kg N ha^{-1} had a higher net cost benefit than the 200 kg N application rate. The strategy for control of mites would be to integrate spraying with the use of fertilizer.

In Assam, India, the 'China' tea variety was more susceptible to damage than the 'Assam' variety, and pest infestation was reduced most when bushes were pruned rather than 'skiffed' (Gope and Das, 1992).

BIOLOGICAL CONTROL. Unfortunately, biological control alone does not solve the problem of infestation by *B. phoenicis s.l.* Albuquerque *et al.* (2000) reported that the fungi *Metarhizium anisopliae* (Metschnikoff) Sorokin, *Aschersonia* sp. and *Paecilomyces* sp. (Clavicipitaceae) and *Verticillium lecanii* (Zimmerman) Viégas (Ascomycota) did not produce reliable control, while Magalhães *et al.* (2005), using *M. anisopliae* var. *acridum*, observed 90% adult mortality after 8 days. *Hirsutella thompsonii* Fisher (Clavicipitaceae) gave a good response (Rosas Acevedo and Sampedro Rosas, 2000), and some isolates caused high mortality with indices higher than 90% (Rossi Zalaf and Alves, 2006), but there is no information on their practical application. Other trials done later showed that concentration and time to death differed between treatments and control. Applications of *H. thompsonii* resulted in a reduction in adult survival and the number of eggs of *B. phoenicis s.l.* (Rossi-Zalaf *et al.*, 2008).

Similar results have been observed with predatory mites in various countries. Gravena *et al.* (1994) investigated predation by the phytoseiid *Euseius citrifolius* (Denmark *et* Muma) on *B. phoenicis s.l.* on oranges. In Israel, *Galendromus occidentalis* (Nesbitt) has been introduced from California (Swirski and Dorzia, 1969). In Brazil, *E. citrifolius* and *E. concordis* (Chant) are the most frequent phytoseiid species on coffee leaves. Significant correlations were observed between the population dynamics of *E. concordis* and *B. phoenicis*, between *Zetzellia malvinae* Matioli, Ueckermann *et* Oliveira (Stigmaeidae) and *B. phoenicis s.l.*, and between *Z. malvinae* and *E. concordis*. Significant correlations were also found between the number of domatia and the population densities of *B. phoenicis s.l.*, *E. concordis*, *Lorryia* sp. and *Z. malvinae* (Mineiro *et al.*, 2008).

On leaves of coffee under laboratory conditions, the females of *A. compositus* Denmark *et* Muma were the most efficient consumers of *B. phoenicis s.l.* in all development phases. The predatory activity and the oviposition of *A. compositus* increase in a positive and highly significant manner according to the increase in prey density. Regression analysis suggests a type II functional response with maximum predation between 30 and 35 *B. phoenicis* cm^{-2} per female (Reis *et al.*, 2007). Also under laboratory conditions on coffee leaves, Silva *et al.* (2006) found that the first instar larvae of *Chrysoperla externa* (Hagen) (Chrysopidae) exhibited better predatory capacity when compared to those of the second and third instars, regardless of the mite developmental stage. The preference of first instar larvae of *C. externa* was for eggs, followed by larvae, nymphs and finally of adults of *B. phoenicis s.l.* These results indicate that larvae of *C. externa* can act as auxiliary predators in regulating mite populations in coffee tree agroecosystems.

Prey consumption by *Agistemus brasiliensis* Matioli, Ueckermann *et* Oliveira was studied at different densities (5, 10, 20, 40 and 60) of *B. phoenicis s.l.* females per cage (3 cm in diameter) on citrus fruits at 29°C. The maximum number of prey (7.6 females a day), was consumed at a density of 20 *B. phoenicis s.l.* At densities of >40 prey per cage, *A. brasiliensis* laid 4.7 eggs a day compared with 2.5 eggs per day at 20 mites per cage (Matioli and Oliveira, 2007).

CHEMICAL CONTROL. The control of *B. phoenicis s.l.* on citrus in Central and South America is based on spray applications of

acaricides. The use of organochlorines, organotin, petroleum oils and sulfur, either applied alone or in various mixtures, is well known on citrus (Vacante, 2010). For this mite, it is also necessary to rationalize the use of acaricides to prevent the selection of resistant strains (Omoto, 1998; Alves *et al.*, 2000a,b; Campos and Omoto, 2002, 2006). Thus, applications of fenbutatin oxide reduce the numbers of *B. phoenicis s.l.* on citrus by 75–100% for as long as 4 months post-treatment (Oliveira *et al.*, 1998), but the large number of sprays required for this purpose leads to the development of resistance to organochlorines and to phenylpyrazoles. Abamectin in various formulations is effective on citrus (Andrade *et al.*, 2010a), and spray applications of spirodiclofen on adults produced maximum mortality of ~60% under semi-field conditions (Poletti and Omoto, 2003).

On coffee, spirodiclofen was selective for predatory mites, but azocyclotin was less selective. In the field, both acaricides reduced all post-embryonic stages of *B. phoenicis s.l.*, mainly on the leaves (Reis *et al.*, 2005). Increasing spray volumes by duplicating the lances (nozzle branches) resulted in a significant increase in control efficiency compared with conventional lance and in the control plot, without dependence on the spray volume (Fernandes *et al.*, 2010).

On tea (cv. 'TRFK 11/4'), spray applications of propargite are recorded as giving the highest control of the pest, followed by horticultural mineral oil and petroleum oil. The mean yield of the crop was highest with the application of propargite and lowest with the application of petroleum oil (Sudoi, 2004). Some of the above acaricides are not permitted under the regulations of different countries.

Various botanical acaricides have been tested under laboratory conditions and in the field on citrus. Under laboratory conditions, the most effective extracts (giving a control of >50%) were from the leaves of *Luffa aegyptiaca* Mill., *Eucalyptus citriodora* Hook. and *Hedera helix* L., and the fruits of *Melia azedarach* L. (Guirado and Ambrosano, 2000). One-year-old seedlings of 'Pera' orange were evaluated under field conditions in São Paulo, Brazil. Extracts of *L. cylindrica* M. Roem. (= *L. aegyptiaca*), *Allium sativum* L., *H. helix* and *Datura metel* L. were active in order decreasing efficiency. *A. sativum* extract acted as an antifeedant, while the extracts of *L. cylindrica*, *H. helix* and *D. metel* prevented inoculation of the associated virus into the host plant (Guirado *et al.*, 2001). The application of distilled and decanted pyroligneous extract preparations of *Eucalyptus* sp. did not produce significant repellence to *B. phoenicis*. The protective action recognized at low doses that are normally recommended is not due to mortality or repellence by direct contact of these products with mites (Alves *et al.*, 2007). Under laboratory conditions, neem oil was effective in the control of the mite (Justiniano *et al.*, 2009).

INTEGRATED PEST MANAGEMENT. In Brazil, the need to limit the number of chemicals used, and their associated disadvantages, has suggested that control should move towards the use of IPM (Gravena, 1998), while respecting the technical characteristics of the main acaricides registered for citrus and the detailed procedures of a laboratory bioassay conducted to evaluate the efficacy of the acaricides against the citrus leprosis mite (Gravena *et al.*, 2005).

The Genus *Tenuipalpus* Donnadieu

Diagnostic characteristics

The anterior margin of the prodorsum has a broad flat projection over the gnathosoma, and coxae I–II are strongly forked medially. The palpus is 1–3 segmented. The opisthosomal setae c_2, d_2 and e_2 are absent; the setae c_1, and f_2 are present/absent, and the setae h_2 are elongate. The ventral and genital plates are developed, frequently fused and slightly sclerotized. The anal plates are not developed, and 2 pairs of setae *ps* are present. Both the claws and empodium are pad like and provided with tenent hairs (Beard *et al.*, 2013).

Economic importance

This genus has a worldwide distribution, and includes about 300 species, reported for the African (91 species), Oriental (69 species), Neotropical (about 64 species) and Western Palaearctic (29 species) regions. In particular, *Tenuipalpus* spp. have been reported from Brazil, China, France, India, Iran, Iraq, Mexico, Pakistan, Philippines, South Africa and the USA. Host plants belong to various botanical families, such as the Anacardiaceae, Arecaceae, Asteraceae, Combretaceae, Euphorbiaceae, Fabaceae, Meliaceae, Mimosaceae, Myrtaceae, Orchidaceae, Rosaceae and Rubiaceae (Beard *et al.*, 2013). *T. pacificus* is the most important species. The main features of other *Tenuipalpus* spp. known as occasional or secondary pests are summarized below.

T. caudatus (Dugès) was described from specimens collected in France on laurestinus (*Viburnum tinus* L.) (Dugès, 1834). The mite has also been collected on *Acacia*, *Calligonum*, *Citrus*, *Laurus*, *Malus* and *Olea* (Ehara and Masaki, 2001), and it has a worldwide distribution (Gerson, 2003). The bio-ecology of *T. caudatus* is unknown, and neither is there information on the damage caused by its feeding. *T. crassulus* Baker *et* Tuttle was reported on *Crassula lycopodioides* Lamarck (Crassulaceae) from the Netherlands (Baker and Tuttle, 1972). A new species currently being described by Welbourn, Beard, Bauchan and Ochoa (in preparation) was recently observed feeding on Crassulaceae in Florida, and was initially identified as the morphologically similar species, *T. crassulus*. Feeding by this mite on *Aloe* causes dark blotches that often turn dark red/maroon. Feeding on crassulaceous host plants starts along the stems and distal leaf tips. As the population increases, infested leaves turn a rusty colour before complete darkening of the leaf surface occurs, resulting in dry cracked lesions. The mite feeds on and lays its eggs along the stems and leaf tips, but mites are often observed in high numbers on the lower leaf surface as well (Beard *et al.*, 2013). *T. granati* Sayed has been described on pomegranate (*Punica granatum* L.) in Egypt (Sayed, 1946d), where it usually lives on the lower leaf surface. Occasionally, it has been reported as a pest of grapes (*Vitis* sp.) in various countries, such as Azerbaijan, Egypt, Georgia, Greece, Kazakhstan, etc. (Jeppson *et al.*, 1975). *T. punicae* Pritchard *et* Baker causes defoliation on pomegranate and sometimes also spots on fruits, causing fruit cracking (Jeppson *et al.*, 1975). The mite has been

described on specimens from Israel (Pritchard and Baker, 1958), but it currently has a wide range of distribution, and has been intercepted in South Africa on persimmon (*Diospyros kaki* Thunb.) originating from Japan (Meyer Smith, 1979).

Tenuipalpus pacificus Baker (see Beard *et al.*, 2013)

Common name

Unknown.

Diagnostic characteristics

FEMALE. The body is red. The palptarsus has a long straight solenidion and a short, curved tapered seta. The anterior margin of the prodorsum has a forked projection, variable in shape and length, with two central projections flanked by two reduced projections. The prodorsum is mostly smooth, with some weak areolae centrally and a series of longitudinal oblique folds laterally. The dorsal opisthosoma is mostly smooth, with a few wrinkles; the setae c_2, d_2 and e_2 are missing, and the setae f_2, f_3 and h_1 are thick and curved. The ventral plate is not developed; the cuticle posterior to setae $4a_{1-2}$ is transversely striate, forming an oval shape. The genital plate is weakly developed; the cuticle between the setae *ag* and g_{1-2} is usually broken striate, forming patterns of short 'dashes', or sometimes with unbroken striae. The ventral lateral cuticle anterior to coxae III and lateral to setae $3a_{1-2}$ has a distinct pebble-like pattern (Beard *et al.*, 2013).

MALE. The male is similar to the female (Beard *et al.*, 2013).

Geographical distribution

Australia, Barbados, Brazil, England, Germany, Japan, Malaysia, New Zealand, the Netherlands, Panama, Philippines, Puerto Rico, Siam, Java, Thailand, the USA (California, Florida) and Venezuela (Baker, 1945; Pritchard and Baker, 1958; Collyer, 1973; Beard *et al.*, 2013).

Host plants

Host plants belong to the following families: Davalliaceae – *Davallia fejeensis* Hook., *Platycerium* sp. (Beard *et al.*, 2013); Orchidaceae – *Aerides* sp., *Calanthe* sp., *Cattleya* sp., *Cypripedium* sp., *Dendrobium* sp., *Epidendrum* sp., *Grammatophyllum* sp., *Oncidium* sp., *Phalaenopsis stuartiana* Rchb.f., *Phalaenopsis* sp., *Saccolabium* sp. (Baker, 1945; Pritchard and Baker, 1958; Beard *et al.*, 2013).

Bio-ecology

T. pacificus was first described on *Phalaenopsis stuartiana* in Panama (Baker, 1945), and was later reported as a pest of orchids in various countries (Pritchard, 1951b; Denmark, 2006). In Europe (England, Germany, the Netherlands, etc.), it is known to damage orchids in greenhouses. Egg development takes 18–23 days, and the developmental stages each require 14–15 days. The life cycle takes about 64 days (Dosse, 1954). The mites anchor themselves to the leaf surface by inserting their stylets directly into the leaf epidermis (Beard *et al.*, 2012).

Symptomatology and damage

The feeding activity of *T. pacificus* causes dark spots on the leaves, with pits on the upper leaf surface. High population densities cause severe injuries that result in necrosis of the tissue (Dosse, 1954; Jeppson *et al.*, 1975).

Control

Horticultural oil can be mixed with an adjuvant (Silwet® L-77) to improve coverage and control. The mixture is considered to have only a low toxicity, may be less expensive than some pesticides, and may conserve natural enemies, making it a potential addition to an integrated pest management programme for orchids (Cating *et al.*, 2010).

The Genus *Cenopalpus* Pritchard *et* Baker

Diagnostic characteristics

The palpus is 4 segmented (rarely 3 segmented). The anterior margin of the prodorsum has a broad flat projection extending over coxae I–II and the gnathosoma (similar to *Brevipalpus*), and can be reduced in some species. It often has a characteristic collicular pattern on the dorsal cuticle, although this can be mostly smooth or weakly wrinkled. The opisthosomal setae c_2 are present, and the setae d_2, and e_2 are absent; the setae f_2 are present or absent; when present, they are inserted in a lateral position (as in *Brevipalpus*); the setae h_2 are not elongate. The ventral plate is distinct, surrounded by a characteristic cuticle pattern. The genital and anal plates are well developed, and 2 pairs of setae *ps* are present. The solenidia on tarsi I–II are long and slender (Beard *et al.*, 2013).

Economic importance

The genus includes over 60 species, most of which are recorded from the western Palaearctic (30 species) and Oriental (17 species) regions. *Cenopalpus* spp. are reported from Crete, Egypt, Greece, Iran, Israel, Italy, Spain, Turkey and Ukraine, and inhabit host plants belonging to several botanical families, including several Rosaceae, followed by the Oleaceae and Pinaceae (Beard *et al.*, 2003). The most important species are *C. lanceolatisetae* (Attiah), which infests deciduous trees, and *C. pulcher* (Canestrini *et* Fanzago), an occasional pest of deciduous fruit trees and reported as a pest of quince in Egypt and Turkey (Pritchard and Baker, 1958). *C. spinosus* (Donnadieu) is an occasional pest of primrose, blackberries, raspberries, roses and dogwood (*Cornus*) in some European countries (Baker, 1949). *C. lineola* (Canestrini *et* Fanzago) infests pine trees in various countries, including the Netherlands, Italy and Georgia (USSR) (Jeppson *et al.*, 1975). *C. lanceolatisetae* and *C. pulcher* are further described below.

Cenopalpus lanceolatisetae (Attiah) (see Beard *et al.*, 2013)

Common name

Unknown.

Diagnostic characteristics

FEMALE. The palp femur lacks setae, the genu has 1 dorsal seta, the tibia 2 setae and the tarsus 3 setae. The prodorsum is reticulate, with large regular cells centrally and medium cells sublaterally; the prodorsal setae are long, broadly lanceolate and heavily barbed. The central opisthosoma between the setae c_1–d_1 has even reticulation, with large regular cells; between the setae d_1–e_1 the reticulation becomes a series of short transverse folds, and between the setae e_1–h_1 a series of short transverse bands, becoming reticulate towards the setae h_1. The sublateral opisthosoma is reticulate with regular medium cells, and smaller cells laterally. The length of the dorsal setae follows the order c_2, c_3, d_3, e_3, f_2, f_3 > c_1 > d_1, h_2 > e_1, h_1, with d_1 and e_1 often of the same length. Ventrally, the cuticle between the setae *3a* to *4a* is more or less smooth, with extremely fine transverse striae; the cuticle between the setae *4a* to the ventral plate has even large rounded cells laterally; the cuticle is smooth centrally, with extremely fine transverse striae. The anterior cuticle of the ventral plate is smooth, and the posterior cuticle has medium cells. The genital plate has large rounded cells (Beard *et al.*, 2013).

Geographical distribution

Cyprus, Egypt, England, Greece, Iran, Israel, Jordan, Lebanon, Libya and Portugal (Attiah, 1956; Pritchard and Baker, 1958; Jeppson *et al.*, 1975; Meyer Smith and Gerson, 1981; Beard *et al.*, 2013).

Host plants

Host plants belong to the following families: Punicaceae – pomegranate (*Punica granatum* L.); Rosaceae – quince (*Cydonia oblonga* Mill.), apple (*Malus domestica* Borkh.), European crab apple (*M. sylvestris* (L.) Mill.), apricot (*Prunus armeniaca* L.), plum (*P. domestica* L.), peach (*P. persica* (L.) Batsch)), pear (*Pyrus communis* L.) (Attiah, 1956; Pritchard and Baker, 1958; Jeppson *et al.*, 1975; Beard *et al.*, 2013).

Bio-ecology

C. lanceolatisetae lives on the young shoots and buds, and on both leaf surfaces of species of the Rosaceae and Punicaceae. Among these, specimens collected on apple, European crab apple, quince and plum in England, Jordan and Lebanon were recently confirmed (Beard *et al.*, 2013). The mite may be found on the lower leaf surface only during favourable summer weather conditions. In Egypt, populations reach higher densities during July, August and September, and remain high through the autumn. In the field, the mite is often confused with *C. pulcher* (Zaher and El-Badry, 1964).

Symptomatology and damage

The symptoms of and damage caused by attack of *C. lanceolatisetae* on various host plants need to be better described.

Control

This is not usually necessary.

Cenopalpus pulcher (Canestrini *et* Fanzago) (see Beard *et al.*, 2013)

Common name

Flat scarlet mite.

Diagnostic characteristics

FEMALE. The body is about 320 μm long, 160 μm wide, and intense scarlet in colour (Jeppson *et al.*, 1975). The palpfemur lacks setae; the genu has 1 dorsal seta, the tibia bears 2 setae and the tarsus has 3 setae. The prodorsum has even reticulation, and large regular cells. The prodorsal setae are longer than the opisthosomal setae and narrowly lanceolate. The central opisthosoma between the setae c_1–d_1 has even reticulation and large regular cells; between the setae d_1–e_1 the reticulation becomes a series of short transverse folds, and between the setae e_1–h_1, a series of short transverse bands become an even reticulation towards the setae h_1. The sublateral opisthosoma has an even reticulation of large rounded and convex cells. The length of the dorsal setae follows the order c_1, c_2, c_3, d_3, e_3 > d_1, f_3 > e_1, f_2, h_1, h_2, with f_2 sometimes as long as f_3. Ventrally, the cuticle from the setae *3a* to *4a* is more or less smooth, with extremely fine transverse striae, and from the setae *4a* to the ventral plate, it is variable. The ventral plate has large rounded cells, often transversely elongate. The genital plate has large cells (Beard *et al.*, 2013).

MALE. The male is similar to the female, except that the dorsolateral setae are much longer, and the posterior body is striate (Jeppson *et al.*, 1975).

Geographical distribution

Algeria, Afghanistan, Argentina, Austria, Bulgaria, China, Croatia, Cyprus, Denmark, Egypt, France, Germany, Greece, India, Iran, Israel, Italy, Lebanon, Morocco, the Netherlands, Pakistan, Portugal, Romania, former Soviet Union/USSR, Syria, Turkey, the UK (England, Wales) the USA (Oregon) and Yugoslavia (Pritchard and Baker, 1958; Jeppson *et al.*, 1975; Meyer-Smith, 1979; Bajwa *et al.*, 2001; Beard *et al.*, 2013).

Host plants

Host plants belong to the following families: Amaranthaceae – *Amaranthus* sp.; Araliaceae – ivy leaves (*Hedera* sp.); Juglandaceae – English walnut (*Juglans regia* L.); Moraceae – fig (*Ficus carica* L.); Oleaceae – lilac (*Syringa vulgaris* L.); Punicaceae – pomegranate (*Punica granatum* L.); Rosaceae – quince (*Cydonia oblonga* Mill.), loquat (*Eriobotrya japonica* (Thunb.) Lindl.), apple (*Malus domestica* Borkh.), almond (*Prunus amygdalus* Batsch), apricot (*P. armeniaca*

L.), cherry (*P. avium* (L.) L.), plum (*P. domestica* L.), *P. insititia* L., peach (*P. persica* (L.) Batsch), pear (*Pyrus communis* L.); Rutaceae – lemon (*Citrus limon* (L.) Burm.f.); Salicaceae – willow (*Salix* sp.); Vitaceae – grape (*Vitis vinifera* L.) (Pritchard and Baker, 1958; Jeppson *et al.*, 1975; Meyer Smith, 1979; Bajwa *et al.*, 2001; Beard *et al.*, 2013).

Bio-ecology

C. pulcher is an occasional pest of deciduous fruit plants, walnut, quince and ornamentals. Recently, Beard *et al.* (2013) confirmed specimens collected on almond, apple, loquat, quince and pear in Argentina, Croatia, France, Greece, India, Iran, Italy, Morocco, Pakistan, Portugal, Romania, Turkey, the UK (England, Wales), the USA (Oregon) and Yugoslavia.

Populations of the flat scarlet mite are usually sedentary and live mainly in groups on the lower leaf surfaces along the midrib and veins. Females overwinter in cracks of the bark of various host plants, and in Europe become active early in the spring from April, but do not move completely to the leaves until May. The first eggs are laid on wood late in April. Later, oviposition starts on the leaves, along the midrib, where the eggs are protected by the leaf hairs, and continues until mid-July. In Egypt, the peak populations are observed in August, and population densities remain high until the end of December. During winter, when the trees are bare, the mite populations move to the buds. Egg development takes about 55 days, and the eggs begin to hatch near the end of June; the next generation of adults occurs in late July, with males initially more numerous. Mating occurs in August and September; later, the males die and the females begin to overwinter. Dosse (1953) observed one generation a year in Europe, but Sepasgosarian (1970) reported three generations a year in Iran. Overwintering populations can survive temperatures as low as −30°C.

Symptomatology and damage

The symptoms of and damage caused by the mite on various host plants needs to be better documented.

Control

The need for control of *C. pulcher* is unclear, and probably varies in different countries. Jeppson *et al.* (1975) reported that the mite was a pest of neglected deciduous fruit plants. However, trials on the control of the mite have been done in Egypt, where the temperature/toxicity relationships were investigated of various pyrethroids (bioallethrin, d-phenothrin, fenvalerate and cypermethrin) on *C. pulcher* individuals collected from quince (Eesa and Moursy, 1990).

The Genus *Dolichotetranychus* Sayed

Diagnostic characteristics

The palpus is 3 segmented. The body is distinctly longer than wide, and usually with a distinct 'granular' cuticle. The anterior margin of the prodorsum lacks a notch or projection. The dorsal setae d_2, e_1, e_2 and f_2 are missing, and the setae h_2 are not very elongate. The ventral, genital and anal plates are membranous and not developed; 1–2 pairs of setae *ps* are present. Genua III–IV are nude, and the true claws are uncinate or pad like. The males have a distinctly shaped posterior opisthosoma (Beard *et al.*, 2013).

Economic importance

This worldwide genus includes about 23 species, mostly recorded from Nearctic (eight species) and Oriental (five species) regions for a restricted number of host families, such as the Poaceae, Bromeliaceae and Asteraceae. Known species are reported from Australia, Egypt, India, Indonesia, Iraq, Israel, New Zealand, South Africa, Sri Lanka and the USA (Meyer Smith and Gerson, 1981; Beard *et al.*, 2013). *Dolichotetranychus floridanus* (Banks) plays the most important economic role.

Dolichotetranychus floridanus (Banks) (see Beard *et al.*, 2013)

Common name

Pineapple flat mite, orange mite of pineapple.

Diagnostic characteristics

FEMALE. Adults, nymphs and eggs are bright orange in colour. The palp tarsus has 1 solenidion and 1 tiny eupathidium set distally, and 1 dorsal seta. The ventral infracapitular seta (*m*) is present. The striae on the prodorsum are mostly smooth, sometimes with a few tubercles centrally. The striae on the genital flap lack granular texture; there are 2 pairs of genital setae (g_{1-2}) that are shorter than the distance between them, and 2 pairs of setae *ps*. Femur I has a long dorsal seta, reaching the tarsus, and femur II has a short dorsal seta, about the length of the segment; coxa II bears 2 setae (*2b*, *2c*) (Beard *et al.*, 2013).

MALE. The palp tarsus has 1 solenidion and 1 eupathidium distally, and 1 dorsal seta. The setae ps_1 are moderately long (Beard *et al.*, 2013).

Geographical distribution

Australia, Brazil, Costa Rica, Cuba, Honduras, Indonesia (Java), Japan, Mexico, Panama, Philippines, Puerto Rico, South Africa (Eastern Cape Province, Natal, northern Transvaal, Swaziland, Zululand) and the USA (Florida, Hawaii) (Pritchard and Baker, 1958; Jeppson *et al.*, 1975; Meyer Smith, 1979, 1981a; Beard *et al.*, 2013).

Host plants

Host plants belong to the following families: Bromeliaceae – pineapple (*Ananas comosus* (L.) Merr.) (Pritchard and Baker, 1958; Jeppson *et al.*, 1975; Meyer-Smith, 1979, 1981a; Beard *et al.*, 2013); Poaceae – grasses, giant reed (*Arundo donax* L.) (Wafa *et al.*, 1968/1969).

Bio-ecology

D. floridanus is a pest of pineapple in various tropical countries, and large populations have been collected under the leaf sheaths of grasses and *A. donax* in Egypt (Wafa *et al.*, 1968/1969). The mite lives most commonly on the lower leaf surfaces, near the midribs and veins. High population densities occur at the base of the leaves near the leaf stem, and between the bases of the tightly overlapping leaves (Jeppson *et al.*, 1975; Poli, 1991). Heavy infestations may develop on detached crowns stored for later planting as well. High populations of mites can be observed in the soil at the base of infested pineapples or on the pineapple stem itself, from which the mite probably migrates to the soil to avoid extremes in climate. It is particularly prevalent in drier areas, where its massive development may require control. Feeding mites exhibit a typical posture, standing on the gnathosoma and with the posterior part of the body raised in the air (Poli, 1991).

One phytoseiid mite (*Amblyseius*) was found to be associated with *D. floridanus*, but its efficacy as a natural enemy is unknown (Meyer Smith, 1981a).

Symptomatology and damage

The pineapple flat mite infests plants of all ages, and young plants or freshly planted crowns can be severely damaged or killed. However, when plants are not killed, fruit yield is reduced. Feeding by *D. floridanus* on the soft white tissues at the base of the plant induces the epidermal tissue to dry and crack, allowing fungi and bacteria to enter into the plant and to induce the tissue to rot. The dry lesions lead to scarring and tissue malformation (Pritchard and Baker, 1958; Jeppson *et al.* 1975; Sanches and Zem, 1978; Poli, 1991). In South Africa, the mite does not usually play an economic role (Meyer-Smith, 1981a).

Control

Control is provided by the use of systemic insecticides/acaricides and biological means. Prevention plays a fundamental role.

PREVENTION. It is necessary to keep young plants free of mite infestation. Planting infested plant material spreads infestation into clean plantations (Meyer-Smith, 1981a).

PLANT RESISTANCE. The chemical basis of resistance of the pineapple plant to *D. floridanus* was investigated in India. Among the chemical constituents of the leaves of five varieties of pineapple, zinc and silicon contents showed a positive relationship with the population density of *D. floridanus* in the varieties 'Queen', 'Giant Kew' and 'Kew' (Das *et al.*, 2000).

BIOLOGICAL CONTROL. The fungus *Hirsutella thompsoni* Fisher has been successfully used to control this mite (Ochoa *et al.*, 1994).

CHEMICAL CONTROL. Jeppson *et al.* (1975) suggested that insecticides and fungicides should be applied to the buds, depending on rain and dew to carry the chemicals down into the buds. The use of systemic chemicals is also suggested because the mites are well protected by the bases of old leaves and young leaves in the buds. Spray applications must be effected early in the morning, so that the active ingredient, together with the dew, runs off into the plant. Young plants should also be dipped into an acaricide before planting in the field (Meyer-Smith, 1981a).

Several acaricides were tested for the control of *D. floridanus* on pineapples in Queensland (Australia). The fumigation of pineapple tops with methyl bromide at 32 and 48 g m^{-2} gave complete control without plant damage; conversely, methyl bromide at 16 g m^{-2} was unsatisfactory. Field spraying and/or dipping treatment of plant tops with a number of acaricides/insecticides, such as bifenthrin, 'Carter's mixture' (diesel fuel and water mixed with bentonite), chlorpyrifos, clofentezine, dicofol, dimethoate, fenbutatin oxide, fluvalinate, hexythiazox, monocrotophos, propargite, sulfur or white oil gave no control. The phytoseiid mite *A. benjamini* Schicha was adversely affected by chlorpyrifos and dimethoate (Elder, 1988). In Brazil, four applications of acephate, vamidothion, methamidophos and monocrotophos applied at intervals of 30 days were tested in the field, with the first application 150 days after planting. The applications were made immediately after plant samples were collected for population counts. The production was weighed at harvest and five population counts were done during the growing cycle. The results showed that vamidothion was best for control of the mite up to 270 days after planting, and did not interfere with productivity; acephate was inefficient against the mite (Ferreira and Veiga, 2000). The use of some of the above active ingredients is not allowed in various countries.

The Genus *Raoiella* Hirst

Diagnostic characteristics

The palpus is 2 segmented. The anterior margin of the prodorsum is smoothly rounded, without a notch or ventral projection. There is a full complement of dorsal setae; the setae h_2 are not elongate; the dorsal setae are spatulate, the lateral setae are often obviously longer than the central setae, with droplets of liquid present when alive or frozen. The genital and anal plates are membranous, and not developed; two pairs of the setae *ps* are present. The setae ps_1 of the males are modified, and are usually thicker and/or longer than ps_2 (Beard *et al.*, 2013).

Economic importance

Currently, this genus includes only five species, but several new species are being described by Beard *et al.* (in preparation). Known species are *R. australica* Womersley, collected on *Eucalyptus* spp. in Australia (Womersley, 1940), *R. eugeniae* (Mohanasundaram), reported on *Eugenia* sp. in India (Mohanasundaram, 1996), the red palm mite, *R. indica* Hirst, described on coconut (*Cocos nucifera* L.) in India (Hirst, 1924), *R. macfarlanei* Pritchard *et* Baker, found on olive in Greece (Pritchard and Baker, 1958) and *R. shimapana* Meyer on *Cassine*

transvaalensis (Burtt Davy) Codd (Meyer Smith, 1979). Taking into account the work of Beard *et al.* (in preparation), most species occur in Australia, while others are known from Greece, India and South Africa. The most common families of host plants are the Myrtaceae (Ochoa *et al.*, 2011; Dowling *et al.*, 2012), followed by the Arecaceae, Celastraceae, Musaceae and Oleaceae. From an economic point of view, the red palm mite is the most important species of this genus; this is a palm specialist, infesting mainly arecaceous species, and also other monocotyledonous hosts including *Musa* spp. (Musaceae). *R. indica* has recently spread to the Nearctic, Neotropical and African regions, where it has caused severe damage to coconut palms (Dowling *et al.*, 2012).

All known species in this genus produce and carry droplets of an unknown liquid at the tips of their dorsal setae. A droplet is also usually observed on the tip of the stipe on the egg. All known species of *Raoiella* feed via the stomata of their host plants (Ochoa *et al.*, 2011; Beard *et al.*, 2012). The molecular characteristics of the COI mitochondrial and 28S ribosomal regions show a basal split between *R. indica* and all other *Raoiella* spp., which indicates Africa or the Middle East as the most probable origin of the genus. Moreover, the COI data suggest that at least eight new species are represented among the 20 Australian populations that have been examined (Dowling *et al.*, 2012).

Only the most important species known, the red palm mite, is further described below.

Raoiella indica Hirst (see Beard *et al.*, 2013)

Common name

Coconut mite, coconut red mite, red palm mite, leaflet false spider mite, frond crimson mite, scarlet mite.

Diagnostic characteristics

FEMALE. The body is oval, about 245 μm long and 182 μm wide, and reddish in colour. After feeding, dark markings appear on the dorsum (Hoy and Peña, 2006). The dorsal setae c_1, d_1 and e_1 are weakly spatulate (e_1 is often tapered), and the setae decrease in length in the order $c_1 > d_1 > e_1$; the setae h_1 are subequal in length to h_2, which are setiform, and with a finely tapered tip, and the setae f_2 are shorter than f_3. Coxae III–IV are nude, femur II bears 4 setae (*d*, *l′*, *bv′*, *v′*), genua I–II have 3 setae (*d*, *l′*, *l″*) and tarsi I–II have companion setae that are obviously longer than the solenidia; the dorsal setae on tibiae I–II are setiform, with a finely tapered tip. The palp tibiotarsus has 1 solenidion and 1 tapered seta; the setiform eupathidium is set distally and 1 seta is set dorsally. The setae ps_1 and ps_2 are inserted distant to one another (Beard *et al.*, 2013).

MALE. The posterior opisthosoma of the male is distinctively shaped. The setae ps_1 are short. The femurs II and genua I—II are as in the female (Beard *et al.*, 2013).

Geographical distribution

Antigua, Aruba, Barbados, Benin, Brazil, Cambodia, Colombia, Cuba, Dominica, Dominican Republic, Egypt, Granada, Guadeloupe, Haiti, Kenya, India, Iran, Iraq, Israel, Jamaica, Martinique, Mauritius, Mexico, Oman, Pakistan, Panama, Philippines, Puerto Rico, Réunion, Saint Lucia, Saint Martin, Saint Vincent and Saint Thomas, Saudi Arabia, Sri Lanka, Sudan, Thailand, Trinidad and Tobago, Tunisia, Turks and Caicos Islands, United Arab Emirates, the USA (Florida), Venezuela and the (US) Virgin Islands (Beard *et al.*, 2013).

Host plants

The list of host plans of *R. indica* currently includes 91 monocotyledonous species from the orders Arecales (Arecaceae), Zingiberales (Heliconiaceae, Musaceae, Strelitziaceae, Zingiberaceae) and Pandanales (Pandanaceae). Most are palms belonging to the family Arecaceae that originated in areas of the eastern hemisphere, and about a quarter are native to the New World (Carrillo *et al.*, 2012a). The main host plants reported here are listed by family: Arecaceae – *Areca* spp., *Caryota mitis* Lour., coconut palm (*Cocos nucifera* L.), *Dypsis lutescens* (H. Wendl.) Beentje and J. Dransf. (= *Chrysalidocarpus lutescens* H. Wendl.), African oil palm (*Elaeis guineensis* Jacq.), *Livistonia chinensis* (Jacq.) R. Br. ex Mart., *Phoenix* spp., date palm (*P. dactylifera* L.), bamboo palm (*Rhapis excelsa* (Thunb.) A. Henry ex Rehder), fan palm (*Washingtonia robusta* H. Wendl.); Heliconiaceae – *Heliconia* spp.; Musaceae – *Musa* spp.; Strelitziaceae – bird of paradise (*Strelitzia reginae* Banks); Zingiberaceae – red ginger (*Alpinia purpurata* Vieill.) (Beard *et al.*, 2013).

Bio-ecology

The red palm mite was first described in India on leaves of coconut palm by Hirst (1924). Later, it was reported in Egypt, Israel, Mauritius, Réunion, Sudan, Iran, Oman, Pakistan and the United Arab Emirates (Sayed, 1942; Moutia, 1958; Meyer Smith, 1979; Gerson *et al.*, 1983; Elwan, 2000; Gassouma, 2005). In 2004, *R. indica* was detected in Martinique, Dominica, Guadeloupe, St Martin, St Lucia and Trinidad and Tobago, in the Caribbean (Flechtmann and Etienne, 2004, 2005; Etienne and Flechtmann, 2006). In 2006, the red palm mite was found in Puerto Rico (Rodrigues *et al.*, 2007). In 2007, it was reported in Florida (Hoy and Peña, 2006) and also in Cuba (Rodrigues *et al.*, 2007), where significant infestations have been found damaging *Musa* spp., raising fears that it could become a key pest because bananas have greater economic importance than coconuts (Ramos Lima *et al.*, 2010). In 2008, *R. indica* was reported for the first time in South America, where it was causing severe damage on coconut and banana leaves in north-eastern Venezuela (Vasquez *et al.*, 2008b).

The red palm mite has now spread rapidly through the Neotropical region, where it causes both economic and ecological damage (Carrillo *et al.*, 2012a). The spread of the mite in this region is highly dynamic, and is encouraged by favourable ecological factors, both biotic and abiotic. Six years after its detection in the Caribbean, *R. indica* has expanded its host plant range to 27 new reproductive hosts, a 30% of increase on previous host plant records. It can be realistically assumed that as this pest mite continues to spread in the Neotropical region, a great diversity of plants will potentially be affected (Carrillo *et al.*, 2012a). In the Caribbean, in addition to several ornamental

and fruit-producing palms, such as coconut, areca palm, banana and plantain, the mite also damages ornamental palms, such as hurricane or princess palm (*Dictyosperma album* (Bory) H. Wendl. and Drude ex Scheff.) and Christmas palm (*Veitchia merrillii* (Becc.) H.E. Moore) (Arecaceae), heliconia (Heliconiaceae), bird of paradise (Strelitziaceae), and red ginger and *Etlingera elatior* (Jack) R.M. Sm. (Zingiberaceae). However, it is not known whether the mite appears just randomly on these plants after moving from coconuts and other palms, where its populations reach very high levels of density.

In any case, the red palm mite is expected to cause economic damage to tropical and subtropical agriculture, and to urban and indoor environments where palms are grown in the western hemisphere. Palm nurseries, landscape palms and horticultural gardens will all be affected by this new pest mite (Hoy and Peña, 2006). The incidence of *R. indica* in Cuba might have serious consequences for coconut, plantain crops and ornamentals, and the development of a management strategy to mitigate the possible impact of the red palm mite and of a comprehensive public awareness programme for farmers and the public in general is very important (Rodrigues *et al.*, 2007). In Benin and Tanzania, *R. indica* represents more than 70% of the total number of pest mites recorded on coconut leaves and is one of the most important mite pests in term of damage living on coconut leaves in both countries (Zannou *et al.*, 2010). Despite the above reports, Cocco and Hoy (2009) found that coconut leaf discs and trees were better hosts for rearing the red palm mite in quarantine than banana, plantain varieties or the native palms that were tested in rearing studies in Florida. Moreover, quarantine tests and field observations suggested that the host range of *R. indica* might not be as broad as some reports have indicated because plants from which the red palm mite adults and/or eggs had been collected might not be suitable for the establishment of a multi-generational colony. This indicates that more field research is needed to evaluate the ability of *R. indica* to establish and spread on native and ornamental palms in natural landscapes (in this case) in Florida.

R. indica does not produce webbing. Males and females are sexually mature when they emerge, and males actively seek out females, probably because of the involvement of a sex pheromone. A male settles close to a female deutonymph in the quiescent stage, and waits for up to 2 days for her to moult. The male becomes active when the female deutonymph begins to moult, and moves under her body, bending his posterior up and forwards to mate. The mating posture lasts for about 16 min. The life cycle from egg to adult takes 23–28 days for females and 20–22 days for males. Under laboratory conditions, mated females have a 5- to 6-day pre-oviposition period and oviposit for 47 days. Unmated females lay an average of 18 eggs after a 2-day pre-oviposition period, oviposit for 40 days and live for about 48 days. Males originate from unmated females, and live for about 22 days (Hoy and Peña, 2006).

At 25.42 ± 1.21°C and 57.54 ± 6.54% RH on leaf discs of *A. catechu* L. under laboratory conditions, the average duration of the life cycle was 31 days. The egg stage was the longest, while that of protonymphs was the shortest. The pre-oviposition period averaged 4 days, while the oviposition period could last 17 days, with an average of 11 days and female longevity of 30 days. The sex ratio was 56% females. The mortality of immature stages was 14.6% for larvae, 7.3% for protonymphs and 17.1% for deutonymphs, with 39% total mortality (Flores-Galano *et al.*, 2010). On coconut and banana leaves, eggs hatched after an average of 8.5 and 8.1 days, respectively. The larvae developed to protonymphs in 8.7–8.8 days. Deutonymphs developed to adults in 6.6–9.2 days. The whole period from egg to adult lasted 31.4 ± 3.31 days on coconut and 33.4 ± 4.76 on banana leaves. The oviposition period, longevity and total oviposition of females were higher on coconut than on banana (Ramos Lima *et al.*, 2010).

R. indica was observed in large populations on the lower surface of median leaves of coconut palms along the coastline and also on ornamental palms, ornamental plants and bananas in Guadeloupe, and on coconut leaves in St Martin (Etienne and Flechtmann, 2006). The largest populations of *R. indica* occur on coconut palms, where they range from 300 to more than 4000 individuals per colony, usually with multiple colonies per leaflet and millions of mites per tree (Pons and Bliss, 2007). However, *R. indica* does not feed on palm species in the genus *Sabal*, commonly known as palmetto palms (Cocco and Hoy, 2009), one of the most common and iconic genera of palms in the Caribbean basin. In India, populations of *R. indica* are negatively affected by rainfall and high RH, and populations are highest during hot, sunny and dry conditions on palms, so it is likely that populations will be found more readily during hot, dry periods in the western hemisphere. The mite is easily spread by wind currents and during the movement of infested plants through nursery stock and via cut branches of plants (Hoy and Peña, 2006). Field observations made monthly between December and July in Kerala, India, during one non-monsoon season (when *R. indica* populations usually increase) showed that there were significantly higher red palm mite densities on areca in the peak season (May/June) than on coconut; although significantly more coconut sites were infested with the mite than areca sites.

Although no one climatic factor was significantly related to *R. indica* numbers, interactions were found between temperature, humidity and rainfall, and the partitioning of host plant locality showed that where conditions were warmer and drier, *R. indica* densities were significantly higher. On coconut, red palm mite densities were significantly related to the combined interaction between site temperature, site humidity and phytoseiid densities (Taylor *et al.*, 2012). In Cuba, the *R. indica* population showed an exponential increase, reaching a peak in the dry season, and here, *A largoensis* (Muma) was the only predator species recorded in association with the red palm mite. In general, there was a correspondence between prey and predator population trends; a forecasting model helped to conceptualize how the various processes affected mite populations (Ramos Lima *et al.*, 2010).

Natural enemies include some fungi, and predatory insects and mites (see Table 21.2).

Symptomatology and damage

A study on the specialized feeding behaviour of *R. indica* indicated that host plant use by this species could be affected by complex stomatal morphology (Beard *et al.*, 2012). Infested leaves initially became yellow and later developed more or less extensive areas of brown and dark necrotic tissue (Etienne and Flechtmann, 2006). The mites are easily seen against the background

Table 21.2. *Raoiella indica* Hirst. Natural enemies (fungi, insects and mites).

Families and species	References
FUNGI	
Clavicipitaceae	
Hirsutella thompsonii Fisher	Rodrigues and Colon, in Carrillo *et al*., 2012b
Lecanicillium lecanii (Zimmerm.)	
Cordycipitaceae	
Simplicillium sp.	Rodrigues and Colon, in Carrillo *et al*., 2012b
INSECTS	
Cecidomyiidae	
Arthrocnodax sp.	Carrillo, in Carrillo *et al*., 2012b
Chrysopidae	
Ceraeochrysa claveri (Navás)	Carrillo and Peña, 2010; Carrillo, in Carrillo *et al*., 2012b
Chrysoperla spp.	Peña *et al*., 2009a,b
Chrysopodes collaris (Scheider)	Carrillo, in Carrillo *et al*., 2012b
Coccinellidae	
Chilorus cacti L.	Peña, in Carrillo *et al*., 2012b
Jauravia soror (Weise)	Puttarudriah and ChannaBasavanna, 1956
Stethorus keralicus Kapur	Kapur, 1961; Puttaswamy and Rangaswamy, 1976; Carrillo *et al*., 2012b
Stethorus parcempunctatus Puttarudrian *et* ChannaBasavanna	Puttarudriah and ChannaBasavanna, 1956; Gupta, 2001
Stethorus pauperculus Weise	Yadav Babu and Manjunatha, 2007
Stethorus tetranychi Kapur	Puttarudriah and ChannaBasavanna, 1956; Gupta, 2001
Stethorus utilis (Horn)	Peña *et al*., 2009a,b; Carrillo and Peña, 2010
Telsimia ephippiger Chapin	Gallego *et al*., 2003; Carrillo *et al*., 2012b
Phlaeothripidae	
Aleurodothrips fasciapennis (Franklin)	Roda *et al*., 2008; Peña *et al*., 2009a,b; Carrillo and Peña, 2010
Staphylinidae	
Oligota sp.	Somchoudhury and Sarkar, 1987; Yadav Babu and Manjunatha, 2007
MITES	
Ascidae	
Lasioseius sp.	Sheeja and Ramani, 2009
Bdellidae	
Bdella sp.	Roda *et al*., 2008; Peña *et al*., 2009b
Bdella distincta (Barker *et* Bullock)	Peña *et al*., 2009a,b; Carrillo and Peña, 2010
Cheyletidae	
Cheletomimus sp.	Roda *et al*., 2008; Peña *et al*., 2009b
Hemicheyletia bakeri (Ehara)	Carrillo, in Carrillo *et al*., 2012b
Mexecheles sp.	Welbourn, in Carrillo *et al*., 2012b
Cunaxidae	
Armascirus taurus Kramer	Gallego *et al*., 2003
Phytoseiidae	
Amblyseius sp.	Somchoudhury and Sarkar, 1987
Amblyseius caudatus Berlese	Moutia, 1958; Carrillo *et al*., 2012b
Amblyseius channabasavanni Gupta *et* Daniel	Daniel, 1981; Carrillo *et al*., 2012b
Amblyseius largoensis (Muma)	Ueckermann, 2004; Peña *et al*., 2009a,b; Carrillo and Peña, 2010; Carrillo *et al*., 2010, 2012b; Bowman and Hoy, 2012; Domingos *et al*., 2012
Amblysieus raoiellus Denmark *et* Muma	Denmark and Muma, 1989
Amblyseius swirski Athias-Henriot	Peña *et al*., 2009a
Amblyseius tamatavensis Blommers	Taylor *et al*., 2012
Neoseiulus longispinosus (Evans)	Gupta, 2001; Roda *et al*., 2008
Phytoseius sp.	Somchoudhury and Sarkar, 1987
Typhlodromips tetranychivorus Gupta	Jagadish and Nageshachandra, 1981; Nangia and ChannaBasavanna, 1989
Xenocaligonellidae	
Xenocaligonellidus sp.	Welbourn, in Carrillo *et al*., 2012b

of green leaves. Heavy infestations typically occur on the lower leaf surfaces, and feeding damage shows as yellow speckles and blotches on the leaves. Yellowing of the leaves may often be severe. Palms attacked by *R. indica* exhibit symptoms from scattered yellow spots on both leaflet surfaces to a strong yellowish coloration on the entire leaflet, but most of the leaflets are affected in the middle area of the leaf. Leaves of coconut palms that are heavily infested by *R. indica* become entirely yellow, especially on the lower third portion of the plant. It is not known whether this alteration results only from mite feeding in combination with the dry season or is also the result of the presence of a plant pathogen transmitted by the mite, because there is no information as to whether this mite is a disease vector (Moutia, 1958; Jeppson *et al.*, 1975; Hoy and Peña, 2006). Coconut palms can be severely attacked from both the very young to the very old (more than 50 feet tall), and show most of their lower leaves yellow. Young coconut plants may be the most damaged by the feeding of the mite. The yellow colour of the leaflets is followed by the abortion of the flowers or small nuts. The mite is spreading to other exotic and ornamental palms. In Dominica, where coconut is grown with bananas, *R. indica* has moved to this new host at numbers close to 200 mites cm^{-2}. The lower leaves of banana and plantain turn yellow with small patchy green–yellow areas. Most of the mites inhabit the lower leaf surfaces of the host plant, and the presence of yellow coloration on the upper surface of the leaf corresponds to the presence of a mite population on the lower leaf surface (Hoy and Peña, 2006).

Control

Control of the red palm mite needs the application of rigid preventive measures, and the use of chemical and biological means. Monitoring, plant resistance and cultural practices may be important.

MONITORING. Mites can be monitored by visual examination of the individual plant or plant part under a dissecting microscope or collected by beating the plants over a sieve screen fitted to a plastic funnel with a vial attached to it. Both methods allow tenuipalpids as well as predatory phytoseiid mites to be collected (Hoy and Peña, 2006).

PREVENTION. Importation of palms into the Neotropical region requires strict supervision if infestation by this new pest is to be limited (Flechtmann and Etienne, 2005).

PLANT RESISTANCE. Although significant differences in the morphological characters of coconut leaflets of different varieties are known, no relationship has been seen to occur between these characters, singly or in combination, and the rate of infestation by *R. indica*. Conversely, the chemical composition of the leaves (nitrogen, crude protein, calcium and phosphorus, and moisture contents) sometimes differs significantly among varieties and shows a positive correlation with the population increase of *R. indica* (Sarkar and Somchoudhury, 1989b).

The banana and plantain industries have been negatively affected to different degrees by *R. indica* infestation throughout the Caribbean. Genetic resistance in the host plant and the proximity of natural sources of mite infestation seem to be two of the main factors affecting *R. indica* densities in *Musa* plantations. Studies on the effect of coconut palm proximities and planting densities on *R. indica* populations infesting *Musa* plants, and the evaluation of susceptibility/resistance to colonization by *R. indica* populations of 14 *Musa* spp. hybrid accessions, showed differences in mite population increases in both the density and germplasm accession evaluations. The results of this research have implications for how the mite pest can be managed on bananas and plantains (Rodrigues and Irish, 2012).

BIOLOGICAL CONTROL. Twenty-eight species of predators, including mites and insects, have been found associated with *R. indica* in Asia, Africa and the Neotropics, and pathogenic fungi develop on *R. indica* in the Caribbean. Among the predators, the phytoseiid mites *Amblyseius caudatus* Berlese, *A. channabasavanni* Gupta and *A. largoensis* (Muna) are reported as important natural enemies of *R. indica*, and despite each area possessing a different natural enemy complex, *A. largoensis* is reported in all the geographical areas (Carrillo *et al.*, 2012b).

The biology of *A. largoensis* was investigated on various food sources (*R. indica*, *Tetranychus gloveri* Banks, *Aonidiella orientalis* (Newstead), *Nipaecocus nipae* (Maskell) and oak (*Quercus virginiana* Mill.) pollen) at 26.5 ± 1°C, 70 ± 5% RH and under a 12L:12D photophase to evaluate the development, survivorship, oviposition rate, sex ratio and longevity of the predator. *A. largoensis* completes its life cycle and reproduces when fed exclusively on *R. indica*. The development of immature stages of the predator was faster and the fecundity and survivorship were higher when fed on *R. indica* or *T. gloveri* than on the other test food sources. The intrinsic rate of natural increase of *A. largoensis* was higher when fed on *R. indica* than on other diets. These results suggest that *A. largoensis* could play a role in controlling *R. indica* (Carrillo *et al.*, 2010; Carrillo and Peña, 2010).

On excised *A. catechu* leaves infested with all the stages of *R. indica* under laboratory conditions at 25.42 ± 1.21°C and 57.54 ± 6.54% RH, the number of prey killed by each gravid starving female of *A. largoensis* was 1.30, there were 39.39% of successful attacks and the mean time spent on feeding on each prey was 74.65 s. *A. largoensis* fed on the egg, larva, nymph stages and female adults, but not on the males. It spent most of the time on prey foraging, grooming and resting (Rodríguez *et al.*, 2010). Under no-choice conditions, in which eggs, larvae, (proto) nymphs and adult females of *R. indica* were offered individually over 2 days to *A. largoensis*, the predator consumed more eggs than other stages of *R. indica*. In choice tests, in which different combinations of all four stages were offered in different arenas, the predator exhibited a significant preference for *R. indica* eggs over the other prey stages. *A. largoensis* displayed a type II functional response, showing an increase in the number of prey killed with an increase in prey population density. The consumption of prey stabilized at approximately 45 eggs daily, the level at which oviposition by the predator was maximized (2.36 ± 0.11 eggs daily; mean ± SEM). The results of this study suggest that *A. largoensis* can play a role in controlling *R. indica* populations even when prey densities are low (Carrillo and Peña, 2012).

Further studies by Carrillo *et al.* (2012c) investigated variations in the response of *A. largoensis* to *R. indica* using four different populations of *A. largoensis* with different previous

feeding experiences on *R. indica*. The results show that *A. largoensis* is likely to accept and consume high numbers of *R. indica* eggs regardless of its previous feeding experience, although all of the populations consumed relatively fewer *R. indica* larvae than the other prey tested (*T. gloveri*). Predators previously exposed to *R. indica* were more likely to consume *R. indica* larvae. By contrast, predators not previously exposed to *R. indica* showed the lowest likelihood of choosing to feed on this prey item. Plasticity in the response of *A. largoensis* to *R. indica* larvae could be associated with selection, learning or a combination of both (Carrillo *et al.*, 2012c).

Comparison of the biology of a population of *A. largoensis* that had been in association with *R. indica* for a long time on Réunion Island in the Indian Ocean, with a population from Roraima State in northern Brazil, where *R. indica* was first found only recently, showed no significant differences between populations in relation to the duration of different immature stages or total survivorship. However, the oviposition period, prey consumption and net reproductive rate were significantly higher for the Réunion population, warranting further investigation to determine whether that population should be released in Roraima to control the pest (Domingos *et al.*, 2012).

The analysis of high-fidelity RAPD (random amplified polymorphic DNA)-PCR markers, in addition to Bayesian and sequence divergence analyses of 12S ribosomal RNA sequences, shows that *A. largoensis* populations from Mauritius and south Florida differ genetically, but whether these are species or population differences is unknown. The degenerate EF-Iα (nuclear elongation factor Iα) primers used to survey the phytoseiids amplified two different elongation factor sequences with distinct amino acid translations, the putative EF-Iα and an unknown elongation factor. Variability within the 12S gene has been used to develop population-specific primers for identifying the Mauritius phytoseiids in the event they are released in south Florida (Bowman and Hoy, 2012). Molecular detection of predation by identifying prey markers in the digestive tract of predators has developed into a powerful tool for assessing predator–prey systems in which diet identification is too time-consuming or impossible. This technique needs refinement specific for this system, but the results obtained here confirm that it could turn into a very useful tool for assessing aspects of this predator–prey system (Rivera-Rivera *et al.*, 2012). However, despite the positive results of initial approaches to evaluating the action of *A. largoensis*, there is no information yet available on the practical use of this predator in the field.

Another biological control agent, *Bacillus thuringiensis* Berliner, was judged to be unsatisfactory against females and eggs of *R. indica* (Ramos Lima *et al.*, 2010).

CHEMICAL CONTROL. If infested palm trees are very tall, chemical control is difficult and expensive. However, in the absence of or without the adequate presence of natural enemies, spray applications of chemicals may be needed to control the red palm mite in palm nurseries. At present, it is unclear which chemicals are appropriate, although some common acaricides, such as oil and sulfur, should be further tested for their effectiveness (Hoy and Peña, 2006).

A number of active ingredients have been applied to coconut palms, bananas and other economic plants in various countries. These include abamectin, acequinocyl (Peña and Rodrigues, 2010; Rodrigues and Peña, 2012), bifenazate, chlorfenapyr (Peña and Rodrigues, 2010), cypermethrin, demeton-*S*-methyl (Senapati and Biswas, 1990), dicofol (Sarkar and Somchoudhury, 1988; Senapati and Biswas, 1990; Jayaraj *et al.*, 1991; Peña and Rodrigues, 2010; Ramos Lima *et al.*, 2010; Rodrigues and Peña, 2012), dimethoate (Jalaluddin and Mohanasundaram, 1990), endosulfan (Sarkar and Somchoudhury, 1988; Senapati and Biswas, 1990; Jayaraj *et al.*, 1991), ethion (Sarkar and Somchoudhury, 1988; Jayaraj *et al.*, 1991), etotaxole (Peña and Rodrigues, 2010; Rodrigues and Peña, 2012), fluvalinate (Senapati and Biswas, 1990), lime sulfur (Mathurin *et al.*, 2010), methomyl (Jalaluddin and Mohanasundaram, 1990), milbemectin (Peña and Rodrigues, 2010; Rodrigues and Peña, 2012), monocrotophos (Sarkar and Somchoudhury, 1988; Senapati and Biswas, 1990; Jalaluddin and Mohanasundaram, 1990; Jayaraj *et al.*, 1991), oil (Peña and Rodrigues, 2010), phosalone (Jayaraj *et al.*, 1991), phosphamidon (Jalaluddin and Mohanasundaram, 1990), pyridaben (Peña and Rodrigues, 2010; Rodrigues and Peña, 2012), quinalphos (Sarkar and Somchoudhury, 1988), safe oil (Mathurin *et al.*, 2010), spiromesifen (Peña and Rodrigues, 2010; Rodrigues and Peña, 2012), sulfur (Jayaraj *et al.*, 1991; Peña and Rodrigues, 2010; Mathurin *et al.*, 2010; Rodrigues and Peña, 2012) and tetradifon (Jayaraj *et al.*, 1991).

In West Bengal, India, monocrotophos was the most effective on coconuts (Sarkar and Somchoudhury, 1988), and on arecanuts during April–May a water spray of monocrotophos was also effective (Senapati and Biswas, 1990). In Tamil Nadu, India, topical sprays of ethion, endosulfan or phosalone were suggested on young coconut palms, and stem injection or root feeding on older plants (Jayaraj *et al.*, 1991). On coconut palms and bananas, the mixture of natural sugar esters plus acequinocyl or natural sugar esters plus spiromesifen, or natural sugar esters alone, provided satisfactory control 3–12 days after treatment (Peña and Rodrigues, 2010). Other trials on palms and bananas indicated the effectiveness of spiromesifen, dicofol and acequinocyl in reducing the population of *R. indica* on coconut in Puerto Rico, and spray treatments with etoxanole, abamectin, pyridaben, milbemectin and sulfur gave mite control in Florida. In addition, the acaricides acequinocyl and spiromesifen were able to reduce the population of *R. indica* in banana trials (Rodrigues and Peña, 2012). On coconut palms in St Lucia, sulfur and lime sulfur gave the best control, while that of a mixture of cottonseed oil, clove oil and garlic oil was ineffective (Mathurin *et al.*, 2010).

22 The Family Tarsonemidae Canestrini *et* Fanzago

Morphological Characteristics, Systematics and Bio-ecology

The Tarsonemidae are small mites, from 100 to 300 μm long, translucent, pale or whitish, but with their colour influenced by the food they ingest. These phytophagous species may appear greenish, depending on the green leaves that they have attacked. They are usually dimorphic, with the males different from the females in size and morphological characteristics. The body of the female is usually ovoid, and sometimes elongate, such as in *Steneotarsonemus* spp., while the male is smaller than the female and has a characteristic subterminal 'genital capsule' and a strong leg IV, particularly conformed and with a single claw. The body consists of a gnathosoma, prodorsum and opisthosoma (Fig. 22.1). The gnathosoma includes parts, such as the stylophore, palpi, infracapitulum and chelicerae (Fig. 2.7G). The stylophore is derived from the merged bases of the chelicerae and, integrated with the infracapitulum, forms the gnathosomal capsule. The palpi are 2-articulate and usually closely appressed, and the chelicerae are retractable and of different lengths in different species, with the movable digit as a fine stylet. The prodorsum of the female has a sclerotized dorsal shield that is variously ornate, one pair of stigmata, two pairs of setae (v_1, sc_2), one pair of bothridia supporting the pseudostigmatic organs and one pair of pits (v_2) (Figs 2.17A, 22.1). The prodorsal shield of the male has vertical (v_1, v_2) and scapular setae (sc_1, sc_2) that are variously expressed.

The opisthosoma has five dorsal tergites (C, D, EF, H and Ps) that are related to one another in a telescopic manner. Tergite C of the female is usually entire, well separated from the next tergite D, and bears the setae c_1 and c_2. Tergites C and D of the male are fused to form tergite CD. Tergite D of the female is usually entire, with one pair of setae (*d*) and cupules (ia). In the female, tergite EF derives from the fused tergites E and F, and bears two pairs of setae (*e*, *f*) and one pair of cupules (im). In the male, tergite EF normally has one pair of setae and one pair of pits, except in some species such as the broad mite, *Polyphagotarsonemus latus* (Banks), which has two setae (Fig. 22.2). Tergite H bears the setae *h* and the cupules ih. The pseudanal tergite Ps is a small structure, caudal or ventrocaudal, little sclerified, and with one pair of caudal setae (*ps*) in the adult female (Fig. 22.1) and no setae in the male (Fig. 22.2).

Ventrally, the female has an expanded and consistent aggenital shield which sometimes has one pair of aggenital setae; the shield is absent in the male. The ventral podosoma consists of four different coxisternal plates that are generally not ornate. The first two plates are fused to each other in the middle position and internally they form a prosternal apodema, and apodemes 1 and 2; they stretch around the edge of the rear trochanters I and II in the dorsolateral position above the legs, and they have one pair of setae (*1a*, *2a*) with a pit each (*1b*, *2b*). The coxisternal metapodosomal plates III and IV have three apodemes (3, 4 and poststernal); those of the fourth pair of apodemes are normally associated with the poststernal, which is unequal and median. Two pairs of setae (*3c*, *3b*) are set near apodemes 3 and 4 (Fig. 22.1).

The legs have six articles. Tibia and tarsus I of the female are fused into a single article, while in the male they are always separated. In most species, leg I of the male and female has a single claw, which in the female may exhibit profound changes and is sessile, large and uncinate. Pretarsus I, unlike pretarsi II and III, lacks a true empodium (Figs 2.11D, 22.2). Trochanter III of the female is long, subelliptic and fixed; femur III is fused with the genu. Leg IV of the male is usually 4-segmented, with the trochanter short and femur and patella fused. Leg IV of the female has three segments, with the trochanter short, the second segment resulting from the fusion of the femur and genu, and the last from the fusion of the tibia with the tarsus; it ends with a long seta and is devoid of a distal claw. The chaetotaxy and setal notation are typical; for an extensive discussion of the topic it is suggested that Lindquist (1986) is consulted. An example is given in Fig. 22.2B for leg I of the male of *P. latus*.

The life cycle of tarsonemids develops through the egg, larva and adult stages, with a quiescent nymph stage inside the larval cuticle. The absence of active nymphs reduces the length of the life cycle, which depends on the host plant, temperature, RH and other ecological factors. The reproduction is fundamentally arrhenotokous parthenogenesis, with the involvement of a haplodiploid sex mechanism. Males usually originate from haploid eggs and females mostly from diploid eggs, and the basic chromosome numbers are $n = 2$ and $2n = 4$ (Flechtmann and Flechtmann, 1984; Helle *et al.*, 1984). Garman (1917) observed only females of *Phytonemus pallidus* (Banks) for several generations, but Karl (1965a) found that unfertilized females of the same species generated offspring of both sexes. These results show that the tarsonemid reproduces more often by amphitoky than by thelytoky. Moreover, Karl (1965b) reported that most populations of *Tarsonemus confusus* Ewing reproduced by arrhenotoky, but that one population was thelytokous for over 30 generations. Schaarschmidt (1959) observed amphitoky in *T. fusarii* Cooreman. However, in free-living arrhenotokous tarsonemids, unfertilized adult females generate only males, although if they subsequently mate with one or more male offspring, they generate mostly female progeny (Schaarschmidt, 1959; Karl, 1965b; Flechtmann and Flechtmann, 1984).

Males find pharate females still contained within the larval cuticle, and affix the genital capsule to the body of the female, which they carry for as long as 24 h (this behaviour is called

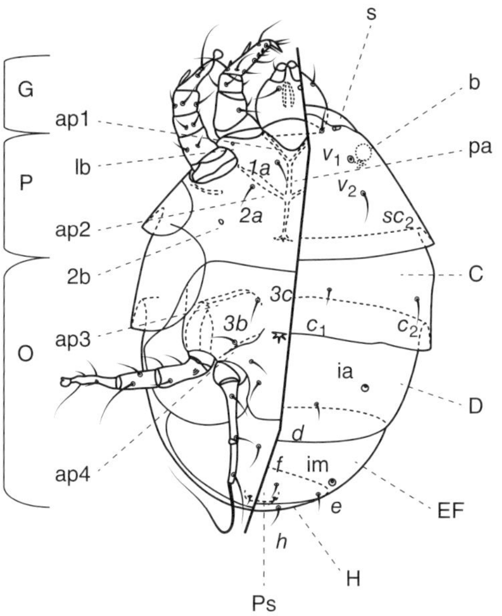

Fig. 22.1. *Polyphagotarsonemus latus* (Banks). Female, dorsal (right) and ventral (left) views (from Nucifora and Vacante, 2004). Key: ap 1–4, apodemes 1–4; b, bothridium; C, D, EF, H, Ps, opisthosomal tergites; G, gnathosoma; ia, im, cupules; O, opisthosoma; P, prodorsum; pa, prosternal apodeme; s, stigma. The rest of the labels designate setae, and an explanation of the setal notation is given in the text.

'precopulation'); they make a true copulation as soon as the females emerge (Nucifora, 1963; Suski, 1972). Dispersal occurs by adult females, which are usually fertilized and adopt different strategies, such as an association with their host (parasitic species) (Rennie, 1921; Lindquist, 1969, 1970), walking for a short distance or by phoresy with flying insects for long distances, as occurs in *P. latus* (Flechtmann *et al.*, 1990; Parker and Gerson, 1994; Palevsky *et al.*, 2001).

Members of this family have various feeding habits, including algivorous, fungivorous, insectophilous, nidicolous and phytophagous, and the symptoms that result from feeding vary according to the species and are sometimes unclear or in doubt (Walter *et al.*, 2009). Only a restricted number of species in the genera *Polyphagotarsonemus* Beer *et* Nucifora, *Hemitarsonemus* Ewing, *Steneotarsonemus* Beer, *Phytonemus* Lindquist and *Tarsonemus* Canestrini *et* Fanzago are plant feeders. Some species (*Steneotarsonemus* spp.) initiate fungal infection. The most important pests are *P. latus*, which causes severe damage to a variety of economic plants (ornamentals, vegetables, citrus, etc.), and *P. pallidus*, which damages strawberries and several ornamental flowers and shrubs (Van Eyndhoven and Groenewold, 1959; Nucifora, 1961; Jeppson *et al.*, 1975; Vacante, 2000, 2010; Zhang, 2003; Vacante and Benuzzi, 2007). Other plant-feeding species belong to the genus *Steneotarsonemus*.

Of the *Steneotarsonemus* spp., the pineapple leathery pocket mite, *S. ananas* (Tryon), causes brown spots on the white tissues at the base of the leaves of pineapple, especially in young plants (Meyer Smith, 1981a). In South Africa, the presence of

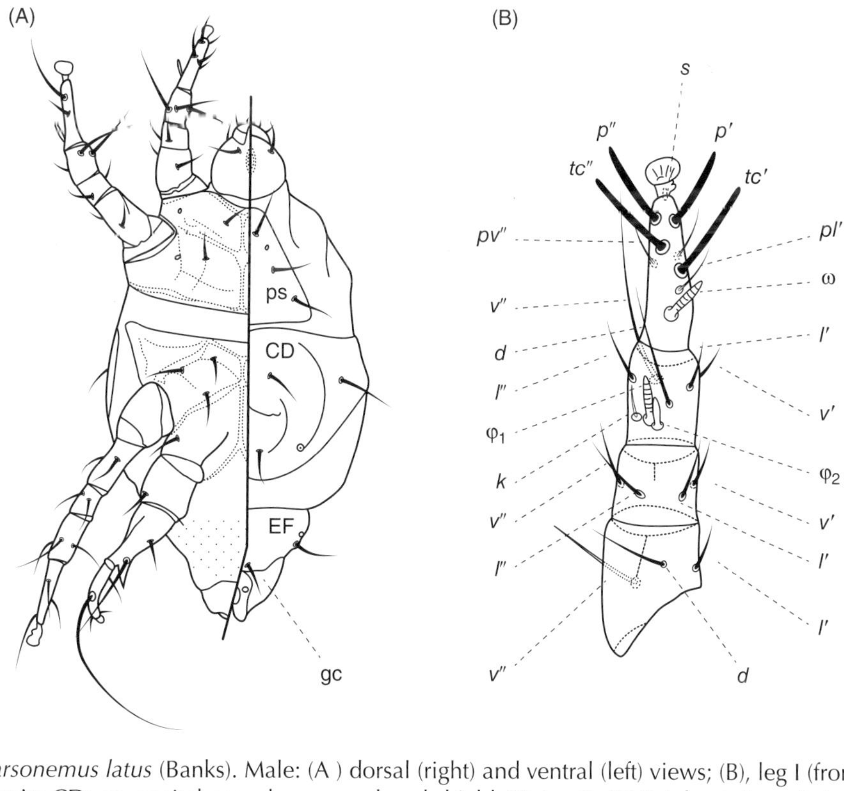

Fig. 22.2. *Polyphagotarsonemus latus* (Banks). Male: (A) dorsal (right) and ventral (left) views; (B), leg I (from Nucifora and Vacante, 2004); CD, tergite CD; gc, genital capsule; ps, prodorsal shield; EF, tergite EF. Setal notation: *d*, dorsal seta; *k*, tibial euphathidium; *l'*, *l''*, lateral setae; *p'*, *p''*, proral eupathidia; *pl'*, primilateral seta; *pv''*, primiventral seta; *s*, subunguinal seta; *tc'*, *tc''*, tectal eupathidia; *v'*, *v''*, ventral setae; φ_1, φ_2, tibial solenidia; ω, tarsal solenidion.

this mite is associated with 'leathery pocket disease' (Petty, 1975), which is probably due primarily to the fungus *Penicillium funiculosum* Thom (Rohrbach, 1979, in Meyer Smith, 1981a). In the West Indies, feeding of the sugarcane mite, *S. bancrofti* (Michael), causes transparent, brown, crater-like depressions on the young sugarcane stalk, and the plant tissues appear scabby or scarred; later, as the cane grows, these areas become opaque, brown and rough (Holloway, 1936; Butani, 1959). *S. furcatus* De Leon causes severe distortion of leaves and shortened internodes on ornamental grasses (*Paspalum* spp.) and on *Maranta leuconeura* Morr, resulting in stunted plants (Jeppson *et al.*, 1975). The bulb scale mite, *S. laticeps* (Halbert), feeds mainly on the epidermal surface of the scales of bulbs of *Narcissus* and other Amaryllidaceae. On developing shoots, it causes distortion, stunting and death of leaves and flowers. Scar-like tissues on leaves and flower stems cause bronze streaks and transverse cracks. Infested bulbs have yellowish brown areas (Doucette, 1936; Jeppson *et al.*, 1975). The rice panicle mite, *S. spinki* Smiley, feeds on rice (*Oryza sativa* L.), on the invasive *O. latifolia* Desv. and other *Oryza* spp. (USDA NRCS, 2007); it has a wide range of distribution, including regions of the Americas; its spread to these regions has recently been discussed by Návia *et al.* (2010b). *S. spirifex* (Marchal) causes spiral malformations of the rachis, blind spikelets and incompletely emerged inflorescences of cereals (Evans *et al.*, 1961). The fern mite, *Hemitarsonemus tepidariorum* (Warburton), feeds on various species of ferns in greenhouses in the USA and the UK (Cameron, 1925; Pritchard, 1951a; Beer, 1954). *T. myceliophagus* Hussey is reported as a serious pest in mushroom culture (Hussey, 1963; Hussey and Gurney, 1967). Moreover, tarsonemid mites have been reported to carry spores of pathogenic fungi in sporotheca on their bodies (Bridges and Moser, 1983; Moser, 1985; Blackwell *et al.*, 1986, 1989; Levieux *et al.*, 1989; Ochoa *et al.*, 1991b, 1994; Moser *et al.*, 1995).

The Tarsonemidae include about 530 described species ascribed to 40 genera worldwide (Lindquist, 1986; Lin and Zhang, 2002) and included in three subfamilies (Pseudotarsonemoidinae Lindquist, Acarapinae Schaarschmidt and Tarsoneminae Canestrini *et* Fanzago) and seven tribes. The genus *Polyphagotarsonemus* belongs to the subfamily Pseudotarsonemoidinae and the tribe Pseudotarsonemoidini Lindquist. The genus *Hemitarsonemus* Ewing is included in the tribe Hemitarsonemini Lindquist, the genera *Steneotarsonemus* and *Phytonemus* in the tribe Steneotarsonemini Lindquist, and the genus *Tarsonemus* in the tribe Tarsonemini Canestrini *et* Fanzago (Lindquist, 1986). The key to the identification of subfamilies and genera (adults) that is given below is based on Lindquist (1986).

Selected websites

Acari: Plant-Feeding Mites. Systematic Entomology Laboratory, US Department of Agriculture Agricultural Research Service, Beltsville, Maryland. Available at: http://www.sel.barc.usda.gov/acari/frames/plantfeed.html

New Zealand Biosecurity – Taxonomy, Tarsonemidae, from PaDIL (High quality images and Information tools designed for Biosecurity and Biodiversity). An initiative of the Australian Government Department of Agriculture, in collaboration with Museum Victoria, Plant Health Australia, the Department of Agriculture and Food Western Australia and the Plant Biosecurity Cooperative Research Centre. Available at: http://www.padil.gov.au/maf-border/search?queryText1=&queryType1=All&sortType=ScientificName&viewType=Details&pageSize=10&page=1&taxonomy=Acari:Prostigmata:Tarsonemidae

Key for the Identification of Genera (Adults)

1. • The metapodosomal venter of both sexes has 3 or 4 pairs of setae, including 1 pair of setae between the bases of legs IV in the female and on the coxal plates in the male (subfamily **Pseudotarsonemidae** Lindquist); the metapodosomal venter of both sexes has 4 pairs of setae; the prodorsal shield of the female does not cover the stigmata, which open on the dorsolateral edge of the shield, slightly anteriad of the vertical setae; the dorsal setae are slender, and the scapular setae are longer than the tergital pairs; leg IV of the female has 2 setae on the tibiotarsus; leg IV of the male has a pointed, triangular flange set distally of the posterolateral surface of the femorogenu, and a reduced button-like claw distally...... ....genus ***Polyphagotarsonemus*** Beer *et* Nucifora
   • The metapodosomal venter of both sexes has 2 pairs of setae, none of which are set between the bases of legs IV in the female nor on coxal plates IV in the male; the prodorsum of the female is usually provided with a pair of capitate trichobothria, rarely setiform, vestigial or absent; the prodorsum of the male has 4 or rarely 3 pairs of setae... ....subfamily **Tarsoneminae** Canestrini *et* Fanzago....2
2. • Femur I with 3 setae; the prodorsal shield of the female is not hood like over the gnasthosoma, and has the stigmata enlarged and aligned between the vertical and scapular setae; ventral apodemes 3 not extending laterad of trochanter III; apodemes 3 extending posterolaterad of setae *3b* to bases of trochanters IV; tarsi II and III have paired claws that are symmetrically developed; the male has the plate CD with the setae slightly barbed; leg IV has the femorogenual flange rounded apically, and set subdistally; the genital capsule is elevated well above the bases of legs IV in unmounted specimens........................*genus* **Hemitarsonemus** *Ewing*
   • Femur I with 2–4 setae...3
3. • Idiosoma of female usually elongated–ellipsoidal, with relatively short legs, and leg II usually not longer than leg IV; the femorogenu of leg IV of the male usually with a prominent, rounded flange whose distal margin forms an acute angle where it is attached to the surface of the legs, and rarely lacking a flange; apodeme 2 of the female is obliquely slanted, and not bifurcate anterolaterally; the coxisternal plates III and IV are continuous medially; male with scapular setae sc_2 usually much shorter than sc_1............................... ....genus ***Steneotarsonemus*** Beer
   • Idiosoma of female usually not elongated, with moderately long legs, and leg II usually longer than leg IV; femorogenu of leg IV of male with or without flange; when present and rounded, the flange has the distal margin

forming an obtuse angle where it is attached to the surface of the leg..4

4. • Male and female without a sejugal apodeme; female with apodemes 4 extending posterolaterad of the bases of the setae *3b* though not reaching to the bases of legs IV; leg IV of male with prominent flange on femorogenu, and with fused tibiotarsus......................genus *Phytonemus* Lindquist

• Male and female with sejugal apodeme fully developed or reduced only medially; female with apodemes 4 not extending posterolaterad of bases of setae *3b*; leg IV of male with or without flange on femorogenu, and with tibia and tarsus separate or fused..........genus *Tarsonemus* Canestrini *et* Fanzago

The Genus *Polyphagotarsonemus* Beer *et* Nucifora

Diagnostic characteristics

The prodorsal shield of the female is not enlarged and does not cover the stigmata; the dorsal idiosomal setae are slender, and the setae sc_2 are longer than the other setae. Femur I and tarsus II of the female lack seta *l″* and spine-like seta *pl″*, respectively; the female has a gently curved, sessile claw and reduced, contiguous unguinal setae on leg I; the female usually has tarsal setae *pv″* enlarged, spine like on leg I, and only 2 setae on the tibiotarsus of leg IV (Lindquist, 1986).

Economic importance

The genus *Polyphagotarsonemus* has only two species, *P. latus* (Banks) and *P. beeri* (Smiley). *P. latus* is commonly called the broad mite, and feeds on the tender tissues of young leaves, flowers and the immature fruits of a large variety of economic plants in subtropical and tropical countries, and in greenhouses of temperate regions, causing severe damage under appropriate conditions (Nucifora, 1961; Jeppson *et al.* 1975; Gerson, 1992; Vacante, 2000, 2010; Vacante and Benuzzi, 2007). *P. beeri* is reported on *Ficus* in Congo (Smiley, 1964), but its bio-ecology and damage are not well documented.

Polyphagotarsonemus latus (Banks) (Figs. 22.1–22.2)

Common name

Broad mite, citrus silver mite, yellow tea mite, white mite or tropical mite.

Diagnostic characteristics

FEMALE. The body is oval, about 200 μm long and white amber to yellow or pale greenish coloured, sometimes with a white median spot, intense and with faint contours. The body setae are short, the pseudostigmatic organs are spherical and the ventral metapodosoma has 4 pairs of setae. The tibiotarsus of leg I has one claw.

MALE. The body is about 170 μm long, and similar in colour to that of the female. The prodorsum has 3 pairs of setae and the ventral metapodosoma 4 pairs of setae; the tibia and tarsus IV are fused into a tibiotarsus ending with a knob-like claw; coxae III and IV are intimately associated.

Geographical distribution

Worldwide. Common in tropical and subtropical areas and in greenhouses of temperate and subtropical countries (Jeppson *et al.*, 1975).

Host plants

The list of host plants includes wild and cultivated species belonging to over 57 families (Gerson, 1992). The most important economic plants are beans, citrus, cotton, peppers, potatoes, rubber, tobacco (Mann *et al.*, 1920; Kulkarni, 1923; Hambleton, 1938; Jeppson *et al.*, 1975), cowpea (Karmakar *et al.*, 1996a), sesame (Ahuja, 2000; Dhooria, 2005a), mulberry (Chauhan *et al.*, 2002), green gram (Dhooria, 2005b), coastal morning glory (Maia and Buzzi, 2006), grapevines (Ferreira *et al.*, 2006), watermelon (Kousik *et al.*, 2007) and jute (Karmakar *et al.*, 1996a; Somchoudhury *et al.*, 2008). Ornamental plants include the African violet, ageratum, azalea, begonia, chrysanthemums, cyclamen, dahlia, gerbera, gloxinia, ivy, jasmine, Japanese camellia, impatiens, lantana, marigold, peperomia, pittosporum, snapdragon, verbena and zinnia (Johnson and Lyon, 1991; Baker, 1997; Pérez Otero *et al.*, 2007).

Bio-ecology

P. latus was described by Banks (1904b) from the terminal buds of mango in a greenhouse in Washington, DC. It has a worldwide distribution and is one of the most important phytophagous mites as it is injurious to many economic plants in tropical and subtropical areas, and in greenhouses in temperate and subtropical climates.

The broad mite feeds on tender tissues of new leaves, blossoms and young fruits. Females oviposit in hollows of the lower leaf surface and on younger fruit. Reproduction is by arrhenotokous parthenogenesis and development typically requires tropical climatic conditions. Experimental studies using lemons show that optimum development occurs at 25°C and an RH between 90 and 100% (Jones and Brown, 1983). Li (1990) reported that at 80% RH, the lethal thermal threshold was 43.6°C and at 25°C the intrinsic rate of increase (r_m) of the population was 0.3283. Several studies noted that there was no active nymphal stage, and that males emerged earlier than females and were commonly seen congregating around a quiescent female nymph (Nucifora, 1963; Dhooria, 2005b; Manjit Singh and Dhooria, 2006). The broad mite is active throughout the year and its reproductive activity slows during the coldest months

(Jeppson *et al.*, 1975; Schwartz, 1977c). The broad mite exhibits similar behaviour in cold Mediterranean greenhouses (Vacante and Benuzzi, 2007). The males carry the quiescent nymphs, also known as pharate females, from older leaves on the new sprouts (Lavoipierre, 1940; Moutia, 1958). The host plant generally influences the different life stages of *P. latus*.

Under laboratory conditions, the longest life cycle (from egg to egg) was observed on tea (190.50 h) and the lowest (67.88 h) on cowpea (*Vigna unguiculata* (L.) Walp.). The longest oviposition period (241.25 h) was on jute (*Corchorus* sp.) and the shortest (181.25 h) on black nightshade (*Solanum nigrum* L.). The post-oviposition period (67.00 h) was longest on *Ludwigia parviflora* Roxb. and shortest (33.00 h) on cowpea. The longest lifespan (356.88 h) was observed on *L. parviflora* and the shortest (309.50 h) on *S. nigrum*. The greatest fecundity (64.25 eggs per female) was observed on chilli (*Capsicum annuum* L.) and the lowest fecundity (20.75 eggs per female) on *L. parviflora*. During an oviposition period of 8–10 days, the broad mite laid 62.20–71.19% of its eggs within the first 5 days (Karmakar *et al.*, 1996a).

Many studies on the bio-ecology of the broad mite have been done on chilli both in the field and under laboratory conditions. On chilli plants (cv. 'LCA-305'), the mean durations of the egg, larval and quiescent stages of males and females were 36.3, 19.8 and 17.0 h, and 38.8, 21.5 and 19.3 h, respectively. The pre-oviposition, oviposition and post-oviposition periods of the female lasted 1.15, 6.30 and 1.17 days, respectively. The longevity of males and females was 6.41 and 7.92 days, respectively (Srinivasulu *et al.*, 2002a). On pepper at 25°C and 85-90% RH, the intrinsic rate of increase (r_m), finite rate of increase (λ), net reproduction rate (R_0), mean generation time and double population time were 0.2948, 1.3429, 25.4944, 13.8509 days and 2.3512 days, respectively. In the experimental population, the age distribution was stable at 70.1781% of immature stages and 29.821% of mature stages (Zhi, 2002). In rearing studies on chilli leaves, at 29.5°C and 85% RH, the maximum female birth (m_x = 3.919) in the life cycle was observed on the third day of oviposition (11th day of pivotal age). The net reproductive rate (R_0) was 17.329 females per female a day in a precise generation time of 11.78 days. The innate capacity for increase (r_m), finite rate of increase (λ), weekly multiplication, doubling time and hypothetical (F_2) female population were 0.2430 females per female a day, 1.28 females per female a day, 5.63 times, 2.85 days and 300.29, respectively. On reaching stable age distribution, the contributions of the egg, larval and adult stages were 66.49, 18.2 and 15.49%, respectively (Rai and Solanki, 2003). In another study on chilli leaves, the average incubation period was 58.39 h at 22.5°C and 47.77 h at 27.5°C. Larval development took 45.25 h at 22.5°C and 37.02 h at 27.5°C. At 22.5 and 27.5°C, the mean oviposition period was 5.93 and 5.35 days, respectively; females lived longer than males at both temperatures. At 22.5 and 27.5°C, the female laid on average 7.16 eggs and 5.61 eggs, respectively; the sex ratio was in favour of females at both temperatures (Manjit Singh and Dhooria, 2006). On chilli in the field in India, the incidence of mites was highest in the 40th meteorological week when the prevailing maximum and minimum temperatures, morning and evening RH, rainfall and bright sunshine hours were 35.8°C, 18.0°C, 76%, 34%, 0.00 mm and 11 h, respectively. The mite population was negatively correlated with rainfall and positively correlated with bright sunshine hours. Regression analysis showed that the population decreased by 0.055 and increased by 0.368 per unit increase in rainfall and bright sunshine hours, respectively (Bhede *et al.*, 2008).

On potato leaves at 25 ± 1°C, 70 ± 10% RH and with a 14L:10D photoperiod, the developmental stages of egg hatching, nymphal, quiescent and pre-oviposition lasted 1.32, 1.1, 1.12 and 1.3 days, respectively. Male and female longevity was 7.9 and 9.76 days, and mean fecundity was 30.53 eggs (Namvar and Arbabi, 2007). Also on potato leaves, but at different temperatures (15, 20, 25, 30 and 35 ± 1°C), 50–60% RH and with a 16L:8D photoperiod, the duration of incubation, active and quiescent larval stages, pre-oviposition, oviposition, longevity, life cycle and the generation period was 6.25, 3.50, 2.38, 3.13, 4.69, 9.06, 12.13 and 14 days at 15°C for mated females and 6.44, 3.13 (active larval period only), 2.13, 3.25, 4.88, 9.19 and 11.69 days for unmated (parthenogenetic) females, respectively. The daily egg deposition was highest (5.00 and 2.25 for mated and parthenogenetic females, respectively) at 35°C and lowest (1.36 for both types of female) at 15°C. The highest fecundity was at 25°C for fertilized females (11.57) and 30°C (8.86) for unmated females. The hatching rate was highest (93.06 and 86.67% for mated and unmated females, respectively) at 35°C. The percentages of females and males reached 84.33 and 15.67% at 30°C, and 60.20 and 39.80% at 20°C, respectively. The sex ratio (female:male) was highest (6.5:1) at 30°C and lowest (2.1:1) at 20°C (Al-Ani and Al-Jboory, 2008).

In acclimatized chambers on grapevines (*Vitis vinifera* L., cv. 'Italy'), the development of *P. latus* from egg to adult took 3.4 and 6.8 days for males and 3.5 and 7.4 days for females at 32 and 18°C, respectively. At 18, 25 and 32°C, the female laid 16.5, 44.3 and 13.3 eggs, respectively. The thermal thresholds for the egg, larva and pupa and egg to adult period were 11.23, 9.45, 12.19 and 9.71°C, and the thermal constant was 28.51, 14.59, 8.33 and 62.73 degree-days. At 18, 25 and 32°C, the mean duration of one generation (T) was 25.6, 10.8 and 8.2 days, respectively. The net reproductive rate (R_0) was highest at 25°C, corresponding to an increase of 30.12 times at each generation. At 18, 25 and 32°C, the intrinsic rate of population increase (r_m) was 0.10, 0.31 and 0.12, and the finite rate of population increase (λ) was 1.10, 1.36 and 1.13. According to these experimental mean temperature values, it is concluded that in the municipal districts of Petrolina (Pernambuco) and Juazeiro (Bahia) in Brazil, *P. latus* can develop 95 and 99 generations a year, respectively (Ferreira *et al.*, 2006).

On green gram (*Vigna radiata* L.) under laboratory conditions, females laid eggs singly, preferably on the ventral surface of the leaves. The incubation period lasted 1–3 days and the larval period 2–3 days. The larvae then entered a quiescent stage that later transformed into an adult male or female. Females laid 3–11 eggs on green gram leaf discs. Male longevity varied from 1 to 3 days, while female longevity was around 8 days (Dhooria, 2005b).

In an experimental study on lemon (*Citrus limon* Burm) at 27.5 ± 0.5°C, 67.6 ± 1.3% RH and under a constant photoperiod, the development of the young stages of female and male *P. latus* took about 3.7 ± 0.1 and 3.6 ± 0.1 days, respectively, with 100%

survival. Pre-oviposition lasted 1 ± 0.2 days and the female laid 5.6 ± 0.5 eggs a day for 10.5 ± 0.9 days. The longevity was 13.4 ± 1.0 days for females and 12.0 ± 2.4 days for males. The intrinsic rate of increase was 0.359, the finite rate of increase 1.42 individuals per female a day, the average time of a generation was 10.34 days and the net reproductive rate was 41.0 (Vieira and Chiavegato, 1999). On Persian limes (*C. latifolia* (Yu. Tanaka) Tanaka) in the laboratory at 22.43°C and 63.0% RH, the development from egg to adult stage took 3.95 days and mortality was 17% (Ramos and Alvarez, 1987).

P. latus was observed attacking mulberry (*Morus* spp.) leaves in Nilgiris, Tamil Nadu, India. It was observed that the development of the life cycle took 100 h (Chauhan *et al.*, 2002).

In Punjab, India, on green gram sown in the field in the first week of July, broad mites were first detected in the first half of August, and the population increased thereafter, reaching a peak in the first week of September. The mite population then decreased, reaching a minimum at the end of September when the crop was mature. Besides green gram, the broad mite was found feeding and breeding on pulses, vegetables, ornamentals and fruits (Dhooria, 2005b). The incidence of the mite on sesame in Rajasthan (India) was found to be influenced by abiotic factors. Correlation between RH and incidence in both the morning and evening was positive and significant. Maximum temperature showed a negative and significant correlation with the broad mite population. However, the correlation between minimum temperature and rainfall and the number of mites was negative and non-significant. Symptoms of damage due to the mite started appearing 1 month after sowing in August and continued until harvest. The peak broad mite populations were recorded during mid-September, when the maximum and minimum temperatures ranged from 30.1 to 33.0°C and 22.6 to 25.2°C, respectively. The morning and evening RH recorded was 86–92% and 67–76%, respectively (Ahuja, 2000). In Punjab, India, peak populations of mites on chilli were observed during July–September (Manjit Singh and Dhooria, 2006). On jute (*Corchorus olitorius* L. and *C. capsularis* L.) in West Bengal (India), the peak population of mites was observed during the third week of June irrespective of cultivar and date of sowing. No relationship was found between population build-up and age of the crop; rather it was dependent upon the prevailing weather conditions (Hath and Jaydeb Ghosh, 2001). On the same host plant, and also in West Bengal, there was a positive correlation between temperature and the mite population per leaf. The maximum mite population was recorded during July, when the temperature ranged between 27.8 and 29.2°C (Somchoudhury *et al.*, 2008).

P. latus disperses through the males, which carry the quiescent nymphs (pharate females) from older leaves on to new sprouts, and also by wind and via farmers and (phoretic) insects. Notable among the latter are the tobacco whitefly, *Bemisia tabaci* (Gennadius) (Flechtmann *et al.*, 1990), and the greenhouse whitefly, *Trialeurodes vaporariorum* (Westwood) (Parker and Gerson, 1994). Palevsky *et al.* (2001) determined the specificity of the association between the mite and its phoretic hosts, and observed that it also attached to a citrus pest, the whitefly *Dialeurodes citri* (Asmehad) and to other whiteflies. The attachment to the carrier occurs equally well in the dark and the light, and broad mites do not differentiate between the sexes of their phoretic host *B. tabaci*. Waxy particles of the whitefly affect broad mite behaviour, and the broad mite is attracted specifically to whitefly waxy particle-treated leaf areas and also to leaf trichomes in this area, to which they become attached. The olfactory cues from the whitefly waxy particles play a crucial role in the recognition process of the phoretic host and/or the induction of the attachment behaviour to whitefly legs or leaf trichomes (Soroker *et al.*, 2003). Free-moving mites of all stages and both sexes preferred young cucumber (*Cucumis sativus* L., cv. 'Kfir') leaves to old cucumber leaves, and leaves of young cucumber rather than young tomato (*Solanum lycopersicum* L., cvs 'M82' and 'Moneymaker') plants, indicating that broad mites might choose their host actively. As for phoretic mated females, although most of the mites abandon the phoretic vector (*B. tabaci*), the rate of detachment from the whitefly vector is host dependent and correlated with the fitness of the mite for the particular host. In general, the host preference of phoretic female mites resembles that of the free-moving females, but the cues used by mites for the host selection remain to be explored (Alagarmalai *et al.*, 2009).

The natural enemies of broad mite include fungi and mites (Table 22.1).

Symptomatology and damage

The broad mite feeds on tender tissues and is common on young leaves, flowers and immature fruits of a large variety of economic plants. On infested gerbera, the lower leaf surface becomes rigid and rolled under at the edges, turns bronze and the flower rays are distorted and discoloured. In general, the feeding activity of the broad mite causes curling and crinkling of leaves and blister patches. High population densities stop plant growth and cause death (Smith, 1939a; Vacante and Benuzzi, 2007).

Feeding by *P. latus* on potato plants causes blackish spots on the lower leaf surfaces of young leaves, which turn gradually reddish. The plant becomes rosetted, showing prominent leaf hairs, with the leaves wrinkled at the edges. The plants wither and the buds are killed (Mann *et al.*, 1920). Masoud Arbabi *et al.* (2001) observed that during autumn, the twisting terminal and lateral sides of young potato leaves might exhibit browning or bronzing on the lower surfaces; high infestation densities decreased the length of stems, total plant growth, tuber size and crop yield.

On pepper seedlings, infestation by the broad mite prevents flower and fruit development, later causing flower dropping and silvering of the fruits (Kulkarni, 1923; Vacante and Benuzzi, 2007); similar damage occurs on aubergine plants, both in the field and in greenhouses (Vacante and Benuzzi, 2007).

The stems of terminal shoots and the lower surfaces of the young leaves of tomato become shiny bronze or brownish. The first injury is browning of the epidermal cells, but later the cells of developing tissues collapse and die. The young expanding leaves become narrow, stiff, twisted, or crumpled, and fail to develop, and they dry. The stems of young plants appear swollen, roughened or russeted and greyish green in colour. On cotton, the first symptoms occur on the lower leaf surfaces, and the leaves appear rigid and rolled under at the

Table 22.1. *Polyphagotarsonemus latus* (Banks). Natural enemies (fungi and mites).

Families and species	References
FUNGI	
Ascomycota	
Verticillium lecanii (Zimmermann) Viégas	Smitha and Giraddi, 2006a
Clavicipitaceae	
Beauveria bassiana (Balsamo) Vuill.	Peña *et al*., 1996
Hirsutella thompsonii Fisher	
Paecilomyces fumosoroseus (Wise) Brown and Smith	Peña *et al*., 1996; Smitha and Giraddi, 2006a
MITES	
Bdellidae	
Bdellid mites	Umeh *et al*., 2007
Phytoseiidae	
Amblyseius largoensis Muma	Rodriguez and Ramos, 2000b, 2004
Euseius hibisci (Chant)	
Euseius nicholsi (Ehara *et* Lee)	Wu, 1984
Euseius ovalis (Evans)	Moutia, 1958; Manjunatha *et al*., 2001
Euseius stipulatus (Athias Henriot)	Brown and Jones, 1983
Euseius victoriensis (Womersley)	Smith and Papacek, 1985
Neoseiulus agrestis (Karg)	Kolodochka and Prutenskaya, 1987
Neoseiulus barkeri Hughes	Fan and Petitt, 1994
Neoseiulus californicus (McGregor)	Peña and Osborne, 1996
Neoseiulus cucumeris (Oudemans)	Prabaningrum *et al*., 1999; Cantliffe *et al*., 2004
Neoseilus longispinosus (Evans)	Hariyapa and Kulkarni, 1988
Typhlodromalus peregrinus (Muma)	Peña, 1992
Typhlodromus athiasae Porath *et* Swirski	Karut *et al*., 1998

edges. Mature leaves may split or crack open, and have various shapes (Hambleton, 1938).

Infested cucumber plants (cvs 'M450' and 'Kfir') produce more ethylene than non-infested plants, and have smaller and harder leaves. The mites feed on epidermal cells, which disappear or collapse. Infested leaves are thicker, and display additional layers of more compact cells which do not undergo typical mesophyll differentiation. Feeding by the broad mite induces both the salicylic acid (SA) and the jasmonic acid (JA) pathways (Grinberg *et al*., 2004). The growth of infested plants is inhibited and there is a decrease in leaf numbers and leaf area. There is also an increase in the firmness of the infested leaves, as measured by a texture analyser. Although the mite feeds on the epidermis, structural and ultrastructural studies show that there are aberrations in the whole of the leaf tissue. Severe infestation leads to a complete loss of the epidermis and an increase in mesophyll cell size and number. The entire epidermal tissue appears to have collapsed, and the mesophyll cell walls appear thick and distorted. A Northern blot analysis of infested leaves showed the induction of genes related to the JA/ethylene and SA pathways, and an induction of oxidative stress-responsive gene transcripts, such as peroxidase (*PRX*). Transcript levels of aminocyclopropanecarboxylate (ACC) oxidase (*ACO*), which participate in ethylene biosynthesis, remained relatively constant. Broad mite feeding then induces strong morphological, structural and ultrastructural changes, along with an induction of genes involved in defence pathways. However, further studies are needed to evaluate whether these changes in leaf structure and ultrastructure affect the mite, and how the induction of the defence pathways affects the susceptibility of plants to *P. latus* infestations (Grinberg *et al.*, 2005).

Injury by the broad mite to watermelon mainly occurs on the growing terminals and the tender apical leaves, which appear bronzed, grow poorly and, in some cases, become distorted and curl upward (Kousik *et al.*, 2007). On mulberry, the broad mites develop on the abaxial surface of the tender and medium leaves, causing severe damage to the leaves of the crop by sucking their sap (Chauhan *et al.*, 2002).

Sesame plants infested by broad mites develop typical downward curling of the leaf margins. Necrosis of vegetative and floral buds occurs, resulting in general stunted growth of mite-infested plants. In Punjab (India), serious mite infestation was observed during July–September. The top canopy of the plants was most infested, and shady conditions were more preferred for mite infestation than sunny conditions. Leaves affected by phyllody were less infested by the mite, and floral and vegetative buds were more preferred than green pods containing seeds. Delaying sowing time by 1 week did not affect mite infestation in any way (Dhooria, 2005a). Also in Punjab, infested leaves of green gram had a deep green colour with a coppery red ventral surface. Downward rolling sometimes occurred on older leaves. Necrosis of growing points, particularly in young plants, is also noticed. Mite damage is quite high if the climatic conditions are hot and humid during the cropping season (Dhooria, 2005b).

P. latus damages young leaves and fruits of citrus, especially of lemon. Feeding by the mite causes distortions and curlings of leaves, sometimes so pronounced as to prevent the development of the sprout. The lower surface of the young leaves has a shining silver colour. Fruits are infested as soon as they appear and until they reach the size of a walnut. Young fruits in the shade and inside the apical part of the canopy are attacked most frequently. Infested fruits show a silvering of the pericarp with irregular contours and sometimes over the entire fruit surface; as the infestation and fruit develop, the alteration of the pericarp evolves from russeting with an irregular shape and variable colour starting from a whitish silver and then turning brown (Nucifora, 1961; Jeppson *et al.*, 1975; Vacante, 2010). The damage to lemon fruits is similar to that caused by *Phyllocoptruta oleivora* (Ashmead)

(see Chapter 16), and may result in a reduction of commercial value of the fruit of up to 100%. In South Africa, lemon losses of 30–70% have been reported (Meyer Smith and Schwartz, 1998e). In Nigeria, it has been observed that mite populations and the damage from them on sweet orange varieties increases to fruit maturity. The 'Washington Navel', 'Valencia Late', 'Agege' and 'Lue Gin Gong' varieties were most susceptible to mite attack and recorded higher fruit damage than other varieties (Umeh *et al.*, 2007).

The broad mite also attacks young seedling of apples, mainly the tender shoots, and these become thin, making the tree weak and stopping its growth (Zhao *et al.*, 2005).

Control

Because of their small size, broad mites are not noticeable until they cause serious damage to apical leaves in pepper seedlings (Jovicich *et al.*, 2004), and in other crops. Sometimes, it is possible to confuse broad mite damage with herbicide damage, nutritional deficiencies (e.g. of boron), physiological disorders or virus infections. Much effort has been devoted to the study of plant resistance and the biological control of the broad mite. However, there are no current reports on the extensive application of biological control. The correct use of chemical means may be a part of integrated pest management (IPM) programmes, taking into account their application at the first symptoms of the attack. For some crops, e.g. citrus and chilli, decision tools for control are well documented. Prevention is a fundamental part of any control strategy.

PREVENTION. Broad mites are active throughout the year and require a food source for survival, so implementing sanitation practices, such as cleaning greenhouses before introducing or transplanting new plants, and disinfecting benches, help to reduce the risk of outbreaks (Vacante, 2000; Vacante and Benuzzi, 2007; Cloyd, 2010). All-year activity also occurs under field conditions. Manjit Singh and Dhooria (2006) reported 31 host plants of broad mite, and an additional 13 plant species were found suitable for feeding and breeding of mite through artificial releases in the field. The Abbotsford Agriculture Centre (2010) of British Columbia suggests that infested floral plants should be identified quickly and any found in greenhouses carefully removed before the mites spread. Imported cuttings must be inspected to prevent introduction into a non-infested greenhouse. The presence of broad mite problems in one year means that growers must be vigilant in the next crop-growing season to avoid a carry-over population.

DECISION TOOLS

Citrus. In Argentina, chemical treatment of lemon fruits infested by 10 mites cm^{-2} was successful and provided good-quality fruit using just one acaricide spray (Leonel *et al.*, 1999). In India, treatments were suggested when 5% of fruits were infested (Dhooria *et al.*, 2005). Curvilinear and simple linear regressions represent the relationship between broad mite days and the percentage of fruit surface damage, and between mite days and percentage damaged fruits on the tree. Economic injury levels of 9–15 broad mite days causing yield losses were established (Peña, 1990). In Florida, an economic injury level per lime tree is 42 and 45 mite days in the spring and summer, respectively (Peña *et al.*, 2002). In Nigeria, a progressive increase in the mite population on sweet orange suggests the need to initiate control of phytophagous mites in July before the peak population period of September (Umeh *et al.*, 2007). In the Mediterranean region, Vacante and Bonsignore (2012) suggested an action threshold of 2–3% of infested fruits.

Chilli. Transplants with no symptoms, which are a potential source of infestation in greenhouse-grown crops, become infested at the end of the transplant production cycle (Jovicich *et al.*, 2004). Rodriguez Morell *et al.* (2008) reported that 65 leaves were enough to realize a satisfactory estimate of the total broad mite population in a protected crop of pepper. In Cuba, population counts done on 100 randomly selected leaves of chilli in 16 serial samplings made weekly showed that the maximum capacity of increment of *P. latus* was up to 227.8 individuals per sampling, with an intrinsic ratio of 0.36. It was demonstrated that density stability can in fact change to severe dynamic movements with balance points when the densities are higher, thus showing that in the absence of an effective control, the populations of *P. latus* can rise considerably. A control measure may be applied when the mite populations are higher than 56 individuals per each 100 leaf sample (Miranda *et al.*, 2009). On chilli and other crops in the greenhouse, Vacante (2000) and Vacante and Benuzzi (2007) suggested an action threshold of detection of the first attack.

PLANT RESISTANCE. Studies on plant resistance to *P. latus* have involved various economic plants, including the potato, aubergine, chilli, cucurbits, cotton, jute and sesame. The main studies that have been done on these plants are summarized below.

Relatively few studies have been done on potato. Techniques for potato germplasm evaluation for broad mite resistance were investigated in the Philippines (Eusebio and Bernardo, 1999), and the response of the main varieties was investigated in Iran, where maximum damage on potato was recorded on 'Arinda', followed by 'Maradona' and 'Diamond', while the lowest damage was observed on 'Sante' (Namvar and Arbabi, 2007).

Field experiments in China to determine the resistance to *P. latus* of 27 aubergine cultivars with different leaf stomatal densities showed that broad mite populations were positively correlated with the stomatal density of the lower leaf surface, but negatively correlated with the stomatal density of the upper leaf surface (Gui *et al.*, 2001a). The cultivars 'Fengyan 1', 'Zhongduwandunzao', 'Zhongdu special-culture', 'Chengdumo', 'Yuzhao 2', 'Xiangyan 2' and 'Yonguan' were the most resistant to the mite, while 'Xi'an Green', '8819', 'Zhongduhuangtaizi' and 'Xiangyanzao' were the most susceptible (Gui *et al.*, 2001b).

Screenings on 24 genotypes of chilli in Karnataka, India, showed that the percentage leaf curl index (LCI) (the curl symptom is called murda) resulting from mite infestation ranged from 19.6 to 40.4%. The chilli cultivars 'Pant C-1', 'LCA-304' and 'LCA-312' were promising sources of resistance (Tatagar *et al.*, 2001). Further field experiments conducted in Andhra Pradesh (India) on 77 different chilli cultivars enabled them to be classified into three groups , according to infestation by the broad mite, and the resulting injury grade and damage

index. In particular, nine cultivars ('LCA 235', 'LCA 330', 'EC 128946', 'Cluster mutant', 'LIC 19', 'LCA 312', 'Yellow anther mutant', 'LIC 13' and 'LIC 45') were resistant, whereas 31 cultivars were susceptible and 37 were highly susceptible (Khalid Ahmed *et al.*, 2001). The influence of plant morphological characters on the incidence of the broad mite on chilli was investigated in 21 accessions grown in the field in Tamil Nadu (India). Accession '45' recorded the least population of mites compared with the susceptible accession '43' and the variety 'K2'. The petiole length (alone) had a significant positive relationship with the mite population (Rajaram and Ramamurthy, 2001). Based on the mite leaf curl reaction, five germplasm accessions ('EC-378630', 'EC-378633', 'EC-391082', 'IC-214991' and 'NIC-23897') of chilli were found to be resistant in screening carried out in Andhra Pradesh; these accessions had less than 1% disease incidence, while incidence ranged from 2.1 to 5.2% in the rest of the entries. Most of the accessions reacted independently to leaf curl caused by thrips and mites, a finding that was specifically evident in the reaction of lines 'EC-348630' and 'IC-214991', which showed nil damage due to mites and heavy leaf curl due to thrips. However, one exotic entry ('EC-391082', a paprika type) was found to be resistant to leaf curl caused by both thrips and mites. Mixed reactions of resistance (resistant to tolerant) to both leaf curls were observed in different entries ('PBC-613', 'EC-378633', 'EC-378634', 'EC-391090', 'IC-214990', 'NIC-23897' and 'NIC-23906') (Babu *et al.*, 2002). In Brazil, significant statistical differences between chilli hybrids cultivated in the greenhouse when compared in terms of percentage attack and recovery at 45, 65 and 85 days after transplantation was found. The hybrid 'Nathalie' had the lowest percentage of attacked plants (13–28%) in all evaluations, while the hybrid 'Martha' had the highest percentage (38–59%) of attacked plants (Pereira *et al.*, 2007).

Screenings of cotton varieties in Brazil indicated no significant differences between the times taken for development from egg to adult on the different varieties tested, as well as no significant differences between the number of eggs per female (daily and in total), longevity and percentage spread of females on the different varieties (Vieira and Castro, 1999). Under laboratory conditions in Brazil that evaluated the resistance mechanisms of non-preference and antibiosis to the broad mite in 19 cotton cultivars, the cultivars 'CNPA 7H', 'IPR 94' and 'IPR 96' seemed to be less preferred for feeding and oviposition than the others. 'Coodetec 401' was the least suitable for increase of the *P. latus* population, showing the presence of antibiosis factors, and 'ITA 96' was the most adequate for the increase of the broad mite population (Vieira *et al.*, 2002). Evaluation of 36 cotton genotypes for injury caused by *P. latus* in the field in the state of São Paulo, Brazil, indicated that the varieties 'IAC 24' and 'Fibermax 966' were the least affected by the pest among the cultivars, whereas 'Deltaopal' and 'Makina' were the most injured (Goulart *et al.*, 2009).

In the USA, nine plant introductions (PIs) of watermelon out of a total of 219 did not show visible broad mite injury in the field (in South Carolina) compared with the commercial cultivar 'Mickey Lee', which was severely injured. After further assessment in the greenhouse, PIs in accessions belonging to *Citrullus lanatus* (Thunb.) Matsum. and Nakai (var. *lanatus*) (PI 357708), *C. lanatus* var. *citroides* ('PI 500354'), *C. colocynthis* (L.) Schrad. ('PI 386015', 'PI 386016', 'PI 525082') and *Praecitrullus fistulosus* (Stocks) Pangalo ('PI 449332') showed lower broad mite injury ratings and counts compared with Mickey Lee and other susceptible PIs (Kousik *et al.*, 2007).

The distribution pattern of *P. latus* populations on different leaves of early sown crops of three varieties each of *Corchorus capsularis* L. ('JRC 321', 'JRC 212' and 'JRC 7447') and *C. olitorius* L. ('JRO 632', 'JRO 524' and 'JRO 7835') jute in West Bengal (India) was studied based on five apical leaves of the plant. The highest number of broad mites was observed on the second leaf from the top in both types of jute. Variety 'JRO 52' showed some degree of tolerance while there was no significant variation in the abundance of mites among *capsularis* varieties (Hath, 2000). Other studies on jute showed that the mite population was always higher on cultivars of *C. olitorius* than those of *O. capsularis*. Among the six cultivars evaluated ('JRC 212', 'JRC 7447' and 'JRC 321' of *C. capsularis* and 'JRO 7835', 'JRO 632' and 'JRO 524' of *C. olitorius*), 'JRO 7835' was the most susceptible cultivar and 'JRC 321' was the least susceptible (Hath and Jaydeb Ghosh, 2001). Field experiments in West Bengal to evaluate the susceptibility of different jute cultivars to the broad mite showed that the maximum population *P. latus* occurred on the cultivar 'PRO-4', followed by 'CO-114' and 'NJD-PL-1'. The 'JRO-524' cultivar showed the least susceptibility, followed by 'CO-105' and 'BC-50'. There was a non-significant negative correlation between leaf length and mite population as well as between leaf breadth and mite population (Jana *et al.*, 2007). Further studies in India on eight *C. capsularis* jute cultivars, along with standard check cultivars ('JRC-212' and 'JRC-321'), showed that the cultivar 'NDC-2005-7' was significantly superior and most promising in resistance to the broad mite, with an average infestation of 7.9% and significantly superior to the standard check varieties (Yadav and Yadav, 2010).

Germplasm of sesame (*Sesamum indicum* L.) was screened in Madhya Pradesh (India) against the broad mite, and resistance to *P. latus* was observed in the line 'UCR/82/Mo/3' (Ahuja and Kalyan, 2001).

CULTURAL PRACTICES. The incidence of the broad mite on different cultivars of jute (*C. olitorius* and *C. capsularis*), sown at different dates during the crop growing season in West Bengal (India), showed that the mite population was higher on early sown crops but decreased gradually with delay in sowings. The mite population per five leaves decreased on crops sown in the second and fourth week of April and the second week of May, respectively (Hath and Jaydeb Ghosh, 2001). In a later study that also evaluated the effect of sowing date (third week of April, first week of May and third week of May) and cultivar on the incidence of broad mite on three *C. capsularis* cultivars ('JRC 212', 'JRC 321' and 'JRC 698') and three *C. olitorius* cultivars ('JRC 524', 'JRO 632' and 'JRO 66'), early sowing resulted in high yellow mite infestation for all six cultivars, which gradually decreased with delay in the sowing date. However, a delay in sowing date also resulted in lower fibre yield. In both jute species, the highest yield was obtained when the crop was sown in the third week of April (Hath, 2004).

BIOLOGICAL CONTROL. Different mycopathogens and predatory mites were tested against the broad mite both in the field and

under laboratory conditions. However, no current strategy seems to be superior at a practical level. The main results are reported below.

The first tests on the responses of *Beauveria bassiana* (Balsamo) Vuill., *Hirsutella thompsonii* Fisher and *Paecilomyces fumosoroseus* (Wise) Brown and Smith (Clavicipitaceae) showed that *B. bassiana* appeared to offer the best results (Peña *et al.*, 1996; Ihsan and Yusof, 2004; Smitha and Giraddi, 2006b). The action of several other fungi was evaluated later: *Verticillium lecanii* (Zimmerman) Viégas (Ascomycota) (Smitha and Giraddi, 2006a,b), *Fusarium* spp. (Mikunthan and Manjunatha, 2006), *F. semitectum* Berk. *et* Ravenel (Nectriaceae) (Mikunthan and Manjunatha, 2008) and *Metarhizium anisopliae* (Metchnikoff) Sorokin ('CKM-048') (Clavicipitaceae) (Maketon *et al.*, 2008). *F. semitectum* can be induced to develop and build up further by enhancing the microclimate, and the mixed cropping of chilli–sorghum, chilli–cotton–chilli and chilli–red gram seemed the best cropping systems in terms of total chilli yield (Mikunthan and Manjunatha, 2008). *M. anisopliae* 'CKM-048' appeared to be a virulent strain in controlling both larvae and adults of the broad mite on mulberry, and could reduce the broad mite population within 4 days of treatment. However, after 7 days, its efficacy decreased significantly (Maketon *et al.*, 2008). Despite these encouraging results, there is no information available on the practical use in the field of these mycopathogens.

A number of experiments have been done in various countries using phytoseiid mites applied by inoculative or inundative releases for control of the broad mite, and these are outlined here. *Euseius stipulatus* (Athias Henriot) has been introduced on to citrus in California (USA), where it gave more convincing control, especially in the spring, than did *E. hibisci* (Chant) (Brown and Jones, 1983). *E. stipulatus*, also reported as *Neoseiulus californicus* (McGregor), together with other indigenous predatory mites, has lowered the population of *P. latus* on lime fruits to below the levels of economic damage (Peña and Osborne, 1996). In southern China, *E. nicholsi* (Ehara *et* Lee) preyed on 10.5 *P. latus* a day and gave good control of the pest on citrus (Wu, 1984). In Florida, the action of *Typhlodromalus peregrinus* (Muma) has been evaluated, with promising results (Peña, 1992). Equally interesting were the actions of *N. barkeri* Hughes (Fan and Petitt, 1994) and *N. cucumeris* (Oudemans) (Prabaningrum *et al.*, 1999; Cantliffe *et al.*, 2004; Mizobe and Tamura, 2004). The predatory action of *E. ovalis* (Evans) has been known for a long time (Moutia, 1958), and if the predator is employed in a ratio of 1 mite:20 prey, it provides good control of the broad mite (Manjunatha *et al.*, 2001). The action of *E. ovalis* in India has been reviewed, together with that of *N. longispinosus* (Evans) (Smitha and Giraddi, 2006a), which seems to be promising for practical use (Hariyapa and Kulkarni, 1988), while *N. agrestis* (Karg) is less satisfactory (Kolodochka and Prutenskaya, 1987). In Queensland (Australia), *E. victoriensis* (Womersley) has provided effective pest control (Smith and Papacek, 1985). In Cuba, Rodriguez and Ramos (2000a) assessed the contribution of *Amblyseius largoensis* (Muma), and *Typhlodromus athiasae* Porath *et* Swirski was positively evaluated in the Czech Republic (Karut *et al.*, 1998). The role of *Typhlodromips swirskii* (Athias Henriot) was examined in the Canary Islands (Hernández-Suárez *et al.*, 2006). Among the predatory mites discussed above, *A. largoensis*, *E. ovalis*, *N. cucumeris* and *N. californicus* have been extensively investigated, and there has been some investigation of *N. barkeri*.

A. largoensis was reared on excised potato leaves (*Solanum tuberosum* L.) and the detached leaf culture method was evaluated as the best for rearing this predator, resulting in a population increase of 17.8 times in 10 days (Rodriguez and Ramos, 2000a). No significant differences were observed in the duration of the life history and pre-oviposition periods of *A. largoensis* reared on *P. latus* fed with different substrates (*S. tuberosum*, *Capsicum annuum* L. and *Citrus latifolia* (Yu. Tanaka) Tanaka). The fecundity and average number of eggs laid per female a day by *A. largoensis* were significantly higher on *P. latus* fed with *C. latifolia* than on *P. latus* fed with *S. tuberosum*. The *C. annuum* substrate yielded results on a par with intermediate between the two other substrates (Rodriguez and Ramos, 2003). *A. largoensis* fed on all stages of development of the prey, but it had less difficulty feeding on the immature life stages than on the adult mites. The phytoseiid spent most of the time actively walking and feeding. The biological and predation data indicate that the broad mite is a suitable prey for *A. largoensis*, and the results suggest that this predatory mite is a potential biological control agent for *P. latus* in tropical conditions (Rodriguez and Ramos, 2004).

The interaction between *E. ovalis* (which has also been called *A. ovalis* (Evans)) and the broad mite were investigated both under laboratory conditions and in the field. Predator:prey ratios of 1:40, 1:20 and 1:100 were investigated in relation to the efficacy of *E. ovalis* in controlling *P. latus* infesting chillies under laboratory conditions. Predators were released on to the infested plants at 5, 10, 15 and 20 mites per plant, and at recommended packages of practice (RPP). At a predator:prey ratio of 1:20, there was a rapid decline in the adult prey population 7 days after the start of the experiment. At ratios of 1:40 and 1:100, the peak in the prey population was seen 4 and 8 days after the start of the experiment, respectively. The population of *P. latus* after release of the predator was lowest under RPP at 15 and 30 days after release. At 45 days after release, RPP and a rate of 20 mites per plant resulted in the lowest pest population (Manjunatha *et al.*, 2001). Srinivasulu *et al.* (2002c) examined the seasonal occurrence of *P. latus* with reference to biotic and abiotic factors on *C. annuum* (cv. 'LCA-305') in India. Predatory mite activity started again in the first week of January and progressively increased, reaching a peak in the third week of February (1 week after the peak in the population of the broad mite.

The population life tables of *N. cucumeris* (which has also been called *A. cucumeris* (Oudemans)) with *P. latus* as the prey were constructed at various temperatures (23, 25, 28, 30 and 31°C), and 25°C was estimated to be the optimum temperature for the development and reproduction of *A. cucumeris* (Li *et al.*, 2003). Predation tests showed that *N. cucumeris* fed on 8.8, 15.2, 27.0 and 30.3 individuals of *P. latus* a day at temperatures of 15, 20, 25 and 30°C, respectively. At 25°C, the larva, protonymph, deutonymph and adults – male and female – of *N. cucumeris* fed on 0.5, 1.6, 5.2, 3.8 and 27 individuals of adult *P. latus* a day, respectively. The functional response curve of adult *P. latus* to the densities of *N. cucumeris* showed a Holling's Type II curve. The consumption of prey by *N. cucumeris* increased

with increase in prey density, but the consumption rate decreased. The critical ratio of predator to prey to effectively (>98%) suppress the population of *P. latus* by releasing *N. cucumeris* was 1:40 on red pepper (La *et al.*, 2009). From a practical point of view, releases of *N. cucumeris* successfully controlled the broad mite on two varieties of greenhouse-grown sweet peppers. A survey of pre-plant seedlings showed that nurseries were a source of infestation for the broad mite. The predatory mites were released twice (on day 1 and 5, or 15 days later) on each plant, and on every second plant or every fourth plant. Broad mite populations were evaluated by sampling young leaves from the top of the plant. The effect of the broad mite on plant height, dry mass and yield was evaluated. All three release rates of *N. cucumeris* significantly controlled broad mite populations, but when the predatory mites were released only on every fourth plant, the overall height and yield of the plants were adversely affected by broad mites. Releasing of *N. cucumeris* on to each or every second plant was as efficacious at controlling broad mites as sulfur treatments in terms of plant height, dry mass and yield. Plants treated with sulfur, however, had significantly higher thrip populations and fruit damage (Weintraub *et al.*, 2003).

Timing of the release of *N. californicus* was investigated in scenarios of pest and predator presence to simulate preventive and curative control strategies. The seedlings were not damaged when two predators per seedling were released preventively. Curative control strategies were not effective for producing undamaged or broad mite-free transplants. Further studies were carried out to evaluate the effectiveness of *N. californicus* on pepper seedlings that hosted a few broad mites and were transplanted into the production greenhouse. *N. californicus* released at different times and densities, and conventional control methods (micronized sulfur sprays), were evaluated for pest control, plant damage and fruit yield. Yields of red fruits from plants with four predators per plant released at any of the tested release times (6 days before, at, or 4 days after transplanting) or with sulfur sprays were not significantly different from those of an uninfested control treatment. Releases of two predatory mites per plant led to yields similar to the uninfested control only when introductions were a week before transplanting, while later introductions led to serious plant damage and significantly lower yields. Early broad mite infestations in pepper crops grown in greenhouses might then be effectively controlled with *N. californicus* released at early seedling developmental stages in the nursery and at transplanting in the fruit production greenhouse (Jovicich *et al.*, 2009).

Cloyd (2010) reported that *N. barkeri* had been successfully used in suppressing broad mite populations, as well as *N. cucumeris* and *N. californicus*, and might be utilized against broad mite populations on certain greenhouse-grown crops, including vegetables. It is important to apply predatory mites early in the crop production cycle before broad mites become established.

PHYSICAL CONTROL. Exposure times of 15–45 min at 40–43°C are effective in killing broad mites. Plants have to be immersed in hot water long enough to allow penetration into areas such as the meristematic tissues where the mites live, and at the same time not damage the plants. This practice may be implemented, but it is a short-term solution lacking residual effect, and after transplantation, a reinfestation may occur (Cloyd, 2010).

CHEMICAL CONTROL. In the last 30 years, several active ingredients (acaricides, insecticides and fungicides) have been tested and proposed, alone or mixed. Vacante (2010) listed the main active ingredients reported in a bibliography for citrus up until 2010. There is a similarly long list of active ingredients that have been use for control of the broad mite on other economic plants, such as chilli, jute, sesame and cotton. They are as follows:

Abamectin (Bellettini *et al.*, 1999; Santos *et al.*, 1999; Srinivasulu *et al.*, 2002b; Sarkar *et al.*, 2005, 2008; Zhao *et al.*, 2005; Rodriguez Morell *et al.*, 2008; Venzon *et al.*, 2008), acephate (Sharma and Sharma, 2005), acetamiprid (Reddy *et al.*, 2005), bromopropilate (Banu *et al.*, 2007b), buprofezin (Smitha and Giraddi, 2006b), carbaryl (Reddy *et al.*, 2005, 2007), carbofuran, carbosulfan (Panda and Sarangi, 2004), chlorfenapyr (Bellettini *et al.*, 1999, Santos *et al.*, 1999), chlorpyrifos (Karmakar *et al.*, 1996b; Reddy *et al.*, 2005), clofentezine (Sarkar *et al.*, 2005, 2007, 2008), cypermethrin (Karmakar *et al.*, 1996b; Laskar and Ghosh, 2005), dichlorvos (Karmakar *et al.*, 1996b; Reddy *et al.*, 2005), dicofol (Karmakar *et al.*, 1996b; Srinivasulu *et al.*, 2002b; Panda and Sarangi, 2004; Sarkar *et al.*, 2005, 2007; Sharma and Sharma, 2005; Laskar and Ghosh, 2005; Reddy *et al.*, 2005, 2007; Rodriguez Morell *et al.*, 2008), dimethoate (Reddy *et al.*, 2005; Sharma and Sharma, 2005), endosulfan (Karmakar *et al.*, 1996b; Srinivasulu *et al.*, 2002b; Ahuja and Kalyan, 2003; Reddy *et al.*, 2005, 2007), ethion (Panda and Sarangi, 2004; Sharma and Sharma, 2005; Smitha and Giraddi, 2006b; Mahalingappa *et al.*, 2008), fenazaquin (Sarkar *et al.*, 2005, 2008), fenpropathrin (Jeyarani and Chandrasekaran, 2006), fenpyroxymate (Smitha and Giraddi, 2006b; Somchoudhury *et al.*, 2008), fipronil (Reddy *et al.*, 2005, 2007; Venkateshalu *et al.*, 2009), flufenzin (Sarkar *et al.*, 2007), fluvalinate (Karmakar *et al.*, 1996b), imidacloprid (Srinivasulu *et al.*, 2002b; Panda and Sarangi, 2004; Reddy *et al.*, 2005; Bhede *et al.*, 2008), indoxacarb (Reddy *et al.*, 2005, 2007), monocrotophos (Karmakar *et al.*, 1996b; Sharma and Sharma, 2005; Reddy *et al.*, 2005), oxydemeton-methyl (Srinivasulu *et al.*, 2002b; Panda and Sarangi, 2004; Sharma and Sharma, 2005), phosalone (Srinivasulu *et al.*, 2002b; Reddy *et al.*, 2005, 2007), phosphamidon (Sharma and Sharma, 2005; Bhede *et al.*, 2008), profenofos (Reddy *et al.*, 2005, 2007; Mahalingappa *et al.*, 2008), propargite (Smitha and Giraddi, 2006b), quinalphos (Reddy *et al.*, 2005, 2007; Sharma and Sharma, 2005), spiromesifen (Kavitha *et al.*, 2006), thiamethoxam (Panda and Sarangi, 2004; Reddy *et al.*, 2005), triazophos (Panda and Sarangi, 2004; Laskar and Ghosh, 2005; Reddy *et al.*, 2005, 2007; Mahalingappa *et al.*, 2008).

Many of the above active ingredients exhibit no activity or a modest activity against the broad mite, and only a restricted number are labelled, for example, as for citrus in California (USA), including abamectin, chlorpyrifos, dicofol, fenproximate narrow-range oil, spirodiclofen, wettable sulfur (UC IPM Online, 2009g), or for ornamentals and vegetables in greenhouses, e.g. abamectin, chlorfenapyr, fenpyroximate, pyridaben and spiromesifen (Cloyd, 2010). In addition, some compounds, e.g. oxydemeton-methyl and imidacloprid, caused resurgence of the broad mite after spraying (Srinivasulu *et al.*, 2002b). Among the labelled active ingredients, sulfur is well known and widely tested for the control of the broad mite (Cranham *et al.*, 1962; Jeppson *et al.*, 1975; Brown and Jones, 1983; Buergo *et al.*, 1986; Karmakar *et al.*, 1996a,b; Srinivasulu *et al.*, 2002b; Banu *et al.*, 2007b; Venzon *et al.*, 2008), and probably represents the best

acaricide in the control of this pest mite in Mediterranean greenhouses (Vacante, 2000; Vacante and Benuzzi, 2007). Lime sulfur has been positively used on apple (Zhao *et al.*, 2005). Other fungicides that have been tested unsuccessfully against the broad mite are dinocap (Sharma and Sharma, 2005), mancozeb (Srinivasulu *et al.*, 2002b) and ziram (Karmakar *et al.*, 1996b).

There have been a number of experiences with botanical acaricides. Among these, azadirachtin (from neem, *Azadirachta indica* A. Juss.) is probably the most widely tested (Palaniswamy and Ragini, 2000; Srinivasulu *et al.*, 2002b; Ahuja and Kalyan, 2003; Smitha and Giraddi, 2003, 2006b; Uraisakul, 2003; Panda and Sarangi, 2004; Reddy *et al.*, 2005; Laskar and Ghosh, 2005; Banu *et al.*, 2007c; Venzon *et al.*, 2008). Leaf extracts of *Adathoda vasica* Nees, *Argemone mexicana* L., *Aristolochia bracteata* Retz., *Cissus quadrangularis* L., *Lippia nodiflora* (L.) Michx., neem, *Sansevieria* sp. and *Vitex negundo* L., and *Aloe* sp. as 5% aqueous extracts were sprayed on chillies and reduced the *P. latus* populations in all treatments (Palaniswamy and Ragini, 2000). The effects of extracts of *Andrographis paniculata* (Burm.f.) Wall. ex Nees, *Annona squamosa* L., *Chromolaena odorata* (L.) R.M. King and H. Rob., *Cymbopogon nardus* (L.) Rendle, *Derris elliptica* (Wall.) Benth., *Eucalyptus*, *Hyptis suaveolens* (L.) Poit., *Melaleuca bracteata* F. Muell., *Nicotiana tabacum* L., *Stemona tuberosa* Lour. and *Vernonia squarrosa* (D. Don) Lessing, alone or in combination, were investigated on chilli in Thailand. Treatment with an *Annona* suspension was efficient for controlling broad mites, killing 100% of the eggs and larvae, and 80% of the adults in the laboratory. However, it did not harm the eggs and adults of the phytoseiid mite *Amblyseius longicaudatus* (Uraisakul, 2003). Plant oils (castor, *Hibiscus cannabinus* L., neem, *Pongamia* sp. and sesame), combined with dicofol were tested on chilli (Smitha and Giraddi, 2003). Aqueous leaf extract of *Clerodendrum inerme* (L.) Gaertn. and *V. negundo* were tested on chilli in India, but exhibited a moderate efficacy that was not comparable with that of synthetic acaricides (Smitha and Giraddi, 2006b). More recently, extracts of garlic paste and turmeric powder were evaluated in the greenhouse on jute (Banu *et al.*, 2007a,c). Castor oil-based soft soaps with potassium hydroxide and naturally available materials such as botanicals and preservatives that varied from soap to soap have been evaluated on chilli in India (David *et al.*, 2009).

INTEGRATED PEST MANAGEMENT. Selective acaricides and other substances such as sulfur (as dust, lime or wettable) are used. This active ingredient is effective against adult and immature stages of the broad mite, and can be applied at least twice a year in cases where no other chemicals are used (Krause *at al.*, 1996; Meyer Smith and Schwartz, 1998e). Dhooria *et al.* (2005) suggested that infested plants could be dusted with sulfur 2–3 times at 5-day intervals. However, the application of sulphur negatively influences populations of natural enemies (McCoy, 1977a; Smith and Papacek, 1991), although it does have only a feeble persistence. Sulfur should not be used at high temperatures or mixed with mineral oils.

The evaluation of different modules for the IPM of *P. latus* on sweet pepper in greenhouse studies in Bangalore, India, showed that spray application of abamectin followed by ethion and abamectin or spray application of abamectin followed by profenofos and abamectin were more effective than spraying dicofol followed by pongamia oil and neem seed kernal extract. Spray application of dicofol followed by release of the phytoseiid mite *A. tetranychivorus* (Gupta) and spraying with *V. lecanii*, and spray application of dicofol followed by release of *A. tetranychivorus* and a spray of pongamia oil was marginally effective. However, in March–September, the use of abamectin followed by dicofol ('module' M_1), dicofol–fenazaquin (M_2), fenazaquin–pongamia oil (M_3) and the organic module of oxymetrin-neem soap (M_4) were effective in the control of the broad mite (Reddy and Kumar, 2006).

The Genus *Hemitarsonemus* Ewing

Diagnostic characteristics

Femur I has 3 setae. The prodorsal shield of the female is not hood like over the gnathosoma; the stigmata are enlarged and aligned between the vertical and scapular setae. The ventral apodemes 3 do not extend laterad of trochanter III; apodemes 4 extend posterolaterad of setae *3b* to bases of trochanters IV. Tarsi II and III have symmetrically developed paired claws. The setae of plate CD of the male are slightly barbed; leg IV bears a femorogenual flange rounded apically and set subdistally; the genital capsule is elevated well above the bases of legs IV in unmounted specimens (Lindquist, 1986).

Economic importance

This genus has only four species. The type species, *H. tepidariorum* (Warburton), commonly called the fern mite, feeds on various species of ferns in greenhouses in the USA (Ewing, 1939; Pritchard, 1951a; Beer, 1954), UK (Warburton, 1904; Cameron, 1925) and Costa Rica (USDA ARS SEL, 2005). Two of the three remaining species were described from China: *H. biconvexa* Lin *et* Zhang, collected on *Cibotium barometz* (L.) J. Sm. (Lin and Zhang, 1994) and *H. furcalis* Lin *et* Zhang on *Pteris esquirolii* Christ (Lin and Zhang, 1995). The fourth species, *H. ganeo* Magowski, was described from ferns (*Athyrium filix-femina* (L.) Roth, *Dryopteris dilatata* (Hoffm.) A. Gray and *D. filix-mas* (L.) Schott) outside in Poland (Magowski, 2012). Only the fern mite has some economic importance.

Hemitarsonemus tepidariorum (Warburton) (Fig. 22.3)

Common name

Fern mite.

Diagnostic characteristics

FEMALE. The body is oval, and yellowish brown in colour. The stigmata are large, round and obliquely aligned with the prodorsal setae v_1 and sc_1, and related internally with enlarged, bulbous

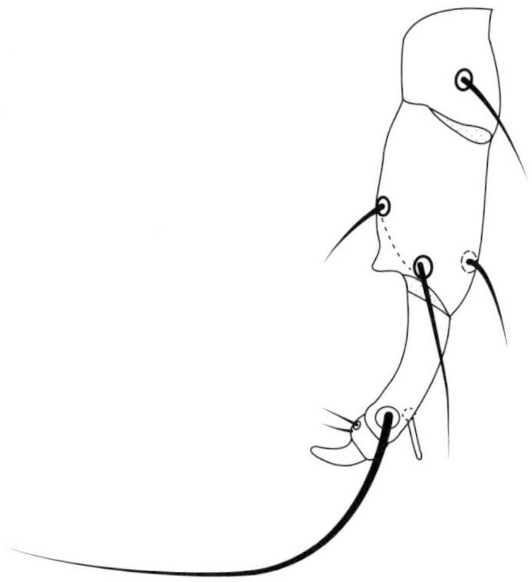

Fig. 22.3. *Hemitarsonemus tepidariorum* (Warburton). Male, leg IV (drawing by Vacante).

postatrial structures. The anterior margin of coxisternal plates III is strongly arched medially. The femoral seta *d* of leg I is widened, serrate–lamellate; the tibial solenidion φ_2 is elongated, rod like and set dorsoproximally, well apart from φ_1 and *k* (the tibial eupathidium) on the tibiotarsus (Lindquist, 1986).

MALE. The male is smaller than the female, suboval and yellow in colour. The posterior half of the idiosoma is strongly elevated dorsally, and the genital capsule is well removed above the bases of legs IV. Legs IV have a femorogenual flange that is narrowly rounded in shape and set in a subapical position, with elongated tibia and a blade-like tarsal claw. On leg I, the solenidion φ_2 is well apart from φ_1 and *k*, though not elongated or displaced proximally (Lindquist, 1986).

Geographical distribution

California (Pritchard, 1951a), Costa Rica (USDA ARS SEL, 2005), England (Warburton, 1904; Cameron, 1925; Lindquist, 1986) and Minnesota (Ewing, 1939; Pritchard, 1951a; Beer, 1954).

Host plants

These include the following families and species: Aspleniaceae – mother fern (*Asplenium bulbiferum* G. Forst.) (Warburton, 1904; Cameron, 1925); Athyriaceae – *Athyrium procumbens*, *Diplazium procorum* (Lindquist 1986); Dryopteridaceae – holly fern (*Polystichum* sp.) (Ewing, 1939; Pritchard, 1951a; Beer, 1954), leather fern (*Rumohra adiantiformis* (G. Forst.) Ching) (USDA ARS SEL, 2005); Pteridaceae – *Pteris argyraea* Moore, Cretan brake (*Pteris cretica* L., cultivars 'Alexandrae', 'Mayii', 'Ouvardii', 'Parkeri', 'Wilsonii', 'Wimsettii'), slender brake (*Pteris ensiformis* Burm.f., cv. victoriae) (Pritchard, 1951a).

Bio-ecology

H. tepidariorum was first described from ferns in England by Warburton (1904). Cameron (1925) collected additional specimens from ferns in greenhouses in England, and gave a detailed account of the life history of the mite. Later, Ewing (1939) described the genus *Hemitarsonemus* and based his description of *H. tepidariorum* on both Warburton's original description and other specimens collected from holly fern (*Polystichum* sp.) in Minnesota. Additional references indicate that the fern mite lives in association with a number of different fern species grown in commercial greenhouses (Pritchard, 1951a; Beer, 1954). Lindquist (1986) studied specimens collected in 1955 on *Athyrium procumbens* from Malaya and *Diplazium procorum* from Sri Lanka housed in the propagating houses of the Royal Botanical Gardens at Kew, England. Taking into account that almost none of the host plants, with the probable exception of *Rumohra adiantiformis*, are native to areas where the fern mite is reported, and that the majority of host plants of fern mite originate from tropical and subtropical areas of the Old World, it is possible to hypothesize that the primary geographical distribution of the fern mite coincides with these areas, e.g. southern India, South-east Asia and eastern Australia (Magowski, 2012).

The fern mite feeds on the younger portions of ferns and avoids the old fronds. It generally prefers grooves and hollows on the pinnae or the stalk, but also inhabits the scale leaves and scales at the base of the plant. On larger leaflets, the fen mite populations are more numerous around the main vein. The females lay eggs in protected surfaces within the distorted pinnae, on the upper surface of the leaf stalks near the tips of fronds or between the young scales near the tips of fronds, or between the young scales near the base of the plant. The eggs hatch in 2–4 days, and the larval stage feeds on the fern leaflets for 1–2 weeks and later becomes quiescent. The adults emerge after 3 or 4 days and lay eggs after 2 days. Daily oviposition ranges from one to two eggs. Adult longevity ranges from 2 to 16 days. The fern mite overwinters as eggs and adults, mainly females. Dispersal occurs by the adults walking on greenhouse benches, but mainly through the movement of infested plants (Cameron, 1925). No natural enemies are known.

Symptomatology and damage

These mites feed on the leaf tissue of ferns, and produce depressions on the frond surface, stunting leaf growth and deforming the leaflets. Severely affected leaflets have a reduced size and rounded borders, and older leaflets show a bronze or rusty colouring on the ventral surface. Severe infestations induce the arrest of plant growth and sometimes kill the plants (Cameron, 1925; Pritchard, 1951a; Beer 1954).

Control

Control measures are mainly sanitation and prevention of the infestation of ferns.

The Genus *Steneotarsonemus* Beer

Diagnostic characteristics

Both sexes have a spine-like subunguinal seta on tarsus I and the unguinal seta *u′* of tarsi I and II is not recurved, commonly bifid

distally but lacking a spur midventrally. The idiosoma of the female is usually elongated–ellipsoidal, with apodeme 2 obliquely slanted, not bifurcate anterolaterally, and coxisternal plates III and IV continuous medially; the legs are relatively short, and leg II is usually not longer than leg IV. The male has the scapular setae sc_2 usually much shorter than sc_1; the femorogenu of leg IV usually has a prominent, rounded flange, whose distal margin forms an acute angle where it is attached to the surface of the legs, and rarely lacks a flange (Lindquist, 1986).

Economic importance

This genus has worldwide distribution, and includes over 70 species. *Steneotarsonemus* spp. are mainly phytophagous on monocotyledonous plants, including grasses, sedges, bamboos, pineapple and liliaceous species; some species are found in ground-level or subterranean nests containing grasses, e.g. the nests of mammals (Lindquist, 1986). The phytophagous habit is well documented for different *Steneotarsonemus* spp. (Beer, 1954; Jeppson *et al.*, 1975), some of which are host specific, such, the pineapple leathery pocket mite, *S. ananas* (Tryon), the rice panicle mite, *S. spinki* Smiley, and the West Indian sugarcane mite, *S. bancrofti* (Michael). Other species may be restricted to plants belonging to one genus, such as *S. phyllophorus* (Ewing), which lives on bamboo, or to one family, such as the bulb scale mite, *S. laticeps* (Halbert), which lives on bulbs of *Narcissus* and other Amaryllidaceae, or *S. spirifex* (Marchal), which lives on Poaceae, causing spiral malformations of the rachis, blind spikelets and incompletely emerged inflorescences of cereals (Evans *et al.*, 1961). *S. furcatus* De Leon causes distortion of the leaves and shortened internodes, resulting in stunted plants on ornamental grass (*Paspalum* spp.) and on *Maranta leuconeura* Morr (Jeppson *et al.*, 1975). *S. phyllophorus*, *S. spirifex* and *S. furcatus* are occasional pests.

Four species are further described below: *S. ananas*, *S. spinki*, *S. bancrofti* and *S. laticeps*.

Steneotarsonemus ananas (Tryon) (Fig. 22.4)

Common name

Pineapple leathery pocket mite.

Diagnostic characteristics

FEMALE. The body is elongate, and cream coloured. The pseudostigmatic organs are rounded. Apodemes 3 and 4 are weak and not contiguous. Legs I–IV are well separate; the patella II has a stout seta and 2 normal setae; ventrally, the setae *1a* of the coxisternal metapodosomal plate I lie behind apodemes 1; the terminal seta of tarsus IV is much longer than the subterminal setae (Jeppson *et al.*, 1975).

MALE. The prodorsum has 4 pairs of dorsal setae; the third pair is longer than the others and the first pair is the shortest. The inner flange on tarsus IV is evenly rounded; all 3 setae of the tarsus are short and subequal in length (Jeppson *et al.*, 1975).

Geographical distribution

Australia (Tryon, 1898; Waite, 1993), Brazil (Flechtmann, 1967c), Hawai (Beer, 1954; Hughes, 1959), Indonesia (Fauzan, 1999, in Petty *et al.*, 2002), Ivory Coast (Adico *et al.*, 1999, in Petty *et al.*, 2002), Réunion (Vuillaume, 1982, in Peña *et al.*, 2002), South Africa (Petty, 1975; Meyer Smith, 1981a) and Swaziland (Dodson, 1969).

Host plants

Bromeliaceae: pineapple (*Ananas comosus* (L.) Merr.) (Jeppson *et al.*, 1975).

Bio-ecology

The pineapple leathery pocket mite, *S. ananas*, is host specific to pineapples in their areas of cultivation. The mites usually live in the basal portions of the leaves, where they are well protected and from which they can invade the seed cavities during blossom time. The warm, moist conditions (95–100% RH) and low light intensity of the seed cavities and nectary ducts of the fresh fruits of pineapples favour the development of high levels of *S. ananas* populations (Meyer Smith, 1981a).

Symptomatology and damage

The feeding activity of the pineapple leathery pocket mite causes brown spots on the white tissues at the base of the leaves of pineapple, especially in young plants (Meyer Smith, 1981a).

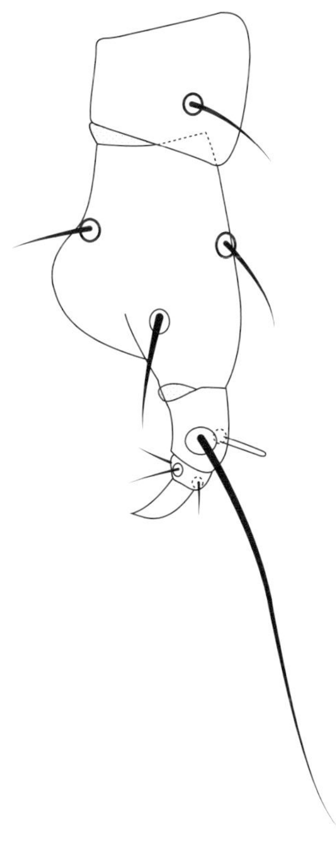

Fig. 22.4. *Steneotarsonemus ananas* (Tryon). Male, leg IV (drawing by Vacante).

The presence of the mite is associated with the 'leathery pocket disease', in which the walls of the seed cavities and nectary ducts of the pineapple fruitlet harden (Petty, 1975); the disease is probably primarily due to the fungus *Penicillium funiculosum* Thom (Rohrbach, 1979, in Meyer Smith, 1981a). The fruitlets of severely infested fruits appear externally depressed. Seed cavities affected at an early stage are smaller, and early damage results in more extensive lesions. Affected fruits have delayed ripening and green fruitlets, especially in 'Queen' pineapple (Petty, 1975). In South Africa, leathery pocket disease usually occurs in the Eastern Cape during late summer and early autumn. Curiously, although the disease is also known in Hawai, Australia and Brazil, in these countries it has no economic importance, while in some areas of South Africa (Eastern Cape, etc.), up to 16% of the fruit may be rejected on account of the disease (Petty, 1977).

Control

Spray applications of endosulfan reduced the incidence of leathery pocket disease in South Africa (Petty, 1977). The hypothesis that the fungus responsible for the disease is transmitted by the mites during the blossom period of pineapples in October, November and December, when the flowers open, suggests that the mites and the fungus should be treated before the formation of the buds (Meyer Smith, 1981a). Clearly, it is only possible to use chemicals that are labelled for both mites and crops.

Steneotarsonemus spinki **Smiley (Fig. 22.5)**

Common name

Panicle rice mite.

Diagnostic characteristics

FEMALE. The idiosoma is elongate, finely punctate, 222–350 µm long and 84–108 µm wide. The prodorsal shield is not projected beyond the basal part of the gnathosoma. The stigmata are set on the margins of the prodorsal shield, posterolateral to setae v_1, which are stout and finely pilose; the setae sc_2 are filiform and posterolateral to the setae sc_1. The bothridial setae sc_1 are leaf like, finely pilose and with rounded bases, and not covered by the prodorsal shield. The setae c_1 are barbed, shorter than setae c_2, and filiform. The setae c_2 and c_1 are transversely aligned. The setae *d* are setiform, stout and barbed. The setae *e* are as long as the setae *f*, setiform, stout and barbed. The setae *e* and *f* of the tergite EF are transversely aligned. The setae *h* are elongate, stout and barbed. Apodemes 1 are short and united with the prosternal apodeme; apodemes 2 are conspicuous and curved, seam-ripper like, and connected with the prosternal apodeme. The prosternal apodeme has a strong triangular ending, with the sejugal apodeme inconspicuous. Apodemes 3 are shorter than apodemes 2, conspicuous, not united with the poststernal apodeme and not extending posteriad of trochanters III. Apodemes 4 are blade like distally and not united with the poststernal apodeme, but extend posteromediad of setae *3b*. The setae *1a* are two times longer than setae *2a*, and setae *3a* are long and filiform. The pseudanal setae are present. The subunguinal seta of tarsus I is spine like and bifid distally. Tibia I has a complete sensory cluster (Hummel *et al.*, 2009).

MALE. The idiosoma is smooth, finely punctate and about 217 µm long and 121 µm wide. The prodorsal shield is weakly sclerotized and subtriangular. The ventral hysterosoma is finely punctate. The genital capsule is subcircular in dorsal view, as long as it is wide, with the caudal setae *h* setiform; the accessory copulatory structures ps_1 are spine like and pointed distally. Leg IV has a strong, long smooth flange and the ventral trochanter setae are longer than the genual setae $v'G$ (Hummel *et al.*, 2009).

Geographical distribution

Arkansas (NAPPO, 2007), California (UC Davis, 2009), China (Ou *et al.*, 1977), Colombia (ICA, 2005), Costa Rica (Sanabria and Aguilar, 2005), Cuba (Ramos and Rodríguez, 1998, 2001), Dominican Republic (Pellerano, 1998, in Hummel *et al.*, 2009; Ramos *et al.*, 2001), Guatemala (Castro *et al.*, 2006), Haiti (Herrera, 2005), Honduras (Castro *et al.*, 2006), India, Kenya (Rao and Das, 1977), Korea (Cho *et al.*, 1999; Schall, 2004), Japan (Shikata *et al.*, 1984), Louisiana (Smiley, 1967; Hummel *et al.*, 2007), Mexico (Arriaga, 2007), New York (NAPPO, 2007), Nicaragua (Aguilar and Murillo, 2008), Ohio (Ochoa, 2004, in Hummel *et al.*, 2009), Panama (Almaguel Rojas and Botta Ferret, 2005), Philippines (Sogawa, 1977), Puerto Rico (Texas Department of Agriculture, 2007), Sri Lanka (Cabrera *et al.*, 2002a), Taiwan (Lo and Ho, 1977), Texas (Texas Department of Agriculture, 2007), Thailand (Cheng and Chiu, 1999) and Venezuela (Aguilar and Murillo, 2008).

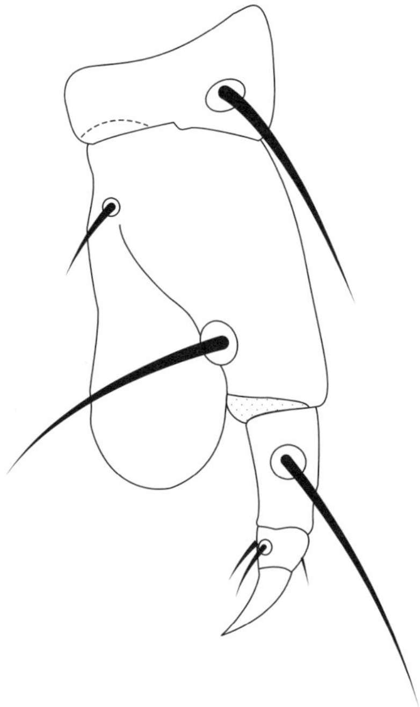

Fig. 22.5. *Steneotarsonemus spinki* Smiley. Male, leg IV (drawing by Vacante).

Host plants

These include the following families and species: Cyperaceae – rice flatsedge (*Cyperus iria* L.) (Central Rice Research Institute, 2006), *Schoenoplectus articulatus* (L.) Palla (Rao and Prakash, 2002); Poaceae – Bermuda grass (*Cynodon dactylon* (L.) Pers.) (Rao and Prakash, 1996), American wild rice (*Oryza latifolia* Desv.) (Sanabria and Aguilar, 2005), rice (*O. sativa* L.) (Ho and Lo, 1979; Jiang *et al.*, 1994). In addition, different biological stages of *S. spinki* were collected in various countries on other plant species, mainly belonging to Poaceae, although it is not known whether these plants are able to support the reproduction of *S. spinki* or whether they are accidental hosts (Hummel *et al.*, 2009).

The male holotype and a female paratype were also collected on a planthopper (*Sogata orizicola* Muir) (Smiley, 1967).

Bio-ecology

S. spinki has a probable Asian origin and was first described by Smiley (1967) from a planthopper in Louisiana. The mite can be misidentified with two congeneric species, *S. konoi* Smiley *et* Emmanouel and *S. madecassus* Gutierrez, but there are some distinctive characteristics that enable the three species to be separated. The female of *S. spinki* has the hysterosomal setae c_1 and *d* barbed, and the ventral apodeme 2 is strong and seam-ripper like. The male of *S. spinki* differs from the male of *S. madecassus* in having the setae sc_1 longer than the setae sc_2 and apodemes 3 and 4 united distally; it differs from the male of *S. konoi* in having no ventral ornamentation, leg IV having a strong, long smooth flange and the ventral trochanter setae being longer than the genual setae *v'G* (Hummel *et al.*, 2009). Moreover, *S. konoi* is associated with Bermuda grass, its presence on rice is accidental and no reproductive populations have been found on rice. Another tarsonemid, *S. furcatus* DeLeon, causes direct damage to rice in Brazil (Návia *et al.*, 2006c), and has been found associated with *S. spinki* on rice in Colombia, Costa Rica, Cuba and India (Návia *et al.*, 2005c, 2006a; Sanabria and Aguilar, 2005; Karmakar, 2007). However, a rice survey conducted by US Department of Agriculture in 2007 found only low numbers of *S. furcatus* in the USA, which could either indicate the presence of only low-level populations of this mite on rice or its association with another host plant near the rice crops (McDonald, in Hummel *et al.*, 2009).

The preferred host plant of *S. spinki* is rice (Ho and Lo, 1979; Jiang *et al.*, 1994). In Taiwan, Ho and Lo (1979) observed over 70 plant species, including weed species, growing in or near rice paddies, but detected *S. spinki* only on rice; of the 70 plants, 44 species were poaceous and one species belonged to the Oryzae tribe. In China, Jiang *et al.* (1994) examined plants belonging to nine families and found that the preferred host for *S. spinki* was only rice. Conversely, Sanabria and Aguilar (2005) reported American wild rice as an alternate host for *S. spinki* in Costa Rica and Panama. In addition, Bermuda grass (Rao and Prakash, 1996), rice flatsedge (Central Rice Research Institute, 2006) and *S. articulatus* are regarded as alternate hosts for *S. spinki* in India (Rao and Prakash, 2002). Notwithstanding these reports, *S. spinki* is largely known in areas of rice production, and represents a severe problem in tropical and subtropical rice-growing areas (Rao and Das, 1977; Sogawa, 1977; Ho and Lo, 1979; Jiang *et al.*, 1994; Ramos and Rodríguez, 1998, 2001; Ramos *et al.* 2001; Schall, 2004; Herrera, 2005; ICA, 2005), while it should not be a major problem in temperate rice-growing regions of the world, e.g. the continental USA, as it is probably unable to overwinter in such regions (Schall, 2004). The accidental introduction of *S. spinki* is seriously feared in North and South America, and strict quarantine measures are needed to prevent its accidental introduction into these regions (Návia *et al.*, 2010b), and also into mainland USA from countries such as Puerto Rico (USDA, 2009).

At low population densities, *S. spinki* usually inhabits the internal surfaces of the leaf sheaths of rice plants, but at high densities it can spread on to all plant parts, including the kernels (Chow *et al.*, 1980; Ramos and Rodríguez, 2001). Large populations, e.g. greater than 450 mites cm^{-2}, induce typical symptoms of attack on rice grains and leaf sheaths (Ramos and Rodríguez, 2001). The panicle rice mite reproduces by arrhenotokous parthenogenesis, and unmated females generate male offspring. Under laboratory conditions at temperatures from 24.1 to 35.3°C, unmated females generated an average of 79.4 adult mites in 17 days. The male offspring mate with their mothers to generate both female and male offspring (Xu *et al.*, 2001). The female:male sex ratio is 22:1, 32:1 and 8:1 at 32 and 28°C, and under field temperature conditions, respectively (Zhang, 1984). Under laboratory conditions at temperatures ranging from 24.5 to 35.4°C, the fecundity of mated females was 55.5 eggs and the time to produce a new generation was 8.5–13.6 days at 25–30°C (Xu *et al.*, 2001). In Cuba, it took 8–17 days to produce a new generation at 25°C (Ramos and Rodríguez, 2000). Castro *et al.* (2006) reported that the developmental period was 11.3 days at 20°C, 7.8 days at 23.9°C and 4.9 days at 33.9°C, and that the mite developed from 48 to 55 generations a year under an appropriate climate. When adults of *S. spinki* were reared on sections of leaf sheaths at constant temperatures of 37, 39 and 41°C, there was no mortality until 48 h, when 21.1% mortality was observed; by 96 h, the recorded mortality was 97.7%; at 39°C, mortality at 24 h was 15.57%, increased further at 36 h, and by 72 h, all the mites were dead; at 41°C, 50% of the mites were dead after 36 h, and there was 100% mortality at 60 h. The females of *S. spinki* were more tolerant to high temperatures than the males, and the first female mortality began after the last males had died (Xu *et al.*, 2002). High humidities favoured mite development, and the egg hatch was strongly influenced by RH and temperature during development. When eggs were maintained for 7 days at 25°C and under 70, 95 and 100% RH, the hatch was 0, 65 and 96%, respectively. The hatch of eggs maintained for 7 days at constant RH and at 18, 25 and 36°C was 19.2, 92 and 50%, respectively (Liang, 1980). A similar relationship between RH and percentage egg hatch was found by Chen *et al.* (1979).

In China and Taiwan, *S. spinki* infests rice during different phenological stages (flowering, milk, soft dough, medium dough and hard dough grain maturity), and its population density varies according to the phenological stage of the host plant. During the early stage of rice growth, *S. spinki* lives on the lower leaves, and its density gradually increases as the

season progresses. Late-planted rice has higher mite densities than early-season planted rice. The seedling beds are usually mite free. At tillering, there are low population densities, but during flowering the mite density increases, as it also does during the milk stage (by mid-June). After grain maturation, the density of the population decreases, and after rice harvesting, the mites move to new and alternative sources of moisture and sustenance, such as rice stubble, regenerated rice, seedlings from fallen seeds and other poaceous hosts where they can survive. In China, the panicle rice mite overwinters on rice stubble and ratoons, and tillers that emerge from previously harvested plants. In Taiwan, the population of *S. spinki* peaked in May on the first rice crop (grown from January to May); on the second crop (grown from mid-May to October), it peaked between August and October; the mite population crashed later on, depending on the (warm) temperature and (high) humidity (Lo and Ho, 1977, 1980; Ou and Fang, 1978; Tseng, 1984; Jiang *et al.*, 1994).

Changes in the density of *S. spinki* during the season were also observed in Cuba. Panicle rice mite populations were low during the tillering stage, and multiplied by 24× at the green ring stage and then another three times during the blooming stage. The population reaches a maximum at grain filling and then decreases as the grain progresses from the milk to soft dough stages of maturation (Ramos and Rodríguez, 2001). Also in Cuba, Leyva Fernández *et al.* (2003) found that rice planted during December to May escaped higher infestations by *S. spinki*, while rice planted from August through to October had much greater amounts of infestation and mite damage. Under laboratory conditions, Xu *et al.* (2002) showed the ability of *S. spinki* to survive in a flooded rice field, with no plant material. *S. spinki* has also been observed floating and mating on the surface of water (Pellerano, 1998, in Hummel *et al.*, 2009).

The dispersal of *S. spinki* is via wind (Jiang *et al.*, 1994), water (Xu *et al.*, 2002), insects (Tseng, 1984), green rice seed (Rao *et al.*, 2000) and contaminated rice harvesting equipment. Ou and Fang (1978) observed no soil-borne or water-borne dispersal, and nor was *S. spinki* found on the bees foraging in the crop, but an important mode of dispersal was found via the clothes of people. There is only one report of dispersal via seed (Rao *et al.*, 2000). In Texas, *S. spinki* was noted floating in the water and 'swimming' to find females, and females were reported purposely falling from the plant into the water to find males. Large numbers of mites were observed near the water doors (gates) that allow the flood waters to rise and recede in rice fields ('Biology and the status of *Steneotarsonemus spinki*'. Personal communication to S.M. Kubilus on 14 August 2007, from R. Ochoa, US Department of Agriculture Agricultural Research Service Systematic Entomology Laboratory, Baltimore, Maryland).

Natural enemies of *S. spinki* include mycopathogens and and mites (Table 22.2).

Symptomatology and damage

Symptoms and damage caused by the feeding of *S. spinki* have been discussed by various authors in the different rice-growing areas. Chow *et al.* (1980) reported that feeding by the panicle rice mite caused brown necrotic areas on the upper surface of the leaf sheath and on the hull of grains that were similar to those produced by the pathogen *Sarocladium oryzae* (Sawada) W. Gams and D. Hawkswon (Sawada, 1922). Feeding by *S. spinki* on rice has also been associated with black lesions of the leaf sheath, discoloured grains, grains that are partially to completely chaffy and grain malformations (Chein, 1980; Reissig *et al.*, 1986; Rao and Prakash, 1992; Rao *et al.*, 1993). High mite

Table 22.2. *Steneotarsonemus spinki* Smiley. Natural enemies (fungi and mites).

Families and species	References
FUNGI	
Clavicipitaceae	
Hirsutella nodulosa Petch	Cabrera *et al.*, 2002a, 2005a
Hypocreaceae	
Cephalosporium spp.	
Davidiellaceae	
Cladosporium spp.	
Trichomaceae	
Penicillium spp.	
MITES	
Ascidae	
Aceodromus asternalis Lindquist *et* Chant	Ramos and Rodríguez, 1998; Cabrera *et al.*, 2003; Ramos *et al.*, 2005
Asca pineta De Leon	
Cheiroseius napalensis (Evans *et* Hyatt)	Tseng, 1984
Cheiroseius serratus (Halbert)	
Lasioseius sp.	Ramos and Rodríguez, 1998; Cabrera *et al.*, 2003; Ramos *et al.*, 2005
Lasioseius parberlesei Bhattacharyya	Lo and Ho, 1979
Lasioseius (≈ *tridentis*)	Ramos and Rodríguez, 1998; Cabrera *et al.*, 2003; Ramos *et al.*, 2005
Proctolaelaps bickleyi Bram	
Laelapidae	
Hypoaspis sp.	
Phytoseiidae	
Amblyseius taiwanicus Ehara	Lo and Ho, 1979
Galendromus sp.	Ramos and Rodríguez, 1998; Cabrera *et al.*, 2003; Ramos *et al.*, 2005
Galendromus alveolaris (De Leon)	
Galendromus longipilus (Nesbitt)	
Neoseiulus baraki Athias Henriot	
Neoseiulus paraibensis (Moraes *et* McMurtry)	
Neoseiulus paspalivorus De Leon	
Proprioseiopsis asetus (Chant)	
Typhlodromus sp.	

populations have been held responsible for necrotic bands and spots on rice grains and leaf sheaths as well (Ramos and Rodríguez, 2001). Hummel *et al.* (2009) reported that damage symptoms on rice plants infested with *S. spinki* included empty panicles, discoloured hulls and an orange discoloration of the leaf sheath. Mite feeding on developing panicles at the boot stage causes the most significant crop damage, resulting in sterility and partial infertility of the plants, although the mechanism of induction of the sterility is not yet understood.

Different factors might be involved in the damage caused by *S. spinki*, such as the association of the mite with pathogens, the toxins injected by mites during feeding, and seed and sheath damage caused by mite feeding that makes it easier for bacteria to enter the seed and cause sterility. The main aspects of these three possible mechanisms are discussed below.

A complex of bacteria, fungi, spiroplasmas and virus-like particles has been reported in association with *S. spinki* (Hsieh *et al.*, 1977; Chow *et al.*, 1980; Shikata *et al.*, 1984; Rao *et al.*, 1993; Almaguel *et al.*, 2003; Rao and Prakash, 2003; Sanabria and Aguilar, 2005), and in some countries, e.g. Cuba, the *S. spinki*–fungus complex is now considered to be the greatest economic threat to rice production (Perez, 2002). In Taiwan, the infestation of rice plants by *S. spinki* was associated with 'sterile grain syndrome', which is characterized by empty or partially filled grains with diseased brown spots (Chen *et al.*, 1979). In India, grains infested with *S. spinki* were discoloured, and pathogenic fungi and bacteria, such as *Alternaria padwickii* (Ganguly), *Burkholderia* (*Pseudomonas*) *glumae* (Kurita *et* Tabei), *Cochliobolus lunatus* R.R. Nelson *et* Haasis, *Gibberella zeae* (Schwein) Petch and *Fusarium moniliforme* J. Sheld, were isolated from infested plants (Rao and Prakash, 2003). However, sampling of specimens of *S. spinki* on plants that were pathogen free suggested that the grain discoloration might depend on a chemical reaction to the toxic saliva injected by mite. Moreover, in Taiwan, it was observed that the panicle rice mite carried spores of *S. oryzae* on its body, suggesting that plant symptoms could depend on a combination of mite damage and disease (Chen *et al.*, 1979). In Cuba, *S. oryzae* was isolated from 70% of rice mite specimens, confirming the hypothesis of the role of the mite as a vector of this pathogen (Sanabria, 2005). In the Dominican Republic, *S. spinki* was found in all the discoloured rice grains and in association with *Fusarium* spp. and *Helminthosporium* spp., and *Sclerophthora* spp., *Fusarium* spp. and *Rhizoctonia* spp. were most commonly associated with the mite in rice stems. In closed panicles, the rice mite was found most often associated with *Sarocladium* spp. and *Sclerophthora* spp. (Pellerano, 1998, in Hummel *et al.*, 2009).

In addition to typical mite damage symptoms, rice plants infested by *S. spinki* in Louisiana also exhibited symptoms of 'bacterial panicle blight' damage, such as empty hulls and panicles standing erect. Discoloured areas on the flag leaf sheaths of rice plants were associated with *S. spinki* feeding inside the leaf sheath, and spores (type unknown) were found attached to the body of the mite (Hummel *et al.*, 2009). Rice fields infested by *S. spinki* in Honduras contained grains with 'parrot-beaking' symptoms (Castro *et al.*, 2006). In Panama, Nandakumar *et al.* (2007) reported the association of the bacteria *B. glumae* and *B. gladioli* (Severini) with panicle blight symptoms on rice. These observations and results from other research suggest that the majority of the damage attributed to *S. spinki* is actually caused by bacteria (CIAT, ICA and FEDEARROZ, 2005). Nandakumar *et al.* (2007) noted that Bernal *et al.* (2002) attributed similar symptoms in Cuba, Haiti and the Dominican Republic to damage from *S. spinki* in association with *S. oryzae*. An unidentified virus was isolated from *S. spinki* on greenhouse-grown rice in Japan, and in rice tissues infested with *S. spinki* (Shikata *et al.*, 1984). *S. spinki* may be injecting the virus-like particles into the plant tissue during feeding, but further research is needed to determine its vector role.

The common association of *S. spinki* with diseases could depend on the damage caused to the seed and sheath by mite feeding, which makes it easier for bacteria to enter the seed and cause sterility. The association of the mite with a wide number of diseases of rice is of particular interest. For instance, of the 12 plant diseases associated with *S. spinki*, eight already occur in rice fields in Louisiana (Hummel *et al.*, 2007).

Losses caused by the panicle rice mite, directly and/or indirectly, are alarming. In Taiwan, 60% of crop damage by *S. spinki* was registered in 1974, and about 20–60% of harvested grains were empty in 1976, corresponding to a loss of 20,000 t and valued at US$9.2 million (Cheng and Chiu, 1999). In southern China, a 5–20% yield reduction in both early- and late-season rice crops was attributed to infestation by *S. spinki* (Jiang *et al.*, 1994). In India, rice sterility has been correlated with the levels of the *S. spinki* population present in the field, with a significant negative correlation between *S. spinki* density and grain yield and a significant positive correlation between mite density and percentage of chaffy grains found (Rao and Prakash, 2003; Karmakar, 2007). In some areas, e.g. the state of Orissa, 4–90% sterility was associated with 7–600 mites per tiller; in Ghandari District, Andhra Pradesh, 15–50% sterility was associated with *S. spinki* densities of 150–900 mites per tiller, and in Jharkhand, 19–28% sterility was associated with a density of only 3–7 mites per tiller (Rao and Prakash, 2003). Similar losses have been reported from Central America. In Cuba, crop losses in 1997 were 30–70% (Ramos and Rodríguez, 1998, 2000). In 1998, in some rice-producing regions of the Dominican Republic, more than 40% of grains were empty or stained (Díaz *et al.*, 1999; Ramos *et al.*, 2001). In Haiti, 60% yield losses were caused by *S. spinki* (Almaguel Rojas and Botta Ferret, 2005), and 40–60% yield losses were estimated in Panama (García, 2005). In Costa Rica, yield losses of up to 45% were calculated for 2004, corresponding to a loss of US$10.96 million (Barquero, 2004). It is expected that the introduction of *S. spinki* into Brazil may cause 30–70% losses, corresponding to 3.8–8.9 million t year^{-1} (Návia *et al.*, 2005c).

Control

Control of the panicle rice mite requires the rigid application of prevention measures, the use of resistant varieties, appropriate cultural practices and spray applications of pesticides.

PREVENTION. In cultivated and neighbouring areas, destruction of rice crop residues and the removal of invasive plant species suppress the probable source of infestation (Ho and Lo, 1979). *S. spinki* could be introduced into new areas on propagated rice, as the mite is thought to be transmitted from seed to seedling

(Rao *et al.*, 2000), so the use of mite-free seeds is suggested. Before planting, it is necessary to clean the new cropping area surfaces in order to remove mites. Similarly, it is necessary to clean machinery and other equipment when it is used by different farmers or in different areas in order to avoid the dissemination of the mite from an infested crop to an uninfested crop (Hummel *et al.*, 2009).

Quarantine programmes are also fundamental to prevention. The US Department of Agriculture has published the '*Panicle Rice Mite Program Manual*' (USDA, 2009), which is intended primarily for Plant Protection and Quarantine (PPQ) officers, staff officers and state and federal cooperators who are involved in carrying out the Panicle Rice Mite Program. This is a guide for designing a programme to detect, monitor, control, contain or eradicate an infestation of the panicle rice mite. If *S. spinki* is detected in the USA, the PPQ will produce a site-specific action plan based on the Program Manual developed through discussion, consultation or agreement with staff at the USDA Agricultural Research Service (ARS), universities, industries and state departments of agriculture.

PLANT RESISTANCE. Research in Cuba has shown the potential of plant resistance in combating the rice panicle mite. Densities of up to 141 *S. spinki* per stem were observed on the rice variety 'J-104' compared with 58 *S. spinki* per stem on the 'iaCuba28' variety. Moreover, peak infestation in variety 'iaCuba28' occurred from August to September, while the peak for variety 'J-104' was in September and December (Leyva Fernández *et al.*, 2003). In the Dominican Republic, the variety 'ISA-40' was more susceptible to *S. spinki* infestation at the tillering and panicle initiation stages than the varieties 'JUMA-57', 'Prosedoca' and 'Prosequisa', with 100% infestation observed in 'ISA-40'. As plants reached the blooming stage, the differences between the varieties decreased (Ramos *et al.*, 2001). In Taiwan, the variety 'Tainan 5' was highly susceptible to *S. spinki*, and in response to the severe crop losses registered, the cultivation of this variety was reduced from 400,000 ha in 1975 to 80,000 ha in 1981 (Cheng and Chiu, 1999). *S. spinki* was not reported as a pest in subsequent years, thus emphasizing the importance of varietal resistance.

CULTURAL PRACTICES. Various cultural practices reduce the populations of *S. spinki*, delay its spread into the crop and diminish yield losses and production costs. Hence, between two consecutive crop cycles, it a good practice to leave the fields fallow for at least 2 weeks. It is also necessary to use certified seed to plant fields, as well as to plant rice varieties resistant to *S. spinki*, and to use the lowest sowing rate possible for the variety planted, as a high rice planting density and high RH induce high populations of *S. spinki*. Staggered plantings should be avoided in infested areas, as should planting rice in adjacent areas during harvest time or in areas that have recently been harvested. The wind direction should be considered as well, because planting rice downwind from an infested field must be avoided. Finally, it is good practice to divide the doses of nitrogen fertilizers into different treatments, and to use a shallow depth of permanent flood water (Ho and Lo, 1979; Lo and Ho, 1980; Cabrera *et al.*, 1998; Ramos *et al.*, 2001; Hernández *et al.*, 2003, 2005; Romero *et al.*, 2003; Sanabria and Aguilar, 2005).

BIOLOGICAL CONTROL. The mycopathogen *Hirsutella nodulosa* Petch caused 71% mortality in treated mites (Cabrera *et al.*, 2005a). However, there is no information on its practical use, or that of predatory mites, such as *Amblyseius taiwanicus* Ehara.

CHEMICAL CONTROL. Chemical control depends on the strict monitoring of mites in the field. Crop monitoring 15 days after planting, especially in fields downwind of infested areas, ensures early detection of *S. spinki* and the implementation of control measures using acaricides. However, the use of chemical control is hampered by the adaptation of the mite to live behind the leaf sheath or inside the hulls of developing grains. In addition, the association of rice plants with water and nearby wetlands limits the availability of active ingredients labelled for this crop in several countries. Furthermore, various systemic pesticides are ineffective against *S. spinki* (Chow *et al.*, 1980; Almaguel *et al.*, 2000). This has resulted in the intensive use of chemicals, which are the primary causes of mite outbreaks in Taiwan (Ou and Fang, 1978). As for other pest mites, the use of chemicals does not represent a productive choice for managing *S. spinki* (Cabrera *et al.*, 1998), and chemical control must be applied only in outbreak treatments (Ramos and Rodríguez, 1998, 2000; Cheng and Chiu, 1999; Almaguel *et al.*, 2000). Active ingredients effective in the control of *S. spinki* include insecticides such as dichlorvos (Jiang *et al.*, 1994), dimethoate (Ghosh *et al.*, 1998), isoprocarb (Jiang *et al.*, 1994), parathion (Lo *et al.*, 1981), thiophanate (Jiang *et al.*, 1994) and triazophos (Cabrera *et al.*, 1999, 2002b). High mortality of adult *S. spinki* was obtained under laboratory conditions with bromopropilate, diafenthiuron, dicofol and edifenphos (Cabrera *et al.*, 2005b). Lo *et al.* (1981) reported the good action of dicofol, which was later confirmed in field trials by Bhanu *et al.* (2006).

Currently, the integrated management programme for *S. spinki* suggested for the Caribbean and other Central American countries provides for the use of abamectin, biomite (geraniol, citronellol, nerolidol, farnesol), dicofol, endosulfan, ethoprophos and triazophos (Almaguel *et al.*, 2005). García *et al.* (2002) suggested treating seeds with benomyl plus thiram to reduce the percentage of sterile and spotted grains and increase yield in rice-producing regions infested with *S. spinki*. Daravath *et al.* (2013) found spiromesifen plus propiconazole (fungicide), diafenthiuron plus propiconazole and fenpyroxymate alone effective against the panicle mite and associated grain discoloration. Beard (2007) reported some success in controlling grain mites with phosphine fumigation, and this technique could represent an alternative treatment for rice seed protection.

Steneotarsonemus bancrofti (Michael) (Fig. 22.6)

Common name

West Indian sugarcane mite, sugarcane stalk mite.

Diagnostic characteristics

FEMALE. The colour is opalescent light green. General features are as for the genus (Jeppson *et al.*, 1975).

MALE. The gnathosoma is small. The first pair of prodorsal setae are shorter than the second and fourth pairs. Coxae III are not ornamented. Femur IV is relatively small, with a bag-like flange, which reaches to the distal margin of the femur, and is indented distally between the femur and the flange; the tibial seta is strong and about as long as the length of the femur (Jeppson *et al.*, 1975).

Geographical distribution

Worldwide, in all countries (with the exception of southern India) where sugarcane is grown (Jeppson *et al.*, 1975).

Host plants

Poaceae: sugarcane (*Saccharum officinarum* L.) (Holloway, 1936; Butani, 1959; Jeppson *et al.*, 1975).

Bio-ecology

S. bancrofti was first described by Bancroft (1877) in Queensland (Australia), but was not named until 1890, when Michael (1890) found the mite on sugarcane in Barbados and proposed the name '*Tarsonemus bancrofti*' for it. The sugarcane stalk mite is host specific for sugarcane in all areas of sugarcane cultivation except southern India. It lives in cane tops under the leaf sheaths, and does not occur on the older part of the cane. A few individuals live on the youngest portions of the cane that have just been damaged by mite feeding (Jeppson *et al.*, 1975).

Symptomatology and damage

Feeding by the sugarcane stalk mite causes transparent, brown, crater-like depressions on the young sugarcane stalk; the plant tissues appear scabby or scarred, and later, as the cane grows, these areas become opaque, brown and rough. At cutting, the canes may appear covered with roughened areas. The feeding activity of *S. bancrofti* makes it easy for spores of red rot, *Colletotrichum*, to enter, thus causing red spots that give a reddened appearance to the leaves (Holloway, 1936; Butani, 1959).

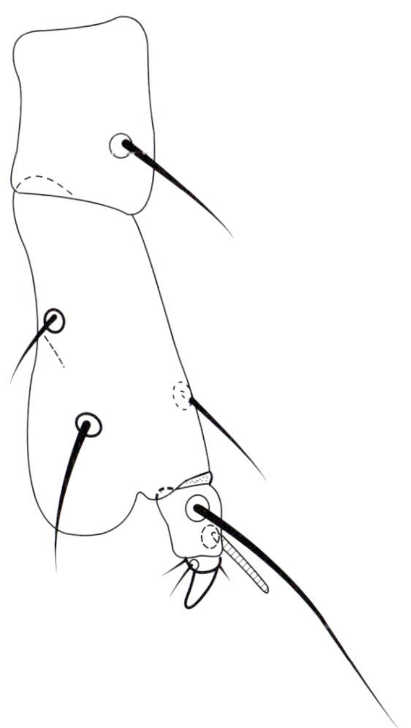

Fig. 22.6. *Steneotarsonemus bancrofti* (Michael). Male, leg IV (drawing by Vacante).

Control

The need for control is not well documented. If it is done, acaricides labelled for sugarcane must be used, favouring when possible the use of sulfur.

Steneotarsonemus laticeps (Halbert) (Fig. 22.7)

Common name

Bulb scale mite.

Diagnostic characteristics

FEMALE. The body is oval, with a bright semi-translucent appearance, colourless in the young and pale pink in older specimens. The pseudostigmatic organs are acuminate (Meyer Smith, 1981a). General features are as for the genus (Jeppson *et al.*, 1975).

MALE. The third pair of prodorsal setae are $^2/_3$ as long as the width of the gnathosoma, and the other setae are in a linear

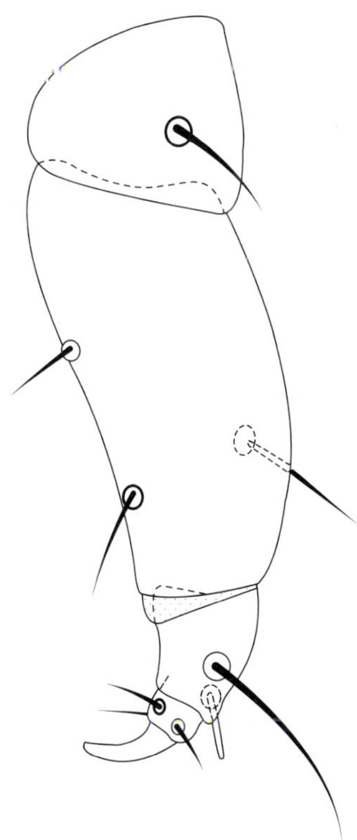

Fig. 22.7. *Steneotarsonemus laticeps* (Halbert). Male, leg IV (drawing by Vacante).

arrangement. Femur IV lacks a flange, and the tactile seta of tibia IV is short and not as long as the femoral segment (Jeppson *et al.*, 1975).

Geographical distribution

Bohemia (former Czechoslovakia) (Blattny, 1933), Germany (Hodson, 1934; Smith, 1934; Karnkowski, 1995), Hungary (Timár *et al.*, 2004), Irish Republic (Halbert, 1923), Poland (Łabanowski *et al.*, 1990; Łabanowski and Jaworski, 1992), the Netherlands (Doucette, 1929; Karnkowski, 1995; Messelink and van Holstein-Saj, 2006), South Africa (Wiese and Meyer Smith, 1967, in Meyer Smith, 1981a; Meyer Smith, 1981a), Sweden (Vitzthum, 1929b), the UK (Massee, 1933a; Hodson, 1934; Hussey *et al.*, 1969; Lane, 1979; Lynch, 1993; Karnkowski, 1995), Ukraine (Livshits *et al.*, 1981) and the USA (West Coast) (Ewing, 1929b; Doucette, 1929; Jeppson *et al.*, 1975).

Host plants

These belong to one family, the Amaryllidaceae: *Amaryllis* (Smith, 1934; Hussey *et al.*, 1969; Meyer Smith, 1981a; Lane, 1984a), *Eucharis* (Lane, 1984a; Alford, 2012), garlic (Timár *et al.*, 2004), *Hippeastrum* (Blattny, 1933; Fox-Wilson, 1939; Lane, 1984a; Łabanowski *et al.*, 1990; Labanowski and Jaworski, 1992; Lynch, 1993; Alford, 2012), *Narcissus* (Halbert, 1923; Hussey *et al.*, 1969; Meyer Smith, 1981a; Lane, 1984a; Łabanowski *et al.*, 1990; Lynch, 1993; Alford, 2012), *Pancratium* (Lane, 1984a), *Sprekelia* (Alford, 2012), *Vallotta purpurea* (Aiton) Herb. (Lane, 1984a; Alford, 2012).

Bio-ecology

S. laticeps infests bulbs of various Amaryllidaceae and is a major pest of bulbs of narcissus. The mites feed on the epidermal surfaces of the scales of the bulbs, mainly in the neck area, extending upwards from the point where the leaves separate, usually at the level of the soil surface; from here they move into the growth region within the bulb. During spring, the growth of the bulb induces a swelling that may crush many mites between the scales, and vigorous bulbs show nearly complete mite mortality. Many mites congregate in the necks of forced bulbs, and as the levels of the population increase, they start to move towards the leaves and flowers that have just developed, and begin to produce the typical damage (Doucette, 1936). In early summer (late April and early May in temperate climates), the higher temperatures stimulate the increase of mites, which lay large numbers of eggs on the leaves. At this time, the mites disperse over the leaves, and in early June they move from one plant to another (Hodson, 1934). As the foliage dies back, the mites move from dying leaves to the bulbs, and by mid-June the greater part of the *S. laticeps* populations is found within the bulb scales. Within the bulbs, eggs and immature forms of the bulb scale mite are present throughout the year, but populations fluctuate significantly. At lifting, usually occurring in temperate climate in late July, the bulb scale mite populations are numerous in the neck area. The bulbs are fully turgid and solid, and the mites cannot penetrate between the adpressed scales. Later, during storage in August and September, when the bulbs lose moisture and shrink, the mites spread along the crevices and reach the soft tissues, where they produce the yellowish strips of injured tissue. By October and November, the *S. laticeps* populations decline and the mites slow their activity. During this period, the populations consist mostly of females, with about 5% males. In the new crop, as the root action starts, the bulbs assume full turgor and the swelling scales crush many mites, with only the mites inhabiting the neck region surviving (Hodson, 1934). The death of mites by crushing in bulbs that do not swell uniformly is less marked. Moreover, at lifting, some mites remain on the outside of the bulbs or in the soil, and may later infest bulbs that are being forced (Doucette, 1936; Hussey *et al.*, 1969; Jeppson *et al.*, 1975).

Mite fecundity ranges from 5 to 28 eggs; the eggs hatch in about 11 days, and the first larval development takes about 15 days. The developmental time from the egg to adult takes about 7 weeks in the field. High temperatures within forcing houses facilitate mite reproduction, so that low infestations within the dormant bulbs may increase, damaging leaves and flowers. Outdoors, the growth of narcissus appears normal, and the basal portions of the leaves and flower stems show few symptoms, although severely affected leaves show a 'saw-tooth' alteration (Doucette, 1936; Hussey *et al.*, 1969; Jeppson *et al.*, 1975). Lynch and Bedi (1994) reported that the duration of the life cycle varied according to temperature and season. The minimum developmental time from egg to adult is 15 days at 20°C, increasing to 51 days at 10°C. The optimal temperature for population growth is 10–20°C, and a negative response to temperature occurs above 25°C. These data are helpful in understanding the low bulb scale mite populations and damage to bulbs grown in the open, and contrasting the severe damage caused in glasshouse-forced bulbs, where the high temperatures and humidities used in forcing favour mite population growth (Doucette, 1936; Winfield, 1964). The increase in the use of forcing as a common practice to maximize production has allowed to bulb scale mite to become a major pest compared with field production (Lynch, 1993).

Mites may move from bulb to bulb by themselves, either in the field or when harvested bulbs are stored, or they may be transported passively by winds or insects, such as the large narcissus fly (*Merodon equestris* (Fabricius)) (Collier *et al.*, 2009). Dispersal of the mite over a long distance involves commercial activities and the transport of infested bulbs (Doucette, 1936).

Natural enemies are some phytoseiid mites (*Amblyseius andersoni* (Chant), *Neoseiulus barkeri* Hughes, *N. cucumeris* (Oudemans)) and laelapid mites (*Hypoaspis miles* Berlese) that have been reported preying on *S. laticeps* (Messelink and van Holstein-Saj, 2006).

Symptomatology and damage

The bulb scale mite feeds mainly on epidermal surfaces of the scales of bulbs of *Narcissus* and other Amaryllidaceae. On developing shoots, the mite causes distortion, stunting and death of leaves and flowers. The scar-like tissues produced by the mite feeding on leaves and flower stems results in longitudinal bronze streaks and transverse cracks.

Infested bulbs are recognized at the time of lifting and in storage by their unusual dryness, loose consistency and lightness in weight. The dried outer scales adhere tightly to the bulb

and show yellowish brown areas where the mite has fed that are visible only by cutting the bulb open from the neck to the base and separating the scales and shoots. Small particles of brown and discoloured tissue are exposed, mainly at the angular points of the scales. The neck area may harbour many mites. The single scales show yellow or brown marks extending as narrow scars from the apex to the base. However, mites can occur in bulbs without symptoms (Hodson, 1934).

Forced *Narcissus* bulbs exhibit the most serious damage, resulting in loss yield and quality of flowers (Lynch, 1993). The high temperatures used for forcing induce a rapid increase of mite populations. The bulb scale mites congregate at the bases of leaves, in the area of the neck, and their populations appear as greyish dust. Mites contaminate the flower buds as they emerge and spread on to the leaves where egg deposition occurs on both leaf surfaces (Hodson, 1934). Feeding by *S. laticeps* removes the surface wax from the leaves, and the leaf tissue becomes bright green. The leaves may become distorted and sickle shaped, and the leaf edges appear scarred, producing a 'saw-edge' effect. A yellow flecking of leaves, similar to symptoms caused by the '*Yellow stripe virus*', may occur. Lunate-shaped areas, light in colour and located below the tip, affect the two pairs of innermost leaves (Lynch, 1993).

The congregation of mites at the neck of bulbs usually occurs at the time of protrusion of the flower buds. At high population densities, the flower bud and the stems may be killed, but at low levels of attack, although the bud is killed, the flower stem extends bearing the blind bud. Low numbers of mites also impair bud growth, and the buds then develop weak and malformed flowers. Plants infested for successive seasons produce bulbs without vigour, growing poorly and producing stunted plants, with the foliage senescing prematurely (Doucette, 1936, Hussey *et al.*, 1969, Jeppson *et al.*, 1975, Lynch, 1993). On *Narcissus*, feeding damage promotes secondary infections, such as attack by the fungal pathogen smoulder (*Botrytis narcissicola* Kleb.), leading to further crop loss (Gray and Sheil, 1987).

Symptoms are usually less severe on outdoor *Narcissus* bulbs, but a long-term cumulative effect may intervene. Infested bulbs usually have normal growth up to the time of flowering. However, mild climate conditions during winter and early spring may induce, especially in warmer areas, more severe damage. Outdoor plants show feeding symptoms as yellow streaks and blotches, often streaked with brown, on the lower portions of both the foliage and flower stems, which in the UK is more evident during April and May according to the rise in temperature and increase of mite populations. These symptoms may result in wilting and death of leaves, and in a reduction in the size of bulbs. The slow plant growth induces the mites to concentrate on the young leaves before the flower buds develop sufficiently. In these cases, leaf growth may be stopped, but the flower stems emerge intact and produce small and imperfect blooms (Lynch, 1993).

The mite causes red streaks and spots on leaves of amaryllis as they emerge from the neck region of the bulb. This reddening appears intensified on the upper part of both sides of the bulb scales, and may extend down into the bulb. These lesions become elongated as the leaves and the flower stems develop, and may result in a scar several inches long. High levels of attack induce the flowers to shrivel or the flower stem to be short (Blattny, 1933; Hussey *et al.*, 1969). Fox-Wilson (1939) reported that the mite caused reddening of the leaf tissues before they emerged from the bulbs.

Control

The inspection of bulbs used for forcing is a fundamental practice in the control strategy for this mite, as well as that of the bulb storage conditions. The adaptation of the bulb scale mite to live in the heart of the bulb makes control difficult, and regular inspection of bulbs in storage is needed, as well as the use of mite-free bulbs. Control methods include the use of both physical and chemical means.

INSPECTION OF BULBS FOR FORCING. Inspection tends to detect infestation at an early stage or before its development and aims to minimize the risk of damage. Healthy bulbs are usually solid, while bulbs that are soft and small indicate the risk of mite presence. Such bulbs must be cut and examined to detect the symptomatology of the attack, e.g. yellow or brown streaks. When these symptoms are not evident, it is a good practice to observe the scales under the stereomicroscope to detect the presence of mites, especially in the neck area. Bulbs that are severely infested must be discarded, while those with only a little infestation can be used for flowers if they are appropriately treated (Lynch, 1993).

STORAGE CONDITIONS. It is necessary to improve the hygiene conditions of storage, avoiding the proximity of bulbs that have been treated to highly infested bulbs (Lynch, 1993). The management of storage conditions, mainly after hot water treatment of the bulbs, helps to prevent mite reinfestation. Boxes and pallets used to carry untreated stocks must be dipped in cresylic acid (Lane, 1984a). Vehicles for transporting stock must be cleaned, and areas of possible contamination need to be avoided. The storage of narcissus bulbs in warm conditions improves the action of the hot water treatment (HWT) (see below), while the storage of amaryllis bulbs at low temperatures and humidity seems to reduce the development of severe infestation (Hussey *et al.*, 1969).

BIOLOGICAL CONTROL. The first efforts to develop integrated pest management (IPM) in amaryllis were started in 2004 with a project funded by the Dutch Product Board for Horticulture. High densities of the phytoseiid mite *N. cucumeris* were introduced into two amaryllis nurseries. Despite repeated releases of high rates of mites (up to 1000 m^{-2}), the phytoseiid never became established (Messelink and van Holstein-Saj, 2006).

A survey of naturally occurring predatory mites with the potential to control *S. laticeps* was conducted in different commercial amaryllis greenhouses in the Netherlands. The most abundant species was *N. barkeri*, which was found in almost half of the nurseries sampled to be frequently associated with *S. laticeps*. Laboratory predation tests indicated *N. barkeri* to be a promising candidate for biological control of the bulb scale mite in amaryllis (Messelink and van Holstein-Saj, 2006), so an assessment was made of the ability of *N. barkeri* and

A. andersoni to control or restrict the bulb scale mite in amaryllis in a greenhouse experiment. Phytoseiid mites were released at rates of 1000 mites m^{-2} by sprinkling them on the soil between the bulbs. Later, an extra release of 1000 mites m^{-2} was added for the *A. andersoni* treatment. The number of damaged bulbs and leaves was significantly lower in the predator treatments, with *N. barkeri* performing significantly better than *A. andersoni*. In the *N. barkeri* treatment, only 4% of the leaves were damaged compared with 57% in the control. However, 45% of the bulbs were still colonized by the bulb scale mite. Neither *N. barkeri* nor *A. andersoni* appeared to be able to completely eliminate a bulb scale mite infection in amaryllis (Messelink and van Holstein-Saj, 2007). Despite these results, it seems that no biological control programmes for *S. laticeps* have been planned.

PHYSICAL CONTROL. Hot water treatment (HWT) of dormant bulbs has been applied from time to time in the control of *S. laticeps* and has sometimes provided conflicting results. Blattny (1933) suggested the immersion of narcissus bulbs in hot water for 30 min at 40°C, and Hodson (1934) reported that 1 h at 43.3°C produced 100% mite mortality; practical requirements suggest that this treatment should be repeated every 2 years. Doucette (1926, 1929) found HWT at 43°C for 1 h, or vapour treatment at 43°C for 2 h, to be sufficient to reduce injury to flowers. Winfield (1958) stated that the immersion of the bulbs in water heated to 43.3°C for 4 h, or to 44.4°C for 3 h, might eradicate mite infestation on narcissus bulbs. Moreover, HWT is also used in narcissus to target the stem nematode, for which the standard treatment is 3 h at 44.4°C (Winfield, 1971). However, the treatment has consequences on the following season's flower growth, and there is a fine line to be drawn between eradicating the pest and damaging the bulbs (Winfield, 1967; Hussey *et al.*, 1969; Lane, 1984b). HWT is therefore unsuitable for bulbs destined for forcing or sale. Additionally, it is thought that only the outer scales of the bulb reach the required kill temperature, possibly leaving mite infestations within the bulb unharmed. This could explain the recurrence in mite numbers post HWT. These reports indicate that there is clear discordance on the duration of treatment, and the solution to the problem seems to be given by Wallis (1967) and Gratwick and Southey (1986), who reported that warm storage of bulbs before the hot water treatment reduced the bulb damage caused by the treatment. In addition, this technique improves growth and flowering. Bulbs stored in this way have a lower rate of respiration, which makes the developing flowers, leaves and roots more tolerant to the hot water treatment. During warm seasons, this condition may be reached naturally, usually during July–August. This occurs from 4 to 6 weeks after lifting, when the flowers of most varieties are fully formed (Tompsett *et al.*, 1977). According to the variety, locality and season, the safest time in south-west England is mid-July, and in eastern counties it is late July to mid-August (Lynch, 1993). Despite the above reports on the control of bulb scale mites in lifted amaryllis (*Hippeastrum*) bulbs, hot-water treatment is not always satisfactory, even though growers apply a great variety of time × temperature combinations. Doorduin *et al.* (1997) investigated the influence of bulb size, water temperature and initial bulb temperature on the time until 43°C was reached in the bulb. Larger bulbs, low initial bulb temperature and low water temperature increased the time until the temperature of 43°C was reached. The effects are described in a model. Growers underestimate the influence on the duration of the initial bulb temperature and bulb size, and overestimate the effect of the water temperature (Doorduin *et al.*, 1997).

An alternative treatment is the application of the controlled atmosphere temperature treatment (CATT), which could provide good results, but lacks experimental verification (see *Phytonemus pallidus* (Banks) later in the chapter, and Chapter 9).

CHEMICAL CONTROL. A number of active ingredients have been tested over the years with varying degrees of success. Hodson (1934, 1948) reported some degree of control with white oil emulsion applied at the centre of the plants when the flower bud appeared. Harrison (1956) investigated the action of systemic insecticides (amiton, fluoracetamide, fluoacetanilide, dimefox, mipafox, schradan (octamethylpyrophosphoramide)) on bulbs before their introduction into the glasshouse, but the results were not very satisfactory. Collingwood (1959b) tested fluoracetamide, endrin and demeton-*S*-methyl on infested bulbs and found none effective, except that endrin suppressed mites feeding on the leaves. Drench sprays of endrin, endosulfan or azinphos-methyl on narcissus plants in the greenhouse gave good control (Winfield, 1964). Wiese and Meyer Smith (1967, in Meyer Smith, 1981a) suggested dipping bulbs in dicofol for about 15 s to prevent reinfestation. The immersion of bulbs for over 2 h in a water emulsion of thionazin produced good control of the bulb scale mite in narcissus (Hussey *et al.*, 1969). Good control of *S. laticeps* was achieved in bulbs of narcissus stored for 6 weeks at 8–9°C and treated 1 or 2 h in water at 44.4°C or with thionazin solution (Winfield, 1971). Gurney (1970, 1972) highlighted the inefficacy of various systemic insecticides (aldicarb, fonofos, phorate, thionazin) as granular formulations applied in the compost on which infested bulbs were grown. Wilkin *et al.* (1976) evaluated the action of phoxim and pirimiphos-methyl under laboratory conditions and in the field, but the positive results obtained in the laboratory were not confirmed by the results of the field tests. Among the pesticides discussed above, endosulfan gave better control of *S. laticeps*. In England, the application of a drench of 0.1% endosulfan has been suggested within a few days of bringing the infested bulb outdoors (Lane, 1984a,b). However, its application to boxes of bulbs pre-forcing has a severe toxic impact on humans and the environment (Pesticide Action Network UK, 2000).

Methyl bromide fumigation was applied successfully for the control of mite on bulbs of narcissus, amaryllis and daffodil (Wiese and Meyer Smith, 1967, in Meyer Smith, 1981a; Hussey *et al.*, 1969; Gurney and Gandy, 1974), as was pirimiphos-methyl fumigation (Gurney and Gandy, 1974). No significant phytotoxicity due to methyl bromide was observed by Powell (1977), except at the largest of the dosages tested. A treatment of 250 mg h^{-1} with 5 h exposure was recommended as being virtually non-phytotoxic and providing satisfactory control of the bulb scale mite, although high doses can be damaging to both bulbs and flowers. The use of these fumigants has now been replaced (for ecotoxicological reasons) by other means.

Despite their effectiveness, the use of the above-mentioned active ingredients is prohibited in different countries, and few acaricides are labelled for mites on bulbs of Amaryllidaceae. Thus, in the UK (and other countries), there are no approved acaricides to control *S. laticeps* in glasshouse *Narcissus*, or when bulbs are stored. Other than hot-water treatment, there is no non-chemical means of control (Collier *et al.*, 2009).

INTEGRATED PEST MANAGEMENT. Physical and chemical approaches to mite control within an IPM strategy include the use of acaricides and high- or low-temperature treatments. Although extreme temperatures may kill mites, it is important that the bulbs themselves are unaffected and that their quality is not impaired. Options for biological control appear fundamental (Collier *et al.*, 2009).

The Genus *Phytonemus* Lindquist

Diagnostic characteristics

Both the male and female are without a sejugal apodeme, and with the posterior margin of the coxisternal plate II emarginated medially. The female has apodemes 4 extending posterolaterad of the bases of the setae *3b*, though not reaching to the bases of legs IV. Leg IV of the male has a prominent flange on the femorogenu, and a fused tibiotarsus (Lindquist, 1986).

Economic importance

The genus includes only *P. pallidus* (Banks), reported in North America, Eurasia, Hawaii, Japan (Ito, 1964, 1977), South America (Flechtmann, 1967c, 1971; Urueta Sandino and Navarro Alzate, 1978) and Africa (Meyer Smith *et al.*, 1973; Meyer Smith and Loots, 1978; Meyer Smith, 1981a) on plants of economic importance, including strawberries, watercress, red pepper and a large variety of flowering plants (Smith and Goldsmith, 1936; Ewing, 1939; Van Eyndhoven and Groenewold, 1959; Jeppson *et al.*, 1975). In cool temperate regions, *P. pallidus* develops both in the greenhouse and outdoors (Fjelddalen and Stenseth, 1958; Stenseth and Nordby, 1976).

Phytonemus pallidus (Banks) (Fig. 22.8)

Common name

Cyclamen mite, strawberry mite.

Diagnostic characteristics

FEMALE. The body is about 250 μm long and yellowish brown. The prodorsal shield does not extend over the gnathosoma and has 2 pairs of setae, the first pair (v_1) of which is shorter than the second pair (sc_2); this last is much longer than the dorsal hysterosomal setae. The trichobotria (sc_1) are capitate, partially or not

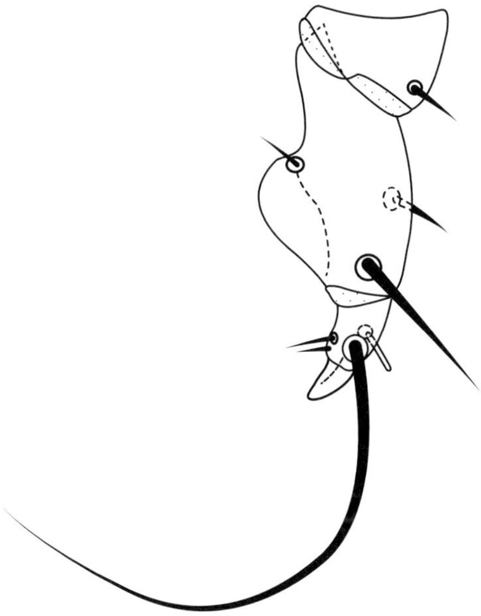

Fig. 22.8. *Phytonemus pallidus* (Banks). Male, leg IV (drawing by Vacante).

covered by the lateral margins of the prodorsal shield, set closely posterolaterad of setae v_1. Ventrally, apodeme 4 extends beyond the bases of setae *3b*.

MALE. The male is smaller than the female. The prodorsum has 4 pairs of setae, the fourth pair (sc_2) of which is shorter than the third pair (sc_1) and set laterad of the line formed by the first 3 pairs of setae (v_1, v_2, sc_1). The tibia and the tarsus of the leg IV are fused, and the femurogenu has a conspicuous and rounded flange.

Geographical distribution

Worldwide.

Host plants

These include the following 26 families and various species (Garman, 1917; Smith and Goldsmith, 1936; Ewing, 1939; van de Vrie, 1957; Van Eyndhoven and Groenewold, 1959; Karl, 1965a; French *et al.*, 1968; Hussey *et al.*, 1969; Jeppson *et al.*, 1975; Meyer Smith, 1981a; Oliveira de Andrade-Bertolo *et al.*, 2013):

Amaranthaceae: pigweed redroot (*Amaranthus retroflexus* L.), *Iresine* sp.

Araliaceae: ivy (*Hedera helix* L.).

Asteraceae: ragweed (*Ambrosia artemisiifolia* L.), *Aster dumosus* L., *Dahlia* sp., *Chrysanthemum* sp., feverfew (*C. parthenium* (L.) Bernh.), *Felicia* sp., galinsoga weed (*Galinsoga parviflora* Cav.), *Gerbera* sp., African daisy (*G. jamesoni* Adlam).

Balsaminaceae: snapweed (*Impatiens sultani* Hook.f.).

Begoniaceae: *Begonia* sp.

Boraginaceae: *Heliotropium* sp.

Brassicaceae: watercress (*Nasturtium officinale* W.T. Aiton).

Caryophyllaceae: carnation (*Dianthus caryophyllus* L.).

Crassulaceae: *Crassula rubicunda* E. Mey. ex Harv., *Kalanchoe coccinea* (H. Perrier) Boiteau.

Ericaceae: *Azalea* sp., *Vaccinium* sp.

Gentianaceae: Cape centaury (*Chironia linoides* L.), Persian violet (*Exacum affine* Balf.f.).

Geraniaceae: *Geranium* sp., *Pelargonium* sp.

Gesneriaceae: *Gloxinia* sp., African violet (*Saintpaulia ionantha* H. Wendl.).

Oleaceae: pikake (*Jasminum sambac* (L.) Aiton).

Onagraceae: *Fuchsia* sp.

Oxalidaceae: *Oxalis* sp.

Plantaginaceae: snapdragon (*Antirrhinum* sp.), plantain (*Plantago major* L.), purslane speedwell (*Veronica peregrina* L.).

Polygonaceae: field sorrel (*Rumex acetosella* L.).

Portulacaceae: common purslane (*Portulaca oleracea* L.).

Primulaceae: *Cyclamen indicum*?

Ranunculaceae: *Delphinium* sp., rocket larkspur (*D. ajacis* L.), *D. belladonna* hort. ex Bergmans.

Rosaceae: strawberry (*Fragaria* × *ananassa* Duchesne), *Geum* sp., cutleaf blackberry (*Rubus laciniatus* Willd.).

Solanaceae: red pepper (*Capsicum annuum* L.), *Petunia* sp.

Urticaceae: stinging nettle (*Urtica dioica* L.).

Verbenaceae: *Lantana* sp., *Verbena* sp.

Vitaceae: Japanese ivy (*Parthenocissus tricuspidata* (Siebold *et* Zucc.) Planch.), grapevine (*Vitis vinifera* L., cv. Merlot).

Bio-ecology

P. pallidus was first described by Banks (1901) as *Tarsonemus pallidus* from leaves of greenhouse chrysanthemums at Jamaica, New York. However, the morphological and ecological traits that it displayed suggested that it may represent a species complex. The possibility of species-level distinction arose from observations made by Massee (1930b, 1933b) that distinguished the species *pallidus* and *fragariae* according to differences in the lengths of the setae on the femorogenu and tibiotarsus of leg IV of the male, and also in host plant preference. Other authors have not encouraged this conclusion (Smith and Goldsmith, 1936; Ewing, 1939; Beer, 1954), while Van Eyndhoven and Groenewold (1959) found a species-level distinction between the European populations living on cyclamen and strawberry, according to differences in length and the reduction of the setae sc_1 of males and sc_2 of females, and, again, in host plant preference. Karl (1965a, 1969) studied specimens from North America and Europe and confirmed the above differences, in addition to other differences in the legs of females; Karl identified three forms in the *pallidus* species complex. However, parthenogenetic reproduction hindered cross-breeding tests between these forms and suggested that they be designated as the subspecies *T. pallidus pallidus* Banks, 1899, *T. pallidus fragariae* Zimmermann, 1905, and *T. pallidus asteris* Karl, 1965a. *T. p. pallidus* lives on several host plants, mainly in greenhouses, and is commonly called the cyclamen mite; *T. p. fragariae* is the strawberry mite, and lives mainly on rosaceous plants, especially strawberries in the field; *T. p. asteris* is the aster mite, living only on *Aster dumosus* L. in the field. At present though, the problem of this species complex seems not to be resolved, and to avoid further confusion, the species is discussed here only as *P. pallidus*, the common cyclamen mite.

The cyclamen mite lives on a wide range of wild and cultivated plants, belonging to almost 26 botanical families. It is a severe pest of strawberries, and of ornamental flowers and shrubs such as cyclamen, gerbera, begonia, African violet, ivy and other plants (Smith and Goldsmith, 1936; Ewing, 1939; Van Eyndhoven and Groenewold, 1959; Jeppson *et al.*, 1975). Recently, the mite has been reported on grapevines in Brazil (Oliveira de Andrade-Bertolo *et al.*, 2013). Populations of *P. pallidus* are negatively phototropic and their development requires high levels of humidity. This adaptation means that the mite prefers to live in sites such as unopened leaflets in the crown of the host plant, between the young leaves in the bud or in the cavities of flower buds. In these sites, the cyclamen mite lays its eggs and reproduces. *P. pallidus* usually feeds on the upper side of young and not completely developed leaves. In temperate climates, *P. pallidus* overwinters as females, but during mild winters or in mild climates, it can continue to reproduce. In Canada, the females overwinter in the crowns within the leaf sheaths of strawberry, and begin to ovideposit by mid-April. Peak populations were observed from the middle of June to late July, with as many as 300 mites per plant. The population density of *P. pallidus* is inversely related to the rate of new leaf growth. Egg deposition generally decreases in August and stops in October, although under greenhouse conditions, the mite may reproduce throughout the year (Dustan and Matthewman, 1932). In central and northern Europe, population peaks were observed in July and August, when the flower buds start to form (Alford, 1972, 1976; Stenseth and Nordby, 1976). In Poland, the population of *P. pallidus* is not so large in May, but it increases rapidly in the summer, reaching a peak in July–August, which extends until the beginning of September in some years. The mite develops 3.5 generations a year, and at the end of August and in September the young females start to search for wintering places (Łabanowska, 2006a). In cool temperate regions, *P. pallidus* develops both in greenhouses and outdoors (Fjelddalen and Stenseth, 1958; Stenseth and Nordby, 1976).

P. pallidus reproduces by haplodiploidy, but unfertilized females of some populations can generate male and female offspring, while other populations reproduce by thelytoky (Karl, 1965a). The eggs are laid in clusters or in masses of two or three units between the young leaves of the bud. Daily oviposition ranges from one to three eggs and fecundity from 12 to 16 eggs. The total developmental time from egg to adult is 1–3 weeks, with egg development lasting from 3 to 7 days, larva development taking 1–4 days and the quiescent nymph stage lasting 2–7 days (Smith and Goldsmith, 1936).

Dispersal of the cyclamen mite within the same field takes place by the crawling of motile forms, the transplantation of daughter infested plants or by wind. Dispersal at long distances is usually via the transport of infested plants, irrigation, animals (bees, birds), and also via pickers and equipment, including strawberry freezer trays (UC IPM Online, 2014d).

Natural enemies include predatory mites such as the phytoseiid mites *Typhlodromus bellinus* Womersley and *T. reticulatus* Oudemans (= *Neoseiulus reticulatus* (Oudemans)); however, their populations increase too slowly to produce economic control. Other predators develop at high population densities of *P. pallidus*, such as the six-spotted thrips, minute

pirate bugs and the western predatory mite, *Galendromus occidentalis* (Nesbitt) (= *Metaseiulus occidentalis* (Nesbitt)) (UC IPM Online, 2014d).

Symptomatology and damage

Infestations of *P. pallidus* on strawberry primarily occur on plants transplanted in the autumn and in second-year plantings, but infested plants can also be transplanted into first-year fields and the damage symptoms appear on leaves as the season progresses. The cyclamen mite usually damages the youngest leaves of the strawberry plant. The upper surface of infested leaves becomes roughened and wrinkled, with irregular folding and fluting of the margins, and bulging of the veins upward as blisters. The petioles do not develop and plants show a central dense appearance. Plants that are severely infested are dwarfed at the crown and have small leaflets that do not extend regularly. The smaller leaflets are at first pale yellowish green with a hard brittle texture, and afterwards turn brown or silvered (if the lower leaf surfaces are exposed to the sun). Infested flowers and young fruits are brown near the inner bases of the sepals, and at high levels of attack may turn black and dry. The fruits of infested plants are dwarfed, and the seeds stand out on the flesh of the berry. When uncontrolled, the cyclamen mite can prevent plants from producing fruits. Infestation generally decreases fruit size, yield and quality (Dustan and Matthewman, 1932; Alford, 1972, 1976; Stenseth and Nordby, 1976; UC IPM Online, 2014d).

Infestations on gerbera cause bronzed patches along the midribs and slight curling of the leaves. Infestation to flowers usually occurs at the bud stage and the main symptom is deformation of the flower rays (Boyle and Haramoto, 1956).

Infestations of cyclamen mite on pikake result in fewer flowers per cluster and the total abortion of the clusters. Buds and flowers may be poorly formed. The leaves are twisted, distorted and smaller than normal. The internodes of infested shoots are elongated, and the leaf buds are smaller at the shoot apex. Severely infested shoots bear leaves that are totally destroyed, with the leaves gradually becoming smaller and more distorted towards the apex (Boyle and Haramoto, 1956).

Control

Control is difficult because the cyclamen mite populations are protected by leaf buds. Prevention, monitoring and the application of biological, physical and chemical means may suppress mite outbreaks.

MONITORING. At the first detection of damage symptoms on strawberry, it is suggested that the whole field be carefully monitored, with the priority on newly unfolding leaves. The detection of one cyclamen mite per ten leaves suggests that the infested crop surface should be treated (UC IPM Online, 2014d). In Poland, the economic threshold level on strawberry should be estimated after the harvest of the strawberries, and it has been specified as six mites per leaf (Łabanowska, 2006a).

PREVENTION. The fumigation of planting stock in closed chambers with methyl bromide or ethylene dibromide has been suggested in the past to prevent the spread of cyclamen mite from one area to another. Regardless of whether these chemicals are used (as they are currently subject to severe restrictions), plant inspection at this time allows the detection of mite presence on young plants before and after transplantation, thus optimizing the application of control strategies, taking into account that the cyclamen mites can easily be transferred from one crop to another by animals, pickers and equipment, including strawberry freezer trays. From this point of view, it is useful to dip trays of long-term cold storage (−2.22°C) transplants into a hot water bath for 7 min before planting. Similarly, it is useful to wash young plants in hot water before transplanting (see Physical control below).

PLANT RESISTANCE. Even though some strawberry cultivars have been reported to be less susceptible to *P. pallidus*, no cultivar has been found to be completely resistant (Alford, 1972). Studies in Poland on 13 strawberry cultivars showed that none of the cultivars tested were completely resistant to strawberry mites. However, the number of strawberry mites found on some cultivars ('Dukat', 'Marmolada', 'Redgauntlet', 'Senga Sengana') was low for three growing seasons, and no control was needed even though the population increased from year to year. Other cultivars were either mostly infested ('Gerida', 'Kama', 'Syriusz') or moderately infested ('Cortina', 'Dana', 'Elsanta', 'Honeoye', 'Kent', 'Real') (Łabanowska, 2004).

CULTURAL PRACTICES. It is good practice not to transplant – where and when possible – into areas that have previously been affected by the presence of the mite. Because infested nursery plants are the major source of *P. pallidus* in annual plantings, it is necessary to use uninfested nursery stock. The removal of infested plants as soon as symptoms are detected also slows the spread of infestations in the crop (UC IPM Online, 2014d).

BIOLOGICAL CONTROL. Different phytoseiid mites (*Amblyseius andersoni* (Chant), *Neoseiulus californicus* (McGregor), *N. cucumeris* (Oudemans), *N. fallacis* (Garman), *N. reductus* (Wainstein) and *N. reticulatus*) have been proposed in various countries for the biological control of *P. pallidus*. Among these, *N. californicus* and *N. cucumeris* showed potential for the biological control of phytophagous mites on strawberry. Both phytoseiids reduced the numbers of *P. pallidus* on potted strawberry plants under glasshouse conditions. Reductions in the range of 71–81% in the numbers of tarsonemid active stages and eggs on plants on to which the predators had been released, compared with non-released plants, were observed in England. A suitable predator:prey ratio at an early stage was showed in an experiment in which an initial ratio of 1 *N. cucumeris*:10 *P. pallidus* produced greater control than ratios of 1:20 or 1:40 (Easterbrook *et al.*, 2001). Results from other research in Sweden showed that releases of *N. cucumeris* reduced cyclamen mite populations on strawberry by up to 50% but that there was no increase in yield. It was difficult to draw clear conclusions from *P. pallidus* dispersal, because mites tended to move more easily along planted rows than between rows (Berglund, 2007).

In Finland, releases of *N. cucumeris* and *N. barkeri* Hughes, on to strawberry gave encouraging results (Tuovinen *et al.*, 2009), but tests with a different set of predatory mites (*Anthoseius*

rhenanus (Oudemans), *Euseius finlandicus* (Oudemans), *N. cucumeris*, *N. barkeri*) indirectly demonstrated that the biological control of cyclamen mite on this crop awaits a definitive solution. In this latter study, *N. cucumeris* and *N. barkeri* replaced *E. finlandicus* and *A. rhenanus*, and were able to form mixed populations when *P. pallidus* was available. However, these phytoseiids did not prevent the development of natural infestation by the two-spotted spider mite, *Tetranychus urticae* Koch, after the initial study was concluded (Tuovinen and Lindqvist, 2010). The volatile emission profile of strawberry plants (cvs 'Polka' and 'Honeoye') damaged by *P. pallidus* was analysed to determine the potential of these strawberry plants to emit herbivore-induced volatiles. The phytoseiids *N. cucumeris*, *N. californicus* and *E. finlandicus* did not prefer *P. pallidus*-damaged plants over intact plants in a Y-tube olfactometer. This behaviour seems to indicate that these predatory mites are not attracted by the release of herbivore-induced volatiles from young strawberry plants (Himanen *et al.*, 2005).

PHYSICAL CONTROL. The immersion of strawberry plants in water at 43.5°C for 30 min gave complete mortality of cyclamen mites, but treatment of the plants with saturated air at 43.5°C for 1 h was more efficient. The plants must be loose on screens or slatted boxes, or stacked, to allow the penetration of gas, but after the treatment the plants should be dried before packing, and planted as soon as possible (Munger, 1933; Smith, 1939a). However, the use of this technique on strawberry may retard its growth (Hussey *et al.*, 1969), so investigations were clearly needed to reduce the risk of damage. It was observed that during the summer, the shortest exposure times to kill all *P. pallidus* adults in hot water at temperatures of 44°C, 46°C and 48°C were 6, 2.5 and 1.25 min, respectively, and hot-water treatment for 6.5 min at 46°C killed all mites on infested runners. This exposure time includes the time needed to raise the temperature of the plant material to that of the water bath. From the middle of September and later, the mites became more heat tolerant as they prepared for hibernation, and longer exposure periods could be necessary to kill mites on frigo (cold-stored) plants (Hellqvist, 2002). The University of California Pest Management Guidelines for the cyclamen mite on strawberry (UC IPM Online, 2012d) recommend dipping trays of long-term cold-stored (−2.22°C) transplants into a hot water bath for 7 min before planting. The plants should be washed to remove all dirt, and then placed in a circulating water bath that is held at a constant temperature of 48.88°C. Afterwards, the plants should be submerged in very cold water and then planted as soon as possible. This treatment is not recommended for freshly dug transplants that have only been stored at 0.55°C.

The ban on the use of methyl bromide to treat mother planting stock for runner production suggested that the controlled atmosphere temperature treatment (CATT) should be optimized to provide a non-chemical and sustainable method of disinfestation. The application of CATT for 48 h at a temperature of 35°C and under 50% CO_2 produces over 99.8% mortality of *P. pallidus*. No harmful irreversible results of CATT on the vitality of mother plants in the field or on runner production were observed. From 2009, CATT has been upscaled to a commercial level and widely applied by Dutch producers of planting stock (van Kruistum *et al.*, 2012).

CHEMICAL CONTROL. Strawberry plants not treated with hot water or CATT before transplanting are usually treated with chemicals. Chemical treatments require the use of a high rate of water per acre (300–500 gal) to soak the folded leaves and immature flower buds located in the crowns. The rapid increase of cyclamen mite populations recommends the use of effective active ingredients. The removal and treatment of infested hot spots with a hand sprayer suppress early infestations. Early season control in nurseries before the plant canopy closes over is important (UC IPM Online, 2014d). The choice of acaricides is primarily led by their ecotoxicological properties and by the phenology of the crop. This constraint is crucial for strawberry, as it needs to take account of the harvesting of the fruits and the spray application of low persistence principles that have an acceptable level of selectivity towards natural enemies and pollinators.

The main acaricides tested or suggested for use on strawberry include: abamectin (Tuovinen *et al.*, 2009; Zalom *et al.*, 2009; Fountain *et al.*, 2010; UC IPM Online, 2014d), acequinocyl (Zalom *et al.*, 2009; Fountain *et al.*, 2010; UC IPM Online, 2014d), amitraz (Łabanowska, 1992), bifenazate (Łabanowska, 2006a,b; Zalom *et al.*, 2009; Fountain *et al.*, 2010), dicofol (Hussey *et al.*, 1969; Jeppson *et al.*, 1975; Zalom *et al.*, 2009), endosulfan (Hussey *et al.*, 1969; Jeppson *et al.*, 1975; Bashkatova *et al.*, 1983; Łabanowska, 2006a,b; UC IPM Online, 2014d), endrin (Hussey *et al.*, 1969; Jeppson *et al.*, 1975), essential oils (Ecotrol™ or GC Mite) (Zalom *et al.*, 2009), etoxazole (Zalom *et al.*, 2009; Fountain *et al.*, 2010), fenazaquin (Łabanowska, 2006a,b), fenpropathrin (Zalom *et al.*, 2009; UC IPM Online, 2014d), fenpyroximate (Łabanowska, 2006a,b; Zalom *et al.*, 2009; Fountain *et al.*, 2010), lambda-cyhalothrin (Raudonis, 2006), milbemectin (Zalom *et al.*, 2009), pirimiphos-methyl (Bashkatova *et al.*, 1983), propargite (Łabanowska, 2006a,b), pyridaben (Labanowska, 1992), spirodiclofen (Łabanowska, 2006a,b; Raudonis, 2006; Fountain *et al.*, 2010), spiromesefen (Zalom *et al.*, 2009; Fountain *et al.*, 2010), spirotetramat (Łabanowska *et al.*, 2013), tebufenpyrad (Fountain *et al.*, 2010), triazophos (Łabanowska, 1992). The use of the surfactant heptamethyltrisiloxane improved the action of some acaricides (Fountain *et al.*, 2010).

Chemicals applied to cyclamen include in addition to dicofol, endosulfan and endrin, spray applications of dienochlor, fenbutatin oxide, oxamil and triazof (Tusnádi and Kerényi-Nemestóthy, 1989), or a drench of schradan (Hussey *et al.*, 1969) on *Saintpaulia* and other plants. However, it is not possible to generalize this information, as various active ingredients cited here have been withdrawn and each choice must be made in relation to the law of the country concerned.

INTEGRATED PEST MANAGEMENT. The harmonization of the above technical possibilities allows the implementation of IPM programmes, which sometimes need different methods to be used as well, e.g. *N. cucumeris* releases and covering with fleece in combination with pyrethrum sprayings in the organic production of strawberry (Berglund *et al.*, 2007). IPM experiences with strawberry in Finland summarize the possible lines of strategy, including the risks or limits of its implementation in the field. In Finland, only methiocarb and abamectin are available for control of the cyclamen mite before flowering and after harvesting, and only biological control is permitted in the

meantime. Pyrethroids and organophosphates are applied to control insect pests, often just before flowering. The phytoseiid mites *N. cucumeris* and *N. barkeri* are released in strawberry fields using variable timing and waiting periods after the preceding insecticide treatments. During the season, from 1 to 1.5 million phytoseiid mites ha^{-1} are released once, and monitoring of the mite populations is provided by leaf samples taken two to six times during the season. The integration of biological control of the cyclamen mite and chemical control of insects appears to be realistic. However, implementation of the strategy requires a clear knowledge of the situation in the field and the use of some precautions to optimize the application of the strategy in the field. Thus, systematic monitoring of pest insects allows earlier timing of insecticide applications, and consequently the earlier releases of predators. Delayed spray applications of insecticides invite the choice of low-persistence principles so as to enable the early release of predators when cyclamen mite populations are low. Late-season leaf samples help in the decision making for acaricide treatments after harvesting (Tuovinen *et al.*, 2009).

The Genus *Tarsonemus* Canestrini *et* Fanzago

Diagnostic characteristics

Males and females have either a fully developed sejugal apodeme or one that is reduced only medially. The female has stigmata spaced posterolaterad from the setae v_1, apodemes 4 do not extend posterolaterad of the bases of setae *3b*, and the tegula is normally rounded and set between the bases of legs IV. Leg IV of the male is with or without a flange on the femorogenu, with the tibia and tarsus separate or fused, and the terminal claw well developed and usually unguiform (Lindquist, 1986).

Economic importance

This genus includes a large number of species. Lindquist (1986) proposed around 200 nominal species, but other species from Palaearctic regions and other countries worldwide were reported later by various authors. In the Palaearctic region, the genus *Tarsonemus* has 77 species, mainly fungivores and with a wide geographical distribution (Kaliszewski, 1993). Mites of this genus are known throughout the world, with the probable exception of the polar regions. They live in soil, litter, in ground-level and subterranean nests of vertebrate and invertebrate animals, in decaying wood, bracket fungi and subcortical habitats associated with insects, in stored grain and food products, laboratory cultures, man-made habitats (houses), and on all sorts of woody and herbaceous plants (monocotyledonous and dicotyledonous) (Lindquist, 1986).

Despite this wide range of habitats, the species of this genus are primarily fungivorous, although facultative fungivory has been reported for various species (Cooreman, 1941; Beer, 1954; Schaarschmidt, 1959; Hussey, 1963; Karl, 1965a; Suski, 1972; Santos and Whitford, 1981), and some of these fungivorous species are serious pests (Thomas, 1942; Hussey, 1963). Among these species, *T. myceliophagus* Hussey was reported as injurious to commercial mushroom cultures (Hussey, 1963; Hussey and Gurney, 1967). Seed-borne fungi causing the deterioration of stored grain were associated with the presence of *T. granarius* Lindquist (Lindquist, 1972; White and Sinha, 1979, 1981). *T. fusarii* Cooreman may contaminate fungal cultures in the laboratory (Schaarschmidt, 1959; Boness, 1968; Lindquist, 1972). *T. waitei* Banks feeds and reproduces on *Alternaria* sp. under laboratory conditions (Nucifora and Vacante, 1986). Other *Tarsonemus* spp. were found associated with fungi in the decomposition of soil litter (Santos and Whitford, 1981). Other feeding habits may occur in other species of this genus, such as phytophagy, algivory, saprophagy, necrophagy and, probably, zoophagy (feeding on living animal exudates or products) (Lindquist, 1986).

Some species, e.g. *T. waitei*, *T. bakeri* Ewing and *T. smithi* Ewing, could be facultatively phytophagous on host plants of economic importance (Karl, 1965b; Jeppson *et al.*, 1975; Lindquist, 1978). Other species reported as pests of economic plants include *T. confusus* Ewing, which lives on plants, soil litter, house dust and birds' nests. It is primarily a fungivorous species, which occasionally becomes a minor pest of some ornamentals and vegetables (Hussey *et al.*, 1969; Heungens and Tirry, 2000). The damage caused by this mite is not well documented, and is probably confused with that of the broad mite, *Polyphagotarsonemus latus* (Banks), and *Phytonemus pallidus* (Banks), which are frequently associated with it in the field. *T. bilobatus* Suski is a worldwide species, living in a wide range of habitats, such as plant species, fungal and bacterial cultures, stored food and products, litter and soil. It is primarily a fungivorous species, but is thought to be responsible for damage to ornamental flowers and cucumber, Chinese cabbage and other host plants in greenhouses; the damage is seen as lustrous, discoloured and deformed leaves with irregular folding of the upper leaf surface (Nemestothy, 1983; Nakao, 1991; Ito, 1993). *T. floricolus* Canestrini *et* Fanzago is primarily a fungivorous species, but is also known as a pest of mushrooms (Zhang, 2003). This mite was a dominant species, among the mites living on house dust in Danish city dwellings, and may be responsible for human allergies to house dust (Korsgaard and Hallas, 1979). In Korea, *T. floricolus* was collected from sputum and produced a positive reaction on a skin test (Ryu *et al.*, 2003).

The aspects that have been briefly discussed above indicate the need for further research on the bio-ecology of these interesting mites. Many reports of their damage are not well documented, and in some case problems of misidentification could be occurring. There is also doubt of the ability of several species to cause definite economic damage.

Only *T. myceliophagus* is discussed here. This is a mite that has in the past been reported as injurious to mushroom cultures (Hussey, 1963; Hussey and Gurney, 1967), but it is currently regarded as a secondary pest that is sporadically reported on mushrooms (White, 1991; Fletcher and Gaze, 2008).

Tarsonemus myceliophagus Hussey (Fig. 22.9)

Common name

Unknown.

Diagnostic characteristics

FEMALE. The body is oval, less than 200 μm long and pale brown in colour. Other characteristics are as for the genus.

MALE. The male is smaller than the female. The fourth pair of prodorsal setae (sc_2) are set laterad from the third pair (sc_1), which is longer than the other prodorsal setae. Tarsus II has a large solenidion (ω). Femur IV is round at the base, and lacking a flange; the tactile seta of tibia IV is as long as femur IV.

Geographical distribution

China (Yang and Liang, 1986), Germany (Greenslade and Clift, 2004), Hungary (Magowski and Beron, 2013), the Netherlands (Greenslade and Clift, 2004), New Zealand (Greenslade and Clift, 2004), Poland (Magowski and Beron, 2013), South Africa (Ueckermann, 2013), the UK (Hussey, 1963; Magowski and Beron, 2013), Ukraine (Magowski and Beron, 2013), the USA (Greenslade and Clift, 2004).

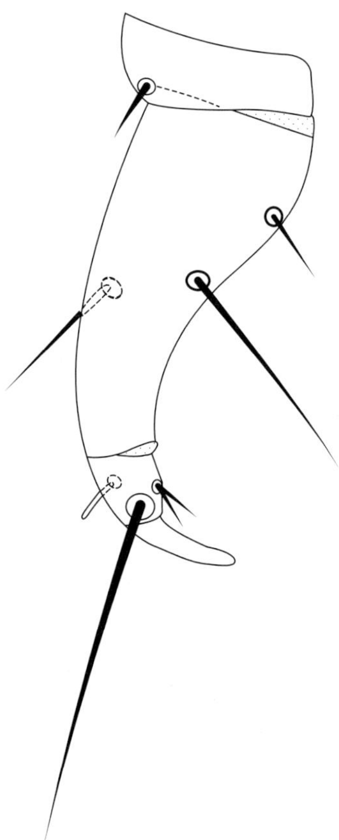

Fig. 22.9. *Tarsonemus myceliophagus* (Hussey). Male, leg IV (drawing by Vacante).

Host plants

Mycelium of mushrooms, weed moulds (*Chaetomium*, *Penicillium* spp., *Trichoderma*) (Hussey, 1963; Hussey and Gurney, 1967; Fletcher and Gaze, 2008).

Bio-ecology

T. myceliophagus was first described from mushroom cultures in England by Hussey (1963), and for some time it was reported as injurious to commercial mushroom cultures (Hussey and Gurney, 1967). The mite has a wide range of distribution and feeds on mushroom mycelium and weed moulds. Although it is potentially able to cause severe damage, it is currently regarded as a minor pest in England (White, 1991) and other countries. It is common on grasses in the Sydney region of New South Wales (Australia), but is rare in mushroom cultivation, as its presence depends on inadequate compost preparation (Greenslade and Clift, 2004). Under spawn-running and cropping conditions at 24°C, the egg to egg developmental time (with mated females) is 8 days, while at 16°C it is 12 days. The mite lays one egg a day for 2–3 weeks. At 16°C, the female longevity is about 2 months, and at 24°C it is 15 days. When the populations are high, they inhabit the mushroom casing, and when the casing dries out, the females stand erect and jump about 5 cm into the air. In this way, they exploit air movement that spreads them towards more favourable sites (Hussey and Gurney, 1967; Fletcher and Gaze, 2008).

Symptomatology and damage

Feeding by *T. myceliophagus* causes reddish brown discoloration and rounding of the bases of mushroom stalks, sometimes completely severing the basal attachments of the mushrooms. Rarely, the discoloration affects the mushrooms extensively. Damage does not usually appear until the third flush, because the mite populations develop slowly. However, if early infestation occurs after spawning, high population densities may develop, causing severe damage (Hussey and Gurney, 1967; Fletcher and Gaze, 2008).

Control

Effective control depends largely on hygiene treatments effected between the crops. Improvement in composting and 'cooking out' in recent years has promoted a decline in the importance of this mite. An effective phase II and a thorough and efficient cook out should be sufficient for control (Fletcher and Gaze, 2008).

In the past, the application of methyl bromide fumigation or 'cooking out' for several hours at 66–71C°, followed by the application of 0.01% dicofol sprays directed at the internal surfaces of the empty mushroom-growing building produced satisfactory control (Hussey and Gurney, 1967). However, methyl bromide fumigation is currently not permitted, and laboratory tests carried out have now shown that adult mites can be killed by exposure to a temperature of 49°C for 20 min in a saturated atmosphere. However, living mites have been found in compost that has been cooked out at 60°C, probably because they were protected by sheltering in cool crevices, especially when the treatment has not been applied uniformly (Fletcher and Gaze, 2008).

Other Tarsonemid Mites

This group includes some species, such as *Xenotarsonemus belemnitoides* (Weis-Fogh), reported from rice, strawberry and ornamentals in some European countries (Schaarschmidt, 1959) and in China (Lin and Zhang, 2002), which has become a pest of azalea in Belgium and the Netherlands. Unfortunately, the damage caused by this mite to economic plants is not well documented, and so the extensive chemical control that is applied in Belgium cannot be justified (Heungens, 1993; Heungens and Tirry, 2000).

23 Minor Families

This chapter includes mites in five families (Dolichocybidae, Siteroptidae, Pygmephoridae, Scutacaridae, Microdispidae) that have in common a capacity to be either directly or indirectly injurious to mushrooms.

The Family Dolichocybidae Mahunka

Morphological characteristics, systematics and bio-ecology

The Dolichocybidae are small mites that are 100–280 μm long. The prodorsal shield of the female lacks anterolateral stigmata and the associated tracheae; it has 2–3 pairs of setiform setae that may or may not have anterolateral bothridia with capitate sensilla (the normal scapular setae in the male). The prodorsum and the tergite C are normal in form, without wing-like expansions. In the genus *Formicomotes* Sevastyanov, the opisthosomal tergites C, D and EF are distinct, while the segment H may be consolidated with the pseudoanal segment Ps, and the aggenital region Ag into the undivided plate HPsAg located ventrally (Magowski, 1988). The female has a small genital aperture, with 1–3 pairs of small or vestigial genital setae (g_{1-3}) (Figs 23.1 and 23.2), which are sometimes absent, and 1 pair of aggenital setae (*ag*). Species of the genus *Dolichocybe* Krantz show tergites C, D, EF and Ps distinctly (Khaustov, 2006a). The tergites C, D, EF and Ps bear the setae c_{1-2}, *d*, *e* and *f* and h_{1-2}, respectively; the cupules ia, and im are located in the tergites D and EF, respectively. The coxisternal plates I–IV have 1–3 setae, and the apodemes I are weakly developed, while apodemes II–IV are easily discernible and of various sizes. Tarsi II–IV have a smooth flexibile empodium located between the claws. Trochanter III lacks setae, and the setation of genua I–IV is maximally 2-1-1-1 (Walter *et al.*, 2009).

Currently, the family includes about 25 species ascribed to five genera (Walter *et al.*, 2009), but with the exception of some genera, such as *Dolichocybe* (Khaustov, 2006a), its systematic organization appears to be unclear in various traits and further insight is needed.

The feeding habits of the Dolichocybidae have been little documented, although the fungivory of some species is confirmed, such as that of *Dolichocybe hippocastani* Rack and *Formicomotes heteromorphus* Magowski, which are both characterized by physogastry during feeding and generate directly as many as 200 adult male and female progeny (Rack, 1967; Magowski, 1988). Other species living in habitats adapted to fungal growth may be fungivorous (Kaliszewski *et al.*, 1995). Adults emerge through the larval skins and a nymphal apoderm as well as the egg chorion (Rack, 1967). Magowski (1988) determined two female morphs of *F. heteromorphus*, the first record of non-sexual dimorphism in this family; this mite causes retardation of mycelial growth of the cultivated mushroom *Auricularia polytricha* (Mont.) Sacc. (Kantaratanakul *et al.*, 1989). *D. shenagarici* Zou *et* Gao, and *D. perniciosa* Zou *et* Gao have been reported on edible fungi in China (Zou and Gao, 1996). Moreover, *D. keiferi* Krantz directly affects mushroom mycelia (Woodhall *et al.*, 2009), and *D. indica* Mahunka damages the black poplar mushroom, *Agrocybe cylindracea* (DC.) Maire (Kantaratanakul *et al.*, 1989). Among these species, only *F. heteromorphus* has any economic importance.

The Genus *Formicomotes* Sevastyanov

Diagnostic characteristics

A hyaline rim partially encompasses the prodorsal shield, and the outline of the idiosoma is nearly round. Mites of this genus are found in association with ants (Sevastianov, 1980).

Economic importance

The genus *Formicomotes* was established on the basis of a single morph, and may be congeneric with *Dolichomotes* Smiley (Magowski, 1988). Currently, it includes only two species, *F. octipes* Sevastyanov and *F. heteromorphus*, of which only the latter has any economic importance and has been reported causing damage to cultivated mushrooms in Thailand (Kantaratanakul *et al.*, 1989).

Formicomotes heteromorphus Magowski (Figs 23.1 and 23.2)

Common name

Unknown.

Diagnostic characteristics

F. heteromorphus has three different morphotypes: the (normal) female or female X-morph, which is physogastric; the heteromorphic female or female Y-morph, which is also physogastric; and the male (Magowski, 1988).

FEMALE X-MORPH. The body is about 145 μm long and 64 μm wide. The gnathosoma is roundish and tapering. The prodorsum lacks stigmata and peritremes. The prodorsal shield bears

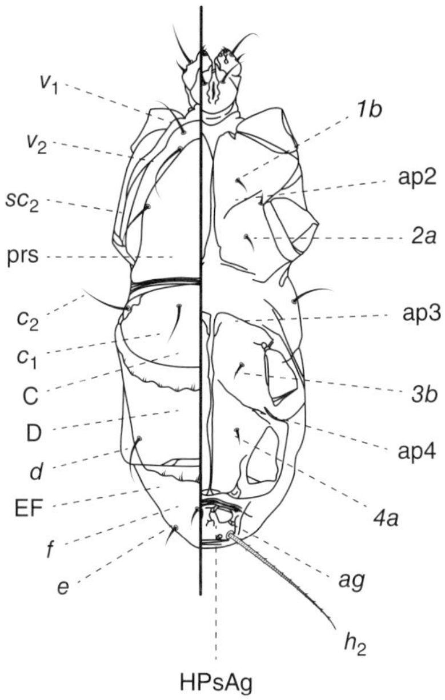

Fig. 23.1. *Formicomotes heteromorphus* Magowski. Female X-morph, dorsal (left) and ventral (right) views (modified from Magowski, 1988). Key: ap 2–4, apodemes 2–4; C, D, EF, tergites; HPsAg, plate consolidated from segments H, Ps and Ag; prs, prodorsal shield. The rest of the labels designate setae, and the setal notation is explained in the text.

3 pairs of setae (v_1, v_2, sc_2) and lacks trichobotria (sc_1). The tergites C and D usually have striae and dentation. The segment H is consolidated with Ps and Ag into the undivided plate HPsAg, which is located ventrally, is shorter than wide, and has 1 pair of large, external papillae (not the genital papillae of other mites). The setae v_2 are similar in length to the setae sc_2, the setae *f* are set anteriorly to the level of the setae *e*, and the coxal setae (1 pair on each pair of coxal plates) are short, not reaching or extending beyond the apodemes or ridges of the coxisternal plates (and about 4 times shorter than the distance between their bases). The solenidia of all the legs adhere tightly, formed as roundish fields, and never clavate, rod or cob like. The empodia of all the legs are sessile and expanded; the claws of leg I can be different in size, while those of legs III–IV are similar in size (Magowski, 1988).

FEMALE Y-MORPH. The body is about 124 μm long and 70 μm wide. The gnathosoma is more robust and wider than in the X-morph. The anterior rim of the prodorsal shield covers at least the base of the gnathosoma and the posterior margins; tergites C and D are striated with dentation. Three pairs of prodorsal setae (v_1, v_2, sc_2) are present, while stigmata, peritremes and the trichobotria sc_1 are absent. The idiosoma is oval, roundish, and no more than 1.5 times longer than wide. A weakly defined hyaline anterior rim of the prodorsal shield is present, the v_2 setae are longer than the v_1 setae and subequal to the sc_2 setae, and the setae *f* are set anteriorly to the setae *e*. Ventrally, there is 1 pair of setae on each pair of coxisternal plates. The segment H is consolidated with Ps and Ag into the undivided plate HPsAg, located ventrally, with a pair of large,

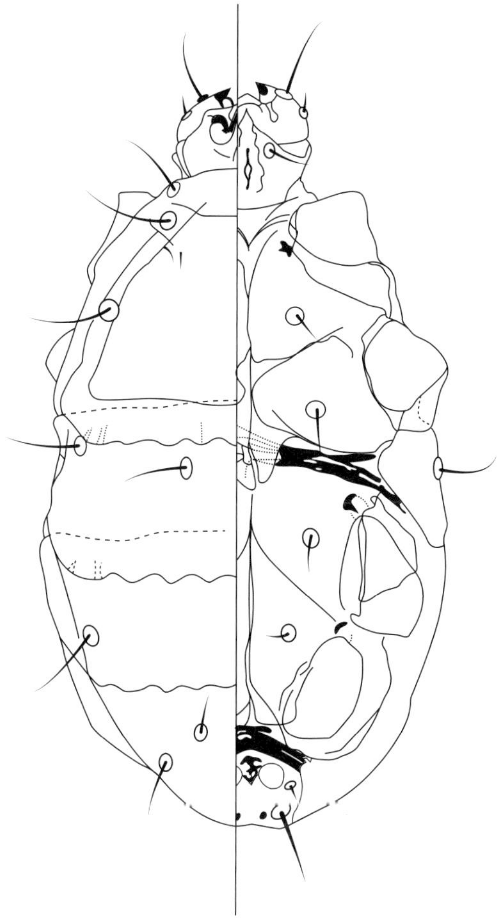

Fig. 23.2. *Formicomotes heteromorphus* Magowski. Female Y-morph, dorsal (left) and ventral (right) views (modified from Magowski, 1988).

external papillae (not the genital papillae of other mites). The solenidia of all legs adhere tightly, formed as roundish fields, and never clavate, rod or cob like. The claws of leg I can be different in size, while those of legs III and IV are similar in size; the empodia of legs II–IV are stalked, expanded beyond the end of the claws, and sucker like (Magowski, 1988).

MALE. The body is about 100 μm long and 66 μm wide. The gnathosoma is much smaller than that of either female and has 3 pairs of setae. The palpi are reduced a great deal, although cheliceral stylets are visible with small levers, probably used only restrictively or even not used. The stigmata, the peritremes and the trichobotria (sc_1) are absent, and only 3 pairs of setae (v_1, v_2, sc_2) are located on the prodorsal shield. There is one pair of setae on each pair of coxisternal plates. The aedeagus is well developed, tubiform, located dorsomedially in the space between parts of divided medially tergite EF near the posterior extremity of the body and directed upward. The posterior margins of the tergites are dentated with weak striae. The solenidia of all legs adhere tightly, formed as roundish fields, and never clavate, rod or cob like. The claws are symmetrically developed, and the empodia are sessile; the claws of legs III and IV are similar in size (Magowski, 1988).

Geographical distribution

Thailand (Magowski, 1988; Kantaratanakul *et al.*, 1989; Kulpiyawat *et al.*, 2009).

Hosts

The mushroom hosts include: Auriculariaceae – Jew's ear mushroom (*Auricularia polytricha* (Mont.) Sacc.) (Magowski, 1988), *A. auricula* (Hook) Underwood; Marasmiaceae – *Flammulina velutipes* (Curtis) Singer (Kulpiyawat *et al.*, 2009); Pleurotaceae – *Pleurotus* sp. (Hungary strain) (Kulpiyawat *et al.*, 2009), maple oyster (*P. cystidiosus* O.K. Miller) (Kulpiyawat *et al.*, 2009), oyster mushroom (*P. ostreatus* (Jacq.) P. Kumm.) (Magowski, 1988; Kulpiyawat *et al.*, 2009).

Bio-ecology

F. heteromorphus was first described from Thailand on cultivated mushrooms such as the Jew's ear mushroom and the oyster mushroom (Magowski, 1988; Kantaratanakul *et al.*, 1989). Studies on the host preferences of *F. heteromorphus* at 27 ± 2°C and 65 ±3% RH in which 11 mushroom species were tested indicated that the mite damaged four of them: *Pleurotus* sp. (Hungary strain), *P. cystidiosus*, *P. ostreatus* and *F. velutipes*. It did not damage seven other species: *A. polytricha*, *Ganoderma lucidum* (Curtis) P. Karst., *Lentinula edodes* (Berk.) Pegler, *Lentinus polychrous* Lév., *L. squarrosulus* Mont., *P. sajor-caju* (Fr.) Singer and *Schizophyllum commune* Fries (Kulpiyawat *et al.*, 2009).

The life cycle of *F. heteromorphus* is characterized by having three morphotypes: the (normal) female or female X-morph, which is physogastric, the heteromorphic female or female Y-morph, which is also physogastric, and the male. The biological reason of this female non-sexual dimorphism is unknown. The Y-morph is shorter and more roundish and could be a phoretomorph that increases the abilities of the mite population to disperse, but it is also possible that the Y-morph is parasitic because it has chelicerae and a gnathosoma that are both stronger than those of an X-morph (Magowski, 1988).

Under laboratory conditions on *A. polytricha* sporophore cultures, the life cycle of *F. heteromorphus* develops through the normal and physogastric females, with a rare presence of males. The active stages last from 18 to 66 h, during which the normal females are more active and exhibit a jumping behaviour. The physogastric females are characterized by their elongated and rod-shaped opisthosoma behind legs IV, which can be up to about 1900 μm long and contains many eggs. These eggs hatch within 2–4 days in the body of the mother and no larvae or copulation have been observed. The post-embryonic development of the male takes from 72 to 120 h, and that of the normal female and gravid female from 18 to 66 h, and from 66 to 228 h, respectively. The life cycle of the male and female takes 4 ± 0.66 and 10 ± 2.06 days, respectively. The number of offspring per female is 5 ± 4.1 males and 198 ± 52.1 females (total number of 203 ± 53.6). The hatching period of the female lasts 3.4 ± 0.7 days (Kantaratanakul *et al.*, 1989).

Studies on the seasonal fluctuation of *F. heteromorphus* in Thailand indicated that populations of the mite occurred during April to December, with an average number ranging from 7.28 to 369.46 mites cm^{-2} of polyethylene bag (Kulpiyawat *et al.*, 2009).

Symptomatology and damage

F. heteromorphus damages the mycelium of Jew's car and oyster mushrooms, especially during mycelial growth in the sawdust spawn. There is no damage to the sporophores, but within a few days of infestation, mycelial growth stops completely and no sporophore production occurs. The newly infested spawn may be brought into the mushroom house for fructification because of the tiny, hyaline nature of the adults of the mite, which are not seen but can spread rapidly because of the jumping behaviour of the normal females, their high reproductive potential and the short life cycle. In severely infested spawn, the active mites appear as light brown dust particles on the upper compost layer, as well as in clusters of elongated rod-shaped gravid females (Kantaratanakul *et al.*, 1989). It has been observed that the sporophore can be harvested only in the first week of fructification if the sawdust spawn is infested by 200 mites during mycelial growth. Mushroom production from infested spawn is significantly less than from uninfested spawn (Kulpiyawat *et al.*, 2009).

Control

The application of amitraz, bromopropylate and pyridaben suppressed the *F. heteromorphus* population with high efficacy on fully grown mycelium of *Pleurotus* sp. (Hungary strain), *P. cystidiosus*, *P. ostreatus* and *F. velutipes*. The use of one tablet of phosphine fumigant for 24, 48 and 72 h or two tablets for 24 h in a 0.5 cubic m^3 volume did not induce mycelium phytotoxicity. Spraying amitraz, pyridaben and tebufenpyrad at 7- or 10- or 14-day intervals on to the cotton pads of the sorghum grain mother spawn in bottles, and on to the cotton pads of the mycelium running stage in polythene bags, prevented the bottles and the bags from being contaminated by the mites (Kulpiyawat *et al.*, 2009).

Among botanical pesticides, the most toxic to *F. heteromorphus* was a hexane extract of *Litsea cubeba* (Lour.) Pers., followed by the methanol and dichloromethane extracts of the same plant (Insung *et al.*, 2008).

The Family Siteroptidae Mahunka

Morphological characteristics, systematics and bio-ecology

The Siteroptidae are small mites, with the body of the female body about 100 to over 500 μm long, and pale yellowish in colour. The female gnathosoma is capsulate, variable in shape and with 1–2 dorsal setae (ch_1–$_2$). The palpi are reduced, and the movable cheliceral digits are stylet like. The male gnathosoma is non-functional. The prodorsal shield of the female has a pair of anterolateral stigmata associated with tracheae, 3 pairs of setiform setae (v_1, v_2, sc_2), and a pair of dorsal or anterolateral bothridia with capitate sensilla (sc_1) (the male has normal scapular setae). The female opisthosoma has 5 dorsal plates (C, D, EF, H, Ps). Of these, plate C bears 2 pairs of setae (c_1, c_2), plate D bears the setae d and the cupules ia, plate EF has the setae f and a pair of cupules (im), plate H bears the setae h_1 and h_2 and the cupules ih. The plate Ps has 3 pairs of setae (ps_{1-3}), with ps_3 longer and set laterally (see Fig. 23.3). Ventrally, the coxisternal plates I have

3 pairs of setae (*1a–c*), and the coxisternal plates II have 2 pairs of setae (*2a–b*); the coxisternal plates I and II are united medially, forming a prosternal apodeme; the coxisternal plates III bear 3 pairs of setae (*3a–c*), and the coxisternal plates IV have 2 pairs of setae (*4b*, *4c*) (see Fig. 23.3); the coxisternal plates III and IV may be united medially and form a poststernal apodeme. The genital opening of the female is small, has a consolidated aggenital plate, and the genital and aggenital setae are usually lacking. The males have the plates C and D consolidated, with the aedeagus enclosed in a caudal genital capsule formed by the consolidated plates H, Ps and Ag. Leg I of the female has a single claw, and legs II–IV each have paired claws. All setae of femur I of the females are slender and attenuate, and the seta *d* is setiform (Walter *et al.*, 2009).

The Siteroptidae have a haplodiploid system of sex determination, with unfertilized eggs that generate haploid males and fertilized eggs that generate diploid females (Cooper, 1937). The life cycle develops through two active post-embryonic instars, the larva and the adult, separated by a quiescent nymph. Only the larvae and adult females exhibit feeding activity; the adult males live only to mate. Females of some *Siteroptes* spp. exhibit a marked hysterosomatic physogastry during feeding, which is usually associated with high fecundity. In *S. cerealium* (Kirchner) each female may generate up to 500 progeny, consisting largely (about 95%) of females. In this species, the progeny of a female develop through the larval stage within the mother's body, where copulation takes place before birth, and the progeny emerge at maturity. The process often involves the bursting open of the mother's body, and the mothers do not survive long after egg laying (Cooper, 1937; Krczal, 1959).

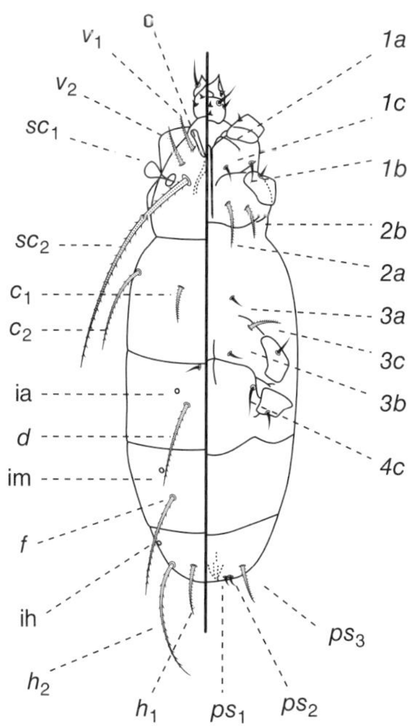

Fig. 23.3. *Siteroptes cerealium* (Kirchner). Female, dorsal (left) and ventral (right) views (modified from Suski, 1973). Key: ia, ih, im, cupules; s, stigma. The rest of the labels designate setae, and the setal notation is explained in the text.

Some species are characterized by a phoretic and non-phoretic female dimorphism called 'phoretomorphy', and this phenomenon makes the recognition of some genera difficult as it involves distinct adult female forms that have been placed by past authors into different families. The normal adult female is a fairly sedentary individual, while phoretomorphs are able to disperse by phoresy and may also differ in morphological features (Walter *et al.*, 2009).

The Siteroptidae are primarily free-living mites, although they sometimes live on plants and display a fungivorous feeding behaviour. Some species live in soil, litter, rotten plants, grasses or cultivated plants, or are phoretic on arthropods (Cross, 1965; Rack, 1974; Kaliszewski, 1988a,b; Livshits *et al.*, 1986). *Siteroptes* spp. have been reported from various habitats, including the leaves of poaceous plants (Rack, 1972; Livshits *et al.*, 1986). Females of *S. cerealium* carry fungal spores on the external surface of their bodies, e.g. the spores of the fungus *Fusarium poae* (Peck) Wollenw., the causal agent of carnation bud rot (Cooper, 1940); the mite is also involved in the 'silver top' of grasses in North America and Eurasia (Hardison, 1959; Suski, 1973, 1984). Other species have paired eversible sporothecae set on the ventral body surface that carry only spores of fungi preferred as food, e.g. *S. avenae* (Müller), which has deeply invaginated sporothecae (see Fig. 23.4) set between coxisternal plates II and III, and carrying conidiospores of species of *Botrytis*, *Fusarium*, *Harzia* and *Nigrospora* (Cooper, 1940; Suski, 1973; Su *et al.*, 1979, 1981).

Walter *et al.* (2009) recognized the superfamilies Pygmephoroidea (including the family Siteroptidae) and Scutacaroidea, but Khaustov and Ermilov (2011) ascribed both superfamilies to the single superfamily Pygmephoroidea. The family Siteroptidae has about eight genera, including 80 species (Walter *et al.*, 2009). The species of economic importance belong to the genus *Siteroptes* Amerling.

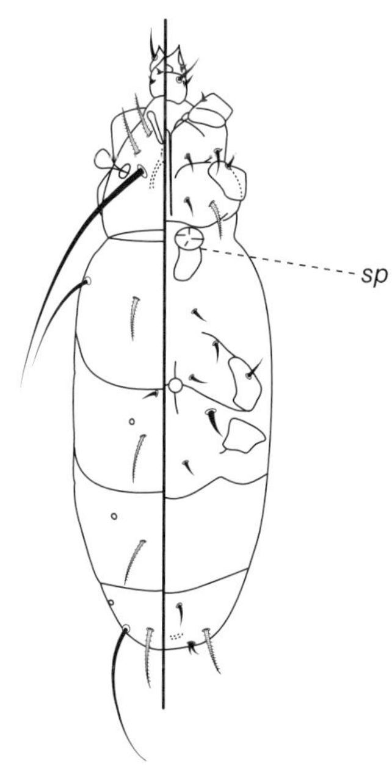

Fig. 23.4. *Siteroptes avenae* (Müller). Female, dorsal (left) and ventral (right) views (modified from Suski, 1973). Key: sp, sporotheca.

Selected websites

Plant and Soil – a 3D Adventure: *Siteroptes graminum*. Vaxten and Marken (1999), Swedish University of Agricultural Sciences, Uppsala, Sweden. Available at: http://www-vaxten.slu.se/marken/markdjur/siteroptes_graminum.htm

Progenesis and Neoteny 311: *Siteroptes graminum*. Available at: http://www.sjgarchive.org/library/text/ontogeny/p0311.htm

The Genus *Siteroptes* Amerling

Diagnostic characteristics

Legs I have 5 segments and a simple pedicellate claw, lacking a thumb. The chelicerae are small and indistinct, and the palpi are small and free. The gravid females distend behind legs II (Jeppson *et al.*, 1975).

Economic importance

This genus includes 14 species (Khaustov and Ermilov, 2011) that transmit spores of phytopathogenic fungi to herbaceous plants, including various economic plants (Kaliszewski, 1987). Two of these species, *S. avenae* and *S. cerealium*, play some economic role as a carrier of phytopathogens. Unfortunately, there is some confusion over their identity. Jeppson *et al.* (1975) report that *S. cerealium* (Kirchner) (Fig. 23.3) is also known in the literature as *S. graminum* (Reuter). Zhang (2003) refers to *S. avenae* (Müller) (Fig. 23.4) as a synonym of *S. graminum* (Reuter) but not of *S. cerealium* (Kirchner), while Walter *et al.* (2009) share the view of Jeppson *et al.* (1975) and consider *S. cerealium* (Kirchner) to be a synomym of *S. graminum* (Reuter). To avoid further confusion, the view of Jeppson *et al.* (1975) is adopted here, but the problem does await a solution. Regardless of their taxonomic features, the economic importance of both mites appears to be restricted to particular environments and the direct damage caused by them is not well documented, with the exception of the risk associated with the transmission of spores of phytopathogenic fungi. In addition, there are inconsistent reports on their control.

S. cerealium is reported by Jeppson *et al.* (1975) as a pest of grasses and cereals, including wheat, barley and rye. On these host plants, the mite feeds within the sheaths on the upper part of the stem, causing 'silver tip' of grasses in North America and Eurasia, in which there is growth retardation associated with partial emergence of inflorescences that appear malformed and silvered (Hardison, 1959; Suski, 1973, 1984). The mites are often associated with the pathogenic fungus *N. oryzae* (Berk. *et* Br.) Petch and play a role in its spread. *S. cerealium* feeds both on healthy plant tissues and on the fungus, and on plant tissue destroyed by the fungus. The mite also spreads spores of *F. poae*, the causal agent of carnation bud rot (Cooper, 1940).

S. avenae has some importance in greenhouses (Zhang, 2003). It is known as a carrier of conidiospores of species of *Botrytis*, *Fusarium*, *F. poae* and *Harzia* (Cooper, 1940; Suski, 1973; Su *et al.*, 1979, 1981). The mite is ovoviviparous, and only the adult stages occur in its life cycle. Ben *et al.* (1978) studied the life cycle of *S. avenae* at different temperatures (10, 15, 20, 25, 30°C) and found that 20°C was the optimum for the population increase, but that within the range of 10–25°C, development was faster with an increase of temperature. The mite seems to complete its development and reproduction successfully only on the mycelium of the fungus *Sporotrichum poae* Peck (= *Fusarium poae*), indicating a specific symbiotic relationship. Young and growing mycelium of the symbiont fungus is a prerequisite for the mite to be physogastric. The mite carries and spreads the spores of the fungus, and its physogastric body fluid provides a superior substrate for and stimulates the growth of the pathogen (Su *et al.*, 1981). The results of field tests based on the examination of diseased ears of cereals supported the specificity of the symbiont (Ben *et al.*, 1978). The mite feeds on fungal spores, and the damage that it causes is mainly by the spread of the fungal spores. Infected carnation buds appear normal for 16 days, but then the calyx appears to be abnormal and the petals and the sepals have rotted. Attacks on carnations in the greenhouse often occur in the autumn, following a spell of hot and dry weather. Reports on control of this mite on carnations are very few, and control of the pathogenic fungi is important (Zhang, 2003). The association between *S. avenae* and *F. poae* has been identified on wheat in South Africa, and the evidence suggests that both the mite and the pathogenic fungus cause 'Fusarium glume spot' of wheat (Kemp *et al.*, 1996).

The Family Pygmephoridae Cross

Morphological characteristics, systematics and bio-ecology

The Pygmephoridae are small mites, largely known only as the females. The body is from 110 to over 500 μm long, and pale yellow in colour. The general morphological features are similar to those of Siteroptidae, with the exception of seta *d* of femur I of the females, which has various shapes (thickened, rigid, blade like, rod like or hooked), but not setiform. The coxisternal plates I–II together have 4–6 pairs of setae, and the prodorsal stigmata of females are round or oval, but not extended into elongate grooves (see Figs 23.5–23.7) (Walter *et al.*, 2009). Like the Siteroptidae, some species of this family exhibit adult female dimorphism, with a typical phoretomorphy, and adult females use insects and mammals for dispersal. The life cycle of pygmephorids develops through an active larval and a female stage. Adult males do not feed, and the females exhibit a variable degree of physogastry. Pygmephorids living in commercial mushroom houses, such as *Pediculaster* sp. (Košir, 1975; Cross and Kaliszewski, 1988) and *Luciaphorus* sp. (Rack, 1983; Kantaratanakul and Jitrat, 1984; Zou *et al.*, 1993a), show a pattern of life cycle characterized by phoretomorphy. The eggs hatch within the body of the mother, which becomes physogastric, and the larvae develop within the body of the mother until they reach the adult stage. Adults mate (or not) within the body of the mother. At least 350 species included in 30 genera are described, the majority of which are based only on a description of the female (Walter *et al.*, 2009).

Pygmephorid mites live in soil, litter, organic substrates and nests of birds and mammals, and also in association with

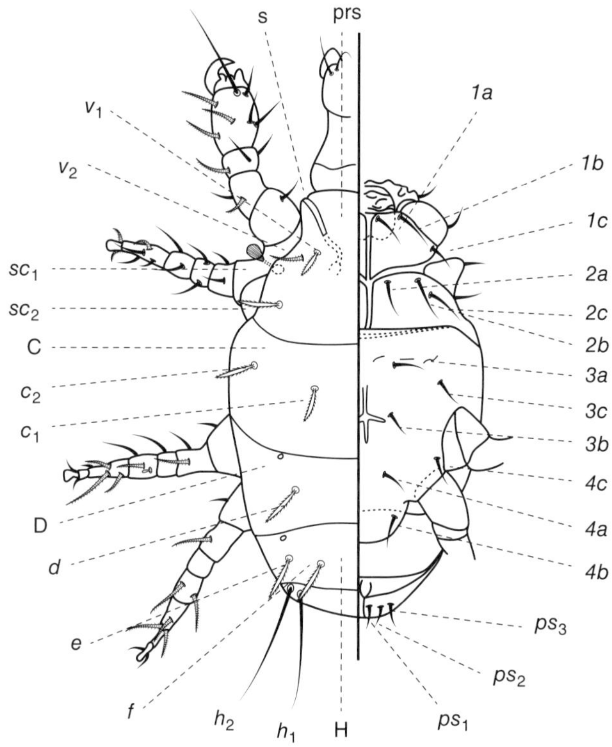

Fig. 23.5. *Luciaphorus perniciosus* Rack. Normal female, dorsal (left) and ventral (right) views (modified from Rack, 1983). Key. C, D, H, tergites; prs, prodorsal shield; s, stigma. The rest of the labels designate setae, and the setal notation is explained in the text.

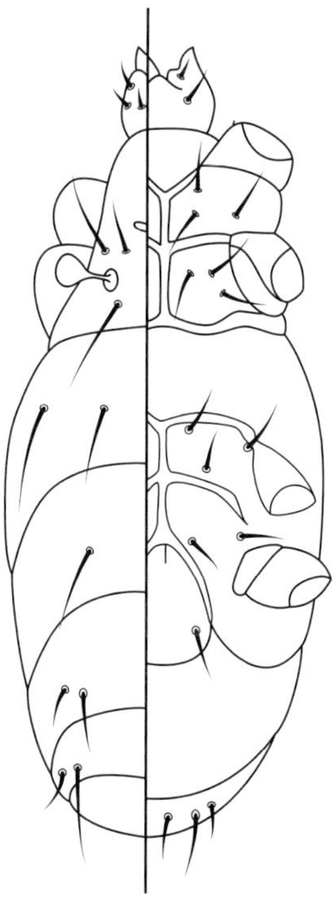

Fig. 23.6. *Pediculaster mesembrinae* (Canestrini). Normal female, dorsal (left) and ventral (right) views (drawing by Vacante).

subsocial and social insects (Krczal, 1959; Cross, 1965, Mahunka, 1969, 1970; Savulkina, 1981). Different species are associated with mushrooms in various countries, and cause direct or indirect damage. The species that are well known are *Luciaphorus perniciosus* Rack (Rack, 1983; Kantaratanakul and Jitrat, 1984; Gao *et al.*, 1990a; Zou *et al.*, 1993a,b; Lin and Zhang, 2000), *Pediculaster flechtmanni* (Wicht) (Wicht, 1970; Cross and Kaliszewski, 1988; Gao *et al.*, 1988; Zou *et al.*, 1988, 1992), and *Pediculaster mesembrinae* (Canestrini) (Košir, 1975; Camerik *et al.*, 2006). With the exception of *L. perniciosus*, which causes direct damage to mushrooms in China and Thailand, other pygmephorid mites feed on weed moulds in mushroom houses, and cause indirect damage. Different phoretic species were collected in compost and found being carried by other invertebrates, such as the red pepper mites, *Pygmephorus athiasae* Wicht, reported from France, and *Pygmephorus kneeboni* Wicht (Wicht, 1970) and *Pygmephorus murphyi* Smiley (Smiley, 1978), both reported from the USA. These mites, as well as *Pediculaster fletchmanni* (or *Pediculaster mesembrinae*), although harmless to the mushroom crop itself (because they feed mainly on weed moulds such as *Trichoderma*), when present in large numbers can lead to crop rejection due to contamination with their brightly coloured bodies. In addition, contact with these mites can induce allergic responses in mushroom pickers (Woodhall *et al.*, 2009). Similarly, *Pygmephorus sellnicki* Krczal feeds only on *Trichoderma* sp. (Hussey *et al.*, 1969). Another pygmephorid, *Pseudopygmephorus quadratus* (Ewing), has been found in association with mushrooms (Krczal, 1959), and is very common in mushroom farms in various Chinese provinces (Zou and Zeng, 1993; Wu, 1994; Lin and Zhang, 2000), where it has caused 20–30% of early mushroom loss in Fujian (Lin *et al.*, 1992). The bio-ecology of this mite has been investigated by Wu (1994), but the scientific name of the mushroom species used to rear it was not reported, which led to the belief that the mushroom was *Agaricus bisporus* (J.E. Lange) Pilát. However, other studies gave different results (Košir, 1975), as it had been observed before that *P. quadratus* fed only on weed moulds (Gao and Zou, 2010). This lack of knowledge on the bio-ecology of pygmephorid mites is widespread, such as in the case of *Pygmephorus allmanni* Krczal, which has been reported on cultivated mushrooms in Australia (Krczal, 1964a) and Turkey (Toros and Çobanoğlu, 1985), although its bio-ecology and any damage that it causes are not documented.

The above account indicates that there is some confusion around the damage caused by these mites, with several species developing only on weed moulds, and others exhibiting a specific symbiotic relationship with mushroom species; only some pygmephorid mites seem to cause direct damage to mushrooms. Paradoxically, some mites are reported as pests of mushrooms without any scientific support. According to Greenslade and Clift (2004), many pest species of mushrooms 'were economically damaging in the past but improved hygiene and growing techniques on farms have reduced their pest status to low or negligible'. In addition, as the systematics of the families Microdispidae, Scutacaridae, Siteroptidae and Pygmephoridae, all belonging to the superfamily Pygmephoroidea and including

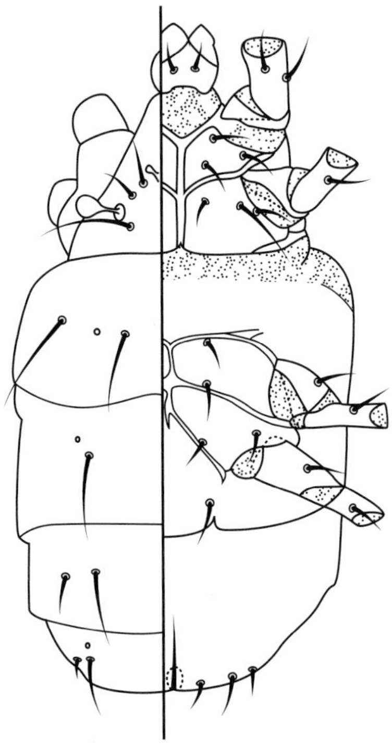

Fig. 23.7. *Pediculaster flechtmanni* (Wicht). Female, dorsal (left) and ventral (right) views (modified from Martin, 1978).

mite species living in association with mushrooms, is currently artificial and unstable at both family and genus level (Kaliszewski and Rack, 1986), the systematic arrangement that has been adopted to date may be provisional (or, indeed, questionable). For example, *Pseudopygmephorus quadratus* has been recently transferred into the family Neopygmephoridae by Khaustov (2010). In spite of the work of the latter author being very appreciable, to avoid further confusion, this species is viewed here as belonging to the family Pygmephoridae until there is a more extensive review of the family.

Key for the identification of genera (adults)

1. • Trichobotria located laterally.. ....genus *Luciaphorus* Mahunka
   • Trichobotria located dorsally.................................... ....genus *Pediculaster* Vitzthum

The Genus *Luciaphorus* Mahunka

Diagnostic characteristics

The trichobotria are set laterally (Mahunka, 1981).

Economic importance

This genus has only two species. The type species, *L. hauseri* Mahunka, is at present only known from the island of St Lucia in the Antilles (West Indies) (Mahunka, 1981). The other species, *L. perniciosus*, was described in Thailand from a commercial mushroom farm near Bangkok (Rack, 1983). A third species, *L. auriculariae* Gao, Zou *et* Jiang, was reported from Jew's ear mushroom, *Auricularia polytricha* (Mont.) Sacc., in Fujian Province, China (Gao *et al.*, 1990a; Zou *et al.*, 1993a,b; Lin and Zhang, 2000), but the name is a junior synonym of *L. perniciosus* (Kaliszewski *et al.*, 1995). The bio-ecology of *L. hauseri* has not been documented, so only *L. perniciosus*, which is known to be injurious to mushrooms in China (Gao and Zou, 2010) and Thailand (Bussaman *et al.*, 2004), is described below.

Luciaphorus perniciosus Rack (Fig. 23.5)

Common name

Unknown.

Diagnostic characteristics

NORMAL FEMALE. The body is 113–142 μm long and 62–80 μm wide. The prodorsum has the stigmata set anterolaterally, and associated with tracheae; there are 3 pairs of prodorsal setae (v_1, v_2, sc_2) and a pair of lateral bothridia with capitate sensilla (sc_1). Tergite C has 2 pairs of setae (c_1, c_2), tergite D has 1 pair of setae (*d*), tergite EF has 2 pairs of setae (*e*, *f*), tergite H has 2 pairs of setae (h_1, h_2), and tergite Ps has 3 pairs of setae (ps_{1-3}). All the dorsal setae are serrate, with h_1 and h_2 about 3 times longer than the other opisthosomal dorsal setae. Ventrally, the coxisternal plates I–VI each have 3 setae (*1a–c*, *2a–c*, *3a–c*, *4a–c*). The setae ps_3 are longer and robust, with ps_2 shorter (Rack, 1983).

MALE. The body of the male is 105–152 μm long and 72–97 μm wide. No mouthparts are present (Rack, 1983).

Geographical distribution

China (Fujian Province) (Gao *et al.*, 1990a; Zou *et al.*, 1993a,b; Lin and Zhang, 2000) and Thailand (Rack, 1983; Kantaratanakul and Jitrat, 1984; Bussaman *et al.*, 2004).

Hosts

The mushroom hosts include: Auriculariaceae – *Auricularia auricula-judae* (Bull. Fr.) Wettst. (Bussaman *et al.*, 2004), Jew's ear mushroom (*A. polytricha* (Mont.) Sacc.) (Rack, 1983; Gao *et al.*, 1990a; Zou *et al.*, 1993a,b; Lin and Zhang, 2000); Marasmiaceae – *Flammulina velutipes* (Curtis) Singer (Bussaman *et al.*, 2004); Polyporaceae – *Lentinus polychrous* Lév., *L. squarrosulus* (Mont.) Singer (Bussaman *et al.*, 2004).

Bio-ecology

L. perniciosus was first described on *A. polytricha* in Thailand (Rack, 1983), where it is considered to be among the most destructive pests of mushroom cultivation, causing severe production losses of *A. auricula-judae*, *F. velutipes*, *L. polychrous* and *L. squarrosulus* mushrooms (Bussaman *et al.*, 2004). The

mite is also the major pest in the commercial production of *A. polytricha* in Fujian Province (China), where it can complete normal development and reproduction when it feeds on mycelia or sporophores of *A. auricula* and *F. velutipes* (Zou *et al.*, 1993a,b).

Unfertilized females produce only male offspring, while fertilized females produce both males and females. The mite exhibits clear phoretomorphy. The pregnant form reaches an average body diameter of 1500 μm, and hosts within its body the eggs and the biological instars that follow. However, copulation occurs both inside and outside the mother's body. The average adult stage of the male lasts 9 days and those of the fertilized and unfertilized females last 12 and 14 days, respectively (Bussaman *et al.*, 2004). The first adults conceal their legs in the ventral sides of their bodies and remain inactive until most of the eggs hatch. They then mate, break open the body and free themselves to feed. Development is fast at 30°C, and a generation develops in 15 days. At 20–30°C, one female can generate from 125 to 163 offspring, but at 35°C, this reduces to 77 offspring (Zou *et al.*, 1993a,b).

RAPD-PCR profiles and information on genetic distance from five populations collected from *L. squarrosulus* compost and basidiocarps from several farms in four provinces in north-east Thailand categorized these mites into three groups, indicating that although all of the mites were *L. perniciosus*, they were genetically diverse (Moukoumbi *et al.*, 2011).

Symptomatology and damage

L. perniciosus is the major pest in the commercial production of *A. polytricha* in Fujian Province (China) (Gao *et al.*, 1990a; Zou *et al.*, 1993a,b; Lin and Zhang, 2000). It infests both mother and original spawns, as well as sawdust spawns encased in plastic bags, and damages both the fungal mycelium and sporophore. The mite may cause 10–50% of total loss of the mushroom crop depending on when infestation occurs (Zou *et al.*, 1993a,b). In Thailand, the mite causes severe production losses of *A. auricula-judae*, *F. velutipes*, *L. polychrous* and *L. squarrosulus* (Bussaman *et al.*, 2004).

Control

There are no reports on the control of *L. perniciosus* from China (Gao and Zou, 2010). In Thailand, few reports are available about the effective measures of control for the mite, and routine horticultural hygiene is the only practice able to reduce the level of infestation. Biological control is not well documented, and consists only of some experimental microbiological applications. Chemical control mainly concerns the study and use of plant extracts.

BIOLOGICAL CONTROL. It has been observed that different parts of a culture of the bacterium *Xenorhabdus stokiae* isolate PB09 had variable effects on mite mortality and fecundity, but application of a cell-free supernatant from an *X. stokiae* culture resulted in both the highest mite mortality (89.00 ± 3.60%) and the lowest mite fecundity (41.33 ± 23.69 eggs per gravid female). The cell-free supernatant of *X. stokiae* isolate PB09 was the most effective biological control agent tested and could conveniently be used in future formulations instead of live bacteria (Bussaman *et al.*, 2012a).

CHEMICAL CONTROL. Mushroom growers use in Thailand a large amount of carbamate and organophosphate insecticides, and even some harmful solvents to manage *L. perniciosus*, but this has had very limited success (Bussaman *et al.*, 2006). Accordingly, research has been done in Thailand on the use of various plant extracts; the results are summarized below.

Insung *et al.* (2008) found that dichloromethane extracts of *Justicia adhatoda* L. and a methanolic extract of *Litsea cubeba* (Lour.) Pers. had the potential to be used as botanical pesticides to control mushroom mites. Extracts of flowers of the clove (*Syzygium aromaticum* (L.) Merr. and L.M. Perry) and the bark of cinnamon (*Cinnamomum bejolghota* (Buchanan-Hamilton) Sweet) caused high mortality of mites (Pumnuan *et al.*, 2008). Essential oils from lemon grass (*Cymbopogon citratus* (DC.) Stapf), citronella grass (*Cymbopogon nardus* (L.) Rendle), lime (*Citrus aurantifolia* (Christm.) Swingle), clove (*Syzygium aromaticum* (L.) Merr. and L.M. Perry), the kernels and seeds of black pepper (*Piper nigrum* L.), turmeric (*Curcuma longa* L.) and cinnamon (*Cinnamomum cassia* auct.) were more effective in controlling *Formicomotes heteromorphus* (Dolichocybidae) than *L. perniciosus*, although the difference was not significant (Pumnuan *et al.*, 2010). Rhizome extracts of *Curcuma xanthorrhiza* Roxb. and *Zingiber montanum* (Koenig) Link ex Dietr. and leaf extracts of *Ocimum sanctum* L. and *Melissa officinalis* L. have a great potential for future development as natural acaricides for controlling *Luciaphorus* sp. (Bussaman et al., 2012b).

The Genus *Pediculaster* Vitzthum

Diagnostic characteristics

The trichobotria are set dorsally, and the coxisternal plates I–II have altogether 6 pairs of setae (Vitzthum, 1931; Mahunka, 1970).

Economic importance

The genus *Pediculaster* Vitzthum includes about 100 species and is the largest genus in the family Pygmephoridae (Khaustov, 2011). Some *Pediculaster* spp. live in mushroom houses, where they feed on weed moulds, but do not cause direct damage. The genus includes *P. mesembrinae* and *P. flechtmanni*, neither of which are known as primary pests of mushrooms. *P. mesembrinae* feeds preferentially on weed moulds and carries spores of *Trichoderma* in a pair of sporotheca located ventrally near the base of legs IV (Terras and Hales, 1995). It commonly lives on baled straw and can survive in compost during pasteurization, but its presence indicates that the compost is not selective for the mushroom mycelium (Gurney and Hussey, 1967; Clift and Terras, 1994). Similarly, *P. flechtmanni* does not feed directly on mushrooms, but on weed moulds. The mite grew well on *Chaetomium cochliodes* Palliser, *C. globosum* Kunze ex Fr., *Monilia* sp., *Nigrospora* sp. and *Trichoderma arzianum*

Rifai, common moulds in mushroom-growing environments (Zou *et al.*, 1988, 1992). Another species living in mushroom compost is *P. sellnicki*, which feeds and reproduces on cultures of *Humicola* sp., *Monilia* sp. and *Trichoderma* sp. (Gurney and Hussey, 1967). Even though none of these mites cause direct damage, their presence in mushroom houses alarms growers, who treat the mushrooms with chemicals, but usually underestimate the importance of prevention of the accidental introduction of carriers of both mites and the application of more stringent hygiene practices. The mites may sometimes cause allergies. Hussey (1964) found that only *P. mesembrinae* swarmed on the first mushrooms produced, while *P. sellnicki* swarmed on the sporophores of the third and later flushes. Only *P. mesembrinae* and *P. flechtmanni* are discussed here. The life history of *P. mesembrinae* is similar to that of *P. sellnicki* (Gurney and Hussey, 1967).

Pediculaster mesembrinae (Canestrini) (Fig. 23.6)

Common name

Red pepper mite, pygmy mite.

Diagnostic characteristics

NORMAL FEMALE. The body is over 260 μm long, with a soft-skinned exoskeleton, and all setae serrate (the phoretic females have the exoskeleton sclerotized and punctate). The prodorsal shield has a pair of anterodorsal oval stigmata associated with tracheae, 3 pairs of setiform setae (v_1, v_2, sc_2), and a pair of dorsal bothridia with capitate sensilla (sc_1). The setae v_1 and v_2 are set on the same transversal line, with the v_1 setae innermost. The setae sc_2 are markedly longer than v_1 and v_2, and reach the setae c_1 and c_2. The six opisthosomal dorsal tergites C, D, EF, H and Ps have 2 pairs of setae (c_1, c_2), 1 pair of setae (d), 2 pairs of setae (e, f), 2 pairs of setae (h_1, h_2) and 3 pairs of setae (ps_{1-3}), respectively. The setae f are two times longer than e, and the setae h_1 are about 4 times longer than the setae h_2. The setae c_2, f and h_1 are longer than the other opisthosomal dorsal setae, and the setae ps_2 are longer than ps_1 and ps_3. Ventrally, the apodemes are sclerotized and complete, and the coxisternal plates I–IV each have 3 smooth setae (*1a–c*, *2a–c*, *3a–c*, *4a–c*). The tegula is roundish.

MALE. The body of the male is about 170 μm long and well sclerotized. The propodosoma and hysterosoma are separated by a disjugal furrow. The prodorsal shield has 4 pairs of setae; the setae v_1 are set anteriorly, and the setae sc_1 and v_2 on the same line, with the setae v_2 innermost. The opisthosomal plate CD has 3 pairs of setae (c_1, c_2, d), and the plate EF bears 2 pairs of setae (e, f), with the setae f longer than the setae e. The tergite H has the setae h_1 and h_2, and the tergite Ps has the setae ps_{1-3}, with ps_2 shorter. Ventrally, the apodemes are sclerotized, and the coxisternal plates I–IV each have 3 smooth setae (*1a–c*, *2a–c*, *3a–c*, *4a–c*).

Geographical distribution

Worldwide.

Hosts

The main mould hosts are Hypocreaceae – *Trichoderma* sp. (Gurney and Hussey, 1967) and *T. harzianum* Rifai (Terras and Hales, 1995; Hernández-Abarca *et al.*, 2005). The mite has also been found on species of Agaricaceae (*Agaricus bisporus* (J.E. Lange) Pilát, *A. bitorquis* (Quél.) Sacc.) (Clift and Tofflon, 1981a,b; Clift 1983, 1986; Nair and Clift, 1993; Clift and Terras, 1994, 1995; Terras and Hales, 1995; Clift *et al.*, 1996), in azalea cultures, and on species of Chaetomiaceae (*Chaetomium elatum* Kuntze), Pleurotaceae (*Pleurotus eryngii* (DC.) Gillet) and Trichosphaeriaceae (*Nigrospora* sp.).

The mite has also been found in manure and in galleries of the bark beetle *Hylurgops palliatus* (Gyll.) (Scolytidae) (Camerik *et al.*, 2006), and reported on gladiolus corms, in mushroom beds, on onions, in mushroom compost and casing, in litter at the forest edge, on garlic, in garden soil, in a pit trap in pasture and on taro (Martin, 1978).

Bio-ecology

P. mesembrinae was first described as a phoretic female collected on the dipteran fly *Mesembrina mystacea* L. (Muscidae) by R. Canestrini (1881). Following Gurney and Hussey (1967), the mite was described by Lombardini (1960) as *Microdispodides fungorum* Lombadini in Italy, and it was later discussed by Bua (1964). Thomas (1942) reports that *P. mesembrinae* inhabits manure and mushroom beds and that after spawning, the levels of its population increase to the point where they suppress mycelial growth in the upper compost layers, causing poor and delayed sporophore production (Gurney and Hussey, 1967). It is the most common mite occurring in mushroom houses, but it is not a primary pest of mushrooms, as it feeds preferentially on weed moulds in the crop and carries spores of *Trichoderma* in a pair of sporotheca located ventrally near the base of legs IV (Terras and Hales, 1995). It is common in baled straw and can survive in compost during pasteurization, but its presence indicates that the compost is not selective for the mushroom mycelium (Clift and Terras, 1994). The life cycle of *P. mesembrinae* develops through the stages of egg, active larva, quiescent larva and adult male and female; the pharate nymph and the adult develop within the larval skin. The mite has two female stases, the normal female, which is sedentary and non-phoretic, and has leg I 5-segmented, and the phoretic female (phoretomorph), which has leg I 4-segmented. The normal females become physogastric after fertilization, while the phoretic females rarely become physogastric in their natural enviroment (Camerik *et al.*, 2006).

Under laboratory conditions at 15, 20, 25 and 30°C and with 60 ± 5% RH, with fermented wheat straw (compost) previously inoculated with *T. harzianum* spores as substrate and food source, the life cycle and the total longevity of *P. mesembrinae* showed an inverse relation to temperature. At 15, 20, 25 and 30°C and with 60 ± 5% RH, the intrinsic rate of increase (r_m) was 0.146, 0.132, 0.170 and 0.204, respectively. Developmental rates were directly and linearly related to temperature; and threshold temperatures ranged between −0.53 and −3.22°C. The development from egg to adult required 152 degree-days (Hernández-Abarca *et al.*, 2005).

Symptomatology and damage

Populations of *P. mesembrinae* can develop while feeding on weed moulds, but they are unable to feed and reproduce on substrate colonized by mushroom mycelium. However, the mites can indirectly damage mushroom cultures by favouring the development of the parasitic fungi *T. harzianum* and *T. viride* Pers. Simultaneous infestation by *P. mesembrinae* and another mite, *Histiostoma feroniarum* (Dufour), resulted in a 20–40% reduction in mushroom yield (Clift and Terras, 1995).

Control

P. mesembrinae is not a primary pest of mushrooms, as it feeds preferentially on weed moulds in the crop and carries spores of *Trichoderma* in a pair of sporotheca located ventrally near the base of legs IV (Terras and Hales, 1995). It is usually found in baled straw and can survive in compost during pasteurization, but its presence indicates that the compost is not selective for the mushroom mycelium (Clift and Terras, 1994) and suggests the need for more scrupulous hygiene and prevention.

Gurney and Hussey (1967) found that in pure culture, species of this genus do not develop on mushroom mycelium, showing that their appearance depends on the presence of unsatisfactory compost. In particular, any compost that is not rapidly colonized by mushroom mycelium allows the development of weed moulds on which the mites feed and develop. The use of high amounts of sugars and ammonium salts in composting showed that the mites developed at high population densities because these compounds influenced the microbial flora of the compost. The mites develop on the mushrooms and bed surface only when the food supply is depleted, and consequently there are few possibilities to apply control. Correct preparation of compost prevents swarms of *P. mesembrinae*, and when the mites do appear, they do not cause severe damage (Gurney and Hussey, 1967).

Pediculaster flechtmanni (Wicht) (Fig. 23.7)

Common name

Red pepper mite, pygmy mite.

Diagnostic characteristics

NORMAL FEMALE. The body is on average about 280 µm long. The prodorsum has a pair of stigmata with associated tracheae, which are close but not contiguous, and 4 pairs of setae (v_1, v_2, sc_1, sc_2). The stigmata, setae v_1, v_2 and a pair of dorsal bothridia with capitate sensilla (sc_1) are located in a straight line between v_1 and sc_1. The idiosomal dorsal setae c_1 are shorter than c_2, the setae *d* are longer than the other dorsal idiosomal setae, and the distance that separates them is almost twice that separating the setae c_1. The setae *e* are half the length of the setae *f* and the shorter idiosomal dorsal setae. The coxisternal plates I–IV each have 3 smooth setae (*1a–c*, *2a–c*, *3a–c*, *4a–c*) (Wicht, 1970).

Geographical distribution

Worldwide.

Hosts

The main mould hosts are: Chaetomiaceae – *Chaetomium cochliodes* Palliser, *C. globosum* Kunze ex Fr.; Hypocreaceae – *Trichoderma harzianum* Rifai; Sclerotiniaceae – *Monilia* (Zou *et al.*, 1992); and Trichosphaeriaceae – *Nigrospora oryzae* (Berk. *et* Br.) Petch (Cross and Kaliszewski, 1988), *Nigrospora* sp. (Zou *et al.*, 1992).

Bio-ecology

P. flechtmanni was collected by Carlos Flechtmann in Brazil on mushrooms and in mushroom houses in 1968 and described by Wicht (1970). The mite sometimes develops high population densities in commercial mushroom houses, where it feeds on weed fungi (Cross and Kaliszewski, 1988). In China, the red pepper mite is common on mushroom farms in various provinces (Beijing, Fujian, Henan, Shanghai) (Zou *et al.*, 1988). The mite does not feed directly on mushrooms, but on weed moulds, and it develops well on the moulds *C. cochliodes*, *C. globosum*, *Monilia* sp., *Nigrospora* sp. and *T. harzianum*, which are common in mushroom houses but not on the mushrooms *Agaricus bisporus* (J.E. Lange) Pilát, *Lentinula edodes* (Berk.) Pegler and *Pleurotus sajor-caju* (Fr.) Singer. Its movement led to the spread of the above moulds as well as the harming of mushroom mycelium, which affected the growth and yield of mushrooms (Zou *et al.*, 1992).

The pre-adult stages are egg, active larva and quiescent larva, and on *N. oryzae* grown on yeast extract agar at 21–27°C, these stages last about 41, 7 and 43 h, respectively. The life cycle from female to female takes 125 h at 21–27°C ambient temperature. The females are physogastric, each producing about 200 eggs within 1–6 oviposition days. The net reproduction rate is approximately 65 females per female each generation. The sex ratios are initially female biased, often becoming male biased as the mothers age. Unmated females, if they reproduce, are arrhenotokous (Cross and Kaliszewski, 1988). On *Monilia* sp. as the food fungus at 15, 20, 25, 30 and 35°C, one generation (from egg to egg) completed in 10.3, 8.0, 5.8, 4.0 and 2.5 days, respectively. At the same temperatures, the fecundity was 292, 296, 269, 249 and 198 eggs, respectively (Zou *et al.*, 1992). On mycelia of *Trichoderma* sp. as substrate and food source at 20 ± 1, 22 ± 1 and 25 ± 1°C, 70% ± 5% RH and under a photoperiod of 16L:8D, the mean developmental time of the egg and the active larva did not differ significantly, but there were significant differences for the quiescent larval stage. The pre-oviposition period ranged from 2.3 to 2.8 days, the ovipositional period increased with temperature, and all females died immediately after oviposition. The development of active larvae was the fastest of all the life stages. The developmental threshold ranged from 5.25 to 14.22°C, the highest value being observed for the quiescent larval development. The immature development required 89.29 degree-days. The values of intrinsic rate of increase (r_m) were 0.229, 0.398 and 0.386 at 20, 22 and 25°C, respectively. The finite rates of increase (λ) increased with increasing temperature from 20 to 25°C; the population doubling time and mean generation time showed significant differences with increasing temperature (Kheradmand *et al.*, 2006a). On *Trichoderma* sp. mycelium at three constant

temperatures (20, 22 and 25°C), 70% RH and under a photoperiod of 16L:8D, the highest gross fecundity and fertility rates were observed at 20°C (277.11 ± 2.86 and 207.67 ± 1.99, respectively), while the highest net fecundity and fertility rates were at 25°C (85.65 ± 1.02 and 83.14 ± 1.31, respectively). The highest mean fertile eggs a day and mean eggs a day were 29.21 ± 1.36 and 33.21 ± 1.14 when rearing was at 22°C. The maximum values of life expectancy were obtained at 25°C (Kheradmand and Zamani, 2008).

P. flechtmanni exhibits a marked sexual dimorphism and phoretic and non-phoretic females, which differ in morphology and behaviour (Cross and Kaliszewski, 1988; Gao *et al.*, 1988). The proportion of phoretomorphs increases as cultures age. This increase is not due to attenuated phoretomorph developmental time or increasing age of the mother, but is probably related to environmental factors associated with habitat deterioration; and such factors could be pheromonal or nutritive, and may act upon any stage in the life cycle (Cross and Kaliszewski, 1988). When larvae fed on abundant, young host mycelium, they developed into normal form females, but when they fed on aged mycelium or were supplied with only scarce food mycelium, they developed phoretomorph form females. When newly emerged females are reared under the same conditions, the numbers of eggs laid by these two types of females are not significantly different (Zou *et al.*, 1992).

Symptomatology and damage

P. fletchmanni and other 'red pepper mites' feed mainly on weed moulds, such as *Trichoderma* sp. and other species. Although harmless to the mushroom crop itself, high populations of the mites lead to crop rejection due to contamination by their brightly coloured bodies. In addition, contact with these mites can induce allergic responses in mushroom pickers (Woodhall *et al.*, 2009). Mite movement leads to the spread of moulds as well as harming the mushroom mycelium, and so affecting the growth and yield of mushrooms (Zou *et al.*, 1992). Comparison of the dry matter, organic matter and mineral elements content kg^{-1} of compost in samples into which different numbers of mites (0, 50, 100 and 250 adults) had been released shows that *P. flechtmanni* can cause qualitative losses by influencing the nutritional values of the button mushroom *A. bisporus* (Kheradmand *et al.*, 2006b).

Control

The preparation of compost of good quality and preventing insects (which may carry these mites) from invading mushroom cultivation houses are important in the control of *P. fletchmanni* (Zou *et al.*, 1992).

The Family Scutacaridae Oudemans

Morphological characteristics, systematics and bio-ecology

The Scutacaridae are primarily free-living mites. The females have a broad body, 120–370 μm long and with an expansive tergite C, which extends completely on to the prodorsum and forms a roof over the gnathosoma. The prodorsum usually bears two pairs of small setiform setae. The idiosoma is oval or rounded, and compact. The distance between the insertions of legs II and III is similar to that between legs III and IV. The seta *d* of femur I is commonly short, stout and branched. The aedeagal tube of the male is annulated or striated (Walter *et al.*, 2009).

The bio-ecology of many of the species of Scutacaridae is not well documented, but males probably exist in many species, as thelytoky has not been observed in any species (Norton *et al.*, 1993). Scutacarid mites are oviparous and the larva is the only juvenile stage. Eggs inside gravid females and freshly laid eggs generally show early stages of development and thus contain no differentiated embryo. However, ovoviviparity or larviparity caused by egg retention have been observed in different species. Egg retention may be an age-related phenomenon or an adaptation to unfavourable environments (Jagersbacher-Baumann and Ebermann, 2013). Non-feeding males may carry pharate females for up to 37 h before female ecdysis, and unfertilized females generate male progeny (Ebermann, 1982). Adult females of various species exhibit a typical dimorphism that is correlated with their phoretic behaviour (Ebermann, 1991a). Gravid females of the fungivorous *Scutacarus baculitarsus* Mahunka show a limited swelling of the body, but not the typical physogastry observed in other families of the Tarsonemina (Binns, 1981).

The feeding habitus of Scutacaridae seem to be primarily fungivorous, and parasitoidism or parasitism are not documented (Ebermann, 1991a; Kaliszewski *et al.*, 1995); known species live in soil, forest litter, humus, manure, compost and nests of animals, and are phoretic on various insects (Karafiat, 1959; Mahunka, 1965; Ebermann, 1991a). Some *Imparipes* spp. use their atrium genitale as sporothecae for carrying fungal spores (Ebermann and Hall, 2004). Some species exhibit a typical thanatosis (apparent death) when they are touched, and remain motionless for up to 6 min (Ebermann, 1991b). In contrast, *I. dispar* (Rack) makes jumps of at least 75 times its own body length in response to some dangers (Ebermann, 1995). The family includes about 750 species ascribed to 30 genera (Walter *et al.*, 2009).

S. australiensis Mahunka and *S. baculitarsus* are reported on mushrooms, but they do not cause direct economic damage. *S. australiensis* is known only in Australia (Halliday, 1998), and is rare in New South Wales (Clift and Terras, 2001). The mite has been found in the compost of cultivated mushrooms. The females are dimorphic, and the phoretic form has claspers distally on the tibiotarsus and is associated with phorid (*Megaselia halterata* (Wood)) and sciarid flies (Binns, 1980; Greenslade and Clift, 2004).

S. baculitarsus has been reported in mushroom houses in various countries. Phoretic forms of this mite have been found in commercial mushroom houses in the USA (Norton and Ide, 1974) and in England (Binns, 1980), where the mite was found clustered on the abdomens of adult males and females of *M. halterata* infesting mushrooms. Several adults of *S. baculitarsus* were taken from the compost of the dry casing layer in one mushroom house that contained the intrusive weed fungus *Trichothecium roseum* Link. Detached mites

were able to reattach both to phorids and to sciarids. The mite develops on unsterilized brewers' yeast on a damp plaster substrate, on which a species of the fungus *Mucor* developed. The life cycle develops through the stages of egg, larva, immobile chrysalis and adult. Physogastric females that developed oviposited near the fungus colony, but the early-emerging males picked up the immobile chrysalids containing phoretomorphic females and carried them away from the fungus colony; it is assumed that mating took place before the adult female emerged. The factors determining the development of phoretomorphic or physogastric females are not yet known (Binns, 1980).

No reports are known of the control of these mites. However, the application of strict hygiene measures reduces the presence of both *S. australiensis* and *S. baculitarsus* in mushroom farms.

The Family Microdispidae Cross

Morphological characteristics, systematics and bio-ecology

The Microdispidae are small mites, mainly known as adult females, the bodies of which are 80 to over 250 μm long. The males are rare and smaller. The pharynx has three pharyngeal pumps, usually well sclerotized and transversely striated. The prodorsum bears dorsally three pairs of setae (v_2, sc_1, sc_2), with v_2 rudimentary or absent, sc_1 clavate trichobothria and sc_2 simple setae. The idiosoma is oval, and the tergite C of the female covers the posterior region of the prodorsum to a variable extent, not covering the entire prodorsum and gnathosoma. Two pairs of cupules are present (ia and ih). The coxisternal plates I and II have two pairs of setae each, *1a*, *1b* and *2a*, *2b*, respectively. The pseudanal segment usually has two pairs of pseudanal setae (ps_1, ps_3), and the setae ps_2 are rarely present. The distance between the insertions of legs II and III is at least twice that between legs III and IV (Fig. 23.8). The femur I has three setae, with the seta *d* attenuated, smooth or barbed. Tibiotarsus I has five eupathidia (*p'* is absent), usually without claws or with small claws (Khaustov, 2006b; Walter *et al.*, 2009).

Microdispid mites live in soil, forest litter, decaying vegetation, rotting wood, moss beds and other similar habitats (Cross, 1965; Mahunka, 1975; Savulkina, 1981). Their feeding habit is primarily fungivorous, but several species exhibit a parasitoid association with insects, such as the army ants (Cross, 1965). Different species are phoretic on ants (Cross, 1965; Mahunka, 1977; Savulkina, 1981), on termites (Cross, 1965) or, unusually, on centipedes (Rack, 1979). *Premicrodispus lambi* (Krczal) is a pest of mushrooms in Australia (Clift and Toffolon, 1981a,b), China (Gao *et al.*, 1986), New Zealand (Krczal, 1964b) and Spain (Ferragut *et al.*, 1997).

The family includes about 100 species ascribed to 16 genera, mainly reported from the tropics (Khaustov, 2006b). The systematics of the Microdispidae is uncertain, and is treated by some authors as a subset either of the Pygmephoridae (Cross, 1965; Savulkina, 1981) or the Scutacaridae (Mahunka and Mahunka-Papp, 1994), as a separate group with uncertain relationships between the Pygmephoridae and the Scutacaridae (Mahunka, 1970; Rack and Kaliszewski, 1985; Ebermann, 1986) or as a sister family of Scutacaridae (Kaliszewski *et al.*, 1995). Similarly, the situation appears not to be well defined at the level of genera.

The Genus *Premicrodispus* Cross

Diagnostic characteristics

The body of the female is elliptical. The stigmata are about four times longer than wide. The dorsal setae are usually thin, smooth or indistinctly barbed. The cupules ia and ih have a rhombic shape. The anterior margin of the posterior sternal plate is without a large tongue-like elongation, and is sometimes slightly convex. The posterior margin of the posterior sternal plate is tripartite. The setae *4a* are usually absent (in some species they are present). The apodemes 4 are short, reaching to the level of the setae *3b*. The apodemes 5 are lacking. There are two pairs of pseudanal setae, and rarely three pairs. Legs I and II are subequal. Tibiotarsus I lacks a claw; tarsi II–IV have a simple claw (Khaustov, 2006b).

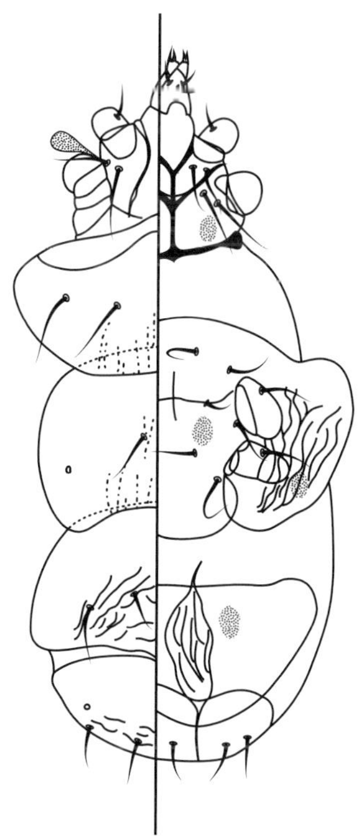

Fig. 23.8. *Premicrodispus lambi* (Krczal). Female, dorsal (left) and ventral (right) views (modified from Kaliszewski and Rack, 1986).

Economic importance

According to Khaustov (2006b), this genus includes at least 14 species, mainly inhabiting soils and litter and distributed worldwide. Of these, only *P. lambi* plays an economic role, causing severe damage to mushrooms in different countries.

Premicrodispus lambi (Krczal) (Fig. 23.8)

Common name

Unknown.

Diagnostic characteristics

FEMALE. The body averages 268 μm long and 75 μm wide. The gnathosoma is quadrangular in outline, and the chelicerae are stylet like with relatively short bases; the palpi are slightly longer than half the length of the gnathosoma, have 2 pairs of solenidia and end with a claw. The prodorsum bears 3 pairs of setae dorsally, 1 pair that are simple setae (sc_2), a second pair (v_2) that are rudimentary, and a third pair that are clavate trichobothria (sc_1); the setae sc_1 are club like, with a long pedicel and many little spines, and set more or less at the same level as the setae v_2 and sc_2. The stigmata are oval and set anterolaterally. The idiosoma has a granulated cuticle, and is divided into tergites (C, D, EF, H). There are 2 pairs of cupules, set on tergites D (ia) and H (ih). The idiosomal dorsal setae are barbed and end sharply. The setae c_1 are slightly shorter than the setae c_2, the setae *d*, *e*, *f* and h_1 are about the same length, and the setae h_2 are shorter. Ventrally, the apodemes are relatively weak; the anteromedian apodeme is connected with the arched apodeme I, II and the sejugal apodeme. The postero-median apodeme is connected with apodemes 4, which are almost transverse to the longitudinal axis of the body. Apodemes 3 and 5 are absent. Epimeres 1 and 2 have 2 pairs of setae each, *1a*, *1b* and *2a*, *2b*, respectively. The setae *4a* are present (Kaliszewski and Rack, 1986).

MALE. The body is about 180 μm long and 71 μm wide. The gnathosoma has a small tubiform structure, with 1 pair of solenidia and 1 pair of normal setae. The tergites CD, EF and H have a typical cuticular structure, and the tergite CD has the cupules ia. Together, the tergites H and Ps form the genital capsule. The setae of tergites CD and EF are slight, ending sharply. The setae h_1 are vestigial, and the setae h_2 are relatively stout. The prodorsal setae v_2 are shorter than the setae sc_1, which are longer than the setae sc_2; the setae c_1 are shorter than the setae c_2 and longer than the setae *d*, *e*, *f*, h_1 and h_2 (Kaliszewski and Rack, 1986).

Geographical distribution

Australia (Clift and Toffolon, 1981a,b), China (Gao *et al.*, 1986), New Zealand (Krczal, 1964b) and Spain (Ferragut *et al.*, 1997b).

Hosts

The mushroom hosts include: Agaricaceae – white mushroom (*Agaricus bisporus* (Lange) Imbach) (Clift and Toffolon, 1981a,b; Wu and Zhang, 1993a; Gao and Zou, 2001; Navarro *et al.*, 2010); Auriculariaceae – Jew's ear (*Auricularia auricula* (Hook. f.) Underw.); Hericiaceae – lion's mane mushroom (*Hericium erinaceus* (Bull.) Persoon); Tremellaceae – *Tremella fuciformis* Berk. (Wu and Zhang, 1993a).

Bio-ecology

P. lambi was first described as *Pygmephorus lambi* from mushrooms in New Zealand (Krczal, 1964b). About two decades later, Kaliszewski and Rack (1986) redescribed the female and described the male, larva and egg of the mite, referring to it as *Microdispus* (*Premicrodispus*) *lambi* (Krczal, 1964b). Later, Ferragut *et al.* (1997) reported the mite in Spain as *Brennandania lambi* (Krczal). Recently, Khaustov (2006b) redescribed the genus *Premicrodispus* Cross and included in it different species that had previously been ascribed to the genera *Microdispus* (*Premicrodispus*) Cross and *Brennandania* Sasa, respectively. This systematic choice is adopted here.

P. lambi is oviparous, and its life cycle includes the egg, larva, quiescent larva (or nymphocrysalis) and adult stages (Gao and Zou, 2001; Hernández-Abarca and Telesforo, 2007). Wu and Zhang (1993a) reported that the mite developed and reproduced on the mycelia of *Agaricus bisporus*, *Auricularia auricula*, *H. erinaceus* and *T. fuciformis*. However, Gao and Zou (2001) found that *P. lambi* completed its development only when fed on *Agaricus bisporus* mycelium growing on mushroom compost, and could not survive on mushroom mycelia of *Auricularia auricula*, *Auricularia polytricha* (Mont.) Sacc., *Ganoderma lucidum* (Curtis) P. Karst., *H. erinaceus*, *Lentinula edodes* (Berk.) Pegler, *Pleurotus ostreatus* (Jacq.) P. Kumm., *P. sajor-caju* (Fr.) Singer and *T. fuciformis*. Wu and Zhang (1993a) reported that *P. lambi* could not develop and reproduce on the mycelia of *L. edodes*, *Flammulina velutipes* (Curtis) Singer, *P. ostreatus* or *P. sajor-caju*, but that 12–20% of the females survived for 15–31 days on these hosts, whereas all mites died after 8 days in the absence of food. When fed the mycelia of *Auricularia polytricha*, *P. lambi* adults could become gravid and reproduce, but active larvae failed to develop to the next stage.

On *Agaricus bisporus* mycelium cultured on mushroom compost, the duration development of the egg and larva, the pre-oviposition and oviposition periods, female longevity and the time to 50% mortality of *P. lambi* declined as the temperature increased from 16 to 28°C. The threshold temperature for female development was 9°C and the thermal constant for the completion of female development 195 degree-days. At 16, 20, 24 and 28°C, the total fecundity was 71, 67, 66 and 57, respectively, and the daily egg-laying rate was 5.6, 8.7, 8.7 and 9.1, respectively. The sex ratio (female:male) ranged from 1.9 to 2.1 at 16–28°C. At 16, 20, 24 and 28°C, the intrinsic rate of natural increase (r_m) was 0.11, 0.18, 0.22 and 0.27, and the population doubling time was 6.1, 3.9, 3.2 and 2.5 days, respectively. All life stages of *P. lambi* died when exposed to 35°C constant temperature for 24 h, to 32°C constant temperature for 12 days or to 31–35°C (average 32.9°C) ambient temperature for 4 days (Gao and Zou, 2001). Under laboratory conditions at 15, 20, 25 and 30°C, and at 60 ± 5% RH, on fermented wheat straw (compost) previously inoculated with *A. bisporus* spores as substrate and food source for *P. lambi*, the intrinsic rate of increase (r_m) was 0.156, 0.167, 0.198 and 0.295, respectively.

Developmental rates then have a direct and linear relationship with temperature, and the threshold temperatures range between 6.01 and 6.74°C. The development from egg to adult requires 214 degree-days (Hernández-Abarca and Telesforo, 2007).

P. lambi is phoretic as mated adult female on various species of Diptera associated with *A. bisporus*. While the mite will attach to almost anything that moves, in a mixed population of adult Diptera, the preferred host in eastern Australia is the phorid *Megaselia halterata* (Wood). *P. lambi* can be found attached to a phorid female invading recently established mushroom crops on commercial farms. After several generations, the mites swarm on to the surface of the mushroom beds at the same time as adult phorids are also emerging from the same beds. The phorids mate and disperse to newly established crops, starting the next generation of phorids and mites (Clift and Terras, 2001).

The mite has a negative binomial distribution pattern on the mushroom bed (Wu *et al.*, 1989). Results of studies carried out in Spain on *Agaricus bisporus* over a period of 18 months, during which 24 consecutive mushroom crop cycles were monitored and a total of 24 spawn and 960 substrate samples were analysed, showed that it was usually the substrates in the growing rooms that were infested, most commonly the compost. In many cases, the mite can be detected when the first flush of the mushroom growth surge (these occur weekly) is harvested, although damage does not become evident until the third flush. Mites were detected at the back of the mushroom growing room and, to a lesser extent, near the access door (Navarro *et al.*, 2010)

Symptomatology and damage

P. lambi is a major pest of mushrooms in Australia as it feeds directly on the mushroom mycelium (Greenslade and Clift, 2004). The microdispid is a severe pest in commercial production of the white mushroom, *Agaricus bisporus*, in different provinces of China, where its feeding activity on mushroom mycelium causes 10–20% of yield losses, and occasionally even a total crop loss on some mushroom farms (Gao *et al.*, 1986; Gao and Zou, 2001; Wu and Zhang, 1993a,b). It has been listed as a quarantine pest of agricultural plants of Zhejiang Province. In Spain, *P. lambi* was first observed in 1996 in mushroom crops (Ferragut *et al.*, 1997b), and is now causing substantial economic losses, particularly in the spring and summer (Navarro *et al.*, 2010).

Control

According to the above discussion, the major infection source of mushroom beds are contaminated spawns, and the most important way to control mite infestation is by producing and sowing mite-free spawns. With this aim, laboratory and field studies were conducted to evaluate freezing as a control measure against *P. lambi* infesting *Agaricus bisporus* in Shanghai (China), primarily through contaminated spawn. Laboratory experiments showed that exposure of contaminated spawn to −10°C for 24 h killed all stages of the mite. The mites probably died due to the freezing of body tissues during sustained exposure at −10°C because the supercooling points for the egg, active larva, quiescent larva, non-gravid adult and gravid adult were, respectively, −11.34, −10.96, −11.43, −7.65 and −11.20°C, whereas the freezing points were −6.68, −6.84. −6.56, −3.96 and −6.75°C. Semi-field experiments showed that compost inoculated with *P. lambi*-infested spawn that had been exposed at −10°C for 24 h had no mite infestation during spawn running, and thus mycelium growth was normal. Laboratory and field experiments from 1989 to spring 1990 showed that freezing the spawn had no effect on mushroom yield. Further field experiments in two farms during autumn 1990 showed no effect of freezing spawn on yield in one farm and an increased yield in the other farm. A field experiment during 1991 to 1992 also indicated that freezing spawn at −10°C for 24–34 h had no effect on mushroom yield (Wu and Zhang, 1993b).

Field surveys of mite damage carried out in the suburbs of Shanghai showed that contaminated spawn was the major source of the mite problem and that losses of the crop averaged 10–20%. Experiments on the effects of temperature on mite survival indicated that exposure of the mites to 50°C for 1 h or to −10°C for 24 h killed all stages of *P. lambi*. Thus, heating treatment during well-executed phase II composting and freezing of mushroom spawn at –10°C for 24 h should eliminate, or at least reduce, the damage to mushrooms by this pest (Wu and Zhang, 1993a).

References

Abad, R., Castanera, P. and Urbaneja, A. (2006) Natural enemies of the spider mites, *Tetranychus urticae* Koch and *Panonychus citri* (McGregor) (Acari: Tetranychidae) in Spanish citrus orchards. *IOBC-WPRS Bulletin* 29(3), 179.

Abawi, G.S. and Lorbeer, J.W. (1971) Populations of *Fusarium oxysporum* f. sp. *cepae* in organic soils in New York. *Phytopathology* 61, 1042–1048.

Abbotsford Agriculture Centre (2010) Mites of Floriculture Crops – Biology and Control. Ministry of Agriculture, Abbotsford, British Columbia, Canada. Available at: http://www.agf.gov.bc.ca/cropprot/mites.htm#broad (accessed 19 July 2014).

Abdel-Samad, M.A., El-Halawany, M.E. and El-Saied, K.M. (1996) Utilizing *Euseius scutalis*, Athias-Henriot to control *Eutetranychus orientalis*, Klein on citrus trees. *Egyptian Journal of Agricultural Research* 74, 671–684.

Abdel-Sater, M.A. and Eraky, S.A. (2001) Bulbs micoflora and their relation with three stored product mites. *Mycopathologia* 153, 33–39.

Abdel-Shaheed, G.A., Hammad, S.M. and El-Sawaf, S.K. (1973) Survey and population density studies on mites found on cotton and corn in Abis, Abou-Hommos localities, El-Beheira Province (Egypt). *Bulletin of the Entomological Society of Egypt* 57, 101–108.

Abdullaev, A.N. and Mugutdinov, N.S. (1976) Brown mite in the orchards of the Dagestan foothills. *Sadovodstvo* 11, 24–25.

Abe, J., Ganaha-Kikumura, T. and Yukawa, J. (2011) Morphological features, distribution, prey mites, and life history traits of *Feltiella acarisuga* (Vallot), (Diptera: Cecidomyiidae) in Japan. *Applied Entomology and Zoology* 46, 271–279.

Abenavoli, L.M., Bonsignore, C.P., Sciarrone, G., Mammoliti, V., Smorto, D., Giametta, F. and Vacante, V. (2009) Mechanical distribution of predators in biological control in protected crops. In: Giametta, G. and Zimbalatti, G. (eds) *Technology and Management to Ensure Sustainable Agriculture, Agro-systems, Forestry and Safety, CIOSTA-CIGR V Conference, 17–19 June 2009, Reggio Calabria, Italy*. Commission International de l'Organisation Scientifique de Travail en Agriculture, "Mediterranea" University of Reggio Calabria, Reggio Calabria, Italy, pp. 651–654.

Abhilash, B. and Sudharma, K. (2003) Effect of pesticides on *Amblyseius longispinosus* (Evans), a predator of red spider mite *Tetranychus ludeni* Zacher. *Entomon* 28, 165–167.

Abo-Korah, S.M. (1978) Mites associated with maize and their predators in Monoufeia Governorate, Egypt. *Bulletin of the Entomological Society of Egypt* 62, 275–278.

Abou-Awad, B.A. (1979) The tomato russet mite, *Aculops lycopersici* (Massee) (Acari: Eriophyidae) in Egypt. *Anzeiger für Schädlingskunde* 52, 153–156.

Abou-Awad, B.A. (1980) New species of genus *Vasates* in Egypt (Acari: Eriophyoidea: Eriophyidae). *Acarologia* 21, 389–391.

Abou-Awad, B.A. (1981a) Ecological and biological studies on the mango bud mite, *Eriophyes mangiferae* (Sayed), with description of immature stages (Eriophyoidea: Eriophyidae). *Acarologia* 22, 145–150.

Abou-Awad, B.A. (1981b) Bionomics of the mango rust mite *Metaculus mangiferae* (Attiah) with description of immature stages (Eriophyoidea: Eriophyidae). *Acarologia* 22, 151–155.

Abou-Awad, B.A. (1983) *Amblyseius gossipi* (Acarina: Phytoseiidae) as a predator of the tomato erineum mite, *Eriophyes lycopersici* (Acarina: Eriophyidae). *Entomophaga* 28, 363–366.

Abou-Awad, B.A. and El-Banhawy, E.M. (1985) Susceptibility of the tomato russet mite, *Aculops lycopersici* (Acari: Eriophyidae), in Egypt to methamidophos, pyridaphenthion, cypermethrin, dicofol and fenarimol. *Experimental and Applied Acarology* 1, 11–15.

Abou-Awad, B.A. and El-Banhawy, E.M. (1991) New mites of the family Eriophyidae from Kenya (Acari: Eriophyoidea). *Acarologia* 32, 329–333.

Abou-Awad, B.A. and Nasr, A.K. (1983) *Heterotergum flacourtiae* n. sp. from Egypt (Acari: Eriophyidae). *International Journal of Acarology* 9, 179–181.

Abou-Awad, B.A., Nasr, A.K., Gomaa, E.A. and Abou Elela, M.M. (1989) Life history of the predatory mite, *Cydnodromella negevi* and the effect of nutrition on its biology (Acari: Phytoseiidae). *Insect Science and Its Application* (now *International Journal of Tropical Insect Science*) 10, 617–623.

Abou-Awad, B.A., El Sherif, A.A., Hassan, M.F. and Abou Elela, M.M. (1998a) Life history and life table of *Amblyseius badryi*, as a specific predator of eriophyid grass mites (Acari: Phytoseiidae: Eriophyidae). *Zeitschrift für Pflanzenkrankheiten und Pflanzenschutz* (now *Journal of Plant Diseases and Plant Protection*) 105, 422–428.

Abou-Awad, B.A., El Sherif, A.A., Hassan, M.F. and Abou Eleila, M.M. (1998b) Studies on development, longevity, fecundity and predation of *Amblyseius olivi* Nasr & Abou Awad (Acari: Phytoseiidae) on various kinds of prey and diets. *Zeitschrift für Pflanzenkrankheiten und Pflanzenschutz* (now *Journal of Plant Diseases and Plant Protection*) 105, 538–544.

Abou-Awad, B.A., El-Sawaf, B.M. and Abdel Kader, A.A. (1999a) Life history and life table of *Pronematus ubiquitous* (McGregor) as a predator of eriophyoid mites in Egypt (Acari: Tydeidae). *Acarologia* 40, 29–32.

Abou-Awad, B.A., El-Sawaf, B.M., Reda, A.S. and Abdel-Khalek, A.A. (1999b) Environmental management and biological aspects of two eriophyoid fig mites in Egypt: *Aceria ficus* and *Rhyncaphytoptus ficifoliae*. *Acarologia* 40, 419–429.

Abou-Awad, B.A., Metwally, A.M. and Al-Azzazy, M.M. (2005) Environmental management and biological aspects of two eriophyid olive mites in Egypt: *Aceria oleae* and *Tegolophus hassani*. *Zeitschrift für Pflanzenkrankheiten und Pflanzenschutz* (now *Journal of Plant Diseases and Plant Protection*) 112, 287–303.

Abou-Awad, B.A., Al-Azzazy, M.M. and El-Sawi, S.A. (2010a) The life-history of the peach silver mite, *Aculus fockeui* (Acari: Eriophyidae) in Egypt. *Archives of Phytopathology and Plant Protection* 43, 384–389.

Abou-Awad, B.A., Metwally, A.S.M. and Al-Azzazy, M.M. (2010b) Effect of different eriophyid and tetranychid mango mite species on development, longevity, fecundity and predation of *Typhlodromus mangiferus* Zaher and El-Brolossy (Acari: Phytoseiidae). *Archives of Phytopathology and Plant Protection* 43, 390–403.

Abou-Awad, B.A., Metwally, A.S.M. and Al-Azzazy, M.M. (2011a) Environmental management and biological aspects of two eriophyid mango mites in Egypt: *Aceria mangiferae* and *Metaculus mangiferae*. *Acarologia* 51, 481–497.

Abou-Awad, B.A., Al-Azzazy, M.M. and Afia, S.I. (2011b) Effect of temperature and relative humidity on the rate of development, fecundity and life table parameters of the red spider mite *Oligonychus mangiferus* (Rahman and Sapra) (Acari: Tetranychidae). *Archives of Phytopathology and Plant Protection* 44, 1862–1866.

Abou-Awad, B.A., Afia, S.I. and Al-Azzazy, M.M. (2012) Ecological studies on the mango red spider mite *Oligonychus mangiferus* (Rahman and Sapra) in mango orchards (Acari: Tetranychidae). *Acarines* 6, 7–13.

Aboul-Nasr, A.E. and Bassal, T.T.M. (1971) Biochemical and physiological study of certain ticks (Ixodoidea). The sugar content and concentration in *Argas* and *Hyalomma* biological fluids. *Journal of Medical Entomology* 8, 521–524.

Abou-Setta, M.M. and Childers, C.C. (1987) Biology of *Euseius mesembrinus* (Acari: Phytoseiidae): life tables on ice plant pollen at different temperatures with notes on behavior and food range. *Experimental and Applied Acarology* 3, 123–130.

Abou-Setta, M.M. and Childers, C.C. (1989) Biology of *Euseius mesembrinus* (Acari: Phytoseiidae): life tables and feeding behavior of tetranychid mites on citrus. *Environmental Entomology* 18, 665–669.

Abraham, E.V., Thirumurthi, S., Ali, K.A. and Subramaniam, T.R. (1973) Some new pests of sesamum. *Madras Agricultural Journal* 60, 593.

Abreu, E., Medina-Gaud, S. and Maldonado-Capriles, J. (1987) New records of mites for Puerto Rico. *Journal of Agriculture of the University of Puerto Rico* 71, 399–402.

Åbro, A. (1988) The mode of attachment of mite larvae (*Leptus* spp.) to harvestmen (opilines). *Journal of Natural History* 22, 123–130.

A'Brook, J. (1975) Epidemiology and effects of grass viruses in relation to their insect vectors. *Annals of Applied Biology* 81, 261–264.

A'Brook, J. and Heard, A.J. (1975) The effect of ryegrass mosaic virus on the yield of perennial ryegrass swards. *Annals of Applied Biology* 80, 163–168.

Achor, D.S., Ochoa, R., Erbe, E.F., Aguilar, H., Wergin, W.P. and Childers, C.C. (2001) Relative advantages of low temperature versus ambient temperature scanning electron microscopy in the study of mite morphology. *International Journal of Acarology* 27, 3–12.

Adango, E., Onzo, A., Hanna, R., Atachi, P. and James, B. (2006) Comparative demography of the spider mite, *Tetranychus ludeni*, on two host plants in West Africa. *Journal of Insect Science* 6:49. Available at: http://jinsectscience.oxfordjournals.org/content/6/1/49 (accessed 2 April 2015).

Adar, E., Inbar, M., Gal, S., Doron, N., Zhang, Z.-Q. and Palevsky, E. (2012) Plant-feeding and non-plant feeding phytoseiids: differences in behavior and cheliceral morphology. *Experimental and Applied Acarology* 58, 341–357.

Afifi, A.-A.M., Mabrouk, A.M. and Asranl, A.A. (2010a) Effect of the entomopathogenic fungus *Beauveria bassiana* on three acarine pests. In: Sabelis, M.W. and Bruin, J. (eds) *Trends in Acarology: Proceedings of the 12th International Congress, 21–26 August 2006*, Amsterdam. Springer, Dordrecht, the Netherlands, pp. 439–440.

Afifi, A.-A.M., El-Laithy, A.Y.M., Shehata, S.A. and El-Saiedy, E.M.A. (2010b) Resistance of strawberry plants against the two-spotted spider mite, *Tetranychus urticae* (Acari: Tetranychidae). In: Sabelis, M.W. and Bruin, J. (eds) *Trends in Acarology: Proceedings of the 12th International Congress, 21–26 August 2006*, Amsterdam. Springer, Dordrecht, the Netherlands, pp. 505–508.

Afifi, A.M., Hassan, M.F. and Nawar, M.S. (1987) Notes on the biology and feeding habits of *Protogamasellus minutus* Hafez, El-Badry and Nasr (Acari: Gamasida: Ascidae). *Bulletin of the Entomological Society of Egypt* 66, 251–259.

Afshari, G.A. (1999) A survey on the ladybirds belonging to the genus *Stethorus* and study of the biology, prey consumption and population dynamics of *Stethorus gilvifrons* in sugarcane farms in Khuzestan, Iran. MSc. thesis, Shahid Chamran University of Ahvaz, Ahvaz, Iran.

Afshari, M., Mossadegh, S., Soleyman-Nejadian, E. and Kamali, K. (2007) Geographical distribution and host plants of *Stethorus gilvifrons* (Mulsant) (Col.: Coccinellidae) and its biology under laboratory conditions in Khuzestan Province. *Journal of Agricultural Sciences and Natural Resources* 14(4), 201–210.

Agarwal, R.A. (1957) The sugar cane mite and its control. *Indian Sugar* 7, 394–399.

Agarwal, R.A. (1969) Morphological characteristics of sugarcane and insect resistance. *Entomologia Experimentalis et Applicata* 12, 767–776.

Agarwal, R.A. and Kandasami, P.A. (1959) Nature of damage caused by eriophyid mite in sugarcane. *Current Science* 28, 297.

Agboton, B.V., Hanna, R. and von Tiedemann, A. (2011) Molecular detection of establishment and geographical distribution of Brazilian isolates of *Neozygites tanajoae*, a fungus pathogenic to cassava green mite, in Benin (West Africa). *Experimental and Applied Acarology* 53, 235–244.

Aghajanzadeh, S. and Malik, B. (2007) Sampling and distribution pattern of citrus rust mite, *Phyllocoptruta oleivora* Ashmead (Acari, Eriophyidae) using adhesive tape method. *International Journal of Agriculture and Biology* 9, 329–332.

Aghajanzadeh, S., Mallik, B. and Chandrashekar, S.C. (2006) Bioefficacy of six isolates of *Hirsutella thompsonii* Fisher against citrus rust mite, *Phyllocoptruta oleivora* Ashmead (Acari: Eriophyidae) and two spotted spider mite, *Tetranychus urticae* Koch (Acari: Tetranychidae). *Pakistan Journal of Biological Sciences* 9, 871–875.

Agrawal, A.A. and Karban, R. (1997) Domatia mediate plant–arthropod mutualism. *Nature* 387, 562–563.

Agriculture Western Australia (2000) *Balaustium* mite *Balaustium medicagoense*. In: PestWeb, Agriculture Western Australia, South Perth, Australia. Available at: http://agspsrv34.agric.wa.gov.au/Ento/pestweb/Query1_1.idc?ID=-1792933947 (accessed 19 July 2014).

Agrios, G.N. (1980) Insect involvement in the transmission of fungal pathogens. In: Harris, K.F. and Maramorosch, K. (eds) *Vectors of Plant Pathogens*. Academic Press, New York, pp. 293–324.

AgroAtlas-Pests (2003–2009a) *Eotetranychus pruni* (Oudemans) – Garden Spider Mite. In: Interactive Agricultural Ecological Atlas of Russia and Neighboring Countries: Economic Plants and their Diseases, Pests and Weeds. Available at: http://www.agroatlas.ru/en/content/pests/Eotetranychus_pruni/ (accessed 11 July 2014).

AgroAtlas-Pests (2003–2009b) *Tetranychus viennensis* Zacher – Hawthorn Red Spider Mite, Fruit Tree Spider Mite. In: Interactive Agricultural Ecological Atlas of Russia and Neighboring Countries: Economic Plants and their Diseases, Pests and Weeds. Available at: http://www.agroatlas.ru/en/content/pests/Tetranychus_viennensis/ (accessed 11 July 2014).

Agudelo-Silva, P. and Silva, P.A. (1986) A species of *Triplosporium* (Zygomycetes: Entomophthorales) infecting *Mononychelus progressivus* (Acari: Tetranychidae) in Venezuela. *Florida Entomologist* 69, 444–446.

Aguilar, H. and Gonzalez, L. (1990) *Tuckerella knorri* Baker and Tuttle agente causal del resquebrajamiento del fruto de los citricos. *Boletín Informativo MIP (Manejo Integrado de Plagas) (CATIE, Turrialba, Costa Rica)* 18, 2–3.

Aguilar, H. and Murillo, P. (2008) Nuevos hospederos y registros de ácaros fitófagos para Costa Rica : periodo 2002–2008. *Agronomía Costarricense* 32(2), 7–28.

Aguilar, H. and Murillo, P. (2012) New hosts and records of plant feeding mites for Costa Rica: interval 2008–2012. *Agronomía Costarricense* 36(2), 11–28.

Aguilar, H. and Ochoa, R. (1988) *Brevipalpus californicus* (Banks) (Acari: Tenuipalpidae), a new pest of cardamoms (*Elettaria cardamomum* Maton) in Costa Rica. *Agronomía Costarricense* 12(2), 251–252.

Aguilar, H. and Salas, L.A. (1989) Biologia y table de vida de *Typhlodromus pilosus* Chant (Acari: Phytoseiidae) en Costa Rica. *Turrialba* 39, 162–169.

Aguilar-Fenollosa, E., Ibañez-Gual, M.V., Pascual-Ruiz, S., Hurtado, M. and Jacas, J.A. (2011) Effect of ground-cover management on spider mites and their phytoseiid natural enemies in clementine mandarin orchards (II): top-down regulation mechanisms. *Biological Control* 59, 171–179.

Aguilar-Fenollosa, E., Pina, T., Gomez-Martinez, M.A., Hurtado, M.A. and Jacas, J.A. (2012) Does host adaptation of *Tetranychus urticae* populations in clementine orchards with a *Festuca arundinacea* cover contribute to a better natural regulation of this pest mite? *Entomologia Experimentalis et Applicata* 144, 181–190.

Aguilera, P.A. (1987) Nuevas localidades para *Stethorus histrio* Chazeau (Coleoptera: Coccinellidae) en Chile. *Revista Chilena de Entomologia* 15, 33–36.

Ahmed, Z.I. and Ahmed, R.F. (1989) Biological studies of the predator *Stethorus gilvifrons* Mulsant (Coleoptera: Coccinellidae) on strawberry mites *Tetranychus turkestani* Ugarov and Nikolski (Acarina: Tetranychidae). *Journal of Biological Science Research* 20, 22–23.

Ahuja, D.B. (2000) Influence of abiotic factors on the population of mite, *Polyphagotarsonemus latus* (Banks) infesting sesame (*Sesamum indicum* L.) in the arid region of Rajasthan (India). *Journal of Entomological Research* 24, 87–89.

Ahuja, D.B. and Kalyan, R.K. (2001) Field screening of genotypes of sesame against leaf webber/capsule borer, *Antigastra catalaunalis* Dup., gallfly, *Asphondylia sesami* Felt., and mite, *Polyphagotarsonemus latus* (Banks). *Pest Management and Economic Zoology* 9, 5–9.

Ahuja, D.B. and Kalyan, R.K. (2003) Evaluation of various plant products against *Antigastra catalaunalis* Dup. and *Polyphagotarsonemus latus* (Banks) on sesame, *Sesamum indicum* L. *Journal of Insect Science (Ludhiana)* 16, 92–94.

AICRP (1994) *Progress Report 1993–94*. AICRP (All India Coordinated Research Project) on Agricultural Acarology, University of Agricultural Sciences, Bangalore, India.

AICRP (1996) *Progress Report 1994–96*. AICRP (All India Coordinated Research Project) on Agricultural Acarology, University of Agricultural Sciences, Bangalore, India.

AICRP (1998) *Progress Report 1996–98*. AICRP (All India Coordinated Research Project) on Agricultural Acarology, University of Agricultural Sciences, Bangalore, India.

Aist, J.R. (1976) Papillae and related wound plugs of plant cells. *Annual Review of Phytopathology* 14, 145–163.

Akimov, I.A., Badanin, I.V. and Wita, I. (2002) Some aspects of trophic specialization in both free-living and parasitic prostigmatid mites (Acari: Actinotrichida) from a morpho-functional view-point. *Acarina* 10, 65–73.

Akinlosotu, T.A. and Leuschner, K. (1981) Outbreak of two new cassava pests (*Mononychellus tanajoa* and *Phenacoccus manihoti*) in southwestern Nigeria. *Tropical Pest Management* 27, 247–250.

Akiyama, M., Sakata, T., Mori, N., Kato, T., Amano, H. and Kuwahara, Y. (1997) Chemical ecology of astigmatid mites. 46. Neryl formate, the alarm pheromone of *Rhizoglyphus setosus* Manson (Acarina: Acaridae) and the common pheromone component among four *Rhizoglyphus* mites. *Applied Entomology and Zoology* 32, 75–79.

Aksoy, H.M., Özman-Sullival, S.K., Ocal, H., Celik, N. and Sullivan, G.T. (2008) The effects of *Pseudomonas putida* biotype B on *Tetranychus urticae* (Acari: Tetranychidae). *Experimental and Applied Acarology* 46, 223–230.

Al-Abbasi, S.H. and Weigle, J.L. (1982) Resistance in New Guinea *Impatiens* species and hybrids to the two-spotted spider mite. *HortScience* 17, 47–48.

Alagarmalai, J., Grinberg, M., Perl-Treves, R. and Soroker, V. (2009) Host selection by the herbivorous mite *Polyphagotarsonemus latus* (Acari: Tarsonemidae). *Journal of Insect Behavior* 22, 375–387.

Alam, M. (1933) Arhar sterility. In: *Proceedings of 20th Annual Meeting of Indian Science Congress, Section Agriculture (Poona, India)* 43, 15–16.

Alan, M.Z. and Wadud, M.A. (1963) On the biology of the litchi mite, *Aceria litchii* (Eriophyidae, Acarina) in East Pakistan. *Pakistan Journal of Science* 15, 231–240.

Al-Ani, L.K. and Al-Jboory, I. (2008) Effect of different temperatures on the biology of the broad mite *Polyphagotarsonemus latus* (Banks) on potato. *Arab Journal of Plant Protection* 26, 95–101.

Al-Azzazy, M.M.A. (2000) Studies on mites associated with olive trees. MSc. thesis, Al-Azhar University, Cairo.

Alba, J.M., Ataide, L.M.S., Villaroel, C., Glasl, J.J., Schimmel, B.C.J., Pappas, M.L., Schuurink, R.C. Sabelis, M.W. and Kant, M.R. (2012a) Functional transcriptomics to study the role of induction and suppression of plant defenses in the chemical ecology of herbivorous mites. In: Schausberger, P., Walzer, A., Strodel, M.A. and Peneder, S. (eds) *7th Symposium of the European Association of Acarologists (EURAAC), July 9–13 2012, Vienna, Austria: Acari in a Changing World. Program, Abstracts, Participants*. University of Natural Resources and Life Sciences, Vienna, Austria, p. 24. Available at: http://euraac.boku.ac.at/SympVienna/files/ProgramBook.pdf (accessed 14 January 2015).

Alba, J.M., Pappas, M.L., Glasl, J.J., Schimmel, B.C.J., Ataide, L.S., Sabelis, M.W., Schuurink, R.C. and Kant, M.R. (2012b) Suppression of host-plant defenses by spider mites affects downstream phytohormone accumulation. In: Schausberger, P., Walzer, A., Strodel, M.A. and Peneder, S. (eds) *7th Symposium of the European Association of Acarologists (EURAAC), July 9–13 2012, Vienna, Austria: Acari in a Changing World. Program, Abstracts, Participants*. University of Natural Resources and Life Sciences, Vienna, Austria, p. 74. Available at: http://euraac.boku.ac.at/SympVienna/files/ProgramBook.pdf (accessed 14 January 2015).

Alberti, G. (1973) Ernährungsbiologie und spinnvermögen der Schnabelmilben (Bdellidae, Trombidiformes). *Zeitschrift für Morphologie und Ökologie der Tiere* 76, 285–338.

Alberti, G. (1979) Fine structure and probable function of genital papillae and Claparède organs of Actinotrichida. In: Rodriguez, G.R. (ed.) *Recent Advances in Acarology, Volume II*. Academic Press, New York, pp. 501–507.

Alberti, G. (1984) The contribution of comparative spermatology to problems of acarine systematics. In: Griffiths, D.A. and Bowman, C.E. (eds) *Acarology VI, Volume 1*. Ellis Horwood, Chichester, UK, pp. 479–490.

Alberti, G. (1998) Fine structure of receptor organs in oribatid mites (Acari). In: Ebermann, E. and Schuster, R. (eds) *Arthropod Biology: Contributions to Morphology, Ecology and Systematics*. Biosystematics and Ecology Series 14, Österreichische Akademie der Wissenschaften, Vienna, pp. 27–77.

Alberti, G. (2000) Chelicerata. In: Jamieson, B.G.M. (ed.) *Reproductive Biology of Invertebrates, Volume IX, Part B, Progress in Male Gamete Ultrastructure and Phylogeny*. Oxford & IBH Publishing., New Delhi/Wiley, New York, pp. 311–388.

Alberti, G. and Bader, C. (1990) Fine structure of external "genital" papillae in the freshwater mite *Hydrovolzia placophora* (Hydrovolziidae, Actinedida, Actinotichida, Acari). *Experimental and Applied Acarology* 8, 115–124.

Alberti, G. and Coons, L.B. (1999) Mites. In: Harrison, F.W. and Foelix, R.F. (eds) *Microscopic Anatomy of Invertebrates, Volume 8C, Chelicerate Arthropoda*. Wiley, New York, pp. 515–1265.

Alberti, G. and Crooker, A.R. (1985) Internal anatomy. In: Helle, W. and Sabelis, M.W. (eds) *Spider Mites: Their Biology, Natural Enemies and Control*. World Crop Pests, Volume 1A. Elsevier Science, Amsterdam, pp. 29–62.

Alberti, G. and Di Palma, A. (2002) Fine structure of the phytoseiid-type sperm access system (Acari, Gamasida, Phytoseiidae). In: Bernini, F., Nannelli, R., Nuzzaci, G. and de Lillo, E. (eds) *Acarid Phylogeny and Evolution: Adaptations in Mites and Ticks. Proceedings of IV Symposium of the European Association of Acarologists*. Kluwer Academic Publishers, Dordrecht, the Netherlands, pp. 241–252.

Alberti, G. and Nuzzaci, G. (1996) SEM and TEM techniques. In: Lindquist, E.E., Sabelis, M.W. and Bruin, J. (eds) *Eriophyoid Mites: Their Biology, Natural Enemies and Control*. World Crop Pests, Volume 6. Elsevier Science, Amsterdam, pp. 399–410.

Alberti, G. and Seeman, O. (2005) Ultrastructural observations on Holothyrida (Acari: Anactinotrichida). In: Weigmann, G., Alberti, G., Wohltmann, A. and Ragusa, S. (eds) *Acarine Biodiversity in the Natural and Human Sphere, 5th Symposium of the European Association of Acarologists, Berlin. Phytophaga* 14 (2004), 103–111.

Alberti, G. and Storch, V. (1974) Über Bau und Funktion der Prosoma-Drüsen von Spinnmilben (Tetranychidae, Trombidiformes). *Zeitschrift für Morphologie und Ökologie der Tiere* 79, 133–153.

Alberti, G., Storch, V. and Renner, H. (1981) Über den feinstrukturellen Aufbau der Milbencuticula (Acari, Arachnida). *Zoologische Jahrbücher. Abteilung für Anatomie und Ontogenie der Tiere* 105, 183–236.

Alberti, G., Moreno, A.I. and Kratzmann, M. (1994) The fine structure of trichobothria in moss mite with special emphasis on *Acrogalumna longipluma* (Berlese, 1904) (Oribatida, Acari, Arachnida). *Acta Zoologica* 75, 57–74.

Alberti, G., Norton, R.A., Adis, J., Fernandez, N.A., Franklin, E., Kratzmann, M., Moreno, A.I., Ribeiro, E.F., Weigmann, G. and Woas, S. (1996) Porose areas and related organs in oribatid mites (Oribatida). In: Mitchell R., Horn, D.J., Needham, G.R. and Welbourn, W.C. (eds) *Acarology IX: Proceedings, Volume 1*. Ohio Biological survey, Columbus, Ohio, pp. 277–283.

Albrigo, L.G. and McCoy, C.W. (1974) Characteristic injury by citrus rust mite to orange leaves and fruit. *Proceedings of the Florida State Horticultural Society* 87, 48–55.

Albrigo, L.G., Childers, C.C. and Syvertsen, J.P. (1983) Structural damage to citrus leaves from spider mite feeding. In: Matsumoto, K. (ed.) *Proceedings of the 4th International Citrus Congress, Tokyo, Japan, 9–12 November 1981, Volume 2*. International Society of Citriculture, Riverside, California, pp. 649–652.

Albuquerque, F.A. de, Arantes, A.M.V.T. and Correia, Ado, C.B. (2000) Patogenicidade de fungos para o acaro da leprose *Brevipalpus phoenicis* (Geijskes, 1939) (Acari: Tenuipalpidae). *Acta Scientiarum, Agronomy* 22, 969–971.

Alcázar, M.D., Aranda, G., Márquez, A.L., Sánchez, L. and Ruiz, C. (2005) *Oligonychus persea* (Acari: Tetranychidae); una nueva plaga en el aguacate en el Sur de España. In: Bento, A., Miranda-Arabolaza, M.J., Pereira, J.A. (eds) *Programa, Resumos IV Congresso Nacional de Entomologia Aplicada, X Jornadas Científicas de la Sociedad Española de Entomología Aplicada, I Jornadas Portuguesas de Entomologia Aplicada*, Bragança, 17–21 de Outubro de 2005. Instituto Politécnico de Bragança, Escola Superior Agrária Bragança, Portugal, p. 213. Available at: http://www.seea.es/pdf/resumenesIVCongreso.pdf (accessed 28 April 2015).

Al-Doghairi, M.A. (2004) Effect of eight acaricides against the date dust mite, *Oligonychus afrasiaticus* (Mcgregor) (Acari: Tetranychidae). *Pakistan Journal of Biological Sciences* 7, 1168–1171.

Aldosari, S.A. (2009) Occurrence of dust mite, *Oligonychus afrasiaticus* Meg. on fruits, leaflets of some date palm trees and evaluation the efficiency of botanical compound, (Biaco) as compared with some acaricides. *Assiut University Bulletin for Environmental Researches* 12(2), 69–77. Available at: http://www.aun.edu.eg/env_enc/oct2009/69-76.pdf (accessed 26 March 2015).

Al-Dosary, N.H. (2010) Evaluate efficiency of some insecticides and sticker color traps to protected date palm fruits infested by dust mite *Oligonychus afrasiaticus* (McGregor) and lesser date moth *Batrachedra amydraula* (Merck). *Basrah Journal of Agricultural Sciences* 23(1), 1–22.

Alfaro, A. (1964) Los acaros de los frutales. *Boletín de Patología Vegetal y Entomología Agrícola* 27, 201–221.

Alford, D.V. (1972) The effect of *Tarsonemus fragariae* Zimmermann (*Acarina: Tarsonemidae*) on strawberry yields. *Annals of Applied Biology* 70, 13–18.

Alford, D.V. (1976) Some observations on the effect of pests on strawberry yields. *Annals of Applied Biology* 81, 110–111.

Alford, D.V. (1979) Chemical control of blackberry mite, *Acalitus essigi* (Hassan). *Plant Pathology* 28, 91–94.

Alford, D.V. (1984) *A Colour Atlas of Fruit Pests, Their Recognition, Biology and Control*. Wolfe Publishing, London.

Alford, D.V. (2012) *Pests of Ornamental Trees, Shrubs and Flowers: A Color Handbook*, 2nd edn. Academic Press, San Diego, California.

Ali, A.G. and Aldosari, S.A. (2007) Susceptibility of date palm fruit cultivars to the natural infestation by *Oligonychus afrasiaticus* (McG.) (Acari: Tetranychidae) in relation to their chemical composition. *Assiut University Bulletin for Environmental Researches* 10(2), 1–7. Available at: http://www.aun.edu.eg/env_enc/oct2007/1-6.pdf (accessed 26 March 2015).

AliNiazee, M.T. (1979) Mite populations on apple foliage in Western Oregon. In: Rodriguez, J.G. (ed.) *Recent Advances in Acarology, Volume I*. Academic Press, New York, pp. 71–76.

AliNiazee, M.T. (1980) *Filbert Insect and Mite Pests*. Station Bulletin 643, Agricultural Experiment Station Oregon State University, Corvallis, Oregon.

AliNiazee, M.T. (1989) Suitability of biphenthrin in IPM programmes on apples. In: ChannaBasavanna, G.P. and Viraktamath, C.A. (eds) *Progress in Acarology, Volume 2*. Oxford & IBH Publishing, New Delhi, pp. 461–465.

AliNiazee, M.T. (1994) Insect pest management in hazelnut orchards of North America. *Acta Horticulturae* 351, 543–549.

AliNiazee, M.T. (1997) Integrated pest management of hazelnut pests: a worldwide perspective. *Acta Horticulturae* 445, 469–476.

AliNiazee, M.T. (1998) Ecology and management of hazelnut pests. *Annual Review of Entomology* 43, 395–419.

Al-Jboory, I.J. (1987) Taxonomic studies of false spider mites (Acari:Tenuipalpidae) in central Iraq. PhD thesis, University of Bonn, Bonn, Germany.

Al-Jboory, I.J. and Al-Suaide, T.M. (2010) Effect of temperature on the life history of the old world date mite, *Oligonychus afrasiaticus* (Acari: Tetranychidae). In: Sabelis, M.W. and Bruin, J. (eds) *Trends in Acarology: Proceedings of the 12th International Congress, 21–26 August 2006*, Amsterdam. Springer, Dordrecht, the Netherlands, pp. 361–363.

Alkan, B. (1959) Haselnussschadlinge in der Turkei. *Zeitschrift für Angewandte Entomologie* 44, 187–202.

Allen, J.C. (1976a) A modified sine wave method for calculating degree-days. *Environmental Entomology* 5, 388–396.

Allen, J.C. (1976b) A model for predicting citrus rust mite damage on Valencia orange fruit. *Environmental Entomology* 5, 1083–1088.

Allen, J.C. (1978) The effect of citrus rust mite damage on citrus fruit drop. *Journal of Economic Entomology* 71, 746–750.

Allen, J.C. (1979) Effect of citrus rust mite damage on citrus fruit growth. *Journal of Economic Entomology* 72, 195–201.

Allen, J.C. (1981) The citrus rust mite game: a simulation model of pest losses. *Environmental Entomology* 10, 171–176.

Allen, J.C. and McCoy, C.W. (1979) The thermal environment of the citrus rust mite. *Agricultural Meteorology* 20, 411–425.

Allen, J.C. and Stamper, J.H. (1979) Frequency distribution of citrus rust mite damage on citrus fruit. *Annals of the Entomological Society of America* 72, 327–330.

Allen, J.C., Yang, Y. and Knapp, J.L. (1995) Temperature effects on development and fecundity of the citrus rust mite (Acari: Eriophyidae). *Environmental Entomology* 24, 996–1004.

Allington, W.B., Staples, R. and Viehmeyer, G. (1968) Transmission of rose rosette virus by the eriophyid mite *Phyllocoptes fructiphilus*. *Journal of Economic Entomology* 61, 1137–1140.

Almaguel, L., Perez, R., Caceres, I., Feito, E. and Sanchez, Y.G. (1986) Desinfeccion de semillas agamicas de ajo con remojado previo al tratamiento quimico contra *Eriophyes* (*Aceria*) *tulipae*. *Ciencia y Técnica en la Agricultura. Protección de Plantas* 9, 57–72.

Almaguel, L., Hernández, J., de la Torre, P.E., Santos, A., Cabrera, R.I., García, A., Rivero, L.E., Báez, L., Cácerez, I. and Ginarte, A. (2000) Evaluación del comportamiento del acaro *Steneotarsonemus spinki* (Acari: Tarsonemidae) en los estúdios de regionalización desarrollados en Cuba. *Fitosanidad* 4(1–2), 15–19.

Almaguel, L., Santos, A., de la Torre, P., Botta, E., Hernández, J., Cáceres, I. and Ginarte, A. (2003) Dinámica de población e indicadores ecológicos del ácaro *Steneotarsonemus spinki* Smiley 1968 (Acari: Tarsonemidae) en arroz de riego en Cuba. *Fitosanidad* 7(1), 23–30.

Almaguel, L.R., Botta, E.F., Hernández, J. and Ginarte, A. (2005) Propuesta de manejo integrado del ácaro del vaneado del arroz para los países de la región de latinoamérica y el Caribe. In: *III Encuentro International del Arroz: El Ácaro del Arroz* Steneotarsonemus spinki *(Tarsonemidae) Retos y Alternativas para América Latina y el Caribe, Hotel Palco Ciudad de La Habana, Cuba, 6 de Junio de 2005. Libro de Resumen*, pp. 53–59.

Almaguel, L. (2002) *Curso Introductorio a la Acarología Aplicada, Honduras 30 Sept. al 4 Oct. del 2002*. Instituto de Investigaciones de Sanidad Vegetal, Havana. Available et: http://www.inisav.cu/OtrasPub/Libro%20curso%20Venezuela%202007.pdf (accessed 28 April 2015).

Almaguel, L. and Botta, E. (2005) *Manejo Integrado de* Steneotarsonemus spinki, *Smiley. Resultados de Cuba y Transferencia para la Región de Latinoamérica y el Caribe. Curso de Postgrado de Acarología, Introducción a la Acarología Agrícola. Caso Estudio:* Steneotarsonemus spinki *Smiley 1967 (Tarsonemidae)*. Instituto de Investigaciones de Sanidad Vegetal, Havana. Available at: http://www.researchgate.net/publication/235933688_Manejo_Integrado_de_Steneotarsonemus_spinki_Smiley (accessed 20 July 2014).

Al-Mallah, N.M. and Mohammad, M.A. (1989) Biology of the fig leaf mite, *Rhyncaphytoptus ficifoliae* (K). (Acariformes: Rhyncaphytoptidae). *Arab Journal of Plant Protection* 7, 23–29 (in Arabic).

Almeida, S.L., Corte, C.R., Morais, A.A., Galhardo, L.C.S., Fekete, T.J. and Mariconi, F.A.M. (1981) Defensivos quimicos e o fungo *Hirsutella thompsonii* pulverizados contra *Phyllocoptruta oleivora* (ácaro da falsa ferrugem dos citros). *O Solo (Piracicaba, Sao Paulo)* 73, 11–17.

Al-Safadi, M.M. (1987) The life cycle of the Acari *Tyrophagus similis*. *Journal of Zoology* 213, 141–146.

Al-Safadi, M.M. (1990) The effect of light on the life cycle of the Acari *Tyrophagus similis* (Astigmata: Acaridae). *Revue d'Ecologie et de Biologie du Sol* 27, 459–465.

Al-Safadi, M.M. (1991) The effect of temperature on the developmental rate of the *Tyrophagus similis* Volgin (Acari: Acaridae). *Revue d'Ecologie et de Biologie du Sol* 28, 341–347.

Alston, D.G. (1994) Effect of apple floor vegetation on density and dispersal of phytophagous and predaceous mites in Utah. *Agriculture, Ecosystems and Environment* 50, 73–84.

Alston, D.G. and Reding, M.E. (2011a) *European Red Mite (*Panonychus ulmi*)*. Utah Pests Fact Sheet ENT-152-05, Utah State University Extension and Utah Plant Pest Diagnostic Laboratory, Logan, Utah. Available at: http://extension.usu.edu/files/publications/publication/euro-red-mites97.pdf (accessed 6 January 2015).

Alston, D.G. and Reding, M.E. (2011b) *Web Spinning Spider Mites: Twospotted Spider Mite (*Tetranychus urticae*); McDaniel Spider Mite (*Tetranychus mcdanieli*)*. Utah Pests Fact Sheet ENT-151-06, Utah State University Extension and Utah Plant Pest Diagnostic Laboratory, Logan, Utah. Available at: http://extension.usu.edu/files/publications/publication/web-spinning-spider-mite97.pdf (accessed 20 July 2014).

Al-Sweedy, T.M., Al-Jboory, I.J. and Ahmad, T.R. (2006) Age-specific fecundity of old world date mite *Oligonychus afrasiaticus* (McGregor) (Acari: Tetranychidae). *Arab Journal of Plant Protection* 24, 14–19 (in Arabic).

Alvarado, D.G. and Fréitez, R.F.P. (1976) *Schizotetranychus paezi* sp.n. y *S. oryzae* (Acarina: Tetranychidae) atacando arroz en Venezuela. *Agronomía Tropical (Maracay)* 26, 159–165.

Álvarez, L.D. and Alarcón, L. (2004) Catálogo de ácaros fitófagos de la provincia de Las Tunas. *Fitosanidad* 8(1), 19–21.

Alves, E.B., Omoto, C. and Franco, C.R. (2000a) Resistencia cruzada entre o dicofol e outros acaricidas em *Brevipalpus phoenicis* (Geijskes) (Acari: Tenuipalpidae). *Anais de Sociedade Entomológica do Brasil* 29, 765–771.

Alves, E.B., Omoto, C. and Franco, C.R. (2000b) Mistura de dicofol com fempiroximato no manejo da resistencia de *Brevipalpus phoenicis* (Geijskes) (Acari: Tenuipalpidae) ao dicofol. *Anais de Sociedade Entomológica do Brasil* 29, 789–797.

Alves, E.B., Casarin, N.F.B. and Omoto, C. (2005a) Mecanismos de dispersao de *Brevipalpus phoenicis* (Geijskes) (Acari: Tenuipalpidae) em pomares de citros. *Neotropical Entomology* 34, 89–96.

Alves, L.F.A., Spongoski, S., Vieira, F.N. da S. and Moraes, G.J. de (2004) Biologia e danos de *Oligonychus yothersi* (McGregor) (Acari: Tetranychidae) em *Ilex paraguariensis*. *Arquivos do Instituto Biológico (São Paulo)* 71, 211–214.

Alves, M., Cazetta, J.O., Nunes, M.A., Oliveira, C.A.L. de and Colombi, C.A. (2007) Action of different pyroligneous extract preparations when applied on *Brevipalpus phoenicis* (Geijskes). *Revista Brasileira de Fruticultura* 29, 382–385.

Alves, S.B., Calafiori, M.H. and Paschoal, A.D. (1972) Eriofiídeos (Acarina: Eriophyidae) prejudiciais à batata – doce. *Revista de Agricultura (Piracicaba)* 47, 139–140.

Alves, S.B., Tamai, M.A., Rossi, L.S. and Castiglioni, E. (2005b) *Beauveria bassiana* pathogenicity to the citrus rust mite *Phyllocoptruta oleivora*. *Experimental and Applied Acarology* 37, 117–122.

Amalin, D.M., Vasquez, E.A. and Vanderzaag, P. (1991) Note: arthropods of sweet potato in the Philippines. *Philippine Agriculturalist* (now *Philippine Agricultural Scientist*) 74, 39–50.

Amano, H. (2001) Species structure and abundance of invertebrate natural enemies in sustainable agroecosystems. In: Shiyomi, M. and Koizumi, H. (eds) *Structure and Function in Agroecosystem Design and Management*. CRC Press, Boca Raton, Florida, pp. 167–182.

Amano, H. and Chant, D.A. (1977) Life history and reproduction of two species of predacious mites, *Phytoseiulus persimilis* Athias-Henriot and *Amblyseius andersoni* (Chant) (Acarina: Phytoseiidae). *Canadian Journal of Zoology* 55, 1978–1983.

Amano, H. and Chant, D.A. (1986) Laboratory studies on the feeding habits, reproduction and development of three phytoseiid species, *Typhlodromus pomi*, *Phytoseius macropilis* and *Amblyseius finlandicus* (Acari: Phytoseiidae), occurring on abandoned apple trees in Ontario, Canada. *Experimental and Applied Acarology* 2, 299–313.

Amano, H. and Chant, D.A. (1990) Species diversity and seasonal dynamics of Acari on abandoned apple trees in southern Ontario, Canada. *Experimental and Applied Acarology* 8, 71–96.

Ambrosi, M. and Lenarduzzi, R. (1959) Prove di lotta control "Acaro giallo" della vite. *Progresso Agricolo* 5, 769–807.

Amer, S.A.A., Refaat, A.M. and Momen, F.M. (2001) Repellent and oviposition-deterring activity of rosemary and sweet marjoram on the spider mites *Tetranychus urticae* and *Eutetranychus orientalis* (Acari: Tetranychidae). *Acta Phytopathologica et Entomologica Hungarica* 36, 155–164.

Amos, J.M., Hatton, R.G., Knight, R.C. and Massee, A.M. (1927) Experiments in the transmission of reversion disease in black currants. In: *Annual Report 1925, East Malling Research Station, II Supplement*. East Malling, UK, pp. 126–150.

Amrine, J.W. Jr (1996) *Phyllocoptes fructiphilus* and biological control of multiflora rose In: Lindquist, E.E., Sabelis, M.W. and Bruin, J. (eds) *Eriophyoid Mites: Their Biology, Natural Enemies and Control*. World Crop Pests, Volume 6. Elsevier Science, Amsterdam, pp. 741–750.

Amrine, J.W. Jr (2002) Multiflora rose. In: Van Driesche, R., Lyon, S., Blossey, B., Hoddle, M. and Reardon, R. (eds) *Biological Control of Invasive Plants in the Eastern United States*. USDA Forest Service Publication FHTET-2002-04, Washington, DC, pp, 265–292.

Amrine, J.W. Jr and Hindal, D.F. (1988) *Rose Rosette: a Fatal Disease of Multiflora Rose*. Circular 147, West Virginia University Agricultural and Forestry Experiment Station, Morgantown, West Virginia.

Amrine, J.W. Jr and Manson, D.C.M. (1996) Preparation, mounting and descriptive study of eriophyoid mites. In: Lindquist, E.E., Sabelis, M.W. and Bruin, J. (eds) *Eriophyoid Mites: Their Biology, Natural Enemies and Control*. World Crop Pests, Volume 6. Elsevier Science, Amsterdam, pp. 383–396.

Amrine, J.W. Jr and Stasny, T.A. (1994) *Catalog of Eriophyoidea (Acarina: Prostigmata) of the World*. Indira Publishing House, West Bloomfield, Michigan.

Amrine, J.W. Jr and Stasny, T.A. (2005) The genus *Epitrimerus* Nalepa, 1898 and the pear rust mite, *Epitrimerus pyri* (Nalepa, 1891) (Prostigmata: Eriophyidae). *International Journal of Acarology* 31, 137–141.

Amrine, J.W. Jr, Hindal, D.F., Stasny, T.A., Williams, R.L. and Coffman, C.C. (1988) Transmission of the rose rosette disease agent to *Rosa multiflora* by *Phyllocoptes fructiphilus* (Acari: Eriophyidae). *Entomological News* 99, 239–252.

Amrine, J.W. Jr, Hindal, D.F., Williams, R., Appel, J., Stasny, T. and Kassar, A. (1990) Rose rosette as a biocontrol of multiflora rose, 1987-1989. In: Champaign, G.A. (ed.) *Proceedings of the 43rd Annual Meeting of the Southern Weed Science Society, 15–17 January 1990*. Southern Weed Science Society, Atlanta, Georgia, pp. 316–320.

Amrine, J.W. Jr, Duncan, G.H., Jones, A.T., Gordon, S.C. and Roberts, I.M. (1994) *Cecidophyopsis* mites (Acari: Eriophyidae) on *Ribes* spp. (Grossulariaceae). *International Journal of Acarology* 20, 139–168.

Amrine, J.W. Jr, Stasny, T.A. and Flechtmann, C.H.W. (2003) *Revised Keys to World Genera of Eriophyoidea (Acari: Prostigmata)*. Indira Publishing House, West Bloomfield, Michigan.

Anand, R.K. and Chauhan, R. (1983) Simultaneous control of red pumpkin beetle (*Raphidopalpa foveicollis* Lucas), mite (*Tetranychus cucurbitae* Rahman & Sapra) and powdery mildew (*Sphaerotheca fuliginea* Schlecht) of bottle-gourd with mixtures of pesticides. *Indian Journal of Agricultural Sciences* 53, 851–854.

Ananda, N., Kotikal, Y.K. and Balikai, R.A. (2009) Sucking insect and mite pests of pomegranate and their natural enemies. *Karnataka Journal of Agricultural Sciences* 22, 781–783.

Anandkumar, V., Nagangoud, A. and Patil, B.V. (2003) Bioefficacy of insecticides and acaricides against brinjal red spider mite. In: *Proceedings of the National Symposium on Frontier Areas of Entomological Research, 5–7 November 2003, IARI, New Delhi*. Indian Agricultural Research Insitute, New Delhi, pp. 283–284.

Ananthakrishnan, N.R. (1962) Experiments on the control of tea mites in southern India (1961–1962 field trials). In: Report, Tea Scientific *Department, United Planters' Association of Southern India, 1961–62*. Coonoor, India, pp. 40–48.

Ananthakrishnan, N.R. (1963) Control of mites on tea in southern India 1962–63 trials. In: *Annual Report, Tea Scientific Department, United Planters' Association of Southern India, 1962–63*. Coonoor, India, pp. 44–54.

Ananthakrishnan, N.R. and Krishnamurthy, H.S. (1964) Field tests with some new acaricides for the control of red spider *Oligonychus coffeae* and purple mite (*Calacarus carinatus*). In: *Annual Report. Tea Scientific Department, United Planters' Association of Southern India, 1963–64*. Coonoor, India, pp. 93–95.

Andersen, P.C. and Mizell III, R.F. (1987) Impact of the peach silver mite, *Aculus cornutus* (Acari: Eriophyidae), on leaf gas exchange of 'Flordaking' and 'June Gold' peach trees. *Environmental Entomology* 16, 660–663.

Andersen, V.S. (1947) Untersuchungen über die Biologie und Bekämpfung der Obstbaumspinnmilbe *Paratetranychus pilosus* Can. et Fanz. Thesis, University of Bonn, Bonn, Germany.

Anderson, L.D. (1954) The tomato russet mite in the United States. *Journal of Economic Entomology* 47, 1001–1005.

Anderson, M.M. (1971) Resistance to gall mite (*Phytoptus ribis* Nal.) in the Eucoreosma section of *Ribes*. *Euphytica* 20, 422–426.

Anderson, N.H. and Morgan, C.V.G. (1958) Life histories of the clover mite, *Bryobia praetiosa* Koch and the brown mite, *B. arborea* Morgan and Anderson, in British Columbia. *Canadian Entomologist* 40, 23–32.

Andrade, D.J. de, Oliveira, C.A.L. de, Assumpcao, R.M. de and Morais, M.R. de (2009) Efeito da abamectina com e sem a adicao de oleo mineral sobre o acaro da ferrugem dos citros *Phyllocoptruta oleivora* (Ashmead) (Acari: Eriophydae). *Revista de Agricultura (Piracicaba)* 84, 62–70.

Andrade, D.J. de, Oliveira, C.A.L. de, Santos, N.C. dos and Morais, M.R. de (2010a) Differential toxicity of abamectin based products over *Brevipalpus phoenicis* (Geijskes, 1939) (Acari: Tenuipalpidae) mite in citrus. *Revista Brasileira de Fruticultura* 32, 82–89.

Andrade, H.B., Silveira, E.C., Reis, P.R., de Toledo, M.A. and Souza-Pimentel, G.C. (2010b) Evaluation of neem extracts and sulfur in the control of *Oligonychus ilicis* (McGregor, 1917) (Acari: Tetranychidae). In: Moraes, G.J. de, Castilho, R.C. and Flechtmann, C.H.W. (eds) *XIII International Congress of Acarology, August 23–27, 2010, Recife-PE, Brazil, Abstract Book*, abstract 21, p. 13. Available at: http://www.acarology.org/ica/ica2010/docs/0910_15_abstractBook-Capa.pdf (accessed 3 February 2015).

André, H.M. (1988) Age-dependent evolution: from theory to practice. In: Humphries C.J. (ed.) *Ontogeny and Systematics*. Columbia University Press, New York, pp. 137–187.

André, H.M. (1989) The concept of stase. In: André, H.M. and Lions J.-C. (eds) *L'Ontogenèse et le Concept de Stase chez les Arthropodes*. AGAR Publishers, Wavre, Belgium, pp. 3–16.

André, H.M. and Fain, A. (2000) Phylogeny, ontogeny and adaptative radiation in the superfamily Tydeoidea (Acari: Actinedida), with a reappraisal of morphological characters. *Zoological Journal of the Linnean Society* 139, 405–448.

André, H.M. and Remacle, C. (1984) Comparative and functional morphology of the gnathosoma of *Tetranychus urticae* (Acari: Tetranychidae). *Acarologia* 25, 179–190.

André, H.M. and Van Impe, G. (2012) The missing stase in spider mites (Acari: Tetranychidae): when the adult is not the imago. *Acarologia* 52, 3–16.

André, H.M., Noti, M.-I. and Lebrun, P. (1994) The soil fauna: the other last biotic frontier. *Biodiversity and Conservation* 3, 45–56.

André, M. (1932a) Contribution à l'étude du "Bou-Faroua" Tétranyque nuisible au dattier en Algérie. *Bulletin de la Société d'Histoire Naturelle d'Afrique du* Nord 23, 301–346.

André, M. (1932b) Note sur un acarien (*Penthaleus major* Dugès) nuisible aux plantes potagères. *Bulletin du Muséum National d'Histoire Naturelle (Série 2)* 4, 284–291.

André, M. (1933a) Sur la biologie des tétranyques tisserands. *Revue de Pathologie Végétale et d'Entomologie Agricole de France* 1, 8–25.

André, M. (1933b) Note sur un tétranyque nuisible au cotonier en Nouvelle-Calédonie. *Bulletin du Muséum National d'Histoire Naturelle (Série 2)* 5, 302–308.

André, M. (1942) L'hivernation chez les tétranyques et ses rapports avec la lutte contre ces acariens phytophages. *Bulletin du Muséum National d'Histoire Naturelle (Série 2)* 14, 57–62.

André, M. (1945) Sur les dommages causés en France aux culture de soja par l'invasion d'un tétranyque. *Compte Rendus Hebdomadaire des Séances de l'Académie d'Agriculture de France* 31, 463–464.

André, M. (1959) Note complémentaire sur *Tetranychus neocaledonicus* André. *Acarologia* 1, 53–55.

Andrewartha, H.G. and Birch, L.C. (1973) The history of insect ecology. In: Smith, R.F., Mittler, T.E. and Smith, C.N. (eds) *History of Entomology*. Annual Reviews, Palo Alto, California, pp. 229–266.

Andrews, E.A. (1915) Entomologist's tour in Cachar and the Duars. *Quarterly Journal of the Indian Tea Association* 4, 113–117.

Andrews, E.A. (1928) Red spider. *Quarterly Journal of the Indian Tea Association* 4, 206–219.

Andrews, K.L. and La Pré, L.F. (1979) Effects of Pacific spider mite on physiological processes of almond foliage. *Journal of Economic Entomology* 72, 651–654.

Andrews, K.L. and Poe, L. (1980) Spider mites of El Salvador, Central America (Acari: Tetranychidae). *Florida Entomologist* 63, 502–505.

Angeli, G., Rizzi, C., Dorigoni, A. and Ioriatti, C. (2007) Population injury levels of the apple rust mite *Aculus schlechtendali* (Nal.) on Golden Delicious and Red Delicious apple fruits. *IOBC-WPRS Bulletin* 30(4), 255–260.

Angeli, G., Rizzi, C., Giuliani, G., Tomasi, C. and Baldessari, M. (2008) Valide alternative per il controllo dell'eriofide del melo. *Informatore Agrario* 64(23), 68–71.

Angsumarn, C. (1994) *Kan Khuapkhum Rai Daeng Thurian* Eutetranychus orientalis *(K.) doi Chiwawithi*. Kasetsart University, Bangkok, Thailand (in Thai).

Angsumarn, C. and Kosol, C. (1988) *Sattru som Lae Kan Pongkan Kamchat*. Kasetsart University, Bangkok, Thailand (in Thai).

Ansaloni, T. and Perring, T.M. (2002) Biology of *Aceria guerreronis* Keifer on queen palms. In: *Program and Abstract Book, 11th International Congress of Acarology, Merida, Yucatan, Mexico*. Universidad Nacional Autonoma de Mexico, Mexico City, abstract 145.

Ansaloni, T. and Perring, T.M. (2004) Biology of *Aceria guerreronis* (Acari: Eriophyidae) on queen palm, *Syagrus romanzoffiana* (Arecaceae). *International Journal of Acarology* 30, 63–70.

Ansaloni, T., Perez Diaz, P., Aucejo Romero, S., Hurtado Ruiz, M., Gomez Cadenas, A. and Jacas Miret, J.A. (2006) Modified performance of *Tetranychus urticae* on NaCl stressed citrus plants. *IOBC-WPRS Bulletin* 29(3), 141.

Ansari, M.A. and Pawar, A.D. (1992) Biology of spider mite, *Tetranychus ludeni*, Zacher (Acari: Tetranychidae) recorded on waterhyacinth. *Plant Protection Bulletin (Faridabad)* 44(3), 28–31.

Anstead, R.D. (1911) Report on a tour in the Nilgiris. *Planters' Chronicle* 6, 189–190.

Anthony, D. (2008) *Draft Import Risk Analysis: Fresh Citrus Fruit (7 Species) from Samoa*. MAF (Ministry of Agriculture and Forestry) Biosecurity New Zealand, Wellington, New Zealand.

Anthony, M., Sattler, R. and Cooney-Sovetts, C. (1983) Morphogenetic potential of *Fraxinus ornus* under the influence of the gall mite *Aceria fraxinivora*. *Canadian Journal of Botany* 61, 1580–1594.

Anthony, M., Westphal, E. and Sattler, R. (1988) Proliférations épiphylles provoquées par l'acarien *Eriophyes cladophthrius* chez le *Solanum lycopersicum* et le *Nicandra physaloides* (Solanaceae). *Canadian Journal of Botany* 66, 1974–1985.

Antonacci, D., Tarricone, L. and Guerra, G.B. (2000) Controllo di *Calepitrymerus vitis* (Nalepa) su vitigni di uva da tavola. *Informatore Fitopatologico* 50(6), 27–33.

Antonelli, A.L. (1992) *Maple Bladder Gall Mite*. Extension Bulletin EB1711, Cooperative Extension, College of Agriculture and Home Economics, Washington State University, Pullman, Washington.

Antonious, G.F. and Snyder, J.C. (2006) Repellency and toxicity of wild tomato leaf exracts to two-spotted spider mite, *Tetranychus urticae* Koch. *Journal of Environmental Science and Health, Part B: Pesticides, Food Contaminants, and Agricultural Wastes* 41, 43–55.

Antonious, G.F., Meyer, J.E. and Synder, J.C. (2006) Toxicity and repellency of hot pepper extracts to spider mite, *Tetranychus urticae* Koch. *Journal of Environmental Science and Health, Part B: Pesticides, Food Contaminants, and Agricultural Wastes* 41, 1383–1391.

Aoki, J. (1973) Soil mites (oribatids) climbing trees. In: Daniel, M. and Rosický, B. (eds) *Proceedings of the 3rd International Congress of Acarology Held in Prague, 31 August–6 September 1971*. Dr W. Junk, The Hague, the Netherlands (reprinted by Springer), pp. 59–65.

APHIS (1979) An eriophyid mite (*Eriophyes vaga*) – California – new North American record. *Cooperative Plant Pest Report* 4(23), 427. Animal and Plant Health Inspection Service, US Department of Agriculture, Hyattsville, Maryland.

Aponte, O. and McMurtry, J.A. (1997a) Damage on 'Hass' avocado leaves, webbing and nesting behaviour of *Oligonychus perseae* (Acari: Tetranychidae). *Experimental and Applied Acarology* 21, 265–272.

Aponte, O. and McMurtry, J.A. (1997b) Biology, life table and mating behavior of *Oligonychus perseae* (Acari: Tetranychidae). *International Journal of Acarology* 23, 199–207.

Appiano, A. (1982) Cytological observations on leaves of figs infected with mosaic. *Caryologia* 35(1), 152 (In: Abstracts of Papers Presented at the 13th Congress).

AQIS (2000) Import risk analysis for the importation of bulk maize (*Zea mays* L.) from the United States of America. Australian Quarantine and Inspection Service, Canberra. Available at: http://www.agriculture.gov.au/SiteCollectionDocuments/ba/memos/2000/plant/rev_dft_ira_maize.doc (accessed 25 February 2015).

Aranda, B.R. (1969) Contribução para o conhecimento de ácaros de plantas do Paraguay. MSc. thesis, Escola Superior de Agricultura "Luiz de Queiroz", Universidade de São Paulo, Piracicaba, São Paulo, Brazil.

Aranda, B.R. (1974) Tetranychoidea (Acari) de uma area de cerrado do estado de São Paulo. PhD thesis, Escola Superior de Agricultura "Luiz de Queiroz", Universidade de São Paulo, São Paulo, Brazil.

Aranda, B.R. and Flechtmann, C.H.W. (1971) A report on the Tetranychidae of Paraguay. *Proceedings of the Entomological Society of Washington* 73, 29–33.

Aratchige, N.S., Lesna, I. and Sabelis, M.W. (2004) Belowground plant parts emit herbivore-induced volatiles: olfactory responses of a predatory mite to tulip bulbs infested by rust mites. *Experimental and Applied Acarology* 33, 21–30.

Aratchige, N.S., Sabelis, M.W. and Leśna, I. (2007) Plant structural changes due to herbivory: do changes in *Aceria*-infested coconut fruits allow predatory mites to move under the perianth? *Experimental and Applied Acarology* 43, 97–107.

Arbabi, M. and Singh, J. (2000) Studies on biological aspects of predaceous mite *Cunaxa setirostris* on *Tetranychus ludeni* at laboratory condition in Varanasi, India. *Journal of Agriculture and Rural Development* 2, 13–23.

Arbabi, M. and Singh, J. (2002) Studies on *Agistemus industani* Gonzalez-Rodriguez (Acarina: Stigmaeidae), an efficient predator of *Tetranychus ludeni* Zacker on mulberry. *Acarina* 10, 85–89.

Arbabi, M., Kamali, H., Mohsenin, A.B. and Baradaran, P. (1999) Eriophyid mites status on fruit trees of Iran. In: *Symposium, Acarological Society of India, 27–30 October, Bangalore, India*.

Arbabi, M., Asgari, M., Lotfiyan, M., Fasihi, M.T., Damghani, M.R. and Golmohammadzadeh-Khiyaban, N. (2006) *Study on Effects of Different Pesticides on Date Palm Spider Mite Population (*Oligonychus afrasiaticus *Massee) in Southern Date Palm Orchards of Iran*. Agricultural Scientific Information and Documentation Centre, Agricultural Research and Education Organization, Tehran. Available as English summary at: http://agris.fao.org/agris-search/search.do?recordID=IR2008000394 (accessed 29 August 2014).

Arbabi, M., Asgari, M., Fasihi, M.T., Golmohammadzadeh-Khiaban, N., Damghani, M.R., Latifiean, M. and Babai, M. (2010) Evaluation of water spray application for organic control of date palm spider mite *Oligonychus afrasiaticus* (McGregor) (Acari: Tetranychidae) of date palm orchards in southern parts of Iran. *Iranian Journal of Entomological Research* 1, 269–277.

Archer, T.L., Bynum, E.D. Jr and Peterson, G.C. (1986) Influence of sorghum maturity on abundance and damage potential of Banks grass mite, *Oligonychus pratensis* (Banks). *Experimental and Applied Acarology* 2, 217–222.

Argov, Y. (1993) Rearing and colonization of predatory mites from Australia to control the citrus rust mite, *Phyllocoptruta oleivora*. *Hassadeh* 73, 893–896 (in Hebrew).

Argov, Y., Amitai, S., Beattie, G.C. and Gerson, U. (2002) Rearing, release and establishment of imported predatory mites to control citrus rust mite in Israel. *BioControl* 47, 399–409.

Arillo, A., Subías, L.S. and Shtanchaeva, U. (2012) A new species of fossil oribatid mite (Acariformes, Oribatida, Trhypochthoniidae) from the Lower Cretaceous amber of San Just (Teruel Province, Spain). *Systematic and Applied Acarology* 17, 106–112.

Arjunan, G., Swamiyappan, R. and Reddy, M.V. (1989) Studies on redgram sterility mosaic disease and its vector. In: *National Seminar on Red Gram Organized by University of Agricultural Sciencec, Dharwad, and Agriculture Research Station Gulbarga, held at Gulbarga from 6–7th February 1987*. University of Agricultural. Sciences, Gulbarga, Karnataka, India, abstract.

Arnaudov, V. (2003) Efficacy of some acaricides for control of hawthorn spider mite *Tetranychus viennensis* Zacher (Acari: Tetranychidae). *Rasteniev'dni Nauki* 40, 165–169 (in Bulgarian).

Arora, P.K., Battu, R.S. and Balwinder Singh (2004) Bioefficacy of ethion against mite, *Eutetranychus orientalis* (Klein) and its residue on pear fruits. *Pesticide Research Journal* 16, 33–35.

Arriaga, J.T. (2007) Detection of the rice tarsonemid mite (*Steneotarsonemus spinki* Smiley) in Palizada, Campeche, Mexico. North American Plant Protection Organization's Phytosanitary Alert System. Available at: http://www.pestalert.org/oprDetail.cfm?oprID=268 (accessed 20 July 2014).

Arru, G.M. (1961) Un acaro del pioppo nuovo per l'Italia. *Informatore Fitopatologico* 11(1), 6–7.

Arruda, G.P. and Aquino, M.L.N. (1970) *Dois Novos Ácaros (Acarina, Eriophyidae) Atacando Fruteiras em Pernambuco*. Boletin Técnico No. 42, Instituo de Pesquisas Agronômicas, Recife, Brazil.

Arthofer, R. (1976) Spider-mites in vine cultivation. *Pflanzenarzt* 29, 73.

Arthur, A.L. and Umina, P.A. (2010) Balaustium *Mite*. Note Number AG1413, Department of Primary Industries, Melbourne, Victoria, Australia.

Arthur, A.L., Weeks, A.R., Umina, P.A. and Hoffmann, A.A. (2010) Survival and reproduction of the pest mites *Balaustium medicagoense* and *Bryobia* spp. on winter grain crops. *Experimental and Applied Acarology* 52, 141–153.

Arzone, A. (1975) L'acaro delle gemme del nocciuolo: *Phytoptus avellanae* Nal. (Acarina, Eriophyidae). Reperti biologici e prove sperimentali di lotta chimica in Piemonte. *Annali della Facolta di Scienze Agrarie della Università degli Studi di Torino* 9, 372–388.

Arzone, A. (1977) Esperimenti di lotta biologica contro *Phytoptus avellanae* Nalepa in Piemonte (Acarina, Eriophyidae). *Informatore Fitopatologico* 27(12), 29–32.

Arzone, A. (1983) Due fitomizi dannosi al nocciolo: l'acaro delle gemme e il Miride degli amenti. In: *Atti del Convegno Internazionale sul Nocciuolo, 22–24 September 1983*, *Avellino, Italy*, pp. 199–204.

Ascerno, M.E., Pfleger, F.L., Morgan, F. and Wilkins, H.F. (1983) Relationship of *Rhizoglyphus robini* (Acari: Acaridae) to root rot control in greenhouses-forced Easter lilies. *Environmental Entomology* 12, 422–425.

Ashburner, M. (1989) Mites. In: Ashburner, M., Golic, K.G. and Hawley, R.S. (eds) *Drosophila*, *A Laboratory Handbook*. Cold Spring Harbor Laboratory Press, New York, pp. 1191–1225.

Ashihara, W. (1987) Infestation and reproduction of the citrus red mite, *Panonychus citri* (McGregor) (Acarina: Tetranychidae) on leguminous plants. *Applied Entomology and Zoology* 22, 512–518.

Ashihara, W. (1996) Effect of grape cultivars and insecticide application on reproduction and occurrence of *Tetranychus kanzawai* Kishida (Acarina: Tetranychidae). *Japanese Journal of Applied Entomology and Zoology* 40, 55–59.

Ashihara, W., Inoue, K., Osakabe, M. and Hamamura, T. (1992) Effectiveness of *Phytoseiulus persimilis* Athias-Henriot (Acarina: Phytoseiidae) as a control agent for the Kanzawa spider mite, *Tetranychus kanzawai* Kishida (Acarina: Tetranychidae) and occurrence of native natural enemies of the spider mite on grapevine in glasshouse. *Bulletin of the Fruit Tree Research Station* 22, 109–130.

Ashihara, W., Kondo, A., Shibao, M., Tanaka, H., Hiehata, K. and Izumi, K. (2004) Ecology and control of eriophyid mites injurious to fruit trees in Japan. *Japan Agricultural Research Quarterly* 38, 31–41.

Ashley, E. and Manson, D.C.M. (1987) The occurrence of *Brevipalpus essigi* and *B. russulus* (Acari: Tenuiplapidae) in New Zealand. *New Zealand Entomologist* 10, 131–133.

Ashok, S. and Mahendra, S. (2001) Effect of nutrients and salinity on the incidence of *Petrobia latens* (Muller) (Prostigmata: Tetranychidae). In: Halliday, R.B., Walter, D.E., Proctor, H.C., Norton, R.A. and Colloff, M.J. (eds) *Acarology: Proceedings of the 10th International Congress of Acarology, 5–10 July 1998, Canberra*. CSIRO Publishing, Collingwood, Australia, pp. 349–351.

Askari, A. and Stern, V.M. (1972) Effect of temperature and photoperiod on *Orius tristicolor* feeding on *Tetranychus pacificus*. *Journal of Economic Entomology* 65, 132–135.

Ataide, L.M.S., Schimmel, B.C.J., Alba, J.M., Pappas, M.L., Sabelis, M.W., Pallini, A., Janssen, A.R.M. and Kant, M.R. (2012) Suppression of plant defenses by *Tetranychus evansi* can backfire during competition with other mite species. In: Schausberger, P., Walzer, A., Strodel, M.A. and Peneder, S. (eds) *7th Symposium of the EURopean Association of Acarologists (EURAAC), July 9–13 2012, Vienna, Austria: Acari in a Changing World. Program, Abstracts, Participants*. University of Natural Resources and Life Sciences, Vienna, Austria, p. 25–26. Available at: http://euraac.boku.ac.at/SympVienna/files/ProgramBook.pdf (accessed 14 January 2015).

Atanasov, N. (1971) Physiological functions of plants as affected by damage caused by *Tetranychus atlanticus* McGregor [*turkestani* (Ugar. & Nik.)]. In: Daniel, M. and Rosický, B. (eds) *Proceedings of the 3rd International Congress of Acarology Held in Prague, 31 August–6 September 1971*. Dr W. Junk, The Hague, the Netherlands (reprinted by Springer), pp. 183–186.

Atanasov, N. (1979) Experiments on biological control of *Tetranychus urticae* Koch and *Tetranychus turkestani* Ugarov et Nikolski in Bulgaria. In: Piffl, E. (ed.) *Proceedings of the 4th International Congress of Acarology, Saalfelden (Austria)*. Akadémiai Kiadó, Budapest, pp. 663–666.

Atchenson, W.C. (1953) An ecological study of three species of mites on American linden. *Journal of Economic Entomology* 46, 705.

Athias Henriot, C. (1959) Acarologie appliquée et agronomie algérienne. *Acarologia* 1, 181–200.

Athias Henriot, C. (1968) L'appareil d'insémination laelapoïde (Acariens Anactinotrichides: Laelapoidea, ♀♀). Premières observations. Possibilié d'emploi à des fins taxonomiques. *Bulletin Scientifique de Bourgogne* 25, 229–274.

Athias Henriot, C. (1969a) Contribution aux peuplements acarologiques des biotopes terrestres humides médioeuropéens: Observations sur le genre *Panteniphis* Willm. (Mésostigmates, Rhodacaride). *Revue d'Ecologie et de Biologie du Sol* 6, 69–78.

Athias Henriot, C. (1969b) Les organes cuticulaires sensoriels et glandulaires des Gamasides. Poroidotaxie et adenotaxie. *Bulletin de la Société Entomologique de France* 94, 485–492.

Athias Henriot, C. (1969c) Notes sur la morphologie externe des gamasides (Acariens Anactinotriches). *Acarologia* 11, 609–629.

Athias-Binche, F. (1994) *La Phoresie chez les Acariens. Aspects Adaptatifs et Évolutifs*. Editions du Castillet, Perpignan, France.

Atkins, E.L., Bailey, J.B., Bellows, T.S., Browner, O.L., Carman, G.E., Elmer, H.S., Ewart, W.H., Hare, J.D., Jeppson, L.R., Morse, J.G. *et al.* (1987) Insects, mites and snails. In: Bailey, J.B. and Morse, J.G. (eds) *Citrus Treatment Guide*. Publication 2903, Cooperative Extension, University of California, Division of Agriculture and Natural Resources, Oakland, California pp. 29–98.

Attiah, H.H. (1955) A new eriophyid mite on mango from Egypt (Acarina). *Bulletin of the Entomological Society of Egypt* 39, 379–383.

Attiah, H.H. (1956) The genus *Brevipalpus* in Egypt (Acarina: Tenuipalpidae). *Bulletin of the Entomological Society of Egypt* 40, 433–448.

Attiah, H.H. (1967a) Two new species of mites on figs, from Egypt. *Bulletin of the Entomological Society of Egypt* 51(1), 1–24.

Attiah, H.H. (1967b) The genus *Eutetranychus* in U.A.R., with description of three new species (Acarina: Tetranychidae). *Bulletin of the Entomological Society of Egypt* 51(11), 1–16.

Attiah, H.H. (1969) The Tetranychini of the U.A.R. 1. The genus *Tetranychus* Dufour (Acarina: Tetranychidae). *Acarologia* 11, 733–741.

Attiah, H.H. (1970) New records of eriophyid mites from Egypt. *Bulletin of the Entomological Society of Egypt* 54, 43–47.

Attiah, H.H. and Boudraux, H.B. (1964) Influence of DDT on egglaying in spider mites. *Journal of Economic Entomology* 57, 50–53.

Attiah, H.H. and Wahba, M.L. (1973) Phosphorus compounds as a cause of an increase in numbers of the flat mite *Brevipalpus californicus* (Banks), on lemon trees in Egypt. In: Daniel, M. and Rosický, B. (eds) *Proceedings of the 3rd International Congress of Acarology Held in Prague, 31 August–6 September 1971*. Dr W. Junk, The Hague, the Netherlands (reprinted by Springer), pp. 193–196.

Attiah, H.H., Wahba, M.L. and Kodirah, S.M. (1973) Chlorobenzilate, an acaricide of wide spectrum against citrus mites. In: Daniel, M. and Rosický, B. (eds) *Proceedings of the 3rd International Congress of Acarology Held in Prague, 31 August–6 September 1971*. Dr W. Junk, The Hague, the Netherlands (reprinted by Springer), pp. 639–643.

Atwa, W.A., Abdel All, H.E. and Afify, E.A. (1987) Response of *Eutetranychus anneckei* Meyer to temperature (Acari: Tetranychidae). *Annals of Agricultural Sciences, Ain Shams University* 32, 785–797.

Atyeo, W.T., Kethley, J.B. and Pérez, T.M. (1984) Paedomorphosis in *Metacheyletia* (Acari: Cheyletidae), with the description of new species. *Journal of Medical Entomology* 21, 125–131.

Aucejo, S., Foo, M., Gimeno, E., Gomez-Cadenas, A., Monfort, R., Obiol, F., Prades, E., Ramis, M., Ripolles, J.L., Tirado, V. *et al.* (2003) Management of *Tetranychus urticae* in citrus in Spain: acarofauna associated to weeds. *IOBC-WPRS Bulletin* 26(6), 213–220.

Auger, P. and Migeon, A. (2007) Les tétranyques des bambous en France. *PHM Revue Horticole* 488, 17–19.

Auger, P., Migeon, A., Ueckerman, E.A., Tiedt, L. and Navajas, M. (2013) Evidence for synonymy between *Tetranychus urticae* and *Tetranychus cinnabarinus* (Acari, Prostigmata, Tetranychidae): review and new data. *Acarologia* 53, 383–415.

Avery, D.J. (1962) The vegetative growth of young plants of brompton plum infested with fruit tree red spider mite. In: *Annual Report 1961, East Malling Research Station*. East Malling, UK, pp. 77–80.

Avery, D.J. and Briggs, J.B. (1968a) The aetiology and development of damage in young fruit trees infested with the spider mite, *Panonychus ulmi* (Koch). *Annals of Applied Biology* 61, 277–288.

Avery, D.J. and Briggs, J.B. (1968b) Damage to leaves caused by fruit tree red spider mite, *Panonychus ulmi* (Koch). *Journal of Horticultural Science* 43, 463–473.

Avidov, Z. and Harpaz, I. (1969) *Plant Pests of Israel*. Israel University Press, Jerusalem.

Avidov, Z., Blumberg, D. and Gerson, U. (1968) *Cheletogenes ornatus* (Acarina: Cheyletidae), a predator of the chaff scale on citrus in Israel. *Israel Journal of Entomology* 3, 77–94.

Ayyappath, R., Witkowski, J.F. and Higley, L.G. (1997) Ovipositional responses of two species of spider mites (Acari: Tetranychidae) to sublethal concentrations of permethrin and methyl parathion on corn. *Environmental Entomology* 26, 489–496.

Azevedo, F.R. de and Vieira, F.V. (2002) Cotton insect pests population in rainfed crop. *Revista Ciência Agronômica* 33, 15–19.

Ba-Angood, S.A. and Basshaih, G.S. (2000) A study on the effect of date palm dust mite *Oligonychus afrasiaticus* (McGregor) (Acarin: Tetranychidae) on the physiochemical characters of three different date varieties in Wadi Hadhramout, Yemen. *Arab Journal of Plant Protection* 18, 82–85.

Ba-Angood, S.A. and Basshaih, G. (2001) On the occurrence and host preference, have the date palm dust mite *Oligonychus afrasiaticus*(McG) on different date palm varieties in Wadi Hadramout – Yemen. In: *Proceedings of the 2nd International Conference on Date Palm (SICDP), March 25–27, 2001*, Al-Ain, *UAE*, pp. 381–390.

Baars, J.R. and Neser, S. (1999) Past and present initiatives on the biological control of *Lantana camara* (Verbenaceae) in South Africa. *African Entomology Memoir* No. 1, 21–33. Available at: http://www.arc.agric.za/arc-ppri/documents/baars-neser.pdf (accessed 29 April 2015).

Babaei, M., Kamali, H. and Vafaei Shoushtari, R. (2010) Faunistic study of eriophyes (Acari: Prostigmata: Eriophyoidea) in orchards of Neishaboor. *Iranian Journal of Entomological Research* 2, 165–178 (in Farsi).

Babar, M.H., Afzal, M., Bashir, M.H. and Ali, M.A. (2005) Chemical control of mushroom mite (Tarsonemidae) at different time intervals and concentrations of dicofol, abamectin and hexythiazox. *Pakistan Journal of Agricultural Sciences* 42(3–4), 41–44.

Babu, A., Selvasundaram, R., Muraleedharan, N. and Sasidhar, R. (2004) A new predator of red spider mites. *Newsletter – UPASI Tea Research Foundation (Valparai, India)* 14(1), 1.

Babu, B.S., Pandravada, S.R., Reddy, K.J., Varaprasad, K.S. and Sreekanth, M. (2002) Field screening of pepper germplasm for sources of resistance against leaf curl caused by thrips (*Scirtothrips dorsalis* Hood) and mites (*Polyphagotarsonemus latus* Banks). *Indian Journal of Plant Protection* 30, 7–12.

Baco, D., Sama, S. and Hasanuddin, A. (1991) Sugarcane pest found in Sulawesi rice fields. *International Rice Research Newsletter* 16(1), 22.

Baco, D., Nasmawati and Sama, S. (1992) Sugarcane mites *Aceria saccharium* Chan., on rice and their control. *Agrikam* 7(19), 1–6.

Bader, C. (1989) *Milben Acari*. Naturhistorisches Museum, Basel, Switzerland.

Badgley, M.A. and Fleschner, C.A. (1956) Biology of *Oligota oviformis* Casey (Coleoptera: Staphylinidae). *Annals of the Entomological Society of America* 44, 501–502.

Badii, M.H. and McMurtry, J.A. (1983) Effect of different foods on development, reproduction and survival of *Phytoseiulus longipes* (Acarina: Phytoseiidae). *Entomophaga* 28, 161–166.

Badii, M.H., McMurtry, J.A. and Johnson, H.G. (1990) Comparative life-history studies on the predaceous mites *Typhlodromus annectens* and *T. porresi* (Acari: Mesostigmata: Phytoseiidae). *Experimental and Applied Acarology* 10, 129–136.

Badii, M.H., Hernández, E., Flores, S., Holling, C.S., Rogers, D., Livdahl, T.P. and Stiven, A.E. (1993) Functional response of *Euseius mesembrinus* (Dean) a function of the density of *Brevipalpus californicus* (Banks) (Acari: Phytoseiidae, Tenuipalpidae). *Southwestern Entomologist* 18, 301–304.

Badii, M.H., Varela, S., Flores, A.E. and Landeros, J. (2003) Temperature based life history and life table parameters of Texas citrus mite on orange (Acari: Tetranychidae). *Systematic and Applied Acarology* 8, 25–38.

Badii, M.H., Cerna, E., Landeros, J. and Guevara, L.P. (2010a) Prey consumption rates of the male predator *Euseius mesembrinus* on different instars of *Oligonychus punicae*: prey density and exposure time effect. In: Moraes, G.J. de, Castilho, R.C. and Flechtmann, C.H.W. (eds) *XIII International Congress of Acarology, August 23–27, 2010, Recife-PE, Brazil, Abstract Book*, abstract 37, pp. 22–23. Available at: http://www.acarology.org/ica/ica2010/docs/0910_15_abstractBookCapa.pdf (accessed 3 February 2015).

Badii, M.H., McMurtry, J.A., Landeros, J. and Ochoa, Y. (2010b) Partial fitness of *Phytoseiulus longipes*: effect of food. In: Moraes, G.J. de, Castilho, R.C. and Flechtmann, C.H.W. (eds) *XIII International Congress of Acarology, August 23–27, 2010, Recife-PE, Brazil, Abstract Book*, abstract 42, pp. 25–26. Available at: http://www.acarology.org/ica/ica2010/docs/0910_15_abstractBookCapa.pdf (accessed 3 February 2015).

Badowska-Czubik, T., Suski, Z.W. and Mercik, T. (1977) Effect of mineral fertilizers applied to raspberries on the population density of the twospotted spider mite *Tetranychus urticae* Koch. *Fruit Science Reports* 4, 45–49.

Badowska-Czubik, T., Kruczynska, D. and Pala, E. (2002) The density of pear rust mite *Epitrimerus piri* (Nalepa) on some pear cultivars in Central Poland. *Acta Horticulturae* 596, 575–578.

Bagdasarian, A.T. (1948) Novyj vid nautinnogo klesca iz Armenii. *Doklady Akademii Nauk Armenia SSR (Erevan)* 9, 129–130 (in Russian).

Bagdasarian, A.T. (1957) *Tetranychoid Mites (Superfamily Tetranychoidea). Fauna of the Armenian S.S.R.* Zoologii Institut, Akademia Nauk Armenia SSR, Erevan (in Russian).

Bagdasarian, A.T. (1970) The tetrapod mites of stone fruit trees of Armenia (Acarina: Eriophyidae). *Zoolosicheskii Sbornik (Erevan)* No. 15, 138–149 (in Russian).

Bagdasarian, A.T. (1972) Eriofiidnyie klesci na fistachke (Eriophyoidea). *Doklady Akademii Nauk Armenia SSR (Erevan)* 15, 123–128 (in Russian).

Baggiolini, M., Guignard, E., Hugi, H. and Epard, S. (1969) Contribution à la connaissance de la biologie de l'érinose de la vigne et nouvelles possibilités de lutte. *Revue Suisse de Viticulture Arboriculture Horticulture* 1, 50–52.

Baggiolini, M., Descoins, C., Baillod, M., Touzeau, J., Simon, J. L. and Schmid, A. (1976) La lutte integrée en viticulture. *Revue Suisse de Viticulture Arboriculture Horticulture* 8, 147–160.

Bagnyuk, I.G. (1984) Novii chetirekhnogii kleshch (Acarina, Tetrapodili) iz pochek pikhti sibirskoi. *Zoologicheskii Zhurnal* 63, 373–301 (in Russian).

Bahadroglu, C. and Avgin, S. (2003) Experiential infestation and seasonal variations in density of *Eriophyes vitis* Pgst. (Acarina: Eriophyidae) in Turkey. Consequences of earliest treatment. *Acarologia* 43, 261–265.

Bai, L.X. and Carey, J.R. (1988) Laboratory studies on demography of Pacific spider mite (*T. pacificus*) on cotton leaf. *Jiangsu Journal of Agricultural Sciences* 4(3), 29–37 (in Chinese).

Bailey, J.S. and Bourne, A.I. (1946) The control of the blueberry bud mite. *Journal of Economic Entomology* 39, 89.

Bailey, S.F. and Keifer, H.H. (1943) The tomato russet mite, *Phyllocoptes destructor* Keifer: its present status. *Journal of Economic Entomology* 36, 702–712.

Baillod, M. and Fiaux, G. (1975) *Description d'une Méthode Simplifiée pour le Contrôle des Oeufs Hivernants de l'Acarien Rouge des Arbres Fruitiers (*P. ulmi *Koch) en Verger de Pommier.* Rapport Interne, Station Fédérale de Recherches Agronomiques de Changins, Switzerland.

Baillod, M. and Guignard, E. (1986) Noveaux dégâts de l'acariose bronzée et du court-noué parasitaire dus à *Calepitrimerus vitis* (Nalepa) (Acari, Eriophyidae) en 1984 et 1985. *Revue Suisse de Viticulture Arboriculture Horticulture* 18, 285–288.

Baillod, M. and Schlaepfer, R. (1982) Technique simplifiée de contrôle pour l'acarien rouge (*P. ulmi* Koch) et les vers de la grape (1re génération). *Revue Suisse de Viticulture Arboriculture Horticulture* 14, 211–215.

Baillod, M., Bassino, J.P. and Piganeau, P. (1979) L'estimation du risque provoqué par l'acarien rouge (*Panonychus ulmi* Koch) et l'acarien des charmilles (*Eotetranychus carpini* Oud.) en viticolture. *Revue Suisse de Viticulture Arboriculture Horticulture* 11, 123–130.

Baillod, M., Guignard, E., Mischler, M. and Vautier, P. (1989a) Controle des oeufs de l'acarien rouge (*Panonychus ulmi* Koch) sur le bois de taille en viticulture et prevision du risque. *Revue Suisse de Viticulture Arboriculture Horticulture* 21, 7–14.

Baillod, M., Staubli, A. and Hohn, H. (1989b) Principaux acariens des petits fruits. *Revue Suisse de Viticulture Arboriculture Horticulture* 21, 103–104.

Baillod, M., Antonin, P. and Mittaz, C. (1989c) Migrations, estimation des populations et nuisibilité de l'acarien jaune commun, *Tetranychus urticae* Koch dans la viticulture valaisanne. *Revue Suisse de Viticulture Arboriculture Horticulture* 21, 179–183.

Baillod, M., Antonin, P., Guignard, E. and Jermini, M. (1989d) Vers une généralisation de la lutte biologique contre les acariens phytophages en vergers de pommiers. *Revue Suisse de Viticulture Arboriculture Horticulture* 21, 279–284.

Baillod, M., Oppikofer, A. and Antonin, P. (1991) Roussissure des poires causeée par l'eriophyide libre du poirier, *Epitrimerus pyri* (Nalepa), en Valais. *Revue Suisse de Viticulture Arboriculture Horticulture* 23, 87–92.

Bailly, X., Migeon, A. and Navajas, M. (2004) Analysis of microsatellite variation in the spider mite pest *Tetranychus turkestani* (Acari: Tetranychidae) reveals population genetic structure and raises questions about related ecological factors. *Biological Journal of the Linnean Society* 82, 69–78.

Bainotti, C. and Vanzetti, L. (2008) Perspectivas para el mejoramiento de trigo en la búsqueda de resistencia a *Wheat streak mosaic virus* (WSMV) en américa del Sur. In: *Libro de Resúmenes. Workshop en Virosis de Cereales Transmitidas por el Ácaro* Aceria tosichella – *Wheat streak mosaic virus y High plain virus – en los Países del Cono Sur: Situación, Detección y Manejo, Balcarce, Argentina.* Instituto de Fitopatología y Fisiología Vegetal (IFFIVE), Instituto Nacional de Tecnología Agropecuaria (INTA), Buenos Aires, p. 9.

Bajwa, W.I., Krantz, G.W. and Kogan, M. (2001) Discovery of *Cenopalpus pulcher* (C. & F.) (Acari: Tenuipalpidae) in the New World. *Proceedings of the Entomological Society of Washington* 103, 754–756.

Baker, A.S. (1990) A survey of external morphology of mites of the superfamily Eupodoidea Banks, 1894 (Acari: Acariformes). *Journal of Natural History* 24, 1227–1261.

Baker, A.S. (1995) A redescription of *Halotydeus destructor* (Tucker) (Prostigmata: Penthaleidae), with a survey of ontogenetic setal development in the superfamily Eupodoidea. *International Journal of Acarology* 21, 261–282.

Baker, A.S. and Beron, P. (2004) Penthaleidae. In: Magowski, W.Ł. (ed.) Acari: Acariformes: Fauna Europaea, version 2.6.2. Available at: http://www.faunaeur.org/full_results.php?id=15763 (accessed 30 April 2015).

Baker, E.W. (1939) The fig mite, *Eriophyes ficus* Cotte, and other mites of the fig tree, *Ficus carica* L. *Bulletin of the California Department of Agriculture* 28(4), 266–275.

Baker, E.W. (1945) Mites of the genus *Tenuipalpus* (Acarina: Trichadenidae). *Proceedings of the Entomological Society of Washington* 47, 33–38.

Baker, E.W. (1949) The Genus *Brevipalpus* (Acarina: Pseudoleptidae). *American Midland Naturalist* 42, 350–402.

Baker, E.W. (1962) *Tetranychus sinhai*, a new species of mite from grasses (Acari: Tetranychidae). *Canadian Entomologist* 94, 956–959.

Baker, E.W. (1968) The genus *Pronematus* Canestrini. *Annals of the Entomological Society of America* 61, 1091–1097.

Baker, E.W. (1975a) Spider mites (Tetranychidae: Acarina) from Southeast Asia and Japan. *US Department of Agriculture Cooperative Economic Insect Report* 25(49–52), 911–921.

Baker, E.W. (1975b) *Plant-feeding Mites of Thailand (Tetranychidae, Tenuipalpidae, and Tuckerellidae)*. Plant Protection Service Technical Bulletin No. 35, Ministry of Agriculture and Co-operatives, Bangkok.

Baker, E.W. (1979a) A note on paedogenesis in *Brevipalpus* sp. (Acari: Tenuipalpidae), the first such record for a mite. *International Journal of Acarology* 5, 355–356.

Baker, E.W. and Pritchard, A.E. (1953a) The family categories of tetranychoid mites, with a review of the new families Linotetranidae and Tuckerellidae. *Annals of the Entomological Society of America* 46, 243–258.

Baker, E.W. and Pritchard, A.E. (1953b) A guide to the spider mites of cotton. *Hilgardia* 22(7), 203–234.

Baker, E.W. and Pritchard, A.E. (1960) The tetranychoid mites of Africa. *Hilgardia* 29(11), 455–574.

Baker, E.W. and Pritchard, A.E. (1962) Arañas rojas de América Central (Acarina: Tetranychidae). *Revista de la Sociedad Mexicana de Historia Natural* 23, 309–340.

Baker, E.W. and Tuttle, D.M. (1972) *New Species and Further Notes on the Tetranychoidea Mostly from the South-western United States (Acarina: Tetranychidae and Tenuipalpidae)*. Smithsonian Contributions to Zoology 116, Smithsonian Institution Press, Washington, DC.

Baker, E.W. and Tuttle, D.M. (1975) A new species of *Tuckerella* (Acarina: Tuckerellidae) from Thailand. *US Department of Agriculture Cooperative Economic Insect Report* 25(17), 337–340.

Baker, E.W. and Tuttle, D.M. (1987) *The False Spider Mites of Mexico (Tenuipalpidae: Acari)*. Technical Bulletin No. 1706, US Department of Agriculture Agricultural Research Service, Washington, DC.

Baker, E.W. and Tuttle, D.M. (1994) *A Guide to the Spider Mites (Tetranychidae) of the United States*. Indira Publishing House, West Bloomfield, Michigan.

Baker, E.W. and Wharton, G.W. (1952) *An Introduction to Acarology*. Macmillan, New York.

Baker, E.W., Camin, J.H., Cunliffe, F., Woolley, T.A. and Yunker, C.E. (1958) *Guide to the Families of Mites*. Contribution No. 3, Department of Zoology, Institute of Acarology, University of Maryland, College Park, Maryland.

Baker, E.W., Tuttle, D.M. and Abbatiello, M.J. (1975) *The False Spider Mites of Northwestern and Central Mexico (Acarina: Tenuipalpidae)*. Smithsonian Contributions to Zoology 194, Smithsonian Institution Press, Washington, DC.

Baker, G.T. (1983) Observations on the biology of *Rhizoglyphus robini* Claparède (Acaridae). *Zeitschrift für Angewandte Entomologie* 95, 162–166.

Baker, J.E. and Connell, W.A. (1961) Mites on soybeans in Delaware. *Journal of Economic Entomology* 54, 1024–1026.

Baker, J.R. (1997) *Cyclamen Mite and Broad Mite*. Ornamentals and Turf, Department of Entomology Insect Note, North Carolina Cooperative Extension, North Carolina State University, Raleigh, North Carolina. Available at: http://www.ces.ncsu.edu/depts/ent/notes/O&T/flowers/note28/note28.html (accessed 21 July 2014).

Baker, J.R. and Neunzig, H.N. (1968) *Hirsutella thompsonii* as a fungus parasite of the blue-berry bud mite. *Journal of Economic Entomology* 61, 1117–1118.

Baker, J.R. and Neunzig, H.H. (1970) Biology of the blueberry bud mite. *Journal of Economic Entomology* 63, 74–79.

Baker, R.A. (1964a) The further development of the hypopus of *Histiostoma feroniarum* (Dufour, 1839) (Acari). *The Annals and Magazine of Natural History (Series 13)* 7(84), 693–695.

Baker, R.A. (1964b) Some preliminary observations on the mite fauna of sewage bacteria beds. *The Annals and Magazine of Natural History (Series 13)* 7(84), 707–709.

Baker, R.T. (1979b) Insecticide resistance in the peach silver mite *Aculus cornutus* (Banks) (Acari: Eriophyidae). *New Zealand Journal of Experimental Agriculture* 7, 405–406.

Balaji, K. and Hemavathy, A.T. (2007) Studies on the seasonal incidence of coconut eriophyid mite *Aceria guerreronis* Keifer (Acari: Eriophyidae). *Asian Journal of Bio Science* 2, 88–91.

Bałazy, S., Mietkiewski, R., Tkaczuk, C., Wegensteiner, R. and Wrzosek, M. (2008) Diversity of acaropathogenic fungi in Poland and other European countries. *Experimental and Applied Acarology* 46, 53–70.

Baldwin, B., Gilchrist, K. and Snare, L. (2007) *Hazelnut Variety Assessment for South-eastern Australia*. Available at: https://rirdc.infoservices.com.au/items/07-062 (accessed 30 April 2015).

Balevski, A. (1960) Experiments to determine the effect of DDT on fecundity of the brown apple mite (*Bryobia redikorzevi* Reck). *Rastitelna Zashtita* 8(6), 59–64 (in Bulgarian).

Balevski, A. (1967) *Tetranychid Mites Occurring on Fruit Crops*. Institut za Zastita na Rastenijata, Gara Kostinbrod, Bulgaria. Izdat. Bulgar. Akad. Nauk (Publishing House of the Bulgarian Academy of Sciences), Sofia (in Bulgarian).

Balevski, A. (1980) How we control the second generation of the codling moth and the yellow apple mite on vines (Bulg.). *Rastitelna Zashtita* 28, 42–43 (in Bulgarian).

Balevski, A.D., Natchev, P. and Simova, S. (1982) Acari of cultivated plants. *Bulletin Lenin* 47, 1–242, Zemizdat, Sofia (in Bulgarian).

Ballard, R.C. (1954) The biology of the predacious mite *Typhlodromus fallacis* (Gar.) (Acarina: Phytoseiidae) at 78°F. *Ohio Journal of Science* 54, 175–179.

Balogh, J. (1972) *The Oribatid Genera of the World*. Akademiai Kiado, Budapest.

Bambara, S. (2003) Clover mites in home lawns. North Carolina Department of Entomology Insect Note ENT/ort-124, North Carolina Cooperative Extension Service, North Carolina State University at Raleigh and North Carolina A&T State University at Greensboro, North Carolina. Available at: http://www.ces.ncsu.edu/depts/ent/notes/O&T/lawn/note124/note124.html (accessed 21 July 2014).

Bameul, F. (1990) Le DMHF: Un excellent milieu de montage en entomologie. *L'Entomologie* 46, 233–239.

Bancroft, J. (1877) Causes of diseases affecting live stock and plants. In: *Votes and Proceedings of the Legislative Assembly of Queensland, Annual Report 2, Volume 3*, pp. 1048–1049.

Banerjee, B. (1969) Shade and manurial experiments at Murmuria: a report on the incidence of mites. *Two and a Bud* 16, 93–96.

Banerjee, B. (1971) The threshold values in mite pest control. *Two and a Bud* 18, 20–21.

Banerjee, B. (1978) Recognition and control of mite pests. *Two and a Bud* 25, 5–7.

Banerjee, B. (1979) The mating speed in *Oligonychus gossypii* (Zacher) (Tetranychidae: Acarina). *International Journal of Invertebrate Reproduction and Development* 1, 201–204.

Banerjee, B. and Cranham, J.E. (1985) Tea. In: Helle, W. and Sabelis, M.W. (eds) *Spider Mites: Their Biology, Natural Enemies and Control.* World Crop Pests, Volume 1B. Elsevier Science, Amsterdam, pp. 371–374.

Banks, N. (1894) Some new american Acarina. *Transactions of the American Entomological Society* 21, 209–222.

Banks, N. (1900) The red spiders of the United States (Tetranychus and Stigmaeus). *US Department of Agriculture, Division of Entomology, Technical Series No.* 8, 65–77.

Banks, N. (1901) *Tarsonemus* in America. *Proceedings of the Entomological Society of Washington* 4, 294–296.

Banks, N. (1904a) A treatise on the Acarina, or mites. *Proceedings of the US National Museum* 28(1382), pp. 1–114.

Banks, N. (1904b) Class III, Arachnida, Order 1, Acarina, four new species of injurious mites. *Journal of the New York Entomological Society* 12, 53–56.

Banks, N. (1908) A new *Tetranychus*. *Proceedings of the Entomological Society of Washington* 10, 36.

Banks, N. (1912) New American Mites. *Proceedings of the Entomological Society of Washington* 14, 96–99.

Banks, N. (1914) New Acarina. *Journal of Entomology and Zoology, Pomona College* 6, 55–66.

Banks, N. (1915) *The Acarina or Mites.* US Department of Agriculture Report 108, Washington, DC.

Banks, N. (1917) New mites, mostly economic (Arach., Acar.). *Entomological News* 28, 193–199.

Banu, H., Sultana, K., Islam, M.N., Polan, M.S. and Haque, S.M.A. (2007a) Use of neem leaf (*Azadirachta indica* A. Juss) for controlling jute yellow mite, *Polyphagotarsonemus latus* (Banks). *International Journal of Sustainable Agriculture and Technology* 3(5), 7–9.

Banu, H., Sultana, K., Islam, M.N., Polan, M.S. and Kamruzzaman, A.S.M. (2007b) Evaluation of some acaricides for controlling jute yellow mite *Polyphagotarsonemus latus* (Bank). *International Journal of Sustainable Agriculture and Technology* 3(5), 15–17.

Banu, H., Islam, M.N., Kamruzzaman, A.S.M., Polan, M.S. and Haque, S.M.A. (2007c) Effect of turmeric powder and garlic paste extract against jute yellow mite, *Polyphagotarsonemus latus* (Banks). *International Journal of Sustainable Agriculture and Technology* 3(5), 23–25.

Banu, K. and ChannaBasavanna, G.P. (1972) Plant feeding mites of India I. A preliminary account of the biology of the spider mite *Eutetranychus orientalis* (Klein) (Acarina: Tetranychidae). *Mysore Journal of Agricultural Sciences* 6, 253–268.

Baptist, B.A. and Ranawerra, D.J.W. (1955) The scarlet mites of the genus *Brevipalpus* as pests of tea in Ceylon. *Tea Quarterly* 26, 127–136.

Baradaran, P. and Arbabi, M. (2005) First record of the eriophyid, *Trisetacus juniperinus* on *Cupressus sempervirens* in Teheran. *Letters of the Entomological Society of Iran* 26, 1.

Baradaran, P., Arbabi, M. and Ranjbar, V. (2002) Comparative population fluctuation of fig spider mite (*Eotetranychus hirsti*) on fig varieties in Saveh Region. *Journal of Entomological Society of Iran* 22(1), 49–61.

Baradaran Anaraki, P. and Daneshvar, H. (1992) Studies on biology and chemical control of tomato russet mite, *Aculops lycopersici* (Acari: Eriophyidae), in Varamin. *Applied Entomology and Phytopathology* 59(1–2), 63–78 (in Persian).

Barajas, B.R. (1998) Control biologico del "acaro de la yema del mango" *Eriophyes mangiferae* Sayed (Eriophyidae: Prostigmata), alternativa para combatir la "escoba de bruja". In: *Proceedings of the National Entomological Congress 33, Acapulco, 24–27 May 1998.* Sociedad Mexicana de Entomologia, Mexico, pp. 249–253.

Barbar, Z., Tixier, M.-S., Cheval, B. and Kreiter, S. (2010) Does agroforestry affect phytoseiid mite communities in vineyards in the South of France? In: Sabelis, M.W. and Bruin, J. (eds) *Trends in Acarology: Proceedings of the 12th International Congress, 21–26 August 2006*, Amsterdam. Springer, Dordrecht, the Netherlands, pp. 409–412.

Barbosa, D.G.F., Gondim Junior, M.G.C., Barros, R. and Oliveira, J.V. de (2003) Diversidade de ácaros em aceroleira (*Malpighia emarginata* A. DC.) na Universidade Federal de Pernanmbuco em Recife, PE. *Neotropical Entomology* 32, 577–583.

Barbosa, D.G.F., Gondim, M.G.C.J., Barros, R. Oliveira, J.V. de and Silva, F.R. da (2004) Compared biology of *Eutetranychus banksi* (McGregor) (Acari: Tetranychidae) on tropical fruits species. *Neotropical Entomology* 33, 403–406.

Bardia y Bardia, R. (1958) Los "badochs" del avellano. *Boletín Agro-pecuario (Obra Social Agrícola de la Caja de Pensiones para la Vejez y de Ahorros de Cataluña y Baleares)* 38, 120–123.

Barindra Talukdar, Somchoudhury, A.K., Ghosh, P. and Sarkar, P.K. (2008a) Laboratory evaluation of buprofezin: a thiadizine group chitin synthesis inhibitor against tea red spider mite *Oligonychus coffeae* (Niet.) (Acari: Tetranychidae). *Environment and Ecology* 26, 148–151.

Barindra Talukdar, Somchoudhury, A.K. and Sarkar, P.K. (2008b) Relative toxicity of three newer molecules and two conventional acaricides against tea red spider mite *Oligonychus coffeae* (Niet.) (Tetranychidae: Prostigmata). *Environment and Ecology* 26, 152–154.

Barké, H.E. and Davis, R. (1971) Some eriophyid mites occurring in Georgia with descriptions of four new species. *Journal of the Georgia Entomological Society* 6, 157–169.

Barké, H.E., Davis, R. and Hunter, P.E. (1972) Studies on the peach silver mite, *Aculus cornutus* (Acarina: Eriophyidae). *Journal of the Georgia Entomological Society* 7, 171–178.

Barker, P.S. (1991) Bionomics of *Cheyletus eruditus* (Schrank) (Acarina: Cheyletidae), a predator of *Lepidoglyphus destructor* (Schrank) (Acarina: Glycyphagidae), at three constant temperatures. *Canadian Journal of Zoology* 69, 2321–2325.

Barnes, M.M. (1958) Relationships among pruning time response, symptoms attributed to grape bud mite, and temporary early season boron deficiency in grapes. *Hilgardia* 28(7), 193–224.

Barnes, M.M. (1970) *Calepitrimerus vitis* (Acarina: Eriophyidae) on grape leaves. *Annals of the Entomological Society of America* 63, 1193–1194.

Barnes, M.M. (1992) Grape erineum mite. In: Flaherty, D.L., Christensen, L.P., Lanini, W.T., Harois, J.J., Philips, P.A. and Wilson, L.T. (eds) *Grape Pest Management*, 2nd edn. University of California, Division of Agriculture and Natural Resources Publication No. 3343, Oakland, California, pp. 262–264.

Barnes, M.M. and Andrews, K.L. (1978) Effects of spider mites on almond tree growth and productivity. *Journal of Economic Entomology* 71, 555–558.

Barquero, M. (2004) Millonaria pérdida en arroz. *La Nación Lunes*, 18 October 2004, Caderno 2 Econ, p. 2. Ed. Lidiette Brenes, San José, Costa Rica.

Barrion, A.T. and Corpuz Raros, L.A. (1975) Studies on citrus mites (Acarina): biology of *Eotetranychus cendanai* Rimando (Tetranychidae) and population trends in *Brevipalpus obovatus* Donnadieu (Tenuipalpidae). *The Philippine Entomologist* 3, 30–45.

Barrios, J.R. (1972) *Reacción de 25 Variedades do Yuca* Manihot esculenta, *al Ataque de Ácaros*. Instituto de Agronomia, Universidad Central de Venezuela, Maracay, Venezuela.

Bartkowski, J., Odindo, M.O. and Otieno, W.A. (1988) Some fungal pathogens of the cassava green spider mite *Mononychellus* spp. (Tetranychidae) in Kenya. *Insect Science and Its Application* (now *International Journal of Tropical Insect Science*) 9, 457–459.

Bartlett, B.R. (1951) The action of certain "inert" dust materials on parasitic Hymenoptera. *Journal of Economic Entomology* 46, 891–896.

Bashkatova, T.V., Prokof'ev, M.A., Ogneva, L.I. and Shamanskaya, L.D. (1983) Tested in orchards in Siberia. *Zashchita Rasteniĭ* No. 8, 1–30 (in Russian).

Baskerville, G.L. and Emin, P. (1969) Rapid estimation of heat accumulation from maximum and minimum temperatures. *Ecology* 50, 514–517.

Bass, J.A. and Hays, S.B. (1976) Predation by the mite, *Tyrophagus putrescentiae* on eggs of the imported fire ant. *Journal of the Georgia Entomological Society* 11, 16.

Bassanezi, R.B. and Laranjeira, F.F. (2007) Spatial patterns of leprosis and its mite vector in commercial citrus groves in Brazil. *Plant Pathology* 56, 97–106.

Bassino, J.P., Blanc, M. and Choppin de Janvry, E. (1973) Estimation rapide du risque que represente l'acarien rouge *Panonychus ulmi* Koch en vergers de pommiers dans une perspective de strategie de lutte. *La Défense des Végétaux* 163, 215–228.

Bastianel, M., Oliveira, A.C. de, Cristofani, M., Freitas Astua, J., Rodrigues, V., Arrivabem, F. and Machado, M.A. (2005) Melhoramento para resistencia a leprose dos citros. *Laranja* 26(2), 387–397.

Bastianel, M., Oliveira, A.C. de, Cristofani, M., Guerriero Filho, O., Freitas Astua, J., Rodrigues, V., Astua Monge, G. and Machado, M.A. (2006) Inheritance and heritability of resistance to citrus leprosis. *Phytopathology* 96, 1092–1096.

Bastianel, M., Novelli, V.M., Kitajima, E.W., Kubo, K.S., Bassanezi, R.B., Machado, M.A. and Freitas-Astúa, J. (2010) Citrus leprosis: centennial of an unusual mite-virus pathosystem. *Plant Disease* 94, 284–292.

Batchelor, G.S. (1952) *Eriophyid Mites of Washington*. Bulletin No. 6, Washington State Agricultural Experiment Station, Pullman, Washington.

Baten, W. and Hutson, R. (1943) Precautions necessary in estimating populations of small animals. *Journal of Economic Entomology* 36, 501–504.

Baum, D. (1984) Control of the grape-bud-mite, *Eriophyes vitis* (Pgst.) in vineyards. *Alon Hanotea* 38, 289–293 (in Hebrew).

Bayan, A. (1988) Mites on plum trees in Lebanon. I. A general survey and a diagnostic key. *Arab Journal of Plant Protection* 6, 1–6 (in Arabic).

Bayan, A. (1989) Mites on plums in Lebanon. II. Fluctuations of their population density and abundance on plum trees in Baalbeck (Bekaa). *Arab Journal of Plant Protection* 7, 30–36 (in Arabic).

Beament, J.W.L. (1951) The structure and formation of the egg of the fruit tree spider mite, *Metatetranychus ulmi* Koch. *Annals of Applied Biology* 38, 1–24.

Beard, J. (2007) FW: Fostoxin. Personal communication to R. Ochoa on August 14, 2007, from J. Beard (Australian Quarantine and Inspection Service). Archived at the PERAL Library, Raleigh, North Carolina.

Beard, J. and Ochoa, R. (2010) Ontogenetic modification in the Tuckerellidae (Acari: Tetranychoidea). *International Journal of Acarology* 36, 169–173.

Beard, J.J. and Walter, D.E. (2005) A new *Tuckerella* (Prostigmata: Tetranychoidea; Tuckerellidae) from Australia with description of all stages and a discussion of the tritonymphal stage. *Acarologia* 45, 49–60.

Beard, J.J., Walter, D.E. and Allsopp, P.G. (2003) Spider mites of sugarcane in Australia: a review of grass-feeding Oligonychus Berlese (Acari: Prostigmata: Tetranychidae). *Australian Journal of Entomology* 42, 51–78.

Beard, J.J., Ochoa, R., Bauchan, G.R., Welbourn, W.C., Pooley, C. and Dowling, A.P.G. (2012) External mouthpart morphology in the Tenuipalpidae (Tetranychoidea): *Raoiella* a case study. *Experimental and Applied Acarology* 57, 227–255.

Beard, J.J., Ochoa, R., Bauchan, G.R., Trice, M.D., Redford, A.J., Walters, T.W. and Mitter, C. (2013) Flat Mites of the World, Edition 2. University of Maryland (UMD) and Identification Technology Program, Center for Plant Health Science and Technology (CPHST), Plant Health and Quarantine (PPQ), Animal and Plant Health Inspection Service (APHIS), US Department of Agriculture, Fort Collins, Colorado. Available at: http://idtools.org/id/mites/flatmites/ (accessed 21 July 2014).

Beattie, G.A.C. (1978) Biological control of citrus mites in New South Wales. In: Larey, P.R. (ed.) *Proceedings of 3rd International Society of Citriculture, Sydney, Australia, 1978*. International Society of Citriculture, Riverside, California, pp. 156–158.

Beattie, G.A.C. and Gellatley, J.G. (1983) *Mite Pests of Citrus*. Agfacts H2.AE 3, Department of Agriculture, New South Wales (now NSW Department of Primary Industries. Orange, New South Wales, Australia).

Beattie, G.A.C., Roberts, E.A., Vanhoff, C.L. and Flack, L.K. (1991) Effects of climate, natural enemies and biocides on three citrus mites in coastal New South Wales. *Experimental and Applied Acarology* 11, 271–295.

Beaulieu, F. and Weeks, A.R. (2007) Free-living mesostigmatic mites in Australia: their roles in biological control and bioindication. *Australian Journal of Experimental Agriculture* 47, 460–478.

Beavers, J.B. and Oldfield, G.N. (1970) Portable platforms for watering leaves in acrylic cages containing leaf feeding arthopods. *Journal of Economic Entomology* 63, 312–313.

Beber, K. (1994) Studies of big bud mite (*Phytoptus avellanae* Nal.) in northeastern Slovenia. *Acta Horticulturae* 351, 625–629.

Becerra, V.C., Miano, J.L., Herrera, M.E. and Dagatti, C.V. (2007) Ensayo control del "eriófido del olivo" *Aceria oleae* N. y *Oxycenus maxwelli* K. *Horticultura Argentina* 26, 61.

Beccari, F. (1971) Contributo alla conoscenza dell'entomofauna dell'Arabia Saudita. Primo elenco di insetti, acari e nematodi. *Rivista di Agricoltura Subtropicale e Tropicale* 65, 243–258.

Beck, S.D. (1980) *Insect Photoperiodism*, 2nd edn. Academic Press, New York.

Bedford., E.C.G. (1968) Biological and chemical control of citrus pests in the Western Transvaal. An integrated spray programme. *South African Citrus Journal* 417, 9, 11, 13, 15, 17, 19, 21–28.

Bedford, E.C.G. (1998) Citrus grey mite, *Calacarus citrifolii* Keifer and concentric ring blotch. In: Bedford, E.C.G. (ed.) *Citrus Pests in the Republic of South Africa*. Science Bulletin No. 391, Department of Agricultural Technical Services, Republic of South Africa. Government Printer, Pretoria, pp. 50–53.

Beer, R.E. (1954) A revision of the Tarsonemidae of the western hemisphere (Order Acarina). *Kansas University Science Bulletin* 36, 1091–1387.

Beer, R.E. (1963) Social parasitism in the Tarsonemidae, with description of a new species of tarsonemid mite involved. *Annals of the Entomological Society of America* 56, 153–160.

Beer, R.E. and Lang, D.S. (1958) The Tetranychidae of Mexico *Kansas University Science Bulletin* 38, 1231–1259.

Beer, R.E. and Nucifora, A. (1965) Revisione dei generi della famiglia Tarsonemidae (Acarina). *Bollettino di Zoologia Agraria e di Bachicoltura (Ser. II)* 7, 19–43.

Beerling, E.A.M. and Van der Geest, L.P.S. (1991a) Microsporidiosis in mass-rearings of the predatory mites *Amblyseius cucumeris* and *A. barkeri* (Acarina: Phytoseiidae). *Proceedings of the Section Experimental and Applied Entomology of the Netherlands Entomological Society* 2, 157–162.

Beerling, E.A.M. and Van der Geest, L.P.S. (1991b) A microsporidium (Microspora: Pleistophoridae) in mass-rearings of the predatory mites *Amblyseius cucumeris* and *A. barkeri* (Acarina: Phytoseiidae): analysis of a problem. *IOBC-WPRS Bulletin* 14(7), 5–8.

Beers, E.H. and Hoyt, S.C. (1993) European red mite. Orchard Pest Management online. Washington Sate Univeristy. Available at: http://jenny.tfrec.wsu.edu/opm/displaySpecies.php?pn=290 (accessed 21 July 2014).

Beers, E.H. and Hull, L.A. (1990) Timing of mite injury affects the bloom and fruit development of apple. *Journal of Economic Entomology* 85, 547–551.

Beers, E.H., Riedl, H. and Dunley, J.E. (1998) Resistance to abamectin and reversion to susceptibility to fenbutation oxide in spider mite (Acari: Tetranychidae) populations in the Pasicif Northwest, *Journal of Economic Entomology* 91, 352–360.

Beevi, S.P., Beena, S., Lyla, K.R., Varma, A.S., Mathew, M.P. and Nadarajan, L. (1999) *Hirsutella thompsonii* var. *Synnematosa* Samson, McCoy & O'Donnell on coconut mite *Aceria* (*Eriophyes*) *guerreronis* (Keifer) – a new report from India. *Journal of Tropical Agriculture* 37, 91–93.

Beglyarov, G.A. (1959) On the bionomics of the hawthorne mite *Tetranychus crataegi* Hirst. *Entomologicheskoe Obozrenie (Entomological Review, USSR)* 38(1), 135–144 (in Russian).

Beglyarov, G.A. and Mitrofanov, V.I. (1973) New species of the genus *Oligonychus* and *Tetranychus* from Vladivostok area. *Biological Revue, Zoology* 7, 23–29 (in Russian).

Begon, M., Harper, J.L. and Townsend, C.R. (1986) *Ecology. Individuals, Populations and Communities*. Blackwell, Oxford, UK.

Behrens, E. (1964) Zur Biologie und Okologie der Johannisbeergallmilbe Eriophyes ribis Nal. Sowie ihrer Bekampfung in Johannisbeerenaubaugebiet Perleberg, Bezirk Schwerin. *Wissenschaftliche Zeitschrift der Wilhelm-Pieck-Universität Rostock. Mathematisch-naturwissenschaftliche Reihe* 13, 279–288.

Behura, S.K. (2006) Molecular marker systems in insects: current trends and future avenues. *Molecular Ecology* 15, 3087–3113.

Behysi, M. Taveras María, Lenin E. González Brito, Pellerano, M., Juan M. Alcántara F. and Jorge Mancebo (2002) Distribución geográfica de los ácaros *Aceria bakkeri* y *Steneotarsonemus spinkii*, en las principales zonas arroceras de República Dominicana. *Esporas* 1(2), 37–44.

Bell, J., Toapanta, M. and Morris, R. (2005) Biological performance of ENVIDOR® 2SC (spirodiclofen) against eriophyid mites on Florida citrus. Bayer CropScience, Research Triangle Park, North Carolina. Available at: https://esa.confex.com/esa/responses/2005/18.pdf (accessed 4 March 2015).

Bellettini, S., Ulbrich, A.V. and Gastaldi, L.F. (1999) Control of white mite *Polyphagotarsonemus latus* Banks, 1904 with different doses of chlorfenapyr on cotton. In: *Anais II Congresso Brasileiro de Algodao: o Algodao no Seculo XX, Perspectivar para o Seculo XXI. Ribeirao Preto, SP, Brasil, 5–10 Setembro 1999*. Empresa Brasileira de Pesquisa Agropecuáia, Embrapa Algodão, Campina Brande, Brazil, pp. 152–154.

Bellini, M.R., Moraes, G.J.d. and Feres, R.J.F. (2005) Acaros (Acari) de dois sistemas de cultivo de seringueira no noroeste do Estado de São Paulo. *Neotropical Entomology* 34, 475–484.

Bellotti, A.C. (1985) Cassava. In: Helle, W. and Sabelis, M.W. (eds) *Spider Mites: Their Biology, Natural Enemies and Control*. World Crop Pests, Volume 1B. Elsevier Science, Amsterdam, pp. 333–338.

Bellotti, A.C., Reyes, J.A., Guerrero, J.M. and Varela, A.M. (1982) The mealybug and green spider mite complexes in South America. The problem and potential for biological control. In: Cock, J.H. and Reyes, J.A. (eds) *Cassava. Research, Production and Utilization*. Centro Internacional de Agricultura Tropical (CIAT), Cali, Colombia, pp. 393–439.

Bellotti, A.C., Smith, L. and Lapointe, S.L. (1999) Recent advances in cassava pest management. *Annual Review of Entomology* 44, 343–370.

Bellotti, A.C., Herrera, C.J., del Pilar Hernández, M., Arias, B., Guerrero, J.M. and Melo, E.L. (2012) Cassava pests in Latin America, Africa and Asia. In: Howeler, R.H. (ed.) *The Cassava Handbook. A Reference Manual based on the Asian Regional Cassava Training Course, Held in Thailand, 2011*. Centro Internacional de Agricultura Tropical (CIAT), Cali, Colombia, pp. 199–257.

Belozerov, V.N. (2010) Seasonal adaptations in the life cycles of mites and ticks: comparative and evolutionary aspects. In: Sabelis, M.W. and Bruin, J. (eds) *Trends in Acarology: Proceedings of the 12th International Congress, 21–26 August 2006*, Amsterdam. Springer, Dordrecht, the Netherlands, pp. 319–326.

Ben, J.-Y., Su, T.-Z. and Chen, S.-D. (1978) Studies on bionomics of three pyemotid mites from Chingai. *Acta Entomologica Sinica* 21, 151–160 (in Chinese).

Ben-Ali, Z., Boursot, P., Said, K., Lagnel, J., Chatti, N. and Navajas, M. (2000) Comparaison of ribosomal ITS regions among *Androctonus* spp. scorpions (Scorpionida: Buthidae) from Tunisia. *Journal of Medical Entomology* 37, 787–790.

Ben Chaaban, S. and Chermiti, B. (2009) Characteristics of date fruit and its influence on population dynamics of *Oligonychus afrasiaticus* McGregor (Acari: Tetranichydae) in the southern of Tunisia. *Acarologia* 49, 29–37.

Ben Chaaban, S., Chermiti, B. and Kreiter, S. (2011a) *Oligonychus afrasiaticus* and phytoseiid predators' seasonal occurrence on date palm *Phoenix dactylifera* (Deglet Noor cultivar) in Tunisian oases. *Bulletin of Insectology* 64, 15–21.

Ben Chaaban, S., Chermiti, B. and Kreiter, S. (2011b) Comparative demography of the spider mite, *Oligonychus afrasiaticus*, on four date palm varieties in southwestern Tunisia. *Journal of Insect Science* 11:136. Available at: http://jinsectscience.oxfordjournals.org/content/jis/11/1/136.full.pdf (accessed 15 March 2015).

Ben Chaaban, S., Chermiti, B. and Kreiter, S. (2012) Effects of host plants on distribution, abundance, developmental time and life table parameters of *Oligonychus afrasiaticus* (McGregor) (Acari: Tetranychidae). *Papéis Avulsos de Zoologia* 52(10), 121–132.

Ben-David, T. (2008) Molecular characterization of Israel's spider mites (Acari: Tetranychidae). PhD thesis, Hebrew University of Jerusalem, Jerusalem.

Ben-David, T., Melamed, S., Gerson, U. and Morin, S. (2007) ITS2 sequences as barcodes for identifying and analyzing spider mites (Acari: Tetranychidae). *Experimental and Applied Acarology* 41, 169–181.

Ben-David, T., Gerson, U. and Morin, S. (2009) Asymmetric reproductive interference between two closely related spider mites: *Tetranychus urticae* and *T. turkestani* (Acari: Tetranychidae). *Experimental and Applied Acarology* 48, 213–227.

Benavides, P., Góngora, C. and Bustillo, A. (2012) Chapter 23: IPM program to control coffee berry borer *Hypothenemus hampei*, with emphasis on highly pathogenic mixed strains of *Beauveria bassiana*, to overcome insecticide resistance in Colombia. In: Perveen, F. (ed.) *Insecticides – Advances in Integrated Pest Management*. InTech, Rijeka, Croatia, pp. 511–540. Available at: http://www.intechopen.com/books/insecticides-advances-in-integrated-pest-management/ipm-program-to-control-coffee-berry-borer-hypothenemus-hampei-with-emphasis-on-highly-pathogenic-mix (accessed 31 March 2015).

Bender, G.S. (1993) A new mite problem in avocados. *California Avocado Society Yearbook* 77, 73–77.

Benfatto, D. (1994) Nuovi insetticidi e acaricidi per la difesa degli agrumi. *Informatore Agrario* 50(36), 65–68.

Benfatto, D. and Lanza, G. (1980) Risultati di una prova di lotta contro il "nuovo Ragno rosso degli agrumi". *Informatore Agrario* 36(48), 13237–13241.

Benfatto, D., Di Franco, F. and Vacante, V. (1992) Spiders of Italian citrus groves. In: *Proceedings of the 7th International Citrus Congress, Acireale, Italy, 8–13 March 1992, Volume 3*. International Society of Citriculture, University of California, Riverside, California, pp. 938–942.

Bennett, F.D. (1974) *Investigations on the Natural Enemies of the Cassava Mite* Mononychellus tanajoa. *Report for* April–September, 1974. Commonwealth Institute of Biological Control, Curepe, Trinidad.

Bennett, F.D. and Yaseen, M. (1975) *Investigation on the Cassava Mite* Mononychellus tanajoa *(Bondar) and its Natural Enemies in the Neotropics. Report for April 1974.* Commonwealth Institute of Biological Control, Curepe, Trinidad.

Bennison, J.A. and Jacobson, R. (1991) Integrated control of *Frankliniella occidentalis* (Pergande) in UK cucumber crops – evaluation of a controlled release system of introducing *Amblyseius cucumeris*. *Mededelingen van de Faculteit Landbouwwetenschappen, Rijksuniversiteit Gent* 56, 251–255.

Bergh, J.C. (1992) Monitoring the emergence and behavior of pear rust mite (Acarina: Eriophyidae) deutogynes using sticky-band traps. *Journal of Economic Entomology* 85, 1754–1761.

Bergh, J.C. (1994) Pear rust mite (Acari: Eriophyidae) fecundity and development at constant temperatures. *Environmental Entomology* 23, 420–424.

Bergh, J.C. and Judd, G.J.R. (1993) Degree-day model for predicting emergence of pear rust mite (Acari: Eriophyidae) deutogynes from overwintering sites. *Environmental Entomology* 22, 1325–1332.

Bergh, J.C. and McCoy, C.W. (1997) Aerial dispersal of citrus rust mite (Acari: Eriophyidae) from Florida citrus groves. *Environmental Entomology* 26, 256–264.

Bergh, J.C. and Weiss, C.R. (1993) Pear rust mite, *Epitrimerus pyri* (Acari: Eriophyidae) oviposition and nymphal development on *Pyrus* and non-*Pyrus* hosts. *Experimental and Applied Acarology* 17, 215–224.

Bergh, J.C., Rugg, D., Jansson, R.K., McCoy, C.W. and Robertson, J.L. (1999) Monitoring the susceptibility of citrus rust mite (Acari: Eriophyidae) populations to abamectin. *Journal of Economic Entomology* 92, 781–787.

Berglund, R. (2007) Organic production of strawberries – focus on practical applications. PhD thesis, Swedish University of Agricultural Sciences, Alnarp, Sweden. Available as: *Acta Universitatis Agriculturae Sueciae* 2007:30.

Berglund, R., Svensson, B. and Nilsson, C. (2007) Evaluation of methods to control *Phytonemus pallidus* and *Anthonomus rubi* in organic strawberry production. *Journal of Applied Entomology* 131, 573–578.

Bergoughoux, F., Germain, E. and Sarraquigne, J.P. (1978) *The Hazel.* Production and Culture. National Institute for the Popularization of Fruits, Vegetable and Mushrooms, Paris.

Bergstrom, J. (1979) Morphology of fossil arthropods as a guide to phylogenetic relationships. In: Gupta, A.P. (ed.) *Arthropod Phylogeny*. Van Nostrand Reinhold, New York, pp. 1–56.

Berlese, A. (1882–1902) *Acari, Myriopoda et Scorpiones, Hucusque in Italia Reperta, Fasc. 1–101.* Portici, Padua, Italy.

Berlese, A. (1886) *Acari Dannosi alle Piante Coltivate*. Tip. F. Sacchetto, Padua, Italy.

Berlese, A. (1889) *Acari, Myriopoda, et Scorpionida, Hucusque in Italia Reperta, Fasc. 56, Dem. 102.* Portici, Padua, Italy.

Berlese, A. (1897) *Acari, Myriopoda et Scorpiones, Hucusque in Italia Reperta, Fasc. 80.* Portici, Padua, Italy.

Berlese, A. (1910) Lista di nuove specie e nuovi generi di acari. *Redia* 6, 242–271.

Berlese, A. (1913) *Acarotheca Italica*. Tipografia M. Ricci, Florence, Italy.

Berlese, A. (1917) *Insetti delle Case e dell'Uomo e Malattie che Diffondono*. Hoepli, Milan, Italy.

Berlese, A. (1925) Gli affini degli insetti. In: *Gli Insetti, loro Organizzazione, Sviluppo, Abitudini e Rapporti coll'Uomo, Volume Secundo*. Società Editrice Libraria, Milan, Italy, pp. 1–151.

Berlinger, M.J., Dahan, R. and Mordechi, S. (1988) Integrated pest management of organically grown greenhouse tomatoes in Israel. *Applied Agricultural Research* 3, 233–238.

Bernal, R. and Piñeiro, C. (1982) *Nuevos Acaros Encontrados sobre Citrus en la Zona de Salto*. Miscelanea No. 41, Estación Experimental de Citricultura Centro de Investigationes Agricolas Alberto Boerger, Montevideo.

Bernal, T.B., Ramírez, I.S., López, M.O. and Porras, Á. (2002) Determinación del medio de cultivo para el crecimiento y esporulación de *Sarocladium oryzae* (Sawada) Gams & Hawks. *Fitosanidad* 6(1), 15–18.

Bernard, C. (1909) Over de Ziecten der Theeplant veroorzaakt door Mijten. *Medeelingen vanhet Proefstation voor Thee (Buitenzorg)* 3, 52–60.

Bernard, M.B., Horne, P.A. and Hoffman, A.A. (2005) Eriophyoid mite damage in *Vitis vinifera* (grapevine) in Australia: *Calepitrimerus vitis* and *Colomerus vitis* (Acari: Eriophyidae) as the common cause of the widespread 'Restricted Spring Growth' syndrome. *Experimental and Applied Acarology* 35, 83–109.

Bernays, E. and Chapman, R.F. (1994) *Host-Plant Selection by Phytophagous Insects*. Chapman and Hall, London.

Bernstein, Z. (1988) The grape bud mite in cv. Perlette. *Alon Hanotea* 43, 109–122 (in Hebrew).

Berry, P. (1959) *Entomología Económica de El Salvador*. Boletín Técnico 24, Servicio Cooperativo Agricola Salvadorēno Americano, Ministerio de Agricultura y Ganadería, Santa Tecla, El Salvador .

Berryman, A.A. (1999) *Principles of Population Dynamics and Their Application*. Samuel Thornes, London.

Bertrand, M., Kreiter, S., McKoy, K., Migeon, A., Navajas, M., Tixier, M.-S. and Vial, L (2008) A quick overview of the story of French acarologists. In: Bertrand, M., Kreiter, S., McKoy, K., Migeon, A., Navajas, M., Tixier, M.S. and Vial, L. (eds) *Integrative Acarology. Proceedings of the 6th European Congress of the European Association of Acarologists, Montpellier, France, 21–25 July 2008*. European Association of Acarologists, Montpellier, France.

Besse, D. and Götz, B. (1963) Der Einfluss der Nährstoffversorgung bei der Rebe *Vitis vinifera* L. auf die Massenentwicklung der Obstbaumspinnmilbe *Paratetranychus pilosus* C. und F. *Wein-Wissenschaft* 18, 577–587.

Beute, M.K. and Benson, D.M. (1979) Relation of small soil fauna to plant disease. *Annual Review of Phytopathology* 17, 485–502.

Beyzavi, G., Ueckermann, E.A., Faraji, F. and Ostovan, H. (2013) A catalog of Iranian prostigmatic mites of superfamilies Raphignathoidea and Tetranychoidea (Acari). *Persian Journal of Acarology* 2(3), 389–474.

Bezbaruah, B., Kumar, B.S.D. and Barthakur, M. (1996) Fungicidal and insect controlling properties of *Proteus* strain RRLJ 16, isolated from tea, *Camellia sinensis* (L) O. Kuntze, plantations. *Indian Journal of Experimental Biology* 34, 706–709.

Bhanderi, G.R. (1991) An ecology of *Tetranychus macfarlanei* Baker and Pritchard (Acari: Tetranychidae) in relation to different okra cultivars and its control. MSc. thesis, Gujarath Agricultural University, Navasari, India.

Bhanu, K.S., Reddy, P.S. and Zaheruddeen, S.M. (2006) Evaluation of some acaricides against leaf mite and sheath mite in rice. *Indian Journal of Plant Protection* 34, 132–133.

Bharadwaj, R.K. and Banerjee, S.K. (1973) Phostoxin for control of *Eriophyes mangiferae* (Acarina: Eriophyidae) associated with malformation disease in mango. *Florida Entomologist* 56, 147–148.

Bhardwaj, S. and Sharma, S. (2010) Weather factors affecting European red mite, *Panonychus ulmi* (Koch), density in apple orchards in India. In: Moraes, G.J. de, Castilho, R.C. and Flechtmann, C.H.W. (eds) *XIII International Congress of Acarology, August 23–27, 2010, Recife-PE, Brazil, Abstract Book*, abstract 57, pp. 32–33. Available at: http://www.acarology.org/ica/ica2010/docs/0910_15_abstractBookCapa.pdf (accessed 3 February 2015).

Bhardwaj, S.P. and Bhardwaj, S. (2010) Response of horticultural mineral oils and some new acaricides to European red mite, *Panonychus ulmi* (Koch), infesting apple in temperate region of India. In: Moraes, G.J. de, Castilho, R.C. and Flechtmann, C.H.W. (eds) *XIII International Congress of Acarology, August 23–27, 2010, Recife-PE, Brazil, Abstract Book*, abstract 58, pp. 33–34. Available at: http://www.acarology.org/ica/ica2010/docs/0910_15_abstractBook-Capa.pdf (accessed 3 February 2015).

Bhatia, S.K., Singh, V.S. and Saxena, J.D. (1974) Relative toxicity of some pesticides to the brown wheat mite. *Indian Journal of Agricultural Sciences* 44, 528–530.

Bhede, B.V., Bhosle, B.B. and More, D.G. (2008) Influence of meteorological factors over the incidence of chilli mite, *Polyphagotarsonemus latus* and its chemical control strategies. *Indian Journal of Plant Protection* 36, 200–203.

Bhuiya, D., Majumder, M.Z:R., Hahar, G., Shahjahan, M. and Khan, M. (1999) Irradiation as a quarantine treatment of cut flowers, ginger and tumeric against mites, thrips and nematodes. In: *Irradiation as a Quarantine Treatment of Arthropod Pests. Proceedings of a final Research Co-ordination Meeting organized by the Joint FAO/IAEA Division of Nuclear Techniques in Food and Agriculture and held in Honolulu, Hawaii, 3–7 November 1997. IAEA Technical Documents (IAEA-TECDOCs) 1082*. International Atomic Eneregy Agency, Vienna, pp. 57–66.

Bhumannavar, B.S. and Singh, S.P. (1986) Studies on population dynamics of oriental red mite of citrus, *Eutetranychus orientalis* (Klein). *Entomon* 11, 223–226.

Bhumannavar, B.S., Singh, S.P. and Sulladmath, V.V. (1988) Evaluation of citrus germplasm for resistance to the oriental red mite, *Eutetranychus orientalis* (Klein) under tropical humid south Indian conditions. *Tropical Pest Management* 34, 193–198, 243, 247.

Biasi, L.A. and Santos, H.R. dos (1988a) Consumo de ovos de *Tetranychus ludeni* Zacher, 1913a (Acarina: Tetranychidae) por *Amblyseius fallacis* Garman, 1948 (Acarina: Phytoseiidae). *Revista do Setor de Ciencias Agrarias, Universidade Federal do Parana* 10, 99–102.

Biasi, L.A. and Santos, H.R. dos (1988b) Susceptibilidade da ervilhaca comum (*Vicia sativa* L.) e do tremoco branco (*Lupinus albus* L.) ao acaro fitofago *Tetranychus ludeni* Zacher, 1913b (Acarina: Tetranychidae). *Revista do Setor de Ciencias Agrarias, Universidade Federal do Parana* 10, 179–182.

Bibby, F.F. and Tuttle, D.M. (1959) Notes on phytophagous and predatory mites of Arizona. *Journal of Economic Entomology* 52, 186–190.

Biddinger, D.J., Weber, D.C. and Hull, L.A. (2009) Coccinellidae as predators of mites: Stethorini in biological control. *Biological Control* 51, 268–283.

Bindra, O.S. and Bakhetia, D.R.C. (1969) Studies on the population dynamics of the mango bud mite, *Aceria mangiferae* Sayed, in relation to the incidence of malformation. *Journal of Research – Punjab Agricultural University* 6, 200–206.

Bindra, O.S. and Goyal, N.P. (1966) Prelimlinary studies in the control of vegetable mite on bhindi. *Journal of Research – Punjab Agricultural University* 3, 147–174.

Bindra, O.S. and Kittur, S.V. (1961) Two new records of tetranychid mites as crop pests in M.P. *Indian Journal of Entomology* 23, 71.

Bindra, O.S. and Varma, G.C. (1966) A study in the control of the fig mite, *Eotetranychus hirsti* Pritchard and Baker. *Journal of Research – Punjab Agricultural University* 3, 417–420.

Binnington, K.C. and Obenchain, F.D. (1982) Structure and function of the circulatory, nervous and neuroendocrine systems in ticks. *Current Themes in Tropical Science* 1, 351–398.

Binns, E.S. (1980) *Scutacarus baculitarsus* Mahunka (Acarina: Scutacaridae) phoretic on the mushroom phorid fly *Megaselia halterata* (Wood). *Acarologia* 21, 91–107.

Binns, E.S. (1981) Physogastry in *Scutacarus baculitarsus* Mahunka (Acarina, Scutacaridae). *Entomologist's Monthly Magazine* 116, 157–159.

Binns, E.S. (1982) Phoresy as migration – some functional aspects of phoresy in mites. *Biological Reviews* 57, 571–620.

Biota Taiwanica (2011) *Acaphylla steinwedeni* Keifer 1943. Available at: http://taibif.org.tw//BiotaTaiwanicaAlt/result_species.php?asc=Acaphylla+steinwedeni&search=1&Submit=Search (accessed 23 April 2015).

Bird, F.T. (1967) A virus disease of the European red mite *Panonychus ulmi* (Koch). *Canadian Journal of Microbiology* 13, 1131.

Bjørnson, S. and Keddie, B.A. (1999) Effects of *Microsporidium phytoseiuli* (Microspora) on the performance of the predatory mite, *Phytoseiulus persimilis* (Acari: Phytoseiidae). *Biological Control* 15, 153–161.

Bjørnson, S., Steiner, M.Y. and Keddie, B.A. (1997) Birefringent crystals and abdominal discoloration of the predatory mite *Phytoseiulus persimilis* (Acari: Phytoseiidae). *Journal of Invertebrate Pathology* 69, 85–91.

Blackwell, M., Bridges, J.R. and Moser, J.C. (1986) Hyperphoretic dispersal of a Pyxidiophora anamorph. *Science* 232, 993–995.

Blackwell, M., Moser, J.C. and Wisniewski, J. (1989) Ascospores of Pyxidiophora on mites associated with beetles in trees and wood. *Mycological Research* 92, 397–403.

Blair, B.W. (1983) *Tetranychus evansi* Baker & Pritchard (Acari: Tetranychidae); a new pest of tobacco in Zimbabwe. CORESTA Phytopathology and Agronomy Study Group, Bergerac, France.

Blair, B.W. (1989) Laboratory screening of acaricides against *Tetranychus evansi* Baker & Pritchard. *Crop Protection* 8, 212–216.

Blanchard, E.E. (1940) Tres ácaros dañinos para los cultivos argentinos. *Revista de la Facultad de Agronomía, Universidad Nacional de La Plata* 24, 11–18.

Blanck, A., Miquel, L. and Motemps, J. (1954) Premiers essais d'efficacité comparée des nouveaux acaricides dans la lutte contre l'acariose bronzée de la tomate. *Phytoma* 7, 5–9.

Blattny, C. (1933) On a "Red Burn" disease of amaryllis caused by a mite, *Tarsonemus hydrocephalus* Vitzthum. *Gartenbauwissenschaft* 7, 425–437.

Blauvelt, W.E. (1945) The internal morphology of the common red spider mite, *Tetranychus telarius* L. *Memoirs of Cornell University Agricultural Experiment Station* No. 270, 1–11.

Bledsoe, M.E. and Minnick, D.R. (1982) Citrus tree snail and suppression of citrus microbiota. *Environmental Entomology* 11, 1091–1095.

Blommers, L.H.M. (1976) Capacities for increase and predation in *Amblyseius bibens* (Acarina: Phytoseiidae). *Zeitschrift für Angewandte Entomologie* 81, 225–244.

Blommers, L.H.M. (1994) Integrated pest management in European apple orchards. *Annual Review of Entomology* 39, 213–241.

Blommers, L.H.M. and Chazeau, J. (1974) Two new species of predator mites of the genus *Amblyseius* Berlese (Acarina: Phytoseiidae) from Madagascar. *Zeitschrift für Angewandte Entomologie* 75, 308–315.

Blommers, L.H.M. and Gutierrez, J. (1975) Les tétranyques vivant sur agrumes et avocatiers dans la région de Tamatave (Madagascar-est) et quelques-uns de leurs prédateurs. *Fruits* 30, 191–200.

Blommers, L.H.M., Alkema, P. and De Reede, R. (1986) The effects of pesticides and other spraying material on the predacious mite *Typhlodromus pyri*. *IOBC-WPRS Bulletin* 9(3), 60–62.

Boaretto, M.A.C. and Chiavegato, L.G. (1994) Transmissao da leprose por acaros *Brevipalpus phoenicis* (Geijskes, 1939) (Acari: Tenuipalpidae) temporariamente mantidos em hospedeiros intermediarios, em condicoes de laboratorio. *Científica (Jaboticabal)* 22, 81–93.

Bobot, T. da Encarnação, Franklin, E., Návia, D., Gasnier, T.R.J., Lofego, A.C. and Oliveira, B.M. de (2011) Mites (Arachnida, Acari) on *Citrus sinensis* L. Osbeck orange trees in the State of Amazonas, Northern Brazil. *Acta Amazonica* 41, 557–566.

Boctor, F.N. (1972) Biochemical and physiological studies of certain ticks (Ixodoidea). Free amino acid in female *Argas* (*Persicargas*) *arboreus* Kaiser, Hoogstraal and Kohls (Argasisae) analyzed by gas-liquid chromatography. *Journal of Medical Entomology* 9, 201–204.

Boczek, J. (1962) *Nalepella haarlovi* n. sp. (Acarina, Eriophyidae). *Entomologiske Meddelelser* 31, 195–197.

Boczek, J. (1964a) Studies on mites (Acarina) living on plants in Poland. IV. *Bulletin of the Polish Academy of Sciences Cl. V* (*Biological Sciences*) 12, 365–369.

Boczek, J. (1964b) Studies on mites (Acarina) living on plants in Poland. V. *Bulletin of the Polish Academy of Sciences Cl. V* (*Biological Sciences*) 12, 391–398.

Boczek, J. (1966) Studies on mites (Acarina) living on plants in Poland. VII. *Bulletin of the Polish Academy of Sciences Cl. V* (*Biological Sciences*) 14, 335–341.

Boczek, J. (1974) *Ecology of Eriophyid Mites on Economic Crops*. Final Report PL-480 Project E21-Ent-24, Fg-Po-425.

Boczek, J. and Chandrapatya, A. (1996) Studies on eriophyoid mites (Acari: Eriophyoidea). XVIII. *Bulletin of the Polish Academy of Sciences, Biological Sciences* 44, 61–70.

Boczek, J. and Chyczewski, J. (1975) Beobachtungen zur Biologie einiger Gallmilbenarten (Eriophyoidea) der Gräser. *Tagungsbericht der Akademie der Landwirtschaftswissenschaften der DDR* 134, 83–90.

Boczek, J. and Davis, R. (1990) Three new species of Eriophyid mites from Georgia Coastal islands (Acari: Eriophyoidea). *Journal of Entomological Science* 25, 125–133.

Boczek, J. and Griffiths, D.A. (1994) Structure and systematics of eriophyid mites (Acari: Eriophyoidea) and their relationship to host plants. In: Williams, M.A.J. (ed.) *Plant Galls: Organisms, Interactions, Populations*. Systematics Association Special Volume 49. Clarendon Press, Oxford, UK, pp. 119–129.

Boczek, J. and Kropczynska, D. (1964) Studies on mites (Acarina) living on plants in Poland I. *Fragmenta Faunistica Warszawa* 11, 161–188 (in Polish).

Boczek, J. and Petanović, R. (1996) Eriophyid mites as agents for the biological control of weeds. In: Moran, V.C. and Hoffmann, J.H. (eds) *Proceedings of the IX International Symposium on Biological Control of Weeds, 19–26 January 1996, University of Cape Town, Stellenbosch, South Africa*, pp. 127–131. Available at: http://www.invasive.org/proceedings/pdfs/9_127-131.pdf (accessed 30 April 2015).

Boczek, J. and Shevchenko, V.G. (1996) Ancient associations: eriophyoid mites on gymnosperms. In: Lindquist, E.E., Sabelis, M.W. and Bruin, J. (eds) *Eriophyoid Mites: Their Biology, Natural Enemies and Control*. World Crop Pests, Volume 6. Elsevier Science, Amsterdam, pp. 217–225.

Boczek, J., Zawadzki, W. and Davis, R. (1984) Some morphological and biological differences in *Aculus fockeui* (Nalepa and Trouessart) (Acari: Eriophyidae) on various host plants. *International Journal of Acarology* 10, 81–87.

Boczek, J., Shevchenko, V.G. and Davis, R. (1989) *Generic Key to World Fauna of Eriophyid Mites*. Warsaw Agricultural University Press, Warsaw (in English and Russian).

Boczek, J., Harding, S., Shi, A. and Bresciani, J. (2002) A new species of *Nalepella* Keifer (Acarina: Eriophyoidea: Phytoptidae) from *Abies* in Denmark. *Acarologia* 42, 53–60.

Bodenheimer, F.S. (1951) *Citrus Entomology in the Middle East*. W. Junk, The Hague, the Netherlands.

Boekhout, T., Theelen, B., Houbraken, J., Robert, V., Scorzetti, G., Gafni, A., Gerson, U. and Sztejnberg, A. (2003) Novel anamorphic mite-associated fungi belonging to the Ustilaginomycetes: *Meira geulakonigii* gen. nov., sp. nov., *Meira argovae* sp. nov. and *Acaromyces ingoldii* gen. nov., sp. nov. *International Journal of Systematic and Evolutionary Microbiology* 53, 1655–1664.

Boesewinkel, H.J. (1977) Walnut diseases. *New Zealand Journal of Agriculture* 135(5), 21–22.

Boethel, D.J. and Ezell, J.E. (1978) Influence of spray programs on seasonal abundance of pecan leaf scorch mite and phytoseiids in Louisiana. *Journal of Economic Entomology* 71, 507–517.

Bogatko, W., Łabanowski, G.S. and Pala, E. (1987) Effect of feeding of the spruce spider mite, *Oligonychus ununguis* (Jacobi), on the growth and quality of dwarf spruces. *Prace Instytutu Sadownictwa i Kwiaciarstwa, Seria B, Rośliny Ozdobne* 11, 285–288.

Boggs, J.F., Young, C.E., Shetlar, D.J., Bloetscher, B., Stone, A.K., Goerig, D.J., Malinich, T.J., Dyke, D.E., Draper, E.A., Bennett, P.J., Gao, G.Y. and Chatfield, J.A. (2004) Insect and mite activity noted in Ohio nurseries and landscapes: 2003. *Special Circular, Ohio Agricultural Research and Development Center* No. 193, pp. 33–53

Bognar, S. (1961) Contributions to the knowledge of the spider mite fauna of Hungary. *Annales Instituti Protectionis Plantarum Hungarici* 8, 261–268.

Boh, B., Kornhauser, A., Krumpak, A., Novosel, F., Pokorny, M., Radez, I. and Skerlava, V. (1995) Vecnamensko fitofarmacevtsko sredstvo s fizikalnim ucinkovanjem. *Sodobno Kmetijstvo* 28(2), 55–59 (in Slovenian).

Bohm, M. (1960) Untersuchungen über Spinnmilbenfinde in Osterreich. *Pflanzenschutzberichte* 25, 23–46.

Boisduval, A. (1867) *Essai sur l'Entomologie Horticole*. Donnaud, Paris. Reprinted in 2009 by Kessinger Publishing, Whitefish, Montana.

Bolay, A., Stäubli, A. Baillod, M., Ducrot, V., Guignard, E., Antonin, P., Neury, G. and Terretaz, R. (1989) Guide des traitments des groseilliers et des cassis. *Revue Suisse de Viticulture Arboriculture Horticulture* 21, 81–87.

Bolckmans, K.J.F. (1999) Commercial aspects of biological pest control. In: Albajes, R., Gullino, M.L., van Lenteren, J.P. and Elad, Y. (eds) *Integrated Pest and Disease Management in Greenhouse Crops*. Kluwer, Dordrecht, the Netherlands, pp. 310–338.

Boldt, P.E. and Sobhian, R. (1993) Release and establishment of *Aceria malherbae* (Acari: Eriophyidae) for control of field bindweed in Texas. *Environmental Entomology* 22, 234–237.

Bolland, H.R. (1983) A description of *Neophyllobius aesculi* n. sp. and its developmental stages (Acari: Camerobiidae). *Entomologische Berichten* 43, 42–47.

Bolland, H.R. (1999) *Verslag Baikalreis 16 Juili t:m 3 Augustus 1999*. Vereniging voor Veldbiologie, Wageningen, the Netherlands.

Bolland, H.R. (2001) *Pyreneënreis 2001, 29/6 t/m 13/7: Mitjen Soorten Gevonden in de Spaanse Pyreneën*. Vereniging voor Veldbiologie, Wageningen, the Netherlands.

Bolland, H.R. and Beron, P. (2004) Tetranychoidea. In: Magowski, W.Ł. (ed.) Acari: Acariformes: Fauna Europaea version 2.6.2. Available at: http://www.faunaeur.org/full_results.php?id=15636 (accessed 30 April 2015).

Bolland, H.R. and Helle, W. (1981) A survey of chromosome complements in the Tenuipalpidae. *International Journal of Acarology* 7, 157–160.

Bolland, H.R., Gutierrez, J. and Helle, W. (1981) Chromosomes in spider mites (Tetranychidae-Acari). *Acarologia* 22, 271–275.

Bolland, H.R., Gutierrez, J. and Flechtmann, C.H.W. (1998) *World Catalogue of the Spider Mite Family (Acari: Tetranychidae)*. Brill., Leiden, the Netherlands.

Bonafos, R., Serrano, E., Auger, P. and Kreiter, S. (2007) Resistance to deltamethrin, lambda-cyhalothrin and chlorpyriphos-ethyl in some populations of *Typhlodromus pyri* Scheuten and *Amblyseius andersoni* (Chant) (Acari: Phytoseiidae) from vineyards in south-west of France. *Crop Protection* 26, 169–172.

Bonato, O. (1999) The effect of temperature on life history parameters of *Tetranychus evansi* (Acari: Tetranychidae). *Experimental and Applied Acarology* 23, 11–19.

Bonato, O. and Gutierrez, J. (1993) Elaboration de plans d'échantillonnage communs à deux ravageurs: exemple de deux acariens phytophages (Acari: Tetranychidae) attaquant le manioc. *ANPP – Troisième Conférence Internationale sur les. Ravageurs Agriculture, Montpellier, 7–8–9 Décembre 1993.* Association Nationale de Protection des Plantes, Paris, pp. 571–578. Available at: http://horizon.documentation.ird.fr/exl-doc/pleins_textes/pleins_textes_6/b_fdi_35-36/41756.pdf (accessed 30 April 2015).

Bonato, O. and Gutierrez, J. (1999) Effect of mating status on the fecundity and longevity of four spider mite species (Acari: Tetranychidae). *Experimental and Applied Acarology* 23, 623–632.

Bonato, O., Cotton, D., Kreiter, S. and Gutierrez, J. (1989) Les principaux paramètres démographiques d'une population de l'acarien jaune de la vigne *Eotetranychus carpini* (Acari: Tetranychidae). In: *Colloque sur les Acariens des Cultures, Montpellier – 24, 25, 26 Octobre 1989. Annales A.N.P.P. No. 2, Vol. 1/1. 1989.* Association Nationale de Protection des Plantes, Paris, pp. 541–551. Available at: http://horizon.documentation.ird.fr/exl-doc/pleins_textes/pleins_textes_6/b_fdi_33-34/38644.pdf (accessed 30 April 2015).

Bonato, O., Baumgärtner, J. and Gutierrez, J. (1994) Impact of *Mononychellus progresivus* and *Oligonychus gossypii* (Acari: Tetranychidae) on cassava growth and yield in Central Africa. *Journal of Horticultural Science* 69, 1089–1094.

Bonato, O., Baumgärtner, J. and Gutierrez, J. (1995a) Comparison of biological and demographic parameters for *Mononychellus progresivus* and *Oligonychus gossypii* on cassava: influence of temperature. *Entomologia Experimentalis et Applicata* 75, 119–125.

Bonato, O., Mapangou-Divassa, S. and Gutierrez, J. (1995b) Influence of relative humidity on life-history parameters of *Mononychellus progresivus* and *Oligonychus gossypii* (Acari: Tetranychidae). *Environmental Entomology* 24, 841–845.

Bonato, O., Baumgärtner, J. and Gutierrez, J. (1995c) Sampling plans for *Mononychellus progresivus* and *Oligonychus gossypii* (Acari : Tetranychidae) on cassava in Africa. *Journal of Economic Entomology* 88, 1296–1300.

Bondar, G. (1930) Insectos daminhoes e molestias dos feijões na Bahia. *Boletim do Laboratório de Patologia Vegetal (Salvador)* 9, 1–83.

Bondar, G. (1938) Notas entomológicas de Bahia III. *Revista Brasileira de Entomologia* 9, 441–445.

Bondarenko, N.V. (1950) The influence of shortened day on the annual cycle of development of the common spider mite. *Doklady Akademii Nauk SSSR* 70, 1077–1080 (in Russian).

Bondarenko, N.V. (1958) Characteristics of the diapause in *Tetranychus urticae* Koch. *Zoologicheskii Zhurnal* 37, 1012–1023 (in Russian).

Boness, M. (1968) Über Milbenauftreten in Pilzkulturen. *Anzeiger für Schädlingskunde* 51, 41–42.

Borden, A.D. (1932) *The Pear Leaf Blister Mite as a Cause of Fruit-Bud Injury*. Circular 324, Agricultural Experiment Station, College of Agriculture, University of California, Berkeley, California.

Borges, P.A.V., Aguiar, A.M.F., Boieiro, M., Carles-Tolra, M. and Serrano, A.R.M. (2008) List of arthropods. In: Borges, P.A.V., Aguiar, A.M.F., Carvalho, P., Jardim, R., Melo, I., Oliveira, P., Sérgio, C., Serrano, A.R.M. and Vieira, P. (eds) *A list of the terrestrial fungi, flora and fauna of Madeira and Selvagens archipelagos*. Direcção Regional do Ambiente da Madeira and Universidade dos Açores, Funchal and Angra do Heroísmo, Portugal, p. 440.

Borgman, H.H. (1950) De "rode vrucht ziekte" bij bramen, veroorzaakt door de galmijt *Eriophyes essigi* Hassan. *Tijdschrift over Plantenziekten* (now *European Journal of Plant Pathology*) 56, 149–160 (in Dutch).

Borgo, M. (1988) Dégats provoqués par *Eotetranychus carpini* Oud. et par *Panonychus ulmi* Koch, sur les vignes en Venetie orientale. In: Cavalloro, R. (ed.) *Plant-protection Problems and Prospects of Integrated Control in Viticulture. Proceedings of a CEC/IOBC International Symposium held at Lisboa-Vila Real, Portugal, 6 to 9 June* 1988. Commission of the European Communities, Luxembourg, pp. 209–217.

Borgo, M. and Giorgessi, F. (1987) Effetti della termoregolazione estiva mediante acqua su *Panonychus ulmi* Koch. *Rivista di Viticoltura e di Enologia (Conegliano)* 40, 45–52.

Borichenko, N. and Manolov, P. (1982) Changes in some physiological indicators and biochemical processes of apple leaves infected to various extents by *Panonychus ulmi* Koch. *Lozrska Nauka (Horticultural and Viticultural Science)* 19, 44–50 (in Bulgarian).

Borthakur, M. (1981) Biological notes on a mite predator of the scarlet mite *Brevipalpis phoenicis* (Geijskes). *Two and a Bud* 28, 18–19.

Borthakur, M. and Das, S.C. (1988) Studies on acarine predators of phytophagous mites on tea in north-east India. *Two and a Bud* 34, 21–24.

Borthakur, M., Handique, R. and Banerjee, B. (1997) Effect of pollen feeding on development and reproduction of *Agisteus hystrix* Gupta (Acarina: Stigmaeidae). *Two and a Bud* 44, 18–20.

Bostanian, N.J. and Vincent, C. (1985) Pre-bloom pesticide treatments for phytophagous mite control in apple orchards. *Journal of Horticultural Science* 60, 459–463.

Bostanian, N.J., Paradis, R.O. and Pitre, D. (1981a) Susceptibility of phytophagous mites to a single summer treatment of acaricides in a Quebec apple orchard. *Phytoprotection* 62, 33–38.

Bostanian, N.J., Paradis, R.O. and Pitre, D. (1981b) Essais de traitements préventifs en verger contre le tetranyque rouge du pommier, *Panonychus ulmi* (Koch) et l'Eriophyidae du pommier, *Aculus schlechtendali* (Nalepa). *Phytoprotection* 62, 53–58.

Bostanian, N.J., Paradis, R.O. and Pitre, D. (1983) Essais additionels de lutte chimique contre le tetranyque rouge du pommier, *Panonychus ulmi* (Koch), et l'eriophyide du pommier, *Aculus schlechtendali* (Nalepa) dans le sud-ouest du Québec. *Annales de la Société Entomologique de Québec* 28, 19–24.

Bostanian, N.J., Racette, G. and Lasnier, J. (2010) Biocontrol of phytophagous mites in Québec apple orchards. In: Sabelis, M.W. and Bruin, J. (eds) *Trends in Acarology: Proceedings of the 12th International Congress, 21–26 August 2006*, Amsterdam. Springer, Dordrecht, the Netherlands, pp. 451–456.

Bostrom, S., Johansson, E., Harfast, B., Lundqvist, L., Backman, I., Rosen, E. von and Hage-Hamsten, M. van. (1997) Characterization of the mite fauna (Acari) in Swedish barn dust. *International Journal of Acarology* 23, 127–132.

Boubou, A., Migeon, A., Auger, P. and Navajas, M. (2010) How much environmental conditions drive the invasive potential of *Tetranychus evansi*. In: Moraes, G.J. de, Castilho, R.C. and Flechtmann, C.H.W. (eds) *XIII International Congress of Acarology, August 23–27, 2010, Recife-PE, Brazil, Abstract Book*, abstract 66, pp. 38–39. Available at: http://www.acarology.org/ica/ica2010/docs/0910_15_abstractBookCapa.pdf (accessed 3 February 2015).

Boubou, A., Migeon, A., Roderick, G.K. and Navajas, M. (2011) Recent emergence and worldwide spread of the red tomato spider mite, *Tetranychus evansi*: genetic variation and multiple cryptic invasions. Recent emergence and multiple cryptic invasions of *Tetranychus evansi*. *Biological Invasions* 13, 81–92.

Boubou, A., Migeon, A., Roderick, G.K., Auger, P., Cornuet, J.M., Magalhães, S. and Navajas, M. (2012) Test of colonisation scenarios reveals complex invasion history of the red tomato spider mite *Tetranychus evansi*. *PLoS ONE* 7(4): e35601.

Bouchelos, T. (1962) *Tetranychus viennensis* (Zacher) (Acarina: Tetranychidae). In: *Annual Report of the Phytopathological Station, Patras, Greece*, pp. 17–18.

Boucias, D.G. and Pendland, J.C. (1998) *Principles of Insect Pathology*. Kluwer, Boston, Massachusetts.

Boudreaux, H.B. (1953) A simple method of collecting spider mites. *Journal of Economic Entomology* 46, 1102–1103.

Boudreaux, H.B. (1956) Revision of the two spotted spider mite complex *Tetranychus telarius*. *Annals of the Entomological Society of America* 49, 43–48.

Boudreaux, H.B. (1958a) The effect of relative humidity on egg-laying, hatching, and survival in various spider mites. *Journal of Insect Physiology* 2, 65–72.

Boudreaux, H.B. (1958b) *Tetranychus tumidus* Banks versus *Tetranychus gloveri* Banks (Acarina, Tetranychidae). *Annals of the Entomological Society of America* 51, 174–177.

Boudreaux, H.B. (1963) Biological aspects of some phytophagous mites. *Annual Review of Entomology* 8, 137–154.

Boudreaux, H.B. (1979) Confusion of names for the spider mites *Tetranychus tumidus* and *T. gloveri*. In: Rodriguez, J.G. (ed.) *Recent Advances in Acarology, Volume II*. Academic Press, New York, pp. 395–398.

Bounfour, M. and McMurtry, J.A. (1987) Biology and ecology of *Euseius scutalis* (Athias-Henriot) (Acarina: Phytoseiidae). *Hilgardia* 55(5), 1–23.

Bounfour, M. and Tanigoshi, L.K. (2001a) Host plant-mediated interactions between *Tetranychus urticae* and *Eotetranychus carpini borealis* (Acari: Tetranychidae). *Experimental and Applied Acarology* 25, 13–24.

Bounfour, M. and Tanigoshi, L.K. (2001b) Effect of temperature on development and demographic parameters of *Tetranychus urticae* and *Eotetranychus carpini borealis* (Acari: Tetranychidae). *Annals of the Entomological Society of America* 94, 400–404.

Bounfour, M., Tanigoshi, L.K., Chen, C., Cameron, S.J. and Klauer, S. (2002) Chlorophyll content and chlorophyll fluorescence in red raspberry leaves infested with *Tetranychus urticae* and *Eotetranychus carpini borealis* (Acari: Tetranychidae). *Environmental Entomology* 31, 215–220.

Bourdouxhe, L. and Collingwood, E.F. (1982) Efficacite de trois pyrethrinoides photostables a l'egard des principaux ordres d'insectes et des acariens nuisibles aux cultures maraicheres au Senegal. *Agronomie Tropicale* 37, 379–388.

Bournier, A., Lacasa, A. and Pivot, Y. (1978) Biologie d'un thips prédateur *Aeolothrips intermedius* (Thys.: Aeolothripidae). *Entomophaga* 23, 403–410.

Bournier, A., Lacasa, A. and Pivot, Y. (1979) Régime alimentaire d'un thrips prédateur *Aeolothrips intermedius* (Thys.: Aeolothripidae). *Entomophaga* 24, 353–361.

Bowman, C.E. and Bartlett, P.W. (1978) *Panonychus citri* (McGregor) (Acari: Tetranychidae) infesting imported *Citrus mitis* in England. *Plant Pathology* 27, 201.

Bowman, H.M. and Hoy, M.A. (2012) Molecular discrimination of phytoseiids associated with the red palm mite *Raoiella indica* (Acari: Tenuipalpidae) from Mauritius and South Florida. *Experimental and Applied Acarology* 57, 395–407.

Box, H.E. (1953) *List of Sugarcane Insects*. Commonwealth Institute of Entomology, London (now CAB International, Wallingford, UK).

Boyce, A.M. and Korsmeier, R.B. (1941) The citrus bud mite, *Eriophyes sheldoni* Ewing. *Journal of Economic Entomology* 34, 745–756.

Boyce, A.M., Korsmeier, R.B. and Persing, C.O. (1942) The citrus bud mite and its control. *California Citrograph* 27, 124, 125, 136–138, 140, 141.

Boyd, D.W. and Alverson, D.R. (2000) Repellency effects of garlic extracts on two spotted spider mite, *Tetranychus urticae* Koch. *Journal of Entomological Science* 35, 86–90.

Boykin, L.S. and Campbell, W.V. (1982) Rate of population increase of the two spotted spider mite (Acari: Tetranychidae) on peanut leaves treated with pesticides. *Journal of Economic Entomology* 75, 966–971.

Boyle,W.W. and Haramoto, F.H. (1956) Cyclamen mite on pikake. *Hawaii Farm Science* 5(2), 1, 7, 8.

Boyne, J.V. and Hain, F.P. (1983) Effects of constant temperature, relative humidity, and simulated rainfall on development and survival of the spruce spider mite (*Oligonychus ununguis*). *Canadian Entomologist* 115, 93–105.

Bozai, J. (1970) Determination key for spider mites. *Növényvédelen* 10, 455–460.

Bozai, J. (1974) Contributions to the knowledge of the tetranychoid fauna of Hungary (Acari). *Folia Entomologica Hungarica* 27, 5–7.

Bradfute, O.E., Whitmoyer, R. and Nault, L.R. (1970) Ultrastructure of plant leaf tissue infected with mite-borne viral-like pathogens.In: Arceneaux, C.J. (ed.) *Proceedings of 28th Annual Meeting of the Electron Microscope Society of America, Houston, Texas*. Claitor's Publishing Division, Baton Rouge, Louisiana, pp. 178–179.

Braga, F.M. (1957) Catálogo dos ácaros que vivem nas plantas cultivadas e nos productos armazenados no Brasil. *Boletim Fitossanitário* 7, 31–44.

Brandenburg, R.L. and Kennedy, G.G. (1982a) Intercrop relationships and spider mite dispersal in a corn/peanut agroecosystem. *Entomologia Experimentalis et Applicata* 32, 269–276.

Brandenburg, R.L. and Kennedy, G.G. (1982b) Relationship of *Neogyzites floridana* (Entomophthorales: Entomophthoraceae) to two-spotted spider mite (Acari: Tetranychidae) populations in field corn. *Journal of Economic Entomology* 75, 691–694.

Braun, A.R., Guerrero, J.M., Bellotti, A.C. and Wilson, L.T. (1987) Relative toxicity of permethrin to *Mononychellus progresivus* Doreste and *Tetranychus urticae* Koch (Acari: Tetranychidae) and their predators *Amblyseius limonicus* Garman & McGregor (Acari: Phytoseiidae) and *Oligota minuta* Cameron (Coleoptera: Staphylinidae): bioassays and field validation. *Environmental Entomology* 16, 545–550.

Braun, A.R., Bellotti, A.C., Guerrero, J.M. and Wilson, L.T. (1989) Effect of predator exclusion on cassava infested with tetranychid mites (Acari: Tetranychidae). *Environmental Entomology* 18, 711–714.

Bravemboer, L. and Dosse, G. (1962) *Phytoseiulus riegeli* Dosse als pradator einiger schadmilben aus der *Tetranychus urticae* gruppe. *Entomologia Experimentalis et Applicata* 5, 291–304.

Breakey, E.P. and Batchelor, G.S. (1950) Phyllody of chrysanthemum and the eriophyid mite, *Paraphytoptus chrysanthemi* Keifer. *Annals of the Entomological Society of America* 43, 492–494.

Breakey, E.P. and Batchelor, G.S. (1957) *Biology and Control of the Dryberry Mite,* Phyllocoptes gracilis *(Nal.)*. Bulletin No. 574, Washington State Agricultural Experiment Station, Pullman, Washington.

Breeuwer, J.A.J. (1997) *Wolbachia* and cytoplasmic incompatibility in the spider mites *Tetranychus urticae* and *T. turkestani*. *Heredity* 79, 41–47.

Breeuwer, J.A.J. and Jacobs, G. (1996) *Wolbachia*: intracellular manipulators of mite reproduction. *Experimental and Applied Acarology* 20, 421–434.

Bremer, K. and Heikinheimo, O. (1979) Problems of the disease of *Ribes* in Finland. *Acta Horticulturae* 95, 87–91.

Brennan, R., Stewart, D. and Russell, J. (2008) Developments and progress in RIBES breeding. *Acta Horticulturae* 777, 49–55.

Brennan, R., Jorgensen, L., Gordon, S., Loades, K., Hackett, C. and Russell, J. (2009) The development of a PCR-based marker linked to resistance to the blackcurrant gall mite (*Cecidophyopsis ribis*) (Acari: Eriophyidae). *Theoretical and Applied Genetics* 118, 205–211.

Brennan, R.M., Robertson, G.W., McNicol, J.W., Fyffe, L. and Hall, J.E. (1992) The use of metabolic profiling in the identification of gall mite (*Cecidophyopsis ribis* Westw.) – resistant black currant (*Ribes nigrum* L.) genotypes. *Annals of Applied Biology* 121, 503–509.

Breukel, L.M. and Post, A. (1959) The influence of the manurial treatment of orchards on the population density of *Metatetranychus ulmi* (Koch). *Entomologia Experimentalis et Applicata* 2, 38–47.

Bridges, R.J. and Moser, J.C. (1983) Role of two phoretic mites in transmission of bluestain fungus, *Ceratocystis minor*. *Ecological Entomology* 8, 9–12.

Briese, D.T. (2005) Translating host-specificity test results into the real world: the need to harmonise the yin and yang of current testing procedures. *Biological Control* 35, 208–214.

Briese, D.T. and Cullen, J.M. (2001) The use and usefulness of mites in weed biological control. In: Halliday, R.B., Walter, D.E., Proctor, H.C., Norton, R.A. and Colloff, M.J. (eds) *Acarology: Proceedings of the 10th International Congress of Acarology, 5–10 July 1998*, Canberra. CSIRO Publishing, Collingwood, Australia, pp. 453–463.

Briones, M.L. and McDaniel, B. (1976) *Eriophyid Plant Mites of South Dakota*. Technical Bulletin No. 43, Agricultural Experiment Station, South Dakota State University, Brookings, South Dakota.

Briones, M.L. and Sill, W.H. Jr (1972) Survival, maturation, and reproduction of *Aceria tulipae* (Keifer), vector of wheat streak mosaic virus, at different temperatures. *Proceedings of the South Dakota Academy of Science* 51, 201–213.

Britto, E.P.J., Gondim, M.G.C., Jr, Návia, D. and Flechtmann, C.H.W. (2008) A new deuterogynous eriophyid mite (Acari: Eriophyidae) with dimorphic males from *Caesalpinia echinata* (Caesalpiniaceae) from Brazil: description and biological observations. *International Journal of Acarology* 34, 307–316.

Britto, E.P.J., Gondim, M.G.C., Jr, Torres, J.B., Fiaboe, K.K.M., Moraes, G.J. and Knapp, M. (2009) Predation and reproductive output of the ladybird beetle *Stethorus tridens* preying on tomato red spider mite *Tetranychus evansi*. *BioControl* 54, 363–368.

Brodeur, J. and Cloutier, C. (1992) A modified leaf disk method for rearing predaceous mites (Acarina: Phytoseiidae). *Phytoprotection* 73, 69–72.

Brodrick, H.T. (1965) The increasing importance of red mite in South African citrus orchards. *South African Citrus Journal* 379, 21–23.

Brody, A.R. (1970) Observations on the fine structure of the developing cuticle of a soil mite *Oppia coloradensis* (Acarina: Cryptostigmata). *Acarologia* 12, 421–431.

Brody, A.R., McGrath, J.C. and Wharton, G.W. (1976) *Dermatophagoides farinae*: the supracoxal glands. *Journal of the New York Entomological Society* 84, 34–47.

Bronner, R. and Westphal, E. (1987) Modifications des noyaux des cellules epidermiques foliaires de Solanacees soumises a l'action parasitaire d'un acerien cecidogene, *Eriophyes cladophthirus*. *Premier Congres de la Societé Française de Phytopathologie, Rennes, 19–20 Novembre 1987. Résumé des Communications*. Societé Française de Phytopathologie, Paris, abstract 39.

Bronner, R., Westphal, E. and Dreger, F. (1989) Chitosan, a component of the compatible interaction between *Solanum dulcamara* L. and the gall mite *Eriophyes cladophthrius* Nal. *Physiological and Molecular Plant Pathology* 34, 117–130.

Bronner, R., Westphal, E. and Dreger, F. (1991a) Pathogenesis-related proteins in *Solanum dulcamara* L. resistant to the gall mite *Aceria cladophthirus* Nalepa (syn. *Eriophyes cladophthirus* Nal.). *Physiological and Molecular Plant Pathology* 38, 93–104.

Bronner, R., Westphal, E. and Dreger, F. (1991b) Enhanced peroxidase activity associated with the hypersensitive response of *Solanum dulcamara* to the gall mite *Aceria cladophthrius* (Acari: Eriophyoidea). *Canadian Journal of Botany* 69, 2192–2196.

Brooking, B.C. Jr (1957) A study of the location and termination of diapause in *Petrobia apicalis* (Banks). MSc. thesis, Louisuansa State University, Baton Rouge, Louisiana.

Broufas, G.D. and Koveos, D.S. (2000) Threshold temperature for post-diapause development and degree-days to hatching of winter eggs of the European red mite (Acari: Tetranychidae) in Northern Greece. *Environmental Entomology* 29, 710–713.

Brown, J.H. and Davidson, D.W. (1977) Competition between seed-eating rodents and ants in desert ecosystems. *Science* 196, 880–882.

Brown, R.D. (1980) Citrus red mite control. *Insecticicide and Acaricide Tests* 5, 43.

Brown, R.D. and Jesser, M.J. (1981) Citrus bud mite control. *Insecticicide and Acaricide Tests* 6, 40–41.

Brown, R.D. and Jesser, M.J. (1982) Control of citrus bud mite and citrus red mite. *Insecticicide and Acaricide Tests* 7, 48–49.

Brown, R.D. and Jones, V.P. (1983) The broad mite on lemons in southern California. *California Agriculture* 37(7, 8), 21–22.

Brown Guedira, G.L., Gill, B.S., Bockus, W.W., Cox, T.S., Hatchett, J.H., Leath, S., Peterson, C.J., Thomas, J.B. and Zwer, P.K. (1996) Evaluation of a collection of wild timopheevi wheat for resistance to disease and arthropod pests. *Plant Disease* 80, 928–933.

Browne, F.G. (1968) *Pest and Diseases of Forest Plantation Trees. An Annotated List of the Principal Species Occuring in the British Commonwealth*. Clarendon Press, Oxford, UK.

Browning, H.W. (1992) Approaches to integrated pest management on processing fruit. In: *Proceedings of the Citrus Integrated Pest Management Short Course*. Publication 503, University of Florida, Gainesville, Florida, pp. 3–11.

Bruin, J. and van der Geest, L. (eds) (2009) *Diseases of Mites and Ticks*. Springer, Dordrecht, Netherlands. Originally published as *Experimental and Applied Acarology* 46.

Brun, L.O., Edge, V.E. and Gutierrez, J. (1983) The occurrence of five *Tetranychus* spp. (Acari: Tetranychidae) in New Caledonia and their responses to acaricides. *Tropical Pest Management* 29, 371–377.

Bu, G.-S., and Li, L.-S. (1998) Taxonomic notes on and key to known species of the genus *Rhizoglyphus* (Acari: Acaridae) from China. *Systematic and Applied Acarology* 3, 179–182.

Bua, G. (1964) Due nuovi acari parassiti delle fungaie. *Informatore Fitopatologico* 14, 117–118.

Buchanan, G.A., Bengston, M. and Exley, E.M. (1980) Population growth of *Brevipalpus lewisi* McGregor (Acarina: Tenuipalpidae) on grapevines. *Australian Journal of Agricultural Research* 31, 957–965.

Buergo, J.A., Gonzalez, E., Ramos, M., Perez, N. and Batista, O. (1986) Evaluacion de plaguicidas para el control quimico contra acaros tetranicos y tarsonemidos en citricos. *Revista de Protección Vegetal* 1, 163–171.

Bugeme, D.M., Maniania, N.K., Knapp, M. and Boga, H.I. (2008) Effect of temperature on virulence of *Beauveria bassiana* and *Metarhizium anisopliae* isolates to *Tetranychus evansi*. *Experimental and Applied Acarology* 46, 275–285.

Bugeme, D.M., Maniania, N.K., Boga, H.I. and Knapp, M. (2010) Efficacy of *Metarhizium anisopliae* for the control of the two-spotted spider mite, *Tetranychus urticae*, attacking bean plants. In: Moraes, G.J. de, Castilho, R.C. and Flechtmann, C.H.W. (eds) *XIII International Congress of Acarology, August 23–27, 2010, Recife-PE, Brazil, Abstract Book*, abstract 71, p. 41. Available at: http://www.acarology.org/ica/ica2010/docs/0910_15_abstractBookCapa.pdf (accessed 3 February 2015).

Bulla, A.D. (1975) *Pear Rust Mite Control on Bartlett Pears*. Progress Report No. 75–20, Colorado State University Agricultural Experiment Station, Fort Collins, Colorado.

Burches, E., Pelaez, A., Morales, C., Braso, J.V., Rochina, A., Lopez, S. and Benito, M. (1996) Occupational allergy due to spider mites: *Tetranychus urticae* (Koch) and *Panonychus citri* (Koch). *Clinical and Experimental Allergy* 26, 1262–1267.

Burditt, A.K. and Reed, D.K. Jr (1963) Damage caused by *Aculops pelekassi* Keifer, a rust mite on citrus. *Proceedings of Florida State Horticultural Society* 76, 41–47.

Burditt, A.K. Jr, Reed, D.K. and Chittenden, C.R. (1963) Observations on the mites, *Phyllocoptruta oleivora* Ashm. and *Aculus pelekassi* Keifer under laboratory conditions. *Florida Entomologist* 46, 1–5.

Burger, H.C. and Ulenberg, S.A. (1990) Quarantine problems and procedures. In: Rosen, D. (ed.) *Armored Scale Insects*. World Crop Pests, Volume 4B. Elsevier Science, Amsterdam, pp. 313–327.

Burgess, J.E. and Thompson, M.M. (1985) Shoot development and bud mite infestation in hazelnut (*Corylus avellana*). *Annals of Applied Biology* 107, 397–408.

Burnett, H. C. (1961) Species of *Ficus* susceptible to the fig mosaic virus. *Proceedings of Florida State Horticultural Society* 73, 316–320.

Burnett, H.C. (1962) Additional hosts of the fig mosaic virus. *Plant Disease Reporter* 46, 693.

Burrell, R.W. and McCormick, W.J. (1964) *Typhlodromus* and *Amblyseius* (Acarina: Phytoseiidae) as predators on orchard mites. *Annals of the Entomological Society of America* 57, 483–487.

Burts, E.C. (1970) Biology of blister mites, *Eriophyes* spp., of pear and apple in the Pacific Northeast. *Melanderia* 4, 41–53.

Buskovskaya, L.M. (1976) The effect of chemicals on the mite *Anystus*. *Zashchita Rasteniĭ* No. 10, 53 (in Russian).

Bussaman, P., Chandrapatya, A., Sermswan, R.W. and Grewal, P.S. (2004) Morphology, biology and behavior of the genus *Pygmephorus* (Acari: Heterostigmata), parasite of economic edible mushrooms in Northeast, Thailand. In: *Abstracts of the 22nd International Congress of Entomology, Brisbane, Australia, August 15–21, 2004*, abstract 595.

Bussaman, P., Sermswan, R.W. and Grewal, P.S. (2006) Toxicity of the entomopathogenic bacteria *Photorhabdus* and *Xenorhabdus* to the mushroom mite (*Luciaphorus* sp.: Acari: Pygmephoridae). *Biocontrol Science and Technology* 16, 245–256.

Bussaman, P., Sa-Uth, C., Rattanasena, P. and Chandrapatya, A. (2012a) Acaricidal activities of whole cell suspension, cell-free supernatant, and crude cell extract of *Xenorhabdus stokiae* against mushroom mite (*Luciaphorus* sp.). *Journal of Zheijang University, Science B* 13(4), 261–266.

Bussaman, P., Sa-Uth, C., Rattanasena, P. and Chandrapatya, A. (2012b) Effect of crude plant extracts on mushroom mite, *Luciaphorus* sp. (Acari: Pygmephoridae). *Psyche* 2012, article ID 150958. Available at: http://www.hindawi.com/journals/psyche/2012/150958/ (accessed 24 July 2014).

Busvine, J.R. (1980) *Recommended Methods for Measurement of Pest Resistance to Pesticides*. FAO Plant Production and Protection Paper No 21, Food and Agriculture Organization of the United Nations, Rome.

Butani, D.K. (1959) Sugarcane mites – a review. *Proceedings of the Indian Academy of Science, Section B* 49, 99–102.

Butani, D.K. (1979a) Insect pests of citrus and their control. *Pesticides* 13(4), 27–33.

Butani, D.K. (1979b) *Insects and Fruits*. Periodical Expert Book Agency, New Delhi.

Butler, G.D. Jr and Abid, M.K. (1965) The biology of *Oligonychus platani* on *Pyracantha*. *Journal of Economic Entomology* 58, 687–688.

Buxton, J.H. (1989) *Tyrophagus longior* (Gervais) (Acarina; Acaridae) as a pest of ornamentals grown under protection. *Plant Pathology* 38, 447–448.

Buxton, P.A. (1920) Insect pests of dates and the date palm in Mesopotamia and elsewhere. *Bulletin of Entomological Research* 11, 287–303.

Bynum, E.D., Xu, W. and Archer, T.L. (2004) Diallel analysis of spider mite resistant maize inbred lines and F1 crosses. *Crop Science* 44, 1535–1541.

Byrne, D.H., Guerrero, J.M., Bellotti, A.C. and Gracen, V.E. (1982a) Behavior and development of *Mononychellus tanajoa* (Acari: Tetranychidae) on resistant and susceptible cultivars of cassava. *Journal of Economic Entomology* 75, 924–927.

Byrne, D.H., Guerrero, J.M., Bellotti, A.C. and Gracen, V.E. (1982b) Yield and plant growth responses of *Mononychellus* mite resistant and susceptible cassava cultivars under protected vs. infested conditions. *Crop Science* 22, 486–490.

Byrne, D.H., Bellotti, A.C. and Guerrero, J.M. (1983) The cassava mites. *Tropical Pest Management* 29, 378–394.

CABI (1963) *Distribution Maps of Plant Pests, June 1963, Map No. 165*. CAB International, Wallingford, UK.

CABI (1965) *Distribution Maps of Pests, December 1965, Map No. 209*. CAB International, Wallingford, UK.

CABI (1974) *Distribution Maps of Plant Pests, June 1974, Map No. 324*. CAB International, Wallingford, UK.

CABI (1988) *Distribution Maps of Plant Pests, June 1988, Map No. 497*. CAB International, Wallingford, UK.

CABI (1990a) *Distribution Maps of Plant Pests, June 1990, Map No. 514*. CAB International, Wallingford, UK.

CABI (1990b) *Listing of Mites Identified by CAB International and Found in Trinidad (October, 1990)*. CAB International, Wallingford, UK.

CABI (1996) *Distribution Maps of Plant Pests, December 1996 (rev.), Map No. 129*. CAB International, Wallingford, UK.

CABI (2000) *Distribution Maps of Plant Pests, December 2000, Map No. 612*. CAB International, Wallingford, UK.

CABI (2002) *Crop Protection Compendium*. CAB International, Wallingford, UK. (CD-ROM).

CABI (2007) *Crop Protection Compendium*, 2007 edn. CAB International, Wallingford, UK. Available at: http://www.cabicompendium.org/cpc/home.asp (accessed 25 July 2014).

CABI (2009) *Distribution Maps of Plant Pests, December 2009, Map No. 727*. CAB International, Wallingford, UK.

Cabral, R.V.G. and Carmona, M.M. (1969) *Aceria guerreronis* Keifer (Acarina: Eriophyidae), uma espécie nova para S. Tomé e Príncipe. *Portugaliae Acta Biologica, Série A* 10, 353–358.

Cabrera, I.M., Ramos, M. and Fernándes, B.M. (2003) Factores que influyen en la abundancia de *Steneotarsonemus spinki* en arroz en Cuba. *Manejo Integrado de Plagas y Agroecología* 69, 34–37.

Cabrera, R.I. (1977) Estudio en Cuba del *Hirsutella thompsonii*, Fisher. Control biológico del ácaro del moho (*Phyllocoptruta oleivora*, Ash. 1879). *Agrotecnia de Cuba* 9(1), 3–11.

Cabrera, R.I. (1978) Presencia de *Hirsutella thompsonii* sobre *Brevipalpus phoenicis*. *Ciencia y Técnica en la Agricultura, Cítricos y otros Frutales* 1(3), 35–38.

Cabrera, R.I. (1984) El acaro *Vasates destructor*, nuevo hospedero del hongo *Hirsutella thompsonii*. *Ciencia y Técnica en la Agricultura, Protección de Plantas* 7(3), 69–79.

Cabrera, R.I. (2004) *Hirsutella thompsonii* Fisher y los plaguicidas químicos en una nueva estrategia para el manejo integrado del ácaro del moho *Phyllocoptruta oleivora* Ashmead (Acarina: Eriophyidae) en cítricos. *Fitosanidad* 8(3), 68.

Cabrera, R.I. and Dominguez, D. (1987a) El hongo Hirsutella nodulosa, nuevo parasito para el acaro del cocotero Eriophyes guerreronis. *Ciencia y Técnica en la Agricultura, Cítricos y otros Frutales* 10(1), 41–51.

Cabrera, R.I. and Dominguez, D. (1987b) *Hirsutella nodulosa* e *Hirsutella kirchneri*: Dos nuevos hongos patogenos del acaro del moho, *Phyllocoptruta oleivora*. *Ciencia y Técnica en la Agricultura, Protección de Plantas* 10(2), 139–142.

Cabrera, R.I., Rivero, N. and Gonzales, R. (1981) Primeros estudios comparativos del control biológico y químico del ácaro del moho *Phyllocoptruta oleivora* en Cuba. In: *Memorias Primer Congreso de Nacional Citricos y Otros Frutales, Vol. 2*. Instituto de Investigaciones de Citricos y otros Frutales, Havana, pp. 63–89.

Cabrera, R.I., Caceres, I. and Dominguez, D. (1987) Estudio de dosespecies de *Hirsutella* y su hospedantes en el cultivo de laguayaba *Psidium guajava*. *Agrotecnia de Cuba* 19(1), 29–34.

Cabrera, R.I., García, A., Almaguel, L. and Ginarte, A. (1998) Microorganismos patógenos del ácaro tarsonémido del arroz *Steneotarsonemus spinki* (Acari: Tarsonemidae). In: *Libro de Resúmenes del I Encuentro Internacional de Arroz, 9–11 de Junio de 1998*. Instituto de Investigaciones del Arroz, Havana, p. 185.

Cabrera, R.I., Ginarte, A. and Hernandéz, J. (1999) Efecto de triazophos en el control del ácaro tarsonémido del arroz *Steneotarsonemus spinki* Smiley (Acari: Tarsonemidae). In: *Libro de Resúmenes, Primer Congreso de Arroz de Riego y Secano del Área del Caribe, Primer Seminario Científico Internacional de Agrotécnia Tropical, 24-26 Mayo 1999*, Camagüey, *Cuba*, p. 24

Cabrera, R.I., Nugaliyadde, L. and Ramos, M. (2002a) Presencia de *Hirsutella nodulosa* sobre el ácaro tarsonémido del arroz *Steneotarsonemus spinki* (Acari: Tarsonemidae) en Sri Lanka. In: *Memórias de II Encuentro Internacional de Arroz, 10–12 Julio* 2002a, *La Habana*, *Cuba*, pp. 186–188.

Cabrera, R.I., Hernández, J.L. and García, A. (2002b) Resultado de las aplicaciones aéreas de triazophos y *Bacillus thuringiensis* para combatir el ácaro *Steneotarsonemus spinki* (Acari: Tarsonemidae) en el cultivo del arroz. In: *Memórias de II Encuentro International de Arroz, 10–12 Julio* 2002b, *La Habana, Cuba*, pp. 206–208.

Cabrera, R.I., Garciá, A., Otero-Colina, G., Almaguel, L. and Ginarte, A. (2005a) *Hirsutella nodulosa* and other fungus species associated to the rice tarsonemid mite *Steneotarsonemus spinki* (Acari: Tarsonemidae) in Cuba. *Folia Entomológica Mexicana* 44, 115–121.

Cabrera, R.I., Hernández, J., Ginarte, A., García, A., Páez, Y., Rivero, L.E. and Vega, M. (2005b) Alternativas de los plaguicidas para el control efectivo de *Steneotarsonemus spinki* Smiley en el cultivo del arroz. In: *Libro de Resumen del III Encuentro International del Arroz: el Ácaro del Arroz* Steneotarsonemus spinki *(Tarsonemidae) Retos y Alternativas para América Latina y el Caribe, 6 de Junio de* 2005a, *La Habana, Cuba*, pp. 31–35.

Caceres, S. and Childers, C.C. (1991) Biology and life tables *Galendromus helveolus* (Acari: Phytoseiidae) on Florida citrus. *Environmental Entomology* 20, 224–229.

Cadogan, B.L. and Laing, J.E. (1977) A technique for rearing the predaceous mite *Balaustium putmani* (Acarina: Erythraeidae), with notes on its biology and life history. *Canadian Entomologist* 109, 1535–1544.

Cadogan, B.L. and Laing, J.E. (1982) A study of *Balaustium putmani* (Acari: Erythraeidae) in apple orchards in southern Ontario. *Proceedings of the Entomological Society of Ontario* 112, 13–22.

Cagle, L.R. (1943) *Life History of the Spider Mite* Tetranychus schoenei *McG*. Technical Bulletin No. 87, Virginia Agricultural Experiment Station, Blacksburg, Virginia.

Cagle, L.R. (1949) *Life History of the Two-spotted spider Mite*. Technical Bulletin No. 113, Virginia Agricultural Experiment Station, Blacksburg, Virginia.

Cagle, L.R. (1956) *Life History of the Spider Mite,* Tetranychus atlanticus *McGregor*. Technical Bulletin No. 124, Virginia Agricultural Experiment Station, Blacksburg, Virginia.

Cagle, L.R. (1962) Biology of a red spider mite, *Panonychus* sp. on raspberry in Virginia. *Annals of the Entomological Society of America* 55, 373–379.

Cai, Z.J., Luo, Y.F., Wen, S.X., Zheng, Y.Q. and Wei Z.P. (2001) Comprehensive study on applying sunspray oil year around to control citrus pests. *Fujian Journal of Agricultural Sciences* 16, 28–32 (in Chinese).

Çakmak, I., Janssen, A. and Sabelis, M.W. (2006) Intraguild interactions between the predatory mites *Neoseiulus californicus* and *Phytoseiulus persimilis*. *Experimental and Applied Acarology* 38, 33–46.

Çakmak, I., Janssen, A., Sabelis, M.W. and Baspinar, H. (2009) Biological control of an acarine pest by single and multiple natural enemies. *Biological Control* 50, 60–65.

Caldwell, S.D. (1967) Cotton, a new host for the spider mite, *Eotetranychus smithi*. *Journal of Economic Entomology* 60, 1169.

California Date Commission (1998) *A Comprehensive Systems Approach for Managing Pest of California Dates Using Low Risk Strategies. A Proposal to the Pest Management Alliance Program, DPR* (*California Department of Pesticide Regulation*). Indio, California.

Calis, J.N.M., Overmeer, W.P.J. and van der Geest, L.P.S. (1988) Tydeids as alternative prey for phytoseiid mites in apple orchards. *Mededelingen van de Faculteit Landbouwwetenschappen, Rijksuniversiteit Gent* 53, 793–798.

Calkin, J.D. (1991) Distribution of *Oligonychus* (*Oligonychus*) *ununguis* (Jacobi) (Acari: Tetranychidae) and predator mite species (Acari: Phytoseiidae) on field-grown Douglas-fir (*Pseudotsuga menziesii* (Mirb) Franco) Christmas trees. MSc. Thesis, Oregon State University, Corvallis, Oregon.

Çalmasur, O., Aslan, I. and Sahin, F. (2006) Insecticidal and acaricidal effect of three Lamiaceae plant essential oils against *Tetranychus urticae* Koch and *Bemisia tabaci* Genn. *Industrial Crops and Products* 23, 140–146.

Calvert, D.J. and Huffaker, C.B. (1974) Predator (*Metaseiulus occidentalis*)–prey (*Pronematus* spp.) interactions under sulphur and cattail pollen application in a noncommercial vineyard. *Entomophaga* 19, 361–369.

Calza, R. and Sauer, H.F.G. (1952) A aranha vermelha dos cafezais. *Biològico (Saõ Paulo)* 18, 201–208.

Camerik, A.M. (2010) Phoresy revisited. In: Sabelis, M.W. and Bruin, J. (eds) *Trends in Acarology: Proceedings of the 12th International Congress, 21–26 August 2006*, Amsterdam. Springer, Dordrecht, the Netherlands, pp. 333–336.

Camerik, A.M., de Lillo, E. and Lalkhan, C. (2006) The neotype of *Pediculaster mesembrinae* (Canestrini, 1881) (Acari: Siteroptidae) and the description of all life stages. *International Journal of Acarology* 32, 45–67.

Cameron, W.P.L. (1925) The fern mite (*Tarsonemus tepidarum*, Warburton). *Annals of Applied Biology* 12, 93–112.

Campbell, C.A.M. and Easterbrook, M.A. (1985) Insecticides/acaricides for control of pear sucker and rust mite. Tests of Agrochemicals and Cultivars, No 6. *Annals of Applied Biology* 106(Supplement), 18–19.

Campbell, C.A.M., Easterbrook, M.A. and Fisher, A.J. (1989) Effect of the plant growth regulators paclobutrazol and chlormequat chloride on pear psyllid (*Cacopsylla pyricola* (Foerster)) and pear rust mite (*Epitrimerus piri* (Nal.)). *Journal of Horticultural Science* 64, 561–564.

Camporese, P. and Duso, C. (1995) Life history and life table parameters of the predatory mite *Typhlodromus talbii*. *Entomologia Experimentalis et Applicata* 17, 149–157.

Campos, F.J. and Omoto, C. (2002) Resistance to hexythiazox in *Brevipalpus phoenicis* (Acari: Tenuipalpidae) from Brazilian citrus. *Experimental and Applied Acarology* 26, 243–251.

Campos, F.J. and Omoto, C. (2006) Estabilidade da resistencia de *Brevipalpus phoenicis* (Geijskes) (Acari: Tenuipalpidae) a hexythiazox em pomares de citros. *Neotropical Entomology* 35, 840–848.

Candolfi, M.P., Wermelinger, B. and Boller, E.F. (1993a) Photosynthesis and transpiration of Riesling × Sylvaner grapevine leaves as affected by the European red mite (*Panonychus ulmi* Koch) (Acari: Tetranychidae) feeding. *Journal of Applied Entomology* 115, 233–239.

Candolfi, M.P., Wermelinger, B. and Boller, E.F. (1993b) Influence of the red mite (*Panonychus ulmi* KOCH) on yield, fruit quality and plant vigour of three *Vitis vinifera* varieties. *Wein-Wissenschaft (Viticulture and Enological Science)* 48, 161–164.

Canerday, T.D. and Arant, F.S. (1964) The effect of late season infestations of the strawberry spider mite, *Tetranychus atlanticus*, on cotton production. *Journal of Economic Entomology* 57, 931–933.

Canestrini, G. (1891a) Abbozzo del sistema acarologico. *Atti del Reale Istituto Veneto di Scienze, Lettere ed Arti, Serie 7*, 2, 699–725.

Canestrini, G. (1891b) Sopra due nuove species di *Phytoptus*. *Atti della Società Veneto-Trentina di Scienze Naturali* 12, 138–141.

Canestrini, G. (1892) Sopra due nuove species di *Phytoptus*. *Bollettino della Società Veneto-Trentina di Scienze Naturali* 5, 79–80.

Canestrini, G. and Fanzago, F. (1876) Nuovi acari Italiana (sec. ser.). *Atti della Accademia di Scienze Veneto-Trentina-Istriana* 5, 130–142.

Canestrini, G. and Fanzago, F. (1877) Intorno agli acari italiani. *Atti del Reale Istituto Veneto di Scienze, Lettere ed Arti, Serie 5*, 4, 69–208.

Canestrini, G. and Massalongo, C. (1894) Nuova specie di fitoptidi italiani. *Atti della Società Veneto-Trentina di Scienze Naturali, Serie 2*, 1(2), 465–466.

Canestrini, R. (1881) Contribuzione allo studio degli acari parassiti degli insetti. *Atti della Società Veneto-Trentina di Scienze Naturali* 7(2), 168–169.

Cantliffe, D.J., Stoffella, P.J. and Shaw, N.L. (2004) Distribution of the predatory mite, *Neoseiulus cucumeris*, in greenhouse pepper. *Acta Horticulturae* 659(1), 281–285.

Cao, D., Pimentel D. and Hart K. (2002) Postharvest crop losses (insects and mites). In: Pimentel, D. (ed.) *Encyclopedia of Pest Management*. Marcel Dekker, New York, pp. 645–647.

Cao, H.G., Liang, X.N. and Yang, Z.Q. (1998) Influence of temperature–humidity on the development of *Tetranychus kanzawai*. *Acta Entomologica Sinica* 41, 153–156.

Cao, Y., Liu, M.-L. and Wang, M.-N. (1993) Evaluation on the resistance of cotton spider mites to three insecticides. *Plant Protection* 19(2), 12–14 (in Chinese).

Capua, S. and Gerson, U (1983) The effects of humidity and temperature on hypopodial molting of *Rhizoglyphus robini*. *Entomologia Experimentalis et Applicata* 34, 96–98.

Carayon, J. and Steffan, J.R. (1959) Observations sur le régime alimentaire des *Orius* et particulièrement d'*Orius pallidicornis* (Reuter) (Heteropetara Anthocotidae). *Cahiers des Naturalistes* 15, 53–63.

Cardenas, M.R. (1983) La arañita roja del cafeto (*Oligonychus yothersi*). *Avances Técnicos Cenicafé* 110, 1–2. Available at: http://www.cenicafe.org/es/publications/avt0110.pdf (accessed 4 May 2015).

Cardenas, R.M., Gonzalez, L., Parra, Y., Rivero, D. and Cruz, A. (2003) Influence of rice grain discoloration (*Oryza sativa* Lin.) on variety J-104, harmfulness and genera of fungi present. *Revista de Protección Vegetal* 18, 124–128.

Carèsche, L.A. and Wapshere, A.J. (1974) Biology and host specificity of the chondrilla gall mite, *Aceria chondrillae* (G. Can.) (Acarina: Eriophyoidea). *Bulletin of Entomological Research* 64, 183–192.

Carew, M.E., Goodisman, M.A.D. and Hoffmann, A.A. (2004) Species status and population genetic structure of grapevine eriophyoid mites. *Entomologia Experimentalis et Applicata* 111, 87–96.

Carew, M.[E.], Schiffer, M., Umina, P., Weeks, A. and Hoffmann, A. (2009) Molecular markers indicate that the wheat curl mite, *Aceria tosichella* Keifer, may be a species complex in Australia. *Bulletin of Entomological Research* 99, 479–486.

Carey, J.R. (1983) Demography of the two-spotted spider mite. *Oecologia* 52, 389–395.

Carey, J.R. and Bradley, J.W. (1982) Developmental rates, vital schedules, sex ratios and life tables for *Tetranychus urticae*, *T. turkestani* and *T. pacificus* (Acarina: Tetranychidae) on cotton. *Acarologia* 23, 333–345.

Carlton, J.T. (1996) Biological invasions and cryptogenic species. *Ecology* 77, 1653–1655.

Carmona, M.M. (1960) Contribução para o conhecimento dos àcaros das plantas cultivadas em Portugal – I. *Agronomia Lusitana* 22, 221–230.

Carmona, M.M. (1964) Contribução para o conhecimento dos àcaros das plantas cultivadas em Portugal – III. *Agronomia Lusitana* 24, 273–288.

Carmona, M.M. (1965) An eriophyoid mite, *Eriophyes spathodeae* n. sp. (Acarina: Eriophyidae) on *Spathodea campanulata* Beauv. (Bignoniaceae). *Agronomia Lusitana* 27, 181–183.

Carmona, M.M. (1966) Acaros das plantas cultivadas IV. *Agronomia Lusitana* 26, 175–203.

Carmona, M.M. (1967) Contribução para o estudo de alguns ácaros fitófagos e depredadores, de Angola. *Agronomia Lusitana* 29, 267–288.

Carmona, M.M. (1969) *Metaculus rapistri*, n. sp. (Acarina: Eriophyidae) from Portugal. *Agronomia Lusitana* 30, 89–92.

Carmona, M.M. (1970) Contribução para o conhecimento dos àcaros das plantas cultivadas em Portugal – V. *Agronomia Lusitana* 31, 137–183.

Carmona, M.M. (1973) Notes on the bionomics of *Calepitrimerus vitis* (Nal.) (Acarina: Eriophyidae). In: Daniel, M. and Rosický, B. (eds) *Proceedings of the 3rd International Congress of Acarology Held in Prague, 31 August–6 September 1971*. Dr W. Junk, The Hague, the Netherlands (reprinted by Springer), pp. 197–199.

Carmona, M.M. (1978) *Calepitrimerus vitis* (Nalepa), responsável pela "Acariose da videira". 1 – Notas sobre a morfologia, biologia e sintomatologia. *Agronomia Lusitana* 39, 29–56.

Carmona, M.M. (1992) Acaros fitofagos e predadores da Ilha da Madeira – II. *Boletín de Sanidad Vegetal, Plagas* 18, 469–482.

Carmona, M.M. and Dias, J.C.S. (1980) O complexo Acarina nas culturas portuguesas. In: *I Congresso Português Fitiatria e Fitofarmacologia, Dezembro 1980, Lisboa, Portugal, Volume* 2, pp. 97–115.

Carner, G.R. and Carnerday, T.D. (1968) Field and laboratory investigations with *Entomophthora fresenii*, a pathogen of *Tetranychus* spp. *Journal of Economic Entomology* 61, 956–959.

Carrillo, D. and Peña, J.E. (2010) Studies on the biology of native predators associated with *Raoiella indica* (Acari: Tenuipalpidae) in Florida, USA: implications on their potential as biological control agents of this exotic species. In: Moraes, G.J. de, Castilho, R.C. and Flechtmann, C.H.W. (eds) *XIII International Congress of Acarology, August 23–27, 2010, Recife-PE, Brazil, Abstract Book*, abstract 79, p. 45. Available at: http://www.acarology.org/ica/ica2010/docs/0910_15_abstractBookCapa.pdf (accessed 3 February 2015).

Carrillo, D. and Peña, J.E. (2012) Prey-stage preferences and functional and numerical responses of *Amblyseius largoensis* (Acari: Phytoseiidae) to *Raoiella indica* (Acari: Tenuipalpidae). *Experimental and Applied Acarology* 57, 361–372.

Carrillo, D., Peña, J.E., Hoy, M.A. and Frank, J.H. (2010) Development and reproduction of *Amblyseius largoensis* (Acari: Phytoseiidae) feeding on pollen, *Raoiella indica* (Acari: Tenuipalpidae), and other microarthropods inhabiting coconuts in Florida, USA. *Experimental and Applied Acarology* 52, 119–129.

Carrillo, D., Amalin, D., Hosein, F., Roda, A., Duncan, R.E. and Peña, J.E. (2012a) Host plant range of *Raoiella indica* (Acari: Tenuipalpidae) in areas of invasion of the New World. *Experimental and Applied Acarology* 57, 271–289.

Carrillo, D., Frank, J.H., Rodrigues, J.C.V. and Peña, J.E. (2012b) A review of the natural enemies of the red palm mite, *Raoiella indica* (Acari: Tenuipalpidae). *Experimental and Applied Acarology* 57, 347–360.

Carrillo, D., de Coss, M.E., Hoy, M.A. and Peña, J.E. (2012c) Variability in response of four populations of *Amblyseius largoensis* (Acari: Phytoseiidae) to *Raoiella indica* (Acari: Tenuipalpidae) and *Tetranychus gloveri* (Acari: Tetranychidae) eggs and larvae. *Biological Control* 60, 39–45.

Carroll, D. P. (1980) Biological notes on the spider of some citrus groves in central and southern California. *Entomological News* 91, 147–154.

Carruthers, R.I. and D'Antonio, C.M. (2005) Science and decision making in biological control of weeds: benefits and risks of biological control. *Biological Control* 35, 181–182.

Carson, C. and Gough, N. (2000) Hibiscus erineum mite. (Publication) H00054, Queensland Horticulture Institute, Department of Primary Industries, Indooroopilly, Queensland, Australia.

Carson, C. and Gough, N. (2007) Hibiscus erinose mite. Avalilable at: http://www.hibiscusworld.com/pests/Erinose.htm (accessed 4 May 2015).

Carter, W. (1973) *Insects in Relation to Plant Disease*. Wiley, New York.

Carvalho, T.M.B. de, Reis, P.R., Oliveira, D.F. de, Carvalho, G.A. and Carvalho, D.A. de (2008) Evaluation of plant extracts in the control of *Oligonychus ilicis* (McGregor, 1917) (Acari: Tetranychidae) on coffee tree. *Coffee Science* 3, 94–103. Available at: http://www.coffeescience.ufla.br/index.php/Coffeescience/article/view/79 (accessed 4 May 2015).

Casey, C.A. and Parrella, M.P. (2005) Evaluation of a mechanical dispenser and interplant bridges on the dispersal and efficacy of the predator, *Phytoseiulus persimilis* (Acari: Phytoseiidae) in greenhouse cut roses. *Biological Control* 32, 130–136.

Castagnoli, M. (1973) Contribution to the knowledge of eriophyid mites living on the genus *Pinus* in Italy. *Redia* 54, 1–22.

Castagnoli, M. (1978) Ricerche sulle cause di deperimento e moria dello *Spartium junceum* L. *Eriophyes genistae* (Nal.) e *E. spartii* (G.Can.) (Acarina: Eriophyoidea): ridescrizione, cenni di biologia e Danni in Italian. *Redia* 61, 539–550.

Castagnoli, M. (1981) Gli acari dell'olivo in Italia e loro importanza fitopatologica. In: *Etat d'Avances des Travaux et Échange d'Information sur les Problèmes Posés par la Lutte Integré en Oléicolture. Actes de Ia Réunion Groupe d'Experts CCE (Commission des Communautés Européennes), INRA (Institut National de la Recherche Agronomiques, France), Antibes, 4–6 Novembre 1981*, pp. 179–187. (See also Castagnoli, M. (1982) *Annali dell'Istituto Sperimentale per la Zoologia Agraria* 7, 85–89).

Castagnoli, M. (1996) Ornamental coniferous and shade trees. In: Lindquist, E.E., Sabelis, M.W. and Bruin, J. (eds) *Eriophyoid Mites: Their Biology, Natural Enemies and Control*. World Crop Pests, Volume 6. Elsevier Science, Amsterdam, pp. 661–671.

Castagnoli, M. and Liguori, M. (1986) Development and oviposition times of *Typhlodromus exhilaratus* Ragusa (Acarina: Phytoseiidae) reared on various types of food. *Redia* 69, 361–368.

Castagnoli, M. and Nannelli, R. (1987) Further observations on population trends of mites in a peach orchard-meadow in central Italy. *Redia* 70, 121–134.

Castagnoli, M. and Simoni, S. (2000) Observations on intraplant distribution and life history of eriophyoid mites (Acari: Eriophyidae, Phytoptidae) inhabiting evergreen cypress (*Cupressus sempervirens* L.). *International Journal of Acarology* 26, 93–99.

Castagnoli, M. and Souliotis, P.P. (1982) Seasonal fluctuations and biology of the eriophyids of olive in Tuscany. *Redia* 65, 329–339.

Castagnoli, M., Liguori, M. and Nannelli, R. (1984) Contributo alla conoscenza degli acari associati al pesco in Toscana e osservazioni sull'andamento delle loro popolazioni. *Redia* 67, 493–504.

Castagnoli, M., Simoni, S., Panconesi, A. and Failla, O. (2002) Susceptibility of cypress seedlings to the eriophyoid mite *Trisetacus juniperinus*. *Experimental and Applied Acarology* 26, 195–207.

Castagnoli, M., Lewandowski, M., Łabanowski, G.S., Simoni, S. and Soika, G.M. (2010) An insight into some relevant aspects concerning eriophyoid mites inhabiting forests, ornamental trees and shrubs. *Experimental and Applied Acarology* 51, 169–189.

Castiglioni, E. and Návia, D. (2010) Presence of the wheat curl mite, *Aceria tosichella* Keifer (Prostigmata: Eriophyidae), in Uruguay. *Agrociencia (Montevideo)* 14(1), 19–26.

Castillo, R. (1990) Damage of *Colomerus vitis* Pgst. on buds in the area of Jerez de la Frontera. *IOBC-WPRS Bulletin* 13(7), 150–153.

Castro, B.A., Ochoa, R. and Cuevas, F.E. (2006) The threat of the panicle rice mite, *Steneotarsonemus spinki* Smiley, to rice production in the United States. In: *Proceedings of the Thirty First Rice Technical Working Group, 26 February–1 March 2006. The Woodlands, Texas*, pp. 97–98.

Castro, D., Espinosa, J. and Vargas, M. (2004) Ionising radiation as a quarantine treatment for controlling *Brevipalpus chilensis* (Acarina: Tenuipalpidae) in Thompson Seedless grapes. In: *IAEA Technical Documents (IAEA-TECDOCs) 1427*. International Atomic Eneregy Agency, Vienna, pp. 143–153.

Castro, E.B., Nuvoloni, F.M. and Feres, R.J.F. (2010a) Population dynamics of phytophagous mites (Acari: Eriophyidae, Tenuipalpidae, Tetranychidae) in three rubber tree clones (*Hevea brasiliensis* Muel Arg., Euphorbiaceae) in southern Bahia State, Brazil. In: Moraes, G.J. de, Castilho, R.C. and Flechtmann, C.H.W. (eds) *XIII International Congress of Acarology, August 23–27, 2010, Recife-PE, Brazil, Abstract Book*, abstract 83, p. 17. Available at: http://www.acarology.org/ica/ica2010/docs/0910_15_abstractBookCapa.pdf (accessed 3 February 2015).

Castro, T.R., Wekesa, V.W., Delalibera, I. Jr and Klingen, I. (2010b) Effect of photoperiod and light intensity on sporulation and germination of *Neozygites floridana* (Entomophthorales: eozygitaceae) isolates from Brazil and Norway. In: Moraes, G.J. de, Castilho, R.C. and Flechtmann, C.H.W. (eds) *XIII International Congress of Acarology, August 23–27, 2010, Recife-PE, Brazil, Abstract Book*, abstract 86, pp. 48–49. Available at: http://www.acarology.org/ica/ica2010/docs/0910_15_abstractBookCapa.pdf (accessed 3 February 2015).

Caterino, M.S., Cho, S. and Sperling, F.A.H. (2000) The current state of insect molecular systematics: a thriving tower of Babel. *Annual Review of Entomology* 45, 1–54.

Catherall, P.L. (1986) Resistance to ryegrass mosaic virus in *Lolium perenne*. *Annals of Applied Biology* 7(Supplement), 146–147.

Catherall, P.L. (1987) Effects of barley yellow dwarf and ryegrass mosaic viruses alone and in combination on the productivity of perennial and Italian ryegrasses. *Plant Pathology* 36, 73–78.

Catherall, P.L. and Chamberlain, J.A. (1975) Occurrence of agropyron mosaic virus in Britain. *Plant Pathology* 24, 155–157.

Cating, R.A., Hoy, M.A. and Palmateer, A.J. (2010) Silwet L-77 improves the efficacy of horticultural oils for control of Boisduval scale *Diaspis boisduvalii* (Hemiptera: Diaspididae) and the flat mite *Tenuipalpus pacificus* (Arachnida: Acari: Tenuipalpidae) on orchids. *Florida Entomologist* 93, 100–106.

Cavalcante, A.C.C., Pallini, A., Visôtto, L.E., Janssen, A., Sarmento, R.A. and Bernardo, A.M.G. (2010) Associated microorganisms affect the performance of *Tetranychus evansi* Baker and Pritchard (Acari: Tetranychidae) on tomato. In: Moraes, G.J. de, Castilho, R.C. and Flechtmann, C.H.W. (eds) *XIII International Congress of Acarology, August 23–27, 2010, Recife-PE, Brazil, Abstract Book*, abstract 87, p. 49. Available at: http://www.acarology.org/ica/ica2010/docs/0910_15_abstractBookCapa.pdf (accessed 3 February 2015).

Cavalloro, R. and Prota, R. (1983) *Integrated Control in Citrus Orchards: Sampling Methodology and Threshold for Intervention Against the Principal Phytophagous Pests: Proceedings of the E.C. Experts' Meeting "Integrated Control in Citrus: Comparison of Results Achieved by Applying a Standardized Methodology", 20–22 October 1982, Siniscola-Muravera*, Italy. Commission of the.European Communities, Luxembourg.

Cave, G.L. and Randall-Schadel, B. (2012) *Pest Risk Assessment for Rice Seed from the Western Hemisphere and Europe, Version 2*. Animal and Plant Health Inspection Service, US Department of Agriculture, Raleigh, North Carolina.

CDPR (2007) Summary of Pesticide Use Report Data – 2006, Indexed by Commodity. California Department of Pesticide Regulation, Sacramento, California. Available at: http://www.cdpr.ca.gov/docs/pur/pur06rep/06com.htm (accessed 3 April 2015)

Central Rice Research Institute (2006) A new alternate host of rice panicle mite. *CRRI Newsletter* 27(3), 10.

Cermeli, L.M. and Aponte, O.L. (1990) El acaro de la caña de azucar, *Abacarus sacchari* ChannaBasavanna 1966 (Acari, Eriophyidae), en Venezuela. *Boletín de Entomología Venezolana* 5(20), 164.

Cermeli, L.M., Doreste, S.E. and Balen, L. van (1982) *Aculops lycopersici* (Massee, 1937) (Acari, Eriophyidae) nueva plaga del cultivo del tomate en Venezuela. *Revista de la Facultad de Agronomía, Universidad Central de Venezuela* 12, 227–234.

Cerna, E., Badii, M.H., Ochoa, Y., Aguirre, L.A.U. and Landeros, J. (2009) Tabla de vida de *Oligonychus punicae* Hirst (Acari: Tetranychidae) en hojas de aguacate (*Persea americana* Mill) variedad Hass, Fuerte y Criollo. *Universidad y Ciencia – Trópico Húmedo* 25, 133–140.

Chaboussou, F. (1966) Nouveaux aspects de la phytiatrie et de la phytopharmacie. Le phénomène de la trophobiose. *Proceedings of the FAO Symposium on Integrated Pest Control, Rome, Italy, 11–15 October 1965, Vol. 1.* Food and Agriculture Organization of the United Nations, Rome, pp. 33–61.

Chaboussou, F. and Bessard, A. (1954) La question des tétranyques ou araignées rouges sur les arbres fruitiers. *Revue de Zoologie Agricole et Appliquée* 53, 1–18.

Chada, H.L. (1956) Biology of the winter grain mite and its control in small grains. *Journal of Economic Entomology* 49, 515–520.

Chagas, C.M., Rossetti, V., Colariccio, A., Lovisolo, O., Kitajima, E.W. and Childers, C.C. (2001) *Brevipalpus mites* (Acari: Tenuipalpidae) as vectors of plant viruses. In: Halliday, R.B., Walter, D.E. Proctor, H.C., Norton, R.A. and Colloff, M.J. (eds) *Acarology: Proceedings of the 10th International Congress of Acarology, 5–10 July 1998*, Canberra. CSIRO Publishing, Collingwood, Australia, pp. 369–375.

Chagas, C.M., Kitajima, E.W. and Rodrigues, J.C.V. (2003) Coffee ringspot virus vectored by *Brevipalpus phoenicis* (Acari: Tenuipalpidae) in coffee. *Experimental and Applied Acarology* 30, 203–213.

Chahine, H. and Michelakis, S. (1994) Bioecological control studies on *Tetranychus urticae* Koch (Acarina, Tetranychidae). *IOBC-WPRS Bulletin* 17(1), 121–124.

Chakrabarti, S. and Mondal, S. (1979) Studies on eriophyid mites (Acarina : Eriophyoidea) of India II. Descriptions of three new species from West Bengal. *Oriental Insects* 3, 47–54.

Chakrabarti, S. and Mondal, S. (1980) Studies on the eriophyid mites (Acarina : Eriophyoidea) of India III. Description of five new species of *Calacarus* from West Bengal with key to Indian species. *Acarologia* 21, 396–407.

Chakrabarti, S. and Mondal, S. (1982) Studies on the eriophyid mites (Acarina: Eriophyoidea) of India. XV. New genus, species and new records from West Bengal. *Oriental Insects* 16, 519–525.

Chakrabarti, S. and Mondal, S. (1983) An account of the genus *Diptilomiopus* Nalepa (Acarina: Eriophyoidea) from India with descriptions of three new species and key to Indian species. *Acarologia* 24, 299–307.

Chakraborty, G., Bhabatosh, S. and Somchoudhury, A.K. (2006) Efficacy of spiromesifen 240SC in the management of tea red spider mite *Oligonychus coffeae* Nietner on tea. *Environment and Ecology* 24(Special 1), 168–170.

Chakraborty, S., Dey, P.K., Bhattacharya, D.K., Chakraborty, G., Sarkar, P.K. and Somchoudhury, A. K. (2005) Laboratory evaluation of a predatory mite, *Amblyseius coccosocius* Ghai and Menon against red spider mite, *Oligonychus coffeae* (Nietner) in tea. *Journal of Entomological Research* 29, 309–312.

Chakraborty, S., Dey, P.K., Bhattacharya, D.K., Sarkar, P.K. and Somchoudhury, A.K. (2007) Age specific life tables for red spider mite, *Oligonychus coffeae* on two different types of tea plant. *Journal of Entomological Research* 31, 213–216.

Chakrov, K. (1973) Control measures against pests of grape vine. *Rastitelna Zashtita* 21(12), 12–14 (in Bulgarian).

Chan, K.L. and Turner, C.E. (1998) Discovery of the gall mite *Aceria genistae* (Nalepa) (Acarina: Eriophyidae) on gorse and French broom in the United States. *The Pan-Pacific Entomologist* 74, 55–57.

Chandler, D., Davidson, G., Pell, J.K., Ball, B.V., Shaw, K. and Sunderland, K.D. (2000) Fungal biocontrol of Acari. *Biocontrol Science and Technology* 10, 357–384.

Chandler, D., Davidson, G. and Jacobson, R.J. (2005) Laboratory and glasshouse evaluation of entomopathogenic fungi against the two-spotted spider mite, *Tetranychus urticae* (Acari: Tetranychidae), on tomato, *Lycopersicon esculentum*. *Biocontrol Science and Technology* 15, 37–54.

Chandramani, P., Murugesan, N. and Chezhiyan, N. (2004) Effect of mite, *Phyllocoptruta oleivora* (Ashm.) infestation on fruit quality of mandarin orange. *Insect Environment* 10, 13–14.

Chandrapatya, A. (1984) Observations on morphological aspects of some eriophyid mites (Prostigmata: Eriophyidae). PhD thesis, Mississippi State University, Starkville, Mississippi.

Chandrapatya, A., Konvipasruang, P., Flechtmann, C.H.W. and Moraes, G.J. (2014) Complementary description of *Colomerus novahebridensis* Keifer (Acari, Eriophyidae), with a discussion about the constitution of the genus and its economic importance, and a tentative key to *Colomerus* Newkirk & Keifer species. *ZooKeys* 434, 17–35.

Chandrasekaran, R. (1980) Acaricidal action of calixin against tea mites. *Pesticides* 14(5), 16–17.

Chang, C.P. and Huang, S.C. (1995) Evaluation of the effectiveness of releasing green lacewing, *Mallada basalis* (Walker) for the control of tetranychid mites on strawberry. *Plant Protection Bulletin (Taipei)* 37, 41–58.

Chang, D.C. and Leu, T.S. (1986) Seasonal population changes of spider mites on carambola and their chemical control. *Plant Protection Bulletin (Taipei)* 28, 263–272.

Chang, H.Y. and Chen, C.N. (1984) Review of the mite problems and their biological control in Taiwan. *National Taiwan University Phytopathologist and Entomologist* 11, 137–156 (in Chinese).

ChannaBasavanna, G.P. (1966) *A Contribution to the Knowledge of Indian Eriophyid Mites (Eriophyoidea: Trombidiformes: Acarina)*. University of Agricultural Sciences, Bangalore, India.

ChannaBasavanna, G.P. (1973) The present status of our knowledge of Indian plant-feeding mites. In: Daniel, M. and Rosický, B. (eds) *Proceedings of the 3rd International Congress of Acarology Held in Prague, 31 August–6 September 1971*. Dr W. Junk, The Hague, the Netherlands (reprinted by Springer), pp. 201–204.

ChannaBasavanna, G.P. (1996) Sugarcane, coffee and tea. In: Lindquist, E.E., Sabelis, M.W. and Bruin, J. (eds) *Eriophyoid Mites: Their Biology, Natural Enemies and Control.* World Crop Pests, Volume 6. Elsevier Science, Amsterdam, pp. 631–640.

ChannaBasavanna, G.P., Nageshandra, B.K. and Mallik, B. (1979) *Bibliography of Indian Plant Feeding Mites.* University of Agricultural Sciences, Bangalore, India.

Chant, D.A. (1961) An experiment in biological control of *Tetranychus telarius* (L.) (Acarina: Tetranychidae) in a greenhouse using the predacious mite *Phytoseiulus persimilis* Athias-Henriot (Phytoseiidae). *Canadian Entomologist* 93, 437–443.

Chant, D.A. and McMurtry, J.A. (2007) *Illustrated Keys and Diagnoses for the Genera and 18 Subgenera of the Phytoseiidae of the World (Acari: Mesostigmata)*. Indira Publishing House, West Bloomfield, Michigan.

Chao, R.S. (1971) An ecological study on yellow rust mite, *Acaphylla steinwedeni* Keifer. *Plant Protection Bulletin (Taipei)* 13, 135–146 (in Chinese).

Chao, S.R.S. and Lo, P.K.C. (1974) Biological studies of the spider mite, *Tetranychus truncatus* and its natural enemies. *Journal of Taiwan Agricultural Research* 23, 126–135 (in Chinese).

Chapot, H. and Delucchi, V.L. (1964) *Maladies, Troubles et Ravageurs des Agrumes au Maroc*. Institut National de Recherche Agronomique, Rabat.

Charanasri, V. and Kongchuensin, M. (2001) Species and population densities of mites on jujube. In: Halliday, R.B., Walter, D.E., Proctor, H.C., Norton, R.A. and Colloff, M.J. (eds) *Acarology: Proceedings of the 10th International Congress of Acarology, 5–10 July 1998, Canberra*. CSIRO Publishing, Collingwood, Australia, pp. 419–422.

Charanasri, V., Sarinkaphaibul, C. and Seripunpanich, C. (1993) Taxonomic study on mite injurious to cassava in Thailand. In: *Insect Taxonomy and Acarology Research, Annual Report 1993*. Department of Agriculture, Bangkok, pp. 1–11 (in Thai).

Charles, J.G. and White, V. (1988) Airborne dispersal of *Phytoseiulus persimilis* (Acarina: Phytoseiidae) from a raspberry garden in New Zealand. *Experimental and Applied Acarology* 5, 47–54.

Charlet, L.D. and McMurtry, J.A. (1977) Systematics and bionomics of predaceous and phytophagous mites associated with pine foliage in California. III. Laboratory studies on the biology of the phytoseiids *Metaseiulus validus* (Chant) and *Typhloseiopsis pini* (Chant). *Hilgardia* 45(7), 223–236.

Charudattan, R. (1986) Integrated control of waterhyacinth (*Eichhornia crassipes*) with a pathogen, insects and herbicides. *Weed Science* 34(Suppl. 1), 26–30.

Chaudhri, W.M., Akbar, S. and Rasool, A. (1974) *Taxonomic Studies of the Mites Belonging to the Families Tenuipalpidae, Tetranychidae, Tuckerellidae, Caligonellidae, Stigmaeidae and Phytoseiidae: PL-480 Project on Mites*. University of Agriculture, Lyallpur, Pakistan.

Chauhan, T.P.S., Shankar Narayan and Vineet Kumar (2002) *Polyphagotarsonemus latus* (Banks): a new pest of mulberry. *Indian Journal of Forestry* 25, 171–176.

Chazeau, J. (1971a) *Stethorus comoriensis*, nouvelle espèce de Coccinellidae de l'archipel des Comores (Col.). *Bulletin de la Société Entomologique de France* 76, 60–63.

Chazeau, J. (1971b) Le genre *Stethorus* à Madagascar et aux Mascareignes (Col. Coccinellidae). *Annales de la Société Entomologique de France* (N.S.) 7, 779–796.

Chazeau, J. (1974) Evaluation de l'action predatrice de *Stethorus madecassus* (Coleoptere Coccinellidae) sur *Tetranychus neocaledonicus* (acarien, Tetranychidae). *Entomophaga* 19, 183–193.

Chazeau, J. (1979) Mise au point sur le genre *Stethorus* en Océanie et description de deux espèces nouvelles de Mélanésie (Col. Coccinellidae). *Entomophaga* 24, 295–303.

Chazeau, J. (1983) Deux prédateurs de Tetranychidae en Nouvelle-Guinée: *Stethorus exspectatus* n.sp. et *Stethorus exsultabilis* n.sp. (Col.: Coccinellidae). *Entomophaga* 28, 373–378.

Chazeau, J. (1985) Predaceous insects. In: Helle, W. and Sabelis, M.W. (eds) *Spider Mites: Their Biology, Natural Enemies and Control*. World Crop Pests, Volume 1B. Elsevier Science, Amsterdam, pp. 211–246.

Chazeau, J. and Gutierrez, J. (1977) *Quelques Données sur les Principaux Prédateurs de Tetranychidae (Acariens) dans la Zone d'Action de la Commission du Pacific Sud. Commission du Pacifique Sud: Conference Régionale de la Protection des Vegetaux, Nouméa, Nouvelle-Calédonie, 14-18 November 1977)*. Document No. SPC/Plant Protection/W.P. 9, 10 Octobre 1977, South Pacific Commission, Nouméa, New Caledonia. Available at: http://horizon.documentation.ird.fr/exl-doc/pleins_textes/pleins_textes_5/b_fdi_16-17/09388.pdf (accessed 4 May 2015).

Chazeau, J., Etienne, J. and Fursch, H. (1974) Les Coccinellidae de l'ile de La Réunion (Insecta Coleoptera). *Bulletin du Muséum National d'Histoire Naturelle (Série 3)*, *Zoologie* 210, 265–297.

Chein, C.C. (1980) Studies on the sheath-rot disease and its relation to the sterility of the rice plants. *Plant Protection Bulletin (Taipei)* 22, 31–39 (in Chinese).

Chen, C.N., Cheng, C.C. and Hsiao, K.C. (1979) Bionomics of *Steneotarsonemus spinki* attacking rice plants in Taiwan. In: Rodriguez, J.G. (ed.) *Recent Advances in Acarology, Volume I*. Academic Press, New York, pp. 111–117.

Chen, C.Y., Xu, Y.X., Fei, D.B. and Deng, G.Q. (1988) A preliminary study on the predaceous mite *Tydeus* sp. of *Phyllocoptruta oleivorus* Ashmead. *Chinese Journal of Biological Control* 4, 181 (in Chinese).

Chen, D.M. and Chen, W.M. (1986) Use of *Hirsutella thompsonii* in *Phyllocoptes oleivorus* control. *Chinese Journal of Biological Control* 3, 115.

Chen, D.M., Chen, W.M. and Li, C.J. (1987) Studies on controlling *Phyllocoptruta oleivora* with *Hirsutella thompsonii*. *Natural Enemies of Insects* 9, 13–16.

Chen, H. (1994) Effects of physical and chemical characteristics of tea leaf surface on the damage caused by pink mite (*Acaphylla theae*). *China Tea* 16(5), 12–13 (in Chinese).

Chen, H.-C., Xu, N., Chen, X.-F., Chen, Z.-M. and Yu, F.-L. (1996a) On the resistance mechanisms of tea clones to pink tea rust mite. *Acta Phytophylacica Sinica* 23, 137–142 (in Chinese).

Chen, H.-C., Xu, N. and Chen, Z.-M. (2000) On the relationship between content of free amino acid in tea shoot and resistance of tea tree to tea pink mite, *Acaphylla theae* Watt. *Acta Phytophylacica Sinica* 27, 338–342 (in Chinese).

Chen, J.L. and Li, L.S. (1985a) Study on the damage of citrus yellow mite, *Eotetranychus kankitus* Ehara on the spring tips of orange trees. *Journal of Southwest Agricultural University* 1985-3, 85–90 (in Chinese).

Chen, J.L. and Li, L.S. (1985b) Analysis of transferring damage by the citrus yellow mite, *Eotetranychus kankitus* Ehara in spring. *Journal of Southwest Agricultural University* 1985-3, 91–97 (in Chinese).

Chen, J.-S. (1990a) An improved method for determining the susceptibility of *Rhizoglyphus robini* and *R. setosus* (Acarina: Acaridae) to pesticides. *Experimental and Applied Acarology* 8, 175–178.

Chen, J.-S. (1990b) The biology, insecticide resistence, and control countermeasures of bulb mites. *Chinese Journal of Entomology, Special Publication* No. 3, 93–106.

Chen, J.-S. and Lo, K.-C. (1989) Susceptibility of two bulb mites, *Rhizoglyphus robini* and *Rhizoglyphus setosus* (Acarina: Acaridae), to some acaricides and insecticides. *Experimental and Applied Acarology* 6, 55–66.

Chen, J.-S. and Lo, K.-C. (1990) Toxicity of organophosphorus insecticides to diazinon-reversely-selected, pesticide-pressure-relaxed and field-resistant strains of bulb mite *Rhizoglyphus robini* (Acari: Acaridae). *Experimental and Applied Acarology* 8, 243–252.

Chen, Q., Conner, R.L., Li, H.J., Graf, R., Laroche, A., Li, H.Y and Ahmad, F. (2003) Genomic characterization of new sources of resistance to both wheat streak mosaic virus and wheat curl mite in wheat–*Thinopyrum* partial amphiploids. *Journal of Genetics and Breeding (Rome)* 57, 155–164.

Chen, S. (2007) Study on the control techniques for pest mites of moso bamboo. *Anhui Agricultural Science Bulletin* 2007-02, 131–132. Available as English summary at: http://en.cnki.com.cn/Article_en/CJFDTOTAL-AHNB200702063.htm (accessed 25 July 2014).

Chen, T.Y., French, J.V., Liu, T.X. and Graca, J.V. da (2006a) Predation of *Galendromus helveolus* (Acari: Phytoseiidae) on *Brevipalpus californicus* (Acari: Tenuipalpidae). *Biocontrol Science and Technology* 16, 753–759.

Chen, W. and Zhao, Z. (1994) Study on predation of *Stethorus chengi* Sasaji to *Panonychus citri* McGregor. *Journal of Southwest Agricultural University* 16, 27–31 (in Chinese).

Chen, W.H., Liu, Y.C. and Ho, C.C. (2002a) The life cycle, distribution and host plants of bulb mite, *Rhizoglyphus setosus* Manson (Acari: Acaridae) in Taiwan. *Plant Protection Bulletin (Taipei)* 44, 341–352.

Chen, W.H., Liu, Y.C., Ho, C.C. and Chang, T. Y. (2002b) A newly recorded mite pest, *Rhizoglyphus setosus* Manson (Acari: Acaridae), of onion in Taiwan. *Plant Protection Bulletin (Taipei)* 44, 249–253.

Chen, W.M., Fu, Y.G., Zhang, F.P. and Peng, Z.Q. (2005a) Effect of different varieties of litchi on the development and reproduction of *Oligonychus biharensis* (Hirst). *Systematic and Applied Acarology* 10, 11–16.

Chen, X., Zhou, Z. and Li, J. (1996b) Studies on the bionomics and control of *Tetranychus truncatus* Ehara. *Scientia Silvae Sinicae* 32(2), 144–149 (in Chinese).

Chen, X., Zhang, Y.-X., Ji, J. and Lin, J.-Z. (2007a) Predation of *Bdellodes japonicus* (Ehara) on four species of spider mites (Acari, Bdellidae & Tetranychidae). *Systematic and Applied Acarology* 12, 195–198.

Chen, X., Zhang, Y.-X., Ji, J. and Lin, J.-Z. (2011) Influence of temperature on development of *Amblyseius swirskii* (Athias-Henriot) with *Tetranychus truncatus* (Ehara) as its prey. *Acta Arachnologica Sinica* 20(1), 52–56 (in Chinese).

Chen, X.-F., Xiao, Q., Jiang, D.Y. and Ni, L.H. (1992) Efficacy of dimilin (diflubenzuron) on pest insects on tea plants. *China Tea* 14(1), 14–16 (in Chinese).

Chen, X.-F., Xiao, Q.-A. and Yao, H.-M. (1993) Efficacy of pyridaben on tea mite pests. *China Tea* 14(6), 6–7 (in Chinese).

Chen, X.-R., Wei, S.-G. and Nan, Z.-B. (2005b) A new species of Phytoptinae (Acari: Phytoptidae) from China. *Acta Zootaxonomica Sinica* 30, 81–83 (in Chinese).

Chen, X.Y., Zhang, M.-R. and Zhang, X. (1999) On the bionomics and ecological control tactics of *Tetranychus truncatus* Ehara in corn field in Shaanxi Province. *Acta Phytophylacica Sinica* 26, 7–12 (in Chinese).

Chen, Y., Lin, Y.-W. and Fan, Q.-H. (2006b) Influences of foodstuffs on the development, reproduction and hypopus induction of bulb mite *Rhizoglyphus echinopus* (Acari: Acaridae). *Entomological Journal of East China* 15, 250–252.

Chen, Y., Ran, C., Xiong, L. and Wang, J.-J. (2006c) Influence of citruses germplasm on the carboxylesterase activity and acaricide susceptibility of *Panonychus citri* McGregor. *Southwest China Journal of Agricultural Sciences* 19, 434—437 (in Chinese).

Chen, Y., Opit, G.P., Jonas, V.M., Williams, K.A., Nechols, J.R. and Margolies, D.C. (2007b) Two spotted spider mite population level, distribution, and damage on ivy geranium in response to different nitrogen and phosphorus fertilization regimes. *Journal of Economic Entomology* 100, 1821–1830.

Chen, Z.H. (1999) Experiment on the control of citrus yellow mite. *South China Fruits* 28(6), 22–23 (in Chinese).

Chen, Z.-Y., Ran, C., Zhang, L., Dou, W. and Wang, J.-J. (2009) Susceptibility and esterase activity in citrus red mite *Panonychus citri* (McGregor) (Acari: Tetranychidae) after selection with phoxim. *International Journal of Acarology* 35, 33–40.

Cheng, C.H. and Chiu, Y.I. (1999) Review of changes involving rice pests and their control measures in Taiwan since 1945. *Plant Protection Bulletin (Taipei)* 41, 9–34 (in Chinese).

Cheng, L.L., Nechols, J.R., Margolies, D.C., Campbell, J.F., Yang, P.S., Chen, C.C. and Lu, C.T. (2009) Foraging on and consumption of two species of papaya pest mites, *Tetranychus kanzawai* and *Panonychus citri* (Acari: Tetranychidae), by *Mallada basalis* (Neuroptera: Chrysopidae). *Environmental Entomology* 38, 715–722.

Cheng, L.L., Nechols, J.R., Margolies, D.C., Campbell, J.F. and Yang, P.S. (2010) Assessment of prey preference by the mass-produced generalist predator, *Mallada basalis* (Walker) (Neuroptera: Chrysopidae), when offered two species of spider mites, *Tetranychus kanzawai* Kishida and *Panonychus citri* (McGregor) (Acari: Tetranychidae), on papaya. *Biological Control* 53, 267–272.

Cheng, L.L., Nechols, J.R., Margolies, D.C., Campbell, J.F., Yang, P.S., Chen, C.C. and Lu, C.T. (2012) Efficacy of the predator *Mallada basalis* (Neuroptera: Chrysopidae) on *Tetranychus kanzawai* and *Panonychus citri* (Acari: Tetranychidae) at different predator:prey release ratios. *Journal of Asia-Pacific Entomology* 15, 142–146.

Cheremushkina, N.P., Aramov, N.K., Makarenkova, A.A. and Golyshin, N.M. (1991) The rust mite of tomato. *Zashchita Rasteniĭ* No. 11, 44–45 (in Rusian).

Cherian, M.C. (1931) South Indian Acarina. *Journal of the Asiatic Society of Bengal* 27, 141–147.

Cherian, M.C. (1933) The Cholam mite, *Paratetranychus indicus*, on sorgum. *Madras Agricultural Journal* 1, 1–6.

Cherian, M.C. (1938) Mites (Acarina) pests of crops in south India and methods of their control. *Agriculture and Livestock in India* 8, 537–540.

Chetverikov, P.E. (2012) Confocal laser scanning microscopy technique for the study of internal genitalia and external morphology of eriophyoid mites (Acari: Eriophyoidea). *Zootaxa* 3453, 56–68.

Chhillar, B.S., Rachna Gulati and Praduman Bhatnagar (2007) *Agricultural Acarology*. Daya Publishing House, New Delhi.

Chiaradia, L.A. and Cruz, F.Z. da (1996) Repercussao de tratamentos acaricidas sobre inimigos naturais de pragas das plantas citricas. *Pesquisa Agropecuária Gaúcha* 2, 173–178.

Chiasson, H.A., Be´langer, A., Bostanian, N., Vincent, C. and Poliquin, A. (2001) Acaricidal properties of *Artemisia absinthium* and *Tanacetum vulgare* (Asteraceae) essential oils obtained by different methods of extraction. *Journal of Economic Entomology* 94, 167–171.

Chiasson, H.A., Bostanian, N. and Vincent, C. (2004) Acaricidal properties of a *Chenopodium*-based botanical. *Journal of Economic Entomology* 97, 1373–1377.

Chiavegato, L.G. (1968) Contribução para o onhecimento de alguns acaros que ocorrem na seringueira (*Hevea brasiliensis* Mull). In: *Anais da I Reunião Anual da Sociedae Brasileira de Entomologia, Piracicaba*. Sociedade Brasileira de Entomologia, Curitiba, Brazil, p. 67.

Chiavegato, L.G. (1986) Biologia do acaro *Brevipalpus phoenicis* em citros. *Pesquisa Agropecuária Brasileira* 21, 813–816.

Chiavegato, L.G. (1988) Biologia do acaro *Panonychus citri* (McGregor, 1916) (Acari: Tetranychidae). Praga dos citros. *Científica (Jaboticabal)* 16, 79–84.

Chiavegato, L.G. and Camargos, A.B. de (1968) Contribução para o conhecimento dos ácaros considerados pragas do morangueiro (*Fragaria hibridos*). In: *Anais da I Reunião Anual da Sociedae Brasileira de Entomologia, Piracicaba*. Sociedade Brasileira de Entomologia, Curitiba, Brazil, pp. 70–71.

Chiavegato, L.G. and Kharfan, P.R. (1993) Comportamento do ácaro da leprose *Brevipalpus phoenicis* G. (Acari: Tenuipalpidae) em citros. *Anais de Sociedade Entomológica do Brasil* 22, 355–359.

Chiavegato, L.G., Bernardi, J.B. and Igue, T. (1968) Prejuizos causados pelo ácaro *Aceria tulipae* Keifer e pelo *Thrips tabaci* Lind., 1888, na cultura de alho (*Allium sativum* L.) cultivar lavinia. In: *Anais da I Reunião Anual da Sociedae Brasileira de Entomologia, Piracicaba*. Sociedade Brasileira de Entomologia, Curitiba, Brazil, pp. 68–69.

Chiavegato, L.G., Batista, G.C. and Igue, T. (1975) Toxicidade de acaricidas para tres especies de acaros fitofagos (Acari: Tetranychidae) obtida atraves de dois diferentes metodos de laboratorio. *Arquivos do Instituto Biológico (São Paulo)* 42, 173–182.

Chigira, A. and Miura, K. (2005) Detection of "Candidatus *Cardinium*" bacteria from the haploid host *Brevipalpus californicus* (Acari: Tenuipalpidae) and effect on the host. *Experimental and Applied Acarology* 37, 107–116.

Childers, C.C. (1985) Citrus rust mite control. *Insecticicide and Acaricide Tests* 10, 57–59.

Childers, C.C. (1988) Citrus rust mite efficacy test in Florida. *Insecticicide and Acaricide Tests* 13, 77–78.

Childers, C.C. (1990) Combination studies of selected acaricides with zineb, mancozeb, or carbamate for mite control on citrus. *International Journal of Acarology* 16, 27–36.

Childers, C.C. (1994a) Feeding injury to 'Robinson' tangerine leaves by *Brevipalpus* mites (Acari: Tenuipalpidae) in Florida and evaluation of chemical control on citrus. *Florida Entomologist* 77, 265–271.

Childers, C.C. (1994b) Effect of different copper formulations tank-mixed with fenbutatin-oxide for control of citrus rust mites (Acari: Eriophyidae) on Florida citrus. *Florida Entomologist* 77, 349–365.

Childers, C.C. (2010) Assessment of an integrated pest mite and disease management program on Florida citrus utilizing horticultural mineral oil foliar sprays. In: Moraes, G.J. de, Castilho, R.C. and Flechtmann, C.H.W. (eds) *XIII International Congress of Acarology, August 23–27, 2010, Recife-PE, Brazil, Abstract Book*, abstract 95, p. 54. Available at: http://www.acarology.org/ica/ica2010/docs/0910_15_abstractBookCapa.pdf (accessed 3 February 2015).

Childers, C.C. and Abou-Setta, M.M. (1999) Yield reduction in 'Tahiti' lime from *Panonychus citri* feeding injury following different pesticide treatment regimes and impact on the associated predacious mites. *Experimental and Applied Acarology* 23, 771–783.

Childers, C.C. and Achor, D.S. (1999) The eriophyoid mite complex on Florida citrus (Acari: Eriophyidae and Diptilomiopidae). *Proceedings of the Florida State Horticultural Society* 112, 79–87.

Childers, C.C. and Enns, W.R. (1975a) Field evaluation of early season fungicide substitutions on tetranychid mites and the predators *Neoseiulus fallacis* and *Agistemus fleschneri* in two Missouri apple orchards. *Journal of Economic Entomology* 68, 719–724.

Childers, C.C. and Enns, W.R. (1975b) Predaceous arthropods associated with spider mites in Missouri apple orchards. *Journal of the Kansas Entomological Society* 48, 453–471.

Childers, C.C. and Konsler, B.A. (1980) Mite control. *Insecticicide and Acaricide Tests* 5, 45–46.

Childers, C.C. and Rock G.C. (1981) Observations on the occurrence and feeding habits of *Balaustium putmani* (Acari: Erythraeidae) in North Carolina apple orchards. *International Journal of Acarology* 7, 63–68.

Childers, C.C. and Rodrigues, J.C.V. (2011) An overview of *Brevipalpus* mites (Acari: Tenuipalpidae) and the plant viruses they transmit. In: Moraes, G.J. de and Proctor, H. (eds) *Acarology XIII: Proceedings of the International Congress. Zoosymposia* 6, 180–192.

Childers, C.C. and Sorrell, R.W. (1982) Citrus rust mite control (CRM). *Insecticicide and Acaricide Tests* 7, 46–47, 53–58.

Childers, C.C. and Sorrell, R.W. (1984) Citrus rust mite control. *Insecticicide and Acaricide Tests* 9, 69, 83–85.

Childers, C.C., Keen, P.M. and Smith, D.A. (1989) Pesticide efficacy against citrus rust mite in Florida. *Insecticicide and Acaricide Tests* 14, 74–75.

Childers, C.C., Abou-Setta, M.M. and Nawar, M.S. (1991) Biology of *Eutetranychus banksi*: life tables on 'Marsh' grapefruit leaves at different temperatures (Acari: Tetranychidae). *International Journal of Acarology* 17, 29–35.

Childers, C.C., Easterbrook, M.A. and Solomon, M.G. (1996) Chemical control of eriophyoid mites. In: Lindquist, E.E., Sabelis, M.W. and Bruin, J. (eds) *Eriophyoid Mites: Their Biology, Natural Enemies and Control*. World Crop Pests, Volume 6. Elsevier Science, Amsterdam, pp. 695–726.

Childers, C.C., Kitajima, E.W., Welbourn, W.C., Rivera, C. and Ochoa, R. (2001) *Brevipalpus* mites on citrus and their status as vectors of citrus leprosis. *Manejo Integrado de Plagas* 60, 66–70.

Childers, C.C., French, J.V. and Rodrigues, J.C.V. (2003a) *Brevipalpus californicus, B. obovatus, B. phoenicis*, and *B. lewisi* (Acari: Tenuipalpidae): a review of their biology, feeding injury and economic importance. *Experimental and Applied Acarology* 30, 5–28.

Childers, C.C., Rodrigues, J.C.V. and Welbourn, W.C. (2003b) Host plants of *Brevipalpus californicus, B. obovatus*, and *B. phoenicis* (Acari: Tenuipalpidae) and their potential involvement in the spread of viral diseases vectored by these mites. *Experimental and Applied Acarology* 30, 29–105.

Childers, C.C., Rodrigues, J.C.V., Derrick, K.S., Achor, D.S., French, J.V., Welbourn, W.C., Ochoa, R. and Kitajima, E.W. (2003c) Citrus leprosis and its status in Florida and Texas: past and present. *Experimental and Applied Acarology* 30, 181–202.

Childers, C.C., McCoy, C.W., Nigg, H.N., Stansly, P.A. and Rogers, M.E. (2007) *2007 Florida Citrus Pest Management Guide: Rust Mites, Spider Mites, and Other Phytophagous Mites*. Publication No. ENY-603, Entomology and Nematology Department, University of Florida/Institute of Food and Agricultural Sciences (UF/IFAS) Extension, Gainesville, Florida. (For updated version see Rogers *et al.*, 2013).

Childs, G.H. (1980) Biology, spatial distribution, and control of *Oligonychus* (*Oligonychus*) *ilicis* (McGregor) (Acarina: Tetranychidae) on *Ilex crenata* "Hetzii". PhD thesis, University of Florida, Gainesville, Florida.

Chinnaiah, C. and Mohanasundaram, M. (1995) Evaluation of the efficacy of neem products and certain pesticides against *Aceria cajani* on pigeonpea. In: *Fifth National Symposium on Acarology, 20–22 September 1995, University of Agricultural Sciences, Bangalore, India*, abstract.

Chmielewski, W. (1979) Mites occurring on greenhouse cucumbers. *Ochrona Roślin* 23(10), 5–7.

Chmielewski, W. (1984) *Tyrophagus longior* (Gerv., 1844) (Acarina, Acaridae) – bioecology, occurrence and harmfulness. *Prace Naukowe Instytutu Ochrony Roślin* 26, 69–87.

Cho, M.R., Kim, D.S. and Im, D.S. (1999) A new record of tarsonemid mite *Steneotarsonemus spinki* (Acari: Tarsonemidae) and its damage on rice in Korea. *Korean Journal of Applied Entomology* 38, 157–164 (in Korean).

Cho, S.H., Kim, Y.K., Oh, H.B., Jung, J.W., Son, J.W., Lee, M.H., Jee, H.S., Kim, Y.Y. and Min, K.U. (2000) Association of HLA-DRB1*07 and DRB1*04 to citrus red mite (*Panonychus citri*) and house dust mite sensitive asthma. *Clinical and Experimental Allergy* 30, 1568–1575.

Choh, Y. and Takabayashi, J. (2010) Herbivore-induced plant volatiles prime two indirect defenses in lima bean. In: Sabelis, M.W. and Bruin, J. (eds) *Trends in Acarology: Proceedings of the 12th International Congress, 21–26 August 2006, Amsterdam*. Springer, Dordrecht, the Netherlands, pp. 255–258.

Choi, D.S. and Kim, K.C. (1998) Population fluctuation, developmental character of *Panonychus citri* and damage degree as its control density on young yuzu (*Citrus junos*). *Korean Journal of Applied Entomology* 37, 193–198 (in Korean).

Chong, J.-H., Ward, N.A., Dunwell, W.C. and Chappell, M.R. (2012) Chapter 11: Oak – *Quercus* spp. In: Fulcher, A.F. and White, S.A. (eds) *IPM for Select Deciduous Trees in Southeastern US Nursery Production*. Southern Nursery IPM Working Group, Knoxville, Tennessee. Available at: http://wiki.bugwood.org/IPM_book (accessed 26 March 2015).

Choudhury, P., Duttal, B.K. and Bhattacharjee, P.C. (2006) Some ecological factors on population dynamics of red spider mite (*Oligonychus coffeae*, Nietner) and their control in the tea agro-ecosystem of Barak Valley, Assam (India). *International Journal of Tea Science* 5(3/4), 29–39.

Chouinard, G., Bellerose, S., Brodeur, C. and Morin, Y. (2006) Effectiveness of *Hyaliodes vitripennis* (Say) (Heteroptera: Miridae) predation in apple orchards. *Crop Protection* 25, 705–711.

Chow, A., Chau, A. and Heinz, K.M. (2009) Reducing fertilization for cut roses: effect on crop productivity and two spotted spider mite abundance, distribution, and management. *Journal of Economic Entomology* 102, 1896–1907.

Chow, Y.S., Tzean, S.S., Chang, C.S. and Wang, C.H. (1980) A morphological approach of the tarsonemid mite *Steneotarsonemus spinki* Smiley (Tarsonemidae) as a rice plant pest. *Acta Arachnologica* 29, 25–42.

Christian, M.L. and Willis, W.G. (1993) Survival of wheat streak mosaic virus in grass hosts in Kansas from wheat harvest to fall wheat emergence. *Plant Disease* 77, 239–242.

Chung, C., Wei, H. and Tieng, Y. (1963) The biology of the round wheat mite (*Penthaleus* sp.), and brown wheat mite, *Petrobia latens* (Müller). *Acta Phytophylacica Sinica* 2, 277–284 (in Chinese).

Ciampolini, M. and Rota, P. (1963) Polifagia e invadenza di *Aculus pelekassi* Keifer (Acarina, Eriophyidae) su agrumi in Puglia. *Bollettino di Zoologia Agraria e di Bachicoltura (Ser. II)* 5, 83–91.

Ciampolini, M., Rota, P.A. and Schulthaus, S.D. (1976) Rugginosità delle mele provocata dall'eriofide *Aculus schlechtendali*. *Informatore Agrario* 32(40), 24243–24245.

Ciampolini, M., Lugaresi, C., Rota, P.A. and Cappella, A. (1985) Un acaro nocivo ad orchidee in serra. *Informatore Agrario* 41(33), 29–33.

CIAT (1976a) Cassava production systems. In: *Annual Report 1974*. International Center for Tropical Agriculture, Cali, Colombia, pp. 54–109.

CIAT (1976b) Sistemas de producción de yuca. In: *Informe Anual 1976a*. Centro Internacional de Agricultura Tropical, Cali, Colombia, pp. 16–21.

CIAT (1977) Cassava production systems. In: *Annual Report 1976b*. International Center for Tropical Agriculture, Cali, Colombia, pp. 1–76.

CIAT, ICA and FEDEARROZ (2005) Steneotarsonemus spinki *Smiley. Acaro del Vaneamiento del Arroz*. International Center for Tropical Agriculture, Cali, Instituto Columbiano Agropecuario, Bogota and Federación Nacional de Arroceros, Bogota, Colombia.

Ciglar, I. and Baric, B. (1998) Pernicious insects and mites fauna in Croatian orhards. *Entomologia Croatica* 4, 63–69.

Ciglar, I. and Baric, B. (2000) Side effect of some fungicides on the mite population in an apple orchard in Croatia. *Anzeiger für Schädlingskunde* 73, 110–112.

Cindea, E. (1978) Present state of research on the control of the pathogens and pests of vegetable crops. *Probleme de Protectia Plantelor* 6, 421–443.

Cinti, S., Storti, C., Vitagliano, A., Nanelli, R. and Paparatti, B. (1995) Prime osservazioni sulla biologia di *Oligonychus bicolor* (Banks) (Acari, Tetranychidae) nel comprensorio castanicolo dei Monti Cimini. *Informatore Fitopatologico* 45(3), 60–62.

Clancy, D.W. and Pollard, H.N. (1952) The effect of DDT on mite and predator populations in apple orchards. *Journal of Economic Entomology* 45, 108–114.

Clancy, G. (1981) Observations of mites associated with the low yielding crops of cultivated *Agaricus bisporus* in Australia. *Mushroom Science* 11, 233–244.

Claparède, E. (1869) Studien an Acariden. *Zeitschrift für Wissenschaftliche Zoologie* 18, 445–546.

Clark, S. and Kowalsick, T. (1992) *Using Growing Degree-days for Insect Pest Management*. Horticulture Diagnostic Laboratory, Cornell University Cooperative Extension of Suffolk County, Riverhead/Great River, New York. Available at: https://s3.amazonaws.com/assets.cce.cornell.edu/attachments/1870/Using-Growing-Degree-Days-for-Insect-Pest-Management.pdf?1408019830 (accessed 5 May 2015).

Clift, A. and Terras, M.A. (2001) A quantitative study of phoresy in *Microdispus Iambi* (Krczal) (Acari: Pygmephoroidea) in Eastern Australia. In: Halliday, R.B., Walter, D.E., Proctor, H.C., Norton, R.A. and Colloff, M.J. (eds) *Acarology: Proceedings of the 10th International Congress of Acarology, 5–10 July 1998, Canberra*. CSIRO Publishing, Collingwood, Australia, pp. 394–398.

Clift, A., Terras, M.A. and Tesoriero, L. (1996) Evaluate options for sustainable pest management in cultivated mushrooms. Unpublished Final Report Project Mu 206, Horticultural Research and Development Corporation, NSW Agriculture (now NSW Department of Primary Industries. Orange, New South Wales, Australia).

Clift, A.D. (1983) *Insect and Mite Pests of Mushrooms*. Agfacts H8.AE.3, Department of Agriculture, New South Wales (now NSW Department of Primary Industries. Orange, New South Wales, Australia).

Clift, A.D. (1986) Development of a pest control programme for Australian mushroom farms. In: Wuest, P.J., Royse, D.J. and Beelman, R.B. (eds) *Cultivating Edible Fungi: International Symposium on Scientific and Technical Aspects of Cultivating Edible Fungi (IMS 86), July 15–17, 1986, Proceedings*. Elsevier Science, Amsterdam, pp. 571–576.

Clift, A.D. and Terras, M.A. (1994) The consequences of mites surviving the composting process. In: Tan Nair, N.G. (ed.) *Proceedings of the Second AMGA/ISMS International Workshop/Seminar on Agaricus Compost, Sydney, 1993*. Australian Mushroom Growers Association/International Society for Mushroom Science. Published by AMGA, Windsor, New South Wales, Australia, pp. 237–245.

Clift, A.D. and Terras, M.A. (1995) Mites as indicator of compost conditioning. In: Elliot, T.J. (ed.) *Proceedings of the Fourteenth International Congress on the Science and Cultivation of Edible Fungi, Volume II*. Balkema, Rotterdam, the Netherlands, pp. 507–513.

Clift, A.D. and Toffolon, R.B. (1981a) Biology, fungal host preferences and economic significance of two pygmephorid mites (Acarina: Pygmephoridae) in cultivated mushrooms, NSW, Australia. *Mushroom Science* 11, 245–253.

Clift, A.D. and Toffolon, R.B. (1981b) Insects and mites associated with mushroom cultivation on three commercial farms near Sydney, NSW, Australia. *Mushroom Science* 11, 537–549.

Clotuche, G., Mailleux, A.-C., Deneubourg, J.-L., Detrain, C., Navajas, M. and Hance, T. (2010) Mass ballooning behaviour in the spider mite *Tetranychus urticae* Koch. In: Moraes, G.J. de, Castilho, R.C. and Flechtmann, C.H.W. (eds) *XIII International Congress of Acarology, August 23–27, 2010, Recife-PE, Brazil, Abstract Book*, abstract 102, p. 57–58. Available at: http://www.acarology.org/ica/ica2010/docs/0910_15_abstractBookCapa.pdf (accessed 3 February 2015).

Cloyd, R.A. (2010) Broad mites. *FloriBytes* 5(3). Available at: http://www.oardc.ohio-state.edu/floriculture/images/10-10_Mites.pdf (accessed 2 February 2015).

Coates, T.J.D. (1974) *The Influence of Some Natural Enemies and Pesticides on Various Populations of* Tetranychus cinnabarinus *(Boisduval)*, T. lombardini *Baker & Pritchard and* T. ludeni *Zacher (Acari: Tetranychidae) with Aspects of their Biologies*. Entomology Memoir 42, Department of Agricultural Technical Services, Republic of South Africa. Government Printer, Pretoria.

Cobanoğlu, S. and Bayram, S. (1998) Mites (Acari) and flies (Insecta: Diptera) from natural edible mushrooms (Morchella: Ascomycetes) in Ankara, Turkey. *Bulletin et Annales de la Societe Royale Belge d'Entomologie* 134, 187–198.

Cocco, A. and Hoy, M.A. (2009) Feeding, reproduction, and development of the red palm mite (Acari: Tenuipalpidae) on selected palms and banana cultivars in quarantine. *Florida Entomologist* 92, 276–291.

Cochran, W.G. (1963) *Sampling Techniques*. Wiley, New York.

Cohic, F. (1963) *Catalogue des Parasites de Plantes Cultivées de la Polyésie Française*. Office de la Recherche Scientifique et Technique Outre-Mer, Nouméa, New Caledonia.

Coineau, Y. (1974a) *Introduction a l'Étude des Microarthropodes du Sol et des ses Annexes*. Documents pour l'Enseignements Pratique de l'Écologie. Doin, Paris.

Coineau, Y. (1974b) *Eléments pour une Monographie Morphologique, Écologique et Biologique des Caeculidae (Acariens)*. Mémoires de Muséum National d'Histoire Naturelle (N.S.) A 81.

Coineau, Y., Theron, P.D. and Valette, C. (1997) Une association de phanères, constituant probablement un système de communication sonore original chez les acariens. *Acarologia* 38, 111–116.

Colburn, R. and Asquith, D. (1970) A cage used to study the finding of a host by the ladybird beetle, *Stethorus punctum*. *Journal of Economic Entomology* 63, 1376–1377.

Cole, L. (1954) The population consequences of life history phenomena. *Quarterly Review of Biology* 29, 103–137.

Colin, J.G. and Rubi, M. (1992) Determinación de la susceptibililad de ocho selecciones de aguacate al ataque de araña roya y trips. In: *Memoria, Centro de Investigaciones Cientificas y Tecnologicas del Aguacate en el Estado de Mexico, Coatepec Harinas, Mexico*, pp. 149–155.

Collier, R., Hanks, G. and Millar, M. (2009) Integrated control of bulb-scale mite (*Steneotarsonemus laticeps*) in *Narcissus*. *IOBC-WPRS Bulletin* 50, 9–10.

Collingwood, C.A. (1959a) The control of black currant gall mite in garden and nursery. *Journal of the Royal Horticultural Society* 84, 133–136.

Collingwood, C.A. (1959b) Control of bulb scale mite with endrin. *Plant Pathology* 8, 98.

Collingwood, C.A. and Brock, A.M. (1959) Ecology of the black currant gall mite (*Phytoptus ribis* Nal.). *Journal of Horticultural Science* 34, 176–182.

Collyer, E. (1951) The separation of *Conwentzia pineticola* End. from *Conwentzia psociformis* (Curt.) and notes on their biology. *Bulletin of Entomological Research* 42, 555–564.

Collyer, E. (1952) Biology of some predatory insects and mites associated with the fruit tree red spider mite (*Metatetranychus ulmi* (Koch)) in south-eastern England. I. Biology of *Blepharidopterus angulatus* (Fall.) (Hemiptera-Heteroptera, Miridae). *Journal of Horticultural Science* 27, 117–129.

Collyer, E. (1953) The biology of some predatory insects and mites associated with the fruit tree red spider mite (*Metatetranychus ulmi* (Koch)) in southeastern England. II. Some important predators of the mite. *Journal of Horticultural Science* 28, 85–97.

Collyer, E. (1964a) Phytophagous mites and their predators in New Zealand orchards. *New Zealand Journal of Agricultural Research* 7, 551–568.

Collyer, E. (1964b) A summary of experiments to demonstrate the role of *Typhlodromus pyri* Scheuten in the control of *Panonychus ulmi* (Koch) in England. *Acarologia* 2, 363–371.

Collyer, E. (1964c) The effect of an alternative food supply on the relationship between two *Typhlodromus* species and *Panonychus ulmi* (Koch) (Acarina). *Entomologia Experimentalis et Applicata* 7, 120–124.

Collyer, E. (1969) Two species of *Tuckerella* (Acarina: Tuckerellidae) from New Zealand. *New Zealand Journal of Science* 12, 811–814.

Collyer, E. (1973) A new species of the genus *Dolichotetranychus* (Acarina: Tenuipalpidae) from New Zealand. *New Zealand Journal of Science* 16, 747–749.

Comins, H.N. (1977) The development of insecticide resistance in the presence of immigration. *Journal of Theoretical Biology* 64, 177–197.

Comroy, H.L. (1958) A preliminary survey of the plant mites of Puerto Rico. *Journal of Agriculture of the University of Puerto Rico* 42, 39–144.

Condit, I.J. and Horne, W.T. (1933) A mosaic of fig in California. *Phytopathology* 23, 887–896.

Cone, W.W. (1979) Pheromones of Tetranychidae. In: Rodriguez, J.G. (ed.) *Recent Advances in Acarology, Volume II*. Academic Press, New York, pp. 309–317.

Cone, W.W. (1985) Mating and chemical communication. In: Helle, W. and Sabelis, M.W. (eds) *Spider Mites: Their Biology, Natural Enemies and Control*. World Crop Pests, Volume 1A. Elsevier Science, Amsterdam, pp. 243–251.

Cone, W.W., McDonough, L.M., Maitlen, J.C. and Burdajewicz, S. (1971a) Pheromone studies of the two-spotted spider mite, *Tetranychus urticae* Koch. I. Evidence of a sex pheromone. *Journal of Economic Entomology* 64, 355–358.

Cone, W.W., Predki, S. and Klostermeyer, E.G. (1971b) Pheromone studies of the two-spotted spider mite, *Tetranychus urticae* Koch. II. Behavioral response of male *T. urticae* to quiescent deutonymphs. *Journal of Economic Entomology* 64, 379–382.

Congdon, B.D. and Logan, J.A. (1983) Temperature effects on development and fecundity of *Oligonychus pratensis* (Acari: Tetranychidae). *Environmental Entomology* 12, 359–362.

Conijn, C. (2006) Gewijzigde luchtsamenstelling bestrijdt ongedierte in bloembollenbewaarcel. *Gewasbescherming* 37, 1–7 (in Dutch).

Conijn, C.G.M., van Aartrijk, J. and Lesna, I. (1996) Flower bulbs. In: Lindquist, E.E., Sabelis, M.W. and Bruin, J. (eds) *Eriophyoid Mites: Their Biology, Natural Enemies and Control*. World Crop Pests, Volume 6. Elsevier Science, Amsterdam, pp. 651–659.

Conijn, C.G.M., Lesna, I. and Altena, K. (1997) Biological control of the bulb mite *Rhizoglyphus robini* by the predatory mite *Hypoaspis aculeifer* on lilies: implementation in practice. *Acta Horticulturae* 430, 619–624.

Connin, R.V. (1956) Oversummering volunteer wheat in the epidemiology of wheat streak mosaic. *Journal of Economic Entomology* 49, 405–406.

Cook, M.T. (1909) Some insect galls of Cuba. In: *2nd Informe Anual de la Estación Agronomica de Cuba*, pp. 143–146.

Cook, S.P. (1992) Influence of monoterpene vapors on spruce spider mite, *Oligonychus ununguis*, adult females. *Journal of Chemical Ecology* 18, 1497–1504.

Coons, L.B. and Alberti, G. (1999) Acari: ticks. In: Harrison, F.W. and Foelix, R.F. (eds) *Microscopic Anatomy of Invetebrates Chelicerate Arthropoda, Volume 8B*. Wiley, New York, pp. 267–514.

Coons, L.B., Roshdy, M.A. and Axtell, R.C. (1974) Fine structure of the central nervous system of *Dermacentor variabilis* (Say), *Amblyomma americanum* (L.), and *Argas arboreus* Kaiser, Hoogstraal, and Kohls (Ixodoidea). *Journal of Parasitology* 60, 687–698.

Cooper, J.P. (ed.) (1975) *Photosynthesis and Productivity in Different Environments*. Cambridge University Press, Cambridge, UK.

Cooper, K.W. (1937) Reproductive behavior and haploid parthenogenesis in the grass mite, *Pediculopsis graminum* (Reut.). *Proceedings of the National Academy of Sciences of the United States of America* 23, 41–44.

Cooper, K.W. (1940) Relations of *Pediculopsis graminum* and *Fusarium poae* to central bud rot of carnations. *Phytopathology* 30, 853–859.

Cooreman, J. (1941) Un tarsonémide mycophage nouveau (Acarien). *Bulletin du Musée Royal d'Histoire Naturelle de Belgique* 17(20), 1–7.

Cooreman, J. (1944) Notes et observations sur les Acariens. *Bulletin du Musée Royal d'Histoire Naturelle de Belgique* 20(8), 1–16.

Copping, L.G. (ed.) (2001) *Biopesticide Manual: A World Compendium*, 2nd edn. British Crop Production Council, Farnham, UK.

Cordo, H.A. and DeLoach, C.J. (1976) Biology of the waterhyacinth mite in Argentina. *Weed Science* 24, 245–249.

Coria-Avalos, V. (1993) *Principales Plagas del Aguacate en Michoacán*. Folleto Para Productores No. 19, Centro de Investigaciones del Pacifico Centro, Instituto Nacional de Investigationes Forestales y Agropecuarias, Michoacán, Mexico.

Cornell University (2012) Crop and Pest Management Guidelines (various), Cornell University Cooperative Extension, New York. Available at: http://ipmguidelines.org/ (accessed 25 July 2014).

Corpuz-Raros, L.A. (1989a) Review of Philippine *Tuckerella* (Tuckerellidae, Acari). *Philippine Entomologist* 7, 557–560.

Corpuz-Raros, L.A. (1989b) Hosts, geographic distribution and predatory mite associations of Philippine phytophagous mites (Acari). *Philippine Agriculturalist* (now *Philippine Agricultural Scientist*) 72, 303–322.

Corpuz-Raros, L.A. and Garcia, R.C. (1995) Philippine predatory mites of the family Cunaxidae (Acari). 1. Genus *Cunaxa* Von Heyden. *Philippine Entomologist* 9, 605–624.

Corriveau, M., Uppstrom, K. and Klompen, H. (2010) Effect of eight storage modes on DNA preservation. In: Sabelis, M.W. and Bruin, J. (eds) *Trends in Acarology: Proceedings of the 12th International Congress, 21–26 August 2006, Amsterdam*. Springer, Dordrecht, the Netherlands, pp. 553–556.

Cortesero, A.M., Stapel, J.O. and Lewis, W.J. (2000) Understanding and manipulating plant attributes to enhance biological control. *Biological Control* 17, 35–49.

Cosenza, G.W., Menezes Sobrinho, J.A. de, Regina, S.M. and Gontijo, V.P.M. (1981) Efeito do expurgo sobre a preservacao do alho armazenado. *Pesquisa Agropecuária Brasileira* 16, 199–203.

Costa, A.S. and Correa, D.M. (1960) Mite species found on cotton crops. *Bragantia* 19, 183.

Costa, A.S. and Gonçalves, R.D. (1950) Cinza ou penugem branca do tomateiro. *Bragantia* 10, 383–385.

Costa, A.S., Rigitano, O., Carvalho, A.M.B. and Netto, J.A.G. (1962) Presence of the peach silver mite in Sao Paulo. *Bragantia* 21(6), 37–39.

Costa, J.M. da (1973) Resultados experimentais no control do acaro da mandioca "*Mononychus tanajoa*" (Bondar, 1938). *Série Pesquisa* l(l), 25–30. Escola de Agronomia/Brascan Nordeste, Universidade Federal da Bahia, Cruz das Almas, Brazil.

Costa, J.M. da (1975) O "tanajoa" damandioca. *Série Pesquisa* 2(1), 15–19. Escola de Agronomia/Brascan Nordeste, Universidade Federal da Bahia, Cruz das Almas, Brazil.

Costantino, G. (1962) Acari vecchi e nuovi, parassiti degli agrumi in Italia. *Atti della Accademia Nazionale Italiana di Entomologia* 10, 154–164.

Costantino, G. (1968) *Aculus pelekassi*, acaro degli agrumi (*Aculus pelekassi*, a mite pest of citrus). *Italia Agricola* 105, 87–97.

Costanza, R., D'Arge, R., De Groot, R., Farber, S., Grasso, M., Hannon, B., Limburg, K., Naeem, S., O'Neill, R.V., Paruelo, J., *et al.* (1997) The value of the world's ecosystem services and natural capital. *Nature* 387, 253–260.

Costello, M.J. (2004) Willamette mite on grapes: response to sulfur and irrigation. Poster D0410. In: Posters from *The 2004 ESA Annual Meeting and Exhibition, Salt Lake City*, Utah. Entomological Socirty of America, Annapolis, Maryland. Available at: http://esa.confex.com/esa/2004/techprogram/paper_16152.htm (accessed 25 July 2014).

Costello, M.J. (2007) Impact of sulfur on density of *Tetranychus pacificus* (Acari: Tetranychidae) and *Galendromus occidentalis* (Acari: Phytoseiidae) in a central California vineyard. *Experimental and Applied Acarology* 42, 197–208.

Costilla, M.A. (1991) El acaro *Aculops lycopersici* (Massee, 1937) (Acari: Eriophyidae) causante del tostado o amorronado del tomate en el noroeste Argentino. *Revista Industrial y Agrícola de Tucumán* 68(1/2), 83–90.

Costilla, M.A., Basco, H.J. and Osores, V.M. (1987) Experiencia de control quimico del acaro de la yema *Aceria sheldoni* (Ewing) en plantas de limoeiro. *Revista Industrial y Agrícola de Tucumán* 64(2), 85–92.

Cotte, J. (1920) Deux parasites de la figue sauvage. *Bulletin de la Société de Pathologie Végetale de France* 7, 26.

Courtin, O., Fauvel, G. and Leclant, F. (2000) Temperature and relative humidity effects on egg and nymphal development of *Aceria tulipae* (K.) (Acari: Eriophyidae) on garlic leaves (*Allium sativum* L.). *Annals of Applied Biology* 137, 207–211.

Coutts, B.A., Strickland, G.R., Kehoe, M.A., Severtson, D.L. and Jones, R.A.C. (2008) The epidemiology of wheat streak mosaic virus in Australia: case histories, gradients, mite vectors, and alternative hosts. *Australian Journal of Agricultural Research* 59, 844–853.

Couturier, G., Tanchiva, E., Inga, H., Vásquez, J. and Riva, R. (1996) Notas sobre los artrópodos que viven en el pijuayo (*Bactris gasipaes* H.B.K.: Palmae) en la Amazonía peruana. *Revista Peruana de Entomología* 39, 135–142.

Cox, H.C. and Lieberman, F.V. (1960) Biology of brown wheat mite. *Journal of Economic Entomology* 53, 704–705.

Craemer, C. (1995) Host specificity, and release in South Africa, of *Aceria malherbae* Nuzzaci (Acari: Eriophyoidea), a natural enemy of *Convolvulus arvensis* L. (Convolvulaceae). *African Entomology* 3, 213–215.

Craemer, C. (1996) Eriophyoidea (Acari) associated with *Lantana camara* L., with descriptions of two new species. *African Plant Protection* 2(1), 59–66.

Craemer, C. and Neser, S. (1990) Mites imported against lantana. *Plant Protection News* 19, 3.

Craemer, C. and Saccaggi, D.L. (2013) Frequent quarantine interception in South Africa of grapevine *Colomerus* species (Trombidiformes: Prostigmata: Eriophyidae): taxonomic and distributional ambiguities. *International Journal of Acarology* 39, 239–243.

Craemer, C., Neser, S. and Meyer Smith, M.K.P. (1996) Eriophyid-myte (Acari: Eriophyoide: Eriophyidae) as moontlike beheeragente van ongewenste uithemse plante in Suid-Africa. *Suid-Afrikaanse Tydskrif vir Natuurwetenskap en Tegnologie* 15, 99–109 (in Afrikaans).

Cranham, J.E. (1966) *Insect and Mite Pests of Tea in Ceylon and their Control*. Monographs on Tea Production in Ceylon No. 6, The Tea Research Institute, Talawakelle, Sri Lanka.

Cranham, J.E. (1972) Influence of temperature on hatching of winter eggs of fruit-tree red spider mite, *Panonychus ulmi* (Koch). *Annals of Applied Biology* 70, 119–137.

Cranham, J.E. (1973) Variation in the intensity of diapause in winter eggs of fruit tree red spider mite, *Panonychus ulmi*. *Annals of Applied Biology* 75, 173–182.

Cranham, J.E. (1974) Resistance to organophosphates in red spider mite, *Tetranychus urticae*, from English hop gardens. *Annals of Applied Biology* 78, 99–111.

Cranham, J.E. (1982) Resistance to binapacryl and tetradifon, and the genetic background, in fruit tree red spider mite, *Panonychus ulmi*, from English apple orchards. *Annals of Applied Biology* 100, 25–38.

Cranham, J.E. and Easterbrook, M.A. (1984) The effects of pesticides on the predacious phytoseiid *Typhlodromus pyri* and its prey *Panonychus ulmi*. *IOBC-WPRS Bulletin* 7(3), 43–46.

Cranham, J.E. and Helle, W. (1985) Pesticide resistance in Tetranychidae. In: Helle, W. and Sabelis, M.W. (eds) *Spider Mites: Their Biology, Natural Enemies and Control*. World Crop Pests, Volume 1B. Elsevier Science, Amsterdam, pp. 405–421.

Cranham, J.E. and Solomon, M.G. (1981) Mite management in commercial apple orchards. In: *Annual Report 1980, East Malling Research Station*. East Malling, UK, pp. 171–172.

Cranham, J.E., Raneweera, D.J.W. and Rajapakse, G.B. (1962) The mite pests of Ceylon tea: recognition and control. *Tea Quarterly* 33, 189–195.

Cranshaw, W., Casey Sclar, D. and Cooper, D. (1996) A review of 1994 pricing and marketing policy by suppliers of organisms for biological control of arthropods in the United States. *Biological Control* 6, 291–296.

Cranshaw, W.S. (2012) *Clover and Other Mites of Turfgrass*. Fact Sheet No 5.505, Insect Series, Home and Garden,. Colorado State University, Cooperative Extension, Fort Collins, Colorado. Available at: http://www.ext.colostate.edu/pubs/insect/05505.html (accessed 5 May 2015).

Crawford, M. (1995) *Hazelnuts. Production and Culture*. Agroforestry Research Trust, Dartington, Totnes, UK.

Creary, S. (2009) Indirect effects of imidacloprid on natural enemies of spider mites in two systems. MSc. thesis, University of Maryland, College Park, Maryland.

Crespan, G. and Cestari, F. (2009) Quando intervenire contro gli eriofidi della vite. *Informatore Agrario* 65(20), 60–62.

Croft, B.A. (1975) *Integrated Control of Apple Mites*. Bulletin E-825, Michigan State University Cooperative Extension Series, East Lansing, Michigan.

Croft, B.A. (1976) Establishing insecticide-resistant phytoseiid mite predators in deciduous tree fruit orchards. *Entomophaga* 21, 383–399.

Croft, B.A. (1990) *Arthropod Biological Control Agents and Pesticides*. Wiley, New York.

Croft, B.A. and Brown, A.W.A. (1975) Responses of arthropods natural enemies to insecticides. *Annual Review of Entomology* 20, 285–335.

Croft, B.A. and Hoying, S.A. (1977) Competitive displacement of *Panonychus ulmi* (Koch) by *Aculus schlechtendali* Nalepa in apple orchards. *Canadian Entomologist* 109, 1025–1034.

Croft, B.A. and Jeppson, L.R. (1970) Comparative studies on four strains of *Typhlodromus occidentalis*. II. Laboratory toxicity of ten compounds common to apple pests control. *Journal of Economic Entomology* 63, 1528–1531.

Croft, B.A. and Jorgensen, C.D. (1977) *Typhlodromus mcgregori* Chant (Acarina: Phytoseiidae) and its potential control of phytophagous mites in southern California and central Utah apple orchards. *Great Basin Naturalist* 37, 366–374.

Croft, B.A. and Luh, H.K. (2004) Phytoseiid mites on unsprayed apple trees in Oregon, and other western states (USA): distributions, life-style types and relevance to commercial orchards. *Experimental and Applied Acarology* 33, 281–326.

Croft, B.A. and McGroarty, D.L. (1977) *The Role of* Amblyseius fallacis *(Acarina: Phytoseiidae) in Michigan Apple Orchards*. Michigan State University Research Report No. 333, East Lansing, Michigan.

Croft, B.A. and McMurtry, J.A. (1972a) Minimum releases of *Typhlodromus occidentalis* to control *Tetranychus mcdanieli* on apple. *Journal of Economic Entomology* 65, 188–191.

Croft, B.A. and McMurtry, J.A. (1972b) Comparative studies on four strains of *Typhlodromus occidentalis* Nesbitt (Acarina: Phytoseiidae). IV. Life history studies. *Acarologia* 13, 460–470.

Croft, B.A. and McRae, I.V. (1993) Biological control of apple mites. Impact of *Zetzellia mali* (Acari, Stigmaeidae) on *Typhlodromus pyri* and *Metaseiulus occidentalis* (Acari, Phytoseiidae). *Environmental Entomology* 22, 865–887.

Croft, B.A. and Nelson, E.E. (1972) An index to predict efficient interactions of *Typhlodromus occidentalis* in control of *Tetranychus mcdanieli* in southern California apple trees. *Journal of Economic Entomology* 65, 310–312.

Croft, B.A. and Slone, D.H. (1998) Perturbation of regulated apple mites: immigration and pesticide effects on outbreaks of *Panonychus ulmi* and associated mites (Acari: Tetranychidae, Eriophyidae, Phytoseiidae and Stigmaeidae). *Environmental Entomology* 27, 1548–1556.

Croft, B.A. and van de Baan, H.E. (1988) Ecological and genetic factors influencing evolution of pesticide resistance in tetranychid and phytoseiid mites. *Experimental and Applied Acarology* 4, 277–300.

Croft, B.A., Welch, S.M. and Dover, M.J. (1976) Dispersion statistics and sample size estimates for populations of the mite species *Panonychus ulmi* and *Amblyseius fallacis* on apple. *Environmental Entomology* 5, 227–234.

Croft, B.A., Monetti, L.N. and Pratt, P.D. (1998a) Comparative life histories and predation types – are *Neoseiulus californicus* and *N. fallacis* (Acari, Phytoseiidae) similar type II selective predators of spider mites? *Environmental Entomology* 27, 531–538.

Croft, B.A., McMurtry, J.A. and Luh, H.-K (1998b) Do literature records of predation reflect food specialization and predation types among phytoseiid mites (Acari: Phytoseiidae)? *Experimental and Applied Acarology* 22, 467–480.

Cromroy, H.L. (1958) A preliminary survey of the plant mites of Puerto Rico. *Journal of Agriculture of the University of Puerto Rico* 42, 39–144.

Cromroy, H.L. (1979) Eriophyoidea in biological control of weeds. In: Rodriguez, J.G. (ed.) *Recent Advances in Acarology, Volume I*. Academic Press, New York, pp. 473–475.

Cromroy, H.L. (1983) Potential use of mites in biological control of terrestrial and aquatic weeds. In: Hoy, M.A., Cunningham, G.L. and Knutson, L. (eds) *Recent Advances in Knowledge of the Phytoseiidae*. Division of Agriculture Special Publication No 3304, University of California, Berkeley, California, pp. 61–66.

Crooker, A.R., Drenth-Diephuis, L.J., Ferwerda, M.A. and Weyda, F. (1985) Histological techniques. In: Helle, W. and Sabelis, M.W. (eds) *Spider Mites: Their Biology, Natural Enemies and Control*. World Crop Pests, Volume 1A. Elsevier Science, Amsterdam, pp. 359–381.

Cross, E.A. (1965) The generic relationships of the family Pyemotidae (Acarina: Trombidiformes). *University of Kansas Science Bulletin* 45, 29–275.

Cross, E.A. and Kaliszewski, M.J. (1988) The life history of a mushroom pest mite, *Pediculaster flechtmanni* (Wicht) (Acari: Pygmephoroidea), with studies of alternate morph formation. *Environmental Entomology* 17, 309–315.

Cross, J.V. (2006) Further investigation of IPM methods for blackcurrant gall mite and leaf midge. *IOBC-WPRS Bulletin* 29(9), 131–138.

Cross, J.V. and Easterbrook, M.A. (1998) Prospects for integrated management of blackcurrant pests. *IOBC-WPRS Bulletin* 21(10), 55–61.

Cross, J.V. and Harris, A. (2003) Development of integrated pest management approaches for blackcurrant crops. *IOBC-WPRS Bulletin* 26(2), 11–18.

Cross, J.V. and Ridout, M.S. (2001) Emergence of blackcurrant gall mite (*Cecidophyopsis ribis*) from galls in spring. *Journal of Horticultural Science and Biotechnology* 76, 311–319.

Csapo, Z. (1992) Eriophyid mites (Acarina-Eriophyoidea) on currants: morphology, taxonomy and ecology. PhD thesis., Warsaw Agricultural University, Warsaw.

CSIRO (2004) Commonwealth Scientific and Industrial Research Organisation, Australian Insect Common Names: *Acaphylla theae* (Watt & Mann). Available at: http://www.ces.csiro.au/aicn/system/c_50.htm (accessed 25 July 2014).

Cuenod, H. (1958) Les acariens nuisibles aux agrumes de Tunisie. In: *Proceedings of the 4th International Congress of Méditerranean Agrumiculture, 20–27 May 1956, Tel-Aviv, Israel*. Citrus Marketing Board, Israel, pp. 299–305.

Cui, Y.-Q. (1989) Spider mites injury bamboo's forest in China. *Kunchong Zhishi (Entomological Knowledge)* 26, 303–304 (in Chinese).

Culen, J. (2001) Culenova pasca – metoda na dostatocnu eliminaciu roztocov na vinici. *Vinohrad* 39(1–2), 28–29 (in Slovak).

Cull, B.W. and Paxton, B.F. (1983) Growing the lychee in Queensland. *Queensland Agricultural Journal* 109, 53–59.

Cullen, J.M. and Moore, A.D. (1983) The influence of three populations of *Aceria chondrillae* on three forms of *Chondrilla juncea*. *Journal of Applied Ecology* 20, 235–243.

Cullen, J.M., Groves, R.H. and Alex, J.F. (1982) The influence of *Aceria chondrillae* on the growth and reproductive capacity of *Chondrilla juncea*. *Journal of Applied Ecology* 19, 529–537.

Cunliffe, N. and Ryle, G.B. (1923) The conifer spinning mite on Sitka spruce (*Oligonychus* (*Paratetranychus*) *ununguis* (Jacobi)). *Queensland Journal of Forestry* 17, 359–362.

Curkovic, T. and Araya, J.E. (2004) Acaricidal action of two detergents against *Panonychus ulmi* (Koch) and *Panonychus citri* (McGregor) (Acarina: Tetranychidae) in the laboratory. *Crop Protection* 23, 731–733.

Curkovic, T.S., Barría, G.P. and González, R. (1994) Evaluación de acaricidas en el control de las arañitas, *Panonychus ulmi* (Koch) y *Brevipalpus chilensis* Baker, y degradación de residuos de chinometionate y pyridaben. *Revista Frutícola* 15, 105–114.

Curry, J.P. (1976) The arthropod fauna of some common grass and weed species of pasture. *Proceedings of the Royal Irish Academy* 76B, 1–35.

Cuthbertson, A.G.S. and Murchie, A.K. (2004) The phenology, oviposition and feeding rate of *Anystis baccarum*, a predatory mite in Bramley apple orchards in Northern Ireland. *Experimental and Applied Acarology* 34, 367–373.

Cuthbertson, A.G.S. and Murchie, A.K. (2006) Environmental impact of an orchard winter wash and early season pesticide applications on both a beneficial and a pest mite species in Bramley apple orchards. *International Journal of Environmental Science and Technology* 3, 333–339.

Cuthbertson, A.G.S. and Murchie, A.K. (2012) *Anystis baccarum* – a predatory mite in UK apple orchards. *Research Journal of Chemistry and Environment* 16(3), 18–21.

Cuthbertson, A.G.S., Bell, A.C. and Murchie, A.K. (2003) Impact of the predatory mite *Anystis baccarum* (Prostigmata: Anystidae) on apple rust mite *Aculus schlechtendali* (Prostigmata: Eriophyidae) populations in Northern Ireland Bramley orchards. *Annals of Applied Biology* 142, 107–114.

Cutright, C.R. (1944) Populations of the European red mite as affected by spray schedules. *Journal of Economic Entomology* 37, 499–502.

Czajkowska, B. and Ewa, P. (2005) Spruce spider mite (*Oligonychus ununguis* Jacobi) and its predators associated with ornamental coniferous plants in Polish nurseries. *IOBC-WPRS Bulletin* 28(1), 75–78.

Czajkowska, B. and Kropczynska, D. (1991) The influence of different host plants on the reproductive potential of *Tyrophagus putrescentiae* (Schrank) and *Tyrophagus neiswanderi* Johnston and Bruce (Acaridae). In: Schuster, R. and Murphy P.W. (eds) *The Acari. Reproduction, Development and Life History Strategies.* Chapman and Hall, London, pp. 313–317.

Czaikowska, B., van de Vrie, M. and Kropczynska, D. (1988) Mites of the genus *Tyrophagus* as pests of ornamentals in greenhouses. *Mededelingen van de Faculteit Landbouwwetenschappen, Rijksuniversiteit Gent* 53(2b), 799–809.

Czajkowska, B., Kielkiewicz, M. and Puchalska, E. (2003) Population parameters as an indicator of susceptibility of ornamental coniferous plants to spruce spider mite. *Journal of Plant Protection Research* 43, 155–162.

Czeczuga, B. (1975) The carotenoid content of galls produced by *Eriophyes tiliae* var. Rudis Nal. (Acarina) on *Tilia cordata* Mill., leaves. *Marcellia* 38, 223–225.

Czermainski, A.B.C., Bassanezi, R.B., Laranjeira, F.F. and Amorim, L. (2007) Temporal dynamics of populations of *Brevipalpus phoenicis* mite and citrus leprosis under natural epidemic conditions. *Fitopatologia Brasileira* 32, 295–303.

Dabert, J., Ehrnsberger, R. and Dabert, M. (2008) *Glaucalges tytonis* sp. n. (Analgoidea, Xolalgidae) from the barn owl *Tyto alba* (Strigiformes, Tytonidae): compiling morphology with DNA barcode data for taxon descriptions in mites (Acari). *Zootaxa* 1719, 41–52.

Dabrowski, Z.T. and Bielak, B. (1978) Effect of some plant chemical compounds on the behaviour and reproduction of spider mites (Acarina: Tetranychidae). *Entomologia Experimentalis et Applicata* 24, 117–126.

Dabrowski, Z.T. and Marczak, Z. (1972) Studies on the relationship of *Tetranychus urticae* Koch and host plants. I. Effect of plant species. *Polskie Pismo Entomologiczne* 42, 821–855.

Dabrowski, Z.T., Rodriguez, J.G. and Chaplin, C.E. (1971) Studies in the resistance of strawberries to mites. IV. Effect of season on preference and non-preference of strawberries to *Tetranychus urticae*. *Journal of Economic Entomology* 64, 806–809.

Dacosta, C.P. and Jones, C.M. (1971) Cucumber beetle resistance and mite susceptibility controlled by the bitter gene in *Cucumis sativus* L. *Science* 172, 1145–1146.

Daftari, A. (1979) Studies on feeding, reproduction and development of *Amblyseius aberrans* (Acarina: Phytoseiidae) on various food substances. *Zeitschrift für Angewandte Entomologie* 88, 449–453.

Dahal, G. and Neupane, K.R. (1991) Incidence and effect of sterility mosaic disease of pigeonpea in Nepal. *International Pigeonpea Newsletter* 13, 23–24.

Daiber, K.C. (1985) *Tomato Pests in South Africa, their Potential Damage and their Control.* Technical Communication No. 2, Horticultural Science Series, Department of Agriculture, Republic of South Africa.

Daly, J.M. (1983) Current concepts of disease resistance in plants. In: Kommendahl, T. and Williams, P.M. (eds) *Challenging Problems in Plant Health.* American Phytopathology Society, St Paul, Minnesota, pp. 311–323.

Dam Quoc Tru (2009) Vietnam plant protection review 2007–2009. In: *Report of the Twenty-sixth Session of the Asia and Pacific Plant Protection Commission, 31 August to 4 September 2009, New Delhi, India.* Food and Agriculture Organization of the United Nations, Regional Office for Asia and the Pacific, Bangkok, pp. 27–28.

Damiano, A. (1964) The mites on citrus in Tripolitania. *Rivista di Agricultura Subtropicale e Tropicale* 58(7–9), 354–361.

Damiano, A. (1968a) The animal pests (including some 20 species of insects and mites) injurious to fruit trees in Tripolitania. *Rivista di Agricultura Subtropicale e Tropicale* 62(7–9), 282–298.

Damiano, A. (1968b) The principal animal pests (some 20 species of insects and mites) injurious to fruit trees in Tripolitania. *Rivista di Agricultura Subtropicale e Tropicale* 62(10–12), 39–370.

Daneshnia, N., Akrami, M.A. and Aleosfoor, M. (2012) Life table parameters of fig mite *Eotetranychus hirsti* Pritchard & Baker, 1955 (Acari: Tetranychidae) under laboratory conditions. In: Murariu, M., Adam, C., Chişamera, G., Iorgu, E., Popa, L.O. and Popa, O.P. (eds) *Annual Zoological Congress of "Grigore Antipa" Museum, 21–23 November 2012, Bucharest, Romania, Book of Abstracts.* "Grigore Antipa" National Museum of Natural History, Bucharest, Romania, p. 83. Available at: http://www.czga.ro/pozepagini/CZGA_2012___Book_of_Abstracts.pdf (accessed 5 May 2015). (Full paper: Daneshnia, N., Akrami, M.A. and Aleosfoor, M. (2013) Life table parameters of fig mite, *Eotetranychus hirsti* (Acari: Tetranychidae) under laboratory conditions. *Persian Journal of Acarology* 2, 277–286. Available at: http://www.acarology.ir/Daneshnia%20et%20al.,%20Vol.%202,%20No.%202,%202013.pdf (accessed 5 May 2015)).

Daneshvar, H. (1987) Some predatory mites from Iran, with descriptions of one new genus and six new species (Acari: Phytoseiidae, Ascidae). *Applied Entomology and Phytopathology* 54(1–2), 13–37.

Daneshvar, H. and Abaii, M.G. (1994) Efficient control of *Tetranychus turkestani* on cotton, soybean and bean by *Phytoseiulus persimilis* (Acari: Tetranychidae) in pest foci. *Applied Entomology and Phytopathology* 61(1–2), 22.

Daneshvar, H. and Khosrowshahi, M. (1994) The activities of *Amblydromella iranesis* and *Phytoseius plumifer* on European red mite and hazelnut bud mite. *Applied Entomology and Phytopathology* 61(1–2), 35–36.

Daneshvar, H., Rodriguez, J.G. and Rodriguez, L.D. (1977) Monoxenic culture of *Caloglyphus berlesei* (Acarina: Acaridae) on fungi. *Journal of Invertebrate Pathology* 30, 175–180.

Daniel, M. (1977) The spider mite *Tetranychus fijiensis* Hirst on the betelnut palm. *Acarologia* 19, 220–221.

Daniel, M. (1979) Survey of the indigenous predators of arecanut phytophagous mites. In: Venkata Ram, C.S. and Chandra Mouli, B. (eds) *PLACROSYM II. Plant Protection (Entomology, Microbiology, Nematology, Plant Pathology and Rodentology): Proceedings of the Second Annual Symposium on Plantation Crops, 1979, Ootacamund*, India. PLACROSYM Standing Committee, India, pp. 227–236.

Daniel, M. (1981) Bionomics of the predaceous mite *Amblyseius channabasavanni* (Acari: Phytoseiidae), predaceous on the palm mite. In: ChannaBasavanna, G.P., Mallik, B. and Ghorpade, K.D. (eds) *Contributions to Acarology in India: Proceedings of the I All India Symposium in Acarology, April 23–25, 1979*, Bangalore. Acarological Society of India, Bangalore, India, pp. 167–173.

Dara, S. (2011) Lewis mite: a potential pest of strawberries and raspberries. *Western Farm Press*, Mar 15, 2011. Available at: http://westernfarmpress.com/management/lewis-mite-potential-pest-strawberries-and-raspberries (accessed 25 July 2014).

Daravath, S., Vemuri, S.B., Ramagopal, N. and Bellamkonda, R. (2013) Efficacy of acaricides alone and in combination with propiconozole against rice panicle/sheath mite, *Steneotarsonemus spinki*. *Journal of Agricultural Science and Technology B (New York)* 3, 107–110. Available at: http://www.davidpublishing.com/DownLoad/?id=11196 (accessed 5 May 2015).

Darrow, G.M., Wilcox, R.B. and Ceckwith, C.S. (1944) *Blueberry Growing*. Farmers' Bulletin of the US Department of Agriculture No. 1951, Washington, DC.

Das, G.M. (1959) Bionomics of tea red spider, *Oligonychus coffeae* (Nietner). *Bulletin of Entomological Research* 50 (2), 265–274.

Das, G.M. (1960) Occurrence of the red spider, *Oligonychus coffeae* (Nietner) on tea in northeast India in relation to pruning and defoliation. *Bulletin of Entomological Research* 51, 415–426.

Das, G.M. (1965) *Pests of Tea in North East India and their Control*. Toklai Experimental Station Memorandum 27, Jorhat, India.

Das, G.M. and Das, S.C. (1967) Effect of temperature and humidity on the development of the tea red spider mite, *Oligonychus coffeae* (Nietner). *Bulletin of Entomological Research* 57, 433–436.

Das, G.M. and Sengupta, N. (1958) Observations on the pink mite, *Acaphylla theae* (Watt) Keifer, of tea in north east India. *Journal of the Zoological Society of India* 10, 39–48.

Das, G.M. and Sengupta, N. (1962) Biology and control of the purple mite, *Calacarus carinatus* (Green), a pest of tea in north-east India. *Journal of the Zoological Society of India* 14, 64–72.

Das, S.C. (1983) Some basic points in mite control on tea. *Two and a Bud* 30, 26–28.

Das, S.C. and Gope, B. (1983) Translaminar effect of dicofol and ethion on three mite pests of tea. *Two and a Bud* 30, 64–66.

Das, T.K., Sarkar, P.K., Dey, P.K. and Somchoudhury, A.K. (2000) The chemical basis of resistance of pineapple plant to *Dolichotetranychus floridanus* Banks (Prostigmata: Tenuipalpidae). *Acarologia* 41, 317–320.

David, A.L. (1959) Host range of the sweet potato rust mite, *Epitrimerus* sp. *Madras Agricultural Journal* 46, 313.

David, P.M.M. and Varadarajan, M.K. (2001) A new glycerine drop trap method to sample eriophyoid mites. *Entomon* 26, 97–100.

David, P.M.M., Rajkumar, K., Elanchezhyan, K., Razak, T.A., Nelson, S.J., Nainar, P. and Muralibaskaran, R.K. (2009) Effect of soft soaps on broad mite, *Polyphagotarsonemus latus* (Banks) on chilli. *Karnataka Journal of Agricultural Sciences* 22, 554–556.

Davidson, N.A., Dibble, J.E., Flint, M.L., Marer, P.J. and Guye, A. (1991) *Managing Insects and Mites with Spray Oils*. Publication 3347, IPM Education and Publications, Statewide IPM Project, Division of Agriculture and Natural Resources, University of California, Davis, California.

Davies, J.T., Allen, G.R. and Williams, M.A. (2001a) Intraplant distribution of *Acalitus essigi* (Acari: Eriophyoidea) on blackberries (*Rubus fruticosus* agg.). *Experimental and Applied Acarology* 25, 625–639.

Davies, J.T., Allen, G.R. and Williams, M.A. (2001b) Dispersal of *Acalitus essigi* to blackberry (*Rubus fruticosus* agg.) fruit. *Entomologia Experimentalis et Applicata* 101, 19–23.

Davies, J.T., Williams, M.A. and Allen, G.R. (2002) A new method for sampling eriophyoid mites from architecturally complex plant structures. *Journal of Applied Entomology* 126, 303–305.

Davies, J.T., Ireson, J.E. and Allen, G.R. (2007) The impact of the gorse spider mite, *Tetranychus lintearius*, on the growth and development of gorse, *Ulex europaeus*. *Biological Control* 41, 86–93.

Davis, D.W. (1952) Influence of population density on *Tetranychus multisetis*. *Journal of Economic Entomology* 45, 652–654.

Davis, H.N. (2008) Impact of cover cropping on arthropods in corn on the western High Plains. MSc. thesis, Kansas State University, Manhattan, Kansas.

Davis, J.J. (1968a) Survey of Tetranychidae. In: *Proceedings of the Field Entomologists' Conference, 25–28 June 1968, Nambour*, Australia. Queensland Department of Primary Industries, Brisbane, Australia, pp. 119–123.

Davis, J.J. (1968b) Studies of Queensland Tetranychidae. 3. Records of the genus *Tetranychus*. *Queensland Journal of Agricultural and Animal Sciences* 25, 27–67.

Davis, J.J. (1968c) Studies of Queensland Tetranychidae. 5. Records of the genus *Panonychus* from Queensland, with a note on the occurrence of *Panonychus citri* (McGregor) in New Guinea. *Queensland Journal of Agricultural and Animal Sciences* 25, 73–75.

Davis, J.J. and Heather, N.W. (1962) Control of red spiders in strawberries. *Queensland Journal of Agricultural and Animal Sciences* 19, 143–148.

Davis, R. (1964a) Some eriophyid mites occurring in Georgia with descriptions of three new species. *Florida Entomologist* 47, 17–27.

Davis, R. (1964b) Autecological studies of *Rhynacus breitlowi* Davis (Acarina: Eriophyidae). *Florida Entomologist* 47, 113–121.

Davis, R., Flechtmann, C.H.W., Boczek, J.H. and Barké, H.E. (1982) *Catalogue of Eriophyid Mites (Acari: Eriophyoidea)*. Warsaw Agricultural University Press, Warsaw.

Day, M., Wiley, C.H., Playford, J. and Zalucki, M.P. (2003) Lantana*: Current Management Status and Future Prospects.* ACIAR Monograph 102, Australian Centre for International Agricultural Research, Canberra.

De, R.K., Dhar, V. and Rathore, Y.S. (1996) Control of mosaic in pigeonpea with insecticides and acaricides. *Indian Journal of Pulses Research* 9, 87–89.

De Angelis, J.D., Larson, K.C., Berry, R.E. and Krantz, G.S. (1982) Effects of spider mite injury on transpiration and leaf water status in peppermint. *Environmental Entomology* 11, 975–978.

De Angelis, J.D., Berry, R.E. and Krantz, G.W. (1983a) Evidence for spider mite (Acari: Tetranychidae) injury-induced leaf water deficits and osmotic adjustments in peppermint. *Environmental Entomology* 12, 336–339.

De Angelis, J.D., Berry, R.E. and Krantz, G.W. (1983b) Photosynthesis, leaf conductance and leaf chlorophyll content in spider mite (Acari: Tetranychidae) injures peppermint leaves. *Environmental Entomology* 12, 345–349.

De Carvalho, J.P. and Cardoso, H.A.L. (1972) *Pragas dos Citrinos de Angola*. Instituto de Investigação Agronómica de Angola, Nova Lisboa, Angola.

de Coss, M., Otero-Colinal, G., Peña, J.E. and Magallanes, R. (2010) Demographic and reproductive parameters of *Polyphagotarsonemus latus* in Carica papaya. In: Sabelis, M.W. and Bruin, J. (eds) *Trends in Acarology: Proceedings of the 12th International Congress, 21–26 August 2006, Amsterdam*. Springer, Dordrecht, the Netherlands, pp. 357–360.

de Klerk, C.A. (1985) Chemical control of the grape bud mite, *Eriophyes vitis* (Pagenstecher). *South African Journal of Enology and Viticulture* 6, 13–16.

de la Torre, P.E. and Martínez, H. (2004) Lista de los ácaros eriofioideos (Acari: Prostigmata: Eriophyoidea) de Cuba. *Revista Ibérica de Aracnología* 9, 123–126.

de la Torre, P.E., Almaguel, L. and Perez, E. (2004) *Eriophyes lantanae* Cook (Acari: Eriophyidae) en *Lantana camara* L. en Cuba. *Fitosanidad* (8)1, 67–68.

De Leon, D. (1955) Hosts of *Tuckerella pavoniformis* (Ewing) and *Tuckerella ornata* (Tucker) in Florida (Acarina: Tuckerellidae). *Florida Entomologist* 38, 89–90.

De Leon, D. (1967) *Some Mites of the Caribbean Area. Part I. Acarina on Plants in Trinidad, West Indies*. Allen Press, Lawrence, Kansas.

de Lillo, E. (1988) Acari Eriofidi (Acari: Eriophyoidea) nuovi per l'Italia. *Entomologica* 23, 13–46.

de Lillo, E. (1991) Preliminary observations of ovoviviparity in the gall-forming mite, *Aceria caulobius* (Nal.) (Eriophyoidea: Eriophyidae). In: Schuster, R. and Murphy, P.W. (eds) *The Acari: Reproduction, Development and Life-history Strategies*. Chapman and Hall, London, pp. 223–229.

de Lillo, E. and Amrine, J.W. Jr (1998) Eriophyoidea (Acari) on a computer database. *Entomologica* 31, 137–146.

de Lillo, E. and Amrine, J.W. Jr (2006) Family Eriophyidae. Eriophyidae species listing. Biology Catalog Web Site, Texas A & M University, Austin, Texas. Available at: http://insects.tamu.edu/research/collection/hallan/acari/Eriophyidae.htm (accessed 26 July 2014).

de Lillo, E. and Beron, P. (2004) Eriophyoidea. In: Magowski, W.Ł. (ed.) Acari: Acariformes: Fauna Europaea version 2.6.2. Available at: http://www.faunaeur.org/full_results.php?id=15628 (accessed 30 April 2015).

de Lillo, E. and Monfreda, R. (2002) La "saliva" degli acari Eriophyoidea: messa a punto di un' indagine e risultati preliminari. In: *Atti XIX Congresso Nazionale Italiano di Entomologia, 10–15 Giugno 2002, Catania*. Accademia Nazionale Italiana di Entomologia, Florence, Italy, pp. 1365–1370.

de Lillo, E. and Monfreda, R. (2004) 'Salivary secretions' of eriophyoids (Acari: Eriophyoidea): first results of an experimental model. *Experimental and Applied Acarology* 34, 291–306.

de Lillo, E., Di Palma, A. and Nuzzaci, G. (2001a) Morphological adaptation of mite chelicerae to different trophic activities (Acari). *Entomologica* 32, 125–180.

de Lillo, E., Guario, A. and Lasorella, V. (2001b) Un semplice metodo di estrazione per il campionamento degli acari eriofidi (Eriophyoidea) (*Vitis vinifera* L.). *Informatore Fitopatologico* 51(10), 57–60.

de Lillo, E., Guario, A., Monfreda, R., Lasorella, V. and Grande, O. (2004) Applicazioni fungicide e acaricide nel controllo di *Calepitrimerus vitis*. *Informatore Agrario* 60(17), 63–66.

de Lillo, E., Bari, G. and Monfreda, R. (2005) Preliminary study on the distribution of *Calepitrimerus vitis* (Nalepa) on tendone trained vineyards in Apulia, southern Italy. *Phytophaga* 14, 605–610.

de Lillo, E., Monfreda, R. and Bari, G. (2007) Efficacia di spirodiclofen nel controllo del *Calepitrimerus vitis* (Nalepa), agente dell'acariosi della vite. *Informatore Agrario* 63(21), 74–76.

De Ong, E.R., Knight, H. and Chamberlain, J.C. (1927) A preliminary study of petroleum oil as an insecticide for citrus trees. *Hilgardia* 2, 351–384.

De Ponti, O.M.B. (1977) Resistance in *Cucumis sativus* L. to *Tetranychus urticae* Koch. 1. The role of plant breeding in integrated control. *Euphytica* 26, 633–640.

De Ponti, O.M.B. (1978) Resistance in *Cucumis sativus* L. to *Tetranychus urticae* Koch. 3. Search for sources of resistance. *Euphytica* 27, 167–176.

De Ponti, O.M.B. (1982) Plant resistance to insects: a challenge to plant breeders and entomologists. In: Visser, J.H. and Minks, A.K. (eds) *Proceedings of the 5th International Symposium on Insect–Plant Relationships*. Pudoc, Wageningen, the Netherlands, pp. 337–347.

De Ponti, O.M.B. (1985) Host plant resistance and its manipulation through plant breeding. In: Helle, W. and Sabelis, M.W. (eds) *Spider Mites: Their Biology, Natural Enemies and Control*. World Crop Pests, Volume 1B. Elsevier Science, Amsterdam, pp. 395–403.

De Ponti, O.M.B. and Garretsen, F. (1980) Resistance in *Cucumis sativus* L. to *Tetranychus urticae* Koch. 7. The inheritance of resistance and bitterness and the relation between these characters. *Euphytica* 29, 513–523.

De Silva, P.H.P.R. and Fernando, L.C.P. (2008) Rearing of coconut mite *Aceria guerreronis* and the predatory mite *Neoseiulus baraki* in the laboratory. *Experimental and Applied Acarology* 44, 37–42.

de Villiers, M. and Pringle, K.L. (2011) The presence of *Tetranychus urticae* (Acari: Tetranychidae) and its predators on plants in the ground cover in commercially treated vineyards. *Experimental and Applied Acarology* 53, 121–137.

De Vis, R.M.J., Moraes, G.J. de and Bellini, M.R. (2006) Initial screening of little known predatory mites in Brazil as potential pest control agents. *Experimental and Applied Acarology* 39, 115–125.

de Wolf, E, and Seifers, D. (2008) Triticum *Mosaic: A New Wheat Disease in Kansas*. Publication EP-145, Kansas State University Agricultural Experiment Station and Cooperative Extension Service, Manhattan, Kansas. Available at: http://www.plantpath.ksu.edu/doc1189.ashx (accessed 5 May 2015).

Dean, F.P. and Newcomer, E.J. (1948) Comparative efficiency of materials for controlling orchard mites. In: *Proceedings of the 44th Annual Meeting of the* Washington State Horticultural Association. Washington State Horticultural Association, Wenatchee, Washington, pp. 131–133.

Dean, H.A. (1959a) Seasonal distribution of mites on Texas grapefruit. *Journal of Economic Entomology* 52, 228–232.

Dean, H.A. (1959b) Quadrant distribution of mites on leaves of Texas grapefruit. *Journal of Economic Entomology* 52, 725–727.

Dean, H.A. (1979) Citrus rust mite control affected by certain pesticides. *Journal of the Rio Grande Valley Horticultural Society* 33, 55–62.

Dean, H.A. (1980) Population differences of texas citrus mite, *Eutetranychus banksi* on leaves of four orange varieties in texas. *Journal of Economic Entomology* 73, 813–816.

Dean, H.A. and Maxwell, N.P. (1967) Spotting of grapefruit associated with false spider mites. *Journal of the Rio Grande Valley Horticultural Society* 21, 35–45.

DeBach, P. (ed.) (1964) *Biological Control of Insect Pests and Weeds*. Chapman and Hall, London.

DeBach, P. and Rosen, D. (1991) *Biological Control by Natural Enemies*, 2nd edn. Cambridge University Press, Cambridge, UK.

DeBach, P., Fleschner, C.A. and Dirtrick, E.J. (1950) Studies on the efficacy of natural enemies of citrus red mite in southern California. *Journal of Economic Entomology* 43, 807–819.

DeBach, P., Rosen, D. and Kennett, C.E. (1971) Biological control of coccids by introduced natural enemies. In: Huffaker, C.B. (ed.) *Biological Control*. Plenum Press, New York, pp. 165–194.

Decou, G.C. (1994) Biological control of the two spotted spider mite (Acarina: Tetranychidae) on commercial strawberries in Florida with *Phytoseiulus persimilis* (Acarina: Phytoseiidae). *Florida Entomologist* 77, 33–41.

DeGeer, C. (1778) *Memoires pour Servir a l'Histoire des Insectes*, Tome Septième. Pierre Hesselberg, Stockholm.

Dekle, G.W. (1957) *Twenty-first Biennial Report for the Period July 1, 1954–June 30,* 1956. Bulletin 11-A, State Plant Board of Florida 2, Gainesville, Florida, pp. 71–75.

Dekle, G.W. (1960) *Twenty-third Biennial Report for the Period July 1, 1958–June 30, 1960*. Bulletin 14, State Plant Board of Florida 2, Gainesville, Florida, pp. 65–70.

Del Fosse, E.S. (1978a) Interaction between the mottled waterhyacinth weevil, *Neochetina eichhorniae* Warner and the waterhyacinth mite, *Orthogalumna terebrantis* Wallwork. In: Freeman, T.E. (ed.) *Proceedings of the IV International Symposium on Biological Control Weeds, University of Florida, Gainesville, Florida, August 30–September 2, 1976*. Center for Environmental Programs, Institute of Food and Agricultural Sciences, University of Florida, Gainesville, Florida, pp. 93–97.

Del Fosse, E.S. (1978b) Effect on waterhyacinth of *Neochetina eichhorniae* (Col.: Curculionidae) combined with *Orthogalumna terebrantis* (Acari: Galumnidae). *Entomophaga* 23, 379–387.

Del Fosse, E.S. (2004) Introduction. In: Coombs, E.M., Clark, J.K., Piper, G.L. and Cofrancesco, A.F. Jr (eds) *Biological Control of Invasive Plants in the United States*. Oregon State University Press, Corvallis, Oregon, pp. 1–11.

Del Guercio, G. (1911) Prima contribuzione alla conoscenza degli eriofiidi delle gemme del nocciuolo e delle foglie del pero e le esperienze tentate per combatterli. *Redia* 7, 1–64.

del Rivero, J.M. and García Marí, F. (1983) El ácaro *Petrobia iatens* (Müller) en España. *Boletín del Servicio de Defensa contra Plagas e Inspección Fitopatológica* 9, 109–126.

del Rosario, M.S. and Sill, W.H. Jr (1958) A method of rearing large colonies of an eriophyid mite, *Aceria tulipae* (Keifer), in pure culture from single eggs or adults. *Journal of Economic Entomology* 51, 303–306.

Delalibera, I. Jr and Mascarin, G.M. (2010) Control of *Tetranychus urticae* (Acari: Tetranychidae) with entomopathogenic fungi. In: Moraes, G.J. de, Castilho, R.C. and Flechtmann, C.H.W. (eds) *XIII International Congress of Acarology, August 23–27, 2010, Recife-PE, Brazil, Abstract Book*, abstract 118, p. 67. Available at: http://www.acarology.org/ica/ica2010/docs/0910_15_abstractBookCapa.pdf (accessed 3 February 2015).

Delalibera, I. Jr, Sosa Gomez, D.R., Moraes, G.J. de, Alencar, J.A. de and Farias Araujo, W. (1992) Infection of *Mononychellus tanajoa* (Acari: Tetranychidae) by the fungus *Neozygites* sp. (Entomophthorales) in northeastern Brazil. *Florida Entomologist* 75, 145–147.

Delalibera, I., Hajek, A.E. and Humber, R.A. (2004) *Neozygites tanajoae* sp. nov., a pathogen of the cassava green mite. *Mycologia* 96, 1002–1009.

Dellei, A. and Szendrey, L. (1991) Különböző szőlőfajták atkafertőzöttségének vizsgálata Heves megye szőlőültetvényeiben (The mite infestation of different grapevine varieties in the vineyards of Heves County). *Növényvédelem* 27, 55–61 (in Hungarian).

Delley, B. (1972) Quatre especes d'eriophyides libres nuisibles aux pruniers cultives. *Revue Suisse de Viticulture Arboriculture Horticulture* 4, 39–44.

Delley, B. (1973) Contribution a l'etude des eriophyides libres du prunier dans le verger Neuchatelois. *Bulletin de la Société Entomologique Suisse/Mitteilungen der Schweizerischen Entomologischen Gesellschaft* 46, 75–118.

Delmas, R. and Rambier, A. (1954) L'invasion des "araignées rouges" sur la vigne. *Progrès Agricole et Viticole* 34/35, 101–104.

Delrio, G. and Monagheddu, M. (1986) Effetti della temperatura sull'incremento delle popolazioni di *Panonychus citri* (McGregor) (Acarina: Tetranychidae). *Redia* 69, 627–650.

Demite, P.R. and Feres, R.J.F. (2005) Influência de vegetação vizinha na distribuição de acaros em seringal (*Hevea brasiliensis* Muell. Arg., Euphorbiaceae) em São José do Rio Preto, SP. *Neotropical Entomology* 34, 829–836.

Demite, P.R., Moraes, G.J. de, McMurtry, J.A., Denmark, H.A. and Castilho, R.C. (2014a) Phytoseiidae Database. Available from: www.lea.esalq.usp.br/phytoseiidae (accessed 19 October 2015).

Demite, P.R., McMurtry, J.A. and Moraes, G.J. de (2014b) Phytoseiidae Database: a website for taxonomic and distributional information on phytoseiid mites (Acari). *Zootaxa* 3795, 571–577.

Demokidov, K.E. (1916) On the life history of the tea moth, *Parametriotes thaea* Kush. *Entomologicheskoe Obozrenie (Entomological Review, USSR)* 15, 618–626.

den Houter, J.G. (1976) The effect of acaricides on active and diapausing females of *Tetranychus urticae* Koch, the two-spotted spider mite. *Zeitschrift für Angewandte Entomologie* 81, 248–252.

Deng, Z., Zhang, N.X. and Jia, X.F. (1990) Colonization of *Metaseiulus occidentalis* Acari Phytoseiidae and its control effect on *Eotetranychus pruni* Acari Tetranychidae in Lanzhou apple orchards. *Chinese Journal of Biological Control* 6, 54–58 (in Chinese).

Denmark, H.A. (1962) *Aculus pelekassi* Keifer, another citrus mite in Florida. *Proceedings of the Florida State Horticultural Society* 75, 25–27.

Denmark, H.A. (1968) *The Southern Red Mite,* Oligonychus ilicis *(McGregor)*. Entomology Circular No. 79:1, Florida Department of Agriculture and Consumer Services, Division of Plant Industry, Gainesville, Florida.

Denmark, H.A. (1970) *A Juniper Gall Mite,* Trisetacus quadrisetus juniperinus *(Nal.)*. Entomology Circular No. 91, Florida Department of Agriculture and Consumer Services, Division of Plant Industry, Gainesville, Florida.

Denmark, H.A. (1973) Tetranychus evansi *Baker and Pritchard in Florida (Acarina: Tetranychidae)*. Entomology Circular No. 1, Florida Department of Agriculture and Consumer Services, Division of Plant Industry, Gainesville, Florida.

Denmark, H.A. (1983) Eriophyes mangiferae *(Sayed) a pest of mango (Acarina: Eriophyidae)*. Entomology Circular No. 254, Florida Department of Agriculture and Consumer Services, Division of Plant Industry, Gainesville, Florida.

Denmark, H.A. (2006) *Featured Creatures: Phalaenopsis Mite,* Tenuipalpus pacificus *Baker (Arachnida: Acari: Tenuipalpidae)*. Publication Number: EENY-377, Institute of Food and Agricultural Sciences, University of Florida, Gainesville, Florida. Available at: http://entnemdept.ufl.edu/creatures/orn/flowers/Tenuipalpus_pacificus.htm (accessed 5 May 2015).

Denmark, H.A. and Keifer, H.H. (1979) *A Mite,* Eriophyes aloinis *(Keifer) (Acarina: Eriophyidae)*. Entomology Circular No. 206, Florida Department of Agriculture and Consumer Services, Division of Plant Industry, Gainesville, Florida.

Denmark, H.A. and Muma, M.H. (1989) *A Revision of the Genus* Amblyseius *Berlese, 1914 (Acari: Phytoseiidae)*. Occasional Papers of the Florida State Collection of Arthropods Volume 4, Florida Department of Agriculture and Consumer Services, Division of Plant Industry, Gainesville, Florida. Available at: http://freshfromflorida.s3.amazonaws.com/occasional-papers-vol-4.pdf (accessed 5 May 2015).

Dennehy, T.J. and Granett, J. (1982) Improved detection of dicofol-resistant spider mites in cotton. *California Agriculture* 36(11, 12), 11–12.

Dennehy, T.J. and Omato, C. (1994) Sustaining the efficacy of dicofol against citrus rust mite: a case history of industrial and academic collaboration. In: *Brighton Crop Protection Conference, Pests and Diseases – 1994. Proceedings of an International Conference Organised by the British Crop Protection Council, Held at the Brighton Centre and the Brighton Metropole Hotel, Brighton, England, 21–24 November 1994, Volume 3.* British Crop Production Council, Farnham [now Alton], UK, pp. 955–962.

Dennill, G.B. (1986) An ecological basis for timing control measures against the grape vine bud mite *Eriophyes vitis* Pgst. *Crop Protection* 5, 12–14.

Dennill, G.B. (1991) A pruning technique for saving vineyards severely infested by the grape vine bud mite *Colomerus vitis* (Pagenstecher) (Eriophyidae). *Crop Protection* 10, 310–314.

Deol, G.S. and Sandhu, G.S. (1974) Note on chemical control of brown wheat mite. *Indian Journal of Agricultural Sciences* 44, 681–682.

Depaoli, J.G., Vivarelli, J.B., Oliveira, L.C. de Jr, Silva, E.A., Plens, E.B., Calafiori, M.H. and Franco, J.F. (1983) Ensaio para controle de tripes, *Thrips tabaci* Lindeman, 1888, microácaro *Eriophyes tulipae* Keifer,1938, em alho (*Allium sativum* L.) (Test on the control of thrips, *Thrips tabaci* Lindeman, 1888, russet mite, *Eriophyes tulipae* Keifer, 1983, on garlic (*Allium sativum* L.)). *Ecossistema* 8, 65–70.

Depew, L.J. (1960) Control of *Oligonychus pratensis* attacking winter wheat in western Kansas. *Journal of Economic Entomology* 53, 1061–1063.

Dermauw, W., Ilias, A., Riga, M., Tsagkarakou, A., Grbić, M., Tirry, L., Van Leeuwen, T. and Vontas, J. (2012) The cys-loop ligand-gated ion channel gene family of *Tetranychus urticae*: implications for acaricide toxicology and a novel mutation associated with abamectin resistance. *Insect Biochemistry and Molecular Biology* 42,455–465.

Deshpande, R.R., Shaw, S.S., Shrivastav, R.C. and Mandloi, K.C. (1988) Relative toxicity of pesticides against red spider mite *Eutetranychus orientalis* (Klein) on Citrus. *Pesticides* 22(3), 45–46.

Devi, K.D., Varatharajan, R. and Loganathan, S. (2010) Bioassay of *Pochonia* (*Verticillium*) *lecanii* against tea aphid and red spider mite and their field density in Manipur. *Annals of Plant Protection Sciences* 18, 17–21.

Dhar, V. and Rathore, Y.S. (1994) Pattern of population distribution of *Aceria cajani* on pigeonpea plant. *Indian Journal of Pulses Research* 7, 137–143.

Dhar, V., Chaudhary, R.G. and Naimuddin (1998) Population studies on sterility mosaic vector (*Aceria cajani*). In: *Annual Report 1996–97, Indian Insititute of Pulse Reasearch, Kanpur, India*, p. 28.

Dharmaraj, P.S., Narayana, Y.D., Kumar, P.L., Waliyar, F. and Jones, A.T. (2004) Pigeonpea sterility mosaic disease: an emerging problem in northern Karnataka, India. *International Chickpea and Pigeonpea Newsletter* 11, 47–49.

Dhooria, M.S. (1981) Feeding behavior of predatory mites, thrips and beetles of the citrus mite *Eutetranychus orientalis*. *Acarology Newsletter* 10, 4–6.

Dhooria, M.S. (1982) Differential response of citrus species/varieties to infestation of *Eutetranychus orientalis*. *Indian Journal of Acarology* 7, 11–14.

Dhooria, M.S. (1983) An outbreak of citrus mite, *Eutetranychus orientalis* (Klein) in Delhi. *Pesticides* 17(11), 36.

Dhooria, M.S. (2005a) Effect of different ecological parameters on infestation of tarsonemid mite, *Polyphagotarsonemus latus* (Banks) (Acari: Tarsonemidae) on sesamum. *Journal of Applied Zoological Research* 16, 137–138.

Dhooria, M.S. (2005b) Tarsonemid mite, *Polyphagotarsonemus latus* (Banks) (Acari: Tarsonemidae) – a serious pest of green gram (*Vigna radiata*) in Punjab. *Insect Environment* 11, 3–4.

Dhooria, M.S. and Butani, D.K. (1984) Citrus mite, *Eutetranychus orientalis* (Klein) and its control. *Pesticides* 18(10), 35–38.

Dhooria, M.S. and Mann, G.S. (1980) Chemical control trials using ultra low volume and high volume sprays against spider mite, *Tetranychus neocaledonicus* Andre infesting brinjal and long-melon crops in Punjab. *Indian Journal of Entomology* 42, 269–271.

Dhooria, M.S. and Sandhu, G.S. (1973) Resistance of different species of citrus to the attack of mite *Eutetranychus orientalis* (Klein). *Punjab Horticultural Journal* 13, 135–136.

Dhooria, M.S., Bhullar, M.B. and Mallik, B. (2005) *Mite Pests of Citrus and their Management in India*. AINP (All India Network Project) on Agricultural Acarology, University of Agricultural Sciences, Bangalore, India.

Di Martino, E. (1985) I parassiti animali. In: Spina, P. and Di Martino, E. (eds) Trattato di Agrumicoltura. Edagricole, Bologna, Italy, pp. 113–195.

Di Martino, E. and Benfatto, D. (1973) Osservazioni etologiche danni ed esiti di lotta sperimentale nei confronti dell'*Aculus pelekassi* K. su limone. *Annali dell'Istituto Sperimentale per l'Agrumicoltura* 6, 181–193.

Di Palma, A. (1995) Morfologia funzionale delle parti boccali di *Penthaleus major* (Dugès) (Eupodoidea: Penthaleidae). *Entomologica* 29, 69–86.

Di Stefano, M. (1971) Contributi alla conoscenza degli Acari Eriophyidae: *Phyllocoptes phloeocoptes* (Nalepa) n. comb. var. n. acaro galligeno del Pesco (*Prunus persica* Stokes). *Marcellia* 37, 59–74.

Díaz, A., Okabe K., Eckenrode, C.J., Villani, M.G. and O'Connor, B.M. (2000) Biology, ecology, and management of the bulb mites of the genus *Rhizoglyphus* (Claparède) (Acari: Acaridae). *Experimental and Applied Acarology* 24, 85–113.

Díaz, F., Abud, A. and González, M. (1999) El ácaro de la panícula del arroz. *Fersan Informa* 24, 11–12.

Díaz, G., Romero, E., Boyero, J.R. and Malpiga, N. (2009) Recognition and quantification of area damaged by *Oligonychus perseae* in avocado leaves. In: Bayro Corrochano, E. and Eklundh, J.-O. (eds) *Progress in Pattern Recognition, Image Analysis, Computer Vision, and Applications: 14th Iberoamerican Conference on Pattern Recognition, CIARP '09, Guadalajara, Jalisco, Mexico, November 2009, Proceedings*. Springer, Berlin/Heidelberg, pp. 667–684.

Dick, G.L. and Buschman, L.L. (1995) Seasonal occurrence of a fungal pathogen, *Neozygites adjarica* (Entomophthorales: Neozygitaceae), infecting Banks grass mites, *Oligonychus pratensis*, and twospotted spider mites, *Tetranychus urticae* (Acari: Tetranychidae), in field corn. *Journal of the Kansas Entomological Society* 68, 425–436.

Dicke, M. (1995) Plant characteristics influence biological control agents: implications for breeding for host plant resistance. *IOBC-WPRS Bulletin* 19(5), 72–80.

Dicke, M. and Grostal, P. (2001) Chemical detection of natural enemies by arthropods: an ecological perspective. *Annual Review of Ecology, Evolution, and Systematics* 32, 1–23.

Dicke, M., De Jong, M., Alers, M.P.T., Stelder, F.C.T., Wundennk, R. and Post, J. (1989) Quality control of mass-reared arthropods: nutritional effects on performance of predatory mites. *Journal of Applied Entomology* 108, 462–475.

Dicke, M., Sabelis, M.W., Takabayashi, J., Bruin, J. and Posthumus, M.A. (1990a) Plant strategies of manipulating predator–prey interactions through allelochemicals: prospects for application in pest control. *Journal of Chemical Ecology* 16, 3091–3118.

Dicke, M., Sabelis, M.W., De Jong, M. and Alers, M.P.T. (1990b) Do phytoseiid mites select the best prey species in tems of reproductive success? *Experimental and Applied Acarology* 8, 302–309.

Dicke, M., Gols, R., Ludeking, D. and Posthumus, M.A. (1999) Jasmonic acid and herbivory differentially induce carnivore-attracting plant volatiles in lima bean plants. *Journal of Chemical Ecology* 25, 1907–1922.

Dicker, G.H.L., Gambrill, R.G. and Easterbrook, M.A. (1972) Chemicals tested for control of black currant gall mite, *Cecidophyopsis ribis* (Westw.). *Journal of Horticultural Science* 47, 535–539.

Dickson, R.C. (1949) Factors governing the induction of diapause in the oriental fruit moth. *Annals of the Entomological Society of America* 42, 511–537.

Dieckmann, M., Frison, E.A. and Putter, T. (eds) (1994) *FAO/IPGRI Technical Guidelines for the Safe Movement of Small Fruit Germplasm, Volume 2*. Food and Agriculture Organization of the United Nations and International Plant Genetic Resources Institute, Rome.

Dippenaar, B.J. (1958a) Concentring ring blotch of citrus. Its cause and control. *South African Journal of Science* 1, 83–106.

Dippenaar, B.J. (1958b) Concentring ring blotch of citrus. I. Description and symptoms. *Farming in South Africa* 34(5), 40–43.

Dippenaar, B.J. (1958c) Concentring ring blotch of citrus. II. *Farming in South Africa* 34(6), 33–35, 40.

Dippenaar Schoeman, A.S. (1998) Spider as predators of citrus pests. In: Bedford, E.C.G., van den Berg, M.A. and de Villiers, E.A. (eds) *Citrus Pests in the Republic of South Africa*, 2nd edn. ARC-Institute for Tropical and Subtropical Crops, Nelspruit, South Africa, pp. 34–35.

Dittrich, V. (1971) Electron-microscopic studies of the respiratory mechanism of spider mite eggs. *Annals of the Entomological Society of America* 64, 1134–1143.

Dixon, R.A., Harrison, M.J. and Lamb, C.J. (1994) Early events in the activation of plant defense responses. *Annual Review of Phytopathology* 32, 479–501.

Dobosz, R., Skorupska, A. and Blaszak, C. (1995) The appearance of spider mites (Tetranychidae) in parks of Poznan. In: Boczek, J. and Ignatowicz, S. (eds) *Materialy z Sympozjum na temat: "Osiagniecia Akarologii w Polsce", Siedlce, 26–27 wresnia 1995* (*Proceedings of the Symposium on "Advances in Acarology in Poland", Siedlce, 26–27 September 1995*). Instytut Ochrony Roślin, Polska Akademia Nauk, Olsztyn, Poland, pp. 39–42 (in Polish).

Dobrivojevic, K. and Petanović, R. (1982) Big bud mite of black currents, *Cecidophyopsis ribis* Westw. (Eriophyidae, Acarina) and its role in the extermination of black currant plantations. *Zastita Bilja* 33, 507–518 (in Serbo-Croatian).

Dobrivojevic, K. and Petanović, R. (1985) Eriofidna grinja lista maline *Phyllocoptes gracilis* (Nal.) (Eriophyoidea, Acarina) malo poznata stetocina u Jugoslaviji. *Zastita Bilja* 36, 247–254 (in Serbo-Croatian).

Dodd, A.P. (1929) *The Progress of Biological Control of Prickly-Pear in Australia*. Government Printer, Brisbane, Australia.

Dodd, A.P. (1940) *The Biological Campaign Against Prickly-Pear*. Commonwealth Prickly Pear Board, Brisbane, Australia.

Dodson, P.G.C. (1969) *Annual Report 1968–1969*. Malkerns Experiment Station, Malkerns, Swaziland.

Dohino, T. and Tanabe, K. (1994) Electron beam irradiation of immature stages and adult males of two spotted spider mite, *Tetranychus urticae* Koch (Acari: Tetranychidae). *Research Bulletin of the Plant Protection Service, Japan* 30, 27–34.

Doidge, E.M. (1925) Citrus diseases. In: Davis, R.E. (ed.) *Citrus Growing in South Africa*. Speciality Press, Johannesburg, South Africa, pp. 217–219.

Dolatyar, S., Jafari, S. and Pakyari, H. (2015) Effect of temperature on the development of *Eotetranychus hirsti* (Tetranychidae) on fig leaves. *Acarologia* 55(3): 247–254

Domes, R. (1957) Zur Biologie der Gallmilbe *Eriophyes gracilis* Nalepa. *Zeitschrift für Angewandte Entomologie* 41, 411–424.

Domes, R. (1998) A new species of the genus *Anthocoptes* (Phyllocoptinae: Eriophyidae) on *Juglans regia* L. *Acarologia* 39, 69–71.

Domingos, C.A., Melo, J.W.S., Oliveira, J.E. de M. and Gondim, M.G.C. Jr (2010) Diversity of mites on Vitaceae in the São Francisco River Valley, Brazil. In: Moraes, G.J. de, Castilho, R.C. and Flechtmann, C.H.W. (eds) *XIII International Congress of Acarology, August 23–27, 2010, Recife-PE, Brazil, Abstract Book*, abstract 130, p. 72. Available at: http://www.acarology.org/ica/ica2010/docs/0910_15_abstractBookCapa.pdf (accessed 3 February 2015).

Domingos, C.A., Oliveira, L.O., Morais, E.G.F. de, Návia, D., Moraes, G.J. de and Gondim, M.G.C. Jr (2012) Comparison of two populations of the pantropical predator *Amblyseius largoensis* (Acari: Phytoseiidae) for biological control of *Raoiella indica* (Acari: Tenuipalpidae). *Experimental and Applied Acarology* 60, 83–93.

Dondale, C.D. (1967) A model outbreak of the mite, *Tetranychus mcdanieli* McGregor in Ontario. *Proceedings of the Entomological Society of Ontario* 98, 29–45.

Dong, H.-F. (1990) Control effect of enhanced liuyangmycin on leaf spider mites and its influence on the beneficials. *Chinese Journal of Biological Control* 6, 97–101 (in Chinese).

Donia, A.R.A., Helal, E.M., El Hamid, M.M.A. and Zakzouk, E.A. (1995) Mass rearing and field evaluation of released predaceous mite, *Euseius scutalis* (Athias-Henriot) on the citrus whitefly, *Aleurotrachelus citri* (Priesner & Hosny). *Alexandria Journal of Agricultural Research* 40, 209–219.

Donnadieu, A.L. (1875) Recherches pour servir à l'histoire des tétranyques. *Annales de la Société Linnéenne de Lyon (n. sér.)* 22, 29–136.

Doorduin, J.C., Lilien, H.K., Borochov, A. and Halevy, A.H. (1997) Factors influencing the duration of hot-water treatment of amaryllis bulbs (*Hippeastrum*). In: Lilien-Kipnis, H., Borochov, A. and Halevy, A.H. (eds) *Seventh International Symposium of Flower Bulbs, March 10–16, 1996, Herzliya, Israel. Acta Horticulturae* 430, 695–698.

Doreste, S.E. (1963) Avances obtenidos en el control del ácaro del ajo (*Aceria tulipae* (K.)), nueva plaga de este cultivo en Venezuela. *Ingeniería Agronómica* 11, 13 18.

Doreste, S.E. (1964) Influence of three food plants on the bionomics of *Panonychus ulmi* (Koch). *Agronomía Tropical (Maracay)* 14, 83–100.

Doreste, S.E. (1965) Control del "ácaro del ajo" (*Aceria tulipae* (K.)), mediante tratamiento a la semilla. In: Marcano, L. and Casseres, E. (eds) *Proceedings of the Caribbean Region, American Society Horticultural Science, XIIth Annual Meeting, October 25–31, 1964, Cagua, Venezuela*, pp. 39–43.

Doreste, S.E. (1966) Control del ácaro del ajo (*Aceria tulipae* (K.)) mediante tratamiento a la "semilla". *Agronomía Tropical (Maracay)* 16, 199–203.

Doreste, S.E. (1968a) El ácaro de la flor del cocotero (*Aceria guerreronis* Keifer) en Venezuela. *Agronomía Tropical (Maracay)* 18, 370 386.

Doreste, S.E. (1968b) Primera lista de acaros de importancia agricola en Venezuela. *Agronomía Tropical (Maracay)* 18 449–460. (Trabajo presentado en la VII Reunion Latinoamericana de Fitotecnía, Maracay, Venezuela, Septiembre, 1967.)

Doreste, S. E. (1979) Dos acaros plagas del durazno (*Prunus persica* L.) en Venezuela. *Revista de la Facultad de Agronomía, Universidad Central de Venezuela* 10, 121–127.

Doreste, S.E. (1981) Acaros del género *Mononychellus* Wainstein (Acari Tetranychidae) asociados con la yuca (*Manihot* spp.) en Venezuela. *Boletín de Entomología Venezolana, N.S.* 1, 119–130.

Doreste, S.E. (1982) El complejo de acaros tetraniquidos, como plaga importante del cultivo de la yuca. *Revista de la Facultad de Agronomía, Universidad Central de Venezuela* 12, 249–254.

Doreste, S.E. (1984a) Informacion sobre el eriofido del mango, *Eriophyes mangiferae* (Sayed), en Venezuela. *Revista de la Facultad de Agronomía, Universidad Central de Venezuela* 13, 91–100.

Doreste, S.E. (1984b) Spider mites as important pest of cassava (*Manihot esculenta* Crantz). In: Griffiths, D.A. and Bowman, C.E. (eds.) *Acarology VI, Volume 2*. Ellis Horwood, Chichester, UK, pp. 761–769.

Doreste, S.E. (1988) *Acarología*. Instituto Interamericano de Cooperación para la Agricultura, Costa Rica.

Doreste, S.E. and Aponte, O. (1979) Efectos de los ataques del complejo de ácaros Tetranychidae en los rendimientos del cultivo de la yuca. *Revista de la Facultad de Agronomía, Universidad Central de Venezuela* 10, 105–119.

Doreste, S.E., Arias, C. and Bellotti, A. (1977) Field evaluations of cassava cultivars for resistance to tetranychid mites. In: Brekelbaum, T., Bellotti, A. and Lozano, J.C. (eds) *Proceedings CASSAVA Protection Workshop, 7–12 November 1977, Cali, Colombia*. CIAT (Centro Internacional de Agricultura Tropical), Cali, Colombia, pp. 161–164.

Dosse, G. (1953) *Tenuipalpus oudemansi* Geijskes, eine fur Deutschland neue spinnmilbenart. *Zeitschrift für Angewandte Entomologie* 34, 587–597.

Dosse, G. (1954) *Tenuipalpus orchidarum* Parfitt num auch in duetschen Gewachschausern. *Zeitschrift für Angewandte Entomologie* 36, 304.

Dosse, G. (1958) Uber einige neue Raubmilbenarten (Acarina: Phytoseiidae). *Pflanzenschutzberichte* 21, 44–61.

Dosse, G. (1960) Über den Einfluss der Raubmilbe *Typhlodromus tiliae* Oud. auf die Obstbaumspinnmilbe *Metatetranychus ulmi* Koch (Acari). *Pflanzenschutzberichte* 24, 113–137.

Dosse, G. (1963) *Bryobia rubrioculus* Scheuten in der nördlichen Bekaa des Libanons (*Acarina: Tetranychidae*). *Zeitschrift für Pflanzenkrankheiten Pflanzenpathologie und Pflanzenschutz* (now *Journal of Plant Diseases and Plant Protection*) 70, 652–666.

Dosse, G. (1964a) Boebachtungen über Entwicklungstandenzen in *T. urticae-cinnabarinus* complex. *Pfänzenschutzberichte* 31, 113–128.

Dosse, G. (1964b) Studies on *Tetranychus cinnabarinus* Boisd. complex in citrus plantations in Lebanon. *Zeitschrift für Angewandte Entomologie* 53, 453–461.

Dosse, G. (1967) Schädmilben des Libanons und ihre Prädatoren. *Zeitschrift für Angewandte Entomologie* 59, 16–48.

Doucette, C.F. (1926) The effect on narcissus bulb pests of immersion in hot water. *Journal of Economic Entomology* 19, 248–251.

Doucette, C.F. (1929) A tarsonemid mite attacking *Narcissus*. *Journal of Economic Entomology* 22, 423–424.

Doucette, C.F. (1936) Observations on bulb scale mite as a major pest of narcissus. *Journal of Economic Entomology* 29, 1103–1105.

Doucette, C.F. (1962) The lewis mite, *Eotetranychus lewisi* on greenhouse poinsettia. *Journal of Economic Entomology* 55, 139–140.

Doudrick, R.L. (1983) Rose rosette. In: *American Rose Annual 1983*. American Rose Society, Shreveport, Louisiana, pp. 119–120.

Dover, M.J., Croft, B.A., Welch, S.M. and Tummula, R.I. (1979) Biological control of *Panonychus ulmi* (Acarina: Tetranychidae) by *Amblyseius fallacis* (Acarina: Phytoseiidae) on apple: a prey-predator model. *Environmental Entomology* 8, 282–292.

Dowling, A.P.G., Ochoa, R., Beard, J.J., Welbourn, W.C. and Ueckermann, E.A. (2012) Phylogenetic investigation of the genus *Raoiella* (Prostigmata: Tenuipalpidae): diversity, distribution, and world invasions. *Experimental and Applied Acarology* 57, 257–269.

Downing, R.S. (1966) Quinoxalines as orchards acaricides in British Columbia. *Canadian Entomologist* 98, 134–138.

Downing, R.S. and Moilliet, T.K. (1967) Relative densities of predacious and phytophagous mites on three varieties of apple tree. *Canadian Entomologist* 99, 738–741.

Downing, R.S. and Moilliet, T.K. (1969) Control of the pear rust mite *Epitrimerus pyri*. *Canadian Entomologist* 101, 1000–1002.

Downing, R.S. and Moilliet, T.K. (1974) Control of the pear leaf blister mite and the pear rust mite (Acarina: Eriophyidae) in British Columbia. *Journal of the Entomological Society of British Columbia* 71, 3–4.

Downing, R.S. and Moilliet, T.K. (1975) Preliminary trials with citrazon – a selective acaricide. *Journal of the Entomological Society of British Columbia* 72, 3–5.

Drapek, R.J., Calkin, J.A. and Fisher, G.C. (1990) A hazelnut pest management expert system. *Acta Horticulturae* 276, 21–25.

Dresner, E. (1949) Culture and use of entomogenous fungi for the control of insect pests. *Contributions from Boyce Thompson Institute* 15, 319–335.

Drishpoun, Y., Goldshmidt, Y. and Yaakov, B. (1995) Acaricides screening for controlling the citrus rust mite (*Phyllocoptruta oleivora*) in citrus groves. *Alon Hanotea* 49, 364–368 (in Hebrew).

Drukker, B., Yaninek, J.S. and Herren, H.R. (1993) A packaging and delivery system for aerial release of phytoseiids for biological control. *Experimental and Applied Acarology* 17, 129–143.

Du, K.-S., Qin, X.-F., Du, L.-Y. and Chai, Li-Y. (2009) Effect of six kinds of pesticide on prevention and control for *Eotetranychus populi* (Koch). *Hubei Agricultural Sciences* 2009-09, 2168–2169, 2210. Available as English summary at: http://en.cnki.com.cn/Article_en/CJFDTOTAL-HBNY200909041.htm (accessed 27 July 2014).

du Fretay, G., Quenin, H., Allemoz, P. and Guillon, M. (2001) Nouveaux methodes de l'utilization du soufre et des formulations de terpenes dans un programme de protection integree en viticulture. *IOBC/WPRS Bulletin* 24(7), 159–163.

du Pasquier, R. (1932) Principales maladies parasitaires du théier et du caféier en Extreme-Orient. *Bulletin Économique de l'Indo-Chine* 35, 689–720.

du Pasquier, R.D. (1933) Principales maladies parasitaires du théier et du caféier en Exteme-Orient, Hanoi, Vietnam. *Bulletin Économique de l'Indo-Chine* 36, 127–131.

Duarte, V.S. and Delalibera, I. Jr (2010c) Impact of natural epizootics of *Neozygites floridana* (Zygomycetes: Entomophthorales) on populations of the tomato red spider mite. In: Moraes, G.J. de, Castilho, R.C. and Flechtmann, C.H.W. (eds) *XIII International Congress of Acarology, August 23–27, 2010, Recife-PE, Brazil, Abstract Book*, abstract 132, p. 73. Available at: http://www.acarology.org/ica/ica2010/docs/0910_15_abstractBookCapa.pdf (accessed 3 February 2015).

Duarte, V.S., Rumenos, N.N. and Delalibera, I. Jr (2010) Molecular characterization of strains of the fungus *Neozygites floridana*, pathogenic to spider mites. In: Moraes, G.J. de, Castilho, R.C. and Flechtmann, C.H.W. (eds) *XIII International Congress of Acarology, August 23–27, 2010, Recife-PE, Brazil, Abstract Book*, abstract 133, pp. 73–74. Available at: http://www.acarology.org/ica/ica2010/docs/0910_15_abstractBookCapa.pdf (accessed 3 February 2015).

Dubitzki, E. (1981) Spider mites (Acari: Tetranychidae) infesting wild plants in Israel. PhD thesis, Hebrew University of Jerusalem, Jerusalem (in Hebrew, with English abstract).

Duffner, K (1999) Untersuchungen zur Biologie, Morphologie und Bekämpfung der Kräuselmilbe *Calepitrimerus vitis* Nalepa 1905 (Acari, Eriophyoidea). PhD thesis, Faculty of Biology, Albert-Ludwigs University, Freiburg im Breisgau, Germany.

Duffner, K. and Schruft, G. (1998) Die klebebandmethode zur erfassung des wanderungsverhaltens von krauselmilben. *Deutsches Weinbau-Jahrbuch* 49, 201–206.

Duffner, K., Schruft, G., Gueggelin, M., Mathys, D., Wirtz, S. and Guggenheim, R. (1998) Low-temperature SEM studies on genital structures and spermatophores of the grape rust mite *Capelitrimerus vitis* Nalepa, 1905 (Acari: Eriophyidae). *Entomologische Mitteilungen aus dem Zoologisches Institut und Zoologisches Museum der Universität Hamburg* 12(157), 277–288.

Duffner, K., Schruft, G. and Guggenheim, R. (2001) Passive dispersal of the grape rust mite *Calepitrimerus vitis* Nalepa 1905 (Acari, Eriophyoidea) in vineyards. *Journal of Pest Science* 74, 1–6.

Dufour, L. (1832) Description et figure du *Tetranychus lintearicus*, arachnide nouvelle de la tribu des Acarides. *Annales des Sciences Naturelles (Paris)* 25, 276.

Dugdales, J.S. (1975) *A Preliminary Report on Entomological Visits to Rarotonga, Cook Is. in March and October* 1975, *under Project RAS 427, U.N.D.P./FAO*. Entomology Division, DSIR (Department of Scientific and Industrial Research), Auckland, New-Zealand.

Dugès, A. (1834) Recherches pour servir a l'ordre des acariens en general et à la famille des Trombidies en particulier. *Annales des Sciences Naturelles (Paris) (Ser. 2)* 1, 5–46.

Dugès, A. (1839) Recherches sur l'ordre des Acariens. *Annales des Sciences Naturelles, Zoologie* 1, 18–63.

Duke, J.A. (1983) *Cajanus cajan* (L.) Millsp. Center for New Crops and Plant Products, Purdue University. Available at: http://www.hort.purdue.edu/newcrop/duke_energy/Cajanus_cajun.html (accessed 26 February 2015).

Dulinafka, G. and Szendrei, L. (1995) Szőlő szaporítóanyag atkamentesítése hőkezeléssel (Application of thermotherapy to obtain grapevine propagating material free of leaf mites). *Növényvédelmi Tudományos Napok* 41, 42 (in Hungarian).

Dulinafka, G., Szendrey, G. and Szegedi, E. (1995) Szőlő szaporítóanyag atkamentesítése hőkezeléssel (Application of thermotherapy to obtain grapevine propagating material free of leaf mites). *Növényvédelem* 31, 381–385 (in Hungarian).

Duncombe, W.G. (1972) Red spider mite on cotton and its control. *Rhodesia Agriculture Journal* 69(1), 7–10.

Dunley, J.E. and Croft, B.A. (1996) Eriophyoids as competitors of other phytophagous mites. In: Lindquist, E.E., Sabelis, M.W. and Bruin, J. (eds) *Eriophyoid Mites: Their Biology, Natural Enemies and Control*. World Crop Pests, Volume 6. Elsevier Science, Amsterdam, pp. 751–755.

Dunlop, J.A. and Bertrand, M. (2011) Fossil labidostomatid mites (Prostigmata: Labidostomatidae) from Baltic amber. *Acarologia* 51, 191–198.

Dunlop, J.A. and Selden, P.A. (2009) Calibrating the chelicerate clock: a paleontological reply to Jeyaprakash and Hoy. *Experimental and Applied Acarology* 48, 183–197.

Duran, J.M., Rosa, A. de la, Sanchez, A., Alvarado, M. and Prats, T. (2006) Control de *Acalitus phloeocoptes* Nalepa (Acarina, Eriophyidae), eriofido de las agallas del ciruelo, en la Vega de Sevilla. *Boletín de Sanidad Vegetal, Plagas* 32, 71–78.

Duso, C. (1997) Distribution, biology and management of important phytophagous mites on temperate fruits in Italy. *Redia* 80, Appendice 1–24.

Duso, C. (2006) Il controllo biologico ed integrato degli acari fitofagi associati alla vite. In: Ragusa, S. and Tsolakis, H. (eds) *La Difesa della Vite Dagli Artropodi Dannosi, Marsala 10–11 Ottobre 2005*. Università degli Studi di Palermo, Palermo, Italy, pp. 171–204.

Duso, C. and de Lillo, E. (1996) Grape. In: Lindquist, E.E., Sabelis, M.W. and Bruin, J. (eds) *Eriophyoid Mites: Their Biology, Natural Enemies and Control*. World Crop Pests, Volume 6. Elsevier Science, Amsterdam, pp. 571–582.

Duso, C. and Fontana, P. (2002) On the identity of *Phytoseius plumifer* (Canestrini & Fanzago, 1876) (Acari: Phytoseiidae). *Acarologia* 42, 127–136.

Duso, C. and Girolami, V. (1986) *Lotta Integrata in Viticoltura*. Centro Regionale IRIPA, Mestre-Venezia. Venice, Italy.

Duso, C. and Vettorazzo, E. (1999) Mite population dynamics on different grape varieties with or without phytoseiids released (Acari: Phytoseiidae). *Experimental and Applied Acarology* 23, 741–763.

Duso, C., Kreiter, S., Tixier, M.-S., Pozzebon, A. and Malagnini, V. (2010) Biological control of mites in European vineyards and the impact of natural vegetation. In: Sabelis, M.W. and Bruin, J. (eds) *Trends in Acarology: Proceedings of the 12th International Congress, 21–26 August 2006*, Amsterdam. Springer, Dordrecht, the Netherlands, pp. 399–408.

Duso, C., Pozzebon, A., Kreiter, S., Tixier, M.-S. and Candolfi, M. (2012) Management of phytophagous mites in European vineyards. In: Bostanian, N.J., Vincent, C. and Isaacs, R. (eds) *Arthropod Management in Vineyards: Pests*, Approach and Future Directions. Springer, Dordrecht, the Netherlands, pp. 191–217.

Dustan, A.G. and Matthewman, W.G. (1932) Some notes on the cyclamen mite, *Tarsonemus pallidus* (Banks), a pest of strawberry plants. *Annual Report of the Entomological Society of Ontario (1931)* 62, pp. 34–37.

Dutcher, J.D. (2007) Impact of predatory mite releases on the abundance of pecan leaf scorch mite (Acari: Tetranychidae) *Journal of Entomological Science* 42, 517–524.

Dutcher, J.D. (2012) Integrated control of pecan leaf scorch mite (*Eotetranychus hicoriae*) with predatory mites and miticides. In: *Entomological Society of America Annual Meeting 2012, ESA Annual Meetings Online Program*, poster D0528.

Dutcher, J.D., Fonsah, G.E. and Hudson, W.G. (2009) Integration of bifenazate and western predatory mite (*Acari: Phytoseiidae*) for control of pecan leaf scorch mite (*Acari: Tetranychidae*) in pecan orchards. *Journal of Entomological Science* 44, 98–110.

Duverney, C. and Ngueye-Ndiaye, A. (2005) Essais préliminaires pour limiter les dégâts de Tetranychidae sur les cultures maraîchères dans le Sine-Saloum (Sénégal). In: *Deuxième Colloque International sur les Acariens des Cultures, 24–25 Octobre, Montpellier, France*. Association Française de Protection des Plantes, Paris.

Düzgünes, Z. (1952) Citrus mites in Turkey. *Bitki Koruma Bülteni* 1, 6–11.

Düzgünes, Z (1954) *Taxonomic and Biological Studies on the Species of the Family Tetranychidae, Damaging Fruit Trees in Central Anatolia Ankara (Entomol. Inst.)*. Zir. Vekfl. Nes. ve Haber. Md. Say 706 (in Turkish).

Düzgünes, Z. (1961) Akdiken akari. *Tetranychus viennensis* Zacher. *Ankara Üniversitesi, Ziraat Fakültesi Yilliği* 4, 389–396 (in Turkish).

Düzgünes, Z. (1963) Türkiyede yeni bulunan akarlar. *Bitki Koruma Bulteni* (N.S.) 3(4), 237–246 (in Turkish).

Düzgünes, Z. (1965a) The phytophagous mites of the family Tenuipalpidae in Turkey. *Ankara Üniversitesi, Ziraat Fakultesi Yilliği* 15(3), 120–148 (in Turkish).

Düzgünes, Z. (1965b) The variation in the peritreme of the genus *Eotetranychus* Oud. *Bollettino di Zoologia Agraria e di Bachicoltura (Ser. II)* 7, 15–18.

Dzhadaibaev, Z.M. and Dyagilev, B.K. (1977) A mite on cotoneaster. *Zashchita Rasteniĭ* No. 7, 32–33 (in Russian).

Dzolkhifli, O. and Khoo, K.C. (1989) Toxicity of eight acaricides against *Tetranychus fijiensis* (Hirst). *Journal of Plant Protection in the Tropics* 6 (1), 25–27.

Easterbrook, M.A. (1978) The life history and bionomics of *Epitrimerus pyri* (Acarina: Eriophyidae) on pear. *Annals of Applied Biology* 88, 13–22.

Easterbrook, M.A. (1979) The life history of the eriophyid mite, *Aculus schlechtendali* on apple in south-east England. *Annals of Applied Biology* 91, 287–296.

Easterbrook, M.A. (1980) The host range of a 'non-gall-forming' eriophyid mite living in buds of *Ribes*. *Journal of Horticultural Science* 55, 1–6.

Easterbrook, M.A. (1984a) Effects of pesticides on the apple rust mite *Aculus schlechtendali* (Nal.) (Eriophyidae). *Journal of Horticultural Science* 59, 51–55.

Easterbrook, M.A. (1984b) The biology and control of the rust mites *Aculus schlechtendali* and *Epitrimerus piri* on apple and pear in England. In: Griffiths, D.A. and Bowman, C.E. (eds) *Acarology VI, Volume 2*. Ellis Horwood, Chichester, UK, pp. 797–803.

Easterbrook, M.A. (1985) Orchard pesticides. In: *Annual Report 1984*, East Malling Research Station. East Malling, UK, p. 147.

Easterbrook, M.A. (1996) Damage and control of eriophyoid mites in apple and pear. In: Lindquist, E.E., Sabelis, M.W. and Bruin, J. (eds) *Eriophyoid Mites: Their Biology, Natural Enemies and Control*. World Crop Pests, Volume 6. Elsevier Science, Amsterdam, pp. 527–541.

Easterbrook, M.A. and Buss, D.S. (1988) Tests of acaricides against apple rust mite (*Aculus schlechtendali*). Tests of Agrochemicals and Cultivars 9. *Annals of Applied Biology* 112(Supplement), 2–3.

Easterbrook, M.A. and Campbell, C.A.M. (1986) Tests of early ovicides against mites and insects on apple and pear. Tests of Agrochemicals and Cultivars, 7. *Annals of Applied Biology* 108(Supplement), 38-39.

Easterbrook, M.A. and Cross, J.V. (1987) Evaluation of acaricides for control of blackcurrant gall mite (*Cecidophyopsis ribis*). Tests of Agrochemicals and Cultivars, 8. *Annals of Applied Biology* 110(Supplement), 4–5.

Easterbrook, M.A. and Fuller, M.M. (1986) Russeting of apples caused by apple rust mite *Aculus schlechtendali* (Acarina: Eriophyidae). *Annals of Applied Biology* 109, 1–9.

Easterbrook, M.A. and Palmer, J.W. (1996) The relationship between early-season leaf feeding by apple rust mite, *Aculus schlechtendali* (Nal.), and fruit set and photosynthesis of apple. *Journal of Horticultural Science* 71, 939–944.

Easterbrook, M.A., Solomon, M.G., Cranham, J.E. and Souter, E.F. (1985) Trials of an integrated pest management programme based on selective pesticides in English apple orchards. *Crop Protection* 4, 215–230.

Easterbrook, M.A., Fitzgerald, J.D. and Solomon, M.G. (2001) Biological control of strawberry tarsonemid mite *Phytonemus pallidus* and two-spotted spider mite *Tetranychus urticae* on strawberry in the UK using species of *Neoseiulus* (*Amblyseius*) (Acari: Phytoseiidae). *Experimental and Applied Acarology* 25, 25–36.

Eastmond, M. and Eastmond, C. (2003) A hazelnut farming story. In: Lifestyleblock.co.nz, Friday, 07 November 2008 17:37. Available at: http://www.lifestyle-block.co.nz/lifestyle-file/running-the-farm/crops/item/629-a-hazelnut-farming-story.html (accessed 6 May 2015).

Ebeling, W. (1934) Observations on a method of dissemination employed by mites. *The Pan-Pacific Entomologist* 10, 89.

Ebeling, W. (1959) *Subtropical Fruit Pests*. University California Press, Berkeley, California.

Ebeling, W. and Pence, R.J. (1949) New pomegranate mite. *California Agriculture* 3(6), 11, 14.

Ebeling, W. and Pence, R.J. (1950) A severe case of an uncommon type of injury by the fig mite. *Bulletin of the California Department of Agriculture* 39, 47–48.

Ebermann, E. (1982) Fortpflanzungbiologische Studien an Scutacariden (Acari, Trombidiformes). *Zoologische Jahrbücher Systematik* 109, 98–116.

Ebermann, E. (1986) To the knowledge of the tarsonemid fauna of Italian islands (Acari: Microdispidae, Scutacaridae). *Redia* 69, 469–480.

Ebermann, E. (1991a) Das Phänomen Polymorphismus in der Milbenfamilie Scutacaridae (Acari, Heterostigmata, Tarsonemina, Scutacaridae). *Zoologica*, 141(141), 1–76.

Ebermann, E. (1991b) Thanatosis or feigning death in mites of the family Scutacaridae. In: Schuster, R. and Murphy, P.W. (eds) *The Acari: Reproduction, Development and Life-history Strategies*. Chapman and Hall, London, pp. 399–401.

Ebermann, E. (1995) Indication of jumping ability in the mite family Scutacaridae (Acari, Tarsonemina). *Entomologische Mitteilungen aus dem Zoologisches Institut und Zoologisches Museum der Universität Hamburg* 11(152), 205–209.

Ebermann, E. and Hall, M. (2004) First record of sporothecae within the mite family Scutacaridae (Acari, Tarsonemina). *Zoologischer Anzeiger* 242, 367–375.

Ebrahim, H.M. (2000) Influence of temperature and relative humidity on the biology and life table parameters of *Phyllocoptruta oleivora* and *Aculops pelekassi* (Acari: Eriophyidae) on "Hamlin" orange in Central Florida. *Egyptian Journal of Agricultural Research* 78, 143–161.

Ebrahim, H.M., Achor, D.S. and Childers, C.C. (1996) A new method for preparation of eriophyid mites (Eriophyidae) for scanning electron microscopy using sputter-coat vacuum-drying. In: Mitchell, R., Horn, D.J., Needham, G.R. and Welbourn, W.C. (eds) *Acarology IX, Proceedings of the 9th International Congress on Acarology, Volume 1*. Ohio Biological Survey, Columbus, Ohio, pp. 269–272.

Ecevit, O., Kececi, S., Tuncer, C., Yanilmaz, A.F. and Isik, M. (1992) Dogu karadeniz bolgesi fndk bahcelerinde zarari Eriophyoidea (Acarina, Actinedida) akaralar uzerinde calsmalar. *Proceedings of the Second Turkish National Congress of Entomology, 28–31 January 1992*, pp. 671–681.

Edgington, S., Fernando, L.C.P. and Jones, K. (2008) Natural incidence and environmental profiling of the mite pathogenic fungus *Hirsutella thompsonii* Fisher for control of the coconut mite in Sri Lanka. *International Journal of Pest Management* 54, 123–127.

Edwards, B.A.B. and Hodgson, P.J. (1973) The toxicity of commonly used orchard chemicals to *Stethorus nigripes* (Coleoptera: Coccinellidae). *Journal of the Australian Entomological Society* 12, 222–224.

Edwards, M.C. and McMullen, M.P. (1988) Variation in tolerance to wheat streak mosaic virus among cultivars of hard red spring wheat. *Plant Disease* 72, 705–707.

Edwards, O.R., Melo, E.L., Smith, L. and Hoy, M.A. (1998) Discrimination of three *Typhlodromalus* species (Acari: Phytoseiidae) using random amplified polymorphic DNA markers. *Experimental and Applied Acarology* 22, 101–109.

Edwards, W.O., Gray, K.W., Wilcox, T. and Mote, D.C. (1935) *The Blackberry Mite in Oregon*. Station Bulletin 337, Agricultural Experiment Station, Oregon State Agricultural College, Corvallis, Oregon.

Eesa, N.M. and Moursy, L.E. (1990) Temperature/toxicity relationships of some pyrethroids on *Cenopalpus pulcher* (Acari: Acaridae). *Experimental and Applied Acarology* 10, 77–80.

Ehara, S. (1954) Two new spider mites parasitic on japanese Conifers. *Annotationes Zoologicae Japonenses* 27, 102–106.

Ehara, S. (1955) On two spider mites parasitic on japanese citrus. *Annotationes Zoologicae Japonenses* 28, 178–182.

Ehara, S. (1956a) Some spider mites from Northern Japan. *Journal of the Faculty of Science, Hokkaido University, Series VI, Zoology* 12, 244–258.

Ehara, S. (1956b) Tetranychoid mites of mulberry in Japan. *Journal of the Faculty of Science, Hokkaido University, Series VI, Zoology* 12, 499–510.

Ehara, S. (1956c) Notes on some tetranychid mites of Japan (with 37 txt-figures). *Japanese Journal of Applied Entomology and Zoology* 21, 139–147.

Ehara, S. (1957) On three spider mites of *Schizotetranychus* from Japan. *Journal of the Faculty of Science, Hokkaido University, Series VI, Zoology* 13, 15–23.

Ehara, S. (1959) Mites of the subfamily Bryobiinae from Japan (Tetranychidae). *Journal of the Faculty of Science, Hokkaido University, Series VI, Zoology* 14, 185–195.

Ehara, S. (1960) On some Japanese tetranychid mites of economic importance. *Japanese Journal of Applied Entomology and Zoology* 4, 234–241.

Ehara, S. (1962) Tetranychoid mites of conifers in Hokkaido. *Journal of the Faculty of Science, Hokkaido University, Series VI, Zoology* 15, 157–175.

Ehara, S. (1963a) Notes on spider mites in the collection of the Zoological Survey of India, Calcutta (Acarina : Tetranychidae). *Records of the Indian Museum* 59, 143–148.

Ehara, S. (1963b) A new mite of *Oligonychus* from rice, with notes on some Japanese spider mites. *Japanese Journal of Applied Entomology and Zoology* 7, 228–231.

Ehara, S. (1964a) The tetranychoid mites in Japan. *Acarologia* 9, 409–414.

Ehara, S. (1964b) Some mites of the families *Phytoseiidae* and *Blattisocidae* from Japan (Acarina: Mesostigmata). *Journal of the Faculty of Science, Hokkaido University, Series VI, Zoology* 15, 378–394.

Ehara, S. (1966a) The tetranychoid mites of Okinawa Island (Acarina: Prostigmata). *Journal of the Faculty of Science, Hokkaido University, Series VI, Zoology* 16, 1–22.

Ehara, S. (1966b) Some mites associated with plants in the State of São Paulo, Brazil, with a list of plant mites of South America. *Japanese Journal* of *Zoology* 15, 129–150.

Ehara, S. (1969a) The tetranychoid mites of Taiwan (Acarina: Prostigmata). *Journal of the Faculty of Education, Tottori University, Natural Science* 20, 79–103.

Ehara, S. (1969b) Three spider mites of the genus *Eotetranychus* infesting fruit trees in Japan (Acarina: Tetranychidae). *Applied Entomology and Zoology* 4, 16–22.

Ehara, S. (1970) Four species of the *Carpini* complex of *Eotetranychus* in Japan (Acarina: Tetranychidae). *Journal of the Faculty of Education, Tottori University, Natural Science* 21, 132–141.

Ehara, S. (1973) Three species of the genus *Schizotetranychus* (Acarina : Tetranychidae). *Annotationes Zoologicae Japonenses* 46, 224–232.

Ehara, S. (1975a) A guide to the tetranychid mites of agricultural importance in Japan. In: Yasumatsu, K. and Mori, H. (eds) *Approaches to Biological Control*. JIBP (Japanese Committee for the International Biological Program) Synthesis Volume 7. University of Tokyo Press, Tokyo, pp. 15–23.

Ehara, S. (1975b) Description of a new species of *Tuckerella* from Japan (Acarina: Tuckerellidae). *International Journal of Acarology* 1, 1–5.

Ehara, S. (1978) A new genus and a new subgenus of the spider mites from northern Japan (Acarina: Tetranychidae). *Journal of the Faculty of Education, Tottori University, Natural Science* 28, 87–93.

Ehara, S. (1980) *Illustrations of the Mites and Ticks of Japan*. Zenkoku Nôson Kyôiku Kyôkai, Tokyo.

Ehara, S. (1999) Revision of the spider mite family Tetranychidae of Japan (Acari, Prostigmata). *Species Diversity* 4, 63–141.

Ehara, S. (2004) A collection of spider mites (Arachnida: Acari: Tetranychidae) from Sumatra. *Species Diversity* 9, 67–76.

Ehara, S. and Gotoh, T. (1987) Notes on the genus *Sasanychus* Ehara, new status, with description of a new species from Hokkaido (Acarina, Tetranychidae). *Zoological* Science 4, 375–378.

Ehara, S. and Gotoh, T. (1990) A new *Tetranychus* closely related to *T. viennensis* Zacher (Acari: Tetranychidae). *International Journal of Acarology* 16, 55–58.

Ehara, S. and Gotoh, T. (1991) A new species of *Panonychus* from dwarf bamboo in Japan (Acari: Tetranychidae). *International Journal of Acarology* 17, 9–12.

Ehara, S. and Gotoh, T. (1992) Descriptions of two *Panonychus* spider mites from Japan, with a key to species of the genus in the world (Acari: Tetranychidae). *Applied Entomology and Zoology* 27, 107–115.

Ehara, S. and Gotoh, T. (1996) Two new species of spider mites occurring in Japan (Acari, Tetranychidae). *Journal of the Acarological Society of Japan* 5, 17–25.

Ehara, S. and Lee, L.H.Y. (1971) Mites associated with plants in Hong Kong. *Journal of the Faculty of Education, Tottori University, Natural Science* 22, 61–78.

Ehara, S. and Masaki, M. (1989) Notes on two Japanese species of *Tetranychus* (Acarina: Tetranychidae). *Acta Arachnologica* 38, 49–54.

Ehara, S. and Masaki, M. (2001) Three species of exotic mites (Acari: Tetranychoidea) recently intercepted at Japanese plant quarantine. *Applied Entomology and Zoology* 36, 251–257.

Ehara, S. and Mori, H. (1969) A new species of the genus *Tetranycopsis* Canestrini from Japan (Acarina : Tetranychidae). *Annotationes Zoologicae Japonenses* 42, 42–48.

Ehara, S. and Ohashi, K. (2002) A new species of *Tetranychus* (Acari: Tetranychidae) from the Kinki District, Japan. *Acta Arachnologica* 51, 19–22.

Ehara, S. and Tho, Y.P. (1988) Spider mites of the Malay Peninsula (Acarina: Tetranychidae). *Journal of the Faculty of Education, Tottori University, Natural Science* 37, 1–24.

Ehara, S. and Wongsiri, T. (1975) The spider mites of Thailand (Acarina: Tetranychidae). *Mushi* 48(13), 149–185.

Ehara, S. and Yamaguchi, K. (2001) Discovery of *Tetranychus neocaledonicus* André (Acari: Tetrnychidae) from Amami-Oshima Island, Japan. *Plant Protection (Tokyo)* 55, 268–272.

Ehara, S. and Yogi, K. (1998) Discovery of *Eutetranychus orientalis* (Klein) (Acari: Tetranychidae) from Okinawa Island, Japan. *Plant Protection (Tokyo)* 52, 530–533.

Ehara, S., Ohashi, K., Gotoh, T. and Tsurusaki, N. (2009) The mite taxa created by S. Ehara and his coauthors with depositories of their type series (Arachnida: Acari). *Bulletin of Tottori Prefectural Museum* 46, 9–48.

Ehler, L.E. (1974) *A Review of the Spider-mite Problem on Grain Sorghum and Corn in West Texas*. Texas Agricultural Experimental Station Bulletin 1149, Texas A&M University, College Station, Texas.

Eichmeier, J. and Guyer, G. (1960) An elevation of the rate of reproduction of the two-spotted mite reared on gibberellin-tretaed bean plants. *Journal of Economic Entomology* 53, 661–664.

Eidit, D.C. (1966) The mite *Nalepella tsugifoliae* (Acarina: Eriophyidae) on balsam fir – a new host record. *Journal of Economic Entomology* 59, 1279.

Elbadry, E.A., Abo Elghar, M.R., Hassan, S.M. and Kilany, S.M. (1969) *Agistemus exsertus* as predator of two tetranychid mites. *Annals of the Entomological Society of America* 62, 660–661.

El-Banhawy, E.M. (1975) Biology and feeding behavior of the predatory mite *Amblyseius brazilli* (Mesostigmata, Phytoseiidae). *Entomophaga* 20, 353–360.

El-Banhawy, E.M., Amer, S.A.A. and Saber, S.A. (2000) Development and reproduction of the predacious mite, *Amblyseius cydnodactylon* on different prey species, effect of plant leaf texture on the behaviour and reproduction of the predator. *Zeitschrift fur Pflanzenkrankheiten Pflanzenpathologie und Pflanzenschutz* (now *Journal of Plant Diseases and Plant Protection*) 107, 218–224.

Elder, R.J. (1988) Pesticide control of *Dolichotetranychus floridanus* (Banks) (Acarina: Tenuipalpidae) on pineapples. *Queensland Journal of Agricultural and Animal Sciences* 45, 179–184.

El-Enany, M.A., Mabrouk, A.M. and El-Hadi, M.M. (1993) Biology and feeding capacity of the phytoseiid mite *Amblyseius gossipi* El-Badry when fed on adult females of *Eutetranychus orientalis* Klein treated with some acaricides. *Egyptian Journal of Agricultural Research* 71, 121–129.

Elhadi, F.A. and Birger, R. (1999) A new approach to the control of the olive mite *Aceria* (*Eriophyes*) *oleae* Nalepa in olive trees. *Acta Horticulturae* 474, 555–557.

El-Hady, M.M. (2004) Susceptibility of the citrus brown mite, *Eutetranychus orientalis* (Klein) to the entomopathogenic fungi *Verticillium lecanii* and *Metahrrizium anisopliae*. *Egyptian Journal of Biological Pest Control* 14, 409–410.

El-Haidari, H. (1965) *A Preliminary List of Mites of Iraq*. Bulletin 110, Directorate General of Agriculture Research and Projects, Ministry of Agriculture, Baghdad.

El-Haidari, H.S. (1981) New records of mites and insects infesting date palms in the Near East and North Africa. *Date Palm Journal* 1(1), 133–136.

El-Halawany, M.E. (1991) Some factors affecting the distribution of *Brevipalpus californicus* (Banks) on citrus trees. *Egyptian Journal of Agricultural Research* 69, 185–192.

El-Halawany, M. (2003) Mites associated with mango trees in Egypt. In: Bayaa, B., Makkouk, K.M., Kumari, S.G. and El-Ghariani, I. (eds) *Proceedings of the Eighth Arab Congress of Plant Protection*, 12–16 October 2003, El-Beida, Libya. Arab Society for Plant Protection, Beirut, p. 29E, abstract M3.

El-Halawany, M.E., Ibrahim, G.A. and Abdel-Samad, M.A. (1990a) Mites inhabiting fig varieties. *Agricultural Research Review (Cairo)* 68(1), 38–48.

El-Halawany, M.E., Abou-El-Ela, R.G. and Esmail, H.M. (1990b) Population dynamics of mites and their natural enemies on apple and apricot trees. *Agricultural Research Review (Cairo)* 68(1), 59–66.

El-Halawany, M.E., Abdel-Samed, M.A., Ibrahim, G.A., Radwan, Z. and Ebrahim, H.M. (1993) Evaluating the efficiency of different levels of *Eusieus scutalis* (Athias-Hernriot) released for the control of *Brevipalpus californicus* (Banks) on Baladi orange trees. *Menofia Journal of Agricultural Research* 18, 2697–2707.

El-Halawany, M.E., Abdel-Samad, M.A. and El-Naggar, M.E. (2001) Mites inhabiting date palms. In: *Proceedings of the Second International Conference on Date Palms, 25–27 March 2001, Al-Ain, United Arab Emirates*, pp. 366–373.

El-Halim, S.M.A., Hanna, M.A. and Ramadan, M.F. (2004) Some factors affecting reproduction of the predaceous mite, *Typhlodromus mangiferus* Zaher & El Borolossy (Gamasida: Phytoseiidae). *Egyptian Journal of Biological Pest Control* 14, 323–326.

El-Jaouani, N. (1988) *Contribution à la Connaissance des Acariens Phytophages au Maroc et Étude Bio-écologique de Tetranychus evansi Baker et Pritchard (Acarina: Tetranychidae)*. Mémoire Diplôme d'Ingénieur, Institut Agronomique et Vétérinaire Hassan II, Rabat.

El-Kady, M.H., Nassar, M.E.S. and Heykal, I.H. (1977) Sensitivity of citrus brown mite *Eutetranychus orientalis* (Klien) to certain pesticides on citrus trees, Tetranychidae. *Agricultural Research Review (Cairo)* 55(1), 99–101.

Elliot, M.S., Cromroy, H.L., Zettler, F.W. and Carpenter, W.R. (1987) A mosaic disease of wax myrtle associated with a new species of eriophyid mite. *HortScience* 22, 258–260.

Ellis, H.C. (2008) BugwoodWiki: *Eotetranychus hicoriae*. Available at: wiki.bugwood.org/Eotetranychus_hicoriae (accessed 29 July 2014).

Ellis, W.N. and Ulenberg, S.A. (2005) *Vasates quadripedes*, een galmijt (niet meer zo) nieuw voor Nederland. *Entomologische Berichten* 65, 52–55 (in Dutch).

Elmer, H.S. (1965a) The Yuma spider mite, *Eotetranychus yumensis*, on citrus. *Journal of Economic Entomology* 58, 534–536.

Elmer, H.S. (1965b) Banks grass mite, *Oligonychus pratensis*, on dates in California. *Journal of Economic Entomology* 58, 531–534.

Elmer, H.S. (1968) Citrus flat mite injury patterns on some citrus varieties. *Journal of Economic Entomology* 61, 1689–1692.

Elmer, H.S. (1969) Yuma spider mite (*Eotetranychus yumensis*) on desert citrus. *California Agriculture* 23(10), 15–16.

Elmer, H.S. and Jeppson, L.R. (1957) Biology and control of citrus flat mite. *Journal of Economic Entomology* 50, 566–570.

El-Titi, A., Boller, E.F. and Gendrier, J.P. (eds) (1995) Commisione "Produzione Integrata: Direttive e Approvazione". Produzione integrata – principi e direttive tecniche. *IOBC-WPRS Bulletin* 18(1, 2).

Elwan, A. (2000) A survey of the insect and mite pests associated with date palm trees in Al-Dakhliya region, Sultanate of Oman. *Egyptian Journal of Agricultural Research* 78, 653–664.

El-Zemaity, M.S., Hafez, S.M., Ebaid, N.M. and Kalif, S.K. (1984) Toxicity of certain pesticides to the adult female of citrus brown mite, *Eutetranychus orientalis* (Klein). *Annals of Agricultural Science, Ain Shams University* 29, 1119–1126.

Emmanouel, N.G. and Papadoulis, G.T. (1987) *Panonychus citri* (MacGregor) (Tetranychidae) and *Eriophyes medicaginis* K. (Eriophyidae): two important phytophagous mites recorded for the first time in Greece. *Entomologia Hellenica* 5(1), 3–6.

Engel, H. (1973) Beobachtungen zur Massenvermehrung und Schädlichkeit von *Eriophyes similis* Nalepa (Gallmilbe) an der Hauszwetsche. *Nachrichtenblatt des Deutschen Pflanzenschutzdienstes* 25, 129–133.

Engel, R. (1990) Alternative prey and other food resources of the phytoseiid mite *Typhlodromus pyri* (Scheuten). *IOBC-WPRS Bulletin* 13(7, 1) 124–127.

English, L.L. and Snetsinger, R. (1957a) The habits and control of the clover mite in dwellings. *Journal of Economic Entomology* 50, 135–141.

English, L.L. and Snetsinger, R. (1957b) The biology and control of *Eotetranychus multidigituli* (Ewing) a spider mite of honey locust. *Journal of Economic Entomology* 50, 784–788.

English, L.L. and Turnipseed, G.F. (1941) The influence of temperature and season on the citrus red mite, *Paratetranychus citri* McGregor. *Journal of Agricultural Research (Washington, DC)* 62, 65–67.

English-Loeb, G.M. (1989) Non-linear responses of spider mites to drought-stressed host plants. *Ecological Entomology* 14, 45–55.

English-Loeb, G.M. (1990) Plant drought stress and outbreaks of spider mites: a field test. *Ecology* 71, 1401–1411.

English-Loeb, G.M. and Karban, R. (1988) Negative interactions between Willamette mites and Pacific mites: possible management strategies for grapes. *Entomologia Experimentalis et Applicata* 48, 269–274.

English-Loeb, G.M., Flaherty, D.L., Wilson, L.T., Barnett, W.W., Leavitt, G.M. and Settle, W.H. (1986) Pest management affects spider mites in vineyards. *California Agriculture* 40(3), 28–30.

English-Loeb, G.M., Karban, R. and Hougen-Eitzman, D. (1993) Direct and indirect competition between spider mites feeding on grapes. *Ecological Applications* 3, 699–707.

English-Loeb, G.M., Karban, R. and Walker, M.A. (1998) Genotypic variation in constitutive and induced resistance in grapes against spider mite (Acari: Tetranychidae) herbivores. *Environmental Entomology* 27, 297–304.

Enigi, M., and Schausberger, P. (2007) Incidence of the endosymbionts *Wolbachia, Cardinium* and *Spiroplasma* in phytoiseiid mites and associated prey. *Experimental and Applied Acarology* 42, 75–85.

Enkegaard, A., Saradar, M.A. and Brødsgaard, H.F. (1997) The predatory mite *Hypoaspis miles*: biological and demographic characteristics on two species, the mushroom sciarid fly, *Lycoriella solani*, and the mould mite, *Tyrophagus putrescentiae*. *Entomologia Experimentalis et Applicata* 82, 135–146.

EPICA (2007) Fw: EPICA Notification Thursday, December 27, 2007. Personal communication to B. Spears on December 27, 2007, from Exotic Pest Information Collection and Analysis. Archived at the PERAL Library, Raleigh, North Carolina.

EPPO (1989) EPPO Crop growth stage keys. *Bulletin OEPP/EPPO Bulletin* 19, 373–384.

EPPO (2000) *EPPO Standards. Guidelines on Good Plant Protection Practice. Mushrooms.* Document PP 2/20(1) English. European and Mediterranean Plant Protection Organization, Paris. Available at: https://archives.eppo.int/EPPOStandards/PP2_GPP/pp2-20-e.doc (accessed 10 February 2015).

EPPO (2006) *Tetranychus evansi* is present in Israel. EPPO Reporting Service 01-2006, Article No. 2006/005. European and Mediterranean Plant Protection Organization, Paris. Available at: https://gd.eppo.int/reporting/Rse-2006-01 (accessed 1 April 2015).

EPPO (2007) Report of a Pest Risk Analysis for *Tetranychus evansi.* Document 08-14562, European and Mediterranean Plant Protection Organization, Paris. Available at: http://www.eppo.int/QUARANTINE/Pest_Risk_Analysis/PRAdocs_insects/08-14562%20PRA%20report%20T%20evansi%20final%20 2008-10%20for%20website.doc (accessed 6 May 2015).

Epstein, A.H. and Hill, J.H. (1995) The biology of rose rosette disease: a mite-associated disease of uncertain aetiology. *Journal of Phytopathology* 143, 353–360.

Epstein, A.H. and Hill, J.H. (1999) Status of rose rosette disease as a biological control for multiflora rose. *Plant Disease* 83, 92–101.

Esau, K. (1960) *Anatomy of Seed Plants.* Wiley, New York.

Escalona, C. and Vásquez, C. (2005) Biología y tabla de vida de *Euseius concordi*s (Chant) (Acari: Phytoseiidae) alimentado con dos dietas. In: *Resúmenes XIX Congreso Venezolano de Entomología, 4–7 July 2005, San Felipe, Yaracuy. Entomotropica* 20, 129–130.

Escudero, L.A. and Ferragut, F. (2005) Life-history of predatory mites *Neoseiulus californicus* and *Phytoseiulus persimilis* (Acari: Phytoseiidae) on four spider mite species as prey, with special reference to *Tetranychus evansi* (Acari: Tetranychidae). *Biological Control* 32, 378–384.

Escudero, L.A., Baldo Gosalvez, M. and Ferragut, F. (2005) Eficacia de los fitoseidos como depredadores de las aranas rojas de cultivos horticolas *Tetranychus urticae, T. turkestani, T. ludeni* y *T. evansi* (Acari: Tetranychidae). *Boletín de Sanidad Vegetal, Plagas* 31, 377–383.

Espinosa Becerril, A. and Carrillo Sanchez, J.L. (1986) El hongo *Hirsutella thompsonii* Fisher en el control del eriofid del cocotero *Eriophyes guerreronis* (Keifer). Agricultura Técnica *en* México 12, 319–323.

Essig, E.O. (1925) *The Blackberry Mite, the Case of Redberry Disease of the Himalaya Blackberry and its Control.* Bulletin 399, Agricultural Experiment Station, College of Agriculture, University of California, Berkeley, California.

Estebanes-Gonzalez, M.L. (1997) Acarofauna of nests of wild birds in Mexico. *Acta Zoológica Mexicana* 71, 1–15.

Estebanes-Gonzalez, M.L. and Baker, E.W. (1968) Arañas rojas de Mexico (Acarina: Tetranychidae). *Anales de la Escuela Nacional de Ciencias Biológicas (México)* 15, 61–133.

Estebanes-Gonzalez, M.L. and Rodriguez-Navarro, S. (1991) Observaciones sobre algunos ácaros de las familias Tetranychidae, Eriophyidae, Acaridae y Tarsonemidae (Acari), en hortalizas de México. *Folia Entomológica Mexicana* 83, 199–212.

Estes, Z.E. and Faust, R.M. (1965) The nucleic acid composition of a virus affecting the citrus red mite, *Panonychus citri* (McGregor). *Journal of Invertebrate Pathology* 7, 259–260.

Eswara Reddy, S.G. (2000) Evaluation of brinjal varieties, selected botanicals and synthetic insecticides against pest complex of brinjal (*Solanum melongena* Linn.). MSc. thesis, University Agricultural Sciences, Bangalore, India.

Etienne, J. and Flechtmann, C.H.W. (2006) First record of *Raoiella indica* (Hirst, 1924) (Acari: Tenuipalpidae) in Guadeloupe and Saint Martin, West Indies. *International Journal of Acarology* 32, 331–332.

European Commission (2014) EU Pesticides Database. Available at: http://ec.europa.eu/sanco_pesticides/public/index.cfm?event=homepage&language=EN (accessed 2 February 2015).

Eusebio, J.E. and Bernardo, E.N. (1999) Techniques of potato germplasm evaluation for broad mite (*Polyphagotarsonemus latus* (Banks)) resistance. *Philippine Entomologist* 13, 35–41.

Evans, G.A., Cromroy, H.L. and Ochoa, R. (1993) The Tenuipalpidae of Honduras (Tenuipalpidae: Acari). *Florida Entomologist* 76, 126–155.

Evans, G.O. (1963) Observations on the chaetotaxy of the legs in the free-living Gamasina (Acari: Mesostigmata). *Bulletin of the British Museum (Natural History), Zoology* 10, 277–303.

Evans, G.O. (1992) *Principles of Acarology*. CAB International, Wallingford, UK.

Evans, G.O. and Browning, E. (1955) Techniques for the preparation of mites for study. *Annals and Magazine of Natural History: Series* 12 8(12) (now *Journal of Natural History: Series 12* 8(92)), 631–635.

Evans, G.O. and Loots, G.C. (1972) Scanning electron microscopy in the study of the gnathosoma of the Acari. *Wetenskaplike Bydraes van die se Universiteit vir* Christelike Hoër Onderwys *Potchefstroom, B: Natuur- Wetenskaplike*, 49, 1–13.

Evans, G.O. and Loots, G.C. (1975) Scanning electron microscope study of the structure of the hypostome of *Phityogamasus*, *Laelaps* and *Ornithonyssus* (Acari: Mesostigmata). *Journal of Zoology* 176, 425–436.

Evans, G.O., Sheals, J.G. and MacFarlane, D. (1961) *The Terrestrial Acari of the British Isles. An Introduction to their Morphology, Biology and Classification*. Trustees of the British Museum, London.

Evans, J.W. (1943) *Insect Pests and Their Control*. Department of Agriculture, Hobart, Tasmania, Australia.

Evans, L.M., Allan, G.J., Shuster, S.M., Woolbright, S.A. and Whitham, T.G. (2008) Tree hybridization and genotypic variation drive cryptic speciation of a specialist mite herbivore. *Evolution* 62, 3027–3040.

Evensen, K.B., Bausher, M.G. and Biggs, R.H. (1980) Rust mite damage increases uptake and effectiveness of an abscission-accelerating chemical on 'Valencia' oranges. *Journal of the American society for Horticultural Science* 105, 167–170.

Ewing, H.E. (1909) A systematic and biological study of the Acarina of Illinois. *University of Illinois Studies* 3(8), 1–120.

Ewing, H.E. (1913) The taxonomic value of characters of the male genital armature in the genus *Tetranychus* Dufour. *Annals of the Entomological Society of America* 6, 453–460.

Ewing, H.E. (1914) *The Common Red Spider or Spider Mite*. Station Bulletin 121, Agricultural Experiment Station, Oregon State Agricultural College, Corvallis, Oregon.

Ewing, H.E. (1917) New species of economic mites. *Journal of Economic Entomology* 10, 497–501.

Ewing, H.E. (1921) New Nearctic spider mites of the family Tetranychidae. *Proceedings of the United States National Museum* 59, 659–666.

Ewing, H.E. (1922) Three new species of peculiar and injurious spider mites. *Proceedings of the Entomological Society of Washington* 24, 104–108.

Ewing, H.E. (1926) Two new spider mites (Tetranychidae) from Death Valley, California (Acarina). *Entomological News* 37, 142–143.

Ewing, H.E. (1929a) *A Manual of External Parasites*. Charles C. Thomas, Baltimore, Maryland.

Ewing, H.E. (1929b) A new variety of *Tarsonemus* (Acarina) from the Pacific Coast. *Proceedings of the Entomological Society of Washington* 31, 31–32.

Ewing, H.E. (1932) Early records and present known distribution in the United States of three recently introduced mites. *Proceedings of the Entomological Society of Washington* 34, 13–16.

Ewing, H.E. (1939) *A Revision of the Mites of the Subfamily Tarsoneminae of North America, the West Indies and Hawaiian Islands*. Technical Bulletin No. 653, US Department of Agriculture, Washington, DC.

EXTOXNET (2015) The Extension Toxicology Network. University of California-Davis, Oregon State University, Michigan State University, Cornell University and the University of Idaho. Oregon State University, Corvallis, Oregon. Available at: http://ace.orst.edu/info/extoxnet/ (accessed 2 February 2015).

Fahl, J.I., Queiroz-Voltan, R.B., Carelli, M.L.C., Schiavinato, M.A., Prado, A.K.S. and Souza, J.C. (2007) Alterations in leaf anatomy and physiology caused by the red mite (*Oligonychus ilicis*) in plants of *Coffea arabica*. *Brazilian Journal of Plant Physiology* 19, 61–68.

Fahmy, I. (1956) The economic importance of mites affecting fruit trees, with special reference to citrus, in Egypt (Acarina). *Bulletin of the Entomological Society of Egypt* 40, 449–450.

Fain, A. (1971) Évolution de certains groupes d'hypopes en fonction du parasitisme (Acarina, Sarcoptiformes). *Acarologia* 13, 98–112.

Fain, A. (1980) A method of remounting of old preparations of acariens without raising or displacing the cover slip. *International Journal of Acarology* 6, 169–170.

Fain, A. (1988) Notes sur le genre *Rhizoglyphus* Claparède, 1869 (Acari, Acaridae) avec description de deux espèces nouvelles. *Acarologia* 29, 53–62.

Fain, A. and Fauvel, G. (1993) *Tyrophagus curvipenis* n. sp. from an orchid cultivation in a greenhouse in Portugal (Acari: Acaridae). *International Journal of Acarology* 19, 95–100.

Fain, A. and Schuster, R. (1983) New astigmatic mites from the coastal area of Bermuda Islands (Acari: Hyadesiidae, Saproglyphidae, Acaridae). *Entomologische Mitteilungen aus dem Zoologisches Institut und Zoologisches Museum der Universität Hamburg* 7(119), 417–434.

Fain, A., Lukoschus, F.S., Louppen, J.M.W. and Mendez, E. (1973) *Echimyopus dasypus*, n. sp., a hypopus from *Dasypus novemcinctus* in Panama (Glycyphagidae, Echimyopidae: Sarcoptiformes). *Journal of Medical Entomology* 10, 552–555.

Faizyar, A. (1971) *List of Plant Pests (including Insects and Mites) and Diseases of Economic Importance in Afghanistan*. Near East Plant Protection Commission of the FAO (Food and Agriculture Organization of the United Nations), Cairo, Egypt.

Falcon, L.A. (1985) Development and use of microbial insecticides. In: Hoy, M.A. and Herzog, D.C. (eds) *Biological Control in Agricultural IPM Systems*. Academic Press, Orlando, Florida, pp. 229–242.

Fan, Q.-H. (2006) Bibliographic analysis of acarological papers published in *Zootaxa* from 2001 to 2005, with a catalogue of described new taxa. *Zootaxa* 1385, 53–66.

Fan, Q.-H. and Zhang, Z.-Q. (2003) *Rhizoglyphus echinopus* and *Rhizoglyphus robini* (Acari: Acaridae) from Australia and New Zealand: identification, host plants and geographical distribution. *Systematic and Applied Acarology Special Publications* No. 16, 1–16.

Fan, Q.-H. and Zhang, Z.-Q. (2004) Revision of *Rhizoglyphus* Claparède (Acari: Acaridae) of Australia and Oceania. *Systematic and Applied Acarology Society*, London.

Fan, Q.-H. and Zhang, Z.-Q. (2007) *Tyrophagus* (Acari: Astigmata: Acaridae). Fauna of New Zealand 56. Manaaki Whenua Press, Lincoln, Canterbury, New Zealand.

Fan, Q.-H., Chen, Y. and Wang, Z.-Q. (2010) Acaridia (Acari: Astigmatina) of China: a review of research progress. In: Zhang, Z.Q., Hong, X.-Y. and Fan, Q.H. (eds) *Xin Jie-Liu Centenary: Progress in Chinese Acarology. Zoosymposia* 4, 225–259.

Fan, Y.Q. and Petitt, F.L. (1994) Biological control of broad mite, *Polyphagotarsonemus latus* (Banks), by *Neoseiulus barkeri* Hughes on pepper. *Biological Control* 4, 390–395.

FAO (2001a) Biosecurity in food and agriculture. Item 8 of the Provisional Agenda. In: *Committee on Agriculture, Sixteenth Session Rome, 26–30 March 2001*. Food and Agriculture Organization of the United Nations, Rome. Available at: http://www.fao.org/docrep/meeting/003/x9181e.htm (accessed 30 July 2014).

FAO (2001b) *Pest Risk Analysis for Quarantine Pests*. International Standard for Phytosanitary Measures, ISPM 11, Secretariat of the International Plant Protection Convention of the Food and Agriculture Organization of the United Nations, Rome. Updated 2003, 2004 and 2013. Latest version available at: https://www.ippc.int/sites/default/files/documents/1367503175_ISPM_11_2013_En_2013-05-02.pdf (accessed 4 February 2015).

FAO (2006) *Glossary of Phytosanitary Terms*. International Standard for Phytosanitary Measures, ISPM 5, Secretariat of the International Plant Protection Convention of the Food and Agriculture Organization of the United Nations, Rome. Available at: ftp://ftp.fao.org/docrep/fao/009/a0450e/a0450e.pdf (accessed 4 February 2015).

Farahbakhsh, G. (1961) *A Checklist of Economically Important Insects and Other Enemies of Plants and Agricultural Products in Iran*. Publication No. 1, Department of Plant Protection, Ministry of Agriculture, Tehran.

Farmer, E.E. (2001) Surface-to-air signals. *Nature* 411, 854–586.

Farrar, K. (2000) *Crop Profile for Dates in California*. California Pesticide Impact Assessment Program, University of California, Davis, California. Available at: http://www.ipmcenters.org/cropprofiles/docs/cadates.pdf (accessed 30 July 2014).

Fashing, N.J. (2010) Two novel adaptations for dispersal in the mite family Histiostomatidae (Astigmata). In: Sabelis, M.W. and Bruin, J. (eds) *Trends in Acarology: Proceedings of the 12th International Congress, 21–26 August 2006*, Amsterdam. Springer, Dordrecht, the Netherlands, pp. 81–84

Fashing, N.J. and O'Connor, B.M. (1984) *Sarracenzopus*- a new genus for histiostomatid mites inhabiting the pitchers of the Sarraceniaceae (Astigmata: Histiostomatidae). *International Journal of Acarology* 10, 217–227.

Fasihi, M.T., Mo'tamedi Niya, B., Baqeri, A.O., Damghani, R., Heydari, A.O., Hassan Pur, A.R. and Arbabi, M. (2006) *Evaluation of Population Dynamism of Date Spider Mite,* Oligonychus afrasiaticus *(McGregor)*. Agricultural Scientific Information and Documentation Centre, Agricultural Research and Education Organization, Tehran. Available at: http://agris.fao.org/agris-search/search.do?recordID=IR2008000177 (accessed 26 March 2015).

Fatemi, H. (1986) Pear leaf blister mites and apricot gall mite in Esfahan and their chemical control. *Entomologie et Phytopathologie Appliquées* (*Applied Entomology and Phytopathology*) 53(1, 2), 29–37.

Fauvel, F., Rambier, A. and Cotton, D. (1975) Activité prédatrice et multiplication d'*Orius* (*Heterorius*) *vicinus* (Het.: Anthocoridae) dans les galles d'*Eriophies fraxinivorus* (Acarina: Eriophyidae). *Entomophaga* 23, 261–270.

Fauvel, G., Audemard, H., Rodolphe, F. and Rambier, A. (1978) Le récensement des oeufs d'hiver de l'acarien rouge *Panonychus ulmi* (Koch) sur le pommier. I. *Annales de Zoologie, Ecologie Animale* 10, 461–482.

Fazolin, M. and Pereira, L.V. (1989) Ocorrência de *Oligonychus gossypii* (Zacher 1921) (Acari:Tetranychidae) em seringueiras cultivadas. *Anais de Sociedade Entomológica do Brasil* 18, 199–202.

Feiertag-Koppen, C.C.M. and Pijnacker, L.P. (1982) Development of the female germ cells and process of internal fertilization in the two-spotted spider mite *Tetranychus urticae* Koch (Acariformes: Tetranychidae). *International Journal of Insect Morphology and Embryology* 11, 271–284.

Feldmann, A.M. (1981) Life table and male mating competitiveness of wild-type and of a chromosome mutation strain of *Tetranychus urticae* in relation to genetic pest control. *Entomologia Experimentalis et Applicata* 29, 125–137.

Fellin, D.G. (1968) Mites collected from Douglas-fir foliage in Montana. *Journal of Economic Entomology* 61, 877–878.

Fellows, H. and Sill, W.H. Jr (1955) Predicting wheat streak mosaic epiphytotics in winter wheat. *Plant Disease Reporter* 39, 291–295.

Fenton, B., Malloch, G., Brennan, R.M., Jones, A.T., Gordon, S.C., Amrine, J.W., McGavin, W.J. and Birch, A.N.E. (1993) Taxonomic evaluation of three reputed species of *Cecidophyopsis* mite on Ribes. *Acta Horticulturae* 352, 535–538.

Fenton, B., Malloch, G., Jones, A.T., Amrine, J.W. Jr, Gordon, S.C., A'Hara, S., McGavin, W.J. and Birch, A.N.E. (1995) Species identification of *Cecidophyopsis* mites (Acari: Eriophyidae) from different *Ribes* species and countries using molecular genetics. *Molecular Ecology* 4, 383–387.

Fenton, B., Jones, A.T., Malloch, G. and Thomas, W.P. (1996) Molecular ecology of some *Cecidophyopsis* mites (Acari: Eriophyidae) on *Ribes* species and evidence for their natural cross colonisation of blackcurrant (*R. nigrum*). *Annals of Applied Biology* 128, 405–414.

Fenton, B., Birch, A.N.E., Lanham, P.G. and Brennan, R.M. (2000) Gall mite molecular phylogeny and its relationship to the evolution of plant host specificity. *Experimental and Applied Acarology* 24, 831–861.

Fenton, F.A. (1951) The brown wheat mite, *Petrobia latens*. *Journal of Economic Entomology* 44, 996.

Feres, R.J.F. (1992) *Allonychus brevipenis*, a new species from *Triplaris surinamensis* Cham. in Brazil with new host records for *A. braziliensis* (McGregor) and *A. reisi* Paschoal (Acari: Tetranychidae). *International Journal of Acarology* 18, 299–301.

Feres, R.J.F. and Flechtmann, C.H.W. (1995) Mites (Acari) associated with bamboo (*Bambusa* sp., Poaceae) in a woody area from northwestern São Paulo State, Brazil. *Revista Brasileira de Zoologia* 12, 533–546.

Feres, R.J.F., Lofego, A.C. and Oliveira, A.R. (2005) Ácaros planticolas (Acari) da "Estação ecológica do Noroeste Paulista", estado de São Paulo, Brasil. *Biota Neotropica* 5(1), 43–56.

Feres, R.J.F., Vieira, R.M., Daud, R.D., Pereira, E.G.J., Oliveira, G.F. and Dourado, C.L. (2009) Ácaros (Arachnida, Acari) de plantas ornamentais na região noroeste do estado de São Paulo, Brasil: inventário e descrição dos sintomas causados pelos fitófagos. *Revista Brasileira de Entomologia* 53, 466–474.

Ferla, N.J. and Moraes, G.J. de (2003) Oviposição dos acaros predadores *Agistemus floridanus* Gonzalez, *Euseius concordis* (Chant) e *Neoseiulus anonymus* (Chant & Baker) (Acari) em resposta a diferentes tipos de alimento. *Revista Brasileira de Zoologia* 20, 153–155.

Fernandes, A.P., Ferreira, M. da C. and Oliveira, C.A.L. de (2010) Efficiency of different spraying lances and spraying volumes on the control of *Brevipalpus phoenicis* in coffee crops. *Revista Brasileira de Entomologia* 54, 130–135.

Fernandez, D.E., Beers, E.H., Brunner, J.F., Doerr, M.D. and Dunley, J.E. (2006) Horticultural mineral oil applications for apple powdery mildew and codling moth, *Cydia pomonella* (L.). *Crop Protection* 25, 585–591.

Fernandez, N.A., Alberti, G. and Kümmel G. (1991) Spermatophores and spermatozoa of oribatid mites (Acari: Oribatida). Part I. Fine structure and histochemistry. *Acarologia* 32, 261–286.

Fernandez, R.V. (1972) Acaros de cítricos en la Provincia de Tucumán. *Revista Agronómica del Noroeste Argentino* 9, 413–526.

Fernandez-Munoz, R., Dominguez, E. and Cuartero, J. (2000) A novel source of resistance to the two-spotted spider mite in *Lycopersicon pimpinellifolium* (Jusl.) Mill.: its genetics as affected by interplot interference. *Euphytica* 111, 169–173.

Fernando, L.C.P., Moraes, G.J. de and Wickramananda, I.R. (eds) (2002a) *Proceedings of the International Workshop on Coconut Mite* (Aceria guerreronis), *Coconut Research Institute, Sri Lanka, 6–8 January 2000*. Coconut Research Institute, Lunuwila, Sri Lanka.

Fernando, L.C.P., Wickramananda, I.R. and Aratchige, N.S. (2002b) Status of coconut mite, *Aceria guerreronis* in Sri Lanka. In: Fernando, L.C.P., Moraes, G.J. de and Wickramananda, I.R. (eds) *Proceedings of the International Workshop on Coconut Mite* (Aceria guerreronis*), Coconut Research Institute, Sri Lanka, 6–8 January 2000*. Coconut Research Institute, Lunuwila, Sri Lanka, pp. 1–8.

Ferragut, F. and Escudero, L.A. (1999) *Tetranychus evansi* Baker & Pritchard (Acari, Tetranychidae), una nueva araña roja en los cultivos horticolas españoles. *Boletín de Sanidad Vegetal, Plagas* 25, 157–164.

Ferragut, F. and Santonja, M.C. (1989) Taxonomía y distribución de los ácaros del género *Tetranychus* Dufour 1832 (Acari: Tetranychidae), en España. *Boletín de Sanidad Vegetal, Plagas* 15, 271–281.

Ferragut, F., García Marí, F., Costa Comelles, J. and Laborda, R. (1987) Influence of food and temperature on development and oviposition of *Euseius stipulatus* and *Typhlodromus phialatus* (Acari: Phytoseiidae). *Experimental and Applied Acarology* 3, 317–329.

Ferragut, F., Costa Comelles, J., García Marí, F., Laborda, R., Roca, D. and Marzal, C. (1988) Dinámica poblacional del fitoseido *Euseius stipulatus* (Athias-Henriot) y su presa *Panonychus citri* (McGregor) (Acari: Phytoseiidae, Tetranychidae), en los cítricos españoles. *Boletín de Sanidad Vegetal, Plagas* 14, 45–54.

Ferragut, F., Laborda, R., Costa Comelles, J. and García Marí, F. (1992) Feeding behavior of *Euseius stipulatus* and *Typhlodromus phialatus* on the citrus red mite *Panonychus citri* (Acari: Phytoseiidae, Tetranychidae). *Entomophaga* 37, 537–543.

Ferragut, F., Escudero, L.A. and Olmeda, T. (1997a) *Tetranychus evansi* Baker & Pritchard (Acari: Tetranychidae), una nueva araña roja en los cultivos hortícolas españoles. In: *Resúmenes de las VI Jornadas de la S.E.E.A.* Edicions Universitat de Lleida, Lleida, Spain, p. 66.

Ferragut, F., Gea, F.J. and García-Morrás, J.A. (1997b) El ácaro del champiñón *Brennandania lambi* (Krczal) (Acari: Pygmephoroidea): introducción en España, importancia económica y separación de especes afines. *Boletín de Sanidad Vegetal, Plagas* 23, 301–311.

Ferragut, F., Gallardo, A., Ocete, R. and Lopez, M.A. (2008) Natural predatory enemies of the erineum strain of *Colomerus vitis* (Pagenstecher) (Acari, Eriophyidae) found on wild grapevine populations from southern Spain (Andalusia). *Vitis* 47, 51–54.

Ferragut, F., Moreno, I.-P., Iraola, V. and Escudero, A. (2009) *Ácaros Depredadores de la Familia Phytoseiidae en las Plantas Cultivadas*. Ediciones Agrotecnicas, Madrid.

Ferragut, F., Návia, D. and Ochoa, R. (2013a) New mite invasions in citrus in the early years of the 21st century. *Experimental and Applied Acarology* 59, 145–164.

Ferragut, F., Garzón-Luque, E. and Pekas, A. (2013b) The invasive spider mite *Tetranychus evansi* (Acari: Tetranychidae) alters community composition and host-plant use of native relatives. *Experimental and Applied Acarology* 60, 321–341.

Ferrari M., Menta A., Marcon E., Montermini A. (1999) *Malattie e Parassiti delle Piante da Fiore, Ornamentali e Forestali, Volume 2*. Edagricole, Bologna, Italy.

Ferreira, J.A.M., Cunha, D.F.S., Pallini, A., Sabelis, M.W. and Janssen, A. (2011) Leaf domatia reduce intraguild predation among predatory mites. *Ecological Entomology* 36, 435–441.

Ferreira, M.A. and Carmona, M.M. (1995) Acarofauna do tomateiro em Portugal. In: *Avances en Entomología Ibérica: VI Congreso Ibérico de Entomología, Madrid*. Museo Nacional de Ciencias Naturales (CSIC) and Universidad Autónoma de Madrid, pp. 385–392.

Ferreira, M.A. and Sousa, M.E. (2011) Hosts and distribution of the spider mite *Tetranychus evansi* (Acari: Tetranychidae) in Portugal. In: Hale, C. (ed.) *XXVIII International Horticultural Congress on Science and Horticulture for People (IHC2010): International Symposium on Plant Protection, Lisbon, Portugal, August 22, 2010. Acta Horticulturae* 917, pp. 133–136.

Ferreira, P.F. and Veiga, A.F. de S.L. (2000) Efficiency of the chemical control of orange mite of pineapple, *Dolichotetranychus floridanus* (Banks, 1900) (Acari: Tenuipalpidae), in Pernambuco State, Brazil. *Revista de Agricultura (Piracicaba)* 75, 153–162.

Ferreira, R.C.F., Oliveira, J.V. de, Haji, F.N.P. and Gondim, M.G.C. Jr (2006) Biology, thermal requirements and fertility life table of the broad mite *Polyphagotarsonemus latus* (Banks) (Acari: Tarsonemidae) in grape (*Vitis vinifera* L.) cv. Italia. *Neotropical Entomology* 35, 126–132.

Ferrero, M., Moraes, G.J. de, Kreiter, S., Tixier, M.-S. and Knapp, M. (2007) Life tables of the predatory mite *Phytoseiulus longipes* feeding on *Tetranychus evansi* at four temperatures (Acari: Phytoseiidae, Tetranychidae). *Experimental and Applied Acarology* 41, 45–53.

Ferrero, M., Calvo, F.J., Atuahiva, T., Tixier, M.-S. and Kreiter, S. (2011) Biological control of *Tetranychus evansi* Baker & Pritchard and *Tetranychus urticae* Koch by *Phytoseiulus longipes* Evans in tomato greenhouses in Spain (Acari: Tetranychidae, Phytoseiidae). *Biological Control* 58, 30–35.

Ferro, D.N. and Southwick, E.E. (1984) Microclimates of small arthropods: estimating humidity within the leaf boundary layer. *Environmental Entomology* 13(4), 926–929.

Fetoh, B.E.S.A. and Al-Shammery, K.A. (2011) Acaricidal ovicial and repellent activities of some plant extracts on the date palm dust mite, *Oligonychus afrasiaticus* Meg. (Acari: Tetranychidae). *International Journal of Environmental Science and Engineering* 2, 45–52.

Fiaboe, K.K.M., Gondim, M.G.C. Jr, Moraes, G.J. de, Ogol, C.K.P.O. and Knapp, M. (2007a) Surveys for natural enemies of the tomato red spider mite *Tetranychus evansi* (Acari: Tetranychidae) in northeastern and southeastern Brazil. *Zootaxa* 1395, 33–58.

Fiaboe, K.K.M., Gondim, M.G.C. Jr, Moraes, G.J. de, Ogol, C.K.P.O. and Knapp, M. (2007b) Bionomics of the acarophagous ladybird beetle *Stethorus tridens* fed *Tetranychus evansi*. *Journal of Applied Entomology* 131, 355–361.

Field, R.P., Webster, W.J. and Morris, D.S. (1979) Mass rearing *Typhlodromus occidentalis* Nesbitt (Acarina: Phytoseiidae) for release in orchards. *Journal of the Australian Entomological Society* 18, 213–215.

Fife, L.C. (1937) Damage to sea island cotton by West Indian blister mite, *Eriophyes gossypii* (Banks). *Journal of Agriculture of the University of Puerto Rico* 21, 169–177.

Figueroa, J. (1919) Cultivation of clover. *Revista Nacional de Agricultura (Bogotá)* 9-12, 560–604.

Finney, G.L. and Fisher, T.W. (1964) Culture of entomophagous insects and their hosts. In: DeBach, P. (ed.) *Biological Control of Insect Pests and Weeds*. Chapman and Hall, London, pp. 329–355.

Fischer, S. (1993) Observation of a new pest of cucumber in western Switzerland, *Tyrophagus neiswanderi* Johnston & Bruce (Acari, Acaridae). *Revue Suisse de Viticulture Arboriculture Horticulture* 25, 103–104.

Fischer, S. and Mourrut-Salesse, J. (2005) Tomato russet mite in Switzerland (*Aculops lycopersici*: Acari, Eriophyidae). *Revue Suisse de Viticulture Arboriculture Horticulture* 37, 227–232.

Fischer, S., Klotzli, F., Falquet, L. and Celle, O. (2005) An investigation on biological control of the tomato russet mite *Aculops lycopersici* (Massee) with *Amblyseius andersoni* (Chant). *IOBC-WPRS Bulletin* 28(1), 99–102.

Fisher, F.E. (1950) Two new species of *Hirsutella* Patouillard. *Mycologia* 42, 290–297.

Fisher, F.E. (1951) An *Entomophthora* attacking citrus red mite. *Florida Entomologist* 34, 83–88.

Fjelddalen, J. and Stenseth, C. (1958) Nye kjemiske midler for bekjempelse av jordbaermidd (*Tarsonemus pallidus* Banks). *Frukt og Baer* 1958, 37–42 (in Norwegian).

Flaherty, D.L. (1969) Ecosystem trophic complexity and densities of the Willamette mite, *Eotetranychus willamettei* Ewing (Acarina: Tetranychidae). *Ecology* 50, 911–916.

Flaherty, D.L. and Hoy, M.A. (1971) Biological control of Pacific mites and Willamette mites in San Joaquin Valley vineyards: Part III. Role of tydeid mites. *Researches on Population Ecology (Kyoto)* 13, 80–96.

Flaherty, D.L. and Huffaker, C.B. (1970) Biological control of Pacific mites and Willamette mites in San Joaquin Valley vineyards. I. Role of *Metaseiulus occidentalis*. II. Influence of dispersion patterns of *Metaseiulus occidentalis*. *Hilgardia* 40(10), 267–330.

Flaherty, D.L., Jensen, F.L., Kasimatis, A.N., Kido, H. and Moller, W.J. (eds) (1981) *Grape Pest Management*. Publication No. 4105, Division of Agricultural Sciences, University of California, Berkeley, California.

Flaherty, D.L., Wilson, L.T., Welter, S., Lynn, C.D. and Hanna, R.A. (1992) Spider mites. In: Flaherty, D.L., Christensen, L.P., Lannini, W.T., Marois, J.J., Phillips, P.A. and Wilson, L.T. (eds) *Grape Pest Management*, 2nd edn. University of California, Division of Agriculture and Natural Resources Publication No. 3343, Oakland, California, pp. 180–192.

Flechtmann, C.H.W. (1967a) *Contribution to Knowledge of the Mites of Plants of Some Regions of the State of São Paulo (as a Systematic Survey Including New Species)*. Escola Superior de Agricultura "Luiz de Queiroz", University of São Paulo, Piracicaba, Brazil.

Flechtmann, C.H.W. (1967b) Acaros encontrados sobre maças de procedência Argentina. *Anais da Escola Superior de Agricultura "Luiz de Queiroz" (Piracicaba)* 24, 83–85.

Flechtmann, C.H.W. (1967c) Introduçao á familia Tarsonemidae Kramer, 1877 (Acarina) no estado de São Paulo. *Anais da Escola Superior de Agricultura "Luiz de Queiroz" (Piracicaba)* 24, 265–272.

Flechtmann, C.H.W. (1968) Acaros fitofagos associados a plantas forrageiras. *Pesquisa Agropecuária Brasileira* 3, 171–172.

Flechtmann, C.H.W. (1970) New records and notes on eriophyid mites from Brazil and Paraguay, with list of Eriophyidae from South America. *Proceedings of the Entomological Society of Washington* 72, 94–98.

Flechtmann, C.H.W. (1971) Alguns Trombidiformes do Brasil e do Paraguai (Acari). PhD thesis, Escola Superior de Agricultura "Luiz de Queiroz", University of São Paulo, Piracicaba, Brazil.

Flechtmann, C.H.W. (1972) *Ácaros de Importância Agrícola* (1st edn). Livraria Nobel, São Paulo, Brazil.

Flechtmann, C.H.W. (1973a) *Ácaros de Importância Médico Veterinária*. Livraria Nobel, São Paulo, Brazil.

Flechtmann, C.H.W. (1973b) On an eriophyid mite (Acari) from *Lantana* from Brazil. *Anais de Sociedade Entomológica do Brasil* 2, 109–110.

Flechtmann, C.H.W. (1975) *Elementos de Acarologia*. Livraria Nobel, São Paulo, Brazil.

Flechtmann, C.H.W. (1976a) A report on the Tetranychidae (Acari) of Brazil: an emendation. *Revista Brasileira de Entomologia* 20, 115–116.

Flechtmann, C.H.W. (1976b) *Oligonychus yothersi* (McGregor), uma praga potencial do cha. *Revista de Agricultura (Piracicaba)* 51, 178.

Flechtmann, C.H.W. (1977a) *Ácaros de Importância Agrícola* (2nd edn). Livraria Nobel, São Paulo, Brazil.

Flechtmann, C.H.W. (1977b) The cassava mite complex: taxonomy and identification. In: Brekelbaum, T., Bellotti, A. and Lozano, J.C. (eds) *Proceedings CASSAVA Protection Workshop, 7–12 November 1977, Cali, Colombia*. CIAT (Centro Internacional de Agricultura Tropical), Cali, Colombia, pp. 143–153.

Flechtmann, C.H.W. (1979) *Tuckerella ornata* (Tucker), um ácaro novo para o Brasil e outros Tetranychoidea (Acari) do estado do Pará. *Anais da Escola Superior de Agricultura "Luiz de Queiroz" (Piracicaba)* 36, 615–619.

Flechtmann, C.H.W. (1981a) The cassava mite complex. II. New records and description of two new species in the genus *Tetranychus* from Asia. *International Journal of Acarology* 7, 81–86.

Flechtmann, C.H.W. (1981b) A small collection of spider mites from Jaguanum Island, Rio de Janeiro, Brazil, with description of *Aponychus insularis* (Prostigmata: Tetranychidae). *International Journal of Acarology* 7, 87–90.

Flechtmann, C.H.W. (1982) The cassava mite complex. III – New distribution records, mainly from Colombia and Africa. References to other plants. *Anais da Escola Superior de Agricultura "Luiz de Queiroz" (Piracicaba)* 39, 809–813.

Flechtmann, C.H.W. (1983) Dois ácaros novos para o eucalipto, com uma lista daqueles ja assinalados para esta planta. *IPEF Notícias* 23, 43–46.

Flechtmann, C.H.W. (1986) *Ácaros em Produtos Amazenados e na Poeira Domiciliar*. Universidade de São Paulo, São Paulo, Brazil.

Flechtmann, C.H.W. (1989a) *Cocos weddelliana* H. Wendl. (Palmae: Arecaceae), a new host plant for *Eriophyes guerreronis* (Keifer, 1965) (Acari: Eriophyidae) in Brazil. *International Journal of Acarology* 15, 241.

Flechtmann, C.H.W. (1989b) Seringueira (*Hevea* sp.) um novo hospedeiro para *Oligonychus gossypii* (Zacher, 1921) (Acari: Tetranychidae). *Boletim do Museu Paraense Emílio Goeldi, Série Zoologia* 5, 27–128.

Flechtmann, C.H.W. (1996) Rediscovery of *Tetranychus abacae* Baker & Pritchard, additional description and notes on South American spider mites (Acari, Prostigmata, Tetranychidae). *Revista Brasileira de Zoologia* 13, 569–578.

Flechtmann, C.H.W. (1998) Three Eriophyoidea (Acari: Prostigmata) new for the Brazilian fauna. *International Journal of Acarology* 24, 75–80.

Flechtmann, C.H.W. (1999) *Tegolophus brunneus* n. sp., a new citrus rust mite from Brazil (Acari: Eriophyidae). *International Journal of Acarology* 25, 265–267.

Flechtmann, C.H.W. (2001a) *Aberoptus cerostructor* n.sp., a deuterogynous species from Brazil (Acari: Eriophyidae). *International Journal of Acarology* 27, 199–204.

Flechtmann, C.H.W. (2001b) Two new species of eriophyid mites from fruit trees in Brazil (Acari: Eriophyidae). *International Journal of Acarology* 27, 205–210.

Flechtmann, C.H.W. (2004) Two new plant feeding mites from *Brachiaria ruziziensis* in citrus groves in São Paulo, Brazil and new distribution records of other plant mites in Brazil. *Zootaxa* 708, 1–11.

Flechtmann, C.H.W. and Abreu, J.M. (1973) Ácaros fitofagos do Estado da Bahia, Brasil. (Notas preliminares). *Ciência e Cultura (São Paulo)* 25, 244–250.

Flechtmann, C.H.W. and Aranda, B.R. (1970) New records and notes on eriopyid mites from Brazil and Paraguay, with a list of Eryophidae from South America. *Proceedings of the Entomological Society of Washington* 72, 94–98.

Flechtmann, C.H.W. and Arleu, R.J. (1984) *Oligonychus coffeae* (Nietner), a red spider mite from the rubber tree, new for Brazil and observations on other mites from this plant. *Ecossistema* 9, 123–125.

Flechtmann, C.H.W. and Arruda, G.P. (1967) Ocorrência de dois acaros vermelhos en mamoeiros em Recife. *Revista de Agricultura (Piracicaba)* 42, 123–124.

Flechtmann, C.H.W. and Baker, E.W. (1970) A preliminary report on the Tetranychidae (Acarina) of Brazil. *Annals of the Entomological Society of America* 63, 156–163.

Flechtmann, C.H.W. and Baker, E.W. (1975) A report on the Tetranychidae (Acari) of Brazil. *Revista Brasileira de Entomologia* 19, 111–122.

Flechtmann, C.H.W. and Berti-Filho, E. (1994) Effect of feeding by two species of eriophyid mites (Acari, Eriophyidae) on the mineral content of their host plants. *International Journal of Acarology* 20, 61–65.

Flechtmann, C.H.W. and Etienne, J. (2001) Plant mites from Guadeloupe and French Guyana, with descriptions of five new species of eriophyid mites (Acari: Eriophyidae, Tenuipalpidae, Tetranychidae). *International Journal of Acarology* 27, 261–270.

Flechtmann, C.H.W. and Etienne, J. (2002) New records of plant mites (Acari, Acaridae, Tetranychidae) from Guadeloupe and Marie Galante with description of five new eriophyid species. *Zootaxa* 47, 1–16.

Flechtmann, C.H.W. and Etienne, J. (2003) Two new species of eriophyid mites (Acari: Eriphyidae) from Guadeloupe and Marie Galante, with records on other plant mites. *Zootaxa* 271, 1–7.

Flechtmann, C.H.W. and Etienne, J. (2004) The red palm mite, *Raoiella indica* Hirst, a threat to palms in the Americas (Acari: Prostigmata: Tenuipalpidae). *Systematic and Applied Acarology* 9, 109–110.

Flechtmann, C.H.W. and Etienne, J. (2005) *Raoiella indica*, new pest of palms in the Carribean. *Phytoma* 584, 10–11.

Flechtmann, C.H.W. and Etienne, J. (2006) Further notes on plant associated mites (Acari) from Guadeloupe and Les Saintes, lesser Antilles. *International Journal of Acarology* 32, 377–382.

Flechtmann, C.H.W. and Flechtmann, C.A.H. (1984) Reproduction and chromosomes in the broad mite, *Polyphagotarsonemus latus* (Banks, 1904) (Acari, Prostigmata, Tarsonemidae). In: Griffiths, D.A. and Bowman, C.E. (eds) *Acarology VI, Volume 1*. Ellis Horwood, Chichester, UK, pp. 445–456.

Flechtmann, C.H.W. and Harley, K.L.S. (1974) Preliminary report on mites (Acari) associated with *Lantana camara* L. in the neotropical region. *Anais da Sociedade Entomológica do Brasil* 3, 69–71.

Flechtmann, C.H.W. and Hunter, P.E. (1971) The spider mites (Prostigmata: Tetranychidae) of Georgia. *Journal of the Georgia Entomological Society* 6, 16–30.

Flechtmann, C.H.W. and Knihinicki, D.K. (2002) New species and new record of *Tetranychus* Dufour from Australia, with a key to the major groups in this genus based on females (Acari: Prostigmata: Tetranychidae). *Australian Journal of Entomology* 41, 118–127.

Flechtmann, C.H.W. and Malavolta, E. (1976) Studies on the mineral nutrition of grain sorghum. I. Populations of the common two spotted spider mite on grain sorghum grown on various levels of N, P and K. *Communications in Soil Science and Plant Analysis* 7, 839–841.

Flechtmann, C.H.W. and McMurtry, J.A. (1992) Studies on how phytoseiid mites feed on spider mites and pollen. *International Journal of Acarology* 18, 157–162.

Flechtmann, C.H.W. and Návia, D. (1998) First record of eriophyid mites (Acari) from Pinaceae in Brazil. *International Journal of Acarology* 24, 167–171.

Flechtmann, C.H.W. and Santana, D.L.Q. (1997) A preliminary note on mites on corn in Brazil with redescriptions of *Catarhinus tricholaenae* and *Oligonychus zeae* (Acari: Diptilomiopidae, Tetranychidae). *Systematic and Applied Acarology* 2, 189–194.

Flechtmann, C.H.W. and Zem, A.C. (1980) Effeitos da aplicação de permetrina sobre o ácaro rajado (*Tetranychus urticae* Koch, 1844). *Ecossistema* 5, 29–32.

Flechtmann, C.H.W., Guerrero, B.J.M., Arroyave, J.A. and Costantino, Ch.L.M. (1990) A little known mode of dispersal of *Polyphagotarsonemus latus* (Banks). *International Journal of Acarology* 16, 181–182.

Flechtmann, C.H.W., Da Costa, C.P. and Maielli, J.A. (1998) *A Residência para o Alérgico – Construção e Adaptação*. Editoria UNIMEP, Piracicaba, Brazil.

Flechtmann, C.H.W., Kreiter, S., Etienne, J. and Moraes, G.J. de (1999a) Plant mites (Acari) of the French Antilles. 1. Tetranychoidea (Prostigmata). *Acarologia* 40, 137–144.

Flechtmann, C.H.W., Kreiter, S., Etienne, J. and Moraes, G.J. de (1999b) Plant mites (Acari) of the French Antilles. 4. Eriophyidae (Prostigmata). *Acarologia* 40, 321–342.

Flechtmann, C.H.W., De Moraes, G. and Barbosa, F.R. (2001) A new species of *Calacarus* Keifer (Acari: Eriophyidae) on papaya in northeastern Brazil. *Zootaxa* 5, 1–5.

Flechtmann, N.N.B. and Flechtmann, C.H.W. (1982) Observações sobre a reprodução do ácaro vermelho do cafeeiro. *II Congresso Brasileiro de Iniciação Cientifica em Ciências Agrárias, 8–10 Setembro*. Escola Superior de Agricultura "Luiz de Queiroz", University of São Paulo, Piracicaba, Brazil, pp. 109–111.

Fleschner, C.A. (1952) Host plant resistence, as a factor influencing population densisty of citrus red mites on orchard trees. *Journal of Economic Entomology* 45, 687–695.

Fleschner, C.A. (1958a) Field approach to population studies of tetranychid mites on citrus and avocado in California. In: Becker, E.C. (ed.) *Proceedings of the 10th International Congress of Entomology, Montreal, Auguat 17–25, 1956, Volume 2*. Mortimer, Ottawa, Ontario, Canada, pp. 669–674.

Fleschner, C.A. (1958b) Natural enemies of tetranychid mites on citrus and avocado in Southern California. In: Becker, E.C. (ed.) *Proceedings of the 10th International Congress of Entomology, Montreal, August 17–25, 1956, Volume 4*. Mortimer, Ottawa, Ontario, Canada, pp. 627–631.

Fleschner, C.A. (1958c) The effect of orchard dust on the biological control of avocado pests. *California Avocado Society Yearbook* 42, 94–98.

Fleschner, C.A. and Ricker, D.W. (1953a) An empidid fly predacious on citrus red mites. *Journal of Economic Entomology* 46, 155.

Fleschner, C.A. and Ricker, D.W. (1953b) Food habits of coniopterygids on citrus in southern California. *Journal of Economic Entomology* 46, 458–461.

Fleschner, C.A., Badgley, M.E., Ricker, D.W. and Hall, J.C. (1956) Air drift of spider mites. *Journal of Economic Entomology* 49, 624–627.

Fletcher, J.T. and Gaze, R.H. (2008) *Mushroom Pest and Disease Control: A Color Handbook*. Manson Publishing, London.

Fleury, P.(1992) Slalom. Acaricide–insecticide vigne. *Phytoma* 435, 69–70.

Flint, M.L. and Van Den Bosch, R. (1981) *Introduction to Integrated Pest Management*. Plenum Press, New York.

Floate, K.D. and Whitham, T.G. (1993) The "hybrid bridge" hypothesis: host shifting via plant hybrid swarms? *American Naturalist* 141, 651–662.

Flock, R.A. and Wallace, J.M. (1955) Transmission of fig mosaic by the eriophyid mite, *Aceria ficus* (Cotte). *Phytopathology* 45, 52–53.

Flores, A.E., Badii, M.H., Aranda, H.E. and Flores, B.S. (1994) Tolerance to dicofol in a field population of *Eutetranychus banski* (McGregor). *Southwestern Entomologist* 19, 161–165.

Flores, A.E., Flores, S., Badii, M.H., Aranda, E. and Hernández, E. (1996) Susceptibilidad del arador de los cítricos, *Phyllocoptruta oleivora* y el ácaro depredador, *Euseius mesembrinus*, al acaricida dicofol. *Southwestern Entomologist* 21, 83–84.

Flores-Galano, G., Montoya, A. and Rodríguez, H. (2010) Biología de *Raoiella indica* Hirst (*Acari: Tenuipalpidae*) sobre *Areca catechu* L. *Revista de Protección Vegetal* 25, 11–16. Available at: http://scielo.sld.cu/scielo.php?pid=S1010-27522010000100004&script=sci_arttext (accessed 30 July 2014).

Flores Suarez, A.E., Badii, Z., Mohammed, H., Flores Breceda, S. and Foroughbakhch, R. (1996) Efectos subletales del dicofol sobre parametros poblacionales de *Eutetranychus banksi* (Acari:Tetranychidae). *Agrociencia (Montecillo)* 30, 111–116.

Foott, W.H. (1962) Competition between two species of mites. I. Experimental results. *Canadian Entomologist* 94, 365–375.

Fornazier, M.J., Carmo, C.A.S. do, Souza, J.L. de and Dessaune Filho, N. (1987) *Comportamento de Diversas Cultivares de Alho ao Ataque do Ácaro do Chochamento*. Pesquisa em Andamento, Empresa Capixaba de Pesquisa Agropecuaria (EMCAPA), Cariacica, Brazil.

Forsberg, J.L. (1965) The relationship of *Pseudomonas marginata*, *Stromatina gladiola*, bulb mites, and chemical control treatments to the occurrence of scab and *Stromatinia* rot of gladiolus. *Phytopathology* 55, 1058.

Foster, D.G., Teetes, G.L., Johnson, J.W., Rosenow, D.T. and Ward, C.R. (1977a) Field evaluation of resistance in sorghums to Banks grass mite. *Crop Science* 17, 821–823.

Foster, D.G., Teetes, G.L., Johnson, J.W. and Ward, C.R. (1977b) Resistance in sorghum to the Banks grass mite. *Journal of Economic Entomology* 70, 259–262.

Fouly, A.H. (1997) Effects of prey mites and pollen on the biology and life tables of *Proprioseiopsis asetus* (Chant) (Acari, Phytoseiidae). *Journal of Applied Entomology* 121, 435–439.

Fountain, M.T., Harris, A.L. and Cross, J.V. (2010) The use of surfactants to enhance acaricide control of *Phytonemus pallidus* (Acari: Tarsonemidae) in strawberry. *Crop Protection* 29, 1286–292.

Fourie, P.F. (1988) Citrus bud mite. *Farming in South Africa* H.16, 1–2.

Fournier, D., Millot, P. and Pralavorio, M. (1985) Rearing and mass production of the predatory mite *Phytoseiulus persimilis*. *Entomologia Experimentalis et Applicata* 38, 97–100.

Fox-Wilson, (1939) Contributions from the Wisley Laboratory LXXXVII. Amaryllis pests. *Journal of the Royal Horticultural Society* 64, 318–326.

Franco, R.A., Reis, P.R., Zacarias, M.S. and Altoe, B.F. (2007) Predation potential of three species of phytoseiids on *Oligonychus ilicis* (McGregor, 1917) (Acari: Tetranychidae). *Coffee Science* 2, 175–182.

Franco, R.A., Reis, P.R., Zacarias, M.S., Altoe, B.F. and Pedro Neto, M. (2008) Population dynamics of *Oligonychus ilicis* (McGregor, 1917) (Acari: Tetranychidae) in coffee plants and of their associated phytoseiids. *Coffee Science* 3, 38–46.

Franco, R.A., Reis, P.R., Zacarias, M.S., Altoe, B.F. and Barbosa, J.P.R.A.D. (2009) Influence of *Oligonychus ilicis* (McGregor, 1917) (Acari: Tetranychidae) infestation on the potential photosynthesis rate of coffee-plant leaves. *Arquivos do Instituto Biológico (São Paulo)* 76, 205–210.

Franco, R.A., Reis, P.R., Zacarias, M.S. and Oliveira, D.C. (2010) Influence of the webbing produced by *Oligonychus ilicis* (McGregor) (Acari: Tetranychidae) on associated predatory phytoseiids. *Neotropical Entomology* 39, 97–100.

Frank, J.H., Bennett, F.D. and Cromroy, H.L. (1992) Distribution and prey records for *Oligota minuta* (Coleoptera: Staphylinidae), a predator of mites. *Florida Entomologist* 75, 376–380.

Franklin, E., Adis, J. and Woas, S. (1997) The oribatid mites. In: Junk, W.J. (ed.) *The Central Amazon Floodplain: The Ecology of a Pulsing System*. Ecological Studies Volume 126. Springer Verlag, Berlin, pp. 331–349.

Frazier, N.W. and Smith, L.M. (1946) The Willamette mite on grapes. *Hilgardia* 17(4), 191–196.

Fréitez, F.R. (1974) *Reconcimiento Preliminar de Ácaros Fitoparasitos de la Familia Tetranychidae de Costa Rica (Acarina)*. Facultad de Agronomía, San José, Universidad de Costa Rica, Costa Rica.

Fréitez, F.R. and Alvarado, D.G. (1976) Nota preliminar sobre acaros tetraniquidos del girasol en Venezuela (Acarina: Tetranychidae). *Agronomía Tropical (Maracay)* 26, 167–169.

Fréitez, F.R. and Alvarado, D.G. (1978a) Contribución al conocimiento de ácaros en Musaceae de Venezuela (Acarina). *Revista de la Facultad de Agronomía, Universidad Central de Venezuela* 26, 197–207.

Fréitez, F.R. and Alvarado, D.G. (1978b) Ácaros en gramineas de los estados Portuguesa y Yaracuy – Venezuela. *Ciarco (Acarigua, Venezuela)* 8, 55–61.

French, J.V. (1975) Control of armored scale on citrus with non-oil scalicides. *Journal of the Rio Grande Valley Horticultural Society* 29, 45–51.

French, J.V. (1982) Citrus rust mite control with avermectin B1. *Insecticicide and Acaricide Tests* 7, 47–48.

French, J.V. (1984) Rust mite control on Texas citrus. *Insecticicide and Acaricide Tests* 9, 71.

French, J.V. and Bruno, P.H. (1996) Evaluation of BAS 300 11I (pyridaben) for mite control on Texas citrus. *Subtropical Plant Science* 48, 10–15.

French, J.V. and Gage, E.V. (1983) Evaluation of carbosulfan for rust mite suppression on Texas citrus. *Journal of the Rio Grande Valley Horticultural Society* 36, 81–85.

French, J.V. and Rakha, M.A. (1994) False spider mite: damage and control on Texas citrus. *Subtropical Plant Science* 46, 16–19.

French, J.V. and Timmer, L.W. (1981) Suppression of citrus rust mite and citrus nematode on Texas grapefruit with aldicarb. In: Matsumoto, K. (ed.) *Proceedings of the 4th International Citrus Congress, Tokyo, Japan, 9–12 November 1981, Volume 2*. International Society of Citriculture, Riverside, California, pp. 672–674.

French, J.V. and Villarreal, J. (1988) Evaluation of avermectin B1 and hexythiazox for spider mite control on Texas citrus. *Journal of the Rio Grande Valley Horticultural Society* 41, 75–81.

French, J.V. and Villarreal, J. (1990a) Rust mite control on Texas citrus. *Insecticide and Acaricide Tests* 15, 66–67.

French, J.V. and Villarreal, J. (1990b) Argi-Mek (abamectin) a new miticide for control of citrus rust mite. *Journal of the Rio Grande Valley Horticultural Society* 43, 9–15.

French, J.V., Swietlik, D. and Gage, E. (1985) Carbosulfan and dicofol: efficacy on mites as affected by spray mix pH. *Journal of the Rio Grande Valley Horticultural Society* 38, 43–50.

French, N., Vernon, J.D.R. and Woodville, H.C. (1968) Cyclamen mite, *Steneotarsonemus pallidus*, on chrysanthemum. *Plant Pathology* 12, 143.

French, R. and Stenger, D.C. (2003) Evolution of wheat streak mosaic virus: dynamics of population growth within plants may explain limited variation. *Annual Review of Phytopathology* 41, 199–214.

Frezzi, M.J. (1940) "Lepra explosiva" of orange. Studies in the pathology laboratory of Bella Vista (Corrientes). *Boletín de Frutas y Hortalizas* 5, 1–15.

Friebe, B., Gill, K.S., Tuleen, N.A. and Gill, B.S. (1996) Transfer of wheat streak mosaic virus resistance from *Agropyron intermedium* into wheat. *Crop Science* 36, 857–861.

Fritzsche, R. (1960) Morphologische, Biologische, und Physiologische Variabilität und ihre Bedeuting für de Epidemiologie und Bekämpfing von *Tetranychus uritcae* Koch. *Biologisches Zentrablatt* 79, 521–576.

Froggatt, W.W. (1921) Orchard and garden mites. No. 2 – Spinning mites (Family Tetranychidae). *Agricultural Gazette of New South Wales* 32, 130–135.

Frost, C.J., Mescher, M.C., Carlson, J.E. and Moraes, C.M. de (2008) Plant defense priming against herbivores: getting ready for a different battle. *Plant Physiology* 146, 818–824.

Frost, W.E. (1993a) Australian record of *Abacarus hystrix* (Nalepa) (Acarina: Eriophyidae) backdated through examination of herbarium material. *Journal of the Australian Entomological Society* 32, 389–390.

Frost, W.E. (1993b) Aspects of the population ecology of *Abacarus hystrix* (Nalepa) (Acarina: Eriophyidae) in South Australian dairy pastures. In: Corey, S., Dall, D. and Milne, W. (eds) *Pest Control and Sustainable Agriculture*. CSIRO (Commonwealth Scientific and Industrial Research Organisation), Canberra, pp. 388–390.

Frost, W.E. (1997) Polyphenic wax production in *Abacarus hystrix* (Acari: Eriophyidae), and implications for migratory fitness. *Physiological Entomology* 22, 37–46.

Frost, W.E. and Ridland, P.M. (1996) Grasses. In: Lindquist, E.E., Sabelis, M.W. and Bruin, J. (eds) *Eriophyoid Mites: Their Biology, Natural Enemies and Control*. World Crop Pests, Volume 6. Elsevier Science, Amsterdam, pp. 619–629.

Froud, K.J., Stevens, P.S., Machin, T. and Jamieson, L.E. (2002) Efficacy of new pesticides against sixspotted mite *Eotetranychus sexmaculatus* (Riley) (Acari: Tetranychidae) on avocados. In: *New Zealand Avocado Growers Association Annual Research Report – 2002. Volume 2*. Tauranga, New Zealand, article 3. Available at: http://www.avocadosource.com/Journals/NZAGA/NZAGA_2002/NZAGA_2002_03.pdf (accessed 21 May 2015).

Fu, Y.-G. and Zhang F.-P. (2002) Effects of temperatures on development and reproduction of *Oligonychus mangiferus* (Acari: Tetranychidae). *Chinese Journal of Tropical Crops* 23(3), 47–52 (in Chinese).

Fulton, B.B. (1940) The blueberry bud mite, a new pest. *Journal of Economic Entomology* 33, 699.

Funayama, K. and Takahashi, Y. (1992) Studies on ecology of apple rust mite *Aculus schlechtendali* (Nalepa) (Acarina: Eriophyidae). *Bulletin of the Akita Fruit-tree Experiment Station* 22, 9–22 (in Japanese).

Furman, D.P. and Sousa, O.E. (1969) Morphology and biology of a nest-producing mite, *Bakericheyla chanayi* (Acarina: Cheyletidae). *Annals of the Entomological Society of America* 62, 858–863.

Furniss, R.L. and Carolin, V.M. (1977) *Western Forest Insects*. Miscellaneous Publication No. 1339, US Department of Agriculture Forest Service, Washington, DC.

Furtado, I.P., Moraes, G.J. de, Kreiter, S. and Knapp, M. (2006) Search for effective natural ennemies of *Tetranychus evansi* in south and southeast Brazil. *Experimental and Applied Acarology* 40, 157–174.

Furtado, I.P., Toledo, S., Moraes, G.J. de, Kreiter, S. and Knapp, M. (2007) Search for effective natural enemies of *Tetranychus evansi* (Acari: Tetranychidae) in northwest Argentina. *Experimental and Applied Acarology* 43, 121–127.

Gäbele, M. (1959) Beiträge zur Kenntnis der Guttung Bryobia. *Zeitschrift für Angewandte Zoologie* 46, 191–247.

Gabi G. (1994) Helyzetkép a szőlő- levélatka (*Calepitrimerus vitis* Nal.) elterjedéséről Tolna megyében (Situation on the distribution of grapevine rust mite (*Calepitrimerus vitis* Nal.) in Tolna county). *Növényvédelem* 30, 225–226 (in Hungarian).

Gabi, G. and Mészáros, Z. (2000) A szőlő-levélatka (*Calepitrimerus vitis* nalepa) népességmozgalmának és telelőhelyeinek vizsgálata a szekszárdi borvidéken (Population dynamics and hibernation shelters of *Calepitrimerus vitis* in the vine-growing region of Szekszárd). *Növényvédelem* 36, 349–355 (in Hungarian).

Gabi, G. and Mészáros, Z. (2001) New data to the knowledge of *Calepitrimerus vitis* Nalepa in the vine-growing region of Szekszárd, Hungary (Acari: Eriophyidae). *Acta Phytopathologica et Entomologica Hungarica* 36, 193–200.

Gabi, G. and Mészáros, Z. (2003) Examination of the development of the deutogynes of *Calepitrimerus vitis* Nalepa in the vine-growing region of Szekszárd, Hungary (Acari: Eriophyidae). *Acta Phytopathologica et Entomologica Hungarica* 38, 369–376.

Gabryś, G. (2000) *Balaustium xerothermicum* sp. nov. from Poland with remarks on all world species of the genus (Acari: Actinedida: Erythraeidae). *Annales Zoologici* 50, 47–56.

Gaede, K. (1991) On the water balance of *Phytoseiulus persimilis* A.-H. and its ecological significance. *Experimental and Applied Acarology* 15, 181–189.

Gagne, R.J. (1995) Revision of tetranychid (Acarina) mite predators of the genus *Feltiella* (Diptera: Cecidomyiidae). *Annals of the Entomological Society of America* 88, 16–30.

Gajek, D. and Boczek, J. (1998) The life-cycle of the blackcurrant gall mite, *Cecidophyopsis ribis* (Westw.) (Acari: Eriophyidae). *General Technical Report NC-199*. North Central Research Station, US Department of Agriculture Forest Service, St Paul, Minnesota, pp. 131–135.

Gajek, D. and Niemczyk, E. (2001) Efficacy of chemical and non-chemical treatments against blackcurrant gall mite (*Cecidophyopsis ribis* (Westw)) and their influence on populations of two spotted spider mite (*Tetranychus urticae* Koch), predatory mites (Phytoseiidae) and aphids (Aphididae). *Journal of Fruit and Ornamental Plant Research* 9, 93–102.

Gajek, D. and Nowacki, J. (1996) Cultivars de cassis resistant au phytopte du groseiller (*Cecidophyopsis ribis* Westw.), element de gestion integree des ravageurs. *IOBC-WPRS Bulletin* 19(4), 349–350.

Gajek, D., Abanowska, B.H., Niemczyk, E., Olszak, R.W. and Sekrecka, M. (2004) Effectiveness of conventional and integrated control of blackcurrant pests. *IOBC-WPRS Bulletin* 27(4), 141–144.

Gallego, C.E., Aterrado, E.D. and Batomalaque, C.G. (2003) Biology of the false spider mite, *Rarosiella cocosae* Rimando, infesting coconut palms in Camiguin, northern Mindanao (Philippines). *Philippine Entomologist* 17, 187.

Gallegos, R. (1983) *Algunos Aspectos del Aguacate y su Producción en Michoacán*. Universidad Autónoma de Chapingo, Grupo Editorial Gaceta, Mexico.

Gamliel-Atinsky, E., Freeman, S., Maymon, M., Belausov, E., Ochoa, R., Bauchan, G., Skoracka, A., Peña, J. and Palevsky, E. (2010) The role of eriophyoids in fungal pathogen epidemiology, mere association or true interaction? *Experimental and Applied Acarology* 51, 191–204.

Gams, W. (1971) Cephalosporium-*artige Schimmelpilze (Hyphomycetes)*. Gustav Fisher, Stuttgart, Germany.

Gantner, M. (2001) Occurrence of hazelnut pests in southeastern Poland. *Acta Horticulturae* 556, 469–477.

Gao, G.-R. and Zou, P. (2010) Dolichocyboidea, Pigmephoroidea, Scutacaroidea and Trochometridioidea of China: a review of progress, with a checklist. In: Zhang, Z.-Q., Hong, X.-Y. and Fan, Q.-H. (eds) *Xin Jie-Liu Centenary: Progress in Chinese Acarology*. *Zoosymposia* 4, 165–174.

Gao, G.-R., Zou, P. and Ma, E.-P. (1988) Description of *Siteroptes flechtmanni* (Wicht, 1970) (Acarina: Pygmephoridae). *Acta Agriculturae Shanghai* 4(1), 21–23 (in Chinese).

Gao, G.-R., Zou, P. and Jiang, H.-F. (1990a) A new species of *Luciaphorus* (Acariformes: Pygmephoridae) injuring Jew's ear mushroom (*Auricularia polytricha*) in China. *Acta Agriculturae Shanghai* 6(3), 68–71 (in Chinese).

Gao, J.-R. and Zou, P. (1994) A new species of the genus *Allonychus* (Acari: Tetranychidae) from the People's Repubic of China. *Acarologia* 35, 353–356.

Gao, J.-R. and Zou, P. (2001) Biology, life table and host specificity of the mushroom pest, *Brennandania lambi* (Acari: Pygmephoroidea). *Experimental and Applied Acarology* 25, 187–202.

Gao, J.-R., Zou, P. and Ma, E.-P. (1986) Two new records of mushrooms pygmephorid mites from China (Acari: Pygmephoridae). *Acta Agriculturae Shanghai* 2(3), 27–32 (in Chinese).

Gao, J.Y., Guo, J., Wang, Z.-R., Zhou, D.-G., Peng, M.-X. and Yue, J.-Q. (2012) Study on insect pest species and occurrence rule of main species in Dehong lemon orchard of Yunnan Province. *Acta Agriculturae Jiangxi* 24(6), 70–73.

Gao, S., Pang, B., Liu, J., Meng, H. and Yi, D. (2005) Studies on the population dynamics of *Tetranychus truncatus* Ehara in greenhouse. *Journal of Inner Mongolia Agricultural University (Natural Science Edition)* 26(1), 5–8 (in Chinese).

Gao, Z.-R., Qiu, F. and Li, Q.-S. (1990b) Study on the infestation of spider mites *Tetranychus* spp. on cotton and their economic threshold. *Acta Phytophylacica Sinica* 17, 331–336 (in Chinese).

Garant, D. and Kruuk, L. (2005) How to use molecular marker data to measure evolutionary parameters in wild populations. *Molecular Ecology* 14, 1843–1859.

García, A., Hernández, J., Almaguel, L., Sandoval, I., Botta, E. and Arteaga, I. (2002) Influencia del ácaro *Steneotarsonemus spinki* Smiley (Acari: Tarsonemidae) y del hongo *Sarocladium oryzae* (Sawada) Gams and Hawks. sobre el vaneado y manchado de los granos de arroz. In: *Memórias, II Encuentro International del Arroz, 10–12 Julio 2002, La Habana, Cuba*, pp. 189–193.

Garcia, E., Marquez, A.L., Orta, S. and Alvarado, P. (2003) Caracterizacion de la presencia de *Eutetranychus banksi* (McGregor) y *Eutetranychus orientalis* (Klein) en el Sur de Espana. *Phytoma España* 153, 90–96.

Garcia, E., Marquez, A.L., Orta, S. and Alvarado, P. (2004) Ácaro de Texas, *Eutetranychus banksi* (McGregor): situación y caracteristicas de la plaga. *Terralia* 44, 78–79.

Garcia, E.A. (1966) A acariose ou no' curto parasitario da videira no Minho. *Agricultura (Lisbon)* 30, 20–23.

García, M.P.G. (2005) Vaneamento y manchado de grano en cultivos de arroz en Panamá. *Revista Arroz* 53, 455.

García Marí, F. (2009) *Guía de Campo. Plagas de Cítricos y sus Enemigos Naturals*. Phytoma España, Valencia, Spain.

García Marí, F. and González-Zamora, J.E. (1999) Biological control of *Tetranychus urticae* (Acari: Tetranychidae) with naturally occurring predators in strawberry plantings in Valencia, Spain. *Experimental and Applied Acarology* 23, 487–495.

García Marí, F., Santaballa, E., Ferragut, F., Marzal, C., Colomer, P. and Costa, J. (1983) El acaro rojo *Panonychus citri* (McGregor): incidencia en la problematica fitosanitaria de nuestros agrios. *Boletín, Servicio de Defensa contra Plagas e Inspección Fitopatológica* 9, 191–218.

García Marí, F., Roca, D., Fonbuena, P., Ferragut, F. and Costa Comelles, J. (1988) Accion de los acaricidas tetradifon y dicofol sobre huevos y adultos de *Panonychus citri* (McGregor) y *Tetranychus urticae* Koch (Acari: Tetranychidae), en citricos. *Boletín de Sanidad Vegetal, Plagas* 14, 163–169.

Gardner, W.A., Oetting, R.D. and Storey, G.K. (1982) Susceptibility of the two-spotted spider mite, *Tetranychus urticae* Koch, to the fungal pathogen *Hirsutella thompsonii* Fisher. *Florida Entomologist* 65, 458–465.

Garman, P. (1917) *Tarsonemus pallidus* Banks, a pest of geraniums. *Bulletin of the Maryland Agricultural Experiment Station* No. 208, 327–342.

Garman, P. (1923) Notes on the life history of the spruce mite (*Paratetranychus ununguis* (Jacobi)). *Bulletin of the Connecticut Agricultural Experiment Station* No. 247, 340–342.

Garman, P. (1937) A study of the bulb mite. *Bulletin of the Connecticut Agricultural Experiment Station* No. 402, 889–907.

Garman, P. (1940) Tetranychidae of Connecticut. *Bulletin of the Connecticut Agricultural Experiment Station* No. 431, 1–88.

Garrett, L.E. and Haramoto, F.H. (1967) A catalog of Hawaiian Acarina. *Proceedings of the Hawaiian Entomological Society* 19, 381–414.

Gärtel, W. (1970) Austriebsschäden und Kümmerwuchs als Folge gleichzeitigen Auftretens von Bormangel und Rebblatt-gallmilben (*Eriophyes vitis* Pgst.) in unbewässerten Weinbau-gebieten Chiles. *Weinberg und Keller* 17, 159–200.

Gärtel, W. (1972) Die Rebenblattmilbe, *E. vitis* Pagst., die Erreger der Pockenkrankheit (Erinose) als Knopfschädling und als Ursache starken Blattrollens. *Weinbau und Keller* 19, 589–614.

Garzón-Hidalgo, A. and Ferragut, F. (2010) Assessing the suitability of the Oriental red mite *Eutetranychus orientalis* (Tetranychidae) as prey for the predatory mite *Euseius stipulatus* (Phytoseiidae). In: Moraes, G.J. de, Castilho, R.C. and Flechtmann, C.H.W. (eds) *XIII International Congress of Acarology, August 23–27, 2010, Recife-PE, Brazil, Abstract Book*, abstract 167, p. 91. Available at: http://www.acarology.org/ica/ica2010/docs/0910_15_abstractBookCapa.pdf (accessed 3 February 2015).

Gasser, R. (1951) Zur Kenntnis der gemeinen Spinnmilbe *Tetranychus urticae* Koch. I. Mitteilung: Morphologie, Anatomie, Biologie and Oekologie. *Mitteilungen der Schweizerischen Entomologischen Gesellschaft/Bulletin de la Société Entomologique Suisse* 24, 217–262.

Gasser, R. and Mathys, G. (1960) II Symposium Européen d'Acarologie à East Malling en 1959. *Acarologia* 2, 149–153.

Gassouma, S.M. (2005) Pests of the date palm (*Phoenix dactylifera*). Presented to: *Proceedings of the regional workshop on date palm development in the GCC countries of the Arabian Peninsula, Abu Dhabi, 29–31 May 2004*. Originally available at: http://www.uae.gov.ae/uaeagricent/palmtree2/chap8.htm; http://www.icarda.org/APRP/Datepalm/Topics/Pest/Pestright.htm. Abstract available at: http://www.cabi.org/isc/abstract/20087207796 (accessed 7 May 2015).

Gaull, K.R. and Ridsdill-Smith, T.J. (1996) The foraging behaviour of redlegged earth mite, *Halotydeus destructor* (Acarina: Penthaleidae), in an annual subterranean clover pasture. *Bulletin of Entomological Research* 86, 247–252.

Gaunt, R.E. (1981) Disease tolerance – an indicator of thresholds? *Phytopathology* 71, 915–916.

Geddes, W.A. (2010) Effects of nitrogen and potassium fertilizer on Willamette spider mite (*Eotetranychus willamettei*) (Acari: Tetranychidae). MSc. thesis, Faculty of California Polytechnic State University, San Luis Obispo, California.

Geier, P. and Baggiolini, M. (1952) *Malacocoris chlorizans* Pz. (Hem. Het. Mirid.), prédateur des acariens phytophages. *Bulletin de la Société Entomologique Suisse/Mitteilungen der Schweizerischen Entomologischen Gesellschaft* 25, 257–259.

Geijskes, D.C. (1939) Beiträge zur Kenntnis der europäischen Spinnmilben (Acari, Tetranychidae), mit besonderer Berücksichtigung der niederlandischen Arten. *Mededelingen Landbouwhogeschool Wageningen* 42(4), 1–68.

Geispitz, K.F. (1968) Genetic aspects of variations in photoperiodic adaptations. In: Danileviska, A.S. (ed.) *Photoperiodic Adaptations in Insects and Acari, Volume 2*. Leningrad State University Press, Leningrad, USSR, pp. 52–79 (in Russian).

Geispitz, K.F. and Orlovskaja, E.E. (1971) Physiological characteristics of two types of diapause in the common spider mite *Tetranychus urticae* Koch (Acarina, Tetranychidae). In: Maavara, V. (ed.) *Cold-hardiness in Insects and Mites, Symposium, Tartu (Estonian SSR), April 19–21, 1971*. Academy of Sciences of the Estonian SSR, Tartu, Estonia, pp. 16–20 (in Russian).

Gemrich, E.G., Lamar Lee, B., Tripp, T.L. and Van de Streek, E. (1976) Relationship between formamidine structure and insecticidal, miticidal and ovicidal activity. *Journal of Economic Entomology* 69, 301–306.

Gencsoylu, I., Liu W., Amin Usmani, K. and Knowles, C.O (1998) Toxicity of acaricides to the bulb mite *Rhizoglyphus echinopus* (Acari: Acaridae). *Experimental and Applied Acarology* 22, 343–351.

Genevest, R., Migeon, A., Porter, W., Navajas, M. and Meynard, C.N. (2012) Modeling the potential distribution of several Tetranychidae species under current and future climatic conditions. In: Schausberger, P., Walzer, A., Strodel, M.A. and Peneder, S. (eds) *7th Symposium of the European Association of Acarologists (EURAAC), July 9–13 2012, Vienna, Austria: Acari in a Changing World. Program, Abstracts, Participants*. University of Natural Resources and Life Sciences, Vienna, Austria, p. 38. Available at: http://euraac.boku.ac.at/SympVienna/files/ProgramBook.pdf (accessed 14 January 2015).

Genini, M. and Baillod, M. (1987) Introduction de souches résistantes de *Typhlodromus pyri* (Scheuten) et *Amblyseius andersoni* Chant (Acari: Phytoseiidae) en vergers de pommiers. *Revue Suisse de Viticulture Arboriculture Horticulture* 19, 115–123.

Gentry, C.R., Dutcher, J.D., Littrell, R.H. and Worley, R.E. (1982) Low-pressure trunk injection of dicrotophos and solubilized benomyl for pest control on mature pecan trees. *Journal of Economic Entomology* 75, 611–615.

Genty, P. and Reyes, E. (1977) A new oil-plam mite: the eriophyid *Retracrus elaeis* Keifer (biology, connection with orange spotting, effect on production, treatments). *Oléagineux* 32, 255–262.

Georgala, M.B. (1955) *The Biology of Orchard Mites in the Western Cape Province*. Science Bulletin No. 360, Department of Agricultural Technical Services, Republic of South Africa. Government Printer, Pretoria.

Georgala, M.B. (1958) *A Contribution to the Biology of the Mite,* Bryobia praetiosa *Koch*. Science Bulletin No. 367, Department of Agricultural Technical Services, Republic of South Africa. Government Printer, Pretoria.

Georghiou, G.P. (1959) Plant-feeding mites of Cyprus. *FAO Plant Protection Bulletin* 7(12), 153–160.

Georghiou, G.P. and Saitō, T. (eds) (1983) *Pest Resistance to Pesticides*. Plenum Press, New York.

Georghiou, G.P. and Taylor, C.E. (1977a) Genetic and biological influences in the evolution of insecticide resistance. *Environmental Entomology* 70, 319–323.

Georghiou, G.P. and Taylor, C.E. (1977b) Operational influences in the evolution of insecticide resistance. *Environmental Entomology* 70, 653–658.

Georgis, R., Abdul Wahab, W. and El-Haidari, H.S. (1976) Observations on biology of *Stethorus gilvifrons* Muls., a predator of *Tetranychus atlanticus* McG. *Yearbook of Plant Protection Research, Iraq Ministry of Agriculture and Agrarian Reform* No. 1, 9.

Gerdeman, B.S. and Tanigoshi, L.K. (2004) *Bamboo Spider Mites: Biology and Management in the Pacific Northwest*. Extension Bulletin 1992E, College of Agricultural, Human, and Natural Resource Sciences, Washington State University, Pullman, Washington.

Gerding, P.M. (1992) *Acalitus essigi* (Hassan) (Acarina: Eriophyidae) presente en moras cultivadas y silvestres (*Rubus* spp), en Chile. *Agricultura Técnica (Santiago)* 52, 336–337.

Gergerich, R.C. and Kim, K.S. (1983) A description of the causal agent of rose rosette disease. *Arkansas Farm Research*, May–June 1983, p. 7.

Gergerich, R.C., Kim, K.S. and Kitijima, E.W. (1983) A particle of unique morphology associated with a disease of rose in northwest Arkansas. *Phytopathology* 73, 501–502.

Gerini, V. (1971) Contributo alla conoscenza della entomofauna nociva della Turchia. *Rivista di Agricultura Subtropicale e Tropicale* 65(10–12), 1–48.

Gerson, U. (1971) The mites associated with citrus in Israel. *Israel Journal of Entomology* 6, 5–22.

Gerson, U. (1979) Silk production in *Tetranychus* (Acari: Tetranychidae). In: Rodriguez, J.G. (ed.) *Recent Advances in Acarology, Volume I*. Academic Press, New York, pp. 177–188.

Gerson, U. (1985a) Other predaceous mites and spiders. In: Helle, W. and Sabelis, M.W. (eds) *Spider Mites. Their Biology, Natural Enemies and Control*. World Crop Pests, Volume 1A. Elsevier Science, Amsterdam, pp. 205–210.

Gerson, U. (1985b) Webbing. In: Helle, W. and Sabelis, M.W. (eds) *Spider Mites: Their Biology, Natural Enemies and Control*. World Crop Pests, Volume 1A. Elsevier Science, Amsterdam, pp. 223–232.

Gerson, U. (1986) The spider mite *Oligonychus mangiferus* found on mango in Israel. *Phytoparasitica* 14, 148.

Gerson, U. (1992) Biology and control of the broad mite, *Polyphagotarsonemus latus* (Banks) (Acari: Tarsonemidae). *Experimental and Applied Acarology* 13, 163–178.

Gerson, U. (2003) Acarine pests of citrus: overview and non chemical control. *Systematic and Applied Acarology* 8, 3–12.

Gerson, U. (2008) The Tenuipalpidae: an under-explored family of plant-feeding mites. *Systematic and Applied Acarology* 13, 83–101.

Gerson, U. (2014) Pest control by mites (Acari): present and future. *Acarologia* 54, 371–394.

Gerson, U. and Aronowitz, A. (1980) Feeding of the carmine spider mite on seven host plant species. *Entomologia Experimentalis et Applicata* 28, 109–115.

Gerson, U. and Aronowitz, A. (1981a) Spider mite webbing. Part IV. The effect of acaricides on spinning by the carmine spider mite *Tetranychus cinnabarinus* (Boisduval). *Pesticide Science* (now *Pesticide Management Science*) 12, 211–214.

Gerson, U. and Aronowitz, A. (1981b) Spider mite webbing. V. The effect of various host plants. *Acarologia* 22, 277–281.

Gerson, U. and Cohen, E. (1989) Resurgences of spider mites (Acari: Tetranychidae) induced by synthetic pyrethroids. *Experimental and Applied Acarology* 6, 29–46.

Gerson, U. and Smiley, R.L. (1990) *Acarine Biocontrol Agents, An Illustrated Key and Manual*. Chapman and Hall, London.

Gerson, U. and Vacante, V. (2012) Acari. In. Vacante, V. and Gerson, U. (eds) *Integrated Control of Citrus Pests in the Mediterranean Region*. Bentham Science, Sharja, United Arab Emirates, pp. 88–108.

Gerson, U. and van de Vrie, M. (1979) The potential of mites in the biological control of mite and insect hosts. In: Piffl, E. (ed.) *Proceedings of the 4th International Congress of Acarology, Saalfelden (Austria)*. Akadémiai Kiadó, Budapest, pp. 629–635.

Gerson, U. and Weintraub, P.G. (2007) Mites for the control of pests in protected cultivation. *Pest Management Science* 63, 658–676.

Gerson, U., Kenneth, R. and Muttath, T.I. (1979) *Hirsutella thompsonii*, a fungal pathogen of mites. II. Host–pathogen interactions. *Annals of Applied Biology* 91, 29–40.

Gerson, U., Venezian, A. and Blumberg, D. (1983) Phytophagous mites on date palms in Israel. *Fruits* 38, 133–135.

Gerson, U., Yathom, S., Capua, S. and Thorens, D. (1985) *Rhizoglyphus robini* Claparède (Acari: Astigmata: Acaridae) as a soil mite. *Acarologia* 26, 371–380.

Gerson, U., Smiley, R.L. and Ochoa, R. (2003) *Mites (Acari) for Pest Control*. Blackwell Publishing, Oxford, UK.

Geurs, M., Bongers, J. and Brussard, L. (1991) Variations of the heptane flotation method for improved efficiency of collecting microarthropods from a sandy loam soil. *Agriculture, Ecosystems and Environment* 34, 213–221.

Ghai, S. and Shenhmar, M. (1984) A review of the world fauna of Tenuipalpidae (Acarina: Tetranychoidea). *Oriental Insects* 18, 99–172.

Ghanekar, A.M., Sheila, V.K., Beniwal, S.P.S., Reddy, M.V. and Nene, Y.L. (1992) Sterility mosaic of pigeonpea. In: Singh, U.S., Mukhopadhyay, A.N., Kumar, J. and Chaube, H.S. (eds) *Plant Diseases of International Importance, Volume I (Diseases of Cereals and Pulses)*. Prentice Hall, Englewood Cliffs, New Jersey, pp. 415–428.

Ghasemi Moghadam, S., Ahadiyat, A., Sarafrazi, A. and Rahmani, H. (2012) Fauna of Tetranychoidea (Acari: Prostigmata) and their predator mites (Acari: Mesostigmata, Phytoseiidae), and survey of infestation rate on trees and shrubs in parks of Teheran, Iran. In: Murariu, D., Adam, C., Chişamera, G., Iorgu, E., Popa, L.O. and Popa, O.P. (eds) *Annual Zoological Congress of "Grigore Antipa" Museum, 21–23 November 2012, Bucharest, Romania, Book of Abstracts*. "Grigore Antipa" National Museum of Natural History, Bucharest, p. 135. Available at: http://www.czga.ro/pozepagini/CZGA_2012___Book_of_Abstracts.pdf (accessed 5 May 2015).

Ghazy, N.A. and Amano, H. (2014) Rapid cold hardening response in the predatory mite *Neoseiulus californicus*. *Experimental and Applied Acarology* 63, 535–544.

Gheibi, M. and Hesami, S. (2011) Life table and reproductive table parameters of *Scolothrips longicornis* (Thysanoptera: Thripidae) as a predator of two-spotted spider mite, *Tetranychus turkestani* (Acari: Tetranychidae). *World Academy of Science, Engineering and Technology* 5(12): 13183. Available at: http://waset.org/publications/13183/life-table-and-reproductive-table-parameters-of-scolothrips-longicornis-thysanoptera-thripidae-as-a-predator-of-two-spotted-spider-mite-tetranychus-turkestani-acari-tetranychidae- (accessed 7 May 2015).

Ghobrial, A. and Dittrich, V. (1980) Early and late pest complexes on cotton, their control by aerial and ground application of insecticides and side-effects on the predator fauna. *Zeitschrift für Angewandte Entomologie* 90, 306–313.

Ghosh, C.C. (1940) *Insect Pests of Burma*. Supt. Government Printer and Stationer, Rangoon, Burma.

Ghosh, S.K., Prakash, A. and Rao, J. (1998) Efficacy of some chemical pesticides against rice tarsonemid mite *Steneotarsonemus spinki* Smiley (Acari: Tarsonemidae) under controlled conditions. *Environment and Ecology* 16, 913–915.

Ghoshal, S., Gupta, S.K. and Mukherjee, B. (2006) Life cycle of *Tetranychus neocaledonicus* André on *Rhizophora mucronata* Lamk, under laboratory condition at controlled temperature. *Records of the Zoological Survey of India* 106(Part 2), 61–65.

Ghoshal, S., Barman, S. and Saha, M. (2011) Biochemical changes of some important organic, mineral and inorganic compounds in the leaves of mango (*Mangifera Indica*) due to the infestation of *Oligonychus mangiferus* (Rahman and Sapra). *IUP Journal of Environmental Sciences* 5(2) 59–63.

Giannotti, O. (1971) O uso dos insecticidas e acaricidas. *Boletín de la Sociedad Entomológica del Perú* 6, 85–103.

Gibson, F.A. (1967) Mites of Camellia. *Agricultural Gazette of New South Wales* 78(11), 663–664.

Gibson, R.W. (1974) Studies on the feeding behavior of the eriophyid mite *Abacarus hystrix*, a vector of grass virus. *Annals of Applied Biology* 78, 213–217.

Gibson, R.W. (1975) Measurement of eriophyid mite populations on ryegrass using ultrasonic radiation. *Transactions of the Royal Entomological Society of London* 127, 31–32.

Gibson, R.W. (1976a) Effects of cutting height on the abundance of the eriophyid mite *Abacarus hystrix* (Nalepa) and the incidence of ryegrass mosaic virus in ryegrass. *Plant Pathology* 25, 152–156.

Gibson, R.W. (1976b) Infection of ryegrass plants with ryegrass mosaic virus decreases numbers of the mite vector. *Annals of Applied Biology* 83, 485–488.

Gibson, R.W. (1980) A comparison of the relative importance of between- and within-crop spread of ryegrass mosaic virus by eriophyid mites. In: Plumb, R.T. (ed.) *Proceedings of the 3rd Conference on Virus Diseases of Gramineae in Europe, Rothamsted, 28–30 May 1980*. Rothamsted Experimental Station, Harpenden, UK, pp. 99–102.

Gibson, R.W. (1981) Rapid spread by mites of ryegrass mosaic virus from old sward to seedling ryegrass and its prevention by aldicarb. *Plant Pathology* 30, 25–29.

Gibson, R.W. and Heard, A.J. (1974) Ryegrass mosaic virus, a possible relation between RMV incidence and age of ryegrass swards. In: Hirst, J.M. (ed.) *Rothamsted Experimental Station Report, Part 1*. Rothamsted, UK, p. 232.

Gibson, R.W. and Plumb, R.T. (1976) The transmission and effect on yield of ryegrass mosaic virus in a filtered air environment. *Annals of Applied Biology* 82, 79–84.

Gibson, W.W. and Painter, R.H. (1957) Transportation by aphids of the wheat curl mite, *Aceria tulipae* (K.), a vector of wheat streak mosaic virus. *Journal of the Kansas Entomological Society* 30, 147–153.

Gigja, G. (1964) Pests in stockfish. Department of Agriculture Reports, Series B. N.18, 5–41. University Research Institute, Reykjavik, Iceland.

Giles, D.K., Gardner, H.E. and Studer, H.E. (1995) Mechanical release of predacious mites for biological pest control in strawberries. *Transactions of the American Society of Agricultural Engineers* 38, 1289–1296.

Gillespie, D.J. (1993) Redlegged earth mite (*Halotydeus destructor*) resistance in annual pasture legumes. In: Delfosse, E. (ed.) *Pests of Pastures: Weed, Invertebrate and Disease Pests of Australian Sheep Pastures*. CSIRO (Commonwealth Scientific and Industrial Research Organisation), Melbourne, Australia, pp. 211–213.

Gillespie, D.J. (1994) Identification and development of annual legume varieties with resistance to redlegged earth mite. In: McDonald, G. and Hoffmann, A.A. (eds) *Proceedings of the Second National Workshop on Redlegged Earth Mite, Lucerne Flea and Blue Oat Mite, October 1994, Rutherglen, Victoria, Australia*. Department of Agriculture, Rutherglen, Australia, pp. 1–4.

Gillespie, D.R., Roitberg, B., Basalyga, E., Johnstone, M., Opit, G., Rodgers, J. and Sawyer, N. (1998) *Biology and Application of* Feltiella acarisuga *(Vallot) (Diptera: Cecidomyiidae) for Biological Control of Twospotted Spider Mites on Greenhouse Vegetable Crops*. Technical Report 145, Pacific Agri-food Research Centre (Agassiz), British Columbia, Canada.

Gillespie, D.R., Opit, G. and Roitberg, B. (2000) Effects of temperature and relative humidity on development, reproduction, and predation in *Feltiella acarisuga* (Vallot) (Diptera: Cecidomyiidae). *Biological Control* 17, 132–138.

Gillespie, R.L., Roberts, D.E. and Bentley, E.M. (1997) Population dynamics and dispersal of wheat curl mites (Acari: Eriophyidae) in north central Washington. *Journal of the Kansas Entomological Society* 70, 361–364.

Gillet, J.L., HansPetersen, H.N., Leppla, N.C. and Thomas, D.D. (2006) *Grower's IPM guide for Florida Tomato and Pepper Production*. IPM Florida, IFAS (Institute of Food and Agricultural Sciences) Extension, University of Florida, Gainesville, Florida. Available at: http://ipm.ifas.ufl.edu/resources/success_stories/T&PGuide/index.shtml (accessed 2 August 2014).

Gilliat, F.C. (1935) Some predators of the European red mite, *Paratetranychus pilosus* C. and F. in Nova Scotia. *Canadian Journal of Research* 13, 19–38.

Gilmer, R.M. and McEwen, F.L. (1958) Chlorotic fleck, an eriophyid mite injury of myrobalan plum. *Journal of Economic Entomology* 5, 335–341.

Gilmore. J.E. (1965) Preliminary field evaluation of a noninclusion virus for control of the citrus red mite. *Journal of Economic Entomology* 58, 113–1140.

Gilrein, D. (2006) Lewis mites on poinsettia. *GPN (Greenhouse Product News)*, August 2006, pp. 63–66. Available at: http://www.gpnmag.com/sites/default/files/lewismites.pdf (accessed 2 August 2014).

Gilstrap, F.E. and Friese, D.D. (1985) The predatory potential of *Phytoseiulus persimilis*, *Amblyseius californicus*, and *Metaseiulus occidentalis* (Acarina: Phytoseiidae). *International Journal of Acarology* 11, 163–168.

Gilstrap, F.E. and Oatman, E.R. (1976) The bionomics of *Scolothrips sexmaculatus* (Pergande) (Thysanoptera: Thripidae), an insect predator of spider mites. *Hilgardia* 44(2), 26–60.

Gilstrap, F.E., Friese, D.D., Summy, K.R. and Armstrong, A.A. (1977) Persistence of *Amblyseius californicus*, *Phytoseiulus persimilis* and *Typhlodromus occidentalis* released in field populations of Banks grass mite on corn. *Southwestern Entomologist* 2, 159–163.

Girija, V.K., Umamaheswaran, K. and Paul, A. (2001) Cellotape embedding technique for assessment of population of coconut eriophyid mite, *Aceria* (*Eriophyes*) *guerreronis* Keifer. *Insect Environment* 7, 35.

Girling, D.J. (1977) *Report on the Possibilities of Finding Natural Enemies to Control Gorse,* Ulex europaeus L. *in New Zealand.* Report of Commonwealth Institute of Biological Control, London.

Girling, D.J., Bennett, F.D. and Yaseen, M. (1978) Biological control of the green cassava mite, *Mononychellus tanajoa* (Bondar) (Acarina: Tetranychidae) in Africa. In: Brekelbaum, T., Bellotti, A. and Lozano, J.C. (eds) *Proceedings CASSAVA Protection Workshop, 7–12 November 1977, Cali, Colombia.* CIAT (Centro Internacional de Agricultura Tropical), Cali, Colombia, pp. 165–170.

Girolami, V. (1981) Danni, soglie d'intervento, controllo degli acari della vite. *Atti del Convegno "La Difesa Integrata della Vite", 3–4 Dicembre 1981, Latina, Italy*, pp. 111–143.

Girolami, V. and Mozzi, A. (1984) Distribution, economic threshold and sampling methods of *Panonychus ulmi* (Koch). In: Cavalloro, R. (ed.) *Statistical and Mathematical Methods in Population Dynamics and Pest Control. Proceedings of a Meeting of the EC Experts' Group/Parma 26–28 October 1983.* A.A. Balkema, Rotterdam, the Netherlands, pp. 90–101.

Girolami, V., Duso, C., Refatti, E. and Osler, R. (1989) *Lotta Integrata in Viticoltura. Malattie della Vite.* IRIPA (Istituto Regionale d'Istruzione Professionale in Agricoltura de Veneto)/Federazione Regionale Coldiretti Veneto, Mestre, Italy.

Gispert, C., Perring, T.M. and Oldfield, G.N. (1997) Rearing *Eriophyes insidiosus* Keifer and Wilson (Acari: Eriophyoidea), a fastidiosus bud mite. *International Journal of Acarology* 23, 227–231.

Gispert, C., Oldfield, G.N., Perring, T.M. and Craemer, R. (1998) Biology of the transmission of peach mosaic virus by *Eriophyes insidiosus* (Acari: Eriophyidae). *Plant Disease* 82, 1371–1374.

Gispert, C., Farrar, C. and Perring, T.M. (2001) Seasonal abundance of the Banks grass mite *Oligonychus pratensis* (Banks) (Prostigmata: Tetranychidae) and a predatory mite, and their response to sulfur treatment on commercial date palms *Phoenyx dactilifera* L. in Southern California. In: Halliday, R.B., Walter, D.E., Proctor, H.C., Norton, R.A. and Colloff, M.J. (eds) *Acarology: Proceedings of the 10th International Congress of Acarology, 5–10 July 1998*, Canberra. CSIRO Publishing, Collingwood, Australia, pp. 403–408.

Gispert Galván, M.C., Perring, T.M., Zarate de Lara, G. and Llanderal Cazares, C. (1989) Efecto del riego en la fluctuación poblacional del ácaro del tomate (*Aculops lycopersici* Massee). *Agrociencia* 76, 153–165.

Giustina, W.D. (1981) The pests of the aerial parts of cucumber grown under cover. *Phytoma* 328, 9–12.

Glancey, B.M. (1958) Studies on embryonic diapause in the legume mite, *Petrobia apicalis* (Banks). MSc. thesis, Louisiana State University, Baton Rouge, Louisiana.

Glavendekić, M., Mihajlović, L. and Petanović, R. (2005) Introduction and spread of invasive mites and insects in Serbia and Montenegro. In: *Plant Protection and Plant Health in Europe: Introduction and Spread of Invasive Species. Symposium Held at Humboldt University, Berlin, Germany, 9–11 June 2005.* Symposium Proceedings No. 81, Humbolt University, Berlin, pp. 229–230.

Glinyanaya, E.I. (1972) The role of photoperiod in the reactivation of arthropods with summer and winter diapause. In: Goryshin N.I. (ed.) *Problems of Photoperiodism and Diapause in Insects.* Leningard State Univeristy, Leningrad, USSR, pp. 88–101 (in Russian).

Godse, D.B. and Patil, R.B. (1977) A species of *Cladosporium* infecting *Tetranychus ludeni* Zacher (Acarina: Tetranychidae). *Acarology Newsletter* 4, 7.

Goebel, O., Vergnet, C. and Heller, J.J. (2001) Lutte contre l'erinose et l'acariose de la vigne. Interet des traitements de debut de saison au soufre. *Phytoma — La Défense des Végétaux* 537, 39–41.

Goeden, R.D. (1983) Critique and revision of Harris' scoring system for selection of insects agents in biological control of weeds. *Protection Ecology* 5, 287–301.

Goeden, R.D. (1988) A capsule history of biological control of weeds. *Biocontrol News and Information* 9, 55–61.

Goes, A. de, Barros, J.C. da S.M. and Alves, R.C.P. (1985) *A Leprose dos Citros: o Ácaro e seu Controle.* Comunicado Técnico No. 148, Pesagro Rio (Empresa de Pesquisa Agropecuária do Estado do Rio de Janeiro), Rio de Janeiro.

Goh, K.S. and Lange, W.H. (1989) Microarthropods associated with insecticide-treated and untreated artichoke fields in California. *Journal of Economic Entomology* 82, 621–625.

Goka, K. (1998) Mode of inheritance of resistance to three new acaricides in the Kanzawa spider mite, *Tetranychus kanzawai* Kishida (Acari: Tetranychidae). *Experimental and Applied Acarology* 22, 699–708.

Goka, K. and Takafuji, A. (1995) Allozyme variations among populations of the two-spotted spider mite, *Tetranychus urticae* Koch, in Japan. *Applied Entomology and Zoology* 30, 567–579.

Göksu, M.E. (1968) Studies on the bionomics, control, distribution and food-plants of the hawthorn mite, *Tetranychus viennensis* Zacher, in the Marmara region. *Bitki Koruma Bülteni* 8, 194–213 (in Turkish).

Goldarazena, A., Aguilar, H., Kutuk, H. and Childers, C.C. (2004) Biology of three species of *Agistemus* (Acari: Stigmaeidae): life table parameters using eggs of *Panonychus citri* or pollen of *Malephora crocea* as food. *Experimental and Applied Acarology* 32, 281–291.

Goldsmid, J.M. (1962) *The Mites of the Federation of Rhodesia and Nyasaland.* Bulletin No. 2162, Federal Ministry of Agriculture, Federation of Rhodesia and Nyasaland, Salisbury, Rhodesia.

Goldstein, D.B. and Schlötterer, C. (1999) *Microsatellites, Evolution and Applications.* Oxford University Press, Oxford, UK.

Golik, Z. (1975) A study of the destructiveness of the fruit tree red spider mite, *Panonychus ulmi* (Koch) on apple. *Zeszyty Problemowe Postępów Nauk Rolniczych* 171, 15–34 (in Polish).

Golpayegani, A.Z., Saboori, A., Nowzari, J. and Kamali, K. (2003) Biology of *Amphitetranychus viennensis* (Zacher) (Acari: Tetranychidae) in Baraghan Region of Karaj, Iran. *Acarologia* 14, 69–71.

Gólya, G. and Kozma, E. (2001) Abiotikus tényezők szerepe az almatermésűek levélatkájának (*Aculus schlechtendali* (Nalepa)) populációdinamikájában (Role of abiotic factors in population dynamics of apple rust mite (*Aculus schlechtendali* (Nalepa)). *Növényvédelem* 37, 345–350 (in Hungarian).

Gólya, G., Kozma, E. and Kovács, A. (2001) New data to the knowledge on the eriophyoid fauna on woody plants in Hungary (Acari: Eriophyoidea). *Acta Phytopathologica et Entomologica Hungarica* 36, 89–91.

Gomez, C. and Mizell, R.F. III (2008) *Featured Creatures: Clover Mite,* Bryobia praetiosa *Koch (Arachnida: Acari: Tetranychidae).* Publication Number EENY-437, Institute of Food and Agricultural Sciences, University of Florida, Gainesville, Florida. Available at : http://entnemdept.ufl.edu/creatures/orn/mites/clover_mite.htm (accessed 8 May 2015).

Gomez, D.R.S. and Nasca, A.J. (1983) Primera cita del hongo patógeno de ácaros, *Hirsutella thompsonii* (Fisher, 1950) para la República Argentina. *CIRPON – Revista de Investigación* 1, 137–141.

Gomez-Moya, C.A. and Ferragut, F. (2007) Predator–prey dynamics of the spider mites and predatory phytoseiid mites (Acari: Tetranychidae, Phytoseidae) in horticultural crops. PhD thesis, Universidad Politécnica de Valencia, Valencia, Spain.

Gonçalves, L.D., Maluf, W.R., Graças Cardoso, M. das, Vilela de Resende, J.T., Castro, E.A. de, Macedo Santos, N., Rodrigues do Nascimento, I. and Ventura Faria, M. (2006) Relação entre zingibereno, tricomas foliares e repelência de tomateiros a *Tetranychus evansi*. *Pesquisa Agropecuária Brasileira* 41, 267–273.

Gonçalves, M., Cavaco, M., Fernandes, J.E., Soares, C. and Ramos, N. (2002) *Ácaro do texas (*Eutetranychus banksi *McGregor, 1914): nova especie de ácaro na cultura dos citrinos no Algarve*. Direção Geral de Proteção das Culturas, Oeiras, Portugal.

Gonçalves, M.E.C., Oliveira, J.V., Barros, R. and Torres, J.B. (2001) Efeito de extratos vegetais sobre estágios imaturos e fêmeas adultas de *Mononychellus tanajoa* (Bondar) (Acari: Tetranychidae). *Neotropical Entomology* 30, 305–309.

Gonçalves, M.I.F., Maluf, W., Gomes, L.A.A. and Barbosa, L.V. (1998) Variation of 2-tridecanone level in tomato plant leaflets and resistance to two mite species (*Tetranychus* sp). *Euphytica* 104, 33–38.

Gondim, M.G.C. Jr and Moraes G.J. de (2003) Life cycle of *Retracrus johnstoni* Keifer (Acari: Phytoptidae). *Neotropical Entomology* 32, 197–201.

Gondim, M.G.C. Jr, Flechtmann, C.H.W. and Moraes, G.J. de (2000) Mite (Arthropoda: Acari) associates of palms (Arecaceae) in Brazil. IV. Descriptions of four new species in the Eriophyoidea. *Systematic and Applied Acarology* 5, 99–110.

Gontijo, L.M., Nechols, J.R, Margolies, D.C. and Cloyd, R.A. (2012) Plant architecture and prey distribution influence foraging behavior of the predatory mite *Phytoseiulus persimilis* (Acari: Phytoseiidae). *Experimental and Applied Acarology* 56, 23–32.

González, H.R. (2001) Joint chemical management of the western flower thrips *Frankliniella occidentalis* (Pergande) and the grape flat mite *Brevipalpus chilensis* Baker on table grapes. *Revista Frutícola* 22, 51–61.

González, M.I., Alvarado, M., Duran, J.M., Rosa, A. de la and Serrano, A. (2000) Los eriofidos (Acarina, Eriophidae) del olivar de la provincia de Sevilla. Problematica y control. *Boletín de Sanidad Vegetal, Plagas* 26, 203–214.

González, R.H. (1961) *Contribución al Conocimiento de los Ácaros del Manzano en Chile Central*. Boletin Técnico No. 11, Estación Experimental Agronómica, Universidad de Chile, Santiago.

González, R.H. (1968a) Acaros plantícolas del género *Tenuipalpus* en Chile (Acarina: Tenuipalpidae). *Revista Chilena de Entomologia* 6, 37–46.

González, R.H. (1968b) *Biologia y Control de la Falsa Aranita de la Vid,* Brevipalpus chilensis *Baker*. Boletin Técnico No. 1, Estación Experimental Agronómica, Universidad de Chile, Santiago.

González, R.H. (1969) Biological control of citrus pest in Chile. In: Chapman, H.D. (ed.) *Proceedings of the International Citrus Symposium, Riverside, California, March 16–26, 1968: in Celebration of the Centennial of the University of California, Volume 2*. University of California, Riverside, California, pp. 839–847.

González, R.H. (1975) Revision of the *Brevipalpus phoenicis* 'complex' with descriptions of new species from Chile and Thailand (Acarina, Tenuiplapidae). *Acarologia* 17, 81–91.

González, R.H. (1983a) *Manejo de Plagas de la Vid*. Publicaciones en Ciencias Agricolas 13, Universidad de Chile, Santiago.

González, R.H. (1983b) La falsa aranita de la vid *Brevipalpus chilensis* Baker (Acarina, Tenuipalpidae). *Revista Frutícola* 4, 61–65.

González, R.H. (1985) Acaros eriofidos del manzano y peral en Chile (Acarina: Eriophyidae). *Revista Chilena de Entomologia* 12, 77–84.

González, R.H. (1989) *Insectos y Ácaros de Importancia Agrícola y Cuarantenaria en Chile*. Facultad de Ciencias Agrarias y Forestales, Universidad de Chile, Santiago.

González, R.H. (2006) Biology, quarantine risks and control alternatives of the grape flat mite, *Brevipalpus chilensis* Baker (*Acarina: Tenuipalpidae*). *Revista Frutícola* 27, 77–88.

González, R.H. and Barria, G. (1999) Abamectina: insecticida–acaricida de origen biológico en el control de la falsa aranita de la vid y del trips de California y degradación de residuos en nectarinos. *Revista Frutícola* 20, 5–10.

González, R.H. and Flechtmann, C.H.W. (1977) Revision de los ácaros fitófagos en el Perú y descripción de un nuevo genero de Tetranychidae (Acari). *Revista Peruana de Entomología* 20, 67–71.

González-Zamora, J., López, C. and Avilla, C. (2011) Population studies of arthropods on *Melia azedarach* in Seville (Spain), with special reference to *Eutetranychus orientalis* (Acari: Tetranychidae) and its natural enemies. *Experimental and Applied Acarology* 55, 389–400.

Goodwin, S. (1982) Chemical control of bunch mite, *Brevipalpus lewisi* McGregor (Tenuipalpidae), on grapes in New South Wales. *General and Applied Entomology* 14, 41–44.

Goonewardene, H.F. and Kwolek, W.F. (1975) Sampling dormant apple twigs to estimate the density of eggs of the European red mite. *Environmental Entomology* 4, 923–928.

Gopal, M., Mukherjee, I., Roy, N.K., Das, S.C. and Gope, B. (1987) Efficacy of fluvalinate against pests of tea and its residue studies. *Pesticides (Bombay)* 28, 33–36.

Gope, B. and Borthakur, M. (1988) Ovicidal action of some acaricides on the egg of pink mite of tea. *Two and a Bud* 35, 52.

Gope, B. and Das, S.C. (1992) Biology and effect of pruning and skiffing on the distribution of the scarlet mite, *Brevipalpus phoenicis* (Geijskes), on tea in North-East India. *Two and a Bud* 39, 23.

Gordon, R.D. (1982) New species and new synonymy in neotropical *Stethorus* Weise (Coleoptera: Coccinellidae). *Coleopterists Bulletin* 36, 121–126.

Gordon, S.C. (1981) *Raspberry Leaf and Bud Mite*. Leaflet No. 790, Ministry of Agriculture, Fisheries and Food, London.

Gordon, S.C. and Taylor, C.E. (1976) Some aspects of the biology of the raspberry leaf and bud mite (*Phyllocoptes* (*Eriophyes*) *gracilis* Nal.) (Eriophyidae) in Scotland. *Journal of Horticultural Science* 51, 501–508.

Gordon, S.C. and Taylor, C.E. (1977) Chemical control of the raspberry leaf and bud mite *Phyllocoptes gracilis* (Nal.) (Eriophyidae). *Journal of Horticultural Science* 52, 517–523.

Gotoh, T. (1987a) Life-history parameters of two species of the genus *Eotetranychus* (Acarina: Tetranychidae) on deciduous trees. *Applied Entomology and Zoology* 22, 45–51.

Gotoh, T. (1987b) Annual life cycle of the spider mite, *Eotetranychus uncatus* Garman (Acarina: Tetranychidae). *Applied Entomology and Zoology* 22, 52–58.

Gotoh, T. (1997) Annual life cycles of populations of the two-spotted spider mite, *Tetranychus urticae* Koch (Acari: Tetranychidae) in four Japanase pear orchards. *Applied Entomology and Zoology* 32, 207–216.

Gotoh, T. and Gomi, K. (2000) Population dynamics of *Tetranychus kanzawai* (Acari: Tetranychidae) on hydrangea. *Experimental and Applied Acarology* 24, 337–350.

Gotoh, T. and Gomi, K. (2003) Life-history traits of the Kanzawa spider mite *Tetranychus kanzawai* (Acari: Tetranychidae). *Applied Entomology and Zoology* 38, 7–14.

Gotoh, T. and Higo, Y. (1997) Differences in host range and reproductive compatibility among populations of *Panonychus mori* Yokoyama (Acari: Tetranychidae). *International Journal of Acarology* 23, 119–125.

Gotoh, T. and Nagata, T. (2001) Development and reproduction of *Oligonychus coffeae* (Acari: Tetranychidae) on tea. *International Journal of Acarology* 27, 293–298.

Gotoh, T. and Shida, T. (2007) Life cycles and interactions in spider mites (Acari: Tetranychidae) on dwarf bamboo, *Sasa senanensis* (F. & S.) (Poaceae), in Japan. *International Journal of Acarology* 33, 259–273.

Gotoh, T., Abe, T., Kurihara, A. and Suzuki, M. (1995) Genetic incompatibility in local populations of the spider mite, *Tetranychus quercivorus* Ehara *et* Gotoh (Acari: Tetranychidae). *Applied Entomology and Zoology* 30, 361–368.

Gotoh, T., Kamoto, T. and Oku, H. (1996a) Termination of embryonic diapause by physical and chemical stimuli in the spider mite *Panonychus bambusicola* Ehara and Gotoh (Tetranychidae). In: Mitchell, R., Horn, D.J., Needham, G.R. and Welbourn, W.C. (eds) *Acarology IX, Proceedings of the 9th International Congress on Acarology*. Ohio Biological Survey, Columbus, Ohio, pp. 667–669.

Gotoh, T., Takafuji, A. and Gomi, K. (1996b) Tetranychid mites of Okinawa Island (Acari: Tetranychidae). *Journal of the Acarological Society of Japan* 5, 89–97.

Gotoh, T., Gomi, K. and Nagata, T. (1999) Incompatibility and host plant differences among populations of *Tetranychus kanzawai* Kishida (Acari: Tetranychidae). *Applied Entomology and Zoology* 34, 551–561.

Gotoh, T., Kitashima, Y., Goka, K. and Nagata, T. (2001) Susceptibility of the red spider mite, *Oligonychus coffeae* (Acari: Tetranychidae), to acaricides on tea plants in Japan. *International Journal of Acarology* 27, 303–307.

Gotoh, T., Noda, H. and Hong, X.Y. (2003a) *Wolbachia* distribution and cytoplasmic incompatibility based on a survey of 42 spider mite species (Acari: Tetranychidae) in Japan. *Heredity* 91, 208–216.

Gotoh, T., Ishikawa, Y. and Kitashima, Y. (2003b) Life-history traits of the six *Panonychus* species from Japan (Acari: Tetranychidae). *Experimental and Applied Acarology* 29, 241–252.

Gotoh, T., Kitashima, Y. and Adachi, I. (2004a) Geographic variation of susceptibility to acaricides in two spider mite species, *Panonychus osmanthi* and *P. citri* (Acari: Tetranychidae) in Japan. *International Journal of Acarology* 30, 55–61.

Gotoh, T., Yamaguchi, K. and Mori, K. (2004b) Effect of temperature on life history of the predatory mite *Amblyseius* (*Neoseiulus*) *californicus* (Acari: Phytoseiidae). *Experimental and Applied Acarology* 32, 15–30.

Gotoh, T., Noda, H. and Ito, S. (2007a) *Cardinium* symbionts cause cytoplasmic incompatibility in spider mites. *Heredity* 98, 13–20.

Gotoh, T., Kitashima, Y., Goka, K. and Nagata, T. (2007b) Miticide susceptibility of the red spider mite, *Oligonycus coffeae* (Acari: Tetranychidae). In: Morales-Malacara, J.B., Behan-Pelletier, V., Ueckermann, E., Pérez, T.M., Estrada-Venegas, E.G. and Badii, M. (eds) *Acarology XI: Proceedings of the International Congress, 8–13 September 2002, Mérida, Universidad Nacional Autonoma de México*. Sociedad Latinoamericana de Acarología, México, pp. 529–533.

Gotoh, T., Kitashima, Y. and Goka, K. (2007c) Tetranychus mite species identification using esterase and phosphoglucomutase zymograms. *Applied Entomology and Zoology* 42, 579–585.

Gotoh, T., Sugasawa, J., Noda, H. and Kitashima, Y. (2007d) *Wolbachia*-induced cytoplasmic incompatibility in Japanese populations of *Tetranychus urticae* (Acari: Tetranychidae). *Experimental and Applied Acarology* 41, 1–16.

Gotoh, T., Korenaga, T., Ikejima, S. and Hoshino, T. (2008) Acaricide mediated suppression of *Panonychus citri* (McGregor) (Acari: Tetranychidae) in pear orchards in Japan. *International Journal of Acarology* 34, 393–402.

Gotoh, T., Araki, R., Boubou, A., Migeon, A., Ferragut, F. and Navajas, M. (2009) Evidence of co-specificity between *Tetranychus evansi* and *Tetranychus takafujii* (Acari: Prostigmata, Tetranychidae): comments on taxonomic and agricultural aspects. *International Journal of Acarology* 35, 485–501.

Gotoh, T., Sugimoto, N., Pallini, A., Knapp, M., Hernández-Suarez, E., Ferragut, F., Ho, C.-C., Migeon, A., Navajas, M. and Nachman, G. (2010) Reproductive performance of seven strains of the tomato red spider mite *Tetranychus evansi* (Acari: Tetranychidae) at five temperatures. *Experimental and Applied Acarology* 52, 239–259.

Gotoh, T., Fujiwara, S. and Kitashima, Y. (2011) Susceptibility to acaricides in nine strains of the tomato red spider mite *Tetranychus evansi* (Acari: Tetranychidae). *International Journal of Acarology* 37, 93–102.

Goulart, C., Cia, E., Fuzatto, M.G., Luders, R.R. and Lourencao, A.L. (2009) Evaluation of cotton genotypes to injury caused by *Polyphagotarsonemus latus* Banks (Acari: Tarsonemidae). *Arquivos do Instituto Biológico (São Paulo)* 76, 61–65.

Gould, F. (1978a) Resistance of cucumber varieties to *Tetranychus urticae*: genetic and environmental determinants. *Journal of Economic Entomology* 71, 680–683.

Gould, F. (1978b) Predicting the future of resistance of crop varieties to pest populations: a case study of mites and cucumbers. *Environmental Entomology* 7, 622–626.

Gould, F. (1979) Rapid host range evolution in a population of the phytophagous mite *Tetranychus urticae* Koch. *Evolution* 33, 791–802.

Gould, H.J. and Kingham, H.G. (1965) Observations on diapause in *Tetranychus urticae* infesting cucumbers. *Plant Pathology* 14, 174–178.

Gould, H.J., Hussey, N.W. and Parr, N.J. (1969) Large scale control of *Tetranychus urticae* Koch on cucumbers by the predator *Phytoseiulus persimilis* A.H. In: Evans, G.O. (ed.) *Proceedings of the 2nd International Congress of Acarology, 19–25 July, 1967, Sutton Bonington, UK*. Akadémiaei Kiadó, Budapest, pp. 383–388.

Gould, S.J. (1981) Palaeontology plus ecology as palaebiology. In: May, R.M. (ed.) *Theoretical Ecology: Principles and Applications*, 2nd edn. Blackwell Scientific Publications, Oxford, UK, pp. 295–317.

Gould, S.J. (1989) *Wonderful Life. The Burgess Shale and the Nature of History*. W.W. Norton, New York.

Gouvea, A. de, Zanella, C.F. and Alves, L.F.A. (2004) Dinâmica populacional do ácaro *Oligonychus yothersi* (McGregor), 1914, (Acari: Tetranychidae) em plantas de erva-mate *Ilex paraguariensis* St. Hil. (Aquifoliacea), em dois vizinhos. *Scientia Agraria Paranaensis* 3, 35–48.

Gouvea, A. de, Boaretto, L.C., Zanella, C.F. and Alves, L.F.A. (2006) Populational dynamics of mites (Acari) in the mate-tea tree (*Ilex paraguariensis* St. Hil.: Aquifoliaceae). *Neotropical Entomology* 35, 101–111.

Gouvea, A. de, Bertoldo, G. and Alves, L.F.A. (2007) Presence–absence sampling plan for *Oligonychus yothersi* (McGregor) (Acari: Tetranychidae) on mate-tea orchard. *Neotropical Entomology* 36, 583–586.

Gowariker, V., Krishnamurthy, V. N., Gowariker, S. and Paranjape, K. (2014) *The Pesticide Encyclopedia*. CAB International, Wallingford, UK.

Gower, G.M.C., Hoffmann, A.A. and Weeks, A.R. (2008) Effectiveness of spring spraying targeting diapause egg production for controlling redlegged earth mites and other pests in pasture. *Australian Journal of Experimental Agriculture* 48, 1118–1125.

Goyal, M., Sadana, G.L. and Sharma, N.K. (1985) Influence of temperature on the development of *Brevipalpus obovatus* (Acarina: Tenuipalpidae). *Entomon* 10, 125–129.

GPDD (2007) Global Pest and Disease Database. NSF Center for Integrated Pest Management, North Carolina State University, Raleigh, North Carolina. Available at: https://www.gpdd.info/ (accessed 3 August 2014).

Grafton Cardwell, E.E., Ouyang, Y. and Striggow, R.A. (1997) Predaceous mites (Acari: Phytoseiidae) for control of spider mites (Acari: Tetranychidae) in nursery citrus. *Environmental Entomology* 26, 121–130.

Grahl, A. and Leuprecht, B. (1998) Untersuchungen zur Biologie der Tomatenrostmilbe *Aculus lycopersici* undihrer biologischen Bekämpfung. *Mitteilungen der Deutschen Phytomedizinischen Gesellschaft* 28, 41–42.

Grandjean, F. (1935a) Observations sur les acariens (1e séries). *Bulletin du Muséum National d'Histoire Naturelle (Paris)* 7(2), 119–126.

Grandjean, F. (1935b) Observations sur les oribates (8e série). *Bulletin du Muséum National d'Histoire Naturelle (Paris)* 8(2), 237–244.

Grandjean, F. (1935c) Les poils et les organes sensitifs portés par les pattes et le palpe chez les oribates. Première partie. *Bulletin de la Société Zoologique de France* 60, 6–39.

Grandjean, F. (1938) Sur l'ontogénie des Acariens. *Comptes Rendus de l'Académie des Sciences (Paris)* 206D, 146–150.

Grandjean, F. (1939a) Quelques genres d'acariens appartenent au groupe des Endeostigmata. *Annales des Sciences Naturelles, Zoologie* 11, 1–122.

Grandjean, F. (1939b) La chaetotaxie des pattes chez les Acaridiae. *Bulletin de la Société Zoologique de France* 64, 50–60.

Grandjean, F. (1939c) Les segments post-larvaires de l'hystérosoma chez les Oribates (Acariens). *Bulletin de la Société Zoologique de France* 64, 273–284.

Grandjean, F. (1940) Les poils et les organes sensitifs portés par les pattes et le palpe chez les Oribates. Deuxième partie. *Bulletin de la Société Zoologique de France* 65, 32–44.

Grandjean, F. (1941) La chaetotaxie comparée des pattes chez les Oribates (1re série). *Bulletin de la Société Zoologique de France* 66, 33–50.

Grandjean, F. (1942) La chaetotaxie comparée des pattes chez les Oribates (2re série). *Bulletin de la Société Zoologique de France* 67, 40–53.

Grandjean, F. (1946a) Les poils et les organes sensitifs portés par les pattes et le palpe chez les Oribates. Troisième partie. *Bulletin de la Société Zoologique de France* 71, 10–29.

Grandjean, F. (1946b) Observations sur les Acariens. *Mémoires du Muséum National d'Histoire Naturelle (Paris)* 18, 337–344.

Grandjean, F. (1947) L'origine pileuse des mors et la chaetotaxie de la mandibule chez les Acariens actinochitineux. *Comptes Rendus de l'Académie des Sciences (Paris)* 224, 1251–1254.

Grandjean, F. (1949) Observation et conservation des trés petit arthropodes. *Bulletin du Muséum National d'Histoire Naturelle (Paris)* 21(2), 363–370.

Grandjean, F. (1954) Etude sur les Palaeacaroides (Acariens, Oribates). *Mémoires du Muséum National d'Histoire Naturelle (Paris) (n.s.), Série A., Zoologie* 7, 179–274.

Grandjean, F. (1958) Sur le comportement et la notation des poils accessoires postérieurs aux tarses des Nothroïdes et d'autres acariens. *Archives de Zoologie Expérimentale et Générale* 96, 277–308.

Grandjean, F. (1970) Stase. Actinopiline. Rappel de ma classification des Acariens en trois groupes majeurs. *Acarologia* 11, 796–827.

Graniti, A. (1954) Ricerche sulle anomalie fogliari dell'olivo in Sardegna. I. Studio sulle alterazioni indotte da *Eriophyes oleae* Nalepa alle foglie d'olivo. *Annali della Sperimentazione Agraria* viii, 709–715.

Gratwick, M. and Southey, J.F. (eds) (1986) *Hot Water Treatment of Plant Material*, 3rd edn. ADAS/MAFF (Agricultural Development and Advisory Service/ Ministry of Agriculture, Fisheries and Food) Reference Book 201. Her Majesty's Stationery Office, London.

Gravena, S. (1998) Manejo ecológico de pragas dos citros – aspectos práticos. *Laranja* 19(1), 61–77.

Gravena, S., Coletti, A. and Yamamoto, P.T. (1993) Influence of green cover with *Ageratum conyzoides* and *Eupatorium pauciflorum* on predatory and phytophagous mites in citrus. *IOBC/WPRS Bulletin* 16(7), 104–114.

Gravena, S., Benetoli, I., Moreira, P.H.R. and Yamamoto, P.T. (1994) *Euseius citrifolius* Denmark & Muma predation on citrus leprosis mite *Brevipalpus phoenicis* (Geijskes) (Acari: Phytoseiidae: Tenuipalpidae). *Anais de Sociedade Entomológica do Brasil* 23, 209–218.

Gravena, S., Benvenga, S.R., Amorim, L.C.S., Silva, J.L. da and Araujo Junior, N. (2005) Manejo prático da resistência do ácaro da leprose dos citros. *Laranja* 26(1), 11–24.

Gray, E.G. and Sheil, R.S. (1987) *Narcissus* smoulder. A review of the disease and its association with bulb scale mite infestation. *Notes from the Royal Botanic Garden, Edinburgh* 44, 541–547.

Gray, E.G., Shaw, M.W. and Shiel, R.S. (1975) The role of mites in the transmission of smoulder in narcissus. *Plant Pathology* 24, 105–107.

Grbić, M., Van Leeuwen, T., Clark, R.M., Rombauts, S., Rouzé, P., Grbić, V., Osborne, E.J., Dermaw, W., Ngoc, P.C., Ortego, F., *et al.* (2011) The genome of *Tetranychus urticae* reveals herbivorous pest adaptations. *Nature* 479, 487–492.

Greco, N.M., Liljesthröm, G.G. and Sánchez, N.E. (1999) Spatial distribution and coincidence of *Neoseiulus californicus* and *Tetranychus urticae* (Acari: Phytoseiidae, Tetranychidae) on strawberry. *Experimental and Applied Acarology* 23, 567–580.

Green, E.E. (1890) *Insect Pests of Tea Plant*. Independent Press, Colombo, Ceylon.

Greenslade, P. and Clift, A. (2004) Review of pest arthropods recorded from commercial mushroom farms in Australia. *Australasian Mycologist* 23(3), 77–93.

Greenup, L.R. (1967) The blue oat mite, *Penthaleus major* (Dugès). *Agricultural Gazette of New South Wales* 78, 410–411.

Grewal, J.S. (1993) Biochemical factors responsible for susceptibility or resistance of various plants against the scarlet mite, *Brevipalpus phoenicis*: I. Amino acids analysis. *International Journal of Environmental Studies* 43, 293–295.

Gribble, G.W. (1999) The diversity of naturally occurring organobromine compounds. *Chemical Society Reviews* 28, 335–346.

Griffith, R. (1984) The problem of the coconut mite, *Eriophyes guerreronis*, in the coconut groves of Trinidad and Tobago. In: Webb, R., Knausenberger, W. and Yntema, L. (eds) *Proceedings of the Caribbean Food Crops Society 20th Annual Meeting, St. Crois, US Virgin Islands, October 21–26, 1984. Small Farm Systems in the Caribbean*. Eastern Caribbean Center, College of the Virgin Islands and Caribbean Food Crops Society, Mayaguez, Puerto Rico, pp. 128–132.

Griffiths, D.A. (1960) Some field habitats of mites of stored food products. *Annals of Applied Biology* 48, 134–144.

Griffiths, D.A. (1970) A further systematic study of the genus *Acarus* L., 1758 (Acaridae, Acarina), with a key to species. *Bulletin of the British Museum (Natural History), Zoology* 19, 85–118.

Griffiths, D.A. (1979) The morpho-species and its relationship to the biological species in the genus *Tyrophagus* (Acaridae, Acarina). In: Rodriguez, J.G. (ed.) *Recent Advances in Acarology, Volume I*. Academic Press, New York, pp. 199–212.

Griffiths, D.A. and Southgate, B.J. (1967) Mites of the genus *Tyrophagus* as pests of cucumber. In: *Pest Infestation Research 1967. The Report of the Pest Infestation Laboratory, Agricultural Research Council, UK*. Her Majesty's Stationery Office, London, p. 3.

Griffiths, D.A., Atyeo, W.T., Norton, R.A. and Lynch, C.A. (1990) The idiosomal chaetotaxy of astigmatid mites. *Journal of Zoology* 220, 1–32.

Griffiths, J.T. and Thompson, W.L. (1957) *Insects and Mites found in Florida Citrus*. Bulletin No. 591, University of Florida, Gainesville, Florida.

Griffiths, M. (1994) *Index of Garden Plants*. Timber Press, Portland, Oregon.

Grimm, M., Michael, P., Hyder, M. and Doyle, P. (1995) Effects of pasture pest damage and grazing management on efficiency of animal production. *Plant Protection Quarterly* 10, 62–64.

Grinberg, M., Soroker, V., Palevsky, E., Shomer, I. and Perl-Treves, R. (2004) Response of cucumber to the broad mite (*Polyphagotarsonemus latus*). In: Lebeda, A. and Paris, H.S. (eds) *Progress in Cucurbit Genetics and Breeding Research. Proceedings of Cucurbitaceae 2004, the 8th EUCARPIA Meeting on Cucurbit Genetics and Breeding*, Olomouc, Czech Republic, 12–17 July 2004. Palacký University in Olomouc, Czech Republic, pp. 319–324.

Grinberg, M., Perl-Treves, R., Palevsky, E., Shomer, I. and Soroker, V. (2005) Interaction between cucumber plants and the broad mite, *Polyphagotarsonemus latus*: from damage to defense gene expression. *Entomologia Experimentalis et Applicata* 115, 135–144.

Grissa, K.L. and Sahraoui, H. (2007) Demographic traits of two phytophagous mites (*Tetranychus cinnabarinus* and *Aculops lycopersici*) and biological control on tomato. *Acta Horticulturae* 758, 81–87.

Groot, T.V.M. and Breeuwer, J.A.J. (2006) *Cardinium* symbionts induce haploid thelytoky in most clones of three closely related *Brevipalpus* species. *Experimental and Applied Acarology* 39, 257–271.

Groot, T.V.M., Janssen, A., Pallini, A. and Breeuwer, J.A.J. (2005) Adaptation in the asexual false spider mite *Brevipalpus phoenicis*: evidence for frozen niche variation. *Experimental and Applied Acarology* 36, 165–176.

Grosscurt, A.C., Wixley, R.A.J. and Haar, M. ter (1994) Cross-resistance between flucycloxuron, clofentezine and hexythiazox in *Panonychus ulmi* (fruit tree red spider mite). *Experimental and Applied Acarology* 18, 445–458.

Grostal, P. and Dicke, M. (2000) Recognising one's enemies: a functional approach to risk assessment. *Behavioral Ecology and Sociobiology* 47, 258–264.

Grostal, P. and O'Dowd, D.J. (1994) Plants, mites and mutualism: leaf domatia and the abundance and reproduction of mites on *Viburnum tinus* (Caprifoliaceae). *Oecologia* 97, 308–315.

Grout, T.G. (1998) Predacious mites as biocontrol agents on citrus in Southern Africa. In: Bedford, E.C.G., van den Berg, M.A. and Villiers, E.A. (eds) *Citrus Pests in the Republic of South Africa*, 2nd edn. ARC-Institute for Tropical and Subtropical Crops, Pretoria, South Africa, pp. 29–33.

Grout, T.G. and Richards, G.I. (1992a) The dietary effect of windbreak pollens on lengevity and fecundity of a predacious mite *Euseius addoensis addoensis* (Acari: Phytoseiidae) found in citrus orchards in South Africa. *Bulletin of Entomological Research* 82, 317–320.

Grout, T.G. and Richards, G.I. (1992b) Susceptibility of *Euseius addoensis addoensis* (Acari: Phytoseiidae) to field-weathered residues of insecticides used on citrus. *Experimental and Applied Acarology* 15, 1–13.

Grout, T.G. and Schutte, G.C. (2004) Citrus grey mite: causing necrotic spot. *South African Fruit Journal* 3(2), 32–34.

Grout, T.G. and Stephen, P.R. (2002) Citrus rust mite: develops resistance to mancozeb. *South African Fruit Journal* 1(2), 40–43.

Grout, T.G., Richards, G.I. and Stephen, P.R. (1997) Further non-target effects of citrus pesticides on *Euseius addoensis* and *Euseius citri* (Acari: Phytoseiidae). *Experimental and Applied Acarology* 21, 171–177.

Gu, Q.-Y., Chen, W.-B., Wang, L.-J., Shen, J. and Zhang, J.-P. (2010) Effects of sublethal dosage of abamectin and pyridaben on life table of laboratory populations of *Tetranychus turkestani* (Acari: Tetranychidae). *Acta Entomologica Sinica* 53, 876–883.

Guagliumi, P. (1967) *Insects and Arachnids of the Common Plants in Venezuela reported in the Period 1938–1963*. Relazioni e Monografie Agrarie Subtropicali e Tropicali (N.S.) No. 86, L'Istituto Agronomico per l'Oltremare, Florence, Italy.

Gualaccini, F. (1950) Alterazioni istologiche prodotte dall'*Eriophyes pistaciae* Nal. su *Pistacia terebinthus* L. e presenza di *Uromyces terebinthi* (D.C.) Winter sulla stessa pianta. *Bollettino della Stazione di Patologia Vegetale, Seria 3* 8, 221–233.

Guanilo, A.D. and Martinez, N. (2007) Predadores asociados a *Panonychus citri* McGregor (Acari: Tetranychidae) en la costa central del Peru. *Ecología Aplicada, Universidad Nacional Agraria La Molina* 6, 119–129.

Guanilo, A.D. and Martinez, N. (2009) Biology and behavior of *Amblyseius chungas* Denmark & Muma (Acari: Phytoseiidae) as a predator of *Panonychus citri* (McGregor) (Acari: Tetranychidae). *Ecología Aplicada, Universidad Nacional Agraria La Molina* 8, 15–25.

Guanilo, A.D., Moraes, G.J. de and Knapp, M. (2008) Phytoseiid mites (Acari: Phytoseiidae) of the subfamily Amblyseiinae Muma from Peru, with descriptions of four new species. *Zootaxa* 1880, 1–47.

Guanilo, A.D., Moraes, G.J. de, Toledo, S. and Knapp, M. (2010) New records of *Tetranychus evansi* and associated natural enemies in northern Argentina. *Systematic and Applied Acarology* 15, 3–20.

Guanilo, A.D., Moraes, G.J. de, Flechtmann, C.H.W. and Knapp, M. (2012) Phytophagous and fungivorous mites (Acari: Prostigmata, Astigmata) from Peru. *International Journal of Acarology* 38, 120–134.

Gudleifsson, B.E., Hallas, T.E., Ólafsson, S. and Sveinsson, T. (2002) Chemical control of *Penthaleus major* (Acari: Prostigmata) in hayfields in Iceland. *Journal of Economic Entomology* 95, 307–312.

Guedes, J.V.C., Návia, D., Lofego, A.C. and Dequech, S.T.B. (2007) Ácaros associados à cultura da soja no Rio Grande do Sul, Brasil. *Neotropical Entomology* 36, 288–293.

Gueguen, F. (1998) Turkoise: acaricide de la vigne. *Phytoma – La Défense des Végétaux* 508, 42–43.

Guenaoui, Y. (2010) *Tetranychus evansi* (Baker & Pritchard) (Acari: Tetranychidae) invasive mite reported on tomato crops in Mostaganem in the north-west of Algeria. *OEPP/EPPO Bulletin* 40, 193–195.

Guerere, P. and Gonzalez, M.Q. de (2000) Assessment of degree of feeding damage caused by the scarlet mite, *Brevipalpus phoenicis* (Geijskes) (Tenuipalpidae), on guava fruits (*Psidium guajava* L.). *Revista de la Facultad de Agronomía, Universidad de Zulia* 17, 471–481.

Guerra, A., Bertelli, L., Cantoni, M. and Gollo, M. (2002) Spirodiclofen (Envidor®): un nuovo acaricida per frutticoltura e viticoltura. In: *Atti, Giornate Fitopatologiche 2002, Baselga di Pinè (Trento) 7–11 Aprile 2002*. Fitogest®, Image Line Network, Faenza, Italy, pp. 371–376.

Guerrero, J.M. and Bellotti, A.C. (1980) Contribución al conocimiento de algunos acaros fitofagos encontrados en el cultivo de la yuca *Manihot esculenta* Crantz, en Colombia. *Revista Colombiana de Entomología* 6, 85–98.

Guessoum, M. (1986) Approach of a bio-ecological study of the mite *Oligonychus afrasiaticus* (Boufaroua) on date palms. *Annales de l'Institut National Agronomique (El Harrach)* 10, 153–166.

Gui, L.-Y., Gong, X.-W., Meng, G.-L. and Xiong, S.-H. (2001a) The relationship between the stoma density of eggplant leaves and their resistance to *Polyphagotarsonemus latus*. *Acta Horticulturae Sinica* 28, 170–172 (in Chinese).

Gui, L.-Y., Meng, G.-L., Gong, X.-W. and Xiong, S.-H. (2001b) The cluster analysis of resistance to *Polyphagotarsonemus latus* (Banks) for eggplant varieties. *Scientia Agricultura Sinica* 34, 506–510 (in Chinese).

Guichou, S., Tixier, M.-S. and Kreiter, S. (2010) Assessment of morphological and molecular variation among strains of *Neoseiulus californicus* (Acari: Phytoseiidae). In: Sabelis, M.W. and Bruin, J. (eds) *Trends in Acarology: Proceedings of the 12th International Congress, 21–26 August 2006*, Amsterdam. Springer, Dordrecht, the Netherlands, pp. 271–278.

Guido, M., Battisti, A. and Roques, A. (1995) A contribution to the study of cone and seed pests of the evergreen cypress (*Cupressus sempervirens* L.) in Italy. *Redia* 78, 211–227.

Guirado, N. and Ambrosano, E.J. (2000) Mitcide effect of natural products on *Brevipalpus phoenicis* in the laboratory. *Revista de Agricultura (Piracicaba)* 75, 387–393.

Guirado, N., Nogueira, N. de L., Ambrosano, E.J. and Francozo, M. (2001) Vegetable extracts effect on *Brevipalpus phoenicis* vector activity. *Summa Phytopatholica* 27, 343–347.

Gunson, F.A. and Hutchins, R.F.N. (1982) Absence of farnesol in strawberry and hop foliage. *Journal of Chemical Ecology* 8, 785–796.

Günthart, E. and Günthart, H. (1959) Uber das Auftreten von Spinnmilben (Tetranychidae) und Raubmilben (Phytoseiinae, Acari) in der Schweiz. In: *Sonderdruck Verh IV Internationalen Pflanzenschutz-Kongress, Hamburg* 1957, *Volume 2*. Braunschweig, Germany, pp. 927–928.

Guo, Y.J., Zuo, G.S., Zhao, J.H., Wang, N.Y. and Jiang, J.W. (1993a) A laboratory test on the toxicity of thuringiensin to *Tetranychus urticae* (Acari: Tetranychidae) and *Phytoseiulus persimilis* (Acari: Phytoseiidae). *Chinese Journal of Biological Control* 9, 151–155 (in Chinese).

Guo, Y.-L., Jiao, X.-D., Xu, J.-J., Yang, S., Duan, X.-K. and Zhang, J.-P. (2013) Growth and reproduction of *Tetranychus turkestani* and *Tetranychus truncatus* (Acari: Tetranychidae) on cotton and corn. *Systematic and Applied Acarology* 18(19), 89–98.

Guo, Z., Weston, P.A. and Snyder, J.C. (1993b) Repellency to two-spotted spider mite, *Tetranychus urticae* Koch, as related to leaf surface chemistry of *Lycopersicon hirsutum* accessions. *Journal of Chemical Ecology* 19, 2965–2979.

Gupta, S.K. (1976) Contribution to our knowledge of tetranychid mites (Acarina) with descriptions of three new species from India. *Oriental Insects* 10, 327–351.

Gupta, S.K. (1985) *Handbook: Plant Mites of India*. Zoological Survey of India, Calcutta, India.

Gupta, S.K. (1991) The mites of agricultural importance in India with remarks on their economic status. *Modern Acarology* 1, 509–522.

Gupta, S.K. (1992) Arachnida: plant mites (Acari). In: Ghosh, A.K. (ed.) *State Fauna Series 3: Fauna of West Bengal, Part 3*. Zoological Survey of India, Calcutta, India, pp. 61–211.

Gupta, S.K. (1995) Plant mites (Acari). In: Ghosh, A.K. (ed.) *State Fauna Series 4: Fauna of Meghalaya, Part 2*. Zoological Survey of India, Calcutta, India, pp. 17–50.

Gupta, S.K. (2001) A conspectus of natural enemies of phytophagous mites and mites as potential biocontrol agents of agricultural pests in India. In: Halliday, R.B., Walter, D.E., Proctor, H.C., Norton, R.A. and Colloff. M.J. (eds) *Acarology: Proceedings of the 10th International Congress of Acarology, 5–10 July 1998, Canberra*. CSIRO Publishing, Collingwood, Australia, pp. 484–497.

Gupta S.K. (2003) Mite pests of agricultural crops in India, their management and identification. In: Yadav, P.R., Chauhan, R., Putatunda, B.N. and Chhillar, B.S (eds) *Mites, their Identification and Management*. CCS-HAU (Chaudhary Charan Singh Haryana Agricultural University), Hisar, India, pp. 48–61.

Gupta, S.K. and Chatterjee, K. (1997) Acari: plant mites. In: Alfred, J.R.B. (ed.) *State Fauna Series 6: Fauna of Delhi*. Zoological Survey of India, Calcutta, India, pp. 485–532.

Gupta, S.K. and Gupta, Y.N. (1994) *A Taxonomic Review of Indian Tetranychidae (Acari: Prostgmata) with Description of New Species, Redescriptions of Known Species and Keys to Genera and Species*. Memoirs of the Zoological Survey of India No. 18, Calcutta, India.

Gupta, S.K. and Krishna Karmakar (2011) Diversity of mites (Acari) on medicinal and aromatic plants in India. In: Moraes, G.J. de and Proctor, H. (eds) *Acarology XIII: Proceedings of the International Congress. Zoosymposia* 6, 56–61.

Gupta, S.K., Dhooria, M.S. and Sidhu, A.S. (1971) A note on predators of citrus mites in Punjab. *Science and Culture* 37, 484–485.

Gupta, S.K., Sidhu, A.S., Singh, G. and Dhooria, M.S. (1972) Evaluation of new pesticides for the control of *Tetranychus cucurbitae* Rahman & Sapra, infesting brinjal in Punjab. *Indian Journal of Agricultural Sciences* 42, 921–942.

Gupta, S.K., Dhooria, M.S. and Sidhu, A.S. (1975) Effect of food and temperature on the development, longevity and fecundity of sugarcane red mite, *Oligonychus indicus* (Hirst). *Acarologia* 16, 436–440.

Gupta, Y.N. (1980) Some spider mites (Acarina: Tetranychidae) from Andaman and Nicobar Islands with descriptions of three new species. *Records of the Zoological Survey of India* 77, 111–117.

Gupta, Y.N. (1981) On a collection of tetranychid mites from Jammu and Kashmir with description of one new species of *Eotetranychus* (Acarina: Tetranychidae). *Bulletin of the Zoological Survey of India* No. 3, 235–238.

Gupta, Y.N. (1984) On a collection of tetranychoid mites from Tamil Nadu with description of a new species of *Aponychus* (Acari: Tetranychidae). *Bulletin of the Zoological Survey of India* No. 6, 237–245.

Gupta, Y.N. and Gupta, S.K. (1976) Notes on a collection of tetranychoid mites from Gujarat (Acarina: Trombidiformes). *Acarology Newsletter* 3, 6–7.

Gupta, Y.N. and Gupta, S.K. (1978) On a collection of tetranychoid mites from Madhya Pradesh, India, with a description of a new *Eotetranychus* (Acari: Trombidiformes). *Indian Journal of Acarology* 3, 87–91.

Gurney, B. (1970) Bulb-scale mite. *Annual Report of the Glasshouse Crops Research Institute for 1969*. Littlehampton, UK.

Gurney, B. (1972) Bulb-scale mite. *Annual Report of the Glasshouse Crops Research Institute for 1971*. Littlehampton, UK.

Gurney, B. and Gandy, D.G. (1974) Methyl bromide for control of bulb scale mite, *Steneotarsonemus laticeps* Halb. *Plant Pathology* 23, 17–19.

Gurney, B. and Hussey, N.W. (1967) *Pygmephorus* species (Acarina: Pyemotidae) associated with with cultivated mushrooms. *Acarologia* 9, 353–358.

Gushina, N.I., Gar, K.A., Gulidova, L.A. and Abelentseva, N.V. (1976) Tests on the control of the gall mite. *Zashchita Rasteniĭ* No. 7, 37 (in Russian).

Gutierrez, J. (1966) Cinq nouvelles espèces de Tetranychidae de Madagascar. *Acarologia* 8, 594–610.

Gutierrez, J. (1967) Huit nouvelles espèces du genre *Eotetranychus* Oudemans (Acariens: Tetranychidae) de Madagascar. *Acarologia* 9, 370–394.

Gutierrez, J. (1968) Note sur quelques acariens phytophages de l'Ile de la Réunion avec description d'une nouvelle espèce du genre *Eotetranychus* Oudemans (Tetranychidae). *Acarologia* 10, 444–446.

Gutierrez, J. (1974) Les espèces du genre *Tetranychus* Dufour (Acariens: Tetranychidae) ayant une incidence économique à Madagascar et dans les Iles voisines. Compétition entre les complexes *Tetranychus neocaledonicus* André et *Tetranychus urticae* Koch. *Acarologia* 16, 258–270.

Gutierrez, J. (1975) *Rapport d'une Mission Effectuée aux Nouvelles-Hébrides du 16 au 27 Juin 1975*. ORSTOM Laboratoire de Zoologie Appliquée et de Lutte Biologique, Nouméa, New Caledonia.

Gutierrez, J. (1976a) *Report on a Mission to American Samoa from 20th May to 1st June 1976 to Investigate an Outbreak of* Spodoptera litura *(F.) (Lepidoptera: Noctuidae) Affecting Taro*. ORSTOM Laboratoire de Zoologie Appliquée et de Lutte Biologique, Nouméa, New Caledonia.

Gutierrez, J. (1976b) *Etude Biologique et Ecologique de* Tetranychus neocaledonicus *André (Acariens: Tetranychidae)*. Travaux et Documents de l'ORSTOM No. 57, Office de la Recherche Scientifique et Technique d'Outre-Mer, Montpellier, Paris.

Gutierrez, J. (1977) Phytophagous mites of economic significance in the area of the South Pacific Commission. In: *Regional Meeting on Plant Protection, Nouméa, Nouvelle-Calédonie, 14–18 November 1977*. South Pacific Commission, Noumea, New Caledonia.

Gutierrez, J. (1978) *Biological Control of the Coconut Hispid (Brontispa longissima) and a Survey of Fruit Flies in American Samoa. Nouméa, Nouvelle-Calédonie.* ORSTOM Laboratoire de Zoologie Appliquée et de Lutte Biologique, Nouméa, New Caledonia.

Gutierrez, J. (1979) *Enquête sur les Acariens Phytophages ayant une Incidence Economique en Polynésie Française.* ORSTOM, Laboratoire de Zoologie Appliquée et de Lutte Biologique, Nouméa, New Caledonia.

Gutierrez, J (1980) Acarina of the genus *Tetranychus* injurious to economic plants in the tropical Indo-Pacific area. In: *XVI International Congress of Entomology 3–9 August 1980, Kyoto, Japan*, pp. 1–11.

Gutierrez, J. (1981) *Actualisation des Données sur l'Entomologie Économique à Wallis et Futuna.* ORSTOM Laboratoire de Zoologie Appliquée et de Lutte Biologique, Nouméa, New Caledonia.

Gutierrez, J. (1982) Deux acariens phytophages vivant sur canne à sucre à la Réunion: *Oligonychus etienni* n. sp. (Tetranychidae) et *Abacarus sacchari* (Eriophyidae). *Agronomie Tropicale* 37, 389–392.

Gutierrez, J. (1983) Rôle de l'homme dans la dispersion et l'établissement des acariens phytophages à travers le domaine insulaire de l'Indo-Pacifique. In: *La Nature Insulaire Tropicale: Spécificité et Protection: 8 Colloque de la SEPANRIT (Société pour l'Etude, la Protection et l'Aménagement de la Nature dans les Régions Inter-tropicales), Talence, Nouméa, 1981/11/02-03.* Bulletin de Liaison – SEPANRIT No. 12, Nouméa, New Caledonia, pp. 161–166.

Gutierrez, J. (1985a) Systematics. In: Helle, W. and Sabelis, M.W. (eds) *Spider Mites: Their Biology, Natural Enemies and Control.* World Crop Pests, Volume 1A. Elsevier Science, Amsterdam, pp. 75–90.

Gutierrez, J. (1985b) Mounting techniques. In: Helle, W. and Sabelis, M.W. (eds) *Spider Mites: Their Biology, Natural Enemies and Control.* World Crop Pests, Volume 1A. Elsevier Science, Amsterdam, pp. 351–353.

Gutierrez, J. (1987a) *Rapport d'une Mission Effectuée en République Populaire du Congo et en République Centrafricaine à Propos des Pullulations d'Acariens Phytophages dans les Plantations de Manioc, 21 September–October 1987.* ORSTOM, Office de la Recherche Scientifique et Technique d'Outre-Mer, Montpellier.

Gutierrez, J. (1987b) The cassava green mite in Africa: one or two species? (Acari: Tetranychidae). *Experimental and Applied Acarology* 3, 163–168.

Gutierrez, J. (1992) Les acariens déprédateurs du cotonnier. *Coton et Fibres Tropicales* 47, 153–172.

Gutierrez, J. and Bonato, O.A.D. (1994) Les acariens Tetranychidae attaquant le manioc au Congo et quelques uns de leurs predateurs. *Journal of African Zoology* 108, 191–200.

Gutierrez, J. and Chazeau, J. (1972) Cycles de developpement et tables de vie de *Tetranychus neocaledonicus* Andre (Acariens: Tetranychidae) et d'un de ses principaux predateurs a Madagascar *Stethorus madecassus* Chazeau (Coccinellidae). *Entomophaga* 17, 275–295.

Gutierrez, J. and Etienne, J. (1981a) Une nouvelle espèce du genre *Oligonychus* (Acariens: Tetranychidae) attaquant le riz au Sénégal. *Agronomie Tropicale* 36, 389–390.

Gutierrez, J. and Etienne, J. (1981b) Quelques données sur les acariens Tetranychidae attaquant les plantes cultivées au Sénégal. *Agronomie Tropicale* 36, 391–394.

Gutierrez, J. and Etienne, J. (1986) Les Tetranychidae de l'île de la Réunion et quelques-uns de leurs predateurs. *Agronomie Tropicale* 41, 84–91.

Gutierrez, J. and Etienne, J. (1988) *Quelques Données sur les Acariens Phytophages de la Martinique et de la Guadeloupe. Journées de la Protection des Végétaux dans les Départements d'Outre-Mer, Montpellier, France, 5 October 1988,* ORSTOM, Montpellier.

Gutierrez, J. and Helle, W. (1971) Deux nouvelles espèces du genre *Eutetranychus* Banks (Acariens: Tetranychidae) vivant sur plantes cultivèes à Madagascar. *Entomologische Berichten* 31, 45–60.

Gutierrez, J. and Helle, W. (1981) *Eotetranychus uncatus* Garman, a pest of apple new to the Netherlands (Acari: Tetranychidae). *Entomologische Berichten* 41, 102–103.

Gutierrez, J. and Helle, W. (1983) *Eotetranychus rubiphilus* espèce nouvelle pour la France et l'Europe occidentale: redescription et caryotype (Acari: Tetranychidae). *Entomologische Berichten* 43, 109–112.

Gutierrez, J. and Helle, W. (1985) Evolutionary changes in the Tetranychidae. In: Helle, W. and Sabelis, M.W. (eds) *Spider Mites: Their Biology, Natural Enemies and Control.* World Crop Pests, Volume 1A. Elsevier Science, Amsterdam, pp. 91–107.

Gutierrez, J. and Schicha, E. (1983) The spider mite family Tetranychidae (Acari) in New South Wales. *International Journal of Acarology* 9, 99–116.

Gutierrez, J. and Schicha, E. (1984) Phytoseiidae and Tetranychoidea in Fiji and other South Pacific islands (Acari). *International Journal of Entomology* 26, 386–388.

Gutierrez, J. and Schicha, E. (1985) Notes on the occurrence of *Oligonychus milleri* (McGregor) and *O. coffeae* (Nietner) (Acari: Tetranychidae) in Australia. *Australian Entomological Magazine* 12, 69–70.

Gutierrez, Z.M. and Gonzalez, R.M. (1993) Weeds as wild hosts to mites and their relation to economically important crops. *Centro-Agrícola (Santa Clara)* 20(1), 89–91.

Gutue, M., Gutue, C. and Rosca, I. (2010) Mites of parks and ornamental gardens in Bucharest urban area. In: Moraes, G.J. de, Castilho, R.C. and Flechtmann, C.H.W. (eds) *XIII International Congress of Acarology, August 23–27, 2010, Recife-PE, Brazil, Abstract Book*, abstract 186, pp. 102–103. Available at: http://www.acarology.org/ica/ica2010/docs/0910_15_abstractBookCapa.pdf (accessed 3 February 2015).

Gyenis, K., Penzes, B. and Hegyi, T. (2005) Fitofág és ragadozó atkafajok vadgesztenyén (Phytophagous and predatory mites on the horse chestnut tree). *Növényvédelem* 41, 143–148.

Gyorffy, S. (2006) Őszibarack-ültetvény és a környező vegetáció atkapopulációinak lehetséges kölcsönhatása (Possible relationship between the mite populations of a peach orchard and the surrounding vegetation). *Növényvédelem* 42, 195–204 (in Hungarian).

Györffyné, J.M. (1992) Új levélatkafaj a hazai faunában (New leaf mite species in the Hungarian fauna). *Növényvédelem* 28, 320–322 (in Hungarian).

Györffyné, J.M. (1994) Új atkaölő szer a szőlő növényvédelmében (New acaricide in grapevine protection). *Növényvédelem* 30, 428–431 (in Hungarian).

Györffyné, J.M., Dulinafka, G. and Dobi, D. (1994) Uj atkaoloszer a szolo novenyvedelmeben. *Növényvédelmi Tudomanyos Napok* 40, 85 (in Hungarian).

Hadwiger, L.A. (1999) Host–parasite interactions: elicitation of defense responses in plants with chitosan. In: Jolles, P. and Muzarelli, R.A.A. (eds) *Chitin and Chitinases. Experientia* Supplementum 87. Birkhauser, Basel, Switzerland, pp. 185–200.

Hafez, S.M. and Maksoud, M.A. (1984) Control of *Eriophyes tulipae* K. attacking garlic (Acari: Eriophyidae) I. Chemical control. *Egyptian Journal of Horticulture* 11, 93–97.

Hage-Hamsten, M. Van, Johansson, S.G.O. and Zetterstrom, O. (1985) Predominance of mites allergy over allergy to pollens and animal danders in a farming population. *Clinical Allergy* 17, 417–423.

Hagen, K.S. (1987) Nutritional ecology of terrestrial insect predators. In: Slansky, J. and Rodriguez, J.G. (eds) *Nutritional Ecology of Insects, Mites, Spiders and Related Invertebrates.* Wiley, New York, pp. 533–577.

Haji, F.N.P., Moraes, G.J. de, Lacerda, C.A. de and Neto, R.S. (1988) Controle quimico do ácaro do bronzeamento do tomateiro *Aculops lycopersici* (Massee, 1937). *Anais de Sociedade Entomológica do Brasil* 17, 437–442.

Hajiqanbar, H.R., Irani-Nejad, K.H. and Chaichi, P.T. (2002) Records of astigmatic mites of family Acaridae from sugarbeet fields in Miandoab Plain. *Journal of Agricultural Science, University of Tabriz* 12(2), 1–10.

Hajizadeh, J. and Hosseini, R. (2003a) *Proceedings of the Applied-Scientific Seminar on Flowers and Ornamental Plants, 2nd, 15–16 October 2003, Mahallat, Iran* (in Persian).

Hajizadeh, J. and Hosseini, R. (2003b) Introducing of privet rust mite *Aculus ligustri* (Acar.: Eriophyidae) and its some biological aspect in Guilan Province. *Iranian Journal of Forest and Range Protection Research* 1(1), 25–35 (in Persian).

Hajizadeh, J. and Hosseini, H. (2004) Introducing of eight species of family Eriophyidae from forest plants in Guilan Province. In: *Proceedings of the16th Plant Protection Congress Iran, Tabriz, Iran*, p. 279.

Hajizadeh, J., Kamali, G.K. and Assadi, H.B. (1993) Investigations on the functional response and populations fluctuations of *Stethorus gilvifrons* on red spider mite, *Panonychus ulmi* (Koch) in Karaj vicinity (Iran). *Applied Entomology and Phytopathology (Tehran)* 61, 32–34 (in Persian).

Halawa, A.M. and El-Sebaay, M.M. (2010) Biological studies on the mango red mite, *Oligonychus mangiferus* R and S on two mango cultivars. *Egyptian Journal of Agricultural Research* 88, 377–380.

Halbert, J.N. (1923) Notes on Acari, with descriptions of new species. *Zoological Journal of the Linnean Society* 35, 363–392.

Halbert, S.E. and Mead, F.W. (1995) Entomology section. *Tri-ology* 34(3), 4–9. Available at: http://www.freshfromflorida.com/Divisions-Offices/Plant-Industry/Plant-Industry-Publications/Tri-ology-FDACS-DPI/Triology-Vol.-34-No.-3 (accessed 22 May 2015).

Hall, C.C. Jr (1958) A new eriophyid mite from Kansas. *Journal of the Kansas Entomological Society* 31, 233–235.

Hall, C.C. Jr (1967) The Eriophyoidea of Kansas. *Kansas University Science Bulletin* 47, 601–675.

Hall, D.G. and Simms, M.K. (2003) Damage by infestations of Texas citrus mite (Acari: Tetranychidae) and its effect on the life of 'Valencia' leaves in an irrigated citrus grove. *Florida Entomologist* 86, 15–28.

Hall, D.G., Childers, C.C. and Eger, J.E. (1991) Estimating citrus rust mite (Acari: Eriophyidae) levels on fruit in individual citrus trees. *Environmental Entomology* 20, 383–390.

Hall, D.G., Childers, C.C. and Eger, J.E. (1994) Spatial dispersion and sampling of citrus rust mite (Acari: Eriophyidae) on fruit in "Hamlin" and "Valencia" orange groves in Florida. *Journal of Economic Entomology* 87, 687–689.

Hall, D.G., Childers, C.C. and Eger, J.E. (2005) Effects of reducing sample size on density estimates of citrus rust mite (Acari: Eriophyidae) on citrus fruit: simulated sampling. *Journal of Economic Entomology* 98, 1048–1057.

Hall, D.G., Childers, C.C. and Eger, J.E. (2007) Binomial sampling to estimate citrus rust mite (Acari: Eriophyidae) densities on orange fruit. *Journal of Economic Entomology* 100, 233–240.

Hall, F.R. and Ferree, D.C. (1975) Influence of two-spotted spider mite populations on photosynthesis of apple leaves. *Journal of Economic Entomology* 68, 517–520.

Hall, I.M., Hunter, D.K. and Arakawa, K.Y. (1971) The effect of the β-exotoxin fraction of *Bacillus thuringiensis* on the citrus red mite. *Journal of Invertebrate Pathology* 18, 359–362.

Hall, R.A. and Espinosa, B.A. (1981) The coconut mite, *Eriophyes guerreronis*, with special reference to the problem in Mexico. In: *Proceedings of the Brighton Crop Protection Conference: Pests and Diseases – 1981*. British Crop Production Council, Croydon (now Alton), UK, pp. 113–120.

Hall, R.A., Hussey, N.W. and Mariau, D. (1980) Results of a survey of biological control agents of the coconut mite, *Eriophyes guerreronis*. *Oléagineux* 35, 395–400.

Hall, W.J. (1954) *Acaphylla steinwedeni*: a first record. *FAO Plant Protection Bulletin* 3, 125–127.

Hallas, T.E. and Gudleifsson, B.E. (2008) Summer occurrence of *Tyrophagus similis* Volgin (Acari, Acaridae) in hayfields in northern Iceland may include a facultative deutonymph. *Journal of the Acarological Society of Japan* 17, 101–106.

Hallas, T.E. and Solberg, H. (1989) Mites of stored hay on the Faroe Islands (Acari). *Entomologiske Meddelelser* 57, 151–155.

Halliday, R.B. (1991) Taxonomic background of the redlegged earth mite *Halotydeus destructor* (Tucker) (Acarina: Penthaleidae). *Plant Protection Quarterly* 6, 162–165.

Halliday, R.B. (1998) *Mites of Australia: A Checklist and Bibliography*. CSIRO Publishing, Melbourne, Australia.

Halliday, R.B. (2000) Additions and corrections to mites of Australia: a checklist and bibliography. *Australian Journal of Entomology* 39, 233–235.

Halliday, R.B. (2001a) 250 years of Australian Acarology. In: Halliday, R.B., Walter, D.E., Proctor, H.C., Norton, R.A. and Colloff, M.J. (eds) *Acarology: Proceedings of the 10th International Congress of Acarology, 5–10 July 1998, Canberra*. CSIRO Publishing, Collingwood, Australia, pp. 3–16.

Halliday, R.B. (2001b) Systematics and biology of the Australian species of *Balaustium* von Heyden (Acari: Erythraeidae). *Australian Journal of Zoology* 40, 326–330.

Halliday, R.B. (2004) The occurrence of *Aceria tulipae* (Keifer) and *Aceria tosichella* Keifer in Australia (Acari: Eriophyidae). *International Journal of Acarology* 30, 113–118.

Halliday, R.B. (2005a) Predatory mites from crops and pastures in South Africa: potential natural enemies of redlegged earth mite *Halotydeus destructor* (Acari: Penthaleidae). *Zootaxa* 1079, 11–64.

Halliday, R.B. (2005b) Systematics and biology of *Penthaleus tectus* sp. n. (Acari: Penthaleidae), a recently discovered pest of grain crops in eastern Australia. *Australian Journal of Entomology* 44, 144–149.

Halliday, R.B. (2010) Taxonomic confusion surrounding mite pests of sugarcane and rice (Acari: Eriophyidae). *Systematic and Applied Acarology* 15, 257–262.

Halliday, R.B. (2013) Higher Taxon Acari: Mites. Available at: http://www.environment.gov.au/biodiversity/abrs/online-resources/fauna/afd/taxa/Acari (accessed 3 August 2014).

Halliday, R.B. and Paull, C. (2004) Assessment of *Chaussieria capensis* (Acari: Anystidae) as predator of *Halotydeus destructor* (Acari: Penthaleidae). *African Entomology* 12, 286–290.

Hambleton, E.J. (1938) A occorrencia do acaro tropical *Tarsonemus latus* Banks. Causador da rasgadura das folhas nos algodoais de S. Paulo. *Arquivos do Instituto Biológico (São Paulo)* 9, 201–209.

Hambleton, E.J. and Sauer, H.F.G. (1938) Observacoes sobre as prágas da cultura algodoeira no nordeste e norte do Brasil (Observations on cotton pests in north-eastern and northern Brazil). *Arquivos do Instituto Biológico (São Paulo)* 9, 319–330.

Hamilton, A. (1948) Why the blackberries fail to ripen. *New Zealand Science Review* 6, 33–34.

Hamilton, A. (1950) The blackberry mite (*Aceria essigi*). *New Zealand Journal of Science and Technology (A)* 31, 42–45.

Hamilton, C.C. (1926) *Insect Pests of Boxwood*. New Jersey Agricultural Experiment Station Circular No. 179, New Brunswick, New Jersey.

Hamstead, E.O. (1970) Greenhouse integrated control studies of the two-spotted spider mite on lima beans with a predaceous mite, *Typhlodromus fallacis*, and insecticides. *Journal of Economic Entomology* 63, 1027–1028.

Han, J., Choi, B.-R., Lee, S.-G., Kim, S. I., and Ahn, Y.-J. (2010) Toxicity of plant essential oils to acaricide-susceptible and -resistant *Tetranychus urticae* (Acari: Tetranychidae) and *Neoseiulus californicus* (Acari: Phytoseiidae). *Journal of Economic Entomology* 103, 1293–1298.

Han, K.P. (1969) Studies on mites (II). On some mites of the apple trees. *Korean Journal of Plant Protection* 8, 29–35 (in Korean).

Han, K.P. (1970) Studies on mites (III). Mites on persimmon and citrus. *Korean Journal of Plant Protection* 9, 33–35 (in Korean).

Hanchinal, S.G. and Manjunatha, M. (2000) Effect of *Metarhizium anisopliae* (Metsch.) Sor. on *Tetranychus neocaledonicus* Andre on bhendi (okra) under field conditons. *Karnataka Journal of Agricultural Sciences* 13, 1000–1002.

Handoko (2004) Preference of *Stethorus gilvifrons* Mulsant (Coleoptera: Coccinellidae) for spider mites infesting orange (*Citrus* spp.), apple (*Malus* spp.) and rose (*Rosa* spp.) plants. MSc. thesis, Philippines University of Los Baños, Laguna, Philippines.

Hänel, H. (1986) Effect of juvenile hormone (III) from the host *Apis mellifera* (Insecta: Hymenoptera) on neurosecretion of the parasitic mite *Varroa jacobsoni* (Acari: Mesostigmata). *Experimental and Applied Acarology* 2, 257–271.

Haney, P.B., Luck, R.F. and Moreno, S.S. (1987) Increases in densities of the citrus red mite *Panonychus citri* (Acarina: Tetranychidae) in association with the Argentine ant *Iridomyrmex humilis* (Hymenoptera: Formicidae) in southern California citrus. *Entomophaga* 21, 49–57.

Hanna, M.A., Wahba, M.L. and Hafez, M.A.A. (1977) Effects of insecticide–acaricide mixtures on populations of the flat mite, *Brevipalpus californicus*, on lemon. *Agricultural Research Review (Cairo)* 55(1), 163–166.

Hanna, R. and Toko, M. (2001) Cassava green mite biological control in Africa: project overview and summary of progress. In: Hanna, R. and Toko, M. (eds) *Proceedings of the Regional Meeting of the Cassava Green Mite Biocontrol Project, 15–17 November 2000, Dar es Salaam, Tanzania*. International Institute for Tropical Agriculture, Cotonou, Benin, pp. 4–22.

Hanna, R. and Wilson, L.T. (1991) Prey preference by *Metaseiulus occidentalis* (Acari: Phytoseiidae) and the role of prey aggregation. *Biological Control* 1, 51–58.

Hanna, R., Wilson, L.T., Zalom, F.G., Flaherty, D.L. and Leavitt, G.M. (1996) Spatial and temporal dynamics of spider mites (Acari: Tetranychidae) in "Thompson seedless" vineyards. *Environmental Entomology* 25, 370–382.

Hanna, R., Wilson, L.T., Zalom, F.G. and Flaherty, D.L. (1997a) Effects of predation and competition on the population dynamics of *Tetranychus pacificus* on grapevines. *Journal of Applied Ecology* 34, 878–888.

Hanna, R., Zalom, F.G., Wilson, L.T. and Leavitt, G.M. (1997b) Sulfur can suppress mite predators in vineyards. *California Agriculture* 51(1), 19–21.

Hansen, A.J. (1978) Natural occurrence of cherry mottle leaf in apricot. *Plant Disease Reporter* 50, 902–904.

Hansen, D.L., Brodsgaard, H.F. and Enkegaard, A. (1999) Life table characteristics of *Macrolophus caliginosus* preying upon *Tetranychus urticae*. *Entomologia Experimentalis et Applicata* 93, 267–273.

Hanson, A.J. (1930) The redberry disease of blackberries. *Proceedings of the Washington State Horticultural Association* 26, 199–201.

Hanson, A.J. (1933) *The Blackberry Mite and its Control (*Eriophyes essigi *Hassan)*. Bulletin No. 279, Washington Agricultural Experiment Station, Pullman, Washington.

Hao, H., Cheng, L., Li, P. and Wang, H. (2009) Determination of toxicities of 8 insecticides to *Eotetranychus sexmaculatus* Riley. *Chinese Journal of Tropical Agriculture* 2009 (3), 1–3.

Hao, H.H., Wang, H.L., Lee, W.T. and Lo, K.C. (1996) Studies on biological control of spider mites on papaya. *Journal of Agricultural Research of China* 45, 411–421 (in Chinese).

Haq, M.A. (2001) Culture and rearing of *Aceria guerreronis* and its predators. *Entomon* 26, 297–302.

Haq, M.A. (2007) Coconut mite tragedy in peninsular India. In: Morales-Malacara, J.B., Behan-Pelletier, V., Ueckermann, E., Pérez, T.M., Estrada-Venegas, E.G. and Badii, M. (eds) *Acarology XI: Proceedings of the International Congress, 8–13 September 2002, Mérida, Universidad Nacional Autonoma de México.* Sociedad Latinoamericana de Acarología, México, pp. 427–432.

Haq, M.A. and Sobhal, T.R. (2010) Weight loss of copra due to infestation by *Aceria guerreronis*. In: Sabelis, M.W. and Bruin, J. (eds) *Trends in Acarology: Proceedings of the 12th International Congress, 21–26 August 2006*, Amsterdam. Springer, Dordrecht, the Netherlands, pp. 509–512.

Haq, M.A. and Sumangala, K. (2003) Acarine regulators of water hyacinth in Kerala (India). *Experimental and Applied Acarology* 29, 27–33.

Haq, M.A., Sumangala, K. and Ramani, N. (2002) Coconut mite invasion, injury and distribution. In: Fernando, L.C.P., Moraes, G.J. de and Wickramananda, I.R. (eds) *Proceedings of the International Workshop on Coconut Mite* (Aceria guerreronis)*, Coconut Research Institute, Sri Lanka, 6–8 January 2000.* Coconut Research Institute, Lunuwila, Sri Lanka, pp. 41–49.

Haq, M.I. and Ali, M.A. (1973) Pink mite of tea and its control. *Tea Journal of Bangladesh* 9(2), 5–11.

Haque, M., Wahab, A., Naher, N. and Begum, A. (2007) Developmental stages of the red spider mite, *Oligonychus coffeae* Neitner (Acari: Tetranychidae) infesting rose. *University Journal of Zoology, Rajshahi University* 26, 71–72.

Haque, M.M. and Kawai, A. (2003) Effect of temperature on development and reproduction of the tomato russet mite, *Aculops lycopersici* (Massee) (Acari: Eriophyidae). *Applied Entomology and Zoology* 38, 97–101.

Hara, A., Tsuda, D., Tavares, J., Yogi, J. and Hensley, D. (1996) *Hibiscus Erineum Mite*. Instant Information Series No. 18, Cooperative Extension Service, University of Hawaii at Manoa, Honolulu, Hawaii.

Hara, A., Tsuda, D., Tavares, J., Yogi, J. and Hensley, D. (2001) *Hibiscus Erineum Mite*. CTAHR Insect Pests 2001, IP-7, Cooperative Extension Service/College of Tropical Agriculture and Human Resources (CTAHR), University of Hawaii at Manoa, Honolulu, Hawaii. Available at: http://www.ctahr.hawaii.edu/oc/freepubs/pdf/ip-7.pdf (accessed 11 May 2015).

Haramoto, F.H. (1969) *Biology and Control of* Brevipalpus phoenicis *(Geijskes) (Acarina: Tenuipalpidae)*. Technical Bulletin No. 68, Hawaii Agricultural Experiment Station, Honolulu, Hawaii.

Harbans, S.K. and Sudhu, A.S. (1961) Sugarcane mite and its control. *Indian Journal of Sugarcane Research* 5, 92–95.

Hardison, J.R. (1959) Evidence against *Fusarium poae* and *Siteroptes graminum* as causal agents of silver top of grasses. *Mycologia* 51, 712–728.

Hardman, J.M., Franklin, J.L., Moreau, D.L. and Bostanian, N.J. (2003) An index for selective toxicity of miticides to phytophagous mites and their predators based on orchard trials. *Pest Management Science* 59, 1321–1332.

Hardouin, R. (1934) Un acarien très nuisible à l'agriculture *Tetranychus ludeni* Zacher. *Bulletin de la Société Entomologique de France* 39, 123–125.

Hare, J.D. (1992) Effects of plant variation on herbivore–natural enemy interactions. In: Fritz, R.S. and Simms, E.L. (eds) *Plant Resistance to Herbivores and Pathogens: Ecology, Evolution and Genetics*. University of Chicago Press, Chicago, Illinois, pp. 278–298.

Hare, J.D. and Phillips, P.A. (1992) Economic effect of the citrus red mite (Acari: Tetranychidae) on southern California coastal lemons. *Journal of Economic Entomology* 85, 1926–1932.

Hare, J.D., Rettig, J.M. and Pehrson, J.E. (1989) Egg production and population growth of the citrus red mite (Acari: Tetranychidae) on differentially irrigated citrus trees. *Environmental Entomology* 18, 651–659.

Hare, J.D., Pehrson, J.E., Clemens, T., Menge, J.A., Coggins, C.W. Jr, Embleton, T.W. and Meyer, J.L. (1992) Effect of citrus red mite (Acari: Tetranychidae) and cultural practices on total yield, fruit size, and crop value of 'Navel' orange: years 3 and 4. *Journal of Economic Entomology* 85, 486–495.

Hare, J.D., Rakha, M. and Phillips, P.A. (1999) Citrus bud mite (Acari: Eriophyidae): an economic pest of California lemons? *Journal of Economic Entomology* 92, 663–675.

Hariyapa, A.S. and Kulkarni, K.A. (1988) Biology and feeling efficiency of the predatory mite *Amblyseius longispinus* (Evans) on chilli mite *Polyphagotarsonemus latus* (Banks). *Journal of Biological Control (Bangalore)* 2, 131–132.

Harley, K.L.S. (1990) The role of biological control in the management of water hyacinth, *Eichhornia crassipes*. *Biocontrol News and Information* 11, 11–22.

Harman, J., Paul, P., Craig, R., Cox-Foster, D., Medford, J. and Mumma, R.O. (1996) Development of a mite bioassay to evaluate plant resistance and its use in determining regeneration of spider mite resistance. *Entomologia Experimentalis et Applicata* 81, 301–305.

Harpaz, I. (1955) Notes on the eriophyid mites of Israel. *Bulletin of the Research Council of Israel* 5, 61–66.

Harpaz, I. and Bernstein, Z. (1960) Occurrence of the bud mite strain of *Eriophyes vitis* (Pgst.) in the Old World and the nature of its damage to grape vines. In: *Verhandlungen XI Internationaler Kongress für Entomologie, Wien, 17 bis 25 August 1960, Band II*. Internationalen Kongresses für Entomologie, Vienna, pp. 47–48.

Harpaz, I. and Rosen, D. (1971) Development of integrated control programs for crop pests in Israel. In: Huffaker, C.B. (ed.) *Biological Control. Proceedings of an AAAS Symposium on Biological Control, Held at Boston, Massachusetts December 30–31, 1969*. Plenum Press, New York, pp. 458–469.

Harranger, J. (1983) Les phytoptes du prunier. *Phytoma* 344, 22–23.

Harries, F.H. (1963) Effects of some antibiotics and other compounds on fertility and mortality of orchard mites. *Journal of Economic Entomology* 56, 438–441.

Harris, K.M. (1982) *Lestodiplosis oomeni* sp.n. (Diptera: Cecidomyiidae), a predator on the carinate tea mite, *Calacarus carinatus* (Green) (Acarina: Eriophyidae) and on other mites on tea plants in Indonesia. *Entomologische Berichten* 42, 20–23.

Harris, P. (1973) The selection of effective agents for the biological control of weeds. *Canadian Entomologist* 105, 1495–1503.

Harrison, I.R. (1956) Control of bulb-scale mite in *Narcissus*. *Plant Pathology* 5, 127–129.

Harrison, R.A. and Smith, A.G. (1961) The influence of temperature and relative humidity on the development of eggs and on the effectiveness of ovicides against *Tetranychus telarius* (L.). *New Zealand Journal of Science* 4, 540–549.

Hart, B.J. and Fain, A. (1987) A new technique for isolation of mites exploiting the difference in density between ethanol and saturated NaCl: qualitative and quantitative studies. *Acarologia* 28, 251–254.

Hartley, S.E. (1998) The chemical composition of plant galls: are levels of nutrients and secondary compounds controlled by the gall-former? *Oecologia* 113, 492–501.

Harty, A., Dooling, W. and Little, A. (2004) Producing world class navel oranges in New Zealand. Part 3: pest and disease control research. *The Orchardist* (formerly *The Orchardist of New Zealand*) 77(6), 57–62.

Harvey, T.L. and Martin, T.J. (1980) Effects of wheat pubescence on infestations of wheat curl mite and incidence of wheat streak mosaic. *Journal of Economic Entomology* 73, 225–227.

Harvey, T.L. and Martin, T.J. (1988) Sticky-tape method to measure cultivar effect on wheat curl mite (Acari: Eriophyidae) populations in wheat spikes. *Journal of Economic Entomology* 81, 731–734.

Harvey, T.L. and Seifers, D.L. (1991) Transmission of wheat streak mosaic virus to sorghum by the wheat curl mite (Acari: Eriophyidae). *Journal of the Kansas Entomological Society* 64, 18–22.

Harvey, T.L., Martin, T.J. and Thompson, C.A. (1979) Controlling wheat curl mite and wheat streak mosaic virus with systemic insecticide. *Journal of Economic Entomology* 72, 854–855.

Harvey, T.L., Martin, T.J. and Seifers, D.L. (1990) Wheat curl mite and wheat streak mosaic in moderate trichome density wheat cultivars. *Crop Science* 30, 534–536.

Harvey, T.L., Martin, T.J. and Seifers, D.L. (1994) Importance of plant resistance to insect and mite vectors in controlling virus diseases of plants: resistance to the wheat curl mite (Acari: Eriophyidae). *Journal of Agricultural Entomology* 11, 271–277.

Harvey, T.L., Martin, T.J., Seifers, D.L. and Sloderbeck, P.E. (1995) Adaptation of wheat curl mite (Acari: Eriophyidae) to resistant wheat in Kansas. *Journal of Agricultural Entomology* 12, 119–125.

Harvey, T.L., Seifers, D.L. and Martin, T.J. (1998) Effect of imidacloprid seed treatment on infestations of wheat curl mite (Acari: Eriophyidae) and the incidence of wheat streak mosaic virus. *Journal of Agricultural and Urban Entomology* 15, 75–81.

Harvey, T.L., Martin, T.J. and Seifers, D.L. (2002) Wheat yield reduction due to Wheat Curl Mite (Acari: Eriophyidae) infestations. *Journal of Agricultural and Urban Entomology* 19, 9–13.

Harvey, T.L., Seifers, D.L., Martin, T.J. and Michaud, J.P. (2005) Effect of resistance to wheat streak mosaic virus on transmission efficiency of wheat curl mites. *Journal of Agricultural and Urban Entomology* 22, 1–6.

Hassan, A.S. (1928) The biology of the Eriophyidae with special reference to *Eriophyes tristriatus* (Nalepa). *University of California Publications in Entomology* 4, 341–394.

Hassan, A.S. (1944) Notes on *Eriophyes mangiferae* sp. n. (Acarina). *Bulletin de la Société Fouad 1er d'Entomologie* 28, 179–182.

Hassan, E.F.O. and Keifer, H.H. (1978) The mango leaf-coating mite, *Cisaberoptus kenyae* K. (Eriophyidae, Aberoptinae). *The Pan-Pacific Entomologist* 54, 185–193.

Hassan, M.F. (1989) Influence of food quantity on development and female fecundity of *Eutogenes punctata* Zaher and Soliman (Cheyletidae: Acari). *Annals of Agricultural Science (Moshtohor)* 27, 587–591.

Hassan, M.F., Afifi, A.M. and Nawar, M.S. (1987) *Dendrolaelaps sayedi*, a new digamasellid mite (Acari: Gamasida) from Egypt, with notes on its biology. *Bulletin of the Entomological Society of Egypt* 66, 227–235.

Hastings, C. (1994) *Penthaleus major*. An occasional little-known pest. *Phytoma* 465, 43–44.

Hatcher, P.E. and Paul, N.D. (2001) Plant pathogen–herbivore interactions and their effects on weeds. In: Jeger, M.J. and Spence, N.J. (eds) *Biotic Interactions in Plant–Pathogen Associations*. CAB International, Wallingford, UK, pp. 193–218.

Hath, T.K. (2000) Distribution of yellow mite (*Polyphagotarsonemus latus* Banks) population on leaves of different jute varieties. *Environment and Ecology* 18, 578–580.

Hath, T.K. (2004) Effect of sowing and jute varieties on the incidence of yellow mite (*Polyphagotarsonemus latus* Banks) and jute semilooper (*Anomis sabulifera* Guen.) under terai region of West Bengal. *Journal of Entomological Research* 28, 311–315.

Hath, T.K. and Jaydeb Ghosh (2001) Influence of date of sowing on the incidence of yellow mite, *Polyphagotarsonemus latus* Banks on different varieties of jute under terai agroecology of West Bengal. *Journal of Entomological Research* 25, 87–92.

Hatzinikolis, E.N. (1968a) A new mite, *Aculus benakii* n. sp. from *Olea europea* L. *Acarologia* 10, 450–452.

Hatzinikolis, E.N. (1968b) Liste d'acariens de la sous-famille Bryobiinae trouvés sur les plantes cultivées en Grèce. *Annales de l'Institut Phytopathologique Benaki* 8, 162–163.

Hatzinikolis, E.N. (1968c) Liste d'acariens de la sous-famille Tetranychinae trouvés sur les plantes cultivées en Grèce. *Annales de l'Institut Phytopathologique Benaki* 8, 193–194.

Hatzinikolis, E.N. (1969a) Acariens Eriophyoidea signalès sur des plantes cultivées en Grèce. *Annales de l'Institut Phytopathologique Benaki* 9, 54–56.

Hatzinikolis, E.N. (1969b) Acariens phytophages signalés en Grèce sur l'olivier (*Olea europea* L.). In: *Report of the 8th FAO ad hoc Conference on the Control of Olive Pests and Diseases, Held in Athens, Greece, 8–12 May 1969/ 8° Congrès ad hoc FAO sur la Lutte Contre le Ravageurs et les Maladies de l'Olivier, Athènes.* Document de Travail No. 14, Food and Agriculture Organization of the United Nations, Rome, p. 5.

Hatzinikolis, E.N. (1969c) Preliminary notes on tetranychoid and eriophyid mites infesting cultivated plants in Greece. In: Evans, G.O. (ed.) *Proceedings of the 2nd International Congress of Acarology, 19–25 July, 1967, Sutton Bonington*, UK. Akadémiaei Kiadó, Budapest, pp. 161–167.

Hatzinikolis, E.N. (1970) Nine species of mites recorded for the first time in Greece. *Annales de l'Institut Phytopathologique Benaki* 9, 238–241.

Hatzinikolis, E.N. (1971) A contribution to the study of *Aceria oleae* (Nalepa, 1900) (Acarina: Eriophyidae). In: Daniel, M. and Rosicky, B. (eds) *Proceedings of the 3rd Internatuonal Congress of Acarology, 31 August–6 September 1971, Prague.* Dr W. Junk, The Hague, the Netherlands, pp. 221–224.

Hatzinikolis, E.N. (1972) La pathogénie et l'écologie de *Tegonotus hassani* Keifer, 1959 sur l'olivier (Acarina, Eriophyidae). *Zeszyty Problemowe Postępów Nauk Rolniczych* 129, 186–191.

Hatzinikolis, E.N. (1979) Studies on the biology and ecology of *Aculus benakii* Hatzinikolis, 1968 (Acarina: Eriophyidae). In: Piffl, E. (ed.) *Proceedings of the 4th International Congress of Acarology, Saalfelden, Austria.* Akadémiai Kiadó, Budapest, pp. 189–191.

Hatzinikolis, E.N. (1982) The mites of olive trees in Greece. In: *II Réunion du Groupe d'Experts CCE. sur "Etat d'Avancement des Travaux et Echange d'Informations sur les Problèmes Posés par la Lutte Intégrée en Oléiculture", Antibes, 4–6 Novembre 1981.* INRA (Institut National de la Recherche Agronomique), Paris, pp. 188–197.

Hatzinikolis, E.N. (1984) Description of *Panonychus caricae* new species on fig-trees in Greece (Acari: Tetranychidae). *Entomologia Hellenica* 2, 55–61.

Hatzinikolis, E.N. (1986) The genus *Tetranychus* Dufour, 1832, in Greece (Acari: Tetranychidae). *Biologia Gallo-Hellenica* 12, 383–388.

Hatzinikolis, E.N. and Emmanouel, N.G. (1991) A revision of the genus *Bryobia* in Greece (Acari: Tetranychidae). *Entomologia Hellenica* 9, 21–34.

Hatzinikolis, E.N. and Kolovos, A. (1985) Tenuipalpid mites of olive trees in the Mediterranean area (Acarida-Tenuipalpidae). In: *Comptes-rendus: 2e Congrès International sur La Zoogéographie et l'Écologie de la Grèce et des Régions Avoisinantes, Tenu à Athènes du 7 au 12 septembre 1981.* Biologia Gallo-hellenica Volume 10, Groupe Franco-hellénique de Recherches Biologiques, Toulouse, France, pp. 201–212.

Hatzinikolis, E.N. and Papanou, A.H. (1975) Les acariens phytophages et leurs hôtes en Grèce. *Nea Agrotike Epitheoresis* 1975, 325–327.

Haug, T., Herren, H.R., Nadel, D.J. and Akinwumi, J.B. (1987) Technologies for the mass-rearing of cassava mealybugs, cassava green mites and their natural enemies. *Insect Science and its Application* (now *International Journal of Tropical Insect Science*) 8, 879–881.

Hazan, A., Gerson, U. and Tahori, A.S. (1973) Life history and life tables of the carmine spider mite. *Acarologia* 15, 414–440.

Hazan, A., Gerson, U. and Tahori, A.S. (1974) Spider mite webbing. I. The production of webbing under various environmental conditions. *Acarologia* 16, 68–84.

Hazan, A., Gerson, U. and Tahori, A.S. (1975) Spider mite webbing. II. The effect of webbing on egg hatchability. *Acarologia* 17, 270–273.

Headley, J.C. (1983) Economic analysis of navel orangeworm control in almonds. *California Agriculture* 27(5, 6), 26–29.

Headley, J.C. and Hoy, M.A. (1986) The economics of integrated mite management in almonds. *California Agriculture* 40(1), 28–30.

Headley, J.C. and Hoy, M.A. (1987) Benefit/cost analysis of an integrated mite management program for almonds. *Journal of Economic Entomology* 80, 555–559.

Heard, A.J. and Chapman, P.E. (1986) A field study of the pattern of local spread of ryegrass mosaic virus in mown grassland.*Annals of Applied Biology* 108, 341–345.

Heikal, I.H., Iskander, N.G. and Sedrak, R.A. (1990) Feeding efficiency of *Scolothrips longicornis* Priesner females on different stages of *Tetranychus urticae* Koch and *Eutetranychus orientalis* (Klein). *Bulletin of the Entomological Society of Egypt* 69, 69–72.

Heitmans, W.R.B., Overmeer, W.P.J. and van der Geest, L.P.S. (1986) The role of *Orius vicinus* Ribaut (Heteroptera, Anthocoridae) as a predator of phytophagous and predaceous mites in a Dutch orchard. *Journal of Applied Entomology* 102, 391–402.

Helal, E.M., El Hamid, M.M.A., Donia, A.R.A. and Zakzouk, E.A. (1995) Toxicity of some pesticides to the citrus brown mite *Eutetranychus orientalis* Klein and its predatory mite *Euseius scutalis* (Athias Henriot). *Alexandria Journal of Agricultural Research* 40, 221–230.

Helle, W. (1962) Genetic of resistence to organophosphorus compounds and its relation to diapause in *Tetranychus urticae* Koch. *Tijdschrift over Plantenziekten* (now *European Journal of Plant Pathology*) 68, 155–195.

Helle, W. (1965a) Population genetics of arrhenotokous mites. *Bollettino di Zoologia Agraria e di Bachicoltura (Ser. II)* 7, 219–225.

Helle, W. (1965b) Resistance in the Acarina: mites. In: Naegele, J.A. (ed.) *Advances in Acarology, Volume 2.* Cornell University Press/Comstock Publishing Associates , Ithaca, New York, pp. 71–93.

Helle, W. (1967) Fertilization in the two spotted spider mite (*Tetranychus urticae*: Acari). *Entomologia Experimentalis et Applicata* 10, 103–110.

Helle, W. (1968) Genetic variability of photoperiodic response in arrhenotokous mite (*Tetranychus urticae*). *Entomologia Experimentalis et Applicata* 11, 101–113.

Helle, W. (1985) Genetics. In: Helle, W. and Sabelis, M.W. (eds) *Spider Mites: Their Biology, Natural Enemies and Control.* World Crop Pests, Volume 1B. Elsevier Science, Amsterdam, pp. 185–192.

Helle, W. and Bolland, H.R. (1967) Karyotypes and sex-determination in spider mites (Tetranychidae). *Genetica* 38, 43–53.

Helle, W. and Bolland, H.R. (1972) Artificial induction of males in a thelytokous mite species by means of X-rays. *Entomologia Experimentalis et Applicata* 15, 395–396.

Helle, W. and Gutierrez, J. (1983) Karyotypes of tetranychid species of New Caledonia. *International Journal of Acarology* 9, 123–126.

Helle, W. and Overmeer, W.J.P. (1985a) Rearing techniques. In: Helle, W. and Sabelis, M.W. (eds) *Spider Mites: Their Biology, Natural Enemies and Control*. World Crop Pests, Volume 1A. Elsevier Science, Amsterdam, pp. 331–335.

Helle, W. and Overmeer, W.J.P. (1985b) Toxicological test methods. In: Helle, W. and Sabelis, M.W. (eds) *Spider Mites: Their Biology, Natural Enemies and Control*. World Crop Pests, Volume 1A. Elsevier Science, Amsterdam, pp. 391–395.

Helle, W. and Pijnacker, L.P. (1985) Parthenogenesis, chromosomes and sex. In: Helle, W. and Sabelis, M.W. (eds) *Spider Mites: Their Biology, Natural Enemies and Control*. World Crop Pests, Volume 1A. Elsevier Science, Amsterdam, pp. 129–139.

Helle, W. and Sabelis, M.W. (eds) (1985) *Spider Mites. Their Biology, Natural Enemies and Control*. World Crop Pests, Volume 1B. Elsevier Science, Amsterdam.

Helle, W. and van de Vrie, M. (1974) Problems with spider mites. *Outlook on Agriculture* 8, 119–125.

Helle, W. and Wysoki, M. (1983) The chromosomes and sex-determination of some actinotrichid taxa (Acari), with special reference to Eriophyidae. *International Journal of Acarology* 9, 67–71.

Helle, W. and Wysoki, M. (1996) Arrhenotokous parthenogenesis. In: Lindquist, E.E., Sabelis, M.W. and Bruin, J. (eds) *Eriophyoid Mites: Their Biology, Natural Enemies and Control*. World Crop Pests, Volume 6. Elsevier Science, Amsterdam, pp. 169–172.

Helle, W., Gutierrez, J and Bolland, H.R. (1970) A study on sex-determination and karyotypic evolution in Tetranychidae. *Genetica* 41, 21–32.

Helle, W., Bolland, H.R., van Arrendok, R., DeBoer, R., Schulten, G.G.M. and Russel, V. (1978) Genetic evidence for biparental males in haplo-dipolid predator mites (Acarina: Phytoseiidae). *Genetica* 49, 165–171.

Helle, W., Bolland, H.R. and Heitmans, W.R.B. (1980) Chromosomes and types of parthenogenesis in the false spider mites (Acari: Tenuipalpidae). *Genetica* 54, 45–50.

Helle, W., Bolland, H.R. and Heitmans, W.R.B. (1981) A survey of chromosome complements in the Tetranychidae. *International Journal of Acarology* 7, 147–156.

Helle, W., Tempelaar, M.J. and Drenth-Diephuis, L.J. (1983) DNA-contents in spider mites (Tetranychidae) in relation to karyotype evolution. *International Journal of Acarology* 9, 127–129.

Helle, W., Bolland, H.R., Jeurissen, S.H.M. and Van Seventer, G.A. (1984) Chromosome data on the Actinedida, Tarsonemida and Oribatida. In: Griffiths, D.A. and Bowman, C.E. (eds) *Acarology VI, Volume 1*. Ellis Horwood, Chichester, UK, pp. 449–454.

Hellemans, A. and Goossens, F. (2000) Control of box spider mite (*Eurytetranychus buxi*). *Verbondsnieuws* 44 (15), 27–29.

Hellqvist, S. (2002) Heat tolerance of strawberry tarsonemid mite *Phytonemus pallidus*. *Annals of Applied Biology* 141, 67–71.

Hely, P.C. (1968) *The Entomology of Citrus in New South Wales*. Australian Entomological Society Miscellaneous Publication 1, East Melborne, Victoria, Australia.

Henderson, C.F. (1960) A sampling technique for estimating populations of small arthropods in soil and vegetation. *Journal of Economic Entomology* 53, 115–121.

Henderson, C.F. and Holloway, J.K. (1942) Influence of leaf age and feeding injury on the citrus red mite. *Journal of Economic Entomology* 35, 683–686.

Henderson, C.F. and McBurnie, H.V. (1943) *Sampling Techniques for Determining Populations of the Citrus Red Mite and Its Predators*. Circular No. 671, US Department of Agriculture Circular No. 671, Washington, DC.

Henderson, R.C. (2001) Technique for positional slide-mounting of Acari. *Systematic and Applied Acarology Special Publications* 7, 1–4.

Hendrichs, J., Franz, G. and Rendon, P. (1994) Use of sterile male-only releases as "biological pesticides" for control of Mediterranean fruit flies in commercial fruit production. *IOBC-WPRS Bulletin* 17(6), 212–224.

Henry, J.M., Wegulo, S.N., Gibeault, V.A. and Autio, R. (2005) Turfgrass performance with reduced irrigation and nitrogen fertilization. *International Turfgrass Society Research Journal* 10, 93–101.

Herbert, H.J. (1959) Note on feeding ranges of six species of predaceous mites (Acarina: Phytoseiidae) in the laboratory. *Canadian Entomologist* 91, 812.

Herbert, H.J. (1962) Influence of *Typhlodromus* (*T.*) *pyri* Scheuten on the development of *Bryobia arborea* M. & A. populations in the greenhouse. *Canadian Entomologist* 94, 870–873.

Herbert, H.J. (1965) The brown mite, *Bryobia arborea* Morgan and Anderson, on apple in Nova Scotia. *Canadian Entomologist* 97, 1303–1318.

Herbert, H.J. (1974) Notes on the biology of the apple rust mite, *Aculus schlechtendali* (Prostigmata: Eriophyidae), and its density on several cultivars of apple in Nova Scotia. *Canadian Entomologist* 106, 1035–1038.

Herbert, H.J. (1979) Population trends and behavior of the pear rust mite, *Epitrimerus pyri* (Prostigmata: Eriophyoidea) on pears in Nova Scotia. *Canadian Entomologist* 111, 955–957.

Herbert, H.J. (1981) Biology, life tables, and intrinsic rate of increase of the European red mite, *Panonychus ulmi* (Acarina: Tetranychidae). *Canadian Entomologist* 113, 65–71.

Herbert, H.J. (1982) Biology, life tables, and innate capacity for increase of the twospotted spider mite, *Tetranychus urticae* (Acarina: Tetranychidae). *Canadian Entomologist* 113, 371–378.

Herbert, H.J. and Butler, K.P. (1973a) The effects of European red mite, *Panonychus ulmi* (Acarina: Tetranychidae) infestations on N, P and K concentrations in apple foliage throughout the season. *Canadian Entomologist* 105, 263–269.

Herbert, H.J. and Butler, K.P. (1973b) Sampling systems for European red mite, *Panonychus ulmi* (Acarina: Tetranychidae), eggs on apple in Nova Scotia. *Canadian Entomologist* 105, 1519–1523.

Herbert, H.J. and McRae, K.B. (1982) Predicting eclosion of overwintering eggs of European red mite, *Panonychus ulmi* (Acarina: Tetranychidae), in Nova Scotia. *Canadian Entomologist* 114, 703–712.

Herbert, H.J. and Sanford, K.H. (1969) The influence of spray programs on the fauna of apple orchards in Nova Scotia. XIX. Apple rust mite, *Vasates schlechtendali*, a food source for predators. *Canadian Entomologist* 101, 62–67.

Heriot, A.D. (1935) Notes on the blister mite, *Eriophyes pyri*. *Proceedings of the Entomological Society of British Columbia* 31, 41–42.

Herms, D.A. and Mattson, W.J. (1992) The dilemma of plants: to grow or to defend. *Quarterly Review of Biology* 67, 283–335.

Hernandes, F.A. and Feres, R.J.F. (2006) Diversidade e sazonalidade de ácaros (Acari) em seringal (*Hevea brasiliensis*, Muell. Arg.) no Noroeste do Estato de São Paulo. *Neotropical Entomology* 35, 523–535.

Hernandes, F.A. and Feres, R.J.F. (2013) Phylogeny and taxonomic revision of the spider mite genera *Aponychus*, *Paraponychus* and *Stylophoronychus* using morphology (Acari: Tetranychidae). *Invertebrate Systematics* 27, 265–281.

Hernández, C. and Escobio, J. (1988) Efecto de diferentes formas de aplicación con asperjadoras de ventilador axial sobre el combate de *Mycosphaerella citri* y *Phyllocoptruta oleivora* en el cultivo de la toronja. *Ciencia y Tecnica en la Agricultura. Proteccion de Plantas* 11(3), 33–41.

Hernández, E.N., Vargas, H.A. and Zuluaga, J.I. (1988) Efecto de algunos insecticidas sobre el comportamiento de dispersión de *Tetranychus mexicanus* (Acari: Tetranychidae) y observaciones sobre la acción de varios acaricidas para su control. *Revista Colombiana de Entomologia* 14(2), 15–24.

Hernández, J.L., Ginarte, A., Gómez, P.J., Cabrera, I., Galano, R., Vieira, R., Duany, A. and Veja, M. (2003) Recomendaciones agronómicas para el manejo del ácaro *Steneotarsonemus spinki* Smiley em el cultivo de arroz. In: *Memórias del I Forum Ramal del Cultivo del Arroz, Instituto de Investigaciones del Arroz, 11–12 December 2003, Camagüey, Cuba*, pp. 19–23.

Hernández, J.L., Gómez, P., Galano, R., Botta, E., Duany, A., Ginarte, A. and Berbén, T. (2005) Influencia del marco y densidad de siembra la fertilización nitrogenadada y el manejo de agua sobre la población y daños de *S. spinki* y los patógenos a él asociados. In: *III Encuentro International del Arroz: El Ácaro del Arroz* Steneotarsonemus spinki *(Tarsonemidae) Retos y Alternativas para América Latina y el Caribe, Hotel Palco Ciudad de La Habana, Cuba, 6 de Junio de 2005. Libro de Resumen*, pp. 129–131.

Hernández-Abarca, J.T. and Telesforo, J. (2007) Tablas de vida y fecundidad y constantes térmicas de *Pediculaster mesembrinae* (Canestrini) (Acari: Siteroptidae) y *Brennandania lambi* (Kreczal), (Acari: Pygmephoridae), en champiñon *Agaricus bisporus* (Lange) Imbach. PhD thesis, Colegio de Postgraduados, Montecillo, Mexico. Available at: http://www.biblio.colpos.mx:8080/xmlui/handle/10521/1503 (accessed 5 August 2014).

Hernández-Abarca, J.T., Otero-Colina, G., Vera-Graziano, J., Dionicio Alvarado-Rosales, J.L. and Aranda-Escobar, E. (2005) Life tables and thermal constants of *Pediculaster mesembrinae* (Acari: Siteroptidae) fed on the fungus, *Trichoderma harzianum* Rifai. *International Journal of Acarology* 31, 165–170.

Hernández Riesco, S., Cabaleiro Sobrino, C., Jacas Miret, J. and Martín López, B. (2002) El empleo de aceites minerales, vegetales y de pescado en el control integrado de plagas y enfermedades del vinedo. *Boletín de Sanidad Vegetal, Plagas* 28, 223–237.

Hernández Roque, F. (1977) Combate quimico del eriofido del cocotero *Aceria (Eriophyes) guerreronis* (K.) en la costa de Guerrero. *Agricultura Técnica en México* 4, 23–38.

Hernández-Suárez, E., Velásquez, M. del C., Paz-González, M.I., González, F.J., Carnero, A. and Ferragut, F.J. (2006) Effect of different types of prey on fecundity of the phytoseiid mite *Typhlodromips swirskii*, a potential biological control agent for horticultural greenhouse pests in the Canary Islands. *IOBC-WPRS Bulletin* 29(4), 137 (abstract).

Hernández Suárez, E., Torres Luis, E., López González, A.A., Perera González, S. and Saavedra Oliva, O.L. (2010) *Evaluación de Eficacia de Acaricidas y Sueltas Inundativas de* Neoseiulus californicus *para el Control de la Araña Cristalina* Oligonychus perseae *en Aguacate*. Cabildo Insular de Tenerife, Santa Cruz de Tenerife and Instituto Canario de Investigaciones Agrarias, La Laguna, Tenerife.

Herne, D.H.C. (1968) Some responses of European red mites, *Panonychus ulmi* (Koch) to immersion in water. *Canadian Entomologist* 100, 540–541.

Herne, D.H.C., Cranham, J.E. and Easterbrook, M.A. (1979) New acaricides to control resistant mites. In: Rodriguez, J.G. (ed.) *Recent Advances in Acarology, Volume I*. Academic Press, New York, pp. 95–104.

Hernera-Villamil, G. (1958) Biologia y control de la falsa aranita roja de la vid. (*Brevipalpus chilensis* Baker). *Agricultura Tecníca (Casilla)* 18, 35–42.

Herr, R. (1986) Untersuchungen über die Resistenzmechanismen der Gattung *Ribes* gegen die Johannisbeergallmilbe *Cecidophyopsis ribis*. *Mitteilungen aus der Biologischen Bundesanstalt für Land- und Forstwirtschaft, Berlin-Dahlem* No. 232, 359 (abstract).

Herr, R. (1988) Untersuchungen zum Resistenzmechanismus der Gattung *Ribes* gegen die Johannisbeergallmilbe *Cecidophyopsis ribis*. *Mitteilungen der Deutschen Gesellschaft für Allgemeine und Angewandte Entomologie* 6, 17–21.

Herr, R. (1991) Untersuchungen zur Resistenz der Gattung *Ribes* gegen die Johannisbeergallmilbe, *Cecidophyopsis ribis* (Westw.) (Acari, Eriophyidae). Infektionsversuche und Biotests. *Journal of Applied Entomology* 112, 181–193.

Herrera, L.A.R. (2005) *Ácaro del Vaneamiento del Arroz* – Steneotarsonemus spinki *Smiley (Prostigmata: Tarsonemidae)*. Ciencia Tecnología (Plegable Informativo), Fondo LatinoAmericana, Cali, Colombia.

Herron, G.A. and Rophail, J. (2000) Potential insecticides for control of *Oligonychus ilicis* (McGregor) (Acari: Tetranychidae), a new threat to Australian horticulture. *Australian Journal of Entomology* 39, 86–88.

Herron, G.A., Edge, V. and Rophail, J. (1993) Clofentezine and hexythiazox resistance in *Tetranychus urticae* Koch in Australia. *Experimental and Applied Acarology* 17, 433–440.

Herron, G.A., Beattie, G.A.C., Parkes, R.A. and Barchia, I. (1995) Potter spray tower bioassay of selected citrus pests to petroleum spray oil. *Journal of the Australian Entomological Society* 34, 255–263.

Herron, G.A., Edge, V.E., Wilson, L.J. and Rophail, J. (1998) Organophosphate resistance in spider mites (Acari: Tetranychidae) from cotton in Australia. *Experimental and Applied Acarology* 22, 17–30.

Hesse, M. (1971) Über Mehrkernigkeit und Polyploidisierung der Nahrgewebe einiger Milbengallen. *Österreichische Botanische Zeitschrift* 119, 74–93.

Hessein, N.A. and Perring, T.M. (1986) Feeding habits of the Tydeidae with evidence of *Homeopronematus anconai* (Acari: Tydeidae) predation of *Aculops lycopersici* (Acari: Eriophyidae). *International Journal of Acarology* 12, 215–221.

Hessein, N.A. and Perring, T.M. (1988) *Homeopronematus anconai* (Baker) (Acari: Tydeidae) predation on citrus flat mite, *Brevipalpus lewisi* McGregor (Acari: Tenuipalpidae). *International Journal of Acarology* 14, 89–90.

Hetenyi, E. (1954) A gyapoton é lö takacsatkak (*Tetranychus urticae* Koch) és az ellen ük valo védekezés. *Növényvédelem* 1, 11–19 (in Hungarian).

Heungens, A. (1993) Chemical control of the soft mite *Tarsonemus belemnitoides* Weis Fogh (Tarsonemidae) in azalea culture. *Parasitica* 49, 3–9.

Heungens, A. and Tirry, L. (2000) Curative chemical control of the tarsonemid mites *Tarsonemus confusus* and *Tarsonemoides belemnitoides* in azalea culture. *Parasitica* 56, 123–130.

Higgins, T.J. and Ridsdill-Smith, T.J. (1994) Preliminary testing of proteins with potential for the design of genetically engineered subterranean clovers with protection against redlegged earth mite. In: McDonald, G. and Hoffmann, A.A. (eds) *Proceedings of the Second National Workshop on Redlegged Earth Mite, Lucerne Flea and Blue Oat Mite, October 1994, Rutherglen, Victoria, Australia*. Department of Agriculture, Rutherglen, Australia, p. 8.

Hightower, B.C. and Martin, D.P. (1956) Ecological studies of spider mites attacking cotton in central Texas. *Journal of Economic Entomology* 49, 424–425.

Higley, L.G. and Pedigo, L.P. (1996) *Economic Thresholds for Integrated Pest Management*. University of Nebraska Press, Lincoln, Nebraska.

Hildebrand, E.M. (1945) Myrobalan mottle and asteriod spot. *Phytopathology* 35, 47–50.

Hill, A. and Deahl, K.L. (1978) Description and life cycle of a new species of *Histiostoma* (Acari: Histiostomidae) associated with commercial mushroom production. *Proceedings of the Entomological Society of Washington* 80, 317–329.

Hill, R.L. and O'Donnell, D.J. (1991a) Reproductive isolation between *Tetranychus lintearius* and two related species, *T. urticae* and *T. turkestani* (Acarina: Tetranychidae). *Experimental and Applied Acarology* 11, 241–251.

Hill, R.L. and O'Donnell, D.J. (1991b) The host range of *Tetranychus lintearius* (Acarina: Tetranychidae). *Experimental and Applied Acarology* 11, 253–269.

Hill, R.L. and Stone, C. (1985) Spider mites as control agents for weeds. In: Helle, W. and Sabelis, M.W. (eds) *Spider Mites: Their Biology, Natural Enemies and Control*. World Crop Pests, Volume 1B. Elsevier Science, Amsterdam, pp. 443–448.

Hill, R.L., Gourlay, A.H. and Wigley, P.J. (1989) The introduction of gorse spider mite, *Tetranychus lintearius*, for biological control of gorse. In: Popay, A.J. (ed.) *Proceedings of the Forty Second New Zealand Weed and Pest Control Conference, 8–10 August 1989, Taranki Country Lodge, New Plymouth, New Zealand*. New Zealand Weed and Pest Control Society, Palmerston North, New Zealand, pp. 137–139.

Himanen, S., Vuorinen, T., Tuovinen, T. and Holopainen, J.K. (2005) Effects of cyclamen mite (*Phytonemus pallidus*) and leaf beetle (*Galerucella tenella*) damage on volatile emission from strawberry (*Fragaria* × *ananassa* Duch.) plants and orientation of predatory mites (*Neoseiulus cucumeris*, *N. californicus*, and *Euseius finlandicus*). *Journal of Agricultural and Food Chemistry* 53, 8624–8630.

Hinomoto, N. and Takafuji, A. (2001) Genetic diversity and phylogeny of the Kanzawa spider mite, *Tetranychus kanzawai*, in Japan. *Experimental and Applied Acarology* 25, 355–370.

Hinomoto, N. and Takafuji, A. (2004) Evaluation of mitochondrial cytochrome oxydase subunit I sequences in *Tetranychus kanzawai* Kishida (Acari: Tetranychidae) for phylogeographic studies. *Journal of the Acarological Society of Japan* 13, 47–55.

Hinomoto, N., Osakabe, M., Gotoh, T. and Takafuji, A. (2001) Phylogenetic analysis of green and red forms of the two-spotted spider mite, *Tetranychus urticae* Koch (Acari: Tetranychidae), in Japan, based on mitochondrial cytochrome oxidase subunit I sequences. *Applied Entomology and Zoology* 36, 459–464.

Hinomoto, N., Dinh Pha, T., Anh Tuan, P., Thi Bao Ngoc, L., Tajima, R., Ohashi, K., Osakabe, M. and Takafuji, A. (2007) Identification of spider mites (Acari: Tetranychidae) by DNA sequences: a case study in northern Vietnam. *International Journal of Acarology* 33, 53–60.

Hirata, H., Sakamaki, Y. and Tsuda, K. (2007) A simple method to monitor the population density of the tomato russet mite, *Aculops lycopersici* on tomato plants. *Kyushu Plant Protection Research* 53, 82–85.

Hirata, K., Kudo, M., Miyake, T., Kawamura, Y. and Ogura, T. (1988) NC-129, a new acaricide. In: *Proceedings of the Brighton Crop Protection Conference: Pests and Diseases – 1988, Volume 1*. BCPC Publications, Farnham (now Alton), UK, pp. 41–48.

Hirst, S. (1920) Revision of the English species of red spider (Genera *Tetranychus* and *Oligonychus*). *Proceedings of the Zoological Society of London* 90, 49–60.

Hirst, S. (1922) On some new parasitic mites. *Proceedings of the Zoological Society of London* 92, 769–802.

Hirst, S. (1923a) On some arachinid remains from the Old Red Sandstone (Rhynie Chert Bed, Aberdeenshire). *The Annals and Magazine of Natural History (Series 9)* 12(70), 455–474.

Hirst, S. (1923b) On some new or little-known species of Acari. *Proceedings of the Zoological Society of London* 93, 971–1000.

Hirst, S. (1924) LV. On some new species of red spider. *The Annals and Magazine of Natural History (Series 9)* 14(83), 522–527.

Hirst, S. (1925) 4. Descriptions of new Acari, mainly parasitic on rodents. *Proceedings of the Zoological Society of London* 95, 49–69.

Hirst, S. (1926) Descriptions of new mites including four new species of "red spider". *Proceedings of the Zoological Society of London* 96, 825–841.

Hislop, R.G. and Jeppson, L.R. (1976) Morphology of the mouthparts of several species of phytophagous mites. *Annals of the Entomological Society of America* 69, 1125–1135.

Hislop, R.G. and Prokopy, R.J. (1981) Mite predator responses to prey and predator-emitted stimuli. *Journal of Chemical Ecology* 7, 895–904.

Hislop, R.G., Auditore, P.J., Weeks, B.L. and Prokopy, R.J. (1981) Repellency of pesticides to the mite predator *Amblyseius fallacis*. *Protection Ecology* 3, 253–257.

Hluchy, M. (1993) Zur biologischen Bekämpfung der Kräuselmilbe *Calepitrimerus vitis* Nalepa (Acari, Eriophyidae) auf der Weinrebe durch die Raubmilbe *Typhlodromus pyri* Scheuten (Acari, Phytoseiidae). *Journal of Applied Entomology* 116, 449–458.

Hluchy, M. and Pospisil, Z. (1992) Damage and economic injury levels of eriophyid and tetranychid mites on grapes in Czechoslovakia. *Experimental and Applied Acarology* 14, 95–106.

Hluchy, M., Pospisil, Z. and Zacharda, M. (1991) Phytophagous and predatory mites (Acari: Tetranychidae, Eriophyidae, Phytoseiidae, Stigmaeidae) in south Moravian vineyards, Czechoslovakia, treated with various types of chemicals. *Experimental and Applied Acarology* 13, 41–52.

Hıncal, P., Yasaraknc, N. and Cnarlı, İ. (2002) İzmir ilinde domates pas akar (*Aculops lycopersici* Massee) (Acarina: Eriophyidae) 'n popülasyon seyri, doğal düşmanları ve kimyasal mücadelesi üzerinde araştırmalar. *Bitki Koruma Bülteni* 42(1/4), 9–22 (in Turkish).

Ho, C.-C. (1990) A preliminary study on the biological control of *Tetranychus kanzawai* in tea field by *Amblyseius fallacis* and *Phytoseiulus persimilis* (Acarina: Tetranychidae, Phytoseiidae). *Journal of Agricultural Research of China* 39, 133–140 (in Chinese).

Ho, C.-C. (2000) Spider-mite problems and control in Taiwan. *Experimental and Applied Acarology* 24, 453–462.

Ho, C.-C. (2006) Monitoring on two exotic spider mites in Taiwan. Available at: http://www.agnet.org/htmlarea_file/activities/20110826121346/paper-246339602.pdf (accessed 24 March 2015).

Ho, C.–C. and Chen, J.S. (1987) A new record of bulb mite, *Rhizoglyphus setosus* Manson, (Acarina: Acaridae) from Taiwan. *Journal of Agricultural Research of China* 36, 237–238 (in Chinese).

Ho, C.-C. and Chen, W.-H. (1992a) Control of phytoseiids in a spider mite mass-rearing system (Acari: Phytoseiidae, Tetranychidae). *Experimental and Applied Acarology* 13, 287–293.

Ho, C.-C. and Chen, W.H. (1992b) Species survey of spider mites and seasonal occurrence of *Thrips palmi, Chlorita biguttula biguttula*, and *Tetranychus cinnabarinus* on eggplant. *Chinese Journal of Entomology* 12, 259–268 (in Chinese).

Ho, C.-C. and Chen, W.-H. (1994) Infestation of grape berry by spider mites. *Plant Protection Bulletin (Taipei)* 36, 251–255 (in Chinese).

Ho, C.-C. and Chen, W.H. (1998) Life history, food consumption, and seasonal occurrence of *Feltiella minuta* (Diptera: Cecidomyiidae) on eggplant. *Chinese Journal of Entomology* 18, 27–37 (in Chinese).

Ho, C.-C. and Chen, W.H. (2001) Evaluation of feeding and ovipositing responses of *Scolothrips indicus* (Thysanoptera: Aeolothripidae) to amounts of Kanzawa spider mite eggs (Acari: Tetranychidae). *Plant Protection Bulletin (Taipei)* 43, 165–172.

Ho, C.-C. and Chen, W.H. (2002) Evaluation of feeding and ovipositing responses of *Oligota flavicornis* (Coleoptera: Staphylinidae) to amounts of Kanzawa spider mite eggs (Acari: Tetranychidae). *Plant Protection Bulletin (Taipei)* 44, 15–20.

Ho, C.-C. and Lo, K.-C. (1979) The survey of the host range of *Steneotarsonemus spinki* (Acari: Tarsonemidae). *National Science Council Monthly (Tapei)* 7, 1022–1028 (in Chinese).

Ho, C.-C. and Shih, S.-P. (2004) *Eotetranychus lewisi*, a new pest of poinsettia from Taiwan. *Plant Protection Bulletin (Taipei)* 46, 173–176.

Ho, C.-C. and Wang, S.C. (2007) Addendum on the records of *Tetranychus evansi* in Taiwan. *Plant Protection Bulletin (Taipei)* 49, 167–168.

Ho, C.-C., Lo, K.-C. and Chen, W.-H. (1997) Spider mite (Acari: Tetranychidae) on various crops in Taiwan. *Journal of Agricultural Research of China* 46, 333–346.

Ho, C.-C., Wang, S.-C. and Chien, Y.-L. (2004) Field observation on 2 newly recorded spider mites in Taiwan. *Plant Protection Bulletin (Taiwan)* 47, 391–402.

Hobson, G.W. (1977) Characterization of acaricidal resistance of the Banks grass mite, *Oligonychus pratensis* (Banks) in west Texas. MSc. thesis, Faculty of Texas Tech University, Lubbock, Texas.

Hobza, R.F. (1970) Interrelationships of temperature and humidity on development of citrus rust mite, *Phyllocoptruta oleivora* (Ashm.). MSc. thesis, University of California, Riverside, California.

Hoda, F.M., El Naggar, M.E., Taha, A.H. and Ibrahim, G.A. (1987) Effect of different types of food on fecundity of predacious mite *Amblyseius swirskii* Athias Henriot (Acari: Phytoseiidae). *Bulletin of the Entomological Society of Egypt* 66, 113–116.

Hoddle, M.S. (1998) Biology and management of the persea mite. *California Avocado Society Yearbook* 82, pp. 75–86.

Hoddle, M.S. and Kerguelen, V. (1999a) Biological control of persea mite, *Oligonychus perseae* Tuttle, Baker & Abatiello (Acari: Tetranychidae) and investigation of host plant resistance. *Proceedings of the 99 California Avocado Research Symposium, Riverside, California*. California Avocado Society and University of California, Riverside, California, pp. 47–51. Available at: http://www.avocadosource.com/ARAC/SUM_1999/SYMP_1999_PG_47-51.pdf (accessed 11 May 2015).

Hoddle, M.S. and Kerguelen, V. (1999b) Determining the susceptibility of avocado cultivars to feeding by the persea mite, *Oligonychus perseae* (Acari: Tetranychidae). *Revista Chapingo, Serie Horticultura* 5, 301–308.

Hoddle, M.S., Aponte, O., Kerguelen, V. and Heraty, J. (1999) Biological control of *Oligonychus perseae* (Acari: Tetranychidae) on avocado: I. Evaluating release timings, recovery and efficacy of six commercially available phytoseiids. *International Journal of Acarology* 25 (3), 211–219.

Hoddle, M.S., Robinson, L. and Virzi, J. (2000a) Biological control of *Oligonychus perseae* (Acari: Tetranychidae) on avocado: III. Evaluating the efficacy of varying release rates and release frequency of *Neoseiulus californicus* (Acari: Phytoseiidae). *International Journal of Acarology* 26(3), 203–214.

Hoddle, M.S., Jones, J., Oishi, K. and Robinson, L. (2000b) The suitability of different diets for the development and reproduction of *Franklinothrips* n. sp. (Thysanoptera: Aeolothripidae). In: Hoddle, M.S. (ed.) *California Conference on Biological Control II, The Historic Mission Inn, Riverside, 11–12 July 2000*. Center for Biological Control, College of Natural Resources, University of California, Berkeley, California, pp. 139–142.

Hodek, I., Van Emden, H.F. and Honek, A. (eds) (2012) *Ecology and Behaviour of the Ladybird Beetles (Coccinellidae)*. Wiley-Blackwell, Chichester, UK.

Hodel, D.R. and Johnson, D.V. (2007) *Imported and American Varieties of Dates in the United States*. Agriculture and Natural Resources Publication 3498, University of California, Oakland, California.

Hodgkiss, H.E. (1930) *The Eriophyidae of New York. II. The Maple Mites*. Technical Bulletin No. 163, New York State Agricultural Experiment Station, Geneva, New York.

Hodson, W.E.H. (1934) The bionomics of the bulb scale mite, *Tarsonemus approximatus* Banks var. *narcissi Ewing*. *Bulletin of Entomological Research* 25, 177–185.

Hodson, W.E.H. (ed.) (1948) *Narcissus Pests*, 3rd edn. MAFF (Ministry of Agriculture and Fisheries) Bulletin No. 51. Her Majesty's Stationery Office, London.

Hoffmann, A.A., Porter, S. and Kovacs, I. (1997) The response of the major crop and pasture pest, the red-legged earth mite (*Halotydeus destructor*) to pesticides: dose-response curves and evidence for tolerance. *Experimental and Applied Acarology* 21, 151–162.

Hoffmann, A.A., Weeks, A.R., Nash, M.A., Mangano, G.P. and Umina P.A. (2008) The changing status of invertebrate pests and the future of pest management in the Australian grains industry. *Australian Journal of Experimental Agriculture* 48, 1481–1493.

Hoffmann, D., Vierheilig, H. and Schausberger, P. (2011) Mycorrhiza-induced trophic cascade enhances fitness and population growth of an acarine predator. *Oecologia* 166, 141–149.

Höhn, H. and Höpli, H.U. (1990) Die Apfelrostmilbe-oft überschätzt, aber kaum prognostizierbar! *Schweizer Zeitschrift für Obst- und Weinbauu* 126, 259–266.

Holanda, A.C.A. de and Oliveira, J.V. de (1992) Population fluctuations of *Polyphagotarsonemus latus* (Banks, 1904) and *Tetranychus desertorum* Banks, 1900 (Acari: Tetranychidae) on *Phaseolus vulgaris* L. and *Vigna unguiculata* (L.) Walp. *Anais de Sociedade Entomológica do Brasil* 21, 5–14.

Holdaway, F.G. (1941) *Insects of Tomato and their Control*. Circular 116, Agricultural Extension Service, University of Hawaii, Honolulu, Hawaii.

Holighaus, F. and Dahlbender, W. (1992) Birnenpockenmilbenbekämpfung. In: *Jahresbericht des Landespflanzenschutzdienstes Rheinland-Pfalz 1992*. Rheinland-Pfalz Ministerium für Landwirtschaft, Weinbau und Forsten, Mainz, Germany, p. 107.

Hollingsworth, C.S. and Berry, R.E. (1982) Two-spotted spider mite (Acari: Tetranychidae) in peppermint: population dynamics and influence of cultural practices. *Environmental Entomology* 11, 1280–1284.

Holloway, T.E. (1936) Insect pests of sugarcane XIV. The Java leaf mite. *F.S.S.: Facts About Sugar* (now *Sugar*) 31(6), 216–217.

Holmes, S.J.I. (1979) Effect of ryegrass mosaic virus on the quality of perennial ryegrass. *Annals of Applied Biology* 91, 75–79.

Holmes, S.J.I. (1980) Field studies on the effects of ryegrass mosaic virus on ryegrass yield. In: Plumb, R.T. (ed.) *Proceedings of the 3rd Conference on Virus Diseases of Gramineae in Europe, Rothamsted, 28–30 May 1980*. Rothamsted Experimental Station, Harpenden, UK, pp. 103–108.

Honarparvar, N., Khanjani, M. and Ueckermann, E.A. (2010a) Effect of temperature on development and fecundity of the brown mite, *Bryobia rubrioculus* Scheuten (Acari: Tetranychidae), on sweet cherry. In: Moraes, G.J. de, Castilho, R.C. and Flechtmann, C.H.W. (eds) *XIII International Congress of Acarology, August 23–27, 2010, Recife-PE, Brazil, Abstract Book*, abstract 197, pp. 108–109. Available at: http://www.acarology.org/ica/ica2010/docs/0910_15_abstractBookCapa.pdf (accessed 3 February 2015).

Honarparvar, N., Khanjani, M. and Ueckermann, E.A. (2010b) Temperature-dependent demographic parameters of the brown mite, *Bryobia rubrioculus* (Scheuten) (Acari: Tetranychidae), under laboratory conditions. In: Moraes, G.J. de, Castilho, R.C. and Flechtmann, C.H.W. (eds) *XIII International Congress of Acarology, August 23–27*, 2010, *Recife-PE, Brazil, Abstract Book*, abstract 198, p. 109. Available at: http://www.acarology.org/ica/ica2010/docs/0910_15_abstractBookCapa.pdf (accessed 3 February 2015).

Hong, C. and Hansen, M.A. (2012) *Rose Rosette Disease*. Publication No. 450-620 (VT/0812/PPWS-10NP), Virginia Cooperative Extension, Virginia Polytechnic Institute and State University, Blacksburg, Virginia. Available at: https://pubs.ext.vt.edu/450/450-620/450-620_pdf.pdf (accessed 28 January 2015).

Hong, X.-Y. and Kuang, H.-Y. (1989) Three new genera and seven new species of the subfamily Phyllocoptinae (Acari: Eriophyidae) from China. *International Journal of Acarology* 15, 145–152.

Hong, X. and Zhang, Z.Q. (1996a) A cladistic analysis of the Eriophyoidea (Acari: Prostigmata): tests of monophyly of families. *Systematic and Applied Acarology* 1, 107–122.

Hong, X.-Y. and Zhang, Z.-Q. (1996b) *Eriophyoid Mites of China: Illustrated Catalogue and Identification Keys*. Memoir of Entomology International, Volume 7. The Associated Publishers, Gainesville, Florida.

Hong, X.-Y., Ji, J., Ma, J.-X. and Dong, H.-Q. (1998) Tests of host plant suitability of the tea pink mite, *Acaphylla theae* (Eriophyoidea: Eriophyidae), at different temperatures. *Systematic and Applied Acarology* 3, 63–68.

Hong, X.-Y., Wang, D.-S. and Zhang, Z.-Q. (2006) Distribution and damage of recent invasive eriophyoid mites (Acari: Eriophyoidea) in mainland China. *International Journal of Acarology* 32, 227–240.

Hong, X.-Y., Xue, X.-F. and Song, Z.-W. (2010a) Eriophyoidea of China: a review of progress with a checklist. In: Zhang, Z.Q., Hong, X.-Y. and Fan, Q.H. (eds) *Xin Jie-Liu Centenary: Progress in Chinese Acarology*. *Zoosymposia* 4, 57–93.

Hong, X.-Y., Zhang, Z.-Q. and Li, G.-Q. (2010b) Tetranychidae of China: a review of progress with a checklist. In: Zhang, Z.Q., Hong, X.-Y. and Fan, Q.H. (eds) *Xin Jie-Liu Centenary: Progress in Chinese Acarology*. *Zoosymposia* 4, 133–150.

Horn, T.B., Johann, L. and Ferla, N.J. (2011) Ecological interactions between phytophagous and predaceous mites in citrus agroecosystems in Taquari Valley, Rio Grande do Sul, Brazil. *Systematic and Applied Acarology* 16, 133–144.

Horsburgh, R.L. and Asquith, D. (1968) Initial survey of arthropod predators of the European red-mite in south-central Pennsylvania. *Journal of Economic Entomology* 61, 1752–1754.

Horsfall, J.G. and Barratt, R.W. (1945) An improved grading system for measuring plant disease. *Phytopathology* 35, 655.

Horton, D.R., Broers, D.A., Hinojosa, T., Lewis, T.M., Miliczky, E.R. and Lewis, R.R. (2002) Diversity and phenology of predatory arthropods overwintering in cardboard bands placed in pear and apple orchards of central Washington State. *Annals of the Entomological Society of America* 95, 469–480.

Horváth, A. (1989) A szőlő-levélatka (*Calepitrimerus vitis* Nal.) elleni fajtaspecifikus védekezés lehetőségei [Possibility of a specific control for *Calepitrimerus vitis* Nal.]. *Növényvédelem* 25, 171–176 (in Hungarian).

Hoskins, W.M., Boyce, A.M. and Lamiman, J.F. (1938) The use of selenium in sprays for the control of mites on citrus and grapes. *Hilgardia* 12(2), 115–175.

Hossain, S.M. (1992) Comparison of sampling techniques for the European red mite, *Panonychus ulmi* (Koch) (Acari: Tetranychidae) and the apple rust mite, *Aculus schlechtendali* (Nalepa) (Acari: Eriophyidae). *Acta Agriculturae Scandinavica, Section B — Soil and Plant Science* 42, 128–132.

Hosseini, M., Hatami, B., Saboori, A., Allahyari, H. and Ashouri, A. (2005) Predation by *Allothrombium pulvinum* on the spider mites *Tetranychus urticae* and *Amphitetranychus viennensis*: predation rate, prey preference and functional response. *Experimental and Applied Acarology* 37, 173–181.

Hotta, M. (1928) A study on a Japanese tea spider mite. *Transactions of the Shizuoka Prefectural Agricultural Association* 32, 1–107 (in Japanese).

Houard, C. (1903) Caractères morphologiques des *Pleurocecidies caulinaires. Comptes Rendus de l'Académie des Sciences (Paris)* 136, 1338–1340.

Houck, M.A. (1989) Isozyme analysis of *Hemisarcoptes* and its beetle associate *Chilocorus. Entomologia Experimentalis et Applicata* 52, 167–172.

Houck, M.A. (1994) Adaptation and transition into parasitism from commensalism: a phoretic model. In: Houck, M.A. (ed.) *Mites: Ecological and Evolutionary Analyses of Life-history Patterns* (1st edn). Chapman and Hall, New York, pp. 252–281.

Houck, M.A. and O'Connor, B.M. (1990) Ontogeny and life history of *Hemisarcoptes cooremani* (Acari: Hemisarcoptidae). *Annals of the Entomological Society of America* 83, 869–886.

Hougen-Eitzman, D. and Karban, R. (1995) Mechanisms of interspecific competition that result in successful control of Pacific mites following inoculations of Willamette mites on grapevines. *Oecologia* 103, 157–161.

Hountondji, F.C.C., Lomer, C.J., Hanna, R., Cherry, A.J. and Dara, S.K. (2002) Field evaluation of Brazilian isolates of *Neozygites floridana* (Entomophthorales: Neozygitaceae) for the microbial control of cassava green mite in Benin, West Africa. *Biocontrol Science and Technology* 12, 361–370.

Hountondji, F.C.C., Sabelis, M.W. and Hanna, R. (2010) The role of infochemicals in the interaction between cassava green mite and its fungal pathogen *Neozygites tanajoae*. In: Sabelis, M.W. and Bruin, J. (eds) *Trends in Acarology: Proceedings of the 12th International Congress, 21–26 August 2006*, Amsterdam. Springer, Dordrecht, the Netherlands, pp. 249–254.

Houseman, R.M. (2000) *Clover Mites.* MU Extension Publication G07358, University of Missouri Extension, Columbia, Missouri. Available at: http://extension.missouri.edu/explorepdf/agguides/pests/g07358.pdf (accessed 11 May 2015).

Houston, K.J. (1980) A revision of the Australian species of *Stethorus* Weise (Coleoptera: Coccinellidae). *Journal of the Australian Entomological Society* 19, 81–91.

Hovasse, R. (1930) Les insectes nuisibles au noisetier en Turquie. *Revue de Pathologie Végétale et d'Entomologie Agricole* 17, 407–412.

Howard, F.W. and Abreu, R.E. (1991) Tightness of the perianth of coconuts in relation to infestation by coconut mites. *Florida Entomologist* 74, 358–361.

Howard, F.W. and Moore, D. (2008) *A Coconut Mite,* Aceria guerreronis *Keifer (Arachnida: Acari: Eriophyidae)*. Publication Number EENY-397, Institute of Food and Agricultural Sciences, University of Florida, Gainesville, Florida. Available at: http://edis.ifas.ufl.edu/IN709 (accessed 5 August 2014).

Howard, F.W., Abreu-Rodriguez, E. and Denmark, H.A. (1990) Geographical and seasonal distribution of the coconut mite, *Aceria guerreronis* (Acari: Eriophyidae), in Puerto Rico and Florida, USA. *Journal of Agriculture of the University of Puerto Rico* 74, 237–251.

Howard, F.W., Moore, D., Giblin-Davis, R.M. and Abad, R.G. (2001) *Insects on Palms.* CAB International, Wallingford, UK.

Howell, A.D. and Daugovish, O. (2013) Biological control of *Eotetranychus lewisi* and *Tetranychus urticae* (Acari: Tetranychidae) on strawberry by four phytoseiids (Acari: Phytoseiidae). *Journal of Economic Entomology* 106, 80–85.

Hoy, C.W., Feldman, J., Gould, F., Kennedy, G.G., Reed, G. and Wyman, J.A. (1998) Naturally occurring biological controls in genetically engineered crops. In: Barbosa, P. (ed.) *Conservation Biological Control.* Academic Press, San Diego, California, pp. 185–205.

Hoy, M.A. (1982a) Aerial dispersal and field efficacy of a genetically improved strain of the spider mite predator *Metaseiulus occidentalis. Entomologia Experimentalis et Applicata* 32, 205–212.

Hoy, M.A. (1982b) Genetic improvement of *Metaseiulus occidentalis*: implementing pesticide resistant strains – progress and problems. In: Cameron, P.J., Wearing, C.H. and Kain, W.M. (eds) *Proceedings of the Australasian Workshop Development and Implementation of IPM, Mt Albert Research Centre, Auckland, New Zealand, 20–22 July 1982.* Entomology Division, Department of Scientific and Industrial Research, Auckland, New Zealand, pp. 18–28.

Hoy, M.A. (1984) *Managing Mites in Almonds: An Integrated Approach.* UC IPM Publication 1, University of California, Davis, California.

Hoy, M.A. (1985) Almonds: integrated mite management for California almond orchards. In: Helle, W. and Sabelis, M.W. (eds) *Spider Mites: Their Biology, Natural Enemies, and Control.* World Crop Pests, Volume 1B. Elsevier Science, Amsterdam, pp. 299–310.

Hoy, M.A. (1990) Genetic improvement of natural enemies of armored scale insects. In: Rosen D. (ed.) *Armored Scale Insects.* World Crop Pests, Volume 4B. Elsevier Science, Amsterdam, pp. 441–451.

Hoy, M.A. (1992) Criteria for release of genetically-improved phytoseiids: an examination of the risks associated with release of biological agents. *Experimental and Applied Acarology* 14, 393–416.

Hoy, M.A. (1998) Myths, models and mitigation of resistance to pesticides. *Philosophical Transactions of the Royal Society B: Biological Sciences* 353, 1787–1795.

Hoy, M.A. (2000) Transgenic arthropods for pest management: risks and realities. *Experimental and Applied Acarology* 24, 463–495.

Hoy, M.A. (2011) *Agricultural Acarology: Introduction to Integrated Mite Management.* CRC Press (Taylor & Francis Group), Boca Raton, Florida.

Hoy, M.A. and Conley, J. (1987) Selection for abamectin resistance in *Tetranychus urticae* and *T. pacificus* (Acari: Tetranychidae). *Journal of Economic Entomology* 80, 221–225.

Hoy, M.A. and Jeyaprakash, A. (2008) Symbionts, including pathogens, of the predatory mite *Metaseiulus occidentalis*: current and future analysis. *Experimental and Applied Acarology* 46, 329–347.

Hoy, M.A. and Jeyaprakash, A. (2009) Symbionts, including pathogens, of the predatory mite *Metaseiulus occidentalis*: Current and future analysis methods. In: Bruin, S. and van der Geest, L.P.S. (eds) *Diseases of Mites and Ticks.* Springer, Dordrecht, the Netherlands, pp. 329–347. Originally published in *Experimental and Applied Acarology* 46, 329–347.

Hoy, M.A. and Ouyang, Y.L. (1986) Selectivity of the acaricides clofentezine and hexythiazox to the predator *Metaseiulus occidentalis* (Acari: Phytoseiidae). *Journal of Economic Entomology* 79, 1377–1380.

Hoy, M.A. and Peña, J. (2006) *Featured Creatures: Red Palm Mite,* Raoiella indica *Hirst (Arachnida: Acari: Tenuipalpidae)*. Publication Number EENY-397, Institute of Food and Agricultural Sciences, University of Florida, Gainesville, Florida. Available at: http://entnemdept.ufl.edu/creatures/orn/palms/red_palm_mite.htm (accessed 5 August 2014).

Hoy, M.A. and Smilanick, J. (1979) A sex pheromone produced by immature and adult females of the predatory mite, *Metaseiulus occidentalis* (Acarina: Phytoseiidae). *Entomologia Experimentalis et Applicata* 26, 291–300.

Hoy, M.A. and Smilanick, J. (1981) Non-random prey location by the phytoseiid predator *Metaseiulus occidentalis*: differential responses to several spider mites species. *Entomologia Experimentalis et Applicata* 29, 241–253.

Hoy, M.A. and Smith, K.B. (1982) Evaluation of *Stethorus nigripes* (Col: Coccinellidae) for biological control of spider mites in California almond orchards. *Entomophaga* 27, 301–310.

Hoy, M.A. and Standow, K.A. (1981) Resistance to sulfur in a vineyard spider mite predator. *California Agriculture* 35(5, 6), 8–10.

Hoy, M.A. and Standow, K.A. (1982) Inheritance of resistance to sulfur in the spider mite predator *Metaseiulus occidentalis*. *Entomologia Experimentalis et Applicata* 31, 316–323.

Hoy, M.A., Castro, D. and Cahn, D. (1982a) Two methods for large scale production of pesticide-resistant strains of the spider mite predator *Metaseiulus occidentalis* (Nesbitt) (Acarina, Phytoseiidae). *Zeitschrift für Angewandte Entomologie* 94, 1–9.

Hoy, M.A., Barnett, W.W., Reil, W.O., Castro, D., Cahn, D., Hendricks, L.C., Coviello, R. and Bentley, W.J. (1982b) Large-scale releases of pesticide-resistant spider mite predators. *California Agriculture* 36(1, 2), 8–10.

Hoy, M.A., Cunningham, G.L. and Knutson, L. (1983) *Biological Control of Pests by Mites*. Special Publication No. 3304, University of California, Berkeley, California.

Hoyt, S.C. (1962) New materials for the control of the apple rust mite. *Journal of Economic Entomology* 55, 639–641.

Hoyt, S.C. (1969a) Integrated chemical control of insects and biological control of mites on apples in Washington. *Journal of Economic Entomology* 62, 74–86.

Hoyt, S.C. (1969b) Populations studies of five mite species on apple trees in Washington. In: Evans, G.O. (ed.) *Proceedings of the 2nd International Congress of Acarology, 19–25 July, 1967, Sutton Bonington, UK*. Akadémiaei Kiadó, Budapest, pp. 117–133.

Hoyt, S.C. and Beers, E.H. (1993) McDaniel spider mite. Orchard Pest Management Online, Tree Fruit Research and Extension Center, Washington Sate University, Wenatchee, Washington. Available at: http://jenny.tfrec.wsu.edu/opm/displayspecies.php?pn=270 (accessed 6 August 2014).

Hoyt, S.C. and Tanigoshi, L.K. (1983) Economic injury levels for apple mite pests. In: Croft, B.A. and Hoyt, S.C. (eds) *Integrated Management of Insect Pests of Pome and Stone Fruit*. Wiley, New York, pp. 203–218.

Hoyt, S.C., Westigard, P.H. and Burts, E.C. (1978) Effects of 2 synthetic pyrethroids on the codling moth, pear psylla and various mite species in northwest apple and pear orchards. *Journal of Economic Entomology* 71, 431–434.

Hoyt, S.C., Tanigoshi, L.K. and Browne, R.W. (1979) Economic injury level studies in relation to mites on apple. In: Rodriguez, J.G. (ed.) *Recent Advances in Acarology, Volume I*. Academic Press, New York, pp. 3–12.

Hrncic, S., Perovic, T., Radonjic, S., Petanović, R. and Vidovic, B. (2009) Pojava, intenzitet napada i suzbijanje dvije nove vrste eriofida, *Aceria oleae, Shevtchenkella barensis* na sadnicama masline u Crnoj Gori. *Pesticidi i Fitomedicina* 24, 303–308 (in Serbian).

Hsieh, S.P.Y., Liang, W.J. and Chang, S.Y. (1977) Studies on the sterility of rice plant. I. Association of sheath rot fungus *Acrocylindrium oryzae* Sawada, with sterile rice plants. *Plant Protection Bulletin (Taipei)* 19, 30–36 (in Chinese).

Hu, C.C. and Wang, L.C. (1965) A study of the annual life cycle of the tea red spider mite, *Oligonychus coffeae* (Nietner). *Journal of the Agricultural Association of China* 50, 1–11 (in Chinese).

Hu, D. and Krantz, G.W. (1991) A new species of *Nalepella* Keifer (Acari: Eriophyoidea: Nalepellidae) from conifers in Oregon, USA. *International Journal of Acarology* 17, 5–8.

Hu, J.-H., Li, H.-J., Lei, H.-D., Ran, C., Ling, B.-M., Zhang, Q.-B., Tian, W.-H. and Qian, K.-M. (2004a) A performance investigation on resistance of citruses germplasm to the citrus red mite, *Panonychus citri* Mcg. *Southwest China Journal of Agricultural Sciences* 17, 460–464 (in Chinese).

Hu, M.-Y., Liu, X.Q., Zhou, L., Lo, X.-M., Hou, R.-H. and Weng, Q.-F. (2004b) Irradiation as a quarantine treatment against citrus rust mite (*Phyllocoptruta oleivora*). In: *IAEA Technical Documents (IAEA-TECDOCs) 1427*. International Atomic Enerегy Agency, Vienna, pp. 127–132.

Hu, S.Q. and Chen, X.W. (1994) Two new species of the *genus Eurytetranychus* from China (Acari: Tetranychidae). *Acta Zootaxonomica Sinica* 19, 302–305.

Hu, Y., Wang, C., Shen, Z. and Li, B. (1998) Study on the control of pyridaben against *Tetranychus truncatus*. *Forest Pest and Disease* 3, 37–38 (in Chinese).

Huang, H.X. (2005) The occurrence and control of citrus mite *Phyllocoptes oleivorus* in southern part of Fujian province. *China Fruits* 4, 58–59 (in Chinese).

Huang, K.-W. (2008) *Aceria* (Acarina: Eriophyoidea) in Taiwan: five new species and plant abnormalities caused by sixteen species. *Zootaxa* 1829, 1–30.

Huang, M.-D., Mai, S.-W., Li, S.-X and Situ, J. (1983) Biological control of citrus red mite, *Panonychus citri* (McG.) in Guangdong Province. In: Matsumoto, K. (ed.) *Proceedings of the 4th International Citrus Congress, Tokyo, Japan, 9–12 November 1981, Volume 2*. International Society of Citriculture, Riverside, California, pp. 643–646.

Huang, N., Enkegaard, A., Osborne, L.S., Ramakers, P.M.J., Messelink, G.J., Pijnakker, J. and Murphy, G. (2011) The banker plant method in biological control. *Critical Reviews in Plant Sciences* 30, 259–278.

Huang, P.K., Luo, X.N. and Song, S.L. (1988) Biological control of *Panonychus citri* by releasing *Stethorus siphonulus* Kapur (Coleoptera: Coccinellidae) in south-eastern China. *Acta Phytophylacica Sinica* 15, 1–6 (in Chinese).

Huang, T. (1967) A study on morphological features of erinose mite of litchi (*Eriophyes litchii* Keifer) and an observation on the conditions of its damage. *Plant Protection Bulletin (Taipei)* 9, 35–46 (in Chinese).

Huang, T. (1971) Records of ten eriophyid mites associated with plants in Japan. *Journal of the Faculty of Science, Hokkaido University, Series VI, Zoology* 18, 256–276.

Huang, T. (1974) Record of six eriophyid mites associated with economic plants in Taiwan. *Journal of Agriculture and Forestry (Nong Lin Xuebao)* 23, 75–88.

Huang, T. (1978) Main injurious eriophyid mites and their infestations in Taiwan. *Institute of Zoology Academica Sinica (Taiwan), Supplement Publication* No. 3, 255–279.

Huang, T. (1980) *Acaphylla steinwedeni*. In: Ehara, S. (ed.) *Illustrations of the Mites and Ticks of Japan*. Zenkoku Noson Kyoiku Kyokai, Tokyo (in Japanese).

Huang, T. and Wang, C.F. (1997) A new-record species of eriophyid mite on lemon from Taiwan (Acarina: Eriophyidae). *Chinese Journal of Entomology* 17, 269–274.

Huang, T., Huang, K.W. and Horng, I.J. (1990) Two species of eriophyid mites injurious to litchi trees in Taiwan. *Chinese Journal of Entomology, Special Publication* No. 3, 57–64 (in Chinese).

Huang, Z.Y., Liu, X., Li, L.H., Qi, J.X. and Liang, A.G. (1994) Study on the biology of *Aculus fockeui*. *Entomological Knowledge* 31, 15–17 (in Chinese).

Hubbard, H.G. (1885) *Insects Affecting the Orange: Report on the Insects Affecting the Culture of the Orange and Other Plants of the Citrus Family, with Practical Suggestions for their Control or Extermination*. Miscellaneous Document No. 40, Division of Entomology, US Department of Agriculture. Government Printing Office, Washington, DC.

Hueck, H.J. (1951) Influence of light upon the hatching of winter eggs of the fruit tree red spider. *Nature* 167, 933–944.

Huffaker, C.B. (1958) Experimental studies on predation: dispersion factors and predator–prey oscillations. *Hilgardia* 27(14), 343–383.

Huffaker, C.B. and Flaherty, D.L. (1966) Potential of biological control of two-spotted spider mites on strawberries in California. *Journal of Economic Entomology* 59, 786–792.

Huffaker, C.B. and Kennett, C.E. (1953) Differential tolerance to parathion in two *Typhlodromus* predatory on cyclamen mite. *Journal of Economic Entomology* 46, 707–708.

Huffaker, C.B. and Kennett, C.E. (1956) Experimental studies on predation: predation and cyclamen-mite populations on strawberries in California. *Hilgardia* 26(4), 191–222.

Huffaker, C.B., van de Vrie, M. and McMurtry, J.S. (1969) The ecology of tetranychid mites. *Annual Review of Entomology* 14, 125–173.

Huffaker, C.B., van de Vrie, M. and McMurtry, J.S. (1970) Ecology of tetranychid mites and their natural enemies: a review. II. Tetranychid populations and their possible control by predators: an evaluation. *Hilgardia* 40(11), 391–458.

Hughes, A.M. (1961) *The Mites of Stored Food*. MAFF (Ministry of Agriculture, Fisheries and Food), Technical Bulletin No. 9. Her Majesty's Stationery Office, London.

Hughes, A.M. (1976) *The Mites of Stored Food and Houses*. MAFF (Ministry of Agriculture, Fisheries and Food), Technical Bulletin No. 9 (rev. edn). Her Majestry's Stationery Office, London.

Hughes, R.D. and Jackson, C.G. (1958) A review of the Anoetidae (Acari). *Virginia Journal of Science, New Series* 9, 1–198.

Hughes, T.E. (1953) The functional morphology of the mouth parts of the mite *Anoetus sapromyzarum* Dufour 1839, compared with those of the more typical Sarcoptiformes. *Proceedings of the Academy of Sciences, Amsterdam* 56C, 2, 278–287.

Hughes, T.E. (1954) Some histological changes which occur in the gut epithelium of *Ixodes ricinus* females during gorging and up to oviposition. *Annals of Tropical Medicine and Parasitology* 48, 397–404.

Hughes, T.E. (1959) *Mites or the Acari*. Athlone Press, London.

Hukusima, S. (1958) The effect of varying nitrogen levels upon the arthropod fauna in young apple trees, with special reference to the increase of the mite and aphid populations. *Bulletin of the Faculty of Agriculture, Hirosaki University* 4, 72–79.

Hull, L.A. and Beers, E.H. (1985) Ecological selectivity: modifying chemical control practices to preserve natural enemies. In: Hoy, M.A. and Herzog, D.C. (eds) *Biological Control in Agricultural IPM Systems*. Academic Press, Orlando, Florida, pp. 103–122.

Hull, L.A., Asquith, D. and Mowery, P.D. (1977) The mite searching ability of *Stethorus punctum* within an apple orchard. *Environmental Entomology* 6, 684–688.

Hull, V.J. and Groves, R.H. (1973) Variation in *Chondrilla juncea* L. in Australia. *Proceedings of the Ecological Society of Australia* 10, 113–135.

Humber, R.A. (1992) *Collection of Entomopathogenic Fungi: Catalog of Strains 1992*. Agricultural Research Service Publications ARS 110, US Department of Agriculture, Washington, DC.

Humber, R.A., Moraes, G.J. and dos Santos, J.M. (1981) Natural infection of *Tetranychus evansi* (Acarina: Tetranychidae) by a *Triplosporium* sp. (Zygomycetes: Entomophthorales) in northeastern Brazil. *Entomophaga* 26, 421–425.

Hummel, N.A., Castro, B.A., Stout, M.J. and Saichuk, J.K. (2007) *The Panicle Rice Mite* Steneotarsonemus spinki Smiley. Rice Pest Notes, Pest Management and Insect Identification Series, Publication 3023, Louisiana State University Agricultural Center, Baton Rouge, Louisiana and Texas Cooperative Extension, College Station, Texas.

Hummel, N.A., Castro, B.A., McDonald, E.M., Pellerano, M.A. and Ochoa, R. (2009) The panicle rice mite, *Steneotarsonemus spinki* Smiley, a re-discovered pest of rice in the United States. *Crop Protection* 28, 547–560.

Humphreys, W.F. (1979) Production and respiration in animal populations. *Journal of Animal Ecology* 48, 427–454.

Hunter, M.D. (2002a) A breath of fresh air: beyond laboratory studies of plant volatile–natural enemy interactions. *Agricultural and Forest Entomology* 4, 81–86.

Hunter, M.D. (2002b) Ecological causes of pest outbreaks. In: Pimentel, D. (ed.) *Encyclopedia of Pest Management*. Marcel Dekker, New York, pp. 214–217.

Hunter, M.D. and Price, P.W. (1998) Cycles in insect populations: delayed density dependence or exogenous driving variables. *Ecological Entomology* 23, 216–222.

Hürková, J. and Weyda, F. (1982) Response to pesticides of diapausing females of *Tetranychus urticae* (Acarina, Tetranychidae). *Věstník Československé Společnosti Zoologické* 46, 92–99.

Hurtado, M.A., Ansaloni, T., Cros-Arteil, S., Jacas, J.A. and Navajas, M. (2008) Sequence analysis of the ribosomal internal transcribed spacers region in spider mites (Prostigmata: Tetranychidae) occurring in citrus orchards in eastern Spain: use for species discrimination. *Annals of Applied Biology* 153, 167–174.

Hussain, A.A. (1969) Biology of *Paratetranychus afrasiaticus* (McGregor) (Acari: Tetranychidae) infesting date palms in Iraq. *Bulletin of the Entomological Society of Egypt* 53, 221–225.

Hussey, N.W. (1963) A new species of *Tarsonemus* (Acarina: Tarsonemidae) from cultivated mushrooms. *Acarologia* 5, 540–544.

Hussey, N.W. (1964) Mites as pests of cultivated mushrooms. *Annual Report of the Glasshouse Crops Research Institute for 1963*. Little Hampton, UK, pp. 114–117.

Hussey, N.W. (1985) Cucumbers. In: Hussey, N.W. and Scope, N.E.A. (eds) *Biological Pest Control*. The Glasshouse Experience. Blandford Press, Poole, UK, pp. 175–179.

Hussey, N.W. and Gurney, B. (1967) Bionomics and control of *Tarsonemus myceliophagus* Hussey in mushroom composts. *Entomologia Experimentalis et Applicata* 10, 287–294.

Hussey, N.W. and Parr, W.J. (1963a) Dispersal of the glasshouse red spider mite *Tetranychus urticae* Koch (Acarina: Tetranychidae). *Entomologia Experimentalis et Applicata* 6, 207–214.

Hussey, N.W. and Parr, W.J. (1963b) The effect of glasshouse red spider mite (*Tetranychus urticae* Koch) on the yield of cucumbers. *Journal of Horticultural Science* 38, 255–263.

Hussey, N.W. and Scopes, N.E.A. (1985) Greenhouse vegetables (Britain). In: Helle, W. and Sabelis, M.W. (eds) *Spider Mites: Their Biology, Natural Enemies and Control*. World Crop Pests, Volume 1B. Elsevier Science, Amsterdam, pp. 285–197.

Hussey, N.W., Read, W.H. and Hesling, J.J. (1969) *The Pest of Protected Cultivations: The Biology and Control of Glasshouse and Mushroom Pests.* Edward Arnold, London.

Huston, M. (1979) A general hypothesis of species diversity. *American Naturalist* 113, 81–101.

Iacob, N. (1978) Clasa Arachnida–Acari. In: Iacob, N., Romaşcu, E., Grossu, A.V., Manolache, C. and Boguleanu, G.G. (eds) *Tratat de Zoologie Agricolă: Dăunătorii Plantelor Cultivate: Vol. 1: Nematoda (Secernentea, Adenophorea); Annelida (Oligochaeta); Mollusca (Gastropoda); Arthropoda (Crustacea, Arachnida–Acari, Myriapoda) (Treatise of Agricultural Zoology. Pests of Cultivated Plants: Vol. 1: Nematoda (Secernentea, Adenophorea) Annelida (Oligochaeta) Mollusca (Gastropoda) Arthropoda (Crustacea, Arachnida–Acari, Myriapoda)).* Editura Academiei Republicii Socialiste România, Bucharest, pp. 205–329 (in Romanian).

Iacob, N., Novac, N., Balutescu, I. and Nereuta, C. (1977) Influenta unor factori ecologici si de combatere asupra acarianului *Aculus* (*Phyllocoptes*) *fockeui* Nal. et Trt. (Influence of some ecological and control factors on the mite *Aculus* (*Phyllocoptes*) *fockeui* Nal. et Trt). *Analele Institutului de Cercetări pentru Protecţia Plantelor* 13, 143–155 (in Romanian).

Ibrahim, G.A. (1992) Effect of different fungicides (dithane, comaprop, compazin, numspor, sulfur) on the citrus rust mite *Phyllocoptruta oleivora* (Ashmead) and *Brevipalpus californicus* (Banks) with their associated predator mites. *Egyptian Journal of Agricultural Research* 70, 207–213.

Ibrahim, G.A., El-Ghobashy, M.S., El-Sayed, K.M. and Shoeib, A.A. (2005) Biological control of citrus brown mite *Eutetranychus orientals* using predatory mite, *Neoseiulus californicus* (McGregor) (Acari: Tetranychidae & Phytoseiidae) on citrus trees. *Egyptian Journal of Agricultural Research* 83, 131–139.

Ibrahim, S.M. (1994) Influence of three mineral oils and other compounds on the population of the citrus rust mite, *Phyllocoptruta oleivora* Ashmead and the citrus brown mite, *Eutetranychus orientalis*, Klein. *Egyptian Journal of Agricultural Research* 72, 401–407.

ICA (2005) *Resolución No. 001195 de 2005. Diario Oficial, Edición 45.892. Miércoles 27 abril de 2005.* Instituto Colombiano Agropecuario, Bogota, Colombia.

ICAR (1994–1996) *Indian Council of Agricultural Research Progress Report 1994–96, AICRP (All India Coordinated Research Project) on Agricultural Acarology.* University of Agricultural Sciences, Bangalore, India.

ICAR (1996–1998) *Indian Council ofAgricultural Research Progress Report 1996–98, AICRP (All India Coordinated Research Project) on Agricultural Acarology.* University of Agricultural Sciences, Bangalore, India.

Idder, M.A. and Pintureau, B. (2008) Effectiveness of the ladybird *Stethorus punctillum* (Weise) as a predator of the mite *Oligonychus afrasiaticus* (McGregor) in the palm plantations of the area of Wargla in Algeria. *Fruits* 63, 85–92.

Ido, T., Kumkiri, M., Lao, L.-M., Yano, Y. and Takada, N. (2004) Dermatitis caused by *Balaustium murorum*. *Acta Dermato-Venereologica* 84, 80–81.

Iglinsky, W.E. and Rainwater, C.F. (1950) *Orius insidiosus*, an enemy of a spider mite on cotton. *Journal of Economic Entomology* 43, 567–568.

Iglinsky, W.E. and Rainwater, C.R. (1954) Life history and habitus of *Tetranychus desertorum* and *bimaculatus* on cotton. *Journal of Economic Entomology* 47, 1084–1086.

Ihsan, N. and Yusof, I. (2004) Laboratory bioassay of some entomopathogenic fungi against broad mite (*Polyphagotarsonemus latus* Banks). *International Journal of Agriculture and Biology* 6, 223–225.

IITA (1976) *Annual Report for 1975.* International Institute of Tropical Agriculture, Ibadan, Nigeria.

Ikemiya, J., Ueda, Y., Tanaka, H. and Shibao, M. (2007) Presence–absence density estimation of pink citrus rust mite, *Aculops pelekassi* (Keifer) on Satsuma mandarin fruit. *Annual Report of the Kansai Plant Protection Society* 49, 77–78 (in Japanese).

Ikeshima, K., Sakamaki, Y., Miyagi, A., Ohno, S., Fukuda, T. and Tsuda, K. (2009) Occurrence of the spider mite *Tetranychus takafujii* Ehara & Ohashi in Kyushu and Okinawa. *Kyushu Plant Protection Research* 55, 136–140.

Imani, Z. and Shishehbor, P. (2009) Effect of temperature on life history and life tables of *Eutetranychus orientalis* (Klein) (Acari: Tetranychidae). *Systematic and Applied Acarology* 14, 11–18.

Imani, Z., Shishehbor, P. and Sohrabi, F. (2009) The effect of *Tetranychus turkestani* and *Eutetranychus orientalis* (Acari: Tetranychidae) on the development and reproduction of *Stethorus gilvifrons* (Coleoptera: Coccinellidae). *Journal of Asia-Pacific Entomology* 12, 213–216.

Immelman, D. W. (1983) *Annual Report of the Director General: Agriculture for the Period 1 April 1981 to 31 March 1982.* Department of Agriculture, Pretoria.

Ingram, W.R. and Green, S.M. (1972) Sequential sampling for bollworms on rain grown cotton in Botswana. *Cotton Growing Review* 49, 265–275.

Injac, M., Dulic, K. and Petanović, R. (1988) Pojava *Epitrimerus pyri* Nal. (Acaridida: Eriophyidae) i rezultati ogleda suzbijanja. *Zastita Bilja* 39, 125–132 (in Serbian).

Injac, M., Almasi, R. and Almasi, S. (1995) Suzbijanje *Calepitrimerus vitis* (Nal.) (Acarina: Eriophydae) pred odlazak deutoginih zenki na prezimljavanje. *Savremena Poljoprivreda* 43(5–6), 47–52 (in Serbian).

Inoue, K. (1980) Relationship between dicofol resistance and fitness in the citrus red mite, *Panonychus citri* (McG.). *Journal of Pesticicide Science (Tokyo)* 5, 165–175 (in Japanese).

Inoue, K. and Tanaka, M. (1983) Biological characteristics of *Agistemus terminalis* (Quayle) (Acarina: Stigmaeidae) as a predator of citrus red mite, *Panonychus citri* (McGregor). *Japanese Journal of Applied Entomology and Zoology* 27, 280–288 (in Japanese).

Inserra, R. (1970) Osservazioni morfologiche ed appunti di biologia su *Zetzellia graeciana* Gonzalez (Acarina: Stigmaeidae). *Bollettino di Zoologia Agraria e di Bachicoltura (Ser. II)* 10, 85–119.

Insung, A. (2001) Effect of some medicinal plant extracts on *Tetranychus truncatus* (Ehara). *King Mongkut's Agricultural Journal* 19(3), 15–22 (in Thai).

Insung, A., Pumnuan, J. and Chandrapatya, A. (2008) Acaricidal activities of wild plant extracts against *Luciaphorus perniciosus* Rack (Acari: Pygmephoridae) and *Formicomotes heteromorphus* Magowski (Acari: Dolichocybidae). *Systematic and Applied Acarology* 13, 188–194.

Ioachim, E. and Bobarnac, B. (1997) Research on the hazelnut pests in Romania. *Acta Horticulturae* 445, 527–536.

Ioffe, I.D. (1965) Distribution of neurosecretory cells in the central nervous apparatus of *Dermacentor pictus* Herm. (Acarina, Chelicerata). *Doklady Akademii Nauk SSSR* 154, 25–29.

IPM Practitioner (2000) 2001 Directory of least-toxic pest control products. *The IPM Practitioner* 21(11–12), 1–48.

IPM Thailand (2005) IPM Thailand. Available at: http://thailand.ipm-info.org/ (accessed 6 August 2014).

IPPC Secretariat (2005) *Identification of Risks and Management of Invasive Alien Species Using the IPPC Framework. Proceedings of the Workshop on Invasive Alien Species and the International Plant Protection Convention, Braunschweig, Germany, 22–26 September 2003.* Food and Agriculture Organization of the United Nations, Rome.

Ippolito, R. and Triggiani, O. (1988) Prove di efficacia di alcuni acaricidi nei riguardi di *Tyrophagus similis* Volgin (Acaridida-Acaridae) su spinacio. *Informatore Fitopatologico* 38(2), 43–48.

IRAC (2012) *Acaricide Mode of Action Classification: A Key to Effective Acaricide Resistance Management.* Insecticide Resistance Action Committee, www.irac-online.org. Available at: http://www.irac-online.org/documents/mites-moa-poster/?ext=pdf (accessed 2 February 2015).

Ireson, J.E. and Webb, W.R. (1996) Redistribution and establishment of *Neomolgus capillatus* (Kramer) (Acarina: Bdellidae) for the biological control of *Sminthurus viridis* (L.) (Collembola: Sminthuridae) in Tasmania. *Journal of the Australian Entomological Society* 35, 243–246.

Ireson, J.E., Gourlay, A.H., Kwong, R.M., Holloway, R.J. and Chatterton, W.S. (1999) Progress on the rearing, release and establishment of the gorse spider mite, *Tetranychus lintearius* Dufour, for the biological control of gorse in Australia. In: Bishop, A.C., Boersma, M. and Barnes, C.D. (eds) *Proceedings of the XIIth Australian Weeds Conference, 12–16 September 1999, Hobart, Tasmania, Australia*. Council of Australasian Weeds Societies (http://www.caws.org.au/), pp. 320–324.

Ireson, J.E., Gourlay, A.H., Kwong, R.M., Holloway, R.J. and Chatterton, W.S. (2003) Host specificity, release, and establishment of the gorse spider mite, *Tetranychus lintearius* Dufour (Acarina: Tetranychidae), for the biological control of gorse, *Ulex europaeus* L. (Fabaceae), in Australia. *Biological Control* 26, 117–127.

Isaacs, R. and Gajek, D. (2003) Abundance of blueberry bud mite (*Acalitus vaccinii*) in Michigan blueberries, and variation in infestation among common highbush blueberry varieties. *IOBC-WPRS Bulletin* 26(2), 127–132.

Isaacs, R., Morrone, V. and Gajek, D. (2004) Potential acaricides for the management of blueberry bud mite in Michigan blueberries. *HortTechnology* 14, 188–191.

Isac, M., Preda, S. and Marcu, M. (1998) Aphid species-vectors of plum pox virus. *Acta Virologica* 42, 233–234.

Isenhour, D.J. and Yeargan, K.V. (1981) Effect of temperature on the development of *Orius insidiosus*, with notes on laboratory rearing. *Annals of the Entomological Society of America* 74, 114–116.

Ishaaya, I. and Sternlicht, M. (1969) Growth accelerators and inhibitors in lemon buds infested with *Aceria sheldoni* (Ewing) (Acarina: Eriophyidae). *Journal of Experimental Botany* 20, 796–804.

Ishaaya, I. and Sternlicht, M. (1971) Oxidative enzymes, ribonuclease, and amylase in lemon buds infested with *Aceria sheldoni* (Ewing) (Acarina: Eriophyidae). *Journal of Experimental Botany* 22, 146–152.

Ishiwari, H., Suzuki, T. and Maeda, T. (2007) Essential compounds in herbivore-induced plant volatiles that attract the predatory mite *Neoseiulus womersleyi*. *Journal of Chemical Ecology* 33, 1670–1681.

Iskander, N.G. (1993) Chemical control of the rust mite *Phyllocoptruta oleivora* Ashmead, flat mite *Brevipalpus californicus* Banks and its side effect on *Amblyseius scutalis* (Athias-Henriot) on citrus trees. *Egyptian Journal of Agricultural Research* 71, 463–472.

Iskander, N.G. and El-Atrouzy, N.A. (1991) The latent effects of flufenoxuron on *Brevipalpus californicus* (Banks). *Bulletin of the Entomological Society of Egypt, Economic Series* 17, 199–206.

Iskander, N.G., Ibrahim, S.M. and Heikal, I.H. (1993) Effect of certain chemicals on different development stages of some phytophagous and predaceous mite species. *Egyptian Journal of Agricultural Research* 71, 453–461.

Ismail, M.A. (1971) Seasonal variation in bonding force and abscission of citrus fruit in response to ethylene, ethephon, and cycloheximide. *Proceedings of the Florida State Horticultural Society* 84, 77–81.

Isman, M.B. (2006) Botanical insecticides, deterrents, and repellents in modern agriculture and an increasingly regulated world. *Annual Review of Entomology* 51, 45–66.

Isman, M.B., Machial, C.M., Miresmailli, S. and Bainard, L.D. (2007) Essential oil-based pesticides: new insights from old chemistry. In: Ohkawa, H., Miyagawa, H. and Lee, P. (eds) *Pesticide Chemistry*. Wiley-VCH Verlag, Weinheim, Germany, pp. 201–209.

Israel Ministry of Agriculture (1975) Mango. In: *Scientific Activities 1971–1974*. Institute of Plant Protection, Ministry of Agriculture, Bet Dagan, Israel.

Itimu, O.A. and Sauti, R.F.N. (1982) *Cassava Green Spider Mite (G.S.M.) (*Mononychellus tanajoa *(Bondar)): Its History and Possible Control Measures in Malawi*. Lunyangwa Agricultural Research Station, Ministry of Agriculture, Mzuzu, Malawi.

Ito, K. and Saitō, Y. (2006) Effects of host-plant species on diapause induction of the Kanzawa spider mite, *Tetranychus kanzawai*. *Entomologia Experimentalis et Applicata* 121, 177–184.

Ito, Y. (1964) Descriptions of eight tarsonemid mites from Japan. *Japanese Journal of Applied Entomology and Zoology* 8, 34–44.

Ito, Y. (1977) A taxonomy and key to Japanese tarsonemid mites. In: Sasa, M. and Aoki, J. (eds) Contributions to Acarology in Japan. Zukan-no-Hokuryukan, Tokyo, pp. 230–240 (in Japanese).

Ito, Y. (1993) *Tarsonemus bilobatus* Suski. In: Ehara, S. (ed.) *Plant Mites of Japan in Colors*. Zenkoku Noson Kyoiku Kyokai, Tokyo, pp. 34–35.

Ivanov, V.P. and Leonovich, S.A. (1983) Sensory organs. In: Balashov, Y.S. (ed.) *An Atlas of Ixodid Tick Ultrastructure.* English Translation from Russian (1979). Special Publication of the Entomological Society of America, Lanham, Maryland., pp. 191–220.

Izquierdo, J., Mansanet, V., Sanz, J.V. and Puiggros, J.M. (2002) Development of EnvidorReg. for the control of spider mites in Spanish citrus production. *Pflanzenschutz-Nachrichten Bayer* 55, 255–266.

Jäckel, B., Balder, H., Schneider, K. and Pradel, B. (2000) *Stethorus punctillum*, an important antagonist of spidermites in the city. *Mitteilungen aus der Biologischen Bundesanstalt für Land- und Forstwirtschaft, Berlin-Dahlem* No. 370, 291 (abstract).

Jackson, P.R. and Hunter, P.E. (1983) Effects of nitrogen fertilizer level on development and populations of the pecan leaf scorch mite (Acari: Tetranychidae). *Journal of Economic Entomology* 76, 432–435.

Jackson, P.R., Hunter, P.E. and Payne, J.A. (1983) Biology of the pecan leaf scorch mite (Acari: Tetranychidae). *Environmental Entomology* 12, 55–59.

Jacob, H. (1976) Untersuchungen zur Übertragung des virösen Atavismus der Schwarzen Johannisbeere (*Ribes nigrum* L.) durch die Gallmilbe *Cecidophyes ribis* Westw. *Zeitschrift für Pflanzenkrankheiten und Pflanzenschutz* (now *Journal of Plant Diseases and Plant Protection*) 83, 448–458.

Jacobi, A. (1905) Eine Spinnmilbe (*Tetranychus ununguis* n. sp.) als Koniferenschädling. *Naturwissenschaftliche Zeitschrift für Land- und Forstwirtschaft* 3, 269–247.

Jacot, A.P. (1925) Phylogeny in the Oribatoidea. *American Naturalist* 59, 272–279.

Jadue, Y., Vargas, C., Rubio, T. and Araya, J.E. (1996) Effects of cold storage on the false grape mite, *Brevipalpus chilensis* Baker. *Zeitschrift für Pflanzenkrankheiten und Pflanzenschutz* (now *Journal of Plant Diseases and Plant Protection*) 103, 403–408.

Jadue, Y., Vargas, C., Araya, J.E. and Rubio, T. (1997) Preliminary evaluation of ionizing radiation as a quarantine treatment for the false grape mite, *Brevipalpus chilensis*. *Zeitschrift für Pflanzenkrankheiten und Pflanzenschutz* (now *Journal of Plant Diseases and Plant Protection*) 104, 222–230.

Jaenike, J. (1990) Host specialization in phytophagous insects. *Annual Review of Ecology and Systematics* 21, 243–273.

Jagadish, P.S. and ChannaBasavanna, G.P. (1983) Toxicity of different pesticides to *Tetranychus ludeni* in the laboratory. *Acarology Newsletter* 12, 8–9.

Jagadish, P.S. and Nageshachandra, B.K. (1981) Biology of *Typhlodromips tetranychivorous* (Acari: Phytoseiidae) on red palm mite *Raoiella indica* (Acari: Tenuipalpidae). In: ChannaBasavanna, G.P., Mallik, B. and Ghorpade, K.D. (eds) *Contributions to Acarology in India: Proceedings of the I All India Symposium in Acarology, April 23–25, 1979, Bangalore*. Acarological Society of India, Bangalore, India, pp. 173–179.

Jagadish, P.S., ChannaBasavanna, G.P. and Nangia, N. (1994) Biology of the predatory mite *Euseius concordis* (Chant) (Acari: Phytoseiidae) an effective predator on *Tetranychus neocaledonicus* Andre (Acari: Tetranychidae). *Mysore Journal of Agricultural Sciences* 28, 244–248.

Jagersbacher-Baumann, J. and Ebermann, E. (2013) Ovoviviparity in scutacarid mites (Scutacaridae, Heterostigmatina)? *International Journal of Acarology* 39, 347–350.

Jaime de Herrero, A. (1984) *Phytoseiulus macropilis* (Banks) (Acarina-Phytoseiidae) un nuevo ácaro benefico para los citricos de Tucuman. *CIRPON – Revista de Investigación* 2, 27–35.

Jalaluddin, S.M. and Mohanasundaram, M. (1990) Control of the coconut red mite *Raoiella indica* Hirst (Tenuipalpidae: Acari) in the nursery. *Indian Coconut Journal* 21(6), 7–8.

Jalaluddin, S.M., Ravi, G. and Jebaraj, S. (2008) Suppression of rice leaf mite pest. Plant Protection India Blog, Thursday, November 13, 2008. Available at: http://cropcareindia.blogspot.it/2008_11_01_archive.html (accessed 6 August 2014).

James, D. and Mukerji, S. (1993) Mechanical transmission, identification, and characterization of a virus associated with mottle leaf in cherry. *Plant Disease* 77, 271–275.

James, D.G. (1989) Influence of diet on development, survival and oviposition in an Australian phytoseiid, *Amblyseius victoriensis* (Acari: Phytoseiidae). *Experimental and Applied Acarology* 6, 1–10.

James, D.G. (1990) Biological control of *Tetranychus urticae* Koch (Acari: Tetranychidae) in southern New South Wales peach orchards: the role of *Amblyseius victoriensis* (Acarina: Phytoseiidae). *Australian Journal of Zoology* 37, 645–655.

James, D.G. (1993) Pollen, mould mites and fungi: improvements to mass rearing of *Typhlodromus doreenae* and *Amblyseius victoriensis*. *Experimental and Applied Acarology* 17, 271–276.

James, D.G. (1995) Biological control of earth mites in pasture using endemic natural enemies. *Plant Protection Quarterly* 10, 58–59.

James, D.G. (1996) Integrated management of redlegged earth mite in pasture. In: Virgona, J. and Michalk, D. (eds) *The Art of Grazing Management: Proceedings of the Eleventh Annual Conference of the Grassland Society of New South Wales; Invited Papers, Contributed Papers and Poster Summaries from the Conference Held at the Wagga Wagga Leagues Club, Gurwood St., Wagga Wagga, NSW, 2650, July 10–11, 1996*. The Grassland Society of New South Wales, Orange, Australia, pp. 76–83.

James, D.G. (1997) Imidacloprid increases egg production in *Amblyseius victoriensis* (Acari: Phytoseiidae). *Experimental and Applied Acarology* 21, 75–82.

James, D.G. (2000) Abundance and phenology of earth mites (Acari: Penthaleidae) and predatory mites in pesticide-treated and pesticide-free grassland habitats in southern New South Wales, Australia. *International Journal of Acarology* 26, 363–369.

James, D.G. (2003) Synthetic herbivore-induced plant volatiles as field attractants for beneficial insects. *Environmental Entomology* 32, 977–992.

James, D.G. and O'Malley, K.J. (1991) Oversummering of eggs of *Halotydeus destructor* (Tucker) (Acarina: Penthaleidae): diapause termination and mortality. *Australian Entomological Magazine* 18, 35–41.

James, D.G. and O'Malley. K.J. (1993) Phenology of egg production and diapause in *Halotydeus destructor* Tucker and *Penthaleus major* Dugès (Acari: Penthaleidae) in southern New South Wales during 1988/89. *General and Applied Entomology* 24, 33–38.

James, D.G. and Price, T.S. (2002) Fecundity in two-spotted spider mite (Acari: Tetranychidae) is increased by direct and systemic exposure to imidacloprid. *Journal of Economic Entomology* 95, 729–732.

James, D.G. and Whitney, J. (1991) Biological control of grapevine mites in inland south-eastern Australia. *Australian and New Zealand Wine Industry Journal* 6, 210–214.

James, D.G. and Whitney, J. (1993) Mite populations on grapevines in south-eastern Australia: implications for biological control of grapevine mites (Acarina: Tenuipalpidae, Eriophyidae). *Experimental and Applied Acarology* 17, 259–270.

James, D.G., Warren, G.N. and Whitney, J. (1992) Phytoseiid mite populations of dormant grapevines: extraction using a microwave oven. *Experimental and Applied Acarology* 14, 175–178.

James, D.G., O'Malley, K. and Rayner, M. (1995) Effect of alphacypermethrin and bifenthrin on the survival of five acarine predators of *Halotydeus destructor* (Acari: Penthaleidae). *Experimental and Applied Acarology* 19, 647–654.

James, D.G., Price, T.S., Wright, L.C. and Perez, J. (2002) Abundance and phenology of mites, leafhoppers and thrips on pesticide-treated and untreated wine grapes in southcentral Washington. *Journal of Agricultural and Urban Entomology* 19, 45–54.

James, D.G., Castle Del Conte, S., Price, T.S., Grasswitz, T.R., Wright, L.C., Reyna, V., Garcia, O. and Perez, J. (2010) Manipulating plant–arthropod conversations to improve conservation biological control of mites In: Sabelis, M.W. and Bruin, J. (eds) *Trends in Acarology: Proceedings of the 12th International Congress 21–26 August 2006, Amsterdam*. Springer, Dordrecht, the Netherlands, pp. 413–418.

Jamieson, L.E. and Stevens, P.S. (2007) Development rates, longevity and fecundity of six-spotted mite (*Eotetranychus sexmaculatus*) at constant temperatures. *New Zealand Plant Protection* 60, 72–77.

Jamieson, L.E. and Stevens, P.S. (2009) Miticides against citrus red mites (*Panonychus citri*). *New Zealand Plant Protection* 62, 302–309.

Jamieson, L.E., Charles, J.G., Stevens, P.S., McKenna, C.E. and Bawden, R. (2005) Natural enemies of citrus red mite (*Panonychus citri*) in citrus orchards. *New Zealand Plant Protection* 58, 299–305.

Jamieson, L.E., Chhagan, A. and Charles, J.G. (2008) Predation of citrus red mite (*Panonychus citri*) by *Stethorus* sp. and *Agistemus longisetus*. *New Zealand Plant Protection* 61, 317–321.

Jana, S.K. and Somchoudhury, A.K. (2005) Bioefficacy of fenpropathrin against tea red spider mite, *Oligonychus coffeae* Nietner at Terai, Darjeeling (West Bengal). *Journal of Entomological Research* 29, 209–210.

Jana, S.K., Nair, N., Sarkar, P.K. and Somchoudhury, A.K. (2007) Varietal susceptibility of jute varieties against yellow mite, *Polyphagotarsonemus latus* Banks (Acari: Tarsonemidae). *Journal of Entomological Research* 31, 29–32.

Janarthanan, R., Navaneethan, G., Subramanian, K.S. and Sathiabalan, S.G. (1971) A method for assessment of eriophyid mites on pigeon pea leaves. *Madras Agricultural Journal* 59, 437.

Janarthanan, R., Navaneethan, G., Subramanian, K.S., and Samuel, G.S. (1972) Some observations on the transmission of sterility mosaic of pigeonpea. *Current Science* 41, 645–647.

Janarthanan, R., Samuel, G.S., Subramanian, K.S., Navaneethan, G. and Kandaswamy, T.K. (1973) A report on the survey of sterility mosaic disease incidence on redgram in Tamil Nadu. *Madras Agricultural Journal* 60, 41–44.

Janssen, A., Sarmento, R.A. and Sabelis, M.W. (2012) Plant and predator-mediated competition between invasive and endemic spider mites on tomato: does this explain invasiveness of *T. evansi* in the Old World? In: Schausberger, P., Walzer, A., Strodel, M.A. and Peneder, S. (eds) *7th Symposium of the*

European Association of Acarologists (EURAAC), July 9–13 2012, Vienna, Austria: Acari in a Changing World. Program, Abstracts, Participants. University of Natural Resources and Life Sciences, Vienna, Austria, p. 44. Available at: http://euraac.boku.ac.at/SympVienna/files/ProgramBook.pdf (accessed 14 January 2015).

Jara, B., Acosta, C. and Cardona, C. (1981) Efecto de cinco variedades de frijol sobre la biologia y la fecundidad de la arañita roja, *Tetranychus desertorum* Banks (Acari, Tetranychidae). *Revista Colombiana de Entomología* 7, 33–39.

Jayaraj, J., Natarajan, K. and Ramasubramanian, G.V. (1991) Control of *Raoiella indica* Hirst. (Tenuipalpidae: Acari) on coconut with pesticides. *Indian Coconut Journal* 22(8), 7–9.

Jaydeb, G., Mukherjee, A.B. and Sarkar, P.K. (1996) Assessment of loss of bhendi against red spider mite. *Environment and Ecology* 14, 480–481.

Jedlicková, J. (1992) Increasing the effectiveness of rearing and release of the predatory mite *Phytoseiulus persimilis*. *Bulletin OEPP/EPPO Bulletin* 22, 479–482.

Jedlicková, J. (1997) Modification of a leaf-washing apparatus for the recovery of mites. *Experimental and Applied Acarology* 21, 273–277.

Jedlicková, J. and Kolodochka, L.A. (1994) Faunistical notes: Acari, Mesostigmata, Phytoseiidae. *Biologia* 49, 696.

Jensen, A. and Mingochi, D.S. (1988) Chemical control of red spider mite (*Tetranychus urticae* Koch) on tomatoes in Zambia. *Acta Horticulturae* 218, 275–280.

Jensen, S.G., Lane, L.C. and Seifers, D.L. (1996) A new disease of maize and wheat in the high plains. *Plant Disease* 80, 1387–1390.

Jeppson, L.R. (1951) New acaricides for control of citrus red mite. 1948–50. *Journal of Economic Entomology* 44, 823–32.

Jeppson, L.R. (1977) Bionomics and control of mites attacking citrus. In: Grierson, W. (ed.) *Proceedings of the 2nd International Citrus Congress, Orlando, Florida, 1977, Volume 2*. International Society of Citriculture, University of California, Riverside, California, pp. 445–451.

Jeppson, L.R. (1989) Biology of citrus insects, mites and molluscs. In: Reuther, W., Calavan, E.C. and Carman, G.E. (eds) *The Citrus Industry, Volume 5*. University of California, Oakland, California, pp. 1–87.

Jeppson, L.R., Jesser, M.J. and Complin, J.O. (1955) Control of mites on citrus with chlorobenzilate. *Journal of Economic Entomology* 48, 375–377.

Jeppson, L.R., Jesser, M.J. and Complin, J.O. (1958) Factors affecting populations of the citrus bud mite in southern California lemon orchards and acaricide treatments for control of this eriophyid. *Journal of Economic Entomology* 51, 657–662.

Jeppson, L.R., Complin, J.O. and Jesser, M.J. (1962) Effects of application programs on citrus red mite control and development of resistance to acaricides. *Journal of Economic Entomology* 55, 17–22.

Jeppson, L.R., Complin, J.O. and Jesser, M.J. (1966) Cycloheximide derivatives and mite control, with special reference to mites on citrus. *Journal of Economic Entomology* 59, 15–19.

Jeppson, L.R., Keifer, H.H. and Baker, E.W. (1975) *Mites Injurious to Economic Plants*. University of California Press, Berkeley and Los Angeles, California.

Jesionowska, K. (1991) Morphology of the Eupodidae (Actinotrichida, Actinedida, Eupodoidea). In: Dusbabek, F. and Bukva, V. (eds) *Modern Acarology, Volume 2*. Academia, Prague and SPB Academic Publishing, The Hague, the Netherlands, pp. 265–278.

Jesiotr, J. (1980) The influence of host plants on the reproduction potential of the two-spotted spider mite, *Tetranychus urticae* Koch (Acarina: Tetranychidae) IV. Changes within different populations affected by a new species of host plant. *Ekologia Polska* (now *Polish Journal of Ecology*) 28, 633–647.

Jesiotr, J. and Suski, W. (1976) The influence of the host plants on the reproduction potential of the two-spotted spider mite, *Tetranychus urticae* Koch (Acarina: Tetranychidae). *Ekologia Polska* (now *Polish Journal of Ecology*) 24, 407–411.

Jesse, L.C., Moloney, K.A. and Obrycki, J.J. (2006) Abundance of arthropods on the branch tips of the invasive plant, *Rosa multiflora* (Rosaceae). *Weed Biology and Management* 6, 204–211.

Jeyaprakash, A. and Hoy, M.A. (2009) First divergence time estimate of spiders, scorpions, mites and ticks (subphylum: Chelicerata) inferred from mitochondrial phylogeny. *Experimental and Applied Acarology* 47, 1–18.

Jeyarani, S. and Chandrasekaran, M. (2006) Bioefficacy of certain acaricides against chilli mite, *Polyphagotarsonemus latus* (Banks). *Agricultural Science Digest* 26, 132–134.

Jhansi Lakshmi, V., Krishnaiah, N.V., Pasalu, I.C. and Katti, G. (2008) Bio-ecology and management of rice mites – a review. *Agricultural Reviews* 29, 31–39.

Ji, J., Zhang, Y.-X., Chen, X. and Lin, J.-Z. (2005a) Life table of a laboratory population of *Oligonychus biharensis* (Hirst) (Acari: Tetranychidae) at different temperatures. *Acta Arachnologica Sinica* 14(1), 37–41.

Ji, J., Zhang, Y., Chen, X. and Lin, J. (2005b) Responses of *Amblyseius cucumeris* (Acari: Phytoseiidae) to *Tetranychus turkestani* (Acari: Tetranychidae) on sweet potato leaves. *Entomological Journal of East China* 3, 224–228.

Ji, J., Zhang, Y.-X., Chen, X. and Lin, J.-Z. (2008) Responses to stimuli from *Oligonychus biharensis* Hirst (Acari: Tetranychidae) on loquat leaves by *Neoseiulus cucumeris* (Oudemans) (Acari: Phytoseiidae). *International Journal of Acarology* 34, 175–181.

Jia, C.-L., Cui, W.-J., Li, S.-H. and Li, A.-R. (2002) *Rhizoglyphus echinopus* severely damages *Allium fistulosum*. *Plant Protection Technology and Extension* 22(10), 45 (in Chinese).

Jiang, J., Friebe, B., Dhaliwal, H.S., Martin, T.J. and Gill, B.S. (1993) Molecular cytogenetic analysis of *Agropyron elongatum* chromatin in wheat germplasm specifying resistance to wheat streak mosaic virus. *Theoretical and Applied Genetics* 86, 41–48.

Jiang, P.Z., Xie, X.J., Chen, W.X., Cao, S.Y. and Liang, Z.H. (1994) Regularity of incidence of *Steneotarsonemus spinki*. *Guandong Agricultural Sciences* 5, 37–40 (in Chinese).

Jiang, W., Garrett, K.A., Peterson, D.E., Harvey, T.L., Bowden, R.L. and Fang, L. (2005) The window of risk for emigration of wheat streak mosaic virus varies with host eradication method. *Plant Disease* 89, 853–858.

Jiang, Y. and Ridsdill-Smith, T.J. (1996) Examination of the involvement of mechanical strength in antixenotic resistance of subterranean clover cotyledons to the redlegged earth mite *Halotydeus destructor* (Acarina: Penthaleidae). *Bulletin of Entomological Research* 86, 263–270.

Jiang, Y., Ridsdill-Smith, T.J. and Ghisalberti, E.L. (1996) Assays for the effects of volatile compounds from artificially damaged cotyledons of subterranean clover on the redlegged earth mite, *Halotydeus destructor*. *Experimental and Applied Acarology* 20, 61–72.

Jiang, Z.-T. and Ge, C.-H. (1994) Studies on the life history and reproductive behaviour of *Rhizoglyphus callae*. *Journal of Nanchang University (Natural Science)* 18, 117–122 (in Chinese).

Jin, D.Y. (2006) Biological characteristics of *Oligonychus ununguis*. *Journal of Northeast Forestry University* 34(6), 52–53.

Jin, D.Y., Lu, L.S. and Li, L.G. (2002) Effect of 16 types of insecticide on the experimental population of *Tetranychus truncatus* Ehara. *Journal of Jilin Agricultural Sciences* 27(1), 35–37, 46 (in Chinese).

Jing, Y., Huang, J. and Han, J.C. (2003) Effects of alternative prey *Tetranychus fijiensis* on the experimental population of *Delphastus catalinae* (Horn) (Coleoptera: Coccinellidae). *Acta Arachnologica Sinica* 12, 100–105 (in Chinese).

Johann, L., Klock, C.L., Ferla, N.J. and Botton, M. (2009) Acarofauna (Acari) associada à videira (*Vitis vinifera* L.) no Estado do Rio Grande do Sul. *Biociências* 17, 1–19.

Johanowicz, D.L. and Hoy, M.A. (1996) *Wolbachia* in a predator–prey system: 16S ribosomal DNA analysis of two phytoseiids (Acari: Phytoseiidae) and their prey (Acari: Tetranychidae). *Annals of the Entomological Society of America* 89, 435–441.

Johanowicz, D.L. and Hoy, M.A. (1998) Experimental induction and termination of non-reciprocal reproductive incompatibilities in a parahaploid mite. *Entomologia Experimentalis et Applicata* 87, 51–58.

Johansen, T.J. (1986) Middskader i eng i Nord-Norge. *Norden* 90, 17, 34 (in Norwegian).

Johnson, C.G. (1969) *Migration and Dispersal of Insects by Flight*. Methuen, London.

Johnson, D.T. and Croft, B.A. (1981) Dispersal of *Amblyseius fallacis* (Acarina: Phytoseiidae) in an apple ecosystem. *Environmental Entomology* 10, 313–319.

Johnson, P.C. (1958) *Spruce Spider Mite Infestations in Northern Rocky Mountain Douglas-fir Forests*. Research Paper No. 55, US Department of Agriculture Forest Service Intermountain Forest and Range Experiment Station, Ogden, Utah.

Johnson, W.T. and Lyon, H.H. (1976) *Insects that Feed on Trees and Shrubs*. Cornell University Press, Ithaca, New York.

Johnson, W.T. and Lyon, H.H. (1988) *Insects that Feed on Trees and Shrubs* (rev. edn). Cornell University Press, Ithaca, New York.

Johnson, W.T. and Lyon, H.H. (1991) *Insects that Feed on Trees and Shrubs*, 2nd edn. Cornell University Press, Ithaca, New York.

Johnston, D.E. (1982) Acari. In: Parker, S.P. (ed.) *Synopsis and Classification of Living Organisms, Volume 2*. McGraw-Hill, New York, pp. 111–169.

Johnston, D.E. and Bruce, W.A. (1965) Tyrophagus neiswanderi, *a New Acarid Mite of Agricultural Importance*. Research Bulletin No. 977, Ohio Agricultural Experiment Station, Wooster, Ohio.

Johnston, R.B., King, J.R. and McBribe, J.J. Jr (1957) Zineb controls citrus rust mite. *Proceedings of the Florida State Horticultural Society* 70, 38–48.

Jones, A.T. (2000) Black currant reversion disease-the probable causal agent, eriophyid mite vectors, epidemiology and prospects for control. *Virus Research* 71, 71–84.

Jones, A.T., Gordon, S.C. and Jennings, D.L. (1984) A leaf-blotch disorder of tayberry associated with the leaf and bud mite (*Phyllocoptes gracilis*) and some effects of three aphid-borne viruses. *Journal of Horticultural Science* 59, 523–528.

Jones, A.T., Kumar, P.L., Saxena, K.B., Kulkarni, N.K., Muniyappa, V. and Waliyar, F. (2004) Sterility mosaic disease – the "green plague" of pigeonpea: advances in understanding the etiology, transmission and control of a major virus disease. *Plant Disease* 88, 436–445.

Jones, L.S. and Prendergast, D.T. (1937) Method of obtaining an index to density of field populations of citrus red mite. *Journal of Economic Entomology* 30, 934–940.

Jones, L.S. and Wilson, N.S. (1952) Peach mosaic spreads throughout the growing season. *Bulletin of the California Department of Agriculture* 40(3), 117–118.

Jones, L.S., Wilson, N.S., Burr, W. and Barnes, M.M. (1970) Restriction of peach mosaic virus spread through control of the vector mite *Eriophyes insidiosus*. *Journal of Economic Entomology* 63, 1551–1552.

Jones, M.B., Heard, A.J., Woledge, J., Leafe, E.L. and Plumb, R.T. (1977) The effect of ryegrass mosaic virus on carbon assimilation and growth of ryegrasses. *Annals of Applied Biology* 87, 393–405.

Jones, O.T. (1987) The use of behaviour modifying chemicals in the integrated management of selected fruit fly species. In: Economopoulos, A.P. (ed.) *Proceedings of the 2nd International Symposium on Fruit Flies, 16–21 September 1986, Colymbari, Crete*. Tsiveriotis, Athens, distributed by Elsevier Science, Amsterdam and New York, pp. 451–458.

Jones, R.K. (1979) The effectiveness of some insecticides against blemish pests on covered and uncovered banana bunches. *Citrus and Subtropical Fruit Journal* 547, 18–19.

Jones, V.P. (1990) Sampling and dispersion of the twospotted spider mite (Acari: Tetranychidae) and the western orchard predatory mite (Acari: *Phytoseiidae*) on tart cherry. *Journal of Economic Entomology* 83, 1376–1380.

Jones, V.P. and Brown, R.D. (1983) Reproductive responses of the broad mite, *Polyphagotarsonemus latus* (Banks) (Acari: Tarsonemidae) to constant temperature-humidity regimes. *Annals of the Entomological Society of America* 76, 466–469.

Jones, V.P. and Parrella, M.P. (1984a) Greenhouse and laboratory studies on the sub-lethal effects of selected insecticides on *Panonychus citri* (McGregor). *Canadian Entomologist* 116, 1033–1040.

Jones, V.P. and Parrella, M.P. (1984b) Presence–absence sampling of citrus red mite on lemons. *California Agriculture* 38(9), 31–32.

Jones, V.P., Toscano, N.C., Johnson, M.W., Welter, S.C. and Youngman, R.R. (1986) Pesticide effects on plant physiology: integration into a pest management program. *Bulletin of the Entomological Society of America* 32, 103–109.

Jonge, J.T. de (1988) Remarkable mites and insects in and around buildings in 1986. *Entomologische Berichten* 48, 18–19.

Joos, J.L., Berlowitz, A. and Davis, C.S. (1974) Controlling pear rust mite. *California Agriculture* 28(11), 15.

Jorgensen, C.D. (1967) A new species of *Tuckerella* (Acarina: Tuckerellidae) from Nevada. *Entomological News* 78, 141–146.

Jose, V.T. and Shah, A.H. (1989a) Bionomics of the spider mite, *Tetranychus macfarlanei* injurious to cotton in Gujarat, India. In: Channabasavanna, G.P. and Viraktamath, C.A. (eds) *Progress in Acarology, Volume 2*. Oxford & IBH Publishing, India, pp. 23–28.

Jose, V.T. and Shah, A.H. (1989b) Carry over of spider mite, *Tetranychus macfarlanei* through alternate host plants in cotton-growing areas of south and central Gujarat, India. In: Channabasavanna, G.P. and Viraktamath, C.A. (eds) *Progress in Acarology, Volume 2*. Oxford & IBH Publishing, New Delhi, pp. 29–31.

Jovicich, E., Cantliffe, D.J., Osborne, L.S. and Stoffella, P.J. (2004) Mite population and damage caused by broad mites (*Polyphagotarsonemus latus* (Banks)) infesting bell pepper (*Capsicum annuum* L.) at different seedling developmental stages. *Acta Horticulturae* 659(1), 339–344.

Jovicich, E., Cantliffe, D.J., Osborne, L.S., Stoffella, P.J. and Simonne, E.H. (2009) Release of *Neoseiulus californicus* on pepper transplants to protect greenhouse-grown crops from early broad mite (*Polyphagotarsonemus latus*) infestations. In: Mason, P.G., Gillespie, D.R. and Vincent, C. (eds) *Proceedings of the 3rd International Symposium on the Biological Control of Arthropods, 8–13 February 2009, Christchurch, New Zealand*. Publication FHTET-2008-06, US Department of Agriculture Forest Service Forest Health Technology Enterprise Team, Morgantown, Virginia, pp. 347–353.

Julia, J.F. and Mariau, D. (1979a) New research on the coconut mite, *Eriophyes guerreronis* K., in the Ivory Coast. *Oléagineux* 34, 181–189.

Julia, J.F. and Mariau, D. (1979b) Nouvelles recherches en Côte d'Ivoire sur *Eryophyes guerreronis* K. acarien ravageur des noix de cocotier. In: *Congrès sur la Lutte Contre les Insectes en Milieu Tropical. Chambre de Commerce et d'Industrie de Marseille, 13–16 Mars 1979. Compte Rendu des Travaux. Premiere partie. Cultures Tropicales. Deuxieme partie. Sante Humaine et Animale/Congress on the Control of Insects in the Tropical Environment. Chamber of Commerce and Industry of Marseilles, 13–16 March 1979. Report of the Proceedings. Part I. Tropical crops. Part II. Human and Animal Health*. Chamber of Commerce and Industry, Marseilles, France, pp. 373–387.

Jung, C., El Taj, H. and Mitsuhiro, K. (2010) Biological control of spider mites with the introduced *Neoseiulus californicus* in Korean apple orchard. In: Moraes, G.J. de, Castilho, R.C. and Flechtmann, C.H.W. (eds) *XIII International Congress of Acarology, August 23–27, 2010, Recife-PE, Brazil, Abstract Book*, abstract 214, p. 117. Available at: http://www.acarology.org/ica/ica2010/docs/0910_15_abstractBookCapa.pdf (accessed 3 February 2015).

Jung, Y.-C., Mizuki, E., Akao, T. and Côté, J. C. (2007) Isolation and characterization of a novel *Bacillus thuringiensis* strain expressing a novel crystal protein with cytocidal activity against human cancer cells. *Journal of Applied Microbiology* 103, 65–79.

Jupp, P.W. and Cullen, J.M. (1996) Expected and observed effects of the mite *Aculus hyperici* on St John's wort, *Hypericum perforatum*, in Australia. In: Moran, V.C. and Hoffmann, J.H. (eds) *Proceedings of the IX International Symposium on Biological Control of Weeds, 19–26 January 1996, University of Cape Town, Stellenbosch, South Africa*, pp. 365–370. Available at: http://www.invasive.org/proceedings/pdfs/9_365-370.pdf (accessed 12 May 2015).

Jupp, P.W., Briese, D.T. and Cullen, J.M. (1997) Evidence for resistance in *Hypericum perforatum* to a biological control agent, the eriophyid mite *Aculus hyperici*. *Plant Protection Quarterly* 12, 67–70.

Justiniano, W., Pereira, M.F.A., Amorim, L.C. de S. and Maciel, C.D. de G. (2009) Efficiency of neem oil on citrus leprosis mite *Brevipalpus phoenicis* (Geijskes, 1939) control. *Pesquisa Agropecuária Tropical* 39, 38–42.

Jyotika, K.G. and Mandeep, K. (2003) Biochemical basis of differential susceptibility of citrus cultivars to infestation with the citrus mite, *Tetranychus orientalis* (Klein). *Annals of Biology (Hissar)* 19, 235–240.

Kadono, F. (1981) Two new eriophyid mites from Japan (Acarina, Eriophyidae). *Applied Entomology and Zoology* 16, 419–422.

Kadono, F. (1985) Three species of eriophyid mites injurious to fruit trees in Japan (Acarina: Eriophyidae). *Applied Entomology and Zoology* 20, 458–464.

Kadono, F. (1992) A new species of eriophyid mite injurious to tea plants in Japan (Acari: Eriophyidae). *Acta Arachnologica* 41, 149–152.

Kadono, F. and Endo, M. (1996) Seasonal population trend and control of *Tyrophagus neiswanderi* Johnston et Bruce on Phalaenopsis in the greenhouse. *Proceedings of the Kanto-Tosan Plant Protection Society* 43, 247–252.

Kadzhaya, G.S. (1988) On the taxonomy of the genus *Mycetoglyphus* (Acariformes, Acaroidea). *Zoologicheskii Zhurnal* 67, 141–142 (in Russian).

Kaestner, A. (1980) *Invertebrate Zoology. Volume II. Arachnids and Myriapods.* Translated from Russian by Levi, H.W. and Levi, L.R. Krieger, Huntington, New York.

Kafil, M., Allahyari, H. and Saboori, A. (2007) Effect of host plants on developmental time and life table parameters of *Amphitetranychus viennensis* (Acari: Tetranychidae). *Experimental and Applied Acarology* 42, 273–281.

Kahu, K., Janes, H., Luik, A. and Klaas, L. (2009) Yield and fruit quality of organically cultivated blackcurrant cultivars. *Acta Agriculturae Scandinavica, Section B – Soil and Plant Science* 59, 63–69.

Kaimal, S.G. (2013) Life table of *Oligonychus biharensis* (Hirst) (Acari: Tetranychidae) – a pest on *Vigna unguiculata* (L.) Walp. *Global Journal of Bio-Science and Biotechnology* 2(4), 576–579.

Kaimal, S.G. and Ramani, N. (2007a) Chlorophyll loss due to infestation by *Tetranychus neocaledonicus* André (Acari: Tetranychidae) on *Moringa oleifera* Lam. *Journal of Acarology* 17(1–2), 12–14.

Kaimal, S.G. and Ramani, N. (2007b) Biological studies of *Tetranychus neocaledonicus* (Andre) (Acari: Tetranychidae) infesting *Moringa oleifera* Lam. *Bulletin of Pure and Applied Sciences – Zoology*, Jul. 1, 2007. Available at: http://www.thefreelibrary.com/Biological+studies+of+Tetranychus+neocaledonicus+(Andre)+(Acari%3A...-a0199539298 (accessed 21 August 2014).

Kaimal, S.G. and Ramani, N. (2011a) Biology of *Oligonychus biharensis* (Hirst) (Acari: Tetranychidae) on cassava. *Journal of Experimental Zoology India* 14, 27–30.

Kaimal, S.G. and Ramani, N. (2011b) Studies on feeding characteristics of *Oligonychus biharensis* (Hirst) (Acari: Tetranychidae) infesting cassava. *Biological Forum – An International Journal* 3(2), 9–13. Available at: http://www.researchtrend.net/biological_forum_3%282%29_2011/2%20SANGEETA.pdf (accessed 12 May 2015).

Kaimal, S.G. and Ramani, N. (2011c) Biology of *Tetranychus ludeni* Zacher (Acari: Tetranychidae) – a pest of velvet bean. *Indian Journal of Fundamental and Applied Life Sciences* 1(3), 1–6.

Kaimal, S.G., Sheeja, U.M. and Ramani, N. (2011) Ultrastructural elucidation of leaf damage on cassava induced by *Oligonychus biharensis* (Hirst) (Acari: Tetranychidae). *International Journal of Acarology* 37(Supplement 1), 108–113.

Kalaisekar, A., Naidu, V.G. and Rao, N.V. (2000) Citrus rust mite, *Phyllocoptruta oleivora* (Ashmead) (Eriophyidae: Acarina): effect of its damage on fruit quality and its chemical control. *Indian Journal of Plant Protection* 28, 132–134.

Kalaisekar, A., Naidu, V.G. and Rao, N.V. (2003) Efficacy of some pesticides against citrus rust mite, *Phyllocoptruta oleivora*. *Indian Journal of Entomology* 65, 308–310.

Kaliszewski, M. (1987) *Siteroptes longisomus* sp. n. from Poland, with remarks on the genus and key to the species (Acari, Pygmephoroidea). *Entomologische Mitteilungen aus dem Zoologisches Institut und Zoologisches Museum der Universität Hamburg* 9(130), 21–36.

Kaliszewski, M. (1988a) *Diroptes* gen. n. (Acari, Pygmephoroidea) with a key to the species. *Entomologische Mitteilungen aus dem Zoologisches Institut und Zoologisches Museum der Universität Hamburg* 9(132), 115–122.

Kaliszewski, M. (1988b) *Pseudobakerdania* gen. n. (Acari, Pygmephoroidea) with the description of two new species and some remarks on the phylogenetic position of the genus. *Entomologische Mitteilungen aus dem Zoologisches Institut und Zoologisches Museum der Universität Hamburg* 9(132), 123–134.

Kaliszewski, M. (1993) *Key to Palearctic (Species) of the Genus Tarsonemus: Acari, Tarsonemidae*. Seria: Zoologia 14, Wydawnictwo Naukowe, Uniwersytetu im. Adama Mickiewicza, Poznań, Poland (in Polish).

Kaliszewski, M. and Rack, G. (1986) A systematic study of the mite *Microdispus* (*Premicrodispus*) *lambi* Krczal, 1964 (Acari: Pygmephoroidea), a mushroom pest in Australia. *Mitteilungen aus dem Hamburgischen Zoologischen Museum und Institut* 83, 231–241.

Kaliszewski, M.J., Athias Binche, F. and Lindquist, E.E. (1995) Parasitism and parasitoidism in Tarsonemina (Acari: Heterostigmata) and evolutionary considerations. *Advances in Parasitology* 35, 335–367.

Kalshoven, L.G.E. (1951) Acarina. In: Kalshoven, L.G.E. *De Plagen van de Cultuurgewassen in Indonesië (2 Vols)*. W. Van Hoeve's, Gravenhage/Bandung, Indonesia, pp. 68–90.

Kamali, K. (1990) A checklist of plant mites (Acari) of Khuzestan, southwestern Iran. *Scientific Journal of Agriculture* 13(13), 83–73 (in Arabic).

Kamali, K., Ostovan, H. and Atamehr, A. (2001) *A Catalogue of Mites and Ticks (Acari) of Iran*. Scientific Publication Center, Islamic Azad University, Tehran.

Kamath, M.K. (1979) A review of biological control of insect pests and noxious weeds in Fiji (1969–1978). *Fiji Agricultural Journal* 41, 55–72.

Kamau, A.W. (1977) Effect of eriophyid mite, *Aculops lycopersici* Massee, (Acarina: Eriophyidae) on tomato. *Kenya Entomological News*. 5, 4.

Kamau, A.W., Mueke, J.M. and Khaemba, B.M. (1992) Resistance of tomato varieties to the tomato russet mite, *Aculops lycopersici* (Massee) (Acarina: Eriophyidae). *Insect Science and Its Application* (now *International Journal of Tropical Insect Science*) 13, 351–356.

Kanavel, R.F. and Selhime, A.G. (1967) Biological studies on *Paracheyletia bakeri* (Acarina: Cheyletidae). *Florida Entomologist* 50, 107–113.

Kanazawa, M., Sahara, K. and Saitō, Y. (2010) Silk threads function as an 'adhesive cleaner' for nest space in a social spider mite. *Proceedings of the Royal Society B: Biological Sciences* 278, 1653–1660.

Kanda, K. and Hirai, Y. (1990) Outbreak of the winter grain mite, *Penthaleus major* (Dugès) in forage grasses in Nasu district. *Japanese Journal of Applied Entomology and Zoology* 34, 79–81.

Kandulna A., Singh, J. and Prasad, R. (2005) Management of eriophyid mite *Aceria cajani*, as transmitting agent of sterility mosaic disease of pigeonpea. *Shashpa* 12, 51–53.

Kane, N.A., Jones, C.S. and Vuorisalo, T. (1997) Development of galls of *Alnus glutinosa* and *Alnus incana* (Betulaceae) caused by the eriophyid mite *Eriophyes laevis* (Nalepa). *International Journal of Plant Sciences* 158, 13–23.

Kanervo, V. and Makinen, K.L. (1967) Herukan äkämäpumkin torojunta. *Maatalous Koetoim* 21, 148–158 (in Finnish).

Kang, S.G., Koo, B.J., Lee, E.T. and Chang, M.U. (2007) Allexivirus transmitted by eriophyid mites in garlic plants. *Journal of Microbiology and Biotechnology* 17, 1833–1840.

Kannaiyan, J., Nene, Y.L., Reddy, M.V. and Raju, T.N. (1981) Prevalence of pigeonpea wilt and sterility mosaic in India 1975-80. *International Pigeonpea Newsletter* 1, 25.

Kannaiyan, J., Nene, Y.L., Reddy, M.V., Ryan, J.G. and Raju, T.N. (1984) Prevalence of pigeonpea diseases and crop losses in Asia, Africa and the America. *Tropical Pest Management* 30, 62–71.

Kanta, S., Rai, B.K. and Rattan, L. (1963) Evaluation of the toxicity of some pesticides to the fig mite, *Eotetranychus hirsti* Pritchard and Baker (Tetranychidae: Acarina). *Indian Journal of Entomology*, 26–32.

Kantack, E.J. and Knutson, H. (1958) Chemical control studies of the wheat curl mite. *Journal of Economic Entomology* 51, 68–72.

Kantaratanakul, S, and Jitrat, S. (1984) Bionomics of the pyemotid mite, *Luciaphorus hauseri* infesting cultivated Jew's ear fungus *Auricularia polytrica* in Thailand. In: Griffiths, D.A. and Bowman, C.E. (eds) *Acarology VI, Volume 1*. Ellis Horwood, Chichester, UK, pp. 622–628.

Kantaratanakul, S., Sombatsiri, K. and Tauthong, P. (1979) Effectiveness of arthropods in the control of carmine mite. In: Rodriguez, J.G. (ed.) *Recent Advances in Acarology, Volume I*. Academic Press, New York, pp. 499–505.

Kantaratanakul, S., Chaiwat, W. and Prayoonrat, P. (1989) Bionomics of the dolichocybid mite, *Formicotes* sp., as a pest of cultivated mushroom in Thailand. In: ChannaBasavanna, G.P. and Viraktamath, C.A. (eds) *Progress in Acarology, Volume 2*. Oxford & IBH Publishing, New Delhi, pp. 223–227.

Kapaxidi, E.V. Markoyiannaki-Printziou, D. and Papaioanou-Souliotis, P. (2008) First record of *Aceria cynodoniensis* feeding on Bermuda grass in Greece. *Hellenic Plant Protection Journal* 1, 89–91.

Kapetanakis, E.G., Warman, T.M. and Cranham, J.E. (1986) Effects of permethrin sprays on the mite fauna of apple orchards. *Annals of Applied Biology* 108, 21–32.

Kappers, I.F., Verstappen, F.W.A., Luckerhoff, L.L.P., Bouwmeester, H.J. and Dicke, M. (2010) Genetic variation in jasmonic acid- and spider mite-induced plant volatile emission of cucumber accessions and attraction of the predator *Phytoseiulus persimilis*. *Journal of Chemical Ecology* 36, 500–512.

Kapur, A.P. (1948) On the Old World species of the genus *Stethorus* Weise (Coleoptera, Coccinellidae). *Bulletin of Entomological Research* 39, 297–320.

Kapur, A.P. (1961) A new species of *Stethorus* Weise (Coleoptera: Coccinellidae), feeding on arecanut palm mites in Kerala, southern India. *Entomophaga* 6, 35–38.

Kapur Ghai, J. and Mandeep Kaur (2003a) Differential susceptibility of various cultivars to infestation with the citrus brown mite, *Eutetranychus orientalis*. *Annals of Agri Bio Research* 8, 247–253.

Kapur-Ghai, J. and Mandeep Kaur (2003b) Biochemical basis of differential susceptibility of citrus cultivars to infestation with the citrus mite, *Eutetranychus orientalis* (Klein). *Annals of Biology (Hissar)* 19, 235–240.

Karafiat, H. (1959) Systematik und Ökologie der Scutacariden. In: Stammer, H.J. (ed.) *Beiträge zur Systematik und Ökologie Mitteleuropäischer Acarina, Bund 1, Abschnitt 4*. Akademische Verlagsgesellschaft Geest & Portig, Leipzig, Germany, pp. 627–712.

Karami-Jamour, T. and Shishehbor, P. (2012) Development and life table parametersof *Tetranychus turkestani* (Acarina: Tetranychidae) at different constant temperatures. *Acarologia* 52, 113–122.

Karban, R. (1986) Induced immunelike resistance to spider mites in cotton. *California Agriculture* 40(11, 12), 13–15.

Karban, R. and Carey, J.R. (1984) Induced resistance of cotton seedlings to mites. *Science* 225, 53–54.

Karban, R. and English-Loeb, G.M. (1990) A "vaccination" of Willamette spider mites (Acari: Tetranychidae) to prevent large populations of Pacific spider mites on grapevines. *Journal of Economic Entomology* 83, 2252–2257.

Karban, R. and Meyers, J.H. (1989) Induced plant responses to herbivory. *Annual Review of Ecology and Systematics* 20, 331–348.

Karg, W. (1970) *Tyrophagus cucumeris* n. sp. eine neue Milbenart an Gurkenkulturen in Gewächshäusern. *Archiv für Pflanzenschutz* 6, 481–486.

Karg, W. (1972) Untersuchungen über die Korrelation zwischen dominierenden Raubmilbenarten und ihr möglichen Beute in Apfelanlagen. *Archiv für Pflanzenschutz* 8, 29–52.

Karg, W. (1994) *Raubmilben, nützliche Regulatoren im Naturhaushalt*. Die Neue Brehm-Bücherei, Bd. 624, Westarp Wissenschaften, Magdeburg, Germany.

Karg, W., Burth, U. and Ramson, A. (1973) Der Einfluss von Fungiziden auf das Auftreten von Spinnmilben und anderen blattbewohnenden Milbengruppen in Apfelanlagen. *Nachrichten Pflanzenschutz DDR* 27, 169–175.

Karl, E. (1965a) Untersuchungen zur Morphologie und Ökologie von Tarsonemiden gartnerischer Kulturpflanzen. I. *Tarsonemus pallidus* Banks. *Biologisches Zentralblatt* 84, 47–80.

Karl, E. (1965b) Untersuchungen zur Morphologie und Ökologie von Tarsonemiden gartnerischer Kulturpflanzen. II. *Hemitarsonemus latus* (Banks), *Tarsonemus confusus* Ewing, *T. talpae* Schaarschmidt, *T. setifer* Ewing, *T. smithi* Ewing und *Tarsonemoides belemnitoides* Weis-Fogh. *Biologisches Zentralblatt* 84, 331–357.

Karl, E. (1969) Untersuchungsergebnisse zur Klarung der Artfrage bei *Tarsonemus pallidus* Banks. In: Fankhänel, H. (ed.) *Der Berichte üer die 10. Wanderversammlung Deutscher Entomologen, 15 bis 19 September 1965 in Dresden*. *Tagungsberichte Nr* 80. Deutsche Akademie der Landwirtschaftswissenschaften zu Berlin, DDR, pp. 567–574.

Karmakar, K. (2007) *Steneotarsonemus spinki* Smiley (Acari: Tarsonemidae)-a silent yield reducing mite of rice in West Bengal, India. *International Journal of Acarology* 22, 1–5.

Karmakar, K. and Saha, G. (2005) Population dynamics of false spider mite, *Brevipalpus phoenicis* (Giejskes) (Acari:Tenuipalpidae) on *Mikania micrantha* Kunth. in relation to weather parameters. *Journal of Crop and Weed* 1(2), 68–70.

Karmakar, K. and Soma Dey (2006) Studies on seasonal incidence of phytophagous mite species on selected germplasms of banana in West Bengal. *Indian Journal of Crop Science* 1(1–2), 138–139.

Karmakar, K., Sarkar, P.K., Somchoudhury, A.K. and Mukherjee, A.B. (1996a) Influence of host plants on different life stages of *Polyphagotarsonemus latus* (Banks) (Acari: Tarsonemidae). *Annals of Entomology (New Delhi)* 14(2), 41–45.

Karmakar, K., Sarkar, P.K., Somchoudhury, A.K. and Mukherjee, A.B. (1996b) Effectiveness of some modern pesticides against different stages of yellow mite, *Polyphagotarsonemus latus* (Banks) (Acari: Tarsonemidae) infesting chilli. *Annals of Entomology (New Delhi)* 14(2), 47–54.

Karnkowski, W. (1995) Method for examination of Narcissus bulbs to detect the bulb scale mite (*Steneotarsonemus laticeps* Helbert) (Acarina: Tarsonemidae). In: Boczek, J. and Ignatowicz, S. (eds) *Materialy z Sympozjum na temat: "Osiagniecia Akarologii w Polsce", Siedlce, 26–27 wresnia 1995* (*Proceedings of the Symposium on "Advances in Acarology in Poland", Siedlce, 26–27 September 1995*). Instytut Ochrony Roślin, Polska Akademia Nauk, Olsztyn, Poland, pp. 106–110.

Kartasheva, T.T. and Lesteva, E.E. (1979) The role of acarophages in the regulation of numbers of fruit spider mites in orchards in Kirgizia. *Entomologicheskie Issledovaniya v Kirgizii* 13, 82–85 (in Russian).

Karuppuchamy, P. (1994) Studies on the management of pests of pomegranate with special reference to fruit borer, *Virachola isocrates* (Fab.). PhD thesis, Tamil Nadu Agricultural University, Coimbatore, India.

Karuppuchamy, P. and Mohanasundaram, M. (1987) New species and records of tetranychid mites (Tetranychidae: Acarina) from India. *Entomon* 12, 89–94.

Karut, K., Kasap, I., Kazak, C., Yildiz, S. and Sekeroglu, E. (1998) Biological control of *Polyphagotarsonemus latus* (Acarina, Tarsonemidae) using the predatory mite, *Typhlodromus athiasae* (Acarina, Phytoseiidae) in greenhouse. In: Brunnhofer, V. and Soldán, T. (eds) *Book of Abstracts. Proceedings of the 6th European Congress of Entomology, Ceske Budejovice, Czech Republic, 23–29 August 1998*. Institute of Entomology (CAS), Ceske Budejovice, Czech Republic, pp. 2, 332, 562.

Kasap, I. (2002) Biology and life tables of the twospotted spider mite, *Tetranychus urticae* Koch (Acari: Tetranychidae) on three different host plants in laboratory conditions. *Turkiye Entomoloji Dergisi* 26, 257–266 (in Turkish).

Kasap, I. (2003) Life history of hawthorn spider mite *Amphitetranychus viennensis* (Acarina: Tetranychidae) on various apple cultivars and at different temperatures. *Experimental and Applied Acarology* 31, 79–91.

Kasap, I. (2009) The biology and fecundity of the citrus red mite *Panonychus citri* (McGregor) (Acari: Tetranychidae) at different temperatures under laboratory conditions. *Turkish Journal of Agriculture and Forestry* 33, 593–600.

Kasap, I. (2010) Seasonal population development of spider mites (Acari: Tetranychidae) and their predators in sprayed and unsprayed apple orchards in Van, Turkey. In: Moraes, G.J. de, Castilho, R.C. and Flechtmann, C.H.W. (eds) *XIII International Congress of Acarology, August 23–27, 2010, Recife-PE, Brazil, Abstract Book*, abstract 221, p. 120. Available at: http://www.acarology.org/ica/ica2010/docs/0910_15_abstractBookCapa.pdf (accessed 3 February 2015).

Kasap, I. (2011) Biological control of the citrus red mite *Panonychus citri* by the predator mite *Typhlodromus athiasae* on two citrus cultivars under greenhouse conditions. *BioControl* 56, 327–332.

Kasap, I. and Sekeroglu, E. (2004) Life history of *Euseius scutalis* feeding on citrus red mite *Panonychus citri* at various temperatures. *BioControl.* 49, 645–654.

Kasap, I., Karut, K. and Sekeroglu, E. (2004) Determination of sequential sampling plan of the citrus red mite *Panonychus citri* (McGregor) (Acarina: Tetranychidae) for Washington and Valencia orange cultivars. *Turkiye Entomoloji Dergisi* 28, 3–14.

Kashio, T. and Tanaka, M. (1979) Effects of fungicides on the predaceous mite, *Amblyseius deleoni* Muma & Denmark (Acarina: Phytoseiidae). *Proceedings of the Association for Plant Protection of Kyushu* 25, 153–156.

Kassar, A. and Amrine, J.W. Jr (1990) Rearing and development of *Phyllocoptes fructiphilus*. *Entomological News* 101, 276–282.

Kast, W.K., Rupp, D., Schiefer, H.C. and Trankle, L. (2004) Statistische Beziehungen zwischen Witterungsdaten und dem Auftreten von Krankheiten und Schadlingen im Weinbaugebiet Wurttemberg/Deutschland. *Mitteilungen Klosterneuburg, Rebe und Wein, Obstbau und Früchteverwertung* 54, 239–248.

Kasuga, S. and Amano, H. (2000a) Influence of temperature on the life history parameters of *Tyrophagus similis* Volgin (Acari: Acaridae). *Applied Entomology and Zoology* 35, 237–244.

Kasuga, S. and Amano, H. (2000b) Survey for the genus *Tyrophagus* and its damage in spinach crop with view to develop its pest management strategy. *Journal of the Acarological Society of Japan* 9, 31–42 (in Japanese).

Kasuga, S. and Amano, H. (2002) Susceptibility of 18 Japanese populations of *Tyrophagus similis* Volgin (Acari: Acaridae) to DDVP EC. *Japanese Journal of Applied Entomology and Zoology* 46, 99–101.

Kasuga, S. and Amano, H. (2003) Seasonal prevalence and susceptibility to agrochemicals of *Tyrophagus similis* (Acari: Acaridae) in spinach buds and agricultural soil under greenhouse conditions. *Experimental and Applied Acarology* 30, 279–288.

Kasuga, S. and Amano, H. (2005) Spatial distribution of *Tyrophagus similis* (Acari: Acaridae) in agricultural soils under greenhouse conditions. *Applied Entomology and Zoology* 40, 507–511.

Kasuga, S. and Amano, H. (2006) Infestation of *Tyrophagus similis* Volgin (Acari: Acaridae) on spinach during the seed germination period. *Journal of the Acarological Society of Japan* 15, 69–73.

Kasuga, S. and Honda, K. (2006a) High-temperature tolerance of *Tyrophagus similis*, and its comparison with *T. neiswanderi*, *T. putrescentiae* and *Rhizoglyphus robini*. *Japanese Journal of Applied Entomology and Zoology* 50, 19–23.

Kasuga, S. and Honda, K. (2006b) Infestation of *Tyrophagus similis* Volgin (Acari: Acaridae) on spinach during the seed germination period. *Journal of the Acarological Society of Japan* 15, 69–73.

Kasuga, S. and Honda, K. (2006c) Suitability of organic matter, fungi and vegetables as food for *Tyrophagus similis* (Acari: Acaridae). *Applied Entomology and Zoology* 41, 227–231.

Kasuga, S., Honda, K., Kawai, A. and Amano, H. (2005) A bait trap for monitoring acarid mites (Acari: Acaridae) in agricultural soils. *Applied Entomology and Zoology* 40, 303–308.

Kasuga, S., Kanno, H. and Amano, H. (2006) Development, oviposition, and predation of *Hypoaspis aculeifer* (Acari: Laelapidae) feeding on *Tyrophagus similis* (Acari: Acaridae). *Journal of the Acarological Society of Japan* 15, 139–143.

Katayama, H., Masui, S., Tsuchiya, M., Tatara, A., Doi, M., Kaneko, S. and Saitō, T. (2006) Density suppression of the citrus red mite *Panonychus citri* (Acari: Tetranychidae) due to the occurrence of *Neoseiulus californicus* (McGregor) (Acari: Phytoseiidae) on Satsuma mandarin. *Applied Entomology and Zoology* 41, 679–684.

Katayama, M., Mizutani, K., Kishino, H., Yabuta, S., Tomita, I., Mizuno, T., Yoneda, M., Soma, Y. and Kawakami, F. (2001) Mortality tests for Kanzawa spider mite, six-spotted mite, tropical citrus aphid and citrus psylla on Satsuma mandarins by methyl bromide fumigation. *Research Bulletin of the Plant Protection Service, Japan* 37, 27–33.

Kato, T., Kadono, F. and Amano, H. (1993) Insecticide susceptibility of the robin bulb mite, *Rhizoglyphus robini* Claparède (Acari: Acaridae), from the Welsh onion fields in Chiba Prefecture. *Proceedings of the Kanto-Tosan Plant Protection Society* 40, 29–31

Kato, T. (1977) Qualitative change of Satsuma mandarin fruits infested with the Japanese citrus mite, *Aculops pelekassi* (Keifer). *Bulletin of the Yamaguchi Agricultural Experiment Station* 28, 129–137 (in Japanese).

Kats, L.B. and Dill, L.M. (1998) The scent of death: chemosensory assessment of predation risk by prey animals. *Ecoscience* 5, 361–194.

Kavadas, D. (1927) *Phytoptus oleae*. Un acarien de l'olivier (*Phytoptus oleae*, a mite parasitic to olive trees). *Georgiko Deltio Elaioparagogon Piliou* 66, 36–38 (in Greek).

Kavitha, J., Kuttalam, S. and Chandrasekaran, S. (2006) Evaluation of spiromesifen 240 SC against chilli mite, *Polyphagotarsonemus latus* (Banks). *Annals of Plant Protection Sciences* 14, 52–55.

Kavousi, A., Chi, H., Talebi, K., Bandani, A., Ashouri, A. and Naveh, V.H. (2009) Demographic traits of *Tetranychus urticae* (Acari: Tetranychidae) on leaf discs and whole leaves. *Journal of Economic Entomology* 102, 595–601.

Kawai, A. and Haque, M.M. (2004) Distribution pattern of *Aculops lycopersici* (Massee) (Acari: Eriophyidae) in tomato leaf and estimation method for the population density on leaf. *Journal of the Acarological Society of Japan* 13, 31–39.

Kawashima, M. and Jung, C. (2010) Artificial ground shelters for overwintering phytoseiid mites in orchards. *Experimental and Applied Acarology* 52, 35–47.

Kawashima, M., Chung, B.K. and Jung, C.L. (2008) Herbivorous and predacious mites on persimmon trees, *Diospyros kaki* Thunb., in Korea. *International Journal of Acarology* 34, 167–174.

Kay, I.R. (1986) Tomato russet mite: a serious pest of tomatoes. *Queensland Agricultural Journal* 112, 231–232.

Kay, I.R. and Shepherd, R.K. (1988) Chemical control of the tomato russet mite on tomatoes in the dry tropics of Queensland. *Queensland Journal of Agricultural and Animal Sciences* 45, 1–8.

Kazhdaya, G.S. (1996) Some data on acaroid mites in Saarland (with description of variability of *Forcellinia diamesa*, Acaroidea, Acariformes). *Zoologicheskii Zhurnal* 75, 620–624.

Kazimi, S.K. and Ghani, M.A. (1969) *Resumé of Work on Spider Mites and Their Natural Enemies in West Pakistan*. Commonwealth Institute of Biological Control, Pakistan Station, Rawalpindi, Pakistan.

Keep, E. (1985) The mildew resistance gene *Sph* in relation to gall mite resistance in the black currant. *Euphytica* 34, 865–868.

Keetch, D.P. (1968) Some aspects of the ecology of the citrus red mite, *Panonychus citri* (McGr.) in the Eastern Cape Province. PhD thesis, Rhodes University, Grahamstown, South Africa.

Keetch, D.P. (1972) Ecology of the citrus red mite, *Panonychus citri* (McGregor), (Acarina: Tetranychidae) in South Africa. 3. The influence of the predacious mite, *Amblyseius* (*Typhlodromus*) *addoensis* van der Merwe & Ryke. *Journal of the Entomological Society of South Africa* 35, 69–79.

Keifer, H.H. (1938a) Eriophyid studies. *Bulletin of the California Department of Agriculture* 27(2), 181–206.

Keifer, H.H. (1938b) Eriophyid studies II. *Bulletin of the California Department of Agriculture* 27(3), 301–323.

Keifer, H.H. (1939a) Eriophyid studies III. *Bulletin of the California Department of Agriculture* 28(2), 144–163.

Keifer, H.H. (1939b) Eriophyid studies IV. *Bulletin of the California Department of Agriculture* 28(3), 223–239.

Keifer, H.H. (1939c) Eriophyid studies V. *Bulletin of the California Department of Agriculture* 28(5), 328–345.

Keifer, H.H. (1939d) Eriophyid studies VI. *Bulletin of the California Department of Agriculture* 28(6), 416–426.

Keifer, H.H. (1939e) Eriophyid studies VII. *Bulletin of the California Department of Agriculture* 28(7, 8, 9), 484–505.

Keifer, H.H. (1940a) Eriophyid studies VIII. *Bulletin of the California Department of Agriculture* 29(1), 21–46.

Keifer, H.H. (1940b) Eriophyid studies IX. *Bulletin of the California Department of Agriculture* 29(2), 112–117.

Keifer, H.H. (1940c) Eriophyid studies X. *Bulletin of the California Department of Agriculture* 29(3), 160–179.

Keifer, H.H. (1941) Eriophyid studies XI. *Bulletin of the California Department of Agriculture* 30(2), 196–216.

Keifer, H.H. (1942) Eriophyid studies XII. *Bulletin of the California Department of Agriculture* 31(3), 117–129.

Keifer, H.H. (1943) Eriophyid studies XIII. *Bulletin of the California Department of Agriculture* 32(3), 212–222.

Keifer, H.H. (1944) Eriophyid studies XIV. *Bulletin of the California Department of Agriculture* 33(1), 18–38.

Keifer, H.H. (1945) Eriophyid studies XV. *Bulletin of the California Department of Agriculture* 34(3), 137–140.

Keifer, H.H. (1946) Eriophyid studies XVI. *Bulletin of the California Department of Agriculture* 35(1), 39–48.

Keifer, H.H. (1951) Eriophyid studies XVII. *Bulletin of the California Department of Agriculture* 40(3), 93–104.

Keifer, H.H. (1952a) The eriophyid mites of California (Acarina: Eriophyidae). *Bulletin of the California Insect Survey* 2(1), 1–128.

Keifer, H.H. (1952b) Eriophyid studies XVIII. *Bulletin of the California Department of Agriculture* 41(1), 31–42.

Keifer, H.H. (1952c) Eriophyid studies XIX. *Bulletin of the California Department of Agriculture* 41(2), 65–74.

Keifer, H.H. (1952d) Eriophyid studies XX. *Bulletin of the California Department of Agriculture* 41(3), 147–149.

Keifer, H.H. (1953) Eriophyid studies XXI. *Bulletin of the California Department of Agriculture* 42(2), 65–79.

Keifer, H.H. (1954) Eriophyid studies XXII. *Bulletin of the California Department of Agriculture* 43(3), 121–131.

Keifer, H.H. (1955a) Eriophyid studies XXIII. *Bulletin of the California Department of Agriculture* 44(3), 126–130.

Keifer, H.H. (1955b) Eriophyid mites-notes and new species (Acarina). *Pan-Pacific Entomologist* 31, 109–116.

Keifer, H.H. (1956) Eriophyid studies XXIV. *Bulletin of the California Department of Agriculture* 44(4), 159–163.

Keifer, H.H. (1957) Eriophyid studies XXV. *Bulletin of the California Department of Agriculture* 46(3), 242–248.

Keifer, H.H. (1959a) Eriophyid studies, XXVI. *Bulletin of the California Department of Agriculture* 47(4), 271–281.

Keifer, H.H. (1959b) *Eriophyid Studies, XXVII*. Occasional Papers No. 1, California Department of Agriculture, Sacramento, California.

Keifer, H.H. (1959c) *Eriophyid Studies, XXVIII*. Occasional Papers No. 2, California Department of Agriculture, Sacramento, California.

Keifer, H.H. (1959d) New eriophyid mites. *Annals of the Entomological Society of America* 52, 649–657.

Keifer, H.H. (1960) *Eriophyid Studies B-1*. Special Publication of the Bureau of Entomology, California Department of Agriculture, Sacaramento, California.

Keifer, H.H. (1961a) *Eriophyid Studies B-2*. Special Publication of the Bureau of Entomology, California Department of Agriculture, Sacaramento, California.

Keifer, H.H. (1961b) *Eriophyid studies B-4*. Special Publication of the Bureau of Entomology, California Department of Agriculture, Sacaramento, California.

Keifer, H.H. (1962a) *Eriophyid Studies B-5*. Special Publication of the Bureau of Entomology, California Department of Agriculture, Sacaramento, California.

Keifer, H.H. (1962b) *Eriophyid Studies B-6*. Special Publication of the Bureau of Entomology, California Department of Agriculture, Sacaramento, California.

Keifer, H.H. (1962c) *Eriophyid Studies B-8*. Special Publication of the Bureau of Entomology, California Department of Agriculture, Sacaramento, California.

Keifer, H.H. (1963a) *Eriophyid Studies B-9*. Special Publication of the Bureau of Entomology, California Department of Agriculture, Sacaramento, California.

Keifer, H.H. (1963b) *Eriophyid Studies B-10*. Special Publication of the Bureau of Entomology, California Department of Agriculture.

Keifer, H.H. (1964) *Eriophyid Studies B-12*. Special Publication of the Bureau of Entomology, California Department of Agriculture, Sacaramento, California.

Keifer, H.H. (1965a) *Eriophyid Studies B-13*. Special Publication of the Bureau of Entomology, California Department of Agriculture, Sacaramento, California.

Keifer, H.H. (1965b) *Eriophyid Studies B-14*. Special Publication of the Bureau of Entomology, California Department of Agriculture, Sacaramento, California.

Keifer, H.H. (1965c) *Eriophyid Studies B-15*. Special Publication of the Bureau of Entomology, California Department of Agriculture, Sacaramento, California.

Keifer, H.H. (1965d) *Eriophyid Studies B-16*. Special Publication of the Bureau of Entomology, California Department of Agriculture, Sacaramento, California.

Keifer, H.H. (1966a) *Eriophyid Studies B-18*. Special Publication of the Bureau of Entomology, California Department of Agriculture, Sacaramento, California.

Keifer, H.H. (1966b) *Eriophyid Studies B-20*. Special Publication of the Bureau of Entomology, California Department of Agriculture, Sacaramento, California.

Keifer, H.H. (1969a) *Eriophyid Studies C-1*. Agricultural Research Service, US Department of Agriculture, Washington, DC.

Keifer, H.H. (1969b) *Eriophyid Studies C-2*. Agricultural Research Service, US Department of Agriculture, Washington, DC.

Keifer, H.H. (1969c) *Eriophyid Studies C-3*. Agricultural Research Service, US Department of Agriculture, Washington, DC.

Keifer, H.H. (1970) *Eriophyid Studies C-4*. Agricultural Research Service, US Department of Agriculture, Washington, DC.

Keifer, H.H. (1972) *Eriophyid Studies C-6*. Agricultural Research Service, US Department of Agriculture, Washington, DC.

Keifer, H.H. (1973) *Eriophyid Studies C-8*. Agricultural Research Service, US Department of Agriculture, Washington, DC.

Keifer, H.H. (1975a) *Eriophyid Studies C-10*. Agricultural Research Service, US Department of Agriculture, Washington, DC.

Keifer, H.H. (1975b) *Eriophyid Studies C-11*. Agricultural Research Service, US Department of Agriculture, Washington, DC.

Keifer, H.H. (1977) *Eriophyid Studies C-14*. Agricultural Research Service, US Department of Agriculture, Washington, DC.

Keifer, H.H. (1978) *Eriophyid Studies C-15*. Agricultural Research Service, US Department of Agriculture, Washington, DC.

Keifer, H.H. and Denmark, H.A. (1976) Eriophyes lantanae *Cook (Acarina: Eriophyidae) in Florida*. Entomology Circular No. 166, Florida Department of Agriculture and Consumer Services, Division of Plant Industry, Gainesville, Florida.

Keifer, H.H. and Knorr, L.C. (1978) *Eriophyid Mites of Thailand*. Technical Bulletin No. 38, Plant Protection Service, Department of Agriculture, Ministry of Agriculture and Co-operatives, Bangkok.

Keifer, H.H. and Wilson, N.S. (1955) A new species of eriophyid responsible for vection of peach mosaic virus. *Bulletin of the California Department of Agriculture* 44(4), 145–146.

Keifer, H.H., Baker, E.W., Kono, T., Delfinado, M. and Styer, W. (1982) *An Illustrated Guide to Plant Abnormalities Caused by Eriophyid Mites in North America*. Agriculture Handbook No. 573, US Department of Agriculture, Washington, DC.

Keizer, M. and Zuurbier, J. (2001) Red spider mite. *Namibian Crop Pests* 37, 21–26.

Keller, S. and Wuest, J. (1983) Observations sur trois espèces de *Neozygites* (Zygomycetes: Entomophthoraceae). *Entomophaga* 28, 123–134.

Kemp, G.H.J., Pretorius, Z.A. and Wingfield, M.J. (1996) Fusarium glume spot of wheat: a newly recorded mite-associated disease in South Africa. *Plant Disease* 80, 48–51.

Kendall, J. (1930) The structure and development of certain eriophyid galls. *Zeitschrift für Parasitenkunde* 2, 477–501.

Kennedy, G.G. and Smitley, D.R. (1985) Dispersal. In: Helle, W. and Sabelis, M.W. (eds) *Spider Mites: Their Biology, Natural Enemies and Control*. World Crop Pests, Volume 1A. Elsevier Science, Amsterdam, pp. 233–242.

Kennedy, J.S., Impe, G. van, Hance, T. and Lebrun, P. (1996) Demecology of the false spider mite, *Brevipalpus phoenicis* (Geijskes) (Acari, Tenuipalpidae). *Journal of Applied Entomology* 120, 493–499.

Kenneth, R., Wallis, G., Olmert, Y. and Halperin, J. (1971) A list of entomophagous fungi of Israel. *Israel Journal of Agricultural Research* 21, 63–66.

Kenneth, R., Wallis, G., Gerson, U. and Plaut, H.N. (1972) Observations and experiments on *Triplosporium floridanum* attacking spider mites in Israel. *Journal of Invertebrate Pathology* 19, 366–369.

Kerfoot, W.C. (1982) A question of taste: crypsis and warning coloration in freshwater zooplankton communities. *Ecology* 63, 538–554.

Kerguelen, V. and Hoddle, M.S. (1999) Measuring mite *Oligonychus perseae* feeding damage on avocado leaves with automated image analysis software. *Florida Entomologist* 82, 119–122.

Kerguelen, V. and Hoddle, M.S. (2000) Comparison of the susceptibility of several cultivars of avocado to the persea mite, *Oligonychus perseae* (Acari: Tetranychidae). *Scientia Horticulturae* 84, 101–114.

Kerns, D. (2006) Efficacy of envidor on mites of citrus, 2005. *Arthropod Management Tests* 2006, 31, D5. Available at: http://amt.oxfordjournals.org/content/amt/31/1/D5.full.pdf (accessed 12 May 2015).

Kerns, D., Wright, G. and Loghry, J. (2001) Citrus flat mite (*Brevipalpus lewisi*). (Part of *Citrus Arthropod Pest Management in Arizona*). Cooperative Extension, College of Agriculture and Life Sciences, University of Arizona. Available at: http://ag.arizona.edu/crops/citrus/insects/flatmite.pdf (accessed 8 August 2014).

Kerns, D., Wright, G. and Loghry, J. (2004) Yuma spider mite (*Eotetranychus yumensis*). (Part of *Citrus Arthropod Pest Management in Arizona*). Cooperative Extension, College of Agriculture and Life Sciences, University of Arizona. Available at: http://cals.arizona.edu/crop/citrus/insects/yumamite.pdf (accessed 8 August 2014).

Kethley, J.B. (1991) A procedure for extraction of microarthropods from bulk samples of sandy soils. *Agriculture, Ecosystems and Environment* 34, 193–200.

Khajehzadeh, Y. and Latifian, M. (2011) Suceptibility of date palm cultivars to natural infestation by *Oligonychus afrasiaticus* in Behbehan. *Plant Protection (Scientific Journal of Agriculture, University of Shahid Chamran, Iran)* 34(1), 67–79. English abstract available at: http://en.journals.sid.ir/ViewPaper.aspx?ID=256273 (accessed 26 March 2015).

Khalid Ahmed, Rao, V.H. and Rao, P.P. (2001) Resistance in chilli cultivars to yellow mite, *Polyphagotarsonemus latus* Banks. *Indian Journal of Agricultural Research* 35, 95–99.

Khalil-Manesh, B. (1973) Phytophagous mite fauna of Iran. *Entomologie et Phytopathologie Appliquées (Tehran)* (*Applied Entomology and Phytopathology*) 35(1, 2), 30–38.

Khalilov, B.B. (1972) Pests of grape vines in Azerbaidzhan. *Zashchita Rastenii* No. 17, 36–40 (in Russian).

Khan, B.S. and Muhammad, A. (2005) Comparison of life cycle of stigmaeid mite, *Agistemus buntex* on three host tetranychid species. *Pakistan Entomologist* 27(1), 9–11.

Khan, I.A. and Fent, M. (2005) Prey preference and consumption by two polyphagous predators *Typhiodromus pyri* Scheuten (Acari: Phytoseiidae) and *Chrysoperla carnea* (Stephens) (Neuroptera: Chrysophidae) of different mite pest species of apple orchards. *Sarhad Journal of Agriculture* 21(1), 89–96.

Khan, M.Q. and Murty, D.V. (1956) *Dicanthium annulatum* Stapf, an important alternate host plant of jowar and sugarcane mites. *Indian Journal of Entomology* 18, 198–199.

Khan, R.M., Doval, S.L. and Joshi, H.C. (1969) Biology of brown wheat mite. *Petrobia latens* (Müller). *Indian Journal of Entomology* 31, 258–264.

Khan, R.M., Singh, V.S., Saxena, J.D. and Bhatia, S.K. (1977) A simplified technique for sampling the brown wheat mite. *Acarologia* 19, 414–416.

Khanjani, M. and Haddad Irani-Nejad, K. (2006) *Injurious Mites of Agricultural Crops in Iran*. Bu-Ali Sina University Press Center, Iran (in Persian).

Khaustov, A.A. (2006a) A new species of the mite genus *Dolichocybe* (Acari, Heterostigmata, Dolichocybidae) from the Crimea. *Zoologicheskii Zhurnal* 85, 772–774 (in Russian).

Khaustov, A.A. (2006b) A review of the genera *Premicrodispus* Cross, 1965 and *Dolichodispus* gen. nov. (Acari: Microdispidae) of Crimea. *Acarina* 14, 155–174.

Khaustov, A.A. (2010) A new species of mites of the genus *Pseudopygmephorus* (Acari: Heterostigmata: Neopygmephoridae) associated with *Aphodius fimetarius* (Coleoptera: Scarabeidae) from Crimea. *Acarina* 18, 261–264.

Khaustov, A.A. (2011) New species and new records of mites of the genus *Pediculaster* (Acari, Heterostigmata, Pygmephoridae) from Ukraine. *Vestnik Zoologii* 45, 265–268.

Khaustov, A.A. and Ermilov, S.G. (2011) A new species of the genus *Siteroptes* (Acari, Heterostigmata, Pygmephoridae) from European Russia. *Zoologicheskii Zhurnal* 90, 756–760 (in Russian).

Kheradmand, K. and Zamani, A.A. (2008) Life table and reproductive parameters of two mushroom pests, *Tyrophagus putrescentiae* (Acari: Astigmata) and *Pediculaster flechtmanni* (Acari: Prostigmata). *Journal of Entomological Research* 32, 99-102.

Kheradmand, K., Kamali, K., Fathipour, Y., Goltapeh, E.M. and Camerik, A.M. (2006a) Biology and life table parameters of the mushroom pest, *Pediculaster fletchmanni* (Acari: Siteroptidae), at three constant temperatures. *Insect Science* 13, 375–380.

Kheradmand, K., Kamali, K., Fathipour, Y., Barzegar, M. and Mohammadi Goltapeh, E. (2006b) Effect of pigmy mite *Pediculaster fletchmanni* (Acari: Siteroptidae) on mineral elements of button mushroom *Agaricus bisporous*. *Pakistan Journal of Biological Sciences* 9, 2177–2180.

Kheradmand, K., Kamali, K., Fathipour, Y., Goltapeh, E.M. and Ueckermann, E.A. (2007) Thermal requirement for development of *Sancassania rodionovi* (Acari: Acaridae) on mushrooms. *Journal of Economic Entomology* 100, 1098–1103.

Khodayari, S., Kamali, K. and Fathipour, Y. (2009) Comparison of population density and spatial distribution of *Tydeus longisetosus* El-Bagoury & Momen, *Tetranychus turkestani* (Ugarov & Nikolskii) and their predator *Zetzellia mali* (Ewing) on two apple varieties in Maragheh region. *Iranian Journal of Entomological Research* 1, 197–208 (in Persian).

Khot, N.S. and Patel, G.A. (1956) Life history of the gladiolus mite, *Tetranychus equatorius* McGregor. *Indian Journal of Entomology* 18, 149–164.

Khromova, L.M. (2001) A dangerous mite on tomatoes in open soil. *Zashchita i Karantin Rasteniĭ* 6, 34.

Kido, H. (1981) Grape erineum mite. In: Flaherty, D.L., Jensen, F.L., Kasimatis, A.N., Kido, H. and Moller, W.J. (eds) *Grape Pest Management*. Publication No. 4105, Division of Agricultural Sciences, University of California, Berkeley, California, pp. 217–220.

Kido, H. and Stafford, E.M. (1955) The biology of the grape bud mite *Eriophyes vitis* (Pgst.). *Hilgardia* 24(6), 119–142.

Kielczewski, B. (1966) Common mite species of forest biotypes occurring on trees and accompanying certain insects. Translated from the original (*Zeszyty Problemowe Postępów Nauk Rolniczych* 65, 217–223) as Publication No. TT73-54038, by US Department of Commerce, National Technical Information Service, Springfield, Virginia.

Kielkiewicz, M. (1981) Physiological, anatomical and cytological changes in leaves of two strawberry varieties (*Fragaria grandiflora* Duch.) resulting from feeding by the two-spotted spider mite (*Tetranychus urticae* Koch). PhD thesis, Agricultural University of Warsaw, Warsaw (in Polish).

Kielkiewicz, M. and Czajkowska, B. (2002) Seasonal population dynamics of spruce spider mite (*Oligonychus ununguis* Jacobi) and accompanied predatory fauna on ornamental coniferous plants in Botanical Garden in Powsin. *Progress in Plant Protection/Postępy w Ochronie Roślin* 42, 435–438.

Kielkiewicz, M. and Witul, A. (1995) Effectiveness of integrated control of spider mites on gerbera. *Materialy Sesji Instytutu Ochrony Roślin* 35, 133–137.

Kiekiewicz, M., Puchalska, E. and Czajkowska, B. (2005) Changes in biochemical composition of needles of ornamental dwarf spruce (*Picea glauca* 'Conica') induced by spruce spider mite (*Oligonychus ununguis* Jacobi, Acari: Tetranychidae) feeding. *Acta Physiologiae Plantarum* 27, 163–171.

Kikuta, H. and Takagi, R. (2001) Isolation and acaricidal activity of microorganisms from dead *Tyrophagus similis* Volgin. *Journal of Rakuno Gakuen University, Natural Science* 25, 271–276 (in Japanese).

Kim, C.M. (2004) Trigynaspida (Acari: Mesostigmata): new diagnosis, classification, and phylogeny. *Acarologia* 44, 157–194.

Kim, D.G., Park, D.G., Kim, S.H., Park, I.S. and Choi, S.K. (2002a) Morphology, biology and chemical control of tomato russet mite, *Aculops lycopersici* Massee (Acari: Eriophyidae) in Korea. *Korean Journal of Applied Entomology* 41, 255–261 (in Korean).

Kim, D.H., Kim, K.S., Hyun, J.W. and Jeong, S.K. (2002b) Comparison of predation rates of three phytoseiid mite species on citrus red mite (*Panonychus citri* McGregor) on citrus tree. *Korean Journal of Applied Entomology* 41, 55–60.

Kim, D.H., Kim, K.S., Hyun, J.W. and Lim, H.C. (2003) Release level of *Amblyseius fallacis* Garman (Acarina: Phytoseiidae) for biological control of *Panonychus citri* McGregor (Acari: Tetranychidae) on citrus. *Korean Journal of Applied Entomology* 42, 233–240.

Kim, D.H., Kim, S.S., Kim, K.S. and Hyun, J.W. (2006) Characteristics of predation of *Neoseiulus fallacis* (Acarina: Phytoseiidae) on *Panonychus citri* (Acari: Tetranychidae). *Korean Journal of Applied Entomology* 45, 145–152.

Kim, G.J. and Hwang, C.Y. (1996) An investigation of insect pest on the mushroom (*Lentinus edodes*, *Pleurotus ostreatus*) in south region of Korea. *Korean Journal of Applied Entomology* 35, 45–45 (in Korean).

Kim, K.C. and Choi, D.S. (2000) Natural enemies of citrus red mite, *Panonychus citri* McGregor, and seasonal occurrence of major predators on yuzu tree (*Citrus junos*). *Korean Journal of Applied Entomology* 39, 13–19.

Kim, N.S., Whelan, E.D.P., Fedak, G. and Armstrong, K. (1992) Identification of a *Triticum–Lophopyrum* noncompensating translocation line and detection of *Lophopyrum* DNA using a wheatgrass specific molecular probe. *Genome* 35, 541–544.

Kim, N.S., Armstrong, K.C., Fedak G., Fominaya, A. and Whelan, E.W.P. (1993a) Cytological and molecular characterization of a chromosome interchange and addition lines in Cadet involving chromosome 5B of wheat and 6Ag of *Lophopyrum ponticum*. *Theoretical and Applied Genetics* 86, 827–832.

Kim, S.H., Kim, Y.K., Lee, M.H., Hong, S.C., Bae, J.M., Min, K.U., Kim, Y.Y. and Cho, S.H. (2002c) Relationship between sensitization to citrus red mite (*Panonychus citri*) and the prevalence of atopic diseases in adolescents living near citrus orchards. *Clinical and Experimental Allergy* 32, 1054–1058.

Kim, S.S. and Paik, C.H. (1996a) Comparative toxicity of fenpyroximate to the predatory mite, *Amblyseius womersleyi* Schicha and the Kanzawa spider mite, *Tetranychus kanzawai*, Kishida (Acarina: Phytoseiidae, Tetranychidae). *Applied Entomology and Zoology* 31, 369–377.

Kim, S.S. and Paik, C.H. (1996b) Comparative toxicity of abamectin to the spider mites, *Tetranychus urticae* Koch and *Tetranychus kanzawai* Kishida (Acarina: Tetranychidae) and the predatory mite, *Amblyseius womersleyi* Schicha (Acarina: Phytoseiidae). *Korean Journal of Applied Entomology* 35, 164–172.

Kim, S.S. and Paik, C.H. (1999) Predation of five species of phytoseiid mites on *Panonychus citri* and *Aculops pelekassi*. *Korean Journal of Entomology* 29, 261–264 (in Korean).

Kim, S.S., Paik, C.H., Kim, D.I., Park, J.D. and Lee, S.C. (1993b) Some ecological characteristics of tea red spider mite, *Tetranychus kanzawai* Kishida (Acarina, Tetranychidae). *Korean Journal of Entomology* 23, 261–266 (in Korean).

King, A.B.S. and Saunders, J.L. (1984) *The Invertebrate Pests of Annual Food Crops in Central America*. Overseas Development Administration, London.

King, C.B.R. (1937) Notes on the life history of *Eriophyes carinatus* Green. *Bulletin of Entomological Research* 28, 311–314.

Kinn, D.N. and Doutt, R.L. (1972) Natural control of spider mites on wine grape varieties in northern California. *Environmental Entomology* 1, 513–518.

Kinn, D.N., Joos, J.L., Doutt, R.L., Sorensen, J.T. and Foskett, M.J. (1974) Effects of *Tetranychus pacificus* and irrigation practices on yield and quality of grapes in northern coast vineyards of California. *Environmental Entomology* 3, 601–606.

Kirby, A.H.M. and Tew, R.P. (1952) Agricultural acaricides. *Reports on the Progress of Applied Chemistry* 37, 236–276.

Kishida, K. (1927) Notes on *Tetranychus kanzawai* n. sp. a new tetranychid mite injurious to leaves of the mulberry tree in Japan. *Zoological Magazine (Tokyo)* 39, 105–107.

Kishimoto, H. (2002) Species composition and seasonal occurrence of spider mites (Acari: Tetranychidae) and their predators in Japanese pear orchards with different agrochemical spraying programs. *Applied Entomology and Zoology* 37, 603–615.

Kishimoto, H. and Adachi, I. (2006) Effects of predacious insects on the population dynamics of three spider mite species (Acari: Tetranychidae) on Japanese pear. *Journal of the Acarological Society of Japan* 15, 47–54.

Kishimoto, H. and Adachi, I. (2008) Predation and oviposition by predatory *Stethorus japonicus*, *Oligota kashmirica benefica*, and *Scolothrips takahashii* in egg patches of various spider mite species. *Entomologia Experimentalis et Applicata* 128, 294–302.

Kishimoto, H. and Takagi, K. (2001) Evaluation of predation on *Panonychus citri* (McGregor) (Acari: Tetranychidae) from feeding traces on eggs. *Applied Entomology and Zoology* 36, 91–95.

Kitajima, E.W., Muller, G.W., Costa, A.S. and Yuki, W. (1972) Short, rod-like particles associated with citrus leprosis. *Virology* 50, 254–258.

Kitajima, E.W., Rezende, J.A.M., Rodrigues, J.C.V., Chiavegato, L.G., Piza, C.T. Jr and Morozini, W. (1997) Green spot of passion fruit, a possible viral disease associated with infestation by the mite *Brevipalpus phoenicis*. *Fitopatologia Brasileira* 22, 555–559.

Kitajima, E.W., Rezende, J.A.M. and Rodrigues, J.C.V. (2003a) Passion fruit green spot virus vectored by *Brevipalpus phoenicis* (Acari: Tenuipalpidae) on passion fruit in Brazil. *Experimental and Applied Acarology* 30, 225–231.

Kitajima, E.W., Chagas, C.M. and Rodrigues, J.C.V. (2003b) *Brevipalpus*-transmitted plant virus and virus-like diseases: cytopathology and some recent cases. *Experimental and Applied Acarology* 30, 1345–1160.

Kitajima, E.W., Groot, T.V.M., Novelli, V.N., Freitas-Astua, J., Alterti, G. and Moraes, G.J. de (2007) *In situ* observation of the *Cardinium* symbionts of *Brevipalpus* (Acari: Tenuipalpidae) by electron microscopy. *Experimental and Applied Acarology* 42, 263–271.

Kitajima, E.W., Rodrigues, J.C.V. and Freitas-Astua, J. (2010) An annotated list of ornamentals naturally found infected by *Brevipalpus* mite-transmitted viruses. *Scientia Agricola* 67, 348–371.

Kitamura, T. and Honda, K.I. (2006) Susceptibility of tomato russet mite *Aculops lycopersici* (Massee) to sulfur fumigation. *Annual Report of the Kansai Plant Protection Society* 48, 101–103 (in Japanese).

Kitamura, T. and Kawai, A. (2006) Difference of susceptibility to damage from tomato russet mite, *Aculops lycopersici* (Massee) (Acari: Eriophidae), among varieties within and between species in genus *Lycopersicon*. *Japanese Journal of Applied Entomology and Zoology* 50, 57–61 (in Japanese).

Kitashima, Y. and Gotoh, T. (1995) Host range difference and reproductive incompatibility among five populations of the citrus red mite, *Panonychus citri* (McGregor) (Acari: Tetranychidae). *Journal of the Acarological Society of Japan* 4, 91–101.

Kitashima, Y. and Gotoh, T. (1997) Electrophoretic differentiation of F1 hybrids of the sibling species *Panonychus osmanthi* Ehara et Gotoh and *P. citri* (McGregor) (Acari: Tetranychidae). *Applied Entomology and Zoology* 32, 635–637.

Klapperich, B.J. (1957) Control of the oriental red mite on citrus seedlings in Jordan. *FAO Plant Protection Bulletin* 5, 57–58.

Klein, H.Z. (1936) Contributions to the knowledge of the red spiders in Palestine. I. The oriental red spider, *Anychus orientalis* Zacher. II. The common red spider, *Epitetranychus althea* v. Hainstein. *Bulletin of the Agricultural Research Station, Rehovot* 21, 3–36, 37–63.

Klement'eva, K.V. (1979) Evaluation of resistance: of raspberry to the mite. *Zashchita Rasteniĭ* No. 10, 29–30 (in Russian).

Klimov, P.B. and O'Connor, B.M. (2003) Phylogeny, historical ecology and systematics of some mushroom-associated mites of the genus *Sancassania* (Acari: Acaridae), with new generic synonymies. *Invertebrate Systematics* 17, 469–514.

Klimov, P.B. and O'Connor, B.M. (2009) Conservation of the name *Tyrophagus putrescentiae*, a medically and economically important mite species (Acari: Acaridae). *International Journal of Acarology* 35, 95–114.

Klimov, P.B. and O'Connor, B.M. (2010) *Acarus putrescentiae* Schrank, 1781 (currently *Tyrophagus putrescentiae*; Acariformes, Acaridae): proposed conservation of usage by designation of a replacement neotype. *Bulletin of Zoological Nomenclature* 67, 24–27.

Klimov, P.B. and Tolstikov, A.V. (2011) Acaroid mites of northern and eastern Asia (Acari: Acaroidea). *Acarina* 19, 252–264.

Klompen, H. (2010) From sequence to phoresy – molecular biology in acarology. In: Sabelis, M.W. and Bruin, J. (eds) *Trends in Acarology: Proceedings of the 12th International Congress, 21–26 August 2006, Amsterdam*. Springer, Dordrecht, the Netherlands, pp. 1–8.

Klostermeyer, E.C. and Rasmussen, W.B. (1953) The effect of soil insecticide treatment on mite population and damage. *Journal of Economic Entomology* 46, 910–912.

Klostermeyer, E.C. and Rasmussen, W.B. (1956) A counting plate for sampling mite populations. *Journal of Economic Entomology* 49, 705–706.

Knapp, J.L. (1981) Integrated pest management in Florida citrus. *Miscellaneous Publications of the Entomological Society of America* 12, 91–98.

Knapp, J.L. (1991) Effect of different petroleum oils on defoliation, fruit quality, and pest control of Florida citrus. *Proceedings of the Florida State Horticultural Society* 103, 1–4.

Knapp, J.L., Nigg, H.N. and Anderson, H.E. (2001) Update on petroleum spray oil for citrus rust mite control. *Proceedings of the Florida State Horticultural Society* 114, 46–51.

Knapp, M. (2007) Important mite crop pest in Africa. In: Morales-Malacara, J.B., Behan-Pelletier, V., Ueckermann, E., Pérez, T.M., Estrada-Venegas, E.G. and Badii, M. (eds) *Acarology XI: Proceedings of the International Congress, 8–13 September 2002, Mérida, Universidad Nacional Autonoma de México*. Sociedad Latinoamericana de Acarología, México, abstract, p. 120.

Knapp, M., Saunyama, I.G.M., Sarr, I. and Moraes, G.J. de (2003) *Tetranychus evansi* in Africa – status, distribution, damage and control options. In: Wollny, C., Deininger, A., Bhandari, N., Maass, B., Manig, W., Muuss, U., Brodbeck, F. and Howe, I. (eds) *Deutcher Tropentag 1003, International Research on Food Security, Natural Resource Management and Rural Development: Technological and Institutional Innovations for Sustainable Rural Development. Georg-August-Universität Göttingen October 8–10, 2003. Book of Abstracts*. Centre for Tropical and Subtropical Agriculture and Forestry, Göttingen, Germany, p. 102. Available at: http://www.tropentag.de/2003/pdf/proceedings.pdf (accessed 13 May 2015).

Knapp, M., Hoogerbrugge, H., van Houten, Y.M., Calvo, J.C. and Bolckmans, K.J. (2010) Can the use of *Phytoseiulus longipes* improve the biological control of two-spotted spider mite on greenhouse grown tomatoes? In: Moraes, G.J. de, Castilho, R.C. and Flechtmann, C.H.W. (eds) *XIII International Congress of*

Acarology, August 23–27, 2010, Recife-PE, Brazil, Abstract Book, abstract 234, pp. 126–127. Available at: http://www.acarology.org/ica/ica2010/docs/0910_15_abstractBookCapa.pdf (accessed 3 February 2015).

Knight, R.L., Keep, E. Briggs, J.B. and Parker, J.H. (1974) Transference of resistance to black currant gall mite, *Cecidophyopsis ribis*, from gooseberry to black currant. *Annals of Applied Biology* 76, 123–130.

Knihinicki, D., Fletcher, M. and Nash, J. (1999) *Southern Red Mite Identified in NSW*. Agnote DPI/241, NSW Agriculture, Orange, New South Wales, Australia.

Knihinicki, D.K. (1999) *A New Spider Mite for Australia, the Southern Red Mite*. Agnote DPI/238. NSW Agriculture, Orange, New South Wales, Australia.

Knihinicki, D.K. and Boczek, J. (2002) New eriophyoid mites (Acari: Eriophyoidea) from Australia. *International Journal of Acarology* 28, 241–249.

Knisley, C.B. and Swift, P.C. (1972) Qualitative study of mite fauna associated with apple foliage in New Jersey. *Journal of Economic Entomology* 65, 445–448.

Knorr, L.C. (1968) Studies on the etiology of leprosis in citrus. In: Childs, J.F.L. (ed.) *Proceedings of the 4th Conference of the International Organization of Citrus Virologists (IOCV), Gainesville, Florida*, pp. 332–340.

Knorr, L.C. and Denmark, H.A. (1970) Injury to *Citrus* by the mite *Brevipalpus phoenicis*. *Journal of Economic Entomology* 63, 1996–1998.

Knorr, L.C. and Malaguti, G. (1960) Halo scab – result of an association of *Elsinoe fawcetti* and *Brevipalpus phoenicis* destructive to sour-orange seedlings. *Plant Disease Reporter* 44, 662–663.

Knorr, L.C., Webster, B.N. and Malaguti, G. (1960) Injuries in citrus attributed to brevipalpus mites, including brevipalpus gall, a newly reported disorder in sour-orange seedlings. *FAO Plant Protection Bulletin* 8(12), 35–42.

Knorr, L.C., Denmark, H.A. and Burnett, H.C. (1968) Occurrence of *Brevipalpus* mites, leprosis and false leprosis on citrus in Florida. *Florida Entomologist* 51, 12–17.

Knorr, L.C., Phatak, H.C. and Keifer, H.H. (1976) Web-spinning eriophyid mites. *Journal of the Washington Academy of Sciences* 66, 228–234.

Knowles, C.O. (1997) Mechanisms of resistance to acaricides. In: Sjut, V. (ed.) *Molecular Mechanisms of Resistance to Agrochemicals*. Volume 13: Chemistry of Plant Protection. Springer, Heidelberg, Germany, pp. 57–77.

Koch, C.L. (1838) *Deutschlands Crustaceen, Myriapoden, und Arachniden: ein beitrag zur Deutschen Fauna, Fasc. 17*. F. Pustet, Regensburg, Germany.

Koleva, R. and Jenser, G. (1996) A csonkázás hatása a szőlő-levélatka (*Calepitrimerus vitis* Nalepa) egyedsűrűségére. (The effect of summer pruning of shoots on the population density of *Calepitrimerus vitis* Nalepa). *Növényvédelem* 32, 277–279 (in Hungarian).

Koleva, R., Ferenczy, A. and Jenser, G. (1996) Effect of various plant protection programs on mite populations in vineyards. *Horticultural Science* 28, 79–82.

Koller, M., Knapp, M. and Schausberger, P. (2007) Direct and indirect adverse effects of tomato on the predatory mite *Neoseiulus californicus* feeding on the spider mite *Tetranychus evansi*. *Entomologia Experimentalis et Applicata* 125, 297–305.

Kolodochka, L.A. and Prutenskaya, M.D. (1987) Ability of the phytoseiid mite *Amblyseius agrestis* to feed on the tarsonemid mite *Hemitarsonemus latus*. *Vestnik Zoologii* 2, 80 (in Russian).

Kolodziej, A., Kropczyńska, D. and Poskuta, J. (1975) The effect of two spotted spider mite (*Tetranychus urticae* Koch) injury on carbon metabolism in *Chrysanthemum morifolium* L. In: *Proceedings of the 8th International Congress on Plant Protection, Moscow, Part II*, pp. 217–229.

Kolodziej, A., Kropczyńska, D. and Poskuta, J. (1979) Comparative studies on carbon dioxide exchange rates of strawberry and chrysanthemum plants infested with *Tetranychus urticae* Koch. In: Piffl, E. (ed.) *Proceedings of the 4th International Congress of Acarology, Saalfelden (Austria)*. Akadémiai Kiadó, Budapest, pp. 209–214.

Kolokytha, P.D., Fantinou, A.A. and Papadoulis, G.T. (2011) Effect of several different pollens on the bio-ecological parameters of the predatory mite *Typhlodromus athenas* Swirski and Ragusa (Acari: Phytoseiidae). *Environmental Entomology* 40, 597–604.

Kondo, A. and Hiramatsu, T. (1999a) Analysis of peach tree damage caused by peach silver mite, *Aculus fockeui* (Nalepa *et* Trouessart) (Acari: Eriophyidae). *Japanese Journal of Applied Entomology and Zoology* 43, 189–193 (in Japanese).

Kondo, A. and Hiramatsu, T. (1999b) Resurgence of the peach silver mite, *Aculus fockeui* (Nalepa *et* Trouessart) (Acari: Eriophyidae), induced by a synthetic pyrethroid fluvalinate. *Applied Entomology and Zoology* 34, 531–535.

Kondo, H., Maeda, T. and Tamada, T. (2003) Orchid fleck virus: *Brevipalpus californicus* mite transmission, biological properties and genome structure. *Experimental and Applied Acarology* 30, 215–223.

Kondo, H., Takanori, M., Shirako, Y. and Tamada, T. (2006) Orchid fleck virus is a rhabdovirus with an unusual bipartite genome. *Journal of General Virology* 87, 2413–2421.

Kong, C.-H., Hu, F., Xu, X.-H., Zhang, M.-X. and Liang, W.-J. (2005) Volatile allelochemicals in the *Ageratum conyzoides* intercropped citrus orchard and their effects on mites *Amblyseius newsami* and *Panonychus citri*. *Journal of Chemical Ecology* 31, 2193–2203.

Kongchuensin, M. (1983) Geographic variation and combining abilities of sex ratio and offspring number in the two-spotted spider mite, *Tetranychus urticae* Koch. PhD thesis, Ohio State University, Columbus, Ohio.

Kongchuensin, M., Charanasri, V. and Takafuji, A. (2005) Geographic distribution of *Neoseiulus longispinosus* (Evans) and its habitat plants in Thailand. *Journal of the Acarological Society of Japan* 14, 1–11.

Konno, T., Kuriyama, F., Hamaguchi, H. and Kajihara, O. (1990) Fenpyroximate (NNI-850), a new acaricide. In: *Proceedings of the Brighton Crop Protection Conference: Pests and Diseases – 1990, Volume 1*. BCPC Publications, Farnham (now Alton), UK, pp. 71–78.

Korsgaard, J. and Hallas, T.E. (1979) Tarsonemid mites in Danish house dust. *Allergy* 34, 225–232.

Koschier, E. (1997) Einfluss verschiedener Baumschnittmassnahmen auf die Populationsdynamik der Obstbaumspinnmilbe (*Panonychus ulmi* Koch) und der Apfelrostmilbe (*Aculus schlechtendali* Nalepa) auf Apfelbaumen. *Mitteilungen Klosterneuburg, Rebe und Wein, Obstbau und Früchteverwertung* 47, 44–55.

Košir, M. (1975) Ernährung und Entwicklung von *Pygmephorus mesembrinae* und *P. quadratus* (Pygmephoridae, Tarsonemini, Acari) und Bemerkungen über drei weitere Arten. *Pedobiologia* 15, 313–329.

Kostiainen, T.S. and Hoy, M.A. (1996) *The Phytoseiidae as Biological Control Agents of Pest Mites and Insects: A Bibliography (1960–1994)*. Monograph 17, University of Florida Publications, Gainesville, Florida.

Kotze, J.M., Putterill, J.F., Labuschangne, N. and Wehner, F.C. (1987) Occurrence of a spiroplasma-like organism in lesions of concentric ring blotch on citrus. *Phytophylactica* 19, 363–364.

Kousik, C.S., Shepard, B.M., Hassell, R., Levi, A. and Simmons, A.M. (2007) Potential sources of resistance to broad mites (*Polyphagotarsonemus latus*) in watermelon germplasm. *HortScience* 42, 1539–1544.

Koveos, D.S. and Veerman, A. (1996) Involvement of the circadian system in photoperiodic termination of diapause in the spider mite *Tetranychus urticae*. *Journal of Insect Physiology* 42, 681–691.

Kowalsick, T. (2003) *Spruce Spider Mites*. Cornell University Cooperative Extension of Schenectady County, Schenectady, New York.

Kowalsick, T. and Clark, S. (2010) *Using Growing Degree-Days for Insect Pest Management*. Cornell University Cooperative Extension of Suffolk County, Riverhead, New York. Available at: https://s3.amazonaws.com/assets.cce.cornell.edu/attachments/1870/Using-Growing-Degree-Days-for-Insect-Pest-Management.pdf?1408019830 (accessed 26 March 2015).

Kozłowski, J. (1979) Zbadan nad szpecielami (Eriophyoidea) wystepujacymi na jabloniach i sliwach w sadach wielkopolski. *Prace Naukowe Instytutu Ochrony Roślin* 21, 137–148 (in Polish).

Kozłowski, J. (1980) Researches on the occurrence and noxiousness of apple leaf mite *Aculus schlechtendali* (Nal.). *Prace Naukowe Instytutu Ochrony Roślin* 22, 155–162.

Kozłowski, J. (1983) Observations on the overwintering of gall mites (Acarina: Eriophyoidea) on fruit trees. *Prace Naukowe Instytutu Ochrony Roślin* 25, 99–109 (in Polish).

Kozłowski, J. (1993) Obserwacje nad wystepowaniem, biologia i szkodliwoscia szpecieli (Eriophyoidea) na winorosli. *Prace Naukowe Instytutu Ochrony Roślin* 33, 56–70 (in Polish).

Kozłowski, J. (1995) The acceptance of leaves of various apple varieties by the apple rust mite-*Aculus schlechtendali* (Nal.). *Roczniki Nauk Rolniczych. Seria E, Ochrona Roślin* 24, 39–44.

Kozłowski, J. (1998) *Factors Determining Susceptibility and Response of Apple Tree Cultivars to the Applerust Mite* Aculus schlechtendali *(Nalepa)*. Zeszyt 2, Rozprawy Naukowe, Institutu Ochrony Roślin, Poznań, Poland.

Kozłowski, J. (2001) Occurrence of eriophyoids (Acari; Eriophyoidea) on wild and cultivated grasses. *Journal of Plant Protection Research* 41, 402–412.

Kozłowski, J. and Boczek, J. (1987a) Density and host plants of the apple rust mite-*Aculus schlechtendali* (Nal.) (Acarina: Eriophyoidea). *Prace Naukowe Instytutu Ochrony Roślin* 29, 39–50.

Kozłowski, J. and Boczek, J. (1987b) Overwintering of the apple rust mite *Aculus schlechtendali* (Nal.) (Acarina: Eriophyoidea). *Prace Naukowe Instytutu Ochrony Roślin* 29, 51–62.

Kozłowski, J. and Boczek, J. (1989) Life-cycle of apple rust mite, *Aculus schlechtendali* (Nal.). *Prace Naukowe Instytutu Ochrony Roślin* 31, 49–66.

Kozłowski, J. and Zielinska, L. (1997) Cytological changes in apple leaves infested with the apple rust mite-*Aculus schlechtendali* (Nal.) (Acarina: Eriophyoidea). *Journal of Plant Protection Research* 37, 72–84.

Krainacker, D.A. and Carey, J.R. (1989) Reproductive limits and heterogeneity of male two-spotted spider mites. *Entomologia Experimentalis et Applicata* 50, 209–214.

Kramer, D.A. and Hain, F.P. (1989) Effect of constant-and variable-humidity and temperature regimes on the survival and developmental periods of *Oligonychus ununguis* (Acarina: Tetranychidae) and *Neoseiulus fallacis* (Acarina: Phytoseiidae). *Environmental Entomology* 18, 741–746.

Kramer, K. and Cranshaw, W.S. (2009) Effects of supplemental irrigation on populations of clover mite, *Bryobia praetiosa* Koch (Acari: Tetranychidae), and other arthropods in a Kentucky bluegrass lawn. *Southwestern Entomologist* 34(1), 69–74.

Kramer, P. (1877) Grundzüge zur Systematik der Milben. *Archiv für Naturgeschichte* 43(1), 215–247.

Kramer, W., Schirmer, U., Jeschke, P. and Witschel, M. (eds) (2012) *Modern Crop Protection Compounds, Volume 1, Herbicides*, 2nd edn. Wiley-VCH Verlag, Weinheim, Germany.

Krantz, G.W. (1973) Observations on the morphology and behaviour of the filbert rust mite *Aculus comatus* (Prostigmata: Eriophyoidea) in Oregon. *Annals of the Entomological Society of America* 66, 709–717.

Krantz, G.W. (1978) *A Manual of Acarology*. Oregon State University, Corvallis, Oregon.

Krantz, G.W. (1979) The role of *Phytocoptella avellanae* (Nal.) and *Cecidophyopsis vermiformis* (Nal.) (Eriophyoidea) in big bud of filbert. In: Piffl, E. (ed.) *Proceedings of the 4th International Congress of Acarology, Saalfelden (Austria)*. Akadémiai Kiadó, Budapest, pp. 201–208.

Krantz, G.W. (2009a) Form and function. In: Krantz, G.W. and Walter, D.E. (eds) *A Manual of Acarology*. Texas Tech University Press, Lubbock, Texas, pp. 5–53.

Krantz, G.W. (2009b) Introduction. In: Krantz, G.W. and Walter, D.E. (eds) *A Manual of Acarology*. Texas Tech University Press, Lubbock, Texas, pp. 1–2.

Krantz, G.W. (2009c) Origins and phylogenetic relationships. In: Krantz, G.W. and Walter, D.E. (eds) *A Manual of Acarology*. Texas Tech University Press, Lubbock, Texas, pp. 3–4.

Krantz, G.W. (2009d) Habits and habitats. In: Krantz, G.W. and Walter, D.E. (eds) *A Manual of Acarology*. Texas Tech University Press, Lubbock, Texas, pp. 64–82.

Krantz, G.W. and Ehrensing, D.T. (1990) Deuterogyny in the skeleton weed mite, *Aceria chondrillae* (G. Can.) (Acari: Eriophyidae). *International Journal of Acarology* 16, 129–133.

Krantz, G.W. and Walter, D.E. (eds) (2009) *A Manual of Acarology*, 3rd edn. Texas Tech University Press, Lubbock, Texas.

Krause, M., Nel, A. and Van Zyl, K. (1996) *A Guide to the Use of Pesticides and Fungicides in the Republic of South Africa*, 37th rev. edn. National Department of Agriculture, Pretoria.

Krczal, H. (1959) Systematik und Ökologie der Pyemotiden. In: Stammer, H.J. (ed.) *Beiträge zur Systematik und Ökologie Mitteleuropäischer Acarina, Bund 1, Abschnitt 3*. Akademische Verlagsgesellschaft Geest & Portig, Leipzig, Germany, pp. 385–625.

Krczal, H. (1963) Uber einen Bekampfungsversuch gegen die Hsealnussgallmibe *Phytoptus avellanae* Nal. mit Thiodan. *Nachrichtenblatt des Deutschen Pflanzenschutzdienstes* 15(2), 27–28.

Krczal, H. (1964a) *Pygmephorus allmanni,* eine neue Pyemotide aus Australien. *Acarologia* 6, 669–671.

Krczal, H. (1966) Untersuchungen zur Bekämpfung der Brombeergallmilbe *Eriophyes essigi* Hassan. *Bad Obst-und Gartenbauverein* 12, 53–54.

Krczal, H. (1969) Untersuchungen zur Biologie und Bekämpfung der Brombeergallmilbe *Eriophyes essigi* Hassan. *Erwerbsobstbau* 11, 239–242.

Krczal, H.T. (1964b) *Pygmephorus lambi,* eine neue Pyemotide aus Champignonkulturen. *Zoologischer Anzeiger* 172, 318–322.

Kreiter, S. and Planas, R. (1987) L'acariose n'a pas fini de faire parler d'elle. *Phytoma* 387, 24–29.

Kreiter, S. and Sentenac, G. (2004) *Méthode CEB No. 167: Méthode d'Étude des Effets Non Intentionnels à Moyen Terme sur les Phytoseiidae de la Vigne des Préparations Phytopharmaceutiques Utilisés en Traitement des Parties Aériennes*. CEB (Commission des essais Biologiques), AFPP (Association Française de Protection des Plantes), Paris.

Kreiter, S. and Tixier, M.-S. (2010) A new genus and species of phytoseiid mites (Acari: Mesostigmata) from the Brazilian Atlantic forest. *Acarologia* 50, 197–205.

Kreiter, S., Tixier, M.-S., Auger, P., Muckensturm, N., Sentenac, G., Doublet, B. and Weber, M. (2000) Phytoseiid mites in vineyards in France (Acari: Phytoseiidae). *Acarologia* 41, 77–96.

Kreiter, S., Auger, P., Lebdi-Grissa, K., Tixier, M.-S., Chermiti, B. and Dali, M. (2002) Plant inhabiting mites (Acari: Prostigmata & Mesostigmata) of some northern Tunisian crops. *Acarologia* 42, 389–402.

Kreiter, S., Tixier, M.-S. and Barbar, Z. (2006) The importance of the vegetation surrounding the agrosystems on predatory mites associated to vineyards. In: Ragusa, S. and Tsolakis, H. (eds) *La Difesa della Vite Dagli Artropodi Dannosi, Marsala 10–11 Ottobre 2005*. Università degli Studi di Palermo, Palermo, Italy, pp. 97–147.

Kreiter, S., Auger, P. and Bonafos, R. (2010) Side effects of pesticides on phytoseiid mites in French vineyards and orchards: laboratory and field trials. In: Sabelis, M.W. and Bruin, J. (eds) *Trends in Acarology: Proceedings of the 12th International Congress, 21–26 August 2006*. Amsterdam. Springer, Dordrecht, the Netherlands, pp. 457–464.

Kreiter, S., Mailloux, J., Tixier, M.-S., Le Bellec, F., Douin, M., Guichou, S. and Etienne, J. (2013) New phytoseiid mites of French West Indies, with description of a new species, and new records (Acari: Mesostigmata). *Acarologia* 53, 285–303.

Kremer, F.W. (1956) Studies on the biology, epidemiology, and control of *Bryobia praetiosa* Koch. *Höfchenbriefe* 9(4), 189–256.

Krieg, A. (1968) Effectiveness of *Bacillus thuringiensis* exotoxin on *Tetranychus telarius* (Acarina: Tetranychidae). *Journal of Invertebrate Pathology* 12, 478.

Krischik, V. and Davidson, J. (2013a) IV. Pests of trees and shrubs. Hemlock rust mite. In: Krischik, V. and Davidson, J. (eds) *IPM (Integrated Pest Management) of Midwest Landscapes*. University of Minnesota, Minneapolis, Minnesota, p. 146. Available at: http://cues.cfans.umn.edu/old/Web/146HemlockRustMite.pdf (accessed 9 March 2015).

Krischik, V. and Davidson, J. (2013b) IV. Pests of trees and shrubs. Honeylocust spider mite. In: Krischik, V. and Davidson, J. (eds) *IPM (Integrated Pest Management) of Midwest Landscapes*. University of Minnesota, Minneapolis, Minnesota, pp. 151–152. Available at: http://cues.cfans.umn.edu/old/Web/151HoneylocustSpiderMite.pdf (accessed 9 March 2015).

Krischik, V. and Davidson, J. (2013c) IV. Pests of trees and shrubs. Spruce spider mite. In: Krischik, V. and Davidson, J. (eds) *IPM (Integrated Pest Management) of Midwest Landscapes*. University of Minnesota, Minneapolis, Minnesota, pp. 204–205. Available at: http://cues.cfans.umn.edu/old/Web/204SpruceSpiderMite.pdf (accessed 9 March 2015).

Krishnaiah, K. and Tandon, P.L. (1975) Chemical control of *Tetranychus neocaledonicus* Andre (*cucurbitae* Rahman and Sapra) infesting okra and brinjal. *Current Research, University of. Agricultural Sciences (Bangalore)* 4, 87–89.

Krishnamoorthy, A. (1983) A new record of phytoseiid mite from Karnataka. *Acarology Newsletter* 13, 2.

Kruel, W. (1963) Gallmilben an Kiefer (Acarina: Eriophyidae). *Beiträge zur Entomologie* 13, 566–576.

Krüssman, G. (1986) *Manual of Cultivated Broad-leaved Trees & Shrubs, Volume 3, PRU–Z*. Timber Press, Portland, Oregon.

Kuang, H.-Y. (1983) Bionomics and population dynamics of avocado brown mite *Oligonychus punicae* (Hirst) on camphor tree. *Acta Entomologica Sinica* 26, 63–68 (in Chinese).

Kuang, H.-Y. (1995) *Economic Insect Fauna of China, Fasc. 44 (Acari Eriophyoidea) (1)*. Science Press, Beijing (in Chinese).

Kuang, H.-Y. (1998) Four new eriophyid mites from forest plants in China (Acari: Eriophyoidea). *Acta Entomologica Sinica* 41, 300–304 (in Chinese).

Kuang, H.-Y. and Zhao, J. (1988) Biology and population dynamics of the orange camellia rust mite. *Journal of Nanjing Agricultural University* 11(2), 48–53 (in Chinese).

Kuang, H.-Y., Li, Y.X. and Shen, B.Y. (1994) A new genus and three new species from conifers in China (Acari: Eriophyoidea). *Acta Zootaxonomica Sinica* 19, 175–180.

Kuang, H.-Y., Luo, G.-H. and Wang, A.-W. (2005) *Fauna of Eriophyid Mites from China (II) (Acari: Eriophyoidea)*. China Forestry Publishing House, Beijing (in Chinese).

Kulkarni, G.S. (1923) The "murda" disease of chilli (*Capsicum*). *Agricultural Journal of India* 17, 51–54.

Kulkarni, N.K. (2003) Studies on pigeonpea sterility mosaic disease; transmission, virus–vector relationships and identification of resistant sources. PhD thesis, University of Agricultural Sciences, Bangalore, India.

Kulkarni, N.K. and Kumar, P.L. (2002) Transmission of Pigeon pea sterility mosaic virus by the eriophyid mite, *Aceria cajani* (Acari: Arthropoda). *Plant Disease* 86, 1297–1302.

Kulpiyawat, T., Payapanon, A., Vugsilabutr, W., Kongchuensin, M., Chaowattanawong, P. and Konvipasruang, P. (2009) Dolichocybid mite, *Formicomotes heteromorphus* Magowski on mushrooms and their control. *Thai Agricultural Research Journal* 27, 2–25 (in Thai).

Kumar, J. and Beniwal, S.P.S. (1992a) Mango malformation. In: Kumar, J., Chawbe, H.S., Singh, U.S. and Mukhopadhyay, A.N. (eds) *Plant Diseases of International Importance*, 3rd edn. Prentice Hall, Englewood Cliffs, New Jersey, pp. 357–393.

Kumar, J. and Beniwal, S.P.S. (1992b) Role of Fusarium species in the etiology of mango malformation. In: *Procedings of the 4th International Mango Symposium, 5–10 July, Miami, Florida*, p. 17.

Kumar, P.L., Fenton, B., Duncan, G.H., Jones, A.T., Sreenivasulu, P. and Reddy, D.V.R. (2001) Assessment of variation in *Aceria cajani* using analysis of rDNA ITS regions and scanning electron microscopy: implications for the variability observed in host plant resistance to pigeonpea sterility mosaic disease. *Annals of Applied Biology* 139, 61–73.

Kumar, P.L., Jones, A.T. and Reddy, D.V.R. (2003a) A novel mite-transmitted virus with a divided RNA genome closely associated with pigeonpea sterility mosaic disease. *Phytopathology* 93, 71–81.

Kumar, P.L., Latha, T.K.S., Kulkarni, N.K., Raghavendra, N., Saxena, K.B., Waliyar, F., Rangaswamy, K.T., Muniyappa, V., Doriswamy, S. and Jones, A.T. (2005) Broad-based resistance to pigeonpea sterility mosaic disease in wild relatives of pigeonpea (*Cajanus*: Phaseoleae). *Annals of Applied Biology* 146, 371–379.

Kumar, S., Jain, P.C. and Kumawat, S.R. (1989) Field efficacy of various insecticides against brinjal mite (*Tetranychus macfarlanei* Baker and Prichard). *Indian Journal of Entomology* 51, 253–255.

Kumar, S., Prasad, S. and Singh, R.N. (2003b) Resurgence of spider mite *Tetranychus ludeni* Zacher (Acarina: Tetranychidae) against acaricides and botanical pesticides on cowpea. *Resistant Pest Management Newsletter* 12(2), 32–35.

Kumar, V. and Sharma, D.D. (1993) Bio ecology and chemical control of spider mite, *Tetranychus ludeni* Zecher on okra. *Indian Journal of Plant Protection* 21, 68–71.

Kumari, M. and Sadana, G.L. (1991) Influence of temperature and relative humidity on the development of *Amblyseius alstoniae* (Acari: Phytoseiidae). *Experimental and Applied Acarology* 11, 199–203.

Kumashiro, B. (1998) Minutes for 1994–1995. November 1995. Notes and Exhibitions. New Hawaii state, island, and host records. *Proceedings of the Hawaiian Entomological Society* 33, 4.

Kumral, N.A., Çobanoğlu, S. and Yalcin, C. (2009) Acaricidal, repellent and oviposition deterrent activities of *Datura stramorium* L. against adult *Tetranychus urticae* Koch. *Journal of Pest Science* 83, 173–180.

Kunugi, Y., Terai, Y. and Kato, S. (1993) Seasonal occurrence of peach silver mite, *Aculus fockeui* (Nalepa *et* Trouessart), on peach leaves and their effects on fruit development. *Proceedings of the Kanto Tosan-Plant Protection Society* 40, 269–271 (in Japanese).

Kurosa, K. and Nakao, H. (1993) *Tyrophagus neiswanderi* Johnston *et* Bruce. In: Ehara, S. (ed.) *Plant Mites of Japan in Colors*. Zenkoku Noson Kyoiku Kyokai, Tokyo, pp. 156–157.

Küster, E. (1911) *Die Gallen der Pflanzen. Ein Lehrbuch für Botaniker und Entomologen*. S. Hirzel, Leipzig, Germany.

Kuwahara, M. (1977) The development and inheritance of resistance in the Kanzawa spider mite, *Tetranychus kanzawai* Kishida, selected with chlordimeform, dicofol and phenthoate. *Japanese Journal of Applied Entomology and Zoology* 21, 163–168.

Kuwahara, M. (1986) Resistance of the bulb mite, *Rhizoglyphus robini* Claparède (Acarina: Acaridae), to insecticides. I. Resistance patterns to organophosphorus insecticides. *Japanese Journal of Applied Entomology and Zoology* 30, 290–295 (in Japanese).

Kuwahara, Y. (2010) How astigmatic mites control the emission of two or even three types of pheromones from the same gland. In: Sabelis, M.W. and Bruin, J. (eds) *Trends in Acarology: Proceedings of the 12th International Congress, 21–26 August 2006, Amsterdam*. Springer, Dordrecht, the Netherlands, pp. 241–248.

Kuwahara, Y., Akimoto, K., Leal, W.S., Nakao, H. and Suzuki, T. (1987) Isopiperitenone: a new alarm pheromone of the acarid mite, *Tyrophagus similis* (Acarina, Acaridae). *Agricultural and Biological Chemistry* 51, 3441–3442.

Kuwahara, Y., Leal, W.S., Nakano, Y., Kaneko, Y., Nakao, H. and Suzuki, T. (1989) Pheromone study on astigmatid mites XXIII. Identification of the alarm pheromone on the acarid mite, *Tyrophagus neiswanderi* and species specificities of alarm pheromones among four species of the same genus. *Applied Entomology and Zoology* 24, 424–429.

Kwee, L.T. and Chong, K.K. (1990) *Guava in Malaysia, Production, Pests, and Diseases*. Tropical Press, Kuala Lumpur.

La, S.Y., Paik, C.H., Lee, G.H. and Kim, D.H. (2009) Biological control of *Polyphagotarsonenmus latus* (Acari: Tasonemidae) by *Amblyseius cucumeris* (Acari: Phytoseiidae). *Korean Journal of Applied Entomology* 48, 347–354.

Labandeira, C.C., Phillips, T.L. and Norton, R.A. (1997) Oribatid mites and the decomposition of plants tissues in Paleozoic coal-swamp forests. *Palaios* 12, 319–353.

Łabanowska, B.H. (1992) Effectiveness of acaricides in the control of strawberry mite (*Phytonemus pallidus* ssp. *fragariae* Zimm.). *Fruit Science Reports* 19, 137–146.

Łabanowska, B.H. (2004) Spread of the strawberry mite (*Phytonemus pallidus* ssp. *fragariae* Zimm.) on thirteen strawberry cultivars. *Journal of Fruit and Ornamental Plant Research* 12, 105–111.

Łabanowska, B.H. (2006a) Potential agents for controlling the strawberry mite (*Phytonemus pallidus* ssp. *fragariae* Zimm.) after the withdrawal of endosulfan. *Journal of Fruit and Ornamental Plant Research* 14(Suppl. 3), 67–72.

Łabanowska, B.H. (2006b) Efficiency of new-generation acaricides in controlling the strawberry mite *Phytonemus pallidus* ssp. *fragariae* Zimm. on strawberry. *Biological Letters* 43, 335–340.

Łabanowska, B.H. and Gajek, D. (1999) Monitoring i zwalczanie najwazniejszych szkodnikow porzeczki czarnej. *Progress in Plant Protection/Postępy w Ochronie Roślin* 39, 305–311 (in Polish).

Łabanowska, B.H. and Suski, Z.W. (1990) Preliminary observations on the occurrence and control of the blackberry mite – *Acalitus essigi* (Hassan). *Prace Instytutu Sadownictwa i Kwiaciarstwa w Skierniewicach. Seria A, Prace Doswiadczalne z Zakresu Sadownictwa* 29, 97–114 (in Polish).

Łabanowska, B.H., Korzeniowski, M., Gasparski, T. and Piotrowski, W. (2013) Efficacy of spirotetramat (Movento 100 SC) in the control of strawberry mite (*Phytonemus pallidus*) on strawberry plantations in Poland. In: *Future IPM in Europe, 19–21 March 2013*, Riva del Garda, Italy, Book of Abstracts, p. 28 Available at: http://futureipm.eu/content/download/2496/17830/file/book%20of%20abstracts.pdf (accessed 13 May 2015).

Łabanowski, G. S. (1993) The box spider mite, *Eurytetranychus buxi* on *Buxus sempervirens*, in Poland. *Entomologia Hellenica* 8, 45–51.

Łabanowski, G.S. and Jaworski, A. (1992) Control of the bulb scale mite-*Steneotarsonemus laticeps* (Halbert) on *Hippeastrum*. *Prace Instytutu Sadownictwa i Kwiaciarstwa w Skierniewicach. Seria B, Rośliny Ozdobne* 17, 179–188.

Łabanowski, G.S. and Soika, G.M. (1995) Lesser known and unknown mites (Acarina)-pests of ornamental trees and shrubs of nurseries in Poland. In: Boczek, J. and Ignatowicz, S. (eds) *Materialy z Sympozjum na temat: "Osiagniecia Akarologii w Polsce", Siedlce, 26–27 Wresnia 1995* (*Proceedings of the Symposium on "Advances in Acarology in Poland", Siedlce, 26–27 September 1995*). Instytut Ochrony Roślin, Polska Akademia Nauk, Olsztyn, Poland, pp. 127-132.

Łabanowski, G.S. and Soika, G.M. (1998) Zagrożenie drzew i krzewów liściastych w szkółkach roślin ozdobnych przez szkodniki. *Progress in Plant Protection/Postępy w Ochronie Roślin* 38, 172–179 (in Polish).

Łabanowski, G.S. and Soika, G.M. (2000) Eriophyoid mites (Acari: Eriophyoidea) on ornamental coniferous plants in Poland. *Journal of Plant Protection Research* 40, 85–93.

Łabanowski, G.S., Łabanowska, B.H. and Suski, Z.W. (1990) New species of mites (Acarina) in the fauna of Poland. *Zeszyty Problemowe Postępów Nauk Rolniczych* 373, 9–17.

Lacasa, A., Torres, J. and Martinez, M.C. (1990a) *Acalitus phloeocoptes* (Napela) (Acarina: Eriophyidae) plaga del ciruelo en el Sureste espanol. *Boletín de Sanidad Vegetal, Plagas* 16, 285–295.

Lacasa, A., Martinez, M.C. and Torres, J. (1990b) Efecto de los tratamientos en el control de las poblaciones del 'acaro de las yemas' (*Aceria sheldoni* Ewing) del limonero. *Boletín de Sanidad Vegetal, Plagas* 16, 305–315.

Laccone, G., Montemurro, P., Piglionica, V. and Viggiani, G. (1993) Programma di difesa integrata per fasi fenologiche. In: Laccone *et al. La Difesa Integrata dell'Agroecosistema Vite*. INEA (Istituto Nazionale di Economia Agraria), Rome, pp. 1–18.

Lacey, L.A., Frutos, R., Kaya, H.K. and Vail, P. (2001) Insect pathogens as biological control agents: do they have a future? *Biological Control* 21, 230–248.

Laffi, F. (1980) Un acaro danno ai semenzai di melone: *Tyrophagus similis* Volgin. *Informatore Fitopatologico* 30(7/8), 17–21.

Laffi, F. (1981) *Acari di Interesse Agrario*. CLUEB (Cooperativa Libraria Universitaria Editrice Bologna), Bologna, Italy.

Laffi, F. (1982) Il ragnetto giallo della vite *Eotetranychus carpini* Oud. f. *vitis* Dosse. *Informatore Fitopatologico* 32(6), 31–34.

Laffi, F. (1983) Prove di lotta contro l'eriofide rugginoso del pero (*Epitrimerus pyri* Nal.). *Informatore Fitopatologico* 33(9), 49–54.

Laffi, F. (1994) L'acariosi della vite: sintomatologia e prove di lotta. *Informatore Agrario* 50(18), 49–52.

Laffi, F. (1996) La difesa dagli acari. *Informatore Agrario* 52(14), 81–83.

Laffi, F. (1999) Attivita' sull' eriofide rugginoso del pero – *Epitrimerus pyri* – dei fungicidi diclofluanide e tolilfluanide. *Informatore Fitopatologico* 49(3), 53–55.

Laffi, F. (2001) Infestazioni dell'acaro *Bryobia praetiosa* su rucola. *Informatore Agrario* 57(10), 83–84.

Laffi, F. and Montermini, A. (1985) Gli eriofidi del noce. *Informatore Fitopatologico* 35(1), 11–14.

Laffi, F. and Ponti, I. (1997) *Acari Dannosi alle Piante*. Edizioni l'Informatore Agrario, Verona, Italy.

Lagerstedt, H.B. (1979) Filberts. In: Jaynes, R.A. (ed.) *Nut Tree Culture in North America*. The Northern Nut Growers Association, New York, pp. 128-147.

Lai, H.S. and Lin, F.C. (2005) Development and population parameters of the Lewis spider mite, *Eotetranychus lewisi*, on poinsettia. *Plant Protection Bulletin (Taipei)* 47, 37–390 (in Chinese).

Laing, J.E. (1968) Life history and life table of *Phytoseiulus persimilis* Athias-Henriot. *Acarologia* 10, 578–588.

Laing, J.E. (1969) Life history and life table of *Tetranychus urticae* Koch. *Acarologia*, 11, 32–42.

Laing, J.E. (1973) Evaluating the effectiveness of *Paracheyletia bakeri* (Acarina: Cheyletidae) as a predator of the two spotted spider mite *Tetranychus urticae*. *Annals of the Entomological Society of America* 66, 641–646.

Laing, J.E., Calvert, D.L. and Huffaker, C.B. (1972) Preliminary studies of effects of *Tetranychus pacificus* on yield and quality of grapes in the San Joaquin Valley, California. *Environmental Entomology* 1, 658–663.

Lake, A.W.H. and Howie, J.H. (1995) Selection for redlegged earth mite resistance in annual *Medicago* species. *Plant Protection Quarterly* 10, 45–46.

Lakshmikantha, B.P. (1992) Studies on *Aceria cajani* ChannaBasavanna (Acari : Eriophyidae), a vector of pigeonpea (*Cajanus cajan*) sterility mosaic disease. MSc. thesis, University of Agricultural Sciences, Bangalore, India.

Lakshmikantha, B.P. and Prabhuswamy, H.P. (2003) Reaction of pigeon pea varieties to mite (*Aceria cajani* Channabasavanna) transmitting sterility mosaic disease. *Plant Protection Bulletin (Faridabad)* 54(3/4), 6–7.

Lal, K. and Pillai, K.S. (1976) Occurrence of new tetranchid spider mites, *Eutetranychus orientalis* (Klein) and *Oligonychus biharensis* (Hirst) on cassava. *Journal of Root Crops* 2, 59–60.

Lal, L. (1977) Studies on the biology of the mite *Eutetranychus orientalis* (Klein) (Tetranychidae: Acarina). *Entomon* 2, 53–57.

Lal, L. and Mukharji, S.P. (1978) Biology of *Eotetranychus uncatus* (Acari: Tetranychidae) in north India. *Indian Journal of Acarology* 3, 38–42.

Lal, L. and Mukharji, S.P. (1979) Observations of the injury symptoms caused by phytophagous mites. *Zoologische Beiträge* 25, 13–17.

Lal, S.S. and Pillai, K.S. (1980) The effect of tetranychid spider mite infestations on tuber yield of cassava. *Indian Journal of Plant Protection* 8, 128–132.

Laminman, J.F. (1935) The Pacific mite, *Tetranychus pacificus* McGregor, in California. *Journal of Economic Entomology* 28, 900–903.

Lampedro, L. and Luis Rosas, J. (1989) Seleccion de cepas de *Hirsutella thompsonii* Fisher para combatir al ácaro del cocotero *Eriophyes guerreronis*. *Revista Mexicana de Micología* 5, 225–231.

Landwehr, V.R. (1977) *Ischyropalpus nitidulus* (Coleoptera: Anthicidae), a predator of mites associated with Monterey pine. *Annals of the Entomological Society of America* 70, 81–83.

Landwehr, V.R. and Koehler, C.S. (1980) *Brevipalpus pini* and eriophyoid mite injury on Monterey pine. *Journal of Economic Entomology* 73, 675–678.

Lane, A. (1979) *Bulb Scale Mite*. Advisory Leaflet 456, Ministry of Agriculture, Fisheries and Food, Alnwick, UK. Her Majesty's Stationery Office, London.

Lane, A. (1984a) *Bulb Scale Mite*. Leaflet No 456 (revised), Ministry of Agriculture, Fisheries and Food, Alnwick, UK. Her Majesty's Stationery Office, London.

Lane, A. (1984b) *Bulb Pests*. ADAS/MAFF (Agricultural Development and Advisory Service/Ministry of Agriculture, Fisheries and Food) Reference Book 51, 7th edn. Her Majesty's Stationery Office, London.

Laney, A.G., Keller, K.E., Martin, R.R. and Tzanetakis, I.E. (2011) A discovery 70 years in the making: characterization of the rose rosette virus. *Journal of General Virology* 92, 1727–1732.

Lange, A.B., Drozdovskii, E.M. and Bushkovskaya, L.M. (1974a) Collection and release of *Anystis*. *Zashchita Rasteniĭ* No. 7, 33–34 (in Russian).

Lange, A.B., Drozdovskii, E.M. and Bushkovskaya, L.M. (1974b) *Ancystis* mites – effective predators of small phytophages. *Zashchita Rasteniĭ* No. 1, 26–28 (in Russian).

Lange, W.H. Jr (1955) *Aceria tulipae* (K.) damaging garlic in California. *Journal of Economic Entomology* 48, 612–613.

Lange, W.H. and Bacon, O.G. (1958) Crown mite damage on spinach. *California Agriculture* 12(2), 9, 16.

Lanza, G., Caruso, A. and Di Martino, E. (1980) Prove invernali di lotta contro il *Panonychus citri* McGregor in Calabria. *Informatore Fitopatologico* 30(6), 15–20.

Lara Benítez, M. and Ocete Rubio, R. (1993) "Erinosis", una constante de las poblaciones españolas de "*Vitis vinifera silvestris*" (Gmelin) Hegi (The grape erineum mite, a constant pest in the Spanish populations of *Vitis vinifera silvestris* (Gmelin) Hegi). *Fruticultura Profesionel* No. 59, 11–16/*Viticultura Enologia Profesionel* No. 29, 11–16. Available as Spanish summary at: http://agris.fao.org/agris-search/search.do?recordID=ES19940087701 (accessed 16 August 2014).

Larew, H.G. (1982) A comparative anatomical study of galls caused by the major cecidogenic groups, with special emphasis on nutritive tissue. PhD thesis, Oregon State University, Cornvallis, Oregon.

Larraín, S.P. (1986) Incidencia del ataque del ácaro de los bulbos *Eriophyes tulipae* Keifer (Acar., Eriophyidae) en el rendimiento y calidad del ajo (*Allium sativum* L.). *Agricultura Técnica (Santiago)* 46, 147–150.

Larson, K.C. (1998) The impact of two gall-forming arthropods on the photosynthetic rates of their hosts. *Oecologia* 115, 161–166.

Laskar, N. and Ghosh, S.K. (2005) Efficacy of pesticides on the incidence of thrips, *Scirtothrips dorsalis* Hood and mite, *Polyphagotarsonemus latus* (Banks) on chilli. *Journal of Plant Protection and Environment* 2, 80–83.

Latha, T.K.S. and Doraiswamy, S. (2004) Interaction studies between pigeonpea sterility mosaic virus and mite vector. *Indian Journal of Plant Protection* 32, 105–110.

Latha, T.K.S. and Doraiswamy, S. (2008) Detection of pigeonpea sterility mosaic virus, the causal agent of sterility mosaic disease of pigeonpea in viruliferous mite vector by DAS-ELISA and DIBA. *Archives of Phytopathology and Plant Protection* 41, 537–541.

Latha, T.K.S., Kumar, P.L., Doraiswamy, S. and Waliyar, F. (2007) Biodiversity of mite vector of pigeonpea sterility mosaic virus in Tamil Nadu, India. *Indian Journal of Plant Protection* 35, 138–139.

Latifian, M. (2012) Voracity and feeding preferences of larvae and adult stages of *Stethorus gilvifrons* Mulsant. (Coleoptera: Coccinellidae) on larvae and adult of *Oligonychus afrasiaticus* McGregor (Acarina: Tetranychidae). *International Journal of Agriculture and Crop Sciences* 4, 540–546.

Latifian, M., Marashi, S.S., Ahmadizadeh, S. and Nikbakht, P. (2007) Host preference of date palm spider mite, *Oligonychus afrasiaticus* (McGregor), to native date palm cultivars of Khuzestan. *Seed and Plant Improvement Journal* 23, 245–255 (in Persian). English abstract available at: http://spij.spii.ir/browse.php?a_id=171&sid=1&slc_lang=en (accessed 26 March 2015).

Latreille, P.A. (1806-1809) *Genera Crustaceorum et Insectorum Secundum Ordinem Naturalem in Familias Disposita, Iconibus Exemplurisque Plurimis Explicata: T.1* (1806); *T.2* (1807); *T.3* (1807); *T.4* (1809). Amand Koenig, Paris et Argentorati [= Strasbourg].

Lattin, J.D. (1999) Bionomics of the Anthocoridae. *Annual Review of Entomology* 44, 207–231.

Lavoipierre, M.M.J. (1940) *Hemitarsonemus latus* (Banks), (Acarina) a mite of economic importance to South Africa. *Journal of the Entomological Society of Southern Africa* 3, 116–123.

Lawson-Balagbo, L.M., Gondim, M.G.C. Jr, Moraes, G.J. de, Hanna, R. and Schausberger, P. (2007a) Life history of the predatory mites *Neoseiulus paspalivorus* and *Proctolaelaps bickleyi*, candidates for biological control of *Aceria guerreronis*. *Experimental and Applied Acarology* 43, 49–61.

Lawson-Balagbo, L.M., Gondim, M.G.C. Jr, Moraes, G.J. de, Hanna, R. and Schausberger, P. (2007b) Refuge use by the coconut mite *Aceria guerreronis*: fine scale distribution and association with other mites under the perianth. *Biological Control* 43, 102–110.

Lawson-Balagbo, L.M., Gondim, M.G.C. Jr, Moraes, G.J. de, Hanna, R. and Schausberger, P. (2008) Exploration of the acarine fauna on coconut palm in Brazil with emphasis on *Aceria guerreronis* (Acari: Eriophyidae) and its natural enemies. *Bulletin of Entomological Research* 98, 83–96.

Lawton, J.H. and Hassell, M.P. (1981) Asymmetrical competition in insects. *Nature* 289, 793–795.

Lazar, M.D., Worrall, W.D., Peterson, G.L., Porter, K.B., Rooney, L.W., Tuleen, N.A., Marshall, D.S., McDaniel, M.E. and Nelson, L.R. (1997) Registration of 'TAM 110' wheat. *Crop Science* 37, 1978–1979.

Leach, W.E. (1815) A tabular view of the external characters of four classes of animals which Linné arranged under Insecta. *Transactions of the Linnean Society of London* 11, 306–400.

Leal, W.S., Nakano, Y., Kuwahara, Y., Nakao, H. and Suzuki, T. (1988) Pheromone study of acarid mites XVII. Identification of 2-hydroxy-6-methyl-benzaldehyde as the alarm pheromone of the acarid mite, *Tyrophagus perniciosus* (Acarina: Acaridae), and its distribution among related mites. *Applied Entomology and Zoology* 23, 422–427.

Learner, M.A. and Chawner, H.A. (1998) Macro-invertebrates associations in sewage filter-beds and their relationship to operational practice. *Journal of Applied Ecology* 35, 720–747.

Leary, R.F. and Allendorf, F.W. (1989) Fluctuating asymmetry as an indicator of stress: implications for conservation biology. *Trends in Ecology and Evolution* 4, 214–216.

Leatherdale, D. (1965) Fungi infecting rust and gall mites (Acarina: Eriophyidae). *Journal of Invertebrate Pathology* 7, 325–328.

Lebdi-Grissa, K., van Impe, G. and Lebrun, P. (2002) Demographic traits of *Eotetranychus pruni* from Belgian and Tunisian orchards, in comparison with *Tetranychus urticae*. *Experimental and Applied Acarology* 26, 209–17.

Lebdi-Grissa, K. and Mbarki, A. (2012) Demographic traits of *Tetranychus evansi* (Acari: Tetranychidae) on different solanaceous crops (tomato, potato, pepper and *Solanum nigrum*). In: Schausberger, P., Walzer, A., Strodel, M.A. and Peneder, S. (eds) *7th Symposium of the European Association of Acarologists (EURAAC), July 9–13 2012, Vienna, Austria: Acari in a Changing World. Program, Abstracts, Participants*. University of Natural Resources and Life Sciences, Vienna, Austria, p. 92. Available at: http://euraac.boku.ac.at/SympVienna/files/ProgramBook.pdf (accessed 14 January 2015).

LeClerg, E.L. (Chairman) *et al.* (1965) *Losses in Agriculture*. Agriculture Handbook No. 291, US Department of Agriculture Agricultural Research Service, Washington, DC.

Ledesma, C., Wong, M.E., Vela, J.M., Jacas, J.A. and Boyero, J.R. (2011) Population dynamics of the citrus oriental mite, *Eutetranychus orientalis* (Klein) (Acari: Tetranychidae), and its mite predatory complex in southern Spain. *IOBC-WPRS Bulletin* 62, 83–92.

Lee, H. and Chung, B. (2011) Occurrences of major pests in Japanese apricot, *Prunus mume* Siebold & Zucc. in Gyeongnam Province. *Korean Journal of Applied Entomology* 50, 21–27.

Lee, J.-H., Jung, S. and Lee, S. (2014) Three newly recorded species of the genera *Acaphylla* Keifer and *Calacarus* Keifer (Prostigmata: Eriophyidae) from *Camellia* spp. (Theaceae) in Korea. *Korean Journal of Applied Entomology* 53, 59–64.

Lee, J.S. (1989) *A Systematic Study on the Tetranychoid Mites (Order: Acarina) in Korea*. Chonbuk University, Chonju, Korea.

Lee, M., Suh, S. and Kwon, Y. (1999) Phylogeny and diagnostic markers of six *Tetranychus* species (Acarina: Tetranychidae) in Korea based on the mitochondrial cytochrome oxidase subunit I. *Journal of Asia-Pacific Entomology* 2, 85–92.

Lee, M.-L. and Leu, L.-S. (1997) Effect of soil fumigation on gladiolus plants grown in the field rotated with *Sesbania javanica* Miq. *Plant Protection Bulletin (Taipei)* 39, 377–382.

Lee, W.K., Lee, B.H., Kim, Y.J. and Kang, D.W. (1987) Taxonomic studies on spider mites (Tetranychidae : Acarina) of Korea. II. Spider mites, parasitic on wild plants. *Korean Journal of Systematic Zoology* 3, 95–116.

Lee, W.-T. and Lo, K.C. (1990) Integrated control of two spotted spider mite on strawberry in Taiwan. *Chinese Journal of Entomology, Special Publication* No. 3, 125–137.

Lee Goff, M. (1986) Spider mites (Acari: Tetranychidae) in the Hawaiian Islands. *International Journal of Acarology* 12, 43–49.

Lees, A.D. (1950) Diapause and photoperiodism in the fruit tree red spider mite (*Metatetranychus ulmi* Koch). *Nature* 166, 874–875.

Lees, A.D. (1953a) Environmental factors controlling the evocation and termination of diapause in the fruit tree red spider mite, *Metatetranychus ulmi* (Koch). *Annals of Applied Biology* 40, 449–486.

Lees, A.D. (1953b) The significance of the light and dark phases in the photoperiodic control of diapause in *Metatetranychus ulmi* (Koch). *Annals of Applied Biology* 40, 487–497.

Lees, A.D. (1955) *The Physiology of Diapause in Arthropods*. Cambridge University Press, Cambridge, UK.

Lees, A.D. (1961) On the structure of the egg shell in the mite *Petrobia latens* Müller (Acarina: Tetranychidae). *Journal of Insect Physiology* 6, 146–151.

Lehman, R.D. (1997) "Arborvitae spider mite", *Platytetranychus thujae* (McGregor). *Regulatory Horticulture* 23(l), 25–26. (From Bureau of Plant Industry, Pennsylvania Department of Agriculture, Harrisburg, Pennsylvania).

Lehman, R.D. (2002) *Spruce Spider Mite,* Oligonychus ununguis *(Jacobi), an Integrated Approach to Management*. Entomology Circular No.190, 1998, rev. 2001/ Regulatory Horticulture 24(1), Bureau of Plant Industry, Pennsylvania Department of Agriculture, Harrisburg, Pennsylvania. Available at: http://ento.psu.edu/extension/christmas-trees/information/pest-fact-sheets/spspidermiteipment190.pdf (accessed 31 March 2015).

Lehman, W.F. (1971) *Alfalfa Varieties and Production in the Low Desert Valley Areas of California – 1971*. Bulletin, University of California Imperial Valley Field Station, El Centro, California.

Lei, H.-D., Li, H.-J., Ran, C., Hu, J.-H., Lin, B.-M., Zhang, Q.-B., Tian, W.-H. and Qian, K.-M. (2003) Application of Sunspray oils for pollution-free control of the main pests in citrus orchards. *Journal of Southwest Agricultural University* 25, 409–412 (in Chinese).

Lei, H.-D., Hu, J.-H., Li, H.-J., Ran, C., Zhang, Q.-B., Lin, B.-M., Tian, W.-H. and Qian, K.-M. (2004) Performances of the citrus red mite, *Panonychus citri* (McGregor) (Acarina: Tetranychidae) on various citrus varieties. *Acta Entomologica Sinica* 47, 607–611.

Leigh, T., Goodell, P., Toscano, N., Burton, V., Bentley, W., Wilson, L.T., Natwick, E.T., Roberts, P., Davis, R.M., Vargas, R., *et al.* (1990) *Cotton. Pest Management Guidelines*. UCPMG Publication (17), University of California Pest Management Guidelines, Division of Agriculture and Natural Resources, Davis, California, pp. 21–24, 37–56.

Leigh, T.F. (1985) Cotton. In: Helle, W. and Sabelis, M.W. (eds) *Spider Mites: Their Biology, Natural Enemies and Control*. World Crop Pests, Volume 1B. Elsevier Science, Amsterdam, pp. 349–358.

Leigh, T.F. and Hyer, A. (1963) Spider mite-resistant cotton. *California Agriculture* 17(2), 6–7.

Leigh, T.F., Maggi, V.L. and Wilson, L.T. (1984) Development and use of a machine for recovery of arthropods from plant leaves. *Environmental Entomology* 77, 271–276.

Leisering, R.O.B. (1960) Beiträge zum Phytopatologischen Wirkungsmechanismus von *Tetranychus urticae*. *Pflanzenschutz* 67, 525–542.

Leite, G.L.D., Picanço, M., Guedes, R.N.C. and Zanuncio, J.C. (1999) Influence of canopy height and fertilization levels on the resistance of *Lycopersicon hirsutum* to *Aculops lycopersici* (Acari: Eriophyidae). *Experimental and Applied Acarology* 23, 633–642.

Leite, G.L.D., Picanco, M., Zanuncio, J.C. and Marquini, F. (2003) Factors affecting mite herbivory on eggplants in Brazil. *Experimental and Applied Acarology* 31, 243–252.

Lemmetty, A., Tikkanen, M., Tuovinen, T. and Lehto, K. (2004) Identification of different *Cecidophyopsis* mites on *Ribes* in Finland. *Acta Horticulturae* 656, 115–118.

Lemos, F., Almeida Sarmento, R., Pallini, A., Rosa Dias, C., Sabelis, M.W. and Janssen, A. (2010) Spider mite web mediates anti-predator behaviour. *Experimental and Applied Acarology* 52, 1–10.

Léon Lobos, O.A. (2003) Estudio de los parámetros de vida de *Oligonychus yothersi* McGregor (Acarina: Tetranychidae) en dos cultivares de palto (*Persea americana* Mill.), Hass y Fuerte. Thesis, Universidad Austral de Chile, Valdivia, Chile.

Leonel, S., Negri, J.D. de, Tonet, R.M. and Cardoso, S.A. de B. (1999) Niveis de controle para o ácaro-branco na cultura do limoeiro 'Siciliano'. *Laranja* 20(2), 345–355.

Leroux, E.J. (1954) Effects of various levels of nitrogen, phosphorus, and potassium in nutrient solution on the fecundity ot the two-spotted mite reared on cucumber. *Canadian Journal of Agricultural Science* 34, 145–151.

Leśna, I., Sabelis, M.W., Bolland, H.R. and Conijn, C.G.M. (1995) Candidate natural enemies for control of *Rhizoglyphus robini* Claparède (Acari: Astigmata) in lily bulbs: exploration in the field and pre-selection in the laboratory. *Experimental and Applied Acarology* 19, 655–669.

Leśna, I., Sabelis, M. and Conijn, C. (1996) Biological control of the bulb mite, *Rhizoglyphus robini*, by the predatory mite, *Hypoaspis aculeifer*, on lilies: predator—prey interaction at various spatial scales. *Journal of Applied Ecology* 33, 369–376.

Leśna, I., Conijn, C.G.M., Sabelis, M.W. and Straalen, N.M. van (2000) Biological control of the bulb mite, *Rhizoglyphus robini*, by the predatory mite, *Hypoaspis aculeifer*, on lilies: predator-prey dynamics in the soil, under greenhouse and field conditions. *Biocontrol Science and Technology* 10, 179–193.

Leśna, I., Conijn, C.G.M. and Sabelis, M.W. (2004) From biological control to biological insight: rust-mite induced change in bulb morphology, a new mode of indirect plant defense? *Phytophaga* 14, 285–291.

Lester, P.J., Dentener, P.R., Bennett, K.V. and Connolly, P.G. (1997) Postharvest disinfestation of diapausing and non-diapausing twospotted spider mite (*Tetranychus urticae*) on persimmons: hot water immersion and cool storage. *Entomologia Experimentalis et Applicata* 83, 189–193.

Letourneau, D.K., Jedlicka, J.A., Bothwell, S.G. and Moreno, C.R. (2009) Effects of natural enemy biodiversity on the suppression of arthropod herbivores in terrestrial ecosystems. *Annual Review of Ecology, Evolution, and Systematics* 40, 573–592.

Leuschner, K. (1980) Pest control for cassava and sweet potato. In: *Root Crops in Eastern Africa: Proceedings of a Workshop, 23–27 November, 1980, Kigali, Rwanda*. International Development Research Centre, Ottawa, pp. 60–64.

Levieux, J., Lieutier, F., Moser, J.C. and Perry, T.J. (1989) Transportation of phytopathogenic fungi by the bark beetle *Ips sexdentatus* Boerner and associated mites. *Journal of Applied Entomology* 108, 1–11.

Lewis, G.C. (1982) Evaluation of fenpropathrin for the control of the mite vector of ryegrass mosaic virus. *Annals of Applied Biology* 100(Supplement 3), 24–25.

Lewis, G.C. and Heard, A.J. (1980) The incidence in ryegrass (*Lolium* spp.) crops of eriophyid mites, vectors of ryegrass mosaic virus. In: Plumb. R.T. (ed.) *Proceedings of the 3rd Conference on Virus Diseases of Gramineae in Europe, Rothamsted, 28–30 May 1980*. Rothamsted Experimental Station, Harpenden, UK, pp. 89–98.

Lewis, G.C., Heard, A.J., Brady, B.L. and Minter, D.W. (1981) Fungal parasitism of the eriophyoid mite vector of ryegrass mosaic virus. In: *Proceedings of the Brighton Crop Protection Conference: Pests and Diseases – 1981*. British Crop Production Council, Croydon (now Alton), UK, pp. 109–111.

Lewis, G.C., Heard, A.J., Gutteridge, R.A., Plumb, R.T. and Gibson, R.W. (1985) The effect of mixing Italian ryegrass (*Lolium multiflorum*) with perennial ryegrass (*L. perenne*) or red clover (*Trifolium pratense*) on the incidence of viruses. *Annals of Applied Biology* 106, 483–488.

Lewis, H.C. (1949) Injury to citrus by *Tenuipalpus* mites. *California Citrograph* 29, 87.

Lewis, H.C., Worthy, W.L. and Barnhart, E.L. (1951) Spider mite attacking lemons. *California Citrograph* 36, 363.

Lewontin, R.C. (1965) Selection for colonizing ability. In: Baker, H.G. and Ledyard Stebbins, G. (eds) *The Genetics of Colonizing Species*. Academic Press, New York, pp. 77–91.

Leyva Fernández, Y., Zamora Zamora, N., Álvarez Álvarez, E. and Jiménez, M. (2003) Resultados preliminares de la dinámica poblacional del ácaro *Steneotarsonemus spinki*. *Granma Ciencia* 7(1), 1–6.

Li, C. (1999) Preliminary study of the acaroid mites breeding in stored Chinese traditional medicinal materials. *Chinese Journal of Parasitic Disease Control* (now *Journal of Pathogen Biology*) 12, 72–73.

Li, D., Janssen, A. and Sabelis, M.W. (2013a) Population dynamics of *Tetranychus urticae* and *Tetranychus evansi* on tomato: differential impact of plant apex removal. In: Schausberger, P., Walzer, A., Strodel, M.A. and Peneder, S. (eds) *7th Symposium of the European Association of Acarologists (EURAAC), July 9–13 2012, Vienna, Austria: Acari in a Changing World. Program, Abstracts, Participants*. University of Natural Resources and Life Sciences, Vienna, Austria, p. 93. Available at: http://euraac.boku.ac.at/SympVienna/files/ProgramBook.pdf (accessed 14 January 2015).

Li, D.-X., Tian, J. and Shen, Z.-R. (2007a) Functional response of the predator *Scolothrips takahashii* to hawthorn spider mite, *Tetranychus viennensis*: effect of age and temperature. *BioControl* 52, 41–61.

Li, D-X., Tian, J., Guo, Y.-L. and Zhang, J.G. (2009) Effect of peach cultivars on developmental time and life table parameters of *Amphitetranychus viennensis* (Acari: Tetranychidae). *Systematic and Applied Acarology* 14, 206–212.

Li, G., Li, H., Guo, Y., Yang, S., Li, J., Xia, W. and Zhang, J. (2013b) Effect of transgenic cotton on life table parameters of *Tetranychus turkestani*. *Chinese Bulletin of Entomology* 2, 376–381.

Li, H., Conner, R.L., Chen, Q., Graf, R.J., Laroche, A., Ahmad, F. and Kuzyk, A.D. (2005) Promising genetic resources for resistance to wheat streak mosaic virus and the wheat curl mite in wheat–*Thinopyrum* partial amphiploids and their derivatives. *Genetic Resources and Crop Evolution* 51, 827–835.

Li, H., Conner, R.L., Liu, Z., Li, Y., Chen, Y., Zhou, Y., Duan, X., Shen, T., Chen, Q., Graf, R.J. and Jia, X. (2007b) Characterization of wheat–triticale lines resistant to powdery mildew, stem rust, stripe rust, wheat curl mite, and limitation on spread of WSMV. *Plant Disease* 91, 368–374.

Li, H.-Y., Hua, T. and Yeh, W.-B. (2010) Amplification of single bulb mites by nested PCR: species-specific primers to detect *Rhizoglyphus robini* and *R. setosus* (Acari: Acaridae). *Journal of Asia-Pacific Entomology* 13, 267–271.

Li, J. and Margolies, D.C. (1991) Factors affecting location of Banks grass mite, *Oligonychus pratensis* (Acari: Tetranychidae), on corn leaves. *Experimental and Applied Acarology* 12, 27–34.

Li, J. and Margolies, D.C. (1994) Barometric pressure influences initiation of aerial dispersal in the two-spotted spider mite. *Journal of the Kansas Entomological Society* 67, 386–393.

Li, J.-B., Li, Y.-X., Sun, J.-T., Xue, X.-F., Xu, X.-N. and Hong, X.-Y. (2012) COI barcoding as a molecular assay for the identification of phytoseiid mites. *Systematic and Applied Acarology* 17, 397–406.

Li, J.-M., Yang, Y.-Y., Qu, Y.-F. and Wu, Q.-H. (2003) Experimental population life table of *Amblyseius cucumeris* with *Polyphagotarsonemus latus* as prey. *Acta Phytophylacica Sinica* 30, 389–395.

Li, L.S. (1990) Recent progress in the study of agricultural mites. *Scientia Agricultura Sinica* 23, 22–30 (in Chinese).

Li, L.S., Huang, F.N. and Chen, J.L. (1989) Ecological characteristics of the citrus rust mite, *Phyllocoptruta oleivora* (Ashm.). *Acta Entomologica Sinica* 32, 184–191 (in Chinese).

Li, Q. and Cai, R.-X. (1996) Occurrence and damage of the apple rust mite *Aculus schlechtendali*. *Plant Protection* 22(3), 16–17 (in Chinese).

Li, Q.-P., Bai, H.-Q. and Luo, Y.-Z. (2008) Threshold of developmental and effective thermal sum of *Rhizoglyphus robini* Claparède. *Journal of Yangzhou University (Agricultural and Life Sciences Edition)* (now *Jiangsu Agricultural Research*) 29(2), 91–94.

Li, X.-J., Wen, X.-J. and Sun, S.-X. (1998) Study on the bionomics of *Tetranychus truncatus* Ehara. *Forest Pest and Disease* 1998-3, 3–4 (in Chinese).

Liang, W. and Huang, M. (1994) Influence of citrus orchard ground cover plants on arthropod communities in China: a review. *Agriculture, Ecosystems and Environment* 50, 29–37.

Liang, W.J. (1980) Rearing method for a tarsonemid mite (*Steneotarsonemus spinki* Smiley) free from the rice sheath rot fungus. *Plant Protection Bulletin (Taipei)* 22, 23–29 (in Chinese).

Lidster, P.D., Sanford, K.H. and McRae, K.B. (1981) Effects of modified atmosphere storage on overwintering populations of the apple rust mite and European red mite eggs. *HortScience* 16, 328–329.

Lienk, S.E. and Chapman, P.J. (1951) Orchard mite studies in 1950. *Journal of Economic Entomology* 44, 301–306.

Lienk, S.E. and Chapman, P.J. (1958a) Effect of the winter temperatures of 1956–57 on the survival rate of European red mite eggs. *Journal of Economic Entomology* 51, 263.

Lienk, S.E. and Chapman, P.J. (1958b) Influence of presence or absence of the European red mite on two-spotted mite abundance. *Journal of Economic Entomology* 44, 623.

Lienk, S.E., Minns, J. and Łabanowska, B.H. (1978) Evaluation of pesticides against the European red mite, apple rust mite and two mite predators in 1976-1977. *New York's Food and Life Sciences Bulletin* 71, 1–12.

Lienk, S.E., Watve, C.M. and Weires, R.W. (1981) Apple mites and their predators. *Agrochemical Age*, April 1981, 32–35.

Light, W.I.ST.G. and Ludlam, F.A.B. (1972) The effects of fruit tree red spider mite (*Panonychus ulmi*) on yield of apple trees in Kent. *Plant Pathology* 21, 175–181.

Liguori, M. (1987) Population trends of phytophagous and predatory mites in two vineyards of the Chianti area, Italy. *Redia* 70, 141–150.

Lin, D., Zeng, X., Li, P. and Huang, Y. (1992) Study on the distribution pattern and sampling method of *Pseudopygmephorus quadratus* Ewing in mushroom beds. *Fujian Journal of Agricultural Sciences* 7(2), 69–73 (in Chinese).

Lin, D.W., Cui, G.C. and Li, J.L. (1987) Investigation of the occurrence and bionomics of *Eriophyes tulipae* Keifer in Tibet. *Plant Protection* 13(5), 23–24 (in Chinese).

Lin, F.-P. and Kuang, H.-Y. (1997) A new genus and three new species of the family Diptilomiopidae from China (Acari: Diptilomiopidae). *Acta Entomologica Sinica* 40, 154–158 (in Chinese).

Lin, J.-Z. and Zhang, Y.-X. (1994) A note on 13 species of tarsonemid mites in Sichuan Province with a description of two new species (Acari: Tarsonemidae). *Journal of Southwest Agricultural University* 16, 525–528 (in Chinese).

Lin, J.-Z. and Zhang, Y.-X. (1995) A new species of genus *Hemitarsonemus* from Fujian Province of China (Acari: Tarsonemidae). *Fujian Journal of Agricultural Sciences* 10(4), 45–47 (in Chinese).

Lin, J.-Z. and Zhang, Y.-Z. (2000) Pygmephoroidea (Acari: Actinedida). In: Wang, B.-K. (ed.) *Fauna of Insects in Fujian Province of China, Vol. 9*. Fujian Science and Technology Press, Fuzhou, China, pp. 121–132.

Lin, J.-Z. and Zhang, Z.-Q. (1999) *Tarsonemidae of China (Acari: Prostigmata): an Annotated and Illustrated Catalogue and Bibliography*. Systematic and Applied Acarology Society, London.

Lin, J.-Z. and Zhang, Z.-Q. (2002) *Tarsonemidae of the World: Key to Genera, Geographical Distribution, Systematic Catalogue and Annotated Bibliograhy*. Systematic and Applied Acarology Society, London.

Lin, J.-Z. and Zhang, Z.-Q. (2010) Tarsonemidae of China: a review of progress on the systematics and biology, with an updated checklist of species. In: Zhang, Z.Q., Hong, X.-Y. and Fan, Q.H. (eds) *Xin Jie-Liu Centenary: Progress in Chinese Acarology*. *Zoosymposia* 4, 175–185.

Lin, J.-Z., Zhang, Z.-Q., Zhang, Y., Liu, Q. and Ji, J. (2000) Checklist of mites from moso bamboo in Fujian, China. *Systematic and Applied Acarology Special Publications* No. 4, 81–92.

Lin, M.Y. and Chen, S.K. (2008) Occurrence of mango red spider mite, *Oligonychus mangiferus*, and its control in the field. *Tainan District Agricultural Research and Extension Station Research Report* 51, 1–18 (in Chinese).

Lin, Y., Yang, G., Wang, H. and Yu, G. (1985) Bionomics and control of *Eotetranychus sexmaculatus* (Riley). *Chinese Journal of Tropical Crops* 1985 (02). Available as English summary at: http://en.cnki.com.cn/Article_en/CJFDTOTAL-RDZX198502013.htm (accessed 11 August 2014).

Lin, Y.-M. (1982) A new species and two new records of *Tuckerella* from China (Acari: Tuckerellidae). *Acta Entomologica Sinica* 25, 448–449 (in Chinese).

Lin, Y.M. and Chen, P.Z. (1984) Preliminary observations on *Stethorus* (*Allostethorus*) *parapauperculus*. *Natural Enemies of Insects* 6, 126–128 (in Chinese).

Lin, Y.-M. and Fu, Y.-G. (1997a) Two new species of the genus *Tuckerella* from Hainan Island, China. *Acta Zootaxonomica Sinica* 22, 150–153 (in Chinese).

Lin, Y.-M. and Fu, Y.-G. (1997b) A new species of the genus *Tuckerella* from China (Acari: Tuckerellidae). *Acta Entomologica Sinica* 40, 311–312 (in Chinese).

Linder, C., Baroffio, C. and Mittaz, C. (2008) Traitement post-recolte de l'eriophyide des framboises *Phyllocoptes gracilis*. *Revue Suisse de Viticulture Arboriculture Horticulture* 40, 105–107.

Linder, C., Jermini, M. and Zufferey, V. (2009) Nuisibilite de l'erinose sur le cepage Muscat. *Revue Suisse de Viticulture Arboriculture Horticulture* 41, 177–181.

Lindo, D.E. and Grenn, H.H. (1968) *Bryobia praetiosa* (clover mite) infestation in a feline. *Canadian Veterinary Journal* 9, 254–256.

Lindquist, E.E. (1969) Review of holarctic tarsonemid mites (Acarina: Prostigmata) parasitizing eggs of pine bark beetles. *Memoirs of the Entomological Society of Canada* 60, 1–111.

Lindquist, E.E. (1970) Review of the genus *Heterotarsonemus* (Acarina: Tarsonemidae). *Canadian Entomologist* 102, 812–829.

Lindquist, E.E. (1972) A new species of *Tarsonemus* from stored grain (Acarina: Tarsonemidae). *Canadian Entomologist* 104, 1699–1708.

Lindquist, E.E. (1975) Associations between mites and other arthropods in forest floor habitats. *Canadian Entomologist* 107, 425–437.

Lindquist, E.E. (1978) On the synonymy of *Tarsonemus waitei* Banks, *T. setifer* Ewing, and *T. bakeri* Ewing, with redescription of species (Acari: Tarsonemidae). *Canadian Entomologist* 110, 1023–1048.

Lindquist, E.E. (1984) Current theories on the evolution of major groups of Acari and on their relationships with other groups of Arachnida, with consequent implications for their classification. In: Griffiths, D.A. and Bowman, C.E. (eds) *Acarology VI, Volume 1.* Ellis Horwood, Chichester, UK, pp. 28–62.

Lindquist, E.E. (1985) External anatomy. In: Helle, W. and Sabelis, M.W. (eds) *Spider Mites. Their Biology, Natural Enemies and Control.* World Crop Pests, Volume 1A. Elsevier Science, Amsterdam, pp. 3–28.

Lindquist, E.E. (1986) The world genera of Tarsonemidae (Acari: Heterostigmata): a morphological, phylogenetic, and systematic revision, with a reclassification of family-group taxa in the Heterostigmata. *Memoirs of the Entomological Society of Canada* 136, 1–517.

Lindquist, E.E. (1996) External anatomy and notation of structures. In: Lindquist, E.E., Sabelis, M.W. and Bruin, J. (eds) *Eriophyoid Mites: Their Biology, Natural Enemies and Control.* World Crop Pests, Volume 6. Elsevier Science, Amsterdam, pp. 3–31.

Lindquist, E.E. (2001) Poising for a new century in acarology. In: Halliday, R.B., Walter, D.E., Proctor, H.C., Norton, R.A. and Colloff, M.J. (eds) *Acarology: Proceedings of the 10th International Congress of Acarology, 5–10 July 1998, Canberra.* CSIRO Publishing, Collingwood, Australia, pp. 17–34.

Lindquist, E.E. and Amrine, J.W. (1996) Systematics, diagnoses for major taxa, and keys to families and genera with species on plants of economic importance. In: Lindquist, E.E., Sabelis, M.W. and Bruin, J. (eds) *Eriophyoid Mites: Their Biology, Natural Enemies and Control.* World Crop Pests, Volume 6. Elsevier Science, Amsterdam, pp. 33–87.

Lindquist, E.E. and Oldfield, G.N. (1996) Evolution of eriophyoid mites in relation to their host plants. In: Lindquist, E.E., Sabelis, M.W. and Bruin, J. (eds) *Eriophyoid Mites: Their Biology, Natural Enemies and Control.* World Crop Pests, Volume 6. Elsevier Science, Amsterdam, pp. 277–300.

Lindquist, E.E., Sabelis, M.W. and Bruin, J. (eds) (1996) *Eriophyoid Mites: Their Biology, Natural Enemies and Control.* World Crop Pests, Volume 6. Elsevier Science, Amsterdam.

Lindquist, E.E., Krantz G.W. and Walter, D.E. (2009) Classification. In: Krantz, G.W. and Walter, D.E. (eds) *A Manual of Acarology,* 3rd edn. Texas Tech University Press, Lubbock, Texas, pp. 97–103.

Lindroth, J.I. (1904) Nya och salsynta finska Eriophyider. *Acta Societatis pro Fauna et Flora Fennica* 26, 3–18.

Linke, W. (1953) Investigation of the biology and epidemiology of the common spider mite, *Tetranychus alteae* von Hanst with particular consideration to hop as the host. *Höfchenbriefe* 4(4), 181–232.

Liotta, G. (1981) Osservazioni bio-etologiche su *Clitostethus arcuatus* (Rossi) (Col. Coccinellidae) in Sicilia. *Redia* 64, 173–185.

Liu, A. and Ridsdill-Smith, T.J. (2000) Cold storage of *Halotydeus destructor* (Acari: Penthaleidae) for use in experiments. *Experimental and Applied Acarology* 24, 123–133.

Liu, C.-Z., Yan, L., Li, H.-R. and Wang, G. (2006a) Effects of predator-mediated apparent competition on the population dynamics of *Tetranychus urticae* on apples. *BioControl* 51, 453–463.

Liu, H., Zhao, Z.-M., Wang, J.-J., Li, L.-S. and Ding, W. (2000a) Temperature-dependent development and reproduction of the spider mite, *Schizotetranychus bambusae* Reck (Acari: Tetranychidae). *Systematic and Applied Acarology* 5, 33–39.

Liu, H., Zhao, Z., Ding, W., Lu, L.,Wang, S. and Tan, X. (2000b) Biology of *Schizotetranychus bambusae* Reck and population dynamics of the spider mites on the bamboo of *Phyllostachys pubescens* Mazel in Changning Bamboo Area. *Journal of Sichuan Forestry Science and Technology* 2000 (03). Available as English summary at: http://en.cnki.com.cn/Article_en/CJFDTOTAL-SCLK200003000.htm (accessed 11 August 2014).

Liu, H., Zhao, Z.-M., Deng, Y.-X., He, L., Wang, J.-J. and Wu, S. Y. (2002) Effect of temperature on experimental population dynamics of *Typhlodromus bambusae* Ehara. *Zoological Research* 23 356–360.

Liu, H., Zhao, Z., Deng, Y., Deng, X., He, L., Wu, S. and Shuai, X. (2004a) Host plant suitability and reproductive isolation of two *Schizotetranychus bambusae* populations from the bamboo species *Phyllostachys pubescens* and *Neosinocalamus affinis*. *Chinese Journal of Applied Ecology*15, 299–302 (in Chinese).

Liu, H., Zhao, Z., Wang, J.-J., He, L. and Wu, S.-Y. (2004b) Effect of different bamboo ages on population parameters of *Schizotetranychus bambusae* Reck (A: Tetranychidae). *Systematic and Applied Acarology* 9, 15–21.

Liu, H., Zhao, Z., Wang, J. and Wu, S. (2006b) Predation of *Typhlodromus bambusae* Ehara on *Schizotetranychus bambusae*. *Chinese Journal of Applied Ecology* 17, 280–284.

Liu, M.-P. and Kuang, H.-Y. (1998) Three new species of the family Diptilomiopidae from China (Acari: Eriophyoidea). *Acta Zootaxonomica Sinica* 23, 25–29 (in Chinese).

Liu, M.Z., Li, X.Y., Zhao, Y.H. and Li, X.H. (1984) Control of *Aculops lycii* Kuang on *Lycium chinense* Mill. *Ningxia Journal of Agriculture and Forestry Science and Technology* 6, 24–25.

Liu, Q.-H., Xu, Z.-Q., Yao, T., Zheng, L.-Y. and Pu, X.-P. (2011) Pests investigation of ancient tree and its control strategies – a case study in Lanzhou Wuquan Mountain Park. *Pratacultural Science* 28, 661–666 (in Chinese).

Liu, T.-S. (1995) Control of bulb mites on gladiolus by soil treatment. *Bulletin of Taichung District Agricultural Improvement Station* 49, 33–40 (in Chinese).

Liu, T.-S. (1996) Effect of organic matters on the occurrence of the bulb mites on bulb crops. *Bulletin of Taichung District Agricultural Improvement Station* 51, 49–57 (in Chinese).

Liu, T.S. and Tzeng, A.K. (1994) Biological and damaging characteristics of bulb mite, *Rhizoglyphus robini* Claparede. *Plant Protection Bulletin (Taipei)* 36, 177–187.

Liu, Y.H. and Tsai, J.H. (1998) Development, survivorship, and reproduction of *Tetranychus tumidus* Banks (Acarina: Tetranychidae) in relation to temperature. *International Journal of Acarology* 24, 245–252.

Livshits, I.Z. (1960a) Brown fruit mite *Bryobia redikorzevi* Reck, 1947 (Taxonomic position, morphology and biology). *Trudy Gosudarstvennogo Nikitskogo Botanicheskogo Sada* 33, 1–76 (in Russian).

Livshits, I.Z. (1960b) Contribution to morphology and biology of fruit mites injuring cultures (New species of tetranychid mites on plum). *Trudy Gosudarstvennogo Nikitskogo Botanicheskogo Sada* 33, 77–156 (in Russian).

Livshits, I.Z. and Mitrofanov, V.I. (1967) Materials to the cognition of the Acariformes: Tenuipalpidaes fauna. *Proceedings of the Nikita Botanical Gardens* 39, 3–72 (in Russian).

Livshits, I.Z. and Mitrofanov, V.I. (1975) *Plant-inhabiting Mites: An Illustrated Key to Families. Trudy Gosudarstvennogo Nikitskogo Botanicheskogo Sada* 66, 1–183 (in Russian).

Livshits, I.Z. and Salinas-Croche, A. (1968) *Preliminares Acerca de los Ácaros "Tetranicos" de Cuba*. Centro National Fitosanid, Cuba, Havana.

Livshits, I.Z., Mitrofanov, V.I. and Sharonov, A.A. (1981) Novye vidy kleshchei roda *Steneotarsonemus* Beer, 1954 so zlakov i mcha (Acariformes, Tarsonemidae) (New mite species of the genus *Steneotarsonemus* Beer, 1954 (Acariformes, Tarsonemidae) from grasses and moss). *Vestnik Zoologii* 1981 (6), 79–82 (in Russian).

Livshits, I.Z., Mitrofanov, V.I. and Sharonov, A.A. (1986) Revision of mites of the family Siteroptidae Mahunka, 1970 (Acari, Tarsonemina). *Pests and Diseases of Fruit, Subtropical, and Ornamental Cultures* 99, 7–31 (in Russian).

Lloyd, R.W. and Krieg, D.R. (1987) Cotton development and yield as affected by insecticides. *Journal of Economic Entomology* 80, 854–858.

Lo, K.-C. (1996) Potential use of mites as biological agents of insect and mite pests. *Plant Protection Bulletin (Taipei)* 38, 99–110 (in Chinese).

Lo, K.-C. and Ho, C.C. (1977) Preliminary studies on rice tarsonemid mite *Steneotarsonemus spinki* (Acarina: Tarsonemidae). *National Science Council Monthly (Taipei)* 5, 274–284 (in Chinese).

Lo, K.-C. and Ho, C.C. (1979) Ecological observations on rice tarsonemid mite, *Steneotarsonemus spinki* (Acarina: Tarsonemidae). *Journal of Agricultural Research of China* 28, 181–192 (in Chinese).

Lo, K.-C. and Ho, C.C. (1980) Studies on the rice tarsonemid mite, *Steneotarsonemus spinki* Smiley. *Plant Protection Bulletin (Taipei)* 22, 1–9 (in Chinese).

Lo, K.-C., Ho, C.C. and Lin, K.C. (1981) Screening of chemicals for the control of rice tarsonemid mite, *Steneotarsonemus spinki*. *Journal of Agricultural Research of China* 30, 303–307 (in Chinese).

Lo, K.-C., Lee, W.T., Wu, T.-K. and Ho, C.-C. (1990) Use of predators to control spider mites (Acarina: Tetranychidae) in the Republic of China on Taiwan. In: Bay-Peterson, J. (ed.) *The Use of Natural Enemies to Control Agricultural Pests*. FFTC Series 40, Food and Fertilizer Technology Center, Asian and Pacific Region, Taipei, pp. 166–177.

Lo, P.K.C. (1968a) Tetranychoid mites infesting tea in Taiwan. *Bulletin of the Sun Yat-Sen Cultural Foundation* 1, 275–285 (in Chinese).

Lo, P.K.C. (1968b) Tetranychoid mites infesting fruit plants in Taiwan. *Bulletin of the Sun Yat-Sen Cultural Foundation* 2, 97–137 (in Chinese).

Lo, P.K.C. (1969) Tetranychoid mites infesting special crops in Taiwan. *Bulletin of the Sun Yat-Sen Cultural Foundation* 4, 43–82 (in Chinese).

Lo, P.K.C. and Hsia, D.N.T. (1968) Tenuipalpid and tetranychid mites infesting citrus in Taiwan and a life history study of the citrus green mite, *Schizotetranychus baltazarae* Rimando. *Bulletin of the Sun Yat-Sen Cultural Foundation* 1, 254–274. (in Chinese).

Lobbes, P. and Schotten, C. (1980) Capacities for increase of the soil mite *Hypoaspis aculeifer* Canestrini (Mesostigmata: Laelapidae). *Zeitschrift für Angewandte Entomologie* 90, 9–22.

Lochar, F.J. (1958) Der enfluss Von dichlordiphenyl trichlormethyl methan (DDT) auf einigen Tetranychiden. *Zeitschrift für Angewandte Zoologie* 45, 201–248.

Loganathan, S. and Muraleedharan, N. (1997) Life table studies on the pink mite, *Acaphylla theae* (Watt) (Eriophyidae: Acarina) infesting tea. *Journal of Plantation Crops* 25, 109–111.

Lombardini, G. (1960) Acari nuovi 41. *Redia* 45, 255–261.

Longhurst, C., Bacci, L., Buendia, J., Hatton, C.J., Petitprez, J. and Tsakonas, P. (1992) Fenazaquin, a novel acaricide for the management of spider mites in a variety of crops. In: *Proceedings of the Brighton Crop Protection Conference: Pests and Diseases – 1992, Brighton, November 23–26, 1992*. British Crop Production Council, Croydon (now Alton), pp. 51–58.

Longworth, J.F. (1987) Biological control of gorse: a case study. *New Zealand Entomologist* 10, 8–13.

Lopes, M.V., Oliveira, C.A.L. de, Barreto, M., Barbosa, J.C. and Rossi, F.M. (2007) Measurement of samples for monitoring the acarid of leprous *Brevipalpus phoenicis* (Geijskes, 1939) in citros. *Revista Brasileira de Fruticultura* 29, 671–676.

Lopez, R., Levi, A., Shepard, B.M., Simmons, A.M. and Jackson, D.M. (2005) Sources of resistance to twospotted spider mite (Acari: Tetranychidae) in *Citrullus* spp. *HortScience* 40, 1661–1663.

Lord, E.T. (1949) The influence of spray programs on the fauna of apple orchards in Nova Scotia. III. Mites and their predators. *Canadian Entomologist* 81, 202–230.

Lord, F.T. (1971) Laboratory tests to compare the predatory value of six mirid species in each stage of development against the winter eggs of the European red mite, *Panonychus ulmi* (Acari: Tetranychidae). *Canadian Entomologist* 103, 1663–1669.

Lord, F.T. and Stewart, D.K.R. (1961) Effects of increasing the nitrogen levels of apple leaves on mite and predator populations. *Canadian Entomologist* 93, 924–927.

Lorenzato, D. (1984) Ensaios de controle aos ácaros nocivos em alho (*Allium sativum* L.) armazenado. *Agronomia Sulriograndense* 20, 153–165.

Lorenzato, D. (1987) Controle biológico de ácaros fitófagos na cultura da macieira no município de Farroupilha, RS. *Agronomia Sulriograndense* 23, 167-183.

Lotka, A.J. (1925) *Elements of Physical Biology*. Williams and Wilkins, Baltimore, Maryland.

Lovisolo, O. (2001) Citrus leprosis virus: properties, diagnosis, agro-ecology and phytosanitary importance. *OEPP/EPPO Bulletin* 31, 79–89.

Löyttyniemi, K. (1969) Akamapunkkilaji (*Nalepella haarlovi* var. *picea-abietis* Löyttyniemi, Acarina, Eriophyidae) kuusen taimien tuholaisena taimitarhoissa. *Silva Fennica* 3, 191–200.

Löyttyniemi, K. (1970a) Zur Biologie der Nadelholzspinnmilbe (*Oligonychus ununguis* (Jacobi), Acarina, Tetranychidae) in Finland. *Acta Entomologica Fennica* 27, 1–64.

Löyttyniemi, K. (1970b) Control of mites *Oligonychus ununguis* and *Nalepella haarlovi* var. *piceae-abietis*. *Folia Forestalia* 93, 1–10.

Löyttyniemi, K. (1971) On the biology of *Nalepella haarlovi* Boczek var. *piceae-abietis* Löyttyniemi (Acarina: Eriophyidae). *Communicationes Instituti Forestalis Fenniae* 73(3), 1–16.

Löyttyniemi, K. (1973) *Eurytetranychus piceus* sp. n. (Acarina, Tetranychidae) with notes on its life history. *Annales Entomologici Fennici* 39, 119–123.

Löyttyniemi, K. (1975) Mass outbreaks of *Epitrimerus pungiscus* Keifer (Acarina, Eriophyidae) on Norway spruce *Picea abies* (L.) Karst. *Annales Entomologici Fennici* 41, 13–15.

Löyttyniemi, K. and Heliövaara, K. (1991) Effect of forest fertilization on the spruce spider mite *Oligonychus ununguis* (Jacobi) (Acarina: Tetranychidae). *Acarologia* 32, 139–143.

Lu, X.-Y., Xu, J. and Li, Q.-Y. (2008) Impact of *Lolium perenne* planted in chestnut orchards and its mowing methods on population densities of *Oligonychus ununguis* and its natural enemy *Amblyseius castaneae*. *Chinese Journal of Biological Control* 24, 108–112.

Lu, S.J. (2000) Experiment of control of citrus rust mite by enhanced abamectin. *South China Fruits* 29(4), 25 (in Chinese).

Lu, T. (1979) New discovery of the male of *Petrobia latens* Müller. Acarina: Tetranychidae. *Acta Entomologica Sinica* 22, 477 (in Chinese).

Lu, W. and Lou, Y. (1995) Investigation on the fluctuation dynamics of tea pink mite (*Acaphylla theae* Watt) and the prediction of occurring date. *Journal of Tea Science (Hangzhou)* 15, 27–32 (in Chinese).

Lu, W.-M. (1993) Mites in tea plantations and their controls. *China Tea* 15(5), 12–13 (in Chinese).

Lucas, H. (1848) Genus *Tetranychus*. Exploration scientifique de l'Algérie pendant les années 1840, 1841, 1842. *Sciences Physiques et Zoologiques (Paris)* 1, 309.

Luckey, T.D. (1968) Insecticide hormoligosis. *Journal of Economic Entomology* 61, 7–12.

Luczynski, A., Nyrop, J.P. and Shi, A. (2008) Pattern of female reproductive age classes in mass-reared populations of *Phytoseiulus persimilis* (Acari: Phytoseiidae) and its influence on population characteristics and quality of predators following cold storage. *Biological Control* 47, 159–166.

Luh, H.-K. and Croft, B.A. (1999) Classification of generalist or specialist life styles of predaceous phytoseiid mites using a computer genetic algorithm, information theory, and life history traits. *Environmental Entomology* 28, 915–923.

Lui, Z.G. and Lui, N.Z. (1986) A preliminary report on *Tetranychus piercei* McGregor. *Insect Knowledge* 23, 18–19 (in Chinese).

Luo, J., Chen, M., Lü, H., Luo, L. and Luo, G. (1994) Spatial patterns of mobile mite of *Eotetranychus pruni* Oudemans (Acari: Tetranychidae) in apple trees. *Acta Agriculturae Boreali-occidentalis Sinica* 1994–04, 32–36. Available at: http://en.cnki.com.cn/Article_en/CJFDTOTAL-XBNX404.007.htm (accessed 11 August 2014).

Lydon, J. and Duke, S.O. (1989) Pesticide effects on secondary metabolism of higher plants. *Pesticide Science* (now *Pesticide Management Science*) 25, 361–373.

Lykouressis, D.P. and Emmanouel, N.G. (1991) Biological aspects on some arthropod pests of lucerne in Greece. *Journal of Applied Entomology* 111, 526–532.

Lynch, S.M.T. (1993) *The Bulb-scale Mite,* Steneotarsonemus laticeps *(Halbert) – A Review*. Report (code) BOF 25, Horticultural Development Council, East Malling, UK.

Lynch, S.M.T. and Bedi, A. (1994) *The Development of an* in vitro *Method for Culturing the Bulb-scale Mite (*Steneotarsonemus laticeps *Halbert) and its Use for Life History Studies*. Report (code) BOF 25, Horticultural Development Council, East Malling, UK.

Lyon, W.F. (1973) A plant-feeding mite *Mononychellus tanajoa* (Bondar) (Acarina:Tetranychidae) new to the African continent threatens cassava (*Manihot esculenta* Crantz) in Uganda, East Africa. *PANS (Pest Articles and News Summaries)* 19, 36–37.

Lyon,W.F. (2008) *Clover Mite*. Ohio State University Extension Fact Sheet, Entomology, HYG 2095-94, Columbus, Ohio. Available at: http://ohioline.osu.edu/hyg-fact/2000/2095.html (accessed 11 August 2014).

Ma, E.-P. and Yuan, Y.L. (1975) Preliminary study on the genus *Tetranychus* Dufour in China (Acarina: Tetranychidae). *Acta Entomologica Sinica* 18, 220–228 (in Chinese).

Ma, E.-P. and Yuan, Y.-L. (1976) On the genus *Oligonychus* Berlese from China (Acarina: Tetranychidae). *Acta Entomologica Sinica* 19, 357–362 (in Chinese).

Ma, E.-P. and Yuan, Y.-L. (1980) New species and new records of tetranychid mites from China I. (Acarina: Tetranychidae). *Acta Zootaxonomica Sinica* 5, 42-45 (in Chinese).

Ma, E.-P. and Yuan, Y.-L. (1981a) Two new species of the genus *Eurytetranychus* from China (Acarina: Tetranychidae). *Zoological Research* 2, 195–198 (in Chinese).

Ma, E.-P. and Yuan, Y.-L. (1981b) New species and new records of tetranychid mites from China II. (Acarina: Tetranychidae). *Zoological Research* 2, 281–287 (in Chinese).

Ma, E.-P. and Yuan, Y.-L. (1992) Tetranychid mites. In: Chen, X.B. and Ma, E.P. (eds) *Researches of Acarology in China*. Chongqing Publishing House, Chongqing, China, pp. 28–32 (in Chinese).

Ma, E.-P., Yuan, Y.-L. and Lin, Y.M. (1979) The spider mites of Hainan Island (Acarina: Tetranychidae). *Journal of Jiangxi University* 3, 39–49 (in Chinese).

Ma, E.-P., Yuan, Y.-L. and Qian, Y.-H. (1984) Spider mites. In: Jiangxi University (ed.) *Agricultural Mites of China* (*Zhong-Guo-Nong-Ye-Man-Lei*). Shanghai Science and Technology Press, Shanghai, China, pp. 88–164 (in Chinese).

MacArthur, R.H. and Wilson, E.O. (1967) *The Theory of Island Biogeography*. Princeton University Press, Princeton, New Jersey.

MacDonald, O.C. and Mills, K.A. (1994) Plant quarantine treatments: are there alternatives to methyl bromide? In: *Brighton Crop Protection Conference, Pests and Diseases – 1994. Proceedings of an International Conference Organised by the British Crop Protection Council, Held at the Brighton Centre and the Brighton Metropole Hotel, Brighton, England, 21–24 November 1994, Volume 1*. British Crop Production Council, Farnham (now Alton), UK , pp. 183–189.

Macfarlane, D. (1984) Key to spider mites (Tetranychidae) recorded on cassava in Africa, with a note on slide preparations. In: Greathead, A.H., Markham, R.H., Murphy, R.J., Murphy, S.T. and Robertson, I.A.D. (eds) *Integrated Pest Management of Cassava Green Mite. Proceedings of a Regional Training Workshop in East Africa, 30 April–4 May 1984*. CIBC (Commonwealth Institute of Biological Control)., Ascot, UK, pp. 31–35.

Macintyre, I.D. and Foster, G.N. (1977) Autumn mortality in the raspberry leaf and bud mite (*Phyllocoptes gracilis* (Nalepa), Eriophyidae). *Horticultural Research (Edinburgh)* 17(1), 47–50.

Mackie, D.B., Steinweden, J.B. and Carter, W.B. (1942) *Methyl Bromide Fumigation 1942 of Narcissus Bulbs for the Control of Bulb Flies and Bulb Mites*. Department of Agriculture, State of California, Sacramento, California.

Maclennan, K.E., McDonald, G. and Ward, S.A. (1998) Soil microflora as hosts of redlegged earth mite (*Halotydeus destructor*). *Entomologia Experimentalis et Applicata* 86, 319–323.

MacLeod, L. (2007) *CSL* (Central Science Laboratory, UK) *Pest Risk Analysis for* Aceria tulipae. Available at: http://webarchive.nationalarchives.gov.uk/20130123162956/http:/www.defra.gov.uk/planth/pra/Aceriatulipae.pdf (accessed 2 February 2015).

MacPhee, A.W. (1961) Mortality of winter eggs of the European red mite *Panonychus ulmi* (Koch) at low temperature, and its ecological significance. *Canadian Journal of Zoology* 39, 229–243.

MacPhee, A.W. (1963) The effect of low temperatures on some predacious phytoseiid mites, and on brown mite, *Bryobia arborea* Morgan and Anderson. *Canadian Entomologist* 95, 41–44.

MacPhee, A.W. (1964) Cold-hardness, habitat and winter survival of some orchard arthropods in Nova Scotia. *Canadian Entomologist* 96, 617–625.

MacRae, I.V. and Croft, B.A. (1993) Influence of temperature on interspecific predation and cannibalism by *Metaseiulus occidentalis* and *Typhlodromus pyri* (Acarina: Phytoseiidae). *Environmental Entomology* 22, 770–775.

Madrigal, C.J.A. (1974) *Ácaros de Importancia Economica en Plantas Cultivadas en Colombia*. Seminario, Universidad Nacional de Colombia, Medellin, Colombia.

Magalhães, B.P., Rodrigues, J.C.V., Boucias, D.G. and Childers, C.C. (2005) Pathogenicity of *Metarhizium anisopliae* var. *acridum* to the false spider mite *Brevipalpus phoenicis* (Acari: Tenuipalpidae). *Florida Entomologist* 88, 195–198.

Magalhães, S., Forbes, M.R., Skoracka, A., Osakabe, M., Chevillon, C. and McCoy, K. (2007) Host race formation in Acari. *Experimental and Applied Acarology* 42, 225–238.

Magowski, W. (1988) Description of a new species of *Formicomotes* Sevastianov, 1980 (Acari: Dolichocybidae) with notes on the female dimorphism within this genus. *Mitteilungen aus dem Hamburgischen Zoologischen Museum und Institut* 85, 163–182.

Magowski, W.Ł. (2012) Two new species and a new subgenus of tarsonemid mites (Acari: Heterostigmatina: Tarsonemidae) from ferns in Poland. *Zoological Studies (Taipei)* 51, 512–525.

Magowski, W.Ł. and Beron, P. (2013) *Tarsonemus (Tarsonemus) myceliophagus* Hussey 1963. In: Magowski, W.Ł. (ed.) Acari: Acariformes: Fauna Europaea version 2.6.2. Available at: http://www.faunaeur.org/full_results.php?id=229464 (accessed 21 April 2015).

Mague, D.L. and Streu, H.T. (1980) Life history and seasonal population growth of *Oligonychus ilicis* infesting Japanese holly in New Jersey. *Environmental Entomology* 9, 420–424.

Mahabaleshwar Hegde and Balikai, R.A. (2011) Management of paddy leaf mite, *Oligonychus oryzae* through acaricides. *International Journal of Agricultural and Statistics Sciences* 7, 613–616.

Mahalingappa, P.B., Reddy, K.D., Reddy, K.N. and Subbaratnam, G.V. (2008) Bio-efficacy of certain insecticides against thrips (*Scirtothrips dorsalis* Hood) and mite (*Polyphagotarsonemus latus* Banks) infesting chillies (*Capsicum annuum* L.). *Journal of Research ANGRAU* 36(1), 11–15.

Mahgoob, A.E.A. (2006) Susceptibility of some mango varieties to the bud mite *Eriophyes mangiferae* (Sayed) and malformation disease and the relation of mite infestation to malformation disease. *Arab Universities Journal of Agricultural Sciences* 14, 491–507.

Mahmood, T., Hein, G.L. and Jensen, S.G. (1998) Mixed infection of hard red winter wheat with high plains virus and wheat streak mosaic virus from wheat curl mites in Nebraska. *Plant Disease* 82, 311–315.

Mahr, F.A., Kwong, R.M., McLaren, D.A. and Jupp, P.W. (1997) Redistribution and present status of the mite *Aculus hyperici* for the control of St. John's wort, *Hypericum perforatum*, in Australia. *Plant Protection Quarterly* 12, 84–88.

Mahunka, S. (1961) Contributions to the tyroglyphid fauna of Hungary (Acari). *Annales Universitatis Scientiarum Budapestinensis de Rolando Eötvös Nominatae, Sectio Biologica* 4, 113–118.

Mahunka, S. (1965) Identification key for the species of the family Scutacaridae (Acari: Tarsonemini). *Acta Zoologica Academiae Scientiarum Hungaricae* 11, 353–401.

Mahunka, S. (1969) The scientific results of the Hungarian soil zoological expedition to South America. 13. Acari: Pygmephoridae and Scutacaridae from the material of the second expedition (Brazil and Bolivia). *Acta Zoologica Academiae Scientiarum Hungaricae* 15, 333–370.

Mahunka, S. (1970) Considerations on the systematics of the Tarsonemina and description of new European taxa (Acari: Trombidiformes). *Acta Zoologica Academiae Scientiarum Hungaricae* 16, 137–174.

Mahunka, S. (1975) Äthiopische Tarsonemiden (Acari: Tarsonemida). I. *Acta Zoologica Academiae Scientiarum Hungaricae* 21, 369–410.

Mahunka, S. (1977) The examination of myrmecophilous tarsonemid mites based on the investigations of Dr. C.W. Rettenmeyer (Acari). I. *Acta Zoologica Academiae Scientiarum Hungaricae* 23, 99–132.

Mahunka, S. (1981) Milben (Acari) aus St. Lucia (Antillen). I. Tarsonemina. Anoetidae. *Acta Zoologica Academiae Scientiarum Hungaricae* 27, 323–353.

Mahunka, S. and Mahunka-Papp, L. (1994) Two new microdispid (Acari, Heterostigmata: Microdispidae) species from Hungary. *Annales Historico-naturales Musei Nationalis Hungarici* 86, 119–124.

Maia, O.M. de A. and Buzzi, Z.J. (2006) Occurrence of *Brevipalpus phoenicis* (Geijskes) (Acari, Tenuipalpidae), *Tetranychus urticae* (Koch) (Acari, Tetranychidae) and *Polyphagotarsonemus latus* (Banks) (Acari, Tarsonemidae) on leaves of *I. cairica* (Linnaeus) Sweet (Solanales, Convolvulaceae). *Revista Brasileira de Zoologia* 23, 886–887.

Maia, O.M.A. and Oliveira, C.A.L. de (2004) Capacidade de colonização de *Brevipalpus phoenicis* (Geijskes) (Acari: Tenuipalpidae) em cercas-vivas, quebra-ventos e plantas invasoras. *Neotropical Entomology* 33, 625–629.

Mailloux, J., Le Bellec, F., Kreiter, S., Tixier, M.-S. and Dubois, P. (2010) Influence of ground cover management on diversity and density of phytoseiid mites (Acari: Phytoseiidae) in Guadeloupean citrus orchards. *Experimental and Applied Acarology* 52, 275–290.

Mailloux, M. (1984) Le réseau d'avertissements phytosanitaires du pommier au Quebec. In: Vincent, C. and Bostanian, N.J. (eds) *La Phytoprotection des Vergers des Pommier au Quebec*. Technical Bulletin No. 19, St. Jean sur Richelieu Research Station, Agriculture Canada, Quebec, Canada, pp. 177–188.

Maixner, M. (1990) Untersuchungen zur Insektizidresistenz der Raubmilbe *Typhlodromus pyri* Scheuten (Acari: Phytoseiidae) an Reben des Weinbaugebiets Mosel-Saar-Ruwer. *Mitteilungen aus der Biologischen Bundesanstalt für Land- und Forstwirtschaft, Berlin-Dahlem* No. 257, 1–118.

Majer, J.D. and and Recher, H.F. (1988) Invertebrate communities on Western Australian eucalypts: a comparison of branch clipping and chemical knockdown procedures. *Australian Journal of Ecology* 13, 269–278.

Majorana, G. (1970) Alterazioni virus – simili causate da un eriofide (*Aceria medicaginis* K.) su sulla. *Informatore Fitopatologico* 20(15/16), 12–17.

Majumder, M.Z.R., Bhuiya, A.D. and Chowdhury, N. (1996) Effects of radiation on mortality, fecundity and sterility of *Oligonychus biharensis* (Hirst) infesting common flower plants in Bangladesh. *Bangladesh Journal of Zoology* 24, 25–32.

Maketon, M., Orosz-Coghlan, P. and Sinprasert, J. (2008) Evaluation of *Metarhizium anisopliae* (Deuteromycota: Hyphomycetes) for control of broad mite *Polyphagotarsonemus latus* (Acari: Tarsonemidae) in mulberry. *Experimental and Applied Acarology* 46, 157–167.

Makundi, R.H. and Kashenge, S. (2002) Comparative efficacy of neem, *Azadirachta indica*, extract formulations and the synthetic acaricide, amitraz (Mitac), against the two spotted spider mites, *Tetranychus urticae* (Acari: Tetranychidae), on tomatoes, *Lycopersicum esculentum*. *Zeitschrift für Pflanzenkrankheiten und Pflanzenschutz* (now *Journal of Plant Diseases and Plant Protection*) 109, 57–63.

Malagnini, V. (2001) Caratterizzazione genetica di artropodi. Thesis, Università di Bologna, Bologna Italy.

Malagnini, V., Navajas, M., Migeon, A. and Duso, C. (2012) Differences between sympatric populations of *Eotetranychus carpini* collected from *Vitis vinifera* and *Carpinus betulus*: insights from host-switch experiments and molecular data. *Experimental and Applied Acarology* 56, 209–219.

Malcolm, D.R. (1955) *Biology and Control of the Timothy Mite,* Paratetranychus pratensis *(Banks)*. Technical Bulletin No. 17, Washington State Agricultural Experiment Station, Pullman, Washington.

Mali, U.R., Shirsat, A.M. and Godbole, G.M. (1977) Occurrence of pigeonpea sterility mosaic in Marathwada. *Research Bulletin, Marathwada Agricultural University* No. 1, 148–149.

Malik, R., Smith. C.M., Brown Guedira, G.L., Harvey, T.L. and Gill, B.S. (2003a) Assessment of *Aegilops tauschii* for resistance to biotypes of wheat curl mite (Acari: Eriophyidae). *Journal of Economic Entomology* 96, 1329–1333.

Malik, R., Brown Guedira, G.L., Smith, C.M., Harvey, T.L. and Gill, B.S. (2003b) Genetic mapping of wheat curl mite resistance genes *Cmc3* and *Cmc4* in common wheat. *Crop Science* 43, 644–650.

Malinoski, M.K., Clement, D.L. and Bosmans, R. (2009) *IPM Series HG 52: Boxwood*. Home and Garden Information Center, University of Maryland Extension, College Park, Maryland. Available at: http://extension.umd.edu/sites/default/files/_images/programs/hgic/Publications/HG52_IPM_Boxwood.pdf (accessed 11 August 2014).

Mallik, B., Srinivasa, N. and Chinnamade Gowda, C. (2003) *Bibliography of Plant Feeding Mites of India*. All India Network Project on Agricultural Acarology, Indian Council of Agricultural Research, University of Agricultural Sciences, Bangalore, India.

Mallikarjunappa, S. (1987) Studies on biology, ecology and control of *Eotetranychus hicoriae* (McGregor) (Acari: Tetranychidae) on guava. *Mysore Journal of Agricultural Sciences* 21, 4–5.

Mallikarjunappa, S. and Nageshchandra, B.K. (1989) Ecology of *Eotetranychus hicoriae* (Acari:Tetranychidae) on guava in South India. In: ChannaBasavanna, G.P. and Viraktamath, C.A. (eds) *Progress in Acarology, Volume 2.* Oxford & IBH Publishing, New Delhi, pp. 33–38.

Mallikarjunappa, S. and Nageshchandra, B.K. (1990) Efficacy of pesticides on *Eotetranychus hicoriae* (Acari: Tetranychidae) infesting guava. *Mysore Journal of Agricultural Sciences* 24, 471–474.

Mallikarjunappa, S., Nageshchandra, B.K. and Kumar, P. (1989) Laboratory evaluation of chemicals for the control of the mite, *Eotetranychus hicoriae* (Acari Tetranychidae) infesting guava. *Current Research, University of. Agricultural Sciences (Bangalore)* 18, 174–175.

Maloney, D., Drummond, F.A. and Alford, R. (2003) *Spider Predation in Agroecosystems: Can Spiders Effectively Control Pest Populations?* Technical Bulletin No. 190, Maine Agricultural and Forest Experiment Station, University of Maine, Orono, Maine.

Maltshenkova, N.I. (1969) The biology and ecology of *Epitrimerus vitis* Nal. (Acarina: Tetrapodili), a pest of the grape vine. In: *Vrednaya i Poleznaya Fauna Bespozvonochnykh Moldavii, Part/Vol. 4–5.* Izdatel'stvo Kartya Moldovenskaya Kishinev, USSR (now Chişinău, Moldova), pp. 205–222 (in Russian).

Malyuganov, V. A. (1990) Protection of new grape varieties on the basis of their resistance to diseases and pests. *Sadovodstvo-i-Vinogradarstvo* 5, 31–33 (in Russian).

Mancia, J.E. and Cortès, M.R. (1975) Lista de insectos clasificados encontrados en el cultivo de frijol, *Phaseolus vulgaris* L. *Siades* 4, 120–136.

Manglitz, G.R. and Cory, E.N. (1953) Biology of *Brevipalpus australis*. *Journal of Economic Entomology* 46, 116–119.

Mani, M.S. (1964) *Ecology of Plant Galls.* W. Junk, The Hague, the Netherlands.

Manita, K. and Vatana, C. (2000) Biology and efficiency of exotic predatory mite, *Phytoseiulus persimilis* Athias Henriot and *Amblyseius californicus* (McGregor) and indigenous predatory mite, *Amblyseius longispinosus* (Evans). *Entomology and Zoology Gazette (Bangkok)* 22, 202–216 (in Thai).

Manjit Singh and Dhooria, M.S. (2006) Biology, host range and population dynamics of tarsonemid mite, *Polyphagotarsonemus latus* (Banks) on chilli (*Capsicum annum*) in Punjab. *Journal of Applied Zoological Research* 17, 18–23.

Manjunatha, M. and Puttaswamy (1989a) A new record of spiders as predators of sorghum spidermite *Oligonychus indicus* (Hirst) (Acari: Tetranychidae). *Current Science* 58, 272.

Manjunatha, M. and Puttaswamy (1989b) Life history of *Tetranychus neocaledonicus* under green house conditions. *Indian Journal of Acarology* 11, 35–40.

Manjunatha, M. and Puttaswamy (1990a) Chemical control of spider mite, *Oligonychus indicus* Hirst on sorghum. *Indian Journal of Plant Protection* 18, 37–41.

Manjunatha, M. and Puttaswamy (1990b) Estimation of crop losses due to the spider mite *Oligonychus indicus* Hirst in sorghum. *Indian Journal of Plant Protection* 18, 245–249.

Manjunatha, M., Hanchinal, S.G. and Kulkarni, S.V. (2001) Interaction beetween *Amblyseius ovalis* and *Polyphagotarsonemus latus* and efficacy of *A. ovalis* on chilli mite and thrips. *Karnataka Journal of Agricultural Sciences* 14, 506–509.

Mann, H.H., Nagpurkar, S.D. and Kulkarni, G.S. (1920) The "tambera" disease of potato. *Agricultural Journal of India* 15, 282–288.

Mann, J. (1969) *Cactus-feeding Insects and Mites.* Bulletin of the United States National Museum No. 256. Government Printing Office, Washington, DC.

Mann, J. (1970) *Cacti Naturalised in Australia and Their Control.* Government Printer, Brisbane, Australia.

Mann, L.K. and Lange, W.H. (1960) Fumigation controls microscopic mite attacking garlic. *California Agriculture* 14(12), 9–10.

Mansilla Vázquez, J.P. (1991a) Caida del boton floral de la camelia provocado por *Cosetacus camelliae* (Keifer) (Acari: Eriophyidae). *Boletín de Sanidad Vegetal, Plagas* 17, 125–132.

Mansilla Vázquez, J.P. (1991b) Detection of "purple mite" (*Calacarus carinatus* Green) in *Camellia japonica* L. *Boletín de Sanidad Vegetal, Plagas* 17, 283–286.

Mansilla Vázquez, J.P., Pérez Otero, R. and López Villarreal, M. (2002) Ácaros eriofidos sobre camelia en Galicia (Espana). *Phytoma España* 138, 25–26.

Mansilla Vázquez, J.P., Pérez Otero, R., Salinero, C., López Villarreal, M.J., Barrote, C.R. and Meira, M. (2003) The distribution of eriophyid mites and its presence on camellias in Galicia and Portugal. *Boletín de Sanidad Vegetal, Plagas* 29, 35–41.

Manson, D.C.M. (1963) Mites of the families Tetranychidae and Tenuipalpidae associated with citrus in South East Asia. *Acarologia* 5, 351–364.

Manson, D.C.M. (1967a) The spider mite family Tetranychidae in New Zealand. I. The genus *Bryobia*. *Acarologia* 9, 76–123.

Manson, D.C.M. (1967b) The spider mite family Tetranychidae in New Zealand. II. The genus *Tetranychus*. *Acarologia* 9, 581–597.

Manson, D.C.M. (1967c) Mites of the families Tenuipalpidae and Tetranychidae intercepted entering New Zealand from overseas. *New Zealand Journal of Science* 10, 664–674.

Manson, D.C.M. (1968) *Panonychus elongatus* Manson (Acarina: Tetranychidae) – a description and comparison with *P. citri* (McG.). *Journal of the Australian Entomological Society* 7, 6–10.

Manson, D.C.M. (1970) Wheat curl mite on garlic. *New Zealand Journal of Agriculture* 131, 61–62.

Manson, D.C.M. (1972) A contribution to the study of the genus *Rhizoglyphus* Claparède (Acarina: Acaridae). *Acarologia* 13, 621–650.

Manson, D.C.M. (1983) List of new mite/host plant records from New Zealand. *New Zealand Entomologist* 7, 41–403.

Manson, D.C.M. (1984a) *Eriophyoidea except Eriophyinae (Arachnida: Acari).* Fauna of New Zealand, No. 4, Department of Science and Industrial Research, New Zealand, Wellington.

Manson, D.C.M. (1984b) *Eriophyinae (Arachnida: Acari: Eriophyoidea).* Fauna of New Zealand, No. 5, Department of Science and Industrial Research, New Zealand, Wellington.

Manson, D.C.M. (1989) New species and records of eriophyid mites from New Zealand. *New Zealand Journal of Zoology* 16, 37–49.

Manson, D.C.M. and Gerson, U. (1996) Web spinning, wax secretion and liquid secretion by eriophyoid mites. In: Lindquist, E.E., Sabelis, M.W. and Bruin, J. (eds) *Eriophyoid Mites: Their Biology, Natural Enemies and Control.* World Crop Pests, Volume 6. Elsevier Science, Amsterdam, pp. 251–258.

Manson, D.C.M. and Oldfield, G.N. (1996) Life forms, deuterogyny, diapause and seasonal development. In: Lindquist, E.E., Sabelis, M.W. and Bruin, J. (eds) *Eriophyoid Mites: Their Biology, Natural Enemies and Control.* World Crop Pests, Volume 6. Elsevier Science, Amsterdam, pp. 173–183.

Mansour, F. and Whitcomb, W.H. (1986) The spiders of a citrus grove in Israel and their role as biological agents of *Ceroplastes floridensis* (Homoptera: Coccidae). *Entomophaga* 31, 269–276.

Mansour, F., Cohen, H. and Shain, Z. (1993) Integrated mite management in apples in Israel: augmentation of a beneficial mite and sensitivity of tetranychid and phytoseiid mites to pesticides. *Phytoparasitica* 21, 39–51.

Mansour, F., Bernstein, E. and Abo-Moch, F. (1995) The potential of spiders of different taxa and a predacious mite to feed on the carmine spider mite, a laboratory study. *Phytoparasitica* 23, 217–220.

Mansour, F.A., Ascher, K.R.S. and Abo-Moch, F. (1997) Effects of Neemgard on phytophagous and predacious mites and on spiders. *Phytoparasitica* 25, 333–336.

Manton, S.M. (1977) *The Arthropoda: Habits, Functional Morphology and Evolution*. Clarendon Press, Oxford, UK.

Manzur, B.E. (1989) Plan de muestreo secuencial para ácaro del tostado *Phyllocoptruta oleivora* (Ashm.) (Acari: Eriophyidae), en citricos. *CIRPON – Revista de Investigación* 7, 55–63.

Maoz, Y., Palevsky, E., Gal, S., Zilberstein, M., Noy, M., Izhar, Y., Gan-Mor, J.A.S. and Coll, M. (2009) Integrated pest management of *Oligonychus perseae*: developing action thresholds and the identification and conservation of natural enemies. *IOBC-WPRS Bulletin* 50, 57–60.

Maoz, Y., Gal, S., Zilberstein, M., Izhar, Y., Alchanatis, V., Coll, M. and Palevsky, E. (2011a) Determining an economic injury level for the persea mite, *Oligonychus perseae*, a new pest of avocado in Israel. *Entomologia Experimentalis et Applicata* 138, 110–116.

Maoz, Y., Gal, S., Argov, Y., Coll, M. and Palevsky, E. (2011b) Biocontrol of persea mite, *Oligonychus perseae*, with an exotic spider mite predator and an indigenous pollen feeder. *Biological Control* 59, 147–157.

Marafeli, P.P., Pedro Neto, M., Reis, P.R. and Souza-Pimentel, G.C. (2010) Acaricidal effect of plant extracts on the coffee pest mite *Oligonychus ilicis*. In: Moraes, G.J. de, Castilho, R.C. and Flechtmann, C.H.W. (eds) *XIII International Congress of Acarology, August 23–27, 2010, Recife-PE, Brazil, Abstract Book*, abstract 262, p. 142. Available at: http://www.acarology.org/ica/ica2010/docs/0910_15_abstractBookCapa.pdf (accessed 3 February 2015).

Marasas, W.F.O., Ploetz, R.C., Wingfield, M.J., Wingfield, B.D. and Steenkamp, E.T. (2006) Mango malformation disease and the associated *Fusarium* species. *Phytopathology* 96, 667–672.

March, R.B. (1958) The chemistry and action of acaricides. *Annual Review of Entomology* 3, 355–376.

Marcic, D., Mutavdzic, S., Medjo, I., Prijovic, M. and Peric, P. (2010) Field and greenhouse evaluation of spirodiclofen (Envidor) against spider mites (Acari: Tetranychidae) in Serbia. In: Moraes, G.J. de, Castilho, R.C. and Flechtmann, C.H.W. (eds) *XIII International Congress of Acarology, August 23–27, 2010, Recife-PE, Brazil, Abstract Book*, abstract 263, pp. 142–143. Available at: http://www.acarology.org/ica/ica2010/docs/0910_15_abstractBookCapa.pdf (accessed 3 February 2015).

Marcone, C., Ragozzino, A., Schneider, B., Lauer, U., Smart, C.D. and Seemüller, E. (1996) Genetic characterization and classification of two phytoplasmas associated with Spartium witches' broom disease. *Plant Disease* 80, 365–371.

Margolies, D.C. and Kennedy, G.G. (1985) Movement of the two spotted spider mite *Tetranychus urticae* Koch (Acari: Tetranychidae), among hosts in a corn–peanut agroecosystem. *Entomologia Experimentalis et Applicata* 37, 55–61.

Mariau, D. (1969) *Aceria guerreronis* Keifer: récent ravageur de la cocoteraie Dahoméenne. *Oléagineux* 24, 269–272.

Mariau, D. (1977) *Aceria* (*Eriophyes*) *guerreronis*: un important ravageur des cocoteraies africaines et americaines. *Oléagineux* 32, 101–111.

Mariau, D. (1986) Comportement de *Eriophyes guerreronis* Keifer à l"egard de différentes variétés de cocotiers. *Oléagineux* 41, 499-505.

Mariau, D. and Genty, P. (1987) Utilisation du soufre dans la lutte contre *Retracrus elaeis* Keifer acarien foliaire du palmier à huile en Colombie. *Compte-rendus du Symposium International du Soufre Élémentaire en Agriculture, 25–27 Mars 1987, Nice, Volume 1*. Syndicat Francais du Soufre, Marseille, France, pp. 281–282.

Mariau, D. and Julia, J.F. (1970) L'acariose à *Aceria guerreronis* (Keifer), ravageur du cocotier. *Oléagineux* 25, 459–464.

Markin, G.P., Yoshioka, E.R. and Conant, P. (1996) Biological control of gorse in Hawaii. In: Moran, V.C. and Hoffmann, J.H. (eds) *Proceedings of the IX International Symposium on Biological Control of Weeds, 19–26 January 1996, University of Cape Town, Stellenbosch, South Africa*, pp. 371–375. Available at: http://www.invasive.org/proceedings/pdfs/9_371-375.pdf (accessed 12 May 2015).

Markkula, M. and Tiittanen, K. (1969) Effect of fertilizers on the reproduction of *Tetranychus telarius* (L.), *Myzus persicae* (Sulz.) and *Acyrthosiphon pisum* Harris. *Annales Agriculturae Fenniae* 8, 9-14.

Markkula, M. and Tiittanen, K. (1976) 'Pest in first' and 'natural infestation' methods in the control of *Tetranychus urticae* Koch with *Phytoseiulus persimilis* Athias-Henriot on glasshouse cuccumbers. *Acta Agriculturae Fenniae* 15, 81-85.

Marquez, A.L., Wong, E., Garcia, E. and Olivero, J. (2006) Efficacy assay of different phytosanitary chemicals for the control of *Eutetranychus orientalis* (Klein) (Oriental Spider Mite) on Fine lemon and Valencia-Late orange crops. *IOBC-WPRS Bulletin* 29(3), 305–310.

Marriott, J., Florentine, S. and Raman, A. (2013) Effects of *Tetranychus lintearius* (Acari: Tetranychidae) on the structure and water potential in the foliage of the invasive *Ulex europaeus* (Fabaceae) in Australia. *International Journal of Acarology* 39, 275–284.

Marsaro, A.L. Jr, Sato, M.E., Mineiro, J.L.C., Návia, D., Aguiar, R.M. and Vieira, G.B. (2009) Ácaros predadores associados ao ácaro hindú dos citros, *Schizotetranychus hindustanicus* (Hirst, 1924) (Acari: Tetranychidae), no estado de Roraima, Brasil. In: *Resumos do XI Simpósio de Controle Biológico, SICONBIOL 2009, 1 a 5 de Junho, Bento Gonçalves-RS*. Sociedade Entomologica do Brasil, Londrina, Brazil, p. 28, abstract 09/093. Available at: http://www.seb.org.br/eventos/SICONBIOL/XISICONBIOL/09%E2%80%93PREDADORES.pdf (accessed 14 May 2015).

Marsaro, A.L. Jr, Návia, D., Sato, M.E., Silva, R.J. Jr, Moreira, G.A.M. and Duarte, O.R. (2010) Host plant range of the citrus hindú mite, *Schizotetranychus hindustanicus* (Hirst) (Tetranychidae), in Brazil. In: Moraes, G.J. de, Castilho, R.C. and Flechtmann, C.H.W. (eds) *XIII International Congress of Acarology, August 23–27, 2010, Recife-PE, Brazil, Abstract Book*, abstract 266, pp. 144–145. Available at: http://www.acarology.org/ica/ica2010/docs/0910_15_abstractBookCapa.pdf (accessed 3 February 2015).

Marshall, V.G. and Lindquist, E.E. (1972) Notes on the genus *Nalepella* (Acarina: Eriophyidae) and the occurrence of *N. halourga* on black spruce in Canada. *Canadian Entomologist* 104, 239–244.

Martelli, G.M. (1964) L'eriofide rugginoso degli agrumi (*Aculus pelekassi* K.). *Italia Agricola* 101, 1241–1246.

Martelli, G.M. and Di Martino, E. (1963) Eriofide del mandarino. *Informatore Fitopatologico* 13, 355.

Martí Fabregat, F. and del Rivero, J.M. (1966) Nuevos ensayos de lucha contra el 'badoc' del avellano. *Boletín de Patología Vegetal y Entomología Agrícola* 29, 257–259.

Martín, B., Hernández, S., Silvarrey, C., Jacas, J.A. and Cabaleiro, C. (2005) Vegetable, fish and mineral oils control grapevine powdery mildew. *Phytopathologia Mediterranea* 44, 169–179.

Martin, H. (1958) Ravageurs and maladies du palmier-dattier en Libye. *FAO Plant Protection Bulletin/FAO Bulletin Phytosanitaire* 6(8), 120–123.

Martin, H. (1959) Ravageurs and maladies du palmier-dattier en Libye. Paper presented at: La *Première Reunion Technique Internationale sur la Production et le Traitement des Dattes, Tripoli, Libye, 5–11 Decembre 1959* (Report published by Food and Agriculture Organization of the United Nations, Rome, in 1960).

Martin, N.A. (1978) *Siteroptes* (*Siteroptoides*) species with *Pediculaster*-like phoretomorphs (Acari: Tarsonemida: Pygmephoridae) from New Zealand and Polynesia. *New Zealand Journal of Zoology* 5, 121–155.

Martin, N.A. and Workman, P. (1985) Pest control in boxes of *Cymbidium* orchid flowers with dichlorvos-impregnated plastic. In: Hartley, M.J. (ed.) *Proceedings of the NZ Weed and Pest Control Conference 38 (1985)*. New Zealand Weed and Pest Control Society, Henderson, Auckland New Zealand, pp. 169–171. Available at: http://www.nzpps.org/journal/38/nzpp_381690.pdf (14 May 2015).

Martin, T.J., Harvey, T.L., Bender, C.G. and Seifers, D.L. (1984) Control of wheat streak mosaic virus with vector resistance in wheat. *Phytopathology* 74, 963–964.

Martínez, A.Z., de la Torre, P.E. and García, S.A. (2004) Principales ácaros detectados en la provincia de Villa Clara. *Fitosanidad* 8(1), 3–17.

Martinez, W., Wilson, K., Hernández, K. and Lopez, Y. (2013) Don't Let Spider Mites Get the Best of your Silage. In: *Forage Seminars, 12 February 2013, Tulare.* World Ag Expo, Tulare, California.

Martínez-Ferrer, M.T., Jacas-Miret, J.A., Ripollés-Moles, J.L. and Aucejo-Romero, S. (2006) Sampling plans for *Tetranychus urticae* (Acari: Tetranychidae) for IPM decisions on clementines in Spain. *IOBC-WPRS Bulletin* 29(3), 303.

Martínez-Villar, E., Sáenz-de-Cabezón, F., Moreno-Grijalba, F., Marco, V. and Pérez-Moreno, I. (2005) Effects of azadirachtin on the two-spotted spider mite, *Tetranychus urticae* (Acari: Tetranychidae). *Experimental and Applied Acarology* 35, 215–222.

Masaki, M. (1991) A list of Acarina intercepted in plant quarantine. *Research Bulletin of the Plant Protection Service Japan* 27, 87–92.

Maslennikov, A.A. (1979) Protection of tea gardens. *Zashchita Rastenii* No. 1, 32–34 (in Russian).

Masoud Arbabi, Payman Namvar, Samed Karmi and Majied Farokhi (2001) First damage record of *Polyphagotarsonemus latus* (Banks., 1904) (Acarina: Tarsonemidae) on potato cultivated in Jhiroft of Iran. *Applied Entomology and Phytopathology* 69(1), 41–42.

Massee, A.M. (1928) The life history of the black currant gall mite, *Eriophyes ribis* (Westw.) Nal. *Bulletin of Entomological Research* 18, 297–309.

Massee, A.M. (1930a) On some gall-mites (Eriophyidae) found on *Corylus avellana* L. *Bulletin of Entomological Research* 21, 165–168.

Massee, A.M. (1930b) The tarsonemid mite of strawberry. *Journal of Pomology and Horticultural Science* 8, 305–308.

Massee, A.M. (1932) Some injurious and beneficial mites on top and soft fruits. *Journal of Pomology and Horticultural Science* 10, 106–129.

Massee, A.M. (1933a) *Tarsonemus approximatus* Banks var. *narcissi* Ewing, a variety of tarsonemids new to the British List. *The Annals and Magazine of Natural History (Series 10)* 101(11), 198–201.

Massee, A.M. (1933b) Further observations on the strawberry tarsonemid mite (*Tarsonemus fragariae* Zimm.). In: *Annual Report 1932, East Malling Research Station*. East Malling, UK, pp. 117–131.

Massee, A.M. (1939) A species of gall mite (Eriophyidae) injurious to tomato. *The Annals and Magazine of Natural History (Series 2)* 3, 617–619.

Massee, A.M. (1954) *The Pests of Fruits and Hops*. Crosby Lockwood, London.

Masui, S. and Ikeda, M. (2003) Regional differences in occurrence of the citrus red mite (*Panonychus citri*) and their predacious insects in Satsuma mandarin orchards in Shizuoka prefecture. *Proceedings of the Kansai Plant Protection Society* 45, 11–16 (in Japanese).

Mathew, J. (1973) Timing of measures for budmite control to reduce blossom malformation. *Fruit World and Market Grower* 74(11), 32–34.

Mathez, F. (1965) Contribution à l'étude morphologique et biologique d'*Eriophyes vitis* Pgst., agent de l'erinose de la vigne. *Bulletin de la Société Entomologique Suisse / Mitteilungen der Schweizerischen Entomologischen Gesellschaft* 37, 233–283.

Mathur, A.C. and Tandon, P.L. (1974) Litchi mite can be controlled. *Indian Horticulture* 19, 11–12.

Mathur, L.M.L. (1969) Bionomics of *Stethorus gilvifrons* Mulsant (Coleoptera: Coccinellidae). *Madras Agricultural Journal* 56, 7–11.

Mathurin, G., George, G. and Lauckner, B. (2010) Field trials to determine the efficacy of three chemicals in the control of the red palm mite (*Raoiella indica* Hirst) on coconut (*Cocos nucifera*) in Saint Lucia. *CARDI Review (Caribbean Agricultural Research and Development Institute)* 10, 12–19.

Mathys, G. (1957a) Étude des possibilitiés d'intervention contre *Phyllocoptes vitis* Nal. agent du court-noué parasitaire de la vigne. *Revue Romande d'Agriculture de Viticulture et d'Arboriculture* (now *Agriculture Romande*) 13, 95–97.

Mathys, G. (1957b) Contribution à la connaissance de la systématique et de la biologie du genre *Bryobia* en Suisse romande. *Bulletin de la Société Entomologique Suisse / Mitteilungen der Schweizerischen Entomologischen Gesellschaft* 30, 189–204.

Mathys, G. (1959) L'acariose ou court-noué parasitaire de la vigne. *Revue Romande d'Agriculture de Viticulture et d'Arboriculture* (now *Agriculture Romande*) 15, 21–23.

Mathys, G. and Hugi, H. (1961) L'érinose de la vigne (*Eriophyes vitis* Pgst.). *Revue Romande d'Agriculture de Viticulture et d'Arboriculture* (now *Agriculture Romande*) 17, 29–30.

Mathys, G. and Tencalla, Y. (1959) L'identification par les praticiens des principaux acariens nuisibles aux arbres fruitiers et à la vigne. *Revue Romande d'Agriculture de Viticulture et d'Arboriculture* (now *Agriculture Romande*) 15, 30–32.

Matioli, A.L. and Oliveira, C.A.L. de (2007) Biology of *Agistemus brasiliensis* Matioli, Ueckermann & Oliveira (Acari: Stigmaeidae) and its predation potential on *Brevipalpus phoenicis* (Geijskes) (Acari: Tenuipalpidae). *Neotropical Entomology* 36, 577–582.

Matioli, A.L., Leite, G.L.D., Pallini Filho, A. and Picanco, M. (1998) Distribuição espacial e temporal e efeito de diferentes tratos culturais em ácaros associados a laranja Pêra-Rio. *Agro-Ciencia* 14, 395–405.

Matons, A. (1920) Els Badochs del Avellaner. *Agricultura: Revista Agricola Catalana* 4, 279–281.

Matsuda, K., Shimizu, N., Nakamura, S., Moriyama, H. and Sakaida, T. (2001) Seasonal occurrence and control of tea rust mite, *Acaphylla theavagrans* Kadono. *Bulletin of the Fukuoka Agricultural Research Center* 20, 27–30 (in Japanese).

Matsuda, T., Hinomoto, N., Singh, R.N. and Gotoh, T. (2012) Molecular-based identification and phylogeny of *Oligonychus* species (Acari: Tetranychidae). *Journal of Economic Entomology* 105, 1043–1050.

Matsuda, T., Fukumoto, C., Hinomoto, N. and Gotoh, T. (2013) DNA-based identification of spider mites: molecular evidence for cryptic species of the genus *Tetranychus* (Acari: Tetranychidae). *Journal of Economic Entomology* 106, 463–472.

Matsumura, F. and Voss, G. (1964) Mechanism of malathion and parathion resistance in the two-spotted spider mite, *Tetranychus urticae*. *Journal of Economic Entomology* 57, 911–917.

Matsumura, M., Nakano, T., Ono, D. and Fukui, T. (2005) Control of acarid mites *Tyrophagus similis* Volgin by soil sterilization. *Annual Report of the Kansai Plant Protection Society* 47, 1–8.

Matsushima, R., Ozawa, R., Uefune, M., Gotoh, T. and Takabayashi, J. (2006) Intraspecies variation in the Kanzawa spider mite differentially affects induced defensive response in lima bean plants. *Journal of Chemical Ecology* 22, 2501–2512.

Matthysse, J.G. (1980) Preliminary report on the mites collected from plants and animals in Nigeria. (I Mites from plants). *Nigerian Journal of Entomology* 1(3), 57–70.

Matthysse, J.G. and Naegele, J.A. (1952) Spruce mite and southern red mite control experiments. *Journal of Economic Entomology* 45, 383–387.

May, P. and Webster, W.J. (1958) The bud strain of *Eriophyes vitis* (Pgst.) in Australia. *Journal of the Australian Institute of Agricultural Science* 24, 163–165.

Maziarz, B. (1984) Opanowanie leszczyny przez wielkopakowca leszczynowego – *Phytoptus avellanae* (Nal.) (Acarina, Phytoptidae). *Polskie Pismo Entomologiczne* 53, 649–655.

Mazid, S., Rajkhowa, R.C. and Kalita, J.C. (2013) Effect of temperature on duration of developmental stages of red spider mite (*Oligonychus Coffeae* Nietner), a serious pest of tea. *Global Research Analysis* 2, 245–246.

McBride, D.K. (1977) *Insect Pests of Evergreens*. Circular E-297, Cooperative Extension Service, North Dakota State University, Fargo, North Dakota.

McClay, A.S., Littlefield, J.L. and Kashefi, J. (1999) Establishment of *Aceria malherbae* (Acari: Eriophyidae) as a biological control agent for field bindweed (Convolvulaceae) in the northern Great Plains. *Canadian Entomologist* 131, 541–547.

McCoy, C.W. (1976) Leaf injury and defoliation caused by the citrus rust mite, *Phyllocoptruta oleivora*. *Florida Entomologist* 59, 403–410.

McCoy, C.W. (1977a) Horticultural practices affecting phytophagous mite populations on citrus. In: Grierson, W. (ed.) *Proceedings of the 2nd International Citrus Congress, Orlando, Florida, 1977, Volume 2*. International Society of Citriculture, University of California, Riverside, pp. 459-462.

McCoy, C.W. (1977b) Resurgence of citrus rust mite populations following application of methidathion. *Journal of Economic Entomology* 70, 748–752.

McCoy, C.W. (1979) Migration and development of citrus rust mite on the spring flush of Valencia orange. *Proceedings of the Florida State Horticultural Society* 92, 48–51.

McCoy, C.W. (1981) Fungi: pest control by *Hirsutella thompsonii*. In: Burges, H.D. (ed.) *Microbial Control of Pests and Plant Diseases 1970–1980*. Academic Press, London, pp. 499–512.

McCoy, C.W. (1985) Citrus: current status of biological control in Florida. In: Hoy, M.A. and Herzog, D.C. (eds) *Biological Control in Agricultural IPM Systems*. Academic Press, Orlando, Florida, pp. 481–490.

McCoy, C.W. (1996a) Pathogens of eriophyoid mites. In: Lindquist, E.E., Sabelis, M.W. and Bruin, J. (eds) *Eriophyoid Mites: Their Biology, Natural Enemies and Control*. World Crop Pests, Volume 6. Elsevier Science, Amsterdam, pp. 481–490.

McCoy, C.W. (1996b) Stylar feeding injury and control of eriophyoid mites in citrus. In: Lindquist, E.E., Sabelis, M.W. and Bruin, J. (eds) *Eriophyoid Mites: Their Biology, Natural Enemies and Control*. World Crop Pests, Volume 6. Elsevier Science, Amsterdam, pp. 513–526.

McCoy, C.W. and Albrigo, L.G. (1975) Feeling injury to the orange caused by the citrus rust mite, *Phyllocoptruta oleivora* (Prostigmata: Eriophyoidea). *Annals of the Entomological Society of America* 68, 289–297.

McCoy, C.W. and Couch, T.L. (1982) Microbial control of the citrus rust mite with the mycoacaricide, Mycar. *Florida Entomologist* 65, 117–126.

McCoy, C.W. and Kanavel, R.F. (1969) Isolation of *Hirsutella thompsonii* from the citrus rust mite, *Phyllocoptruta oleivora*, and its cultivation on various synthetic media. *Journal of Invertebrate Pathology* 14, 386–390.

McCoy, C.W. and Rakha, M.A. (1985) *Euseius mesembrinus* (Acari: Phytoseiidae): a new record on Florida citrus and literature correction. *Florida Entomologist* 68, 347–348.

McCoy, C.W. and Selhime, A.G. (1974) The fungus pathogen *Hirsutella thompsonii* and its potential use for the control of the citrus rust mite in Florida. Carpena, O. (ed.) *Proceedings of the 1st International Citrus Congress, Murcia and Valencia, Spain, 1973, Volume 2*. International Society of Citriculture, University of California, Riverside, pp. 521–527.

McCoy, C.W., Selhime, A.G., Kanavel, R.F. and Hill, A.J. (1971) Suppression of citrus rust mite population with application of fragmented mycelia of *Hirsutella thompsonii*. *Journal of Invertebrate Pathology* 17, 270–276.

McCoy, C.W., Hill, A.J. and Karnavel, R.F. (1975) Large-scale production of the fungus pathogen *Hirsutella thompsonii* in submerged culture and its formulation for application in the field. *Entomophaga* 20, 229–240.

McCoy, C.W., Davis, P.L. and Munroe, K.A. (1976a) Effect of late season fruit injury by the citrus rust mite, *Phyllocoptruta oleivora* (Prostigmata: Eriophyoidea), on the internal quality of Valencia orange. *Florida Entomologist* 59, 335–342.

McCoy, C.W., Brooks, R.F., Allen, J.C., Selhime, A.G. and Wardowski, W.F. (1976b) Effect of reduced pest control programs on yield and quality of 'Valencia' orange. *Proceedings of the Florida State Horticultural Society* 89, 74–77.

McCoy, C.W., Bullock, R.C. and Dybas, R.A. (1982) Avermectin B1: a novel miticide active against citrus mites in Florida. *Proceedings of the Florida State Horticultural Society* 95, 51–56.

McCoy, C.W., Samson, R.A. and Boucias, D.G. (1988) Entomogenous fungi. In: Ignoffo, C.M. (ed.) *Handbook of Natural Pesticides, Volume 5, Microbial Insecticides, Part A, Entomogenous Protozoa and Fungi*. CRC Press, Boca Raton, Florida, pp. 151–236.

McCoy, C.W., Lye, B.H. and Salami, M. (1989) Spray volume and acaricide rate effects on the control of the citrus rust mite. *Proceedings of the Florida State Horticultural Society* 102, 36–40.

McCoy, C.W., Lye, B.H. and Salyani, M. (1990) Effect of spray volume on the residual control of citrus rust mite. *Citrus and Vegetable Magazine* 53(5), 23, 64–65.

McCoy, C.W., Samson, R.A., Boucias, D.G., Osborne, L.S., Peña, J.E. and Buss, L.J. (2009) *Pathogens Infecting Insects and Mites of Citrus*. LLC Friends of Microbes, Winter Park, Florida.

McDaniel, B., Morihara, D. K. and Lewis, J.K. (1975) A new species of *Tuckerella* from South Dakota and a key with illustrations of all known described species. *Acarologia* 17, 274–283.

McDonald, G., Moritz, K., Merton, E. and Hoffmann, A.A. (1995) The biology and behaviour of redlegged earth mite and blue oat mite on crop plants. *Plant Protection Quarterly* 10, 52–55.

McEnroe, W.D. (1961) The control of water loss by the two-spotted spider mite *Tetranychus telarius* (L.). *Annals of the Entomological Society of America* 54, 883–887.

McEnroe, W.D. (1963) The role of the digestive system in the water balance of the two-spotted spider mite. In: Naegele, J.A. (ed.) *Advances in Acarology, Volume 1*. Cornell University Press/Comstock Publishing Associates, Ithaca, New York, pp. 225–231.

McEnroe, W.D. (1969) Spreading and inbreeding in the spider mite. *Journal of Heredity* 60, 343–345.

McEnroe, W.D. and Dronka, K. (1971) Photobehavioral classes of the spider mite, *Tetranychus urticae*. *Entomologia Experimentalis et Applicata* 14, 419–424.

McEnroe, W.D. and Naegele, J.A. (1968) The coadaptive process in an organophosphorous-resistant strain of the two-spotted spider mite, *Tetranychus urticae*. *Annals of the Entomological Society of America* 61, 1055-1059.

McEvoy, P.B. (1996) Host specificity and biological pest control. *Bioscience* 46, 401–405.

McFadyen, R.E.C. (1998) Biological control of weeds. *Annual Review of Entomology* 43, 369–393.

McGregor, E.A. (1913) *The Red Spider on Cotton*. Bureau of Entomology Circular No. 172, US Department of Agriculture, Washington, DC

McGregor, E.A. (1914) Four new tetranychids. *Annals of the Entomological Society of America* 7, 354–364.

McGregor, E.A. (1917) Description of seven new species of red spiders. *Proceedings of the United States National Museum* 51, 581–590.

McGregor, E.A. (1919) The red spiders of America and a few European species likely to be introduced. *Proceedings of the United States National Museum* 56, 641–679.

McGregor, E.A. (1930) A new spinning mite attacking *Asparagus plumosus* in Florida. *Proceedings of the Entomological Society of Washington* 32, 161–163.

McGregor, E.A. (1931) A new spinning mite attacking raspberry in Michigan. *Proceedings of the Entomological Society of Washington* 33, 193–195.
McGregor, E.A. (1934) A new spinning mite on citrus at Yuma, Arizona. *Proceedings of the Entomological Society of Washington* 36, 256–259.
McGregor, E.A. (1936) Two spinning mites attacking incense cedar in California. *Annals of the Entomological Society of America* 29, 770–775.
McGregor, E.A. (1939) The specific identity of the American date mite; description of two new species of *Paratetranychus*. *Proceedings of the Entomological Society of Washington* 41, 247–256.
McGregor, E.A. (1941a) A new spinning mite attacking strawberry on the mid-Atlantic coast. *Proceedings of the Entomological Society of Washington* 43, 26–28.
McGregor, E.A. (1941b) The avocado mite of California, a new species. *Proceedings of the Entomological Society of Washington* 43, 85–88.
McGregor, E.A. (1941c) A new spider mite from Virginia (Acarina: Tetranychidae). *Proceedings of the Entomological Society of Washington* 43, 223–225.
McGregor, E.A. (1943a) A new spider mite on citrus in south-western California (Acarina: Tetranychidae). *Proceedings of the Entomological Society of Washington* 45, 127–129.
McGregor, E.A. (1943b) A new spider mite from Argentina. *Proceedings of the Entomological Society of Washington* 45, 176–178.
McGregor, E.A. (1949) Nearctic mites of the family Pseudoleptidae. *Memoirs of the Southern California Academy of Sciences* 3(2), 1–45.
McGregor, E.A. (1950) Mites of the family Tetranychidae. *American Midland Naturalist* 44, 257–420.
McGregor, E.A. (1954) Two new mites in the genus *Typhlodromus* (Acarina: Phytoseiidae). *Bulletin of the Southern California Academy of Sciences* 53, 89–92.
McGregor, E.A. (1955) Notes on the spider mites of Equador. *Revista Ecuatoriana de Entomologia y Parasitologia* 2, 365–377.
McGregor, E.A. (1956) Mites on citrus trees in southern California. *Memoirs of the Southern California Academy of Sciences* 3(3), 15–42.
McGregor, E.A. and McDonough, F.L. (1917) *The Red Spider Mite on Cotton*. US Department of Agriculture Bulletin No. 416, Washington, DC.
McIntyre, P.J. and Whitham, T.G. (2003) Plant genotype affects long-term herbivore population dynamics and extinction: conservation implications. *Ecology* 84, 311–322.
McKenzie, H.L. (1935) *Biology and Control of Avocado Insects and Mites*. Bulletin 592, Agricultural Experiment Station, College of Agriculture, University of California, Berkeley, California.
McLarty, H.R., Lott, T.B., Milbrath, J.A., Reeves, E.L. and Zeller, S.M. (1951) Mottle leaf. Virus diseases and other disorders with viruslike symptoms of stone-fruits in North America. In: *Agriculture Handbook No. 10*. US Department of Agriculture, Washington, DC, pp.106–111.
McMurtry, J.A. (1969) Biological control of citrus red mite in California. In: Chapman, H.D. (ed.) *Proceedings of the International Citrus Symposium, Riverside, California, March 16–26, 1968: in Celebration of the Centennial of the University of California, Volume 2*. University of California, Riverside, California, pp. 855–862.
McMurtry, J.A. (1970) Some factors of foliage condition limiting population growth of *Oligonychus punicae* (Acarina: Tetranychidae). *Annals of the Entomological Society of America* 63, 406–412.
McMurtry, J.A. (1977a) Biological control of citrus mites. In: Grierson, W. (ed.) *Proceedings of the 2nd International Citrus Congress, Orlando, Florida, 1977, Volume 2*. International Society of Citriculture, University of California, Riverside, pp. 456–459.
McMurtry, J.A. (1977b) Some predaceous mites (Phytoseiidae) on citrus in the Mediterranean region. *Entomophaga* 22, 19–30.
McMurtry, J.A. (1977c) Description and biology of *Typhlodromus persianus*, n. sp., from Iran, with notes on *T. kettanchi* (Acari: Mesostigmata: Phytoseiidae). *Annals of the Entomological Society of America* 70, 563–568.
McMurtry, J.A. (1982) The use of phytoseiids for biological control: progress and future prospects. In: Hoy, M.A. (ed.) *Recent Advances in Knowledge of Phytoseiidae*. Special Publication No. 3284, Division of Agricultural Sciences, University of California, Berkeley, California, pp. 23–48.
McMurtry, J.A. (1985a) Avocado. In: Helle, W. and Sabelis, M.W. (eds) *Spider Mites: Their Biology, Natural Enemies and Control*. World Crop Pests, Volume 1B. Elsevier Science, Amsterdam, pp.327–332.
McMurtry, J.A. (1985b) Citrus. In: Helle, W. and Sabelis, M.W. (eds) *Spider Mites: Their Biology, Natural Enemies and Control*. World Crop Pests, Volume 1B. Elsevier Science, Amsterdam, pp. 339–347.
McMurtry, J.A. (2010) Concepts of classification of the Phytoseiidae: relevance to biological control of mites. In: Sabelis, M.W. and Bruin, J. (eds) *Trends in Acarology: Proceedings of the 12th International Congress, 21–26 August 2006, Amsterdam*. Springer, Dordrecht, the Netherlands, pp. 393–398.
McMurtry, J.A. and Croft, B.A. (1997) Life styles of phytoseiid mites and their roles in biological control. *Annual Review of Entomology* 42, 291–321.
McMurtry, J.A. and Johnson, H.G. (1966) An ecological study of the spider mite, *Oligonychus punicae* (Hirst) and its natural enemies. *Hilgardia* 37(11), 363–402.
McMurtry, J.A. and Scriven, G.T. (1965) Insectary production of phytoseiid mites. *Journal of Economic Entomology* 58, 282–284.
McMurtry, J.A. and Scriven, G.T. (1975) Population increase of *Phytoseiulus persimilis* on different insectary feeding programs. *Journal of Economic Entomology* 68, 319–321.
McMurtry, J.A., Johnson, H.G. and Scriven, G.T. (1969) Experiments to determine the effects of mass releases of *Stethorus picipes* on the level of infestation of the avocado brown mite. *Journal of Economic Entomology* 62, 1216–1221.
McMurtry, J.A., Huffaker, C.B. and van de Vrie, M. (1970) Ecology of tetranychid mites and their natural enemies: a review – I. Tetranychid enemies: their biological characters and the impact of spray practices. *Hilgardia* 40(11), 331–390.
McMurtry, J.A., Scriven, G.T. and Malone, R.S. (1974) Factors affecting oviposition of *Stethorus picipes* (Coleoptera: Coccinellidae), with special reference to photoperiod. *Environmental Entomology* 3, 123–127.
McMurtry, J.A., Shaw, J.G. and Johnson, H.G. (1979) Citrus red mite populations in relation to virus disease and predaceous mites in southern California. *Environmental Entomology* 8, 160–164.
McMurtry, J.A., Johnson, H.G. and Badii, M.H. (1984) Experiments to determine effects of predator releases on populations of *Oligonychus punicae* (Acarina: Tetranychidae) on avocado. *Entomophaga* 29, 11–19.
McMurtry, J.A., Morse, J.G. and Johnson, H.G. (1992) Studies of the impact of *Euseius* species (Acari: Phytoseiidae) on citrus mites using predator exclusion and predator release experiments. *Experimental and Applied Acarology* 15, 233–248.
McMurtry, J.A., Moraes, G.J. de and Sourassou, N.F. (2013) Revision of the lifestyles of phytoseiid mites (Acari: Phytoseiidae) and implications for biological control strategies. *Systematic and Applied Acarology* 18, 297–320.
McNeill, S. and Southwood, T.R.E. (1978) The role of nitrogen in the development of insect plant relationship. In: Harbone, J.B. (ed.) *Biochemical Aspects of Plant and Animal Coevolution*. Academic Press, New York, pp. 77–98.
McSorely, R. and Walter, D.E. (1991) Comparison of soil extraction methods for nematodes and microarthropods. *Agriculture, Ecosystems and Environment* 34, 201–207.
Mead, F.A. (1993) Entomology section. *Triology* 32(4), 3.
Meagher, R.L. Jr and Meyer, J.R. (1990) Influence of ground cover and herbicide treatments on *Tetranychus urticae* populations in peach orchards. *Experimental and Applied Acarology* 9, 149–158.

Meats, A. (1971) The relative importance to population increase of fluctuations in mortality, fecundity and the time variables of the reproductive schedule. *Oecologia* 6, 223–237.

Medina Gaud, S. and Garcia Tuduri, J. (1977) New arthropod records for Puerto Rico. *Journal of Agriculture of the University of Puerto Rico* 61, 409–412.

Meenakshi, N. and Vadivel, E. (2004) Quality characters of tender and mature coconut as affected by eriophyid mite (*Eriophyes guerreronis*). *Progressive Horticulture* 36, 138–141.

Mégevand, B., Klay, A., Gnanvossou, D. and Paraiso, G. (1993) Maintenance and mass rearing of phytoseiid predators of the cassava green mite. *Experimental and Applied Acarology* 17, 115–128.

Mégnin, P. (1876) Memoire sur l'organisation et la distribution zoologique des acariens de la famille des Gamasides. *Journal de l'Anatomie et de la Physiologie Normales et Pathologiques de l'Homme et des Animaux* 12, 288–336.

Mehan, U.K. and Reddy, M.V. (1990) Diseases of pigeonpea and chickpea and source of resistance. Paper presented at: *In-country Training Course on Legumes Production, Sri Lanka, 9–17, July 1990.* ICRISAT (International Crops Research Institute), Patancheru, India.

Mehlenbacher, S.A. (1994) Genetic improvement of the hazelnut. *Acta Horticulturae* 351, 23–38.

Mehrnejad, M.R. and Daneshvar, H. (1991) Two new eriophyid mites of pistachio orchards in the Kerman and Yazd Provinces. *Entomologie et Phytopathologie Appliquées* (*Applied Entomology and Phytopathology*) 58(1, 2), 87–88.

Mehrnejad, M.R. and Ueckermann, E.A. (2001) Mites (Arthropoda, Acari) associated with pistachio trees (Anacardiaceae) in Iran (I). *Systematic and Applied Acarology Special Publications* No. 6, 1–12.

Mejia, O.I., Ramirez, L.M., Escalona, L.E., Messa, N.C., Duque, M.C. and Pantoja, A. (1998) Desarrollo de *Schizotetranychus oryzae* (Rossi de Simons) sobre arroz. *Journal of Agriculture of the University of Puerto Rico* 82, 183–188.

Mellot, J.L. (1968) *Panonychus caglei*, new species, the raspberry red mite (Acarina: Tetranychidae). *Acarologia* 10, 230–244.

Mellot, J.L. and Connell, W.A. (1965) Notes on the life history of *Tetranychus atlanticus*. *Annals of the Entomological Society of America* 58, 379–383.

Mendonça, R.S. de, Návia, D., Diniz, I.R., Auger, P. and Navajas, M. (2010) A critical review of some closely related species of *Tetranychus sensu stricto* Tuttle & Baker (1968) (Prostigmata: Tetranychidae) in the public DNA sequences database. In: Moraes, G.J. de, Castilho, R.C. and Flechtmann, C.H.W. (eds) *XIII International Congress of Acarology, August 23–27, 2010, Recife-PE, Brazil, Abstract Book*, pp. 153–154, abstract 284. Available at: http://www.acarology.org/ica/ica2010/docs/0910_15_abstractBookCapa.pdf (accessed 23 January 2015).

Mendonça, R.S., Návia, D., Diniz, I.R. and Flechtmann, C.H.W. (2011a) South American spider mites: New hosts and localities. *Journal of Insect Science* 11(1):121. Available at: http://jinsectscience.oxfordjournals.org/content/11/1/121 (accessed 23 March 2015).

Mendonça, R.S., Návia, D., Diniz, I.R., Auger, P. and Navajas, M. (2011b) A critical review on some closely related species of *Tetranychus sensu stricto* (Acari: Tetranychidae) in the public DNA sequences databases. *Experimental and Applied Acarology* 55, 1–23.

Meng, H.-S., Wang, K.-Y., Jiang, X.-Y. and Yi, M.-Q. (2000) Selection of resistance to fenpropathrin in *Panonychus citri* and its cross resistance. *Plant Protection* 26(6), 7–9 (in Chinese).

Meng, H.-S., Wang, K.-Y., Jiang, X.-Y. and Yi, M.-Q. (2002) Studies on resistance selection by abamectin and fenpropathrin and activity change of detoxicant enzymes in *Panonychus citri*. *Acta Entomologica Sinica* 45, 58–62 (in Chinese).

Menon, M.G.R. and Ghai, S. (1968) Further records of the distribution of *Petrobia latens* (Mueller) (Acarina, Tetranychidae) a pest of wheat in India together with the description of a new species of predatory mites on the same. *Indian Journal of Entomology* 30, 77–79.

Merabet, H., Dutzmann, S., Haeuser-Hahn, I., Kniehase, U., Ohs, P., Stuebler, D., Stumpf, N., Bylemans, D. and Creemers, P. (2002) Euparen Reg. Multi (tolylfluanid), a broad spectrum fungicide for *Rubus* and *Ribes* fruit crops. *Acta Horticulturae* 585, 381–386.

Meredith, T. (2001) *Bamboo for Gardens*. Timber Press., Portland, Oregon.

Merriam, T.L., Burkart, S.E., von Maltzahn, C., Kyomura, N., Kohyama, Y. and Motojima, S. (1990) Field evaluation of AC 801,757 (MK-239) acaricide. In: *Proceedings of the Brighton Crop Protection Conference: Pests and Diseases – 1990, Volume 1.* BCPC Publications, Farnham (now Alton), UK, pp. 63–69.

Merton, E., McDonald, G. and Hoffmann, A. (1995) Cultural control of redlegged earth mite, blue oat mite and lucerne flea in canola. *Plant Protection Quarterly* 10, 65–66.

Mesa, N.C., Braun, A.R. and Belotti, A.C. (1990) Comparison of *Mononychellus progresivus* and *Tetranychus urticae* as prey for five species of phytoseiid mites. *Experimental and Applied Acarology* 9, 159–168.

Mesa, N.C., Ochoa, R., Welbourn, W.C., Evans, G.A. and Moraes, G.J. de (2009) *A Catalog of the Tenuipalpidae (Acari) of the World with a Key to the Genera. Zootaxa* 2098. Magnolia Press, Auckland, New Zealand.

Mesa Cobo, N.C. and Zuluaga Cardona, J.I. (1983) *Review of the Literature of Agricultural Mites in Colombia*. Facultad de Ciencia Agropecuaria, Universidad Nacional de Colombia, Palmira, Columbia.

Messelink, G.J. and van Holstein-Saj, R. (2006) Potential for biological control of the bulb scale mite (Acari: Tarsonemidae) by predatory mites in amaryllis. *Proceedings of the Netherlands Entomological Society Meeting* 17, 113–118.

Messelink, G.J. and van Holstein-Saj, R. (2007) Biological control of the bulb scale mite *Steneotarsonemus laticeps* (Acari: Tarsonemidae) with *Neoseiulus barkeri* (Acari: Phytoseiidae) in amaryllis. *IOBC-WPRS Bulletin* 30(5), 81–85.

Messiaen, C.M. (1971) Les acariens nuisibles aux solanees maraichères en Guadeloupe. *Nouvelles Maraichères et Vivrières de l'INRA aux Antilles* 2, 1–5.

Messing, R.H. and Croft, B.A. (1996) Pesticide resistance in eriophyoid mites, their competitors and predators. In: Lindquist, E.E., Sabelis, M.W. and Bruin, J. (eds) *Eriophyoid Mites: Their Biology, Natural Enemies and Control*. World Crop Pests, Volume 6. Elsevier Science, Amsterdam, pp. 689–694.

Meyer, M.K.P. and Ryke, P.A.J (1959a) Nine new species of the superfamily Erythraeoidea (Acarina: Trombidiformes) associated with plants in South Africa. *Acarologia* 1, 304–323.

Meyer, M.K.P. and Ryke, P.A.J. (1959b) A revision of the spider mites (Acarina: Tetranychidae) of South Africa, with descriptions of a new genus and new species. *Journal of the Entomological Society of South Africa* 22, 330–366.

Meyer, M.K.P. and Ryke, P.A.J. (1960) Mites of the superfamily Eupodoidea (Acarina: Prostigmata) associated with South African plants. *South African Journal of Agricultural Science* 3, 481–496.

Meyer Smith, M.K.P. (1965) South African Acarina I. Nine species of the subfamily Tetranychidae collected on wild plants. *Koedoe* 8, 82–94.

Meyer Smith, M.K.P. (1968) The grass stunt mite, *Aceria neocynodonis* Keifer in South Africa. *South African Journal of Agricultural Science* 11, 803–804.

Meyer Smith, M.K.P. (1970a) Two important immigrant spider mites new to South Africa (Acari: Tetranychidae) *Tetranychus urticae* Koch and *Oligonychus mangiferus* (Rahm. & Sapra). *Phytophylactica* 2, 147–148.

Meyer Smith, M.K.P. (1970b) South African Acari II. Check-list of mites in our Parks (Part 1). *Koedoe* 13, 29–35.

Meyer Smith, M.K.P. (1974) *A Revision of the Tetranychidae of Africa (Acari) with a Key to the Genera of the World*. Entomology Memoir 36, Department of Agriculture and Water Supply, Republic of South Africa. Government Printer, Pretoria.

Meyer Smith, M.K.P. (1979) *The Tenuipalpidae (Acari) of Africa with Keys to the World Fauna*. Entomology Memoir 50, Department of Agriculture and Water Supply, Republic of South Africa. Government Printer, Pretoria.

Meyer Smith, M.K.P. (1981a) *Mite Pests of Crops in Southern Africa*. Science Bulletin 397, Department of Agriculture and Fisheries, Republic of South Africa, Pretoria.

Meyer Smith, M.K.P. (1981b) South African Eriophyoidea (Acari): The genus *Aceria* Keifer, 1944. *Phytophylactica* 13, 117–126.

Meyer Smith, M.K.P. (1987) *African Tetranychidae (Acari: Prostigmata) – with Reference to the World Genera*. Entomology Memoir 69, Department of Agriculture and Water Supply, Republic of South Africa. Government Printer, Pretoria.

Meyer Smith, M.K.P. (1989a) African Eriophyoidea: on species of the subfamily Aberoptinae (Acari: Eriophyidae). *Phytophylactica* 21, 271–274.

Meyer Smith, M.K.P. (1989b) African Eriophyoidea: the genus *Abacarus* Keifer, 1966 (Acari: Eriophyidae). *Phytophylactica* 21, 421–423.

Meyer Smith, M.K.P. (1990a) African Eriophyoidea: on some related genera: *Neooxycenus* Abou-Awad, *Oxycenus* Keifer, *Neotegonotus* Newkirk & Keifer, *Tegonotus* Nalepa and *Tegolophus* Keifer (Acari: Eriophyidae). *Phytophylactica* 22, 371–386.

Meyer Smith, M.K.P. (1990b) On the African species of *Anthocoptes* Nalepa and *Metaculus* Keifer (Acari: Eriophyidae). *Journal of African Zoology* 104, 558–572.

Meyer Smith, M.K.P. (1992) African Eriophyoidea: on *Heterotergum* Keifer, 1955 and a new genus *Pyelotus* (Acari: Eriophyidae). *Phytophylactica* 24, 157–161.

Meyer Smith, M.K.P. (1996a) Ornamental flowering plants. In: Lindquist, E.E., Sabelis, M.W. and Bruin, J. (eds) *Eriophyoid Mites: Their Biology, Natural Enemies and Control*. World Crop Pests, Volume 6. Elsevier Science, Amsterdam, pp. 641–650.

Meyer Smith, M.K.P. (1996b) *Mite Pests and their Predators on Cultivated Plants in Southern Africa: Vegetables and Berries*. Handbook 6, ARC (Agricultural Research Council)-Plant Protection Research Institute, Pretoria.

Meyer Smith, M.K.P. (1998) Lowveld citrus mite – *Eutetranychus orientalis* (Klein). In: Bedford, E.C.G., van den Berg, M.A. and de Villiers, E.A. (eds) *Citrus Pests in the Republic of South Africa*, 2nd edn. ARC (Agricultural Research Council)-Institute for Tropical and Subtropical Crops, Nelspruit, South Africa, pp. 69–72.

Meyer Smith, M.K.P. and Bolland, H.R. (1984) Tetranychoid mites (Acari: Prostigmata) from Cameroun and a survey of their chromosome complements. *Phytophylactica* 16, 209–220.

Meyer Smith, M.K.P. and Craemer, C. (1999) Mites (Arachnida: Acari) as crop pests in southern Africa: an overview. *African Plant Protection* 5(1), 37–51.

Meyer Smith, M.K.P. and Gerson, U. (1981) Some false spider mites (Prostigmata: Tenuipalpidae) from Israel. *Israel Journal of Entomology* 15, 67–81.

Meyer Smith, M.K.P. and Loots, G.C. (1978) Acari. In: Werger, M.J.A. (ed.) *Biogeography and Ecology of Southern Africa*. W. Junk, The Hague, the Netherlands, pp. 703–718.

Meyer Smith, M.K.P. and Rodrigues, M.C. (1966) Acari associated with cotton in Southern Africa (with reference to other plants). *Garcia de Orta* 13(2), 1–33.

Meyer Smith, M.K.P. and Ryke, P.A.J. (1959) A revision of the spider mites (Acarina: Tetranychidae) of South Africa, with descriptions of a new genus and new species. *Journal of the Entomological Society of Southern Africa* 22, 330–366.

Meyer Smith, M.K.P. and Schwartz, A. (1998a) Citrus silver mite, *Polyphagotarsonemus latus* (Banks). In: Bedford, E.C.G., van den Berg, M.A. and de Villiers, E.A. (eds) *Citrus Pests in the Republic of South Africa* 2nd edn. ARC (Agricultural Research Council)-Institute for Tropical and Subtropical Crops, Nelspruit, South Africa, pp. 78–83.

Meyer Smith, M.K.P. and Schwartz, A. (1998b) Citrus red mite – *Panonychus citri* (McGregor). In: Bedford, E.C.G., van den Berg, M.A. and de Villiers, E.A. (eds) *Citrus Pests in the Republic of South Africa*, 2nd edn. ARC (Agricultural Research Council)-Institute for Tropical and Subtropical Crops, Nelspruit, South Africa, pp. 58–65.

Meyer Smith, M.K.P. and Schwartz, A. (1998c) Citrus flat mite – *Brevipalpus californicus* (Banks). In: Bedford, E.C.G., van den Berg, M.A. and de Villiers, E.A. (eds) *Citrus Pests in the Republic of South Africa*, 2nd edn. ARC (Agricultural Research Council)-Institute for Tropical and Subtropical Crops, Nelspruit, South Africa, pp. 72–74.

Meyer Smith, M.K.P. and Schwartz, A. (1998d) Ornamental flat mite – *Brevipalpus obovatus* Donnadieu. In: Bedford, E.C.G., van den Berg, M.A. and de Villiers, E.A. (eds) *Citrus Pests in the Republic of South Africa*, 2nd edn. ARC (Agricultural Research Council)-Institute for Tropical and Subtropical Crops, Nelspruit, South Africa, pp. 77–78.

Meyer Smith, M.K.P. and Schwartz, A. (1998e) Reddish-black flat mite – *Brevipalpus phoenicis* (Geijskes). In: Bedford, E.C.G., van den Berg, M.A. and de Villiers, E.A. (eds) *Citrus Pests in the Republic of South Africa*, 2nd edn. ARC (Agricultural Research Council)-Institute for Tropical and Subtropical Crops, Nelspruit, South Africa, pp. 74–77.

Meyer Smith, M.K.P. and Ueckermann, E.A. (1992) Four new *Aceria* spp. (Acari: Eriophyidae) from South Africa. *International Journal of Acarology* 18, 293–298.

Meyer Smith, M.K.P. and Ueckermann, E.A. (1997) A review of some species of the families Allochaetophoridae, Linotetranidae and Tuckerellidae (Acari: Tetranychoidae). *International Journal of Acarology* 23, 67–92.

Meyer Smith, M.K.P., Loots, G.C., Van Pletzen, R., Engelbrecht, C.M. and Walker, J.B. (1973) *Acari of the Ethiopian Region*. Entomology Memoir 29, Department of Agricultural Technical Services, Republic of South Africa. Government Printer, Pretoria.

Mezei, I. (1995) Adatok a *Quadrastichus (Myiomisa) sajói* (Szelényi) (Hymenoptera, Chalcidoidea) biológiájához (Data on the biology of *Quadrastichus* (*Myiomisa*) *sajoi* Szelenyi (Hymenoptera: Chalcidoidea)). *Növényvédelem* 31, 211–214 (in Hungarian).

Mezei, I. and Czepo, M. (1988) Damage and control of the bark mite, *Aceria phloecoptes*. *Növényvédelem* 24, 27–30 (in Hungarian).

Michael, A.D. (1884) *British Oribatidae, Vol. 1*. The Ray Society, London.

Michael, A.D. (1890) Report on diseased sugar-cane from Barbados, forwarded by Mr. John R. Bovell. *Bulletin of Miscellaneous Information, Royal Botanic Gardens Kew* 40, 85–86.

Michael, A.D. (1903) *British Tyroglyphidae*. II. The Ray Society, London.

Michael, P.J. (1995) Biological control of redlegged earth mite and lucerne flea by the predators *Anystis wallacei* and *Neomologus capillatus*. *Plant Protection Quarterly* 10, 55–57.

Michael, P.J., Dutch, M.E. and Pekin, C.J. (1991) A review of the predators of redlegged earth mite, blue oat mite and lucerne flea. *Plant Protection Quarterly* 6, 178–180.

Michael, P.J., Dutch, M.E. and Pekin, C.J. (1993) Biological control of redlegged earth mite, *Halotydeus destructor*, blue oat mite, *Penthaleus major*, and lucerne flea, *Sminthurus viridis*. In: Delfosse, E.S. (ed.) *Pests of Pastures: Weed, Invertebrate and Disease Pests of Australian Sheep Pastures. Proceedings of a Conference*. CSIRO Information Services, Melbourne, Australia, pp. 329–331.

Michalska, K. (1997) Reproductive behaviour of some species of eriophyoid mites. PhD thesis, Warsaw University of Life Sciences, Warsaw.

Michalska, K. (1999) Spermatophore deposition and guarding in the free-living eriophyid mite *Vasates robiniae* (Acari). *Behaviour* 136, 899–918.

Michalska, K. (2000) The influence of conspecific males on spermatophore deposition in the eriophyoid mite *Aculus fockeui*. *Experimental and Applied Acarology* 24, 905–911.

Michalska, K. (2005) Spermatophore deposition throughout the day by the plum rust mite, *Aculus fockeui*. *Experimental and Applied Acarology* 35, 111–116.

Michalska, K. and Boczek, J. (1991) Sexual behavior of males attracted to quiescent deutonymphs in the Eriophyoidea (Acari). In: Dusbabek, F. and Bukva, V. (eds) *Modern Acarology, Volume 2*. Academia, Prague and SPB Academic Publishing, The Hague, the Netherlands, pp. 549–553.

Michalska, K. and Mańkowski, D.R. (2006) Population sex ratio in three species of eriophyoid mites differing in degree of sex dissociation. *Biological Letters* 43, 197–207.

Michalska, K. and Shi, A. (2004) A first view on factors influencing spermatophore deposition by the eriophyoid mite *Cecidophyopsis hendersoni* (Keifer). *Phytophaga* 14, 141–148.

Michalska, K., Skoracka, A., Návia, D. and Amrine, J.W. (2010) Behavioural studies on eriophyoid mites: an overview. *Experimental and Applied Acarology* 51, 31–59.

Michelatti, G., Pinoggi, G., Schreiber, G. and Mozzone, G.C. (1994) Experience of rational control of the hazelnut eriophyoid (*Phytocoptella avellanae* Nal.) studied for 5 years in Piemonte. *Acta Horticulturae* 351, 575–581.

Michelbacher, A.E. (1959) Spider mites on walnut in Northern California. *Journal of Economic Entomology* 52, 936–939.

Micic, S., Hoffmann, A.A., Strickland, G., Weeks, A.R., Bellati, J., Henry, K., Nash, M.A. and Umina, P.A. (2008) Pests of germinating grain crops in southern Australia: an overview of their biology and management options. *Australian Journal of Experimental Agriculture* 48, 1560–1573.

Micinski, S., Boethel, D.J. and Boudreaux, H.B. (1979) Influence of temperature and photoperiod on development and oviposition of the pecan leaf scorch mite, *Eotetranychus hicoriae*. *Annals of the Entomological Society of America* 72, 649–654.

Micinski, S., Boethel, D.J. and Boudreaux, H.B. (1981) Life tables and intrinsic rates of increase of the pecan leaf scorch mite. *Journal of Economic Entomology* 74, 612–616.

Micke, W.E., Keser, D.E., Rizzi, A.D. and Carlson, C.V. (1966) Early harvesting of almonds. *California Agriculture* 20(12), 14–16.

Mico, V., Laborda, R., García Marí, F., Soto, T. and Costa Comelles, J. (1992) Distribución de las poblaciones del ácaro rojo *Panonychus citri* McGregor en agrios. *Boletín de Sanidad Vegetal, Plagas* 18, 45–55.

Miętkiewski, R. and Baazy, S. (2003) *Neozygites abacaridis* sp. nov. (Entomophthorales), a new pathogen of phytophagous mites (Acari, Eriophyidae). *Journal of Invertebrate Pathology* 83, 223–229.

Miętkiewski, R., Bałazy, S. and Van der Geest, L.P.S. (1993) Observations on a mycosis of spider mites (Acari: Tetranychidae) caused by *Neozygites floridana* in Poland. *Journal of Invertebrate Pathology* 61, 317–319.

Miętkiewski, R., Bałazy, S. and Tkaczuk, C. (2000) Mycopathogens of mites in Poland – a review. *Biocontrol Science and Technology* 10, 459–465.

Mifsud, D., Monfreda, R. and de Lillo, E. (2006) Eriophyoid mites (Acari: Prostigmata: Eriophyoidea) from the Maltese Islands (Central Mediterranean): new reports and a preliminary check-list. *Entomologica* 40, 117–138.

Migeon, A. (2003) *Eurytetranychus admes*: un nouvel acarien tétranyque sur les cyprès en France. *Phytoma – La Défense des Végétaux* 561, 30–31.

Migeon, A. (2005) Un nouvel acarien ravageur en France: *Tetranychus evansi* Baker et Pritchard. *Phytoma – La Défense des Végétaux* 579, 38–43.

Migeon, A. (2007) Acarien rouge de la tomate: nouvelles observations et perspectives. *PHM Revue Horticole* 488, 20–24.

Migeon, A. and Dorkeld, F. (2006–2013) Spider Mites Web: a comprehensive database for the Tetranychidae. Available at: http://www.montpellier.inra.fr/CBGP/spmweb (accessed 13 August 2014).

Migeon, A. and Flechtmann, C.H.W. (2004) First additions and corrections to the world catalogue of the spider mite family (Acari: Tetranychidae). *International Journal of Acarology* 30, 143–152.

Migeon, A., Malagnini, V., Navajas, M. and Duso, C. (2007) Notes on the genus *Eotetranychus* (Acari: Tetranychidae) in Italy and France with a redescription of *Eotetranychus fraxini* Reck, new record for Italy and Western Europe. *Zootaxa* 1509, 51–60.

Migeon, A., Ferragut, F., Escudero-Colomar, L.A., Fiaboe, K., Knapp, M., Moraes, G.J. de, Ueckermann, E. and Navajas, M. (2009) Modelling the potential distribution of the invasive tomato red spider mite, *Tetranychus evansi* (Acari: Tetranychidae). *Experimental and Applied Acarology* 48, 199–212.

Migeon, A., Meynard, C.N. and Navajas, M. (2012) Current and future potential distribution of *Tetranychus evansi*: environmental niche differentiation between two distinct haplotypes and consequences for climate change. In: Schausberger, P., Walzer, A., Strodel, M.A. and Peneder, S. (eds) *7th Symposium of the European Association of Acarologists (EURAAC), July 9–13 2012, Vienna, Austria: Acari in a Changing World. Program, Abstracts, Participants*. University of Natural Resources and Life Sciences, Vienna, Austria, p. 50. Available at: http://euraac.boku.ac.at/SympVienna/files/ProgramBook.pdf (accessed 14 January 2015).

Mijuskovic, M. (1973a) *The Study of Mites on the Citrus Trees on the Yugoslav Littoral – Final Technical Report 2*. Agriculture Research Institute, Titograd, (former) Yugoslavia (in Serbo Croatian).

Mijuskovic, M. (1973b) Essais de lutte contre l'*Aceria sheldoni* Ewing (Acarina, Eriophyidae). *Poljoprivreda i Šumarstvo* 19, 41–50 (in Serbo Croatian).

Mijuskovic, M. (1981) Harmful mites on citrus fruits. *Poljoprivreda i Šumarstvo* 27, 3–22.

Mijuskovic, M. and Kosac, D. (1972) Control of *Aculops pelekassi* Keifer (Acarina: Eriophyidae) an important damageable mites of citrus in Yugoslavia. *Journal of Yugoslav Pomology* 21/22, 835–842 (in Serbo Croatian).

Mijuskovic, M. and Tomašević, B. (1975) *The Mites on the Citrus Trees on the Yugoslav Littoral, Volume 1*. Society for Science and Art of Montenegro Monographs, Section of Natural Sciences. Podgorica, Montenegro (in Serbo Croatian).

Mike's Garden Guide (2013) Bamboo Mites. Available at: http://mikesgardenguide.wordpress.com/2013/01/22/bamboo-mites/ (accessed 13 August 2014).

Mikulas, I. (2001) Suitability of environmentally friendly vineyard management with the resistant grapevine cultivar 'Viktoria gyongye' in Hungary. *Mitteilungen Klosterneuburg, Rebe und Wein, Obstbau und Früchteverwertung* 51, 254–258.

Mikunthan, G. and Manjunatha, M. (2006) Mycopathogens associated with pests of chilli and their pathogenicity against thrips (*Scirtothrips dorsalis*) and mites (*Polyphagotarsonemus latus*). *Tropical Agricultural Research* 18, 163–172.

Mikunthan, G. and Manjunatha, M. (2008) Impact of habitat manipulation on mycopathogen, *Fusarium semitectum* to control *Scirtothrips dorsalis* and *Polyphagotarsonemus latus* of chilli. *BioControl* 53, 403–412.

Mikunthan, G. and Manjunatha, M. (2010) *Fusarium* species: acaropathogenic fungi as potential control agents against coconut mite, *Aceria guerreronis*. In: Sabelis, M.W. and Bruin, J. (eds) *Trends in Acarology: Proceedings of the 12th International Congress, 21–26 August 2006, Amsterdam*. Springer, Dordrecht, the Netherlands, pp. 445–447.

Miles, P.M. (1964) Chrysanthemum semibud mite in Wales. *Entomologist's Monthly Magazine* 100, 97.

Millar, G.F. (1993) Aggregation and development of the gorse spider mite *Tetranychus lintearius* Dufour (Acari: Tetranychidae). MSc. thesis, University of Canterbury, Christchurch, New Zealand. Available at: http://hdl.handle.net/10092/1495 (accessed 13 August 2014).

Miller, L.W. (1950) Factors influencing diapause in the European red mite. *Nature* 166, 875.

Miller, L.W. (1966) The tetranychid mites of Tasmania. *Papers and Proceedings of the Royal Society of Tasmania* 100, 53–76.

Millet, E.R. (1958) Summary of insect conditions in Lebanon. *Cooperative Economic Insect Report* 8(8), 137–138. Available at: https://archive.org/stream/cooperativeecono88unit/cooperativeecono88unit_djvu.txt (accessed 15 May 2015).

Millet, E.R. (1959) Summary of insect conditions in Tunisia. *Cooperative Economic Insect Report* 9(6), 77–78. Available at: https://archive.org/stream/cooperativeecono96unit/cooperativeecono96unit_djvu.txt (accessed 15 May 2015).

Mills, L.R. (1973) Morphology of glands and ducts in the two-spotted spider mite, *Tetranychus urticae* Koch, 1836. *Acarologia* 15, 218–236.

Milne, M. and Walter, G.H. (1997) The significance of prey in the diet of the phytophagous thrips, *Frankliniella schultzei*. *Ecological Entomology* 22, 74–81.

Mineiro, J.L. de C., Lofego, A.C., Raga, A. and Moraes, G.J. de (2005) Primeiros registros dos ácaros *Amblyseiella setosa* Muma (Phytoseiidae) e *Tuckerella pavoniformis* (Ewing) (Tuckerellidae) no Brasil. *Arquivos do Instituto Biológico (São Paulo)* 72, 395–396.

Mineiro, J.L. de C., Sato, M.E., Raga, A. and Arthur, V. (2008) Population dynamics of phytophagous and predaceous mites on coffee in Brazil, with emphasis on *Brevipalpus phoenicis* (Acari: Tenuipalpidae). *Experimental and Applied Acarology* 44, 277–291.

Minetti, G., Mozzone, G.C. and Pellegrino, S. (1986) Esperienza di lotta guidata all'eriofide del nocciolo (ácaro delle gemme). *Informatore Agrario* 42(21), 71–73.

Minter, D.W., Brady, B.L. and Hall, R.A. (1983) Five hyphomycetes isolated from eriophyid mites. *Transactions of the British Mycological Society* 81, 455–471.

Miranda, I., Montoya, A., Rodriguez, Y., Depestre, T., Ramos, M. and Rodriguez, H. (2009) Density limit for the control of *Polyphagotarsonemus latus* (Banks) (Acari: Tarsonemidae) in protected crop of pepper (*Capsicum annuum* L.). *Revista de Protección Vegetal* 24, 146–151.

Misayo, M., Tomohiko, N., Daigo, O. and Toshio, F. (2005) Control of acarid mites *Tyrophagus similis* Volgin by soil sterilization. *Annual Report of the Kansai Plant Protection Society* 47, 1–8.

Mishra, D.N., Saha, K., Das, T.K., Somchoudhury, A.K. and Sarkar, P.K. (1996) Efficacy of some modern pesticides and plant hormones in the control of mango bud mite and its impact on various growth characters. *Indian Agriculturist* 40, 183–192.

Mishra, R.K. (1980) Gall formation and control of the erinose mite *Eriophyes litchii* Keifer, Acarina: Eriophyidae. In: Russ, K. and Berger, H. (eds) *Proceedings of the International Symposium of IOBC/WPRS on Integrated Control in Agriculture and Forestry, Vienna, 8–12 October, 1979*. International Organisation for Biological Control, West Palearctic Regional Section (http://www.iobc-wprs.org/), pp. 435–436.

Misra, B.C. and Israel, P. (1968) Studies on the bionomics of the paddy mite, *Oligonychus oryzae* Hirst (Acarina: Tetranychidae). *Oryza* 5, 32–37.

Misra, C.S. (1913) The red spider mite on jute (*Tetranychus bioculatus* Wodd Mason). *Agricultural Journal of India* 8, 309–316.

Mitchell, R. (1970) An anlysis of dispersal in mites. *American Naturalist* 104, 425–431.

Mitchell, R. (1972) The sex ratio of the spider mite *Tetranychus urticae*. *Entomologia Experimentalis et Applicata* 15, 299–304.

Mitchell, R.D. and Nadchatram, M. (1969) Schizekenosy: the substitute for defecation in chigger mites. *Journal of Natural History* 3, 121–124.

Mito, T. and Uesugi, T. (2004) Invasive alien species in Japan: the status quo and the new regulation for prevention of their adverse effects. *Global Environmental Research* 8, 171–191.

Mitra, M. (1931) Report of the Imperial Mycologist. In: *Scientific Reports 1929–30*. Imperial Agricultural Research Institute, Pusa, India, pp. 58–71.

Mitrofanov, V.I. (1967) Tetranychoid mites of the fauna of the USSR, damaging conifers. *Trudy Gosudarstvennogo Nikitskogo Botanicheskogo Sada* 39, 111–130 (in Russian).

Mitrofanov, V.I. and Antipov, V.P. (1973) Bud mite, a new grapevine pest in the Crimea. *Byulleten' Gosudarstvennogo Nikitskogo Botanicheskogo Sada* 3(22), 39–43 (in Russian).

Mitrofanov, V.I. and Sharonov, A.A. (1983) A contribution to the fauna of plant-feeding mites (Acariformes, Tetranychoidea) of Crimea. *Zoologicheskii Zhurnal* 62, 947–950 (in Russian).

Mitrofanov, V.I. and Strunkova, Z.I. (1979) *Operdelitl Kleshchei-ploskotelok* (*A Key to False Spider Mites*). Donish, Dushanbe, USSR (in Russian).

Mitrofanov, V.I., Bossenko, L.I. and Bitchevskis, M.Y.A. (1975) A key for determination of tetranychoid mites on coniferous trees. *Latvijas Entomologs* 1975, Supplement 3, 1–41 (in Russian).

Mitrofanov, V.I., Shaponov, A.A. and Livshits, I.Z. (1986) New species of the genus *Dendroptus* (Acariformes, Tarsonemidae) from the Crimea. *Zoologicheskii Zhurnal* 65, 141–149 (in Russian).

Mitrofanov, V.I., Strunkova, Z.I. and Livshits, I.Z. (1987) *Keys to the Tetranychid Mites (Tetranychidae, Bryobiidae) Fauna of the USSR and Adjacent Countries*. SSR, I.o.Z.a.P.E.N.P.T., Donish, Dushanbe, Tajikistan SSR, USSR (in Russian).

Mizell, R.F. III and Schiffhauer, D.E. (1991) Kairomone response, pesticide tolerance and field efficacy of the predatory mite, *Neosieulus collegae* (De Leon). *Journal of Environmental Horticulture* 9, 155–159.

Mizobe, M. and Tamura, I. (2004) Biological control of the broad mite *Polyphagotarsonemus latus* (Banks) by *Amblyseius cucumeris* (Oudemans) on greenhouse sweet pepper. *Kyushu Plant Protection Research* 50, 62–65 (in Japanese).

Mizoguchi, A., Mori, N., Nishida, R. and Kuwahara, Y. (2003) Alpha-acaridial, a female sex pheromone from an alarm pheromone emitting mite, *Rhizoglyphus robini*. *Journal of Chemical Ecology* 29, 1681–1690.

Mizoguchi, A., Murakami, K., Shimizu, N., Mori, N., Nishida, R. and Kuwahara, Y. (2005) *S*-isorobinal as the female sex pheromone from an alarm pheromone emitting mite, *Rhizoglyphus setosus*. *Experimental and Applied Acarology* 36, 107–117.

Mizuta, T. (2000) Effect of temperature on developmental period and oviposition of tea rust mite, *Acaphylla theavagrans* Kadono. *Japanese Journal of Applied Entomology and Zoology* 44, 95–99.

Mizuta, T. (2002) The adult density of the tea rust mite, *Acaphylla theavagrans* Kadono (Acari: Eriophyidae) on different tea leaves and its reproductive rate on different tea varieties. *Japanese Journal of Applied Entomology and Zoology* 46, 7–14 (in Japanese).

Mo, T.L. and Liu, T.X. (2007) Predation and life table of *Feltiella acarisuga* (Diptera: Cecidomyiidae) preying on eggs of *Tetranychus urticae* (Acari: Tetranychidae). *Environmental Entomology* 36, 369–375.

Mochizuki, M. (2002) Control of Kanzawa spider mite, *Tetranychus kanzawai* Kishida (Acari: Tetranychidae) on tea by a synthetic pyrethroid resistant predatory mite, *Amblyseius womersleyi* Schicha (Acari: Phytoseiidae). *Japanese Journal of Applied Entomology and Zoology* 46, 243–251 (in Japanese).

Moeezipoor, M., Kafil, M., Nooeei, S. and Allahyari, H. (2008) Functional response of predatory mite, *Phytoseius plumifer* (Canestrini & Fanzago) on different densities of *Amphitetranychus viennensis* (Zacher) and *Tetranychus urticae* (Koch). *Agricultural Research, Bu-Ali Sina University* 8(3), 107–116. Available at: http://ar.basu.ac.ir/jufile?c2hvd1BERj0yMTcmX2FjdGlvbj1zaG93UERGJmFydGljbGU9MjE3Jl9vYj1iZjU0N2YyYTUxNTVkMmZkN2YzZTkx-ODBiNjJiNjE0OA== (accessed 15 May 2015).

Moffitt, L.J. and Baritelle, J.L. (1990) Economic thresholds. In: Rosen, D. (ed.) *Armored Scale Insects*. World Crop Pests, Volume 4B. Elsevier Science, Amsterdam, the Netherlands, pp. 357–362.

Moghadam, M.M., Ghadamyari, M. and Talebi, K. (2012) Resistance mechanisms to fenazaquin in Iranian populations of two-spotted spider mite, *Tetranychus urticae* Koch (Acari: Tetranychidae). *International Journal of Acarology* 38, 138–145.

Moha, C. (1969) Cytologie des premiers stades de la cécidogenèse et de l'évolution de la galle d'*Eriophyes brevitarsus* Fockeu sur *Alnus glutinosa* Gaertn. *Bulletin de la Société Botanique de France* 116, 145–163.

Mohamed, I.I. (1963) Acarine mites occurring on cotton plants in Egypt. *Bulletin of the Entomological Society of Egypt* 46, 511.

Mohamed, I.I. (1965) Host preference of the citrus brown mite *Eutetranychus banksi* (McGregor). *Bulletin of the Entomological Society of Egypt* 48, 163–170.

Mohamed, I.I. and El-Haidari, H. (1965) A supplementary list of phytophagous mites in Iraq (1964–65). *Ministry of Agriculture Research and Projects Bulletin (Baghdad)* No. 131, 1–25.

Mohamed, I.I. and El-Haidari, H. (1968) Contribution to the knowledge of the mite fauna of Iraq (Acarina). *Bulletin of the Entomological Society of Egypt* 50, 127–128.

Mohamed, M.I., Zaher, M.A. and Hassan, M.F. (1982) Observations on *Cheyletus cacahuamilpensis*, a predator of the tenuipalpid mite *Dolichotetranychus floridanus*. *Entomophaga* 27, 343–348.

Mohanasundaram, M. (1981a) New gall-mites of the subfamily Nothopodinae (Acarina: Eriophyidae) from India. *Oriental Insects* 15, 145–166.

Mohanasundaram, M. (1981b) Four new species of Eriophyid mites (Acari : Eriophidae) from Tamil Nadu, India. *Colemania (Bangalore)* 1, 39–45.

Mohanasundaram, M. (1981c) The significance of the occurrence of thick and thin forms in the sugarcane blister, *Eriophyes saccharini* (Acari: Eriophyidae). In: ChannaBasavanna, G.P., Mallik, B. and Ghorpade, K.D. (eds) *Contributions to Acarology in India: Proceedings of the I All India Symposium in Acarology, April 23–25, 1979, Bangalore*. Acarological Society of India, Bangalore, India, pp. 72–74.

Mohanasundaram, M. (1983) Indian eriophyid studies IV. Record of new phyllocoptine mites (Phyllocoptinae: Eriophyidae: Acarina) from South India. *Acarologia* 24, 13–35.

Mohanasundaram, M. (1984) New eriophyid mites from India (Acarina: Eriophyoidea). *Oriental Insects* 18, 251–283.

Mohanasundaram, M. (1986) Three new species of rhynchaphytoptid mites (Rhynchaphytoptidae: Eriophyoidea) from Tamil Nadu. *Entomon* 11, 47–51.

Mohanasundaram, M. (1991) Further studies on the eriophyid fauna (Eriophyoidea: Acari) of Tamil Nadu. *Entomon* 16, 187–192.

Mohanasundaram, M. (1996) A new tenuipalpid mite, *Neoraoiella eugeniae* gen. *et* sp. nov. from south India (Tenuipalpidae: Acari). *Journal of Entomological Research* 20, 141–143.

Mohanasundaram, M., Parameswaran, S. and Natarajan, K. (1989) A note on the occurrence of *Phyllocoptruta musae* Keifer (Eriophyidae: Acari) on banana in south India. *South Indian Horticulture* 37, 300.

Molins, R.A. (2001) *Food Irradiation: Principles and Applications*. Wiley, New York.

Momen, F. and El-Borolossy, M. (1999) Suitability of the citrus brown mite, *Eutetranychus orientalis* as prey for nine species of phytoseiid mites (Acari: Tetranychidae, Phytoseiidae). *Acarologia* 40, 19–23.

Momen, F.M. and Abdel-Khalek, A. (2008) Effect of the tomato rust mite *Aculops lycopersici* (Acari: Eriophyidae) on the development and reproduction of three predatory phytoseiid mites. *International Journal of Tropical Insect Science* 28, 53–57.

Monfreda, R., Lekveishvili, M., Petanović, R. and Amrine, J.W. Jr (2010) Collection and detection of eriophyoid mites. *Experimental and Applied Acarology* 51, 273–282.

Monkman, K.D. (1992) Tomato russet mite – Acari: Eriophyoidea. *Monthly Bulletin, Department of Agriculture, Fisheries and Parks, Bermuda* 63(3), 19–21.

Montaldo, A. (1972) *La Yuca: Trabajo Sobre este Cultivo, con Especial Referencia a Venezuela*. Oficina de Communicaciones Agricolas, Ministerio de Agricultura y Cria, Maracay, Venezuela.

Montano, C.P.R. (1995) *Control de* Brevipalpus chilensis *Baker. Falsa Aranita de la Vid en* Vitis vinifera L. *Cultivar Cabernet Sauvignon Mediante Aplicaciones de Aceite Mineral*. Escola de Agronomía, Universidad de Chile, Santiago.

Montes, S.M.N.M., Raga, A., Mineiro, J.L.C. and Montes, R.M. (2010) Mite fauna in cassava varieties in Presidente Prudente, State of São Paulo, Brazil. In: Moraes, G.J. de, Castilho, R.C. and Flechtmann, C.H.W. (eds) *XIII International Congress of Acarology, August 23–27, 2010, Recife-PE, Brazil, Abstract Book*, abstract 292, p. 158. Available at: http://www.acarology.org/ica/ica2010/docs/0910_15_abstractBookCapa.pdf (accessed 3 February 2015).

Montesinos, E. (2003) Development, registration and commercialization of microbial pesticides for plant protection. *International Microbiology* 6, 245–252.

Montréal Biodôme (1999) Faune et flore du Biodôme, Tétranyque, Birch spider mite. Available at: http://espacepourlavie.ca/faune-biodome/tetranyque (accessed 25 March 2015).

Montserrat, M. and Guzmán, C. (2010) Is global warming jeopardizing the biological control of invasive species? A case study in avocado agro-ecosystems from Spain. In: Moraes, G.J. de, Castilho, R.C. and Flechtmann, C.H.W. (eds) *XIII International Congress of Acarology, August 23–27, 2010, Recife-PE, Brazil, Abstract Book*, abstract 294, p. 159. Available at: http://www.acarology.org/ica/ica2010/docs/0910_15_abstractBookCapa.pdf (accessed 3 February 2015).

Moore, D. (1986) Bract arrangement in the coconut fruit in relation to attack by the coconut mite *Eriophyes guerreronis* Keifer. *Tropical Agriculture* 63, 285–288.

Moore, D. (2000) Non-chemical control of *Aceria guerreronis* on coconuts. In: Fernando, L.C.P., Moraes, G.J. de and Wickramananda, I.R. (eds) *Proceedings of the International Workshop on Coconut Mite (*Aceria guerreronis*), Coconut Research Institute, Sri Lanka, 6–8 January 2000*. Coconut Research Institute, Lunuwila, Sri Lanka, pp. 63–70.

Moore, D. and Alexander, L. (1987a) Aspects of migration and colonization of the coconut palm by the coconut mite, *Eriophyes guerreronis* (Keifer) (Acari: Eriophyidae). *Bulletin of Entomological Research* 77, 641–650.

Moore, D. and Alexander, L. (1987b) Stem injection of vamidothion for control of coconut mite, *Eriophyes guerreronis* Keifer, in St. Lucia. *Crop Protection* 6, 329–333.

Moore, D. and Alexander, L. (1990) Resistance of coconuts in St. Lucia to attack by the coconut mite *Eriophyes guerreronis* Keifer. *Tropical Agriculture* 67, 33–36.

Moore, D. and Howard, F.W. (1996) Coconuts. In: Lindquist, E.E., Sabelis, M.W. and Bruin, J. (eds) *Eriophyoid Mites: Their Biology, Natural Enemies and Control*. World Crop Pests, Volume 6. Elsevier Science, Amsterdam, pp. 561–570.

Moore, D., Alexander, L. and Hall, R.A. (1989) The coconut mite, *Eriophyes guerreronis* Keifer in St. Lucia: yield losses and attempts to control it with acaricide, polybutene and *Hirsutella* fungus. *Tropical Pest Management* 35, 83–89.

Moore, D., Ridout M.S. and Alexander, L. (1991) Nutrition of coconuts in St. Lucia and relationship with attack by coconut mite *Aceria guerreronis* Keifer. *Tropical Agriculture* 68, 41–44.

Mora Morin, J. (1991) *Stethorus utilis* (Coleoptera: Coccinellidae) pequena cotorrita que se alimenta de acaros tetraniquidos que atacan a los citricos. Comunicacion corta. *Centro Agricola (Villa Clara, Cuba)* 18(2), 83–84.

Moraes, F.H.R., Belo, W.R.F., Moraes, G.J. and Kitajima, E.W. (2006) Ocorrência do vírus da pinta verde em maracujá no Estado do Maranhão, Brasil. *Fitopatologia Brasileira* 31, 100. Available as English and Portuguese abstracts at: http://dx.doi.org/10.1590/S0100-41582006000100019 (accessed 13 August 2014).

Moraes, G.J. de and Delalibera, I. Jr (1992) Specificity of a strain of *Neozygites* sp. (Zygomycetes: Entomophthorales) to *Mononychellus tanajoa* (Acari: Tetranychidae). *Experimental and Applied Acarology* 14, 89–94.

Moraes, G.J. de and Flechtmann, C.H.W. (1981) Phytophagous mites of Northeastern Brazil. *Pesquisa Agropecuária Brasileira* 16, 177–186.

Moraes, G.J. de and Flechtmann, C.H.W. (2007) Phytophagous mites of tropical crops in eastern South America. In: Morales-Malacara, J.B., Behan-Pelletier, V., Ueckermann, E., Pérez, T.M., Estrada-Venegas, E.G. and Badii, M. (eds) *Acarology XI: Proceedings of the International Congress, 8–13 September 2002, Mérida, Universidad Nacional Autonoma de México*. Sociedad Latinoamericana de Acarología, México, pp. 279–288.

Moraes, G.J. de and Flechtmann, C.H.W. (2008) *Manual de Acarologia: Acarologia Básica e Ácaros de Plantas Cultivadas no Brasil*. Holos, Ribeirão Preto, Brazil.

Moraes, G.J. de and Lima, H.C. (1983) Biology of *Euseius concordis* (Chant), a predator of the tomato russet mite. *Acarologia* 24, 251–255.

Moraes, G.J. de and McMurtry, J.A. (1981) Biology of *Amblyseius citrifolius* (Denmark and Muma) (Acarina: Phytoseiidae). *Hilgardia* 49(1), 29.

Moraes, G.J. de and McMurtry, J.A. (1985) Comparison of *Tetranychus evansi* and *T. urticae* (Acari: Tetranychidae) as prey for eight species of phytoseiid mites. *Entomophaga* 30, 393–397.

Moraes, G.J. de and McMurtry, G.A. (1987) Physiological effect of the host plant on the suitability of *Tetranychus urticae* as prey for *Phytoseiulus persimilis* (Acari: Tetranychidae: Phytoseiidae). *Entomophaga* 32, 35–38.

Moraes, G.J. de and Zacarias, M.S. (2002) Use of predatory mites for the control of eriophyid mites. In: Fernando, L.C.P., Moraes, G.J. de and Wickramananda, I.R. (eds) *Proceedings of the International Workshop on Coconut Mite (*Aceria guerreronis*), Coconut Research Institute, Sri Lanka, 6–8 January 2000*. Coconut Research Institute, Lunuwila, Sri Lanka, pp. 78–88.

Moraes, G.J. de, McMurtry, G.A. and Denmark, H.A. (1986) *A Catalog of the Mite Family Phytoseiidae. References to Taxonomy, Synonymy, Distribution and Habitat*. EMBRAPA-DDT (Departamento de Difusão de Tecnologia, Empresa Brasileira de Pesquisa Agropecuária), Brasilia, Brazil.

Moraes, G.J. de, McMurtry, J.A. and Baker, E.W. (1987) Redescription and distribution of the spider mites *Tetranychus evansi* and *T. marianae*. *Acarologia* 28, 333–343.

Moraes, G.J. de, McMurtry, J.A. and Yaninek, J.S. (1989) Some phytoseiid mites (Acari: Phytoseiidae) from tropical Africa with description of a new species. *International Journal of Acarology* 15, 95–102.

Moraes, G.J. de, Alencar, J.A. de, Lima, J.L.S. de, Yaninek, J.S. and Delalibera, I. Jr (1993) Alternative plant habitats for common phytoseiid predators of the cassava green mite (Acari: Phytoseiidae, Tetranychidae) in northeast Brazil. *Experimental and Applied Acarology* 17, 77–90.

Moraes, G.J. de, da Silva, C.A.D. and Moreira, A.N. (1994) Biology of a strain of *Neoseiulus idaeus* (Acari: Phytoseiidae) from northwest Brazil. *Experimental and Applied Acarology* 18, 213–220.

Moraes, G.J. de, Moreira, N.A. and Delalibera, I. Jr (1995a) Growth of the mite *Mononychellus tanajoa* (Acari: Tetranychidae) on alternative plant hosts in north-eastern Brazil. *Florida Entomologist* 78, 350–354.

Moraes, G.J. de, McMurtry, J.A., Denmark, H.A. and Campos, C.B. (2004) *A Revised Catalog of the Mite Family Phytoseiidae. Zootaxa* 434. Magnolia Press, Auckland, New Zealand.

Moraes, L.A.H., de Porto, O. de M. and Braun, J. (1995b) Efeito de três doses de enxofre na população do "ácaro da falsa ferrugem" *Phyllocoptruta oleivora* (Ashmead, 1879) (Acari, Eriophyidae) em citrus. *Pesquisa Agropecuária Gaúcha* 1, 17–20.

Morales-Galvan, O., Bravo-Mojica, H., López-Collado, J., González-Hernández, H. and Villegas-Monter, A. (2003) Desarrollo y validación de un plan de muestreo para *Oligonychus punicae* y *O. perseae* (Acari:Tetranychidae) en aguacate cv. Hass. In: *V Congreso Mundial del Aguacate del 19 al 24 de Octubre de 2003, Granada-Málaga. Actas: Volumen I (Proceedings of the V World Avocado Congress., 19–24 October 2003, Granada-Málaga)*. Consejería de Agricultura y Pesca, Junta de Andalucia. Published by A.G. Novograf, Seville, Spain, pp. 509–514. Available at: http://www.avocadosource.com/WAC5/Papers/WAC5_p509.pdf (accessed 15 May 2015).

Morando, A., Morando, M. and Morando, D. (1998) Confronti tra acaricidi in vigneto ed influenze della difesa sulle caratteristiche quanti-qualitative della produzione. In: *Atti, Giornate Fitopatologiche 1998, Scicli e Ragusa 3–7 Maggio, Volume 1*. Fitogest®, Image Line Network, Faenza, Italy, pp. 241–246.

Moreira, A.N., Oliveira, J.V. de, Haji, F.N.P. and Pereira, J.R. (1999) Efeito de diferentes niveis de NPK na infestacao de *Aculops lycopersici* (Massee) (Acari: Eriophyidae), em tomateiro no Submedio do Vale do Sao Francisco. *Anais de Sociedade Entomológica do Brasil* 28, 275–284.

Moreira, C.S. (1992) Eficiencia do oleo SUN e do oleo OPA-E no controle do ácaro da ferrugem (*Phyllocoptruta oleivora* Ashm.) em frutos de laranjeira. *Revista de Agricultura (Piracicaba)* 67, 237–249.

Morewood, W.D. (1992) Cold storage of *Phytoseiulus persimilis* (Phytoseiidae). *Experimental and Applied Acarology* 13, 231–236.

Morgan, C.V.G. and Anderson, N.H. (1958) Notes on parathion-resistant strains of two phytophagous mites and a predacious mite in British Columbia. *Canadian Entomologist* 90, 92–97.

Morgan, C.V.G. and Arrand, J.C. (1969) *Pear Rust Mite and its Control in British Columbia*. Publication 69–8, British Columbia Department of Agriculture Entomology Branch, Victoria, British Columbia, Canada.

Morgan, C.V.G., Chant, D.A., Anderson, N.H. and Ayre, G.L. (1955) Methods for estimating orchard mite populations, especially with the mite brushing machine. *Canadian Entomologist* 77, 189–200.

Morgan, C.V.G., Yee, P.T. and Brinton, F.E. (1962) Sevin as a systematic miticide for the pear leaf blister mite, *Eriophyes pyri* (Pgst.). Acarina: Eriophyidae. *Canadian Entomologist* 94, 680–686.

Mori, H. (1961) Comparative studies on thermal reaction in four species of spider mites. *Journal of the Faculty of Agriculture, Hokkaido University* 51, 574–591.

Mori, H. (1962a) Seasonal difference of phototactic response in three species of spider mites. *Journal of the Faculty of Agriculture, Hokkaido University* 52, 1–9.

Mori, H. (1962b) The effects of photo-stimuls on the thermal reaction of four species of spider mites. *Journal of the Faculty of Agriculture, Hokkaido University* 52, 10–19.

Mori, H., and Chant, D.A. (1966a) The influence of prey density, relative humidity, and starvation on the predacious behavior of *Phytoseiulus persimilis* Athias-Henriot (Acarina: Phytoseiidae). *Canadian Journal of Zoology* 44, 483–491.

Mori, H., and Chant, D.A. (1966b) The influence of humidity on the activity of *Phytoseiulus persimilis* Athias-Henriot and its prey, *Tetranychus urticae* (C.L. Koch) (Acarina: Phytoseiidae, Tetranychidae). *Canadian Journal of Zoology* 44, 863–871.

Mori, H. and Saitō, Y. (1979) Biological control of *Tetranychus urticae* Koch (Acarina: Tetranychidae) populations by three species of phytoseiid mites (Acarina: Phytoseiidae). *Journal of the Faculty of Agriculture, Hokkaido University* 59, 303–311.

Mori, K. and Saitō, Y. (2005) Variation in social behavior within a spider mite genus, *Stigmaeopsis* (Acari: Tetranychidae). *Behavioral Ecology* 16, 232–238.

Mori, K., Nozawa, M., Arai, K. and Gotoh, T. (2005) Life-history traits of the acaraphagous lady beetle, *Stethorus japonicus* at three constant temperatures. *BioControl* 50, 35–51.

Mori, N., Kuwahara, Y. and Kurosa, K. (1998) Rosefuran: the sex pheromone of an acarid mite. *Journal of Chemical Ecology* 24, 1771–1779.

Morishita, F.S. (1954) Biology and control of *Brevipalpus inornatus* (Banks). *Journal of Economic Entomology* 47, 449–456.

Morishita, M. (2000) Effect of predators on the occurrence of the Kanzawa spider mite, *Tetranychus kanzawai* Kishida (Acari: Tetranychidae), on feral plants, *Clerodendrum trichotomum* and *Akebia quinata*. *Japanese Journal of Applied Entomology and Zoology* 44, 235–239 (in Japanese).

Moritz, K. and McDonald, G. (1995) Developing redlegged earth mite resistance in canola. In: Potter, T.D. (ed.) *Proceedings of the 10th Australian Research Assembly on Brassicas, Struan, South Australia, September 26, 27 and 28, 1995*. South Australian Research and Development Institute, Adelaide, Australia, pp. 30–35.

Morris, O.N. (1961) The development of the clover mite *Bryobia praetiosa* Koch in relation to the nitrogen, phosphorus and potassium nutrition of its host plant. *Annals of the Entomological Society of America* 54, 551–575.

Morros, M.E. and Aponte, O. (1994) Biología y tabla de vida de *Tetranychus ludeni* Zacher en caraota *Phaseolus vulgaris* L. *Agronomía Tropical (Maracay)* 44, 667–677.

Morros, M.E. and Aponte, O. (1995a) Efecto de dos niveles de infestación de *Tetranychus ludeni* Zacher sobre las fases de desarrollo de la caraota I: Nivel de campo. *Agronomía Tropical (Maracay)* 54(2), 189–194.

Morros, M.E. and Aponte, O. (1995b) Efecto de dos niveles de infestación de *Tetranychus ludeni* Zacher sobre las fases de desarrollo de la caraota II: Nivel de invernadero. *Agronomía Tropical (Maracay)* 54(2), 195–202.

Morse, J.G. and Jones, V.P. (1983) Control of the citrus red mite. *Insecticicide and Acaricide Tests* 8, 32.

Moser, J.C. (1985) Use of sporothecae by phoretic *Tarsonemus* mites to transport ascospores of coniferous bluestain fungi. *Transactions of the British Mycological Society* 84, 750–753.

Moser, J.C., Perry, T.J. and Yin, H.F. (1995) Ascospore dispersal of *Ceratocystiopsis ranaculosus*, a mycangial fungus of the southern pine beetle. *Mycologia* 87, 84–86.

Moss, W.W. (1962) Studies on the morphology of the trombidiid mite *Allothrombidium lerouxi* Moss (Acari). *Acarologia* 4, 313–345.

Mostafa, A.R., DeBach, P. and Fisher, T.W. (1975) Anystid mite: citrus thrips predator. *California Agriculture* 29(3), 5.

Mothes, U. and Seitz, K.A. (1981a) Fine structure and function of the prosomal glands of the two-spotted spider mite, *Tetranychus urticae* (Acari, Tetranychidae). *Cell and Tissue Research* 221, 339–349.

Mothes, U. and Seitz, K.A. (1981b) Functional microscopic anatomy of the digestive system of the *Tetranychus urticae* (Acari: Tetranychidae). *Acarologia* 22, 257–270.

Mothes, U. and Seitz, K.A. (1982) Fine structural alternations of bean plant leaves by feeding injury of *Tetranychus urticae* Koch (Acari: Tetranychidae). *Acarologia* 23, 149–157.

Moukoumbi, Y.D., Sie, M., Vodouhe, R., Bonou, W., Toulou, B. and Ahanchede, A. (2011) Genetic diversity of mushroom mite (*Luciaphorus* sp.) infesting cultivated mushrooms in the Northeast of Thailand. *African Journal of Agricultural Research* 6, 5438–5445.

Mound, L.A. (1994) Plant galls. In: Williams, M.A.J. (ed.) *Plant Galls: Organisms, Interactions, Populations*. Systematics Association Special Volume 49. Clarendon Press, Oxford, UK, pp. 131–149.

Mound, L.A. (2011) Species recognition in the genus *Scolothrips* (Thysanoptera, Thripidae), predators of leaf-feeding mites. *Zootaxa* 2797, 45–53.

Moura, M.Z.D., Isaias, R.M. dos S. and Soares, G.L.G. (2005) Ontogenesis of internal secretory cells in leaves of *Lantana camara* (Verbenaceae). *Botanical Journal of the Linnean Society* 148, 427–431.

Moura, M.Z.D., Soares, G.L.G. and Isaias, R.M. dos S. (2008) Species-specific changes in tissue morphogenesis induced by two arthropod leaf gallers in *Lantana camara* L. (Verbenaceae). *Australian Journal of Botany* 56, 153–160.

Moura, M.Z.D., Soares, G.L.G. and Isaias, R.M. dos S. (2009) Ontogênese da folha e das galhas induzidas por *Aceria lantanae* Cook (Acarina: Eriophyidae) em *Lantana camara* L. (Verbenaceae). *Revista Brasileira de Botânica* 32, 271–282.

Mourao, S.A., Zanuncio, J.C., Pallini Filho, A., Guedes, R.N.C. and Camargos, A.B. de (2004) Toxicity of neem extracts (*Azadirachta indica*) to the coffee red mite *Oligonychus ilicis*. *Pesquisa Agropecuária Brasileira* 39, 827–830.

Moutia, L.A. (1958) Contribution to study of some phytophagous Acarina and their predators in Mauritius. *Bulletin of Entomological Research* 49, 59–75.

Mowat, D.J. (1978) The influence of cutting programmes on populations of eriophyid mites in ryegrass. *Scientific Proceedings of the Royal Dublin Society* 6, 351–358.

Mowat, D.J. (1985) The control of *Abacarus hystrix* (Nalepa) on ryegrass by pesticides. *Record of Agricultural Research, Department of Agriculture for Northern Ireland* 33, 5–7.

Mowery, P.D., Hull, L.A. and Asquith, D. (1980) Two new sampling plans for European red mite surveys on apple utilizing the negative binomial distribution. *Environmental Entomology* 9, 159–163.

Mozzone, G.C., Pellegrino, S., Pio, R., Griseri, G. and Fenoglio, P. (1994) Coordination of the technical assistance for hazelnut in Piemonte. *Acta Horticulturae* 351, 159–167.

Mpedi, P.F. and Urban, A.J. (2003) *Host Specificity Report in Relation to an Application for Permission to Release the Central American mite,* Aceria lantanae, *from Quarantine in Pretoria, for Biological Control of the Noxious Weed,* Lantana camara, *in South Africa*. Unpublished report to the National Department of Agriculture, Directorate of Plant Health and Quality. ARC-PPRI (Agricultural Research Council-Plant Protection Research Insititute), Pretoria.

Mpedi, P.F. and Urban, A.J. (2005) *Host Specificity Testing of* Aceria lantanae *(Acari: Eriophyidae), a candidate agent for biological control of* Lantana camara *(Verbenaceae)*. Final report to the Department of Natural Resources and Mines, Alan Fletcher Research Station Brisbane, Queensland, Australia.

Mpkagiannis, G. (1982) Biologische Aspekte und Orientierungsverhalten der raüberischen Blumenwanze *Anthocoris nemorum* L. (Het.: Anthocoridae). PhD thesis, Georg-August-Universität Göttingen, Göttingen, Germany.

Mtambo, C.C. and Hoeschle-Zeledon, I. (2001) The development of integrated control methods for the tomato bed spider mite (*Tetranychus evansi*) in Malawi. In: Phiri, I.M.G., Saka, A.R. and Chilembwe, E.H.C. (eds) *Agricultural Technologies for Sustainable Development in Malawi. Proceedings of the First Annual Scientific Conference Held at the Malawi Institute of Management (MIM), Lilongwe, Malawi, 6–10 November, 2000*. Malawi Institute of Management, Lilongwe, Malawi, pp. 139–147.

Mukerjea, T.D. (1967) Ethion – a new acaricide for control of red spider, scarlet, pink and purple mites. *Two and a Bud* 14, 112–122.

Mukerjea, T.D. (1968) Trithion – a new acaricide. *Two and a Bud* 15, 53–60.

Mukherjee, A.B., Sarkar, P.K., Somchoudhury, A.K. and Roy, M. (1989) Relative incidence of *Oligonychus oryzae* on nine varieties of paddy and its control. In: ChannaBasavanna, G.P. and Viraktamath, C.A. (eds) *Progress in Acarology, Volume 2*. Oxford & IBH Publishing, New Delhi, pp. 53–57.

Mukherjee, I.N. and Singh, J. (1993) Records of phytophagous mites and predatory mites associated with fruit plants in Uttar Pradesh. *Journal of Insect Science (Ludhiana)* 6, 13–136.

Müller, G.F.W. (1957) Morphology, biology, and control of the hawthorn spider mite, *Tetranychus viennensis* Zacher. *Höfchenbriefe* 10(1), 1–60.

Mulligan, T.E. (1960) The transmission by mites, host range, and properties of rye grass mosaic virus. *Annals of Applied Biology* 48, 575–579.

Mullis, K.B. and Faloona, F.A. (1987) Specific synthesis of DNA *in vitro* via a polymerase-catalyzed chain reaction. *Methods in Enzymology* 155, 335–350.

Muma, M.H. (1955) Factors contributing to the natural control of citrus insects and mites in Florida. *Journal of Economic Entomology* 48, 432–438.

Muma, M.H. (1961) *Mites Associated with Citrus in Florida*. Florida Agricultural Experiment Station Bulletin No. 640, University of Florida, Gainesville, Florida.

Muma, M.H. (1967) Biological notes on *Coniopterix vicina* (Neuroptera: Coniopterygidae). *Florida Entomologist* 50, 285–293.

Muma, M.H. (1969) Biological control of mites on subtropical crops. In: Evans, G.O. (ed.) *Proceedings of the 2nd International Congress of Acarology, 19–25 July, 1967, Sutton Bonington, UK*. Akadémiaei Kiadó, Budapest, pp. 373–382.

Muma, M.H. (1970) Natural control potential of *Galendromus floridanus* (Acarina: Phytoseiidae) on Tetranychidae on Florida citrus trees. *Florida Entomologist* 53, 79–88.

Muma, M.H. (1975) *Mites Associated with Citrus in Florida*. Bulletin No. 640A, Florida Agricultural Experiment Station, University of Florida, Gainesville, Florida.

Muma, M.H. and Selhime, A.G. (1971) *Agistemus floridanus* (Acarina: Stigmaeidae), a predatory mite, on Florida Citrus. *Florida Entomologist* 54, 249–258.

Muma, M.H., Selhime, A.G. and Denmark, H.A. (1961) *An Annotated List of Predators and Parasites Associated with Insects and Mites on Florida Citrus*. Technical Bulletin No. 643, University of Florida, Gainesville, Florida.

Muma, M.H., Selhime, A.G. and Denmark, H.A. (1975) *An Annotated List of Predators and Parasites Associated with Insects and Mites on Florida Citrus*. IFAS Bulletin No. 634B, Institute of Food and Agricultural Science, University of Florida, Gainesville, Florida.

Munger, F. (1933) *Investigations in the Control of the Cyclamen Mite (*Tarsonemus pallidus *Banks)*. Technical Bulletin No. 93, Minnesota Agricultural Experiment Station, St Paul, Minnesota.

Munger, F., Gilmore, J.E. and Davis, W.S. (1959) A disease of citrus red mite. *California Citrograph* 44, 190, 216.

Munier, P. (1952) *L'Assaba, Essai Monographique*. Études Mauritaniennes No. 3, Centre IFAN (Institut Fondamental d'Afrique Noire), Mauritanie, Saint-Louis, Senegal.

Munro, J.A. (1954) Entomology problems in Bolivia. A supplemental note to Entomology problems in Bolivia. *FAO Plant Protection Bulletin* 2, 97–101, 150.

Muraleedharan, N. (1981) Entomology. In: *Annual Report for the Period 1st January 1981 to 31st December 1981, Tea Scientific Department, United Planters' Association of Southern India*. Cinchona, India, pp. 88–104.

Muraleedharan, N. and Chandrasekharan, R. (1981) Observations on the seasonal variations of *Acaphylla theae* Keifer and *Calacarus carinatus* (Green) (Acarina: Eriophyidae) in a tea field at the Anamalais (South India). *Pestology* 5, 11–15.

Muraleedharan, N. and Kandasamy, C. (1982) Evaluation of cyhexatin against the pink mite of tea. *Pesticides* 16(12), 23–24.

Muraleedharan, N. and Selvasundaram, R. (1996) Cost benefit in tea pest control. *Bulletin of the United Planters' Association of Southern India* 49, 35–46.

Muraleedharan, N. and Varatharajan, R. (1985) Preliminary observations on the effect of synthetic pyrethroids on the pink mite of tea. *Pesticides* 19(12), 51–56.

Muraleedharan, N. and Varatharajan, R. (1988) Pattern of insecticide usage in the tea plantations of Anamalais (Coimbatore District, Tamil Nadu). *Journal of Entomological Research* 12, 113–116.

Muraleedharan, N., Radhakrishnan, B. and Devadas, V. (1988a) Vertical distribution of three species of eriophyid mites on tea in south India. *Experimental and Applied Acarology* 4, 359–364.

Muraleedharan, N., Selvasundaram, R. and Radhakrishnan, B. (1988b) Natural enemies of certain tea pests occurring in southern India. *Insect Science and Its Application* (now *International Journal of Tropical Insect Science*) 9, 647–654.

Muraleedharan, N., Mohan, M.S., Radhakrishnan, B. and Udayabhanu, K.G. (1994) Seasonal abundance and vertical distribution of purple mite, *Calacarus carinatus* (Green) (Eriophyidae: Acarina) infesting tea. *Entomon* 19, 53–56.

Muraoka, M. and Shiyomi, M. (1986) Simulation on the occurrence of citrus red mite, *Panonychus citri*, using a system model developed for the occurrence in citrus orchard. *Bulletin of the Saga Prefecture Fruit Tree Experiment Station* 9, 103–123 (in Japanese).

Muriithi, L.M., Rangan, T.S. and Waite, B.H. (1982) *In vitro* propagation of fig through shoot tip culture. *HortScience* 17, 86–87.

Murray, A. (1877) *Economic Entomology, Aptera*. Chapman and Hall, London.

Murray, N.D. (1985) Rates of change in introduced organisms. In: Delfosse, E.S. (ed.) *Proceedings of the VI International Symposium on the Biological Control of Weeds, Vancouver, August 19–25, 1984*. Agriculture Canada, Ottawa, pp. 191–199.

Murthy, R.L.N., Rao, G.N. and Chandrasekaran, R. (1979) Recent developments in mite control: to achieve early prevention, greater coverage and longer spraying intervals. In: Venkata Ram, C.S. and Chandra Mouli, B. (eds) *PLACROSYM II. Plant Protection (Entomology, Microbiology, Nematology, Plant Pathology and Rodentology): Proceedings of the Second Annual Symposium on Plantation Crops, 1979, Ootacamund, India*. PLACROSYM Standing Committee, India, pp. 243–257.

Murungi, L.K., Knapp, M., Masinde, P.W., Onyambu, G., Gitonga, L. and Gaya Agong, S. (2009) Host-plant acceptance, fecundity and longevity of *Tetranychus evansi* (Acari: Tetranychidae) on selected tomato accessions. *African Journal of Horticultural Science* 2, 79–91. Available as English summary at: http://www.hakenya.net/ajhs/index.php/ajhs/rt/printerFriendly/33/0 (accessed 14 August 2014).

Murungi, L.K., Knapp, M., Salifu, D., Wesonga, J., Nyende, A., Masinde, P. and Torto, B. (2011) Differential effects of various African nightshade species on the fecundity and movement of *Tetranychus evansi* (Acari: Tetranychidae). *International Journal of Tropical Insect Science* 31, 269–276.

Muthiah, C. and Bhaskaran, R. (2000) Major outbreak of eriophyid mite of coconut in India. *The Planter* 76, 243–246.

Muthiah, C., Bhaskaran, R. and Kannaiyan, S. (2001) Bio-ecology and control of eriophyid mite of coconut-an Indian experience. *The Planter* 77, 255–263.

Muthukrishnan, T.S. (1956) An eriophyid mite on sugarcane in South India. *Current Science* 25, 234–235.

Mwase, E.T. and Baker, A.S. (2006) An annotated checklist of mites (Arachnida: Acrai) of Zambia. *Zootaxa* 1106, 1–124.

Nachev, P. (1976) Studies on eriophyid mites in Bulgaria. IX. Eriophyid mites on pear trees (Acarina, Eriophyoidea). *Rastitelnozashchitna Nauka* 3, 70–76 (in Bulgarian).

Nachev, P. (1978) The study of eriophyid mites in Bulgaria. XII. Eriophyid mites on apple trees (Acarina; Eriophyoidea). *Gradinarska i Lozarska Nauka* 15(5/6), 71–80 (in Bulgarian).

Nachev, P. (1983) Eriophyid mites on plum trees – species and population dynamics. *Zeszyty Problemowe Postępów Nauk Rolniczych* 252, 89–93.

Naegele, J.A., McEnroe, W.D. and Soans, A.B. (1965) Spectral sensitivity and orientation in the two-spotted spider mite, *Tetranychus urticae* K. from 350 mμ to 700 mμ. *Journal of Insect Physiology* 12, 1187–1195.

Nagarajan, K.R. (1957) A short note on *Paratetranychus oryzae* Hirst. *Madras Agricultural Journal* 44, 480.

Nagatomo, S., Atari, N., Onimaru, T. and Nonaka, T. (1991) Studies on the control of Kanzawa spider mite, *Tetranychus kanzawai* Kishida by the predacious mite, *Amblyseius longispinosus* (Evans) in tea (*Camellia sinensis*) fields. *Bulletin of the Kagoshima Tea Experiment Station* 7, 1–76 (in Japanese).

Nair, C.P.R. (2002) Status of coconut eriophyid mite *Aceria guerreronis* Keifer in India. In: Fernando, L.C.P., Moraes, G.J. de and Wickramananda, I.R. (eds) *Proceedings of the International Workshop on Coconut Mite (*Aceria guerreronis*), Coconut Research Institute, Sri Lanka, 6–8 January 2000.* Coconut Research Institute, Lunuwila, Sri Lanka, pp. 9–12. Available at: http://cri.nsf.ac.lk/bitstream/handle/1/2535/05650.PDF?sequence=1&isAllowed=y (accessed 15 May 2015).

Nair, C.P.R. and Koshy, P.K. (2000) Studies on coconut eriophyid mite, *Aceria guerreronis* Keifer in India. In: *Programme of the International Workshop on Coconut Mite (*Aceria guerreronis*), 1, 6–8 January 2000.* Coconut Research Institute, Lunuwilla, Sri Lanka, p. 7.

Nair, C.P.R., Rajan, P. and Mohan Chandrika (2005) Coconut eriophyid mite *Aceria guerreronis* Keifer – an overview. *Indian Journal of Plant Protection* 33, 1–10.

Nair, N.G. and Clift, A.D. (1993) *Integrated Pest and Disease Management.* Horticultural Research and Development Corporation Final Report Project MU/002 for NSW Agriculture. HRDC, Gordon, New South Wales, Australia.

Nakagawa, T. (1984) Growth and prey quantity of *Amblyseius longispinosus* (Evans) as a predator of *Tetranychus kanzawai* Kishida. *Proceedings of the Association for Plant Protection of Kyushu* 30, 158–161 (in Japanese).

Nakamura, Y., Gotoh, T., Imanishi, S., Mita, K., Kurtti, T.J and Noda, H. (2011) Differentially expressed genes in silkworm cell cultures in response to infection by *Wolbachia* and *Cardinium* endosymbionts. *Insect Molecular Biology* 20, 279–289.

Nakao, H. (1989) Studies on acarid mites injurious to vegetable plants (Acari: Astigmata) I. Occurrence of damage to spinach by acarid mites. *Bulletin of Hokkaido Prefectural Agricultural Experiment Stations* 59, 41–47.

Nakao, H. (1991) Studies on acarid mites (Acari: Astigmata) damaging vegetable plants. II. Damage to vegetable seedlings. *Japanese Journal of Applied Entomology and Zoology* 35, 303–309.

Nakao, H. and Kurosa, K. (1988) Description of four species of acarid mites newly recorded from Japan, with reference to the damage caused to crops (Acari: Astigmata). *Japanese Journal of Applied Entomology and Zoology* 32, 135–142.

Nakao, H., Saitō, Y. and Mori, H. (1990) Mass production of predatory mites. In: Bay-Petersen, J. (ed.) *The Use of Natural Enemies to Control Agricultural Pests.* FFTC Series 40, Food and Fertilizer Technology Center, Asian and Pacific Region, Taipei, pp. 184–189.

Nakao, S., Nohara, K. and Nagatomi, A. (1977) Studies on pests and their predators at two groves of 'Kuroshima-mikan', a native citrus on Nagashima, Kagoshima Pref. *Applied Entomology and Zoology* 12, 334–346.

Nakayama, K., Abreu, J.M. de and Ferraz, E.C. de A. (1987) Controle quimico do ácaro mexicano *Tetranychus mexicanus* (McGregor, 1950) em mudas de cacaueiro. *Revista Theobroma* 17, 9–16.

Nalepa, A. (1887) Die Anatomie der Phytopten. *Sitzungsberichte der Kaiserlichen Akademie der Wissenschaften, Mathematisch-naturwissenschaftliche Classe, Abteilung I* 96, 115–165.

Nalepa, A. (1890) Zur Systematik der Gallmilben. *Sitzungsberichte der Kaiserlichen Akademie der Wissenschaften, Mathematisch-naturwissenschaftliche Classe, Abteilung I* 99, 40–69.

Nalepa, A. (1891) Genera und species der familae Phytoptidae. *Sitzungsberichte der Kaiserlichen Akademie der Wissenschaften, Mathematisch-naturwissenschaftliche Classe, Abteilung I* 58, 875–884.

Nalepa, A. (1892) Neue Gallmilben (5. Fortsetzung). *Anzeiger der Akademie der Wissenschaften in Wien, Mathematische-naturwissenchaftliche Klasse, Wien* 29(19), 190–192.

Nalepa, A. (1897) Neue Gallmilben (15. Fortsetzung). *Anzeiger der Akademie der Wissenschaften in Wien, Mathematische-naturwissenchaftliche Klasse, Wien* 34(23), 231-233.

Nalepa, A. (1898) Neue Gallmilben (16. Fortsetzung). *Anzeiger der Akademie der Wissenschaften in Wien, Mathematische-naturwissenchaftliche Klasse, Wien* 35(17), 163–164.

Nalepa, A. (1906) Ueber zwei neve Eriophyiden von den Fidschiinseln. *Journal of Economic Biology* 1, 147–151.

Nalepa, A. (1909a) Eriophyiden. In: Rehniner, K. (ed.) *Botanische und Zoologische Ergebnisse einer Wissenschaftlichen Forschungsreise nach den Samooninseln, dem Neuguinea-Archipel und den Salomoainseln, dem Neuguinea-Archipel und den Salomoninseln, von Maerz dis Dez. 1905. Denkschriften der Kaiserlichen Akademie der Wissenschaften, Mathematisch-naturwissenschaftliche Klasse* 84, 523–536.

Nalepa, A. (1909b) Neue Gallmilben (30. Fortsetzung). *Anzeiger der Akademie der Wissenschaften in Wien, Mathematische-naturwissenchaftliche Klasse, Wien* 46(10), 116–117.

Nalepa, A. (1916) Neue Gallmilben (32. Fortsetzung). *Anzeiger der Akademie der Wissenschaften in Wien, Mathematische-naturwissenchaftliche Klasse, Wien* 53(22), 283–284.

Nalepa, A. (1918) Eriophyiden aus Java (Zweiter Beitrag.). *Verhandlungen der Zoologisch-Botanischen Gesellschaft in Wien* 68, 40–92.

Nalepa, A. (1929) Neuer Katalog der bisher Beschriebenen Gallmilben, ihrer Gallen und Wirtspflanzen. *Marcellia* 25, 67–183.

Namaghi, H.S. (2010) Mites (Acari: Prostigmata & Mesostigmata) inhabiting green plantings in urban environment of north-eastern Iran, including six new records. *Munis Entomology and Zoology Journal* 5, 123–130.

Namvar, P. and Arbabi, M. (2007) Study on biology, population fluctuations and rate of damages of yellow broad mite, *Polyphagotarsonemus latus* (Acari: Tarsonemidae), on different potato varieties in Jiroft. *Applied Entomology and Phytopathology* 74(2, 3–4), 23–43.

Nandakumar, R., Rush, M.C. and Correa, F. (2007) Association of *Burkholderia glumae* and *B. gladioli* with panicle blight symptoms on rice in Panama. *Plant Disease* 91, 767.

Nangia, N. and ChannaBasavanna, G.P. (1983) Effect of sulphur and dicofol on *Tetranychus ludeni* and its predator *Amblyseius tetranychivorus*. In: *Plant Protection for Human Welfare: 10th International Congress on Plant Protection. Proceedings of a Conference Held at Brighton, England, 20–25 November 1983, Volume 3.* British Crop Production Council, Croydon, UK, p. 1028.

Nangia, N. and ChannaBasavanna, G.P. (1989) Feeding potential of *Amblyseius tetranychivorus* (Acari: Phytoseiidae), an indigenous predator on selected tetranychid and tenuipalpid mites. *Indian Journal of Acarology* 10, 75–81.

Nangia, N., Nageshchandra, B.K. and Jagadish, P.S. (1990) Toxicity of some pesticides to two mite pests of mulberry, *Morus alba*. *Mysore Journal of Agricultural Sciences* 24, 349–352.

NAPPO (2007) Detections of Panicle Rice Mite, *Steneotarsonemus spinki*, in Stuttgart, Arkansas and Ithaca, New York – United States. North American Plant Protection Organization's Phytosanitary Alert System, Ottawa. Available at: http://www.pestalert.org/oprDetail.cfm?oprID=283 (accessed 15 August 2014).

Narayan, D.S. (1962) Morphological, biological and ecological studies of the winter grain mite, *Penthaleus major* (Dugès) Pt. 1. *Journal of the Zoological Society of India* 14, 45–63.

Narayanswamy, C.A. (2004) Studies on *Aceria cajani* ChannaBasavanna (Acari: Eriophyidae), mite vector of pigeonpea sterility mosaic disease with special reference to varietal screening and its intraplant distribution. MSc. thesis, University of Agricultural Sciences, Bangalore, India.

Nascimento, A.S. do, Caldas, R.C. and Silva, L.M.S. da (1982) Amostragem do ácaro da ferrugem *Phyllocoptruta oleivora* (Ashmead, 1879) (Acari: Eriophyidae). *Anais de Sociedade Entomológica do Brasil* 11, 157–162.

Nascimento de Vasconcelos, G.J., Moraes, G.J. de, Delalibera, I. Jr and Knapp, M. (2008) Life history of the predatory mite *Phytoseiulus fragariae* on *Tetranychus evansi* and *Tetranychus urticae* (Acari: Phytoseiidae, Tetranychidae) at five temperatures. *Experimental and Applied Acarology* 44, 27–36.

Nash, A.J. (1991) Summary of report on bud mite research. *International Camellia Journal* 23, 87–89.

Nassar, O.A. and Ghai, S. (1981) Taxonomic studies on tetranychoid mites infesting vegetable and fruit crops in Delhi and surrounding areas. *Oriental Insects* 15, 333–396.

Nassar, O.A. and Kandeel, M.M.H. (1986) Biological and morphological studies on the predatory mite *Acaropsellina sollers* (Rohdendrof), preying on different stages of spider mites (Acari: Cheyletidae). *Bulletin of the Entomological Society of Egypt* 66, 177–187.

Natarajan, K. and Beevi, D. (1992) Control of *Aceria cajani* (Eriophyidae: Acarina) on pigeonpea (*Cajanus cajan* (L.) Millsp. *International Pigeonpea Newsletter* 16, 22–24.

National Research Council (2002) *Predicting Invasions of Nonindigenous Plants and Plant Pests*. National Academy Press, Washington, DC.

Nault, L.R. and Styer, W.E. (1969) The dispersal of *Aceria tulipae* and three other grass-infesting eriophyid mites in Ohio. *Annals of the Entomological Society of America* 62, 1446–1455.

Nault, L.R. and Styer, W.E. (1970) Transmission of an eriophyid-borne wheat pathogen by *Aceria tulipae*. *Phytopathology* 60, 1616–1618.

Nault, L.R., Briones, M.L., Williams, L.E. and Barry, B.D. (1967) Relation of the wheat curl mite to kernel red streak of corn. *Phytopathology* 57, 986–989.

Nault, L.R., Styer, W.E., Gordon, D.T., Bradfute, O.E., LaFever, H.N. and Willliams, L.E. (1970) An eriophyid-borne pathogen from Ohio, and its relation to wheat spot mosaic virus. *Plant Disease Reporter* 54, 156–160.

Naumann, I.D. (1993) *CSIRO Handbook of Australian Insect Names. Common and Scientific Names for Insects and Allied Organisms of Economic and Environmental Importance*, 6th edn. CSIRO (Commonwealth Scientific and Industrial Research Organisation), Melbourne, Australia.

Navajas, M. and Fenton, B. (2000) The application of molecular markers in the study of diversity in acarology: a review. *Experimental and Applied Acarology* 24, 751–774.

Navajas, M. and Návia, D. (2010) DNA-based methods for eriophyoid mite studies: review, critical aspects, prospects and challenges. *Experimental and Applied Acarology* 51, 257–271.

Navajas, M., Gutierrez, J., Bonato, O., Bolland, H.R. and Mapangou-Divassa, S. (1994) Intraspecific diversity of the cassava green mite *Mononychellus progresivus* (Acari: Tetranychidae) using comparisons of mitochondrial and nuclear ribosomal DNA sequences and cross-breeding. *Experimental and Applied Acarology* 18, 351–360.

Navajas, M., Gutierrez, J., Lagnel, J. and Boursot, P. (1996) Mitochondrial cytochrome oxidase I in tetranychid mites: a comparison between molecular phylogeny and changes of morphological and life history traits. *Bulletin of Entomological Research* 86, 407–417.

Navajas, M., Gutierrez, J. and Gotoh, T. (1997) Convergence of molecular and morphological data reveals phylogenetic information on *Tetranychus* species and allows the restoration of the genus *Amphitetranychus* (Acari: Tetranychidae). *Bulletin of Entomological Research* 87, 283–288.

Navajas, M., Gutierrez, J., Lagnel, J., Fauvel, G. and Gotoh, T. (1999) DNA sequences and cross-breeding experiments in the hawthorn spider mite *Amphitetranychus viennensis* reveal high genetic differentiation between Japanese and French populations. *Entomologia Experimentalis et Applicata* 90, 113–122.

Navajas, M., Thistlewood, H., Lagnel, J., Marshall, D., Tsagkarakou, A. and Pasteur, N. (2001a) Field releases of the predatory mite *Neoseiulus fallacis* (Acari: Phytoseiidae) in Canada, monitored by pyrethroid resistance and allozyme markers. *Biological Control* 20, 191–198.

Navajas, M., Gutierrez, J., Williams, M. and Gotoh, T. (2001b) Synonymy between two spider mite species, *Tetranychus kanzawai* and *T. hydrangeae* (Acari: Tetranychidae), shown by ribosomal ITS2 sequences and cross-breeding experiments. *Bulletin of Entomological Research* 91, 117–123.

Navarro, M.J., Gea, F.J. and Escudero-Colomar, L.A. (2010) Abundance and distribution of *Microdispus lambi* (Acari: Microdispidae) in Spanish mushroom crops. *Experimental and Applied Acarology* 50, 309–316.

Navasero, M.M., Rejesus, B.M., Hirao, G.A. and Navasero, M.V. (2003) Mass-rearing of *Orius tantillus* (Mots.) (Hemiptera: Anthocoridae). *Philippine Entomologist* 17, 194.

Návia, D. (2004) Ácaros Eriophyoidea (Prostigmata) associados a palmeiras (Arecaceae), com ênfase no ácaro do coqueiro, *Aceria guerreronis* Keifer – espectro de hospedeiros e aspectos biogeográficos. PhD thesis, Escola Superior de Agricultura "Luiz de Queiroz", Univ. São Paulo, Brazil. Available at: http://www.teses.usp.br/teses/disponiveis/11/11146/tde-02072004-142301/ (accessed 15 August 2014).

Návia, D. and Flechtmann, C.H.W. (2000a) Eriophyid mites (Acari: Prostigmata) from mango, *Mangifera indica* L., in Brazil. *International Journal of Acarology* 26, 73–80.

Návia, D. and Flechtmann, C.H.W. (2000b) A new genus and two new species of eriophyoid mites from conifers in Brazil. *International Journal of Acarology* 26, 265–270.

Návia, D. and Flechtmann, C.H.W. (2002) Mite (Arthropoda: Acari) associates of palms (Arecaceae) in Brazil: VI. New genera and new species of Eriophyidae and Phytoptidae (Prostigmata: Eriophyoidea). *International Journal of Acarology* 28, 121–146.

Návia, D. and Flechtmann, C.H.W. (2004) Rediscovery and redescription of *Tetranychus gigas* (Acari, Prostigmata, Tetranychidae). *Zootaxa* 547, 1–8.

Návia, D. and Flechtmann, C.H.W. (2005) A new genus and five new species of Eriophyoidea (Prostigmata) associated with palm trees from Brazilian Amazon. *Zootaxa* 1078, 41–58.

Návia, D. and Flechtmann, C.H.W. (2008) Eriophyoid mites intercepted from plant germplasm in Brazil – addressing questions on new introduction pathways. In: Kreiter, B.M., McCoy, S., Migeon, K.D., Navajas, M.A., Tixier, M.S. and Vial, L. (eds) (2008) *Integrative Acarology: Proceedings 6th European Congress EURAAC (European Association of Acarologists, Montpellier, July, 21–25, 2008* (CD-ROM). EURAAC, France, abstracts, 58.

Návia, D. and Marsaro, A.L. Jr (2010) First report of the citrus hindú mite, *Schizotetranychus hindustanicus* (Hirst) (Prostigmata: Tetranychidae), in Brazil. *Neotropical Entomology* 39, 140–143.

Návia, D., Flechtmann, C.H.W. and Amrine, J.W. Jr (2005a) Supposed ovoviviparity and viviparity in the coconut mite, *Aceria guerreronis* Keifer (Prostigmata: Eriophyidae), as a result of female senility. *International Journal of Acarology* 31, 63–65.

Návia, D., Moraes, G.J. de, Roderick, G. and Navajas, M. (2005b) The invasive coconut mite *Aceria guerreronis* (Acari: Eriophyidae): origin and invasion sources inferred from mitochondrial (16S) and nuclear (ITS) sequences. *Bulletin of Entomological Research* 95, 505–516.

Návia, D., Mendonça, R.S. and Melo, L.A.M.P. de (2005c) *Steneotarsonemus spinki* – an invasive tarsonemid mite threatening rice crops in South America. In: Alford, D.V. and Backhaus, G.F. (eds) *Plant Protection and Plant Health in Europe: Introduction and Spread of Invasive Species, Held at Humboldt University, Berlin, Germany, 9–11 June 2005*. Symposium Proceedings No. 81, British Crop Production Council, Alford, UK, pp. 267–268.

Návia, D., Mendonca, R.S. de and Ochoa, R. (2006a) The rice mite *Steneotarsonemus spinki* Smiley – an invasive species in America. In: *XIIth International Congress on Acarology, 21-26 August 2006, Amsterdam*, abstracts.

Návia, D., Truol, G., Mendonça, R.S. and Sagadín, M. (2006b) *Aceria tosichella* Keifer (Acari: Eriophyidae) from wheat streak mosaic virus-infected wheat plants in Argentina. *International Journal of Acarology* 32, 189–193.

Návia, D., Mendes, M.A.S. and Ochoa, R. (2006c) *Steneotarsonemus furcatus* DeLeon (Prostigmata: Tarsonemidae) infesting rice crops in Brazil. *International Journal of Acarology* 32, 219–222.

Návia, D., Moraes, G.J. de and Querino, R.B. (2006d) Geographic variation in the coconut mite, *Aceria guerreronis* Keifer (Acari: Eriophyidae): a geometric morphometric analysis. *International Journal of Acarology* 32, 301–314.

Návia, D., Moraes, G.J. de and Flechtmann, C.H.W. (2007a) Phytophagous mites as invasive alien species: quarantine procedures. In: Morales-Malacara, J.B., Behan-Pelletier, V., Ueckermann, E., Pérez, T.M., Estrada-Venegas, E.G. and Badii, M. (eds) *Acarology XI: Proceedings of the International Congress, 8–13 September 2002, Mérida, Universidad Nacional Autonoma de México*. Sociedad Latinoamericana de Acarología, México, pp. 307–316.

Návia, D., Mendonça, R.S., Batista, M.F., Truol, G., Pereira, P.R.V.S., Guedes, J.V.C., Castiglioni, E. and Morel, N.N.E. (2007b) The wheat curl mite, *Aceria tosichella* Keifer, and associated viruses, wheat streak mosaic virus and high plain virus – the risks posed to cereal crops in South America In: *Congress Proceedings: XVI International Plant Protection Congress: Glasgow, 15–18 October 2007, Scottish Exhibition & Conference Centre, Glasgow, Scotland, UK, Volume 2*. British Crop Production Council, Alton, UK, pp. 612–613.

Návia, D., Manoel, G.C. Gondim Jr and Moraes, G.J. de (2007c) Eriophyoid mites (Acari: Eriophyoidea) associated with palm trees. *Zootaxa* 1389, 1–30.

Návia, D., Ochoa, R.,Welbourn, C. and Ferragut, F. (2010a) Adventive eriophyoid mites: a global review of their impact, pathways, prevention and challenges. *Experimental and Applied Acarology* 51, 225–255.

Návia, D., Santos de Mendonça, R. and Ochoa, R. (2010b) The rice mite *Steneotarsonemus spinki*, an invasive species in the Americas. In: Sabelis, M.W. and Bruin, J. (eds) *Trends in Acarology: Proceedings of the 12th International Congress, 21–26 August 2006, Amsterdam*. Springer, Dordrecht, the Netherlands, pp. 379–384.

Návia, D., Flechtmann, C.H.W., Lindquist, E.E. and Aguilar, H. (2011) A new species of *Abacarus* (Acari: Prostigmata: Eriophyidae) damaging sugarcane, *Sacharrum officinarum* L., from Costa Rica – the first eriophyoid mite described with a tibial seta on leg II. *Zootaxa* 3025, 51–58.

Návia, D., Marsaro, A.L. Jr, Correa Gondim, M.G. Jr, Santos de Mendonça, R. and Valle da Silva Pereira, P.R. (2013) Recent mite invasions in South America. In: Peña, J. (ed.) *Potential Invasive Pests of Agricultural Crops*. CAB International, Wallingford, UK, pp. 251–287.

Nawar, M.S., Zaher, M.A., El Enany, M.A.M. and Ibrahim, A.A. (2001a) Life tables and food range of *Typhlodromus athiasae* at different temperatures (Acari: Phytoseiidae). *Egyptian Journal of Agricultural Research* 79, 1291–1302.

Nawar, M.S., Zaher, M.A., El Enany, M.A.M. and Ibrahim, A.A. (2001b) Life table parameters of *Phytoseius plumifer* (Canestrini & Fanzago) reared at different temperatures (Acari: Phytoseiidae). *Egyptian Journal of Agricultural Research* 79, 1341–1352.

Nayak, H.G., Hugar, P.S. and Mahabaleshwar Hegde (2008a) Biology of leaf mite, *Oligonychus oryzae* (Hirst.) on paddy. *Annals of Plant Protection Sciences* 16, 81–82.

Nayak, H.G., Hugar, P.S. and Hegde, M.G. (2008b) Chemical control of paddy leaf mite (*Oligonychus oryzae* (Hirst)) on paddy. *Journal of Maharashtra Agricultural Universities* 33, 272–273.

Nazir, Z. (2009) Sustainable management of *Chilo partellus* (Swinhoe) and *Oligonychus pratensis* (Banks) on spring maize crop, in Punjab, Pakistan. PhD thesis, University of Agriculture, Faisalabad, Pakistan.

NBN (2015) National Biodiversity Network, Nottingham, UK. Available at the National Biodiversity Network Gateway: https://data.nbn.org.uk/Taxa/NHMSYS0020190410 (accessed 26 February 2015).

Neal, J.W. Jr (ed.) (1979) *Guidelines for Control of Insect and Mite Pests of Foods, Fibers, Feeds, Ornamentals, Livestock, Forests, and Forest Products*. Agriculture Handbook No. 554, US Department of Agriculture, Washington, DC.

Neelu Nangia, Jagadish, P.S. and Nageshchandra, B.K. (1990) Biology and control of *Oligonychus mangiferus* (Rahman and Sapra) (Acari: Tetranychidae) on *Terminalia spp.*, important host plant of silk worms. *Mysore Journal of Agricultural Sciences* 23, 355–358.

Neena Singla (2001) Effect of weather factors and the predatory mite, *Amblyseius* (*Euseius*) *delhiensis* (Narayanan & Kaur) on populations of *Brevipalpus phoenicis* (Geijskes) and *Eutetranychus orientalis* (Klein) on guava. *Pest Management and Economic Zoology* 9, 135–140.

Neena Singla and Sadana, G.L. (1998) Population development of *Eutetranychus orientalis* (Klein) on citrus: a correlation with weather factors and predators. *Indian Journal of Ecology* 25, 32–35.

Neena Singla and Sadana, G.L. (2000) Population of *Brevipalpus phoenicis* on *Citrus aurantifolia* var. sourlime in relation to seasonal fluctuations and predators. *Annals of Agri Bio Research* 5, 183–187.

Negm, M.W. (2012) Towards the biological control of the Old World date mite (Ghoubar Mite), *Oligonychus afrasiaticus* on date palm using phytoseiid mites from Saudi Arabia (Acari: Tetranychidae, Phytoseiidae) In: *Proceedings of The Third Scientific Conference for Higher Education Students, May 2012, Ministry of Higher Education, Al-Khobar, Saudi Arabia*, pp. 15–16.

Nehru, R.K. and Bhagat, K.C. (2006a) Biology of false spider mite, *Brevipalpus californicus* Banks on sweet orange, *Citrus sinensis* Osbeck cv. Jaffa. *Annals of Plant Protection Sciences* 14, 300–303.

Nehru, R.K. and Bhagat, K.C. (2006b) Life table studies on *Brevipalpus californicus* Banks (Acarina: Tenuipalpidae). *Environment and Ecology* 24(Special 3), 499–501.

Nehru, R.K., Bhagat, K.C., Koul, V.K. and Rajinder, B. (2005) Preferential distribution of false spider mite, *Brevipalpus californicus* Banks on sweet orange, *Citrus sinensis* Osbeck. *Annals of Plant Protection Sciences* 13, 396–399.

Nelson, E.E., Croft, B.A. Hewitt, A.J. and Jones, A.L. (1973) Toxicity of apple orchard pesticides to *Agistemus fleschneri*. *Environmental Entomology* 2, 219–222.

Nemati, A., Saeidi, Z., Soleimannejadian, E., Shishehbor, P. and Kamali, K. (2010a) Effect of temperature on development and reproductive parameters of *Tetranychus turkestani* U. & N. (Acari: Tetranychidae). In: Moraes, G.J. de, Castilho, R.C. and Flechtmann, C.H.W. (eds) *XIII International Congress of Acarology, August 23–27, 2010a, Recife-PE, Brazil, Abstract Book*, abstract 326, p. 177. Available at: http://www.acarology.org/ica/ica2010/docs/0910_15_abstractBookCapa.pdf (accessed 3 February 2015).

Nemati, A., Saeidi, Z., Soleimannejadian, E., Shishehbor, P. and Kamali, K. (2010b) Population dynamics of strawberry spider mite, *Tetranychus turkestani* U. & N. (Acari: Tetranychidae), on aubergine. In: Moraes, G.J. de, Castilho, R.C. and Flechtmann, C.H.W. (eds) *XIII International Congress of Acarology, August 23–27, 2010b, Recife-PE, Brazil, Abstract Book*, abstract 327, pp. 177–178. Available at: http://www.acarology.org/ica/ica2010/docs/0910_15_abstractBookCapa.pdf (accessed 3 February 2015).

Němec, B. (1924) Untersuchung uber Eriophyidengallen. *Studies from the Plant Physiological Laboratory of Charles University (Prague)* 5(2), 47–94.

Nemestothy, K.K. (1983) A Magyarorszagon elofordulo tetuatkak (Acari: Tarsonemidae) (The tarsonemid species occurring in Hungary (Acari: Tarsonemidae)). *Növényvédelem* 19, 198-202 (in Hungarian).

Nemoto, H. (1991) Ecological and morphological studies on the eriophyid and tarsonemid mites injurious to horticultural plants and their control. *Bulletin of the Saitama Horticultural Experiment Station* 3, 47–77.

Nemoto, H. and Aoki, J. (1975) *Entomophthora floridana* (Entomophthorales: Entomophthoraceae) attacking the sugi spider mite, *Oligonychus hondoensis* (Acarina: Tetranychidae), in Japan. *Applied Entomology and Zoology* 10, 90–95.

Nene, Y.L., Beniwal, S.P.S., Reddy, M.V., Deena, E., Sithananthan, S. and Rameswara Rao, V. (1983) Nature and extent of pigeonpea sterility mosaic under field conditions. In: *All India Symposium in Acarology, 26–28 May 1983, Pune, India*, abstracts.

Nene, Y.L., Sheila, V.K. and Sharma, S.B. (1984) *A World List of Chickpea and Pigeonpea Pathogens*. Pulse Pathology, Progress Report 32, ICRISAT (International Crops Research Institute), Patancheru, India.

Nene, Y.L., Hall, S.D. and Sheila, V.K. (eds) (1990) *The Pigeonpea*. CAB International, Wallingford, UK.

Nentwig, W., Frank, T. and Lethmayer, C. (1998) Sown weed strips: artifical ecological compensation areas as an important tool in conservation biological control. In: Barbosa, P. (ed.) *Conservation Biological Control*. Academic Press, San Diego, California, pp. 133–153.

Nestel, B.I. (1976) Introduction. In: *African Cassava Mosaic Report, Kenya*. IDRC (International Development Research Centre), Ottawa, pp. 5–6.

Netto, A.V. (1987) Causas do insucesso no controle do ácaros dos pomares de citros. *Laranja* 1(8), 51–59.

Neunzig, H.H. and Galletta, G.J. (1977) Abundance of the blueberry bud mite (Acarina: Eriophyidae) on various species of blueberry. *Journal of the Georgia Entomological Society* 12, 182–184.

New South Wales Department of Agriculture (1975) Spider mites. *Entomology Branch Insect Pest Bulletin* 37, 1–11.

Newell, I.M. (1947) Quantitative methods in biological and control studies of orchard mites. *Journal of Economic Entomology* 40, 683–689.

Newell, I.M. (1963) Feeding habits in the genus *Balaustium* (Acarina: Erythraeidae), with special reference to attacks on man. *Journal of Parasitology* 49, 498–502.

Newkirk, R.A. and Keifer, H.H. (1971) Revision of types of *Eriophyes* and *Phytoptus*. In: Keifer, H.H. (ed.) *Eriophyid Studies C-5*. Agricultural Research Service, US Department of Agriculture, Washington, DC.

Newkirk, R.A. and Keifer, H.H. (1975) Eriophyoidea: synoptic keys to groups and genera. In: Jeppson, L.R., Keifer, H.H. and Baker, E.W. (eds) *Mites Injurious to Economic Plants*. University of California Press, Berkeley, California, pp. 562–587.

Newman, L.J. (1923) Red legged velvet earth mite. *Western Australia Department of Agriculture Bulletin* No. 106, 1–4.

Nguyen, D.T., Vangansbeke, D. and de Clercq, P. (2014) Artificial and factitious foods support the development and reproduction of the predatory mite *Amblyseius swirskii*. *Experimental and Applied Acarology* 62, 181–194.

Nguyen, T.V. and Thomas Shih, C.-I. (2012) Life-table parameters of *Neoseiulus womersleyi* (Schicha) and *Euseius ovalis* (Evans) (Acari: Phytoseiidae) feeding on six food sources. *International Journal of Acarology* 38, 197–205.

Nicholls, R., Buxton, J., Umpelby, R. and Dennis, E.B. (1986) Protection of black currant against black currant gall mite with the synthetic pyrethroid fenpropathrin. In: *Proceedings of the Brighton Crop Protection Conference: Pests and Diseases – 1986, Volume 1*. British Crop Production Council, Croydon (now Alton), UK, pp. 129–135.

Nickel, J.L. (1958) Agricultural insects of the Paraguayan Chaco. *Journal of Economic Entomology* 51, 633–636.

Nickel, J.L. (1960) Temperature and humidity relationships of *Tetranychus desertorum* Banks with special reference to distribution. *Hilgardia* 30(2), 41–100.

Nicotina, M. and Viggiani, G. (1985) Distribuzione e densità stagionali delle popolazioni degli Acari fitoseidi in niccioleti delle Campania. *Atti XIV Congresso Nazionale Italiano di Entomologia, Palermo-Erice-Bagheria, 28 Maggio–1 Giugno 1985, Vol. 1*. Accademia Nazionale Italiana di Entomologia, Palermo, Italy, pp. 729–732.

Nielsen, F. (1958) Biology of the McDaniel mite, *Tetranychus mcdanieli* McGregor, in Utah. *Journal of Economic Entomology* 51, 588–592.

Nielsen, S.L. (1986) Postharvest sprayings with 2 systemic pesticides against the black currant gall mite (*Cecidophyopsis ribis* Westw.) on black currant (*Ribes nigrum*). *Danish Journal of Plant and Soil Science* 90, 385–388.

Nielsen, S.L. (1987) Pesticides tested for the control of the black currant gall mite (*Cecidophyopsis ribis*, Westw.). *Journal of Horticultural Science* 62, 27–30.

Nienstaedt-Arreaza, B.M. (2007) Estudio de algunos aspectos biológicos y ecológicos del ácaro hindú de los cítricos *Schizotetranychus hindustanicus* (Hirst, 1924) (Acari: Tetranychidae), en Maracay, Venezuela. MSc. thesis, Universidad Central de Venezuela, Maracay, Venezuela.

Nietner, J. (1861) *Observations on the Enemies of the Coffee Tree in Ceylon*. Colombo, Ceylon.

Night, G.N.G., Asiimwe, P., Gashaka, G., Nkezabahizi, D., Legg, J.P., Okao-Okuja, G., Obonyo, R., Nyirahorana, C., Mukakanyana, C., Mukase, F., *et al.* (2011) Occurrence and distribution of cassava pests and diseases in Rwanda. *Agriculture, Ecosystems and Environment* 140, 492–497.

Niklas, K.J., Tiffney, B.H. and Knoll, A.H. (1983) Patterns in vascular land plant diversification. *Nature* 303, 614–616.

Nikolskii, V.V. (1953) Mites injurious to tea shrub in Transcaucasus. *Plant Protection (Leningrad)* 4, 126–128 (in Russian).

Nims, R.C., Halliwell, R.S. and Rosberg, D.W. (1967) Wound healing in cultured tobacco cells following microinjections. *Protoplasma* 64, 305–314.

Nishida, T. and Holdoway, F.G. (1955) *The Erinose Mites of Lychee*. Circular 48, Hawaiian Agricultural Experiment Station, Honolulu.

Nishimura, S., Hinomoto, N. and Takafuji, A. (2010) Development of microsatellite markers for *Tetranychus kanzawai* (Acari: Tetranychidae) and analysis of spatio-temporal gene flow among populations on different host plants. In: Sabelis, M.W. and Bruin, J. (eds) *Trends in Acarology: Proceedings of the 12th International Congress, 21–26 August 2006*, Amsterdam. Springer, Dordrecht, the Netherlands, pp. 349–356.

Noble, W.E., Poe, S.L. and Stall, R.E. (1978) Surface sterilization of bulb mites for phytopathogen transmission studies. *Florida Entomologist* 61, 120.

Noggle, G.R. (1983) The organization of plants. In: Noggle, G.R. and Fritz, G.J. (eds) *Introductory Plant Physiology*. Prentice Hall, Englewood Cliffs, New Jersey, pp. 9–38.

Noguchi, S., Mori, N., Kurosa, K. and Kuwahara, Y. (1998) Chemical ecology of astigmatid mites XLIX. beta-acaridial (2(E)-(4-methyl-3-pentenylidene)-butanedial), the alarm pheromone of *Tyrophagus longior* Gervais (Acarina: Acaridae). *Applied Entomology and Zoology* 33, 53–57.

Nohara, K., Nakao, S. and Nagatomi, A. (2000) A study of the relationship between pesticide treatment and the fauna in citrus groves on Nagashima Island, Kagoshima Prefecture. *Applied Entomology and Zoology* 35, 271–281.

Norambuena, M.H. and Escobar, S.S. (2004) El ácaro *Tetranychus lintearius* Dufour (Acarina: Tetranychidae) exploración en el extranjero e internación a Chile. *Agro Sur* 32, 76–79.

Norambuena, H., Martinez, G., Carrillo, R. and Neira, M. (2007) Host specificity and establishment of *Tetranychus lintearius* (Acari: Tetranychidae) for biological control of gorse, *Ulex europaeus* (Fabaceae) in Chile. *Biological Control* 40, 204–212.

Noronha, A.C.S., Boaventura, V.J. and Alves, A.A.C. (2010) Oviposition rate as a tool for evaluating species of *Manihot* for resistance to cassava green mite (CGM), *Mononychellus tanajoa* (Acari: Tetranychidae). In: Moraes, G.J. de, Castilho, R.C. and Flechtmann, C.H.W. (eds) *XIII International Congress of Acarology, August 23–27, 2010, Recife-PE, Brazil, Abstract Book*, abstract 332, p. 180. Available at: http://www.acarology.org/ica/ica2010/docs/0910_15_abstract-BookCapa.pdf (accessed 3 February 2015).

Norris, K.R. (1948) Seasonal severity of the attack of the red-legged earth mite (*Halotydeus destructor*) on subterranean clover. *Australian Council of Science and Industrial Research Journal* 21, 7–15.

Norton, R.A. (1977) A review of F. Grandjean's system of leg chaetotaxy in the Oribatei and its application to Damaeidae. In: Dindal, D.L. (ed.) *Biology of Oribatid Mites, Volume 33*. SUNY College of Environmental Science and Forestry, Syracuse, New York, pp. 33–62.

Norton, R.A. and Ide, S.G. (1974) *Scutacarus baculitarsus agaricus* n. subsp. (Acarina: Scutacaridae) from commercial mushroom houses, with notes on phoretic behavior. *Journal of the Kansas Entomological Society* 47, 527–534.

Norton, R.A., Bonamo, P.M., Grierson, J.D. and Shear, W.A. (1988) Oribatid mites fossils from a terrestrial Devonian deposit near Gilboa, New York. *Journal of Paleontology* 62, 259–269.

Norton, R.A., Kethley, J.B., Johnston, D.E. and O'Connor, B.M. (1993) Phylogenetic perspectives on genetic systems and reproductive modes in mites. In: Wrensch, D. and Ebbert, M. (eds) *Evolution and Diversity of Sex Ratio in Insects and Mites*. Chapman and Hall, New York, pp. 8–99.

Norton, R.A., Alberti, G., Weigmann, G. and Woas, S. (1997) Porose integumental organs of oribatid mites (Acari, Oribatida). 1, Overview of types and distribution. *Zoologica* 146, 1–31.

NPAG (2008) *Amended NPAG Report* Steneotarsonemus spinki *Smiley: Rice Panicle Mite*. New Pest Advisory Group, Plant Epidemiology and Risk Analysis Laboratory, Center for Plant Health Science and Technology, US Department of Agriculture Animal and Plant Health Inspection Service (APHIS), Raleigh, North Carolina. Available at: http://www.aphis.usda.gov/plant_health/plant_pest_info/rice_mite/downloads/sspinki-npag.pdf (accessed 27 January 2015).

Nucifora, A. (1961) Un nuovo ácaro dannoso agli agrumi in Sicilia (*Hemitarsonemus latus* Banks). *Tecnica Agricola (Catania)* 12, 118–126.

Nucifora, A. (1963) Osservazioni sulla riproduzione di *Hemitarsonemus latus* (Banks) (Acarina: Tarsonemidae). *Atti della Accademia Nazionale Italiana di Entomologia* 10, 142–153.

Nucifora, A. (1967) Un grave pericolo incombe sull'agrumicoltura siciliana: l'eriofide rugginoso degli agrumi (*Aculus pelekassi* Keifer). *Tecnica Agricola (Catania)* 5, 3–14.

Nucifora, A. (1969) L'*Aculus pelekassi* Keifer è in Sicilia. *Tecnica Agricola (Catania)* 4, 1–6.

Nucifora, A. and Vacante, V. (1978) Ricerche sul controllo di *Panonychus ulmi* (Koch) nei meleti dell'Etna: soglia di intervento. In: *Atti, Giornate Fitopatologiche 1978, Catania-Acireale 8–10 Marzo 1978, Volume 1*. Fitogest®, Image Line Network, Faenza, Italy, pp. 215–222.

Nucifora, A. and Vacante, V. (1986) Laboratory observations on the biology of *Tarsonemus waitei* Banks. In: Cavalloro, R. and Di Martino, E. (eds) *Integrated Pest Control in Citrus-groves. Proceedings of the Experts' Meeting, 26–29 March 1985*, Acireale, Italy. Published for the Commission of European Communities by Balkema, Rotterdam, the Netherlands, pp. 203–207.

Nucifora, A. and Vacante, V. (2004) Citrus mites in Italy. VII. The family Tarsonemidae. Species collected and notes on ecology. *Acarologia* 44, 49–65.

Nucifora, A., Calabretta, C. and Vacante, V. (1979a) Aspetti bio-etologici di *Phytoptus avellanae* Nal. (Acarina, Eriophyidae) in Sicilia e gradi diversi di alterazioni delle gemme attaccate. *Tecnica Agricola (Catania)* 31, 309–326.

Nucifora, A., Vacante, V. and Galvagno, S. (1979b) Attuali possibilità di controllo della "Acarosi delle gemme del nocciolo" e risultati preliminari sulla resistenza di alcune cultivar all'acaropatia. *Atti del Convegno Nazionale "Il Miglioramento della Coltura del Mandorlo e del Nocciolo – Aspetti Genetici e Tecnici", Messina e Siracusa, 29–30 Novembre e 1 Dicembre*. La Nuova Grafica, Catania, pp. 221–235.

Nunez, S. and Maeso, D. (1983) Secondary action of fungicides on *Aculops lycopersici* on tomatoes. *Investigaciones Agronómicas (Montevideo)* 4, 3–5.

Nuzzaci, G. (1976) Feeding behavior of eriophyid mites. *Entomologica* 12, 75–80.

Nuzzaci, G. (1979) Contributo alla conoscenza dello gnathosoma degli eriofidi. *Entomologica* 15, 73–101.

Nuzzaci, G. and Alberti, G. (1996) Internal anatomy and physiology. In: Lindquist, E.E., Sabelis, M.W. and Bruin, J. (eds) *Eriophyoid Mites: Their Biology, Natural Enemies and Control*. World Crop Pests, Volume 6. Elsevier Science, Amsterdam, pp. 101–150.

Nuzzaci, G. and de Lillo, E. (1989) Contributo alla conoscenza dello gnatosoma degli Acari Tenuipalpidi (Tetranychoidea: Tenuipalpidae). *Entomologica* 24, 5–32.

Nuzzaci, G. and de Lillo, E. (1991a) Contributo alla conoscenza delle parti boccali di *Penthaleus major* (Dugès) (Acari: Eupodoidea Penthaleidae). *Atti XVI Congresso Nazionale Italiano di Entomologia, 23–28 Settembre 1991, Bari- Martina Franca*. Accademia Nazionale Italiana di Entomologia, Florence, Italy, pp. 265–277.

Nuzzaci, G. and de Lillo, E. (1991b) Fine structure and functions of the mouthparts involved in the feeding mechanisms in *Cenopalpus pulcher* (Canestrini and Fanzago) (Tetranychoidea: Tenuipalpidae). In: Schuster, R. and Murphy, P.W. (eds) *The Acari: Reproduction, Development and Life-history Strategies*. Chapman and Hall, London, pp. 367–376.

Nuzzaci, G. and Monaco, R. (1977) Danni al cipresso da *Trisetacus juniperinus* (Nalepa). *Informatore Fitopatologico* 27(11), 11–14.

Nuzzaci, G. and Vovlas, N. (1976) Osservazione dei caratteri tassinomici degli eriofidi al microscopio elettronico a scansione. In: *Atti XI Congresso Nazionale Italiano di Entomologia, 10–15 Maggio 1976, Portici-Sorrento*. Accademia Nazionale Italiana di Entomologia, Florence, Italy, pp. 117–122.

Nuzzaci, G., Mimmocchi, T. and Clement, S.L. (1985) A new species of *Aceria* (Acari: Eriophyidae) from *Convolvulus arvensis* L. (Convolvulaceae) with notes on other eriophyid associated of convolvulaceous plants. *Entomologica* 20, 81–89.

Nuzzaci, G., Di Palma, A., Magowski, W.L. and Aldini, P. (2002) Moutparts of *Tarsonemus nodosus* Schaarschmidt, 1959 (Acari: Tarsonemidae): fine structure and functional morphology. In: Bernini, F., Nannelli, R., Nuzzaci, G. and de Lillo, E. (eds) *Acarid Phylogeny and Evolution: Adaptation in Mites and Ticks. Proceedings of the IV Symposium of the European Association of Acarologists*. Kluwer Academic Publishers, Dordrecht, the Netherlands, pp. 269–281.

Nyiira, Z.M. (1976) Advances in research on the economic significance of the green cassava mite (*Mononychellus tanajoa*) in Uganda. In: Terry, E.R. and MacIntyre, R. (eds) *The International Exchange and Testing of Cassava Germplasm in Africa. Proceedings 1975.* International Development Research Centre, Ottawa, Canada, pp. 27–29.

Nyiira, Z.M. (1978) *Mononychellus tanajoa* (Bondar): biology, ecology and economic importance. In: Brekelbaum, T., Bellotti, A. and Lozano, J.C. (eds) *Proceedings CASSAVA Protection Workshop, 7–12 November 1977, Cali, Colombia.* CIAT (Centro Internacional de Agricultura Tropical), Cali, Colombia, pp. 155–159.

Nyiira, Z.M. (1982) *Mononychellus tanajoa* (Bondar): implications and approach to its management. In: Okoli, O.O., Enyinnia, T., Igbokwe, M.C., Odurukwe, S.O., Okeke, J.E. and Okereke, H.E. (eds.) *Proceedings of the International Workshop on Control of Cassava Mealybug and Green Spider Mite, 1981, National Root Crops Research Institute, Umudike, Nigeria.* Guinea Printing Press, Umuahia, Nigeria, pp. 47–64.

Nyoni, B.N., Gorman, K., Mzilahowa, T., Williamson, M.S., Navajas, M., Field, L.M. and Bass, C. (2011) Pyrethroid resistance in the tomato red spider mite, *Tetranychus evansi*, is associated with mutation of the *para*-type sodium channel. *Pest Management Science* 67, 891–897.

Nyrop, J.P., Minns, J.C. and Herring, C.P. (1994) Influence of ground cover on dynamics of *Amblyseius fallacis* Garman (Acarina; Phytoseiidae) in New York apple orchards. *Agriculture, Ecosystems and Environment* 50, 61–72.

Oatman, E.R. (1973) An ecological study of arthropod populations on apple in northeastern Wisconsin: population dynamics of mite species on the foliage. *Annals of the Entomological Society of America* 66, 122–131.

Oatman, E.R., Fleschner, C.A. and McMurtry, J.A. (1967) New, highly destructive spider mite present in southern California. *Journal of Economic Entomology* 60, 477–480.

Oatman, E.R., McMurtry, J.A., Gilstrap, F.E. and Voth, V. (1977) Effect of releases of *Amblyseius californicus, Phytoseiulus persimilis*, and *Typhlodromus occidentalis* on the twospotted spider mite on strawberry in southem California. *Journal of Economic Entomology* 70, 45–47.

Oatman, E.R., Wyman, J.A., Browning, H.W. and Voth, V. (1981) Effects of releases and varying infestation levels of the two spotted spider mite on strawberry yield in southern California. *Journal of Economic Entomology* 74, 112–115.

Oatman, E.R., Sances, F.V., LaPré, L.P., Toscano, N.C. and Voth, V. (1982) Effect of different infestation levels of the two-spotted spider mite on strawberry yield in winter plantings in southern California. *Journal of Economic Entomology* 75, 94–96.

Obrycki, J.J. and Kring, T.J. (1998) Predaceous Coccinellidae in biological control. *Annual Review of Entomology* 43, 295–321.

Ocete, R. and Pérez, M.A. (1996) Efectos de la aplicación de extractos de *Daphne gnidium* L. y *Anagyis foetida* L. sobre diversos grupos taxonómicos. *Boletín de Sanidad Vegetal, Plagas* 22, 45–56.

Ocete, R. and Skuhrava, M. (1996) Nota sobre la presencia de *Arthrocnodax vitis* Rubsaamen (Diptera, Cecidomyiidae) en erineos de poblaciones de vid silvestre de la región subbética. *Boletín de Sanidad Vegetal, Plagas* 22, 465–468.

Ochiai, N., Mizuno, M., Mimori, N., Miyake, T., Dekeyser, M., Canlas, L.J. and Takeda, M. (2007) Toxicity of bifenazate and its principal active metabolite, diazene, to *Tetranychus urticae* and *Panonychus citri* and their relative toxicity to the predaceous mites, *Phytoseiulus persimilis* and *Neoseiulus californicus. Experimental and Applied Acarology* 43, 181–197.

Ochieng, R.S., Oloo, G.W. and Amboga, E.O. (1987) An artificial diet for rearing the phytoseiid mite *Amblyseius teke* Pritchard and Baker. *Experimental and Applied Acarology* 3, 169–173.

Ochoa, R. (1989a) A note on paedogenesis in Tetranychoidea. *International Journal of Acarology* 15, 117–118.

Ochoa, R. (1989b) The genus *Tuckerella* in Costa Rica (Acari: Tuckerellidae). *International Journal of Acarology* 15, 205–207.

Ochoa, R. and Aguilar, H. (1989) Combate quimico de la aranita roja (*Tetranychus urticae* Koch) en fresa (*Fragaria* sp.). *Manejo Integrado de Plagas* 11, 51–60.

Ochoa, R., Aguilar, H. and Merino, F.L. (1989) Chemical control of red spider mites (Acari: Tetranychidae) in chayote (*Sechium edule* (Jacq.) SW.). *Manejo Integrado de Plagas* 14, 31–45.

Ochoa, R., Aguilar, H. and Sanabria, C. (1990) Acaros fitoparasitos asociados al cultivo del mango (*Mangifera indica* L.) en Costa Rica. *Manejo Integrado de Plagas* 16, 32–37.

Ochoa, R., Aguilar, H. and Vargas, C. (1991a) *Acaros Fitofagos de America Central: Guia Ilustrada.* Serie Técnica: Manual Técnico No. 6, CATIE (Centro Agronómico Tropical de Investigación y Enseñanza), Turrialba, Costa Rica.

Ochoa, R., Smiley, R.L. and Saunders, J.L. (1991b) The family Tarsonemidae in Costa Rica (Acari: Heterostigmata). *International Journal of Acarology* 17, 41–86.

Ochoa, R., Aguilar, H. and Vargas, C. (1994) *Phytophagous Mites of Central America: An Illustrated Guide*, English edn. Serie Tecnica: Manual Técnico No. 6, CATIE (Centro Agronómico Tropical de Investigación y Enseñanza), Turrialba, Costa Rica.

Ochoa, R., Pettis, J.S., Erbe, E. and Wergin, W.P. (2005) Observations on the honey bee tracheal mite *Acarapis woodi* (Acari: Tarsonemidae) using low-temperature scanning electron microscopy. *Experimental and Applied Acarology* 35, 239–249.

Ochoa, R., Beard, J.J., Bauchan, G.R., Kane, E.C., Dowling, A.P.G. and Erbe, E.F. (2011) Herbivore exploits chink in armour of host. *American Entomologist* 57, 26–29.

O'Connor, B.M. (1998) Keys to North American Genera: Family Acaridae Latreille, 1802. Available at: http://insects.ummz.lsa.umich.edu/PEET/nagenera.html (accessed 18 May 2015).

O'Connor, B.M. (2008) Family Acaridae Latreille, 1802 (Species listing from the 'Biology Catalog' of Texas A&M University). Available at: http://insects.tamu.edu/research/collection/hallan/Acari/Family/Acaridae.txt http://bug.tamu.edu/research/collection/hallan/Acari/Family/Acaridae.txt (accessed 18 May 2015).

O'Connor, B.M. (2009) Cohort Astigmatina. In: Krantz, G.W. and Walter, D.E. (eds.) *A Manual of Acarology*, 3rd edn. Texas Tech University Press, Lubbock, Texas, pp. 565–657.

O'Dowd, D.J. and Willson, M.F. (1989) Leaf domatia and mites on Australian plants: ecological and evolutionary implications. *Biological Journal of the Linnean Society* 37, 191–236.

O'Dowd, D.J. and Willson, M.F. (1997) Leaf domatia and the distribution and abundance of foliar mites in broadleaf deciduous forest in Wisconsin. *American Midland Naturalist* 137, 337–348.

OEPP/EPPO (2006) *Bulletin OEPP/EPPO Bulletin* 36, 161–163.

Ohashi, K., Ehara, S., Kunimoto, Y., Amano, H. and Takafuji, A. (2009) The occurrence of *Schizotetranychus baltazari* Rimando (Acari: Tetranychidae) in Japan. *Journal of the Acarological Society of Japan* 18, 29–31.

Ohkubo, N. (1983) Role of petroleum oil sprays in an integrated pest management system of citrus crops in Japan. In: Matsumoto, K. (ed.) *Proceedings of the 4th International Citrus Congress, Tokyo, Japan, 9–12 November 1981, Volume 2.* International Society of Citriculture, Riverside, California, pp. 611–614.

Ohnishi, R., Kanazaki, S. and Ohmasa, Y. (2008) Relation between the wintering population density and succeeding seasonal prevalence of occurrence of pink citrus rust mite, *Aculops pelekassi* (Keifer) on satsuma mandarin. *Bulletin of the Ehime Fruit Tree Experiment Station* 22, 25–35 (in Japanese).

Ohno, S., Miyagi, A., Ganaha-Kikumura, T., Gotoh, T., Kitashima, Y., Ooishi, T., Ando, T., Kijima, K., Futagami, K., Uesato, T. and Yasuda, K. (2009) Species composition of spider mites (Acari: Tetranychidae) on vegetables in Okinawa, southwestern Japan. *Applied Entomology and Zoology* 44, 627–633.

Ohno, S., Miyagi, A., Ganaha-Kikumura, T., Gotoh, T., Kijima, K., Ooishi, T., Moromizato, C., Haraguchi, D., Yonamine, K. and Uezato, T. (2010) Non-crop host plants of *Tetranychus* spider mites (Acari: Tetranychidae) in the field in Okinawa, Japan: determination of possible sources of pest species and inference on the cause of peculiar mite fauna on crops. *Applied Entomology and Zoology* 45, 465–475.

Ohno, S., Miyagi, A., Gotoh, T., Ganaha-Kikumura, T., Shiromoto, K., Kijima, K. and Ooishi, T. (2011a) Wild host plants of four spider mite species (Acari: Tetranychidae) infesting fruit crops in Okinawa. *Journal of Asia-Pacific Entomology* 14, 281–284.

Ohno, S., Kodama, H. and Ganaha-Kikumura, T. (2011b) Temporal occurrence of the two-spotted spider mite, *Tetranychus urticae* (Acari: Tetranychidae), on Ishigaki Island, Okinawa. *Journal of the Acarological Society of Japan* 20, 37–40.

Ohno, S., Takaki-Kikuchi, M., Kodama, H., Kurima, M., Kijima, K., Ganaha-Kikumura, T., Yamaguchi, A., Shimatani, M. and Miyagi, A. (2012) Range expansion of an invasive pest spider mite, *Tetranychus evansi* (Acari: Tetranychidae), in Okinawa, Southwestern Japan, and the acaricidal effect of several agrochemicals on an Okinawan population of the species. *Japanese Journal of Applied Entomology and Zoology* 56, 74–77.

Oi, D.H., Sanderson, J.P., Youngman, R.R. and Barnes, M.M. (1989) Development times of the Pacific spider mite (Acari: Tetranychidae) on water-stressed almond trees. *Environmental Entomology* 18, 208–212.

Okabe, K. (2006) Pest on commercial mushrooms in Japan. *Bulletin of the Forestry and Forest Products Research Institute, Ibaraki* 5, 119–133 (in Japanese).

Okabe, K. and Amano, H. (1990) Attractancy of alcohols isolated from culture filtrates of *Fusarium* fungi for the robine bulb mite, *Rhizoglyphus robini* Claparede (Acari: Acaridae), in sand. *Applied Entomology and Zoology* 25, 397–404.

Okabe, K. and Amano, H. (1991) Penetration and population growth of the robine bulb mite, *Rhizoglyphus robini* Claparede (Acari: Acaridae), on healthy and *Fusarium*-infected rakkyo bulbs. *Applied Entomology and Zoology* 26, 129–136.

Okasabe, M. and Saitō, Y. (1991) Morphological differences between the diapausing and non-diapausing forms of the citrus red mite, *Panonychus citri* (McGregor) (Acari: Tetranychidae). *Bulletin of the Fruit Tree Research Station* 21, 95–115 (in Japanese).

Okassa, M., Tixier, M.-S., Cheval, B. and Kreiter, S. (2009) Molecular and morphological evidence for a new species status within the genus *Euseius* (Acari: Phytoseiidae). *Canadian Journal of Zoology* 87, 689–698.

Oku, K. (2008) Role of excreta in predator avoidance by the Kanzawa spider mite, *Tetranychus kanzawai* (Acari: Tetranychidae). *European Journal of Entomology* 105, 619–623.

Oku, K. (2010) Do quiescent deutonymph females guarded by a male attract additional males in spider mites? In: Moraes, G.J. de, Castilho, R.C. and Flechtmann, C.H.W. (eds) *XIII International Congress of Acarology, August 23–27, 2010, Recife-PE, Brazil, Abstract Book*, abstract 350, p. 189. Available at: http://www.acarology.org/ica/ica2010/docs/0910_15_abstractBookCapa.pdf (accessed 3 February 2015).

Oku, K., Yano, S. and Takafuji, A. (2003) Spider mite's use of a refuge during the quiescent stage in the presence of a predator. *Entomologia Experimentalis et Applicata* 108, 71–74.

Oku, K., Yano, S., Osakabe, M. and Takafuji, A. (2005) Mating strategies of *Tetranychus kanzawai* (Acari: Tetranychidae) in relation to mating status of females. *Annals of the Entomological Society of America* 98, 625–628.

Oldfield, G.N. (1969) The biology and morphology of *Eriophyes emarginatae*, a *Prunus* finger gall mite, and notes on *E. prunidemissae*. *Annals of the Entomological Society of America* 62, 269–277.

Oldfield, G.N. (1970) Mite transmission of plant viruses. *Annual Review of Entomology* 15, 343–380.

Oldfield, G.N. (1973) Sperm storage in female Eriophyidae (Acarina). *Annals of the Entomological Society of America* 66, 1089–1092.

Oldfield, G.N. (1984) Evidence for conspecificity of *Aculus cornutus* and *A. fockeui* (Acari: Eriophyidae), rust mites of *Prunus* fruit trees. *Annals of the Entomological Society of America* 77, 564–567.

Oldfield, G.N. (1989) Observations on interspecific attraction to spermathophores by species of Eriophyoidae. In: ChannaBasavanna, G.P. and Viraktamath, C.A. (eds) *Progress in Acarology, Volume 1.* Oxford & IBH Publishing, New Delhi, pp. 249–253.

Oldfield, G.N. (1996a) Diversity and host plant specificity. In: Lindquist, E.E., Sabelis, M.W. and Bruin, J. (eds) *Eriophyoid Mites: Their Biology, Natural Enemies and Control.* World Crop Pests, Volume 6. Elsevier Science, Amsterdam, pp. 199–216.

Oldfield, G.N. (1996b) Toxemias and other non-distortive feeding effects. In: Lindquist, E.E., Sabelis, M.W. and Bruin, J. (eds) *Eriophyoid Mites: Their Biology, Natural Enemies and Control.* World Crop Pests, Volume 6. Elsevier Science, Amsterdam, pp. 243–250.

Oldfield, G.N. (1999) Distribution of asymmetrical and symmetrical sperm storage in the Eriophyoidea and its phylogenetic implications. In: Bruin, J., Geest, L.P.S. and Sabelis, M.W. (eds) *Ecology and Evolution of the Acari.* Kluwer Academic Publishers, Dordrecht, the Netherlands, pp. 157–161.

Oldfield, G.N. (2005) Biology of gall-inducing acari. In: Raman, A., Schaefer, C.W. and Withers, T.M. (eds) *Biology, Ecology and Evolution of Gall-inducing Arthropods.* Science Publishers, Portland, Oregon, pp. 35–57.

Oldfield, G.N. and Michalska, K. (1996) Spermatophore deposition, mating behavior and population mating structure. In: Lindquist, E.E., Sabelis, M.W. and Bruin, J. (eds) *Eriophyoid Mites: Their Biology, Natural Enemies and Control.* World Crop Pests, Volume 6. Elsevier Science, Amsterdam, pp. 185–198.

Oldfield, G.N. and Newell, I.M. (1973a) The role of the spermatophore in the reproductive biology of protogynes of *Aculus cornutus* (Acarina : Eriophyoidae). *Annals of the Entomological Society of America* 66, 160–163.

Oldfield, G.N. and Newell, I.M. (1973b) The spermatophore as the source of sperm for deutogynes of *Aculus cornutus* (Acari: Eriophyidae). *Annals of the Entomological Society of America* 66, 223–225.

Oldfield, G.N. and Perring, T.M. (1996) Rearing techiniques. In: Lindquist, E.E., Sabelis, M.W. and Bruin, J. (eds) *Eriophyoid Mites: Their Biology, Natural Enemies and Control.* World Crop Pests, Volume 6. Elsevier Science, Amsterdam, pp. 377–381.

Oldfield, G.N. and Proeseler, G. (1996) Eriophyoid mites as vectors of plant pathogens. In: Lindquist, E.E., Sabelis, M.W. and Bruin, J. (eds) *Eriophyoid Mites: Their Biology, Natural Enemies and Control.* World Crop Pests, Volume 6. Elsevier Science, Amsterdam, pp. 259–275.

Oldfield, G.N., Hobza, R.F. and Wilson, N.S. (1970) Discovery and characterization of spermatophores in the Eriophyoidea (Acari). *Annals of the Entomological Society of America* 63, 520–526.

Oldfield, G.N., Newell, I.M. and Reed, D.K. (1972) Insemination of protogynes of *Aculus cornutus* from spermatophores and description of the sperm cell. *Annals of the Entomological Society of America* 65, 1080–1084.

Oldfield, G.N., Reddy, M.V., Nene, Y.L. and Reed, W. (1981) Preliminary studies of the eriophyid vectors of mosaic. *International Pigeonpea Newsletter* 1, 25–26.

Oldfield, G.N., Creamer, R., Gispert, C., Osorio, F., Rodriguiz, R. and Perring, T.M. (1995) Incidence and distribution of peach mosaic and its vector, *Eriophyes insidiosus* (Acari: Eriophyidae) in Mexico. *Plant Disease* 79, 186–189.

Oliveira, C.A.L. de (2000) Effect of pyrethroid application on the mite *Oligonychus ilicis* (Mc Gregor) (Acari: Tetranychidae) and its predators, in the coffee crop. *Ecossistema* 25, 28–34.

Oliveira, C.A.L. de and Sponchiado, O.J. (1983) Efeito de aplicação de acaricida, inseticida e fungicida sobre o microácaro *Aculops lycopersici* (Massee, 1937) na cultura do tomateiro (*Lycopersicon esculentum* Mill.). *Anais de Sociedade Entomológica do Brasil* 12, 213–219.

Oliveira, C.A.L. de, Mauro, A.O. and Kronka, S.N. (1982) Comparação de metodos para estimativas da população do ácaro *Phyllocoptruta oleivora* (Ashmead, 1879) na cultura do citrus. *Anais de Sociedade Entomológica do Brasil* 11, 101–114.

Oliveira, C.A.L. de, Rigotto, E.L. and Silva, J.R.T. da (1985) Efeitos da aplicação de produtos químicos no controle do ácaro da ferrugem *Phyllocoptruta oleivora* (Ashmead, 1879) na cultura de citros. *Anais de Sociedade Entomológica do Brasil* 14, 147–159.

Oliveira, C.A.L., Campos Neto, R.R. and Fernandes, C.B. (1998) Efeito de diferentes volumes de calda no controle do ácaro da leprose *Brevipalpus phoenicis* (Geijskes) em citros. *Anais de Sociedade Entomológica do Brasil* 27, 117–124.

Oliveira, C.M., Ferreira, J.A.M., Oliveira, R.M., Santos, F.O. and Pallini, A. (2014) *Ricoseius loxocheles*, a phytoseiid mite that feeds on coffee leaf rust. *Experimental and Applied Acarology* 64, 223–233.

Oliveira, R.C. de, Alves, L.F.A. and Neves, P.M.O.J. (2002) Susceptibility of *Oligonychus yothersi* (Acari: Tetranychidae) to the fungus *Beauveria bassiana*. *Scientia Agricola* 59, 187–189.

Oliveira, R.C. de, Neves, P.M.O.J. and Alves, L.F.A. (2004) Entomopathogenic fungi selection to control *Oligonychus yothersi* (McGregor) (Acari: Tetranychidae) in Paraguay tea crops (*Ilex paraguariensis* St. Hill.). *Neotropical Entomology* 33, 347–351.

Oliveira Azevedo, A., Silva Noronha, A.C. da, Tavares Ferreira, C. and Sousa Oliveira, F. (2012) Viabilidade de ovos de *Tetranychus mexicanus* (McGregor) em dendezeiro em diferentes temperaturas. In: *16° Seminário de Iniciação Científica da EMBRAPA, 6 e 17 de agosto de 2012, Embrapa Amazônia Oriental, Belém-PA*. CD-ROM, Embrapa Amazônia Oriental, Belém-PA, Brazil. Available at: http://ainfo.cnptia.embrapa.br/digital/bitstream/item/76490/1/Adria.pdf (accessed 18 May 2015).

Oliveira de Andrade-Bertolo, F. de, Bottini de Moura, R., Matioli, A.L. and Ott, A.P. (2013) New records of mites (Acari: Prostigmata) on *Vitis* sp. (L.) in Brazil. *Comunicata Scientiae* 4, 414–418.

Oliveira Guassu, C.M. de, Bueno, R. and Kenji, U. (1993) Eficiencia do flufenoxuron e teflubenzuron para o controle do ácaro da falsa ferrugem do citros *Phyllocoptruta oleivora* (Ashm., 1879) (Acari Eryophyidae). *UNIMAR Ciências* 1, 52–56.

Olivier, A.D. and Cancienne, E.A. (1980) Status of two species of rust mites as pests on *Camellia japonica* L. in Lousiana. *Journal of the Georgia Entomological Society* 15, 210–214.

Olvera-Fonseca, S. (1986) El acaro causante de la "rona del cocotero" en Veracruz, Mexico (Acarina: Eriophyidae). *Folia Entomológica Mexicana* 67, 45–51.

Omoto, C. (1998) Acaricide resistance management of leprosis mite (*Brevipalpus phoenicis*) in Brazilian citrus. *Pesticide Science* (old series) 52, 189–191.

Omoto, C. and McCoy, C.W. (1998) Toxicity of purified fungal toxin hirsutellin A to the citrus rust mite *Phyllocoptruta oleivora* (Ash.). *Journal of Invertebrate Pathology* 72, 319–322.

Omoto, C., Dennehy, T.J., McCoy, C.W., Crane, S.E. and Long, J.W. (1994) Detection and characterization of the interpopulation variation of citrus rust mite (Acari: Eriophyidae) resistance to dicofol in Florida citrus. *Journal of Economic Entomology* 87, 566–572.

Onillon, J.C., Onillon, J., Brun, P., Franco, E. and Decoux, G. (1986) Effect of fertilizer on the population level of the citrus white-fly *Dialeurodes citri* (Homoptera, Aleyrodidae). In: Cavalloro, R. and Di Martino, E. (eds) *Integrated Pest Control in Citrus-groves. Proceedings of the Experts' Meeting, 26–29 March 1985, Acireale, Italy*. Published for the Commission of European Communities by Balkema, Rotterdam, the Netherlands, pp. 109–120.

Onsager, J.A. (1976) *The Rationale of Sequential Sampling, with Emphasis on its Use in Pest Management*. Technical Bulletin No. 1526, US Department of Agriculture, Washington, DC.

Onucar, A. and Ulu, O. (1988) Notes on the mite species present on chestnut trees. *Turkiye Entomoloji Dergisi* 12, 33–38 (in Turkish).

Onyambus, G.K., Maranga, R.O., Gitonga, L.M. and Knapp, M. (2011) Host plant resistance among tomato accessions to the spider mite *Tetranychus evansi* in Kenya. *Experimental and Applied Acarology* 54, 385–393.

Onzo, A. (2003) Population dynamics of cassava green mite and its predator *Typhlodromalus aripo* in Bénin, West Africa. PhD thesis, University of Amsterdam, Amsterdam.

Onzo, A., Hanna, R. and Sabelis, M.W. (2009) Within-plant migration of the predatory mite *Typhlodromalus aripo* from the apex to the leaves of cassava: response to day–night cycle, prey location and prey density. *Journal of Insect Behavior* 22, 186–195.

Oomen, P.A. (1982) Studies on population dynamics of the scarlet mite, *Brevipalpus phoenicis*, a pest of tea in Indonesia. *Mededelingen Landbouwhogeschool (Wageningen)* 82(1), 1–88.

Opit, G., Nechol, J.R., Margolies, D.C. and Williams, K.A. (2005) Survival, horizontal distribution, and economics of releasing predatory mites (Acari: Phytoseiidae) using mechanical blowers. *Biological Control* 33, 344–351.

Orlob, G.B. (1966) Feeding and transmission characteristics of *Aceria tulipae* Keifer as vector of wheat streak mosaic virus. *Phytopathologische Zeitschrift* 55, 218–238.

Orozco Hoyos, J., Duque Echeverry, M.C. and Mesa Cobo, N.C. (1990) Efecto de la temperatura sobre la tabla de vida de *Oligonychus yothersi* en *Coffea arabica*. *Cenicafé (Manizales (Caldas) Colombia)* 41(1), 5–18.

Ortega, C.A. and Banda Torre, A. (1967) El eriafido del cocotero *Aceria guerreronis* (Keifer) en la Costa de Guerrero. *Folia Entomológica Mexicana* 15–16, pp. 61–63.

Ortega, C.A., Rodriguez, V.J. and Garibay, C.V. (1967) Investigaciones preliminares sobre el eriófido del fruto del cocotero *Aceria guerreronis* Keifer, en la Costa Grande Guerrero (México). *Oleaginéux* 6, 371–372.

Ortiz, J.A. and Trejo, J.A. (1976) Plagas de cafeto. In: *Manual Técnico del Cultivo de Café en El Salvador*. Instituto Salvadoreño de Investigations del Café, Santa Tecla, El Salvador, pp. 141–167.

Orton, D.A. (1989) *Coincide: The Orton System of Plant Management*. Plantsmen's Publications. Flossmoor, Illinois.

Osakabe, M. (1962) On the diapause of the tea red spider mite, *Tetranychus kanzawai* Kishida. *Study of Tea* 26, 12–25 (in Japanese).

Osakabe, M. (1963) On the natural enemy of the tea red spider mite *Tetranychus kanzawai* Kishida. *Study of Tea* 27, 1–11.

Osakabe, M. (1965) Seasonal fluctuations of population density of the tea red spider mite, *Tetranychus kanzawai* Kishida, in the tea plantation. *Japanese Journal of Applied Entomology and Zoology* 9, 206–210.

Osakabe, M. (1967) Biological studies on the tea red spider mite, *Tetranychus kanzawai* Kishida, in tea plantation. *Bulletin of the Tea Research Station (Chagyo Shikenjo Kenhyu Hokoku)* 4, 35–156 (in Japanese).

Osakabe, M. (1974) Acaricide resistance of the Kanzawa spider mite, *Tetranychus kanzawai* Kishida, parasitic on tea plant. *Japan Agricultural Research Quarterly* 8, 28–31.

Osakabe, M.(H.) (2002) Which predatory mite can control both a dominant mite pest, *Tetranychus urticae*, and a latent mite pest, *Eotetranychus asiaticus*, on strawberry? *Experimental and Applied Acarology* 26, 219–230.

Osakabe, M., Inoue, K. and Ashihara, W. (1987) Effect of *Amblyseius sojaensis* Ehara (Acarina: Phytoseiidae) as a predator of *Panonychus citri* (McGregor) and *Tetranychus kanzawai* Kishida (Acarina: Tetranychidae). *Applied Entomology and Zoology* 22, 594–599.

Osakabe, M., Ehara, S. and Adhikari, S. (2000) Damage to young leaves of pear trees by *Bryobia rubrioculus* (Scheuten) (Acari: Tetranychidae) in Nepal. *Journal of the Acarological Society of Japan* 9, 15–22.

Osakabe, M., Hirose, T. and Sato, M. (2002) Discrimination of four Japanese *Tetranychus* species (Acari: Tetranychidae) using PCR-RFLP of the inter-transcribed spacer region of nuclear ribosomal DNA. *Applied Entomology and Zoology* 37, 399–407.

Osakabe, M., Goka, K., Toda, S., Shintaku, T. and Amano, H. (2005) Significance of habitat type for the genetic population structure of *Panonychus citri* (Acari: Tetranychidae). *Experimental and Applied Acarology* 36, 25–40.

Osakabe, M., Uesugi, R., Asahara, M. and Matsuo, T. (2010) An approach to acaricide-resistance developments from perspectives of genetic linkages and population structures in spider mites. In: Moraes, G.J. de, Castilho, R.C. and Flechtmann, C.H.W. (eds) *XIII International Congress of Acarology, August 23–27, 2010, Recife-PE, Brazil, Abstract Book*, abstract 365, pp. 197–198. Available at: http://www.acarology.org/ica/ica2010/docs/0910_15_abstractBookCapa.pdf (accessed 3 February 2015).

Osborne, L.S., Peña, J.E. and Oi, D.H. (1995) Predation by *Tapinoma melanocephalum* (Hymenoptera: Formicidae) on twospotted spider mites (Acari: Tetranychidae) in Florida greenhouses. *Florida Entomologist* 78, 565–570.

Osman, A.A. (1975a) On the population and control of the mite *Eriophyes vitis* (Pgst.) infesting vineyards in Tahreer Province (Acarina: Eriophyidae). *Bulletin of the Entomological Society of Egypt, Economic Series* 9, 111–114.

Osman, A.A. (1975b) Efficiency of some fungicides in the control of the eriophyid mite, *Vasates lycopersici* (Massee) in Egypt (Acarina: Eriophyidae). *Bulletin of the Entomological Society of Egypt, Economic Series* 9, 115–118.

Osman, A.A. (1979) Notes on the control of the bud mite *Aceria mangiferae* (Sayed) in Egypt (Acarina: Eriophyidae). *Bulletin of the Entomological Society of Egypt, Economic Series* 9, 119–126.

Osman, A.A. and Zaki, A.M. (1986) Studies on the predation efficiency of *Agistemus exsertus* Gonzalez (Acarina, Stigmaeidae) on the eriophyid mite *Aculops lycopersici* (Massee). *Anzeiger für Schädlingskunde, Pflanzenschutz, Umweltschutz* 59, 135–136.

Osman, A.A., Abo-Taka, S.M. and Zaki, A.M. (1991) *Agistemus exsertus* Gonzales (Acarina: Stigmaeidae) as a predator of the grapevine mite *Colomerus vitis* (Pgst.) (Acarina: Actinedida). In: Dusbábek, F. and Bukva, V. (eds) *Modern Acarology, Volume II. Proceedings of the VIII International Congress of Acarology, Held in České Budějovice, Czechoslovakia, 6–11 August 1990.* SPB Academic Publishers, The Hague, the Netherlands, co-published with Academia, Publishing House of the Czechoslovak Academy of Sciences, Prague, pp. 689–690.

Osman, M.S. (1997) Petroleum oils as a component of integrated pest management of phytophagous mites. *Arab Gulf Journal of Scientific Research* 15, 125–135.

Osman, M.S., Ibrahim, G.A. and El-Halawany, M.E. (1994) Suitability of petroleum and tar oils as acaricides and ovicides. *Menofia Journal of Agricultural Research* 19, 1123–1130.

Ostojá-Starzewski, J.C. (2000) *Schizotetranychus celarius* (Banks) (Acari: Prostigmata) a mite pest of bamboo; first records for Britain and two new host records. *British Journal of Entomology and Natural History* 13, 95–97.

Ostojá-Starzewski, J.C. and Halstead, A.J. (2006) *Trisetacus chamaecypari* Smith (Acari: Phytoptdae) damaging Leyland cypress in Britain: a new host and country record for this mite. *British Journal of Entomology and Natural History* 19, 201–205.

Ostojá-Starzewski, J.C. and Matthews, L. (2009) *Aceria tulipae* (Keifer) (Acari: Eriophyoidae) on imported onion sets, and present status in Britain. *Entomologist's Monthly Magazine* 145, 61–68.

Ostovan, H. and Kamali, K. (1996) Biology of two bulb mites: *Rhizoglyphus echinopus* (Fumouze and Robin) and *R. robini* Claparede under two different laboratory conditions. *Journal of Agricultural Sciences, Islamic Azad University* 2 (5/6), 47–59.

Othieno, C.O., Kilavuka, C.I., Mbaya, M. and Gitungo, G. (1981) Economics of fertilizer application to smallholder tea farms in Kenya. *Tea* 2(1), 10–19.

Othman, K., Rhouma, A., Belhadj, R., Alimi, E., Fallah, H., Kreiter, P., Lenfant, C. and Brun, J. (2001) Biological control of a date pest, the mite *Oligonychus afrasiaticus*. *Phytoma* 540, 30–31.

Othman, Y. and Zhang, Z.Q. (2003) *Tetranychidae* (Acari: Prostigmata) of Malay Peninsula: Checklist, key to genera and species and description of three new species. *Systematic and Applied Acarology* 8, 149–173.

Otto, J.C. (1999) Revision of the genus *Erythracarus* Berlese (Acarina: Anystidae: Erythracarinae). *Journal of Natural History* 33, 825–909.

Otto, J.C. and Halliday, R.B. (1991) Systematics and biology of a predatory mite (*Anystis* sp.) introduced in Australia for biological control of redlegged mite. *Plant Protection Quarterly* 6, 181–185.

Otto, S.P. and Jarne, P. (2001) Haploids – hapless or happening? *Science* 292, 2441–2443.

Ou, Y.T. and Fang, H.C. (1978) Studies on the rice tarsonemid mite *Steneotarsonemus spinki* Smiley (Acarina: Tarsonemidae) and its mode of transmission in rice plant. *National Science Council Monthly (Taipei)* 6, 773–779 (in Chinese).

Ou, Y.T., Fang, H.C. and Tseng, Y.H. (1977) Studies on *Steneotarsonemus madecassus* Gutierrez of rice. *Plant Protection Bulletin (Taipei)* 19, 21–29 (in Chinese).

Oudemans, A.C. (1897) List of Dutch Acari Latr. Fifth part: Trombidides Leach, with synonymical notes and other remarks, and description of an apparently new, but indeed very old species of *Cheyletus*, *Ch. squamosus* De Geer. *Tijdschrift voor Entomologie* 40, 117–35 (in Dutch).

Oudemans, A.C. (1905) Notes on Acari. XIV Series (Parasitidae and Thrombidiidae). *Tijdschrift voor Entomologie* 48, 221–244.

Oudemans, A.C. (1906) Nieuwe classificatie der Acari. *Entomologische Berichten* 2, 43–46.

Oudemans, A.C. (1910) A short survey of the more important families of Acari. *Bulletin of Entomological Research* 1, 105–119.

Oudemans, A.C. (1915) Notizen über Acari. *Archiv für Naturgeschichte* 81, 1–78.

Oudemans, A.C. (1924) Acarologische Aanteekeningen LXXIV. *Entomologische Berichten* 136, 241–260.

Oudemans, A.C. (1926) *Kritisch Historisch Overzicht der Acarologie, Eerste Gedeelte, 850 v. C. tot 1758. Tijdschrift voor Entomologie* 69(Suppl.).

Oudemans, A.C. (1928) Acarologische Aanteekeningen LXXXIX. *Entomologische Berichten* 7, 285–293.

Oudemans, A.C. (1931a) Acarologische Aanteekeningen CVI. *Entomologische Berichten* 8, 189–204.

Oudemans, A.C. (1931b) Acarologische Aanteekeningen CVII. *Entomologische Berichten* 8, 221–236.

Oudemans, A.C. (1931c) Acarologische Aanteekeningen CIX. *Entomologische Berichten* 8, 272–280.

Oudemans, A.C. (1931d) Acarologische Aanteekeningen CX. *Entomologische Berichten* 8, 289–293.

Oudemans, A.C. (1937) *Kritisch Historisch Overzicht der Acarologie III C.* E.J. Brill, Leiden, the Netherlands, pp. 799–1348.

Overmeer, W.P.J. (1967) Genetics of resistance to tedion in *Tetranychus urticae* C.L. Koch. *Archives Néerlandaises de Zoologie* 17, 295–349.

Overmeer, W.P.J. (1972) Notes on mating behaviour and sex ratio control of *Tetranychus urticae* Koch (Acarina: Tetranychidae). *Entomologische Berichten* 32, 240–244.

Overmeer, W.P.J. (1985a) Rearing and handling. In: Helle, W. and Sabelis, M.W. (eds) *Spider Mites: Their Biology, Natural Enemies and Control.* World Crop Pests, Volume 1B. Elsevier Science, Amsterdam, pp. 161–170.

Overmeer, W.P.J. (1985b) Toxicological methods. In: Helle, W. and Sabelis, M.W. (eds) *Spider Mites: Their Biology, Natural Enemies and Control.* World Crop Pests, Volume 1B. Elsevier Science, Amsterdam, pp. 183–189.

Overmeer, W.P.J. and Harrison, R.A. (1969) Notes on the control of the sex ratio in populations of the two-spotted spider mite, *Tetranychus urticae* Koch (Acarina: Tetranychidae). *New Zealand Journal of Science* 12, 920–928.

Ownley, B.H., Gwinn, K.D. and Vega, F.E. (2010) Endophytic fungal entomopathogens with activity against plant pathogens: ecology and evolution. *BioControl* 55, 113–128.

Ozawa, A. (1995) The susceptibility of the Kanzawa spider mite, *Tetranychus kanzawai* Kishida, collected from tea fields in Chuuen and Ogasa districts, Shizuoka Prefecture. II. Several new type acaricides. *Proceedings of the Kanto-Tosan Plant Protection Society* 42, 239–240 (in Japanese).

Özman, S.K. (2000) Some biological and morphological differences between gall and vagrant forms of *Phytoptus avellanae* Nal. (Acari: Phytoptidae). *International Journal of Acarology* 26, 215–219.

Özman, S.K. and Cobanoğlu, S. (2001) Current status of hazelnut mites in Turkey. *Acta Horticulturae* 556, 479–487.

Özman, S.K. and Toros, S. (1997a) Life cycles of *Phytoptus avellanae* Nal. and *Cecidophyopsis vermiformis* Nal. (Eriophyoidea: Acarina). *Acta Horticulturae* 445, 493–501.

Özman, S.K. and Toros, S. (1997b) Damage caused by *Phytoptus avellanae* Nal. and *Cecidophyopsis vermiformis* Nal. (Eriophyoidea: Acarina) in hazelnut. *Acta Horticulturae* 445, 537–541.

Özman, S.K. and Toros, S. (1997c) Population fluctuations of *Phytoptus avellanae* Nal. and *Cecidophyopsis vermiformis* Nal. (Eriophyoidea: Acarina) in big buds of hazelnut. *Acta Horticulturae* 445, 551–519.

Özman-Sullivan, S.K. (2006) Life history of *Kampimodromus aberrans* as a predator of *Phytoptus avellanae* (Acari: Phytoseiidae, Phytoptidae). *Experimental and Applied Acarology* 38, 15–23.

Özman-Sullivan, S.K. and Akça, İ. (2005) Efficiency of pesticides against big bud mites [*Phytoptus avellanae* Nal. and *Cecidophyopsis vermiformis* Nal. (Acarina: Eriophyoidea)] on hazelnut. *Acta Horticulturae* 686, 393–399.

Özman-Sullivan, S.K., Cobanoğlu, S. and Kaźmierski, A. (2005) Alycina and Eupodina mites in hazelnut orchards in Turkey. *Acta Horticulturae* 686, 401–406.

Özman-Sullivan, S.K, Amrine, J.W. and Walter, D.E. (2006) A new species of eriophyoid mite (Acari: Eriophyidae) on sugarcane in Australia. *International Journal of Acarology* 32, 387–395.

Ozsisli, T. and Cobanoğlu, S. (2011) Mite (Acari) fauna of some cultivated plants from Kahramanmaras, Turkey. *African Journal of Biotechnology* 10, 2149–2155.

Pady, S.M. (1955) The occurrence of the vector of wheat streak mosaic, *Aceria tulipae*, on slides exposed in the air. *Plant Disease Reporter* 39, 296–297.

Pagenstecher, H.A. (1857) Uber Milben, besonders die Gattung *Phytoptus*. *Verhandlungen Natur-historich-medizinischen Vereins zu Heidelberg* 1(2; Bd. 1857–1859), 46–53.

Pagliano, T. (1951) *Les Ennemis des Vergers des Olivettes et des Palmeraies.* Publication de l'Office de l'Expérimentation et de la Vulgarisation Agricoles de Tunisie. Société d'Editions Françaises en Afrique du Nord, Tunis.

Painter, R.H. (1951) *Insect Resistance in Crop Plants.* MacMillan, New York.

Painter, R.H. (1968) *Insect Resistance.* The University Press of Kansas, Lawrence, Kansas.

Palaniswamy, S. and Ragini, J.C. (2000) Influence of certain plant extracts on yellow mite *Polyphagotarsonemus latus* (Banks) on chillies. *Insect Environment* 6, 25 26.

Palevsky, E., Oppenheim, D., Reuveny, H. and Gerson, U. (1996) Impact of European red mite on Golden Delicious and Oregon Spur apples in Israel. *Experimental and Applied Acarology* 20, 343–354.

Palevsky, E., Soroker, V., Weintraub, P., Mansour, F., Abo-Moch, F. and Gerson, U. (2001) How species-specific is the phoretic relationship between the broad mite, *Polyphagotarsonemus latus* (Acari: Tarsonemidae), and its insect hosts? *Experimental and Applied Acarology* 25, 217–224.

Palevsky, E., Argov, Y., David, T.B. and Gerson, U. (2003) Identification and evaluation of potential predators of the citrus rust mite, *Phyllocoptruta oleivora*, in Israel. *Systematic and Applied Acarology* 8, 39–48.

Palevsky, E., Ucko, O., Peles, S., Yablonski, S. and Gerson, U. (2004) Evaluation of control measures for *Oligonychus afrasiaticus* infesting date palm cultivars in the Southern Arava Valley of Israel. *Crop Protection* 23, 387–392.

Palevsky, E., Borochov-Neori, H. and Gerson, U. (2005) Population dynamics of *Oligonychus afrasiaticus* in the southern Arava Valley of Israel in relation to date fruit characteristics and climatic conditions. *Agricultural and Forest Entomology* 7, 283–290.

Palevsky, E., Gal, S. and Ueckermann, E.A. (2009) Phytoseiidae from date palms in Israel with descriptions of two new taxa and a key to the species found on date palms worldwide (Acari: Mesostigmata). *Journal of Natural History* 43, 1715–1747.

Palevsky, E., Lotan, A. and Gerson, U. (2010a) Evaluation of *Eutetranychus palmatus* (Acari: Tetranychidae) as a pest of date palms in Israel. *Israel Journal of Plant Sciences* 58, 43–51.

Palevsky, E., Weintraub, P.G., Zchori-Fein, E., Gal, S., Mozes-Daube, N., Argov, Y., Castagnoli, M., Liguori, M., Nannelli, R., Simoni, S., *et al.* (2010b) Development of an economic rearing and transport system for an arid-adapted strain of the predatory mite, *Neoseiulus californicus*, for spider mite control. In: Sabelis, M.W. and Bruin, J. (eds) *Trends in Acarology: Proceedings of the 12th International Congress, 21–26 August 2006, Amsterdam.* Springer, Dordrecht, the Netherlands, pp. 425–430.

Palevsky, E., Gerson, U. and Zhang, Z.-Q. (2013) Can exotic phytoseiids be considered 'benevolent invaders' in perennial cropping systems? *Experimental and Applied Acarology* 59, 11–26.

Paliwal, Y.C. (1980) Relationship of wheat streak mosaic and barley stripe mosaic viruses to vector and nonvector eriophyid mites. *Archives of Virology* 63, 123–132.

Paliwal, Y.C. and Tremaine, J.H. (1976) Multiplication, purification and properties of ryegrass mosaic virus. *Phytopathology* 66, 406–414.

Palmer, A.R. (1994) Fluctuating asymmetry analyses: a primer. In: Markow, T.A. (ed.) *Development Instability: Its Origins and Evolutionary Implications.* Kluwer, Dordrecht, the Netherlands, pp. 335–364.

Palmer, W.A. and Pullen, K.R. (1995) The phytophagous arthropods associated with *Lantana camara*, *L. hirsuta*, *L. urticifolia* and *L. urticoides* (Verbenaceae) in North America. *Biological Control* 5, 54–72.

Palmieri, M., Donis, I., Salazar, A.L., Blanco, S., Porres, M., Brlansky, R.H., Guerra-Moreno, A.S., Manjunath, K.L. and Lee, R.F. (2007) Leprosis in Guatemala. In: Hilf, M.E., Duran-Vila, N. and Rocha-Peña, M.A. (eds) *Proceedings of the Sixteenth Conference of the International Organization of Citrus Virologists, IOCV 2005.* IOCV, c/o University of California, Riverside, California, p. 510 (abstract).

Paltrinieri, S., Martini, M., Pondrelli, M. and Bertaccini, A. (1998) X-disease-related phytoplasmas in ornamental trees and shrubs with witches' broom and malformation symptoms. *Journal of Plant Pathology (Pisa)* 80, 261.

Pan, W.-G., Luo, P., Fu, R.-B., Gao, P., Long, Z.-F., Xu, F.-Y., Xiao, H.-B. and Liu, S.-G. (2006) Acaricidal activity against *Panonychus citri* of a ginkgolic acid from the external seed coat of *Ginkgo biloba*. *Pest Management Science* 62, 283–287.

Panda, P.K. and Sarangi, P.K. (2004) Chemical management of yellow mite, *Polyphagotarsonemus latus* (Banks) in chilli. *Orissa Journal of Horticulture* 32, 68–70.

Pande, Y. D. and Nandi, S. (1983) Mites associated with tea in Tripura and the effect of shade on their populations. *Science and Culture* 49, 141–142.

Pande, Y.D. and Sharma, B.L. (1984) Record of some new predators of *Tetranychus neocaledonicus* (Acarina: Tetranychidae) from India. *Science and Culture* 50, 175–176.

Pande, Y.D. and Sharma, B.L. (1986) Effect of food and temperature on the biology of red cucurbit mite *Tetranychus neocaledonicus* André (Acarina: Tetranychidae). *Revista Brasileira de Biologia* 46, 597–602.

Pande, Y.D., Carnero, A. and Hernández, M. (1989) Notes on biological observations on some unrecorded species of phytophagous and predatory mites in Canary Islands. *Investigación Agraria, Producción y Protección Vegetales* 4, 275–281.

Pang, B.-P., Zhou, X.-R., Shi, L. and Mu, H.-B. (2004) Performance of *Tetranychus truncatus* Ehara (Acarina: Tetranychidae) reared with different host plants. *Acta Entomologica Sinica* 47, 55–58 (in Chinese).

Pang, B.-P., Liu, J., Liu, M. and Gao, S. (2005a) Population dynamics of *Tetranychus truncatus* in corn fields. *Chinese Journal of Ecology* 24, 1115–1119 (in Chinese).

Pang, B-P., Liu, J., Zhou, X.-R. and Zhang, R. (2005b) Effects of corn cultivar on *Tetranychus truncatus* Ehara (Acarina: Tetranychidae) population parameters. *Chinese Journal of Applied Ecology* 16, 1313–1316 (in Chinese).

Pantoja, A., Mejia, O.I., Ramirez, L.M., Escalona, L.E., Messa, N.C. and Duque, M.C. (1998) Comportamiento de *Schizotetranychus oryzae* sobre nueve variedades de arroz. *Journal of Agriculture of the University of Puerto Rico* 82, 213–215.

Pap, L., Hajimichael, J., Bleicher, E., Botar, S. and Skekely, I. (1994) Chemical and biological evaluation of a new acaricide. In: *Brighton Crop Protection Conference, Pests and Diseases – 1994. Proceedings of an International Conference Organised by the British Crop Protection Council, Held at the Brighton Centre and the Brighton Metropole Hotel, Brighton, England, 21–24 November 1994, Volume 1.* British Crop Production Council, Farnham (now Alton), UK, pp. 75–82.

Papaioannou-Souliotis, P., Markoyiannaki-Printziou, D., Rumbos, I. and Adamopoulos, I. (1999a) Phytoseiid mites associated with vine in various provinces of Greece: a contribution to faunistics and biogeography, with reference to eco-ethological aspects of *Phytoseius finitimus* (Ribaga) (Acari: Phytoseiidae). *Acarologia* 40, 113–125.

Papaioannou-Souliotis, P., Markoyiannaki-Printziou, D. and Tsagkarakou, A. (1999b) Seasonal fluctuations of the citrus bud mite population with an improved spraying programme for lemon trees in view. *Bollettino di Zoologia Agraria e di Bachicoltura (Ser. II)* 31, 67–74.

Parent, B. and Beaulieu, A.A. (1957) Life history of the European red mite. *Canadian Entomologist* 89, 328–333.

Parent, B. and LeRoux, E.J. (1956) Notes on *Mediolata mali* (Ewing) (Acarina: Raphignathidae) as a predator of ERM. *Canadian Entomologist* 88, 487.

Parent, B.J.P.H. (1973) Natural population densities of the European red mite on apple in Quebec. *Environmental Entomology* 2, 1064–1068.

Parent, B.J.P.H. (1979) Principaux acariens des pommeraies du Quebec (Canada). In: Piffl, E. (ed.) *Proceedings of the 4th International Congress of Acarology, Saalfelden (Austria)*. Akadémiai Kiadó, Budapest, pp. 657–661.

Park, C.G., Lee, M.H., Yoo, J.K., Lee, J.O. and Choi, B.R. (1995) Relative toxicity of abamectin to the predatory mite *Amblyseius womersleyi* Schicha (Acari: Phytoseiidae) and twospotted spider mite *Tetranychus urticae* Koch (Acari: Tetranychidae). *Korean Journal of Applied Entomology* 34, 360–367.

Parker, R. and Gerson, U. (1994) Dispersal of the broad mite, *Polyphagotarsonemus latus* (Banks) (Heterostigmata: Tarsonemidae), by the greenhouse whitefly, *Trialeurodes vaporariorum* (Westwood) (Homoptera: Aleyrodidae). *Experimental and Applied Acarology* 18, 581–585.

Parker, W.B. (1913) *The Red Spider Mite on Hops in the Sacramento Valley of California*. Bulletin 117, Bureau of Entomology, US Department of Agriculture, Washington, DC.

Parkinson, C.L. (1990) Population increase and damage by three species of mites on wheat at 20°C and two humidities. *Experimental and Applied Acarology* 8, 179–193.

Parolin, P., Bresch, C., Desneux, N., Brun, R., Bout, A., Boll, R. and Poncet, C. (2012) Secondary plants used in biological control: a review. *International Journal of Pest Management* 58, 91–100.

Parolin, P., Bresch, C., Ruiz, G., Desneux, N. and Poncet, C. (2013) Testing banker plants for biological control of mites on roses. *Phytoparasitica* 41, 249–262.

Parr, W.J. and Hussey, N.W. (1966) Diapause in the glasshouse red spider mite *Tetranychus urticae* Koch. A synthesis of present knowledge. *Horticulture Research* 6, 1–21.

Parrella, M.P., McCaffrey, J.P. and Horsburgh, R.L. (1980) Compatibility of *Leptothrips mali* with *Stethorus punctum* and *Orius insidiosus*. Predators of *Panonychus ulmi*. *Environmental Entomology* 9, 694–696.

Parrella, M.P., McCaffrey, J.P. and Horsburgh, R.L. (1981) Population trends of selected phytophagous arthropods and predators under different pesticide programs in Virginia apple orchards. *Journal of Economic Entomology* 74, 492–498.

Parrella, M.P., Rowe, D.J. and Horsburgh, R.L. (1982) Biology of *Leptothrips mali*, a common predator in Virginia apple orchards. *Annals of the Entomological Society of America* 75, 130–135.

Parrott, P.J., Hodgkiss, H.E. and Schoene, W.J. (1906) The apple and pear mites. *New York Agricultural Experiment Station Bulletin* 283, 281–318.

Parsons, P.A. (1990) Fluctuating asymmetry: an epigenic measure of stress. *Biological Reviews* 65, 131–145.

Paschoal, A.D. (1968a) Especies de ácaros encontradas emplantas no Estado de São Paulo. *Ciência e Cultura (São Paulo)* 20, 258.

Paschoal, A.D. (1968b) *Tetranychus marianae* (Acarina: Tetranychidae): inimigo potencial das Solanaceae. *Anais da Reunião Anual da Sociedade Brasileira de Entomologia (Piracicaba)* 1, 83–84.

Paschoal, A.D. (1970) *Contribuicao ao Conhecimento da Familia Tetranychidae no Brasil (Arachnida: Acarina)*. Escola Superior de Agricultura "Luiz de Queiroz", Universidade de São Paulo, Piracicaba, Brazil.

Paschoal, A.D. (1971a) A new *Schizotetranychus* from the Brazilian fauna (Acarina: Tetranychidae). In: *Anales del Primer Congreso Latinoamericano de Entomologia Cusco, Peru, 12–18 de Abril 1971 Revista Peruana de Entomología* 14, 172–173. Available at: http://www.revperuentomol.com.pe/publicaciones/vol14/SCHIZOTETRANYCHUS-TAQUARAE.pdf (accessed 18 May 2015).

Paschoal, A.D. (1971b) *Allonychus querci* Baker & Pritchard, 1962 (Acarina: Tetranychidae): novo representante da fauna Brasileira. *Revista de Agricultura (Piracicaba)* 46, 115–116.

Paschoal, A.D. and Reis, P.R. (1968) Relacao de ácaros encontrados em plantas. *Revista de Agricultura (Piracicaba)* 43, 137–139.

Pascon, R.C., Kitajima, J.P., Breton, M.C., Assumpção, L., Greggio, C., Zanca, A.S., Okura, V.K., Alegria, M.C., Camargo, M.E., Silva, G.G.C., *et al.* (2006) The complete nucleotide sequence and genomic organization of citrus leprosis associated virus, cytoplasmatic type (CiLV-C). *Virus Genes* 32, 289–298.

Pascual, A. and Ferragut, F. (2003) Influence of climatic conditions on population dynamics of *Tetranychus urticae* and *Euseius stipulatus* (Acari: Tetranychidae, Phytoseiidae) on clementines (*Citrus reticulata*). *IOBC-WPRS Bulletin* 26(6), 211.

Pascual-Ruiz, S., Aguilar-Fenollosa, E., Ibáñez-Gual, M.V., Hurtado-Ruiz, M.A. and Jacas, J.A. (2010) Economic thresholds for *Tetranychus urticae* (Acari: Tetranychidae) on *Citrus clementina*. In: Moraes, G.J. de, Castilho, R.C. and Flechtmann, C.H.W. (eds) *XIII International Congress of Acarology, August 23–27, 2010, Recife-PE, Brazil, Abstract Book*, abstract 372, p. 201. Available at: http://www.acarology.org/ica/ica2010/docs/0910_15_abstractBookCapa.pdf (accessed 3 February 2015).

Pascual-Ruiz, S., Ansaloni, T., Jacas, J.A. and Hurtado, M.A. (2011) Comparison of the genetic structure of two populations of *Tetranychus urticae* Koch (Acari: Prostigmata) in clementine orchards under integrated and organic crop management. *IOBC-WPRS Bulletin* 62, 100.

Pasini, A., Capelo, S.M.J. and Oliveira, R.C. (2003) Ensaios preliminares para avaliação da eficiência de óleo de neem no controle de *Oligonychus yothersi* (Acari: Tetranychidae). *Semina: Ciências Agrárias, Londrina* 24, 315–316.

Pasqualini, E. and Malavolta, C. (1985) Possibilitá di contenimento naturale di *Panonychus ulmi* (Koch) (Acarina: Tetranychidae) su melo in Emilia-Romagna. *Bollettino dell'Istituto di Entomologia dell'Università di Bologna* 39, 221–230.

Patel, J.R., Patel, C.B., Patel, M.B., Desai, H.R., Patel, A.J. and Rai, A.B. (1995) Some ecological studies on eriophyid mite, *Aceria cajani* infesting pigeonpea in south Gujarat. In: *Fifth National Symposium on Acarology, 20–22 September 1995, University of Agricultural Sciences, Bangalore, India*, p. 29–30 (abstract).

Patel, L. and Karmakar, K. (2005) Seasonal abundance of false spider mite, *Brevipalpus phoenicis* (Geijskes) (Acari: Tenuipalpidae), on selected cultivars of pointed gourd (*Trichosanthes dioica* Roxb.). *Environment and Ecology* 23(Special 2), 242–244.

Paternotte, E. (2000) Schadelijke insecten en mijten op houtig kleinfruit. *Fruitteelt Nieuws* 13(9), 34–35 (in Dutch).

Paternotte, E. and Soenen, A. (1982) *Aculus* spec., les phytoptes. *Mededelingen, Faculteit Landbouwkundige en Toegepaste Biologische Wetenschappen, Universiteit Gent* 47(2), 627–633.

Patil, R.S. (2005) Investigations on mite pests of solanaceous vegetables with special reference to brinjal. PhD thesis, University of Agricultural Sciences, Dharwad, India. Abstract available at: http://etd.uasd.edu/abst/th8487.pdf (accessed 2 April 2015).

Patterson, C.G., Thurston, R. and Rodriguez, J.G. (1974) Two-spotted spider mite resistance in *Nicotiana* species. *Journal of Economic Entomology* 67, 341–343.

Patterson, C.G., Knavel, D.E., Kemp, T.R. and Rodriguez, J.G. (1975) Chemical basis for resistance to *Tetranychus urticae* Koch in tomatoes. *Environmental Entomology* 4, 670–674.

Patti, I. and Vacante, V. (1985) Un caso di dermatite umana connessa all'infestazione di *Cheyletiella blakei* Smiley (Acarina, Cheyletiellidae) su gatto persiano. *Bollettino di Zoologia Agraria e di Bachicoltura (Ser. II)* 18(1984–85), 13–22.

Pauriene, B. (1970) Predacious mites on orchard plants in the Lithuanian, SSR. *Acta Entomologica Lituanica* 1, 178–179 (in Russian).

Payandeh, A., Dehghan, A., Kamrani, R., Behzadi, N.M. and Yavarzadeh, M.R. (2013) Population abundance and seasonal activity of *Oligonychus afrasiaticus* (McGregor) and its predator *Stethorus gilvifrons* (Mulsant) in Bam region, Iran. *Ecology, Environment and Conservation* 19(1), 225–231.

Payne, J., Cannon, R.J.C. and Bagley, A.L. (1993) *Bacillus thuringiensis* isolates for controlling acarides. US Patent 5,211,946. US Patent and Trademark Office, Alexandria, Virginia.

Paz, Z., Gerson, U. and Sztejnberg, A. (2007a) Assaying three new fungi against citrus mites in the laboratory, and a field trial. *BioControl* 52, 855–862.

Paz, Z., Burdman, S., Gerson, U. and Sztejnberg, A. (2007b) Antagonistic effects of the endophytic fungus *Meira geulakonigii* on the citrus rust mite *Phyllocoptruta oleivora*. *Journal of Applied Microbiology* 103, 2570–2579.

Peairs, F.B. (2010) *Mites in Wheat*. Fact Sheet No. 5.578. Colorado State University Extension, Fort Collins, Colorado. Available at: http://www.ext.colostate.edu/pubs/insect/05578.html (accessed 18 August 2014).

Pećinar, I.B., Stevanović, B., Rector, G. and Petanović, R. (2007) Anatomical injuries caused by *Leipothrix dipsacivagus* Petanovic & Rector on cut-leaf teasel, *Dipsacus laciniatus* L. (Dipsacaceae). *Archives of Biological Sciences (Belgrade)* 59, 363–367.

Pedro Neto, M., Carvalho, C.F., Reis, P.R., Santa-Cecilia, L.V.C., Souza, B., Alcantra, E. and Silva, R.A. (2008) Biological aspects of *Chrysoperla externa* (Hagen) preying on *Oligonychus ilicis* (McGregor) and *Planococcus citri* (Risso). *Coffee Science* 3, 85–93.

Pedro Neto, M., Reis, P.R., Andrade, H.B. and Zacarias, M.S. (2010) Effect of windbreaks of arboreal legumes on mite population on coffee. In: Moraes, G.J. de, Castilho, R.C. and Flechtmann, C.H.W. (eds) *XIII International Congress of Acarology, August 23–27, 2010, Recife-PE, Brazil, Abstract Book*, abstract 373, p. 202. Available at: http://www.acarology.org/ica/ica2010/docs/0910_15_abstractBookCapa.pdf (accessed 3 February 2015).

Pegazzano, F. (1971) Acari fitofagi dell'olivo in Italia. *Redia* 52, 361–366.

Pellizzari, G. and Duso, C. (2009) Occurence of *Stigmaeopsis nanjingensis* in Europe. *Bulletin of Insectology* 62, 149–151.

Pellizzari, G., Dalla Montà, L. and Vacante, V. (2005) List of alien insect and mite pests introduced to Italy in sixty years (1945–2004). *Plant Protection and Plant Health in Europe: Introduction and Spread of Invasive Species. BCPC Conference, 9–11 June 2005, Humbolt University, Berlin, Germany*. BCPC Symposium Proceedings No. 81, British Crop Production Council, Alton, UK, Poster. Available at: http://www.researchgate.net/publication/267216407_Alien_insect_and_mite_pests_introduced_to_Italy_in_sixty_years_%281945-2004%29 (accessed 18 May 2015).

Peña, J.E. (1990) Relationships of broad mite (Acari: Tarsonemidae) density to lime damage. *Journal of Economic Entomology* 83, 2008–2015.

Peña, J.E. (1992) Predator–prey interactions between *Typhlodromalus peregrinus* and *Polyphagotarsonemus latus*: effects of alternative prey and other food resources. *Florida Entomologist* 75, 241–248.

Peña, J.E. and Baranowski, R.M. (1990) Dispersion indices and sampling plans for the broad mite (Acari: Tarsonemidae) and the citrus rust mite (Acari: Eriophyidae) on limes. *Environmental Entomology* 19, 378–382.

Peña, J.E. and Osborne, L. (1996) Biological control of *Polyphagotarsonemus latus* (Acarina: Tarsonemidae) in greenhouses and field trials using introductions of predacious mites (Acarina: Phytoseiidae). *Entomophaga* 41, 279–285.

Peña, J.E. and Rodrigues, J.C.V. (2010) Alternatives for the chemical control of the red palm mite, *Raoiella indica* (Acari: Tenuipalpidae) on palms and bananas. In: Moraes, G.J. de, Castilho, R.C. and Flechtmann, C.H.W. (eds) *XIII International Congress of Acarology, August 23–27, 2010, Recife-PE, Brazil, Abstract Book*, abstract 374, pp. 202–203. Available at: http://www.acarology.org/ica/ica2010/docs/0910_15_abstractBookCapa.pdf (accessed 3 February 2015).

Peña, J.E., Osborne, L. and Duncan, R.E. (1996) Potential of fungi as biocontrol agents of *Polyphagotarsonemus latus* (Acari: Tarsonemidae). *Entomophaga* 41, 27–36.

Peña, J.E., Duncan, R., Klema, E. and Hunsberger, A. (1999) Evaluation of direct and indirect action of insecticides and acaricides for control of lime and avocado pests. *Proceedings of the Florida State Horticultural Society* 112, 213–217.

Peña, J.E., Sharp, J.L. and Wysoki, M. (2002) *Tropical Fruit Pests and Pollinators. Biology, Economic Importance, Natural Enemies and Control.* CAB International, Walllingford, UK.

Peña, J.E., Palevsky, E., Otero Colina, G., Ochoa, R. and Meister, C.W. (2005) Mango bud mite, *Aceria mangiferae.* Bionomics and control under Florida conditions. *Proceedings of the Florida State Horticultural Society* 118, 228–234.

Peña, J.E., Rodrigues, J.C., Osborne, L. and Roda, A. (2009a) The red palm mite, *Raoiella indica*: effect of resident and commercially produced predators against a recently introduced pest in Florida, USA. In: Mason, P.G., Gillespie, D.R. and Vincent, C. (eds) *Proceedings of the 3rd International Symposium on the Biological Control of Arthropods, 8–13 February 2009, Christchurch, New Zealand.* Publication FHTET-2008-06, US Department of Agriculture Forest Service Forest Health Technology Enterprise Team, Morgantown, Virginia, pp. 108–115.

Peña, J.E., Rodrigues, J.C.V., Roda, A., Carrillo, D. and Osborne, L.S. (2009b) Predator–prey dynamics and strategies for control of the red palm mite (*Raoiella indica*) (Acari: Tenuipalpidae) in areas of invasion in the Neotropics. *IOBC-WPRS Bulletin* 50, 69–79.

Peña, M.A. and Ferragut, F. (1994) Primera cita para Espana de *Eriophyes mangiferae* (Sayed) (Acari: Eriophydae). *Boletín de Sanidad Vegetal, Plagas* 20, 605–609.

Penman, D.R. and Chapman, R.B. (1983) Fenvalerate-induced distributional imbalances of two-spotted spider mite on bean plants. *Entomologia Experimentalis et Applicata* 33, 71–78.

Penman, D.R. and Cone, W.W. (1972) Behaviour of male two-spotted spider mites in response to quiescent female deutonymphs and to web. *Annals of the Entomological Society of America* 65, 1289–1293.

Penman, D.R. and Cone, W.W. (1974) Role of web, tactile stimuli and female sex pheromone in attraction of male two-spotted spider mites to quiescent female deutonymphs. *Annals of the Entomological Society of America* 67, 179–182.

Pennisi, L., Di Giacomo, A. and Giambattista, R. (1975) Experiencias sobre los efectos de los danos del *Aculus pelekassi* K. sobre el rendimiento y la calidad del aceite esencial de limon. *Essenze e dei Derivati Agrumari* 44, 348–352.

Pepper, B.B. (1942) Studies of the mite, *Epitrimerus hystrix*, on timothy. *Journal of Economic Entomology* 35, 201–204.

Peralta, A.L. (1991) Nuevas especies de eriofidos (Acarina: Eriophyidae) presentes en duraznos y ciruelos. *Revista Frutícola* 12, 8–9.

Peralta, A.L. (1998) Contribución al conocimiento de nuevas especies de ácaros eriofidos (Acarina; Eriophyoidea) en olivo (*Olea europea* L.) en Chile. *Agricultura Técnica (Santiago)* 58, 298–304.

Pereira, E.B., Fornazier, M.J., Ventura, J.A., Carmo, C.A.S. do, Souza, J.L. de, Prezotti, L.C. and Dessaune Filho, N. (1987) *Efeito do Composto Orgânico na Produtividae do Alho e Reação a Praga e Patógenos.* Comunicado Técnico No. 48, Empresa Capixaba de Pesquisa Agropecuaria (EMCAPA), Cariacica, Brazil.

Pereira, P.R. Valle da Silva, Halfeld-Vieira, B. de A., Nechet, K. de L. and Mourão Júnior, M. (2007) Occurrence, damage and control of the mite *Polyphagotarsonemus latus* (Banks, 1904) (Acarina: Tarsonemidae) on sweet pepper cultivated in greenhouses. *Revista Academica Ciências Agrarias e Ambientais* 5, 39–46.

Pereira, P.R. Valle da Silva, Návia, D., Salvadori, J.R. and Lau, D. (2009) Occurrence of *Aceria tosichella* in Brazil. *Pesquisa Agropecuária Brasileira* 44, 539–542.

Perez, R. (2002) Experience in CUBA with the system of rice intensification. In: Uphoff, M., Fernandes, E.C.M., Yuan, L., Peng, J., Rafaralahy, S. and Rabenandrasana, J. (eds) *Proceedings of an International Conference on Assessments of the System of Rice Intensification (SRI) Sanya, China, April 1–4, 2002.* Cornell International Institute for Food, Agriculture and Development, Cornell University, Ithaca, New York, pp. 52–55. Available at: http://sri.ciifad.cornell.edu/proc1/SRI_Assessments.pdf (accessed 18 May 2015).

Pérez, R.P., Almaguel, L., Cáceres, I., Feitó, E. and Torre, E. de la (2004) Los depredadores de *Tetranychus tumidus* Banks en Cuba. *Fitosanidad* 8(1), 47–50.

Pérez Alvarez, R.P., Almaguel Rojas, L., Torre Lopéz, E. de la and Dominguez Barrios, M. (1997) Umbral minímo de desarrollo de *Tetranychus tumidus* en el cultivo del platano. *Manejo Integrado de Plagas* 44, 26–28.

Pérez Marín, J.L. and Ortega Sáenz, C. (1994) Posibilidades actuales de lucha contra la acariosis de la vina. *Cuadernos de Fitopatologia* 11(40), 7–11.

Pérez Moreno, I. and Moraza, M.L. (1996) Metodo de muestreo secuencial-enumerativo y binomial para *Calepitrimerus vitis* (Nalepa, 1905) (Acari: Eriophyidae). *Boletín de Sanidad Vegetal, Plagas* 22, 179–187.

Pérez Moreno, I. and Moraza, M.L. (1997) Etude sur le *Typhlodromus pyri* Scheuten en relation avec le *Calepitrimerus vitis* (Nalepa) dans les vignobles de La Rioja (Espagne). *Bulletin de l'OIV* 70(801–802), 832–846.

Pérez Moreno, I. and Moraza-Zorrilla, M.L. (1998) Population dynamics and hibernation shelters of *Calepitrimerus vitis* in the vineyards of Rioja, Spain, with a description of a new eriophyid extraction technique (Acari: Eriophyidae). *Experimental and Applied Acarology* 22, 215–226.

Pérez Otero, R., Mansilla Vézquez, P. and Salinero Corral, M.C. (2004) Efficiency of pesticides on the control of *Cosetacus camelliae* Keifer in Galicia. In: *American Camellia Yearbook 2004.* American Camellia Society, Fort Valley, Georgia, pp. 57–60.

Pérez Otero, R., Mansilla Vázquez, J.P. and Salinero Corral, M.C. (2007) First report of the broad mite *Polyphagotarsonemus latus* banks on *Camellia japonica* in Spain. In: *American Camellia Yearbook 2007.* American Camellia Society, Fort Valley, Georgia, pp. 52–56.

Pérez-Santiago, G., Alcaraz, J., Garcia-Gutierre, C. and Alvarez Zagoya, R. (2007a) Susceptibility of *Eotetranychus lewisi* (Acari: Tetranychidae) to four acaricides. In: Morales-Malacara, J.B., Behan-Pelletier, V., Ueckermann, E., Pérez, T.M., Estrada-Venegas, E.G. and Badii, M. (eds) *Acarology XI: Proceedings of the International Congress*, 8–13 September 2002, Mérida, Universidad Nacional Autonoma de México. Sociedad Latinoamericana de Acarología, México, abstract, 253.

Pérez-Santiago, G., Otero-Colina, G., González Hernández, V.A., Ramírez Guzmán, M.E., González Hernández, H. and López Jiménez, A. (2007b) The population level of *Eotetranychus lewisi* and the concentration of carbohydrates in peach trees. *Experimental and Applied Acarology* 43, 255–263.

Pérez-Sayas, C., Pina, T., Jacas, J.A. and Hurtado-Ruiz, M.A. (2010) Use of PCR-Multiplex technique for prey detection in citrus mites. In: Moraes, G.J. de, Castilho, R.C. and Flechtmann, C.H.W. (eds) *XIII International Congress of Acarology, August 23–27, 2010, Recife-PE, Brazil, Abstract Book*, abstract 382, pp. 206–207. Available at: http://www.acarology.org/ica/ica2010/docs/0910_15_abstractBookCapa.pdf (accessed 3 February 2015).

Perkins, B.D. (1973) Preliminary studies of a strain of the waterhyacinth mite from Argentina. In: Dunn, R.H. (ed.) *Proceedings of the 2nd International Symposium on the Biological Control of Weeds, 4–7 October 1971, Rome.* Commonwealth Agricultural Bureaux, Slough, UK [now CAB International, Wallingford, UK], pp. 180–184.

Perring, T.M. (1983) Influences of temperature, humidity, and corn canopy microenvironment on population dynamics of the Banks grass mite, *Oligonychus pratensis* (Banks). PhD thesis, University of Nebraska-Lincoln, Available from ETD collection for University of Nebraska-Lincoln, Paper AAI8318673 (as English abstract) at: http://digitalcommons.unl.edu/dissertations/AAI8318673 (accessed 18 August 2014).

Perring, T.M. (1996) Vegetables. In: Lindquist, E.E., Sabelis, M.W. and Bruin, J. (eds) *Eriophyoid Mites: Their Biology, Natural Enemies and Control*. World Crop Pests, Volume 6. Elsevier Science, Amsterdam, pp. 593–610.

Perring, T.M. and Farrar, C.A. (1986) Historical perspective and current world status of the tomato russet mite (Acari: Eriophyidae). *Miscellaneous Publications of the Entomological Society of America* 63, 1–19.

Perring, T.M. and McMurtry, J. (1996) Other predatory arthropods. In: Lindquist, E.E., Sabelis, M.W. and Bruin, J. (eds) *Eriophyoid Mites: Their Biology, Natural Enemies and Control*. World Crop Pests, Volume 6. Elsevier Science, Amsterdam, pp. 471–479.

Perring, T.M. and Trumble, J.T. (1984) Impact of insecticides on secondary outbreak of tomato russet mites. *Insecticicide and Acaricide Tests* 9, 197.

Perring, T.M., Holtzer, T.O., Kalisch, J.A. and Norman, J.M. (1984) Temperature and humidity effects on ovipositional rates, fecundity and longevity of adult female Banks grass mites (Acari: Tetranychidae). *Annals of the Entomological Society of America* 77, 581–586.

Perring, T.M., Farrar C.A. and Oldfield, G.N. (1996) Sampling techniques. In: Lindquist, E.E., Sabelis, M.W. and Bruin, J. (eds) *Eriophyoid Mites: Their Biology, Natural Enemies and Control*. World Crop Pests, Volume 6. Elsevier Science, Amsterdam, pp. 367–376.

Perumalsamy, K., Selvasundaram, R., Roobakkumar, A., Rahman, V.J., Babu, A. and Muraleedharan, N.N. (2009) Life table and predation of *Oligota pygmaea* (Coleoptera: Staphylinidae) a major predator of the red spider mite, *Oligonychus coffeae* (Acarina: Tetranychidae) infesting tea. *Biological Control* 51, 96–101.

Pesante, A. (1961) L'acariosi delle gemme del nocciolo. *Informatore Fitopatologico* 11, 198–199.

Pesante, A. (1962) L'acariosi delle gemme del nocciolo. *Bollettino del Laboratorio Sperimental*e *e Regio Osservatorio di Fitopatologia*, Torino 25, 27–74.

Pesic, M., Stamenkovic, S., Papic, S. and Uric, G. (1998) The effect of eriophyid mite *Vasates fockeui* (Nal. *et* Trt.) on growth increment of plum cv. Cacanska Rodna nursery trees. *Acta Horticulturae* 478, 287–291.

Pesticide Action Network UK (2000) Endosulfan. *Pesticides News* No. 47, March 2000, p. 20/Fact Sheet, Pesticide Action Network UK, Brighton, UK. Available at: http://www.pan-uk.org/pestnews/Actives/endosulf.htm (accessed 20 April 2015).

Petanović, R. (1993) Mites (Acari) biological control agents. I. Use in orchards and vineyards. *Zastita Bilja* 44, 5–20 (in Serbian).

Petanović, R. (1997) Allochthonous mite species (Acari) in the fauna of Yugoslavia. *Zastita Bilja* 48, 211–224 (in Serbian).

Petanović, R. (1998) New data to the knowledge on the eriophyoid fauna (Acari: Eriophyoidea) in Yugoslavia. *Acta Entomologica Serbica* 3, 149–157.

Petanović, R. (2001) Stetne grinje u proizvodnji cvetnih kultura u yatvorenom prostoru (4): Acaride (Acaridae). *Biljni Lekar* 29, 155–159 (in Serbian).

Petanović, R (2004) *Štetne Grinje Ukrasnih Biljaka – Atlas* (*Mite Pests of Decorative Plants – Atlas*). Beografik., Belgrade, Serbia (in Serbian).

Petanović, R. and Dobrivojevic, K. (1987) A complex of gall forming eriophyid mites on plum leaves. *Zastita Bilja* 38, 145–156.

Petanović, R. and Filipi-Matutinovic, S. (1988) The history and bibliography on phytophagous and predatory Acarida (Gamasida, Actinedida, Acaridida) in Yugoslavia till 1988. *Pesticidi* 4, 201–206.

Petanović, R. and Kielkiewicz, M. (2010a) Plant–eriophyoid mite interactions: cellular biochemistry and metabolic responses induced in mite-injured plants. Part I. *Experimental and Applied Acarology* 51, 61–80.

Petanović, R. and Kielkiewicz, M. (2010b) Plant-eriophyoid mite interactions: specific and unspecific morphological alterations. Part II. *Experimental and Applied Acarology* 51, 81–91.

Petanović, R. and Stanković, S. (1999) Catalogue of the Eriophyoidea (Acari: Prostigmata) of Serbia and Montenegro. *Acta Entomologica Serbica*, Special Issue, 1–143.

Petanović, R.U., Boczek, J., Jovanovic, S. and Stojnic, B. (1996) Eriophyoidea (Acari: Prostigmata). In: Karaman, G. (ed.) *Fauna Durmitora, Knj. 32, sv. 18*. Crnogorska Akademija Nauka i Umjetnosti, Podgorica, Montenegro, pp. 5–42.

Peter, C. and David, B.V. (1988) Evaluation of some insecticides for the control of major insect pests of okra and their side effects. *Tropical Pest Management* 34, 76–80, 117, 122.

Peter, C., David, B.V. and Zutshi, M.K. (1987) Evaluation of Plictran 50W against the red spider mite *Tetranychus neocaledonicus* Andre (Tetranychidae: Acarina) on lady finger. *Pesticides* 21(6), 34–35.

Petersen, H. (1982) Structure and size of soil animal populations. *Oikos* 39, 306–329.

Peterson, K.D. and Higley, L.G. (2002) Economic decision levels. In: Pimentel, D. (ed.) *Encyclopedia of Pest Management*. Marcel Dekker, New York, pp. 228–230.

Peterson, L.O.T. and Hildahl, V. (1969) *The Spruce Spider Mite in the Prairie Provinces*. Canadian Forest Service Liaison and Service Note L-7, Department of Fisheries and Forestry, Northwest Region, Canada Department of the Environment, Edmonton, Alberta.

Peterson, P.G., McGregor, P.G., Springett, B.P. (1994) Development of *Stethorus bifidus* in relation to temperature: implications for regulation of gorse mite populations. In: Popay, E.J. (ed.) *Proceedings of the 47th New Zealand Plant Protection Conference, 1994*. New Zealand Weed and Pest Control Society, Palmerston North, New Zealand, pp. 103–106.

Petrushov, A.Z. (1987) Results of introduction into the USSR of a population of a predaceous mite, *Metaseiulus occidentalis* (Acarina, Phytoseiidae), resistant to pesticides. *Zoologicheskii Zhurnal* 66, 674–680 (in Russian).

Petty, G.J. (1975) Pineapple mites. *Citrus and Subtropical Fruit Journal* 498, 15–18.

Petty, G.J. (1977) *Leathery Pocket in Pineapples*. Leaflet, Pineapple Series No H. 2, Department of Agricultural Technical Services of the Republic of South Africa. Government Printer, Pretoria.

Petty, G.J., Stirling, G.R. and Bartholomew, D.P. (2002) Pests of pineapple. In: Peña, J.E., Sharp, J. and Wysoki, M. (eds) *Tropical Fruit Pests and Pollinators: Their Economic Importance, Natural Enemies and Control*. CAB International, Wallingford, UK, pp. 157–196.

Pfeiffer, D.G., Ponton, L.F. and Varn, M. (1989) Apple: mite control trials 1986. *Insecticicide and Acaricide Tests* 14, 30–32.

Philipp, W. (1957) Intensified occurrence of the plum gall mite, *Aceria* (*Eriophyes*) *phloeocopte*s (Nal.). *Zeitschrift für Pflanzenkrankheiten Pflanzenpathologie und Pflanzenschutz* 64, 35–38.

Phillips, P.A. and Walker, G.P. (1997) Increase in flower and young fruit abscission caused by citrus bud mite (Acari: Eriophyidae) feeding in the axillary buds of lemon. *Journal of Economic Entomology* 90, 1273–1282.

Pianka, E.R. (1970) On *r*- and *k*-selection. *American Naturalist* 104, 592–597.

Pickett, A.D. (1965) The influence of spray programs on the fauna of apple orchards in Nova Scotia. XIV. Supplement to II. Oystershell scale, *Lepidosaphes ulmi* (L.). *Canadian Entomologist* 97, 816–821.

Pickett, C.H. and Gilstrap, F.E. (1986) Predation of *Oligonychus pratensis* [Aca.: Tetranychidae] by *Phytoseiulus persimilis* and *Amblyseius californicus* [Aca.: Phytoseiidae] under controlled laboratory conditions. *Entomophaga* 31, 205–212.

Pickett, C.H., Gilstrap, F.E., Morrison, R.K. and Bouse, L.F. (1987) Release of predatory mites (Acari: Phytoseiidae) by aircraft for the biological control of spider mites (Acari: Tetranychidae) infesting corn. *Journal of Economic Entomology* 80, 906–910.

Pickett, C.H., Morrison, R.K., Gilstrap, F.E. and Bouse, L.F. (1990) Storage of *Phytoseiulus persimilis* for aerial release (Acari: Phytoseiidae). *International Journal of Acarology* 16, 37–40.

Pielou, D.P. (1960) The effect of DDT on oviposition, and on behavior in the European red mite *Panonychus ulmi* (Koch). *Canadian Journal of Zoology* 38, 1147–1151.

Pierce, W.C. (1953) Studies of mites and their control on pecan in Louisiana. *Journal of Economic Entomology* 46, 561–565.

Pieterse, A.H. (1972) A preliminary investigation on control of water hyacinth by spider mites. *Proceedings of the 11th British Weed Control Conference, Volume 1.* British Crop Production Council, Nottingham, UK, pp. 1–3.

Pijnacker, L.P., Ferwerda, M.A., Bolland, H.R. and Helle, W. (1980) Haploid female parthenogenesis in the false spider mite *Brevipalpus obovatus* (Acari: Tenuipalpidae). *Genetica* 51, 211–214.

Pijnacker, L.P., Ferwerda, M.A. and Helle, W. (1981) Cytological investigations on the female and male reproductive system of the parthenogenetic privet mite *Brevipalpus obovatus* Donnadieu (Phytoptipalpidae, Acari). *Acarologia* 22, 157–163.

Pillai, K.S. and Palaniswami, M.S. (1983) Economics on the control of spidermites on cassava. *Entomon* 8, 373–375.

Pillai, K.S. and Palaniswami, M.S. (1991) Integrated approach for the management of spidermite complex on cassava. *Journal of Root Crops* 17, 44–49.

Pillai, S.V. and Jolly, M.S. (1986) Studies on the population build up and control measures of the red spider mite, *Tetranychus equatorius* (McGr). *Indian Journal of Sericulture* 25, 15–21.

Pimentel, D. (1979) *World Food, Pest Losses and the Environment.* AAAS (American Association for the Advancement of Science) Selected Symposium 13. Westview Press, Boulder, Colorado.

Pimentel, D. (ed.) (2002) *Encyclopedia of Pest Management.* Marcel Dekker, New York.

Pimental, D., Rumsey, M.W. and Streams, F.A. (1960) Rearing tyroglyphid mites on *Neurospora*. *Annals of the Entomological Society of America* 53, 549.

Pinto, R., Ferreira, J.A.M., Pires, E.M. and Zanuncio, J.C. (2012) New record and characteristics of damage caused by *Oligonychus yothersi* on *Eucalyptus urophylla*. *Phytoparasitica* 40, 143–145.

Planes, S., del Rivero, J.M., Marti Fabregat, F. and Carrero, J.M. (1965) Contribution al estudio de la biologia de *Phytoptus avellanae* Nal. y de los medios de lucha contra la plaga. *Boletín de Patología Vegetal y Entomología Agrícola* 28, 27–55.

Plaut, H.N. (1965) On the phenology and control value of *Stethorus punctillum* Weise as a predator of *Tetranychus cinnabarinus* Boisd. in Israel. *Entomophaga* 10, 133–137.

Plavsic, B. and Milicic, D. (1980) Electron microscopy of fig mosaic disease. In: *Proceedings of the Fifth Congress of the Mediterranean Phytopathological Union, Patras, Greece, 21–27 September 1980.* Hellenic Phytopathological Society, Athens, pp. 3–5.

Poderoso, J.C.M., Ribeiro, G.T., Návia, D., Passos, E.M., Gonçalves G.B., Correia-Oliveira, M.E. and Dantas, P.C. (2010) Primeiro registro no Brasil de *Erythrina velutina* Willd. como hospedeira de *Tetranychus neocaledonicus* (Acari: Tetranychidae). *Revista Brasileira de Plantas Medicinais (Botucatu)* 12, 398–401.

Poe, S.L. (1971) Microfaunal populations on gladiolus corms. *Florida Entomologist* 54, 127–133.

Poe, S.L. (1980) Sampling mites on soybean. In: Kogan, M. and Herzog, D.C. (eds) *Sampling Methods in Soybean Entomology*. Springer, New York, pp. 313–323.

Poe, S.L. and Enns, W.R. (1969) Predacious mites (Acarina: Phytoseiidae) associated with Missouri orchards. *Transactions of the Missouri Academy of Science* 3, 69–82.

Poe, S.L. and Enns, W.R. (1970) Effects of inbreeding on closed populations of predaceous mites (Acarina: Phytoseiidae). *Canadian Entomologist* 102, 1222–1229.

Poe, S.L., Collins, H. and Shih, C. (1976) Numeric response of *Oligonychus ilicis* to contact acaricides. *Southern Nurserymen's Association Research Conference* 3, 39–40.

Poe, S.L., Noble, W.E. and Stall, R.E. (1979) Acquisition and retention of *Pseudomonas marginata* by *Anoetus feroniarum* and *Rhizoglyphus robini*. In: Rodriguez, J.G. (ed.) *Recent Advances in Acarology, Volume I.* Academic Press, New York, pp. 119–124.

Poinar, G.O. Jr (1992) *Life in Amber*. Stanford University Press, Redwood City, California.

Poinar, G.O. Jr (1995) First fossil soft ticks, *Ornithodoros antiquus* n. sp. (Acari: Argasidae) in Dominican amber with evidence of their mammalian host. *Experientia* 51, 384–387.

Poinar, G.O. Jr and Brown, A.E. (2003) A new genus of hard ticks in Cretaceous Burmese amber (Acari: Ixodida: Ixodidae). *Systematic Parasitology* 54, 199–205.

Poinar, G.O. Jr and Poinar, R. (1998) Parasites and pathogens of mites. *Annual Review of Entomology* 43, 449–469.

Poinar, G.O. Jr, Pike, E.T.M. and Krantz, G.W. (1993) Animal–animal parasitism. *Science* 361, 307–308.

Poinar, G.O. Jr, Krantz, G.W., Boucout, A.J. and Pike, T.M. (1997) A unique Mesozoic parasitic association. *Naturwissenschaften* 84, 321–322.

Poletti, M. and Omoto, C. (2003) Effect of spirodiclofen on the reproduction and survivorship of *Brevipalpus phoenicis* (Geijskes) (Acari: Tenuipalpidae). *Arquivos do Instituto Biológico (São Paulo)* 70(Suplemento 3), 65.

Poli, R.C.D. (1991) The biology of the false spider mite *Dolichotetranychus floridanus*: a pest of pineapples in Central Queensland. MSc. Thesis, University of Central Queensland, North Rockhampton, Australia.

Policápio da Costa Gonçalves, G.C., Costa Botelho, A., Cardoso Pires, V., Návia Magalhães Ferreira, D., Soares Pires, C.S., Gouveia Fontes, E.A. and Ryoiti Sujii, E. (2008) Ocorrência do ácaro da erinose do algodeiro, *Acalitus gossypii* (Banks) (Acari: Eriophyidae), em *Gossypium barbadense* L. no Districto Federal. In: *Resumos, XXII Congresso Brasileiro de Entomologia, 24 a 29 de agosto 2008, Centro de Convenções-Uberlândia, Minas Gerais.* Sociedade Entomológica do Brasil, Londrina, Brazil, Resumo ID 1599-2. Available at: http://www.seb.org.br/eventos/CBE/XXIICBE/resumos/R1599-2.html (accessed 19 May 2015).

Pollock, D.A. and Michels, G.J. Jr (2002) Distributions of *Stethorus nigripes* Kapur (Coleoptera: Coccinellidae), a predator of Banks grass mite (*Oligonychus pratensis* Banks) in the southern United States. *Southwestern Entomologist* 27, 217–220.

Pollock, D.A. and Michels, G.J. (2003) First records of *Stethorus histrio* Chazeau (Coleoptera: Coccinellidae) from the United States. *Southwestern Entomologist* 28, 221–222.

Pons, L. and Bliss, R.M. (2007) A tiny menace island-hops the Caribbean. *AgResearch (Agricultural Research Magazine, Agricultural Research Service, US Department of Agriculture)* 55(5) May/June, 4–6. Available at: http://agresearchmag.ars.usda.gov/AR/archive/2007/May/island0507.pdf (accessed 19 May 2015).

Popov, N.A. and Khudyakova, O.A. (1989) Development of *Phytoseiulus persimilis* (Acarina, Phytoseiidae) fed on *Tetranychus urticae* (Acarina, Tetranychidae) on various food plants. *Acta Entomologica Fennica* 53, 43–46.

Post, A. (1962) Effect of cultural measures on the population density of the fruit tree red spider mite, *Metatetranychus ulmi* Koch. *Tijdschrift over Plantenziekten* (now *European Journal of Plant Pathology*) 68, 1–110.

Potapenko, A.A. (1985) Breeding black currant for a combination of economically useful characters. In: *Selektsiya i Agrotekhnika Plodovo-yagodnykh i Dekorativnykh Kul'tur*. Novosibirsk, USSR, pp. 15–22 (in Russian).

Potenza, M.R., Takematsu, A.P., Jocys, T., Felicio, J.D.F., Rossi, M.H. and Sakita, M.N. (2003) Acaricide evaluation of extracts for the control of coffee red mite *Oligonychus ilicis* (McGregor) (Acari: Tetranychidae). *Arquivos do Instituto Biológico (São Paulo)* 70(Suplemento 3), 4.

Potenza, M.R., Takematsu, A.P., Jocys, T., Felicio, J.D.F., Rossi, M.H. and Sakita, M.N. (2005) Evaluation of plant extracts for the control of coffee red mite *Oligonychus ilicis* (McGregor) (Acari: Tetranychidae). *Arquivos do Instituto Biológico (São Paulo)* 72, 499–503.

Potter, D.A. and Wrensch, D.L. (1978) Interrupted matings and the effectiveness of second inseminations in the two spotted spider mite. *Annals of the Entomological Society of America* 71, 882–885.

Potter, D.A., Wrensch, D.L. and Johnston, D.E. (1976a) Guarding, aggressive behaviour and mating success in male two-spotted spider mites. *Annals of the Entomological Society of America* 69, 707–711.

Potter, D.A., Wrensch, D.L. and Johnston, D.E. (1976b) Agression and mating success in male spider mites. *Science* 193, 160–161.

Powell, D.F. (1977) The effects on narcissus bulbs of methyl bromide fumigation used to control bulb scale mite. *Plant Pathology* 26, 79–84.

Power, A.G. (2001) The ecology of agriculture. In: Levin, S.A. (ed.) *Encyclopedia of Biodiversity, Volume 2*. Academic Press, New York, pp. 269–276.

Prabaningrum, L., Sastrosiswojo, S. and Rubiati, T. (1999) Efficacy of *Amblyseius cucumeris* in controlling *Thrips parvispinus* and *Polyphagotarsonemus latus* on hot pepper communities. *Jurnal Hortikultura (Jakarta)* 9(3), 220–225 (in Indonesian).

Prabhuswamy, H.P., Srinivasa, N. and Gupta, S.K. (1995) Population fluctuation of eriophyid mite *Aceria cajani*, vector of pigeonpea sterility mosaic disease. In: *Fifth National Symposium of Acarology, 20–22 September 1955, University of Agricultural Sciences, Bangalore*. Acarological Society India/UAS, Bangalore, India, abstract, 29.

Prado, E. (1991) *Artrópodos y sus Enemigos Naturales Asociados a Plantas Cultivadas en Chile*. Boletín Técnico No. 169, Estación Experimental La Platina, Instituto de Investigaciones Agropecuarias, Santiago.

Pralavorio, M. and Almaguel-Rojas, L. (1980) Influence de la temperature et de l'humidite relative sur le developpement et la reproduction de *Phytoseiulus persimilis*. *IOBC-WPRS Bulletin* 3(3), 157–162.

Prameela, H.A., Vishwanath, S. and Anilkumar, T.B. (1990) Effect of single and combined infection of powdery mildew and sterility mosaic on nodulation and biomass of pigeonpea. *Zentralblatt für Mikrobiologie* 145, 37–39.

Prasad, G.S. and Ranganath, H.R. (2000) Occurrence of nut infesting eriophyid mite *Eriophyes guerreronis* (K.) in coconut plantations in Andamans. *Indian Coconut Journal* 31(6), 17.

Prasad, V. (1973) A new species of *Tuckerella* (Acarina: Tuckerellidae) from India. *Acarologia* 15, 237–239.

Prasad, V. (1974) *A Catalogue of Mites of India*. Indira Acarology Publishing House, Ludhiana, India.

Prasad, V. (1988) *Mites: A Bibliography (1926–1966)*. Indira Publishing House, Oak Park (now West Bloomfield), Michigan.

Prasad, V.G. and Bagle, B. (1981) Evaluation of acaricides for the control of litchi mite, *Aceria litchii* Keifer (Acari: Eriophyidae). *Pesticides* 15(1), 22–23.

Prasad, V.G. and Singh, R.K. (1981) Prevalence and control of litchi mite, *Aceria litchii* Keifer in Bihar. *Indian Journal of Entomology* 43, 67–75.

Prasanna, K.P. (2007) Seasonal incidence and management of tetranychid mite in brinjal. MSc. thesis, University of Agricultural Sciences, Dharwad, India. Abstract available at: http://etd.uasd.edu/abst/th9531.pdf (accessed 2 April 2015).

Prat Kricun, S., Belingheri, L.D., Fontana, H.P., Rivera Flores, S.E. and Dehle, R.A. (2011) *Té: Técnicas de Cultivo y Manufactura – Plagas y Enfermedades*. Estación Experimental Agropecuaria Cerro Azul, Instituto Nacional de Tecnología Agropecuaria, Buenos Aires. Available at: http://inta.gob.ar/documentos/te-tecnicas-de-cultivo-y-manufactura-plagas-y-enfermedades/at_multi_download/file/Te_tecnicas_cultivomanuf_plagas_enferm.pdf (accessed 19 August 2014).

Prather, T.S., Callihan, R.H., McCaffrey, J.P. and Shafii, B. (1989) Effects of a gall midge and a gall mite on rush skeletonweed. *Proceedings, Western Society of Weed Science* 42, 284–285. Available at: http://www.wsweedscience.org//wp-content/uploads/proceedings-archive/1989.pdf (accessed 19 May 2015).

Pratt, P.D. and Croft, B.A. (1999) Expanded distribution of the bamboo spider mite, *Schizotetranychus longus* (Acari: Phytoseiidae), and predation by *Neoseiulus fallacis* (Acari: Phytoseiidae). *Acarologia* 40, 191–197.

Pratt, P.D., Coombs, E.M. and Croft, B.A. (2002) Phytoseiid mite fauna on gorse, *Ulex europaeus* L., in western Oregon, USA with new records for *Phytoseiulus persimilis* Athias-Henriot and *Amblyseius graminis* (Chant) (Acari: Phytoseiidae). *The Pan-Pacific Entomologist* 78, 215–218.

Pratt, P.D., Coombs, E.M. and Croft, B.A. (2003) Predation by phytoseiid mites on *Tetranychus lintearius* (Acari: Tetranychidae), an established weed biological control agent of gorse (*Ulex europaeus*). *Biological Control* 26, 40–47.

Pratt, R.M. (1957) Relation between moisture conditions and rust mite infestations. *Proceedings of the Florida State Horticultural Society* 70, 48–51.

Pree, D.J., Bittner, L.A. and Whitty, K.J. (2002) Characterization of resistance to clofentezine in populations of European red mite from orchards in Ontario. *Experimental and Applied Acarology* 27, 18–193.

Prescott, R.T.M. (1934) The pear leaf blister mite, *Eriophyes pyri* (Pgst.). *Journal of the Department of Agriculture, Victoria* 32, 538–539.

Price, D.W. (1976) Passage of *Verticillum albo-atrum* propagules through the alimentary canal of the bulb mite. *Phytopathology* 66, 46–50.

Price, P.W., Bouton, C.E., Gross, P., McPheron, B.A., Thompson, J.N. and Weis, A.E. (1980) Interactions among three trophic levels: influence of plants on interactions between insect herbivores and natural enemies. *Annual Review of Ecology and Systematics* 11, 41–65.

Prieto Trueba, D. (1975) Research on the biology of *Brevipalpus phoenicis* (Geijskes) (Acarina: Tenuipalpidae). In: *VIII International Plant Protection Congress, Moscow, 1975. Reports and Informations: Section II, Progress in the Study of Biology of Pest Organisms and the Development of Forecast Methods*. VIII International Plant Protection Congress, Moscow pp. 414–416.

Pringle, K.L. (1975) The seasonal occurrence of phytophagous mites (Acari) and their predators on apple trees in the Elgin area. In: *Proceedings of the 1st Congress of the Entomological Society of Southern Africa, 1975*. Entomological Society of Southern Africa, Pretoria, pp. 131–140.

Pritchard, A.E. (1951a) The fern mite, a newly recognized pest in California ferns readily controlled by treatment with proper chemicals. *California Agriculture* 5(7), 10.

Pritchard, A.E. (1951b) Control of orchid mites: false spider mites and spider mites must be distinguished for proper control purposes. *California Agriculture* 5(9), 11.

Pritchard, A.E. (1957) New carnation pest found in California. *Florists' Exchange and Horticultural Trade World* 129(1), 12.

Pritchard, A.E. and Baker, E.W. (1952a) A guide to the spider mites of deciduous fruit trees. *Hilgardia* 21(9), 253–287.

Pritchard, A.E. and Baker, E.W. (1952b) The false spider mites of California (Acarina: Phytoptipalpidae). *University of California Publications in Entomology* 9(1), 1–94.

Pritchard, A.E. and Baker, E.W. (1955) *A Revision of the Spider Mite Family Tetranychidae*. Memoirs of the Pacific Coast Entomological Society, Volume 2, Pacific Coast Entomological Society, San Francisco, California.

Pritchard, A.E. and Baker, E.W. (1958) The false spider mites (Acarina: Tenuipalpidae). *University of California Publications in Entomology* 14(3), 175–274.

Proeseler, G. (1969) Transmission of fig mosaic virus by the gall mite *Aceria ficus* (Cotte). *Zentralblatt für Bakteriologie, Parasitenkunde, Infektionskrankheiten und Hygiene, II Abteilung – Naturwissenschaftliche-Mikrobiologie der Landwirtschaft der Technologie* 123, 288–292.

Proeseler, G. (1972a) Relationship between virus, vector, and host plant shown by fig mosaic virus and *Aceria ficus* (Cotte). *Acta Phytopathologica Academiae Scientiarum Hungaricae* 7, 179–186.

Proeseler, G. (1972b) Ecological studies of *Aculus fockeui* Nalepa and Trt. and *Cecidophyopsis ribis* Westw. in the D.D.R. *Zeszyty Problemowe Postępów Nauk Rolniczych* 129, 177–184.

Proeseler, G. (1972c) Ein Beitrag zur Morphologie und Ökologie der Gallmilben (Eriophyoidea) an Gramineen. *Biologisches Zentrablatt* 91, 477–492.

Proeseler, G. (1973) The gall mite *Cecidophyopsis ribis* (Westw.) as pest of currants. *Archiv fur Phytopathologie und Pflanzenschutz* 9, 383–394.

Pruthi, H.S. and Mani, M.S. (1945) *Our Knowledge of the Insect and Mite Pests of Citrus in India and their Control*. Scientific Monograph No. 16, Council of Agricultural Research, Delhi.

Przygoda, E. (2001) The occurrence of spruce spider mite (*Oligonychus ununguis* Jacobi) on ornamental coniferous plants in botanical gardens in Poland. *Annals of Warsaw Agricultural University – Horticulture and Landscape Architecture* 22, 23–29.

Pscheidt, J.W., Peachey, E. and Walton, V. (2014) *Hazelnut 2014 Pest Management Guide for the Willamette Valley*. Oregon State University Extension Service Publication No. EM 8328, Corvallis, Oregon. Available at: http://ir.library.oregonstate.edu/xmlui/bitstream/handle/1957/46564/em8328.pdf (accessed 16 August 2014).

Puchalska, E. (2002) Preliminary studies concerning changes of photosynthetic activity of white spruce (*Picea glauca* 'Conica') infested by spruce spider mite (*Oligonychus ununguis* Jacobi). *Annals of Warsaw Agricultural University – Horticulture and Landscape Architecture* 23, 7–10.

Puchalska, E. (2006) The influence of *Oligonychus ununguis* Jacobi (Acari: Tetranychidae) on photosynthetic activity and needle damage of *Picea glauca* 'Conica'. *Biological Letters* 43, 353–360.

Puchalska, E., Tykarska, T. and Czajkowska, B. (2006) Morphological and anatomical factors responsible for varied susceptibility of some species of spruce to infestation by spruce spider mite (*Oligonychus ununguis* Jacobi). *Acta Physiologiae Plantarum* 28, 599–604.

Puchalska, E., Czajkowska, B. and Kiekiewicz, M. (2008) Morphological, anatomical and chemical characterization of white spruce (*Picea glauca* 'Conica') differently aged needles and hypotheses on their influence on *Oligonychus ununguis* infestation. *Acta Physiologiae Plantarum* 30, 225–232.

Pugh, P.J.A. (1996) Using artifical substrata to monitor how cryptofaunal acari colonize littoral algae on sub-antarctic South Georgia. *Acarologia* 37, 189–200.

Pukall, R., Schumann, P., Schütte, C., Gols, R. and Dicke, M. (2006) *Acaricomes phytoseiuli* gen. nov., sp. nov., isolated from the predatory mite *Phytoseiulus persimilis*. *International Journal of Systematic and Evolutionary Microbiology* 56, 465–469.

Pumnuan, J., Insung, A. and Chandrapatya, A. (2008) Acaricidal effects of herb extracts on the mushroom mites, *Luciaphorus perniciosus* Rack and *Formicomotes heteromorphus* Magowski. *Systematic and Applied Acarology* 13, 33–38.

Pumnuan, J., Chantrasorn, W., Insung, A. and Rongpol, P. (2009) Effect of black pepper (*Piper nigrum* Linn.) essential oil on African red mite (*Eutetranychus africanus* (Tucker)) (Actinedida: Tetranychidae). *Journal of Agriculture, Faculty of Agriculture, Chiang Mai University* 25, 169–176 (in Thai with English abstract).

Pumnuan, J., Rongpol, P. and Insung, A. (2010) Repellent efficiency of essential oils of medicinal plants to mushroom mites, *Formicomotes heteromorphus* Magowski and *Luciaphorus perniciosus* Rack. *Agricultural Science Journal* 41(2, Suppl.), 633–636 (in Thai with English abstract). Available at: http://wireless02.agri.kmitl.ac.th/agri_officer/userfiles/files/53_32.pdf (accessed 19 May 2015)

Purohit, S.D., Ramawat, K.G. and Arya, H.C. (1979) Phenolics, peroxidase and phenolase as related to gall formation in some arid zone plants. *Current Science* 48, 714–716.

Pushpa, V. and Nandihalli, B.S. (2008) Bioefficacy of botanicals in the management of coconut eriophyid mite, *Aceria guerreronis* Keifer. *Journal of Plant Protection and Environment* 5, 90–94.

Putatunda, B.N. and Anupam Tagore (1997) Beneficial phytoseiid mites associated with some crop plants at Hisar, Haryana. *Pest Management and Economic Zoology* 5, 123–126.

Putatunda, B.N., Mathur, R.B. and Mathur, S. (2002) Mites associated with some fruit trees in Hisar, Haryana, India. *Indian Journal of Agricultural Research* 36, 88–95.

Puterka, G.J., Glenn, D.M., Sekutowski, D.G., Unruh, T.R. and Jones, S.K. (2000) Progress toward liquid formulations of particle films for insect and disease control in pear. *Environmental Entomology* 29, 329–339.

Putman, W.L. (1939) The plum nursery mite (*Phyllocoptes fockeui* Nal. and Trt.). *Annual Report of the Entomological Society of Ontario* 70, 33–40.

Putman, W.L. (1955) Bionomics of *Stethorus punctillum* Weise (Coleoptera: Coccinellidae) in Ontario. *Canadian Entomologist* 87, 9–33.

Putman, W.L. (1965) The predaceous thrips *Haplothrips faurei* Hood (Thysanoptera, Phloethripidae) in Ontario peach orchards. *Canadian Entomologist* 97, 1208–1221.

Putman, W.L. (1967) Prevalence of spiders and their importance as predators in Ontario peach orchards. *Canadian Entomologist* 99, 160–170.

Putman, W.L. (1970a) Occurrence and transmission of a virus disease of the European red mite, *Panonychus ulmi*. *Canadian Entomologist* 102, 305–321.

Putman, W.L. (1970b) Life history and behavior of *Balaustium putmani* (Acarina: Erythraeidae). *Annals of the Entomological Society of America* 63, 76–81.

Putman, W.L. and Herne, D.H.C. (1966) The role of predators and other biotic agents in regulating the population density of phytophagous mites in Ontario peach orchards. *Canadian Entomologist* 98, 808–820.

Puttarudriah, M. and ChannaBasavanna, G.P. (1956) Some beneficial coccinellids of Mysore. *Journal of the Bombay Natural History Society* 54, 156–159.

Puttarudriah, M. and ChannaBasavanna, G.P. (1959) A preliminary account of phytophagous mites of Mysore. In: Seshaiya, R.V. and Chauhan, B.S. (eds) *Proceedings of the First All-India Congress of Zoology, 24–28 October 1959, Part 2, Scientific Papers*. Zoological Society of India, Calcutta, India, p. 530–539.

Puttarudriah, M. and Usman, S. (1957) An eriophyid mite injurious to sugarcane. *Current Science* 26, 290.

Puttaswamy and ChannaBasavanna, G.P. (1976) Effect of some important acaricides on the predator *Stethorus pauperculus* Weise (Coleoptera: Coccinellidae) and its host *Tetranychus cucurbitae* Rahman and Sapra. *Mysore Journal of Agricultural Sciences* 10, 636–641.

Puttaswamy and ChannaBasavanna, G.P. (1977) Biology of *Stethorus pauperculus* Weise (Coleoptera: Coccinellidae), a predator of mites. *Mysore Journal of Agricultural Sciences* 11, 81–89.

Puttaswamy and ChannaBasavanna, G.P. (1979a) Effect of temperature and relative humidity on the development and oviposition of *Tetranychus ludeni* (Acari: Tetranychidae). *Indian Journal of Acarology* 4, 31–40.

Puttaswamy and ChannaBasavanna, G.P. (1979b) Life history of *Tetranychus ludeni* (Acari: Tetranychidae) under field conditions. *Indian Journal of Acarology* 4, 41–48.

Puttaswamy and ChannaBasavanna, G.P. (1979c) *Typhlodromips tetranychivorus* (Acari: Phytoseiidae) feeding on *Tetranychus ludeni* (Acari: Tetranychidae) at Bangalore. *Acarology Newsletter* 8, 5.

Puttaswamy and ChannaBasavanna, G.P. (1980) Influence of weather factors and predators on the populations of the spidermite *Tetranychus ludeni* (Acari: Tetranychidae). *Indian Journal of Acarology* 5, 69–79.

Puttaswamy and ChannaBasavanna, G.P. (1981a) Influence of three *Amaranthus* species on the development, fecundity and longevity of *Tetranychus neocaledonicus* (Acari: Tetranychidae). *Colemania (Bangalore)* 1, 35–36.

Puttaswamy and ChannaBasavanna, G.P. (1981b) Influence of host plants on the reproductive biology of *Tetranychus neocaledonicus* (Acari: Tetranychidae). *Indian Journal of Acarology* 6, 72–76.

Puttaswamy and Rangaswamy, H.R. (1976) *Stethorus keralicus* Kapur (Coleoptera: Coccinellidae), a predator of the areca palm mite. *Current Research, University of Agricultural Sciences (Bangalore)* 5, 27–28.

Puttaswamy and Reddy, D.N.R. (1980) *Tetranychus neocaledonicus* (André) (Acari: Tetranychidae), a new pest on cardamom. *Acarology Newsletter* 10, 6–7.

Puttaswamy, C.G.P. (1981) Influence of host plants on the reproductive biology of *Tetranychus neocaledonicus* (Acari: Tetranychidae). *Indian Journal of Acarology* 6, 72–76.

Queensland Department of Agriculture and Fisheries (2012) Strawberry spider mite. Available at: http://www.daff.qld.gov.au/plants/field-crops-and-pastures/broadacre-field-crops/integrated-pest-management/a-z-insect-pest-list/mites-overview/strawberry-spider-mite (accessed 2 April 2015).

Qin, T.K. (1997) Population genetics of redlegged earth mites *Halotydeus destructor* and *H. anthropus* (Acarina: Penthaleidae) from Australia and/or South Africa. *Bulletin of Entomological Research* 87, 289–298.

Qin, T.K. and Halliday, R.B. (1995) Systematic studies of redlegged earth mite *Halotydeus destructor* (Tucker) and related species (Acarina: Penthaleidae). *Plant Protection Quarterly* 10, 50–52.

Qin, T.K. and Halliday, R.B. (1996a) Revision of the Australian and South African species of *Halotydeus* Berlese (Acarina: Penthaleidae). *Bulletin of Entomological Research* 86, 441–450.

Qin, T.K. and Halliday, R.B. (1996b) The Australian species of *Chromotydaeus* and *Penthaleus* C.L. Koch (Acarina: Penthaleidae). *Journal of Natural History* 30, 1833–1848.

Qin, T.K. and Rohitha, B.H. (1996) The astigmatid mite *Histiostoma feroniarum* (Acari: Astigmata: Histiostomatidae) in New Zealand. *New Zealand Entomologist* 19, 65–69.

Qu, S.C., Tashiro, N., Etoh, T. and Sadamatsu, M. (1997) Occurrence of dithiocarbamate resistant strain and its susceptibilities to several chemicals in the pink citrus rust mite, *Aculops pelekassi* (Keifer), in Saga Prefecture. *Proceedings of the Association for Plant Protection of Kyushu* 43, 125–129 (in Japanese).

Quayle, H.J. (1938) *Insects of Citrus and Other Subtropical Fruits*. Cornell University Press/Comstock Publishing, Ithaca, New York.

Quilici, S., Kreiter, S., Ueckermann, E.A. and Vincenot, D. (1997) Predatory mites (Acari) from various crops on Réunion Island. *International Journal of Acarology* 23, 283–291.

Quirós de González, M. (1978) Presencia del ácaro marron del aguacate *Oligonychus punicae* (Hirst) en hojas de platano del distrito Baralt, Estado Zulia. *Revista de la Facultad de Agronomía, Universidad de Zulia* 5, 422–424.

Quirós de González, M. (1981) A systematic revision of the spider mites of Central America (Tetranychidae: Acari). M.Sc. thesis, University of Maryland, College Park, Maryland.

Quirós de González, M. and Dorado, I. (2005) Eficiencia de tres productos comerciales en el control del ácaro hindú de las cítricas *Schizotetranychus hindustanicus* (Hirst), en el laboratorio. In: *Resúmenes del XIX Congreso Venezolano de Entomología, 4–7 July 2005, Maracaibo, Venezuela*. *Entomotropica* 20, 184–185, abstract 183.

Quirós de González, M. and Geraud-Pouey, F. (2007) *Schizotetranychus hindustanicus* (Hirst) (Acari: Tetranychidae), new spider mite pest damaging citrus in Venezuela, South America. In: Morales-Malacara, J.B., Behan-Pelletier, V., Ueckermann, E., Pérez, T.M., Estrada-Venegas, E.G. and Badii, M. (eds) *Acarology XI: Proceedings of the International Congress, 8–13 September 2002, Mérida, Universidad Nacional Autonoma de México*. Sociedad Latinoamericana de Acarología, México, abstract, pp. 255–256.

Quiróz de González, M., Poleo, N. and Petit, M.Y. de (2002) Development of the apical tip damage of guava fruit, *Psidium guajava* L., caused by *Brevipalpus phoenicis* (Geijskes) (Acari: Tenuipalpidae). *Entomotropica* 17, 91–96.

Quirós-Gonzalez, M. (2000) Phytophagous mite populations on Tahiti lime, *Citrus latifolia*, under induced drought conditions. *Experimental and Applied Acarology* 24, 897–904.

Quirós-Gonzalez, M.J. and Baker, E.W. (1984) Idiosomal and leg chaetotaxy in the Tuckerellidae Baker and Pritchard: ontogeny and nomenclature. In: Griffiths, D.A. and Bowman, C.E. (eds) *Acarology VI, Volume 1*. Ellis Horwood, Chichester, UK, pp. 166–173.

Qureshi, S.A., Oatman, E.R. and Fleschner, C.A. (1969) Biology of the spider mite, *Tetranychus evansi*. *Annals of the Entomological Society of America* 62, 898–903.

Rabbinge, R. (1976) *Biological Control of Fruit-tree Red Spider Mite*. Simulation Monographs, Pudoc, Wageningen, the Netherlands.

Rachana, R.R., Manjunatha, M., Gayathridevi, S. and Naik, M.I. (2009) Seasonal incidence of red spider mite *Tetranychus neocaledonicus* André and its natural enemies. *Karnataka Journal of Agricultural Sciences* 21, 213–214.

Rack, G. (1967) Untersuchungen über die Biologie von *Dolichocybe* Krantz, 1957 und Beschreibung von zwei neuen Arten (Acarina, Pyemotidae). *Mitteilungen aus dem Hamburgischen Zoologischen Museum und Institut* 64, 29–42.

Rack, G. (1972) Pyemotiden an Gramineen in schwedichen landwirtschaftlichen Betrieben. Ein Beitrag zur Entwicklung von *Siteroptes graminum* (Reuter, 1900) (Acarina, Pyemotidae). *Zoologischer Anzeiger* 188, 157–174.

Rack, G. (1973) *Balaustium murorum* (Hermann, 1804) (Acarina, Erythraeidae) in Häusern. Eine wenig beachtete sommerliche Milbenplage. *Anzeiger für Schädlingskunde, Pflanzenschutz, Umweltschutz* 46, 129-132.

Rack, G. (1974) Neue und bekannte Milbenarten der Überfamilie Pygmephoroidea aus dem Saalkreis bei Halle (Acarina, Tarsonemida). *Entomologische Mitteilungen aus dem Zoologisches Institut und Zoologisches Museum der Universität Hamburg* 4(87), 499-521.

Rack, G. (1979) *Brennandania scolopendrae* sp. n. von *Scolopendra morsitans* L. (Acarina, Tarsonemida, Pygmephoroidea, Microdispidae). *Revue de Zoologie Africaine* 93, 376–381.

Rack, G. (1983) Beschreibung einer in Thailand an speisepilzen (*Auricularia polytricha*) schadlichen Milbe, *Luciaphorus perniciosus* sp. n. (Acari, Tarsonemina, Pygmephoridae). *Folia Entomologica Hungarica* 44, 139–146.

Rack, G. and Kaliszewski, M. (1985) Beschreibung von *Cochlodispus operosus* sp. n. (Acari, Prostigmata) aus einem Gewächshaus in Gent/Belgien. *Entomologische Mitteilungen aus dem Zoologisches Institut und Zoologisches Museum der Universität Hamburg* 8(124), 81–88.

Radford, C.D. (1950) *Systematic Check List of Mite Genera and Type Species.* Union Internationale des Sciences Biologiques, Série C: Entomologie, No. 1. Secrétariat Général de l'UISB, Paris.

Radhakrishnan, B., Muraleedharan, M., Selvasundaram, R. and Varatharajan, R. (1987) Seasonal abundance of *Acaphylla theae* (Watt), (Eriophyidae: Acarina) in a tea field at the Anamallais. *Journal of Coffee Research (Bangalore)* 17(1), 105–106.

Radhakrishnan, V. and Ramaraju, K. (2009) Development durations, colonization and insecticide efficacy of leaf mite, *Oligonychus oryzae* Hirst on rice. *Tropical Agricultural Research* 21, 30–38.

Rae, D.J., Watson, D.M., Huang, M.D., Cen, Y.J., Wang, B.Z., Beattie, G.A.C., Liang, W.G., Tan, B.L. and Liu, D.G. (2000) Efficacy and phytotoxicity of multiple petroleum oil sprays on sweet orange (*Citrus sinensis*) and pummelo (*C. grandis* (L.)) in southern China. *International Journal of Pest Management* 46, 125–140.

Raga, A., Mineiro, J.L.C., Sato, M.E., Moraes, G.J. and Flechtmann, C.H.W. (2008) Primeiro relato de *Eriophyes litchii* Keifer (Prostigmata: Eriophyidae) em plantas de lichia no Brasil. In: *Anais do XXII Congresso Brasileiro de Entomologia, 24 a 29 de agosto 2008, Centro de Convenções-Uberlândia, Minas Gerais.* Sociedade Entomológica do Brasil, Londrina, Brazil (CD). Also available as abstract at: http://www.seb.org.br/eventos/CBE/XXIICBE/resumos/R1554-1.html (accessed 19 May 2015).

Ragusa, S. (1974) Difesa del nocciolo dagli Artropodi dannosi. VIII. Effetto del lindano e dell'azinphos-metile sugli acari Fitoseidi (Acarina: Mesostigmata). *Bollettino dell'Istituto di Entomologia Agraria e dell'Osservatorio di Fitopatologia di Palermo* 8, 203–214.

Ragusa, S. (1981) Influence of different kinds of food substances on the developmental time in young stages of the predacious mite *Typhlodromus exhilaratus* Ragusa (Acarina: Phytoseiidae). *Redia* 64, 237–243.

Ragusa, S., Vargas, R., Tsolakis, H. and Ashbach, R. (2000) Laboratory studies on the influence of various food substances on some biological and life-table parameters of *Cydnodromus picanus* Ragusa (Parasitiformes, Phytoseiidae) associated with citrus trees in the Chilean desert. *Phytophaga* 10, 11–23.

Rahman, A., Sarmah, M. and Gurusubramanian, G. (2007a) Bioefficacy of neem formulations combined with reduced dosages of acaricides and insecticides against *Oligonychus coffeae* Nietner and *Scirtothrips dorsalis* Hood in tea. *Journal of Plantation Crops* 35, 59–63.

Rahman, A., Sarmah, M., Phukan, A.K. and Gurusubramanian, G. (2007b) Bioactivity of different formulations of propargite 57% EC against red spider mite, *Oligonychus coffeae* in tea. *Research on Crops* 8, 474–480.

Rahman, K.A. and Sapra, A.N. (1940a) Mites of the family Tetranychidae from Lyallpur with descriptions of four new species. *Proceedings of the Indian Academy of Sciences. Section B*, 11, 17–196.

Rahman, K.A. and Sapra, A.N. (1940b) Biology of the mite *Paratetranychus indicus* Hirst, a pest of sugar cane in the Punjab. *Indian Journal of Entomology* 2, 201–212.

Rahman, K.A. and Sapra, A.N. (1945) On the biology of the vegetable mite, *Tetranychus cucurbitae* Rahaman and Sapra. *Indian Journal of Agricultural Sciences* 15, 124–130.

Rai, A.B. and Solanki, V.Y. (2003) The rate of natural increase of tarsonemid mite *Polyphagotarsonemus latus* Banks (Acari: Tarsonemidae) when reared on chilli leaf (*Capsicum annum* L.). *Shashpa* 10, 13–18.

Rai, A.B., Jhala, R.C., Patel, C.B., Patel, V.J., Patel, K.G. and Shah, A. (1991) Ecology of *Oligonychus indicus* (Acarina: Tetranychidae) infesting coconut in South Gujarat, India. *Florida Entomologist* 53, 241.

Rai, B., Shah, A. and Patel, R. (1988) Biology of *Oligonychus mangiferus* (Tetranychidae: Acarina), a pest of mango in Gujarat. *Gujarat Agricultural University Research Journal* 14, 5–10.

Rai, B.K., Verma, S. and Kumar, K. (1966) Evaluation of pesticides for the control of mango bud mite, *Aceria mangiferae* Sayed (Acarina: Eriophyidae). *Indian Journal of Entomology* 28, 176–180.

Raikov, E. and Nachev, P. (1965) Winter control of the vine phytoalpid. *Rastitelna Zashtita* 13(8), 6–8 (in Bulgarian).

Rajagopal, K., Jayaraj, S. and Subramanium, T.R. (1970) Physiological mechanim of resistance in jasmine to blister mite, *Aceria jasmini* C. (Eriophyidae: Acarina). *Indian Journal of Experimental Biology* 8, 44–47.

Rajaram, V. and Ramamurthy, R. (2001) Influence of morphological characters on the incidence of chilli thrips, *Scirtothrips dorsalis* Hood and mite *Polyphagotarsonemus latus* Banks. *Annals of Plant Protection Sciences* 9, 54–57.

Ram, C.S.V. (1978) Aerial spraying of pesticides in tea. 1. Feasibility trials. *Planters' Chronicle* 73, 87–89, 91.

Ramalho, F.S. and Leao Veiga, A.F. de S. (1980) Acaricidas no controle do ácaro do bronzeamento-*Aculops lycopersici* (Massee, 1937) na cultura do tomateiro. *Revista de Agricultura (Piracicaba)* 55, 281–286.

Ramaraju, K. and Rabindra, R.J. (2002) Palmyra, *Borassus flabellifer* L. (Palmae): a host of the coconut eriophyid mite *Aceria guerreronis* Keifer. *Pest Management in Horticultural Ecosystems* 7, 149–151.

Ramaraju, K., Natarajan, K., Sundara Babu, P.C., Palanisamy, S. and Rabindra, R.J. (2002) Studies on coconut eriophyid mite, *Aceria guerreronis* Keifer in Tamil Nadu, India. In: Fernando, L.C.P., Moraes, G.J. de and Wickramananda, I.R. (eds) *Proceedings of the International Workshop on Coconut Mite* (Aceria guerreronis*), Coconut Research Institute, Sri Lanka, 6–8 January 2000.* Coconut Research Institute, Lunuwila, Sri Lanka, pp. 13–31.

Ramaseshiah, G. (1971) Occurrence of an *Entomophthora* on tetranychid mites in India. *Journal of Invertebrate Pathology* 18, 421–424.

Rambier, A. (1954) Un acarien nuisible méconnu: le tétranycque du pommier (*Amphitetranychus viennensis* Zacher, 1920). *Comptes Rendus de l'Académie des Sciences (Paris)* 8, 340–345.

Rambier, A. (1958a) Les acariens nuisibles en cultures fruitières. *Comptes-Rendus des Journées Fruitières et Maraichères d'Avignon*, 59–64.

Rambier, A. (1958b) Les tétranychiques nuisibles à la vigne en France continenental. *Revue de Zoologie Agricole et Appliquée* 57(1–3), 1–20.

Rambier, A. (1964) Essais acaricides réalisés en Bas-Languedoc dans un foyer de *Panonychus ulmi* Koch (Tetranychidae) résistant au Déméton. *Comptes Rendus de l'Académie d'Agriculture de France* 50, 267–278.

Rambier, A. (1965) Etude de quelques populations de *Tetranychus urticae* Koch. *Bollettino di Zoologia Agraria e di Bachicoltura (Ser. II)* 7, 51–59.

Rambier, A. (1979) Ou en est aujourd hui en viticulture, la lutte contre les acariens tetranyques. *Progrès Agricole et Viticole* 96, 236–241.

Rambier, A.M. (1982a) Un acarien, sur vigne en Champagne, nouveau en France: *Tetranychus mcdanieli* McGregor 1931 du groupe *Pacificus*. *Progrès Agricole et Viticole* 99, 261–263.

Rambier, A.M. (1982b) Protection intégrée contre les acariens de la vigne. *Progrès Agricole et Viticole* 99, 267–271.

Ramirez, L.A. and Esponda, O. (1986) *Lasioseius scapulatus* (Acari: Ascidae) nueva especie para Cuba. *Revista de Protección Vegetal* 1, 33–35.

Rammer, I.A., Arias, R.O. and Kurtz, E.A. (1963) The control of the pear leaf blister mite with endosulfan. *Journal of Economic Entomology* 56, 664–666.

Ramos, M. and Alvarez, C.D. (1987) Ciclo biológico de *Polyphagotarsonemus latus* (Acari: Tarsonemidae) en frutos de lima Persa. *Revista de Protección Vegetal* 2, 269–271.

Ramos, M. and Rodríguez, H. (1998) *Steneotarsonemus spinki* Smiley (Acari: Tarsonemidae): Nuevo informe para Cuba. *Revista de Protección Vegetal* 13, 25–28.

Ramos, M. and Rodríguez, H. (2000) Ciclo de desarollo de *Steneotarsonemus spinki* Smiley (Acari: Tarsonemidae) en laboratório. *Revista de Protección Vegetal* 15, 751–752.

Ramos, M. and Rodríguez, H. (2001) Aspectos biológicos y ecológicos de *Steneotarsonemus spinki* en arroz, en Cuba. *Manejo Integrado de Plagas* 61, 48–52.

Ramos, M., Gómez, C. and Cabrera, R.I. (2001) Presencia de *Steneotarsonemus spinki* Smiley (Acari: Tarsonemidae) en cuatro variedades de arroz en le Republica Dominicana. *Revista de Protección Vegetal* 16, 6–9.

Ramos, M., Rodriguéz, H. and Chico, R. (2005) Los ácaros depredadores y su potencial en la regulación de *Steneotarsonemus spinki*. In: *III Encuentro International del Arroz: El Ácaro del Arroz* Steneotarsonemus spinki *(Tarsonemidae) Retos y Alternativas para América Latina y el Caribe, Hotel Palco Ciudad de La Habana, Cuba, 6 de Junio de 2005. Libro de Resumen*, pp. 10–13.

Ramos Lima, M., González, A.I. and González, M. (2010) Management strategy of *Raoiella indica* Hirst in Cuba, based on biology, host plants, seasonal occurrence and use of acaricide. In: Moraes, G.J. de, Castilho, R.C. and Flechtmann, C.H.W. (eds) *XIII International Congress of Acarology, August 23–27, 2010, Recife-PE, Brazil, Abstract Book*, abstract 405, pp. 218–219. Available at: http://www.acarology.org/ica/ica2010/docs/0910_15_abstractBookCapa.pdf (accessed 3 February 2015).

Ramsay, G.W. and Paterson, S.E. (1977) Mites (Acari) from *Rattus* species on Raoul Island. *New Zealand Journal of Zoology* 4, 389–392.

Ramsdell, D.R. and Jubb, G.L. Jr (1979) Seasonal life history of European red mite on concord grapes in Erie County, Pennsylvania. *Environmental Entomology* 8, 1018–1020.

Ran, C., Lei, H.D., Li, H.Y., Hu, J.H., Lin, B.M. and Zhang, Q.B. (2001) Experiment on the use of mancozeb M-45 to control citrus diseases and citrus rust mite. *South China Fruits* 30(6), 14–15 (in Chinese).

Ran, C., Chen, Y., Yuan, M.-L., Wang, J.-J., Liu, H.-Q. and Yao, T.-S. (2008) Susceptibility of *Panonychus citri* field populations to different acaricides. *Acta Phytophylacica Sinica* 35, 537–540 (in Chinese).

Ran, C., Chen, Y. and Wang, J.-J. (2009) Susceptibility and carboxylesterase activity of five field populations of *Panonychus citri* (McGregor) (Acari: Tetranychidae) to four acaricides. *International Journal of Acarology* 35, 115–121.

Rančić, D. and Petanović, R. (2008) Differences in the leaf morphology of sycamore maple infested by two congeneric eriophyid species at Tara National Park in western Serbia. In: Kreiter, B.M., McCoy, S., Migeon, K.D., Navajas, M.A., Tixier, M.S. and Vial, L. (eds) (2008) *Integrative Acarology: Proceedings 6th European Congress EURAAC (European Association of Acarologists, Montpellier, July, 21–25, 2008* (CD-ROM). EURAAC, France, pp. 326–330.

Rangaswamy, K.T., Sulladmath, V.V. and Byregowda, M. (1997) Reaction of certain pigeonpea entries and varieties to sterility disease. In: *Abstracts of the Symposium on Economically Important Diseases of Crop Plants, IPS* (Indian Phytopathological Society) *Southern Chapter, 18–20 December, UAS, Bangalore, India.*

Rao, G.N. (1970) Tea pests in southern India and their control. *PANS (Pest Articles and News Summaries)* 16, 667–672.

Rao, J. and Prakash, A. (1992) Infestation of tarsonemid mite, *Steneotarsonemus spinki* Smiley, in rice in Orissa. *Journal of Applied Zoological Research* 3, 103.

Rao, J. and Prakash, A. (1996) *Cynodon dactylon* (Linn.) Pers. (Graminae): an alternate host of rice tarsonemid mite, *Steneotarsonemus spinki* Smiley. *Journal of Applied Zoological Research* 7, 50–51.

Rao, J. and Prakash, A. (2002) Paddy field weed, *Schoenoplectus articulatus* (Linn.) Palla (Cyperaceae): a new host of tarsonemid mite, *Steneotarsonemus spinki* Smiley and panicle thrips, *Haplothrips ganglbaureri* Schmutz. *Journal of Applied Zoological Research* 13, 174–175.

Rao, J. and Prakash, A. (2003) Panicle mites causing sterility in farmers' paddy fields in India. *Journal of Applied Zoological Research* 14, 212–217.

Rao, J., Prakash, A., Dhanasekharan, S. and Ghosh, S.K. (1993) Observations on rice tarsonemid mite *Steneotarsonemus spinki*, white-tip nematode and sheath-rot fungus interactions deteriorating grain quality in paddy fields. *Journal of Applied Zoological Research* 4, 89–90.

Rao, P.R.M., Bhavani, T.R.M., Rao, T.R.M. and Reddy, P.R. (2000) Notes From the Field: Spikelet sterility/grain discoloration in Andhra Pradesh India. *International Rice Research Notes* 25(3), 40. Available at: http://books.irri.org/IRRN25no3_content.pdf (accessed 19 May 2015).

Rao, V.P., Dutta, B. and Ramaseshiah, G. (1970) *Natural Enemy Complex of the Flushworm and Phytophagous Mites on Tea in India*. Scientific Publications Series 5, Tea Board of India, Kolkata, India.

Rao, Y.S. and Das, P.K. (1977) A new mite pest of rice in India. *International Rice Research Newsletter* 2(1), 8.

Raros, E.S. and Haramoto, F.H. (1974) Biology of *Stethorus siphonulus* Kapur (Coccinellidae: Coleoptera), a predator of spider mites in Hawaii. *Proceedings of the Hawaiian Entomological Society* 21, 457–465.

Raros, L.A.C. (2001) New mite pests and new host records of phytophagous mites (Acari) from the Philippines. *Philippine Agricultural Scientist* 84, 341–351.

Rasmy, A.H. (1969) Responses of populations of phytophagous and predaceous mites on citrus to cessation of spraying. *Canadian Entomologist* 101, 1078–1080.

Rasmy, A.H. (1970a) Management of orchard mite populations with particular reference to citrus mites. *Zeitschrift für Angewandte Entomologie* 66, 174–177.

Rasmy, A.H. (1970b) Relation between predaceous and phytophagous mites on citrus. *Zeitschrift für Angewandte Entomologie* 67, 6–9.

Rasmy, A.H. (1977) Biology of the citrus brown mite, *Eutetranychus orientalis* as affected by some citrus species. *Acarologia* 19, 222–224.

Rasmy, A.H. and El-Banhawy, E.M. (1974) The phytoseiid mite *Phytoseiulus plumifer* as a predator of the eriophyid mite *Aceria ficus* (Acarina). *Entomophaga* 19, 427–430.

Rasmy, A.H. and Hussein, H.E. (1994) Effect of age and mating on release of female sex pheromones and male response in the two-spotted spider mite. *Journal of Applied Entomology* 117, 109–111.

Rasmy, A.H. and Zaher, M.A. (1983) Population density of the citrus brown mite, *E. orientalis* as affected by citrus species. In: *Plant Protection for Human Welfare: 10th International Congress on Plant Protection. Proceedings of a Conference Held at Brighton, England, 20–25 November 1983, Volume 3.* British Crop Production Council, Croydon, UK, p. 1018.

Rasmy, A.H., El-Bagoury, M.E. and Reda, A.S. (1987) A new diet for reproduction of two predaceous mites *Amblyseius gossypii* and *Agistemus exsertus* (Acari: Phytoseiidae, Stigmaeidae). *Entomophaga* 32, 277–280.

Rasmy, A.H., Zaher, M.A., Momen, F.M., Nawar, M.S. and Abou Elella, G.M. (2000) The effect of prey species on biology and predatory efficiency of some phytoseiid mites: I – *Amblyseius deleoni* (Muma & Denmark). *Egyptian Journal of Biological Pest Control* 10, 117–121.

Rasmy, A.H., Momen, F.M., Zaher, M.A. and Abou Elella, G.M. (2003) Influence of diet on life history and predatory capacity of *Amblyseius zaheri* Yousef & El Brolossy (Acari: Phytoseiidae). *Insect Science and Its Application* (now *International Journal of Tropical Insect Science*) 23, 31–34.

Raspotnig, G., Leutgeb, V., Krisper, G. and Leis, H.J. (2011) Discrimination of *Oribotritia* species by oil gland chemistry (Acari, Oribatida). *Experimental and Applied Acarology* 54, 211–24.

Rath, A.C. (1991) Pathogens for the biological control of mite and collembolan pests. *Plant Protection Quarterly* 6, 172–174.

Rather, A.Q. (1983) Controlling mites of maize in Jammu and Kashmir. *Indian Farming* 33(5), 13, 29.

Rather, A.Q. (1984) New species and a new record of the genus *Amblydromella* Muma (Acarina, Phytoseiidae) from India. *Entomologist's Monthly Magazine* 120(1440/1443), 103–107.

Rather, A.Q. (2008) Management of phytophagous and predatory mites in vineyards of Jammu and Kashmir, India. *Acta Horticulturae* 785, 327–333.

Rathi, Y.P.S. (1979) Temik treatment of pigeonpea seeds for prevention of sterility mosaic. *Acta Botanica Indica* 7, 90–91.

Rathi, Y.P.S. (1983) *Studies on Sterility Mosaic Disease of Pigeonpea (*Cajanus cajan *(L.) Millsp.)*. Final Technical Report, G.B. Pant University of Agriculture and Technology, Pantnagar, Uttar Pradesh, India.

Rathman, R.J., Brunner, J.F. and Hulbert, S.J. (1988) Feeding by *Medetera* species (Diptera: Dolichopodidae) on aphids and eriophyid mites on apple, *Malus domestica* (Rosaceae). *Proceedings of the Entomological Society of Washington* 90, 510–512.

Rattan, L., Rai, B.K. and Kanta, S. (1962) Evaluation of ovicidal and residual efficacy of some pesticides against the sugarcane mite, *Schizotetranychus andropogoni* Hirst. *Indian Journal of Agricultural Sciences* 32, 305–308.

Raudonis, L. (2006) Comparative toxicity of spirodiclofen and lambdacihalotrin to *Tetranychus urticae*, *Tarsonemus pallidus* and predatory mite *Amblyseius andersoni* in a strawberry site under field conditions. *Agronomy Research* 4(Special Issue), 317–322.

Raudonis, L., Valiuskaite, A. and Surviliene, E. (2007) Toxicity of abamectin to rust mite, *Aculus schlechtendali* (Acari: Eriophyidae) in apple tree orchard. *Sodininkyste ir Darzininkyste* 26(2), 10–17 (in Lithuanian).

Raut, S.K. and Ruma Sarkar (1991) The influence of temperature on the life-cycle of *Rhizoglyphus robini* Claparède (Acari: Acaridae). *International Journal of Acarology* 17, 145–148.

Raworth, D.A., Fauvel, G. and Auger, P. (1994) Location, reproduction and movement of *Neoseiulus californicus* (Acari: Phytoseiidae) during autumn, winter and spring in orchards in the south of France. *Experimental and Applied Acarology* 18, 593–602.

Razmjou, J., Tavakkoli, H. and Nemati, M. (2009) Life history traits of *Tetranychus urticae* Koch on three legumes (Acari: Tetranychidae). *Munis Entomology and Zoology* 4, 204–211. Available at: http://www.munisentzool.org/yayin/vol4/issue1/204-211.pdf (accessed 7 April 2015).

Razumova, A.P. (1967) Variability of photoperiodic reaction in a number of successive generations of spider mites (Acarina). *Entomologicheskoe Obozrenie (Entomological Review, USSR)* 46, 268–272 (in Russian).

Razvyazkina, G.M. (1966) Biology and feeding specializations of grass mites of the family Eriophyidae. In: Efremenko T.S. (ed.) *Proceedings of the Vth All-Union Conference on Virus Diseases of Plants.* VASKhNIL (All-Union Academy of Agricultural Sciences of the Soviet Union), Moscow, pp. 322–324 (in Russian).

Readshaw, J.L. (1975) The ecology of tetranychid mites in Australian orchards. *Journal of Applied Ecology* 12, 473–495.

Reck, G.F. (1941a) *Mite Pests on Cultivated Plants.* Izdatel'stvo Akademii Nauk Gruzinskoĭ SSR, Tiblisi, USSR (in Russian).

Reck, G.F. (1941b) Eine neue *Schizotetranychus*-Art (Tetranychida, Acari). *Soobshcheniya Akademii Nauk Gruzinskoĭ SSR* 2, 449–452.

Reck, G.F. (1947) New species of tetranychid mites from Georgia (Tetranychidae, Acari). *Soobshcheniya Akademii Nauk Gruzinskoĭ SSR* 8, 471–475 (in Russian).

Reck, G.F. (1948a) Fauna of spider mites (Tetranychidae, Acari) from Georgia. *Trudy Zoologicheskogo Instituta Akademia Nauk Gruzinskoĭ SSR* 8, 175–185 (in Russian).

Reck, G.F. (1948b) Descriptions of species of the genus *Schizotetranychus* (Träg.) from Georgia. *Soobshcheniya Akademii Nauk Gruzinskoĭ SSR* 9, 445–452 (in Russian).

Reck, G.F. (1950) Spider mite fauna from Georgia (Tetranychidae: Acarina). *Trudy Zoologicheskogo Instituta Akademia Nauk Gruzinskoĭ SSR* 9, 117–134 (in Russian).

Reck, G.F. (1953) Research investigation on the fauna of the Tetranychidae in Georgia. *Trudy Zoologicheskogo Instituta Akademia Nauk Gruzinskoĭ SSR* 11, 161–181 (in Russian).

Reck, G.F. (1956) Novye vidy tetranihovyh klescej iz Vostocnoj Gruzii. *Trudy Zoologicheskogo Instituta Akademia Nauk Gruzinskoĭ SSR* 15, 5–28 (in Russian).

Reck, G.F. (1959) *A Key to the Tetranychoid Mites. Fauna Transcaucasica, 1.* Institut Zoologii, Akademii Nauk Gruzinskoĭ SSR, Tbilisi, USSR (in Russian).

Reddy, A.V., Sreehari, G. and Kumar, A.K. (2005) Evaluation of certain new insecticides against chilli thrips (*Scirtothrips dorsalis*) and mites (*Polyphagotarsonemus latus*). *Research on Crops* 6, 625–626.

Reddy, A.V., Srihari, G. and Kumar, A.K. (2007) Evaluation of certain new insecticides against chilli thrips (*Scirtothrips dorsalis*) and mites (*Polyphagotarsonemus latus*). *Asian Journal of Horticulture* 2(2), 8–9.

Reddy, D.N.R. and Jagadish, A. (1977) *Scolothrips indicus* (Thysanoptera: Thripidae) predating on mites in Bangalore, Karnataka. *Acarology Newsletter* 5, 7.

Reddy, G.V.P. (2001) Comparative effectiveness of an integrated pest management system and other control tactics for managing the spider mite *Tetranychus ludeni* (Acari: Tetranychidae) on eggplant. *Experimental and Applied Acarology* 25, 985–992.

Reddy, G.V.P. (2002) Plant volatiles mediate orientation and plant preference by the predator *Chrysoperla carnea* Stephens (Neuroptera: Chrysopidae). *Biological Control* 25, 49–55.

Reddy, G.V.P. and Baskaran, P. (1991) Biology and varietal preference of *Tetranychus ludeni* Zacher (Acari: Tetranychidae) on four varieties of eggplant, *Solanum melongena* L. *Mysore Journal of Agricultural Science* 25, 331–334.

Reddy, G.V.P. and Baskaran, P. (2006) Damage potential of spider mite *Tetranychus ludeni* Zacher (Acari: Tetranychidae) on four varieties of eggplant. *International Journal of Tropical Insect Science* 26, 48–56.

Reddy, G.V.P., Kikuchi, R. and Bautista, J.R. (2013) Threshold-based spraying decision programmes for the red spider mite *Tetranychus marianae* on eggplant. *Journal of Applied Entomology* 137, 429–436.

Reddy, L.S. and Naik, S.L. (2000) Spread of eriophyid mites into the coconut gardens of Chittoor district, Andhra Pradesh. *Indian Coconut Journal* 30(12), 8–9.

Reddy, M.V. and Nene, Y.L. (1979) Ring spot symptom: a genotypic expression of pigeonpea sterility mosaic. *Tropical Grain Legume Bulletin* 5, 27–29.

Reddy, M.V. and Nene, Y.L. (1981) Estimation of yield loses in pigeonpea due to sterility mosaic. In: Nene, Y.L. (ed.) *Proceedings of the International Workshop on Pigeonpeas, 15–19 December 1980, ICRISAT Centre, Patancheru, India, Volume 2.* ICRISAT (International Crops Research Institute for the Semi-Arid Tropics), Patancheru, India, pp. 305-312. Available at: http://oar.icrisat.org/1954/1/Proceedings_of_the_international_workshop_on_pigeonpeasV2.pdf (accessed 19 May 2015).

Reddy, M.V., Beniwal, S.P.S., Sheila, U.K., Sithanantham, S. and Nene, Y.L. (1989) Role of eriophyid mite *Aceria cajani* (Acari: Eriophyidae) in transmission and spread of sterility mosaic of pigeonpea. In : ChannaBasavanna, G.P. and Viraktamath, C.A. (eds) *Progress in Acarology, Volume 2.* Oxford & IBH Publishing, New Delhi, pp. 121–127.

Reddy, M.V., Sharma, S.B. and Nene, Y.L. (1990a) Pigeonpea: disease management. In: Nene, Y.L., Hall, S.D. and Sheila, V.K. (eds) *The Pigeonpea.* CAB International, Wallingford, UK and ICRISAT (International Crops Research Institute for the Semi-Arid Tropics), Patancheru, India, pp. 303–316.

Reddy, M.V., Arjunan, N. and Muniyappa, V. (1990b) Survival of pigeonpea sterility mosaic pathogen and its vector during summer in Southern India. *International Pigeonpea Newsletter* 11, 16–17.

Reddy, M.V., Raju, T.N., Sharma, S.B., Nene, Y.L. and McDonald, D. (1993a) *Handbook of Pigeonpea Diseases*. Information Bulletin 42, ICRISAT (International Crops Research Institute for the Semi-Arid Tropics), Patancheru, India.

Reddy, M.V., Sheila, V.K. and Nene, Y.L. (1993b) *Cajanus scarabaeoides*, an alternative host of pigeonpea sterility mosaic pathogen and its vector *Aceria cajani*. *International Pigeonpea Newsletter* 18, 24.

Reddy, M.V., Raju, T.N. and Lenne, J.M. (1998) Diseases of pigeonpea. In: Allen, D.J. and Lenne, J.M. (eds) *The Pathology of Food and Pasture Legumes*. CAB International, Wallingford, UK and ICRISAT (International Crops Research Institute for the Semi-Arid Tropics), Patancheru, India, pp. 517–558.

Reddy, S.G.E. and Kumar, N.K.K. (2006) Integrated management of the yellow mite, *Polyphagotarsonemus latus* (Banks), on sweet pepper grown under polyhouse. *Journal of Horticultural Sciences* 1, 120–123.

Reddy, S.V., Reddy, M.V., Ghanekar, A.M., Nene, V.L. and Amin, K. S. (1988) Annual recurrence of pigeonpea sterility mosaic in eastern Uttar Pradesh, India. *International Pigeonpea Newsletter* 7, 30.

Redl, H. and Hiebler, A. (1991) Untersuchungen uber die Bekampfung der Krauselmilbe *Calepitrimerus vitis* Nal. in einem integrierten Rebschutzprogramm. *Mitteilungen Klosterneuburg, Rebe und Wein, Obstbau und Früchteverwertung* 41, 56–64.

Reed, D.K. (1974) Effects of temperature on virus–host relationships and on activity of the non-inclusion virus of the citrus red mite, *Panonychus citri*. *Journal of Invertebrate Pathology* 24, 218–223.

Reed, D.K. (1981) Control of mites by non-occluded viruses. In: Burges, H.D. (ed.) *Microbial Control of Pests and Plant Diseases 1970–1980*. Academic Press, London, pp. 427–432.

Reed, D.K. and Desjardins, P.R. (1978) Isometric virus-like particles from citrus red mites, *Panonychus citri*. *Journal of Invertebrate Pathology* 31, 188–193.

Reed, D.K. and Hall, I.M. (1972) Electron microscopy of a rod-shaped non-inclusion virus infecting the citrus red mite. *Journal of Invertebrate Pathology* 20, 272–278.

Reed, D.K., Burditt, A.K. and Crittenden, C.R. (1964) Laboratory methods for rearing rust mites (*Phyllocoptruta oleivora* and *Aculus pelekassi*) on citrus. *Journal of Economic Entomology* 57, 130–133.

Reed, D.K., Hall, I.M., Rich, J.E. and Shaw, J.G. (1972a) Birefringent crystal formation in citrus red mites associated with non-inclusion virus disease. *Journal of Invertebrate Pathology* 20, 170–175.

Reed, D.K., Rich, J.E. and Shaw, J.G. (1972b) A portable apparatus for detection of virus-diseased red mites in the field. *Journal of Economic Entomology* 65, 890–891.

Reed, D.K., Hendrickson, R.M., Rich, J.R. and Shaw, J.G. (1973) Laboratory evaluation of extenders for the non-inclusion virus of the citrus red mite. *Journal of Invertebrate Pathology* 22, 182–185.

Reeves, R.M. (1963) *Tetranychidae Infesting Woody Plants in New York State, and a Life History Study of the Elm Spider Mite,* Eotetranychus matthyssei *n.sp.* Memoir No. 380, Cornell University Agricultural Experiment Station, Ithaca, New York.

Regenfuss, H. (1973) Beinreduktion und Verlagerung des Kopulationsapparates in der Milbenfamilie Podapolipidae, ein Beispiel fur verhaltensgesteuerte Evolution morphologischer Strukturen. *Zeitschrift für Zoologische Systematik und Evolutionsforschung* 11, 173–195.

Regev, S. (1978) Differences in farnesol content in strawberry varieties and their susceptibility to the carmine spider mite *Tetranychus cinnabarinus* (Boisd.) (Acari: Tetranychidae). *Entomologia Experimentalis et Applicata* 24, 22–26.

Regev, S. and Cone, W.W. (1975a) Evidence of farnesol as a male sex attractant of the two-spotted spider mite, *Tetranychus urticae* Koch (Acarina: Tetranychidae). *Environmental Entomology* 4, 307–311

Regev, S. and Cone, W.W. (1975b) Chemical differences in hop varieties vs. scusceptibility to the two-spotted spider mite. *Environmental Entomology* 4, 697–700.

Regev, S. and Cone, W.W. (1976) Evidence of gonadotropic effect of farnesol in the two-spotted spider mite *Tetranychus urticae*. *Environmental Entomology* 5, 517-519.

Regev, S. and Cone, W.W. (1980) The monoterpene citronellol, as male sex attractant of the two-spotted spider mite, *Tetranychus urticae* (Acarina: Tetranychidae). *Environmental Entomology* 9, 50–52.

Regev, U. (2002) Thresholds for pesticide treatment. In: Pimentel, D. (ed.) *Encyclopedia of Pest Management*. Marcel Dekker, New York, pp. 831–833.

Rêgo, A.S., Costa, E.C., Maciel, A.G.S. and Teodoro, A.V. (2010) Comparative biological aspects and growth rate of *Mononychellus tanajoa* (Bondar) (Acari: Tetranychidae) and *Amblyseius largoensis* (Muma) (Acari: Phytoseiidae). In: Moraes, G.J. de, Castilho, R.C. and Flechtmann, C.H.W. (eds) *XIII International Congress of Acarology, August 23–27, 2010, Recife-PE, Brazil, Abstract Book*, abstract 409, pp. 220–221. Available at: http://www.acarology.org/ica/ica2010/docs/0910_15_abstractBookCapa.pdf (accessed 3 February 2015).

Reiff, M. (1949) Physiologische Merkmale bei Spinnmilben (Acari, Tetranychidae) nach Veränderung des Blattstoffwechsels. *Verhandlungen Schweizerwischen Naturforschenden Gesellschaft* 29, 165–166.

Reinecke, D., Korner, H. and Kirchhof, H. (1986) Die Einsatzmoglichkeiten von Filitox im Obst-und Weinbau. *Nachrichtenblatt für den Pflanzenschutz in der DDR* 40, 132–136.

Reinert, J.A. and Neel, P.L. (1977) Tumid spider mite control on parlor palm grown in containers. *HortScience* 12, 339–341.

Reinert, J.A., Dudeck, A.E. and Snyder, G.H. (1978) Resistance in Bermudagrass to the Bermudagrass mite. *Environmental Entomology* 7, 885–888.

Reinert, J.A., Taliaferro, C.M. and McAfee, J.A. (2008) Susceptibility of Bermudagrass (*Cynodon*) varieties to bermduagrass mite (*Eriophyes cynodoniensis*). *Acta Horticulturae* 783, 519–528.

Reis, P.R. and Franco, R.A. (2010) Selectivity of the insecticide rynaxypyr for three species of phytoseiid mites relevant to coffee and citrus in Brazil. In: Moraes, G.J. de, Castilho, R.C. and Flechtmann, C.H.W. (eds) *XIII International Congress of Acarology, August 23–27, 2010, Recife-PE, Brazil, Abstract Book*, abstract 413, pp. 222–223. Available at: http://www.acarology.org/ica/ica2010/docs/0910_15_abstractBookCapa.pdf (accessed 3 February 2015).

Reis, P.R. and Teodoro, A.V. (2000) Effects of copper oxychloride on reproduction of coffee red spider mite, *Oligonychus ilicis* (McGregor, 1917). *Ciência e Agrotecnologia* 24, 347–352.

Reis, P.R., de Camargo, A.H., Igue, T. and Rossetto, C.J. (1970) Comportamento de variedade de mangueira (*Mangifera indica* L.) em relação a *Aceria mangiferae* (Sayed) (Acarina: Eriophyidae). *Revista de Agricultura (Piracicaba)* 45, 145–151.

Reis, P.R., Souza, J.C. de, Sousa, E. de O. and Teodoro, A.V. (2000) Spatial distribution of *Brevipalpus phoenicis* (Geijskes) (Acari: Tenuipalpidae) in coffee plants (*Coffea arabica* L.). *Anais de Sociedade Entomológica do Brasil* 29, 177–183.

Reis, P.R., Sousa, E.O., Teodoro, A.V. and Neto, M.P. (2003) Effect of prey density on the functional and numerical responses of two species of predaceous mites (Acari: Phytoseiidae). *Neotropical Entomology* 32, 461–467.

Reis, P.R., Pedro Neto, M., Franco, R.A. and Teodoro, A.V. (2004) Control of *Brevipalpus phoenicis* (Geijskes, 1939) and *Oligonychus ilicis* (McGregor, 1917) (Acari: Tenuipalpidae, Tetranychidae) in coffee plants and the impact on beneficial mites. I – Abamectin and emamectin. *Ciência e Agrotecnologia* 28, 269–281.

Reis, P.R., Pedro Neto, M. and Franco, R.A. (2005) Control of *Brevipalpus phoenicis* (Geijskes, 1939) and *Oligonychus ilicis* (McGregor, 1917) (Acari: Tenuipalpidae, Tetranychidae) in coffee plants and the impact on beneficial mites. II – Spirodiclofen and azocyclotin. *Ciência e Agrotecnologia* 29, 528–537.

Reis, P.R., Teodoro, A.V. and Neto, M.P. (2007) História de vida de *Amblyseius compositus* Denmark *et* Muma predando *Brevipalpus phoenicis* (Geijskes) (Acari: Phytoseiidae, Tenuiplapidae). *Coffee Science* 2, 150–158.

Reis, P.R., Oliveira, A.F. de and Návia, D. (2011) First record of the olive bud mite *Oxycenus maxwelli* (Keifer) (Acari: Eriophyidae) from Brazil. *Neotropical Entomology* 40, 622–624.

Reissig, W.H., Heinrichs, E.A., Litsinger, J.A., Moody, K., Fiedler, L., Mew, T.W. and Barrion, A.T. (1986) Rice panicle mite (Acarina: Tarsonemidae). In: *Illustrated Guide to Integrated Pest Management in Rice in Tropical Asia*. International Rice Research Institute, Los Baños, Laguna, Philippines, pp. 228–232.

Rennie, J. (1921) Notes on acarine disease. VI. *Bee World* 13, 115–117.

Reuter, E. (1909) Zur Morphologie und Ontogenie der Acariden mit besonderer Berücksichtigung von *Pediculopsis graminum*. *Acta Societatis Scientiarum Fennicae* 36, 1–288.

Reyes, J.C. and Salgado, M. (1994) Tolerancia de algunas selecciones de aguacate al Ataque de araña roja (*O. punicae*) y thrips (*Scirtothrips* sp). In: *Memorias del Centro de Investigaciones Científicas y Tecnológicas del Aguacate en el Estado de México*. Fundación Salvador Sánchez Colín, Coatepec Harinas, México, pp. 175–178.

Reyes-Bello, J.C. and Mesa-Cobo, N.C. (2011) Biología de *Oligonychus yothersi* (McGregor) (Acari: Tetranychidae) sobre aguacate *Persea americana* Mill. cv. Lorena (Lauraceae). *Caldasia* 33, 211–220.

Reynolds, H.T. and Swift, J.E. (1951) Control of *Petrobia latens* in the Imperial Valley of California. *Journal of Economic Entomology* 44, 642–645.

Rezk, H.A. (2001) The false spider mite, *Brevipalpus obovatus* Donnadieu (Acari: Tenuipalpidae): host related biology, seasonal abundance, and control. In: Halliday, R.B., Walter, D.E., Proctor, H.C., Norton, R.A. and Colloff, M.J. (eds) *Acarology: Proceedings of the 10th International Congress of Acarology, 5–10 July 1998, Canberra*. CSIRO Publishing, Collingwood, Australia, pp. 291–294.

Rezk, H.A. and Gadelhak, G.G. (1996) Relationship between phytophagous and predatory mites in citrus orchards and the effect of acaricides on their populations. *Alexandria Journal of Agricultural Research* 41, 217–224.

Rheinland-Pfalz Ministerium für Landwirtschaft, Weinbau und Forsten (1991) Bekämpfung von Rostmilben. In: *Jahresbericht des Landespflanzenschutzdienstes Rheinland-Pfalz 1991*. Mainz, Germany, pp. 121–122.

Rhodes, E.M. and Liburd, O.E. (2006) Evaluation of predatory mites and acramite for control of twospotted spider mites in strawberries in north central Florida. *Journal of Economic Entomology* 99,1291–1298.

Ribeiro, A.E.L., Gondim, M.G.C. Jr, Melo, J.W.S. and Delalibera, I. Jr (2012) *Solanum americanum* as a reservoir of natural enemies of the tomato red spider mite, *Tetranychus evansi* (Acari: Tetranychidae). *International Journal of Acarology* 38, 692–698.

Rice, R.E. (1961) Bionomics of the tomato russet mite, *Vasates lycopersici* (Massee). MSc. thesis, University of California, Davis, California.

Rice, R.E. and Jones, R.A. (1972) Experimental miticides on European red and Pacific mites. *California Agriculture* 26(3), 12–13.

Rice, R.E. and Jones, R.A. (1978) Mites in almonds and stone fruits. *California Agriculture* 32(4), 20–21.

Rice, R.E. and Strong, F.E. (1962) Bionomics of the tomato russet mite, *Vasates lycopersici* (Massee). *Journal of Economic Entomology* 55, 431–435.

Rice, R.E. and Weinberger, G.B. (1981) Citrus flat mite on pistachios in California. *California Agriculture* 35(7), 25–26.

Rice, R.E., Jones, R.A. and Hoffman, M.L. (1976) Seasonal fluctuations in phytophagous and predaceous mite populations on stonefruits in California. *Environmental Entomology* 5, 557–564.

Richardson, R.G. and Hill, R.L. (1998) The biology of Australian weeds. 34. *Ulex europaeus* L. *Plant Protection Quarterly* 13, 46–58.

Richmond, D.S. and Shetlar, D.J. (1996) Eclosion time and spatial distribution of over-wintering spruce spider mite eggs on Colorado spruce. *Journal of Economic Entomology* 89, 447–452.

Richter, P. and Schulze, H.U. (1990) Investigations on the resistance spectrum of a greenhouse population of the two-spotted spider mite, *Tetranychus urticae* Koch from Gross Gaglow (district of Cottbus). *Archiv für Phytopathologie und Pflanzenschutz* 26, 551–556.

Ridland, P.M. (1996) Forage crops. In: Lindquist, E.E., Sabelis, M.W. and Bruin, J. (eds) *Eriophyoid Mites: Their Biology, Natural Enemies and Control*. World Crop Pests, Volume 6. Elsevier Science, Amsterdam, pp. 673–680.

Ridland, P.M. and Halloran, G.M. (1979) The influence of the lucerne bud mite (*Eriophyes medicaginis* Keifer) on the growth of annual and perennial *Medicago* species. *Australian Journal of Agricultural Research* 30, 1027–1033.

Ridland, P.M. and Halloran, G.M. (1980a) The influence of the lucerne bud mite (*Eriophyes medicaginis* Keifer) on the growth of annual and perennial *Trifolium* species. *Australian Journal of Agricultural Research* 31, 713–718.

Ridland, P.M. and Halloran, G.M. (1980b) Population variability in the reaction of alfalfa to alfalfa bud mite. *Crop Science* 20, 755–757.

Ridland, P.M. and Halloran, G.M. (1981) The influence of the lucerne bud mite *(Eriophyes medicaginis* Keifer) on the growth of lucerne. *Australian Journal of Agricultural Research* 32, 773–781.

Ridsdill-Smith, T.J. (1991) A contribution to assessing the economic impact of redlegged earth mite on agricultural production in Australia. *Plant Protection Quarterly* 6, 168–169.

Ridsdill-Smith, T.J. (1997) Biology and control of *Halotydeus destructor* (Tucker) (Acarina: Penthaleidae): a review. *Experimental and Applied Acarology* 21, 195–224.

Ridsdill-Smith, T.J. and Annells, A.J. (1997) Seasonal occurrence and abundance of redlegged earth mite *Halotydeus destructor* (Acari, Penthaleidae) in annual pastures of south western Australia. *Bulletin of Entomological Research* 87, 413–423.

Ridsdill-Smith, T.J. and Pavri, C. (1998) Spring spraying for redlegged earth mites. *Australian Grain* 8 (October–November), Western Focus i–iv.

Ridsdill-Smith, T.J. and Pavri, C.C. (2000) Feeding life style of redlegged earth mite, *Halotydeus destructor* (Acari: Penthaleidae), in pastures and the role of broad-leafed weeds. *Experimental and Applied Acarology* 24, 397–414.

Ridsdill-Smith, T.J. and Pavri, C.C. (2010) TIMERITE®. CSIRO and Australian Wool Innovation Limited, Sydney, Australia. Available at: http://www.timerite.com.au/ (accessed 20 August 2014).

Ridsdill-Smith, T.J., Jiang, Y. and Ghisalberti, E.L. (1995) A method to test compounds for feeding deterrence towards redlegged earth mites (Acarina: Penthaleidae). *Annals of Applied Biology* 127, 593–600.

Riechert, S.E. and Lockley, T. (1984) Spiders as biological control agents. *Annual Review of Entomology* 29, 299–320.

Riedl, H. and Hoying, S.A. (1980) Impact of fenvalerate and diflubenzuron on target and nontarget arthropod species on Bartlett pears in northern California. *Journal of Economic Entomology* 73, 117–122.

Riedl, H. and Shearer, P.W. (1988) Codling moth control with XRD-473 1989. *Insecticicide and Acaricide Tests* 15, 48–49.

Riehl, L.A. (1969) Advances relevant to narrow-range spray oils for citrus pest control. In: Chapman, H.D. (ed.) *Proceedings of the International Citrus Symposium, Riverside, California, March 16–26, 1968: in Celebration of the Centennial of the University of California, Volume 2.* University of California, Riverside, California, pp. 897–907.

Riehl, L.A. (1983) Fundamental consideration and current development in the production and use of petroleum oils. Insects and mites pests. In: Matsumoto, K. (ed.) *Proceedings of the 4th International Citrus Congress, Tokyo, Japan, 9–12 November 1981, Volume 2.* International Society of Citriculture, Riverside, California, pp. 601–607.

Ries, D.T. (1935) A new mite (*Neotetranychus buxi* n. sp. Garman) on boxwood. *Journal of Economic Entomology* 28, 55–62.

Riley, C.V. (1890) The six-spotted spider mite of the orange (*Tetranychus 6-maculatus*, n. sp.). *Insect Life* 2, 225–226.

Rilling, G. (1989) Differential response of grapevine cultivars to European red mite (*Panonychus ulmi* Koch)-elaboration of a screening method. *Vitis* 28, 97–110.

Rimando, L.C. (1962a) *The Tetranychoid Mites of the Philippines.* Technical Bulletin No. 11, College of Agriculture, University of Philippines. Los Baños, Laguna, Philippines.

Rimando, L.C. (1962b) Four new species of spider mites of the genera *Eotetranychus* and *Schizotetranychus* (Tetranychidae, Acarina). *Philippine Agriculturalist* (now *Philippine Agricultural Scientist*) 45, 535–544.

Rimes, G.D. (1968) The pear leaf blister mite. *Journal of Agriculture, Western Australia (Series 4)* 9, 159–160.

Rioja, S.T. and Vargas M.R. (2009) Life table parameters and consumption rate of *Cydnodromus picanus* Ragusa, *Amblyseius graminis* Chant, and *Galendromus occidentalis* (Nesbitt) on avocado red mite *Oligonychus yothersi* (McGregor) (Acari: Phytoseiidae, Tetranychidae). *Chilean Journal of Agricultural Research* 69, 160–170.

Ripa, R. and Rodriguez, F. (1999) *Plagas de Citricos, sue Enemigos Naturals y Manejo.* Instituto de Investigaciones Agropecuarias, Ministerio de Agricultura, Santiago.

Ripa, S.R., Rodriguez, A.F., Larral, D.P. and Luck, R.F. (2006) Evaluation of a detergent based on sodium benzene sulfonate for the control of woolly whitefly *Aleurothrixus floccosus* (Maskell) (Hemiptera: Aleyrodidae) and red citrus red mites *Panonychus citri* (McGregor) (Acarina: Tetranychidae) on oranges and mandarins. *Agricultura Técnica (Santiago)* 66, 115–123.

Ripka, G., Fain, A., Kazmierski, A., Kreiter, S. and Magowski, W. (2002) Recent data to the knowledge of the arboreal mite fauna in Hungary (Acari: Mesostigmata, Prostigmata, and Astigmata). *Acarologia* 42, 271–281.

Ripka, G., Fain, A., Kazmierski, A., Kreiter, S. and Magowski, W. (2005) New data to the knowledge of the mite fauna of Hungary (Acari: Mesostigmata, Prostigmata and Astigmata). *Acta Phytopathologica et Entomologica Hungarica* 40, 159–176.

Ripolles, J.L. and Melia, A. (1980) Primeras observaciones sobre la proliferación de *Conwentzia psociformis* (Curt.) (Neuroptera, Coniopterygidae), en los citricos de Castellon de la Plana. *Boletín del Servicio contra Plagas e Inspección Fitopatológica* 6(1), 61–66.

Ripper, W.E. (1956) Effect of pesticides on balance of arthropod populations. *Annual Review of Entomology* 1, 403–438.

Rivera-Rivera, C., Galindo-Cardona, A. and Rodrigues, J.C.V. (2012) Testing prey DNA fingerprinting on *Amblyseius largoensis* (Acari: Phytoseiidae) predation of *Raoiella indica* (Acari: Tenuipalpidae). *Experimental and Applied Acarology* 57, 373–379.

Rivero, F. and Vásquez, C. (2009) Biologia e tabela de vida de *Tetranychus desertorum* (Acari. Tetranychidae) sobre folhas de feijão (*Phaseolus vulgaris*). *Zoologia* 26, 38–42.

Rivero, G., Quiros, M., Aponte, O., Sanchez, A., Ortega, J., Colmenares, C., Petit, Y. and Poleo, N. (2010) Population fluctuation and distribution of *Brevipalpus phoenicis* (Geijskes) (Acari: Tenuipalpidae) on peduncle, sepals and exocarp of guava (*Psidium guajava* L.) fruits of different ages. *Acta Horticulturae* 849, 277–282.

Rivero-Maldonado, G., Sanchez-Urdaneta, A., Quirós de Gonzales, M., Sanabria, M.E., Colmenares, C. and Ortega, J. (2008) Histological alterations caused by *Brevipalpus phoenicis* (Geijskes) to peduncles and sepals of *Psidium guajava* L. *Revista de la Facultad de Agronomía, Universidad de Zulia* 25, 525–549.

Rizk, G.A., Karaman, G.A. and Ali, M.A. (1983) Population densities of phytophagous and predacious mites on citrus trees in Middle Egypt. *Bulletin of the Entomological Society of Egypt* 62, 97–103.

Robbs, C.F. and Peracchi, A.L. (1965) Sobre a ocorrência de um ácaro prejudicial ao coqueiro (*Cocos nucifera* L.). In: *Anais da Reunião Fitossanitária 9.* Serviço de Defesa Sanitária Vegetal, Ministério da Agricultura, Rio de Janeiro, pp. 65–70.

Roberts, I.M., Jones, A.T. and Amrine, J.W. Jr (1994) Ultrastructure of the black currant gall mite, *Cecidophyopsis ribis* (Acari: Eriophyidae), the vector of the agent of reversion disease. *Annals of Applied Biology* 125, 447–455.

Robertson, J.L., Russell, R.M., Preisler, H.K. and Savin, E. (2007) *Bioassays with Arthropods*, 2nd edn. CRC Press, Boca Raton, Florida.

Robertson, N.L. and Carroll, T.W. (1988) Virus like particles and spider mite intimately associated with a new disease of barley. *Science* 240, 1188–1190.

Robertson, P.L. (1959) A revision of the genus *Tyrophagus*, with a discussion on its taxonomic position in the Acarina. *Australian Journal of Zoology* 7, 146–181.

Robertson, P.L. (1961) A morphological study of variation in *Tyrophagus* (Acarina), with particular reference to populations infesting cheese. *Bulletin of Entomological Research* 52, 501–529.

Robinson, A.G. (1953) Notes on *Stethorus punctum* (Lec.) (Coleoptera: Coccinellidae), a predator of tetranychid mites in Manitoba. In: *Annual Report of the Entomological Society of Ontario 1952.* Guelph, Canada, pp. 24–26.

Robinson, M.T. and Hoffmann, A.A. (2000) Additional tests on the effects of pesticides on cryptic species of blue oat mite (*Penthaleus* spp.) and the redlegged earth mite (*Halotydeus destructor*). *Australian Journal of Experimental Agriculture* 40, 671–678.

Robinson, M.T. and Hoffmann, A.A. (2001) The pest status and distribution of three cryptic blue oat mite species (*Penthaleus* spp.) and redlegged earth mite (*Halotydeus destructor*) in southeastern Australia. *Experimental and Applied Acarology* 25, 699–716.

Robinson, M.T., Weeks, A.R. and Hoffmann, A.A. (2002) Geographic patterns of clonal diversity in the earth mite species *Penthaleus major* with particular emphasis on species margins. *Evolution* 56, 1160–1167.

Roda, A., Dowling, A., Welbourn, C., Peña, J.E., Rodrigues, J.C.V., Hoy, M.A., Ochoa, R., Duncan, R.A. and De Chi, W. (2008) Red palm mite situation in the Caribbean and Florida. *Proceedings of the Caribbean Food Crops Society* 44(1), 80–87.

Roddick, J.G. (1974) The steroidal glycoalkaloid α-tomatine. *Phytochemistry* 13, 9–25.

Rodgers, C.A. and Baltensperger, A.A. (2004) Registration of 'FMC-6' Bermudagrass. *Crop Science* 44, 2262–2263.

Rodrigues, J.C.V. (2006) "Is Sabara sweet orange immune or resistant to citrus leprosis?": addressing Bitancourt's 1938 question. *Laranja* 27(1), 31–39.

Rodrigues, J.C.V. and Irish, B.M. (2012) Effect of coconut palm proximities and *Musa* spp. germplasm resistance to colonization by *Raoiella indica* (Acari: Tenuipalpidae). *Experimental and Applied Acarology* 57, 309–316.

Rodrigues, J.C.V. and Peña, J.E. (2012) Chemical control of the red palm mite, *Raoiella indica* (Acari: Tenuipalpidae) in banana and coconut. *Experimental and Applied Acarology* 57, 317–329.

Rodrigues, J.C.V., Nogueira, N.L., Freitas, D.S. and Prates, H.S. (1997) Virus-like particles associated with *Brevipalpus phoenicis* Geijskes (Acari: Tenuipalpidae), vector of citrus leprosis virus. *Anais de Sociedade Entomológica do Brasil* 26, 391–395.

Rodrigues, J.C.V., Machado, M.A., Kitajima, E.W. and Muller, G.W. (2000) Transmission of citrus leprosis virus by *Brevipalpus phoenicis* (Acari: Tenuipalpidae). In: Graça, J.V. da, Lee, R.F. and Yokomi, R.K. (eds) *Proceedings of the Fourteenth Conference of the International Organization of Citrus Virologists, IOCV 2000*. IOCV, c/o University of California, Riverside, California, pp. 174–178.

Rodrigues, J.C.V., Kitajima, E.W., Childers, C.C. and Chagas, C.M. (2003) Citrus leprosis virus vectored by *Brevipalpus phoenicis* (Acari: Tenuipalpidae) on citrus in Brazil. *Experimental and Applied Acarology* 30, 161–179.

Rodrigues, J.C.V., Ochoa, R. and Kane, E.C. (2007) First report of *Raoiella indica* Hirst (Acari: Tenuipalpidae) and its damage to coconut palms in Puerto Rico and Culebra Island. *International Journal of Acarology* 33, 3–5.

Rodrigues, M.C. (1959) Os ácaros na cultura algodoeira em Mocambique. *Garcia de Orta* 7(4), 715–740.

Rodrigues, M.C. (1968) Acarina de Mocambique catalogo das especies relacionadas com a agricultura. *Agronomia Moçambicana* 2, 215–256.

Rodriguez, H. and Ramos, M. (1998) Muestreo presencia–ausencia de *Panonychus citri* (McGregor) (Acari: Tetranychidae) en viveros de citricos. *Revista de Protección Vegetal* 13, 103–105.

Rodriguez, H. and Ramos, M. (2000a) Evaluación de motodos de cria para *Amblyseius largoensis* (Muma) (Acari: Phytoseiidae) sobre *Polyphagotarsonemus latus* (Banks) (Acari: Tarsonemidae). *Revista de Protección Vegetal* 15, 105–108.

Rodriguez, H. and Ramos, M. (2000b) Rearing of *Amblyseius largoensis* (Muma) (Acari: Phytoseiidae) on *Panonychus citri* (McGregor). *Revista de Protección Vegetal* 15, 149–151.

Rodriguez, H. and Ramos, M. (2003) Biology of *Amblyseius largoensis* (Muma) (Acari: Phytoseiidae) on *Polyphagotarsonemus latus* (Banks) (Acari: Tarsonemidae) reared in different substrates. *Revista de Protección Vegetal* 18, 58–61.

Rodriguez, H. and Ramos, M. (2004) Biology and feeding behavior of *Amblyseius largoensis* (Muma) (Acari: Phytoseiidae) on *Polyphagotarsonemus latus* (Banks) (Acari: Tarsonemidae). *Revista de Protección Vegetal* 19, 73–79.

Rodriguez, H. and Ramos, M. (2006) Feeding behavior of *Amblyseius largoensis* (Muma) and *Phytoseiulus macropilis* (Banks) (Acari: Phytoseiidae) on *Panonychus citri* (McGregor) (Acari: Tetranychidae). *Revista de Protección Vegetal* 21, 114–118.

Rodriguez, H., Montoya, A. and Ramos, M. (2007) *Raoiella indica* Hirst (Acari: Tenuipalpidae): a threat for Cuba. *Revista de Protección Vegetal* 22, 142–153.

Rodriguez, H., Montoya, A. and Flores-Galano, G. (2010) Conducta alimentaria de *Amblyseius largoensis* (Muma) sobre *Raoiella indica* Hirst. *Revista de Protección Vegetal* 25, 26–30.

Rodriguez, J.A., Arias, A., Santiago, R. and Nieto, J. (1987) Observaciones sobre la biologia de *Brevipalpus lewisi* (McGregor) en vinedos de la Comarca de Guarena (Badajoz), 1984–1986. *Boletín de Sanidad Vegetal, Plagas* 13, 249–259.

Rodriguez, J.G. (1951) Mineral nutrition of the two-spotted spider mite, *Tetranychus bimaculatus* Harvey. *Annals of the Entomological Society of America* 44, 511–526.

Rodriguez, J.G. (1958) The comparative NPK nutrition of *Panonychus ulmi* (Koch) and *Tetranychus telarius* (L.) on apple trees. *Journal of Economic Entomology* 51, 369–373.

Rodriguez, J.G. (1960) Nutrition of the host and reaction to pests. In: Reitz, L.P. (ed.) *Biological and Chemical Control of Plant and Animal Pests*. American Association for the Advancement of Science, Washington, DC, 149–163.

Rodriguez, J.G. (1978) The biology and ecology of the tetranychid mite complex in cassava in perspective. In: Brekelbaum, T., Bellotti, A. and Lozano, J.C. (eds) *Proceedings CASSAVA Protection Workshop, 7–12 November 1977, Cali, Colombia*. CIAT (Centro Internacional de Agricultura Tropical), Cali, Colombia, pp. 171–175.

Rodriguez, J.G. and Rodriguez, L.D. (1952) The relation between minerals, B-complex vitamins, and mite populations in tomato foliage. *Annals of the Entomological Society of America* 45, 331–338.

Rodriguez, J.G. and Rodriguez, L.D. (1987) Nutritional ecology of phytophagous mites. In: Slansky, F. and Rodriguez, J.G. (eds) *Nutritional Ecology of Insects, Mites and Spiders*. Wiley, New York, pp. 177–208.

Rodriguez, J.G. and Stepien, Z.A. (1973) Biology and populations dynamics of *Caloglyphus berlsesei* (Michael) (Acarina, Acaridae) in xenic diet. *Journal of the Kansas Entomological Society* 46, 176–183.

Rodriguez, J.G., Knavel, D. and Aina, O.J. (1972) Studies in the resistance of tomatoes to mites. *Journal of Economic Entomology* 65, 50–53.

Rodriguez-Almaraz, G. and Contreras-Fernandez, E. (1993) Vertical distribution of spiders in citrus orchards in Allende, Nuovo Leon, Mexico. *Southwestern Entomologist* 18, 51–56.

Rodriguez Morell, H., Miranda Cabrera, I., Montoya, A., Rodriguez, Y. and Ramos Lima, M. (2008) Population behavior of *Polyphagotarsonemus latus* (Banks) in protected crop of pepper (*Capsicum annuum* L.). *Fitosanidad* 12(4), 215–220.

Roeser, I. (1996) Aktuelle Brombeersorten. *Gartenpraxis* 22(5), 46–49.

Rogers, J.S., McCoy, C.W. and Manners, M.M. (1994) Standardized visual comparison keys for rapid estimations of citrus rust mite (Acari: Eriophyidae) populations. *Journal of Economic Entomology* 87, 1507–1512.

Rogers, M.E., Stansly, P.A., Childers, C.C., McCoy, C.W. and Nigg, H.N. (2013) *2014 Florida Citrus Pest Management Guide: Rust Mites, Spider Mites, and Other Phytophagous Mites*. SP 043 Florida Citrus Pest Management Guide. Publication No. ENY-603, Entomology and Nematology Department, University of Florida/Institute of Food and Agricultural Sciences (UF/IFAS) Extension, Gainesville, Florida. (Originally published in 2007, see Childers *et al.*, 2007.) Available at: http://edis.ifas.ufl.edu/cg002 (accessed 4 March 2015).

Roggia, S. (2007) Spider mite (Prostigmata: Tetranychidae) associates with soybean in Rio Grande do Sul: occurrence, species identification and effect of soybean cultivars and weeds. MSc. thesis, Universidade Federal de Santa Maria, Santa Maria, Brazil.

Roggia, S., Vanderlei Carús Guedes, J., Roppa Kuss, R.C., Arnemann, J.A. and Návia, D. (2008) Spider mites associated to soybean in Rio Grande do Sul, Brazil. *Pesquisa Agropecuária Brasileira* 43, 295–301.

Roggia, S., Guedes, J.V.C., Kuss-Roggia, R.C.R., Vasconcelos, G.J.N. de, Návia, D. and Delalibera Junior, I. (2009) Ácaros predadores e o fungo *Neozygites floridana* associados a tetraniquídeos em soja no Rio Grande do Sul. *Pesquisa Agropecuária Brasileira* 44, 107–110

Rohfritsch, O. (1975) Etude comparative de cellules du tissue naurricier de la jeune galle de l'*Aulax glechomae* L. sur le *Glechoma hederaceum* L. et de cellules du tissue naurricier abandone par le parasite. *Marcellia* 38, 185–196.

Roivainen, H. (1947) Eriophyid news from Finland. *Acta Entomologica Fennica* 3, 1–51.

Roivainen, H. (1950) Eriophyid news from Sweden. *Acta Entomologica Fennica* 7, 1–51.

Roivainen, H. (1951) Contributions to the knowledge of the eriophyids of Finland. *Acta Entomologica Fennica* 8, 1–72.

Roivainen, H. (1953) Some gall mite (Eriophyidae) from Spain. *Archivos del Instituto de Aclimatación* 1, 9–42.

Rojas, E.W. and Klimov P.B. (2007) Mites of the genus *Rhizoglyphus* (Acari: Acaridae) infesting cultivated bulbs in central and southern Chile, with taxonomic notes on *Acarus hyacinthi* Boisduval and *Rhizoglyphus frickorum* Nesbitt. *International Journal of Acarology* 33, 87–90.

Rolfs, P.H. (1907) Tomato diseases. *Florida Agricultural Experiment Station Bulletin* No. 91, 34–35.

Romeih, A.H.M., Hassan, M.F., Rizk, M.A. and Abo Shnaf, R.I.A. (2005) Description and life history of *Typhlodromus citri* Hassan & Romeih sp. n. (Acari: Phytoseiidae). *Egyptian Journal of Biological Pest Control* 15, 49–56.

Romero, L., Pérez, R., Arana, R., Cabrera, R.I., Hernández, J.L., Castillo, D. and Hernández, A.A. (2003) Experiência adquirida de la interrelación entre la comunidade San Paul y el Instituto de Investigaciones del Arroz. In: *Memorias del I Forum Ramal del Cultivo del Arroz Institute de Investigaciones del Arroz, Camagüey Cuba, 11–12 December 2003*, pp. 40–45.

Roobakkumar, A., Subramaniam, M.S.R., Babu, A. and Muraleedharan, N. (2010) Bioefficacy of certain plant extracts against the red spider mite, *Oligonychus coffeae* (Nietner) (Acarina: Tetranychidae) infesting tea in Tamil Nadu, India. *International Journal of Acarology* 36, 255–258.

Roopa, S.P. (2005) Investigations on mite pests of solanoceous vegetables with special references to brinjal. PhD thesis, University Agricultural Science, Dharwad, Karnataka, India.

Root, R.B. (1973) Organization of a plant-arthropod association in simple and diverse habitats: the fauna of collards (*Brassica oleraceae*). *Ecological Monographs* 43, 95–124.

Rosas Acevedo, J.L. and Sampedro Rosas, L. (2000) Control biológico de *Brevipalpus* spp. en *Citrus aurantifolia* en Guerrero, México. *Manejo Integrado de Plagas* 55, 56–59.

Rosas Acevedo, J.L. and Sampedro Rosas, L. (2006) Variabilidad de cepas de *Hirsutella thompsonii*, a partir de ácaros fitofagos en tres sistemas terrestres del estado de Colima, Mexico. *Revista Mexicana de Biodiversida* 77, 7–16.

Rosen, D. (1986a) Methodologies and strategies of pest control in citriculture. In: Cavalloro, R. and Di Martino, E. (eds) *Integrated Pest Control in Citrus-groves. Proceedings of the Experts' Meeting, 26–29 March 1985, Acireale, Italy*. Published for the Commission of European Communities by Balkema, Rotterdam, the Netherlands, pp. 521–530.

Rosen, D. (1986b) The role of taxonomy in effective biological control programs. *Agriculture, Ecosystems and Environment* 15, 121–129.

Rosenheim, J.A., Kaya, H.K., Ehler, L.E., Marois, J.J. and Jaffee, B.A. (1995) Intraguild predation among biological control agents: theory and evidence. *Biological Control* 5, 303–335.

Rosenthal, S.S. (1983) Current status and potential for biological control of field bindweed, *Convolvulus arvensis*, with *Aceria convolvuli*. In: Hoy, M.A., Cunningham, G.L. and Knutson, L. (eds) *Biological Control of Pests by Mites*. Special Publication, No. 3304, University of California, Berkeley, California, pp. 57–60.

Rosenthal, S.S. (1996) *Aceria*, *Epitrimerus* and *Aculus* species and biological control of weeds. In: Lindquist, E.E., Sabelis, M.W. and Bruin, J. (eds) *Eriophyoid Mites: Their Biology, Natural Enemies and Control*. World Crop Pests, Volume 6. Elsevier Science, Amsterdam, pp. 729–739.

Rosenthal, S.S. and Platts, B.E. (1990) Host specificity of *Aceria* (*Eriophyes*) *malherbae* (Acari: Eriophyidae), a biological control agent for the weed, *Convolvulus arvensis* (Convolvulaceae). *Entomophaga* 35, 459–463.

Rosetta, R. (2001) Bamboo mites: biology and integrated management. *Bamboo, the Magazine of the American Bamboo Society* 22, 16–20.

Rosetta, R. (2005) Bamboo Mite IPM, Oregon State University, Corvallis, Oregon. Available at: http://oregonstate.edu/dept/nurspest/Bamboo/bamboomite.htm (accessed 21 July 2014).

Rossem, G. van, Burger, H.C. and Bund, C.F. van de (1961) Verslag over het optreden van enige schadelijke insecten in het jaar 1960. *Entomologische Berichten* 21, 156–163.

Rossetto, C.J. (1972) Acaros eriofiidios pragas de fruteiras e outras plantas no Brasil. *Ciência e Cultura (São Paulo)* 24(91) 817–829.

Rossetto, C.J. and Camargo, L.S. (1966) *Rhizoglyphus* sp. (Acarina – Acaridae) prejudicando raizes de cenoura, em campinas. *Bragantia* 25, 11–17.

Rossetto, C.J., Ribeiro, I.J.A. and Santos, R. (1967) *Aceria mangiferae* Sayed (Acarina Eriophyidae) praga de mangueira em São Paulo. *Agronômico (Campinas)* 19 (9–10), 33–34.

Rossetto, C.J., Ojima, M., Rigitano, O. and Igue, T. (1971) Queda dos frutos do caquizeiro, associda a infestação de *Aceria diospyri* K. (Acarina, Eriophyidae). *Bragantia*, 30, 1–9.

Rossi, E. and Conti, B. (1997) Acaricide and insecticide resistance in some strains of *Tetranychus urticae* Koch (Acarina: Tetranychidae) and of its predator *Phytoseiulus persimilis* Athias-Henriot (Acarina: Phytoseiidae). *Frustula Entomologica* 20, 168–177.

Rossi Simons, N.H. (1961) Lista de las especies de Tetranychidae (Acari) de la Republica Argentina. *Idia* 163, 9–13.

Rossi Simons, N.H. (1966) Descripción de *Schizotetranychus oryzae* sp. n. (Acari – Tetranychidae). *Revista de Investigaciones Agropecuárias (Ser. 5)* 3, 1–10.

Rossi Simons, N.H. (1971) Novedades acarologicas argentinas. *Idia* 278, 22–26.

Rossi-Zalaf, L.S. and Alves, S.B. (2006) Susceptibility of *Brevipalpus phoenicis* to entomopathogenic fungi. *Experimental and Applied Acarology* 40, 37–47.

Rossi-Zalaf, L.S., Alves, S.B. and Vieira, S.A. (2008) Effect of culture media on virulence of *Hirsutella thompsonii* (Fischer) (Deuteromycetes) to control *Brevipalpus phoenicis* (Geijskes) (Acari: Tenuipalpidae). *Neotropical Entomology* 37, 312–320.

Rossouw, D.J. and Smith, A.J. (1963) The relation of *Calacarus citrifolii* Keifer to concentric ring blotch of citrus. *South African Citrus Journal* 354, 7, 9.

Rostami, E., Abassipour, H., Khanjani, M. and Askarianzadeh, A. (2010) Effect of relative humidity on the seasonal fluctuation of *Eupalopsellus hamedaniensis* Khanjani *et al.* (Acari: Eupalopsellidae) in cherry orchards in Hamedan Province of Iran. In: Moraes, G.J. de, Castilho, R.C. and Flechtmann, C.H.W. (eds) *XIII International Congress of Acarology, August 23–27, 2010, Recife-PE, Brazil, Abstract Book*, abstract 429, p. 232. Available at: http://www.acarology.org/ica/ica2010/docs/0910_15_abstractBookCapa.pdf (accessed 3 February 2015).

Rota, P. (1962) Osservazioni sugli acari tetranychidi dannosi alle piante coltivate ed ornamentali in Italia. *Bollettino di Zoologia Agraria e di Bachicoltura (Ser. II)* 2, 31–136.

Rota, P. (1991) Danni da acariosi su uva da tavola in Puglia. *Informatore Agrario* 47(15), 75–78.

Rota, P. and Biraghi, C. (1987) *Oligonychus ilicis* (McGregor): acaro tetranichide nuovo per l'Europa, fitofago su azalee, camelie e rododendri. *Informatore Agrario* 43(15), 105–107.

Rott, A.S. and Ponsonby, D.J. (2000a) The effects of temperature, relative humidity and host plant on the behaviour of *Stethorus punctillum* Weise (Coleoptera: Coccinellidae) as a predator of the twospotted spider mite, *Tetranychus urticae* Koch (Acari: Tetranychidae). *Biocontrol* 45, 155–164.

Rott, A.S. and Ponsonby, D.J. (2000b) Improving the control of *Tetranychus urticae* on edible glasshouse crops using a specialist coccinellid (*Stethorus punctillum* Weise) and a generalist mite (*Amblyseius californicus* McGregor) as biocontrol agents. *Biocontrol Science and Technology* 10, 487–498.

Roush, R.T. (1990) Genetic variation in natural enemies: critical issues for colonization in biological control. In: Mackauer, M., Ehler, L.E. and Roland, J. (eds) *Critical Issues in Biological Control*. Intercept, Andover, UK, pp. 263–288.

Roush, R.T. and Hopper, K.R. (1995) Use of single family lines to preserve genetic variation in laboratory colonies. *Annals of the Entomological Society of America* 88, 713–717.

Roussel, J.S., Webber, J.C., Newsom, L.D. and Smith, C.E. (1951) The effect of infestation by spider mite, *Septanychus tumidus* on growth and yield of cotton. *Journal of Economic Entomology* 44, 523–527.

Roy, I., Aditya, G. and Saha, G.K. (2011) Preliminary assessment of selected botanicals in the control of *Tetranychus neocaledonicus* André (Acari: Tetranychidae). *Proceedings of the Zoological Society* 64, 124–127.

Roy, M., Brodeur, J. and Cloutier, C. (1999) Seasonal abundance of spider mites and their predators on red raspberry in Quebec, Canada. *Environmental Entomology* 28, 735–747.

Roy, M., Brodeur, J. and Cloutier, C. (2002) Relationship between temperature and development rate of *Stethorus punctillum* (Coleoptera: Coccinellidae) and its prey *Tetranychus mcdanieli* (Acarina: Tetranychidae). *Environmental Entomology* 31, 177–187.

Royalty, R.N. (1987) Morphological and physiological damage to tomato caused by tomato russet mite, *Aculops lycopersici* (Massee) (Acari: Eriophyidae). MSc. thesis, University of California, Riverside, California.

Royalty, R.N. and Perring, T.M. (1987) Comparative toxicity of acaricides to *Aculops lycopersici* and *Homeopronematus anconai* (Acari: Eriophyidae, Tydeidae). *Journal of Economic Entomology* 80, 348–351.

Royalty, R.N. and Perring, T.M. (1988) Morphological analysis of damage to tomato leaflets by tomato russet mite (Acari: Eriophyidae). *Journal of Economic Entomology* 81, 816–820.

Royalty, R.N. and Perring, T.M. (1989) Reduction in photosynthesis of tomato leaflets caused by tomato russet mite (Acari: Eriophyidae). *Environmental Entomology* 18, 256–260.

Royalty, R.N. and Perring, T.M. (1996) Nature of damage and its assessment. In: Lindquist, E.E., Sabelis, M.W. and Bruin, J. (eds) *Eriophyoid Mites: Their Biology, Natural Enemies and Control*. World Crop Pests, Volume 6. Elsevier Science, Amsterdam, pp. 493–512.

Royalty, R.N., Hall, F.R. and Taylor, R.A.J. (1990) Effects of thuringiensin on *Tetranychus urticae* (Acari: Tetranychidae) mortality, fecundity, and feeding. *Journal of Economic Entomology* 83, 792–798.

Royalty, R.N., Phelan, P.L. and Hall, F.R. (1992) Arrestment of male two-spotted spider mite caused by female sex pheromone. *Journal of Chemical Ecology* 18, 137–153.

Royalty, R.N., Phelan, P.L. and Hall, F.R. (1993) Quantitative and temporal analysis of effects of two-spotted spider mite (Acari: Tetranychidae) female sex pheromone on male guarding behavior. *Journal of Chemical Ecology* 19, 211–223.

Ruffinelli, A. and Carbonell, C.S. (1954) Second list of insects and other arthropods of economic importance in Uruguay. *Revista de la Asociación de Ingenieros Agrónomos (Montevideo)* 94, 33-82.

Russell, G.E. (1978) *Plant Breeding for Pest and Disease Resistance*. Butterworths, London.

Russo, L.F. (1973) *L'Oxypleurites maxwelli* Keifer (Acarina: Eriophyidae) in Campania. *Bollettino del Laboratorio di Entomologia Agraria Filippo Silvestri, Portici* 30, 165–168.

Ryke, P.A.J. and Meyer, M.K.P. (1960) South African gall mites, rust mites and bud mites (Acarina: Eriophyidae) of economic importance. *South African Journal of Agricultural Science* 3, 231–242.

Ryu, J.S., Ree, H.I., Min, D.Y. and Ahn, M.H. (2003) A human case of house dust mite *Tarsonemus floricolus* collected from sputum. *Korean Journal of Parasitology* 41, 171–173.

Ryu, M.O. and Ehara, S. (1995) Two spider mites from Korea (Acari, Tetranychidae). *Journal of the Acarological Society of Japan* 4, 15–20.

Saba, F. (1964) Report on injurious mites in the Near East. *Anzeiger für Schädlingskunde* 37, 17–19.

Saba, F. (1971) *Tetranychus yusti*, a spider mite of potential economic importance. *Journal of Econonic Entomology* 64, 141–144.

Saba, F. (1973a) Population dynamics of some tetranychids in subtropical Florida. In: Daniel, M. and Rosický, B. (eds) *Proceedings of the 3rd International Congress of Acarology Held in Prague, 31 August–6 September 1971*. Dr W. Junk, The Hague, the Netherlands (reprinted by Springer), pp. 237–240.

Saba, F. (1973b) Les acariens nuisibles aux plantes cultivées au Maroc. *Al Awamia* 49(3), 69–97.

Saba, F. (1974) Life history and population dynamics on *Tetranychus tumidus* in Florida (Acarina: Tetranychidae). *Florida Entomologist* 57, 47–63.

Sabelis, M.W. (1981) *Biological Control of Two-spotted Spider Mites Using Phytoseiid Predators. Part I. Modelling the Predator–Prey Interaction at the Individual Level*. Agricultural Research Reports 910, Pudoc, Wageningen, the Netherlands.

Sabelis, M.W. (1985a) Reproductive strategies. In: Helle, W. and Sabelis, M.W. (eds) *Spider Mites: Their Biology, Natural Enemies and Control*. World Crop Pests, Volume 1A. Elsevier Science, Amsterdam, pp. 265–278.

Sabelis, M.W. (1985b) Sampling techniques. In: Helle, W. and Sabelis, M.W. (eds) *Spider Mites: Their Biology, Natural Enemies and Control*. World Crop Pests, Volume 1A. Elsevier Science, Amsterdam, pp. 337–350.

Sabelis, M.W. (1985c) Sex allocation. In: Helle, W. and Sabelis, M.W. (eds) *Spider Mites: Their Biology, Natural Enemies and Control*. World Crop Pests, Volume 1B. Elsevier Science, Amsterdam, pp. 83–94.

Sabelis, M.W. (1991) Life history evolution of spider mites. In: Schuster, R. and Murphy, P.W. (eds) *The Acari: Reproduction, Development and Life-history Strategies*. Chapman and Hall, London, pp. 23–49.

Sabelis, M.W. (1996) Phytoseiidae. In: Lindquist, E.E., Sabelis, M.W. and Bruin, J. (eds) *Eriophyoid Mites: Their Biology, Natural Enemies and Control*. World Crop Pests, Volume 6. Elsevier Science, Amsterdam, pp. 427–456.

Sabelis, M.W. and Bruin, J. (1996) Evolutionary ecology: life history patterns, food plant choice and dispersal. In: Lindquist, E.E., Sabelis, M.W. and Bruin, J. (eds) *Eriophyoid Mites: Their Biology, Natural Enemies and Control*. World Crop Pests, Volume 6. Elsevier Science, Amsterdam, pp. 329–366.

Sabelis, M.W. and Dicke, M. (1985) Long-rate dispersal and searching behavior. In: Helle, W. and Sabelis, M.W. (eds) *Spider Mites: Their Biology, Natural Enemies and Control*. World Crop Pests, Volume 1B. Elsevier Science, Amsterdam, pp. 141–160.

Sabelis, M.W. and Van de Baan, H.E. (1983) Location of distant spider mite colonies by phytoseiid predators: demonstration of specific kairomones emitted by *Tetranychus urticae* and *Panonychus ulmi*. *Entomologia Experimentalis et Applicata* 33, 303–314.

Sabelis, M.W. and van Rijn, P.C.J. (1996) Eriophyoid mites as alternative prey. In: Lindquist, E.E., Sabelis, M.W. and Bruin, J. (eds) *Eriophyoid Mites: Their Biology, Natural Enemies and Control*. World Crop Pests, Volume 6. Elsevier Science, Amsterdam, pp. 757–764.

Sabelis, M.W., Afman, B. and Slim, P.J. (1984) Location of distant spider mite colonies by *Phytoseiulus persimilis* Athias-Henriot (Acarina: Phytoseiidae): localization and extraction of a kairomone. In: Griffiths, D. and Bowman, C. (eds) *Acarology VI, Volume 1*. Ellis Horwood, Chichester, UK, pp. 431–440.

Sabelis, M.W., van Baalen, M., Bakker, F.M., Bruin, J., Drukker, B., Egas, M., Janssen, A.R.M., Lesna, I.K., Pels, B., Van Rijn, P. and Scutareanu, P. (1999) The evolution of direct and indirect plant defence against herbivorous arthropods. In: Olff, H., Brown, V.K. and Drent, R.H. (eds) *Herbivores: Between Plant and Predators*. Blackwell, Oxford, UK, pp. 109–166.

Saboori, A. (2001) Description of the larva, deutonymph and adult of *Balaustium zhangi* sp. nov. (Acari: Erythraeidae) from Iran. *Systematic and Applied Acarology* 6, 171–178.

Sachin, J.P., Babu, A., Perumalsamy, K., Muraleedharan, N. and Selvasundaram, R. (2008) Red spider mite, *Oligonychus coffeae* (Nietner) induced plant volatiles from tea leaves, an invitation to its predators. *Journal of Plantation Crops* 36, 425–429.

Sadana, G.L. (1985) *Plant Feeding Mites of India (Tetranychoidea)*. Kalyani Publishers, New Delhi.

Sadana, G.L. (1997) *False Spider Mites Infesting Crops in India*. Kalyani Publishers, New Delhi.

Sadana, G.L. and Joshi, R. (1979) Comparative susceptibility of different varieties of citrus to the attack of phytophagous mite, *Brevipalpus californicus*. *Science and Culture* 45, 413–414.

Sadana, G.L. and Kanta, V. (1972) Susceptibility of different varieties of citrus to the citrus mite, *Eutetranychus orientalis* (Klein). *Science and Culture* 38, 525–526.

Sadieva, B.E. (1984) A mite and the pathogen of brown mosaic. *Zashchita Rastenii̮* No. 7, 20 (in Russian).

Sadigov, G.I. (1990) A pest of vines. *Zashchita Rastenii̮* No. 8, 28 (in Russian).

Sagliocco, J.L. (2008) *Biological Control of English Broom with the Broom Gall Mite*. Landcare Notes. Department of Primary Industries, Frankston, Australia.

Sah, K., Somchoudhury, A.K., Sarkar, P.K. and Gupta, S.K. (1999) Evaluation of some pesticides for efficacy against tea red spider mite, *Oligonychus coffeae* (Nietner) and their adverse effects on natural enemies belonging to *Amblyseius* spp., *Agistemus* spp. and *Cunaxa* spp. in Dooars areas of West Bengal, India. *Tea* 20(1), 12–20.

Sahu, B., Nair, N., Dhar, P.P., Sekh, K. and Somchoudhury, A.K. (2008) Relative toxicity of some acaricides to tea red spider mite, *Oligonychus coffeae* (Nietner) and its predator *Amblyseius longispinosus* Evans. *Journal of Entomological Research* 32, 53–56.

Saitō, Y. (1977) Study on the spinning behaviour of the spider mite (Acarina: Tetranychidae). II. The process of secretion and the webbing on the various stages of *Tetranychus urticae* Koch under different environmental conditions. *Japanese Journal of Applied Entomology and Zoology* 21, 150–157 (in Japanese).

Saitō, Y. (1979a) Study on spinning behaviour of spider mites III. Responses of mites to webbing residues and their preferences for particular physical conditions of leaf surfaces (Acarina: Tetranychidae). *Japanese Journal of Applied Entomology and Zoology* 23, 82–91 (in Japanese).

Saitō, Y. (1979b) Comparative studies on life histories of three species of spider mites (Acarina: Tetranychidae). *Applied Entomology and Zoology* 14, 83–94.

Saitō, Y. (1983) The concept of 'life types' in Tetranychinae – an attempt to classify the spinning behaviour of Tetranychinae. *Acarologia* 24, 377–391.

Saitō, Y. (1984) Prey kills predator – counterattack of tetranychid mite against phytoseiid predator. In: *Proceedings of the XVII International Congress of Entomology, 20–26 August 1984, Hamburg*. (Abstract), p. 284. (See also Saitō, Y. (1986) Prey kills predator: counter-attack success of a spider mite against its specific phytoseiid predator. *Experimental and Applied Acarology* 2, 47–62.)

Saitō, Y. (1985) Life types of spider mites. In: Helle, W. and Sabelis, M.W. (eds) *Spider Mites: Their Biology, Natural Enemies and Control*. World Crop Pests, Volume 1A. Elsevier Science, Amsterdam, pp. 253–264.

Saitō, Y. (1986) Biparental defence in a spider mite (Acari: Tetranychidae). *Behavioral Ecology and Sociobiology* 18, 377–386.

Saitō, Y. (1990a) Two new spider mite species of the *Schizotetranychus celarius* complex (Acari: Tetranychidae). *Applied Entomology and Zoology* 25, 389–396.

Saitō, Y. (1990b) Life-history and feeding habit of *Typhlodromus bambusae*, a specific predator of *Schizotetranychus celarius* (Acari: Phytoseiidae: Tetranychidae). *Experimental and Applied Acarology* 10, 45–51.

Saitō, Y. (2010) *Plant Mites and Sociality – Diversity and Evolution*. Springer, Tokyo.

Saitō, Y. and Osakabe, M. (1992) A new fixation method for preparing mite specimens for optical and SEM microscope observations. *Applied Entomology and Zoology* 27, 427–436.

Saitō, Y. and Takahashi, K. (1980) Study on variation of *Schizotetranychus celarius* (Banks). *Japanese Journal of Applied Entomology and Zoology* 24, 62–70.

Saitō, Y. and Takahashi, K. (1982) Study on variation of *Schizotetranychus celarious* (Banks) II. Comparison of mode of life between two sympatric forms (Acarina: Tetranychidae). *Japanese Journal of Ecology* 32, 69–78 (in Japanese).

Saitō, Y. and Ueno, J. (1979) Life history studies on *Schizotetranychus celarius* (Banks) and *Aponychus corpuzae* Rimando as compared with other tetranychid mite species (Acarina: Tetranychidae). *Applied Entomology and Zoology* 14, 445–452.

Saitō, Y., Osakabe, M., Sakagami, Y. and Yasui, Y. (1993) A method for preparing permanent specimens of mites with Canada balsam. *Applied Entomology and Zoology* 28, 593–597.

Saitō, Y., Mori, K., Sakagami, T. and Lin, J. (2004) Reinstatement of the genus *Stigmaeopsis* Banks, with descriptions of two new species (Acari, Tetranychidae). *Annals of the Entomological Society of America* 97, 635–646.

Sakagami, T. (2002) Phylogenic analysis of subfamily Tetranychinae (Acari: Tetranychidae) in Japan, based on 28S ribosomal DNA sequences. Does the *celarius* species group belong to the genus *Schizotetranychus*? PhD thesis, Hokkaido University, Sapporo, Japan.

Sakagami, T., Saitō, Y., Kongchuensin, M. and Sahara, K. (2009) Molecular phylogeny of *Stigmaeopsis*, with special reference to speciation through host plant shift. *Annals of the Entomological Society of America* 102, 360–366.

Sakunwarin, S. (2004) Bionomics, bio-pesticide effects and morphological aspects of *Tetranychus truncatus* Ehara (Prostigmata: Tetranychidae). PhD thesis, Graduate School, Kasetsart University, Bangkok.

Sakunwarin, S., Chandrapatya, A. and Baker, G.T. (2003) Biology and life table of the cassava mite, *Tetranychus truncatus* Ehara (Acari: Tetranychidae). *Systematic and Applied Acarology* 8, 13–24.

Sakurai, H., Inaba, T. and Takeda, S. (1992) Effect of temperature on the development of bulb mite, *Rhizoglyphus echinopus*. *Research Bulletin of the Faculty College of Agriculture Gifu University* 57, 81–90.

Salas, H. and Macian, A. (2003) Efecto del aceite emulsionable y mancozeb sobre las poblaciones de ácaro de la yema en limon. *Avance Agroindustrial* 24(3), 27–29.

Salas, L.A. (1978) Algunas notas sobre las arañitas rojas (Acari: Tetranychidae) halladas en Costa Rica. *Agronomía Costarricense* 2(1), 47–59.

Sale, P. (1988) Mite pests of citrus in New Zealand. *The Orchardist of New Zealand* 61(1), 26–27.

Saleh, M.R. and Hosny, M.M. (1979) Observation on *Oligonychus* spp. occurring on date bunches (Acari: Tetranychidae). *Research Bulletin of the Faculty of Agriculture, Ain Shams University* No. 1114, 1–8.

Salehuzzanian, M. and Wilkins, P.W. (1984) Components of resistance to ryegrass mosaic virus in a clone of *Lolium perenne* and their strain-specificity. *Euphytica* 33, 411–417.

Salinas Anaya, P. and Reséndiz García, B. (1995) Ciclo biológico de la araña cristalina del aguacatero *Oligonychus perseae* (Tuttle, Baker y Abbatiello) (Prostigmata: Tetranychidae). *Revista Chapingo, Serie Horticultura* 2(1), 53–56.

Salm, S.N., Ray, M.E.C. and Wolfson, M.S. (1994) A South African isolate of ryegrass mosaic virus. *Plant Pathology* 43, 708–712.

Samson, R.A., McCoy, C.W. and O'Donnell, K.L. (1980) Taxonomy of the acarine parasite *Hirsutella thompsonii*. *Mycologia* 72, 359–377.

Sanabria, C. (2005) El ácaro del vaneo del arroz (*Steneotarsonemus spinki* L: Tarsonemidae). Actualidad fitosanitaria Mayo–Jun 2005. Boletín Informativo. No. 22, Ministerio de Agricultura y Ganaderia, San José, Costa Rica, pp. 1–3.

Sanabria, C. and Aguilar, H. (2005) *El Ácaro del Vaneo del Arroz (*Steneotarsonemus spinki *L: Tarsonemidae)*. Boletín Fitosanitario, Ministerio de Agricultura y Ganaderia, San José, Costa Rica.

Sances, F.V., Wyman, J.A. and Ting, J.P. (1979a) Morphological responses of strawberry leaves to infestations of the two-spotted spider mite. *Journal of Economic Entomology*, 72, 710–713.

Sances, F.V., Wyman, J.A. and Ting, J.P. (1979b) Physiological responses to spider mite infestation on strawberry. *Environmental Entomology* 8, 711–714.

Sances, F.V., Toscano, N.C., Hoffman, M.P., LaPré, L.F., Johnson, M.W. and Bailey, J.B. (1982a) Physiological responses of avocado leaves to avocado brown mite feeding injury. *Environmental Entomology* 11, 516–518.

Sances, F.V., Toscano, N.C., Oatman, E.R., LaPré, L.F., Johnson, M.W. and Voth, V. (1982b) Reduction in plant processess by *Tetranychus urticae* (Acari: Tetranychidae) feeding on strawberry. *Environmental Entomology* 11, 733–737.

Sanches, N.F. and Zem, A.C. (1978) Action of different miticides in the control of *Dolichotetranychus floridanus* (Banks, 1900) in Perola pineapple 'slips'. *Anais de Sociedade Entomológica do Brasil* 7, 155–161.

Sánchez, C. and Vásquez, C. (2005) Ácaros asociados con *Vitis vinifera* L. en el estado Lara. In: *Resúmenes XIX Congreso Venezolano de Entomología "Dr. Carlos Pereira Núñez", San Felipe Estado Yaracuy, del 4 al 7 de Julio de 2005. Entomotropica* 20, 129, abstract 006.

Sanchez Salas, J.A. and Padron Chavez, J.E. (1985) Poblaciones de arador o negrilla *Phyllocoptruta oleivora* Ashm. y su relación con el dano que causan a la naranja *Citrus sinensis* Osbeck, en General Teran, Nuevo Leon. *Agricultura Técnica en México* 9(1), 3–11.

Sanchez Sanchez, H., Henry, M., Cardenas Soriano, E. and Alvizo Villasana, H.F. (2001) Identification of wheat streak mosaic virus and its vector *Aceria tosichella* in Mexico. *Plant Disease* 85, 13–17.

Sánchez-Ramos, I., Álvarez-Alfageme, F. and Castañera, P. (2007) Reproduction, longevity and life table parameters of *Tyrophagus neiswanderi* (Acari: Acaridae) at constant temperatures. *Experimental and Applied Acarology* 43, 213–226.

Sanders, R.W. (2006) Taxonomy of *Lantana* sect. *Lantana* (Verbenaceae): I. Correct application of *Lantana camara* and associated names. *SIDA, Contributions to Botany* 22, 381–421.

Sandhu, G.S. (1976) Record of bulb mite *Rhizoglyphus echinopus* (Fumouze and Robin) from stored onions in India. *Science and Culture*, 42, 221–222.

Sandhu, G.S., Kaushal, K.K. and Gupta, S.K. (1973) Mites associated with maize and their predators in the Punjab. *Science and Culture* 39, 226–229.

Sandhu, G.S., Singh, B. and Dhooria, M.S. (1975) Effect of rain on the population of *Oligonychus indicus* (Hirst) (Acarina: Tetranychidae) on different varieties of maize (*Zea mays* L.) in the Punjab, India. *International Journal of Acarology* 1, 10–13.

Sandhu, G.S., Batra, R.C., Sohi, A.S. and Singh, S. (1979) A new record of *Brevipalpus phoenicis* (Geijk) (Acarina: Tenuipalpidae) as a pest of citrus in India. *Indian Journal of Entomology* 41, 280.

Sandhu, M.S., Gatoria, G.S., Sandhu, S.S. and Sukhdev, S. (1982) Chemical control of tetranychid mite (*Eutetranychus banksi* McGregor) infesting cotton in Punjab. *Journal of Research – Punjab Agricultural University* 19, 127–129.

Sandoval, M.F., Aponte, O. and Vásquez, C. (2011) Idiosomal setae analysis in *Oligonychus peruvianus* and *Oligonychus perseae* (Acari: Tetranychidae) from different hosts. *Entomotropica* 26, 127–136.

Sandquist, R.E., Overhulser, D.L. and Stein, J.D. (1993) Aerial applications of esfenvalerate to suppress *Contarinia oregonensis* (Diptera: Cecidomyiidae) and *Megastigmus spermotrophus* (Hymenoptera: Torymidae) in Douglas-fir orchards. *Journal of Economic Entomology* 86, 470–474.

Sangsrijun, C. and Chandrapatya, A. (2001) Bionomics of the mango red mite, *Oligonychus mangiferus* Rahman and Sapra, and its control using holy basil, lemon grass and neem extracts. In: *Abstracts Research and Thesis 2001: 5th BRT Annual Conference, Udon Thani (Thailand), 8–11 Oct 2001*. Biodiversity Research and Training Program, Bangkok, p. 223.

Sanjaya, Y., Ocampo, V.R. and Caoili, B.L. (2013) Transmission effect of entomopathogenic fungi on population of *Tetranychus kanzawai* (Kishida) (Tetranychidae: Acarina). *Arthropods (International Academy of Ecology and Environmental Sciences)* 2, 36–41. Available at: http://www.iaees.org/publications/journals/arthropods/articles/2013-2%281%29/transmission-effect-of-entomopathogenic-fungi.pdf (accessed 20 May 2015).

Santaballa, E., Roca, M. and Laborda, R. (1994) Comportamiento de clofentecin frente al ácaro rojo *Panonychus citri* (McGregor) (Acari: Tetranychidae) y su predador *Euseius stipulatus* (Athias -Henriot) (Acari: Phytoseiidae) en los agrios espanoles. *Boletín de Sanidad Vegetal, Plagas* 20, 419–428.

Santamarina, M.P., Sanchis, V., Jimenez, M., Garcia, F. and Hernández, E. (1988) A strain of *Penicillium funiculosum* Thom with activity against insects and mites. *Journal of Applied Entomology* 105, 167–173.

Santana, D.L.Q. and Flechtmann, C.H.W. (1998) Mite (Arthropoda:Acari) associates of palms (Arecaceae) in Brazil. I. Present status and new records. *Revista Brasileira de Zoologia* 15, 959–963.

Santana, D.L.Q., Flechtmann, C.H.W. and Lima, M.F. (1994) *Novos Ácaros do Coqueiro no Brasil*. Comunicado Técnico 3, CPATC (Embrapa Tabuleiros Costeiros), EMBRAPA (Empresa Brasileira de Pesquisa Agropecuária)-CPATC, Aracaju, Brazil.

Santiago, J.L., Mandl, K., Hack, R. and Regner, F. (2009) Quantitative trait loci analysis of sensitivity to erinose (*Eriophyes vitis* Pgst.) and hair density, based on a genetic map of 'Welschriesling' × 'Sirius'. *Acta Horticulturae* 827, 341–346.

Santos, J.C.C. dos, Scarpellini, J.R. and Luswarghi, H.N. (1999) Chlorfenapyr and abamectin for the control of white mite *Polyphagotarsonemus latus* (Banks, 1904) (Acari – Tarsonemidae) and red spider mite *Tetranychus urticae* Koch, 1836 on cotton. In: *Anais II Congreso Brasileiro Algodao: O Algodao no Seculo XX, Perspectivar para o Seculo XXI, 5–10 Setembro 1999, Ribeirao Preto, São Paulo, Brasil*, pp. 189–191.

Santos, J.H.R. and Lima, P.J.B.F. (1976) Study of garlic in state of Ceara. III. Weight losses during storage of bulbs of plants attacked by field mites (*Aceria tulipae*). *Boletim Cearense de Agronomia* 17, 27–30.

Santos, M.A. (1976) Evaluation of *Zetzellia mali* as predator of *Panonychus ulmi* and *Aculus schlechtendali*. *Environmental Entomology* 5, 187–191.

Santos, M.A. (1984) Effects of host plant on the predator–prey cycle of *Zetzellia mali* (Acari: Stigmaeidae) and its prey. *Environmental Entomology* 13, 65–69.

Santos, M.W. and Laing, J.E. (1985) Stigmaeid predators. In: Helle, W. and Sabelis, M.W. (eds) *Spider Mites: Their Biology, Natural Enemies and Control*. World Crop Pests, Volume 1B. Elsevier Science, Amsterdam, pp. 197–203.

Santos, P.F. and Whitford, W.G. (1981) The effects of microarthropods on litter decomposition in a Chihuahuan desert ecosystem. *Ecology* 62, 654–663.

Santos da Silva, L.M., Gravena, S. and Donadio, L.C. (1981) Efeitos do quimicos e o fungo *Hirsutella thompsonii* no ácaro da ferrrugem do citros *Phyllocoptruta oleivora*. *Anais da Estação Experimental de Bouquim* 8, 106–120.

Sao Jose, A.R., Souza, S.E., Vega Pina, A. and Ataide, E.M. (2000) Incidence and severity of mango flower malformation in Bahia State, Brazil. *Acta Horticulturae* 509, 765–767.

Saowanit, M., Angsumarn, C., Chiradej, C. and McCoy, C.W. (1999) Preliminary testing of biomass production of *Hirsutella thompsonii* var. *synematosa*. *Thai Journal of Agricultural Science* 32, 229–239.

Sapozhnikova, F.D. (1982) Photoperiodic reaction of the eriophyid mite, *Aculus schlechtendali* (Nal.) (Acarina: Tetrapodili). *Entomologicheskoe Obozrenie (Entomological Review, USSR)* 61, 637–643 (in Russian).

Sarkar, H., Mahato, S., Somchoudhury, A.K. and Sarkar, P.K. (2008) Management of yellow mite, *Polyphagotarsonemus latus* (Banks) (Acari: Tarsonemidae) infesting chilli (*Capsicum annum* L.) in Gangetic alluvial plains of West Bengal. *Journal of Entomological Research* 32, 127–130.

Sarker, M. and Mukhopadhyay, A. (2006) General esterases of *Oligonychus coffeae* (Acarina: Tetranychidae) occurring in pesticide-treated and untreated tea plantations of Darjeeling Plains, India. *Journal of Applied Zoological Research* 17, 67–71.

Sarkar, P.K. and Somchoudhury, A.K. (1988) Evaluation of some pesticides against *Raoiella indica* Hirst on coconut palm in West Bengal. *Pesticides* 22(10), 21–22.

Sarkar, P.K. and Somchoudhury, A.K. (1989a) Influence of major abiotic factors on the seasonal incidence of *Raoiella indica* and *Tetranychus fijiensis* on coconut. In: ChannaBasavanna, G.P. and Viraktamath, C.A. (eds) *Progress in Acarology, Volume 2*. Oxford & IBH Publishing, New Delhi, pp. 59–65.

Sarkar, P.K. and Somchoudhury, A.K. (1989b) Interrelationship between plant characters and incidence of *Raoiella indica* Hirst. on coconut. *Indian Journal of Entomology* 51, 45–50.

Sarkar, P.K., Sarkar, H., Sarkar, M.A. and Somachoudhury, A.K. (2005) Yellow mite, *Polyphagotarsonemus latus* (Banks): a menace in chilli cultivation and its management options using biorational acaricides. *Indian Journal of Plant Protection* 33, 294–296.

Sarkar, P.K., Sarkar, H., Sarkar, M.A. and Talukdar, B. (2007) Schedule of application of chitin synthesis inhibitors clofentezine and flufenzin of tetrazine group against yellow mite, *Polyphagotarsonemus latus* (Banks) (Acari: Tarsonemidae) in chilli. *Journal of Entomological Research* 31, 307–311.

Sarkar, P.K., Talukdar, B. and Somchoudhury, A.K. (2009) Relative efficacy of diafenthiuron (50 WP) alone and with two different adjuvants against adult and egg stages of tea red spider mite, *Oligonychus coffeae* (Nietner) under laboratory condition. *Journal of Entomological Research* 33, 109–112.

Sarmah, M. and Bhattacharyya, B. (2002) Biology and feeding potential of *Stethorus gilvifrons* Mulsant (Coccinellidae: Coleoptera) on tea red spider mite, *Oligonychus coffeae* Neitner. *Shashpa* 9, 23–26.

Sarmah, M., Basit, A. and Hazarika, L.K. (2000) Effect of *Polygonum hydropiper* L. and *Lantana camara* L. on tea red spider mite, *Oligonychus coffeae*. *Two and a Bud* 46, 20–22.

Sarmah, M., Basit, A., Bhattacharyya, B., Rahman, A. and Gurusubramanian, G. (2007) Ovicidal, acaricidal and growth inhibitory activity of *Xanthium strumarium* L., *Acorus calamus* L., and *Pongamia pinnata* L. (Pierre) against a major pest of tea, *Oligonychus coffeae* Nietner (Tetranychidae: Acarina). *International Journal of Tea Science* 6(3), 1–11.

Sarmah, M., Rahman, A., Phukan, A.K. and Gurusubramanian, G. (2009) Effect of aqueous plant extracts on tea red spider mite, *Oligonychus coffeae*, Nietner (Tetranychidae: Acarina) and *Stethorus gilvifrons* Mulsant. *African Journal of Biotechnology* 8, 417–423.

Sasa, M. (1965) *Mites. An Introduction to Classification, Bionomics and Control of Acarina*. University of Tokyo Press, Tokyo (in Japanese).

Sasso, R. and Viggiani, G. (2001) Notizie preliminari sulla presenza di *Arthrocnodax vitis* Rubsaamen (Diptera: Cecidomyiidae), predatore di *Colomerus vitis* (Pagenstecher) (Acarina: Eriophyidae), in Italia. *Bollettino del Laboratorio di Entomologia Agraria Filippo Silvestri, Portici* 57, 155–157.

Sathiamma, B. (1995) Biological suppression of the spider mite *Oligonychus iseilemae* (Hirst) on coconut foliage. *Entomon* 20, 237–243.

Sathiamma, B., Nair, C.P.R. and Koshy, P.K. (1998) Outbreak of a nut infesting eriophyid mite *Eriophyes guerreronis* (K.) in coconut plantations in India. *Indian Coconut Journal* 29(2), 1–3.

Sato, Y. and Saitō, Y. (2010) Variation in nest sanitation behavior among social spider mites of genus *Stigmaeopsis*. In: Moraes, G.J. de, Castilho, R.C. and Flechtmann, C.H.W. (eds) *XIII International Congress of Acarology, August 23–27, 2010, Recife-PE, Brazil, Abstract Book*, abstract 446, p. 241. Available at: http://www.acarology.org/ica/ica2010/docs/0910_15_abstractBookCapa.pdf (accessed 3 February 2015).

Saunyama, I.G.M. and Knapp, M. (2003) Effect of pruning and trellising of tomatoes on red spider mite incidence and crop yield in Zimbabwe. *African Crop Science Journal* 11, 269–277.

Savage, M.J. (1978) Damage to vines by *Colomerus vitis* (Pgst.) (Prostigmata: Eriophyidae). *Plant Pathology* 27, 147–148.

Savulkina, M.M. (1981) System, ecology and distribution of mites of family Pygmephoridae. *Entomologicheskoe Obozrenie (Entomological Review, USSR)* 60, 434–450 (in Russian).

Savzdarg, E.E. (1957) Ways of freeing berry fruits from mites in relation to peculiarities of their biology and ecology. *Izvestiya Timiryazevskoi Sel'Skokhozyaistvennoi Akademii* 1, 5–19 (in Russian).

Sawada, K. (1922) *Descriptive Catalogue of the Formosan Fungi II*. Report of the Department of Agriculture Government Research Institute of Formosa No. 2, Taihoku, Japan.

Saxena, D.K. and Rawat, R.R. (1969) Chemical control of the brown wheat mite *Petrobia latens* (Müller) (Acariña: Tetranychidae) on wheat. *Plant Protection Bulletin (Faridabad)* 21(3), 7–12.

Sayed, M.T. (1946a) *Aceria mangiferae* nov. spec. (Acarina-Eriophyidae). *Bulletin de la Société Fouad 1er d'Entomologie* 30, 7–10.

Sayed, M.T. (1946b) Contribution to the knowledge of Acarina of Egypt. V. Five new species of Tetranychidae. *Bulletin de la Société Fouad 1er d'Entomologie* 30, 143–148.

Sayed, M.T. (1946c) Three new eriophyid mites from Egypt (Acarina, Eriophyidae). *Bulletin de la Société Fouad 1er d'Entomologie* 30, 149–154.

Sayed, M.T. (1946d) Description of *Tenuipalpus granati* nov. sp. and *Brevipalpus pyri* nov. sp. (Acarina: Trichadenidae). *Bulletin de la Société Fouad 1er d'Entomologie* 30, 99–103.

Sayed, M.T. (1950a) On the taxonomy of Tetranychidae and allied genera. A new family and two new sub-families in Acarina. *Proceedings of the 8th International Congress on Entomology, August 1948, Stockholm, Sweden*, pp. 1012–1017.

Sayed, M.T. (1950b) Mite problems in Egypt. *Proceedings of the 8th International Congress on Entomology, August 1948, Stockholm, Sweden*, pp. 1022–1025.

Sayed, T. (1942) Contribution to the knowledge of Acarina in Egypt: The genus *Raoiella* Hirst (Pseudotetranychinae: Tetranychidae). *Bulletin de la Société Fouad 1er d'Entomologie* 26, 81–91.

Sazo, R.L. and Gerstle, M.C. (1989) Effect of caesium 137 on the mortality of *Naupactus xanthographus* (Germar) (Coleoptera: Curculionidae) and *Brevipalpus chilensis* Baker (Acari: Tenuipalpidae). *Ciencia e Investigación Agraria* 16(1–2), 65–67.

Scalopi, E.J., Vasconcellos, E.F.C. and Nakano, O. (1971) Sintomatologia do ataque de ácaros a variedades de alho. *O Solo (Piracicaba, Sao Paulo)* 63, 37–38.

Scannavini, M., Cavazza, F. and Franceschelli, F. (2008) La difesa della vite nei confronti di *Calepitrimerus vitis* alla luce della revoca all'impiego del bromopropylate. In: *Atti, Giornate Fitopatologiche 2008, Cervia 12–14 Marzo, Volume 1*. Fitogest®, Image Line Network, Faenza, Italy, pp. 221–224.

Scarpellini, J.R. and Clari, A.R. (1999) Controle do ácaro da falsa ferrugem com diflubenzuron sozinho ou associado ao enxofre em citros. *Revista de Agricultura (Piracicaba)* 74, 15–23.

Scarpellini, J.R. and Santos, J.C.C. dos (1997) Efeito de acaricidas no controle do ácaro da falsa ferrugem dos citros *Phyllocoptruta oleivora* Ashmead, 1879 atraves de dois metodos de avaliacao. *Revista de Agricultura (Piracicaba)* 72, 351–361.

Schaarschmidt, L. (1959) Systematik und Okologie der Tarsonemiden. In: Stammer, H.J. (ed.) *Beiträge zur Systematik und Ökologie Mitteleuropäischer Acarina, Bund 1, Abschnitt 5*. Akademische Verlagsgesellschaft Geest & Portig, Leipzig, Germany, pp. 713–823.

Schall, R.A. (2004) *NPAG Data:* Steneotarsonemus spinki, *Rice Tarsonemid Mite*. New Pest Advisory Group, Plant Epidemiology and Risk Analysis Laboratory, Center for Plant Health Science and Technology, US Department of Agriculture Animal and Plant Health Inspection Service (APHIS), Raleigh, North Carolina. See updated version at NPAG (2008).

Schatz, H., Behan-Pelletier, V.M., O'Connor, B.M. and Norton, R.A. (2011) Suborder Oribatida van der Hammen, 1968. In: Zhang, Z.-Q. (ed.) *Animal Biodiversity: An Outline of Higher-level Classification and Survey of Taxonomic Richness. Zootaxa* 3148. Magnolia Press, Auckland, pp. 141–148.

Schausberger, P., Peneder, S., Jurschik, S. and Hoffmann, D. (2012) Mycorrhiza changes plant volatiles to attract spider mite enemies. *Functional Ecology* 26, 441–449.

Schausberger, V.P. (1992) Vergleichende Untersuchungen über den Einfluß unterschiedlicher Nahrung auf die Präimaginalentwicklung und die Reproduction von *Amblyseius aberrans* Oud. und *Amblyseius finlandicus* Oud. (Acarina, Phytoseiidae). *Journal of Applied Entomology* 113, 476–486.

Scheucher, R.I. (1957) Systematik und Ökologie der deutschen Anoetinen. In: Stammer, H.J. (ed.) *Beiträge zur Systematik und Ökologie Mitteleuropäischer Acarina, Bund 1, Abschnitt 1*. Akademische Verlagsgesellschaft Geest & Portig, Leipzig, Germany, pp. 233–284.

Schicha, E. (1987) *Phytoseiidae of Australia and Neighbouring Areas*. Indira Publishing House, Oak Park, Michigan.

Schicha, E. and Gutierrez, J. (1985) Phytoseiidae of Papua New Guinea, with three new species, and new records of Tetranychidae (Acari). *International Journal of Acarology* 11, 173–181.

Schimitschek, E. (1939) Beiträge zur forstentomologie der Turkei. II. Ueber einige sehadlinge der haselkulturen. *Zeitschrift für Angewandte Entomologie* 26, 449–461.

Schlechtendal, D.H.R. von (1911) Eriophyidocecidien, die durch Gallmilben verursachten Pflanzengallen. In: Rübsaamen, H. and Hedicke, H. (eds) Die Zoocecidien, durch Tiere erzeugte Pflanzengallen Deutschland und ihre Bewohner. *Zoologica* 61, 295–498.

Schliesske, J. (1977a) Zum Wirtspflanzenkreis der Gallmilbe *Aculus fockeui* Nal. et Trt. und zum Einfluss der Witterung und Bekampfungsmassnahmen auf die Populationsentwicklung. *Mitteilungen aus der Biologischen Bundesanstalt für Land- und Forstwirtschaft, Berlin-Dahlem* No. 178, 254 (abstract).

Schliesske, J. (1977b) Freilebende Gallmilben (Acari: Eriophyidae) als Schadlinge an Prunus-Arten. *Erwerbsobstbau* 19, 178–180.

Schliesske, J. (1977c) Ausbreitung und Wirtspflanzenkreis von *Aculus fockeui* Nal. et Trt. (Acari: Eriophyidae) und der mit ihr vergesellschafteten Arten. *Mededelingen van de Faculteit Landbouwwetenschappen, Universiteit Gent* 42, 1343–1351.

Schliesske, J. (1981) Investigations on the biology of *Aculus fockeui* (Nalepa and Trt.) (Acari: Eriophyoidea). *Zeitschrift für Angewandte Zoologie* 68, 419–434.

Schliesske, J. (1983) Economic important species of gall mites (Acari: Eriophyoidea) on fruit trees and shrubs. *Anzeiger für Schädlingskunde, Pflanzenschutz, Umweltschutz* 56, 121–125.

Schliesske, J. (1985a) Zur Gallmilbenfauna (Acari: Eriophyoidea) der Geholze in den Wallhecken und Baumschulen Schleswig-Holsteins. *Mitteilungen der Deutschen Gesellschaft für Allgemeine und Angewandte Entomologie* 4, 406–408.

Schliesske, J. (1985b) Zur Biologie and zum Schadauftreten der Gallmilbe *Aculus schlechtendali* (Nalepa) (Acari: Eriophyoidea) an *Malus* spp. *Erwerbsobstbau* 27, 195–197.

Schliesske, J. (1985c) Zur Verbreitung und Ökologie einer neuen ursprünglichen gallmilbenart (Acari: Eriophyoidea) an *Araucaria araucara* (Molina). *Entomologische Mitteilungen aus dem Zoologischen Museum Hamburg* 8(124), 97–106.

Schliesske, J. (1988) Zur Gallmilbenfauna (Acari: Eriophyoidea) von *Cocos nucifera* L. in Costa Rica. *Nachrichtenblatt Deutscher Pflanzenschutzdienst* 40(8–9), 124–127.

Schliesske, J. (1990) On the gall mite fauna (Acari: Eriophyoidea) of *Cocos nucifera* L. in Costa Rica. *Plant Research and Development* 31, 74–81.

Schliesske, J. (1992) The free-living gall mite species (Acari: Eriophyoidea) on pomes and stone fruits and their natural enemies in Northern Germany. *Acta Phytopathologica et Entomologica Hungarica* 27, 583–586.

Schliesske, J. and Billen, W. (1989) Zum Schadauftreten der Gallmilbe *Phyncaphytoptus platani* Keifer (Acari: Eriphyoidea) an Platanen. *Gesunde Pflanzen* 41, 325–326.

Schmidt, A.R., Jancke, S., Lindquist, E.E., Ragazzi, E., Roghi, G., Nascimbene, P.C., Schmidt, K., Wappler, T. and Grimaldi, D.A. (2012) Arthropods in amber from the Triassic Period. *Proceedings of the National Academy of Sciences of the United States of America* 109, 14796–14801.

Schmidt, G. (1976) Der Einfluss der von den Beutetieren hinterlassen Spuren auf Suchverhalten und Sucherfolg von *Phytoseiulus persimilis* A.H. (Acarina: Phytoseiidae). *Zeitschrift für Angewandte Entomologie* 82, 216–218.

Schmidt, R.A. (2014) Leaf structures affect predatory mites (Acari: Phytoseiidae) and biological control: a review. *Experimental and Applied Acarology* 62, 1–17.

Schopf, Y.J.M. (1974) Permo-Triassic extinctions: relation to sea-floor spreading. *Journal of Geology* 82, 129–143.

Schread, J.C. (1951) Mite pests of ornamentals and their control. *Bulletin of the Connecticut Agricultural Experiment Station* No. 591.

Schreiner, U. (1984) Untersuchungen zur Anfälligkeit verschiedener Rebsorten, Pfropfkombinationen und Unterlagen gegenüber *Tetranychus urticae* und *Panonychus ulmi*. PhD thesis, Universität Kaiserslautern, Kaiserslautern, Germany.

Schruft, G. (1962) Beiträge zur Kenntnis der Biologie der Krauselmilben (*Phyllocoptes vitis* Nal. und *Epitrimerus vitis* Nal., Fam. Eriophyidae (Acarina)) and Reben (*Vitis vinifera* L.). *Wein-Wissenschaft* 17, 191–211.

Schruft, G. (1966) Untersuchungen zur Klarung der Arten-Frage beim Krauselmilben-Komplex. *Wein-Wissenschaft* 21, 426–434.

Schruft, G. (1969) Das Vorkommen rauberischer Milben aus den Familien Cunaxidae und Stigmaeidae (Acari) an Reben. IV. Beitrag über Untersuchungen zur Faunistik und Biologie der Milben (Acari) an Kulturreben (*Vitis* spec.). *Wein-Wissenschaft* 24, 320–326.

Schruft, G. (1971) *Haleupalus olivieri* nov. spec., eine Dornpalpenmilbe an Reben (*Vitis* spec.) (Acari: Cunaxidae). *Deutsche Entomologische Zeitschrift* 18, 377–382.

Schruft, G. (1972a) Les tydéidés (Acari) sur vigne. *Bulletin OEPP/EPPO Bulletin* 3, 51–55.

Schruft, G. (1972b) Das Vorkommen von Milben aus der Familie Tydeidae (Acari) an Reben. VI. Beitrag über Untersuchungen zur Faunistik und Biologie der Milben (Acari) an Kulturreben (*Vitis* spec.). *Zeitschrift für Angewandte Entomologie* 71, 124–133.

Schruft, G.A. (1985) Grape. In: Helle, W. and Sabelis, M.W. (eds) *Spider Mites: Their Biology, Natural Enemies and Control.* World Crop Pests, Volume 1B. Elsevier Science, Amsterdam, pp. 359–366.

Schulten, G.G.M. (1968) *Genetics of Organophosphate Resistance in the Two-spotted Mite (*Tetranychus urticae *Koch).* Communication of the Department of Agricultural Research No. 57, Royal Tropical Institute, Amsterdam.

Schulten, G.G.M. (1985) Pseudo-arrhenotoky. In: Helle, W. and Sabelis, M.W. (eds) *Spider Mites: Their Biology, Natural Enemies and Control.* World Crop Pests, Volume 1B. Elsevier Science, Amsterdam, pp. 67–71.

Schuster, D.J., Price, J.F., Martin, F.G., Howard, C.M. and Albregts, E.E. (1980) Tolerance of strawberry cultivars to two-spotted spider mites in Florida. *Journal of Economic Entomology* 73, 52–54.

Schuster, M.F. (1959) Chemical control of *Tetranychus marianae* McGregor on tomatoes in the lower Rio Grande Valley. *Journal of Economic Entomology* 52, 763–764.

Schuster, R. (1972) Spinnvermögen der Tydeiden (Milben). *Naturwissenschaften* 59, 275.

Schütte, C. (2006) A novel bacterial disease of the predatory mite *Phytoseiulus persimilis*: disease syndrome, disease transmission and pathogen isolation. PhD thesis, Wageningen University, Wageningen, the Netherlands.

Schwartz, A. (1970) *Brevipalpus* myt op sitrus. *South African Citrus Journal* 438, 27–28 (in Afrikaans).

Schwartz, A. (1972) Evaluasie van die sitrusknopmyt *Aceria sheldoni* (Ewing) (Acarina: Eriophyidae) probleem op nawellemoene. *Phytophylactica* 4, 41–45 (in Afrikaans).

Schwartz, A. (1975a) Aftermath of a rust mite infestation. *Citrus and Subtropical Fruit Journal* 494, 22.

Schwartz, A. (1975b) Wees gewaarsku teen die gevaar van sitrusknopmyt. *Citrus and Subtropical Fruit Journal* 495, 19–21 (in Afrikaans).

Schwartz, A. (1975c) Simultaneous control of citrus mite. *Citrus and Subtropical Fruit Journal* 499, 5, 7.

Schwartz, A. (1977a) Hoe skakel roesmytbeheer by swartvlekbeheer in op sitrus in die Laeveld? *Citrus and Subtropical Fruit Journal* 520, 17 (in Afrikaans).

Schwartz, A. (1977b) *Flat Mites on Citrus.* Farming in South Africa, Citrus Series H 13, Department of Agricultural Technical Services of the Republic of South Africa. Government Printer, Pretoria.

Schwartz, A. (1977c) *Citrus Silver Mite,* Hemitarsonemus latus *(Banks).* Leaflet, Citrus Series H 14, Department of Agricultural Technical Services of the Republic of South Africa. Government Printer, Pretoria.

Schwartz, A. (1978) Citrus red mite *Panonychus citri* (McGregor). In: Bedford, E.C.G. (ed.) *Citrus Pests in the Republic of South Africa.* Science Bulletin No. 391, Department of Agricultural Technical Services of the Republic of South Africa. Government Printer, Pretoria, pp. 41-46.

Schwartz, A. (1984) Invloed van bemesting ten opsigte van die voorkoms van sitrusrooimyt. *Information Bulletin* No. 138. Citrus and Subtropical Fruit Research Institute, Nelspruit, South Africa, pp. 17–19 (in Afrikaans).

Schwartz, A. (1986) *Leaf Curl Mite in Vineyards.* Viticulture and Oenology, Series E 28, Department of Agriculture and Water Supply, South Africa. Government Printer, Pretoria.

Schwartz, A. (1990) Pesticide effect on populations of *Tetranychus urticae* Koch (Acari: Tetranychidae) and a predaceous mite on table grapes in the Hex River Valley. *South African Journal for Enology and Viticulture* 11, 33–37.

Schwartz, A. and Meyer Smith, M.K.P. (1998) Citrus rust mite, *Phyllocoptruta oleivora* (Ashmead). In: Bedford, E.C.G., van den Berg, M.A. and de Villiers, E.A. (eds) *Citrus Pests in the Republic of South Africa*, 2nd edn. ARC-Institute for Tropical and Subtropical Crops, Nelspruit, South Africa, pp. 54–58.

Schwartz, A. and Riekert, F. J. (1967) Effectiveness of standard and new compounds for the control of bud mite, *Aceria sheldoni* Ewing, on navel oranges. *South African Journal of Agricultural Science* 10, 609–616.

Sclar, D.C., Gerace, D. and Cranshaw, W.S. (1998) Observations of population increases and injury by spider mites (Acari: Tetranychidae) on ornamental plants treated with imidacloprid. *Journal of Economic Entomology* 91, 250–255.

Scopes, N.E.A. (1968) Mass rearing *Phytoseiulus riegeli* Dosse for use in commercial horticture. *Plant Pathology* 17, 123–26.

Scopes, N.E.A. (1985) Red spider mites and their predator *Phytoseilus persimilis.* In: Hussey, N.W. and Scopes, N. (eds) *Biological Pest Control.* Blandford Press, Poole, UK, pp. 43–52.

Scott, D.R. (1965) *Carrot Bud Mite.* University of Idaho Current Information Series No 10, University of Idaho College of Agriculture, Moscow, Idaho.

Scott, J.K., Yeoh, P.B. and Knihinicki, D.K. (2008) Redberry mite, *Acalitus essigi* (Hassan) (Acari: Eriophyidae), an additional biological control agent for *Rubus* species (blackberry) (Rosaceae) in Australia. *Australian Journal of Entomology* 47, 261–264.

Searle, C.M. and Meyer Smith, M.K.P. (1998) Family Eriophyidae: rust, gall and bud mites. In: Bedford, E.C.G., van den Berg, M.A. and de Villiers, E.A. (eds) *Citrus Pests in the Republic of South Africa*, 2nd edn. ARC-Institute for Tropical and Subtropical Crops, Nelspruit, South Africa, pp. 43–58.

Searle, C.M.S.L. (1973) The role of citrus bud mite in biological and integrated control orchards in Southern Africa. In: Carpena, O. (ed.) *Proceedings of the 1st International Citrus Congress, Murcia and Valencia, Spain, 1973, Volume 2.* International Society of Citriculture, University of California, , pp. 481–490.

Searle, C.M.S.L. (1978) Citrus bud mite, *Aceria sheldoni* (Ewing). In: Bedford, E.C.G. (ed.) *Citrus Pests in the Republic of South Africa.* Science Bulletin No. 391, Department of Agricultural Technical Services, Republic of South Africa. Government Printer, Pretoria, pp. 26–32.

Seber, G.A.F. (1973) *The Estimation of Animal Abundance.* Griffin, London.

Sebesta, E.E., Wood, E.A. Jr, Porter, D.R., Webster, J.A. and Smith, E.L. (1995) Registration of Amigo wheat germplasm resistant to greenbug. *Crop Science* 35, 293.

Secord, D. and Kareiva, P. (1996) Perils and pitfalls in the host specificity paradigm. *BioScience* 46, 448–453.

Seeman, O.D. and Beard, J.J. (2011) Identification of exotic pest and Australian native and naturalised species of *Tetranychus* (Acari: Tetranychidae). *Zootaxa* 2961, 1–72.

Šefrová, H. and Laštůvka, Z. (2005) Catalogue of alien animal species in the Czech Republic. *Acta Universitatis Agriculturae* et *Silviculturae Mendelianae Brunensis* 53(4), 151–170.

Seguni, Z. (2002) Incidence, distribution and economic importance of the coconut eriophyid mite, *Aceria guerreronis* Keifer in Tanzanian coconut based cropping systems. In: Fernando, L.C.P., Moraes, G.J. de and Wickramananda, I.R. (eds) *Proceedings of the International Workshop on Coconut Mite* (Aceria guerreronis*), Coconut Research Institute, Sri Lanka, 6–8 January 2000.* Coconut Research Institute, Lunuwila, Sri Lanka, pp. 54–57.

Seiedy, M., Saboori, A., Allahyari, H., Talaei-Hassanloui, R. and Tork, M. (2010) Laboratory investigation on the virulence of two isolates of the entomopathogenic fungus *Beauveria bassiana* against the twospotted spider mite *Tetranychus urticae* (Acari: Tetranychidae). *International Journal of Acarology* 36, 527–532.

Seifers, D.L., Harvey, T.L., Louie, R., Gordon, D.T. and Martin, T.J. (2002) Differential transmission of isolates of the High Plains virus by different sources of wheat curl mites. *Plant Disease* 86, 138–142.

Seifers, D.L., Martin, T.J., Harvey, T.L., Fellers, J.P. and Michaud, J.P. (2009) Identification of the wheat curl mite as the vector of *Triticum* mosaic virus. *Plant Disease* 93, 25–29.

Sejalia, A.S., Rai, A.B., Patel, C.B. and Radadai, G.G. (1993) On the biological aspects of *Tetranychus macfarlanei* (Acari: Tetranychidae) infesting okra (*Abelmoschus aesculentus*) in south Gujarat. *Gujarat Agricultural University Research Journal* 19, 32–37.

Seki, M. (1979) Ecological studies of the pink citrus rust mite, *Aculops pelekassi* (Keifer), with special reference to the life cycle, forecasting of occurrence and chemical control of *A. pelekassi*. *Special Bulletin of Saga Prefectural Fruit Tree Experiment Station* 2, 1-66 (in Japanese).

Seki, M. (1983) Life cycle of the pink citrus mite, *Aculops pelekassi* (Keifer), in Japan. In: Matsumoto, K. (ed.) *Proceedings of the 4th International Citrus Congress, Tokyo, Japan, 9–12 November 1981, Volume 2.* International Society of Citriculture, Riverside, California, pp. 656–658.

Sekulic, D., Peric, P., Stojnic, B., Petanovic, R., Dimic, N., Miletic, N. and Sestovic, M. (1997) Efikasnost fenazakvina, novog akaricida, u suzbijanju *Panonychus ulmi* (Acari: Tetranychidae) na jabuci i *Calepitrimerus vitis* (Acari: Eriophydae) na vinovoj lozi. *Pesticidi* 12(1), 25–29 (in Serbian).

Selhime, A.G. (1984) Oil sprays control citrus rust mite. *Citrus Industry* 65, 24–27, 31.

Selhime, A.G. and Muma, M.H. (1966) Biology of *Entomophthora floridana* attacking *Eutetranychus banksi*. *Florida Entomologist* 49, 161–168.

Selvasundaram, R., Sudarmani, D.N.P. and Muraleedharan, N. (2001) Entomopathogenic fungi for pink mite control in tea. *Newsletter – UPASI Tea Research Foundation* 11, 1, 2.

Senapati, S.K. and Biswas, A.K. (1990) Efficacy of some pesticides against *Raoiella indica* Hirst. (Tenuipalpidae) on arecanut seedlings in Terai region of West Bengal. *Environment and Ecology* 8, 763–765.

Sengonca, C., Khan, I.A. and Blaeser, P. (2003) Prey consumption during development as well as longevity and reproduction of *Typhlodromus pyri* Scheuten (Acari, Phytoseiidae) at higher temperatures in the laboratory. *Anzeiger für Schädlingskunde* 76, 57–64.

Sepasgosarian, H. (1956) Morphologie und Biologie der gelben Apfelspinnmilbe *Eotetranychus pomi* n. sp. (Acari, Tetranychidae). *Zeitschrift für Angewandte Entomologie* 43, 435–491.

Sepasgosarian, H. (1970) *The Red False Spider Mite of Apple,* Cenopalpus pulcher *(Can. and Fanz.).* Bulletin No. 115, Agricultural College, University of Teheran, Karadj, Iran.

Sepasgosarian, H. (1972) Pflanzenspinnmilben und ihre wirtschaftliche Bedeutung im Iran. *Anzeiger für Schädlingskunde, Pflanzenschutz, Umweltschutz* 45, 105–107.

Sepasgosarian, H. (1973) Mites and their economic importance in Iran. In: Daniel, M. and Rosický, B. (eds) *Proceedings of the 3rd International Congress of Acarology Held in Prague, 31 August–6 September 1971.* Dr W. Junk, The Hague, the Netherlands (reprinted by Springer), pp. 241–245.

Sepasgosarian, H. (1975) Neue und bekannte Milben aus Iran. *Anzeiger für Schädlingskunde, Pflanzenschutz, Umweltschutz* 48, 6–8.

Sepasgosarian, H. (1977) The 20 years research of acarology in Iran. *Journal of Engineering Society of Iran* 56, 40–50.

Serafimovski, A. and Thalenhorst, W. (1962) Biologische und Okologische Beobachtungen an der Fitchenspinnmelbe, *Paratetranychus ununguis*. *Anzeiger für Schädlingskunde* 36, 37–42.

Serrano, L., Ramon, J., Segarra, J., Medina, V., Achón, M.A., López, M. and Juárez, M. (2004) New approach in the identification of the causal agent of fig mosaic disease. *Acta Horticulturae* 657, 559–566.

Seth, M.L. (1962) Transmission of pigeon pea sterility by an eriophyid mite. *Indian Phytopathology* 15, 225–227.

Sevastianov, E. (1980) New taxa of the mite family Dolichocybidae (Trombidiformes, Tarsonemina) and phylogenetic relationships of its subfamilies. *Zoologicheskii Zhurnal* 59, 1453–1461 (in Russian).

Severmutlu, S., Rodgers, C., Riordan, T.P. and Shearman, R.C. (2005a) Registration of 'Bowie' buffalograss. *Crop Science* 45, 2120–2122.

Severmutlu, S., Riordan, T.P. and Shearman, R.C. (2005b) Registration of 'Cody' buffalograss. *Crop Science* 45, 2122–2123.

Servicio para el Agricultor (1973) Control de plagas en citricas. *Noticias Agricolas (Venezuela)* 6 (34), 135–138.

Sewify, G.H. and Mabrouk, A.M. (1990) The susceptibility of different stages of citrus brown mite *Eutetranychus orientalis* Oudemans (family: Tetranychidae) to entomopathogenic fungus *Verticillium lecanii*. In: *Proceedings and Abstracts, Vth International Colloquium on Invertebrate Pathology and Microbial Control, 20–24 August 1990, Adelaide, Australia*, p. 395.

Shah, A.H. and Jose, V.T. (1989) The bionomics of *Amblyseius alstoniae* Gupta the predatory mite of cotton spider mite. *Gujarat Agricultural University Research Journal* 14, 53–57.

Shahrokhi, M.B. and Nikkhoo, F. (1990) Apple yellow mite *Eotetranychus pruni* (Oudemans) Acari: (Tetranychidae). *Applied Entomology and Phytopathology* 57(1–2), 23–25.

Shalini, K.V., Manjunatha, S., Lebrun, P., Berger, A., Baudouin, L., Pirany, N., Ranganath, R.M. and Prasad, D.T. (2007) Identification of molecular markers associated with mite resistance in coconut (*Cocos nucifera* L.). *Genome* 50, 35–42.

Shanks, C.H. and Barritt, B.H. (1975) Resistance of strawberry to the two-spotted spider mite. *Journal of Economic Entomology* 68, 7–10.

Sharanabasava, H. and Manjunatha, M. (2001) Predator prey interactions between *Chrysoperla carnea* Stephens (Neuroptera: Chrysopidae) and *Tetranychus neocaledonicus* (Andre) (Acari: Tetranychidae) on okra. In: Halliday, R.B., Walter, D.E., Proctor, H.C., Norton, R.A. and Colloff, M.J. (eds) *Acarology: Proceedings of the 10th International Congress of Acarology, 5–10 July 1998, Canberra*. CSIRO Publishing, Collingwood, Australia, pp. 423–426.

Sharma, A. and Srinivasa, N. (2004) *Brown Wheat Mite* Petrobia latens *(Muller) (Acari: Tetranychidae)*. All India Network Project on Agricultural Acarology, University of Agricultural Sciences, Bangalore, India.

Sharma, B.L. and Pande, Y.D. (1983) Effect of rain on the population of *Tetranychus neocaledonicus* Andre (Acarina: Tetranychidae) on cucurbits. *Indian Agriculturist* 27, 135–141.

Sharma, B.M. and Sharma, A. (2005) Bio efficacy of some pesticides against *Polyphagotarsonemus latus* (Banks) infesting chilli under semi-arid ecosystem. *Udyanika* 11(2), 94–96.

Sharma, D.C., Choudhary, A. and Sharma, D.K. (2005) Maximum residue limit and risk assessment of spiromesifen (BAY BSN 2060; Oberon 240SC) on tea (*Camellia sinensis* (L.) O'Kuntze). *Bulletin of Environmental Contamination and Toxicology* 75, 768–774.

Sharma, D.D. (1985) Major pests of litchi in Bihar. *Indian Farming* 35, 25–26.

Sharma, D.D. and Rahman, M.F. (1982) Control of litchi mite *Aceria litchii* (Keifer) with particular reference to evaluation of prebloom and postbloom application with different insecticides. *Entomon* 7, 55–56.

Sharma, D.D., Singh, S.P. and Akhauri, R.K. (1986) Relationship between the population of *Aceria litchii* Keifer on litchi and weather factors. *Indian Journal of Agricultural Sciences* 56, 59–63.

Sharma, N.K. and Sadana, G.L. (1987) Effect of predator–prey density on the prey consumption and daily rate of egg production of the predatory mite, *Amblyseius finlandicus* (Oudemans) (Acarina: Phytoseiidae). *Entomon* 12, 191–195.

Sharma, S. and Bhardwaj, S.P. (2010) Effect of nitrogen, phosphorus, and potash levels on population fluctuation of European red mite, *Panonychus ulmi*, on apple. In: Sabelis, M.W. and Bruin, J. (eds) *Trends in Acarology: Proceedings of the 12th International Congress, 21–26 August 2006, Amsterdam*. Springer, Dordrecht, the Netherlands, pp. 501–504.

Shaw, J.G., Moffitt, C. and Scriven, G.T. (1967) Biotic potential of phytoseiid mites fed on virus-infected citrus red mites. *Journal of Economic Entomology* 60, 1751–1752.

Shaw, J.G., Chambers, D.L. and Tashiro, H. (1968) Introducing and establishing the non-inclusion virus of the citrus red mite in citrus groves. *Journal of Economic Entomology* 61, 1352–1355.

Shaw, P.W. and Wallis, D.R. (2007) Predator mite application methods for biological control of two-spotted mites in hops. *New Zealand Plant Protection* 60, 89–93.

Sheeja, U.M. and Ramani, N. (2009) Feeding potential of *Lasioseius* sp. (Acari: Mesostigmata), a promising predator of the red palm mite, *Raoiella indica* Hirst (Acari: Prostigmata) ecosystem. *Karnataka Journal of Agricultural Sciences* 22, 698–700.

Sheikholeslam Zadeh, S. and Sadeghi, H. (2010a) First records of four mite species (Acari: Tetranychidae) in Iran. *Applied Entomology and Phytopathology* 78(1), 121–125.

Sheikholeslam Zadeh, S. and Sadeghi Namaghi, H. (2010b) Faunistic survey of injurious mites associated with broad leaf trees in green spaces of Mashhad region. *Quarterly Journal of Plant Protection (Agricultural Science and Technology)* 24, 210–217 (in Persian).

Sheila, V.K., Manohar, S.K. and Nene, Y.L. (1988) Biology and morphology of *Aceria cajani*. *International Pigeonpea Newsletter* 7, 28.

Sheppard, A.W., Hill, R., DeClerck-Floate, R.A., McClay, A., Olckers, T., Quimby, P.C. Jr and Zimmermann, H.G. (2003) A global review of risk–benefit–cost analysis for the introduction of classical biological control agents against weeds: a crisis in the making? *Biocontrol News and Information* 24(4), 91–108.

Sheppard, A.W., van Klinken, R.D. and Heard, T.A. (2005) Scientific advances in the analysis of direct risks of weed biological control agents to nontarget plants. *Biological Control* 35, 215–226.

Sheppard, A.W., Hosking, J.R., Sagliocco, J.L., Thomann, T., Downey, P.O. and Kwong, R.M. (2006) Biological control of brooms in Australia: an update. In: *Managing Weeds in a Changing Climate, 15th Australian Weeds Conference, Papers and Proceedings, Adelaide, South Australia, 24–28 September 2006*, pp. 573–576.

Shevtchenko, V.G. (1957) The life-history of alder gall mite, *Eriophyes laevis* (Nalepa). *Entomologicheskoe Obozrenie (Entomological Review, USSR)* 36, 558–618 (in Russian).

Shevtchenko, V.G. (1961) Peculiarities of the postembryonic development of gall-mites and some notes on the classification of *Eriophyes laevis* (Nalepa). *Zoologicheskii Zhurnal* 60, 1143–1158 (in Russian).

Shevchenko, V.G. (1962) Novii chetirekhnogij kleshch *Trisetacus kirghizorum* sp.n. (Acarina, Eriophyidae) vreditel semjan arci (New four legged mite – *Trisetacus kirghizorum* sp.n. (Acarina: Eriophyidae) – the pest of seeds of Asian mountain juniper). *Trudy Kirghiz Lesnika Opytnaia Stantsia (Proceedings of Kirghyz Forest Experiment Station)* 3, 299–305 (in Russian).

Shevchenko, V.G. and De-Millo, A.P. (1968) Life-cycle of *Trisetacus kirghisorum* (Acarina: Tetrapodili) – pest of *Juniperus semiglobosa* Rgl. *Vestnik LGU* 3, 60–67 (in Russian).

Shevtchenko, V.G. and Silvere, A.P. (1968) Rotovoi aparat chetyrekhnogikh kleshchei (Acarina: Eriophyoidea). *Esti NSV Teaduste Akadeemia Toimetised (Izvestiya Akademii Nauk Estonskoi SSR), Biologia* 17, 248–263 (in Russian).

Shevtchenko, V.G., De-Millo, D.P., Razvyazkina, G.M. and Kapkova, E.A. (1970) Taxonomic separation of similar species of eriophyid mites, *Aceria tulipae* Keirf. and *A. tritici* sp.n. (Acarina, Eriophyoidea) – vectors of the viruses of onions and wheat. *Zoologicheskii Zhurnal* 49, 224–235 (in Russian).

Shevchenko, V.G., Bagnyuk, I.G. and Rinne, V. (1993) *Trisetacus pini* (Nalepa) in some Baltic countries and in Russia (taxonomy, morphology, biology, distribution). *Acarina* 1, 51–71.

Shi, A. and Tomczyk, A. (2001) Impact of feeding of eriophyid mite *Epitrimerus gibbosus* (Nalepa) (Acari: Eriophyoidea) on some biochemical components of blackberry (*Rubus* spp.). *Bulletin of the Polish Academy of Sciences, Biological Sciences* 49(1), 41–47.

Shi, J.-M., Yu, H.-X. and Yang, S.-D. (1992) Studies on the bionomics of three species of bamboo spider mites and their control. *Journal of Zhejiang Forestry Science and Technology* 12(2), 11–14 (in Chinese).

Shi, W.B. and Feng, M.G. (2006) Field efficacy of application of *Beauveria bassiana* formulation and low rate pyridaben for sustainable control of citrus red mite *Panonychus citri* (Acari: Tetranychidae) in orchards. *Biological Control* 39, 210–217.

Shi, W.B., Feng, M.G. and Liu, S.S. (2008) Sprays of emulsifiable *Beauveria bassiana* formulation are ovicidal towards *Tetranychus urticae* (Acari: Tetranychidae) at various regimes of temperature and humidity. *Experimental and Applied Acarology* 46, 247–257.

Shiao, S.N. (1976) An ecological study of the tea purple mite, *Calacarus carinatus* Green. *Plant Protection Bulletin (Taipei)* 18, 183–198.

Shih, C.I.T, Poe, S.L. and Cromroy, H.L. (1976) Biology, life table, and intrinsic rate of increase of *Tetranychus urticae*. *Annals of the Entomological Society of America* 69, 362–364.

Shih, C.I.T., Chang, H.Y., Hsu, P.H. and Hwang, Y.F. (1993) Responses of *Amblyseius ovalis* (Evans) (Acarina: Phytoseiidae) to natural food resources and two artificial diets. *Experimental and Applied Acarology* 17, 503–519.

Shih, C.T. (1979) The influence of age of female *Tetranychus kanzawai* on sex ratio and life cycle of its progeny. In: Rodriguez, J.G. (ed.) *Recent Advances in Acarology, Volume 1*. Academic Press, New York, pp. 511–517.

Shih, C.T. (1985) Biological control of mulberry spider mites, *Tetranychus kanzawai*, with augmentation of *Amblyseius womersleyi*. *Taiwan Agricultural Research Institute Special Publication* No. 19, pp. 21–27.

Shikata, E., Kawano, S., Senboku, T., Tiongco, E.R. and Mayejima, K. (1984) Small virus-like particles isolated from the leaf sheath tissues of rice plant and from the rice tarsonemid mite, *Steneotarsonemus spinki* Smiley (Acarina: Tarsonemidae). *Annals of the Phytopathological Society of Japan* 5, 368–374.

Shimoda, T., Kishimoto, H., Takabayashi, J., Amano, H. and Dicke, M. (2009) Comparison of thread-cutting behavior in three specialist predatory mites to cope with complex webs of *Tetranychus* spider mites. *Experimental and Applied Acarology* 47, 111–120.

Shinkaji, N. (1954) A review of ecological studies on spider mites. *Agriculture and Horticulture* 29, 1365–1368, 1497–1500 (in Japanese).

Shinkaji, N. (1975) Hatching time of the winter eggs and termination of diapause in the common conifer spider mite *Oligonychus ununguis* (Jacobi), on chestnut in relation to temperature (Acarina: Tetranychidae). *Japanese Journal of Applied Entomology and Zoology* 19, 144–148 (in Japanese).

Shinkaji, N. (1979) Geographical distribution of the citrus red mite, *Panonychus citri* and European red mite, *P. ulmi* in Japan. In: Rodriquez, J.G. (ed.) *Recent Advances in Acarology, Volume I*. Academic Press, New York, pp. 81–87.

Shinkaji, N., Okabe, K. and Amano, H. (1986) Rhizoglyphine mites (Acari: Acaridae) infesting rakkyo (*Allium chinense* G. Don) and Chinese chive (*A. tuberosum* Rottler), and their susceptibility to dimethoate and disulfoton. *Japanese Journal of Applied Entomology and Zoology* 30, 285–289.

Shipp, J.L., Gillespie, D.R. and Ferguson, G.M. (2001) *Aculops lycopersici* (Massee), tomato russet mite (Acari: Eriophyidae). In: Mason, P.G. and Huber, J.T. (eds) *Biological Control Programmes in Canada, 1981–2000*. CAB International, Wallingford, UK, pp. 32–33.

Shiraki, T. (1919) Insect pests of the tea-plant in Formosa (preliminary report). In: *Report, Proceedings 3rd Entomology Meeting, Pusa, Calcutta, India*, pp. 629–669.

Short, D.E. and Buss, E.A. (2005) *Bermudagrass Mite*. Publication No. ENY-328, Entomology and Nematology Department, University of Florida/Institute of Food and Agricultural Sciences (UF/IFAS) Extension, Gainesville, Florida. Updated version by Buss, E.S. (2008) available at: http://ufdcimages.uflib.ufl.edu/IR/00/00/28/49/00001/LH03500.pdf (accessed 20 May 1015).

Shrewsbury, P.M. and Hardin, M.R. (2003) Evaluation of predatory mite (Acari: Phytoseiidae) releases to suppress spruce spider mites, *Oligonychus ununguis* (Acari: Tetranychidae), on juniper. *Journal of Economic Entomology* 96, 1675–1684.

Shvanderov, F.A. (1975) Role of phoresy in the migration of Eriophyoidea. *Zoologicheskii Zhurnal* 54, 458–461 (in Russian).

Shvanderov, F.A. (1984) Gall mites on pear. *Zashchita Rastenii* No. 8, 48–49 (in Russian).

Shwetaketu Dev and Chenchaiah, K.C. (2005) Chemical control of red spider mite (*Oligonychus coffeae*) in sub Himalayan Tarai Region of West Bengal. *Orissa Journal of Horticulture* 33, 86–88.

Sibanda, T., Dobson, H.M., Cooper. J.F., Manyangarirwa, W. and Chiimba, W. (2000) Pest management challenges for smallholder vegetable farmers in Zimbabwe. *Crop Protection* 19, 807–815.

Sibbett, G.S., Dibble, J.E. and Babcock, J.D. (1976) Black scale now a major olive pest. *California Agriculture* 30(11), 12–13.

Siddig, M.A. and El-Badry, E. (1971a) Tetranychid mites of the Sudan (Acarina: Tetranychidae). *Sudan Agricultural Journal* 6, 36–39.

Siddig, M.A. and El-Badry, E. (1971b) Biology of the spider mite *Eutetranychus sudanicus*. *Annals of the Entomological Society of America* 64, 806–809.

Sidebottom, J.R. (1996) *The Spruce Spider Mite in Fraser Fir*. Christmas Tree Notes No. 29, North Carolina State Cooperative Extension, Raleigh, North Carolina.

Sidhu, A.S. and Singh, G. (1971) Chemical control of the red vegetable mite, *Tetranychus cucurbitae* Rahman and Sapra, on brinjal. *Punjab Horticultural Journal* 11, 95–97.

Sidorchuk, E. and Bertrand, M. (2013) New fossil labidostomatids (Acari: Labidostomatidae) from Eocene amber and presence of an apustulate species in Europe. *Acarologia* 53, 25–39.

Siewniak, M. (1971) Eine bisher unbekannte Schädigung der Sämlinge von *Picea abies* Karst. durch Gallmilben (Acar. Eriophyidae). *Entomologisches Nachrichtenblatt* 15, 66–67.

Sigsgaard, L. (2006) Biological control of arthropod pests in outdoor crops – the new challenge. In: Hansen, L.S., Enkegaard, A., Steenberg, T., Ravnskov, S. and Larsen, J. (eds) *Implementation of Biocontrol in Practice in Temperate Regions – Present and Near Future: Proceedings of the International Workshop at Research Centre Flakkebjerg, Denmark on November 1 to 3, 2005*. DIAS Report (Plant Production) No. 119, Danish Institute of Agricultural Sciences, Tjele, Denmark, pp.153–167. Available at: http://pure.au.dk/portal/files/318213/119.pdf (accessed 20 May 2015).

Silberman, L. (1983) Aerial dispersal of the two-spotted spider mite, *Tetranychus urticae* (Koch), from its host plant. MSc. thesis, North Carolina State University, Raleigh, North Carolina.

Silva, A.L. da, Veloso, V.R.S., Silva, R.M.C. de and Rocha, M.R. da (1988) Ensaio de controle ao microacaro do tomatreio *Aculops lycopersici* (Massee, 1937) (Acari: Eriophyidae) com um novo acaricida. *Anais da Escola de Agronomia e Veterinária (Goiania)* 18(1), 29–34.

Silva, C.A.D. da (2002) Biologia e exigencias termicas do acaro-vermelho (*Tetranychus ludeni* Zacher) em folhas de algodoeiro. *Pesquisa Agropecuária Brasileira* 37, 573–580.

Silva, E.A., Reis, P.R., Carvalho, T.M.B. and Altoé, B.F. (2009) *Tetranychus urticae* (Acari: Tetranychidae) on *Gerbera jamesonii* Bolus and Hook (Asteraceae). *Brazilian Journal of Biology* 69, 1121–1125.

Silva, F.R. da, Moraes, G.J. de and Knapp, M. (2008) Distribution of *Tetranychus evansi* and its predator *Phytoseiulus longipes* (Acari: Tetranychidae, Phytoseiidae) in southern Brazil. *Experimental and Applied Acarology* 45, 137–145.

Silva, F.R. da, Moraes, G.J. de, Gondim, M.G.C. Jr, Knapp, M., Rouam, S.L., Jefferson, L.A. and Oliveira, P.G.M. de (2010) Efficiency of *Phytoseiulus longipes* Evans as a control agent of *Tetranychus evansi* Baker and Pritchard (Acari: Phytoseiidae: Tetranychidae) on screenhouse tomatoes. *Neotropical Entomology* 39, 991–995.

Silva, M.A., Parra, J.R.P. and Chiavegato, L.G. (1985) Biologia comparada de *Tetranychus urticae* em cultivares de algodoeiro. II. Tabela de vida de fertilidade. *Pesquisa Agropecuária Brasileira* 20, 1015–1019.

Silva, P. (1954) Um novo ácaro nocivo ao tomateiro na Bahia. *Boletim do Instituto Biológico da Bahia* 1, 1–20.

Silva, R.A., Reis, P.R., Carvalho, C.F. and Souza, B. (2006) Predatory capacity of *Chrysoperla externa* (Hagen, 1861) (Neuroptera: Chrysopidae) on *Brevipalpus phoenicis* (Geijskes, 1939) (Acari: Tenuipalpidae). *Coffee Science* 1(1), 50–54.

Silva, R. R. da, Teodoro, A.V., Vasconcelos, J.F., Martins, C.R., Rilho, W. dos Santos S., Carvalho, H. W. L. de and Guzzo, E.C. (2015) Citrus rootstocks influence the population densities of pest mites. *Ciência Rural*. Online. http://dx.doi..org/10.1590/0103-8478cr20150486

Silveira, E.C., Reis, P.R., Souza-Pimentel, G.C., Toledo, M.A. de, Marafeli, P.P. and Pedro Neto, M. (2010) Influence of physiological insecticides used in the control of coffee leaf-miner on the mite *Oligonychus ilicis*. In: Moraes, G.J. de, Castilho, R.C. and Flechtmann, C.H.W. (eds) *XIII International Congress of Acarology, August 23–27, 2010, Recife-PE, Brazil, Abstract Book*, abstract 468, pp. 253–254. Available at: http://www.acarology.org/ica/ica2010/docs/0910_15_abstractBookCapa.pdf (accessed 3 February 2015).

Simanton, W.A. (1976) Occurrence of insect and mite pests in citrus, their predators and parasitism in relation to spraying operations. In: *Proceedings of 6th Tall Timbers Conference on Ecological Animal Control by Habitat Management, 1976, Gainesville, Florida*. Tall Timbers Research Station, Tallahassee, Florida, pp. 135–163.

Simmonds, N.W. (1979) *Principles of Crop Improvement*. Longmans, Harlow, UK.

Simmons, A.T. and Gurr, G.M. (2005) Trichomes of *Lycopersicon* species and their hybrids: effects on pests and natural enemies. *Agricultural and Forest Entomology* 7, 265–276.

Simon, C., Frati, F., Beckenbach, A., Crespi, B., Liu, H. and Flook, P. (1994) Evolution, weighting and phylogenetic utility of mitochondrial gene sequences and a compilation of conserved polymerase chain reaction primers. *Annals of the Entomological Society of America* 87, 651–701.

Simoni, S., Cantini, R., Castagnoli, M. and Battisti, A. (2004) Impact and management of the eriophyoid mite *Trisetacus juniperinus* on the evergreen cypress *Cupressus sempervirens*. *Agricultural and Forest Entomology* 6, 175–180.

Simons, J.N. (1964) Tetranychid mites as defoliators of cotton. *Journal of Economic Entomology* 57, 145–148.

Singer, G. (1967) A comparison between different mounting techniques commonly employed in acarology. *Acarologia* 9, 475–484.

Singh, A.K. (1992) Sterility mosaic of pigeonpea: etiology, epidemiology and management. PhD thesis, G.B. Pant University of Agriculture and Technology, Pantnagar, India.

Singh, A.K. and Rathi, Y.P.S. (1994) Nature of losses caused by sterility mosaic in pigeonpea. *Journal of Applied Biology* 4, 35–37.

Singh, A.K. and Rathi, Y.P.S. (1997) Epidemiology of vector of pigeonpea sterility mosaic virus. *Indian Journal of Virology* 13, 143–145.

Singh, A.K., Rathi, Y.P.S. and Agrawal, K.C. (1999) Sterility mosaic of pigeonpea: a challenge of 20th century. *Indian Journal of Virology* 15, 85–92.

Singh, B.U. (1997) Screening for resistance to sorghum shoot bug and spider mites. In: Sharma, H.C., Faujdar Singh, and Nwanze, K.F. (eds) *Plant Resistance to Insects in Sorghum*. International Crops Research Institute for the Semi-Arid Tropics (ICRISAT), Patancheru, India, pp. 52–59.

Singh, B.U., Rana, B.S. and Rao, N.G.P. (1981a) Host plant resistance to mite (*Oligonychus indicus* Hirst) and its relationship with shoot fly (*Atherigona soccata* Rond.) resistance in sorghum. *Journal of Entomological Research* 5, 25–30.

Singh, G. (1982) Sugarcane mites and their control. *Pesticides* 16(2), 9–16.

Singh, J. and Mukherjee, I.N. (1989) Pest status of phytophagous mites in some northern states of India. In: *Proceedings of the 1st Asia-Pacific Conference of Entomology (APCE)*. Entomology and Zoology Association of Thailand, Bangkok, pp. 169–185.

Singh, J. and Raghuraman, M. (2011) Emerging scenario of important mite pests in north India In: Moraes, G.J. de and Proctor, H. (eds) *Acarology XIII: Proceedings of the International Congress*. *Zoosymposia* 6, 170–179.

Singh, J., Singh, R.N. and Rai, S.N. (2000a) Expanding pest status of phytophagous mites and integrated pest management. In: Upadhyay, R.K., Mukherji, K.G. and Dubey, O.P. (eds) *IPM-System in Agriculture, Animal Pests*. Aditya Books, New Delhi, pp. 1–29.

Singh, L.B., Singh, S.M. and Nirvan, R.S. (1961) Studies on mango malformation. I. Review, symptoms, extent, intensity, and cause. *Horticultural Advances* 5, 197–217.

Singh, M.P. and Singh, K.M. (1999) Some phytophagous and predaceous mites associated with summer vegetables in Manipur. *Insect Environment* 4, 117.

Singh, M.P., Sharma, D.D. and Hameed, S.F. (1988) Post-treatment temperature effect on toxicity of some acaricides to *Tetranychus neocaledonicus* on okra (*Abelmoschus esculentus*). *Indian Journal of Agricultural Sciences* 58, 77–79.

Singh, O.P. (1966) Record of leaf sheath mite (*Eriophyes* sp.) on sugarcane in the Punjab and Uttar Pradesh. *Science and Culture* 32, 51.

Singh, P. (1984) Insect diets. Historical developments, recent advances, and future prospects. In: King, E.G. and Leppla, N.C. (eds) *Advances and Challenges in Insect Rearing*. USDA/ARS (US Department of Agriculture Agricultural Research Service), New Orleans, Louisiana, pp. 32–44.

Singh, P., Beattie, G.A.C., Clift, A.D., Watson, D.M., Furness, G.O., Tesoriero, L., Rajakulendran, V., Parkes, R.A. and Scanes, M. (2000b) Petroleum spray oils and tomato integrated pest and disease management in southern Australia. *General and Applied Entomology* 29, 69–93.

Singh, R.A. (2001) Occurrence of paddy leaf mite, *Oligonychus oryzae* Hirst. (Tetranychidae: Acarina) at Faizabad. *Journal of Applied Zoological Research* 12, 59.

Singh, R.N. (1995) Studies on seasonal abundance of *Tetranychus ludeni* Zacher (Acari: Tetranychidae) on cowpea (*Vigna unguiculata* Savi.). *Journal of Recent Advances in Applied Sciences* 10, 1–2.

Singh, R.P. and Agnihotri, N.P. (1984) Residues of dicofol, endosulfan and malathion on tea, *Camellia sinensis* (L.) O. Kuntze. *Journal of Entomological Research* 8, 14–16.

Singh, S.P., Rao, N.S. and Kumar, K.K. (1981b) Occurrence of vegetable mite, *Tetranychus neocaledonicus* Andre on passion fruit, *Passiflora edulis* Sims and evaluation of certain acaricides against it. *Pestology* 5(1), 25–27.

Singh, S.P., Rao, N.S., Kumar, K.K. and Bhumannavar, B.S. (1983) Field evaluation of citrus germplasm for the incidence of the oriental red mite, *Eutetranychus orientalis* (Klein). *Indian Journal of Plant Protection* 11, 140–142.

Singh, S.P., Rao, N.S. and Bhumannavar, B.S. (1986) Chemical control of oriental red mite, *Eutetranychus orientalis* (Klein) on coorg mandarin. *Pesticides* 20(8), 51–53.

Sinha, R.N. (1964) Ecological relationships of stored-products mites and seed-borne fungi. *Acarologia* 6, 372–389.

Sinha, R.N. and Wallace, H.A.H. (1963) *Tetranychus sinhai* Baker (Acarina: Tetranychidae) a new pest of cereals – varietal reaction of barley. *Canadian Entomologist* 95, 588–596.

Sinha, R.N. and Wallace, H.A.H. (1966) Association of granary mites and seed-borne fungi in stored grain and outdoor and indoor habitats. *Annals of the Entomological Society of America* 59, 1170–1181.

Siqueira, F. (2011) Biologia e flutuação populacional de *Mononychellus planki* (McGregor) (Acari: Tetranychidae) em cultivares de soja *Glycine max* (L.) Merr. e impacto do imidaclopride em aspectos biológicos do adult. PhD thesis, Universidade Federal do Paraná, Curitiba, Brazil.

Sithanantham, S. (1981) Distribution of the eriophyid mite *Aceria sacchari* in sugarcane. In: ChannaBasavanna, G.P., Mallik, B. and Ghorpade, K.D. (eds) *Contributions to Acarology in India: Proceedings of the I All India Symposium in Acarology, April 23–25, 1979, Bangalore*. Acarological Society of India, Bangalore, India, pp. 68–72.

Sithanantham, S. and Velayutham, N. (1981) Susceptibility of different varieties of sugarcane to *Aceria sacchari* (Acari: Eriophyidae). In: ChannaBasavanna, G.P., Mallik, B. and Ghorpade, K.D. (eds) *Contributions to Acarology in India: Proceedings of the I All India Symposium in Acarology, April 23–25, 1979, Bangalore*. Acarological Society of India, Bangalore, India, pp. 74–79.

Sithanantham, S., Muthuswamy, S. and Bhaskaran, T.L. (1972) A note on the role of the eriophyid mite (*Aceria sacchari* ChannaBasavanna, Eriophyoidea: Acarina) in the transmission of streak virus disease of sugarcane. *Science and Culture* 38, 248–249.

Sithanantham, S., Muthuswamy, S. and Bhaskaran, T.L. (1973) Varietal difference in the expression of streak virus disease symptoms in sugarcane as a factor in transmission studies. *Science and Culture* 39, 232–234.

Sithanantham, S., Muthusamy, S. and Dura, D. (1975) Direct effect of infestation by the eriophyid mite, *Aceria sacchari*, on the composition of sugarcane leaf sheat. *Science and Culture* 41, 327–328.

Sitnikova, L.G. (1978) The main evolutionary trends of the Acari and the problem of their monophyletism. *Entomologicheskoe Obozrenie (Entomological Review, USSR)* 57, 431–457 (in Russian).

Sitti, A., D'Ascenzo, D. and Guastamacchia, F. (2008) Attivita' di alcuni principi attivi nei confronti di *Calepitrimerus vitis*, agente dell'acariosi della vite. In: *Atti, Giornate Fitopatologiche 2008, Cervia 12–14 Marzo, Volume 1*. Fitogest®, Image Line Network, Faenza , pp. 219–220.

Skaf, J.S. and Carroll, T.W. (1995) Purification of barley yellow streak mosaic virus and detection by DASELISA and ISEM using polyclonal antibodies. *Plant Disease* 79, 1003–1007.

Skare, J.M., Wijkamp, I., Rezende, J., Michels, G., Rush, C., Scholthof, K.-B.G. and Scholthof, H.B. (2003) Colony establishment and maintenance of the eriophyid wheat curl mite *Aceria tosichella* for controlled transmission studies on a new virus-like pathogen. *Journal of Virological Methods* 108, 133–137.

Skare, J.M., Wijkamp, I., Denham, I., Rezende, J.A.M., Kitajima, E.W., Park, J.-W., Desvoyes, B., Rush, C.M., Michels, G., Scholthof, K.-B.G. and Scholthof, H.B. (2006) A new eriophyid mite-borne membrane-enveloped virus-like complex isolated from plants. *Virology* 347, 343–353.

Skoracka, A. (2006) Host specificity of eriophyoid mites: specialists or generalists? *Biological Letters* 43, 289–298.

Skoracka, A. (2008) Reproductive barriers between populations of the cereal rust mite *Abacarus hystrix* confirm their host specialization. *Evolutionary Ecology* 22, 607–616.

Skoracka, A. (2009) Quackgrass-and ryegrass-adapted populations of the cereal rust mite, *Abacarus hystrix* (Acari: Eriophyidae), differ in their potential for wheat, *Triticum aestivum*, colonization. *Bulletin of Entomological Research* 99, 33–39.

Skoracka, A. and Boczek, J. (2001) Eriophyoid mites (Acari: Eriophyoidea) as the vectors of virus diseases of cereals and grasses. *Postepy Nauk Rolniczych* 5, 79–84 (in Polish).

Skoracka, A. and Dabert, M. (2010) The cereal rust mite *Abacarus hystrix* (Acari: Eriophyoidea) is a complex of species: evidence from mitochondrial and nuclear DNA sequences. *Bulletin of Entomological Research* 100, 263–272.

Skoracka, A. and Kuczynski, L. (2004) Demography of the cereal rust mite *Abacarus hystrix* (Acari: Eriophyoidea) on quack grass. *Experimental and Applied Acarology* 32, 231–242.

Skoracka, A. and Kuczynski, L. (2006a) Is the cereal rust mite, *Abacarus hystrix* really a generalist? – testing colonization performance on novel hosts. *Experimental and Applied Acarology* 38, 1–13.

Skoracka, A. and Kuczynski, L. (2006b) Host related differences in the development and reproduction of the cereal rust mite, *Abacarus hystrix* (Acari: Eriophyidae) in Poland. *International Journal of Acarology* 32, 397–405.

Skoracka, A., Magowski, W. and Kuczynski, L. (2003) Spatial and temporal distribution of cereal rust mite *Abacarus hystrix* (Acari, Eriophyoidea) on *Agropyron repens*. *Biologia* 58, 913–917.

Skoracka, A., Lewandowski, M. and Boczek, J. (2005) *Catalogue of Eriophyoid Mites of Poland.* Catalogus Faunae Poloniae (N.S.), 1, Natura Optima dux Foundation. Museum and Institute of Zoology, Polish Academy of Sciences, Warsaw.

Skoracka, A., Kuczynski, A. and Rector, B.G. (2007) Divergent host acceptance behavior suggests host specialization in populations of the polyphagous mite *Abacarus hystrix* (Acari: Prostigmata: Eriophyidae). *Environmental Entomology* 36, 899–909.

Skoracka, A., Smith, L., Oldfield, G., Cristofaro, M. and Amrine, J.W. (2010) Host-plant specificity and specialization in eriophyoid mites and their importance for the use of eriophyoid mites as biocontrol agents of weeds. *Experimental and Applied Acarology* 51, 93–113.

Skorupska, A. (1976) *Eotetranychus uncatus* Garman, a pest of apple and plum new to Poland. *Prace Naukowe Instytutu Ochrony Roślin* 18, 197–200 (in Polish).

Skorupska, A. (1989) Biology and ecology of *Eotetranychus uncatus* Garman. 1. Spread of *Eotetranychus uncatus* Garman and its host plants. *Prace Naukowe Instytutu Ochrony Roślin* 30, 151–163 (in Polish).

Skorupska, A. (1994) *Tetranycopsis horridus* Can. et Fan. (Acarina, Tetranychidae) – a new mite species for the fauna of Poland occurring on hazel. *Prace Naukowe Instytutu Ochrony Roślin* 35, 40–42.

Skorupska, A. (1998) Morphologico-anatomical structure of leaves and demographic parameters of the hawthorn spider mite, *Tetranychus viennensis* Zacher and the two-spotted spider mite, *Tetranychus urticae* (Koch) (Acarina, Tetranychidae) on selected scab-resistant apple varieties. *Journal of Applied Entomology* 122, 493–496.

Skorupska, A. and Boczek, J. (1984) Host-plant acceptance and preference of hawthorn spider mite (*Tetranychus viennensis* Zacher). *Prace Naukowe Instytutu Ochrony Roślin* 26, 147–165 (In Polish).

Skorupska, A. and Boczek, J. (1985) Effect of the host plants change on the reproduction of hawthorn spider mite (*Tetranychus viennensis* Zacher). *Prace Naukowe Instytutu Ochrony Roślin* 27, 5–22.

Skorupska, A. and Tomczyk, A. (1988) Effect of hawthorn spider mite (*Tetranychus viennensis* Zacher) feeding on some physiological processes in apple-tree seedlings. *Prace Naukowe Instytutu Ochrony Roślin* 29, 47–57.

Slansky, F. Jr and Scriber, J.M. (1985) Food consumption and utilization. In: Kerkut, G.A. and Gilbert, L.I. (eds) *Comprehensive Insect Physiology Biochemistry and Pharmacology, Volume 4.* Pergamon, Oxford, UK, pp. 87–163.

Slepyan, E.I., Landsberg, G.S. and Mal'chenkova, N.I. (1969) The gall of the mite *Eriophyes vitis* Pgt. (Acarina, Eriophyidae). *Entomological Review* 48, 67–74 (English version of *Entomologicheskoe Obozrenie* 48, 125–131).

Slykhuis, J.T. (1955) *Aceria tulipae* Keifer (Acarina: Eriophyidae) in relation to the spread of wheat streak mosaic. *Phytopathology* 45, 116–128.

Slykhuis, J.T. (1956) Wheat spot mosaic, caused by a mite-transmitted virus associated with wheat streak mosaic. *Phytopathology* 46, 682–687.

Slykhuis, J.T. (1958) A survey of virus diseases of grasses in northern Europe. *FAO Plant Protection Bulletin* 6, 129–134.

Slykhuis, J.T. (1963) Mite transmission of plant viruses. In: Naegele, J.A. (ed.) *Advances in Acarology, Volume 1.* Cornell University Press/Comstock Publishing Associates, Ithaca, New York, pp. 326–340.

Slykhuis, J.T. (1965) Mite transmission of plant viruses. In: Smith, K.M. (ed.) *Advances in Virus Research* 11, 97–137.

Slykhuis, J.T. (1967) Methods for experimenting with mite transmission of plant viruses. In: Maramorosch, K. and Kaprowski, H. (eds) *Methods in Virology.* Academic Press, New York, pp. 347–368.

Slykhuis, J.T. (1969) Transmission of Agropyron mosaic virus by the eriophyid mite, *Abacarus hystrix*. *Phytopathology* 59, 29–32.

Slykhuis, J.T. (1980) Mites. In: Harris, K.F. and Maramorosch, K. (eds) *Vectors of Plant Pathogens.* Academic Press, New York, pp. 325–356.

Smalley, E.B. (1956) The production on garlic by an eriophyid mite of symptoms like those produced by viruses. *Phytopathology* 46, 346–347.

Smee, C. (1943) *Report of Entomologist for the Year 1943.* Department of Agriculture, Zomba, Nyasaland (now Malawi).

Smidansky, E.D. (1996) Factors influencing the outcome of barley yellow streak mosaic virus: brown wheat mite–barley interactions. *Plant Disease* 80, 186–193.

Smilanick, J.M. and Zalom, F.G. (1983) Eriophyid mites in relation to Kentucky bluegrass seed production. *Entomologia Experimentalis et Applicata* 33, 31–34.

Smiley, R.L. (1964) Some Tarsonemidae from the Republic of the Congo (Acarina). *Proceedings of the Entomological Society of Washington* 66, 145–150.

Smiley, R.L. (1966) Further descriptions of two erythraeids predaceous upon cotton bollworm eggs (Acari: Erythraeidae). *Proceedings of the Entomological Society of Washington* 68, 25–28.

Smiley, R.L. (1967) Further studies on the Tarsonemidae (Acarina). *Proceedings of the Entomological Society of Washington* 69, 127–146.

Smiley, R.L. (1968) A new genus and three new species of Erythraeoidea (Acarina: Erythraeidae and Smarididae). *Proceedings of the Entomological Society of Washington* 70, 13–21.

Smiley, R.L. (1978) Taxonomic studies of *Pygmephorus* species from the western hemisphere, with a key to females and an overview of the current problems for classification (Acari: Pyemotidae and Pygmephoridae). *International Journal of Acarology* 4, 125–160.

Smiley, R.L. and Baker, E.W. (1995) A report on some tetranychid mites (Acari: Prostigmata) from Yemen. *International Journal of Acarology* 21, 135–164.

Smiley, R.L. and Landwehr, V.R. (1976) A new species of *Tarsonemus* (Acarina: Tarsonemidae), predaceous on tetranychid mite eggs. *Annals of the Entomological Society of America* 69, 1065–1072.

Smiley, R.L., Flechtmann, C.H.W. and Ochoa, R. (1993) A new species of *Steneotarsonemus* (Acari: Tarsonemidae) and an illustrated key to grass-infesting species in the western hemisphere. *International Journal of Acarology* 19, 87–93.

Smissaert, H.R. (1964) Cholinesterase inhibition in spider mites susceptible and resistant to organophosphate. *Science* 143, 129-131.

Smith, A.J. (1953) The citrus red mite. *Farming in South Africa* 28(331), 340, 344.

Smith, B.D. (1960) The behavior of the black currant gall mite, *Phytoptus ribis* Nal. during the free living phase of its life cycle. In: *Annual Report of the Agricultural and Horticultural Research Station, Long Ashton, 1959.* Bristol, UK, pp. 130–136.

Smith, B.D. (1961) Population studies of the blackcurrant gall mite (*Phytoptus ribis* Nalepa). In: *Annual Report of the Agricultural and Horticultural Research Station, Long Ashton, 1960.* Bristol, UK, pp. 120–124.

Smith, C.E. and Webber, J.C. (1954) The legume mite, *Petrobia apicalis* (Banks), a pest of several winter-growing legumes. *Louisiana State University Agricultural Experiment Station Technical Bulletin* No. 493, 1–25.

Smith, D. and Papacek, D.F. (1985) Integrated pest management in Queensland citrus. *Queensland Agricultural Journal* 111, 249–259.

Smith, D. and Papacek, D.F. (1991) Studies on the predatory mite *Amblyseius victoriensis* (Acarina: Phytoseiidae) in citrus orchards in south-east Queensland: control of *Tegolophus australis* and *Phyllocoptruta oleivora* (Acarina: Eriophyidae), effect of pesticides, alternative host plants and augmentative release. *Experimental and Applied Acarology* 12, 195–217.

Smith, D., Smith, N.J. and Smith, K.M. (1998) Effect of abamectin on citrus rust mite *Phyllocoptruta oleivora* and brown citrus rust mite *Tegolophus australis* and the scale natural enemies *Aphytis lingnanensis* and *Chilocorus circumdatus* on oranges. *Plant Protection Quarterly* 13, 136–139.

Smith, F.F. (1934) Insect and mite pests of *Amaryllis* and their control. *American Amaryllis Yearbook* 1, 94–96.

Smith, F.F. (1939a) Control of cyclamen and broad mite on gerbera. *US Department of Agriculture Circular* No. 516, 1–14.

Smith, F.F., Henneberry, T.J. and Boswell, A.L. (1963) The pesticide tolerance of *Typhlodromus fallacis* (Garman) and *Phytoseiulus persimilis* A.H. with some observations on the predator efficiency of *P. persimilis*. *Journal of Economic Entomology* 56, 274–278.

Smith, G.L. (1942) *California Cotton Insects*. Bulletin 660, Agricultural Experiment Station, College of Agriculture, University of California, Berkeley, California.

Smith, I.M. (1977) Two new species of *Trisetacus* (Prostigmata: Eriophyoidea) associated with Cupressaceae. *Canadian Entomologist* 109, 843–847.

Smith, I.M. (1978) Two new species of *Trisetacus* (Prostigmata, Eriophyoidea) from berries of *Juniperus* in North America. *Canadian Entomologist* 110, 1157–1160.

Smith, I.M. (1979) Description of new species of *Trisetacus* (Prostigmata, Eriophyoidea) from Pinaceae in eastern North America. *Canadian Entomologist* 111, 5–129.

Smith, J.C. and Newsom, L.D. (1970) The biology of *Amblyseius fallacis* (Acarina: Phytoseiidae) at various temperature and photoperiod regimes. *Annals of the Entomological Society of America* 63, 460–462.

Smith, J.W. and Furr, R.E. (1975) Spider mites and some natural control agents found in cotton in the Delta area of Mississippi. *Environmental Entomology* 4, 559–560.

Smith, K.M. and Cressman, A.W. (1962) Birefringent crystals in virus diseased citrus red mites. *Journal of Insect Pathology* (now *Journal of Invertebrate Pathology*) 4, 229–236.

Smith, K.M., Hills, G.J., Munger, F. and Gilmore, J.E. (1959) A suspected virus disease of the citrus red mite *Panonychus citri* (McG.). *Nature* 184, 70.

Smith, L., de Lillo, E. and Amrine, J.W. Jr (2010) Effectiveness of eriophyid mites for biological control of weedy plants and challenges for future research. *Experimental and Applied Acarology* 51, 115–149.

Smith, L.M. (1950) Mite populations on grapes. *California Agriculture* 4(3),13–14.

Smith, L.M. and Goldsmith, E.V. (1936) The cyclamen mite, *Tarsonemus pallidus*, and its control on field strawberries. *Hilgardia* 10(3), 53–94.

Smith, L.M. and Schuster, R.O. (1963) The nature and extent of *Eriophyes vitis* injury to *Vitis vinifera* L. *Acarologia* 5, 530–539.

Smith, L.M. and Stafford, E.M. (1948) The bud mite and the erineum mites of grapes. *Hilgardia* 18(7), 317–334.

Smith, L.R. (1980) Development of extension demonstration work and scouting techniques for citrus rust mites. *Journal of the Rio Grande Valley Horticultural Society* 34, 67–69.

Smith, R.H. (1939b) Observations on the ilicis mite, *Paratetranychus ilicis*. *Bulletin of the California Department of Agriculture* 28(5), 412–414.

Smith, W.A. (1955) The tomato mite. *Queensland Agricultural Journal* 81, 90–91.

Smitha, M.S. and Giraddi, R.S. (2003) Increased efficacy of dicofol in combination with different plant oils against yellow mite, *Polyphagotarsonemus latus* (Banks) on chilli. *Pest Management in Horticultural Ecosystems* 9, 67–70.

Smitha, M.S and Giraddi, R.S. (2006a) Safety of pesticidal sprays to natural enemies in chilli (*Capsicum annuum* L.). *Journal of Biological Control (Bangalore)* 20, 7–12.

Smitha, M.S. and Giraddi, R.S. (2006b) Efficacy of new acaricides, botanicals and bioagents against yellow mite *Polyphagotarsonemus latus* (Banks) on chilli. *Indian Journal of Entomology* 68, 31–35.

Smolarz, S. and Pala, E. (1982) Termiczna metoda niszczenia wielkopakowca porzeczkowego w pakach sadzonek porzeczki czarnej (Thermic method of gall mite control in buds of black currants). *Prace Instytutu Sadownictwa I Kwiaciarstwa, Seria A, Prace Doświadczalne z Zakresu Sadownictwa* 23, 131–135.

Snare, L. and Knihinicki, D. (2000) Big bud mite. An issue for Australian Growers? *Australian Nutgrower* 14(1), 17–19.

Snell, T.W. (1978) Fecundity, developmental time, and population growth rate. *Oecologia* 32, 119–125.

Snetsinger, R. (1956) Biology of *Bdella depressa*, a predaceous mite. *Journal of Economic Entomology* 49, 745–746.

Sobhian, R. and Andres, L.A. (1978) The response of the skeletonweed gall midge, *Cystiphora schmidti* (Diptera: Cecidomyiidae), and gall mite, *Aceria chondrillae* (Eriophyidae) to North American strains of rush skeletonweed (*Chondrilla juncea*). *Environmental Entomology* 7, 506–508.

Sogawa, K. (1977) Occurrence of the rice tarsonemid mite at IRRI. *International Rice Research Newsletter* 2(5), 17.

Sohrabi, F. and Shishenbor, P. (2007) Functional and numerical responses of *Stethorus gilvifrons* Mulsant feeding on strawberry spider mite, *Tetranychus turkestani* Ugarov and Nikolski. *Pakistan Journal of Biological Sciences* 10, 4563–4566.

Sohrabi, F. and Shishehbor, P. (2008) Effects of host plant and temperature growth and reproduction on strawberry spider mite *Tetranychus turkestani* Ugarov & Nikolski (Acari: Tetranychidae). *Systematic and Applied Acarology* 13, 26–32.

Soika, G.M. and Kielkiewicz, M. (2004) Occurrence of *Phytoptus tetratrichus* (Nalepa) (Acari: Eriophyoidea) and differences in morphology of leaf galls on two linden species. *Phytophaga* 14, 615–622.

Soika, G.M. and Łabanowski, G.S. (1998a) New species of eriophyid mites (Eriophyoidea) in ornamental nurseries. *Progress in Plant Protection/Postępy w Ochronie Roślin* 38, 416–418 (in Polish).

Soika, G.M. and Łabanowski, G.S. (1998b) Gall mites – pests of London plane in Poland. *Ochrona Roślin* 42(2), 6–7 (in Polish).

Soika, G.M. and Łabanowski, G.S. (1999a) Types of damages caused by eriophyoid mites (Acari: Eriophyoidea) on ornamental trees and shrubs. In: *Materiały XXVI Sympozjum Akarologicznego "Akarologia Polska u Progu Nowego Tysiąclecia", Kazimierz Dolny, 24–26 października 1999 r.* Wydawnictwo SGGW (Szkoła Główna Gospodarstwa Wiejskiego w Warszawie), Warszaw, pp. 371–380.

Soika, G.M. and Łabanowski, G.S. (1999b) Szpeciele – sprawcy wyrośli i zniekształceń na roślinach ozdobnych. *Progress in Plant Protectection/Postępy w Ochronie Roślin* 39, 537–540 (in Polish).

Soika, G.M. and Łabanowski, G.S. (2000) Eriophyoid mites (Acari: Eriophyoidea) on ornamental plants in Poland. Fabaceae family. Descriptions of two new species. *Bulletin of the Polish Academy of Sciences, Biological Sciences* 47, 157–167.

Soika, G.M. and Łabanowski, G.S. (2002) Eriophyoid mites (Acari: Eriophyoidea) living on *Fagus* spp. (Fagaceae) in Poland. *Bulletin of the Polish Academy of Sciences, Biological Sciences* 50, 71–83.

Soika, G.M. and Łabanowski, G.S. (2004) Novel compounds against spider mites on ornamental plants grown in the field and under cover. *Progress in Plant Protection/Postępy w Ochronie Roślin* 44, 387–392.

Soika, G.M. and Łabanowski, G.S. (2006) New records of eriophyoid mites (Acari: Eriophyoidea) on ornamental plants in Poland. In: Gabryś, G. and Ignatowicz, S. (eds) *Proceedings of the Symposium "Advances in Polish Acarology"*. Wydawnictwo SGGW (Szkoła Główna Gospodarstwa Wiejskiego w Warszawie), Warszaw, pp. 385–398.

Solano, D.A., Alvarez-Herrera, J.G. and Rodriguez, J.A. (2008) Spatial distribution of *Brevipalpus phoenicis* vector of citrus leprosis virus in Valencia orange (*Citrus sinensis*) crop in Yopal, Casanare (Colombia). *Agronomia Colombiana* 26, 399–410.

Soliman, Z.R. and Abou-Awad, B.A. (1978) Two new species of the genus *Phyllocoptes* in the A.R.E. (Acarina: Eriophyoidea: Eriophyidae). *Acarologia* 20, 104–108.

Soliman, Z.R. and Mahfood, S.A. (1975) Phytophagous and predaceous mites of People's Republic of Southern Yemen. *Bulletin of the Zoological Society of Egypt* 27, 78–84.

Solomon, M.E. (1937) Experiments on the effects of temperature and humidity on the survival of *Halotydeus destructor* (Dugès). *South Australian Journal of Agriculture* 38, 353–367.

Solomon, M.F. (1969) Experiments on predator–prey interactions of storage mites. *Acarologia* 11, 484–503.

Solomon, M.G. (1975) The colonization of an apple orchard by predators of the fruit tree red spider mite. *Annals of Applied Biology* 80, 119–122.

Solomon, M.G. (1981) Windbreaks as a source of orchard pests and predators. In: Thresh, J.M. (ed.) *Pests, Pathogens and Vegetation*. Pitman, London, pp. 273–283.

Solow, A.R. and Sherman, K. (1997) Testing for stability in a predator–prey system. *Ecology* 78, 2624–2627.

Somchoudhury, A.K. and Sarkar, P.K. (1987) Observations on natural enemies found in association with coconut mite, *Raoiella indica* Hirst. *Bulletin of Entomology (New Delhi)* 28, 104–107.

Somchoudhury, A.K., Chowdhury, A.K. and Mukherjee, A.B. (1985) Mite vectors and their trapping. In: *Use of Traps of Pest/Vector Research and Control. Proceedings of the National Seminar held at Bidhan Chandra Krishi Viswavidyalaya, Mohanpur, West Bengal on March 10–11, 1984 in Collaboration with Department of Agriculture, Government of West Bengal, Department of Science and Technology, Government of India, Indian National Science Academy, New Delhi.* Bidhan Chandra Krishi Viswavidyalaya, Kalyani, India, pp. 41–50.

Somchoudhury, A.K., Jana, S.K., Nair, N., Samanta, A. and Sarkar, P.K. (2008) Studies on the effect of different abiotic factors and bio-efficacy of different chemicals on yellow mite, *Polyphagotarsonemus latus* (Banks) infesting jute. *Journal of Entomological Research* 32, 105–107.

Sømme, L. (1965) Changes in sorbitol content and supercooling points in overwintering eggs of the European red mite (*Panonychus ulmi* (Koch)). *Canadian Journal of Zoology* 43, 881–884.

Sømme, L. (1966) Mortality of winter eggs of *Metatetranychus ulmi* (Koch) during the winter of 1965/66. *Norsk Entomologisk Tidskrift* 13, 420–423.

Sonenshine, D.E. (1985) Pheromones and other semiochemicals of the Acari. *Annual Review of Entomology* 30, 1–28.

Sonenshine, D.E. and Gregson, J.D. (1970) A contribution to the internal anatomy and histology of the bat tick *Ornithodoros kelleyi* Cooley and Kohls, 1941. I, The alimentary system, with notes on the food channel in *Ornithodoros denmarki* Kohls, Sonenshine and Clifford. *Journal of Medical Entomology* 7, 46–64.

Song, J.H., Lee, C.H., Kang, S.H., Kim, D.H., Kang, S.Y. and Riu, K.Z. (2001) Binomial sampling plans for the citrus red mite, *Panonychus citri* (Acari: Tetranychidae) on satsuma mandarin groves in Jeju. *Korean Journal of Applied Entomology* 40, 197–202 (in Korean).

Song, J.H., Kim, S.N. and Riu, K.Z. (2003) Spatial dispersion and sampling of adults of citrus red mite, *Panonychus citri* (McGregor) (Acari: Tetranychidae) in citrus orchard in autumn season. *Korean Journal of Applied Entomology* 42, 29–34 (in Korean).

Sorensen, J.T., Kinn, D.N., Doutt, R.L. and Cate, J.R. (1976) Biology of the mite *Anystis agilis* (Acari: Anystidae): a California vineyard predator. *Annals of the Entomological Society of America* 69, 905–910.

Sorensen, J.T., Kinn, D.N. and Doutt, R.L. (1983) Biological observations on *Bdella longicornis*: a predatory mite in California vineyards (Acari: Bdellidae). *Entomography* 2, 297–305.

Soroker, V., Nelson, D.R., Bahar, O., Reneh, S., Yablonski, S. and Palevsky, E. (2003) Whitefly wax as a cue for phoresy in the broad mite, *Polyphagotarsonemus latus* (Acari: Tarsonemidae). *Chemoecology* 13, 163–168.

Soroosh, M. (2003) Importance of eriophyid mites on olives. In: *1st European Meeting of the IOBC/WPRS Study Group, "Integrated Control in Olives", 20–31 May 2003, Chania (Crete), Greece*, p. 73.

Soto, A., Venzon, M., Oliveira, R.M., Oliveira, H.G. and Pallini, A. (2010) Alternative control of *Tetranychus evansi* Baker & Pritchard (Acari: Tetranychidae) on tomato plants grown in greenhouses. *Neotropical Entomology* 39, 638–44.

Sousa, J.M. de, Gondim, M.G.C. Jr and Lofego, A.C. (2010) Biology of *Tetranychus mexicanus* (McGregor) (Acari: Tetranychidae) on three species of Annonaceae. *Neotropical Entomology* 39, 319–323.

Southcott, R.V. (1961) Studies on the systematics and biology of the Erythraeoidea (Acarina), with a critical revision of the genera and subfamilies. *Australian Journal of Zoology* 9, 367–610.

Southcott, R.V. (1976) Arachnidism and allied syndromes in the Australian region. *Records of. Adelaide Children's Hospital* 1, 97–186.

Spears, A.T. and Yothers, W.W. (1924) Is there an entomogenous fungus attacking the citrus rust mite in Florida?, *Science* 60, 41–42.

Specht, H.B. (1963) *Oligonychus bicolor* (Banks) (Acarina: Tetranychidae) in Nova Scotia. *Canadian Entomologist* 95, 1021–1022.

Speight, M.R., Hunter, D.M. and Watt, A.D. (1999) *The Ecology of Insects: Concepts and Applications.* Blackwell Scientific, Oxford, UK.

Spencer, H. and Selhime, A.G. (1954) New miticides for the citrus red mite (Purple mite). *Proceedings of the Florida State Horticultural Society* 67, 42–44.

Spies, J.J. (1984) A cytotaxonomic study of *Lantana camara* (Verbenaceae) from South Africa. *South African Journal of Botany* 3, 231-250.

Spieser, F., Graf. B., Walther, P. and Noesberger, J. (1998) Impact of apple rust mite (Acari: Eriophyidae) feeding on apple leaf gas exchange and leaf color associated with changes in leaf tissue. *Environmental Entomology* 27, 1149–1156.

Spooner-Hart, R., Tesoriero, L. and Hall, B. (2007) *Field Guide to Olive Pests, Diseases and Disorders in Australia.* Publication No. 07/153, Rural Industries Research and Development Corporation, Australian Government, Barton, Australian Capital Territory, Australia.

Sreenivas, A.G., Ramanujam, B.R., Prashanth-Mohanraj, Nargund, V.B. and Shivaramu, K. (2005) Efficacy of entomopathogenic fungi against red spider mite, *Tetranychus neocalodonicus* Zacher (Acari: Tetranychidae). *Karnataka Journal of Agricultural Sciences* 18, 966–969.

Sreerama Kumar, P. (2010) *Hirsutella thompsonii* as a mycoacaricide for *Aceria guerreronis* on coconut in India: research, development, and other aspects. In: Sabelis, M.W. and Bruin, J. (eds) *Trends in Acarology: Proceedings of the 12th International Congress, 21–26 August 2006, Amsterdam.* Springer, Dordrecht, the Netherlands, pp. 441–444.

Sridhar, Y., Reddy, K.D. and Singh, B.U. (2010) Orientation, settling, and ovipositional responses of spider mite, *Oligonychus indicus* (Hirst) in sorghum. *Indian Journal of Entomology* 72, 1–6.

Srimohanapriya, V., Kandibane, M. and Natarajan, L. (2010) Bioefficacy of insecticides against rice leaf mite, *Oligonychus oryzae* Hirst. *Madras Agricultural Journal* 97, 81–83.

Srinivas, T. and Reddy, M.V. (1995) Pigeonpea lines resistant to old and new strains of sterility mosaic disease. *International Chickpea and Pigeonpea Newsletter* 2, 67–70.

Srinivasa, N. and Sugeetha, G. (1999) Bioeffectiveness of certain botanicals and synthetic pesticides against okra spider mite, *Tetranychus macfarlanei*. *Journal of Acarology (Bangalore)* 15(142), 1–5.

Srinivasa, N., Mallik, B. and Gowda, C.C. (2004) Aceria cajani *(Acari: Eriophyidae) Transmitted Pigeonpea Sterility Mosaic Disease.* AINP (All India Network Project) on Agricultural Acarology, University of Agricultural Sciences, Bangalore, India.

Srinivasa, N., Gowda, C.C., Mallik, B. and Raghavendra, P. (2012) New record of *Tetranychus truncatus* Ehara (Acari: Tetranychidae) as a potential pest from Karnataka. *Indian Journal of Entomology* 74, 379–383.

Srinivasulu, P., Naidu, V.G. and Rao, N.V. (2002a) Biology and bionomics of chilli mite, *Polyphagotarsonemus latus* Banks on chilli. *Journal of Applied Zoological Research* 13, 19–21.

Srinivasulu, P., Naidu, V.G. and Rao, N.V. (2002b) Evaluation of different pesticides for the control of yellow mite, *Polyphagotarsonemus latus* (Banks) on chilli. *Journal of Applied Zoological Research* 13, 71–72.

Srinivasulu, P., Naidu, V.G., Rao, N.V. and Babu, K.H. (2002c) Seasonal occurrence of chilli mite, *Polyphagotarsonemus latus* (Banks) with reference to biotic and abiotic factors. *Journal of Applied Zoological Research* 13, 142–144.

Srivastava, B.K. and Mathur, L.M.L. (1962) Bionomics and control of castor mite. *Indian Journal of Entomology* 24, 229–235.

Srivastava, R.P. (1974) Comparative efficacy of various insecticides against mango bud mite, *Aceria mangiferae* Sayed. *Indian Journal of Entomology* 36, 74–75.

Stables, L.M. (1980) The effectiveness of some recently developed pesticides against stored-product mites. *Journal of Stored Products Research* 16, 143–146.

Stafford, E.M. and Kido, M. (1969) Newer insecticides for the control of grape insect and spider mite pests. *California Agriculture* 23(4), 6–8.

Stamenkovic, S., Pesic, M. and Milenkovic, S. (1994) The possibility of using endosulfan (Thiocid E-35) in the control of *Vasates fockeui* (Nal. *et* Trt) in plum nurseries. *Jugoslovensko Vocarstvo* 28(3/4), 55–60 (in Serbo-Croatian).

Stamenkovic, S., Pesic, M. and Milenkovic, S. (1995) The occurrence of plum rust mite (*Vasates fockeui* (Nal. *et* Trt.)) in the plum nursery and control measures. *Jugoslovensko Vocarstvo* 29(1/2), 103–108 (in Serbo-Croatian).

Stamenkovic, S., Milenkovic, S., Pesic, M. and Mitrovic, M. (1997) Population dynamics, harmfulness and control of *Phytoptus avellanae* (Nalepa) in Western Serbia. *Acta Horticulturae* 445, 521-526.

Stanley, D., Glenn, D.M. and Gary, J. (1998) Particle films – a new kind of plant protectant. *AgResearch (Agricultural Research Magazine, Agricultural Research Service, US Department of Agriculture)* 46(11), 16–19.

Stanley, M.S. (2007) An analysis of the history of marine animal diversity. *Paleobiology* 33(4), 1–55.

Staples, R. and Allington, W.B. (1956) *Streak Mosaic of Wheat in Nebraska and its Control.* Research Bulletin 178, University of Nebraska College of Agriculture Agricultural Experiment Station, Lincoln, Nebraska.

Stavrinides, M.C. and Mills, N.J. (2009) Demographic effects of pesticides on biological control of Pacific spider mite (*Tetranychus pacificus*) by the western predatory mite (*Galendromus occidentalis*). *Biological Control* 48, 267–273.

Stavrinides, M.C. and Mills, N.J. (2011) Influence of temperature on the reproductive and demographic parameters of two spider mite pests of vineyards and their natural predator. *Biocontrol* 56, 315–325.

Stavrinides, M.C., Lara, J.R. and Mills, N.J. (2010a) Comparative influence of temperature on development and biological control of two common vineyard pests (Acari: Tetranychidae). *Biological Control* 55, 126–131.

Stavrinides, M.C., Van Nieuwenhuyse, P., Van Leeuwen, T. and Mills, N.J. (2010b) Development of acaricide resistance in Pacific spider mite (*Tetranychus pacificus*) from California vineyards. *Experimental and Applied Acarology* 50, 243–54.

Stavrinides, M.C., Daane, K. and Mills, N.J. (2012) Inconsistent effects of plant water stress on herbivores: The role of leaf temperature. In: Schausberger, P., Walzer, A., Strodel, M.A. and Peneder, S. (eds) *7th Symposium of the European Association of Acarologists (EURAAC), July 9–13 2012, Vienna, Austria: Acari in a Changing World. Program, Abstracts, Participants.* University of Natural Resources and Life Sciences, Vienna, Austria, p. 66. Available at: http://euraac.boku.ac.at/SympVienna/files/ProgramBook.pdf (accessed 14 January 2015).

Stein, C.P. and Daólio, N. (2012) Biologia de *Tetranychus mexicanus* (McGregor) (Acari: Tetranychidae) em folhas de pupunha (*Bactris gasipaes* Kunth). *Revista Bioikos* 26(1), 23–28.

Steiner, H. (1980) The role of anthocorids and mirids in controlling fruit tree red spider mite in orchards in Baden-Würtemberg. *IOBC-WPRS Bulletin* 3(6), 19.

Steiner, M. (1993) *Quality Control Requirements for Pest Biological Control Agents.* Alberta Government Publication AECV93-R6, Alberta Environmental Centre, Vegrevile, Alberta, Canada.

Steiner, M. and Goodwin, S. (2005) Managing tomato russet mite, *Aculops lycopersici* (Massee) (Acari: Eriophyiidae) in greenhouse tomato crops. *IOBC-WPRS Bulletin* 28(1), 245–248.

Steinhaus, E.A. (1959) Possible virus disease in European red mite. *Journal of Insect Pathology* (now *Journal of Invertebrate Pathology*) 1, 435–437.

Steinhaus, E.A. and Marsh, G.A. (1962) Report of diagnosis of diseased insects 1951–1961. *Hilgardia* 33(9), 349–490.

Stenger, D.C., Hein, G.L. and French, R. (2006) Nested deletion analysis of wheat streak mosaic virus HC-Pro: mapping of domains affecting polyprotein processing and eriophyid mite transmission. *Virology* 350, 465–474.

Stenseth, C. (1973) Bekjempelse av bringebaerbladmidd (*Eriophyes gracilis*). *Gartneryrket* 63(7), 112–114 (in Norwegian).

Stenseth, C. (1979) Effect of temperature and humidity on the development of *Phytoseiulus persimilis* and its ability to regulate populations of *Tetranychus urticae* (Acarina: Phytoseiidae, Tetranychidae). *Entomophaga* 24, 311–317.

Stenseth, C. and Nordby, A. (1976) Damage, and control of the strawberry mite *Steneotarsonemus pallidus* (Acarina: Tarsonemidae), on strawberries. *Journal of Horticultural Science* 51, 49–54.

Stephan, D., Moeller, I., Skoracka, A., Ehrig, F. and Maiss, E. (2008) Eriophyid mite transmission and host range of a brome streak mosaic virus isolate derived from a full-length cDNA clone. *Archives of Virology* 153, 181–185.

Sterk, G. (1994) Control of the fruit tree red spider mite (*Panonychus ulmi*) and the apple rust mite (*A. schlechtendali*) in apple orchards. In: *Brighton Crop Protection Conference, Pests and Diseases – 1994. Proceedings of an International Conference Organised by the British Crop Protection Council, Held at the Brighton Centre and the Brighton Metropole Hotel, Brighton, England, 21–24 November 1994, Volume 2.* British Crop Production Council, Farnham (now Alton), UK, pp. 559–568.

Sterk, G. and Highwood, D.P. (1992) Implementation of IRAC anti-resistance guidelines with IPM programmes for Belgian apple and pear orchards. *Proceedings of the Brighton Crop Protection Conference: Pests and Diseases – 1992, Brighton, November 23–26, 1992.* British Crop Production Council, Croydon (now Alton), pp. 517–526.

Sterk, G., Hassan, S.A., Baillod, M., Bakker, F., Bigler, F., Blümel, S., Bogenschütz, H., Boller, E., Bromand, B., Brun, J., *et al.* (1999) Results of seventh joint pesticide testing programme carried out by the IOBC/WPRS-Working Group "Pesticides and Beneficial Organisms". *BioControl* 44, 99–117.

Sterling, W. (1975) Sequential sampling of cotton insect populations. In: *Proceedings of Beltwide Cotton Production Research Conference, 6–8 January 1975, New Orleans, Louisiana*. National Cotton Council of America, Memphis, Tennessee, pp. 133–135.

Stern, V.M., Smith, R.F., Van Den Bosch, R. and Hagen, K.S. (1959) The integration of chemical and biological control of the spotted alfalfa aphid: the integrated control concept. *Hilgardia* 29(2), 81–101.

Sternlicht, M. (1966) Trials in the control of the citrus bud mite, *Aceria sheldoni* (Ewing) in Israel. *Israel Journal of Agricultural Research* 16, 115–124.

Sternlicht, M. (1967) A method of rearing the citrus bud mite (*Aceria sheldoni* Ewing). *Israel Journal of Agricultural Research* 17, 57–59.

Sternlicht, M. (1969a) A study of fluctuations in the citrus bud mite populations. *Annales de Zoologie, Ecologie Animale* 1, 127–147.

Sternlicht, M. (1969b) Effect of different wave lenghts of light on the behavior of an eriophyid bud mite, *Aceria sheldoni*. *Entomologia Experimentalis et Applicata* 12, 377–382.

Sternlicht, M. (1970) Contribution to the biology of the citrus bud mite *Aceria sheldoni* (Ewing) (Acarina: Eriophyidae). *Annals of Applied Biology* 65, 221–230.

Sternlicht, M. and Goldenberg, S. (1971) Fertilisation, sex ratio and postembryonic stages of the citrus bud mites *Aceria sheldoni* (Ewing) (Acarina, Eriophyidae). *Bulletin of Entomological Research* 60, 391–397.

Sternlicht, M. and Goldenberg, S. (1976) Mango eriophyid mites in relation to inflorescence. *Phytoparasitica* 4, 45–50.

Sternlicht, M. and Griffiths, D.A. (1974) The emission and form of spermatophores and the fine structure of adult *Eriophyes sheldoni* (Ewing) (Acarina: Eriophyoidae). *Bulletin of Entomological Research* 63, 561–565.

Sternlicht, M., Goldenberg, S. and Chen, M. (1973) Development of the plum gall and trials to control its mite *Acalitus phloecoptes* (Eriophyidae: Acarina). *Annales de Zoologie, Ecologie Animale* 5, 365–377.

Steven, D. (2003) Novel chemicals to control six-spotted mite, *Eotetranychus sexmaculatus*. In: *New Zealand Avocado Growers' Association Annual Research Report – 2003, Volume 3.* Tauranga, New Zealand, pp. 1–8. Available at: http://www.avocadosource.com/Journals/NZAGA/NZAGA_2003/NZAGA_2003_01.pdf (accessed 21 May 2015).

Steven, D. (2004) Control of six-spotted mite, *Eotetranychus sexmaculatus*. In: *New Zealand Avocado Growers' Association Annual Research Report – 2004, Volume 4.* Tauranga, New Zealand, pp. 13–24. Available at: http://www.avocadosource.com/Journals/NZAGA/NZAGA_2004/NZAGA_2004_13.pdf (accessed 25 March 2015).

Stevens, P. (2001) *Fact sheet: Six Spotted Mite*. Prepared for Avocado Industry Council, Tauranga, New Zealand.

Stewart, A.J.A. and Wright, A.F. (1995) A new inexpensive suction apparatus for sampling arthropods in grassland. *Ecological Entomology* 20, 98–102.

Stewart, K.E. and Peterson, L.O.T. (1960) *Control of the Spruce Spider Mite*. Publication 1078, Research Branch, Canadian Department of Agriculture, Ottawa.

Stiefel, V.L., Margolies, D.C. and Bramel-Cox, R.J. (1992) Leaf temperature effects of resistance to the Banks grass mite (Acari: Tetranychidae) on drought-resistant grain sorghum. *Journal of Economic Entomology* 85, 2170–2184.

Stiling, P. and Cornelissen, T. (2005) What makes a successful biocontrol agent? A meta-analysis of biological control agent performance. *Biological Control* 34, 236–246.

Stofberg, F.J. (1959) Control of mites on citrus. *Farming in South Africa* 35(9), 33–35.

Stone, B.F. (1968) A formula for determining degree of dominance in cases of monofactorial inheritance of resistance to chemicals. *Bulletin of the World Health Organization* 38, 325–326.

Stone, C. (1981) An electrostatic method for extracting eriophyid mites from leaf galls. *Journal of the Australian Entomological Society* 20, 235–236.

Stone, C. (1982) The taxonomy and biology of *Tetranychus lintearius* Dufour. MSc. thesis, University of London.

Stone, C. (1986) An investigation into the morphology and biology of *Tetranychus lintearius* Dufour (Acari: Tetranychidae). *Experimental and Applied Acarology* 2, 173–186.

Storms, J.J.H. (1969) Observations on the relationships between mineral nutrition of apple rootstocks in gravel culture and the reproduction rate of *Tetranychus urticae*. *Entomologia Experimentalis et Applicata* 12, 297–311.

Storms, J.J.H. (1971) Some physiological effects of spider mite infestations on bean plants. *Netherlands Journal of Plant Pathology* (now *European Journal of Plant Pathology*) 77, 154–167.

Stout, M.J., Workman, J. and Duffey, S.S. (1994) Differential induction of tomato foliar proteins by arthropod herbivores. *Journal of Chemical Ecology* 20, 2575–2594.

Stout, M.J., Workman, K.V. and Duffey, S.S. (1996) Identity, spatial distribution, and variability of induced chemical responses in tomato plants. *Entomologia Experimentalis et Applicata* 79, 255–271.

Strong, D.R. Jr, Lawton, J.H. and Southwood, T.R.E. (1984) *Insects on Plants: Community Patterns and Mechanisms.* Blackwell Scientific Publications, Oxford, UK.

Strunkova, Z.I. (1969) New tetranychid mites from Central Asia (Acarina, Tetranychoidea). *Doklady Akademii Nauk Tadzhikskoj SSR* 12, 67–71 (in Russian).

Strunkova, Z.I. (1971) Phytophagous mites of “Tigrovaya Balka” nursery (Tadzhikistan). *Izvestiya Akademii Nauk Tadzhikskoy SSR (Otdelenie Biologicheskikh Nauk)* 4, 107–109 (in Russian).

Strunkova, Z.I. (1972) *Tetranychid Mites – Pests of Fruit Crops in the Hissar Valley of Tadzhikistan.* Institut Zoologii i Parazitologii Imeni E.N. Pavlovskogo, Akademii Nauk Tadzhikskoy SSR, Dushanbe, Tajikistan (in Russian).

Stubbings, W.A.K. (1947) *Studies in Verband met die Ekologie en Bestryding van die Peerknopmyt (*Eriophyes pyri *Pag.).* Science Bulletin No. 270/Fruit Research Technical Series No. 15, Department of Agriculture, South Africa. Government Pronter, Pretoria (in Afrikaans).

Stubbs, L.L. and Meagher, J.W. (1965) A virosis-like proliferation (witches’broom) of lucerne (*Medicago sativa* L.) caused by an eriophyid mite, *Aceria medicaginis* Keifer. *Australian Journal of Agricultural Research* 16, 125–129.

Stumpf, N. and Nauen, R. (2002) Biochemical markers linked to abamectin resistance in *Tetranychus urticae* (Acari-Tetranychidae). *Pesticide Biochemistry and Physiology* 72, 111–121.

Styer, W.E. and Nault, L.R. (1996) Corn and grain plants. In: Lindquist, E.E., Sabelis, M.W. and Bruin, J. (eds) *Eriophyoid Mites: Their Biology, Natural Enemies and Control.* World Crop Pests, Volume 6. Elsevier Science, Amsterdam, pp. 611–618.

Su, T.-Z., Beng, G.Y. and Chen, X.-D. (1979) Notes on the genus *Siteroptes* in China (Acarina: Pyemotidae). *Acta Zootaxonomica Sinica* 4, 162–169 (in Chinese).

Su, T.-Z., Chen, X.-D. and Beng, G.-Y. (1981) Study of symbiotic relationships between mites and fungi in cereal ear rot. *Acta Phytopathologica Sinica* 11, 19–24 (in Chinese).

Su, U. (1931) Plant disease in Burma. *International Bulletin of Plant Protection* 5, 141–142.

Suárez, A. (1991a) Distribución y evaluación de danos del ácaro *Eriophyes guerreronis* en plantaciones de cocotero de la regiín de Baracoa (Cuba). *Protección de Plantas* 1(1), 75–81.

Suárez, A. (1991b) Mites detected in cocoa cultivation (*Theobroma cacao*) in the province of Guantanamo. *Protección de Plantas* 1(1), 59–65.

Suárez, A. (2004) Catálogo de ácaros de la provincia de Guantánamo. *Fitosanidad* 8(1), 23–31.

Subaharan, K. and Regupathy, A. (2006) Effect of neem formulations on the management of purple mite, *Calacarus carinatus* L. infesting tea. *Journal of Plantation Crops* 34, 435–38.

Subías, L.S. and Arillo, A. (2002) Oribatid mites fossils from the Upper Devonian of South Mountain, New York and the Lower Carboniferous of Country Antrim, Northern Ireland (Acariform, Oribatida). *Estudios del Museo de Ciencias Naturales de Álava* 17, 93–106.

Subirats, F.J. and Self, R.L. (1972) Inciting bud drop and flower deterioration of camellias by a camellia bud mite in Alabama. *Journal of Economic Entomology* 65, 306–307.

Sudharma, K. and Nair, G.M. (1999) Occurrence of phytophagous mites and natural enemies on vegetables in Kerala. *Insect Environment* 5, 15–16.

Sudoi, V. (1990) Remedial application of nitrogen combined with acaricides for control of purple mites *Calacarus carinatus* Green attacking tea. *Tea* 11(2), 101–103.

Sudoi, V. (1992) In search for factors responsible for resistance in tea clones to red spider mite *Oligonychus coffeae* Nietner (Acari – Tetranychidae) and possibility of their future use in screening programmes. *Tea* 13(1), 62–67.

Sudoi, V. (2004) Strategies for control of red crevice mites (*Brevipalpus phoenicis* Geijskes) on tea. *Tea* 25(1), 20–23.

Sudoi, V., Khaemba, B.M. and Wanjala, F.M.E. (1994) Screening of Kenyan tea clones for their resistance to *Brevipalpus phoenicis* Geijskes (Acari: Tenuipalpidae) attack. *Tea* 15(2), 105–109.

Sudoi, V., Wanyoko, J.K., Owuor, P.O. and Lang’at, J.K. (2001a) Prospects for NPKS 25:5:5:5 fertilizer as a component of *Brevipalpus phoenicis* Geiskes (Acari: Tenuipalpidae) mite pest management. *Tea* 22(1), 13–19.

Sudoi, V., Khaemba, B.M. and Wanjala, F.M.E. (2001b) Nitrogen fertilization and yield losses of tea to red crevice mite (*Brevipalpus phoenicis* Geijskes) in the Eastern Highlands of Kenya. *International Journal of Pest Management* 47, 207–210.

Sugiura, N., Toda, S., Tuchida, M. and Sakaki, H. (2008) Chemical control and management of the pink citrus rust mite, *Aculops pelekassi* (Keifer) strain with a decreased susceptibility for mancozeb. *Research Bulletin of the Kumamoto Prefectural Agricultural Research Center* 15, 120–130 (in Japanese).

Sujatha, A., Kumar, D.A. and Srinivasulu, B. (2003) Efficacy of certain pesticides against coconut eriophyid mite, *Aceria guerreronis* (K.) through root feeding. *Journal of Applied Zoological Research* 14, 51–53.

Sukumalanand, P., Buranapanichpan, S., Tayutivutikul, J., Tepsuwan, C., Ratanabhumma, S., Kasetsoontorn, R. and Chanbang, Y. (1999) African Red Mite, *Eutetranychus africanus* (Tucker), on tangerines in Chiang Mai. *Journal of Agriculture, Faculty of Agriculture, Chiang Mai University* 15, 39–46.

Sulaimanov, B., Rashidov, M.I. and Kimsanbaev, K.K. (2000) Exotoxin against the rust mite on solanaceous crops. *Zashchita i Karantin Rastenĭ* 3, 46 (in Russian).

Sumangala, K. and Haq, M.A. (1995) Nutritional diversity of Acari infesting *Eichhornia crassipes*. *Journal of Ecobiology* 7, 289–297.

Sumangala, K. and Haq, M.A. (2005) Diurnal periodicity and dispersal of coconut mite, *Aceria guerreronis* Keifer. *Journal of Entomological Research* 29, 303–307.

Summers, F.M. (1950) Brown almond mites. *California Agriculture* 4(7), 6.

Summers, F.M. and Baker, G.A. (1952) A procedure for determining relative densities of brown almond mite populations on almond trees. *Hilgardia* 21(13), 369–382.

Summers, F.M. and Stocking, C.R. (1972) Some immediate effects on almond leaves of feeding by *Bryobia rubrioculus* (Scheuten). *Acarologia* 14, 170–178.

Summers, F.M., Gonzales-Rodriguez, R.H. and Witt, R.L. (1973) The mouthparts of *Bryobia rubrioculus* (Sch.) (Acarina: Tetranychidae). *Proceedings of the Entomological Society of Washington* 75, 96–111.

Sun, X.-G. and Yin, S.Y. (2002) Studies on the interrelationship among *Oligonychus ununguis* host plants and *Amblyseius finlandicus*. II. The effects of volatile on the correlation of host plants – *Oligonychus ununguis – Amblyseius finlandicus*. *Scientia Silvae Sinicae* 38(2), 73–77 (in Chinese).

Sun, X.-G., Zhou, C.-G., Liu, Y.-M., Liang, H. and Cui, Z.-P. (1996) Studies on bionomics and effective accumulated temperature of *Eotetranychus populi* (Koch) (Acariformes: Tetranychidae). *Acta Entomologica Sinica* 39, 166–172 (in Chinese).

Sun, X.-G., Zhou, C.-G., Liu, Y.-M., You, C.-X. and Yu, B. (1997a) A preliminary study on the bionomics of *Schizotetranychus bambusae* Reck. *Scientia Silvae Sinicae* 33(3), 274–278 (in Chinese).

Sun X.-G., Xu, C.-Q., Guo, H.-L. and Gai, Y.-P. (1997b) Infestation of *Eotetranychus populi* Koch in Taian of Shandong. *Plant Protection* 23(3), 14-16 (in Chinese).

Sun, X.-G., Xu, C.-Q., Zhou, C.-G., Yin, S.-Y. and Zhen, W.-Q. (2000) Performances and reproductive isolation of different populations of *Oligonychus ununguis* (Jacobi) on conifer and broadleaf trees. *Acta Entomologica Sinica* 43, 52–57 (in Chinese).

Sun, X.-G., Yin, S.-Y. and Li, B. (2002) Studies on the interrelationship among *Oligonychus ununguis* host plants and *Amblyseius finlandicus* I. The olfactory responses of *Oligonychus ununguis* to host plants and *Amblyseius finlandicus* to preys. *Scientia Silvae Sinicae* 38(1), 76–81 (in Chinese).

Sundaram, K.M.S. and Sloane, L. (1995) Effects of pure and formulated azadirachtin, a neem-based biopesticide, on the phytophagous spider mite, *Tetranychus urticae*. *Journal of Environmental Science and Health, Part B: Pesticides, Food Contaminants, and Agricultural Wastes* 30, 801–814.

Sunnucks, P. (2000) Efficient genetic markers for population biology. *Trends in Ecology and Evolution* 15, 199–203.

Supkoff, D.M., Joley, D.B. and Marois, J.J. (1988) Effect of introduced biological control organisms on the density of *Chondrilla juncea* in Californa. *Journal of Applied Ecology* 25, 1089–1095.

Sureya, M. (1933) Turkey: plant enemies observed in Anatolia during 1933. *International Bulletin of Plant Protection* 7, 247–248.

Susi, P. (2004) Black currant reversion virus, a mite-transmitted nepovirus. *Molecular Plant Pathology* 5, 167–173.

Suski, Z.W. (1972) Tarsonemid mites on apple trees in Poland. X. Laboratory studies on the biology of certain mite species of the family Tarsonemidae (Acarina, Heterostigmata). *Zeszyty Problemowe Postępów Nauk Rolniczych* 129, 111–137.

Suski, Z.W. (1973) A revision of *Siteroptes cerealium* (Kirchner) complex (Acarina, Heterostigmata, Pyemotidae). *Annales Zoologici* 30, 509–535.

Suski, Z.W. (1984) On the identity of pyemotid mites associated with the silver-top disease of grasses. In: Griffiths, D.A. and Bowman, C.E. (eds) *Acarology VI, Volume 1*. Ellis Horwood, Chichester, UK, pp. 174–179.

Suski, Z.W. and Badowska, T. (1975) Effect of the host plant nutrition on the population of the two spotted spider mite, *Tetranychus urticae* Koch (Acarina, Tetranychidae). *Ekologia Polska* (now *Polish Journal of Ecology*) 23, 185–209.

Suski, Z.W. and Naegele, J.A. (1963) Light response in the two-spotted spider mite. II: Behavior of the "sedentary" and "dispersal" phases. In: Naegele, J.A. (ed.) *Advances in Acarology, Volume 1*. Cornell University Press/Comstock Publishing Associates, Ithaca, New York, pp. 445–453.

Suski, Z.W. and Naegele, J.A. (1968*) Environmental Determinants of White Light Response in the Two-spotted Spider Mite,* Tetranychus urticae *K*. Bulletin No. 571, Agricultural Experiment Station, College of Agriculture, University of Massachusetts, Amherst, Massachusetts.

Svetov, V. G. (1985) The grapevine bud mite. *Zashchita Rasteniĭ* No. 12, 29–30 (in Russian).

Swaddle, J.P., Witter, M.S. and Cuthill, I.C. (1994) The analysis of fluctuating asymmetry. *Animal Behaviour* 48, 986–989.

Swaine, G. (1971) *Agricultural Zoology in Fiji*. Overseas Research Publication 18, Foreign and Commonwealth Office, Overseas Development Administration, London.

Swaminathan, M.S. (2007) *Vegetable Crops*. Horticulture Science, Series, 4. New India Publishing Agency, New Delhi.

Swan, D.C. (1934) The red-legged earth mite, *Halotydeus destructor* (Tucker) in South Australia; with remarks upon *Penthaleus major* (Dugès). *South Australian Journal of Agriculture* 38, 365–367.

Swezey, O.H. (1923) The Java sugar cane leaf-mite in Hawaii. *Hawaiian Planter's Record* 27, 4–7.

Swirski, E. (1962) Contribution to the knowledge of the fluctuations in population of the citrus rust mite (*Phyllocoptruta oleivora* Ashm.) in the coastal plain of Israel. *Israel Journal of Agricultural Research* 12, 175–187.

Swirski, E. and Amitai, S. (1958) Contribution to the biology of the citrus rust mite (*Phyllocoptruta oleivora* Ashm.) A. Development, adult longevity and life cycle. *Katavim* 8, 189–207.

Swirski, E. and Dorzia, N. (1969) Laboratory studies on the feeding, development and fecundity of the predaceous mite *Typhlodromus occidentalis* Nesbitt (Acarina: Phytoseiidae) on various kinds of food substances. *Israel Journal of Agricultural Research* 19, 143–145.

Swirski, E., Kehat, M., Greunberg, S., Dorzia, N. and Amitai, S. (1967) Trials for control of the citrus rust mite (*Phyllocoptruta oleivora* Ashm.). *Israel Journal of Agricultural Research* 17, 121–126.

Swirski, E., Amitai, S. and Dorzia, N. (1970) Laboratory studies on the feeding habits, post-embryonic survival and oviposition of the predaceous mites *Amblyseius chilenensis* Dosse and *Amblyseius hibisci* (Chant) on various kinds of food substances. *Entomophaga* 15, 93–106.

Szabó, A. and Pénzes, B. (2013) A new method for the release of *Amblyseius andersoni* (Acari: Phytoseiidae) in young apple orchards. *European Journal of Entomology* 110, 477–482.

Szabo, G. (1992) Elorejelzes es vedekezesi kiserletek a szolo levelatkaval (*Calepitrimerus vitis* Nalepa) kapcsolatban. *Kertgazdasag* 24(5), 21–30 (in Hungarian).

Szczepaniec, A., Creary, S.F., Laskowski, K.L., Nyrop, J.P. and Raupp, M.J. (2011) Neonicotinoid insecticide imidacloprid causes outbreaks of spider mites on elm trees in urban landscapes. *PLoS ONE* 6(5): e20018.

Szczepaniec, A. and Raupp, M.J. (2013) Direct and indirect effects of imidacloprid on fecundity and abundance of *Eurytetranychus buxi* (Acari: Tetranychidae) on boxwoods. *Experimental and Applied Acarology* 59, 307–318.

Szendrey, G., Dulinafka, G. and Szegedi, E. (1995) Elimination of mites from the buds of dormant grapevine cuttings by hot water treatment. *Vitis* 34, 65–66.

Szendrey, G., Ilovai, Z. and Lucza, Z. (2003) Damage caused by blackberry mite (*Acalitus essigi* Hassan) and the role of natural biological control agents in integrated blackberry production system in Hungary. *IOBC-WPRS Bulletin* 26(2), 133–138.

Sztejnberg, A., Doron-Shloush, S. and Gerson, U. (1997) The biology of the acaropathogenic fungus *Hirsutella kirchneri*. *Biocontrol Science and Technology* 7, 577–590.

Sztejnberg, A., Paz, Z., Boekhout, T., Gafni, A. and Gerson, U. (2004) A new fungus with dual biocontrol capabilities: reducing the numbers of phytophagous mites and powdery mildew disease damage. *Crop Protection* 23, 1125–1129.

Szulc, W. (1966) Gall mites (Eriophyidae) of Lodz Upland. *Uniwersytet Lodski, Zeszyty Naukowe (Series 2)* 21, 27–55.

Tabashnik, B.E. (1990) Modeling and evaluation of resistance management tactics. In: Roush, R.T. and Tabashnik, B.E. (eds) *Pesticide Resistance in Arthropods*. Chapman & Hall, New York, pp. 153–182.

Tabashnik, B.E. and Croft, B.A. (1982) Managing pesticide resistance in crop–arthropod complexes: interactions between biological and operational factors. *Environmental Entomology* 11, 1137–1144.

Taffin, G. de, Zakra, N. and Bonny, C.P. (1991) Dwarf × Tall coconut hybrid performance under commercial conditions in Côte-d'Ivoire. *Oléagineux* 46, 187–195.

Tagaki, K. (1981) The injuries on the citrus fruit caused by green leafhoppers, *Empoasca* spp., tea yellow thrips, *Scirtothrips dorsalis* Hood, citrus red mite, *Panonychus citri* (McGregor) and citrus rust mite, *A. pelekassi*. *Bulletin of the Fruit Tree Research Station, Series D (Kuchinotsu)* 3, 101–112 (in Japanese).

Takabayashi, J., Dicke, M. and Posthumus, M. (1991) Variation in composition of predator-attracting allelochemicals emitted by herbivore-infested plants: relative influence of plant and herbivore. *Chemoecology* 2, 1–6.

Takabayashi, J., Dicke, M., Takahashi, S., Posthumus, M.A. and van Beek, T.A. (1994a) Leaf age affects composition of herbivore-induced synomones and attraction of predatory mites. *Journal of Chemical Ecology* 20, 373–386.

Takabayashi, J., Dicke, M. and Posthumus, M.A. (1994b) Volatile herbivore-induced terpenoids in plant mite interactions: variation caused by biotic and abiotic factors. *Journal of Chemical Ecology* 20, 1329–1354.

Takada, Y. and Kashio, T. (2004) Biological studies of *Amblyseius californicus* (McGregor) and its control effect for *Tetranychus kanzawai* on greenhouse strawberries. *Kyushu Plant Protection Research* 50, 49–54 (in Japanese).

Takafuji, A. (1977) The effect of the rate of successful dispersal of a phytoseiid mite, *Phytoseiulus persimilis* Athias-Henriot on the persistence in the interactive system between the prey and its predator. *Researches on Population Ecology (Kyoto)* 18, 210-222.

Takafuji, A. and Chant, D.A. (1976) Comparative studies of two species of predacious phytoseiid mites (Acarina: Phytoseiidae), with special reference to their responses to the density of their prey. *Researches on Population Ecology (Kyoto)* 17, 255–310.

Takafuji, A. and Hinomoto, N. (2008) The distribution and geographic variation in diapause capacity among populations of two *Tetranychus* species (Acari: Tetranychidae) in East and South-east Asia. *Journal of the Acarological Society of Japan* 17, 1–15.

Takafuji, A. and Ishii, T. (1989) Inheritance of sex ratio in the Kanzawa spider mite, *Tetranychus kanzawai* Kishida. *Researches on Population Ecology (Kyoto)* 31, 123–128.

Takafuji, A. and Morimoto, N. (1983) Diapause attributes and seasonal occurrences of two populations of the citrus red mite, *Panonychus citri* (McGregor) on pear (Acarina: Tetranychidae). *Applied Entomology and Zoology* 18, 525–532.

Takafuji, A., Santoso, S. and Hinomoto, N. (2001) Host-related differences in diapause characteristics of different geographical populations of the Kanzawa spider mite, *Tetranychus kanzawai* Kishida (Acari: Tetranychidae), in Japan. *Applied Entomology and Zoology* 36, 177–184.

Takafuji, A., Santoso, S., Hinomoto, N., Shih, C.I.T., Ho, C.C. and Gotoh, T. (2003) Diapause characteristics of two species of tetranychid mites (Acari: Tetranychidae) in southern Japan and Taiwan. *Applied Entomology and Zoology* 38, 267–270.

Takahashi, F., Inoue, M. and Takafuji, A. (1998) Management of the spider-mite population in a vinylhouse vinery by releasing *Phytoseiulus persimilis* Athias-Henriot onto the ground cover. *Japanese Journal of Applied Entomology and Zoology* 42, 71–76.

Takano-Lee, M. and Hoddle, M.S. (2001) Biological control of *Oligonychus perseae* (Acari: Tetranychidae) on avocado: IV. Evaluating the efficacy of a modified mistblower to mechanically dispense *Neoseiulus californicus* (Acari: Phytoseiidae). *International Journal of Acarology* 27, 157–169.

Takano-Lee, M. and Hoddle, M.S. (2002a) *Oligonychus perseae* (Acari: Tetranychidae) population responses to cultural control attempts in an avocado orchard. *Florida Entomologist* 85, 216–226.

Takano-Lee, M. and Hoddle, M. (2002b) Predatory behaviors of *Neoseiulus californicus* and *Galendromus helveolus* (Acari: Phytoseiidae) attacking *Oligonychus perseae* (Acari: Tetranychidae). *Experimental and Applied Acarology* 26, 13–26.

Takeyama, K., Mori, N. and Osakabe, M. (2006) Effect of cytochrome P450 inhibitor, piperonyl butoxide, on survival of *Panonychus citri* (McGregor) (Acari: Tetranychidae) on citrus leaves. *Applied Entomology and Zoology* 41, 487–491.

Taleb, M.A. and Sardar, M.A. (2007) Predation efficiency of insect predators in suppressing red mite, *Tetranychus bioculatus* attacking marigold. *Journal of Agriculture and Rural Development* 5, 98–104.

Talhouk, A.S. (1961) Additional records of insect pests from Lebanon and Syria. *Bulletin of the Entomological Society of Egypt* 45, 405–406.

Talhouk, A.S. (1963) *Aceria phloeocoptes* (Nal.), a serious eriophyid pest of almond in Lebanon and Syria. *Anzeiger für Schädlingskunde* 36, 129–132.

Talhouk, A.S. (1971) The role of systemic insecticides in Middle East horticulture. In: Gur'eva, E.L. and Kryzhanovskii, O.L. (eds) *Proceedings of the XIII International Congress of Entomology, Moscow, 2–9 August 1968, Volume 2*. Nauka, Leningrad, USSR, pp. 282–283.

Talhouk, A.S. (1977) Contribution to the knowledge of almond pests in East Mediterranean countries. *Zeitschrift für Angewandte Entomologie* 83, 248–257.

Tamai, M.A., Moraes, G.J. de Silva, C.A.D. da and Nunes, A.M. (1997) Suitability of *Brevipalpus obovatus* as prey to *Neoseiulus idaeus* (Acari: Tenuipalpidae, Phytoseiidae) on cassava. *Systematic and Applied Acarology* 2, 101–106.

Tan, F.M. and Ward, C.R. (1977) Laboratory studies on the biology of the Banks grass mite. *Annals of the Entomological Society of America* 70, 534–536.

Tanaka, H. and Shibao, M. (2003) A pattern of occurrence and dispersal of the tomato russet mite, *Aculops lycopersici* (Massee) in a tomato greenhouse. *Proceedings of the Kansai Plant Protection Society* 45, 23–27 (in Japanese).

Tanaka, M. and Kashio, T. (1977) Biological studies on *Amblyseius largoensis* Muma (Acarina: Phytoseiidae) as predator of the citrus red mite, *Panonychus citri* (McGregor) (Acarina: Tetranychidae). *Bulletin of the Fruit Tree Research Station, Series D (Kuchinotsu)* 1, 49–67.

Tandon, P. (1982) Metabolism of auxin in *Eriophyes*-incited *Zizyphus* gall grown in culture. In: Fujiwara, A. (ed.) *Proceedings of the 5th International Congress on Plant Tissue Cell Culture, Tokyo and Lake Yamanake, Japan*. Japanese Association for Plant Tissue Culture, Tokyo, pp. 203–204.

Tandon, P. and Arya, H.C. (1980) Auxin autotrophy and hyperauxinity of *Eriophyes* induced *Zizyphus* stem galls in culture. *Biochemie und Physiologie der Pflanzen* 175, 537–541.

Tandon, P. and Arya, H.C. (1982) Association of auxin protectors, peroxidase, indoleacetic acid oxidase and polyphenol oxidase in *Zizyphus* gall and normal stem tissues grown in culture. *Biochemie und Physiologie der Pflanzen* 177, 114–124.

Tanigoshi, L.K. (1982) Advances in knowledge of the biology of the Phytoseiidae. In: Hoy, M.A: (ed.) *Recent Advances in Knowledge of the Phytoseiidae*. Division of Agriculture Special Publication No 3284, University of California, Berkeley, California, pp. 1–22.

Tanigoshi, L.K. and Davis, R.W. (1978) An ultrastructural study of *Tetranychus mcdanieli* feeding injury to the leaves of 'Red Delicious' apple (Acari: Tetranychidae). *International Journal of Acarology* 4, 47–56.

Tanigoshi, L.K. and McMurtry, J.A. (1977) The dynamics of predation of *Stethorus picipes* (Coleoptera: Coccinellidae) and *Typhlodromus floridanus* on the prey *Oligonychus punicae* (Acarina: Phytoseiidae, Tetranychidae). Part I. Comparative life history and life table studies. *Hilgardia* 45(8), 237–261.

Tanigoshi, L.K., Hoyt, S.C., Browne, R.W. and Logan, J.A. (1975a) Influence of temperature on population increase of *Metaseiulus occidentalis* (Acarina: Phytoseiidae). *Annals of the Entomological Society of America* 68, 979–986.

Tanigoshi, L.K., Browne, R.W. and Hoyt, S.C. (1975b) A study on the dispersion pattern and foliage injury by *Tetranychus mcdanieli* (Acarina: Tetranychidae) in simple apple ecosystems. *Canadian Entomologist* 107, 439–446.

Tanigoshi, L.K., Fargerlund, J. and Nishio Wong, J.Y. (1981) Significance of temperature and food resources to the developmental biology of *Amblyseius hibisci* (Chant) (Acarina, Phytoseiidae). *Zeitschrift für Angewandte Entomologie* 92, 409–419.

Tanigoshi, L.K., Bahdousheh, M. and Babcock, J.M. (1990) *Euseius scutalis* (Athias-Henriot) a predator of *Eutetranychus orientalis* (Klein) (Acari: Phytoseiidae, Tetranychidae) in Jordan: toxicity of some acaricides to *E. orientalis*. *Arab Journal of Plant Protection* 8, 114–120.

Tanigoshi, L.K., Martin, N.A., Osborne, L.S. and Peña, J.E. (2004) Biological control of spider mites on ornamental crops. In: Heinz, K.M., Van Driesche, R.G. and Parrella, M.P. (eds) *Biocontrol in Protected Culture*. Ball Publishing, Batavia, Illinois, pp. 185–199.

Tanzini, M.R., Alves, S.B., Tamai, M.A., Moraes, G.J. de and Ferla, N.J. (2000) An epizootic of *Calacarus heveae* (Acari: Eriophyidae) by *Hirsutella thompsonii* on rubber trees. *Experimental and Applied Acarology* 24, 141–144.

Tashiro, H. (1967) Self-watering acrylic cages for confining insects and mites on leaves. *Journal of Economic Entomology* 60, 354–356.

Tatagar, M.H., Prabhu, S.T. and Jagadeesha, R.C. (2001) Screening chilli genotypes for resistance to thrips, *Scirtothrips dorsalis* Hood and mite, *Polyphagotarsonemus latus* (Banks). *Pest Management in Horticultural Ecosystems* 7, 113–116.

Tatara, A., Kosugi, Y. and Onikubo, T. (1999) Effectiveness of pesticide application in early spring for control of *Tetranychus kanzawai* Kishida in tea fields. *Proceedings of the Kansai Plant Protection Society* 41, 73–74 (in Japanese).

Tauber, M.J., Tauber, C.A., Daane, K.M. and Hagen, K.S. (2000) Commercialization of predators, recent lessons from green lacewings (Neuroptera: *Chrysoperla*). *American Entomologist* 46, 26–38.

Tautz, D., Arctander, P., Minelli, A., Thomas, R.H. and Vogler, A.P. (2003) A plea for DNA taxonomy. *Trends in Ecology and Evolution* 18, 70–74.

Tavamaishvili, L.E. (1990) The chief pests of hazelnut in subtropical western Georgia and their control. *Subtropicheskie Kul'tury* 5, 119–125.

Tawfik, M.F.S. and Ata, A.M. (1973a) The life-history of *Orius albidipennis* (Reut.) (Hemiptera-Heteroptera; Anthocoridae*). Bulletin of the Entomological Society of Egypt* 57, 117–126.

Tawfik, M.F.S. and Ata, A.M. (1973b) The life-history of *Orius laevigatus* (Fieber) (Hemiptera: Anthocoridae). *Bulletin of the Entomological Society of Egypt* 57, 145–151.

Taylor, A.M. (1909) Descriptions and life histories of two new parasites of the black currant mite, *Eriophyes ribis* (Nal.). *Journal of Economic Biology* 4, 1–8.

Taylor, B., Rahman, P.M., Murphy, S.T. and Sudheendrakumar, V.V. (2012) Within-season dynamics of red palm mite (*Raoiella indica*) and phytoseiid predators on two host palm species in south-west India. *Experimental and Applied Acarology* 57, 331–345.

Taylor, C.E. and Georghiou, G.P. (1979) Suppression of insecticide resistance by alteration of dominance and migration. *Journal of Economic Entomology* 72, 105–109.

Taylor, K.B. (1970) The Eriophyoidea of the Coniferales of the New York area. PhD thesis, Cornell University, Ithaca, New York.

Tedders, W.L. (1983) Insect management in deciduous orchard ecosystems: habitat manipulation. *Environmental Management* 7, 29–34.

Telford, M.J. and Thomas, R.H. (1998) Expression of homeobox genes show chelicerate arthropods retain their deuterocerebral segment. *Proceedings of the National Academy of Sciences of the United States of America* 95, 10671–10675.

Telliev, L. and Geshev, T. (1980) Efficient control of the yellow apple mite on grape vine. *Rastitelna Zashtita* 28(5), 18–19 (in Bulgarian).

Teng, B., Qiu, Z. X. and Wang, H. (1988) Investigation of mite infestation of respiratory tract in Chinese herb workers. *Chinese Journal of Parasitic Disease Control* (now *Journal of Pathogen Biology*) 1, 38–39, 77 (in Chinese).

Teodoro, A.V. and Reis, P.R. (2006) Reproductive performance of the mite *Brevipalpus phoenicis* (Geijskes, 1939) on citrus and coffee, using life table parameters. *Brazilian Journal of Biology* 66, 899–905.

Teodoro, A.V., Fadini, M.A.M., Lemos, W.P., Guedes, R.N.C. and Pallini, A. (2005) Lethal and sub-lethal selectivity of fenbutatin oxide and sulfur to the predator *Iphiseiodes zuluagai* (Acari: Phytoseiidae) and its prey, *Oligonychus ilicis* (Acari: Tetranychidae), in Brazilian coffee plantations. *Experimental and Applied Acarology* 36, 61–70.

Teodoro, A.V., Rêgo, A.S., Maciel, A.G.S. and Costa, E.C. (2010) Mechanisms explaining the population dynamics of the cassava green mite, *Mononychellus tanajoa* (Bondar) (Acari: Tetranychidae In: Moraes, G.J. de, Castilho, R.C. and Flechtmann, C.H.W. (eds) *XIII International Congress of Acarology, August 23–27, 2010, Recife-PE, Brazil, Abstract Book*, abstract 506, p. 274. Available at: http://www.acarology.org/ica/ica2010/docs/0910_15_abstractBook-Capa.pdf (accessed 3 February 2015).

Terho, E.O., Husman, K., Vohlonen I., Rautalahti, M. and Tukiainen, H. (1985) Allergy to storage mites of cow dander as a cause of rhinitis among Finnish dairy farmers. *Allergy* 40, 23–26.

Terras, M.A. and Hales, D.F. (1995) Red pepper mites are vectors of *Trichoderma*. In: Elliott, T.J. (ed.) *Proceedings of the Fourteenth International Congress on the Science and Cultivation of Edible Fungi, Volume II*. Balkema, Rotterdam, the Netherlands, pp. 485–490.

Tew, R.P. and Groves, J.R. (1957) Some observations on the effects of formulations on the persistence of DDT and its toxicity to certain orchard mites and insects. In: *44th Report, East Malling Research Station, 1955–1956*. East Malling, UK, pp. 152–160.

Texas Department of Agriculture (2007) Emergency action notification ordered to stop movement of rice products from Texas research facility. TDA Press Release July 2007. Austin, Texas.

Texas Invasive Species Institute (2011) *Brevipalpus californicus*: Scarlet Mite/Omnivorous Flat Mite. Sam Houston State University, Huntsville, Texas. Available at: http://www.tsusinvasives.org/database/B_californicus.html#sthash.cOltd84u.dpuf (accessed 6 August 2014).

Thaker, P.J. and Desa, M.H. (1929) Woolly-mites (*Eriophyes gossypii*) on cotton. Its status and economic importance in Gujarat. *Agricultural Journal of India* 24, 175–182.

Thakur, M., Dinabandhoo, C.L. and Chauhan, U. (2010) Host range, distribution, and morphometrics of predatory mites associated with phytophagous mites of fruit crops in Himachal Pradesh, India. In: Sabelis, M.W. and Bruin, J. (eds) *Trends in Acarology: Proceedings of the 12th International Congress, 21–26 August 2006, Amsterdam*. Springer, Dordrecht, the Netherlands, pp. 431–434.

Thalenhorst, W. (1963) The occurrence of *Oligonychus ununguis* (Jacobi) on a fertilizer test plot. *Mitteilungen aus der Biologischen Bundesanstalt für Land- und Forstwirtschaft, Berlin-Dahlem* No. 108, 39–42.

Theaker, T.L. and Tonks, N.V. (1977) A method for rearing the predaceous mite *Phytoseiulus persimilis* (Acarina: Phytoseiidae). *Journal of the Entomological Society of British Columbia* 74, 8–9.

Thewke, S.E. and Enns, W.R. (1969) *The Spider-mite Complex (Acarina: Tetranychoidae) in Missouri*. Museum Contributions, Monograph No. 1, University of Missouri, Columbia, Missouri.

Thirugnanasuntharan, K. and Amarasinghe, L.D. (1990) Clonal susceptibility and population dynamics of tea red spider mite, *Oligonychus coffeae* Nietner (Acarina: Tetranychidae) under laboratory conditions. *Sri Lanka Journal of Tea Science* 59, 9–15.

Thistlewood, H.M.A., Clements, D.R. and Harmsen, R. (1996) Stigmaeidae. In Lindquist, E.E., Sabelis, M.W. and Bruin, J. (eds) *Eriophyoid Mites: Their Biology, Natural Enemies and Control*. World Crop Pests, Volume 6. Elsevier, Amsterdam, the Netherlands, pp. 457–470.

Thomas, C. and Rajotte, E. (2011) *Integrated Pest Management for Christmas Tree Production. A Guide for Pennsylvania Growers*. Ag Communications and Marketing, The Pennsylvania State University, University Park, Pennsylvania.

Thomas, C.A. (1942) *Mushroom Insects: their Biology and Control*. Bulletin No. 419, Pennylvania State College of Agriculture Experiment Station, State College, Pennylvania.

Thomas, C.E. (1969) Transmission of tobacco ringspot virus by *Tetranychus* sp. *Phytopathology* 59, 633–636.

Thomas, F. (1896) Uber die Lebensweise der Stachelbeermilbe *Bryobia ribis*, deren Verbreitung in Deutschlande. *Zeitschrift für Pflanzenkrankheiten und Pflanzenschutz* (now *Journal of Plant Diseases and Plant Protection*) 6, 80–84.

Thomas, J., Chen, Q. and Talbert, L. (1997) Genetic segregation and the detection of spontaneous wheat–alien translocations. In: Braun, H.J., Altay, F., Kronstad, W.E., Beniwal, S.P.S. and McNab, A. (eds) *Wheat: Prospects for Global Improvement. Proceedings of the 5th International Wheat Conference, 10–14 June 1996, Ankara, Turkey*. Developments in Plant Breeding. Kluwer Academic Publishers (later Springer), Dordrecht, the Netherlands, pp. 327–333.

Thomas, J.A, and Hein, G.L. (2003) Influence of volunteer wheat plant condition on movement of the wheat curl mite, *Aceria tosichella*, in winter wheat. *Experimental and Applied Acarology* 31, 253–268.

Thomas, J.B. and Conner, R.L. (1986) Resistance to colonization by the wheat curl mite in *Aegilops squarrosa* and its inheritance after transfer to common wheat. *Crop Science* 26, 527–530.

Thomas, J.B. and Whelan, E.D.P. (1991) Genetics of wheat curl mite resistance in wheat: recombination of Cmc1 with the 6D centromere. *Crop Science* 31, 936–938.

Thomas, J.B., Conner, R.L. and Graf, R.J. (2004) Comparison of different sources of vector resistance for controlling wheat streak mosaic in winter wheat. *Crop Science* 44, 125–130.

Thomas, W.P. and Walker, J.T.S. (1989) *Tetranychus urticae* Koch, two spotted spider mite (Acari: Tetranychidae). In: Cameron, P.J., Hill., R.L., Bain, J. and Thomas, W.P. (eds) *A Review of Biological Control of Invertebrate Pests and Weeds in New Zealand 1874 to* 1987. CAB International, Wallingford, UK, pp. 245–250.

Thompson, M.M. (1977) Inheritance of big bud mite susceptibility in filbert. *Journal of the American Society for Horticultural Science* 102, 39–42.

Thomsen, J. (1975) Development and histology of galls on *Tilia platyphylla* caused by *Eriophyes tiliae tiliae*. *Botanisk Tidsskrift* 69, 262–270.

Thomsen, J. (1976) Morphology and biology of the gall mite *Eriophyes tiliae tiliae* Pgst. (Acarina, Trombidiformes, Eriophyidae). *Entomologiske Meddelelser* 44, 9–17.

Thomsen, J. (1988) Feeding behaviour of *Eriophyes tiliae tiliae* Pgst. and suction track in the nutrittive cells of the galls caused by the mites. *Entomologiske Meddelelser* 56, 73–78.

Thongtab, T., Chandrapatya, A. and Baker, G.T. (2002) Bionomics of the citrus yellow mite, *Eotetranychus cendanai* Rimando. *Systematic and Applied Acarology* 7, 57–68.

Thor, S. (1929) Über die Phylogenie und Systematik der Acarina, mit Beiträgen zu ersten Entwicklungsgeschichte einzelner Gruppen. *Nyt Magazin for Naturvidenskaberne* 67, 145–210.

Thor, S. and Willmann, C. (1941) Eupodidae, Penthalodidae, Penthaleidae, Rhagidiidae, Pachygnathidae, Cunaxidae. In: Thor, S. and Willmann, C. (eds) *Das Tierreich, Eine Zusammenstellung und Kennzeichung der rezenten Tierformen, Lieferung 71a*. Walter De Gruyter, Leipzig, Germany, pp. 164–175.

Thresh, J.M. (1963) Black currant reversion disease. *Annual Report 1962, East Malling Research Station*. East Malling, UK, pp. 184–189.

Thresh, J.M. (1964a) Increased susceptibility to mite vector (*Phytoptus ribis* Nal.) caused by infection with black currant reversion virus. *Nature* 202, 1028.

Thresh, J.M. (1964b) Association between black currant reversion virus and its gall mite vector (*Phytoptus ribis* Nal.). *Nature* 202, 1085–1087.

Thresh, J.M. (1964c) Warm water treatments to eliminate the gall mite *Phytoptus ribis* Nal. from black currant cuttings. In: *Annual Report 1963a, East Malling Research Station*. East Malling, UK, pp. 131–132.

Thresh, J.M. (1964d) The chemical control of black currant reversion virus and its gall mite vector (*Phytoptus ribis* Nal.). In: *Annual Report 1963, East Malling Research Station*. East Malling, UK, pp. 152–157.

Thresh, J.M. (1967) Increased susceptibility of black currant bushes to the gall-mite vector (*Phytoptus ribis* Nal.) following infection with reversion virus. *Annals of Applied Biology* 60, 455–467.

Thulsiram (1991) Bionomics and chemical control of *Tetranychus macfarlanei* on cotton. MSc. thesis, University of Agricultural Sciences, Dharwad, India.

Thwaite, W.G. (1991) Resistance to clofentezine and hexythiaxoz in *Panonychus ulmi* from apples in Australia. *Experimental and Applied Acarology* 11, 73–80.

Tian, M. (1995) Analysis on control effects of two natural enemies *Amblyseius nicholsi* and *Stethorus punctillum* on *Panonychus citri* by removing method. *Chinese Journal of Biological Control* 11, 153–155 (in Chinese).

Tibilova, A.A. (1932) *Biology of the Red Spider Mite in the Tashkent District*. Byulleten' Nauchno-Issledovatel'skogo Instituta po Khlopkovodstvu, Volume 1. Tashkent, Uzbekistan (former USSR).

Timár, E., Bozai, J. and Bürgés, G. (2004) Adalékok a fokhagymán élő atkák ismeretéhez (Additions to the knowledge of mites living on garlic). *Növényvédelem* 40, 17–25 (in Hungarian).

Timmer, L.W. and Duncan, L.W. (eds) (1999) *Citrus Health Management*. American Phytopathological Society Press, St Paul, Minnesota.

Timms, S., Ferro, D.N. and Emberson, R.M. (1981) General biology and nomenclature of *Sancassania berlesei* (Michael). *Acarologia* 22, 385–390.

Tixier, M.-S., Kreiter, S., Auger, P. and Weber, M. (1998) Colonization of Languedoc vineyards by phytoseiid mites (Acari: Phytoseiidae): influence of wind and crop enviroment. *Experimental and Applied Acarology* 22, 523–542.

Tixier, M.-S., Kreiter, S., Allam, L., Ouahbi, A. and Hmimina, M. (2003) Phytoseiid and tetranychid mites (Acari: Mesostigmata, Prostigmata) of some Moroccan crops. *Acarologia* 43, 87–97.

Tixier, M.-S., Kreiter, S., Cheval, B., Guichou, S., Auger, P. and Bonafos, R. (2006) Immigration of phytoseiid mites from surrounding uncultivated areas into a newly planted vineyard. *Experimental and Applied Acarology* 39, 227–242.

Tixier, M.-S., Ferrero, M., Okassa, M., Guichou, S. and Kreiter, S. (2010a) On the specific identity of specimens of *Phytoseiulus longipes* Evans (Mesostigmata: Phytoseiidae) showing different feeding behaviours: morphological and molecular analyses. *Bulletin of Entomological Research* 100, 569–579.

Tixier, M.-S., Guichou, S. and Kreiter, S. (2010b) Assessment of the usefulness of eight DNA fragments for phylogenetic studies within the family Phytoseiidae. In: Sabelis, M.W. and Bruin, J. (eds) *Trends in Acarology: Proceedings of the 12th International Congress, 21–26 August 2006, Amsterdam*. Springer, Dordrecht, the Netherlands, pp. 41–48.

Tixier, M.-S., Hernandes, F.A., Guichou, S. and Kreiter, S. (2011) The puzzle of DNA sequences of Phytoseiidae (Acari: Mesostigmata) in the public GenBank database. *Invertebrate Systematics* 25, 389–406.

Tixier, M.-S., Kreiter, S., Douin, M. and Moraes, G.J. (2012) Rates of description of Phytoseiidae (Acari: Mesostigmata): space, time and body size variations. *Biodiversity and Conservation* 21, 993–1013.

Tna, D.J. (1986) Field experiments on the integrated control of citrus mites. *Chinese Journal of Biological Control* 2, 181 (in Chinese).

TNAU (2011) Paddy. Agro Advisory Service, Tamil Nadu Agricultural University, Coimbatore, India. Available at: http://agritech.tnau.ac.in/govt_schemes_services/aas/paddy_ex1.html (accessed 1 September 2014).

Tobacco Research Board (1982) *Annual Report and Accounts for the Year Ended 30th June, 1982*. Harare, Zimbabwe.

Toda, S., Osakabe, M. and Komazaki, S. (2000) Interspecific diversity of mitochondrial COI sequences in Japanese *Panonychus* species (Acari: Tetranychidae). *Experimental and Applied Acarology* 24, 821–829.

Toko, M., Yaninek, J.S. and O'Neil, R.J. (1996) Response of *Mononychellus tanajoa* (Acari: Tetranychidae) to cropping systems, cultivars and pest interventions. *Environmental Entomology* 25, 237–249.

Toledo, M.A. de, Reis, P.R., Andrade, H.B., Marafeli, P.P. and Pedro Neto, M. (2010a) Selectivity of plant extracts to the predaceous mite *Amblyseius herbicolus* (Chant) (Acari: Phytoseiidae). In: Moraes, G.J. de, Castilho, R.C. and Flechtmann, C.H.W. (eds) *XIII International Congress of Acarology, August 23–27, 2010, Recife-PE, Brazil, Abstract Book*, abstract 512, pp. 277–278. Available at: http://www.acarology.org/ica/ica2010/docs/0910_15_abstractBookCapa.pdf (accessed 3 February 2015).

Toledo, M.A. de, Reis, P.R., Souza-Pimentel, G.C., Andrade, H.B., Silveira, E.C. and Marafeli, P.P. (2010b) Evaluation of plant extracts in the control of *Oligonychus ilicis* (McGregor) and *Tetranychus urticae* Koch (Acari: Tetranychidae). In: Moraes, G.J. de, Castilho, R.C. and Flechtmann, C.H.W. (eds) *XIII International Congress of Acarology, August 23–27, 2010, Recife-PE, Brazil, Abstract Book*, abstract 513, p. 278. Available at: http://www.acarology.org/ica/ica2010/docs/0910_15_abstractBookCapa.pdf (accessed 3 February 2015).

Tomašević, B. (1964) The yellow poplar mite, *Eotetranychus populi* Koch (Acarina, Tetranychidae). *Zastita Bilja* 15, 687–693 (in Serbo Croatian).

Tomczyk, A. and Boczek, J. (2006) Physiological and biochemical changes in tree leaves infested by some eriophyoid mites (Acari: Eriophyoidea). In: Gabryś, G. and Ignatowicz, S. (eds) *Proceedings of the Symposium "Advances in Polish Acarology"*. Wydawnictwo SGGW (Szkoła Główna Gospodarstwa Wiejskiego w Warszawie), Warszaw, pp. 413–419.

Tomczyk, A. and Burda, W. (2005) Behaviour and activity of *Phytoseiulus persimilis* (A.-H.) on mite infested cucumber plants cultivated in the presence of plant growth promoting rhizobacteria (PGPR). *IOBC-WPRS Bulletin* 28(1), 267–270.

Tomczyk, A. and Kropczyńska, D. (1985) Effects on the host plant. In: Helle, W. and Sabelis, M.W. (eds) *Spider Mites: Their Biology, Natural Enemies and Control*. World Crop Pests, Volume 1A. Elsevier Science, Amsterdam, pp. 317–329.

Tomczyk, A. and van de Vrie, M. (1982) Physiological and biochemical changes in three cultivars of chrysanthemum after feeding by *Tetranychus urticae*. In: Visser, J.H. and Minks, A.K. (eds) *Proceedings of the 5th International Symposium on Insect–Plant Relationships*. Pudoc, Wageningen, the Netherlands, pp. 391–392.

Tomerlin, A.T. (2005) Economic threshold decision rules as an integrated pest management tool in Florida citrus production. PhD thesis, Universiy of Florida, Gainesville, Florida.

Tomkins, A.R. (2002) Sustainable management of sixspotted spider mite (*Eotetranychus sexmaculatus* (Riley)) on avocados. In: *New Zealand Avocado Growers Association Annual Research Report – 2002. Volume 2*. Tauranga, New Zealand, article 4. Available at: http://www.avocadosource.com/Journals/NZAGA/NZAGA_2002/nzaga_2002_04.pdf (accessed 21 May 2015).

Tomlin, C.D.S. (ed.) (2009) *The Pesticide Manual*, 15th edn. CAB International, Wallingford, UK.

Tomlinson, W.E. Jr (1950) Summer oil sprays to control blueberry bud mite. *Journal of Economic Entomology* 43, 727.

Tompsett, A.A., Roberstson, J.S. and Wallis, L.W. (ed.) (1977) *Hot Water Treatment of* Narcissus *Bulbs*. Short Term Leaflet no. 21, ADAS/MAFF (Agricultural Development and Advisory Service/Ministry of Agriculture, Fisheries and Food) Publications, Pinner, UK. Her Majesty's Stationery Office, London. Available at: http://dafflibrary.org/wp-content/uploads/UK-Ministry-of-Ag-Hot-Water-Treatment-of-Narcissus-Bulbs-1977.pdf (accessed 2 February 2015).

Tono, T., Fujita, S. and Yamaguchi, S. (1978) Effect of infestation by citrus rust mite, *Aculops pelekassi* Keifer, on the development of Satsuma mandarin fruit and availability for juice processing from damaged fruit. *Bulletin of the Faculty of Agriculture, Yamaguchi University* 44, 57–66 (in Japanese).

Toroitich, F., Navajas, M. and Knapp, M. (2008) Molecular analysis of the genetic diversity of *Tetranychus evansi* Baker and Pritchard from Kenya and Tanzania. In: Bertrand, M., Kreiter, S., McCoy, K.D., Migeon,A., Navajas, M., Tixier, M.-S. and Vial, L. (eds) (2008) *Integrative Acarology: Proceedings 6th European Congress EURAAC (European Association of Acarologists, Montpellier, July, 21–25, 2008* (CD-ROM), poster.

Toroitich, F.J., Ueckermann, E.A., Theron, P.D. and Knapp, M. (2009) The tetranychid mites (Acari: Tetranychidae) of Kenya and a redescription of the species *Peltanobia erasmusi* Meyer (Acari: Tetranychidae) based on males. *Zootaxa* 2176, 33–47.

Toros, S. and Çobanoğlu, S. (1985) *Kültür Mantarlarında Zararlı Akarlar*. Ankara Üniversitasi Ziraat Fakültesi Yayınları No. 921/Bilimsel Araştırma ve İncelemeler No. 541, Ankara (in Turkish).

Torres, E., Botti, S., Paltrinieri, S., Martin, M.P. and Bertaccini, A. (2002) First report of *Spartium* witches' broom disease in Spain. *New Disease Reports, British Society for Plant Pathology* 5(14). Available at: http://www.ndrs.org.uk/article.php?id=005014 (accessed 25 August 2014).

Torres, E., Hernández-Suárez, E. , López, A., Rolo, P. and Ferragut, F. (2010) The invasive persea mite, *Oligonychus perseae* (Tetranychidae), in Canary Islands: conservative or augmentative biological control? In: Moraes, G.J. de, Castilho, R.C. and Flechtmann, C.H.W. (eds) *XIII International Congress of Acarology, August 23–27, 2010, Recife-PE, Brazil, Abstract Book*, abstract 516, pp. 279–280. Available at: http://www.acarology.org/ica/ica2010/docs/0910_15_abstractBookCapa.pdf (accessed 3 February 2015).

Touzeau, J. (1973) La répartition des oeufs d'hiver de *Panonychus ulmi* sur arbres fruitiers et propositions concernant la prognose hivernale. *ACTA Note d'Information, Le Réseau des Instituts des Filières Animales et Végétales, Paris* 11, 5–9.

Townsend, L. (2007) Clover mites. EntFact-627, Entomology Department, College of Agriculture, Food and Environment, University of Kentucky, Lexington, Kentucky. Available at: http://www2.ca.uky.edu/entomology/entfacts/ef627.asp (accessed 25 August 2014).

Trägårdh, I. (1946) Outlines of a new classification of the Mesostigmata (Acarina) based on comparative morphological data. *Lunds Universitets Årsskrift N.F.* 57, 3–37.

Trammel K. (1965) Properties of petroleum oils in relation to performance as citrus tree sprays in Florida. PhD thesis, University of Florida Citrus Experiment Station, Lake Alfred, Florida.

Travis, B.V. (1968) Glyptal, a useful slide ringing compound. *Journal of Medical Entomology* 5, 24.

Tribe, H.T. (1972) Sealing of lactophenol mounts. *Transactions of the British Mycological Society* 58, 341.

Trichilo, P.J. and Leigh, T.F. (1986) Predation on spider mite eggs by the western flower thrips, *Frankliniella occidentalis* (Thysanoptera: Thripidae), an opportunist in a cotton agroecosystem. *Environmental Entomology* 15, 821–825.

Trindade, M.L.B. and Chiavegato, L.G. (1990) Colonização por *Brevipalpus obovatus* Donnadieu, 1875, *Brevipalpus californicus* (Banks, 1904) e *Brevipalpus phoenicis* (Geijskes, 1939) (Acari: Tenuipalpidae) em variedades citricas. *Laranja* 11(1), 227–240.

Trindade, M.L.B. and Chiavegato, L.G. (1994) Caracterização biologica dos ácaros *Brevipalpus obovatus* D., *B. californicus* B. and *B. phoenicis* G. (Acari: Tenuipalpidae). *Anais de Sociedade Entomológica do Brasil* 23, 189–195.

Trottin-Caudal, Y., Fournier, C. and Leyre, J.M. (2005) Protection biologique contre *Aculops lycopersici* (Massee) a l'aide d'acariens prédateurs *Neoseiulus californicus* McGregor et *Neoseiulus cucumeris* (Oudemans) en culture de tomate sous serre. Paper presented to: AFPP *7ème Conférence Internationale sur les*

Ravageurs en Agriculture, 26 et 27 Octobre 2005 et AFPP 2eme Colloque sur les Acariens des Cultures, 24 et 25 October 2005, Montpellier, France. Association Française de Protection des Plantes, Alfortville, France.

Trouessart, T. (1891) Considerations génerales sur la classification des acariens suivies d'un essai de classification nouvelle. *Revue des Sciences Naturelles de l'Ouest* 1, 289–308.

Truol, G. and Sagadin, M. (2008a) Presencia de High Plains virus (HPV) nueva enfermedad del cultivo de trigo en La Província de Buenos Aires. In: *VII Congreso Nacional de Trigo, 2, 3 y 5 de Julio de 2008, Santa Rosa, Argentina, Resúmenes*, 4.

Truol, G. and Sagadin, M. (2008b) Virosis de los cereales detectadas en Argentina. In: *Libro de Resúmenes. Workshop en Virosis de Cereales Transmitidas por el Ácaro* Aceria tosichella – Wheat streak mosaic virus y High plain virus *– en los Países del Cono Sur: Situación, Detección y Manejo, Balcarce, Argentina*. Instituto de Fitopatología y Fisiología Vegetal (IFFIVE), Córdoba/Buenos Aires and Instituto Nacional de Tecnología Agropecuaria (INTA), Buenos Aires, pp. 21–22.

Truol, G., French, R., Sagadin, M. and Arneodo, J. (2004) First report of wheat streak mosaic virus infecting wheat in Argentina. *Australasian Plant Pathology* 33, 137–138.

Truol, G., Sagadin, M., Melchiorre, G. and Izaurralde, J. (2008) *Enfermedades Virales Asociadas al Cultivo de Trigo en Argentina: Virus Transmitidos por Semilla y de Importancia en el Marco de las Exportaciones. Detección en Argentina de Sitios Geográficos con Presencia de* Wheat streak mosaic virus *(WSMV)*. Informe No. 6, INTA-IFFIVE (Instituto Nacional de Tecnología Agropecuaria- Instituto de Fitopatología y Fisiología Vegetal), Córdoba, Argentina.

Tryon, H. (1898) Fruitlet core-rot of pineapple. *Queensland Agricultural Journal* 3, 462–467.

Tsagkarakou, A., Cros-Arteil, S. and Navajas, M. (2007) First record of the invasive mite *Tetranychus evansi* in Greece. *Phytoparasitica* 35, 519–522.

Tsai, S.M., Kung, K.S. and Shih, C.I. (1989) The effect of temperature on life history and population parameters of Kanzawa spider mite, *Tetranychus kanzawai* Kishida (Acarina: Tetranychidae), on tea. *Plant Protection Bulletin (Taipei)* 31, 119–130 (in Chinese).

Tseng, Y.-H. (1974) Systematics and distribution of phytophagous and predatory mites on grape in Taiwan. Part I. Phytophagous mites. *Journal of the Agricultural Association of China* 88, 56–73 (in Chinese).

Tseng, Y.-H. (1979) Studies on the mites infesting stored food products on Taiwan. In: Rodrigues, J.G. (ed.) *Recent Advances in Acarology, Volume 1*. Academic Press, New York, pp. 311–316.

Tseng, Y.-H. (1980) Mites associated with the conifers of Taiwan, (1) Tetranychoidea, with a checklist of the world. *Journal of Entomology (Chung Hsien University)* 15, 145–168.

Tseng, Y.-H. (1984) Mites associated with weeds, paddy rice and upland rice fields in Taiwan. In: Griffthis, D.A. and Bowman, C.E. (eds) *Acarology VI, Volume 2*. Ellis Horwood, Chichester, UK, pp. 770–780.

Tseng, Y.-H. (1990) *A Monograph of the Mite Family Tetranychidae (Acarina: Trombidiformes) from Taiwan*. Taiwan Museum Special Publication Series No. 9, Taipei.

Tsolakis, H., Ragusa di Chiara, E. and Ragusa di Chiara, S. (2000) Distribution of phytoseiid mites (Parasitiformes, Phytoseiidae) on hazelnut at two different altitudes in Sicily (Italy). *Environmental Entomology* 29, 1251–1257.

Tsuchiya, M. (2002) Systematization of integrated pest and disease control technologies for citrus 1. Systematization of integrated pest control technologies for citrus (3). Establishment of effective use of pesticides. Effect of insecticide on *Aculops pelekassi* Keifer in Ito city. In: *Kaju ni Kansuru Shiken Seisekisho. Byogaichiu, Hassei Yosatsuhen* 2001, 66–67 (in Japanese).

Tsugawa, C., Yamada, M. and Shirasaki, S. (1961) Forecasting the outbreak of destructive insects in apple orchards. III. Forecasting the initial date of hatch in respect of the overwintering eggs of the European red mite, *Panonychus ulmi* (Koch) in Aomori prefecture. *Japanese Journal of Applied Entomology and Zoology* 5, 167–173.

Tsugawa, C., Yamada, M., Shirasaki, S. and Oyama, N. (1966) Forecasting the outbreak of destructive insects in apple orchards. VII. Termination of diapause in hibernating eggs of *Panonychus ulmi* (Koch) in relation to temperature. *Japanese Journal of Applied Entomology and Zoology* 10, 174–180.

Tsuji, N., Chittenden, A.R., Ogawa, T., Zhang, Y.-X., Takada, T. and Saito, Y. (2011) The possibility of sustainable pest management by introducing bio-diversity: simulations of pest mite outbreak and regulation. *Sustainability Science* 6, 97–107.

Tubeuf, C. von (1910) Teratologische Bilder. *Naturwissenschaftliche Zeitschrift für Forst- und Landwirtschaft* 8, 263.

Tucker, M.A. (1926) Some South African mites. Mainly Tetranychidae and Eriophyidae. *Entomology Memoirs, Department of Agriculture, Pretoria, Union of South Africa* 5, 3–15.

Tucker, R.W.E. (1925) The black sand mite: *Penthaleus destructor* n. sp. *Entomology Memoirs, Department of Agriculture, Pretoria, Union of South Africa* 3, 21–36.

Tuft, T.O. and Anderson, L.D. (1953) Acaricides for control of tomato russet mite in California. *Journal of Economic Entomology* 46, 502–504.

Tulashvili, N. (1930) Beobachtungen über die Schädlinge des Teestrauches und der Citrusgewächse (Zitronen, Apfelsinen) am Strandgebiet Batum im Laufe von 1927–1928 (Observations on pests of tea and *Citrus* on the Batum Coast during 1927–1928). *Mitteilungen PflSchAbt. Volhskom. Landw. SSR Georg. (Tiflis)* 11, 189–230 (cited at http://www.cabi.org/isc/abstract/19310500581).

Tulisalo, U. (1972) Resistance to the two-spotted spider mite, *Tetranychus urticae* Koch (Acarina, Tetranychidae) in the genera *Cucumis* and *Citrullus* (Cucurbitaceae). *Annales Entomologici Fennici* 38, 60–64.

Tulisalo, U. (1974) Control of the two spotted spider mite *Tetranychus urticae* Koch by high air humidity or direct contact with water. *Annales Entomologici Fennici* 40, 158–162.

Tuovinen, T. and Lindqvist, I. (2010) Maintenance of predatory phytoseiid mites for preventive control of strawberry tarsonemid mite *Phytonemus pallidus* in strawberry plant propagation. *Biological Control* 54, 119–125.

Tuovinen, T., Parikka, P. and Lemmetty, A. (2008) Plant protection in currant production in Finland. *Acta Horticulturae* 777, 333–337.

Tuovinen, T., Lindqvist, I., Kauppinen, S. and Kivijärvi, P. (2009) Integration of biological mite control and conventional insect control in strawberry. *Acta Horticulturae* 842(2), 661–666.

Turchin, P. (1990) Rarity of density dependence or population regulation with lags. *Nature* 344, 660–663.

Türk, E. and Türk, F. (1957) Systematic und Ökologie der Tyrogliphiden Mitteleuropas. In: Stammer, H.J. (ed.) *Beiträge zur Systematik und Ökologie Mitteleuropäischer Acarina, Bund 1, Abschnitt 1*. Akademische Verlagsgesellschaft Geest & Portig, Leipzig, Germany, pp. 1–231.

Turner, P. (2011) *Magister®/Magus®. Technical Review: Fenazaquin, 2011 IR-4 Food Use Workshop*. Gowan Company, Yuma, Arizona. Available at: http://ir4.rutgers.edu/FoodUse/FUWorkshop/2011FUWPresentations/2011FUWGowan.pdf (accessed 26 August 2014).

Tusnádi, C.K. and Kerényi-Nemestóthy, K. (1989) Újabb növényvédő szerek szamócaatka (*Tarsonemus pallidus sensu stricto* Banks; Acari: Tarsonemidae) elleni hatásának vizsgálata afrikai ibolyán (New acaricides against the mite *Tarsonemus pallidus sensu stricto* Banks (Acari: Tarsonemidae) on *Saintpaulia*). *Növényvédelem* 25, 393–399 (in Hungarian).

Tuttle, D.M. and Baker, E.W. (1964) *The Spider Mites of Arizona (Acarina: Tetranychidae)*. Technical Bulletin No. 158, Agricultural Experiment Station, University of Arizona, Tucson, Arizona.

Tuttle, D.M. and Baker, E.W. (1968) *Spider Mites of Southwestern United States and a Revision of the Family Tetranychidae*. The University of Arizona Press, Tucson, Arizona.

Tuttle, D.M. and Baker, E.W. (1976) New records and species of Tetranychidae and Tenuipalpidae (Acarina) from Utah and Idaho. *Great Basin Naturalist* 36, 57–64.

Tuttle, D.M. and Butler, G.D. (1961) A new eriophyid mite infesting Bermuda grass. *Journal of Economic Entomology* 54, 836–838.

Tuttle, D.M. and Mullis, C.H. (1981) Citrus red mite control on lemon. *Insecticicide and Acaricide Tests* 6, 42.

Tuttle, D.M., Baker, E.W. and Abbatiello, M. (1974) *Spider Mites from Northwestern and North Central Mexico (Acarina: Tetranychidae)*. Smithsonian Contributions to Zoology 171, Smithsonian Institution Press, Washington, DC.

Tuttle, D.M., Baker, E.W. and Abbatiello, M. (1976) Spider mites of Mexico (Acarina : Tetranychidae). *International Journal of Acarology* 2, 1–102.

Tuttle, D.M., Baker, E.W. and Sales, F.M. (1977) Spider mites (Tetranychidae: Acarina) of the state of Ceará, Brazil. *International Journal of Acarology* 3, 1–8.

Tzanetakis, I.E., Gergerich, R.C. and Martin, R.R. (2006) A new *Ilarvirus* found in rose. *Plant Pathology* 55, 568.

Ubertalli, J.A. (1955) Life history of *Eotetranychus uncatus* Garman. *Journal of Economic Entomology* 48, 47–49.

UC Davis (2009) News and Information, February 24, 2009: UC Davis cleanses greenhouses to eliminate rice pest. Available at: http://news.ucdavis.edu/search/news_detail.lasso?id=9020 (accessed 17 April 2015).

UC IPM Online (2003a) UC IPM, Phenology Model Database, European Red Mite. Available at: http://www.ipm.ucdavis.edu/PHENOLOGY/ma-european_red_mite.html (accessed 26 August 2014).

UC IPM Online (2003b) UC IPM, Phenology Model Database, Pacific Spider Mite. Available at: http://www.ipm.ucdavis.edu/PHENOLOGY/ma-pacific_spider_mite.html (accessed 26 August 2014).

UC IPM Online (2003c) UC IPM, Phenology Model Database, Twospotted Spider Mite. Available at: http://www.ipm.ucdavis.edu/PHENOLOGY/ma-2spotted_spider_mite.html (accessed 26 August 2014).

UC IPM Online (2007a) UC IPM, Sixspotted Spider Mite on Plumeria Available at: http://www.ipm.ucdavis.edu/PMG/PESTNOTES/pn74141.html (accessed 26 August 2014).

UC IPM Online (2007b) UC Pest Management Guidelines, Avocado, Avocado Brown Mite. Available at: http://www.ipm.ucdavis.edu/PMG/r8400111.html (accessed 26 August 2014).

UC IPM Online (2009a) How to Manage Pests. UC IPM Pest Management Guidelines, Olive, Olive Mite. UC ANR Publication 3452. Available at: http://www.ipm.ucdavis.edu/PMG/r583400111.html (accessed 3 March 2015).

UC IPM Online (2009b) UC Pest Management Guidelines, Apple, Webspinning Spider Mites. Available at: http://www.ipm.ucdavis.edu/PMG/r4400211.html (accessed 26 August 2014).

UC IPM Online (2009c) UC Pest Management Guidelines, Apple. Available at: http://www.ipm.ucdavis.edu/PMG/selectnewpest.apples.html (accessed 26 August 2014).

UC IPM Online (2009d) UC Pest Management Guidelines, Corn, Spider Mites. Available at: http://www.ipm.ucdavis.edu/PMG/r113400111.html (accessed 26 August 2014).

UC IPM Online (2009e) UC Pest Management Guidelines, Prune, Webspinning Spider Mites. Available at: http://www.ipm.ucdavis.edu/PMG/r606400411.html (accessed 26 August 2014).

UC IPM Online (2009f) UC Pest Management Guidelines, Walnut, European Red Mite. Available at: www.ipm.ucdavis.edu/PMG/r881400211.html (accessed 26 August 2014).

UC IPM Online (2009g) UC Pest Management Guidelines, Citrus, Broad mite. Available at: http://www.ipm.ucdavis.edu/PMG/r107400311.html (accessed 26 August 2014).

UC IPM Online (2010) UC Pest Management Guidelines, Avocado, Sixspotted Mite. Available at: http://www.ipm.ucdavis.edu/PMG/r8400311.html (accessed 26 August 2014).

UC IPM Online (2012a) How to Manage Pests, Degree-Days. Available at: http://www.ipm.ucdavis.edu/WEATHER/ddconcepts.html (accessed 26 August 2014).

UC IPM Online (2012b) UC Pest Management Guidelines, Almond. Available at: http://www.ipm.ucdavis.edu/PMG/selectnewpest.almonds.html (accessed 26 August 2014).

UC IPM Online (2012c) UC Pest Management Guidelines, Almond, Webspinning Spider Mites. Available at: http://www.ipm.ucdavis.edu/PMG/r3400211.html (accessed 26 August 2014).

UC IPM Online (2012d) UC Pest Management Guidelines, Cucurbits. Available at: http://www.ipm.ucdavis.edu/PMG/r116400111.html (accessed 26 August 2014).

UC IPM Online (2012e) UC Pest Management Guidelines, Pear. Available at: http://www.ipm.ucdavis.edu/PMG/selectnewpest.pears.html (accessed 26 August 2014).

UC IPM Online (2013a) UC Pest Management Guidelines, Almond, Dormant Spur or First-Year Twig Sampling and Treatment Guidelines. Available at: www.ipm.ucdavis.edu/PMG/r3900211.html (accessed 26 August 2014).

UC IPM Online (2013b) UC Pest Management Guidelines, Citrus, Yuma Spider Mite. Available at: http://www.ipm.ucdavis.edu/PMG/r107400611.html (accessed 26 August 2014).

UC IPM Online (2013c) UC Pest Management Guidelines, Cotton, Webspinning Spider Mites. Available at: http://www.ipm.ucdavis.edu/PMG/r114400111.html (accessed 26 August 2014).

UC IPM Online (2014a) UC Pest Management Guidelines, Avocado, Persea Mite. Available at: http://www.ipm.ucdavis.edu/PMG/r8400211.html (accessed 26 August 2014).

UC IPM Online (2014b) UC Pest Management Guidelines, Grape, Webspinning Spider Mites. Available at: http://www.ipm.ucdavis.edu/PMG/r302400111.html (accessed 26 August 2014).

UC IPM Online (2014c) UC Pest Management Guidelines, Strawberry, Spider Mites. Available at: http://www.ipm.ucdavis.edu/PMG/r734400111.html (accessed 26 August 2014).

UC IPM Online (2014d) UC Pest Management Guidelines. Strawberry, Cyclamen Mite. Available at: http://www.ipm.ucdavis.edu/PMG/r734400211.html (accessed 26 August 2014).

Uchida, M. (1980) Appearance time of diapausing females and termination of diapause in the two-spotted spider mite, *Tetranychus urticae* Koch and the Kanzawa spider mite, *Tetranychus kanzawai* Kishida on pear tree in Tottori district (Acarina: Tetranychidae). *Japanese Journal of Applied Entomology and Zoology* 24, 175–183 (in Japanese).

Ueckermann, E.A. (2004) Taxonomic research in acarology. In: Baret, S., Rouget, M., Nänni, I. and Le Bourgeois, T. (eds) *Proceedings of a Workshop on Biodiversity Dynamics on La Réunion Island, 29 November–5 December 2004, Saint Pierre-Saint Denis*. UMR PVBMT (Unité Mixte de Recherche – Peuplements

Végétaux et Bioaggresseurs en Milieu Tropical), Région Réunion de CIRAD (Centre de Coopération Internationale en Recherche Agronomique pour le Développement), St Pierre and St Denis, Réunion, pp. 12–13.

Ueckermann, E.A. (ed.) (2010) Special Issue. Eriophyoid Mites: Progress and Prognoses. *Experimental and Applied Acarology* 51(1–3).

Ueckermann, E.A. (2013) New mushroom pest. *Plant Protection News* 97, 6.

Ugarov, A.A. (1937) Spider mites from Central Asia. *Soz. Nauka i Tekhnika* 5(9), 26–40 (in Russian).

Ugarov, A.A. and Nikolskii, V.V. (1937) Systematic study of spider mites from Central Asia. *Trudy Vsesojuznoj Sredneaziatskoi Stantsii Zashchaty Rastenii* 2, 26–64 (in Russian).

Ullah, M.S. and Gotoh, T. (2013) Laboratory-based toxicity of some acaricides to *Tetranychus macfarlanei* and *Tetranychus truncatus* (Acari: Tetranychidae). *International Journal of Acarology* 39, 244–251.

Ullah, M.S., Moriya, D., Kongchuensin, M., Konvipasruang, P. and Gotoh, T. (2011) Comparative toxicity of acaricides to *Tetranychus merganser* Boudreaux and *Tetranychus kanzawai* Kishida (Acari: Tetranychidae). *International Journal of Acarology* 37, 535–543.

Ullah, M.S., Haque, M.A., Nachman, G. and Gotoh, T. (2012) Temperature-dependent development and reproductive traits of *Tetranychus macfarlanei* (Acari: Tetranychidae). *Experimental and Applied Acarology* 56, 327–344.

Umeh, V., Amih, C., Kolade, J. and Onukwu, D. (2007) Damage on sweet orange fruits by the mite *Polyphagotarsonemus latus* (Banks) and the population of its predatory mites in Southwestern Nigeria. *Fruits* 62, 149–156.

Umina, P.A. (2008) *Redlegged Earth Mite*. Agriculture Note AG0414, Department of Primary industries, State of Victoria, Australia.

Umina, P.A. and Hoffmann, A.A. (1999) Tolerance of cryptic species of blue oat mites (*Penthaleus* spp.) and the redlegged earth mite (*Halotydeus destructor*) to pesticides. *Australian Journal of Experimental Agriculture* 39, 621–628.

Umina, P.A. and Hoffmann, A.A. (2003) Diapause and implications for control of *Penthaleus* species and *Halotydeus destructor* (Acari: Penthaleidae) in southeastern Australia. *Experimental and Applied Acarology* 31, 209–223.

Umina, P.A. and Hoffmann, A.A. (2004) Plant host associations of *Penthaleus* species and *Halotydeus destructor* (Acari: Penthaleidae) and implications for integrated pest management. *Experimental and Applied Acarology* 33, 1–20.

Umina, P.A., Hoffmann, A.A. and Weeks, A.R. (2004) Biology, ecology and control of the *Penthaleus* species complex (Acari: Penthaleidae). *Experimental and Applied Acarology* 34, 211–237.

Undurraga, J.M. and Dybas, R.A. (1988) Abamectin, novel naturally-derived insecticide/acaricide for integral key pest controls on pears and tomatoes. In: *Proceedings of the Brighton Crop Protection Conference: Pests and Diseases – 1988, Volume 1*. BCPC Publications, Farnham (now Alton), UK, pp. 205–210.

Upton, M.S. (1991) *Methods for Collecting, Preserving and Studying Insects and Related Forms*, 4th edn. Australian Entomological Society, Brisbane, Australia.

Upton, M.S. (1993) Aqueous gum-chloral slide mounting media: an historical review. *Bulletin of Entomological Research* 83, 267–274.

Uraisakul, K. (2003) *Annona* seed extract and some herb extracts on chilli yield and control broad mite (*Polyphagotarsonemus latus* (Banks)) and some key pests in chilli. *Proceedings of 41st Kasetsart University Annual Conference, 3–7 February 2003. Subject: Plants and Agricultural Extension and Communication*. Kasetsart University, Bangkok, pp. 354–361.

Uraisakul, K., Kanok-Uraisakul and Oates, C.G. (1999) Effectiveness of some botanical extracts on vegetative growth of papaya and egg hatching inhibition of African red spider mite (*Eutetranychus africanus* (Tucker)). In: *Proceedings of 37th Kasetsart University Annual Conference, 3–5 February 1999*. Text and Journal Publications, Bangkok, pp. 11–18.

Urban, A., Day, M., Ellison, C. and den Breeÿen, A. (2001a) Global lantana biocontrol initiative. *Biocontrol News and Information* 22, 56–58.

Urban, A.J., Mpedi, P.F., Neser, S. and Craemer, C. (2001b) Potential of the flower gall mite, *Aceria lantanae* (Cook) (Acari: Eriophyidae), for biocontrol of the noxious weed, *Lantana camara* L. (Verbenaceae). In: Olckers, T. and Brothers, D.J. (eds) *Proceedings of the 13th Entomological Congress, 2–5 July 2001, Pietermaritzburg*. Entomological Society of South Africa, Hatfield, South Africa, pp. 67–68.

Urban, A.J., Simelane, D.O., Mpedi, P.F., Day, M.D., Neser, S. and Craemer, C. (2004) Varietal resistance in lantana: fact or fiction? In: Cullen, J.M., Briese, D.T., Kriticos, D.J., Lonsdale, W.M., Morin, L. and Scott, J.K. (eds) *Proceedings of the XI International Symposium on Biological Control of Weeds, Canberra, Australia, 27 April–2 May 2003*. CSIRO Entomology, Canberra, p. 232 (abstract).

Urisakul, G. (1982) Kan suksa buangton thang chiwawitthaya khong raidaeng mansampalang *Tetranychus truncatus* Ehara (Acarina: Tetranychidae) lae khong rai-tua-ham *Amblyseius* (*Amblyseius*) *longispinosus* (Evans) (Acarina: Phytoseiidae). MSc. thesis, AIBA-SEARCA (Agricultural Information Bank for Asia-Southeast Asian Regional Center for Graduate Study), College, Laguna, Philippines (in Thai).

Urueta, E.J. (1970) *Mononychellus planki* (McGregor) a potential pest to manihot in Colombia. *Tropical Root and Tuber Crops Newsletter* 3, 14.

Urueta, E.J. (1975) Aranas rojas (Acarina: Tetranychidae) del Departamento de Antioquia. *Revista Colombiana de Entomología* 1, 1–14.

Urueta, E.J. (1980) Control del ácaro *Retracrus elaeis* Keifer mediante el hongo *Hirsutella thompsonii* Fisher e inhibición de este por dos fungicidas. *Revista Colombiana de Entomología* 6, 3–9.

Urueta Sandino, E.J. and Navarro Alzate, R. (1978) El ácaro *Steneotarsonemus pallidus* nueva plaga de la fresa en Colombia y estudio sobre su control quimico y diseminación a otros cultivos. *Revista Colombiana de Entomología* 4, 70–75.

US EPA (1986) *Methyl Bromide*. Pesticide Fact Sheet No. 98, Office of Pesticide Programs, US Environmental Protection Agency, Washington, DC.

US EPA (2014) Pesticides. US Environmental Protection Agency, Washington, DC. Available at: http://www.epa.gov/pesticides/ (accessed 26 August 2014).

USDA (1981) New North American record. *Plant Pest News* 1(2), 2.

USDA (2007) *Detections of Panicle Rice Mite*, Steneotarsonemus spinki, *in Lajas, Puerto Rico*. Publication No. DA-2007-28, US Department of Agriculture, Washington, DC.

USDA (2009) *Panicle Rice Mite Program Manual* (updated 25 April 2012). US Department of Agriculture Animal and Plant Health Inspection Service (APHIS) and Cooperating State Departments of Agriculture, Washington, DC. Available at: http://www.aphis.usda.gov/import_export/plants/manuals/domestic/downloads/panicle_rice_mite.pdf (accessed 20 April 2015).

USDA ARS SEL (2005) Acari: Fern Mites, Tarsonemidae. In: *Acari: Plant-Feeding Mites*. Systematic Entomology Laboratory, US Department of Agriculture Agricultural Research Service, Beltsville, Maryland. Available at: http://www.sel.barc.usda.gov/acari/content/fernmites.htm (accessed 16 August 2011).

USDA NRCS (2007) The PLANTS Database. National Plant Data Center. US Department of Agriculture Natural Resources Conservation Service, Washington, DC. Available at: http://plants.usda.gov/ (accessed 27 January 2015).

Uysal, C., Cobanoğlu, S. and Okten, M.E. (2001) Determination of Tetranychoidea (Acarina: Prostigmata) species harmful in the park areas of Ankara. *Turkiye Entomoloji Dergisi – Turkish Journal of Entomology* 25, 147–160.

Vacante, V. (1981) Notizie sulla presenza di *Therodiplosis persicae* Kieffer (Diptera, Cecidomyiidae) in serra su piante orticole e floreali, attaccate da *Tetranychus urticae* Koch (Acarina, Tetranychidae). *Tecnica Agricola (Catania)* 5, 303–312.

Vacante, V. (1982a) La difesa del pomodoro in serra dall'eriofide rugginoso (*Aculops lycopersici* (Massee)) (Acarina, Eriophyidae). *Colture Protette* 11(6), 29–34.

Vacante, V. (1982b) Indagini preliminari sull'impiego combinato di *Phytoseiulus persimilis* A.H. (Acarina, Phytoseiidae) e *Therodiplosis persicae* Kieffer (Diptera, Cecidomyiidae) contro *Tetranychus urticae* Koch (Acarina, Tetranychidae) su melone in serra. In: *Atti della Giornata di Studio su "Aspetti e Problemi delle Successioni Colturali in Orticoltura", Bari, 2 Maggio 1982*, pp. 1–10.

Vacante, V. (1983) Lotta chimica con chinometionato contro *Tetranychus urticae* Koch in coltura protetta. *Colture Protette* 12(8–9), 29–33.

Vacante, V. (1984) Olive tree mites in Sicily. A first paper. In: Cavalloro, R. and Crovetti, A. (eds) *Integrated Pest Control in Olive-groves. Proceedings of the CEC/FAO/IOBC International Joint Meeting, Pisa, 3–6 April 1984*. Commission of the European Communities, Brussels, Belgium, Directorate-General for Agriculture/Food and Agriculture Organization of the United Nations, Rome/International Organization for Biological Control, Paris, West Palaearctic Regional Section. Balkema, Rotterdam, the Netherlands, pp. 285–299.

Vacante, V. (1985a) Prima raccolta di acari Tetranichidi in Sicilia. *Phytophaga* 1, 41–114.

Vacante, V. (1985b) The current state of control of phytophagous mites in protected crops in Sicily. In: *Meeting, Working Group "Integrated Control in Glasshouses"*, 14–16 May 1984, Catania, Italy. *IOBC-WPRS Bulletin* 8(1), 43–50.

Vacante, V. (1985c) Acari presenti nelle colture orticole e floricole nelle serre del ragusano e considerazioni sulle possibilità di lotta chimica, biologica ed integrata. *Tecnica Agricola (Catania)* 37, 299–322.

Vacante, V. (1986) Influence of white mineral oil treatments on the population dynamics of some mites in a lemon orchard in Eastern Sicily. In: Cavalloro, R. and Di Martino, E. (eds) *Integrated Pest Control in Citrus-groves. Proceedings of the Experts' Meeting, 26–29 March 1985, Acireale, Italy*. Published for the Commission of European Communities by Balkema, Rotterdam, the Netherlands, pp. 423–431.

Vacante, V. (1988) Note su *Tyrophagus neiswanderi* Johnston & Bruce (Acarina, Acaridae). I. Descrizione morfologica degli adulti. In: *Atti XV Congresso Nazionale Italiano di Entomologia, L'Aquila, 13–17 Giugno 1988*, pp. 213–222.

Vacante V., (1989a) Gli Acari degli agrumi in Italia. III. Gli Acarididi (Hemisarcoptidae Oudemans, Acaridae Ewing & Nesbitt, Saproglyphidae Oudemans e Glyciphagidae Berlese). *Bollettino di Zoologia Agraria e di Bachicoltura (Ser. II)* 21, 45–84.

Vacante, V. (1989b) I Coleotteri Curculionidi della fragola in Sicilia. I Contributo. *La Difesa delle Piante* 12, 107–118.

Vacante, V. (1995) Eriofidi e Ragnetti rossi. In: Viggiani, G. (ed.) *La difesa Integrata dell'Agroecosistema Agrumi*. INEA (Istituto Nazionale di Economia Agraria,), Rome, pp. 61–71.

Vacante, V. (1997a) Note sugli acari della polvere delle case in Italia. I. *Pyrogliphidae* Cunliffe. *Rivista di Parassitologia* 16(58)(3), 25–44.

Vacante, V. (1997b) Acarofauna delle colture protette mediterranee. *Notiziario delle Piante* 7, 43–52.

Vacante, V. (2000) Animali dannosi alle ortive da serra. In: Baccetti, B., Barbagallo, S., Süss, L. and Tremblay, E. (eds) *Manuale di Zoologia Agraria*. Antonio Delfino Editore, Rome, pp. 429–448.

Vacante, V. (2009a) Generalità sui mezzi e sulle strategie di lotta contro i fitofagi degli agrumi. In: Vacante ,V. and Calabrese, F. (eds) *Citrus: Trattato di Agrumicoltura*. Edagricole, Milan, Italy, pp. 289–311.

Vacante, V. (2009b) Gli acari fitofagi degli agrumi. In: Vacante, V. and Calabrese, F. (eds) *Citrus, Trattato di Agrumicoltura*. Edagricole, Bologna (now Milan), Italy, pp. 333–344.

Vacante, V. (2010) *Citrus Mites: Identification, Bionomy and Control*. CAB International, Wallingford, UK.

Vacante, V. and Benuzzi, M. (2004) *Produzione Massale e Impiego dei Principali Ausiliari nella lotta Biologica in Orticoltura e Agrumicoltura*. Aracne Editrice, Rome.

Vacante, V. and Benuzzi, M. (2007) *Difesa delle Colture in Serra*. Edagricole, Bologna (now Milan), Italy.

Vacante, V. and Bonsignore, C. (2009) L'acaro della ruggine degli agrumi (*Aculops pelekassi* (Keifer)). *Protezione delle Colture* 2, 47–51.

Vacante, V. and Bonsignore, C. (2012) Implementation of IPM in citriculture. In: Vacante, V. and Gerson, U. (eds) *Integrated Control of Citrus Pests in the Mediterranean Region*. Bentham Science, Sharja, United Arab Emirates, pp. 28–55.

Vacante, V. and Calabretta, C. (1980) Interventi primaverili contro *Panonychus ulmi* (Koch) e loro influenza sulla dinamica delle popolazioni in meleti del massiccio etneo. In: *Atti, Giornate Fitopatologiche 1980, Siusi, Bolzano 22–24 Gennaio 1980, Volume 1*. Fitogest®, Image Line Network, Faenza, Italy, pp. 337–344.

Vacante, V. and Di Martino, E. (1987) Reperti sull'acarofauna degli agrumi in Libia. *Atti del Convegno "Il Recente Contributo della Ricerca allo Sviluppo dell'Agrumicoltura Italiana", 29 aprile–30 maggio 1986, Cagliari, Italy*. Antonio Delfino Editore, Rome, pp. 559–563.

Vacante, V. and Gerson, U. (1987) Three species of *Eryngiopus* (Acari: Stigmaeidae) from Italy, with key to species and summary of habitats. *Redia* 70, 385–401.

Vacante, V. and Gerson, U. (eds) (2012) *Integrated Control of Citrus Pests in the Mediterranean Region*. Bentham Science Publishers, Sharja, United Arab Emirates.

Vacante, V. and Nucifora, A. (1984) L'acaro delle meraviglie (*Aceria sheldoni* (Ew.)): soglia d'intervento e trattamenti, I contributo. In: *Atti, Giornate Fitopatologiche 1984, Sorrento, Napoli 26–29 Marzo 1984, Volume 2*. Fitogest®, Image Line Network, Faenza, Italy, pp. 525-533.

Vacante, V. and Nucifora, A. (1985) Gli Acari degli agrumi in Italia. I. Specie rinvenute e chiave per il riconoscimento degli ordini, dei sottordini e delle famiglie. *Bollettino di Zoologia Agraria e di Bachicoltura (Ser. II)* 18, 115–166.

Vacante, V. and Nucifora, A. (1986) A first list of the mites in citrus orchards in Italy. In: Cavalloro, R. and Di Martino, E. (eds) *Integrated Pest Control in Citrus-groves. Proceedings of the Experts' Meeting, 26–29 March 1985, Acireale, Italy*. Published for the Commission of European Communities by Balkema, Rotterdam, the Netherlands, pp. 177–188.

Vacante, V. and Nucifora, A. (1987) Gli Acari degli agrumi in Italia. II. I Gamasidi (Phytoseiidae Berlese e Ascidae Voigts & Oudemans): chiave per il riconoscimento dei generi e delle specie e note di biologia. *Bollettino di Zoologia Agraria e di Bachicoltura (Ser. II)* 19, 145–184.

Vacante, V. and Tropea Garzia, G. (1987a) Reperti dell'acarofauna del melo in Sicilia. In: *Atti del Convegno "La Coltura del Melo verso Gli Anni 90", 18–20 Dicembre 1986, Cordenons, Pordenone, Italy*. Chiandetti Editore, Reana del Rojale, Udine, Italy, pp. 609–616.

Vacante, V. and Tropea Garzia, G. (1987b) Grape mites in Sicily. I Contribution. In: Cavalloro, R. (ed.) *Proceedings of a Meeting of the EC Experts' Group, Portoferraio, 26–28 September 1985. Integrated Pest Control in Viticulture*. Published for the Commission of European Communities by Balkema, Rotterdam, the Netherlands/Brookfield, Vermont, pp. 207–215.

Vacante, V. and Tropea Garzia, G. (1990) Note sul rinvenimento dell'eriofiide della tomentosità fogliare del carciofo ("*Eriophyes neocynarae*" Keifer). *Bollettino di Zoologia Agraria e di Bachicoltura (Ser. II)* 22, 209–214.

Vacante, V. and Tropea Garzia, G. (1993) Impiego programmato di *Amblyseius cucumeris* (Oud.) contro *Frankliniella occidentalis* (Perg.) su peperone in serra fredda. *Colture Protette* 22(Suppl. No. 1), 23–32.

Vacante, V. and Van Eyndhoven, G.L. (1989) Note sui Briobiini Italiani. II. Descrizione di *Bryobia cavalloroi* sp.n. (Acarina, Tetranychidae). *Frustula Entomologica (Nuova serie)* 9, 141–150.

Vacante, V., Longo, S. and Benfatto, D. (1980) Prove sperimentali di lotta contro *Panonychus citri* (McGregor). *Tecnica Agricola (Catania)* 32, 223–228.

Vacante, V., Tropea Garzia, G. and Calabrò, M. (1989a) Mass rearing of beneficial organisms in biological and integrated control of protected crops. I. *Phytoseiulus persimilis* Athias-Henriot. In: Cavalloro, R. and Pelerents, C. (eds) *Proceedings of the CEC-IOBC Experts' Group Meeting, Cabrils, 27–29 May 1987, "Integrated Pest Management in Protected Vegetable Crops"*. Published for the Commission of European Communities by Balkema, Rotterdam, the Netherlands/Brookfield, Vermont, pp.157–166.

Vacante, V., Nucifora, A. and Tropea Garzia, G. (1989b) Citrus mites in the Mediterranean area. In: Goren, R. and Mendel, K. (eds) *Proceedings of the 6th International Citrus Congress, 6–11 March 1988, Tel Aviv, Volume 3*. Published for the International Society of Citriculture, University of California, Riverside by Balaban Publishers, Rehovot, Israel, pp. 1325–1334.

Vacante, V., Bonsignore, C. and Federico, S. (2007) Indagini sul ruolo ecologico di *Aceria sheldoni* su limone. *Italus Hortus* 14(6), 103–105.

Valentine, J.W. (1970) How many marine invertebrate fossil species? *Journal of Paleontology* 44, 410–415.

Van Brussel, E.W. (1975) *Interrelations between Citrus Rust Mite,* Hirsutella thompsonii *and Greasy Spot on Citrus in Surinam*. Bulletin No. 98, Landbouwproefstation in Suriname, Paramaribo, Suriname.

van de Vrie, M. (1957) Observations on the biology and control of the tarsonemid mite on strawberries. *Mededelingen TuinbVoorlDienst* 20, 433–439.

van de Vrie, M. (1964) The distribution of phytophagous and predacious mites on leaves and shoots of apples trees. *Entomophaga* 9, 233–238.

van de Vrie, M. (1965) Problems and prospects in the integrated control of phytophagous mites. *Bollettino di Zoologia Agraria e di Bachicoltura (Ser. II)* 7, 275–283.

van de Vrie, M. (1966) Population sampling for integrated control. In: *Proceedings of the FAO Symposium on Integrated Control, Rome, Italy, 11–15 October 1965, Volume 2*. Food and Agriculture Organization of the United Nations, Rome, pp. 57–75.

van de Vrie, M. (1967) De levenswije en de bestrijding van de rondknopmijt van zwarte bes *Cecidophyopsis ribis*. *Netherlands Journal of Plant Pathology* (now *European Journal of Plant Pathology*) 73, 170–180.

van de Vrie, M. (1985) Apple. In: Helle, W. and Sabelis, M.W. (eds) *Spider Mites: Their Biology, Natural Enemies and Control*. World Crop Pests, Volume 1B. Elsevier Science, Amsterdam, pp. 311–325.

van de Vrie, M. and Kropczyńska, D. (1965) The influence of predatory mites on the population development of *Panonychus ulmi* (Koch) on apple. *Bollettino di Zoologia Agraria e di Bachicoltura (Ser. II)* 7, 119–130.

van de Vrie, M., McMurtry, J.A. and Huffaker, C.B. (1972) Ecology of tetranychid mites and their natural enemies. A review. III. Biology, ecology and pest status, and host plant relations of tetranychids. *Hilgardia* 41(13), 343–432.

Van den Berg, H. and Markham, R.H. (1987) Prey preference and reproductive success of the predatory mite *Neoseiulus idaeus* on the prey species *Mononychellus tanajoa* and *Tetranychus lombardinii*. *Insect Science and Its Application* (now *International Journal of Tropical Insect Science*) 8, 867–869.

Van Den Bosch, R., Messenger, P.S. and Gutierrez, A.P. (1982) *An Introduction to Biological Control*. Plenum Press, New York.

van den Bruel, W.E. (1940) Un ravageur de l'épinard d'hiver: *Tyrophagus dimidiatus* Herm. (*longior* Gerv.). *Bulletin des Recherches Agronomiques de Gembloux* 9, 81–99.

Van der Geest, L.P.S. (1985) Pathogens of spider mites. In: Helle, W. and Sabelis, M.W. (eds) *Spider Mites: Their Biology, Natural Enemies and Control*. World Crop Pests, Volume 1B. Elsevier Science, Amsterdam, pp. 247–258.

Van der Geest, L.P.S. (2010) IPM potentials of microbial pathogens and diseases of mites. In: Ciancio, A. and Mukerji, K.G. (eds) *Integrated Management of Arthropod Pests and Insect Borne Diseases, Volume 5*. Springer, Dordrecht, the Netherlands, pp. 249–308.

Van der Geest, L.P.S., Elliot, S.L., Breeuwer, J.A.J. and Beerling, E.A.M. (2000) Diseases of mites. *Experimental and Applied Acarology* 24, 497–560.

Van der Geest, L.P.S., Moraes, G.J. de, Návia, D. and Tanzini, M.R. (2002) New records of pathogenic fungi in mites (Arachnida: Acari) from Brazil. *Neotropical Entomology* 31, 493–495.

Van der Hammen, L. (1961) Description of *Holothyrus grandjeani* nov. spec., and notes on the classification of the mites. *Nova Guinea (n.s.) Zoology* 9, 173–194.

Van der Hammen, L. (1970) La segmentation primitive des acariens. *Acarologia* 12, 3–10.

Van der Hammen, L. (1972–1976) *F. Grandjean, Complete Acarological Works, Vol. I–VII*. W. Junk, The Hague, the Netherlands.

Van der Hammen, L. (1980) *Glossary of Acarological Terminology. Volume 1, General Terminology*. W. Junk, The Hague, the Netherlands.

Van der Hammen, L. (1989) *An Introduction to Comparative Arachnology*. SPB Academic Publishing, The Hague, the Netherlands.

van der Merwe, G.G. and Coates, T.J. (1965) Biological study of the grey mite *Calacarus citrifolii* Keifer. *South African Journal of Science* 8, 817–824.

van Dijk, P., Verbeek, M. and Bos, L. (1991) Mite-borne virus isolates from cultivated *Allium* species, and their classification into two rymoviruses in the family Potyviridae. *Netherlands Journal of Plant Pathology* (now *European Journal of Plant Pathology*) 97, 381–399.

van Dinther, J.B.M. (1951) *Phyllocoptes gracilis* as a cause of yellow leaf spots on raspberry. Tijdschrift *over Plantenziekten* (now *European Journal of Plant Pathology*) 57, 81–94.

van Dinther, J.B.M. (1952) Bestrijdingsproeven tegen *Eriophyes avellanae* Nal. en *Eriophyes gracilis* Nal. *Tijdschrift over Plantenziekten* (now *European Journal of Plant Pathology*) 58, 96–103.

Van Driesche, R., Hoddle, M. and Center, T.D. (2008) *Control of Pests and Weeds by Natural Enemies. An Introduction to Biological Control*. Blackwell, Oxford, UK.

van Epenhuijsen, C.W. (1981) Vruchtboomgalmijt (*Aculus schlechtendali* Nal.) een niet te onderschatten plaag in de appelbomen. *De Fruitteelt* 8, 238–241.

Van Eyndhoven, G.L. (1954) Invasie van Spintmijten (Acar., Tetran.). *Entomologische Berichten* 15, 221–222.

Van Eyndhoven, G.L. (1955) *Bryobia* from Hedera, apple and pear (Acar., Tetran.) – Notulae ad Tetranychidas 1. *Entomologische Berichten* 13, 340–347.

Van Eyndhoven, G.L. (1967a) *Tetranychus lintearius* Dufour, 1832, is a valid species (Acar.) – Notulae ad Tetranychidas 11. *Entomologische Berichten* 27, 90–100.

Van Eyndhoven, G.L. (1967b) The red currant gall mite, *Cecidophyopsis selachodon* n. sp. *Entomologische Berichten* 27, 149–151.

Van Eyndhoven, G.L. and Groenewold, H. (1959) On the morphology of *Steneotarsonemus pallidus* and *S. fragariae* (Acarina, Tarsonemidae). *Entomologische Berichten* 19, 123–124.

Van Eyndhoven, G.L. and Vacante, V. (1985) The *Berlesei*-group of the genus *Bryobia* Koch (Acari, Tetranychidae). *Redia* 68, 377–437.

Van Hall, C.J.J. (1924) Ziekten en plagen der cultuurgewassen in Nederlandsch-Indië in 1923. *Mededeelingenvan het Instituut voor Plantenziekten* 64, 1–47.

Van Haren, R.J.F., Steenhuis, M.M., Sabelis, M.W. and De Ponti, O.M.B. (1987) Tomato stem trichomes and dispersal success of *Phytoseiulus persimilis* relative to its prey *Tetranychus urticae*. *Experimental and Applied Acarology* 3, 115–121.

van Kruistum, G., Hoek, H., Verschoor, J. and Molendijk, L. (2012) Controlled atmosphere temperature treatment as sustainable alternative to control strawberry tarsonemid mites and plant parasitic nematodes in strawberry plants. In: Mezzetti, B. and Brás de Oliveira, P. (eds) *XXVIII International Horticultural Congress on Science and Horticulture for People (IHC2010): International Symposium on Berries: From Genomics to Sustainable Production, Quality and Health. Acta Horticulturae* 926, 601–608.

van Lenteren, J.P. (ed.) (2005) *IOBC Internet Book of Biological Control.* International Organisation of Biological Control (http://www.iobc-global.org). Version 6, February 2012. Available at: http://www.iobc-global.org/publications_iobc_internet_book_of_biological_control.html (accessed 27 August 2014).

van Lenteren, J.P. (2012) The state of commercial augmentative biological control: plenty of natural enemies, but a frustrating lack of uptake. *BioControl* 57, 1–20.

van Lenteren, J.P. and Woets, J. (1988) Biological and integrated pest control in greenhouses. *Annual Review of Entomology* 33, 239–269.

van Lenteren, J.P., Roskam, M.M. and Timmer, R. (1997) Commercial mass production and pricing of organisms for biological control of pests in Europe. *Biological Control* 10, 143–149.

Van Nieuwenhuyse, P., Vanholme, B., Tirry, L. and Van Leeuwen, T. (2010) An overview of cytochrome b point mutations conferring resistance to bifenazate and acequinocyl in field strains of spider mites (Acari: Tetranychidae). In: Moraes, G.J. de, Castilho, R.C. and Flechtmann, C.H.W. (eds) *XIII International Congress of Acarology, August 23–27, 2010, Recife-PE, Brazil, Abstract Book*, abstract 525, pp. 284–285. Available at: http://www.acarology.org/ica/ica2010/docs/0910_15_abstractBookCapa.pdf (accessed 3 February 2015).

Van Opijnen, T. and Breeuwer, J.A.J. (1999) High temperatures eliminate *Wolbachia*, a cytoplasmic incompatibility inducing endosymbiont, from the two-spotted spider mite. *Experimental and Applied Acarology* 23, 871–881.

Van Rooyen, L.J. (1966) The biology and ecology of *Panonychus citri* (McGr.). MSc. thesis, University of Pretoria, South Africa.

Vanderlip, R.L. (1972) *How a Sorghum Plant Develops.* Publication No. C-477, Cooperative Extension Service, Kansas State University, Manhattan, Kansas.

Vanek, G. and Ackermann, P. (1996) Kuceravost' akarinoza vinica a pokrokove metody chemickej ochrany. *Vinohrad* 34(4), 77–79 (in Slovakian).

Varandi, H.B., Kamali, K. and Abai, M. (2005) Biology of *Oligonychus ununguis* (Acari: Tetranychidae) on *Picea abies* in Kelar-Dasht, Mazandaran Province. *Journal of Entomological Society of Iran* 24(2), 1–18.

Varela, L.G. and Elkins, R.B. (2008) Conversion from use of organophosphate insecticides to codling moth mating disruption in California pear orchards. *Acta Horticulturae* 800, 955–960.

Vargas, C., Merayo, A. and Aguilar, H. (1996a) Identificacion de ácaros en coberturas vivas y malezas en Costa Rica. *Manejo Integrado de Plagas* 40, 40–43.

Vargas, C., Ochoa, R. and Sanabria, C. (1996b) Nuevos representantes de la familia Tetranychidae (Acari) en Costa Rica. *Manejo Integrado de Plagas* 41, 45–48.

Vargas, M.R., Olivares, P.N. and Cardemil, O.A. (2005) Postembryonic development and life table parameters of *Typhlodromus pyri* Scheuten, *Cydnodromus californicus* (McGregor) (Acarina: Phytoseiidae) and *Brevipalpus chilensis* Baker (Acarina: Tenuipalpidae). *Manejo Integrado de Plagas* 65, 147–156.

Vargas, R. (1989) *Control Biológico de Plagas y Enfermedades de la Agricultura. Proyecto Fundación Fondo de Investagicación Agropecuaria, Santiago.* Instituto de Investigaciones Agropecuarias, Santiago.

Varner, M., Mattedi, L., Lucin, R., Forno, F., Feichter, M., Mescalchin, E. and Fellin, F. (2006) Andamenti climatici anomali favoriscono l'acariosi della vite. *Informatore Agrario* 62(17), 85–89.

Vasanthakumar, D., Roobakkumar, A., Subramaniam, M.S.R., Kumar, P., Sundaravadivelan, C. and Babu, A. (2012) Evaluation of certain leaf extracts against red spider mite, *Oligonychus coffeae* Nietner (Acarina: Tetranychidae) infesting tea. *International Journal of Acarology* 38, 135–137.

Vashisth, K.S. and Nagaich, B.B. (1965) *Morus indica*, an additional host of fig mosaic. *Indian Phytopathology* 18, 315.

Vashisth, K.S. and Nagaich, B.B. (1968) *Aceria ficus* (Cotte) as a vector of fig mosaic in India. *Indian Journal of Entomology* 30, 322.

Vásquez, C., Aponte, O., Morales, J., Sanabria, M.E. and García, G. (2008a) Biological studies of *Oligonychus punicae* (Acari: Tetranychidae) on grapevine cultivars. *Experimental and Applied Acarology* 45, 59–69.

Vasquez, C., Quirós de Gonzalez, M., Aponte, O. and Sandoval, D.M.F. (2008b) First report of *Raoiella indica* Hirst (Acari: Tenuipalpidae) in South America. *Neotropical Entomology* 37, 739–740.

Vásquez, C., Mondragón, A., Dávila, M. and Aponte, O. (2009) Phytophagous mites (Tetranychoidea: Tetranychidae, Tenuipalpidae) from natural vegetations in Lara, Venezuela. *Biota Neotropica* 9(4), 55–58.

Vásquez, C., Morales-Sánchez, J., da Silva, F.R. and Sandoval, M.F. (2012) Biological studies and pest management of phytophagous mites in South America. In: Soloneski, S. (ed.) *Integrated Pest Management and Pest Control – Current and Future Tactics.* InTech, Rijeka, Shanghai, China, pp. 353–376.

Vatana, C., Chutchai, S., Manita, K. and Auranuj, K. (2001) Taxonomy and biology of mites associated with passion fruit. *Entomology and Zoology Gazette (Bangkok)* 23, 135–157 (in Thai).

Vaz Nunes, M. and Veerman, A. (1982) Photoperiodic time measurement in the spider mite *Tetranychus urticae*: a novel concept. *Journal of Insect Physiology* 28, 1041–1053.

Vázquez-Rojas, I.M. and Estrada-Venegas, E.G. (2010) Ereynetid mites (Tydeoidea: Ereynetidae) associated with garlic crops in Guanajuato, Mexico. In: Sabelis, M.W. and Bruin, J. (eds) *Trends in Acarology: Proceedings of the 12th International Congress, 21–26 August 2006, Amsterdam.* Springer, Dordrecht, the Netherlands, pp. 207–210.

Veerman, A. (1974) Carotenoid metabolism in *Tetranychus urticae* Koch (Acari: Tetranychidae). *Comparative Biochemistry and Physiology – Part B: Comparative Biochemistry* 47, 101–116.

Veerman, A. (1977a) Photoperiodic termination of diapause in spider mites. *Nature* 266, 526–527.

Veerman, A. (1977b) Aspects of the induction of diapause in a laboratory strain of the mite *Tetranychus urticae*. *Journal of Insect Physiology* 23, 703–711.

Veerman, A. (1980a) Induction and termination of diapause in the spider mite *Tetranychus urticae*. In: Minks, A.K. and Gruys, P. (eds) *Integrated Control of Insect Pests in the Netherlands.* Pudoc, Wageningen, the Netherlands, pp. 129–132.

Veerman, A. (1980b) Functional involvement of carotenoids in photoperiodic induction of diapause in the spider mite, *Tetranychus urticae*. *Physiological Entomology* 5, 291–300.

Veerman, A. (1985) Diapause. In: Helle, W. and Sabelis, M.W. (eds) *Spider Mites: Their Biology, Natural Enemies and Control.* World Crop Pests, Volume 1A. Elsevier Science, Amsterdam, pp. 279–316.

Veerman, A. (1992) Diapause in phytoseiid mites: a review. *Experimental and Applied Acarology* 14, 1–60.

Veerman, A. and Helle, W. (1978) Evidence for the functional involvement of carotenoids in the photoperiodic reaction of spider mites. *Nature* 275, 234.

Veerman, A. and Vaz Nunes, M. (1984) Photoperiod reception in spider mites: photoreceptor, clock and counter. In: Porter, R. and Collins, G.M. (eds) *Photoperiodic Regulation of Insect and Molluscan Hormones.* Ciba Foundation Symposium No 104. Pitman, London, pp. 48–64.

Veiga, A.F. de S.L. (1985) Aspectos bioecologicos e alternativas de controle do ácaro verde da mandioca *Mononychellus tanajoa* (Bondar, 1938) (Acarina, Tetranychidae) no Estado de Pernambuco. PhD thesis, ESALQ/USP (Luiz de Queiroz College of Agriculture, University of São Paulo), Piracicaba, São Paulo, Brazil.

Veire, M. van de and Degheele, D. (1992) Twospotted spider mite, *Tetranychus urticae*, control with fenpyroximate and its possible use in IPM in glasshouse tomatoes and cucumbers. *Mededelingen van de Faculteit Landbouwwetenschappen, Rijksuniversiteit Gent* 57(3A), 925–929.

Velusamy, M., Alice, D. and Subramanian, M. (1987) Reaction of rice varieties to the mite *Oligonychus oryzae*. *International Rice Research Newsletter* 12(3), 15.

Venables, E.P. and Dennys, A.A. (1941) A new method of counting orchard mites. *Journal of Economic Entomology* 34, 324.

Venables, E.P. and Heriot, A.D. (1937) *The Blister Mite of Apple and Pear*. Publication No. 577, Canada Department of Agriculture, Ottawa.

Venkateshalu, V., Srinivas, A.G., Sushila Nadagouda and Hanumantharaya, L. (2009) Bio-efficacy of plant product, Stanza against chilli thrips, *Scirtothrips dorsalis* Hood and chilli mite, *Polyphagotarsonemus latus* (Banks). *Karnataka Journal of Agricultural Sciences* 22, 559–560.

Venkatraman, T.V. and Sharma, G.R. (1960) A note on the natural control of the sugarcane mite *Oligonychus indicus* (Hirst). *Current Science* 29, 313–314.

Venzon, M., Rosado, M.C., Molina-Rugama, A.J., Duarte, V.S., Dias, R. and Pallini, A. (2008) Acaricidal efficacy of neem against *Polyphagotarsonemus latus* (Banks) (Acari: Tarsonemidae). *Crop Protection* 27, 869–872.

Venzon, M., Lemos, F., Almeida Sarmento, R., Consolação Rosado, M. and Pallini, A. (2009) Predação por coccinelídeos e crisopídeo influenciada pela teia de *Tetranychus evansi*. *Pesquisa Agropecuária Brasileira* 44, 1086–1091.

Vercammen-Grandjean, P.H. (1975) Les organes de Claparède et les papilles genitals de certains acariens sont-ils des organes respiratoires? *Acarologia* 17, 624–630.

Vergani, A.R. (1945) Transmision y naturaleza de la "lepra explosiva" del naranjo. *Ministerio de Agricultura de la Nacion, Instituto Sanidad Vegetal, Buenos Aires, Serie A* 5, 1–11.

Vergani, A.R. (1953) Ácaros de interes economico para nuestros cultivos. *Revista de Investigaciones Agrícolas (Buenos Aires)* 7, 213–252.

Verlinden, A. (1995) De teelt van bramen onder glas in containers (Rillaar 1994). *Fruitteelt Nieuws* 8(7), 22–23.

Verma, B.R. and Kapur, M. (1990) Eriophyid mites (Acari: Eriophyoidea) of arid zone fruit crops and quarantine. *Indian Journal of Entomology* 52, 249–252.

Vidal-Barraquer, R., De Mora, M., De Sivatte, M. and Miquel, J. (1966) *Phytoptus avellanae* Nal. y otros eriofidos del avellano. Boletín de *Patología Vegetal* y *Entomología Agrícola (Madrid)* 29, 133–235.

Vieira, M.R. and Castro, T.M.M.G. de (1999) Effect of cotton varieties on the biology of *Polyphagotarsonemus latus* (Banks, 1904) (Acari: Tarsonemidae). In: *Anais II Congresso Brasileiro de Algodao: o Algodao no Seculo XX, Perspectivar para o Seculo XXI. Ribeirao Preto, SP, Brasil, 5–10 Setembro 1999*. Empresa Brasileira de Pesquisa Agropecuáia, Embrapa Algodão, Campina Brande, Brazil, pp. 186–188.

Vieira, M.R. and Chiavegato, L.G. (1999) Biology of *Polyphagotarsonemus latus* (Banks) (Acari: Tarsonemidae) on lemon (*Citrus limon* Burm). *Anais de Sociedade Entomológica do Brasil* 28, 27–33.

Vieira, M.R., Campos, A.R., Castro, T.M.M.G. de, Silva, H.A. de S. da, Figueira, J.C. and Monteverde, M. de S. (2002) Cotton cultivars resistance to broad mite *Polyphagotarsonemus latus* (Banks) (Acari: Tarsonemidae). 1. Laboratory experiments. *Revista de Oleaginosas e Fibrosas* 6, 545–555.

Vierbergen, G. (1990) The spider mites of the Netherlands and their economic significance (Acarina: Tetranychidae). *Proceedings of the Section Experimental and Applied Entomology of the Netherlands Entomological Society* 1, 158–164.

Vierbergen, G. (1994) Entomology. *Panonychus citri*: a pest in the subtropics, found in the Netherlands in the open. In: *Annual Report, Diagnostic Centre, Plant Protection Service, Wageningen*. *Medelingen* 173, 28–30.

Viggiani, G. and Bianco, M. (1974) Osservazioni ed esperienze per una lotta chimica razionale contro *Phytoptus avellanae* Nal. (Acarina: Eriophyidae). *Bollettino del Laboratorio di Entomologia Agraria Filippo Silvestri, Portici* 31, 30–53.

Vilela de Resende, J.T., Maluf, W.R., Graças Cardoso, M. das, Ventura Faria, M., Gonçalves, L.D. and Rodrigues do Nascimento, I. (2008) Resistance of tomato genotypes with high level of acylsugars to *Tetranychus evansi* Baker and Pritchard. *Scientia Agricola* 65, 31–35.

Villacarlos, L.T. and Vasquez, E.A. (1988) Arthropod pests of cassava and their control. In: *Crops State of the Art and Abstract Bibliography Series* No. 15/1988. PCARRD (Philippine Council for Agriculture and Resources Research and Development), Los Baños, Philippines, pp. 31–36.

Villanueva, R.T. and Childers, C.C. (2007) Development of *Iphiseiodes quadripilis* (Banks) (Acari: Phytoseiidae) on pollen or mite diets and predation on *Aculops pelekassi* (Keller) (Acari: Eriophyidae) in the laboratory. *Environmental Entomology* 36, 9–11.

Villaneva, R.T. and Harmsen, R. (1996) Ecological interactions of tarsonemid mites in apple orchards: predation of apple rust mite and use of *Phyllonorycter blancardella* mines. *Proceedings of the Entomological Society of Ontario* 127, 99–106.

Villanueva, R.T., Michaud, J.P. and Childers, C.C. (2004) Ladybeetles as predators of pest and predacious mites in citrus. *Journal of Entomological Science* 39, 23–29.

Villanueva, R.T., Rodrigues, J.C.V. and Childers, C.C. (2005) Larval *Cryptothelea gloveri* (Lepidoptera: Psycidae), an arthropod predator and herbivore on Florida citrus. *Experimental and Applied Acarology* 36, 83–92.

Villanueva, R.T., Gagne, R. and Childers, C.C. (2006) Two species of Cecidomyiidae predacious on citrus rust mite, *Phyllocoptruta oleivora*, on Florida citrus. *Florida Entomologist* 89, 161–167.

Villaronga, P., Cosiallas, J.R. and Bonet, J. (1993) Mite fauna associated to peach orchards in Lleida (Spain). *IOBC-WPRS Bulletin* 16(4), 14–21.

Vis, R.M.J. de and Moraes, G.J. de (2002) A new species of *Panonychus* (Acari: Tetranychidae) from Peru. *Zootaxa* 48, 1–6.

Vis, R.M.J. de, Moraes, G.J. de and Bellini, M.R. (2006) Mites (Acari) of rubber trees (*Hevea brasiliensis* Muell. Arg., Euphorbiaceae) in Piracicaba, state of São Paulo, Brazil. *Neotropical Entomology* 35, 112–120.

Vitzthum, G.H. (1929a) Ordnung: Milben, Acari. In: Brohmer, P., Ehrmann, P. and Ulmer, G. (eds) *Die Tierwelt Mitteleuropas. Spinnentiere, Band 3, Lief 3*. Quelle and Meyer, Leipzig, Germany.

Vitzthum, H.G. (1929b) *Tarsonemus hydrocephalus* n.sp. *Entomologisk Tidskrift* 50, 97–102.

Vitzthum, H.G. (1940–1943) Acarina. In: Brohmer, P., Ehrmann, P. and Ulmer, G. (eds) *Dr H.G. Bronn's Klassen und Ordnung des Tierreichs, Band 5, Abteilung 5, Teil 2*. Quelle and Meyer, Leipzig, Germany.

Vitzthum, V. (1931) Resultats scientifiques du Voyage aux Indes Orientales Néerlandaises. Acarina. *Mémoires du Musée Royal d'Histoire Naturelle de Beligique* 3(5), 31–36.

Vogel, W. and Bachmann, F. (1956) *Die Astproben-Untersuchung ais Prognose-methode für das Auftreten wichtiger schadlinge im Obstbau, 1.* Mittelungen der Eidgenössischen Versuchsanstalt für Obst-, Wein- und Gartenbau, No. 57. Wädenswil, Germany.

Voigt, E. (1998) Pest control of the grapevine in landscape conservation areas. *IOBC-WPRS Bulletin* 21(2), 95–97.

Voigt, E., Szendrey, G. and Mikulas, J. (1998) Integrated pest management in grapevines in Hungary. *IOBC-WPRS Bulletin* 21(2), 107–109.

Voigt, P. (1990) Damage caused by mould mites in cucumbers. *Nachrichtenblatt Pflanzenschutz* 44, 246–247.

Volkonsky, M. (1940) *Podapolipus diander* n. sp. acarien hétérostigmatique parasite du criquet migrateur (*Locusta migratoria* L.). *Archives de l'Institut Pasteur d'Algérie* 18, 321–340.

Volterra, V. (1926) Variazioni e fluttuazioni del numero d'individui in specie animali conviventi. *Memoria della Regia Accademia Nazionale dei Lincei* (*Series 6*) 2, 31–113.

von Palm, G. (1985) Die freilebende Gallmilbe *Aculus schlechtendali* (Nal.) und ihre Bedeutung für unseren Pflanzenschutz im Apfelanbau. *Mitteilungen des Obstbauversuchsringes des Alten Landes* 40, 140–150.

von Scheller, H.D. (1962) On the bionomics and injuriousness of *Oligonychus ununguis* and of *Rhopalosiphum abietinum*. Part I. *O. ununguis*. *Zeitschrift für Angewandte Entomologie* 51, 69–85.

von Siebold, C.T.E. (1851) Ueber *Eriophyes*. *Jahresbericht der Schlesichen Gessellschaft fur Vaterldndische Cultur* 28, 88–89.

Vukovits, G. (1961) Beobachtungen und Untersuchungen über die an Prunus – Arten vorkommende Sternflecken- (Krausel-) Krankheit. *Pflanzenschutz Berichte* 26, 1–17.

Vuorisalo, T., Walls, M., Niemela, P. and Kuitunen, H. (1989) Factors affecting mosaic distribution of galls of an eriophyid mite, *Eriophyes laevis*, in alder, *Alnus glutinosa*. *Oikos* 55, 370–374.

Wafa, A.K., Soliman, Z.R. and El-Kady, M.H. (1967) Mites associated with pear trees in Egypt (Acarina). *Bulletin of the Entomological Society of Egypt* 51, 125–130.

Wafa, A.K., Zaher, A. and Yousef, A. (1968/1969) Survey of the tenuipalpid mites in U.A.R. *Bulletin de la Société Zoologique d'Égypte* 22, 52–59.

Wafa, A.K., Maher Ali, A., Zaher, M.A. and Rasmy, A.H. (1969) Supervised control of citrus mites (Acarina). *Bulletin of the Entomological Society of Egypt, Economic Series* 3, 251–256.

Wafa, A.K., Zaher, M.A. and Soliman, Z.R. (1970) Life history of the predator mite *Acaropsis aegyptiaca* Wafa and Soliman (Acarina: Cheyletidae). *Bulletin of the Entomological Society of Egypt* 54, 155–158.

Wahba, M.L., Doss, S.A. and Farrag, A.M.I. (1985) Source of re-infestation by *Eriophyes tulipae* K. for garlic plant with some biological aspects. *Bulletin of the Entomological Society of Egypt* 65, 179–182.

Wahba, M.L., El-Enany, M.A.M. and Farrag, A.M.I. (1986) Five mango varieties as affected by malformation phenomenon and bud mite infestation in Egypt. *Agricultural Research Review (Cairo)* 61(1), 193–201.

Wainstein, B.A. (1954) Fauna of spider mites injurious to orchards in South Kazakhstan. *Zoologicheskii Zhurnal* 33, 561–564 (in Russian).

Wainstein, B.A. (1956a) Material on the fauna of tetranychid mites of Kazakhstan. *Trudy Respublik, Stantsii Zashchity Rastenii Kazfilial Vaskhnil* 3, 70–83 (in Russian).

Wainstein, B.A. (1956b) New findings of spider mites in South Kazakhstan. *Zoologicheskii Zhurnal* 35, 1146–1151 (in Russian).

Wainstein, B.A. (1956c) Fauna of tetranychoid mites in South Kazakhstan. *Zoologicheskii Zhurnal* 35, 384–391 (in Russian).

Wainstein, B.A. (1958) Notes on the biology of fruit mites. *Trudy Nauchno-Issledovatel'skogo Instituta Zashchita Rasteniĭ, Kazakhskaja Akademiia Sel'skogo* 4, 172–188 (in Russian).

Wainstein, B.A. (1960) Tetranychoid Mites of Kazakhstan (with Revision of the Family). *Kazakhskaja Akademiia Sel'skogo Instituta Zashchity Rastenii/Trudy Nauchno-Issledovatel'skogo Instituta Zashchita Rasteniĭ Kazakh* 5, 1–276 (in Russian).

Waite, G.K. (1988) Integrated control of *Tetranychus urticae* in strawberries in south-east Queensland. *Experimental and Applied Acarology* 5, 23–32.

Waite, G.K. (1999) New evidence further incriminates honey-bees as vectors of lychee erinose mite *Aceria litchii* (Acari: Eriophyidae). *Experimental and Applied Acarology* 23, 145–147.

Waite, G.K. and Gerson, U. (1994) The predator guild associated with *Aceria litchii* (Acari:Eriophyidae) in Australia and China. *Entomophaga* 39, 275–280.

Waite, G.K. and McAlpine, J.D. (1992) Honey bees as carriers of lychee erinose mite *Eriophyes litchii* (Acari: Eriophyiidae). *Experimental and Applied Acarology* 15, 299–302.

Waite, G.R. (1993) Pests. In: Broadley, R.H., Wassman, R.C. and Sinclair, E. (eds) *Pineapple Pests and Disorders*. Queensland Department of Primary Industries, Brisbane, Australia, pp. 21–32.

Wakefield, M.E. and Dunn J.A. (2005) Effectiveness of the BT mite trap for detecting the storage mite pests, *Acarus siro*, *Lepidoglyphus destructor* and *Tyrophagus longior*. *Experimental and Applied Acarology* 35, 17–28.

Wald, A. (1945) *Sequential Analysis in Inspection and Experimentation*. Columbia University Press, New York.

Walde, S.J. (1995) How quality of host plant affects a predator–prey interaction in biological control. *Ecology* 76, 1206–1219.

Walde, S.J., Hardman, J.M. and Magagula, C.N. (1997) Direct and indirect species interactions influencing within-season dynamics of apple rust mite, *Aculus schlechtendali* (Acari: Eriophyidae). *Experimental and Applied Acarology* 21, 587–614.

Walenta, D.L. (2006) Insecticide/miticide efficacy for Banks grass mite control in established Kentucky bluegrass grown for seed in the Grande Ronde Valley of northeastern Oregon. In: Young, W.C. III (ed.) *2005 Seed Production Research at Oregon State University USDA-ARS Cooperating*. Department of Crop and Soil Science, Corvallis, Oregon, pp. 82–83.

Walgenbach, J.F. (1993) *Biological Control of Mites on Apples*. Publication AG-483, North Carolina Cooperative Extension Service, Raleigh, North Carolina.

Walker, G.P. and Aitken, D.C.G. (1996) Non-target effect of sprays to control California red scale (*Aonidiella aurantii* (Maskell), Hom., Diaspididae) on citrus red mite (*Panonychus citri* (McGregor), Acari, Tetranychidae). *Journal of Applied Entomology* 120, 175–180.

Walker, G.P., Voulgaropoulos, A.L. and Phillips, P.A. (1992) Distribution of citrus bud mite (Acari: Eriophyidae) within lemon trees. *Journal of Economic Entomology* 85, 2389–2398.

Wall, R. and Begon, M. (1985) Competition and fitness. *Oikos* 44, 356–360.

Wallace, H.A.H. and Sinha, R.H. (1961) Note on a new mite disease of barley and other cereals. *Canadian Journal of Plant Science* 41, 871.

Wallace, H.A.H. and Sinha, R.N. (1965) Reaction of wheat varieties to *Tetranychus sinhai* Baker (Acarina: Tetranychidae). *Canadian Entomologist* 97, 838–841.

Wallace, M.M.H. (1961) Pasture burning and its effect on the aestivating eggs of *Halotydeus destructor* (Tucker). *Australian Journal of Experimental Agriculture and Animal Husbandry* 1, 109–111.

Wallace, M.M.H. (1970a) Diapause in the aestivating egg of *Halotydeus destructor*. *Australian Journal of Zoology* 18, 295–313.

Wallace, M.M.H. (1970b) The influence of temperature on the post-diapausal development and survival in the aestivating eggs of *Halotydeus destructor* (Acari: Eupodidae). *Australian Journal of Zoology* 18, 315–329.

Wallace, M.M.H. and Mahon, J.A. (1971) The distribution of *Halotydeus destructor* and *Penthaleus major* in Australia in relation to climate and land use. *Australian Journal of Zoology* 19, 65–76.

Wallace, M.M.H. and Mahon, J.A. (1972) The taxonomy and biology of Australian Bdellidae (Acari). I. Subfamilies Bdellinae, Spinibdellinae and Cytinae. *Acarologia* 14, 544–580.

Wallis, L.W. (1967) Warm storage of *Narcissus* bulbs before hot water treatment. *Experimental Horticulture* 17, 27–37.

Wallwork, J.A. (1959) The distribution and dynamics of some forest soil mites. *Ecology* 40, 557–563.

Walter, D.E. (1992) Leaf surface structure and the distribution of *Phytoseius* mites (Acarina: Phytoseiidae) in south-eastern Australian forests. *Australian Journal of Zoology* 40, 593–603.

Walter, D.E. (1996) Living on leaves: mites, tomenta, and leaf domatia. *Annual Review of Entomology* 41, 101–114.

Walter, D.E. (2009) Reproduction and embryogenesis. In: Krantz, G.W. and Walter, D.E. (eds) *A Manual of Acarology*. Texas Tech University Press, Lubbock, Texas, pp. 54–56.

Walter, D.E. and Campbell, N.J.H. (2003) Exotic vs endemic biocontrol agents: would the real *Stratiolaelaps miles* (Berlese) (Acari: Mesostigmata: Laelapidae), please stand up? *Biological Control* 26, 253–269.

Walter, D.E. and Kaplan, D.T. (1991) Observations on *Coleoscirus simplex* (Acarina: Prostigmata), a predatory mite that colonizes greenhouse cultures of rootknot nematode (*Meloidogyne* spp.), and a review of feeding behavior in the Cunaxidae. *Experimental and Applied Acarology* 12, 47–59.

Walter, D.E. and Krantz, G.W. (2009) Collecting, rearing, and preparing specimens. In: Krantz, G.W. and Walter, D.E. (eds) *A Manual of Acarology*. Texas Tech University Press, Lubbock, Texas, pp. 83–96.

Walter, D.E. and O'Dowd, D.J. (1992) Leaf morphology and predators: effect of leaf domatia on the abundance of predatory mites. *Environmental Entomology* 21, 478–484.

Walter, D.E. and Proctor, H.C. (1999) *Mites: Ecology, Evolution and Behaviour*. University of New South Wales Press, Sydney, Australia and CAB International, Wallingford, UK.

Walter, D.E., Hudgens, R.A. and Freckman, D.W. (1986) Consumption of nematodes by fungivorous mites, *Tyrophagus* spp. (Acarina: Astigmata: Acaridae). *Oecologia* 70, 357–361.

Walter, D.E., Kethley, J. and Moore, J.C. (1987) A heptane flotation method for recovering microarthropods from arid grassland soils, with comparisons to the Merchant–Crossley high-gradient extraction method and estimates of microarthropods biomass. *Pedobiologia* 30, 221–232.

Walter, D.E., Halliday, R.B. and Smith, D. (1995) The oriental red mite, *Eutetranychus orientalis* (Klein) (Acarina: Tetranychidae), in Australia. *Journal of the Australian Entomological Society* 34, 307–308.

Walter, D.E., Seeman, O., Rodgers, D. and Kitching, R.L. (1998) Mites in the mist: how unique is a rainforest canopy knock-down fauna? *Australian Journal of Ecology* 23, 501–508.

Walter, D.E., Lindiquist, E.E., Smith, I.M., Cook, D.R. and Krantz, G.W. (2009) Order Trombidiformes. In: Krantz, G.W. and Walter, D.E. (eds) *A Manual of Acarology*. Texas Tech University Press, Lubbock, Texas, pp. 233–420.

Walters, P.J. (1976) Susceptibility of three *Stethorus* spp. (Coleoptera: Coccinellidae) to selected chemicals used in N.S.W. apple orchards. *Journal of the Australian Entomological Society* 15, 49–52.

Walton, W., Dreves, A.J., Gent, D.H., James, D.G., Martin, R.R., Chambers, U. and Skinkis, P.A. (2007) Relationship between rust mites *Calepitrimerus vitis* (Nalepa), bud mites *Colomerus vitis* (Pagenstecher) (Acari: Eriophyidae) and short shoot syndrome in Oregon vineyards. *International Journal of Acarology* 33, 307–318.

Walzer, A., Blümel, S. and Schausberger, P. (2001) Population dynamics of interacting predatory mites, *Phytoseiulus persimilis* and *Neoseiulus californicus*, held on detached bean leaves. *Experimental and Applied Acarology* 25, 731–743.

Wan, S.-Q., Liu, X.-F., Feng, G.-Z. and Pan, D.-J. (2006) Repellent activity of extracts of wild rice species against *Panonychus citri* and *Aphis citricola* associated with esterase isoenzyme in insects. *Rice Science* 13, 146–148.

Wang, B.J., Yuan, M.L., Wei, D.D., Niu, J.Z., Nan, G.Y. and Wang, J.J. (2012) High divergence levels of *Panonychus citri* populations on Rutaceae and Oleaceae as indicated by internal transcribed spacer 1 (ITS1) sequences. *International Journal of Acarology* 38, 66–73.

Wang, C.-L. (1983) The infestation and control of bulb mite (*Rhizoglyphus robini*) on gladiolus. *Journal of Agricultural Research of China* 32, 75–82 (in Chinese).

Wang, C.-L. and Lin, R.-T. (1986) Studies on soil treatments for the control of the bulb mite (*Rhizoglyphus robini* Claparède) on *Gladiolus*. *Journal of Agricultural Research of China* 35, 230–234 (in Chinese).

Wang, C.S. (1964) A new blister mite on sugar cane in Taiwan, *Aceria sacchari* Wang. *Report of the Taiwan Sugar Experiment Station* 33, 83–91 (in Japanese).

Wang, H., Varma, R.V. and Tiansen, X. (1998) *Insect Pests of Bamboos in Asia – An Illustrated Manual*. INBAR Technical Report No. 13, International Network for Bamboo and Rattan (INBAR), Beijing, Eindhoven and New Delhi. c/o International Development Research Centre, New Delhi. Available at: http://www.inbar.int/wp-content/plugins/download-monitor/download.php?id=inbar_technical_report_no13_01.pdf (accessed 22 May 2015).

Wang, H.-F. (1977) *Eurytetranychus* from China with a new species (Acarina: Tetranychidae). *Acta Entomologica Sinica* 20, 339–341 (in Chinese).

Wang, H.-F. (1980) Notes on the genus *Eotetranychus* in China with descriptions of four new species (Acarina: Tetranychidae). *Acta Entomologica Sinica* 23, 216–223 (in Chinese).

Wang, H.-F. (1981a) *Acariformes: Tetranychoidea. Economic Insect Fauna of China, Fasc. 23*. Science Press, Beijing (in Chinese).

Wang, H.-F. (1981b) *Schizotetranychus* from China with a new species (Acarina: Tetranychidae). *Sinozoology* 1, 113–116 (in Chinese).

Wang, H.-F. (1981c) Some predatory species of Stigmaeidae from Chinese orchards. *Insect Knowledge* 18, 81–82, 95 (in Chinese).

Wang, H.-F. and Cui, Y.Q. (1992) Acari; Tetranychoidea. *Insects of the Hengduan Mountains Region* 1(2), 1431–1440 (in Chinese).

Wang, H.-F., Cui, Y.Q. and Zhang, S.Y. (1981) Tetranychid and tenuipalpid mites infesting fruit trees in the north of China. *Acta Phytophylacica Sinica* 8, 9–16 (in Chinese).

Wang, H.-F., Zhang, X.M. and Zhang, K.C. (1985) A preliminary survey of tetranychid mites in Fujian with description of a new species (Acariformes: Tetranychoidea). *Wuyi Science Journal* 5, 89–93 (in Chinese).

Wang, K.H. (1985) A study on the estimation of the population density of the citrus yellow mite (*Eotetranychus kankitus*) in the field. *Journal of Southwest Agricultural College* 3, 124–134 (in Chinese).

Wang, M.-Y., Wang, D.-S., Yuan, Y.-D. and Hong, X.Y. (2008a) Development of the tomato russet mite, *Aculops lycopersici* (Massee) (Acari: Eriophyidae) on various tomato lines. *Acta Entomologica Sinica* 51, 839–843 (in Chinese).

Wang, M.-Y., Wang, D.-S., Hong, X.-Y., Yuan, Y.-D. and Gan, J.-H. (2008b) Relationship between physical and chemical differences of tomato leaf and the resistance to *Aculops lycopersici*. *Chinese Bulletin of Entomology* 45, 904–908 (in Chinese).

Wang, W.J. and Liu, T.S. (1993) Control effect of some common miticides against Kanzawa spider mite, *Tetranychus kanzawai* Kishida. *Bulletin of Taichung District Agricultural Improvement Station* 40, 1–8 (in Chinese).

Wang, Y.T., Guo, W. and He, J. (2013) Functional response of *P. persimilis* to predation of *T. turkestani*, *T. truncatus* and *T. cinnabarinus*. *Xinjiang Agricultural Sciences* 5, 839–844.

Wanibuchi, K. and Saitō, Y. (1983) The process of population increase and patterns of resource utilization of two spider mites, *Oligonychus ununguis* (Jacobi) and *Panonychus citri* (McGregor), under experimental conditions (Acari: Tetranychidae). *Researches on Population Ecology (Kyoto)* 25, 116–129.

Warburton, C. (1904) Annual report for 1904 of the zoologist. Mites of the genus *Tarsonemus* with a description of two new species. *Journal of the Royal Agricultural Society of England* 65, 273–287.

Ward, C.R., Huddleston, E.W., Owens, J.C., Hills, T.M., Richardson, L.G. and Ashdown, D. (1972) Control of the Banks grass mite attacking grain sorghum and corn in west Texas. *Journal of Economic Entomology* 65, 523–529.

Waterhouse, D.F. (1994) *Biological Control of Weeds: Southeast Asian Prospects*. Australian Centre for International Agricultural Research, Canberra.

Watson, T.F. (1964) Influence of host plant condition on population increase of *Tetranychus telarius* (L.). *Hilgardia* 35(11), 273–322.

Watt, G. and Mann, H.H. (1903) *The Pests and Blights of the Tea Plant*. Government of India. Office of Superintendent Government Printing, Calcutta, India.

Wcislo, H. (1977) Observations on leaf galls of *Tilia cordata* Mill. induced by *Eriophyes tiliae*. *Acta Biologica Cracoviensia Series Botanica* 20, 147–152.

Wearing, C.H. and Colhoun, K.J. (1999) Development of *Orius vicinus* (Ribaut) (Heteroptera: Anthocoridae) on different prey. *Biocontrol Science and Technology* 9, 327–334.

Webber, J.D. (2007) Phenology of hazelnut big bud mites in Canterbury and implications for management. MSc. thesis, Lincoln University, Lincoln, New Zealand.

Webber, J.D., Chapman, R.B. and Worner, S.P. (2008) Forecasting emergence and movement of overwintering hazelnut big bud mites from big buds. *Experimental and Applied Acarology* 45, 39–51.

Wedding, R.T., Rihel, L.A. and Jeppson, L.R. (1958) Red mite on citrus, experiments designed to measure damage give bases for further studies. *California Agriculture* 12(8), 9, 10, 12.

Weeks, A.R. and Breeuwer, J.A.J. (2001) *Wolbachia*-induced parthenogenesis in a genus of phytophagous mites. *Proceedings of the Royal Society B: Biological Sciences* 268, 2245–2251.

Weeks, A.R. and Hoffmann, A.A. (1999) The biology of *Penthaleus* species in southeastern Australia. *Entomologia Experimentalis et Applicata* 92, 179–189.

Weeks, A.R. and Hoffmann, A.A. (2000) Competitive interactions between two pest species of earth mites, *Halotydeus destructor* and *Penthaleus major* (Acarina: Penthaleidae). *Journal of Economic Entomology* 93, 1183–1191.

Weeks, A.R. and Stouthamer, R. (2003) Increased fecundity associated with infection by a *Cytophaga*-like intracellular bacterium in the predatory mite, *Metaseiulus occidentalis*. *Proceedings of the Royal Society B: Biological Sciences* 271(Suppl. 4), S193–S195.

Weeks, A.R., Fripp, Y.J. and Hoffmann, A.A. (1995) Genetic structure of *Halotydeus destructor* and *Penthaleus major* populations in Victoria (Acari: Penthaleidae). *Experimental and Applied Acarology* 19, 633–646.

Weeks, A.R., Turelli, M. and Hoffmann, A.A. (2000a) Dispersal patterns of pest earth mites (Acari: Penthaleidae) in pastures and crops. *Journal of Economic Entomology* 93, 1415–1423.

Weeks, A.R., van Opijnen, T. and Breeuwer, J.A.J. (2000b) AFLP fingerprinting for assessing intraspecific variation and genome mapping in mites. *Experimental and Applied Acarology* 24, 775–793.

Weeks, A.R., Marec, F. and Breeuwer, J.A.J. (2001) A mite species that consists entirely of haploid female. *Science* 292, 2479–2482.

Wegner Kib, G. (2003) Die Krauselmilbe (*Calepitrimerus vitis*), ein ernst zu nehmender Schadling im Weinbau. *Schweizer Zeitschrift für Obst- und Weinbau* 139(10), 9–12.

Wei, Y.-L., Chen, X.-R. and Lin, S.-J. (1991) Occurrence and control of the apricot bud gall. *Journal of Gansu Agricultural University* 26, 66–74 (in Chinese).

Weidhaas, J.A. (1979) Spider mites and other Acarina on tree and shrubs. *Journal of Arboriculture* 5, 9–15.

Weidhaas, J.A. and Reeves, R.M. (1963) The occurrence and importance of tetranychid and eriophyid mites on woody plants in New York. *Advances in Acarology* 1, 25–29.

Weinman, J.J., Djordjevic, M.A., Creaser, E.H., Lawson, C.G.R., Broderick, K., Mathesius, U., Gärtner, E., Pittock, C., Majnik, J. de and Rolfe, B.G. (1995) Preparing subterranean clovers for future biotechnology: molecular analysis of genes and proteins involved in stress and defence reactions and the construction of transgenic plants. *Plant Protection Quarterly* 10, 47–49.

Weintraub, P. and Palevsky, E. (2008) Evaluation of the predatory mite, *Neoseiulus californicus*, for spider mite control on greenhouse sweet pepper under hot arid field conditions. *Experimental and Applied Acarology* 45, 29–37.

Weintraub, P.G., Kleitman, S., Mori, R., Shapira, N. and Palevsky, E. (2003) Control of the broad mite (*Polyphagotarsonemus latus* (Banks) on organic greenhouse sweet peppers (*Capsicum annuum* L.) with the predatory mite, *Neoseiulus cucumeris* (Oudemans). *Biological Control* 27, 300–309.

Weintraub, P.G., Kleitman, S., Mori, R., Gan-Mor, S., Gannot, L. and Palevsky, E. (2009) Novel application of pollen to augment the predator *Amblyseius swirskii* on greenhouse sweet pepper. *IOBC-WPRS Bulletin* 50, 119–124.

Weiser, J. (1956) *Nosema steinhausi* n.sp., nova mikrosporidie z rostoâe *Tyrophagus noxius* (Acarina, Tyroglyphidae). *Ceskoslovenska Parasitologie* 11, 187–192 (in Czech).

Weiser, J. (1968) *Triplosporium tetranychi* sp. n. (Phycomycetes, Entomophthoraceae), a fungus infecting the red mite *Tetranychus alteae* Hanst. *Folia Parasitologica* 15, 115–122.

Weiser, J. and Muma, M.H. (1966) *Entomophthora floridana* n. sp. (Phycomycetes: Entomophthoraceae), a parasite of the Texas citrus mite *Tetranychus banksi*. *Florida Entomologist* 49, 155–159.

Weiss, M.J. and McDonald, G. (1998) European earwig, *Forficula auricularia* L. (Dermaptera: Forficulidae), as a predator of the redlegged earth mite, *Halotydeus destructor* (Tucker) (Acarina: Penthaleidae). *Australian Journal of Entomology* 37, 183–185.

Wekesa, V.W., Maniania, N.K., Knapp, M. and Boga, H.I. (2005) Pathogenicity of *Beauveria bassiana* and *Metarhizium anisopliae* to the tobacco spider mite *Tetranychus evansi*. *Experimental and Applied Acarology* 36, 41–50.

Wekesa, V.W., Knapp, M., Maniania, N.K. and Boga, H.I. (2006) Effects of *Beauveria bassiana* and *Metarhizium anisopliae* on mortality, fecundity and egg fertility of *Tetranychus evansi*. *Journal of Applied Entomology* 130, 155–159.

Wekesa, V.W., Knapp, M. and Delalibera, I. Jr (2008) Side-effects of pesticides on the life cycle of the mite pathogenic fungus *Neozygites floridana*. *Experimental and Applied Acarology* 46, 287–297.

Wekesa, V.W., Vital, S., Silva, R.A., Ortega, E.M.M., Klingen, I. and Delalibera, I. (2011) The effect of host plants on *Tetranychus evansi*, *Tetranychus urticae* (Acari: Tetranychidae) and on their fungal pathogen *Neozygites floridana* (Entomophthorales: Neozygitaceae). *Journal of Invertebrate Pathology* 107, 139–145.

Welbourn, W.C. (1995) Balaustium *spp. in Florida (Acari: Erythraeidae)*. Entomology Circular No. 368, Florida Department of Agriculture and Consumer Services, Division of Plant Industry, Gainesville, Florida.

Welbourn, W.C. and Jennings, D.T. (1991) Two new species of Erythraeidae (Acari: Prostigmata) associated with the spruce budworm, *Choristoneura fumiferana* (Clemens) (Lepidoptera: Tortricidae), in Maine. *Canadian Entomologist* 123, 567–580.

Welbourn, W.C. and Young, O.P. (1987) New genus and species of Erythraeinae (Acari: Erythraeidae) from Mississipi with a key to the genera of North American Erythraeidae. *Annals of the Entomological Society of America* 80, 230–242.

Welbourn, W.C., Ochoa, R., Kane, E.C. and Erbe, E.F. (2003) Morphological observations on *Brevipalpus phoenicis* (Acari: Tenuipalpidae) including comparisons with *B. californicus* and *B. obovatus*. *Experimental and Applied Acarology* 30, 107–133.

Welbourn, W.C., Rodrigues, J.C. and Peña, J.E. (2008) *The Hibiscus Erineum Mite,* Aceria hibisci *(Acari: Eriophyidae) A New Introduction in the Caribbean and a Potential Threat to Florida's Hibiscus.* Publication No. ENY-852, University of Florida/Institute of Food and Agricultural Sciences (UF/IFAS) Extension, Gainesville, Florida. Available at: http://ufdcimages.uflib.ufl.edu/IR/00/00/13/47/00001/IN77700.pdf (accessed 26 May 2015)

Welter, S.C., Barnes, M.M., Ting, I.P. and Hayashi, J.T. (1984) Impact of various levels of late-season spider mite (Acari: Tetranychidae) feeding damage on almond growth and yield. *Environmental Entomology* 13, 52–55.

Welter, S.C., McNally, P.S. and Farnham, D.S. (1989a) Effect of Willamette mite (Acari: Tetranychidae) on grape productivity and quality: a reappraisal. *Environmental Entomology* 18, 408–411.

Welter, S.C., Farnham, D.S., McNally, P.S. and Freeman, R. (1989b) Effect of Willamette mite and Pacific spider mite (Acari: Tetranychidae) on grape photosynthesis and stomatal conductance. *Environmental Entomology* 18, 953–957.

Welter, S.C., Freeman, R. and Farnham, D.S. (1991) Recovery of Zinfandel grapevines from feeding damage by Willamette spider mite (Acari: Tetranychidae)-implications for economic injury level studies in perennial crops. *Environmental Entomology* 20, 104–109.

Welty, C. (1995) Survey of predators associated with European red mite (*Panonychus ulmi*; Acari: Tetranychidae) in Ohio apple orchards. *Great Lake Entomologist* 28, 171–184.

Wen, H.C. and Lee, H.S. (1981) Seasonal occurrence and chemical control of the carmine spider mite (*Tetranychus cinnabarinus* (Boisduval)) on papaya. *Plant Protection Bulletin (Taipei)* 23, 131–135 (in Chinese).

Wen, H.C., Lee, H.S. and Lin, C.C. (1991) Field studies on litchi erineum mite (*Eriophyes litchii* Keifer) in southern Taiwan. *Journal of Agricultural Research of China* 40, 298–304 (in Chinese).

Wen, H.C., Hao, H.H. and Lin, C.C. (1993) Seasonal occurrence and chemical control of the citrus red mite (*Panonychus citri* (McGregor)) on Indian jujube. *Journal of Agricultural Research of China* 42, 303–308 (in Chinese).

Wene, G.P. (1956) *Tetranychus marianae* McG., a new pest of tomatoes. *Journal of Economic Entomology* 49, 712.

Wergin, W.P., Ochoa, R., Erbe, E.F. and Joy, D.C. (2000a) Oscillating trichobothria in the low temperature SEM: biological capacitors or charging artifacts. *Scanning* 22, 140–141.

Wergin, W.P., Ochoa, R., Erbe, E.F., Craemer, C. and Raina, A.K. (2000b) Use of low-temperature field emission scanning electron microscopy to examine mites. *Scanning* 22, 145–155.

Wermelinger, B. and Delucchi, V. (1990) Effect of sex-ratio on multiplication of the twospotted spider mite as affected by leaf nitrogen. *Experimental and Applied Acarology* 9, 11–18.

Wermelinger, B., Oertli, J.J. and Delucchi, V. (1985) Effect of host plant nitrogen fertilization on the biology of the two-spotted spider mite, *Tetranychus urticae*. *Entomologia Experimentalis et Applicata* 38, 23–28.

Wermelinger, B., Oertli, J.J and Baumgärtner, J. (1991) Environmental factors influencing the life-tables of *Tetranychus urticae* (Acari: Tetranychidae). III. Host-plant nutrition. *Experimental and Applied Acarology* 12, 259–274.

Wermelinger, B., Candolfi, M.P. and Baumgärtner, J. (1992) A model of the European red mite (Acari: Tetranychidae) population dynamics and its linkage to grapevine growth and development. *Journal of Applied Entomology* 114, 155–166.

West, K.J. and De Angelis, J.D. (1993) *Typhlodromus americanus* Chant and Yoshida-Shaul, a predator of the spruce spider mite. *International Journal of Acarology* 19, 193–194.

Westigard, P.H. (1969) Timing and evaluation of pesticides for control of the pear rust mite. *Journal of Economic Entomology* 62, 1158–1161.

Westigard, P.H. (1975) Population injury levels and sampling of the pear rust mite on pears in southern Oregon. *Journal of Economic Entomology* 68, 786–790.

Westigard, P.H. and Berry, D.W. (1964) Control of the pear rust mite *Epitrimerus pyri*. *Journal of Economic Entomology* 57, 953–955.

Westigard, P.H. and Hoyt, S.C. (1965) The rust mite apple and pear species, *Aculus schlechtendali* Nalepa and *Epitrimerus pyri* Nal. differ but look alive. *Western Fruit Grower* 19(4), 31.

Westigard, P.H., Lombard, P.B. and Allen, R.B. (1979) Effects of overtree irrigation on density and damage of pear pests. *Journal of Economic Entomology* 72, 839–840.

Westigard, P.H., Gut, L.J. and Liss, W.J. (1986) Selective control program for the pear pest complex in southern Oregon. *Journal of Economic Entomology* 79, 250–257.

Westphal, E. (1970) Quelques aspects structuraux cytologiques et histochimiques de la galle en bourse d'*Eriophyes ulmicola* Nal. sur les feuilles d'*Ulmus campestris* L. *Marcellia* 36, 199–215.

Westphal, E. (1972) Traces de succion parasitaire laissées par quelques ériophyides cécidogènes. Aspect histochimique et observations ultrastructurales. *Marcellia* 37, 53–69.

Westphal, E. (1974) Cécidogenèse et aspects ultrastructuraux de la galle en bourse de l'*Eriophyes padi* Nal. sur la feuille de *Prunus padus* L. *Marcellia* 38, 77–93.

Westphal, E. (1975) Observations sur le developpement et l'ultrastructure de quelques erinoses. I. Erinoses provoquée par l'*Eriophyes leiosoma* (Nalepa) sur le *Tilia intermedia* D.C. *Marcellia* 38, 197–209.

Westphal, E. (1977a) Morphogenèse, ultrastructure et étiologie de quelques galles d'Eriophyides (acariens). *Marcellia* 39, 193–375.

Westphal, E. (1977b) Sequence of early changes in the leaf epidermis of *Prunus padus* L. under the influence of brief cecidogenic activity of eriophyid mites. *Marcellia* 40, 151–157.

Westphal, E. (1980) Responses of some Solanaceae to attack by the gall mite *Eriophyes cladophthyrus*. *Plant Disease* 64, 406–409.

Westphal, E. (1982) Modification du pH vacuolaire des cellules épidermiques foliaires de *Solanum dulcamara* soumises à l'action d'un acarien cécidogène. *Canadian Journal of Botany* 60, 2882–2888.

Westphal, E. (1983) Adaptation of gall mites (Acari, Eriophyidae) to live in galls. In: Margaris, N.S., Arianoutsou-Faraggitaki, M. and Reiter, R.J. (eds) *Adaptations to Tetrestrial Environments*. Plenum Press, New York, pp. 69–75.

Westphal, E. (1985) Potentialités morphogènes de l'épiderme foliaire de *Solanum dulcamara* parasité par *Eriophyes lycopersici*. *Beiträge zur Biologie der Pflanzen* 60, 475–481.

Westphal, E. (1992) Cecidogenesis and resistance phenomena in mite-induced galls. In: Shorthouse, J. and Rohfritsch, O. (eds) *Biology of Insect-induced Galls*. Oxford University Press, Oxford, UK, pp. 141–156.

Westphal, E. and Manson, D.C.M. (1996) Feeding effects on host plants: gall formation and other distortions. In: Lindquist, E.E., Sabelis, M.W. and Bruin, J. (eds) *Eriophyoid Mites: Their Biology, Natural Enemies and Control.* World Crop Pests, Volume 6. Elsevier Science, Amsterdam, pp. 231–242.

Westphal, E., Bronner, R. and Le Ret, M. (1981) Changes in leaves of susceptible and resistant *Solanum dulcamara* infested by the gall mite *Eriophyes cladophthirus* (Acarina, Eriophioidea). *Canadian Journal of Botany* 59, 875–882.

Westphal, E., Bronner, R. and Dreger, F. (1989) Résistance par hypersensibilité de *Solanum dulcamara* L. à l'attaque d'un eriophyide. In: Institut Agronomique Mediterraneen (ed.) *Colloque sur les Acariens des Cultures, 24–26 October 1989. Montpellier. Annales ANPP(Association Nationale pour la Protection des Plantes)* No. 2(1/1), pp. 219–226.

Westphal, E., Dreger, F. and Bronner, R. (1990) The gall mite *Aceria cladophthirus* (Nalepa). I. Life cycle, survival outside the gall and symptoms' expression on susceptible and resistant *Solanum dulcamara* L. plants. *Experimental and Applied Acarology* 9, 183–200.

Westphal, E., Dreger, F. and Bronner, R. (1991) Induced resistance in *Solanum dulcamara* triggered by the gall mite *Aceria cladophthirus* (Acari: Eriophyoidea). *Experimental and Applied Acarology* 12, 111–118.

Westphal, E., Perrot-Minnot, M.J., Kreiter, S. and Gutierrez, J. (1992) Hypersensitive reaction of *Solanum dulcamara* to the gall mite *Aceria cladophthirus* causes an increased susceptibility to *Tetranychus urticae*. *Experimental and Applied Acarology* 15, 15–26.

Westphal, E., Bronner, R. and Dreger, F. (1996) Host plant resistance. In: Lindquist, E.E., Sabelis, M.W. and Bruin, J. (eds) *Eriophyoid Mites: Their Biology, Natural Enemies and Control.* World Crop Pests, Volume 6. Elsevier Science, Amsterdam, pp. 681–688.

Weygoldt, P. (1998) Evolution and systematics of the Chelicerata. In: *Proceedings EURAAC III Symposium, Amsterdam. Experimental and Applied Acarology* 22, 63–79.

Weygoldt, P. and Paulus, H.F. (1979a) Untersuchungen zur Morphologie, Taxonomie and Phylogenie der Chelicerata. I, Morphologische Untersuchungen. *Zeitschrift für Zoologische Systematik und Evolutionsforschung* 17, 85–116.

Weygoldt, P. and Paulus, H.F. (1979b) Untersuchungen zur Morphologie, Taxonomie and Phylogenie der Chelicerata. II, Cladogramme und die Entfaltung der Chelicerata. *Zeitschrift für Zoologische Systematik und Evolutionsforschung* 17, 177–200.

Whalon, M.E., Mota-Sanchez, D., Hollingworth, R.M. and Duynslager, L. (2010) Arthropod Pesticide Resistance Database. Insecticide Resistance Action Committee, Michigan State University, East Lansing, Michigan. Available at: www.pesticideresistance.com (accessed 28 August 2014).

Wharton, G.W. and Brody, A.R. (1972) A peritrophic membrane of the mite *Dermatophagoides farinae*: Acariformes. *Journal of Parasitology* 58, 801–804.

Wheeler, A.G., Colburn, R.B. and Lehman, R.D. (1973) *Stethorus punctillum* associated with spruce spider mite on ornamentals. *Environmental Entomology* 4, 718–720.

Whelan, E.D.P. and Conner, R.L. (1989) Registration of LRS-7-50 wheat germplasm. *Crop Science* 29, 838.

Whelan, E.D.P. and Lukow, O.M. (1990) The genetics and gliadin protein characteristics of a wheat–alien translocation that confers resistance to colonization by the wheat curl mite. *Genome* 33, 400–404.

White, L.D. (1964) Laboratory studies on the effects of temperature, relative humidity, and host variations of *Tetranychus canadensis* (McGregor) in Utah. PhD thesis, Utah State University, Logan, Utah.

White, L.D. (1965) *Plant Feeding Mites of South Dakota.* Technical Bulletin No. 27, Entomology–Zoology Department, Agricultural Experiment Station, South Dakota State University, Brookings, South Dakota.

White, N.D.G. and Laing, J.E. (1977) Field observations of *Zetzellia mali* (Ewing) (Acarina: Stigmaeidae) in southern Ontario apple orchards. *Proceedings of the Entomological Society of Ontario* 108, 23–30.

White, N.D.G. and Sinha, R.N. (1979) Natural regulation of *Tarsonemus granarius* numbers in stored wheat ecosystems: a multivariate assessment. In: Rodriguez, J.R. (ed.) *Recent Advances in Acarology, Volume 1.* Academic Press, New York, pp. 291–298.

White, N.D.G. and Sinha, R.N. (1981) Life history and population dynamics of the mycophagous mite *Tarsonemus granarius* Lindquist (Acarina: Tarsonemidae). *Acarologia* 22, 353–360.

White, P.F. (1991) *The Control of Mushroom Pests Using Biological Control Agents and Environmentally Acceptable Products.* The Horticultural Development Council, Petersfield, UK.

White, S.M., Gillham, M.C., Norton, J.A., Starner, V.R. and Dybas, R.A. (1996) The suitability of abamectin 3.4 EC for mite management programs in European apple orchards. *Faculteit Landbouwkundige en Toegepaste Biologische Wetenschappen, Universiteit Gent* 61(3a), 877–885.

Whitehead, V.B., Rust, D.J., Pringle, K.A. and Albertse, G. (1978) The bud-infesting strain of the grape leaf blister mite, *Eriophyes vitis* (Pgst.), on vines in the Western Cape Province. *Journal of the Entomological Society of Southern Africa* 41, 9–15.

Wicht, M.C. Jr (1970) Three new species of pyemotid mites associated with commercial mushrooms. *Acarologia* 12, 262–268.

Wickramananda, I.R., Silva, P.H.P.R. de, Sarathchandra, S.R., Kumara, A.D.N.T. and Rajapaksha, R.H.S. (2005) A method for *in vitro* culture of coconut mite, *Aceria guerreronis* Keifer (Acari: Eriophyidae) in immature leaf tissues of coconut (*Cocos nucifera*). In: Peiris, T.S.G. and Ranasinghe, C.S. (eds) *Proceedings of the International Conference to Mark the 75th Anniversary of the Coconut Research Institute of Sri Lanka, Part II Review Papers and Guest Presentation, (Contributed papers).* Coconut Research Institute of Sri Lanka, Lunuwila, Sri Lanka, pp. 229–236.

Wiese, M.V. (1987) *Compendium of Wheat Diseases.* APS Press, St Paul, Minnesota.

Wilkin, D.R., Murdoch, G. and Woodville, H.C. (1976) The chemical control of mites infesting *Freesia* corms and *Narcissus* bulbs. *Annals of Applied Biology* 82, 186–189.

Wilkins, P.W. (1987) Transfer of polygenic resistance to ryegrass mosaic virus from perennial to Italian ryegrass by backcrossing. *Annals of Applied Biology* 11, 409–413.

Wilkins, P.W. and Hides, D.H. (1976) Tolerance to ryegrass mosaic virus, its assessment and effect on yield. *Annals of Applied Biology* 83, 399–405.

Wille, J.E., Simón, J.E. and González, J.E. (1955) *Los Insectos e Insecticidas en la Campaña Algodonera 1954–55.* Informe Mensual, Estacion Experimental Agricola de 'La Molina', Lima, Peru.

Wilson, L.J. (1994) Plant-quality effect on life-history parameters of the twospotted spider mite (Acari: Tetranychidae), on cotton. *Journal of Economic Entomology* 87, 1665–1673.

Wilson, L.J., Smilanick, J.M., Hoffmann, M.P., Flaherty, D.L. and Ruiz, S.M. (1988) Leaf nitrogen and position in relation to population parameters of Pacific spider mite, *Tetranychus pacificus* (Acari: Tetranychidae) on grapes. *Environmental Entomology* 17, 964–968.

Wilson, L.J., Bauer, L.R. and Walter, G.H. (1996) "Phytophagous" thrips are facultative predators of twospotted spider mites (Acari: Tetranychidae) on cotton in Australia. *Bulletin of Entomological Research* 86, 297–305.

Wilson, L.T. and Barnett, W.W. (1983) Degree-days: an aid in crop and pest management. *California Agriculture* 37(1), 4–7.

Wilson, L.T., Gonzalez, D., Leigh, T.F., Maggi, V., Foristiere, C. and Goodell, P. (1983) Within-plant distribution of spider mites (Acari: Tetranychidae) on cotton: a developing implementable monitoring program. *Environmental Entomology* 12, 128–134.

Wilson, L.T., Hoy, M.A., Zalom, F.G. and Smilanick, J.M. (1984) Sampling mites in almonds. I. Within-tree distribution and clumping pattern of mites with comments on predatory-prey interactions. *Hilgardia* 52(7), 1–13.

Wilson, L.T., Pickett, C.H., Leigh, T.F. and Carey, J.R. (1987) Spider mite (Acari: Tetranychidae) infestation foci: cotton yield reduction. *Environmental Entomology* 16, 614–617.

Wilson, M.F. and O'Dowd, D.J. (1990) The relationship of leaf size and shoot length in *Prunus americana* to leaf-galling by mites. *American Midland Naturalist* 123, 408–413.

Wilson, N.L. and Oliver, A.D. (1969) Evaluation of some acaricides for control of spider mites on three woody ornamentals in Louisiana. *Journal of Economic Entomology* 62, 1400–1401.

Wilson, N.S. (1965) A new species of blister forming mite on pear. *Annals of the Entomological Society of America* 58, 327–330.

Wilson, N.S. (1970) Some eriophyid species from Mexico. *Annals of the Entomological Society of America* 63, 1125–1136.

Wilson, N.S. and Cochran, L.C. (1952) Yellow spot, and eriophyid mite injury on peach. *Phytopathology* 42, 443–447.

Wilson, N.S. and Oldfield, G.N. (1966) New species of eriophyid mites from western North America, with a discussion of eriophyid mites on *Populus*. *Annals of the Entomological Society of America* 59, 585–599.

Wilson, N.S., Jones, L.S. and Cochran, L.C. (1955) An eriophyid mite vector of the peach mosaic virus. *Plant Disease Reporter* 39, 889–892.

Winfield, A.L. (1958) Bulb scale mite. In: Hodson, W.E.H. (ed.) *Narcissus Pests*, 4th edn. MAFF (Ministry of Agriculture and Fisheries) Bulletin No. 51. Her Majesty's Stationery Office, London, pp. 24–27

Winfield, A.L. (1964) Chemical control of bulb scale mite on forced narcissus. *Experimental Horticulture* 11, 69–77.

Winfield, A.L. (1967) Experiments on controlling bulb scale mite of *Narcissus*, *Steneotarsonemus laticeps* Halbt. *Horticulture Research* 7, 34–43.

Winfield, A.L. (1971) Control of bulb scale mite and stem nematode of *Narcissus*, and reclaiming forced bulbs. *Plant Pathology* 20, 10–13.

Winston, P.W. (1964) The physiology of water balance in Acarina. *Acarologia* (Facsimile Hors Série), 307–314.

Winterton, S. and de Freitas, S. (2006) Molecular phylogeny of the green lacewings (Neuroptera: Chrysopidae). *Australian Journal of Entomology* 45, 235–243.

Wipada, V. (1997) Kan suksa prasitthiphap kan kin rai sattru som khong maengmum yai phaen *Hylyphantes graminicola* Sundevall. *Entomology and Zoology Gazette (Bangkok)* 19, 144–156 (in Thai).

Witalinski, W. (2000) *Aclerogamasus stenocornis* sp.n., a fossil mite from the Baltic amber (Acari: Gamasida: parasitidae). *Genus (Warsaw)* 11, 619–626.

Witalinski, W., Liana, M. and Alberti, G. (2002) Fine structure and probable function of ring organs in the mite *Histiostoma feroniarum* (Acari: Actinotrichida: Acaridida: Histiostomatidae). *Journal of Morphology* 253, 255–263.

Witters, J., Casteels, H. and De Bondt, G. (2003) Diagnostic acarological research at the Department of Crop Protection in 2002–2003. *Parasitica* 59, 107–111.

Witul, A. and Kielkiewicz, M. (1999) Life-history parameters of two closely related forms of the *Tetranychus urticae* complex on differents host plants. In: Bruin, J., Van der Geest, L.P.S. and Sabelis, M.W. (eds) *Ecology of the Acari*. Kluwer Academic Publishers, Dordrecht, the Netherlands, pp. 399–404.

Wolffenstein, O. (1879) *Phytoptus lycopersici* W. *Monatsschrift* 22, 42–426.

Womersley, H. (1933) A possible biological control of the clover springtail or lucerne flea (*Sminthurus viridis*) in Western Australia. *Journal of the Council for Scientific and Industrial Research (Melbourne)* 6, 83–91.

Womersley, H. (1940) Studies in Australian Acarina Tetranychidae and Trichadenidae. *Transactions of the Royal Society of South Australia* 64, 233–265.

Womersley, H. (1941) Studies in Australian Acarina. (2). Tyroglyphidae (*s. l.*). *Records of the South Australian Museum* 6, 451–488.

Womersley, H. (1942) Miscellaneous additions to the acarine fauna of Australia. *Transactions of the Royal Society of South Australia* 66, 85–92.

Wongkopras, A. (1992) Phytophagous mites of economic crops. In: *Phytophagous Mites and Spiders. Monograph of the 9th Training Course "Insect and Animal Pests and Their Controls", 24 March–4 April 1997*. Entomology and Zoology Division, Department of Agriculture, Bangkok, Thailand, pp. 47–52 (in Thai).

Wongthong, P. and Pimsamarn, S. (2007) Toxicity of *Derris elliptica* Benth extracts on *Tetranychus truncatus* Ehara. *Agricultural Science Journal* 38(6, Suppl.), 75–78 (in Thai with English abstract). Available at: http://www.kmutt.ac.th/CRDC_symposium/data/75-78.pdf (accessed 22 May 2015).

Wood, B.W. and Reilly, C.C. (2000) Pest damage to pecan is affected by irrigation, nitrogen application, and fruit load. *HortScience* 35, 669–672.

Wood, E.A. Jr, Sebesta, E.E., Webster, J.A. and Porter, D.R. (1995) Resistance to wheat curl mite (Acari: Eriophyidae) in greenbug-resistant 'Gaucho' triticale and 'Gaucho' X wheat crosses. *Journal of Economic Entomology* 88, 1032–1036.

Woodburn, A.T. (1995) The market for agrochemicals present and future. In: *Proceedings of the Brighton Crop Protection Conference: Weeds – 1995, Volume 1*. BCPC Publications, Farnham (now Alton), pp. 121–128.

Wooddy, M.W. and Fashing N.J. (1993) The ability of *Rhizoglyphus robini* Claparède (Astigmata: Acaridae) to subsist solely on a diet of filter paper. *International Journal of Acarology* 19, 345–348

Woodhall, J.W., Smith, J.E., Mills, P.R. and Sansford, C.E. (2009) *A UK Commodity Pest Risk Analysis for the Cultivated Mushroom,* Agaricus bisporus. Commodity PRA for Mushrooms CSL (Central Science Laboratory)/Warwick HRI (formerly Horticulture Research International), UK, 18 December 2007. Available at: https://secure.fera.defra.gov.uk/phiw/riskRegister/plant-health/documents/agaricusBisporus.pdf (accessed 22 May 2015)

Woodring, J.P. (1969a) Environmental regulation of andropolymorphism in tyroglyphids (Acari). In: Evans, G.O. (ed.) *Proceedings of the 2nd International Congress of Acarology, 19–25 July, 1967, Sutton Bonington, UK*. Akadémiaei Kiadó, Budapest, pp. 433–440.

Woodring, J.P. (1969b) Observations on the biology of six species of acarid mites. *Annals of the Entomological Society of America* 62, 102–108.

Woodring, J.P. (1973) Comparative morphology, functions and homologies of the coxal glands in oribatid mites (Arachnida: Acari). *Journal of Morphology* 139, 407–429.

Woodring, J.P. and Galbraith, C.A. (1976) The anatomy of the adult uropodid *Fuscuropoda agitans* (Arachnida: Acari), with comparative observations on other Acari. *Journal of Morphology* 150, 19–58.

Woolley, T.A. (1961) A review of the phylogeny of mites. *Annual Review of Entomology* 6, 263–284.

Woolley, T.A. (1988) *Acarology: Mites and Human Welfare*. Wiley, New York.

Workman, P.J. and Martin, N.A. (2002) Effect of pesticides on *Cymbidium* orchid pollen-cap mite and its predator *Hypoaspis* sp. *New Zealand Plant Protection* 55, 380–384.

Wrensch, D.L. (1979) Components of reproductive success in spider mites. In: Rodriguez, J.G. (ed.) *Recent Advances in Acarology, Volume 1*. Academic Press, New York, pp. 15–164.

Wrensch, D.L. (1985) Reproductive parameters. In: Helle, W. and Sabelis, M.W. (eds) *Spider Mites: Their Biology, Natural Enemies and Control.* World Crop Pests, Volume 1A. Elsevier Science, Amsterdam, pp. 165–170.

Wrensch, D.L. and Young, S.S.Y. (1975) Effects of quality of resource and fertilization status on some fitness traits in the two-spotted spider mite, *Tetranychus urticae* Koch. *Oecologia* 18, 259–267.

Wrensch, D.L. and Young, S.S.Y. (1978) Effects of density and host quality on rate of development, survivorship and sex ratio in the carmine spider mite. *Environmental Entomology* 7, 499–501.

Wrensch, D.L. and Young, S.S.Y. (1983) Relationships between primary and tertiary sex ratio in the two-spotted spider mite (Acarina: Tetranychidae). *Annals of the Entomological Society of America* 76, 786–789.

Wrensch, D.L., Kethley, J. B. and Norton, R.A. (1994) Cytogenetics of holokinetic chromosomes and inverted meiosis: keys to the evolutionary success of mites, with generalizations on eukaryotes. In: Houck, M.A. (ed.) *Mites: Ecological and Evolutionary Analyses of Life-history Patterns* (1st edn). Chapman and Hall, New York, pp. 282–343.

Wright, W.E. (1961) Red-legged earth mites. *Agricultural Gazette of New South Wales* 72, 213–215.

Wu, G.-R. (1994) Studies on the biology and distribution pattern of *Pseudopygmephorus quadratus* (Ewing). *Journal of Zhejiang Agricultural University* 20, 543–546 (in Chinese).

Wu, J.-F. and Zhang, Z.-Q. (1993a) Host feeding, damage and control of the mushroom pest, *Brennandania lambi* (Acari: Pygmephoroidea) in China. *Experimental and Applied Acarology* 17, 233–240.

Wu, J.-F. and Zhang, Z.-Q. (1993b) Control of *Brennandania lambi* (Acari: Pygmephoroidea) by freezing: evaluation of its efficacy and effects on mushroom growth and yield. *Experimental and Applied Acarology* 17, 531–540.

Wu, J.-F., Liu, L.-D. and Mae, E.-P. (1989) Studies on spatial distribution patterns of *Brennandania lambi* (Krczal) adults on mushroom bed and their application. *Acta Agriculturae Shanghai* 5(3), 15–19 (in Chinese).

Wu, S.-X., Jiang, H., Chen, X.-F., Xiao, Q. and Yao, H.-M. (1995) Efficacy of endosulfan, a newly developed pesticide, on tea pests. *China Tea* 17(2), 18–19 (in Chinese).

Wu, T.K. (1995) Integrated control of *Phyllocnistis citrella*, *Panonychus citri*, and *Phyllocoptruta oleivora* with periodic releases of *Mallada basalis* and pesticide applications. *Chinese Journal of Entomology* 15, 113–123 (in Chinese).

Wu, W. (1984) Notes on the genus *Amblyseius* Berlese with description of two new species fron citrus orchards in south China (Acarina: Phytoseiidae). In: Griffiths, D.A. and Bowman, C.E (eds) *Acarology VI, Volume 1.* Ellis Horwood, Chichester, UK, pp. 222–227.

Wurzell, B. (2002) The maple bladder gall of the gall mite *Vasates quadripedes* new to Britain. *Cecidology* 17, 31–35.

Wyman, D. (1986) *Wyman's Gardening Encyclopedia.* Macmillan, New York.

Wysoki, M. (1997) Present status of arthropod fauna in mango orchards in Israel. *Acta Horticulturae* 455, 805–811.

Xia, X.-L., Yang, G.-X. and Sui, W. (2007) The damage of black currant gall mite to black currant and its control. *China Fruits* 1, 38–39.

Xin, J. (1985) Current status and perspective of mites as biological control agents of insect pests. *Chinese Journal of Biological Control* 1, 40–43 (in Chinese).

Xin, J., Lu, J. and Zhang, Z.-Q. (1998) *Predatory Mites. Their Biology and Roles in Biological Control.* Systematic and Applied Acarology Society, London (in Chinese).

Xu, D.J., Deng, D.F., Xia, X.Y., Li,Y. and Luo, Q.H. (1994) Method for rearing and observing *Polyphagotarsonemus latus* (Banks) in the laboratory. *Yangtze River Vegetable* 1994 (2), 16–17 (in Chinese).

Xu, G.L., Wu, H.J., Huan, Z.L., Mo, G. and Wan, M. (2001) Study on the reproductive characteristics of rice mite, *Steneotarsonemus spinki* Smiley (Acari: Tarsonemidae). *Systematic and Applied Acarology* 6, 45–49.

Xu, G.L., Wu, H.J. and Tong, X.L. (2002) Studies on stress resistance of *Steneotarsonemus spinki* Smiley. *Plant Protection* 28(5), 18–21 (in Chinese).

Xu, J.-H. and Li, X.-Z. (1996) The population dynamics and the bionomics of *Eriophyes litchii* Keifer. *Journal of Fujian Agricultural University* 25, 458–460 (in Chinese).

Xu, N., Chen, X. and Chen, H. (1996) Morphological and biochemical parameters of tea varieteis resistant to pink mite (*Acaphylla theae* Watt). *Journal of Tea Science (Hangzhou)* 16, 125–130 (in Chinese).

Xu, X., Li, L.-Y., Wang, D.-S., Hong, X.-Y., Wu, J., Yuan, Y.-D. and Xie, X.-C. (2006) Effect of temperature and relative humidity on development and reproduction of the tomato russet mite, *Aculops lycopersici* (Massee) (Acarina, Eriophyidae). *Acta Entomologica Sinica* 49, 816–821 (in Chinese).

Xu, X., Li, L.-Y., Hong, X.-Y., Yuan, Y.-D. and Wang, D.-S. (2008) Study on spatial distribution pattern and sampling technique of *Aculops lycopersici. Acta Agriculturae Shanghai* 24(3), 72–75 (in Chinese).

Xue, X.-F. and Hong, X.-Y. (2005) A new genus and eight new species of Phyllocoptini (Acari: Eriophyoidea: Phyllocoptinae) from China. *Zootaxa* 1039, 1–17.

Xue, X.-F., Sadeghi, H. and Hong, X.-Y (2009a) Eriophyoid mites (Acari: Eriophyoidea) from Iran, with descriptions of three new species, one new record and a checklist. *International Journal of Acarology* 35, 461–483.

Xue, X.-F, Wang, Z., Song, Z.-W. and Hong, X.-Y. (2009b) Eriophyoid mites on Fagaceae with descriptions of seven new genera and eleven new species (Acari: Eriophyoidea). *Zootaxa* 2253, 1–95.

Xue, X.-F., Han, X., Song, Z.-W. and Hong, X.-Y (2012) Eriophyoid mite fauna of Shaanxi Province, China, with descriptions of five new species (Acari: Eriophyoidea). *Zootaxa* 3292, 1–71.

Yadav, G.S., Anand, R.K. and Yadav, P.R. (1987) Bioefficacy of some chemicals against red spider mite on brinjal. *Indian Journal of Entomology* 49, 582–584.

Yadav, S.K., Ashok Sharma and Veer Singh (2000) Varietal susceptibility of ber to the seasonal incidence of *Eutetranychus orientalis* (Klein). *Current Agriculture* 24, 97–99.

Yadav, S.K., Ashok Sharma and Yadav, L.N. (2003a) Biology of mite, *Eutetranychus orientalis* (Klein) at ambient temperature. *Annals of Agri Bio Research* 8, 73–76.

Yadav, S.K., Ashok Sharma and Veer Singh (2003b) Relative efficacy of some insecticides/acaricides for the management of mite, *Eutetranychus orientalis* (Klein), on ber under field conditions. *Annals of Biology* 19, 221–223.

Yadav, U.S. and Yadav, A. (2010) Performance of jute cultivars against *Apion corchori*, *Anomis sabulifera* and *Polyphagotarsonemus latus. Annals of Plant Protection Sciences* 18, 98–100.

Yadav Babu, R.K. and Manjunatha, M. (2007) Seasonal incidence of mite population in arecanut. *Karnataka Journal of Agricultural Sciences* 20, 401–402.

Yamada, T., Kaeriyama, M., Matsui, N. and Yoneda, H. (1987) Development of a new miticide, hexythiazox. *Journal of Pesticicide Science (Tokyo)* 12, 327–335 (in Japanese).

Yamamoto, A., Yoneda, H., Hatano, R. and Asada, M. (1995) Genetic analysis of hexythiazox resistance in the citrus red mite, *Panonychus citri* (McGregor). *Journal of Pesticide Science (Tokyo)* 20, 513–519.

Yamamoto, P.T. and Gravena, S. (2001) Influência da dieta na oviposição de *Iphiseiodes zuluagai* Denmark & Muma (Acari: Phytoseiidae). *Revista Brasileira de Fruticultura* 23, 82–86.

Yan, H.M., Wang, H.Q. and Yang, H.M. (1992) Natural control of citrus pests by spiders in China. In: *Living with Insects: Proceedings XIX International Congress of Entomology, Beijing, China, June 28–July 4, 1992: Abstracts*. ICE Council, Beijing, 347.

Yanar, D. (2012) Life-history parameters of *Eotetranychus uncatus* Garman (Acari: Tetranychidae) on Red Delicious apple. *Pakistan Journal of Zoology* 44, 129–132.

Yanar, D. and Ecevit, O. (2005) Plant injurious and predatory mites species in apple (*Malus communis* L.) orchards in Tokat Province. *OMÜ Ziraat Fakültesi Dergisi* 20, 18–23 (in Turkish).

Yanar, D. and Ecevit, O. (2008) Species composition and seasonal occurrence of spider mites and their predators in sprayed and unsprayed apple orchards in Tokat, Turkey. *Phytoparasitica* 36, 491–501.

Yanar, D., Kadıoğlu, I. and Gökçe, A. (2011) Acaricidal effects of different plant parts extracts on two-spotted spider mite (*Tetranychus urticae* Koch). *African Journal of Biotechnology* 10, 11745–11750.

Yang, G.R. (1987) A preliminary study of mating habit of *Eotetranychus sexmaculatus* (Riley). *Insect Knowledge* 24, 96–98 (in Chinese).

Yang, H.-Z., Hu, J.-H., Li, Q., Li, H.-J., Liu, H.-Q., Yao, T.-S. and Lei, H.-D. (2007) Primary studies of acarocidal activity of twenty plants extracts against *Panonychus citri*. *Southwest China Journal of Agricultural Sciences* 20, 1012–1015.

Yang, H.-Z., Hu, J.-H., Li, Q., Li, H.-J., Liu, H.-Q., Yao, T.-S., Ran, C. and Lei, H.-D. (2009) Acaricidal activity of *Boenninghausenia sessilicarpa* against *Panonychus citri*. *Agricultural Sciences in China* 8, 1097–1102.

Yang, Q.-S. and Liang, L.-R. (1986) *Knowledge of Acari (Piman Zishi)*. Science Press, Beijing (in Chinese).

Yang, R. (1998) Reasons for the prevalence of citrus yellow mite and its control. *South China Fruits* 27(5), 19–20 (in Chinese).

Yang, S., Zhao, B.-M., Li, G.-Y., Hu, S.-L., Guo, Y.-L. and Zhang, J.-P. (2013a) Effects of brief exposure to high temperature on *Tetranychus turkestani* and *T. truncatus* (Acari: Tetranychidae). *Acta Entomologica Sinica* 56, 276–285.

Yang, S., He, Y., He, H., Li, T., Wei, Q., Feng, P. and Zhang, J. (2013b) Comparison of the predation of *Stethorus punctillum* on *Tetranychus turkestani* and *Tetranychus truncatus*. *Journal of Shihezi University (Natural Science)* 1, 10–13.

Yang, X., Margolies, D., Zhu, K.Y. and Buschman, L.L. (2001) Host plant-induced changes in detoxification enzymes and susceptibility to pesticides in the two-spotted spider mite (Acari: Tetranychidae). *Journal of Economic Entomology* 94, 381–387.

Yang, Y., Allen, J.C., Knapp, J.L. and Stansly, P.A. (1994) Citrus rust mite (Acari: Eriophyidae) damage effects on 'Hamlin' orange fruit growth and drop. *Environmental Entomology* 23, 244–247.

Yang, Y., Allen, J.C., Knapp, J.L. and Stansly, P.A. (1995) Frequency distribution of citrus rust mite (Acari: Eriophyidae) damage on fruit in 'Hamlin' orange trees. *Environmental Entomology* 24, 1018–1023.

Yang, Z.Q., Cao, H.G. and Chen, F.Y. (1991) A preliminary study on *Tetranychus kanzawai*. *Acta Agriculturae Universitatis Jiangxiensis* 13, 129–133 (in Chinese).

Yaninek, J.S. (2007) Biological control of the cassava green mite in Africa: overcoming challenges to implementation. In: Vincent, C., Goettel, M.S. and Lazarovits, G. (eds) *Biological Control: A Global Perspective*. CAB International, Wallingford, UK, pp. 28–37.

Yaninek, J.S. and Herren, H.R. (1988) Introduction and spread of the cassava green mite, *Mononychellus tanajoa* (Bondar) (Acari: Tetranychidae), an exotic pest in Africa and the search for appropriate control methods: a review. *Bulletin of Entomological Research* 78, 1–13.

Yaninek, J.S., Herren, H.R. and Gutierrez, A.P. (1989) Dynamics of *Mononychellus tanajoa* (Acari: Tetranychidae) in Africa: seasonal factors affecting phenology and abundance. *Environmental Entomology* 18, 625–632.

Yaninek, J.S., Mègevand, B., Moraes, G.J. de, Bakker, F., Braun, A. and Herren, H.R. (1992) Establishment of the Neotropical predator *Amblyseius idaeus* (Acari: Phytoseiidae) in Benin, West Africa. *Biocontrol Science and Technology* 1, 323–330.

Yaninek, J.S., Saizonou, S., Onzo, A., Zannou, I. and Gnanvossou, D. (1996) Seasonal and habitat variability in the fungal pathogens, *Neozygites* cf. *floridana* and *Hirsutella thompsonii*, associated with cassava green mites in Benin, West Africa. *Biocontrol Science and Technology* 6, 23–33.

Yaninek, J.S., Megevand, B., Ojo, B., Cudjoe, A.R., Abole, E., Onzo, A. and Zannou, I. (1998) Establishment and spread of *Typhlodromalus manihoti* (Acari: Phytoseiidae), an introduced phytoseiid predator of *Mononychellus tanajoa* (Acari: Tetranychidae) in Africa. *Environmental Entomology* 276, 1496–1505.

Yaninek, S. and Hanna, R. (2003) Cassava green mite in Africa – a unique example of successful classical biological control of a mite pest on a continental scale. In: Neuenschwander, P., Borgemeister, C. and Langewald, J. (eds) *Biological Control in IPM Systems in Africa*. CAB International, Wallingford, UK, pp. 61–75.

Yaseen, M. and Bennett, F.D. (1977) Distribution, biology and population dynamics of the green cassava mites in the neotropics. In: Cock, J., MacIntyre, R. and Graham, M. (eds) *Proceedings of the Fourth Symposium of the International Society for Tropical Root Crops, Held at CIAT, Cali, Colombia, 1–7 August 1976*. Publication IDRC-080e, International Development Research Centre, Ottawa, pp. 197–202.

Yaseen, M. and Bennett, F.D. (1978) Investigations on the natural enemies of cassava mites and mealybugs. In: *III Congreso Latino Americano de Entomología, V. Congreso Brasileiro de Entomología, Julho 1978, Ileus-Itabuna, Bahía, Brazil: Resumes.*

Yaseen, M., Bennett, F.D. and Ingram, W.R. (1982) *Investigation on the Cassava Mite* Mononychellus tanajoa *(Bondar) and its Natural Enemies in the Neotropics and East Africa, 1979–1982*. Cassava Mites, Final Report, Commonwealth Institute of Biological Control, Curepe, Trinidad.

Yasen, S., Wu, K. and Yushanjiang, T. (2007) Influence of host plants on development and reproduction of *Tetranychus turkestani* (Ugarov & Nikoiskii). *Plant Protection* 33(1), 99–102.

Yeh, C.C. and Huang, T. (1975) An ecological study on the tea eriophyid mite, *Acaphylla steinwedeni* Keifer (Acarina: Eriophyidae). *Journal of Agriculture and Forestry, Taiwan Provincial College of Agriculture* 24, 73–95.

Yeh, W.B. (1999) Identification of mites by molecular methods. *Chinese Journal of Entomology, Special Publication* No. 12, 49–62.

Yen, H. (1974) Isolation of the filamentous fungus *Hirsutella thompsonii* from *Phyllocoptruta oleivora*. *Acta Entomologica Sinica* 17, 225–226.

Yiğit, A. and Uygun, N. (1982) Investigations on the population dynamics of hawthorn mite, *Tetranychus viennensis* Zacher (Acarina: Tetranychidae) and its predators in apple orchards. *Çukurova Üniversitesi Ziraat Fakültesi Yıllığı* 13, 64–69.

Yiğit, A. and Uygun, N. (1986) Studies on interactions between hawthorn mite (*Tetranychus viennensis* Zacher) (Acarina: Tetranychidae) and its predator, *Stethorus punctillum* Weise (Col.: Coccinellidae). In: *Proceedings of the First Turkish National Congress on Biological Control, 12–14 February 1986, Adana, Turkey*, pp. 406–422 (in Turkish).

Yin, K.-S., Tang, M.-J. and Xiong, X.-P. (1995) Studies on 3-grade and 4-grade methods of investigating the pink tea mite (*Acaphylla theae* Watt). *Journal of Tea Science (Hangzhou)* 15, 99–104 (in Chinese).

Yin, K.-S., Tang, M.-J., Xiong, X.-P. and Chen, H.-C. (2003) Studies on the effect of ecological factors on the population dynamics of tea pink mite (*Acaphylla theae*). *Journal of Tea Science (Hangzhou)* 23, 53–57 (in Chinese).

Yin, S.G., Lu, C.J. and Lan, W.H. (1987) A new species of genus *Eurytetranychus* from China (Acarina: Tetranychidae). *Acta Entomologica Sinica* 30, 109–112.

Yin, S.-Y. and Sun, X.-G. (2002) Studies on the interrelationship among *Oligonychus ununguis*, host plants and *Amblyseius finlandicus*. III. Relationship between chemical component of host plants and development of *Oligonychus ununguis*. *Scientia Silvae Sinicae* 38(4), 105–110.

Ying, X.Y., Lin, M., Xu, W.Z., Huang, M.J. and Pan, X.J. (2000) Effects of Dithane M-45 on the control of citrus rust mite and on fruit appearance and quality. *South China Fruits* 29(3), 14 (in Chinese).

Yoder, J.A., Benoit, J.B., Rellinger, E.J., Ark, J.T., Halloran, M.C. and Gribbins, K.M. (2006) Structure and function of the urnulae in *Balaustium* sp. (Parasitengona: Erythraeidae) featuring secretion of a defensive allomone and alarm pheromone. *International Journal of Acarology* 32, 3–12.

Yokoyama, K. (1929) *Saishin Nippon Sangyo Gaichu Zensho (New Textbook of Sericultural Insect Pests*). Meibundo, Tokyo (in Japanese).

Yokoyama, K. (1932) New tetranychid mites attacking the mulberry leaf. (I). Bionomics and external structures of the *Tetranychus suginamensis* n. sp. *Bulletin of the Imperial Agricultural Experimental Station* 8, 229–287 (in Japanese).

Yoldas, Z., Madanlar, N., Gul, A. and Onogur, E. (1999) Investigations on integrated control practices in vegetable glasshouses in Izmir. *Acta Horticulturae* 491, 453–460.

Yothers, W.W. (1915) Spraying scheme for the control of insect pests on citrus trees in Florida. *Journal of Economic Entomology* 7, 161–164.

Yothers, W.W. and Mason, A.C. (1930) *The Citrus Rust Mite and its Control*. USDA Technical Bulletin No. 176, US Department of Agriculture, Washington, DC.

Yothers, W.W. and Miller, R.L. (1934) Methods for determining rust mite abundance. *Proceedings of the Florida Horticultural Society* 47, 53–55.

Young, B.A., Hein, G.L., French, R. and Stenger, D.C. (2007) Substitution of conserved cysteine residues in wheat streak mosaic virus HC-Pro abolishes virus transmission by the wheat curl mite. *Archives of Virology* 152, 2107–2111.

Young, J.E.B., Talbot, G.A. and Balderston, M.E. (1990) Evaluation of acaricides against apple rust mite. Tests of agrochemicals and cultivars 11. *Annals of Applied Biology* 116(Supplement), 2–3.

Young, R.A. and Haun, J.R. (1961) *Bamboo in The United States: Description, Culture and Utilization*. Agriculture Handbook No. 193, US Department of Agriculture, Washington, DC.

Young, S.S.Y. and Wrensch, D.L. (1981) Relative influence of fitness components on total fitness of the two-spotted spider mite in different environments. *Environmental Entomology* 10, 1–5.

Young, S.S.Y., Wrensch, D.L. and Kongschuensin, M. (1986) Control of sex ratio by female spider mites. *Entomologia Experimentalis et Applicata* 40, 53–60.

Young, S.S.Y., Yehling, D.M., Sastrodihardjo, S. and Wrensch, D.L. (1995) Sampling population density of *Brevipalpus phoenicis* (Acari: Tenuipalpidae) and assessment of feeding damage level on Indonesia tea. *Journal of Economic Entomology* 88, 283–287.

Youngman, R.R., Sanderson, J.P. and Barnes, M.M. (1988) Life history parameters of *Tetranychus pacificus* McGregor (Acari: Tetranychidae) on almonds under differential water stress. *Environmental Entomology* 17, 488–495.

Yousef, A.A. (1970) Mites associated with vine trees in the U.A.R. (Acarina). *Zeitschrift für Angewandte Entomologie* 67, 1–6.

Yousef, A.A., Zaher, M.A. and Kandil, M.M. (1982) Effect of prey and temperature on the development and biology of *Cheyletus malaccensis* Oudemans (Acari: Cheyletidae). *Zeitschrift für Angewandte Entomologie* 93, 39–42.

Yu, H.-X. and Shi, J.-M. (1991) Studies on *Schizotetranychus nanjinensis*. *Journal of Bamboo Research* 10(2), 61–67 (in Chinese).

Yu, J.-G., Wang, D.-Y. and Ma, M.-M. (2000) Developmental threshold temperature and effective thermal summation of *Tetranychus turkestani* (Ugarov & Nikolskii). *Chinese Bulletin of Entomology* 37, 203–205 (in Chinese).

Yu, S.J. (2008) *The Toxicology and Biochemistry of Insecticides*. CRC Press, Boca Raton, Florida.

Yue, B. and Childers, C.C. (1994) Effects of temperature on life table parameters of *Agistemus exsertus* Gonzalez (Acari: Stigmaeidae) and its attack rate on *Panonychus citri* eggs. *International Journal of Acarology* 20, 109–113.

Yue, B. and Tsai, J.H. (1995) *Agistemus exsertus* Gonzalez (Acari: Stigmaeidae) as a predator of citrus red mite [*Panonychus citri* (McGregor)]. *Journal of the New York Entomological Society* 103, 107–113.

Yüksel, B. (1999) A new harmful, the pine needle sheath mite (*Trisetacus pini* (Nal.)) on Crimean Pine. In: Tatlı, A., Ölçer, H., Bingöl, N. and Akan, H. (eds) *Proceedings of the 1st International Symposium on Protection of Natural Environment and Erhami Karaçam (*Pinus nigra *Arnold ssp.* pallasiana *(Lamb.) Holmboe var.* pyramidata *(Acat.) Yaltırık)*, 23–25 September 1999, Kütahya, Turkey. Environmental Protection and Management Research Center, Dumlupınar University, Kütahya, Turkey, pp. 950–957 (in Turkish).

Zacharda, M., Pultar, O. and Muska, J. (1988) Washing technique for monitoring mites in apple orchards. *Experimental and Applied Acarology* 5, 181–183.

Zacher, F. (1913) Untersuchungen über Spinnmilben. *Mitteilungen der Kaiserlichen Biologischen Anstalt für Land- und Forstwirtschaft, Berlin-Dahlem* 14, 37–41.

Zacher, F. (1920) *Vorläufige Diagnosen einiger neuen Spinnmilbenarten, Reihe 1*. Privatdruck, Berlin.

Zacher, F. (1921a) Neue und wenig bekannte Spinnmilben. *Zeitschrift für Angewandte Entomologie* 7, 181–187.

Zacher, F. (1921b) Untersuchungen uber Spinnmilben. *Mitteilungen aus der Biologischen Reichsanstalt für Land- und Forstwirtschaft, Berlin-Dahlem* 21, 91–100.

Zacher, F. (1949) Arachnoidea, Spinnentiere. In: Sorauer, B.v.P. (ed.) *Handbuch der Pflanzenkrankheiten Band 4*. Paul Parey, Berlin, 139–207.

Zaher, M.A. (2007) Acarology in Egypt: a brief history. *Acarines* 1, 1–4.

Zaher, M.A. and Abou-Awad, B.A. (1979) A new strain and new record of some eriophyid mites in Egypt (Eriophyoidea: Eriophyidae). *Acarologia* 21, 61–64.

Zaher, M.A. and Abou-Awad, B.A. (1980) Two new eriophyid species infesting olive trees in Egypt (Eriophyoidea: Eriophyidae). *Acarologia* 21, 65–69.

Zaher, M.A. and El-Badry, E.A. (1961) Life-history of the predator mite, *Agistemus fleschneri* Summers, and effect of nutrition on its biology. *Bulletin of the Entomological Society of Egypt* 45, 375–385.

Zaher, M.A. and El-Badry, E.A. (1962) Abundance of mites and insects predaceous on tetranychid and tenuipalpid mites in Giza (Acarina). *Bulletin of the Entomological Society of Egypt* 46, 429–441.

Zaher, M.A. and El-Badry, E.A. (1964) Survey and population studies of red and false spider mites. In: *Proceedings of the First International Congress on Acarology, 1963, Fort Collins, Colorado, 2-7 September 1963*. *Acarologia* 6(Hors Serie), 425–435.

Zaher, M.A. and Osman, A.A. (1970) Population studies on mites associated with mango trees in Egypt. *Bulletin of the Entomological Society of Egypt* 54, 141–148.

Zaher, M.A. and Rasmy, H.A. (1969) A new species of the genus *Tuckerella* from UAR (Acarina: Tuckerellidae). *Acarologia* 11, 730–732.

Zaher, M.A. and Soliman, Z.R. (1971) Life history of the predatory mite, *Chiapacheylus macrocorneus* Zaher and Soliman (Acarina: Cheyletidae). *Bulletin of the Entomological Society of Egypt* 55, 137–140.

Zaher, M.A., Wafa, A.K. and Yousef, A.A. (1970) Biology of *Brevipalpus phoenicis* (Geijskes) in Egypt. *Bulletin of the Entomological Society of Egypt* 54, 177–183.

Zaher, M.A., Rasmy, A.H. and Abou-Awad, B.A. (1971) Ecological studies on mites infesting deciduous fruit trees in Lower Egypt. *Zeitschrift für Angewandte Entomologie* 69, 59–64.

Zaher, M.A., Soliman, Z.R. and El Bishlawy, S.M. (1975) Feeding habits of the predaceous mite, *Cunaxa capreolus* (Acarina: Cunaxidae). *Entomophaga* 20, 209–212.

Zaher, M.A., Gomaa, E.A. and El-Enany, M.A. (1982) Spider mites of Egypt (Acari: Tetranychidae). *International Journal of Acarology* 8, 91–114.

Zaher, M.A., Soliman, Z.R., Hanna, M.A., Abo-Awad, B.A. and El-Enany, A.M. (1984) *Survey and Ecological Studies on Phytophagous, Predacious and Soil Mites in Egypt. I – Phytophagous Mites in Egypt (Nile Valley and Delta).* Final Report of PL-480 Programme, USA, Project No. EG-ARS 30, Grant No. FG-EG-139.

Zakhvatkin, A.A. (1941) *A Translation of Fauna of USSR, Arachnoidea, Tyroglyphoidea (Acari), Volume 6, No. 1.* Translated and edited by Ratclyffle, A. and Hughes, A.M.. American Institute of Biological Sciences, Washington, DC.

Zakhvatkin, A.A. (1952) Division of the Acarina into orders and their position in the system of the Chelicerata. *Parazitologiya Sbornik* 14, 5–46 (in Russian).

Zalom, F.G., Goodell, P.B., Wilson, L.T., Barnett, W.W. and Bentley, W.J. (1983) *Degree-Days: The Calculation and Use of Heat Units in Pest Management.* Leaflet 21373, Division of Agriculture and Natural Resources, University of California, Davis, California.

Zalom, F.G., Hoy, M.A., Wilson, L.T. and Barnett, W.W. (1984a) Sampling mites in almonds. II. Presence–absence sequential sampling for *Tetranychus* mite species. *Hilgardia* 51(7), 14–24.

Zalom, F.G., Wilson, L.T., Hoy, M.A., Barnett, W.W. and Smilanick, J.M. (1984b) Sampling *Tetranychus* spider mites in almonds. *California Agriculture* 38(5), 17-19. Ch. 20

Zalom, F.G., Kitzmiller, J., Wilson, L.T. and Gutierrez, A.P. (1986a) Observation of tomato russet mite (Acari: Eriophyidae) damage symptoms in relation to tomato plant development. *Journal of Economic Entomology* 79, 940-942.

Zalom, F.G., Wilson, L.T., Kennett, C.E., O'Connell, N.V., Flaherty, D.L. and Morse, J.G. (1986b) Presence–absence sampling of citrus red mite. *California Agriculture* 40(3, 4), 15–16.

Zalom, F.G., Thompson, P.B. and Nicola, N. (2009) Cyclamen mite, *Phytonemus pallidus* (Banks), and other tarsonemid mites in strawberries. *Acta Horticulturae* 842, 243–246.

Zamorra, J.A. and Nasca, A.J. (1985) Influencia de la fertilización en citricos con dos niveles de N y P, sobre las poblaciones de *Phyllocoptruta oleivora* (Ashm.). *CIRPON – Revista de Investigación* 3, 35–45.

Zamora, J.A., Nasca, A.J. and Sosa, G.D.R. (1989) Eficiencia del aceite mineral en aplicacion otonl, en el control de *Phyllocoptruta oleivora* (Ashm.). *CIRPON – Revista de Investigación* 7, 65-71.

Zannou, I.D., Negloh, K., Hanna, R., Houadakpode, S. and Sabelis, M.W. (2010) Mite diversity in coconut habitat in West and East Africa. In: Moraes, G.J. de, Castilho, R.C. and Flechtmann, C.H.W. (eds) *XIII International Congress of Acarology, August 23–27, 2010, Recife-PE, Brazil, Abstract Book*, abstract 547, p. 295. Available at: http://www.acarology.org/ica/ica2010/docs/0910_15_abstractBookCapa.pdf (accessed 3 February 2015).

Zappalà, L., Kreiter, S., Russo, A., Tropea Garzia, G. and Auger, P. (2015). First record of the Persea Mite Oligonychus perseae (Acari: Tetranychidae) in Italy with a review of the literature, *International Journal of Acarology*, DOI: 10.1080/01647954.2015.1014415

Zawadzki, W. (1975) Wstepne obserwacje nad szkodliwoscia szpeciela pordzewiacza sliwowego *Aculus fockeui* (Nal., Trt.). *Zeszyty Problemowe Postępów Nauk Rolniczych* 171, 157–166 (in Polish).

Zdárková, E. (1991) Application of the bio-preparation 'Cheyletin' in empty stores. In: Dusbábek, F. and Bukva, V. (eds) *Modern Acarology, Volume I. Proceedings of the VIII International Congress of Acarology, Held in České Budějovice, Czechoslovakia, 6–11 August 1990.* SPB Academic Publishing, The Hague, the Netherlands, co-published with Academia, Publishing House of the Czechoslovak Academy of Sciences, Prague, pp. 607–610.

Zeck, E.H. (1955) Grape vine mites. *Agricultural Gazette of New South Wales* 66, 155–157.

Zein-Eldin, E.A. (1956) Studies on the legume mite, *Petrobia apicalis*. *Journal of Economic Entomology* 49, 291–296.

Zemek, R. (1993a) Characteristics of development and reproduction in *Typhlodromus pyri* on *Tetranychus urticae* and *Cecidophyopsis ribis*. I. Overwintered females. *Experimental and Applied Acarology* 17, 405–421.

Zemek, R. (1993b) Characteristics of development and reproduction in *Typhlodromus pyri* on *Tetranychus urticae* and *Cecidophyopsis ribis*. I. Progeny of overwintered females. *Experimental and Applied Acarology* 17, 847–858.

Zhang, B.D. (1984) Preliminary observations on the biological characteristics of *Steneotarsonemus spinki* Smiley. *Insect Knowledge* 18, 55–56.

Zhang, F.-P., Zhang, S.-L. and You, M.-S. (2007a) Composition and structure of arthropod community in *Phyllostachys heterocycla* cv. *pubescens* forest disturbed by lindane. *Acta Ecologica Sinica* 27, 1026–1027.

Zhang, F.-P., Fu, Y.-G., Jin, Q.-A. and Zhang, J.-B. (2007b) Development and fecundity of *Oligonychus biharensis* on three southern fruit crops. *Journal of Fruit Science* 24, 185–188.

Zhang, Y. and Lin, J. (1991) Studies of *Gnorimus chaudhrii* (Acari: Phytoseiidae), an important predator of tarsonemid mites in rice fields. *Chinese Journal of Biological Control* 7, 163–165 (in Chinese).

Zhang, Y.-K., Zhang, K.-J., Sun, J.-T., Yang, X.-M., Ge, C. and Hong, X.-Y. (2013) Diversity of *Wolbachia* in natural populations of spider mites (genus *Tetranychus*): evidence for complex infection history and disequilibrium distribution. *Microbial Ecology* 65, 731–739.

Zhang, Y.-Q., Ding, W., Tian, L. and Zhao, Z.-M. (2009a) Acaricidal bioactivity of *Artemisia annua* extracts against *Panonychus citri* (Acari: Tetranychidae). *Scientia Agricultura Sinica* 42, 2217–2222.

Zhang, Y.-X., Lin, J.-Z., Chi, Y.-B., Chen, W. and Lin, S. (1996a) Biocontrol of *Tetranychus kanzawai* (Acari: Tetranychidae) using *Phytoseiulus persimilis* (Acari: Phytoseiidae) in open-air strawberry gardens. *Systematic and Applied Acarology* 1, 29–34.

Zhang, Y.-X., Lin, J.-Z., Chi, Y.-B., Chen, W. and Lin, S. (1996b) Using *Phytoseiulus persimilis* to control *Tetranychus kanzawai* on strawberry. *Chinese Journal of Biological Control* 4, 188–189 (in Chinese).

Zhang, Y.-X., Liu, Q.-Y., Lin, J.-Z., Song, M.-G., Cai, Q.-J. and He, X.-Y. (1997) Studies on species distribution and damage by spider mites in bamboo trees in Fujian Province. *Journal of Fujian Academy of Agricultural Sciences* 12(3), 11–15.

Zhang, Y.-X., Liu, Q.-Y., Lin, J.-Z., Song, M.-G. and He, X.-Y. (1998) Influence on the physiology and biochemistry of bamboo damage by the *Schizotetranychus nanjingensis* Ma *et* Yuan. *Entomological Journal of East China* 7, 65–70 (in Chinese).

Zhang, Y.-X., Zhang, Z.-Q., Liu, Q.-Y. and Lin, J.-Z. (1999a) Biology of *Schizotetranychus nanjingensis* (Acari: Tetranychidae) with reference to effects of temperature on and importance of nests to its survival and development. *Systematic and Applied Acarology* 4, 75–82.

Zhang, Y.-X., Zhang, Z.-Q., Lin, J.-Z. and Ji, J. (2000a) Potential of *Amblyseius cucumeris* (Acari: Phytoseiidae) as a biocontrol agent against *Schizotetranychus nanjingensis* (Acari: Tetranychidae) in Fujian, China. *Systematic and Applied Acarology Special Publications* No. 4, 109–124.

Zhang, Y.-X., Ji, J., Zhang, Z.-Q. and Lin, J.-Z. (2000b) Arrestment response of the predatory mite *Amblyseius longispinosus* to *Schizotetranychus nanjingensis* webnests on bamboo leaves (Acari: Phytoseiidae, Tetranychidae). *Experimental and Applied Acarology* 24, 227–233.

Zhang, Y.X., Zhang, Z.Q., Lin, J.Z., Ji, J. and Hou, A.P. (2001a) Observations on the life history of *Schizotetranychus bambusae* Reck (Acari: Tetranychidae) infesting bamboo leaves in Fujian, China. *Systematic and Applied Acarology Special Publications* No. 6, 13–20.

Zhang, Y.X., Zhang, Z.Q., Lin, J.Z. and Ji, J. (2001b) Observations on the behaviour and life history of *Schizotetranychus tenuinidus* (Acari: Tetranychidae) inhabiting bamboo leaves in Fujian, China. *Systematic and Applied Acarology Special Publications* No. 8, 13–20.

Zhang, Y.X., Zhang, Z.Q., Lin, J.Z., Ji, J. and Hou, A.P. (2001c) Oviposition and survival of *Schizotetranychus bambusae* females (Acari: Tetranychidae) feeding on young and old bamboo leaves. *Systematic and Applied Acarology Special Publications* No. 9, 1–9.

Zhang, Y.-X., Zhang, Z.-Q., Lin, J.-Z., Ji, J. and Tong, L.-X. (2001d) Larvae and protonymphs of the predator *Typhlodromus bambusae* (Acari: Phytoseiidae) attacked and killed by adult males of their prey, *Schizotetranychus nanjingensis* (Acari: Tetranychidae). *Systematic and Applied Acarology Special Publications* No. 9, 11–21.

Zhang, Y.-X., Zhang, Z.-Q., Zhang, X.-J., Liu, Q.-Y. and Lin, J.-Z. (2001e) Population dynamics of phytophagous and predatory mites (Acari: Tetranychidae, Eriophyidae, Phytoseiidae) on bamboo plants in Fujian. *Experimental and Applied Acarology* 25, 383–391.

Zhang, Y.-X., Zhang, Z.-Q., Chen, C.-P., Lin, J.-Z. and Chen, X. (2001f) *Amblyseius cucumeris* (Acari: Phytoseiidae) as a biocontrol agent against *Panonychus citri* (Acari: Tetranychidae) on citrus in China. *Systematic and Applied Acarology* 6, 35–44.

Zhang, Y.-X., Lin, J.Z., Ji, J., Chen, C.P. and Lin, Z. (2003a) Studies on *Neoseiulus cucumeris* for controlling pest mites on citrus trees. *Plant Protection* 29(5), 31–33 (in Chinese).

Zhang, Y.-X., Saitō, Y., Lin, J.-Z., Chittenden, A.R., Ji, J. and Sato, Y. (2003b) Ambulatory migration in mites (Acari: Tetranychidae, Phytoseiidae) to new leaves of moso bamboo shoots. *Experimental and Applied Acarology* 31, 59–70.

Zhang, Y.-X., Lin, J.-Z., Ji, J., Saitō, Y. and Zhang, Z.-Q. (2004a) Studies on the potential of *Typhlodromus bambusae* (natural) and *Amblyseius cucumeris* (introduced) (Acari: Phytoseiidae) as a biocontrol agent against pest mites of moso bamboo. *Scientia Silvae Sinicae* 40(5), 132–137 (in Chinese).

Zhang, Y.-X., Ji, J., Lin, J.-Z., Song, M.-G., Tong, L.-X. and Liu, Q.-Y. (2004b) Release of *Amblyseius cucumeris* (Acari: Phytoseiidae) to control pest mites of moso bamboo. *Fujian Journal of Agricultural Sciences* 2004 (02), 73–74. Available as English summary at: http://en.cnki.com.cn/Article_en/CJFDTOTAL-FJNX200402002.htm (accessed 29 August 2014).

Zhang, Y.-X., Lin, J.-Z., Ji, J., Kang, Y.-M. and Chen, X. (2004c) Analysis of numerical responses and main life parameters for determining the suppression of *Amblyseius cucumeris* on *Panonychus citri*. *Scientia Agricultura Sinica* 37, 1866–1873 (in Chinese).

Zhang, Y.-X., Lin, J.-Z., Ji, J., Chen, X. and Kang, Y.-M. (2005) Analyses of numerical responses and main life parameters for determining the suppression of *Amblyseius cucumeris* on *Panonychus citri*. *Agricultural Sciences in China* 4, 368–375.

Zhang, Y.-X., Ji, J.,Wang, F.-T., Chen, X. and Chen, F. (2006a) The fecundity potential of *Tetranychus turkestani* Ugarov *et* Nikolskii (Acari:Tetranychidae). *Acta Phytophylacica Sinica* 33, 379–383.

Zhang, Y.-X., Wang, F.-T., Ji, J., Chen, F.,Yi, Z.-B.,Weng, X.-M. and Chen, X. (2006b) Evaluation of *Amblyseius cucumeris* Oudemans for control of pest mites of koerle pear and strategy for its practical application. *Scientia Agricultura Sinica* 3, 518–524 (in Chinese).

Zhang, Y.-X., Lin, J.-Z., Ji, J. and Chen, X. (2009b) Studies on the killing effect of predatory of *Neoseiulus cucumeris* on *Oligonychus biharensis*. *Journal of Fruit Science* 2009 (05), 683–686. Available as English summary at: http://en.cnki.com.cn/Article_en/CJFDTotal-GSKK200905022.htm (accessed 29 August 2014).

Zhang, Y.-Z., Zhang, Z.-Q., Liu, Q.-Y. and Lin, J.-Z. (1999b) Biology of *Typhlodromus bambusae* (Acari: Phytoseiidae), a predator of *Schizotetranychus nanjingensis* (Acari: Tetranychidae) injurious to bamboo in Fujian (China). *Systematic and Applied Acarology* 4, 57–62.

Zhang, Y.-Z., Zhang, Z.-Q., Liu, Q.-Y., Lin, J.-Z. and Ji, J. (2000c) An overview on occurrence, development and damage of bamboo mites and their integrated management in Fujian, China. *Systematic and Applied Acarology Special Publications* No. 4, 9–17.

Zhang, Y.-Z., Zhang, Z.-Q., Tong, L.-X., Liu, Q.-Y. and Song, M.-G. (2000d) Causes of mite pest outbreaks in bamboo forests in Fujian, China: analyses of mite damage in monoculture versus polyculture stands. *Systematic and Applied Acarology Special Publications* No. 4, 93–108.

Zhang, Y.-Z., Zhang, Z.-Q., Saitō, Y., Liu, Q.-Y. and Ji, J. (2004d) On the causes of mite pest outbreaks in mono and poly-cultured moso bamboo forests. *Journal ofApplied Ecology (Ying Yong Sheng Tai Xue Bao Bian)* 17, 1161–1165 (in Chinese).

Zhang, Z.P., Wu, W.G. and Zhang, W.Q. (1990) Studies on the occurrence and control of injurious mites in mulberry plantations of Guangdong Province. *Journal of South China Agricultural University* 11(4), 17–24 (in Chinese).

Zhang, Z.-Q. (1992) Functional response of *Allothrombidium pulvinum* deutonymphs (Acari: Trombidiidae) on two-spotted mites (Acari: Tetranychidae). *Experimental and Applied Acarology* 15, 249–257.

Zhang, Z.-Q. (2002) Taxonomy of *Tetranychus ludeni* (Acari: Tetranychidae) in New Zealand and its ecology on *Sechium edule*. *New Zealand Entomologist* 25, 27–34.

Zhang, Z.-Q. (2003) *Mites of Greenhouses: Identification, Biology and Control.* CAB International, Wallingford, UK.

Zhang, Z.-Q. (2010) Tenuipalpidae of China: a review of progress with a checklist. In: Zhang, Z.Q., Hong, X.-Y. and Fan, Q.H. (eds) *Xin Jie-Liu Centenary: Progress in Chinese Acarology*. *Zoosymposia* 4, 151–157.

Zhang, Z.-Q. (ed.) (2011) *Animal Biodiversity: An Outline of Higher-Level Classification and Taxonomic Richness*. *Zootaxa* 3148. Magnolia Press, Auckland, New Zealand.

Zhang, Z.-Q. and Croft, B.A. (1994) A comparative life history study of immature *Amblyseius fallacis*, *Amblyseius andersoni*, *Typhlodormus occidentalis* and *Typhlodromus pyri* (Acari: Phytoseiidae) with a review of larval feeding patterns in the family. *Experimental and Applied Acarology* 18, 631–657.

Zhang, Z.-Q. and Hong, X.-Y. (2010) Xin centenary and progress in Chinese acarology: an introduction. In: Zhang, Z.Q., Hong, X.-Y. and Fan, Q.H. (eds) *Xin Jie-Liu Centenary: Progress in Chinese Acarology*. *Zoosymposia* 4, 5–10.

Zhang, Z.-Q. and Liang, L.-R. (1997) *An Illustrated Guide to Mites of Agricultural Importance*. Tongji University Press, Shanghai, China.

Zhang, Z.-Q., Zhang, Y.-X. and Lin, J.-Z. (2000e) Mites of *Schizotetranychus* (Acari: Tetranychidae) from moso bamboo in Fujian, China. *Systematic and Applied Acarology Special Publications* No. 4, 19–35.

Zhang, Z.-Q., Fan, Q.-H., Pesic, V., Smit, H., Bochkov, A.V., Khaustov, A.A., Baker, A., Wohltmann, A., Wen, T., Amrine, J.W., *et al.* (2011) Order Trombidiformes Reuter, 1909. In: Zhang, Z.-Q. (ed.) *Animal Biodiversity: An Outline of Higher-level Classification and Survey of Taxonomic Richness*. *Zootaxa* 3148. Magnolia Press, Auckland, New Zealand, pp. 129–138.

Zhao, D.-X., Chen, D.-S., Ge, C., Gotoh, T. and Hong, X.-Y. (2013) Multiple infections with *Cardinium* and two strains of *Wolbachia* in the spider mite *Tetranychus phaselus* Ehara: revealing new forces driving the spread of *Wolbachia*. *PLoS ONE* 8(1): e54964 (accessed 29 August 2014).

Zhao, G., Jiang, S., Xue, L. and Liu, B. (1997) A preliminary report on the medicine effect of the extract of *Peganum harmla* to phytophage spider mites. *Acta Universitatis Agriculturalis Borealis-Occidentalis* 25(4), 111–114 (in Chinese).

Zhao, J. and Kuang, H.-Y. (1996) Two new species and two new records of the genus *Aceria* in China (Acari: Eriophyidae). *Journal of Nanjing Agricultural University* 19(2), 31–33 (in Chinese).

Zhao, J.-Y., Li, S.-H., Niu, T.-Q., Zhao, R.-Y., Dai, Z.-W. and Tian, G.-L. (2005) Damages of apple seedlings caused by attack of *Polyphagotarsonemus latus* under green house conditions. *China Fruits* 3, 8–9 (in Chinese).

Zhao, S. (2000) Study of the dispersal and diversity of eriophyoid mites (Acari: Eriophyoidea). PhD thesis, West Virginia University, Morgantown, West Virginia.

Zhao, S. and Amrine, J.W. Jr (1997a) Investigation of snowborne mites (Acari) and relevancy to dispersal. *International Journal of Acarology* 23, 209–213.

Zhao, S. and Amrine, J.W. Jr (1997b) A new method for studying aerial dispersal behavior of eriophyoids (Acari: Eriophyoidea). *Systematic and Applied Acarology* 2, 107–110.

Zhao, Y.F. and Hou, J.W. (1993) Investigation and feeding of natural enemies of *Acaphylla theae* Watt. *Journal of Tea* 19, 35–37 (in Chinese).

Zhi, J.R. (1997) A study on *Eotetranychus kankitus* and its important natural enemy, *Amblyseius nicholsi*. *Journal of Guizhou Agricultural College* 16(3) 22–26 (Also given as *Journal of Mountain Agriculture and Biology*). Available as English summary at: http://en.cnki.com.cn/Article_en/CJFDTOTAL-SDNS199703005.htm (accessed 24 March 2015).

Zhi, J.R. (2002) The reproductive life table of *Polyphagotarsonemus latus*. *Entomological Knowledge* 39, 199–202 (in Chinese).

Zhi, J.R., Guo, Z.Z. and Xiong, J.W. (1994) A study on the predation of *Amblyseius nicholsi* Ehara et Lee on *Eotetetranychus kankitus* Ehara. *Entomological Knowledge* 31, 19–22 (in Chinese).

Zhi, J.R., Guo, Z.Z. and Xiong, J.W. (1998) The biological and ecological characteristics of *Amblyseius nicholsi* (Acari: Phytoseiidae). *Systematic and Applied Acarology* 3, 35–42.

Zhimo, Z. and McMurtry, J.A. (1990) Development and reproduction of three *Euseius* (Acari: Phytoseiidae) species in the presence and absence of supplementary foods. *Experimental and Applied Acarology* 8, 233–242.

Zhou, C.-G., Zhang, W.-G., Qiao, L.-Q., Sun, X.-G. and Wang, Z.-Y. (2006) The biological characteristics, effective accumulated temperature and occurrence regulation of *Eutetranychus orientalis*. *Scientia Silvae Sinicae* 42(5), 89–93.

Zhou, L., Yue, B. and Zou, F. (1999) Life table studies of *Eotetranychus kankitus* (Acari: Tetranychidae) at different temperatures. *Systematic and Applied Acarology* 4, 69–73.

Zhou, L., Yue, B.-S. and Zou, F.-D. (2000) Influence of temperature–humidity on the generation period and the number of eggs laid of *Eotetranychus kankitus* Ehara. *Journal of Sichuan University (Natural Science Edition)* 37(1), 98–102 (in Chinese).

Zhou, Q.-A. and Li, H.-S. (2001) Geostatistical analysis of the spatial structure of gall mite in litchi orchards. *Acta Horticulturae* 558, 429–433.

Zivanovic, V. (1973) Pregalj *Eriophyes essigi* Hassan-uzrocnik nedozrevanja plodova kupine. *Jugoslovensko Vocarstvo* 7(23), 63–64 (in Serbo Croatian).

Zote, K.K., Mali, V.R., Mayee, C.D. and Kulkarni, S.V. (1991) Outbreak of sterility mosaic disease of pigeonpea in Marathwada region of Maharashtra, India. *International Pigeonpea Newsletter* 14, 19–20.

Zou, F.-D., Yue, B.-S., Xu, L., Zhang, Y.-Z. and Zeng, Z.-Y. (2002) Use of a new method to study karyotypes of *Eotetranychus kankitus* and *Panonychus citri*. *Acta Entomologica Sinica* 45, 662–665 (in Chinese).

Zou, P. and Gao, J. (1996) Two new species of the genus *Dolichocybe* from edible fungi in China (Acariformes: Pyemotoidea). *Acta Entomologica Sinica* 39, 430–433 (in Chinese).

Zou, P. and Zeng, X.-S. (1993) *Pseudopygmephorus quadratus*. In: Huang, N.-L. (ed.) *China Edible Fungi Encyclopedia*. China Agricultural Publishing House, Beijing (in Chinese).

Zou, P., Gao, J.-R. and Ma, E.P. (1988) Study on pyemotid and pygmephorid mites associated with edible fungi from China (Acari: Pygmephoroidea). *Journal of Shanghai Agricultural College* 6, 221–226, 240 (in Chinese).

Zou, P., Gao, J.-R., Wang, J.-M. and Tan, Q. (1992) Study on *Siteroptes flechtmanni* infesting commercially cultivated mushrooms. In: *Proceedings of Youth Satellite Symposium on life Science and Biotech, Shanghai First Academic Annual Meeting of Youth Cast*. Shanghai Science and Technology Press, Shanghai, China, pp. 240–244 (in Chinese).

Zou, P., Gao, J.-R. and Ma, E.P. (1993a) Preliminary studies on the biology of the pest mite and *Luciaphorus auriculariae* (Acari: Pygmephoridae) infesting Jew's ear mushroom *Auricularia polytricha* in China. *Experimental and Applied Acarology* 17, 225–232.

Zou, P., Gao, J.-R. and Wu, S.-Z. (1993b) Study on a destructive mite, *Luciaphorus auriculariae*. *Edible Fungi* 1993a (5), 35–36 (in Chinese).

Zriki, G., Shaabo, A. and Boubou, A. (2015) A preliminary survey of the spider mites (Acari: Tetranychidae) in Latakia governatorate of Syria. *Acarologia* 55(3), 303–309.

Zukova, V.P. (1963) Feeding mechanisms of the spider mite *Tetranychus telarius*. *Trudy Nauchno-Issledovatel'skogo Instituta Zashchita Rasteniĭ Uzbekistan SSR* 6, 13–18 (in Russian).

Zuluaga Cardona, I. (1971) Lista preliminas de ócaros de importancia en Colombia. *Acta Agronómica, Universidad Nacional de Colombia* 21, 119–132.

Zuluaga Cardona, I. and Sánchez Potes, A. (1971) La roña o escoriación de los frutos del cocotero (*Cocos nucifera* L.) en Colombia. *Acta Agronómica, Universidad Nacional de Colombia* 21, 133–139.

Zurawicz, E., Pluta, S., Masny, A. and Danek, J. (2009) Polish small fruit breeding program – aims and achievements. *Journal of Jilin Agricultural University* 31, 486–491.

Zwick, R.W. and Fields, G.J. (1978) Field and laboratory evaluations of fenvalerate against several insect and mite pests of apple and pear in Oregon. *Journal of Economic Entomology* 71, 793–796.

Zwick, R.W., Fields, G.J. and Mellenthin, W.M. (1976) Effects of mite population density on 'Newton' and 'Golden Delicious' apple tree performance. *Journal of the American Society for Horticultural Science* 101, 123–125.

Zwolfer, H. (1964) Ulex europaeus *Project: Investigations for New Zealand*. Report No 1, Report of Commonwealth Insitute of Biological Control, European Station, Délémont, Switzerland.

Taxonomic Index

Pest mites are given under both scientific and common names. Page numbers in **bold** type refer to illustrations and tables.

Subject Index

Page numbers in **bold** type refer to figures and tables.